FOR STUDENTS

- Career Opportunities
- Career Fitness Program
- Becoming an Electronics Technician
- Free On-Line Study Guides (Companion web sites)

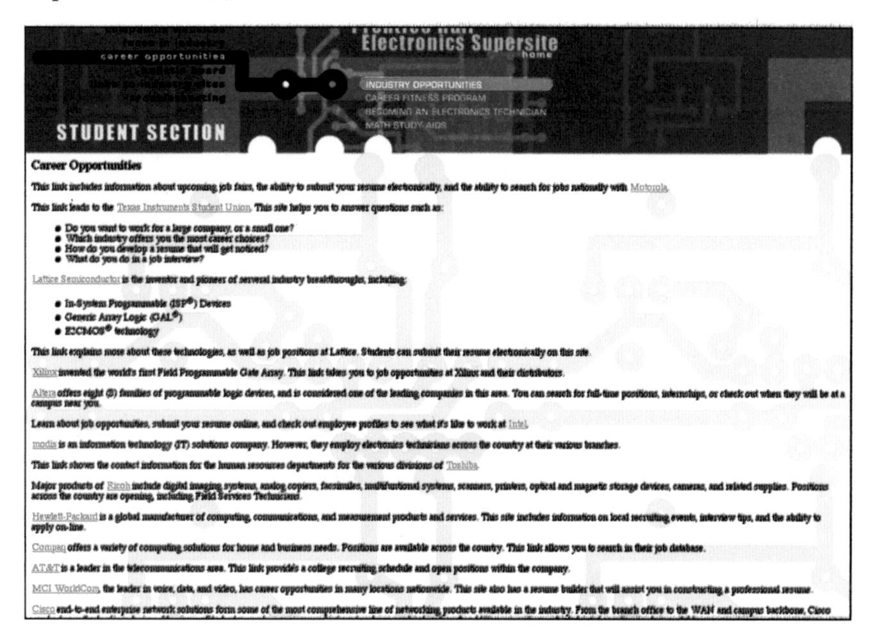

FOR FACULTY*

- Supplements
- On-Line Product Catalog
- Electronics Technology Journal
- Prentice Hall Book Advisor

* *Please contact your Prentice Hall representative for passcode*

FIFTH EDITION

ELECTRONICS FUNDAMENTALS

Circuits, Devices, and Applications

Thomas L. Floyd

Upper Saddle River, New Jersey
Columbus, Ohio

Library of Congress Cataloging in Publication Data

Floyd, Thomas L.
Electronics fundamentals : circuits, devices, and applications / Thomas L. Floyd.–5th ed.
p. cm.
Includes index.
ISBN 0-13-085236-8 (alk. paper)
1. Electronics. 2. Electronic circuits. 3. Electronic apparatus and appliances. I. Title.

TK7816.F57 2001
621.381—dc21

00-029849

Vice President and Publisher: Dave Garza
Editor in Chief: Stephen Helba
Acquisitions Editor: Scott J. Sambucci
Associate Editor: Katie E. Bradford
Production Editor: Rex Davidson
Design Coordinator: Karrie Converse-Jones
Text Design: Seventeenth Street Studios
Cover Designer: Allen Bumpus
Cover Art: Marjory Dressler
Illustrations: Jane Lopez and Steve Botts
Photographs: Part-opening and chapter-opening photographs by EyeWire, Inc.
Copyeditor: Lois Porter
Production Manager: Pat Tonneman
Marketing Manager: Ben Leonard

This book was set in Times Roman and Highlander by The Clarinda Company. It was printed and bound by R. R. Donnelley & Sons Company. The cover was printed by Phoenix Color Corp.

10 9 8 7 6 5 4 3 2
ISBN: 0-13-085236-8

Brief Contents

Once Again, To Sheila
With Love

Contents

PART II ■ AC CIRCUITS

APPENDICES

Preface

This fifth edition of *Electronics Fundamentals: Circuits, Devices, and Applications* provides a comprehensive coverage of basic electrical and electronic concepts, practical applications, and troubleshooting. The organization has been improved for a smoother and more logical flow of the material in certain areas. In this edition, many topics have been strengthened and improved, and some new topics and features have been added. Also, a completely new text design and layout enhance the text's appearance and useability.

This textbook is divided into three parts: DC Circuits in Chapters 1 through 7, AC Circuits in Chapters 8 through 15, and Devices in Chapters 16 through 21.

NEW FEATURES AND IMPROVEMENTS

Engineering Notation Chapter 1 includes an expanded coverage of engineering notation and the use of the calculator (TI-86) in scientific and engineering notation.

Electrical Safety Chapter 2 introduces electrical safety. It is supplemented by a feature called *Safety Point* found throughout portions of the text. Safety Points are identified by a special logo and design treatment.

Troubleshooting An expanded coverage of troubleshooting begins in Section 3-8 with an introduction to troubleshooting. A systematic approach called the APM (analysis, planning, and measurement) method is introduced and used in many of the troubleshooting sections and examples. A new logo identifies troubleshooting features.

Circuit Simulation Tutorials A website tutorial associated with most chapters can be downloaded for student use. These tutorials introduce students to elements of Electronics Workbench™, as needed, on a chapter-by-chapter basis. These tutorials may be found at **http://www.prenhall.com/floyd.**

Circuit Simulation Problems A new set of problems at the end of most chapters reference circuits simulated with both Electronics Workbench and CircuitMaker® on the CD-ROM that accompanies the text. Many of these circuits have hidden faults that the student must locate using troubleshooting skills. Results are provided in a password-protected file on the CD-ROM. Circuit simulation problems and exercises on the CD-ROM are indicated by a special logo.

Hands-On Tips Called HOTips for short, this feature provides useful and practical information interspersed throughout the book. They generally relate to the text coverage but can be skipped over without affecting an understanding of chapter material. HOTips are identified by a special logo and design treatment.

Biographies Brief biographies of those after whom major electrical and magnetic units have been named are located near the point where the unit is introduced. Each biography is indentified by a special design treatment.

Key Terms Terms identified as the most important in each chapter are listed as key terms on the chapter opener. Within the chapter, key terms are highlighted in color and with a special icon. Each Key Term is also defined in the Glossary.

Chapter Reorganization Several chapters in the AC part of the text have been rearranged to provide a smoother and more logical flow of topics. The new chapter sequence is as follows: Chapter 9: Capacitors, Chapter 10: *RC* Circuits, Chapter 11: Inductors, Chapter 12: *RL* Circuits, Chapter 13: *RLC* Circuits and Resonance, and Chapter 14: Transformers.

ADDITIONAL FEATURES

- Full-color format
- A two-page chapter opener for each chapter with an introduction, chapter outline, chapter objectives, key terms, and application assignment preview
- An introduction and list of objectives at the beginning of each section within a chapter keyed to the chapter objectives
- An Application Assignment at the end of each chapter (except Chapter 1)
- Many high-quality illustrations
- Numerous worked examples
- A Related Problem in each worked example with answers at the end of the chapter
- An Electronics Workbench/CircuitMaker simulation on CD-ROM for selected worked examples
- An Electronics Workbench/CircuitMaker exercise in selected Application Assignments
- Section Reviews with answers at the end of the chapter
- Troubleshooting section in many of the chapters
- Self-test in each chapter with answers at the end of the chapter
- Problem set at the end of each chapter divided by chapter sections and organized into basic and advanced categories. Answers to odd-numbered problems are provided at the end of the book.
- A comprehensive Glossary at the end of the book. Terms that appear boldface or in color in the text are defined in the glossary.
- Standard resistor and capacitor values are used throughout.

ACCOMPANYING STUDENT RESOURCES

New—*Student Workbook* by James K. Gee. Features step-by-step explanations of textbook material, additional examples with solutions, explanatory tables, reminders, and a Problem Set for every textbook section. Odd-numbered answers to Problem Set questions are included at the end of the Student Workbook. Gee's Student Workbook is tied section by section to the Floyd text, thus enabling students to easily locate specific sections with which they are having difficulty or would like additional practice. (ISBN 0–13–019392–5)

New—StudyWizard e-tutorial CD-ROM. Students can enhance their understanding of each chapter by answering the review questions and testing their knowledge of the terminology with StudyWizard. This program is available separately from the text. Contact your local bookstore for more information.

New—Electronics Workbench/CircuitMaker CD-ROM. Packaged with each text, this software includes simulation circuits for selected examples and end-of-chapter problems and a Student Version of CircuitMaker. Electronics Workbench software can be obtained through your local bookstore, or by contacting Electronics Workbench at 800-263-5552, or through their website at www.electronicsworkbench.com.

Experiments in Electronics Fundamentals and Electric Circuits Fundamentals Fifth Edition, by David Buchla. (ISBN 0–13–017002–X)

Companion Website (www.prenhall.com/floyd). This website offers students a free online study guide that they can check for conceptual understanding of key topics. It includes simulation tutorials in Electronics Workbench.

Electronics Supersite (www.prenhall.com/electronics). Students will find additional troubleshooting exercises, links to industry sites, an interview with an electronics professional, and more.

INSTRUCTOR RESOURCES

New—PowerPoint CD-ROM. Contains slides featuring all figures from the text, of which 150 selected slides contain explanatory text to elaborate on the presented graphic. This CD-ROM also includes innovative PowerPoint slides for the lab manual by Dave Buchla. (ISBN 0–13–019386–0)

Companion Website (www.prenhall.com/floyd). For the professor, this website offers the ability to post your syllabus online with our Syllabus Builder. This is a great solution for classes taught online, self-paced, or in any computer-assisted manner.

Electronics Supersite (www.prenhall.com/electronics). Instructors will find the *Prentice Hall Electronics Technology Journal,* extra classroom resources, and all of the supplements for this text available online for easy access. Contact your local Prentice Hall sales representative for your "User Name" and "Passcode."

Online Course Support. If your program is offering your electronics course in a distance learning format, please contact your local Prentice Hall sales representative for a list of product solutions.

Instructor's Resource Manual. Includes solutions to chapter problems, solutions to Application Assignments, a section relating SCANS objectives to textbook coverage, and a CEMA skills list. (ISBN 0–13–019387–9)

Lab Solutions Manual. Includes worked-out lab results for the Lab Manual by Buchla. (ISBN 0–13–019391–7)

Test Item File. This edition of the Test Item File has been checked for accuracy and features 166 new questions. (ISBN 0–13–019388–7)

Prentice Hall Test Manager. This is a CD-ROM version of the Test Item File. (ISBN 0–13–019389–5)

ILLUSTRATION OF CHAPTER FEATURES

Chapter Opener Each chapter begins with a two-page spread, as shown in Figure P–1. The left page includes the chapter number and title, a chapter introduction, and a list of sections in the chapter. The right page has a list of chapter objectives, a list of key terms, an application assignment preview, and a website reference for circuit simulation tutorials and other helpful material.

Section Opener Each section in a chapter begins with a brief introduction that includes a general overview and section objectives as related to the chapter objectives. An example is shown in Figure P–2.

Section Review Each section in a chapter ends with a review consisting of questions or exercises that emphasize the main concepts presented in the section. This is also shown in Figure P–2. The answers to the Section Reviews are at the end of the chapter.

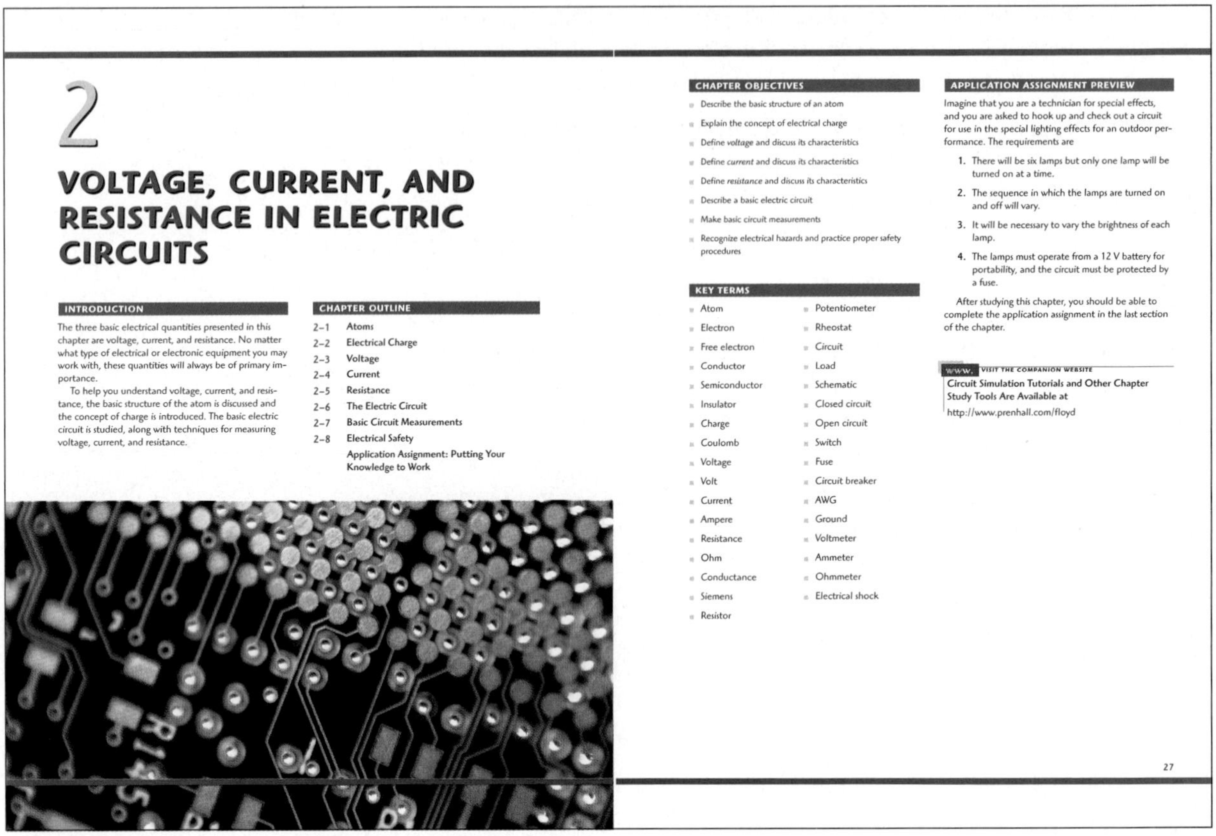

2

VOLTAGE, CURRENT, AND RESISTANCE IN ELECTRIC CIRCUITS

INTRODUCTION

The three basic electrical quantities presented in this chapter are voltage, current, and resistance. No matter what type of electrical or electronic equipment you may work with, these quantities will always be of primary importance.

To help you understand voltage, current, and resistance, the basic structure of the atom is discussed and the concept of charge is introduced. The basic electric circuit is studied, along with techniques for measuring voltage, current, and resistance.

CHAPTER OUTLINE

2–1 Atoms
2–2 Electrical Charge
2–3 Voltage
2–4 Current
2–5 Resistance
2–6 The Electric Circuit
2–7 Basic Circuit Measurements
2–8 Electrical Safety
Application Assignment: Putting Your Knowledge to Work

CHAPTER OBJECTIVES

- Describe the basic structure of an atom
- Explain the concept of electrical charge
- Define *voltage* and discuss its characteristics
- Define *current* and discuss its characteristics
- Define *resistance* and discuss its characteristics
- Describe a basic electric circuit
- Make basic circuit measurements
- Recognize electrical hazards and practice proper safety procedures

KEY TERMS

- Atom
- Electron
- Free electron
- Conductor
- Semiconductor
- Insulator
- Charge
- Coulomb
- Voltage
- Volt
- Current
- Ampere
- Resistance
- Ohm
- Conductance
- Siemens
- Resistor
- Potentiometer
- Rheostat
- Circuit
- Load
- Schematic
- Closed circuit
- Open circuit
- Switch
- Fuse
- Circuit breaker
- AWG
- Ground
- Voltmeter
- Ammeter
- Ohmmeter
- Electrical shock

APPLICATION ASSIGNMENT PREVIEW

Imagine that you are a technician for special effects, and you are asked to hook up and check out a circuit for use in the special lighting effects for an outdoor performance. The requirements are

1. There will be six lamps but only one lamp will be turned on at a time.
2. The sequence in which the lamps are turned on and off will vary.
3. It will be necessary to vary the brightness of each lamp.
4. The lamps must operate from a 12 V battery for portability, and the circuit must be protected by a fuse.

After studying this chapter, you should be able to complete the application assignment in the last section of the chapter.

WWW. VISIT THE COMPANION WEBSITE
Circuit Simulation Tutorials and Other Chapter Study Tools Are Available at
http://www.prenhall.com/floyd

27

▲ **FIGURE P–1**

Chapter opener.

Worked Examples, Related Problems, and EWB/CircuitMaker Exercise Numerous worked examples help illustrate and clarify basic concepts or specific procedures. Each example ends with a Related Problem that reinforces or expands on the example by requiring the student to work through a problem similar to the example. Selected examples contain an EWB/CircuitMaker exercise keyed to the CD-ROM. A typical worked example with a related problem and an EWB/CircuitMaker exercise is shown in Figure P–3. Answers to Related Problems are at the end of the chapter.

Troubleshooting Sections Many chapters include a troubleshooting section that relates to the topics covered in the chapter and emphasizes logical thinking as well as a structured approach called APM (analysis, planning, and measurement). Particular troubleshooting methods such as *half-splitting* are applied.

Application Assignment: Putting Your Knowledge to Work Application Assignments are located at the end of each chapter (except Chapter 1) and are identified by a special photographic logo and colored background design. A practical application of the material covered in the chapter is presented. In a series of steps, the student is required to compare circuit layouts with a schematic, analyze circuits using concepts and theories learned in the chapter, and evaluate and/or troubleshoot circuits. A typical Application Assignment is shown in Figure P–4. The Application Assignments are optional and skipping over them does not affect any other coverage.

Although they are not intended or designed for use as laboratory projects (except the laboratory of the mind), many of the application assignments use

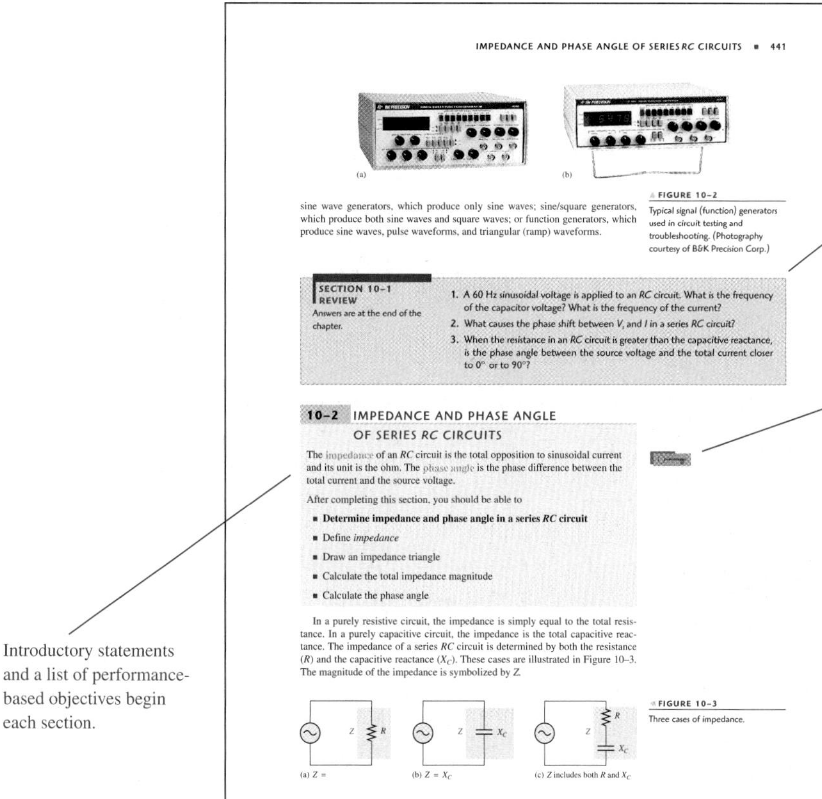

IMPEDANCE AND PHASE ANGLE OF SERIES *RC* CIRCUITS ■ 441

(a) (b)

sine wave generators, which produce only sine waves; sine/square generators, which produce both sine waves and square waves; or function generators, which produce sine waves, pulse waveforms, and triangular (ramp) waveforms.

FIGURE 10–2
Typical signal (function) generators used in circuit testing and troubleshooting. (Photography courtesy of B&K Precision Corp.)

SECTION 10–1 REVIEW
Answers are at the end of the chapter.

1. A 60 Hz sinusoidal voltage is applied to an *RC* circuit. What is the frequency of the capacitor voltage? What is the frequency of the current?
2. What causes the phase shift between V_s and I in a series *RC* circuit?
3. When the resistance in an *RC* circuit is greater than the capacitive reactance, is the phase angle between the source voltage and the total current closer to 0° or to 90°?

10–2 IMPEDANCE AND PHASE ANGLE OF SERIES *RC* CIRCUITS

The impedance of an *RC* circuit is the total opposition to sinusoidal current and its unit is the ohm. The phase angle is the phase difference between the total current and the source voltage.

After completing this section, you should be able to

- **Determine impedance and phase angle in a series *RC* circuit**
- Define *impedance*
- Draw an impedance triangle
- Calculate the total impedance magnitude
- Calculate the phase angle

In a purely resistive circuit, the impedance is simply equal to the total resistance. In a purely capacitive circuit, the impedance is the total capacitive reactance. The impedance of a series *RC* circuit is determined by both the resistance (R) and the capacitive reactance (X_C). These cases are illustrated in Figure 10–3. The magnitude of the impedance is symbolized by Z.

FIGURE 10–3
Three cases of impedance.

(a) $Z = R$ (b) $Z = X_C$ (c) Z includes both R and X_C

Review questions end each section.

A key icon indicates a key term that is in color.

Introductory statements and a list of performance-based objectives begin each section.

FIGURE P–2

Section review and section opener.

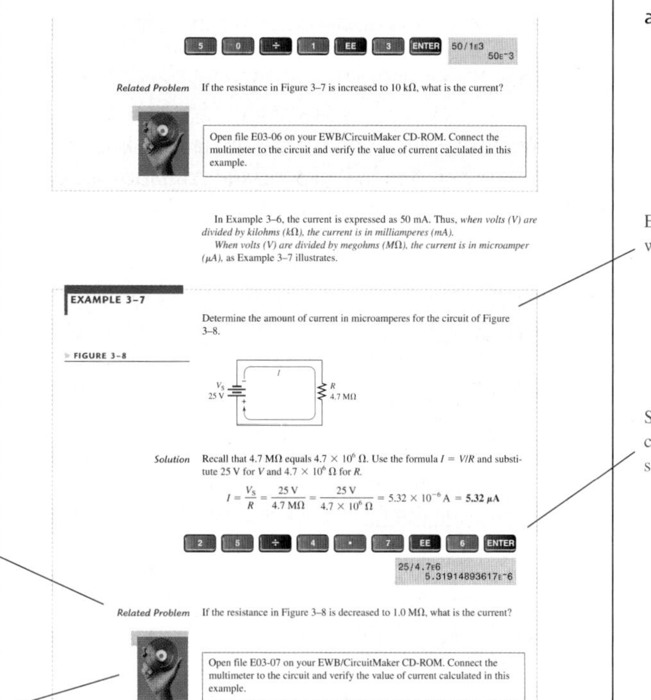

86 ■ OHM'S LAW, ENERGY, AND POWER

5 0 ÷ 1 EE 3 ENTER
50/1E3
50E-3

Related Problem If the resistance in Figure 3–7 is increased to 10 kΩ, what is the current?

Open file E03-06 on your EWB/CircuitMaker CD-ROM. Connect the multimeter to the circuit and verify the value of current calculated in this example.

In Example 3–6, the current is expressed as 50 mA. Thus, *when volts (V) are divided by kilohms (kΩ), the current is in milliamperes (mA).*
When volts (V) are divided by megohms (MΩ), the current is in microamperes (μA), as Example 3–7 illustrates.

EXAMPLE 3–7

Determine the amount of current in microamperes for the circuit of Figure 3–8.

FIGURE 3–8

Solution Recall that 4.7 MΩ equals 4.7×10^6 Ω. Use the formula $I = V/R$ and substitute 25 V for V and 4.7×10^6 Ω for R.

$$I = \frac{V_S}{R} = \frac{25\text{ V}}{4.7\text{ M}\Omega} = \frac{25\text{ V}}{4.7 \times 10^6\ \Omega} = 5.32 \times 10^{-6}\text{ A} = \mathbf{5.32\ \mu A}$$

2 5 ÷ 4 • 7 EE 6 ENTER
25/4.7E6
5.31914893617E-6

Related Problem If the resistance in Figure 3–8 is decreased to 1.0 MΩ, what is the current?

Open file E03-07 on your EWB/CircuitMaker CD-ROM. Connect the multimeter to the circuit and verify the value of current calculated in this example.

Examples are contained within a colored box.

Selected examples contain calculator key sequences and display.

Each example contains a problem related to the example.

Selected examples contain an EWB/CircuitMaker exercise keyed to the CD-ROM.

FIGURE P–3

An example with a related problem and an EWB/CircuitMaker exercise.

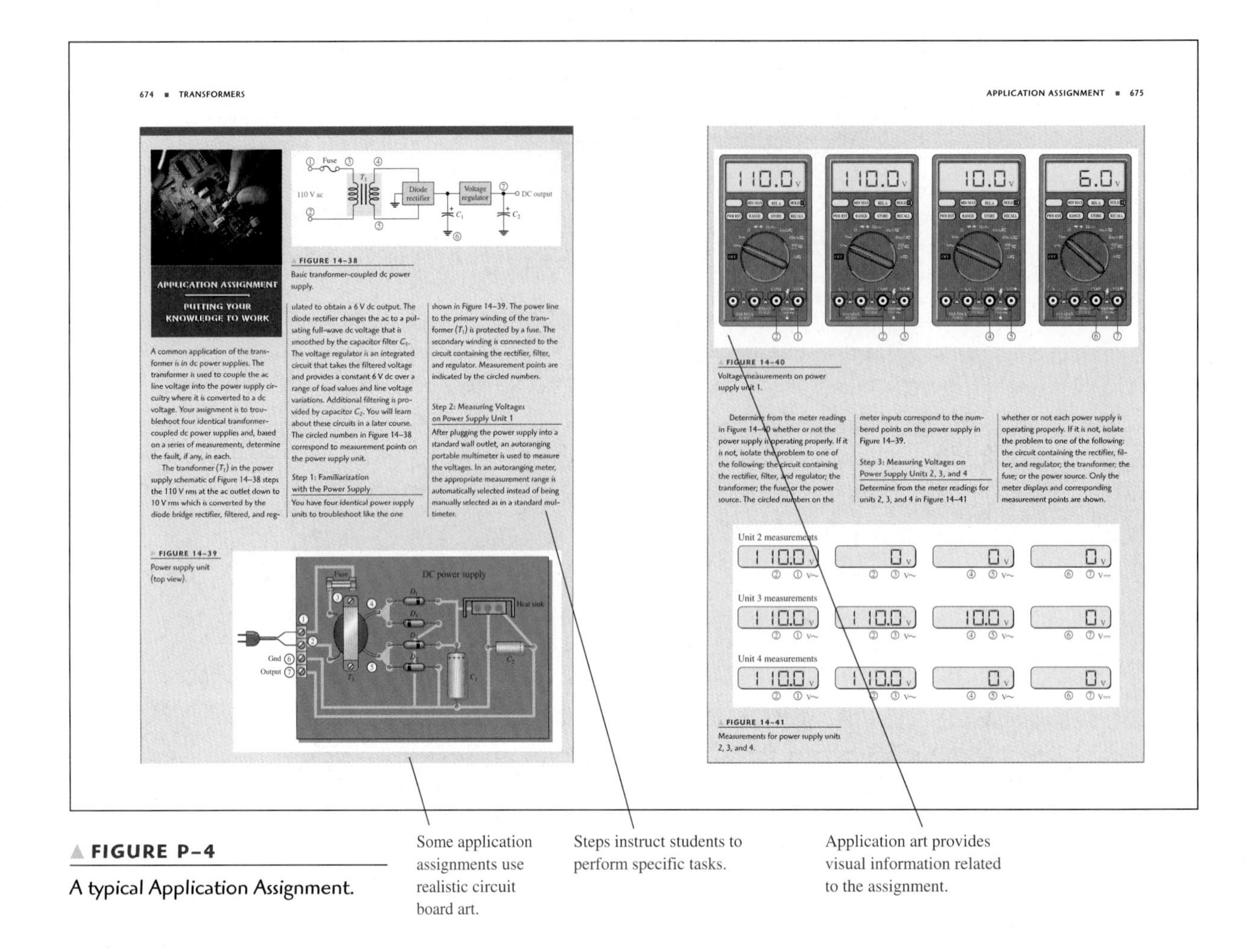

674 ■ TRANSFORMERS

APPLICATION ASSIGNMENT

PUTTING YOUR KNOWLEDGE TO WORK

A common application of the transformer is in dc power supplies. The transformer is used to couple the ac line voltage into the power supply circuitry where it is converted to a dc voltage. Your assignment is to troubleshoot four identical transformer-coupled dc power supplies and, based on a series of measurements, determine the fault, if any, in each.

The transformer (T_1) in the power supply schematic of Figure 14–38 steps the 110 V rms at the ac outlet down to 10 V rms which is converted by the diode bridge rectifier, filtered, and regulated to obtain a 6 V dc output. The diode rectifier changes the ac to a pulsating full-wave dc voltage that is smoothed by the capacitor filter C_1. The voltage regulator is an integrated circuit that takes the filtered voltage and provides a constant 6 V dc over a range of load values and line voltage variations. Additional filtering is provided by capacitor C_2. You will learn about these circuits in a later course. The circled numbers in Figure 14–38 correspond to measurement points on the power supply unit.

Step 1: Familiarization with the Power Supply

You have four identical power supply units to troubleshoot like the one shown in Figure 14–39. The power line to the primary winding of the transformer (T_1) is protected by a fuse. The secondary winding is connected to the circuit containing the rectifier, filter, and regulator. Measurement points are indicated by the circled numbers.

Step 2: Measuring Voltages on Power Supply Unit 1

After plugging the power supply into a standard wall outlet, an autoranging portable multimeter is used to measure the voltages. In an autoranging meter, the appropriate measurement range is automatically selected instead of being manually selected as in a standard multimeter.

FIGURE 14–38
Basic transformer-coupled dc power supply.

FIGURE 14–39
Power supply unit (top view).

APPLICATION ASSIGNMENT ■ 675

FIGURE 14–40
Voltage measurements on power supply unit 1.

Determine from the meter readings in Figure 14–40 whether or not the power supply is operating properly. If it is not, isolate the problem to one of the following: the circuit containing the rectifier, filter, and regulator; the transformer; the fuse; or the power source. The circled numbers on the meter inputs correspond to the numbered points on the power supply in Figure 14–39.

Step 3: Measuring Voltages on Power Supply Units 2, 3, and 4

Determine from the meter readings for units 2, 3, and 4 in Figure 14–41 whether or not each power supply is operating properly. If it is not, isolate the problem to one of the following: the circuit containing the rectifier, filter, and regulator; the transformer; the fuse; or the power source. Only the meter displays and corresponding measurement points are shown.

FIGURE 14–41
Measurements for power supply units 2, 3, and 4.

FIGURE P–4

A typical Application Assignment.

representations based on realistic printed circuit boards and instruments. Results and answers for the steps in the Application Assignments are provided in the Instructor's Resource Manual.

Chapter End Matter The following pedagogical features are found at the end of each chapter:

- Summary
- Equations
- Self-Test
- Basic Problems
- Advanced Problems
- Electronics Workbench/CircuitMaker Troubleshooting Problems (keyed to CD-ROM)
- Answers to Section Reviews
- Answers to Related Problems for Examples
- Answers to Self-Test

SUGGESTIONS FOR USING THIS TEXTBOOK

As mentioned before, this book is divided into three parts: DC Circuits, AC Circuits, and Devices. The text can be used to accommodate a variety of scheduling and program requirements. Some suggestions follow:

Option 1 A three-term dc/ac/devices sequence should allow sufficient time to cover all or most of the topics in the book. Chapters 1 through 7 can be covered in the first term, Chapters 8 through 15 in the second term, and Chapters 16 through 21 in the third term.

Option 2 A modification of Option 1 is to add the coverage of capacitors through Section 9-5 and inductors through Section 11-5 to the first term dc course.

Option 3 Yet another modification to Option 1 for those who prefer to cover reactive components before covering reactive circuits is to cover Chapter 11 on inductors immediately after Chapter 9 on capacitors. Then follow with Chapter 10, Chapter 12, and so on.

Option 4 A two-term dc/ac/devices sequence in which ac is introduced in the first term by covering Part 1 and Chapter 8. Then, in the second term, cover the remaining chapters in Part 2 and Part 3. Obviously, this approach will require selective and faster-paced coverage of much of the material. Since program requirements vary greatly, it is difficult to make specific suggestions for selective coverage.

Option 5 A two-term dc/ac sequence which omits the devices coverage in Part 3. If this is the option you prefer, it is recommended that the companion text *Electric Circuits Fundamentals* be used because it is identical to *Electronics Fundamentals: Circuits, Devices, and Applications* except that it does not include Part 3.

TO THE STUDENT

Any career training requires hard work, and electronics is no exception. The best way to learn new material is by reading, thinking, and doing. This text is designed to help you along the way by providing an overview and objectives for each section, numerous worked-out examples, practice exercises, and review questions with answers.

Don't expect every concept to be crystal clear after a single reading. Read each section of the text carefully and think about what you have read. Work through the example problems step-by-step before trying the related problem that goes with the example. Sometimes more than one reading of a section will be necessary. After each section, check your understanding by answering the section review questions.

Review the chapter summary and equation list. Take the multiple-choice self-test. Finally, work the problems at the end of the chapter. Check your answers to the self-test at the end of the chapter and the odd-numbered problems against those provided at the end of the book. Working problems is the most important way to check your comprehension and solidify concepts.

CAREERS IN ELECTRONICS

The field of electronics is very diverse, and career opportunities are available in many areas. Because electronics is currently found in so many different applications and new technology is being developed at a fast rate, its future appears limitless. There is hardly an area of our lives that is not enhanced to some degree by electronics technology. Those who acquire a sound, basic knowledge of electrical and electronic principles and are willing to continue learning will always be in demand.

The importance of obtaining a thorough understanding of the basic principles contained in this text cannot be overemphasized. Most employers prefer to hire people who have both a thorough grounding in the basics and the ability and eagerness to grasp new concepts and techniques. If you have a good training in the basics, an employer will train you in the specifics of the job to which you are assigned.

There are many types of job classifications for which a person with training in electronics technology may qualify. A few of the most common job functions are discussed briefly in the following paragraphs.

Service Shop Technician Technical personnel in this category are involved in the repair or adjustment of both commercial and consumer electronic equipment that is returned to the dealer or manufacturer for service. Specific areas include TVs, VCRs, CD players, stereo equipment, CB radios, and computer hardware. This area also offers opportunities for self-employment.

Industrial Manufacturing Technician Manufacturing personnel are involved in the testing of electronic products at the assembly-line level or in the maintenance and troubleshooting of electronic and electromechanical systems used in the testing and manufacturing of products. Virtually every type of manufacturing plant, regardless of its product, uses automated equipment that is electronically controlled.

Laboratory Technician These technicians are involved in breadboarding, prototyping, and testing new or modified electronic systems in research and development laboratories. They generally work closely with engineers during the development phase of a product.

Field Service Technician Field service personnel service and repair electronic equipment—for example, computer systems, radar installations, automatic banking equipment, and security systems—at the user's location.

Engineering Assistant/Associate Engineer Personnel in this category work closely with engineers in the implementation of a concept and in the basic design and development of electronic systems. Engineering assistants are frequently involved in a project from its initial design through the early manufacturing stages.

Technical Writer Technical writers compile technical information and then use the information to write and produce manuals and audiovisual materials. A broad knowledge of a particular system and the ability to clearly explain its principles and operation are essential.

Technical Sales Technically trained people are in demand as sales representatives for high-technology products. The ability both to understand technical concepts and to communicate the technical aspects of a product to a potential customer is very valuable. In this area, as in technical writing, competency in expressing yourself orally and in writing is essential. Actually, being able to communicate well is very important in any technical job category because you must be able to record data clearly and explain procedures, conclusions, and actions taken so that others can readily understand what you are doing.

MILESTONES IN ELECTRONICS

Before you begin your study of electronics fundamentals, let's briefly look at some of the important developments that led to the electronics technology we have today. The names of many of the early pioneers in electricity and electromagnetics still live on in terms of familiar units and quantities. Names such as Ohm, Ampere, Volta, Farad, Henry, Coulomb, Oersted, and Hertz are some of the better known examples. More widely known names such as Franklin and Edison are also significant in the history of electricity and electronics because of their tremendous contributions. Short biographies of some of these pioneers, like shown here, are located throughout the text.

BIOGRAPHY

James Prescott Joule
1818–1889
Joule, a British physicist, is known for his research in electricity and thermodynamics. He formulated the relationship that states that the amount of heat energy produced by an electrical current in a conductor is proportional to the conductor's resistance and the time. The unit of energy is named in his honor. (Photo credit: Library of Congress.)

The Beginning of Electronics Early experiments with electronics involved electric currents in vacuum tubes. Heinrich Geissler (1814–1879) removed most of the air from a glass tube and found that the tube glowed when there was current through it. Later, Sir William Crookes (1832–1919) found the current in vacuum tubes seemed to consist of particles. Thomas Edison (1847–1931) experimented with carbon filament bulbs with plates and discovered that there was a current from the hot filament to a positively charged plate. He patented the idea but never used it.

Other early experimenters measured the properties of the particles that flowed in vacuum tubes. Sir Joseph Thompson (1856–1940) measured properties of these particles, later called *electrons*.

Although wireless telegraphic communication dates back to 1844, electronics is basically a 20th century concept that began with the invention of the vacuum tube amplifier. An early vacuum tube that allowed current in only one direction was constructed by John A. Fleming in 1904. Called the Fleming valve, it was the forerunner of vacuum tube diodes. In 1907, Lee deForest added a grid to the vacuum tube. The new device, called the audiotron, could amplify a weak signal. By adding the control element, deForest ushered in the electronics revolution. It was with an improved version of his device that made transcontinental telephone service and radios possible. In 1912, a radio amateur in San Jose, California, was regularly broadcasting music!

In 1921, the secretary of commerce, Herbert Hoover, issued the first license to a broadcast radio station; within two years over 600 licenses were issued. By the end of the 1920s radios were in many homes. A new type of radio, the superheterodyne radio, invented by Edwin Armstrong, solved problems with high-frequency communication. In 1923, Vladimir Zworykin, an American researcher, invented the first television picture tube, and in 1927 Philo T. Farnsworth applied for a patent for a complete television system.

The 1930s saw many developments in radio, including metal tubes, automatic gain control, "midgit sets," directional antennas, and more. Also started in this decade was the development of the first electronic computers. Modern computers trace their origins to the work of John Atanasoff at Iowa State University. Beginning in 1937, he envisioned a binary machine that could do complex mathematical work. By 1939, he and graduate student Clifford Berry had constructed a binary machine called ABC, (for Atanasoff-Berry Computer) that used vacuum tubes for logic and condensers (capacitors) for memory. In 1939, the magnetron, a microwave oscillator, was invented in Britain by Henry Boot and John Randall. In the same year, the klystron microwave tube was invented in America by Russell and Sigurd Varian.

During World War II, electronics developed rapidly. Radar and very high-frequency communication were made possible by the magnetron and klystron. Cathode ray tubes were improved for use in radar. Computer work continued during the war. By 1946, John von Neumann had developed the first stored program computer, the Eniac, at the University of Pennsylvania. The decade ended with one of the most important inventions ever, the transistor.

Solid-State Electronics The crystal detectors used in early radios were the forerunners of modern solid-state devices. However, the era of solid-state electronics began with the invention of the transistor in 1947 at Bell Labs. The inventors were Walter Brattain, John Bardeen, and William Shockley. PC (printed circuit) boards were introduced in 1947, the year the transistor was invented. Commercial manufacturing of transistors began in Allentown, Pennsylvania, in 1951.

The most important invention of the 1950s was the integrated circuit. On September 12, 1958, Jack Kilby, at Texas Instruments, made the first integrated circuit. This invention literally created the modern computer age and brought about sweeping changes in medicine, communication, manufacturing, and the entertainment industry. Many billions of "chips"—as integrated circuits came to be called—have since been manufactured.

The 1960s saw the space race begin and spurred work on miniaturization and computers. The space race was the driving force behind the rapid changes in electronics that followed. The first successful "op-amp" was designed by Bob Widlar at Fairchild Semiconductor in 1965. Called the μA709, it was very successful but suffered from "latch-up" and other problems. Later, the most popular op-amp ever, the 741, was taking shape at Fairchild. This op-amp became the industry standard and influenced design of op-amps for years to come.

By 1971, a new company that had been formed by a group from Fairchild introduced the first microprocessor. The company was Intel and the product was the 4004 chip, which had the same processing power as the Eniac computer. Later in the same year, Intel announced the first 8-bit processor, the 8008. In 1975, the first personal computer was introduced by Altair, and Popular Science magazine featured it on the cover of the January, 1975, issue. The 1970s also saw the introduction of the pocket calculator and new developments in optical integrated circuits.

By the 1980s, half of all U.S. homes were using cable hookups instead of television antennas. The reliability, speed, and miniaturization of electronics continued throughout the 1980s, including automated testing and calibrating of PC boards. The computer became a part of instrumentation and the virtual instrument was created. Computers became a standard tool on the workbench.

The 1990s saw a widespread application of the Internet. In 1993, there were 130 websites, and now there are millions. Companies scrambled to establish a home page and many of the early developments of radio broadcasting had parallels with the Internet. In 1995, the FCC allocated spectrum space for a new service called Digital Audio Radio Service. Digital television standards were adopted in 1996 by the FCC for the nation's next generation of broadcast television.

ACKNOWLEDGMENTS

Many capable people have been part of this revision for the fifth edition of *Electronics Fundamentals: Circuits, Devices, and Applications*. It has been thoroughly reviewed and checked for both content and accuracy. Those at Prentice Hall who have contributed greatly to this project throughout the many phases of development and production include Rex Davidson, Katie Bradford, and Scott Sambucci. Lois Porter, whose attention to details is unbelievable, has once again done an outstanding job editing the manuscript. Jane Lopez has again provided the excellent illustrations and beautiful graphics work used in the text. As with the previous edition, Gary Snyder has checked the manuscript for accuracy. Also, Gary created the circuit files for the Electronics Workbench features in this edition and Karen Dickson at Protel Technology created the CircuitMaker files. A thorough line-by-line review was done by my colleague Dave Buchla and many of the improvements and changes found in this edition were a result of his recommendations.

I wish to express my appreciation to those already mentioned as well as the reviewers who provided many valuable suggestions and constructive criticism that greatly influenced this edition. These reviewers are Jerry L. Wilson, Lamar Institute of Technology; James K. Gee, ITT Technical Institute; William R. Kist, New England Institute of Technology; Peter Westray, Los Angeles Valley College; and Carl Jensen, DeVry Institute of Technology.

Tom Floyd

PART ONE

DC CIRCUITS

1

COMPONENTS, QUANTITIES, AND UNITS

INTRODUCTION

The topics in this chapter present a basic introduction to the field of electronics. An overview of electrical and electronic components and instruments gives you a preview of the types of things you will study throughout this book.

You must be familiar with the units used in electronics and know how to express electrical quantities in various ways using metric prefixes. Scientific notation and engineering notation are indispensable tools whether you use a computer, a calculator, or do computations the old-fashioned way.

CHAPTER OUTLINE

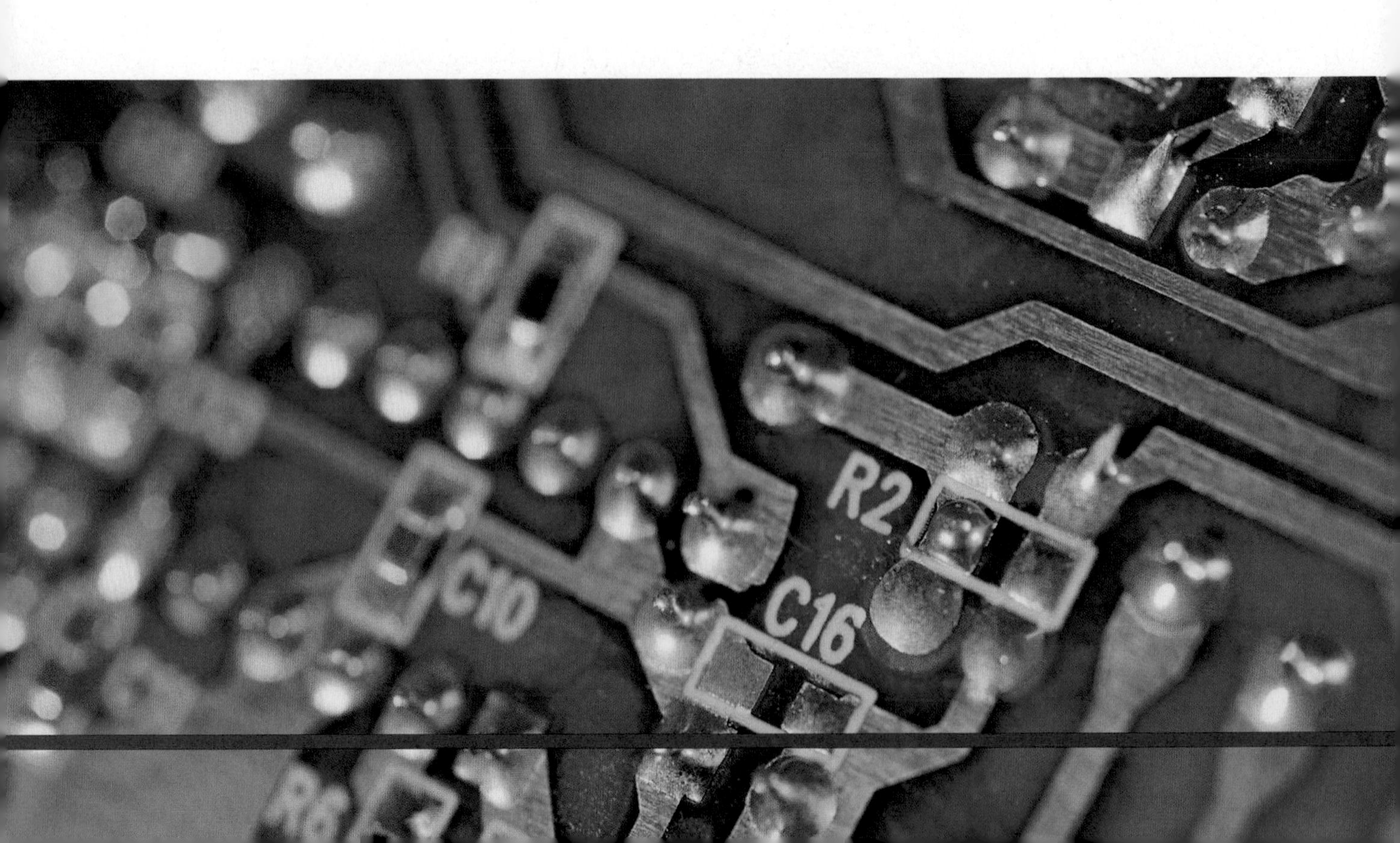

CHAPTER OBJECTIVES

- Recognize some common electrical components and measuring instruments
- State basic electrical and magnetic quantities and their units
- Use scientific notation (powers of ten) to express quantities
- Use engineering notation and metric prefixes to express large and small quantities
- Convert from one metric-prefixed unit to another

KEY TERMS

- Resistor
- Capacitor
- Inductor
- Transformer
- DC power supply
- Function generator
- Digital multimeter
- Oscilloscope
- Scientific notation
- Power of ten
- Exponent
- Engineering notation
- Metric prefix

APPLICATION ASSIGNMENT PREVIEW

At the beginning of each chapter starting with Chapter 2, you will find an application assignment preview that relates to that chapter. The application assignments described by the previews present a variety of practical job situations that a technician might encounter in industry.

As you study each chapter, think about how to approach the application assignment that appears as the last section in each chapter. When you have completed each chapter, you should have a sufficient knowledge of the topics covered to enable you to complete the assignment.

WWW. VISIT THE COMPANION WEBSITE

Circuit Simulation Tutorials and Other Chapter Study Tools Are Available at

http://www.prenhall.com/floyd

1–1 ELECTRICAL COMPONENTS AND MEASURING INSTRUMENTS

A thorough background in dc and ac circuit fundamentals provides the foundation for understanding electronic devices and circuits. In this book, you will study many types of electrical components and measuring instruments. A preview of the basic types of electrical and electronic components and instruments that you will be studying in detail in this and in other courses is provided in this section.

After completing this section, you should be able to

- **Recognize some common electrical components and measuring instruments**
- State the basic purpose of a resistor
- State the basic purpose of a capacitor
- State the basic purpose of an inductor
- State the basic purpose of a transformer
- List some basic types of electronic measuring instruments

Resistors

Resistors resist, or limit, electrical current in a circuit. Several common types of resistors are shown in Figure 1–1 through Figure 1–4.

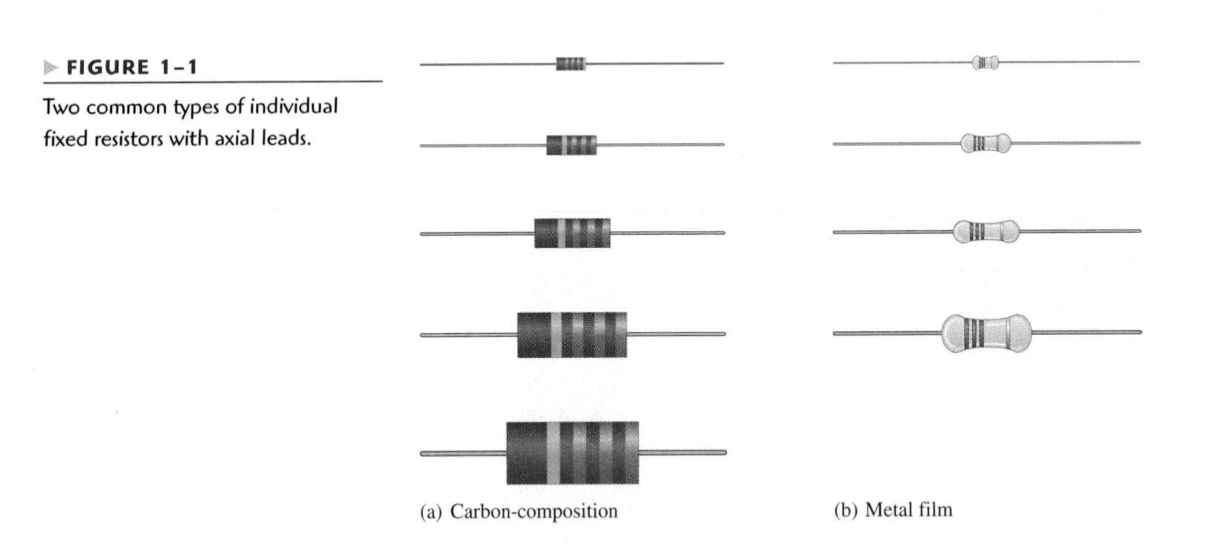

FIGURE 1–1

Two common types of individual fixed resistors with axial leads.

(a) Metal film chip resistor

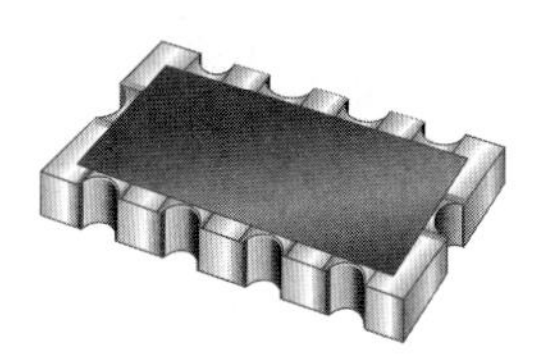
(b) Chip resistor array

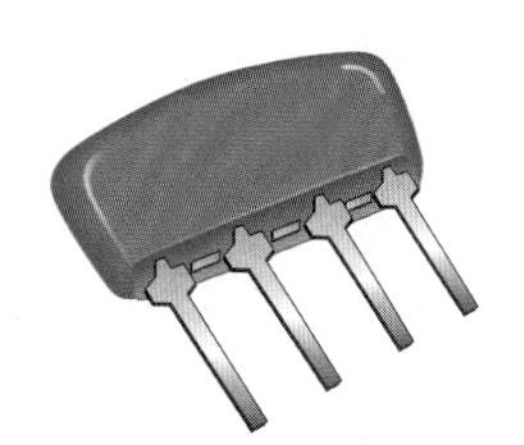
(c) Resistor network (simm)

(d) Resistor network (surface mount)

▲ **FIGURE 1–2**

Chip resistor and resistor networks.

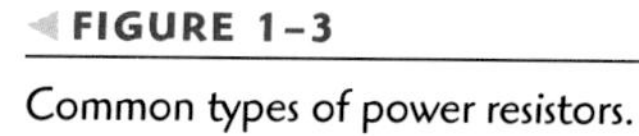
◀ **FIGURE 1–3**

Common types of power resistors.

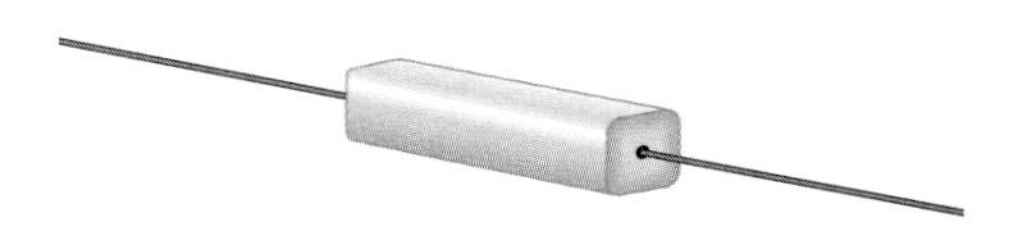
(a) Axial-lead wirewound

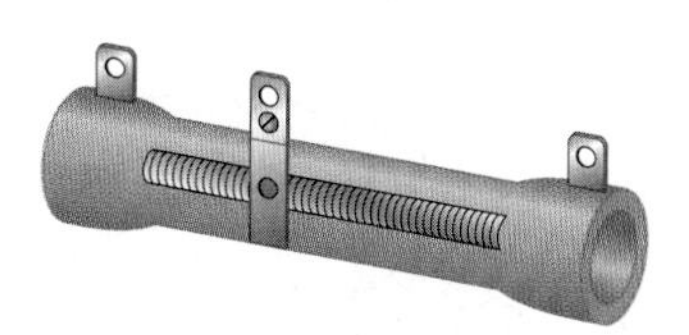
(b) Adjustable wirewound

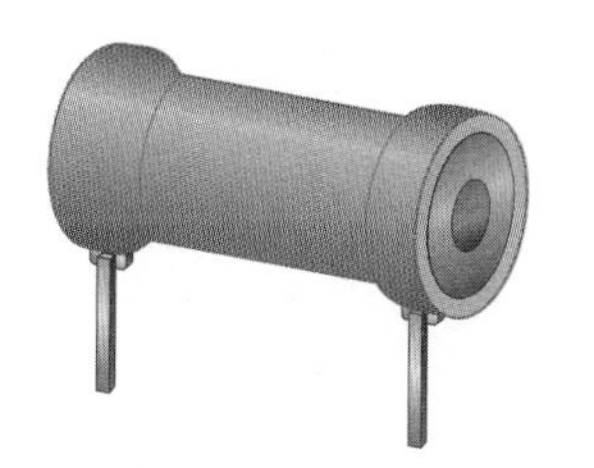
(c) Radial-lead for PC board insertion

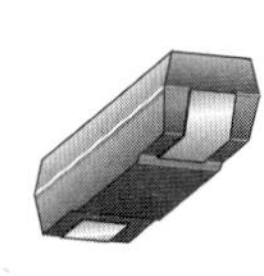
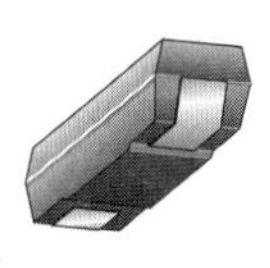
(d) Surface mount

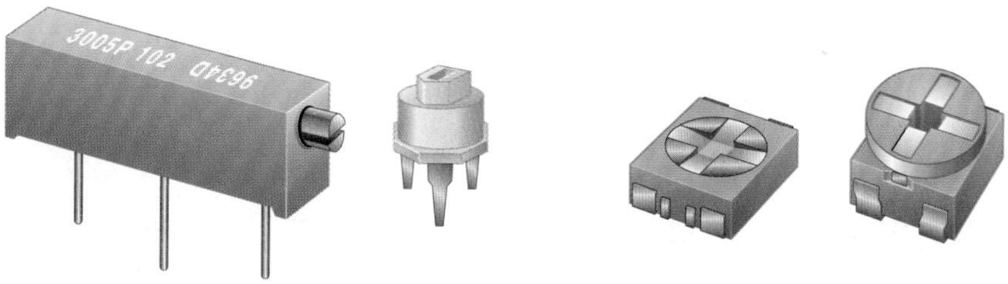

◀ **FIGURE 1–4**

Common types of variable resistors.

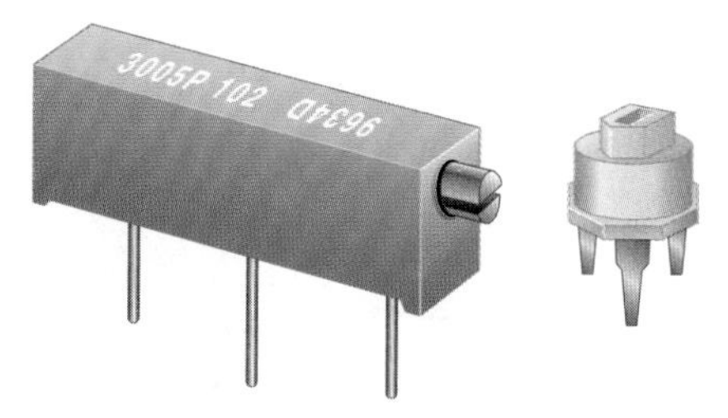

(a) Lead mounted

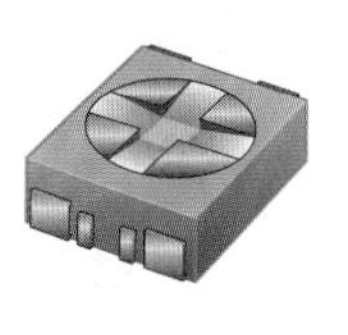
(b) Surface mounted

Capacitors

Capacitors store electrical charge and are used to block dc and pass ac. Figure 1–5 and Figure 1–6 show several typical capacitors.

Inductors

Inductors, also known as *coils,* are used to store energy in an electromagnetic field; they serve many useful functions in an electrical circuit. Figure 1–7 shows several typical inductors.

FIGURE 1–5

Common types of fixed capacitors.

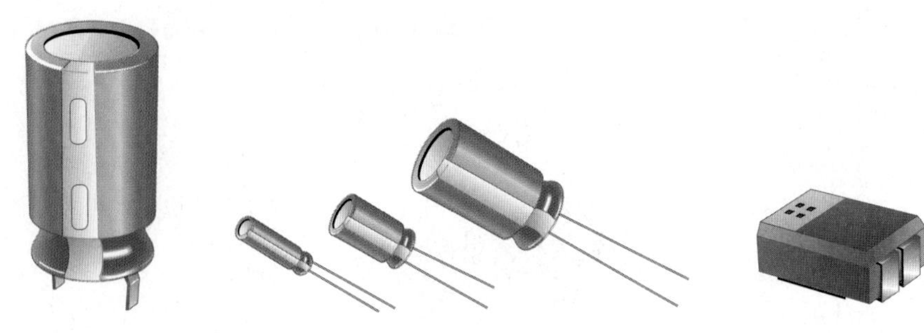

(a) Electrolytic, axial-lead, and surface mount

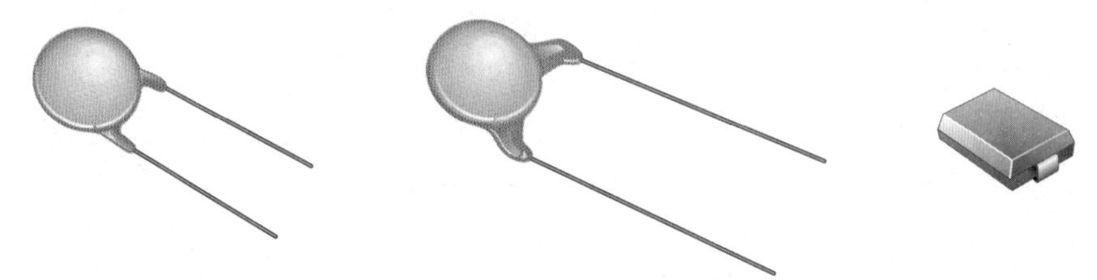

(b) Ceramic, axial-lead, and surface mount

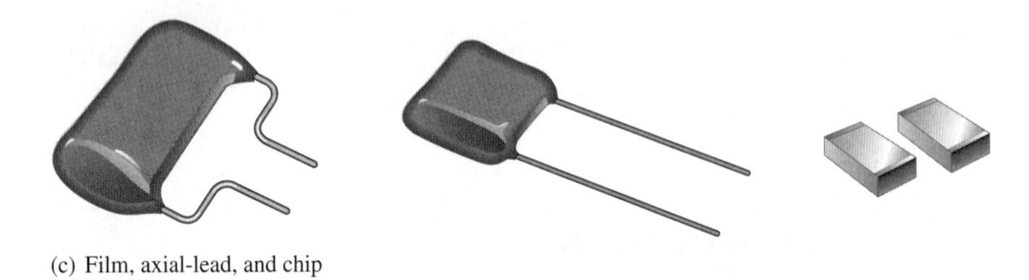

(c) Film, axial-lead, and chip

FIGURE 1–6

Typical variable capacitors.

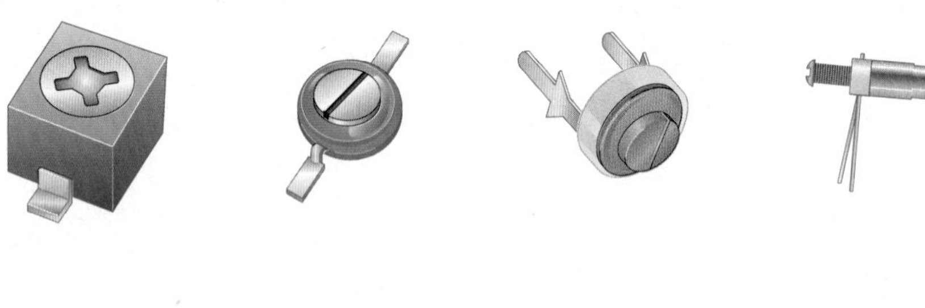

FIGURE 1–7

Some fixed and variable inductors.

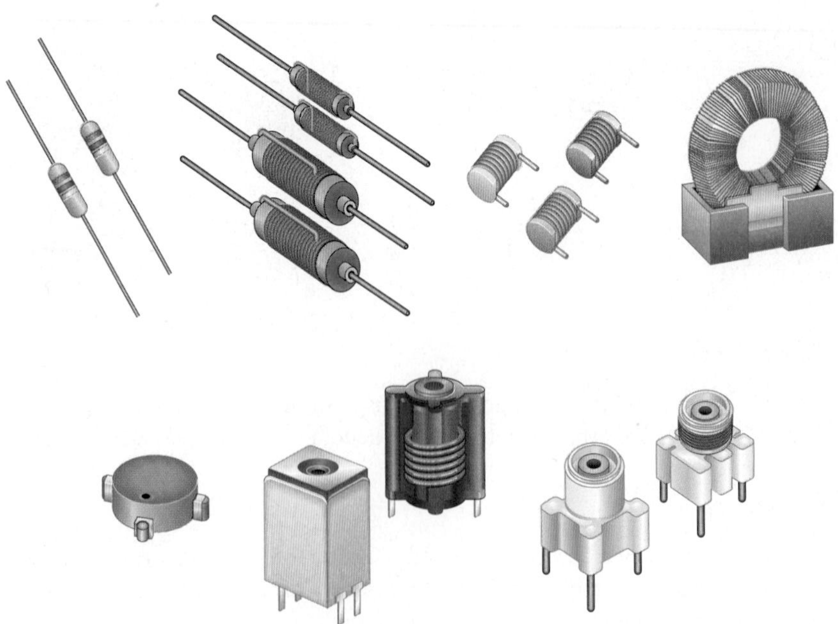

Transformers

Transformers are used to couple ac voltages from one point in a circuit to another, or to increase or decrease the ac voltage. Several types of transformers are shown in Figure 1–8.

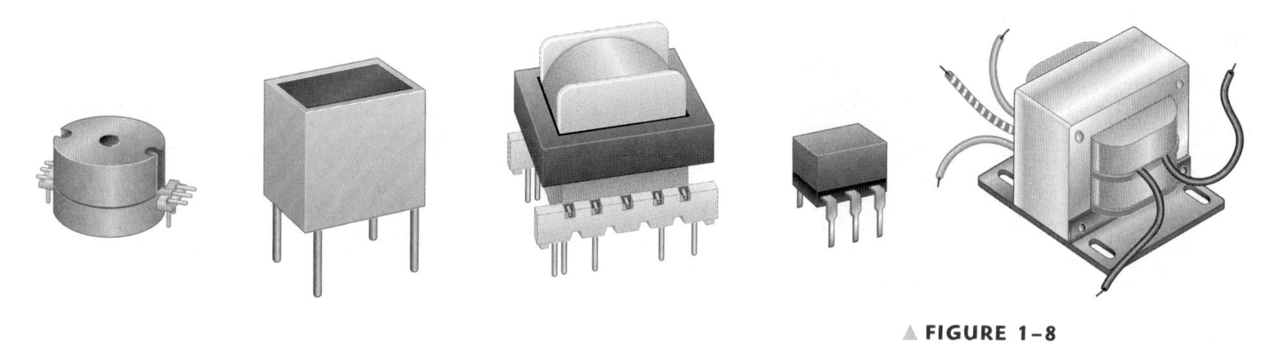

▲ **FIGURE 1–8**

Typical transformers.

Semiconductor Devices

Several varieties of diodes, transistors, and integrated circuits are shown in Figure 1–9. A wide selection of packages are used for semiconductor devices, depending on the function and the power requirements.

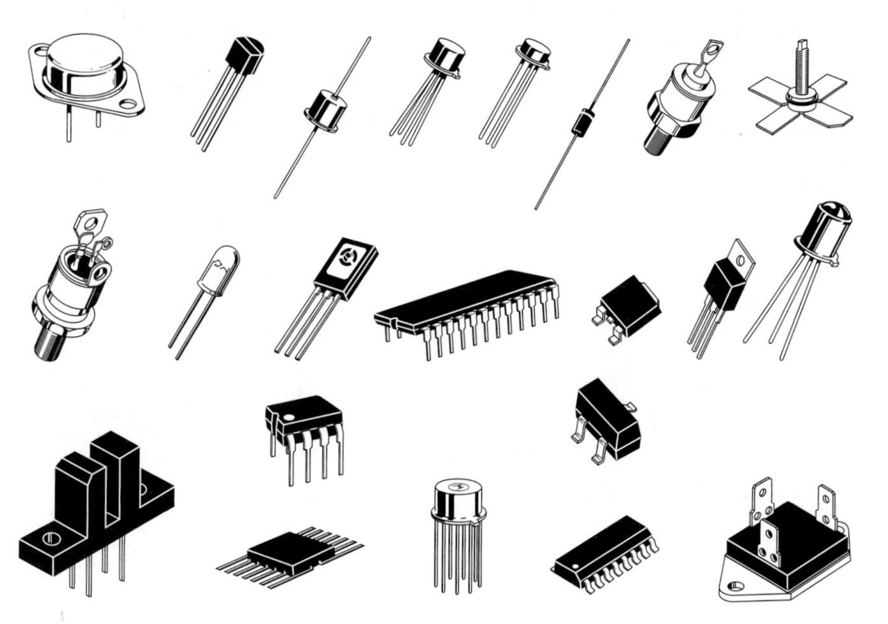

◄ **FIGURE 1–9**

An assortment of semiconductor devices.

Electronic Instruments

Figure 1–10 shows four basic electronic instruments found on the typical laboratory workbench and which will be discussed throughout the book. These instruments include the dc power supply for providing current and voltage to power electronic circuits, the function generator for providing electronic signals, the digital multimeter (DMM) with its voltmeter, ammeter, and ohmmeter functions for measuring voltage, current, and resistance, respectively, and the oscilloscope for observing and measuring ac voltages.

(a)

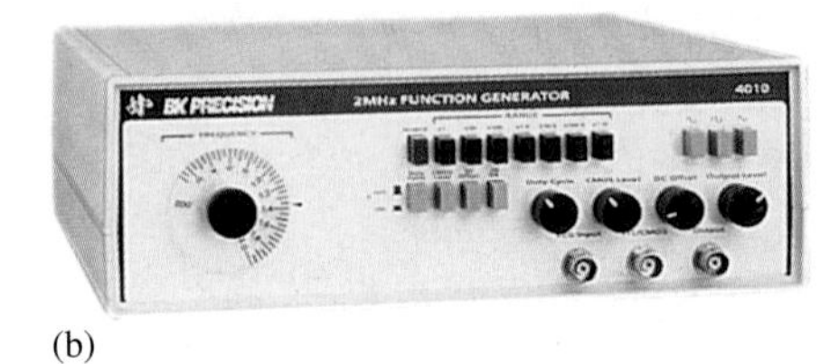

(b)

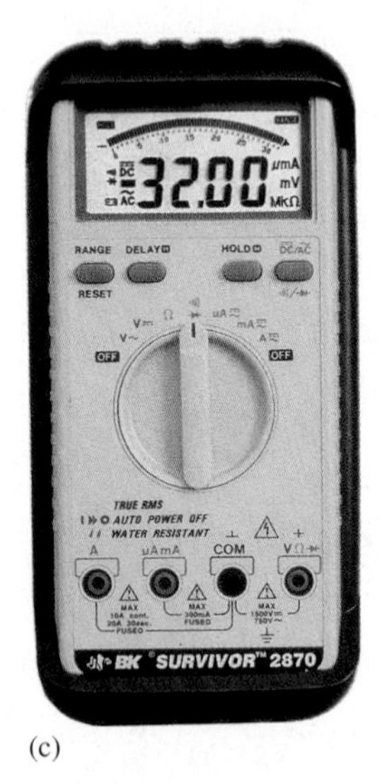

(c)

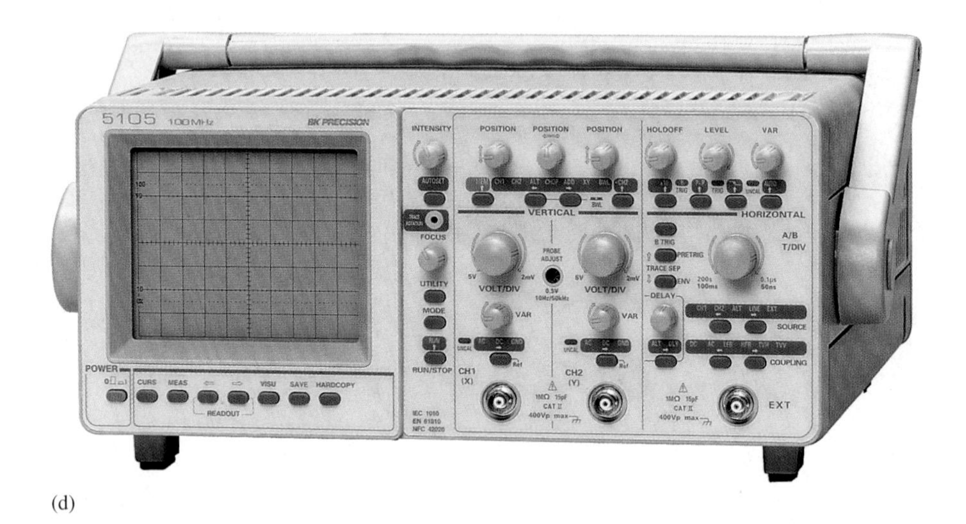

(d)

FIGURE 1–10

Typical instruments. (a) DC power supply; (b) Function generator; (c) Digital multimeter; (d) Digital storage oscilloscope. (Photography courtesy of B&K Precision Corp.)

SECTION 1–1 REVIEW

Answers are at the end of the chapter.

1. Name four types of common electrical components.
2. What instrument is used for measuring electrical current?
3. Name two instruments used for measuring voltage.
4. What instrument is used for measuring resistance?
5. What is a multimeter?

1–2 ELECTRICAL AND MAGNETIC UNITS

In electronics, you must deal with measurable quantities. For example, you must be able to express how many volts are measured at a certain test point in a circuit, how much current there is through a conductor, or how much power a certain amplifier delivers. In this section, you are introduced to the units and symbols for most of the electrical and magnetic quantities that are used throughout the book. Definitions of these and other quantities are presented as they are needed in later chapters.

After completing this section, you should be able to

- **State basic electrical and magnetic quantities and their units**
- Specify the symbol for each quantity
- Specify the symbol for each unit

Letter symbols are used in electronics to represent both quantities and their units. One symbol is used to represent the name of the quantity, and another is used to represent the unit of measurement of that quantity. For example, *P* stands for *power,* and W stands for *watt,* which is the unit of power. Table 1–1 lists the most important electrical quantities, along with their SI units and symbols. These will be used throughout the text. The term *SI* is the French abbreviation for *International System (Système International* in French). Table 1–2 lists the magnetic quantities with their SI units and symbols. These will be used primarily in Chapter 7.

QUANTITY	SYMBOL	UNIT	SYMBOL
capacitance	*C*	farad	F
charge	*Q*	coulomb	C
conductance	*G*	siemens	S
current	*I*	ampere	A
energy	*W*	joule	J
frequency	*f*	hertz	Hz
impedance	*Z*	ohm	Ω
inductance	*L*	henry	H
power	*P*	watt	W
reactance	*X*	ohm	Ω
resistance	*R*	ohm	Ω
voltage	*V*	volt	V

TABLE 1–1

Electrical quantities and their corresponding units with SI symbols.

QUANTITY	SYMBOL	UNIT	SYMBOL
flux density	*B*	tesla	T
magnetic flux	ϕ	weber	Wb
magnetizing force	*H*	ampere-turns/meter	At/m
magnetomotive force	F_m	ampere-turn	At
permeability	μ	webers/ampere-turns-meter	Wb/Atm
reluctance	$\mathcal{R}$	ampere-turns/weber	At/Wb

TABLE 1–2

Magnetic quantities and their corresponding units with SI symbols.

SECTION 1–2 REVIEW

1. What does *SI* stand for?
2. Without referring to Table 1–1, list as many electrical quantities as possible, including their symbols, units, and unit symbols.
3. Without referring to Table 1–2, list as many magnetic quantities as possible, including their symbols, units, and unit symbols.

1–3 SCIENTIFIC NOTATION

In the electrical and electronics fields, you will find both very small and very large quantities. For example, electrical current can range from hundreds of amperes in power applications to a few thousandths or millionths of an ampere in many electronic circuits. For resistive quantities, a wire may have less than one ohm of resistance whereas resistance values of several million ohms are common in circuit applications. This range of values is typical of many other electrical quantities also.

After completing this section, you should be able to

- **Use scientific notation (powers of ten) to express quantities**
- Express any number using a power of ten
- Do calculations with powers of ten

Scientific notation provides a convenient method for expressing large and small numbers and for performing calculations involving such numbers. In scientific notation, a quantity is expressed as a product of a number between 1 and 10 (one digit to the left of the decimal point) and a power of ten. For example, the quantity 150,000 is expressed in scientific notation as 1.5×10^5, and the quantity 0.00022 is expressed as 2.2×10^{-4}.

Powers of Ten

Table 1–3 lists some powers of ten, both positive and negative, and the corresponding decimal numbers. The **power of ten** is expressed as an *exponent* of the *base* 10 in each case.

Base ↘ ↙ Exponent

$$10^x$$

The **exponent** indicates the number of places that the decimal point is moved to the right or left to produce the decimal number. If the power of ten is positive, the decimal point is moved to the right to get the equivalent decimal number. As an example, for an exponent of 4,

$$10^4 = 1 \times 10^4 = 1.0000. = 10{,}000.$$

If the power of ten is negative, the decimal point is moved to the left to get the equivalent decimal number. As an example, for an exponent of -4,

$$10^{-4} = 1 \times 10^{-4} = .0001. = 0.0001$$

The negative exponent does not make a number negative; it simply moves the decimal point to the left.

$10^6 = 1{,}000{,}000$	$10^{-6} = 0.000001$
$10^5 = 100{,}000$	$10^{-5} = 0.00001$
$10^4 = 10{,}000$	$10^{-4} = 0.0001$
$10^3 = 1{,}000$	$10^{-3} = 0.001$
$10^2 = 100$	$10^{-2} = 0.01$
$10^1 = 10$	$10^{-1} = 0.1$
$10^0 = 1$	

TABLE 1–3

Some positive and negative powers of ten.

EXAMPLE 1–1

Express each number in scientific notation:

(a) 200 **(b)** 5000 **(c)** 85,000 **(d)** 3,000,000

Solution In each case, the decimal point is moved an appropriate number of places to the left to determine the positive power of ten.

(a) $200 = \mathbf{2 \times 10^2}$ **(b)** $5000 = \mathbf{5 \times 10^3}$

(c) $85{,}000 = \mathbf{8.5 \times 10^4}$ **(d)** $3{,}000{,}000 = \mathbf{3 \times 10^6}$

*Related Problem** Express 750,000,000 in scientific notation.

*Answers are at the end of the chapter.

EXAMPLE 1–2

Express each number in scientific notation:

(a) 0.2 **(b)** 0.005 **(c)** 0.00063 **(d)** 0.000015

Solution In each case, the decimal point is moved an appropriate number of places to the right to determine the negative power of ten.

(a) $0.2 = \mathbf{2 \times 10^{-1}}$ **(b)** $0.005 = \mathbf{5 \times 10^{-3}}$

(c) $0.00063 = \mathbf{6.3 \times 10^{-4}}$ **(d)** $0.000015 = \mathbf{1.5 \times 10^{-5}}$

Related Problem Express 0.00000093 in scientific notation.

EXAMPLE 1–3

Express each of the following numbers as a normal decimal number:

(a) 1×10^5 **(b)** 2×10^3 **(c)** 3.2×10^{-2} **(d)** 2.5×10^{-6}

Solution The decimal point is moved to the right or left a number of places indicated by the positive or the negative power of ten respectively.

(a) $1 \times 10^5 = \mathbf{100{,}000}$ **(b)** $2 \times 10^3 = \mathbf{2000}$

(c) $3.2 \times 10^{-2} = \mathbf{0.032}$ **(d)** $2.5 \times 10^{-6} = \mathbf{0.0000025}$

Related Problem Express 8.2×10^8 as a normal decimal number.

Calculations Using Powers of Ten

The advantage of scientific notation is in addition, subtraction, multiplication, and division of very small or very large numbers.

Addition The steps for adding numbers in powers of ten are as follows:

1. Express the numbers to be added in the same power of ten.
2. Add the numbers without their powers of ten to get the sum.
3. Bring down the common power of ten, which is the power of ten of the sum.

EXAMPLE 1–4

Add 2×10^6 and 5×10^7 and express the result in scientific notation.

Solution

1. Express both numbers in the same power of ten: $(2 \times 10^6) + (50 \times 10^6)$
2. Add $2 + 50 = 52$.
3. Bring down the common power of ten (10^6) and the sum is $52 \times 10^6 = \mathbf{5.2 \times 10^7}$.

Related Problem Add 4.1×10^3 and 7.9×10^2.

Subtraction The steps for subtracting numbers in powers of ten are as follows:

1. Express the numbers to be subtracted in the same power of ten.
2. Subtract the numbers without their powers of ten to get the difference.
3. Bring down the common power of ten, which is the power of ten of the difference.

EXAMPLE 1–5

Subtract 2.5×10^{-12} from 7.5×10^{-11} and express the result in scientific notation.

Solution

1. Express each number in the same power of ten: $(7.5 \times 10^{-11}) - (0.25 \times 10^{-11})$
2. Subtract $7.5 - 0.25 = 7.25$.
3. Bring down the common power of ten (10^{-11}) and the difference is $\mathbf{7.25 \times 10^{-11}}$.

Related Problem Subtract 3.5×10^{-6} from 2.2×10^{-5}.

Multiplication The steps for multiplying numbers in powers of ten are as follows:

1. Multiply the numbers directly without their powers of ten.
2. Add the powers of ten algebraically (the powers do not have to be the same).

EXAMPLE 1–6

Multiply 5×10^{12} by 3×10^{-6} and express the result in scientific notation.

Solution Multiply the numbers, and algebraically add the powers.

$$(5 \times 10^{12})(3 \times 10^{-6}) = 15 \times 10^{12+(-6)} = 15 \times 10^{6} = \mathbf{1.5 \times 10^{7}}$$

Related Problem Multiply 1.2×10^{3} by 4×10^{2}.

Division The steps for dividing numbers in powers of ten are as follows:

1. Divide the numbers directly without their powers of ten.
2. Subtract the power of ten in the denominator from the power of ten in the numerator (the powers do not have to be the same).

EXAMPLE 1–7

Divide 5.0×10^{8} by 2.5×10^{3} and express the result in scientific notation.

Solution The division problem is written with a numerator and denominator.

$$\frac{5.0 \times 10^{8}}{2.5 \times 10^{3}}$$

Divide the numbers and subtract 3 from 8.

$$\frac{5.0 \times 10^{8}}{2.5 \times 10^{3}} = 2 \times 10^{8-3} = \mathbf{2 \times 10^{5}}$$

Related Problem Divide 8×10^{-6} by 2×10^{-10}.

Scientific Notation on a Calculator

Throughout this textbook, the calculator examples are based on the TI-85/TI-86 calculator. Other calculators are similar but may differ in some functions.

Entering a Number in Scientific Notation There are two ways to enter a number in scientific notation.

1. *Mode screen* Select **Sci** on the Mode screen. When you enter a number, it is automatically converted to scientific notation.
2. *EE key* Enter the number with one digit to the left of the decimal point, press EE, and enter the power of ten. This method requires that the power of ten be determined before entering the number.

EXAMPLE 1–8

Enter the number 23,560 using the Sci mode.

Solution Call up the Mode screen with the following key sequence and use the right arrow key to select **Sci**:

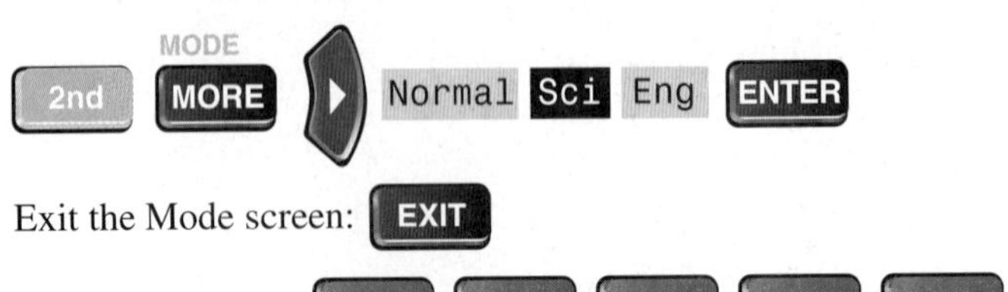

Exit the Mode screen: EXIT

Enter the number: 2 3 5 6 0

Press ENTER

23560
2.356E4

Scientific notation

Related Problem Use the Sci mode to enter the number 150,968 in scientific notation.

EXAMPLE 1–9

Enter 23,560 (same number as in Example 1–8) in scientific notation using the EE key.

Solution Move the decimal point four places to the left so that it comes after the digit 2. This results in the number expressed in scientific notation as

$$2.3560 \times 10^4$$

Enter this number on your calculator as follows:

2 • 3 5 6 0 EE 4 2.3560E4

Related Problem Enter the number 573,946 using the EE key.

SECTION 1–3 REVIEW

1. Scientific notation uses powers of ten. (True or False)
2. Express 100 as a power of ten.
3. Express the following numbers in scientific notation:
   (a) 4350 (b) 12,010 (c) 29,000,000
4. Express the following numbers in scientific notation:
   (a) 0.760 (b) 0.00025 (c) 0.000000597
5. Do the following operations:
   (a) $(1 \times 10^5) + (2 \times 10^5)$ (b) $(3 \times 10^6)(2 \times 10^4)$
   (c) $(8 \times 10^3) \div (4 \times 10^2)$ (d) $(2.5 \times 10^{-6}) - (1.3 \times 10^{-7})$
6. Enter the numbers in Problem 3 into your calculator using the Sci mode.
7. Enter the numbers in Problem 3 into your calculator using the EE key.
8. Repeat the operations in Problem 5 using your calculator.

1–4 ENGINEERING NOTATION AND METRIC PREFIXES

Engineering notation is a specialized form of scientific notation. It is used widely in technical fields to express large and small quantities. In electronics, engineering notation is used to express values of voltage, current, power, resistance, capacitance, inductance, and time, to name a few. Metric prefixes are used in conjunction with engineering notation as a "short hand" for the certain powers of ten that are used.

After completing this section, you should be able to

- **Use engineering notation and metric prefixes to express large and small quantities**
- List the metric prefixes
- Change a power of ten in engineering notation to a metric prefix
- Use metric prefixes to express electrical quantities
- Enter numbers in engineering notation into your calculator
- Convert one metric prefix to another

Engineering Notation

Engineering notation is similar to scientific notation. However, in **engineering notation** a number can have from one to three digits to the left of the decimal point and the power-of-ten exponent must be a multiple of three. For example, the number 33,000 expressed in engineering notation is 33×10^3. In scientific notation, it is expressed as 3.3×10^4. As another example, the number 0.045 is expressed in engineering notation as 45×10^{-3}. In scientific notation, it is expressed as 4.5×10^{-2}.

EXAMPLE 1–10

Express the following numbers in engineering notation:

(a) 82,000 **(b)** 243,000 **(c)** 1,956,000

Solution In engineering notation,

(a) 82,000 is expressed as $\mathbf{82 \times 10^3}$.

(b) 243,000 is expressed as $\mathbf{243 \times 10^3}$.

(c) 1,956,319 is expressed as $\mathbf{1.956 \times 10^6}$.

Related Problem Express 36,000,000,000 in engineering notation.

EXAMPLE 1–11

Convert each of the following numbers to engineering notation:

(a) 0.0022 **(b)** 0.000000047 **(c)** 0.00033

Solution In engineering notation,

(a) 0.0022 is expressed as $\mathbf{2.2 \times 10^{-3}}$.

(b) 0.000000047 is expressed as $\mathbf{47 \times 10^{-9}}$.

(c) 0.00033 is expressed as $\mathbf{330 \times 10^{-6}}$.

Related Problem Express 0.0000000000056 in engineering notation.

Metric Prefixes

Metric prefixes are symbols that represent each of the most commonly used powers of ten in engineering notation. These metric prefixes are listed in Table 1–4 with their designations.

▶ **TABLE 1–4**

Metric prefixes with their symbols and corresponding powers of ten and values.

METRIC PREFIX	SYMBOL	POWER OF TEN	VALUE
pico	p	10^{-12}	one-trillionth
nano	n	10^{-9}	one-billionth
micro	μ	10^{-6}	one-millionth
milli	m	10^{-3}	one-thousandth
kilo	k	10^{3}	one thousand
mega	M	10^{6}	one million
giga	G	10^{9}	one billion
tera	T	10^{12}	one trillion

A metric prefix is used to replace the power of ten in a number that is expressed in engineering notation. Metric prefixes are used only with numbers that have a unit of measure, such as volts, amperes, and ohms, and are placed preceding the unit symbol. For example, 0.025 amperes can be expressed as 25×10^{-3} A. This quantity is expressed using a metric prefix as 25 mA, which is read 25 milliamps. Note that the metric prefix *milli* has replaced 10^{-3}. As another example, 100,000,000 ohms can be expressed as $100 \times 10^{6}\ \Omega$. This quantity is expressed using a metric prefix as 100 MΩ, which is read 100 megohms. The metric prefix *mega* has replaced 10^{6}.

EXAMPLE 1–12

Express each quantity using an appropriate metric prefix:

(a) 50,000 V **(b)** 25,000,000 Ω **(c)** 0.000036 A

Solution **(a)** $50{,}000 \text{ V} = 50 \times 10^3 \text{ V} = \mathbf{50\ kV}$

(b) $25{,}000{,}000\ \Omega = 25 \times 10^6\ \Omega = \mathbf{25\ M\Omega}$

(c) $0.000036 \text{ A} = 36 \times 10^{-6} \text{ A} = \mathbf{36\ \mu A}$

Related Problem Express using appropriate metric prefixes:

(a) 56,000,000 Ω **(b)** 0.000470 A

Engineering Notation on a Calculator

As previously mentioned, the calculator examples in this textbook are based on the TI-85/TI-86 calculator. Other calculators are similar but may differ in some functions.

Entering a Number in Engineering Notation There are two ways to enter a number in engineering notation, similar to the methods for scientific notation.

1. *Mode screen* Select **Eng** on the Mode screen. When you enter a number, it is automatically converted to engineering notation.
2. *EE key* Enter the number with one, two, or three digits to the left of the decimal point, press EE, and enter the power of ten that is a multiple of three. This method requires that the appropriate power of ten be determined before entering the number.

EXAMPLE 1–13

Enter the number 75,200 using the Eng mode.

Solution Call up the Mode screen with the following key sequence and use the right arrow key to select **Eng:**

Exit the Mode screen: EXIT

Enter the number: 7 5 2 0 0

Press ENTER

75200
75.2E3

Engineering notation

Related Problem Use the Eng mode to enter the number 6,481,000.

EXAMPLE 1–14

Enter 51,200,000 in engineering notation using the EE key.

Solution Move the decimal point six places to the left so that it comes after the digit 1. This results in the number expressed in engineering notation as

$$51.2 \times 10^6$$

Enter this number on your calculator as follows:

Related Problem Enter the number 273,900 in engineering notation using the EE key.

SECTION 1–4 REVIEW

1. Express the following numbers in engineering notation:
   (a) 0.0056 (b) 0.0000000283
   (c) 950,000 (d) 375,000,000,000
2. Enter the numbers in Problem 1 into your calculator using engineering notation.
3. List the metric prefix for each of the following powers of ten: 10^6, 10^3, 10^{-3}, 10^{-6}, 10^{-9}, and 10^{-12}
4. Use an appropriate metric prefix to express 0.000001 A.
5. Use an appropriate metric prefix to express 250,000 W.

1–5 METRIC UNIT CONVERSIONS

It is often necessary or convenient to convert a quantity from one metric-prefixed unit to another, such as from milliamperes (mA) to microamperes (μA). A metric prefix conversion is accomplished by moving the decimal point in the number an appropriate number of places to the left or to the right, depending on the particular conversion.

After completing this section, you should be able to

- **Convert from one metric-prefixed unit to another**
- Convert between milli, micro, nano, and pico
- Convert between kilo and mega

The following basic rules apply to metric unit conversions:

1. When converting from a larger unit to a smaller unit, move the decimal point to the right.
2. When converting from a smaller unit to a larger unit, move the decimal point to the left.
3. Determine the number of places that the decimal point is moved by finding the difference in the powers of ten of the units being converted.

For example, when converting from milliamperes (mA) to microamperes (μA), move the decimal point three places to the right because there is a three-place difference between the two units (mA is 10^{-3} A and μA is 10^{-6} A). The following examples illustrate a few conversions.

EXAMPLE 1–15

Convert 0.15 milliampere (0.15 mA) to microamperes (μA).

Solution Move the decimal point three places to the right.

$$0.15 \text{ mA} = 0.15 \times 10^{-3} \text{ A} = 150 \times 10^{-6} \text{ A} = \mathbf{150\ \mu A}$$

Related Problem Convert 1 mA to microamperes.

EXAMPLE 1–16

Convert 4500 microvolts (4500 μV) to millivolts (mV).

Solution Move the decimal point three places to the left.

$$4500\ \mu\text{V} = 4500 \times 10^{-6} \text{ V} = 4.5 \times 10^{-3} \text{ V} = \mathbf{4.5\ mV}$$

Related Problem Convert 1000 μV to millivolts.

EXAMPLE 1–17

Convert 5000 nanoamperes (5000 nA) to microamperes (μA).

Solution Move the decimal point three places to the left.

$$5000 \text{ nA} = 5000 \times 10^{-9} \text{ A} = 5 \times 10^{-6} \text{ A} = \mathbf{5\ \mu A}$$

Related Problem Convert 893 nA to microamperes.

EXAMPLE 1–18

Convert 47,000 picofarads (47,000 pF) to microfarads (μF).

Solution Move the decimal point six places to the left.

$$47{,}000 \text{ pF} = 47{,}000 \times 10^{-12} \text{ F} = 0.047 \times 10^{-6} \text{ F} = \mathbf{0.047\ \mu F}$$

Related Problem Convert 0.0022 μF to picofarads.

EXAMPLE 1–19

Convert 0.00022 microfarad (0.00022 μF) to picofarads (pF).

Solution Move the decimal point six places to the right.

$$0.00022\ \mu\text{F} = 0.00022 \times 10^{-6}\ \text{F} = 220 \times 10^{-12}\ \text{F} = \mathbf{220\ pF}$$

Related Problem Convert 10,000 pF to microfarads.

EXAMPLE 1–20

Convert 1800 kilohms (1800 kΩ) to megohms (MΩ).

Solution Move the decimal point three places to the left.

$$1800\ \text{k}\Omega = 1800 \times 10^{3}\ \Omega = 1.8 \times 10^{6}\ \Omega = \mathbf{1.8\ M\Omega}$$

Related Problem Convert 2.2 kΩ to megohms.

When adding (or subtracting) quantities with different metric prefixes, first convert one of the quantities to the same prefix as the other as the next example shows.

EXAMPLE 1–21

Add 15 mA and 8000 μA and express the result in milliamperes.

Solution Convert 8000 μA to 8 mA and add.

$$15\ \text{mA} + 8000\ \mu\text{A} = 15\ \text{mA} + 8\ \text{mA} = \mathbf{23\ mA}$$

Related Problem Add 2873 mA to 10,000 μA.

SECTION 1–5 REVIEW

1. Convert 0.01 MV to kilovolts (kV).
2. Convert 250,000 pA to milliamperes (mA).
3. Add 0.05 MW and 75 kW and express the result in kW.
4. Add 50 mV and 25,000 μV and express the result in mV.

SUMMARY

- Resistors limit electrical current.
- Capacitors store electrical charge.
- Inductors store energy in their electromagnetic field.
- Inductors are also known as *coils*.

- Transformers magnetically couple ac voltages.
- Semiconductor devices include diodes, transistors, and integrated circuits.
- Power supplies provide current and voltage.
- The voltmeter function on a DMM is used to measure voltage.
- The ammeter function on a DMM is used to measure current.
- The ohmmeter function on a DMM is used to measure resistance.
- A multimeter includes a voltmeter, ammeter, and ohmmeter combined into one instrument.
- Scientific notation is a method for expressing very large and very small numbers as a number between one and ten (one digit to left of decimal point) times a power of ten.
- Engineering notation is a form of scientific notation in which quantities are expressed with one, two, or three digits to the left of the decimal point times a power of ten that is a multiple of three.
- Metric prefixes are symbols used to represent powers of ten that are multiples of three.

SELF-TEST

Answers are at the end of the chapter.

1. Which of the following is not an electrical quantity?
   (a) current **(b)** voltage **(c)** time **(d)** power
2. The unit of current is
   (a) volt **(b)** watt **(c)** ampere **(d)** joule
3. The unit of voltage is
   (a) ohm **(b)** watt **(c)** volt **(d)** farad
4. The unit of resistance is
   (a) ampere **(b)** henry **(c)** hertz **(d)** ohm
5. Hertz is the unit of
   (a) power **(b)** inductance **(c)** frequency **(d)** time
6. The quantity 4.7×10^3 is the same as
   (a) 470 **(b)** 4700 **(c)** 47,000 **(d)** 0.0047
7. The quantity 56×10^{-3} is the same as
   (a) 0.056 **(b)** 0.560 **(c)** 560 **(d)** 56,000
8. The number 3,300,000 can be expressed in engineering notation as
   (a) 3300×10^3 **(b)** 3.3×10^{-6} **(c)** 3.3×10^6 **(d)** either (a) or (c)
9. Ten milliamperes can be expressed as
   (a) 10 MA **(b)** 10 μA **(c)** 10 kA **(d)** 10 mA
10. Five thousand volts can be expressed as
    (a) 5000 V **(b)** 5 MV **(c)** 5 kV **(d)** either (a) or (c)
11. Twenty million ohms can be expressed as
    (a) 20 mΩ **(b)** 20 MW **(c)** 20 MΩ **(d)** 20 $\mu\Omega$
12. 15,000 W is the same as
    (a) 15 mW **(b)** 15 kW **(c)** 15 MW **(d)** 15 μW

PROBLEMS

Answers to odd-numbered problems are at the end of the book.

BASIC PROBLEMS

SECTION 1–3 **Scientific Notation**

1. Express each of the following numbers in scientific notation:
(a) 3000 **(b)** 75,000 **(c)** 2,000,000

2. Express each fractional number in scientific notation:
(a) 1/500 **(b)** 1/2000 **(c)** 1/5,000,000

3. Express each of the following numbers in scientific notation:
(a) 8400 **(b)** 99,000 **(c)** 0.2×10^6

4. Express each of the following numbers in scientific notation:
(a) 0.0002 **(b)** 0.6 **(c)** 7.8×10^{-2}

5. Express each of the following as a regular decimal number:
(a) 2.5×10^{-6} **(b)** 5.0×10^2 **(c)** 3.9×10^{-1}

6. Express each number in regular decimal form:
(a) 4.5×10^{-6} **(b)** 8×10^{-9} **(c)** 4.0×10^{-12}

7. Add the following numbers:
(a) $(9.2 \times 10^6) + (3.4 \times 10^7)$ **(b)** $(5 \times 10^3) + (8.5 \times 10^{-1})$
(c) $(5.6 \times 10^{-8}) + (4.6 \times 10^{-9})$

8. Perform the following subtractions:
(a) $(3.2 \times 10^{12}) - (1.1 \times 10^{12})$ **(b)** $(2.6 \times 10^8) - (1.3 \times 10^7)$
(c) $(1.5 \times 10^{-12}) - (8 \times 10^{-13})$

9. Perform the following multiplications:
(a) $(5 \times 10^3)(4 \times 10^5)$ **(b)** $(1.2 \times 10^{12})(3 \times 10^2)$
(c) $(2.2 \times 10^{-9})(7 \times 10^{-6})$

10. Divide the following:
(a) $(1.0 \times 10^3) \div (2.5 \times 10^2)$ **(b)** $(2.5 \times 10^{-6}) \div (5.0 \times 10^{-8})$
(c) $(4.2 \times 10^8) \div (2 \times 10^{-5})$

SECTION 1–4 **Engineering Notation and Metric Prefixes**

11. Express each of the following numbers in engineering notation:
(a) 89,000 **(b)** 450,000 **(c)** 12,040,000,000,000

12. Express each number in engineering notation:
(a) 2.35×10^5 **(b)** 7.32×10^7 **(c)** 1.333×10^9

13. Express each number in engineering notation:
(a) 0.000345 **(b)** 0.025 **(c)** 0.00000000129

14. Express each number in engineering notation:
(a) 9.81×10^{-3} **(b)** 4.82×10^{-4} **(c)** 4.38×10^{-7}

15. Add the following numbers and express each result in engineering notation:
(a) $2.5 \times 10^{-3} + 4.6 \times 10^{-3}$ **(b)** $68 \times 10^6 + 33 \times 10^6$
(c) $1.25 \times 10^6 + 250 \times 10^3$

16. Multiply the following numbers and express each result in engineering notation:

(a) $(32 \times 10^{-3})(56 \times 10^{3})$ **(b)** $(1.2 \times 10^{-6})(1.2 \times 10^{-6})$

(c) $100(55 \times 10^{-3})$

17. Divide the following numbers and express each result in engineering notation:

(a) $50 \div (2.2 \times 10^{3})$ **(b)** $(5 \times 10^{3}) \div (25 \times 10^{-6})$

(c) $560 \times 10^{3} \div (660 \times 10^{3})$

18. Express each number in Problem 11 in ohms using a metric prefix.

19. Express each number in Problem 13 in amperes using a metric prefix.

20. Express each of the following as a quantity having a metric prefix:

(a) 31×10^{-3} A **(b)** 5.5×10^{3} V **(c)** 20×10^{-12} F

21. Express the following using metric prefixes:

(a) 3×10^{-6} F **(b)** 3.3×10^{6} Ω **(c)** 350×10^{-9} A

22. Express each quantity with a power of ten:

(a) 5 μA **(b)** 43 mV **(c)** 275 kΩ **(d)** 10 MW

SECTION 1–5 **Metric Unit Conversions**

23. Perform the indicated conversions:

(a) 5 mA to microamperes **(b)** 3200 μW to milliwatts

(c) 5000 kV to megavolts **(d)** 10 MW to kilowatts

24. Determine the following:

(a) The number of microamperes in 1 milliampere

(b) The number of millivolts in 0.05 kilovolt

(c) The number of megohms in 0.02 kilohm

(d) The number of kilowatts in 155 milliwatts

25. Add the following quantities:

(a) 50 mA + 680 μA **(b)** 120 kΩ + 2.2 MΩ

(c) 0.02 μF + 3300 pF

26. Do the following operations:

(a) 10 kΩ ÷ (2.2 kΩ + 10 kΩ) **(b)** 250 mV ÷ 50 μV

(c) 1 MW ÷ 2 kW

ANSWERS

SECTION REVIEWS

SECTION 1–1 **Electrical Components and Measuring Instruments**

1. Resistors, capacitors, inductors, and transformers

2. Ammeter function of a DMM (digital multimeter)

3. Voltmeter function of a DMM and oscilloscope

4. Ohmmeter function of a DMM

5. A multimeter is an instrument that measures voltage, current, and resistance.

SECTION 1–2 Electrical and Magnetic Units

1. SI is the abbreviation for Système International.
2. Refer to Table 1–1 after you have compiled your list.
3. Refer to Table 1–2 after you have compiled your list.

SECTION 1–3 Scientific Notation

1. True
2. 10^2
3. **(a)** 4.35×10^3 **(b)** 1.201×10^4 **(c)** 2.9×10^7
4. **(a)** 7.6×10^{-1} **(b)** 2.5×10^{-4} **(c)** 5.97×10^{-7}
5. **(a)** 3×10^5 **(b)** 6×10^{10} **(c)** 2×10^1 **(d)** 2.37×10^{-6}
6. Select Sci mode, enter the digits and press ENTER.
7. Enter the digits, press EE, and enter the power of ten.
8. Use either the Sci mode or the EE key. Enter the first number, press the appropriate arithmetic operation key, enter the second number, and press ENTER.

SECTION 1–4 Engineering Notation and Metric Prefixes

1. **(a)** 5.6×10^{-3} **(b)** 28.3×10^{-9} **(c)** 950×10^3 **(c)** 375×10^9
2. Select Eng mode, enter the digits, and press ENTER; or enter the digits, press EE, and enter the power of ten.
3. Mega (M), kilo (k), milli (m), micro (μ), nano (n), and pico (p)
4. 1 μA (one microampere)
5. 250 kW (250 kilowatts)

SECTION 1–5 Metric Unit Conversions

1. 0.01 MV = 10 kV
2. 250,000 pA = 0.00025 mA
3. 125 kW
4. 75 mV

RELATED PROBLEMS FOR EXAMPLES

1–1 7.5×10^8

1–2 9.3×10^{-7}

1–3 820,000,000

1–4 4.89×10^3

1–5 1.85×10^{-5}

1–6 4.8×10^5

1–7 4×10^4

1–8 Select the Sci mode, enter 150968, press ENTER.

1–9 Enter 5.73946, press EE, enter 5.

1–10 36×10^9

1–11 5.6×10^{-12}

1–12 **(a)** 56 MΩ **(b)** 470 μA

1–13 Select Eng mode, enter 6481000, press ENTER.

1–14 Enter 273.9, press EE, enter 3.

1–15 1000 μA

1–16 1 mV

1–17 0.893 μA

1–18 2200 pF

1–19 0.01 μF

1–20 0.0022 MΩ

1–21 2883 mA

SELF-TEST

1. (c) **2.** (c) **3.** (c) **4.** (d) **5.** (c) **6.** (b) **7.** (a)

8. (c) **9.** (d) **10.** (d) **11.** (c) **12.** (b)

2

VOLTAGE, CURRENT, AND RESISTANCE IN ELECTRIC CIRCUITS

INTRODUCTION

The three basic electrical quantities presented in this chapter are voltage, current, and resistance. No matter what type of electrical or electronic equipment you may work with, these quantities will always be of primary importance.

To help you understand voltage, current, and resistance, the basic structure of the atom is discussed and the concept of charge is introduced. The basic electric circuit is studied, along with techniques for measuring voltage, current, and resistance.

CHAPTER OUTLINE

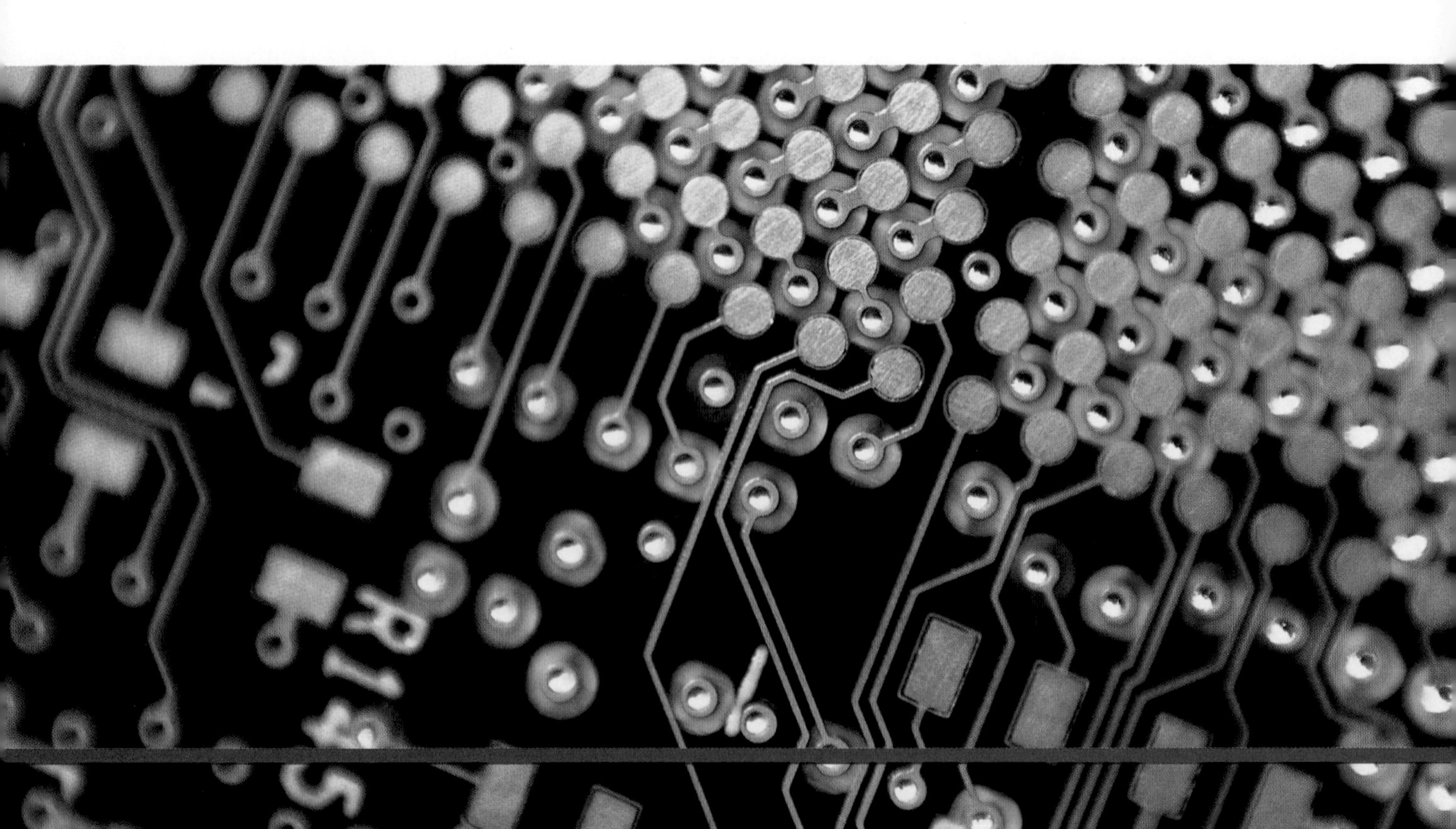

CHAPTER OBJECTIVES

- Describe the basic structure of an atom
- Explain the concept of electrical charge
- Define *voltage* and discuss its characteristics
- Define *current* and discuss its characteristics
- Define *resistance* and discuss its characteristics
- Describe a basic electric circuit
- Make basic circuit measurements
- Recognize electrical hazards and practice proper safety procedures

KEY TERMS

- Atom
- Electron
- Free electron
- Conductor
- Semiconductor
- Insulator
- Charge
- Coulomb
- Voltage
- Volt
- Current
- Ampere
- Resistance
- Ohm
- Conductance
- Siemens
- Resistor
- Potentiometer
- Rheostat
- Circuit
- Load
- Schematic
- Closed circuit
- Open circuit
- Switch
- Fuse
- Circuit breaker
- AWG
- Ground
- Voltmeter
- Ammeter
- Ohmmeter
- Electrical shock

APPLICATION ASSIGNMENT PREVIEW

Imagine that you are a technician for special effects, and you are asked to hook up and check out a circuit for use in the special lighting effects for an outdoor performance. The requirements are

1. There will be six lamps but only one lamp will be turned on at a time.
2. The sequence in which the lamps are turned on and off will vary.
3. It will be necessary to vary the brightness of each lamp.
4. The lamps must operate from a 12 V battery for portability, and the circuit must be protected by a fuse.

After studying this chapter, you should be able to complete the application assignment in the last section of the chapter.

WWW. VISIT THE COMPANION WEBSITE

Circuit Simulation Tutorials and Other Chapter Study Tools Are Available at

http://www.prenhall.com/floyd

2–1 ATOMS

All matter is made of atoms; and all atoms are made of electrons, protons, and neutrons. In this section, you will learn about the structure of the atom, electron orbits and shells, valence electrons, ions, and types of materials used in electronics. Semiconductive material such as silicon or germanium is important because the configuration of certain electrons in an atom is the key factor in determining how a given material conducts electric current.

After completing this section, you should be able to

- **Describe the basic structure of atoms**
- Define *nucleus, proton, neutron,* and *electron*
- Define *atomic number*
- Define *shell*
- Explain what a valence electron is
- Describe ionization
- Explain what a free electron is
- Define *conductor, semiconductor,* and *insulator*

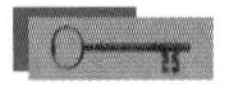

An **atom** is the smallest particle of an **element** that retains the characteristics of that element. Each of the known 109 elements has atoms that are different from the atoms of all other elements. This gives each element a unique atomic structure. According to the classic Bohr model, atoms have a planetary type of structure that consists of a central nucleus surrounded by orbiting electrons, as illustrated in Figure 2–1. The **nucleus** consists of positively charged particles called **protons** and uncharged particles called **neutrons.** The basic particles of negative charge are called **electrons**.

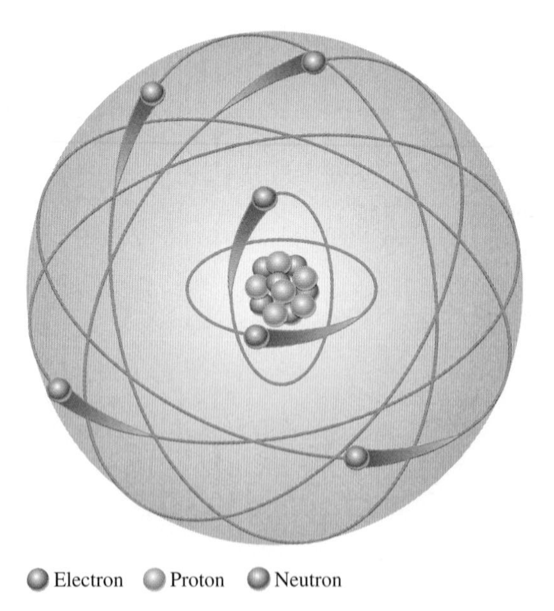

FIGURE 2–1

The Bohr model of an atom showing electrons in orbits around the nucleus. The "tails" on the electrons indicate they are moving.

Each type of atom has a certain number of electrons and protons that distinguishes it from the atoms of all other elements. For example, the simplest atom is that of hydrogen, which has one proton and one electron, as pictured in Figure

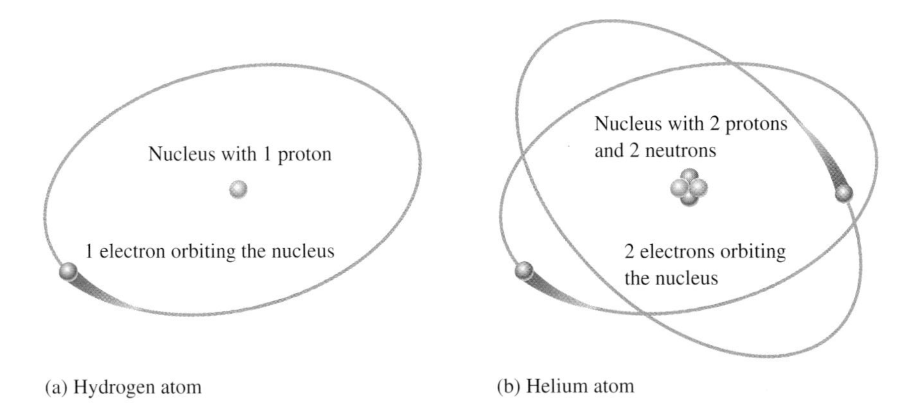

FIGURE 2–2

The two simplest atoms, hydrogen and helium.

2–2(a). As another example, the helium atom, shown in Figure 2–2(b), has two protons and two neutrons in the nucleus and two electrons orbiting the nucleus.

Atomic Number

All elements are arranged in the periodic table of the elements in order according to their **atomic number.** The atomic number equals the number of protons in the nucleus, which is the same as the number of electrons in an electrically balanced (neutral) atom. For example, hydrogen has an atomic number of 1 and helium has an atomic number of 2. In their normal (or neutral) state, all atoms of a given element have the same number of electrons as protons; the positive charges cancel the negative charges, and the atom has a net charge of zero.

Electron Shells and Orbits

Electrons orbit the nucleus of an atom at certain distances from the nucleus. Electrons near the nucleus have less energy than those in more distant orbits. It is known that only discrete (separate and distinct) values of electron energies exist within atomic structures. Therefore, electrons must orbit only at discrete distances from the nucleus.

Energy Levels Each discrete distance (orbit) from the nucleus corresponds to a certain energy level. In an atom, the orbits are grouped into energy bands known as **shells.** A given atom has a fixed number of shells. Each shell has a fixed maximum number of electrons at permissible energy levels (orbits). The differences in energy levels within a shell are much smaller than the difference in energy between shells. The shells are designated 1, 2, 3, and so on, with 1 being closest to the nucleus. This energy band concept is illustrated in Figure 2–3, which shows the 1st shell with one energy level and the 2nd shell with two energy levels. Additional shells may exist in other types of atoms, depending on the element.

The number of electrons in each shell follows a predictable pattern according to the formula, $2N^2$, where N is the number of the shell. The first shell of any atom can have up to 2 electrons, the second shell up to 8 electrons, the third shell up to 18 electrons, and the fourth shell up to 32 electrons.

Valence Electrons

Electrons that are in orbits farther from the nucleus have higher energy and are less tightly bound to the atom than those closer to the nucleus. This is because the force of attraction between the positively charged nucleus and the negatively charged electron decreases with increasing distance from the nucleus. Electrons with the highest energy levels exist in the outermost shell of an atom and are

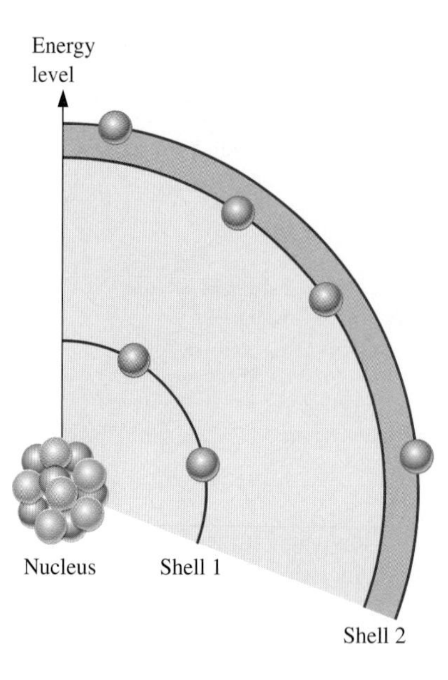

▶ **FIGURE 2–3**

Energy levels increase as the distance from the nucleus increases.

relatively loosely bound to the atom. This outermost shell is known as the **valence** shell and electrons in this shell are called **valence electrons.** These valence electrons contribute to chemical reactions and bonding within the structure of a material and determine its electrical properties.

Ionization

When an atom absorbs energy from a heat source or from light, for example, the energy levels of the electrons are raised. The valence electrons possess more energy and are more loosely bound to the atom than inner electrons, so they can easily jump to higher orbits within the valence shell when external energy is absorbed.

If a valence electron acquires a sufficient amount of energy, it can actually escape from the outer shell and the atom's influence. The departure of a valence electron leaves a previously neutral atom with an excess of positive charge (more protons than electrons). The process of losing a valence electron is known as **ionization,** and the resulting positively charged atom is called a *positive ion.* For example, the chemical symbol for hydrogen is H. When a neutral hydrogen atom loses its valence electron and becomes a positive ion, it is designated H^+. The escaped valence electron is called a free electron. When a free electron loses energy and falls into the outer shell of a neutral hydrogen atom, the atom becomes negatively charged (more electrons than protons) and is called a *negative ion,* designated H^-.

The Copper Atom

Because copper is the most commonly used metal in **electrical** applications, let's examine its atomic structure. The copper atom has 29 electrons that orbit the nucleus in four shells, as shown in Figure 2–4. Notice that the fourth or outermost shell, the valence shell, has only 1 valence electron. When the valence electron in the outer shell of the copper atom gains sufficient thermal energy, it can break away from the parent atom and become a free electron. In copper at room temperature, a "sea" of these free electrons is present. These electrons are not bound to a given atom, but are free to move in the copper material. Free electrons make copper an excellent conductor and make electrical current possible.

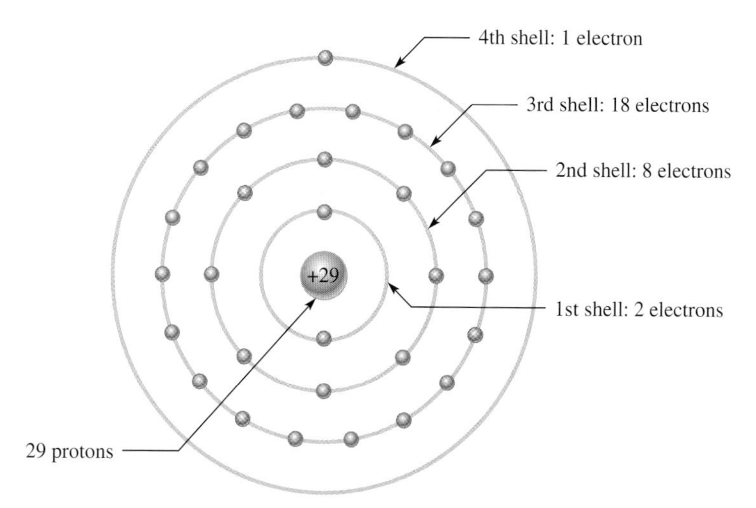

FIGURE 2–4

The copper atom.

Categories of Materials

Three categories of materials are used in electronics: conductors, semiconductors, and insulators.

Conductors Conductors are materials that readily allow current. They have a large number of free electrons and are characterized by one to three valence electrons in their structure. Most metals are good conductors. Silver is the best conductor, and copper is next. Copper is the most widely used conductive material because it is less expensive than silver. Copper wire is commonly used as a conductor in electric circuits.

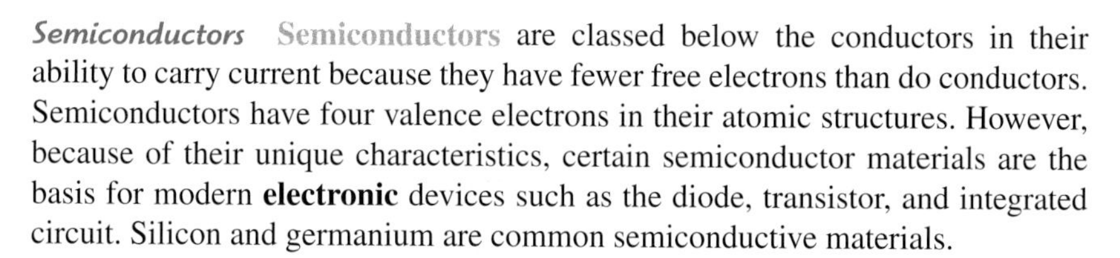

Semiconductors Semiconductors are classed below the conductors in their ability to carry current because they have fewer free electrons than do conductors. Semiconductors have four valence electrons in their atomic structures. However, because of their unique characteristics, certain semiconductor materials are the basis for modern **electronic** devices such as the diode, transistor, and integrated circuit. Silicon and germanium are common semiconductive materials.

Insulators Insulating materials are poor conductors of electric current. In fact, insulators are used to prevent current where it is not wanted. Compared to conductive materials, insulators have very few free electrons and are characterized by more than four valence electrons in their atomic structures.

SECTION 2–1 REVIEW

Answers are at the end of the chapter.

1. What is the basic particle of negative charge?
2. Define *atom.*
3. What does a typical atom consist of?
4. Define *atomic number.*
5. Do all elements have the same types of atoms?
6. What is a free electron?
7. What is a shell in the atomic structure?
8. Name two conductive materials.

2–2 ELECTRICAL CHARGE

As you learned in the last section, the two types of **charge** are the positive charge and the negative charge. The electron is the smallest particle that exhibits negative electrical charge. When an excess of electrons exists in a material, there is a net negative electrical charge. When a deficiency of electrons exists, there is a net positive electrical charge.

After completing this section, you should be able to

- **Explain the concept of electrical charge**
- Name the unit of charge
- Name the types of charge
- Describe the forces between charges
- Determine the amount of charge on a given number of electrons

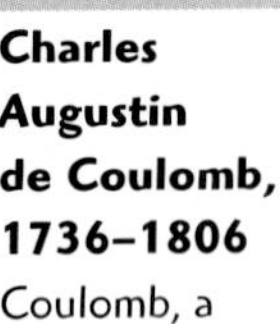

BIOGRAPHY

Charles Augustin de Coulomb, 1736–1806 Coulomb, a Frenchman, spent many years as a military engineer. When bad health forced him to retire, he devoted his time to scientific research. He is best known for his work on electricity and magnetism due to his development of the inverse square law for the force between two charges. The unit of electrical charge is named in his honor. (Photo credit: Courtesy of the Smithsonian Institution. Photo number 52,597.)

The charge of an electron and that of a proton are equal in magnitude. Electrical charge, which is a fundamental characteristic of electrons and protons, is symbolized by Q. Static electricity is the presence of a net positive or negative charge in a material. Everyone has experienced the effects of static electricity from time to time, for example, when attempting to touch a metal surface or another person or when the clothes in a dryer cling together.

Materials with charges of opposite polarity are attracted to each other; materials with charges of the same polarity are repelled, as indicated symbolically in Figure 2–5. A force acts between charges, as evidenced by the attraction or repulsion. This force, called an *electric field,* consists of invisible lines of force as shown in Figure 2–6.

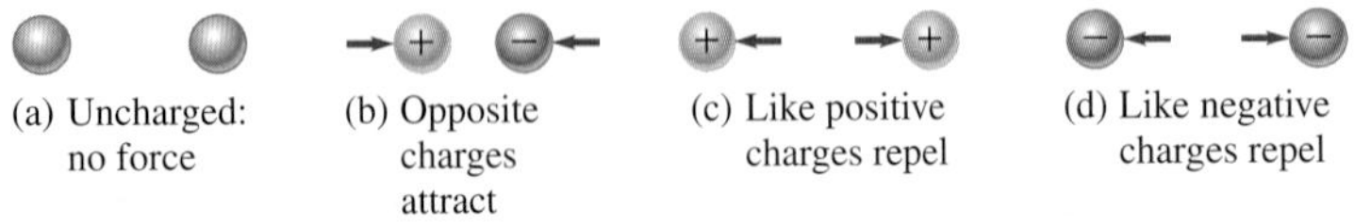

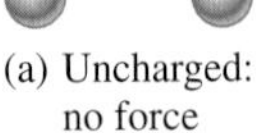

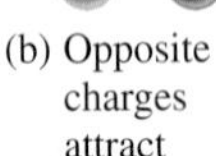

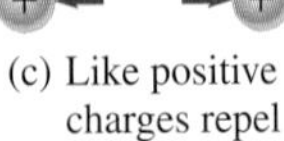

▲ **FIGURE 2–5**

Attraction and repulsion of electrical charges.

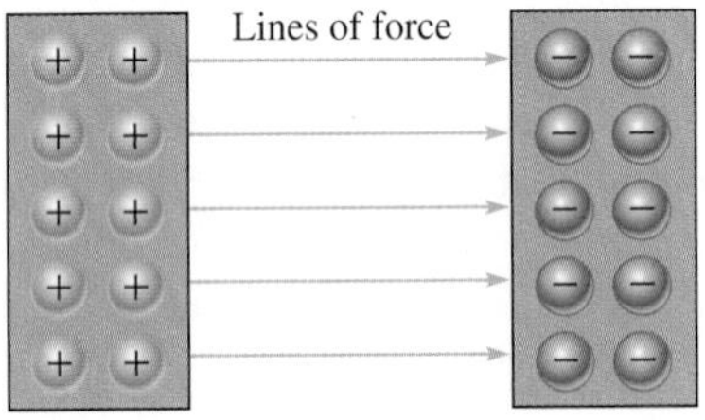

Lines of force

▲ **FIGURE 2–6**

Electric field between oppositely charged surfaces.

Coulomb: The Unit of Charge

Electrical charge is measured in **coulombs**, abbreviated C.

One coulomb is the total charge possessed by 6.25×10^{18} electrons.

A single electron has a charge of 1.6×10^{-19} C. The total charge Q, expressed in coulombs, for a given number of electrons is found by the following formula:

Equation 2–1

$$Q = \frac{\text{number of electrons}}{6.25 \times 10^{18} \text{ electrons/coulomb}}$$

Positive and Negative Charge

Consider a neutral atom—that is, one that has the same number of electrons and protons and thus has no net charge. If a valence electron is pulled away from the atom by the application of energy, the atom is left with a net positive charge (more protons than electrons) and becomes a positive ion. If an atom acquires an extra electron in its outer shell, it has a net negative charge and becomes a negative ion.

The amount of energy required to free a valence electron is related to the number of electrons in the outer shell. An atom can have up to eight valence electrons. The more complete the outer shell, the more stable the atom and thus the more energy is required to release an electron. Figure 2–7 illustrates the creation of a positive and a negative ion when a hydrogen atom gives up its single valence electron to a chloride atom, forming gaseous hydrogen chloride (HCl). When the gaseous HCl is dissolved in the water, hydrochloric acid is formed.

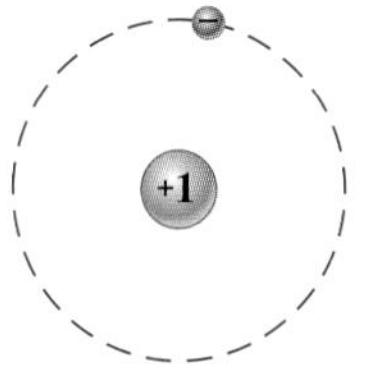

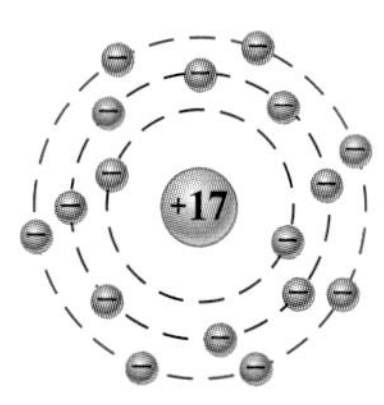

Hydrogen atom (1 proton, 1 electron)

Chloride atom (17 protons, 17 electrons)

(a) The neutral hydrogen atom has a single valence electron.

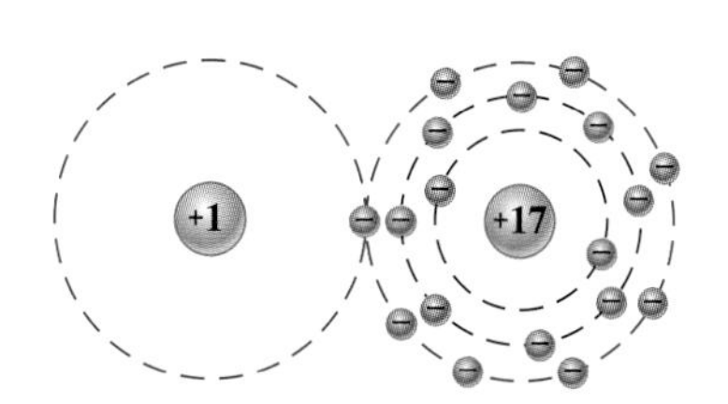

(b) The atoms combine by sharing the valence electron to form gaseous hydrogen chloride (HCl).

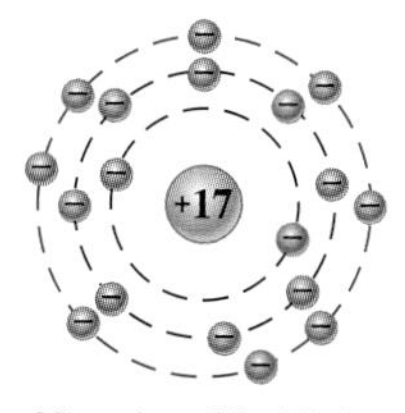

Positive hydrogen ion (1 proton, no electrons)

Negative chloride ion (17 protons, 18 electrons)

(c) When dissolved in water, hydrogen chloride gas separates into positive hydrogen ions and negative chloride ions. The chloride atom retains the electron given up by the hydrogen atom, forming both positive and negative ions in the same solution.

FIGURE 2–7

Example of the formation of positive and negative ions.

EXAMPLE 2–1

How many coulombs of charge do 93.8×10^{16} electrons represent?

Solution

$$Q = \frac{\text{number of electrons}}{6.25 \times 10^{18}\ \text{electrons/C}} = \frac{93.8 \times 10^{16}\ \text{electrons}}{6.25 \times 10^{18}\ \text{electrons/C}} = 15 \times 10^{-2}\ \text{C} = \mathbf{0.15\ C}$$

Related Problem* How many electrons does it take to have 3 C of charge?

* Answers are at the end of the chapter.

SECTION 2–2 REVIEW

1. What is the symbol used for charge?
2. What is the unit of charge, and what is the unit symbol?
3. What are the two types of charge?
4. How much charge, in coulombs, is there in 10×10^{12} electrons?

2–3 VOLTAGE

As you have seen, a force of attraction exists between a positive and a negative charge. A certain amount of energy must be exerted in the form of work to overcome the force and move the charges a given distance apart. All opposite charges possess a certain potential energy because of the separation between them. The difference in potential energy of the charges is the potential difference or voltage. Voltage is the driving force in electric circuits and is what establishes current.

After completing this section, you should be able to

- **Define *voltage* and discuss its characteristics**
- State the formula for voltage
- Name and define the unit of voltage
- Describe the basic sources of voltage

Voltage (V) is expressed as energy (W) divided by charge (Q).

Equation 2–2

$$V = \frac{W}{Q}$$

where W is expressed in **joules** (J) and Q is in coulombs (C).

As a simple analogy, you can think of voltage as corresponding to the pressure difference created by a pump that causes the water to flow through the pipe in a water system.

Volt: The Unit of Voltage

The unit of voltage is the volt, symbolized by V.

One volt is the potential difference (voltage) between two points when one joule of energy is used to move one coulomb of charge from one point to the other.

EXAMPLE 2–2

If 50 J of energy are available for every 10 C of charge, what is the voltage?

Solution

$$V = \frac{W}{Q} = \frac{50 \text{ J}}{10 \text{ C}} = \mathbf{5\ V}$$

Related Problem How much energy is used to move 50 C from one point in a circuit to another when the voltage between the two points is 12 V?

Sources of Voltage

Batteries A voltage source is a source of electrical potential energy or *electromotive force,* more commonly known as voltage. A **battery** is a type of voltage source that converts chemical energy into electrical energy. A battery consists of one or more electrochemical cells that are electrically connected. A cell consists of four basic components: a positive electrode, a negative electrode, electrolyte, and a separator. The *positive electrode* has a deficiency of electrons due to chemical reaction, the *negative electrode* has a surplus of electrons due to chemical reaction, the *electrolyte* provides a mechanism for charge flow between positive and negative electrodes, and the *separator* electrically isolates the positive and negative electrodes. A basic diagram of a battery cell is shown in Figure 2–8.

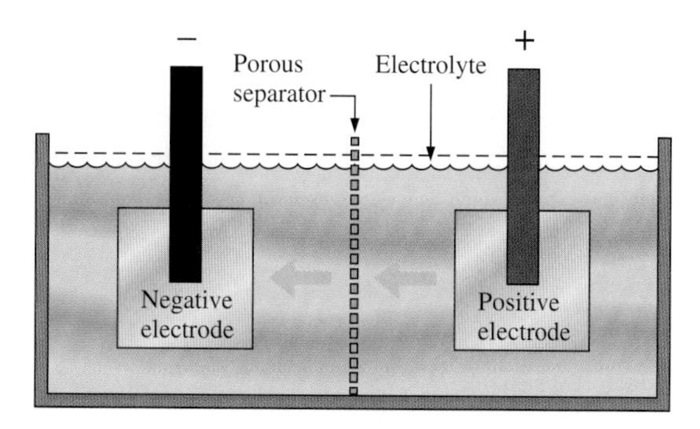

FIGURE 2–8

Diagram of a battery cell.

The voltage of a battery cell is determined by the materials used in it. The chemical reaction at each of the electrodes produces a fixed potential at each electrode. For example, in a lead-acid cell, a potential of −1.685 V is produced at the positive electrode and a potential of +0.365 V is produced at the negative electrode. This means that the voltage between two electrodes of a cell is 2.05 V, which is the standard lead-acid electrode potential. Factors such as acid concentration will affect this value to some degree so that the typical voltage of a commercial lead-acid cell is 2.15 V. The voltage of any battery cell depends on the cell chemistry. Nickel-cadmium cells are about 1.2 V and lithium cells can be as high as almost 4 V.

Although the voltage of a battery cell is fixed by its chemistry, the capacity is variable and depends on the quantity of materials in the cell. Essentially, the *capacity* of a cell is the number of electrons that can be obtained from it and is measured by the amount of current (defined in Section 2–4) that can be supplied over time.

Batteries normally consist of multiple cells that are electrically connected together internally. The way that the cells are connected and the type of cells determine the voltage and capacity of the battery. If the positive electrode of one cell is connected to the negative electrode of the next and so on, as illustrated in Figure 2–9(a) the battery voltage is the sum of the individual cell voltages. This is called a series connection. To increase battery capacity, the positive electrodes of several cells are connected together and all the negative electrodes are connected together, as illustrated in Figure 2–9(b). This is called a parallel connection. Also, by using larger cells, which have a greater quantity of material, the ability to supply current can be increased but the voltage is not affected.

Batteries are divided into two major classes, primary and secondary. Primary batteries are used once and discarded because their chemical reactions are irreversible. Secondary batteries can be recharged and reused many times because

BIOGRAPHY

Alessandro Volta 1745–1827

Volta, an Italian, invented a device to generate static electricity and he discovered methane gas. Volta investigated reactions between dissimilar metals and developed the first battery in 1800. Electrical potential, more commonly known as voltage, is named in his honor as is the unit of voltage, the volt. (Photo credit: AIP Emilio Segrè Visual Archives, Lande Collection.)

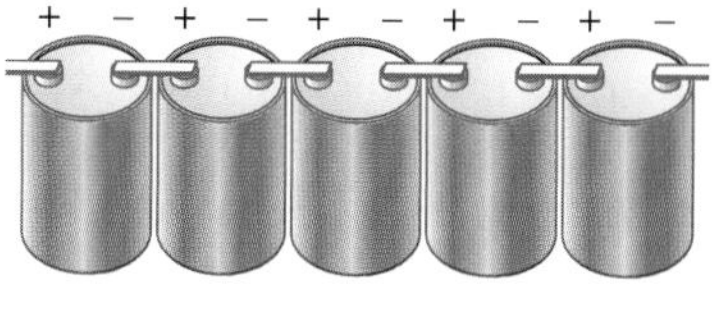

(a) Series-connected battery

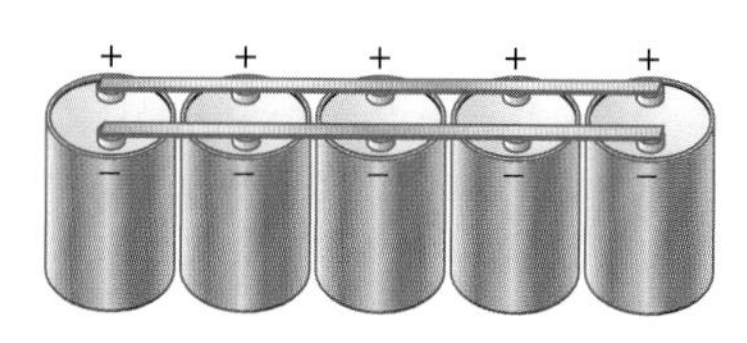

(b) Parallel-connected battery

FIGURE 2–9

Cells connected to form batteries.

they are characterized by reversible chemical reactions. A further discussion of batteries is given in Appendix B.

Solar Cells The operation of solar cells is based on the photovoltaic effect, which is the process whereby light energy is converted directly into electrical energy. A basic solar cell consists of two layers of different types of semiconductive materials joined together to form a junction. When one layer is exposed to light, many electrons acquire enough energy to break away from their parent atoms and cross the junction. This process forms negative ions on one side of the junction and positive ions on the other, and thus a potential difference (voltage) is developed. Figure 2–10 shows the construction of a basic solar cell.

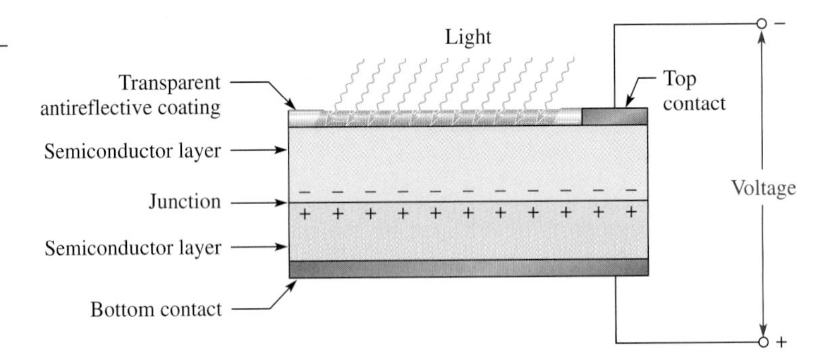

FIGURE 2–10

Construction of a basic solar cell.

Generator Electrical **generators** convert mechanical energy into electrical energy using a principle called *electromagnetic induction* (see Chapter 7). A conductor is rotated through a magnetic field, and a voltage is produced across the conductor. A typical generator is pictured in Figure 2–11.

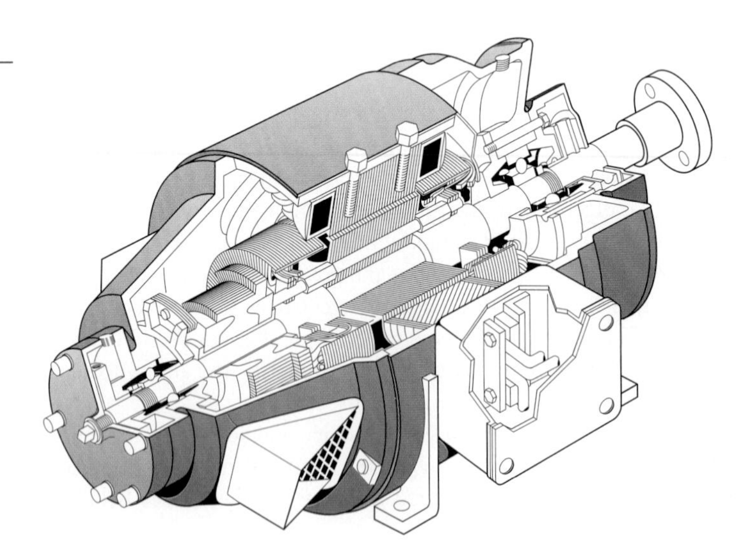

FIGURE 2–11

Cutaway view of a dc generator.

The Electronic Power Supply Electronic power supplies do not produce electrical energy from some other form of energy. They simply convert the ac voltage from the wall outlet to a constant (dc) voltage that is available across two terminals, as indicated in Figure 2–12(a). Typical commercial power supplies are shown in Figure 2–12(b).

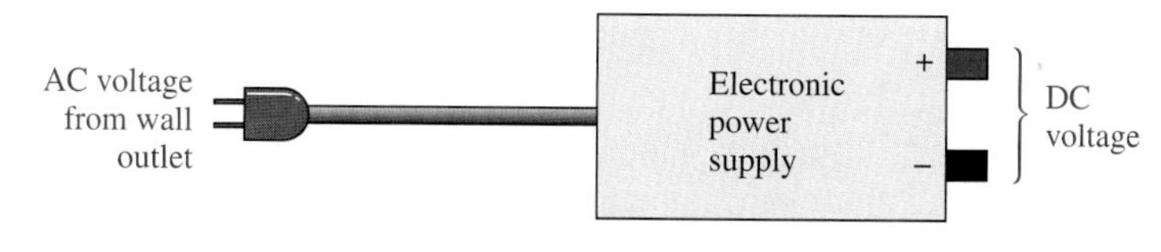

(a)

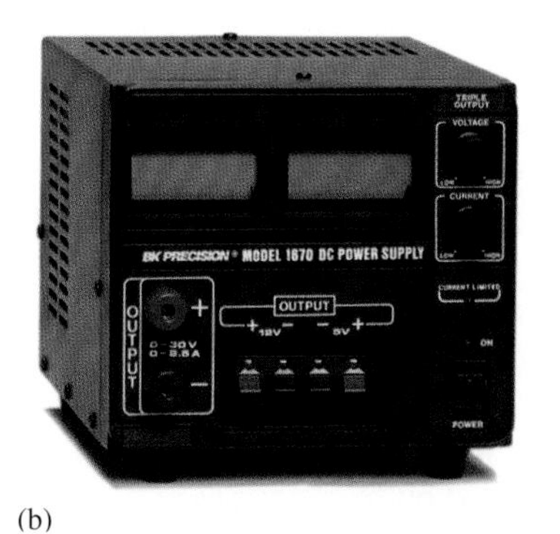

(b)

◀ **FIGURE 2–12**

Electronic power supplies. (Photography courtesy of B&K Precision Corp.)

SECTION 2–3 REVIEW

1. Define *voltage.*
2. What is the unit of voltage?
3. What is the voltage when there are 24 J of energy for 10 C of charge?
4. List four sources of voltage.

2–4 CURRENT

Voltage provides energy to electrons that allows them to move through a circuit. This movement of electrons is the current, which results in work being done in an electric circuit.

After completing this section, you should be able to

- **Define *current* and discuss its characteristics**
- Explain the movement of electrons
- State the formula for current
- Name and define the unit of current

As you have learned, free electrons are available in all conductive and semiconductive materials. These electrons drift randomly in all directions, from atom to atom, within the structure of the material, as indicated in Figure 2–13.

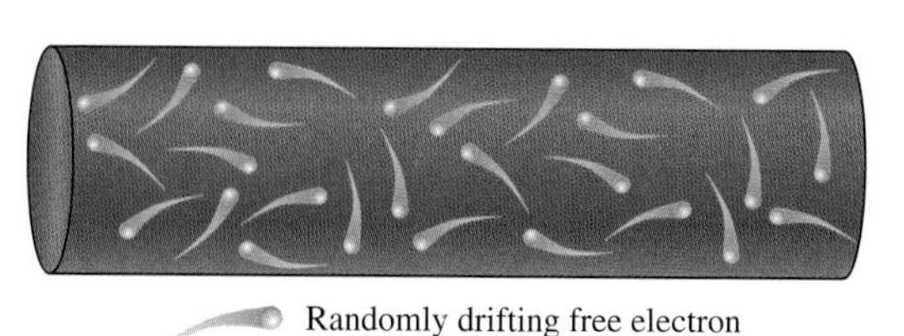

◀ **FIGURE 2–13**

Random motion of free electrons in a material.

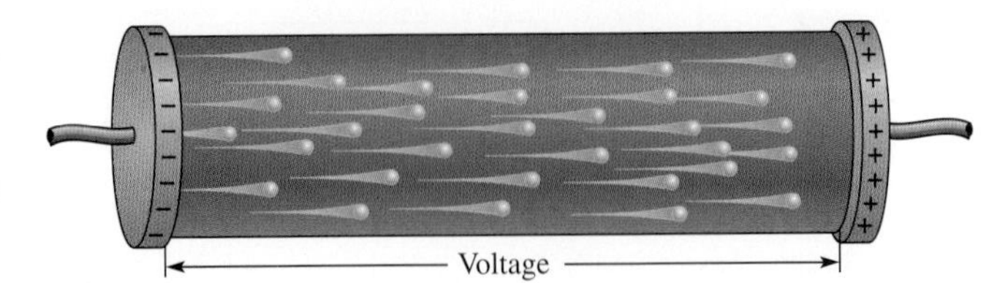

FIGURE 2–14

Electrons flow from negative to positive when a voltage is applied across a conductive or semiconductive material.

Now, if a voltage is placed across the conductive or semiconductive material, one end becomes positive and the other negative, as indicated in Figure 2–14. The repulsive force produced by the negative voltage at the left end causes the free electrons (negative charges) to move toward the right. The attractive force produced by the positive voltage at the right end pulls the free electrons to the right. The result is a net movement of the free electrons from the negative end of the material to the positive end, as shown in Figure 2–14.

The movement of the free electrons from the negative end of the material to the positive end is the electrical current, symbolized by I.

Electrical current is the rate of flow of charge.

Current in a conductive material is measured by the number of electrons (amount of charge, Q) that flow past a point in a unit of time.

Equation 2–3

$$I = \frac{Q}{t}$$

where I is current in amperes, Q is the charge of the electrons in coulombs, and t is the time in seconds.

As a simple analogy, you can think of current as corresponding to the water flowing through the pipe in a water system when pressure (corresponding to voltage) is applied by the pump (corresponding to a voltage source). *Voltage causes current.*

Ampere: The Unit of Current

Current is measured in a unit called the ampere or *amp* for short, symbolized by A.

One ampere (1 A) is the amount of current that exists when a number of electrons having a total charge of one coulomb (1 C) move through a given cross-sectional area in one second (1 s).

See Figure 2–15. Remember, one coulomb is the charge carried by 6.25×10^{18} electrons.

BIOGRAPHY

André Marie Ampère 1775–1836 In 1820 Ampère, a Frenchman, developed a theory of electricity and magnetism that was fundamental for 19th century developments in the field. He was the first to build an instrument to measure charge flow (current). The unit of electrical current is named in his honor. (Photo credit: AIP Emilio Segrè Visual Archives.)

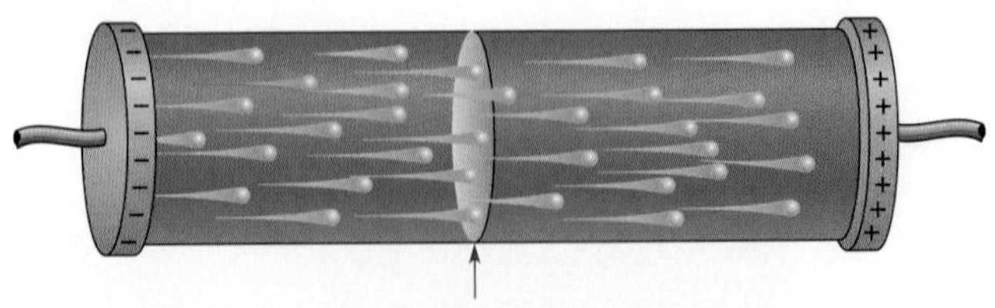

FIGURE 2–15

Illustration of 1 A of current (1 C/s) in a material.

EXAMPLE 2–3

Ten coulombs of charge flow past a given point in a wire in 2 s. What is the current in amperes?

Solution

$$I = \frac{Q}{t} = \frac{10\text{ C}}{2\text{ s}} = \mathbf{5\ A}$$

Related Problem If there are 8 A of direct current through the filament of a light bulb, how many coulombs have moved through the filament in 1.5 s?

SECTION 2–4 REVIEW

1. Define *current* and state its unit.
2. How many electrons make up one coulomb of charge?
3. What is the current in amperes when 20 C flow past a point in a wire in 4 s?

2–5 RESISTANCE

When there is current in a material, the free electrons move through the material and occasionally collide with atoms. These collisions cause the electrons to lose some of their energy, and thus their movement is restricted. The more collisions, the more the flow of electrons is restricted. This restriction varies and is determined by the type of material. The property of a material that restricts the flow of electrons is called resistance, designated with an R.

After completing this section, you should be able to

- **Define *resistance* and discuss its characteristics**
- Name and define the unit of resistance
- Describe the basic types of resistors
- Determine resistance value by color code or labeling

Resistance is the opposition to current.

The schematic symbol for resistance is shown in Figure 2–16.

When there is current through any material that has resistance, heat is produced by the collisions of electrons and atoms. Therefore, wire, which typically has a very small resistance, can become warm or even hot when there is sufficient current through it.

As a simple analogy, you can think of a resistor as corresponding to a partially open valve in a closed water system that restricts the amount of water flowing through the pipe. If the valve is opened more (corresponding to less resistance), the water flow (corresponding to current) increases. If the valve is closed a little (corresponding to more resistance), the water flow (corresponding to current) decreases.

▲ **FIGURE 2–16**

Resistance/resistor symbol.

Ohm: The Unit of Resistance

Resistance, R, is expressed in the unit of ohms and is symbolized by the Greek letter omega (Ω).

One ohm (1 Ω) of resistance exists when there is one ampere (1 A) of current in a material with one volt (1 V) applied across the material.

Conductance The reciprocal of resistance is **conductance**, symbolized by G. It is a measure of the ease with which current is established. The formula is

Equation 2–4

$$G = \frac{1}{R}$$

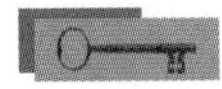

The unit of conductance is the **siemens**, abbreviated S. For example, the conductance of a 22 kΩ resistor is $G = 1/22\text{ k}\Omega = 45.5\ \mu\text{S}$. Occasionally, the obsolete unit of mho is still used for conductance.

Resistors

Components that are specifically designed to have a certain amount of resistance are called **resistors**. The principal applications of resistors are to limit current, divide voltage, and, in certain cases, generate heat. Although there are a variety of different types of resistors that come in many shapes and sizes, they can all be placed in one of two main categories: fixed or variable.

BIOGRAPHY

Georg Simon Ohm 1787–1854 Ohm was born in Bavaria and struggled for years to gain recognition for his work in formulating the relationship of current, voltage, and resistance. This mathematical relationship is known today as Ohm's law and the unit of resistance is named in his honor. (Photo credit: Library of Congress, LC-USZ62-40943.)

Fixed Resistors Fixed resistors are available with a large selection of resistance values that are set during manufacturing and cannot be changed easily. Fixed resistors are constructed using various methods and materials. Figure 2–17 shows several common types.

One common fixed resistor is the carbon-composition type, which is made with a mixture of finely ground carbon, insulating filler, and a resin binder. The ratio of carbon to insulating filler sets the resistance value. The mixture is formed into rods which are cut into short lengths, and lead connections are made. The entire resistor is then encapsulated in an insulated coating for protection.

The chip resistor is another type of fixed resistor and is in the category of SMT (surface-mount technology) components. It has the advantage of a very small size for compact assemblies. Figure 2–18(a) shows the construction of a typical carbon-composition resistor and Figure 2–18(b) shows the construction of a chip resistor.

Other types of fixed resistors include carbon film, metal film, metal-oxide film, and wirewound. In film resistors, a resistive material is deposited evenly

FIGURE 2–17

Typical fixed resistors.

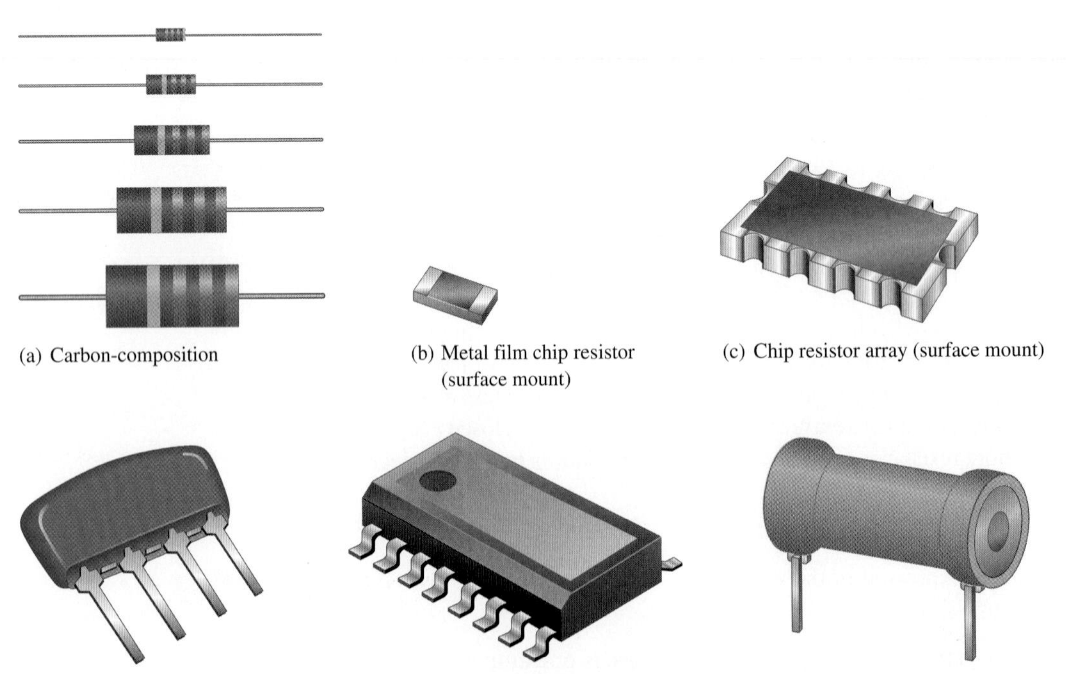

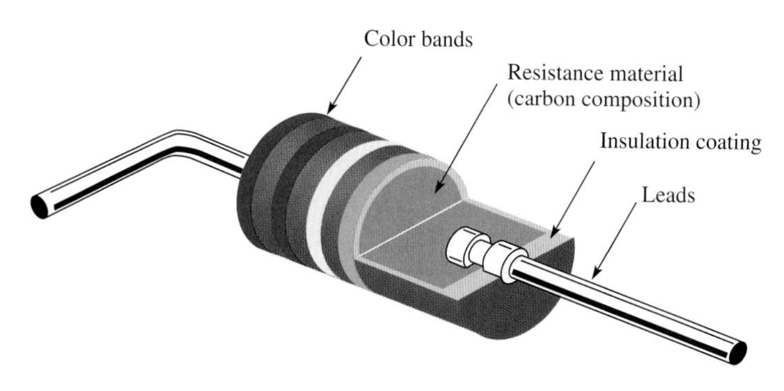

(a) Cutaway view of a carbon-composition resistor

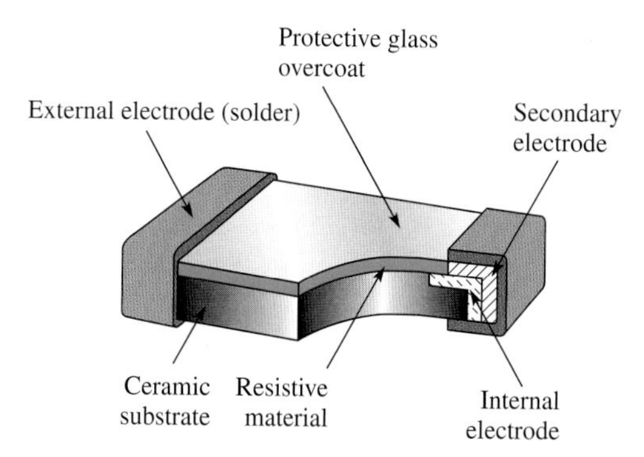

(b) Cutaway view of a tiny surface-mount chip resistor

▲ **FIGURE 2–18**

Two types of fixed resistors (not to scale).

onto a high-grade ceramic rod. The resistive film may be carbon (carbon film) or nickel chromium (metal film). In these types of resistors, the desired resistance value is obtained by removing part of the resistive material in a helical pattern along the rod using a spiraling technique, as shown in Figure 2–19(a). Very close **tolerance** can be achieved with this method. Film resistors are also available in the form of resistor networks, as shown in Figure 2–19(b).

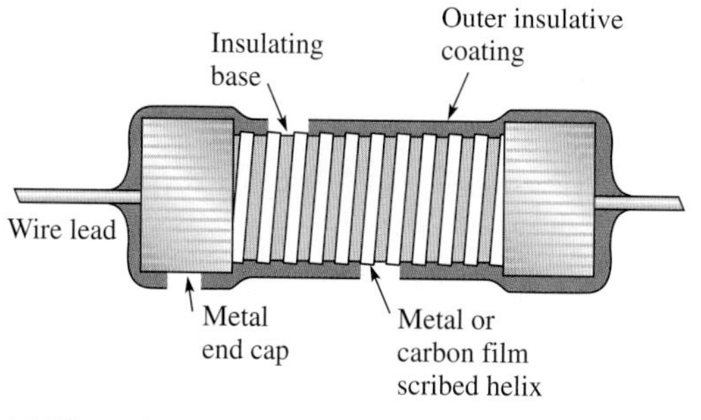

(a) Film resistor showing spiraling technique

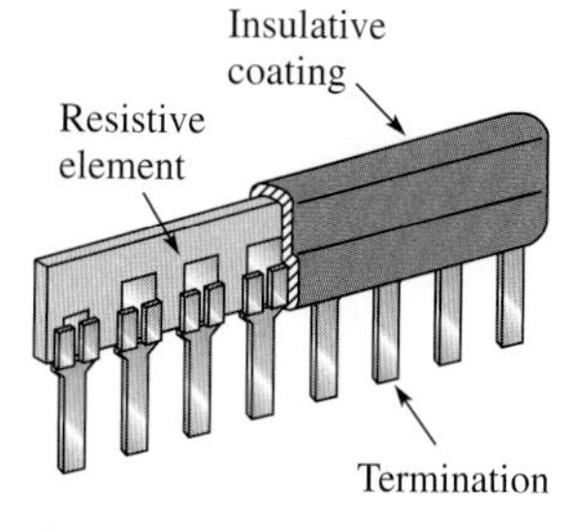

(b) Resistor network

▲ **FIGURE 2–19**

Construction views of typical film resistors.

Wirewound resistors are constructed with resistive wire wound around an insulating rod and then sealed. Normally, wirewound resistors are used because or their relatively high power ratings. Some typical wirewound resistors are shown in Figure 2–20.

BIOGRAPHY

Ernst Werner von Siemens 1816–1892 Siemens was born in Prussia. While in prison for acting as a second in a duel, he began experiments with chemistry which eventually led to his invention of the first electroplating system. Later, he began making improvements in the early telegraph and contributed greatly to the developments in that field. His brother, an engineer, founded the company which today bears the family name. The unit of conductance is named in their honor. (Photo credit: AIP Emilio Segrè Visual Archives, E. Scott Barr Collection.)

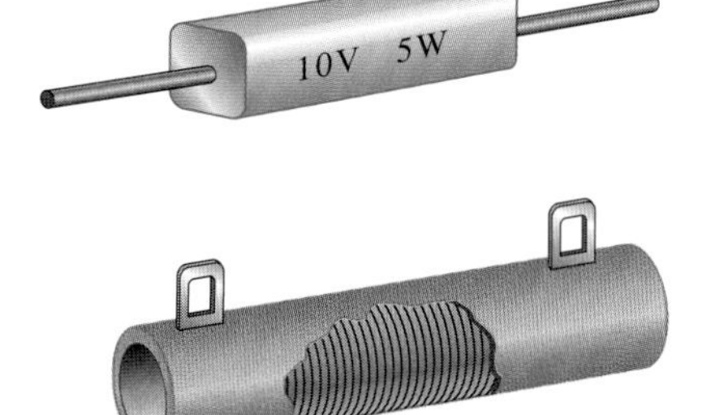

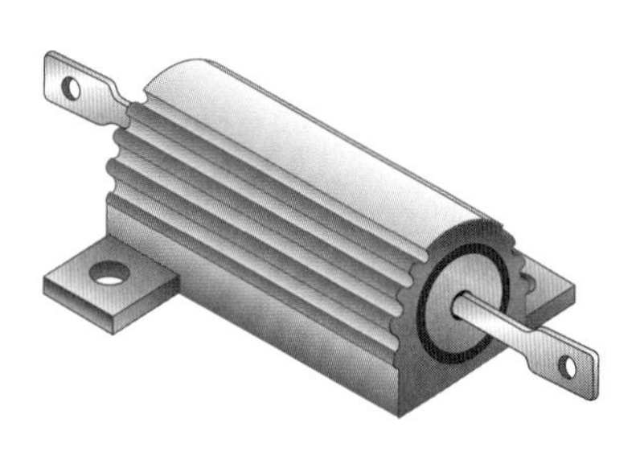

◀ **FIGURE 2–20**

Typical wirewound power resistors.

Resistor Color Codes Some types of fixed resistors with value tolerances of 5% or 10% are color coded with four bands to indicate the resistance value and the tolerance. This color-code band system is shown in Figure 2–21, and the **color code** is listed in Table 2–1. The bands are always located closer to one end.

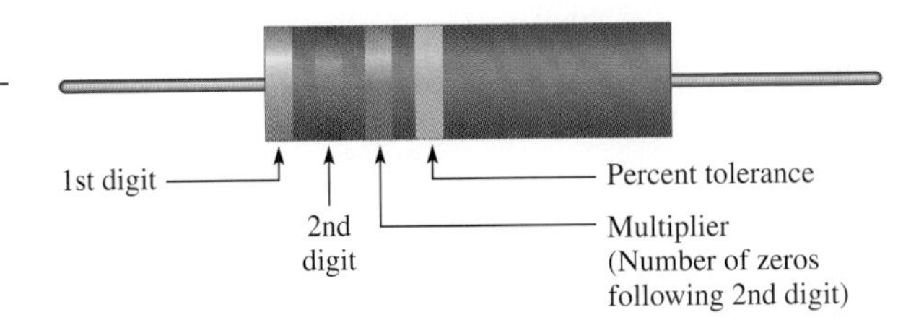

FIGURE 2–21

Color-code bands on a 4-band resistor.

TABLE 2–1

Resistor 4-band color code.

	Digit	Color
Resistance value, first three bands: First band—1st digit; Second band—2nd digit; *Third band—multiplier (number of zeros following the 2nd digit)	0	Black
	1	Brown
	2	Red
	3	Orange
	4	Yellow
	5	Green
	6	Blue
	7	Violet
	8	Gray
	9	White
Fourth band—tolerance	±5%	Gold
	±10%	Silver

* For resistance values less than 10 Ω, the third band is either gold or silver. Gold is for a multiplier of 0.1 and silver is for a multiplier of 0.01.

The 4-band color code is read as follows:

1. Start with the band closest to one end of the resistor. The first band is the first digit of the resistance value. If it is not clear which is the banded end, start from the end that does not begin with a gold or silver band.
2. The second band is the second digit of the resistance value.
3. The third band is the number of zeros following the second digit, or the *multiplier.*
4. The fourth band indicates the percent tolerance and is usually gold or silver.

For example, a 5% tolerance means that the *actual* resistance value is within ±5% of the color-coded value. Thus, a 100 Ω resistor with a tolerance of ±5% can have acceptable values as low as 95 Ω and as high as 105 Ω.

As indicated in the table, for resistance values less than 10 Ω, the third band is either gold or silver. Gold in the third band represents a multiplier of 0.1, and silver represents 0.01. For example, a color code of red, violet, gold, and silver represents 2.7 Ω with a tolerance of ±10%.

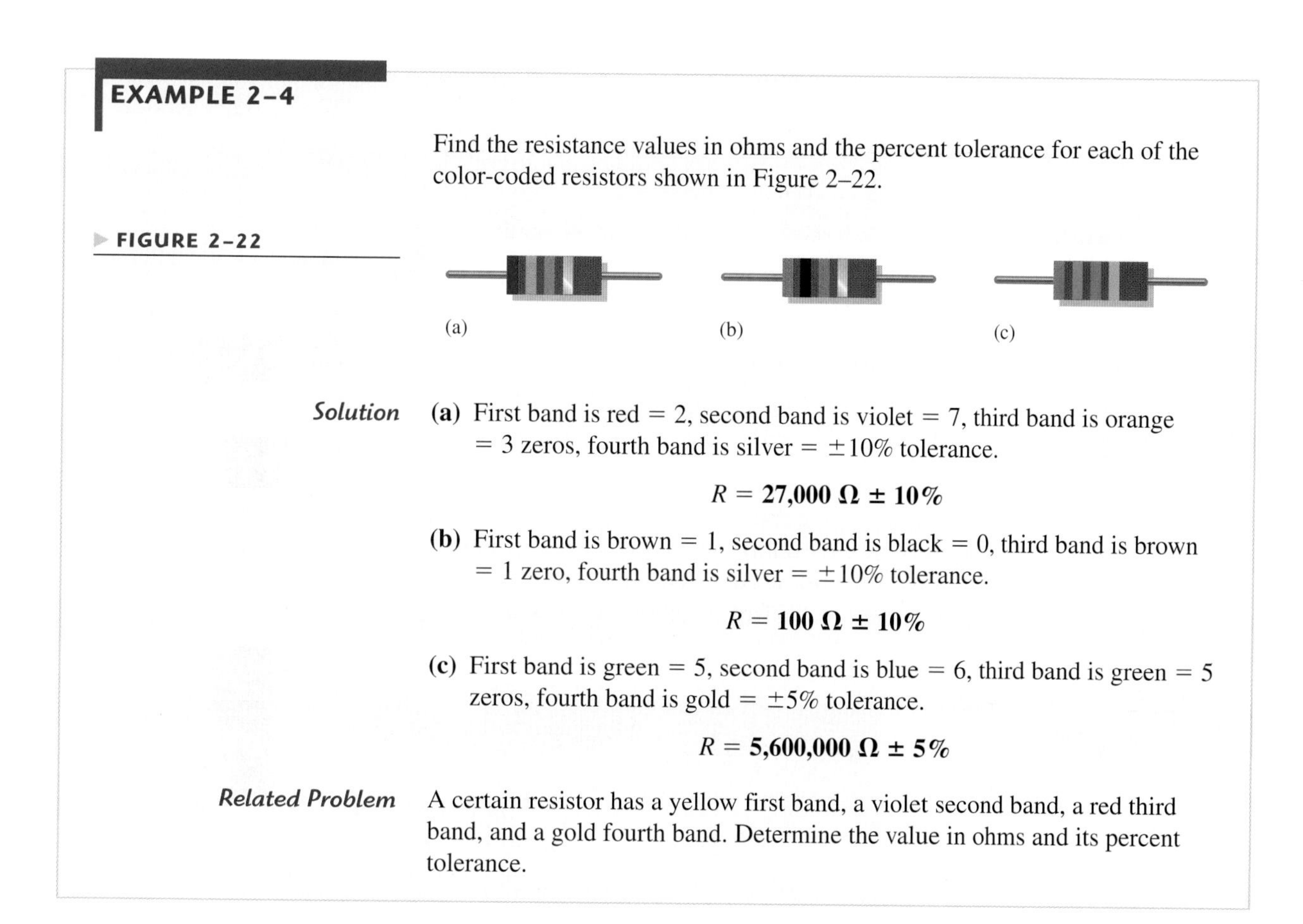

EXAMPLE 2–4

Find the resistance values in ohms and the percent tolerance for each of the color-coded resistors shown in Figure 2–22.

▶ **FIGURE 2–22**

(a) (b) (c)

Solution

(a) First band is red = 2, second band is violet = 7, third band is orange = 3 zeros, fourth band is silver = ±10% tolerance.

$$R = \mathbf{27{,}000\ \Omega \pm 10\%}$$

(b) First band is brown = 1, second band is black = 0, third band is brown = 1 zero, fourth band is silver = ±10% tolerance.

$$R = \mathbf{100\ \Omega \pm 10\%}$$

(c) First band is green = 5, second band is blue = 6, third band is green = 5 zeros, fourth band is gold = ±5% tolerance.

$$R = \mathbf{5{,}600{,}000\ \Omega \pm 5\%}$$

Related Problem A certain resistor has a yellow first band, a violet second band, a red third band, and a gold fourth band. Determine the value in ohms and its percent tolerance.

Five-Band Color Code Certain precision resistors with tolerances of 2%, 1%, or less are generally color coded with five bands, as shown in Figure 2–23. Begin at the band closest to one end. The first band is the first digit of the resistance value, the second band is the second digit, the third band is the third digit, the fourth band is the multiplier (number of zeros after the third digit), and the fifth band indicates the tolerance. Table 2–2 shows the 5-band color code.

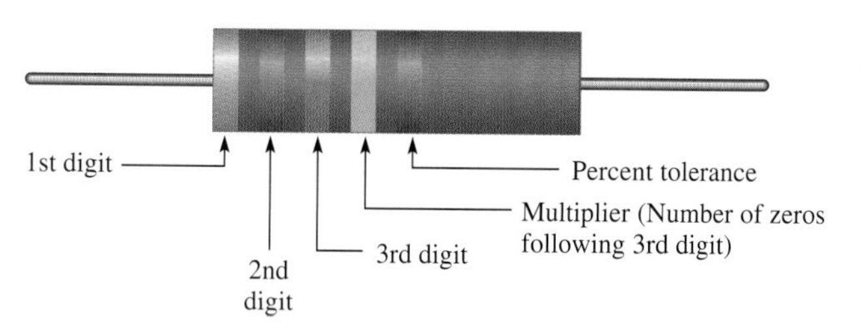

◀ **FIGURE 2–23**

Color-code bands on a 5-band resistor.

TABLE 2–2

Resistor 5-band color code.

	DIGIT	COLOR
Resistance value, first three bands:	0	Black
	1	Brown
First band—1st digit	2	Red
Second band—2nd digit	3	Orange
Third band—3rd digit	4	Yellow
Fourth band—multiplier	5	Green
(number of zeros	6	Blue
following 3rd digit)	7	Violet
	8	Gray
	9	White
Fourth band—multiplier	0.1	Gold
	0.01	Silver
Fifth band—tolerance	±2%	Red
	±1%	Brown
	±0.5%	Green
	±0.25%	Blue
	±0.1%	Violet

Resistor Reliability Band An extra band on some color-coded resistors indicates the resistor's reliability in percent of failures per 1000 hours (1000 h) of use. The reliability color code is listed in Table 2–3. For example, a brown fifth band on a 4-band color-coded resistor means that if a group of like resistors is operated under standard conditions for 1000 h, 1% of the resistors in that group will fail.

TABLE 2–3

Reliability color code.

COLOR	FAILURES DURING 1000 h OF OPERATION
Brown	1.0%
Red	0.1%
Orange	0.01%
Yellow	0.001%

EXAMPLE 2–5

Find the resistance value in ohms and the percent tolerance for each of the color-coded resistors shown in Figure 2–24.

FIGURE 2–24

(a) (b) (c)

Solution **(a)** First band is red = 2, second band is violet = 7, third band is black = 0, fourth band is gold = ×0.1, fifth band is red = ±2% tolerance.

$$R = 270 \times 0.1 = \mathbf{27\ \Omega \pm 2\%}$$

(b) First band is yellow = 4, second band is black = 0, third band is red = 2, fourth band is black = 0, fifth band is brown = ±1% tolerance.

$$R = \mathbf{402\ \Omega \pm 1\%}$$

(c) First band is orange = 3, second band is orange = 3, third band is red = 2, fourth band is orange = 3, fifth band is green = ±0.5% tolerance.

$$R = \mathbf{332{,}000\ \Omega \pm 0.5\%}$$

Related Problem A certain resistor has a yellow first band, a violet second band, a green third band, a gold fourth band, and a red fifth band. Determine its value in ohms and its percent tolerance.

Resistor Label Codes

Not all types of resistors are color coded. Many, including surface-mount resistors, use typographical marking to indicate the resistance value and tolerance. These label codes consist of either all numbers (numeric) or a combination of numbers and letters (alphanumeric). In some cases when the body of the resistor is large enough, the entire resistance value and tolerance are stamped on it in standard form. For example, a 33,000 Ω resistor may be labeled as 33kΩ.

Numeric Labeling This type of marking uses three digits to indicate the resistance value, as shown in Figure 2–25 using a specific example. The first two digits give the first two digits of the resistance value, and the third digit gives the multiplier or number of zeros that follow the first two digits. This code is limited to values of 10 Ω or greater.

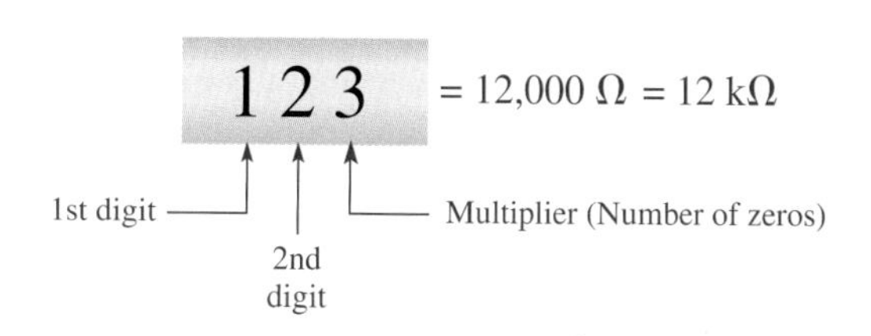

FIGURE 2–25

Example of three-digit labeling for a resistor.

Alphanumeric Labeling Another common type of marking is a three- or four-character label that uses both digits and letters. This type of label typically consists of only three digits or two or three digits and one of the letters R, K, or M. The letter is used to indicate the multiplier, and the position of the letter indicates the decimal point placement. The letter R indicates a multiplier of 1 (no zeros after the digits), the K indicates a multiplier of 1000 (3 zeros after the digits), and the M indicates a multiplier of 1,000,000 (6 zeros after the digits). In this format, values from 100 to 999 consist of three digits and no letter to represent the three digits in the resistance value. Figure 2–26 shows three examples of this type of resistor label.

FIGURE 2–26

Examples of the alphanumeric resistor label.

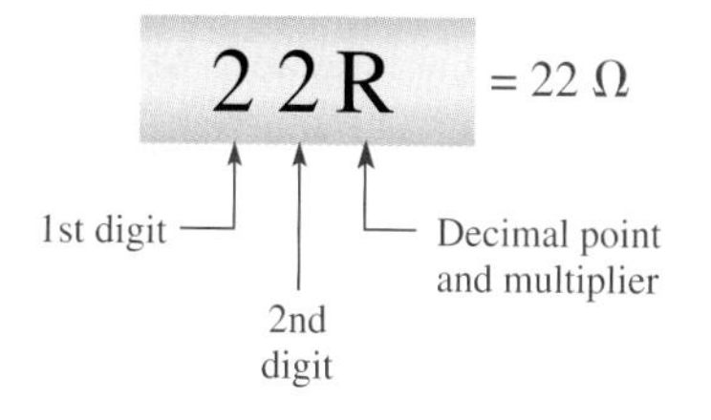

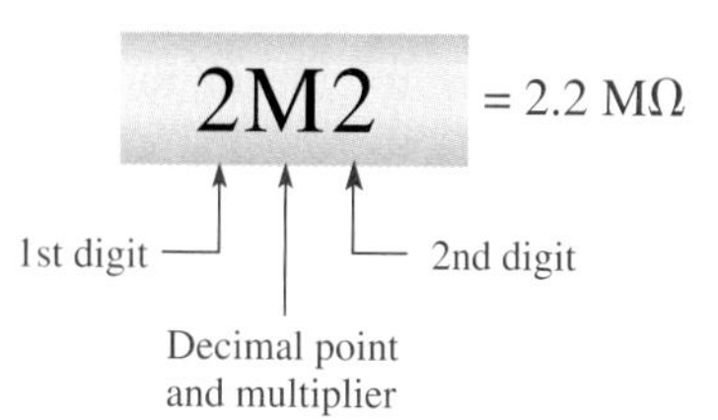

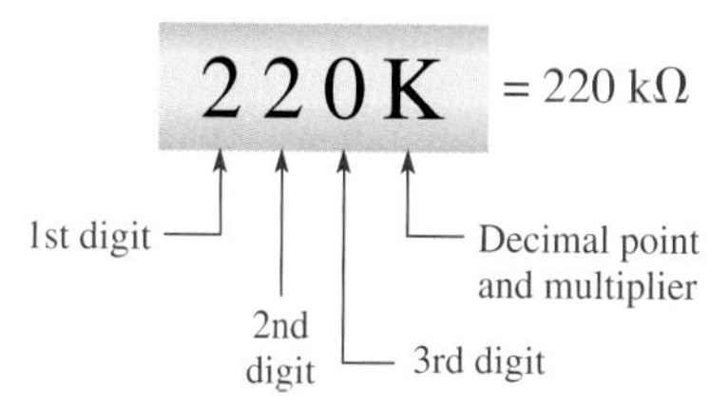

EXAMPLE 2–6

Interpret the following alphanumeric resistor labels:

(a) 470 **(b)** 5R6 **(c)** 68K **(d)** 10M **(e)** 3M3

Solution **(a)** 470 = **470 Ω** **(b)** 5R6 = **5.6 Ω** **(c)** 68K = **68 kΩ**
(d) 10M = **10 MΩ** **(e)** 3M3 = **3.3 MΩ**

Related Problem What is the resistance indicated by 1K25?

One system of labels for resistance tolerance values uses the letters F, G, and J as follows:

$$F = \pm 1\% \qquad G = \pm 2\% \qquad J = \pm 5\%$$

For example, 620F indicates a 620 Ω resistor with a tolerance of ±1%, 4R6G is a 4.6 Ω ±2% resistor, and 56KJ is a 56 kΩ ±5% resistor.

Variable Resistors Variable resistors are designed so that their resistance values can be changed easily with a manual or an automatic adjustment.

Two basic uses for variable resistors are to divide voltage and to control current. The variable resistor used to divide voltage is called a **potentiometer**. The variable resistor used to control current is called a **rheostat**. Schematic symbols for these types are shown in Figure 2–27. The potentiometer is a three-terminal device, as indicated in part (a). Terminals 1 and 2 have a fixed resistance between them, which is the total resistance. Terminal 3 is connected to a moving contact (**wiper**). You can vary the resistance between 3 and 1 or between 3 and 2 by moving the contact.

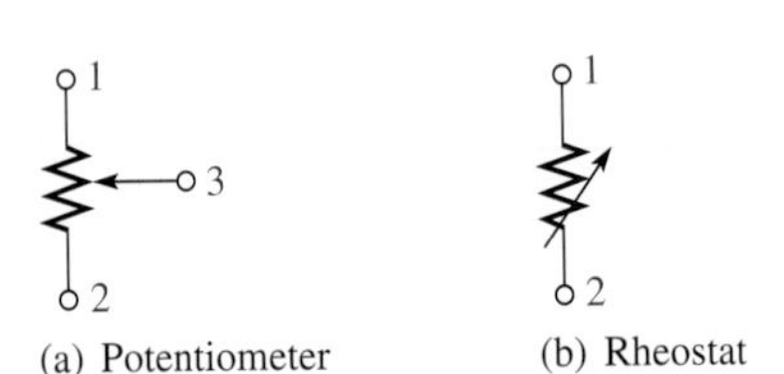

(a) Potentiometer (b) Rheostat

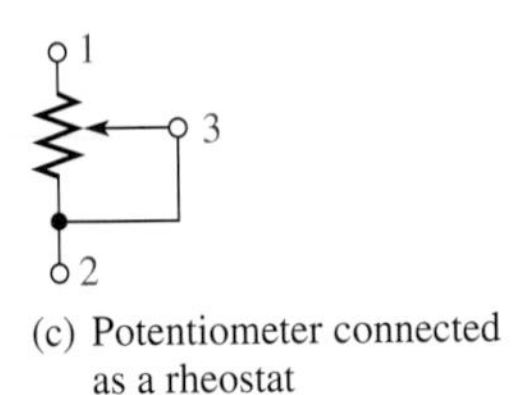

(c) Potentiometer connected as a rheostat

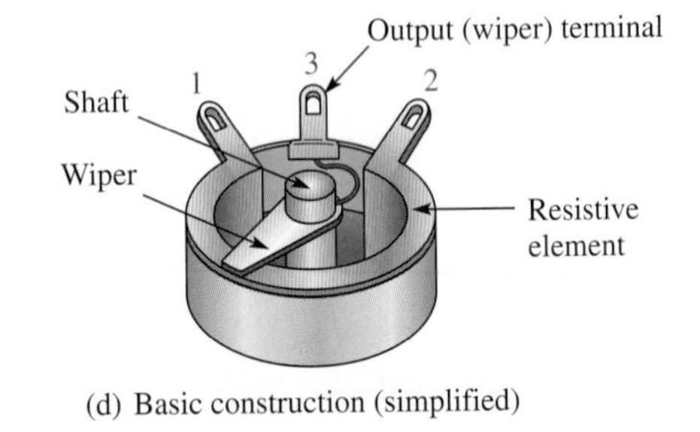

(d) Basic construction (simplified)

FIGURE 2–27

Potentiometer and rheostat symbols and basic construction of one type of potentiometer.

Figure 2–27(b) shows the rheostat as a two-terminal variable resistor. Part (c) shows how you can use a potentiometer as a rheostat by connecting terminal 3 to either terminal 1 or terminal 2. Parts (b) and (c) are equivalent symbols. Part (d) shows a simplified construction diagram of a potentiometer. Some typical potentiometers are pictured in Figure 2–28.

Potentiometers and rheostats can be classified as linear or tapered, as shown in Figure 2–29, where a potentiometer with a total resistance of 100 Ω is used as an example. As shown in part (a), in a linear potentiometer, the resistance between either terminal and the moving contact varies linearly with the position of the moving contact. For example, one-half of a turn results in one-half the total resistance. Three-quarters of a turn results in three-quarters of the total resistance between the moving contact and one terminal, or one-quarter of the total resistance between the other terminal and the moving contact.

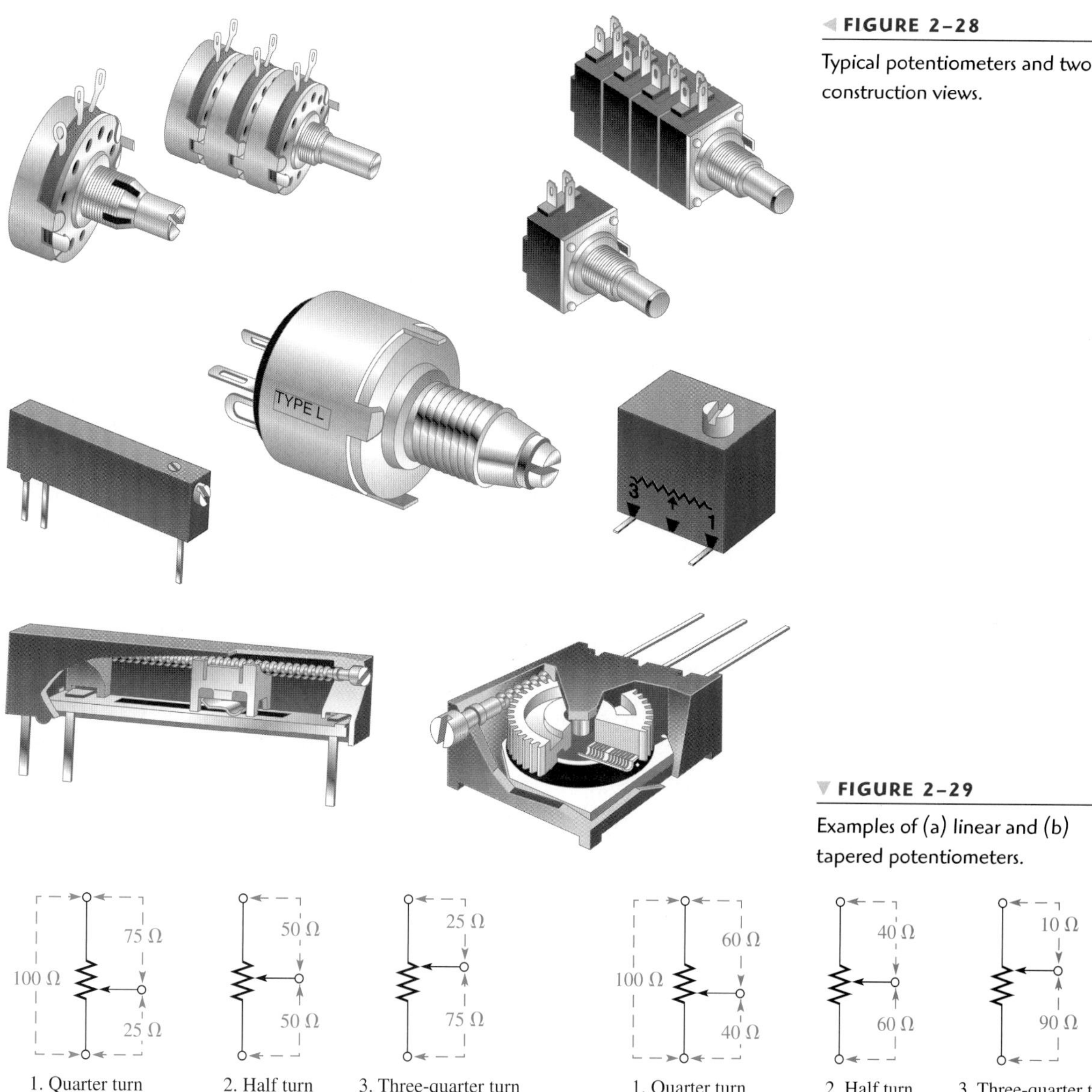

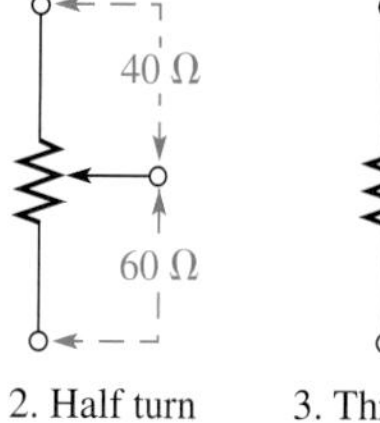

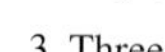

FIGURE 2–28

Typical potentiometers and two construction views.

FIGURE 2–29

Examples of (a) linear and (b) tapered potentiometers.

In the **tapered** potentiometer, the resistance varies nonlinearly with the position of the moving contact, so that one-half of a turn does not necessarily result in one-half the total resistance. This concept is illustrated in Figure 2–29(b), where the nonlinear values are arbitrary.

The potentiometer is used as a voltage-control device. When a fixed voltage is applied across the end terminals, a variable voltage is obtained at the wiper contact with respect to either end terminal. The rheostat is used as a current-control device; the current can be changed by changing the wiper position.

Two Types of Automatically Variable Resistors A **thermistor** is a type of variable resistor that is temperature-sensitive. When its temperature coefficient is negative, the resistance changes inversely with temperature. When its temperature coefficient is positive, the resistance changes directly with temperature.

The resistance of a **photoconductive cell** changes with a change in light intensity. This cell also has a negative temperature coefficient, which means that

HANDS ON TIP

A special type of resistor is used for sensing forces ranging from the pressure of a finger tip on a keyboard to the weight of a truck on a scale. These specialized resistors are known as force sensing resistors (FSRs) and are a type of transducer. The FSR exhibits a change in resistance in response to an applied force. The resistance change is used to indicate the amount of force.

FIGURE 2–30

Symbols for resistive devices with sensitivities to temperature and light.

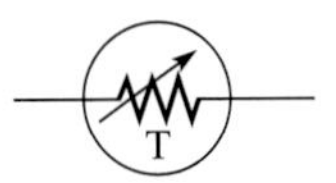

(a) Thermistor

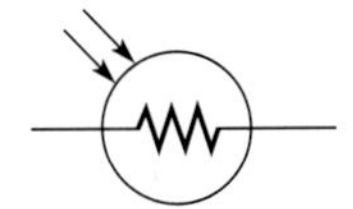

(b) Photoconductive cell

resistance decreases with increasing temperature. Symbols for both of these devices are shown in Figure 2–30.

SECTION 2–5 REVIEW

1. Define *resistance* and name its unit.
2. What are the two main categories of resistors? Briefly explain the difference between them.
3. In the 4-band resistor color code, what does each band represent?
4. Determine the resistance and percent tolerance for each of the following color codes:
   (a) yellow, violet, red, gold (b) blue, red, orange, silver
   (c) brown, gray, black, gold (d) red, red, blue, red, green
5. What resistance value is indicated by each alphanumeric label:
   (a) 33R (b) 5K6 (c) 900 (d) 6M8
6. What is the basic difference between a rheostat and a potentiometer?
7. What is a thermistor?

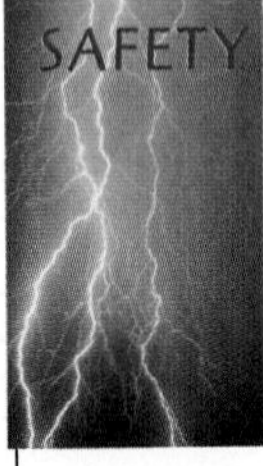

SAFETY POINT

To avoid electrical shock, never touch a circuit while it is connected to a voltage source. If you need to handle a circuit, remove a component, or change a component, first make sure the voltage source is disconnected.

2–6 THE ELECTRIC CIRCUIT

A basic electric circuit is an arrangement of physical components that use voltage, current, and resistance to perform some useful function.

After completing this section, you should be able to

- **Describe a basic electric circuit**
- Relate a schematic to a physical circuit
- Define *open circuit* and *closed circuit*
- Describe various types of protective devices
- Describe various types of switches
- Explain how wire sizes are related to gauge numbers
- Define *ground*

Basically, an electric **circuit** consists of a voltage source, a load, and a path for current between the source and the load. The **load** is a device on which work is done by the current through it. Figure 2–31 shows an example of a simple electric circuit: a battery connected to a lamp with two conductors (wires). The battery is the voltage source, the lamp is the load on the battery because it draws current from the battery, and the two wires provide the current path from the negative terminal of the battery to the lamp and back to the positive terminal of the battery, as indicated by the red arrows. There is current through the filament of the lamp (which has a resistance), causing it to become hot enough to emit visible light.

Current through the battery is produced by chemical action. In many practical cases, one terminal of the battery is connected to a ground point. For example, in automobiles, the negative battery terminal is generally connected to the metal chassis of the car. The chassis is the ground for the automobile electrical system and provides a current path for the circuit. The concept of *ground* is covered later in this chapter.

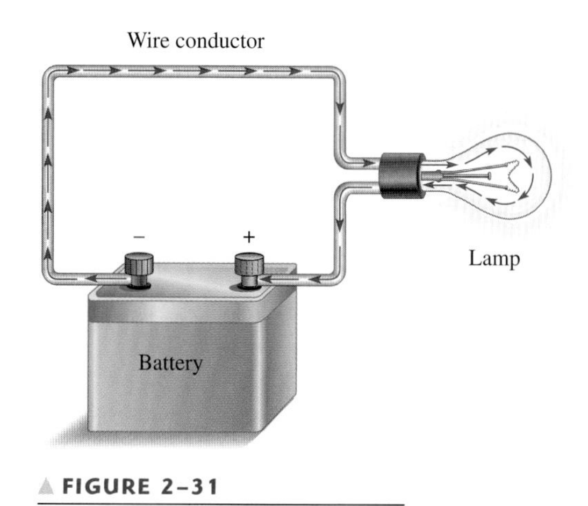

▲ FIGURE 2–31

A simple electric circuit.

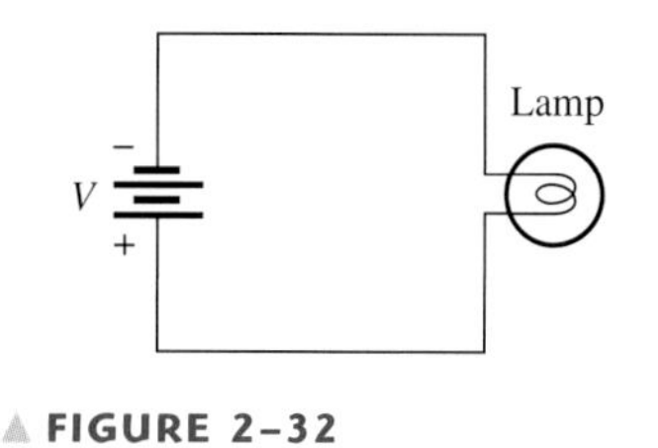

▲ FIGURE 2–32

A schematic for the circuit in Figure 2–31.

The Electric Circuit Schematic

An electric circuit can be represented by a schematic, a diagram that shows the interconnection of components, using standard symbols for each element, as shown in Figure 2–32 for the simple circuit in Figure 2–31. A schematic shows, in an organized manner, how the various components in a given circuit are interconnected so that the operation of the circuit can be determined.

Closed and Open Circuits

The simple circuit in Figure 2–31 illustrates a closed circuit—that is, a circuit in which the current has a complete path. An open circuit is a circuit in which the current path is broken so that there is no current. An open circuit is considered to have infinite resistance (infinite means immeasurably large).

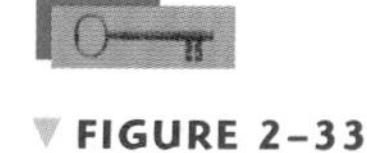

Switches Switches are commonly used for controlling the opening or closing of circuits by either mechanical or electronic means. For example, a switch is used to turn a lamp on or off as illustrated in Figure 2–33. Each circuit pictorial is shown with its associated schematic. The type of switch indicated is a *single-pole–single-throw* (SPST) toggle switch. The term *pole* refers to the movable arm

▼ FIGURE 2–33

Illustration of closed and open circuits using an SPST switch for control.

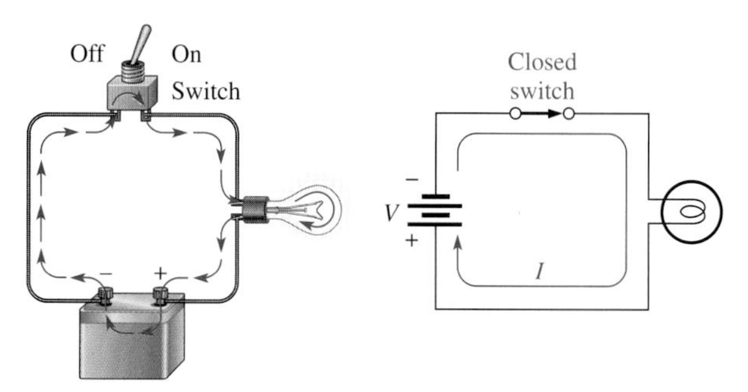

(a) There is current in a *closed* circuit because there is a complete current path (switch is ON or in the *closed* position). Current is almost always indicated by a red arrow in this text.

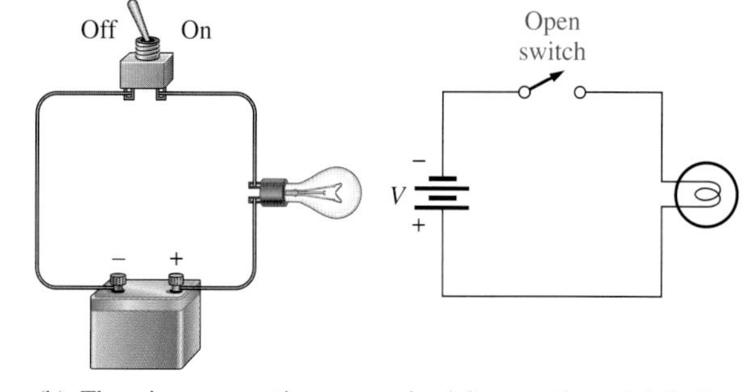

(b) There is no current in an *open* circuit because the path is broken (switch is OFF or in the *open* position).

in a switch, and the term *throw* indicates the number of contacts that are affected (either opened or closed) by a single switch action (a single movement of a pole).

Figure 2–34 shows a somewhat more complicated circuit using a *single-pole–double-throw* (SPDT) type of switch to control the current to two different lamps. When one lamp is on, the other is off, and vice versa, as illustrated by the two schematics that represent each of the switch positions.

FIGURE 2–34

An example of an SPDT switch controlling two lamps.

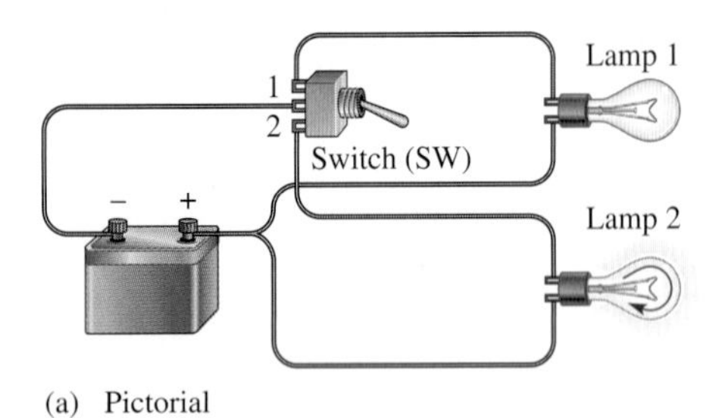

(a) Pictorial

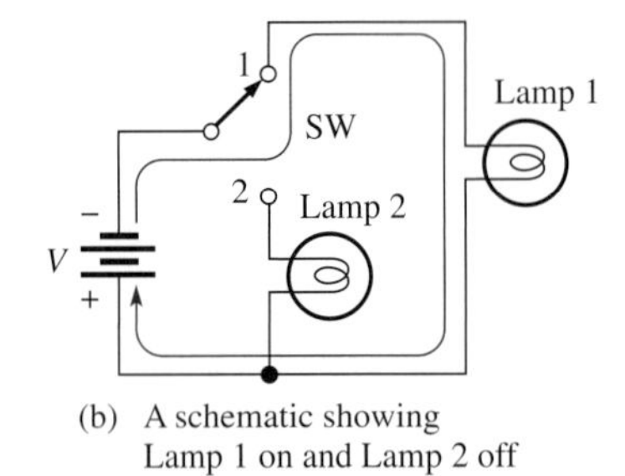

(b) A schematic showing Lamp 1 on and Lamp 2 off

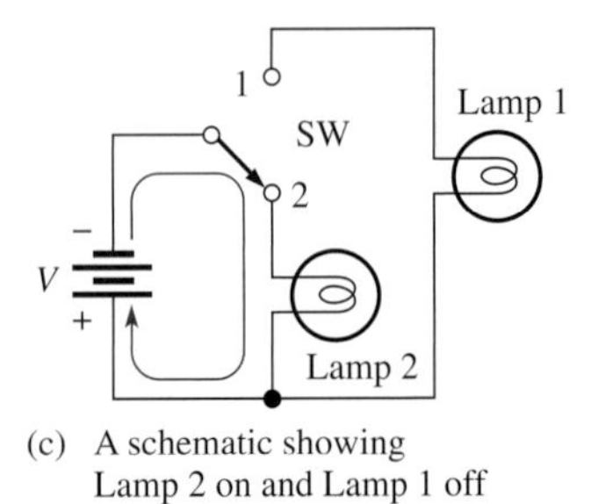

(c) A schematic showing Lamp 2 on and Lamp 1 off

In addition to the SPST and the SPDT switches (symbols are shown in Figure 2–35(a) and (b)), the following other types of switches are also important:

- *Double-pole–single-throw (DPST).* The DPST switch permits simultaneous opening or closing of two sets of contacts. The symbol is shown in Figure 2–35(c). The dashed line indicates that the contact arms are mechanically linked so that both move with a single switch action.
- *Double-pole–double-throw (DPDT).* The DPDT switch provides connection from one set of contacts to either of two other sets. The schematic symbol is shown in Figure 2–35(d).
- *Push button (PB).* In the normally open push-button switch (NOPB), shown in Figure 2–35(e), connection is made between two contacts when the button is depressed, and connection is broken when the button is released. In the normally closed push-button switch (NCPB), shown in Figure 2–35(f), connection between the two contacts is broken when the button is depressed.
- *Rotary.* In a rotary switch, a knob is turned to make a connection between one contact and any one of several others. A symbol for a simple six-position rotary switch is shown in Figure 2–35(g).

Figure 2–36 shows several varieties of switches, and Figure 2–37 shows a construction view of a typical toggle switch.

FIGURE 2–35

Switch symbols.

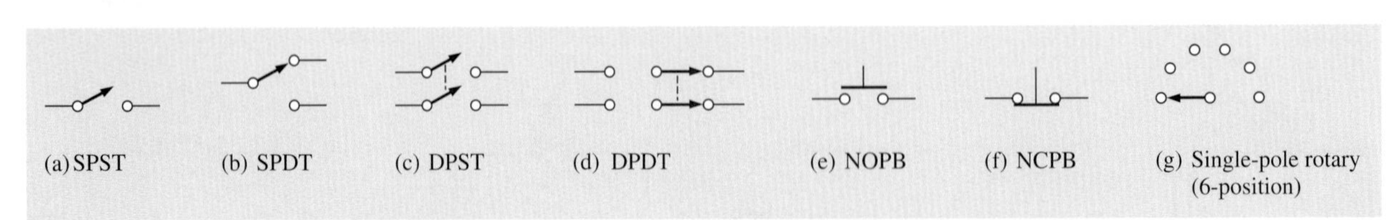

Protective Devices **Fuses** and **circuit breakers** are placed in the current path and are used to deliberately create an open circuit when the current exceeds a specified number of amperes due to a malfunction or other abnormal condition in a circuit. For example, a fuse or circuit breaker with a 20 A rating will open a circuit when the current exceeds 20 A.

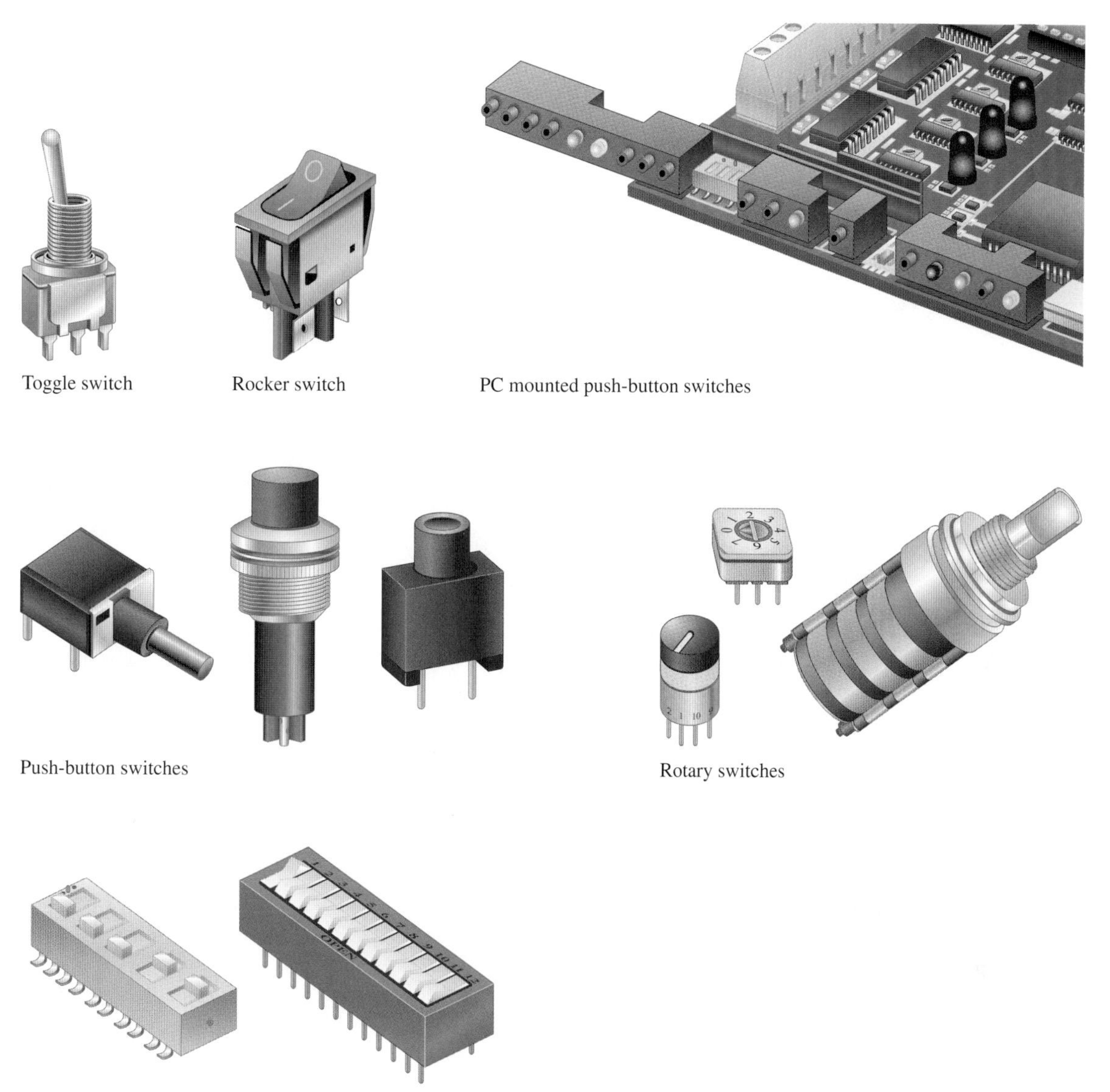

▲ **FIGURE 2–36**

Typical mechanical switches.

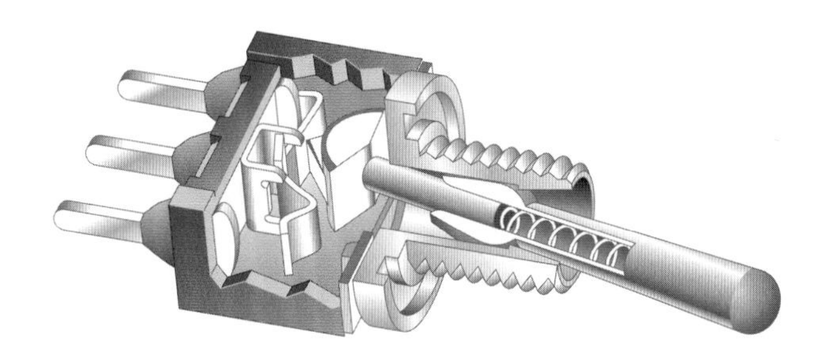

◄ **FIGURE 2–37**

Construction view of a typical toggle switch.

The basic difference between a fuse and a circuit breaker is that when a fuse is "blown," it must be replaced; but when a circuit breaker opens, it can be reset and reused repeatedly. Both of these devices protect against damage to a circuit due to excess current or prevent a hazardous condition created by the overheating of wires and other components when the current is too great. Because fuses cut off excess current more quickly than circuit breakers, fuses are used whenever delicate electronic equipment needs to be protected. Several typical fuses and circuit breakers, along with their schematic symbols, are shown in Figure 2–38.

FIGURE 2–38

Typical fuses and circuit breakers and their symbols.

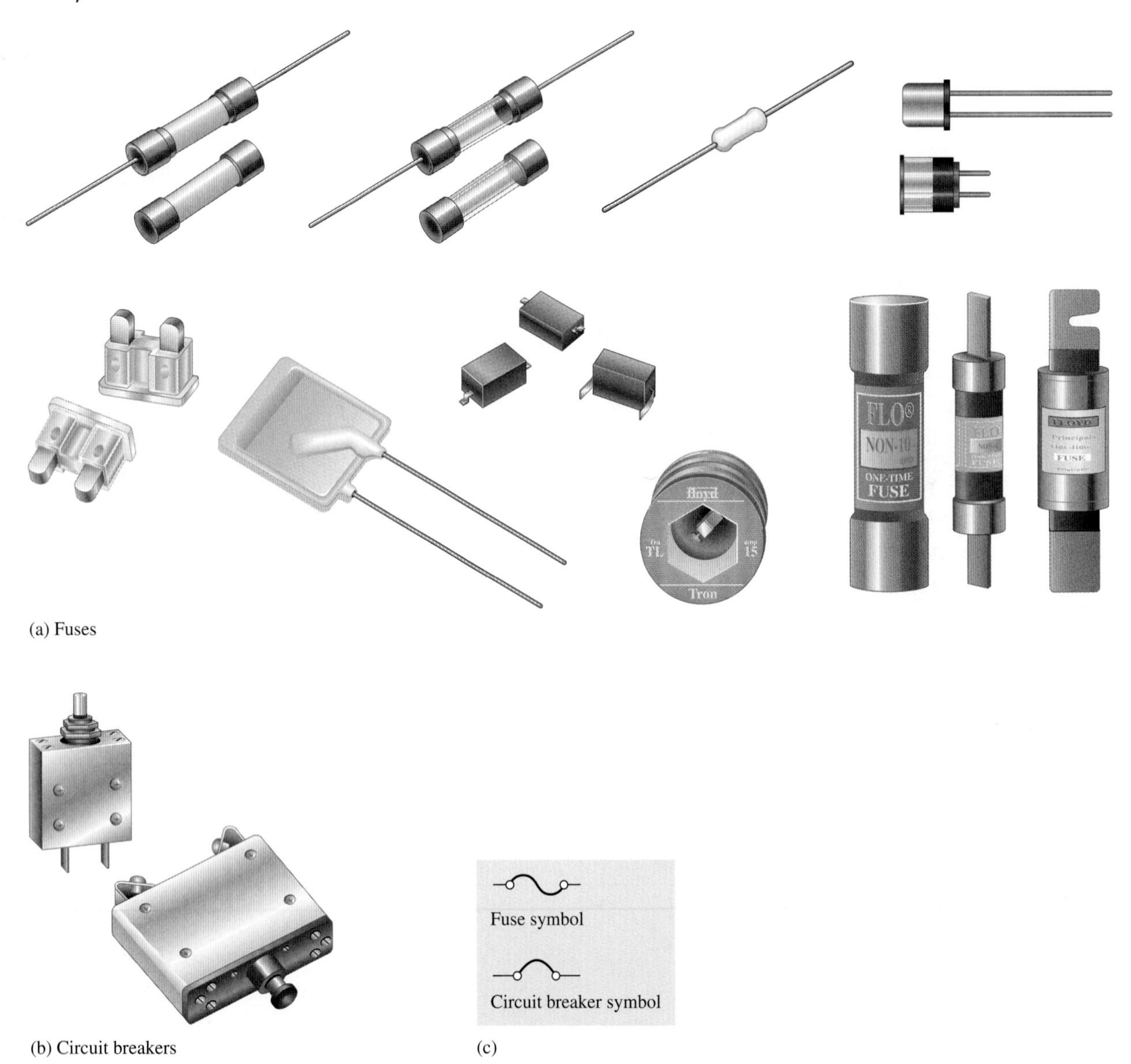

Wires

Wires are the most common form of conductive material used in electrical applications. They vary in diameter size and are arranged according to standard gauge numbers, called AWG (American wire gauge) sizes. The larger the gauge number is, the smaller the wire diameter is. The size of a wire is also specified in terms of its cross-sectional area, as illustrated in Figure 2–39. The unit of cross-sectional

FIGURE 2–39

Cross-sectional area of a wire.

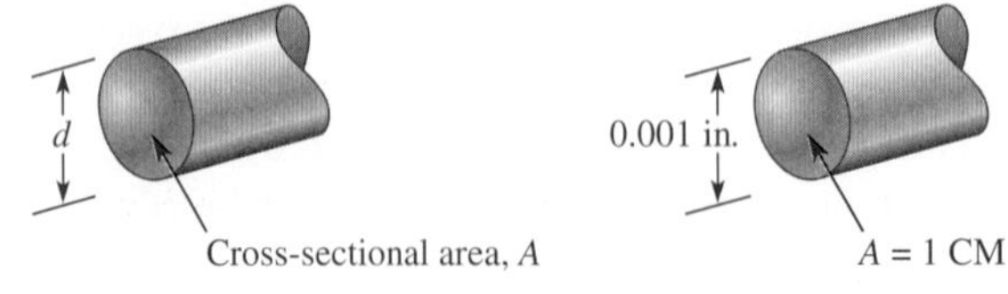

area is the **circular mil,** abbreviated CM. One circular mil is the area of a wire with a diameter of 0.001 inch (0.001 in., or 1 mil). You can find the cross-sectional area in circular mils by expressing the diameter in thousandths of an inch (mils) and squaring it, as follows:

$$A = d^2$$

Equation 2–5

where A is the cross-sectional area in circular mils and d is the diameter in mils. Table 2–4 lists the AWG sizes with their corresponding cross-sectional area and resistance in ohms per 1000 ft at 20°C.

TABLE 2–4

American Wire Gauge (AWG) sizes and resistances for solid round copper.

AWG #	AREA (CM)	RESISTANCE (Ω/1000 FT AT 20°C)	AWG #	AREA (CM)	RESISTANCE (Ω/1000 FT AT 20°C)
0000	211,600	0.0490	19	1,288.1	8.051
000	167,810	0.0618	20	1,021.5	10.15
00	133,080	0.0780	21	810.10	12.80
0	105,530	0.0983	22	642.40	16.14
1	83,694	0.1240	23	509.45	20.36
2	66,373	0.1563	24	404.01	25.67
3	52,634	0.1970	25	320.40	32.37
4	41,742	0.2485	26	254.10	40.81
5	33,102	0.3133	27	201.50	51.47
6	26,250	0.3951	28	159.79	64.90
7	20,816	0.4982	29	126.72	81.83
8	16,509	0.6282	30	100.50	103.2
9	13,094	0.7921	31	79.70	130.1
10	10,381	0.9989	32	63.21	164.1
11	8,234.0	1.260	33	50.13	206.9
12	6,529.0	1.588	34	39.75	260.9
13	5,178.4	2.003	35	31.52	329.0
14	4,106.8	2.525	36	25.00	414.8
15	3,256.7	3.184	37	19.83	523.1
16	2,582.9	4.016	38	15.72	659.6
17	2,048.2	5.064	39	12.47	831.8
18	1,624.3	6.385	40	9.89	1049.0

EXAMPLE 2–7

What is the cross-sectional area of a wire with a diameter of 0.005 inch?

Solution

$$d = 0.005\text{ in.} = 5\text{ mils}$$

$$A = d^2 = (5\text{ mils})^2 = \mathbf{25\ CM}$$

Related Problem What is the cross-sectional area of a 0.0201 in. diameter wire? What is the AWG # for this wire from Table 2–4?

Wire Resistance

Although copper wire conducts current extremely well, it still has some resistance, as do all conductors. The resistance of a wire depends on four factors: (a) type of material, (b) length of wire, (c) cross-sectional area, and (d) temperature.

Each type of conductive material has a characteristic called its *resistivity*, which is represented by the Greek letter ρ. For each material, ρ is a constant value at a given temperature. The formula for the resistance of a wire of length l and cross-sectional area A is

Equation 2–6

$$R = \frac{\rho l}{A}$$

This formula shows that resistance increases with resistivity and length and decreases with cross-sectional area. For resistance to be calculated in ohms, the length must be in feet (ft), the cross-sectional area in circular mils (CM), and the resistivity in CM-Ω/ft.

EXAMPLE 2–8

Find the resistance of a 100 ft length of copper wire with a cross-sectional area of 810.1 CM. The resistivity of copper is 10.37 CM-Ω/ft at 20°C.

Solution

$$R = \frac{\rho l}{A} = \frac{(10.37 \text{ CM-}\Omega/\text{ft})(100 \text{ ft})}{810.1 \text{ CM}} = \mathbf{1.280\ \Omega}$$

Related Problem Determine the resistance of a 1000 ft length of #22 AWG copper wire.

As mentioned, Table 2–4 lists the resistance of the various standard wire sizes in ohms per 1000 ft at 20°C. For example, a 1000 ft length of 14-gauge copper wire has a resistance of 2.525 Ω. A 1000 ft length of 22-gauge wire has a resistance of 16.14 Ω. For a given length, the smaller-gauge wire has more resistance. Thus, for a given voltage, larger-gauge wires can carry more current than smaller ones.

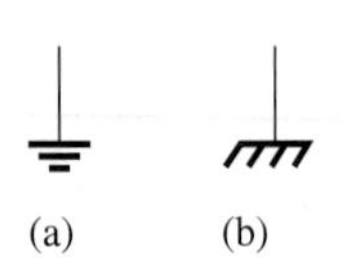

▲ FIGURE 2–40

Symbols for ground.

Ground

The term *ground* comes from the method used in ac power distribution where one side of the power line is neutralized by connecting it to a metal rod driven into the ground. This method of grounding is called *earth ground.*

In electrical and electronic systems, the metal chassis that houses the assembly or a large conductive area on a printed circuit board is used as the electrical reference point and is called *chassis ground* or **circuit ground.** Circuit ground may or may not be connected to earth ground. For example, the negative terminal of the battery and one side of all the electrical circuits in most cars are connected to the metal body chassis, which is isolated from earth ground by the tires.

Ground is a reference point in electric circuits and has a potential of 0 V with respect to other points in the circuit. All of the ground points in a circuit are *electrically* the same and are therefore common points. Two ground symbols are shown in Figure 2–40. The symbol in part (a) is commonly used in schematic drawings to represent a reference ground, but the one in part (b) is also sometimes used. The symbol in part (a) will be used throughout this textbook.

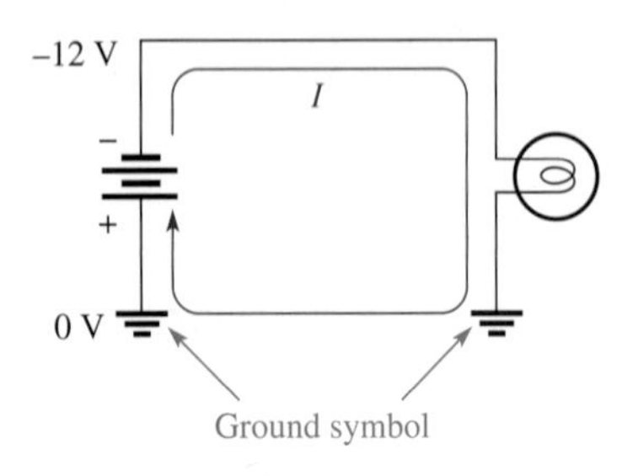

▲ FIGURE 2–41

A simple circuit with ground connections.

Figure 2–41 illustrates a simple circuit with ground connections. The current is from the negative terminal of the 12 V source, through the wire to the lamp, through the lamp, and back to the positive terminal of the source through the common ground connection. **Ground** provides a return path for the current back to the source because all of the ground points are electrically the same point and provide a zero resistance (ideally) current path. The voltage at the top of the circuit is −12 V with respect to ground. You can think of all the ground points in a circuit as being connected together by a conductor.

SECTION 2–6 REVIEW

1. What are the basic elements of an electric circuit?
2. Define *open circuit.*
3. Define *closed circuit.*
4. What is the resistance of an open switch? Ideally, what is the resistance of a closed switch?
5. What is the purpose of a fuse?
6. What is the difference between a fuse and a circuit breaker?
7. Which wire is larger in diameter, AWG #3 or AWG #22?
8. What is ground in an electric circuit?

2–7 BASIC CIRCUIT MEASUREMENTS

In electronics technology, you cannot function without knowing how to measure voltage, current, and resistance.

After completing this section, you should be able to

- **Make basic circuit measurements**
- Properly measure voltage in a circuit
- Properly measure current in a circuit
- Properly measure resistance
- Set up and read basic meters

Voltage, current, and resistance measurements are commonly required in electronics work. Recall that measuring instruments were introduced in Chapter 1. The instrument used to measure voltage is a **voltmeter**, the instrument used to measure current is an **ammeter**, and the instrument used to measure resistance is an **ohmmeter**. Commonly, all three instruments are combined into a single instrument known as a multimeter, in which you can choose the specific quantity to measure by selecting the appropriate function with a switch.

Typical portable multimeters are shown in Figure 2–42. Part (a) shows an analog meter with a needle pointer, and part (b) shows a digital multimeter (DMM), which provides a digital readout of the measured quantity. Many digital multimeters also include a bar graph display.

Meter Symbols

Throughout this book, certain symbols will be used in circuits to represent meters, as shown in Figure 2–43. You may see any of four types of symbols for voltmeters, ammeters, and ohmmeters, depending on which symbol most effectively conveys the information required. The digital meter symbol is used when specific values are to be indicated in a circuit. The bar graph meter symbol and sometimes

▶ **FIGURE 2–42**

Typical portable multimeters. (Photography courtesy of B&K Precision Corp.)

(a)

(b)

▶ **FIGURE 2–43**

Examples of meter symbols used in this book. V, A, or Ω labels indicate voltmeter, ammeter, or ohmmeter respectively for each symbol. Any of the three labels can be used with each meter symbol to indicate its function.

(a) Digital

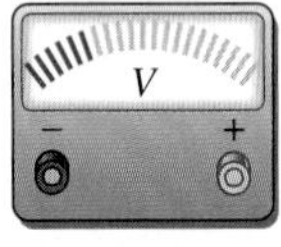

(b) Analog bar graph

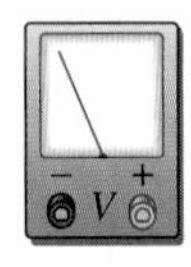

(c) Analog needle

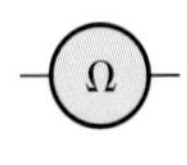

(d) Generic

the needle meter symbol are used to illustrate the operation of a circuit when *relative* measurements or changes in quantities, rather than specific values, need to be depicted. A changing quantity may be indicated by an arrow in the display showing an increase or decrease. The generic symbol is used to indicate placement of meters in a circuit when no values or value changes need to be shown.

Measuring Current

Figure 2–44 illustrates how to measure current with an ammeter. Part (a) shows the simple circuit in which the current through the resistor is to be measured. Connect the ammeter in the current path by first opening the circuit, as shown in part (b). Then insert the meter as shown in part (c). Such a connection is a *series* connection. The polarity of the meter must be such that the current is in at the negative terminal and out at the positive terminal.

SAFETY POINT

Never wear rings or any type of metallic jewelry while working on a circuit. These items may accidentally come in contact with the circuit, causing shock and/or damage to the circuit.

Measuring Voltage

To measure voltage, connect the voltmeter across the component for which the voltage is to be found. Such a connection is a *parallel* connection. The negative terminal of the meter must be connected to the negative side of the circuit, and the positive terminal of the meter to the positive side of the circuit. Figure 2–45 shows a voltmeter connected to measure the voltage across the resistor.

Measuring Resistance

To measure resistance, connect the ohmmeter across the resistor. *The resistor must first be removed or disconnected from the circuit.* This procedure is shown in Figure 2–46.

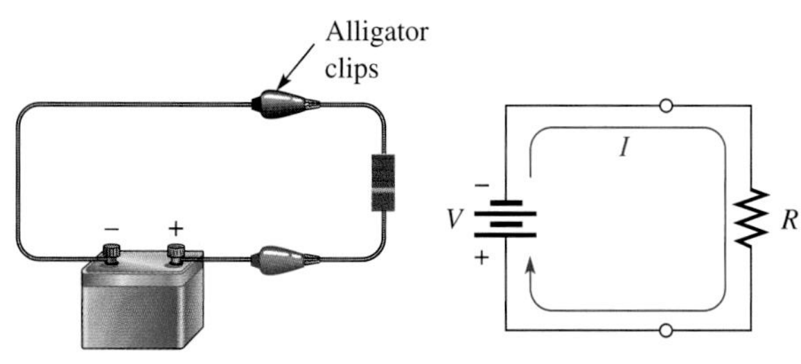

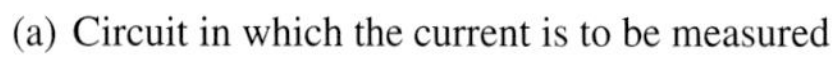

(a) Circuit in which the current is to be measured

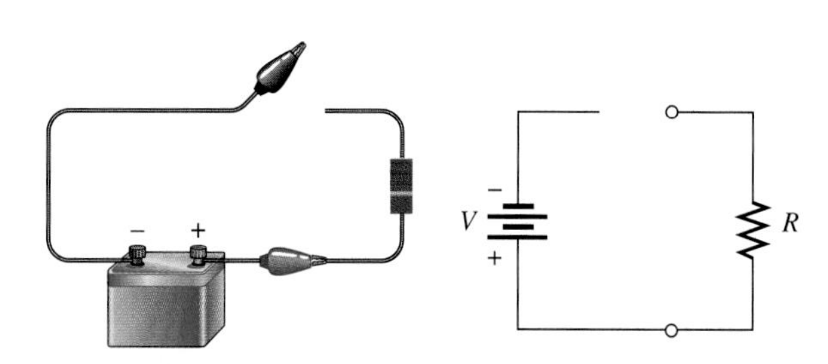

(b) Open the circuit either between the resistor and the positive terminal or between the resistor and the negative terminal of source.

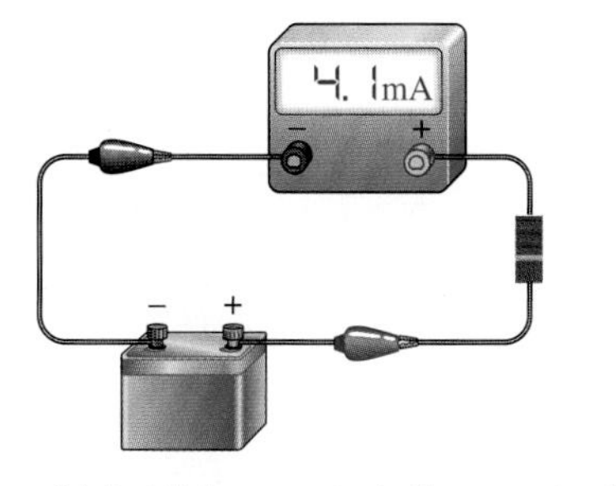

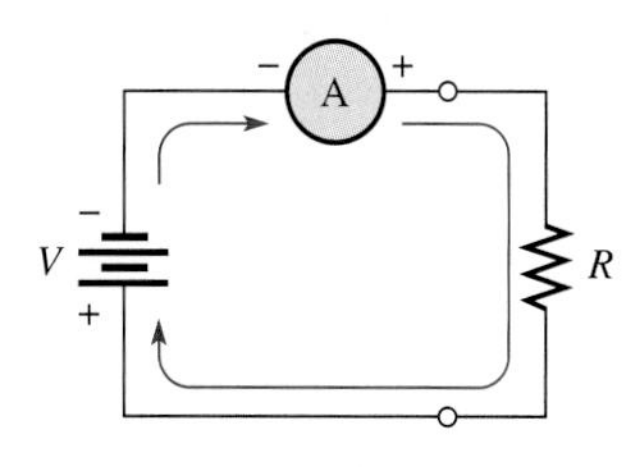

(c) Install the ammeter in the current path with polarity as shown (negative to negative, positive to positive).

FIGURE 2–44

Example of an ammeter connection to measure current in a simple circuit.

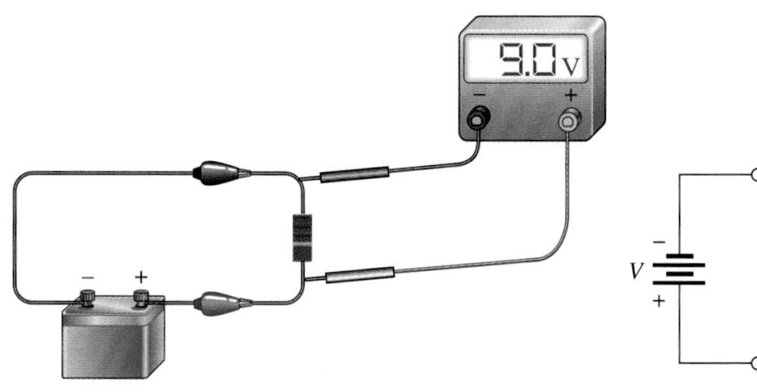

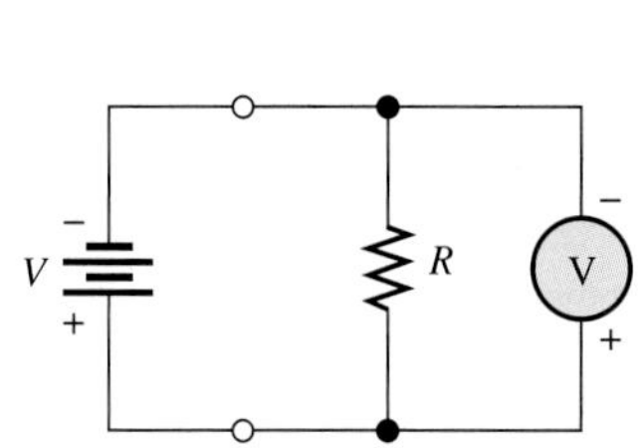

FIGURE 2–45

Example of a voltmeter connection to measure voltage in a simple circuit.

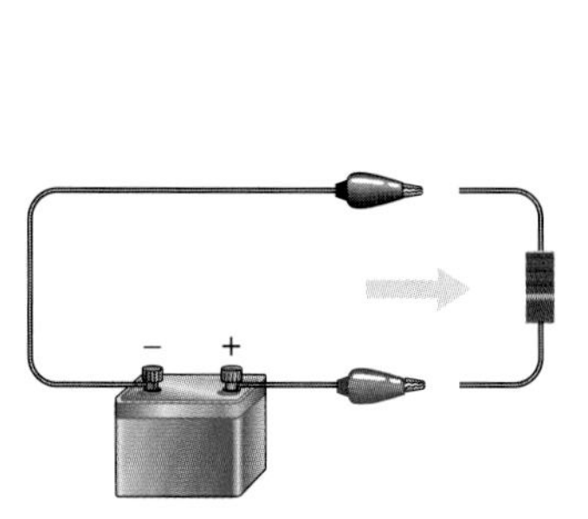

(a) Disconnect the resistor from the circuit to avoid damage to the meter and/or incorrect measurement.

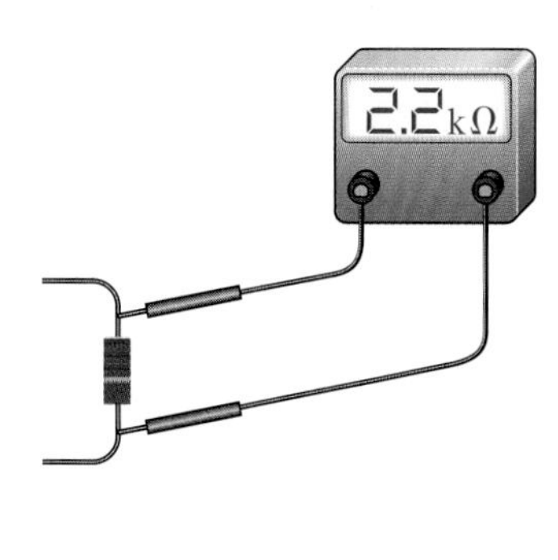

(b) Measure the resistance. (Polarity is not important.)

FIGURE 2–46

Example of using an ohmmeter to measure resistance.

Digital Multimeters (DMMs)

DMMs are the most widely used type of electronic measuring instrument. Generally, DMMs provide more functions, better accuracy, greater ease of reading, and greater reliability than do analog meters, which are covered next. Analog meters have at least one advantage over DMMs, however. They can track short-term variations and trends in a measured quantity that many DMMs are too slow to respond to. Typical DMMs are shown in Figure 2–47. Many DMMs are autoranging types in which the proper range is automatically selected by internal circuitry.

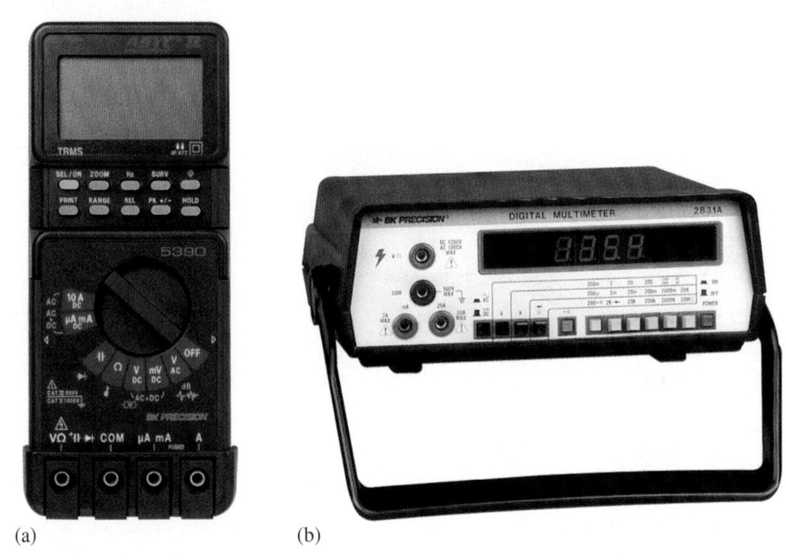

(a) (b)

▶ **FIGURE 2–47**

Typical digital multimeters (DMMs). (Photography courtesy of B&K Precision Corp.)

DMM Functions The basic functions found on most DMMs are

- Ohms
- DC voltage and current
- AC voltage and current

Some DMMs provide additional functions such as analog bar graph displays, transistor or diode tests, power measurement, and decibel measurement for audio amplifier tests.

DMM Displays DMMs are available with either LCD (liquid-crystal display) or LED (light-emitting diode) readouts. The LCD is the most commonly used readout in battery-powered instruments because it requires only very small amounts of current. A typical battery-powered DMM with an LCD readout operates on a 9 V battery that will last from a few hundred hours to 2000 hours and more. The disadvantages of LCD readouts are that (a) they are difficult or impossible to see in low-light conditions and (b) they are relatively slow to respond to measurement changes. LEDs, on the other hand, can be seen in the dark and respond quickly to changes in measured values. LED displays require much more current than LCDs, and, therefore, battery life is shortened when they are used in portable equipment.

Both LCD and LED DMM displays are in a seven-segment format. Each digit in the display consists of seven separate segments, as shown in Figure 2–48(a). Each of the ten decimal digits is formed by activation of appropriate segments, as illustrated in Figure 2–48(b). In addition to the seven segments, there is also a decimal point.

Resolution The **resolution** of a DMM is the smallest increment of a quantity that the DMM can measure. The smaller the increment, the better the resolution. One factor that determines the resolution of a meter is the number of digits in the display.

(a) (b)

FIGURE 2–48

Seven-segment display.

Because many DMMs have 3½ digits in their display, we will use this case for illustration. A 3½-digit multimeter has three digit positions that can indicate from 0 through 9, and one digit position that can indicate only a value of 1. This latter digit, called the *half-digit,* is always the most significant digit in the display. For example, suppose that a DMM is reading 0.999 V, as shown in Figure 2–49(a). If the voltage increases by 0.001 V to 1 V, the display correctly shows 1.000 V, as shown in part (b). The "1" is the half-digit. Thus, with 3½ digits, a variation of 0.001 V, which is the resolution, can be observed.

Now, suppose that the voltage increases to 1.999 V. This value is indicated on the meter as shown in Figure 2–49(c). If the voltage increases by 0.001 V to 2 V, the half-digit cannot display the "2," so the display shows 2.00. The half-digit is blanked and only three digits are active, as indicated in part (d). With only three digits active, the resolution is 0.01 V rather than 0.001 V as it is with 3½ active digits. The resolution remains 0.01 V up to 19.99 V. The resolution goes to 0.1 V for readings of 20.0 V to 199.9 V. At 200 V, the resolution goes to 1 V, and so on.

FIGURE 2–49

A 3½-digit DMM illustrates how the resolution changes with the number of digits in use.

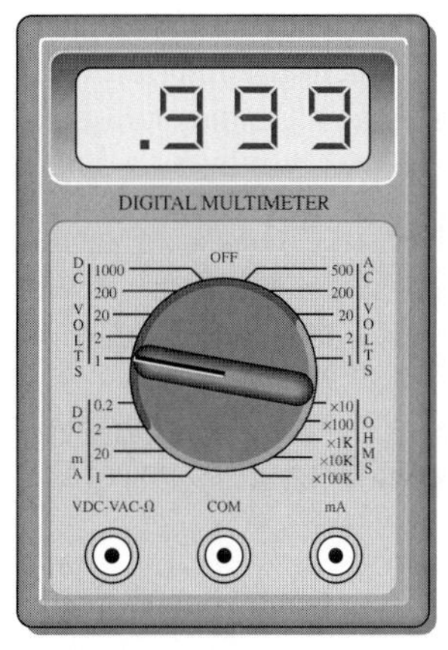

(a) Resolution: 0.001 V

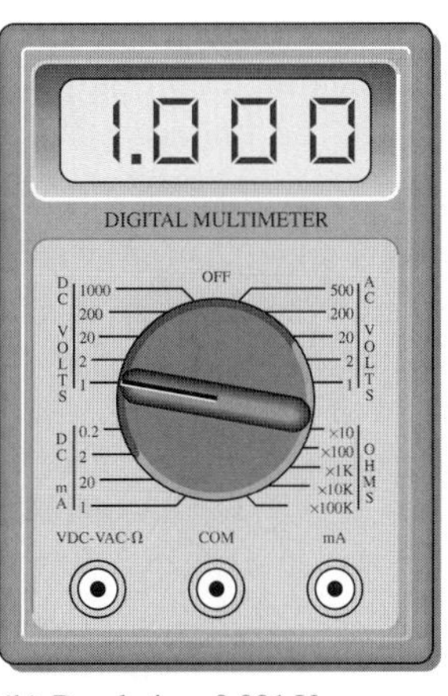

(b) Resolution: 0.001 V

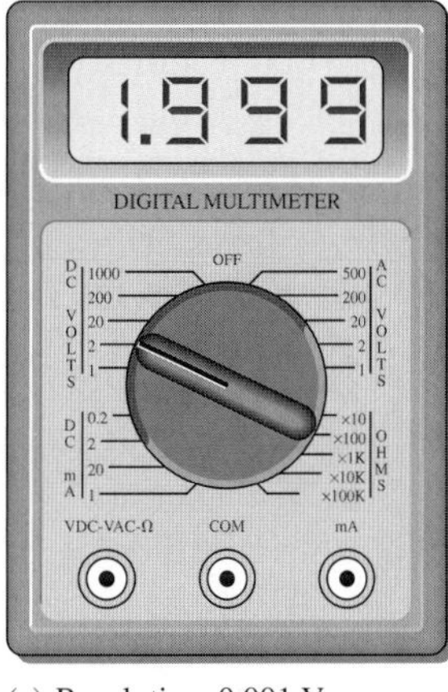

(c) Resolution: 0.001 V

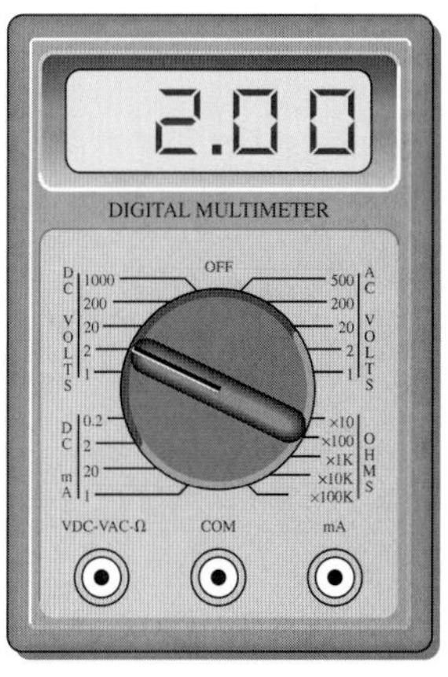

(d) Resolution: 0.01 V

The resolution capability of a DMM is also determined by the internal circuitry and the rate at which the measured quantity is sampled. DMMs with displays of 4½ through 8½ digits are also available.

Accuracy The accuracy is the degree to which a measured value represents the true or accepted value of a quantity. The accuracy of a DMM is established strictly by its internal circuitry and calibration. For typical meters, accuracies range from 0.01% to 0.5%, with some precision laboratory-grade meters going to 0.002%.

HANDS ON TIP

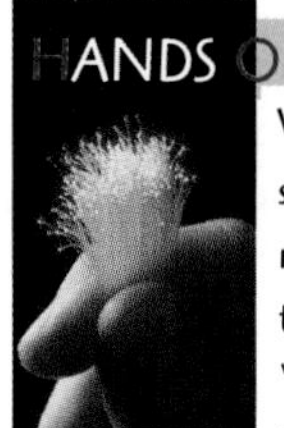

When reading the scales on an analog meter, always view the scale and "needle" directly from the front and not from an angle. This practice will avoid *parallax,* which is an apparent change in position of the needle relative to the meter scale and results in inaccurate readings.

Reading Analog Multimeters

Functions The face of a typical analog needle-type multimeter is represented in Figure 2–50. This particular instrument can be used to measure both direct current (dc) and alternating current (ac) quantities as well as resistance values. It has four selectable functions: dc volts (DC VOLTS), dc milliamperes (DC mA), ac volts (AC VOLTS), and OHMS. Many analog multimeters are similar to this one, although range selections and scales may vary.

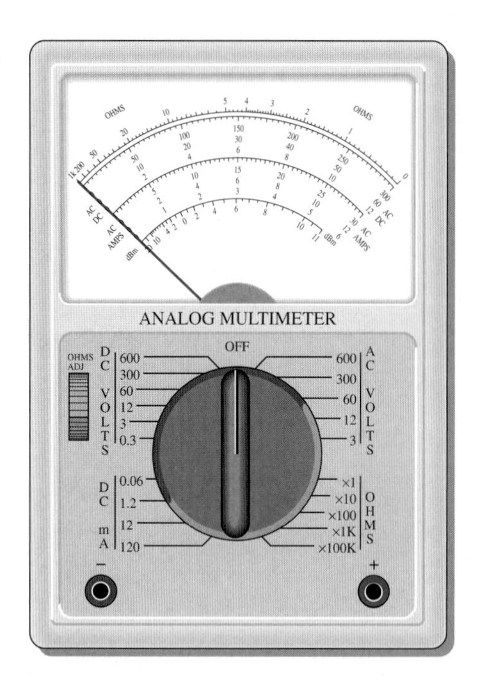

FIGURE 2–50

A typical analog multimeter.

Ranges Within each function there are several ranges, as indicated by the brackets around the selector switch. For example, the DC VOLTS function has 0.3 V, 3 V, 12 V, 60 V, 300 V, and 600 V ranges. Thus, dc voltages from 0.3 V full-scale to 600 V full-scale can be measured. On the DC mA function, direct currents from 0.06 mA full-scale to 120 mA full-scale can be measured. On the ohm scale, the range settings are ×1, ×10, ×100 ×1000, and ×100,000.

The Ohm Scale Ohms are read on the top scale of the meter. This scale is nonlinear; that is, the values represented by each division (large or small) vary as you go across the scale. In Figure 2–50, notice how the scale becomes more compressed as you go from right to left.

To read the actual value in ohms, multiply the number on the scale as indicated by the pointer by the factor selected by the switch. For example, when the switch is set at ×100 and the pointer is at 20, the reading is $20 \times 100 = 2000\ \Omega$.

As another example, assume that the switch is at ×10 and the pointer is at the seventh small division between the 1 and 2 marks, indicating 17 Ω (1.7×10). Now, if the meter remains connected to the same resistance and the switch setting is changed to ×1, the pointer will move to the second small division between the 15 and 20 marks. This, of course, is also a 17 Ω reading, illustrating that a given resistance value can often be read at more than one switch setting. However, the meter should be *zeroed* each time the range is changed by touching the leads together and adjusting the needle.

The AC-DC Scales The second, third, and fourth scales from the top (labeled "AC" and "DC") are used in conjunction with the DC VOLTS, DC mA and AC VOLTS functions. The upper ac-dc scale ends at the 300 mark and is used with the range settings that are multiples of three, such as 0.3, 3, and 300. For example, when the switch is at 3 on the DC VOLTS function, the 300 scale has a full-scale value of 3 V. At the range setting of 300, the full-scale value is 300 V, and so on.

The middle ac-dc scale ends at 60. This scale is used in conjunction with range settings that are multiples of 6, such as 0.06, 60, and 600. For example, when the switch is at 60 on the DC VOLTS function, the full-scale value is 60 V.

The lower ac-dc scale ends at 12 and is used in conjunction with switch settings that are multiples of 12, such as 1.2, 12, and 120. The remaining scales are for ac current and for decibels (covered in Chapter 13).

HANDS ON TIP

When you are using a multimeter, such as the analog multimeter illustrated here, where you manually select the voltage and current ranges, it is good practice to always set the multimeter on the maximum range before you measure an unknown voltage or current. You can then reduce the range until you get an acceptable reading.

EXAMPLE 2–9

In Figure 2–51, determine the quantity (voltage, current, or resistance) that is being measured and its value.

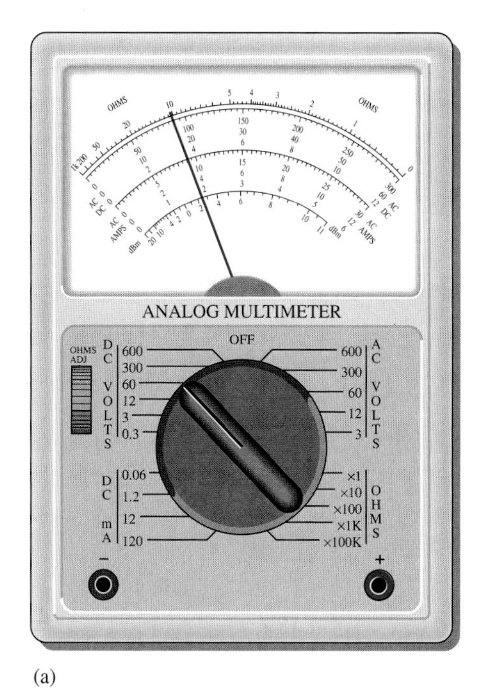

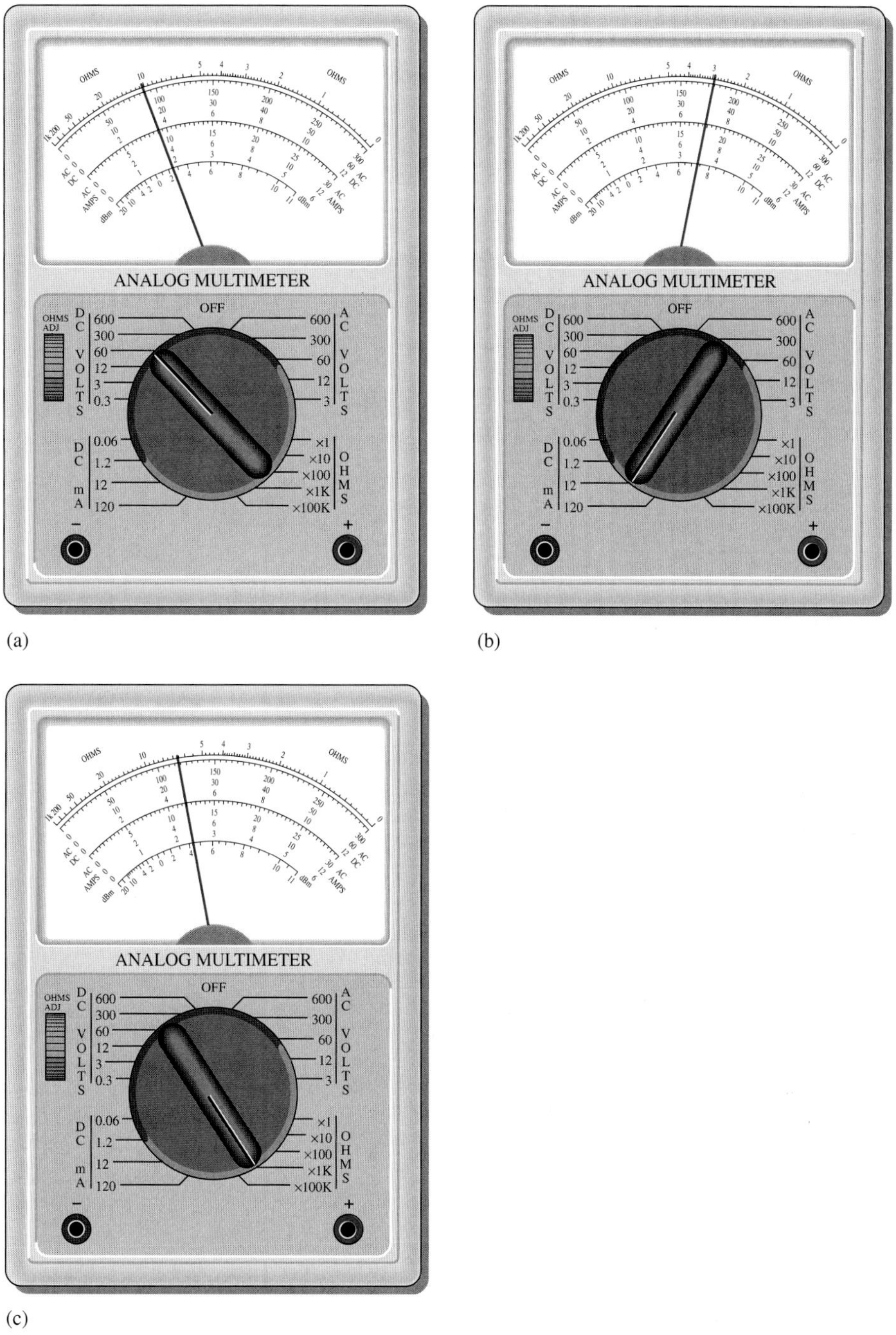

(a)

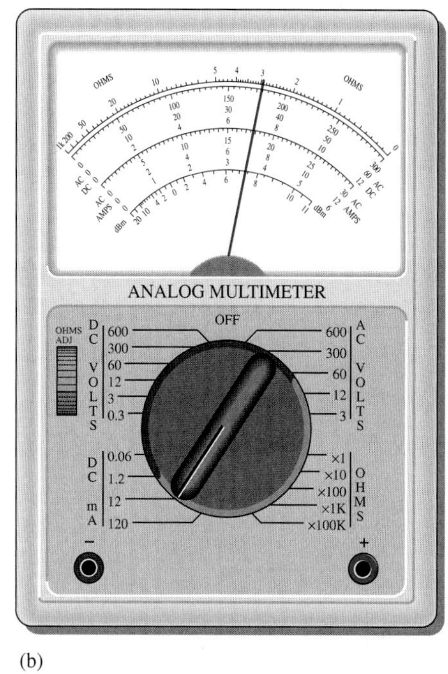

(b)

(c)

▲ **FIGURE 2–51**

Solution **(a)** The switch in Figure 2–51(a) is set on the DC VOLTS function and the 60 V range. The reading taken from the middle dc scale is 18 V.

(b) The switch in Figure 2–51(b) is set on the DC mA function and the 12 mA range. The reading taken from the lower dc scale is 7.2 mA.

(c) The switch in Figure 2–51(c) is set on the ohm (OHMS) function and the ×1000 range. The reading taken from the ohm scale (top scale) is 6.5 kΩ.

Related Problem In Figure 2–51(c) the switch is moved to the ×100 setting. Assuming that the same resistance is being measured, what will the needle do?

SECTION 2–7 REVIEW

1. Name the multimeter functions for measuring (a) current, (b) voltage, and (c) resistance.
2. Show how to place two ammeters in the circuit of Figure 2–34 to measure the current through either lamp (be sure to observe the polarities). How can the same measurements be accomplished with only one ammeter?
3. Show how to place a voltmeter to measure the voltage across lamp 2 in Figure 2–34.
4. List two common types of DMM displays, and discuss the advantages and disadvantages of each.
5. Define *resolution* in a DMM.
6. The multimeter in Figure 2–50 is set on the 3 V range to measure dc voltage. Assume the pointer is at 150 on the upper ac-dc scale. What voltage is being measured?
7. How do you set up the multimeter in Figure 2–50 to measure 275 V dc, and on what scale do you read the voltage?
8. If you expect to measure a resistance in excess of 20 kΩ with the multimeter in Figure 2–50, where do you set the switch?

2–8 ELECTRICAL SAFETY

Safety is a major concern when working with electricity. The possibility of an electric shock or a burn is always present, so caution should always be used. You provide a current path when voltage is applied across two points on your body, and current produces electrical shock. Electrical components often operate at high temperatures, so you can sustain skin burns when you come in contact with them. Also, the presence of electricity creates a potential fire hazard. In this section, safety facts and some precautions are covered.

After completing this section, you should be able to

- **Recognize electrical hazards and practice proper safety procedures**
- Describe the cause of electrical shock
- List the groups of current paths through the body
- Discuss the effects of current on the human body
- List the safety precautions that should be observed when working with electricity

Electric Shock

Current through your body, not the voltage, is the cause of electrical shock. Of course, it takes voltage across a resistance to produce current. When a point on your body comes in contact with a voltage and another point comes in contact with a different voltage or with ground, such as a metal chassis, there will be current through your body from one point to the other. The path of the current depends on the points across which the voltage occurs. The severity of the resulting electrical shock depends on the amount of voltage and the path that the current takes through your body.

The current path through the body determines which tissues and organs will be affected. The current paths can be placed into three groups which are referred to as *touch potential, step potential,* and *touch/step potential.* These are illustrated in Figure 2–52.

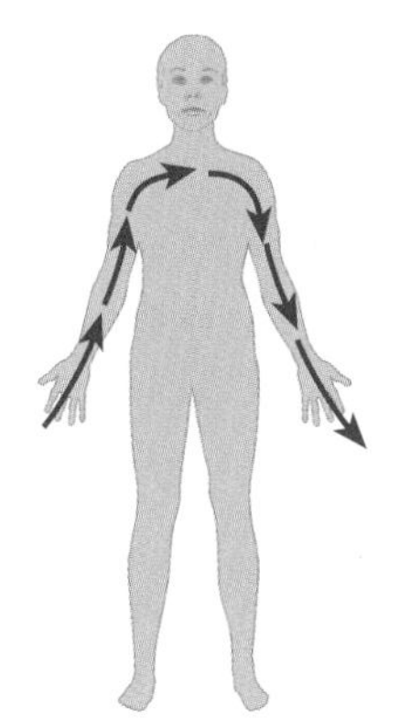

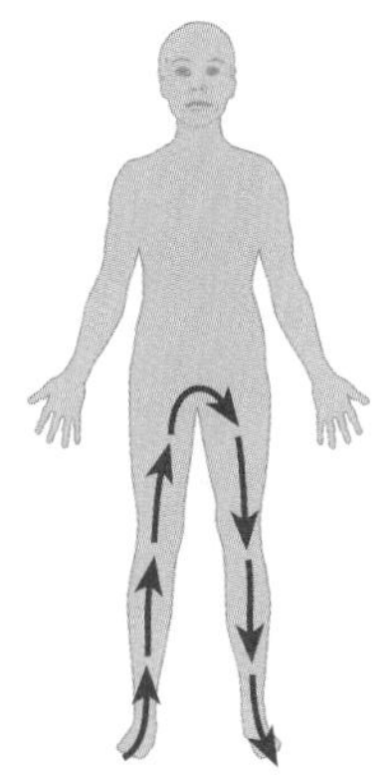

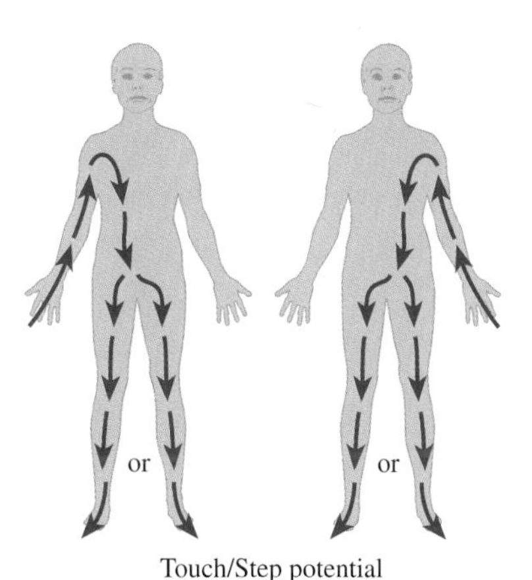

FIGURE 2–52

Shock hazard in terms of three basic current path groups.

Effects of Current on the Human Body The amount of current is dependent on voltage and resistance. The human body has resistance that depends on many factors, which include body mass, skin moisture, and points of contact of the body with a voltage potential. Table 2–5 shows the effects for various values of current in milliamperes.

TABLE 2–5

Physical effects of electrical current. Values vary depending on body mass.

CURRENT (mA)	PHYSICAL EFFECT
0.4	Slight sensation
1.1	Perception threshold
1.8	Shock, no pain, no loss of muscular control
9	Painful shock, no loss of muscular control
16	Painful shock, let-go threshold
23	Severe painful shock, muscular contractions, breathing difficulty
75	Ventricular fibrillation, threshold
235	Ventricular fibrillation, usually fatal for duration of 5 seconds or more
4,000	Heart paralysis (no ventricular fibrillation)
5,000	Tissue burn

Body Resistance Resistance of the human body is typically between 10 kΩ and 50 kΩ and depends on the two points between which it is measured. The moisture of the skin also affects the resistance between two points. The resistance determines the amount of voltage required to produce each of the effects listed in Table 2–5. For example, if you have a resistance of 10 kΩ between two given points on your body, 90 V across those two points will produce enough current (9 mA) to cause painful shock.

Safety Precautions

There are many practical things that you should do when you work with electrical and electronic equipment. Some important precautions are listed here.

- Avoid contact with any voltage source. Turn power off before you work on circuits when touching circuit parts is required.
- Do not work alone.
- Do not work when tired or taking medications that make you drowsy.
- Remove rings, watches, and other metallic jewelry when you work on circuits.
- Do not work on equipment until you know proper procedures and are aware of potential hazards.
- Use equipment with three-wire power cords (three-prong plug).
- Make sure power cords are in good condition and grounding pins are not missing or bent.
- Keep your tools properly maintained. Make sure the insulation on metal tool handles is in good condition.
- Handle tools properly and maintain a neat work area.
- Wear safety glasses when appropriate, particularly when soldering and clipping wires.
- Always shut off power and discharge capacitors before you touch any part of a circuit with your hands.
- Know the location of the emergency power-off switch and emergency exits.
- Never try to override or tamper with safety devices such as an interlock switch.
- Always wear shoes and keep them dry. Do not stand on metal or wet floors.
- Never handle instruments when your hands are wet.
- Never assume that a circuit is off. Double check it with a reliable meter before handling.
- Set the limiter on electronic power supplies to prevent currents larger than necessary to supply the circuit under test.
- Some devices such as capacitors can store a lethal charge for long periods after power is removed. They must be properly discharged before you work with them.
- When making circuit connections, always make the connection to the point with the highest voltage as your last step.
- Avoid contact with the terminals of power supplies.

- Always use wires with insulation and connectors or clips with insulating shrouds.
- Keep cables and wires as short as possible.
- Report any unsafe condition.
- Be aware of and follow all workplace and laboratory rules.
- If another person cannot let go of an energized conductor, switch the power off immediately. If that is not possible, use any available nonconductive material to try to separate the body from the contact.

SECTION 2–8 REVIEW

1. What causes physical pain and/or damage to the body when electrical contact is made?
2. It's OK to wear a ring when working on an electrical circuit. (T or F)
3. Standing on a wet floor presents no safety hazard when working with electricity. (T or F)
4. A circuit can be rewired without removing the power if you are careful. (T or F)
5. Electrical shock can be extremely painful or even fatal. (T or F)

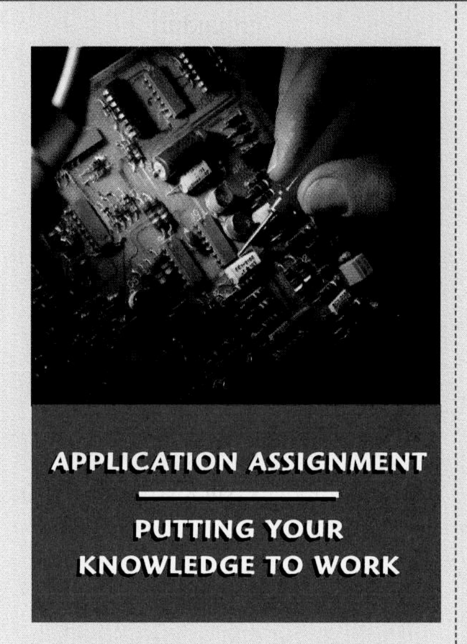

APPLICATION ASSIGNMENT

PUTTING YOUR KNOWLEDGE TO WORK

You are asked to hook up and check out a circuit for use in the special lighting effects for an outdoor performance. You will go through a step-by-step process of working through this assignment using things that you've learned in this chapter. Recall that the requirements from the application assignment preview were

1. The lighting system will consist of six lamps that are controlled so that only one is turned on at a time.
2. The sequence can be in any order.
3. A method must be provided for varying the brightness of the lamp that is on.
4. The system must be fused and operate from a 12 V battery.

Step 1: The Circuit

Select the circuit that meets the requirements from the four configurations in Figure 2–53. Explain why the rejected circuits will not meet the requirements. Explain the purpose of each component in your circuit.

Step 2: Hook Up the Circuit

From the selection of components in Figure 2–54, specify the quantity of each component required for the lighting circuit. Hook up the lighting circuit by developing a point-to-point (From–To) wiring list using component designations and pin numbers.

Step 3: Determine the Current Rating of the Fuse

The fuse protects the circuit by burning open when there is excessive current. The type of lamp to be used draws approximately 5 A when 12 V are applied. Choose a fuse based on its ampere rating that is the most appropriate for your circuit from the following ratings: 1 A, 2 A, 3 A, 4 A, 5 A, 6 A, 7 A, 8 A, 9 A, 10 A, 20 A, and 30 A. Explain your choice.

Step 4: Determine the Capacity of the Battery

A battery can provide a certain amount of current for a certain period of time at its rated voltage. The amount of amps that a battery can deliver to a load for a given number of hours at its rated voltage is its *ampere-hour rating.* Determine the minimum ampere-hour rating of the battery if the lighting circuit must operate for four hours before the battery is recharged. If necessary, see Section 3–7 for a discussion of the ampere-hour rating.

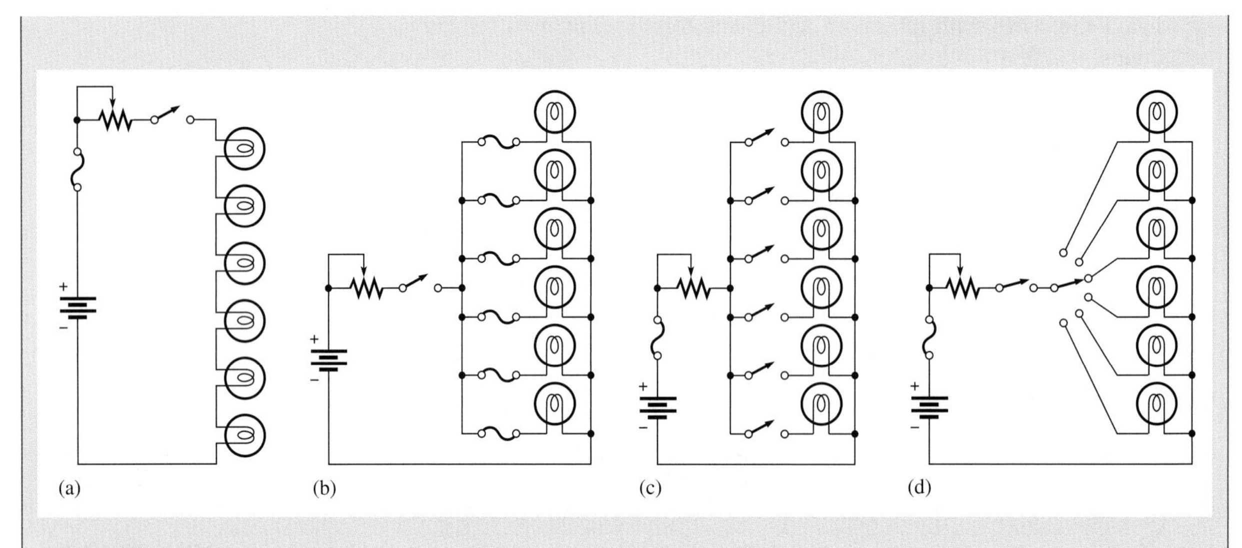

FIGURE 2–53

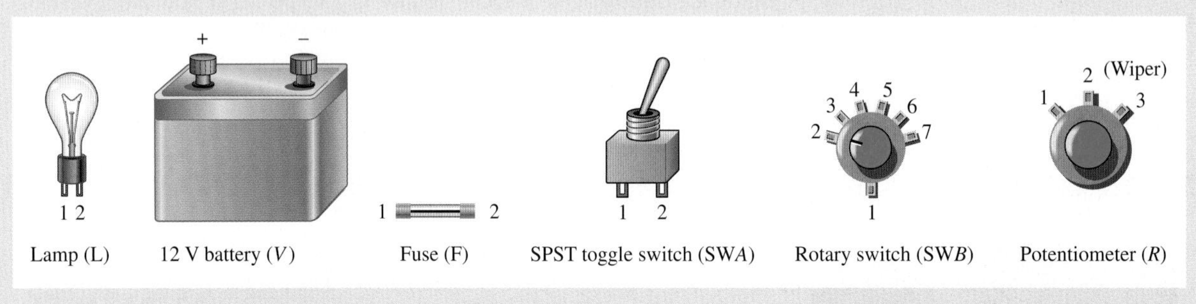

FIGURE 2–54

Component selection.

Step 5: Troubleshoot the Circuit

Once you have connected all the components, it is time to check out the operation to verify that everything is functioning properly.

- Write a brief description of the procedure you would use to thoroughly test the circuit. This is called a *test procedure.*

- For each of the following cases where the circuit is not working properly, determine the possible fault or faults and specify how you would correct it:

1. Each lamp can be turned on except one.
2. None of the lamps can be turned on.
3. Each lamp is too dim and cannot be brightened by adjusting the rheostat.
4. Each lamp is too dim; however, the amount of light can be varied with the rheostat, but not to full brightness.

APPLICATION ASSIGNMENT REVIEW

1. If the resistance of the rheostat in your lighting circuit is reduced, does the brightness of the selected lamp increase or decrease?
2. Does removing any one of the lamps affect the operation of any of the other lamps?
3. What is the current from the battery when no lamps are on?
4. What will happen if you try to turn on two lamps at the same time? Explain.

SUMMARY

- An atom is the smallest particle of an element that retains the characteristics of that element.
- The electron is the basic particle of negative electrical charge.
- The proton is the basic particle of positive charge.
- An ion is an atom that has gained or lost an electron and is no longer neutral.
- When electrons in the outer orbit of an atom (valence electrons) break away, they become free electrons.
- Free electrons make current possible.
- Like charges repel each other, and opposite charges attract each other.
- Voltage must be applied to a circuit before there can be current.
- Resistance limits the current.
- Basically, an electric circuit consists of a source, a load, and a current path.
- An open circuit is one in which the current path is broken.
- A closed circuit is one which has a complete current path.
- An ammeter is connected in line (series) with the current path to measure current.
- A voltmeter is connected across (parallel) the current path to measure voltage.
- An ohmmeter is connected across a resistor to measure resistance. The resistor must be disconnected from the circuit.
- Figure 2–55 shows the electrical symbols introduced in this chapter.
- One coulomb is the charge on 6.25×10^{18} electrons.

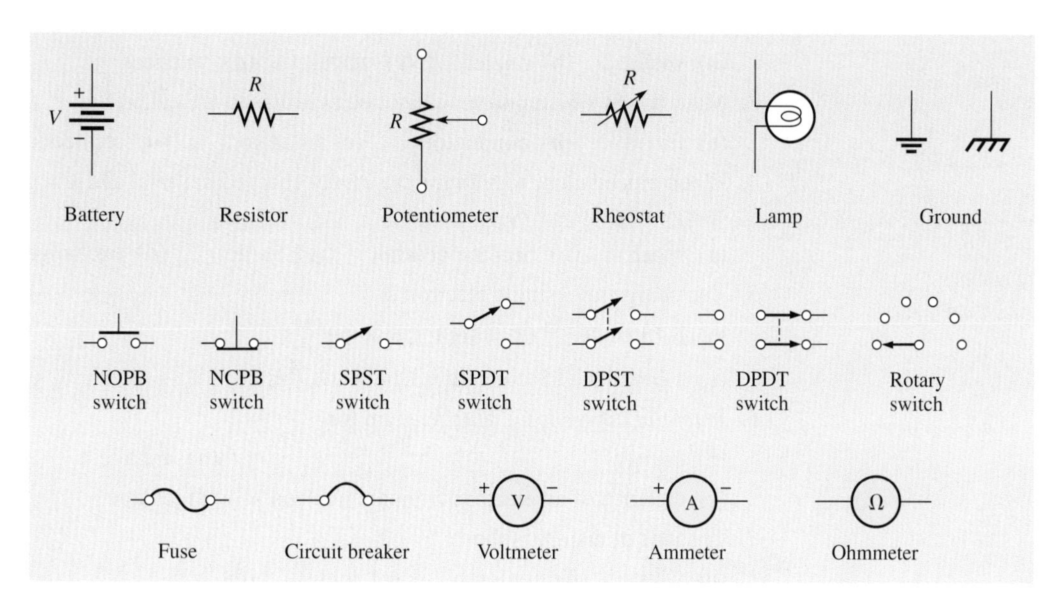

▲ FIGURE 2–55

- One volt is the potential difference (voltage) between two points when one joule of energy is used to move one coulomb of charge from one point to the other.
- One ampere is the amount of current that exists when one coulomb of charge moves through a given cross-sectional area of a material in one second.
- One ohm is the resistance when there is one ampere of current in a material with one volt applied across the material.

EQUATIONS

2–1	$Q = \dfrac{\textbf{number of electrons}}{\textbf{6.25} \times \textbf{10}^{18}\ \textbf{electrons/Coulomb}}$	Charge
2–2	$V = \dfrac{W}{Q}$	Voltage in volts equals energy in joules divided by charge in coulombs.
2–3	$I = \dfrac{Q}{t}$	Current in amperes equals charge in coulombs divided by time in seconds.
2–4	$G = \dfrac{1}{R}$	Conductance in siemens is the reciprocal of resistance in ohms.
2–5	$A = d^2$	Cross-sectional area in circular mils equals the diameter in mils squared.
2–6	$R = \dfrac{\rho l}{A}$	Resistance in resistivity in CM Ω/ft times length in feet divided by cross-sectional area in circular mils.

SELF-TEST

Answers are at the end of the chapter.

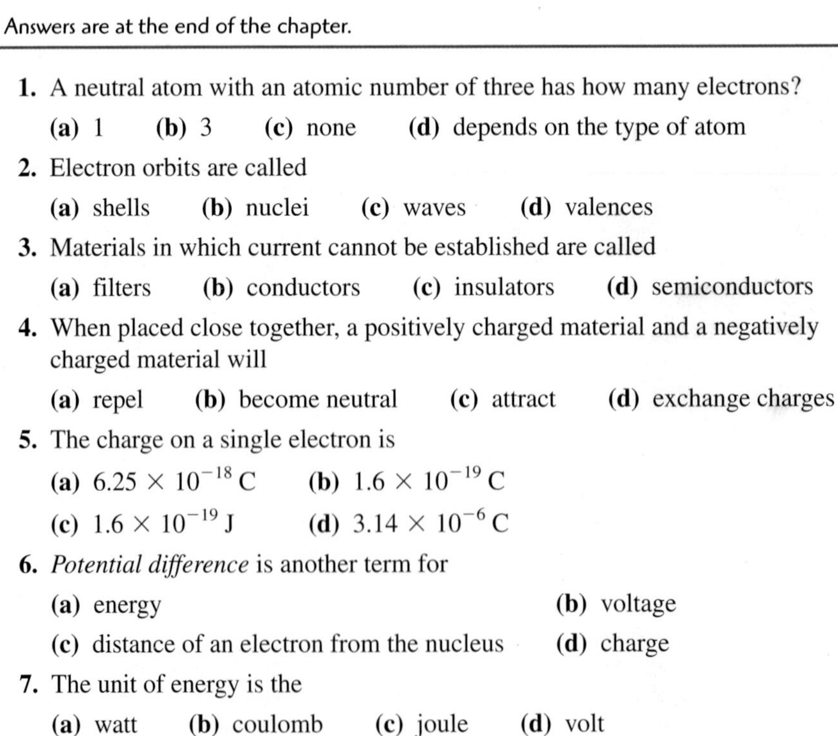

1. A neutral atom with an atomic number of three has how many electrons?
   (a) 1 **(b)** 3 **(c)** none **(d)** depends on the type of atom
2. Electron orbits are called
   (a) shells **(b)** nuclei **(c)** waves **(d)** valences
3. Materials in which current cannot be established are called
   (a) filters **(b)** conductors **(c)** insulators **(d)** semiconductors
4. When placed close together, a positively charged material and a negatively charged material will
   (a) repel **(b)** become neutral **(c)** attract **(d)** exchange charges
5. The charge on a single electron is
   (a) 6.25×10^{-18} C **(b)** 1.6×10^{-19} C
   (c) 1.6×10^{-19} J **(d)** 3.14×10^{-6} C
6. *Potential difference* is another term for
   (a) energy **(b)** voltage
   (c) distance of an electron from the nucleus **(d)** charge
7. The unit of energy is the
   (a) watt **(b)** coulomb **(c)** joule **(d)** volt
8. Which one of the following is not a type of energy source?
   (a) battery **(b)** solar cell **(c)** generator **(d)** potentiometer

9. Which one of the following is generally not a possible condition in an electric circuit?
   (a) voltage and no current (b) current and no voltage
   (c) voltage and current (d) no voltage and no current
10. Electrical current is defined as
    (a) free electrons
    (b) the rate of flow of free electrons
    (c) the energy required to move electrons
    (d) the charge on free electrons
11. There is no current in a circuit when
    (a) a series switch is closed
    (b) a series switch is open
    (c) there is no source voltage
    (d) both (a) and (c) (e) both (b) and (c)
12. The primary purpose of a resistor is to
    (a) increase current (b) limit current
    (c) produce heat (d) resist current change
13. Potentiometers and rheostats are types of
    (a) voltage sources (b) variable resistors
    (c) fixed resistors (d) circuit breakers
14. The current in a given circuit is not to exceed 22 A. Which value of fuse is best?
    (a) 10 A (b) 25 A (c) 20 A (d) a fuse is not necessary

PROBLEMS

Answers to odd-numbered problems are at the end of the book.

BASIC PROBLEMS

SECTION 2–2 **Electrical Charge**

1. How many coulombs of charge do 50×10^{31} electrons possess?
2. How many electrons does it take to make 80 μC of charge?

SECTION 2–3 **Voltage**

3. Determine the voltage in each of the following cases:
   (a) 10 J/C (b) 5 J/2 C (c) 100 J/25 C
4. Five hundred joules of energy are used to move 100 C of charge through a resistor. What is the voltage across the resistor?
5. What is the voltage of a battery that uses 800 J of energy to move 40 C of charge through a resistor?
6. How much energy does a 12 V battery in your car use to move 2.5 C through the electrical circuit?

SECTION 2–4 **Current**

7. Determine the current in each of the following cases:
   (a) 75 C in 1 s (b) 10 C in 0.5 s (c) 5 C in 2 s
8. Six-tenths coulomb passes a point in 3 s. What is the current in amperes?

9. How long does it take 10 C to flow past a point if the current is 5 A?

10. How many coulombs pass a point in 0.1 s when the current is 1.5 A?

SECTION 2–5 **Resistance**

11. Figure 2–56(a) shows color-coded resistors. Determine the resistance value and the tolerance of each.

12. Find the minimum and the maximum resistance within the tolerance limits for each resistor in Figure 2–56(a).

13. **(a)** If you need a 270 Ω resistor, what color bands should you look for?

(b) From the selection of resistors in Figure 2–56(b), choose the following values: 330 Ω, 2.2 kΩ, 39 kΩ, 56 kΩ, and 100 kΩ.

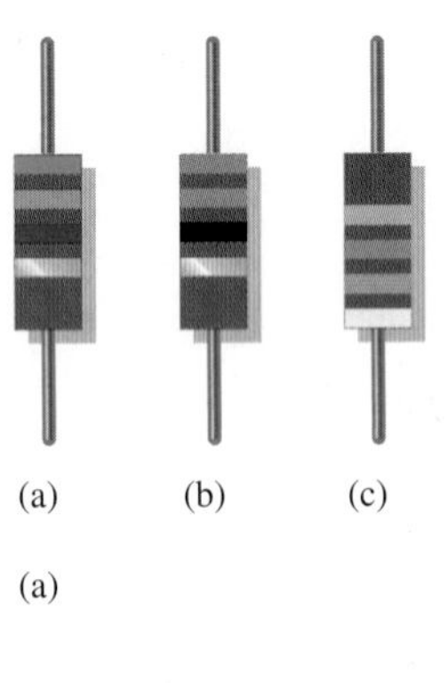

(a)

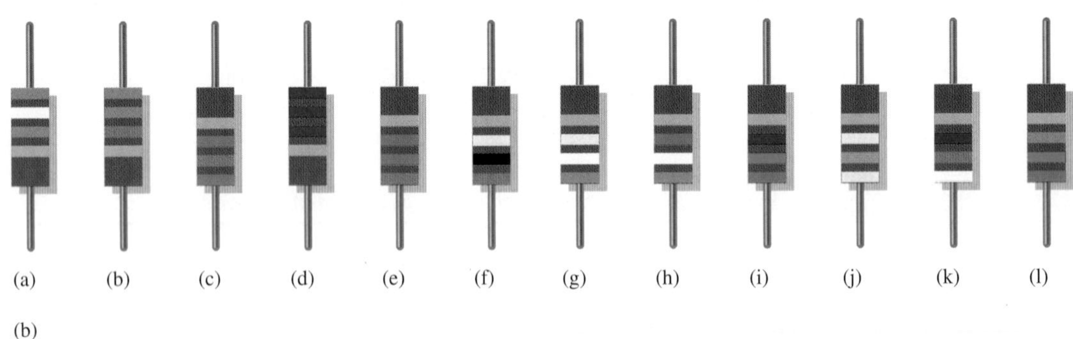

(b)

▲ FIGURE 2–56

14. Determine the resistance value and tolerance for each resistor in Figure 2–57.

▶ FIGURE 2–57

15. Determine the resistance value represented by the numeric labels in parts (a), (b), and (c) and the alphanumeric labels in parts (d), (e), and (f).

(a) 220 **(b)** 472 **(c)** 823

(d) 3K3 **(e)** 560 **(f)** 10M

16. The adjustable contact of a linear potentiometer is set at the mechanical center of its adjustment. If the total resistance is 1000 Ω, what is the resistance between each end terminal and the adjustable contact?

SECTION 2–6

The Electric Circuit

17. Trace the current path in the lamp circuit of Figure 2–34(a) with the switch making contact between the middle and lower pins.

18. With the switch in either position, redraw the lamp circuit in Figure 2–34(b) with a fuse connected to protect the circuit against excessive current.

SECTION 2–7

Basic Circuit Measurements

19. Show the placement of an ammeter and a voltmeter to measure the current and the source voltage in Figure 2–58.

20. Show how you would measure the resistance of R_2 in Figure 2–58.

21. In Figure 2–59 what does each voltmeter indicate when the switch (SW) is in position 1? In position 2?

22. In Figure 2–59, show how to connect an ammeter to measure the current from the voltage source regardless of the switch (SW) position.

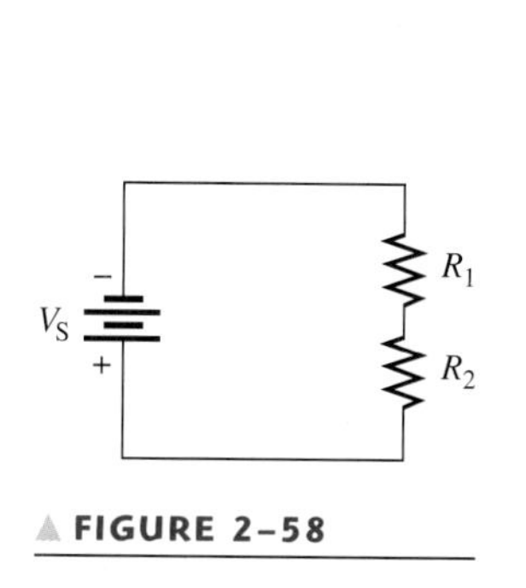

▲ FIGURE 2–58

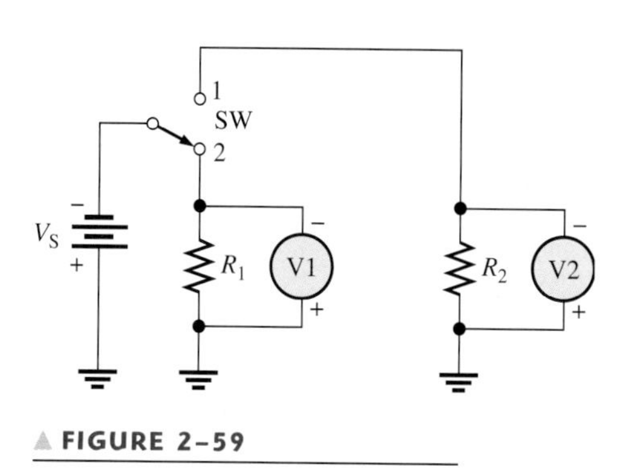

▲ FIGURE 2–59

23. What are the voltage readings of the meters in Figures 2–60(a) and 2–60(b)?

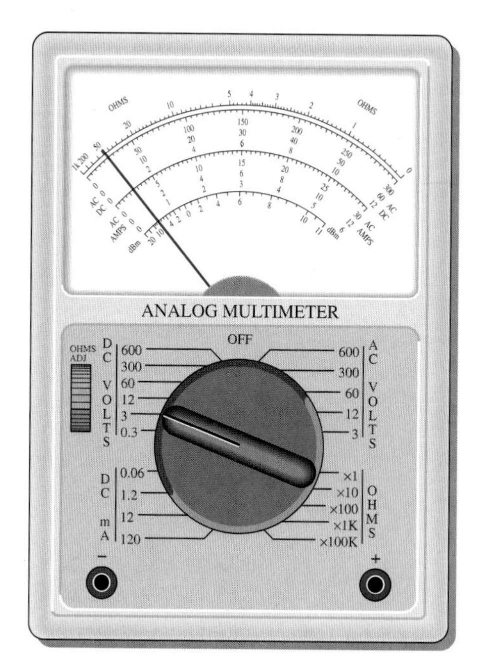

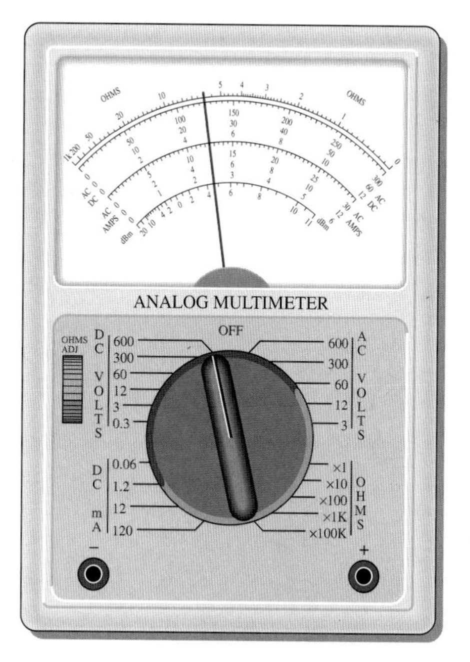

▲ FIGURE 2–60

24. How much resistance is the meter in Figure 2–61 measuring?

FIGURE 2–61

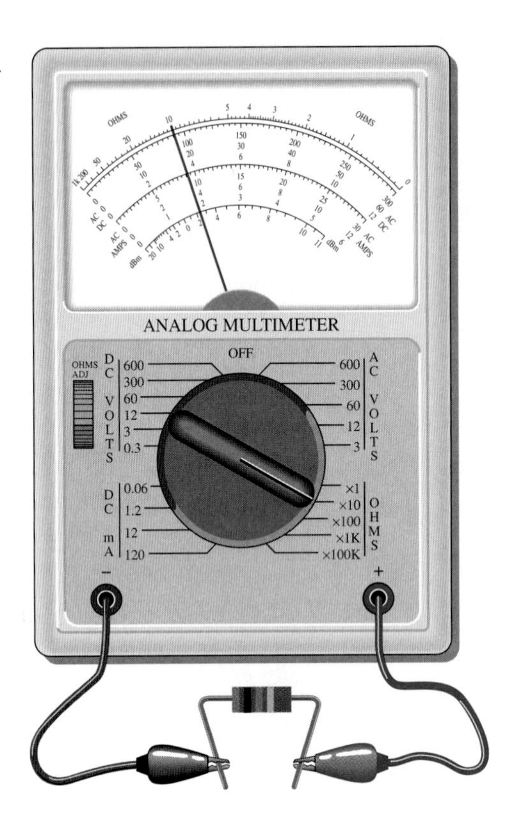

25. Determine the resistance indicated by each of the following ohmmeter readings and range settings:

(a) pointer at 2, range setting at R ×100

(b) pointer at 15, range setting at R ×10M

(c) pointer at 45, range setting at R ×100

26. A multimeter has the following ranges: 1 mA, 10 mA, 100 mA; 100 mV, 1 V, 10 V; R ×1, R ×10, R ×100. Indicate schematically how you would connect the multimeter in Figure 2–62 to measure the following quantities:

(a) I_{R1} **(b)** V_{R1} **(c)** R_1

In each case indicate the *function* to which you would set the meter and the *range* that you would use.

FIGURE 2–62

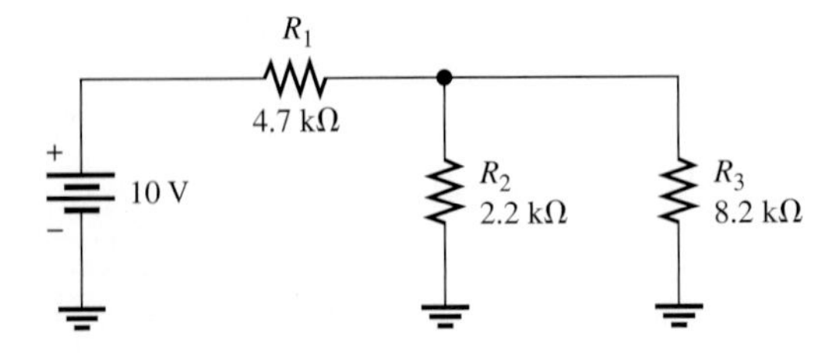

ADVANCED PROBLEMS

27. A resistor with a current of 2 A through it in an amplifier circuit converts 1000 J of electrical energy to heat energy in 15 s. What is the voltage across the resistor?

28. If 574×10^{15} electrons flow through a speaker wire in 250 ms, what is the current in amperes?

29. A 120 V source is to be connected to a 1500 Ω resistive load by two lengths of wire as shown in Figure 2–63. The voltage source is to be located 50 ft from the load. Using Table 2–4, determine the gauge number of the *smallest* wire that can be used if the total resistance of the two lengths of wire is not to exceed 6 Ω.

▶ **FIGURE 2–63**

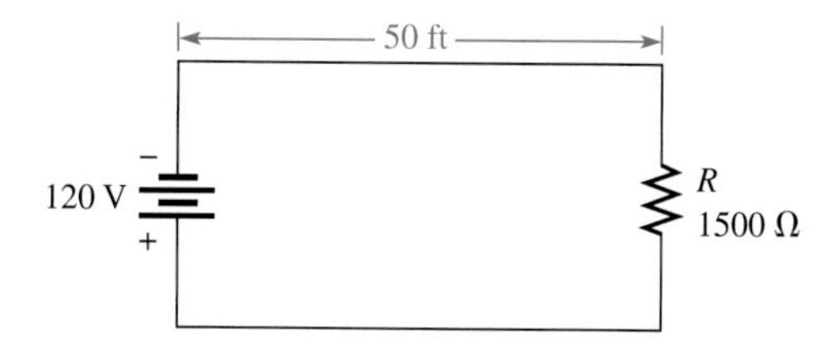

30. Determine the resistance and tolerance of each resistor labeled as follows:

(a) 4R7J **(b)** 560KF **(c)** 1M5G

31. There is only one circuit in Figure 2–64 in which it is possible to have all lamps on at the same time. Determine which circuit it is.

▶ **FIGURE 2–64**

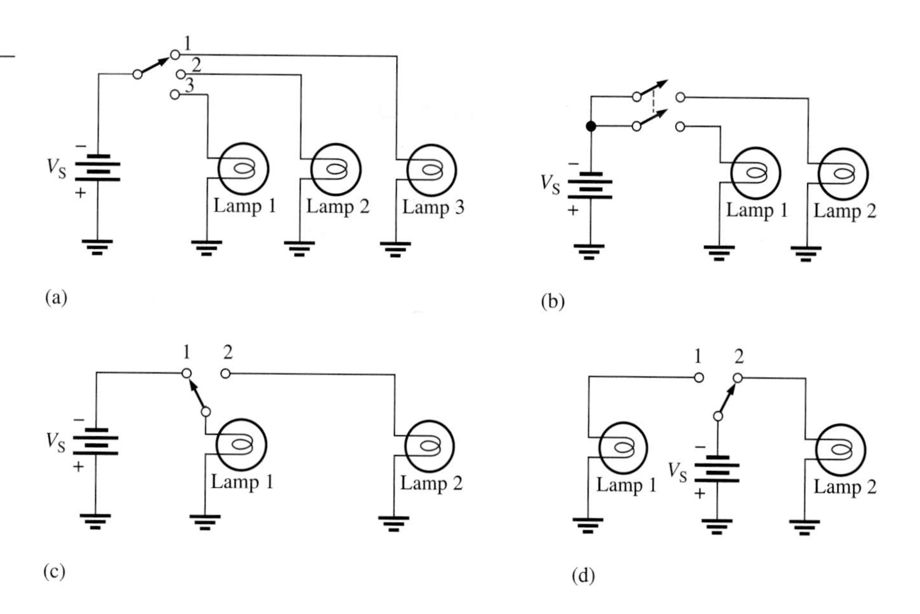

32. Through which resistor in Figure 2–65 is there always current, regardless of the position of the switches?

▶ **FIGURE 2–65**

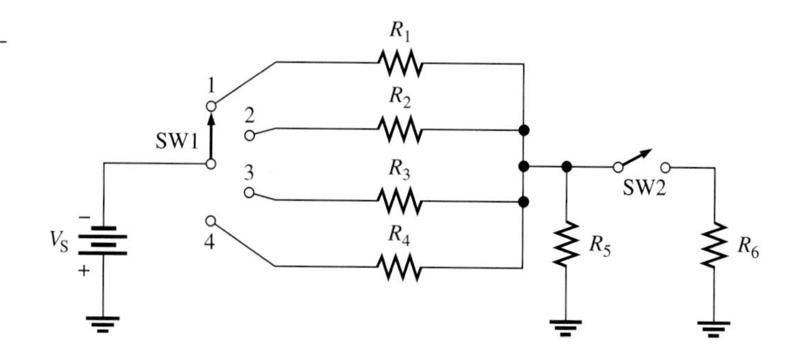

33. In Figure 2–65, show the proper placement of ammeters to measure the current through each resistor and the current out of the battery.

34. Show the proper placement of voltmeters to measure the voltage across each resistor in Figure 2–65.

35. Devise a switch arrangement whereby two voltage sources (V_{S1} and V_{S2}) can be connected simultaneously to either of two resistors (R_1 and R_2) as follows:

V_{S1} connected to R_1 and V_{S2} connected to R_2

V_{S1} connected to R_2 and V_{S2} connected to R_1

36. The different sections of a stereo system are represented by the blocks in Figure 2–66. Show how a single switch can be used to connect the phonograph, the CD (compact disk) player, the tape deck, the AM tuner, or the FM tuner to the amplifier by a single knob control. Only one section can be connected to the amplifier at any time.

FIGURE 2–66

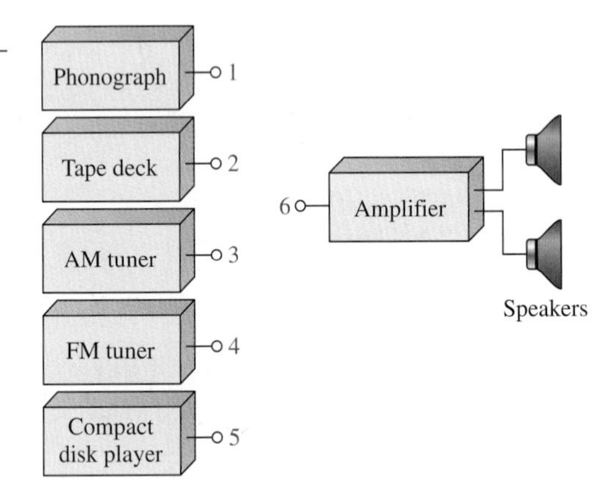

ANSWERS

SECTION REVIEWS

SECTION 2–1 **Atoms**

1. The electron is the basic particle of negative charge.
2. An atom is the smallest particle of an element that retains the unique characteristics of the element.
3. An atom is a positively charged nucleus surrounded by orbiting electrons.
4. Atomic number is the number of protons in a nucleus.
5. No, each element has a different type of atom.
6. A free electron is an outer-shell electron that has drifted away from the parent atom.
7. Shells are energy bands in which electrons orbit the nucleus of an atom.
8. Copper and silver

SECTION 2–2 **Electrical Charge**

1. Q is the symbol for charge.
2. Coulomb is the unit of charge, and C is the symbol for Coulomb.

3. The two types of charge are positive and negative.

4. $Q = \dfrac{10 \times 10^{12}\text{ electrons}}{6.25 \times 10^{18}\text{ electrons/C}} = 1.6 \times 10^{-6}\text{ C} = 1.6\ \mu\text{C}$

SECTION 2–3
Voltage

1. Voltage is energy per unit charge.

2. The unit of voltage is the volt.

3. $V = W/Q = 24\text{ J}/10\text{ C} = 2.4\text{ V}$

4. Battery, power supply, solar cell, and generator are voltage sources.

SECTION 2–4
Current

1. Current is the rate of flow of charge; the unit of current is the ampere (A).

2. There are 6.25×10^{18} electrons in one coulomb.

3. $I = Q/t = 20\text{ C}/4\text{ s} = 5\text{ A}$

SECTION 2–5
Resistance

1. Resistance is opposition to current and its unit is the ohm (Ω).

2. Two resistor categories are fixed and variable. The value of a fixed resistor cannot be changed, but that of a variable resistor can.

3. *First band:* first digit of resistance value.
Second band: second digit of resistance value.
Third band: number of zeros following the 2nd digit.
Fourth band: percent tolerance.

4. **(a)** Yellow, violet, red, gold = 4700 Ω ± 5%
(b) Blue, red, orange, silver = 62,000 Ω ± 10%
(c) Brown, gray, black, gold = 18 Ω ± 5%
(d) Red, red, blue, red, green = 22.5 kΩ ± 0.5%

5. **(a)** 33R = 33 Ω **(b)** 5K6 = 5.6 kΩ
(c) 900 = 900 Ω **(d)** 6M8 = 6.8 MΩ

6. A rheostat has two terminals; a potentiometer has three terminals.

7. A thermistor is a temperature-sensitive resistor.

SECTION 2–6
The Electric Circuit

1. A basic electric circuit consists of source, load, and current path between source and load.

2. An open circuit is one that has no path for current.

3. A closed circuit is one that has a complete path for current.

4. $R = \infty$ (infinite); $R = 0\ \Omega$

5. A fuse protects a circuit against excessive current.

6. A fuse must be replaced once blown. A circuit breaker can be reset once tripped.

7. AWG #3 is larger than AWG #22.

8. Ground is the reference point with zero voltage with respect to other points.

SECTION 2–7
Basic Circuit Measurements

1. **(a)** Ammeter measures current. **(b)** Voltmeter measures voltage.
(c) Ohmmeter measures resistance.

2. See Figure 2–67.

FIGURE 2–67

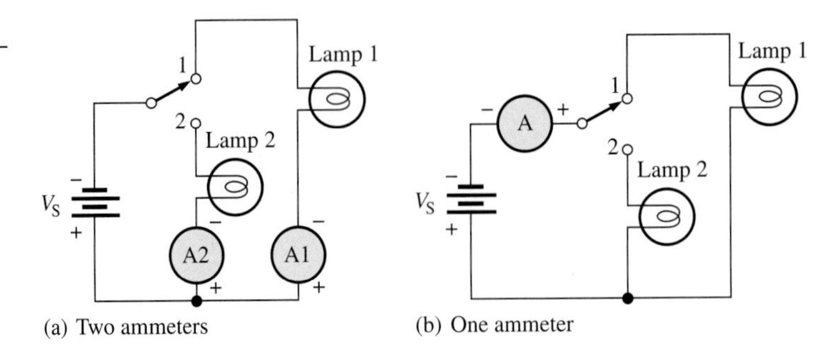

(a) Two ammeters (b) One ammeter

3. See Figure 2–68.

FIGURE 2–68

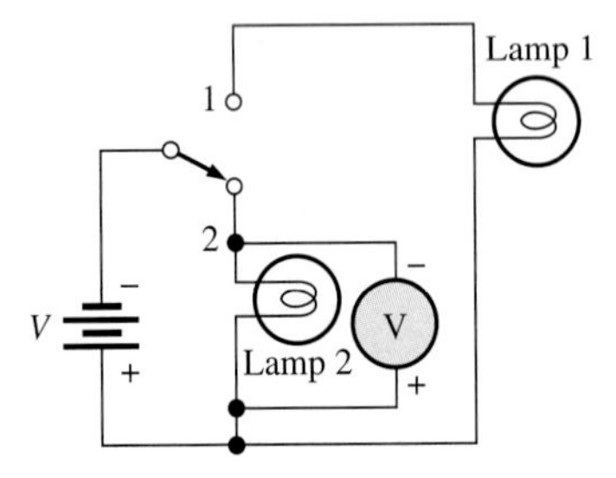

4. Two types of DMM displays are LED and LCD. The LCD requires little current, but it is difficult to see in low light and is slow to respond. The LED can be seen in the dark, and it responds quickly; however, it requires much more current than does the LCD.
5. Resolution is the smallest increment of a quantity that the meter can measure.
6. The voltage being measured is 1.5 V.
7. DC VOLTS, 600 setting; 275 V is read on the 60 scale near midpoint.
8. OHMS ×1000

SECTION 2–8 **Electrical Safety**

1. Current
2. F
3. F
4. F
5. T

■ Application Assignment

1. The brightness increases.
2. No
3. There is no current when all lights are off.
4. The total current would increase, possibly causing the fuse to blow.

RELATED PROBLEMS FOR EXAMPLES

2–1 1.88×10^{19} electrons

2–2 600 J

2–3 12 C

2–4 4700 Ω ± 5%

2–5 47.5 Ω ± 2%

2–6 1.25 kΩ

2–7 404.01 CM; #24

2–8 16.14 Ω

2–9 The needle will move left to the “65” mark on the top scale.

SELF-TEST

1. (b)	**2.** (a)	**3.** (c)	**4.** (c)	**5.** (b)	**6.** (b)	**7.** (c)
8. (d)	**9.** (b)	**10.** (b)	**11.** (e)	**12.** (b)	**13.** (b)	**14.** (c)

3

OHM'S LAW, ENERGY, AND POWER

INTRODUCTION

Georg Simon Ohm (1787–1854) experimentally found that voltage, current, and resistance are all related in a specific way. This basic relationship, known as *Ohm's law,* is one of the most fundamental and important laws in the fields of electricity and electronics. In this chapter, Ohm's law is examined, and its use in practical circuit applications is discussed and demonstrated by numerous examples.

In addition to Ohm's law, the concepts and definitions of energy and power in electric circuits are introduced and the Watt's law power formulas are given. A general approach to troubleshooting using the analysis, planning, and measurement (APM) method is also introduced.

CHAPTER OUTLINE

CHAPTER OBJECTIVES

- Explain Ohm's law
- Use Ohm's law to determine voltage, current, or resistance
- Define *energy* and *power*
- Calculate power in a circuit
- Properly select resistors based on power consideration
- Explain energy conversion and voltage drop
- Discuss power supplies and their characteristics
- Describe a basic approach to troubleshooting

KEY TERMS

- Ohm's law
- Linear
- Energy
- Power
- Joule
- Watt
- Kilowatt-hour
- Watt's law
- Power rating
- Voltage drop
- Ampere-hour rating
- Efficiency
- Troubleshooting
- Half-splitting

APPLICATION ASSIGNMENT PREVIEW

As a technician working in the engineering development lab of an electronics company, you are assigned the task of modifying an existing test fixture for use in a new application. The test fixture is a resistance box with an array of switch-selectable resistors of various values. Your job will be to determine and specify the changes to be made in the existing circuit and develop a procedure to test the box once the modifications have been made. To complete the assignment, you will

1. Determine color-coded resistor values.
2. Determine resistor power ratings.
3. Draw the schematic of the existing circuit.
4. Draw the schematic to meet the new requirements.
5. Use Watt's law power formulas to determine required power ratings.
6. Use Ohm's law to determine resistor values to meet specifications.

After studying this chapter, you should be able to complete the application assignment.

WWW. VISIT THE COMPANION WEBSITE

Circuit Simulation Tutorials and Other Chapter Study Tools Are Available at

http://www.prenhall.com/floyd

3–1 OHM'S LAW

Ohm's law describes mathematically how voltage, current, and resistance in a circuit are related. Ohm's law can be written in three equivalent forms; the formula you use depends on the quantity you need to determine. In this section, you will learn each of these forms.

After completing this section, you should be able to

- **Explain Ohm's law**
- Describe how voltage (V), current (I), and resistance (R) are related
- Express I as a function of V and R
- Express V as a function of I and R
- Express R as a function of V and I

Ohm determined experimentally that *if the voltage across a resistor is increased, the current through the resistor will increase;* and, likewise, *if the voltage is decreased, the current will decrease.* For example, if the voltage is doubled, the current will double. If the voltage is halved, the current will also be halved. This relationship is illustrated in Figure 3–1, with meter indications of voltage and current.

FIGURE 3–1

Effect on the current of a change in the voltage with the resistance at a constant value.

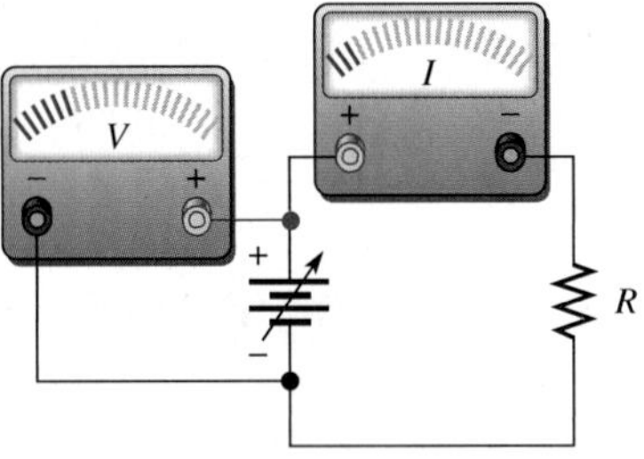

(a) Less V, less I

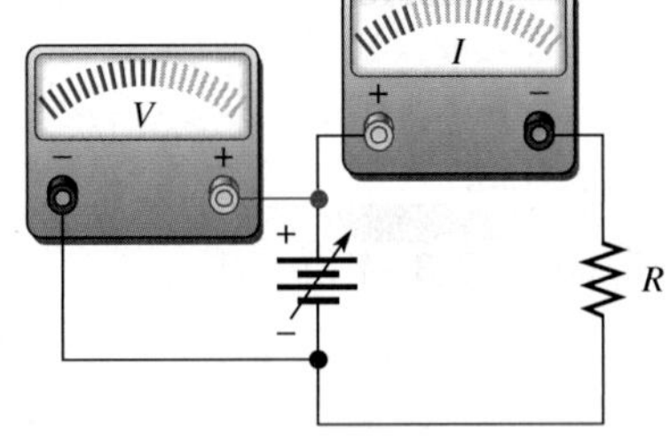

(b) More V, more I

Ohm's law also shows that *if the voltage is kept constant, less resistance results in more current, and more resistance results in less current.* For example, if the resistance is halved, the current doubles. If the resistance is doubled, the current is halved. This concept is illustrated by the meter indications in Figure 3–2, where the resistance is increased and the voltage is held constant.

FIGURE 3–2

Effect on the current of a change in the resistance with the voltage at a constant value.

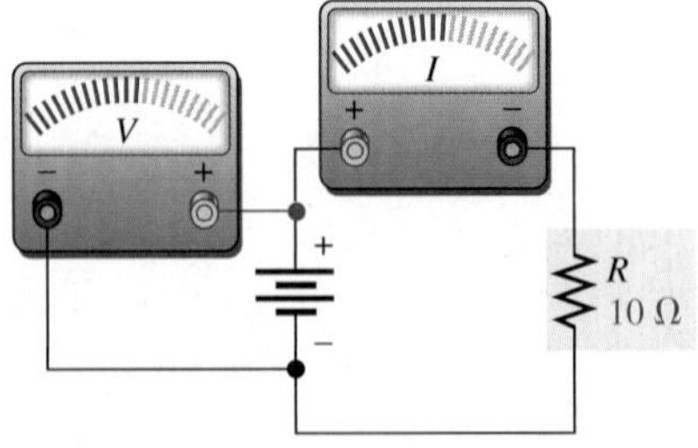

(a) Less R, more I

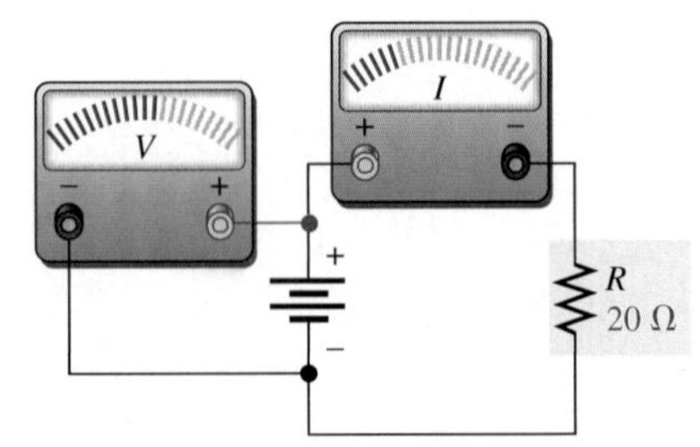

(b) More R, less I

Formula for Current

Ohm's law can be stated as follows:

Equation 3–1

$$I = \frac{V}{R}$$

This formula describes the relationship illustrated by the action in the circuits of Figures 3–1 and 3–2.

For a constant resistance, if the voltage applied to a circuit is increased, the current will increase; and if the voltage is decreased, the current will decrease.

$I = \frac{V}{R}$ Increase V, I increases

$I = \frac{V}{R}$ Decrease V, I decreases

R constant

For a constant voltage, if the resistance in a circuit is increased, the current will decrease; and if the resistance is decreased, the current will increase.

$I = \frac{V}{R}$ Increase R, I decreases

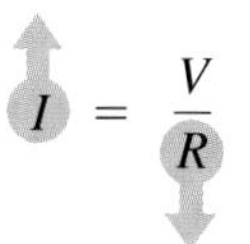

$I = \frac{V}{R}$ Decrease R, I increases

V constant

Using Equation 3–1, you can calculate the current in amperes if the values of voltage in volts and resistance in ohms are known.

EXAMPLE 3–1

Using the Ohm's law formula in Equation 3–1, verify that the current through a 10 Ω resistor increases when the voltage is increased from 5 V to 20 V.

Solution The following calculations show that the current increases from 0.5 A to 2 A. The calculator sequence for each calculation is also shown, based on the TI-85/TI-86 set in the ENG mode.

For $V = 5$ V,

$$I = \frac{V}{R} = \frac{5\text{ V}}{10\ \Omega} = \mathbf{0.5\ A}$$

For $V = 20$ V,

$$I = \frac{V}{R} = \frac{20\text{ V}}{10\ \Omega} = \mathbf{2\ A}$$

*Related Problem** Show that the current decreases when the resistance is increased from 5 Ω to 20 Ω and the voltage is a constant 10 V.

*Answers are at the end of the chapter.

Formula for Voltage

Ohm's law can also be stated another equivalent way. By multiplying both sides of Equation 3–1 by R and transposing terms, you obtain an equivalent form of Ohm's law, as follows:

Equation 3–2

$$V = IR$$

With this formula, you can calculate voltage in volts if the current in amperes and resistance in ohms are known.

EXAMPLE 3–2

Use the Ohm's law formula in Equation 3–2 to calculate the voltage across a 100 Ω resistor when the current is 2 A.

Solution

$$V = IR = (2\ \text{A})(100\ \Omega) = \mathbf{200\ V}$$

Related Problem Find the voltage across a 1.0 kΩ resistor when the current is 1 mA.

Formula for Resistance

There is a third equivalent way to state Ohm's law. By dividing both sides of Equation 3–2 by I and transposing terms, you obtain the following formula:

Equation 3–3

$$R = \frac{V}{I}$$

This form of Ohm's law is used to determine resistance in ohms if the values of voltage in volts and current in amperes are known.

The three formulas—Equations 3–1, 3–2, and 3–3—are all equivalent. They are simply three different ways of expressing Ohm's law.

EXAMPLE 3–3

Use the Ohm's law formula in Equation 3–3 to calculate the resistance in a circuit when the voltage is 12 V and the current is 0.5 A.

Solution

$$R = \frac{V}{I} = \frac{12\ \text{V}}{0.5\ \text{A}} = \mathbf{24\ \Omega}$$

Related Problem Find the resistance when the voltage is 9 V and the current is 10 mA.

The Linear Relationship of Current and Voltage

In resistive circuits, current and voltage are linearly proportional. **Linear** means that if one is increased or decreased by a certain percentage, the other will increase or decrease by the same percentage, assuming that the resistance is constant in value. For example, if the voltage across a resistor is tripled, the current will triple. If the voltage is reduced by half, the current will decrease by half.

EXAMPLE 3–4

Show that if the voltage in the circuit of Figure 3–3 is increased to three times its present value, the current will triple in value.

FIGURE 3–3

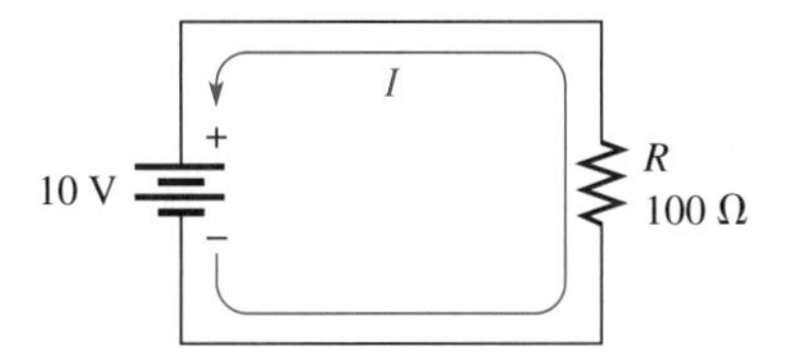

Solution With 10 V, the current is

$$I = \frac{V}{R} = \frac{10\text{ V}}{100\ \Omega} = 0.1\text{ A}$$

If the voltage is increased to 30 V, the current will be

$$I = \frac{V}{R} = \frac{30\text{ V}}{100\ \Omega} = 0.3\text{ A}$$

The current went from 0.1 A to 0.3 A when the voltage was tripled to 30 V.

Related Problem If the voltage in Figure 3–3 is quadrupled, will the current also quadruple?

A Graph of Current Versus Voltage

Let's take a constant value of resistance, for example, 10 Ω, and calculate the current for several values of voltage ranging from 10 V to 100 V. The current values obtained are shown in Figure 3–4(a). The graph of the I values versus the V values is shown in Figure 3–4(b). Note that it is a straight line graph. This graph shows that a change in voltage results in a linearly proportional change in current. No

V	I
10 V	1 A
20 V	2 A
30 V	3 A
40 V	4 A
50 V	5 A
60 V	6 A
70 V	7 A
80 V	8 A
90 V	9 A
100 V	10 A

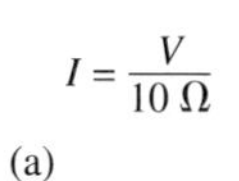

(a)

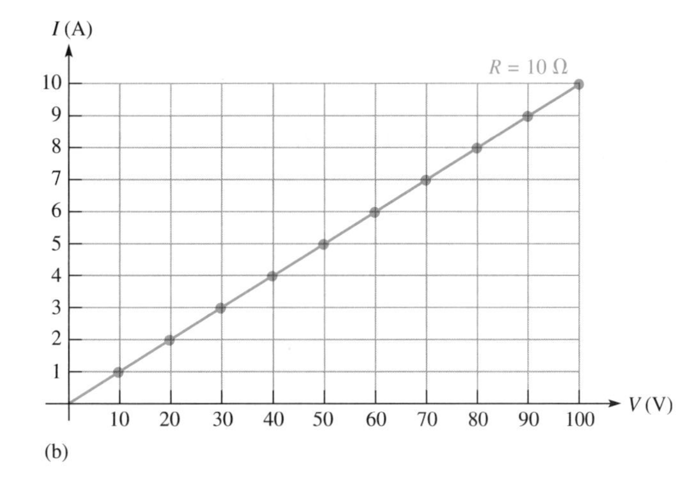

(b)

FIGURE 3–4

Graph of current versus voltage for $R = 10\ \Omega$.

matter what value R is, assuming that R is constant, the graph of I versus V will always be a straight line.

A Graphic Aid for Ohm's Law

You may find the graphic aid in Figure 3–5 helpful for applying Ohm's law. It is a way to remember the formulas.

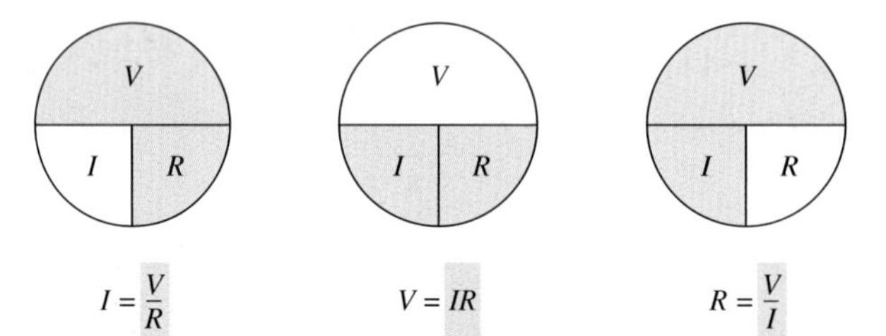

FIGURE 3–5

A graphic aid for the Ohm's law formulas.

SECTION 3–1 REVIEW

Answers are at the end of the chapter.

1. Briefly state Ohm's law in words.
2. Write the Ohm's law formula for calculating current when voltage and resistance are known.
3. Write the Ohm's law formula for calculating voltage when current and resistance are known.
4. Write the Ohm's law formula for calculating resistance when voltage and current are known.
5. If the voltage across a resistor is tripled, does the current increase or decrease? By how much?
6. There is a fixed voltage across a variable resistor, and you measure a current of 10 mA. If you double the resistance, how much current will you measure?
7. What happens to the current in a linear circuit where both the voltage and the resistance are doubled?

3–2 APPLICATION OF OHM'S LAW

This section provides examples of the application of Ohm's law for calculating voltage, current, and resistance in electric circuits. You will also see how to use quantities expressed with metric prefixes in circuit calculations.

After completing this section, you should be able to

- **Use Ohm's law to determine voltage, current, or resistance**
- Use Ohm's law to find current when you know voltage and resistance
- Use Ohm's law to find voltage when you know current and resistance
- Use Ohm's law to find resistance when you know voltage and current
- Use quantities with metric prefixes

Determining I When You Know V and R

In these examples you will learn to determine current values when you know the values of voltage and resistance. In these problems, the formula $I = V/R$ is used.

In order to get current in amperes, you must express the value of V in volts and the value of R in ohms.

EXAMPLE 3–5

How many amperes of current are in the circuit of Figure 3–6?

FIGURE 3–6

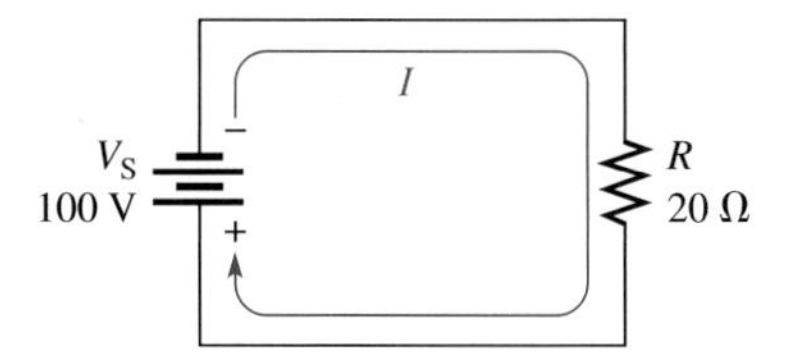

Solution Use the Ohm's law formula $I = V/R$ to find the current. Substitute the values for the source voltage and resistance shown in the figure.

$$I = \frac{V_S}{R} = \frac{100\text{ V}}{20\ \Omega} = \mathbf{5\ A}$$

There are 5 A of current in this circuit.

Related Problem What is the current in Figure 3–6 if the voltage is reduced to 50 V?

Open file E03-05 on your EWB/CircuitMaker CD-ROM. Connect the multimeter to the circuit and verify the value of current calculated in this example.

In electronics, resistance values of thousands or millions of ohms are common. As you learned in Chapter 1, large values of resistance are indicated by the metric system prefixes *kilo* (k) and *mega* (M). Thus, thousands of ohms are expressed in kilohms (kΩ), and millions of ohms are expressed in megohms (MΩ). The following examples illustrate how to use kilohms and megohms when you use Ohm's law to calculate current.

EXAMPLE 3–6

Calculate the current in milliamperes for the circuit of Figure 3–7.

FIGURE 3–7

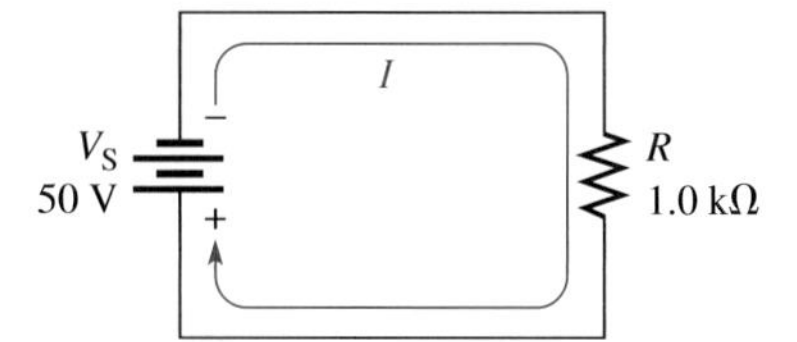

Solution Remember that 1.0 kΩ is the same as $1.0 \times 10^3\ \Omega$. Use the formula $I = V/R$ and substitute 50 V for V and $1.0 \times 10^3\ \Omega$ for R.

$$I = \frac{V_S}{R} = \frac{50\text{ V}}{1.0\text{ k}\Omega} = \frac{50\text{ V}}{1.0 \times 10^3\ \Omega} = 50 \times 10^{-3}\text{ A} = \mathbf{50\ mA}$$

5 0 ÷ 1 EE 3 ENTER

50/1E3
50E⁻3

Related Problem If the resistance in Figure 3–7 is increased to 10 kΩ, what is the current?

Open file E03-06 on your EWB/CircuitMaker CD-ROM. Connect the multimeter to the circuit and verify the value of current calculated in this example.

In Example 3–6, the current is expressed as 50 mA. Thus, *when volts (V) are divided by kilohms (kΩ), the current is in milliamperes (mA).*

When volts (V) are divided by megohms (MΩ), the current is in microamperes (μA), as Example 3–7 illustrates.

EXAMPLE 3–7

Determine the amount of current in microamperes for the circuit of Figure 3–8.

▶ FIGURE 3–8

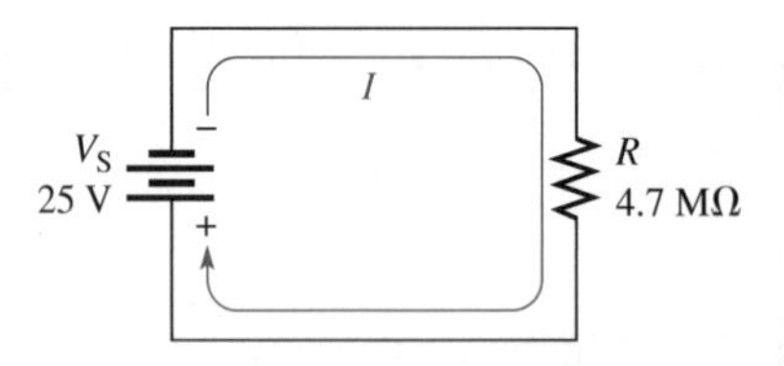

Solution Recall that 4.7 MΩ equals 4.7×10^6 Ω. Use the formula $I = V/R$ and substitute 25 V for V and 4.7×10^6 Ω for R.

$$I = \frac{V_S}{R} = \frac{25\text{ V}}{4.7\text{ M}\Omega} = \frac{25\text{ V}}{4.7 \times 10^6\ \Omega} = 5.32 \times 10^{-6}\text{ A} = \mathbf{5.32\ \mu A}$$

25/4.7E6
5.31914893617E⁻6

Related Problem If the resistance in Figure 3–8 is decreased to 1.0 MΩ, what is the current?

Open file E03-07 on your EWB/CircuitMaker CD-ROM. Connect the multimeter to the circuit and verify the value of current calculated in this example.

Small voltages, usually less than 50 V, are common in electronic circuits. Occasionally, however, large voltages are encountered. For example, the high-voltage supply in a television receiver is around 20,000 V (20 kV). Transmission voltages generated by the power companies may be as high as 345,000 V (345 kV).

EXAMPLE 3–8

How much current in microamperes is there through a 100 MΩ resistor when 50 kV are applied across it?

Solution Divide 50 kV by 100 MΩ to get the current. Substitute 50×10^3 V for 50 kV and $100 \times 10^6\ \Omega$ for 100 MΩ in the formula for current. V_R is the voltage across the resistor.

$$I = \frac{V_R}{R} = \frac{50\text{ kV}}{100\text{ M}\Omega} = \frac{50 \times 10^3\text{ V}}{100 \times 10^6\ \Omega} = 0.5 \times 10^{-3}\text{ A} = 500 \times 10^{-6} = \mathbf{500\ \mu A}$$

Related Problem How much current is there through 10 MΩ when 2 kV are applied?

Determining *V* When You Know *I* and *R*

In these examples you will see how to determine voltage values when you know the current and resistance using the formula $V = IR$. To obtain voltage in volts, you must express the value of I in amperes and the value of R in ohms.

EXAMPLE 3–9

In the circuit of Figure 3–9, how much voltage is needed to produce 5 A of current?

▶ **FIGURE 3–9**

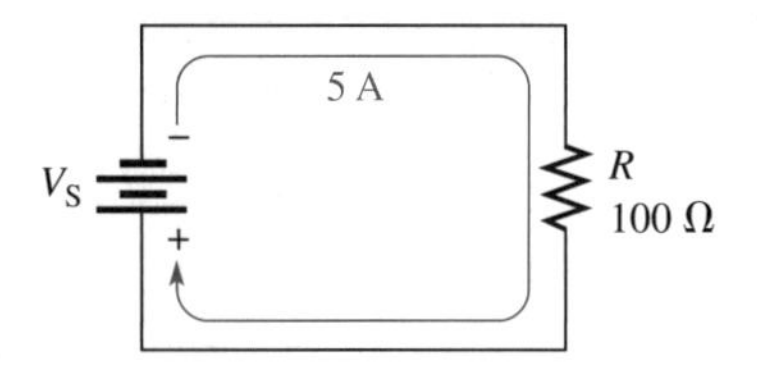

Solution Substitute 5 A for I and 100 Ω for R into the formula $V = IR$.

$$V_S = IR = (5\text{ A})(100\ \Omega) = \mathbf{500\ V}$$

Thus, 500 V are required to produce 5 A of current through a 100 Ω resistor.

Related Problem How much voltage is required to produce 8 A in the circuit of Figure 3–9?

EXAMPLE 3–10

How much voltage will be measured across the resistor in Figure 3–10?

▶ FIGURE 3–10

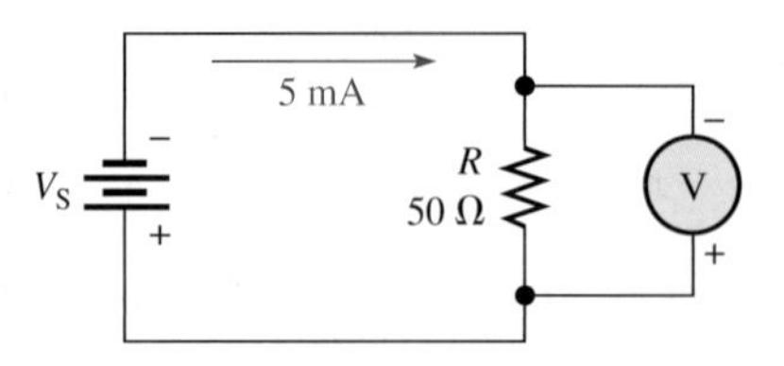

Solution Note that 5 mA equals 5×10^{-3} A. Substitute the values for I and R into the formula $V = IR$.

$$V_R = IR = (5\text{ mA})(50\ \Omega) = (5 \times 10^{-3}\text{ A})(50\ \Omega) = \mathbf{250\ mV}$$

When milliamperes are multiplied by ohms, the result is millivolts.

Related Problem Change the resistor in Figure 3–10 to 22 Ω and determine the voltage required to produce 10 mA.

EXAMPLE 3–11

The circuit in Figure 3–11 has a current of 10 mA. What is the source voltage?

▶ FIGURE 3–11

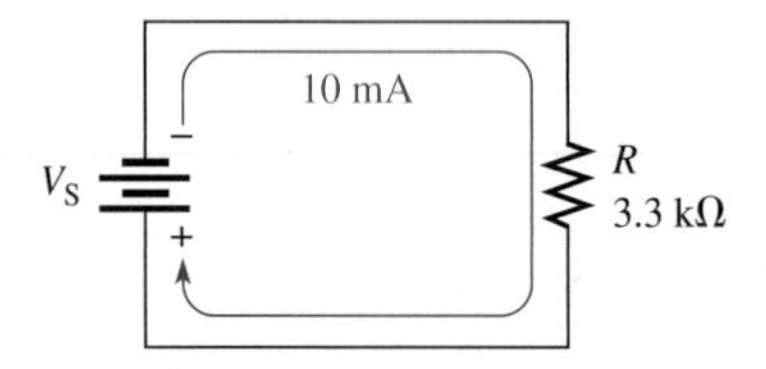

Solution Note that 10 mA equals 10×10^{-3} A and that 3.3 kΩ equals 3.3×10^{3} Ω. Substitute these values into the formula $V = IR$.

$$V_S = IR = (10\text{ mA})(3.3\text{ k}\Omega) = (10 \times 10^{-3}\text{ A})(3.3 \times 10^{3}\ \Omega) = \mathbf{33\ V}$$

When milliamperes and kilohms are multiplied, the result is volts.

1 0 EE (−) 3 × 3 • 3 EE 3 ENTER

10E-3*3.3E3
33E0

Related Problem What is the voltage in Figure 3–11 if the current is 5 mA?

Determining R When You Know V and I

In these examples you will see how to determine resistance values when you know the voltage and current using the formula $R = V/I$. To find resistance in ohms, you must express the value of V in volts and the value of I in amperes.

EXAMPLE 3–12

In the circuit of Figure 3–12 how much resistance is needed to draw 3 A of current from the battery?

▶ FIGURE 3–12

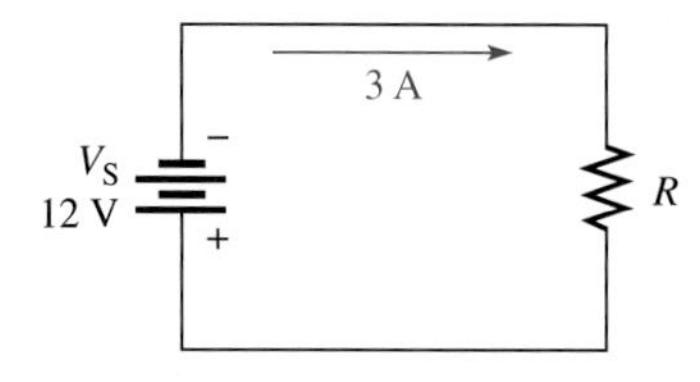

Solution Substitute 12 V for V and 3 A for I into the formula $R = V/I$.

$$R = \frac{V_S}{I} = \frac{12\text{ V}}{3\text{ A}} = \mathbf{4\ \Omega}$$

Related Problem How much resistance is required to draw 3 mA from the battery in Figure 3–12?

EXAMPLE 3–13

The ammeter in Figure 3–13 indicates 5 mA of current and the voltmeter reads 150 V. What is the value of R?

▶ FIGURE 3–13

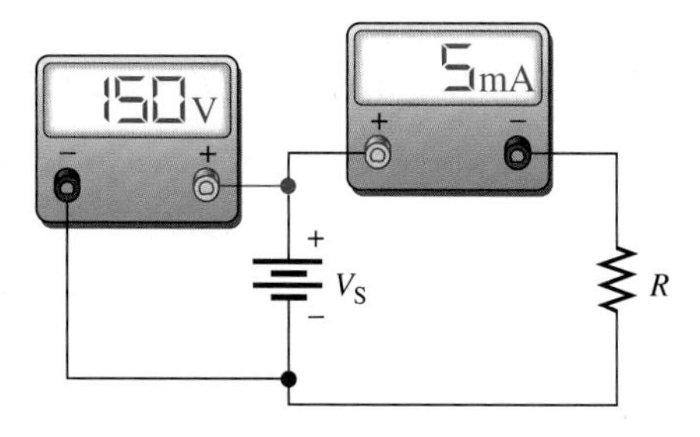

Solution Note that 5 mA equals 5×10^{-3} A. Substitute the voltage and current values into the formula $R = V/I$.

$$R = \frac{V_S}{I} = \frac{150\text{ V}}{5\text{ mA}} = \frac{150\text{ V}}{5 \times 10^{-3}\text{ A}} = 30 \times 10^{3}\ \Omega = \mathbf{30\ k\Omega}$$

Thus, if volts are divided by milliamperes, the resistance will be in kilohms.

Related Problem Determine the value of R if $V_S = 50$ V and $I = 500$ mA in Figure 3–13.

SECTION 3–2 REVIEW

1. $V = 10$ V and $R = 4.7\ \Omega$. Find I.
2. If a 4.7 MΩ resistor has 20 kV across it, how much current is there?
3. How much current will 10 kV across a 2 kΩ resistance produce?
4. $I = 1$ A and $R = 10\ \Omega$. Find V.
5. What voltage do you need to produce 3 mA of current in a 3 kΩ resistance?
6. A battery produces 2 A of current into a 6 Ω resistive load. What is the battery voltage?
7. $V = 10$ V and $I = 2$ A. Find R.
8. In a stereo amplifier circuit there is a resistor across which you measure 25 V, and your ammeter indicates 50 mA of current through the resistor. What is the resistor's value in kilohms? In ohms?

BIOGRAPHY

James Prescott Joule 1818–1889

Joule, a British physicist, is known for his research in electricity and thermodynamics. He formulated the relationship that states that the amount of heat energy produced by an electrical current in a conductor is proportional to the conductor's resistance and the time. The unit of energy is named in his honor. (Photo credit: Library of Congress.)

3–3 ENERGY AND POWER

When there is current through a resistance, electrical energy is converted to heat or other form of energy, such as light. A common example of this is a light bulb that becomes too hot to touch. The current through the filament that produces light also produces unwanted heat because the filament has resistance. Power is a measure of how fast energy is being used; electrical components must be able to dissipate a certain amount of energy in a given period of time.

After completing this section, you should be able to

- **Define energy and power**
- Express power in terms of energy
- State the unit of power
- State the common units of energy
- Perform energy and power calculations

- **Energy is the fundamental capacity to do work.**
- **Power is the rate at which energy is used.**

In other words, power, symbolized by P, is a certain amount of energy, symbolized by W, used in a certain length of time (t), expressed as follows:

Equation 3–4

$$P = \frac{W}{t}$$

Energy is measured in joules (J), time is measured in seconds (s), and power is measured in watts (W). Note that an italic W is used to represent energy in the form of work and a nonitalic W is used for watts, the unit of power.

Energy in joules divided by time in seconds gives power in watts. For example, if 50 J of energy are used in 2 s, the power is 50 J/2 s = 25 W. By definition,

One watt is the amount of power when one joule of energy is used in one second.

Thus, the number of joules used in 1 s is always equal to the number of watts. For example, if 75 J are used in 1 s, the power is 75 W.

EXAMPLE 3–14

An amount of energy equal to 100 J is used in 5 s. What is the power in watts?

Solution

$$P = \frac{\text{energy}}{\text{time}} = \frac{W}{t} = \frac{100\text{ J}}{5\text{ s}} = \mathbf{20\ W}$$

Related Problem If 100 W of power occurs for 30 s, how much energy in joules is used?

Amounts of power much less than one watt are common in certain areas of electronics. As with small current and voltage values, metric prefixes are used to designate small amounts of power. Thus, milliwatts (mW) and microwatts (μW) are commonly found in some applications.

In the electrical utilities field, kilowatts (kW) and megawatts (MW) are common units. Radio and television stations also use large amounts of power to transmit signals.

EXAMPLE 3–15

Express the following powers using appropriate metric prefixes:

(a) 0.045 W **(b)** 0.000012 W **(c)** 3500 W **(d)** 10,000,000 W

Solution

(a) $0.045\text{ W} = 45 \times 10^{-3}\text{ W} = \mathbf{45\ mW}$

(b) $0.000012\text{ W} = 12 \times 10^{-6}\text{ W} = \mathbf{12\ \mu W}$

(c) $3500\text{ W} = 3.5 \times 10^{3}\text{ W} = \mathbf{3.5\ kW}$

(d) $10{,}000{,}000\text{ W} = 10 \times 10^{6}\text{ W} = \mathbf{10\ MW}$

Related Problem Express the following amounts of power in watts without metric prefixes:

(a) 1 mW **(b)** 1800 μW **(c)** 3 MW **(d)** 10 kW

The Kilowatt-hour (kWh) Unit of Energy

Since power is the rate at which energy is used, power utilized over a period of time represents energy consumption. If you multiply power in watts and time in seconds, you have energy in joules, symbolized by *W*.

$$W = Pt \qquad \text{Equation 3–5}$$

The joule has been defined as the unit of energy. However, there is another way to express energy. Since power is expressed in watts and time can be expressed in hours, a unit of energy called the kilowatt-hour (kWh) can be used.

When you pay your electric bill, you are charged on the basis of the amount of energy you use. Because power companies deal in huge amounts of energy, the most practical unit is the **kilowatt-hour**. *You use a kilowatt-hour of energy when you use the equivalent of 1000 W of power for 1 h.* For example, a 100 W light bulb burning for 10 h uses 1 kWh of energy.

$$W = Pt = (100\text{ W})(10\text{ h}) = 1000\text{ Wh} = 1\text{ kWh}$$

EXAMPLE 3–16

Determine the number of kilowatt-hours (kWh) for each of the following energy consumptions:

(a) 1400 W for 1 hr **(b)** 2500 W for 2 h **(c)** 100,000 W for 5 h

Solution

(a) 1400 W = 1.4 kW
$W = Pt = (1.4\text{ kW})(1\text{ h}) = \mathbf{1.4\text{ kWh}}$

(b) 2500 W = 2.5 kW
Energy = (2.5 kW)(2 h) = **5 kWh**

(c) 100,000 W = 100 kW
Energy = (100 kW)(5 h) = **500 kWh**

Related Problem How many kilowatt-hours of energy are used by a 250 W light bulb burning for 8 h?

Table 3–1 lists typical power rating in watts for several household appliances. You can determine the maximum kWh for various appliances by using the power rating in Table 3–1 converted to kilowatts times the number of hours it is used.

TABLE 3–1

APPLIANCE	POWER RATING (WATTS)
Air conditioner	860
Blow dryer	1300
Clock	2
Clothes dryer	4800
Dishwasher	1200
Heater	1322
Microwave oven	800
Range	12,200
Refrigerator	1800
Television	250
Washing machine	400
Water heater	2500

EXAMPLE 3–17

During a typical 24-hour period, you use the following appliances for the specified lengths of time:

air conditioner: 15 hours
blow dryer: 10 minutes
clock: 24 hours
clothes dryer: 1 hour
dishwasher: 45 minutes
microwave oven: 15 minutes
refrigerator: 12 hours
television: 2 hours
water heater: 8 hours

Determine the total kilowatt-hours and the electric bill for the time period. The rate is 10 cents per kilowatt-hour.

Solution Determine the kWh for each appliance used by converting the watts in Table 3–1 to kilowatts and multiplying by the time in hours:

air conditioner:	$0.860 \text{ kW} \times 15 \text{ h} = 12.9 \text{ kWh}$
blow dryer:	$1.3 \text{ kW} \times 0.167 \text{ h} = 0.217 \text{ kWh}$
clock:	$0.002 \text{ kW} \times 24 \text{ h} = 0.048 \text{ kWh}$
clothes dryer:	$4.8 \text{ kW} \times 1 \text{ h} = 4.8 \text{ kWh}$
dishwasher:	$1.2 \text{ kW} \times 0.75 \text{ h} = 0.9 \text{ kWh}$
microwave:	$0.8 \text{ kW} \times 0.25 \text{ h} = 0.2 \text{ kWh}$
refrigerator:	$1.8 \text{ kW} \times 12 \text{ h} = 21.6 \text{ kWh}$
television:	$0.25 \text{ kW} \times 2 \text{ h} = 0.5 \text{ kWh}$
water heater:	$2.5 \text{ kW} \times 8 \text{ h} = 20 \text{ kWh}$

Now, add up all the kilowatt-hours to get the total energy for the 24-hour period.

$$\text{Total energy} = (12.9 + 0.217 + 0.048 + 4.8 + 0.9 + 0.2 + 21.6 + 0.5 + 20) \text{ kWh} = \mathbf{61.165 \text{ kWh}}$$

At 10 cents/kilowatt-hour, the cost of energy to run the appliances for the 24-hour period is

$$\text{Energy cost} = 61.165 \text{ kWh} \times 0.1 \text{ \$/kWh} = \mathbf{\$6.12}$$

Related Problem In addition to the appliances, suppose you used two 100 W light bulbs for 2 hours and one 75 W bulb for 3 hours. Calculate your cost for the 24-hour period for both appliances and lights.

SECTION 3–3 REVIEW

1. Define *power.*
2. Write the formula for power in terms of energy and time.
3. Define *watt.*
4. Express each of the following values of power in the most appropriate units:
   (a) 68,000 W (b) 0.005 W (c) 0.000025 W
5. If you use 100 W of power for 10 h, how much energy (in kilowatt-hours) have you used?
6. Convert 2000 W to kilowatts.
7. How much does it cost to run a heater (1322 W) for 24 hours if energy cost is 9¢ per kilowatt-hour?

3–4 POWER IN AN ELECTRIC CIRCUIT

The generation of heat, which occurs when electrical energy is converted to heat energy, in an electric circuit is often an unwanted by-product of current through the resistance in the circuit. In some cases, however, the generation of heat is the primary purpose of a circuit as, for example, in an electric resistive heater. In any case, you must always deal with power in electrical and electronic circuits.

After completing this section, you should be able to

- **Calculate power in a circuit**
- Determine power when you know I and R
- Determine power when you know V and I
- Determine power when you know V and R

When there is current through a resistance, the collisions of the electrons as they move through the resistance give off heat, resulting in a conversion of electrical energy to thermal energy as indicated in Figure 3–14. There is always a certain amount of power dissipated in an electrical circuit, and it is dependent on the amount of resistance and on the amount of current, expressed as follows:

Equation 3–6

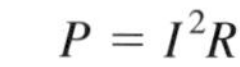

$$P = I^2R$$

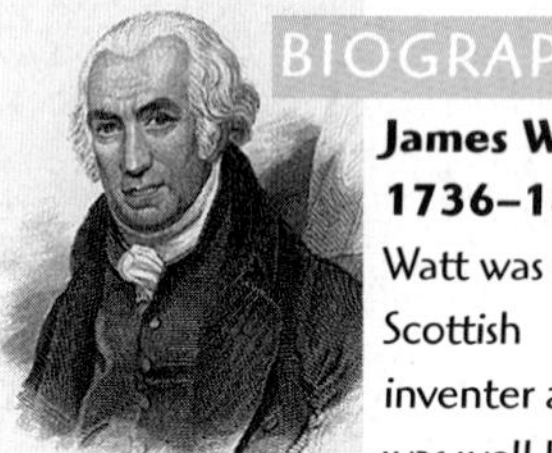

BIOGRAPHY

James Watt
1736–1819
Watt was a Scottish inventer and was well known for his improvements to the steam engine, which made it practical for industrial use. Watt patented several inventions, including the rotary engine. The unit of power is named in his honor. (Photo credit: Library of Congress.)

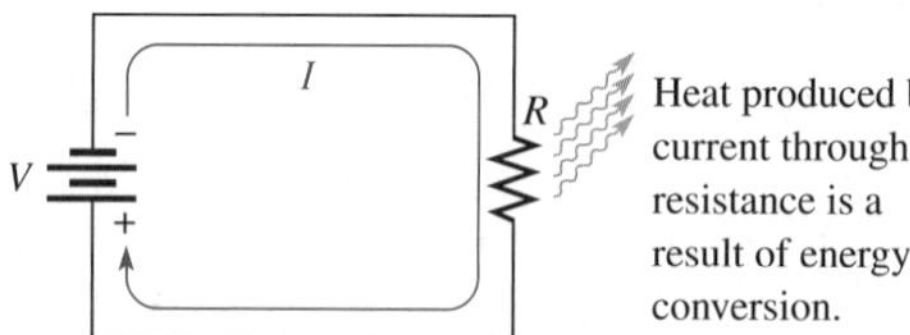

▲ **FIGURE 3–14**

Power dissipation in an electric circuit is seen as heat given off by the resistance. The power dissipation is equal to the power produced by the voltage source.

You can get an equivalent expression for power in terms of voltage and current by substituting $I \times I$ for I^2 and V for IR.

$$P = I^2R = (I \times I)R = I(IR) = (IR)I$$

Equation 3–7

$$P = VI$$

where P is in watts when V is in volts and I is in amperes.

You obtain another equivalent expression by substituting V/R for I (Ohm's law).

$$P = VI = V\left(\frac{V}{R}\right)$$

Equation 3–8

$$P = \frac{V^2}{R}$$

The three power formulas in Equations 3–6, 3–7, and 3–8 are also known as Watt's law. An aid for using both Ohm's law and Watt's law is found in the summary, Figure 3–27, at the end of the chapter.

Using the Appropriate Power Formula

To calculate the power in a resistance, you can use any one of the three equivalent Watt's law power formulas, depending on what information you have. For example, assume that you know the values of current and voltage; in this case, calculate the power with the formula $P = VI$. If you know I and R, use the formula $P = I^2R$. If you know V and R, use the formula $P = V^2/R$.

EXAMPLE 3–18

Calculate the power in each of the three circuits of Figure 3–15.

▶ **FIGURE 3–15**

The ground symbols in these diagrams were introduced in Section 2–6.

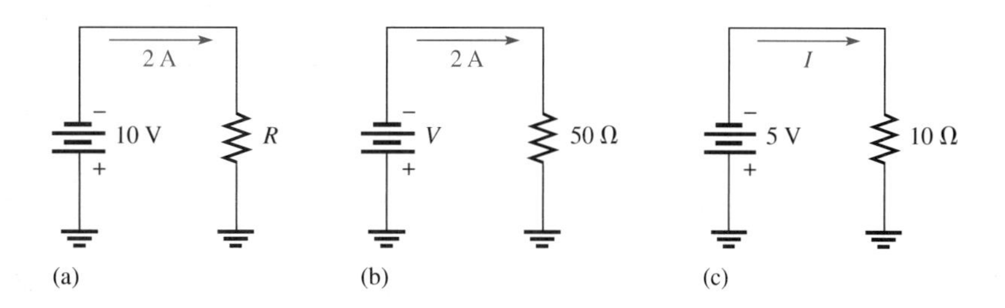

Solution In circuit (a), V and I are known. The power is determined as follows with calculator sequences also shown:

$$P = VI = (10\text{ V})(2\text{ A}) = \mathbf{20\text{ W}}$$

In circuit (b), I and R are known. Therefore,

$$P = I^2R = (2\text{ A})^2(50\ \Omega) = \mathbf{200\text{ W}}$$

In circuit (c), V and R are known. Therefore,

$$P = \frac{V^2}{R} = \frac{(5\text{ V})^2}{10\ \Omega} = \mathbf{2.5\text{ W}}$$

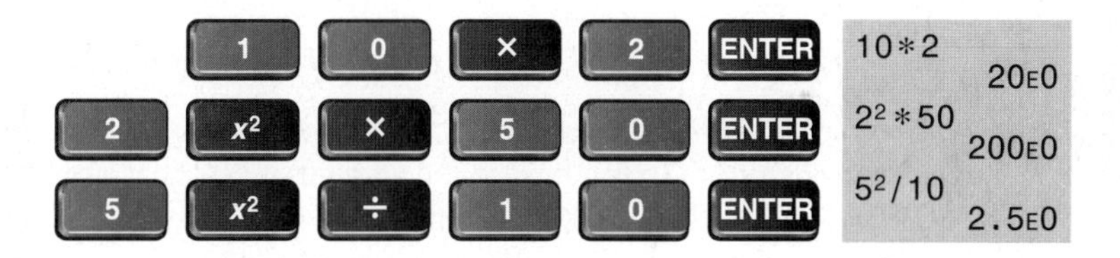

Related Problem Determine the power in each circuit of Figure 3–15 for the following changes: In circuit (a), R is halved; in circuit (b), R is doubled and V is doubled; in circuit (c), V halved with R the same.

EXAMPLE 3–19

A 100 W light bulb operates on 120 V. How much current does it require?

Solution Use the formula $P = VI$ and solve for I by first transposing the terms to get I on the left side of the equation.

$$VI = P$$

Divide both sides of the equation by V to get I by itself.

$$\frac{\not{V}I}{\not{V}} = \frac{P}{V}$$

The V's cancel on the left, leaving

$$I = \frac{P}{V}$$

Substitute 100 W for P and 120 V for V.

$$I = \frac{P}{V} = \frac{100\ \text{W}}{120\ \text{V}} = 0.833\ \text{A} = \mathbf{833\ mA}$$

Related Problem A light bulb draws 545 mA from a 110 V source. What is the power dissipated?

SECTION 3–4 REVIEW

1. If there are 10 V across a resistor and a current of 3 A through it, what is the power dissipated?
2. If there is a current of 5 A through a 47 Ω resistor, what is the power dissipated?
3. How much power is produced by 20 mA through a 5.1 kΩ resistor?
4. Five volts are applied to a 10 Ω resistor. What is the power dissipated?
5. How much power does a 2.2 kΩ resistor with 8 V across it produce?
6. What is the resistance of a 55 W bulb that draws 0.5 A?

SAFETY POINT

Resistors can become very hot in normal operation. To avoid a burn, do not touch a circuit component while the power is connected to the circuit; and after power has been turned off, allow time for the components to cool down.

3–5 THE POWER RATING OF RESISTORS

As you know, a resistor gives off heat when there is current through it. There is a limit to the amount of heat that a resistor can give off, which is specified by its power rating.

After completing this section, you should be able to

- **Properly select resistors based on power consideration**
- Define *power rating*
- Explain how physical characteristics of resistors determine their power rating
- Check for resistor failure with an ohmmeter

The power rating is the maximum amount of power that a resistor can dissipate without being damaged by excessive heat buildup. The power rating is not related to the ohmic value (resistance) but rather is determined mainly by the physical composition, size, and shape of the resistor. All else being equal, the larger the surface area of a resistor, the more power it can dissipate. *The surface area of a cylindrically shaped resistor is equal to the length (l) times the circumference (c)*, as indicated in Figure 3–16. The area of the ends is not included.

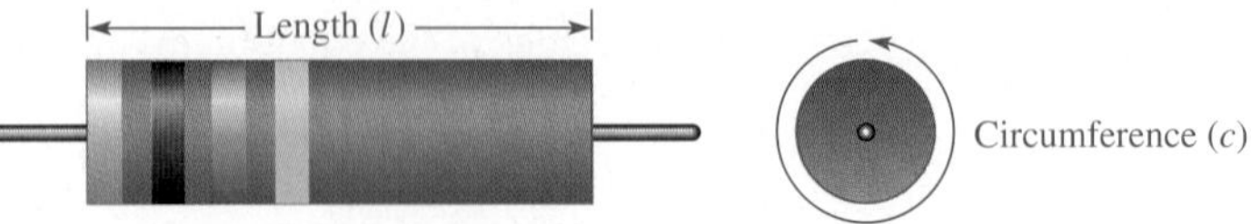

FIGURE 3–16

The power rating of a resistor is directly related to its surface area.

Metal-film resistors are available in standard power ratings from ⅛ W to 1 W, as shown in Figure 3–17. Available power ratings for other types of resistors vary. For example, wirewound resistors have ratings up to 225 W or greater. Figure 3–18 shows some of these resistors.

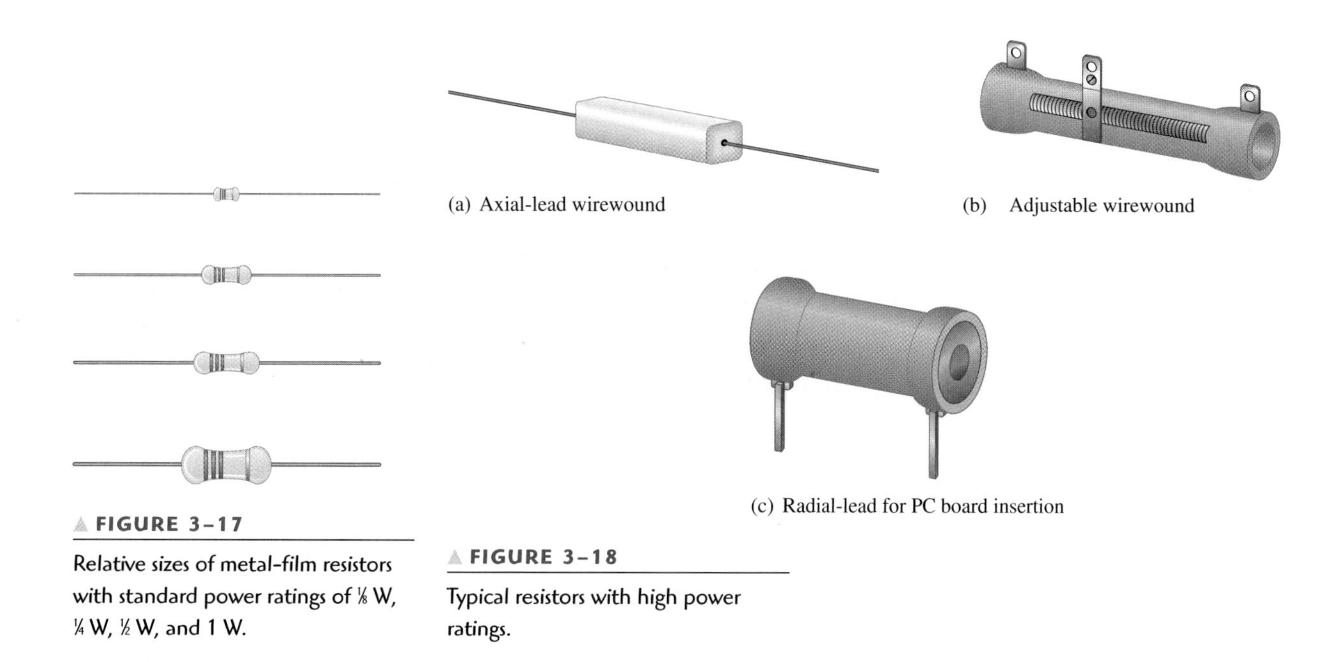

▲ FIGURE 3–17

Relative sizes of metal-film resistors with standard power ratings of ⅛ W, ¼ W, ½ W, and 1 W.

▲ FIGURE 3–18

Typical resistors with high power ratings.

Selecting the Proper Power Rating for an Application

When a resistor is used in a circuit, its power rating should be greater than the maximum power that it will have to handle. Generally, the next higher standard value is used. For example, if a metal-film resistor is to dissipate 0.75 W in a circuit application, its rating should be the next higher standard value which is 1 W.

EXAMPLE 3–20

Choose an adequate power rating for each of the metal-film resistors represented in Figure 3–19 (⅛ W, ¼ W, ½ W, or 1 W).

▶ FIGURE 3–19

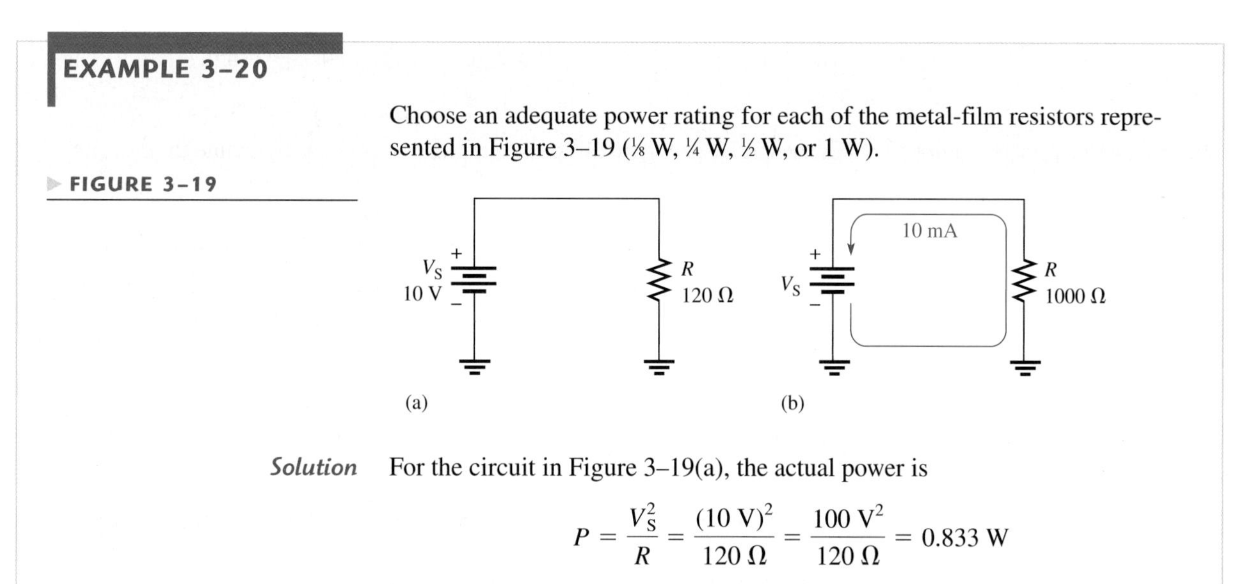

Solution For the circuit in Figure 3–19(a), the actual power is

$$P = \frac{V_S^2}{R} = \frac{(10\text{ V})^2}{120\ \Omega} = \frac{100\text{ V}^2}{120\ \Omega} = 0.833\text{ W}$$

Select a resistor with a power rating higher than the actual power dissipation. In this case, a **1 W resistor** should be used.

For the circuit in Figure 3–19(b), the actual power is

$$P = I^2R = (10\text{ mA})^2(1000\ \Omega) = 0.1\text{ W}$$

A ⅛ W **(0.125 W) resistor** should be used in this case.

Related Problem A certain resistor is required to dissipate 0.25 W (¼ W). What standard rating should be used?

Resistor Failures

When the power dissipated in a resistor is greater than its rating, the resistor will become excessively hot. As a result, either the resistor will burn open or its resistance value will be greatly altered.

A resistor that has been damaged because of overheating can often be detected by the charred or altered appearance of its surface. If there is no visual evidence, a resistor that is suspected of being damaged can be checked with an ohmmeter for an open or increased resistance value. Recall that a resistor should be disconnected from the circuit to measure resistance.

EXAMPLE 3–21

Determine whether the resistor in each circuit of Figure 3–20 has possibly been damaged by overheating.

FIGURE 3–20

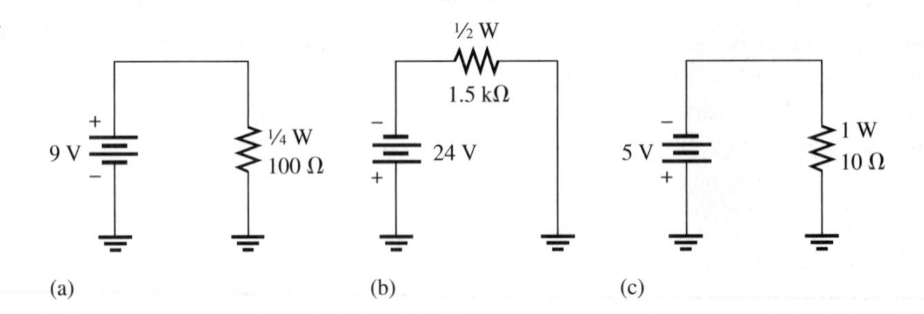

Solution For the circuit in Figure 3–20(a),

$$P = \frac{V^2}{R} = \frac{(9\text{ V})^2}{100\ \Omega} = 0.81\text{ W}$$

The rating of the resistor is ¼ W (0.25 W), which is insufficient to handle the power. The resistor has been overheated and *may* be burned out, making it an open.

For the circuit in Figure 3–20(b),

$$P = \frac{V^2}{R} = \frac{(24\text{ V})^2}{1.5\text{ k}\Omega} = 0.384\text{ W}$$

The rating of the resistor is ½ W (0.5 W), which is sufficient to handle the power.

For the circuit in Figure 3–20(c),

$$P = \frac{V^2}{R} = \frac{(5\text{ V})^2}{10\ \Omega} = 2.5\text{ W}$$

The rating of the resistor is 1 W, which is insufficient to handle the power. The resistor has been overheated and *may* be burned out, making it an open.

Related Problem A ¼ W, 1.0 kΩ resistor is connected across a 12 V battery. Will it overheat?

Checking a Resistor with an Ohmmeter

A typical digital multimeter and an analog multimeter are shown in Figure 3–21(a) and 3–21(b), respectively. For the digital meter in Figure 3–21(a), you use the round function switch to select ohms (Ω). You do not have to manually select a range because this particular meter is autoranging and you have a direct digital readout of the resistance value. The large round switch on the analog meter is called a *range switch.* Notice the resistance (OHMS) settings on both meters.

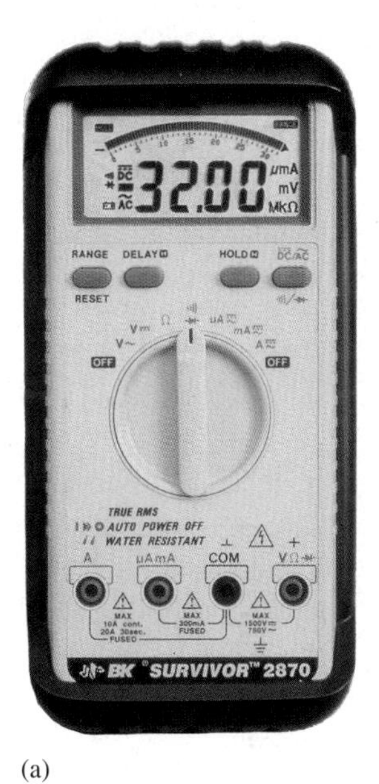

(a)

(b)

FIGURE 3–21

Typical portable multimeters. (Photography courtesy of B&K Precision Corp.)

For the analog meter in part (b), each setting indicates the amount by which the ohms scale (top scale) on the meter is to be multiplied. For example, if the pointer is at 50 on the ohms scale and the range switch is set at ×10, the resistance being measured is $50 \times 10\ \Omega = 500\ \Omega$. *If the resistor is open, the pointer will stay at full left scale (∞ means infinite) regardless of the range switch setting.*

SECTION 3–5 REVIEW

1. Name two important parameters associated with a resistor.
2. How does the physical size of a resistor determine the amount of power that it can handle?
3. List the standard power ratings of metal-film resistors.
4. A resistor must handle 0.3 W. What standard size metal-film resistor should be used to dissipate the energy properly?
5. If the pointer indicates 8 on the ohmmeter scale and the range switch is set at R × 1k, what resistance is being measured?

3–6 ENERGY CONVERSION AND VOLTAGE DROP IN A RESISTANCE

As you have seen, when there is current through a resistance, electrical energy is converted to heat energy. This heat is caused by collisions of the free electrons within the atomic structure of the resistive material. When a collision occurs, heat is given off; and the electron gives up some of its acquired energy as it moves through the material.

After completing this section, you should be able to

- **Explain energy conversion and voltage drop**
- Discuss the cause of energy conversion in a circuit
- Define *voltage drop*
- Explain the relationship between energy conversion and voltage drop

Figure 3–22 illustrates electrons flowing from the negative terminal of a battery, through a circuit, and back to the positive terminal. As they emerge from the negative terminal, the electrons are at their highest energy level. The electrons flow through each of the resistors that are connected together to form a current path (this type of connection is called series, as you will learn in Chapter 4). As the electrons flow through each resistor, some of their energy is given up in the form of heat. So, the electrons have more energy when they enter a resistor than when they exit the resistor, as illustrated in the figure by the decrease in the intensity of the red color. When they have traveled through the circuit back to the positive terminal of the battery, the electrons are at their lowest energy level.

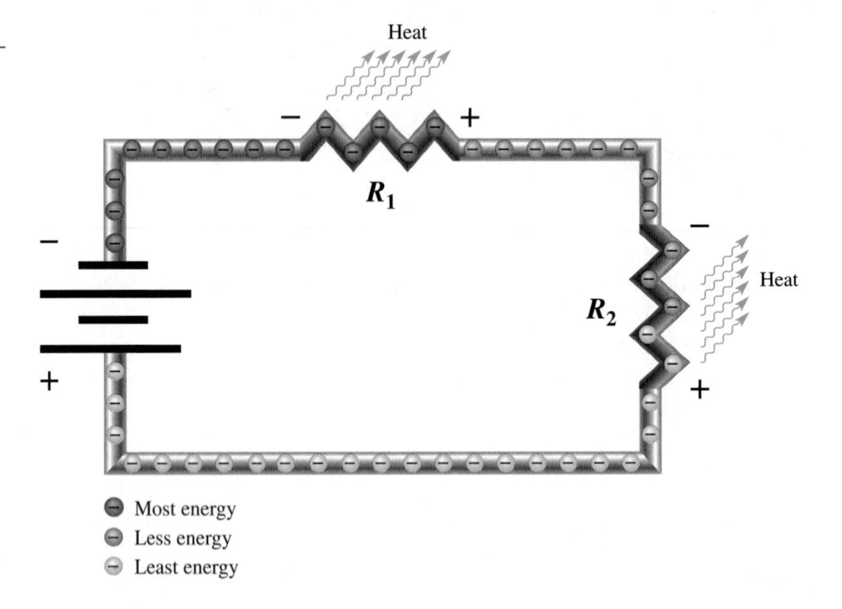

FIGURE 3–22

A loss of energy by electrons as they flow through a resistance creates a voltage drop because voltage equals energy divided by charge.

Recall that voltage equals energy per charge ($V = W/Q$) and charge is a property of electrons. So, based on the voltage of the battery, a certain amount of energy is imparted to all of the electrons that flow out of the negative terminal. The same number of electrons flow at each point throughout the circuit, but their energy decreases as they move through the resistance of the circuit. Electrons exiting R_1 have less energy than they did when they entered R_1.

In Figure 3–22, the voltage at the left end of R_1 is equal to W_{enter}/Q, and the voltage at the right end of R_1 is equal to W_{exit}/Q. The same number of electrons that enter R_1 also exit R_1, so Q is constant. However, the energy W_{exit} is less than W_{enter}, so the voltage at the right end of R_1 is less than the voltage at the left end. This is a voltage drop across the resistor. The voltage at the right end of R_1 is less negative (or more positive) than the voltage at the left end. So, the voltage drop is indicated by − and + signs (the + implies a less negative or more positive voltage).

The electrons have lost some energy in R_1 and now they enter R_2 with a reduced energy level. As they flow through R_2, they lose more energy, resulting in another voltage drop across R_2.

SECTION 3–6 REVIEW

1. What is the basic reason for energy conversion in a resistor?
2. What is a voltage drop?
3. What is the polarity of a voltage drop in relation to current direction?

3–7 POWER SUPPLIES

A power supply is a device that provides power to a load. Recall that a load is any electrical device or circuit that is connected to the output of the power supply and draws current from the supply.

After completing this section, you should be able to

- **Discuss power supplies and their characteristics**
- Define *ampere-hour rating*
- Discuss power supply efficiency

HANDS ON TIP

An electronic power supply provides both output voltage and current. You should make sure the voltage range is sufficient for your applications. Also, you must have sufficient current capacity to assure proper circuit operation. The current capacity is the maximum current that a power supply can provide to a load at a given voltage.

Figure 3–23 shows a block diagram of a power supply with a loading device connected to its output. The load can be anything from a light bulb to a computer. The power supply produces a voltage across its two output terminals and provides current through the load, as indicated in the figure. The product IV_{OUT} is the amount of power produced by the supply and consumed by the load. For a given output voltage (V_{OUT}), more current drawn by the load means more power from the supply.

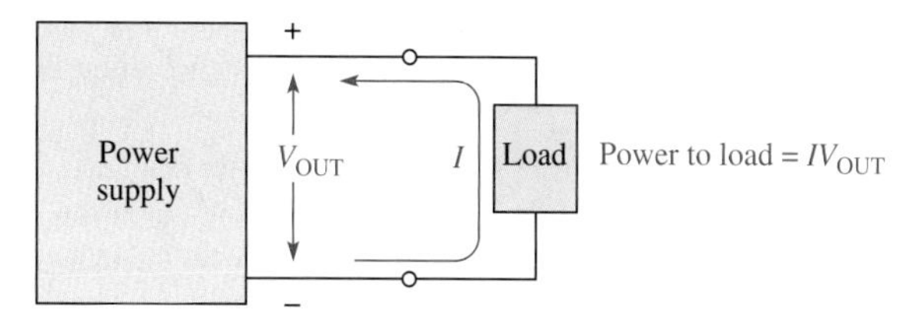

FIGURE 3–23

Block diagram of power supply and load.

Power supplies range from simple batteries to regulated electronic circuits where an accurate output voltage is automatically maintained. A battery is a dc power supply that converts chemical energy into electrical energy. Electronic power supplies normally convert 110 V ac (alternating current) from a wall outlet into a regulated dc (direct current) voltage at a level suitable for electronic components. A regulated voltage is one that remains essentially constant with changes in input voltage or load.

Ampere-hour Ratings of Batteries

Batteries convert chemical energy into electrical energy. Because of their limited source of chemical energy, batteries have a certain capacity that limits the amount of time over which they can produce a given power level. This capacity is measured in ampere-hours (Ah). The ampere-hour rating determines the length of time that a battery can deliver a certain amount of current to a load at the rated voltage.

A rating of one ampere-hour means that a battery can deliver one ampere of current to a load for one hour at the rated voltage output. This same battery can deliver two amperes for one-half hour. The more current the battery is required to deliver, the shorter the life of the battery. In practice, a battery usually is rated for a specified current level and output voltage. For example, a 12 V automobile battery may be rated for 70 Ah at 3.5 A. This means that it can supply 3.5 A for 20 h at the rated voltage.

EXAMPLE 3–22

For how many hours can a battery deliver 2 A if it is rated at 70 Ah?

Solution The ampere-hour rating is the current times the number of hours, x.

$$70 \text{ Ah} = (2 \text{ A})(x \text{ h})$$

Solving for the number of hours, x, yields

$$x = \frac{70 \text{ Ah}}{2 \text{ A}} = \mathbf{35\ h}$$

Related Problem A certain battery delivers 10 A for 6 h. What is its minimum Ah rating?

Power Supply Efficiency

An important characteristic of electronic power supplies is efficiency. Efficiency is the ratio of the output power, P_{OUT}, to the input power, P_{IN}.

Equation 3–9

$$\text{Efficiency} = \frac{P_{OUT}}{P_{IN}}$$

Efficiency is often expressed as a percentage. For example, if the input power is 100 W and the output power is 50 W, the efficiency is $(50 \text{ W}/100 \text{ W}) \times 100\% = 50\%$.

All electronic power supplies are energy converters and require that power be put into them in order to get power out. For example, an electronic dc power supply might use the ac power from a wall outlet as its input. Its output is usually regulated dc voltage. The output power is *always* less than the input power because some of the total power is used internally to operate the power supply circuitry. This internal power dissipation is normally called the *power loss.* The output power is the input power minus the power loss.

Equation 3–10

$$P_{OUT} = P_{IN} - P_{LOSS}$$

High efficiency means that very little power is dissipated in the power supply and there is a higher proportion of output power for a given input power.

EXAMPLE 3–23

A certain electronic power supply requires 25 W of input power. It can produce an output power of 20 W. What is its efficiency, and what is the power loss?

Solution

$$\text{Efficiency} = \left(\frac{P_{OUT}}{P_{IN}}\right)100\% = \left(\frac{20\ \text{W}}{25\ \text{W}}\right)100\% = \mathbf{80\%}$$

$$P_{LOSS} = P_{IN} - P_{OUT} = 25\ \text{W} - 20\ \text{W} = \mathbf{5\ W}$$

Related Problem A power supply has an efficiency of 92%. If P_{IN} is 50 W, what is P_{OUT}?

SECTION 3–7 REVIEW

1. When a loading device draws an increased amount of current from a power supply, does this change represent a greater or a smaller load on the supply?
2. A power supply produces an output voltage of 10 V. If the supply provides 0.5 A to a load, what is the power output?
3. If a battery has an ampere-hour rating of 100 Ah, how long can it provide 5 A to a load?
4. If the battery in Question 3 is a 12 V device, what is its power output for the specified value of current?
5. An electronic power supply used in the lab operates with an input power of 1 W. It can provide an output power of 750 mW. What is its efficiency?

3–8 INTRODUCTION TO TROUBLESHOOTING

Technicians must be able to diagnose and repair malfunctioning circuits and systems. In this section, you learn a general approach to troubleshooting using a simple example. Troubleshooting coverage is an important part of this textbook, so you will find a troubleshooting section in many of the chapters as well as troubleshooting problems for skill building.

After completing this section, you should be able to

- **Describe a basic approach to troubleshooting**
- List three steps in troubleshooting
- Explain what is meant by half-splitting
- Discuss and compare the three basic measurements of voltage, current, and resistance

Troubleshooting is the application of logical thinking combined with a thorough knowledge of circuit or system operation to correct a malfunction. The basic approach to troubleshooting consists of three steps: *analysis, planning,* and *measuring.* We will refer to this 3-step approach as APM.

Analysis

When troubleshooting a circuit, the first step is to analyze clues or symptoms of the failure. The analysis can begin by determining the answer to certain questions:

1. Has the circuit ever worked?
2. If the circuit once worked, under what conditions did it fail?
3. What are the symptoms of the failure?
4. What are the possible causes of failure?

Planning

After analyzing the clues, the second step in the troubleshooting process is formulating a logical plan of attack. Much time can be saved by proper planning. A working knowledge of the circuit is a prerequisite to a plan for troubleshooting. If you are not certain how the circuit is supposed to operate, take time to review circuit diagrams (schematics), operating instructions, and other pertinent information. A schematic with proper voltages marked at various test points is particularly useful. Although logical thinking is perhaps the most important tool in troubleshooting, it rarely can solve the problem by itself.

Measuring

The third step is to narrow the possible failures by making carefully thought out measurements. These measurements usually confirm the direction you are taking in solving the problem, or they may point to a new direction that you should take. Occasionally, you may find a totally unexpected result.

An Example

The thought process that is part of the APM approach is best illustrated with a simple example. Suppose you have a string of 8 decorative lamps connected in series to a 120 V source, as shown in Figure 3–24. Assume that this circuit worked properly at one time but stopped working after it was moved to a new location. When plugged in at the new location, the lamps fail to turn on. How do you go about finding the trouble?

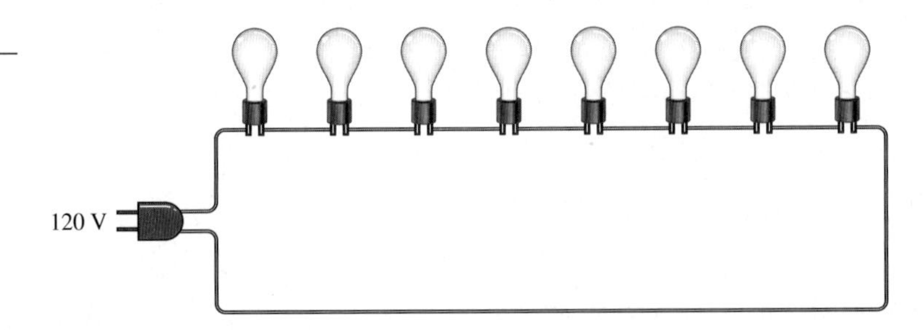

FIGURE 3–24
A string of bulbs connected to a voltage source.

The Analysis Thought Process You may think like this as you proceed to analyze the situation:

- Since the circuit worked before it was moved, the problem could be that there is no voltage at the new location.
- Perhaps the wiring was loose and pulled apart when moved.
- It is possible that a bulb is burned out or loose in its socket.

With this reasoning, you have considered possible causes and failures that may have occurred. The thought process continues:

- The fact that the circuit once worked eliminates the possibility that the original circuit was improperly wired.
- If the fault is due to an open path, it is very unlikely that there is more than one break which could be either a bad connection or a burned out bulb.

You have now analyzed the problem and are ready to plan the process of finding the fault in the circuit.

The Planning Thought Process The first part of your plan is to measure for voltage at the new location. If the voltage is present, then the problem is in the string of lights. If voltage is not present, check the circuit breaker at the distribution box in the house. Before resetting breakers, you should think about why a breaker may be tripped. Let's assume that you find the voltage is present. This means that the problem is in the string of lights.

The second part of your plan is to measure either the resistance in the string of lights or to measure voltages across the bulbs. The decision whether to measure resistance or voltage is a toss-up and can be made based on the ease of making the test. Seldom is a troubleshooting plan developed so completely that all possible contingencies are included. You will frequently need to modify the plan as you go along.

The Measurement Process You proceed with the first part of your plan by using a multimeter to check the voltage at the new location. Assume the measurement shows a voltage of 120 V. Now you have eliminated the possibility of no voltage. You know that, since you have voltage across the string and there is no current because no bulb is on, there must be an open in the current path. Either a bulb is burned out, a connection at the lamp socket is broken, or the wire is broken.

Next, you decide to locate the break by measuring resistance with your multimeter. Applying logical thinking, you decide to measure the resistance of each half of the string instead of measuring the resistance of each bulb. By measuring the resistance of half the bulbs at once, you can possibly reduce the effort required to find the open. This technique is a common troubleshooting procedure called **half-splitting**.

Once you have identified the half in which the open occurs, as indicated by an infinite resistance, you use half-splitting again on the faulty half and continue until you narrow the fault down to a faulty bulb or connection. This process is shown in Figure 3–25, assuming for purposes of illustration that the seventh bulb is burned out.

As you can see in the figure, the half-splitting approach in this particular case takes five measurements to identify the open bulb. If you had decided to measure

▼ **FIGURE 3–25**

Illustration of the half-splitting troubleshooting process. The numbered steps indicate the sequence in which the multimeter is moved from one position to another.

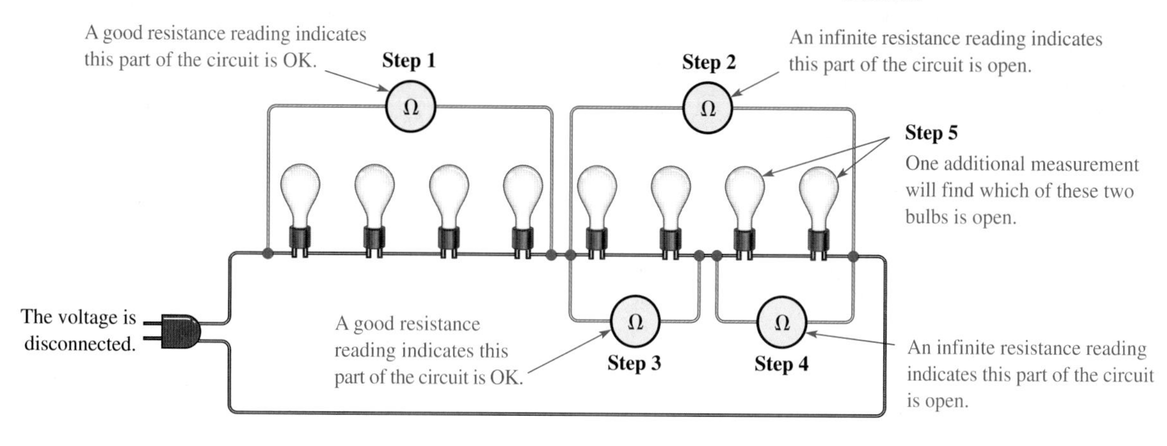

each bulb individually and had started at the left, you would have needed seven measurements. So, sometimes half-splitting saves steps; sometimes it doesn't. The number of steps required depends on where you make your measurements and in what sequence.

Unfortunately, most troubleshooting is more difficult than this example. However, analysis and planning are essential for effective troubleshooting in any situation. As measurements are made, the plan is often modified; the experienced troubleshooter narrows the search by fitting the symptoms and measurements into a probable cause.

Comparison of *V*, *R*, and *I* Measurements

As you know from Section 2–7, you can measure voltage, current, or resistance in a circuit. To measure voltage, the voltmeter is placed in parallel across the component; that is, one lead is placed on each side of the component. This makes voltage measurements the easiest of the three types of measurements.

To measure resistance, the ohmmeter is also connected across a component; however, the voltage must be first disconnected, and sometimes the component itself must be removed from the circuit. Therefore, resistance measurements are generally more difficult than voltage measurements.

To measure current, the ammeter must be placed in series with the component; that is, it must be in line with the current path. To do this a component lead or a wire must be disconnected before the ammeter can be connected. This usually makes a current measurement the most difficult to perform.

SECTION 3–8 REVIEW

1. Name the three steps in the APM approach to troubleshooting.
2. Explain the basic idea of the half-splitting technique.
3. Why are voltages easier to measure than currents in a circuit?

APPLICATION ASSIGNMENT

PUTTING YOUR KNOWLEDGE TO WORK

In this assignment, an existing resistance box that is to be used as part of a test setup in the lab is to be checked out and modified. Your task is to modify the circuit so that it will meet the requirements of the new application. You will have to apply your knowledge of Ohm's law and Watt's law in order to complete this assignment.

The specifications are as follows:

1. Each resistor is switch selectable and only one resistor is selected at a time.
2. The lowest resistor value is to be 10 Ω.
3. Each successively higher resistance in the switch sequence must be a decade (10 times) increase over the previous one.
4. The maximum resistor value must be 1.0 MΩ.
5. The maximum voltage across any resistor in the box will be 4 V.
6. Two additional resistors are required, one to limit the current to 10 mA ± 10% with a 4 V drop and the other to limit the current to 5 mA ± 10% with a 4 V drop.

Step 1: Inspect the Existing Resistance Box

The existing resistance box is shown in both top and bottom views in Figure 3–26. The switch is a rotary type. The resistors all have a power rating of 1 W.

Step 2: Draw the Schematic

From Figure 3–26, determine the resistor values and draw the schematic for the existing circuit so that you will

(a) Top view

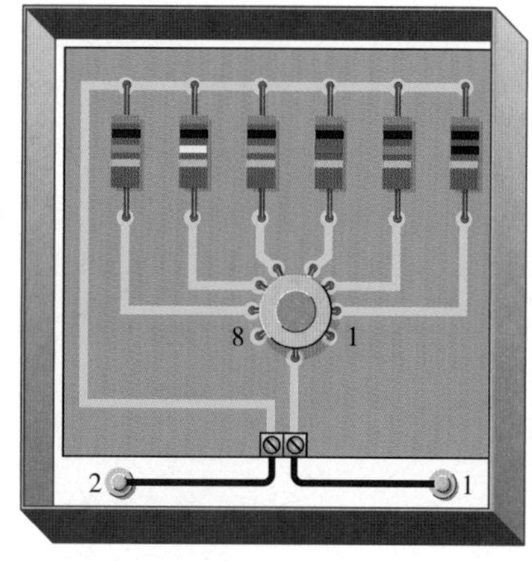

(b) Bottom view

▲ FIGURE 3–26

know what you have to work with. Determine the resistor numbering from the *R* labels on the top view.

Step 3: Modify the Schematic to Meet New Requirements

Change the schematic from Step 2 so the circuit will accomplish the following:

1. One resistor at a time is to be connected by the switch between terminals 1 and 2 of the box.
2. Provide switch selectable resistor values beginning with 10 Ω and increasing in decade increments to 1.0 MΩ.
3. Each of the resistors from Step 1 must be selectable by a sequence of adjacent switch positions in ascending order.
4. In addition to the resistors in Step 1, there must be two switch-selectable resistors, one is in switch position 1 (shown in Figure 3–26, bottom view) and must limit the current to 10 mA ± 10% with a 4 V drop and the other is in switch position 8 and must limit the current to 5 mA ± 10% with a 4 V drop.
5. All the resistors must be standard values with 10% tolerance and have a sufficient power rating for the 4 V operating requirement. See Appendix A for standard resistor values.

Step 4: Modify the Circuit

State the modifications that must be made to the existing box to meet the specifications and develop a detailed list of the changes including resistance values, power ratings, wiring, and new components. You should number each point in the schematic for easy reference.

Step 5: Develop a Test Procedure

After the box has been modified to meet the new specifications, it must be tested to see if it is working properly. Determine how you would test the resistance box and what instruments you would use and then detail your test procedure in a step-by-step format.

Step 6: Troubleshoot the Circuit

When an ohmmeter is connected across terminals 1 and 2 of the resistance box, determine the most likely fault in each of the following cases:

1. The ohmmeter shows an infinitely high resistance when the switch is in position 3.
2. The ohmmeter shows an infinitely high resistance in all switch positions.
3. The ohmmeter shows an incorrect value of resistance when the switch is in position 6.

APPLICATION ASSIGNMENT REVIEW

1. Explain how you applied Watt's law to this application assignment.
2. Explain how you applied Ohm's law to this application assignment.
3. How did you determine the power ratings of the resistors in the existing box?

SUMMARY

- Voltage and current are linearly proportional.
- Ohm's law gives the relationship of voltage, current, and resistance.
- Current is directly proportional to voltage.
- Current is inversely proportional to resistance.
- A kilohm (kΩ) is one thousand ohms.
- A megohm (MΩ) is one million ohms.
- A microampere (μA) is one-millionth of an ampere.
- A milliampere (mA) is one-thousandth of an ampere.
- Use $I = V/R$ to calculate current.
- Use $V = IR$ to calculate voltage.
- Use $R = V/I$ to calculate resistance.
- One watt equals one joule per second.
- Watt is the unit of power, and joule is the unit of energy.
- The power rating of a resistor determines the maximum power that it can handle safely.
- Resistors with a larger physical size can dissipate more power in the form of heat than smaller ones.
- A resistor should have a power rating as high or higher than the maximum power that it is expected to handle in the circuit.
- Power rating is not related to resistance value.
- A resistor usually opens when it overheats and fails.
- Energy is equal to power multiplied by time.
- The kilowatt-hour is a unit of energy.
- An example of one kilowatt-hour is one thousand watts used for one hour.
- A power supply is an energy source used to operate electrical and electronic devices.
- A battery is one type of power supply that converts chemical energy into electrical energy.
- An electronic power supply converts commercial energy (ac from the power company) to regulated dc at various voltage levels.
- The output power of a power supply is the output voltage times the load current.
- A load is a device that draws current from the power supply.
- The capacity of a battery is measured in ampere-hours (Ah).
- One ampere-hour equals one ampere used for one hour, or any other combination of amperes and hours that has a product of one.
- An electronic power supply with a high efficiency has a smaller percentage power loss than one with a lower efficiency.
- The formula wheel in Figure 3–27 gives the Ohm's law and Watt's law relationships.

FIGURE 3–27

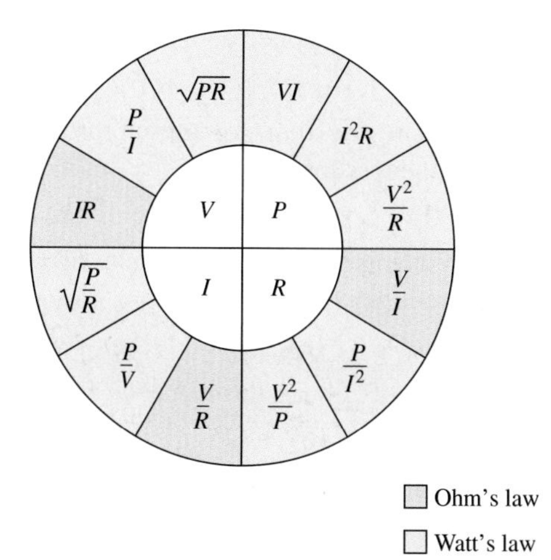

- APM (analysis, planning, and measurement) provides a logical approach to troubleshooting.
- The half-splitting method of troubleshooting generally results in fewer measurements.

EQUATIONS

3–1	$I = \dfrac{V}{R}$	Ohm's law
3–2	$V = IR$	Ohm's law
3–3	$R = \dfrac{V}{I}$	Ohm's law
3–4	$P = \dfrac{W}{t}$	Power equals energy divided by time.
3–5	$W = Pt$	Energy equals power multiplied by time.
3–6	$P = I^2R$	Power equals current squared times resistance.
3–7	$P = VI$	Power equals voltage times current.
3–8	$P = \dfrac{V^2}{R}$	Power equals voltage squared divided by resistance.
3–9	$\text{Efficiency} = \dfrac{P_{OUT}}{P_{IN}}$	Power supply efficiency
3–10	$P_{OUT} = P_{IN} - P_{LOSS}$	Output power

SELF-TEST

Answers are at the end of the chapter.

1. Ohm's law states that
   (a) current equals voltage times resistance
   (b) voltage equals current times resistance
   (c) resistance equals current divided by voltage
   (d) voltage equals current squared times resistance
2. When the voltage across a resistor is doubled, the current will
   (a) triple **(b)** halve **(c)** double **(d)** not change

3. When 10 V are applied across a 20 Ω resistor, the current is

(a) 10 A **(b)** 0.5 A **(c)** 200 A **(d)** 2 A

4. When there are 10 mA of current through a 1.0 kΩ resistor, the voltage across the resistor is

(a) 100 V **(b)** 0.1 V **(c)** 10 kV **(d)** 10 V

5. If 20 V are applied across a resistor and there are 6.06 mA of current, the resistance is

(a) 3.3 kΩ **(b)** 33 kΩ **(c)** 330 Ω **(d)** 3.03 kΩ

6. A current of 250 μA through a 4.7 kΩ resistor produces a voltage drop of

(a) 53.2 V **(b)** 1.175 mV **(c)** 18.8 V **(d)** 1.175 V

7. A resistance of 2.2 MΩ is connected across a 1 kV source. The resulting current is approximately

(a) 2.2 mA **(b)** 455 μA **(c)** 45.5 μA **(d)** 0.455 A

8. Power can be defined as

(a) energy **(b)** heat
(c) the rate at which energy is used **(d)** the time required to use energy

9. For 10 V and 50 mA, the power is

(a) 500 mW **(b)** 0.5 W
(c) 500,000 μW **(d)** answers (a), (b), and (c)

10. When the current through a 10 kΩ resistor is 10 mA, the power is

(a) 1 W **(b)** 10 W **(c)** 100 mW **(d)** 1 mW

11. A 2.2 kΩ resistor dissipates 0.5 W. The current is

(a) 15.1 mA **(b)** 227 μA **(c)** 1.1 mA **(d)** 4.4 mA

12. A 330 Ω resistor dissipates 2 W. The voltage is

(a) 2.57 V **(b)** 660 V **(c)** 6.6 V **(d)** 25.7 V

13. The power rating of a resistor that is to handle up to 1.1 W should be

(a) 0.25 W **(b)** 1 W **(c)** 2 W **(d)** 5 W

14. A 22 Ω half-watt resistor and a 220 Ω half-watt resistor are connected across a 10 V source. Which one(s) will overheat?

(a) 22 Ω **(b)** 220 Ω **(c)** both **(d)** neither

15. When the needle of an analog ohmmeter indicates infinity, the resistor being measured is

(a) overheated **(b)** shorted **(c)** open **(d)** reversed

PROBLEMS

Answers to odd-numbered problems are at the end of the book.

BASIC PROBLEMS

SECTION 3–1 **Ohm's Law**

1. The current in a circuit is 1 A. Determine what the current will be when

(a) the voltage is tripled
(b) the voltage is reduced by 80%
(c) the voltage is increased by 50%

2. The current in a circuit is 100 mA. Determine what the current will be when
   (a) the resistance is increased by 100%
   (b) the resistance is reduced by 30%
   (c) the resistance is quadrupled
3. The current in a circuit is 10 mA. What will the current be if the voltage is tripled and the resistance is doubled?

SECTION 3–2 **Application of Ohm's Law**

4. Determine the current in each case.
   (a) $V = 5\text{ V}, R = 1.0\ \Omega$ (b) $V = 15\text{ V}, R = 10\ \Omega$
   (c) $V = 50\text{ V}, R = 100\ \Omega$ (d) $V = 30\text{ V}, R = 15\text{ k}\Omega$
   (e) $V = 250\text{ V}, R = 4.7\text{ M}\Omega$
5. Determine the current in each case.
   (a) $V = 9\text{ V}, R = 2.7\text{ k}\Omega$ (b) $V = 5.5\text{ V}, R = 10\text{ k}\Omega$
   (c) $V = 40\text{ V}, R = 68\text{ k}\Omega$ (d) $V = 1\text{ kV}, R = 2\text{ k}\Omega$
   (e) $V = 66\text{ kV}, R = 10\text{ M}\Omega$
6. A 10 Ω resistor is connected across a 12 V battery. How much current is there through the resistor?
7. A resistor is connected across the terminals of a dc voltage source in each part of Figure 3–28. Determine the current in each resistor.

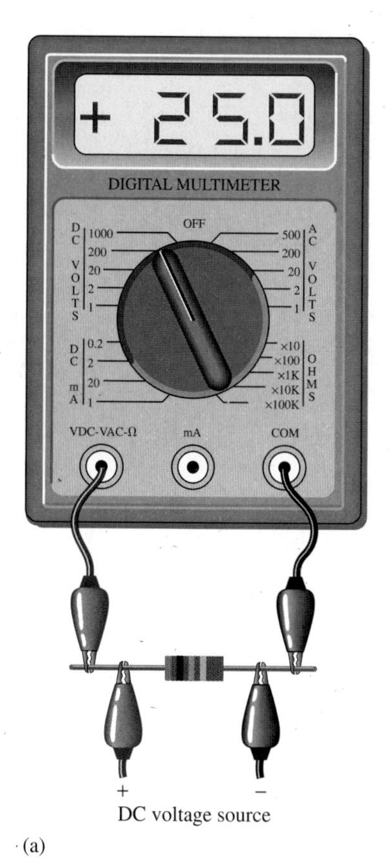

(a)

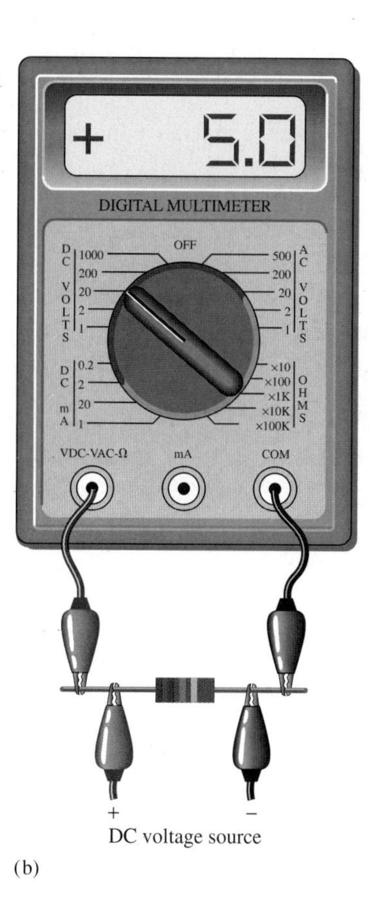

(b)

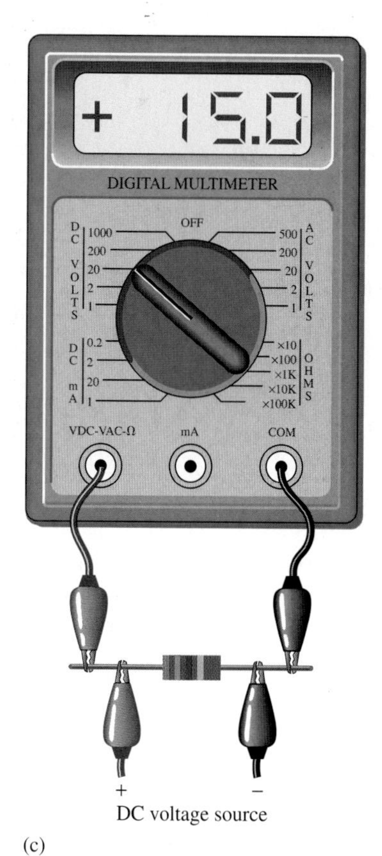

(c)

▲ FIGURE 3–28

8. Calculate the voltage for each value of I and R.

(a) $I = 2\text{ A}, R = 18\ \Omega$ **(b)** $I = 5\text{ A}, R = 47\ \Omega$

(c) $I = 2.5\text{ A}, R = 620\ \Omega$ **(d)** $I = 0.6\text{ A}, R = 47\ \Omega$

(e) $I = 0.1\text{ A}, R = 470\ \Omega$

9. Calculate the voltage for each value of I and R.

(a) $I = 1\text{ mA}, R = 10\ \Omega$ **(b)** $I = 50\text{ mA}, R = 33\ \Omega$

(c) $I = 3\text{ A}, R = 4.7\text{ k}\Omega$ **(d)** $I = 1.6\text{ mA}, R = 2.2\text{ k}\Omega$

(e) $I = 250\ \mu\text{A}, R = 1.0\text{ k}\Omega$ **(f)** $I = 500\text{ mA}, R = 1.5\text{ M}\Omega$

(g) $I = 850\ \mu\text{A}, R = 10\text{ M}\Omega$ **(h)** $I = 75\ \mu\text{A}, R = 47\ \Omega$

10. Three amperes of current are measured through a 27 Ω resistor connected across a voltage source. How much voltage does the source produce?

11. Assign a voltage value to each source in the circuits of Figure 3–29 to obtain the indicated amounts of current.

▶ **FIGURE 3–29**

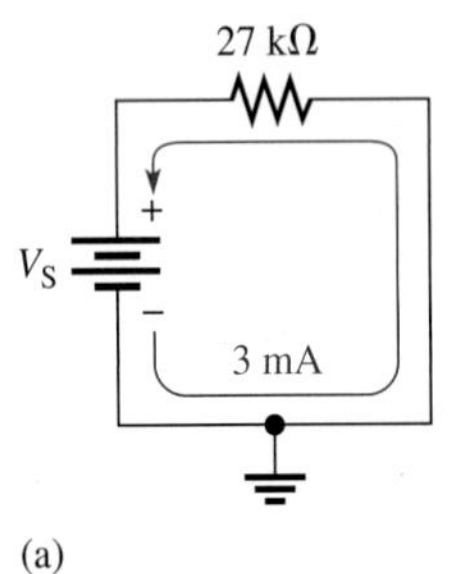

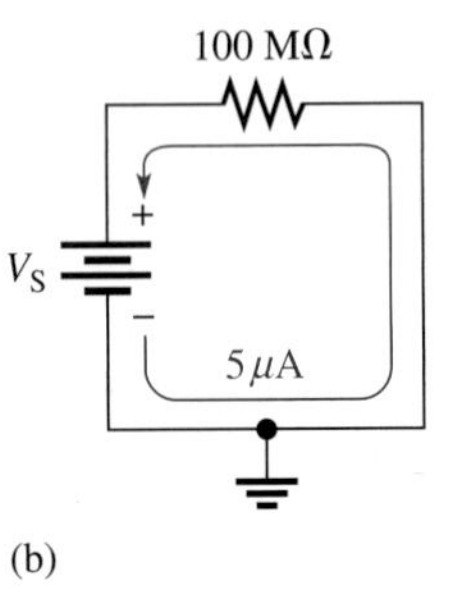

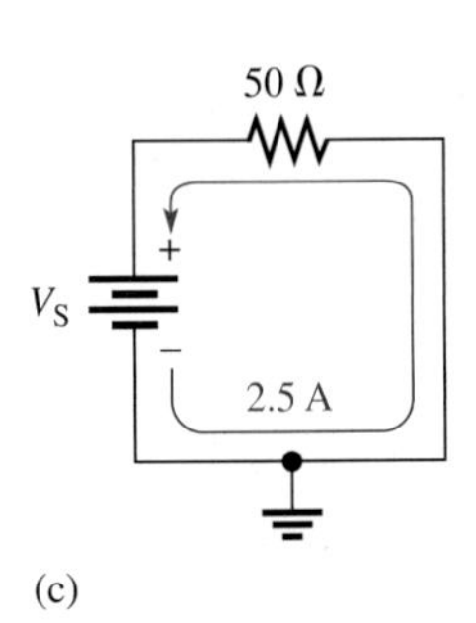

12. Calculate the resistance for each value of V and I.

(a) $V = 10\text{ V}, I = 2\text{ A}$ **(b)** $V = 90\text{ V}, I = 45\text{ A}$

(c) $V = 50\text{ V}, I = 5\text{ A}$ **(d)** $V = 5.5\text{ V}, I = 10\text{ A}$

(e) $V = 150\text{ V}, I = 0.5\text{ A}$

13. Calculate R for each set of V and I values.

(a) $V = 10\text{ kV}, I = 5\text{ A}$ **(b)** $V = 7\text{ V}, I = 2\text{ mA}$

(c) $V = 500\text{ V}, I = 250\text{ mA}$ **(d)** $V = 50\text{ V}, I = 500\ \mu\text{A}$

(e) $V = 1\text{ kV}, I = 1\text{ mA}$

14. Six volts are applied across a resistor. A current of 2 mA is measured. What is the value of the resistor?

15. Choose the correct value of resistance to get the current values indicated in each circuit of Figure 3–30.

▶ **FIGURE 3–30**

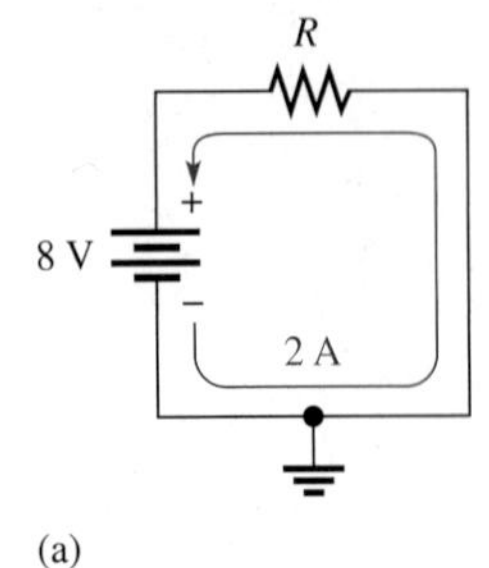

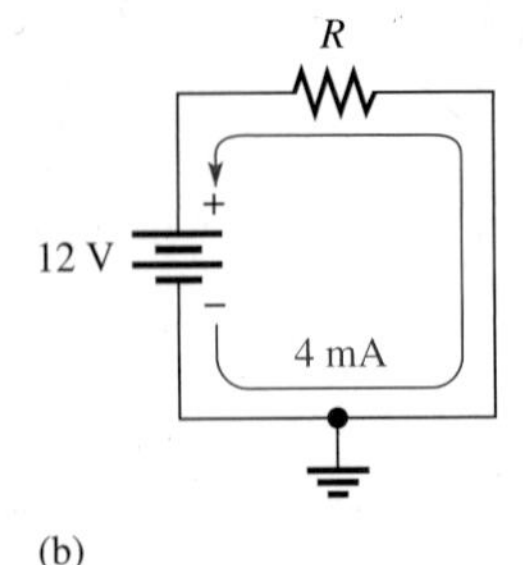

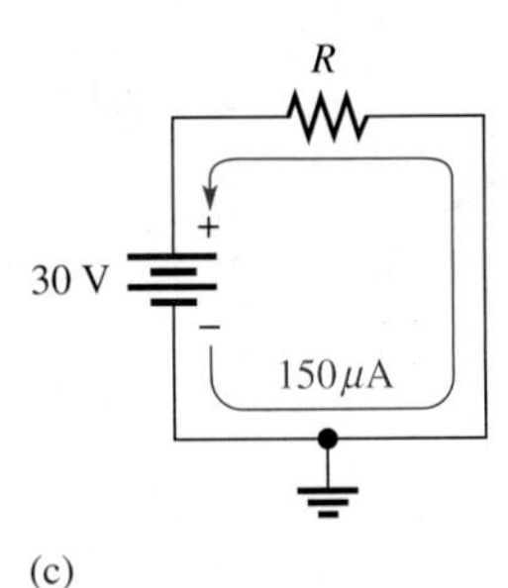

SECTION 3–3 **Energy and Power**

16. What is the power when energy is used at the rate of 350 J/s?
17. What is the power in watts when 7500 J of energy are used in 5 h?
18. Convert the following in kilowatts:
    (a) 1000 W (b) 3750 W (c) 160 W (d) 50,000 W
19. Convert the following to megawatts:
    (a) 1,000,000 W (b) 3×10^6 W (c) 15×10^7 W (d) 8700 kW
20. Convert the following to milliwatts:
    (a) 1 W (b) 0.4 W (c) 0.002 W (d) 0.0125 W
21. Convert the following to microwatts:
    (a) 2 W (b) 0.0005 W (c) 0.25 mW (d) 0.00667 mW
22. Convert the following to watts:
    (a) 1.5 kW (b) 0.5 MW (c) 350 mW (d) 9000 μW

SECTION 3–4 **Power in an Electric Circuit**

23. If a resistor has 5.5 V across it and 3 mA through it, what is the power dissipation?
24. An electric heater works on 115 V and draws 3 A of current. How much power does it use?
25. How much power is produced by 500 mA of current through a 4.7 kΩ resistor?
26. Calculate the power handled by a 10 kΩ resistor carrying 100 μA.
27. If there are 60 V across a 620 Ω resistor, what is the power dissipation?
28. A 56 Ω resistor is connected across the terminals of a 1.5 V battery. What is the power dissipation in the resistor?
29. If a resistor is to carry 2 A of current and handle 100 W of power, how many ohms must it be? Assume that the voltage can be adjusted to any required value.
30. Convert 5×10^6 watts used for 1 minute to kWh.
31. Convert 6700 watts used for 1 second to kWh.
32. How many kilowatt-hours do 50 W use for 12 h equal?

SECTION 3–5 **The Power Rating of Resistors**

33. A 6.8 kΩ resistor has burned out in a circuit. You must replace it with another resistor with the same resistance value. If the resistor carries 10 mA, what should its power rating be? Assume that you have available resistors in all the standard power ratings.
34. A certain type of power resistor comes in the following ratings: 3 W, 5 W, 8 W, 12 W, 20 W. Your particular application requires a resistor that can handle approximately 8 W. Which rating would you use? Why?

SECTION 3–6

Energy Conversion and Voltage Drop in a Resistance

35. For each circuit in Figure 3–31, assign the proper polarity for the voltage across the resistor.

FIGURE 3–31

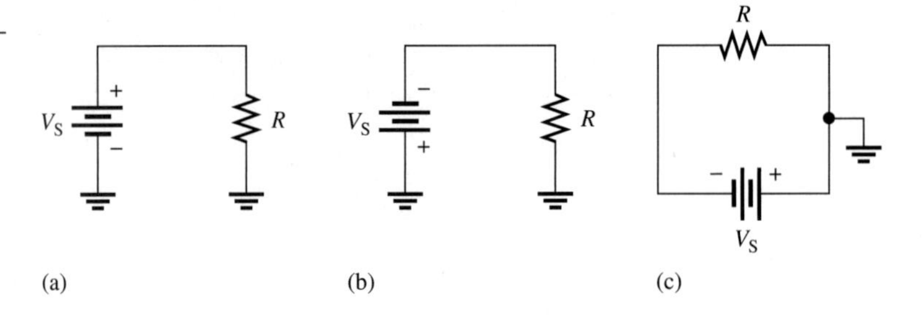

SECTION 3–7

Power Supplies

36. A 50 Ω load consumes 1 W of power. What is the output voltage of the power supply?

37. A battery can provide 1.5 A of current for 24 h. What is its ampere-hour rating?

38. How much continuous current can be drawn from an 80 Ah battery for 10 h?

39. If a battery is rated at 650 mAh, how much current will it provide for 48 h?

40. If the input power is 500 mW and the output power is 400 mW, what is the power loss? What is the efficiency of this power supply?

41. To operate at 85% efficiency, how much output power must a source produce if the input power is 5 W?

SECTION 3–8

Introduction to Troubleshooting

42. In the light circuit of Figure 3–32, identify the faulty bulb based on the series of ohmmeter readings shown.

FIGURE 3–32

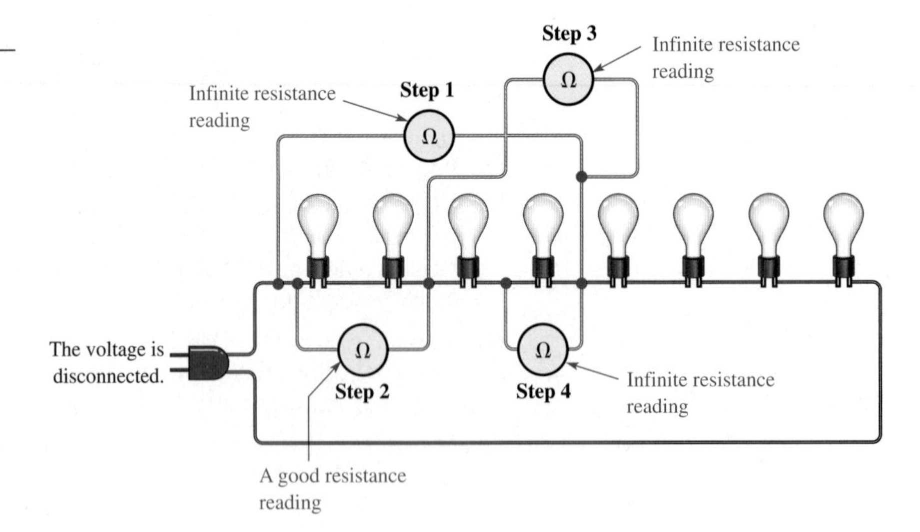

43. Assume you have a 32-light string and one of the bulbs is burned out. Using the half-splitting approach and starting in the left half of the circuit, how many resistance measurements will it take to find the faulty bulb if it is seventeenth from the left. Remember, you don't know which bulb is faulty.

ADVANCED PROBLEMS

44. A certain power supply provides a continuous 2 W to a load. It is operating at 60% efficiency. In a 24 h period, how many kilowatt-hours does the power supply use?

45. The filament of a light bulb in the circuit of Figure 3–33(a) has a certain amount of resistance, represented by an equivalent resistance in Figure 3–33(b). If the bulb operates with 120 V and 0.8 A of current, what is the resistance of its filament?

FIGURE 3–33

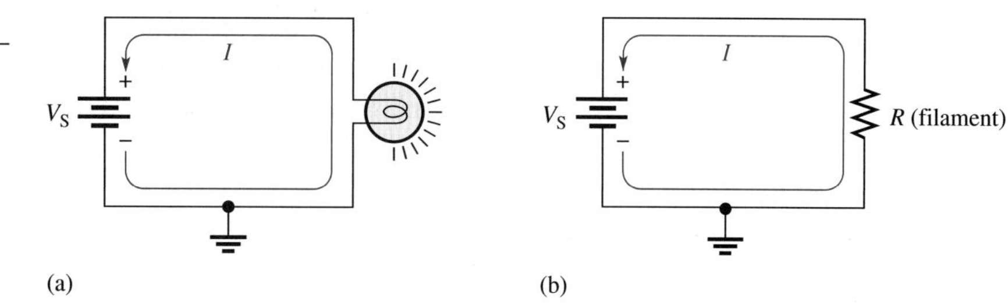

46. A certain electrical device has an unknown resistance. You have available a 12 V battery and an ammeter. How would you determine the value of the unknown resistance? Draw the necessary circuit connections.

47. A variable voltage source is connected to the circuit of Figure 3–34. Start at 0 V and increase the voltage in 10 V steps up to 100 V. Determine the current at each voltage value, and plot a graph of V versus I. Is the graph a straight line? What does the graph indicate?

FIGURE 3–34

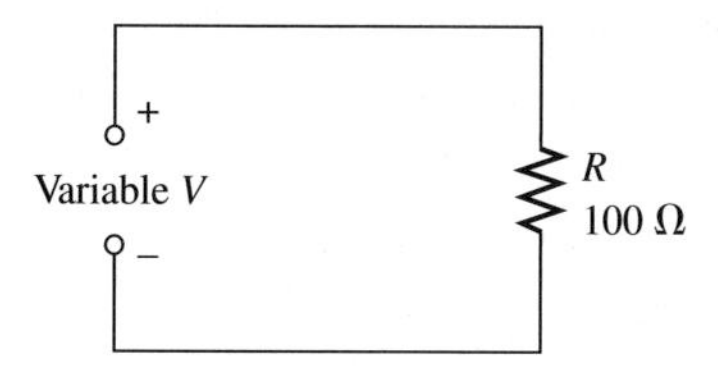

48. In a certain circuit, $V_S = 1$ V and $I = 5$ mA. Determine the current for each of the following voltages in a circuit with the same resistance:

(a) $V_S = 1.5$ V **(b)** $V_S = 2$ V **(c)** $V_S = 3$ V

(d) $V_S = 4$ V **(e)** $V_S = 10$ V

49. Figure 3–35 is a graph of current versus voltage for three resistance values. Determine R_1, R_2, and R_3.

FIGURE 3–35

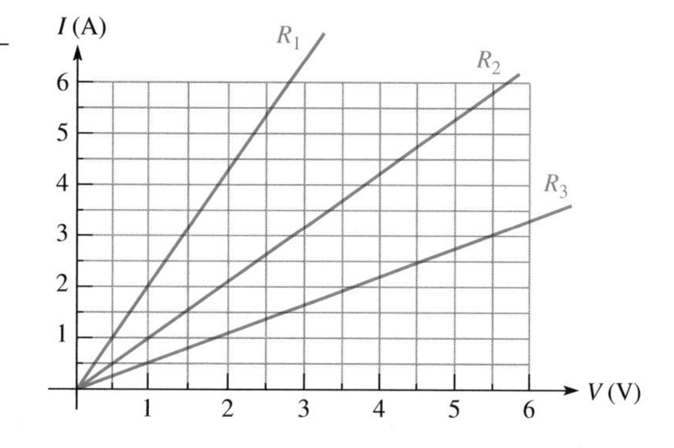

50. You are measuring the current in a circuit that is operated on a 10 V battery. The ammeter reads 50 mA. Later, you notice that the current has dropped to 30 mA. Eliminating the possibility of a resistance change, you must conclude that the voltage has changed. How much has the voltage of the battery changed, and what is its new value?

51. If you wish to increase the amount of current in a resistor from 100 mA to 150 mA by changing the 20 V source, by how many volts should you change the source? To what new value should you set it?

52. By varying the rheostat (variable resistor) in the circuit of Figure 3–36, you can change the amount of current. The setting of the rheostat is such that the current is 750 mA. What is the resistance value of this setting? To adjust the current to 1 A, to what resistance value must you set the rheostat? The rheostat must never be adjusted to 0 Ω. Why?

FIGURE 3–36

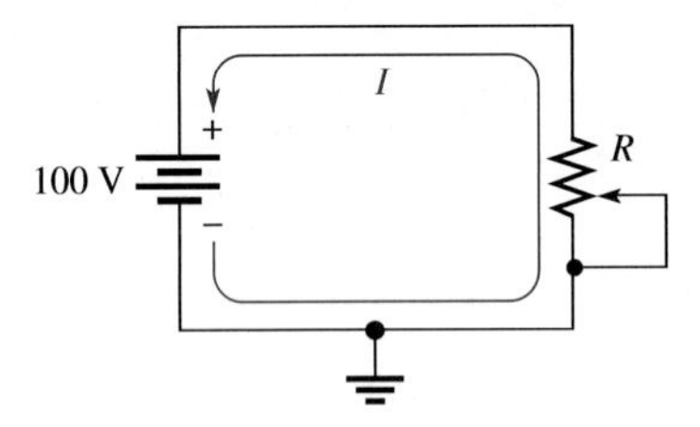

53. A certain resistor has the following color code: orange, orange, red, gold. Determine the maximum and minimum currents you should expect to measure when a 12 V source is connected across the resistor.

54. A 6 V source is connected to a 100 Ω resistor by two 12 ft lengths of 18-gauge copper wire. Refer to Table 2–4 to determine the following:

(a) current **(b)** resistor voltage **(c)** voltage across each length of wire

55. A certain appliance uses 300 W. If it is allowed to run continuously for 30 days, how many kilowatt-hours of energy does it use?

56. At the end of a 31-day period, your utility bill shows that you have used 1500 kWh. What is the average daily power?

57. A certain type of power resistor comes in the following ratings: 3 W, 5 W, 8 W, 12 W, 20 W. Your particular application requires a resistor that can handle approximately 10 W. Which rating would you use? Why?

58. A 12 V source is connected across a 10 Ω resistor for 2 min.

(a) What is the power dissipation?

(b) How much energy is used?

(c) If the resistor remains connected for an additional minute, does the power dissipation increase or decrease?

59. The rheostat in Figure 3–37 is used to control the current to a heating element. When the rheostat is adjusted to a value of 8 Ω or less, the heating element can burn out. What is the rated value of the fuse needed to protect the circuit if the voltage across the heating element at the point of maximum current is 100 V?

FIGURE 3–37

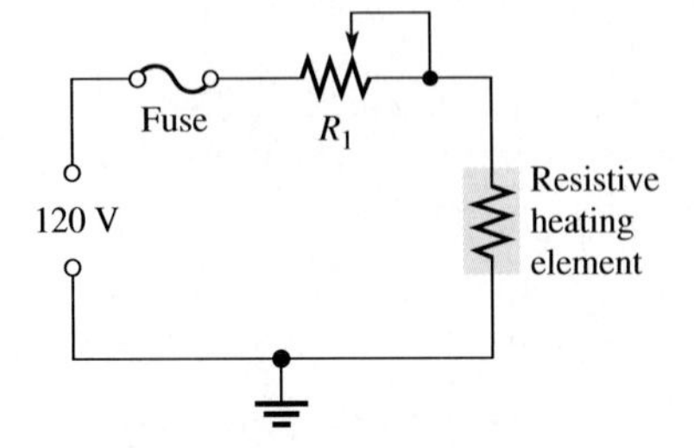

ELECTRONICS WORKBENCH/CIRCUITMAKER TROUBLESHOOTING PROBLEMS

CD-ROM file circuits are shown in Figure 3–38.

60. Open file P03-60 on your CD-ROM. Determine whether or not the circuit is operating properly. If not, find the faulty part.

61. Open file P03-61 on your CD-ROM. Determine whether or not the circuit is operating properly. If not, find the faulty part.

62. Open file P03-62 on your CD-ROM. Determine whether or not the circuit is operating properly. If not, find the faulty part.

63. Open file P03-63 on your CD-ROM. Determine whether or not the circuit is operating properly. If not, find the faulty part.

64. Open file P03-64 on your CD-ROM. Determine whether or not the circuit is operating properly. If not, find the faulty part.

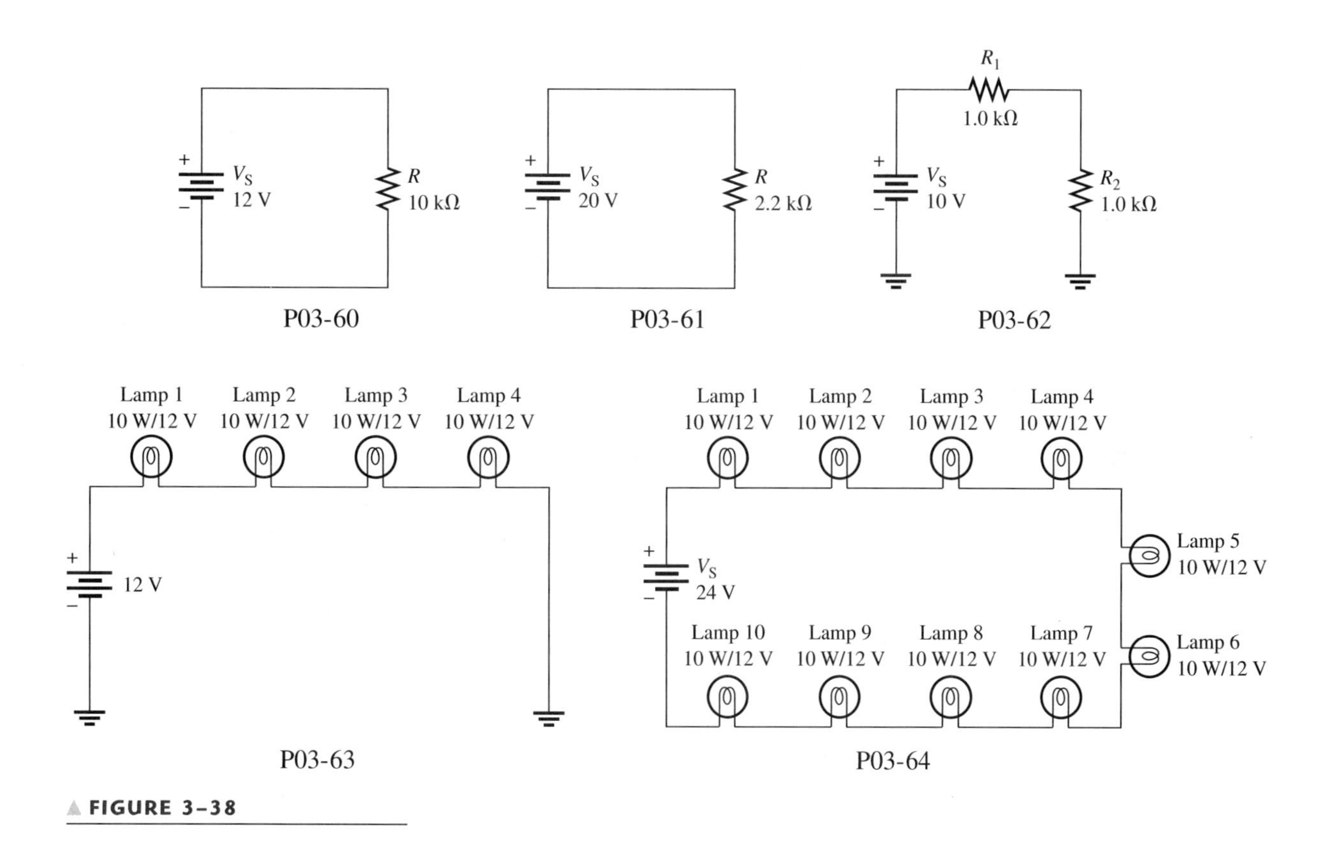

▲ **FIGURE 3–38**

ANSWERS

SECTION REVIEWS

SECTION 3–1 **Ohm's Law**

1. Ohm's law states that current varies directly with voltage and inversely with resistance.

2. $I = V/R$ **3.** $V = IR$ **4.** $R = V/I$

5. The current increases; three times when V is tripled
6. Doubling R cuts I in half to 5 mA.
7. No change in I if V and R are doubled.

SECTION 3–2 **Application of Ohm's Law**

1. $I = 10\text{ V}/4.7\ \Omega = 2.13\text{ A}$
2. $I = 20\text{ kV}/4.7\text{ M}\Omega = 4.26\text{ mA}$
3. $I = 10\text{ kV}/2\text{ k}\Omega = 5\text{ A}$
4. $V = (1\text{ A})(10\ \Omega) = 10\text{ V}$
5. $V = (3\text{ mA})(3\text{ k}\Omega) = 9\text{ V}$
6. $V = (2\text{ A})(6\ \Omega) = 12\text{ V}$
7. $R = 10\text{ V}/2\text{ A} = 5\ \Omega$
8. $R = 25\text{ V}/50\text{ mA} = 0.5\text{ k}\Omega = 500\ \Omega$

SECTION 3–3 **Energy and Power**

1. Power is the rate at which energy is used.
2. $P = W/t$
3. Watt is the unit of power. One watt is the power when 1 J of energy is used in 1 s.
4. **(a)** 68,000 W = 68 kW **(b)** 0.005 W = 5 mW
   (c) 0.000025 = 25 μW
5. (100 W)(10 h) = 1 kWh
6. 2000 W = 2 kW
7. (1.322 kW)(24 h) = 31.73 kWh; (0.09 $/kWh)(31.73 kWh) = $2.86

SECTION 3–4 **Power in an Electric Circuit**

1. $P = (10\text{ V})(3\text{ A}) = 30\text{ W}$
2. $P = (5\text{ A})^2(47\ \Omega) = 1175\text{ W}$
3. $P = (20\text{ mA})^2(5.1\text{ k}\Omega) = 2.04\text{ W}$
4. $P = (5\text{ V})^2/10\ \Omega = 2.5\text{ W}$
5. $P = (8\text{ V})^2/2.2\text{ k}\Omega = 29.1\text{ mW}$
6. $R = 55\text{ W}/(0.5\text{ A})^2 = 220\ \Omega$

SECTION 3–5 **The Power Rating of Resistors**

1. Two resistor parameters are resistance and power rating.
2. A larger resistor physical size dissipates more energy.
3. Standard ratings of metal-film resistors are 0.125 W, 0.25 W, 0.5 W, and 1 W.
4. At least a 0.5 W rating to handle 0.3 W
5. $8 \times 1\text{k} = 8\text{ k}\Omega$

SECTION 3–6 **Energy Conversion and Voltage Drop in a Resistance**

1. Energy conversion in a resistor is caused by collisions of free electrons with the atoms in the material.
2. Voltage drop is the difference in the voltage at two points due to energy conversion.
3. Voltage drop is negative to positive in the direction of current.

SECTION 3–7 **Power Supplies**

1. An increased amount of current represents a greater load.
2. $P_{OUT} = (10\text{ V})(0.5\text{ A}) = 5\text{ W}$
3. 100 Ah/5 A = 20 h

4. $P_{OUT} = (12\text{ V})(5\text{ A}) = 60\text{ W}$
5. Efficiency = (750 mW/1000 mW)100% = 75%

SECTION 3–8

Introduction to Troubleshooting

1. Analysis, Planning, and Measuring
2. Half-splitting identifies the fault by successively isolating half of the remaining circuit.
3. Voltage is measured across a component. Current is measured in series with the component.

■ Application Assignment

1. Watt's law was used to find power ratings ($P = V^2/R$).
2. Ohm's law was used to find the two additional resistors ($R = V/I$).
3. Power ratings were determined by the relative sizes of the resistors, given the range of values.

RELATED PROBLEMS FOR EXAMPLES

3–1 $I_1 = 10\text{ V}/5\ \Omega = 2\text{ A};\ I_2 = 10\text{ V}/20\ \Omega = 0.5\text{ A}$

3–2 1 V

3–3 900 Ω

3–4 Yes

3–5 2.5 A

3–6 5 mA

3–7 25 μA

3–8 200 μA

3–9 800 V

3–10 220 mV

3–11 16.5 V

3–12 4 kΩ

3–13 100 Ω

3–14 3000 J

3–15 **(a)** 0.001 W **(b)** 0.0018 W **(c)** 3,000,000 W **(d)** 10,000 W

3–16 2 kWh

3–17 \$6.12 + \$.06 = 6.18

3–18 **(a)** 40 W **(b)** 400 W **(c)** 0.625 W

3–19 60 W

3–20 0.5 W (½ W)

3–21 No

3–22 60 Ah

3–23 46 W

SELF-TEST

1. (b) **2.** (c) **3.** (b) **4.** (d) **5.** (a) **6.** (d) **7.** (b)
8. (c) **9.** (d) **10.** (a) **11.** (a) **12.** (d) **13.** (c) **14.** (a)
15. (c)

SERIES CIRCUITS

INTRODUCTION

Resistive circuits can be of two basic forms: series or parallel. In this chapter we discuss series circuits. Parallel circuits are covered in Chapter 5, and combinations of series and parallel circuits are examined in Chapter 6. In this chapter you will see how Ohm's law is used in series circuits; and you will study another important law, Kirchhoff's voltage law. Also, several important applications of series circuits are presented.

CHAPTER OUTLINE

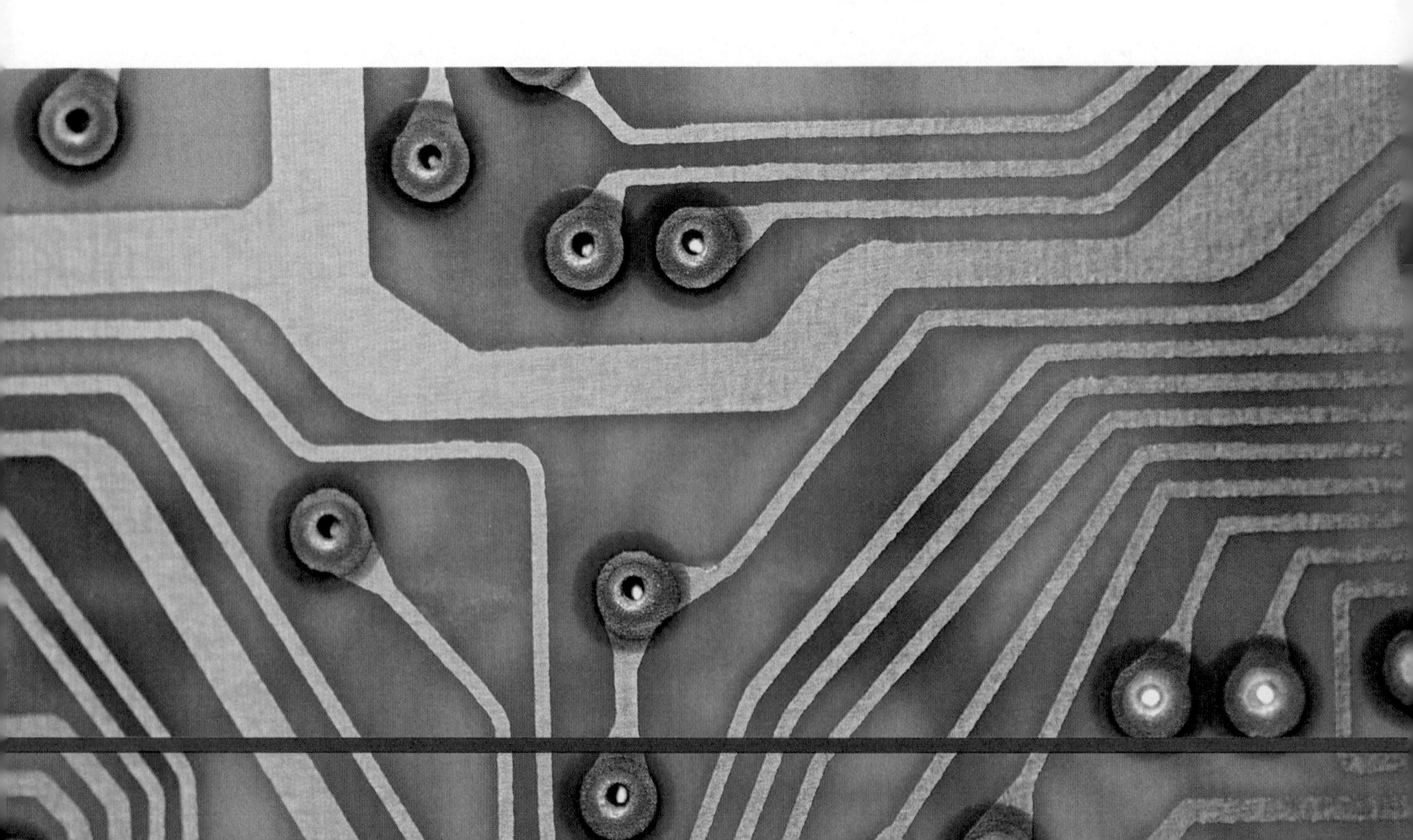

CHAPTER OBJECTIVES

- Identify a series circuit
- Determine the current in a series circuit
- Determine total series resistance
- Apply Ohm's law in series circuits
- Determine the total effect of voltage sources in series
- Apply Kirchhoff's voltage law
- Use a series circuit as a voltage divider
- Determine power in a series circuit
- Determine and identify ground in a circuit
- Troubleshoot series circuits

KEY TERMS

- Series
- Series-aiding
- Series-opposing
- Kirchhoff's voltage law
- Voltage divider
- Circuit ground
- Open
- Short

APPLICATION ASSIGNMENT PREVIEW

For the application assignment in this chapter, suppose your supervisor has given you a voltage-divider board to evaluate and modify if necessary. You will use it to obtain five different voltage levels from a 12 V battery. The voltage divider will be used to provide positive reference voltages to an electronic circuit in an analog-to-digital converter. Your task will be to check the circuit to see if it provides the required voltages and, if not, you will modify it so that it does. Also, you must make sure that the power ratings of the resistors are adequate for the application and determine how long the battery will last with the voltage divider connected to it. After studying this chapter, you should be able to complete the application assignment.

WWW. VISIT THE COMPANION WEBSITE

Circuit Simulation Tutorials and Other Chapter Study Tools Are Available at

http://www.prenhall.com/floyd

4–1 RESISTORS IN SERIES

When connected in series, resistors form a "string" in which there is only one path for current.

After completing this section, you should be able to

- **Identify a series circuit**
- Translate a physical arrangement of resistors into a schematic

Figure 4–1(a) shows two resistors connected in series between point A and point B. Part (b) shows three resistors in series, and part (c) shows four in series. Of course, there can be any number of resistors in series.

FIGURE 4–1

Resistors in series.

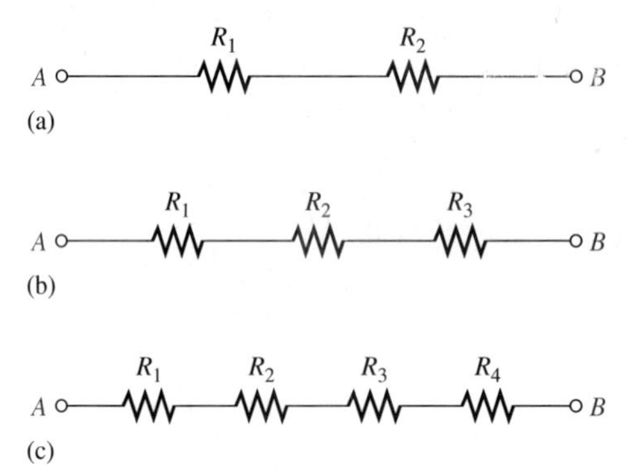

When a voltage source is connected between point A and point B, the only way for electrons to get from one point to the other in any of the connections of Figure 4–1 is to go through each of the resistors. A series circuit is identified as follows:

A series circuit provides only one path for current between two points so that the current is the same through each series resistor.

Identifying Series Circuits

In an actual circuit diagram, a series circuit may not always be as easy to identify as those in Figure 4–1. For example, Figure 4–2 shows series resistors drawn in other ways with a voltage applied. Remember, if there is only one current path between two points, the resistors between those two points are in series, no matter how they appear in a diagram.

FIGURE 4–2

Some examples of series connections of resistors. Notice that the current must be the same at all points because the current has only one path.

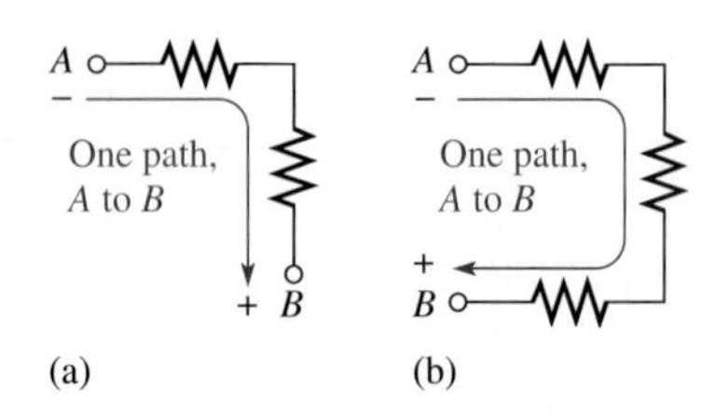

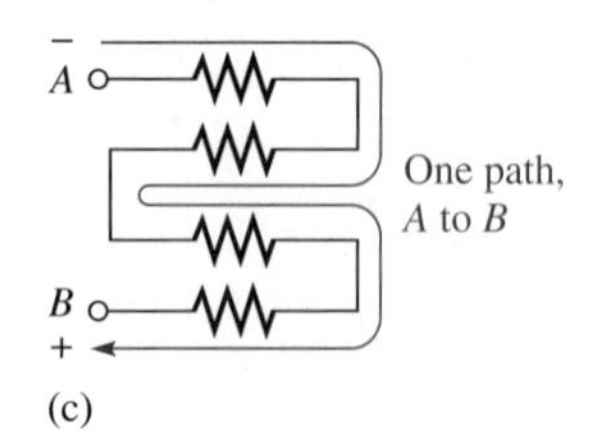

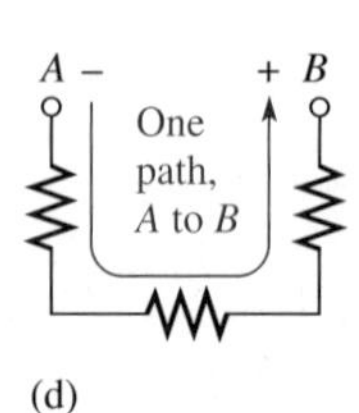

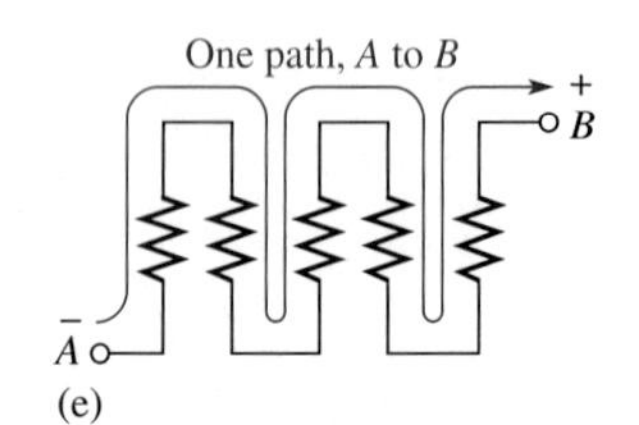

EXAMPLE 4–1

Five resistors are positioned on a circuit board as shown in Figure 4–3. Wire them together in series so that, starting from the negative (−) terminal, R_1 is first, R_2 is second, R_3 is third, and so on. Draw a schematic showing this connection.

FIGURE 4–3

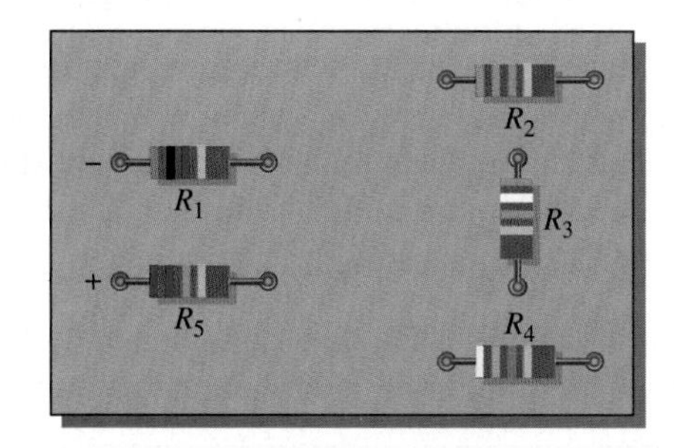

Solution The wires are connected as shown in Figure 4–4(a), which is the assembly diagram. The schematic is shown in Figure 4–4(b). Note that the schematic does not necessarily show the actual physical arrangement of the resistors as does the assembly diagram. The schematic shows how components are connected electrically; the assembly diagram shows how components are arranged and interconnected physically.

FIGURE 4–4

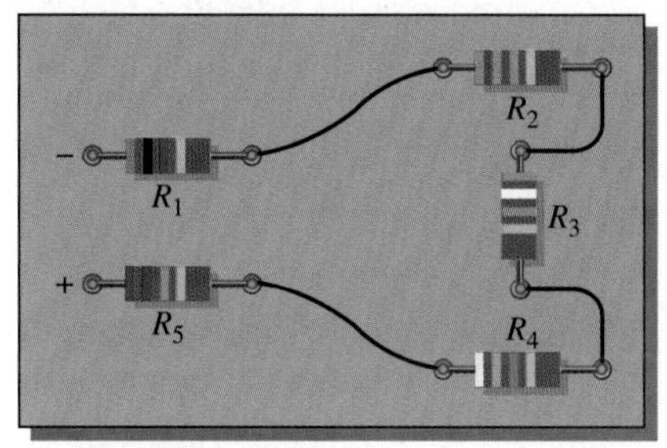

(a) Assembly diagram

(b) Schematic

*Related Problem** **(a)** Show how you would rewire the circuit board in Figure 4–4(a) so that all the odd-number resistors come first followed by the even-numbered ones.
(b) Determine the value of each resistor.

*Answers are at the end of the chapter.

EXAMPLE 4–2

Describe how the resistors on the printed circuit (PC) board in Figure 4–5 are related electrically. Determine the resistance value of each resistor.

FIGURE 4–5

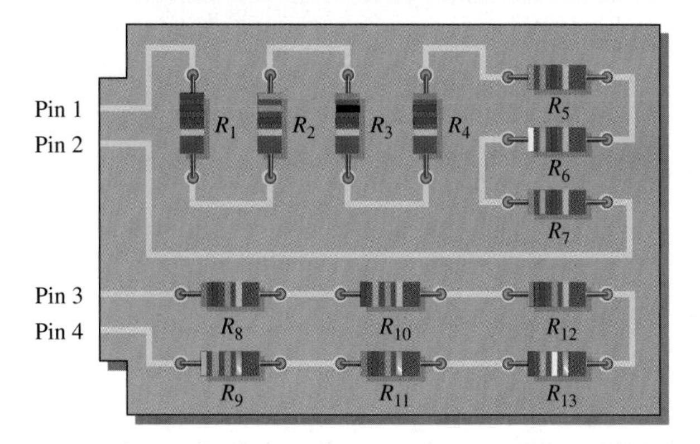

Solution Resistors R_1 through R_7 are in series with each other. This series combination is connected between pins 1 and 2 on the PC board.

Resistors R_8 through R_{13} are in series with each other. This series combination is connected between pins 3 and 4 on the PC board.

The values of the resistors are $R_1 = 2.2\ \text{k}\Omega$, $R_2 = 3.3\ \text{k}\Omega$, $R_3 = 1.0\ \text{k}\Omega$, $R_4 = 1.2\ \text{k}\Omega$, $R_5 = 3.3\ \text{k}\Omega$, $R_6 = 4.7\ \text{k}\Omega$, $R_7 = 5.6\ \text{k}\Omega$, $R_8 = 12\ \text{k}\Omega$, $R_9 = 68\ \text{k}\Omega$, $R_{10} = 27\ \text{k}\Omega$, $R_{11} = 12\ \text{k}\Omega$, $R_{12} = 82\ \text{k}\Omega$, and $R_{13} = 270\ \text{k}\Omega$.

Related Problem How is the circuit in Figure 4–5 changed when pin 2 and pin 3 are connected?

SECTION 4–1 REVIEW

Answers are at the end of the chapter.

1. How are the resistors connected in a series circuit?
2. How can you identify a series circuit?
3. Complete the schematics for the circuits in each part of Figure 4–6 by connecting the resistors in series in numerical order from *A* to *B*.
4. Now connect each group of series resistors in Figure 4–6 in series.

FIGURE 4–6

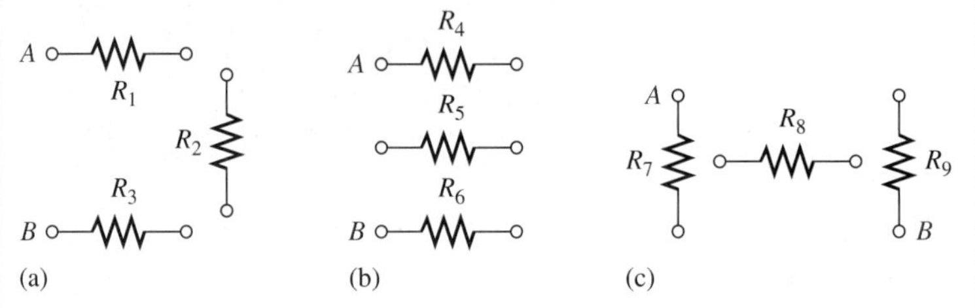

4–2 CURRENT IN A SERIES CIRCUIT

The current is the same through all points in a series circuit. The current through each resistor in a series circuit is the same as the current through all the other resistors that are in series with it.

After completing this section, you should be able to

- **Determine the current in a series circuit**
- Show that the current is the same at all points in a series circuit

Figure 4–7 shows three resistors connected in series to a voltage source. *At any point in this circuit, the current into that point must equal the current out of that point,* as illustrated by the current directional arrows. Notice also that the current out of each of the resistors must equal the current in because there is no place where part of the current can branch off and go somewhere else. Therefore, the current in each section of the circuit is the same as the current in all other sections. It has only one path going from the negative (−) side of the source to the positive (+) side.

In Figure 4–8, the battery supplies 1 A of current to the series resistors. There is one ampere of current out of the battery's negative terminal. As shown, one ampere of current is measured at several points in the series circuit.

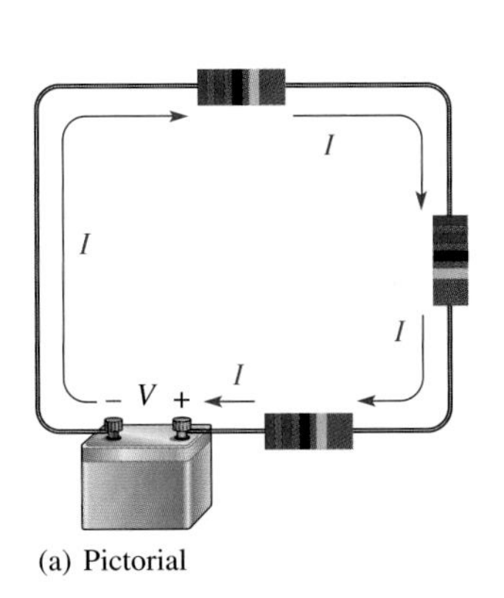

(a) Pictorial

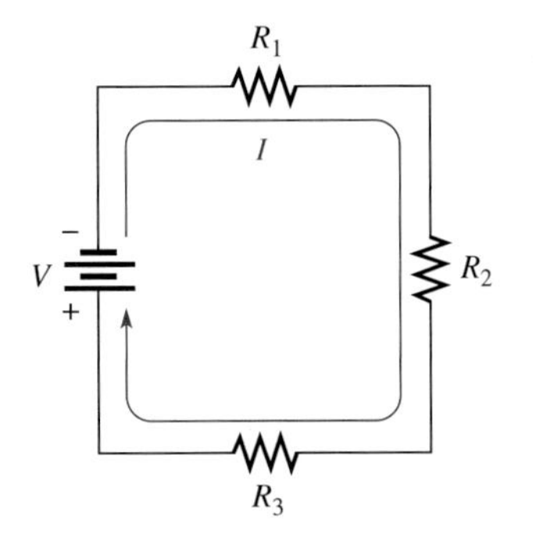

(b) Schematic

FIGURE 4–7

Current entering any point in a series circuit is the same as the current leaving that point.

FIGURE 4–8

Current is the same at all points in a series circuit.

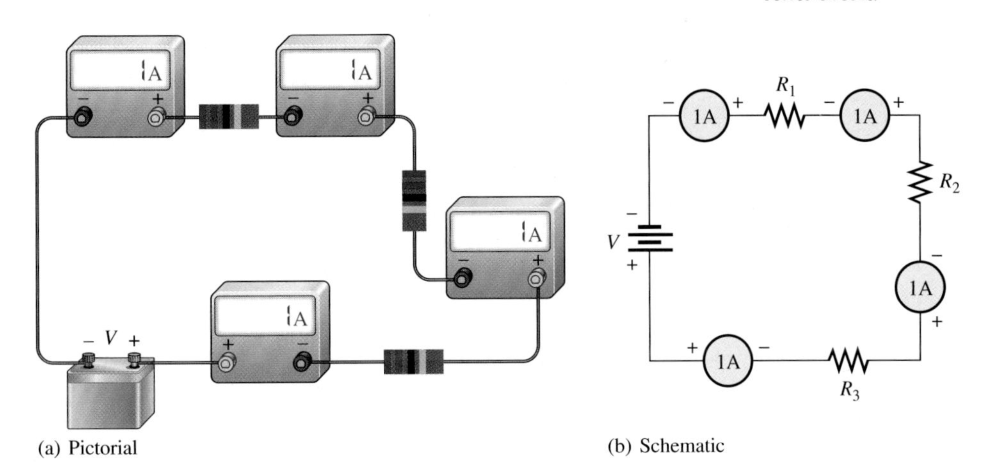

(a) Pictorial

(b) Schematic

SECTION 4–2 REVIEW

1. What statement can you make about the amount of current at any point in a series circuit?
2. In a series circuit with a 10 Ω and a 4.7 Ω resistor in series, there are 2 A through the 10 Ω resistor. How much current is there through the 4.7 Ω resistor?
3. A milliammeter is connected between points *A* and *B* in Figure 4–9. It measures 50 mA. If you move the meter and connect it between points *C* and *D*, how much current will it indicate? Between *E* and *F*?
4. In Figure 4–10, how much current does ammeter 1 indicate? Ammeter 2?

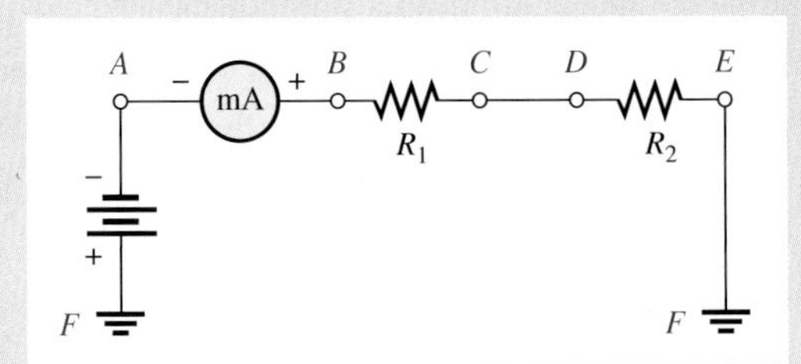

FIGURE 4–9

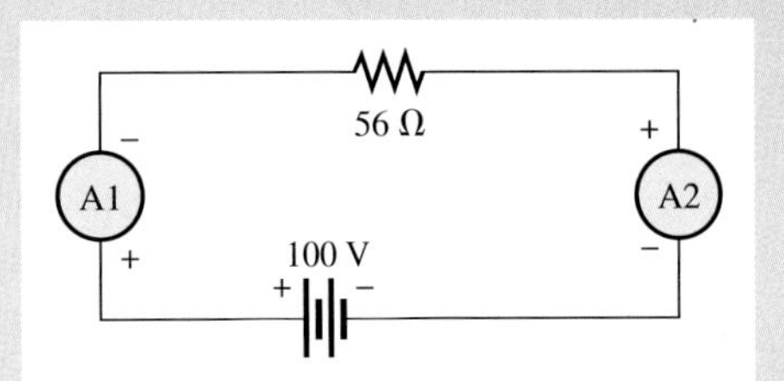

FIGURE 4–10

4–3 TOTAL SERIES RESISTANCE

The total resistance of a series circuit is equal to the sum of the resistances of each individual resistor.

After completing this section, you should be able to

- **Determine total series resistance**
- Explain why resistance values add when resistors are connected in series
- Apply the series resistance formula

Series Resistor Values Add

When resistors are connected in series, the resistor values add because each resistor offers opposition to the current in direct proportion to its resistance. A greater number of resistors connected in series creates more opposition to current. More opposition to current implies a higher value of resistance. So for every resistor that is added in series, the total resistance increases.

Figure 4–11 illustrates how series resistances add together to increase the total resistance. Part (a) of the figure has a single 10 Ω resistor. Part (b) shows another 10 Ω resistor connected in series with the first one, making a total resistance of 20 Ω. If a third 10 Ω resistor is connected in series with the first two, as shown in part (c), the total resistance becomes 30 Ω.

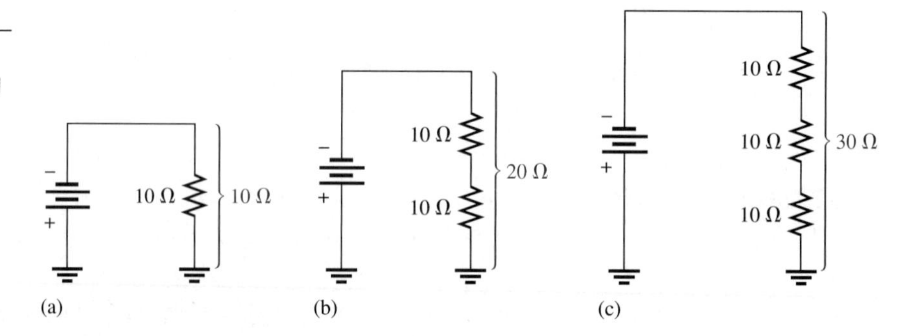

▶ **FIGURE 4–11**

Total resistance increases with each additional series resistor. The ground symbol used here was introduced in Section 2–6.

Series Resistance Formula

For any number of individual resistors connected in series, the total resistance is the sum of each of the individual values.

Equation 4–1

$$R_T = R_1 + R_2 + R_3 + \cdots + R_n$$

where R_T is the total resistance and R_n is the last resistor in the series string (n can be any positive integer equal to the number of resistors in series). For example, if there are four resistors in series ($n = 4$), the total resistance formula is

$$R_T = R_1 + R_2 + R_3 + R_4$$

If there are six resistors in series ($n = 6$), the total resistance formula is

$$R_T = R_1 + R_2 + R_3 + R_4 + R_5 + R_6$$

To illustrate the calculation of total series resistance, let's determine R_T of the circuit of Figure 4–12, where V_S is the source voltage. This circuit has five resistors in series. To find the total resistance, simply add the values.

$$R_T = 56\ \Omega + 100\ \Omega + 27\ \Omega + 10\ \Omega + 47\ \Omega = 240\ \Omega$$

Note in Figure 4–12 that the order in which the resistances are added does not matter. Also, you can physically change the positions of the resistors in the circuit without affecting the total resistance or the current.

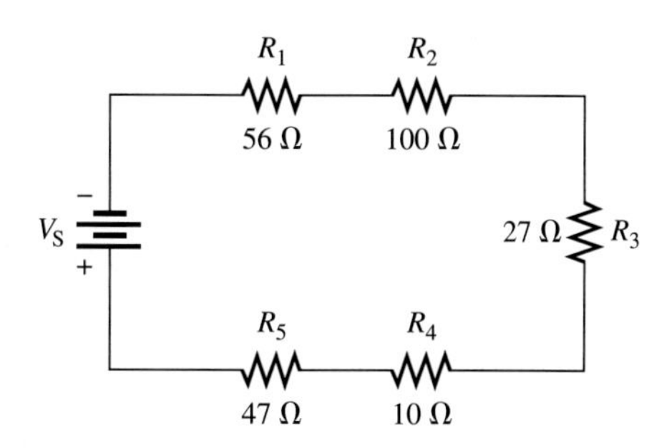

FIGURE 4–12

Example of five resistors in series. V_S stands for source voltage.

EXAMPLE 4–3

Connect the resistors on the protoboard in Figure 4–13 in series, and determine the total resistance, R_T.

FIGURE 4–13

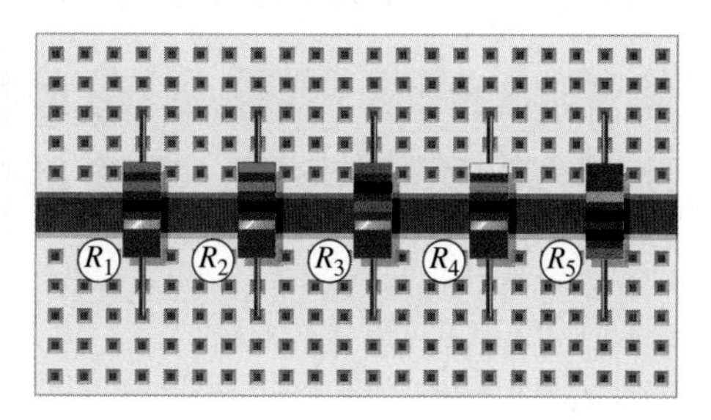

Solution The resistors are connected as shown in Figure 4–14. Find the total resistance by adding all the values.

$$R_T = R_1 + R_2 + R_3 + R_4 + R_5 = 33\ \Omega + 68\ \Omega + 100\ \Omega + 47\ \Omega + 10\ \Omega = \mathbf{258\ \Omega}$$

FIGURE 4–14

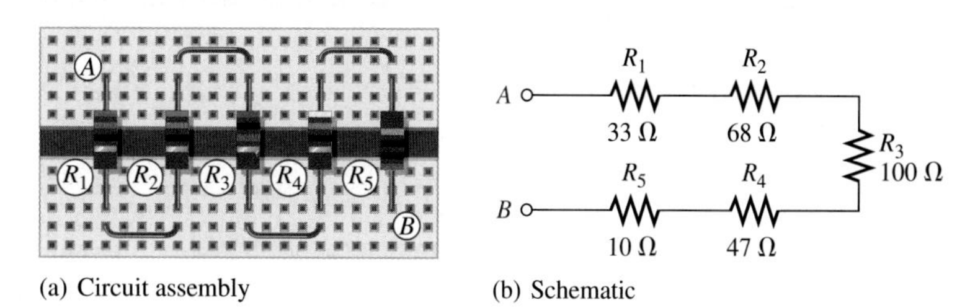

(a) Circuit assembly (b) Schematic

Related Problem Determine the total resistance in Figure 4–14(a) if the positions of R_2 and R_4 are interchanged.

EXAMPLE 4–4

Calculate the total resistance, R_T, for each circuit in Figure 4–15.

▶ FIGURE 4–15

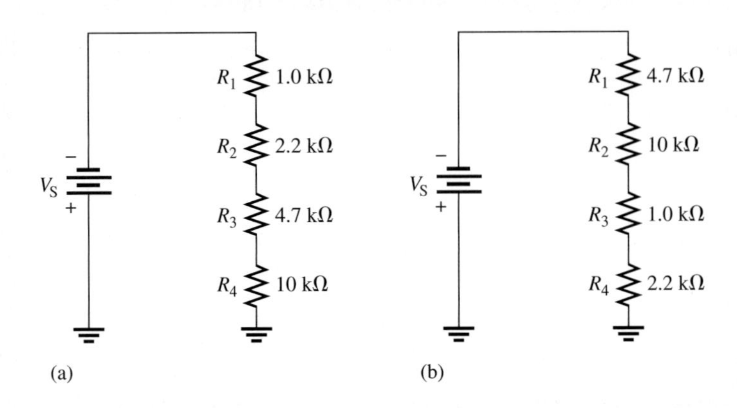

Solution For circuit (a),

$$R_T = 1.0\ \text{k}\Omega + 2.2\ \text{k}\Omega + 4.7\ \text{k}\Omega + 10\ \text{k}\Omega = \mathbf{17.9\ k\Omega}$$

For circuit (b),

$$R_T = 4.7\ \text{k}\Omega + 10\ \text{k}\Omega + 1.0\ \text{k}\Omega + 2.2\ \text{k}\Omega = \mathbf{17.9\ k\Omega}$$

Notice that the total resistance does not depend on the position of the resistors. Both circuits are identical in terms of total resistance.

The calculator sequence and display for R_T in circuit (a) are

1E3+2.2E3+4.7E3+10E3
17.9E3

Related Problem What is the total resistance for the following series resistors: 1.0 kΩ, 2.2 kΩ, 3.3 kΩ, and 5.6 kΩ?

EXAMPLE 4–5

Determine the value of R_4 in the circuit of Figure 4–16.

Solution From the ohmmeter reading, $R_T = 146$ kΩ.

$$R_T = R_1 + R_2 + R_3 + R_4$$

Solving for R_4 yields

$$R_4 = R_T - (R_1 + R_2 + R_3) = 146\ \text{k}\Omega - (10\ \text{k}\Omega + 33\ \text{k}\Omega + 47\ \text{k}\Omega) = \mathbf{56\ k\Omega}$$

FIGURE 4–16

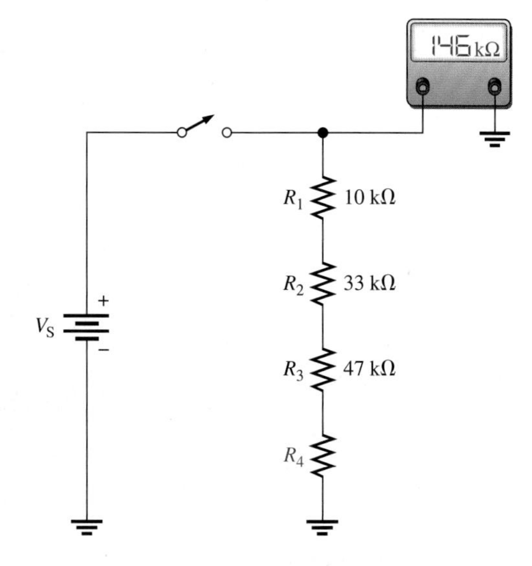

Related Problem Determine the value of R_4 in Figure 4–16 if the ohmmeter reading is 112 kΩ.

Equal-Value Series Resistors

When a circuit has more than one resistor of the same value in series, there is a shortcut method to obtain the total resistance: Simply multiply the resistance value of the resistors having the same value by the number of equal-value resistors that are in series. This method is essentially the same as adding the values. For example, five 100 Ω resistors in series have an R_T of 5(100 Ω) = 500 Ω. In general, the formula is expressed as

$$R_T = nR$$

Equation 4–2

where n is the number of equal-value resistors and R is the resistance value.

EXAMPLE 4–6

Find the R_T of eight 22 Ω resistors in series.

Solution Find R_T by adding the values.

$$R_T = 22\ \Omega + 22\ \Omega + 22\ \Omega + 22\ \Omega + 22\ \Omega + 22\ \Omega + 22\ \Omega + 22\ \Omega = \mathbf{176\ \Omega}$$

However, it is much easier to multiply.

$$R_T = 8(22\ \Omega) = \mathbf{176\ \Omega}$$

Related Problem Find R_T for three 1.0 kΩ and two 680 Ω resistors in series.

SECTION 4–3 REVIEW

1. Calculate R_T between points *A* and *B* for each circuit in Figure 4–17.
2. The following resistors are in series: one 100 Ω, two 47 Ω, four 12 Ω, and one 330 Ω. What is the total resistance?
3. Suppose that you have one resistor each of the following values: 1.0 kΩ, 2.7 kΩ, 3.3 kΩ, and 1.8 kΩ. To get a total resistance of 10 kΩ, you need one more resistor. What should its value be?
4. What is the R_T for twelve 47 Ω resistors in series?

FIGURE 4–17

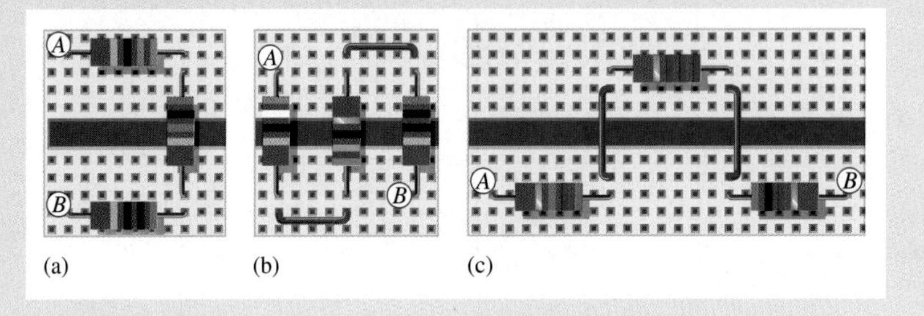

4–4 OHM'S LAW IN SERIES CIRCUITS

The application of Ohm's law and the basic concepts of series circuits are presented in several examples.

After completing this section, you should be able to

- **Apply Ohm's law in series circuits**
- Find the current in a series circuit
- Find the voltage across each resistor in series

Here are several key points to remember when you analyze series circuits.

1. Current through one of the series resistors is the same as the current through each of the other resistors and is the total current.
2. If you know the total voltage and the total resistance, you can determine the total current by using

$$I_T = \frac{V_T}{R_T}$$

3. If you know the voltage drop across one of the series resistors, you can determine the current by using

$$I = \frac{V_R}{R}$$

4. If you know the total current, you can find the voltage drop across any of the series resistors by using

$$V_R = I_T R$$

5. The polarity of a voltage drop across a resistor is positive at the end of the resistor that is closest to the positive terminal of the voltage source.
6. The resistor current is in a direction from the negative end of the resistor to the positive end.

7. An open in a series circuit prevents current; and, therefore, there is zero voltage drop across each series resistor. The total voltage appears across the points between which there is an open.

Now let's look at several examples that involve using Ohm's law.

EXAMPLE 4–7

Find the current in the circuit of Figure 4–18.

▶ FIGURE 4–18

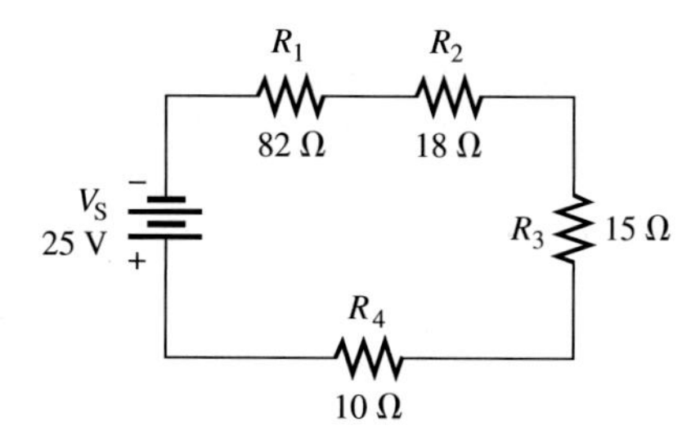

Solution The current is determined by the voltage and the total resistance. First, calculate the total resistance.

$$R_T = R_1 + R_2 + R_3 + R_4 = 82\ \Omega + 18\ \Omega + 15\ \Omega + 10\ \Omega = 125\ \Omega$$

Next, use Ohm's law to calculate the current.

$$I = \frac{V_S}{R_T} = \frac{25\text{ V}}{125\ \Omega} = 0.2\text{ A} = \mathbf{200\ mA}$$

Remember, the same current exists at all points in the circuit. Thus, each resistor has 200 mA through it.

Related Problem What is the current in Figure 4–18 if R_4 is changed to 100 Ω?

Open file E04-07 on your EWB/CircuitMaker CD-ROM. Connect the multimeter and verify the value of current calculated in this example. Change the value of R_4 to 100 Ω and verify the value of current calculated in the related problem.

EXAMPLE 4–8

The current in the circuit of Figure 4–19 is 1 mA. For this amount of current, what must the source voltage V_S be?

Solution In order to calculate V_S, first determine R_T.

$$R_T = 1.2\text{ k}\Omega + 5.6\text{ k}\Omega + 1.2\text{ k}\Omega + 1.5\text{ k}\Omega = 9.5\text{ k}\Omega$$

Now use Ohm's law to get V_S.

$$V_S = IR_T = (1\text{ mA})(9.5\text{ k}\Omega) = \mathbf{9.5\ V}$$

▶ FIGURE 4–19

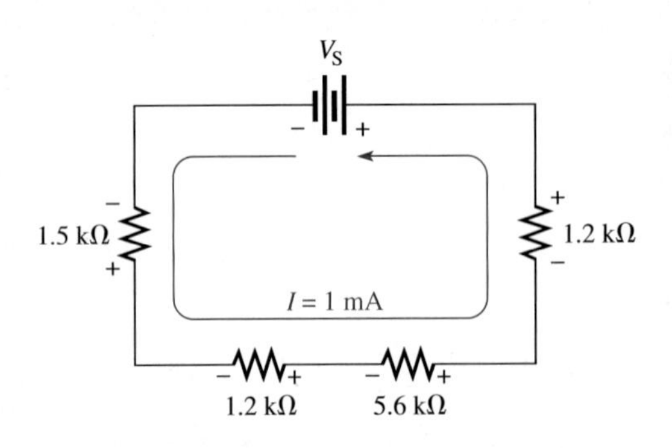

Related Problem If the 5.6 kΩ resistor is changed to 3.9 kΩ, what value of V_S is necessary to keep the current at 1 mA?

EXAMPLE 4–9

Calculate the voltage drop across each resistor in Figure 4–20, and find the value of V_S. To what value can V_S be raised before the 5 mA fuse blows?

▶ FIGURE 4–20

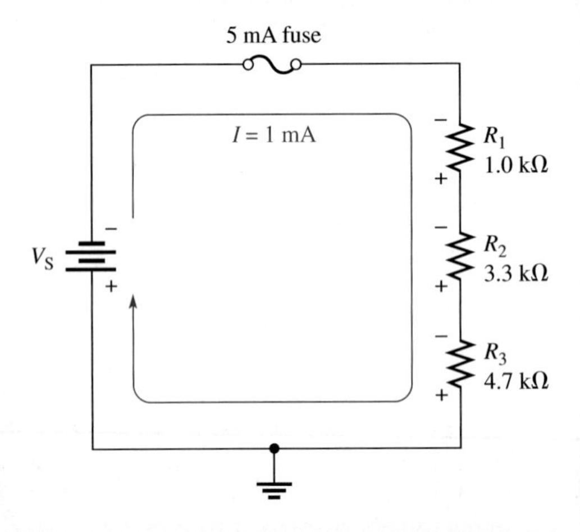

Solution By Ohm's law, the voltage drop across each resistor is equal to its resistance multiplied by the current through it. Use the Ohm's law formula $V = IR$ to determine the voltage drop across each of the resistors. Keep in mind that there is the same current through each series resistor. The voltage drop across R_1 (designated V_1) is

$$V_1 = IR_1 = (1\ \text{mA})(1.0\ \text{k}\Omega) = \mathbf{1\ V}$$

The voltage drop across R_2 is

$$V_2 = IR_2 = (1\ \text{mA})(3.3\ \text{k}\Omega) = \mathbf{3.3\ V}$$

The voltage drop across R_3 is

$$V_3 = IR_3 = (1\ \text{mA})(4.7\ \text{k}\Omega) = \mathbf{4.7\ V}$$

To calculate the value of V_S, first determine the total resistance.

$$R_T = 1.0\ \text{k}\Omega + 3.3\ \text{k}\Omega + 4.7\ \text{k}\Omega = 9\ \text{k}\Omega$$

The source voltage V_S is equal to the current times the total resistance.

$$V_S = IR_T = (1\text{ mA})(9\text{ k}\Omega) = \mathbf{9\text{ V}}$$

Notice that if you add the voltage drops of the resistors, they total 9 V, which is the same as the source voltage.

The fuse can handle a maximum current of 5 mA. The maximum value of V_S is therefore

$$V_{S(max)} = IR_T = (5\text{ mA})(9\text{ k}\Omega) = \mathbf{45\text{ V}}$$

Related Problem Repeat the calculations for V_1, V_2, V_3, V_S, and $V_{S(max)}$ if $R_3 = 2.2\text{ k}\Omega$ and I is maintained at 1 mA.

Open file E04-09 on your EWB/CircuitMaker CD-ROM. Using the multimeter, verify that the voltages across the resistors agree with the values calculated in this example.

EXAMPLE 4–10

Some resistors are not color coded with bands but have the values stamped on the resistor body. When the portion of the circuit board shown in Figure 4–21 was assembled, someone mounted the resistors with the labels turned down, and there is no documentation showing the resistor values. Assume that a voltmeter, ammeter, and power supply are available, but no ohmmeter. Without removing the resistors from the board, use Ohm's law to determine the resistance of each one.

▶ FIGURE 4–21

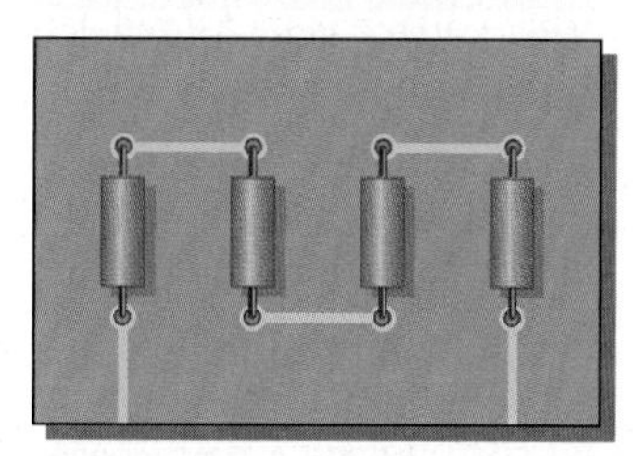

Solution The resistors are all in series, so the current is the same through each one. Measure the current by connecting a 12 V source (arbitrary value) and an ammeter as shown in Figure 4–22. Measure the voltage across each resistor by placing the voltmeter across the first resistor (R_1). Then repeat this measurement for the other three resistors. For illustration, the voltage values indicated are assumed to be the measured values.

Determine the resistance of each resistor by substituting the measured values of current and voltage into the Ohm's law formula.

$$R_1 = \frac{V_1}{I} = \frac{2.5\text{ V}}{25\text{ mA}} = \mathbf{100\ \Omega}$$

$$R_2 = \frac{V_2}{I} = \frac{3\text{ V}}{25\text{ mA}} = \mathbf{120\ \Omega}$$

$$R_3 = \frac{V_3}{I} = \frac{4.5\text{ V}}{25\text{ mA}} = \mathbf{180\ \Omega}$$

$$R_4 = \frac{V_4}{I} = \frac{2\text{ V}}{25\text{ mA}} = \mathbf{80\ \Omega}$$

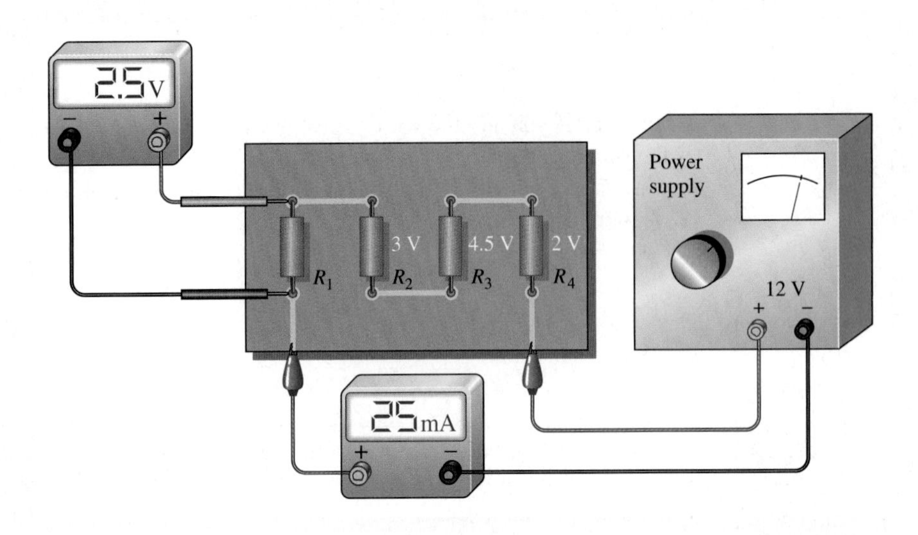

▲ **FIGURE 4–22**

The voltages measured across each resistor are indicated beside the resistor.

The calculator sequence and display for R_1 are

Related Problem If R_2 is open, what voltage is measured across each resistor?

SECTION 4–4 REVIEW

1. A 10 V battery is connected across three 100 Ω resistors in series. What is the current through each resistor?
2. How much voltage is required to produce 5 mA through the circuit of Figure 4–23?
3. How much voltage is dropped across each resistor in Figure 4–23 with 5 mA?
4. Four equal-value resistors are connected in series with a 5 V source. The measured current is 4.63 mA. What is the value of each resistor?

▶ **FIGURE 4–23**

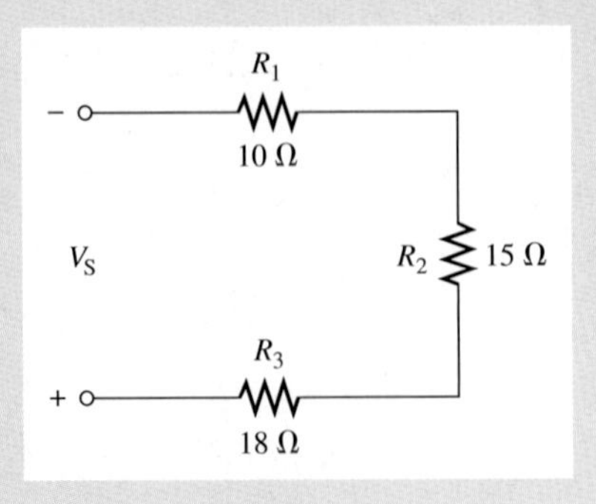

4–5 VOLTAGE SOURCES IN SERIES

A voltage source is an energy source that provides a constant voltage to a load. Batteries and power supplies are practical examples of dc voltage sources. When two or more voltage sources are in series, the total voltage is equal to the algebraic sum of the individual source voltages.

After completing this section, you should be able to

- **Determine the total effect of voltage sources in series**
- Determine the total voltage of series sources with the same polarities
- Determine the total voltage of series sources with opposite polarities

When batteries are placed in a flashlight, they are connected in a series-aiding arrangement to produce a larger voltage, as illustrated in Figure 4–24. In this example, three 1.5 V batteries are placed in series to produce a total voltage ($V_{S(tot)}$).

$$V_{S(tot)} = V_{S1} + V_{S2} + V_{S3} = 1.5\text{ V} + 1.5\text{ V} + 1.5\text{ V} = 4.5\text{ V}$$

FIGURE 4–24

Example of series-aiding voltage sources.

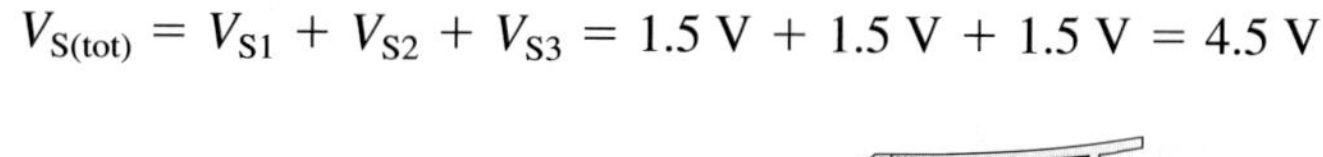

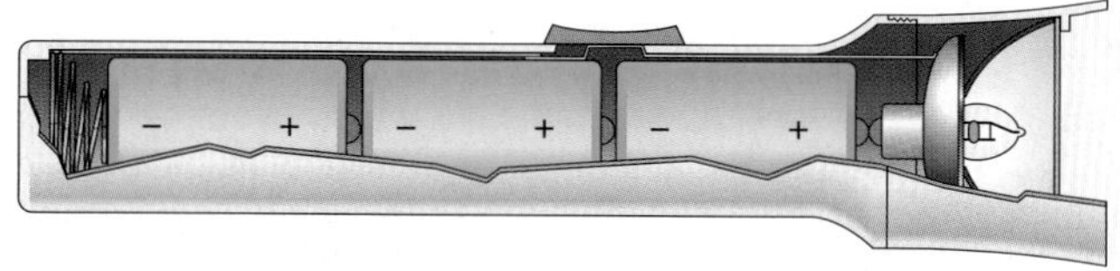

(a) Flashlight with series batteries

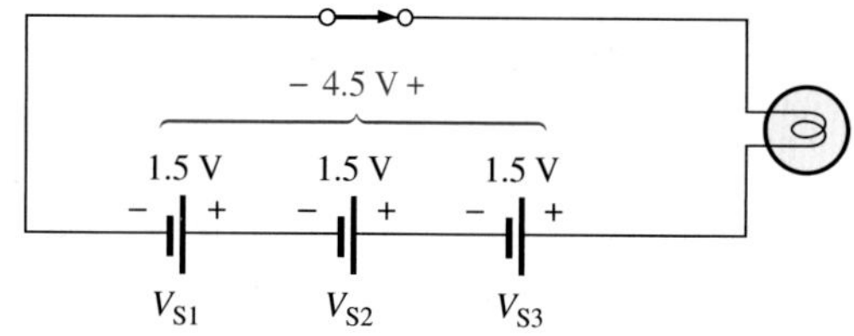

(b) Schematic of flashlight circuit

Series voltage sources (batteries in this case) are added when their polarities are in the same direction, or series-aiding, and are subtracted when their polarities are in opposite directions, or series-opposing. For example, if one of the batteries in the flashlight is turned around, as indicated in the schematic of Figure 4–25, its voltage subtracts because it has a negative value and reduces the total voltage.

$$V_{S(tot)} = V_{S1} - V_{S2} + V_{S3} = 1.5\text{ V} - 1.5\text{ V} + 1.5\text{ V} = 1.5\text{ V}$$

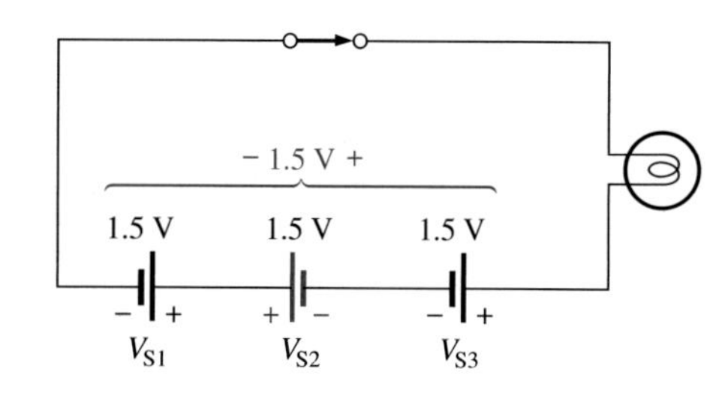

FIGURE 4–25

When batteries are connected in opposite directions, their voltages subtract.

EXAMPLE 4–11

What is the total source voltage ($V_{S(tot)}$) in Figure 4–26?

FIGURE 4–26

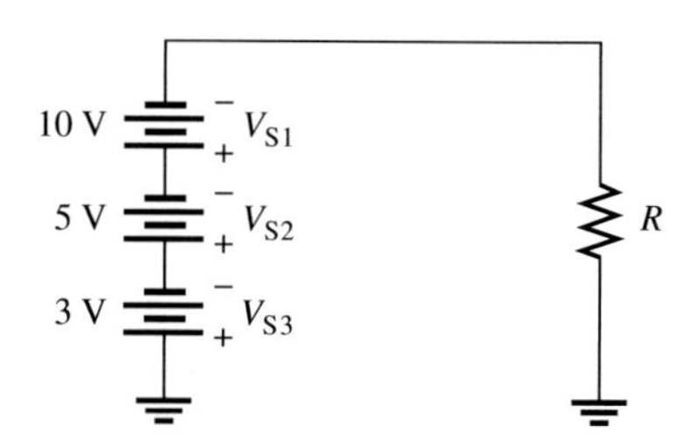

Solution The polarity of each source is the same (the sources are connected in the same direction in the circuit). Therefore, add the three voltages to get the total.

$$V_{S(tot)} = V_{S1} + V_{S2} + V_{S3} = 10\text{ V} + 5\text{ V} + 3\text{ V} = \mathbf{18\text{ V}}$$

The three individual sources can be replaced by a single equivalent source of 18 V with its polarity as shown in Figure 4–27.

▶ FIGURE 4–27

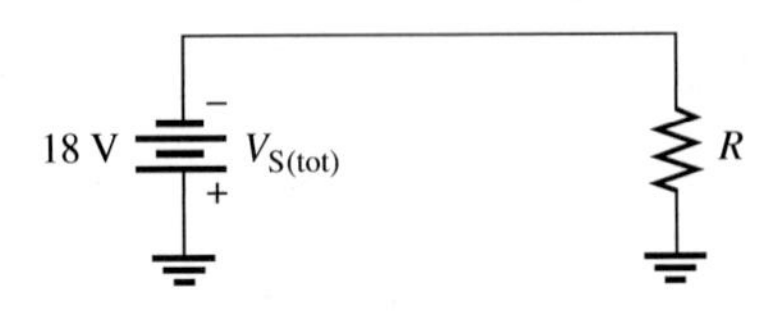

Related Problem If the battery V_{S3} in Figure 4–26 is reversed, what is the total voltage?

EXAMPLE 4–12

Determine $V_{S(tot)}$ in Figure 4–28.

Solution These sources are connected in opposing directions. If you go clockwise around the circuit, you go from plus to minus through V_{S1}, and minus to plus through V_{S2}. The total voltage is the difference of the two source voltages (algebraic sum of oppositely signed values). The total voltage has the same polarity as the larger-value source. Here we will choose V_{S2} to be positive.

$$V_{S(tot)} = V_{S2} - V_{S1} = 25\text{ V} - 15\text{ V} = \mathbf{10\text{ V}}$$

The two sources in Figure 4–28 can be replaced by a 10 V equivalent one with polarity as shown in Figure 4–29.

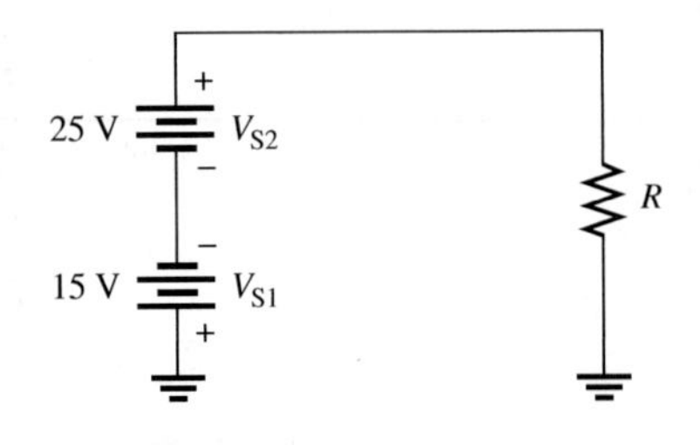

▲ FIGURE 4–28

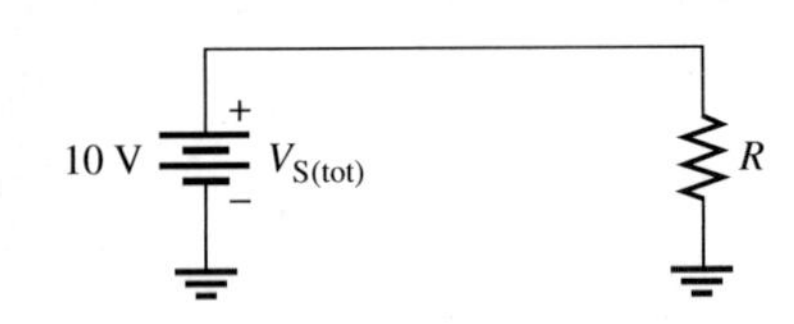

▲ FIGURE 4–29

Related Problem If an 8 V source in the same direction as V_{S1} is connected in series in Figure 4–28, what is the total voltage?

SECTION 4–5 REVIEW

1. How many 12 V batteries must be connected in series to produce 60 V? Sketch a schematic that shows the battery connections.
2. Four 1.5 V flashlight batteries are connected in series positive to negative. What is the total voltage across the bulb?

3. The resistive circuit in Figure 4–30 is used to bias a transistor amplifier. Show how to connect the resistors to two 15 V power supplies in order to get 30 V across the series resistors.
4. Determine the total source voltage in each circuit of Figure 4–31.
5. Sketch the equivalent single source circuit for each circuit of Figure 4–31.

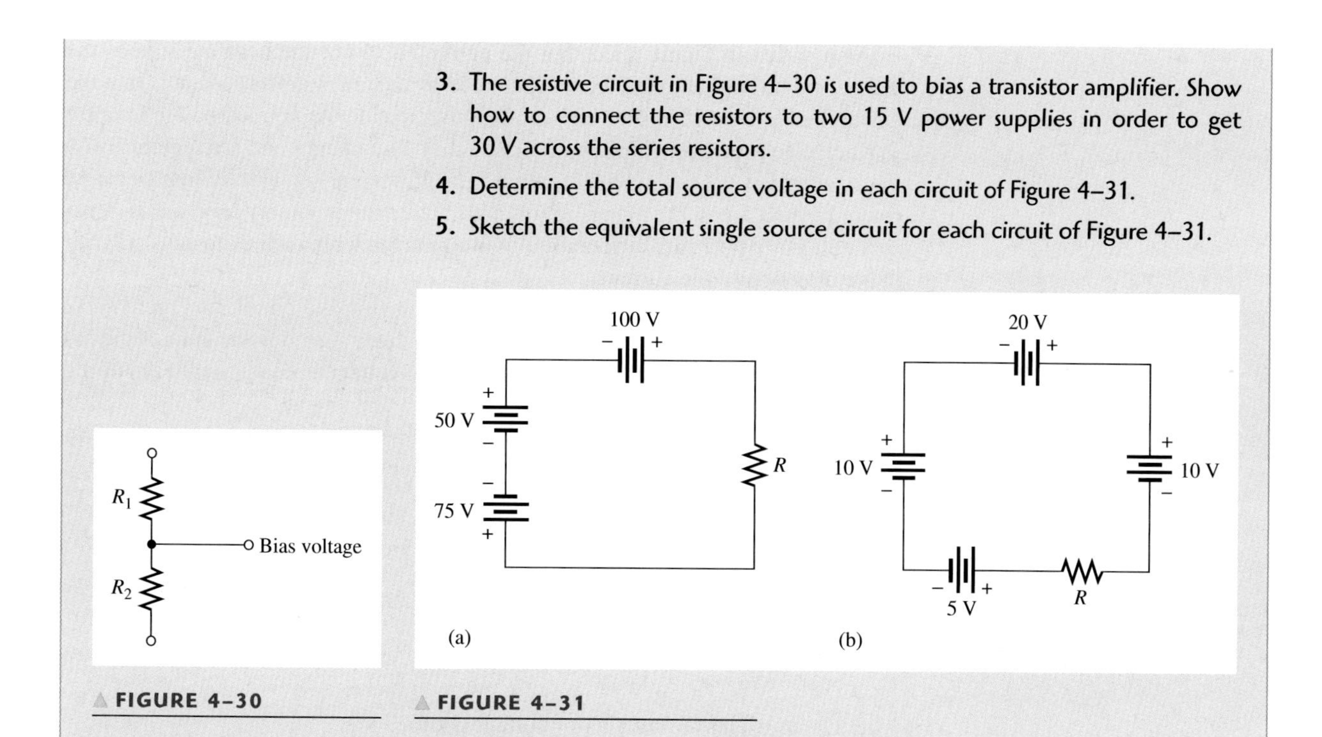

▲ FIGURE 4–30

▲ FIGURE 4–31

4–6 KIRCHHOFF'S VOLTAGE LAW

Kirchhoff's voltage law is a fundamental circuit law that states that the algebraic sum of all the voltages around a closed path is zero or, in other words, the sum of the voltage drops equals the total source voltage.

After completing this section, you should be able to

- **Apply Kirchhoff's voltage law**
- State Kirchhoff's voltage law
- Determine the source voltage by adding the voltage drops
- Determine an unknown voltage drop

In an electric circuit, the voltages across the resistors (voltage drops) *always* have polarities opposite to the source voltage polarity. For example, in Figure 4–32, follow a clockwise loop around the circuit and note that the source polarity is plus-to-minus and each voltage drop is minus-to-plus.

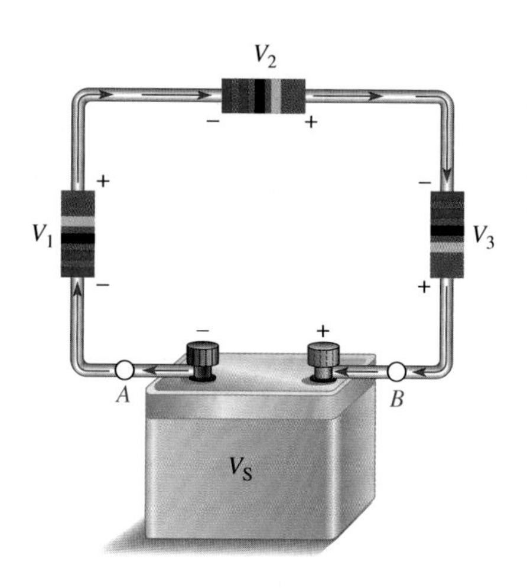

◀ FIGURE 4–32

Kirchhoff's voltage law: The sum of the voltage drops around a single closed path equals the source voltage.

Also notice in Figure 4–32 that the current is out of the negative side of the source and through the resistors as indicated by the arrows. The current is into the negative side of each resistor and out the positive side. As you learned in Chapter 3, when electrons flow through a resistor, they lose energy and are therefore at a lower energy level when they emerge. The lower energy side is less negative (more positive) than the higher energy side. The drop in energy level across a resistor creates a potential difference, or voltage drop, with a minus-to-plus polarity in the direction of the current.

Notice that the voltage from point A to point B in the circuit of Figure 4–32 equals the source voltage, V_S. Also, the voltage from A to B is the sum of the series resistor voltage drops. Therefore, the source voltage is equal to the sum of the three voltage drops.

This discussion is an example of **Kirchhoff's voltage law**, which is generally stated as follows:

The sum of all the voltage drops around a single closed path in a circuit is equal to the total source voltage in that closed path.

The general concept of Kirchhoff's voltage law is illustrated in Figure 4–33 and expressed by Equation 4–3.

Equation 4–3

$$V_S = V_1 + V_2 + V_3 + \cdots + V_n$$

where n represents the number of voltage drops.

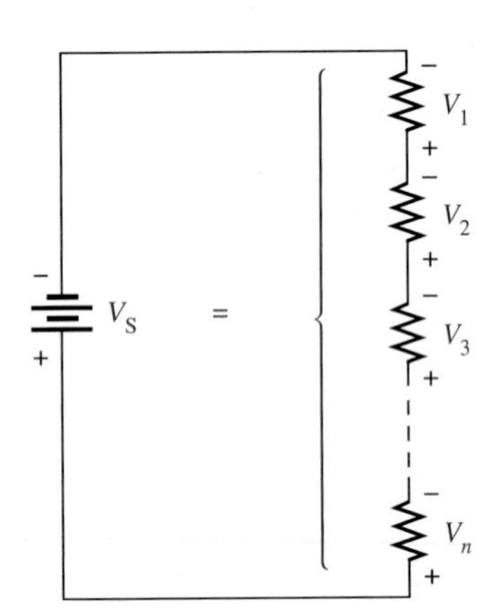

FIGURE 4–33

Sum of n voltage drops equals the source voltage.

Another Way to State Kirchhoff's Voltage Law

If all the voltage drops around a closed path are added and then this total is subtracted from the source voltage, the result is zero. This result occurs because the sum of the voltage drops always equals the source voltage.

The algebraic sum of all voltages (both source and drops) around a closed path is zero.

Therefore, another way of expressing Kirchhoff's voltage law in equation form is as follows:

Equation 4–4

$$V_S - V_1 - V_2 - V_3 - \cdots - V_n = 0$$

You can verify Kirchhoff's voltage law by connecting a circuit and measuring each resistor voltage and the source voltage as illustrated in Figure 4–34. When the resistor voltages are added together, their sum will equal the source voltage but their polarities are opposite. Any number of resistors can be added.

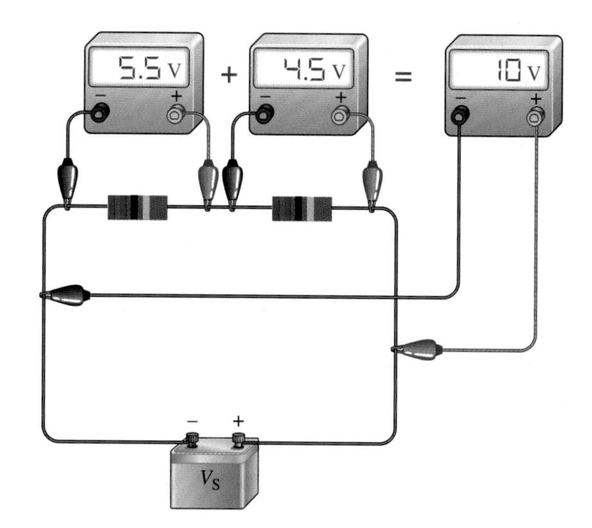

FIGURE 4–34
Illustration of a verification of Kirchhoff's voltage law.

EXAMPLE 4–13

Determine the source voltage, V_S, in Figure 4–35 where the two voltage drops are given.

FIGURE 4–35

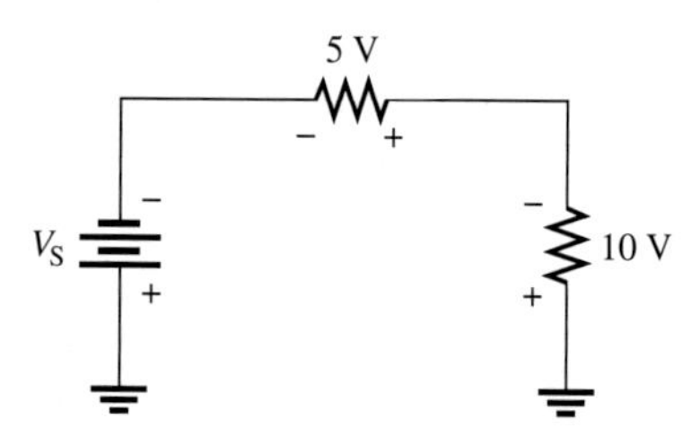

Solution By Kirchhoff's voltage law (Eq. 4–3), the source voltage (applied voltage) must equal the sum of the voltage drops. Adding the voltage drops gives the value of the source voltage.

$$V_S = 5\text{ V} + 10\text{ V} = \mathbf{15\text{ V}}$$

Related Problem If V_S is increased to 30 V in Figure 4–35, what are the two voltage drops?

EXAMPLE 4–14

Determine the unknown voltage drop, V_3, in Figure 4–36.

FIGURE 4–36

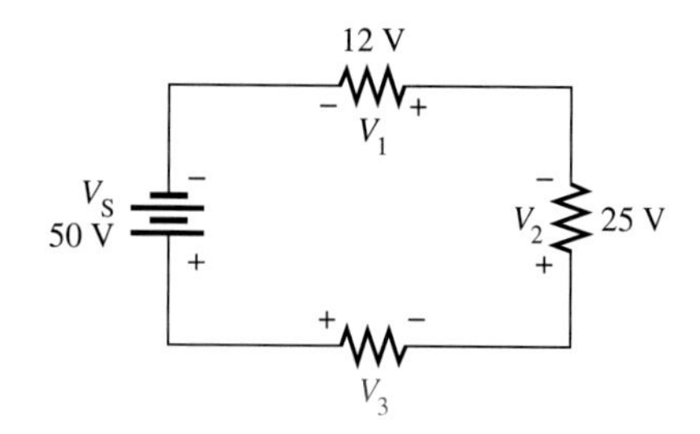

Solution By Kirchhoff's voltage law (Eq. 4–4), the algebraic sum of all the voltages around the circuit is zero (the signs of the voltage drops are opposite the sign of the source).

$$V_S - V_1 - V_2 - V_3 = 0$$

The value of each voltage drop except V_3 is known. Substitute these values into the equation.

$$50\text{ V} - 12\text{ V} - 25\text{ V} - V_3 = 0\text{ V}$$

Next, combine the known values. Transpose 13 V to the right side of the equation, and cancel the minus signs.

$$13\text{ V} - V_3 = 0\text{ V}$$
$$-V_3 = -13\text{ V}$$
$$V_3 = \mathbf{13\text{ V}}$$

The voltage drop across R_3 is 13 V, and its polarity is as shown in Figure 4–36.

Related Problem Determine V_3 in Figure 4–36 if the source is changed to 25 V.

EXAMPLE 4–15

Find the value of R_4 in Figure 4–37.

▶ FIGURE 4–37

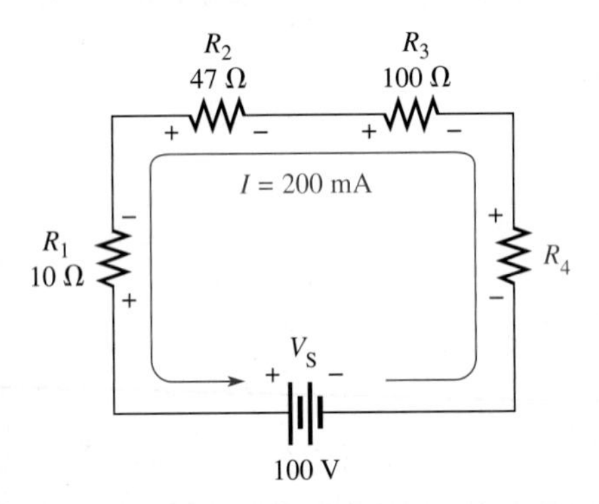

Solution In this problem you will use both Ohm's law and Kirchhoff's voltage law. First, use Ohm's law to find the voltage drop across each of the known resistors.

$$V_1 = IR_1 = (200\text{ mA})(10\ \Omega) = 2\text{ V}$$

$$V_2 = IR_2 = (200\text{ mA})(47\ \Omega) = 9.4\text{ V}$$

$$V_3 = IR_3 = (200\text{ mA})(100\ \Omega) = 20\text{ V}$$

Next, use Kirchhoff's voltage law to find V_4, the voltage drop across the unknown resistor.

$$V_S - V_1 - V_2 - V_3 - V_4 = 0\text{ V}$$
$$100\text{ V} - 2\text{ V} - 9.4\text{ V} - 20\text{ V} - V_4 = 0\text{ V}$$
$$68.6\text{ V} - V_4 = 0\text{ V}$$
$$V_4 = 68.6\text{ V}$$

Equation 4–5 is the general voltage-divider formula. It can be stated as follows:

The voltage drop across any resistor or combination of resistors in a series circuit is equal to the ratio of that resistance value to the total resistance, multiplied by the source voltage.

The following three examples illustrate the use of the voltage-divider formula.

EXAMPLE 4–16

Determine V_1 (the voltage across R_1) and V_2 (the voltage across R_2) in the circuit in Figure 4–40.

▶ FIGURE 4–40

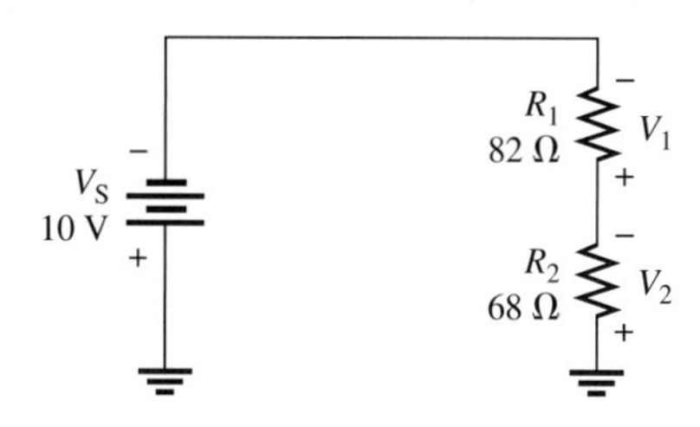

Solution Use the voltage-divider formula, $V_x = (R_x/R_T)V_S$. In this problem you are looking for V_1; so $V_x = V_1$ and $R_x = R_1$. The total resistance is

$$R_T = R_1 + R_2 = 82\ \Omega + 68\ \Omega = 150\ \Omega$$

R_1 is 82 Ω and V_S is 10 V. Substitute these values into the voltage-divider formula.

$$V_1 = \left(\frac{R_1}{R_T}\right)V_S = \left(\frac{82\ \Omega}{150\ \Omega}\right)10\ \text{V} = \mathbf{5.47\ V}$$

There are two ways to find the value of V_2: Kirchhoff's voltage law or the voltage-divider formula. If you use Kirchhoff's voltage law ($V_S = V_1 + V_2$), substitute the values for V_S and V_1 and solve for V_2.

$$V_2 = V_S - V_1 = 10\ \text{V} - 5.47\ \text{V} = \mathbf{4.53\ V}$$

A second way is to use the voltage-divider formula where $x = 2$.

$$V_2 = \left(\frac{R_2}{R_T}\right)V_S = \left(\frac{68\ \Omega}{150\ \Omega}\right)10\ \text{V} = \mathbf{4.53\ V}$$

Related Problem Find the voltage drops across R_1 and R_2 if R_2 is changed to 180 Ω in Figure 4–40.

Open file E04-16 on your EWB/CircuitMaker CD-ROM. Use the multimeter to verify the calculated values of V_1 and V_2. Change R_2 to 180 Ω, measure the voltages across the resistors, and compare to the values calculated in the related problem.

EXAMPLE 4–17

Calculate the voltage drop across each resistor in the circuit of Figure 4–41.

FIGURE 4–41

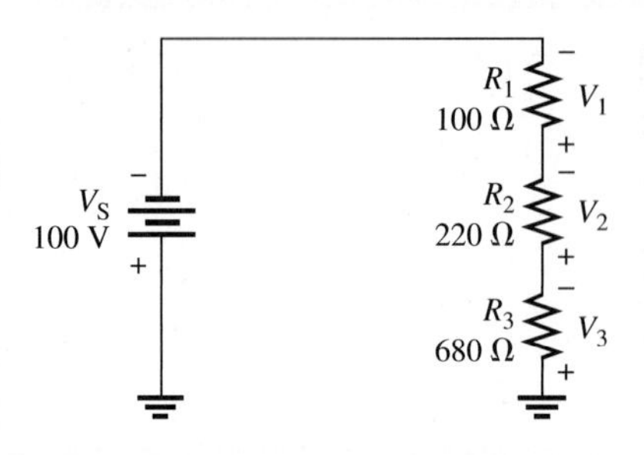

Solution Look at the circuit for a moment and consider the following: The total resistance is 1000 Ω. Ten percent of the total voltage is across R_1 because it is 10% of the total resistance (100 Ω is 10% of 1000 Ω). Likewise, 22% of the total voltage is dropped across R_2 because it is 22% of the total resistance (220 Ω is 22% of 1000 Ω). Finally, R_3 drops 68% of the total voltage (680 Ω is 68% of 1000 Ω).

Because of the convenient values in this problem, it is easy to figure the voltage drops mentally ($V_1 = 0.10 \times 100\text{ V} = 10\text{ V}$, $V_2 = 0.22 \times 100\text{ V} = 22\text{ V}$, and $V_3 = 0.68 \times 100\text{ V} = 68\text{ V}$). Such is not always the case, but sometimes a little thinking will produce a result more efficiently and eliminate some calculating.

Although you have already reasoned through this problem, the calculations are

$$V_1 = \left(\frac{R_1}{R_T}\right)V_S = \left(\frac{100\ \Omega}{1000\ \Omega}\right)100\text{ V} = \mathbf{10\ V}$$

$$V_2 = \left(\frac{R_2}{R_T}\right)V_S = \left(\frac{220\ \Omega}{1000\ \Omega}\right)100\text{ V} = \mathbf{22\ V}$$

$$V_3 = \left(\frac{R_3}{R_T}\right)V_S = \left(\frac{680\ \Omega}{1000\ \Omega}\right)100\text{ V} = \mathbf{68\ V}$$

Notice that the sum of the voltage drops is equal to the source voltage, in accordance with Kirchhoff's voltage law. This check is a good way to verify your results.

Related Problem If R_1 and R_2 are both changed to 680 Ω in Figure 4–41, what are all the voltage drops?

Open file E04-17 on your EWB/CircuitMaker CD-ROM. Verify the values of V_1, V_2, and V_3. Change R_1 and R_2 to 680 Ω, measure the voltages across the resistors, and compare to the results of the related problem.

EXAMPLE 4–18

Determine the voltages between the following points in the circuit of Figure 4–42:

(a) *A* to *B* **(b)** *A* to *C* **(c)** *B* to *C* **(d)** *B* to *D* **(e)** *C* to *D*

FIGURE 4–42

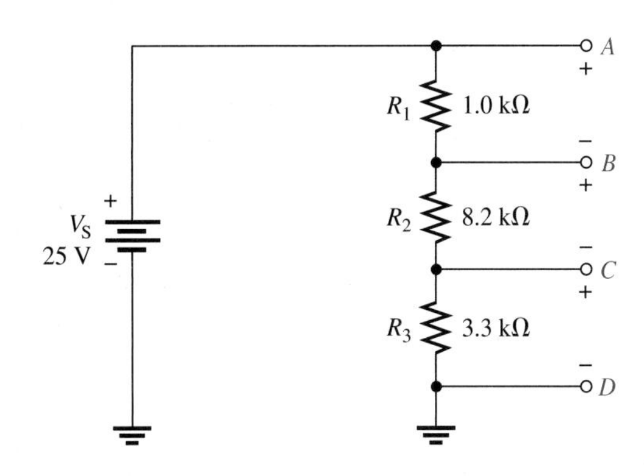

Solution First determine R_T.

$$R_T = 1.0\ \text{k}\Omega + 8.2\ \text{k}\Omega + 3.3\ \text{k}\Omega = 12.5\ \text{k}\Omega$$

Now apply the voltage-divider formula to obtain each required voltage.

(a) The voltage A to B is also the voltage drop across R_1.

$$V_{AB} = \left(\frac{R_1}{R_T}\right)V_S = \left(\frac{1.0\ \text{k}\Omega}{12.5\ \text{k}\Omega}\right)25\ \text{V} = \mathbf{2\ V}$$

(b) The voltage from A to C is the combined voltage drop across both R_1 and R_2. In this case, R_x in the general formula given in Equation 4–5 is $R_1 + R_2$.

$$V_{AC} = \left(\frac{R_1 + R_2}{R_T}\right)V_S = \left(\frac{9.2\ \text{k}\Omega}{12.5\ \text{k}\Omega}\right)25\ \text{V} = \mathbf{18.4\ V}$$

(c) The voltage from B to C is the voltage drop across R_2.

$$V_{BC} = \left(\frac{R_2}{R_T}\right)V_S = \left(\frac{8.2\ \text{k}\Omega}{12.5\ \text{k}\Omega}\right)25\ \text{V} = \mathbf{16.4\ V}$$

(d) The voltage from B to D is the combined voltage drop across both R_2 and R_3. In this case, R_x in the general formula is $R_2 + R_3$.

$$V_{BD} = \left(\frac{R_2 + R_3}{R_T}\right)V_S = \left(\frac{11.5\ \text{k}\Omega}{12.5\ \text{k}\Omega}\right)25\ \text{V} = \mathbf{23\ V}$$

(e) Finally, the voltage from C to D is the voltage drop across R_3.

$$V_{CD} = \left(\frac{R_3}{R_T}\right)V_S = \left(\frac{3.3\ \text{k}\Omega}{12.5\ \text{k}\Omega}\right)25\ \text{V} = \mathbf{6.6\ V}$$

If you connect this voltage divider in the lab, you can verify each of the calculated voltages by connecting a voltmeter between the appropriate points in each case.

Related Problem Determine each of the previously calculated voltages if V_S is doubled.

Open file E04-18 on your EWB/CircuitMaker CD-ROM. Verify the values of V_{AB}, V_{AC}, V_{BC}, V_{BD}, and V_{CD}. Change V_S to 50 V, measure the voltages, and compare to the values calculated in the related problem.

The Potentiometer as an Adjustable Voltage Divider

Recall from Chapter 2 that a potentiometer is a variable resistor with three terminals. A potentiometer connected to a voltage source is shown in Figure 4–43. Notice that the two end terminals are labeled 1 and 2. The adjustable terminal or wiper is labeled 3. The potentiometer acts as a voltage divider, which can be illustrated by separating the total resistance into two parts as shown in Figure 4–43(c). The resistance between terminal 1 and terminal 3 (R_{13}) is one part, and the resistance between terminal 3 and terminal 2 (R_{32}) is the other part. So this potentiometer actually is a two-resistor voltage divider that can be manually adjusted.

FIGURE 4–43

The potentiometer as a voltage divider.

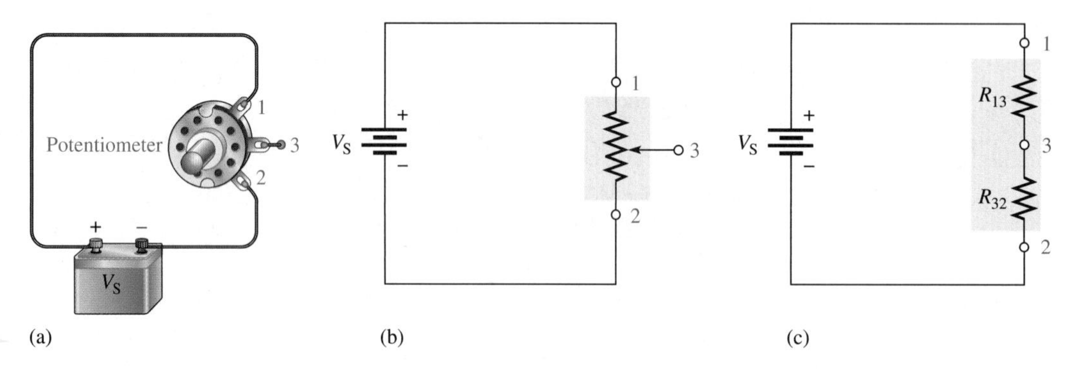

Figure 4–44 shows what happens when the wiper contact (3) is moved. In part (a), the wiper is exactly centered, making the two resistances equal. If you measure the voltage across terminals 3 to 2 as indicated, you have one-half of the total source voltage. When the wiper is moved up, as in part (b), the resistance between terminals 3 and 2 increases, and the voltage across it increases proportionally. When the wiper is moved down, as in part (c), the resistance between terminals 3 and 2 decreases, and the voltage decreases proportionally.

FIGURE 4–44

Adjusting the voltage divider.

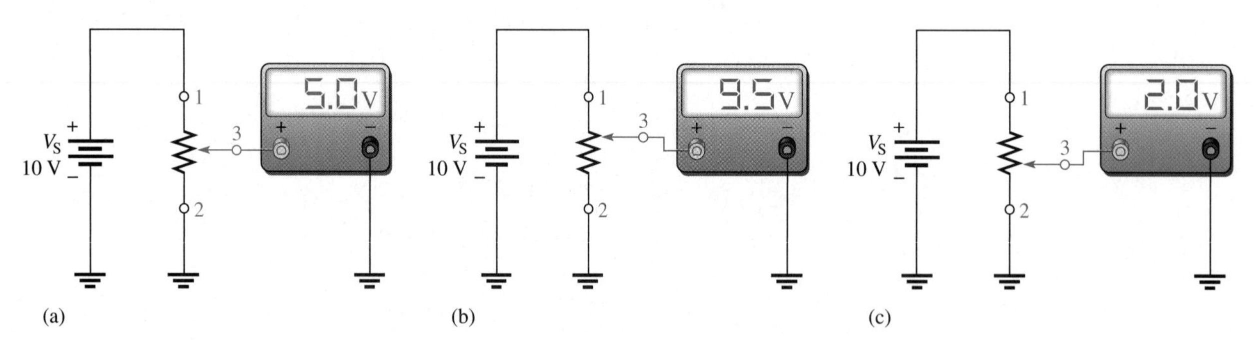

Applications of Voltage Dividers

The volume control of radio or TV receivers is a common application of a potentiometer used as a voltage divider. Since the loudness of the sound is dependent on the amount of voltage associated with the audio signal, you can increase or decrease the volume by adjusting the potentiometer, that is, by turning the knob of the volume control on the set. The block diagram in Figure 4–45 shows how a potentiometer can be used for volume control in a typical receiver.

Another application of a voltage divider is illustrated in Figure 4–46, which depicts a potentiometer voltage divider as a fuel-level sensor in an automobile gas

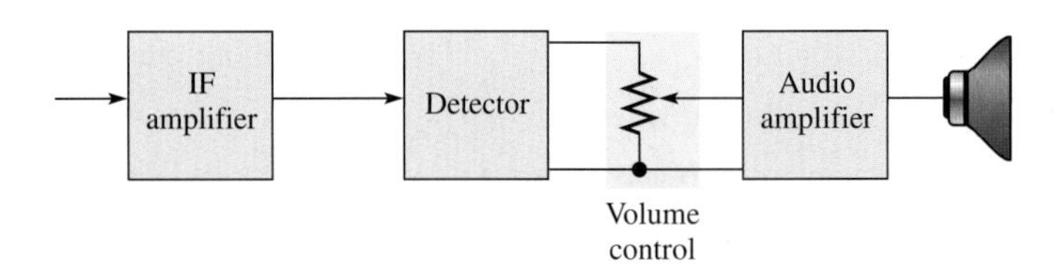

FIGURE 4–45

A voltage divider used for volume control.

tank. As shown in part (a), the float moves up as the tank is filled and moves down as the tank empties. The float is mechanically linked to the wiper arm of a potentiometer, as shown in part (b). The output voltage varies proportionally with the position of the wiper arm. As the fuel in the tank decreases, the sensor output voltage also decreases. The output voltage goes to the indicator circuitry, which controls the fuel gauge or digital readout to show the fuel level. The schematic of this system is shown in part (c).

FIGURE 4–46

A potentiometer voltage divider used as an automotive fuel-level sensor.

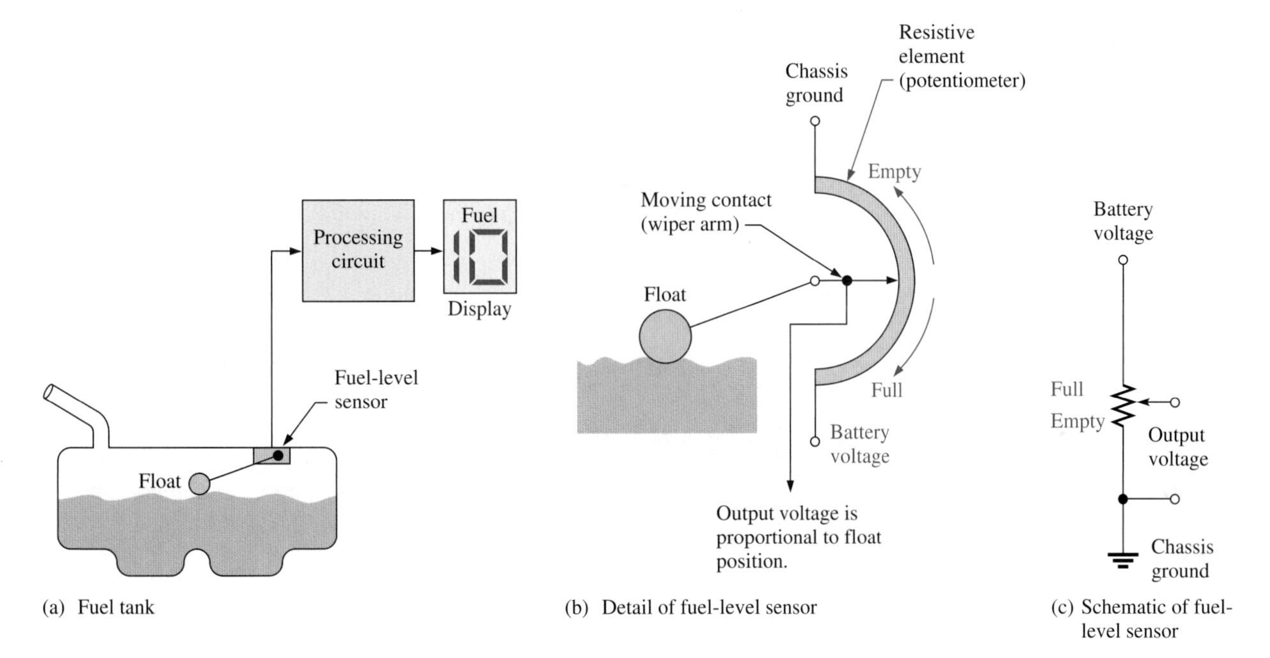

(a) Fuel tank (b) Detail of fuel-level sensor (c) Schematic of fuel-level sensor

Still another application for voltage dividers is in setting the dc operating voltage (bias) in transistor amplifiers. Figure 4–47 shows a voltage divider used for this purpose. You will study transistor amplifiers and biasing later, so it is important that you understand the basics of voltage dividers at this point.

These examples are only three out of many possible applications of voltage dividers.

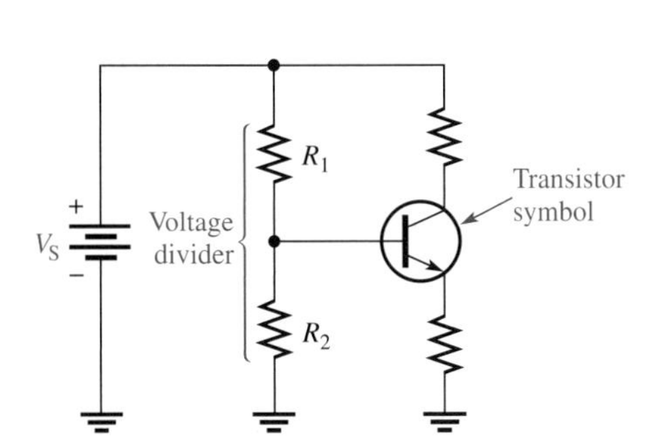

FIGURE 4–47

The voltage divider as a bias circuit for a transistor amplifier.

SECTION 4–7 REVIEW

1. What is a voltage divider?
2. How many resistors can there be in a series voltage-divider circuit?
3. Write the general voltage-divider formula.
4. If two series resistors of equal value are connected across a 20 V source, how much voltage is there across each resistor?
5. A 56 Ω resistor and an 82 Ω resistor are connected as a voltage divider. The source voltage is 100 V. Draw the circuit, and determine the voltage across each of the resistors.
6. The circuit of Figure 4–48 is an adjustable voltage divider. If the potentiometer is linear, where would you set the wiper in order to get 5 V from *B* to *A* and 5 V from *C* to *B*?

▶ FIGURE 4–48

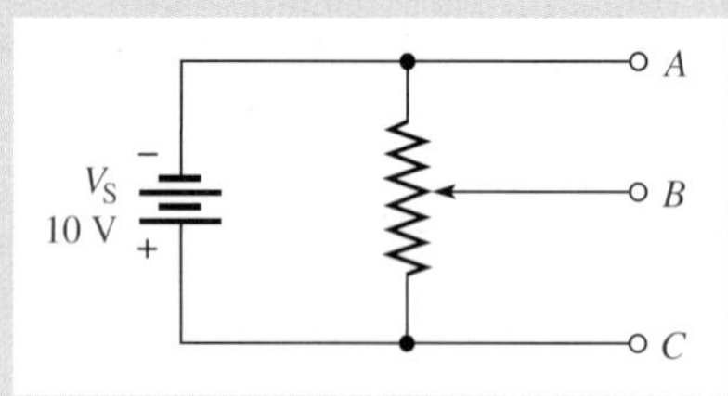

4–8 POWER IN A SERIES CIRCUIT

The power dissipated by each individual resistor in a series circuit contributes to the total power in the circuit. The individual powers are additive.

After completing this section, you should be able to

- **Determine power in a series circuit**

The total amount of power in a series resistive circuit is equal to the sum of the powers in each resistor in series.

Equation 4–6

$$P_T = P_1 + P_2 + P_3 + \cdots + P_n$$

where n is the number of resistors in series, P_T is the total power, and P_n is the power in the last resistor in series. In other words, the powers are additive.

The power formulas that you learned in Chapter 3 are, of course, directly applicable to series circuits. Since each resistor in series has the same current through it, the following formulas are used to calculate the total power:

$$P_T = V_S I$$

$$P_T = I^2 R_T$$

$$P_T = \frac{V_S^2}{R_T}$$

where V_S is the source voltage across the series circuit, R_T is the total resistance, and I is the current in the circuit. Example 4–19 illustrates how to calculate total power in a series circuit.

EXAMPLE 4–19

Determine the total amount of power in the series circuit in Figure 4–49.

FIGURE 4–49

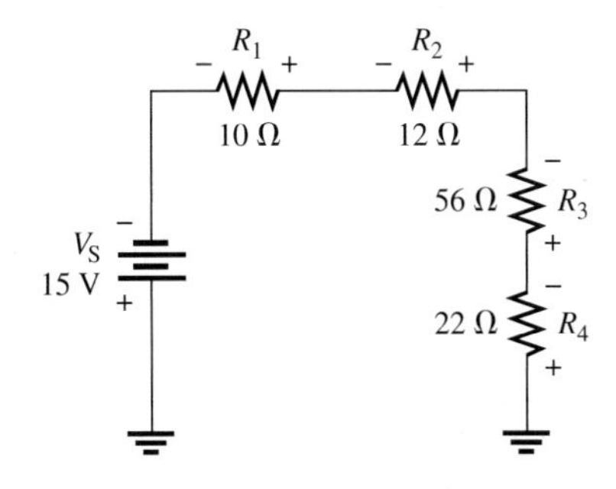

Solution The source voltage is 15 V. The total resistance is

$$R_T = 10\ \Omega + 12\ \Omega + 56\ \Omega + 22\ \Omega = 100\ \Omega$$

The easiest formula to use is $P_T = V_S^2/R_T$ since you know both V_S and R_T.

$$P_T = \frac{V_S^2}{R_T} = \frac{(15\ \text{V})^2}{100\ \Omega} = \frac{225\ \text{V}^2}{100\ \Omega} = 2.25\ \text{W}$$

If you determine the power of each resistor separately and add all these powers, you obtain the same result. Another calculation will illustrate. First, find the current.

$$I = \frac{V_S}{R_T} = \frac{15\ \text{V}}{100\ \Omega} = 150\ \text{mA}$$

Next, calculate the power for each resistor using $P = I^2R$.

$$P_1 = (150\ \text{mA})^2(10\ \Omega) = 225\ \text{mW}$$

$$P_2 = (150\ \text{mA})^2(12\ \Omega) = 270\ \text{mW}$$

$$P_3 = (150\ \text{mA})^2(56\ \Omega) = 1.26\ \text{W}$$

$$P_4 = (150\ \text{mA})^2(22\ \Omega) = 495\ \text{mW}$$

Now, add these powers to get the total power.

$$P_T = 225\ \text{mW} + 270\ \text{mW} + 1.260\ \text{W} + 495\ \text{mW} = \mathbf{2.25\ W}$$

This result shows that the sum of the individual powers is equal to the total power as determined by the formula $P_T = V_S^2/R_T$.

Related Problem What is the total power in the circuit of Figure 4–49 if V_S is increased to 30 V?

SECTION 4–8 REVIEW

1. If you know the power in each resistor in a series circuit, how can you find the total power?
2. The resistors in a series circuit have the following powers: 1 W, 2 W, 5 W, and 8 W. What is the total power in the circuit?
3. A circuit has a 100 Ω, a 330 Ω, and a 680 Ω resistor in series. There is a current of 1 mA through the circuit. What is the total power?

4–9 CIRCUIT GROUND

Voltage is relative. That is, the voltage at one point in a circuit is always measured relative to another point. For example, if we say that there are +100 V at a certain point in a circuit, we mean that the point is 100 V more positive than some reference point in the circuit. This reference point in a circuit is usually the ground point.

After completing this section, you should be able to

- **Determine and identify ground in a circuit**
- Measure voltage with respect to ground
- Define the term *circuit ground*

The concept of *ground* was introduced in Chapter 2. In most electronic equipment, a large conductive area on a printed circuit board or the metal chassis that houses the assembly is used as the common or reference point, called the **circuit ground** or *chassis ground,* as illustrated in Figure 4–50.

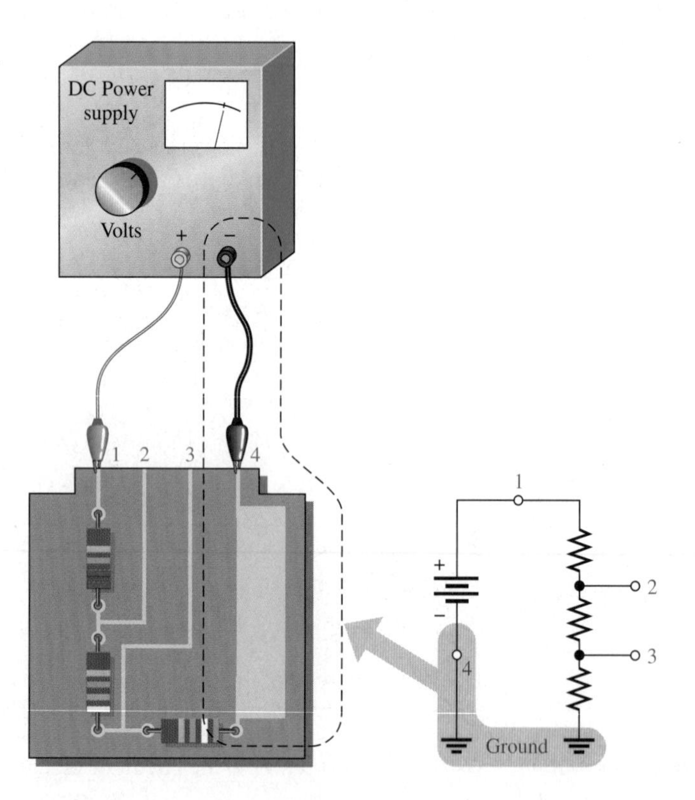

FIGURE 4–50

Simple illustration of circuit ground.

Ground has a potential of zero volts (0 V) with respect to all other points in the circuit that are referenced to it, as illustrated in Figure 4–51. In part (a), the negative side of the source is grounded, and all voltages indicated are positive with respect to ground. In part (b), the positive side of the source is grounded. The voltages at all other points are therefore negative with respect to ground.

Measuring Voltages with Respect to Ground

When voltages are measured with respect to ground in a circuit, one meter lead is connected to the circuit ground, and the other to the point at which the voltage is to be measured. In a negative-ground circuit, the negative meter terminal is con-

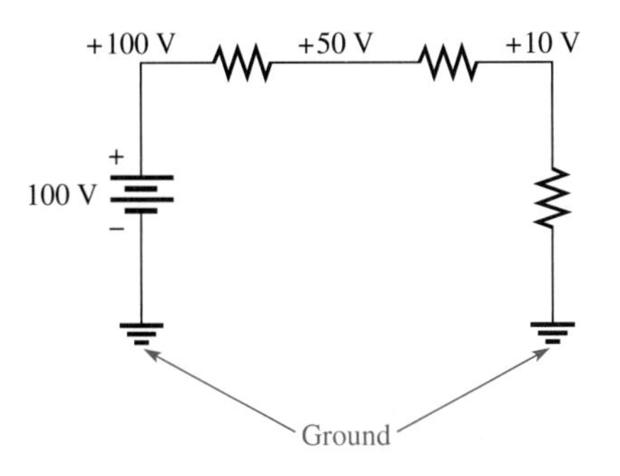

(a) Negative ground

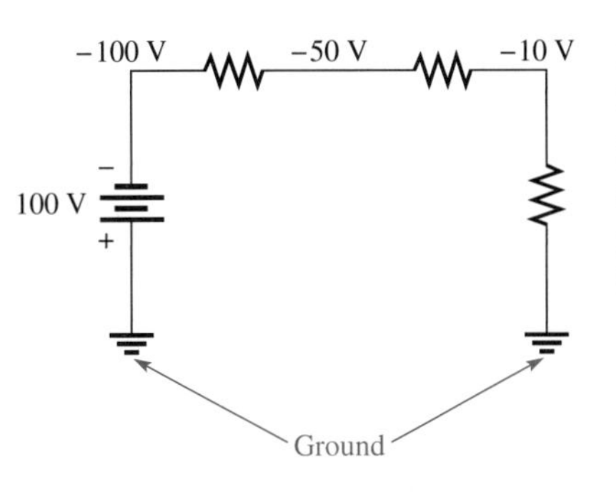

(b) Positive ground

FIGURE 4–51

Examples of negative and positive grounds. Multiple ground symbols actually represent the same electrical point, so you can think of them as being connected together.

nected to the circuit ground. The positive terminal of the voltmeter is then connected to the positive voltage point. Measurement of positive voltage is illustrated in Figure 4–52, where the meter reads the voltage at point A with respect to ground.

For a circuit with a positive ground, the positive analog voltmeter lead is connected to ground, and the negative lead is connected to the negative voltage point, as indicated in Figure 4–53. Here the meter reads the voltage at point A with respect to ground. When using a digital voltmeter, you can connect it either way because a digital meter can display both positive and negative voltages.

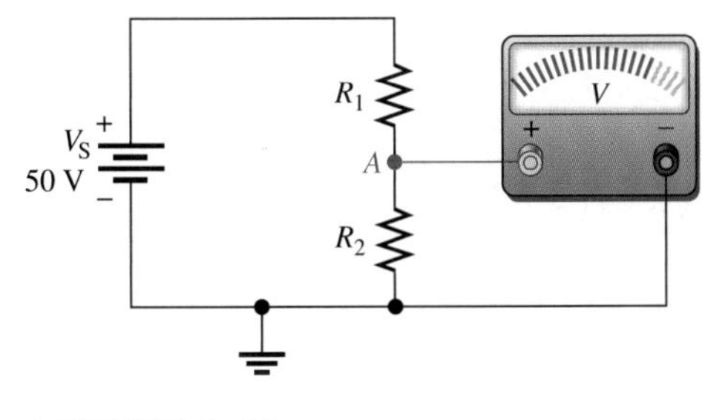

FIGURE 4–52

Measuring a voltage with respect to negative ground.

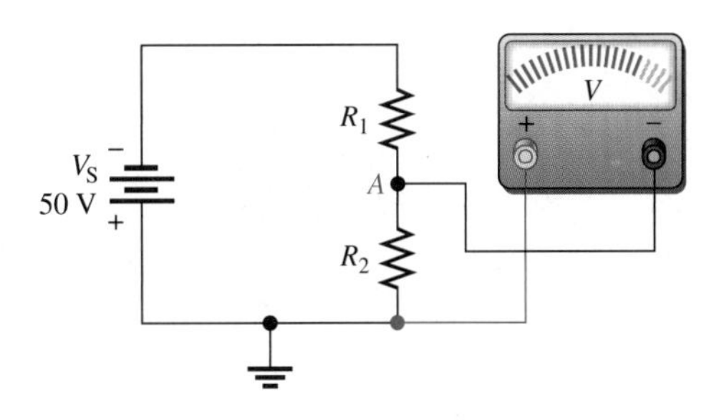

FIGURE 4–53

Measuring a voltage with respect to positive ground.

When voltages must be measured at several points in a circuit, the ground lead can be clipped to ground at one point in the circuit and left there. The other lead is then moved from point to point as the voltages are measured. This method is illustrated in Figure 4–54 and in equivalent schematic form in Figure 4–55.

Measuring Voltage across an Ungrounded Resistor

Voltage can normally be measured across a resistor, as shown in Figure 4–56, even though neither side of the resistor is connected to circuit ground.

Another method can be used, as illustrated in Figure 4–57. The voltages on each side of the resistor (R_2) are measured with respect to ground. The difference of these two measurements is the voltage drop across the resistor.

$$V_{R2} = V_{AB} = V_A - V_B$$

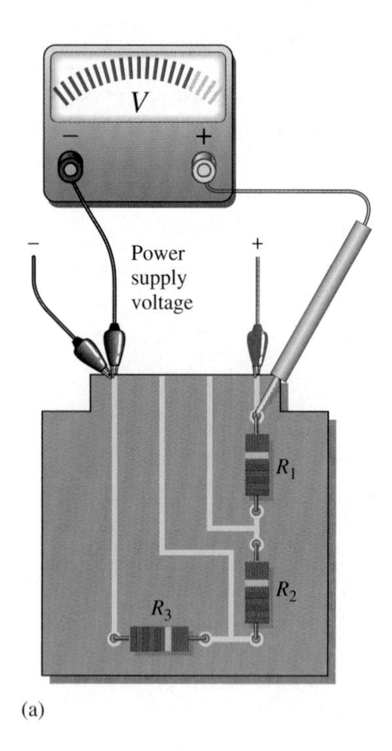

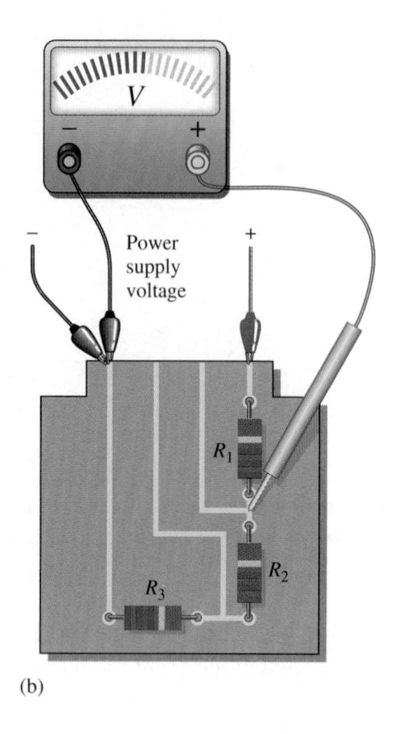

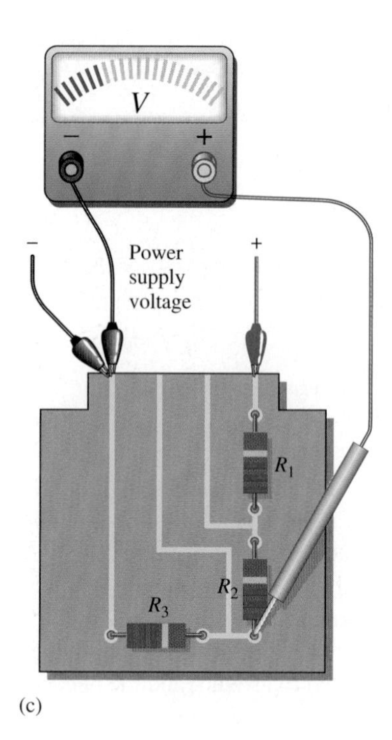

▲ **FIGURE 4–54**

Measuring voltages at several points in a circuit with respect to ground.

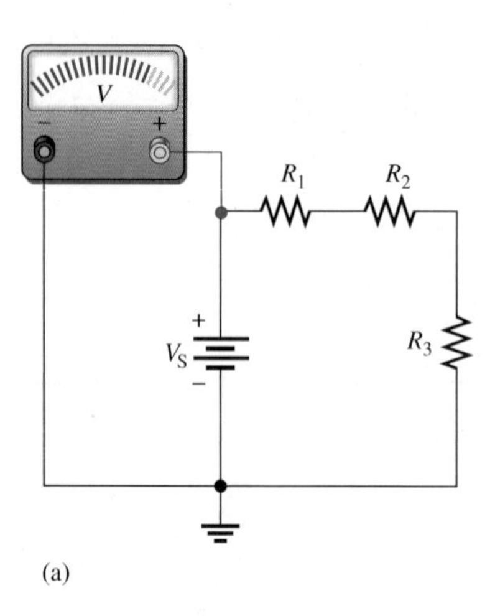

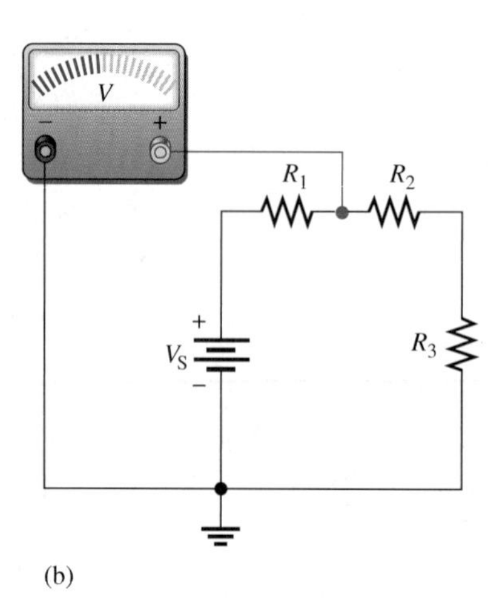

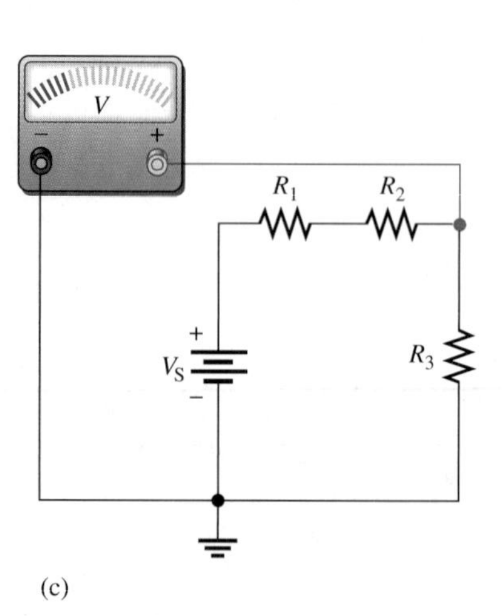

▲ **FIGURE 4–55**

Equivalent schematics for Figure 4–54.

► **FIGURE 4–56**

Measuring voltage directly across a resistor.

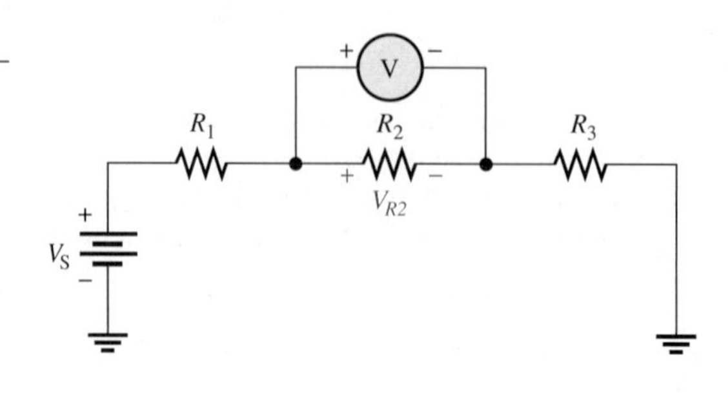

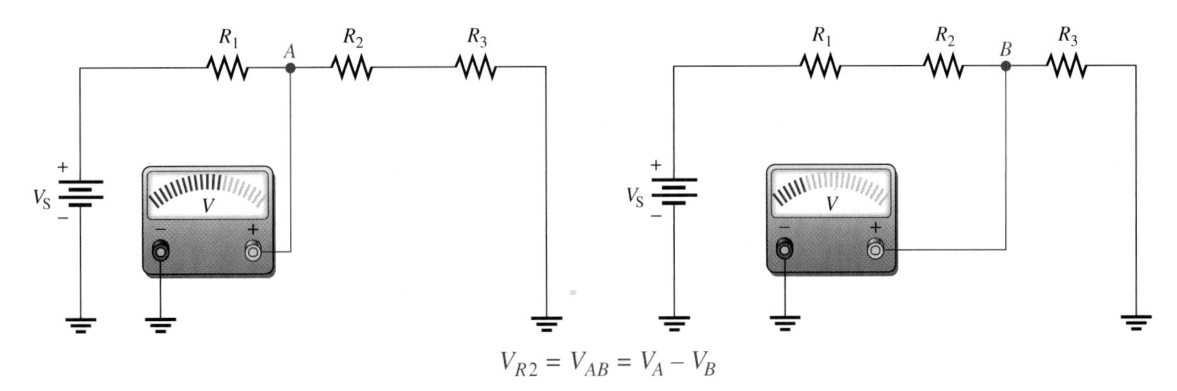

▲ **FIGURE 4–57**

Measuring voltage across R_2 with two separate measurements to ground.

EXAMPLE 4–20

Determine the voltages at each of the indicated points in each circuit of Figure 4–58. Since each of the four resistors has the same value, 25 V are dropped across each one.

▼ **FIGURE 4–58**

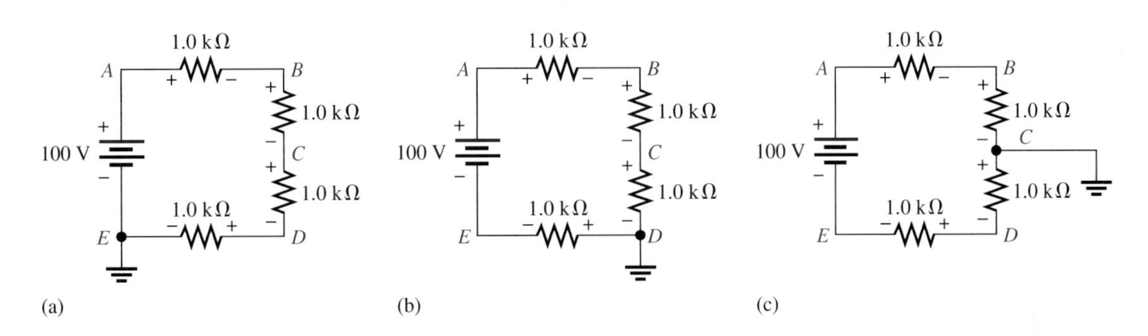

Solution For the circuit in Figure 4–58(a), the voltage polarities are as shown. Point E is ground. Single-letter subscripts denote voltage at a point with respect to ground. The voltages with respect to ground are as follows:

$$V_E = \mathbf{0\ V},\quad V_D = \mathbf{+25\ V},\quad V_C = \mathbf{+50\ V},\quad V_B = \mathbf{+75\ V},\quad V_A = \mathbf{+100\ V}$$

For the circuit in Figure 4–58(b), the voltage polarities are as shown. Point D is ground. The voltages with respect to ground are as follows:

$$V_E = \mathbf{-25\ V},\quad V_D = \mathbf{0\ V},\quad V_C = \mathbf{+25\ V},\quad V_B = \mathbf{+50\ V},\quad V_A = \mathbf{+75\ V}$$

For the circuit in Figure 4–58(c), the voltage polarities are as shown. Point C is ground. The voltages with respect to ground are as follows:

$$V_E = \mathbf{-50\ V},\quad V_D = \mathbf{-25\ V},\quad V_C = \mathbf{0\ V},\quad V_B = \mathbf{+25\ V},\quad V_A = \mathbf{+50\ V}$$

Related Problem If the ground is moved to point A in Figure 4–58(a), what are the voltages at each of the other points with respect to ground?

Open file E04-20 on your EWB/CircuitMaker CD-ROM. For each circuit, verify the values of the voltages at each point with respect to ground. Move the ground to point A and measure the voltages at the other points with respect to ground. Compare the results with those of the related problem.

SECTION 4–9 REVIEW

1. What is the common reference point in a circuit called?
2. Voltages in a circuit are generally referenced to ground. (True or False)
3. The housing or chassis can be used as circuit ground. (True or False)
4. What is the symbol for ground?

4–10 TROUBLESHOOTING

Open components or contacts and shorts between conductors are common problems in all circuits. An open produces an infinite resistance. A short produces a zero resistance.

After completing this section, you should be able to

- **Troubleshoot series circuits**
- Check for an open circuit
- Check for a short circuit
- Identify primary causes of shorts

Open Circuit

The most common failure in a series circuit is an open. For example, when a resistor or a lamp burns out, it causes a break in the current path and creates an open circuit as illustrated in Figure 4–59.

An open in a series circuit prevents current.

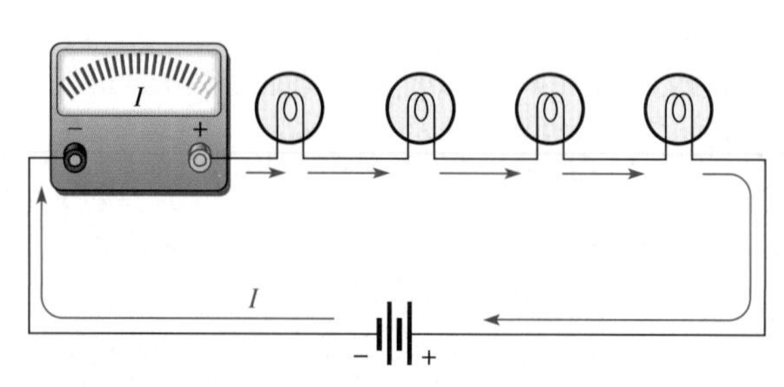

(a) A complete series circuit has current.

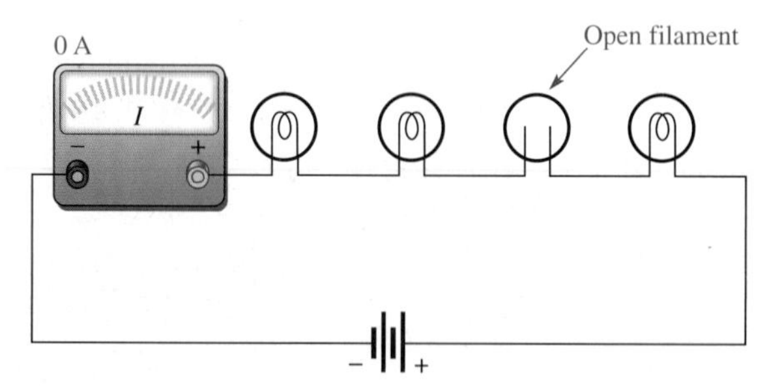

(b) An open series circuit has no current.

▲ **FIGURE 4–59**

An open circuit prevents current.

Troubleshooting an Open In Chapter 3, you were introduced to the analysis, planning, and measurement (APM) approach to troubleshooting. You also learned about the half-splitting method and saw an example using an ohmmeter. Now, the same principles will be applied using voltage measurements instead of resistance measurements. As you know, voltage measurements are generally the easiest to make because you do not have to disconnect anything.

As a beginning step, prior to analysis, it is a good idea to make a visual check of the faulty circuit. Occasionally, a charred resistor, a broken lamp filament, a loose wire, or a loose connection can be found this way. However, it is possible (and probably more common) for a resistor or other component to open without showing visible signs of damage. When a visual check reveals nothing, then proceed with the APM approach.

When an open occurs in a series circuit, all of the source voltage appears across the open. The reason for this is that the open condition prevents current through the series circuit. With no current, there can be no voltage drop across any of the other resistors (or other component). Since $IR = (0\text{ A})R = 0\text{ V}$, the voltage on each end of a good resistor is the same. Therefore, the voltage applied across a series string also appears across the open component because there are no other voltage drops in the circuit, as illustrated in Figure 4–60. The source voltage will appear across the open resistor in accordance with Kirchhoff's voltage law as follows:

$$V_S = V_1 + V_2 + V_3 + V_4 + V_5 + V_6$$

$$\begin{aligned} V_4 &= V_S - V_1 - V_2 - V_3 - V_5 - V_6 \\ &= 10\text{ V} - 0\text{ V} - 0\text{ V} - 0\text{ V} - 0\text{ V} - 0\text{ V} \end{aligned}$$

$$V_4 = V_S = 10\text{ V}$$

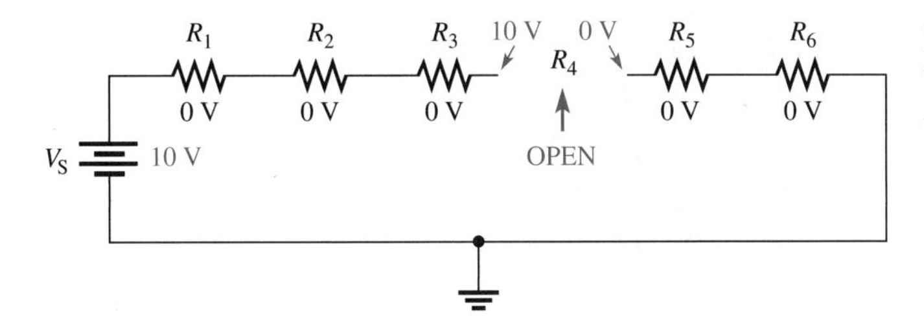

FIGURE 4–60

The source voltage appears across the open series resistor.

Example of Half-Splitting Using Voltage Measurements Suppose a circuit has four resistors in series. You have determined by *analyzing* the symptoms (there is voltage but no current) that one of the resistors is open, and you are *planning* to find the open resistor using a voltmeter for *measuring* by the half-splitting method. A sequence of measurements for this particular case is illustrated in Figure 4–61.

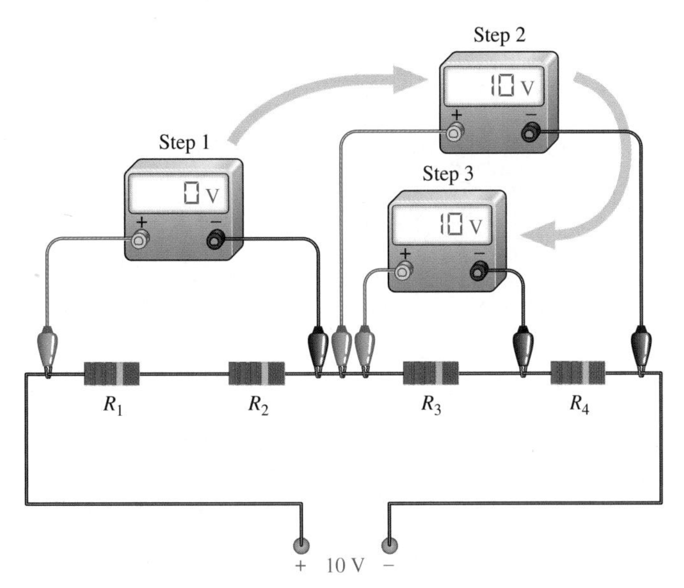

FIGURE 4–61

Troubleshooting a series circuit for an open using half-splitting.

First measure across R_1 and R_2 (the left half of the circuit). A 0 V reading indicates that neither of these resistors is open. Then measure across R_3 and R_4; the reading is 10 V. This indicates there is an open in the right half of the circuit, so either R_3 or R_4 is the faulty resistor (assume no bad connections). As shown in the figure as Step 3, a measurement of 10 V across R_3 identifies it as the open resistor. If you had measured across R_4, it would have indicated 0 V. This would have also

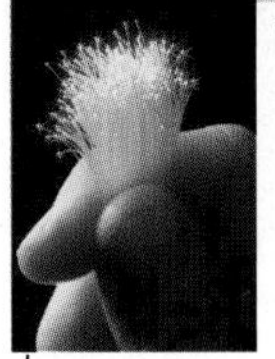

When measuring a resistance, make sure that you do not touch the meter leads or the resistor leads. If you hold both ends of a high-value resistor in your fingers along with the meter probes, the measurement will be inaccurate because your body resistance can affect the measured value. When body resistance is placed in parallel with a high-value resistor, the measured value will be less than the actual value of the resistor.

identified R_3 as the faulty component because it would have been the only one left that could have 10 V across it.

Short Circuit

Sometimes an unwanted short circuit occurs when two conductors touch or a foreign object such as solder or a wire clipping accidentally connects two sections of a circuit together. This situation is particularly common in circuits with a high component density. Three potential causes of short circuits are illustrated on the PC board in Figure 4–62.

FIGURE 4–62

Examples of shorts on a PC board.

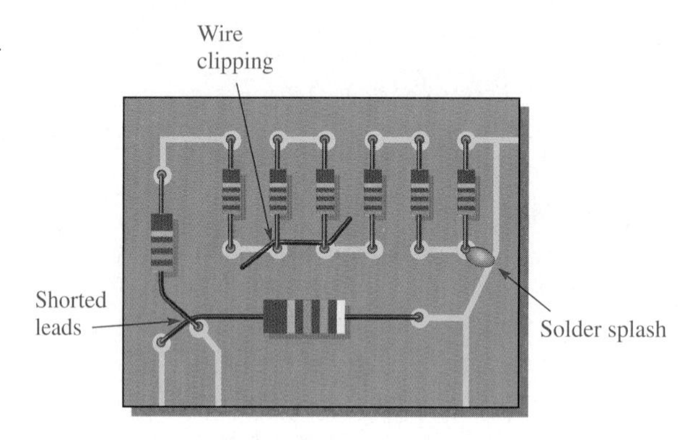

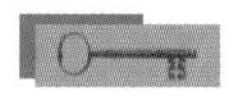

When there is a **short**, a portion of the series resistance is bypassed (all of the current goes through the short), thus reducing the total resistance as illustrated in Figure 4–63. Notice that the current increases as a result of the short.

A short in a series circuit causes more current.

FIGURE 4–63

The effect of a short in a series circuit.

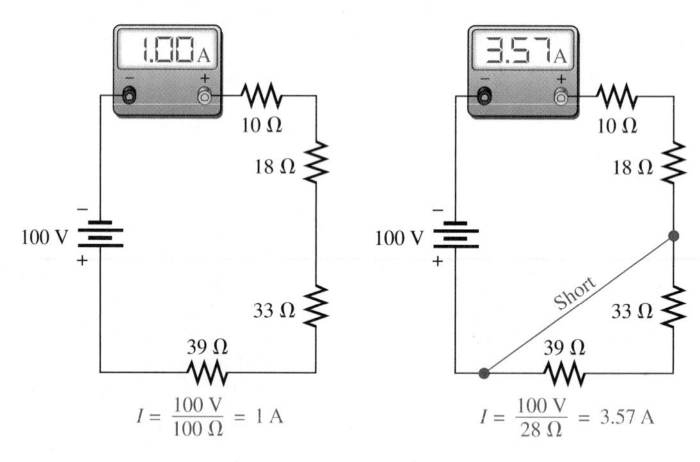

(a) Before short

(b) After short

Troubleshooting a Short As in any troubleshooting situation, it is a good idea to make a visual check of the faulty circuit. In the case of a short in the circuit, a wire clipping, solder splash, or touching leads is often found to be the culprit. In terms of component failure, shorts are less common than opens in many types of components. However, a short is generally more difficult to troubleshoot. Furthermore, a short in one part of a circuit can cause overheating in another part due to the higher current caused by the short. As a result two failures, an open and a short, may occur together.

When a short occurs in a series circuit, there is essentially no voltage across the shorted part. A short has zero or near zero resistance, although shorts with significant resistance values can occur from time to time. These are called *resistive shorts.* For purposes of illustration, zero resistance is assumed for all shorts.

In order to troubleshoot a short, the voltage across each resistor can be measured until you get a reading of 0 V. This is the straight-forward approach and does not use half-splitting. In order to apply the half-splitting method, the correct voltage at each point in the circuit must be known and used for comparison to measured voltages. An example will illustrate using half-splitting to find a short.

EXAMPLE 4–21

Assume you have determined that there is a short in a circuit with four series resistors because the current is higher than it should be. You know that the voltage at each point in the circuit should be as shown in Figure 4–64 if the circuit is working properly. The voltages are shown relative to the negative terminal of the source. Find the location of the short.

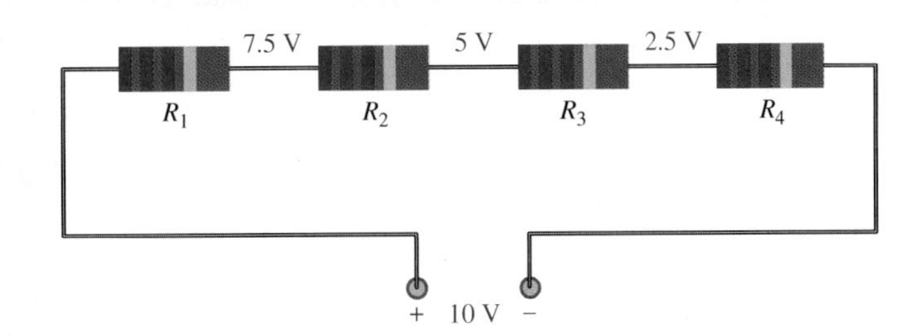

FIGURE 4–64

Series circuit (without a short) with correct voltages marked.

Solution Use the half-splitting method to troubleshoot the short. First, measure across R_1 and R_2. The meter shows a reading of 6.67 V, which is higher than the normal voltage (it should be 5 V). Look for a voltage that is lower than normal because a short will make the voltage less across that part of the circuit.

Next, measure across R_3 and R_4; the reading of 3.33 V is incorrect and lower than normal (it should be 5 V). This shows that the short is in the right half of the circuit and that either R_3 or R_4 is shorted. Measure across R_3. A reading of 3.3 V across R_3 tells you that R_4 is shorted because it must have 0 V across it. Figure 4–65 illustrates this troubleshooting technique.

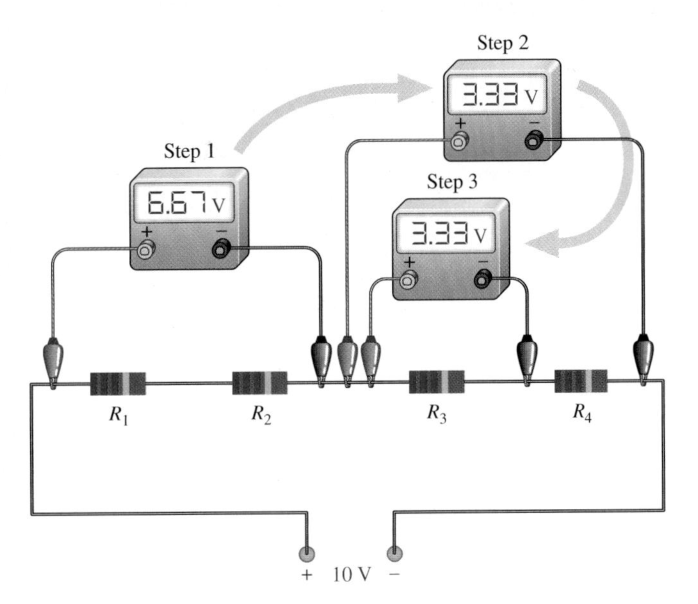

FIGURE 4–65

Troubleshooting a series circuit for a short using half-splitting.

Related Problem Assume that R_1 is shorted in Figure 4–65. What would the Step 1 measurement be?

SECTION 4–10 REVIEW

1. Define *open.*
2. Define *short.*
3. What happens when a series circuit opens?
4. Name two general ways in which an open circuit can occur in practice. How can a short circuit occur?
5. When a resistor fails, it will normally open. (True or False)
6. The total voltage across a string of series resistors is 24 V. If one of the resistors is open, how much voltage is there across it? How much is there across each of the good resistors?
7. Explain why the voltage measured in Step 1 of Figure 4–65 is higher than normal.

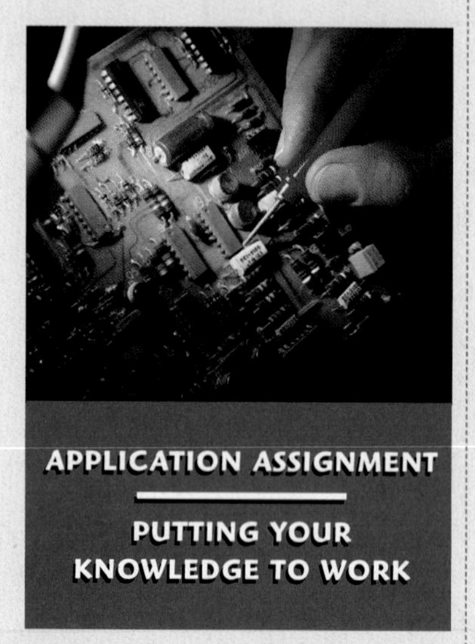

APPLICATION ASSIGNMENT

PUTTING YOUR KNOWLEDGE TO WORK

For this assignment, suppose your supervisor has given you a voltage-divider board to evaluate and modify if necessary. It will be used to obtain five different voltage levels from a 12 V battery that has a 6.5 Ah rating. The voltage divider is to be used to provide positive reference voltages to an electronic circuit in an analog-to-digital converter. Your job will be to check the circuit to see if it provides the following voltages within a tolerance of ±5% with respect to the negative side of the battery: 10.4 V, 8.0 V, 7.3 V, 6.0 V, and 2.7 V. If the existing circuit does not provide the specified voltages, you will modify it so that it does. Also, you must make sure that the power ratings of the resistors are adequate for the application and determine how long the battery will last with the voltage divider connected to it.

Step 1: Draw the Schematic of the Circuit

Use Figure 4–66 to determine the resistor values and draw the schematic of the voltage-divider circuit so you will know what you are working with. All the resistors on the board are 0.25 W.

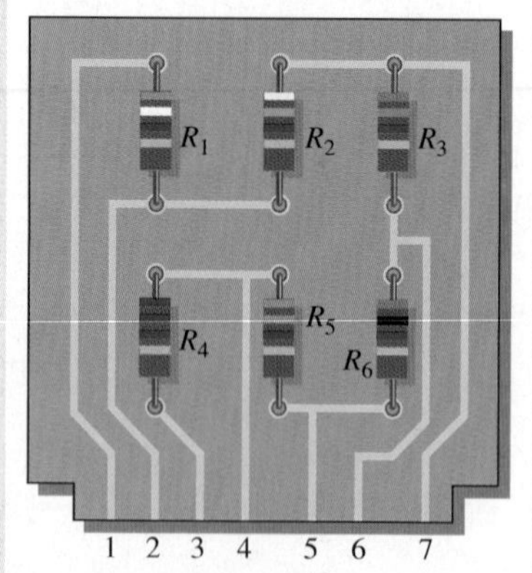

▲ FIGURE 4–66

Step 2: Determine the Voltages

Determine each output voltage on the existing circuit board when the positive side of the 12 V battery is connected to pin 3 and the negative side is connected to pin 1. Compare the existing output voltages to the following specifications:

Pin 1: negative terminal of 12 V battery

Pin 2: 2.7 V ± 5%

Pin 3: positive terminal of 12 V battery

Pin 4: 10.4 V ± 5%

Pin 5: 8.0 V ± 5%

Pin 6: 7.3 V ± 5%

Pin 7: 6.0 V ± 5%

Step 3: Modify the Existing Circuit (if necessary)

If the output voltages of the existing circuit are not the same as those stated in the specifications of Step 2, make the necessary changes in the circuit to meet the specifications. Draw a schematic of the modified circuit showing resistor values and adequate power ratings.

Step 4: Determine the Life of the Battery

Find the total current drawn from the 12 V battery when the voltage-divider circuit is connected and determine how many days the 6.5 Ah battery will last.

Step 5: Develop a Test Procedure

Determine how you would test the voltage-divider board and what instruments you would use. Then detail your test procedure in a step-by-step format.

Step 6: Troubleshoot the Circuit

Determine the most likely fault for each of the following cases (voltages are with respect to the negative battery terminal (pin 1 on the circuit board):

1. No voltage at any of the pins on the circuit board
2. 12 V at pins 3 and 4. All other pins have 0 V.
3. 12 V at all pins except 0 V at pin 1
4. 12 V at pin 6 and 0 V at pin 7
5. 3.3 V at pin 2

Electronics Workbench/CircuitMaker Analysis

1. Using Electronics Workbench or CircuitMaker, connect the circuit based on the schematic from Step 1 and verify the output voltages specified in Step 2.
2. Insert faults determined in Step 6 and verify the resulting voltage measurements.

APPLICATION ASSIGNMENT REVIEW

1. What is the total power dissipated by the voltage-divider circuit with a 12 V battery?
2. What are the output voltages from the voltage divider if a 6 V battery is used?
3. When the voltage-divider board is connected to the electronic circuit to which it is providing positive reference voltages, which pin on the board should be connected to the ground of the electronic circuit?

SUMMARY

- The current is the same at all points in a series circuit.
- The total series resistance is the sum of all resistors in the series circuit.
- The total resistance between any two points in a series circuit is equal to the sum of all resistors connected in series between those two points.
- If all of the resistors in a series circuit are of equal value, the total resistance is the number of resistors multiplied by the resistance value of one resistor.
- Voltage sources in series add algebraically.
- Kirchhoff's voltage law: The sum of the voltage drops equals the total source voltage, or equivalently, the algebraic sum of all the voltages around a closed path is zero.
- The voltage drops in a circuit are always opposite in polarity to the total source voltage.
- Current is out of the negative side of a source and into the positive side.
- Current is into the negative side of each resistor and out of the positive side.
- A voltage divider is a series arrangement of resistors.
- A voltage divider is so named because the voltage drop across any resistor in the series circuit is divided down from the total voltage by an amount proportional to that resistance value in relation to the total resistance.

- A potentiometer can be used as an adjustable voltage divider.
- The total power in a resistive circuit is the sum of all the individual powers of the resistors making up the series circuit.
- All voltages in a circuit are referenced to ground unless otherwise specified.
- Ground is zero volts with respect to all points referenced to it in the circuit.
- *Negative ground* is the term used when the negative side of the source is grounded.
- *Positive ground* is the term used when the positive side of the source is grounded.
- The voltage across an open series component equals the source voltage.
- The voltage across a shorted series component is 0 V.

EQUATIONS

4–1	$R_T = R_1 + R_2 + R_3 + \cdots + R_n$	Total resistance of n resistors in series
4–2	$R_T = nR$	Total resistance of n equal-value resistors in series
4–3	$V_S = V_1 + V_2 + V_3 + \cdots + V_n$	Kirchhoff's voltage law
4–4	$V_S - V_1 - V_2 - V_3 - \cdots - V_n = 0$	Kirchhoff's voltage law stated another way
4–5	$V_x = \left(\frac{R_x}{R_T}\right)V_S$	Voltage-divider formula
4–6	$P_T = P_1 + P_2 + P_3 + \cdots + P_n$	Total power

SELF-TEST

Answers are at the end of the chapter.

1. Five equal-value resistors are connected in series and there is a current of 2 A into the first resistor. The amount of current out of the second resistor is
   (a) 2 A **(b)** 1 A **(c)** 4 A **(d)** 0.4 A
2. To measure the current out of the third resistor in a circuit consisting of four series resistors, an ammeter can be placed
   (a) between the third and fourth resistors
   (b) between the second and third resistors
   (c) at the positive terminal of the source
   (d) at any point in the circuit
3. When a third resistor is connected in series with two series resistors, the total resistance
   (a) remains the same
   (b) increases
   (c) decreases
   (d) increases by one-third

4. When one of four series resistors is removed from a circuit and the circuit reconnected, the current

(a) decreases by the amount of current through the removed resistor

(b) decreases by one-fourth

(c) quadruples

(d) increases

5. A series circuit consists of three resistors with values of 100 Ω, 220 Ω, and 330 Ω. The total resistance is

(a) less than 100 Ω **(b)** the average of the values

(c) 550 Ω **(d)** 650 Ω

6. A 9 V battery is connected across a series combination of 68 Ω, 33 Ω, 100 Ω, and 47 Ω resistors. The amount of current is

(a) 36.3 mA **(b)** 27.6 A **(c)** 22.3 mA **(d)** 363 mA

7. While putting four 1.5 V batteries in a flashlight, you accidentally put one of them in backward. The light will be

(a) brighter than normal **(b)** dimmer than normal

(c) off **(d)** the same

8. If you measure all the voltage drops and the source voltage in a series circuit and add them together, taking into consideration the polarities, you will get a result equal to

(a) the source voltage

(b) the total of the voltage drops

(c) zero

(d) the total of the source voltage and the voltage drops

9. There are six resistors in a given series circuit and each resistor has 5 V dropped across it. The source voltage is

(a) 5 V **(b)** 30 V

(c) dependent on the resistor values **(d)** dependent on the current

10. A series circuit consists of a 4.7 kΩ, a 5.6 kΩ, and a 10 kΩ resistor. The resistor that has the most voltage across it is

(a) the 4.7 kΩ **(b)** the 5.6 kΩ

(c) the 10 kΩ **(d)** impossible to determine from the given information

11. Which of the following series combinations dissipates the most power when connected across a 100 V source?

(a) one 100 Ω resistor **(b)** two 100 Ω resistors

(c) three 100 Ω resistors **(d)** four 100 Ω resistors

12. The total power in a certain circuit is 10 W. Each of the five equal-value series resistors making up the circuit dissipates

(a) 10 W **(b)** 50 W **(c)** 5 W **(d)** 2 W

13. When you connect an ammeter in a series resistive circuit and turn on the source voltage, the meter reads zero. You should check for

(a) a broken wire **(b)** a shorted resistor

(c) an open resistor **(d)** both (a) and (c)

14. While checking out a series resistive circuit, you find that the current is higher than it should be. You should look for

(a) an open circuit **(b)** a short

(c) a low resistor value **(d)** both (b) and (c)

PROBLEMS

Answers to odd-numbered problems are at the end of the book.

BASIC PROBLEMS

SECTION 4–1

Resistors in Series

1. Connect each set of resistors in Figure 4–67 in series between points A and B.

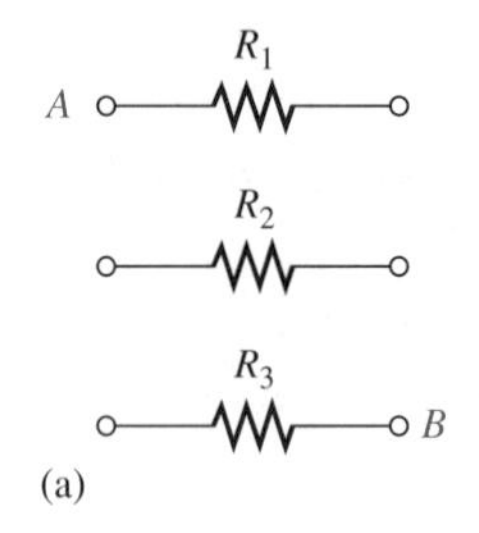

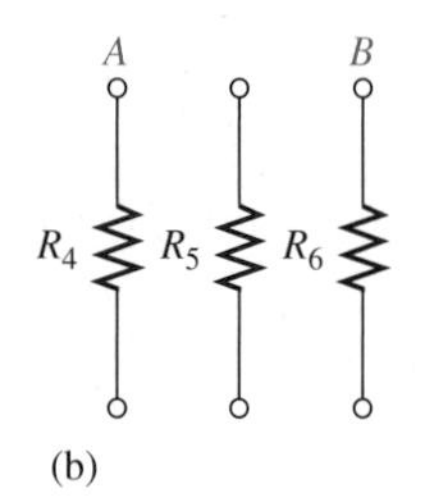

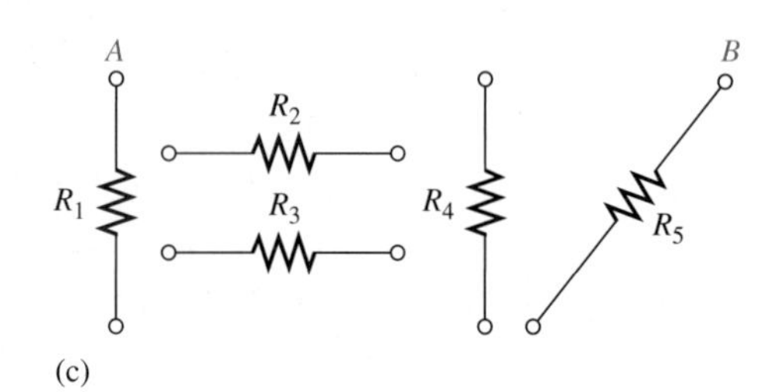

▲ FIGURE 4–67

2. Determine which resistors in Figure 4–68 are in series. Show how to interconnect the pins to put all the resistors in series.

▶ FIGURE 4–68

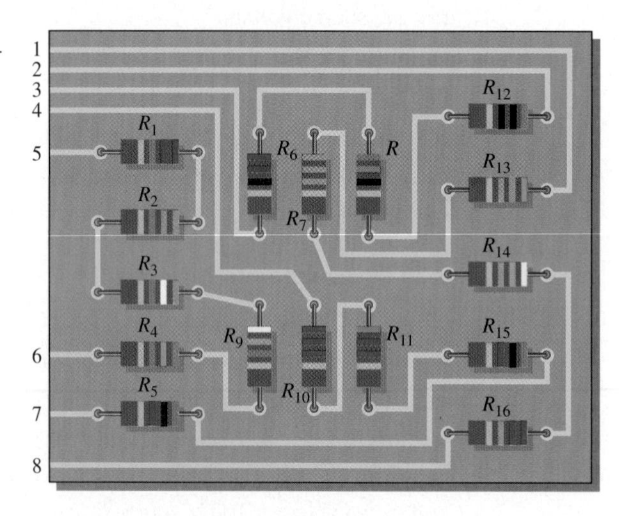

SECTION 4–2

Current in a Series Circuit

3. What is the current through each of four resistors in a series circuit if the source voltage is 12 V and the total resistance is 120 Ω?
4. The current from the source in Figure 4–69 is 5 mA. How much current does each milliammeter in the circuit indicate?

▶ FIGURE 4–69

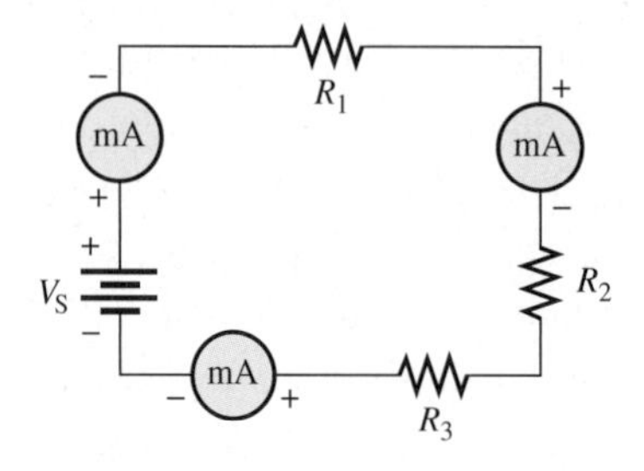

SECTION 4–3 **Total Series Resistance**

5. An 82 Ω resistor and a 56 Ω resistor are connected in series. What is the total resistance?

6. Find the total resistance of each group of series resistors shown in Figure 4–70.

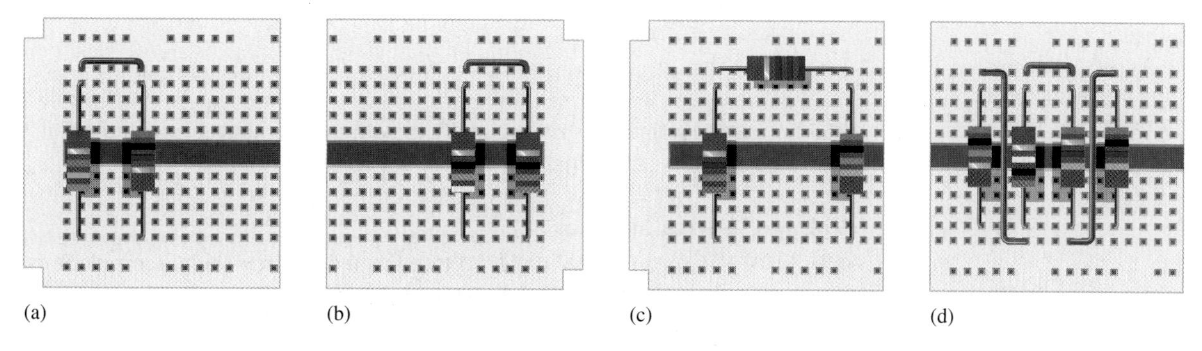

FIGURE 4–70

7. Determine R_T for each circuit in Figure 4–71. Show how to measure R_T with an ohmmeter.

8. What is the total resistance of twelve 5.6 kΩ resistors in series?

9. Six 47 Ω resistors, eight 100 Ω resistors, and two 22 Ω resistors are in series. What is the total resistance?

10. The total resistance in Figure 4–72 is 20 kΩ. What is the value of R_5?

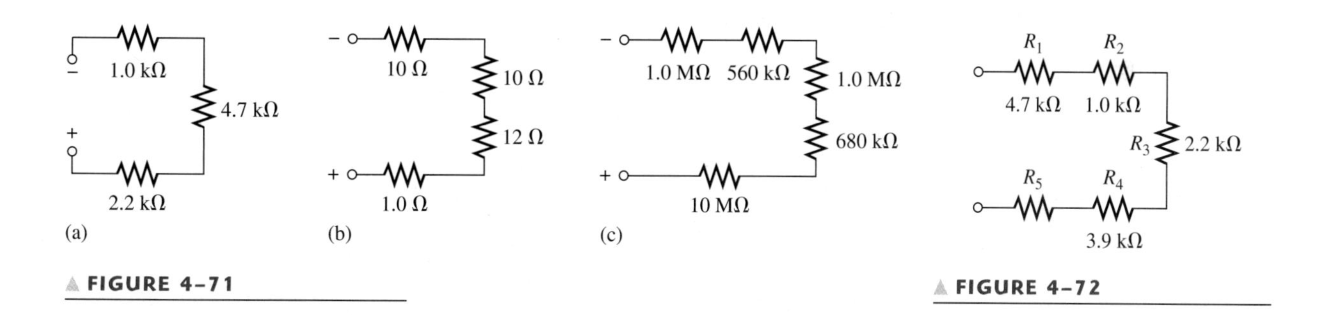

FIGURE 4–71

FIGURE 4–72

11. Determine the resistance between each of the following sets of pins on the PC board in Figure 4–68.
    (a) pin 1 and pin 8 **(b)** pin 2 and pin 3
    (c) pin 4 and pin 7 **(d)** pin 5 and pin 6

12. If all the resistors in Figure 4–68 are connected in series, what is the total resistance?

SECTION 4–4 **Ohm's Law in Series Circuits**

13. What is the current in each circuit of Figure 4–73? Show how to connect an ammeter in each case.

14. Determine the voltage across each resistor in Figure 4–73.

15. Three 470 Ω resistors are in series with a 500 V source. How much current is there?

FIGURE 4–73

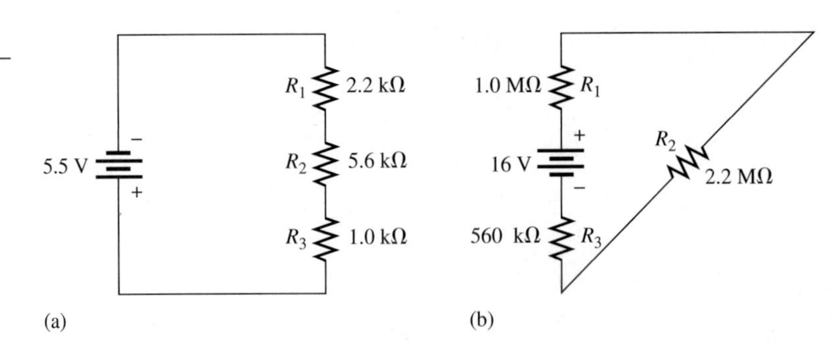

16. Four equal-value resistors are in series with a 5 V source, and a current of 1 mA is measured. What is the value of each resistor?

SECTION 4–5

Voltage Sources in Series

17. A 5 V battery and a 9 V battery are connected in series with their polarities in the same direction. What is the total voltage?

18. Determine the total source voltage in each circuit of Figure 4–74.

FIGURE 4–74

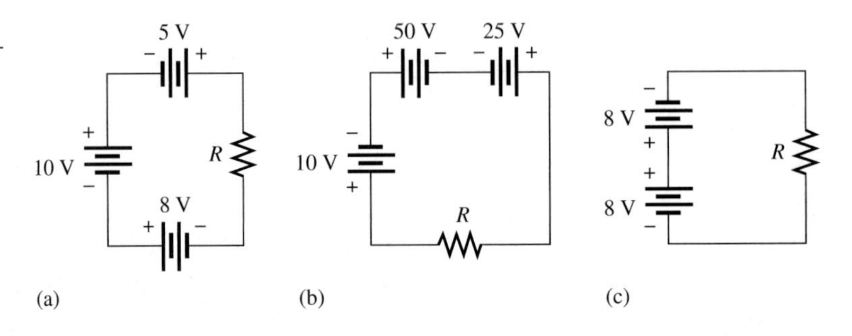

SECTION 4–6

Kirchhoff's Voltage Law

19. The following voltage drops are measured across each of three resistors in series: 5.5 V, 8.2 V, and 12.3 V. What is the value of the source voltage to which these resistors are connected?

20. Five resistors are in series with a 20 V source. The voltage drops across four of the resistors are 1.5 V, 5.5 V, 3 V, and 6 V. How much voltage is across the fifth resistor?

21. Determine the unspecified voltage drop(s) in each circuit of Figure 4–75. Show how to connect a voltmeter to measure each unknown voltage drop.

FIGURE 4–75

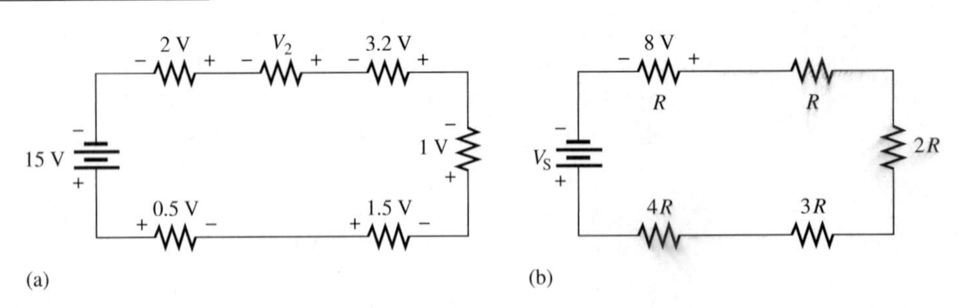

SECTION 4–7

Voltage Dividers

22. The total resistance of a series circuit is 500 Ω. What percentage of the total voltage appears across a 22 Ω resistor in the series circuit?

23. Find the voltage between *A* and *B* in each voltage divider of Figure 4–76.

24. What is the voltage across each resistor in Figure 4–77? *R* is the lowest value and all others are multiples of that value as indicated.

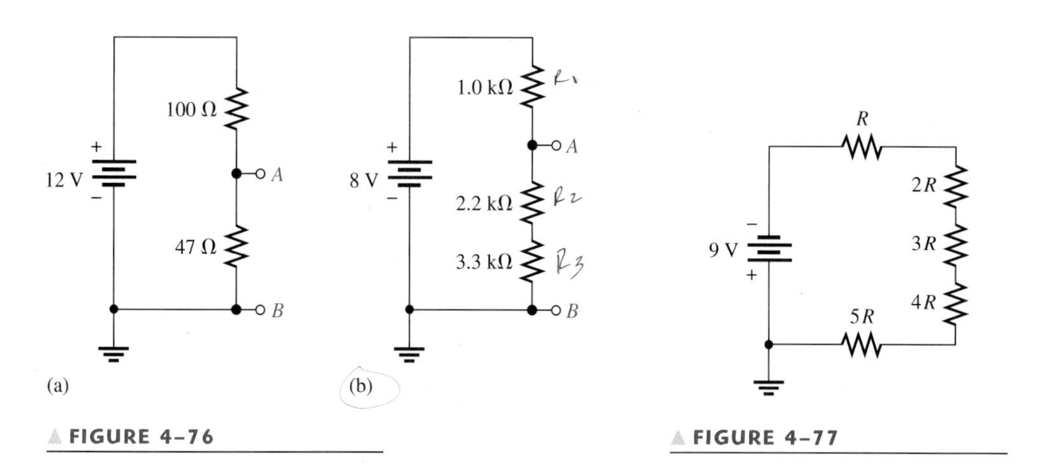

▲ FIGURE 4–76

▲ FIGURE 4–77

25. What is the voltage across each resistor in Figure 4–78(b)?

SECTION 4–8

Power in a Series Circuit

26. Five series resistors each dissipate 50 mW of power. What is the total power?

27. Use the results of Problem 25 to find the total power in Figure 4–78.

▶ FIGURE 4–78

(a) Meter with leads going to protoboard

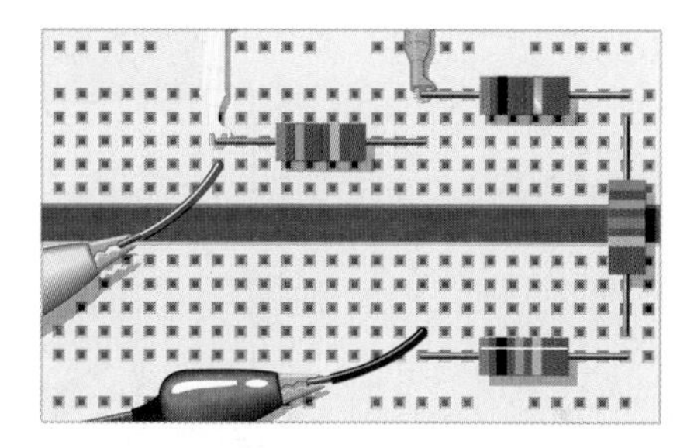

(b) Protoboard with meter leads (yellow and green) and power supply leads (red and gray) connected

SECTION 4–9

Circuit Ground

28. Determine the voltage at each point with respect to ground in Figure 4–79.

29. In Figure 4–80, how would you determine the voltage across R_2 by measurement, without connecting a meter directly across the resistor?

30. Determine the voltage at each point with respect to ground in Figure 4–80.

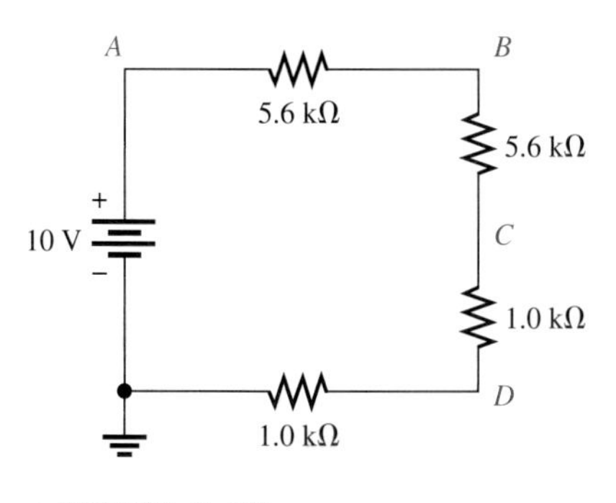

▲ FIGURE 4–79

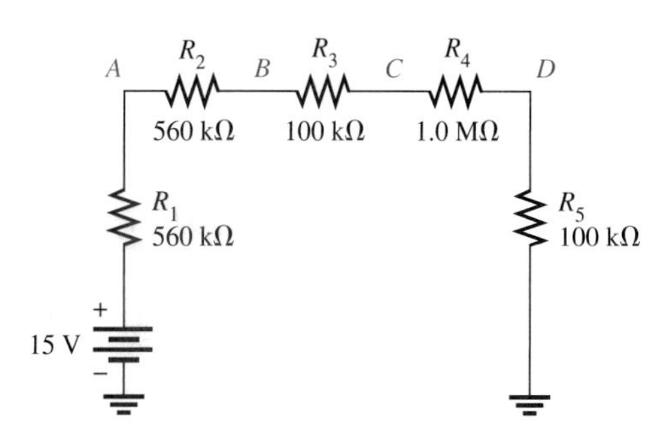

▲ FIGURE 4–80

SECTION 4–10

Troubleshooting

31. By observing the meters in Figure 4–81, determine the types of failures in the circuits and which components have failed.

FIGURE 4–81

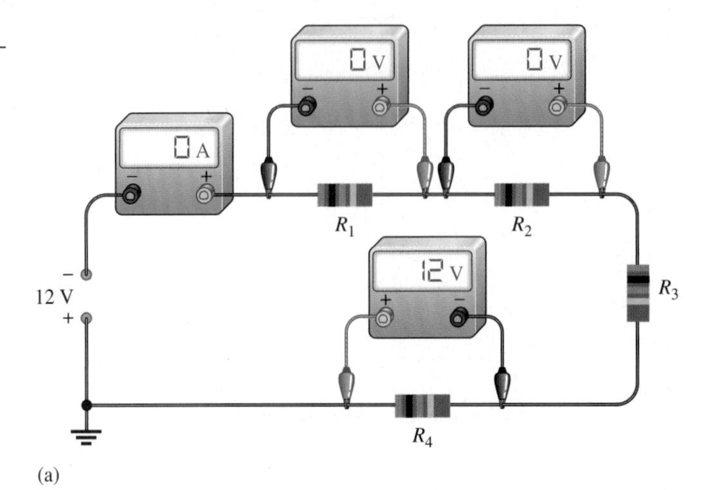

(a)

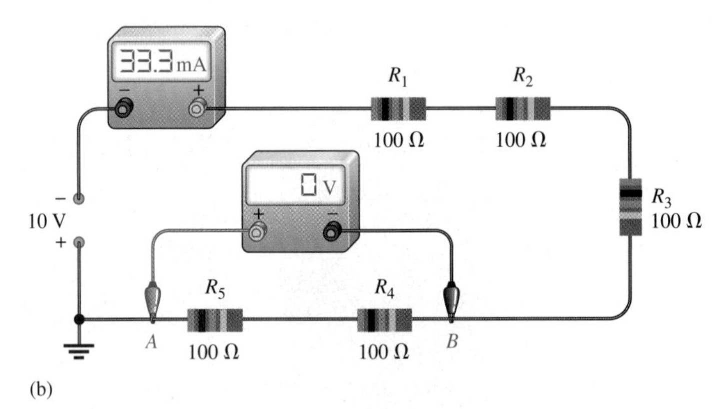

(b)

32. Is the multimeter reading in Figure 4–82 correct? If not, what is wrong?

ADVANCED PROBLEMS

33. Determine the unknown resistance (R_3) in the circuit of Figure 4–83.

(a) Meter with leads going to protoboard

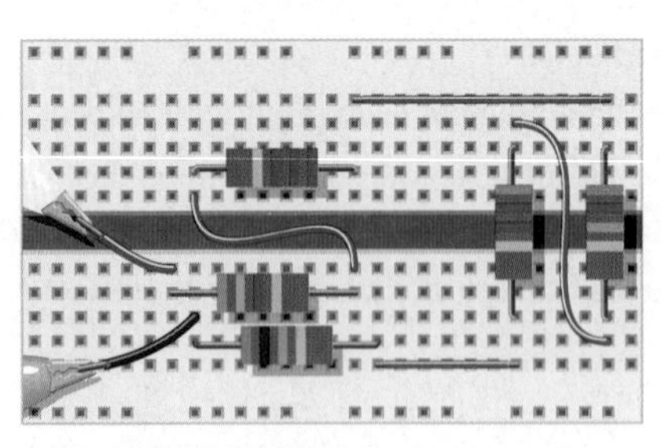

(b) Protoboard with meter leads connected

FIGURE 4–82

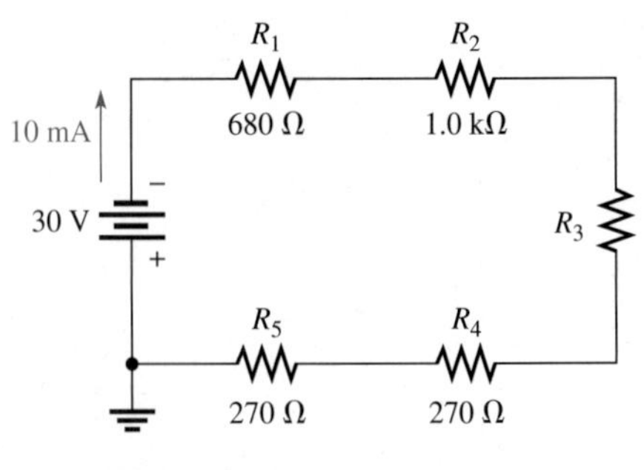

FIGURE 4–83

34. You have the following resistor values available to you in the lab in unlimited quantities: 10 Ω, 100 Ω, 470 Ω, 560 Ω, 680 Ω, 1.0 kΩ, 2.2 kΩ, and 5.6 kΩ. All of the other standard values are out of stock. A project that you are working on requires an 18 kΩ resistance. What combination of available values can you use to obtain the needed value?

35. Determine the voltage at each point in Figure 4–84 with respect to ground.

36. Find all the unknown quantities (appearing in color) in Figure 4–85.

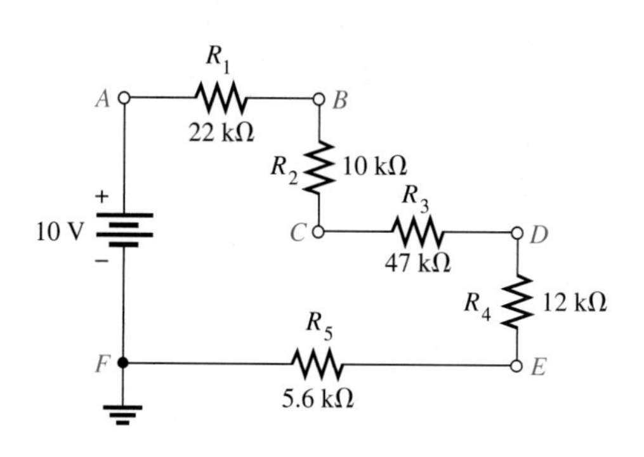

FIGURE 4–84

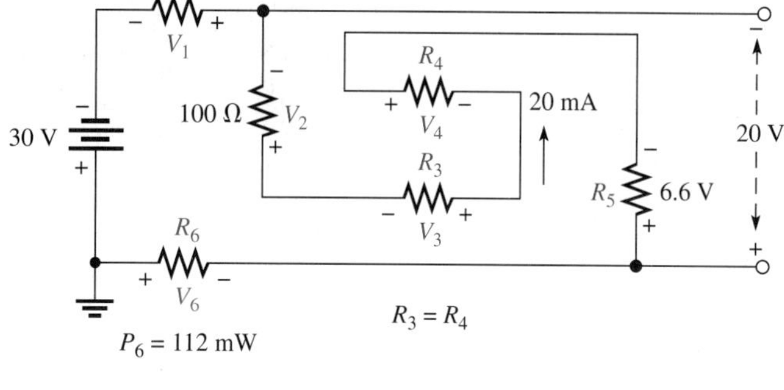

FIGURE 4–85

37. There are 250 mA in a series circuit with a total resistance of 1.5 kΩ. The current must be reduced by 25%. Determine how much resistance must be added in order to accomplish this reduction in current.

38. Four ½ W resistors are in series: 47 Ω, 68 Ω, 100 Ω, and 120 Ω. To what maximum value can the current be raised before the power rating of one of the resistors is exceeded? Which resistor will burn out first if the current is increased above the maximum?

39. A certain series circuit is made up of a ⅛ W resistor, a ¼ W resistor, and a ½ W resistor. The total resistance is 2400 Ω. If each of the resistors is operating at its maximum power level, determine the following:

(a) I **(b)** V_S **(c)** the value of each resistor

40. Using 1.5 V batteries, a switch, and three lamps, devise a circuit to apply 4.5 V across one lamp, two lamps in series, or three lamps in series with a single control switch. Draw the schematic.

41. Develop a variable voltage divider to provide output voltages ranging from a minimum of 10 V to a maximum of 100 V using a 120 V source. The maximum voltage must be at the maximum resistance setting of the potentiometer. The minimum voltage must be at the minimum resistance (zero ohms) setting. The maximum current is to be 10 mA.

42. Using the standard resistor values given in Appendix A, design a voltage divider to provide the following approximate voltages with respect to the negative terminal of a 30 V source: 8.18 V, 14.73 V, and 24.55 V. The current drain on the source must be limited to no more than 1 mA. The number of resistors, their resistance values, and their power ratings must be specified. Draw a schematic showing the circuit with all resistor values indicated.

43. On the double-sided PC board in Figure 4–86, identify each group of series resistors and determine its total resistance. Note that many of the interconnections feed through the board from the top side to the bottom side.

FIGURE 4–86

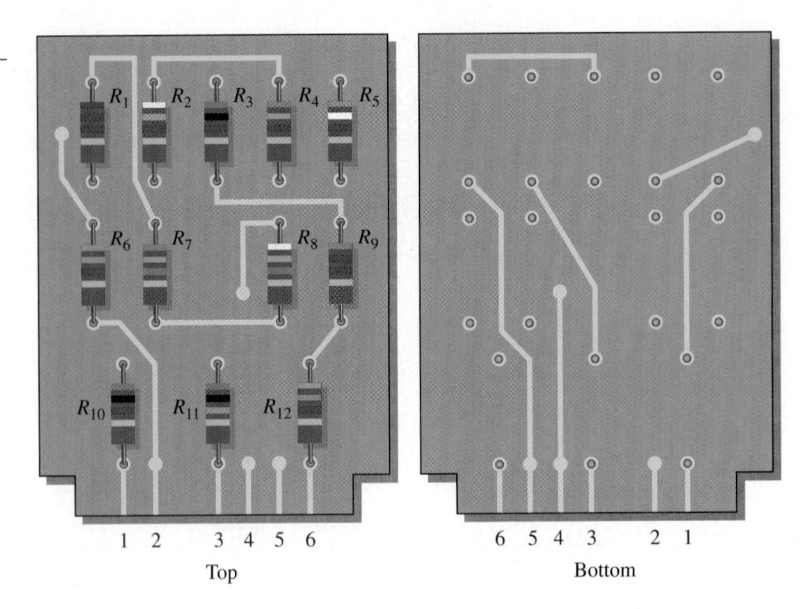

44. What is the total resistance from A to B for each switch position in Figure 4–87?

45. Determine the current measured by the meter in Figure 4–88 for each switch position.

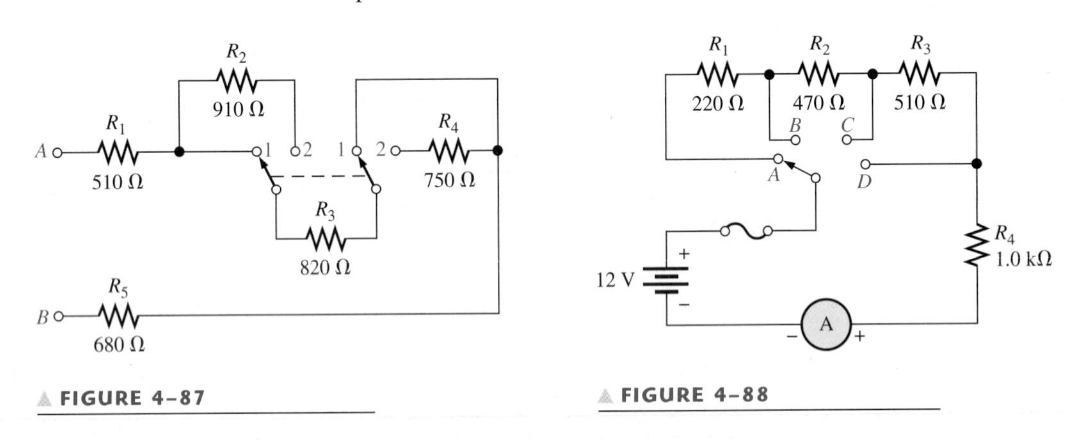

FIGURE 4–87

FIGURE 4–88

46. Determine the current measured by the meter in Figure 4–89 for each position of the ganged switch.

47. Determine the voltage across each resistor in Figure 4–90 for each switch position if the current through R_5 is 6 mA when the switch is in the D position.

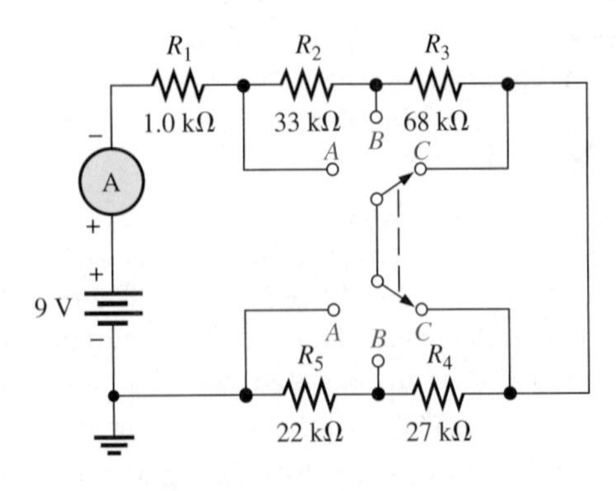

FIGURE 4–89

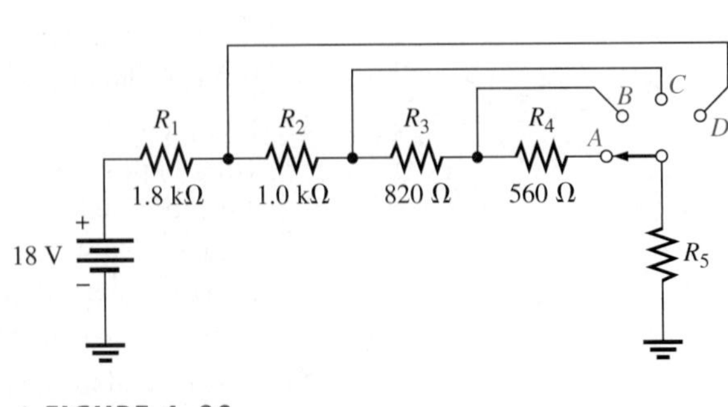

FIGURE 4–90

48. Table 4–1 shows the results of resistance measurements on the PC circuit board in Figure 4–86. Are these results correct? If not, identify the possible problems.

TABLE 4–1

BETWEEN PINS	RESISTANCE
1 and 2	∞
1 and 3	∞
1 and 4	4.23 kΩ
1 and 5	∞
1 and 6	∞
2 and 3	23.6 kΩ
2 and 4	∞
2 and 5	∞
2 and 6	∞
3 and 4	∞
3 and 5	∞
3 and 6	∞
4 and 5	∞
4 and 6	∞
5 and 6	19.9 kΩ

49. You measure 15 kΩ between pins 5 and 6 on the PC board in Figure 4–86. Does this indicate a problem? If so, identify it.

50. In checking out the PC board in Figure 4–86, you measure 17.83 kΩ between pins 1 and 2. Also, you measure 13.6 kΩ between pins 2 and 4. Does this indicate a problem on the PC board? If so, identify the fault.

51. The three groups of series resistors on the PC board in Figure 4–86 are connected in series with each other to form a single series circuit by connecting pin 2 to pin 4 and pin 3 to pin 5. A voltage source is connected across pins 1 and 6 and an ammeter is placed in series. As you increase the source voltage, you observe the corresponding increase in current. Suddenly, the current drops to zero and you smell smoke. All resistors are ½ W.

(a) What has happened?

(b) Specifically, what must you do to fix the problem?

(c) At what voltage did the failure occur?

ELECTRONICS WORKBENCH/CIRCUITMAKER TROUBLESHOOTING PROBLEMS

CD-ROM file circuits are shown in Figure 4–91.

52. Open file P04-52 on your CD-ROM. Determine if there is a fault and, if so, identify the fault.

53. Open file P04-53 on your CD-ROM. Determine if there is a fault and, if so, identify the fault.

54. Open file P04-54 on your CD-ROM. Determine if there is a fault and, if so, identify the fault.

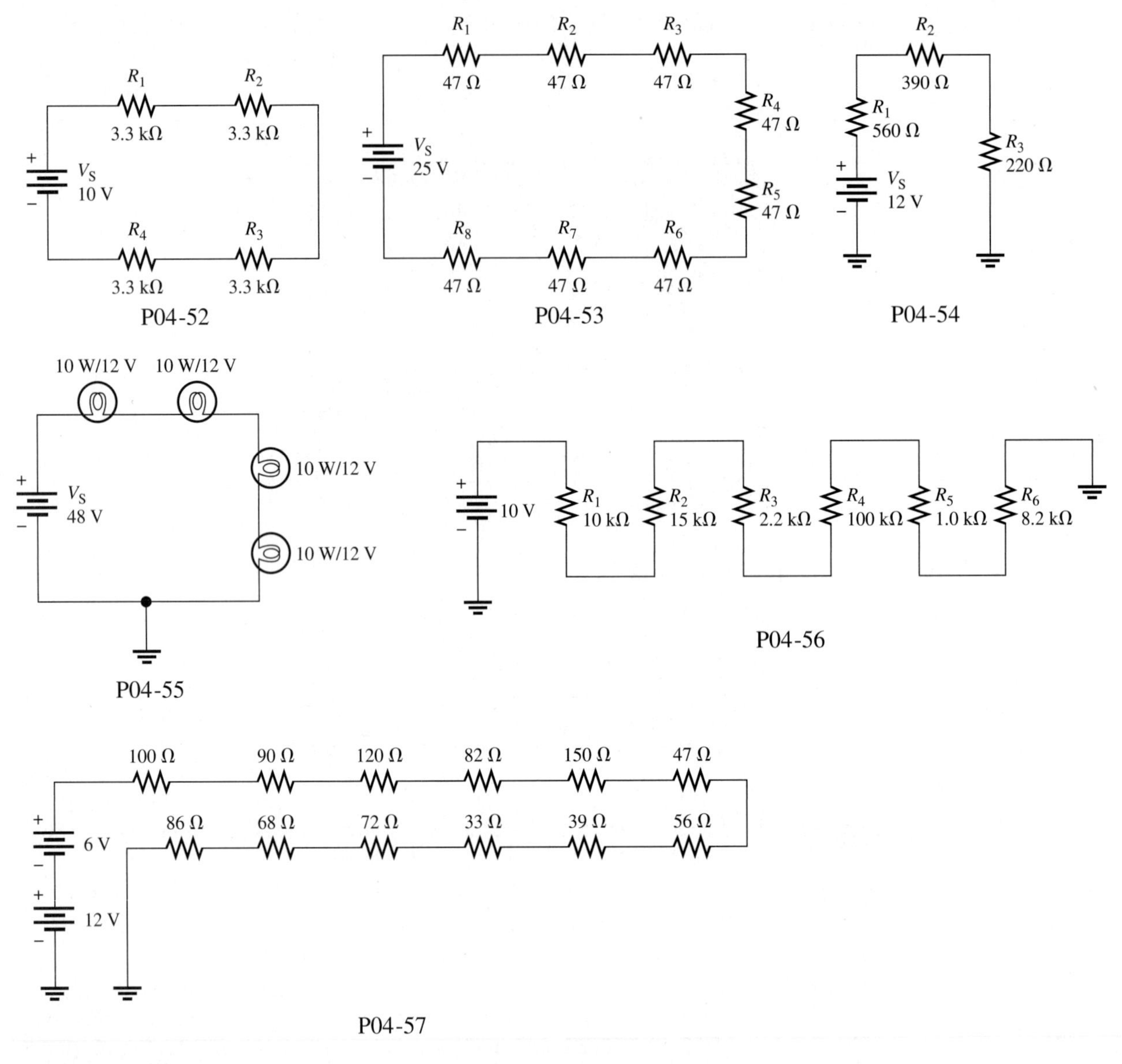

▲ **FIGURE 4–91**

55. Open file P04-55 on your CD-ROM. Determine if there is a fault and, if so, identify the fault.

56. Open file P04-56 on your CD-ROM. Determine if there is a fault and, if so, identify the fault.

57. Open file P04-57 on your CD-ROM. Determine if there is a fault and, if so, identify the fault.

ANSWERS

SECTION REVIEWS

SECTION 4–1 Resistors in Series

1. Series resistors are connected end-to-end in a "string."

2. There is a single current path in a series circuit.

3. See Figure 4–92.

4. See Figure 4–93.

FIGURE 4–92

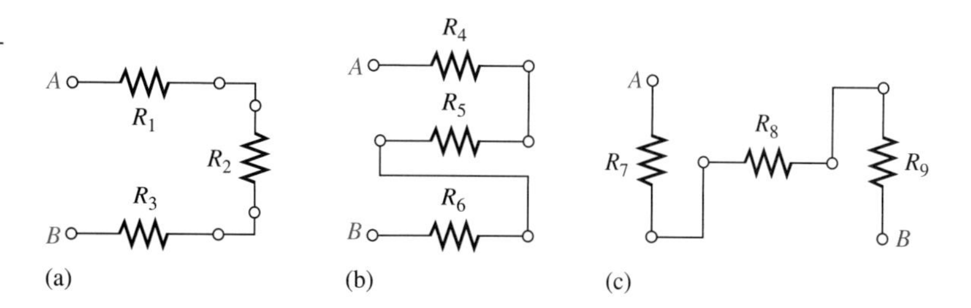

FIGURE 4–93

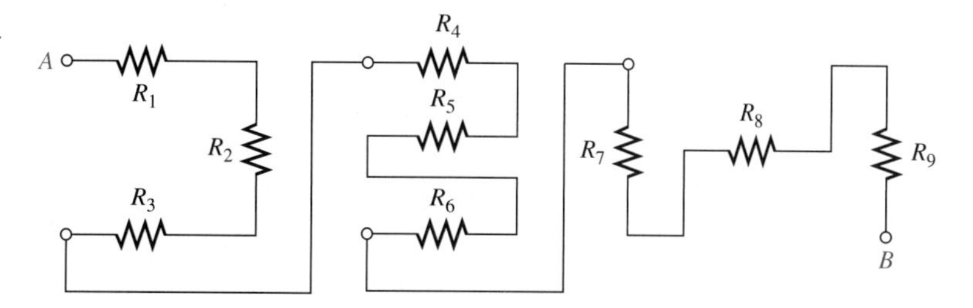

SECTION 4–2

Current in a Series Circuit

1. The current is the same at all points in a series circuit.
2. There are 2 A through the 4.7 Ω resistor.
3. 50 mA between C and D, 50 mA between E and F
4. Ammeter 1 indicates 1.79 A; Ammeter 2 indicates 1.79 A.

SECTION 4–3

Total Series Resistance

1. **(a)** $R_T = 33\ \Omega + 100\ \Omega + 10\ \Omega = 143\ \Omega$
   (b) $R_T = 39\ \Omega + 56\ \Omega + 10\ \Omega = 105\ \Omega$
   (c) $R_T = 820\ \Omega + 2200\ \Omega + 1000\ \Omega = 4020\ \Omega$
2. $R_T = 100\ \Omega + 2(47\ \Omega) + 4(12\ \Omega) + 330\ \Omega = 572\ \Omega$
3. $10\ k\Omega - 8.8\ k\Omega = 1.2\ k\Omega$
4. $R_T = 12(47\ \Omega) = 564\ \Omega$

SECTION 4–4

Ohm's Law in Series Circuits

1. $I = 10\ V/300\ \Omega = 0.033\ A = 33\ mA$
2. $V = (5\ mA)(43\ \Omega) = 125\ mV$
3. $V_1 = (5\ mA)(10\ \Omega) = 50\ mV$, $V_2 = (5\ mA)(15\ \Omega) = 75\ mV$, $V_3 = (5\ mA)(18\ \Omega) = 90\ mV$
4. $R = 1.25\ V/4.63\ mA = 270\ \Omega$

SECTION 4–5

Voltage Sources in Series

1. 60 V/12 V = 5; see Figure 4–94.
2. $V_T = (4)(1.5\ V) = 6.0\ V$
3. See Figure 4–95.

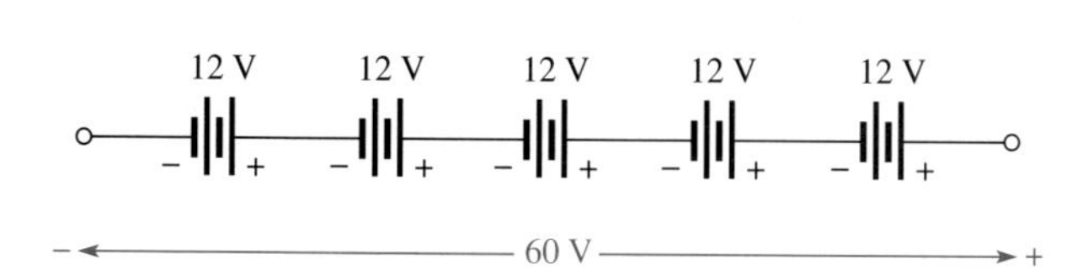

FIGURE 4–94

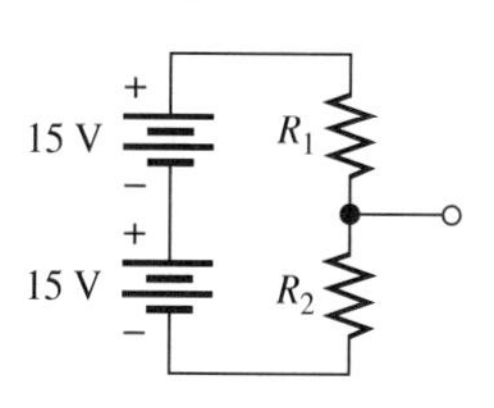

FIGURE 4–95

4. **(a)** $V_{S(tot)} = 100\text{ V} + 50\text{ V} - 75\text{ V} = 75\text{ V}$
   (b) $V_{S(tot)} = 20\text{ V} + 10\text{ V} - 10\text{ V} - 5\text{ V} = 15\text{ V}$
5. See Figure 4–96.

FIGURE 4–96

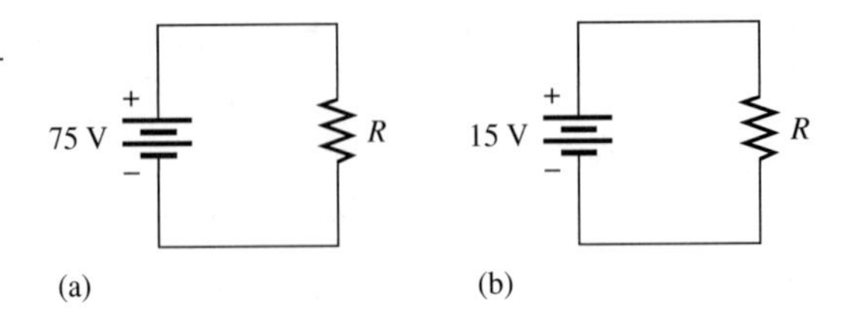

SECTION 4–6

Kirchhoff's Voltage Law

1. Kirchhoff's voltage law states:
   (a) The sum of the voltages around a closed path is zero.
   (b) The sum of the voltage drops equals the total source voltage.
2. $V_{R(tot)} = V_S = 50\text{ V}$
3. $V_{R1} = V_{R2} = 10\text{ V}/2 = 5\text{ V}$
4. $V_{R3} = 25\text{ V} - 5\text{ V} - 10\text{ V} = 10\text{ V}$
5. $V_S = 1\text{ V} + 3\text{ V} + 5\text{ V} + 7\text{ V} + 8\text{ V} = 24\text{ V}$

SECTION 4–7

Voltage Dividers

1. A voltage divider is a series circuit with two or more resistors with an output taken across one or more of the resistors.
2. Two or more resistors form a voltage divider.
3. $V_x = (R_x/R_T)V_S$ is the general voltage-divider formula.
4. $V_R = 20\text{ V}/2 = 10\text{ V}$
5. $V_{56\Omega} = (56\ \Omega/138\ \Omega)100\text{ V} = 40.6\text{ V}$, $V_{82\Omega} = (82\ \Omega/138\ \Omega)100\text{ V} = 59.4\text{ V}$. See Figure 4–97.

FIGURE 4–97

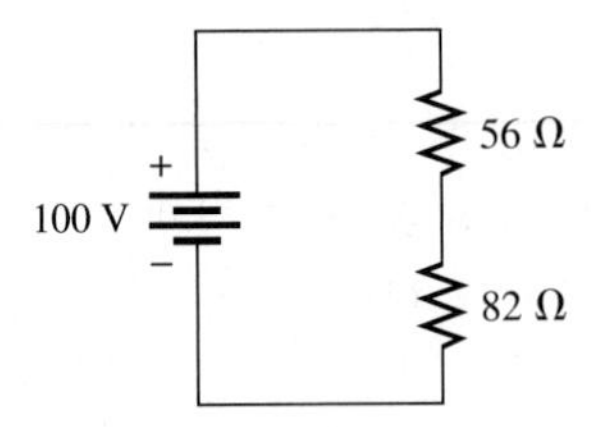

6. Set the potentiometer at the midpoint.

SECTION 4–8

Power in a Series Circuit

1. Add the powers in each resistor to get total power.
2. $P_T = 1\text{ W} + 2\text{ W} = 5\text{ W} + 8\text{ W} = 16\text{ W}$
3. $P_T = (1\text{ mA})^2(100\ \Omega + 330\ \Omega + 680\ \Omega) = 1.11\text{ mW}$

SECTION 4–9

Circuit Ground

1. Ground 2. True 3. True 4. See Figure 4–98.

FIGURE 4–98

SECTION 4–10 **Troubleshooting**

1. An open is a break in the current path.
2. A short is a zero resistance path that bypasses a portion of a circuit.
3. Current ceases when a series circuit opens.
4. An open can be created by a switch or a component failure. A short can be created by a switch or by wire clippings, solder splashes, etc.
5. True
6. 24 V across the open *R;* 0 V across the other resistors.
7. Because R_4 is shorted, more voltage is dropped across the other resistors than normal. The total voltage is divided across three equal-value resistors.

Application Assignment

1. $P_T = (12\text{ V})^2/16.6\text{ k}\Omega = 8.67\text{ mW}$
2. Pin 2: 1.41 V; Pin 6: 3.65 V; Pin 5: 4.01 V; Pin 4: 5.20 mV; Pin 7: 3.11 V
3. Pin 3 connects to ground.

RELATED PROBLEMS FOR EXAMPLES

4–1 **(a)** Left end of R_1 to "minus," right end of R_1 to top end of R_3, bottom end of R_3 to right end of R_5, left end of R_5 to left end of R_2, right end of R_2 to right end of R_4, left end of R_4 to "plus"

(b) $R_1 = 1.0\text{ k}\Omega$, $R_2 = 33\text{ k}\Omega$, $R_3 = 39\text{ k}\Omega$, $R_4 = 470\ \Omega$, $R_5 = 22\text{ k}\Omega$

4–2 The two series circuits are connected in series, so all the resistors on the board are in series.

4–3 258 Ω (No change)

4–4 12.1 kΩ

4–5 22 kΩ

4–6 4.36 kΩ

4–7 116 mA

4–8 7.8 V

4–9 $V_1 = 1\text{ V}$, $V_2 = 3.3\text{ V}$, $V_3 = 2.2\text{ V}$, $V_S = 6.5\text{ V}$, $V_{S(max)} = 32.5\text{ V}$

4–10 $V_1 = 0\text{ V}$, $V_2 = 12\text{ V}$, $V_3 = 0\text{ V}$, $V_4 = 0\text{ V}$

4–11 12 V

4–12 2 V

4–13 10 V, 20 V

4–14 6.5 V

4–15 593 Ω

4–16 $V_1 = 3.13\text{ V}$; $V_2 = 6.87\text{ V}$

4–17 $V_1 = V_2 = V_3 = 33.3\text{ V}$

4–18 $V_{AB} = 4\text{ V}$; $V_{AC} = 36.8\text{ V}$; $V_{BC} = 32.8\text{ V}$; $V_{BD} = 46\text{ V}$; $V_{CD} = 13.2\text{ V}$

4–19 9 W

4–20 $V_A = 0\text{ V}$; $V_B = -25\text{ V}$; $V_C = -50\text{ V}$; $V_D = -75\text{ V}$; $V_E = -100\text{ V}$

4–21 3.33 V

SELF-TEST

1. (a) **2.** (d) **3.** (b) **4.** (d) **5.** (d) **6.** (a) **7.** (b)
8. (c) **9.** (b) **10.** (c) **11.** (a) **12.** (d) **13.** (d) **14.** (d)

5

PARALLEL CIRCUITS

INTRODUCTION

In this chapter, parallel circuits are introduced. Parallel circuits are found in many applications, such as lighting systems, sound systems, appliances, and most electronic equipment.

CHAPTER OUTLINE

CHAPTER OBJECTIVES

- Identify a parallel circuit
- Determine the voltage across each parallel branch
- Apply Kirchhoff's current law
- Determine total parallel resistance
- Apply Ohm's law in a parallel circuit
- Use a parallel circuit as a current divider
- Determine power in a parallel circuit
- Troubleshoot parallel circuits

KEY TERMS

- Parallel
- Branch
- Kirchhoff's current law
- Junction
- Current divider

APPLICATION ASSIGNMENT PREVIEW

As a technician in an electronic instruments company, your job is to troubleshoot any defective instrument that fails the routine check when it comes off the assembly line. In this particular assignment, you must determine the problem(s) with a defective five-range milliammeter so that it can be repaired. The knowledge of parallel circuits and of basic ammeters that you will acquire in this chapter plus your understanding of Ohm's law, current dividers, and the resistor color code will be put to good use. After studying this chapter, you should be able to complete the application assignment.

WWW. VISIT THE COMPANION WEBSITE

Circuit Simulation Tutorials and Other Chapter Study Tools Are Available at

http://www.prenhall.com/floyd

5–1 RESISTORS IN PARALLEL

When two or more resistors are individually connected between the same two points, they are in **parallel** with each other. A parallel circuit provides more than one path for current.

After completing this section, you should be able to

- **Identify a parallel circuit**
- Translate a physical arrangement of parallel resistors into a schematic

Each parallel path in a circuit is called a **branch**. Two resistors connected in parallel are shown in Figure 5–1(a). As shown in part (b), the current out of the source divides when it gets to point *A*. Part of it goes through R_1 and part through R_2. The two currents come back together at point *B*. If additional resistors are connected in parallel with the first two, more current paths are provided, as shown in Figure 5–1(c). All points along the top shown in blue are electrically the same as point *A*, and all the points along the bottom shown in green are electrically the same as point *B*.

Identifying Parallel Circuits

In Figure 5–1, it is obvious that the resistors are connected in parallel. However, in actual circuit diagrams, the parallel relationship often is not as clear. It is im-

FIGURE 5–1

Resistors in parallel.

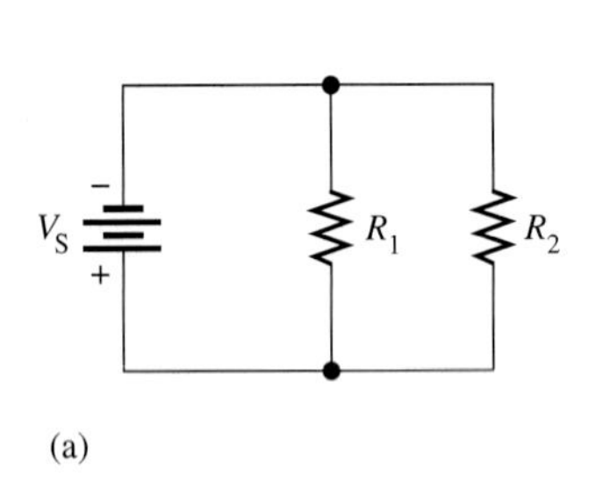

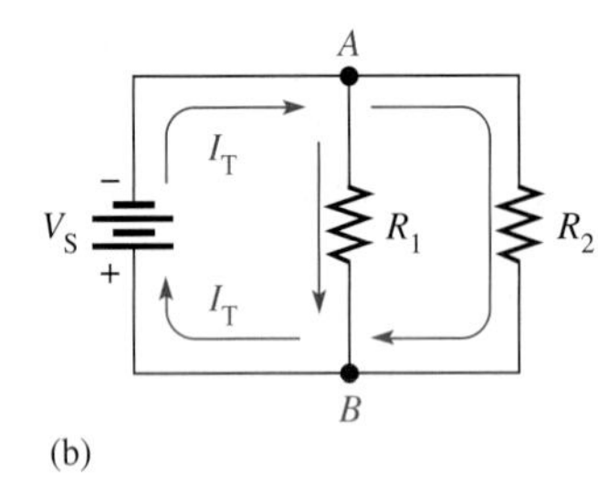

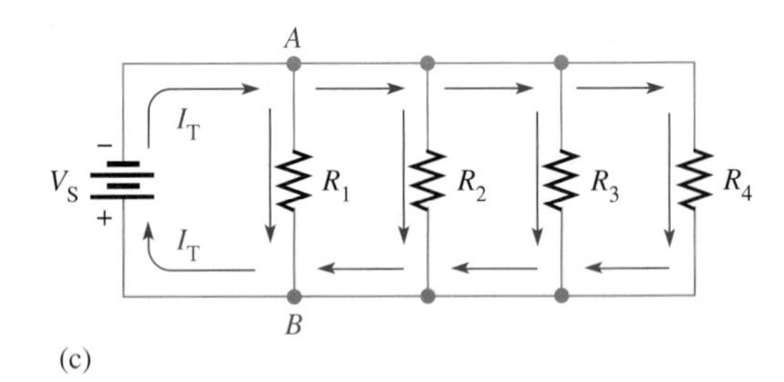

FIGURE 5–2

Examples of circuits with two parallel paths.

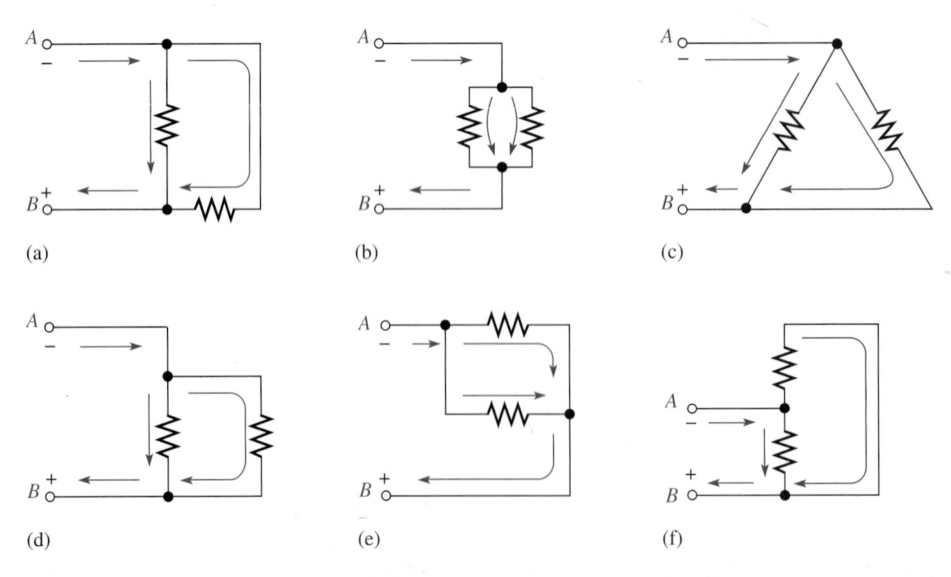

portant that you learn to recognize parallel circuits regardless of how they may be drawn.

A rule for identifying parallel circuits is as follows:

If there is more than one current path (branch) between two points, and if the voltage between those two points also appears across each of the branches, then there is a parallel circuit between those two points.

Figure 5–2 shows parallel resistors drawn in different ways between two points labeled *A* and *B*. Notice that in each case, the current "travels" two paths going from *A* to *B*, and the voltage across each branch is the same. Although these figures show only two parallel paths, there can be any number of resistors in parallel.

EXAMPLE 5–1

Five resistors are positioned on a circuit board as shown in Figure 5–3. Show the wiring required to connect all the resistors in parallel. Draw a schematic and label each of the resistors with its value.

FIGURE 5–3

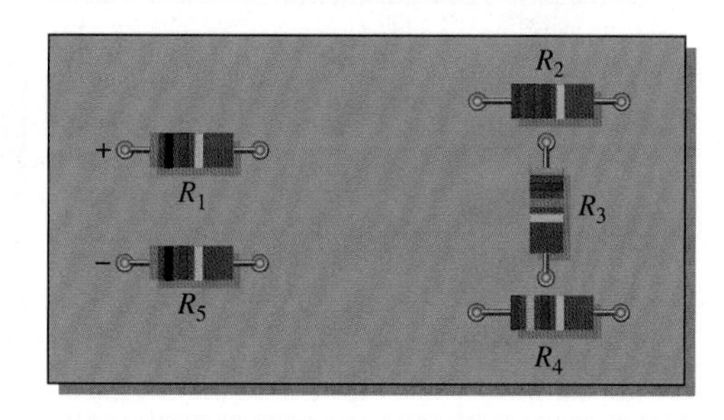

Solution Wires are connected as shown in the assembly wiring diagram of Figure 5–4(a). The schematic is shown in Figure 5–4(b) with resistance values from the color codes. Again, note that the schematic does not necessarily have to show the actual physical arrangement of the resistors. The schematic shows how components are connected electrically.

FIGURE 5–4

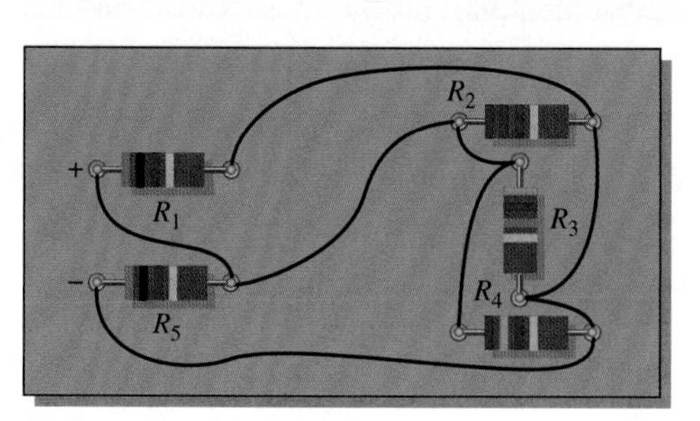

(a) Assembly wiring diagram

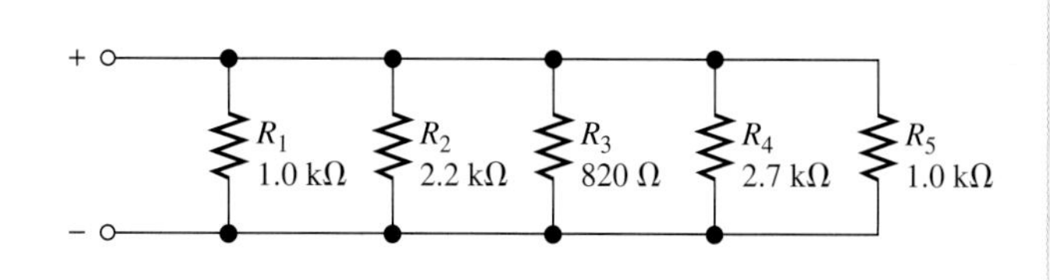

(b) Schematic

Related Problem* Would the circuit have to be rewired if R_2 is removed?

*Answers are at the end of the chapter.

EXAMPLE 5–2

Determine the parallel groupings in Figure 5–5 and the value of each resistor.

▶ FIGURE 5–5

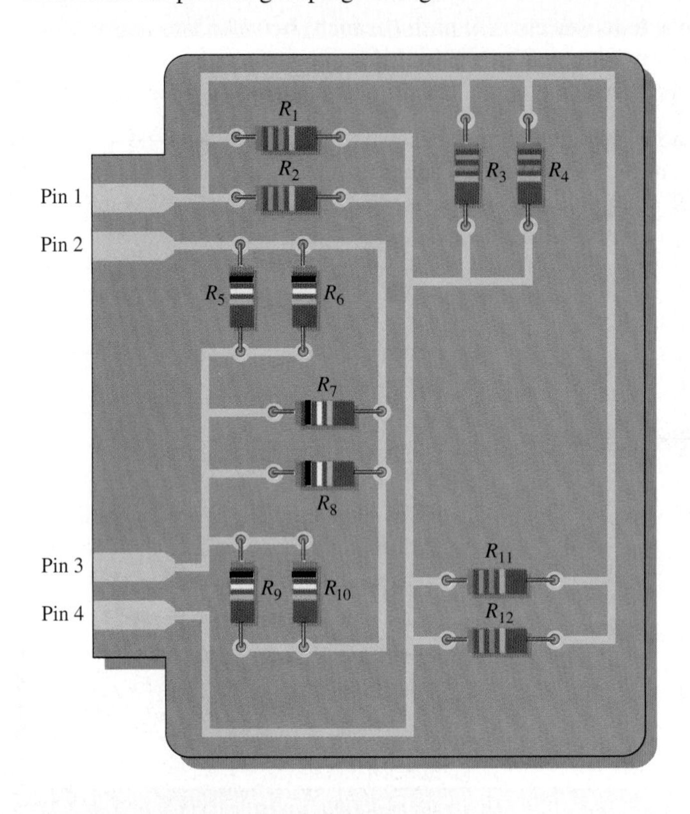

Solution Resistors R_1 through R_4 and R_{11} and R_{12} are all in parallel. This parallel combination is connected to pins 1 and 4. Each resistor in this group is 56 kΩ.

Resistors R_5 through R_{10} are all in parallel. This combination is connected to pins 2 and 3. Each resistor in this group is 100 kΩ.

Related Problem How would you connect all the resistors on the PC board in parallel?

SECTION 5–1 REVIEW

Answers are at the end of the chapter.

1. How are the resistors connected in a parallel circuit?
2. How do you identify a parallel circuit?
3. Complete the schematics for the circuits in each part of Figure 5–6 by connecting the resistors in parallel between points *A* and *B*.
4. Connect each group of parallel resistors in Figure 5–6 in parallel with each other.

▶ FIGURE 5–6

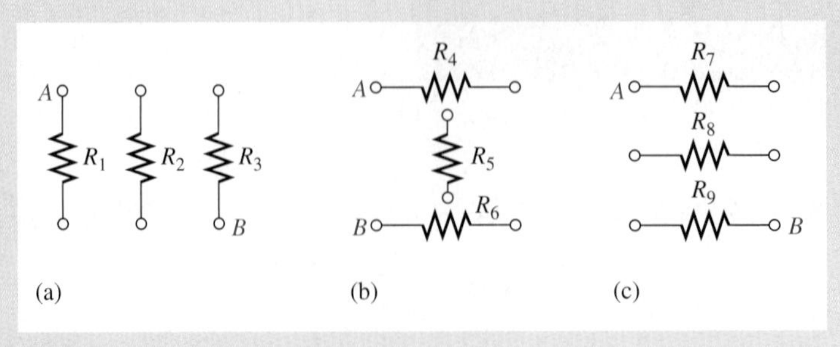

5–2 VOLTAGE IN PARALLEL CIRCUITS

As mentioned in the previous section, each current path in a parallel circuit is called a branch. The voltage across any given branch of a parallel circuit is equal to the voltage across each of the other branches in parallel.

After completing this section, you should be able to

- **Determine the voltage across each parallel branch**
- Explain why the voltage is the same across all parallel resistors.

To illustrate voltage in a parallel circuit, let's examine Figure 5–7(a). Points *A*, *B*, *C*, and *D* along the left side of the parallel circuit are electrically the same point because the voltage is the same along this line. You can think of all of these points as being connected by a single wire to the negative terminal of the battery. The points *E*, *F*, *G*, and *H* along the right side of the circuit are all at a voltage equal to that of the positive terminal of the source. Thus, voltage across each parallel resistor is the same, and each is equal to the source voltage.

Figure 5–7(b) is the same circuit as in part (a), drawn in a slightly different way. Here the left side of each resistor is connected to a single point, which is the negative battery terminal. The right side of each resistor is connected to a single point, which is the positive battery terminal. The resistors are still all in parallel across the source.

In Figure 5–8, a 12 V battery is connected across three parallel resistors. When the voltage is measured across the battery and then across each of the resistors, the readings are the same. As you can see, the same voltage appears across each branch in a parallel circuit.

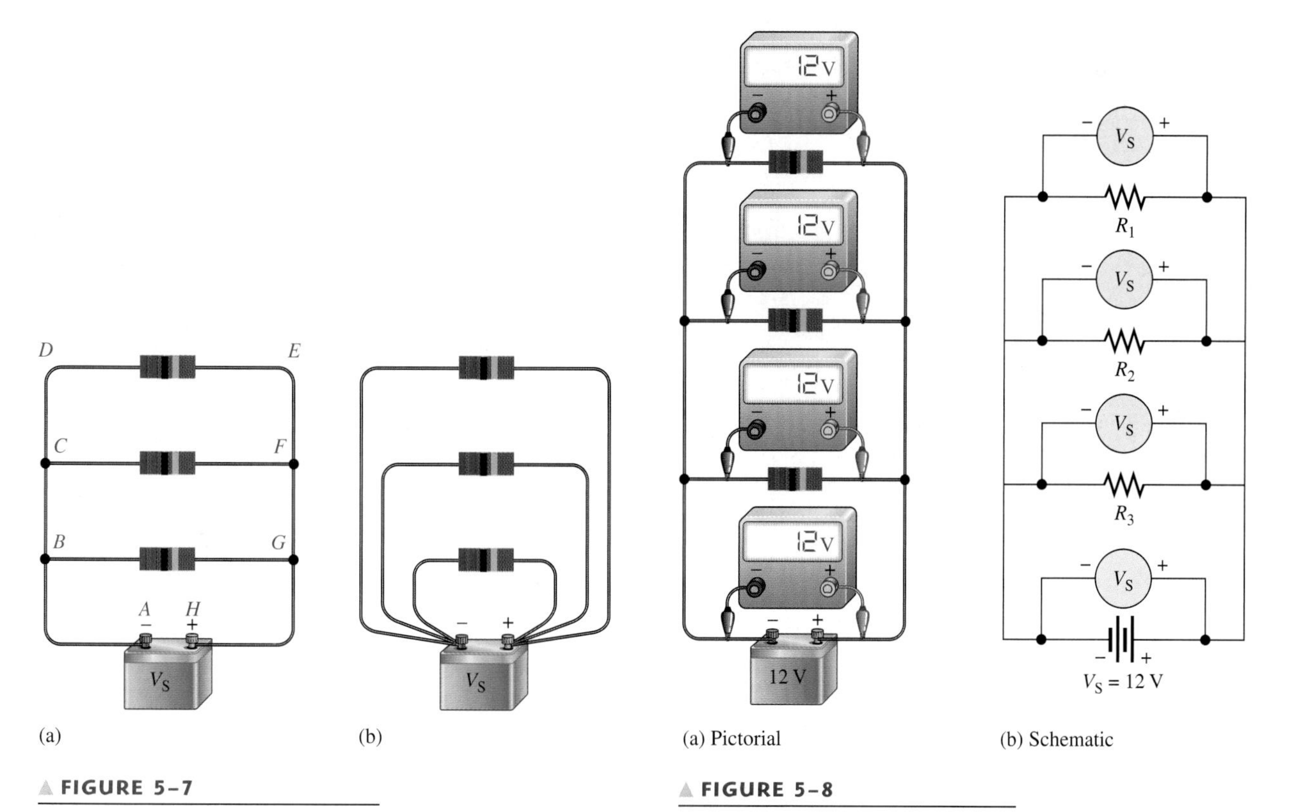

▲ **FIGURE 5–7**

Voltage across parallel branches is the same.

▲ **FIGURE 5–8**

The same voltage appears across each resistor in parallel.

EXAMPLE 5–3

Determine the voltage across each resistor in Figure 5–9.

▶ FIGURE 5–9

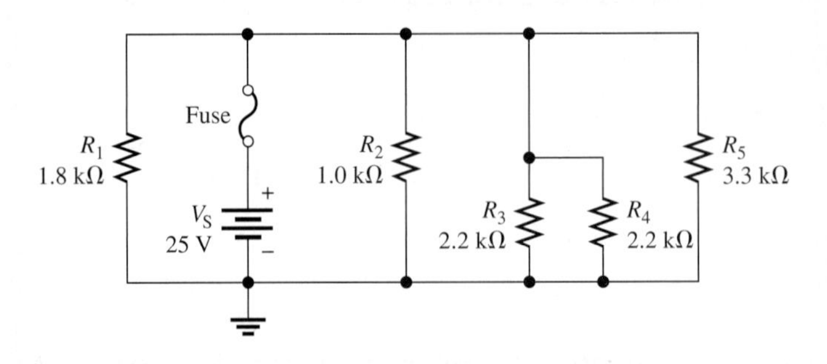

Solution The five resistors are in parallel; so the voltage across each one is equal to the source voltage, V_S. There is no voltage across the fuse.

$$V_1 = V_2 = V_3 = V_4 = V_5 = V_S = \mathbf{25\ V}$$

Related Problem If R_4 is removed from the circuit, what is the voltage across R_3?

SECTION 5–2 REVIEW

1. A 10 Ω and a 22 Ω resistor are connected in parallel with a 5 V source. What is the voltage across each of the resistors?
2. A voltmeter connected across R_1 in Figure 5–10 measures 118 V. If you move the meter and connect it across R_2, how much voltage will it indicate? What is the source voltage?
3. In Figure 5–11, how much voltage does voltmeter 1 indicate? Voltmeter 2?
4. How are voltages across each branch of a parallel circuit related?

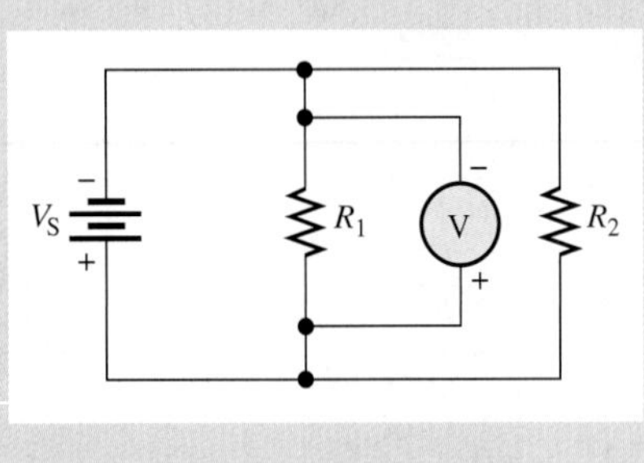

▲ FIGURE 5–10

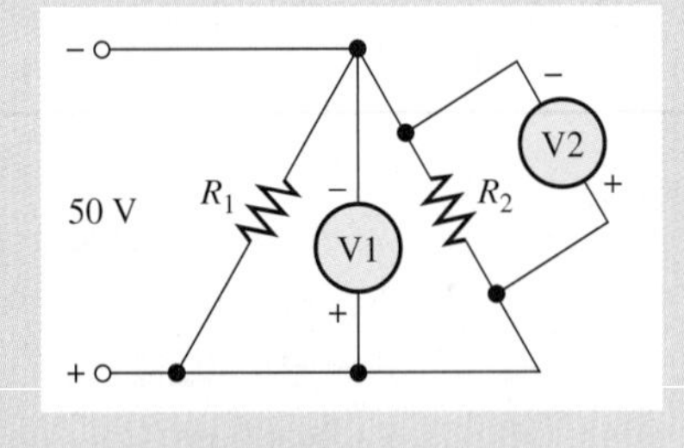

▲ FIGURE 5–11

5–3 KIRCHHOFF'S CURRENT LAW

In the last chapter, you learned Kirchhoff's voltage law that dealt with voltages in a closed series circuit. Now, you will learn Kirchhoff's current law that deals with currents in a parallel circuit.

After completing this section, you should be able to

- **Apply Kirchhoff's current law**
- State Kirchhoff's current law

- Determine the total current by adding the branch currents
- Determine an unknown branch current

Kirchhoff's current law, often abbreviated KCL, is stated as follows:

The sum of the currents into a junction (total current in) is equal to the sum of the currents out of that junction (total current out).

A junction is any point in a circuit where two or more components are connected. In a parallel circuit, a junction is a point where the parallel branches come together. For example, in the circuit of Figure 5–12, point A is one junction and point B is another. Let's start at the negative terminal of the source and follow the current. The total current I_T from the source is *into* the junction at point A. At this point, the current splits up among the three branches as indicated. Each of the three branch currents (I_1, I_2 and I_3) is *out of* junction A. Kirchhoff's current law says that the total current into junction A is equal to the total current out of junction A; that is,

$$I_T = I_1 + I_2 + I_3$$

Now, following the currents in Figure 5–12 through the three branches, you see that they come back together at point B. Currents I_1, I_2, and I_3 are into junction B, and I_T is out of junction B. Kirchhoff's current law formula at this junction is therefore the same as at junction A.

$$I_1 + I_2 + I_3 = I_T$$

General Formula for Kirchhoff's Current Law

The previous discussion used a specific case to illustrate Kirchhoff's current law. Figure 5–13 shows a generalized circuit junction where a number of branches are connected to a point in the circuit. Currents $I_{IN(1)}$ through $I_{IN(n)}$ are into the junction (n can be any number). Currents $I_{OUT(1)}$ through $I_{OUT(m)}$ are out of the junction (m can be any number but not necessarily equal to n).

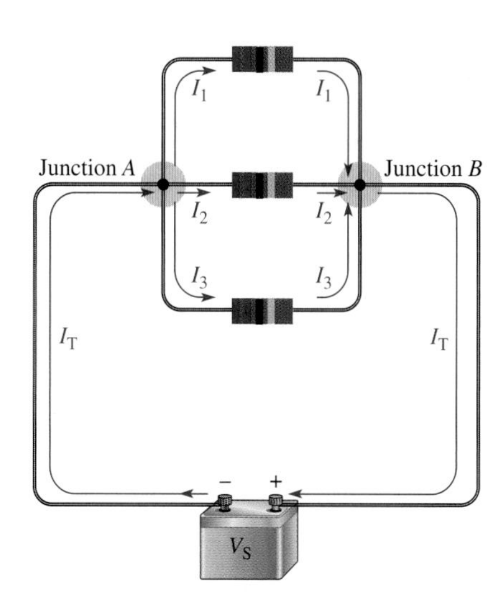

▲ **FIGURE 5–12**

Kirchhoff's current law: The current into a junction equals the current out of that junction.

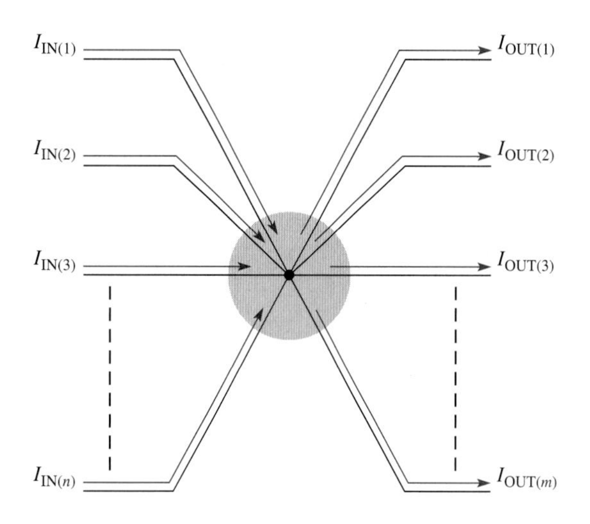

$$I_{IN(1)} + I_{IN(2)} + I_{IN(3)} + \cdots + I_{IN(n)} = I_{OUT(1)} + I_{OUT(2)} + I_{OUT(3)} + \cdots I_{OUT(m)}$$

▲ **FIGURE 5–13**

Generalized circuit junction illustrates Kirchhoff's current law.

By Kirchhoff's current law, the sum of the currents into a junction must equal the sum of the currents out of the junction. With reference to Figure 5–13, the general formula for Kirchhoff's current law is

Equation 5–1

$$I_{IN(1)} + I_{IN(2)} + I_{IN(3)} + \cdots + I_{IN(n)} = I_{OUT(1)} + I_{OUT(2)} + I_{OUT(3)} + \cdots + I_{OUT(m)}$$

If all of the terms on the right side of Equation 5–1 are brought over to the left side, their signs change to negative, and a zero is left on the right side.

$$I_{IN(1)} + I_{IN(2)} + I_{IN(3)} + \cdots + I_{IN(n)} - I_{OUT(1)} - I_{OUT(2)} - I_{OUT(3)} - \cdots - I_{OUT(m)} = 0$$

Kirchhoff's current law can also be stated in this way:

The algebraic sum of all the currents entering and leaving a junction is equal to zero.

You can verify Kirchhoff's current law by connecting a circuit and measuring each branch current and the total current from the source, as illustrated in Figure 5–14. *When the branch currents are added together, their sum will equal the total current.* This rule applies for any number of branches.

The following three examples illustrate use of Kirchhoff's current law.

▶ **FIGURE 5–14**

Illustration of a verification of Kirchhoff's current law.

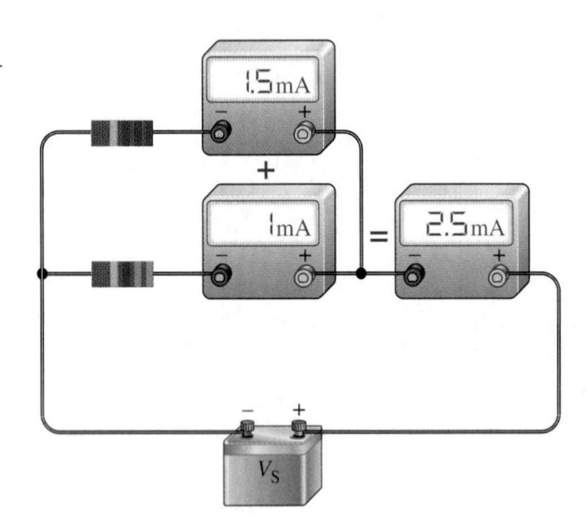

EXAMPLE 5–4

You know the branch currents in the circuit of Figure 5–15. Determine the total current entering junction *A* and the total current leaving junction *B*.

▶ **FIGURE 5–15**

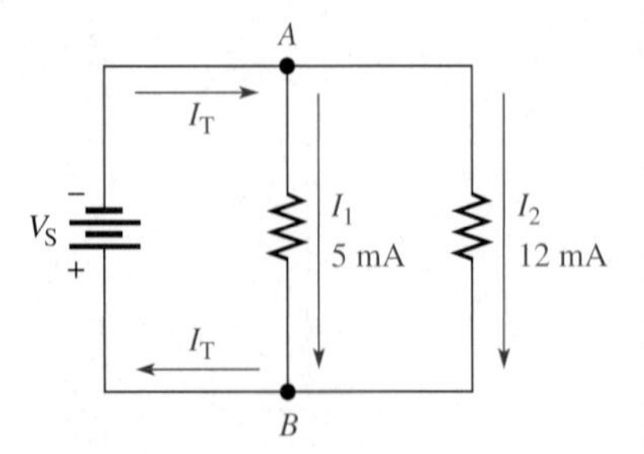

Solution The total current out of junction *A* is the sum of the two branch currents. So the total current into *A* is

$$I_T = I_1 + I_2 = 5\text{ mA} + 12\text{ mA} = \mathbf{17\text{ mA}}$$

The total current entering point B is the sum of the two branch currents. So the total current out of B is

$$I_T = I_1 + I_2 = 5\text{ mA} + 12\text{ mA} = \mathbf{17\text{ mA}}$$

5 EE (−) 3 + 1 2 EE (−) 3 ENTER

5E⁻3+12E⁻3
17E⁻3

Related Problem If a third resistor is connected in parallel to the circuit of Figure 5–15, and its current is 3 mA, what is the total current into junction A and out of junction B?

EXAMPLE 5–5

Determine the current through R_2 in Figure 5–16.

▶ FIGURE 5–16

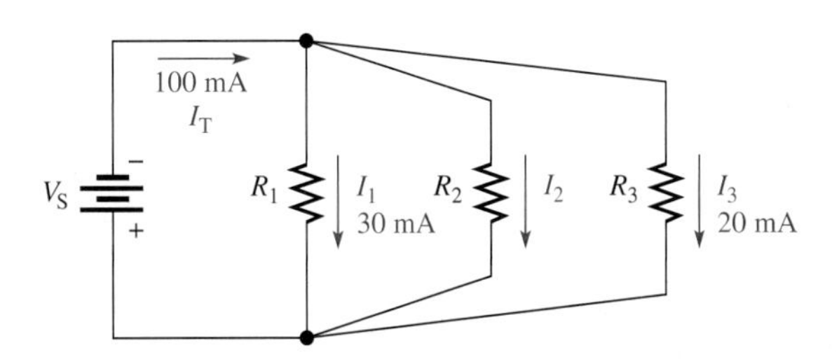

Solution The total current into the junction of the three branches is $I_T = I_1 + I_2 + I_3$. From Figure 5–16, you know the total current and the branch currents through R_1 and R_3. Solve for I_2.

$$I_2 = I_T - I_1 - I_3 = 100\text{ mA} - 30\text{ mA} - 20\text{ mA} = \mathbf{50\text{ mA}}$$

Related Problem Determine I_T and I_2 if a fourth branch is added to the circuit in Figure 5–16 and it has 12 mA through it.

EXAMPLE 5–6

Use Kirchhoff's current law to find the current measured by ammeters A3 and A5 in Figure 5–17.

▶ FIGURE 5–17

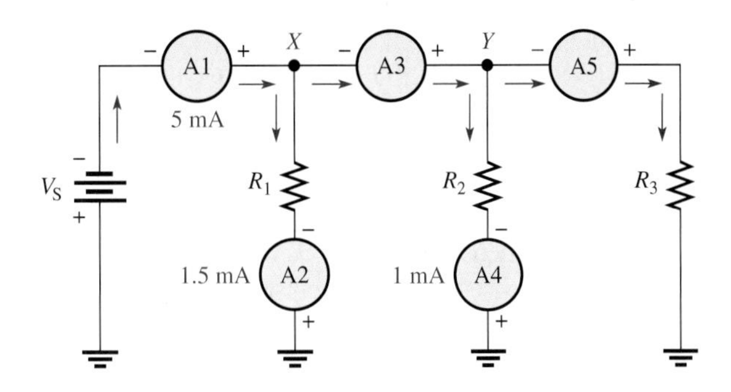

Solution The total current into junction X is 5 mA. Two currents are out of junction X: 1.5 mA through resistor R_1 and the current through A3. Kirchhoff's current law at junction X is

$$5 \text{ mA} = 1.5 \text{ mA} + I_{A3}$$

Solving for I_{A3} yields

$$I_{A3} = 5 \text{ mA} - 1.5 \text{ mA} = \mathbf{3.5 \text{ mA}}$$

The total current into junction Y is $I_{A3} = 3.5$ mA. Two currents are out of junction Y: 1 mA through resistor R_2 and the current through A5 and R_3. Kirchhoff's current law applied to junction Y gives

$$3.5 \text{ mA} = 1 \text{ mA} + I_{A5}$$

Solving for I_{A5} yields

$$I_{A5} = 3.5 \text{ mA} - 1 \text{ mA} = \mathbf{2.5 \text{ mA}}$$

Related Problem How much current will an ammeter measure when it is placed in the circuit in Figure 5–17 right below R_3? Below the positive battery terminal?

SECTION 5–3 REVIEW

1. State Kirchhoff's current law in two ways.
2. A total current of 2.5 A is into the junction of three parallel branches. What is the sum of all three branch currents?
3. In Figure 5–18, 100 mA and 300 mA are into the junction. What is the amount of current out of the junction?
4. Determine I_1 in the circuit of Figure 5–19.
5. Two branch currents enter a junction, and two branch currents leave the same junction. One of the currents entering the junction is 1 A, and one of the currents leaving the junction is 3 A. The total current entering and leaving the junction is 8 A. Determine the value of the unknown current entering the junction and the value of the unknown current leaving the junction.

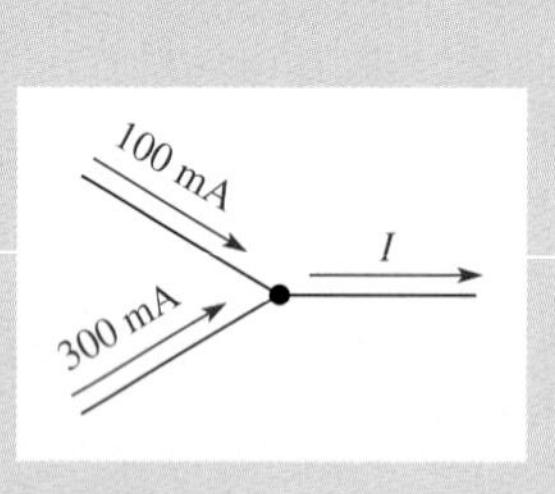

FIGURE 5–18

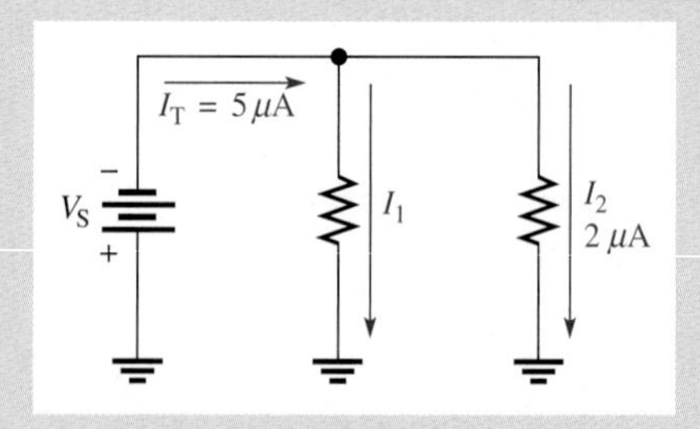

FIGURE 5–19

5–4 TOTAL PARALLEL RESISTANCE

When resistors are connected in parallel, the total resistance of the circuit decreases. The total resistance of a parallel circuit is always less than the value of the smallest resistor. For example, if a 10 Ω resistor and a 100 Ω resistor are connected in parallel, the total resistance is less than 10 Ω.

Calculator Solution The parallel-resistance formula is easily solved on a calculator. The general procedure is to enter the value of R_1 and then take its reciprocal by pressing the 2nd x^{-1} keys (x^{-1} is not a secondary function on some calculators). The notation x^{-1} means $1/x$. Next press the + key; then enter the value of R_2 and take its reciprocal. Repeat this procedure until all of the resistor values have been entered and the reciprocal of each has been added. The final step is to press the 2nd x^{-1} keys to convert $1/R_T$ to R_T. The total parallel resistance is now on the display.

EXAMPLE 5–8

Show the steps required for a calculator (TI-85/TI-86) solution of Example 5–7.

Solution

Step 1: Enter 100. Display shows 100.

Step 2: Press 2nd x^{-1} keys. Display shows 100^{-1} (which is 1/100).

Step 3: Press + key. Display shows 100^{-1} +.

Step 4: Enter 47. Display shows $100^{-1} + 47$.

Step 5: Press 2nd x^{-1} keys. Display shows $100^{-1} + 47^{-1}$.

Step 6: Press + key. Display shows $100^{-1} + 47^{-1}$ +.

Step 7: Enter 22. Display shows $100^{-1} + 47^{-1} + 22$.

Step 8: Press 2nd x^{-1} keys. Display shows $100^{-1} + 47^{-1} + 22^{-1}$.

Step 9: Press ENTER key. Display shows $76.7311411992\text{E}^{-3}$.

Step 10: Press 2nd x^{-1} ENTER keys. Display shows 13.0325182758E0.

The number displayed in Step 10 is the total resistance in ohms. Round it to **13.0 Ω**.

Related Problem Show the additional calculator steps for R_T when a 33 Ω resistor is placed in parallel in the circuit of Example 5–7.

The Case of Two Resistors in Parallel Equation 5–4 is a general formula for finding the total resistance of any number of resistors in parallel. It is often useful to consider only two resistors in parallel because this situation occurs commonly in practice. Also, any number of resistors in parallel can be broken down into pairs as an alternate way to find the R_T.

Derived from Equation 5–4, the formula for two resistors in parallel is

$$R_T = \frac{R_1 R_2}{R_1 + R_2}$$

Equation 5–5

Equation 5–5 states *the total resistance of two resistors in parallel is equal to the product of the two resistors divided by the sum of the two resistors.* This equation is sometimes referred to as the "product over the sum" formula. Example 5–9 illustrates how to use it.

EXAMPLE 5–9

Calculate the total resistance connected to the voltage source of the circuit in Figure 5–23.

▶ FIGURE 5–23

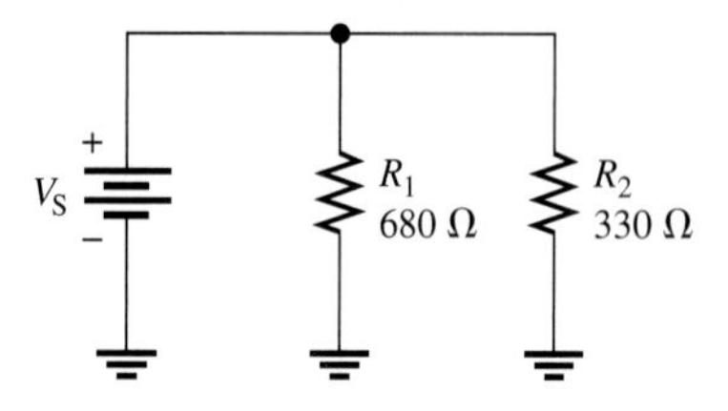

Solution Use Equation 5–5.

$$R_T = \frac{R_1 R_2}{R_1 + R_2} = \frac{(680\ \Omega)(330\ \Omega)}{680\ \Omega + 330\ \Omega} = \frac{224{,}000\ \Omega^2}{1{,}010\ \Omega} = \mathbf{222\ \Omega}$$

680∗330/(680+330)
222.178217822E0

Related Problem Find R_T if a 220 Ω resistor replaces R_1 in Figure 5–23.

The Case of Equal-Value Resistors in Parallel Another special case of parallel circuits is the parallel connection of several resistors having the same value. The following is a shortcut method of calculating R_T when this case occurs:

Equation 5–6

$$R_T = \frac{R}{n}$$

Equation 5–6 says that when any number of resistors (n), all having the same resistance (R), are connected in parallel, R_T is equal to the resistance divided by the number of resistors in parallel. Examples 5–10 and 5–11 show how to use this formula.

EXAMPLE 5–10

Find the total resistance between points A and B in Figure 5–24.

▶ FIGURE 5–24

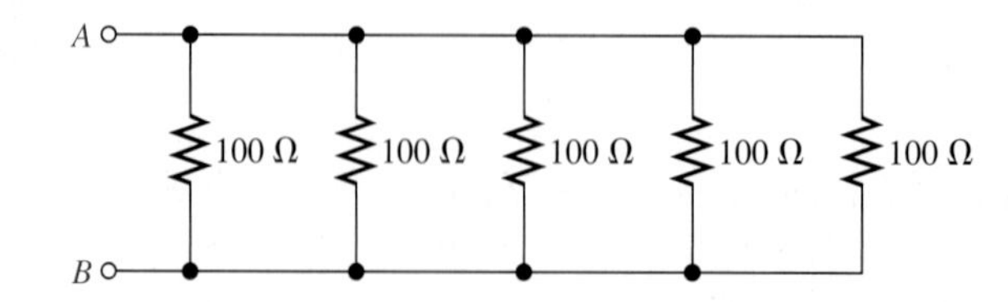

Solution There are five 100 Ω resistors in parallel. Use Equation 5–6.

$$R_T = \frac{R}{n} = \frac{100\ \Omega}{5} = \mathbf{20\ \Omega}$$

Related Problem Find R_T for three 100 kΩ resistors in parallel.

EXAMPLE 5–11

A stereo amplifier drives two 8 Ω speakers in parallel from each channel as shown in Figure 5–25. What is the total resistance across the output terminals of each channel of the amplifier?

FIGURE 5–25

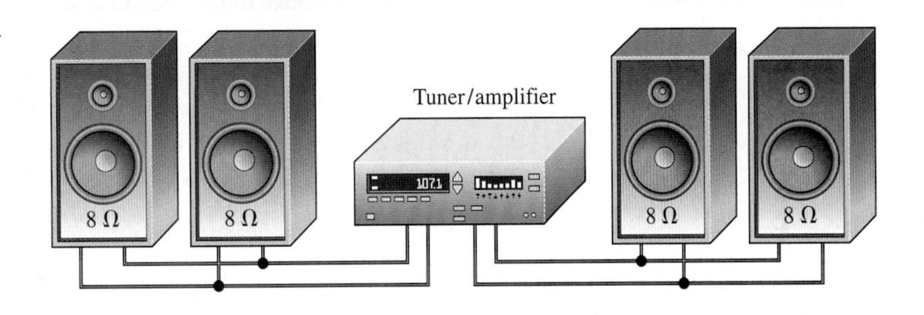

Solution The total parallel resistance across each channel output with the two 8 Ω speakers is

$$R_T = \frac{R}{n} = \frac{8\ \Omega}{2} = \mathbf{4\ \Omega}$$

Related Problem If two additional 8 Ω speakers are added in parallel to each channel, what is the resistance across each channel output?

Notation for Parallel Resistors Sometimes, for convenience, parallel resistors are designated by two parallel vertical marks. For example, R_1 in parallel with R_2 can be written as $R_1 \| R_2$. Also, when several resistors are in parallel with each other, this notation can be used. For example,

$$R_1 \| R_2 \| R_3 \| R_4 \| R_5$$

indicates that R_1 through R_5 are all in parallel.

This notation is also used with resistance values. For example,

$$10\ \text{k}\Omega \| 5\ \text{k}\Omega$$

means that a 10 kΩ resistor is in parallel with a 5 kΩ resistor.

Applications of a Parallel Circuit

Automotive One advantage of a parallel circuit over a series circuit is that when one branch opens, the other branches are not affected. For example, Figure 5–26 shows a simplified diagram of an automobile lighting system. When one headlight on your car goes out, it does not cause the other lights to go out because they are all in parallel.

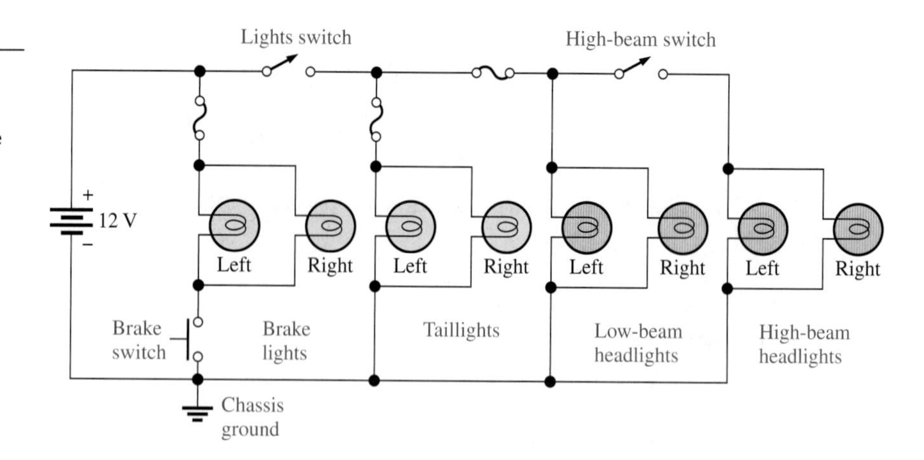

▶ **FIGURE 5–26**

Simplified diagram of the exterior light system of an automobile. All lights are off when the switches are in the positions shown.

Notice that the brake lights are switched on independently of the headlights and taillights. They come on only when the driver closes the brake light switch by depressing the brake pedal. When the lights switch is closed, both low-beam headlights and both taillights are on. The high-beam headlights are on only when both the lights switch and the high-beam switch are closed. If any one of the lights burns out (opens), there is still current in each of the other lights.

Residential Another common use of parallel circuits is in residential electrical systems. All the lights and appliances in a home are wired in parallel. Figure 5–27(a) shows a typical room wiring arrangement with two switch-controlled lights and three wall outlets in parallel.

Figure 5–27(b) shows a simplified parallel arrangement of four heating elements in an electric range. The four-position switches in each branch allow the user to control the amount of current through the heating elements by selecting the appropriate series-limiting resistor. The lowest resistor value (H setting) allows the highest amount of current for maximum heat. The highest resistor value (L setting) allows the least amount of current for minimum heat; M designates the medium settings.

▶ **FIGURE 5–27**

Examples of parallel circuits in residential wiring and appliances.

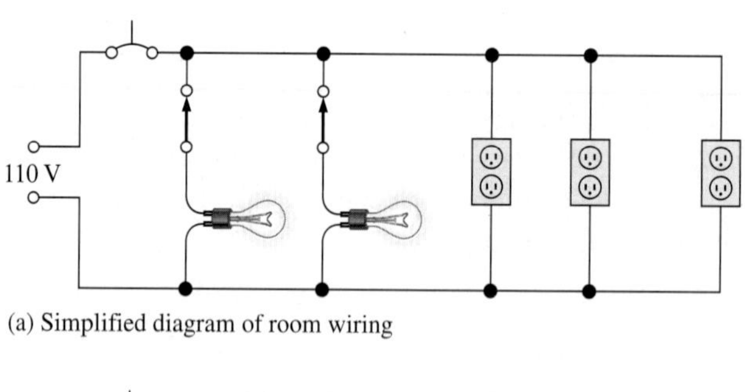

(a) Simplified diagram of room wiring

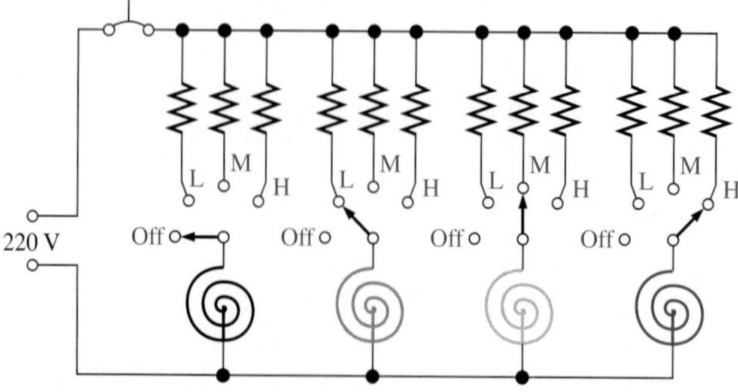

(b) Simplified diagram of a four-burner range

SECTION 5–4 REVIEW

1. Does the total resistance increase or decrease as more resistors are connected in parallel?
2. The total parallel resistance is always less than ______.
3. Determine R_T (between pin 1 and pin 4) for the circuit in Figure 5–28. Note that pins 1 and 2 are connected and pins 3 and 4 are connected.

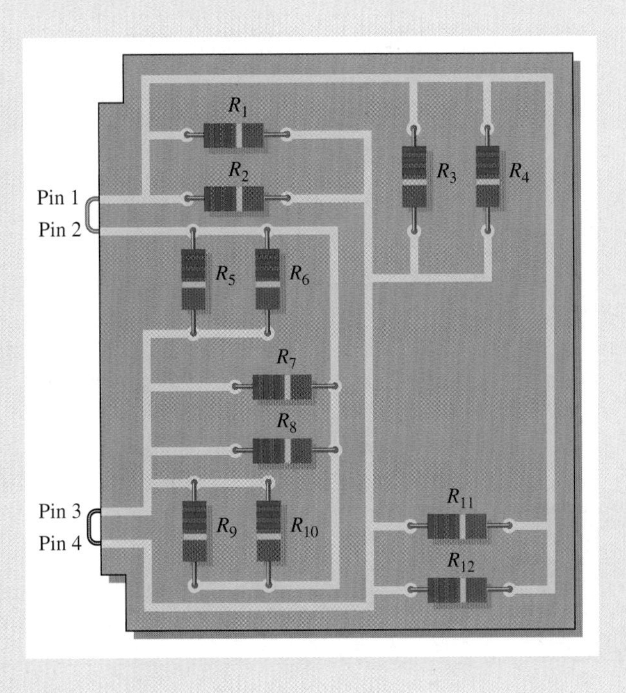

▶ FIGURE 5–28

5–5 OHM'S LAW IN PARALLEL CIRCUITS

In this section, you will see how Ohm's law can be applied to parallel circuit analysis.

After completing this section, you should be able to

- **Apply Ohm's law in a parallel circuit**
- Find the total current in a parallel circuit
- Find each branch current in a parallel circuit
- Find the voltage across a parallel circuit
- Find the resistance of a parallel circuit

The following examples illustrate how to apply Ohm's law to determine the total current, branch currents, voltage, and resistance in parallel circuits.

EXAMPLE 5–12

Find the total current produced by the battery in Figure 5–29.

▶ FIGURE 5–29

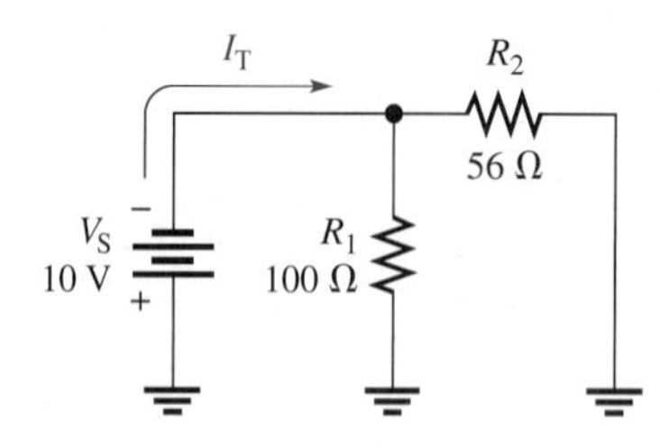

Solution The battery "sees" a total parallel resistance that determines the amount of current that it generates. First, calculate R_T.

$$R_T = \frac{R_1R_2}{R_1 + R_2} = \frac{(100\ \Omega)(56\ \Omega)}{100\ \Omega + 56\ \Omega} = \frac{5600\ \Omega^2}{156\ \Omega} = 35.9\ \Omega$$

The battery voltage is 10 V. Use Ohm's law to find I_T.

$$I_T = \frac{V_S}{R_T} = \frac{10\ \text{V}}{35.9\ \Omega} = \mathbf{279\ mA}$$

Related Problem Find the currents through R_1 and R_2 in Figure 5–29. Show that the sum of the currents through R_1 and R_2 equals the total current.

Open file E05-12 on your EWB/CircuitMaker CD-ROM. Using the multimeter, verify the calculated values of total current and the branch currents.

EXAMPLE 5–13

Determine the current through each resistor in the parallel circuit of Figure 5–30.

▶ FIGURE 5–30

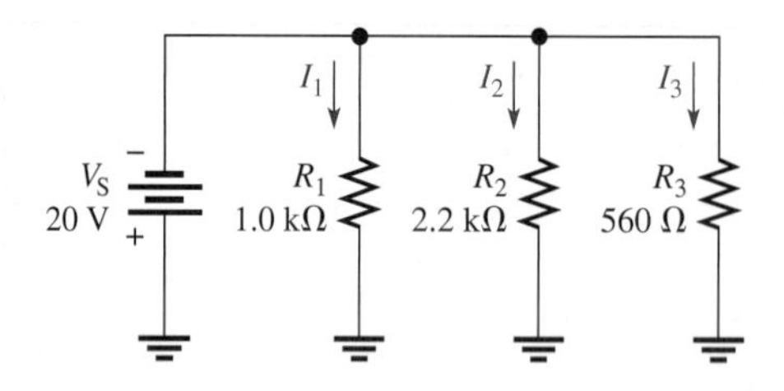

Solution The voltage across each resistor (branch) is equal to the source voltage. That is, the voltage across R_1 is 20 V, the voltage across R_2 is 20 V, and the voltage across R_3 is 20 V. The current through each resistor is determined as follows:

$$I_1 = \frac{V_S}{R_1} = \frac{20\ \text{V}}{1.0\ \text{k}\Omega} = \mathbf{20.0\ mA}$$

$$I_2 = \frac{V_S}{R_2} = \frac{20\ \text{V}}{2.2\ \text{k}\Omega} = \mathbf{9.09\ mA}$$

$$I_3 = \frac{V_S}{R_3} = \frac{20\ \text{V}}{560\ \Omega} = \mathbf{35.7\ mA}$$

Related Problem If an additional resistor of 910 Ω is connected in parallel to the circuit in Figure 5–30, determine all the branch circuits.

Open file E05-13 on your EWB/CircuitMaker CD-ROM. Measure the current through each resistor. Connect a 910 Ω resistor in parallel with the other resistors and measure the branch currents. How much does the total current from the source change when the new resistor is added?

EXAMPLE 5–14

Find the voltage across the parallel circuit found in Figure 5–31.

FIGURE 5–31

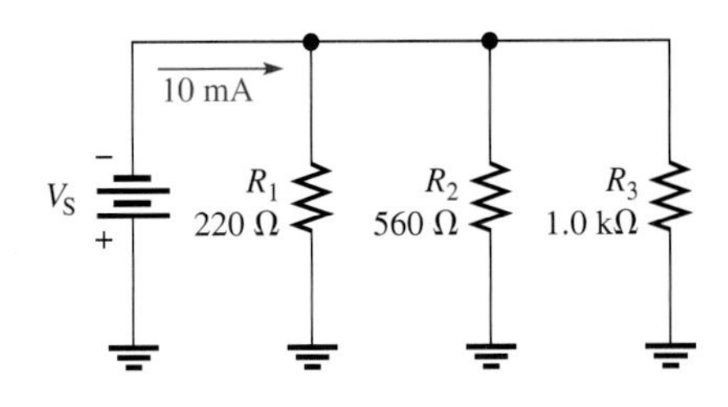

Solution The total current into the parallel circuit is 10 mA. If you know the total resistance, then you can apply Ohm's law to get the voltage. The total resistance is

$$R_T = \frac{1}{\frac{1}{R_1} + \frac{1}{R_2} + \frac{1}{R_3}} = \frac{1}{\frac{1}{220\ \Omega} + \frac{1}{560\ \Omega} + \frac{1}{1.0\ \text{k}\Omega}}$$

$$= \frac{1}{4.55\ \text{mS} + 1.79\ \text{mS} + 1\ \text{mS}} = \frac{1}{7.34\ \text{mS}} = 136\ \Omega$$

Therefore, the source voltage and the voltage across each branch is

$$V_S = I_T R_T = (10\ \text{mA})(136\ \Omega) = \mathbf{1.36\ V}$$

Related Problem Find the total current if R_3 opens in Figure 5–31. Assume V_S remains the same.

EXAMPLE 5–15

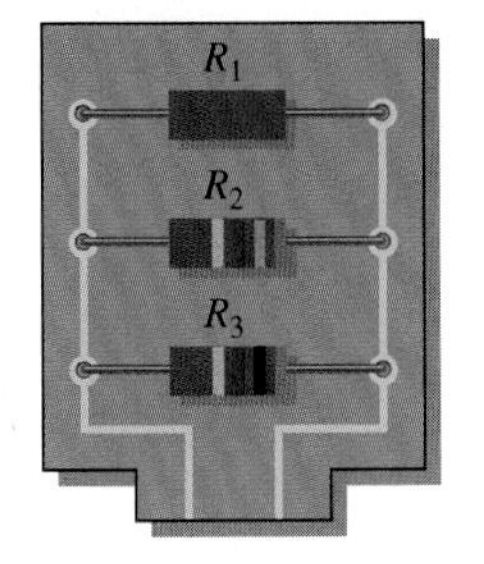

FIGURE 5–32

The circuit board in Figure 5–32 has three resistors in parallel used for bias modification in an instrumentation amplifier. The values of two of the resistors are known from the color codes, but the top resistor is not clearly marked. Determine the value of the unknown resistor without disconnecting it from the circuit board or without using an ohmmeter.

▲ FIGURE 5–33

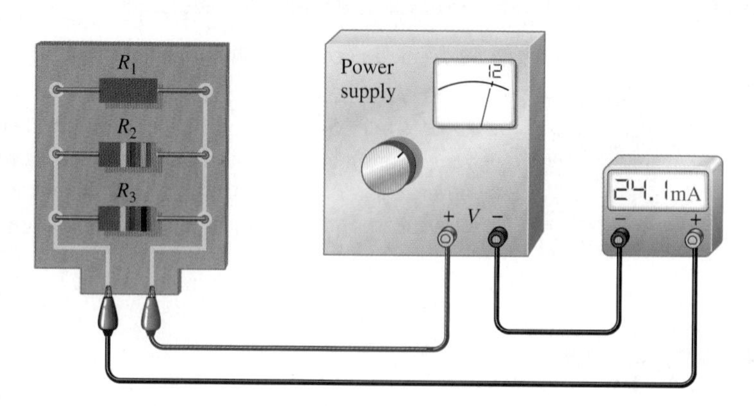

Solution If you can determine the total resistance of the three resistors in parallel, then you can use the parallel-resistance formula to calculate the unknown resistance. You can use Ohm's law to find the total resistance if voltage and total current are known.

In Figure 5–33, a 12 V source (arbitrary value) is connected across the resistors and the total current is measured. Using these measured values, find the total resistance.

$$R_T = \frac{V}{I_T} = \frac{12\text{ V}}{24.1\text{ mA}} = 498\ \Omega$$

Use Equation 5–2 to find the unknown resistance.

$$\frac{1}{R_T} = \frac{1}{R_1} + \frac{1}{R_2} + \frac{1}{R_3}$$

$$\frac{1}{R_1} = \frac{1}{R_T} - \frac{1}{R_2} - \frac{1}{R_3} = \frac{1}{498\ \Omega} - \frac{1}{1.8\text{ k}\Omega} - \frac{1}{1.0\text{ k}\Omega} = 452\ \mu\text{S}$$

$$R_1 = \frac{1}{452\ \mu\text{S}} = \mathbf{2.21\ k\Omega}$$

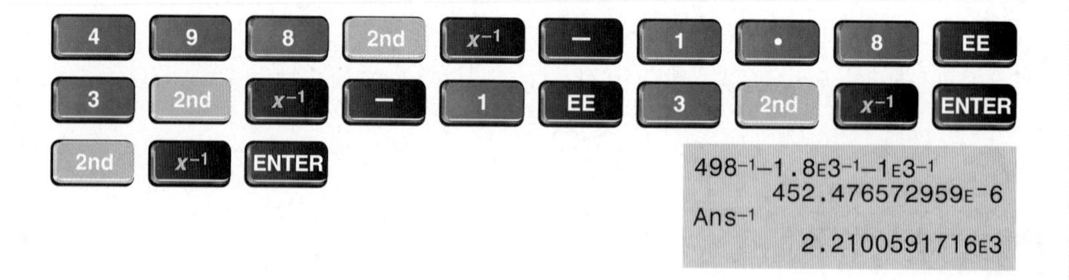

Related Problem Explain how to determine the value of R_1 using an ohmmeter and without removing R_1 from the circuit.

Open file E05-15 on your EWB/CircuitMaker CD-ROM. This is a schematic of the circuit board in Figure 5–33. Connect the circuit as shown and verify the current measurement. Remove R_1 from the circuit and verify its value with the ohmmeter.

SECTION 5–5 REVIEW

1. A 10 V battery is connected across three 68 Ω resistors that are in parallel. What is the total current from the battery?
2. How much voltage is required to produce 20 mA of current through the circuit of Figure 5–34?
3. How much current is there through each resistor of Figure 5–34?
4. There are four equal-value resistors in parallel with a 12 V source, and 6 mA of current from the source. What is the value of each resistor?
5. A 1.0 kΩ and a 2.2 kΩ resistor are connected in parallel. There is a total of 100 mA through the parallel combination. How much voltage is dropped across the resistors?

FIGURE 5–34

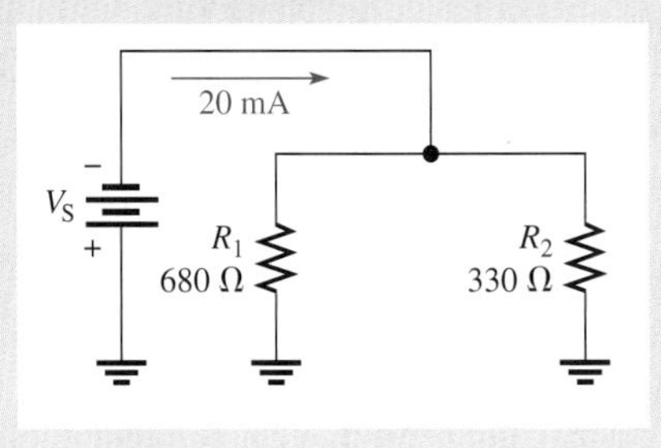

5–6 CURRENT DIVIDERS

A parallel circuit acts as a current divider because the current entering the junction of parallel branches "divides" up into several individual branch currents.

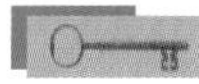

After completing this section, you should be able to

- **Use a parallel circuit as a current divider**
- Apply the current-divider formula
- Determine an unknown branch current

In a parallel circuit, the total current into the junction of the parallel branches divides among the branches. Thus, a parallel circuit acts as a current divider. This current divider principal is illustrated in Figure 5–35 for a two-branch parallel circuit in which part of the total current I_T goes through R_1 and part through R_2.

Since the same voltage is across each of the resistors in parallel, the branch currents are inversely proportional to the values of the resistors. For example, if the value of R_2 is twice that of R_1, then the value of I_2 is one-half that of I_1. In other words,

The total current divides among parallel resistors into currents with values inversely proportional to the resistance values.

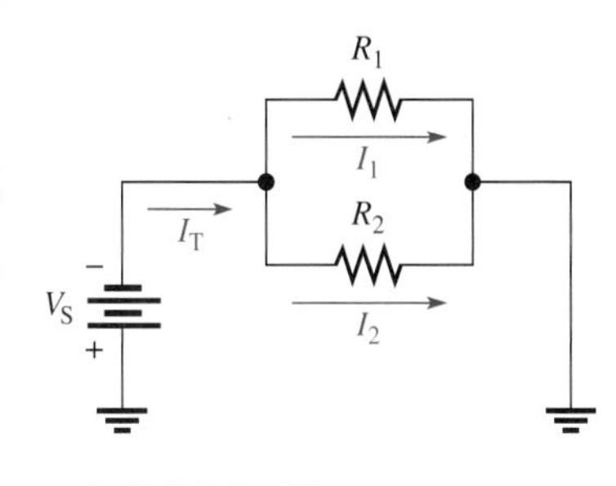

FIGURE 5–35

Total current divides between the two branches.

The branches with higher resistance have less current, and the branches with lower resistance have more current, in accordance with Ohm's law. If all the branches have the same resistance, the branch currents are all equal.

Figure 5–36 shows specific values to demonstrate how the currents divide according to the branch resistances. Notice that in this case the resistance of the upper branch is one-tenth the resistance of the lower branch, but the upper branch current is ten times the lower branch current.

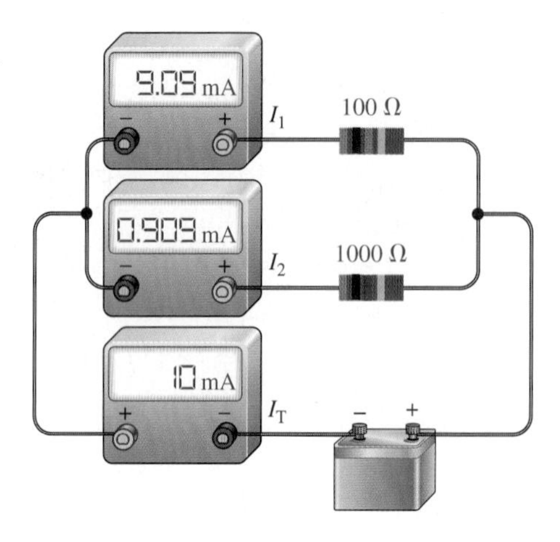

FIGURE 5–36

The branch with the lowest resistance has the most current, and the branch with the highest resistance has the least current.

Current-Divider Formulas for Two Branches

You already know how to use Ohm's law ($I = V/R$) to determine the current in any parallel branch when you know the voltage and resistance. When the voltage is not known but the total current is, you can find both of the branch currents (I_1 and I_2) by using the following formulas:

Equation 5–7

$$I_1 = \left(\frac{R_2}{R_1 + R_2}\right)I_T$$

Equation 5–8

$$I_2 = \left(\frac{R_1}{R_1 + R_2}\right)I_T$$

These formulas show that the current in either branch is equal to the opposite branch resistance divided by the sum of the two resistances and then multiplied by the total current.

EXAMPLE 5–16

Find I_1 and I_2 in Figure 5–37.

FIGURE 5–37

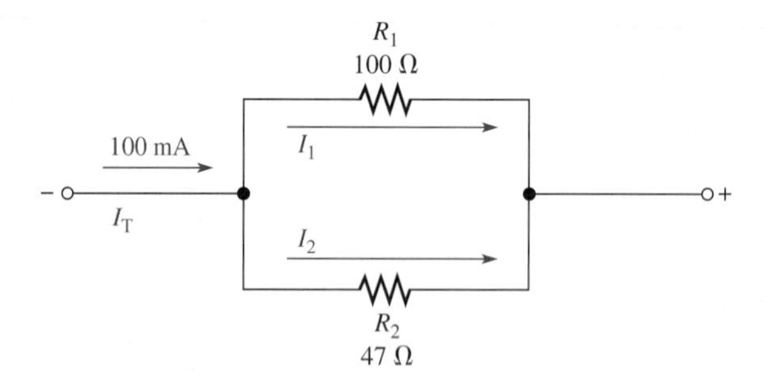

Solution Use Equation 5–7 to determine I_1.

$$I_1 = \left(\frac{R_2}{R_1 + R_2}\right)I_T = \left(\frac{47\ \Omega}{147\ \Omega}\right)100\text{ mA} = \mathbf{32.0\ mA}$$

Use Equation 5–8 to determine I_2.

$$I_2 = \left(\frac{R_1}{R_1 + R_2}\right)I_T = \left(\frac{100\ \Omega}{147\ \Omega}\right)100\text{ mA} = \mathbf{68.0\ mA}$$

The calculator solution for I_2 is

Related Problem If $R_1 = 56\ \Omega$ and $R_2 = 82\ \Omega$ in Figure 5–37 and I_T stays the same, what will each branch current be?

General Current-Divider Formula for Any Number of Parallel Branches

A generalized parallel circuit with n branches is shown in Figure 5–38 (n can represent any number).

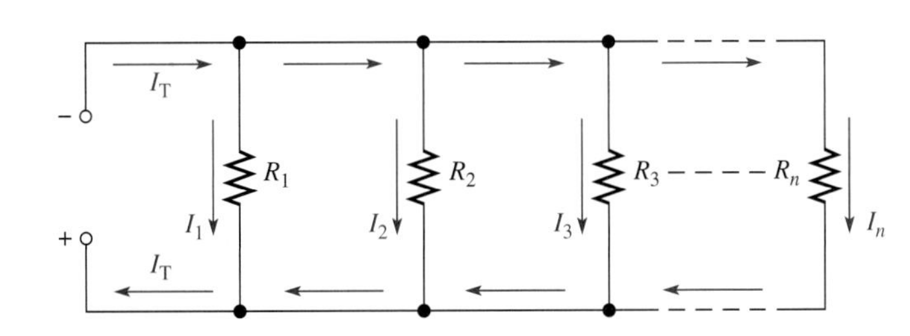

FIGURE 5–38

Generalized parallel circuit with n branches.

The current in any branch can be determined with the following formula:

$$I_x = \left(\frac{R_T}{R_x}\right)I_T$$

Equation 5–9

where I_x represents any branch current (I_1, I_2, and so on) and R_x represents any resistance (R_1, R_2, and so on). For example, the formula for current in the second branch in Figure 5–38 is

$$I_2 = \left(\frac{R_T}{R_2}\right)I_T$$

Equation 5–9 can be used for a parallel circuit with any number of branches. Notice that you must determine R_T in order to use Equation 5–9 to find a branch current.

EXAMPLE 5–17

Determine the current through each resistor in the circuit of Figure 5–39.

FIGURE 5–39

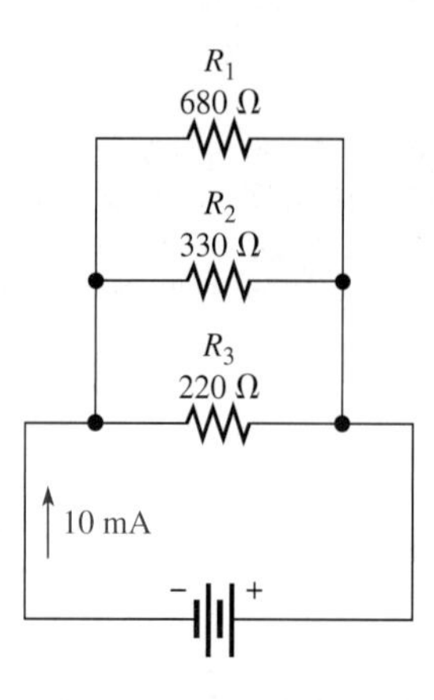

Solution First, calculate the total parallel resistance.

$$R_T = \frac{1}{\frac{1}{R_1} + \frac{1}{R_2} + \frac{1}{R_3}} = \frac{1}{\frac{1}{680\ \Omega} + \frac{1}{330\ \Omega} + \frac{1}{220\ \Omega}} = 110.5\ \Omega$$

The total current is 10 mA. Use Equation 5–9 to calculate each branch current.

$$I_1 = \left(\frac{R_T}{R_1}\right)I_T = \left(\frac{110.5\ \Omega}{680\ \Omega}\right)10\text{ mA} = \mathbf{1.63\ mA}$$

$$I_2 = \left(\frac{R_T}{R_2}\right)I_T = \left(\frac{110.5\ \Omega}{330\ \Omega}\right)10\text{ mA} = \mathbf{3.35\ mA}$$

$$I_3 = \left(\frac{R_T}{R_3}\right)I_T = \left(\frac{110.5\ \Omega}{220\ \Omega}\right)10\text{ mA} = \mathbf{5.02\ mA}$$

Related Problem Determine the current through R_1 and R_2 in Figure 5–39 if R_3 is removed. Assume the source voltage remains the same.

SECTION 5–6 REVIEW

1. A circuit has the following resistors in parallel with a voltage source: 220 Ω, 100 Ω, 68 Ω, 56 Ω, and 22 Ω. Which resistor has the most current through it? The least current?
2. Determine the current through R_3 in Figure 5–40.
3. Find the currents through each resistor in the circuit of Figure 5–41.

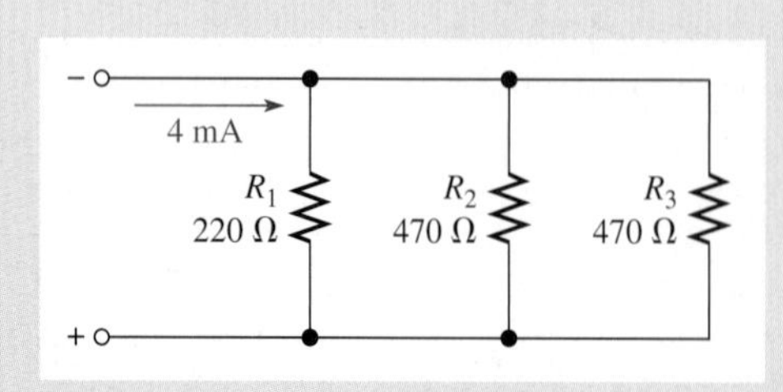

FIGURE 5–40

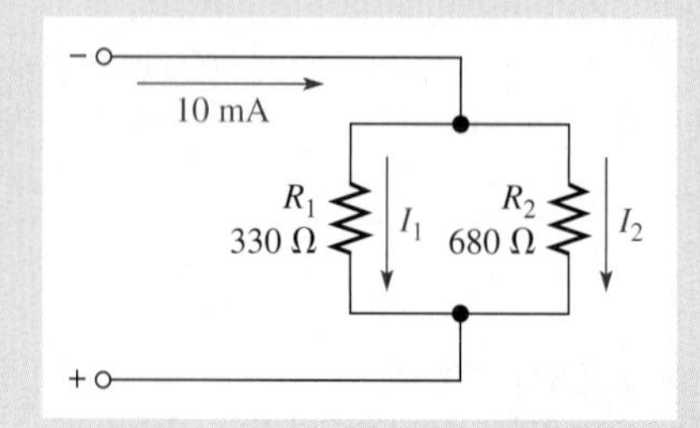

FIGURE 5–41

5–7 POWER IN PARALLEL CIRCUITS

Total power in a parallel circuit is found by adding up the powers of all the individual resistors, the same as you did for series circuits.

After completing this section, you should be able to

- **Determine power in a parallel circuit**

Equation 5–10 states the formula for total power with any number of resistors in parallel.

$$P_T = P_1 + P_2 + P_3 + \cdots + P_n$$ **Equation 5–10**

where P_T is the total power and P_n is the power in the last resistor in parallel. As you can see, the power losses are additive, just as in the series circuit.

The power formulas from Chapter 3 are directly applicable to parallel circuits. The following formulas are used to calculate the total power P_T:

$$P_T = V_S I_T$$

$$P_T = I_T^2 R_T$$

$$P_T = \frac{V_S^2}{R_T}$$

where V_S is the voltage across the parallel circuit, I_T is the total current into the parallel circuit, and R_T is the total resistance of the parallel circuit. Example 5–18 shows how total power can be calculated in a parallel circuit.

EXAMPLE 5–18

Determine the total amount of power in the parallel circuit in Figure 5–42.

FIGURE 5–42

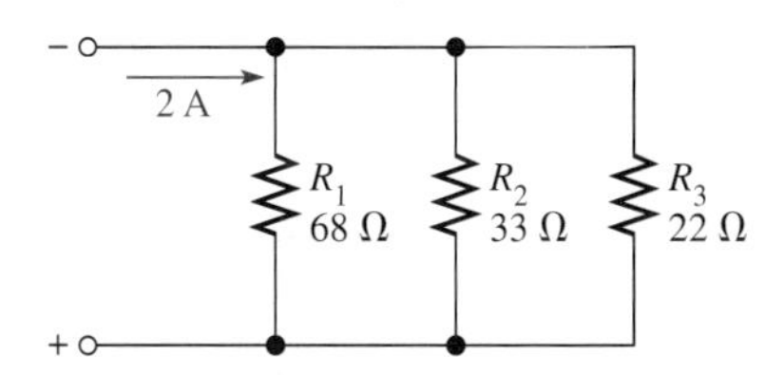

Solution The total current is 2 A. The total resistance is

$$R_T = \frac{1}{\dfrac{1}{68\ \Omega} + \dfrac{1}{33\ \Omega} + \dfrac{1}{22\ \Omega}} = 11.05\ \Omega$$

The easiest formula to use is $P_T = I_T^2 R_T$ since you know both I_T and R_T. Thus,

$$P_T = I_T^2 R_T = (2\text{ A})^2(11.05\ \Omega) = \mathbf{44.2\ W}$$

To demonstrate that if the power in each resistor is determined and if all of these values are added together, you get the same result, let's work through another calculation. First, find the voltage across each branch of the circuit.

$$V_S = I_T R_T = (2\text{ A})(11.05\ \Omega) = 22.1\text{ V}$$

Remember that the voltage across all branches is the same.

Next, use $P = V_S^2/R$ to calculate the power for each resistor.

$$P_1 = \frac{(22.1\ \text{V})^2}{68\ \Omega} = 7.19\ \text{W}$$

$$P_2 = \frac{(22.1\ \text{V})^2}{33\ \Omega} = 14.8\ \text{W}$$

$$P_3 = \frac{(22.1\ \text{V})^2}{22\ \Omega} = 22.4\ \text{W}$$

Now, add these powers to get the total power.

$$P_T = 7.19\ \text{W} + 14.8\ \text{W} + 22.2\ \text{W} = \mathbf{44.2\ W}$$

This calculation shows that the sum of the individual powers is equal to the total power as determined by one of the power formulas.

Related Problem Find the total power in Figure 5–42 if the total current is doubled.

EXAMPLE 5–19

The amplifier in one channel of a stereo system as shown in Figure 5–43 drives four parallel speakers as shown. If the maximum voltage to the speakers is 15 V, how much power must the amplifier be able to deliver to the speakers?

▶ **FIGURE 5–43**

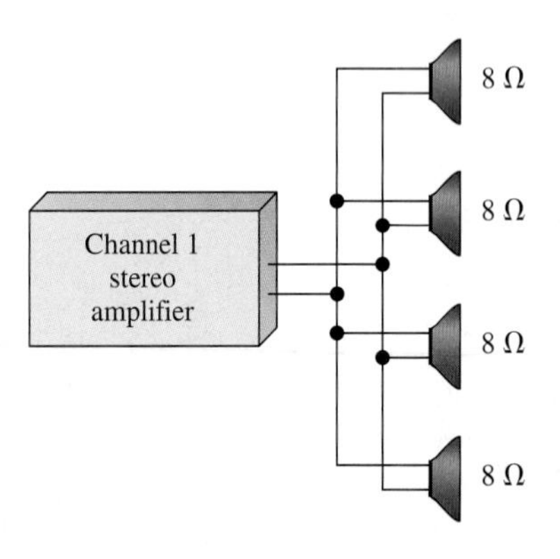

Solution The speakers are connected in parallel to the amplifier output, so the voltage across each is the same. The maximum power to each speaker is

$$P_{max} = \frac{V_{max}^2}{R} = \frac{(15\ \text{V})^2}{8\ \Omega} = 28.1\ \text{W}$$

The total power that the amplifier must be capable of delivering to the speaker system is four times the power in an individual speaker because the total power is the sum of the individual powers.

$$P_{T(max)} = P_{max} + P_{max} + P_{max} + P_{max} = 4P_{max} = 4(28.1\ \text{W}) = \mathbf{112\ W}$$

Related Problem If the amplifier can produce a maximum of 18 V, what is the maximum total power to the speakers?

SECTION 5–7 REVIEW

1. If you know the power in each resistor in a parallel circuit, how can you find the total power?
2. The resistors in a parallel circuit dissipate the following powers: 1 W, 2 W, 5 W, and 8 W. What is the total power in the circuit?
3. A circuit has a 1.0 kΩ, a 2.7 kΩ, and a 3.9 kΩ resistor in parallel. There is a total current of 1 mA into the parallel circuit. What is the total power?

5–8 TROUBLESHOOTING

Recall that an open circuit is one in which the current path is interrupted and there is no current. In this section, you will see how an open in a parallel branch affects the parallel circuit.

After completing this section, you should be able to

- **Troubleshoot parallel circuits**
- Check for an open circuit

Open Branches

If a switch is connected in a branch of a parallel circuit, as shown in Figure 5–44, an open or a closed path can be made by the switch. When the switch is closed, as in Figure 5–44(a), R_1 and R_2 are in parallel. The total resistance is 50 Ω (two 100 Ω resistors in parallel). Current is through both resistors. If the switch is opened, as in Figure 5–44(b), R_1 is effectively removed from the circuit, and the total resistance is 100 Ω. There is still the same voltage across R_2 and the same current through it although the total current from the source is reduced by the amount of the R_1 current.

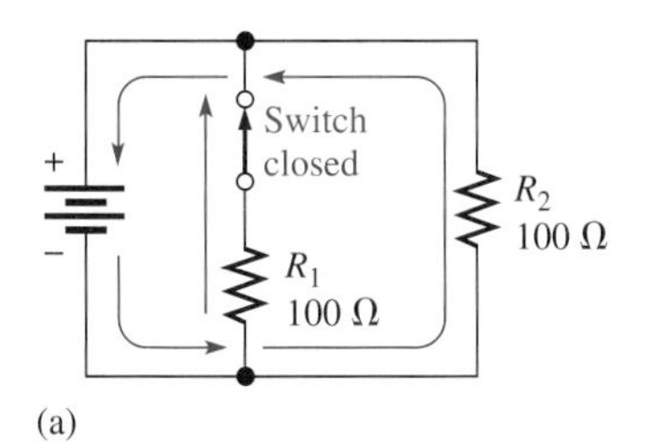

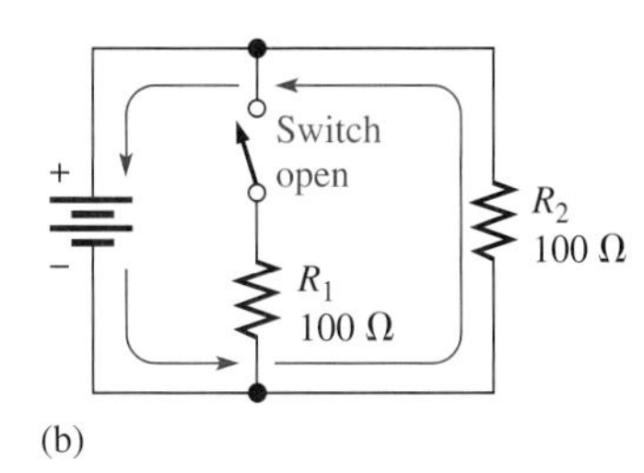

FIGURE 5–44

When the switch opens, total current decreases and current through R_2 remains unchanged.

In general,

When an open circuit occurs in a parallel branch, the total resistance increases, the total current decreases, and the same current continues through each of the remaining parallel paths.

Consider the lamp circuit in Figure 5–45. There are four bulbs in parallel with a 120 V source. In part (a), there is current through each bulb. Now suppose that one of the bulbs burns out, creating an open path as shown in Figure 5–45(b). This light will go out because there is no current through the open path. Notice, however, that current continues through all the other parallel bulbs, and they continue to glow. The open branch does not change the voltage across the remaining parallel branches; it remains at 120 V, and the current through each branch remains the same.

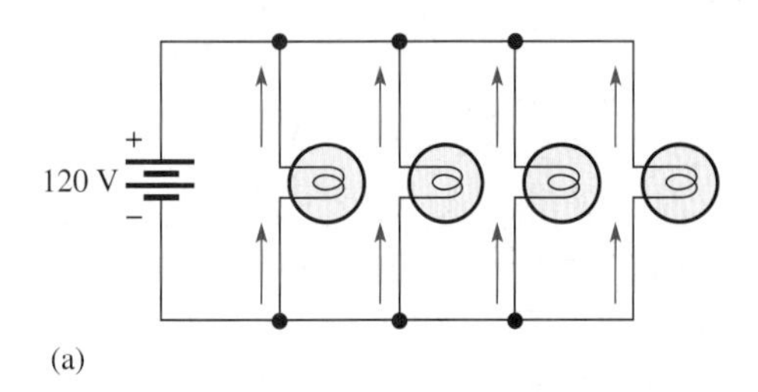

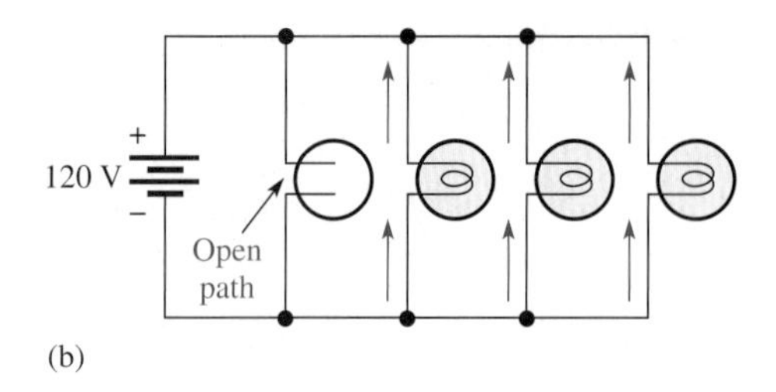

▲ **FIGURE 5–45**

When a lamp filament opens, total current decreases by the amount of current in the lamp that opened. The other branch currents remain unchanged.

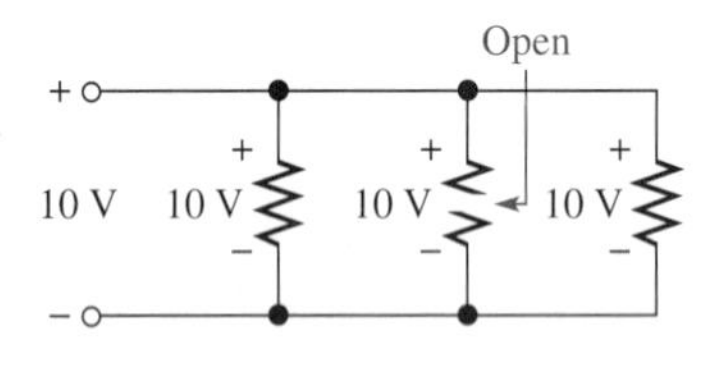

▲ **FIGURE 5–46**

All parallel branches (open or not) have the same voltage.

You can see that a parallel circuit has an advantage over a series connection in lighting systems because if one or more of the parallel bulbs burn out, the others will stay on. In a series circuit, when one bulb goes out, all of the others go out also because the current path is completely interrupted.

When a resistor in a parallel circuit opens, the open resistor cannot be located by measurement of the voltage across the branches because the same voltage exists across all the branches. Thus, there is no way to tell which resistor is open by simply measuring voltage (the half-splitting method is not applicable). The good resistors will always have the same voltage as the open one, as illustrated in Figure 5–46 (note that the middle resistor is open).

If a visual inspection does not reveal the open resistor, it can be located by resistance or current measurements. In practice, measuring resistance or current is more difficult than measuring voltage because you must disconnect a component to measure the resistance and you must insert an ammeter in series to measure the current. Thus, a wire or a printed circuit connection must usually be cut or disconnected, or one end of a component must be lifted off the circuit board, in order to connect a DMM to measure resistance and current. This procedure, of course, is not required when voltage measurements are made because the meter leads are simply connected across a component.

HANDS ON TIP

An alternate way to measure current without having to break a circuit to connect an ammeter is to place, as part of the original circuit, a 1 Ω series resistor in each line through which current is to be measured. This small "sense" resistor will generally not affect the total resistance. By measuring the voltage across the sense resistor, you automatically have the current reading since

$$I = \frac{V}{R} = \frac{V}{1\ \Omega} = V$$

Finding an Open Branch by Current Measurement

In a parallel circuit, the total current should be measured. *When a parallel resistor opens, I_T is always less than its normal value.* Once I_T and the voltage across the branches are known, a few calculations will determine the open resistor when all the resistors are of different values.

Consider the two-branch circuit in Figure 5–47(a). If one of the resistors opens, the total current will equal the current in the good resistor. Ohm's law quickly tells you what the current in each resistor should be.

$$I_1 = \frac{50\ \text{V}}{560\ \Omega} = 89.3\ \text{mA}$$

$$I_2 = \frac{50\ \text{V}}{100\ \Omega} = 500\ \text{mA}$$

If R_2 is open, the total current is 89.3 mA, as indicated in Figure 5–47(b). If R_1 is open, the total current is 500 mA, as indicated in Figure 5–47(c).

▼ **FIGURE 5–47**

Finding an open path by current measurement.

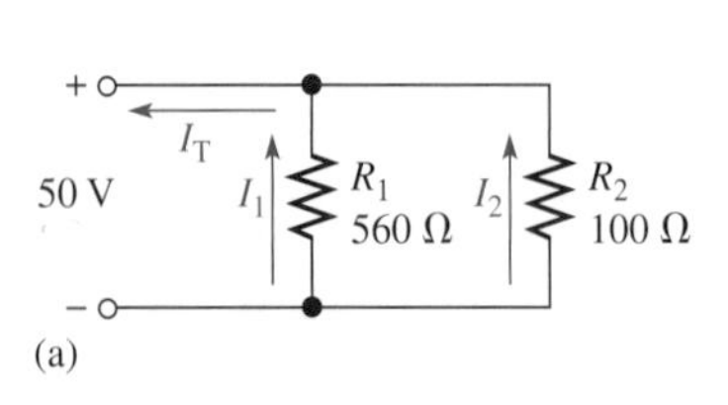

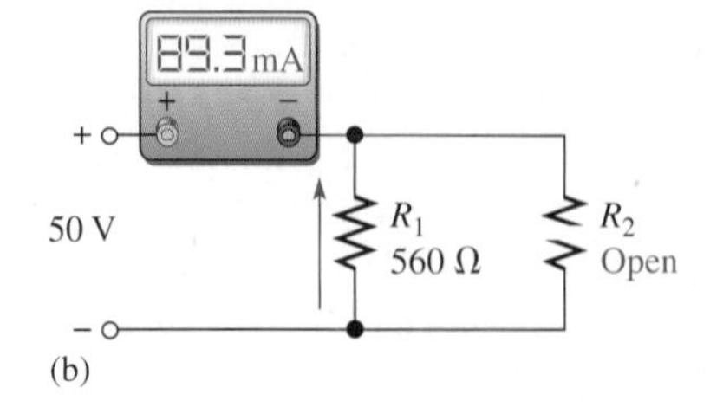

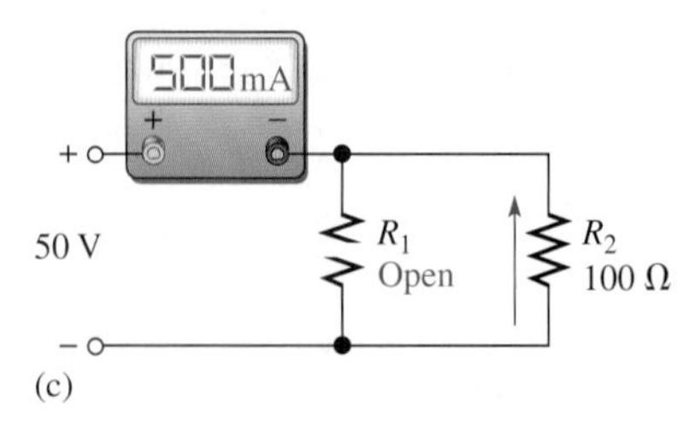

This procedure can be extended to any number of branches having unequal resistances. If the parallel resistances are all equal, the current in each branch must be checked until a branch is found with no current.

EXAMPLE 5–20

In Figure 5–48, there is a total current of 31.09 mA, and the voltage across the parallel branches is 20 V. Is there an open resistor, and if so, which one is it?

▶ **FIGURE 5–48**

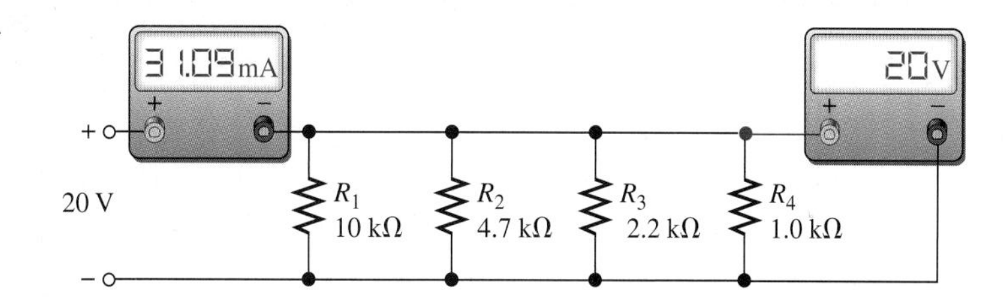

Solution Calculate the current in each branch.

$$I_1 = \frac{V}{R_1} = \frac{20\ \text{V}}{10\ \text{k}\Omega} = 2\ \text{mA}$$

$$I_2 = \frac{V}{R_2} = \frac{20\ \text{V}}{4.7\ \text{k}\Omega} = 4.26\ \text{mA}$$

$$I_3 = \frac{V}{R_3} = \frac{20\ \text{V}}{2.2\ \text{k}\Omega} = 9.09\ \text{mA}$$

$$I_4 = \frac{V}{R_4} = \frac{20\ \text{V}}{1.0\ \text{k}\Omega} = 20\ \text{mA}$$

The total current should be

$$I_T = I_1 + I_2 + I_3 + I_4 = 2\ \text{mA} + 4.26\ \text{mA} + 9.09\ \text{mA} + 20\ \text{mA} = 35.35\ \text{mA}$$

The actual measured current is 31.09 mA, as stated, which is 4.26 mA less than normal, indicating that the branch with 4.26 mA is open. Thus, **R_2 must be open.**

Related Problem What is the total current measured in Figure 5–48 if R_4 and not R_2 is open?

Open file E05-20 on your EWB/CircuitMaker CD-ROM. Measure the total current and the current in each resistor. There are no faults in the circuit.

Finding an Open Branch by Resistance Measurement

If the parallel circuit to be checked can be disconnected from its voltage source and from any other circuit to which it may be connected, a measurement of the total resistance can be used to locate an open branch.

Recall that conductance, G, is the reciprocal of resistance ($1/R$) and its unit is the siemens (S). The total conductance of a parallel circuit is the sum of the conductances of all the resistors.

$$G_T = G_1 + G_2 + G_3 + \cdots + G_n$$

To locate an open branch, do the following steps:

1. Calculate what the total conductance should be using the individual resistor values.

$$G_{T(calc)} = \frac{1}{R_1} + \frac{1}{R_2} + \frac{1}{R_3} + \cdots + \frac{1}{R_n}$$

2. Measure the total resistance and calculate the total measured conductance.

$$G_{T(meas)} = \frac{1}{R_{T(meas)}}$$

3. Subtract the measured total conductance (Step 2) from the calculated total conductance (Step 1). The result is the conductance of the open branch, and the resistance is obtained by taking its reciprocal ($R = 1/G$).

$$R_{open} = \frac{1}{G_{T(calc)} - G_{T(meas)}}$$

EXAMPLE 5–21

Your ohmmeter measures 402 Ω between pin 1 and pin 4 in Figure 5–49. Check the printed circuit board between these two pins for open branches.

▶ FIGURE 5–49

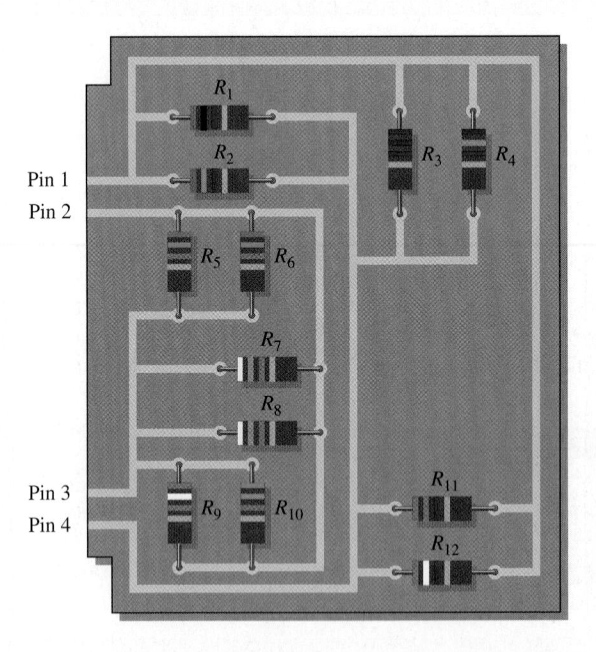

Solution The circuit between pin 1 and pin 4 is checked as follows:

1. Calculate what the total conductance should be using the individual resistor values.

$$G_{T(calc)} = \frac{1}{1.0\text{ k}\Omega} + \frac{1}{1.8\text{ k}\Omega} + \frac{1}{2.2\text{ k}\Omega} + \frac{1}{2.7\text{ k}\Omega} + \frac{1}{3.3\text{ k}\Omega} + \frac{1}{3.9\text{ k}\Omega} = 2.94\text{ mS}$$

2. Calculate the total measured conductance.

$$G_{T(meas)} = \frac{1}{402\ \Omega} = 2.49\text{ mS}$$

3. Subtract the measured total conductance (Step 2) from the calculated total conductance (Step 1). The result is the conductance of the open branch and the resistance is obtained by taking its reciprocal.

$$G_{open} = G_{T(calc)} - G_{T(meas)} = 2.94\text{ mS} - 2.49\text{ mS} = 0.45\text{ mS}$$

$$R_{open} = \frac{1}{G_{open}} = \frac{1}{0.45\text{ mS}} = 2.2\text{ k}\Omega$$

Resistor $\boldsymbol{R_3}$ **is open** and must be replaced.

Related Problem Your ohmmeter indicates 9.6 kΩ between pin 2 and pin 3 on the PC board in Figure 5–49. Determine if this is correct and, if not, which resistor is open.

SECTION 5–8 REVIEW

1. If a parallel branch opens, what changes can be detected in the circuit's voltage and the currents, assuming that the parallel circuit is across a constant voltage source?
2. What happens to the total resistance if one branch opens?
3. If several light bulbs are connected in parallel and one of the bulbs opens (burns out), will the others continue to glow?
4. There is one ampere of current in each branch of a parallel circuit. If one branch opens, what is the current in each of the remaining branches?
5. A three-branch circuit normally has the following branch currents: 1 A, 2.5 A, and 1.2 A. If the total current measures 3.5 A, which branch is open?

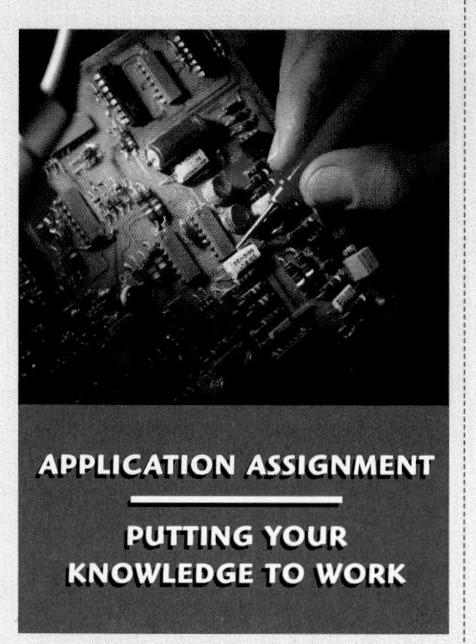

APPLICATION ASSIGNMENT

PUTTING YOUR KNOWLEDGE TO WORK

A five-range milliammeter has just come off the assembly line, but it has failed the line check. For any current in excess of 1 mA, the needle "pegs" (goes off scale to the right) on all the range settings except 1 mA and 50 mA. In these two ranges, the meter seems to work correctly. A multiple-range meter circuit is based on resistance placed in parallel with the meter movement to achieve the specified full-scale deflection of the needle. Your job is to troubleshoot the meter and determine what repairs are necessary.

General Theory of Operation

Parallel circuits are an important part of the operation of this type of ammeter because they allow the user to select various ranges in order to measure many different current values.

The mechanism in an analog ammeter that causes the pointer to move in proportion to the current is called the *meter movement,* which is based on a magnetic principle that you will learn in Chapter 7. Right now, all you need to know is that a meter movement has a certain resistance and a maximum current. This maximum current, called the *full-scale deflection current,* causes the pointer to go all the way to the end of the scale. For example, a certain meter movement has a 50 Ω resistance and a full-scale deflection current of 1 mA. A meter with this particular movement can measure currents of 1 mA or less. Currents greater than 1 mA will cause the pointer to "peg" (or stop) at full scale. Figure 5–50 illustrates a 1 mA meter.

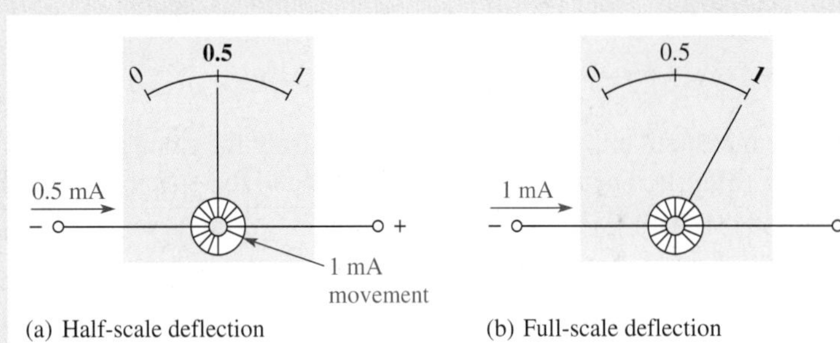

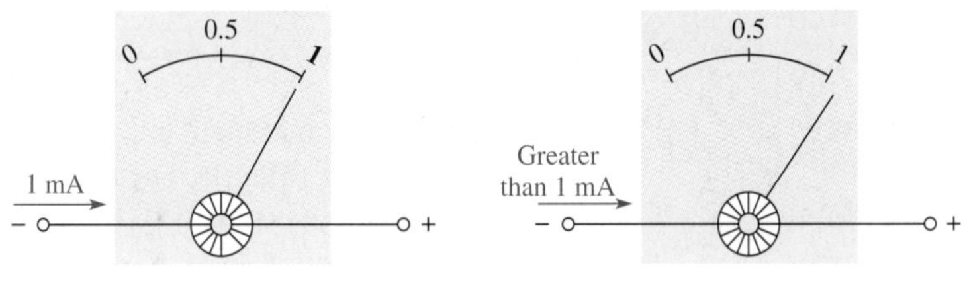

FIGURE 5–50

A 1 mA meter.

Figure 5–51 shows a simple ammeter with a resistor in parallel with the meter movement; this resistor is called a *shunt resistor.* Its purpose is to bypass any current in excess of 1 mA around the meter movement to extend the range of currents that can be measured. The figure specifically shows 9 mA through the shunt resistor and 1 mA through the meter movement. Thus, up to 10 mA can be measured. To find the actual current value, simply multiply the reading on the scale by 10.

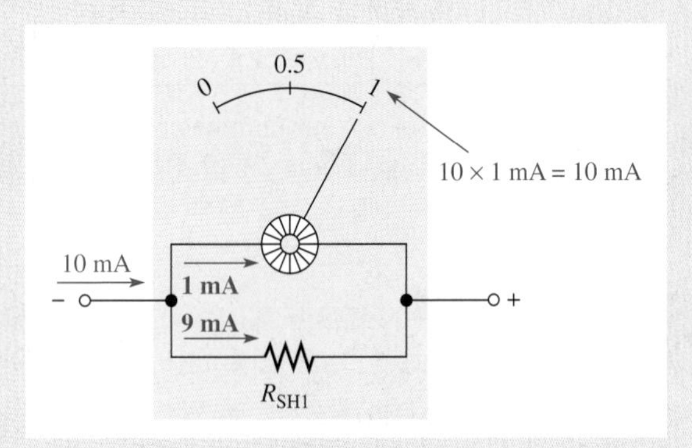

FIGURE 5–51

A 10 mA meter.

A practical ammeter has a range switch that permits the selection of several full-scale current settings. In each switch position, a certain amount of current is bypassed through a parallel resistor as determined by the resistance value. In our example, the current through the movement is never greater than 1 mA.

Figure 5–52 illustrates a meter with three ranges: 1 mA, 10 mA, and 100 mA. When the range switch is in the 1 mA position, all of the current coming into the meter goes through the meter movement. In the 10 mA setting, up to 9 mA goes through R_{SH1} and up to 1 mA through the movement. In the 100 mA setting, up to 99 mA goes through R_{SH2} and the movement can still have only 1 mA for full-scale.

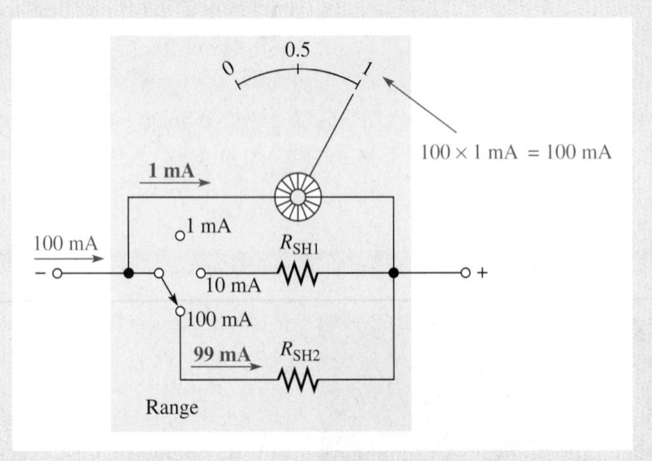

FIGURE 5–52

A milliammeter with three ranges.

For example, in Figure 5–52, if 50 mA of current are being measured, the needle points to the 0.5 mark on the scale; you must multiply 0.5 by 100 to find the current value. In this situation, 0.5 mA is through the movement (half-scale deflection), and 49.5 mA are through R_{SH2}.

Step 1: Open the Meter and Examine the Circuit

The meter face is shown in Figure 5–53(a). The circuitry is exposed by removing the front of the meter as shown in part (b). The precision resistors are 1% tolerance with a five-band color code. Recall from Chapter 2 that the first three bands indicate resistance value. In the fourth band, gold indicates a multiplier of 0.1 and silver a multiplier of 0.01. The brown fifth band is for 1% tolerance.

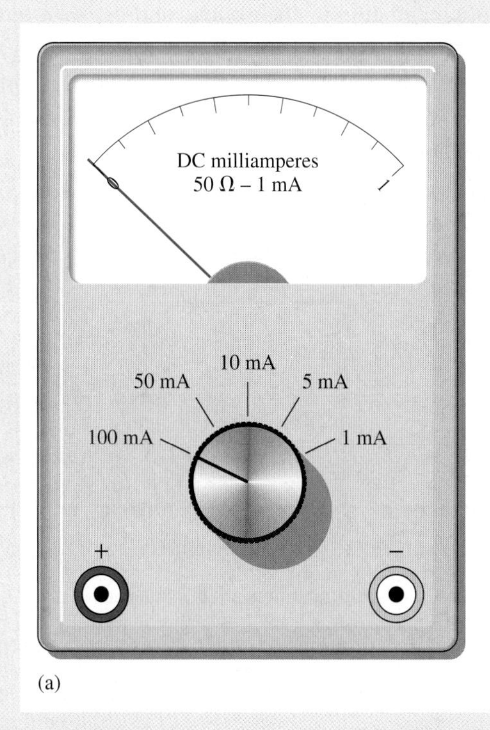

(a)

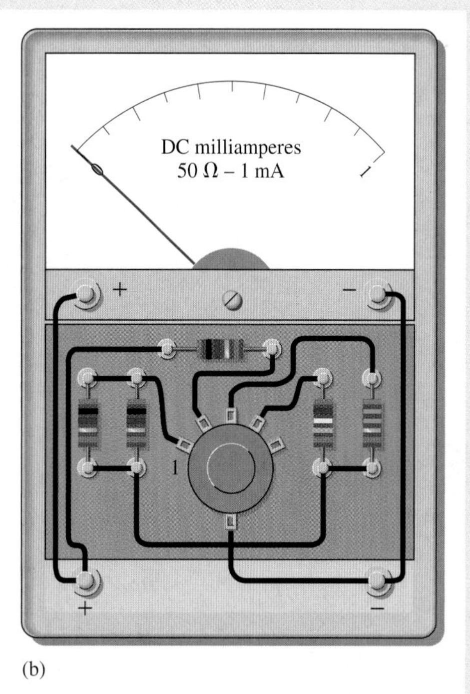

(b)

▲ **FIGURE 5–53**

Step 2: Compare the Meter Circuit to the Schematic

The meter schematic is shown in Figure 5–54. Carefully check the circuit connections in Figure 5–53(b) to make sure they agree with the schematic. If there is a wiring problem, specify how to repair it. Resistors R_1, R_2, etc. can be identified by the way they are wired in the circuit. The 100 mA switch position is labeled "1". Resistance values are not shown in the schematic.

Step 3: Check for Additional Problems

There was a wiring problem in the meter circuit which, let's assume, you have located and corrected. Now, assume that when you connect the meter to the test instrument to check for proper operation you find the meter works properly on the 1 mA, 5 mA, 10 mA, and 50 mA ranges but the needle still "pegs" on the 100 mA range setting for a current of 50 mA or greater.

Step 4: Analyze the Circuit to Determine the Additional Problem

Since the needle pegs on the 100 mA range, too much current is going through the meter movement. Recall that the parallel (shunt) resistor carries most of the current being measured so that the current through the meter movement does not exceed its 1 mA maximum. For this problem to occur, one of the shunt resistors for this range is either not connected, failed open, or has a resistance value that is much too high. Check the circuit in Figure 5–53 again for another wiring error and check the color-coded values of R_4 and R_5. For some reason, the schematic does not show the resistor values so you will have to calculate what they should be using your knowledge of Ohm's law and current dividers. Refer to the table of standard resistor values in Appendix A.

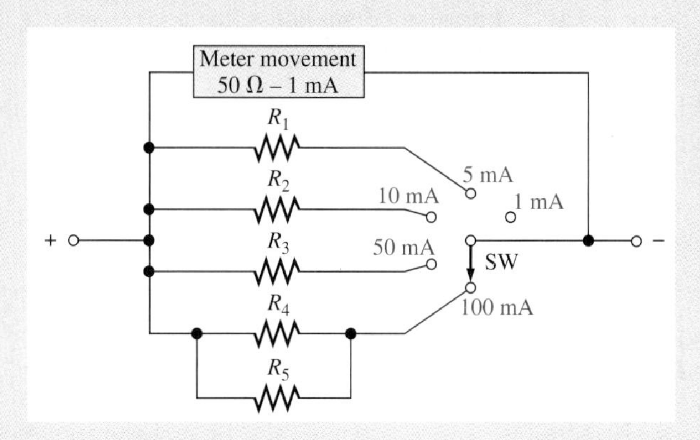

▲ **FIGURE 5–54**

APPLICATION ASSIGNMENT REVIEW

1. What is the total resistance between the positive and negative terminals of the milliammeter in Figure 5–53 when the range switch is in the 1 mA position? The 5 mA position? The 10 mA position?
2. When the switch is in the 5 mA position and the needle points to the middle mark on the scale, how much current is being measured?
3. When the meter is on the 5 mA range and is measuring the amount of current determined in Question 2, how much of the total current is through the meter movement and how much is through the shunt resistor?

SUMMARY

- Resistors in parallel are connected across the same two points in a circuit.
- A parallel circuit provides more than one path for current.
- The number of current paths equals the number of resistors in parallel.
- The total parallel resistance is less than the lowest-value parallel resistor.
- The voltages across all branches of a parallel circuit are the same.
- Kirchhoff's current law: The sum of the currents into a junction (total current in) equals the sum of the currents out of the junction (total current out) or, equivalently, the algebraic sum of all the currents entering and leaving a junction is zero.
- A parallel circuit is a current divider, so called because the total current entering the parallel junction divides up into each of the branches.
- If all of the branches of a parallel circuit have equal resistance, the currents through all of the branches are equal.
- The total power in a parallel-resistive circuit is the sum of all of the individual powers of the resistors making up the parallel circuit.
- The total power for a parallel circuit can be calculated with the power formulas using values of total current, total resistance, or total voltage.
- If one of the branches of a parallel circuit opens, the total resistance increases, and therefore the total current decreases.
- If a branch of a parallel circuit opens, there is still the same current through the remaining branches.

EQUATIONS

5–1 $I_{IN(1)} + I_{IN(2)} + I_{IN(3)} + \cdots + I_{IN(n)} = I_{OUT(1)} + I_{OUT(2)} + I_{OUT(3)} + \cdots + I_{OUT(m)}$ Kirchhoff's current law

5–2 $\frac{1}{R_T} = \frac{1}{R_1} + \frac{1}{R_2} + \frac{1}{R_3} + \cdots + \frac{1}{R_n}$ Reciprocal for total parallel resistance

5–3 $G_T = G_1 + G_2 + G_3 + \cdots + G_n$ Total conductance

5–4	$R_T = \dfrac{1}{\dfrac{1}{R_1} + \dfrac{1}{R_2} + \dfrac{1}{R_3} + \cdots + \dfrac{1}{R_n}}$	Total parallel resistance
5–5	$R_T = \dfrac{R_1R_2}{R_1 + R_2}$	Special case for two resistors in parallel
5–6	$R_T = \dfrac{R}{n}$	Special case for n equal-value resistors in parallel
5–7	$I_1 = \left(\dfrac{R_2}{R_1 + R_2}\right)I_T$	Two-branch current-divider formula
5–8	$I_2 = \left(\dfrac{R_1}{R_1 + R_2}\right)I_T$	Two-branch current-divider formula
5–9	$I_x = \left(\dfrac{R_T}{R_x}\right)I_T$	General current-divider formula
5–10	$P_T = P_1 + P_2 + P_3 + \cdots + P_n$	Total power

SELF-TEST

Answers are at the end of the chapter.

1. In a parallel circuit, each resistor has
(a) the same current **(b)** the same voltage
(c) the same power **(d)** all of the above

2. When a 1.2 kΩ resistor and a 100 Ω resistor are connected in parallel, the total resistance is
(a) greater than 1.2 kΩ
(b) greater than 100 Ω but less than 1.2 kΩ
(c) less than 100 Ω but greater than 90 Ω
(d) less than 90 Ω

3. A 330 Ω resistor, a 270 Ω resistor, and a 68 Ω resistor are all in parallel. The total resistance is approximately
(a) 668 Ω **(b)** 47 Ω **(c)** 68 Ω **(d)** 22 Ω

4. Eight resistors are in parallel. The two lowest-value resistors are both 1.0 kΩ. The total resistance
(a) cannot be determined **(b)** is greater than 1.0 kΩ
(c) is less than 1.0 kΩ **(d)** is less than 500 Ω

5. When an additional resistor is connected across an existing parallel circuit, the total resistance
(a) decreases **(b)** increases
(c) remains the same **(d)** increases by the value of the added resistor

6. If one of the resistors in a parallel circuit is removed, the total resistance
(a) decreases by the value of the removed resistor
(b) remains the same **(c)** increases **(d)** doubles

7. The currents into a junction are along two paths. One current is 5 A and the other is 3 A. The total current out of the junction is
(a) 2 A **(b)** unknown **(c)** 8 A **(d)** the larger of the two

8. The following resistors are in parallel across a voltage source: 390 Ω, 560 Ω, and 820 Ω. The resistor with the least current is

(a) 390 Ω **(b)** 560 Ω

(c) 820 Ω **(d)** impossible to determine without knowing the voltage

9. A sudden decrease in the total current into a parallel circuit may indicate.

(a) a short **(b)** an open resistor

(c) a drop in source voltage **(d)** either (b) or (c)

10. In a four-branch parallel circuit, there are 10 mA of current in each branch. If one of the branches opens, the current in each of the other three branches is

(a) 13.33 mA **(b)** 10 mA **(c)** 0 A **(d)** 30 mA

11. In a certain three-branch parallel circuit, R_1 has 10 mA through it, R_2 has 15 mA through it, and R_3 has 20 mA through it. After measuring a total current of 35 mA, you can say that

(a) R_1 is open **(b)** R_2 is open

(c) R_3 is open **(d)** the circuit is operating properly

12. If there are a total of 100 mA into a parallel circuit consisting of three branches and two of the branch currents are 40 mA and 20 mA, the third branch current is

(a) 60 mA **(b)** 20 mA **(c)** 160 mA **(d)** 40 mA

13. A complete short develops across one of five parallel resistors on a PC board. The most likely result is

(a) the smallest-value resistor will burn out

(b) one or more of the other resistors will burn out

(c) the fuse in the power supply will blow

(d) the resistance values will be altered

14. The power dissipation in each of four parallel branches is 1 W. The total power dissipation is

(a) 1 W **(b)** 4 W **(c)** 0.25 W **(d)** 16 W

PROBLEMS

Answers to odd-numbered problems are at the end of the book.

BASIC PROBLEMS

SECTION 5–1 **Resistors in Parallel**

1. Connect the resistors in Figure 5–55 in parallel across the battery.

▶ **FIGURE 5–55**

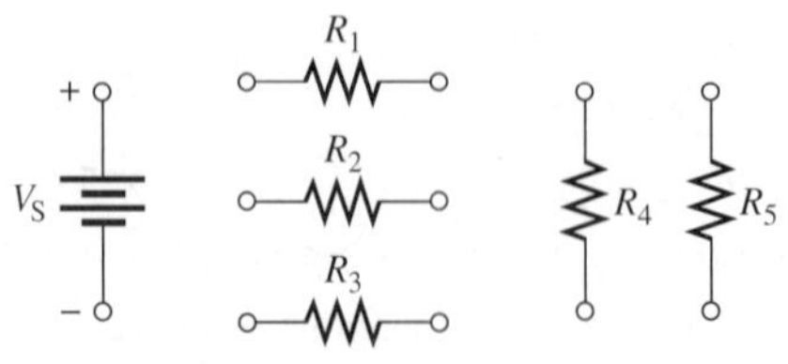

+ V_S − R_1 R_2 R_3 R_4 R_5

2. Determine whether or not all the resistors in Figure 5–56 are connected in parallel on the printed circuit board. Draw the schematic including resistor values.

FIGURE 5–56

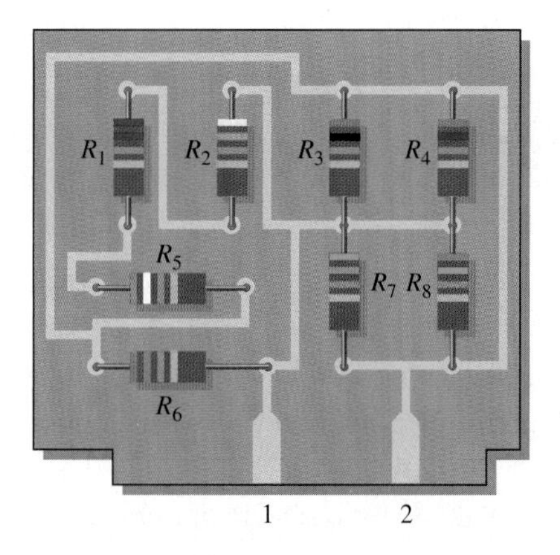

SECTION 5–2

Voltage in Parallel Circuits

3. Determine the voltage across and the current through each parallel resistor if the total voltage is 12 V and the total resistance is 600 Ω. There are four resistors, all of equal value.
4. The source voltage in Figure 5–57 is 100 V. How much voltage does each of the three meters read?

FIGURE 5–57

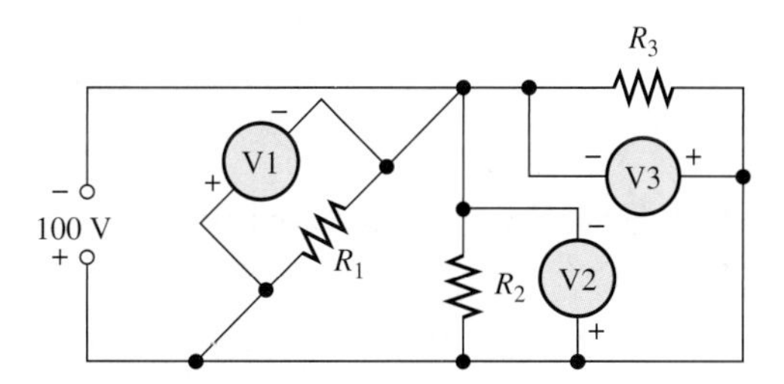

SECTION 5–3

Kirchhoff's Current Law

5. The following currents are measured in the same direction in a three-branch parallel circuit: 250 mA, 300 mA, and 800 mA. What is the value of the current into the junction of these three branches?
6. There is a total of 500 mA of current into five parallel resistors. The currents through four of the resistors are 50 mA, 150 mA, 25 mA, and 100 mA. What is the current through the fifth resistor?
7. How much current is through R_2 and R_3 in Figure 5–58 if R_2 and R_3 have the same resistance? Show how to connect ammeters to measure these currents.

FIGURE 5–58

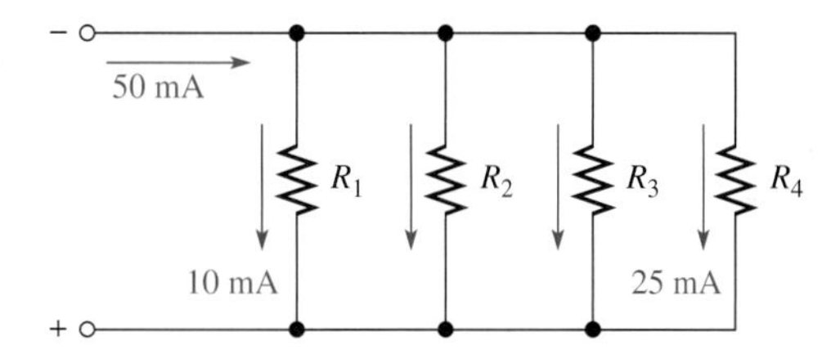

SECTION 5–4

Total Parallel Resistance

8. The following resistors are connected in parallel: 1.0 MΩ, 2.2 MΩ, 4.7 MΩ, 12 MΩ, and 22 MΩ. Determine the total resistance.

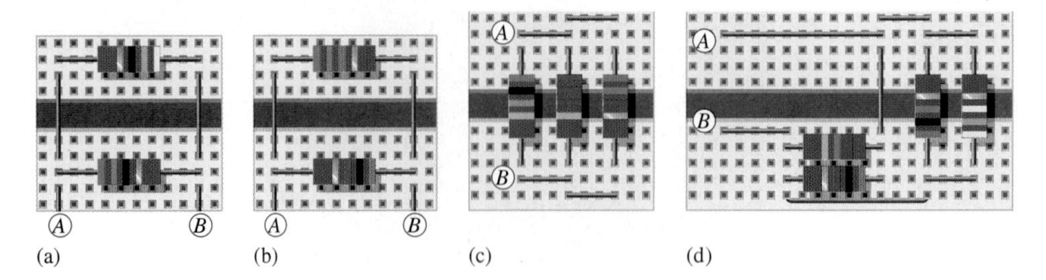

▲ FIGURE 5–59

9. Find the total resistance between points *A* and *B* for each group of parallel resistors in Figure 5–59.
10. Calculate R_T for each circuit in Figure 5–60.

▶ FIGURE 5–60

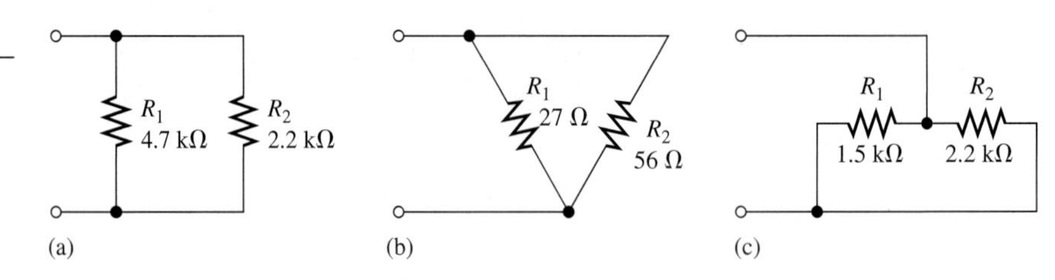

11. What is the total resistance of eleven 22 kΩ resistors in parallel?
12. Five 15 Ω, ten 100 Ω, and two 10 Ω resistors are all connected in parallel. What is the total resistance?

SECTION 5–5 **Ohm's Law in Parallel Circuits**

13. What is the total current I_T in each circuit of Figure 5–61?

▶ FIGURE 5–61

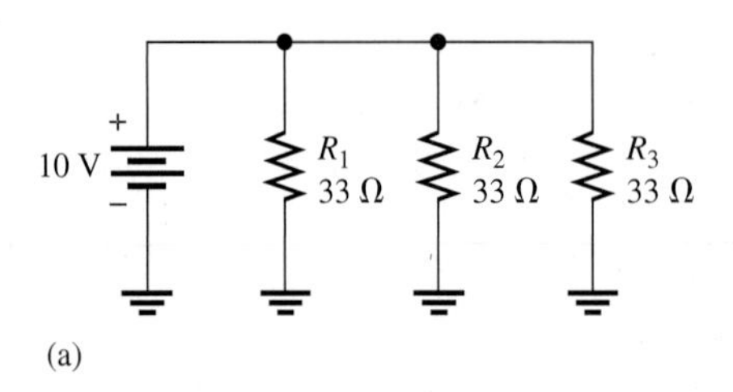

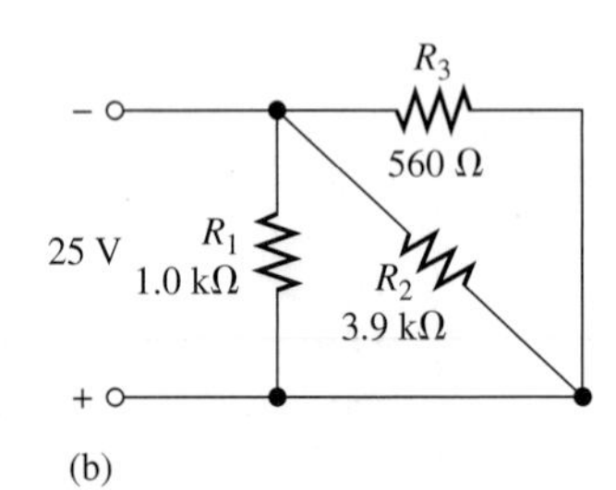

14. Three 33 Ω resistors are connected in parallel with a 110 V source. What is the current from the source?
15. Which circuit of Figure 5–62 has more total current?

▶ FIGURE 5–62

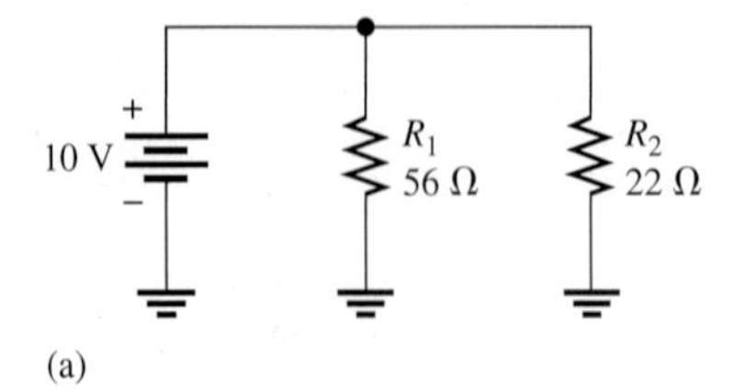

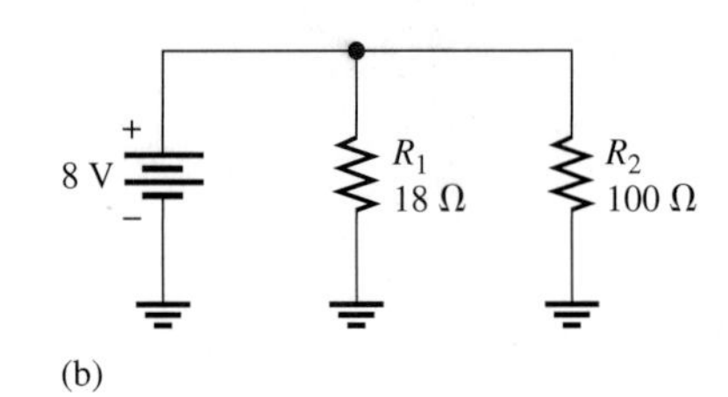

16. Four equal-value resistors are connected in parallel. Five volts are applied across the parallel circuit, and 2.5 mA are measured from the source. What is the value of each resistor?

SECTION 5–6 **Current Dividers**

17. How much branch current should each meter in Figure 5–63 indicate?

18. Using the current-divider formula, determine the current in each branch of the circuits of Figure 5–64.

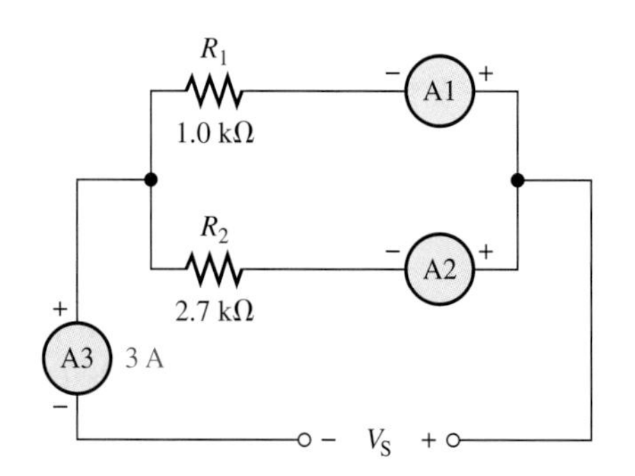

▲ FIGURE 5–63

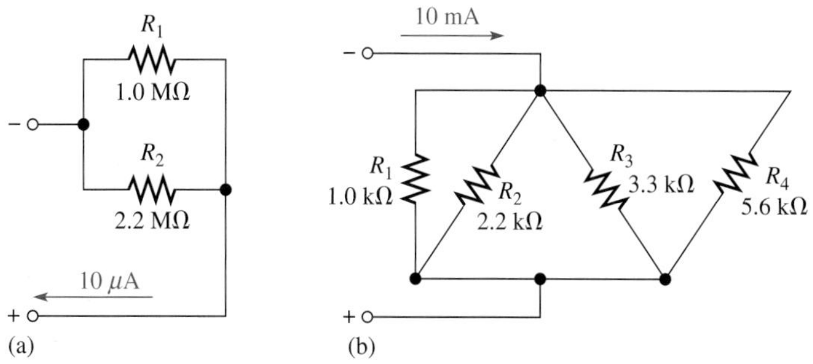

▲ FIGURE 5–64

SECTION 5–7 **Power in Parallel Circuits**

19. Five parallel resistors each handle 40 mW. What is the total power?

20. Determine the total power in each circuit of Figure 5–64.

21. Six light bulbs are connected in parallel across 110 V. Each bulb is rated at 75 W. How much current is through each bulb, and what is the total current?

SECTION 5–8 **Troubleshooting**

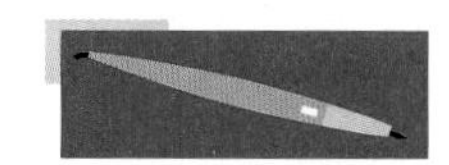

22. If one of the bulbs burns out in Problem 21, how much current will be through each of the remaining bulbs? What will be the total current?

23. In Figure 5–65, the current and voltage measurements are indicated. Has a resistor opened, and, if so, which one?

▶ FIGURE 5–65

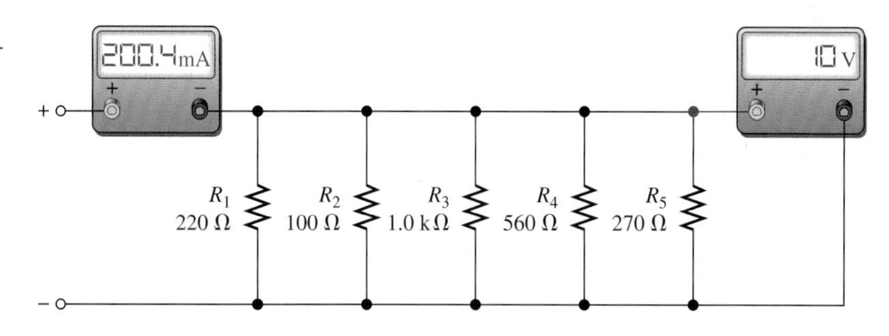

24. What is wrong with the circuit in Figure 5–66?

25. Find the open resistor in Figure 5–67.

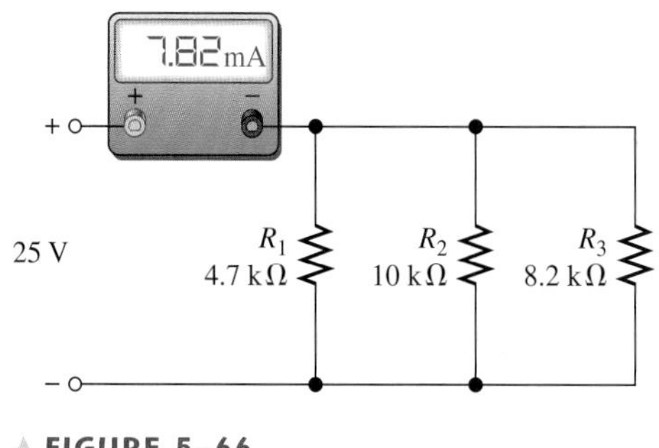

▲ FIGURE 5–66

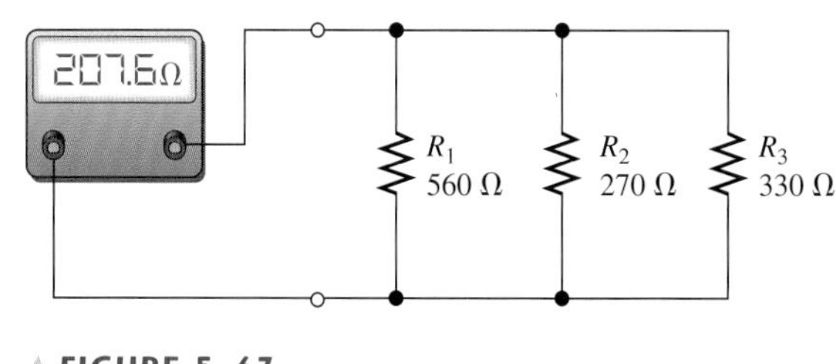

▲ FIGURE 5–67

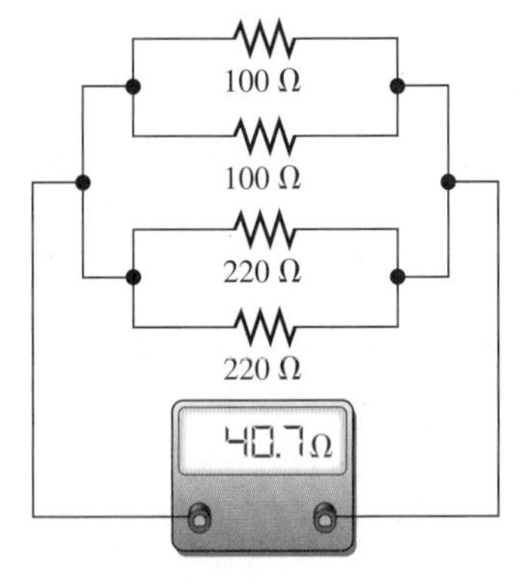

▲ FIGURE 5–68

26. Determine if there is a resistor open in Figure 5–68. If so, identify it.

ADVANCED PROBLEMS

27. In the circuit of Figure 5–69, determine resistances R_2, R_3, and R_4.
28. The total resistance of a parallel circuit is 25 Ω. If the total current is 100 mA, how much current is through a 220 Ω resistor that makes up part of the parallel circuit?
29. What is the current through each resistor in Figure 5–70? R is the lowest-value resistor, and all others are multiples of that value as indicated.
30. A certain parallel network consists of only ½ W resistors each having the same resistance. The total resistance is 1 kΩ, and the total current is 50 mA. If each resistor is operating at one-half its maximum power level, determine the following:
    - **(a)** the number of resistors
    - **(b)** the value of each resistor
    - **(c)** the current in each branch
    - **(d)** the applied voltage

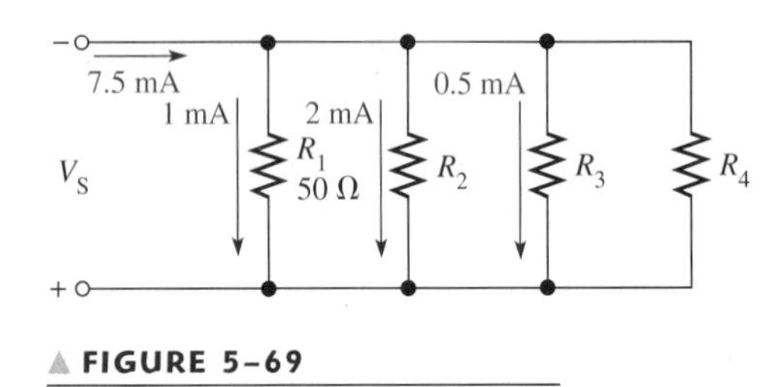

▲ FIGURE 5–69

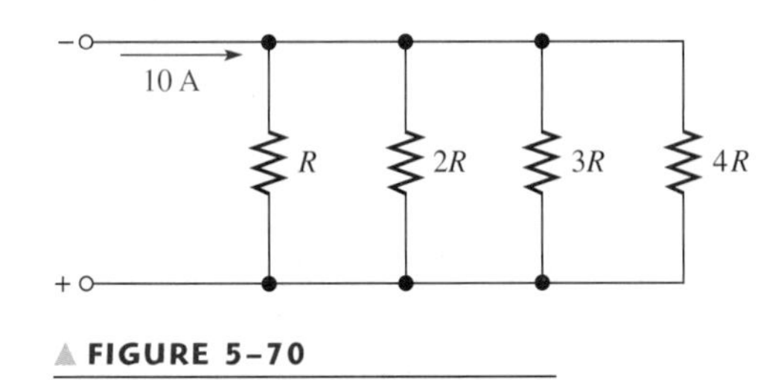

▲ FIGURE 5–70

31. Find the values of the unspecified labeled quantities (shown in color) in each circuit of Figure 5–71.

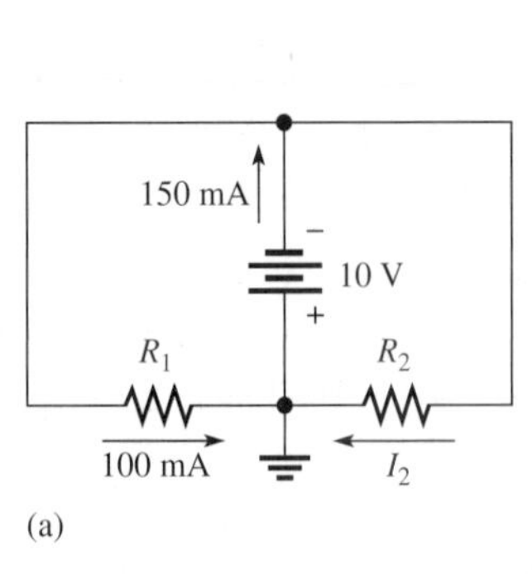

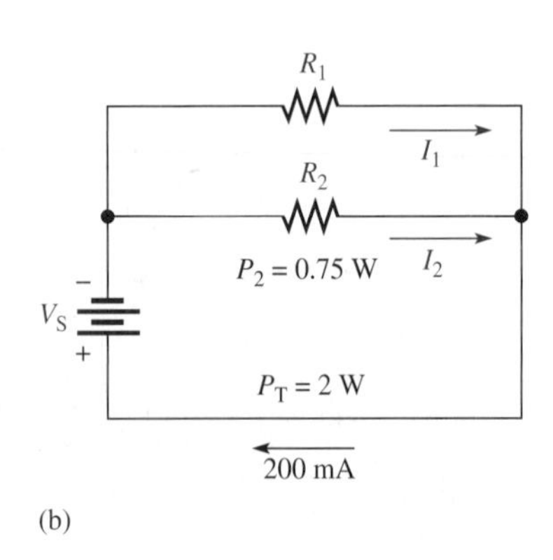

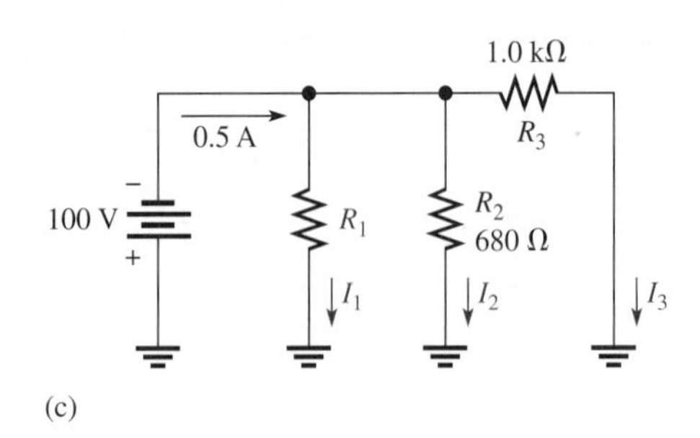

▲ FIGURE 5–71

32. What is the total resistance between point A and ground in Figure 5–72 for the following conditions?
    - **(a)** SW1 and SW2 open
    - **(b)** SW1 closed, SW2 open
    - **(c)** SW1 open, SW2 closed
    - **(d)** SW1 and SW2 closed
33. What value of R_2 in Figure 5–73 will cause excessive current?

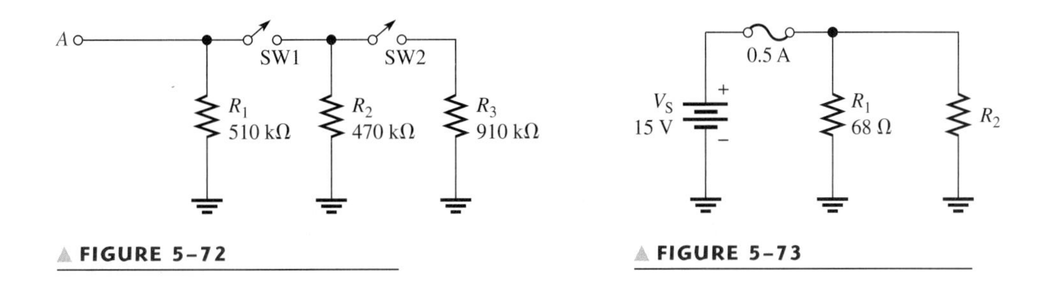

▲ FIGURE 5–72

▲ FIGURE 5–73

34. Determine the total current from the source and the current through each resistor for each switch position in Figure 5–74.

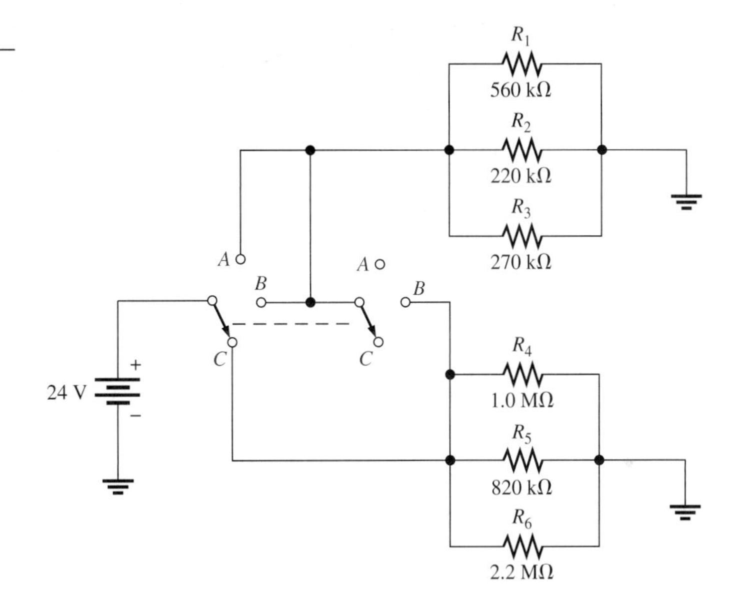

▶ FIGURE 5–74

35. The electrical circuit in a room has a ceiling lamp that draws 1.25 A and four wall outlets. Two table lamps that each draw 0.833 A are plugged into two outlets, and a TV set that draws 1 A is connected to the third outlet. When all of these items are in use, how much current is in the main line serving the room? If the main line is protected by a 5 A circuit breaker, how much current can be drawn from the fourth outlet? Draw a schematic of this wiring.

36. The total resistance of a parallel circuit is 25 Ω. What is the current through a 220 Ω resistor that makes up part of the parallel circuit if the total current is 100 mA?

37. Identify which groups of resistors are in parallel on the double-sided PC board in Figure 5–75. Determine the total resistance of each group.

FIGURE 5–75

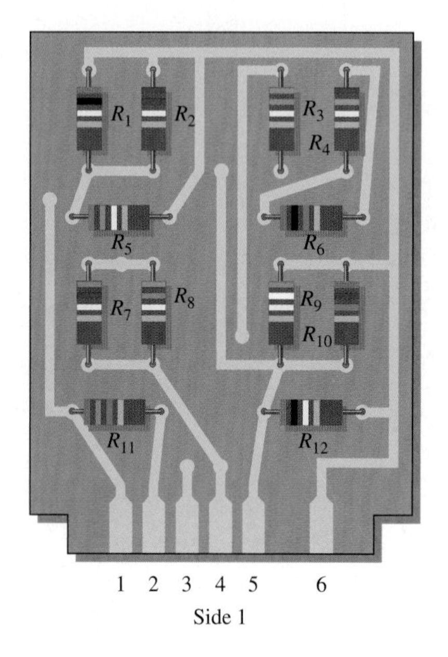

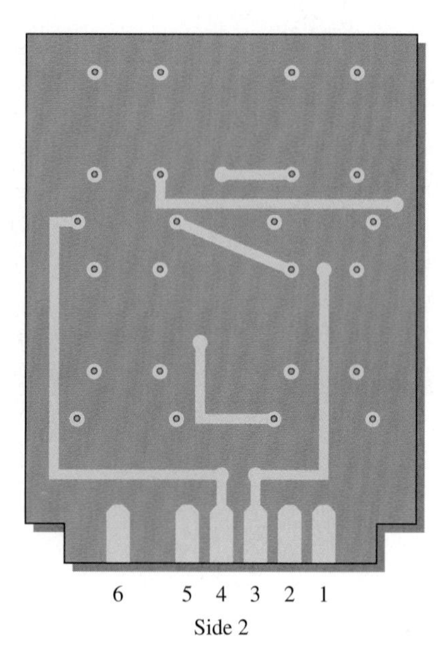

38. If the total resistance in Figure 5–76 is 200 Ω, what is the value of R_2?

FIGURE 5–76

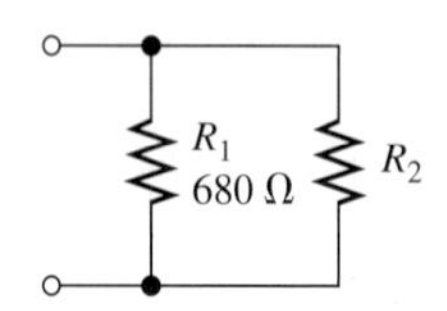

39. Determine the unknown resistances in Figure 5–77.

40. There is a total of 250 mA into a parallel circuit with a total resistance of 1.5 kΩ. The current must be increased by 25%. Determine how much resistance to add in parallel to accomplish this increase in current.

FIGURE 5–77

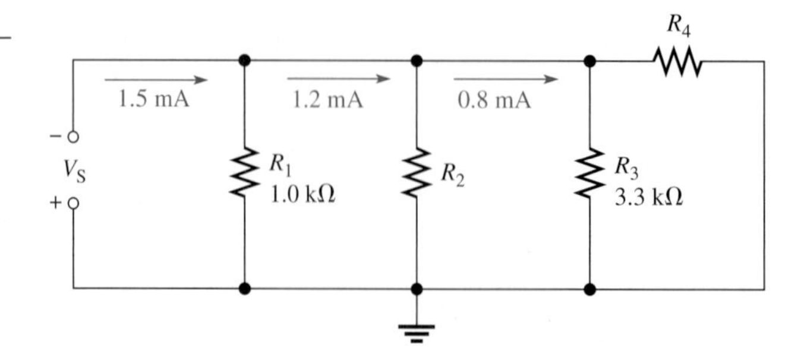

41. Sketch the schematic for the setup in Figure 5–78 and determine what is wrong with the circuit if 25 V are applied across the red and black leads.

FIGURE 5–78

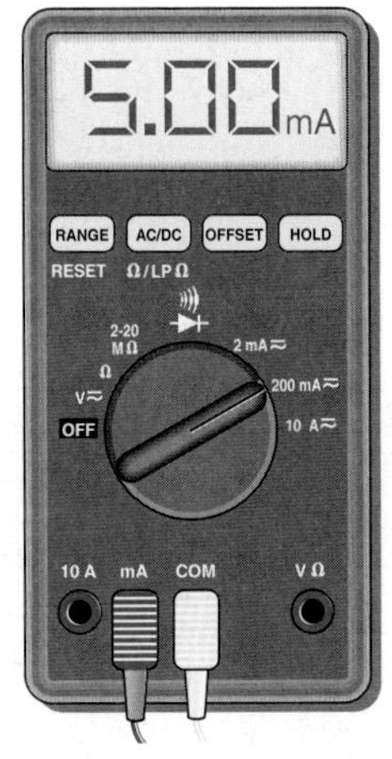

(a) Meter with yellow lead going to protoboard and red lead going to the positive terminal of the 25 V power supply.

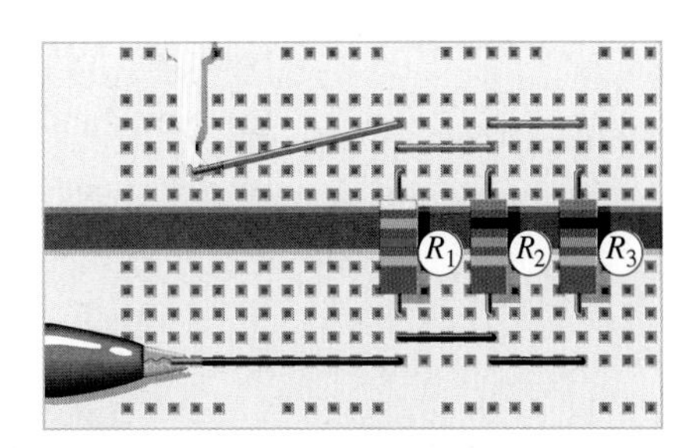

(b) Protoboard with leads connected. Yellow lead is from meter and gray lead is from 25 V power supply ground. The red meter lead goes to +25 V.

42. Develop a test procedure to check the circuit in Figure 5–79 to make sure that there are no open components. You must do this test without removing any component from the board. List the procedure in a detailed step-by-step format.

43. A certain parallel circuit consists of five ½ W resistors with the following values: 1.8 kΩ, 2.2 kΩ, 3.3 kΩ, 3.9 kΩ and 4.7 kΩ. As you slowly increase the voltage across the parallel circuit, the total current slowly increases. Suddenly, the total current drops to a lower value.

(a) Excluding a power supply failure, what happened?

(b) What is the maximum voltage you should have applied?

(c) Specifically, what must be done to repair the fault?

FIGURE 5–79

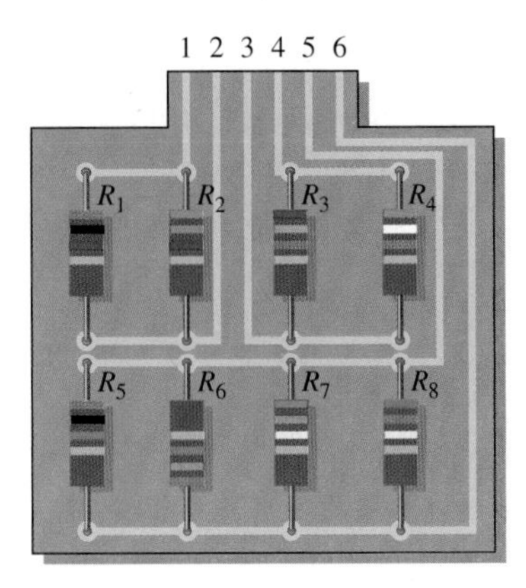

44. For the circuit board shown in Figure 5–80, determine the resistance between the following pins if there is a short between pins 2 and 4.

(a) pins 1 and 2 **(b)** pins 2 and 3

(c) pins 3 and 4 **(d)** pins 1 and 4

45. For the circuit board shown in Figure 5–80, determine the resistance between the following pins if there is a short between pins 3 and 4.

(a) pins 1 and 2 **(b)** pins 2 and 3

(c) pins 2 and 4 **(d)** pins 1 and 4

FIGURE 5–80

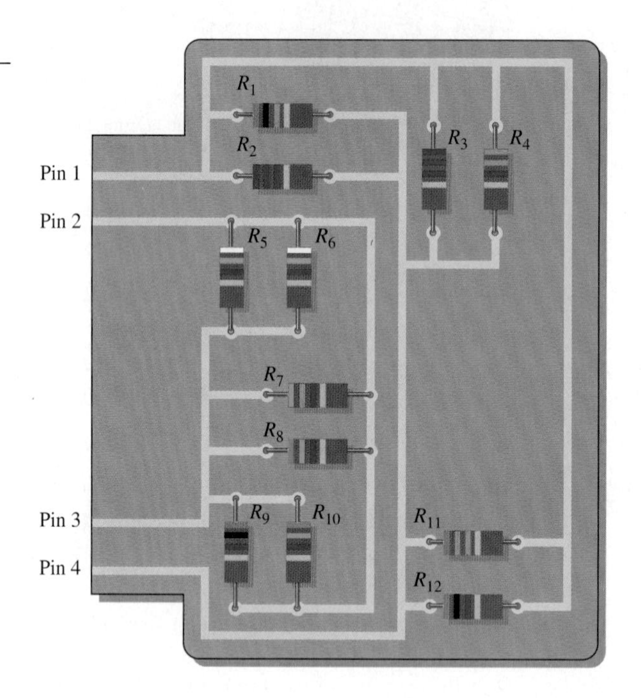

ELECTRONIC WORKBENCH/CIRCUITMAKER TROUBLESHOOTING PROBLEMS

CD-ROM file circuits are shown in Figure 5–81.

46. Open file P05-46 on your CD-ROM. Using current measurements, determine if there is a fault in the circuit. If so, identify the fault.

47. Open file P05-47 on your CD-ROM. Using current measurements, determine if there is a fault in the circuit. If so, identify the fault.

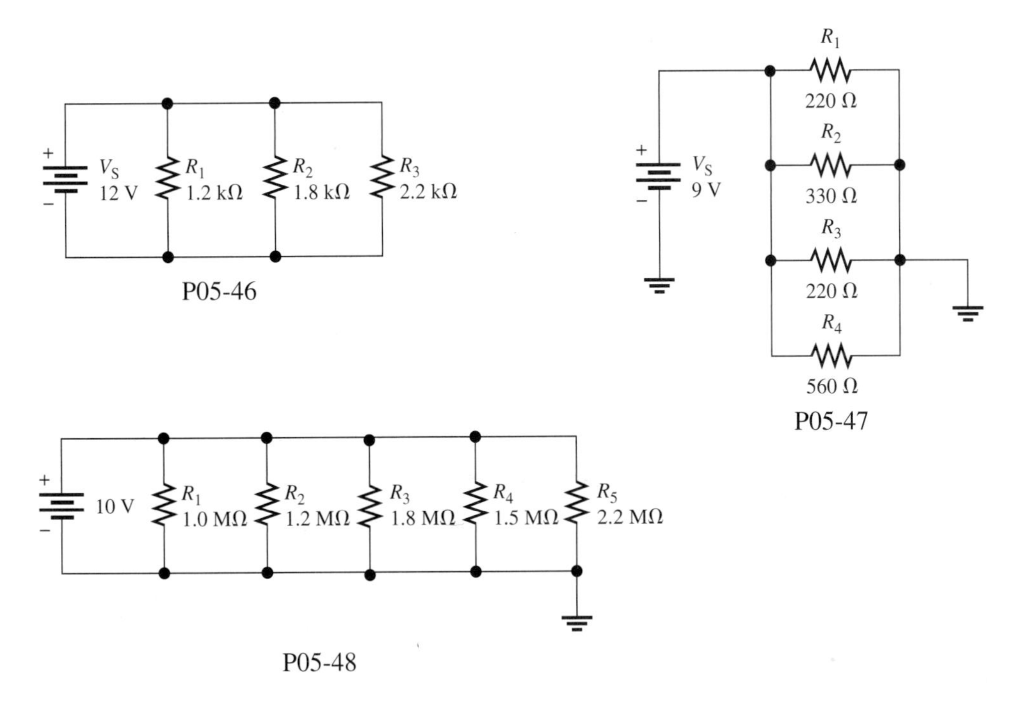

▲ FIGURE 5–81

48. Open file P05-48 on your CD-ROM. Using resistance measurements, determine if there is a fault in the circuit. If so, identify the fault.
49. Using EWB or CircuitMaker, construct the circuits on the circuit board shown in Figure 5–49. Measure the total resistance of each circuit and compare the calculated values.

ANSWERS

SECTION REVIEWS

SECTION 5–1 **Resistors in Parallel**

1. Parallel resistors are connected between the same two points.
2. A parallel circuit has more than one current path between two given points.
3. See Figure 5–82.

▶ FIGURE 5–82

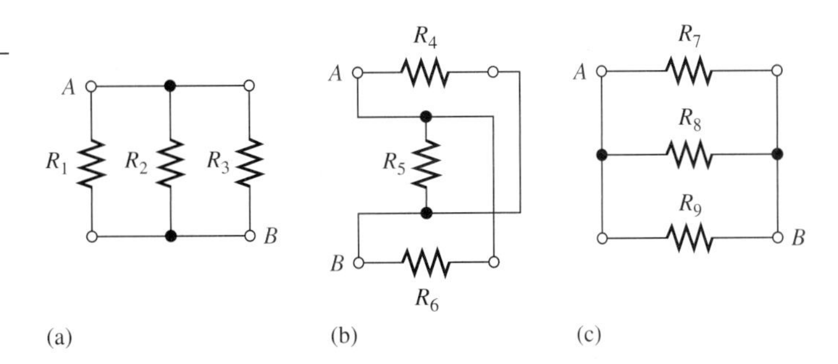

FIGURE 5–83

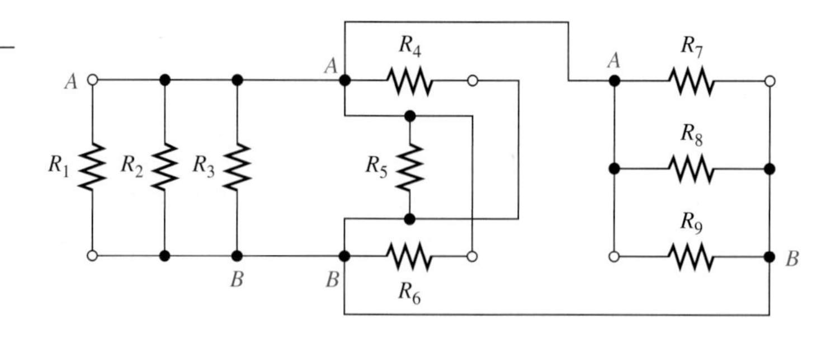

4. See Figure 5–83.

SECTION 5–2

Voltage in Parallel Circuits

1. 5 V
2. $V_{R2} = 118$ V; $V_S = 118$ V
3. $V_1 = 50$ V; $V_2 = 50$ V
4. Voltage is the same across all parallel branches.

SECTION 5–3

Kirchhoff's Current Law

1. Kirchhoff's current law: The algebraic sum of all currents at a junction is zero; The sum of the currents entering a junction equals the sum of all the currents leaving that junction.
2. $I_1 + I_2 + I_3 = 2.5$ A
3. 100 mA + 300 mA = 400 mA
4. $I_1 = 5\ \mu\text{A} - 2\ \mu\text{A} = 3\ \mu\text{A}$
5. 8 A − 1 A = 7 A entering, 8 A − 3 A = 5 A leaving

SECTION 5–4

Total Parallel Resistance

1. As more resistors are connected in parallel, the total resistance decreases.
2. R_T is always less than the smallest resistance value.
3. $R_T = 2.2\ \text{k}\Omega/12 = 183\ \Omega$

SECTION 5–5

Ohm's Law in Parallel Circuits

1. $I_T = 10\ \text{V}/(68\ \Omega/3) = 441$ mA
2. $V_S = 20\ \text{mA}(680\ \Omega \parallel 330\ \Omega) = 4.44$ V
3. $I_1 = 4.44\ \text{V}/680\ \Omega = 6.53$ mA; $I_2 = 4.44\ \text{V}/330\ \Omega = 13.5$ mA
4. $R_T = 4(12\ \text{V}/6\ \text{mA}) = 8\ \text{k}\Omega$
5. $V = (1.0\ \text{k}\Omega \parallel 2.2\ \text{k}\Omega)100\ \text{mA} = 68.8$ V

SECTION 5–6

Current Dividers

1. The 22 Ω has most current; the 200 Ω has least current.
2. $I_3 = (R_T/R_3)4\ \text{mA} = (113.6\ \Omega/470\ \Omega)4\ \text{mA} = 967\ \mu\text{A}$
3. $I_2 = (R_T/680\ \Omega)10\ \text{mA} = 3.27$ mA; $I_1 = (R_T/330\ \Omega)10\ \text{mA} = 6.73$ mA

SECTION 5–7

Power in Parallel Circuits

1. Add the powers in each resistor to get P_T.
2. $P_T = 1\ \text{W} + 2\ \text{W} + 5\ \text{W} + 8\ \text{W} = 16$ W
3. $P_T = (1\ \text{mA})^2 R_T = 615\ \mu\text{W}$

SECTION 5–8 **Troubleshooting**

1. When a parallel branch opens, there is no change in voltage and the total current decreases.
2. When the total resistance is greater than it should be, a branch is open.
3. Yes, all other bulbs continue to glow.
4. The current in each nonopen branch remains 1 A.
5. 1 A + 2.5 A = 3.5 A; therefore, the 1.2 A branch is open.

■ Application Assignment

1. 50 Ω; 9.94 Ω; 5.08 Ω
2. 2.5 mA
3. I_M = 0.5 mA; I_{R1} = 2 mA

RELATED PROBLEMS FOR EXAMPLES

5–1 No rewiring is necessary if terminals remain on the PC board.

5–2 Connect pin 1 to pin 2 and pin 3 to pin 4.

5–3 25 V **5–4** 20 mA

5–5 I_T = 112 mA; I_2 = 50 mA

5–6 2.5 mA; 5 mA

5–7 9.34 Ω

5–8 Replace Step 9 with "Press [+] key."
Display shows $100^{-1} + 47^{-1} + 22^{-1}+$.

Replace Step 10 with "Enter 33."
Display shows $100^{-1} + 47^{-1} + 22^{-1} + 33$.

Step 11: Press [2nd] [x^{-1}] keys.
Display shows $100^{-1} + 47^{-1} + 22^{-1} + 33^{-1}$.

Step 12: Press [ENTER] key. Display shows 107.034171502E^{-3}.

Step 13: Press [2nd] [x^{-1}] [ENTER] keys. Display shows 9.34281067406E0.

5–9 132 Ω

5–10 33.3 kΩ

5–11 2 Ω

5–12 I_1 = 100 mA; I_2 = 179 mA; 100 mA + 179 mA = 279 mA

5–13 I_1 = 20.0 mA; I_2 = 9.09 mA; I_3 = 35.7 mA; I_4 = 22.0 mA

5–14 8.62 mA

5–15 Use an ohmmeter to measure R_T and then calculate R_1 using the parallel-resistance formula.

5–16 I_1 = 59.4 mA; I_2 = 40.6 mA

5–17 I_1 = 1.63 mA; I_2 = 3.35 mA

5–18 177 W

5–19 162 W

5–20 15.4 mA

5–21 Not correct; R_{10} (68 kΩ) is open.

SELF-TEST

1. (b) **2.** (c) **3.** (b) **4.** (d) **5.** (a) **6.** (c) **7.** (c)
8. (c) **9.** (d) **10.** (b) **11.** (a) **12.** (d) **13.** (c) **14.** (b)

SERIES-PARALLEL CIRCUITS

INTRODUCTION

Various combinations of both series and parallel resistors are often found in electronic circuits. In this chapter, examples of such series-parallel arrangements are examined and analyzed. An important circuit called the *Wheatstone bridge* is introduced, and you will learn how complex circuits can be simplified using Thevenin's theorem. Also, circuits with more than one voltage source will be analyzed in simple steps using the superposition theorem. Troubleshooting series-parallel circuits for shorts and opens is also covered.

CHAPTER OUTLINE

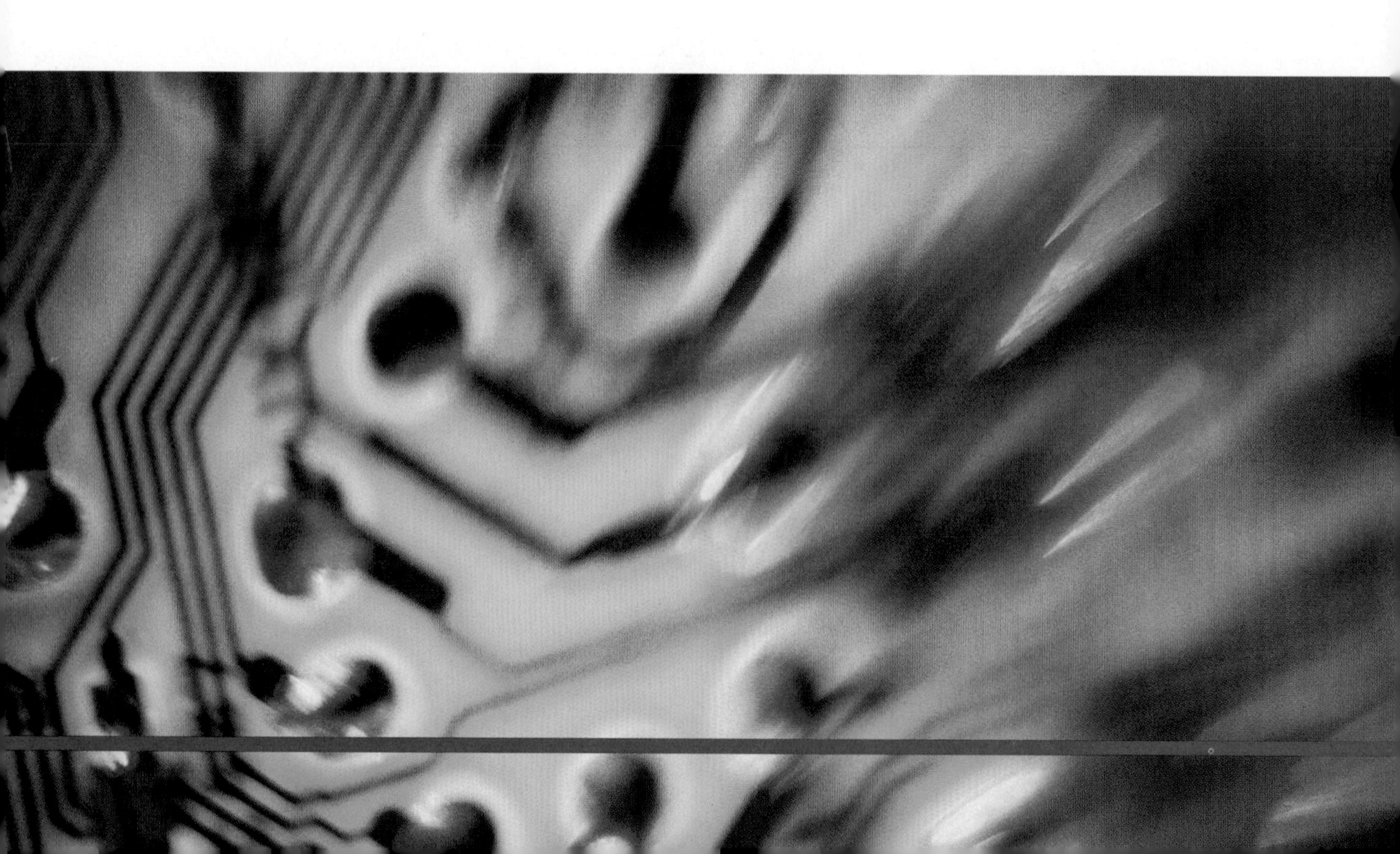

CHAPTER OBJECTIVES

- Identify series-parallel relationships
- Analyze series-parallel circuits
- Analyze loaded voltage dividers
- Determine the loading effect of a voltmeter on a circuit
- Analyze a Wheatstone bridge circuit
- Apply Thevenin's theorem to simplify a circuit for analysis
- Apply the superposition theorem to circuit analysis
- Troubleshoot series-parallel circuits

KEY TERMS

- Loading
- Load current
- Bleeder current
- Wheatstone bridge
- Balanced bridge
- Unbalanced bridge
- Thevenin's theorem
- Terminal equivalency
- Superposition

APPLICATION ASSIGNMENT PREVIEW

Your application assignment in this chapter is to evaluate a voltage-divider circuit board used in a portable power supply by applying your knowledge of loaded voltage dividers gained in this chapter as well as skills developed in previous chapters. The voltage divider in this application is designed to provide reference voltages to three different instruments that act as loads on the circuit. You will also be required to troubleshoot the circuit board for various common faults. After studying this chapter, you should be able to complete the application assignment.

WWW. VISIT THE COMPANION WEBSITE

Circuit Simulation Tutorials and Other Chapter Study Tools Are Available at

http://www.prenhall.com/floyd

6–1 IDENTIFYING SERIES-PARALLEL RELATIONSHIPS

A series-parallel circuit consists of combinations of both series and parallel current paths. It is important to be able to identify how the components in a circuit are arranged in terms of their series and parallel relationships.

After completing this section, you should be able to

- **Identify series-parallel relationships**
- Recognize how each resistor in a given circuit is related to the other resistors
- Determine series and parallel relationships on a PC board

HANDS ON TIP

When building a circuit on a protoboard from a schematic, it is easier to check the circuit by orienting the resistors on the protoboard in the same way that they appear on the schematic. In other words, the actual resistors on the board should be located so that the circuit has the same layout as the schematic.

Figure 6–1(a) shows an example of a simple series-parallel combination of resistors. Notice that the resistance from point A to point B is R_1. The resistance from point B to point C is R_2 and R_3 in parallel ($R_2 \| R_3$). The resistance from point A to point C is R_1 in series with the parallel combination of R_2 and R_3, as indicated in Figure 6–1(b).

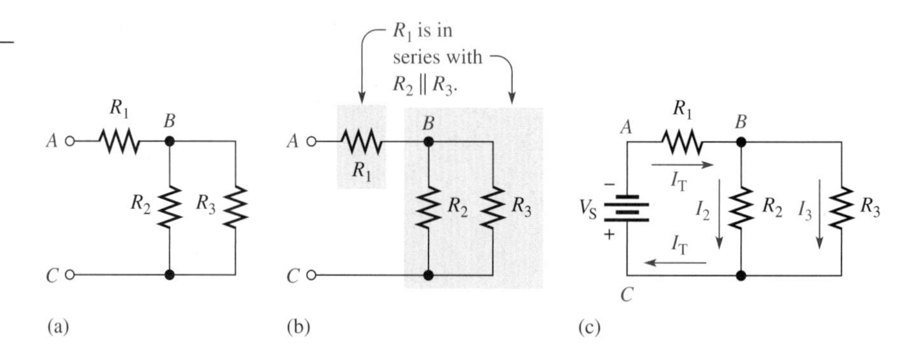

FIGURE 6–1

A simple series-parallel circuit.

When the circuit of Figure 6–1(a) is connected to a voltage source as shown in part (c), the total current through R_1 divides at point B into the two parallel paths. These two branch currents then recombine, and the total current is into the positive source terminal as shown.

Now, to illustrate series-parallel relationships, let's increase the complexity of the circuit in Figure 6–1(a) step-by-step. In Figure 6–2(a), another resistor (R_4) is connected in series with R_1. The resistance between points A and B is now $R_1 + R_4$, and this combination is in series with the parallel combination of R_2 and R_3, as illustrated in Figure 6–2(b).

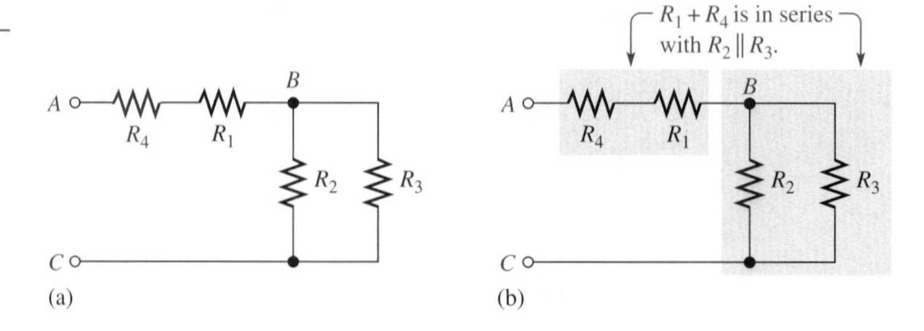

FIGURE 6–2

R_4 is added to the circuit in series with R_1.

In Figure 6–3(a), R_5 is connected in series with R_2. The series combination of R_2 and R_5 is in parallel with R_3. This entire series-parallel combination is in series with the $R_1 + R_4$ combination, as illustrated in Figure 6–3(b).

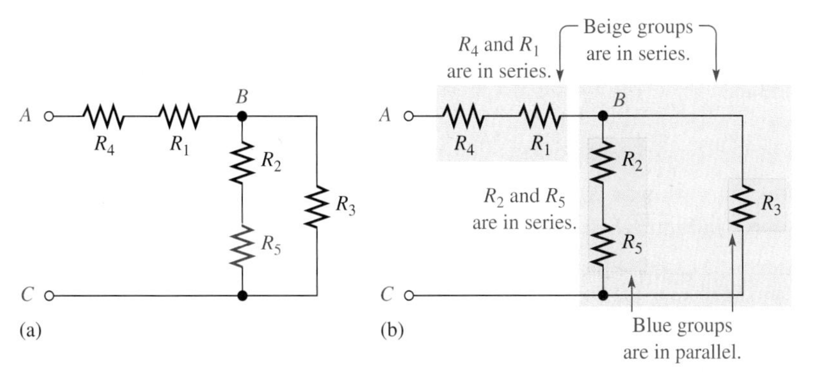

FIGURE 6–3

R_5 is added to the circuit in series with R_2.

In Figure 6–4(a), R_6 is connected in parallel with the series combination of R_1 and R_4. The series-parallel combination of R_1, R_4, and R_6 is in series with the series-parallel combination of R_2, R_3, and R_5, as indicated in Figure 6–4(b).

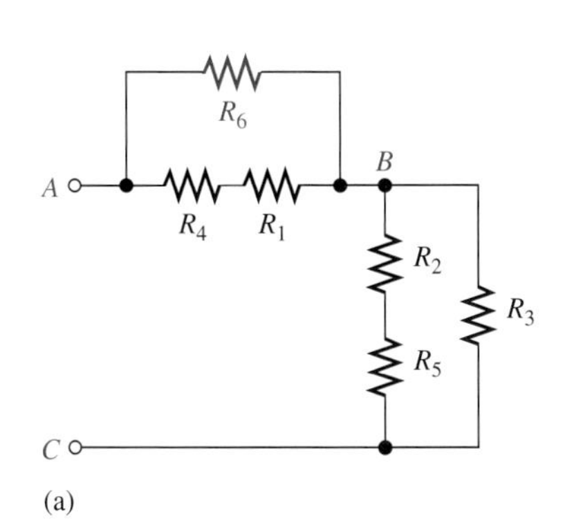

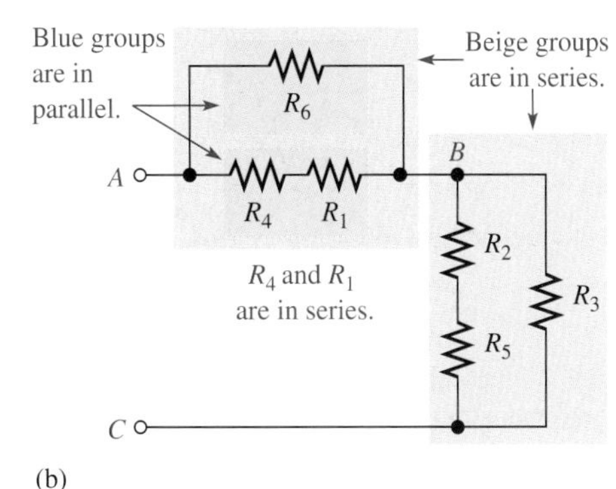

FIGURE 6–4

R_6 is added to the circuit in parallel with the series combination of R_1 and R_4.

EXAMPLE 6–1

Identify the series-parallel relationships in Figure 6–5.

Solution Starting at the negative terminal of the source, follow the current paths. All the current produced by the source must go through R_1, which is in series with the rest of the circuit.

The total current takes two paths when it gets to point A. Part of it is through R_2, and part of it is through R_3. Resistors R_2 and R_3 are in parallel with each other, and this parallel combination is in series with R_1.

At point B, the currents through R_2 and R_3 come together again into a single path. Thus, the total current is through R_4. Resistor R_4 is in series with both R_1 and the parallel combination of R_2 and R_3. The currents are shown in Figure 6–6, where I_T is the total current.

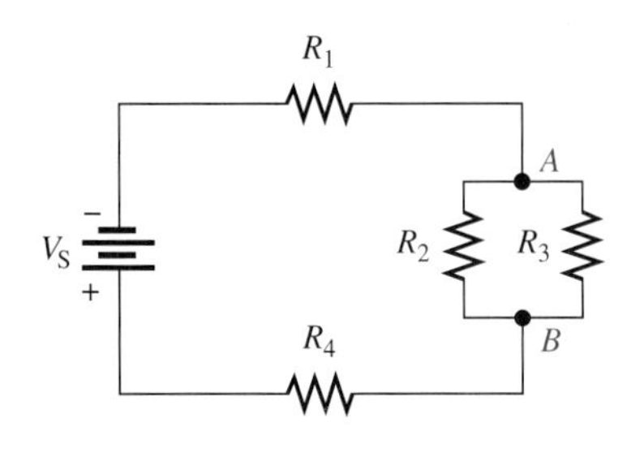

FIGURE 6–5

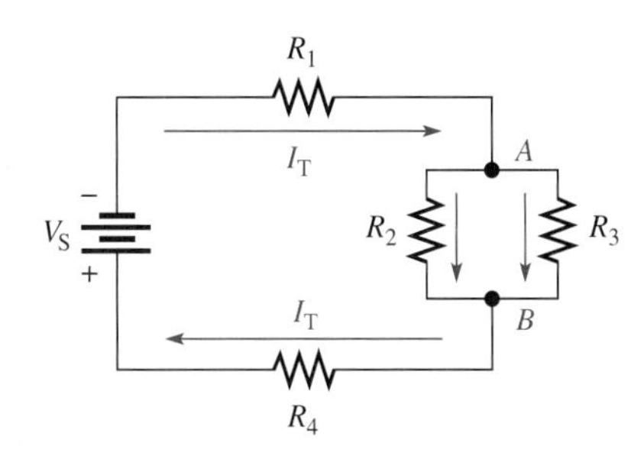

FIGURE 6–6

In summary, R_1 and R_4 are in series with the parallel combination of R_2 and R_3.

$$R_1 + R_2 \parallel R_3 + R_4$$

Related Problem* If another resistor, R_5, is connected from point A to the positive side of the source in Figure 6–6, what is its relationship to the other resistors?

*Answers are at the end of the chapter.

EXAMPLE 6–2

Describe the series-parallel combination between points A and D in Figure 6–7.

FIGURE 6–7

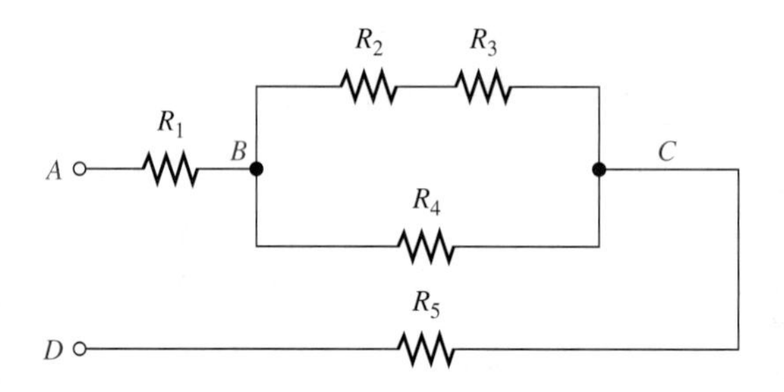

Solution Between points B and C, there are two parallel paths. The lower path consists of R_4, and the upper path consists of a series combination of R_2 and R_3. This parallel combination is in series with both R_1 and R_5.

In summary, R_1 and R_5 are in series with the parallel combination of R_4 and $(R_2 + R_3)$.

$$R_1 + R_5 + R_4 \parallel (R_2 + R_3)$$

Related Problem If a resistor is connected from point C to point D in Figure 6–7, describe its relationship in the circuit.

EXAMPLE 6–3

Describe the total resistance between each pair of points in Figure 6–8.

FIGURE 6–8

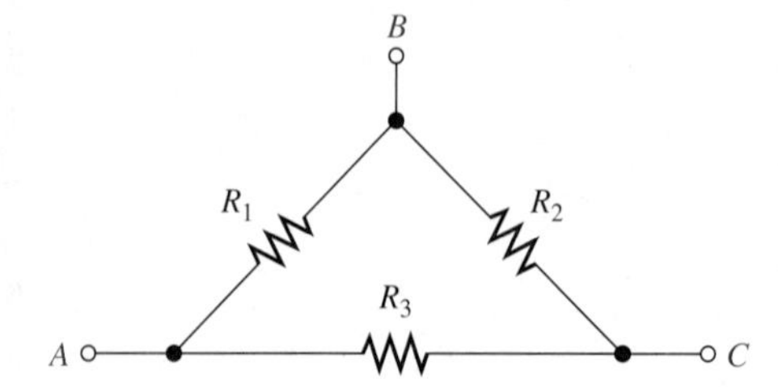

Solution

1. From point *A* to point *B*: R_1 is in parallel with the series combination of R_2 and R_3.
2. From point *A* to point *C*: R_3 is in parallel with the series combination of R_1 and R_2.
3. From point *B* to point *C*: R_2 is in parallel with the series combination of R_1 and R_3.

Related Problem In Figure 6–8, describe the total resistance between each point and an added ground if a new resistor, R_4, is connected from point *C* to ground. None of the existing resistors connect directly to the ground.

Redrawing a Schematic to Determine the Series-Parallel Relationships

Sometimes it is difficult to see the series-parallel relationships on a schematic because of the way in which it is drawn. In such a situation, it helps to redraw the diagram so that the relationships become clear. This is illustrated in Example 6–4.

EXAMPLE 6–4

Identify the series-parallel relationships in Figure 6–9.

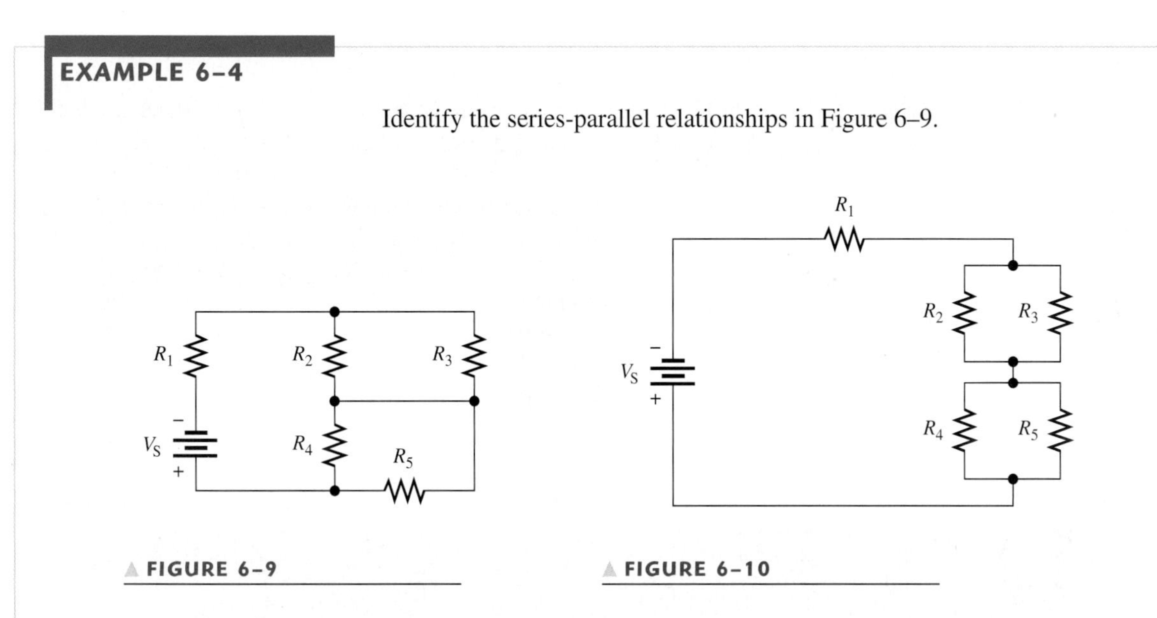

▲ FIGURE 6–9

▲ FIGURE 6–10

Solution The circuit schematic is redrawn in Figure 6–10 to better illustrate the series-parallel relationships. Now you can see that R_2 and R_3 are in parallel with each other and also that R_4 and R_5 are in parallel with each other. Both parallel combinations are in series with each other and with R_1.

$$R_1 + R_2 \parallel R_3 + R_4 \parallel R_5$$

Related Problem If a resistor is connected from the bottom end of R_3 to the top end of R_5 in Figure 6–10, what effect does it have on the circuit? Explain.

Determining Relationships on a Printed Circuit Board

Usually, the physical arrangement of components on a PC board bears no resemblance to the actual electrical relationships. By tracing out the circuit on the

PC board and rearranging the components on paper into a recognizable form, you can determine the series-parallel relationships. This is illustrated in Example 6–5.

EXAMPLE 6–5

Determine the relationships of the resistors on the PC board in Figure 6–11.

FIGURE 6–11

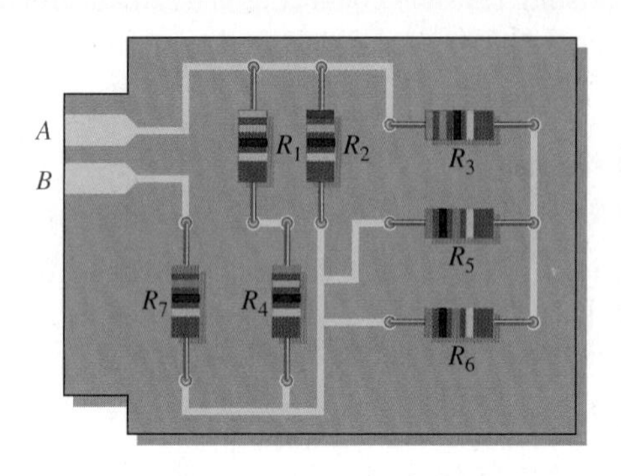

Solution In Figure 6–12(a), the schematic is drawn approximately in the same arrangement as that of the resistors on the board. In part (b), the resistors are reoriented so that the series-parallel relationships are more obvious. Resistors R_1 and R_4 are in series; $R_1 + R_4$ is in parallel with R_2; R_5 and R_6 are in parallel and this combination is in series with R_3. The R_3, R_5, and R_6 series-parallel combination is in parallel with both R_2 and the $R_1 + R_4$ combination. This entire series-parallel combination is in series with R_7. Figure 6–12(c) illustrates these relationships. Summarizing in equation form,

$$R_{AB} = (((R_5 \parallel R_6 + R_3 \parallel R_2)) \parallel (R_1 + R_4)) + R_7$$

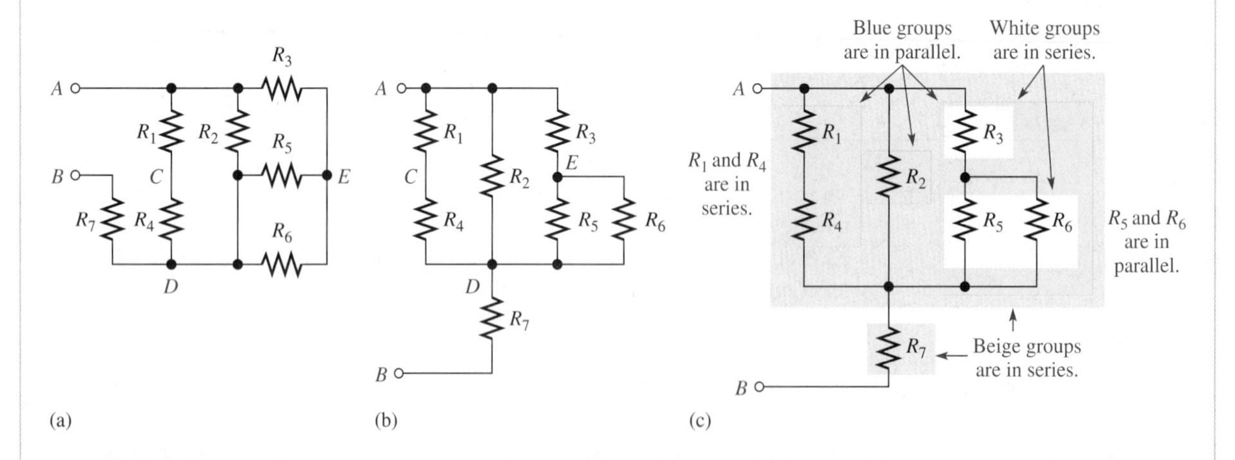

FIGURE 6–12

Related Problem What happens if there is an open connection between R_1 and R_4 on the printed circuit board in Figure 6–11?

SECTION 6-1 REVIEW

Answers are at the end of the chapter.

1. A certain series-parallel circuit is described as follows: R_1 and R_2 are in parallel. This parallel combination is in series with another parallel combination of R_3 and R_4. Sketch the circuit.
2. In the circuit of Figure 6–13, describe the series-parallel relationships of the resistors.
3. Which resistors are in parallel in Figure 6–14?

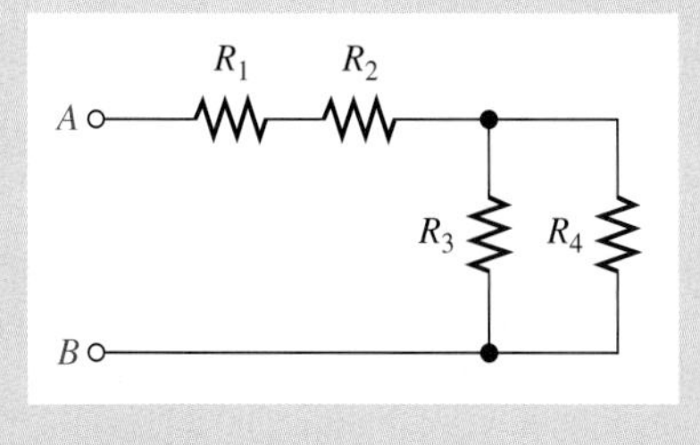

FIGURE 6–13

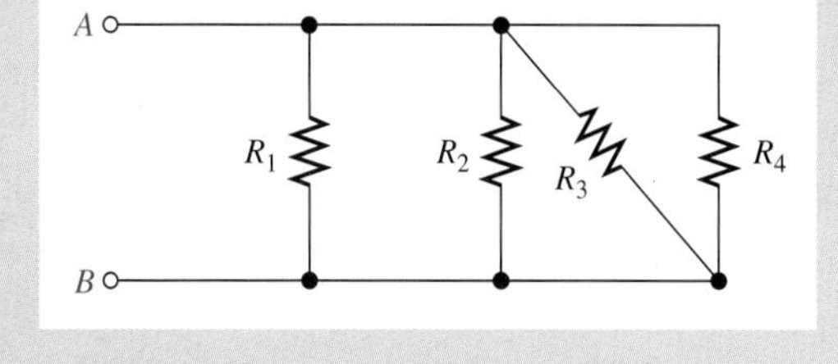

FIGURE 6–14

4. Identify the parallel resistors in Figure 6–15.
5. Are the parallel combinations in Figure 6–15 in series?

FIGURE 6–15

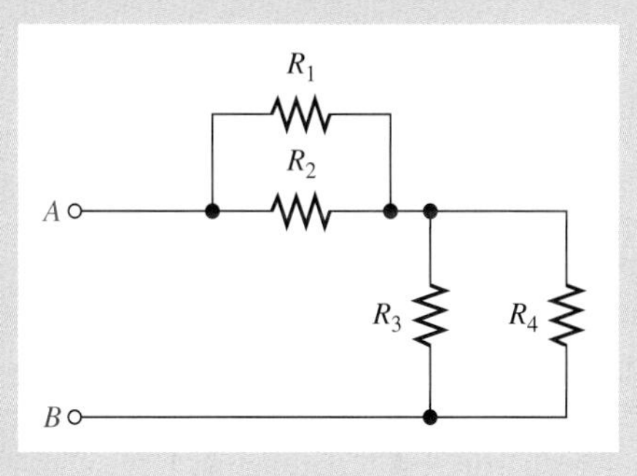

6–2 ANALYSIS OF SERIES-PARALLEL CIRCUITS

Several quantities are important when you have a circuit that is a series-parallel configuration of resistors. Analysis methods learned in the previous chapters can be applied to determine resistances, currents, and voltages.

After completing this section, you should be able to

- **Analyze series-parallel circuits**
- Determine total resistance
- Determine all the currents
- Determine all the voltage drops

Total Resistance

In Chapter 4, you learned how to determine total series resistance. In Chapter 5, you learned how to determine total parallel resistance. To find the total resistance (R_T) of a series-parallel combination, first identify the series and parallel relationships, and then apply what you have previously learned. The following two examples illustrate the general approach.

EXAMPLE 6–6

Determine R_T between points A and B of the circuit in Figure 6–16.

▶ FIGURE 6–16

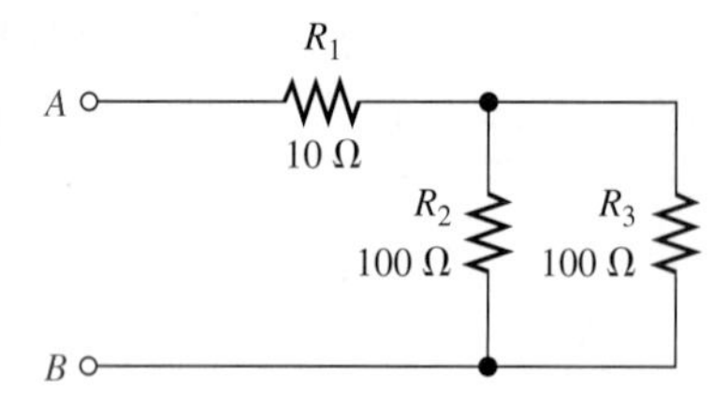

Solution Resistors R_2 and R_3 are in parallel, and this parallel combination is in series with R_1. First find the parallel resistance of R_2 and R_3. Since R_2 and R_3 are equal in value, divide the value by 2.

$$R_{2\|3} = \frac{R}{n} = \frac{100\ \Omega}{2} = 50\ \Omega$$

Now, since R_1 is in series with $R_{2\|3}$, add their values.

$$R_T = R_1 + R_{2\|3} = 10\ \Omega + 50\ \Omega = \mathbf{60\ \Omega}$$

Related Problem Determine R_T in Figure 6–16 if R_3 is changed to 82 Ω.

Open file E06-06 on your EWB/CircuitMaker CD-ROM. Using the multimeter, verify the calculated value of total resistance. Change R_1 to 18 Ω, R_2 to 82 Ω, and R_3 to 82 Ω and measure the total resistance.

EXAMPLE 6–7

Find R_T of the circuit in Figure 6–17.

▶ FIGURE 6–17

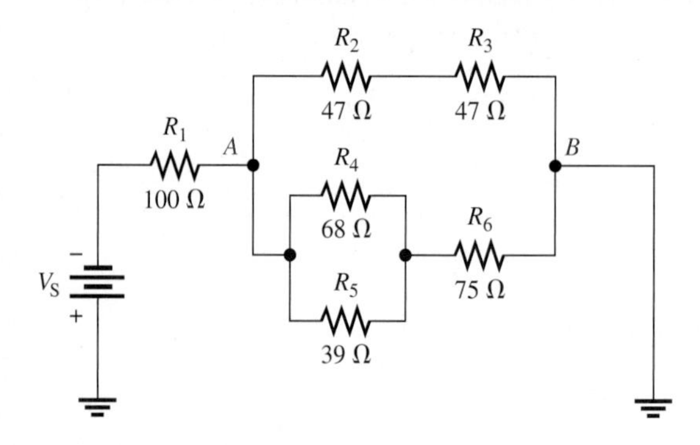

Solution **Step 1:** In the upper branch between points A and B, R_2 is in series with R_3. The series combination is designated $R_{2\|3}$ and is equal to $R_2 + R_3$.

$$R_{2\|3} = R_2 + R_3 = 47\ \Omega + 47\ \Omega = 94\ \Omega$$

Step 2: In the lower branch, R_4 and R_5 are in parallel with each other. This parallel combination is designated $R_{4\|5}$.

$$R_{4\|5} = \frac{R_4R_5}{R_4 + R_5} = \frac{(68\ \Omega)(39\ \Omega)}{68\ \Omega + 39\ \Omega} = 24.8\ \Omega$$

Step 3: Also in the lower branch, the parallel combination of R_4 and R_5 is in series with R_6. This series-parallel combination is designated $R_{4\|5+6}$.

$$R_{4\|5+6} = R_6 + R_{4\|5} = 75\ \Omega + 24.8\ \Omega = 99.8\ \Omega$$

Figure 6–18 shows the original circuit in a simplified equivalent form.

FIGURE 6–18

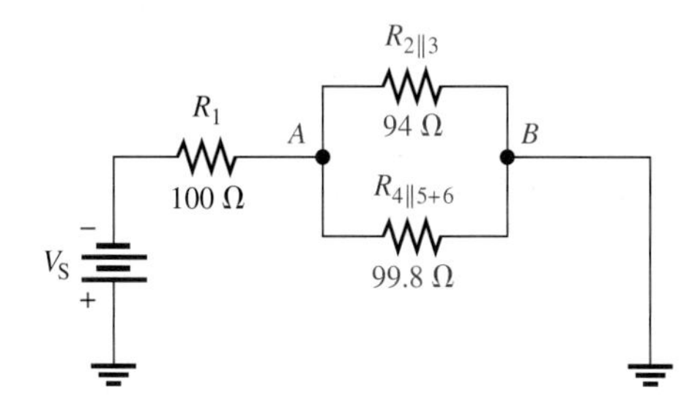

Step 4: Now you can find the resistance between points A and B. It is $R_{2\|3}$ in parallel with $R_{4\|5+6}$. The equivalent resistance is calculated as follows:

$$R_{AB} = \frac{1}{\dfrac{1}{R_{2\|3}} + \dfrac{1}{R_{4\|5+6}}} = \frac{1}{\dfrac{1}{94\ \Omega} + \dfrac{1}{99.8\ \Omega}} = 48.4\ \Omega$$

Step 5: Finally, the total circuit resistance is R_1 in series with R_{AB}.

$$R_T = R_1 + R_{AB} = 100\ \Omega + 48.4\ \Omega = \mathbf{148.4\ \Omega}$$

Related Problem Determine R_T if a 68 Ω resistor is connected to the circuit in Figure 6–17 from point A to point B.

Open file E06-07 on your EWB/CircuitMaker CD-ROM. Verify the total calculated resistance. Remove R_5 from the circuit and measure the total resistance. Then check your measured value by calculating the total resistance.

Total Current

Once the total resistance and the source voltage are known, you can find total current in a circuit by applying Ohm's law. Total current is the total source voltage divided by the total resistance.

$$I_T = \frac{V_S}{R_T}$$

For example, let's find the total current in the circuit of Example 6–7 (Figure 6–17). Assume that the source voltage is 30 V. The calculation is

$$I_T = \frac{V_S}{R_T} = \frac{30\text{ V}}{148.4\ \Omega} = 202\text{ mA}$$

Branch Currents

Using the *current-divider formula, Kirchhoff's current law, Ohm's law,* or combinations of these, you can find the current in any branch of a series-parallel circuit. In some cases, it may take repeated application of the formula to find a given current.

EXAMPLE 6–8

Determine the current through R_4 in Figure 6–19 if $V_S = 50$ V.

▶ FIGURE 6–19

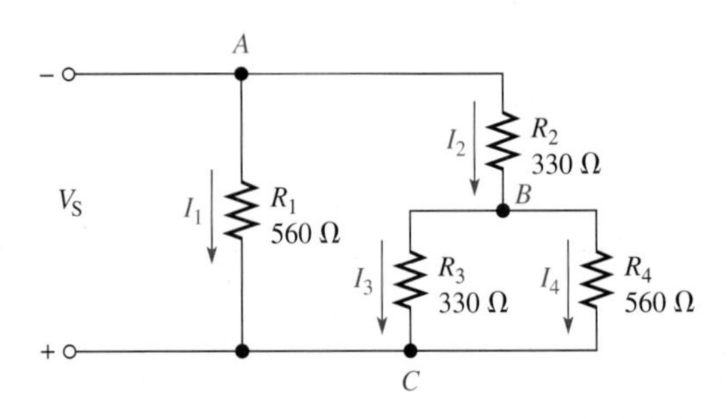

Solution First, find the current (I_2) into the junction of R_3 and R_4 at point B. Once you know this current, you can use the current-divider formula to find I_4, the current through R_4.

Notice that there are two main branches in the circuit. The left-most branch consists of only R_1. The right-most branch has R_2 in series with the parallel combination of R_3 and R_4. The voltage across both of these main branches is the same and equal to 50 V. Find the current (I_2) into the junction of R_3 and R_4 by calculating the equivalent resistance ($R_{2+3\|4}$) of the right-most main branch and then applying Ohm's law; I_2 is the total current through this main branch. Thus,

$$R_{2+3\|4} = R_2 + \frac{R_3R_4}{R_3 + R_4} = 330\ \Omega + \frac{(330\ \Omega)(560\ \Omega)}{890\ \Omega} = 538\ \Omega$$

$$I_2 = \frac{V_S}{R_{2+3\|4}} = \frac{50\text{ V}}{538\ \Omega} = 93\text{ mA}$$

Use the two-resistor current-divider formula to calculate I_4.

$$I_4 = \left(\frac{R_3}{R_3 + R_4}\right)I_2 = \left(\frac{330\ \Omega}{890\ \Omega}\right)93\text{ mA} = \mathbf{34.5\text{ mA}}$$

You can find the current through R_3 by applying Kirchhoff's law ($I_2 = I_3 + I_4$) and subtracting I_4 from I_2 ($I_3 = I_2 - I_4$) or by applying the two-resistor current-divider formula again.

You can find the current through R_1 by using Ohm's law ($I_1 = V_S/R_1$).

The calculator solution for I_4 is

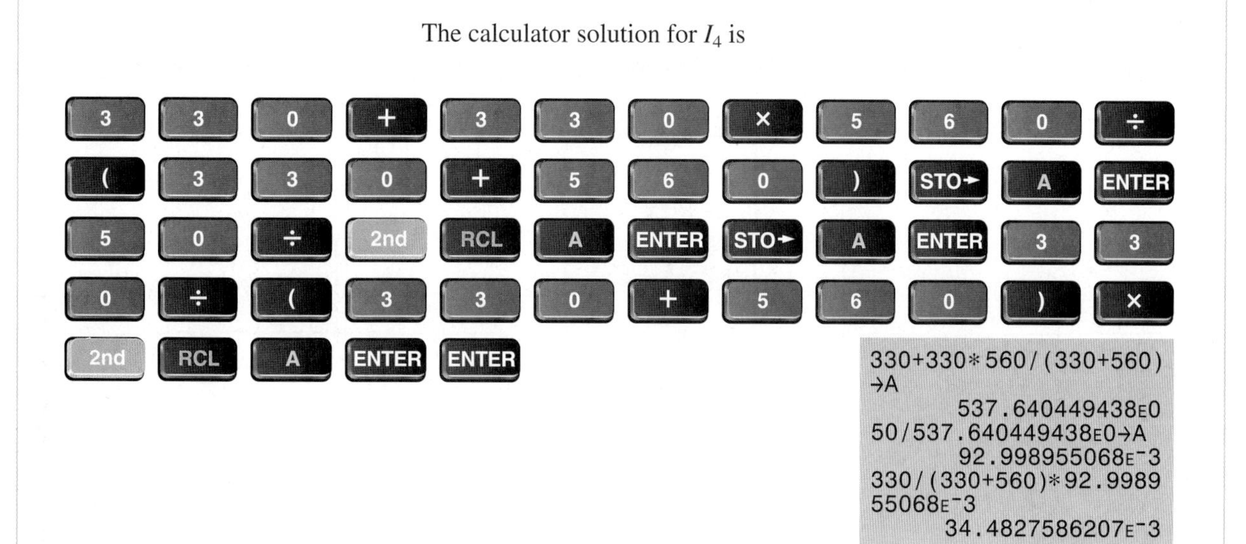

Related Problem Find I_1, I_3, and I_T in Figure 6–19.

Open file E06-08 on your EWB/CircuitMaker CD-ROM. Measure the current in each resistor. Compare the measurements with the calculated values.

Voltage Relationships

The circuit in Figure 6–20 illustrates voltage relationships in a series-parallel circuit. Voltmeters are connected to measure each of the resistor voltages, and the readings are indicated.

Some general observations about Figure 6–20 are as follows:

1. V_{R1} and V_{R2} are equal because R_1 and R_2 are in parallel. (Recall that voltages across parallel branches are the same.) V_{R1} and V_{R2} are the same as the voltage from A to B.

2. V_{R3} is equal to $V_{R4} + V_{R5}$ because R_3 is in parallel with the series combination of R_4 and R_5. (V_{R3} is the same as the voltage from B to C.)

3. V_{R4} is about one-third of the voltage from B to C because R_4 is about one-third of the resistance $R_4 + R_5$ (by the voltage-divider principle).

4. V_{R5} is about two-thirds of the voltage from B to C because R_5 is about two-thirds of $R_4 + R_5$.

5. $V_{R1} + V_{R3}$ equals V_S because, by Kirchhoff's voltage law, the sum of the voltage drops around a single closed path must equal the source voltage.

Example 6–9 will verify the meter readings in Figure 6–20.

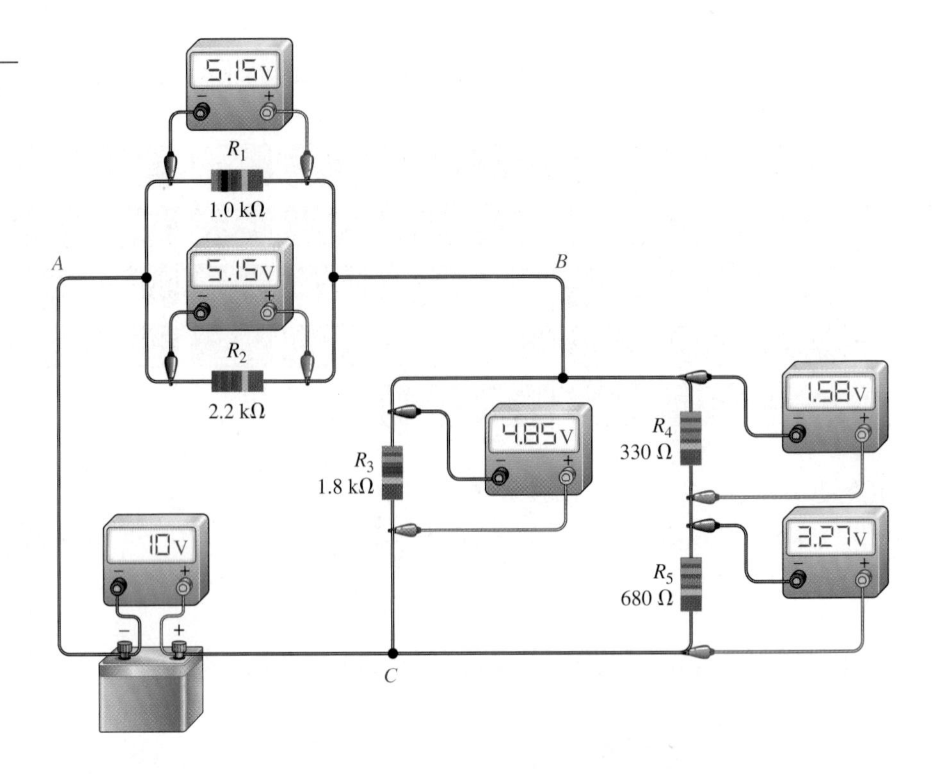

▶ **FIGURE 6–20**

Illustration of voltage relationships.

EXAMPLE 6–9

Verify that the voltmeter readings in Figure 6–20 are correct. The circuit is redrawn as a schematic in Figure 6–21.

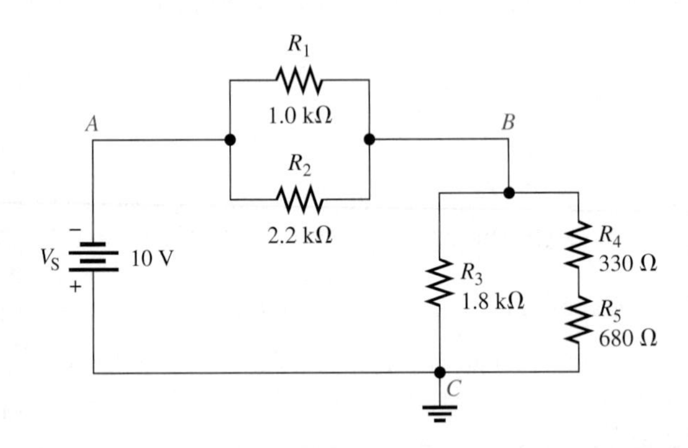

▶ **FIGURE 6–21**

Solution The resistance from A to B is the parallel combination of R_1 and R_2.

$$R_{AB} = \frac{R_1R_2}{R_1 + R_2} = \frac{(1.0\text{ k}\Omega)(2.2\text{ k}\Omega)}{3.2\text{ k}\Omega} = 688\ \Omega$$

The resistance from B to C is R_3 in parallel with the series combination of R_4 and R_5.

$$R_4 + R_5 = 330\ \Omega + 680\ \Omega = 1010\ \Omega = 1.01\text{ k}\Omega$$

$$R_{BC} = \frac{R_3(R_4 + R_5)}{R_3 + R_4 + R_5} = \frac{(1.8\text{ k}\Omega)(1.01\text{ k}\Omega)}{2.81\text{ k}\Omega} = 647\ \Omega$$

The resistance from A to B is in series with the resistance from B to C, so the total circuit resistance is

$$R_T = R_{AB} + R_{BC} = 688\ \Omega + 647\ \Omega = 1335\ \Omega$$

Using the voltage-divider principle, you can calculate the voltages.

$$V_{AB} = \left(\frac{R_{AB}}{R_T}\right)V_S = \left(\frac{688\ \Omega}{1335\ \Omega}\right)10\ \text{V} = 5.15\ \text{V}$$

$$V_{BC} = \left(\frac{R_{BC}}{R_T}\right)V_S = \left(\frac{647\ \Omega}{1335\ \Omega}\right)10\ \text{V} = 4.85\ \text{V}$$

$$V_{R1} = V_{R2} = V_{AB} = \mathbf{5.15\ V}$$

$$V_{R3} = V_{BC} = \mathbf{4.85\ V}$$

$$V_{R4} = \left(\frac{R_4}{R_4 + R_5}\right)V_{BC} = \left(\frac{330\ \Omega}{1010\ \Omega}\right)4.85\ \text{V} = \mathbf{1.58\ V}$$

$$V_{R5} = \left(\frac{R_5}{R_4 + R_5}\right)V_{BC} = \left(\frac{680\ \Omega}{1010\ \Omega}\right)4.85\ \text{V} = \mathbf{3.27\ V}$$

Related Problem Determine each voltage drop in Figure 6–21 if the source voltage is doubled.

Open file E06-09 on your EWB/CircuitMaker CD-ROM. Measure the voltage across each resistor and compare to the calculated values. Verify by measurement that each voltage drop doubles if the source voltage is doubled and is halved if the source voltage is halved.

EXAMPLE 6–10

Determine the voltage drop across each resistor in Figure 6–22.

▶ **FIGURE 6–22**

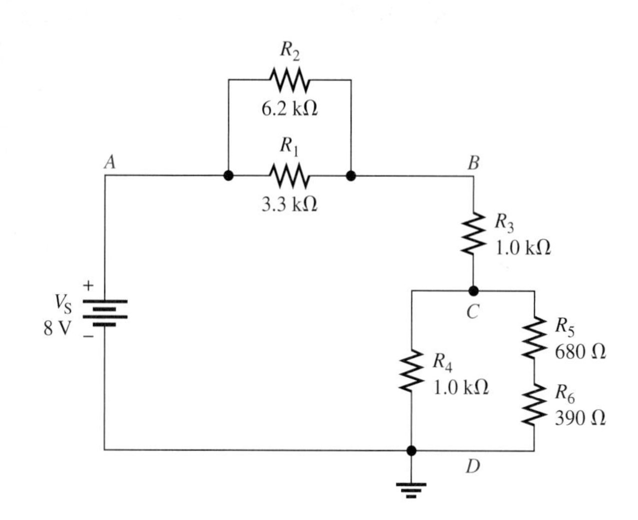

Solution Because you know the total voltage, you can solve this problem using the voltage-divider formula. First reduce each parallel combination to an equivalent resistance. Since R_1 and R_2 are in parallel between points A and B, combine their values.

$$R_{AB} = \frac{R_1 R_2}{R_1 + R_2} + \frac{(3.3\text{ k}\Omega)(6.2\text{ k}\Omega)}{9.5\text{ k}\Omega} = 2.15\text{ k}\Omega$$

Since R_4 is in parallel with the series combination of R_5 and R_6 between points C and D, combine these values.

$$R_{CD} = \frac{R_4(R_5 + R_6)}{R_4 + R_5 + R_6} + \frac{(1.0\text{ k}\Omega)(1.07\text{ k}\Omega)}{2.07\text{ k}\Omega} = 517\ \Omega$$

The equivalent circuit is drawn in Figure 6–23. The total circuit resistance is

$$R_T = R_{AB} + R_3 + R_{CD} = 2.15\text{ k}\Omega + 1.0\text{ k}\Omega + 517\ \Omega = 3.67\text{ k}\Omega$$

FIGURE 6–23

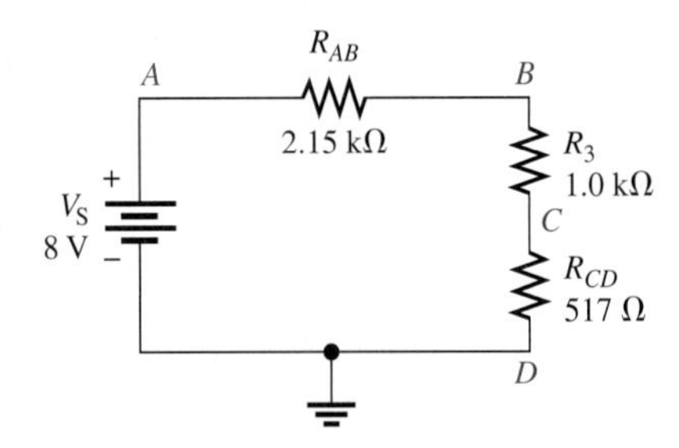

Apply the voltage-divider formula to solve for the voltages in the equivalent circuit.

$$V_{AB} = \left(\frac{R_{AB}}{R_T}\right)V_S = \left(\frac{2.15\text{ k}\Omega}{3.67\text{ k}\Omega}\right)8\text{ V} = 4.69\text{ V}$$

$$V_{BC} = \left(\frac{R_3}{R_T}\right)V_S = \left(\frac{1.0\text{ k}\Omega}{3.67\text{ k}\Omega}\right)8\text{ V} = 2.18\text{ V}$$

$$V_{CD} = \left(\frac{R_{CD}}{R_T}\right)V_S = \left(\frac{517\ \Omega}{3.67\text{ k}\Omega}\right)8\text{ V} = 1.13\text{ V}$$

Refer to Figure 6–22. V_{AB} equals the voltage across both R_1 and R_2.

$$V_{R1} = V_{R2} = V_{AB} = \mathbf{4.69\text{ V}}$$

V_{BC} is the voltage across R_3.

$$V_{R3} = V_{BC} = \mathbf{2.18\text{ V}}$$

V_{CD} is the voltage across R_4 and also across the series combination of R_5 and R_6.

$$V_{R4} = V_{CD} = \mathbf{1.13\text{ V}}$$

Now apply the voltage-divider formula to the series combination of R_5 and R_6 to get V_{R5} and V_{R6}.

$$V_{R5} = \left(\frac{R_5}{R_5 + R_6}\right)V_{CD} = \left(\frac{680\ \Omega}{1070\ \Omega}\right)1.13\text{ V} = \mathbf{718\text{ mV}}$$

$$V_{R6} = \left(\frac{R_6}{R_5 + R_6}\right)V_{CD} = \left(\frac{390\ \Omega}{1070\ \Omega}\right)1.13\text{ V} = \mathbf{412\text{ mV}}$$

Related Problem Determine the current and power in each resistor in Figure 6–22.

Open file E06-10 on your EWB/CircuitMaker CD-ROM. Measure the voltage across each resistor and compare to the calculated values. If R_4 is increased to 2.2 kΩ, specify which voltage drops increase and which ones decrease. Verify this by measurement.

As you have seen in this section, the analysis of series-parallel circuits can be approached in many ways, depending on what information you need and what circuit values you know. The examples in this section do not represent an exhaustive coverage, but they give you an idea of how to approach series-parallel circuit analysis.

If you know Ohm's law, Kirchhoff's laws, the voltage-divider formula, and the current-divider formula, and if you know how to apply these laws, you can solve most resistive circuit analysis problems. The ability to recognize series and parallel combinations is, of course, essential. There is no standard "cookbook" approach that can be applied to all situations. Logical thought is the most powerful tool you can apply to problem solving.

SECTION 6–2 REVIEW

1. Find the total resistance between *A* and *B* in the circuit of Figure 6–24.
2. Find the current through R_3 in Figure 6–24.
3. Find V_{R2} in Figure 6–24.
4. Determine R_T and I_T in Figure 6–25.

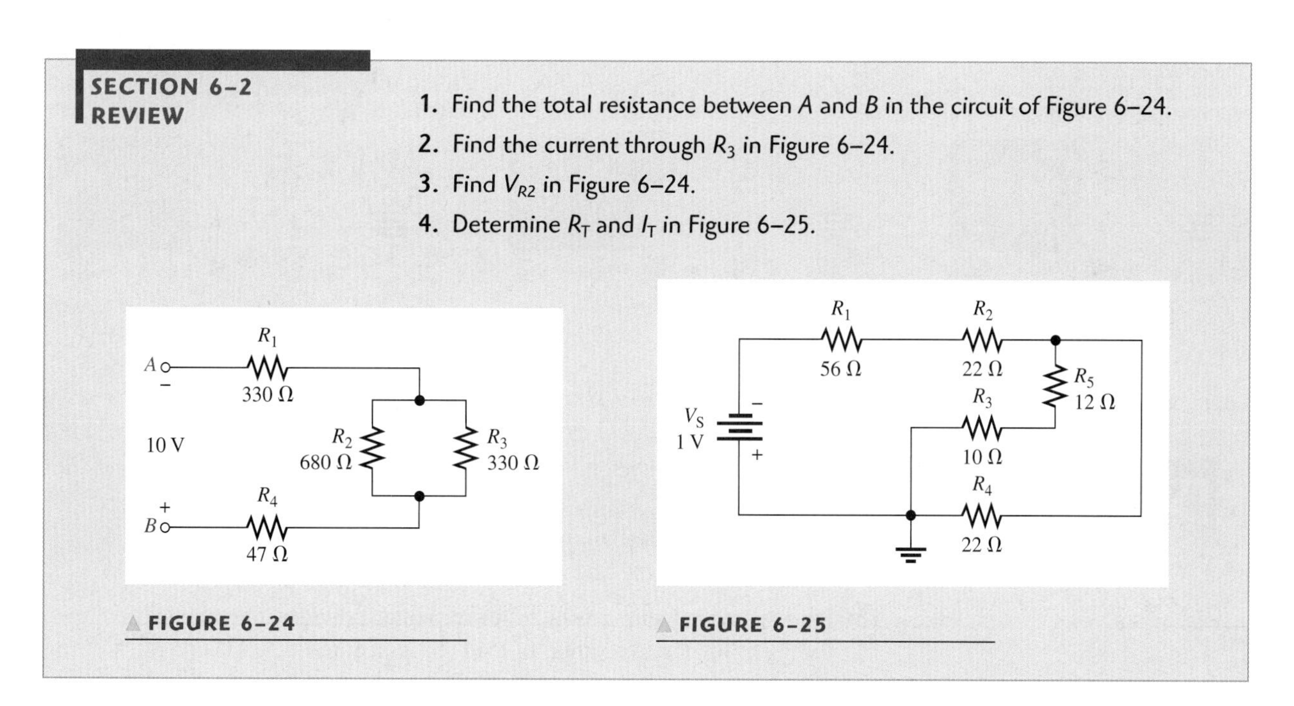

FIGURE 6–24

FIGURE 6–25

6–3 VOLTAGE DIVIDERS WITH RESISTIVE LOADS

Voltage dividers were introduced in Chapter 4. In this section, you will learn how resistive loads affect the operation of voltage-divider circuits.

After completing this section, you should be able to

- **Analyze loaded voltage dividers**
- Determine the effect of a resistive load on a voltage-divider circuit
- Define *bleeder current*

The voltage divider in Figure 6–26(a) produces an output voltage (V_{OUT}) of 5 V because the input is 10 V and the two resistors are of equal value. This voltage is the unloaded output voltage. When a load resistor, R_L, is connected from the output to ground as shown in Figure 6–26(b), the output voltage is reduced by an amount that depends on the value of R_L. This effect is called **loading**. The load resistor is in parallel with R_2, reducing the resistance from point A to ground and, as a result, also reducing the voltage across the parallel combination. This is one effect of loading a voltage divider. Another effect of a load is that more current is drawn from the source because the total resistance of the circuit is reduced.

FIGURE 6–26

A voltage divider with both unloaded and loaded outputs.

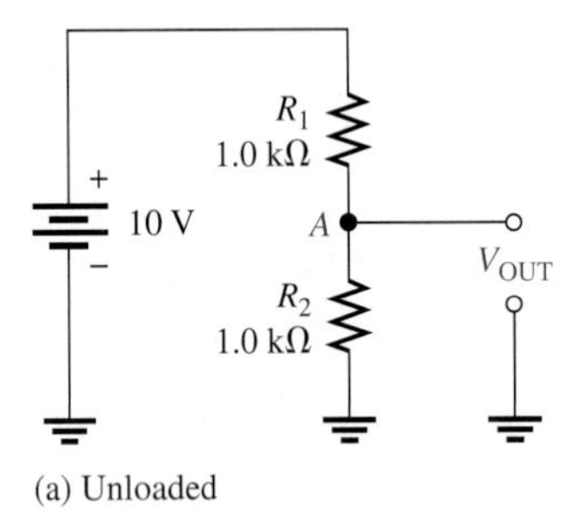

(a) Unloaded

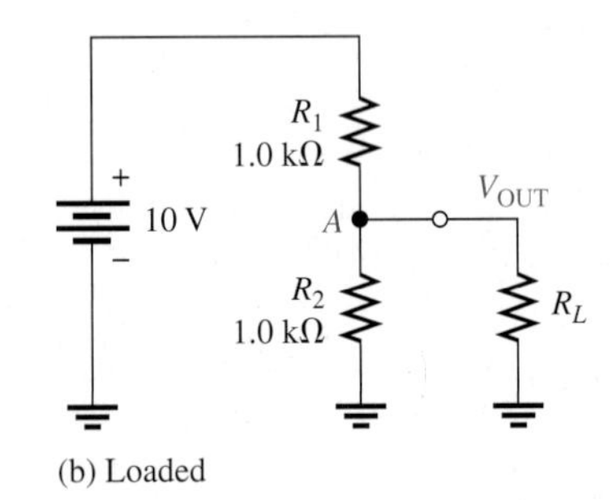

(b) Loaded

The larger R_L is, compared to R_2, the less the output voltage is reduced from its unloaded value, as illustrated in Figure 6–27. When two resistors are connected in parallel and one of the resistors is much greater than the other, the total resistance is close to the value of the smaller resistance.

FIGURE 6–27

The effect of a load resistor.

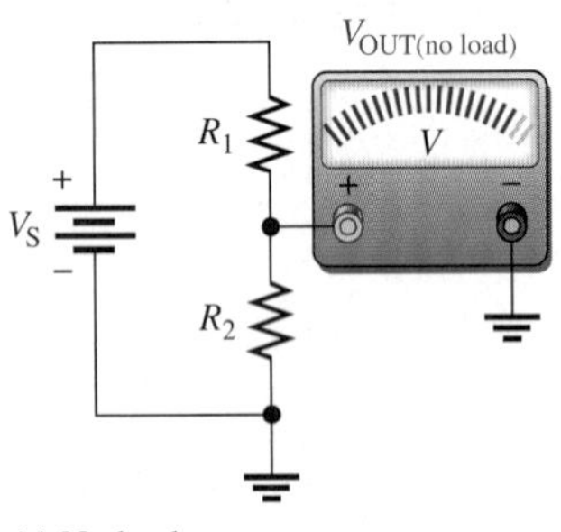

(a) No load

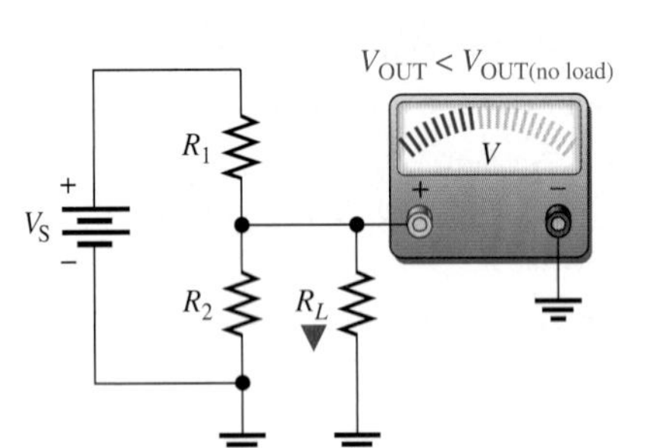

(b) R_L not significantly greater than R_2

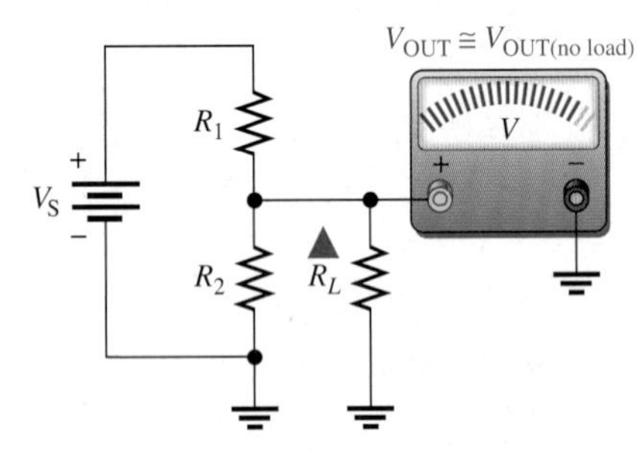

(c) R_L much greater than R_2

EXAMPLE 6–11

(a) Determine the unloaded output voltage of the voltage divider in Figure 6–28.

(b) Find the loaded output voltages of the voltage divider in Figure 6–28 for the following two values of load resistance: $R_L = 10\ \text{k}\Omega$ and $R_L = 100\ \text{k}\Omega$.

FIGURE 6–28

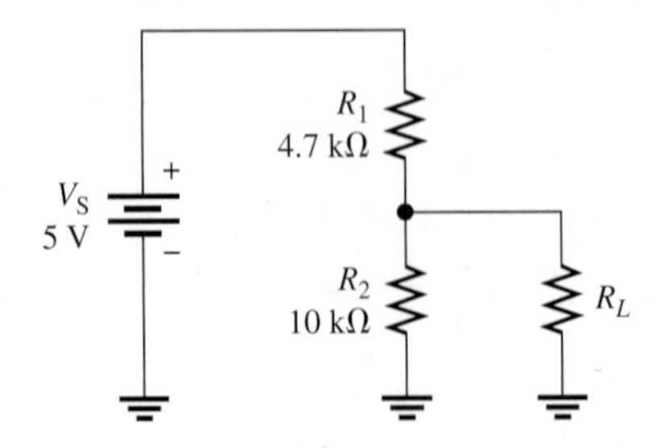

Solution **(a)** The unloaded output voltage is

$$V_{OUT(unloaded)} = \left(\frac{R_2}{R_1 + R_2}\right)V_S = \left(\frac{10\text{ k}\Omega}{14.7\text{ k}\Omega}\right)5\text{ V} = \mathbf{3.40\text{ V}}$$

(b) With the 10 kΩ load resistor connected, R_L is in parallel with R_2, which gives

$$R_2 \parallel R_L = \frac{R_2R_L}{R_2 + R_L} = \frac{(10\text{ k}\Omega)(10\text{ k}\Omega)}{20\text{ k}\Omega} = 5.0\text{ k}\Omega$$

The equivalent circuit is shown in Figure 6–29(a). The loaded output voltage is

$$V_{OUT(loaded)} = \left(\frac{R_2 \parallel R_L}{R_1 + R_2 \parallel R_L}\right)V_S = \left(\frac{5.0\text{ k}\Omega}{9.7\text{ k}\Omega}\right)5\text{ V} = \mathbf{2.58\text{ V}}$$

FIGURE 6–29

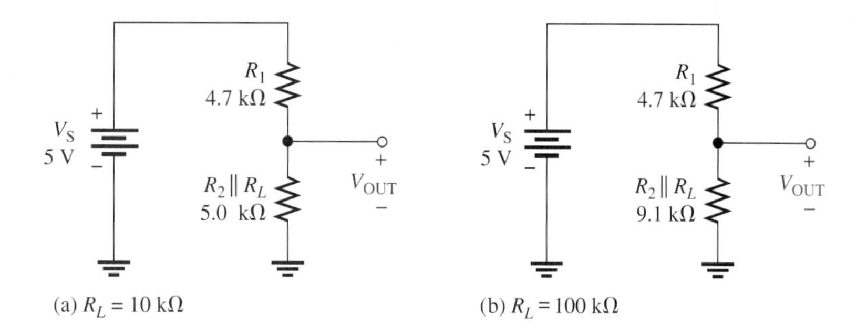

With the 100 kΩ load, the resistance from output to ground is

$$R_2 \parallel R_L = \frac{R_2R_L}{R_2 + R_L} = \frac{(10\text{ k}\Omega)(100\text{ k}\Omega)}{110\text{ k}\Omega} = 9.1\text{ k}\Omega$$

The equivalent circuit is shown in Figure 6–29(b). The loaded output voltage is

$$V_{OUT(loaded)} = \left(\frac{R_2 \parallel R_L}{R_1 + R_2 \parallel R_L}\right)V_S = \left(\frac{9.1\text{ k}\Omega}{13.8\text{ k}\Omega}\right)5\text{ V} = \mathbf{3.30\text{ V}}$$

For the smaller value of R_L, the reduction in V_{OUT} is

$$3.40\text{ V} - 2.58\text{ V} = 0.82\text{ V}$$

For the larger value of R_L, the reduction in V_{OUT} is

$$3.40\text{ V} - 3.30\text{ V} = 0.10\text{ V}$$

This illustrates the loading effect of R_L on the voltage divider.

Related Problem Determine V_{OUT} in Figure 6–28 for a 1.0 MΩ load resistance.

Open file E06-11 on your EWB/CircuitMaker CD-ROM. Measure the voltage at the output terminal with respect to ground. Connect a 10 kΩ load resistor from the output to ground and measure the output voltage. Change the load resistor to 100 kΩ and measure the output voltage. Do these measurements agree closely with the calculated values?

Load Current and Bleeder Current

In a multiple-tap loaded voltage-divider circuit, the total current drawn from the source consists of currents through the load resistors, called **load currents**, and the divider resistors. Figure 6–30 shows a voltage divider with two voltage outputs or two taps. Notice that the total current, I_T, through R_1 enters point A where the current divides into I_{RL1} through R_{L1} and into I_2 through R_2. At point B, the current I_2 divides into I_{RL2} through R_{RL2} and into I_3 through R_3. Current I_3 is called the **bleeder current**, *which is the current left after the total load current is subtracted from the total current into the circuit.*

Equation 6–1

$$I_{BLEEDER} = I_T - I_{RL1} - I_{RL2}$$

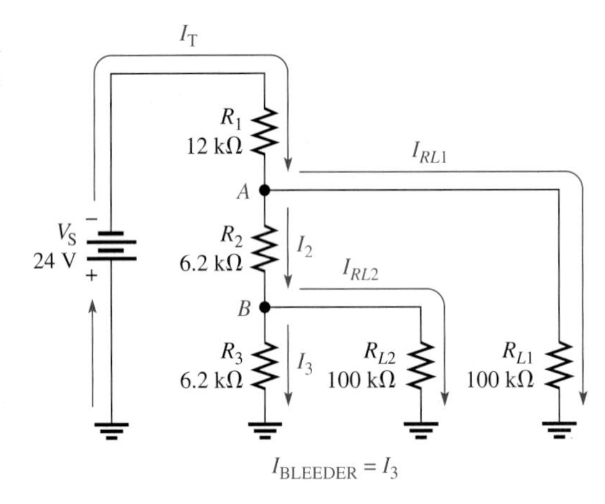

FIGURE 6–30

Currents in a two-tap loaded voltage divider.

EXAMPLE 6–12

Determine the load currents I_{RL1} and I_{RL2} and the bleeder current I_3 in the two-tap loaded voltage divider in Figure 6–30.

Solution The equivalent resistance from point A to ground is the 100 kΩ load resistor R_{L1} in parallel with the combination of R_2 in series with the parallel combination of R_3 and R_{L2}. Determine the resistance values first. R_3 in parallel with R_{L2} is designated R_B. The resulting equivalent circuit is shown in Figure 6–31(a).

$$R_B = \frac{R_3 R_{L2}}{R_3 + R_{L2}} = \frac{(6.2\text{ k}\Omega)(100\text{ k}\Omega)}{106.2\text{ k}\Omega} = 5.84\text{ k}\Omega$$

R_2 in series with R_B is designated R_{2+B}. The resulting equivalent circuit is shown in Figure 6–31(b).

$$R_{2+B} = R_2 + R_B = 6.2\text{ k}\Omega + 5.84\text{ k}\Omega = 12.0\text{ k}\Omega$$

R_{L1} in parallel with R_{2+B} is designated R_A. The resulting equivalent circuit is shown in Figure 6–31(c).

$$R_A = \frac{R_{L1} R_{2+B}}{R_{L1} + R_{2+B}} = \frac{(100\text{ k}\Omega)(12.0\text{ k}\Omega)}{112\text{ k}\Omega} = 10.7\text{ k}\Omega$$

R_A is the total resistance from point A to ground. The total resistance for the circuit is

$$R_T = R_A + R_1 = 10.7\text{ k}\Omega + 12\text{ k}\Omega = 22.7\text{ k}\Omega$$

The voltage across R_{L1} is determined as follows, using the equivalent circuit in Figure 6–31(c):

$$V_{RL1} = V_A = \left(\frac{R_A}{R_T}\right)V_S = \left(\frac{10.7\ \text{k}\Omega}{22.7\ \text{k}\Omega}\right)24\ \text{V} = 11.3\ \text{V}$$

The load current through R_{L1} is

$$I_{RL1} = \frac{V_{RL1}}{R_{L1}} = \left(\frac{11.3\ \text{V}}{100\ \text{k}\Omega}\right) = \mathbf{113\ \mu A}$$

The voltage at point B is determined by using the equivalent circuit in Figure 6–31(a), and the voltage at point A.

$$V_B = \left(\frac{R_B}{R_2 + R_B}\right)V_A = \left(\frac{5.84\ \text{k}\Omega}{12.0\ \text{k}\Omega}\right)11.3\ \text{V} = 5.50\ \text{V}$$

The load current through R_{L2} is

$$I_{RL2} = \frac{V_{RL2}}{R_{L2}} = \frac{V_B}{R_{L2}} = \frac{5.50\ \text{V}}{100\ \text{k}\Omega} = \mathbf{55\ \mu A}$$

The bleeder current is

$$I_3 = \frac{V_B}{R_3} = \frac{5.50\ \text{V}}{6.2\ \text{k}\Omega} = \mathbf{887\ \mu A}$$

FIGURE 6–31

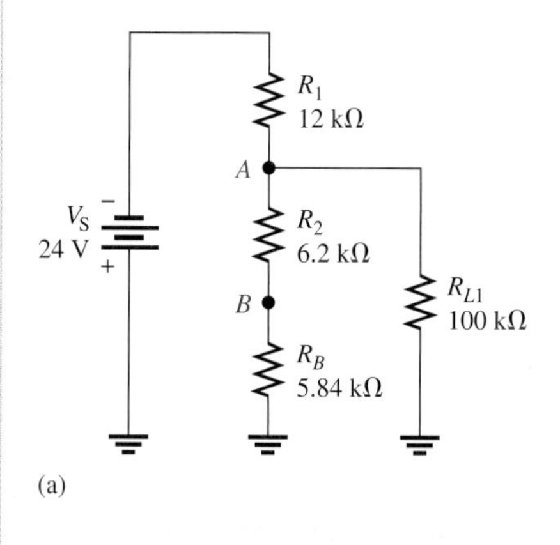

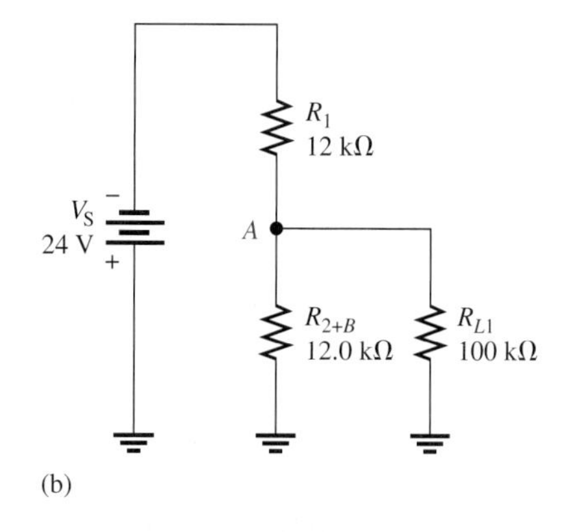

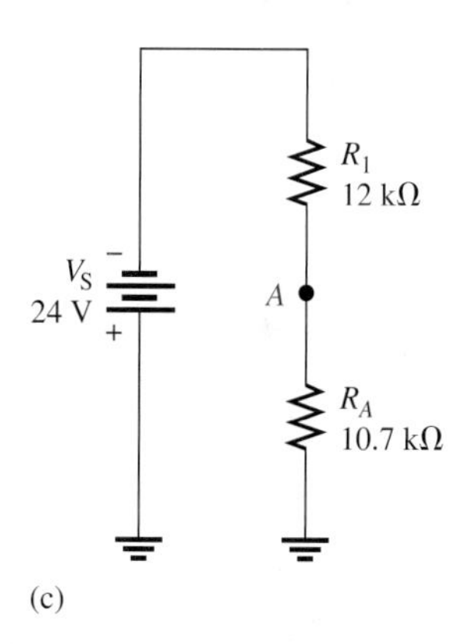

Related Problem How can the bleeder current in Figure 6–30 be reduced without affecting the load currents?

Open file E06-12 on your EWB/CircuitMaker CD-ROM. Measure the voltage across and the current through each load resistor, R_{L1} and R_{L2}.

Bipolar Voltage Dividers

An example of a voltage divider that produces both positive and negative voltages from a single source is shown in Figure 6–32. Notice that neither the positive nor the negative terminal of the source is connected to ground. The voltages at points A and B are positive with respect to ground, and the voltages at points C and D are negative with respect to ground.

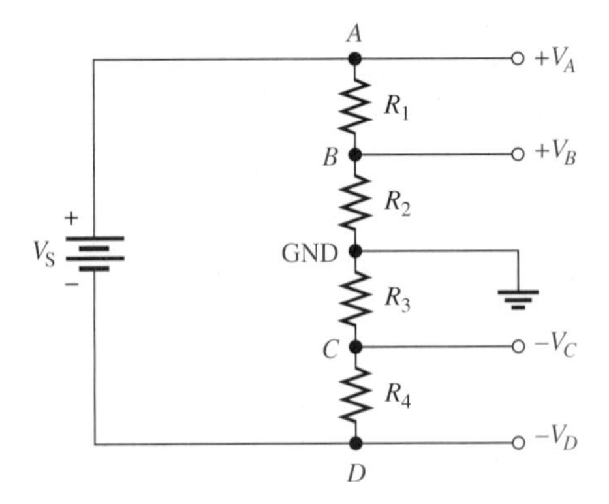

FIGURE 6–32

A bipolar voltage divider. The positive and negative voltages are with respect to ground.

SECTION 6–3 REVIEW

1. A load resistor is connected to an output on a voltage divider. What effect does the load resistor have on the output voltage?
2. A larger-value load resistor will cause the output voltage of a voltage divider to change less than a small-value one will. (True or False)
3. For the voltage divider in Figure 6–33, determine the unloaded output voltage. Also determine the output voltage with a 10 MΩ load resistor connected from the output to ground.

FIGURE 6–33

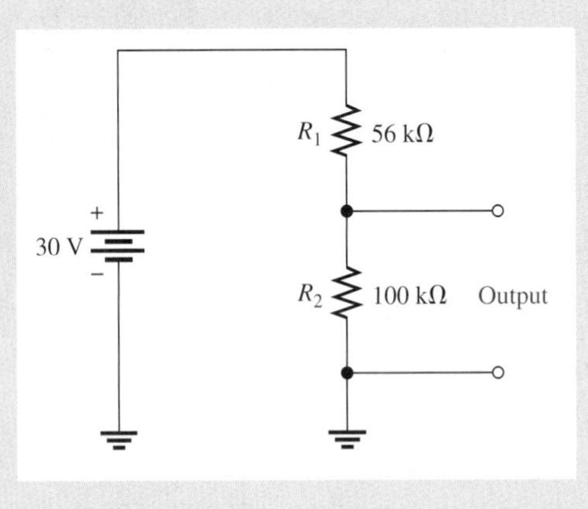

6–4 LOADING EFFECT OF A VOLTMETER

As you have learned, voltmeters must be connected in parallel with a resistor in order to measure the voltage across the resistor. Because of its internal resistance, a voltmeter puts a load on the circuit and will affect, to a certain extent, the voltage that is being measured. Until now, we have ignored the loading effect because the internal resistance of a voltmeter is very high, and normally it has negligible effect on the circuit that is being measured. However, if the internal resistance of the voltmeter is not sufficiently greater than the circuit resistance across which it is connected, the loading effect will cause the measured voltage to be less than its actual value.

After completing this section, you should be able to

- **Determine the loading effect of a voltmeter on a circuit**
- Explain why a voltmeter can load a circuit
- Discuss the internal resistance of a voltmeter

Why a Voltmeter Can Load a Circuit

When a voltmeter is connected to a circuit as shown, for example, in Figure 6–34(a), its internal resistance appears in parallel with R_3, as shown in part (b). The resistance from point A to point B is altered by the loading effect of the voltmeter's internal resistance, R_M, and is equal to $R_3 \parallel R_M$, as indicated in part (c).

FIGURE 6–34

The loading effect of a voltmeter.

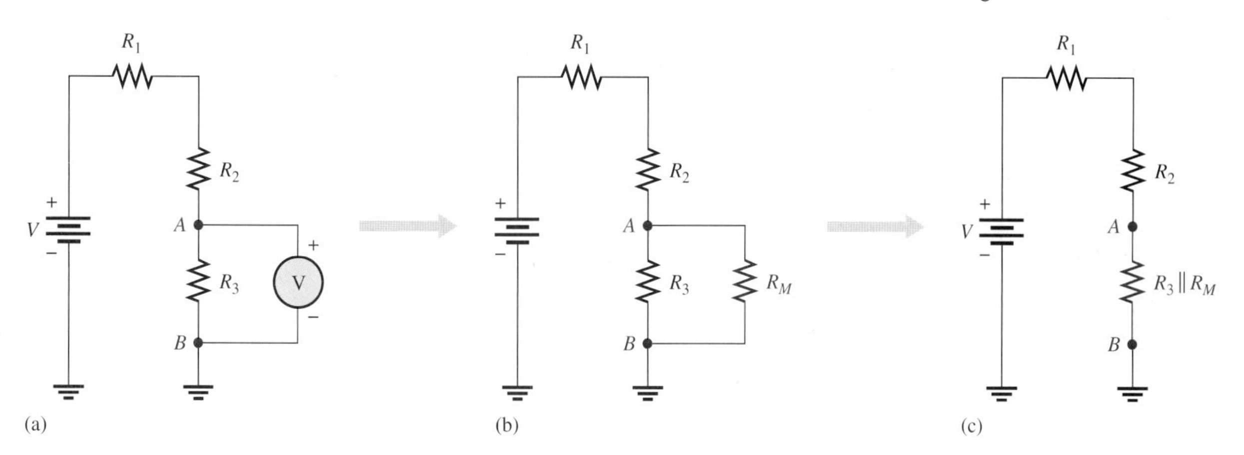

If R_M is much greater than R_3, the resistance from A to B changes very little, and the meter reading is very close to the actual voltage. If R_M is not sufficiently greater than R_3, the resistance from A to B is reduced significantly, and the voltage across R_3 is altered by the loading effect of the meter. A good rule of thumb is that *if the meter resistance is at least ten times greater than the resistance across which it is connected, the loading effect can be neglected (measurement error is less than 10%).*

Internal Resistance of a Voltmeter

The basic categories of voltmeters are the electromagnetic (needle-type) analog voltmeter, whose internal resistance is determined by its **sensitivity factor,** and the digital voltmeter, whose internal resistance is typically at least 10 MΩ. The digital voltmeter presents fewer loading problems than the electromagnetic type because its internal resistance is much higher.

EXAMPLE 6–13

How much does the digital voltmeter affect the voltage being measured for each circuit indicated in Figure 6–35? Assume the meter has an input resistance (R_M) of 10 MΩ.

Solution **(a)** Refer to Figure 6–35(a). The unloaded voltage across R_2 in the voltage-divider circuit is

$$V_{R2} = \left(\frac{R_2}{R_1 + R_2}\right)V = \left(\frac{100\ \Omega}{280\ \Omega}\right)15\ \text{V} = 5.357\ \text{V}$$

The meter's resistance in parallel with R_2 is

$$R_2 \parallel R_M = \left(\frac{R_2 R_M}{R_2 + R_M}\right) = \frac{(100\ \Omega)(10\ \text{M}\Omega)}{10.0001\ \text{M}\Omega} = 99.999\ \Omega$$

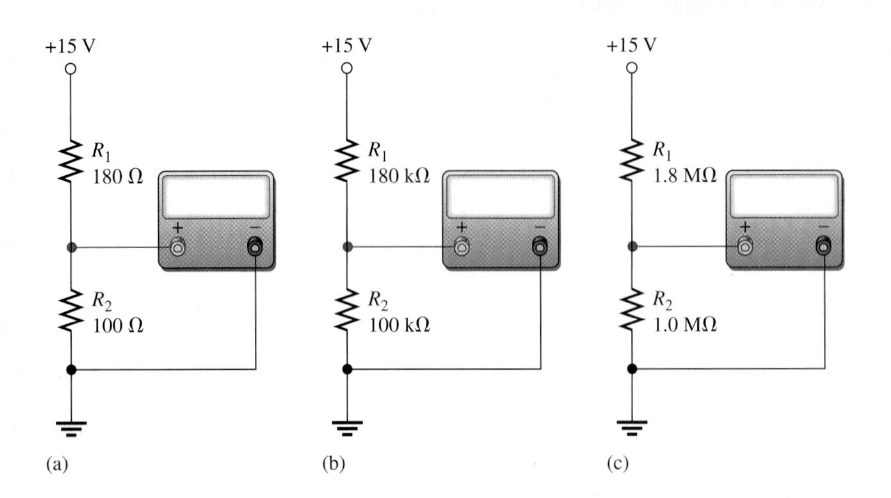

▲ **FIGURE 6–35**

The voltage actually measured by the meter is

$$V_{R2} = \left(\frac{R_2 \parallel R_M}{R_1 + R_2 \parallel R_M}\right)V = \left(\frac{99.999\ \Omega}{279.999\ \Omega}\right)15\ \text{V} = 5.357\ \text{V}$$

The voltmeter has no measurable loading effect.

(b) Refer to Figure 6–35(b).

$$V_{R2} = \left(\frac{R_2}{R_1 + R_2}\right)V = \left(\frac{100\ \text{k}\Omega}{280\ \text{k}\Omega}\right)15\ \text{V} = 5.357\ \text{V}$$

$$R_2 \parallel R_M = \frac{R_2 R_M}{R_2 + R_M} = \frac{(100\ \text{k}\Omega)(10\ \text{M}\Omega)}{10.1\ \text{M}\Omega} = 99.01\ \text{k}\Omega$$

The voltage actually measured by the meter is

$$V_{R2} = \left(\frac{R_2 \parallel R_M}{R_1 + R_2 \parallel R_M}\right)V = \left(\frac{99.01\ \text{k}\Omega}{279.01\ \text{k}\Omega}\right)15\ \text{V} = 5.323\ \text{V}$$

The loading effect of the voltmeter reduces the voltage by a very small amount.

(c) Refer to Figure 6–35(c).

$$V_{R2} = \left(\frac{R_2}{R_1 + R_2}\right)V = \left(\frac{1.0\ \text{M}\Omega}{2.8\ \text{M}\Omega}\right)15\ \text{V} = 5.357\ \text{V}$$

$$R_2 \parallel R_M = \frac{R_2 R_M}{R_2 + R_M} = \frac{(1.0\ \text{M}\Omega)(10\ \text{M}\Omega)}{11\ \text{M}\Omega} = 909.09\ \text{k}\Omega$$

The voltage actually measured is

$$V_{R2} = \left(\frac{R_2 \parallel R_M}{R_1 + R_2 \parallel R_M}\right)V = \left(\frac{909.09\ \text{k}\Omega}{2.709\ \text{M}\Omega}\right)15\ \text{V} = 5.034\ \text{V}$$

The loading effect of the voltmeter reduces the voltage by a noticeable amount. As you can see, the higher the resistance across which a voltage is measured, the more the loading effect.

Related Problem Calculate the voltage across R_2 in Figure 6–35(c) if the meter resistance is 20 MΩ.

SECTION 6–4 REVIEW

1. Explain why a voltmeter can potentially load a circuit.
2. If a voltmeter with a 10 MΩ internal resistance is measuring the voltage across a 1.0 kΩ resistor, should you normally be concerned about the loading effect?
3. If a voltmeter with a 10 MΩ resistance is measuring the voltage across a 3.3 MΩ resistor, should you be concerned about the loading effect?

6–5 THE WHEATSTONE BRIDGE

The Wheatstone bridge circuit is widely used to precisely measure resistance. Also, the bridge is used in conjunction with transducers to measure physical quantities such as strain, temperature, and pressure. Transducers are devices that sense a change in a physical parameter and convert that change into an electrical quantity such as a change in resistance. For example, a strain gauge exhibits a change in resistance when it is exposed to mechanical factors such as force, pressure, or displacement. A thermistor exhibits a change in its resistance when it is exposed to a change in temperature. The Wheatstone bridge can be operated in a balanced or an unbalanced condition. The condition of operation depends on the type of application.

After completing this section, you should be able to

- **Analyze a Wheatstone bridge circuit**
- Determine when a bridge is balanced
- Determine an unknown resistance with a balanced bridge
- Determine when a bridge is unbalanced
- Discuss measurements using an unbalanced bridge

A Wheatstone bridge circuit is shown in its most common "diamond" configuration in Figure 6–36(a). It consists of four resistors, a dc voltage source connected across the top and bottom points of the "diamond." The output voltage is taken across the left and right points of the "diamond" between A and B. In part (b), the circuit is drawn in a slightly different way to more clearly show its series-parallel configuration.

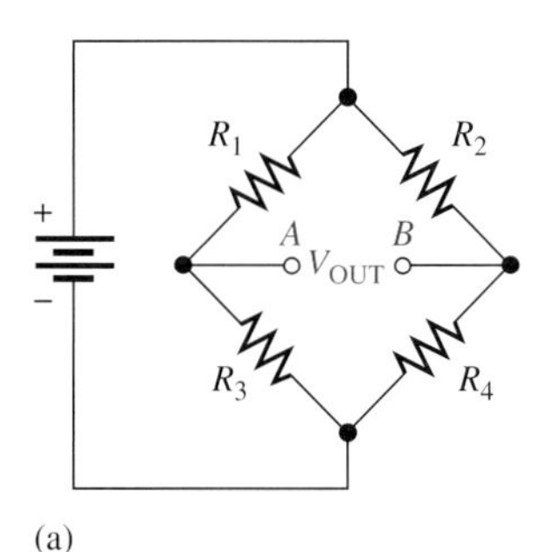

(a)

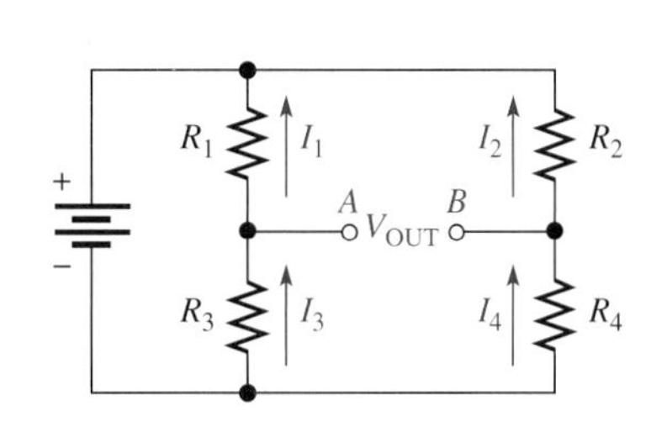

(b)

FIGURE 6–36

Wheatstone bridge.

The Balanced Wheatstone Bridge

The Wheatstone bridge in Figure 6–36 is in the balanced condition when the output voltage (V_{OUT}) between terminals A and B is equal to zero.

$$V_{OUT} = 0\text{ V}$$

When the bridge is balanced, the voltages across R_1 and R_2 are equal ($V_1 = V_2$) and the voltages across R_3 and R_4 are equal ($V_3 = V_4$). Therefore, the voltage ratios can be written as

$$\frac{V_1}{V_3} = \frac{V_2}{V_4}$$

Substituting IR for V by Ohm's law gives

$$\frac{I_1R_1}{I_3R_3} = \frac{I_2R_2}{I_4R_4}$$

Since $I_1 = I_3$ and $I_2 = I_4$, all the current terms cancel, leaving the resistor ratios.

$$\frac{R_1}{R_3} = \frac{R_2}{R_4}$$

Solving for R_1 results in the following formula:

$$R_1 = R_3\left(\frac{R_2}{R_4}\right)$$

This formula allows you to find the value of resistor R_1 in terms of the other resistor values when the bridge is balanced. You can also find the value of any other resistor in a similar way.

Using the Balanced Wheatstone Bridge to Find an Unknown Resistance Assume that R_1 in Figure 6–36 has an unknown value, which we call R_X. Resistors R_2 and R_4 have fixed values so that their ratio, R_2/R_4, also has a fixed value. Since R_X can be any value, R_3 must be adjusted to make $R_1/R_3 = R_2/R_4$ in order to create a balanced condition. Therefore, R_3 is a variable resistor, which we will call R_V. When R_X is placed in the bridge, R_V is adjusted until the bridge is balanced as indicated by a zero output voltage. Then, the unknown resistance is found as

Equation 6–2

$$R_X = R_V\left(\frac{R_2}{R_4}\right)$$

The ratio R_2/R_4 is the scale factor.

Commonly, a measuring instrument called a galvanometer is connected between the output terminals A and B to detect a balanced condition. The galvanometer is essentially a very sensitive ammeter that senses current in either direction. It differs from a regular ammeter in that the midscale point is zero. So if the current through it is in one direction, the needle deflects toward the positive side. If the current is in the other direction, the needle deflects toward the negative side. When the needle is at midpoint, the bridge is balanced because the voltage from A to B is zero and the current through the galvanometer is zero, as illustrated in Figure 6–37. A voltmeter can also be used to indicate when the bridge is balanced.

From Equation 6–2, the value of R_V at balance multiplied by the scale factor R_2/R_4 is the actual resistance value of R_X. If $R_2/R_4 = 1$, then $R_X = R_V$, if $R_2/R_4 = 0.5$, then $R_X = 0.5R_V$, and so on. In a practical bridge circuit, the position of the R_V adjustment can be calibrated to indicate the actual value of R_X on a scale or with some other method of display.

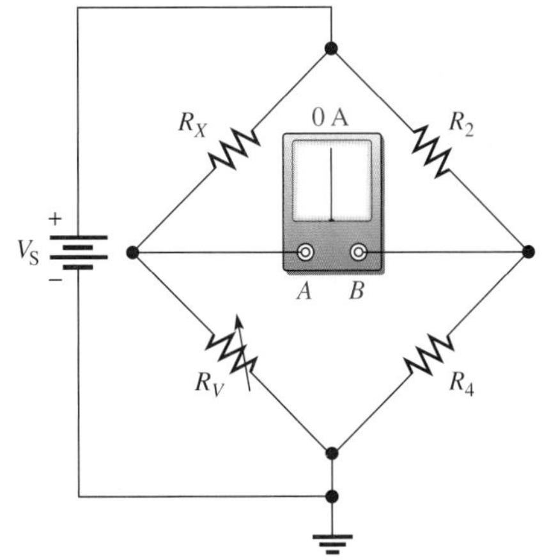

FIGURE 6–37

The Wheatstone bridge is balanced when the galvanometer shows 0 A.

EXAMPLE 6–14

Determine the value of R_X in the balanced bridge shown in Figure 6–38.

FIGURE 6–38

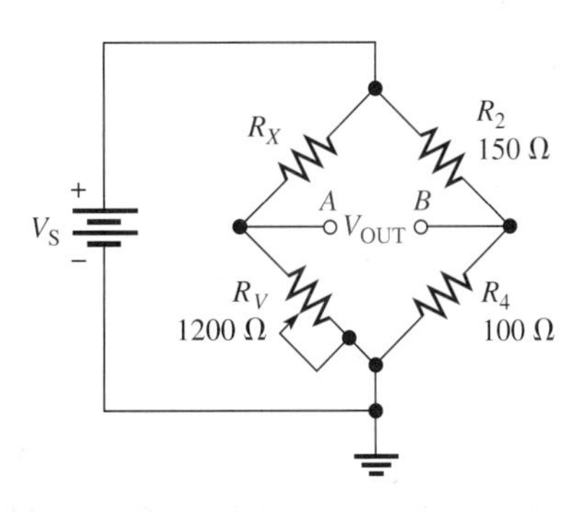

Solution The scale factor is

$$\frac{R_2}{R_4} = \frac{150\ \Omega}{100\ \Omega} = 1.5$$

The bridge is balanced ($V_{OUT} = 0$ V) when R_V is set at 1200 Ω, so the unknown resistance is

$$R_X = R_V\left(\frac{R_2}{R_4}\right) = (1200\ \Omega)(1.5) = \mathbf{1800\ \Omega}$$

Related Problem If R_V must be adjusted to 2.2 kΩ to balance the bridge in Figure 6–38, what is R_X?

The Unbalanced Wheatstone Bridge

The unbalanced bridge, when V_{OUT} is not equal to zero, is used to measure several types of physical quantities such as mechanical strain, temperature, or pressure. This can be done by connecting a transducer in one leg of the bridge, as shown in Figure 6–39. The resistance of the transducer changes proportionally to the changes in the parameter that it is measuring. If the bridge is balanced at a known point, then the amount of deviation from the balanced condition, as indicated by the output voltage, indicates the amount of change in the parameter

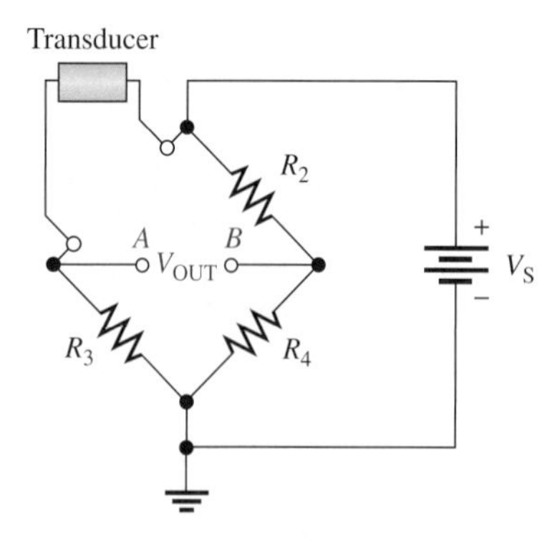

FIGURE 6–39

A bridge circuit for measuring a physical parameter using a transducer.

being measured. Therefore, the value of the parameter being measured can be determined by the amount that the bridge is unbalanced.

A Bridge Circuit for Measuring Temperature If temperature is to be measured, the transducer can be a thermistor, which is a temperature-sensitive resistor. The thermistor resistance changes in a predictable way as the temperature changes. A change in temperature causes a change in thermistor resistance, which causes a corresponding change in the output voltage of the bridge as it becomes unbalanced. The output voltage is proportional to the temperature; therefore, either a voltmeter connected across the output can be calibrated to show the temperature or the output voltage can be amplified and converted to digital form to drive a readout display of the temperature.

A bridge circuit used to measure temperature is designed so that it is balanced at a reference temperature and becomes unbalanced at a measured temperature. For example, let's say the bridge is to be balanced at 25°C. A thermistor will have a known value of resistance at 25°C. For simplicity, let's assume the other three bridge resistors are equal to the thermistor resistance at 25°C, so $R_{therm} = R_2 = R_3 = R_4$. For this particular case, the change in output voltage (ΔV_{OUT}) can be shown to be related to the change in R_{therm} by the following formula:

Equation 6–3

$$\Delta V_{OUT} = \Delta R_{therm}\left(\frac{V_S}{4R}\right)$$

The Δ (Greek letter delta) in front of a variable means a change in the variable. This formula applies only to the case where all resistances in the bridge are equal when the bridge is balanced. Keep in mind that the bridge can be initially balanced without having all the resistors equal as long as $R_1 = R_2$ and $R_3 = R_4$ (see Figure 6–36), but the formula for ΔV_{OUT} would be more complicated.

EXAMPLE 6–15

Determine the output voltage of the temperature-measuring bridge circuit in Figure 6–40 if the thermistor is exposed to a temperature of 50°C and its resistance at 25°C is 1.0 kΩ. Assume the resistance of the thermistor decreases to 900 Ω at 50°C.

Solution

$$\Delta R_{therm} = 1.0\text{ k}\Omega - 900\ \Omega = 100\ \Omega$$

$$\Delta V_{OUT} = \Delta R_{therm}\left(\frac{V_S}{4R}\right) = 100\ \Omega\left(\frac{12\text{ V}}{4\text{ k}\Omega}\right) = 0.3\text{ V}$$

▶ FIGURE 6–40

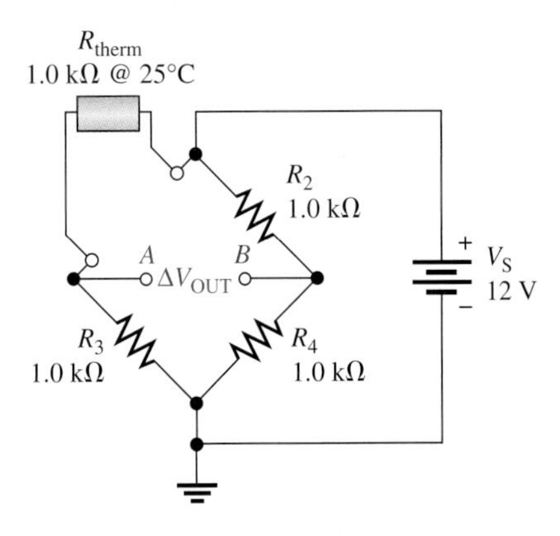

Since $V_{OUT} = 0$ V when the bridge is balanced at 25°C and it changes 0.3 V, then

$$V_{OUT} = \mathbf{0.3\ V}$$

when the temperature is 50°C.

Related Problem If the temperature is increased to 60°C, causing the thermistor resistance in Figure 6–40 to decrease to 850 Ω, what is V_{OUT}?

Other Unbalanced Wheatstone Bridge Applications A Wheatstone bridge with a strain gauge can be used to measure certain forces. A strain gauge is a device that exhibits a change in resistance when it is compressed or stretched by the application of an external force. As the resistance of the strain gauge changes, the previously balanced bridge becomes unbalanced. This unbalance causes the output voltage to change from zero, and this change can be measured to determine the amount of strain. In strain gauges, the resistance change is extremely small. This tiny change unbalances a Wheatstone bridge because of its high sensitivity. For example, Wheatstone bridges with strain gauges are commonly used in precision weight scales.

A Wheatstone bridge can be used in a similar way for measuring pressure using a pressure transducer. The application of pressure changes the resistance of the transducer, causing the bridge to become unbalanced by an amount proportional to the pressure.

SECTION 6–5 REVIEW

1. Draw a basic Wheatstone bridge circuit.
2. Under what condition is a bridge balanced?
3. What is the unknown resistance in Figure 6–37 when $R_V = 3.3$ kΩ, $R_2 = 10$ kΩ, and $R_4 = 2.2$ kΩ?
4. How is a Wheatstone bridge used in the unbalanced condition?

6–6 THEVENIN'S THEOREM

Thevenin's theorem provides a method for simplifying a circuit to a standard equivalent form. In many cases, this theorem can be used to simplify the analysis of series-parallel circuits.

After completing this section, you should be able to

- **Apply Thevenin's theorem to simplify a circuit for analysis**
- Describe the form of a Thevenin equivalent circuit
- Obtain the Thevenin equivalent voltage source
- Obtain the Thevenin equivalent resistance
- Explain terminal equivalency in the context of Thevenin's theorem
- Thevenize a portion of a circuit
- Thevenize a Wheatstone bridge

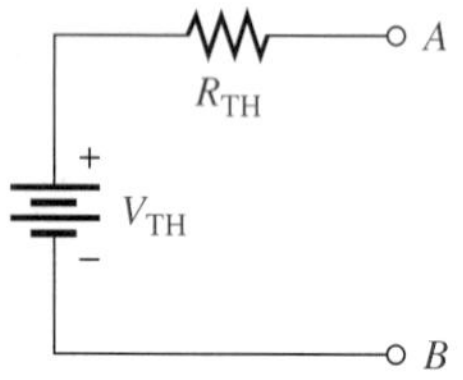

The Thevenin equivalent form of any two-terminal resistive circuit consists of an equivalent voltage source (V_{TH}) and an equivalent resistance (R_{TH}), arranged as shown in Figure 6–41. The values of the equivalent voltage and resistance depend on the values in the original circuit. Any two-terminal resistive circuit can be simplified to a Thevenin equivalent regardless of its complexity.

▲ **FIGURE 6–41**

The general form of a Thevenin equivalent circuit is a voltage source in series with a resistance.

Thevenin's Equivalent Voltage (V_{TH}) and Equivalent Resistance (R_{TH})

The equivalent voltage, V_{TH}, is one part of the complete Thevenin equivalent circuit. The other part is R_{TH}.

The Thevenin equivalent voltage (V_{TH}) is the open circuit (no load) voltage between two specified output terminals in a circuit.

Any component connected between these two terminals effectively "sees" V_{TH} in series with R_{TH}. As defined by **Thevenin's theorem**,

The Thevenin equivalent resistance (R_{TH}) is the total resistance appearing between two specified output terminals in a circuit with all sources replaced by their internal resistances (which for an ideal voltage source is zero).

What Equivalency Means in Thevenin's Theorem

Although a Thevenin equivalent circuit is not of the same form as the original circuit, *it acts the same in terms of the output voltage and current.* Consider the following demonstration as illustrated in Figure 6–42. A resistive circuit of any complexity is placed in a box with only the output terminals exposed. The Thevenin equivalent of that circuit is placed in an identical box with, again, only the output terminals exposed. Identical load resistors are connected across the output terminals of each box. Next, a voltmeter and an ammeter are connected to measure the voltage and current for each load as shown in the figure. The measured values will be identical (neglecting tolerance variations), and you will not be able to determine which box contains the original circuit and which contains the Thevenin equivalent of the original circuit. That is, in terms of what you can observe by electrical measurement, both circuits are the same. This condition is sometimes known as **terminal equivalency** because both circuits look the same from the "viewpoint" of the two output terminals.

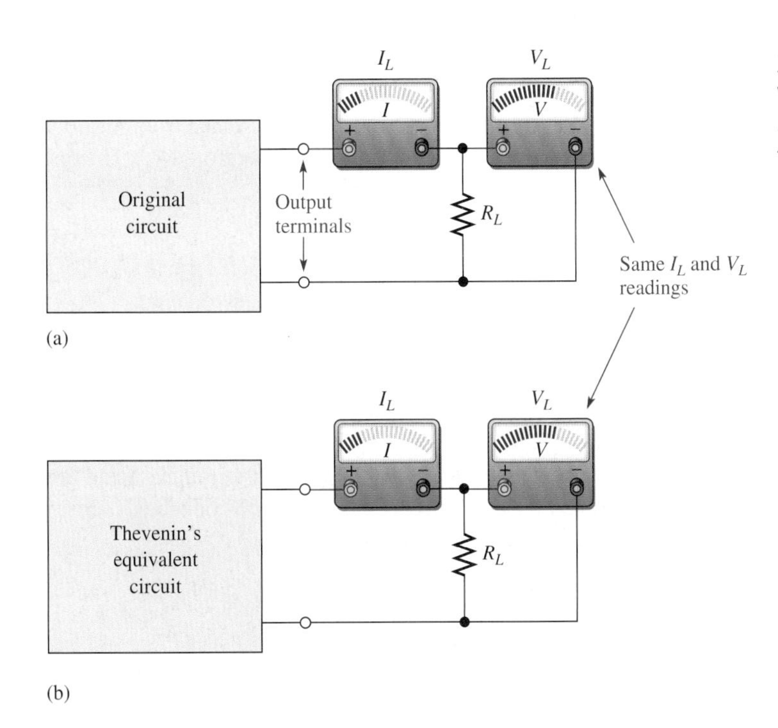

FIGURE 6–42

Which box contains the original circuit and which contains the Thevenin equivalent circuit? You cannot tell by observing the meters because the circuits have terminal equivalency.

The Thevenin Equivalent of a Circuit

To find the Thevenin equivalent of any circuit, you must determine the equivalent voltage, V_{TH}, and the equivalent resistance, R_{TH}. For example, the Thevenin equivalent for the circuit between output terminals A and B is developed in Figure 6–43.

FIGURE 6–43

Example of the simplification of a circuit by Thevenin's theorem.

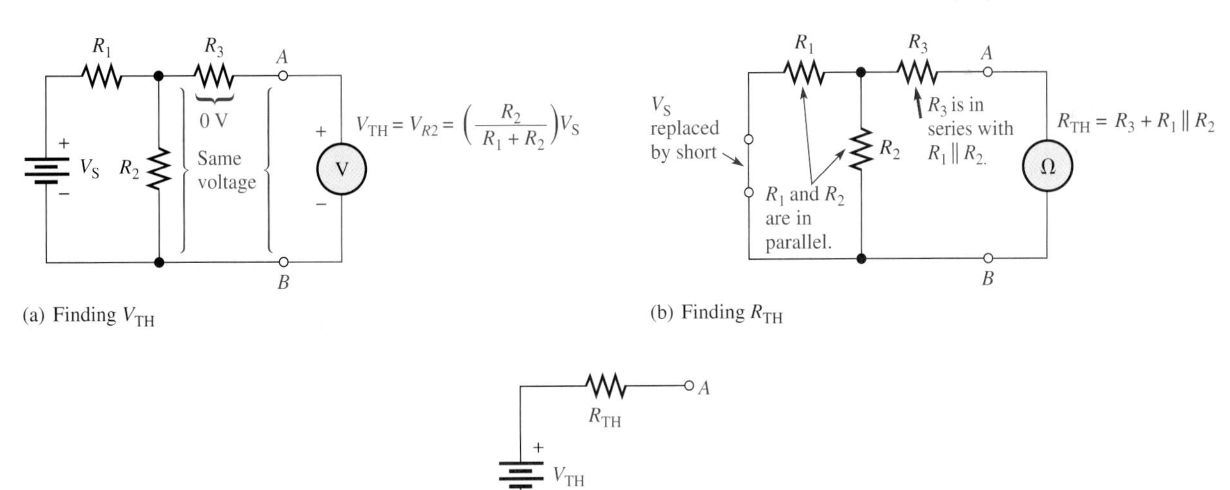

(c) Thevenin equivalent circuit

In Figure 6–43(a), the voltage across the designated terminals A and B is the Thevenin equivalent voltage. In this particular circuit, the voltage from A to B is the same as the voltage across R_2 because there is no current through R_3 and, therefore, no voltage drop across it. V_{TH} is expressed as follows for this particular example:

$$V_{TH} = \left(\frac{R_2}{R_1 + R_2}\right)V_S$$

In Figure 6–43(b), the resistance between terminals A and B with the source replaced by a short to represent its zero internal resistance is the Thevenin equivalent resistance. In this particular circuit, the resistance from A to B is R_3 in series with the parallel combination of R_1 and R_2. Therefore, R_{TH} is expressed as follows:

$$R_{TH} = R_3 + \frac{R_1R_2}{R_1 + R_2}$$

The Thevenin equivalent circuit is shown in Figure 6–43(c).

EXAMPLE 6–16

Find the Thevenin equivalent between the output terminals A and B of the circuit in Figure 6–44. If there were a load resistance connected across terminals A and B, it would first have to be removed.

FIGURE 6–44

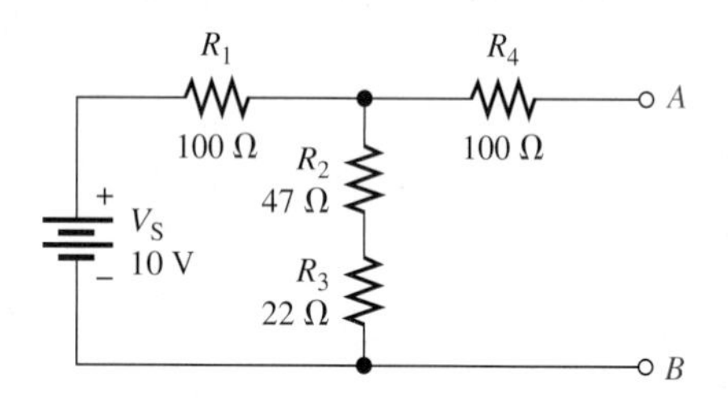

Solution Since there is no voltage drop across R_4, V_{AB} equals the voltage across $R_2 + R_3$ and $V_{TH} = V_{AB}$, as shown in Figure 6–45(a). Use the voltage-divider principle to find V_{TH}.

$$V_{TH} = \left(\frac{R_2 + R_3}{R_1 + R_2 + R_3}\right)V_S = \left(\frac{69\ \Omega}{169\ \Omega}\right)10\ \text{V} = \mathbf{4.08\ V}$$

FIGURE 6–45

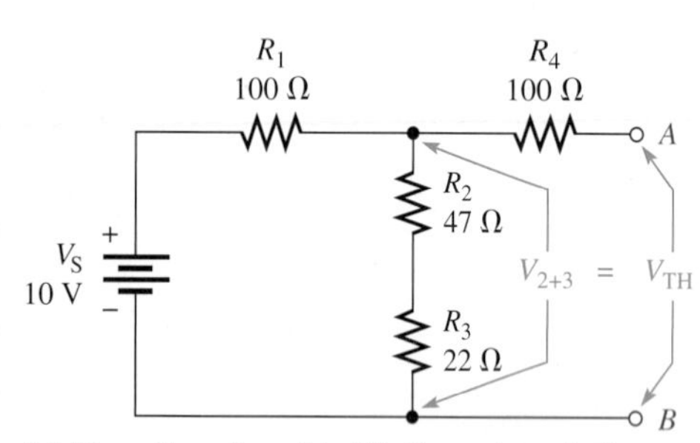

(a) The voltage from A to B is V_{TH} and equals V_{2+3}.

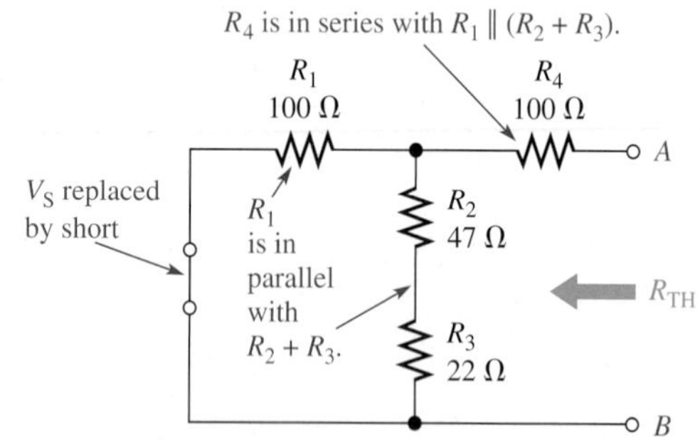

(b) Looking from terminals A and B, R_4 appears in series with the combination of R_1 in parallel with $(R_2 + R_3)$.

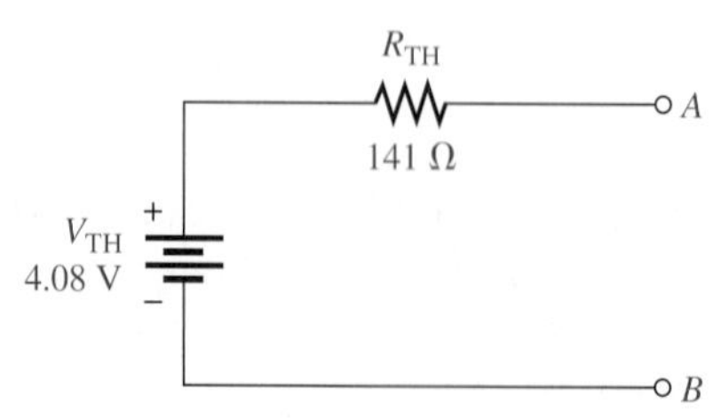

(c) Thevenin equivalent circuit

To find R_{TH}, first replace the source with a short to simulate a zero internal resistance. Then R_1 appears in parallel with $R_2 + R_3$, and R_4 is in series with the series-parallel combination of R_1, R_2, and R_3 as indicated in Figure 6–45(b).

$$R_{TH} = R_4 + \frac{R_1(R_2 + R_3)}{R_1 + R_2 + R_3} = 100\ \Omega + \frac{(100\ \Omega)(69\ \Omega)}{169\ \Omega} = \mathbf{141\ \Omega}$$

The resulting Thevenin equivalent circuit is shown in Figure 6–45(c).

Related Problem Determine V_{TH} and R_{TH} if a 56 Ω resistor is connected in parallel across R_2 and R_3 in Figure 6–44.

Thevenin Equivalency Depends on the Viewpoint

The Thevenin equivalent for any circuit depends on the location of the two output terminals from between which the circuit is "viewed." In Figure 6–44, you viewed the circuit from between the two terminals labeled *A* and *B*. Any given circuit can have more than one Thevenin equivalent, depending on how the output terminals are designated. For example, if you view the circuit in Figure 6–46 from between terminals *A* and *C*, you obtain a completely different result than if you view it from between terminals *A* and *B* or from between terminals *B* and *C*.

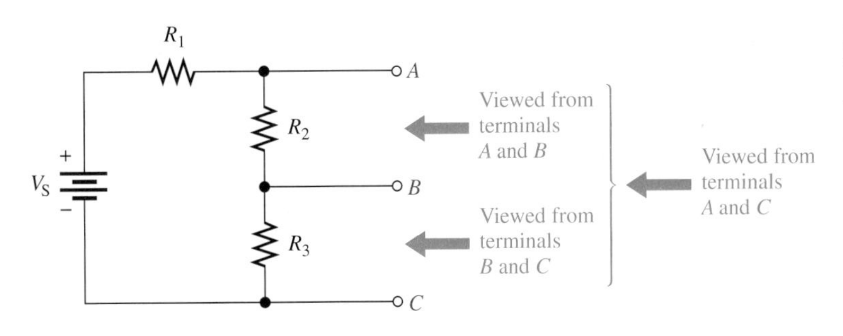

FIGURE 6–46

Thevenin's equivalent depends on the output terminals from which the circuit is viewed.

In Figure 6–47(a), when viewed from between terminals *A* and *C*, V_{TH} is the voltage across $R_2 + R_3$ and can be expressed using the voltage-divider formula as

$$V_{TH(AC)} = \left(\frac{R_2 + R_3}{R_1 + R_2 + R_3}\right)V_S$$

Also, as shown in Figure 6–47(b), the resistance between terminals *A* and *C* is $R_2 + R_3$ in parallel with R_1 (the source is replaced by a short) and can be expressed as

$$R_{TH(AC)} = R_1 \parallel (R_2 + R_3) = \frac{R_1(R_2 + R_3)}{R_1 + R_2 + R_3}$$

The resulting Thevenin equivalent circuit is shown in Figure 6–47(c).

When viewed from between terminals *B* and *C* as indicated in Figure 6–47(d), $V_{TH(BC)}$ is the voltage across R_3 and can be expressed as

$$V_{TH(BC)} = \left(\frac{R_3}{R_1 + R_2 + R_3}\right)V_S$$

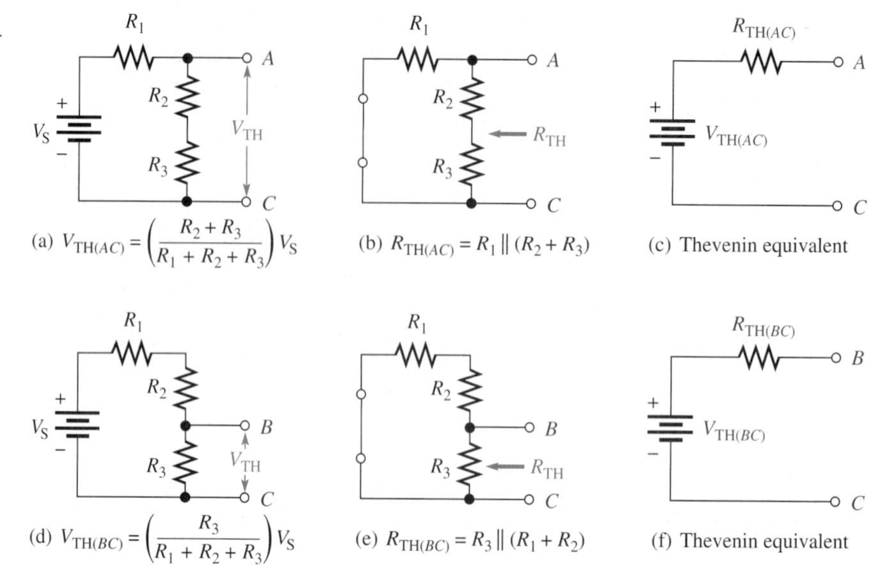

▶ **FIGURE 6–47**

Example of a circuit thevenized from two different sets of terminals. Parts (a), (b), and (c) illustrate one set of terminals and parts (d), (e), and (f) illustrate another set of terminals. (The V_{TH} and R_{TH} values are different for each case.)

As shown in Figure 6–47(e), the resistance between terminals B and C is R_3 in parallel with the series combination of R_1 and R_2.

$$R_{TH(BC)} = R_3 \| (R_1 + R_2) = \frac{R_3(R_1 + R_2)}{R_1 + R_2 + R_3}$$

The resulting Thevenin equivalent is shown in Figure 6–47(f).

Thevenizing a Bridge Circuit

The usefulness of Thevenin's theorem can be illustrated when it is applied to a Wheatstone bridge circuit. For example, consider the case when a load resistor is connected to the output terminals of a Wheatstone bridge, as shown in Figure 6–48. The bridge circuit is very difficult to analyze because it is not a straightforward series-parallel arrangement when a resistance, such as the very small resistance of a galvanometer, is connected between the output terminals A and B. There are no resistors that are in series or in parallel with another resistor.

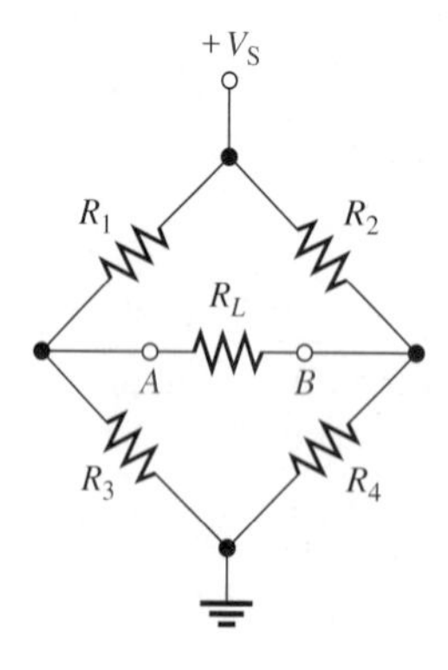

▶ **FIGURE 6–48**

A Wheatstone bridge with a load resistor connected between the output terminals is not a straightforward series-parallel circuit.

Using Thevenin's theorem, you can simplify the bridge circuit to an equivalent circuit viewed from the load resistor as shown step-by-step in Figure 6–49. Study carefully the steps in this figure. Once the equivalent circuit for the bridge is found, the voltage and current for any value of load resistor can easily be determined by Ohm's law.

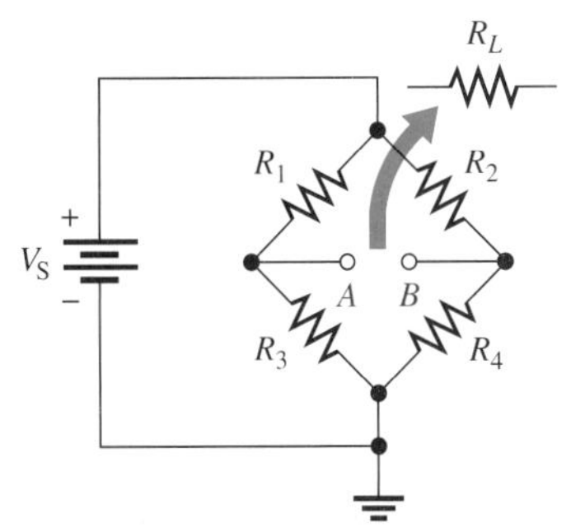

(a) Remove R_L to create an open circuit between the output terminals A and B.

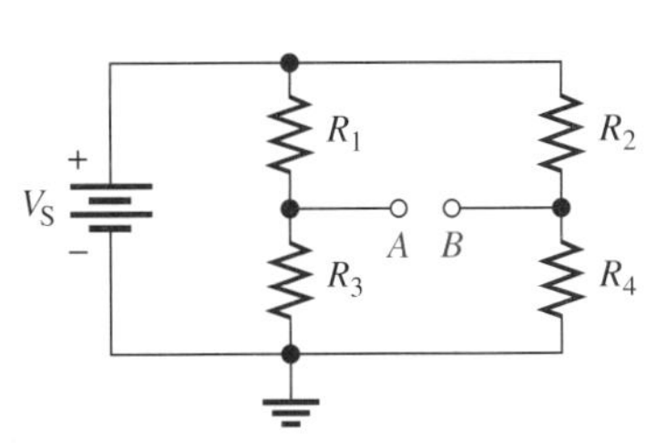

(b) Redraw (if you wish).

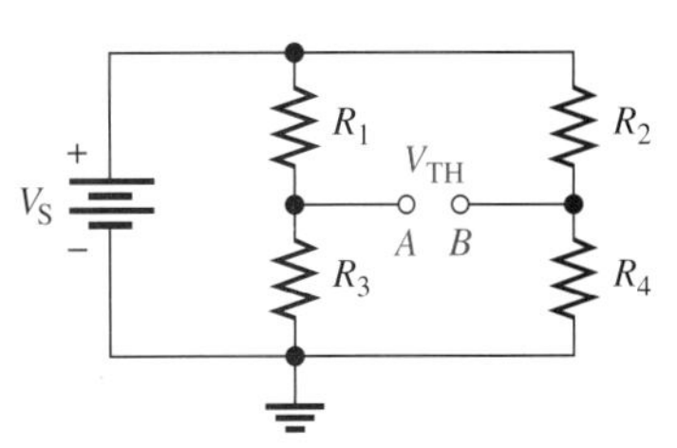

(c) Find V_{TH}:

$$V_{TH} = V_A - V_B = \left(\frac{R_3}{R_1 + R_3}\right)V_S - \left(\frac{R_4}{R_2 + R_4}\right)V_S$$

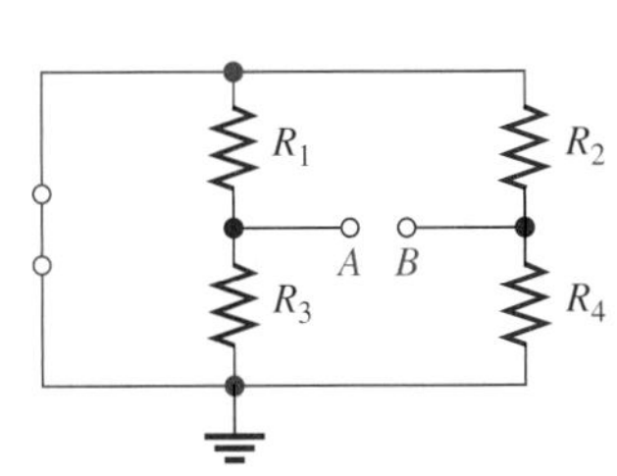

(d) Replace V_S with a short to represent its zero internal resistance. *Note*: The red lines represent the same electrical point as the red lines in Part (e).

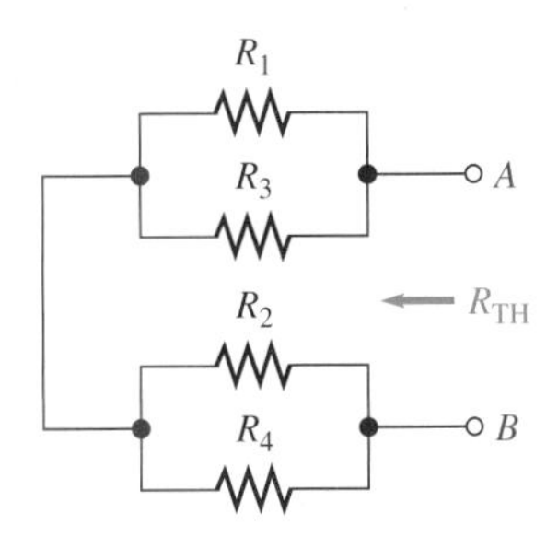

(e) Redraw (if you wish) and find R_{TH}:

$$R_{TH} = R_1 \| R_3 + R_2 \| R_4$$

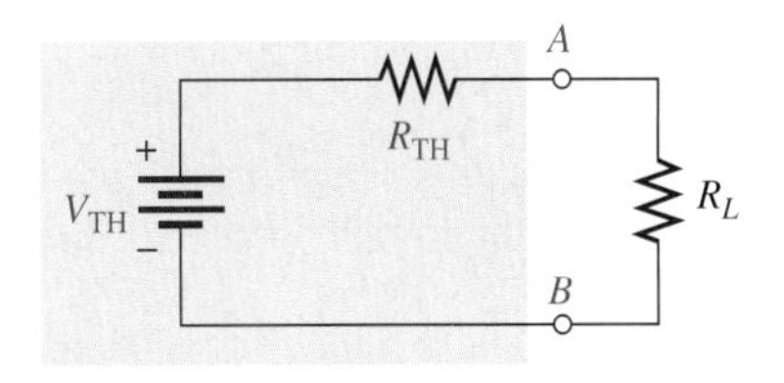

(f) Thevenin's equivalent circuit (beige block) with R_L reconnected

FIGURE 6–49

Simplifying a Wheatstone bridge with Thevenin's theorem.

EXAMPLE 6–17

Determine the voltage and current for the load resistor, R_L, in the bridge circuit of Figure 6–50.

FIGURE 6–50

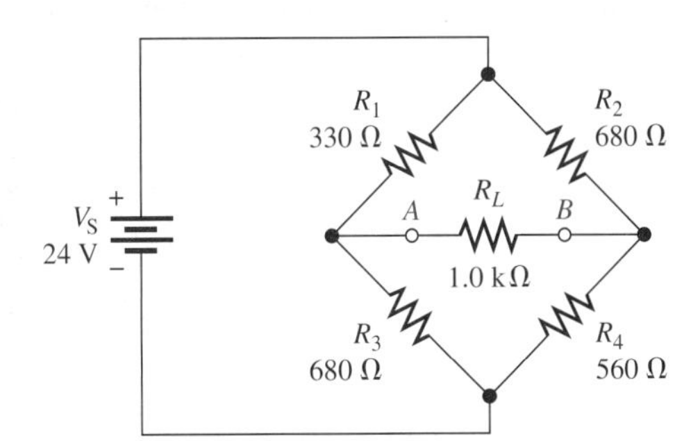

Solution **Step 1:** Remove R_L to create an open circuit between A and B.

Step 2: To thevenize the bridge as viewed from between terminals A and B, as was shown in Figure 6–49, first determine V_{TH}.

$$V_{TH} = V_A - V_B = \left(\frac{R_3}{R_1 + R_3}\right)V_S - \left(\frac{R_4}{R_2 + R_4}\right)V_S$$

$$= \left(\frac{680\ \Omega}{1010\ \Omega}\right)24\ \text{V} - \left(\frac{560\ \Omega}{1240\ \Omega}\right)24\ \text{V} = 16.16\ \text{V} - 1084\ \text{V} = 5.32\ \text{V}$$

Step 3: Determine R_{TH}.

$$R_{TH} = \frac{R_1R_3}{R_1 + R_3} + \frac{R_2R_4}{R_2 + R_4}$$

$$= \frac{(330\ \Omega)(680\ \Omega)}{1010\ \Omega} + \frac{(680\ \Omega)(560\ \Omega)}{1240\ \Omega} = 222\ \Omega + 307\ \Omega = 529\ \Omega$$

Step 4: Place V_{TH} and R_{TH} in series to form the Thevenin equivalent circuit.

Step 5: Connect the load resistor from terminals A to B of the equivalent circuit, and determine the load voltage and current as illustrated in Figure 6–51.

$$V_L = \left(\frac{R_L}{R_L + R_{TH}}\right)V_{TH} = \left(\frac{1.0\ \text{k}\Omega}{1.529\ \text{k}\Omega}\right)5.32\ \text{V} = \mathbf{3.48\ V}$$

$$I_L = \frac{V_L}{R_L} = \frac{3.48\ \text{V}}{1.0\ \text{k}\Omega} = \mathbf{3.48\ mA}$$

▶ **FIGURE 6–51**

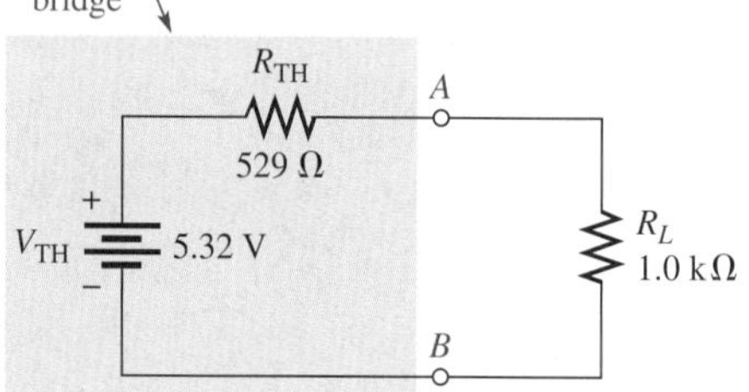

Related Problem Calculate I_L for $R_1 = 2.2\ \text{k}\Omega$, $R_2 = 3.9\ \text{k}\Omega$, $R_3 = 3.3\ \text{k}\Omega$, and $R_4 = 2.7\ \text{k}\Omega$ in Figure 6–50.

Open file E06-17 on your EWB/CircuitMaker CD-ROM. Determine the voltage and current for R_L using a multimeter. Change the resistor values to those specified in the related problem and measure the voltage and current for R_L.

Determining Load Power Using Thevenin's Theorem

Another useful application of Thevenin's theorem is in determining the power delivered to various values of load resistance by a circuit. When a circuit with a given source resistance is connected to a load resistance, the **maximum power transfer** theorem states that *maximum power is delivered to a load when the load resistance and the source resistance of the circuit are equal.* Power to a load decreases when the load resistance is greater or less than the source resistance. The maximum power transfer theorem is discussed again in Chapter 14.

The source resistance, R_S, of a circuit is the equivalent Thevenin resistance as viewed from the output terminals across which the load is connected. Using the Thevenin equivalent circuit for a more complex circuit greatly simplifies the calculation of power in different values of load resistance because you have a simple series circuit, as shown in Figure 6–52.

▲ **FIGURE 6–52**

Maximum power is transferred to the load when $R_L = R_S$.

EXAMPLE 6–18

The Thevenin equivalent circuit in Figure 6–53 represents a more complex circuit and has an equivalent resistance of 75 Ω. Determine the load power for each of the following values of the variable load resistance:

(a) 25 Ω **(b)** 50 Ω **(c)** 75 Ω **(d)** 100 Ω **(e)** 125 Ω

Draw a graph showing the load power versus the load resistance.

▶ **FIGURE 6–53**

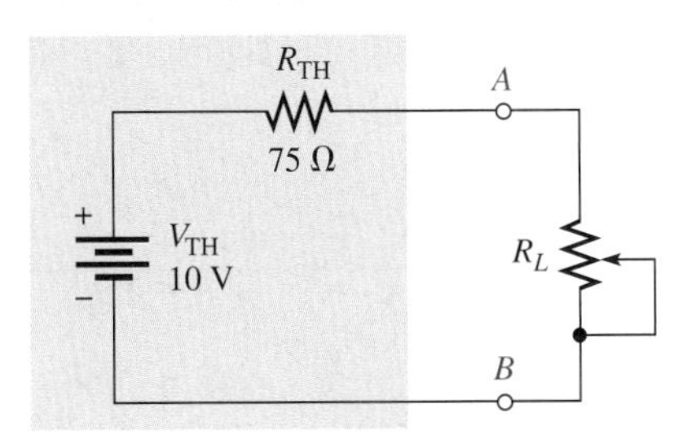

Solution Use Ohm's law ($I = V/R$) and the power formula ($P = I^2R$) to find the load power, P_L, for each value of load resistance.

(a) For $R_L = 25\ \Omega$,

$$I = \frac{V_{TH}}{R_{TH} + R_L} = \frac{10\ \text{V}}{75\ \Omega + 25\ \Omega} = 100\ \text{mA}$$

$$P_L = I^2R_L = (100\ \text{mA})^2(25\ \Omega) = \mathbf{250\ mW}$$

(b) For $R_L = 50\ \Omega$,

$$I = \frac{V_{TH}}{R_{TH} + R_L} = \frac{10\ \text{V}}{125\ \Omega} = 80\ \text{mA}$$

$$P_L = I^2R_L = (80\ \text{mA})^2(50\ \Omega) = \mathbf{320\ mW}$$

(c) For $R_L = 75\ \Omega$,

$$I = \frac{V_{TH}}{R_{TH} + R_L} = \frac{10\ \text{V}}{150\ \Omega} = 66.7\ \text{mA}$$

$$P_L = I^2R_L = (66.7\ \text{mA})^2(75\ \Omega) = \mathbf{334\ mW}$$

(d) For $R_L = 100\ \Omega$,

$$I = \frac{V_{TH}}{R_{TH} + R_L} = \frac{10\ \text{V}}{175\ \Omega} = 57.1\ \text{mA}$$

$$P_L = I^2R_L = (57.1\ \text{mA})^2(100\ \Omega) = \mathbf{326\ mW}$$

(e) For $R_L = 125\ \Omega$,

$$I = \frac{V_{TH}}{R_{TH} + R_L} = \frac{10\ \text{V}}{200\ \Omega} = 50\ \text{mA}$$

$$P_L = I^2R_L = (50\ \text{mA})^2(125\ \Omega) = \mathbf{313\ mW}$$

Notice that the load power is greatest when $R_L = R_{TH} = 75\ \Omega$. When the load resistance is less than or greater than this value, the power drops off, as the curve in Figure 6–54 graphically illustrates.

FIGURE 6–54

Curve showing that the load power is maximum when $R_L = R_{TH}$.

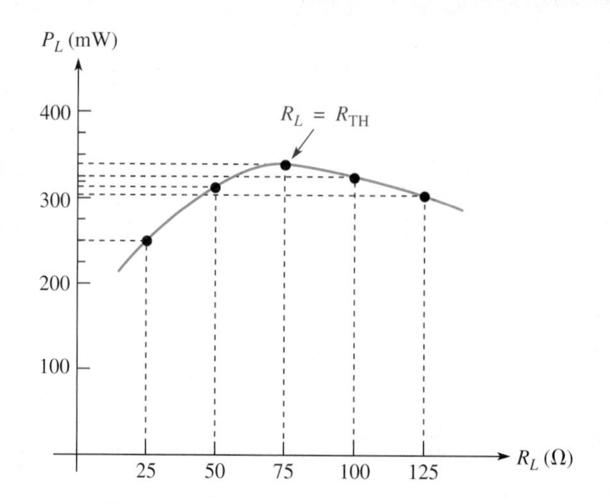

Related Problem If the source resistance in Figure 6–53 is 600 Ω, what is the maximum power that can be delivered to a load?

The Thevenin resistance of a circuit can be measured by connecting a variable resistance on the output of the circuit and adjusting the resistance until the output voltage of the circuit is one-half of its open circuit value. Now, if you remove the variable resistance and measure it, the value equals the Thevenin equivalent resistance of the circuit.

Summary of Thevenin's Theorem

Remember, the Thevenin equivalent circuit for any resistive circuit is *always* an *equivalent voltage source* in series with an *equivalent resistance* regardless of the original circuit that it replaces. The significance of Thevenin's theorem is that the equivalent circuit can replace the original circuit as far as any external load is concerned. Any load resistor connected between the terminals of a Thevenin equivalent circuit will have the same current through it and the same voltage across it as if it were connected to the terminals of the original circuit.

A summary of steps for applying Thevenin's theorem is as follows:

Step 1: Open the two terminals (remove any load) between which you want to find the Thevenin equivalent circuit.

Step 2: Determine the voltage (V_{TH}) across the two open terminals.

Step 3: Determine the resistance (R_{TH}) between the two terminals with all sources replaced by their internal resistance. (An ideal voltage source is replaced by a short.)

Step 4: Connect V_{TH} and R_{TH} in series to produce the complete Thevenin equivalent for the original circuit.

Step 5: Replace the load removed in Step 1 across the terminals of the Thevenin equivalent circuit. The load current and load voltage can now be calculated using only Ohm's law, and they have the same value as the load current and load voltage in the original circuit.

Two additional theorems are sometimes used in circuit analysis. One of these is Norton's theorem, which is similar to Thevenin's theorem except that it deals with current sources instead of voltage sources. The other is Millman's theorem, which deals with parallel voltage sources. See Appendix D for a coverage of the current source, Norton's theorem, and Millman's theorem.

SECTION 6–6 REVIEW

1. What are the two components of a Thevenin equivalent circuit?
2. Draw the general form of a Thevenin equivalent circuit.
3. Define V_{TH}.
4. Define R_{TH}.
5. For the original circuit in Figure 6–55, determine the Thevenin equivalent circuit as viewed by R_L.

▶ FIGURE 6–55

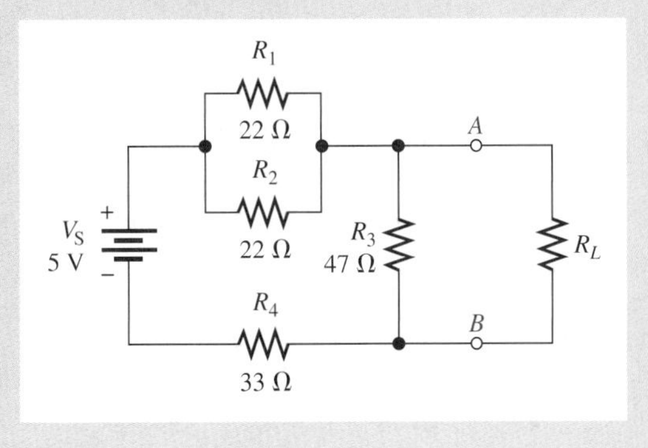

6–7 THE SUPERPOSITION THEOREM

Some circuits require more than one voltage source. For example, certain types of amplifiers require both a positive and a negative dc voltage source for proper operation as well as an ac signal source. In this section, a general method is presented for evaluating series-parallel circuits having multiple voltage sources. The method is illustrated with two-voltage-source circuits. However, once you understand the basic principle, you can extend this method to circuits with any number of sources.

After completing this section, you should be able to

- **Apply the superposition theorem to circuit analysis**
- State the superposition theorem
- List the steps in applying the theorem

The superposition theorem is a way to determine currents and voltages in a linear circuit that has multiple sources by taking one source at a time. The other sources are replaced by their internal resistances. Recall that the ideal voltage source has a zero internal resistance. All voltage sources will be treated as ideal in order to simplify the coverage.

A general statement of the superposition theorem is as follows:

The current in any given branch of a multiple-source linear circuit can be found by determining the currents in that particular branch produced by each source acting alone, with all other sources replaced by their internal resistances. The total current in the branch is the algebraic sum of the individual source currents in that branch.

The steps in applying the superposition theorem are as follows:

Step 1: Take one voltage source at a time and replace each of the other voltage sources with a short, which represents zero internal resistance.

Step 2: Determine the particular current or voltage that you want to find just as if there were only one source in the circuit. This is a component of the total current or voltage that you are looking for.

Step 3: Take the next source in the circuit and repeat Steps 1 and 2 for each source.

Step 4: To find the actual current in a given branch, add or subtract the component currents due to each individual source. If the currents are in the same direction, add them. If the currents are in opposite directions, subtract them with the direction of the resulting current the same as the larger of the individual currents. Once the current is found, voltage can be determined by Ohm's law.

An example of the approach to superposition is illustrated in Figure 6–56 for a series-parallel circuit with two voltage sources. Study the steps in this figure. Examples 6–19 and 6–20 will clarify this procedure.

FIGURE 6–56

Illustration of the superposition theorem.

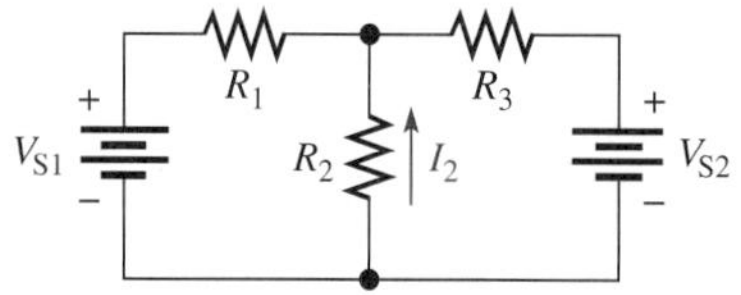

(a) Problem: Find I_2.

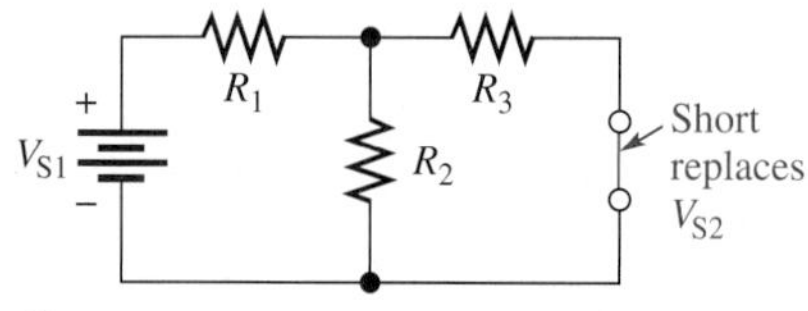

(b) Replace V_{S2} with zero resistance (short).

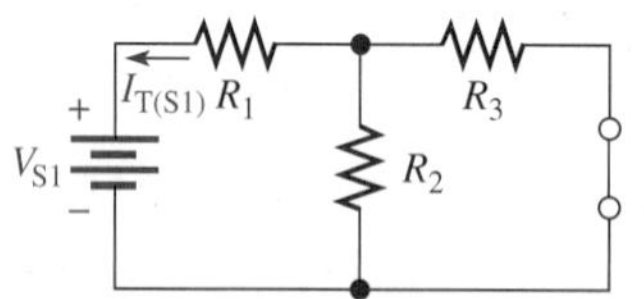

(c) Find R_T and I_T looking from V_{S1}:
$R_{T(S1)} = R_1 + R_2 \| R_3$
$I_{T(S1)} = V_{S1}/R_{T(S1)}$

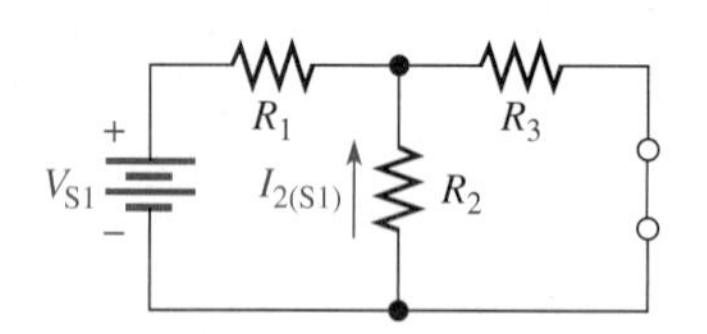

(d) Find I_2 due to V_{S1}:

$$I_{2(S1)} = \left(\frac{R_3}{R_2 + R_3}\right) I_{T(S1)}$$

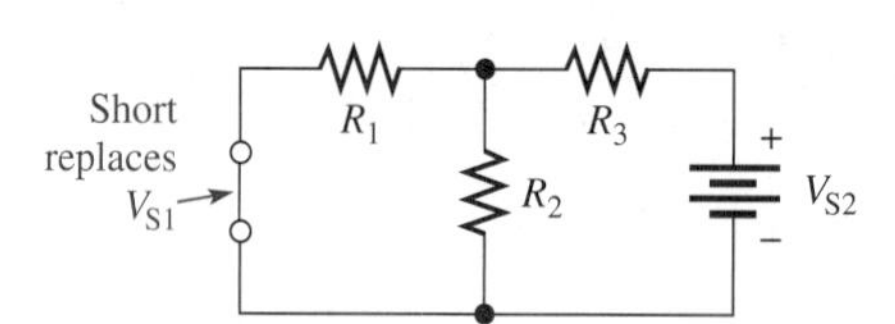

(e) Replace V_{S1} with zero resistance (short).

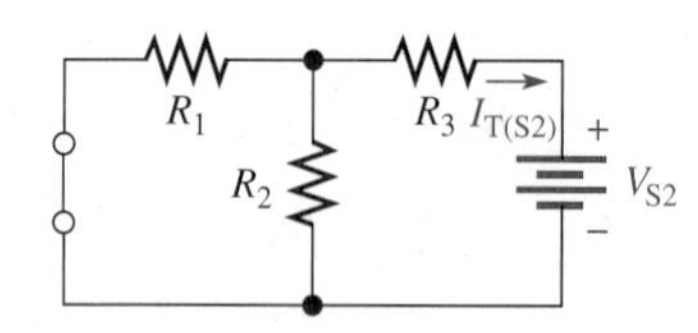

(f) Find R_T and I_T looking from V_{S2}:
$R_{T(S2)} = R_3 + R_1 \| R_2$
$I_{T(S2)} = V_{S2}/R_{T(S2)}$

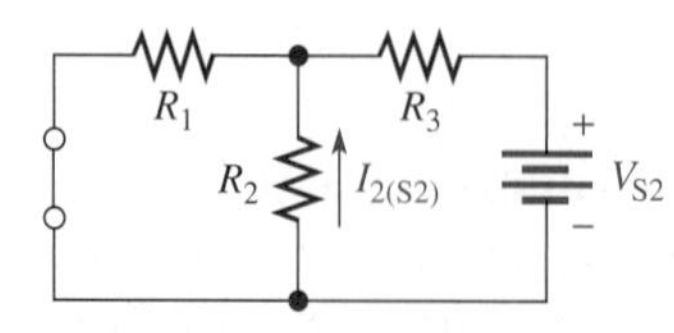

(g) Find I_2 due to V_{S2}:

$$I_{2(S2)} = \left(\frac{R_1}{R_1 + R_2}\right) I_{T(S2)}$$

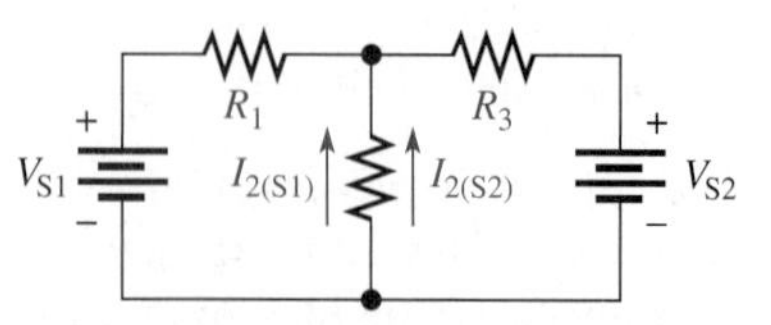

(h) Restore the original sources. Add $I_{2(S1)}$ and $I_{2(S2)}$ to get the actual I_2 (they are in same direction):
$I_2 = I_{2(S1)} + I_{2(S2)}$

EXAMPLE 6–19

Find the current in R_2 and the voltage across it in Figure 6–57 by using the superposition theorem.

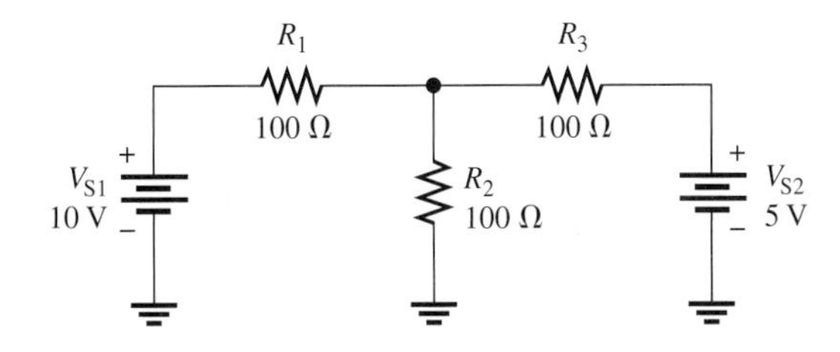

▶ FIGURE 6–57

Solution **Step 1:** Replace V_{S2} with a short to represent its zero internal resistance and find the current in R_2 due to voltage source V_{S1}, as shown in Figure 6–58. To find I_2, use the current-divider formula. Looking from V_{S1},

$$R_{T(S1)} = R_1 + \frac{R_3}{2} = 100\ \Omega + 50\ \Omega = 150\ \Omega$$

$$I_{T(S1)} = \frac{V_{S1}}{R_{T(S1)}} = \frac{10\text{ V}}{150\ \Omega} = 66.7\text{ mA}$$

The component of the total current in R_2 due to V_{S1} is

$$I_{2(S1)} = \left(\frac{R_3}{R_2 + R_3}\right)I_{T(S1)} = \left(\frac{100\ \Omega}{200\ \Omega}\right)66.7\text{ mA} = 33.3\text{ mA}$$

Note that this current is upward through R_2.

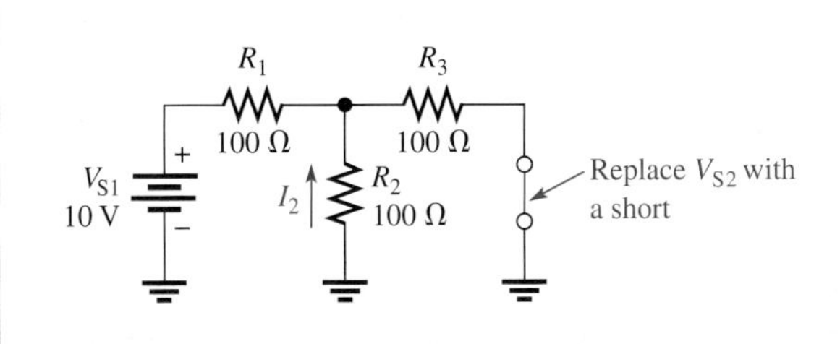

▲ FIGURE 6–58

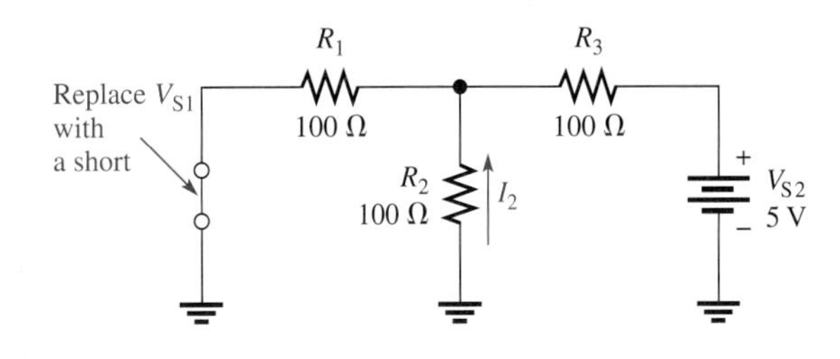

▲ FIGURE 6–59

Step 2: Find the current in R_2 due to voltage source V_{S2} by replacing V_{S1} with a short, as shown in Figure 6–59. Looking from V_{S2},

$$R_{T(S2)} = R_3 + \frac{R_1}{2} = 100\ \Omega + 50\ \Omega = 150\ \Omega$$

$$I_{T(S2)} = \frac{V_{S2}}{R_{T(S2)}} = \frac{5\text{ V}}{150\ \Omega} = 33.3\text{ mA}$$

The component of the total current in R_2 due to V_{S2} is

$$I_{2(S2)} = \left(\frac{R_1}{R_1 + R_2}\right)I_{T(S2)} = \left(\frac{100\ \Omega}{200\ \Omega}\right)33.3\text{ mA} = 16.7\text{ mA}$$

Note that this current is upward through R_2.

Step 3: Both component currents are upward through R_2, so they have the same algebraic sign. Therefore, add the values to get the total current through R_2.

$$I_{2(tot)} = I_{2(S1)} + I_{2(S2)} = 33.3\text{ mA} + 16.7\text{ mA} = \mathbf{50\text{ mA}}$$

The voltage across R_2 is

$$V_{R2} = I_{2(tot)}R_2 = (50\text{ mA})(100\ \Omega) = \mathbf{5\text{ V}}$$

Related Problem Determine the total current through R_2 if the polarity of V_{S2} in Figure 6–57 is reversed.

EXAMPLE 6–20

Find the total current through and voltage across R_3 in Figure 6–60.

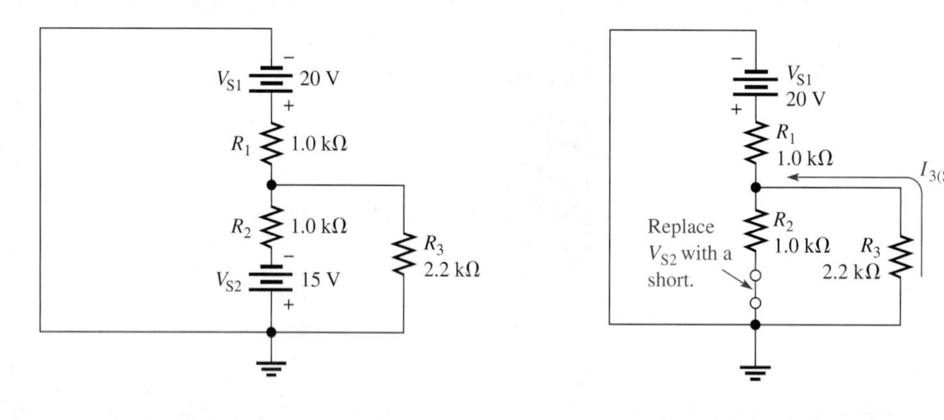

▲ FIGURE 6–60

▲ FIGURE 6–61

Solution **Step 1:** Find the current through R_3 due to source V_{S1} by replacing source V_{S2} with a short to represent its zero internal resistance, as shown in Figure 6–61. Looking from V_{S1},

$$R_{T(S1)} = R_1 + \frac{R_2R_3}{R_2 + R_3} = 1.0\text{ k}\Omega + \frac{(1.0\text{ k}\Omega)(2.2\text{ k}\Omega)}{3.2\text{ k}\Omega} = 1.69\text{ k}\Omega$$

$$I_{T(S1)} = \frac{V_{S1}}{R_{T(S1)}} = \frac{20\text{ V}}{1.69\text{ k}\Omega} = 11.8\text{ mA}$$

Now apply the current-divider formula to get the current through R_3 due to source V_{S1}.

$$I_{3(S1)} = \left(\frac{R_2}{R_2 + R_3}\right)I_{T(S1)} = \left(\frac{1.0\text{ k}\Omega}{3.2\text{ k}\Omega}\right)11.8\text{ mA} = 3.69\text{ mA}$$

Notice that this current is upward through R_3.

Step 2: Find the current through R_3 due to source V_{S2} by replacing source V_{S1} with a short, as shown in Figure 6–62. Looking from V_{S2},

$$R_{T(S2)} = R_2 + \frac{R_1R_3}{R_1 + R_3} = 1.0\text{ k}\Omega + \frac{(1.0\text{ k}\Omega)(2.2\text{ k}\Omega)}{3.2\text{ k}\Omega} = 1.69\text{ k}\Omega$$

$$I_{T(S2)} = \frac{V_{S2}}{R_{T(S2)}} = \frac{15\text{ V}}{1.69\text{ k}\Omega} = 8.88\text{ mA}$$

Now apply the current-divider formula to find the current through R_3 due to source V_{S2}.

$$I_{3(S2)} = \left(\frac{R_1}{R_1 + R_3}\right)I_{T(S2)} = \left(\frac{1.0\text{ k}\Omega}{3.2\text{ k}\Omega}\right)8.88\text{ mA} = 2.78\text{ mA}$$

Notice that this current is downward through R_3.

FIGURE 6–62

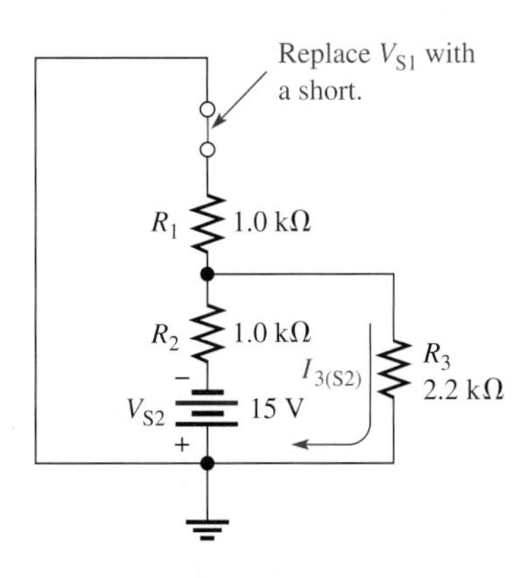

Step 3: Calculate the total current through R_3 and the voltage across it.

$$I_{3(tot)} = I_{3(S1)} - I_{3(S2)} = 3.69\text{ mA} - 2.78\text{ mA} = 0.91\text{ mA} = \mathbf{910\ \mu A}$$

$$V_{R3} = I_{3(tot)}R_3 = (910\ \mu\text{A})(2.2\text{ k}\Omega) \cong \mathbf{2\ V}$$

The current is upward through R_3.

Related Problem Find the total current through R_3 in Figure 6–60 if V_{S1} is changed to 12 V and its polarity reversed.

SECTION 6–7 REVIEW

1. State the superposition theorem.
2. Why is the superposition theorem useful for analysis of multiple-source linear circuits?
3. Why is a voltage source shorted when the superposition theorem is applied?

FIGURE 6–63

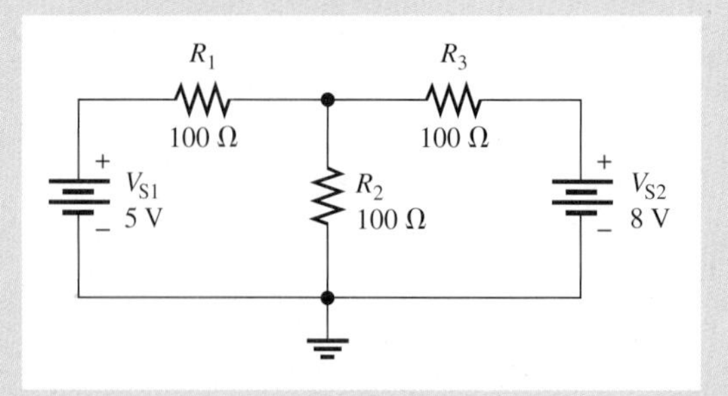

4. Using the superposition theorem, find the current through R_1 in Figure 6–63.
5. If, as a result of applying the superposition theorem, two component currents are in opposing directions through a branch of a circuit, in which direction is the net current?

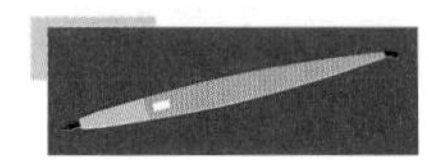

6–8 TROUBLESHOOTING

Troubleshooting is the process of identifying and locating a failure or problem in a circuit. Some troubleshooting techniques and the application of logical thought have already been discussed in relation to both series circuits and parallel circuits. A basic premise of troubleshooting is that you must know what to look for before you can successfully troubleshoot a circuit. These concepts are now applied to series-parallel circuits.

After completing this section, you should be able to

- **Troubleshoot series-parallel circuits**
- Determine the effects of an open circuit
- Determine the effects of a short circuit
- Locate opens and shorts

Opens and shorts are typical problems that occur in electric circuits. As mentioned in Chapter 4, if a resistor burns out, it will normally produce an open circuit. Bad solder connections, broken wires, and poor contacts can also be causes of open paths. Pieces of foreign material, such as solder splashes, broken insulation on wires, and so on, can often lead to shorts in a circuit. A short is considered to be a zero resistance path between two points.

In addition to complete opens or shorts, partial opens or partial shorts can develop in a circuit. A partial open would be a much higher than normal resistance, but not infinitely large. A partial short would be a much lower than normal resistance, but not zero.

The following three examples illustrate troubleshooting series-parallel circuits.

EXAMPLE 6–21

From the indicated voltmeter reading in Figure 6–64, determine if there is a fault. If there is a fault, identify it as either a short or an open.

▶ FIGURE 6–64

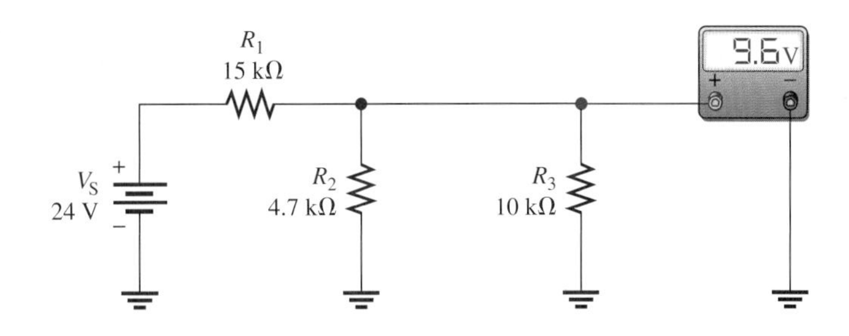

Solution **Step 1:** Analysis

You must know what you expect to measure before you can determine if the measurement is incorrect. First determine what the voltmeter should be indicating as follows. Since R_2 and R_3 are in parallel, their combined resistance is

$$R_{2\|3} = \frac{R_2 R_3}{R_2 + R_3} = \frac{(4.7\text{ k}\Omega)(10\text{ k}\Omega)}{14.7\text{ k}\Omega} = 3.20\text{ k}\Omega$$

The voltage across the parallel combination is determined by the voltage-divider formula.

$$V_{2\|3} = \left(\frac{R_{2\|3}}{R_1 + R_{2\|3}}\right)V_S = \left(\frac{3.2\text{ k}\Omega}{18.2\text{ k}\Omega}\right)24\text{ V} = 4.22\text{ V}$$

This calculation shows that 4.22 V is the voltage reading that you should get on the meter. However, the meter reads 9.6 V across $R_{2\|3}$. This value is incorrect, and, because it is higher than it should be, either R_2 or R_3 is probably open. Why? Because if either of these two resistors is open, the resistance across which the meter is connected is larger than expected. A higher resistance will drop a higher voltage in this circuit.

Step 2: Planning

Start trying to find the open resistor by assuming that R_2 is open. If it is, the voltage across R_3 is

$$V_3 = \left(\frac{R_3}{R_1 + R_3}\right)V_S = \left(\frac{10\text{ k}\Omega}{25\text{ k}\Omega}\right)24\text{ V} = 9.6\text{ V}$$

Since the measured voltage is also 9.6 V, this calculation shows that R_2 is probably open.

Step 3: Measurement

Disconnect power and remove R_2. Measure its resistance to verify it is open. If it is not, inspect the wiring, solder, or connections around R_2, looking for the open.

Related Problem What would be the voltmeter reading if R_3 were open in Figure 6–64? If R_1 were open?

Open file E06-21 on your EWB/CircuitMaker CD-ROM. Determine if a fault exists in the circuit and, if so, isolate the fault to a single component.

EXAMPLE 6–22

Suppose that you measure 24 V with the voltmeter in Figure 6–65. Determine if there is a fault, and, if there is, identify it.

FIGURE 6–65

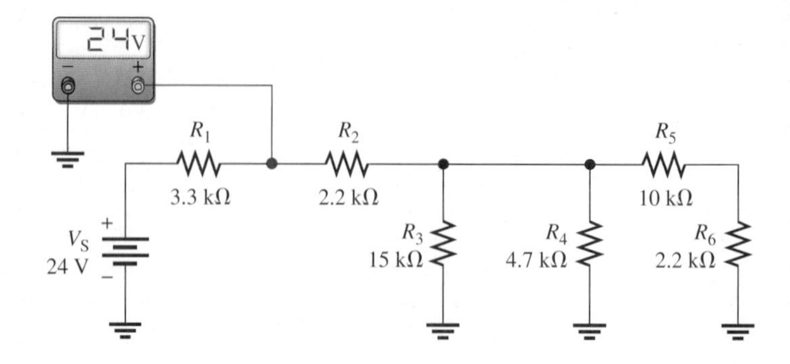

Solution **Step 1:** Analysis

There is no voltage drop across R_1 because both sides of the resistor are at +24 V. Either there is no current through R_1 from the source, which tells you that R_2 is open in the circuit, or R_1 is shorted.

Step 2: Planning

The most probable failure is an open R_2. If it is open, then there will be no current from the source. To verify this, measure across R_2 with the voltmeter. If R_2 is open, the meter will indicate 24 V. The right side of R_2 will be at zero volts because there is no current through any of the other resistors to cause a voltage drop across them.

Step 3: Measurement

The measurement to verify that R_2 is open is shown in Figure 6–66.

FIGURE 6–66

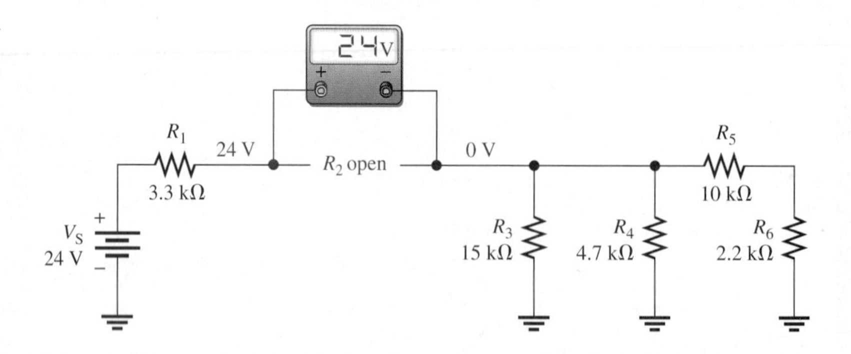

Related Problem What would be the voltage across an open R_5 in Figure 6–65 assuming no other faults?

Open file E06-22 on your EWB/CircuitMaker CD-ROM. Determine if a fault exists in the circuit and, if so, isolate the fault to a single component.

EXAMPLE 6–23

The two voltmeters in Figure 6–67 indicate the voltages shown. Apply logical thought and your knowledge of circuit operation to determine if there are any opens or shorts in the circuit and, if so, where they are located.

FIGURE 6–67

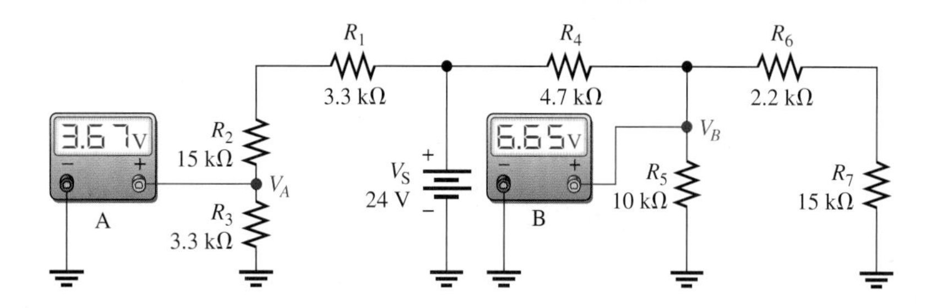

Solution First, determine if the voltmeter readings are correct. R_1, R_2, and R_3 act as a voltage divider. The voltage (V_A) across R_3 is calculated as follows:

$$V_A = \left(\frac{R_3}{R_1 + R_2 + R_3}\right)V_S = \left(\frac{3.3\text{ k}\Omega}{21.6\text{ k}\Omega}\right)24\text{ V} = 3.67\text{ V}$$

The voltmeter A reading is correct. This indicates that R_1, R_2, and R_3 are connected and are not faulty.

Now see if the voltmeter B reading is correct. $R_6 + R_7$ is in parallel with R_5. The series-parallel combination of R_5, R_6, and R_7 is in series with R_4. The resistance of the R_5, R_6, and R_7 combination is calculated as follows:

$$R_{5\|(6+7)} = \frac{R_5(R_6 + R_7)}{R_5 + R_6 + R_7} = \frac{(10\text{ k}\Omega)(17.2\text{ k}\Omega)}{27.2\text{ k}\Omega} = 6.32\text{ k}\Omega$$

$R_{5\|(6+7)}$ and R_4 form a voltage divider, and voltmeter B is measuring the voltage across $R_{5\|(6+7)}$. Is it correct? Check as follows:

$$V_B = \left(\frac{R_{5\|(6+7)}}{R_4 + R_{5\|(6+7)}}\right)V_S = \left(\frac{6.32\text{ k}\Omega}{11\text{ k}\Omega}\right)24\text{ V} = 13.8\text{ V}$$

Thus, the actual measured voltage (6.65 V) at this point is incorrect. Further logical thought will help to isolate the problem.

R_4 is not open, because if it were, the meter would read 0 V. If there were a short across it, the meter would read 24 V. Since the actual voltage is much less than it should be, $R_{5\|(6+7)}$ must be less than the calculated value of 6.32 kΩ. The most likely problem is a short across R_7. If there is a short from the top of R_7 to ground, R_6 is effectively in parallel with R_5. In this case,

$$R_5 \| R_6 = \frac{R_5 R_6}{R_5 + R_6} = \frac{(10\text{ k}\Omega)(2.2\text{ k}\Omega)}{12.2\text{ k}\Omega} = 1.80\text{ k}\Omega$$

Then V_B is

$$V_B = \left(\frac{1.80\text{ k}\Omega}{6.5\text{ k}\Omega}\right)24\text{ V} = 6.65\text{ V}$$

This value for V_B agrees with the voltmeter B reading. So there is a short across R_7. If this were an actual circuit, you would try to find the physical cause of the short.

Related Problem If the only fault in Figure 6–67 is that R_2 is shorted, what will voltmeter A read? What will voltmeter B read?

Open file E06-23 on your EWB/CircuitMaker CD-ROM. Determine if a fault exists in the circuit and, if so, isolate the fault to a single component.

SECTION 6–8 REVIEW

1. Name two types of common circuit faults.
2. For the following faults in Figure 6–68, determine what voltage would be measured at point *A*:
   (a) no faults **(b)** R_1 open **(c)** short across R_5
   (d) R_3 and R_4 open **(e)** R_2 open

FIGURE 6–68

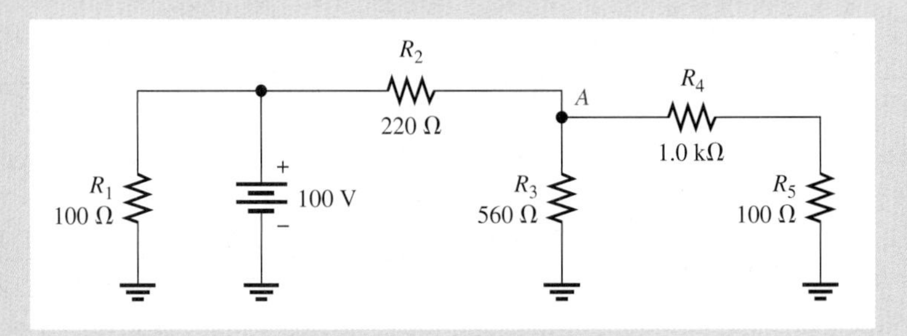

3. In Figure 6–69, one of the resistors in the circuit is open. Based on the meter reading, determine which is the open resistor.

FIGURE 6–69

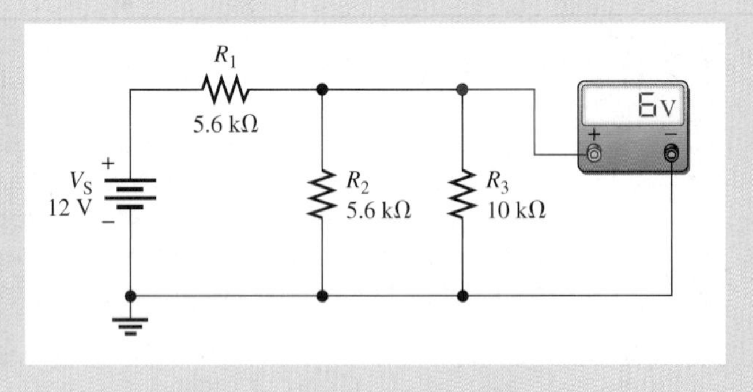

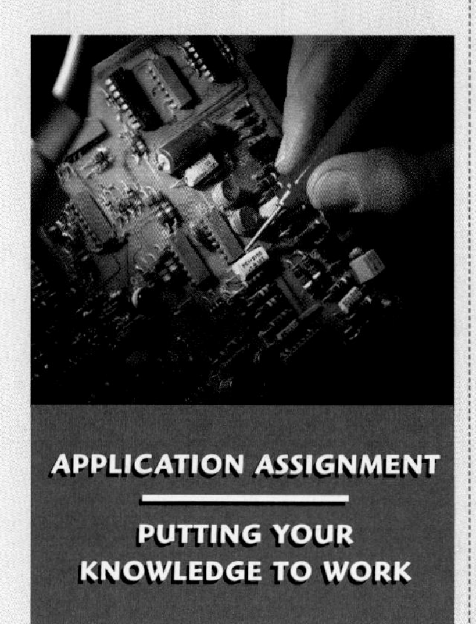

APPLICATION ASSIGNMENT

PUTTING YOUR KNOWLEDGE TO WORK

A voltage divider with three output voltages has been designed and constructed on a PC board. The voltage divider is to be used as part of a portable power supply unit for supplying up to three different reference voltages to measuring instruments in the field. The power supply unit contains a battery pack combined with a voltage regulator that produces a constant +12 V to the voltage-divider circuit board. In this assignment, you will apply your knowledge of loaded voltage dividers, Kirchhoff's laws, and Ohm's law to determine the operating parameters of the voltage divider in terms of voltages and currents for all possible load configurations. You will also troubleshoot the circuit for various malfunctions.

Step 1: Draw the Schematic of the Voltage Divider

Draw the schematic and label the resistor values for the circuit board in Figure 6–70.

Step 2: Connect the 12 V Power Supply

Specify how to connect a 12 V power supply to the circuit board so that all resistors are in series and pin 2 has the highest output voltage.

Step 3: Determine the Unloaded Output Voltages

Calculate each of the output voltages with no loads connected. Add these voltage values to a copy of the table in Figure 6–71.

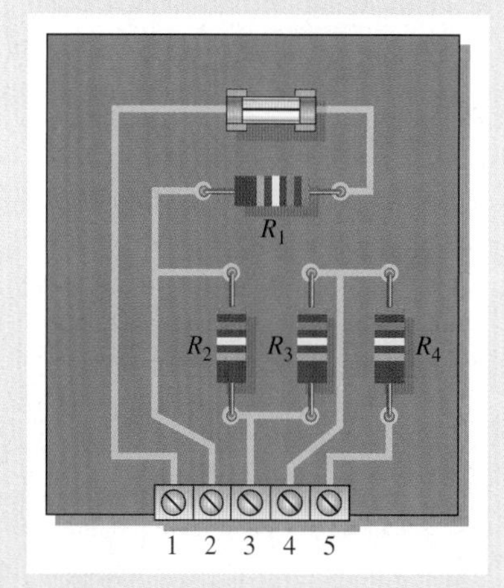

FIGURE 6–70

Voltage-divider circuit board.

FIGURE 6–71

Table of operating parameters for the power supply voltage divider.

10 MΩ Load	$V_{OUT(2)}$	$V_{OUT(3)}$	$V_{OUT(4)}$	% Deviation	$I_{LOAD(2)}$	$I_{LOAD(3)}$	$I_{LOAD(4)}$
None							
Pin 2 to ground							
Pin 3 to ground							
Pin 4 to ground							
Pin 2 to ground Pin 3 to ground				2 3			
Pin 2 to ground Pin 4 to ground				2 4			
Pin 3 to ground Pin 4 to ground				3 4			
Pin 2 to ground Pin 3 to ground Pin 4 to ground				2 3 4			

Step 4: Determine the Loaded Output Voltages

The instruments to be connected to the voltage divider each have a 10 MΩ input resistance. This means that when an instrument is connected to a voltage-divider output there is effectively a 10 MΩ resistor from that output to ground (negative side of source). Calculate the output voltage across each load for each of the following load combinations and add these voltage values to a copy of the table in Figure 6–71.

1. A 10 MΩ load connected from pin 2 to ground.
2. A 10 MΩ load connected from pin 3 to ground.
3. A 10 MΩ load connected from pin 4 to ground.
4. 10 MΩ loads connected from pin 2 and pin 3 to ground.
5. 10 MΩ loads connected from pin 2 and pin 4 to ground.
6. 10 MΩ loads connected from pin 3 and pin 4 to ground.
7. 10 MΩ loads connected from pin 2, pin 3, and pin 4 to ground.

Step 5: Determine the Percent Deviation of the Output Voltages

Calculate how much each loaded output voltage deviates from its unloaded value for each of the load configurations listed in Step 4 and express as a percentage using the following formula:

$$\text{Percent deviation} = \left(\frac{V_{OUT(unloaded)} - V_{OUT(loaded)}}{V_{OUT(unloaded)}}\right)100\%$$

Add the values to the table in Figure 6–71.

Step 6: Determine the Load Currents

Calculate the current to each 10 MΩ load for each of the load configurations listed in Step 4. Add these values to the table in Figure 6–71.

Step 7: Specify a Minimum Value for the Fuse

Step 8: Troubleshoot the Circuit Board

The voltage-divider circuit board is connected to a 12 V power supply and to the three instruments to which it provides reference voltages, as shown in Figure 6–72. Voltages at each of the numbered test points are measured with a voltmeter in each of eight different cases. For each case, determine the problem indicated by the voltage measurements.

FIGURE 6–72

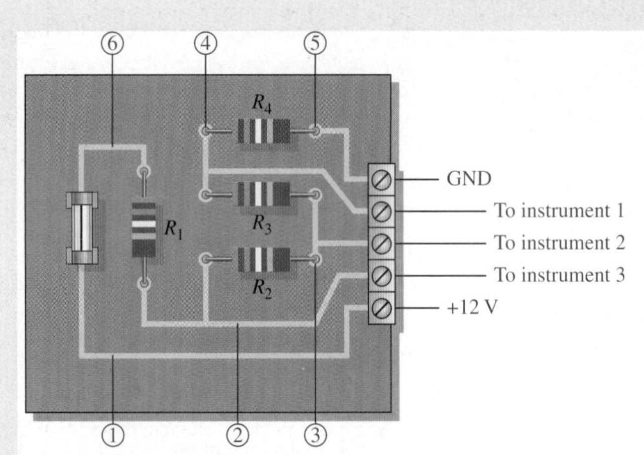

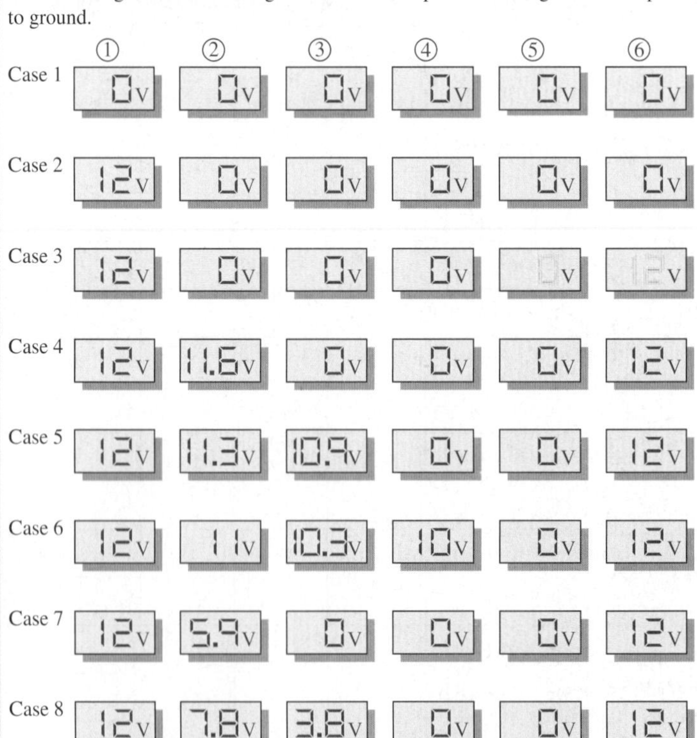

Electronics Workbench/CircuitMaker Analysis and Troubleshooting

1. Using Electronics Workbench or CircuitMaker, connect the circuit shown by the schematic from Step 1 and verify the unloaded output voltages determined in Step 3.
2. Measure the load currents calculated in Step 6.
3. Verify the faults determined in Step 8 for each case by inserting the fault in the circuit and checking the voltage measurements at each point.

APPLICATION ASSIGNMENT REVIEW

1. If the portable unit covered in this section is to supply reference voltages to all three instruments, how many days can a 100 mAh battery be used before recharging?
2. Can ⅛ W resistors be used on the voltage-divider board?
3. If ⅛ W resistors are used, will an output shorted to ground cause any of the resistors to overheat due to excessive power?

SUMMARY

- A series-parallel circuit is a combination of both series current paths and parallel current paths.
- To determine total resistance in a series-parallel circuit, identify the series and parallel relationships, and then apply the formulas for series resistance and parallel resistance from Chapters 4 and 5.
- To find the total current, divide the total voltage by the total resistance.
- To determine branch currents, apply the current-divider formula, Kirchhoff's current law, or Ohm's law. Consider each circuit problem individually to determine the most appropriate method.
- To determine voltage drops across any portion of a series-parallel circuit, use the voltage-divider formula, Kirchhoff's voltage law, or Ohm's law. Consider each circuit problem individually to determine the most appropriate method.
- When a load resistor is connected across a voltage-divider output, the output voltage decreases.
- The load resistor should be large compared to the resistance across which it is connected, in order that the loading effect may be minimized. A *10-times* value is sometimes used as a rule of thumb, but the value depends on the accuracy required for the output voltage.
- To find any current or voltage in a circuit with two or more voltage sources, take the sources one at a time using the superposition theorem.
- A balanced Wheatstone bridge can be used to measure an unknown resistance.
- A bridge is balanced when the output voltage is zero. The balanced condition produces zero current through a load connected across the output terminals of the bridge.

- An unbalanced Wheatstone bridge can be used to measure physical quantities using transducers.
- Any two-terminal resistive circuit, no matter how complex, can be replaced by its Thevenin equivalent.
- The Thevenin equivalent circuit is made up of an equivalent resistance (R_{TH}) in series with an equivalent voltage source (V_{TH}).
- Open circuits and short circuits are typical faults.
- Resistors normally open when they burn out.

EQUATIONS

6–1	$I_{BLEEDER} = I_T - I_{RL1} - I_{RL2}$	Bleeder current
6–2	$R_X = R_V\left(\frac{R_2}{R_4}\right)$	Unknown resistance in a Wheatstone bridge
6–3	$\Delta V_{OUT} = \Delta R_{therm}\left(\frac{V_S}{4R}\right)$	Thermister bridge output

SELF-TEST

Answers are at the end of the chapter.

1. Which of the following statements are true concerning Figure 6–73?
   (a) R_1 and R_2 are in series with R_3, R_4, and R_5.
   (b) R_1 and R_2 are in series.
   (c) R_3, R_4, and R_5 are in parallel.
   (b) The series combination of R_1 and R_2 is in parallel with the series combination of R_3, R_4, and R_5.
   (e) answers (b) and (d)

FIGURE 6–73

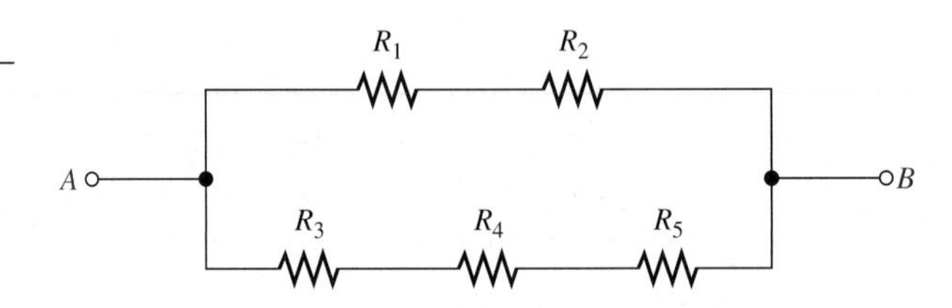

2. The total resistance of Figure 6–73 can be found with which of the following formulas?
   (a) $R_1 + R_2 + R_3 \| R_4 \| R_5$ (b) $R_1 \| R_2 + R_3 \| R_4 \| R_5$
   (c) $(R_1 + R_2) \| (R_3 + R_4 + R_5)$ (d) none of these answers
3. If all of the resistors in Figure 6–73 have the same value, when voltage is applied across terminals A and B, the current is
   (a) greatest in R_5 (b) greatest in R_3, R_4, and R_5
   (c) greatest in R_1 and R_2 (d) the same in all the resistors
4. Two 1.0 kΩ resistors are in series and this series combination is in parallel with a 2.2 kΩ resistor. The voltage across one of the 1.0 kΩ resistors is 6 V. The voltage across the 2.2 kΩ resistor is
   (a) 6 V (b) 3 V (c) 12 V (d) 13.2 V

5. The parallel combination of a 330 Ω resistor and a 470 Ω resistor is in series with the parallel combination of four 1.0 kΩ resistors. A 100 V source is connected across the circuit. The resistor with the most current has a value of

(a) 1.0 kΩ **(b)** 330 Ω **(c)** 470 Ω

6. In the circuit described in Question 5, the resistor(s) with the most voltage has a value of

(a) 1.0 kΩ **(b)** 470 Ω **(c)** 330 Ω

7. In the circuit described in Question 5, the percentage of the total current through any single 1.0 kΩ resistor is

(a) 100% **(b)** 25% **(c)** 50% **(d)** 31.25%

8. The output of a certain voltage divider is 9 V with no load. When a load is connected, the output voltage

(a) increases **(b)** decreases

(c) remains the same **(d)** becomes zero

9. A certain voltage divider consists of two 10 kΩ resistors in series. Which of the following load resistors will have the most effect on the output voltage?

(a) 1.0 MΩ **(b)** 20 kΩ **(c)** 100 kΩ **(d)** 10 kΩ

10. When a load resistance is connected to the output of a voltage-divider circuit, the current drawn from the source

(a) decreases **(b)** increases

(c) remains the same **(d)** is cut off

11. The output voltage of a balanced Wheatstone bridge is

(a) equal to the source voltage

(b) equal to zero

(c) dependent on all of the resistor values in the bridge

(d) dependent on the value of the unknown resistor

12. The primary method of analyzing a circuit with two or more voltage sources is usually

(a) Thevenin's theorem **(b)** Ohm's law

(c) superposition **(d)** Kirchhoff's law

13. In a certain two-source circuit, one source acting alone produces 10 mA through a given branch. The other source acting alone produces 8 mA in the opposite direction through the same branch. With both sources, the total current through the branch is

(a) 10 mA **(b)** 8 mA **(c)** 18 mA **(d)** 2 mA

14. A Thevenin equivalent circuit consists of

(a) a voltage source in series with a resistance

(b) a voltage source in parallel with a resistance

(c) a current source in parallel with a resistance

(d) two voltage sources and a resistance

15. You are measuring the voltage at a given point in a circuit that has very high resistance values and the measured voltage is a little lower than it should be. This is possibly because of

(a) one or more of the resistance values being off

(b) the loading effect of the voltmeter

(c) the source voltage is too low

(d) all of these answers

PROBLEMS

Answers to odd-numbered problems are at the end of the book.

BASIC PROBLEMS

SECTION 6–1 **Identifying Series-Parallel Relationships**

1. Identify the series and parallel relationships in Figure 6–74 as seen from the source terminals.

▶ FIGURE 6–74

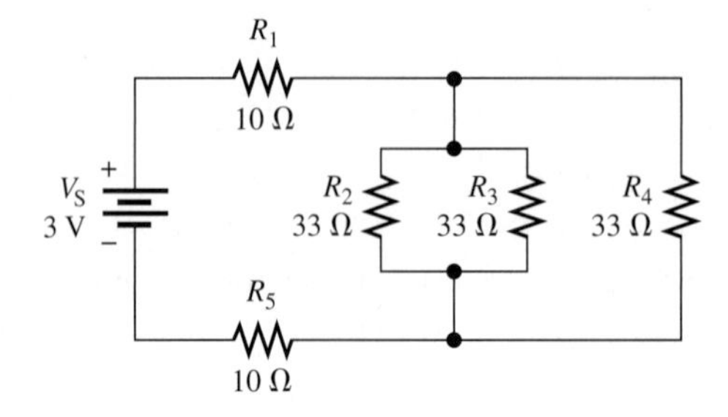

2. Visualize and sketch the following series-parallel combinations:
   (a) R_1 in series with the parallel combination of R_2 and R_3.
   (b) R_1 in parallel with the series combination of R_2 and R_3.
   (c) R_1 in parallel with a branch containing R_2 in series with a parallel combination of four other resistors
3. Visualize and sketch the following series-parallel circuits:
   (a) a parallel combination of three branches, each containing two series resistors
   (b) a series combination of three parallel circuits, each containing two parallel resistors
4. In each circuit of Figure 6–75 identify the series and parallel relationships of the resistors viewed from the source.

▶ FIGURE 6–75

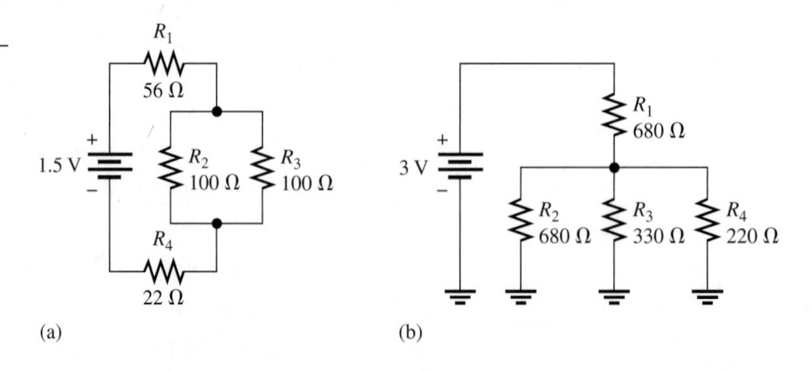

SECTION 6–2 **Analysis of Series-Parallel Circuits**

5. A certain circuit is composed of two parallel resistors. The total resistance is 667 Ω. One of the resistors is 1.0 kΩ. What is the other resistor?
6. For the circuit in Figure 6 –76, determine the total resistance between points *A* and *B*.

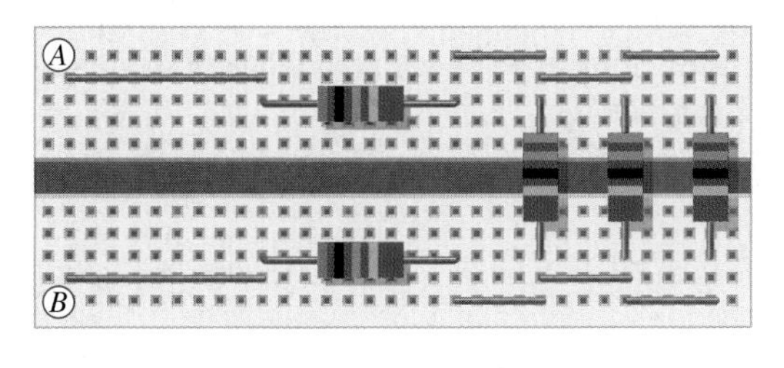

▲ FIGURE 6–76

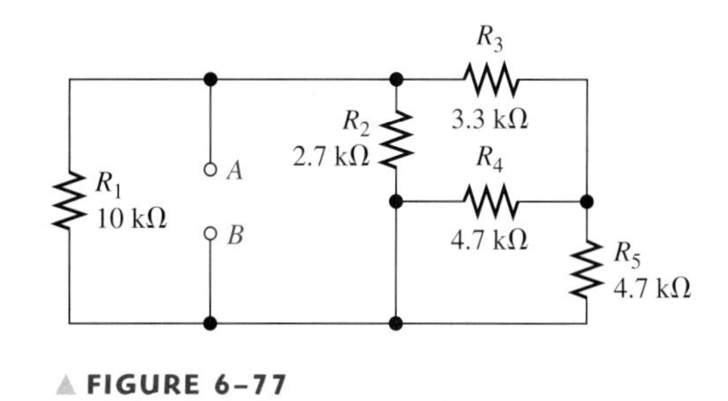

▲ FIGURE 6–77

7. Determine the total resistance for each circuit in Figure 6–75.
8. Determine the current through each resistor in Figure 6–74; then calculate each voltage drop.
9. Determine the current through each resistor in both circuits of Figure 6–75; then calculate each voltage drop.
10. In Figure 6–77, find the following:
    (a) total resistance between terminals *A* and *B*
    (b) total current drawn from a 6 V source connected from *A* to *B*
    (c) current through R_5
    (d) voltage across R_2

SECTION 6–3 **Voltage Dividers with Resistive Loads**

11. A voltage divider consists of two 56 kΩ resistors and a 15 V source. Calculate the unloaded output voltage taken across one of the 56 kΩ resistors. What will the output voltage be if a load resistor of 1.0 MΩ is connected across the output?
12. A 12 V battery output is divided down to obtain two output voltages. Three 3.3 kΩ resistors are used to provide the two outputs with only one output at a time loaded with 10 kΩ. Determine the output voltages in both cases.
13. Which will cause a smaller decrease in output voltage for a given voltage divider, a 10 kΩ load or a 56 kΩ load?
14. In Figure 6–78, determine the current drain on the battery with no load on the output terminals. With a 10 kΩ load, what is the current from the battery?

▶ FIGURE 6–78

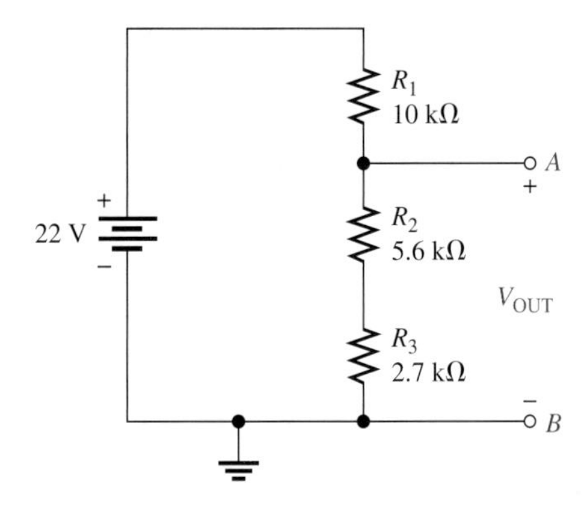

SECTION 6–4 **Loading Effect of a Voltmeter**

15. Across which one of the following resistances will a voltmeter with a 10 MΩ internal resistance present the minimum load on a circuit?
    (a) 100 kΩ (b) 1.2 MΩ (c) 22 kΩ (d) 8.2 MΩ

16. A certain voltage divider consists of three 1.0 MΩ resistors connected in series to a 100 V source. Determine the voltage across one of the resistors measured by a 10 MΩ voltmeter.

17. What is the difference between the measured and the actual unloaded voltage in Problem 16?

18. By what percentage does the voltmeter in Problem 16 alter the voltage which it measures?

SECTION 6–5 **The Wheatstone Bridge**

19. A resistor of unknown value is connected to a Wheatstone bridge circuit. The bridge parameters for a balanced condition are set as follows: R_V = 18 kΩ and $R_2/R_4 = 0.02$. What is R_X?

20. A bridge network is shown in Figure 6–79. To what value must R_V be set in order to balance the bridge?

21. Determine the value of R_X in the balanced bridge in Figure 6–80.

22. Determine the output voltage of the unbalanced bridge in Figure 6–81 for a temperature of 65° C. The thermistor has a nominal resistance of 1 kΩ at 25° C. Assume that its resistance changes 5 Ω for each C° change in temperature.

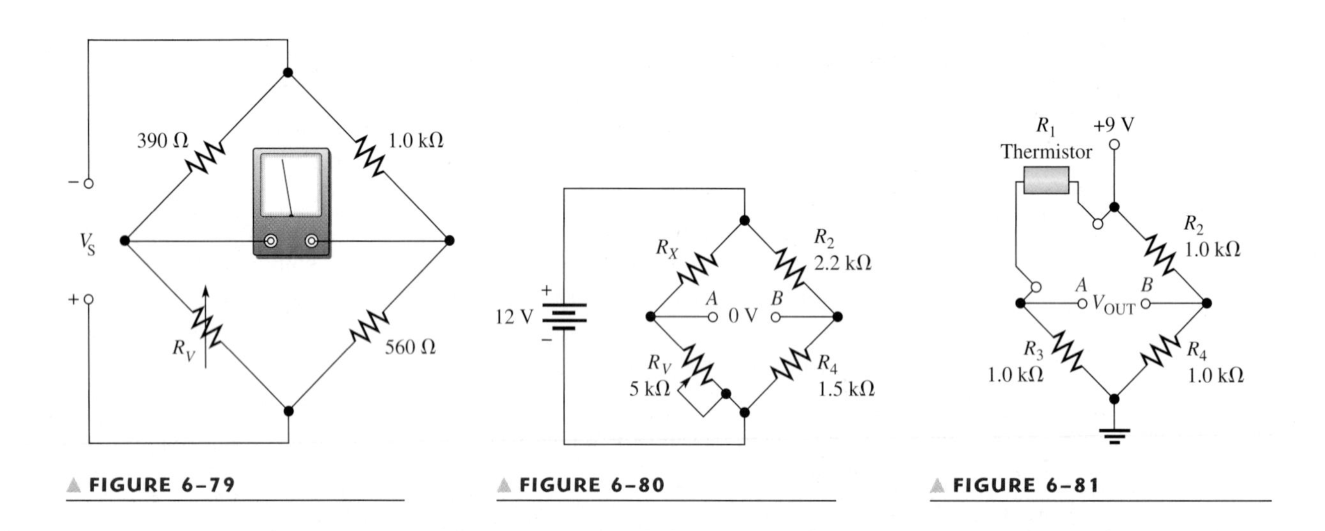

▲ FIGURE 6–79

▲ FIGURE 6–80

▲ FIGURE 6–81

SECTION 6–6 **Thevenin's Theorem**

23. Reduce the circuit in Figure 6–82 to its Thevenin equivalent as viewed from terminals *A* and *B*.

▶ FIGURE 6–82

R_1 100 kΩ

15 V

R_2 22 kΩ

A

B

24. For each circuit in Figure 6–83, determine the Thevenin equivalent as seen by R_L.

▼ **FIGURE 6–83**

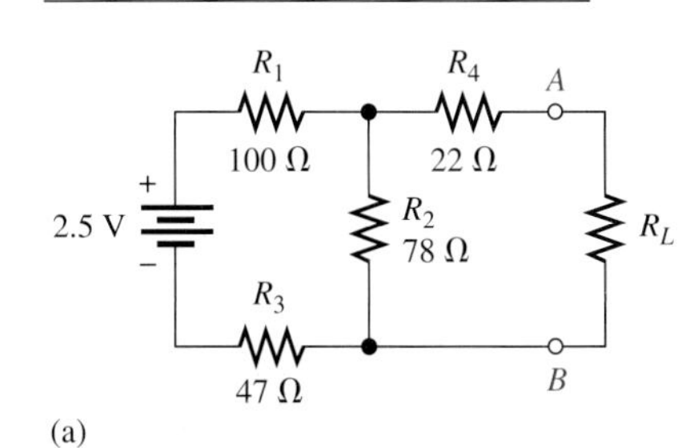

(a)

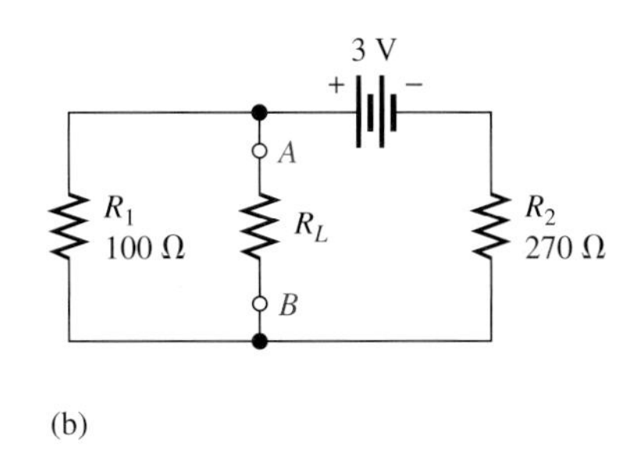

(b)

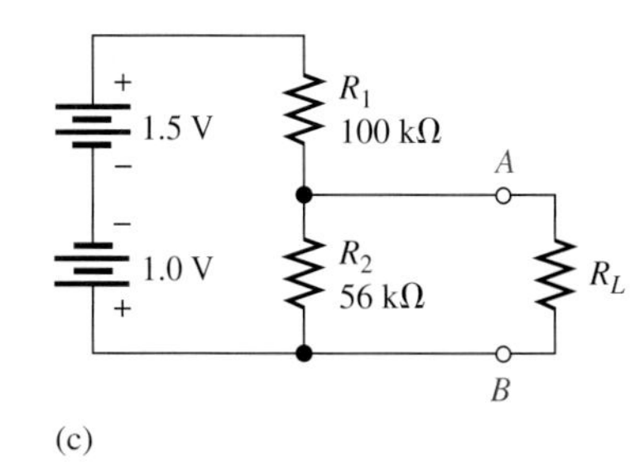

(c)

25. Determine the voltage and current for R_L in Figure 6–84.

26. Determine the value of R_L in Figure 6–83(a) for which R_L dissipates maximum power.

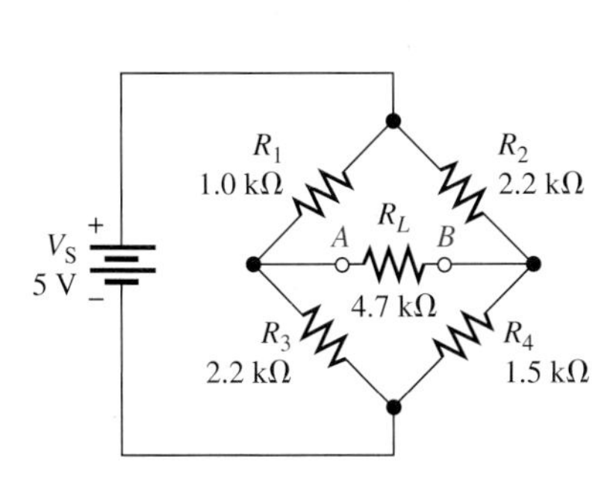

▲ **FIGURE 6–84**

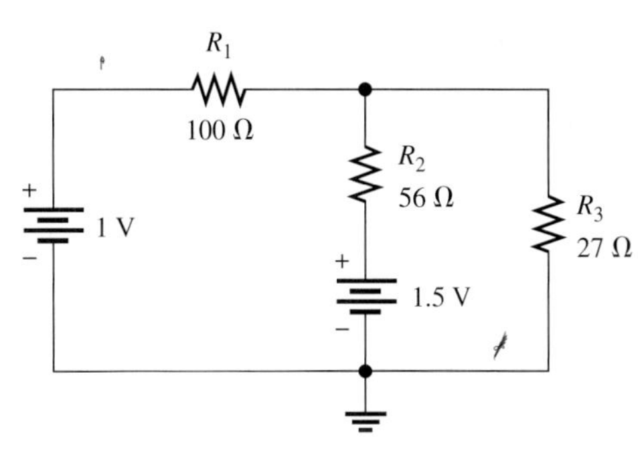

▲ **FIGURE 6–85**

SECTION 6–7 **The Superposition Theorem**

27. In Figure 6–85, use the superposition theorem to find the current in R_3.

28. In Figure 6–85, what is the current through R_2?

SECTION 6–8 **Troubleshooting**

29. Is the voltmeter reading in Figure 6–86 correct? If not, what is the problem?

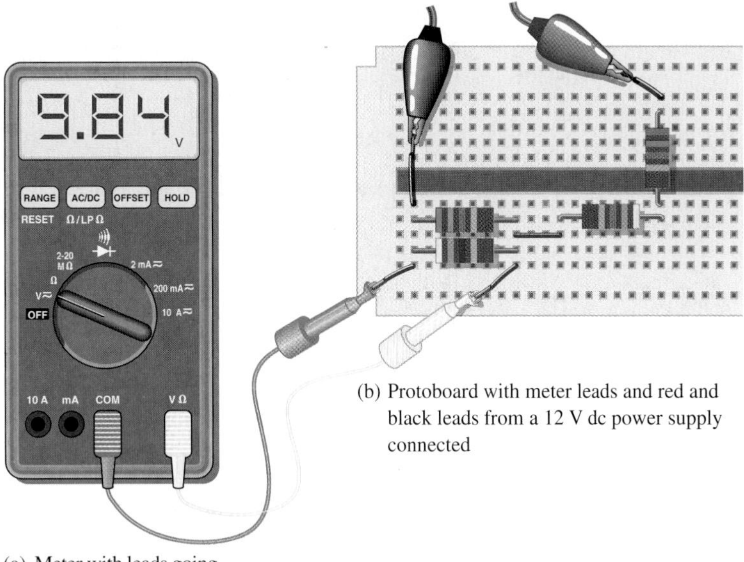

(a) Meter with leads going to protoboard

(b) Protoboard with meter leads and red and black leads from a 12 V dc power supply connected

▶ **FIGURE 6–86**

30. If R_2 in Figure 6–87 opens, what voltages will be read at points *A, B,* and *C?*

31. Check the meter readings in Figure 6–88 and locate any fault that may exist.

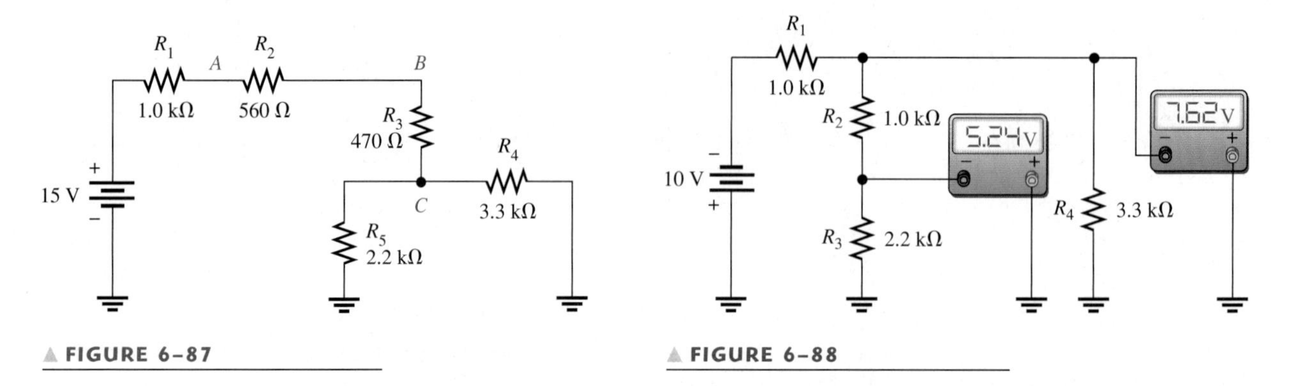

▲ **FIGURE 6–87**

▲ **FIGURE 6–88**

32. Determine the voltage you would expect to measure across each resistor in Figure 6–87 for each of the following faults. Assume the faults are independent of each other.

(a) R_1 open **(b)** R_3 open **(c)** R_4 open

(d) R_5 open **(e)** point *C* shorted to ground

33. Determine the voltage you would expect to measure across each resistor in Figure 6–88 for each of the following faults:

(a) R_1 open **(b)** R_2 open **(c)** R_3 open **(d)** a short across R_4

ADVANCED PROBLEMS

34. In each circuit of Figure 6–89, identify the series and parallel relationships of the resistors viewed from the source.

▶ **FIGURE 6–89**

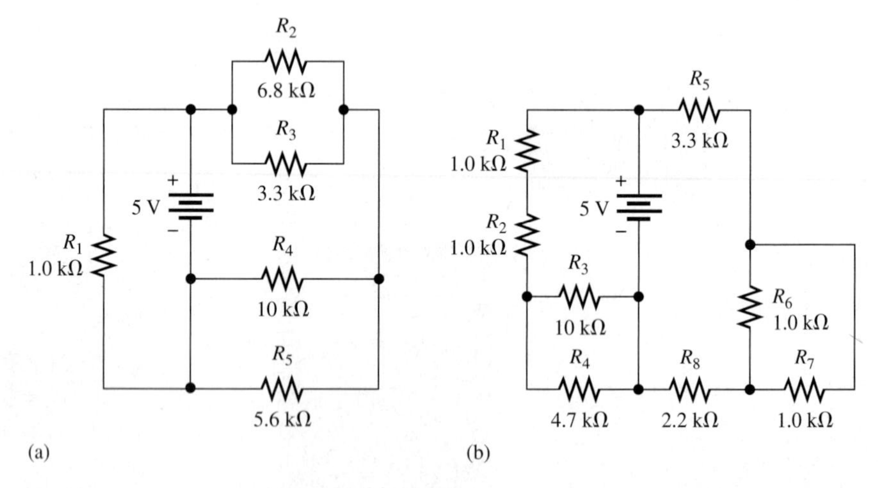

35. Draw the schematic of the PC board layout in Figure 6–90 showing resistor values and identify the series-parallel relationships. Which resistors, if any, can be removed with no effect on R_T?

36. For the circuit shown in Figure 6–91, calculate the following:

(a) total resistance across the source

(b) total current from the source

(c) current through the 910 Ω resistor

(d) voltage from point *A* to point *B*

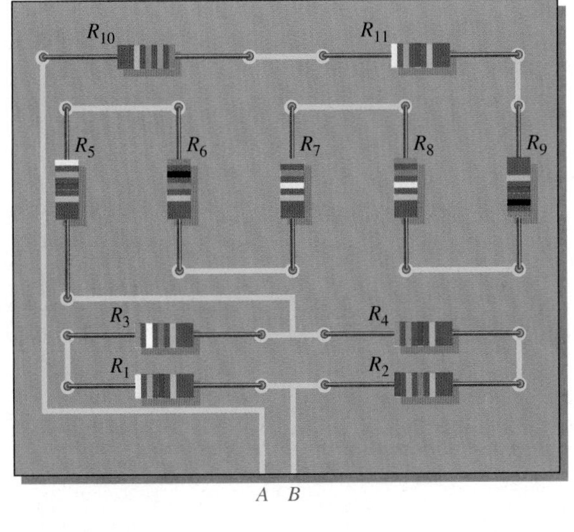

▲ FIGURE 6–90

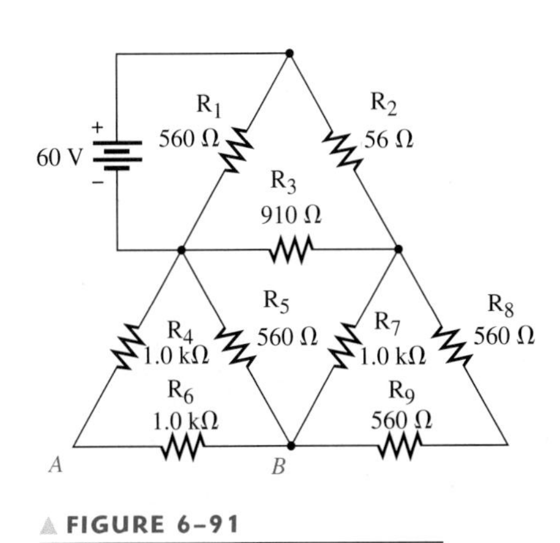

▲ FIGURE 6–91

37. Determine the total resistance and the voltage at points A, B, and C in the circuit of Figure 6–92.

▶ FIGURE 6–92

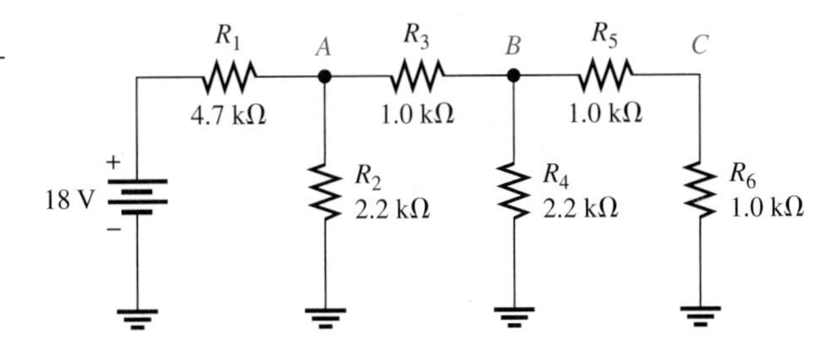

38. Determine the total resistance between terminals A and B of the circuit in Figure 6–93. Also calculate the current in each branch with 10 V between A and B.

▶ FIGURE 6–93

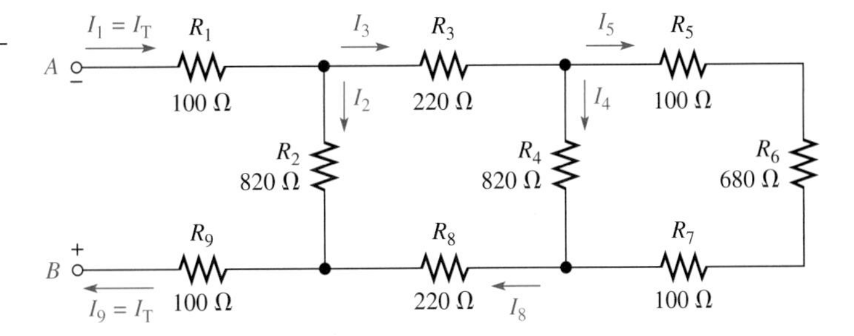

39. What is the voltage across each resistor in Figure 6–93? There are 10 V between A and B.

40. Determine the voltage, V_{AB}, in Figure 6–94.

41. Find the value of R_2 in Figure 6–95.

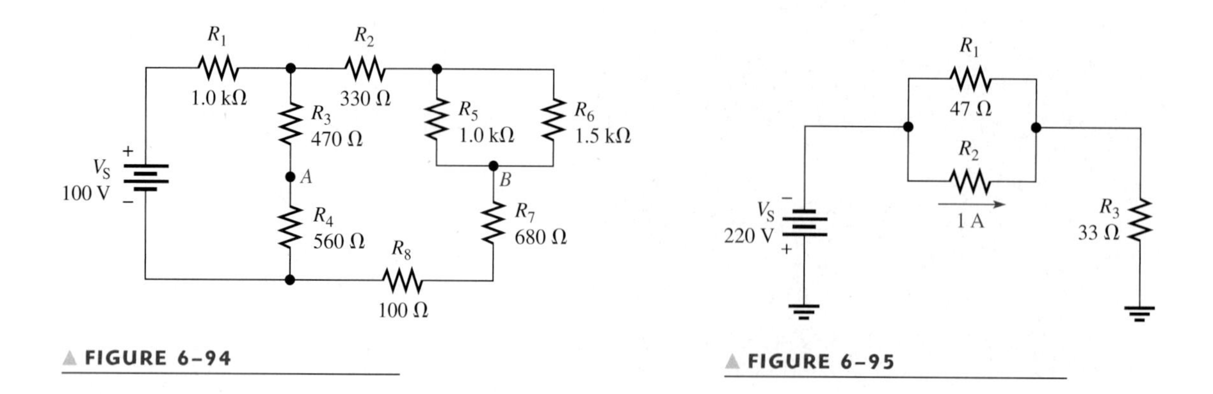

▲ FIGURE 6–94

▲ FIGURE 6–95

42. Determine the total resistance and the voltage at points *A*, *B*, and *C* in the circuit of Figure 6–96.

43. Develop a voltage divider to provide a 6 V output with no load and a minimum of 5.5 V across a 1.0 kΩ load. The source voltage is 24 V, and the unloaded current is not to exceed 100 mA.

44. Determine the resistance values for a voltage divider that must meet the following specifications: The current under an unloaded condition is not to exceed 5 mA. The source voltage is to be 10 V. A 5 V output and a 2.5 V output are required. Sketch the circuit. Determine the effect on the output voltages if a 1.0 kΩ load is connected to each output.

45. Using the superposition theorem, calculate the current in the right-most branch of Figure 6–97.

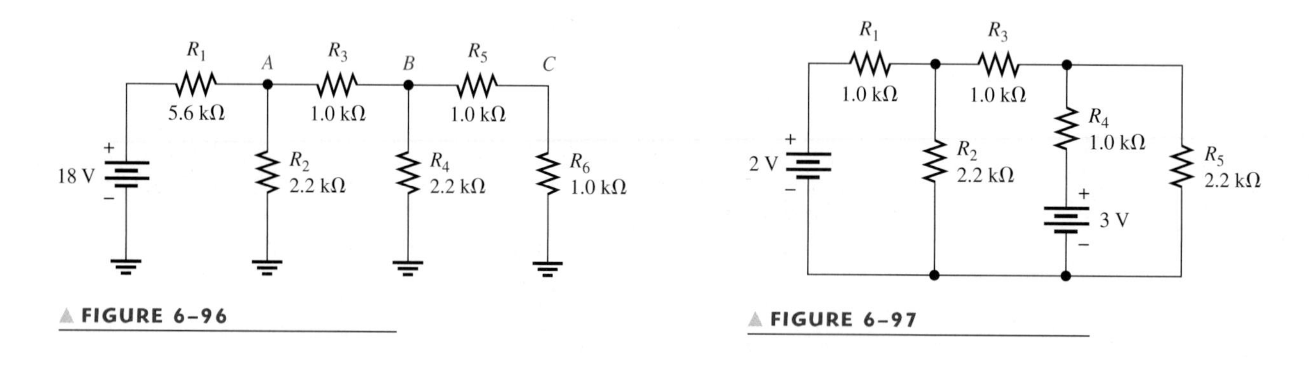

▲ FIGURE 6–96

▲ FIGURE 6–97

46. Find the current through R_L in Figure 6–98.

▶ FIGURE 6–98

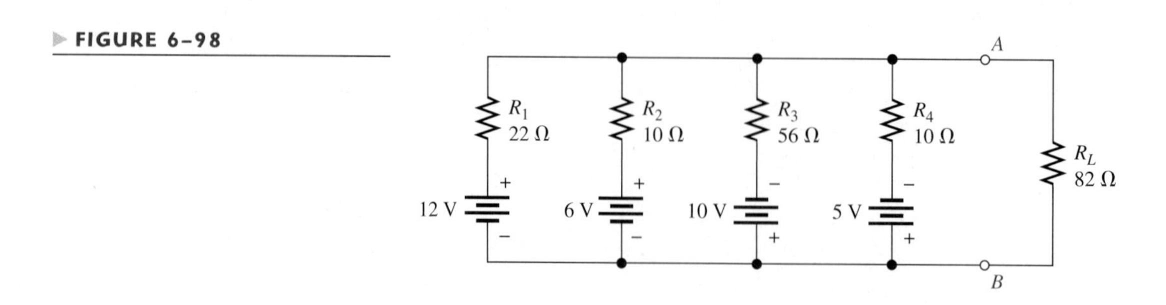

47. Using Thevenin's theorem, find the voltage across R_4 in Figure 6–99.

48. Determine V_{OUT} for the circuit in Figure 6–100 for the following conditions:

(a) Switch SW2 connected to +12 V and the rest to ground

(b) Switch SW1 connected to +12 V and the rest to ground

FIGURE 6–99

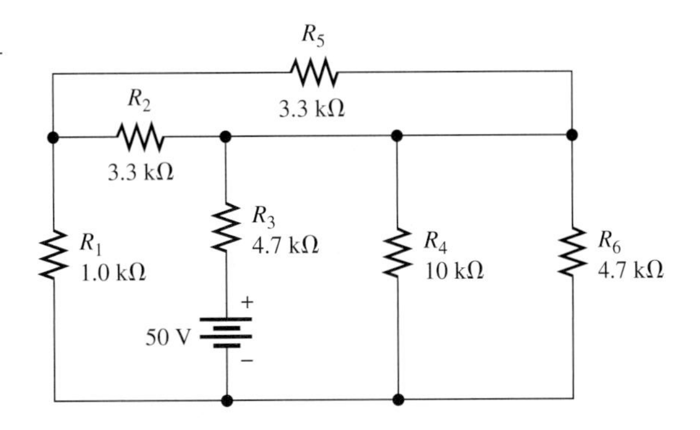

FIGURE 6–100

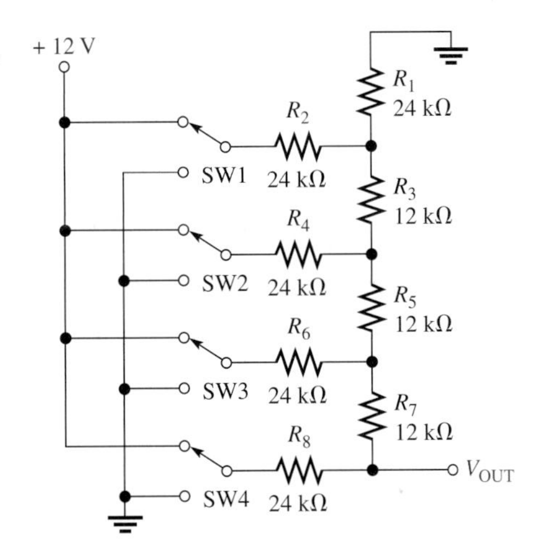

49. Develop a schematic for the double-sided PC board in Figure 6–101 and label the resistor values.

FIGURE 6–101

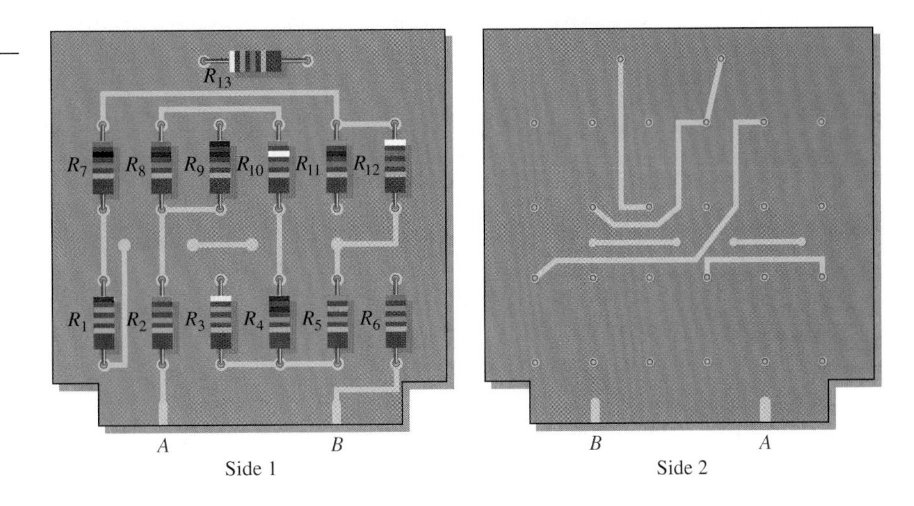

50. Lay out a PC board for the circuit in Figure 6–89(b). The battery is to be connected external to the board.

51. The voltage divider in Figure 6–102 has a switched load. Determine the voltage at each tap (V_1, V_2, and V_3) for each position of the switch.

52. Figure 6–103 shows a dc biasing arrangement for a filed-effect transistor amplifier. Biasing is a common method for setting up certain dc voltage levels required for proper amplifier operation. Although you may not be familiar with transistor amplifiers at this point, the dc voltages and currents in the circuit can be determined using methods that you already know.

(a) Find V_G and V_S with respect to ground

(b) Determine I_1, I_2, I_D, and I_S

(c) Find V_{DS} and V_{DG}

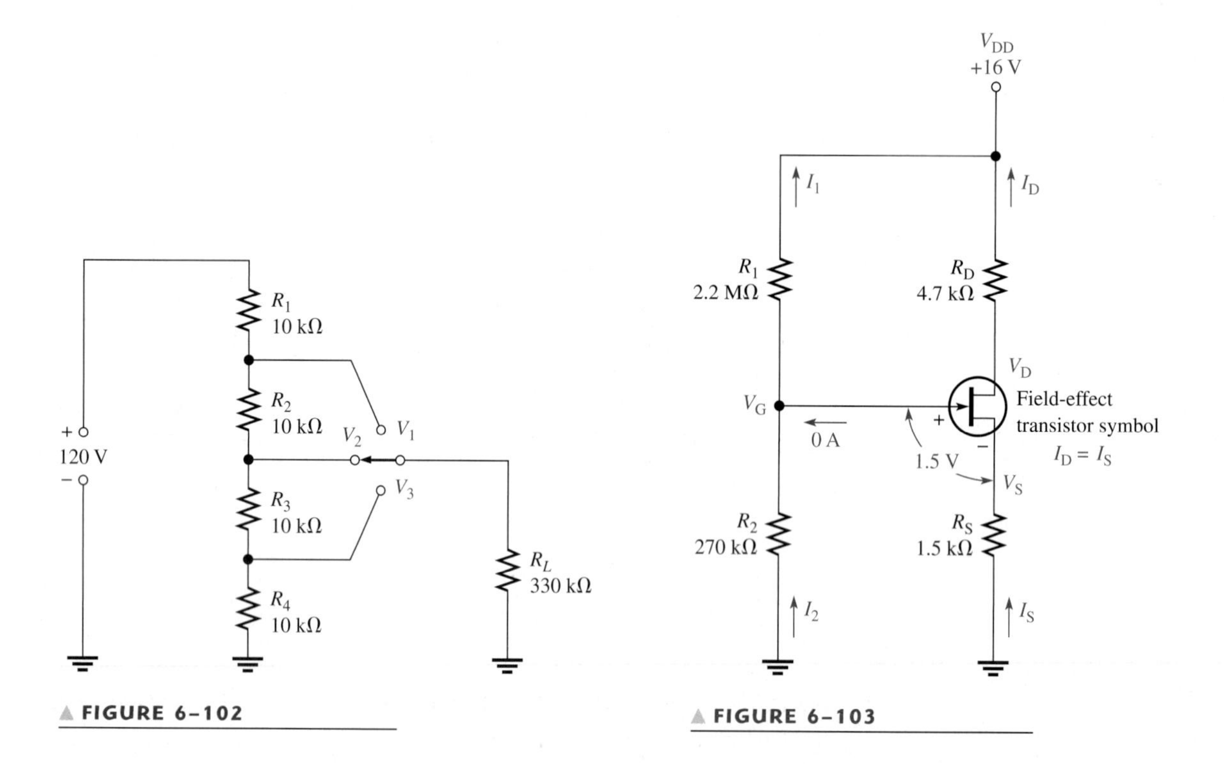

▲ FIGURE 6–102

▲ FIGURE 6–103

53. Look at the voltmeters in Figure 6–104 and determine if there is a fault in the circuit. If there is a fault, identify it.

▶ FIGURE 6–104

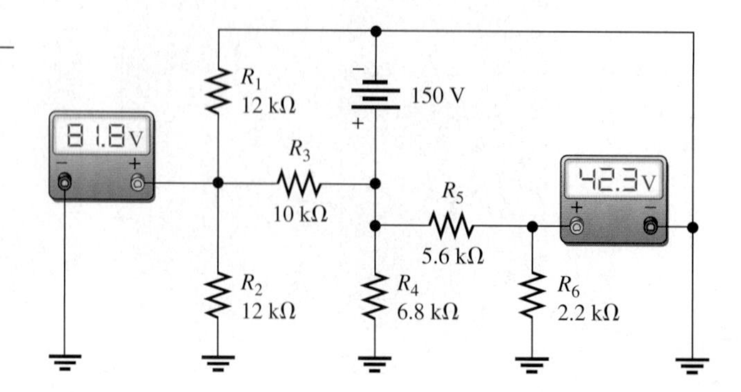

54. Are the voltmeter readings in Figure 6–105 correct?

FIGURE 6–105

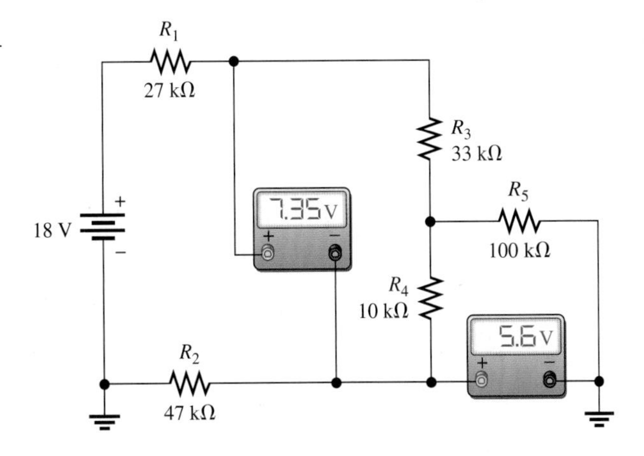

55. There is one fault in Figure 6–106. Based on the voltmeter indications, determine what the fault is.

FIGURE 6–106

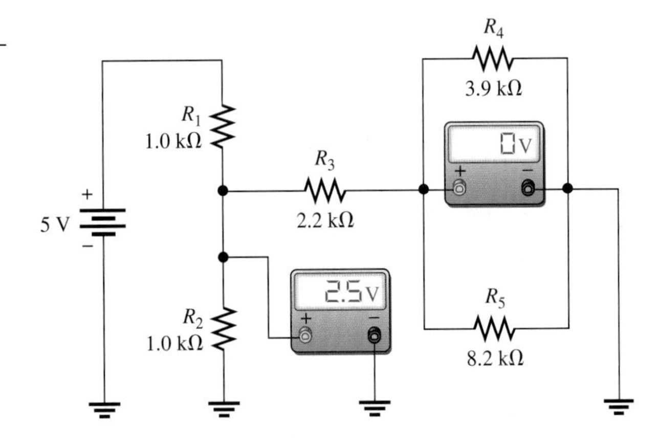

56. Look at the voltmeters in Figure 6–107 and determine if there is a fault in the circuit. If there is a fault, identify it.

57. Determine the voltmeter readings in Figure 6–107 if the 4.7 kΩ resistor opens.

FIGURE 6–107

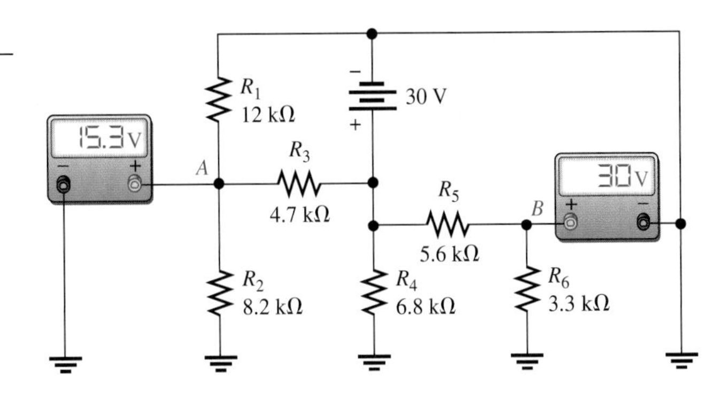

ELECTRONICS WORKBENCH/CIRCUITMAKER TROUBLESHOOTING PROBLEMS

CD-ROM file circuits are shown in Figure 6–108.

58. Open file P06-58 and determine if there is a fault in the circuit. If so, identify the fault.

59. Open file P06-59 and determine if there is a fault in the circuit. If so, identify the fault.

60. Open file P06-60 and determine if there is a fault in the circuit. If so, identify the fault.

61. Open file P06-61 and determine if there is a fault in the circuit. If so, identify the fault.

62. Open file P06-62 and determine if there is a fault in the circuit. If so, identify the fault.

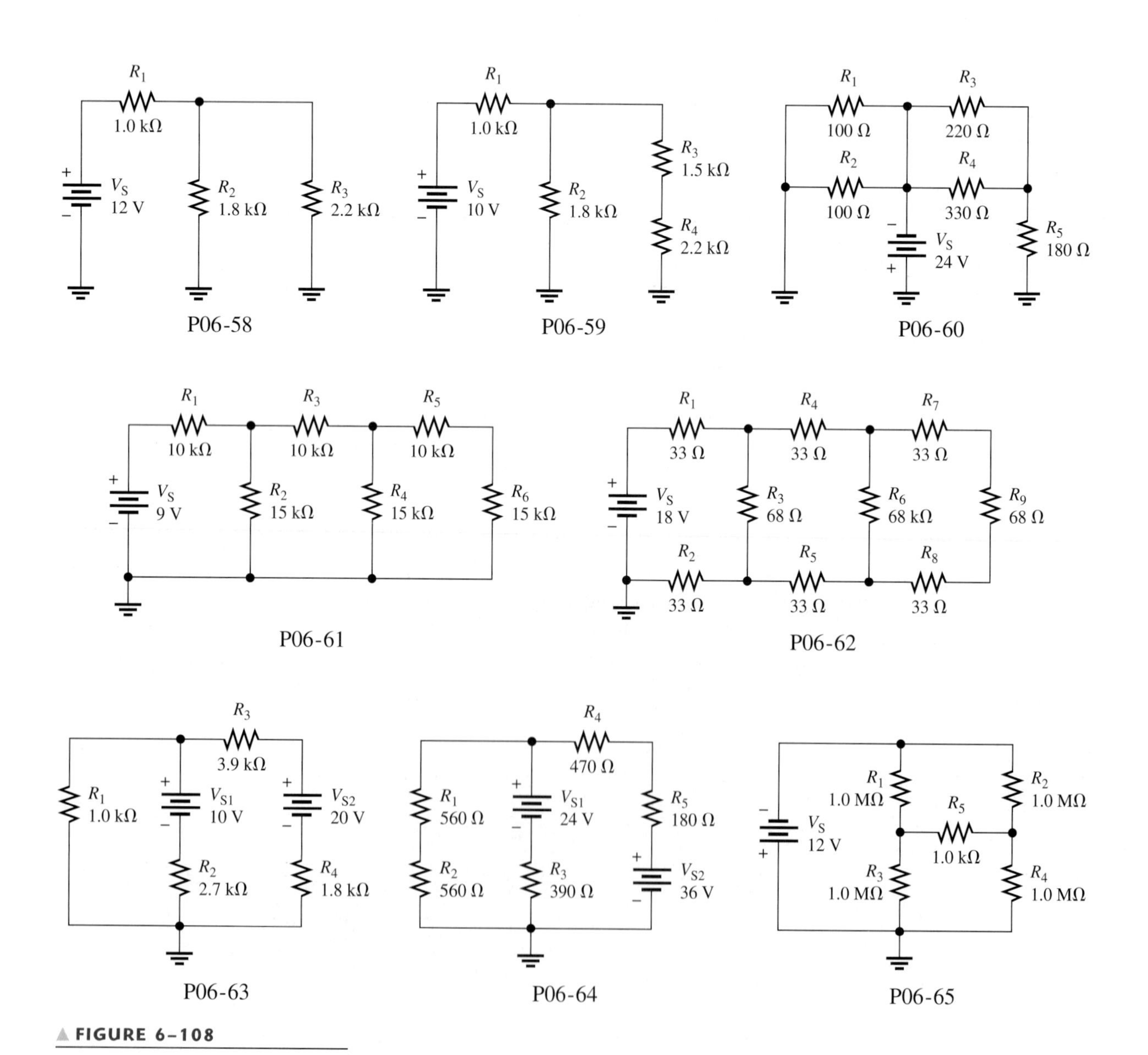

▲ **FIGURE 6–108**

63. Open file P06-63 and determine if there is a fault in the circuit. If so, identify the fault.
64. Open file P06-64 and determine if there is a fault in the circuit. If so, identify the fault.
65. Open file P06-65 and determine if there is a fault in the circuit. If so, identify the fault.

ANSWERS

SECTION REVIEWS

SECTION 6–1

Identifying Series-Parallel Relationships

1. See Figure 6–109.

FIGURE 6–109

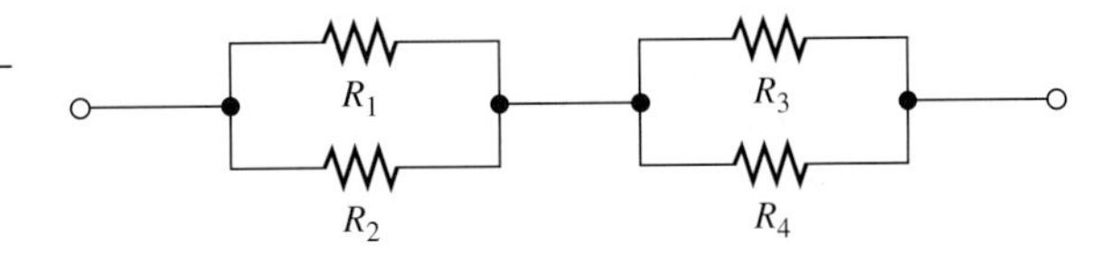

2. R_1 and R_2 are in series with the parallel combination of R_3 and R_4.
3. All resistors are in parallel.
4. R_1 and R_2 are in parallel; R_3 and R_4 are in parallel.
5. Yes, the two parallel combinations are in series with each other.

SECTION 6–2

Analysis of Series-Parallel Circuits

1. $R_T = R_1 + R_4 + R_2 \| R_3 = 599\ \Omega$
2. $I_3 = 11.2$ mA
3. $V_{R2} = I_2R_2 = 3.7$ V
4. $R_T = 89\ \Omega$; $I_T = 11.2$ mA

SECTION 6–3

Voltage Dividers with Resistive Loads

1. The load resistor decreases the output voltage.
2. True
3. $V_{OUT(unloaded)} = 19.23$ V, $V_{OUT(loaded)} = 19.16$ V

SECTION 6–4

Loading Effect of a Voltmeter

1. A voltmeter loads a circuit because the internal resistance of the meter appears in parallel with the circuit resistance across which it is connected, reducing the resistance between those two points of the circuit and drawing current from the circuit.
2. No, because the meter resistance is much larger than 1.0 kΩ.
3. Yes.

SECTION 6–5 **The Wheatstone Bridge**

1. See Figure 6–110.
2. A bridge is balanced when the output voltage is zero.
3. $R_X = 15\ \text{k}\Omega$
4. An unbalanced bridge is used to measure transducer-sensed quantities.

▶ FIGURE 6–110

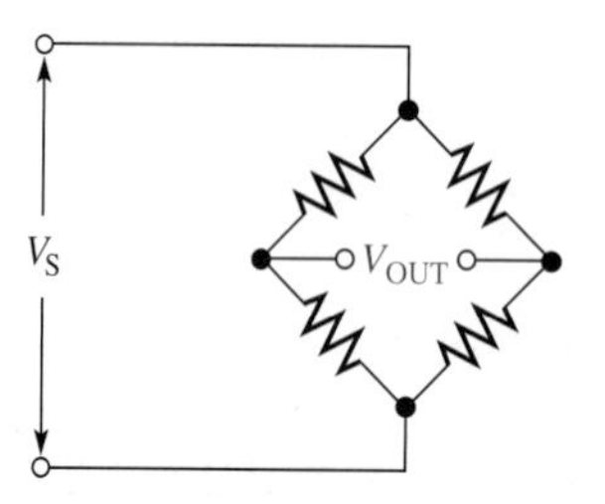

SECTION 6–6 **Thevenin's Theorem**

1. A Thevenin equivalent circuit consists of V_{TH} and R_{TH}.
2. See Figure 6–111.
3. V_{TH} is the open circuit voltage between two points in a circuit.
4. R_{TH} is the resistance as viewed from two terminals in a circuit, with all sources replaced by their internal resistances.
5. See Figure 6–112.

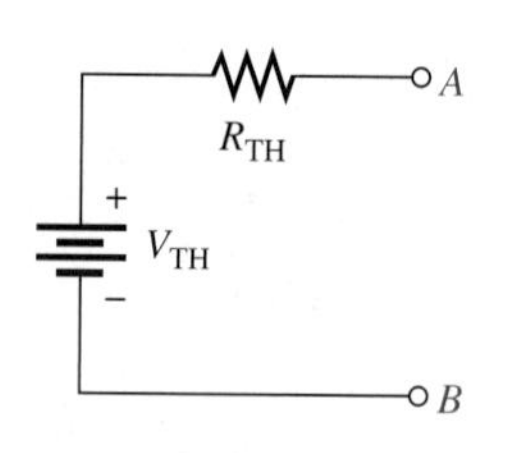

▲ FIGURE 6–111

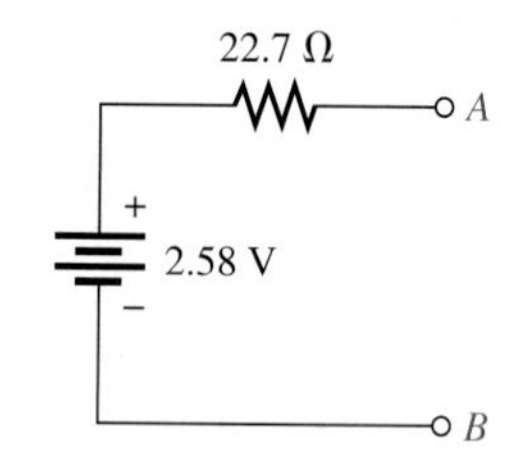

▲ FIGURE 6–112

SECTION 6–7 **The Superposition Theorem**

1. The total current in any branch of a multiple-source linear circuit is equal to the algebraic sum of the currents due to the individual sources acting alone, with the other sources replaced by their internal resistances.
2. The superposition method allows each source to be treated independently.
3. A short simulates the zero internal resistance of an ideal voltage source.
4. $I_{R1} = 6.67\ \text{mA}$
5. The net current is in the direction of the larger current.

SECTION 6–8 **Troubleshooting**

1. Opens and short are two common faults.
2. **(a)** 62.8 V **(b)** 62.8 V **(c)** 62 V **(d)** 100 V **(e)** 0 V
3. The 10 kΩ resistor is open.

■ Application Assignment

1. The battery will last 413 days.
2. Yes, ⅛ W resistors can be used.
3. No, none of the resistors will overheat.

RELATED PROBLEMS FOR EXAMPLES

6–1 The added resistor would be in parallel with the combination of R_4 in series with R_2 and R_3 in parallel.

6–2 The added resistor is in parallel with R_5.

6–3 A to gnd: $R_T = R_3 \| (R_1 + R_2) + R_4$

B to gnd: $R_T = R_2 \| (R_1 + R_3) + R_4$

C to gnd: $R_T = R_4$

6–4 None. The new resistor will be shorted by the existing connection between those points.

6–5 R_1 and R_4 are effectively removed from the circuit.

6–6 55.1 Ω

6–7 128.3 Ω

6–8 $I_1 = 89.3$ mA; $I_3 = 58.5$ mA; $I_T = 182.3$ mA

6–9 $V_1 = V_2 = 10.3$ V; $V_3 = 9.70$ V; $V_4 = 3.16$ V; $V_5 = 6.54$ V

6–10 $I_1 = 1.42$ mA, $P_1 = 6.67$ mW; $I_2 = 756$ μA, $P_2 = 3.55$ mW;
$I_3 = 2.18$ mA, $P_3 = 4.75$ mW; $I_4 = 1.13$ mA, $P_4 = 1.28$ mW;
$I_5 = 1.06$ mA, $P_5 = 758$ μW; $I_6 = 1.06$ mA, $P_6 = 435$ μW

6–11 3.39 V

6–12 Increase the values of R_1, R_2, and R_3 proportionally.

6–13 5.19 V

6–14 3.3 kΩ

6–15 0.45 V

6–16 2.36 V; 124 Ω

6–17 1.17 mA

6–18 41.7 mW

6–19 16.6 mA

6–20 5 mA

6–21 5.73 V, 0 V

6–22 9.46 V

6–23 $V_A = 12$ V; $V_B = 13.8$ V

SELF-TEST

1. (e) **2.** (c) **3.** (c) **4.** (c) **5.** (b) **6.** (a) **7.** (b)
8. (b) **9.** (d) **10.** (b) **11.** (b) **12.** (c) **13.** (d) **14.** (a)
15. (d)

7

MAGNETISM AND ELECTROMAGNETISM

INTRODUCTION

This chapter is somewhat of a departure from the previous six chapters because two new concepts are introduced: magnetism and electromagnetism. The operation of many types of electrical devices is based partially on magnetic or electromagnetic principles.

CHAPTER OUTLINE

CHAPTER OBJECTIVES

- Explain the principles of the magnetic field
- Explain the principles of electromagnetism
- Describe the principle of operation for several types of electromagnetic devices
- Explain magnetic hysteresis
- Discuss the principle of electromagnetic induction
- Describe some applications of electromagnetic induction

KEY TERMS

- Magnetic field
- Lines of force
- Magnetic flux
- Weber
- Tesla
- Gauss
- Electromagnetic field
- Permeability
- Reluctance
- Magnetomotive force
- Ampere-turn
- Solenoid
- Relay
- Speaker
- Magnetizing force
- Hysteresis
- Induced voltage
- Induced current
- Electromagnetic induction
- Faraday's law
- Lenz's law

APPLICATION ASSIGNMENT PREVIEW

For the application assignment, you will apply what you learn in this chapter about relays and other devices to a simple burglar alarm system. You will determine how to connect the components to form a complete system and specify a check-out procedure to make sure the system is working properly. After studying this chapter, you should be able to complete the application assignment.

WWW. VISIT THE COMPANION WEBSITE

Circuit Simulation Tutorials and Other Chapter Study Tools Are Available at

http://www.prenhall.com/floyd

7–1 THE MAGNETIC FIELD

A permanent magnet has a magnetic field surrounding it. The **magnetic field** consists of **lines of force** that radiate from the north pole to the south pole and back to the north pole through the magnetic material.

After completing this section, you should be able to

- **Explain the principles of the magnetic field**
- Define *magnetic flux*
- Define *magnetic flux density*
- Discuss how materials are magnetized
- Explain how a magnetic switch works

BIOGRAPHY

Wilhelm Eduard Weber 1804–1891 Weber was a German physicist who worked closely with Gauss, whose biography appears later. Independently, he established a system of absolute electrical units and also performed work that was crucial to the later development of the electromagnetic theory of light. The unit of magnetic flux is named in his honor. (Photo credit: Courtesy of the Smithsonian Institution. Photo No. 52,604.)

A permanent magnet, such as the bar magnet shown in Figure 7–1, has a magnetic field surrounding it. All magnetic fields originate from charges in motion, and most often the charges in motion are electrons.

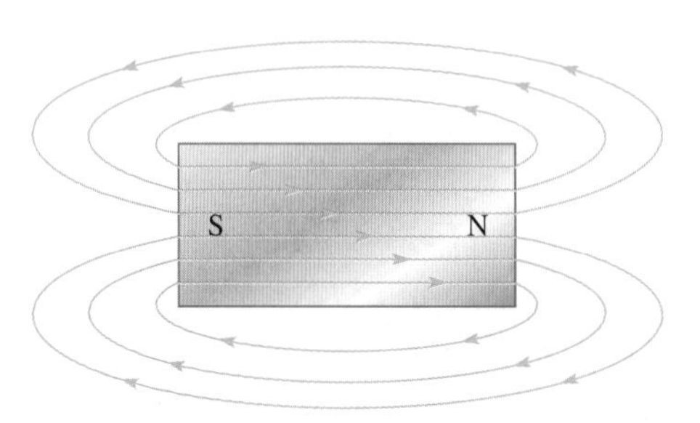

Blue lines represent only a few of the many magnetic lines of force in the magnetic field.

▲ **FIGURE 7–1**

Magnetic lines of force around a bar magnet.

BIOGRAPHY

Nikola Tesla 1856–1943 Tesla was born in Croatia (then Austria-Hungary). He was an electrical engineer who invented the ac induction motor, polyphase ac systems, the Tesla coil transformer, wireless communications, and fluorescent lights. He worked for Edison when he first came to the U.S. in 1884 and later for Westinghouse. The SI unit of magnetic flux density is named in his honor. (Photo credit: Courtesy of the Nikola Tesla Museum, Belgrade, Yugoslavia.)

The magnetic field consists of lines of force, or flux lines, that radiate from the north pole (N) to the south pole (S) and back to the north pole through the magnetic material. For clarity, only a few lines of force are shown in Figure 7–1. Imagine, however, that many lines surround the magnet in three dimensions. The lines shrink to the smallest possible size and blend together, although they do not touch. This effectively forms a continuous magnetic field surrounding the magnet.

Attraction and Repulsion of Magnetic Poles

When unlike poles of two permanent magnets are placed close together, an attractive force is produced by the magnetic fields, as indicated in Figure 7–2(a). When two like poles are brought close together, they repel each other, as shown in part (b).

Altering a Magnetic Field

When a nonmagnetic material such as paper, glass, wood, or plastic is placed in a magnetic field, the lines of force are unaltered, as shown in Figure 7–3(a). However, when a magnetic material such as iron is placed in the magnetic field, the lines of force tend to change course and pass through the iron rather than through

(a) Unlike poles attract.

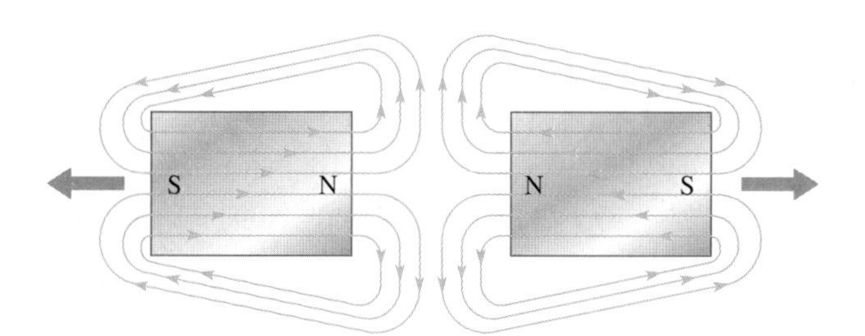

(b) Like poles repel.

FIGURE 7–2

Magnetic attraction and repulsion.

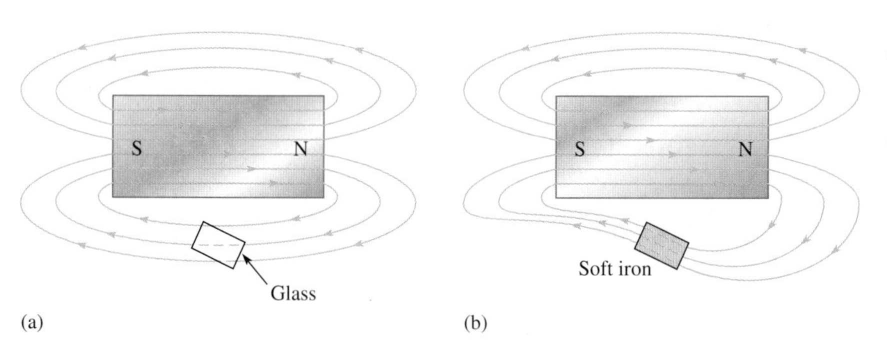

FIGURE 7–3

Effect of (a) nonmagnetic and (b) magnetic materials on a magnetic field.

the surrounding air. They do so because the iron provides a magnetic path that is more easily established than that of air. Figure 7–3(b) illustrates this principle. The fact that magnetic lines of force follow a path through iron or other materials is a consideration in the design of shields that prevent stray magnetic fields from affecting sensitive circuits.

Magnetic Flux (ϕ)

The group of force lines going from the north pole to the south pole of a magnet is called the magnetic flux, symbolized by ϕ (the lowercase Greek letter phi). The number of lines of force in a magnetic field determines the value of the flux. The more lines of force, the greater the flux and the stronger the magnetic field.

The unit of magnetic flux is the weber (Wb). One weber equals 10^8 lines. The weber is a very large unit; thus, in most practical situations, the microweber (μWb) is used. One microweber equals 100 lines of magnetic flux.

Magnetic Flux Density (B)

The **magnetic flux density** is the amount of flux per unit area perpendicular to the magnetic field. Its symbol is B, and its unit is the tesla (T). One tesla equals one weber per square meter (Wb/m^2). The following formula expresses the flux density:

Equation 7–1

$$B = \frac{\phi}{A}$$

where ϕ is the flux and A is the cross-sectional area in square meters (m^2) of the magnetic field.

The Gauss Although the tesla (T) is the SI unit for flux density, another unit called the gauss, from the CGS (centimeter-gram-second) system, is sometimes used (10^4 gauss = 1 T). In fact, the instrument used to measure flux density is the gaussmeter.

EXAMPLE 7–1

Find the flux density in a magnetic field in which the flux in 0.1 m^2 is 800 μWb.

Solution

$$B = \frac{\phi}{A} = \frac{800\ \mu\text{Wb}}{0.1\ \text{m}^2} = \mathbf{8000\ \mu T}$$

*Related Problem** Calculate ϕ if $B = 4700\ \mu$T and $A = 0.05$ m^2.

*Answers are at the end of the chapter.

BIOGRAPHY

Karl Friedrich Gauss
1777–1855
Gauss, a great German mathematician, disproved many 18th century mathematical theories. Later, he worked closely with Weber on a worldwide system of stations for systematic observations of terrestrial magnetism. The most important result of their work in electromagnetism was the later development of telegraphy by others. The CGS unit of magnetic flux density is named in his honor. (Credit: Illustration by Steven S. Nau.)

How Materials Become Magnetized

Ferromagnetic materials such as iron, nickel, and cobalt become magnetized when placed in the magnetic field of a magnet. We have all seen a permanent magnet pick up paper clips, nails, or iron filings. In these cases, the object becomes magnetized (that is, it actually becomes a magnet itself) under the influence of the permanent magnetic field and becomes attracted to the magnet. When removed from the magnetic field, the object tends to lose its magnetism.

Ferromagnetic materials have minute magnetic domains created within their atomic structure by the orbital motion and spin of electrons. These domains can be viewed as very small bar magnets with north and south poles. When the material is not exposed to an external magnetic field, the magnetic domains are randomly oriented, as shown in Figure 7–4(a). When the material is placed in a magnetic field, the domains align themselves, as shown in part (b). Thus, the object itself effectively becomes a magnet.

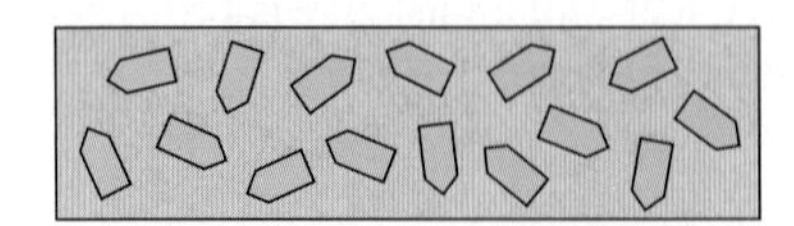

(a) The magnetic domains (N ◁ S) are randomly oriented in the unmagnetized material.

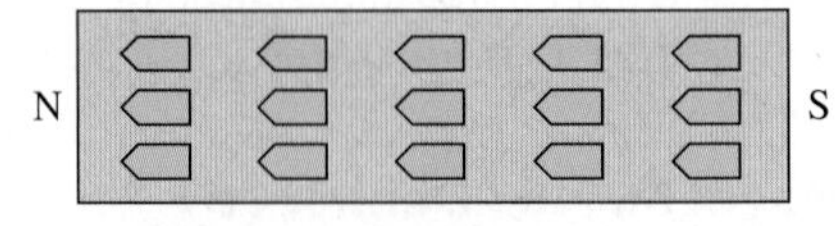

(b) The magnetic domains become aligned when the material is magnetized.

FIGURE 7–4

Ferromagnetic domains in (a) an unmagnetized and (b) a magnetized material.

Application Example

Permanent magnets have numerous applications, one of which is presented here as an illustration. Figure 7–5 shows a typical magnetically operated, normally closed (NC) switch. When the magnet is near the switch mechanism, the metallic arm is held in its NC position, as shown in part (a). When the magnet is moved away, the spring pulls the arm up, breaking the contact as shown in part (b).

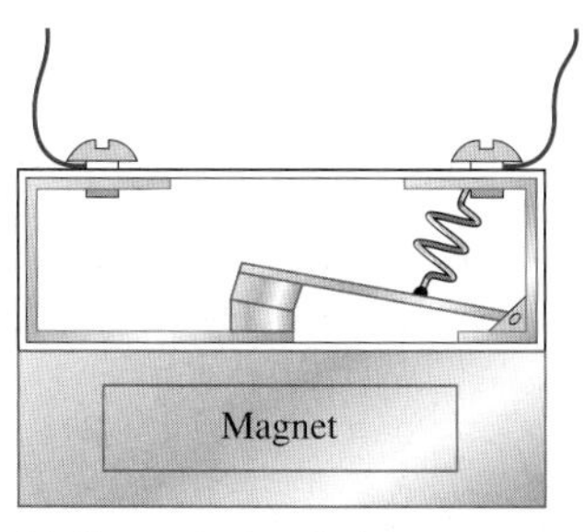

(a) Contact is closed when magnet is near.

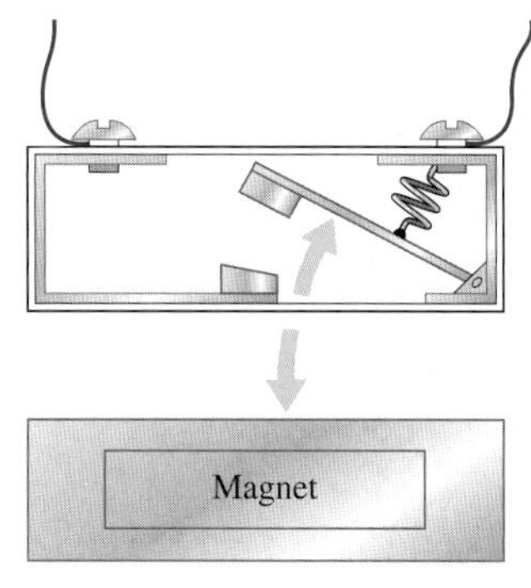

(b) Contact opens when magnet is moved away.

FIGURE 7–5 Operation of a magnetic switch.

Switches of this type are commonly used in perimeter alarm systems to detect entry into a building through windows or doors. As Figure 7–6 shows, several openings can be protected by magnetic switches wired to a common transmitter. When any one of the switches opens, the transmitter is activated and sends a signal to a central receiver and alarm unit.

FIGURE 7–6 Connection of a typical perimeter alarm system.

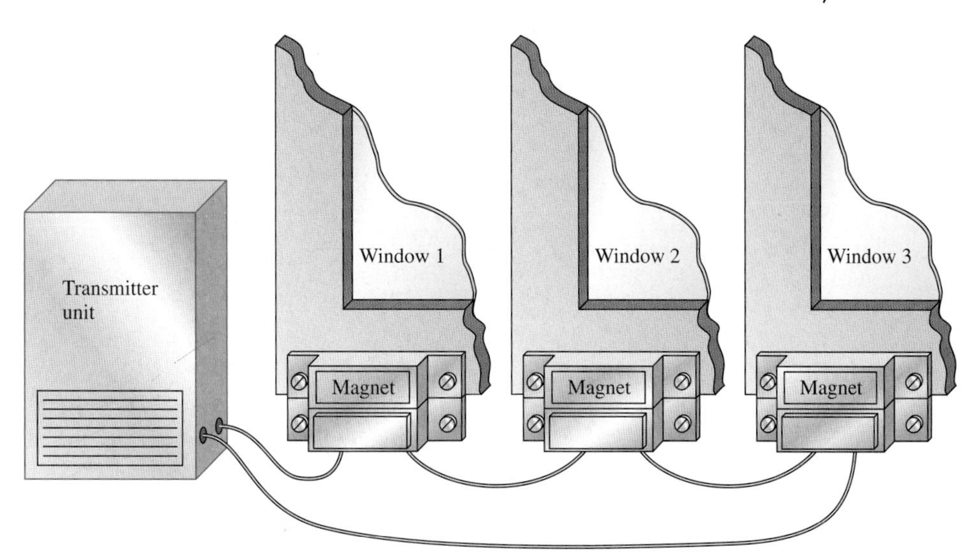

SECTION 7–1 REVIEW
Answers are at the end of the chapter.

1. When the north poles of two magnets are placed close together, do they repel or attract each other?
2. What is magnetic flux?
3. What is the flux density when $\phi = 4.5\ \mu\text{Wb}$ and $A = 5 \times 10^{-3}\ \text{m}^2$?

7–2 ELECTROMAGNETISM

Electromagnetism is the production of a magnetic field by current in a conductor. Many types of useful devices such as tape recorders, electric motors, speakers, solenoids, and relays are based on electromagnetism.

After completing this section, you should be able to

- **Explain the principles of electromagnetism**
- Determine the direction of the magnetic lines of force
- Define *permeability*
- Define *reluctance*
- Define *magnetomotive force*
- Describe a basic electromagnet

Current produces a magnetic field, called an **electromagnetic field**, around a conductor, as illustrated in Figure 7–7. The invisible lines of force of the magnetic field form a concentric circular pattern around the conductor and are continuous along its length.

▼ **FIGURE 7–7**

Magnetic field around a current-carrying conductor.

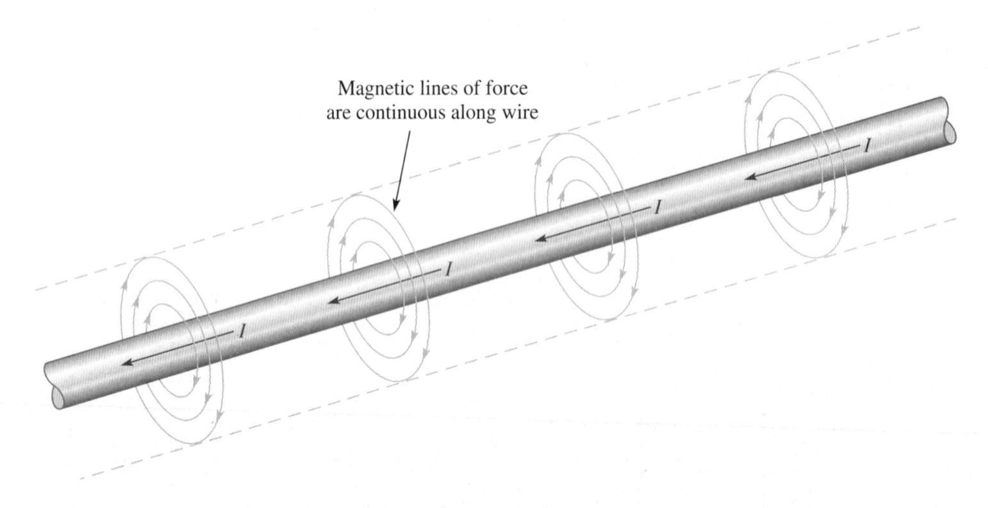

Although the magnetic field cannot be seen, it is capable of producing visible effects. For example, if a current-carrying wire is inserted through a sheet of paper in a perpendicular direction, iron filings placed on the surface of the paper arrange themselves along the magnetic lines of force in concentric rings, as illustrated in Figure 7–8(a). Part (b) of the figure illustrates that the north pole of a compass placed in the electromagnetic field will point in the direction of the lines of force. The field is stronger closer to the conductor and becomes weaker with increasing distance from the conductor.

Direction of the Lines of Force

The direction of the lines of force surrounding the conductor are indicated in Figure 7–9. When the direction of current is right to left, as in part (a), the lines are in a clockwise direction. When current is left to right, as in part (b), the lines are in a counterclockwise direction.

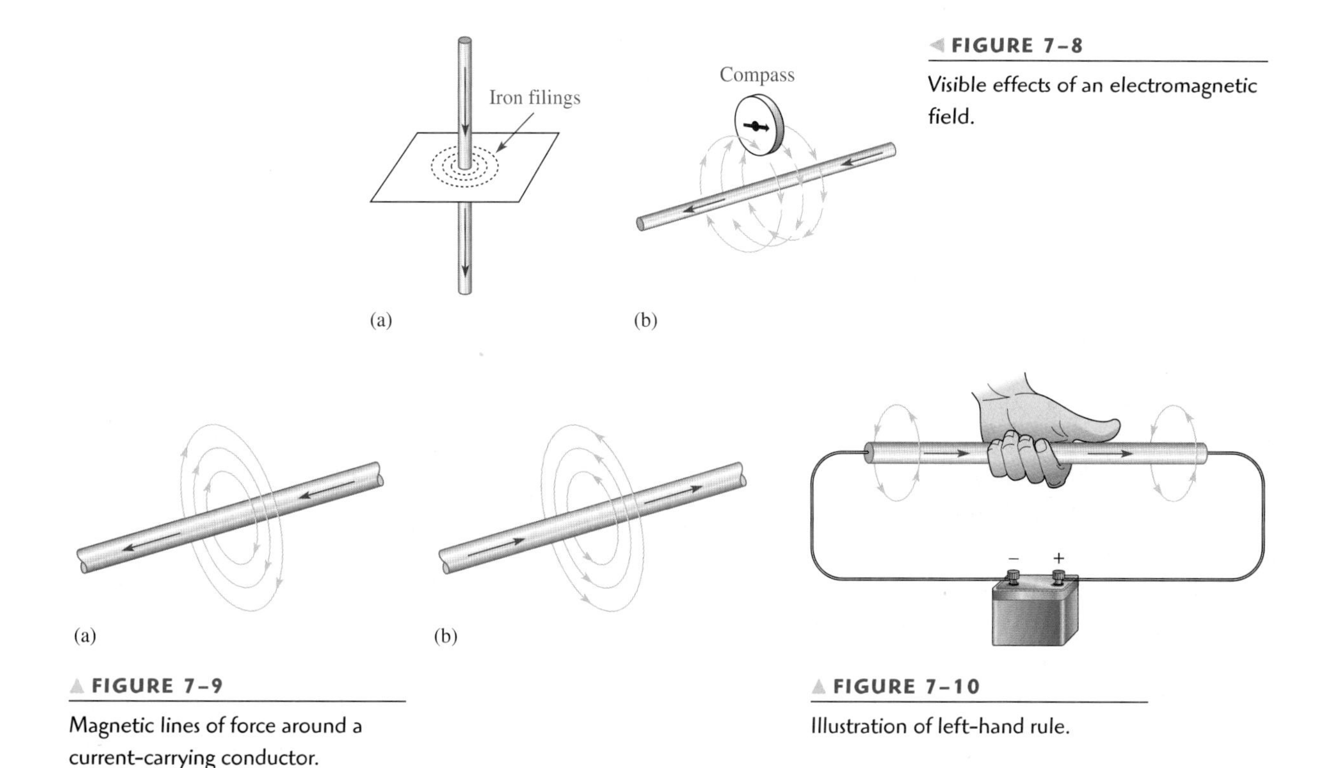

FIGURE 7–8

Visible effects of an electromagnetic field.

FIGURE 7–9

Magnetic lines of force around a current-carrying conductor.

FIGURE 7–10

Illustration of left-hand rule.

Left-Hand Rule An aid to remembering the direction of the lines of force is illustrated in Figure 7–10. Imagine that you are grasping the conductor with your left hand, with your thumb pointing in the direction of current. Your fingers indicate the direction of the magnetic lines of force.

Electromagnetic Properties

Several important properties relating to electromagnetic fields are now discussed.

Permeability (μ) The ease with which a magnetic field can be established in a given material is measured by the permeability of that material. The higher the permeability, the more easily a magnetic field can be established.

The symbol of permeability is μ (the Greek letter mu), and its value varies depending on the type of material. The permeability of a vacuum (μ_0) is $4\pi \times 10^{-7}$ Wb/At·m (weber/ampere-turn·meter) and is used as a reference. Ferromagnetic materials typically have permeabilities hundreds of times larger than that of a vacuum, indicating that a magnetic field can be set up with relative ease in these materials. Ferromagnetic materials include iron, steel, nickel, cobalt, and their alloys.

The *relative permeability* (μ_r) of a material is the ratio of its absolute permeability (μ) to the permeability of a vacuum (μ_0). Since μ_r is a ratio, it has no units.

$$\mu_r = \frac{\mu}{\mu_0}$$

Equation 7–2

Reluctance ($\mathcal{R}$) The opposition to the establishment of a magnetic field in a material is called reluctance ($\mathcal{R}$). The value of reluctance is directly proportional to the length (l) of the magnetic path, and inversely proportional to the permeability (μ) and to the cross-sectional area (A) of the material, as expressed by the following equation:

$$\mathcal{R} = \frac{l}{\mu A}$$

Equation 7–3

Reluctance in magnetic circuits is analogous to resistance in electric circuits. The unit of reluctance can be derived using l in meters, A (area) in square meters, and μ in Wb/At·m as follows:

$$\mathcal{R} = \frac{l}{\mu A} = \frac{\cancel{\text{m}}}{(\text{Wb/At}\cdot\cancel{\text{m}})(\text{m}^{\cancel{2}})} = \frac{\text{At}}{\text{Wb}}$$

At/Wb is ampere-turns/weber.

EXAMPLE 7–2

What is the reluctance of a material that has a length of 0.05 m, a cross-sectional area of 0.012 m^2, and a permeability of 3500 μWb/At·m?

Solution

$$\mathcal{R} = \frac{l}{\mu A} = \frac{0.05\text{ m}}{(3500 \times 10^{-6}\text{ Wb/At}\cdot\text{m})(0.012\text{ m}^2)} = \mathbf{1190\text{ At/Wb}}$$

Related Problem What happens to the reluctance of the material in this example if l is doubled and A is halved?

Magnetomotive Force (mmf) As you have learned, current in a conductor produces a magnetic field. The force that produces the magnetic field is called the **magnetomotive force** (mmf). The unit of mmf, the **ampere-turn** (At), is established on the basis of the current in a single loop (turn) of wire. The formula for mmf is

Equation 7–4

$$F_m = NI$$

where F_m is the magnetomotive force, N is the number of turns of wire, and I is the current in amperes.

Figure 7–11 illustrates that a number of turns of wire carrying a current around a magnetic material creates a force that sets up flux lines through the magnetic path. The amount of flux depends on the magnitude of the mmf and on the reluctance of the material, as expressed by the following equation:

Equation 7–5

$$\phi = \frac{F_m}{\mathcal{R}}$$

Equation 7–5 is known as the *Ohm's law for magnetic circuits* because the flux (ϕ) is analogous to current, the mmf (F_m) is analogous to voltage, and the reluctance ($\mathcal{R}$) is analogous to resistance.

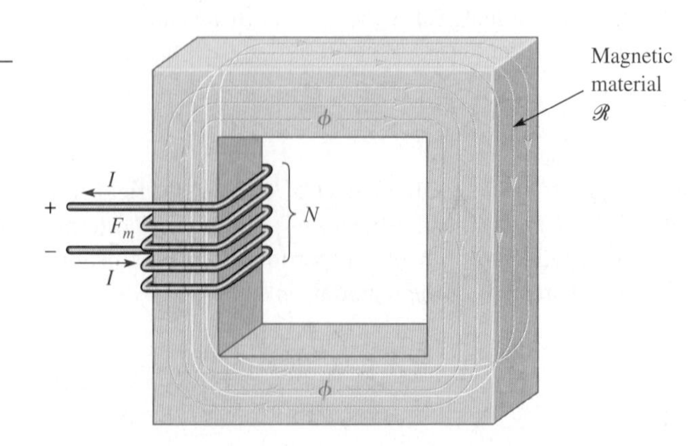

▶ **FIGURE 7–11**

A basic magnetic circuit.

EXAMPLE 7–3

How much flux is established in the magnetic path of Figure 7–12 if the reluctance of the material is 0.28×10^5 At/Wb? Five turns is a small number for a realistic application and is used here only for purposes of illustration.

▶ FIGURE 7–12

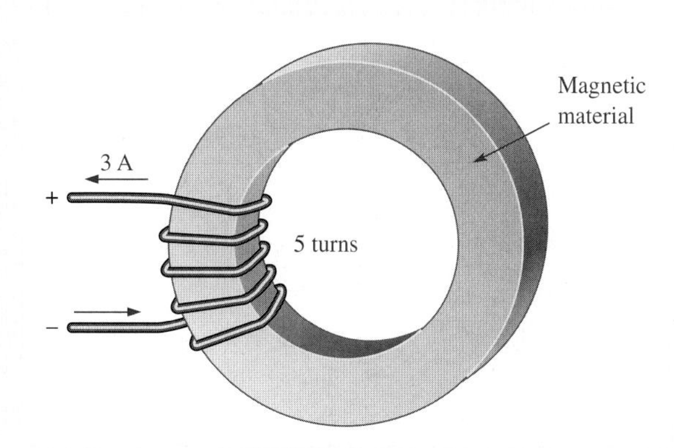

Solution

$$\phi = \frac{F_m}{\mathcal{R}} = \frac{NI}{\mathcal{R}} = \frac{(5\text{ t})(3\text{ A})}{0.28 \times 10^5\text{ At/Wb}} = 5.36 \times 10^{-4}\text{ Wb} = \mathbf{536\ \mu Wb}$$

Related Problem How much flux is established in the magnetic path of Figure 7–12 if the reluctance is 7.5×10^3 At/Wb, the number of turns is 30, and the current is 1.8 A?

EXAMPLE 7–4

There are two amperes of current through a wire with 5 turns.

(a) What is the mmf?

(b) What is the reluctance of the circuit if the flux is 250 μWb?

Solution **(a)** $N = 5$ and $I = 2$ A

$$F_m = NI = (5\text{ t})(2\text{ A}) = \mathbf{10\ At}$$

(b) $$\mathcal{R} = \frac{F_m}{\phi} = \frac{10\text{ At}}{250\ \mu\text{Wb}} = 0.04 \times 10^6\text{ At/Wb} = \mathbf{4.0 \times 10^4\ At/Wb}$$

Related Problem Rework the example for $I = 850$ mA, $N = 50$, and $\phi = 500$ μWb.

The Electromagnet

An electromagnet is based on the properties that you have just learned. A basic electromagnet is simply a coil of wire wound around a core material that can be easily magnetized.

The shape of the electromagnet can be designed for various applications. For example, Figure 7–13 shows a U-shaped magnetic core. When the coil of wire is connected to a battery and there is current, as shown in part (a), a magnetic field is established as indicated. If the current is reversed, as shown in part (b), the direction of the magnetic field is also reversed. The closer the north and south

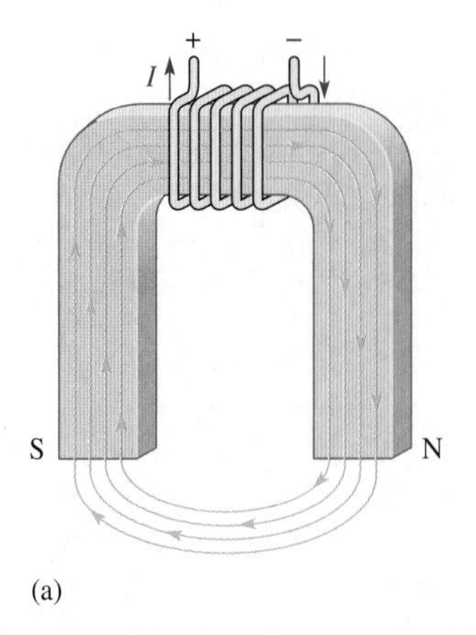

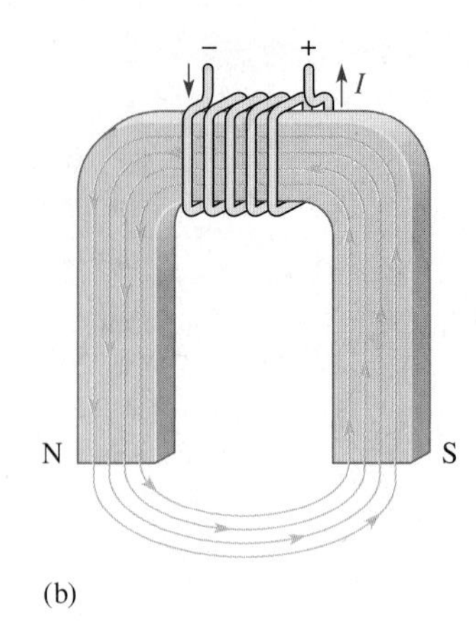

FIGURE 7–13

Reversing the current in the coil causes the electromagnetic field to reverse.

poles are brought together, the smaller the air gap between them becomes, and the easier it becomes to establish a magnetic field, because the reluctance is reduced.

Application Examples

Magnetic Disk/Tape Read/Write Head A simplified diagram of the magnetic disk or tape surface read/write operation is shown in Figure 7–14. A data bit (1 or 0) is written on the magnetic surface by the magnetization of a small segment of the surface as it moves by the write head. The direction of the magnetic flux lines is controlled by the direction of the current pulse in the winding, as shown in Figure 7–14(a) for the case of a positive pulse. At the air gap in the write head, the magnetic flux takes a path through the surface of the storage device. This magnetizes a small spot on the surface in the direction of the field. A magnetized spot of one polarity represents a binary 1, and one of the opposite polarity represents a binary 0. Once a spot on the surface is magnetized, it remains until written over with an opposite magnetic field.

When the magnetic surface passes a read head, the magnetized spots produce magnetic fields in the read head, which induce voltage pulses in the winding. The polarity of these pulses depends on the direction of the magnetized spot and indicates whether the stored bit is a 1 or a 0. This process is illustrated in Figure 7–14(b). Often the read and write heads are combined into a single unit.

FIGURE 7–14

Read/write function on a magnetic surface.

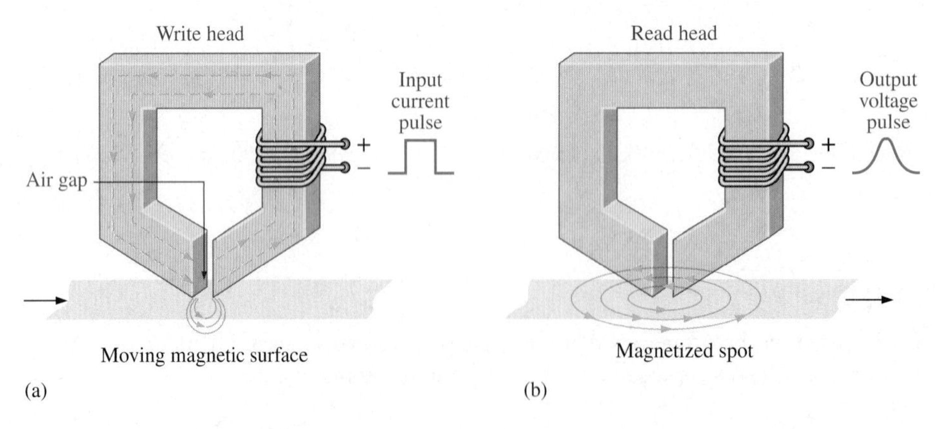

The Magneto-Optical Disk The magneto-optical disk uses an electromagnet and laser beams to read and write (record) data on a magnetic surface. Magneto-optical disks are formatted in tracks and sectors similar to magnetic floppy disks and hard disks. However, because of the ability of a laser beam to be precisely directed to an extremely small spot, magneto-optical disks are capable of storing much more data than standard magnetic hard disks.

Figure 7–15(a) illustrates a small cross-sectional area of a disk before recording, with an electromagnet positioned above it. Tiny magnetic particles, represented by the arrows, are all magnetized in the same direction.

Writing (recording) on the disk is accomplished by applying an external magnetic field opposite to the direction of the magnetic particles as indicated in Figure 7–15(b) and then directing a high-power laser beam to heat the disk at a precise point where a binary 1 is to be stored. The disk material, a magneto-optic alloy, is highly resistant to magnetization at room temperature; but at the spot where the laser beam heats the material, the inherent direction of magnetism is reversed by the external magnetic field produced by the electromagnet. At points where binary 0s are to be stored, the laser beam is not applied and the inherent upward direction of the magnetic particle remains.

As illustrated in Figure 7–15(c), reading data from the disk is accomplished by turning off the external magnetic field and directing a low-power laser beam at a spot where a bit is to be read. Basically, if a binary 1 is stored at the spot (reversed magnetization), the low-power laser beam is reflected and its polarization is shifted; but if a binary 0 is stored, the polarization of the reflected laser beam is unchanged. A detector senses the difference in the polarity of the reflected laser beam to determine if the bit being read is a 1 or a 0.

Figure 7–15(d) shows that the disk is erased by restoring the original magnetic direction of each particle by reversing the external magnetic field and applying the high-power laser beam.

FIGURE 7–15

Basic concept of the magneto-optical disk.

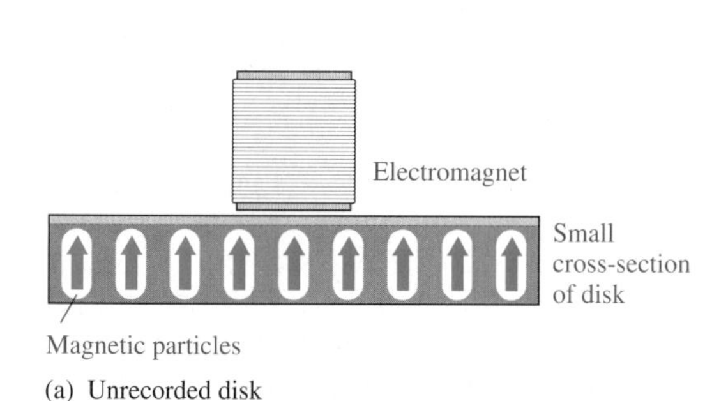

(a) Unrecorded disk

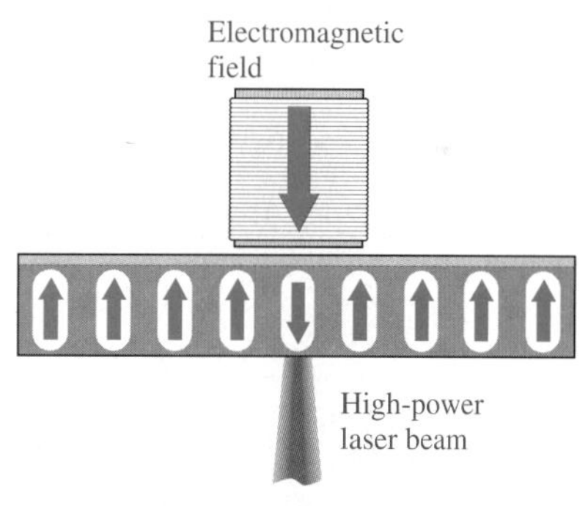

(b) Writing on the disk. High-power laser beam heats the spot and the magnetic particle aligns with the external electromagnetic field.

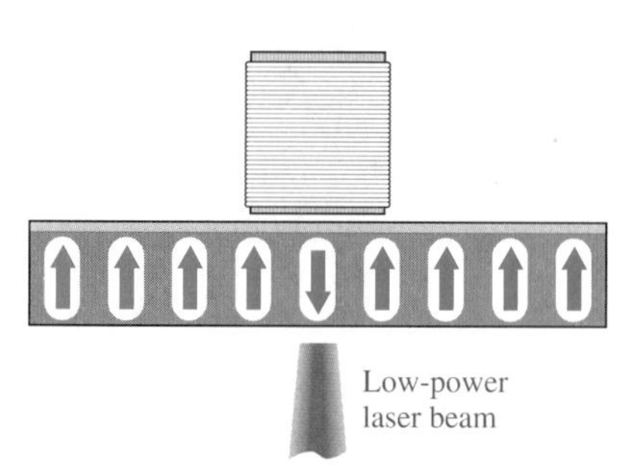

(c) Reading from the disk. Low-power laser beam reflects off of reversed polarity particle and its polarization shifts. If the particle is not reversed, the polarization of the reflected beam is unchanged.

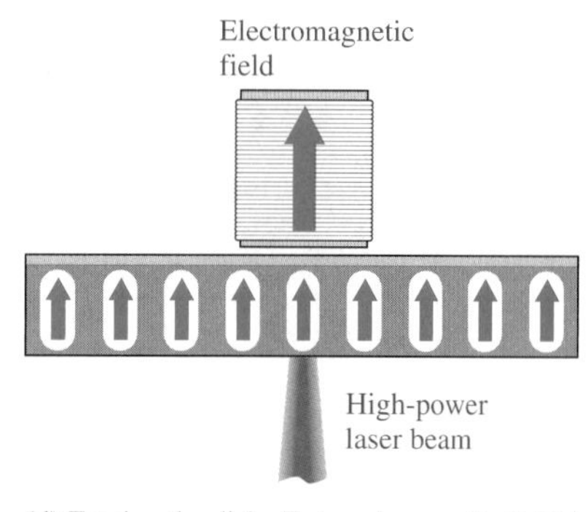

(d) Erasing the disk. External magnetic field is reversed and the high-power laser beam causes the magnetic particle to align with the external beam restoring its original polarity.

SECTION 7–2 REVIEW

1. Explain the difference between magnetism and electromagnetism.
2. What happens to the magnetic field in an electromagnet when the current through the coil is reversed?
3. State Ohm's law for magnetic circuits.
4. Compare each quantity in Question 3 to its electrical counterpart.

7–3 ELECTROMAGNETIC DEVICES

In the last section, you learned that the recording head is a type of electromagnetic device. Now, several other common devices are introduced.

After completing this section, you should be able to

- **Describe the principle of operation for several types of electromagnetic devices**
- Discuss how a solenoid works
- Discuss how a relay works
- Discuss how a speaker works
- Discuss the basic analog meter movement

The magnetic disk/tape read/write head and the magneto-optical disk discussed in the last section are examples of electromagnetic devices. The transformer is another important example and will be covered in Chapter 14. Several other examples are now presented.

The Solenoid

The solenoid is a type of electromagnetic device that has a movable iron core called a *plunger*. The movement of this iron core depends on both an electromagnetic field and a mechanical spring force. The basic structure of a solenoid is shown in Figure 7–16. It consists of a cylindrical coil of wire wound around a nonmagnetic hollow form. A stationary iron core is fixed in position at the end of the shaft and a sliding iron core is attached to the stationary core with a spring.

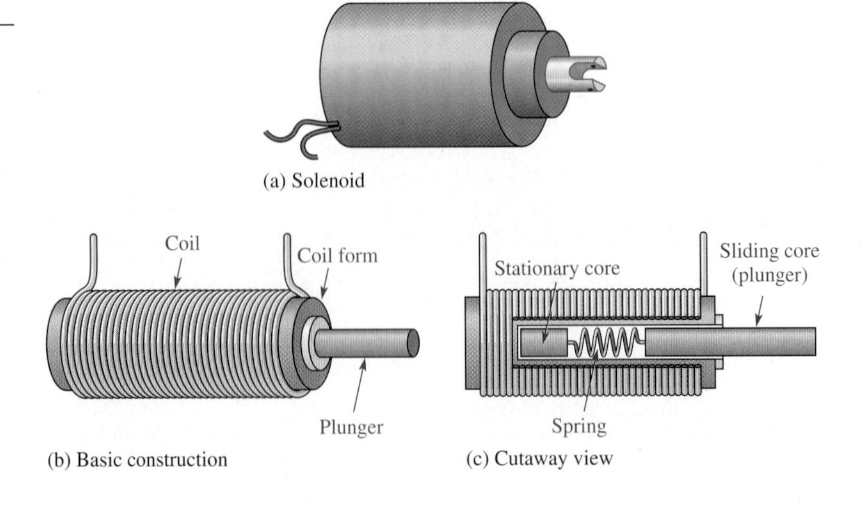

FIGURE 7–16

Basic solenoid structure.

In the at-rest (or unenergized) state, the plunger is extended as shown in Figure 7–17(a). The solenoid is energized by current through the coil, as shown in part (b). The current sets up an electromagnetic field that magnetizes both iron cores as indicated. The south pole of the stationary core attracts the north pole of the movable core, which causes it to slide inward, thus retracting the plunger and compressing the spring. As long as there is coil current, the plunger remains retracted by the attractive force of the magnetic fields. When the current is cut off, the magnetic fields collapse and the force of the compressed spring pushes the plunger back out. The solenoid is used for applications such as opening and closing valves and automobile door locks.

FIGURE 7–17

Basic solenoid operation.

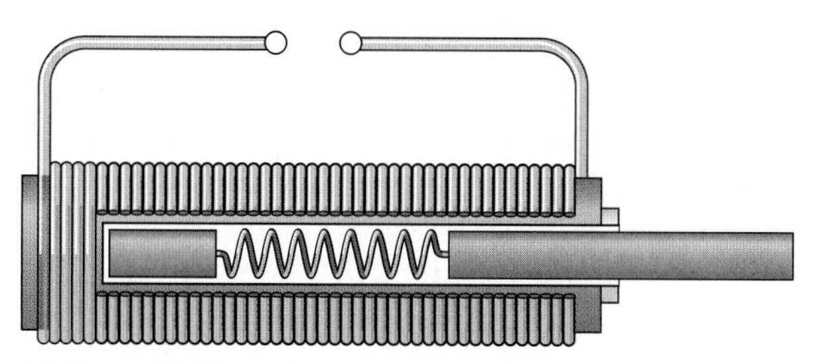

(a) Unenergized (no voltage or current)— plunger extended

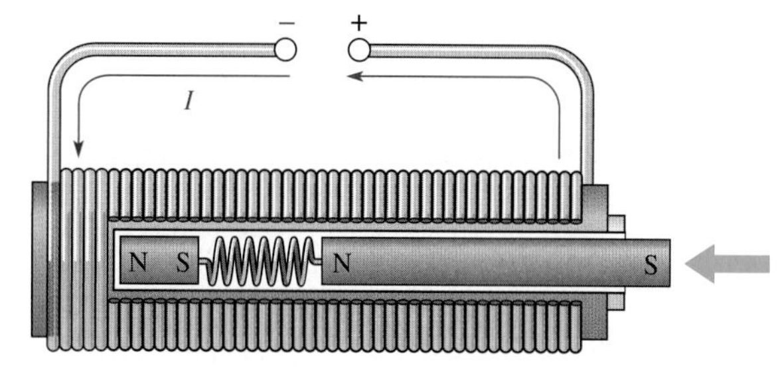

(b) Energized — plunger retracted

The Relay

Relays differ from solenoids in that the electromagnetic action is used to open or close electrical contacts rather than to provide mechanical movement. Figure 7–18 shows the basic operation of a relay with one normally open (NO) contact and one normally closed (NC) contact (single pole–double throw). When there is no coil current, the armature is held against the upper contact by the spring, thus providing continuity from terminal 1 to terminal 2, as shown in Figure 7–18(a). When energized with coil current, the armature is pulled down by the attractive force of the electromagnetic field and makes connection with the lower contact to provide continuity from terminal 1 to terminal 3, as shown in Figure 7–18(b).

FIGURE 7–18

Basic structure of a single-pole–double-throw relay.

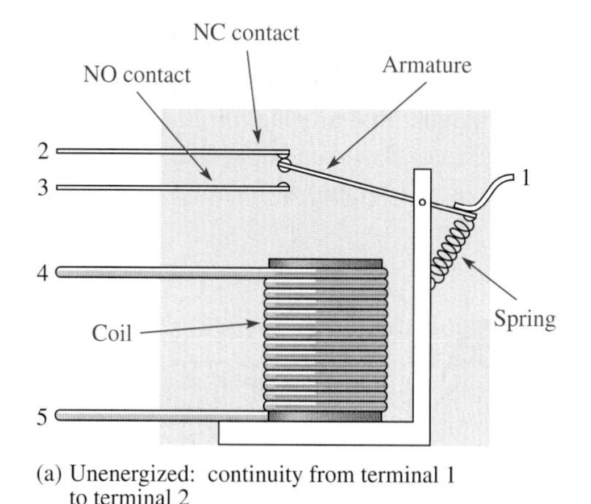

(a) Unenergized: continuity from terminal 1 to terminal 2

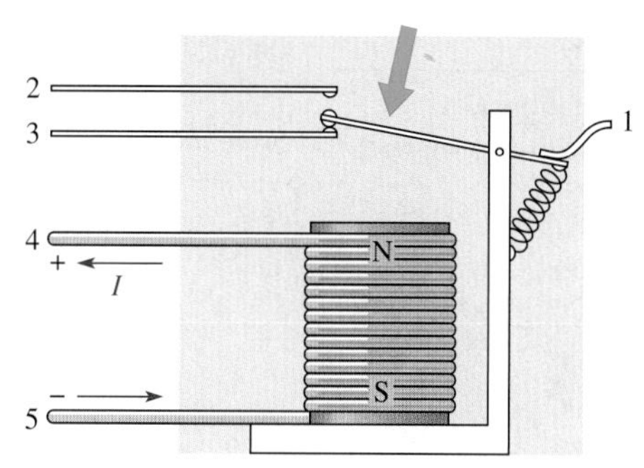

(b) Energized: continuity from terminal 1 to terminal 3

A typical relay and its schematic symbol are shown in Figure 7–19.

Another widely used type of relay is the *reed relay,* which is shown in Figure 7–20. The reed relay, like the armature relay, uses an electromagnetic coil. The contacts are thin reeds of magnetic material and are usually located inside the

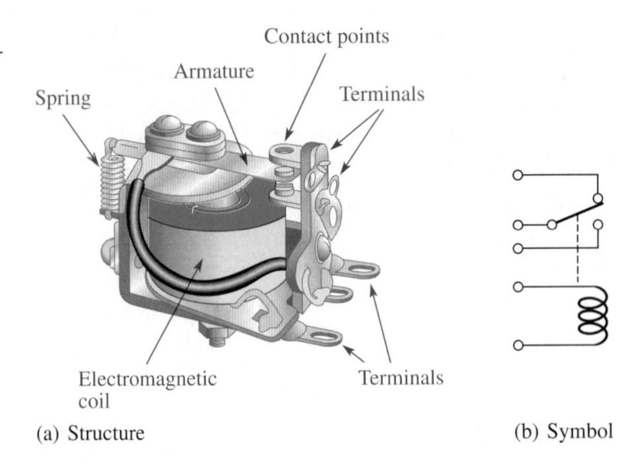

(a) Structure (b) Symbol

▶ FIGURE 7–19

A typical relay.

coil. When there is no coil current, the reeds are in the open position as shown in Figure 7–20(b). When there is current through the coil, the reeds make contact because they are magnetized and attract each other as shown in part (c).

Reed relays are superior to armature relays in that they are faster, more reliable, and produce less contact arcing. However, they have less current-handling capability than armature relays and are more susceptible to mechanical shock.

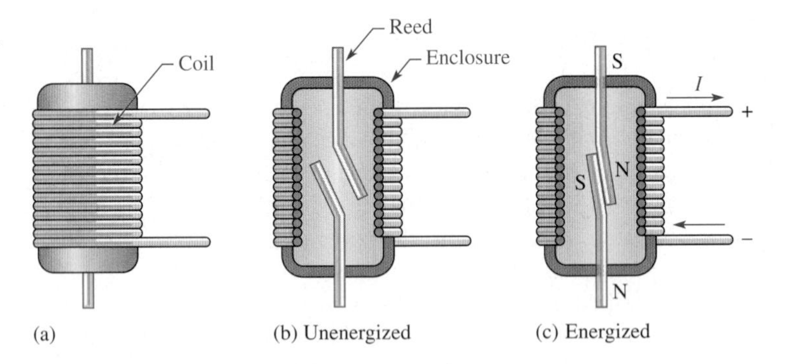

(a) (b) Unenergized (c) Energized

▶ FIGURE 7–20

Basic structure of a reed relay.

The Speaker

Permanent-magnet **speakers** are commonly used in stereos, radios, and TVs, and their operation is based on the principle of electromagnetism. A typical speaker is constructed with a permanent magnet and an electromagnet, as shown in Figure 7–21(a). The cone of the speaker consists of a paper-like diaphragm to which is attached a hollow cylinder with a coil around it, forming an electromagnet. One

▼ FIGURE 7–21

Basic speaker operation with movement exaggerated to illustrate the principle.

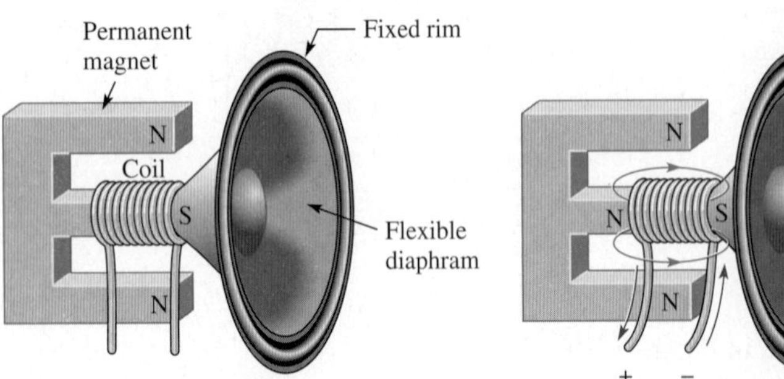

(a) Basic speaker construction

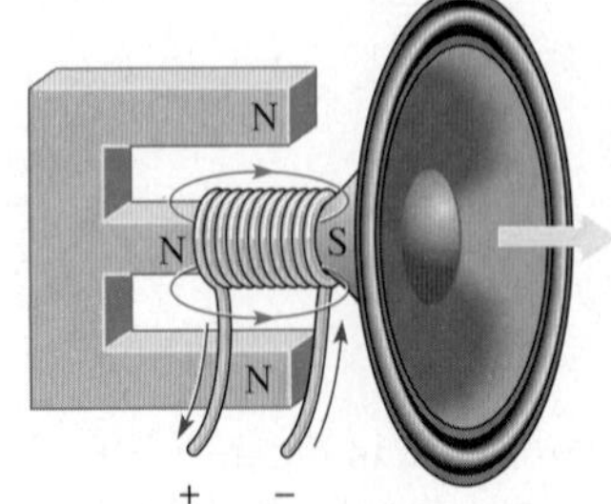

(b) Coil current producing movement of cone to the right

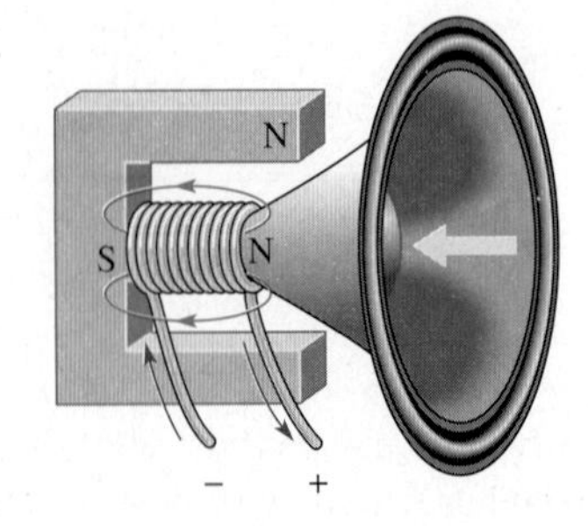

(c) Coil current producing movement of cone to the left

of the poles of the permanent magnet is positioned within the cylindrical coil. When there is current through the coil in one direction, the interaction of the permanent magnetic field with the electromagnetic field causes the cylinder to move to the right, as indicated in Figure 7–21(b). Current through the coil in the other direction causes the cylinder to move to the left, as shown in Figure 7–21(c).

The movement of the coil cylinder causes the flexible diaphragm also to move in or out, depending on the direction of the coil current. The amount of coil current determines the intensity of the magnetic field, which controls the amount that the diaphragm moves.

As shown in Figure 7–22, when an audio signal voltage (voice or music) is applied to the speaker coil, the current varies proportionally in both direction and amount. In response, the diaphragm will vibrate in and out by varying amounts and at varying rates. Vibration in the diaphragm causes the air that is in contact with it to vibrate in the same manner. These air vibrations move through the air as sound waves.

FIGURE 7–22

The speaker converts audio signal voltages into sound waves.

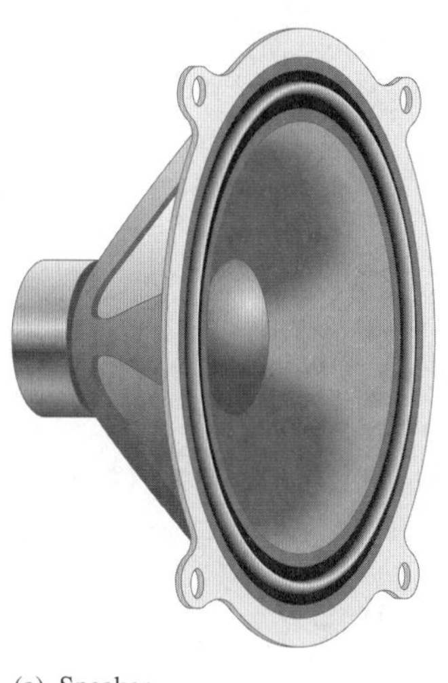

(a) Speaker

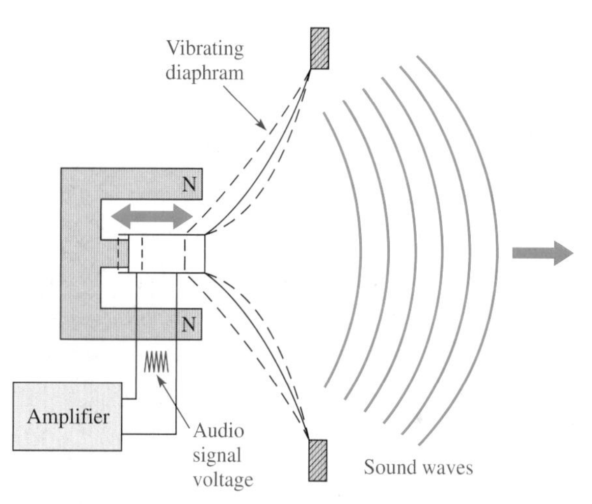

(b) How the speaker produces sound

Meter Movement

The d'Arsonval meter movement is the most common type used in analog multimeters. In this type of meter movement, the pointer is deflected in proportion to the amount of current through a coil. Figure 7–23 shows a basic d'Arsonval meter movement. It consists of a coil of wire wound on a bearing-mounted assembly that is placed between the poles of a permanent magnet. A pointer is attached to the moving assembly. With no current through the coil, a spring mechanism keeps the pointer at its left-most (zero) position. When there is current through the coil, electromagnetic forces act on the coil, causing a rotation to the right. The amount of rotation depends on the amount of current.

Figure 7–24 illustrates how the interaction of magnetic fields produces rotation of the coil assembly. The current is inward at the "cross" and outward at the "dot" in the single winding shown. The inward current produces a counterclockwise electromagnetic field that reinforces the permanent magnetic field below it. The result is an upward force on the left side of the coil as shown. A downward force is developed on the right side of the coil, where the current is outward. These forces produce a clockwise rotation of the coil assembly and are opposed by a spring mechanism. The indicated forces and the spring force are balanced at the value of the current. When current is removed, the spring force returns the pointer to its zero position.

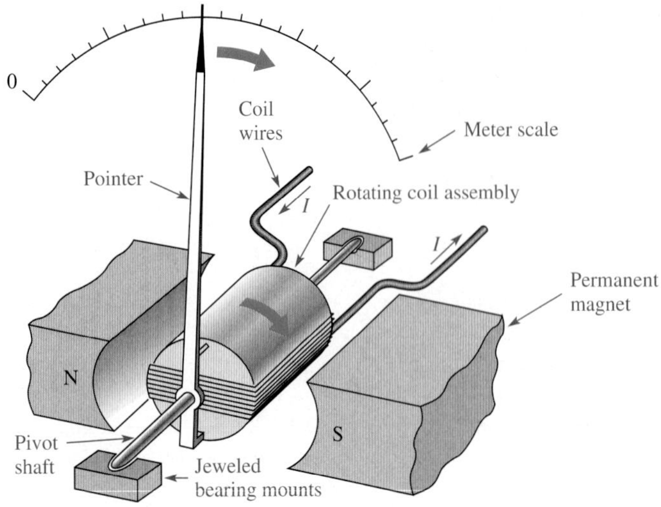

▲ FIGURE 7–23

The basic d'Arsonval meter movement.

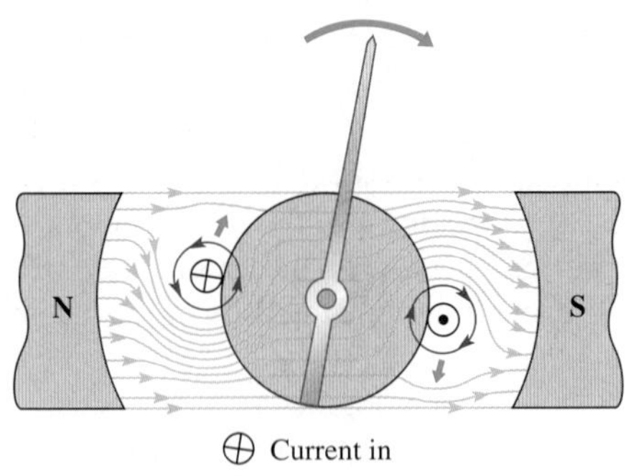

▲ FIGURE 7–24

When the electromagnetic field interacts with the permanent magnetic field, forces are exerted on the rotating coil assembly, causing it to move clockwise and thus deflecting the pointer.

SECTION 7–3 REVIEW

1. Explain the difference between a solenoid and a relay.
2. What is the movable part of a solenoid called?
3. What is the movable part of a relay called?
4. Upon what basic principle is the d'Arsonval meter movement based?

7–4 MAGNETIC HYSTERESIS

When a magnetizing force is applied to a material, the magnetic flux density in the material changes in a certain way, which we will now examine.

After completing this section, you should be able to

- **Explain magnetic hysteresis**
- State the formula for magnetizing force
- Discuss a hysteresis curve
- Define *retentivity*

Magnetizing Force (*H*)

The magnetizing force in a material is defined to be the magnetomotive force (F_m) per unit length (l) of the material, as expressed by the following equation. The unit of magnetizing force (H) is ampere-turns per meter (At/m).

Equation 7–6

$$H = \frac{F_m}{l}$$

where $F_m = NI$. Note that the magnetizing force depends on the number of turns (N) of the coil of wire, the current (I) through the coil, and the length (l) of the material. It does not depend on the type of material.

Since $\phi = F_m/\mathcal{R}$, as F_m increases, the flux increases. Also, the magnetizing force (H) increases. Recall that the flux density (B) is the flux per unit cross-sectional area ($B = \phi/A$), so B is also proportional to H. The curve showing how these two quantities (B and H) are related is called the B-H curve, or the hysteresis curve. The parameters that influence both B and H are illustrated in Figure 7–25.

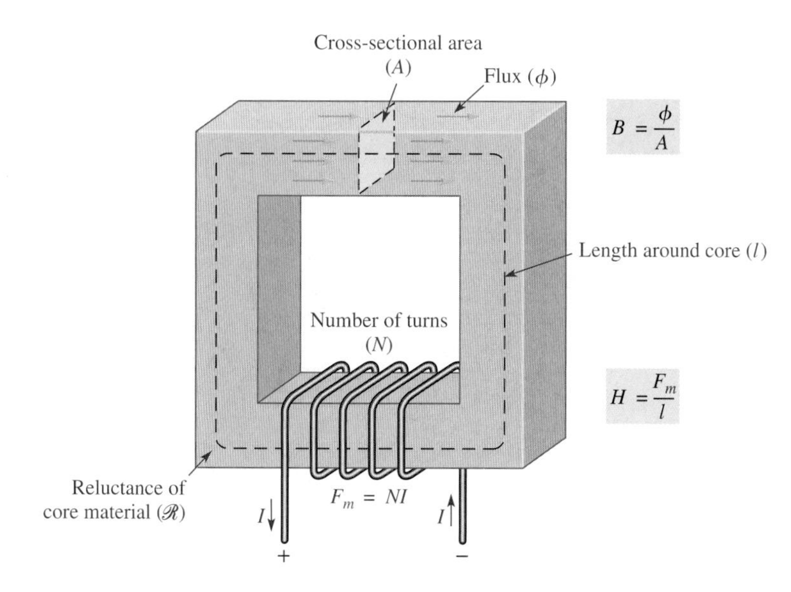

FIGURE 7–25

Parameters that determine the magnetizing force (H) and the flux density (B).

The Hysteresis Curve and Retentivity

Hysteresis is a characteristic of a magnetic material whereby a change in magnetization lags the application of a magnetizing force. The magnetizing force (H) can be readily increased or decreased by varying the current through the coil of wire, and it can be reversed by reversing the voltage polarity across the coil.

Figure 7–26 illustrates the development of the hysteresis curve. Let's start by assuming a magnetic core is unmagnetized so that $B = 0$. As the magnetizing force (H) is increased from zero, the flux density (B) increases proportionally as indicated by the curve in Figure 7–26(a). When H reaches a certain value, the value of B begins to level off. As H continues to increase, B reaches a saturation value (B_{sat}) when H reaches a value (H_{sat}), as illustrated in Figure 7–26(b). Once saturation is reached, a further increase in H will not increase B.

Now, if H is decreased to zero, B will fall back along a different path to a residual value (B_R), as shown in Figure 7–26(c). This indicates that the material continues to be magnetized even with the magnetizing force removed ($H = 0$). The ability of a material, once magnetized, to maintain a magnetized state without the presence of a magnetizing force is called **retentivity.** The retentivity of a material is indicated by the ratio of B_R to B_{sat}.

Reversal of the magnetizing force is represented by negative values of H on the curve and is achieved by reversing the current in the coil of wire. An increase in H in the negative direction causes saturation to occur at a value ($-H_{sat}$) where the flux density is at its maximum negative value, as indicated in Figure 7–26(d).

When the magnetizing force is removed ($H = 0$), the flux density goes to its negative residual value ($-B_R$), as shown in Figure 7–26(e). From the $-B_R$ value, the flux density follows the curve indicated in part (f) back to its maximum positive value when the magnetizing force equals H_{sat} in the positive direction.

The complete B-H curve is shown in Figure 7–26(g) and is called the *hysteresis curve.* The magnetizing force required to make the flux density zero is called the coercive force, H_C.

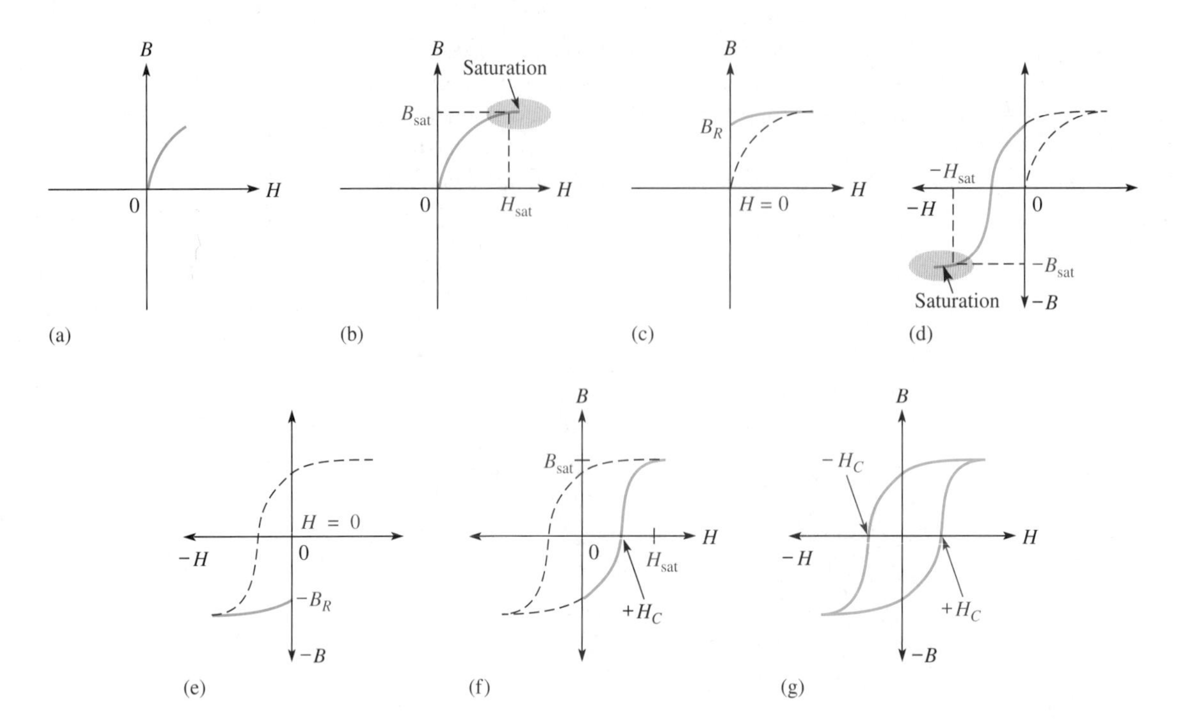

▲ **FIGURE 7–26**

Development of a magnetic hysteresis curve.

Materials with a low retentivity do no retain a magnetic field very well while those with high retentivities exhibit values of B_R very close to the saturation value of B. Depending on the application, retentivity in a magnetic material can be an advantage or a disadvantage. In permanent magnets and memory cores, for example, high retentivity is required. In ac motors, retentivity is undesirable because the residual magnetic field must be overcome each time the current reverses, thus wasting energy.

SECTION 7–4 REVIEW

1. For a given wirewound core, how does an increase in current through the coil affect the flux density?
2. Define *retentivity.*

7–5 ELECTROMAGNETIC INDUCTION

When there is relative motion between a conductor and a magnetic field, a voltage is produced across the conductor. This principle is known as electromagnetic induction and the resulting voltage is an induced voltage. The principle of electromagnetic induction is what makes transformers, electrical generators, electrical motors, and many other devices possible.

After completing this section, you should be able to

- **Discuss the principle of electromagnetic induction**
- Explain how voltage is induced in a conductor in a magnetic field
- Determine polarity of an induced voltage
- Discuss forces on a conductor in a magnetic field
- State Faraday's law
- State Lenz's law

Relative Motion

When a wire is moved across a magnetic field, there is a relative motion between the wire and the magnetic field. Likewise, when a magnetic field is moved past a stationary wire, there is also relative motion. In either case, this relative motion results in an **induced voltage** (v_{ind}) in the wire, as Figure 7–27 indicates. The lowercase v stands for instantaneous voltage. The amount of the induced voltage depends on the rate at which the wire and the magnetic field move with respect to each other: The faster the relative motion, the greater the induced voltage.

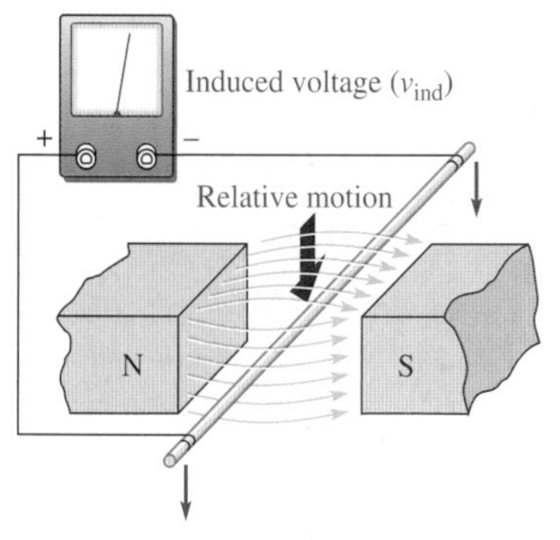

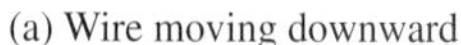
(a) Wire moving downward

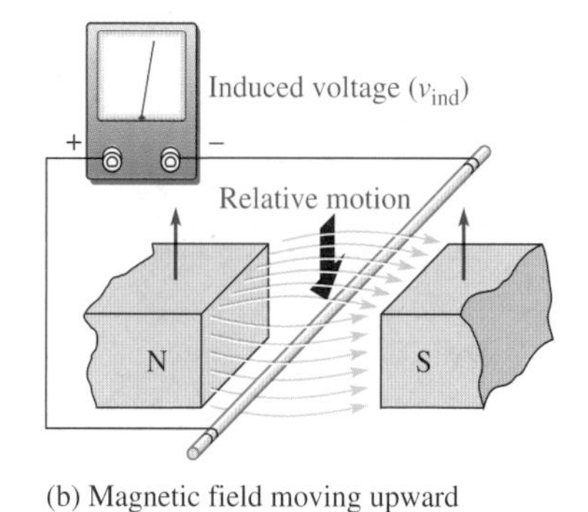

(b) Magnetic field moving upward

FIGURE 7–27

Relative motion between a wire and a magnetic field.

Polarity of the Induced Voltage

If the conductor in Figure 7–27 is moved first one way and then another in the magnetic field, a reversal of the polarity of the induced voltage will be observed. As the wire is moved downward, a voltage is induced with the polarity indicated in Figure 7–28(a). As the wire is moved upward, the polarity is as indicated in part (b) of the figure.

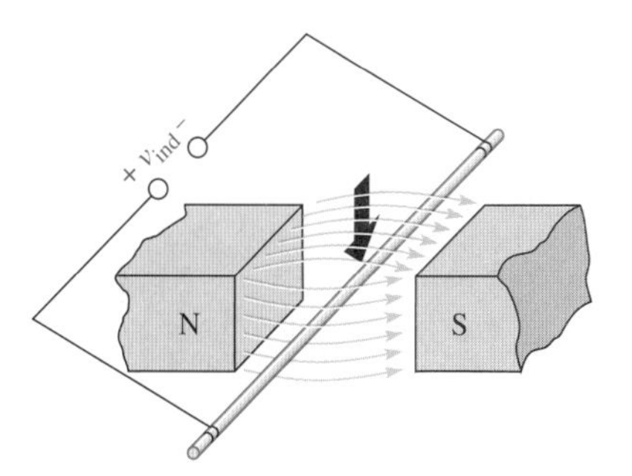

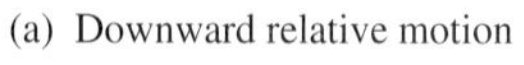
(a) Downward relative motion

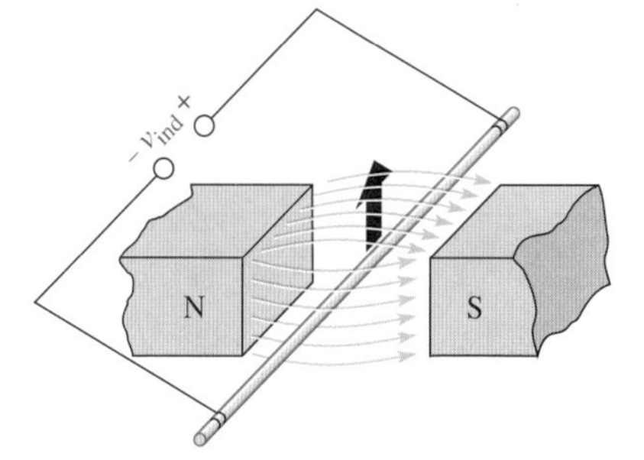

(b) Upward relative motion

FIGURE 7–28

Polarity of induced voltage depends on direction of motion.

Induced Current

When a load resistor is connected to the wire in Figure 7–28, the voltage induced by the relative motion in the magnetic field will cause a current in the load, as shown in Figure 7–29. This current is called the **induced current** (i_{ind}). The lowercase i stands for instantaneous current.

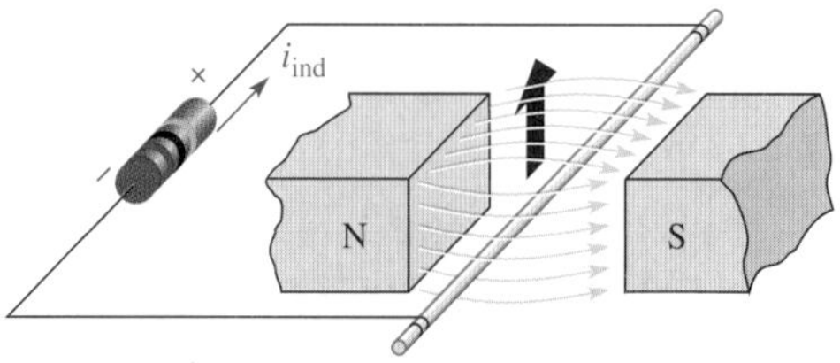

FIGURE 7–29

Induced current (i_{ind}) in a load as the wire moves through the magnetic field.

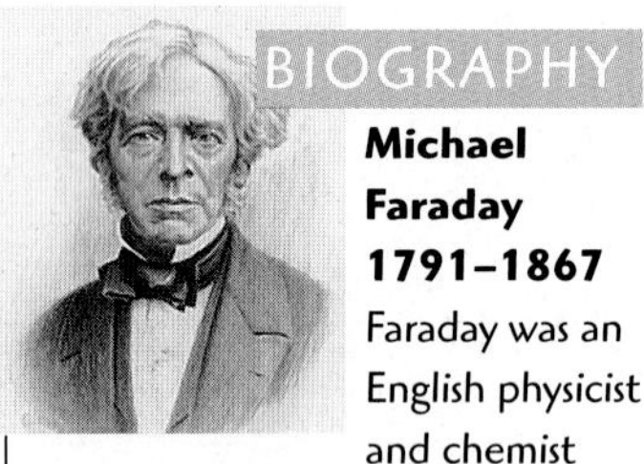

BIOGRAPHY

Michael Faraday 1791–1867

Faraday was an English physicist and chemist who is best remembered for his contribution to the understanding of electromagnetism. He discovered that electricity could be produced by moving a magnet inside a coil of wire, and he was able to build the first electric motor. He later built the first electromagnetic generator and transformer. The statement of the principles of electromagnetic induction is known today as Faraday's law. Also, the unit of capacitance, the *farad,* is named after him. (Photo credit: Library of Congress.)

The action of producing a voltage and a current in a load by moving a conductor across a magnetic field is the basis for electrical generators. Also, the existence of a conductor in a moving magnetic field is fundamental to the concept of inductance in an electric circuit.

Forces on a Current-Carrying Conductor in a Magnetic Field (Motor Action)

Figure 7–30(a) shows current inward through a wire in a magnetic field. The electromagnetic field set up by the current interacts with the permanent magnetic field; as a result, the permanent lines of force above the wire tend to be deflected down under the wire because they are opposite in direction to the electromagnetic lines of force. Therefore, the flux density above is reduced, and the magnetic field is weakened. The flux density below the conductor is increased, and the magnetic field is strengthened. An upward force on the conductor results, and the conductor tends to move toward the weaker magnetic field.

Figure 7–30(b) shows the current outward, resulting in a force on the conductor in the downward direction. These forces are the basis for electric motors.

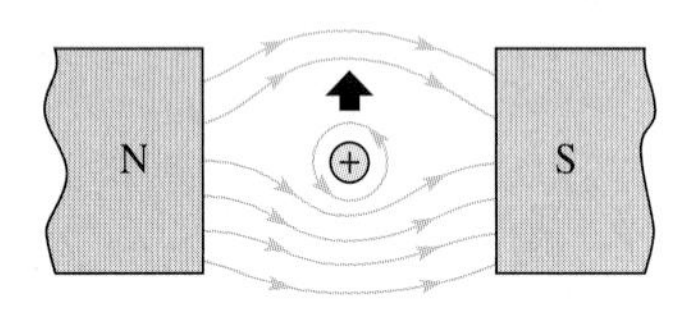

(a) Upward force: weak field above, strong field below.

(b) Downward force: strong field above, weak field below

⊕ Current in
⊙ Current out

▶ **FIGURE 7–30**

Forces on a current-carrying conductor in a magnetic field.

Faraday's Law

Michael Faraday discovered the principle of electromagnetic induction in 1831. He found that moving a magnet through a coil of wire induced a voltage across the coil; he also found that when a complete path was provided, the induced voltage caused an induced current, as you have learned. Faraday's two observations are stated as follows:

1. The amount of voltage induced in a coil is directly proportional to the rate of change of the magnetic field with respect to the coil.
2. The amount of voltage induced in a coil is directly proportional to the number of turns of wire in the coil.

Faraday's first observation is demonstrated in Figure 7–31, where a bar magnet is moved through a coil, thus creating a changing magnetic field. In part (a) of the figure, the magnet is moved at a certain rate, and a certain induced voltage is produced as indicated. In part (b), the magnet is moved at a faster rate through the coil, creating a greater induced voltage.

Faraday's second observation is demonstrated in Figure 7–32. In part (a), the magnet is moved through the coil and a voltage is induced as shown. In part (b), the magnet is moved at the same rate through a coil that has a greater number of turns. The greater number of turns creates a greater induced voltage.

Faraday's law is stated as follows:

The voltage induced across a coil of wire equals the number of turns in the coil times the rate of change of the magnetic flux.

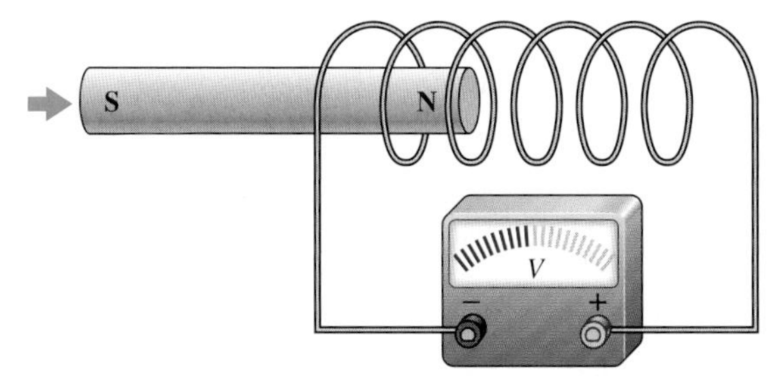

(a) As the magnet moves slowly to the right, its magnetic field is changing with respect to coil, and a voltage is induced.

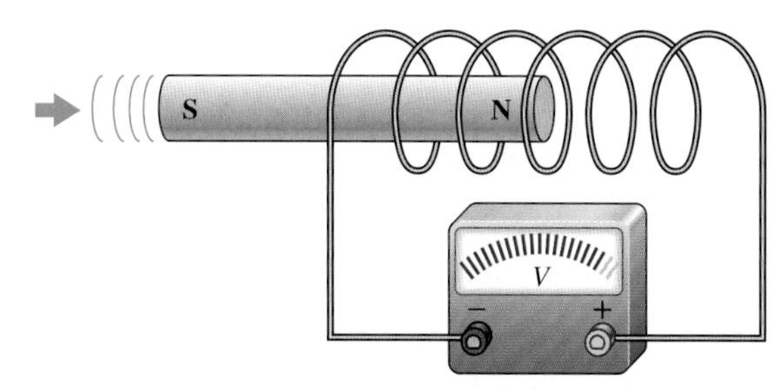

(b) As the magnet moves more rapidly to the right, its magnetic field is changing more rapidly with respect to coil, and a greater voltage is induced.

FIGURE 7–31

A demonstration of Faraday's first observation: The amount of induced voltage is directly proportional to the rate of change of the magnetic field with respect to the coil.

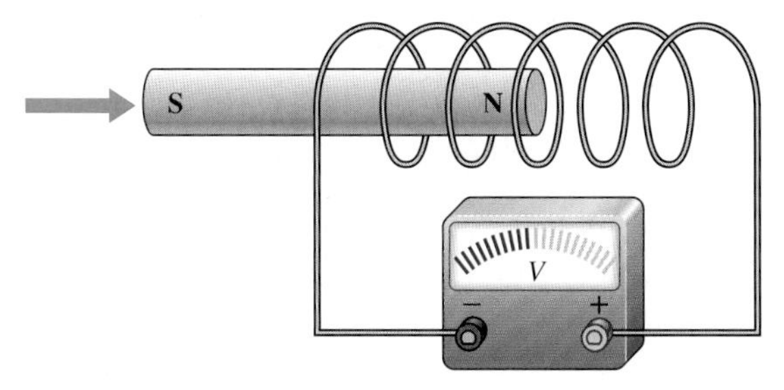

(a) Magnet moves through a coil and induces a voltage.

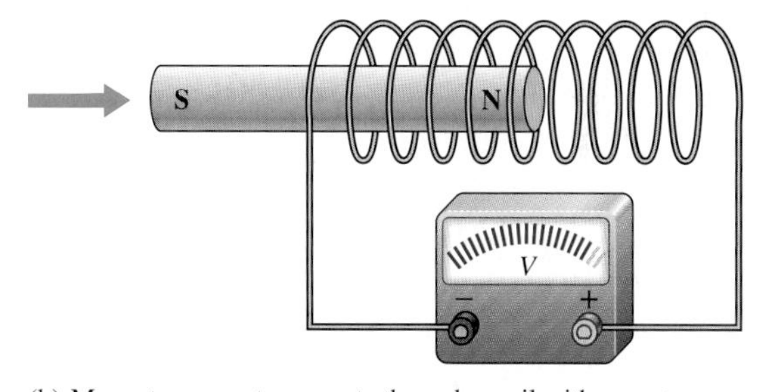

(b) Magnet moves at same rate through a coil with more turns (loops) and induces a greater voltage.

FIGURE 7–32

A demonstration of Faraday's second observation: The amount of induced voltage is directly proportional to the number of turns in the coil.

BIOGRAPHY

Heinrich F. E. Lenz 1804–1865 Lenz was born in Estonia (then Russia) and became a professor at the University of St. Petersburg. He carried out many experiments following Faraday's lead and formulated the principle of electromagnetism, which defines the polarity of induced voltage in a coil. The statement of this principle is named in his honor. (Photo credit: AIP Emilio Segrè Visual Archives, E. Scott Barr Collection.)

Lenz's Law

You have learned that a changing magnetic field induces a voltage in a coil that is directly proportional to the rate of change of the magnetic field and the number of turns in the coil. **Lenz's law** defines the polarity or direction of the induced voltage.

When the current through a coil changes, the polarity of the induced voltage created by the changing magnetic field is such that it always opposes the change in current that caused it.

SECTION 7–5 REVIEW

1. What is the induced voltage across a stationary conductor in a stationary magnetic field?
2. When the rate at which a conductor is moved through a magnetic field is increased, does the induced voltage increase, decrease, or remain the same?
3. When there is current through a conductor in a magnetic field, what happens?

7–6 APPLICATIONS OF ELECTROMAGNETIC INDUCTION

In this section, two interesting applications of electromagnetic induction are discussed—an automotive crankshaft position sensor and a dc generator. Although there are many varied applications, these two are representative.

After completing this section, you should be able to

- **Describe some applications of electromagnetic induction**
- Explain how a crankshaft position sensor works
- Explain how a dc generator works

Automotive Crankshaft Position Sensor

An interesting automotive application is a type of engine sensor that detects the crankshaft position directly using electromagnetic induction. The electronic engine controller in many automobiles uses the position of the crankshaft to set ignition timing and, sometimes, to adjust the fuel control system. Figure 7–33 shows the basic concept. A steel disk is attached to the engine's crankshaft by an extension rod; the protruding tabs on the disk represent specific crankshaft positions.

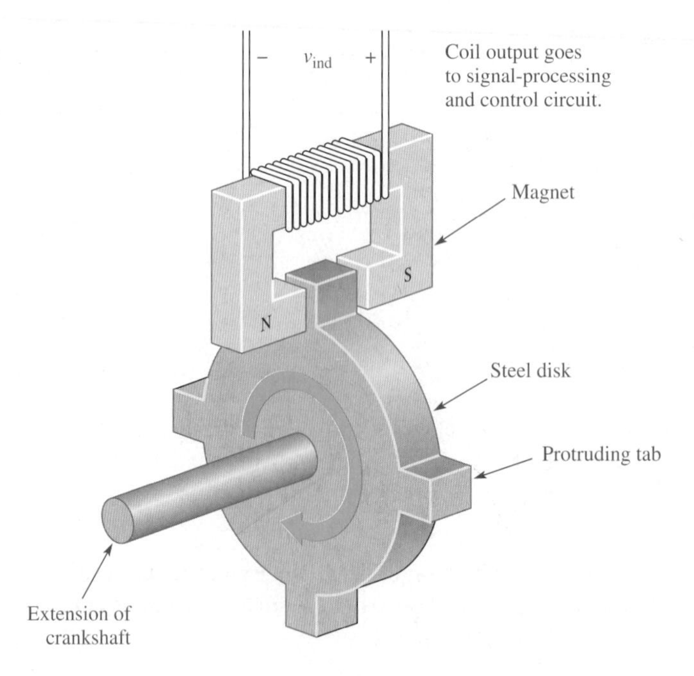

FIGURE 7–33

A crankshaft position sensor that produces a voltage when a tab passes through the air gap of the magnet.

As illustrated in Figure 7–33, as the disk rotates with the crankshaft, the tabs periodically pass through the air gap of the permanent magnet. Since steel has a much lower reluctance than does air (a magnetic field can be established in steel much more easily than in air), the magnetic flux suddenly increases as a tab comes into the air gap, causing a momentary voltage to be induced across the coil. This process is illustrated in Figure 7–34. The electronic engine control circuit uses the induced voltage as an indicator of the crankshaft position.

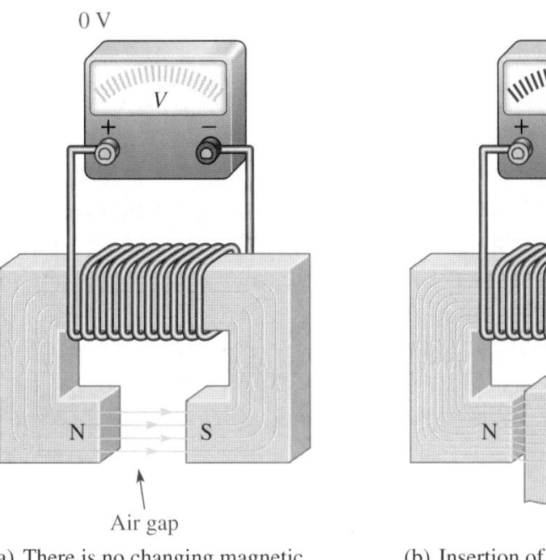

(a) There is no changing magnetic field, so there is no induced voltage.

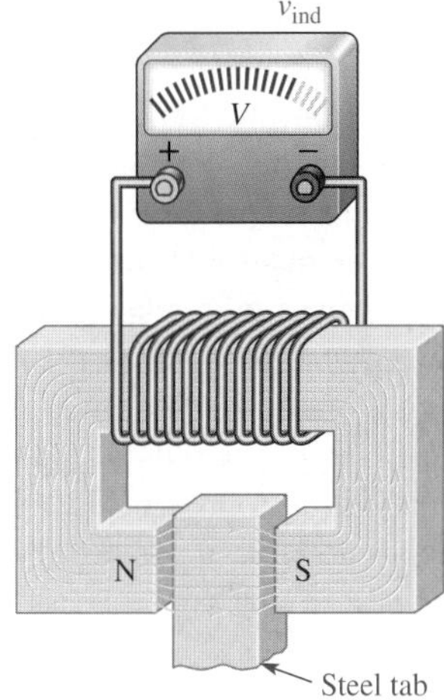

(b) Insertion of the steel tab reduces the reluctance of the air gap, causing the magnetic flux to momentarily increase and thus inducing a momentary voltage.

FIGURE 7–34

As the tab passes through the air gap of the magnet, the coil senses a change in the magnetic field, and a voltage is induced.

DC Generator

Figure 7–35 shows a greatly simplified dc generator consisting of a single loop of wire in a permanent magnetic field. Notice that each end of the loop is connected to a split-ring arrangement. This conductive metal ring is called a *commutator.* As the loop is rotated in the magnetic field, the split commutator ring also rotates. Each half of the split ring rubs against the fixed contacts, called *brushes,* and connects the loop to an external circuit.

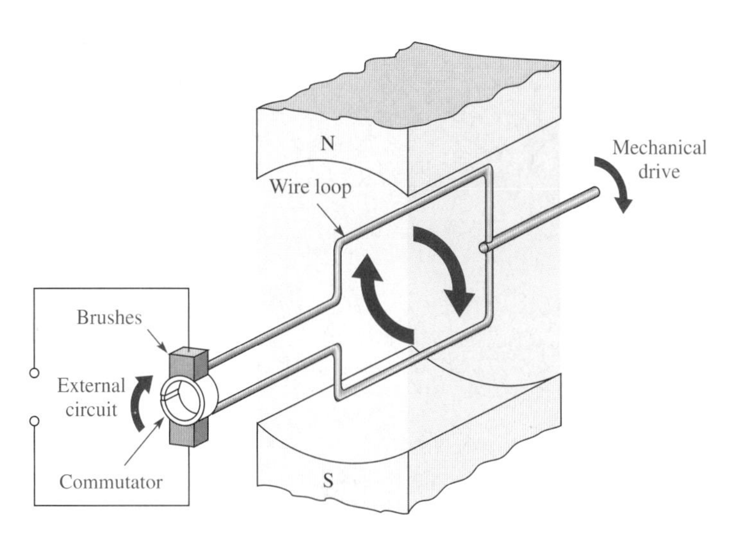

FIGURE 7–35

A basic dc generator.

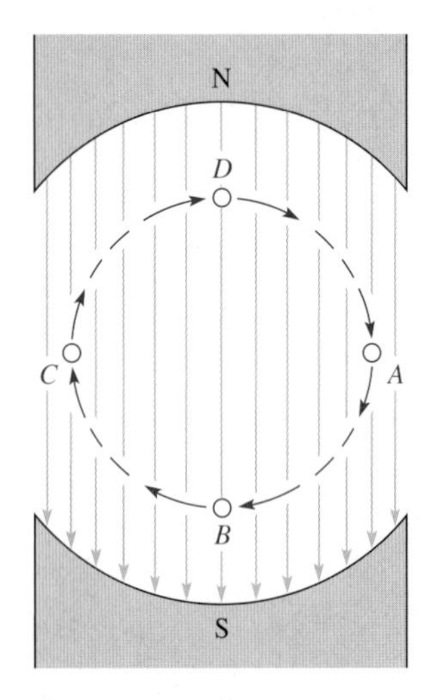

▲ **FIGURE 7–36**

End view of loop cutting through the magnetic field.

As the loop rotates through the magnetic field, it cuts through the flux lines at varying angles, as illustrated in Figure 7–36. At position *A* in its rotation, the loop of wire is effectively moving parallel with the magnetic field. Therefore, at this instant, the rate at which it is cutting through the magnetic flux lines is zero. As the loop moves from position *A* to position *B,* it cuts through the flux lines at an increasing rate. At position *B,* it is moving effectively perpendicular to the magnetic field and thus is cutting through a maximum number of lines. As the loop rotates from position *B* to position *C,* the rate at which it cuts the flux lines decreases to minimum (zero) at *C.* From position *C* to position *D,* the rate at which the loop cuts the flux lines increases to a maximum at *D* and then back to a minimum again at *A.*

As you previously learned, when a wire moves through a magnetic field, a voltage is induced, and by Faraday's law, the amount of induced voltage is proportional to the number of loops (turns) in the wire and the rate at which it is moving with respect to the magnetic field. Now you know that the angle at which the wire moves with respect to the magnetic flux lines determines the amount of induced voltage because the rate at which the wire cuts through the flux lines depends on the angle of motion.

Figure 7–37 illustrates how a voltage is induced in the external circuit as the single loop rotates in the magnetic field. Assume that the loop is in its instantaneous horizontal position, so the induced voltage is zero. As the loop continues in

▼ **FIGURE 7–37**

Basic operation of a dc generator.

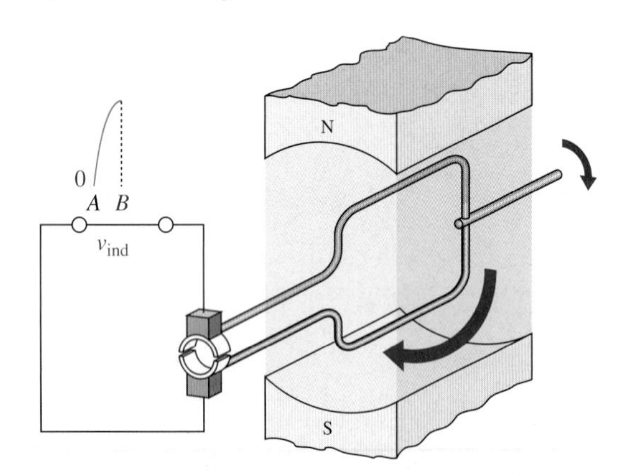

(a) Position *B*: Loop is moving perpendicular to flux lines, and voltage is maximum.

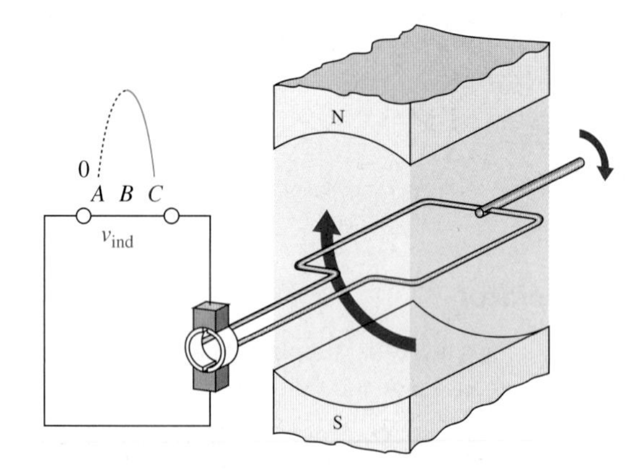

(b) Position *C*: Loop is moving parallel with flux lines, and voltage is zero.

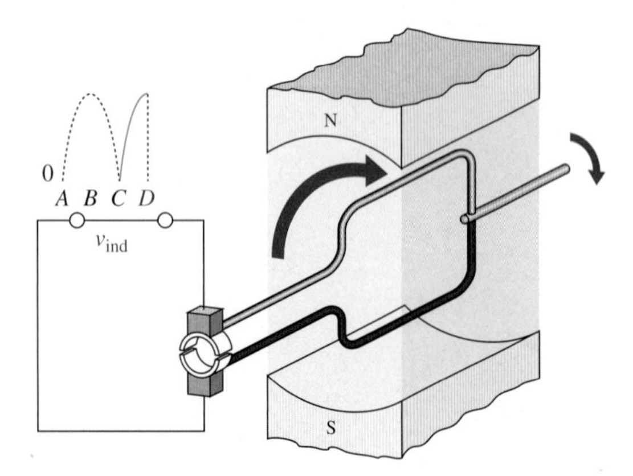

(c) Position *D*: Loop is moving perpendicular to flux lines, and voltage is maximum.

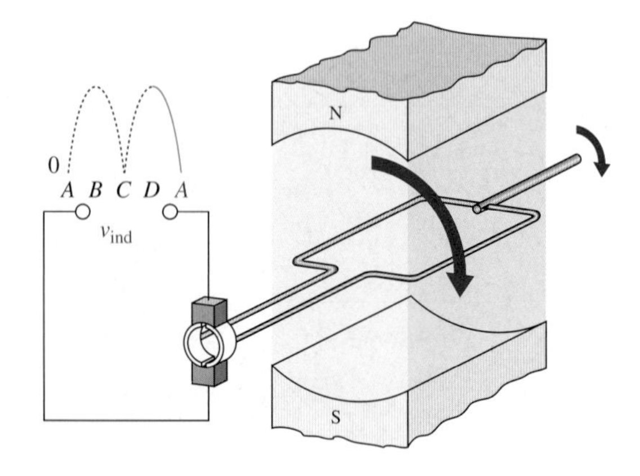

(d) Position *A*: Loop is moving parallel with flux lines, and voltage is zero.

its rotation, the induced voltage builds up to a maximum at position *B*, as shown in part (a) of the figure. Then, as the loop continues from *B* to *C*, the voltage decreases to zero at *C*, as shown in part (b).

During the second half of the revolution, shown in Figure 7–37(c) and (d), the brushes switch to opposite commutator sections, so the polarity of the voltage remains the same across the output. Thus, as the loop rotates from position *C* to *D* and then back to *A*, the voltage increases from zero at *C* to a maximum at *D* and back to zero at *A*.

Figure 7–38 shows how the induced voltage varies as the loop goes through several rotations (three in this case). This voltage is a dc voltage because its polarity does not change. However, the voltage is pulsating between zero and its maximum value.

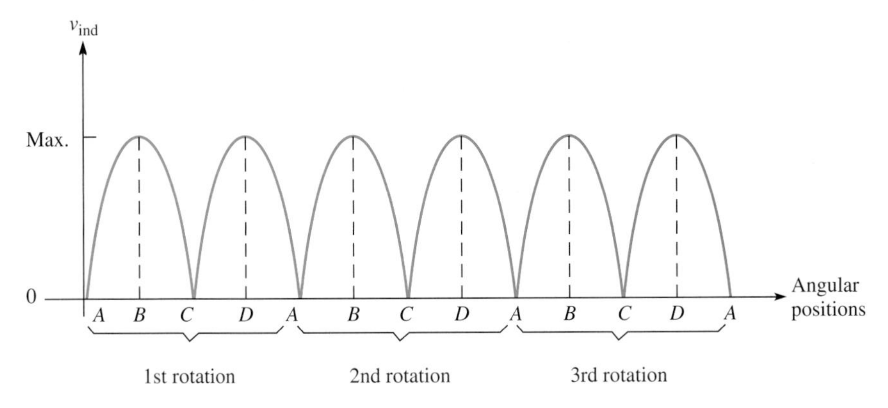

FIGURE 7–38

Induced voltage over three rotations of the loop.

When more loops are added, the voltages induced across each loop are combined across the output. Since the voltages are offset from each other, they do not reach their maximum or zero values at the same time. A smoother dc voltage results, as shown in Figure 7–39 for two loops. The variations can be further smoothed out by filters to achieve a nearly constant dc voltage. (Filters are discussed later in Chapter 13.)

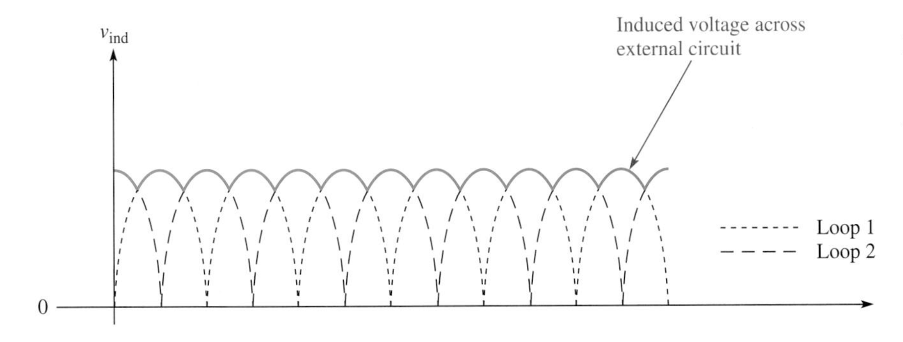

FIGURE 7–39

The induced voltage for a two-loop generator. There is much less variation in the induced voltage.

SECTION 7–6 REVIEW

1. If the steel disk in the crankshaft position sensor has stopped with a tab in the magnet's air gap, what is the induced voltage?
2. What happens to the induced voltage if the loop in the basic dc generator suddenly begins rotating at a faster speed?

APPLICATION ASSIGNMENT

PUTTING YOUR KNOWLEDGE TO WORK

The relay is a common type of electromagnetic device that is used in many types of control applications. With a relay, a lower voltage, such as from a battery, can be used to switch a much higher voltage, such as the 110 V from an ac outlet. In this section, you will see how a relay can be used in a basic burglar alarm system.

The schematic in Figure 7–40 shows a simplified intrusion alarm system that uses a relay to turn on an audible alarm (siren) and lights. The system operates from a 9 V battery so that even if power to the house is off, the audible alarm will still work.

The magnetic detection switches are normally open (NO) switches that are parallel connected and located in the windows and doors. The relay is a triple pole–double-throw device that operates with a coil voltage of 9 V dc and draws approximately 50 mA. When an intrusion occurs, one of the switches closes and allows current from the battery through the relay coil, which energizes the relay and causes the three sets of normally open contacts to close. Closure of contact *A* turns on the audible alarm, which draws 2 A from the battery. Closure of contact *C* turns on a light circuit in the house. Closure of contact *B* latches the relay and keeps it energized even if the intruder closes the door or window through which entry was made. If not for contact *B* in parallel with the detection switches, the alarm and lights would go off as soon as the window or door was shut behind the intruder.

The relay contacts are not physically remote in relation to the coil as the schematic indicates. The schematic is drawn this way for functional clarity. The entire triple-pole–double-throw relay is housed in the package shown in Figure 7–41. Also shown are the pin diagram and internal schematic for the relay.

Step 1: Connect the System

Develop a connection diagram and point-to-point wire list for interconnecting the components in Figure 7–42 to create the alarm system shown in the schematic of Figure 7–40. The connection points on the components are indicated by letters.

Step 2: Test the System

Develop a step-by-step procedure to check out the completely wired burglar alarm system.

FIGURE 7–40

Simplified burglar alarm system.

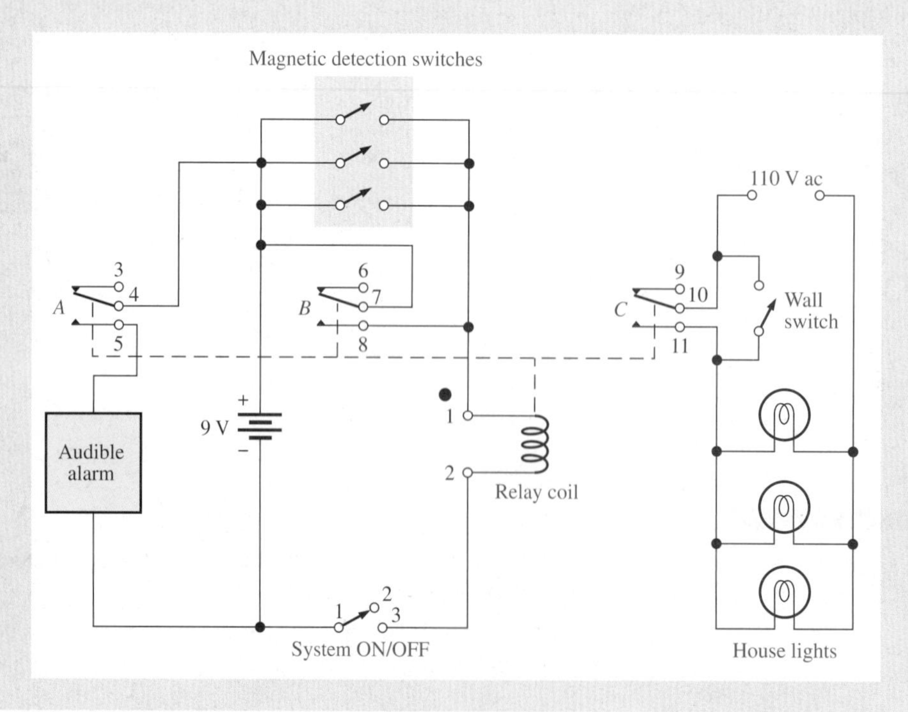

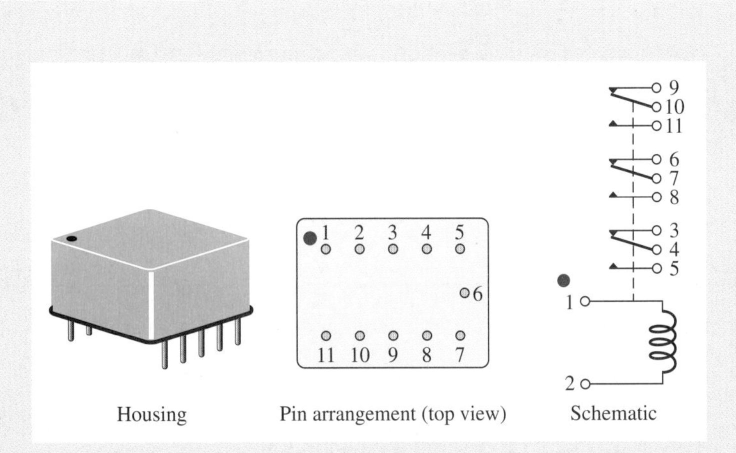

FIGURE 7–41

Triple-pole–double-throw relay.

FIGURE 7–42

Array of burglar alarm components.

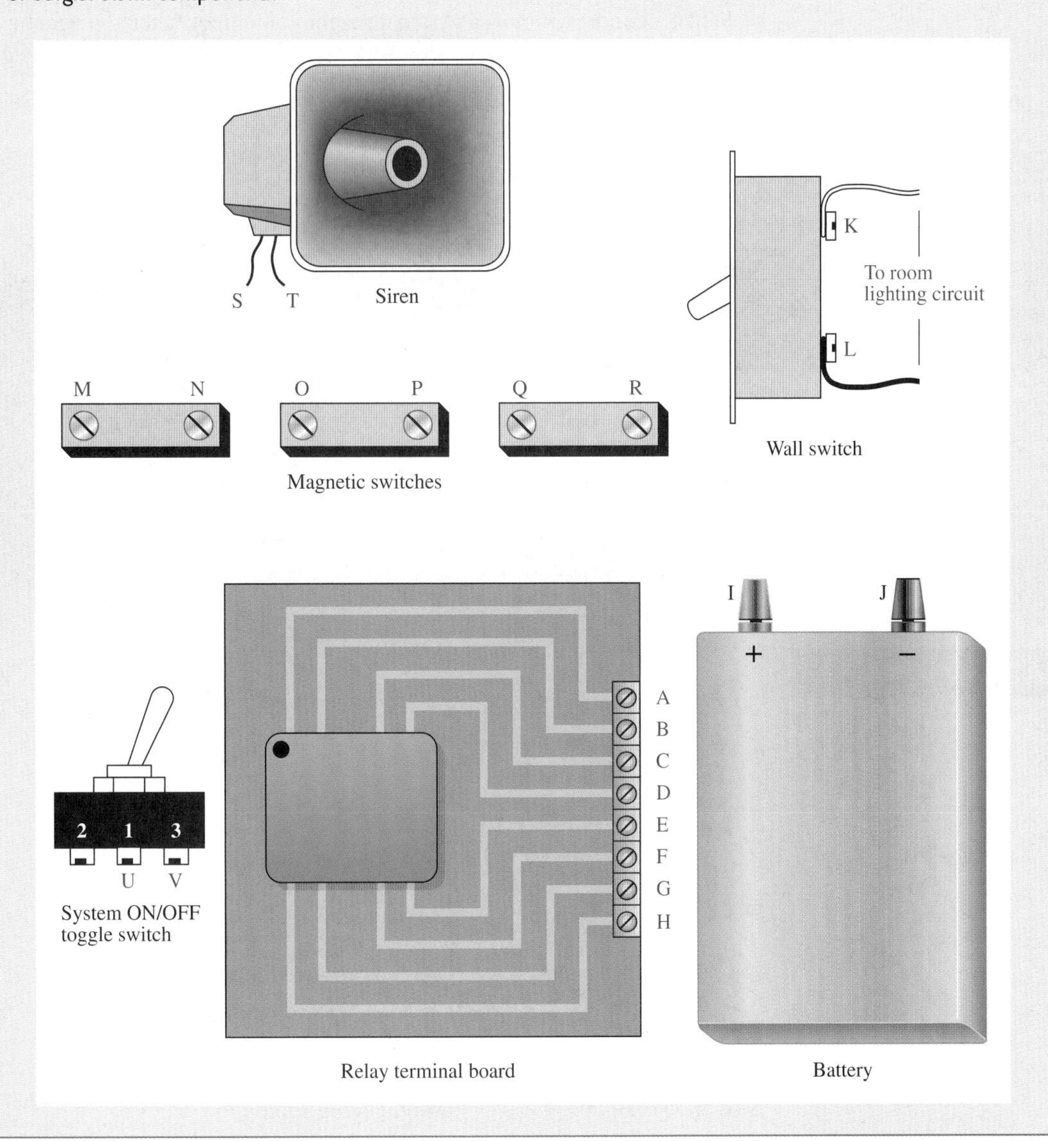

APPLICATION ASSIGNMENT REVIEW

1. How many magnetic detection switches in Figure 7–40 must be closed to activate the system?
2. What is the purpose of each of the three relay contacts?

SUMMARY

- Unlike magnetic poles attract each other, and like poles repel each other.
- Materials that can be magnetized are called *ferromagnetic.*
- When there is current through a conductor, it produces an electromagnetic field around the conductor.
- You can use the left-hand rule to establish the direction of the electromagnetic lines of force around a conductor.
- An electromagnet is basically a coil of wire around a magnetic core.
- When a conductor moves within a magnetic field, or when a magnetic field moves relative to a conductor, a voltage is induced across the conductor.
- The faster the relative motion between a conductor and a magnetic field, the greater is the induced voltage.
- Table 7–1 summarizes the magnetic quantities.

TABLE 7–1

SYMBOL	QUANTITY	SI UNIT
B	Magnetic flux density	Tesla (T)
ϕ	Flux	Weber (Wb)
μ	Permeability	Weber/ampere-turn·meter (Wb/At·m)
$\mathcal{R}$	Reluctance	At/Wb
F_m	Magnetomotive force (mmf)	Ampere-turn (At)
H	Magnetizing force	At/m

EQUATIONS

7–1 $$B = \frac{\phi}{A}$$ Magnetic flux density

7–2 $$\mu_r = \frac{\mu}{\mu_0}$$ Relative permeability

7–3 $$\mathcal{R} = \frac{l}{\mu A}$$ Reluctance

7–4 $$F_m = NI$$ Magnetomotive force

7–5 $$\phi = \frac{F_m}{\mathcal{R}}$$ Magnetic flux

7–6 $$H = \frac{F_m}{l}$$ Magnetizing force

SELF-TEST

Answers are at the end of the chapter.

1. When the south poles of two bar magnets are brought close together, there will be
   (a) a force of attraction (b) a force of repulsion
   (c) an upward force (d) no force
2. A magnetic field is made up of
   (a) positive and negative charges (b) magnetic domains
   (c) flux lines (d) magnetic poles
3. The direction of a magnetic field is from
   (a) north pole to south pole (b) south pole to north pole
   (c) inside to outside the magnet (d) front to back
4. Reluctance in a magnetic circuit is analogous to
   (a) voltage in an electric circuit (b) current in an electric circuit
   (c) power in an electric circuit (d) resistance in an electric circuit
5. The unit of magnetic flux is the
   (a) tesla (b) weber (c) gauss (d) ampere-turn
6. The unit of magnetomotive force is the
   (a) tesla (b) weber (c) ampere-turn (d) electron-volt
7. The unit of magnetic flux density is the
   (a) tesla (b) weber (c) ampere-turn
   (d) gauss (e) answer (a) or (d)
8. The electromagnetic activation of a movable shaft is the basis for
   (a) relays (b) circuit breakers
   (c) magnetic switches (d) solenoids
9. When there is current through a wire placed in a magnetic field,
   (a) the wire will overheat
   (b) the wire will become magnetized
   (c) a force is exerted on the wire
   (d) the magnetic field will be cancelled
10. A coil of wire is placed in a changing magnetic field. If the number of turns in the coil is increased, the voltage induced across the coil will
    (a) remain unchanged (b) decrease
    (c) increase (d) be excessive
11. If a conductor is moved back and forth at a constant rate in a constant magnetic field, the voltage induced in the conductor will
    (a) remain constant (b) reverse polarity
    (c) be reduced (d) be increased
12. In the crankshaft position sensor in Figure 7–33, the induced voltage across the coils is caused by
    (a) current in the coil
    (b) rotation of the disk
    (c) a tab passing through the magnetic field
    (d) acceleration of the disk's rotational speed

PROBLEMS

Answers to odd-numbered problems are at the end of the book.

BASIC PROBLEMS

SECTION 7–1 **The Magnetic Field**

1. The cross-sectional area of a magnetic field is increased, but the flux remains the same. Does the flux density increase or decrease?
2. In a certain magnetic field the cross-sectional area is 0.5 m^2 and the flux is 1500 μWb. What is the flux density?
3. What is the flux in a magnetic material when the flux density is 2500×10^{-6} T and the cross-sectional area is 150 cm^2?

SECTION 7–2 **Electromagnetism**

4. What happens to the compass needle in Figure 7–8 when the current through the conductor is reversed?
5. What is the relative permeability of a ferromagnetic material whose absolute permeability is 750×10^{-6} Wb/At·m?
6. Determine the reluctance of a material with a length of 0.28 m and a cross-sectional area of 0.08 m^2 if the absolute permeability is 150×10^{-7} Wb/At·m.
7. What is the magnetomotive force in a 500 turn coil of wire with 3 A through it?

SECTION 7–3 **Electromagnetic Devices**

8. Typically, when a solenoid is activated, is the plunger extended or retracted?
9. **(a)** What force moves the plunger when a solenoid is activated?
   (b) What force causes the plunger to return to its at-rest position?
10. Explain the sequence of events in the circuit of Figure 7–43 starting when switch 1 (SW1) is closed.

▶ FIGURE 7–43

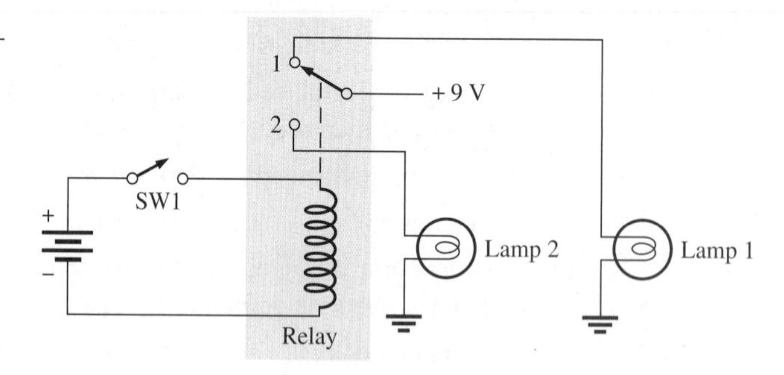

11. What causes the pointer in a d'Arsonval meter movement to deflect when there is current through the coil?

SECTION 7–4 **Magnetic Hysteresis**

12. What is the magnetizing force in Problem 7 if the length of the core is 0.2 m?

13. How can the flux density in Figure 7–44 be changed without altering the physical characteristics of the core?
14. In Figure 7–44, determine the following if the winding has 100 turns:
    (a) H **(b)** ϕ **(c)** B

▶ **FIGURE 7–44**

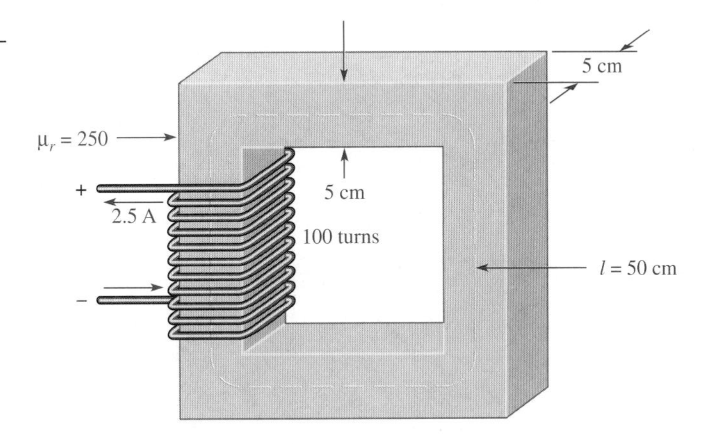

15. Determine from the hysteresis curves in Figure 7–45 which material has the most retentivity.

▶ **FIGURE 7–45**

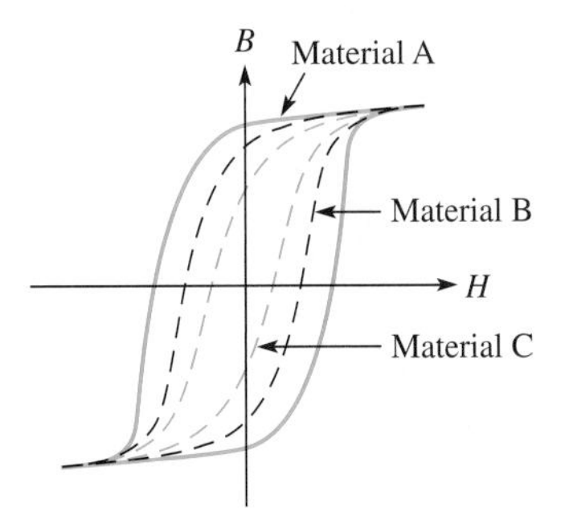

SECTION 7–5 **Electromagnetic Induction**

16. According to Faraday's law, what happens to the induced voltage across a given coil if the rate of change of magnetic flux doubles?
17. The voltage induced across a certain coil is 100 mV. A 100 Ω resistor is connected to the coil terminals. What is the induced current?

SECTION 7–6 **Applications of Electromagnetic Induction**

18. In Figure 7–33, why is there no induced voltage when the steel disk is not rotating?
19. Explain the purpose of the commutator and brushes in Figure 7–35.

ADVANCED PROBLEMS

20. A basic one-loop dc generator is rotated at 60 rps. How many times each second does the dc output voltage peak (reach a maximum)?
21. Assume that another loop, 90° from the first loop, is added to the dc generator in Problem 20. Make a graph of voltage versus time to show how the output voltage appears. Let the maximum voltage be 10 V.

ELECTRONICS WORKBENCH/CIRCUITMAKER TROUBLESHOOTING PROBLEMS

CD-ROM file circuits are shown in Figure 7–46.

22. Open file P07-22 and determine if there is a fault in the circuit. If so, identify the fault.

23. Open file P07-23 and determine if there is a fault in the circuit. If so, identify the fault.

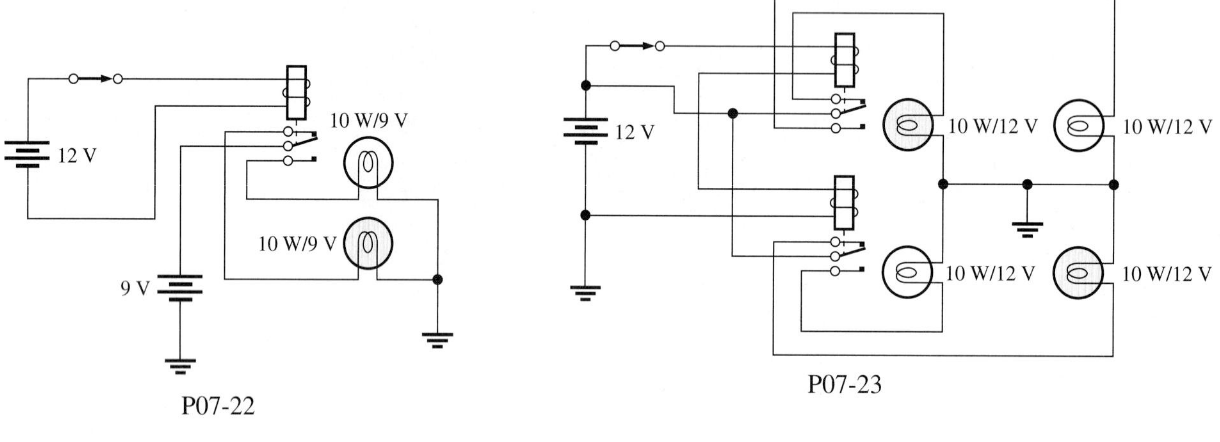

▲ **FIGURE 7–46**

ANSWERS

SECTION REVIEWS

SECTION 7–1

The Magnetic Field

1. The north poles repel.
2. Magnetic flux is the group of lines of force that make up a magnetic field.
3. $B = \phi/A = 900\ \mu T$

SECTION 7–2

Electromagnetism

1. Electromagnetism is produced by current through a conductor. An electromagnetic field exists only when there is current. A magnetic field exists independently of current.
2. The direction of the magnetic field also reverses when the current is reversed.
3. Flux equals magnetomotive force divided by reluctance.
4. Flux is analogous to current, mmf is analogous to voltage, and reluctance is analogous to resistance.

SECTION 7–3

Electromagnetic Devices

1. A solenoid provides mechanical movement of a shaft. A relay provides an electrical contact closure.
2. The movable part of a solenoid is the plunger.
3. The movable part of a relay is the armature.
4. The d'Arsonval meter movement is based on the interaction of magnetic fields.

SECTION 7–4 **Magnetic Hysteresis**

1. In a wirewound core, an increase in current increases the flux density.
2. Retentivity is the ability of a material to remain magnetized after removal of the magnetizing force.

SECTION 7–5 **Electromagnetic Induction**

1. The induced voltage is zero.
2. The induced voltage increases.
3. A force is exerted on the current-carrying conductor in a magnetic field.

SECTION 7–6 **Applications of Electromagnetic Induction**

1. The induced voltage is zero.
2. The induced voltage increases.

■ Application Assignment

1. One or more magnetic detection switches must be closed.
2. Contact *A* activates the audible alarm, contact *B* latches the relay, contact *C* turns on house lights.

RELATED PROBLEMS FOR EXAMPLES

7–1 235 μWb

7–2 Reluctance increases to 4762 At/Wb (quadruples).

7–3 7.2 mWb

7–4 $F_m = 42.5$ At; $\mathcal{R} = 8.5 \times 10^4$ At/Wb

SELF-TEST

1. (b) **2.** (c) **3.** (a) **4.** (d) **5.** (b) **6.** (c) **7.** (e)
8. (d) **9.** (c) **10.** (c) **11.** (b) **12.** (c)

PART TWO

AC CIRCUITS

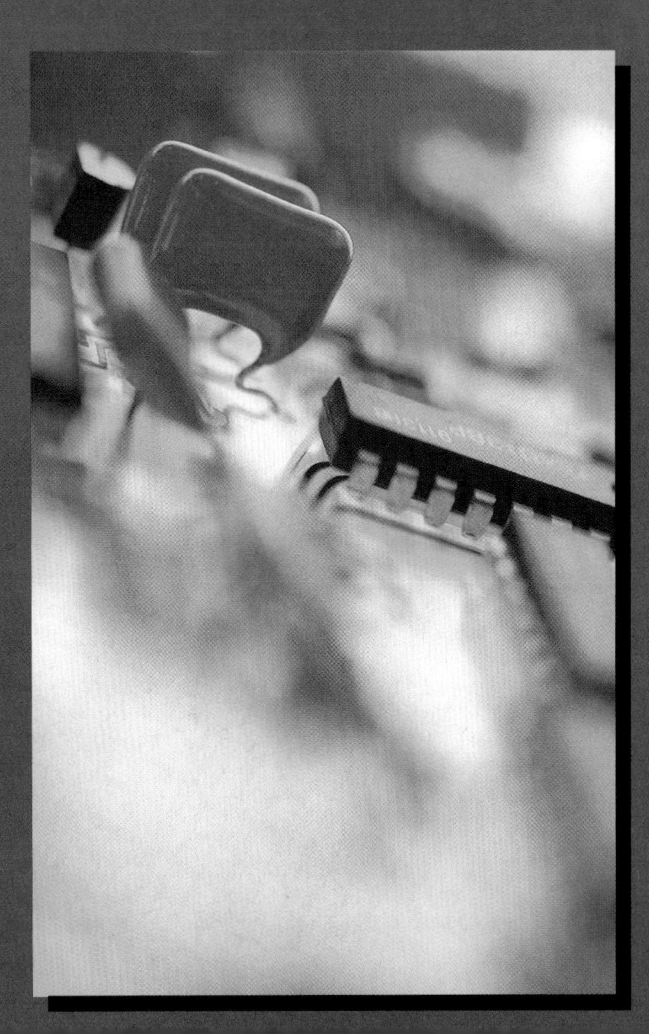

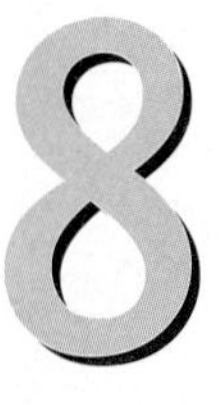

INTRODUCTION TO ALTERNATING CURRENT AND VOLTAGE

INTRODUCTION

This chapter provides an introduction to alternating current (ac) circuits. Alternating voltages and currents fluctuate with time and periodically change polarity and direction according to certain patterns called *waveforms*. Particular emphasis is given to the sine wave because of its basic importance in ac circuits. Other types of waveforms are also introduced, including pulse, triangular, and sawtooth. The use of the oscilloscope for displaying and measuring waveforms is discussed.

CHAPTER OUTLINE

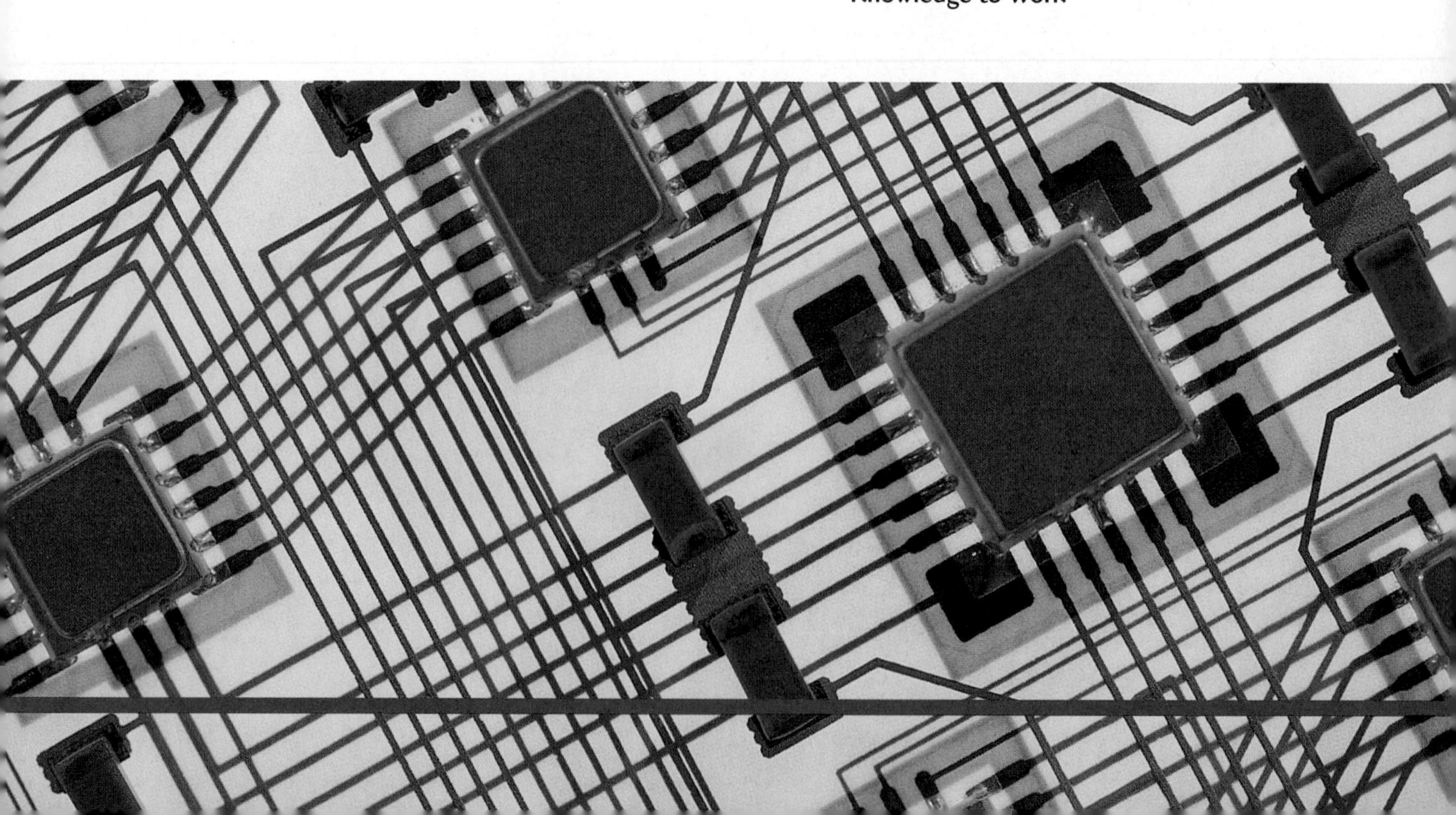

CHAPTER OBJECTIVES

- Identify a sinusoidal waveform and measure its characteristics
- Describe how sine waves are generated
- Determine the voltage and current values of sine waves
- Describe angular relationships of sine waves
- Mathematically analyze a sinusoidal waveform
- Apply the basic circuit laws to ac resistive circuits
- Determine total voltages that have both ac and dc components
- Identify the characteristics of basic nonsinusoidal waveforms
- Use the oscilloscope to measure waveforms

KEY TERMS

- Sine wave
- Alternating current
- Cycle
- Period
- Frequency
- Hertz
- Oscillator
- Instantaneous value
- Peak value
- Peak-to-peak value
- RMS value
- Average value
- Degree
- Radian
- Phase
- Pulse
- Pulse width
- Rise time
- Fall time
- Periodic
- Duty cycle
- Ramp
- Triangular waveform
- Sawtooth waveform
- Harmonics
- Oscilloscope
- Probe

APPLICATION ASSIGNMENT PREVIEW

Imagine, as an electronics technician working for a company that designs and manufactures laboratory instruments, you have been assigned to the testing of a new function generator that produces various types of time-varying voltages with adjustable parameters. The project supervisor has assigned you to measure and record the limits of operation of a prototype model. After studying this chapter, you should be able to complete the application assignment.

WWW. VISIT THE COMPANION WEBSITE

Circuit Simulation Tutorials and Other Chapter Study Tools Are Available at

http://www.prenhall.com/floyd

8–1 THE SINE WAVE

The **sine wave** is the fundamental type of **alternating current** (ac) and alternating voltage. It is also referred to as a sinusoidal wave, or, simply, sinusoid. The electrical service provided by the power companies is in the form of sinusoidal voltage and current. In addition, other types of repetitive waveforms are composites of many individual sine waves called harmonics.

After completing this section, you should be able to

- **Identify a sinusoidal waveform and measure its characteristics**
- Define and determine the period
- Define and determine the frequency
- Relate the period and the frequency

▲ **FIGURE 8–1**

Symbol for a sinusoidal voltage source.

Sine waves, or sinusoidals, are produced by two types of sources: rotating electrical machines (ac generators) or electronic oscillator circuits, which are used in instruments commonly known as electronic signal generators. Figure 8–1 shows the symbol used to represent either source of sinusoidal voltage.

Figure 8–2 is a graph showing the general shape of a sine wave, which can be either an alternating current or an alternating voltage. Voltage (or current) is displayed on the vertical axis and time (t) is displayed on the horizontal axis. Notice how the voltage (or current) varies with time. Starting at zero, the voltage (or current) increases to a positive maximum (peak), returns to zero, and then increases to a negative maximum (peak) before returning again to zero, thus completing one full cycle.

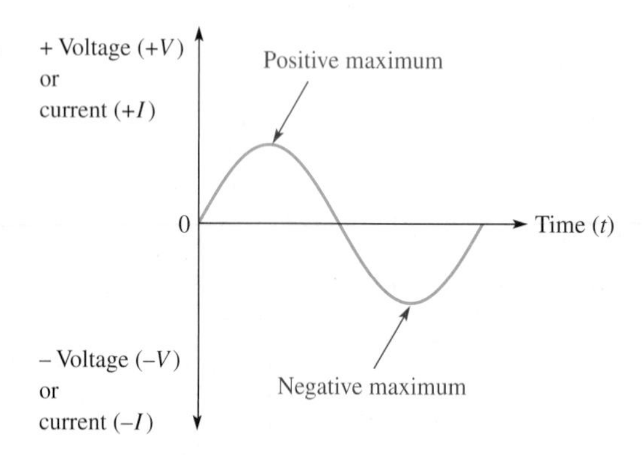

▶ **FIGURE 8–2**

Graph of one cycle of a sine wave.

Polarity of a Sine Wave

As you have seen, a sine wave changes polarity at its zero value; that is, it alternates between positive and negative values. When a sinusoidal voltage source (V_s) is applied to a resistive circuit, as in Figure 8–3, an alternating sinusoidal current results. When the voltage changes polarity, the current correspondingly changes direction as indicated.

During the positive alternation of the source voltage V_s, the current is in the direction shown in Figure 8–3(a). During a negative alternation of the source voltage, the current is in the opposite direction, as shown in Figure 8–3(b). The combined positive and negative alternations make up one **cycle** of a sine wave.

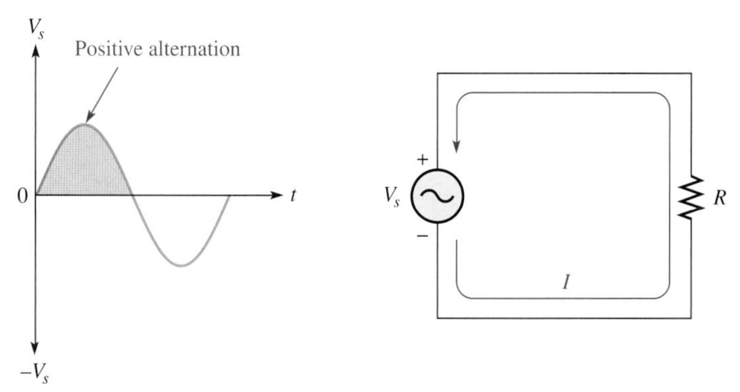

(a) Positive voltage: current direction as shown

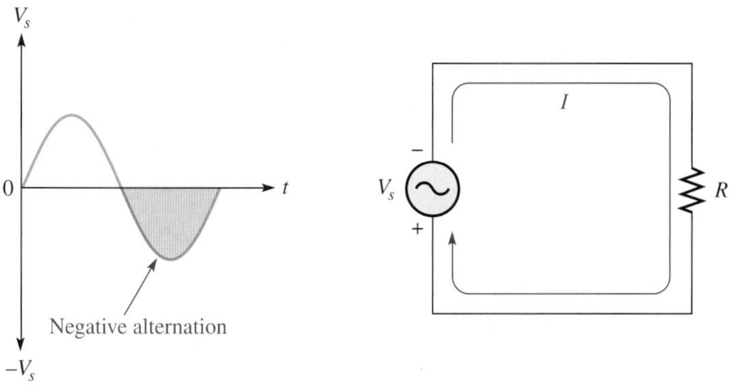

(b) Negative voltage: current reverses direction

FIGURE 8–3

Alternating current and voltage.

Period of a Sine Wave

A sine wave varies with time (t) in a definable manner.

> **The time required for a given sine wave to complete one full cycle is called the period (T).**

Figure 8–4(a) illustrates the period of a sine wave. Typically, a sine wave continues to repeat itself in identical cycles, as shown in Figure 8–4(b). Since all cycles of a repetitive sine wave are the same, the period is always a fixed value for a given sine wave. The period of a sine wave can be measured from a zero crossing to the next corresponding zero crossing, as indicated in Figure 8–4(a). The period can also be measured from any peak in a given cycle to the corresponding peak in the next cycle.

FIGURE 8–4

The period of a given sine wave is the same for each cycle.

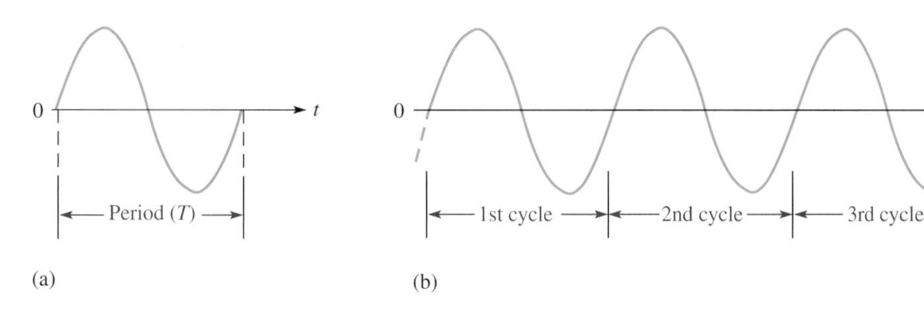

EXAMPLE 8–1

What is the period of the sine wave in Figure 8–5?

▶ FIGURE 8–5

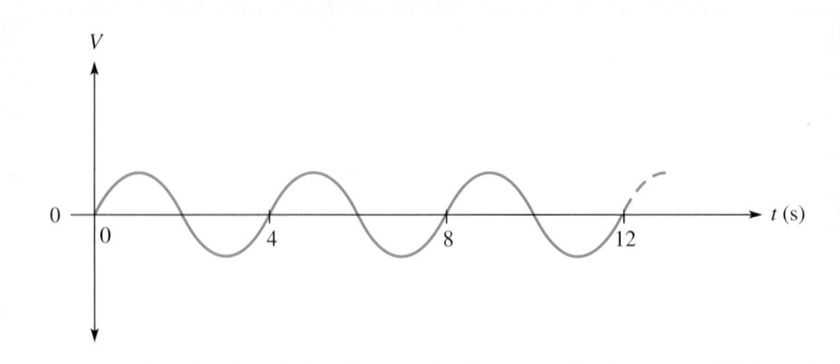

Solution As shown in Figure 8–5, it takes twelve seconds (12 s) to complete three cycles. Therefore, to complete one cycle it takes four seconds (4 s), which is the period.

$$T = \mathbf{4\ s}$$

*Related Problem** What is the period if a given sine wave goes through five cycles in 12 s?

*Answers are at the end of the chapter.

EXAMPLE 8–2

Show three possible ways to measure the period of the sine wave in Figure 8–6. How many cycles are shown?

▶ FIGURE 8–6

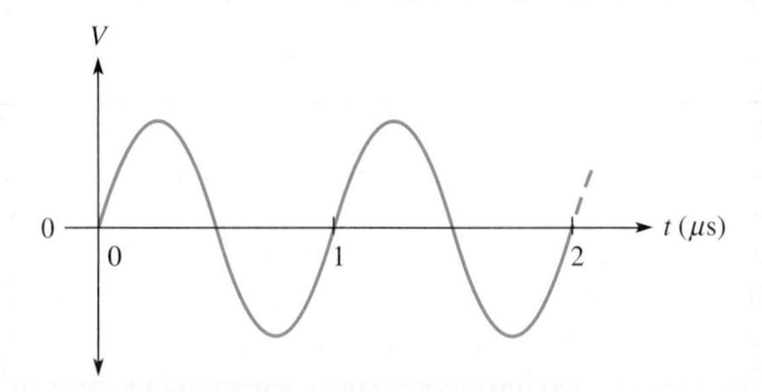

Solution

1. The period can be measured from one zero crossing to the corresponding zero crossing in the next cycle (the slope must be the same at the corresponding zero crossings).
2. The period can be measured from the positive peak in one cycle to the positive peak in the next cycle.
3. The period can be measured from the negative peak in one cycle to the negative peak in the next cycle.

These measurements are indicated in Figure 8–7, where **two cycles** of the sine wave are shown. Keep in mind that you obtain the same value for the period no matter which corresponding peaks or corresponding zero crossings on the waveform you use.

FIGURE 8–7

Measurement of the period of a sine wave.

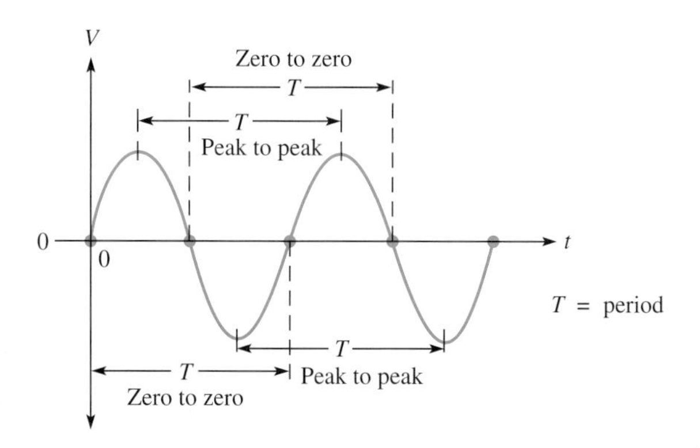

Related Problem If a positive peak occurs at 1 ms and the next positive peak occurs at 2.5 ms, what is the period?

Frequency of a Sine Wave

Frequency is the number of cycles that a sine wave completes in one second.

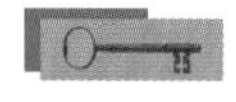

The more cycles completed in one second, the higher the frequency. **Frequency** (f) is measured in units of **hertz**. One hertz (Hz) is equivalent to one cycle per second; for example, 60 Hz is 60 cycles per second. Figure 8–8 shows two sine waves. The sine wave in part (a) completes two full cycles in one second. The one in part (b) completes four cycles in one second. Therefore, the sine wave in part (b) has twice the frequency of the one in part (a).

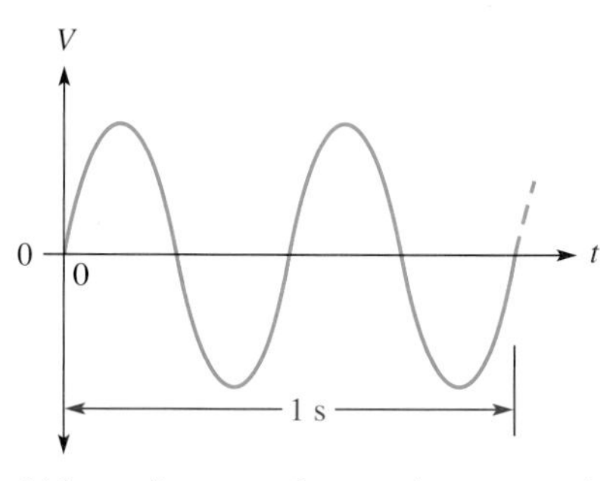

(a) Lower frequency: fewer cycles per second

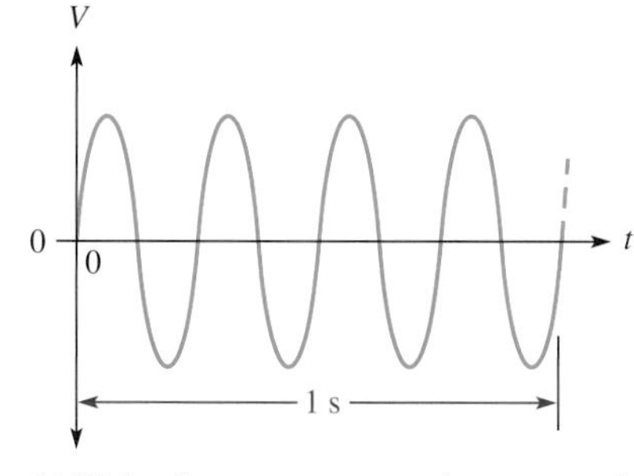

(b) Higher frequency: more cycles per second

FIGURE 8–8

Illustration of frequency.

BIOGRAPHY

Heinrich Rudolf Hertz 1857–1894 Hertz, a German physicist, was the first to broadcast and receive electromagnetic (radio) waves. He produced electromagnetic waves in the laboratory and measured their parameters. Hertz also proved that the nature of the reflection and refraction of electromagnetic waves was the same as that of light. The unit of frequency is named in his honor. (Photo credit: Deutsches Museum, courtesy AIP Emilio Segrè Visual Archives.)

Relationship of Frequency and Period

The relationship between frequency and period is important. The formulas for this relationship are as follows:

$$f = \frac{1}{T}$$ **Equation 8–1**

$$T = \frac{1}{f}$$ **Equation 8–2**

There is a reciprocal relationship between f and T. Knowing one, you can calculate the other with the [2nd] [x^{-1}] keys on your calculator. On some calculators, the reciprocal key is not a secondary function. This inverse relationship indicates that a sine wave with a longer period goes through fewer cycles in one second than one with a shorter period.

EXAMPLE 8–3

Which sine wave in Figure 8–9 has the higher frequency? Determine the period and the frequency of both waveforms.

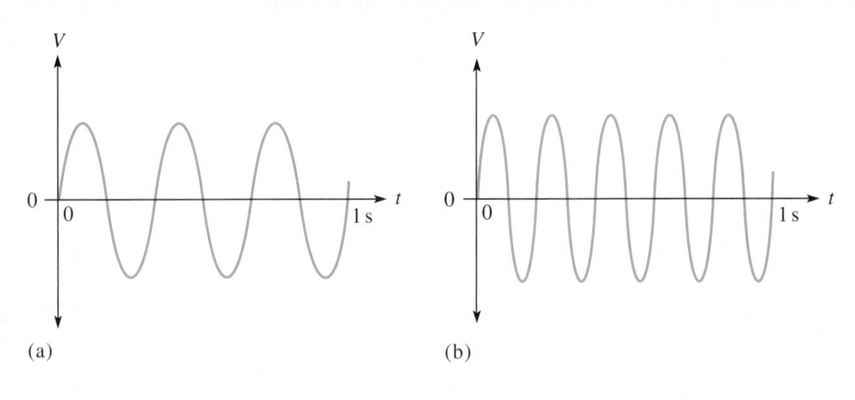

▲ FIGURE 8–9

Solution The sine wave in Figure 8–9(b) has the higher frequency because it completes more cycles in 1 s than does the one in part (a).

In Figure 8–9(a), three cycles take 1 s. Therefore, one cycle takes 0.333 s (one-third second), and this is the period.

$$T = 0.333\text{ s} = \mathbf{333\text{ ms}}$$

The frequency is

$$f = \frac{1}{T} = \frac{1}{333\text{ ms}} = \mathbf{3\text{ Hz}}$$

In Figure 8–9(b), five cycles take 1 s. Therefore, one cycle takes 0.2 s (one-fifth second), and this is the period.

$$T = 0.2\text{ s} = \mathbf{200\text{ ms}}$$

The frequency is

$$f = \frac{1}{T} = \frac{1}{200\text{ ms}} = \mathbf{5\text{ Hz}}$$

Related Problem If the time interval between consecutive negative peaks of a given sine wave is 50 μs, what is the frequency?

EXAMPLE 8–4

The period of a certain sine wave is 10 ms. What is the frequency?

Solution Use Equation 8–1.

$$f = \frac{1}{T} = \frac{1}{10\text{ ms}} = \frac{1}{10 \times 10^{-3}\text{ s}} = \mathbf{100\text{ Hz}}$$

Related Problem A certain sine wave goes through four cycles in 20 ms. What is the frequency?

EXAMPLE 8–5

The frequency of a sine wave is 60 Hz. What is the period?

Solution Use Equation 8–2.

$$T = \frac{1}{f} = \frac{1}{60\text{ Hz}} = \mathbf{16.7\text{ ms}}$$

Related Problem If $f = 1$ kHz, what is T?

SECTION 8–1 REVIEW

Answers are at the end of the chapter.

1. Describe one cycle of a sine wave.
2. At what point does a sine wave change polarity?
3. How many maximum points does a sine wave have during one cycle?
4. How is the period of a sine wave measured?
5. Define *frequency,* and state its unit.
6. Determine f when $T = 5\ \mu s$.
7. Determine T when $f = 120$ Hz.

8–2 SINUSOIDAL VOLTAGE SOURCES

As mentioned in the last section, two basic methods of generating sinusoidal voltages are electromagnetic and electronic. Sine waves are produced electromagnetically by ac generators and electronically by oscillator circuits.

After completing this section, you should be able to

- **Describe how sine waves are generated**
- Discuss the basic operation of an ac generator
- Discuss factors that affect frequency in ac generators
- Discuss factors that affect voltage in ac generators

AC Generator

Figure 8–10 is a cutaway view of the electromechanical ac generator. To illustrate the basic operation, Figure 8–11 shows a greatly simplified ac generator consisting

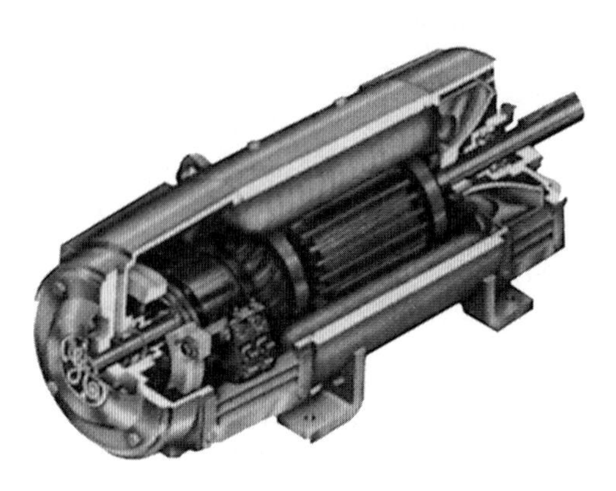

FIGURE 8–10

The electromechanical ac generator, cutaway view (courtesy of General Electric).

▶ FIGURE 8–11

Basic ac generator operation.

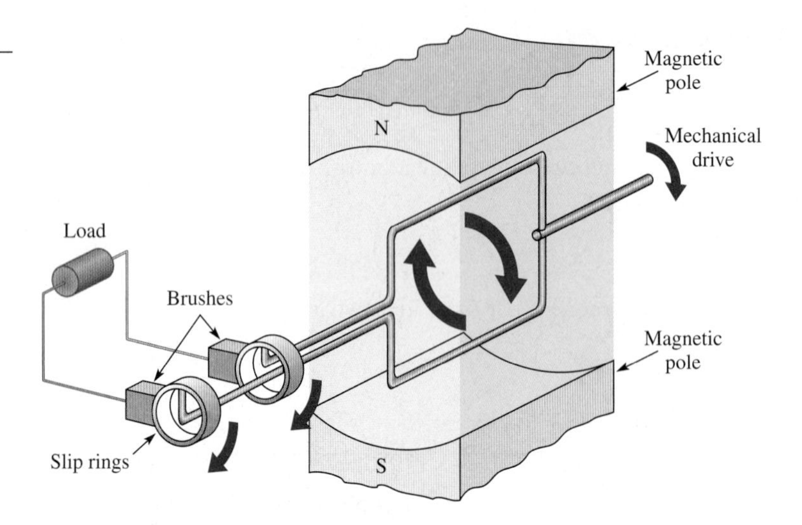

of a single loop of wire in a permanent magnetic field. Notice that each end of the wire is connected to a separate solid conductive ring called a *slip ring*. As the wire loop rotates in the magnetic field, the slip rings also rotate and rub against the brushes that connect the wire to an external load. Compare this generator to the basic dc generator in Figure 7–35, and note the difference in the ring and brush arrangements.

As you learned in Chapter 7, when a conductor moves through a magnetic field, a voltage is induced. Figure 8–12 illustrates how a sinusoidal voltage is pro-

▼ FIGURE 8–12

One revolution of the wire loop generates one cycle of the sinusoidal voltage.

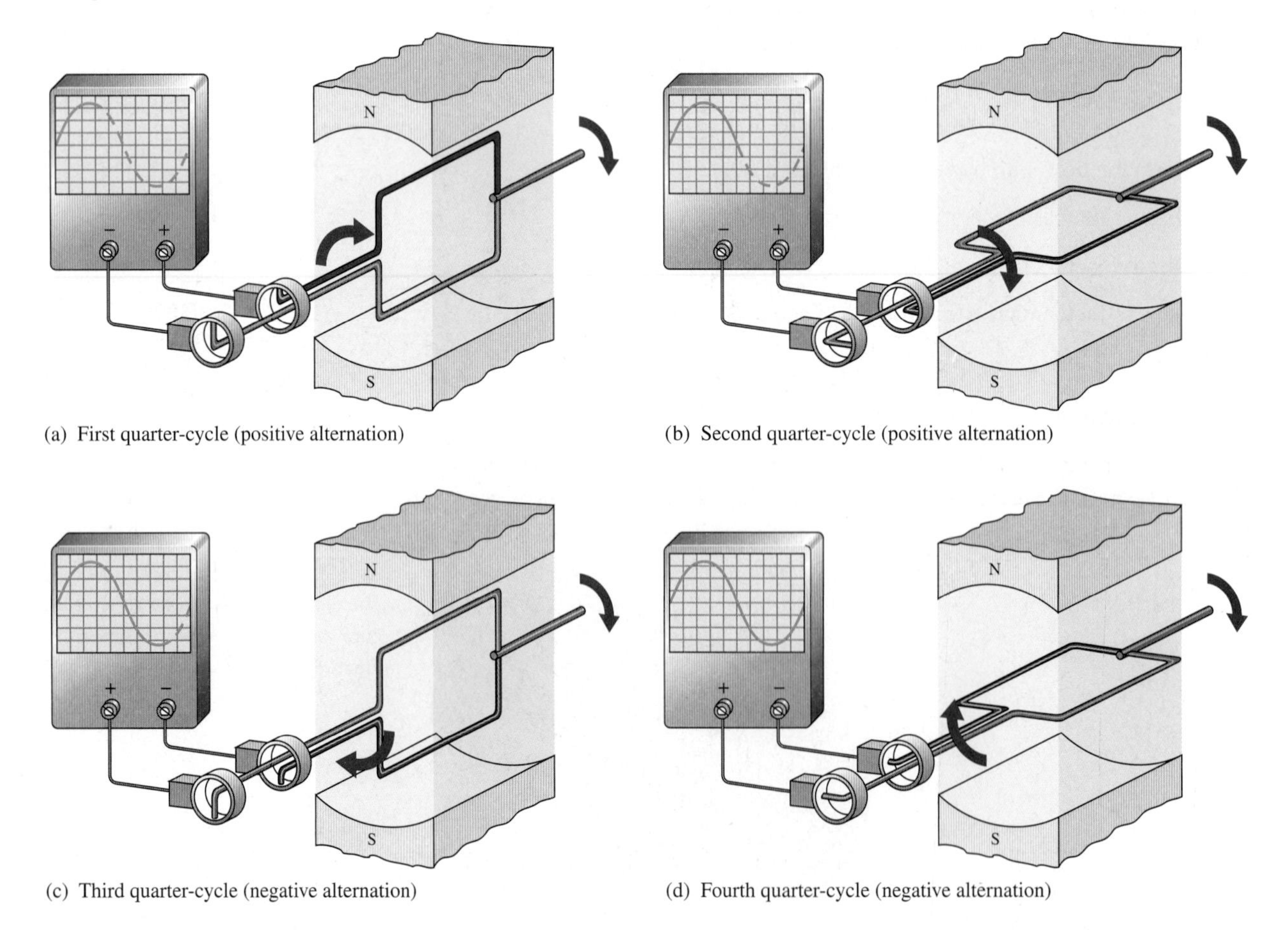

duced by the basic ac generator as the wire loop rotates through the magnetic field. An oscilloscope is used to display the voltage waveform.

To begin, Figure 8–12(a) shows the wire loop rotating through the first quarter of a revolution. It goes from an instantaneous horizontal position, where the induced voltage is zero, to an instantaneous vertical position, where the induced voltage is maximum. At the horizontal position, the loop is instantaneously moving parallel with the flux lines, which exist between the north (N) and south (S) poles of the magnet; thus, no lines are being cut and the voltage is zero. As the loop rotates through the first quarter-cycle, it cuts through the flux lines at an increasing rate until it is instantaneously moving perpendicular to the flux lines at the vertical position and cutting through them at a maximum rate. Thus, the induced voltage increases from zero to a peak during the quarter-cycle. As shown on the display in part (a), this part of the rotation produces the first quarter of the sine wave cycle as the voltage builds up from zero to its positive maximum.

Figure 8–12(b) shows the wire loop completing the first half of a revolution. During this part of the rotation, the voltage decreases from its positive maximum back to zero.

During the second half of the revolution, illustrated in Figure 8–12(c) and 8–12(d), the wire loop is cutting through the magnetic field in the opposite direction, so the voltage produced has a polarity opposite to that produced during the first half of the revolution. After one complete revolution of the loop, one full cycle of the sinusoidal voltage has been produced. As the loop continues to rotate, repetitive cycles of the sine wave are generated.

Frequency You have seen that one revolution of the conductor through the magnetic field in the basic ac generator (also called an *alternator*) produces one cycle of induced sinusoidal voltage. It is obvious that the rate at which the conductor is rotated determines the time for completion of one cycle. For example, if the conductor completes 60 revolutions in one second (rps), the period of the resulting sine wave is 1/60 s, corresponding to a frequency of 60 Hz. The faster the conductor rotates, the higher the resulting frequency of the induced voltage, as illustrated in Figure 8–13.

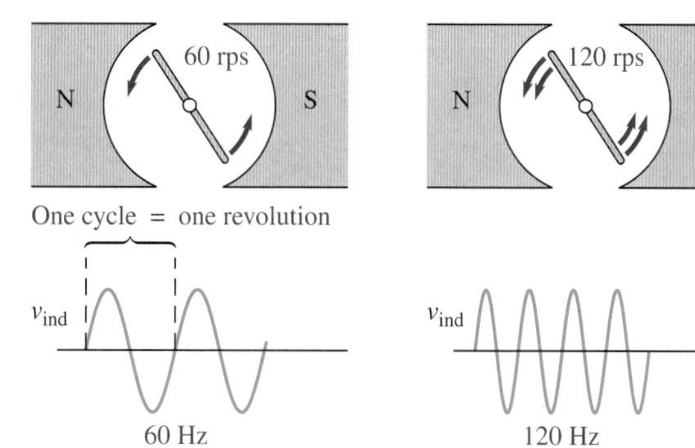

FIGURE 8–13

Frequency is directly proportional to the rate of rotation of the wire loop in an ac generator.

Another way of achieving a higher frequency is to increase the number of magnetic poles. In the previous discussion, two magnetic poles were used to illustrate the ac generator principle. During one revolution, the conductor passes under a north pole and a south pole, thus producing one cycle of a sine wave. When four magnetic poles are used instead of two, as shown in Figure 8–14, one cycle is generated during one-half of a revolution. This doubles the frequency for the same rate of rotation.

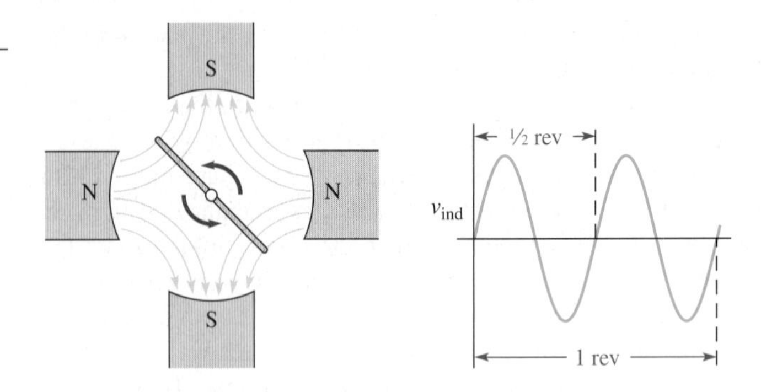

FIGURE 8–14

Four poles achieve a higher frequency than two poles for the same rps.

An expression for frequency in terms of the number of pole pairs and the number of revolutions per second (rps) is

Equation 8–3

$$f = (\text{number of pole pairs})(\text{rps})$$

EXAMPLE 8–6

A four-pole generator has a rotation speed of 100 rps. Determine the frequency of the output voltage.

Solution

$$f = (\text{number of pole pairs})(\text{rps}) = 2(100) = \mathbf{200\ Hz}$$

Related Problem If the frequency of the output of a four-pole generator is 60 Hz, what is the rps?

Voltage Amplitude Recall from Chapter 7 that the amount of voltage induced in a conductor depends on the number of turns (N) and the rate of change with respect to the magnetic field. Therefore, when the speed of rotation of the conductor is increased, not only does the frequency of the induced voltage increase—so also does the amplitude. Since the frequency value normally is fixed, the most practical method of increasing the amount or amplitude of induced voltage is to increase the number of wire loops.

Electronic Signal Generators

The signal generator is an instrument that electronically produces sine waves for use in testing or controlling electronic circuits and systems. There are a variety of signal generators, ranging from special-purpose instruments that produce only one type of waveform in a limited frequency range, to programmable instruments that produce a wide range of frequencies and a variety of waveforms. All signal generators consist basically of an oscillator, which is an electronic circuit that produces sinusoidal voltages or other types of waveforms whose amplitude and frequency can be adjusted. Typical signal generators are shown in Figure 8–15.

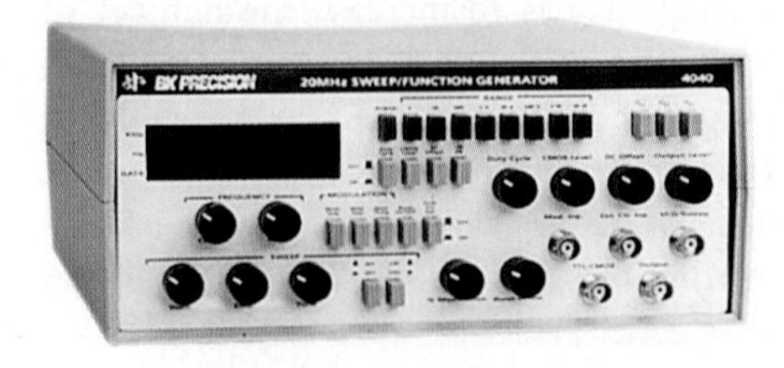

FIGURE 8–15

Typical signal generators. (Photography courtesy of B&K Precision Corp.)

SECTION 8-2 REVIEW

1. List two types of sine wave sources.
2. How are sine waves generated electromagnetically?
3. Upon what is the frequency of a mechanical generator's output voltage dependent?
4. Upon what is the amplitude of a mechanical generator's output voltage dependent?

8-3 VOLTAGE AND CURRENT VALUES OF SINE WAVES

There are five ways to express and measure the value of a sine wave in terms of its voltage or its current magnitude. These are instantaneous, peak, peak-to-peak, rms, and average values.

After completing this section, you should be able to

- **Determine the voltage and current values of sine waves**
- Find the instantaneous value at any point
- Find the peak value
- Find the peak-to-peak value
- Define *rms* and find the rms value
- Explain why the average value of an alternating sine wave is always zero over a complete cycle
- Find the half-cycle average value

Instantaneous Value

Figure 8–16 illustrates that at any point in time on a sine wave, the voltage (or current) has an **instantaneous value**. This instantaneous value is different at different points along the curve. Instantaneous values are positive during the positive

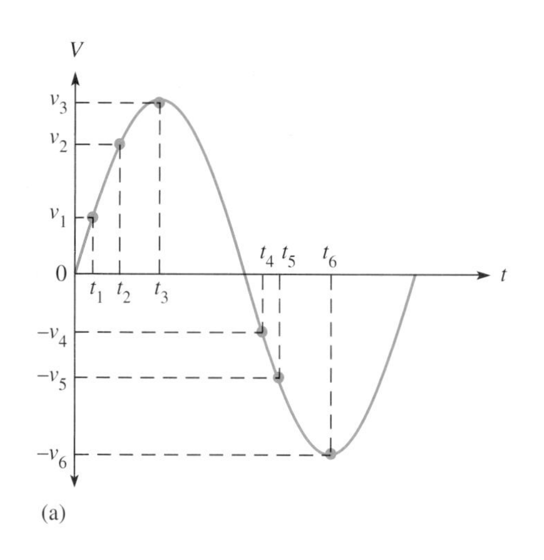

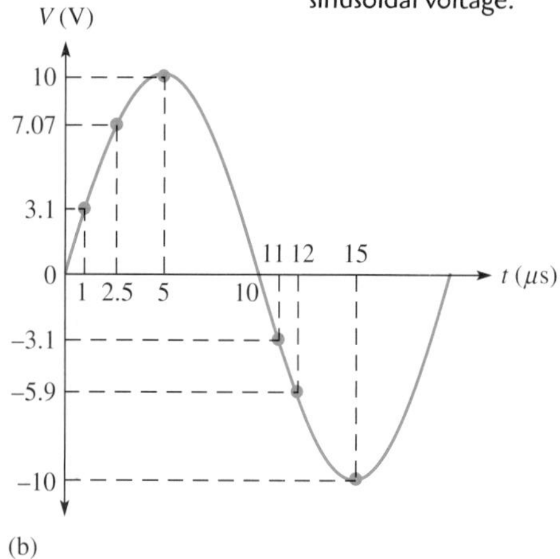

FIGURE 8-16

Example of instantaneous values of a sinusoidal voltage.

alternation and negative during the negative alternation. Instantaneous values of voltage and current are symbolized by lowercase v and i, respectively, as shown in part (a). The curve is shown for voltage only, but it applies equally for current when v's are replaced with i's. An example is shown in part (b), where the instantaneous voltage is 3.1 V at 1 μs, 7.07 V at 2.5 μs, 10 V at 5 μs, 0 V at 10 μs, −3.1 V at 11 μs, and so on.

Peak Value

The peak value of a sine wave is the value of voltage (or current) at the positive or the negative maximum (peaks) with respect to zero. Since the peaks are equal in **magnitude,** a sine wave is characterized by a single peak value, as is illustrated in Figure 8–17. For a given sine wave, the peak value is constant and is represented by V_p or I_p. In the figure, the peak value is 8 V.

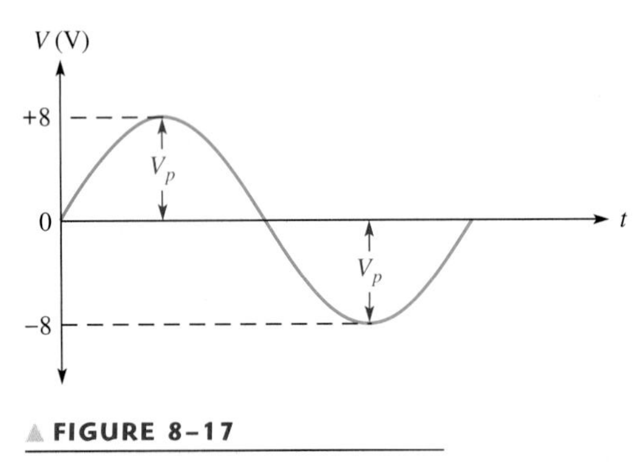

FIGURE 8–17

Peak values.

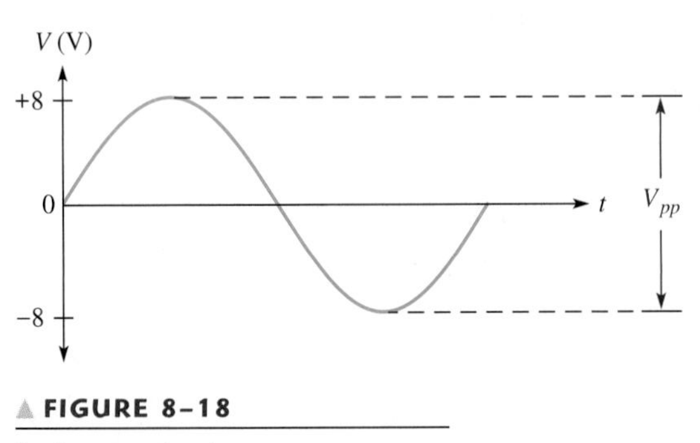

FIGURE 8–18

Peak-to-peak value.

Peak-to-Peak Value

The peak-to-peak value of a sine wave, as illustrated in Figure 8–18, is the voltage (or current) from the positive peak to the negative peak. It is always twice the peak value as expressed in the following equations. Peak-to-peak values are represented by V_{pp} or I_{pp}.

Equation 8–4
$$V_{pp} = 2V_p$$

Equation 8–5
$$I_{pp} = 2I_p$$

In Figure 8–18, the peak-to-peak value is 16 V.

RMS Value

The term *rms* stands for *root mean square.* It refers to the mathematical process by which this value is derived. The rms value is also referred to as the **effective value.** Most ac voltmeters display the rms value of a voltage. The 110 V at your wall outlet is an rms value.

The rms value of a sinusoidal voltage is actually a measure of the heating effect of the sine wave. For example, when a resistor is connected across an ac (sinusoidal) voltage source, as shown in Figure 8–19(a), a certain amount of heat is generated by the power in the resistor. Figure 8–19(b) shows the same resistor connected across a dc voltage source. The value of the ac voltage can be adjusted

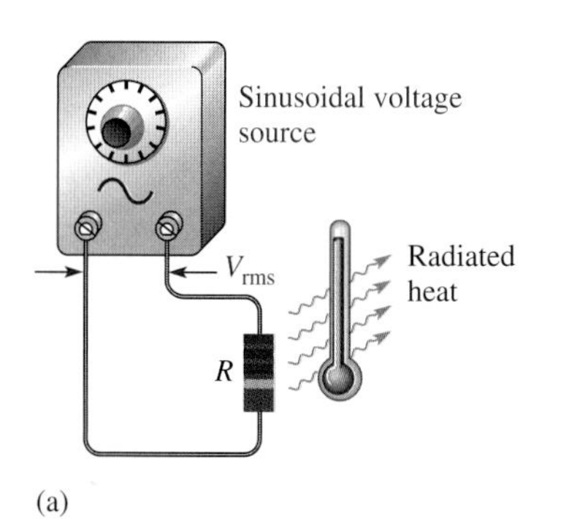

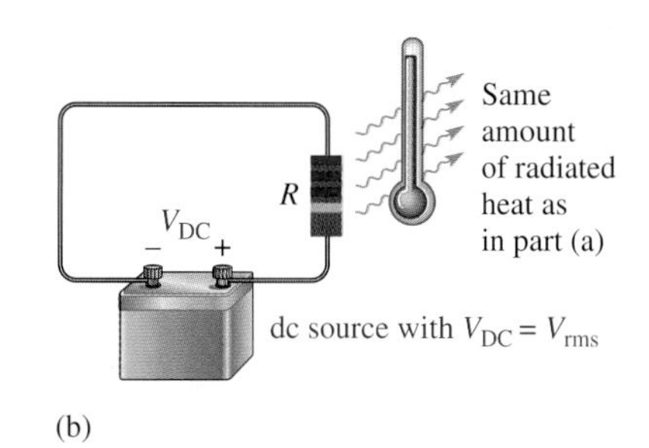

FIGURE 8–19

When the same amount of heat is being produced in both setups, the sinusoidal voltage has an rms value equal to the dc voltage.

so that the resistor gives off the same amount of heat as it does when connected to the dc source.

The rms value of a sinusoidal voltage is equal to the dc voltage that produces the same amount of heat in a resistance as does the sinusoidal voltage.

The peak value of a sine wave can be converted to the corresponding rms value using the following relationships for either voltage or current:

$$V_{rms} = 0.707V_p \quad \text{Equation 8–6}$$

$$I_{rms} = 0.707I_p \quad \text{Equation 8–7}$$

Using these formulas, you can also determine the peak value knowing the rms value.

$$V_p = \frac{V_{rms}}{0.707}$$

$$V_p = 1.414V_{rms} \quad \text{Equation 8–8}$$

Similarly,

$$I_p = 1.414I_{rms} \quad \text{Equation 8–9}$$

To find the peak-to-peak value, simply double the peak value.

$$V_{pp} = 2.828V_{rms} \quad \text{Equation 8–10}$$

and

$$I_{pp} = 2.828I_{rms} \quad \text{Equation 8–11}$$

Average Value

The average value of a sine wave when taken over one complete cycle is always zero because the positive values (above the zero crossing) offset the negative values (below the zero crossing).

To be useful for comparison purposes and in determining the average value of a rectified voltage such as found in power supplies, the **average value** of a sine wave is defined over a half-cycle rather than over a full cycle. The average value is expressed in terms of the peak value as follows for both voltage and current sine waves:

$$V_{avg} = 0.637V_p \quad \text{Equation 8–12}$$

$$I_{avg} = 0.637I_p \quad \text{Equation 8–13}$$

FIGURE 8–20

Half-cycle average value.

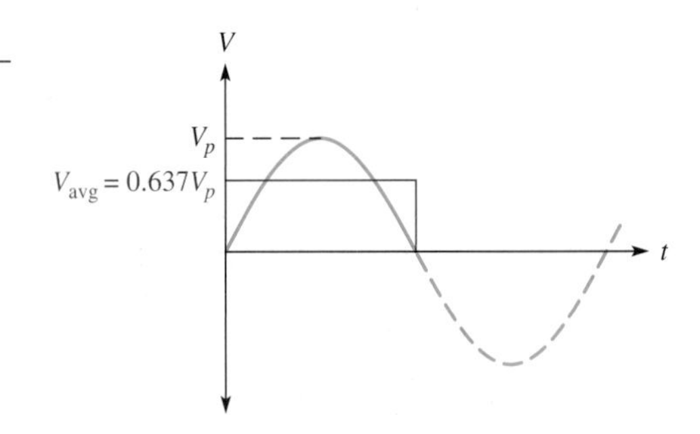

The half-cycle average value of a voltage sine wave is illustrated in Figure 8–20.

EXAMPLE 8–7

Determine V_p, V_{pp}, V_{rms}, and the half-cycle V_{avg} for the sine wave in Figure 8–21.

FIGURE 8–21

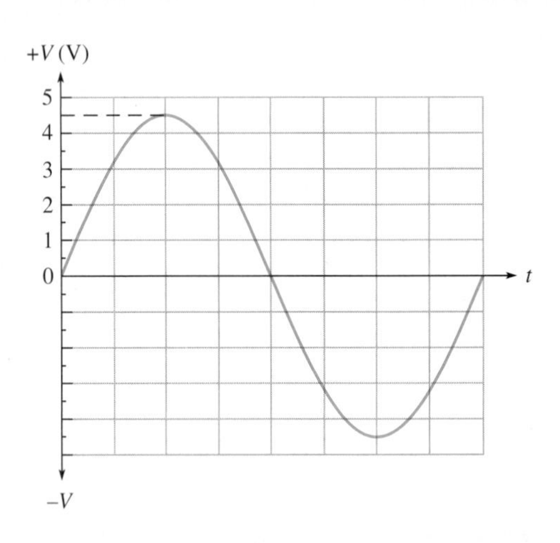

Solution As taken directly from the graph, V_p = **4.5 V**. From this value, calculate the other values.

$$V_{pp} = 2V_p = 2(4.5\text{ V}) = \mathbf{9\ V}$$

$$V_{rms} = 0.707V_p = 0.707(4.5\text{ V}) = \mathbf{3.18\ V}$$

$$V_{avg} = 0.637V_p = 0.637(4.5\text{ V}) = \mathbf{2.87\ V}$$

Related Problem If $V_p = 25$ V, determine V_{pp}, V_{rms}, and V_{avg} for a sine wave.

SECTION 8–3 REVIEW

1. Determine V_{pp} in each case when
   (a) $V_p = 1$ V (b) $V_{rms} = 1.414$ V (c) $V_{avg} = 3$ V
2. Determine V_{rms} in each case when
   (a) $V_p = 2.5$ V (b) $V_{pp} = 10$ V (c) $V_{avg} = 1.5$ V
3. Determine the half-cycle V_{avg} in each case when
   (a) $V_p = 10$ V (b) $V_{rms} = 2.3$ V (c) $V_{pp} = 60$ V

8–4 ANGULAR MEASUREMENT OF A SINE WAVE

As you have seen, sine waves can be measured along the horizontal axis on a time basis; however, since the time for completion of one full cycle or any portion of a cycle is frequency-dependent, it is often useful to specify points on the sine wave in terms of an angular measurement expressed in degrees or radians. Angular measurement is independent of frequency.

After completing this section, you should be able to

- **Describe angular relationships of sine waves**
- Show how to measure a sine wave in terms of angles
- Define *radian*
- Convert radians to degrees
- Determine the phase of a sine wave

As previously discussed, sinusoidal voltage can be produced electromagnetically by rotating electromechanical machines. As the rotor of the ac generator goes through a full 360° of rotation, the resulting output is one full cycle of a sine wave. Thus, the angular measurement of a sine wave can be related to the angular rotation of a generator, as shown in Figure 8–22.

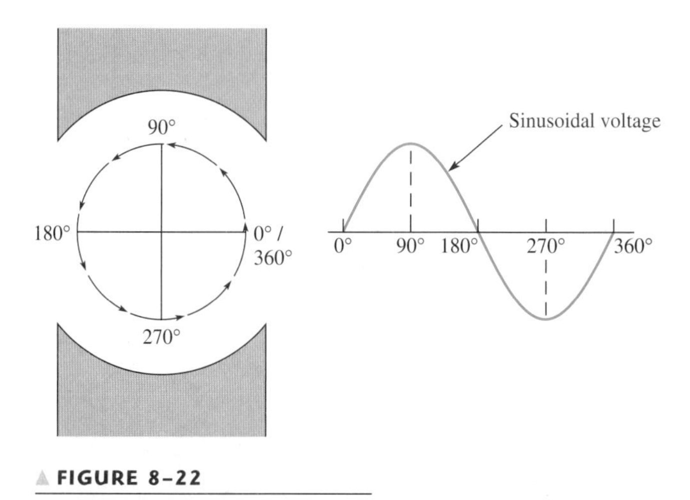

FIGURE 8–22

Relationship of a sine wave to the rotational motion in an ac generator.

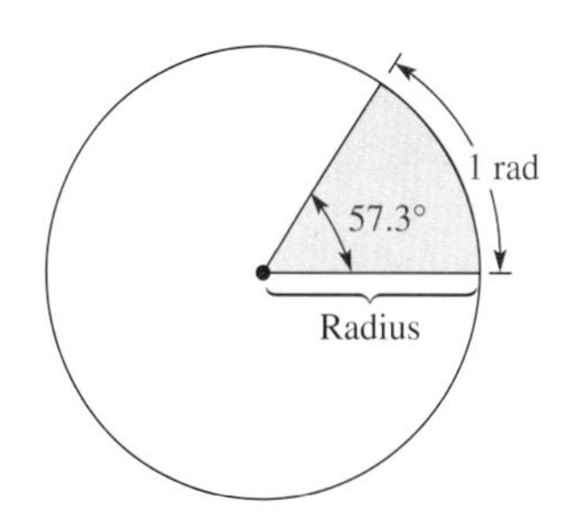

FIGURE 8–23

Angular measurement showing relationship of the radian to degrees.

Angular Measurement

A **degree** is an angular measurement corresponding to 1/360 of a circle or a complete revolution. A **radian** (rad) is the angle formed when the distance along the circumference of a circle is equal to the radius of the circle. One radian is equivalent to 57.3°, as illustrated in Figure 8–23. In a 360° revolution, there are 2π radians.

The Greek letter π (pi) represents the ratio of the circumference of any circle to its diameter and has a constant value of approximately 3.1416.

Scientific calculators have a π key so that the actual numerical value does not have to be entered.

Table 8–1 lists several values of degrees and the corresponding radian values. These angular measurements are illustrated in Figure 8–24.

▼ TABLE 8–1

DEGREES (°)	RADIANS (RAD)
0	0
45	$\pi/4$
90	$\pi/2$
135	$3\pi/4$
180	π
225	$5\pi/4$
270	$3\pi/2$
315	$7\pi/4$
360	2π

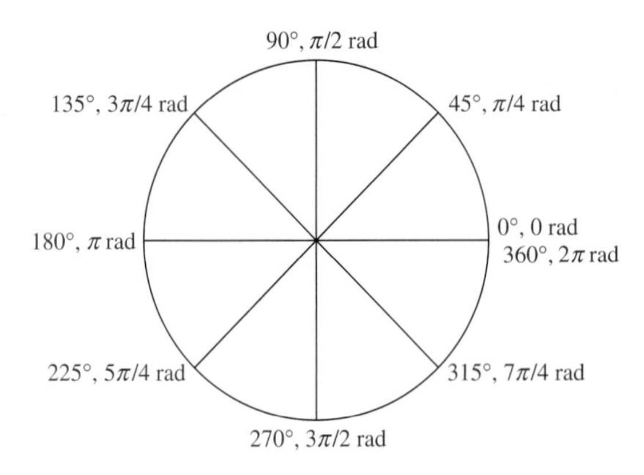

▲ FIGURE 8–24

Angular measurements starting at 0° and going counterclockwise.

Radian/Degree Conversion

Degrees can be converted to radians using Equation 8–14.

Equation 8–14

$$\text{rad} = \left(\frac{\pi\ \text{rad}}{180^\circ}\right) \times \text{degrees}$$

Similarly, radians can be converted to degrees with Equation 8–15.

Equation 8–15

$$\text{degrees} = \left(\frac{180^\circ}{\pi\ \text{rad}}\right) \times \text{rad}$$

EXAMPLE 8–8

(a) Convert 60° to radians. **(b)** Convert $\pi/6$ radian to degrees.

Solution

(a) $\text{Rad} = \left(\frac{\pi\ \text{rad}}{180^\circ}\right)60^\circ = \mathbf{\frac{\pi}{3}\ rad}$

(b) $\text{Degrees} = \left(\frac{180^\circ}{\pi\ \text{rad}}\right)\left(\frac{\pi}{6}\ \text{rad}\right) = \mathbf{30^\circ}$

Related Problem **(a)** Convert 15° to radians. **(b)** Convert 2π radians to degrees.

Sine Wave Angles

The angular measurement of a sine wave is based on 360° or 2π rad for a complete cycle. A half-cycle is 180° or π rad; a quarter-cycle is 90° or $\pi/2$ rad; and so on. Figure 8–25(a) shows angles in degrees over a full cycle of a sine wave; part (b) shows the same points in radians.

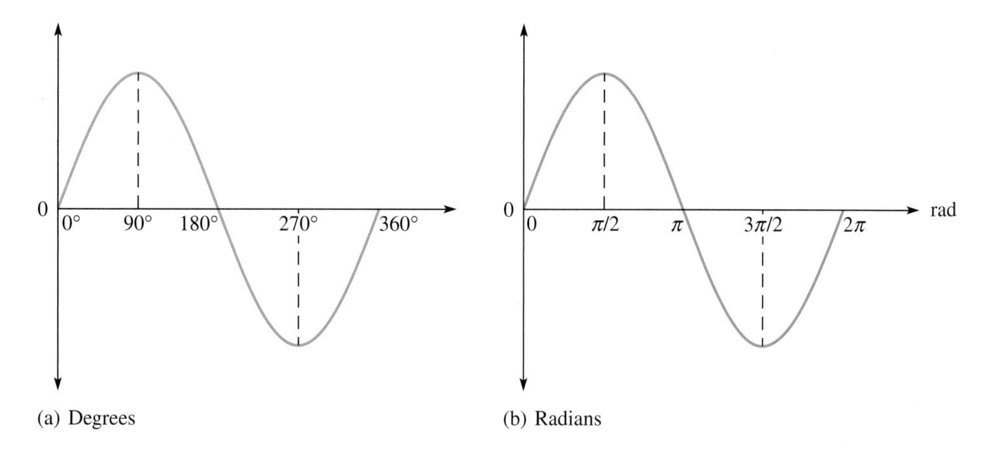

▲ FIGURE 8–25

Sine wave angles.

Phase of a Sine Wave

The phase of a sine wave is an angular measurement that specifies the position of that sine wave relative to a reference. Figure 8–26 shows one cycle of a sine wave to be used as the reference. Note that the first positive-going crossing of the horizontal axis (zero crossing) is at 0° (0 rad), and the positive peak is at 90° (π/2 rad). The negative-going zero crossing is at 180° (π rad), and the negative peak is at 270° (3π/2 rad). The cycle is completed at 360° (2π rad). When the sine wave is shifted left or right with respect to this reference, there is a phase shift.

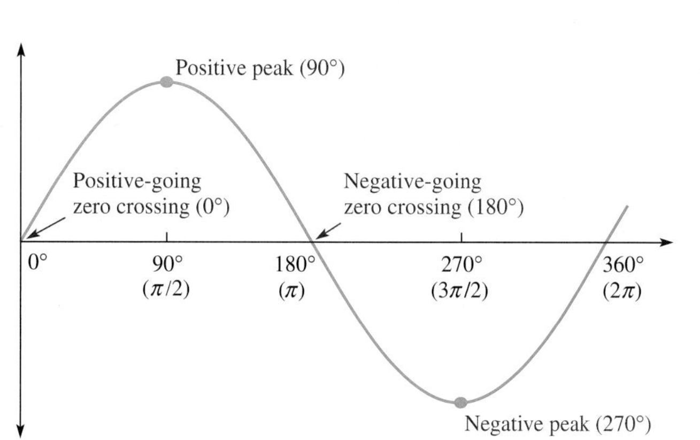

◀ FIGURE 8–26

Phase reference.

Figure 8–27 illustrates phase shifts of a sine wave. In part (a), sine wave *B* is shifted to the right by 90° (π/2 rad). Thus, there is a phase angle of 90° between sine wave *A* and sine wave *B*. In terms of time, the positive peak of sine wave *B* occurs later than the positive peak of sine wave *A* because time increases to the right along the horizontal axis. In this case, sine wave *B* is said to **lag** sine wave *A* by 90° or π/2 radians. In other words, sine wave *A* leads sine wave *B* by 90°.

▼ FIGURE 8–27

Illustration of a phase shift.

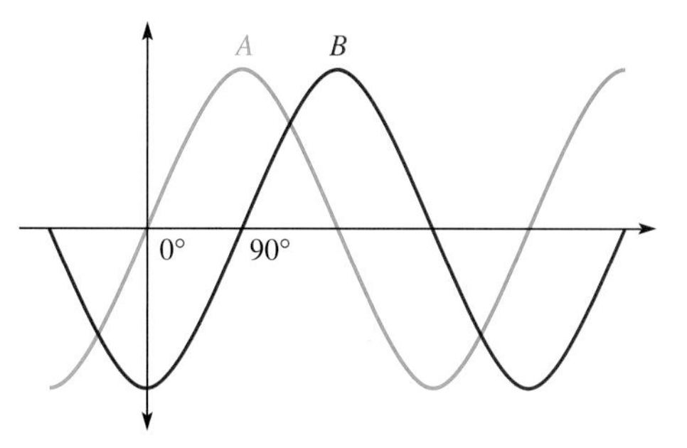

(a) *A* leads *B* by 90°, or *B* lags *A* by 90°.

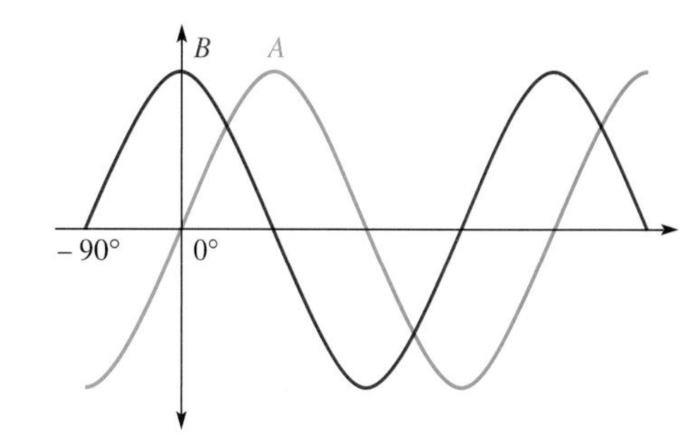

(b) *B* leads *A* by 90°, or *A* lags *B* by 90°.

In Figure 8–27(b), sine wave B is shown shifted left by 90°. Thus, again there is a phase angle of 90° between sine wave A and sine wave B. In this case, the positive peak of sine wave B occurs earlier in time than that of sine wave A; therefore, sine wave B is said to **lead** sine wave A by 90°. In both cases there is a 90° phase angle between the two waveforms.

EXAMPLE 8–9

What are the phase angles between the two sine waves in Figure 8–28(a) and 8–28(b)?

FIGURE 8–28

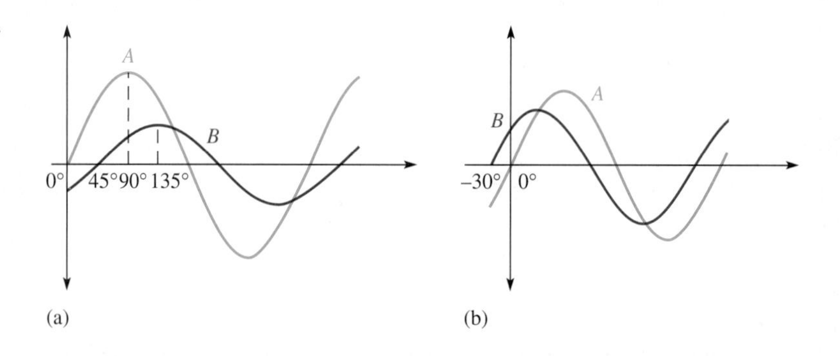

Solution In Figure 8–28(a), the zero crossing of sine wave A is at 0°, and the corresponding zero crossing of sine wave B is at 45°. There is a **45°** phase angle between the two waveforms with **sine wave *A* leading.**

In Figure 8–28(b) the zero crossing of sine wave B is at −30°, and the corresponding zero crossing of sine wave A is at 0°. There is a **30°** phase angle between the two waveforms with **sine wave *B* leading.**

Related Problem If the positive-going zero crossing of one sine wave is at 15° relative to the 0° reference and that of the second sine wave is at 23° relative to the 0° reference, what is the phase angle between the sine waves?

SECTION 8–4 REVIEW

1. When the positive-going zero crossing of a sine wave occurs at 0°, at what angle does each of the following points occur?
   (a) positive peak (b) negative-going zero crossing
   (c) negative peak (d) end of first complete cycle
2. A half cycle is completed in ________ degrees, or ________ radians.
3. A full cycle is completed in ________ degrees, or ________ radians.
4. Determine the phase angle between the sine waves B and C in Figure 8–29.

FIGURE 8–29

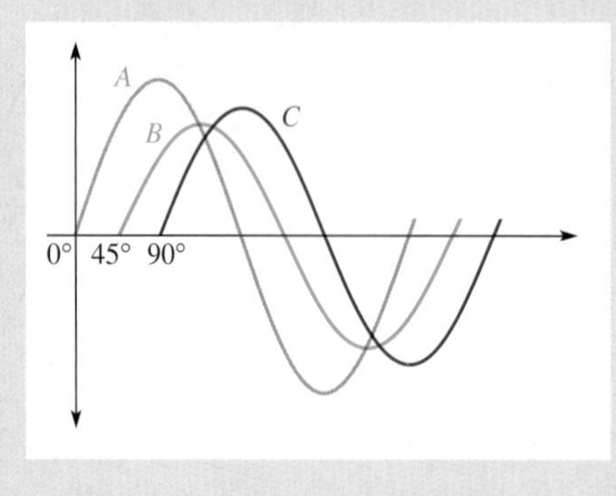

8–5 THE SINE WAVE FORMULA

A sine wave can be graphically represented by voltage or current values on the vertical axis and by angular measurement (degrees or radians) along the horizontal axis. This graph can be expressed mathematically, as you will see.

After completing this section, you should be able to

- **Mathematically analyze a sinusoidal waveform**
- State the sine wave formula
- Find instantaneous values using the sine wave formula

A generalized graph of one cycle of a sine wave is shown in Figure 8–30. The sine wave **amplitude,** A, is the maximum value of the voltage or current on the vertical axis; angular values run along the horizontal axis. The variable y is an instantaneous value representing either voltage or current at a given angle, θ. The symbol θ is the Greek letter *theta.*

All electrical sine waves follow a specific mathematical formula. The general expression for the sine wave curve in Figure 8–30 is

$$y = A \sin \theta$$

Equation 8–16

This formula states that any point on the sine wave, represented by an instantaneous value y, is equal to the maximum value A times the sine (sin) of the angle θ at that point. For example, a certain voltage sine wave has a peak value of 10 V. The instantaneous voltage at a point 60° along the horizontal axis can be calculated as follows, where $y = v$ and $A = V_p$:

$$v = V_p \sin \theta = (10\text{ V})\sin 60° = (10\text{ V})0.866 = 8.66\text{ V}$$

Figure 8–31 shows this particular instantaneous value on the curve. You can find the sine of any angle on your calculator by first entering the value of the angle and then pressing the SIN key. Use the Mode screen to verify that your calculator is in the degree mode.

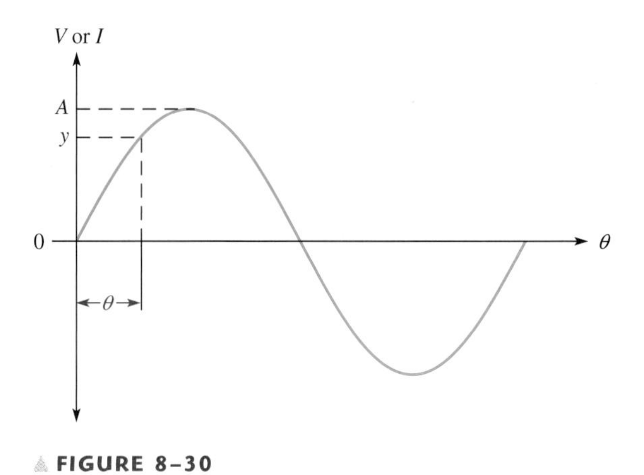

FIGURE 8–30

One cycle of a generic sine wave showing amplitude and phase.

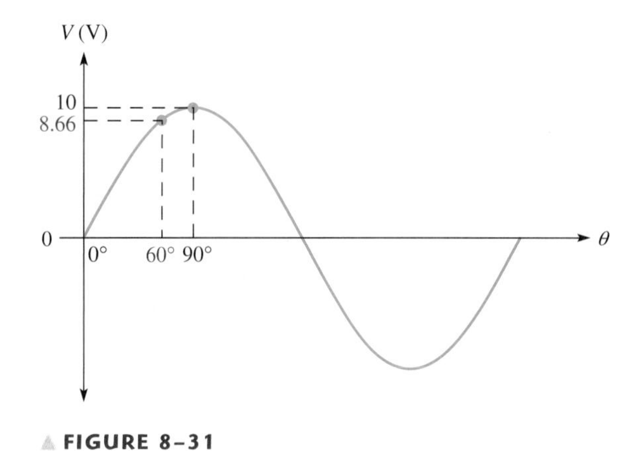

FIGURE 8–31

Illustration of the instantaneous value of a voltage sine wave at $\theta = 60°$.

Derivation of the Sine Wave Formula

As you move along the horizontal axis of a sine wave, the angle increases and the magnitude (height along the y axis) varies. At any given instant, the magnitude of a sine wave can be described by the values of the phase angle and the amplitude (maximum height) and can, therefore, be represented as a **phasor** quantity. *A phasor quantity is one that has both magnitude and direction (phase angle).* A phasor is represented graphically as an arrow that rotates around a fixed point. The length of a sine wave phasor is the peak value (amplitude), and its position as it rotates is the phase angle. One full cycle of a sine wave can be viewed as the rotation of a phasor through 360°.

Figure 8–32 illustrates a phasor rotating counterclockwise through a complete revolution of 360°. If the tip of the phasor is projected over to a graph with the phase angles running along the horizontal axis, a sine wave is "traced out," as shown in the figure. At each angular position of the phasor, there is a corresponding magnitude value. As you can see, at 90° and at 270°, the amplitude of the sine wave is maximum and equal to the length of the phasor. At 0° and at 180°, the sine wave is equal to zero because the phasor lies horizontally at these points.

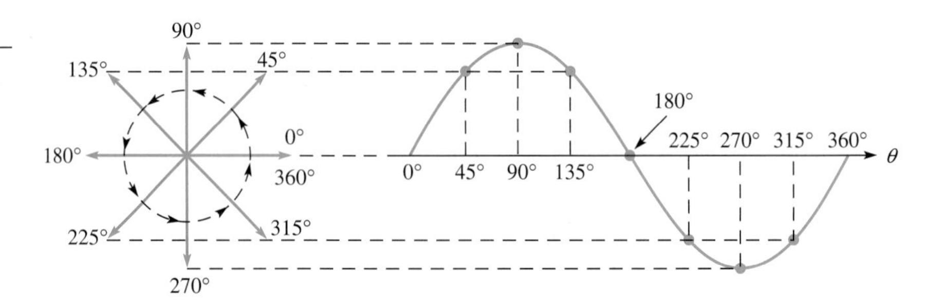

FIGURE 8–32

Sine wave represented by a rotating phasor.

Let's examine a phasor representation at one specific angle. Figure 8–33 shows the voltage phasor at an angular position of 45° and the corresponding point on the sine wave. The instantaneous value, v, of the sine wave at this point is related to both the position (angle) and the length (amplitude) of the phasor. The vertical distance from the phasor tip down to the horizontal axis represents the instantaneous value of the sine wave at that point.

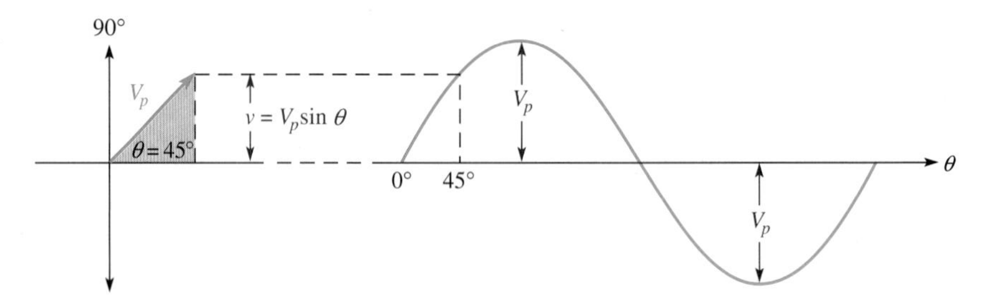

FIGURE 8–33

Right triangle derivation of sine wave formula, $v = V_p \sin \theta$.

Notice that when a vertical line is drawn from the phasor tip down to the horizontal axis, a **right triangle** is formed, as shown in Figure 8–33. The length of the phasor is the **hypotenuse** of the triangle, and the vertical projection is the opposite side. From trigonometry, *the opposite side of a right triangle is equal to the hypotenuse times the sine of the angle θ*. In this case, the length of the phasor is the peak value of the voltage sine wave, V_p. Thus, the opposite side of the triangle, which is the instantaneous value, can be expressed as

Equation 8–17

$$v = V_p \sin \theta$$

This formula also applies to a current sine wave.

$$i = I_p \sin \theta$$

Equation 8–18

Expressions for Phase-Shifted Sine Waves

When a sine wave is shifted to the right of the reference (lagging) by a certain angle, ϕ, as illustrated in Figure 8–34(a), the general expression is

$$y = A \sin(\theta - \phi)$$

Equation 8–19

where y represents instantaneous voltage or current, and A represents the peak value (amplitude). When a sine wave is shifted to the left of the reference (leading) by a certain angle, ϕ, as shown in Figure 8–34(b), the general expression is

$$y = A \sin(\theta + \phi)$$

Equation 8–20

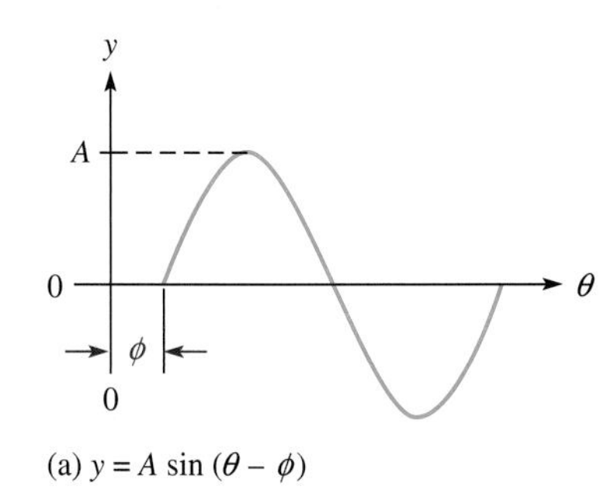

(a) $y = A \sin(\theta - \phi)$

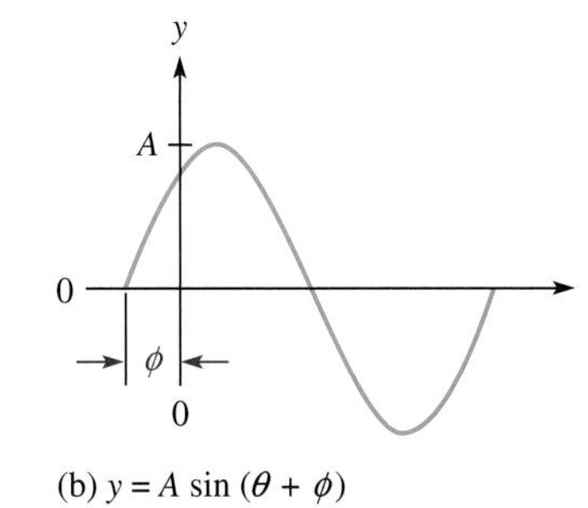

(b) $y = A \sin(\theta + \phi)$

FIGURE 8–34

Shifted sine waves.

EXAMPLE 8–10

Determine the instantaneous value at the 90° reference point on the horizontal axis for each voltage sine wave in Figure 8–35.

FIGURE 8–35

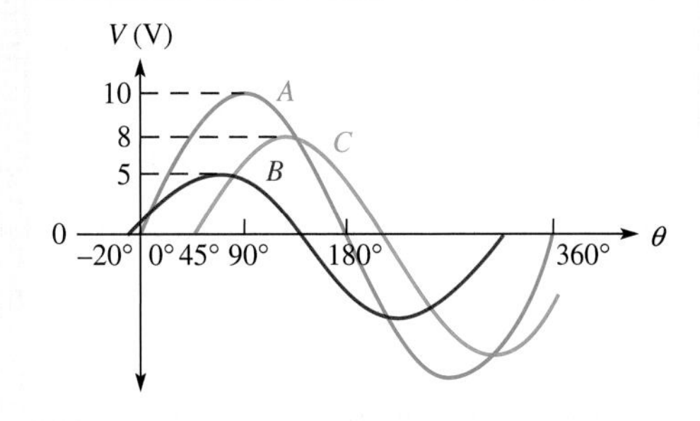

Solution Sine wave A is the reference. Sine wave B is shifted left 20° with respect to A, so B leads. Sine wave C is shifted right 45° with respect to A, so C lags.

$$v_A = V_p \sin \theta = (10 \text{ V})\sin 90^\circ = \mathbf{10 \text{ V}}$$

$$v_B = V_p \sin(\theta + \phi_B) = (5 \text{ V})\sin(90^\circ + 20^\circ) = (5 \text{ V})\sin 110^\circ = \mathbf{4.70 \text{ V}}$$

$$v_C = V_p \sin(\theta - \phi_C) = (8 \text{ V})\sin(90^\circ - 45^\circ) = (8 \text{ V})\sin 45^\circ = \mathbf{5.66 \text{ V}}$$

The calculator solutions for v_A, v_B, and v_C are

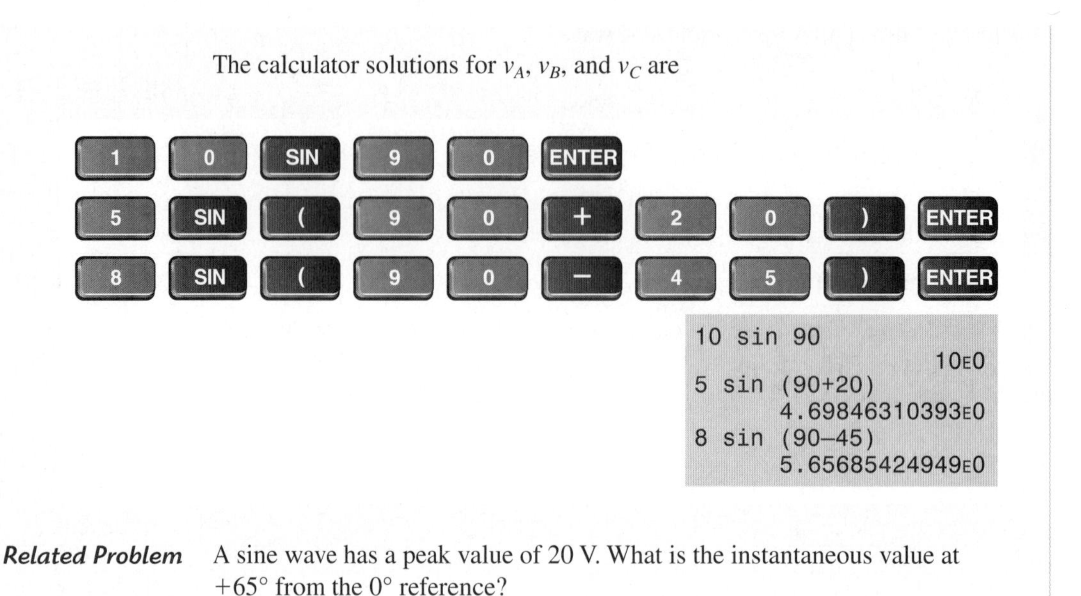

Related Problem A sine wave has a peak value of 20 V. What is the instantaneous value at +65° from the 0° reference?

SECTION 8–5 REVIEW

1. Determine the sine of the following angles:
   (a) 30° (b) 60° (c) 90°
2. Calculate the instantaneous value at 120° for the sine wave in Figure 8–31.
3. Determine the instantaneous value at the 45° point on the reference axis of a sine wave shifted 10° to the left of the zero reference (V_p = 10 V).

8–6 OHM'S LAW AND KIRCHHOFF'S LAWS IN AC CIRCUITS

When time-varying ac voltages such as a sinusoidal voltage are applied to a circuit, the circuit laws that you studied earlier still apply. Ohm's law and Kirchhoff's laws apply to ac circuits in the same way that they apply to dc circuits.

After completing this section, you should be able to

- **Apply the basic circuit laws to ac resistive circuits**
- Apply Ohm's law to resistive circuits with ac sources
- Apply Kirchhoff's voltage law and current law to resistive circuits with ac sources

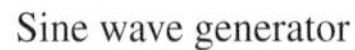
Sine wave generator

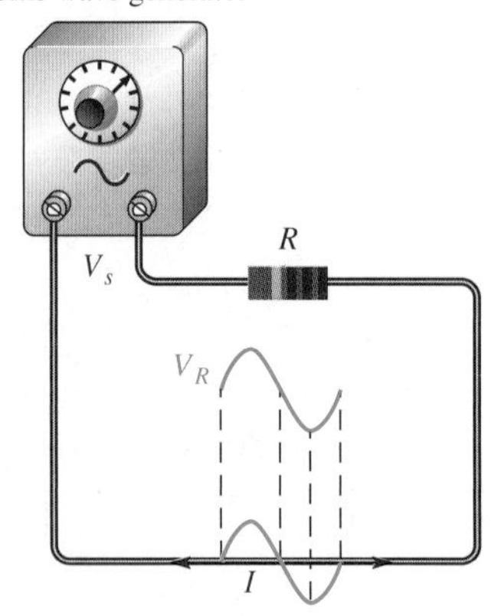

▲ **FIGURE 8–36**

A sinusoidal voltage produces a sinusoidal current.

If a sinusoidal voltage is applied across a resistor, as shown in Figure 8–36, there is a sinusoidal current. The current is zero when the voltage is zero and is maximum when the voltage is maximum. When the voltage changes polarity, the current reverses direction. As a result, the voltage and current are said to be in phase with each other.

When using Ohm's law in ac circuits, remember that both the voltage and the current must be expressed consistently, that is, both as peak values, both as rms values, both as average values, and so on.

EXAMPLE 8–11

Determine the rms voltage across each resistor and the rms current in Figure 8–37. The source voltage is given as an rms value.

▶ **FIGURE 8–37**

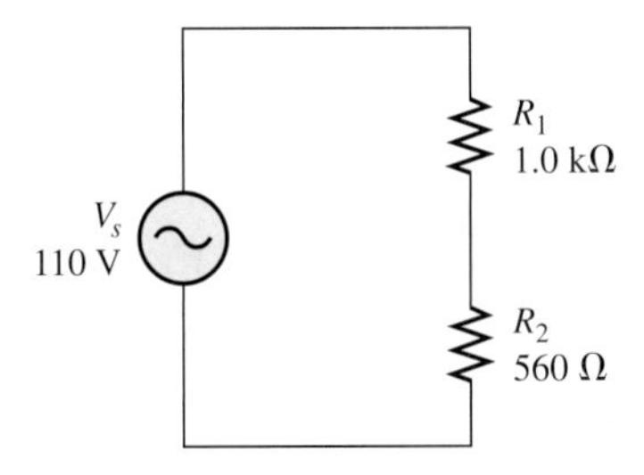

Solution The total resistance of the circuit is

$$R_T = R_1 + R_2 = 1.0\text{ k}\Omega + 560\ \Omega = 1.56\text{ k}\Omega$$

Use Ohm's law to find the rms current.

$$I_{rms} = \frac{V_{s(rms)}}{R_T} = \frac{110\text{ V}}{1.56\text{ k}\Omega} = \mathbf{70.5\text{ mA}}$$

The rms voltage drop across each resistor is

$$V_{1(rms)} = I_{rms}R_1 = (70.5\text{ mA})(1.0\text{ k}\Omega) = \mathbf{70.5\text{ V}}$$

$$V_{2(rms)} = I_{rms}R_2 = (70.5\text{ mA})(560\text{ k}\Omega) = \mathbf{39.5\text{ V}}$$

Related Problem Repeat this example for a source voltage of 10 V peak.

Open file E08-11 on your EWB/CircuitMaker CD-ROM. Measure the rms voltage across each resistor and compare to the calculated values. Change the source voltage to a peak value of 10 V, measure each resistor voltage, and compare to your calculated values.

Kirchhoff's voltage and current laws apply to ac circuits as well as to dc circuits. Figure 8–38 illustrates Kirchhoff's voltage law in a resistive circuit that has a sinusoidal voltage source. As you can see, the source voltage is the sum of all the voltage drops across the resistors, just as in a dc circuit.

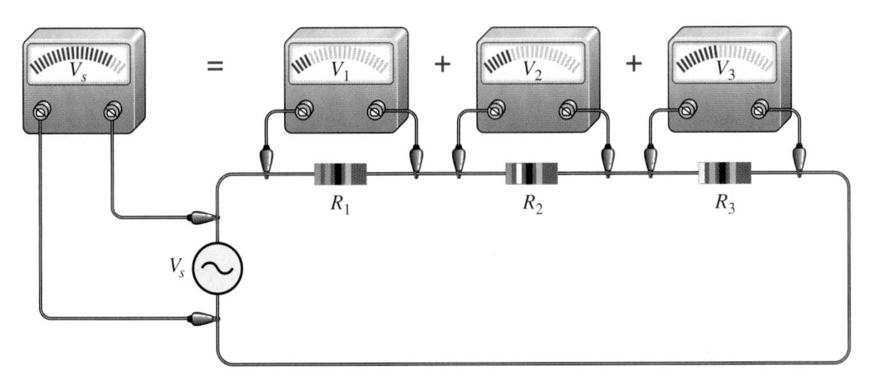

◀ **FIGURE 8–38**

Illustration of Kirchhoff's voltage law in an ac circuit.

EXAMPLE 8–12

All values in Figure 8–39 are given in rms.

(a) Find the unknown peak voltage drop in Figure 8–39(a).

(b) Find the total rms current in Figure 8–39(b).

► FIGURE 8–39

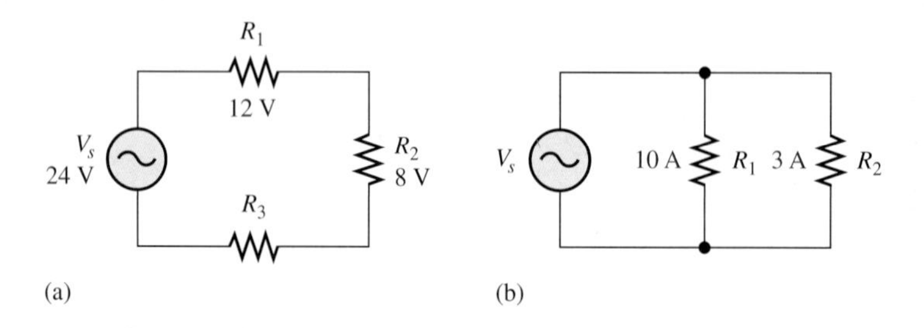

Solution **(a)** Use Kirchhoff's voltage law to find V_3.

$$V_s = V_1 + V_2 + V_3$$

$$V_{3(\text{rms})} = V_{s(\text{rms})} - V_{1(\text{rms})} - V_{2(\text{rms})} = 24\text{ V} - 12\text{ V} - 8\text{ V} = 4\text{ V}$$

Convert rms to peak.

$$V_{3(p)} = 1.414V_{3(\text{rms})} = 1.414(4\text{ V}) = \mathbf{5.66\text{ V}}$$

(b) Use Kirchhoff's current law to find I_{tot}.

$$I_{tot(\text{rms})} = I_{1(\text{rms})} + I_{2(\text{rms})} = 10\text{ A} + 3\text{ A} = \mathbf{13\text{ A}}$$

Related Problem A series circuit has the following voltage drops: $V_{1(\text{rms})} = 3.50$ V, $V_{2(p)} = 4.25$ V, $V_{3(\text{avg})} = 1.70$ V. Determine the peak-to-peak source voltage.

SECTION 8–6 REVIEW

1. A sinusoidal voltage with a half-cycle average value of 12.5 V is applied to a circuit with a resistance of 330 Ω. What is the peak current in the circuit?
2. The peak voltage drops in a series resistive circuit are 6.2 V, 11.3 V, and 7.8 V. What is the rms value of the source voltage?

8–7 SUPERIMPOSED DC AND AC VOLTAGES

In many practical circuits, you will find both dc and ac voltages combined. An example of this is in amplifier circuits where ac signal voltages are superimposed on dc operating voltages.

After completing this section, you should be able to

- **Determine total voltages that have both ac and dc components**

Figure 8–40 shows a dc source and an ac source in series. These two voltages will add algebraically to produce an ac voltage "riding" on a dc level, as measured across the resistor.

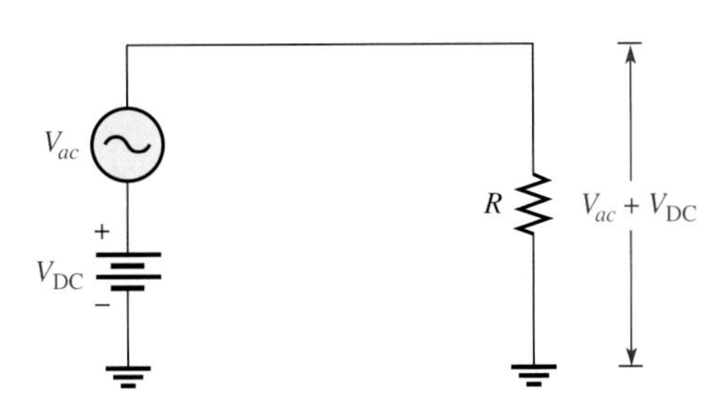

FIGURE 8–40

Superimposed dc and ac voltages.

If V_{DC} is greater than the peak value of the sinusoidal voltage, the combined ac and dc voltage is a sine wave that never reverses polarity and is therefore nonalternating. That is, the sine wave is riding on a dc level, as shown in Figure 8–41(a). If V_{DC} is less than the peak value of the sine wave, the sine wave will be negative during a portion of its lower half-cycle, as illustrated in Figure 8–41(b), and is therefore alternating. In either case, the sine wave will reach a maximum voltage equal to $V_{DC} + V_p$, and it will reach a minimum voltage equal to $V_{DC} - V_p$.

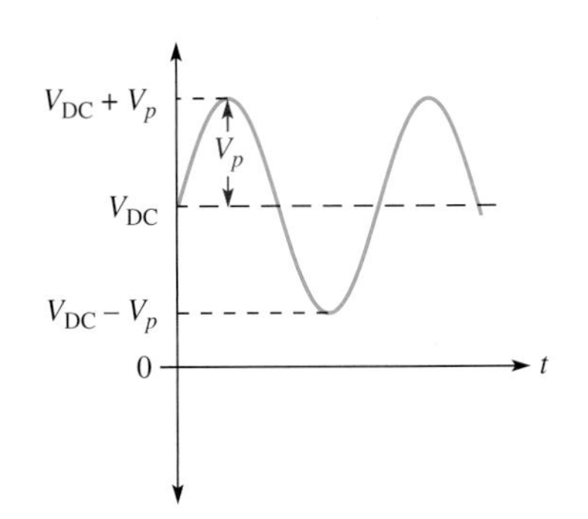

(a) $V_{DC} > V_p$. The sine wave never goes negative.

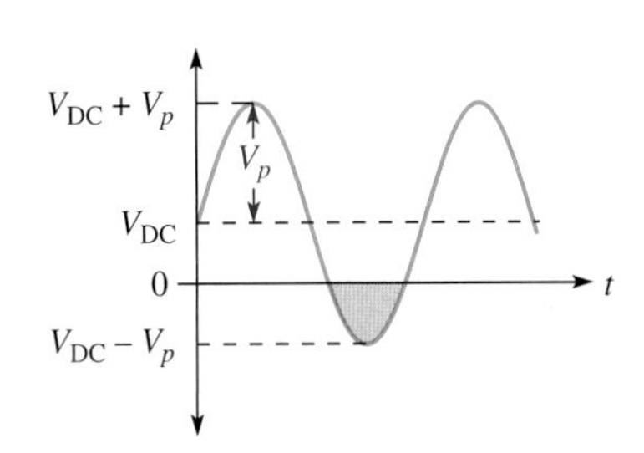

(b) $V_{DC} < V_p$. The sine wave reverses polarity during a portion of its cycle.

FIGURE 8–41

Sine waves with dc levels.

EXAMPLE 8–13

Determine the maximum and minimum voltages across the resistor in each circuit of Figure 8–42 and show the resulting waveforms.

FIGURE 8–42

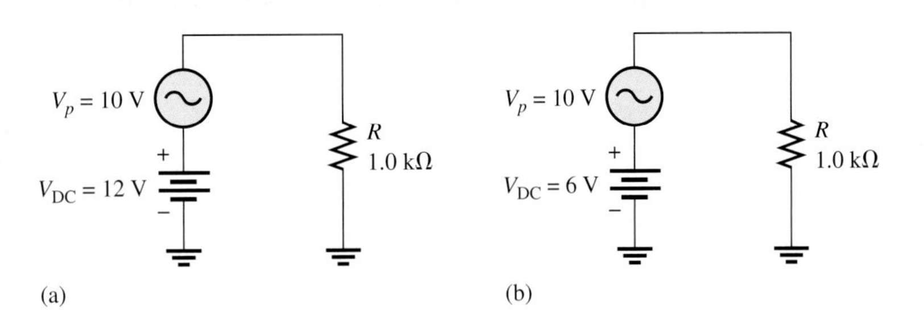

Solution In Figure 8–42(a), the maximum voltage across R is

$$V_{max} = V_{DC} + V_p = 12\text{ V} + 10\text{ V} = \mathbf{22\text{ V}}$$

The minimum voltage across R is

$$V_{min} = V_{DC} - V_p = 12\text{ V} - 10\text{ V} = \mathbf{2\text{ V}}$$

Therefore, $V_{R(tot)}$ is a nonalternating sine wave that varies from +22 V to +2 V, as shown in Figure 8–43(a).

In Figure 8–42(b), the maximum voltage across R is

$$V_{max} = V_{DC} + V_p = 6\text{ V} + 10\text{ V} = \mathbf{16\text{ V}}$$

The minimum voltage across R is

$$V_{min} = V_{DC} - V_p = \mathbf{-4\text{ V}}$$

Therefore, $V_{R(tot)}$ is an alternating sine wave that varies from +16 V to −4 V, as shown in Figure 8–43(b).

FIGURE 8–43

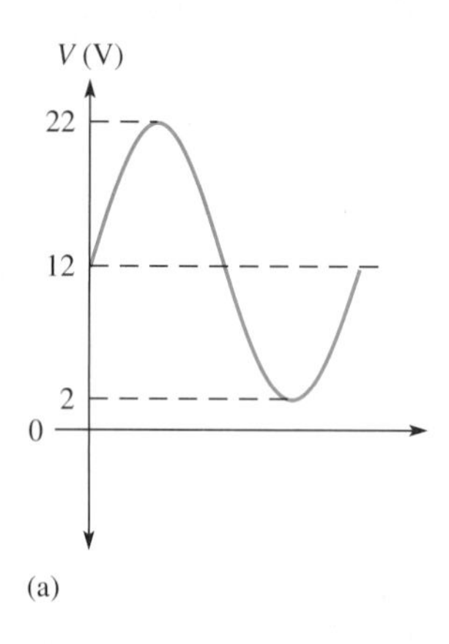

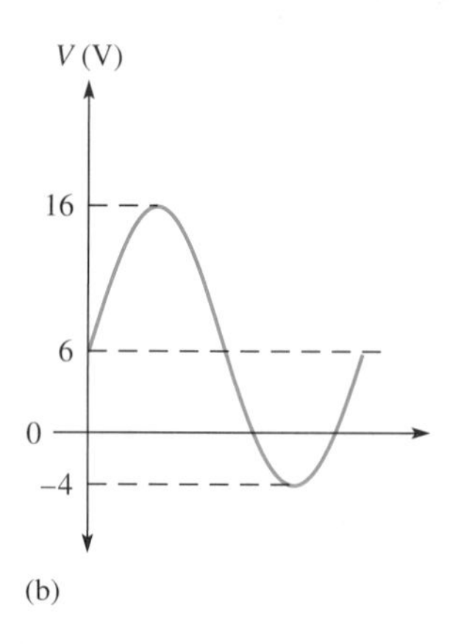

Related Problem Explain why the waveform in Figure 8–43(a) is nonalternating but the waveform in part (b) is considered to be alternating.

SECTION 8–7 REVIEW

1. What is the maximum positive value of the resulting total voltage when a sine wave with $V_p = 5$ V is added to a dc voltage of +2.5 V?
2. Will the resulting voltage in Question 1 alternate polarity?
3. If the dc voltage in Question 1 is −2.5 V, what is the maximum positive value of the resulting total voltage?

8–8 NONSINUSOIDAL WAVEFORMS

Sine waves are important in electronics, but they are not the only type of ac or time-varying waveform. Two other major types of waveforms, the pulse waveform and the triangular waveform, are discussed next.

After completing this section, you should be able to

- **Identify the characteristics of basic nonsinusoidal waveforms**
- Discuss the properties of a pulse waveform
- Define *duty cycle*
- Discuss the properties of triangular and sawtooth waveforms
- Discuss the harmonic content of a waveform

Pulse Waveforms

Basically, a pulse can be described as a very rapid transition (**leading edge**) from one voltage or current level (**baseline**) to another level, and then, after an interval of time, a very rapid transition (**trailing edge**) back to the original baseline level. The transitions in level are called *steps*. An ideal pulse consists of two opposite-going steps of equal amplitude. When the leading or trailing edge is positive-going, it is called a **rising edge.** When the leading or trailing edge is negative-going, it is called a **falling edge.**

Figure 8–44(a) shows an ideal positive-going pulse consisting of two equal but opposite instantaneous steps separated by an interval of time called the pulse width. Figure 8–44(b) shows an ideal negative-going pulse. The height of the pulse measured from the baseline is its voltage (or current) amplitude.

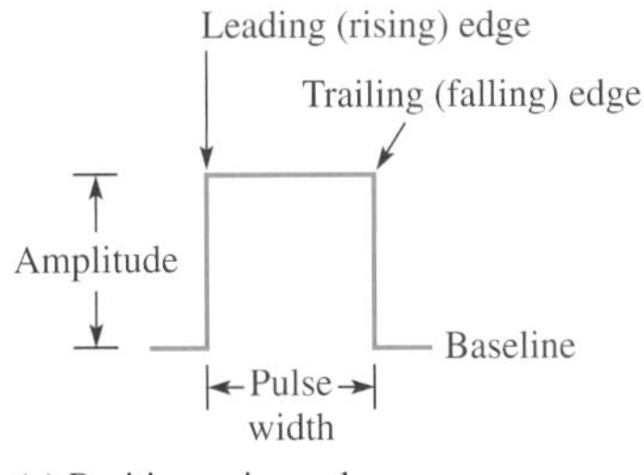

(a) Positive-going pulse

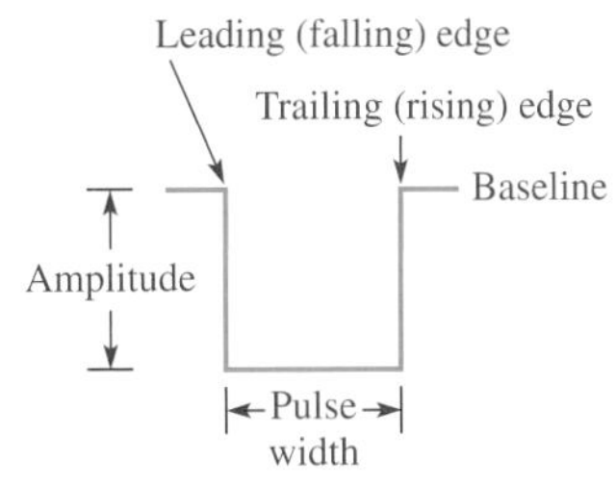

(b) Negative-going pulse

FIGURE 8–44

Ideal pulses.

In many applications, analysis is simplified by treating all pulses as ideal (composed of instantaneous steps and perfectly rectangular in shape). Actual pulses, however, are never ideal. All pulses possess certain characteristics that cause them to be different from the ideal.

In practice, pulses cannot change from one level to another instantaneously. Time is always required for a transition (step), as illustrated in Figure 8–45(a). As you can see, there is an interval of time during which the pulse is rising from its lower value to its higher value. This interval is called the rise time, t_r.

Rise time is the time required for the pulse to go from 10% of its amplitude to 90% of its amplitude.

The interval of time during which the pulse is falling from its higher value to its lower value is called the fall time, t_f.

Fall time is the time required for the pulse to go from 90% of its amplitude to 10% of its amplitude.

Pulse width (t_W) also requires a precise definition for the nonideal pulse because the leading and trailing edges are not vertical.

Pulse width is the time between the point on the leading edge where the value is 50% of the amplitude and the point on the trailing edge where the value is 50% of the amplitude.

Pulse width is shown in Figure 8–45(b).

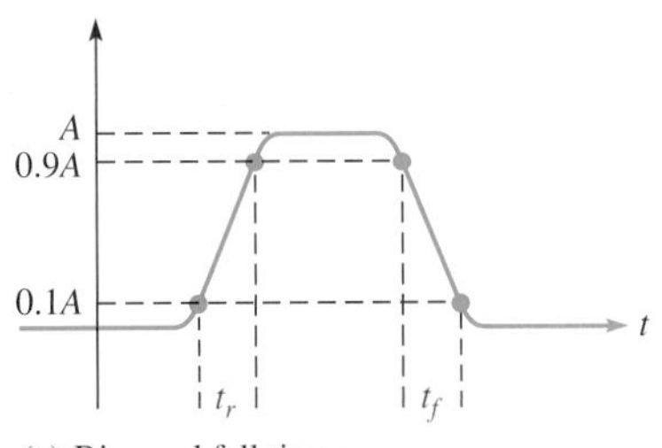

(a) Rise and fall times

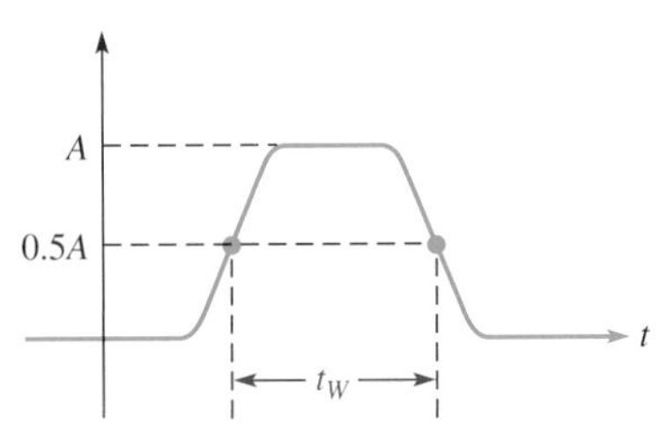

(b) Pulse width

FIGURE 8–45

Nonideal pulse.

Repetitive Pulses

Any waveform that repeats itself at fixed intervals is periodic. Figure 8–46 shows some examples of periodic pulse waveforms. Notice that in each case, the pulses repeat at regular intervals. The rate at which the pulses repeat is the **pulse repetition frequency,** which is the fundamental frequency of the waveform. The frequency can be expressed in hertz or in pulses per second. The time from one pulse to the corresponding point on the next pulse is the period (T). The relationship between frequency and period is the same as with the sine wave, $f = 1/T$.

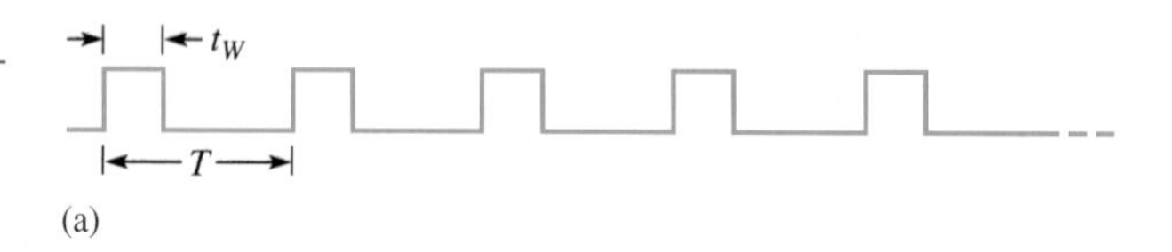

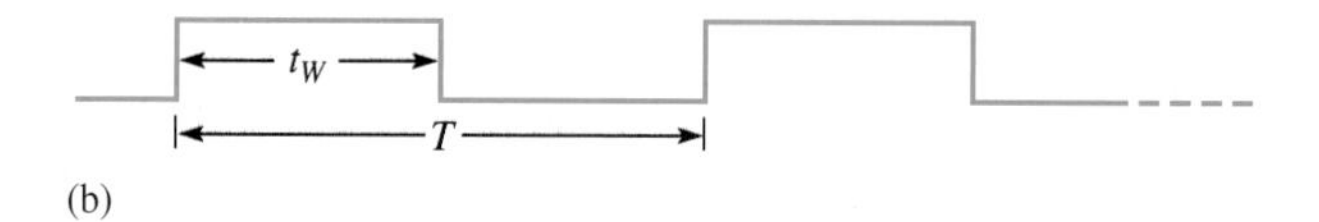

t_W

T

(c)

FIGURE 8–46

Repetitive pulse waveforms.

An important characteristic of periodic pulse waveforms is the duty cycle.

The duty cycle is the ratio of the pulse width (t_W) to the period (T) and is usually expressed as a percentage.

Equation 8–21

$$\text{Percent duty cycle} = \left(\frac{t_W}{T}\right)100\%$$

EXAMPLE 8–14

Determine the period, frequency, and duty cycle for the pulse waveform in Figure 8–47.

FIGURE 8–47

Solution As indicated in Figure 8–47, the period is

$$T = \mathbf{10\ \mu s}$$

Use Equations 8–1 and 8–21 to determine the frequency and duty cycle.

$$f = \frac{1}{T} = \frac{1}{10\ \mu\text{s}} = \mathbf{100\ kHz}$$

$$\text{Percent duty cycle} = \left(\frac{t_W}{T}\right)100\% = \left(\frac{1\ \mu\text{s}}{10\ \mu\text{s}}\right)100\% = \mathbf{10\%}$$

Related Problem A certain pulse waveform has a frequency of 200 kHz and a pulse width of 0.25 μs. Determine the duty cycle expressed as a percentage.

Square Waves

A square wave is a pulse waveform with a duty cycle of 50%. Thus, the pulse width is equal to one-half of the period. A square wave is shown in Figure 8–48.

FIGURE 8–48

Square wave.

The Average Value of a Pulse Waveform

The average value (V_{avg}) of a pulse waveform is equal to its baseline value plus the product of its duty cycle and its amplitude. The lower level of a positive-going waveform or the upper level of a negative-going waveform is taken as the baseline. The formula is as follows:

$$V_{avg} = \text{baseline} + (\text{duty cycle})(\text{amplitude})$$

Equation 8–22

The following example illustrates the calculation of the average value of pulse waveforms.

EXAMPLE 8–15

Determine the average value of each of the positive-going waveforms in Figure 8–49.

FIGURE 8–49

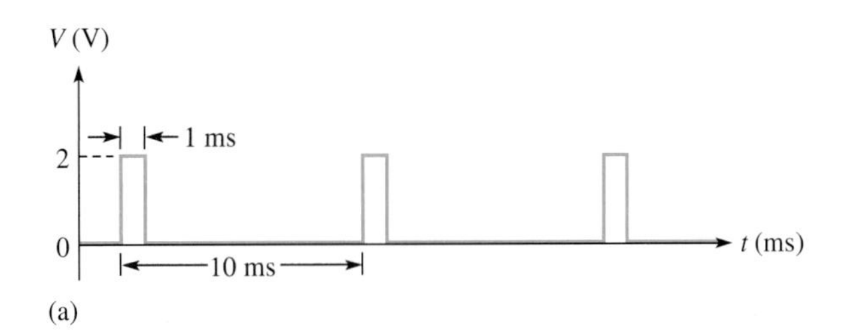

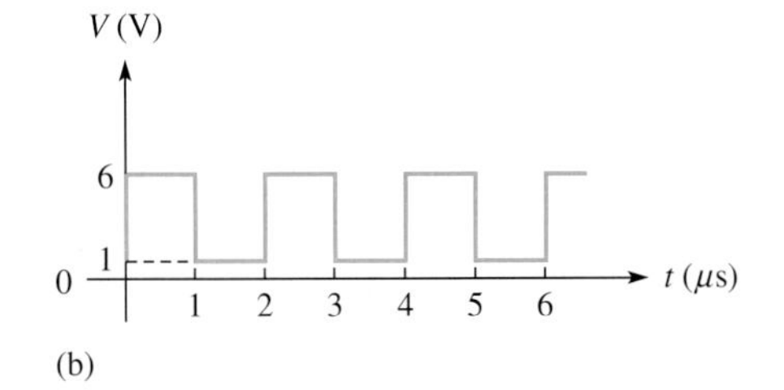

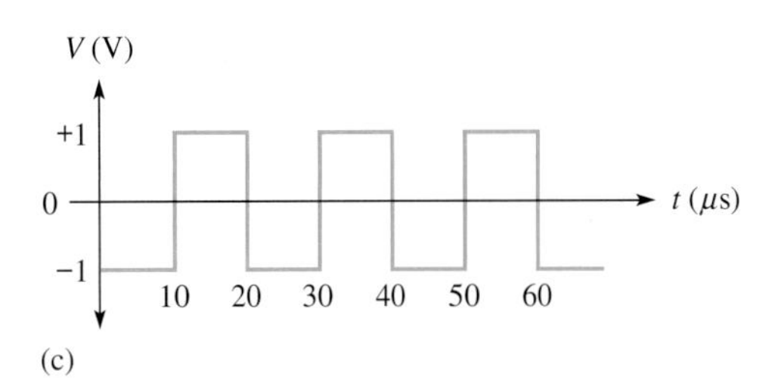

Solution In Figure 8–49(a), the baseline is at 0 V, the amplitude is 2 V, and the duty cycle is 10%. The average value is

$$\begin{aligned} V_{avg} &= \text{baseline} + (\text{duty cycle})(\text{amplitude}) \\ &= 0\text{ V} + (0.1)(2\text{ V}) = \mathbf{0.2\text{ V}} \end{aligned}$$

The waveform in Figure 8–49(b) has a baseline of +1 V, an amplitude of 5 V, and a duty cycle of 50%. The average value is

$$\begin{aligned} V_{avg} &= \text{baseline} + (\text{duty cycle})(\text{amplitude}) \\ &= 1\text{ V} + (0.5)(5\text{ V}) = 1\text{ V} + 2.5\text{ V} = \mathbf{3.5\text{ V}} \end{aligned}$$

The waveform in Figure 8–49(c) is a square wave with a baseline of -1 V and an amplitude of 2 V. The duty cycle is 50%. The average value is

$$V_{avg} = \text{baseline} + (\text{duty cycle})(\text{amplitude})$$
$$= -1\text{ V} + (0.5)(2\text{ V}) = -1\text{ V} + 1\text{ V} = \mathbf{0\text{ V}}$$

This is an alternating square wave, and, like an alternating sine wave, it has an average value of zero over a full cycle.

Related Problem If the baseline of the waveform in Figure 8–49(a) is shifted to $+1$ V, what is the average value?

Triangular and Sawtooth Waveforms

Triangular and sawtooth waveforms are formed by voltage or current ramps. A **ramp** is a linear increase or decrease in the voltage or current. Figure 8–50 shows both positive- and negative-going ramps. In part (a) the ramp has a positive slope; in part (b), the ramp has a negative slope. The slope of a voltage ramp is $\pm V/t$ and is expressed in units of V/s. The slope of a current ramp is $\pm I/t$ and is expressed in units of A/s.

FIGURE 8–50

Ramps.

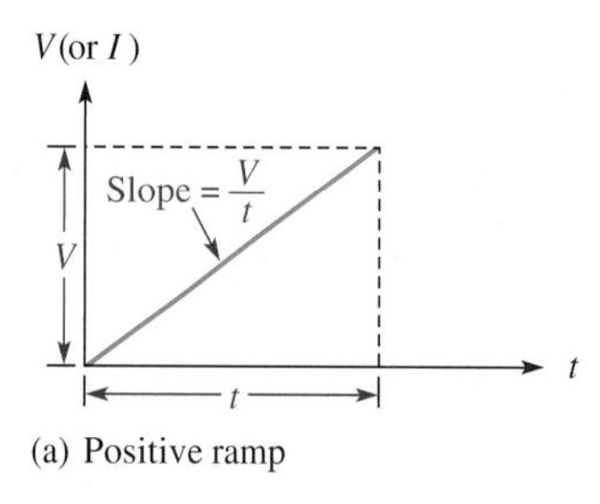

(a) Positive ramp

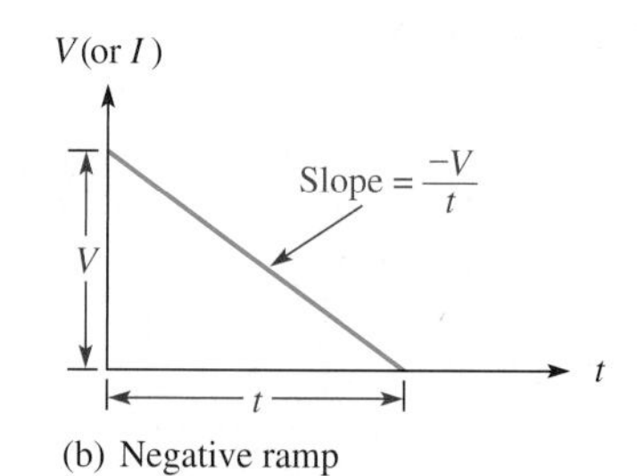

(b) Negative ramp

EXAMPLE 8–16

What are the slopes of the voltage ramps in Figure 8–51?

FIGURE 8–51

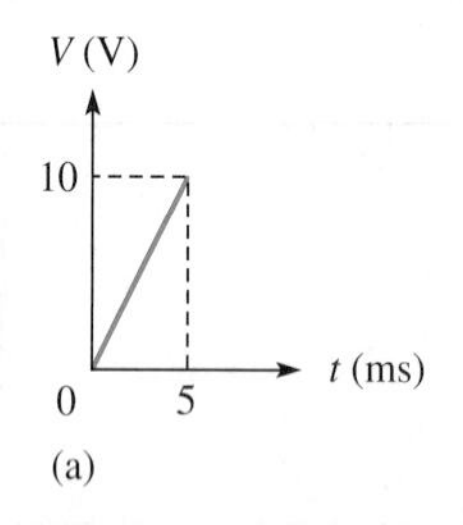

(a)

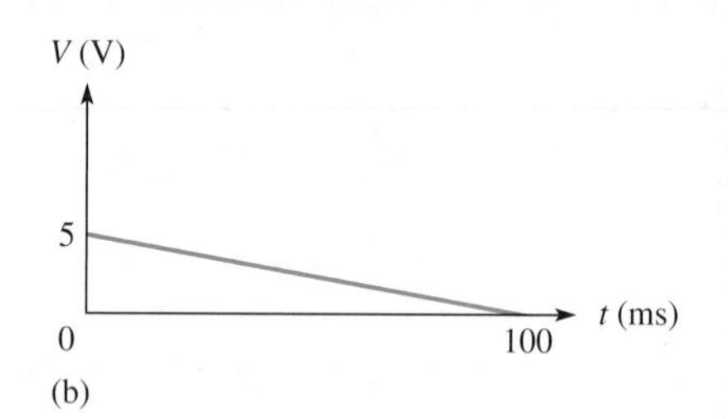

(b)

Solution In Figure 8–51(a), the voltage increases from 0 V to $+10$ V in 5 ms. Thus, $V = 10$ V and $t = 5$ ms. The slope is

$$\frac{V}{t} = \frac{10\text{ V}}{5\text{ ms}} = \mathbf{2\text{ V/ms}}$$

In Figure 8–51(b), the voltage decreases from $+5$ V to 0 V in 100 ms. Thus, $V = -5$ V and $t = 100$ ms. The slope is

$$\frac{V}{t} = \frac{-5\text{ V}}{100\text{ ms}} = -0.05\text{ V/ms} = \mathbf{-50\text{ V/s}}$$

Related Problem A certain voltage ramp has a slope of $+12$ V/μs. If the ramp starts at zero, what is the voltage at 0.01 ms?

Triangular Waveforms Figure 8–52 shows that a **triangular waveform** is composed of positive-going and negative-going ramps having equal slopes. The period of this waveform can be measured from one peak to the next corresponding peak, as illustrated. This particular triangular waveform is alternating and has an average value of zero.

Figure 8–53 depicts a triangular waveform with a nonzero average value. The frequency for triangular waves is determined in the same way as for sine waves, that is, $f = 1/T$.

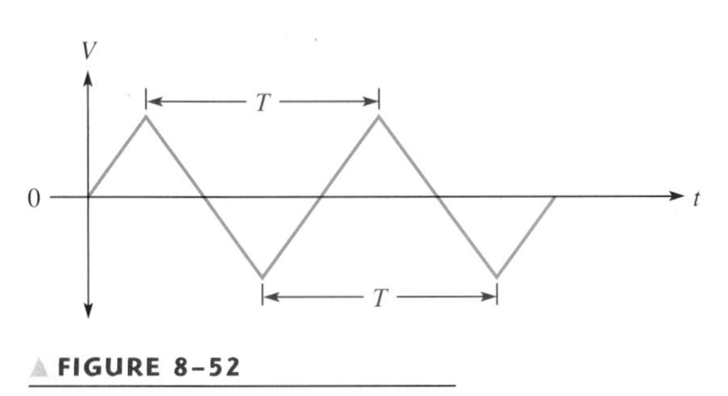

▲ FIGURE 8–52

Alternating triangular waveform.

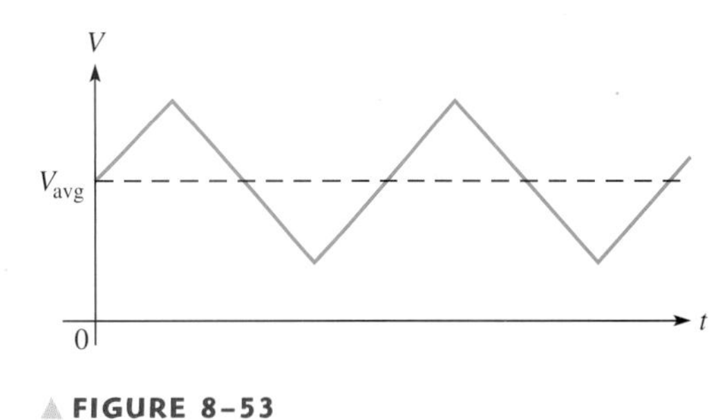

▲ FIGURE 8–53

Nonalternating triangular waveform.

Sawtooth Waveforms The **sawtooth waveform** is actually a special case of the triangular waveform consisting of two ramps, one of much longer duration than the other. Sawtooth waveforms are used in many electronic systems. For example, the electron beam that sweeps across the screen of your TV receiver, creating the picture, is controlled by sawtooth voltages and currents. One sawtooth wave produces the horizontal beam movement, and the other produces the vertical beam movement. A sawtooth voltage is sometimes called a *sweep voltage.*

Figure 8–54 is an example of a sawtooth waveform. Notice that it consists of a positive-going ramp of relatively long duration, followed by a negative-going ramp of relatively short duration.

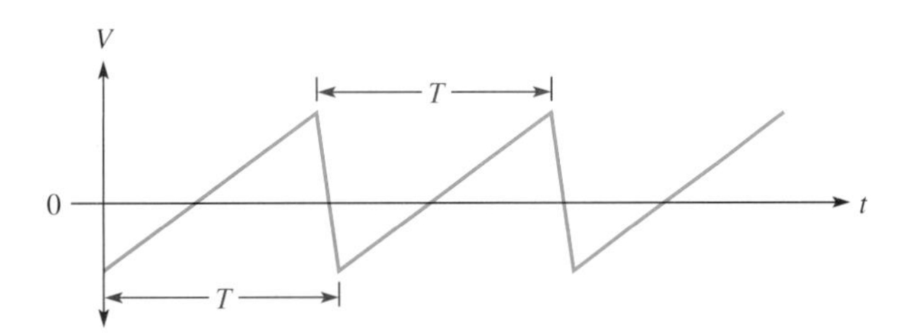

◀ FIGURE 8–54

Alternating sawtooth waveform.

Harmonics

A repetitive nonsinusoidal waveform is composed of a fundamental frequency and harmonic frequencies. The **fundamental frequency** is the repetition rate of the waveform, and the **harmonics** are higher-frequency sine waves that are multiples of the fundamental.

Odd Harmonics *Odd harmonics* are frequencies that are odd multiples of the fundamental frequency of a waveform. For example, a 1 kHz square wave consists of a fundamental of 1 kHz and odd harmonics of 3 kHz, 5 kHz, 7 kHz, and so on. The 3 kHz frequency in this case is the third harmonic; the 5 kHz frequency is the fifth harmonic; and so on.

Even Harmonics *Even harmonics* are frequencies that are even multiples of the fundamental frequency. For example, if a certain wave has a fundamental of 200 Hz, the second harmonic is 400 Hz, the fourth harmonic is 800 Hz, the sixth harmonic is 1200 Hz, and so on. These are even harmonics.

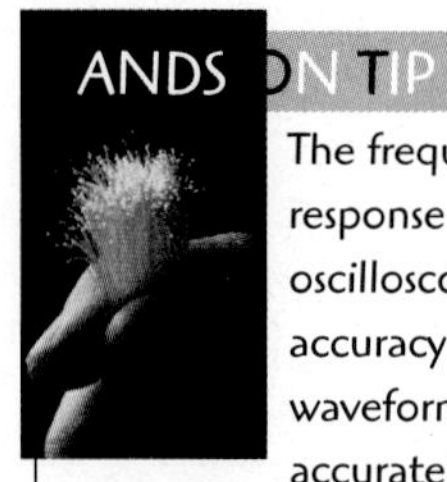

The frequency response of an oscilloscope limits the accuracy with which waveforms can be accurately displayed. To view pulse waveforms, the frequency response must be high enough for all significant harmonics of the waveform. For example, a 100 MHz oscilloscope distorts a 100 MHz pulse waveform because the third, fifth, and higher harmonics are greatly attenuated.

Composite Waveform Any variation from a pure sine wave produces harmonics. A nonsinusoidal wave is a composite of the fundamental and the harmonics. Some types of waveforms have only odd harmonics, some have only even harmonics, and some contain both. The shape of the wave is determined by its harmonic content. Generally, only the fundamental and the first few harmonics are of significant importance in determining the waveshape.

A square wave is an example of a waveform that consists of a fundamental and only odd harmonics. When the instantaneous values of the fundamental and each odd harmonic are added algebraically at each point, the resulting curve will have the shape of a square wave, as illustrated in Figure 8–55. In part (a), the fundamental and the third harmonic produce a waveshape that begins to resemble a square wave. In part (b), the fundamental, third, and fifth harmonics produce a closer resemblance. When the seventh harmonic is included, as in part (c), the resulting waveshape becomes even more like a square wave. As more harmonics are included, a square wave is approached.

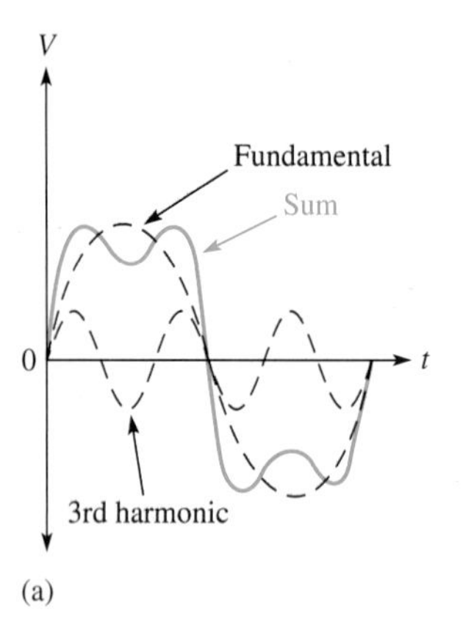

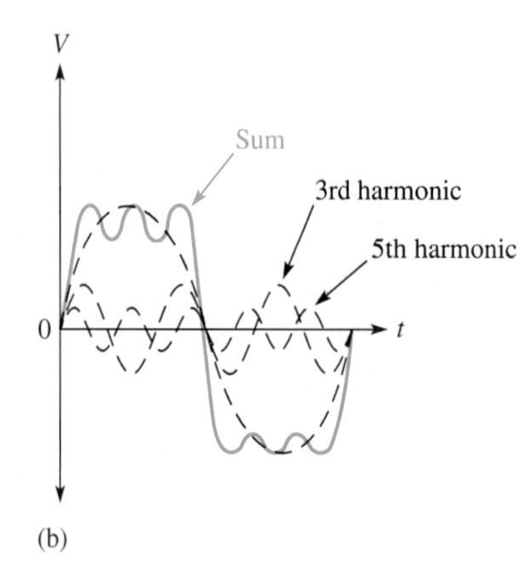

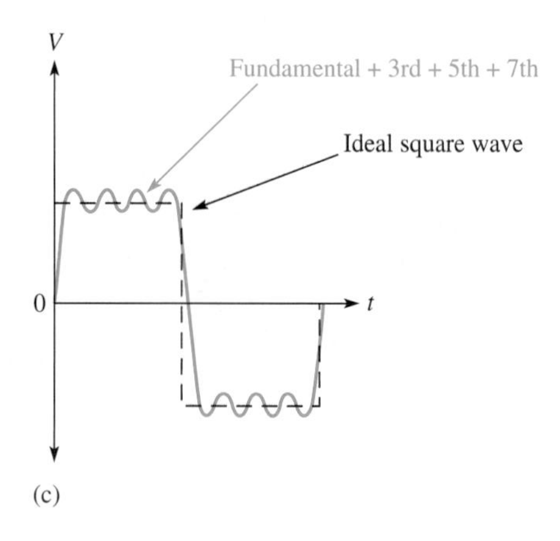

▲ **FIGURE 8–55**

Odd harmonics combine to produce a square wave.

SECTION 8–8 REVIEW

1. Define the following parameters:
   (a) rise time (b) fall time (c) pulse width
2. In a certain repetitive positive-going pulse waveform, the pulses are 200 μs wide and occur once every millisecond. What is the frequency of this waveform?
3. Determine the duty cycle, amplitude, and average value of the waveform in Figure 8–56(a).
4. What is the period of the triangular wave in Figure 8–56(b)?
5. What is the frequency of the sawtooth wave in Figure 8–56(c)?
6. Define *fundamental frequency.*
7. What is the second harmonic of a fundamental frequency of 1 kHz?
8. What is the fundamental frequency of a square wave having a period of 10 μs?

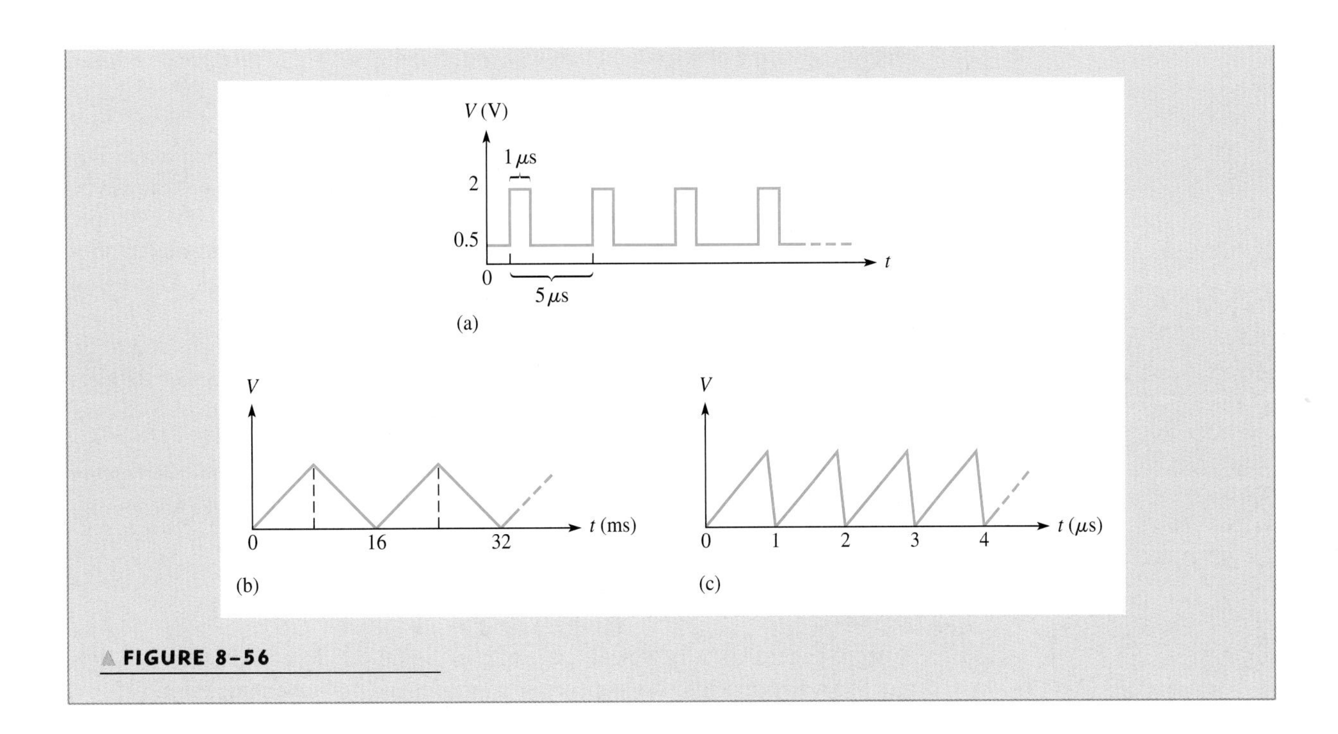

FIGURE 8–56

8–9 THE OSCILLOSCOPE

The oscilloscope, or scope for short, is one of the most widely used and versatile test instruments. It displays on a screen the actual shape of voltages that are changing with time so that waveform measurements can be made and one waveform can be compared to another.

After completing this section, you should be able to

- **Use the oscilloscope to measure waveforms**
- Identify basic oscilloscope controls
- Explain how to measure amplitude
- Explain how to measure period and frequency

Figure 8–57 shows two typical oscilloscopes. The one in part (a) is a digital oscilloscope and the one in part (b) is analog. You can think of the **oscilloscope** as essentially a "graphing machine" that graphs out a voltage and shows you how it varies with time. The shape of the sine wave that you have seen throughout this chapter is simply a graph of voltage versus time. Although oscilloscopes can

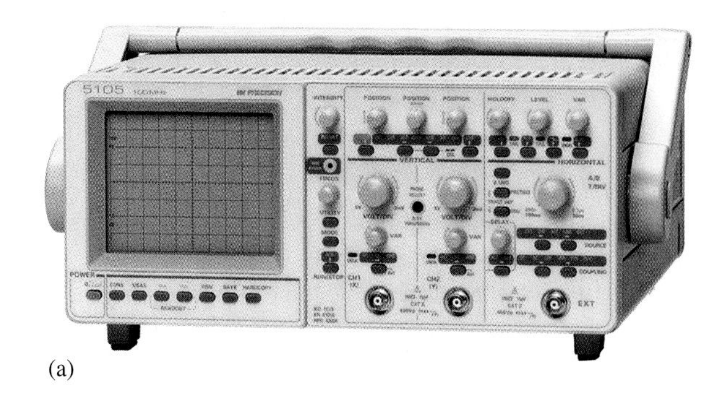
(a)

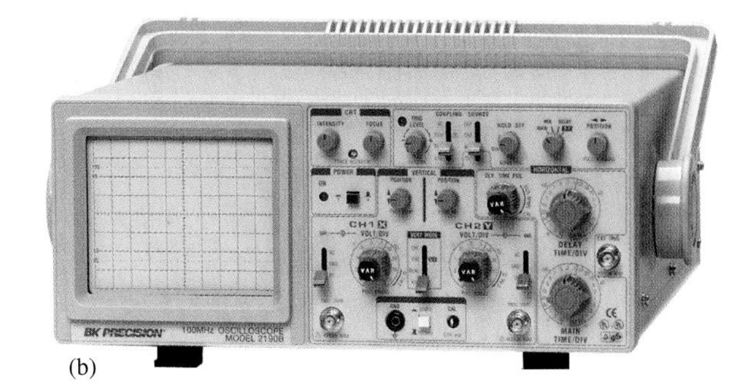
(b)

FIGURE 8–57

Oscilloscopes. (Photography courtesy of B&K Precision Corp.)

display any type of waveform, we will use the sine wave to illustrate some basic concepts.

Digital and analog oscilloscopes are distinguished from each other by the way in which the input signal is processed prior to being displayed on the screen. The principal front panel controls are functionally the same for both types; however, digital scopes can generally do more automated operations than an analog scope.

A digital scope samples the signal voltage and uses an analog-to-digital converter to convert the voltage being measured into digital information. The digital information is then used to reconstruct the waveform on the screen.

An analog scope works by applying the signal voltage being measured to an electron beam that is sweeping across the screen. The measured voltage deflects the beam up and down proportional to the amount of voltage, immediately tracing the waveform on the screen.

Although there are still some analog scopes being manufactured and many older ones that are still in use, the digital scope is widely used and is becoming the instrument of choice to replace older analog instruments.

Digital Oscilloscope

A front-panel view of a typical dual-channel digital oscilloscope is shown in Figure 8–58. Instruments vary depending on model and manufacturer, but most have certain common features. Notice that the two Vertical sections and the Horizontal section each contain a Position control, a channel Menu button, and a Volts/Div or Sec/Div control. Most digital scopes have automated measurements, menu selections, screen displays of parameter settings, and other features not found on analog scopes.

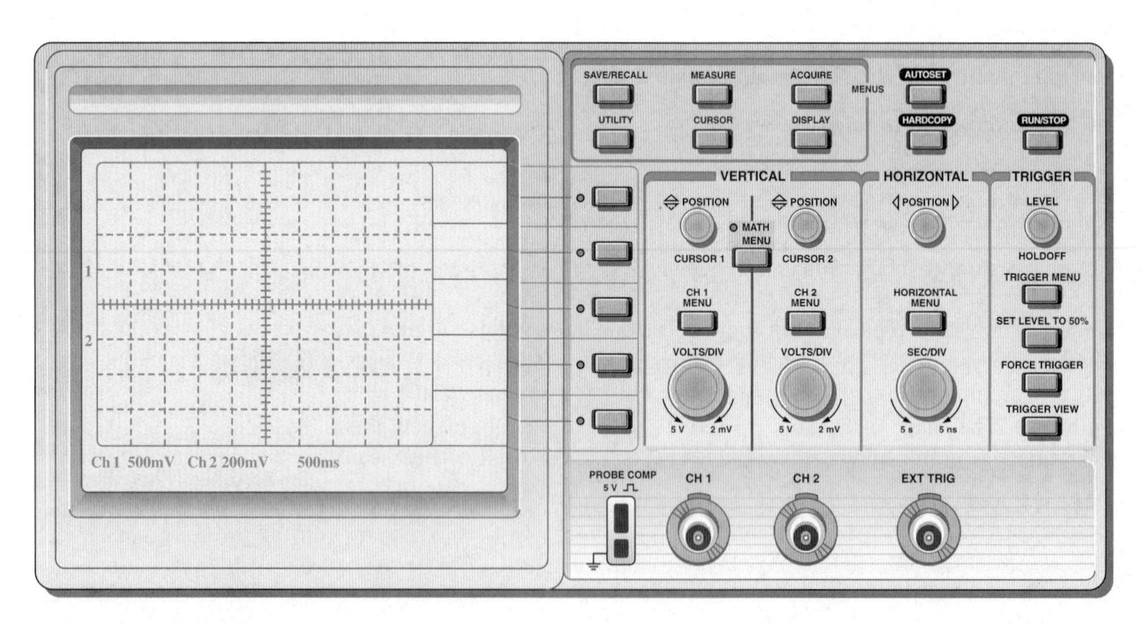

▲ **FIGURE 8–58**

A typical dual-channel digital oscilloscope.

We will discuss some of the controls here. Refer to the user manual for a complete coverage of the details of your particular oscilloscope.

In the Vertical section, there are identical controls for each of the two channels (CH1 and CH2). The Position control lets you move a displayed waveform up or down vertically on the screen. The channel Menu button provides for the selection of several items that appear on the screen, such as the coupling (dc, ac, or ground), course or fine adjustment for the Volts/Div, probe attenuation (1×, 10×, for example), and other parameters. The Volts/Div control adjusts the number of volts represented by each vertical division on the screen. For this type of scope, the Volts/Div setting for each channel is displayed on the screen instead of having to read it from a dial setting, as on many analog scopes. The Math Menu button provides a selection of operations that can be performed on the input waveforms, such as subtraction, addition, and inversion.

In the Horizontal section, the controls apply to both channels. The Position control lets you move a displayed waveform left or right horizontally on the screen. The Horizontal Menu button provides for the selection of several items that appear on the screen, such as main time base, expanded view of a portion of a waveform, and other parameters. The Sec/Div control adjusts the time represented by each horizontal division or main time base. For this type of scope, the Sec/Div setting is displayed on the screen instead of having to read it from a dial setting, as on many analog scopes.

In the Trigger section, the Level control determines the point on the triggering waveform where triggering occurs to initiate the display sweep for the input channel waveforms. The Trigger Menu button provides for the selection of several items that appear on the screen, including edge or slope triggering, trigger source, trigger mode, and other parameters.

There are four input connectors on the front panel. The two marked CH1 and CH2 are for connecting the input signals for display on the screen using voltage probes. The EXT TRIG connector is for connecting an external trigger source. The Probe Comp is used to electrically match the voltage probe to the input circuit.

Probe Attenuation All scope probes have an attenuation factor that indicates by how much they reduce the input signal. A 1× probe does not reduce the input signal at all, while a 10× probe reduces the signal by a factor of 10. The larger the attenuation factor, the less the oscilloscope will load the circuit under test. Less loading means more accuracy. When you use a voltage probe to measure voltage in a circuit, you should always make sure that it is compatible with the scope in terms of the attenuation factor. Use the channel menu in the Vertical section to select the proper probe attenuation to match your probe. For example, if you are using a 10× probe, the 10× attenuation should be selected from the menu.

Probe Compensation A probe can distort a waveform if it is not properly compensated. A scope should always be compensated the first time it is attached to any scope, and compensation should be periodically checked. The Probe Comp output on the scope provides an accurate waveform for checking compensation.

The probe connector should be connected to a channel input (CH1 or CH2), the probe tip should be connected to the Probe Comp connector output, and the probe ground lead should be connected to the grounded pin on the Probe Comp connector, as shown in Figure 8–59. The compensation waveform will appear on the screen having one of the three general shapes shown in Figure 8–60(a). If the waveform is either undercompensated or overcompensated, your probe must be adjusted as indicated in Figure 8–60(b) until it is compensated correctly.

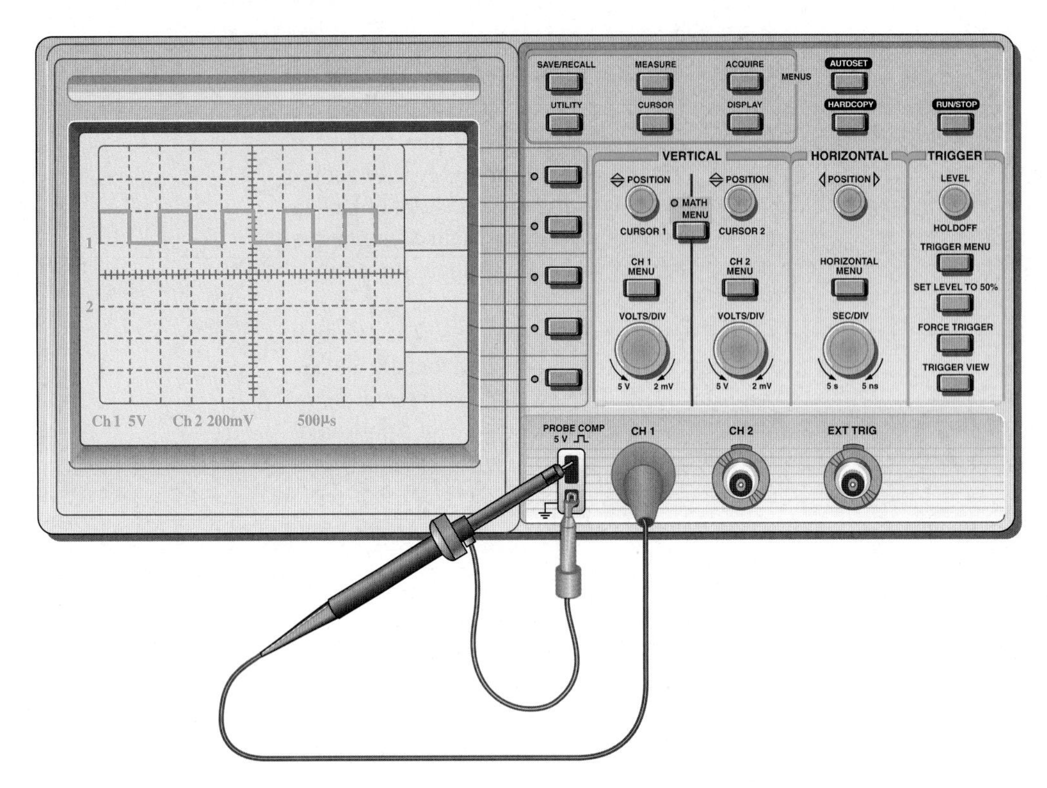

▲ **FIGURE 8–59**

Connection for voltage probe compensation.

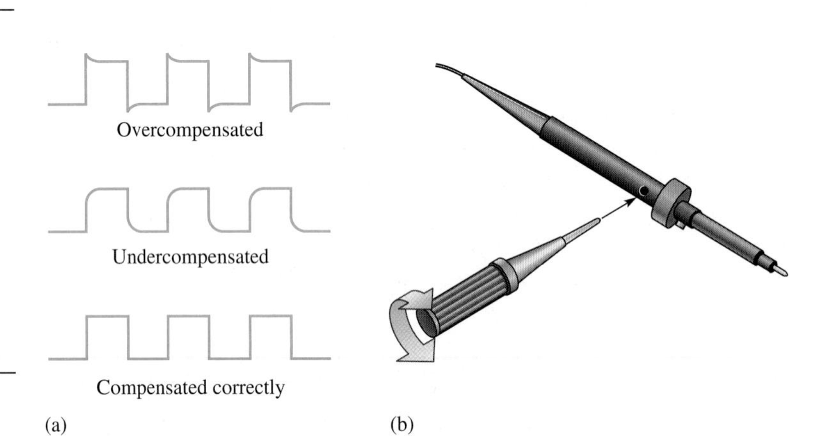

▶ **FIGURE 8–60**

Compensation waveforms and typical voltage probe adjustment.

EXAMPLE 8–17

Determine the peak-to-peak value and period of each sine wave in Figure 8–61 from the digital scope screen displays and the settings for Volts/Div and Sec/Div, which are indicated under the screens.

Solution Looking at the vertical scale in Figure 8–61(a),

$$V_{pp} = 6 \text{ divisions} \times 0.5 \text{ V/division} = \mathbf{3.0\ V}$$

From the horizontal scale (one cycle covers ten divisions),

$$T = 10 \text{ divisions} \times 2 \text{ ms/division} = \mathbf{20\ ms}$$

Looking at the vertical scale in Figure 8–61(b),

$$V_{pp} = 5 \text{ divisions} \times 50 \text{ mV/division} = \mathbf{250\ mV}$$

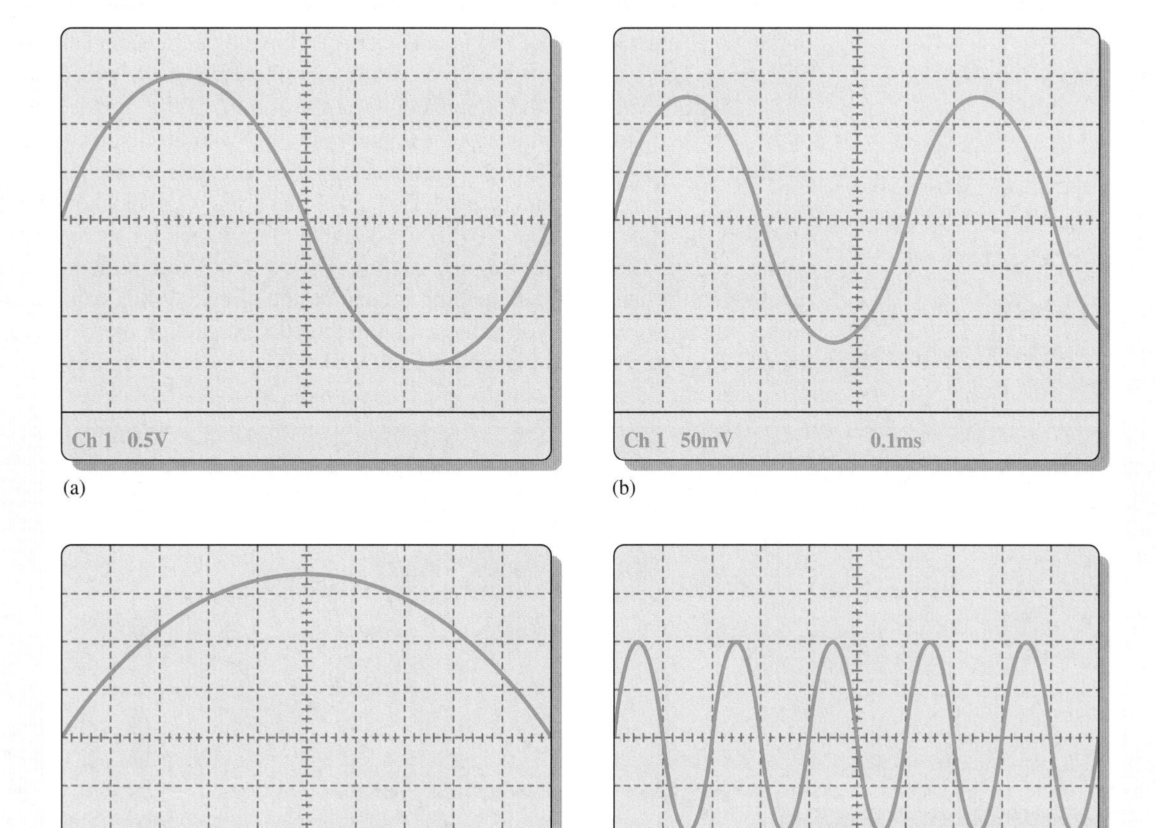

(a) (b) (c) (d)

Sine waves are centered vertically on the screen.

▲ **FIGURE 8–61**

From the horizontal scale (one cycle covers six divisions),

$$T = 6 \text{ divisions} \times 0.1 \text{ ms/division} = 0.6 \text{ ms} = \mathbf{600\ \mu s}$$

Looking at the vertical scale in Figure 8–61(c), the sine wave is centered vertically.

$$V_{pp} = 6.8 \text{ divisions} \times 2 \text{ V/division} = \mathbf{13.6\ V}$$

From the horizontal scale (one-half cycle covers ten divisions),

$$T = 20 \text{ divisions} \times 10\ \mu\text{s/division} = \mathbf{200\ \mu s}$$

Looking at the vertical scale in Figure 8–61(d),

$$V_{pp} = 4 \text{ divisions} \times 5 \text{ V/division} = \mathbf{20\ V}$$

From the horizontal scale (one cycle covers two divisions),

$$T = 2 \text{ divisions} \times 2\ \mu\text{s/division} = \mathbf{4\ \mu s}$$

Related Problem Determine the rms value and the frequency for each waveform displayed in Figure 8–61.

Analog Oscilloscope

Figure 8–62 shows a generic front panel view of a typical dual-channel analog oscilloscope. There are two Volts/Div controls, one for each of the two channels. These controls determine the attenuation or amplification of the signal, and each of the settings on the dial indicates the number of volts (V) or millivolts (mV) represented by each vertical division on the screen. Some analog oscilloscopes display these settings on the screen. Typically, there are two brackets on the Volts/Div dial, one for a times 10 (10×) voltage probe and one for a times 1 (1×) voltage probe. The most commonly used probe is a 10×, which attenuates the input signal by ten to reduce loading effects on the circuit. When using a 10× probe, you must read the Volts/Div setting in the 10× bracket on the type of analog oscilloscope shown in Figure 8–62.

FIGURE 8–62

A typical dual-channel analog oscilloscope.

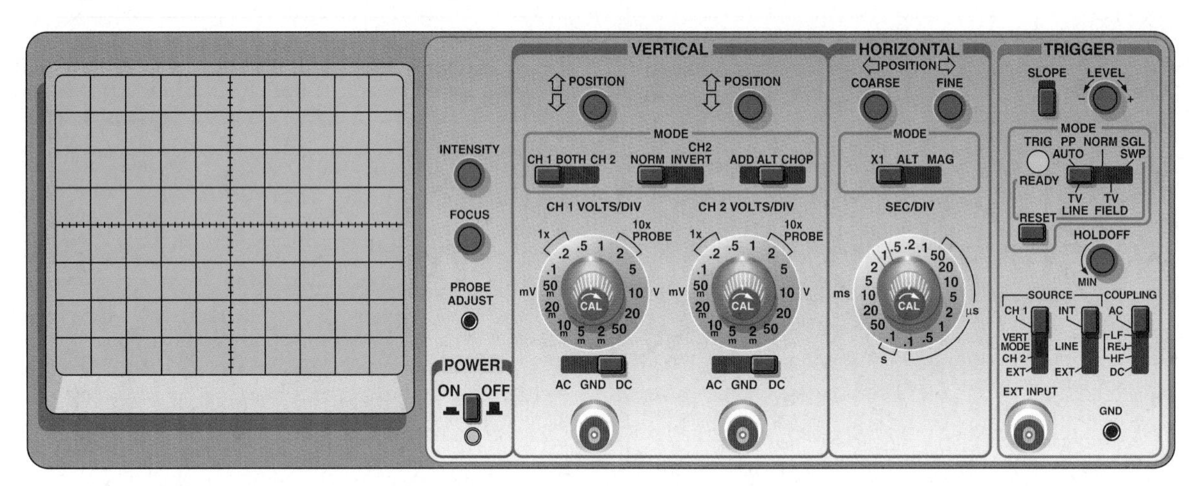

The time base or Sec/Div control sets the number of second(s), milliseconds (ms), or microseconds (μs) represented by each horizontal division as indicated by the dial settings. You can measure the period of a waveform on the horizontal scale and then calculate the frequency.

EXAMPLE 8–18

Based on the control dial settings, determine the peak-to-peak amplitude and the period of the waveform on the screen of an analog oscilloscope as shown in Figure 8–63. Calculate the frequency. Assume a 10× probe is used.

FIGURE 8–63

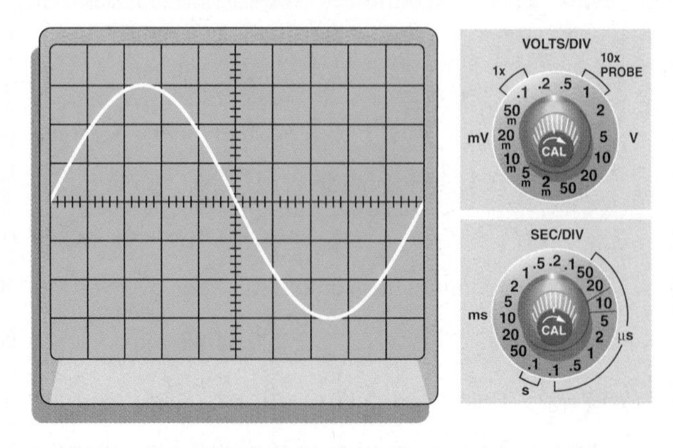

Solution The Volts/Div setting is 1 V (read on the 10× side of the dial). The sine wave is six divisions high from the negative peak to the positive peak and, since each division represents 1 V, the peak-to-peak value of the sine wave is

$$\text{Amplitude} = (6 \text{ divisions})(1 \text{ V/division}) = \mathbf{6\ V}$$

The Sec/Div setting is 10 μs. A full cycle of the waveform covers ten divisions; therefore, the period is

$$\text{Period} = (10 \text{ divisions})(10\ \mu\text{s/division}) = \mathbf{100\ \mu s}$$

The frequency is calculated as

$$f = \frac{1}{100\ \mu\text{s}} = \mathbf{10\ kHz}$$

Related Problem For a Volts/Div setting of 5 V and a Sec/Div setting of 2 ms, determine the peak value, period, and frequency of the sine wave shown on the screen in Figure 8–63.

Other Basic Controls The following descriptions refer to the analog scope in Figure 8–62.

Intensity The Intensity control varies the brightness of the screen trace. Caution should be used so that the intensity is not left too high for an extended time, especially when the electron beam forms a motionless dot on the screen.

Focus This control focuses the electron beam so that it converges to a tiny point at the screen. An out-of-focus condition results in a fuzzy trace.

Horizontal and vertical positions These controls are used to move a trace horizontal across the screen and up or down for easier viewing or measurement.

AC-GND-DC switch This switch allows the input signal to be ac coupled, dc coupled, or grounded. The ac coupling eliminates any dc component on the input signal by inserting a capacitor in series with the signal. The dc coupling permits dc values to be displayed. The ground position disconnects the input and allows a 0 V reference to be established on the screen. For digital signals, dc coupling should be used to correctly show the baseline value.

Signal inputs The signal voltages to be displayed are connected into the channel 1 (CH1) and/or channel 2 (CH2). As mentioned previously, these input connections are done with an attenuator probe that minimizes the loading effect of the scope on a circuit being measured. The common type of oscilloscope probe attenuates (reduces) the signal voltage by a factor of ten and is called a *times 10* (10×) probe. Times 1 (1×) probes are also sometimes used when measuring very small voltages; they lose accuracy for voltage measurements at higher frequencies and tend to load the circuit under test more.

Mode switches These switches provide for displaying either or both channel input signals, inverting channel 2 signal (Invert), adding two waveforms (Add), and selecting between alternate (Alt) and chopped (Chop) mode of sweep. *As a rule, Alt is selected when the frequency is above 100 Hz; otherwise, Chop is used.*

Trigger controls The **trigger** controls allow the scope to be triggered from various selected sources. Triggering causes the trace to begin its sweep across the screen. Triggering can occur from an internal signal, an external signal, or from the line voltage. Typical trigger modes are *normal, auto, single-sweep,* and *TV.* In

HANDS ON TIP

Proper grounding is very important when you set up to take measurements or work on a circuit because proper grounding protects you from shock and assures accurate measurements. Grounding the oscilloscope simply means to connect it to earth ground by plugging the three-prong power cord plug into a grounded outlet. Also, for accurate measurements, make sure that the ground in the circuit you are testing is the same as the scope ground. This can be done by connecting the ground lead on the scope probe to a known ground point in the circuit, such as the metal chassis or a ground contact on the circuit board. You can also connect the circuit ground to the GND jack found on the front of many oscilloscopes.

the auto mode, sweep occurs in the absence of an adequate trigger signal. In the normal mode, a trigger signal must be present for the sweep to occur. The TV mode allows stable triggering on the TV field or TV line signals by placing a low-pass filter in the triggering circuit. The *slope* switch sets the trigger to occur on either the positive-going or the negative-going slope of the trigger waveform. The Level control selects the voltage level on the trigger signal at which the triggering occurs. Basically, the trigger controls provide for synchronization of the horizontal sweep waveform in the scope and the input signals. As a result, the display of the input signal is stable on the screen, rather than appearing to drift, as shown in Figure 8–64.

FIGURE 8–64

Proper triggering stabilizes a repeating waveform.

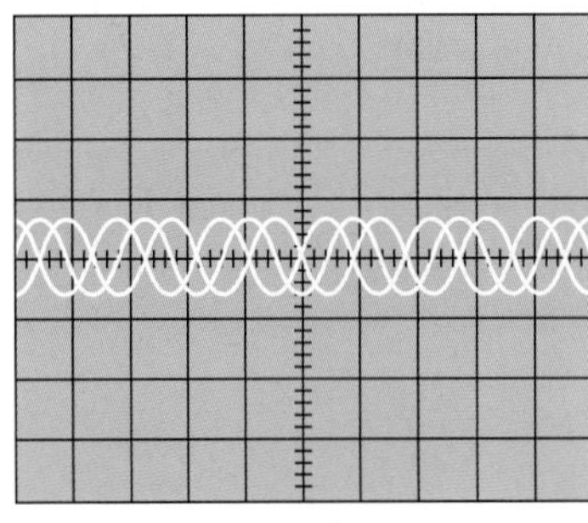

(a) Untriggered display

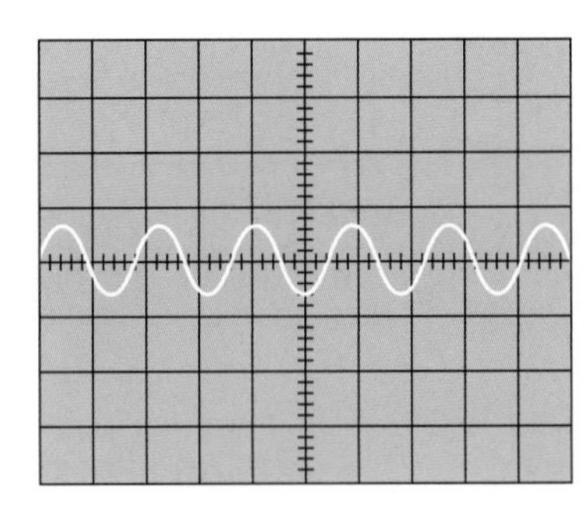

(b) Triggered display

SECTION 8–9 REVIEW

1. Explain the primary difference between a digital and an analog oscilloscope.
2. On an oscilloscope, voltage is measured (horizontally, vertically) on the screen and time is measured (horizontally, vertically).
3. What can an oscilloscope do that a multimeter cannot?
4. What is the advantage of a 10× probe?

Assume you are assigned to check out a new function generator that produces sine wave, triangular wave, and pulse outputs. For each type of output, you will use the oscilloscope to measure the minimum and maximum frequencies, the minimum and maximum amplitudes, the maximum positive and negative dc offsets, and the minimum and maximum duty cycles for the pulse waveform. You will record your measurements in a logical format.

Step 1: Familiarization with the Function Generator

The function generator is shown in Figure 8–65. Each control is marked by a circled number and a description follows.

1. *Power on/off switch* Press this push-button switch to turn the power on. Press again to release the switch and turn the power off.
2. *Function switches* Press one of the switches to select either a sinusoidal, triangular, or pulse output. Only one switch can be engaged at a time.
3. *Frequency range switches* These push-button switches are used in conjunction with the frequency adjustment control (4). Press to select the appropriate range in decade increments from 1 Hz to 1 MHz.
4. *Frequency adjustment control* Turn this dial to set a specific frequency in the range selected by

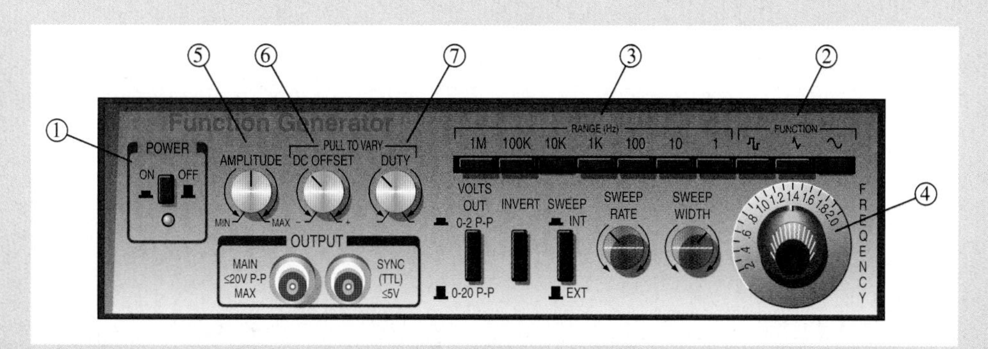

FIGURE 8–65

Function generator.

the push-button switch (3). For example, if you press the 100 Hz frequency range switch and set the dial to 1.4, the output should have a frequency of 140 Hz (1.4×100).

5. *Amplitude control* This knob adjusts the voltage amplitude of the output. Fully counterclockwise is the minimum and fully clockwise is the maximum. It can be adjusted for any value of amplitude between minimum and maximum.
6. *DC offset control* This knob adjusts the dc level of the ac output. You can add a positive or a negative dc level to a waveform.
7. *Duty control* This knob adjusts the duty cycle of the pulse waveform output. The sine wave or triangular wave outputs are not affected by this control.

Step 2: Measure the Sinusoidal Output

The output of the function generator is connected to the channel 1 (CH1) input of the scope and the sinusoidal output function is selected. The scope is set to dc coupling.

In Figure 8–66(a), assume the amplitude and frequency of the function generator are set to their minimum values. Measure and record these values. Express amplitude in both peak and rms.

In Figure 8–66(b), assume the amplitude and frequency are set to their maximum values. Measure and record these values. Express amplitude in both peak and rms.

Step 3: Measure the DC Offset

Assume the sinusoidal amplitude and frequency of the function generator are set to arbitrary values for dc offset measurement. The scope is set to dc coupling.

In Figure 8–67(a), the dc offset of the function generator is adjusted to the maximum positive value. Measure and record this value.

In Figure 8–67(b), the dc offset is adjusted to the maximum negative value. Measure and record this value.

Step 4: Measure the Triangular Output

The triangular output of the function generator is selected. The scope is set to ac coupling.

In Figure 8–68(a), the amplitude and frequency of the function generator are set to the minimum values. Measure and record these values.

In Figure 8–68(b), the amplitude and frequency are set to the maximum values. Measure and record these values.

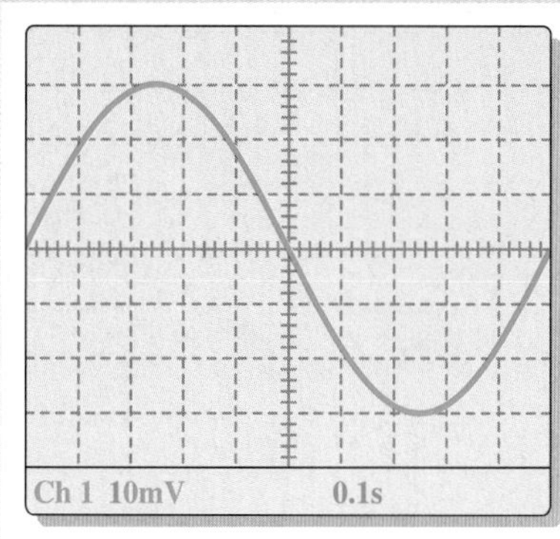

(a) The horizontal axis is 0 V.

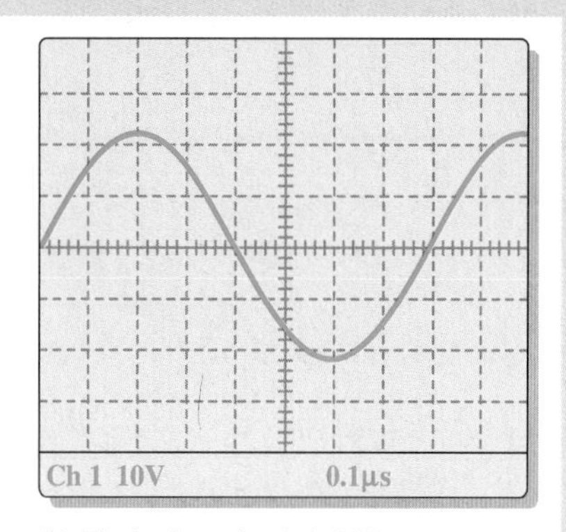

(b) The horizontal axis is 0 V.

FIGURE 8–66

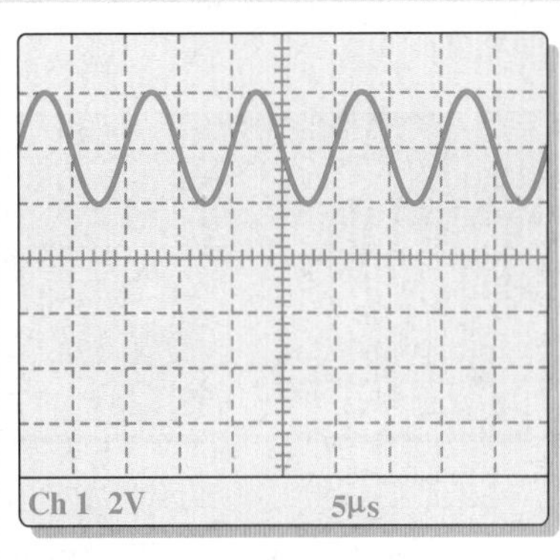

(a) The horizontal axis is 0 V.

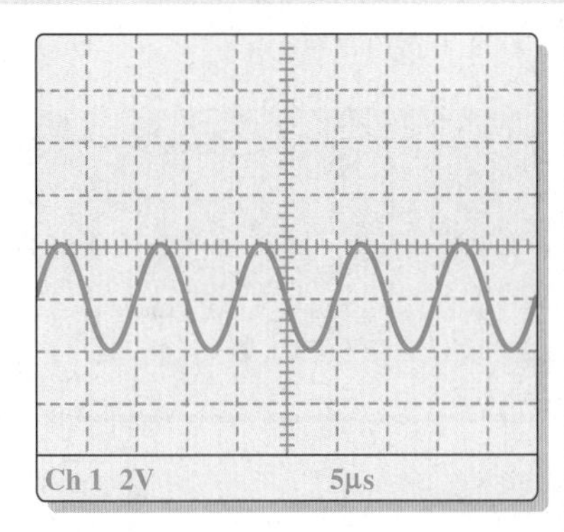

(b) The horizontal axis is 0 V.

FIGURE 8–67

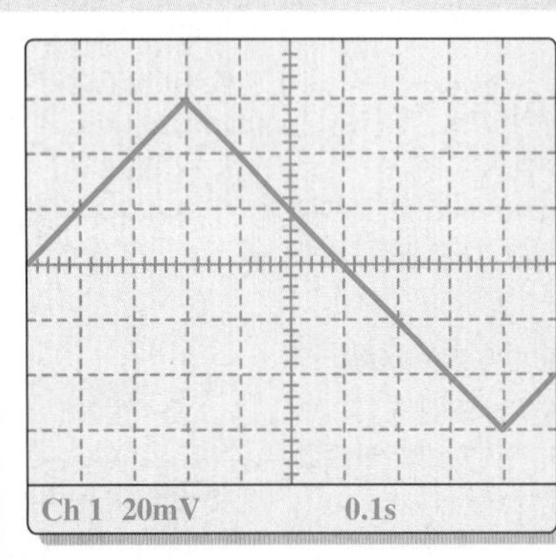

(a) The horizontal axis is 0 V.

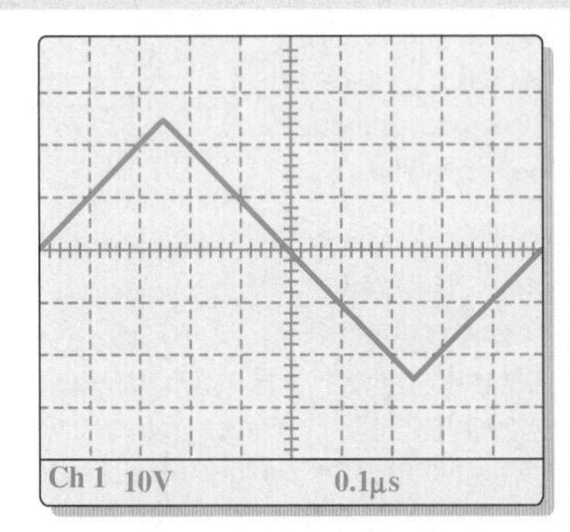

(b) The horizontal axis is 0 V.

FIGURE 8–68

Step 5: Measure the Pulse Output

The pulse output of the function generator is selected. The scope is set to dc coupling.

In Figure 8–69(a), the amplitude from the baseline and frequency of the function generator are set to the minimum values. The duty cycle is adjusted to minimum. Measure and record these values. Express the duty cycle as a percentage.

In Figure 8–69(b), the amplitude and frequency are set to the maximum values. The duty cycle is adjusted to maximum. Measure and record these values. Express the duty cycle as a percentage.

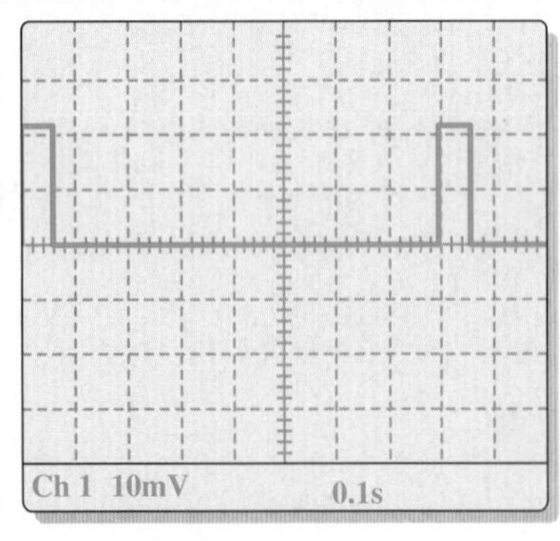

(a) The horizontal axis is 0 V.

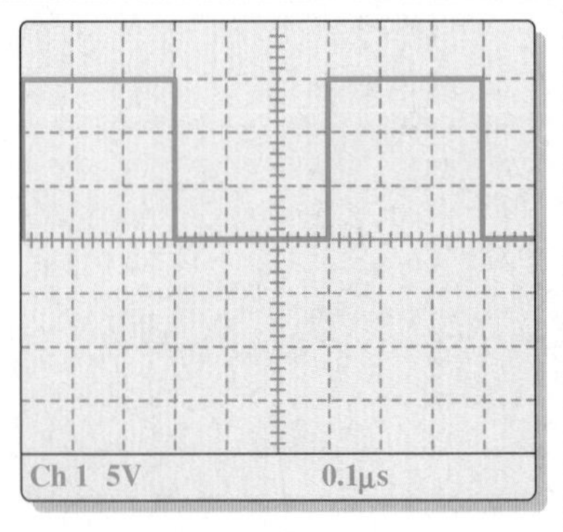

(b) The horizontal axis is 0 V.

FIGURE 8–69

APPLICATION ASSIGNMENT REVIEW

1. Generally, what setting of the Sec/Div control on a scope should be used to obtain the most accurate measurement of frequency?
2. Generally, what setting of the Volts/Div control on a scope should be used to obtain the most accurate measurement of amplitude?
3. Explain the purpose of each position of the AC-GND-DC switch on each channel of the analog scope.

SUMMARY

- The sine wave is a time-varying, periodic waveform.
- The sine wave is the fundamental type of alternating current (ac) and alternating voltage.
- Alternating current changes direction in response to changes in the polarity of the source voltage.
- One cycle of an alternating sine wave consists of a positive alternation and a negative alternation.
- A sine wave voltage can be generated by a conductor rotating in a magnetic field.
- Two common sources of sine waves are the electromagnetic ac generator and the electronic oscillator circuit.
- A full cycle of a sine wave is 360°, or 2π radians. A half-cycle is 180°, or π radian. A quarter-cycle is 90° or $\pi/2$ radians.
- Phase angle is the difference in degrees (or radians) between two sine waves or between a sine wave and a reference wave.
- The angular position of a phasor represents the angle of a sine wave, and the length of the phasor represents the amplitude.
- Voltages and currents must all be expressed with consistent units when you apply Ohm's or Kirchhoff's laws in ac circuits.
- Power in a resistive ac circuit is determined using rms voltage and/or current values.
- A pulse consists of a transition from a baseline level to an amplitude level, followed by a transition back to the baseline level.
- A triangular or sawtooth wave consists of positive-going and negative-going ramps.
- Harmonic frequencies are odd or even multiples of the repetition rate (or fundamental frequency) of a nonsinusoidal waveform.
- Conversions of sine wave values are summarized in Table 8–2.

TABLE 8–2

TO CHANGE FROM	TO	MULTIPLY BY
Peak	rms	0.707
Peak	Peak-to-peak	2
Peak	Average	0.637
rms	Peak	1.414
Peak-to-peak	Peak	0.5
Average	Peak	1.57

EQUATIONS

8–1	$f = \frac{1}{T}$	Frequency
8–2	$T = \frac{1}{f}$	Period
8–3	$f = \text{(number of pole pairs)(rps)}$	Output frequency of a generator
8–4	$V_{pp} = 2V_p$	Peak-to-peak voltage (sine wave)
8–5	$I_{pp} = 2I_p$	Peak to peak current (sine wave)
8–6	$V_{\text{rms}} = 0.707V_p$	Root-mean-square voltage (sine wave)
8–7	$I_{\text{rms}} = 0.707I_p$	Root-mean-square current (sine wave)
8–8	$V_p = 1.414V_{\text{rms}}$	Peak voltage (sine wave)
8–9	$I_p = 1.414I_{\text{rms}}$	Peak current (sine wave)
8–10	$V_{pp} = 2.828V_{\text{rms}}$	Peak-to-peak voltage (sine wave)
8–11	$I_{pp} = 2.828I_{\text{rms}}$	Peak-to-peak current (sine wave)
8–12	$V_{\text{avg}} = 0.637V_p$	Half-cycle average voltage (sine wave)
8–13	$I_{\text{avg}} = 0.637I_p$	Half-cycle average current (sine wave)
8–14	$\text{rad} = \left(\frac{\pi \text{ rad}}{180°}\right) \times \text{degrees}$	Degrees to radian conversion
8–15	$\text{degrees} = \left(\frac{180°}{\pi \text{ rad}}\right) \times \text{rad}$	Radian to degrees conversion
8–16	$y = A \sin \theta$	General formula for a sine wave
8–17	$v = V_p \sin \theta$	Sinusoidal voltage
8–18	$i = I_p \sin \theta$	Sinusoidal current
8–19	$y = A \sin(\theta - \phi)$	Lagging sine wave
8–20	$y = A \sin(\theta + \phi)$	Leading sine wave

8–21	**Percent duty cycle** $= \left(\frac{t_W}{T}\right)100\%$	Duty cycle
8–22	V_{avg} **= baseline + (duty cycle)(amplitude)**	Average value of pulse waveform

SELF-TEST

Answers are at the end of the chapter.

1. The difference between alternating current (ac) and direct current (dc) is
(a) ac changes value and dc does not
(b) ac changes direction and dc does not
(c) both (a) and (b) **(d)** neither (a) nor (b)

2. During each cycle, a sine wave reaches a peak value
(a) one time **(b)** two times
(c) four times **(d)** a number of times depending on the frequency

3. A sine wave with a frequency of 12 kHz is changing at a faster rate than a sine wave with a frequency of
(a) 20 kHz **(b)** 15,000 Hz **(c)** 10,000 Hz **(d)** 1.25 MHz

4. A sine wave with a period of 2 ms is changing at a faster rate than a sine wave with a period of
(a) 1 ms **(b)** 0.0025 s **(c)** 1.5 ms **(d)** 1000 μs

5. When a sine wave has a frequency of 60 Hz, in 10 s it goes through
(a) 6 cycles **(b)** 10 cycles **(c)** 1/16 cycle **(d)** 600 cycles

6. If the peak value of a sine wave is 10 V, the peak-to-peak value is
(a) 20 V **(b)** 5 V **(c)** 100 V **(d)** none of these

7. If the peak value of a sine wave is 20 V, the rms value is
(a) 14.14 V **(b)** 6.37 V **(c)** 7.07 V **(d)** 0.707 V

8. The average value of a 10 V peak sine wave over one complete cycle is
(a) 0 V **(b)** 6.37 V **(c)** 7.07 V **(d)** 5 V

9. The average half-cycle value of a sine wave with a 20 V peak is
(a) 0 V **(b)** 6.37 V **(c)** 12.74 V **(d)** 14.14 V

10. One sine wave has a positive-going zero crossing at 10° and another sine wave has a positive-going zero crossing at 45°. The phase angle between the two waveforms is
(a) 55° **(b)** 35° **(c)** 0° **(d)** none of these

11. The instantaneous value of a 15 A peak sine wave at a point 32° from its positive-going zero crossing is
(a) 7.95 A **(b)** 7.5 A **(c)** 2.13 A **(d)** 7.95 V

12. If the rms current through a 10 kΩ resistor is 5 mA, the rms voltage drop across the resistor is
(a) 70.7 V **(b)** 7.07 V **(c)** 5 V **(d)** 50 V

13. Two series resistors are connected to an ac source. If there is 6.5 V rms across one resistor and 3.2 V rms across the other, the peak source voltage is
(a) 9.7 V **(b)** 9.19 V **(c)** 13.72 V **(d)** 4.53 V

14. A 10 kHz pulse waveform consists of pulses that are 10 μs wide. Its duty cycle is
(a) 100% **(b)** 10% **(c)** 1% **(d)** not determinable

15. The duty cycle of a square wave
    (a) varies with the frequency (b) varies with the pulse width
    (c) both (a) and (b) (d) is 50%

PROBLEMS

Answers to odd-numbered problems are at the end of the book.

BASIC PROBLEMS

SECTION 8–1

The Sine Wave

1. Calculate the frequency for each of the following periods:
   (a) 1 s (b) 0.2 s (c) 50 ms
   (d) 1 ms (e) 500 μs (f) 10 μs
2. Calculate the period for each of the following frequencies:
   (a) 1 Hz (b) 60 Hz (c) 500 Hz
   (d) 1 kHz (e) 200 kHz (f) 5 MHz
3. A sine wave goes through 5 cycles in 10 μs. What is its period?
4. A sine wave has a frequency of 50 kHz. How many cycles does it complete in 10 ms?

SECTION 8–2

Sinusoidal Voltage Sources

5. The conductive loop on the rotor of a simple two-pole, single-phase generator rotates at a rate of 250 rps. What is the frequency of the induced output voltage?
6. A certain four-pole generator has a speed of rotation of 3600 rpm. What is the frequency of the voltage produced by this generator?
7. At what speed of rotation must a four-pole generator be operated to produce a 400 Hz sinusoidal voltage?

SECTION 8–3

Voltage and Current Values of Sine Waves

8. A sine wave has a peak value of 12 V. Determine the following voltage values:
   (a) rms (b) peak-to-peak (c) half-cycle average
9. A sinusoidal current has an rms value of 5 mA. Determine the following current values:
   (a) peak (b) half-cycle average (c) peak-to-peak
10. For the sine wave in Figure 8–70, determine the peak, peak-to-peak, rms, and average values.

▶ FIGURE 8–70

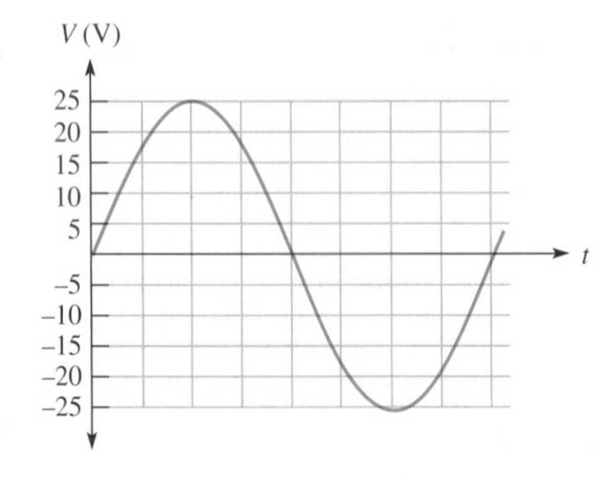

SECTION 8–4

Angular Measurement of a Sine Wave

11. Sine wave *A* has a positive-going zero crossing at 30° from a reference. Sine wave *B* has a positive-going zero crossing at 45° from the same reference. Determine the phase angle between the two signals. Which signal leads?

12. One sine wave has a positive peak at 75°, and another has a positive peak at 100°. How much is each sine wave shifted in phase from the 0° reference? What is the phase angle between them?

13. Make a sketch of two sine waves as follows: Sine wave *A* is the reference, and sine wave *B* lags *A* by 90°. Both have equal amplitudes.

14. Convert the following angular values from degrees to radians:

(a) 30° **(b)** 45° **(c)** 78°
(d) 135° **(e)** 200° **(f)** 300°

15. Convert the following angular values from radians to degrees:

(a) $\pi/8$ **(b)** $\pi/3$ **(c)** $\pi/2$
(d) $3\pi/5$ **(e)** $6\pi/5$ **(f)** 1.8π

SECTION 8–5

The Sine Wave Formula

16. A certain sine wave has a positive-going zero crossing at 0° and an rms value of 20 V. Calculate its instantaneous value at each of the following angles:

(a) 15° **(b)** 33° **(c)** 50° **(d)** 110°
(e) 70° **(f)** 145° **(g)** 250° **(h)** 325°

17. For a particular 0° reference sinusoidal current, the peak value is 100 mA. Determine the instantaneous value at each of the following points:

(a) 35° **(b)** 95° **(c)** 190°
(d) 215° **(e)** 275° **(f)** 360°

18. For a 0° reference sine wave with an rms value of 6.37 V, determine its instantaneous value at each of the following points:

(a) $\pi/8$ rad **(b)** $\pi/4$ rad **(c)** $\pi/2$ rad **(d)** $3\pi/4$ rad
(e) π rad **(f)** $3\pi/2$ rad **(g)** 2π rad

19. Sine wave *A* lags sine wave *B* by 30°. Both have peak values of 15 V. Sine wave *A* is the reference with a positive-going crossing at 0°. Determine the instantaneous value of sine wave *B* at 30°, 45°, 90°, 180°, 200°, and 300°.

20. Repeat Problem 19 for the case when sine wave *A* leads sine wave *B* by 30°.

SECTION 8–6

Ohm's Law and Kirchhoff's Laws in AC Circuits

21. A sinusoidal voltage is applied to the resistive circuit in Figure 8–71. Determine the following:

(a) I_{rms} **(b)** I_{avg} **(c)** I_p **(d)** I_{pp} **(e)** i at the positive peak

▶ **FIGURE 8–71**

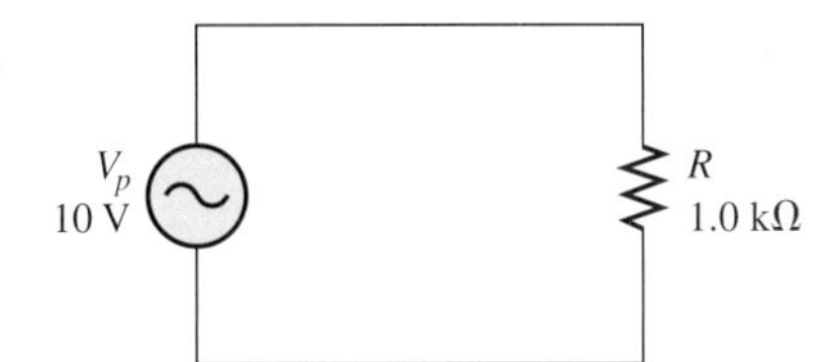

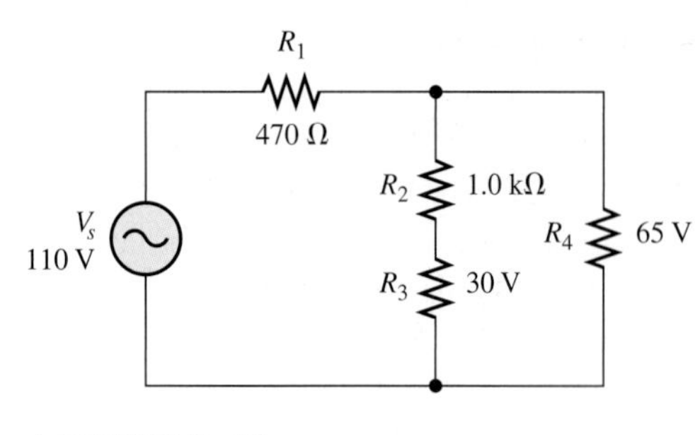

▲ FIGURE 8–72

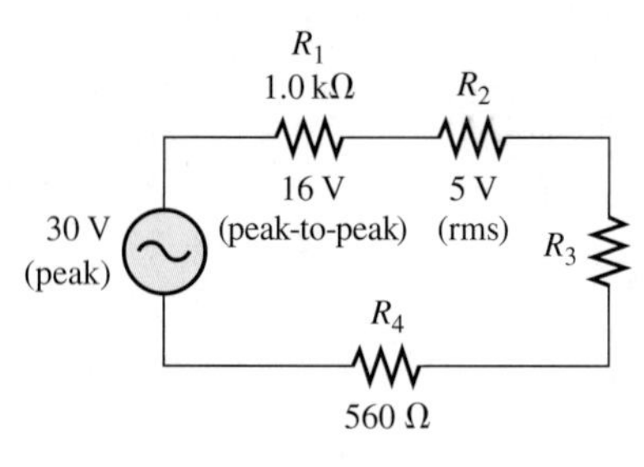

▲ FIGURE 8–73

22. Find the half-cycle average values of the voltages across R_1 and R_2 in Figure 8–72. All values shown are rms.

23. Determine the rms voltage across R_3 in Figure 8–73.

SECTION 8–7 **Superimposed DC and AC Voltages**

24. A sine wave with an rms value of 10.6 V is riding on a dc level of 24 V. What are the maximum and minimum values of the resulting waveform?

25. How much dc voltage must be added to a 3 V rms sine wave in order to make the resulting voltage nonalternating (no negative values)?

26. A 6 V peak sine wave is riding on a dc voltage of 8 V. If the dc voltage is lowered to 5 V, how far negative will the sine wave go?

SECTION 8–8 **Nonsinusoidal Waveforms**

27. From the graph in Figure 8–74, determine the approximate values of t_r, t_f, t_W, and amplitude.

▶ FIGURE 8–74

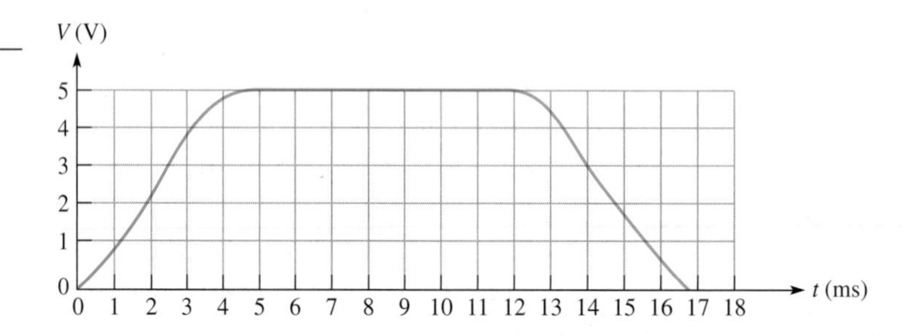

28. Determine the duty cycle for each pulse waveform in Figure 8–75.

29. Find the average value of each positive-going pulse waveform in Figure 8–75.

30. What is the frequency of each waveform in Figure 8–75.

▶ FIGURE 8–75

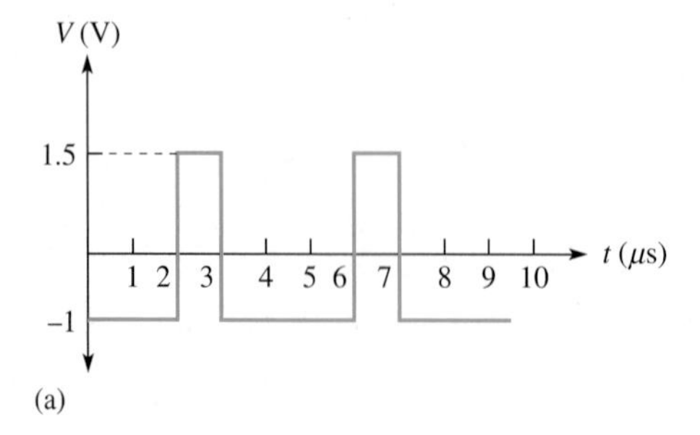

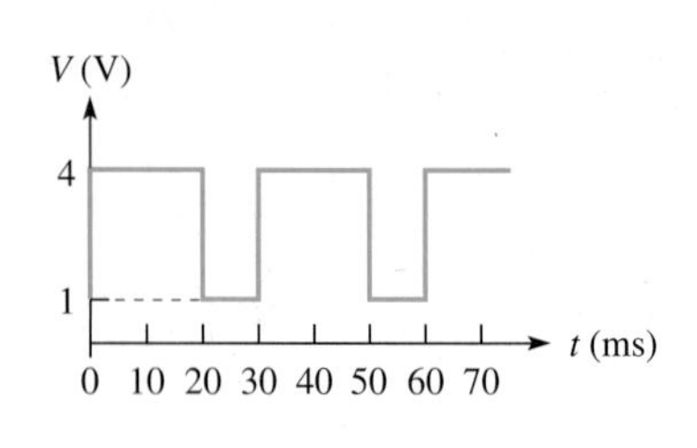

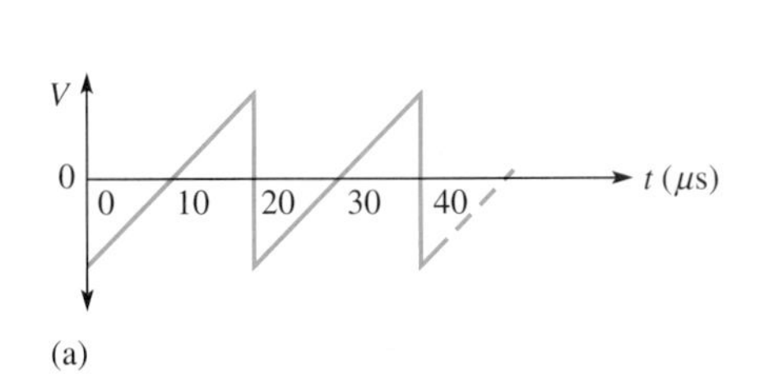

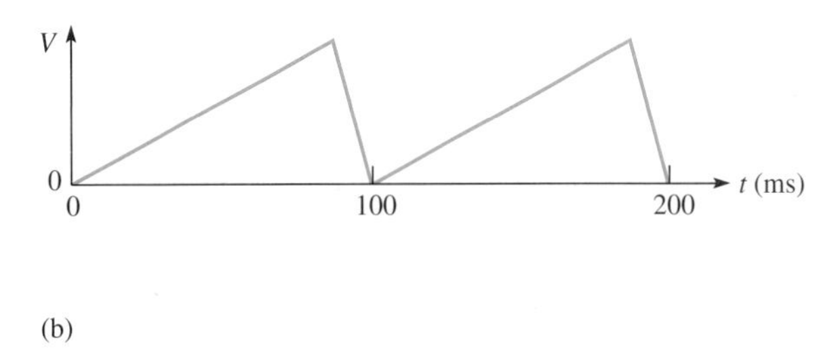

▲ FIGURE 8–76

31. What is the frequency of each sawtooth waveform in Figure 8–76?

32. A square wave has a period of 40 μs. List the first six odd harmonics.

33. What is the fundamental frequency of the square wave mentioned in Problem 32?

SECTION 8–9 **The Oscilloscope**

34. Determine the peak value and the period of the sine wave displayed on the scope screen in Figure 8–77. The horizontal axis is 0 V.

▶ FIGURE 8–77

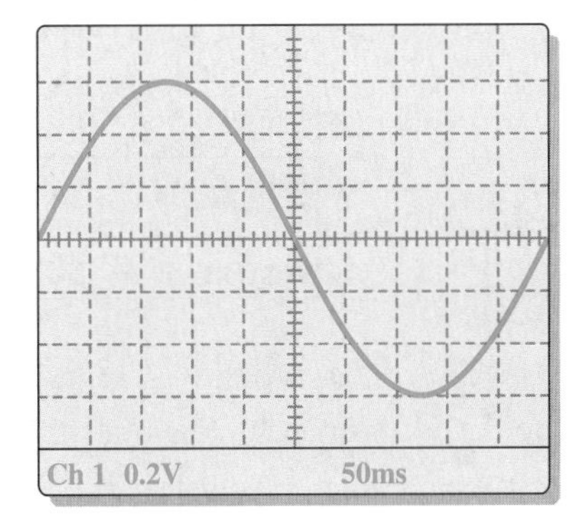

35. Determine the rms value and the frequency of the sine wave displayed on the scope screen in Figure 8–77.

36. Determine the rms value and the frequency of the sine wave displayed on the analog scope screen in Figure 8–78. The horizontal axis is 0 V.

37. Find the amplitude, pulse width, and duty cycle for the pulse waveform displayed on the scope screen in Figure 8–79. The horizontal axis is 0 V.

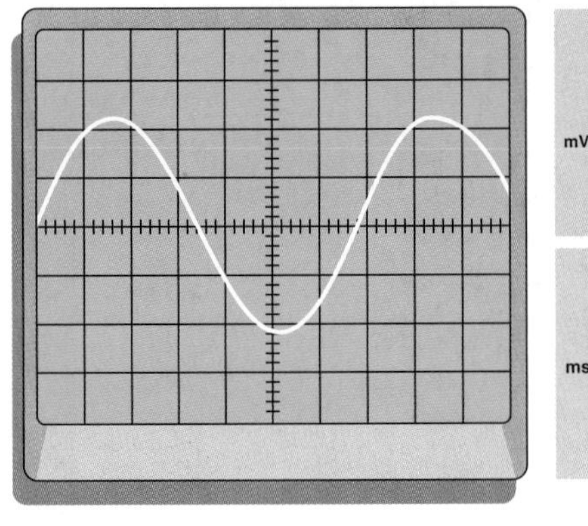

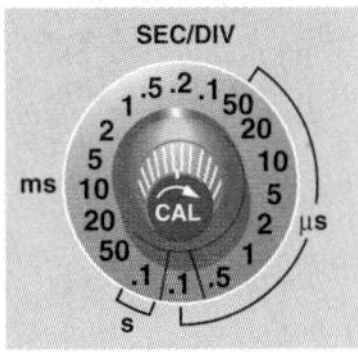

▲ FIGURE 8–78

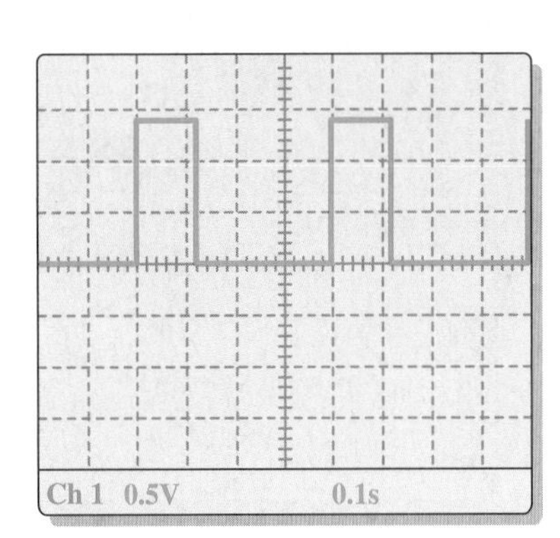

▲ FIGURE 8–79

ADVANCED PROBLEMS

38. A certain sine wave has a frequency of 2.2 kHz and an rms value of 25 V. Assuming a given cycle begins (zero crossing) at $t = 0$ s, what is the change in voltage from 0.12 ms to 0.2 ms?

39. Figure 8–80 shows a sinusoidal voltage source in series with a dc source. Effectively, the two voltages are superimposed. Sketch the voltage across R_L. Determine the maximum current through R_L and the average voltage across R_L.

40. A nonsinusoidal waveform called a *stairstep* is shown in Figure 8–81. Determine its average value.

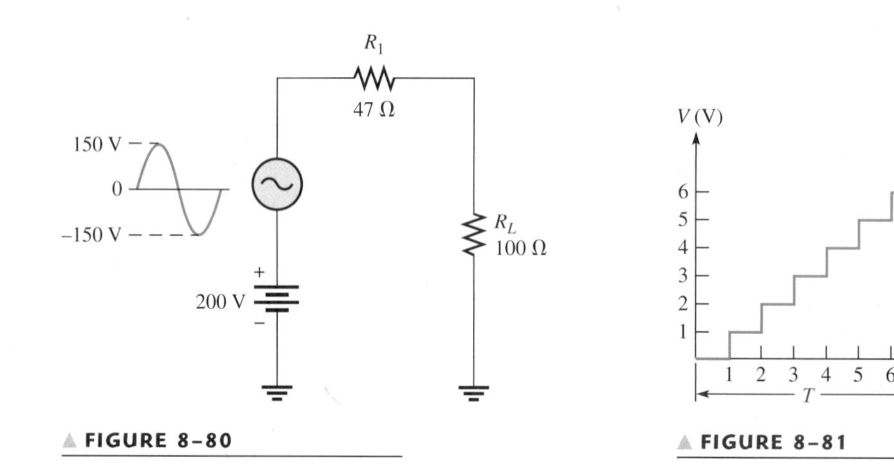

▲ FIGURE 8–80

▲ FIGURE 8–81

41. Refer to the oscilloscope in Figure 8–82.

(a) How many cycles are displayed?

(b) What is the rms value of the sine wave?

(c) What is the frequency of the sine wave?

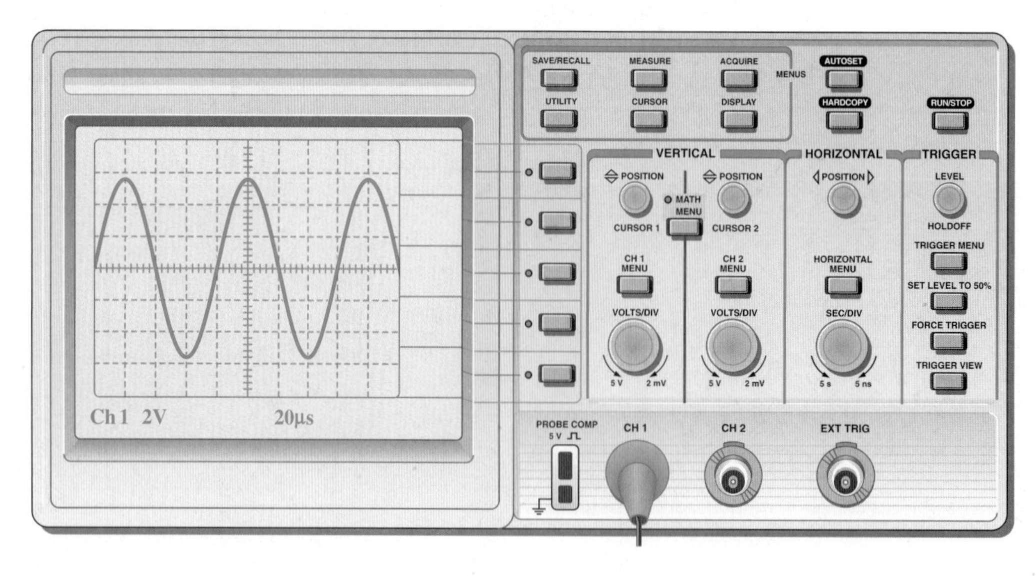

▲ FIGURE 8–82

42. Accurately draw on a grid representing the scope screen in Figure 8–82 how the sine wave will appear if the Volts/Div control is moved to a setting of 5 V.

43. Accurately draw on a grid representing the scope screen in Figure 8–82 how the sine wave will appear if the Sec/Div control is moved to a setting of 10 μs.

44. Based on the instrument settings and an examination of the scope display and the circuit board in Figure 8–83, determine the frequency and peak value of the input signal and output signal. The waveform shown is channel 1. Draw the channel 2 waveform as it would appear on the scope with the indicated settings. After working this problem, open file P08-44 and run the setup to observe the results.

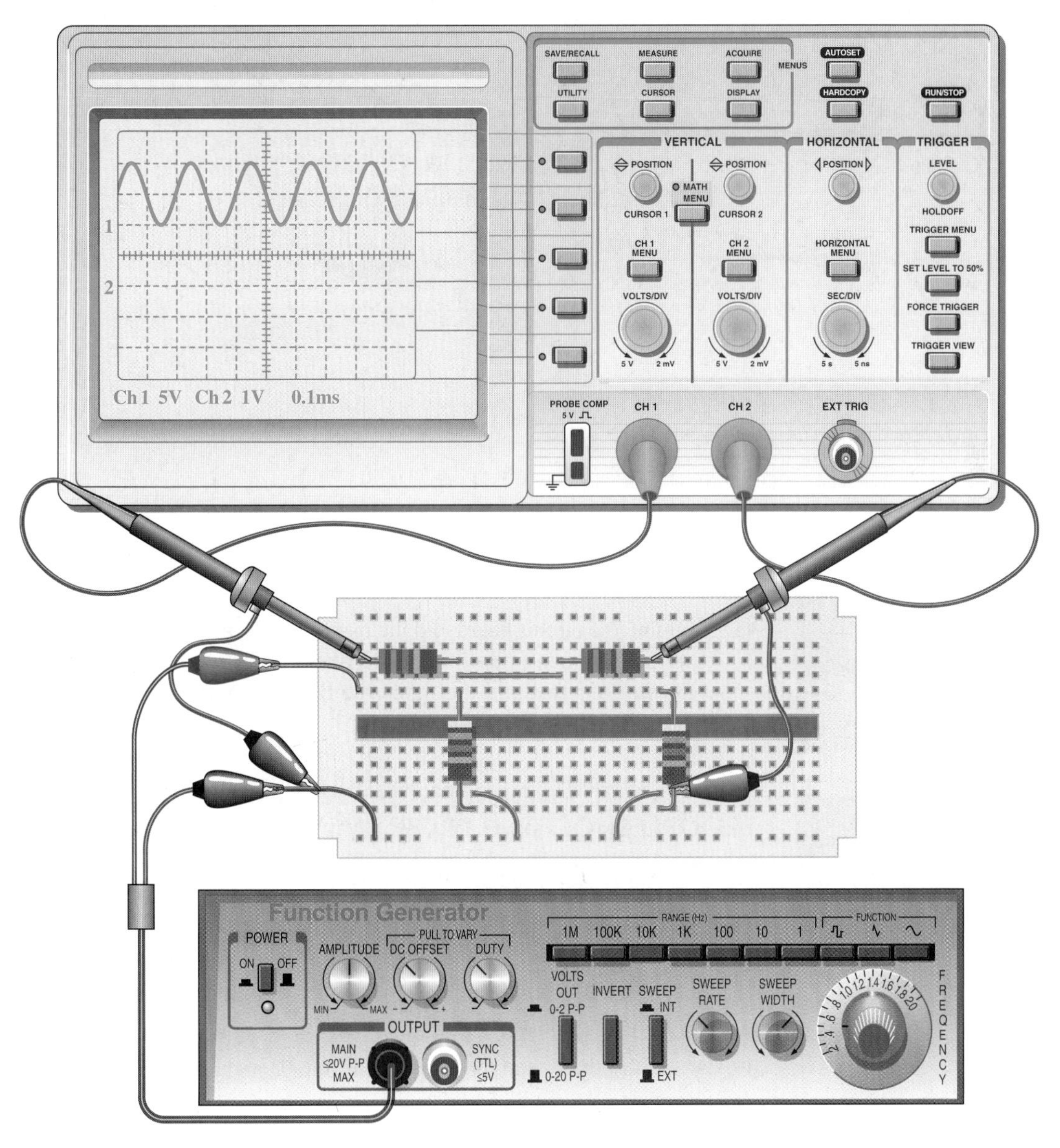

▲ **FIGURE 8–83**

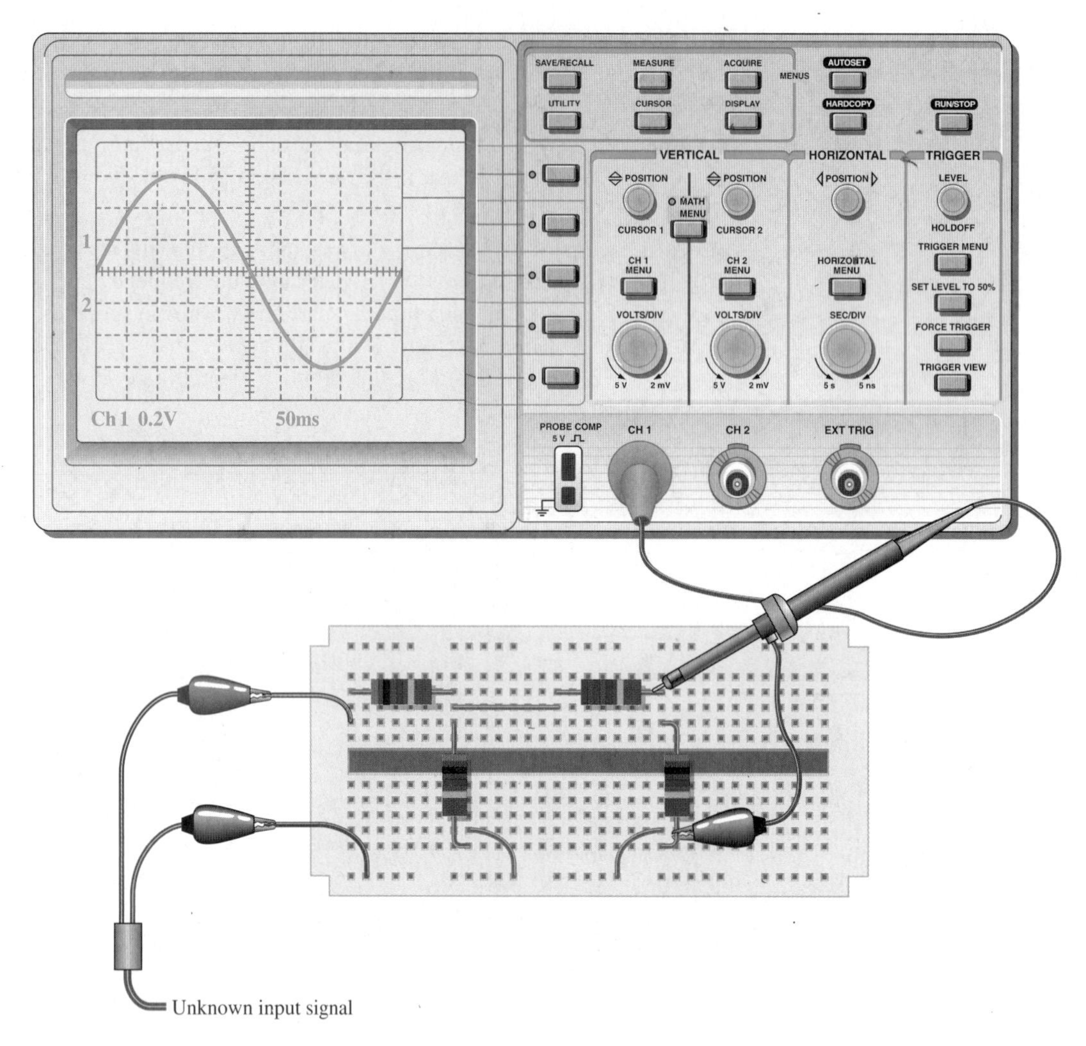

▲ **FIGURE 8–84**

45. Examine the circuit board and the oscilloscope display in Figure 8–84 and determine the peak value and the frequency of the unknown input signal. After working this problem, open file P08-45 and run the setup to observe the results.

ELECTRONICS WORKBENCH/CIRCUITMAKER TROUBLESHOOTING PROBLEMS

CD-ROM file circuits are shown in Figure 8–85.

46. Open file P08-46 and measure the peak value and period of the voltage sine wave using the oscilloscope.

47. Open file P08-47 and determine if there is a fault. If so, identify it.

48. Open file P08-48 and determine if there is a fault. If so, identify it.

49. Open file P08-49 and measure the amplitude and period of the pulse waveform using the oscilloscope.

50. Open file P08-50 and determine if there is a fault. If so, identify it.

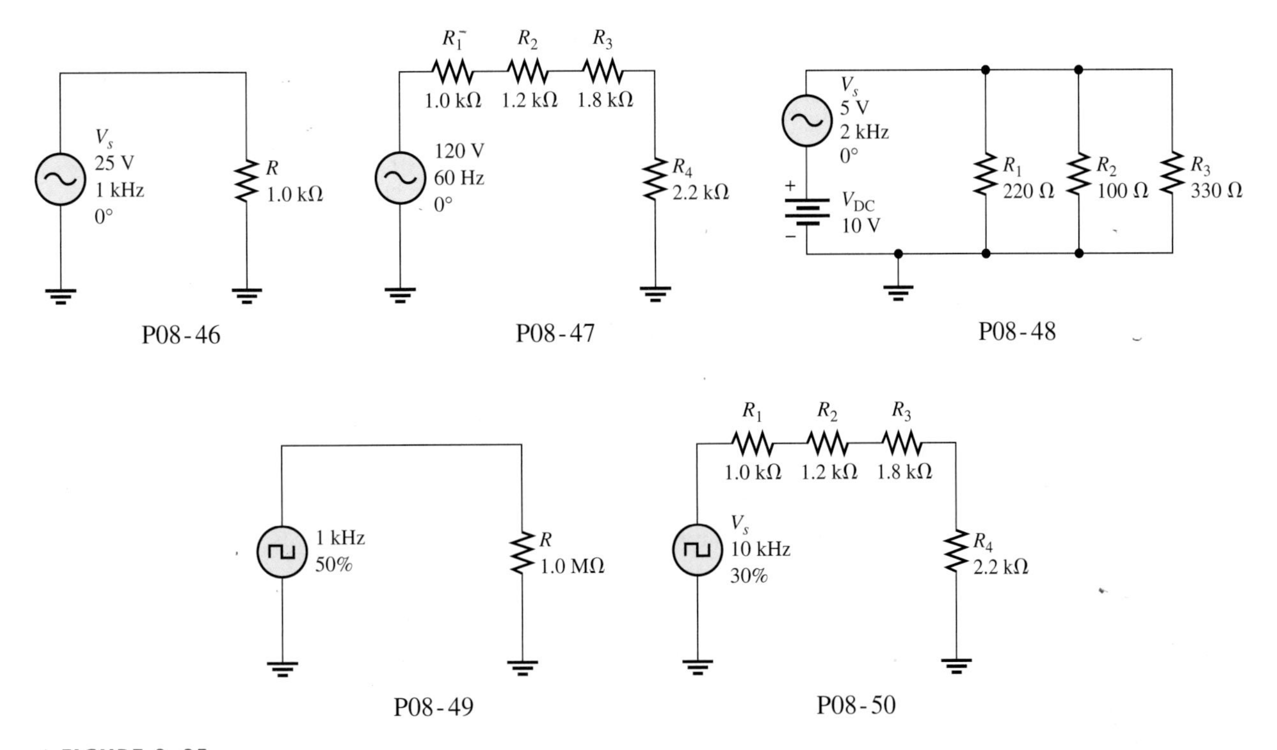

▲ **FIGURE 8–85**

ANSWERS

SECTION REVIEWS

SECTION 8–1 The Sine Wave

1. One cycle goes from the zero crossing through a positive peak, then through zero to a negative peak and back to the zero crossing.
2. A sine wave changes polarity at the zero crossings.
3. A sine wave has two maximum points (peaks) in one cycle.
4. The period is measured from one zero crossing to the next corresponding zero crossing, or from one peak to the next corresponding peak.
5. Frequency is the number of cycles completed in one second and its unit is the hertz.
6. $f = 1/5\ \mu\text{s} = 200$ kHz
7. $T = 1/120$ Hz $= 8.33$ ms

SECTION 8–2 Sinusoidal Voltage Sources

1. Two types of sine wave sources are electromagnetic and electronic.
2. Sine waves are generated electromagnetically by rotating a conductor in a magnetic field.
3. Frequency is dependent on the number of poles and the rps.
4. Amplitude is dependent on the number of turns and the rps.

SECTION 8–3 Voltage and Current Values of Sine Waves

1. **(a)** $V_{pp} = 2(1\text{ V}) = 2\text{ V}$ **(b)** $V_{pp} = 2(1.414)(1.414\text{ V}) = 4\text{ V}$
   (c) $V_{pp} = 2(1.57)(3\text{ V}) = 9.42\text{ V}$
2. **(a)** $V_{rms} = (0.707)(2.5\text{ V}) = 1.77\text{ V}$
   (b) $V_{rms} = (0.5)(0.707)(10\text{ V}) = 3.54\text{ V}$
   (c) $V_{rms} = (0.707)(1.57)(1.5\text{ V}) = 1.66\text{ V}$
3. **(a)** $V_{avg} = (0.637)(10\text{ V}) = 6.37\text{ V}$
   (b) $V_{avg} = (0.637)(1.414)(2.3\text{ V}) = 2.07\text{ V}$
   (c) $V_{avg} = (0.637)(0.5)(60\text{ V}) = 19.1\text{ V}$

SECTION 8–4 Angular Measurements of a Sine Wave

1. **(a)** The positive peak is at 90°.
   (b) The negative-going zero crossing is at 180°.
   (c) The negative peak is at 270°.
   (d) The cycle ends at 360°.
2. A half-cycle is completed in 180° or π radians.
3. A full cycle is completed in 360° or 2π radians.
4. $\theta = 90° - 45° = 45°$

SECTION 8–5 The Sine Wave Formula

1. **(a)** $\sin 30° = 0.5$ **(b)** $\sin 60° = 0.866$ **(c)** $\sin 90° = 1$
2. $v = 10 \sin 120° = 8.66\text{ V}$ **3.** $v = 10 \sin(45° + 10°) = 8.19\text{ V}$

SECTION 8–6 Ohm's Law and Kirchhoff's Laws in AC Circuits

1. $I_p = V_p/R = (1.57)(12.5\text{ V})/330\ \Omega = 59.5\text{ mA}$
2. $V_{s(rms)} = (0.707)(25.3\text{ V}) = 17.9\text{ V}$

SECTION 8–7 Superimposed DC and AC Voltages

1. $+V_{max} = 5\text{ V} + 2.5\text{ V} = 7.5\text{ V}$ **2.** Yes, it will alternate.
3. $+V_{max} = 5\text{ V} - 2.5\text{ V} = 2.5\text{ V}$

SECTION 8–8 Nonsinusoidal Waveforms

1. **(a)** Rise time is the time interval from 10% to 90% of the amplitude;
   (b) Fall time is the time interval from 90% to 10% of the amplitude;
   (c) Pulse width is the time interval from 50% of the leading pulse edge to 50% of the trailing pulse edge.
2. $f = 1/1\text{ ms} = 1\text{ kHz}$
3. d.c. = (1/5)100% = 20%; Ampl. 1.5 V; $V_{avg} = 0.5\text{ V} + 0.2(1.5\text{ V}) = 0.8\text{ V}$
4. $T = 16\text{ ms}$
5. $f = 1/T = 1/1\ \mu\text{s} = 1\text{ MHz}$
6. Fundamental frequency is the repetition rate of the waveform.
7. 2nd harm.: 2 kHz
8. $f = 1/10\ \mu\text{s} = 100\text{ kHz}$

SECTION 8–9 The Oscilloscope

1. Digital: Signal is converted to digital for processing and then reconstructed for display.
   Analog: Signal drives display directly.

2. Voltage is measured vertically; time is measured horizontally.
3. An oscilloscope can display time-varying quantities.
4. A 10× probe provides less loading for the circuit being tested.

Application Assignment

1. Lowest Sec/Div setting that allows frequency to be measured
2. Lowest Volts/Div setting that allows voltage to be measured
3. AC: Couples ac voltage only; GND: Grounds input so trace is at 0 V; DC: Couples ac and dc voltages

RELATED PROBLEMS FOR EXAMPLES

8–1 2.4 s

8–2 1.5 ms

8–3 20 kHz

8–4 200 Hz

8–5 1 ms

8–6 30 rps

8–7 $V_{pp} = 50$ V; $V_{\text{rms}} = 17.7$ V; $V_{\text{avg}} = 15.9$ V

8–8 **(a)** $\pi/12$ rad **(b)** 360°

8–9 8°

8–10 18.1 V

8–11 $I_{\text{rms}} = 4.53$ mA; $V_{1(\text{rms})} = 4.53$ V; $V_{2(\text{rms})} = 2.54$ V

8–12 23.7 V

8–13 The waveform in part (a) never goes negative. The waveform in part (b) goes negative for a portion of its cycle.

8–14 5%

8–15 1.2 V

8–16 120 V

8–17 **(a)** $V_{\text{rms}} = 1.06$ V; $f = 50$ Hz
(b) $V_{\text{rms}} = 88.4$ mV; $f = 1.67$ kHz
(c) $V_{\text{rms}} = 4.81$ V; $f = 5$ kHz
(d) $V_{\text{rms}} = 7.07$ V; $f = 250$ kHz

8–18 15 V, 20 ms, 50 Hz

SELF-TEST

1. (b) **2.** (b) **3.** (c) **4.** (b) **5.** (d) **6.** (a) **7.** (a)
8. (a) **9.** (c) **10.** (b) **11.** (a) **12.** (d) **13.** (c) **14.** (b)
15. (d)

CAPACITORS

INTRODUCTION

The capacitor is a device that can store electrical charge, thereby creating an electric field which in turn stores energy. The measure of the charge-storing ability is called *capacitance*.

In this chapter, the basic capacitor is introduced and its characteristics are studied. The physical construction and electrical properties of various types of capacitors are discussed. Series and parallel combinations are analyzed, and the basic behavior of capacitors in both dc and ac circuits is studied. Representative applications and methods of testing capacitors also are discussed.

CHAPTER OUTLINE

CHAPTER OBJECTIVES

- Describe the basic structure and characteristics of a capacitor
- Discuss various types of capacitors
- Analyze series capacitors
- Analyze parallel capacitors
- Describe how a capacitor operates in a dc switching circuit
- Describe how a capacitor operates in an ac circuit
- Discuss some capacitor applications
- Test a capacitor

KEY TERMS

- Capacitor
- Dielectric
- Capacitance
- Farad
- Coulomb's law
- Dielectric strength
- Temperature coefficient
- Dielectric constant
- Charging
- *RC* time constant
- Exponential
- Capacitive reactance
- Instantaneous power
- True power
- Reactive power
- VAR
- Coupling
- Decoupling
- Bypassing
- Filter

APPLICATION ASSIGNMENT PREVIEW

Capacitors are used in many applications. The application assignment in this chapter focuses on the use of capacitors in amplifier circuits to couple an ac voltage from one point to another while blocking dc voltage. You will determine if voltages measured on an amplifier circuit board with an oscilloscope are correct; and if they are not correct, you will troubleshoot the problem. After studying this chapter, you should be able to complete the application assignment.

WWW. **VISIT THE COMPANION WEBSITE**

Circuit Simulation Tutorials and Other Chapter Study Tools Are Available at

http://www.prenhall.com/floyd

9–1 THE BASIC CAPACITOR

In this section, the basic construction and characteristics of capacitors are examined.

After completing this section, you should be able to

- **Describe the basic structure and characteristics of a capacitor**
- Explain how a capacitor stores charge
- Define *capacitance* and state its unit
- State Coulomb's law
- Explain how a capacitor stores energy
- Discuss voltage rating and temperature coefficient
- Explain capacitor leakage
- Specify how the physical characteristics affect the capacitance

Basic Construction

In its simplest form, a **capacitor** is an electrical device constructed of two parallel conductive plates separated by an insulating material called the **dielectric**. Connecting leads are attached to the parallel plates. A basic capacitor is shown in Figure 9–1(a), and the schematic symbol is shown in part (b).

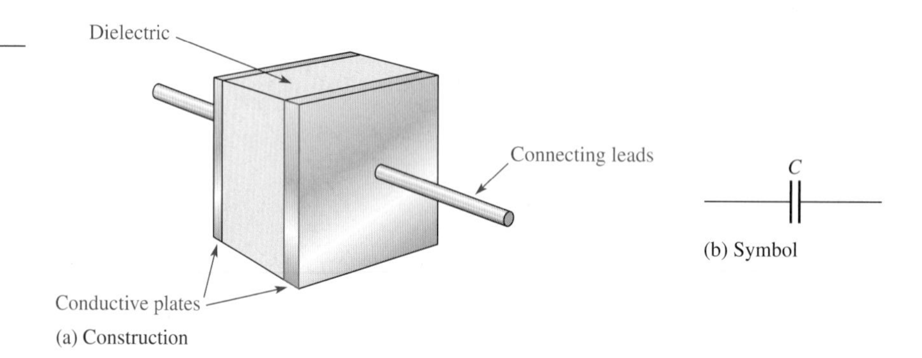

FIGURE 9–1
The basic capacitor.

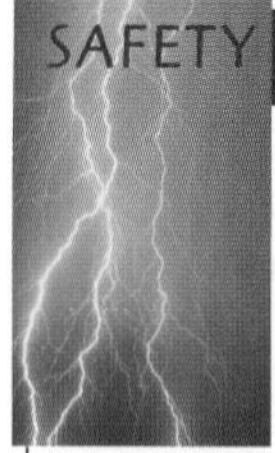

SAFETY POINT

Capacitors are capable of storing electrical charge for a long time after power has been turned off in a circuit. Be careful when touching or handling capacitors in or out of a circuit. If you touch the leads, you may be in for a shock as the capacitor discharges through you! It is usually good practice to discharge a capacitor using a shorting tool with an insulated grip of some sort before handling the capacitor.

How a Capacitor Stores Charge

In the neutral state, both plates of a capacitor have an equal number of free electrons, as indicated in Figure 9–2(a). When the capacitor is connected to a dc voltage source through a resistor, as shown in part (b), electrons (negative charge) are removed from plate *A*, and an equal number are deposited on plate *B*. As plate *A* loses electrons and plate *B* gains electrons, plate *A* becomes positive with respect to plate *B*. During this charging process, electrons flow only through the connecting leads and the source. No electrons flow through the dielectric of the capacitor because it is an insulator. The movement of electrons ceases when the voltage across the capacitor equals the source voltage, as indicated in Figure 9–2(c). If the capacitor is disconnected from the source, it retains the stored charge for a long period of time (the length of time depends on the type of capacitor) and still has the voltage across it, as shown in Figure 9–2(d). Actually, the charged capacitor can be considered as a temporary battery.

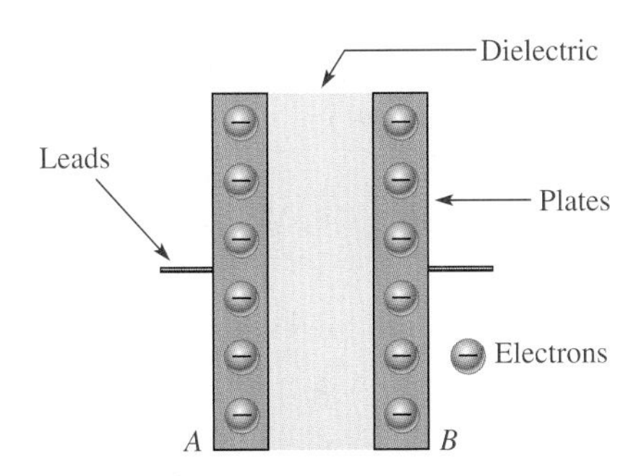

(a) Neutral (uncharged) capacitor (same charge on both plates)

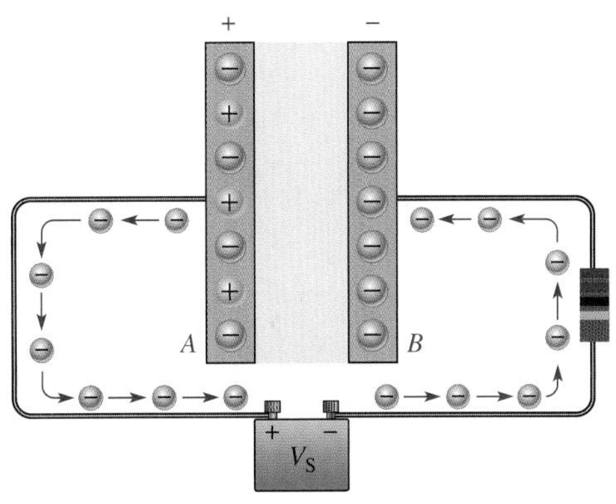

(b) Electrons flow from plate A to plate B as capacitor charges.

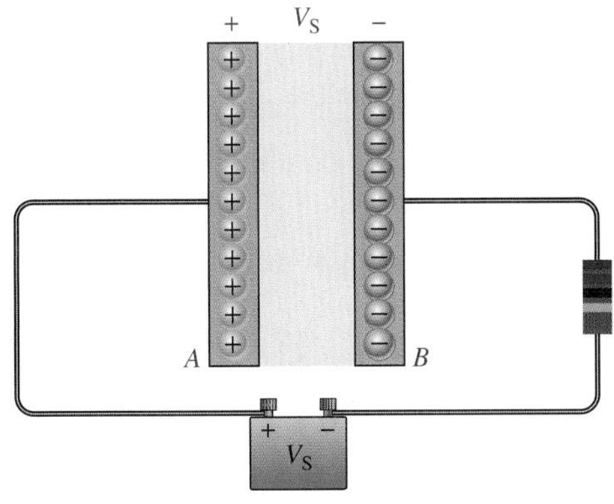

(c) Capacitor charged to V_S. No more electrons flow.

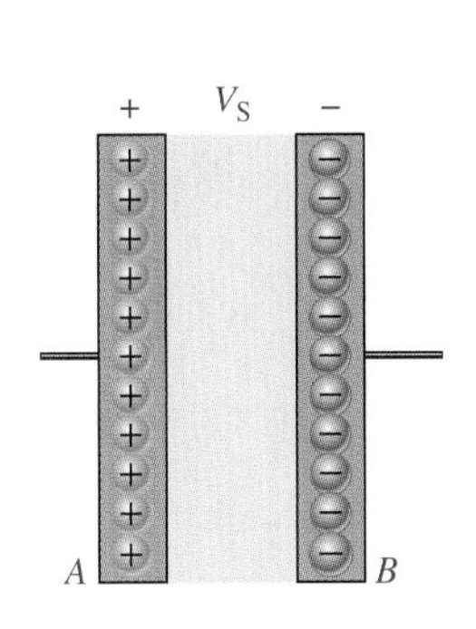

(d) Capacitor retains charge when disconnected from source.

▲ **FIGURE 9–2**

Illustration of a capacitor storing charge.

Capacitance

The amount of charge that a capacitor can store per unit of voltage across its plates is its **capacitance**, designated C. That is, capacitance is a measure of a capacitor's ability to store charge. The more charge per unit of voltage that a capacitor can store, the greater its capacitance, as expressed by the following formula:

Equation 9–1

$$C = \frac{Q}{V}$$

where C is capacitance, Q is charge, and V is voltage.

By rearranging Equation 9–1, you can obtain two other formulas:

Equation 9–2

$$Q = CV$$

Equation 9–3

$$V = \frac{Q}{C}$$

The Unit of Capacitance The **farad** (F) is the basic unit of capacitance. Recall that the coulomb (C) is the unit of electrical charge.

One farad is the amount of capacitance when one coulomb of charge is stored with one volt across the plates.

Most capacitors that are used in electronics work have capacitance values in microfarads (μF) and picofarads (pF).

A microfarad is one-millionth of a farad (1 μF = 1 × 10^{-6} F).

A picofarad is one-trillionth of a farad (1 pF = 1 × 10^{-12} F).

Conversions for farads, microfarads, and picofarads are given in Table 9–1.

TABLE 9–1

Conversions for farads, microfarads, and picofarads.

TO CONVERT FROM	TO	MOVE THE DECIMAL POINT
Farads	Microfarads	6 places to right ($\times 10^6$)
Farads	Picofarads	12 places to right ($\times 10^{12}$)
Microfarads	Farads	6 places to left ($\times 10^{-6}$)
Microfarads	Picofarads	6 places to right ($\times 10^6$)
Picofarads	Farads	12 places to left ($\times 10^{-12}$)
Picofarads	Microfarads	6 places to left ($\times 10^{-6}$)

EXAMPLE 9–1

(a) A certain capacitor stores 50 microcoulombs (50 μC) when 10 V are applied across its plates. What is its capacitance?

(b) A 2 μF capacitor has 100 V across its plates. How much charge does it store?

(c) Determine the voltage across a 100 pF capacitor that is storing 2 μC of charge.

Solution **(a)** $C = \frac{Q}{V} = \frac{50\ \mu\text{C}}{10\ \text{V}} = \mathbf{5\ \mu F}$

(b) $Q = CV = (2\ \mu\text{F})(100\ \text{V}) = \mathbf{200\ \mu C}$

(c) $V = \frac{Q}{C} = \frac{2\ \mu\text{C}}{100\ \text{pF}} = 0.02 \times 10^6 = 20 \times 10^3\ \text{V} = \mathbf{20\ kV}$

The calculator solutions for C, Q, and V are

Related Problem* Determine V if $C = 1000$ pF and $Q = 10\ \mu$C.

*Answers are at the end of the chapter.

EXAMPLE 9–2

Convert the following values to microfarads:

(a) 0.00001 F **(b)** 0.0047 F **(c)** 1000 pF **(d)** 220 pF

Solution **(a)** $0.00001\ \text{F} \times 10^6 = \mathbf{10\ \mu F}$ **(b)** $0.0047\ \text{F} \times 10^6 = \mathbf{4700\ \mu F}$

(c) $1000\ \text{pF} \times 10^{-6} = \mathbf{0.001\ \mu F}$ **(d)** $220\ \text{pF} \times 10^{-6} = \mathbf{0.00022\ \mu F}$

Related Problem Convert 47,000 pF to microfarads.

EXAMPLE 9–3

Convert the following values to picofarads:

(a) 0.1×10^{-8} F **(b)** 0.000027 F **(c)** 0.01 μF **(d)** 0.0047 μF

Solution

(a) $0.1 \times 10^{-8}\ \text{F} \times 10^{12} = \mathbf{1000\ pF}$

(b) $0.000027\ \text{F} \times 10^{12} = \mathbf{27 \times 10^6\ pF}$

(c) $0.01\ \mu\text{F} \times 10^{6} = \mathbf{10{,}000\ pF}$

(d) $0.0047\ \mu\text{F} \times 10^{6} = \mathbf{4700\ pF}$

Related Problem Convert 100 μF to picofarads.

How a Capacitor Stores Energy

A capacitor stores energy in the form of an electric field that is established by the opposite charges on the two plates. The electric field is represented by lines of force between the positive and negative charges and concentrated within the dielectric, as shown in Figure 9–3.

Coulomb's law states

A force exists between two point-source charges that is directly proportional to the product of the two charges and inversely proportional to the square of the distance between the charges.

This relationship is expressed as

$$F = \frac{kQ_1Q_2}{d^2}$$

Equation 9–4

where F is the force in newtons, Q_1 and Q_2 are the charges in coulombs, d is the distance between the charges in meters, and k is a proportionality constant equal to 9×10^9 newton-meter²/coulomb².

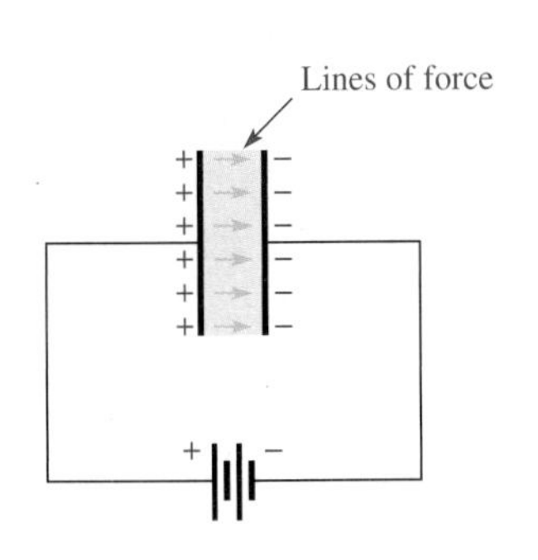

▲ **FIGURE 9–3**

The electric field stores energy in a capacitor.

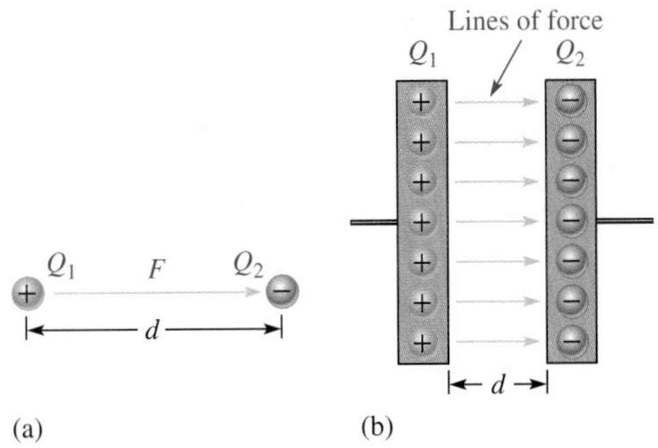

▲ **FIGURE 9–4**

Lines of force are created by opposite charges.

Figure 9–4(a) illustrates the line of force between a positive and a negative charge. Figure 9–4(b) shows that many opposite charges on the plates of a capacitor create many lines of force, which form an electric field between the plates that stores energy within the dielectric.

The greater the forces between the charges on the plates of a capacitor, the more energy is stored. Therefore, the amount of energy stored is directly proportional to the capacitance because, from Coulomb's law, the more charge stored, the greater the force.

Also, from the equation $Q = CV$, the amount of charge stored is directly related to the voltage as well as the capacitance. Therefore, the amount of energy stored is also dependent on the square of the voltage across the plates of the capacitor. The formula for the energy stored by a capacitor is

Equation 9–5

$$W = \frac{1}{2}CV^2$$

When capacitance (C) is in farads and voltage (V) is in volts, the energy (W) is in joules.

Voltage Rating

Every capacitor has a limit on the amount of voltage that it can withstand across its plates. The voltage rating specifies the maximum dc voltage that can be applied without risk of damage to the device. If this maximum voltage, commonly called the *breakdown voltage* or *working voltage,* is exceeded, permanent damage to the capacitor can result.

Both the capacitance and the voltage rating must be taken into consideration before a capacitor is used in a circuit application. The choice of capacitance value is based on particular circuit requirements. The voltage rating should always be above the maximum voltage expected in a particular application.

Dielectric Strength The breakdown voltage of a capacitor is determined by the **dielectric strength** of the dielectric material used. The dielectric strength is expressed in V/mil (1 mil = 0.001 in.). Table 9–2 lists typical values for several materials. Exact values vary depending on the specific composition of the material.

TABLE 9–2

Some common dielectric materials and their typical dielectric strengths.

MATERIAL	DIELECTRIC STRENGTH (V/MIL)
Air	80
Oil	375
Ceramic	1000
Paper (paraffined)	1200
Teflon®	1500
Mica	1500
Glass	2000

The dielectric strength can best be explained by an example. Assume that a certain capacitor has a plate separation of 1 mil and that the dielectric material is ceramic. This particular capacitor can withstand a maximum voltage of 1000 V because its dielectric strength is 1000 V/mil. If the maximum voltage is exceeded, the dielectric may break down and conduct current, causing permanent damage to the capacitor. If the ceramic capacitor has a plate separation of 2 mils, its breakdown voltage is 2000 V.

Temperature Coefficient

The **temperature coefficient** indicates the amount and direction of a change in capacitance value with temperature. A positive temperature coefficient means that the capacitance increases with an increase in temperature or decreases with a decrease in temperature. A negative coefficient means that the capacitance decreases with an increase in temperature or increases with a decrease in temperature.

Temperature coefficients typically are specified in parts per million per Celsius degree (ppm/°C). For example, a negative temperature coefficient of 150 ppm/°C for a 1 μF capacitor means that for every Celsius degree rise in temperature, the capacitance decreases by 150 pF (there are one million picofarads in one microfarad).

Leakage

No insulating material is perfect. The dielectric of any capacitor will conduct some very small amount of current. Thus, the charge on a capacitor will eventually leak off. Some types of capacitors have higher leakages than others. An equivalent circuit for a nonideal capacitor is shown in Figure 9–5. The parallel resistor R_{leak} represents the extremely high resistance (several hundred kΩ or more) of the dielectric material through which there is leakage current.

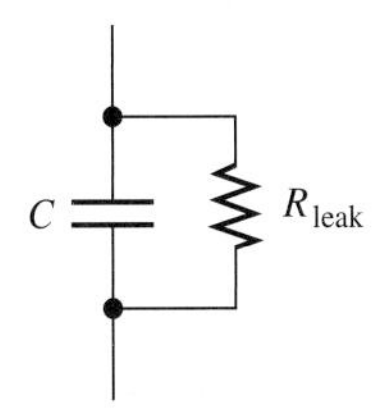

▲ **FIGURE 9–5**

Equivalent circuit for a nonideal capacitor.

Physical Characteristics of Capacitors

The following parameters are important in establishing the capacitance and the voltage rating of a capacitor: plate area, plate separation, and dielectric constant.

Plate Area *Capacitance is directly proportional to the physical size of the plates as determined by the plate area, A.* A large plate area produces a larger capacitance, and a smaller plate area produces a smaller capacitance. Figure 9–6(a) shows that the plate area of a parallel plate capacitor is the area of one of the plates. If the plates are moved in relation to each other, as shown in Figure 9–6(b), the overlapping area determines the effective plate area. This variation in effective plate area is the basis for a certain type of variable capacitor.

Plate Separation *Capacitance is inversely proportional to the distance between the plates.* The plate separation is designated *d*, as shown in Figure 9–7. A greater separation of the plates produces a lesser capacitance, as illustrated in the figure. The breakdown voltage is directly proportional to the plate separation. The further the plates are separated, the greater the breakdown voltage.

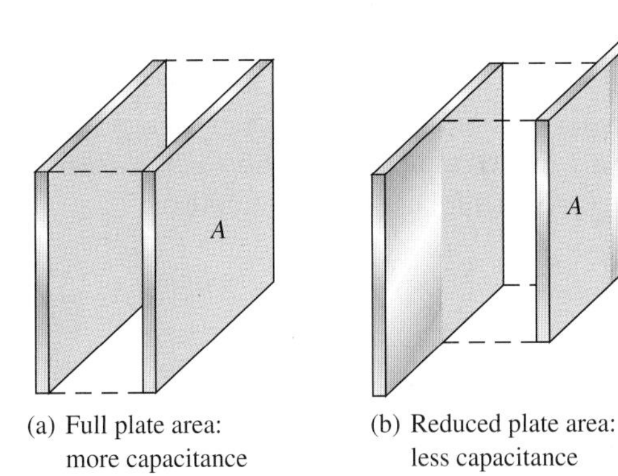

▲ **FIGURE 9–6**

Capacitance is directly proportional to plate area (A).

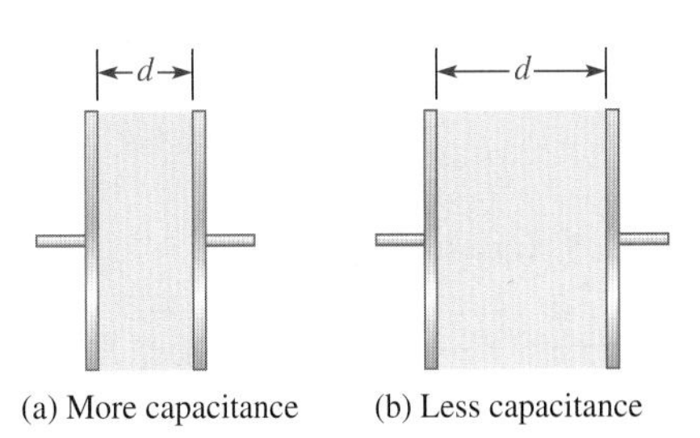

▲ **FIGURE 9–7**

Capacitance is inversely proportional to the distance *d* between the plates.

Dielectric Constant As you know, the insulating material between the plates of a capacitor is called the *dielectric.* Every dielectric material has the ability to concentrate the lines of force of the electric field existing between the oppositely charged plates of a capacitor and thus increase the capacity for energy storage. The measure of a material's ability to establish an electric field is called the

dielectric constant or *relative permittivity,* symbolized by ϵ_r (the Greek letter epsilon).

Capacitance is directly proportional to the dielectric constant. The dielectric constant of a vacuum is defined as 1, and that of air is very close to 1. These values are used as a reference, and all other materials have values of ϵ_r specified with respect to that of a vacuum or air. For example, a material with $\epsilon_r = 5$ can have a capacitance five times greater than that of air, with all other factors being equal.

Table 9–3 lists several common dielectric materials and typical dielectric constants for each. Values can vary because they depend on the specific composition of the material.

TABLE 9–3

Some common dielectric materials and their typical dielectric constants.

MATERIAL	TYPICAL ϵ_r VALUES
Air (vacuum)	1.0
Teflon®	2.0
Paper (paraffined)	2.5
Oil	4.0
Mica	5.0
Glass	7.5
Ceramic	1200

The dielectric constant (relative permittivity) is dimensionless because it is a relative measure. It is a ratio of the absolute permittivity of a material, ϵ, to the absolute permittivity of a vacuum, ϵ_0, as expressed by the following formula:

Equation 9–6

$$\epsilon_r = \frac{\epsilon}{\epsilon_0}$$

The value of ϵ_0 is 8.85×10^{-12} F/m (farads per meter).

Formula for Capacitance in Terms of Physical Parameters

You have seen how capacitance is directly related to plate area, A, and the dielectric constant, ϵ_r, and inversely related to plate separation, d. An exact formula for calculating the capacitance in terms of these three quantities is

Equation 9–7

$$C = \frac{A\epsilon_r(8.85 \times 10^{-12}\ \text{F/m})}{d}$$

where A is in square meters (m^2), d is in meters (m), and C is in farads (F).

EXAMPLE 9–4

Determine the capacitance in pF of a parallel plate capacitor having a plate area of 0.01 m^2 and a plate separation of 0.02 m. The dielectric is mica, which has a dielectric constant of 5.0.

Solution Use Equation 9–7.

$$C = \frac{A\epsilon_r(8.85 \times 10^{-12}\ \text{F/m})}{d} = \frac{(0.01\ \text{m}^2)(5.0)(8.85 \times 10^{-12}\ \text{F/m})}{0.02\ \text{m}} = \mathbf{22.1\ pF}$$

• 0 1 × 5 × 8 • 8 5
EE (–) 1 2 ÷ • 0 2 ENTER

```
.01*5*8.85E-12/.02
        22.125E-12
```

Related Problem Determine C in pF where $A = 0.005\ \text{m}^2$, $d = 0.008$ m, and ceramic is the dielectric.

SECTION 9–1 REVIEW

Answers are at the end of the chapter.

1. Define *capacitance.*
2. (a) How many microfarads in one farad?
   (b) How many picofarads in one farad?
   (c) How many picofarads in one microfarad?
3. Convert 0.0015 μF to picofarads. To farads.
4. How much energy in joules is stored by a 0.01 μF capacitor with 15 V across its plates?
5. (a) When the plate area of a capacitor is increased, does the capacitance increase or decrease?
   (b) When the distance between the plates is increased, does the capacitance increase or decrease?
6. The plates of a ceramic capacitor are separated by 10 mils. What is the typical breakdown voltage?
7. A ceramic capacitor has a plate area of 0.02 m^2. The thickness of the dielectric is 0.0005 m. What is the capacitance in μF?

9–2 TYPES OF CAPACITORS

Capacitors normally are classified according to the type of dielectric material. The most common types of dielectric materials are mica, ceramic, plastic-film, and electrolytic (aluminum oxide and tantalum oxide). In this section, the characteristics and construction of each of these types of capacitors and the variable capacitors are examined.

After completing this section, you should be able to

- **Discuss various types of capacitors**
- Describe the characteristics of mica, ceramic, plastic-film, and electrolytic capacitors
- Describe types of variable capacitors
- Identify capacitor labeling

Fixed Capacitors

Mica Capacitors Two types of mica capacitors are stacked-foil and silver-mica. The basic construction of the stacked-foil type is shown in Figure 9–8. It consists of alternate layers of metal foil and thin sheets of mica. The metal foil forms the plate, with alternate foil sheets connected together to increase the plate area. More layers are used to increase the plate area, thus increasing the capacitance. The mica/foil stack is encapsulated in an insulating material such as Bakelite®, as shown in Figure 9–8(b). The silver-mica capacitor is formed in a similar way by stacking mica sheets with silver electrode material screened on them.

FIGURE 9–8

Construction of a typical radial-lead mica capacitor.

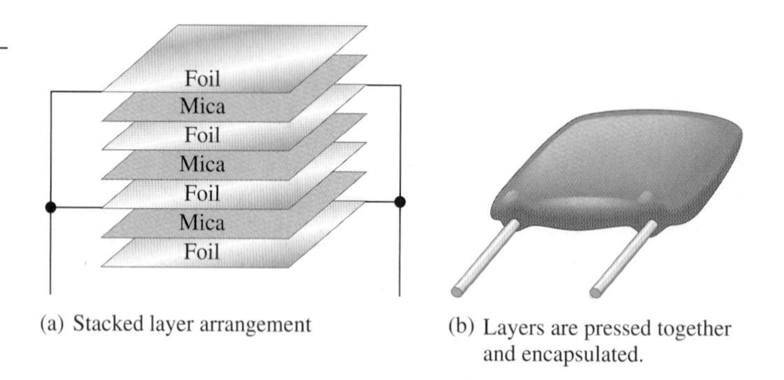

(a) Stacked layer arrangement

(b) Layers are pressed together and encapsulated.

Mica capacitors are generally available with capacitance values ranging from 1 pF to 0.1 μF and voltage ratings from 100 V dc to 2500 V dc and higher. Common temperature coefficients range from -20 ppm/°C to $+100$ ppm/°C. Mica has a typical dielectric constant of 5.

Ceramic Capacitors Ceramic dielectrics provide very high dielectric constants (1200 is typical). As a result, comparatively high capacitance values can be achieved in a small physical size. Ceramic capacitors are commonly available in a ceramic disk form, as shown in Figure 9–9; in a multilayer radial-lead configuration, as shown in Figure 9–10; or in a leadless ceramic chip, as shown in Figure 9–11.

Ceramic capacitors typically are available in capacitance values ranging from 1 pF to 2.2 μF with voltage ratings up to 6 kV. A typical temperature coefficient for ceramic capacitors is 200,000 ppm/°C.

FIGURE 9–9

A ceramic disk capacitor and its basic construction.

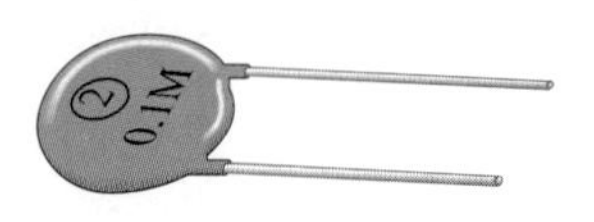

(a)

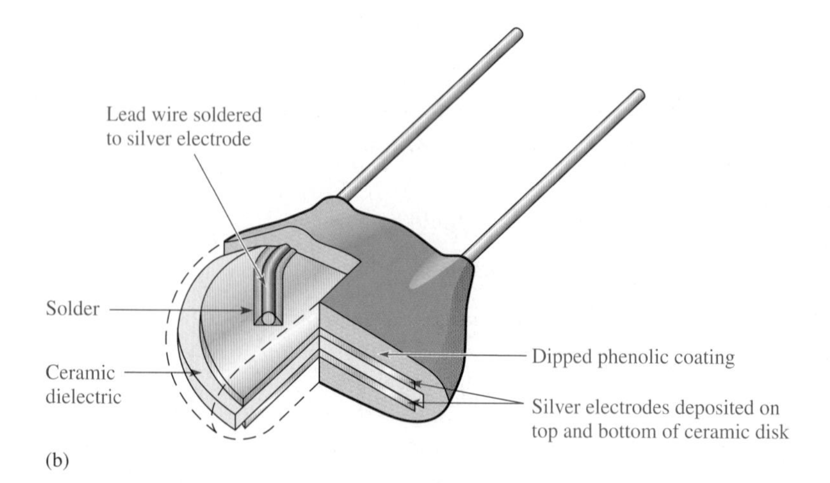

(b)

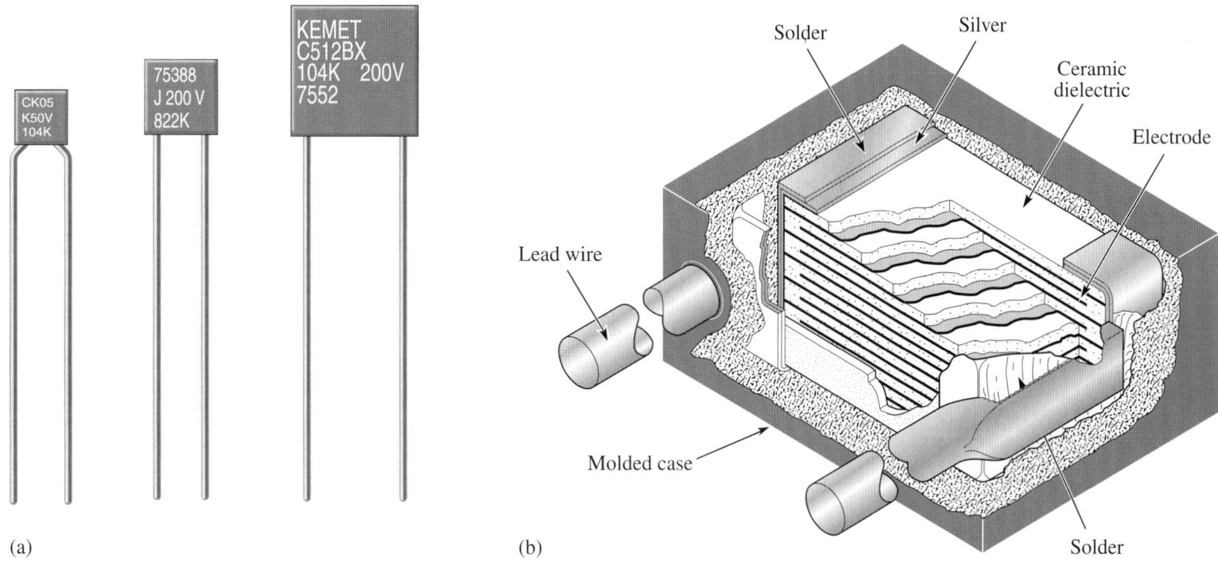

FIGURE 9–10

(a) Typical ceramic capacitors with radial leads. (b) Construction view.

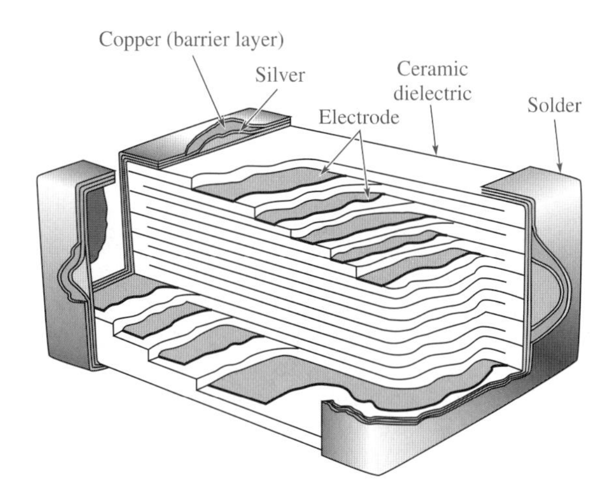

FIGURE 9–11

Construction view of a typical ceramic chip capacitor used for surface mounting on PC boards.

Plastic-Film Capacitors There are several types of plastic-film capacitors. Polycarbonate, propylene, polyester, polystyrene, polypropylene, and mylar are some common dielectric materials used. Some of these types have capacitance values up to 100 μF.

Figure 9–12 shows a basic construction used in many plastic-film capacitors. A thin strip of plastic-film dielectric is sandwiched between two thin metal strips that act as plates. One lead is connected to the inner plate and one to the outer plate as indicated. The strips are then rolled in a spiral configuration and encapsulated in a molded case. Thus, a large plate area can be packaged in a relatively small physical size, thereby achieving large capacitance values. Another method uses metal deposited directly on the film dielectric to form the plates.

Figure 9–13(a) shows typical plastic-film capacitors. Figure 9–13(b) shows a construction view for one type of an axial-lead plastic-film capacitor.

Electrolytic Capacitors Electrolytic capacitors are polarized so that one plate is positive and the other negative. These capacitors are generally used for high capacitance values from 1 μF up to over 200,000 μF, but they have relatively low

HANDS ON TIP

Chip capacitors are for surface mounting on a printed circuit board and have conductive terminals plated on the ends. These capacitors will withstand the molten solder temperatures encountered in reflow and wave soldering used in automated circuit board assembly. Chip capacitors are in great demand because of the move toward miniaturization.

FIGURE 9–12

Basic construction of axial-lead tubular plastic-film dielectric capacitors.

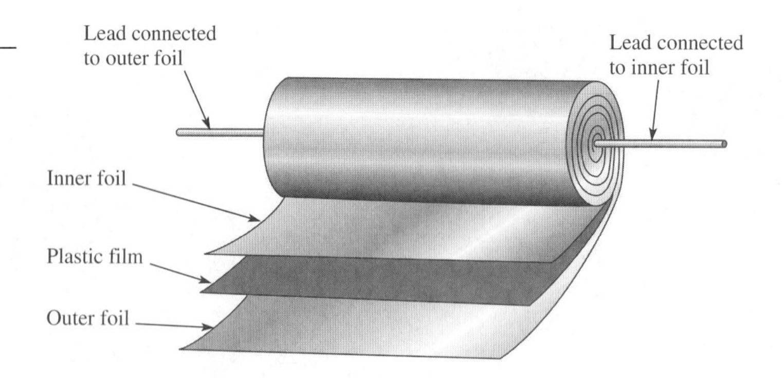

FIGURE 9–13

(a) Several typical plastic-film capacitors. (b) Construction view of an axial-lead capacitor.

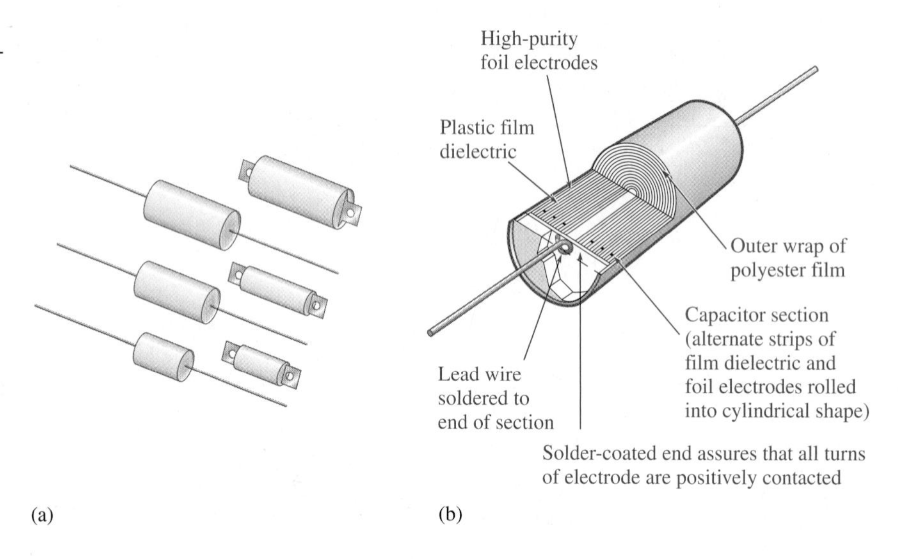

breakdown voltages (350 V is a typical maximum but higher voltages are occasionally found) and high amounts of leakage.

Electrolytic capacitors offer much higher capacitance values than mica or ceramic capacitors, but their voltage ratings are typically lower. Aluminum electrolytic capacitors are probably the most commonly used type. While other capacitors use two similar plates, the electrolytic capacitor consists of one plate of aluminum foil and another plate made of a conducting electrolyte applied to a material such as plastic film. These two "plates" are separated by a layer of aluminum oxide which forms on the surface of the aluminum plate. Figure 9–14(a) illustrates the basic construction of a typical aluminum electrolytic capacitor with axial leads. Other electrolytic capacitors with radial leads are shown in Figure 9–14(b); the symbol for an electrolytic capacitor is shown in part (c).

Tantalum electrolytic capacitors can be in either a tubular configuration similar to Figure 9–14 or "tear drop" shape as shown in Figure 9–15. In the tear drop configuration, the positive plate is actually a pellet of tantalum powder rather than a sheet of foil. Tantalum pentoxide forms the dielectric and manganese dioxide forms the negative plate.

Because of the process used for the insulating pentoxide dielectric, the metallic (aluminum or tantalum) plate is always positive with respect to the electrolyte plate, and, thus all electrolytic capacitors are polarized. The metal plate (positive lead) is usually indicated by a plus sign or some other obvious marking and must always be connected in a dc circuit where the voltage across the capacitor does not change polarity. Reversal of the polarity of the voltage can result in complete destruction of the capacitor.

Be extremely careful with electrolytic capacitors because it does make a difference which way they are connected. Always observe the proper polarity. If a polarized capacitor is connected backwards, it may explode and cause injury.

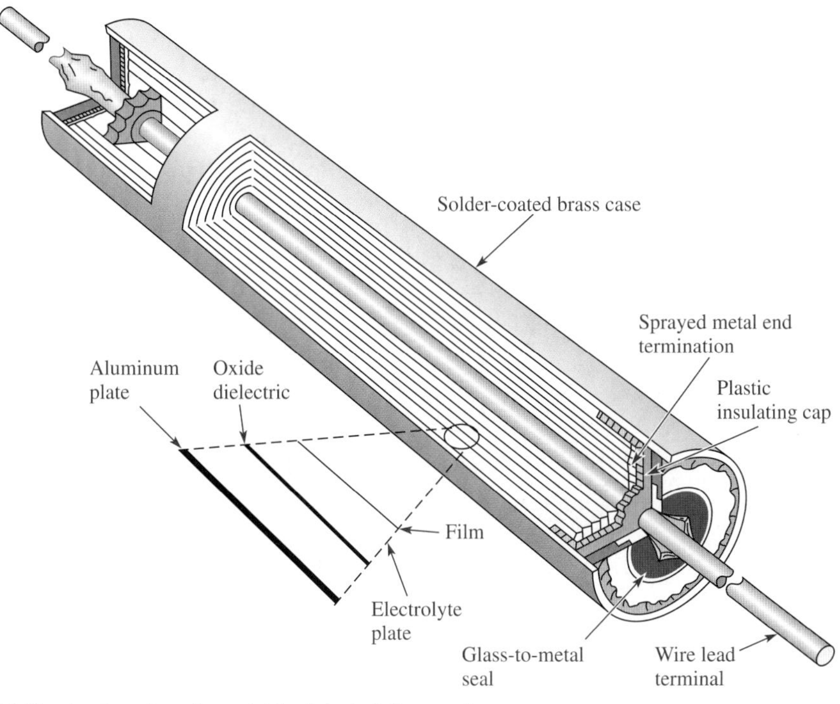

(a) Construction view of an axial-lead electrolytic capacitor

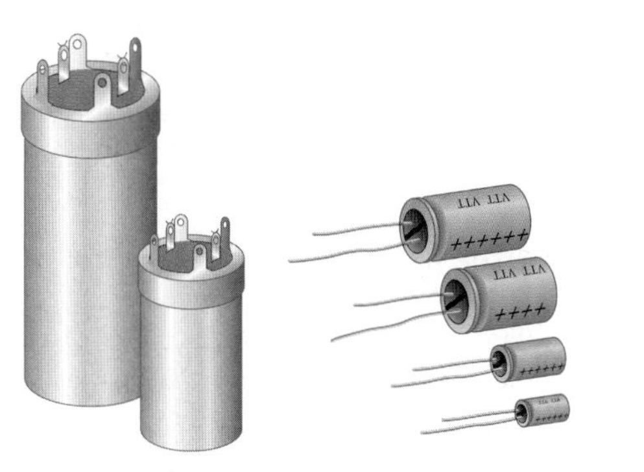

(b) Typical radial-lead electrolytics

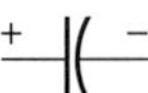

(c) Symbol for an electrolytic capacitor. The straight plate is positive and the curved plate is negative, as indicated.

FIGURE 9–14

Electrolytic capacitors.

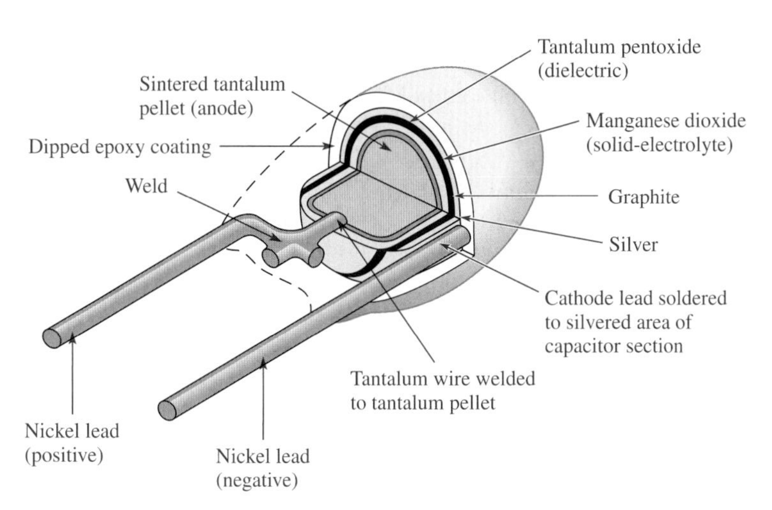

FIGURE 9–15

Construction view of a typical "tear drop" shaped tantalum electrolytic capacitor.

▲ FIGURE 9–16

Schematic symbol for a variable capacitor.

Variable Capacitors

Variable capacitors are used in a circuit when there is a need to adjust the capacitance value either manually or automatically, for example, in radio or TV tuners. The schematic symbol for a variable capacitor is shown in Figure 9–16.

Adjustable capacitors that normally have slotted screw-type adjustments and are used for very fine adjustments in a circuit are called **trimmers.** Ceramic or mica is a common dielectric in these types of capacitors, and the capacitance usually is changed by adjusting the plate separation. Figure 9–17 shows some typical devices.

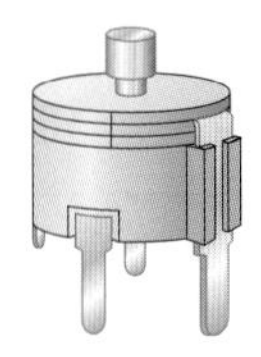

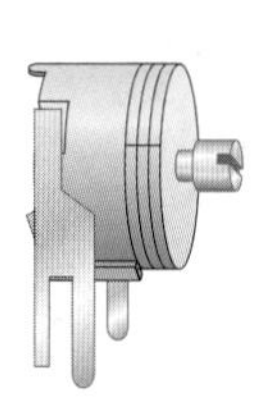

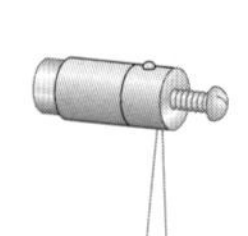

▶ FIGURE 9–17

Trimmer capacitors.

The varactor is a semiconductive device that exhibits a capacitance characteristic that is varied by changing the voltage across its terminals. This device usually is covered in detail in a course on electronic devices.

Capacitor Labeling

Capacitor values are indicated on the body of the capacitor either by numerical or alphanumerical labels and sometimes by color codes. Capacitor labels indicate various parameters such as capacitance, voltage rating, and tolerance.

Some capacitors carry no unit designation for capacitance. In these cases, the units are implied by the value indicated and are recognized by experience. For example, a ceramic capacitor marked .001 or .01 has units of microfarads because picofarad values that small are not available. As another example, a ceramic capacitor labeled 50 or 330 has units of picofarads because microfarad units that large normally are not available in this type. In some cases a 3-digit designation is used. The first two digits are the first two digits of the capacitance value. The third digit is the multiplier or number of zeros after the second digit. For example, 103 means 10,000 pF. In some instances, the units are labeled as pF or μF; sometimes the microfarad unit is labeled as MF or MFD.

A voltage rating appears on some types of capacitors with WV or WVDC and is omitted on others. When it is omitted, the voltage rating can be determined from information supplied by the manufacturer. The tolerance of the capacitor is usually labeled as a percentage, such as $\pm 10\%$. The temperature coefficient is indicated by a *parts per million* marking. This type of label consists of a P or N followed by a number. For example, N750 means a negative temperature coefficient of 750 ppm/°C, and P30 means a positive temperature coefficient of 30 ppm/°C. An NP0 designation means that the positive and negative coefficients are zero; thus the capacitance does not change with temperature. Certain types of capacitors are color coded. Refer to Appendix C for additional capacitor labeling and color code information.

SECTION 9–2 REVIEW

1. How are capacitors commonly classified?
2. What is the difference between a fixed and a variable capacitor?
3. What type of capacitor normally is polarized?
4. What precautions must be taken when a polarized capacitor is installed in a circuit?
5. An electrolytic capacitor is connected between a negative supply voltage and ground. Which capacitor lead should be connected to ground?

9–3 SERIES CAPACITORS

In this section, you will see why the total capacitance of a series connection of capacitors is less than any of the individual capacitances.

After completing this section, you should be able to

- **Analyze series capacitors**
- Determine total capacitance
- Determine capacitor voltages

When capacitors are connected in series, the total capacitance is less than the smallest capacitance value because the effective plate separation increases. The formula for total series capacitance is similar to the formula for total resistance of parallel resistors (Equation 5–4).

To start, we will use two capacitors in series to show how the total capacitance is determined. Figure 9–18 shows two capacitors, which initially are uncharged, connected in series with a dc voltage source. When the switch is closed, as shown in part (a), current begins.

Recall that current is the same at all points in a series circuit and that current is defined as the rate of flow of charge ($I = Q/t$). In a certain period of time, a certain amount of charge moves through the circuit. Since current is the same everywhere in the circuit of Figure 9–18(a), the same amount of charge is moved from the negative side of the source to plate A of C_1, and from plate B of C_1 to

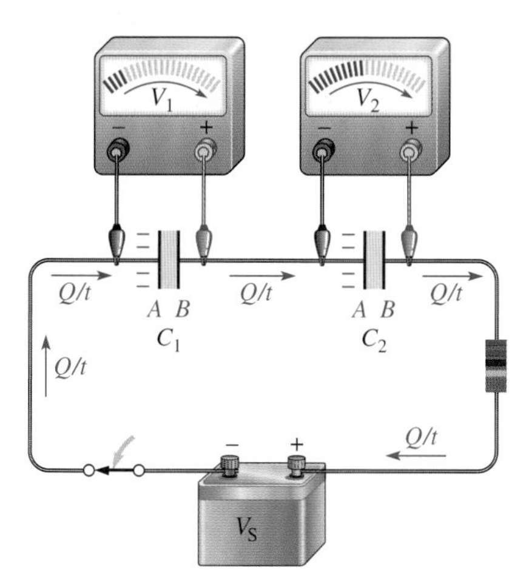

(a) While charging, $I = Q/t$ is the same at all points.

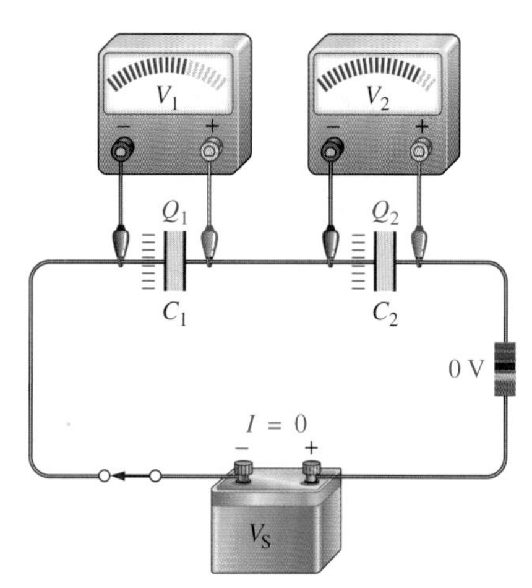

(b) Both capacitors store the same amount of charge ($Q_T = Q_1 = Q_2$).

FIGURE 9–18

Capacitors in series produce a total capacitance that is less than the smallest value.

plate A of C_2, and from plate B of C_2 to the positive side of the source. As a result, of course, the same amount of charge is deposited on the plates of both capacitors in a given period of time, and the total charge (Q_T) moved through the circuit in that period of time equals the charge stored by C_1 and also equals the charge stored by C_2.

$$Q_T = Q_1 = Q_2$$

As the capacitors charge, the voltage across each one increases as indicated.

Figure 9–18(b) shows the capacitors after they have been completely charged and the current has ceased. Both capacitors store an equal amount of charge (Q), and the voltage across each one depends on its capacitance value ($V = Q/C$). By Kirchhoff's voltage law, which applies to capacitive circuits as well as to resistive circuits, the sum of the capacitor voltages equals the source voltage.

$$V_S = V_1 + V_2$$

Using the formula $V = Q/C$, you can substitute into the formula for Kirchhoff's law and get the following relationship (where $Q = Q_T = Q_1 = Q_2$):

$$\frac{Q}{C_T} = \frac{Q}{C_1} + \frac{Q}{C_2}$$

The Q can be factored out of the right side of the equation and canceled with the Q on the left side as follows:

$$\frac{\not{Q}}{C_T} = \not{Q}\left(\frac{1}{C_1} + \frac{1}{C_2}\right)$$

Thus, you can have the following relationship for two capacitors in series:

Equation 9–8

$$\frac{1}{C_T} = \frac{1}{C_1} + \frac{1}{C_2}$$

Taking the reciprocal of both sides of Equation 9–8 gives the formula for the total capacitance for two capacitors in series.

Equation 9–9

$$C_T = \frac{1}{\dfrac{1}{C_1} + \dfrac{1}{C_2}}$$

EXAMPLE 9–5

Find the total capacitance C_T in Figure 9–19.

▶ FIGURE 9–19

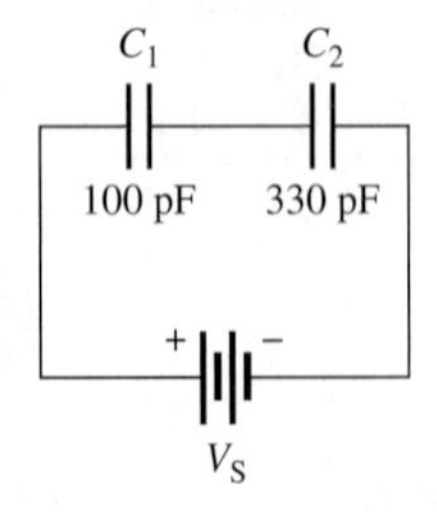

Solution

$$C_T = \frac{1}{\dfrac{1}{C_1} + \dfrac{1}{C_2}} = \frac{1}{\dfrac{1}{100\text{ pF}} + \dfrac{1}{330\text{ pF}}} = \mathbf{76.7\text{ pF}}$$

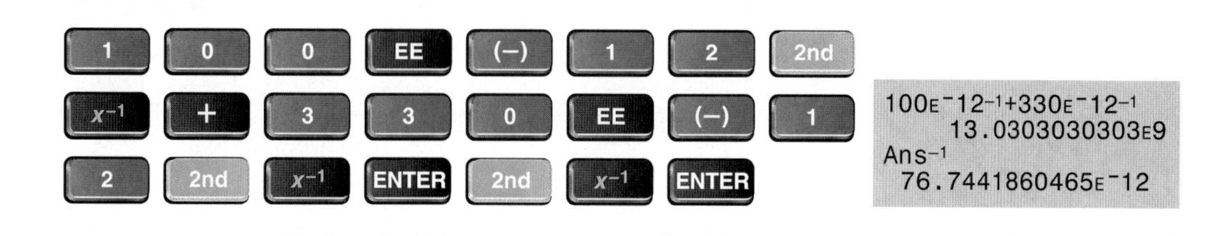

Related Problem Determine C_T if $C_1 = 470$ pF and $C_2 = 680$ pF in Figure 9–19.

General Formula for Total Series Capacitance

Equations 9–8 and 9–9 can be extended to any number of capacitors in series, as shown in Figure 9–20.

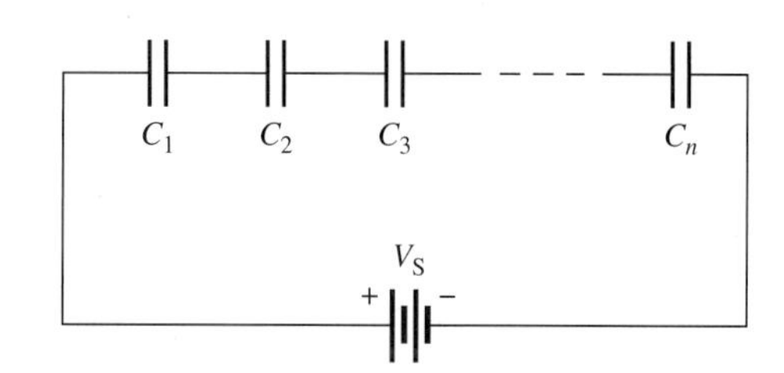

FIGURE 9–20

General circuit with n capacitors in series.

The formulas for total capacitance for any number of capacitors in series are as follows; the subscript n can be any number.

$$\frac{1}{C_T} = \frac{1}{C_1} + \frac{1}{C_2} + \frac{1}{C_3} + \cdots + \frac{1}{C_n}$$

Equation 9–10

$$C_T = \frac{1}{\dfrac{1}{C_1} + \dfrac{1}{C_2} + \dfrac{1}{C_3} + \cdots + \dfrac{1}{C_n}}$$

Equation 9–11

Remember,

The total series capacitance is always less than the smallest capacitance.

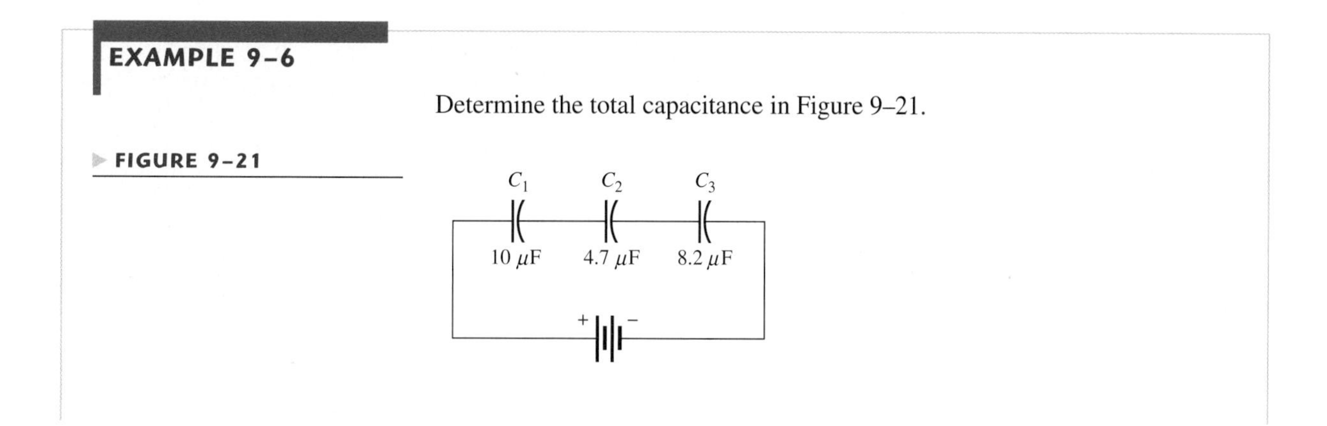

EXAMPLE 9–6

Determine the total capacitance in Figure 9–21.

FIGURE 9–21

Solution

$$\frac{1}{C_T} = \frac{1}{C_1} + \frac{1}{C_2} + \frac{1}{C_3} = \frac{1}{10\ \mu F} + \frac{1}{4.7\ \mu F} + \frac{1}{8.2\ \mu F}$$

Taking the reciprocal of both sides yields

$$C_T = \frac{1}{\frac{1}{10\ \mu F} + \frac{1}{4.7\ \mu F} + \frac{1}{8.2\ \mu F}} = \mathbf{2.30\ \mu F}$$

Related Problem If another 4.7 μF capacitor is connected in series with the three existing capacitors in Figure 9–21, what is the value of C_T?

Capacitor Voltages

The voltage across each capacitor in a series connection depends on its capacitance value according to the formula $V = Q/C$. The voltage across any individual capacitor in series can be determined using the following formula:

Equation 9–12

$$V_x = \left(\frac{C_T}{C_x}\right)V_S$$

where C_x is any capacitor in series, such as C_1, C_2, and C_3, and so on, and V_x is the voltage across C_x.

The largest-value capacitor in series will have the smallest voltage. The smallest capacitance value will have the largest voltage.

EXAMPLE 9–7

Find the voltage across each capacitor in Figure 9–22.

▶ FIGURE 9–22

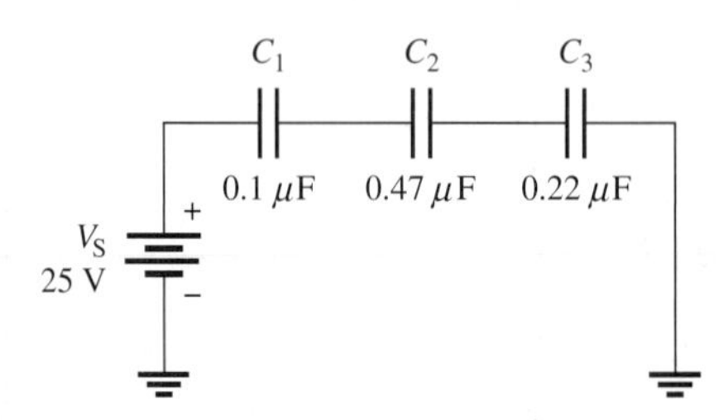

Solution Calculate the total capacitance.

$$\frac{1}{C_T} = \frac{1}{C_1} + \frac{1}{C_2} + \frac{1}{C_3} = \frac{1}{0.1\ \mu F} + \frac{1}{0.47\ \mu F} + \frac{1}{0.22\ \mu F}$$

$$C_T = 0.06\ \mu F$$

The voltages are as follows:

$$V_1 = \left(\frac{C_T}{C_1}\right)V_S = \left(\frac{0.06\ \mu F}{0.1\ \mu F}\right)25\ V = \mathbf{15.0\ V}$$

$$V_2 = \left(\frac{C_T}{C_2}\right)V_S = \left(\frac{0.06\ \mu F}{0.47\ \mu F}\right)25\ V = \mathbf{3.19\ V}$$

$$V_3 = \left(\frac{C_T}{C_3}\right)V_S = \left(\frac{0.06\ \mu F}{0.22\ \mu F}\right)25\ V = \mathbf{6.82\ V}$$

Related Problem Another 0.47 μF capacitor is connected in series with the existing capacitors in Figure 9–22. Determine the voltage across the new capacitor, assuming all the capacitors are initially uncharged.

Open file E09-07 on your EWB/CircuitMaker CD-ROM. Measure the voltage across each of the capacitors and compare with the calculated values. Connect another 0.47 μF capacitor in series with the other three and measure the voltage across the new capacitor. Also measure the voltage across C_1, C_2, and C_3 and compare with the previous voltages. Did the values increase or decrease after the fourth capacitor was added? Why?

SECTION 9–3 REVIEW

1. Is the total capacitance of series capacitors less than or greater than the value of the smallest capacitor?
2. The following capacitors are in series: 100 pF, 220 pF, and 560 pF. What is the total capacitance?
3. A 0.01 μF and a 0.015 μF capacitor are in series. Determine the total capacitance.
4. Determine the voltage across the 0.01 μF capacitor in Question 3 if 10 V are connected across the two series capacitors.

9–4 PARALLEL CAPACITORS

In this section, you will see why capacitances add when they are connected in parallel.

After completing this section, you should be able to

- **Analyze parallel capacitors**
- Determine total capacitance

When capacitors are connected in parallel, the total capacitance is the sum of the individual capacitances because the effective plate area increases. The formula for total parallel capacitance is similar to the formula for total series resistance (Equation 4–1).

Figure 9–23 shows two parallel capacitors connected to a dc voltage source. When the switch is closed, as shown in part (a), current begins. A total amount of charge (Q_T) moves through the circuit in a certain period of time. Part of the total charge is stored by C_1 and part by C_2. The portion of the total charge that is stored by each capacitor depends on its capacitance value according to the relationship $Q = CV$.

Figure 9–23(b) shows the capacitors after they have been completely charged and the current has stopped. Since the voltage across both capacitors is the same, the larger capacitor stores more charge. If the capacitors are equal in value, they store an equal amount of charge. The charge stored by both of the capacitors together equals the total charge that was delivered from the source.

$$Q_T = Q_1 + Q_2$$

Since $Q = CV$, Equation 9–2, you can use substitution to get the following relationship:

$$C_T V_S = C_1 V_S + C_2 V_S$$

Because all the V_S terms are equal, they can be canceled. Therefore, the total capacitance for two capacitors in parallel is

Equation 9–13

$$C_T = C_1 + C_2$$

▶ **FIGURE 9–23**

Capacitors in parallel produce a total capacitance that is the sum of the individual capacitances.

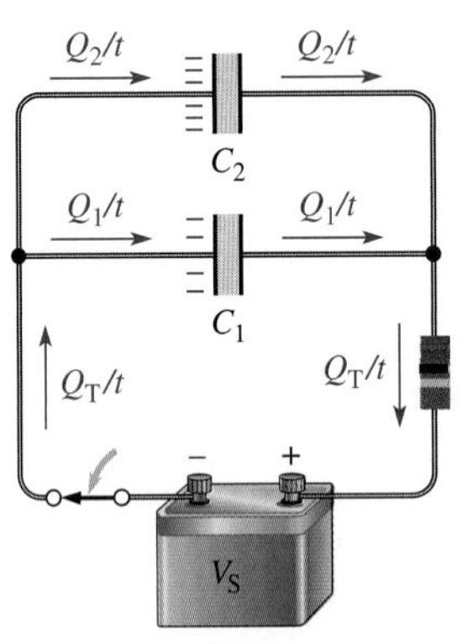

(a) The amount of charge on each capacitor is directly proportional to its capacitance value.

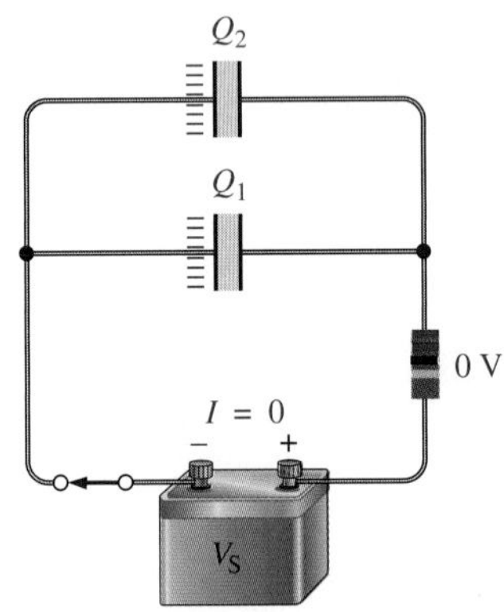

(b) $Q_T = Q_1 + Q_2$

EXAMPLE 9–8

What is the total capacitance in Figure 9–24? What is the voltage across each capacitor?

▶ **FIGURE 9–24**

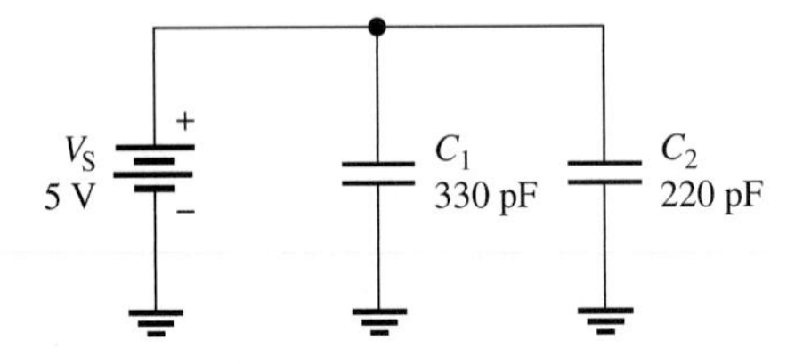

Solution The total capacitance is

$$C_T = C_1 + C_2 = 330 \text{ pF} + 220 \text{ pF} = \mathbf{550 \text{ pF}}$$

The voltage across each capacitor in parallel is equal to the source voltage.

$$V_S = V_1 = V_2 = \mathbf{5 \text{ V}}$$

Related Problem What is C_T if a 100 pF capacitor is connected in parallel with C_1 and C_2 in Figure 9–24?

General Formula for Total Parallel Capacitance

Equation 9–13 can be extended to any number of capacitors in parallel, as shown in Figure 9–25. The expanded formula is as follows; the subscript n can be any number.

Equation 9–14

$$C_T = C_1 + C_2 + C_3 + \cdots + C_n$$

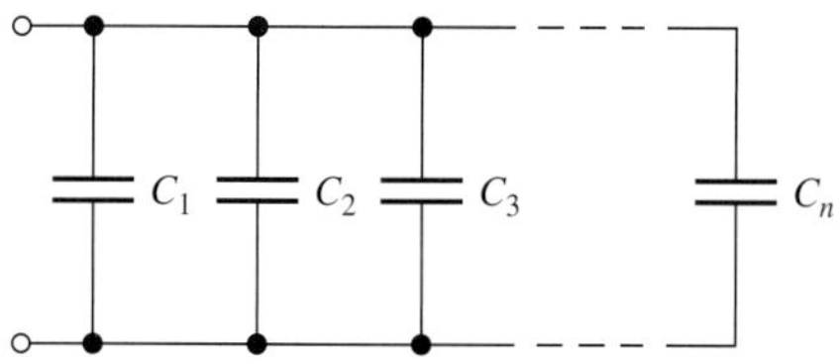

FIGURE 9–25

General circuit with n capacitors in parallel.

EXAMPLE 9–9

Determine C_T in Figure 9–26.

FIGURE 9–26

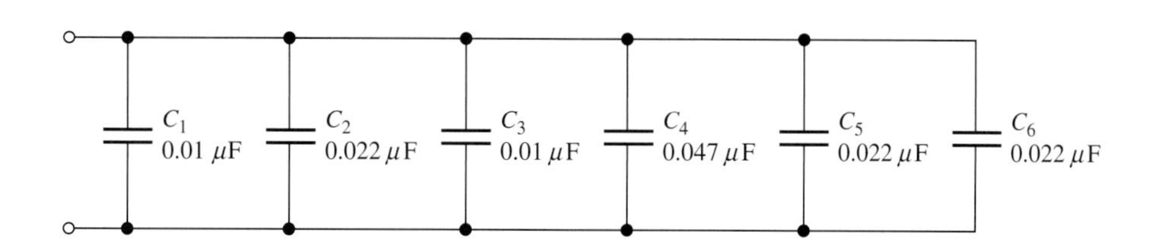

Solution

$$C_T = C_1 + C_2 + C_3 + C_4 + C_5 + C_6$$
$$= 0.01\ \mu\text{F} + 0.022\ \mu\text{F} + 0.01\ \mu\text{F} + 0.047\ \mu\text{F} + 0.022\ \mu\text{F} + 0.022\ \mu\text{F}$$
$$= \mathbf{0.133\ \mu F}$$

Related Problem If three more 0.01 μF capacitors are connected in parallel in Figure 9–25, what is C_T?

SECTION 9–4 REVIEW

1. How is total parallel capacitance determined?
2. In a certain application, you need 0.05 μF. The only capacitor value available is 0.01 μF, and this capacitor is available in large quantities. How can you get the total capacitance that you need?
3. The following capacitors are in parallel: 10 pF, 56 pF, 33 pF, and 68 pF. What is C_T?

9–5 CAPACITORS IN DC CIRCUITS

A capacitor will charge up when it is connected to a dc voltage source. The buildup of charge across the plates occurs in a predictable manner that is dependent on the capacitance and the resistance in a circuit.

After completing this section, you should be able to

- **Describe how a capacitor operates in a dc switching circuit**
- Describe the charging and discharging of a capacitor
- Define *RC time constant*
- Relate the time constant to charging and discharging
- Write equations for the charging and discharging curves
- Explain why a capacitor blocks constant dc

Charging a Capacitor

A capacitor charges when it is connected to a dc voltage source, as shown in Figure 9–27. The capacitor in part (a) is uncharged; that is, plate *A* and plate *B* have equal numbers of free electrons. When the switch is closed, as shown in part (b), the source moves electrons away from plate *A* through the circuit to plate *B* as the arrows indicate. As plate *A* loses electrons and plate *B* gains electrons, plate *A* becomes positive with respect to plate *B*. As this charging process continues, the voltage across the plates builds up rapidly until it is equal to the source voltage, V_S, but opposite in polarity, as shown in part (c). When the capacitor is fully charged, there is no current.

A capacitor blocks constant dc.

FIGURE 9–27

Charging a capacitor.

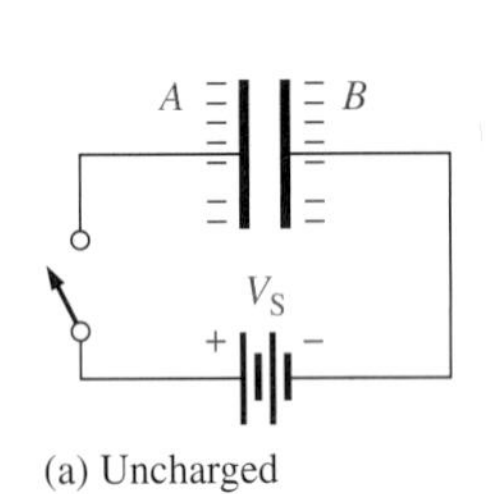

(a) Uncharged

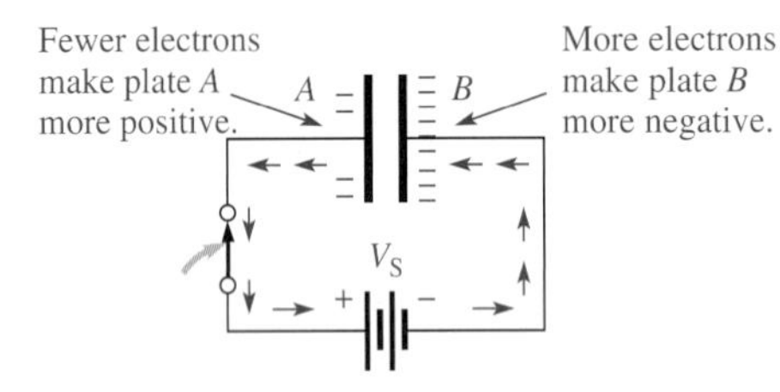

(b) Charging

(c) Fully charged

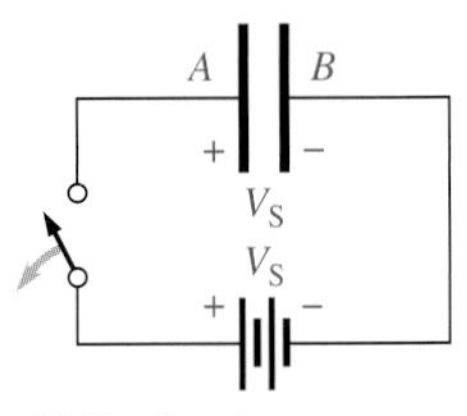

(d) Retains charge

When the charged capacitor is disconnected from the source, as shown in Figure 9–27(d), it remains charged for long periods of time, depending on its leakage resistance, and can cause severe electrical shock. The charge on an electrolytic capacitor generally leaks off more rapidly than in other types of capacitors.

Discharging a Capacitor

When a conductor is connected across a charged capacitor, as shown in Figure 9–28, the capacitor will discharge. In this particular case, a very low resistance path (the conductor) is connected across the capacitor with a switch. Before the switch is closed, the capacitor is charged to 50 V, as indicated in part (a). When the switch is closed, as shown in part (b), the excess electrons on plate *B* move through the circuit to plate *A* (indicated by the arrows); as a result of the current through the low resistance of the conductor, the energy stored by the capacitor is dissipated in the resistance of the conductor. The charge is neutralized when the numbers of free electrons on both plates are again equal. At this time, the voltage across the capacitor is zero, and the capacitor is completely discharged, as shown in part (c).

FIGURE 9–28

Discharging a capacitor.

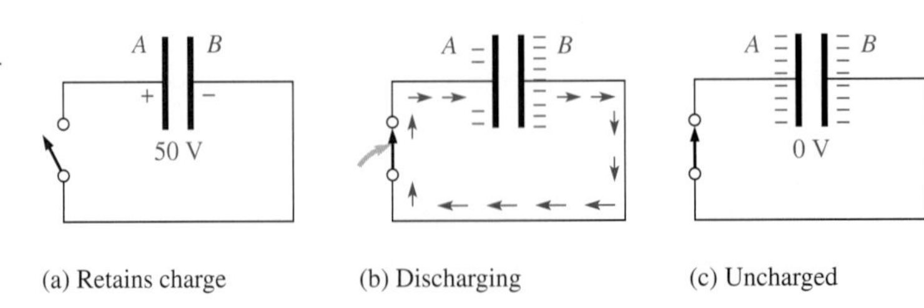

(a) Retains charge (b) Discharging (c) Uncharged

Current and Voltage During Charging and Discharging

Notice in Figures 9–27 and 9–28 that the direction of the current during discharge is opposite to that of the charging current. It is important to understand that *there is no current through the dielectric of the capacitor during charging or discharging because the dielectric is an insulating material.* There is current from one plate to the other only through the external circuit.

Figure 9–29(a) shows a capacitor connected in series with a resistor and a switch to a dc voltage source. Initially, the switch is open and the capacitor is uncharged with zero volts across its plates. At the instant the switch is closed, the current jumps to its maximum value and the capacitor begins to charge. The current is maximum initially because the capacitor has zero volts across it and, therefore, effectively acts as a short; thus, the current is limited only by the resistance. As time passes and the capacitor charges, the current decreases and the voltage across the capacitor (V_C) increases. The resistor voltage is proportional to the current during this charging period.

▼ **FIGURE 9–29**

Current and voltage in a charging and discharging capacitor.

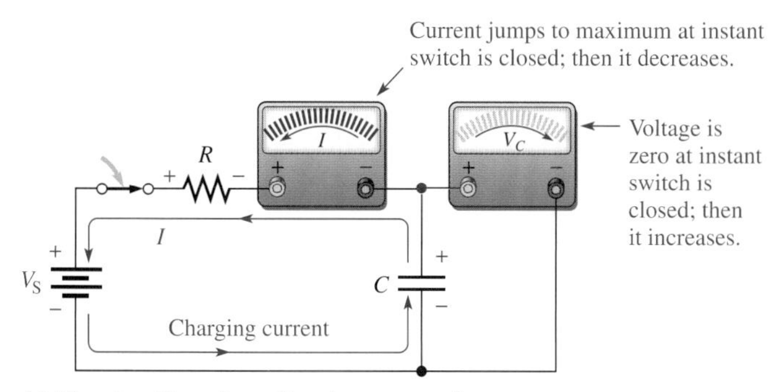

(a) Charging: Capacitor voltage increases as the current and resistor voltage decrease.

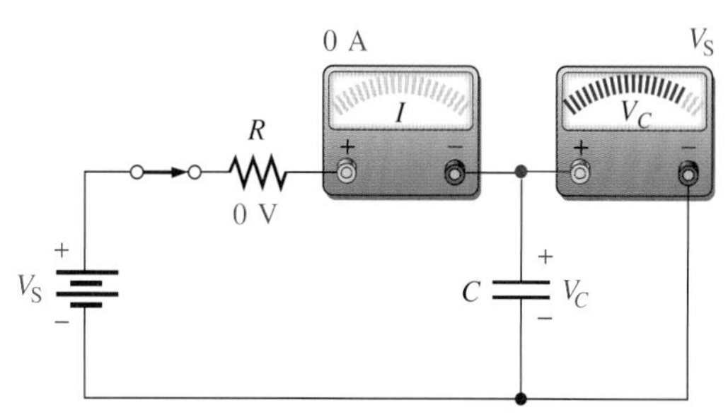

(b) Fully charged: Capacitor voltage equals source voltage. The current is zero.

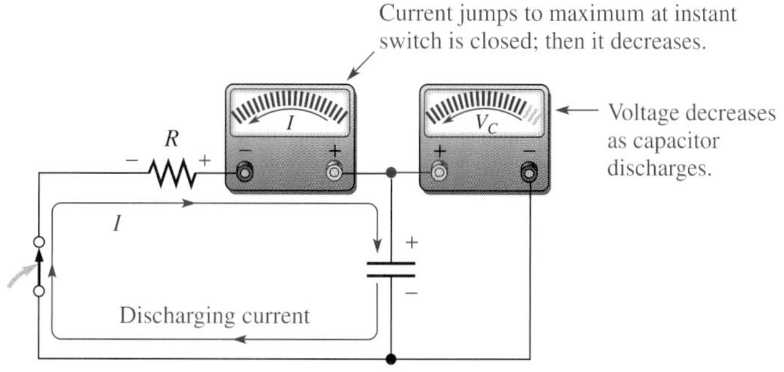

(c) Discharging: Capacitor voltage, resistor voltage, and the current decrease from their initial maximum values. Note that the discharge current is opposite to the charge current.

After a certain period of time, the capacitor reaches full charge. At this point, the current is zero and the capacitor voltage is equal to the dc source voltage, as shown in Figure 9–29(b). If the switch were opened now, the capacitor would retain its full charge (neglecting any leakage).

In Figure 9–29(c), the voltage source has been removed. When the switch is closed, the capacitor begins to discharge. Initially, the current jumps to a maximum but in a direction opposite to its direction during charging. As time passes, the current and capacitor voltage decrease. The resistor voltage is always proportional to the current. When the capacitor has fully discharged, the current and the capacitor voltage are zero.

Remember the following rules about capacitors in dc circuits:

1. Voltage across a capacitor *cannot* change instantaneously.
2. Current in a capacitive circuit *can* ideally change instantaneously.
3. A fully charged capacitor appears as an *open* to unchanging current.
4. An uncharged capacitor appears as a *short* to an instantaneous change in current.

Now let's examine in more detail how the voltage and current change with time in a capacitive circuit.

The *RC* Time Constant

In a practical situation, there cannot be capacitance without some resistance in a circuit. It may simply be the small resistance of a wire, or it may be a designed-in resistance. Because of this, the charging and discharging characteristics of a capacitor must always be considered with the associated series resistance included. The resistance introduces the element of *time* in the charging and discharging of a capacitor.

When a capacitor charges or discharges through a resistance, a certain time is required for the capacitor to charge fully or discharge fully. The voltage across a capacitor cannot change instantaneously because a finite time is required to move charge from one point to another. The rate at which the capacitor charges or discharges is determined by the **RC time constant** of the circuit.

The time constant of a series *RC* circuit is a time interval that equals the product of the resistance and the capacitance.

The time constant is expressed in units of seconds when resistance is in ohms and capacitance is in farads. It is symbolized by τ (Greek letter tau), and the formula is

Equation 9–15

$$\tau = RC$$

Recall that $I = Q/t$. The current depends on the amount of charge moved in a given time. When the resistance is increased, the charging current is reduced, thus increasing the charging time of the capacitor. When the capacitance is increased, the amount of charge increases; thus, for the same current, more time is required to charge the capacitor.

EXAMPLE 9–10

A series *RC* circuit has a resistance of 1.0 MΩ and a capacitance of 4.7 μF. What is the time constant?

Solution

$$\tau = RC = (1.0 \times 10^6\ \Omega)(4.7 \times 10^{-6}\ \text{F}) = \mathbf{4.7\ s}$$

Related Problem A series *RC* circuit has a 270 kΩ resistor and a 3300 pF capacitor. What is the time constant in μs?

When a capacitor is charging or discharging between two voltage levels, the charge on the capacitor changes by approximately 63% of the difference in the levels in one time constant. An uncharged capacitor charges to approximately 63% of its fully charged voltage in one time constant. When a fully charged capacitor is discharging, its voltage drops to approximately 100% − 63% = 37% of its initial voltage in one time constant. This change also corresponds to a 63% change.

The Charging and Discharging Curves

A capacitor charges and discharges following a nonlinear curve, as shown in Figure 9–30. In these graphs, the approximate percentage of full charge is shown at each time-constant interval. This type of curve follows a precise mathematical

formula and is called an **exponential** curve. The charging curve is an increasing

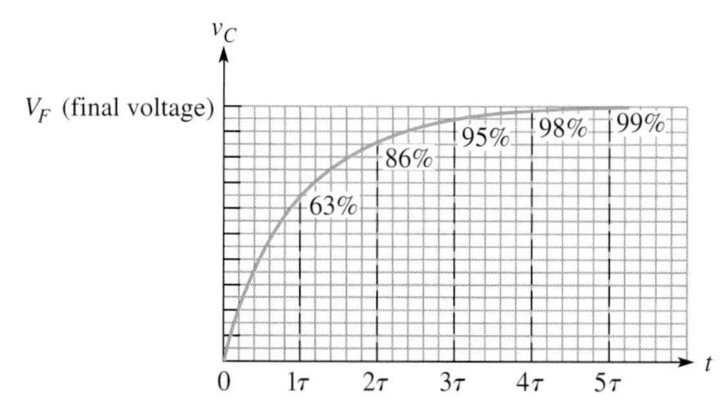

(a) Charging curve with percentages of final voltage

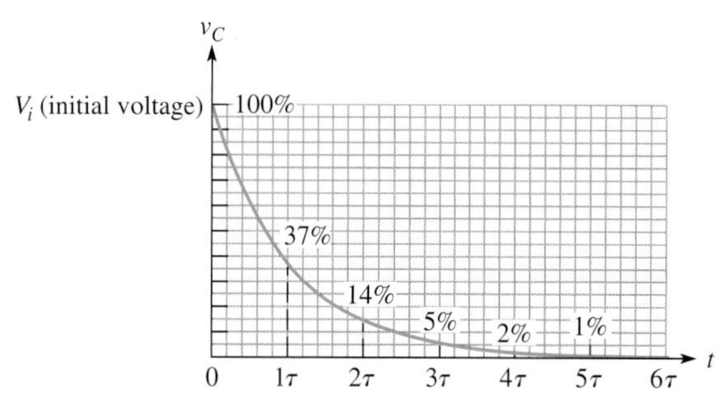

(b) Discharging curve with percentages of initial voltage

▲ **FIGURE 9–30**

Charging and discharging exponential curves for the capacitor voltage in an *RC* circuit.

exponential, and the discharging curve is a decreasing exponential. It takes five time constants to reach 99% (considered 100%) of the final voltage. This five time-constant interval is generally accepted as the time to fully charge or discharge a capacitor and is called the *transient time.*

General Formula The general expressions for either increasing or decreasing exponential curves are given in the following equations for both instantaneous voltage and current:

$$v = V_F + (V_i - V_F)e^{-t/\tau}$$

Equation 9–16

$$i = I_F + (I_i - I_F)e^{-t/\tau}$$

Equation 9–17

where V_F and I_F are the final values of voltage and current and V_i and I_i are the initial values of voltage and current. The lowercase letters v and i are the instantaneous values of the capacitor voltage and current at time t, and e is the base of natural logarithms with a value of 2.71828182846. The [2nd] [e^x] and [LN] keys on the TI-85/TI-86 calculator make it easy to work with this exponential term.

Charging from Zero The formula for the special case in which an increasing exponential voltage curve begins at zero ($V_i = 0$), as shown in Figure 9–30(a), is given in Equation 9–18. It is developed as follows, starting with the general formula, Equation 9–16.

$$v = V_F + (V_i - V_F)e^{-t/\tau} = V_F + (0 - V_F)e^{-t/RC} = V_F - V_Fe^{-t/RC}$$

$$v = V_F(1 - e^{-t/RC})$$

Equation 9–18

Using Equation 9–18, you can calculate the value of the charging voltage of a capacitor at any instant of time if it is initially uncharged.

EXAMPLE 9–11

In Figure 9–31, determine the capacitor voltage 50 μs after the switch is closed if the capacitor initially is uncharged. Draw the charging curve.

▶ **FIGURE 9–31**

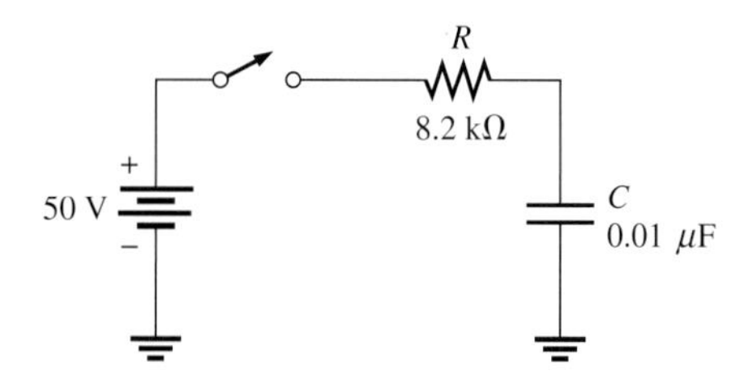

Solution The time constant is $RC = (8.2\ \text{k}\Omega)(0.01\ \mu\text{F}) = 82\ \mu\text{s}$. The voltage to which the capacitor will fully charge is 50 V (this is V_F). The initial voltage is zero. Notice that 50 μs is less than one time constant; so the capacitor will charge less than 63% of the full voltage in that time.

$$v_C = V_F(1 - e^{-t/RC}) = (50\ \text{V})(1 - e^{-50\mu\text{s}/82\mu\text{s}}) = \mathbf{22.8\ V}$$

The charging curve for the capacitor is shown in Figure 9–32.

▶ FIGURE 9–32

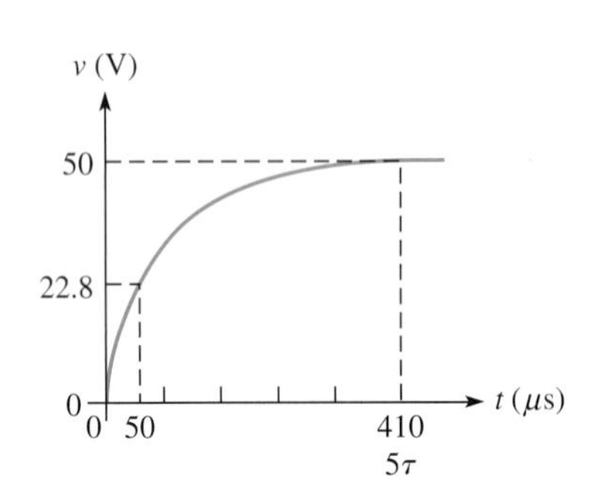

Related Problem Determine the capacitor voltage 15 μs after the switch is closed in Figure 9–31.

Discharging to Zero The formula for the special case in which a decreasing exponential voltage curve ends at zero ($V_F = 0$), as shown in Figure 9–30(b), is derived from the general formula as follows:

$$v = V_F + (V_i - V_F)e^{-t/\tau} = 0 + (V_i - 0)e^{-t/RC}$$

Equation 9–19

$$v = V_i e^{-t/RC}$$

where V_i is the voltage at the beginning of the discharge. You can use this formula to calculate the discharging voltage at any instant, as Example 9–12 illustrates.

EXAMPLE 9–12

Determine the capacitor voltage in Figure 9–33 at a point in time 6 ms after the switch is closed. Draw the discharging curve.

Solution The discharge time constant is

$$RC = (10\ \text{k}\Omega)(2.2\ \mu\text{F}) = 22\ \text{ms}$$

The initial capacitor voltage is 10 V. Notice that 6 ms is less than one time constant, so the capacitor will discharge less than 63%. Therefore, it will have a voltage greater than 37% of the initial voltage at 6 ms.

$$v_C = V_i e^{-t/RC} = (10\ \text{V})e^{-6\text{ms}/22\text{ms}} = \mathbf{7.61\ V}$$

The discharging curve for the capacitor is shown in Figure 9–34.

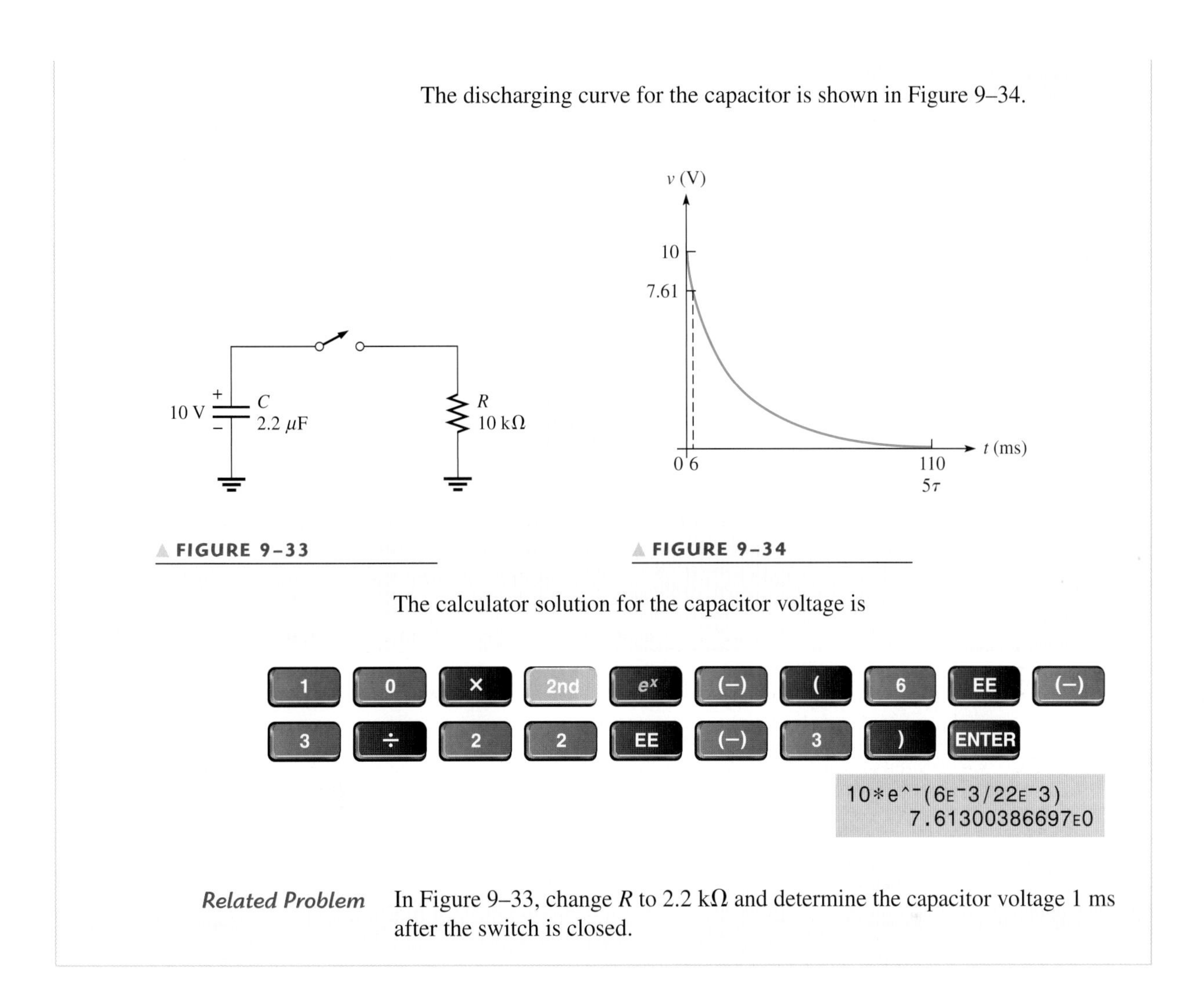

FIGURE 9–33

FIGURE 9–34

The calculator solution for the capacitor voltage is

Related Problem In Figure 9–33, change R to 2.2 kΩ and determine the capacitor voltage 1 ms after the switch is closed.

Graphical Method Using Universal Exponential Curves The universal curves in Figure 9–35 provide a graphic solution of the charge and discharge of capacitors. Example 9–13 illustrates this graphical method.

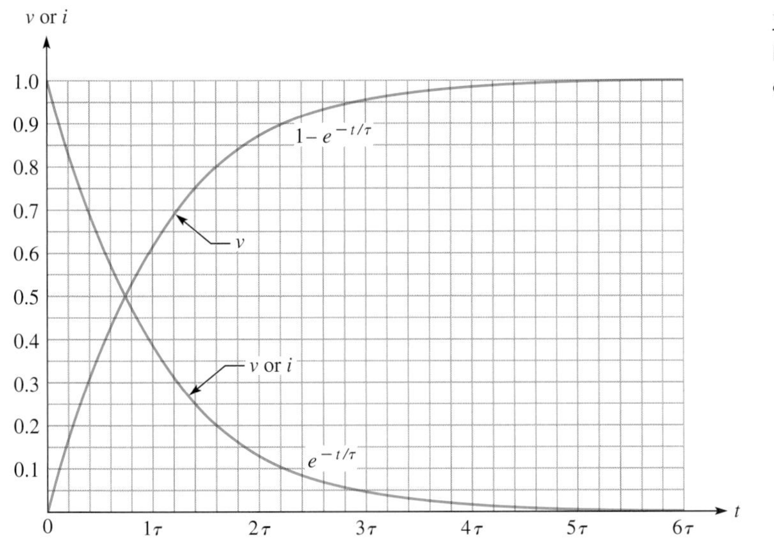

FIGURE 9–35

Normalized universal exponential curves.

EXAMPLE 9–13

How long will it take the capacitor in Figure 9–36 to charge to 75 V? What is the capacitor voltage 2 ms after the switch is closed? Use the normalized universal curves in Figure 9–35 to determine the answers.

FIGURE 9–36

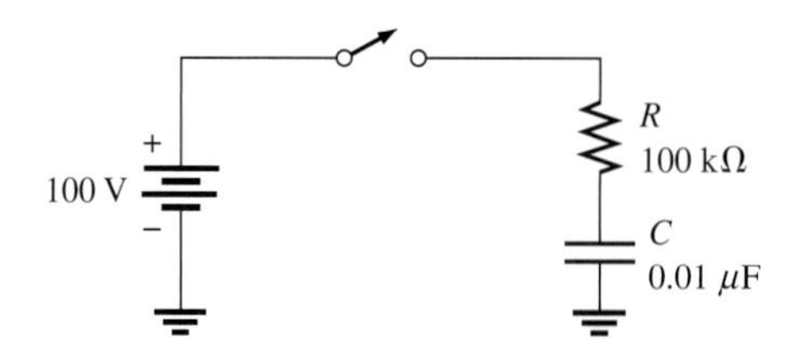

Solution The final voltage is 100 V, which is the 100% level (1.0) on the graph. The value 75 V is 75% of the maximum, or 0.75 on the graph. You can see that this value occurs at 1.4 time constants. One time constant is $RC = (100\ \text{k}\Omega)(0.01\ \mu\text{F}) = 1$ ms. Therefore, the capacitor voltage reaches 75 V at 1.4 ms after the switch is closed.

The capacitor is at approximately 86 V in 2 ms. These graphic solutions are shown in Figure 9–37.

FIGURE 9–37

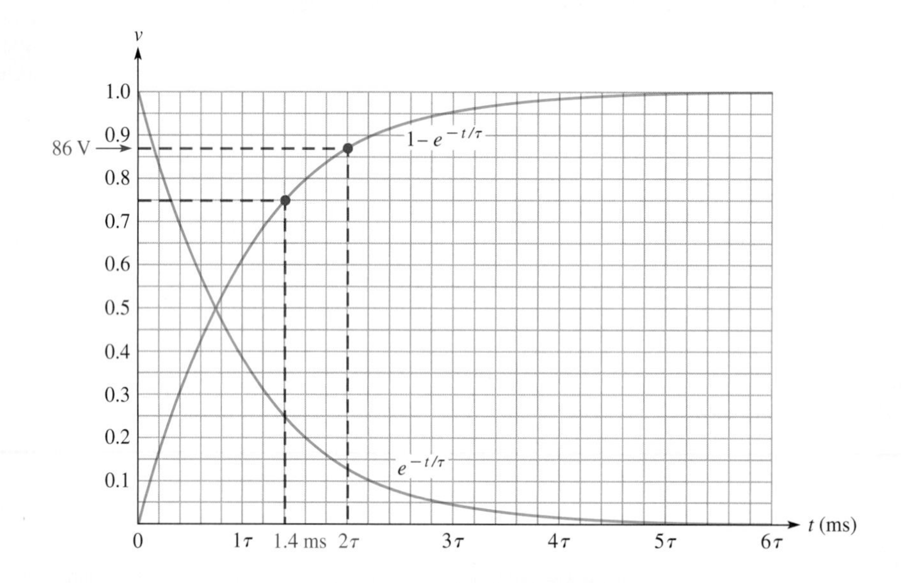

Related Problem Using the normalized universal exponential curves, determine how long it will take the capacitor in Figure 9–36 to charge to 50 V from zero. What is the capacitor voltage 3 ms after switch closure?

SECTION 9–5 REVIEW

1. Determine the time constant when $R = 1.2\ \text{k}\Omega$ and $C = 1000$ pF.
2. If the circuit in Question 1 is charged with a 5 V source, how long will it take the initially uncharged capacitor to reach full charge? At full charge, what is the capacitor voltage?

3. A certain circuit has a time constant of 1 ms. If it is charged with a 10 V battery, what will the capacitor voltage be at each of the following intervals: 2 ms, 3 ms, 4 ms, and 5 ms? The capacitor is initially uncharged.
4. A capacitor is charged to 100 V. If it is discharged through a resistor, what is the capacitor voltage at one time constant?

9–6 CAPACITORS IN AC CIRCUITS

As you saw in the last section, a capacitor blocks constant dc. You will learn in this section that a capacitor passes ac with an amount of opposition that depends on the frequency of the ac.

After completing this section, you should be able to

- **Describe how a capacitor operates in an ac circuit**
- Define *capacitive reactance*
- Determine the value of capacitive reactance in a given circuit
- Explain why a capacitor causes a phase shift between voltage and current
- Discuss instantaneous, true, and reactive power in a capacitor

Capacitive Reactance, X_C

In Figure 9–38, a capacitor is shown connected to a sinusoidal voltage source. When the source voltage is held at a constant amplitude value and its frequency is increased, the amplitude of the current increases. Also, when the frequency of the source is decreased, the current amplitude decreases.

When the frequency of the voltage increases, its rate of change also increases. This relationship is illustrated in Figure 9–39, where the frequency is doubled. Now, if the rate at which the voltage is changing increases, the amount of charge moving through the circuit in a given period of time must also increase. More

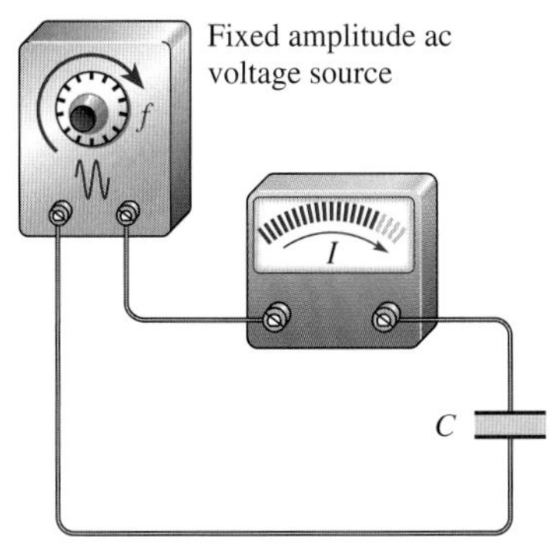

(a) Current increases when the frequency increases.

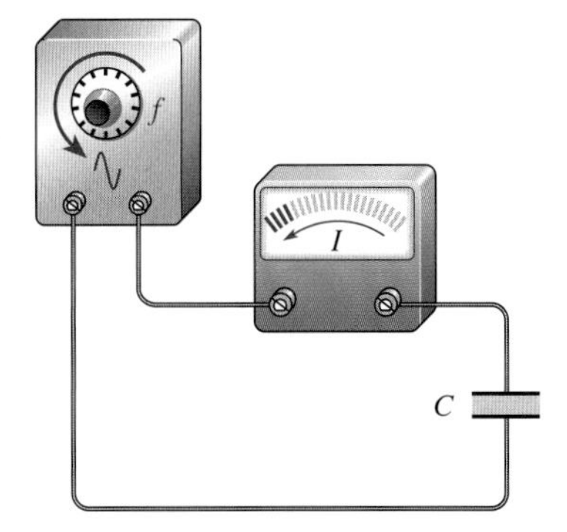

(b) Current decreases when the frequency decreases.

FIGURE 9–38

The current in a capacitive circuit varies directly with the frequency of the source voltage.

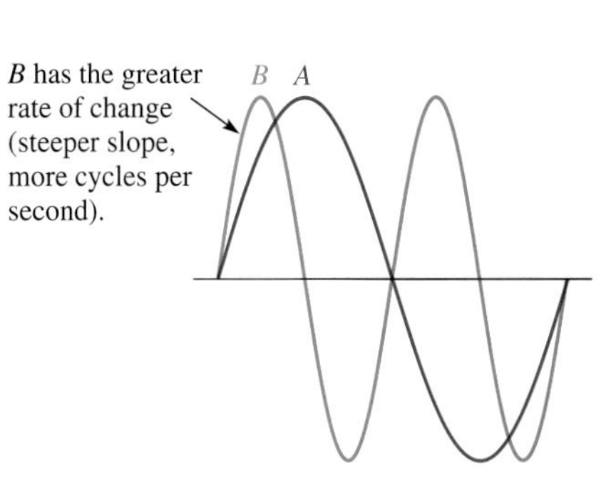

FIGURE 9–39

Rate of change of a sine wave increases when frequency increases.

charge in a given period of time means more current. For example, a tenfold increase in frequency means that the capacitor is charging and discharging 10 times as much in a given time interval. Therefore, since the rate of charge movement has increased 10 times, the current must increase by 10 because $I = Q/t$.

An increase in the amount of current with a fixed amount of voltage indicates that opposition to the current has decreased. Therefore, the capacitor offers opposition to current, and that opposition varies *inversely* with frequency.

The opposition to sinusoidal current in a capacitor is called capacitive reactance.

The symbol for capacitive reactance is X_C, and its unit is the ohm (Ω).

You have just seen how frequency affects the opposition to current (capacitive reactance) in a capacitor. Now let's see how the capacitance (C) itself affects the reactance. Figure 9–40(a) shows that when a sinusoidal voltage with a fixed amplitude and fixed frequency is applied to a 1 μF capacitor, there is a certain amount of alternating current. When the capacitance value is increased to 2 μF, the current increases, as shown in Figure 9–40(b). Thus, when the capacitance is increased, the opposition to current (capacitive reactance) decreases. Therefore, not only is the capacitive reactance *inversely* proportional to frequency, but it is also *inversely* proportional to capacitance. This relationship can be stated as follows:

$$X_C \text{ is proportional to } \frac{1}{fC}.$$

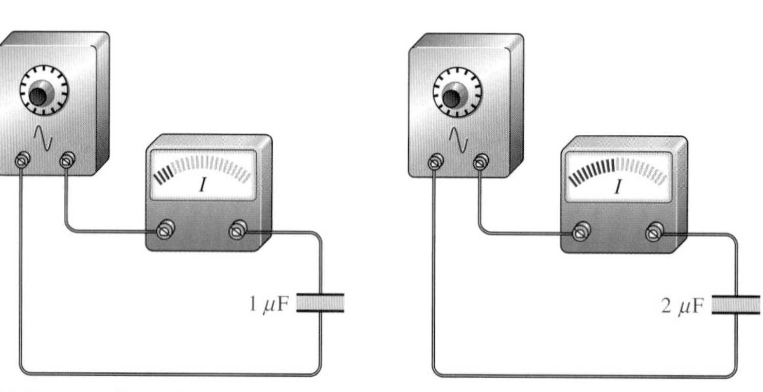

FIGURE 9–40

For a fixed voltage and fixed frequency, the current varies directly with the capacitance value.

It can be proven that the constant of proportionality that relates X_C to $1/fC$ is $1/2\pi$. Therefore, the formula for capacitive reactance (X_C) is

Equation 9–20

$$X_C = \frac{1}{2\pi fC}$$

X_C is in ohms when f is in hertz and C is in farads. The 2π term comes from the fact that, as you learned in Chapter 8, a sine wave can be described in terms of rotational motion, and one revolution contains 2π radians.

EXAMPLE 9–14

A sinusoidal voltage is applied to a capacitor, as shown in Figure 9–41. The frequency of the sine wave is 1 kHz. Determine the capacitive reactance.

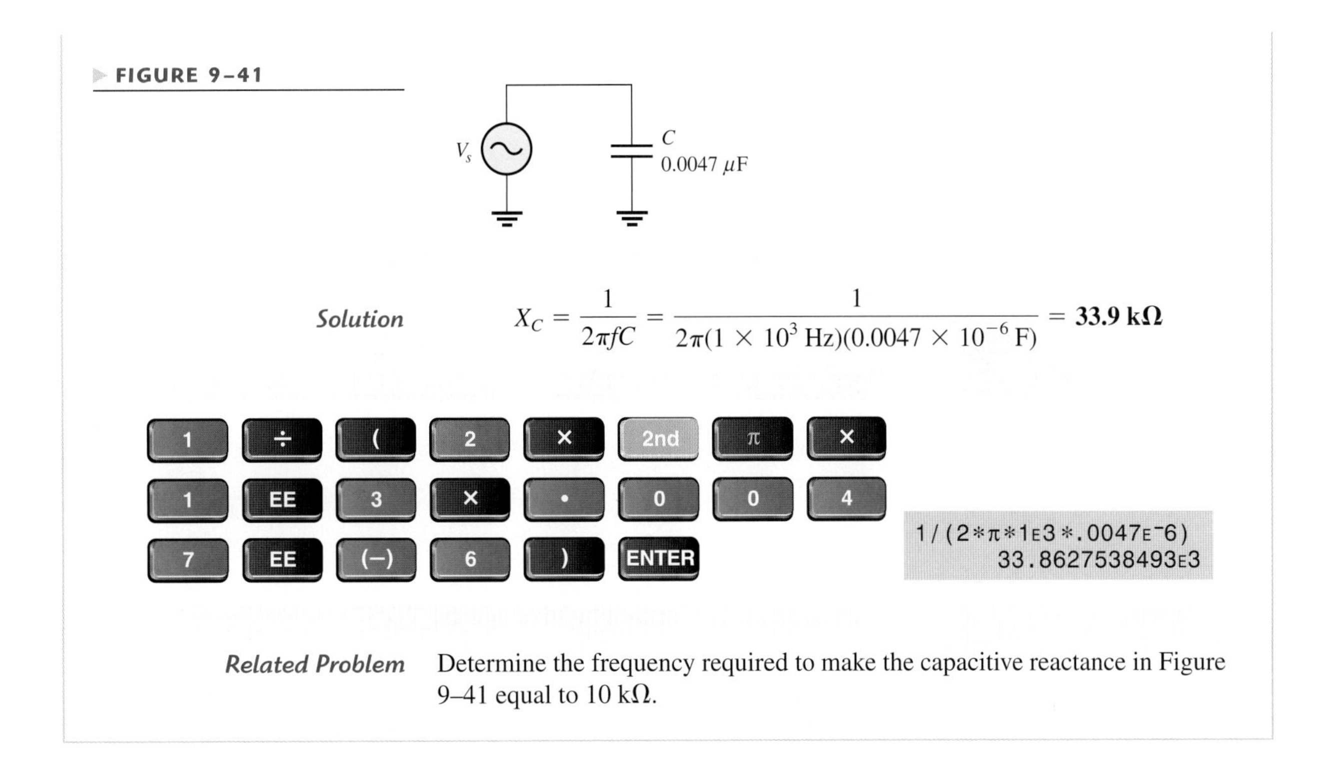

FIGURE 9–41

Solution

$$X_C = \frac{1}{2\pi fC} = \frac{1}{2\pi(1 \times 10^3\ \text{Hz})(0.0047 \times 10^{-6}\ \text{F})} = \mathbf{33.9\ k\Omega}$$

Related Problem Determine the frequency required to make the capacitive reactance in Figure 9–41 equal to 10 kΩ.

Ohm's Law in Capacitive AC Circuits

The reactance of a capacitor is analogous to the resistance of a resistor. In fact, both are expressed in ohms. Since both R and X_C are forms of opposition to current, Ohm's law applies to capacitive circuits as well as to resistive circuits and is stated as follows for capacitive reactance:

$$V = IX_C$$

Equation 9–21

When applying Ohm's law in ac circuits, you must express both the current and the voltage in the same way, that is, both in rms, both in peak, and so on.

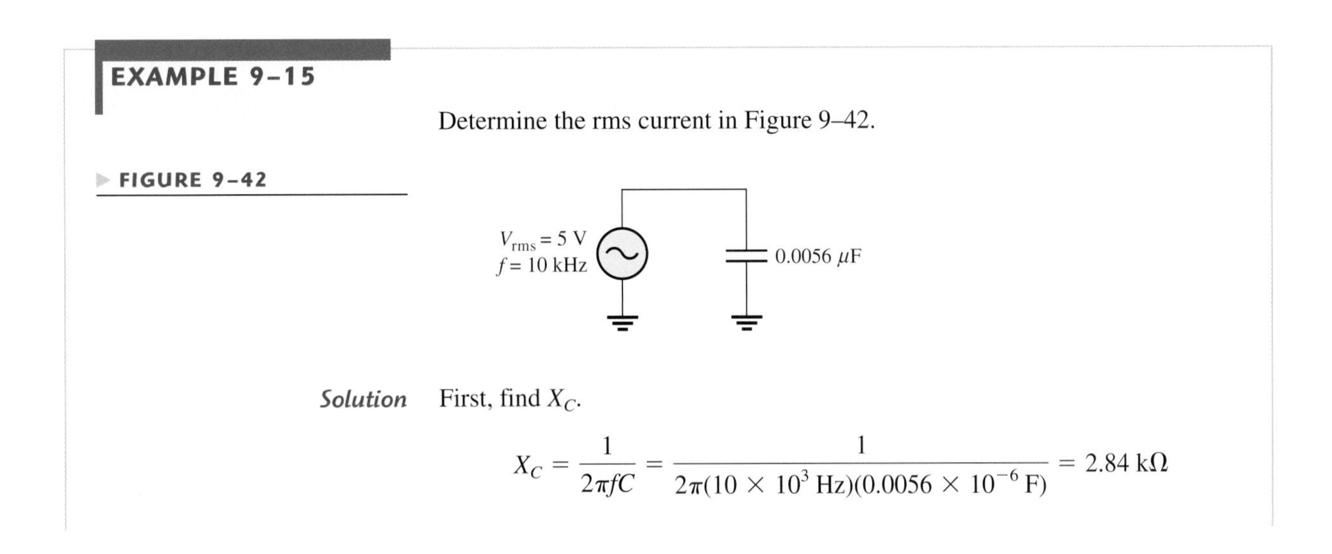

EXAMPLE 9–15

Determine the rms current in Figure 9–42.

FIGURE 9–42

Solution First, find X_C.

$$X_C = \frac{1}{2\pi fC} = \frac{1}{2\pi(10 \times 10^3\ \text{Hz})(0.0056 \times 10^{-6}\ \text{F})} = 2.84\ \text{k}\Omega$$

Then, apply Ohm's law.

$$V_{\text{rms}} = I_{\text{rms}}X_C$$

$$I_{\text{rms}} = \frac{V_{\text{rms}}}{X_C} = \frac{5\text{ V}}{2.84\text{ k}\Omega} = \mathbf{1.76\text{ mA}}$$

Related Problem Change the frequency in Figure 9–42 to 25 kHz and determine the rms value of the current.

Open file E09-15 on your EWB/CircuitMaker CD-ROM. Measure the rms current and compare with the calculated value. Change the frequency of the source voltage to 25 kHz and measure the current.

Current Leads Capacitor Voltage by 90°

A sinusoidal voltage is shown in Figure 9–43. Notice that the rate at which the voltage is changing varies along the sine wave curve, as indicated by the "steepness" of the curve. At the zero crossings, the curve is changing at a faster rate than anywhere else along the curve. At the peaks, the curve has a zero rate of change because it has just reached its maximum and is at the point of changing direction.

The amount of charge stored by a capacitor determines the voltage across it. Therefore, the rate at which the charge is moved ($Q/t = I$) from one plate to the other determines the rate at which the voltage changes. When the current is changing at its maximum rate (at the zero crossings), the voltage is at its maximum value (peak). When the current is changing at its minimum rate (zero at the peaks), the voltage is at its minimum value (zero). This phase relationship is illustrated in Figure 9–44. As you can see, the current peaks occur a quarter of a cycle before the voltage peaks. Thus, the current leads the voltage by 90°.

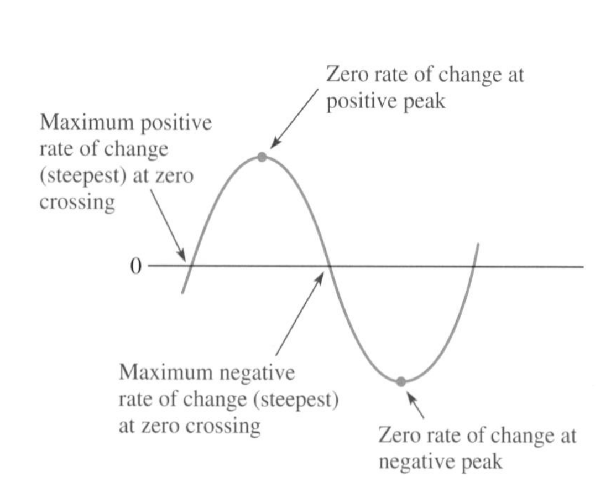

FIGURE 9–43

The rates of change of a sine wave.

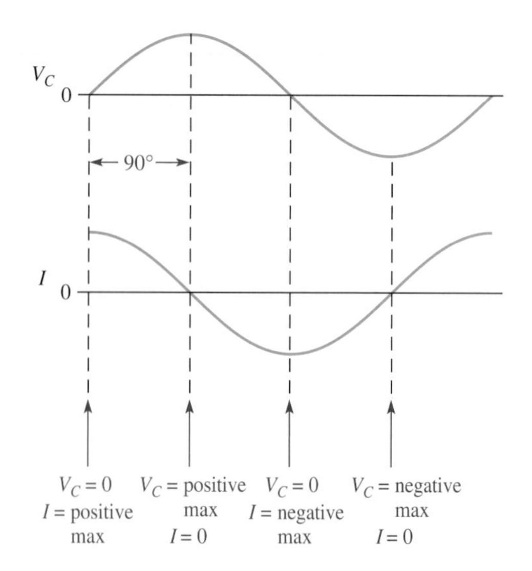

FIGURE 9–44

Current is always leading the capacitor voltage by 90°.

Power in a Capacitor

As discussed earlier in this chapter, a charged capacitor stores energy in the electric field within the dielectric. An ideal capacitor does not dissipate energy; it only stores energy temporarily. When an ac voltage is applied to a capacitor, energy is stored by the capacitor during one-quarter of the voltage cycle. Then the stored energy is returned to the source during another quarter of the cycle. There is no net energy loss. Figure 9–45 shows the power curve that results from one cycle of capacitor voltage and current.

FIGURE 9–45

Power curve for a capacitor.

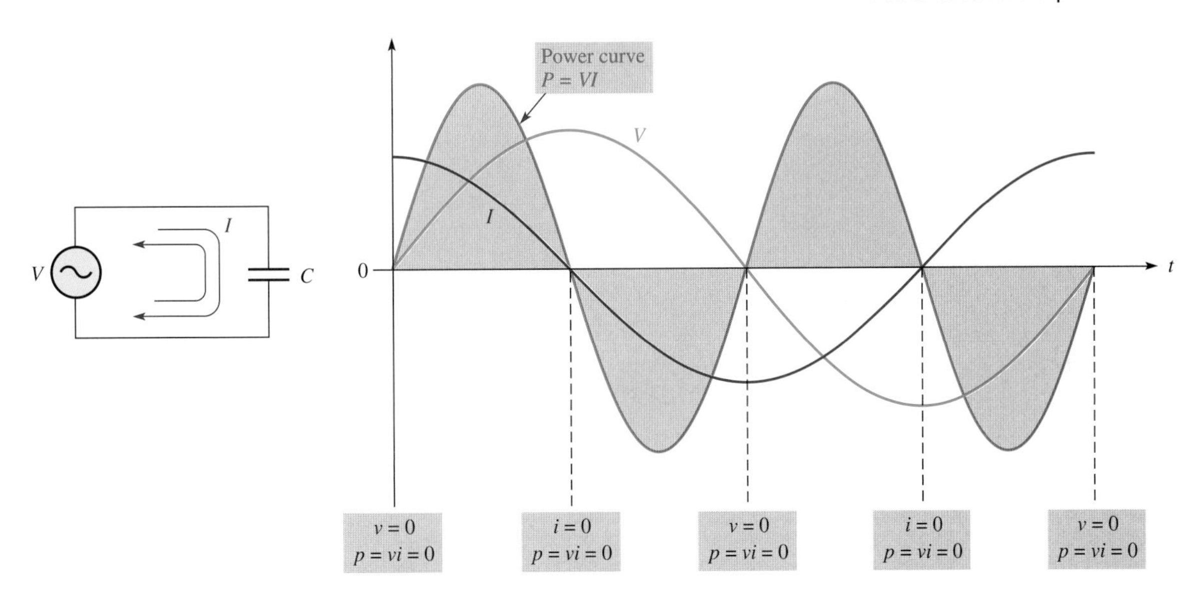

Instantaneous Power (p) The product of instantaneous voltage, v, and instantaneous current, i, gives instantaneous power, p. At points where v or i is zero, p is also zero. When both v and i are positive, p is also positive. When either v or i is positive and the other negative, p is negative. When both v and i are negative, p is positive. As you can see, the power follows a sinusoidal-shaped curve. Positive values of power indicate that energy is stored by the capacitor. Negative values of power indicate that energy is returned from the capacitor to the source. Note that the power fluctuates at a frequency twice that of the voltage or current, as energy is alternately stored and returned to the source.

True Power (P_{true}) Ideally, all of the energy stored by a capacitor during the positive portion of the power cycle is returned to the source during the negative portion. There is no net energy conversion in the capacitor, so the true power is zero. Actually, because of leakage and foil resistance in a practical capacitor, a small percentage of the total power is dissipated in the form of true power.

Reactive Power (P_r) The rate at which a capacitor stores or returns energy is called its reactive power, P_r. The reactive power is a nonzero quantity, because at any instant in time, the capacitor is actually taking energy from the source or returning energy to it. Reactive power does not represent an energy loss. The following formulas apply:

$$P_r = V_{rms}I_{rms}$$ Equation 9–22

$$P_r = \frac{V_{rms}^2}{X_C}$$ Equation 9–23

$$P_r = I_{rms}^2 X_C$$ Equation 9–24

Notice that these equations are of the same form as those introduced in Chapter 3 for true power in a resistor. The voltage and current are expressed as rms values. The unit of reactive power is **VAR** (volt-ampere reactive).

EXAMPLE 9–16

Determine the true power and the reactive power in Figure 9–46.

▶ **FIGURE 9–46**

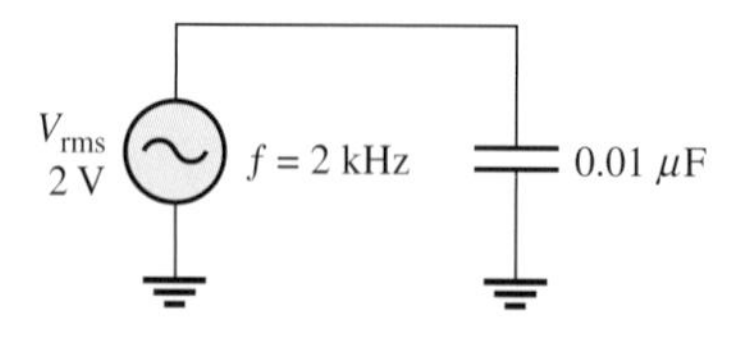

Solution The true power P_{true} is always zero for an ideal capacitor.

$$P_{true} = \mathbf{0\ W}$$

The reactive power is determined by first finding the value for the capacitive reactance and then using Equation 9–23.

$$X_C = \frac{1}{2\pi fC} = \frac{1}{2\pi(2 \times 10^3\text{ Hz})(0.01 \times 10^{-6}\text{ F})} = 7.96\text{ k}\Omega$$

$$P_r = \frac{V_{rms}^2}{X_C} = \frac{(2\text{ V})^2}{7.96\text{ k}\Omega} = 503 \times 10^{-6}\text{ VAR} = \mathbf{503\ \mu VAR}$$

Related Problem If the frequency is doubled in Figure 9–46, what are the true power and the reactive power?

SECTION 9–6 REVIEW

1. Calculate X_C for $f = 5$ kHz and $C = 47$ pF.
2. At what frequency is the reactance of a 0.1 μF capacitor equal to 2 kΩ?
3. Calculate the rms current in Figure 9–47.
4. State the phase relationship between current and voltage in a capacitor.
5. A 1 μF capacitor is connected to an ac voltage source of 12 V rms. What is the true power?
6. In Question 5, determine reactive power at a frequency of 500 Hz.

▶ **FIGURE 9–47**

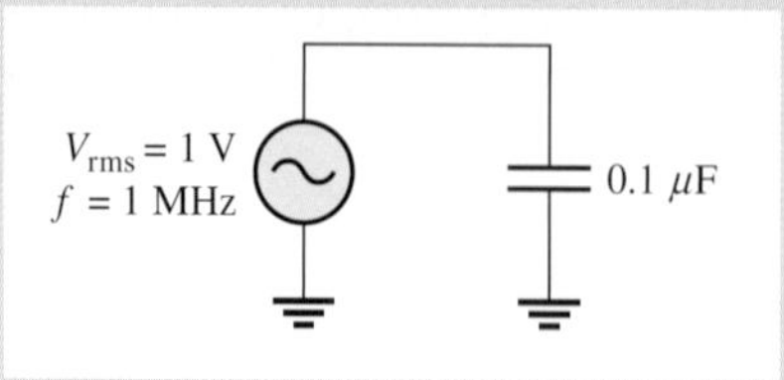

9–7 CAPACITOR APPLICATIONS

Capacitors are widely used in electrical and electronic applications. A few applications are discussed in this section to illustrate the usefulness of this component.

After completing this section, you should be able to

- **Discuss some capacitor applications**
- Describe a power supply filter
- Explain the purpose of coupling and bypass capacitors
- Discuss the basics of capacitors applied to tuned circuits, timing circuits, and computer memories

If you pick up any circuit board, open any power supply, or look inside any piece of electronic equipment, chances are you will find capacitors of one type or another. These components are used for a variety of reasons in both dc and ac applications.

Electrical Storage

One of the most basic applications of a capacitor is as a backup voltage source for low-power circuits such as certain types of semiconductor memories in computers. This particular application requires a very high capacitance value and negligible leakage.

The storage capacitor is connected between the dc power supply input to the circuit and ground. When the circuit is operating from its normal power supply, the capacitor remains fully charged to the dc power supply voltage. If the normal power source is disrupted, effectively removing the power supply from the circuit, the storage capacitor temporarily becomes the power source for the circuit.

The capacitor provides voltage and current to the circuit as long as its charge remains sufficient. As current is drawn by the circuit, charge is removed from the capacitor and the voltage decreases. For this reason, the storage capacitor can only be used as a temporary power source. The length of time that the capacitor can provide sufficient power to the circuit depends on the capacitance and the amount of current drawn by the circuit. The smaller the current and the higher the capacitance, the longer the time.

Power Supply Filtering

A basic dc power supply consists of a circuit known as a **rectifier** followed by a filter. The rectifier converts the 110 V, 60 Hz sinusoidal voltage available at a standard outlet to a pulsating dc voltage that can be either a half-wave rectified voltage or a full-wave rectified voltage, depending on the type of rectifier circuit. As shown in Figure 9–48(a), a half-wave rectifier removes each negative half-cycle of the sinusoidal voltage. As shown in Figure 9–48(b), a full-wave rectifier actually reverses the polarity of the negative portion of each cycle. Both half-wave and full-wave rectified voltages are dc because, even though they are changing, they do not alternate polarity.

To be useful for powering electronic circuits, the rectified voltage must be changed to constant dc voltage because all circuits require constant power. When connected to the rectifier output, the filter nearly eliminates the fluctuations in the rectified voltage and provides a smooth constant-value dc voltage to the load which is the electronic circuit, as indicated in Figure 9–49.

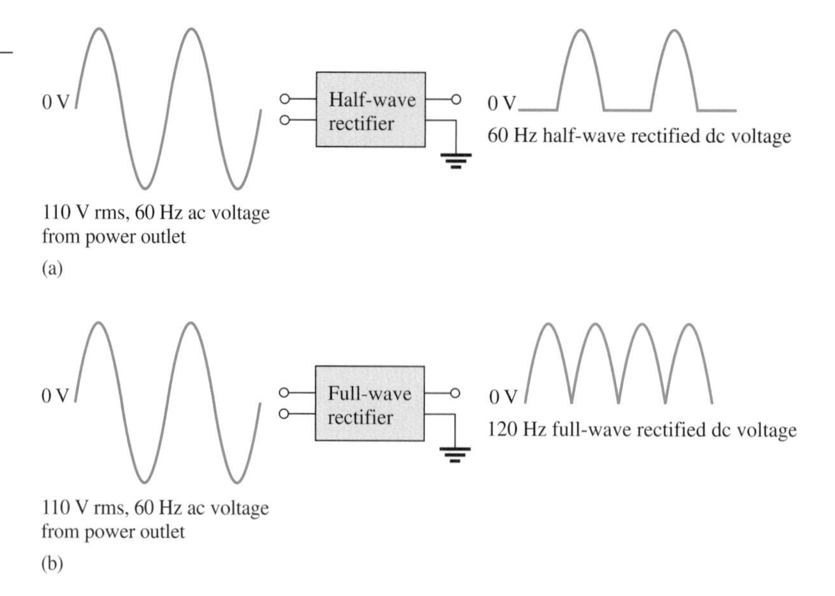

▶ **FIGURE 9–48**

Half-wave and full-wave rectifier operation.

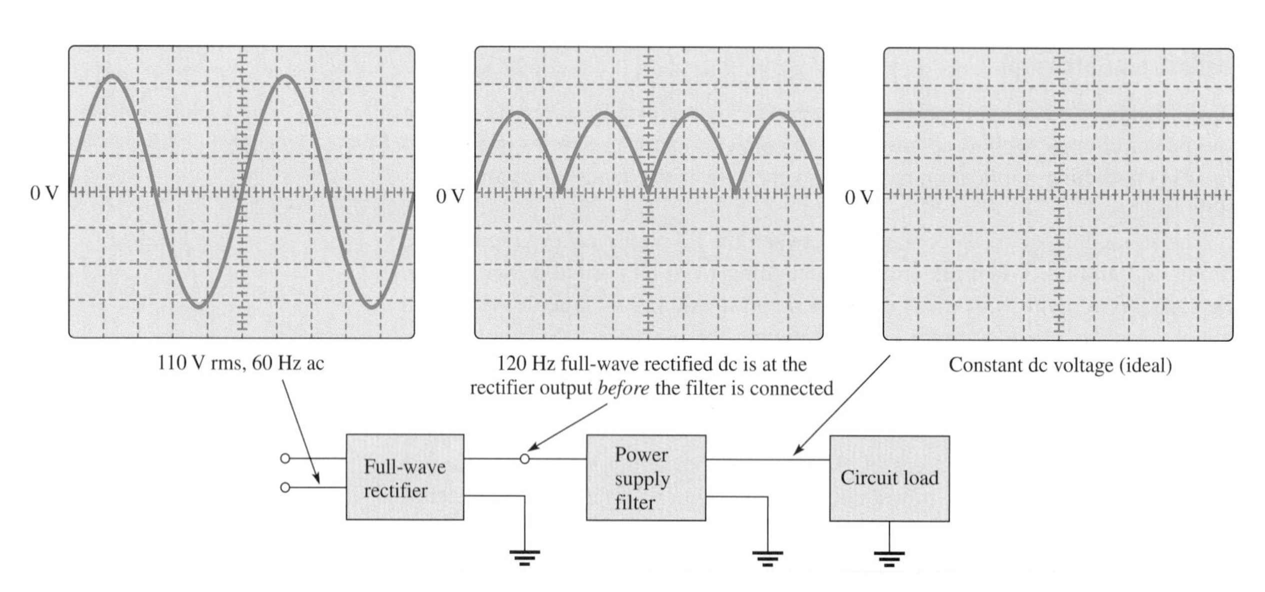

▲ **FIGURE 9–49**

Basic block diagram and operation of a dc power supply.

The Capacitor as a Power Supply Filter Capacitors are used as filters in dc power supplies because of their ability to store electrical charge. Figure 9–50(a) shows a dc power supply with a full-wave rectifier and a capacitor filter. The operation can be described from a charging and discharging point of view as follows. Assume the capacitor is initially uncharged. When the power supply is first turned on and the first cycle of the rectified voltage occurs, the capacitor will quickly charge through the low resistance of the rectifier. The capacitor voltage will follow the rectified voltage curve up to the peak of the rectified voltage. As the rectified voltage passes the peak and begins to decrease, the capacitor will begin to discharge very slowly through the high resistance of the load circuit, as indicated in Figure 9–50(b). The amount of discharge is typically very small and is exaggerated in the figure for purposes of illustration. The next cycle of the rectified voltage will recharge the capacitor back to the peak value by replenishing the small amount of charge lost since the previous peak. This pattern of a small amount of charging and discharging continues as long as the power is on.

A rectifier is designed so that it allows current only in the direction to charge the capacitor. The capacitor will not discharge back through the rectifier but will

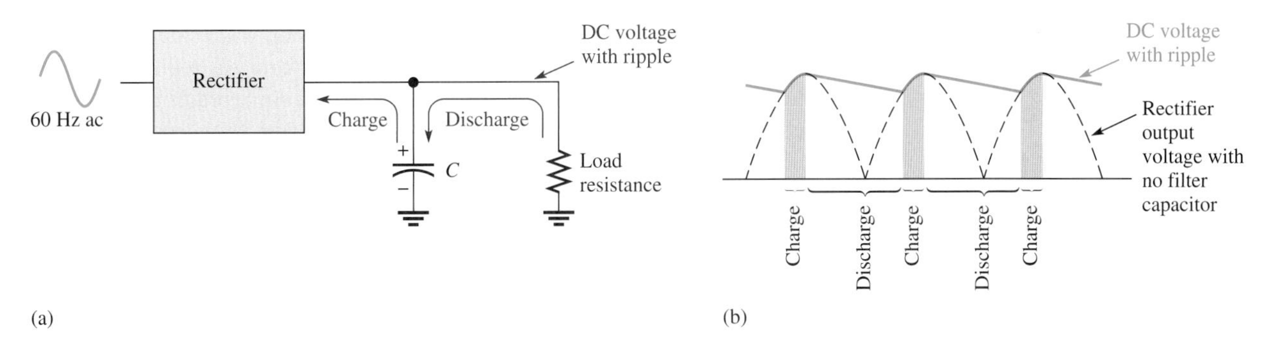

▲ FIGURE 9–50

Basic operation of a power supply filter capacitor.

only discharge a small amount through the relatively high resistance of the load. The small fluctuation in voltage due to the charging and discharging of the capacitor is called the **ripple voltage.** A good dc power supply has a very small amount of ripple on its dc output. Since the discharge time constant of a power supply filter capacitor depends on its capacitance and the resistance of the load, the higher the capacitance value, the longer the discharge time and, therefore, the smaller the ripple voltage.

DC Blocking and AC Coupling

Capacitors are commonly used to block the constant dc voltage in one part of a circuit from getting to another part. As an example of this, a capacitor is connected between two stages of an amplifier to prevent the dc voltage at the output of stage 1 from affecting the dc voltage at the input of stage 2, as illustrated in Figure 9–51. Assume that, for proper operation, the output of stage 1 has a zero dc voltage and the input to stage 2 has a 3 V dc voltage. The capacitor prevents the 3 V dc at stage 2 from getting to the stage 1 output and affecting its zero value, and vice versa.

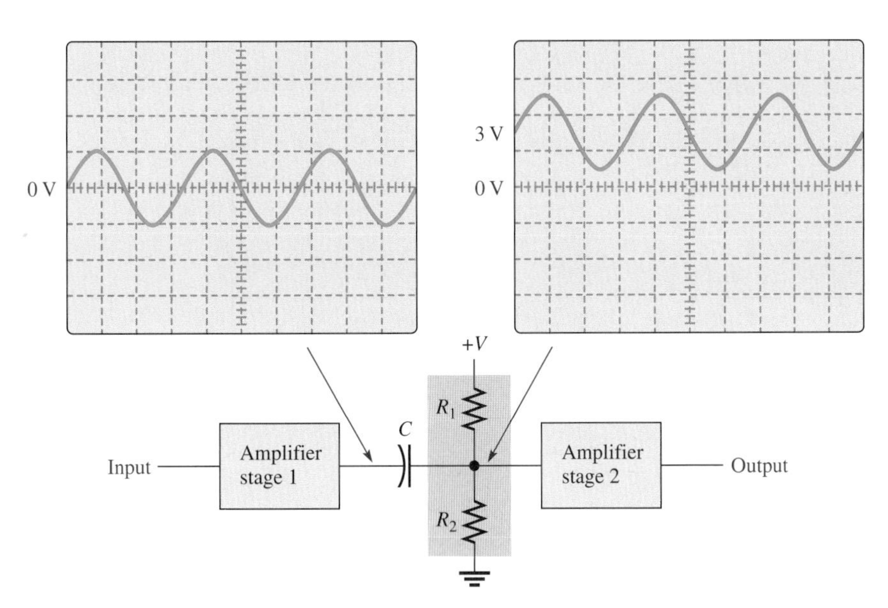

◀ FIGURE 9–51

An application of a capacitor used to block dc and couple ac in an amplifier.

If a sinusoidal signal voltage is applied to the input to stage 1, the signal voltage is increased (amplified) and appears on the output of stage 1, as shown in Figure 9–51. The amplified signal voltage is then coupled through the capacitor to the input of stage 2 where it is superimposed on the 3 V dc level and then again amplified by stage 2. In order for the signal voltage to be passed through the capacitor without being reduced, the capacitor must be large enough so that its reactance at the frequency of the signal voltage is negligible. In this type of

application, the capacitor is known as a coupling capacitor, which ideally appears as an open to dc and as a short to ac. As the signal frequency is reduced, the capacitive reactance increases and, at some point, the capacitive reactance becomes large enough to cause a significant reduction in ac voltage between stage 1 and stage 2.

Power Line Decoupling

Capacitors connected from the dc supply voltage line to ground are used on circuit boards to decouple unwanted voltage transients or spikes that occur on the dc supply voltage because of fast switching digital circuits. A voltage transient contains higher frequencies that may affect the operation of the circuits. These transients are shorted to ground through the very low reactance of the decoupling capacitors. Decoupling capacitors are often used at various points along the supply voltage line on a circuit board, particularly near integrated circuits (ICs).

Bypassing

Another capacitor application is in bypassing an ac voltage around a resistor in a circuit without affecting the dc voltage across the resistor. In amplifier circuits, for example, dc voltages called *bias voltages* are required at various points. For the amplifier to operate properly, certain bias voltages must remain constant and, therefore, any ac voltages must be removed. A sufficiently large capacitor connected from a bias point to ground provides a low reactance path to ground for ac voltages, leaving the constant dc bias voltage at the given point. This bypass application is illustrated in Figure 9–52. At lower frequencies, the bypass capacitor becomes less effective because of its increased reactance.

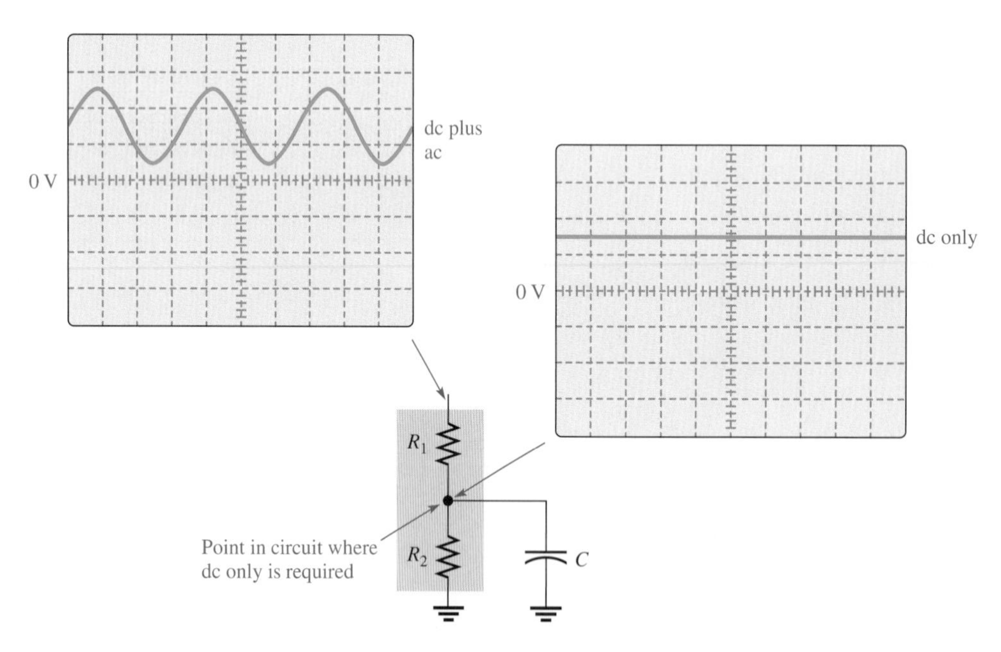

FIGURE 9–52

Example of the operation of a bypass capacitor.

Signal Filters

Capacitors are essential to the operation of a class of circuits called filters that are used for selecting one ac signal with a certain specified frequency from a wide range of signals with many different frequencies or for selecting a certain band of frequencies and eliminating all others. A common example of this application is in radio and television receivers where it is necessary to select the signal transmitted from a given station and eliminate or filter out the signals transmitted from all the other stations in the area.

When you tune your radio or TV, you are actually changing the capacitance in the tuner circuit (which is a type of filter) so that only the signal from the station or channel you want passes through to the receiver circuitry. Capacitors are used in conjunction with resistors, inductors (covered in Chapter 11), and other components in these types of filters.

The main characteristic of a filter is its frequency selectivity, which is based on the fact that the reactance of a capacitor depends on frequency ($X_C = 1/2\pi fC$).

Timing Circuits

Another important area in which capacitors are used is in timing circuits that generate specified time delays or produce waveforms with specific characteristics. Recall that the time constant of a circuit with resistance and capacitance can be controlled by selecting appropriate values for *R* and *C*. The charging time of a capacitor can be used as a basic time delay in various types of circuits. An example is the circuit that controls the turn indicators on your car where the light flashes on and off at regular intervals.

Computer Memories

Dynamic computer memories use capacitors as the basic storage element for binary information, which consists of two digits, 1 and 0. A charged capacitor can represent a stored 1 and a discharged capacitor can represent a stored 0. Patterns of 1s and 0s that make up binary data are stored in a memory that consists of an array of capacitors with associated circuitry. You will study this topic in a computer or digital fundamentals course.

SECTION 9–7 REVIEW

1. Explain how half-wave or full-wave rectified dc voltages are smoothed out by a filter capacitor.
2. Explain the purpose of a coupling capacitor.
3. How large must a coupling capacitor be?
4. Explain the purpose of a decoupling capacitor.
5. Discuss how the relationship of frequency and capacitive reactance is important in frequency-selective circuits such as signal filters.
6. What characteristic of a capacitor is most important in time-delay applications?

9–8 TESTING CAPACITORS

Capacitors are very reliable devices but their useful life can be extended significantly by operating them well within the voltage rating and at moderate temperatures. In this section, the basic types of failures are discussed and methods for checking for them are introduced.

After completing this section, you should be able to

- **Test a capacitor**
- Perform an ohmmeter check
- Explain what an *LCR* meter is

Capacitor failures can be categorized into two areas: catastrophic and degradation. The catastrophic failures are usually a short circuit caused by dielectric breakdown or an open circuit caused by connection failure. Degradation usually results in a gradual decrease in leakage resistance, hence an increase in leakage current or an increase in equivalent series resistance or dielectric absorption.

Ohmmeter Check

When there is a suspected problem, the capacitor can be removed from the circuit and checked with an analog ohmmeter. First, to be sure that the capacitor is discharged, short its leads, as indicated in Figure 9–53(a). Connect the meter—set on a high ohms range such as R × 1M—to the capacitor, as shown in Figure 9–53(b), and observe the needle. It should initially indicate near zero ohms. Then it should begin to move toward the high-resistance end of the scale as the capacitor charges from the ohmmeter's battery, as shown in Figure 9–53(c). When the capacitor is fully charged, the meter will indicate an extremely high resistance, as shown in Figure 9–53(d).

FIGURE 9–53

Checking a capacitor with an analog ohmmeter. This check shows a good capacitor.

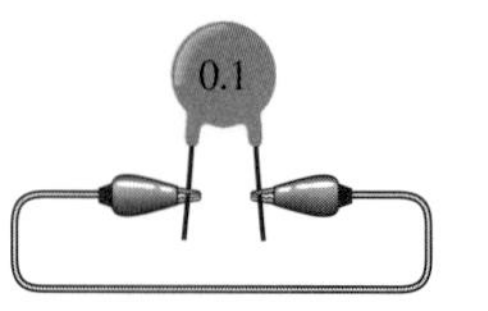

(a) Discharge by shorting leads

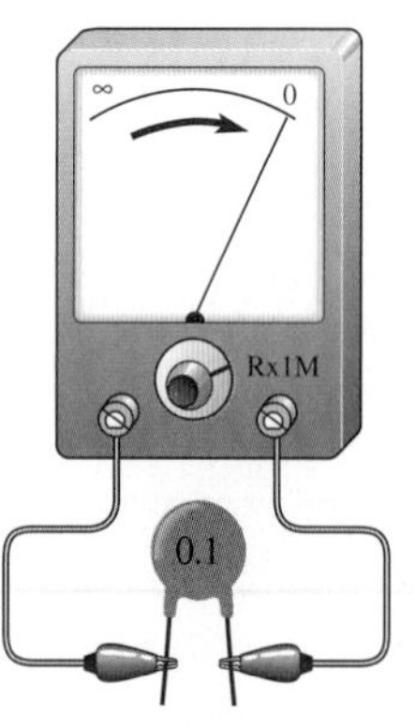

(b) Initially, when the meter is first connected, the pointer jumps immediately to zero.

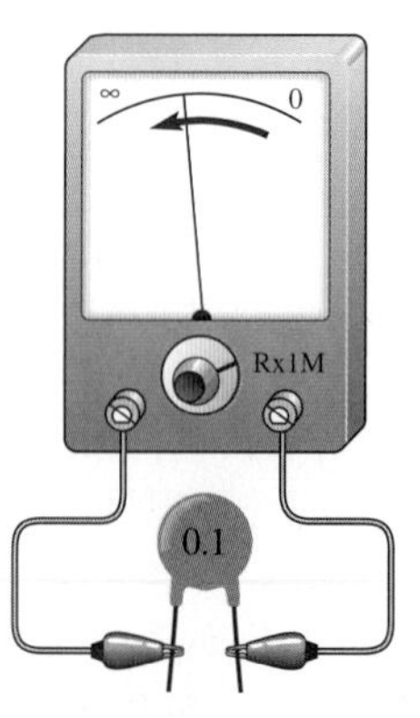

(c) The needle moves slowly toward infinity as the capacitor charges from the ohmmeter's battery.

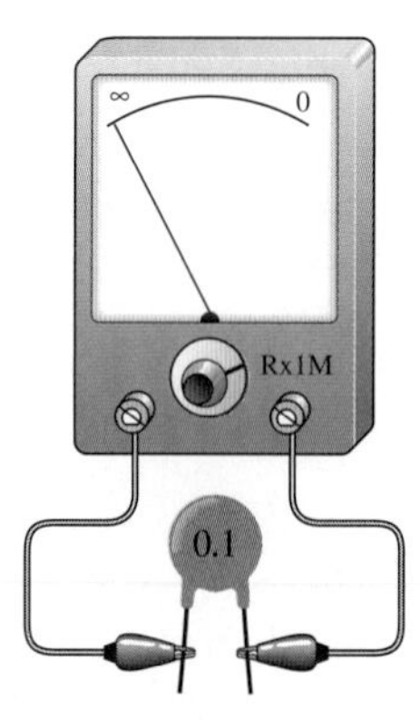

(d) When the capacitor is fully charged, the pointer is at infinity.

As mentioned, the capacitor charges from the internal battery of the ohmmeter, and the meter responds to the charging current. The larger the capacitance value, the more slowly the capacitor will charge, as indicated by the needle movement. For small pF values, the meter response may be too slow to indicate the fast charging action.

If the capacitor is internally shorted, the meter will go to zero and stay there. If it is leaky, the final meter reading will be much less than normal. Most capacitors have a resistance of several hundred megohms. The exception is the electrolytic capacitor, which may normally have less than one megohm of leakage resistance. If the capacitor is open, no charging action will be observed, and the meter will indicate an infinite resistance. Faults such as breakdown at a certain voltage level cannot be detected with this test.

Testing for Capacitance Value and Other Parameters with an *LCR* Meter

An *LCR* meter, such as the one shown in Figure 9–54, can be used to check the value of a capacitor. All capacitors change value over a period of time, some more than others. Ceramic capacitors, for example, often exhibit a 10% to 15% change in the value during the first year. Electrolytic capacitors are particularly subject to value change due to drying of the electrolytic solution. In other cases, capacitors may be labeled incorrectly or the wrong value was installed in the circuit. Although a value change represents less than 25% of defective capacitors, a value check should be made to quickly eliminate this as a source of trouble when troubleshooting a circuit. Values from 200 pF to 20 mF can be measured by simply connecting the capacitor, setting the switch and reading the value of the display.

FIGURE 9–54

A typical *LCR* meter. (Photography courtesy of B&K Precision Corp.)

Some *LCR* meters can check for leakage current in capacitors. In order to check for leakage, a sufficient voltage must be applied across the capacitor to simulate operating conditions. This is automatically done by the test instrument. Over 40% of all defective capacitors have excessive leakage current and electrolytic capacitors are particularly susceptible to this problem.

The problem of dielectric absorption occurs mostly in electrolytic capacitors when they do not completely discharge during use and retain a residual charge. Approximately 25% of defective capacitors have exhibited this condition.

Another defect sometimes found in capacitors is excessive equivalent series resistance. This problem may be caused by a defective lead to plate contacts, resistive leads, or resistive plates and shows up only under ac conditions. This is the least common capacitor defect and occurs in less than 10% of all defective capacitors.

SECTION 9–8 REVIEW

1. How can a capacitor be discharged after removal from the circuit?
2. Describe how the needle of an analog ohmmeter responds when a good capacitor is checked.
3. List four common capacitor defects.

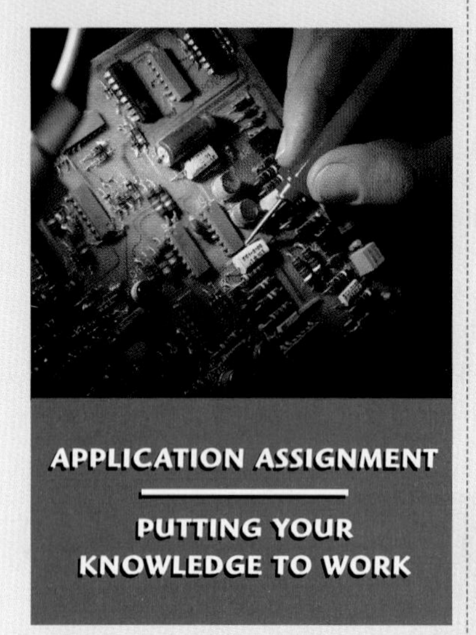

APPLICATION ASSIGNMENT

PUTTING YOUR KNOWLEDGE TO WORK

As you learned in this chapter, capacitors are used in certain types of amplifiers for coupling ac signals and blocking dc voltages. In this assignment, an amplifier circuit board contains two coupling capacitors. Your assignment is to check certain voltages on three identical amplifier circuit boards to determine if the capacitors are working properly. A knowledge of amplifier circuits is not necessary for this assignment.

All amplifier circuits contain transistors that require dc voltages to establish proper operating conditions for amplifying ac signals. These dc voltages are referred to as bias voltages. As indicated in Figure 9–55(a), a common type of dc bias circuit used in amplifiers is the voltage divider formed by R_1 and R_2, which sets up the proper dc voltage at the input to the amplifier.

When an ac signal voltage is applied to the amplifier, the input coupling capacitor, C_1, prevents the internal resistance of the ac source from changing the dc bias voltage. Without the capacitor, the internal source resistance would appear in parallel with R_2 and drastically change the value of the dc voltage.

The coupling capacitance is chosen so that its reactance (X_C) at the frequency of the ac signal is very small compared to the bias resistor values. The coupling capacitance, therefore, efficiently couples the ac signal from the source to the input of the amplifier. On the source side of the input coupling capacitor there is only ac but on the amplifier side there is ac plus dc (the signal voltage is riding on the dc bias voltage set by the voltage divider), as indicated in Figure 9–55(a). Capacitor C_2 is the output coupling capacitor which couples the amplified ac signal to another amplifier stage that would be connected to the output.

You will check three amplifier boards like the one in Figure 9–55(b) for the proper input voltages using an oscilloscope. If the voltages are incorrect, you will determine the most likely fault. For all measurements, assume the amplifier has no dc loading effect on the voltage-divider bias circuit.

Step 1: Compare the Printed Circuit Board with the Schematic

Check the printed circuit board in Figure 9–55(b) to make sure it agrees with the amplifier schematic in part (a).

Step 2: Test the Input to Board 1

The oscilloscope probe is connected from channel 1 to the board as shown in Figure 9–56.

The input signal from a sinusoidal voltage source is connected to the board and set to a frequency of 5 kHz with an amplitude of 1 V rms. Determine if the voltage and frequency displayed on the scope in Figure 9–56 are correct. If the scope measurement is incorrect, specify the most likely fault in the circuit.

FIGURE 9–55

Capacitively coupled amplifier.

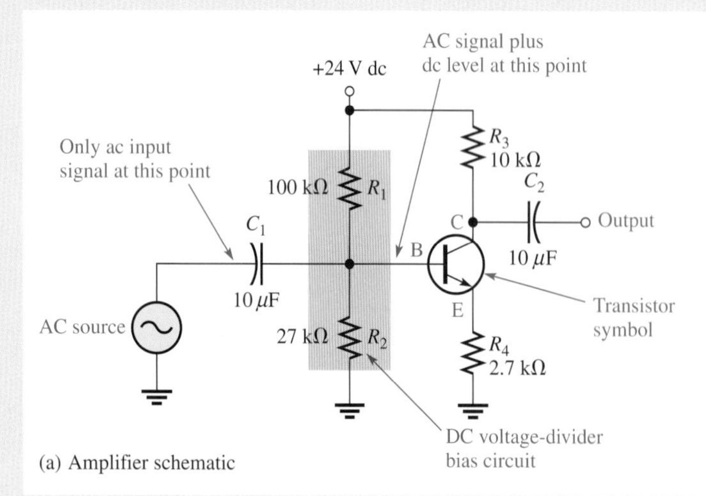

(a) Amplifier schematic

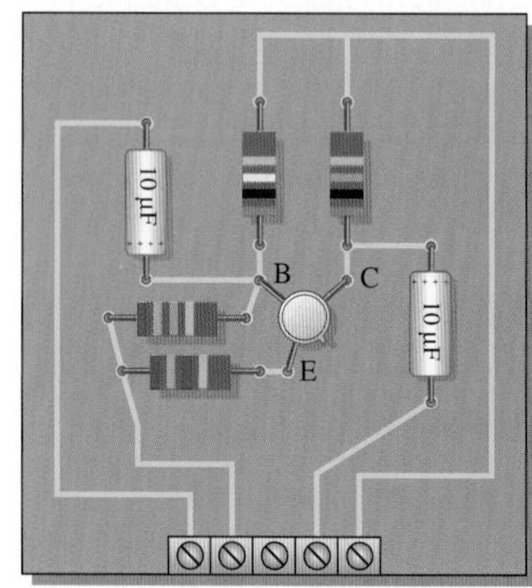

(b) Amplifier board

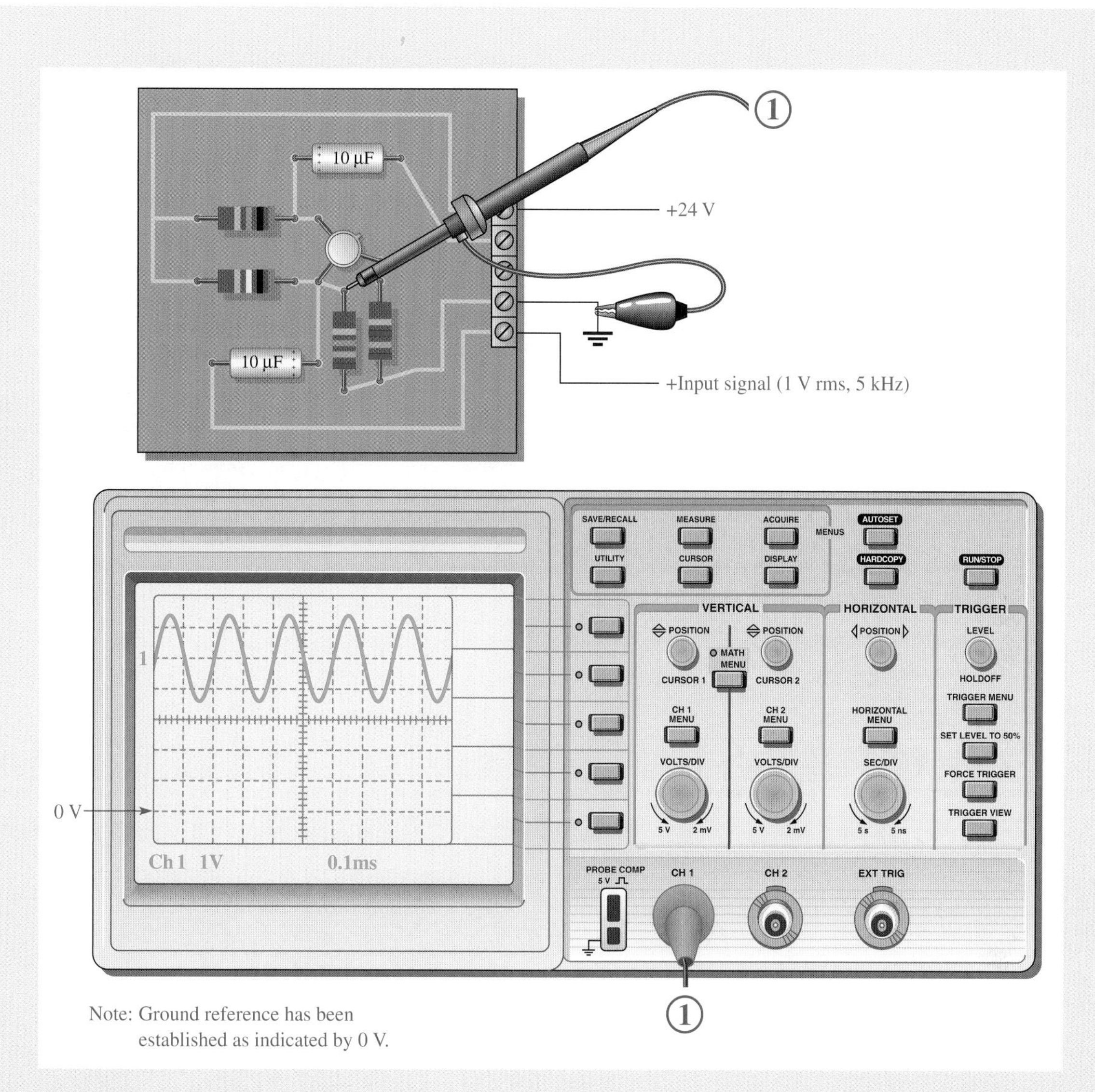

FIGURE 9–56

Testing board 1.

Step 3: Test the Input to Board 2

The oscilloscope probe is connected from channel 1 to board 2 the same as was shown in Figure 9–56 for board 1.

The input signal from the sinusoidal voltage source is the same as Step 2. Determine if the scope display in Figure 9–57 is correct. If the scope measurement is incorrect, specify the most likely fault in the circuit.

Step 4: Test the Input to Board 3

The oscilloscope probe is connected from channel 1 to board 3 the same as was shown in Figure 9–56 for board 1.

The input signal from the sinusoidal voltage source is the same as Step 3. Determine if the scope display in Figure 9–58 is correct. If the scope measurement is incorrect, specify the most likely fault in the circuit.

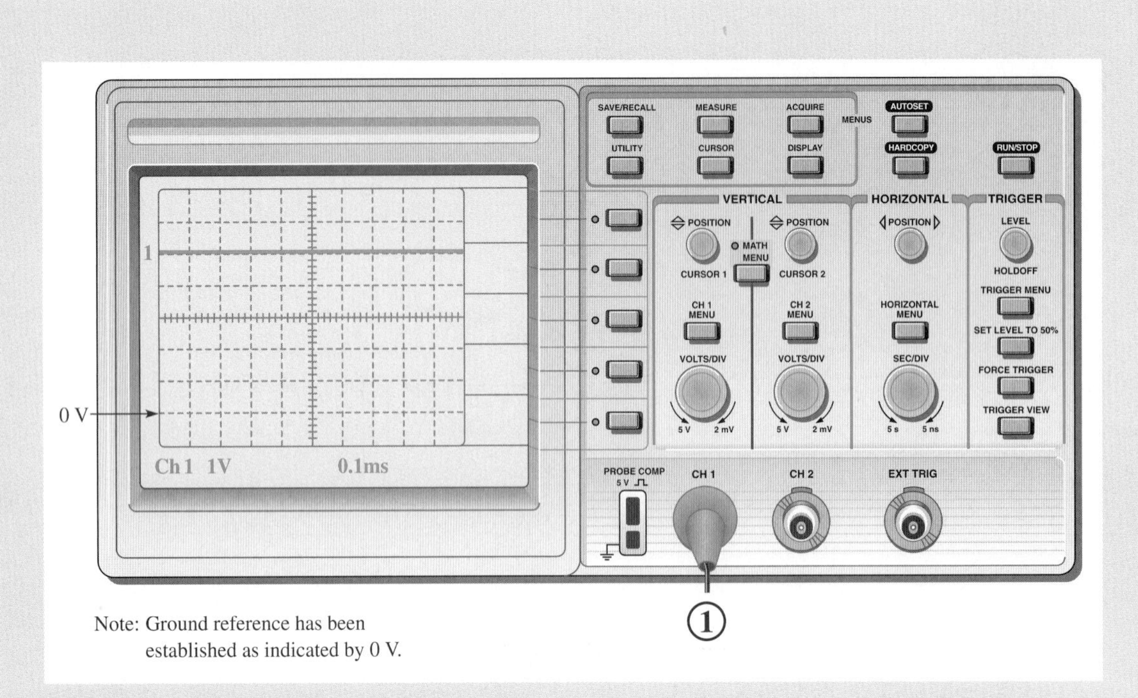

FIGURE 9–57

Testing board 2.

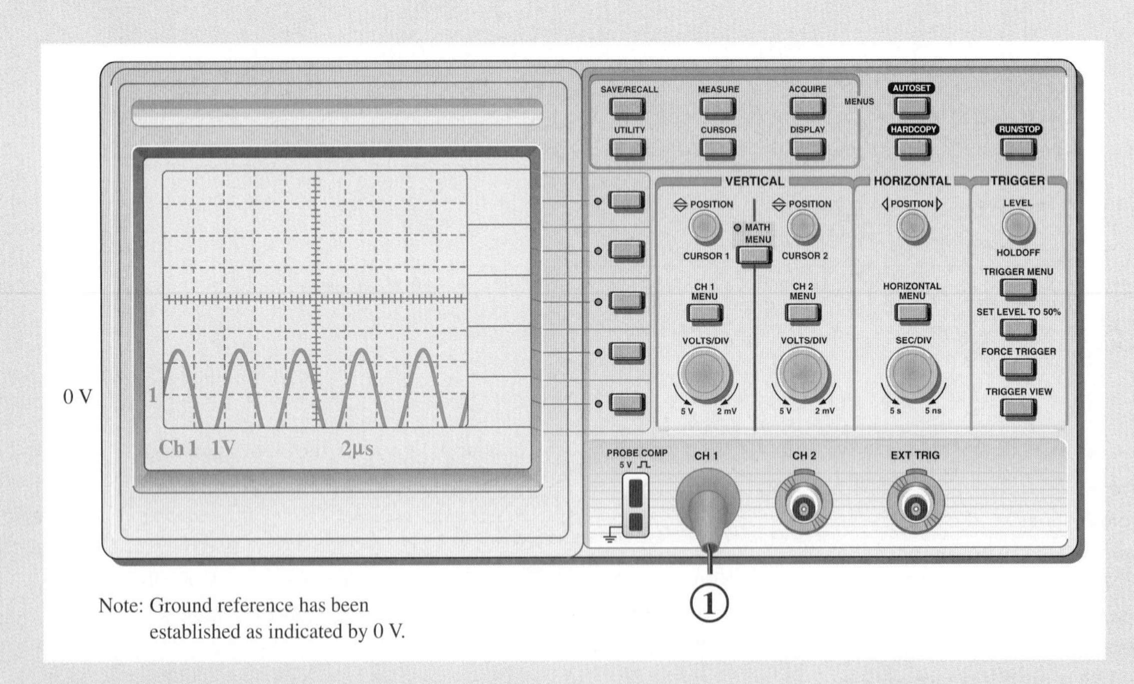

FIGURE 9–58

Testing board 3.

APPLICATION ASSIGNMENT REVIEW

1. Explain why the input coupling capacitor is necessary when connecting an ac source to the amplifier.
2. Capacitor C_2 in Figure 9–55 is an output coupling capacitor. Generally, what would you expect to measure at the point labeled C at the output of the circuit when an input signal is applied to the amplifier?

SUMMARY

- A capacitor is composed of two parallel conductive plates separated by an insulating material called the *dielectric.*
- A capacitor stores electrical charge on its plates.
- Energy is stored in a capacitor by the electric field created between the charged plates in the dielectric.
- Capacitance is measured in units of farads (F).
- Capacitance is directly proportional to the plate area and the dielectric constant and inversely proportional to the distance between the plates (the dielectric thickness).
- The dielectric constant is an indication of the ability of a material to establish an electric field.
- The dielectric strength is one factor that determines the breakdown voltage of a capacitor.
- Capacitors are commonly classified according to the dielectric material. Typical materials are mica, ceramic, plastic, and electrolytic (aluminum oxide and tantalum oxide).
- The total capacitance of series capacitors is less than the smallest capacitance.
- The total capacitance of parallel capacitors is the sum of all the capacitances.
- A capacitor blocks constant dc.
- The time constant determines the charging and discharging time of a capacitor with resistance in series.
- In an *RC* circuit, the voltage and current during charging and discharging make an approximate 63% change during each time-constant interval.
- Five time constants are required for a capacitor to fully charge or discharge. This is called the *transient time.*
- The approximate percentage of final charge after each charging time-constant interval is given in Table 9–4.
- The approximate percentage of initial charge after each discharging time-constant interval is given in Table 9–5.
- Alternating current in a capacitor leads the voltage by 90°.
- A capacitor passes ac to an extent that depends on its reactance and the resistance in the rest of the circuit.
- Capacitive reactance is the opposition to ac, expressed in ohms.

TABLE 9–4

NUMBER OF TIME CONSTANTS	APPROXIMATE % OF FINAL CHARGE
1	63
2	86
3	95
4	98
5	99 (considered 100%)

TABLE 9–5

NUMBER OF TIME CONSTANTS	APPROXIMATE % OF INITIAL CHARGE
1	37
2	14
3	5
4	2
5	1 (considered 0)

- Capacitive reactance (X_C) is inversely proportional to the frequency and to the capacitance value.
- Ideally, there is no energy loss in a capacitor and, thus, the true power (watts) is zero. However, in most capacitors there is some small energy loss due to leakage resistance.

EQUATIONS

9–1	$C = \frac{Q}{V}$	Capacitance in terms of charge and voltage
9–2	$Q = CV$	Charge in terms of capacitance and voltage
9–3	$V = \frac{Q}{C}$	Voltage in terms of charge and capacitance
9–4	$F = \frac{kQ_1Q_2}{d^2}$	Coulomb's law (force between two charged bodies)
9–5	$W = \frac{1}{2}CV^2$	Energy stored by a capacitor
9–6	$\epsilon_r = \frac{\epsilon}{\epsilon_0}$	Dielectric constant (relative permittivity)
9–7	$C = \frac{A\epsilon_r(8.85 \times 10^{-12}\text{ F/m})}{d}$	Capacitance in terms of physical parameters
9–8	$\frac{1}{C_T} = \frac{1}{C_1} + \frac{1}{C_2}$	Reciprocal of total series capacitance (two capacitors)
9–9	$C_T = \frac{1}{\frac{1}{C_1} + \frac{1}{C_2}}$	Total series capacitance (two capacitors)

9–10	$\frac{1}{C_T} = \frac{1}{C_1} + \frac{1}{C_2} + \frac{1}{C_3} + \cdots + \frac{1}{C_n}$	Reciprocal of total series capacitance (general)
9–11	$C_T = \frac{1}{\frac{1}{C_1} + \frac{1}{C_2} + \frac{1}{C_3} + \cdots + \frac{1}{C_n}}$	Total series capacitance (general)
9–12	$V_x = \left(\frac{C_T}{C_x}\right)V_S$	Voltage across series capacitor
9–13	$C_T = C_1 + C_2$	Two capacitors in parallel
9–14	$C_T = C_1 + C_2 + C_3 + \cdots + C_n$	n capacitors in parallel
9–15	$\tau = RC$	RC time constant
9–16	$v = V_F + (V_i - V_F)e^{-t/\tau}$	Exponential voltage (general)
9–17	$i = I_F + (I_i - I_F)e^{-t/\tau}$	Exponential current (general)
9–18	$v = V_F(1 - e^{-t/RC})$	Increasing exponential voltage beginning at zero
9–19	$v = V_i e^{-t/RC}$	Decreasing exponential voltage ending at zero
9–20	$X_C = \frac{1}{2\pi fC}$	Capacitive resistance
9–21	$V = IX_C$	Ohm's law for a capacitor
9–22	$P_r = V_{rms}I_{rms}$	Reactive power in a capacitor
9–23	$P_r = \frac{V_{rms}^2}{X_C}$	Reactive power in a capacitor
9–24	$P_r = I_{rms}^2 X_C$	Reactive power in a capacitor

SELF-TEST

Answers are at the end of the chapter.

1. Which of the following statement(s) accurately describe(s) a capacitor?
   (a) The plates are conductive.
   (b) The dielectric is an insulator between the plates.
   (c) Constant dc flows through a fully charged capacitor.
   (d) A practical capacitor stores charge indefinitely when disconnected from the source.
   (e) none of the above answers
   (f) all the above answers
   (g) only answers (a) and (b)
2. Which one of the following statements is true?
   (a) There is current through the dielectric of a charging capacitor.
   (b) When a capacitor is connected to a dc voltage source, it will charge to the value of the source.
   (c) An ideal capacitor can be discharged by disconnecting it from the voltage source.

3. A capacitance of 0.01 μF is larger than
   (a) 0.00001 F (b) 100,000 pF
   (c) 1000 pF (d) all of these answers
4. A capacitance of 1000 pF is smaller than
   (a) 0.01 μF (b) 0.001 μF
   (c) 0.00000001 F (d) answers (a) and (c)
5. When the voltage across a capacitor is increased, the stored charge
   (a) increases (b) decreases
   (c) remains constant (d) fluctuates
6. When the voltage across a capacitor is doubled, the stored charge
   (a) stays the same (b) is halved
   (c) increases by four (d) doubles
7. The voltage rating of a capacitor is increased by
   (a) increasing the plate separation (b) decreasing the plate separation
   (c) increasing the plate area (d) answers (b) and (c)
8. The capacitance value is increased by
   (a) decreasing plate area (b) increasing plate separation
   (c) decreasing plate separation (d) increasing plate area
   (e) answers (a) and (b) (f) answers (c) and (d)
9. A 1 μF, a 2.2 μF, and a 0.05 μF capacitor are connected in series. The total capacitance is less than
   (a) 1 μF (b) 2.2 μF (c) 0.05 μF (d) 0.001 μF
10. Four 0.022 μF capacitors are in parallel. The total capacitance is
    (a) 0.022 μF (b) 0.088 μF (c) 0.011 μF (d) 0.044 μF
11. An uncharged capacitor and a resistor are connected in series with a switch and a 12 V battery. At the instant the switch is closed, the voltage across the capacitor is
    (a) 12 V (b) 6 V (c) 24 V (d) 0 V
12. In Question 11, the voltage across the capacitor when it is fully charged is
    (a) 12 V (b) 6 V (c) 24 V (d) −6 V
13. In Question 11, the capacitor will reach full charge in a time equal to approximately
    (a) RC (b) $5RC$ (c) $12RC$ (d) cannot be predicted
14. A sinusoidal voltage is applied across a capacitor. When the frequency of the voltage is increased, the current
    (a) increases (b) decreases (c) remains constant (d) ceases
15. A capacitor and a resistor are connected in series to a sine wave generator. The frequency is set so that the capacitive reactance is equal to the resistance and, thus, an equal amount of voltage appears across each component. If the frequency is decreased,
    (a) $V_R > V_C$ (b) $V_C > V_R$ (c) $V_R = V_C$ (d) $V_C < V_R$
16. An ohmmeter is connected across a discharged capacitor and the needle stabilizes at approximately 50 kΩ. The capacitor is
    (a) good (b) charged (c) too large (d) leaky

PROBLEMS

Answers to odd-numbered problems are at the end of the book.

BASIC PROBLEMS

SECTION 9–1 **The Basic Capacitor**

1. **(a)** Find the capacitance when $Q = 50\ \mu C$ and $V = 10$ V.
   (b) Find the charge when $C = 0.001\ \mu F$ and $V = 1$ kV.
   (c) Find the voltage when $Q = 2$ mC and $C = 200\ \mu F$.
2. Convert the following values from microfarads to picofarads:
   (a) 0.1 μF **(b)** 0.0025 μF **(c)** 5 μF
3. Convert the following values from picofarads to microfarads:
   (a) 1000 pF **(b)** 3500 pF **(c)** 250 pF
4. Convert the following values from farads to microfarads:
   (a) 0.0000001 F **(b)** 0.0022 F **(c)** 0.0000000015 F
5. What size capacitor is capable of storing 10 mJ of energy with 100 V across its plates?
6. A mica capacitor has a plate area of 0.04 m^2 and a dielectric thickness of 0.008 m. What is its capacitance?
7. An air capacitor has plates with an area of 0.1 m^2. The plates are separated by 0.01 m. Calculate the capacitance.
8. At ambient temperature (25°C), a certain capacitor is specified to be 1000 pF. It has a negative temperature coefficient of 200 ppm/°C. What is its capacitance at 75°C?
9. A 0.001 μF capacitor has a positive temperature coefficient of 500 ppm/°C. How much change in capacitance will a 25°C increase in temperature cause?

SECTION 9–2 **Types of Capacitors**

10. In the construction of a stacked-foil mica capacitor, how is the plate area increased?
11. What type of capacitor has the higher dielectric constant, mica or ceramic?
12. Show how to connect an electrolytic capacitor across points A and B in Figure 9–59.
13. Determine the value of the typographically labeled ceramic disk capacitors in Figure 9–60.

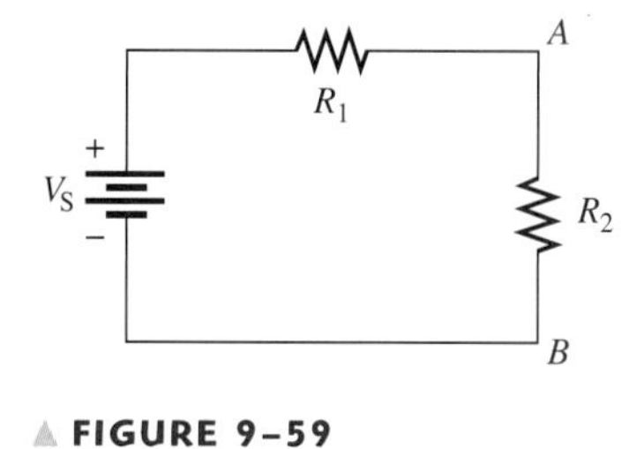

▲ FIGURE 9–59

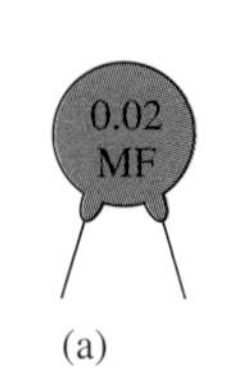

▲ FIGURE 9–60

14. Name two types of electrolytic capacitors. How do electrolytics differ from other capacitors?

15. Identify the parts of the ceramic disk capacitor shown in the cutaway view of Figure 9–61 by referring to Figure 9–9(b).

FIGURE 9–61

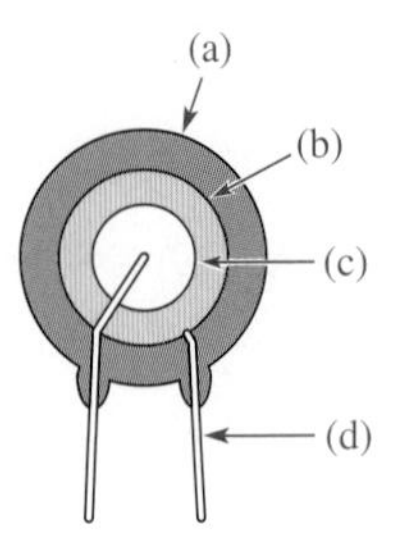

SECTION 9–3

Series Capacitors

16. Five 1000 pF capacitors are in series. What is the total capacitance?

17. Find the total capacitance for each circuit in Figure 9–62.

FIGURE 9–62

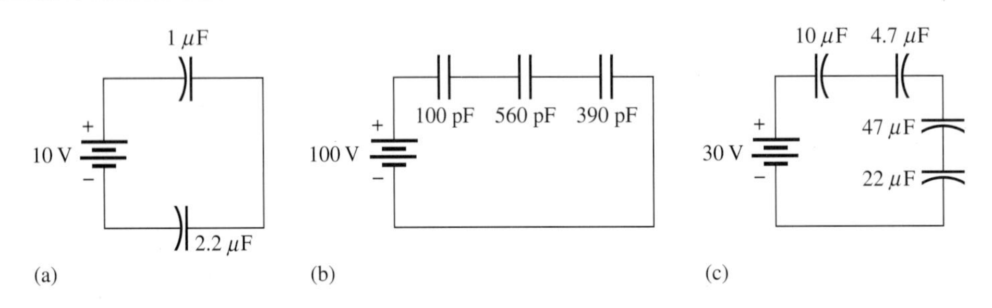

18. For each circuit in Figure 9–62, determine the voltage across each capacitor.

19. The total charge stored by the series capacitors in Figure 9–63 is 10 μC. Determine the voltage across each of the capacitors.

FIGURE 9–63

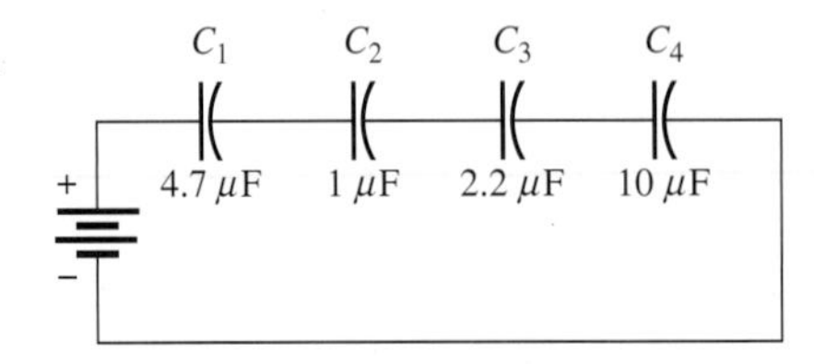

SECTION 9–4

Parallel Capacitors

20. Determine C_T for each circuit in Figure 9–64.

FIGURE 9–64

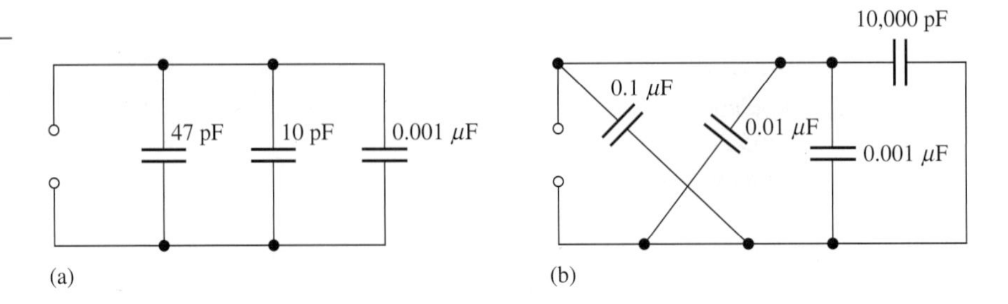

21. Determine C_T for each circuit in Figure 9–65.

22. What is the voltage between points A and B in each circuit in Figure 9–65?

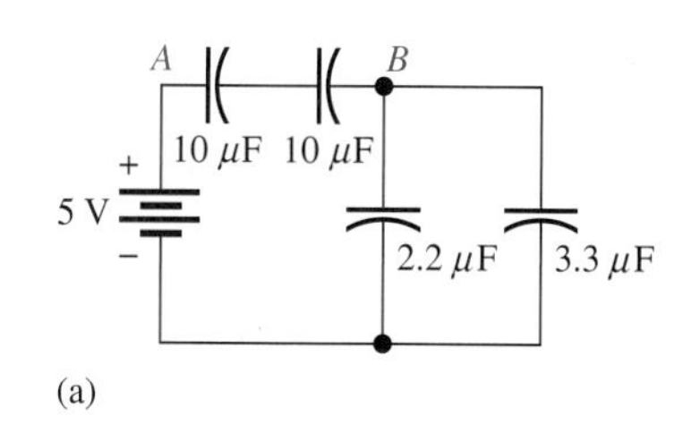

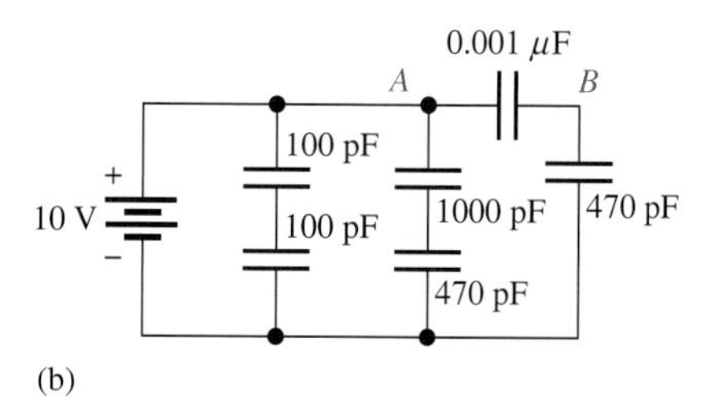

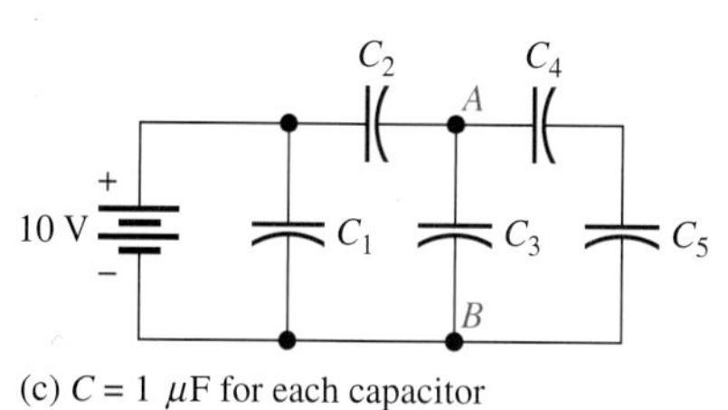

FIGURE 9–65

SECTION 9–5 **Capacitors in DC Circuits**

23. Determine the time constant for each of the following series RC combinations:

(a) $R = 100\ \Omega$, $C = 1\ \mu F$ **(b)** $R = 10\ M\Omega$, $C = 56\ pF$
(c) $R = 4.7\ k\Omega$, $C = 0.0047\ \mu F$ **(d)** $R = 1.5\ M\Omega$, $C = 0.01\ \mu F$

24. Determine how long it takes the capacitor to reach full charge for each of the following combinations:

(a) $R = 47\ \Omega$, $C = 47\ \mu F$ **(b)** $R = 3300\ \Omega$, $C = 0.015\ \mu F$
(c) $R = 22\ k\Omega$, $C = 100\ pF$ **(d)** $R = 4.7\ M\Omega$, $C = 10\ pF$

25. In the circuit of Figure 9–66, the capacitor initially is uncharged. Determine the capacitor voltage at the following times after the switch is closed:

(a) 10 μs **(b)** 20 μs **(c)** 30 μs **(d)** 40 μs **(e)** 50 μs

26. In Figure 9–67, the capacitor is charged to 25 V. Find the capacitor voltage at the following times when the switch is closed:

(a) 1.5 ms **(b)** 4.5 ms **(c)** 6 ms **(d)** 7.5 ms

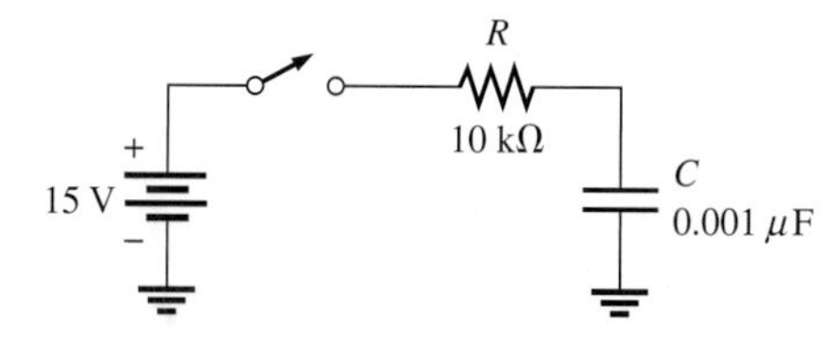

FIGURE 9–66

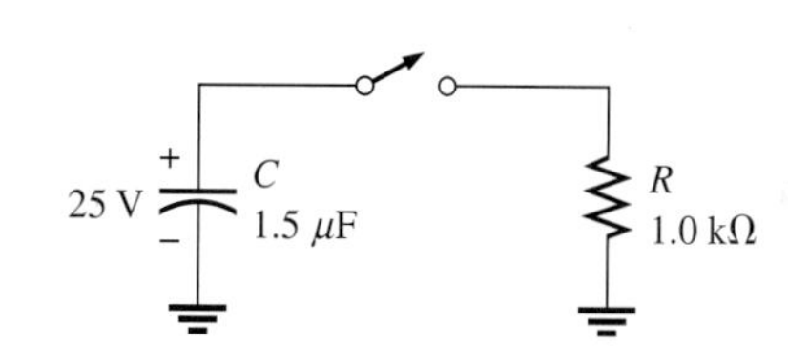

FIGURE 9–67

27. Repeat Problem 25 for the following time intervals:

(a) 2 μs **(b)** 5 μs **(c)** 15 μs

28. Repeat Problem 26 for the following time intervals:

(a) 0.5 ms **(b)** 1 ms **(c)** 2 ms

SECTION 9–6 **Capacitors in AC Circuits**

29. Determine X_C for a 0.047 μF capacitor at each of the following frequencies:

(a) 10 Hz **(b)** 250 Hz **(c)** 5 kHz **(d)** 100 kHz

30. What is the value of the total capacitive reactance in each circuit in Figure 9–68?

FIGURE 9–68

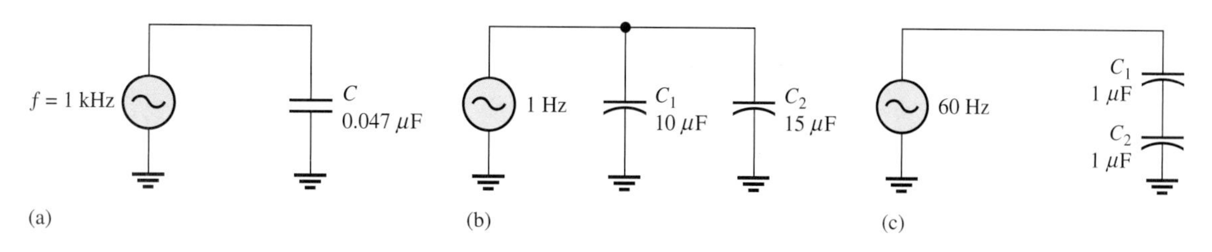

31. In Figure 9–65, each dc voltage source is replaced by a 10 V rms, 2 kHz ac source. Determine the reactance in each case.

32. In each circuit of Figure 9–68, what frequency is required to produce an X_C of 100 Ω? An X_C of 1 kΩ?

33. A sinusoidal voltage of 20 V rms produces an rms current of 100 mA when connected to a certain capacitor. What is the reactance?

34. A 10 kHz voltage is applied to a 0.0047 μF capacitor, and 1 mA of rms current is measured. What is the rms value of the voltage?

35. Determine the true power and the reactive power in Problem 34.

SECTION 9–7 **Capacitor Applications**

36. If another capacitor is connected in parallel with the existing capacitor in the power supply filter of Figure 9–50, how is the ripple voltage affected?

37. Ideally, what should the reactance of a bypass capacitor be in order to eliminate a 10 kHz ac voltage at a given point in an amplifier circuit?

SECTION 9–8 **Testing Capacitors**

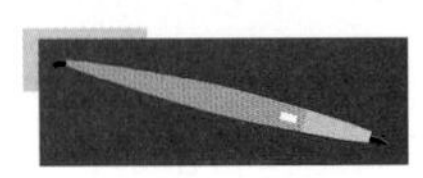

38. Assume that you are checking a capacitor with an analog ohmmeter, and when you connect the leads across the capacitor, the pointer does not move from its left-end scale position. What is the problem?

39. In checking a capacitor with the analog ohmmeter, you find that the pointer goes all the way to the right end of the scale and stays there. What is the problem?

ADVANCED PROBLEMS

40. Two series capacitors (one 1 μF, the other of unknown value) are charged from a 12 V source. The 1 μF capacitor is charged to 8 V, and the other to 4 V. What is the value of the unknown capacitor?

41. How long does it take C to discharge to 3 V in Figure 9–67?

42. How long does it take C to charge to 8 V in Figure 9–66?

43. Determine the time constant for the circuit in Figure 9–69.

44. In Figure 9–70, the capacitor initially is uncharged. At $t = 10\ \mu s$ after the switch is closed, the instantaneous capacitor voltage is 7.2 V. Determine the value of R.

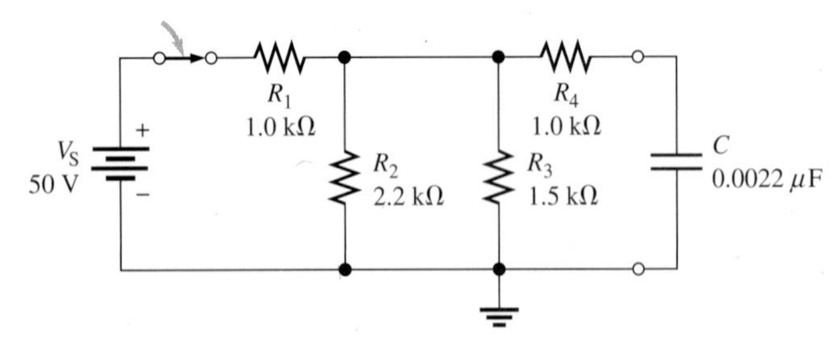

▲ FIGURE 9–69

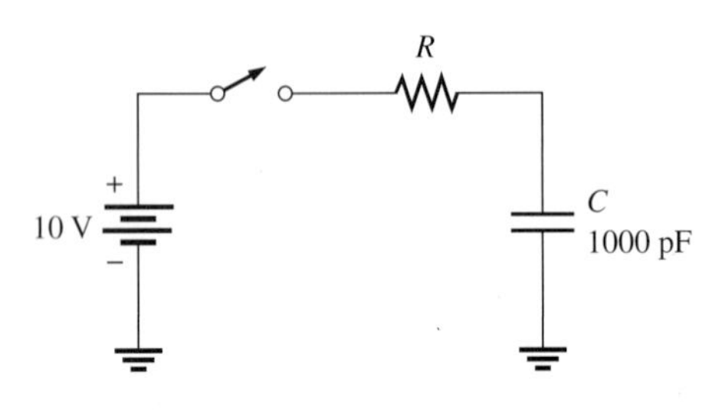

▲ FIGURE 9–70

45. **(a)** The capacitor in Figure 9–71 is uncharged when the switch is thrown into position 1. The switch remains in position 1 for 10 ms and then is thrown into position 2, where it remains indefinitely. Sketch the complete waveform for the capacitor voltage.

(b) If the switch is thrown back to position 1 after 5 ms in position 2, and then is left in position 1, how will the waveform appear?

46. Determine the ac voltage across each capacitor and the current in each branch of the circuit in Figure 9–72.

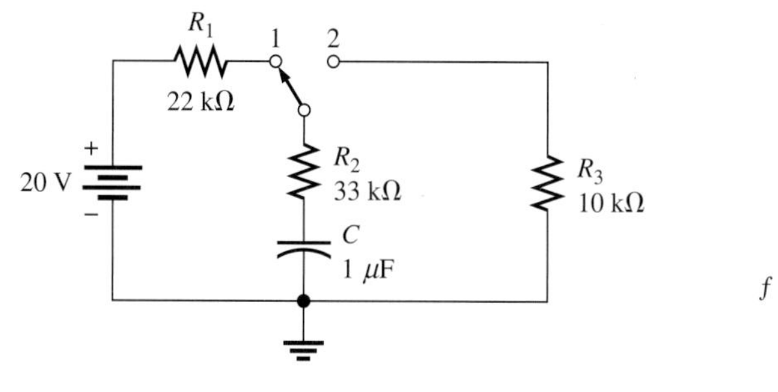

FIGURE 9–71

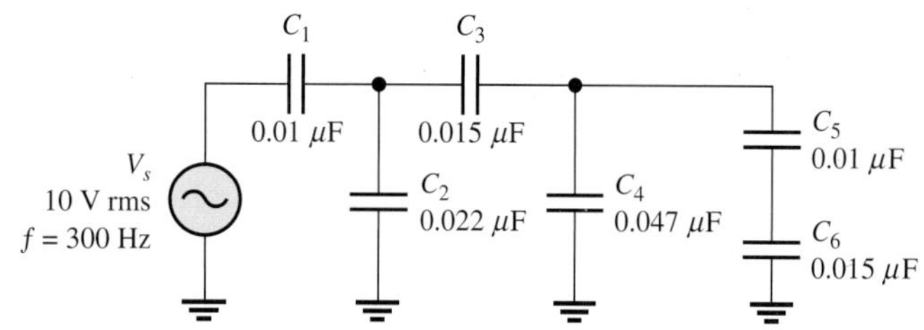

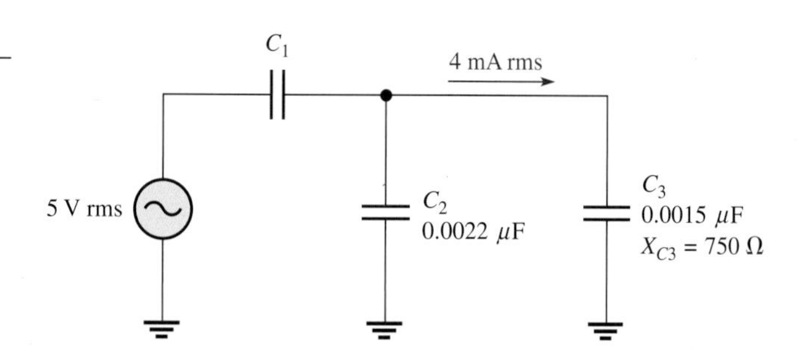

FIGURE 9–72

47. Find the value of C_1 in Figure 9–73.

FIGURE 9–73

48. How much does the voltage across C_5 and C_6 change when the ganged switch is thrown from position 1 to position 2 in Figure 9–74?

FIGURE 9–74

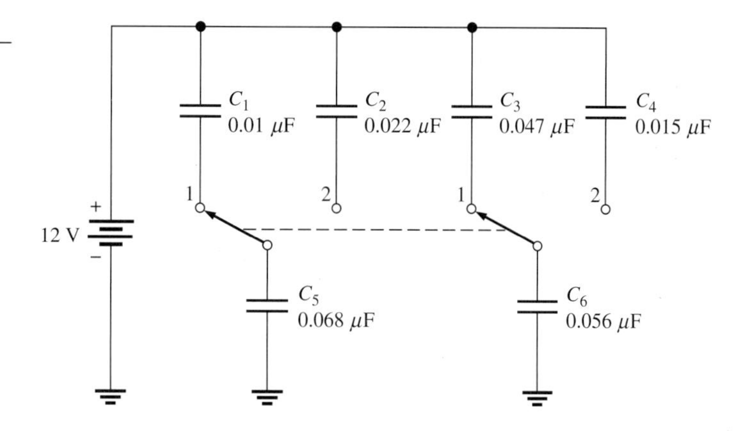

49. If C_4 in Figure 9–72 is open, determine the voltages that would be measured across the other capacitors.

ELECTRONICS WORKBENCH/CIRCUITMAKER TROUBLESHOOTING PROBLEMS

CD-ROM file circuits are shown in Figure 9–75.

50. Open file P09-50 and determine if there is a fault. If so, identify it.

51. Open file P09-51 and determine if there is a fault. If so, identify it.

52. Open file P09-52 and determine if there is a fault. If so, identify it.

53. Open file P09-53 and determine if there is a fault. If so, identify it.

54. Open file P09-54 and determine if there is a fault. If so, identify it.

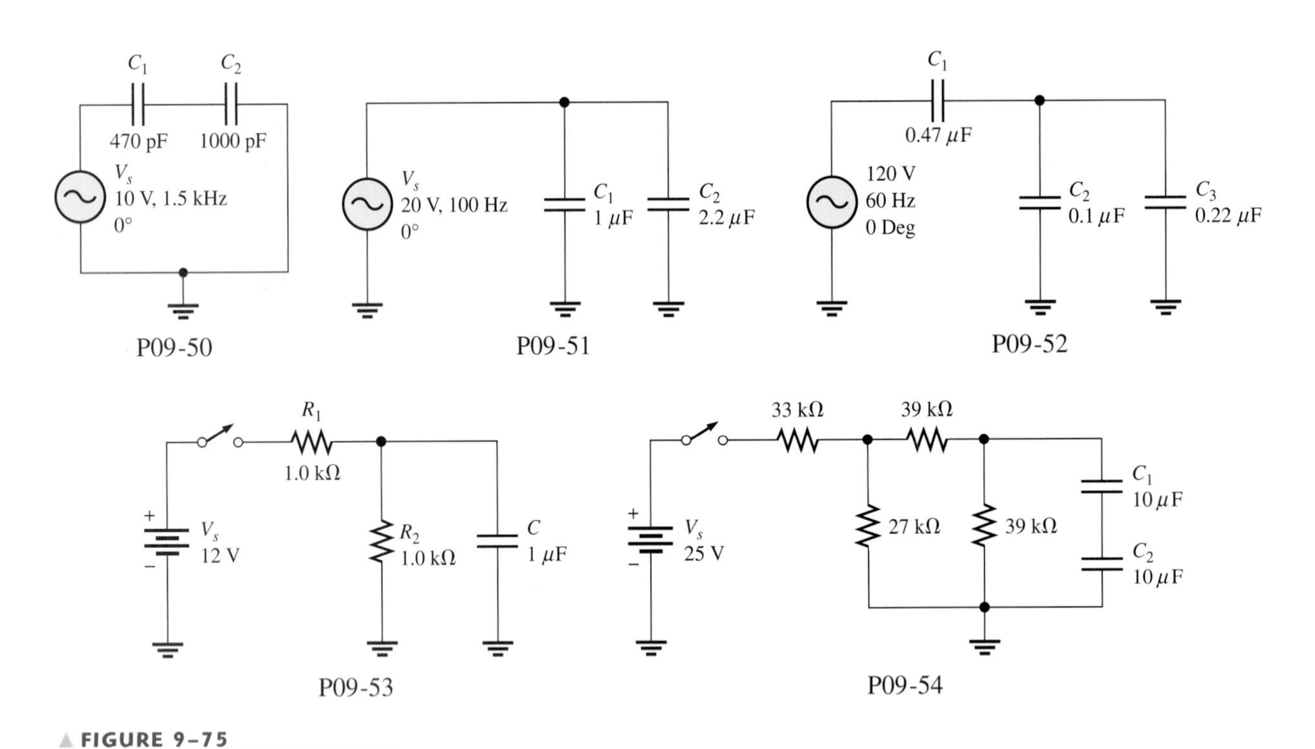

▲ FIGURE 9–75

ANSWERS

SECTION REVIEWS

SECTION 9–1

The Basic Capacitor

1. Capacitance is the ability (capacity) to store electrical charge.

2. **(a)** There are 1,000,000 microfarads in one farad.

(b) There are 1×10^{12} picofarads in one farad.

(c) There are 1,000,000 picofarads in one microfarad.

3. $0.0015\ \mu\text{F} \times 10^6 = 1500\ \text{pF}$; $0.0015\ \mu\text{F} \times 10^{-6} = 0.0000000015\ \text{F}$

4. $W = \frac{1}{2}CV^2 = \frac{1}{2}(0.01\ \mu\text{F})(15\ \text{V})^2 = 1.125\ \mu\text{J}$

5. **(a)** Capacitance increases when plate area is increased.

(b) Capacitance decreases when plate separation is increased.

6. (1000 V/mil)(10 mils) = 10 kV

7. $C = (0.02\ \text{m}^2)(1200)(8.85 \times 10^{-12}\ \text{F/m})/0.0005\ \text{m} = 0.425\ \mu\text{F}$

SECTION 9–2 **Types of Capacitors**

1. Capacitors are commonly classified by the dielectric material.
2. A fixed capacitance cannot be changed; a variable capacitor can.
3. Electrolytic capacitors are polarized.
4. Be sure that the voltage rating is sufficient and connect the positive end to the positive side of the circuit when installing a polarized capacitor.
5. The positive lead should be connected to ground.

SECTION 9–3 **Series Capacitors**

1. C_T of series capacitors is less than the smallest value.
2. $C_T = 61.2$ pF
3. $C_T = 0.006$ μF
4. $V = (0.006\ \mu\text{F}/0.01\ \mu\text{F})10\text{ V} = 6\text{ V}$

SECTION 9–4 **Parallel Capacitors**

1. The individual parallel capacitors are added to get C_T.
2. Use five 0.01 μF capacitors in parallel to get 0.05 μF.
3. $C_T = 167$ pF

SECTION 9–5 **Capacitors in DC Circuits**

1. $\tau = RC = 1.2\ \mu\text{s}$
2. $5\tau = 6\ \mu\text{s}$; V_C is approximately 5 V.
3. $v_{2\text{ms}} = (0.86)10\text{ V} = 8.6\text{ V}$; $v_{3\text{ms}} = (0.95)10\text{ V} = 9.5\text{ V}$; $v_{4\text{ms}} = (0.98)10\text{ V} = 9.8\text{ V}$; $v_{5\text{ms}} = (0.99)10\text{ V} = 9.9\text{ V}$
4. $v_C = (0.37)(100\text{ V}) = 37\text{ V}$

SECTION 9–6 **Capacitors in AC Circuits**

1. $X_C = 1/2\pi fC = 677\text{ k}\Omega$
2. $f = 1/2\pi CX_C = 796\text{ Hz}$
3. $I_{\text{rms}} = 1\text{ V}/1.59\ \Omega = 629\text{ mA}$
4. Current leads voltage by 90°.
5. $P_{\text{true}} = 0\text{ W}$
6. $P_r = (12\text{ V})^2/318\ \Omega = 0.453\text{ VAR}$

SECTION 9–7 **Capacitor Applications**

1. Once the capacitor charges to the peak voltage, it discharges very little before the next peak thus smoothing the rectified voltage.
2. A coupling capacitor allows ac to pass from one point to another, but blocks constant dc.
3. A coupling capacitor must be large enough to have a negligible reactance at the frequency that is to be passed without opposition.
4. A decoupling capacitor shorts power line voltage transients to ground.
5. X_C is inversely proportional to frequency and so is the filter's ability to pass ac signals.
6. The time constant is used in delay applications.

SECTION 9–8 **Testing Capacitors**

1. Short its lead to discharge.
2. Initially, the needle jumps to zero; then it moves to the high-resistance end of the scale when the capacitor is good.
3. Shorted, open, leakage, and dielectric absorption are possible capacitor defects.

Application Assignment

1. The coupling capacitor prevents the source from affecting the dc voltage but passes the ac input signal.
2. An ac voltage superimposed on a dc voltage is at the point labeled C and an ac voltage only at the output.

RELATED PROBLEMS FOR EXAMPLES

9–1 10 kV
9–2 0.047 μF
9–3 100,000,000 pF
9–4 6638 pF
9–5 278 pF
9–6 1.54 μF
9–7 2.83 V
9–8 650 pF
9–9 0.163 μF
9–10 891 μs
9–11 8.36 V
9–12 8.13 V
9–13 0.7 ms; 95 V
9–14 3.39 kHz
9–15 4.40 mA
9–16 0 W; 1.01 mVAR

SELF-TEST

1. (g) **2.** (b) **3.** (c) **4.** (d) **5.** (a) **6.** (d) **7.** (a)
8. (f) **9.** (c) **10.** (b) **11.** (d) **12.** (a) **13.** (b) **14.** (a)
15. (b) **16.** (d)

10

RC CIRCUITS

INTRODUCTION

An *RC* circuit contains both resistance and capacitance. It is one of the basic types of reactive circuits that you will study.

In this chapter, basic series and parallel *RC* circuits and their responses to sinusoidal voltages are covered. Series-parallel combinations are also examined. Power considerations in *RC* circuits are introduced, and practical aspects of power ratings are discussed. Two *RC* circuit applications are presented to give you an idea of how simple combinations of resistors and capacitors can be applied. Troubleshooting common faults in *RC* circuits is also covered.

The methods for analyzing reactive circuits are similar to those you studied in dc circuits. Reactive circuit problems can be solved at only one frequency at a time, and phasor math must be used.

CHAPTER OUTLINE

CHAPTER OBJECTIVES

- Describe the relationship between current and voltage in an *RC* circuit
- Determine impedance and phase angle in a series *RC* circuit
- Analyze a series *RC* circuit
- Determine impedance and phase angle in a parallel *RC* circuit
- Analyze a parallel *RC* circuit
- Analyze series-parallel *RC* circuits
- Determine power in *RC* circuits
- Discuss some basic *RC* applications
- Troubleshoot *RC* circuits

KEY TERMS

- Impedance
- Phase angle
- Capacitive susceptance
- Admittance
- Reactive power
- Apparent power
- Power factor
- *RC* lag network
- *RC* lead network
- Filter
- Frequency response
- Cutoff frequency
- Bandwidth

APPLICATION ASSIGNMENT PREVIEW

Your assignment is to run frequency response measurements on a capacitively coupled amplifier used in a certain communications system. You will be concentrating on the *RC* input circuit of the amplifier and how it responds to different frequencies. The results of your measurements will be graphically documented with a frequency response curve. After studying this chapter, you should be able to complete the application assignment.

WWW. VISIT THE COMPANION WEBSITE

Circuit Simulation Tutorials and Other Chapter Study Tools Are Available at

http://www.prenhall.com/floyd

10–1 SINUSOIDAL RESPONSE OF *RC* CIRCUITS

When a sinusoidal voltage is applied to any type of *RC* circuit, each resulting voltage drop and the current in the circuit are also sinusoidal and have the same frequency as the source voltage. The capacitance causes a phase shift between the voltage and current that depends on the relative values of the resistance and the capacitive reactance.

After completing this section, you should be able to

- **Describe the relationship between current and voltage in an *RC* circuit**
- Discuss voltage and current waveforms
- Discuss phase shift
- Describe types of signal generators

As shown in Figure 10–1, the resistor voltage (V_R), the capacitor voltage (V_C), and the current (I) are all sine waves with the frequency of the source. Phase shifts are introduced because of the capacitance. As you will learn, the resistor voltage and current are in phase with each other and lead the source voltage in phase. The capacitor voltage lags the source voltage. The phase angle between the current and the capacitor voltage is always 90°. These generalized phase relationships are indicated in Figure 10–1.

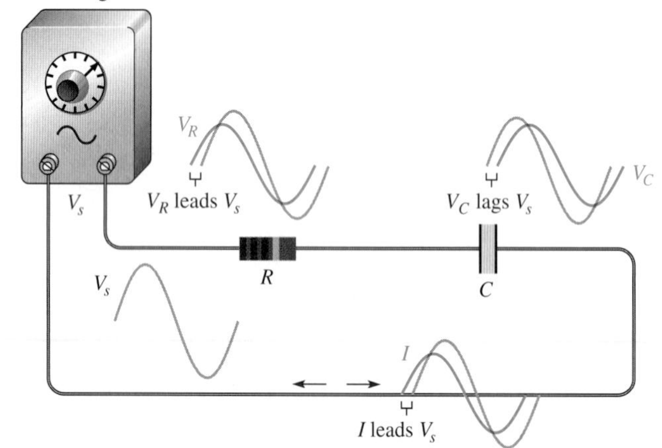

FIGURE 10–1

Illustration of sinusoidal response with general phase relationships of V_R, V_C, and I relative to the source voltage. V_R and I are in phase; V_R leads V_s; V_C lags V_s; and V_R and V_C are 90° out of phase.

The amplitudes and the phase relationships of the voltages and current depend on the values of the resistance and the **capacitive reactance.** When a circuit is purely resistive, the phase angle between the source voltage and the total current is zero. When a circuit is purely capacitive, the phase angle between the source voltage and the total current is 90°, with the current leading the voltage. When there is a combination of both resistance and capacitive reactance in a circuit, the phase angle between the source voltage and the total current is somewhere between zero and 90°, depending on the relative values of the resistance and the capacitive reactance.

Signal Generators

When a circuit is hooked up for a laboratory experiment or for troubleshooting, a signal generator similar to those shown in Figure 10–2 is used to provide the source voltage. These instruments, depending on their capability, are classified as

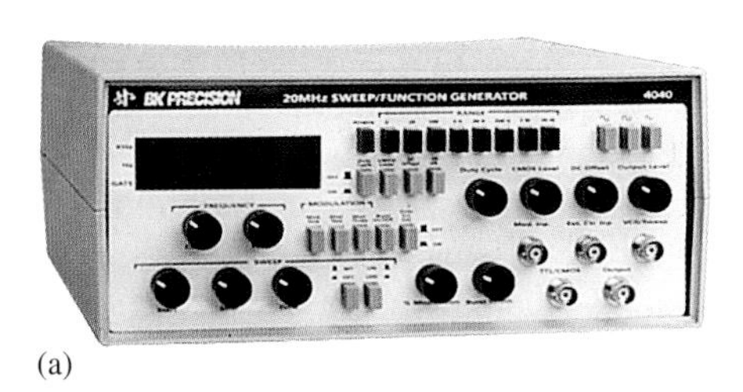
(a)

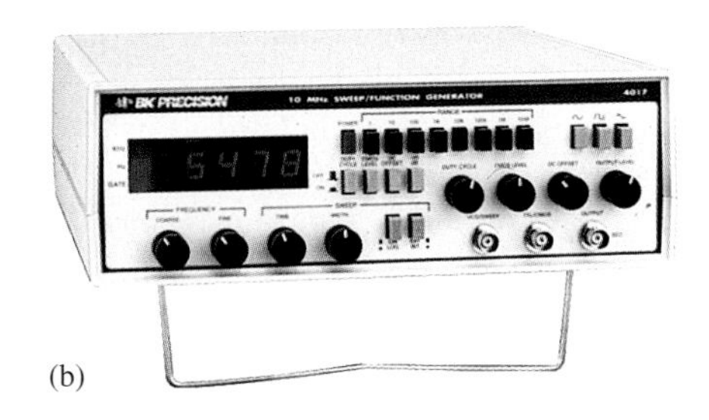
(b)

▲ **FIGURE 10–2**

Typical signal (function) generators used in circuit testing and troubleshooting. (Photography courtesy of B&K Precision Corp.)

sine wave generators, which produce only sine waves; sine/square generators, which produce both sine waves and square waves; or function generators, which produce sine waves, pulse waveforms, and triangular (ramp) waveforms.

SECTION 10–1 REVIEW

Answers are at the end of the chapter.

1. A 60 Hz sinusoidal voltage is applied to an *RC* circuit. What is the frequency of the capacitor voltage? What is the frequency of the current?
2. What causes the phase shift between V_s and I in a series *RC* circuit?
3. When the resistance in an *RC* circuit is greater than the capacitive reactance, is the phase angle between the source voltage and the total current closer to 0° or to 90°?

10–2 IMPEDANCE AND PHASE ANGLE OF SERIES *RC* CIRCUITS

The **impedance** of an *RC* circuit is the total opposition to sinusoidal current and its unit is the ohm. The **phase angle** is the phase difference between the total current and the source voltage.

After completing this section, you should be able to

- **Determine impedance and phase angle in a series *RC* circuit**
- Define *impedance*
- Draw an impedance triangle
- Calculate the total impedance magnitude
- Calculate the phase angle

In a purely resistive circuit, the impedance is simply equal to the total resistance. In a purely capacitive circuit, the impedance is the total capacitive reactance. The impedance of a series *RC* circuit is determined by both the resistance (R) and the capacitive reactance (X_C). These cases are illustrated in Figure 10–3. The magnitude of the impedance is symbolized by Z.

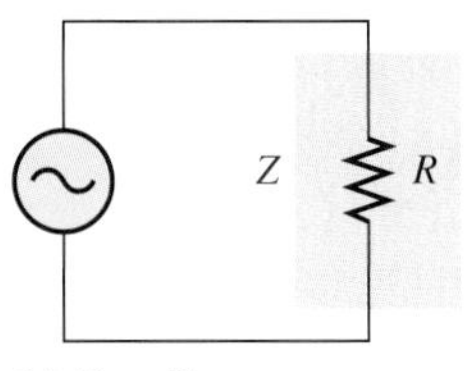

(a) $Z = R$

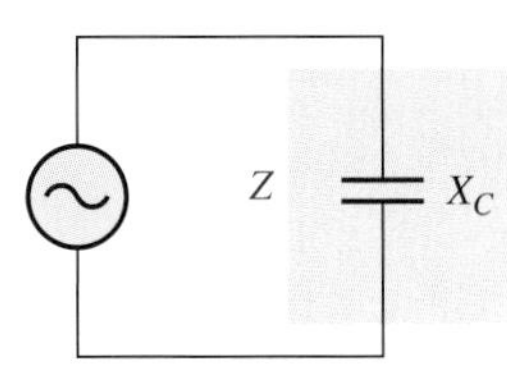

(b) $Z = X_C$

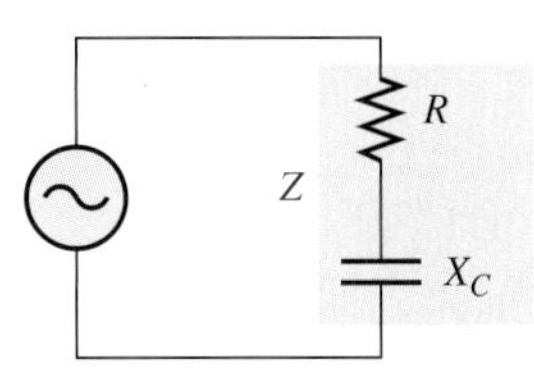

(c) Z includes both R and X_C

◀ **FIGURE 10–3**

Three cases of impedance.

The Impedance Triangle

In ac analysis, both R and X_C are treated as phasor quantities, as shown in the phasor diagram of Figure 10–4(a), with X_C appearing at a $-90°$ angle with respect to R. This relationship comes from the fact that the capacitor voltage in a series RC circuit lags the current, and thus the resistor voltage, by 90°. Since Z is the phasor sum of R and X_C, its phasor representation is as shown in Figure 10–4(b). A repositioning of the phasors, as shown in Figure 10–4(c), forms a right triangle, which is called the *impedance triangle.* The length of each phasor represents the magnitude in ohms, and the angle θ (the Greek letter theta) is the phase angle of the RC circuit and represents the phase difference between the source voltage and the current.

FIGURE 10–4

Development of the impedance triangle for a series RC circuit.

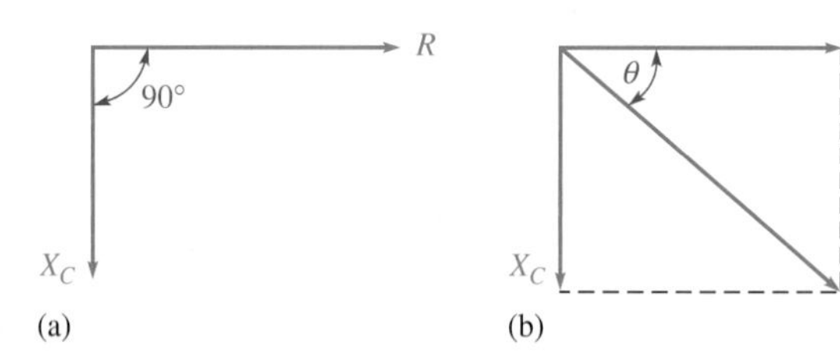

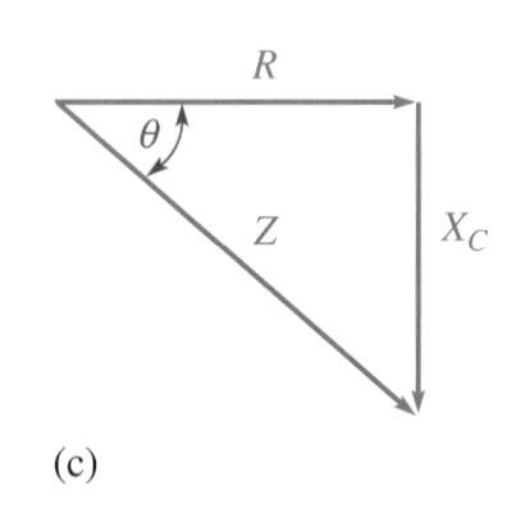

From right-angle trigonometry (Pythagorean theorem), the magnitude of the impedance can be expressed in terms of the resistance and capacitive reactance.

Equation 10–1

$$Z = \sqrt{R^2 + X_C^2}$$

The magnitude of the impedance (Z), as shown in the RC circuit in Figure 10–5, is expressed in ohms.

FIGURE 10–5

Impedance of a series RC circuit.

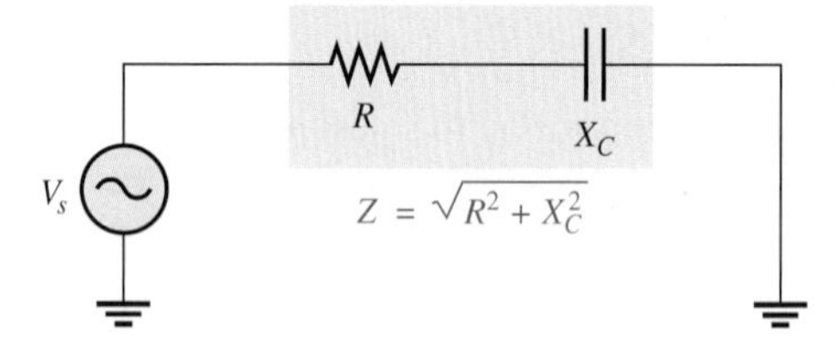

The phase angle, θ, is expressed as

Equation 10–2

$$\theta = \tan^{-1}\left(\frac{X_C}{R}\right)$$

The symbol $\tan^{-1}$ stands for *inverse tangent* and can be found on the calculator by pressing 2nd TAN⁻¹. Another term for inverse tangent is *arctangent* (arctan).

EXAMPLE 10–1

Determine the impedance and the phase angle of the RC circuit in Figure 10–6. Draw the impedance triangle.

FIGURE 10–6

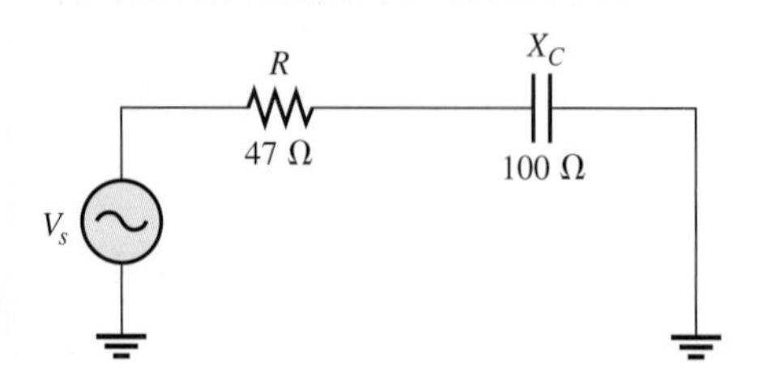

Solution The impedance is

$$Z = \sqrt{R^2 + X_C^2} = \sqrt{(47\ \Omega)^2 + (100\ \Omega)^2} = \mathbf{110\ \Omega}$$

The phase angle is

$$\theta = \tan^{-1}\left(\frac{X_C}{R}\right) = \tan^{-1}\left(\frac{100\ \Omega}{47\ \Omega}\right) = \tan^{-1}(2.13) = \mathbf{64.8^\circ}$$

The source voltage lags the current by 64.8°.

The impedance triangle is shown in Figure 10–7.

▶ **FIGURE 10–7**

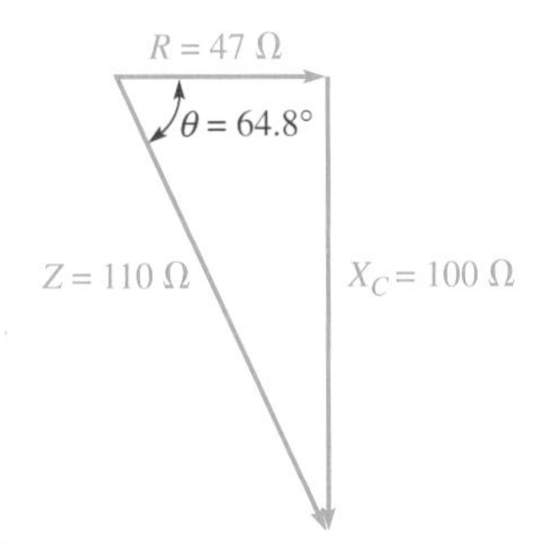

The calculator solutions for Z and θ are

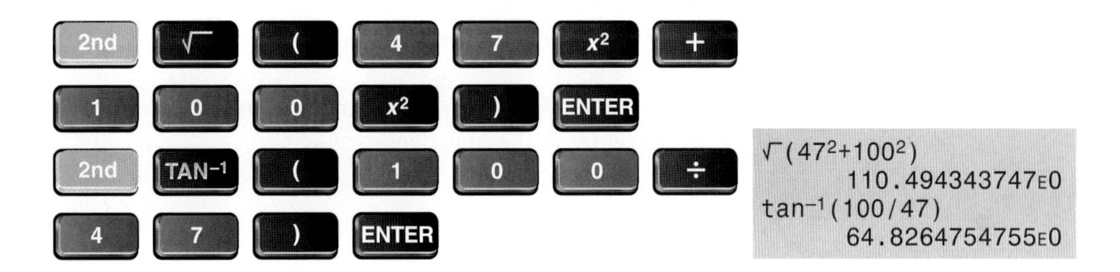

*Related Problem** Find Z and θ for $R = 1.0\ \text{k}\Omega$ and $X_C = 2.2\ \text{k}\Omega$ in Figure 10–6.

*Answers are at the end of the chapter.

SECTION 10–2 REVIEW

1. Define *impedance.*
2. Does the source voltage lead or lag the current in a series *RC* circuit?
3. What causes the phase angle in the *RC* circuit?
4. A series *RC* circuit has a resistance of 33 kΩ and a capacitive reactance of 50 kΩ. What is the value of the impedance? What is the phase angle?

10–3 ANALYSIS OF SERIES *RC* CIRCUITS

In Section 10–2, you learned how to express the impedance of a series *RC* circuit. Now, Ohm's law and Kirchhoff's voltage law are used in the analysis of *RC* circuits.

After completing this section, you should be able to

- **Analyze a series *RC* circuit**
- Apply Ohm's law and Kirchhoff's voltage law to series *RC* circuits
- Determine the phase relationships of the voltages and current
- Show how impedance and phase angle vary with frequency

Ohm's Law

The application of Ohm's law to series *RC* circuits involves the use of the quantities of *Z*, *V*, and *I*. The three equivalent forms of Ohm's law are as follows:

Equation 10–3
$$V = IZ$$

Equation 10–4
$$I = \frac{V}{Z}$$

Equation 10–5
$$Z = \frac{V}{I}$$

The following two examples illustrate the use of Ohm's law.

EXAMPLE 10–2

If the current in Figure 10–8 is 0.2 mA, determine the source voltage and the phase angle. Draw the impedance triangle.

FIGURE 10–8

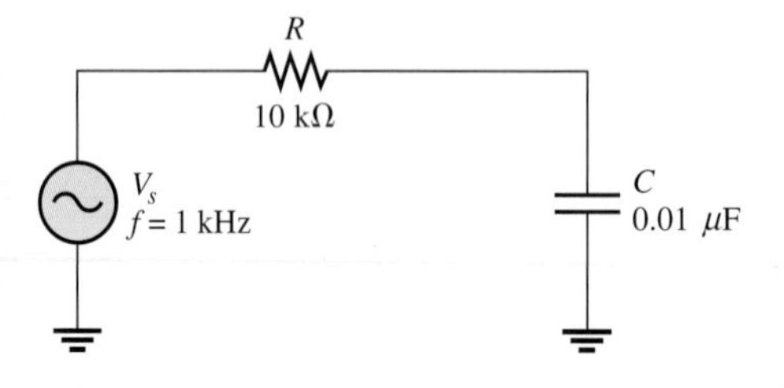

Solution The capacitive reactance is

$$X_C = \frac{1}{2\pi fC} = \frac{1}{2\pi(1000\text{ Hz})(0.01\ \mu\text{F})} = 15.9\text{ k}\Omega$$

The impedance is

$$Z = \sqrt{R^2 + X_C^2} = \sqrt{(10\text{ k}\Omega)^2 + (15.9\text{ k}\Omega)^2} = 18.8\text{ k}\Omega$$

Applying Ohm's law yields

$$V_s = IZ = (0.2\text{ mA})(18.8\text{ k}\Omega) = \mathbf{3.76\ V}$$

The phase angle is

$$\theta = \tan^{-1}\left(\frac{X_C}{R}\right) = \tan^{-1}\left(\frac{15.9\text{ k}\Omega}{10\text{ k}\Omega}\right) = \mathbf{57.8^\circ}$$

The source voltage has a magnitude of 3.76 V and lags the current by 57.8°.

The impedance triangle is shown in Figure 10–9.

FIGURE 10–9

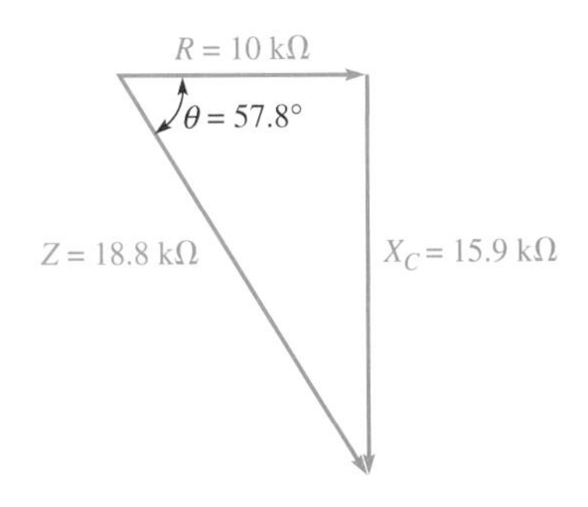

The calculator solution for V_s is

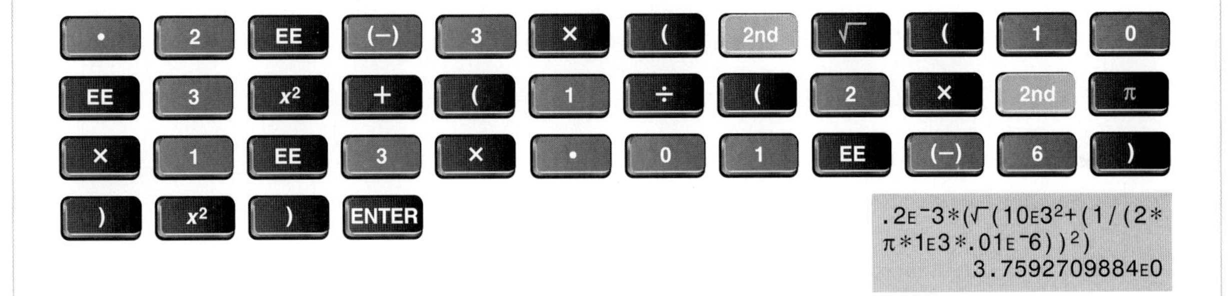

Related Problem Determine V_s in Figure 10–8 if $f = 2$ kHz and $I = 200\ \mu$A.

EXAMPLE 10–3

Determine the current in the *RC* circuit of Figure 10–10.

FIGURE 10–10

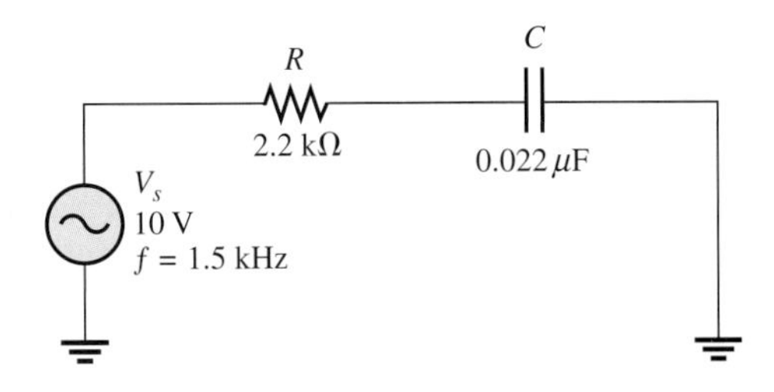

Solution The capacitive reactance is

$$X_C = \frac{1}{2\pi fC} = \frac{1}{2\pi(1.5\text{ kHz})(0.022\ \mu\text{F})} = 4.82\text{ k}\Omega$$

The impedance is

$$Z = \sqrt{R^2 + X_C^2} = \sqrt{(2.2\text{ k}\Omega)^2 + (4.82\text{ k}\Omega)^2} = 5.30\text{ k}\Omega$$

Applying Ohm's law yields

$$I = \frac{V}{Z} = \frac{10\text{ V}}{5.30\text{ k}\Omega} = \mathbf{1.89\text{ mA}}$$

Related Problem Determine the phase angle between V_s and I in Figure 10–10.

Open file E10-03 on your EWB/CircuitMaker CD-ROM. Measure the current and the voltages across the resistor and the capacitor.

Phase Relationships of the Current and Voltages

In a series *RC* circuit, the current is the same through both the resistor and the capacitor. Thus, the resistor voltage is in phase with the current, and the capacitor voltage lags the current by 90°. Therefore, there is a phase difference of 90° between the resistor voltage, V_R, and the capacitor voltage, V_C, as shown in the waveform diagram of Figure 10–11.

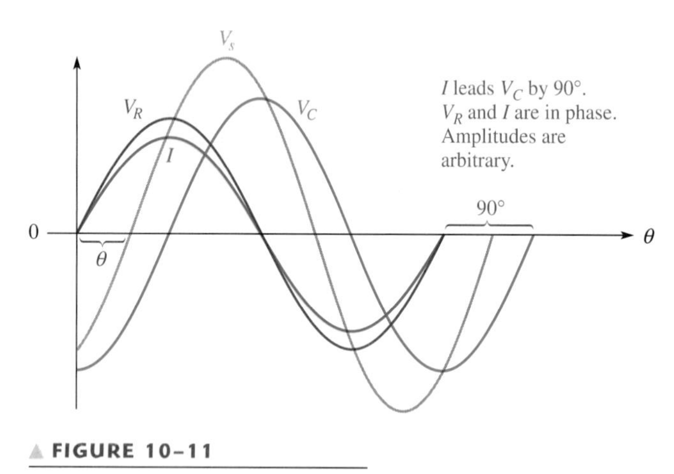

▲ **FIGURE 10–11**

Phase relation of the voltages and current in a series *RC* circuit.

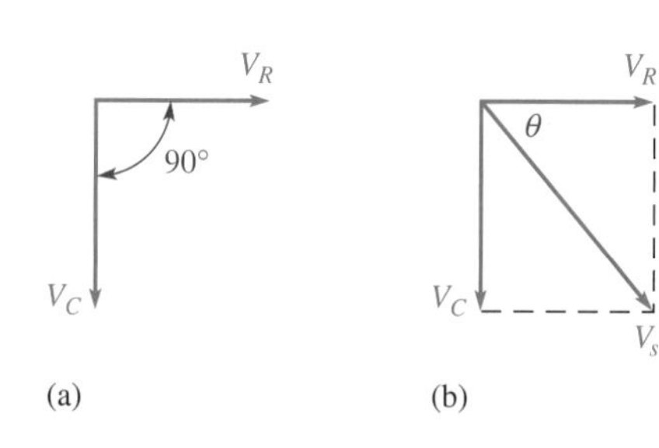

▲ **FIGURE 10–12**

Voltage phasor diagram for the waveforms in Figure 10–11.

You know from Kirchhoff's voltage law that the sum of the voltage drops must equal the source voltage. However, since V_R and V_C are 90° out of phase with each other, they must be added as phasor quantities, with V_C lagging V_R, as shown in Figure 10–12(a). As shown in Figure 10–12(b), V_s is the phasor sum of V_R and V_C, as expressed in the following equation:

Equation 10–6

$$V_s = \sqrt{V_R^2 + V_C^2}$$

The phase angle between the resistor voltage and the source voltage can be expressed as

Equation 10–7

$$\theta = \tan^{-1}\left(\frac{V_C}{V_R}\right)$$

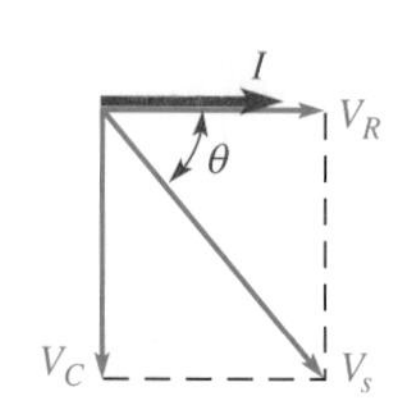

▲ **FIGURE 10–13**

Voltage and current phasor diagram for the waveforms in Figure 10–11.

Since the resistor voltage and the current are in phase, θ in Equation 10–7 also represents the phase angle between the source voltage and the current and is equivalent to $\tan^{-1}(X_C/R)$.

Figure 10–13 shows the voltage and current phasor diagram representing the waveform diagram of Figure 10–11.

EXAMPLE 10–4

Determine the source voltage and the phase angle in Figure 10–14. Draw the voltage phasor diagram.

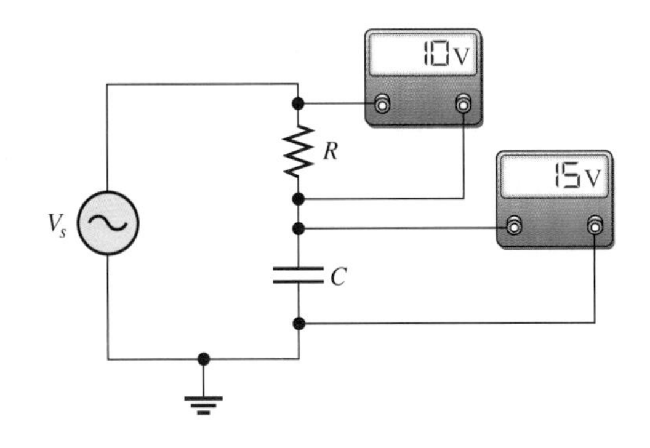

FIGURE 10–14

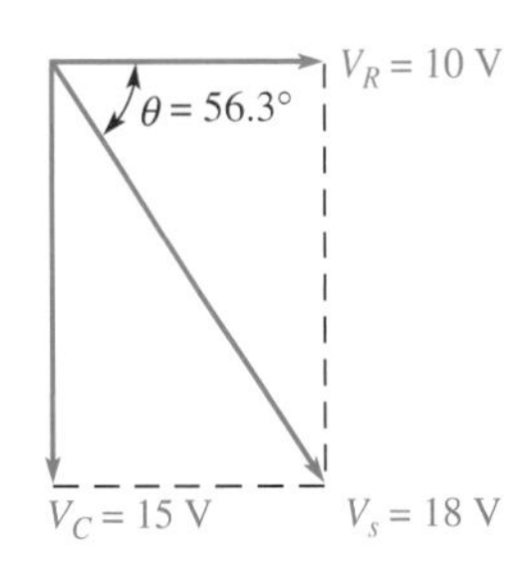

FIGURE 10–15

Solution Since V_R and V_C are 90° out of phase, they cannot be added directly. The source voltage is the phasor sum of V_R and V_C.

$$V_s = \sqrt{V_R^2 + V_C^2} = \sqrt{(10\text{ V})^2 + (15\text{ V})^2} = \mathbf{18\text{ V}}$$

The phase angle between the current and the source voltage is

$$\theta = \tan^{-1}\left(\frac{V_C}{V_R}\right) = \tan^{-1}\left(\frac{15\text{ V}}{10\text{ V}}\right) = \mathbf{56.3°}$$

The voltage phasor diagram is shown in Figure 10–15.

Related Problem In a certain series *RC* circuit, $V_s = 10$ V and $V_R = 7$ V. Find V_C.

Variation of Impedance with Frequency

As you know, capacitive reactance varies inversely with frequency. Since $Z = \sqrt{R^2 + X_C^2}$, you can see that when X_C increases, the entire term under the square root sign increases and thus the total impedance also increases; and when X_C decreases, the total impedance also decreases. Therefore, *in an RC circuit, Z is inversely dependent on frequency.*

Figure 10–16 illustrates how the voltages and current in a series *RC* circuit vary as the frequency increases or decreases, with the source voltage held at a constant value. Part (a) shows that as the frequency is increased, X_C decreases; so less voltage is dropped across the capacitor. Also, *Z* decreases as X_C decreases, causing the current to increase. An increase in the current causes more voltage across *R*.

Figure 10–16(b) shows that as the frequency is decreased, X_C increases; so more voltage is dropped across the capacitor. Also, *Z* increases as X_C increases, causing the current to decrease. A decrease in the current causes less voltage across *R*.

Changes in *Z* and X_C can be observed as shown in Figure 10–17. As the frequency increases, the voltage across *Z* remains constant because V_s is constant

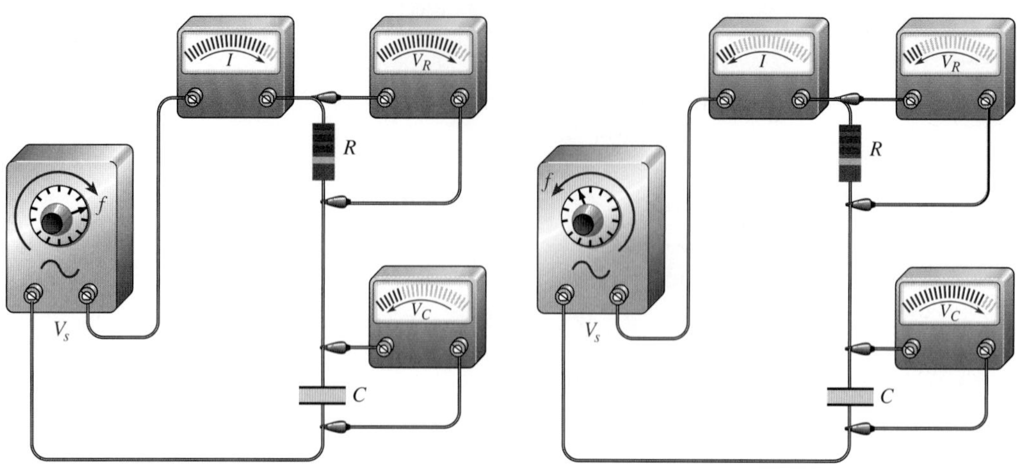

(a) As frequency is increased, I and V_R increase and V_C decreases.

(b) As frequency is decreased, I and V_R decrease and V_C increases.

▲ **FIGURE 10–16**

An illustration of how the variation of impedance affects the voltages and current as the source frequency is varied. The source voltage is held at a constant amplitude.

($V_s = V_Z$). Also, the voltage across C decreases. The increasing current indicates that Z is decreasing. It does so because of the inverse relationship stated in Ohm's law ($Z = V_Z/I$). The increasing current also indicates that X_C is decreasing ($X_C = V_C/I$). The decrease in V_C corresponds to the decrease in X_C.

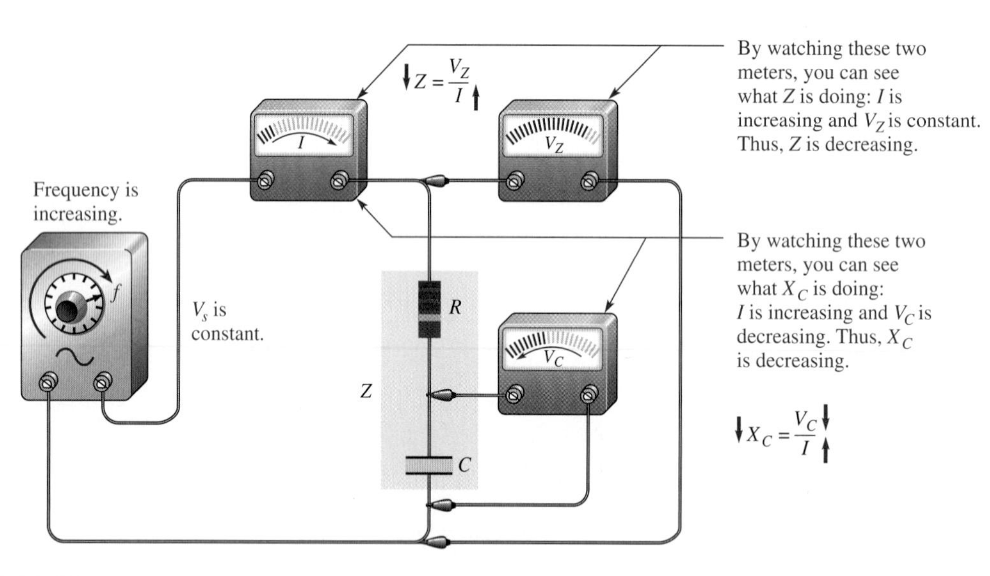

▲ **FIGURE 10–17**

An illustration of how Z and X_C change with frequency.

Variation of the Phase Angle with Frequency

Since X_C is the factor that introduces the phase angle in a series RC circuit, a change in X_C produces a change in the phase angle. As the frequency is increased, X_C becomes smaller, and thus the phase angle decreases. As the frequency is decreased, X_C becomes larger, and thus the phase angle increases. The angle between V_s and V_R is the phase angle of the circuit because I is in phase with V_R.

Figure 10–18 uses the impedance triangle to illustrate the variations in X_C, Z, and θ as the frequency changes. Of course, R remains constant. The key point is that because X_C varies inversely with the frequency, so also do the magnitude of the total impedance and the phase angle. Example 10–5 illustrates this.

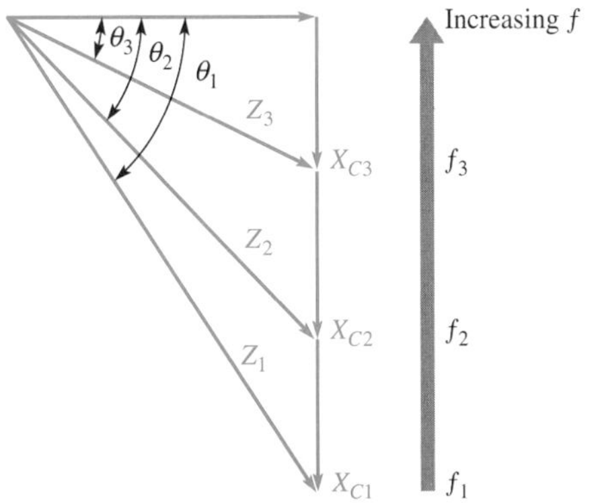

FIGURE 10–18

As the frequency increases, X_C decreases, Z decreases, and θ decreases. Each value of frequency can be visualized as forming a different impedance triangle.

EXAMPLE 10–5

For the series *RC* circuit in Figure 10–19, determine the impedance and phase angle for each of the following values of frequency:

(a) 10 kHz **(b)** 20 kHz **(c)** 30 kHz

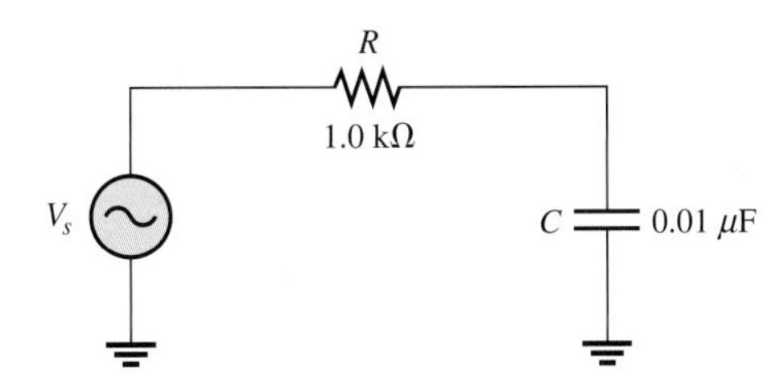

FIGURE 10–19

Solution **(a)** For $f = 10$ kHz, calculate the impedance as follows:

$$X_C = \frac{1}{2\pi fC} = \frac{1}{2\pi(10\text{ kHz})(0.01\ \mu\text{F})} = 1.59\text{ k}\Omega$$

$$Z = \sqrt{R^2 + X_C^2} = \sqrt{(1.0\text{ k}\Omega)^2 + (1.59\text{ k}\Omega)^2} = \mathbf{1.88\text{ k}\Omega}$$

The phase angle is

$$\theta = \tan^{-1}\left(\frac{X_C}{R}\right) = \tan^{-1}\left(\frac{1.59\text{ k}\Omega}{1.0\text{ k}\Omega}\right) = \mathbf{57.8^\circ}$$

(b) For $f = 20$ kHz,

$$X_C = \frac{1}{2\pi(20\text{ kHz})(0.01\ \mu\text{F})} = 796\ \Omega$$

$$Z = \sqrt{(1.0\text{ k}\Omega)^2 + (796\ \Omega)^2} = \mathbf{1.28\text{ k}\Omega}$$

$$\theta = \tan^{-1}\left(\frac{796\ \Omega}{1.0\text{ k}\Omega}\right) = \mathbf{38.5^\circ}$$

(c) For $f = 30$ kHz,

$$X_C = \frac{1}{2\pi(30\text{ kHz})(0.01\ \mu\text{F})} = 531\ \Omega$$

$$Z = \sqrt{(1.0\text{ k}\Omega)^2 + (531\ \Omega)^2} = \mathbf{1.13\text{ k}\Omega}$$

$$\theta = \tan^{-1}\left(\frac{531\ \Omega}{1.0\text{ k}\Omega}\right) = \mathbf{28.0^\circ}$$

Notice that as the frequency increases, X_C, Z, and θ decrease.

Related Problem Find the total impedance and phase angle in Figure 10–19 for $f = 1$ kHz.

SECTION 10–3 REVIEW

1. In a certain series *RC* circuit, $V_R = 4$ V and $V_C = 6$ V. What is the magnitude of the source voltage?
2. In Question 1, what is the phase angle?
3. What is the phase difference between the capacitor voltage and the resistor voltage in a series *RC* circuit?
4. When the frequency of the source voltage in a series *RC* circuit is increased, what happens to each of the following?
   (a) the capacitive reactance (b) the impedance (c) the phase angle

10–4 IMPEDANCE AND PHASE ANGLE OF PARALLEL *RC* CIRCUITS

In this section, you will learn how to determine the impedance and phase angle of a parallel *RC* circuit. Also conductance (G), capacitive susceptance (B_C), and total admittance (Y_{tot}) are discussed because of their usefulness in parallel circuit analysis.

After completing this section, you should be able to

- **Determine impedance and phase angle in a parallel *RC* circuit**
- Express total impedance in a product-over-sum form
- Express the phase angle in terms of R and X_C
- Determine the conductance, capacitive susceptance, and admittance
- Convert admittance to impedance

Figure 10–20 shows a basic parallel *RC* circuit.

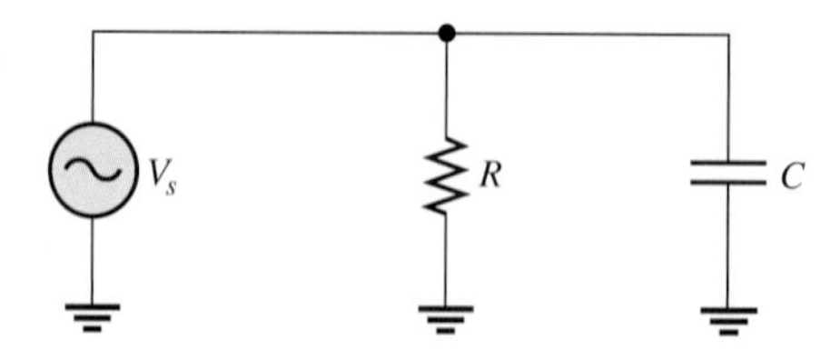

FIGURE 10–20

Parallel *RC* circuit.

The expression for the impedance in Equation 10–8 is given in a product-over-sum form similar to the way two resistors in parallel can be expressed. In this case, the denominator is the phasor sum of R and X_C.

Equation 10–8

$$Z = \frac{RX_C}{\sqrt{R^2 + X_C^2}}$$

The phase angle between the source voltage and the total current can be expressed in terms of R and X_C as shown in Equation 10–9.

$$\theta = \tan^{-1}\left(\frac{R}{X_C}\right)$$

Equation 10–9

This formula is derived from an equivalent formula using branch currents, which is introduced in Section 10–5.

EXAMPLE 10–6

For each circuit in Figure 10–21, determine the impedance and the phase angle.

▶ FIGURE 10–21

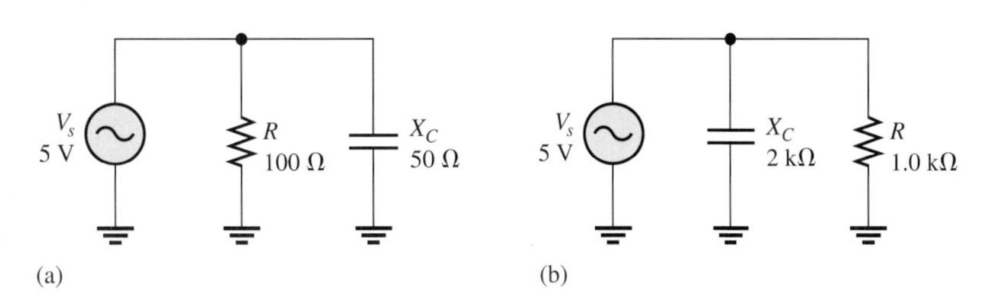

Solution For the circuit in Figure 10–21(a), the impedance and phase angle are

$$Z = \frac{RX_C}{\sqrt{R^2 + X_C^2}} = \frac{(100\ \Omega)(50\ \Omega)}{\sqrt{(100\ \Omega)^2 + (50\ \Omega)^2}} = \mathbf{44.7\ \Omega}$$

$$\theta = \tan^{-1}\left(\frac{R}{X_C}\right) = \tan^{-1}\left(\frac{100\ \Omega}{50\ \Omega}\right) = \mathbf{63.4^\circ}$$

For the circuit in Figure 10–21(b),

$$Z = \frac{(1.0\ \text{k}\Omega)(2\ \text{k}\Omega)}{\sqrt{(1.0\ \text{k}\Omega)^2 + (2\ \text{k}\Omega)^2}} = \mathbf{894\ \Omega}$$

$$\theta = \tan^{-1}\left(\frac{1.0\ \text{k}\Omega}{2\ \text{k}\Omega}\right) = \mathbf{26.6^\circ}$$

The calculator solutions for Z and θ in Figure 10–21(a) are

```
100*50/√(100²+50²)
        44.72135955E0
tan⁻¹(100/50)
        63.4349488229E0
```

Related Problem Determine Z in Figure 10–21(a) if the frequency is doubled.

Conductance, Susceptance, and Admittance

Recall that **conductance** (G) is the reciprocal of resistance and is expressed as

Equation 10–10
$$G = \frac{1}{R}$$

Two new terms are now introduced for use in parallel RC circuits. Susceptance is the reciprocal of reactance; therefore, **capacitive susceptance** (B_C) is the reciprocal of capacitive reactance and is expressed as

Equation 10–11
$$B_C = \frac{1}{X_C}$$

Admittance (Y) is the reciprocal of impedance and is expressed as

Equation 10–12
$$Y = \frac{1}{Z}$$

The unit of each of these three quantities is the siemens (S), which is the reciprocal of the ohm.

In working with parallel circuits, it is often easier to use conductance (G), capacitive susceptance (B_C), and admittance (Y) rather than resistance (R), capacitive reactance (X_C) and impedance (Z). In a parallel RC circuit, as shown in Figure 10–22, the total admittance is the phasor sum of the conductance and the capacitive susceptance.

Equation 10–13
$$Y_{tot} = \sqrt{G^2 + B_C^2}$$

FIGURE 10–22

Admittance in a parallel RC circuit.

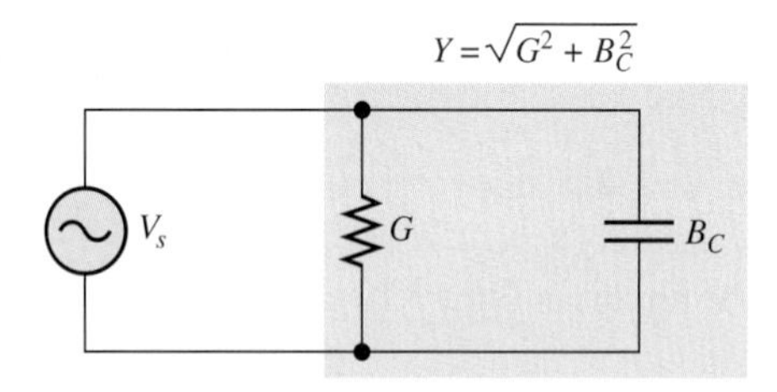

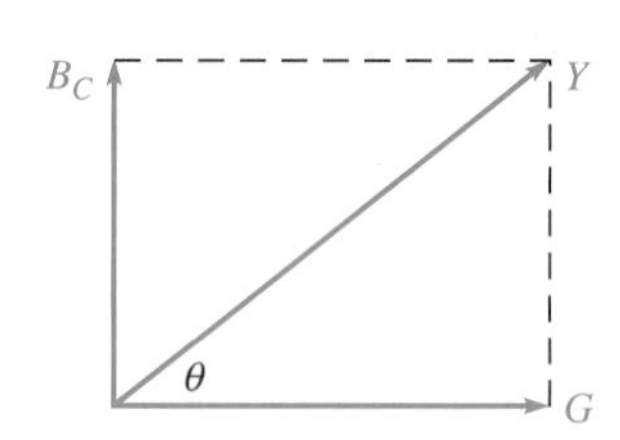

EXAMPLE 10–7

Determine the total admittance in Figure 10–23, and then convert it to impedance.

FIGURE 10–23

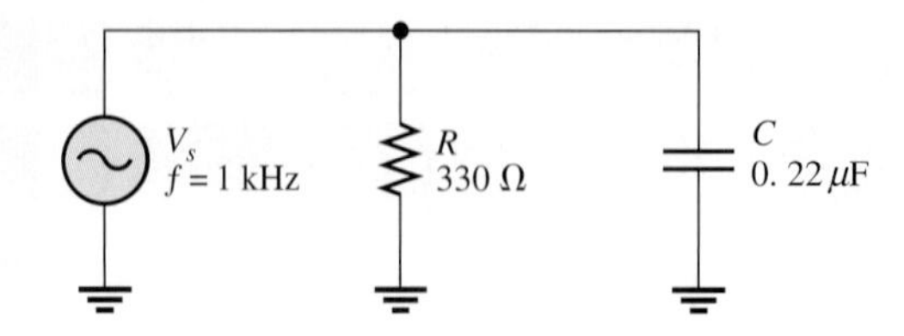

Solution To determine Y, first calculate the values for G and B_C. Since $R = 330\ \Omega$,

$$G = \frac{1}{R} = \frac{1}{330\ \Omega} = 3.03\text{ mS}$$

The capacitive reactance is

$$X_C = \frac{1}{2\pi fC} = \frac{1}{2\pi(1000 \text{ Hz})(0.22 \ \mu\text{F})} = 723 \ \Omega$$

The capacitive susceptance is

$$B_C = \frac{1}{X_C} = \frac{1}{723 \ \Omega} = 1.38 \text{ mS}$$

Therefore, the total admittance is

$$Y_{tot} = \sqrt{G^2 + B_C^2} = \sqrt{(3.03 \text{ mS})^2 + (1.38 \text{ mS})^2} = \mathbf{3.33 \text{ mS}}$$

Convert to impedance.

$$Z = \frac{1}{Y_{tot}} = \frac{1}{3.33 \text{ mS}} = \mathbf{300 \ \Omega}$$

Related Problem Calculate the admittance in Figure 10–23 if f is increased to 2.5 kHz.

SECTION 10–4 REVIEW

1. Determine Z if a 1.0 kΩ resistance is in parallel with a 650 Ω capacitive reactance.
2. Define *conductance, capacitive susceptance,* and *admittance.*
3. If $Z = 100 \ \Omega$, what is the value of Y?
4. In a certain parallel *RC* circuit, $R = 50 \ \Omega$ and $X_C = 75 \ \Omega$. Determine Y.

10–5 ANALYSIS OF PARALLEL *RC* CIRCUITS

In the previous section, you learned how to express the impedance of a parallel *RC* circuit. Now, Ohm's law and Kirchhoff's current law are used in the analysis of *RC* circuits. Current and voltage relationships in a parallel *RC* circuit are examined.

After completing this section, you should be able to

- **Analyze a parallel *RC* circuit**
- Apply Ohm's law and Kirchhoff's current law to parallel *RC* circuits
- Show how impedance and phase angle vary with frequency
- Convert from a parallel circuit to an equivalent series circuit

For convenience in the analysis of parallel circuits, the Ohm's law formulas using impedance—Equations 10–3, 10–4, and 10–5—can be rewritten for admittance using the relation $Y = 1/Z$.

$$V = \frac{I}{Y} \qquad \text{Equation 10–14}$$

$$I = VY \qquad \text{Equation 10–15}$$

$$Y = \frac{I}{V} \qquad \text{Equation 10–16}$$

EXAMPLE 10–8

Determine the total current and the phase angle in Figure 10–24.

▶ FIGURE 10–24

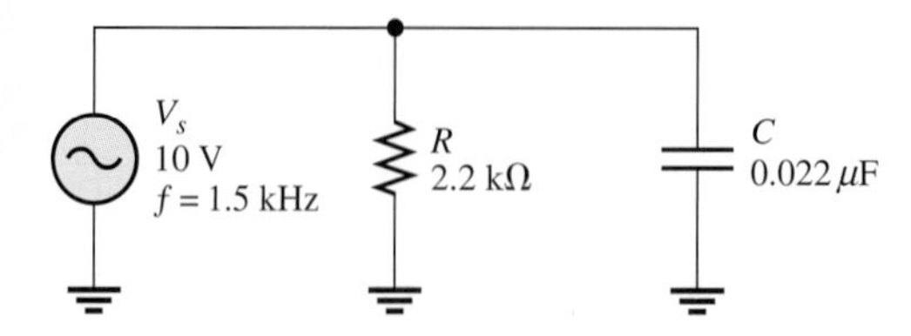

Solution First, determine the total admittance. The capacitive reactance is

$$X_C = \frac{1}{2\pi fC} = \frac{1}{2\pi(1.5\text{ kHz})(0.022\ \mu\text{F})} = 4.82\text{ k}\Omega$$

The conductance is

$$G = \frac{1}{R} = \frac{1}{2.2\text{ k}\Omega} = 455\ \mu\text{S}$$

The capacitive susceptance is

$$B_C = \frac{1}{X_C} = \frac{1}{4.82\text{ k}\Omega} = 207\ \mu\text{S}$$

Therefore, the total admittance is

$$Y_{tot} = \sqrt{G^2 + B_C^2} = \sqrt{(455\ \mu\text{S})^2 + (207\ \mu\text{S})^2} = 500\ \mu\text{S}$$

Next, use Ohm's law to calculate the total current.

$$I_{tot} = VY_{tot} = (10\text{ V})(500\ \mu\text{S}) = \mathbf{5.00\text{ mA}}$$

The phase angle is

$$\theta = \tan^{-1}\left(\frac{R}{X_C}\right) = \tan^{-1}\left(\frac{2.2\text{ k}\Omega}{4.82\text{ k}\Omega}\right) = \mathbf{24.5^\circ}$$

The total current is 5.00 mA, and it leads the source voltage by 24.5°.

Related Problem What is the total current if the frequency is doubled?

Open file E10-08 on your EWB/CircuitMaker CD-ROM. Verify the value of the total current that was calculated. Then measure each branch current. Double the frequency to 3 kHz and measure the total current.

Phase Relationships of the Currents and Voltages

Figure 10–25(a) shows all the currents and voltages in a basic parallel *RC* circuit. As you can see, the source voltage, V_s, appears across both the resistive and the capacitive branches, so V_s, V_R, and V_C are all in phase and of the same magnitude. The total current, I_{tot}, divides at the junction into the two branch currents, I_R and I_C.

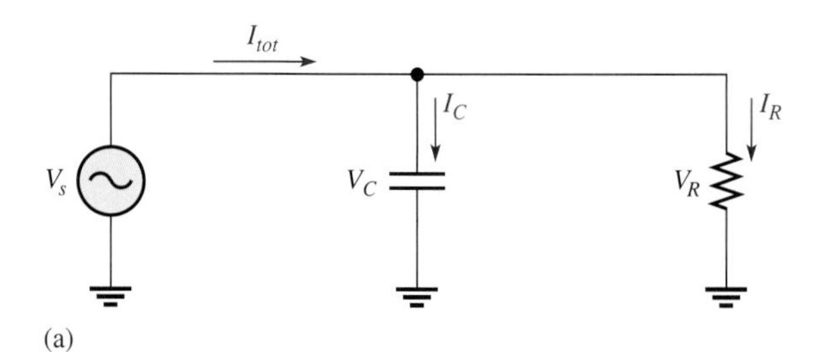

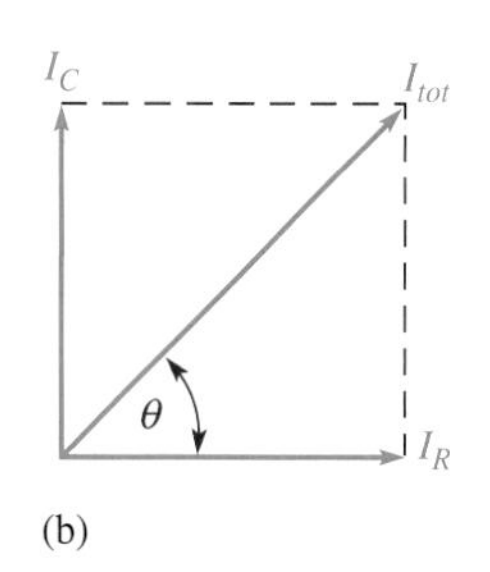

FIGURE 10–25

Currents and voltages in a parallel *RC* circuit. The current directions shown in (a) are instantaneous and, of course, reverse when the source voltage reverses. The current phasors in (b) rotate once per cycle.

The current through the resistor is in phase with the voltage. The current through the capacitor leads the voltage, and thus the resistive current, by 90°. By Kirchhoff's current law, the total current is the phasor sum of the two branch currents, as shown by the phasor diagram in Figure 10–25(b). The total current is expressed as

$$I_{tot} = \sqrt{I_R^2 + I_C^2}$$

Equation 10–17

The phase angle between the resistor current and the total current is

$$\theta = \tan^{-1}\left(\frac{I_C}{I_R}\right)$$

Equation 10–18

Equation 10–18 is equivalent to Equation 10–9, $\theta = \tan^{-1}(R/X_C)$.

Figure 10–26 shows a complete current and voltage phasor diagram. Notice that I_C leads I_R by 90° and that I_R is in phase with the voltage ($V_s = V_R = V_C$).

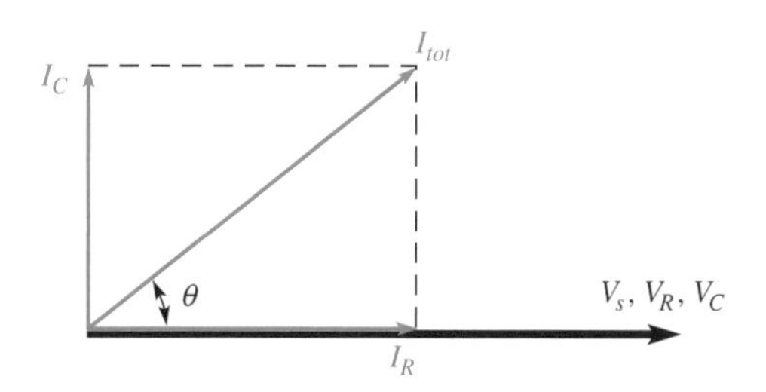

FIGURE 10–26

Current and voltage phasor diagram for a parallel *RC* circuit (amplitudes are arbitrary).

EXAMPLE 10–9

Determine the value of each current in Figure 10–27, and describe the phase relationship of each with the source voltage. Draw the current phasor diagram.

FIGURE 10–27

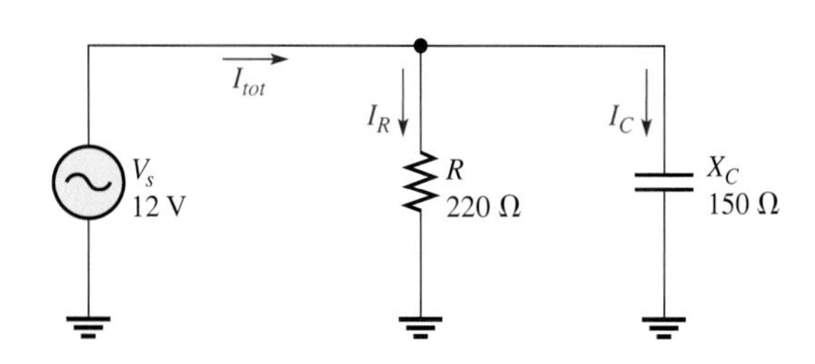

Solution The resistor current, the capacitor current, and the total current are expressed as follows:

$$I_R = \frac{V_s}{R} = \frac{12\text{ V}}{220\ \Omega} = \mathbf{54.5\ mA}$$

$$I_C = \frac{V_s}{X_C} = \frac{12\text{ V}}{150\ \Omega} = \mathbf{80\ mA}$$

$$I_{tot} = \sqrt{I_R^2 + I_C^2} = \sqrt{(54.5\text{ mA})^2 + (80\text{ mA})^2} = \mathbf{96.8\ mA}$$

The phase angle is

$$\theta = \tan^{-1}\left(\frac{I_C}{I_R}\right) = \tan^{-1}\left(\frac{80\text{ mA}}{54.5\text{ mA}}\right) = 55.7°$$

I_R is in phase with the source voltage, I_C leads the source voltage by 90°, and I_{tot} leads the source voltage by 55.7°. The current phasor diagram is shown in Figure 10–28.

FIGURE 10–28

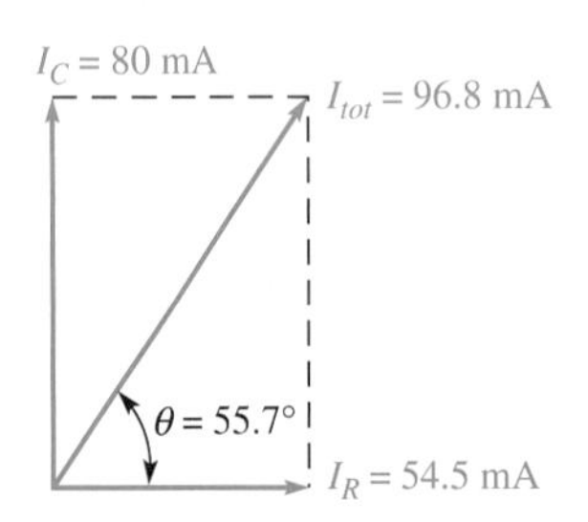

Related Problem In a certain parallel circuit, $I_R = 100$ mA and $I_C = 60$ mA. Determine the total current and the phase angle.

Conversion from Parallel to Series Form

For every parallel *RC* circuit, there is an equivalent series *RC* circuit for any given frequency. Two circuits are equivalent when they both present an equal impedance and phase angle.

To obtain the equivalent series circuit for a given parallel *RC* circuit, first find the impedance and phase angle of the parallel circuit. Then use the values of Z and θ to construct an impedance triangle, shown in Figure 10–29. The vertical and horizontal sides of the triangle represent the equivalent series resistance and capacitive reactance as indicated. These values can be found using the following trigonometric relationships:

Equation 10–19

$$R_{eq} = Z\cos\theta$$

Equation 10–20

$$X_{C(eq)} = Z\sin\theta$$

The cosine (cos) and the sine (sin) functions are available on the TI-85/TI-86 calculator.

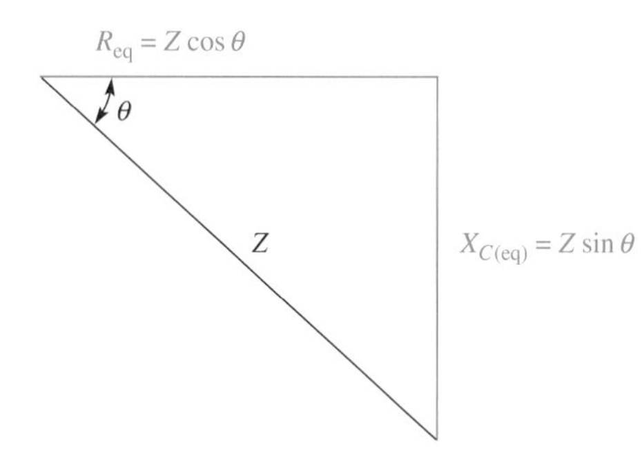

FIGURE 10–29

Impedance triangle for the series equivalent of a parallel *RC* circuit. Z and θ are the known values for the parallel circuit. R_{eq} and $X_{C(eq)}$ are the series equivalent values.

EXAMPLE 10–10

Convert the parallel circuit in Figure 10–30 to an equivalent series form.

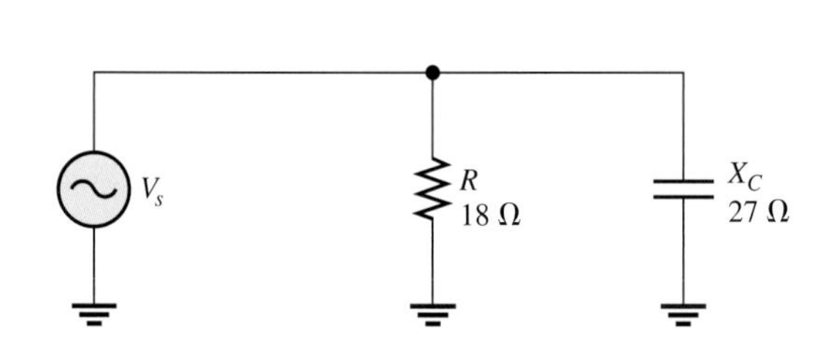

FIGURE 10–30

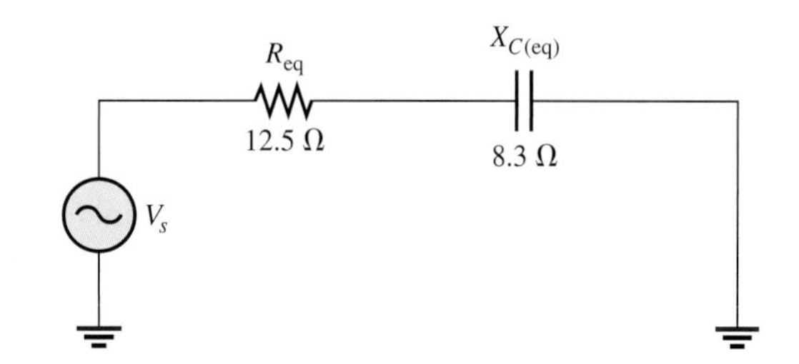

FIGURE 10–31

Solution First, find the total admittance of the parallel circuit as follows:

$$G = \frac{1}{R} = \frac{1}{18\ \Omega} = 55.6\ \text{mS}$$

$$B_C = \frac{1}{X_C} = \frac{1}{27\ \Omega} = 37.0\ \text{mS}$$

$$Y_{tot} = \sqrt{G^2 + B_C^2} = \sqrt{(55.6\ \text{mS})^2 + (37.0\ \text{mS})^2} = 66.8\ \text{mS}$$

Then, the total impedance is

$$Z_{tot} = \frac{1}{Y_{tot}} = \frac{1}{66.8\ \text{mS}} = 15.0\ \Omega$$

The phase angle is

$$\theta = \tan^{-1}\left(\frac{R}{X_C}\right) = \tan^{-1}\left(\frac{18\ \Omega}{27\ \Omega}\right) = 33.6^\circ$$

The equivalent series values are

$$R_{eq} = Z\cos\theta = (15.0\ \Omega)\cos(33.6^\circ) = \mathbf{12.5\ \Omega}$$

$$X_{C(eq)} = Z\sin\theta = (15.0\ \Omega)\sin(33.6^\circ) = \mathbf{8.3\ \Omega}$$

The series equivalent circuit is shown in Figure 10–31. The value of C can only be determined if the frequency is known.

The calculator solutions for R_{eq} and $X_{C(eq)}$ are

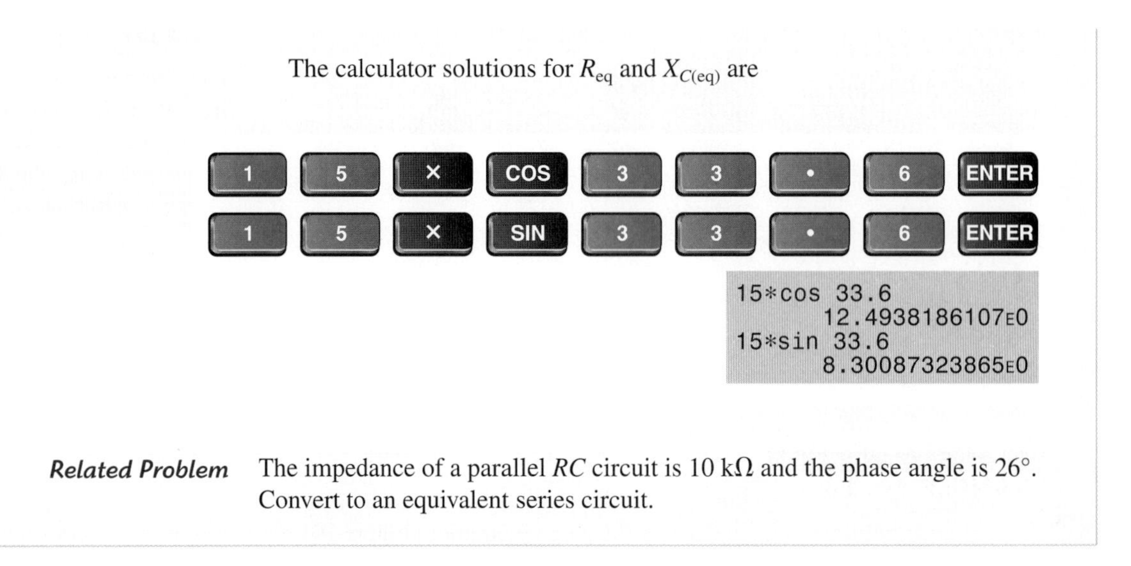

Related Problem The impedance of a parallel *RC* circuit is 10 kΩ and the phase angle is 26°. Convert to an equivalent series circuit.

Note that a parallel *RC* circuit becomes less reactive when X_C is increased. That is, the circuit phase angle becomes smaller. The reason for this effect is that when X_C is increased relative to *R*, less current is through the capacitive branch, and although the in-phase or resistive current does not increase, it becomes a greater percentage of the total current.

SECTION 10–5 REVIEW

1. The admittance of an *RC* circuit is 3.5 mS, and the source voltage is 6 V. What is the total current?
2. In a certain parallel *RC* circuit, the resistor current is 10 mA, and the capacitor current is 15 mA. Determine the phase angle and the total current.
3. What is the phase angle between the capacitor current and the source voltage in a parallel *RC* circuit?

10–6 SERIES-PARALLEL ANALYSIS

In this section, the concepts studied in the previous sections are used to analyze circuits with combinations of both series and parallel *R* and *C* elements.

After completing this section, you should be able to

- **Analyze series-parallel *RC* circuits**
- Determine total impedance
- Calculate currents and voltages
- Measure impedance and phase angle

The following example demonstrates the procedures used to analyze a series-parallel reactive circuit.

EXAMPLE 10–11

In the circuit of Figure 10–32, determine the following:

(a) total impedance **(b)** total current

(c) phase angle by which I_{tot} leads V_s

▶ FIGURE 10–32

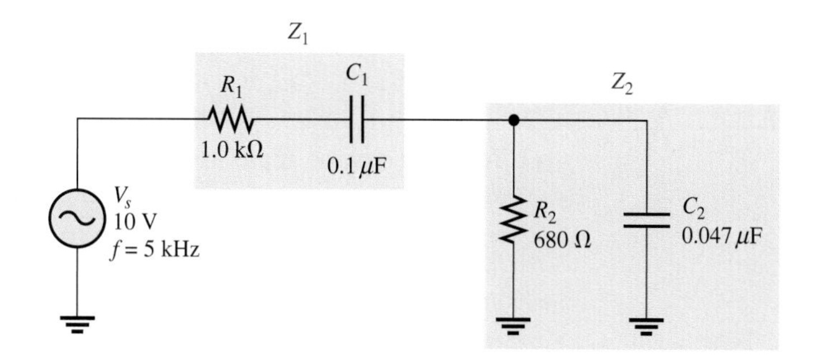

Solution **(a)** First, calculate the magnitudes of the capacitive reactances.

$$X_{C1} = \frac{1}{2\pi(5\text{ kHz})(0.1\ \mu\text{F})} = 318\ \Omega$$

$$X_{C2} = \frac{1}{2\pi(5\text{ kHz})(0.047\ \mu\text{F})} = 677\ \Omega$$

One approach is to find the series equivalent resistance and capacitive reactance for the parallel portion of the circuit; then add the resistances ($R_1 + R_{eq}$) to get total resistance and add the reactances ($X_{C1} + X_{C(eq)}$) to get total reactance. From these totals, you can determine the total impedance.

Find the impedance of the parallel portion (Z_2) by first finding the admittance.

$$G_2 = \frac{1}{R_2} = \frac{1}{680\ \Omega} = 1.47\text{ mS}$$

$$B_{C2} = \frac{1}{X_{C2}} = \frac{1}{677\ \Omega} = 1.48\text{ mS}$$

$$Y_2 = \sqrt{G_2^2 + B_{C2}^2} = \sqrt{(1.47\text{ mS})^2 + (1.48\text{ mS})^2} = 2.09\text{ mS}$$

$$Z_2 = \frac{1}{Y_2} = \frac{1}{2.09\text{ mS}} = 479\ \Omega$$

The phase angle associated with the parallel portion of the circuit is

$$\theta_p = \tan^{-1}\left(\frac{R_2}{X_{C2}}\right) = \tan^{-1}\left(\frac{680\ \Omega}{677\ \Omega}\right) = 45.1°$$

The series equivalent values for the parallel portion are

$$R_{eq} = Z_2\cos\theta_p = (479\ \Omega)\cos(45.1°) = 338\ \Omega$$

$$X_{C(eq)} = Z_2\sin\theta_p = (479\ \Omega)\sin(45.1°) = 339\ \Omega$$

The total circuit resistance is

$$R_{tot} = R_1 + R_{eq} = 1000\ \Omega + 338\ \Omega = 1.34\text{ k}\Omega$$

The total circuit reactance is

$$X_{C(tot)} = X_{C1} + X_{C(eq)} = 318\ \Omega + 339\ \Omega = 657\ \Omega$$

The total circuit impedance is

$$Z_{tot} = \sqrt{R_{tot}^2 + X_{C(tot)}^2} = \sqrt{(1.34\ \text{k}\Omega)^2 + (657\ \Omega)^2} = \mathbf{1.49\ k\Omega}$$

(b) Use Ohm's law to find the total current.

$$I_{tot} = \frac{V_s}{Z_{tot}} = \frac{10\ \text{V}}{1.49\ \text{k}\Omega} = \mathbf{6.71\ mA}$$

(c) To find the phase angle, view the circuit as a series combination of R_{tot} and $X_{C(tot)}$. The phase angle by which I_{tot} leads V_s is

$$\theta = \tan^{-1}\left(\frac{X_{C(tot)}}{R_{tot}}\right) = \tan^{-1}\left(\frac{657\ \Omega}{1.34\ \text{k}\Omega}\right) = \mathbf{26.1^\circ}$$

Related Problem Determine the voltages across Z_1 and Z_2 in Figure 10–32.

Open file E10-11 on your EWB/CircuitMaker CD-ROM. Verify the calculated value of the total current. Measure the current through R_2 and the current through C_2. Measure voltages across Z_1 and Z_2.

Measurement of Total Impedance, Z_{tot}

Now, let's see how the value of Z_{tot} for the circuit in Example 10–11 can be determined by measurement. First, the total impedance is measured as outlined in the following steps and as illustrated in Figure 10–33 (other ways are also possible):

Step 1: Using a sine wave generator, set the source voltage to a known value (10 V) and the frequency to 5 kHz. Check the voltage with an ac voltmeter and the frequency with a frequency counter rather than relying on the marked values on the generator controls.

Step 2: Connect an ac ammeter as shown in Figure 10–33, and measure the total current.

Step 3: Calculate the total impedance by using Ohm's law.

Measurement of Phase Angle, θ

To measure the phase angle, the source voltage and the total current must be displayed on an oscilloscope screen in the proper time relationship. Two basic types of scope probes are available to measure the quantities with an oscilloscope: the voltage probe and the current probe. Although the current probe is a convenient device, it is often not as readily available as a voltage probe. For this reason, we will confine our phase measurement technique to the use of voltage probes in conjunction with the oscilloscope. A typical oscilloscope voltage probe has two points that are connected to the circuit: the probe tip and the ground lead. Thus, all voltage measurements must be referenced to ground.

Since only voltage probes are to be used, the total current cannot be measured directly. However, for phase measurement, the voltage across R_1 is in phase with the total current and can be used to establish the phase angle of the current.

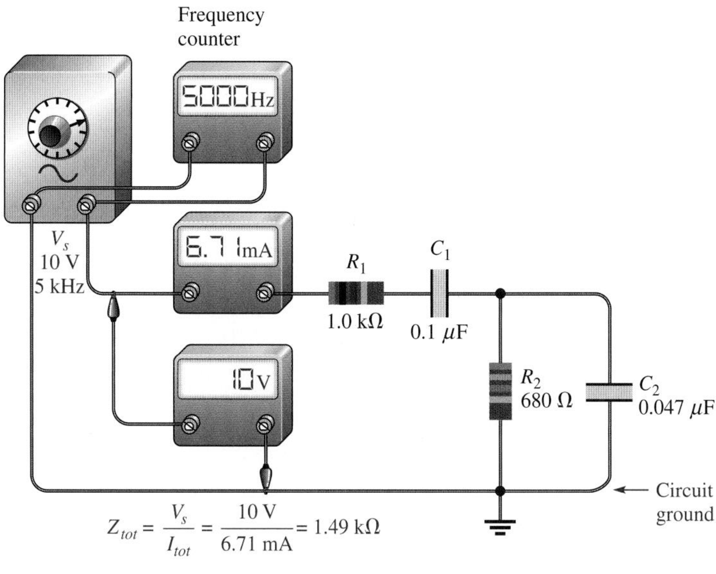

FIGURE 10–33

Determining Z_{tot} by measurement of V_s and I_{tot}.

Before proceeding with the actual phase measurement, there is a problem with displaying V_{R1}. If the scope probe is connected across the resistor, as indicated in Figure 10–34(a), the ground lead of the scope will short point B to ground, thus bypassing the rest of the components and effectively removing them from the circuit electrically, as illustrated in Figure 10–34(b) (assuming that the scope is not isolated from power line ground).

FIGURE 10–34

Effects of measuring directly across a component when the instrument and the circuit are grounded.

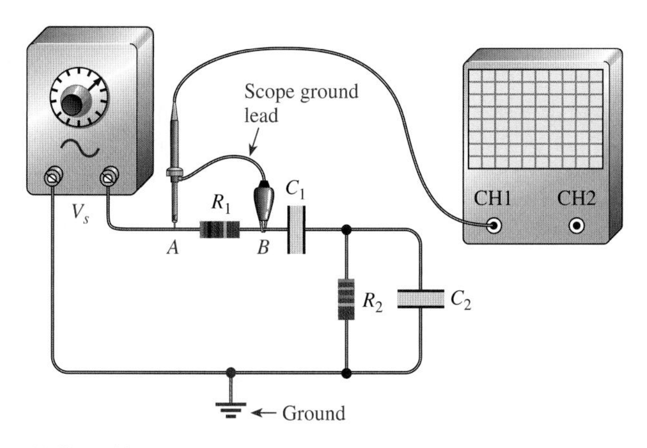

(a) Ground lead on scope probe grounds point B.

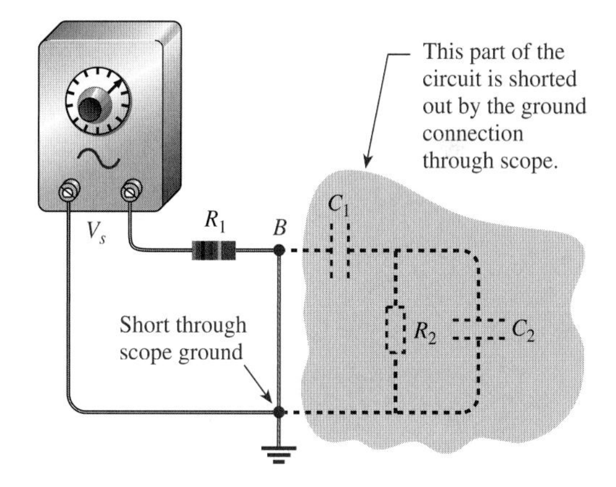

(b) The effect of grounding point B is to short out the rest of the circuit.

To avoid this problem, you can switch the generator output terminals so that one end of R_1 is connected to the ground terminal, as shown in Figure 10–35(a). Now the scope can be connected across it to display V_{R1}, as indicated in Figure 10–35(b). The other probe is connected across the voltage source to display V_s as indicated. Now channel 1 of the scope has V_{R1} as an input, and channel 2 has V_s. The scope should be triggered from the source voltage (channel 2 in this case).

Before connecting the probes to the circuit, you should align the two horizontal lines (traces) so that they appear as a single line across the center of the scope screen. To do so, ground the probe tips and adjust the vertical position knobs to move the traces toward the center line of the screen until they are superimposed. This procedure ensures that both waveforms have the same zero crossing so that an accurate phase measurement can be made.

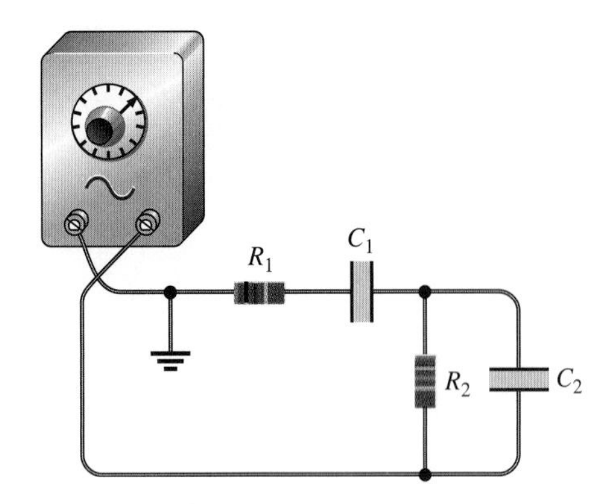

(a) Ground repositioned so that one end of R_1 is grounded.

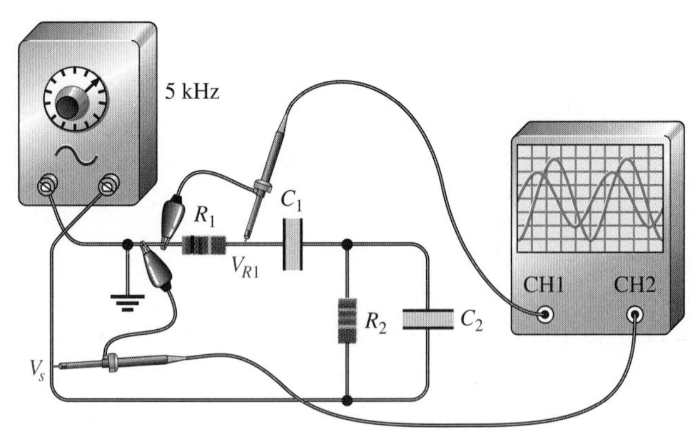

(b) The scope displays V_{R1} and V_s. V_{R1} represents the phase of the total current.

FIGURE 10–35

Repositioning ground so that a direct voltage measurement can be made with respect to ground without shorting out part of the circuit.

Once you have stabilized the waveforms on the scope screen, you can measure the period of the source voltage. Next, use the Volts/Div controls to adjust the amplitudes of the waveforms until they both appear to have the same amplitude. Now, spread the waveforms horizontally by using the Sec/Div control to expand the distance between them. This horizontal distance represents the time between the two waveforms. The number of divisions between the waveforms along any horizontal lines times the Sec/Div setting is equal to the time between them, Δt. Also, you can use the cursors to determine Δt if your oscilloscope has this feature.

Once you have determined the period, T, and the time between the waveforms, Δt, you can calculate the phase angle with the following equation:

Equation 10–21

$$\theta = \left(\frac{\Delta t}{T}\right)360^\circ$$

An example screen display is shown in Figure 10–36. In this illustration, there are 1.5 horizontal divisions between the two waveforms, as indicated, and the Sec/Div control is set at 10 μs. The period of these waveforms is 200 μs and the Δt is

$$\Delta t = 1.5 \text{ divisions} \times 10\ \mu\text{s/division} = 15\ \mu\text{s}$$

The phase angle is

$$\theta = \left(\frac{\Delta t}{T}\right)360^\circ = \left(\frac{15\ \mu\text{s}}{200\ \mu\text{s}}\right)360^\circ = 27^\circ$$

FIGURE 10–36

Determining the phase angle on the oscilloscope.

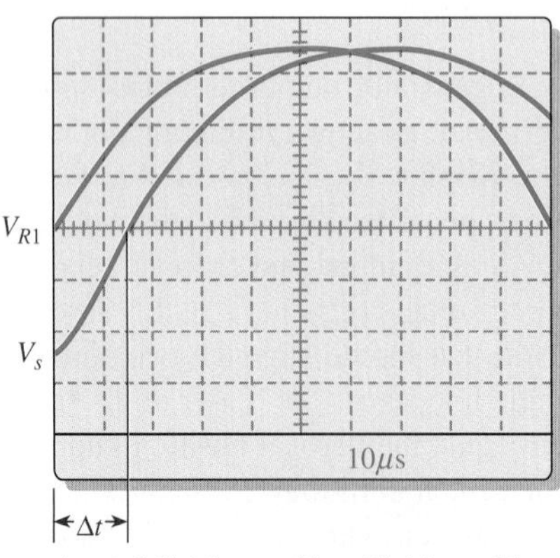

SECTION 10–6 REVIEW

1. Determine the series equivalent circuit for the series-parallel circuit in Figure 10–32.
2. What is the voltage across R_1 in Figure 10–32?

10–7 POWER IN *RC* CIRCUITS

In a purely resistive ac circuit, all of the energy delivered by the source is dissipated in the form of heat by the resistance. In a purely capacitive ac circuit, all of the energy delivered by the source is stored by the capacitor during a portion of the voltage cycle and then returned to the source during another portion of the cycle so that there is no net energy conversion to heat. When there is both resistance and capacitance, some of the energy is alternately stored and returned by the capacitance and some is dissipated by the resistance. The amount of energy converted to heat is determined by the relative values of the resistance and the capacitive reactance.

After completing this section, you should be able to

- **Determine power in *RC* circuits**
- Explain true and reactive power
- Draw the power triangle
- Define *power factor*
- Explain apparent power
- Calculate power in an *RC* circuit

It is reasonable to assume that when the resistance is greater than the capacitive reactance, more of the total energy delivered by the source is dissipated by the resistance than is stored by the capacitance. Likewise, when the reactance is greater than the resistance, more of the total energy is stored and returned than is converted to heat.

The formulas for the power dissipated in a resistor, sometimes called *true power* (P_{true}), and the power in a capacitor, called **reactive power** (P_r), are restated here. The unit of true power is the watt (W), and the unit of reactive power is the VAR (volt-ampere reactive).

$$P_{true} = I_{tot}^2 R \quad \text{Equation 10–22}$$

$$P_r = I_{tot}^2 X_C \quad \text{Equation 10–23}$$

The Power Triangle for *RC* Circuits

The generalized impedance phasor diagram is shown in Figure 10–37(a). A relationship for power can also be represented by a similar diagram because the respective magnitudes of the powers, P_{true} and P_r, differ from R and X_C by a factor of I_{tot}^2, as shown in Figure 10–37(b).

The resultant power, $I_{tot}^2 Z$, represents the **apparent power**, P_a. At any instant in time, P_a is the total power that appears to be transferred between the source and the *RC* circuit. Part of the apparent power is true power, and part of it is reactive power. The unit of apparent power is the volt-ampere (VA). The expression for apparent power is

$$P_a = I_{tot}^2 Z \quad \text{Equation 10–24}$$

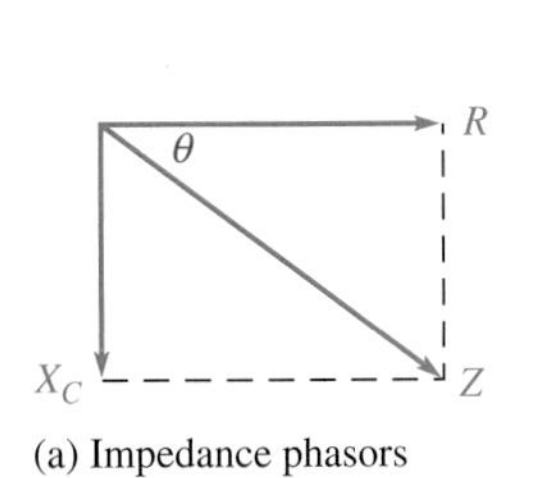

(a) Impedance phasors

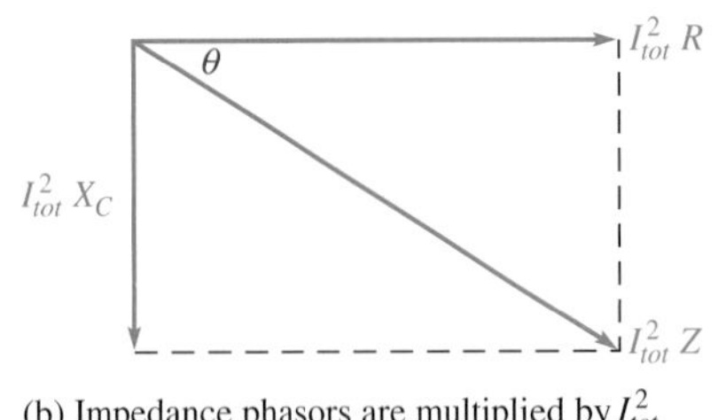

(b) Impedance phasors are multiplied by I^2_{tot} to get power

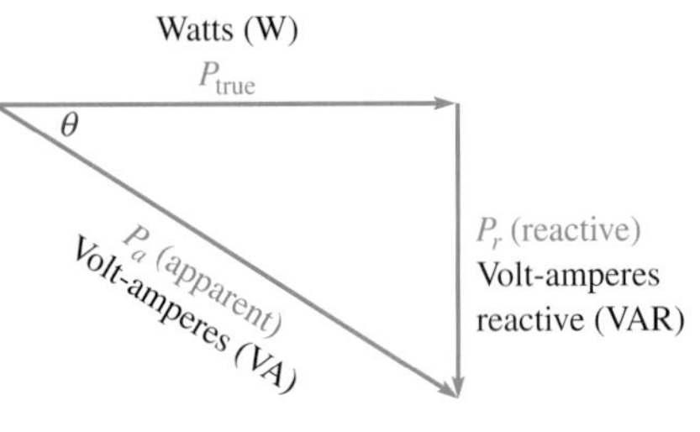

(c) Power triangle

▲ **FIGURE 10–37**

Development of the power triangle for an *RC* circuit.

The diagram in Figure 10–37(b) can be rearranged in the form of a right triangle as shown in Figure 10–37(c), which is called the *power triangle.* Using the rules of trigonometry, P_{true} can be expressed as

$$P_{true} = P_a \cos \theta$$

Since P_a equals $I^2_{tot}Z$ or V_sI_{tot}, the equation for true power can be written as

Equation 10–25

$$P_{true} = V_sI_{tot}\cos \theta$$

where V_s is the source voltage and I_{tot} is the total current.

For the case of a purely resistive circuit, $\theta = 0°$ and $\cos 0° = 1$, so P_{true} equals V_sI_{tot}. For the case of a purely capacitive circuit, $\theta = 90°$ and $\cos 90° = 0$, so P_{true} is zero. As you already know, there is no power dissipated in an ideal capacitor.

Power Factor

The term *cos θ* is called the **power factor** and is stated as

Equation 10–26

$$PF = \cos \theta$$

As the phase angle between the source voltage and the total current increases, the power factor decreases, indicating an increasingly reactive circuit. The smaller the power factor, the smaller the power dissipation.

The power factor can vary from 0 for a purely reactive circuit to 1 for a purely resistive circuit. In an *RC* circuit, the power factor is referred to as a leading power factor because the current leads the voltage.

EXAMPLE 10–12

Determine the power factor and the true power in the *RC* circuit of Figure 10–38.

▶ **FIGURE 10–38**

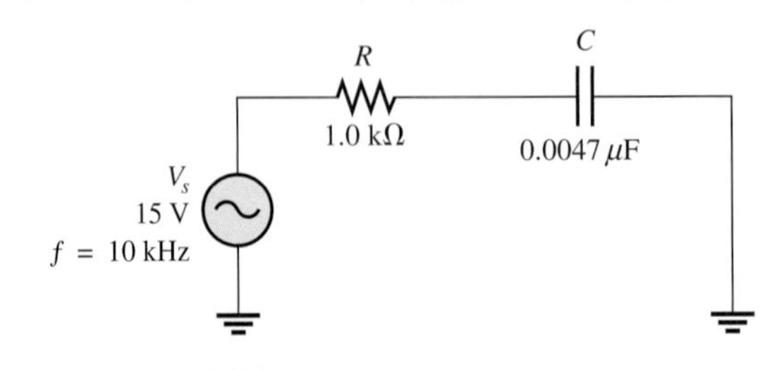

Solution Calculate the capacitive reactance and phase angle.

$$X_C = \frac{1}{2\pi fC} = \frac{1}{2\pi(10\text{ kHz})(0.0047\ \mu\text{F})} = 3.39\text{ k}\Omega$$

$$\theta = \tan^{-1}\left(\frac{X_C}{R}\right) = \tan^{-1}\left(\frac{3.39\text{ k}\Omega}{1.0\text{ k}\Omega}\right) = 73.6^\circ$$

The power factor is

$$PF = \cos\theta = \cos(73.6^\circ) = \mathbf{0.282}$$

The impedance is

$$Z = \sqrt{R^2 + X_C^2} = \sqrt{(1.0\text{ k}\Omega)^2 + (3.39\text{ k}\Omega)^2} = 3.53\text{ k}\Omega$$

Therefore, the current is

$$I = \frac{V_s}{Z} = \frac{15\text{ V}}{3.53\text{ k}\Omega} = 4.25\text{ mA}$$

The true power is

$$P_{\text{true}} = V_s I\cos\theta = (15\text{ V})(4.25\text{ mA})(0.282) = \mathbf{18.0\ mW}$$

Related Problem What is the power factor if *f* is reduced by half in Figure 10–38?

Open file E10-12 on your EWB/CircuitMaker CD-ROM. Measure the current and compare to the calculated value. Measure the voltages across *R* and *C* at 10 kHz, 5 kHz, and 20 kHz. Explain your observations.

The Significance of Apparent Power

Apparent power is the power that *appears* to be transferred between the source and the load, and it consists of two components: a true power component and a reactive power component.

In all electrical and electronic systems, it is the true power that does the work. The reactive power is simply shuttled back and forth between the source and the load. Ideally, in terms of performing useful work, all of the power transferred to the load should be true power and none of it reactive power. However, in most practical situations the load has some reactance associated with it, and therefore you must deal with both power components.

For any reactive load, there are two components of the total current: the resistive component and the reactive component. If you consider only the true power (watts) in a load, you are dealing with only a portion of the total current that the load demands from a source. In order to have a realistic picture of the actual current that a load will draw, you must consider apparent power (in VA).

A source such as an ac generator can provide current to a load up to some maximum value. *If the load draws more than this maximum value, the source can be damaged.* Figure 10–39(a) shows a 120 V generator that can deliver a maximum current of 5 A to a load. Assume that the generator is rated at 600 W and is connected to a purely resistive load of 24 Ω (power factor of 1). The ammeter shows that the current is 5 A, and the wattmeter indicates that the power is 600 W. The generator has no problem under these conditions, although it is operating at maximum current and power.

▶ FIGURE 10–39

The wattage rating of a source is inappropriate when the load is reactive. The rating should be in VA rather than in watts.

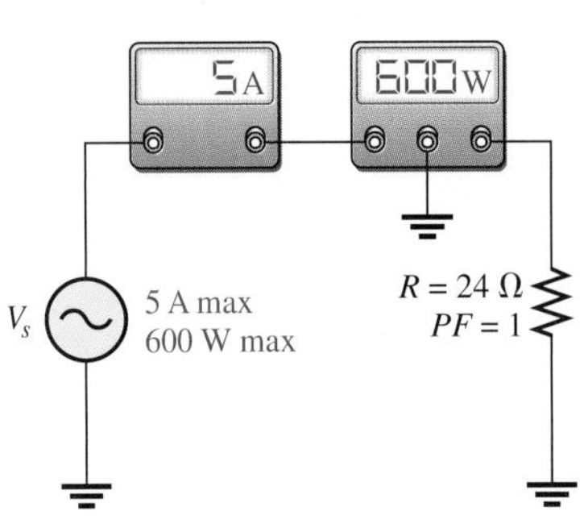

(a) Generator operating at its limits with a resistive load.

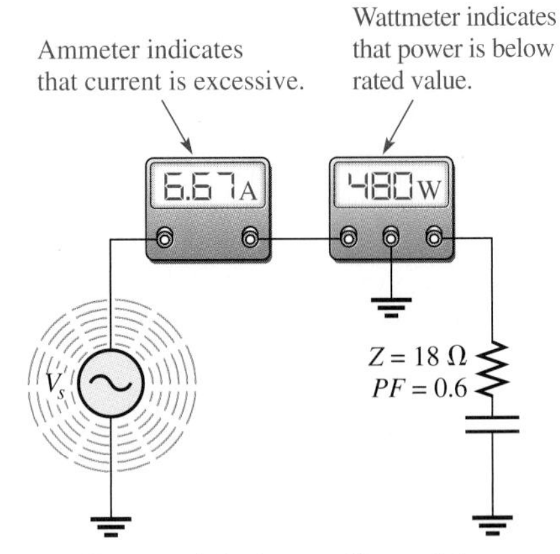

(b) Generator is in danger of internal damage due to excess current, even though the wattmeter indicates that the power is below the maximum wattage rating.

Now, consider what happens if the load is changed to a reactive one with an impedance of 18 Ω and a power factor of 0.6, as indicated in Figure 10–39(b). The current is 120 V/18 Ω = 6.67 A, which exceeds the maximum. Even though the wattmeter reads 480 W, which is less than the power rating of the generator, the excessive current probably will cause damage. This example, shows that a true power rating can be deceiving and is inappropriate for ac sources. This particular ac generator should be rated at 600 VA, a rating the manufacturer would use, rather than 600 W.

EXAMPLE 10–13

For the circuit in Figure 10–40, find the true power, the reactive power, and the apparent power. X_C has been determined to be 2 kΩ.

▶ FIGURE 10–40

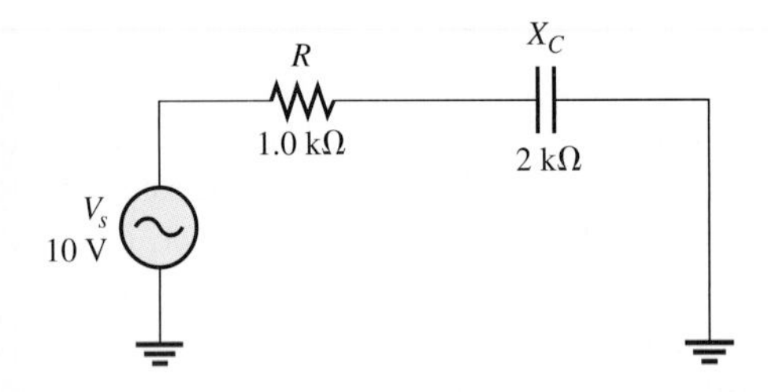

Solution First find the total impedance so that the current can be calculated.

$$Z_{tot} = \sqrt{R^2 + X_C^2} = \sqrt{(1.0\ \text{k}\Omega)^2 + (2\ \text{k}\Omega)^2} = 2.24\ \text{k}\Omega$$

$$I = \frac{V_s}{Z} = \frac{10\ \text{V}}{2.24\ \text{k}\Omega} = 4.46\ \text{mA}$$

The phase angle, θ, is

$$\theta = \tan^{-1}\left(\frac{X_C}{R}\right) = \tan^{-1}\left(\frac{2\ \text{k}\Omega}{1.0\ \text{k}\Omega}\right) = 63.4°$$

The true power is

$$P_{true} = V_sI\cos\theta = (10\text{ V})(4.46\text{ mA})\cos(63.4°) = \mathbf{20\text{ mW}}$$

Note that the same result is realized using the formula $P_{true} = I^2R$.

The reactive power is

$$P_r = I^2X_C = (4.46\text{ mA})^2(2\text{ k}\Omega) = \mathbf{39.8\text{ mVAR}}$$

The apparent power is

$$P_a = I^2Z = (4.46\text{ mA})^2(2.24\text{ k}\Omega) = \mathbf{44.6\text{ mVA}}$$

The apparent power is also the phasor sum of P_{true} and P_r.

$$P_a = \sqrt{P_{true}^2 + P_r^2} = 44.6\text{ mVA}$$

Related Problem What is the true power in Figure 10–40 if $X_C = 10\text{ k}\Omega$?

SECTION 10–7 REVIEW

1. To which component in an *RC* circuit is the power dissipation due?
2. If the phase angle, θ, is 45°, what is the power factor?
3. A certain series *RC* circuit has the following parameter values: $R = 330\ \Omega$, $X_C = 460\ \Omega$, and $I = 2$ A. Determine the true power, the reactive power, and the apparent power.

10–8 BASIC APPLICATIONS

RC circuits are found in a variety of applications, often as part of a more complex circuit. Two major applications, phase shift circuits and frequency-selective circuits (filters), are covered in this section.

After completing this section, you should be able to

- **Discuss some basic *RC* applications**
- Discuss and analyze the *RC* lag network
- Discuss and analyze the *RC* lead network
- Discuss how the *RC* circuit operates as a filter

RC Lag Network

The **RC lag network** is a phase shift circuit in which the output voltage lags the input voltage by a specified angle, ϕ. Phase shift circuits are commonly used in electronic communication systems and in other application areas.

A basic *RC* lag network is shown in Figure 10–41(a). Notice that the output voltage is taken across the capacitor, and the input voltage is the total voltage applied across the circuit. The relationship of the voltages is shown in the phasor diagram in Figure 10–41(b). Keep in mind that the circuit phase angle is the angle between V_{in} and the current. Notice in Figure 10–41(b) that the output voltage, V_{out}, lags V_{in} by an angle, designated ϕ, that is the difference between 90° and the circuit phase angle θ. The angle ϕ is called the phase lag when the output voltage lags the input voltage and the phase lead when the output voltage leads the input.

FIGURE 10–41

The *RC* lag network ($V_{out} = V_C$).

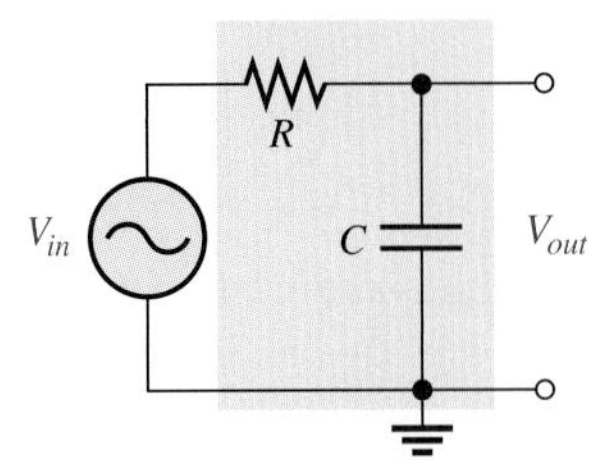

(a) A basic *RC* lag network

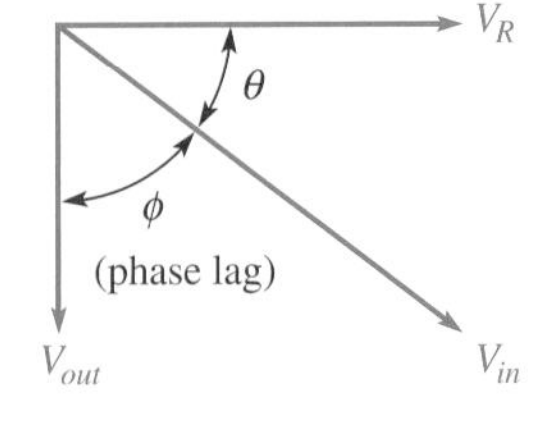

(b) Phasor voltage diagram showing the phase lag between V_{in} and V_{out}

Since $\theta = \tan^{-1}(X_C/R)$, the phase lag, ϕ, can be expressed as

Equation 10–27

$$\phi = 90° - \tan^{-1}\left(\frac{X_C}{R}\right)$$

When the input and output voltage waveforms of the lag network are displayed on an oscilloscope, a relationship similar to that in Figure 10–42 is observed. The exact amount of phase lag between the input and the output depends on the values of the resistance and the capacitive reactance. The magnitude of the output voltage depends on these values also.

FIGURE 10–42

General oscilloscope display of the input and output waveforms of a lag network (V_{out} lags V_{in}).

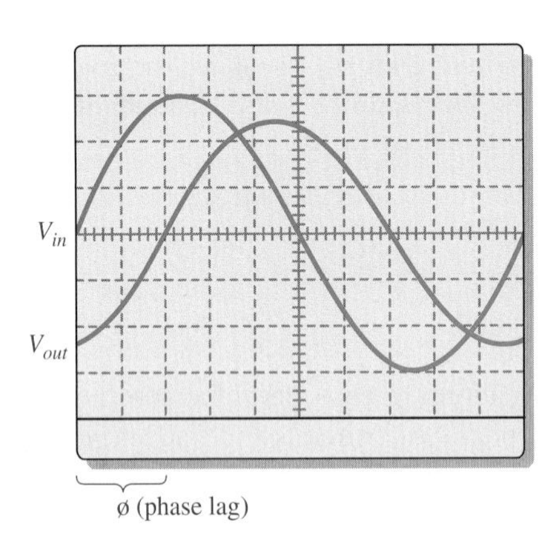

EXAMPLE 10–14

Determine the amount of phase lag from input to output in the lag network in Figure 10–43.

FIGURE 10–43

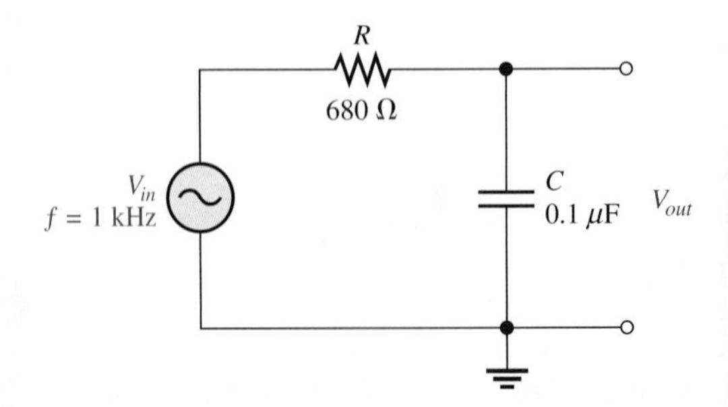

Solution First determine the capacitive reactance.

$$X_C = \frac{1}{2\pi fC} = \frac{1}{2\pi(1\text{ kHz})(0.1\ \mu\text{F})} = 1.59\text{ k}\Omega$$

The phase lag between the output voltage and the input voltage is

$$\phi = 90^\circ - \tan^{-1}\left(\frac{X_C}{R}\right) = 90^\circ - \tan^{-1}\left(\frac{1.59\ \text{k}\Omega}{680\ \Omega}\right) = \mathbf{23.2^\circ}$$

The output voltage lags the input voltage by 23.2°.

Related Problem In a lag network, what happens to the phase lag if the frequency increases?

The phase-lag network can be viewed as a voltage divider with a portion of the input voltage dropped across R and a portion across C. The output voltage can be determined with the following formula:

$$V_{out} = \left(\frac{X_C}{\sqrt{R^2 + X_C^2}}\right)V_{in}$$

Equation 10–28

EXAMPLE 10–15

For the lag network in Figure 10–43 of Example 10–14, determine the output voltage when the input voltage has an rms value of 10 V. Draw the input and output waveforms showing the proper relationships. The values for X_C (1.59 kΩ) and ϕ (23.2°) were found in Example 10–14.

Solution Use Equation 10–28 to determine the output voltage for the lag network in Figure 10–43.

$$V_{out} = \left(\frac{X_C}{\sqrt{R^2 + X_C^2}}\right)V_{in} = \left(\frac{1.59\ \text{k}\Omega}{\sqrt{(680\ \Omega)^2 + (1.59\ \text{k}\Omega)^2}}\right)10\ \text{V} = \mathbf{9.2\ V\ rms}$$

The waveforms are shown in Figure 10–44.

▶ FIGURE 10–44

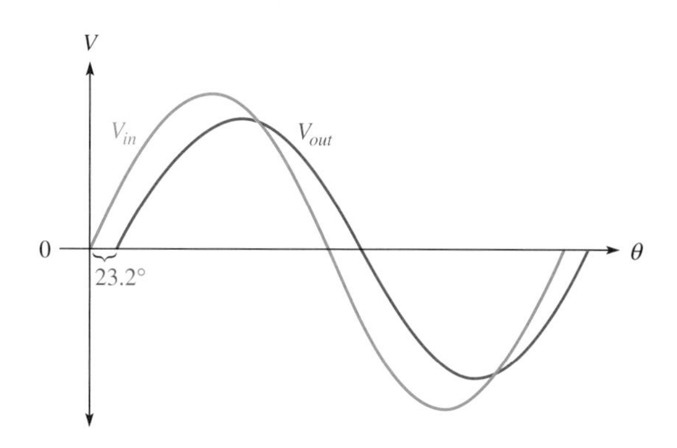

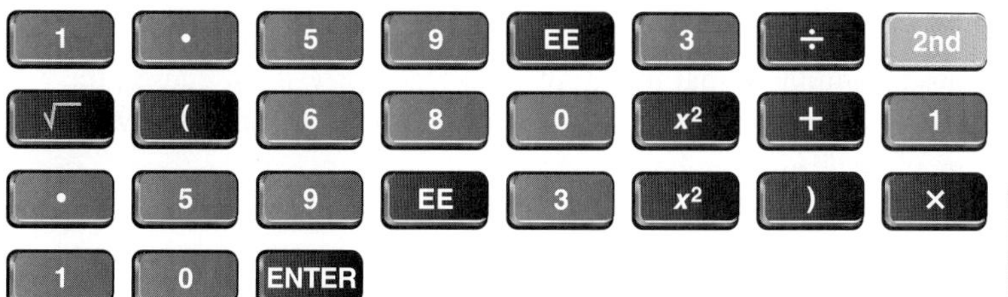

Related Problem In a lag network, what happens to the output voltage if the frequency increases?

Effects of Frequency on the Lag Network Since the circuit phase angle, θ, decreases as frequency increases, the phase lag between the input and the output voltages increases. You can see this relationship by examining Equation 10–27. Also, the magnitude of V_{out} decreases as the frequency increases because X_C becomes smaller and less of the total input voltage is dropped across the capacitor. Figure 10–45 illustrates the changes in phase lag and output voltage magnitude with frequency.

FIGURE 10–45

Illustration of how the frequency affects the phase lag and the output voltage in an *RC* lag network with the amplitude of V_{in} held constant.

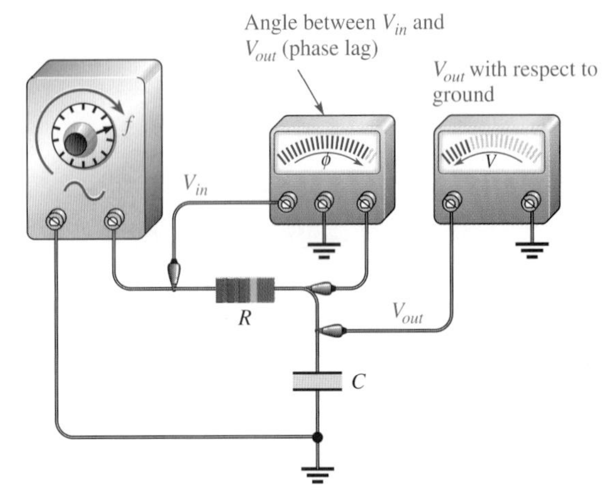

(a) When f increases, the phase lag ϕ increases and V_{out} decreases.

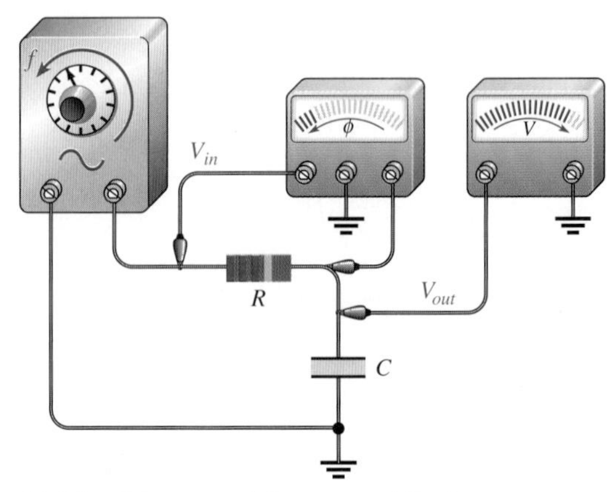

(b) When f decreases, ϕ decreases and V_{out} increases.

RC Lead Network

The *RC* **lead network** is a phase shift circuit in which the output voltage leads the input voltage by a specified angle, ϕ. A basic *RC* lead network is shown in Figure 10–46(a). Notice how it differs from the lag network. Here the output voltage is taken across the resistor. The relationship of the voltages is given in the phasor diagram in Figure 10–46(b). The output voltage, V_{out}, leads V_{in} by an angle that is the same as the circuit phase angle because V_R and I are in phase with each other.

FIGURE 10–46

The *RC* lead network ($V_{out} = V_R$).

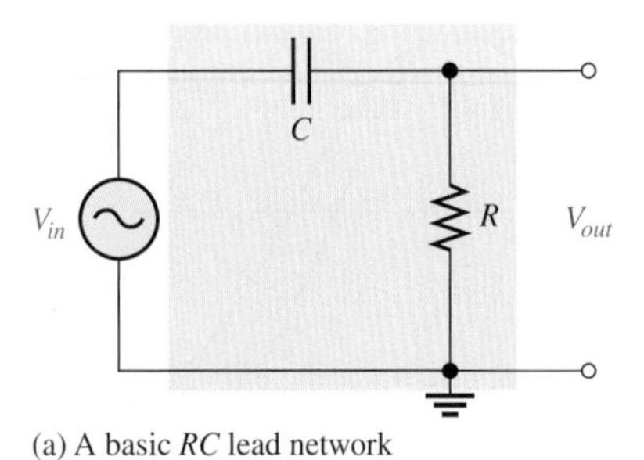

(a) A basic *RC* lead network

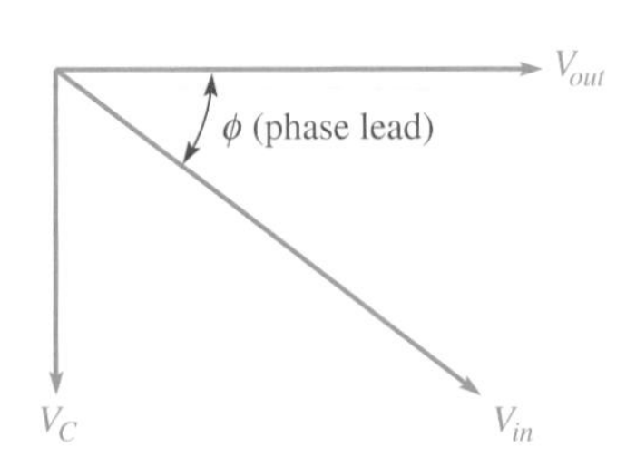

(b) Phasor voltage diagram showing the phase lead between V_{in} and V_{out}

When the input and output waveforms are displayed on an oscilloscope, a relationship similar to that in Figure 10–47 is observed. Of course, the exact amount of phase lead and the output voltage magnitude depend on the values of R and X_C. The phase lead, ϕ, is expressed as

Equation 10–29

$$\phi = \tan^{-1}\left(\frac{X_C}{R}\right)$$

The output voltage is expressed as

Equation 10–30

$$V_{out} = \left(\frac{R}{\sqrt{R^2 + X_C^2}}\right)V_{in}$$

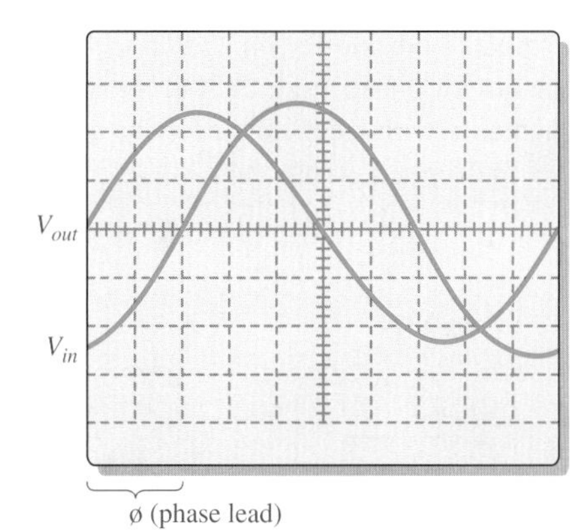

FIGURE 10–47

General oscilloscope display of the input and output waveforms of a lead network (V_{out} leads V_{in}).

EXAMPLE 10–16

Calculate the phase lead and the output voltage for the circuit in Figure 10–48.

FIGURE 10–48

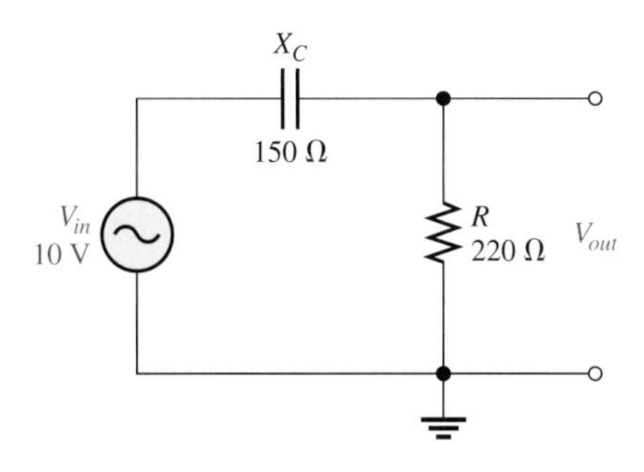

Solution The phase lead is

$$\phi = \tan^{-1}\left(\frac{X_C}{R}\right) = \tan^{-1}\left(\frac{150\ \Omega}{220\ \Omega}\right) = \mathbf{34.3^\circ}$$

The output leads the input by 34.3°.

Use Equation 10–30 to determine the output voltage.

$$V_{out} = \left(\frac{R}{\sqrt{R^2 + X_C^2}}\right)V_{in} = \left(\frac{220\ \Omega}{\sqrt{(220\ \Omega)^2 + (150\ \Omega)^2}}\right)10\text{ V} = \mathbf{8.26\ V}$$

(220/√(220²+150²))*10
8.26227342808E0

Related Problem How does an increase in R affect the phase lead and the output voltage in Figure 10–48?

FIGURE 10–49

Illustration of how the frequency affects the phase lead and the output voltage in an *RC* lead network with the amplitude of V_{in} held constant.

Effects of Frequency on the Lead Network Since the phase lead is the same as the circuit phase angle θ, it decreases as frequency increases. The output voltage increases with frequency because as X_C becomes smaller, more of the input voltage is dropped across the resistor. Figure 10–49 illustrates this relationship.

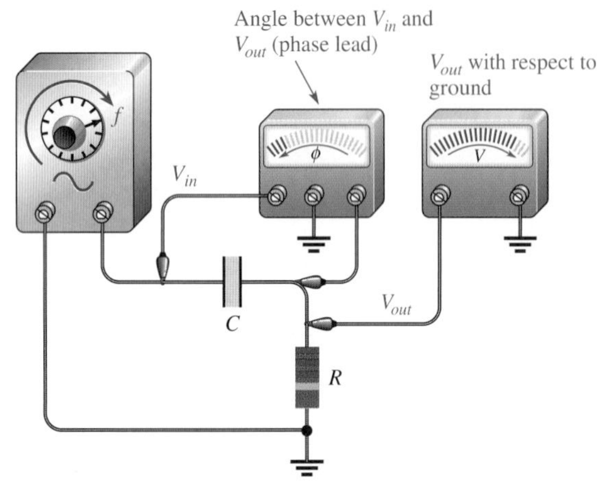

(a) When f increases, the phase lag ϕ decreases and V_{out} increases.

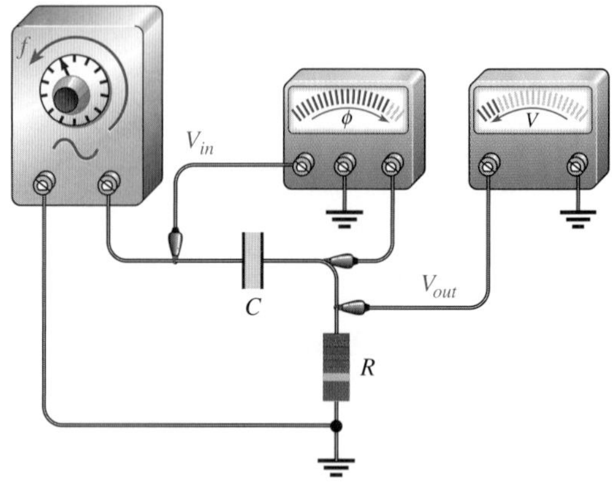

(b) When f decreases, ϕ increases and V_{out} decreases.

The *RC* Circuit as a Filter

Filters are frequency-selective circuits that permit signals of certain frequencies to pass from the input to the output while blocking all others. That is, all frequencies but the selected ones are filtered out.

Series *RC* circuits exhibit a frequency-selective characteristic and therefore act as basic filters. There are two types. The first one that we examine, called a **low-pass filter,** is realized by taking the output across the capacitor, just as in a lag network. The second type, called a **high-pass filter,** is implemented by taking the output across the resistor, as in a lead network.

Low-Pass Filter You have seen what happens to the phase angle and the output voltage in the lag network. In terms of its filtering action, the variation in the magnitude of the output voltage as a function of frequency is important.

Figure 10–50 shows the filtering action of a series *RC* circuit using a specific series of measurements in which the frequency starts at 100 Hz and is increased in increments up to 20 kHz. At each value of frequency, the output voltage is measured. As you can see, the capacitive reactance decreases as frequency increases, thus dropping less voltage across the capacitor while the input voltage is held at a constant 10 V throughout each step. Table 10–1 summarizes the variation of the circuit parameters with frequency.

TABLE 10–1

f (kHz)	X_C (Ω)	Z_{tot} (Ω)	I (mA)	V_{out} (V)
0.1	1,590	≈ 1,590	≈ 6.29	9.98
1	159	188	53.2	8.46
10	15.9	101	99.0	1.57
20	7.96	≈ 100	≈ 100	0.793

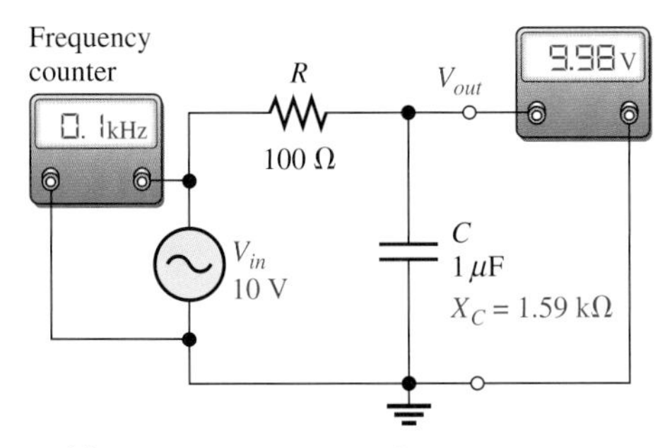

(a) $f = 0.1$ kHz, $X_C = 1.59$ kΩ, $V_{out} = 9.98$ V

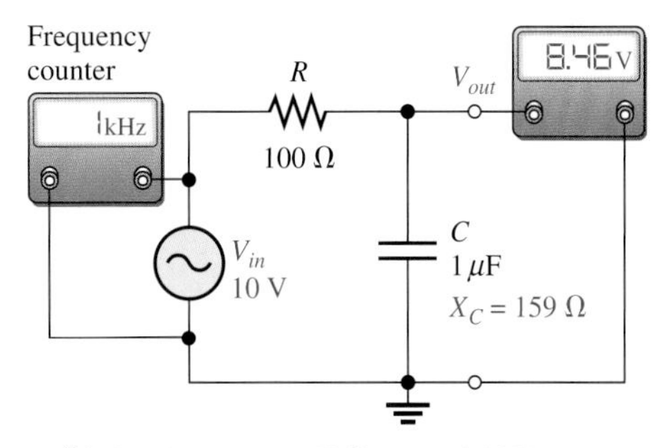

(b) $f = 1$ kHz, $X_C = 159$ Ω, $V_{out} = 8.46$ V

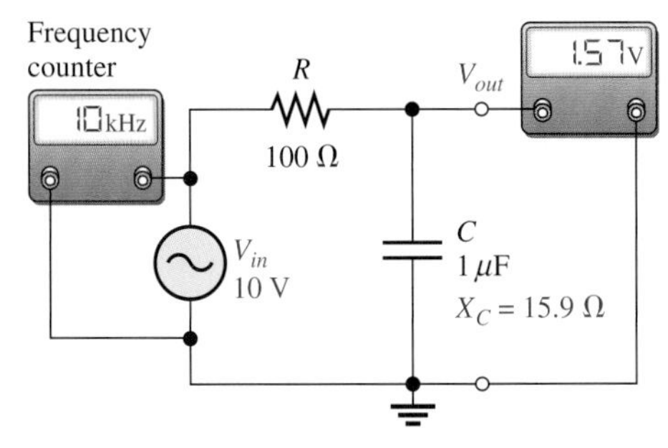

(c) $f = 10$ kHz, $X_C = 15.9$ Ω, $V_{out} = 1.57$ V

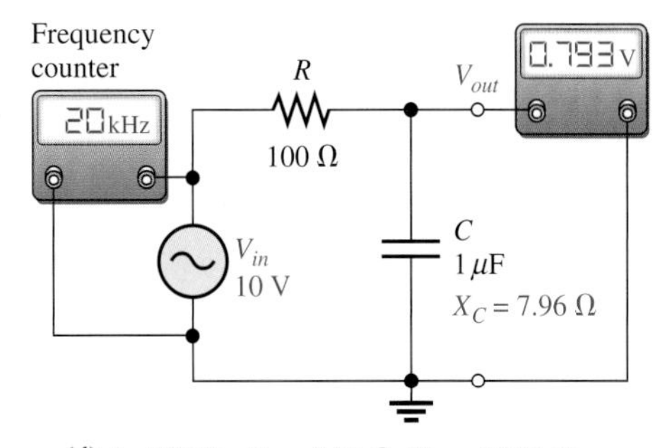

(d) $f = 20$ kHz, $X_C = 7.96$ Ω, $V_{out} = 0.793$ V

▲ **FIGURE 10–50**

Example of low-pass filtering action. As frequency increases, V_{out} decreases.

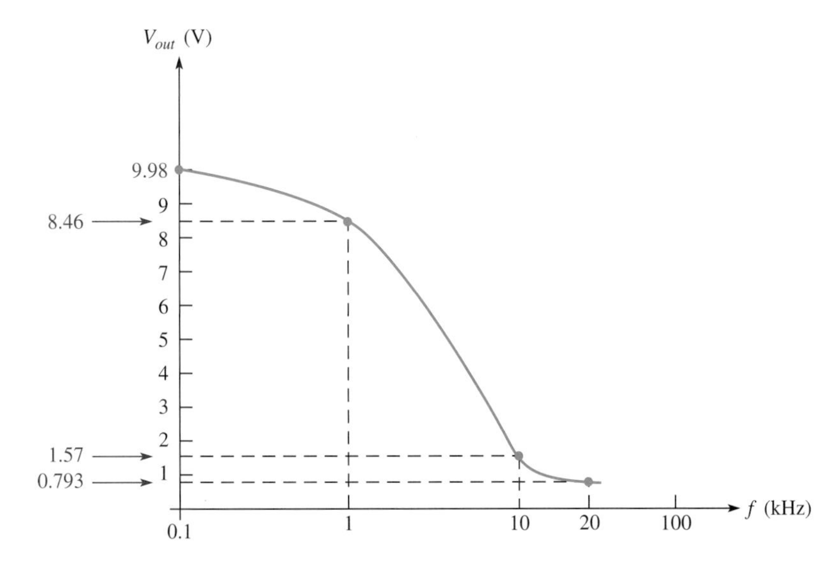

◀ **FIGURE 10–51**

Frequency response curve for the low-pass filter in Figure 10–50.

The frequency response of the low-pass filter in Figure 10–50 is shown in Figure 10–51, where the measured values are plotted on a graph of V_{out} versus f, and a smooth curve is drawn connecting the points. This graph, called a *response curve,* shows that the output voltage is greater at the lower frequencies and decreases as the frequency increases. The frequency scale is logarithmic.

High-Pass Filter To illustrate high-pass filtering action, Figure 10–52 shows a series of specific measurements. The frequency starts at 10 Hz and is increased in increments up to 10 kHz. As you can see, the capacitive reactance decreases as the frequency increases, thus causing more of the total input voltage to be dropped across the resistor. Table 10–2 summarizes the variation of circuit parameters with frequency.

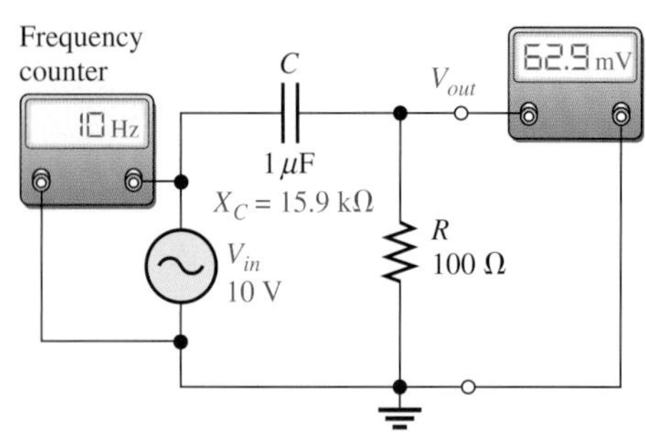

(a) $f = 10$ Hz, $X_C = 15.9$ kΩ, $V_{out} = 62.9$ mV

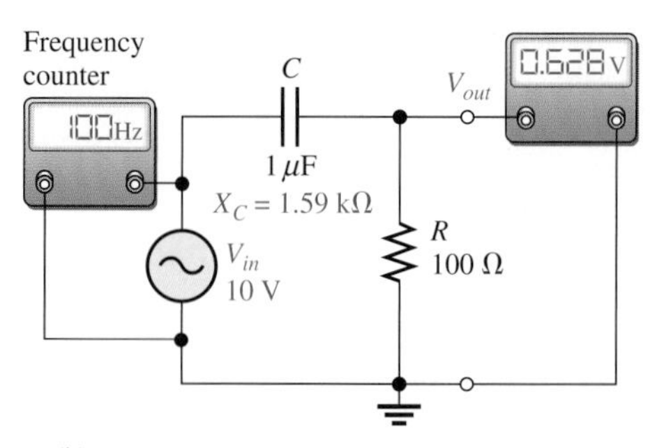

(b) $f = 100$ Hz, $X_C = 1.59$ kΩ, $V_{out} = 0.628$ V

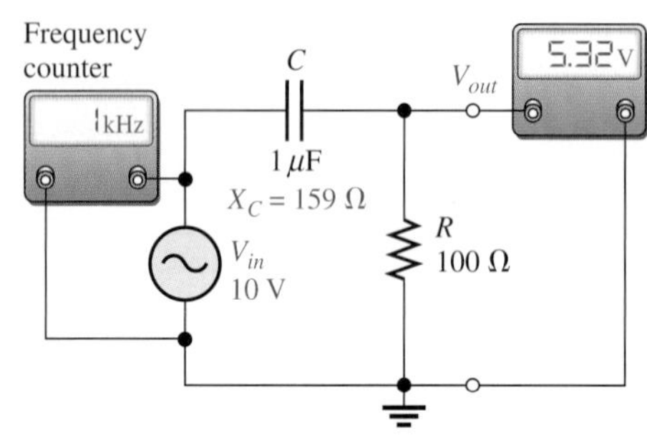

(c) $f = 1$ kHz, $X_C = 159$ Ω, $V_{out} = 5.32$ V

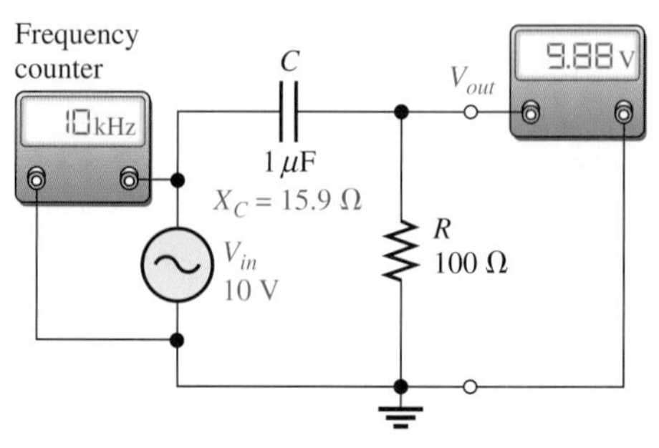

(d) $f = 10$ kHz, $X_C = 15.9$ Ω, $V_{out} = 9.88$ V

▲ **FIGURE 10–52**

Example of high-pass filtering action. As frequency increases, V_{out} increases.

▶ **TABLE 10–2**

f (kHz)	X_C (Ω)	Z_{tot} (Ω)	I (mA)	V_{out} (V)
.01	15,900	≈ 15,900	0.629	0.0629
.1	1590	1593	6.28	0.628
1	159	188	53.2	5.32
10	15.9	101	98.8	9.88

In Figure 10–53, the measured values for the high-pass filter shown in Figure 10–52 have been plotted to produce a response curve for this circuit. As you can see, the output voltage is greater at the higher frequencies and decreases as the frequency is reduced. The frequency scale is logarithmic.

The Cutoff Frequency and the Bandwidth of a Filter The frequency at which the capacitive reactance equals the resistance in a low-pass or high-pass *RC* filter is called the **cutoff frequency** and is designated f_c. This condition is expressed as $1/(2\pi f_c C) = R$. Solving for f_c results in the following formula:

Equation 10–31

$$f_c = \frac{1}{2\pi RC}$$

At f_c, the output voltage of the filter is 70.7% of its maximum value. It is standard practice to consider the cutoff frequency as the limit of a filter's performance in terms of passing or rejecting frequencies. For example, in a high-pass filter, all frequencies above f_c are considered to be passed by the filter, and all those below f_c are considered to be rejected. The reverse is true for a low-pass filter.

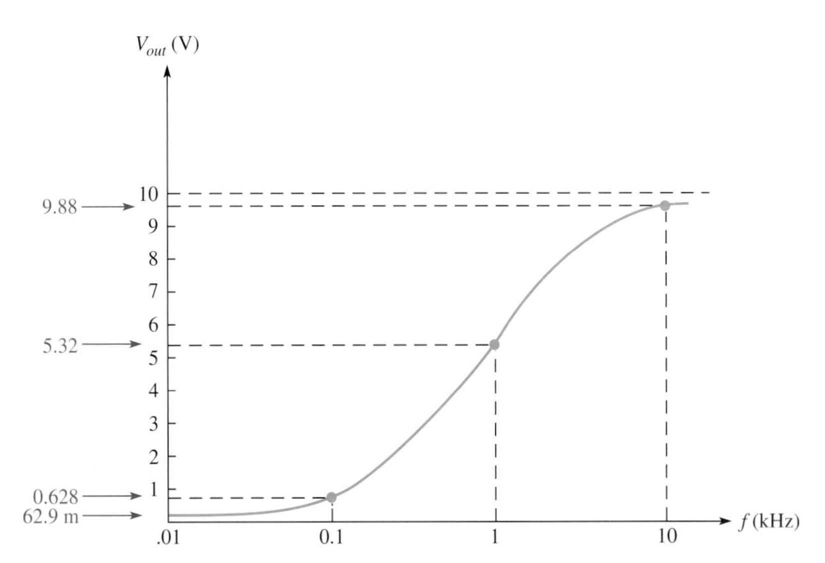

FIGURE 10–53

Frequency response curve for the high-pass filter in Figure 10–52.

The range of frequencies that is considered to be passed by a filter is called the **bandwidth**. Figure 10–54 illustrates the bandwidth and the cutoff frequency for a low-pass filter.

FIGURE 10–54

Normalized general response curve of a low-pass filter showing the cutoff frequency and the bandwidth.

Coupling an AC Signal into a DC Bias Network

Figure 10–55 shows an *RC* circuit that is used to create a dc voltage level with an ac voltage superimposed on it. This type of circuit is commonly found in amplifiers in which the dc voltage is required to **bias** the amplifier to the proper operating point and the signal voltage to be amplified is coupled through a capacitor and superimposed on the dc level. The capacitor prevents the low internal resistance of the signal source from affecting the dc bias voltage.

In this type of application, a relatively high value of capacitance is selected so that for the frequencies to be amplified, the reactance is very small compared to the resistance of the bias network. When the reactance is very small (ideally zero), there is practically no phase shift or signal voltage dropped across the capacitor. Therefore, all of the signal voltage passes from the source to the input to the amplifier.

FIGURE 10–55

Amplifier bias and signal-coupling circuit.

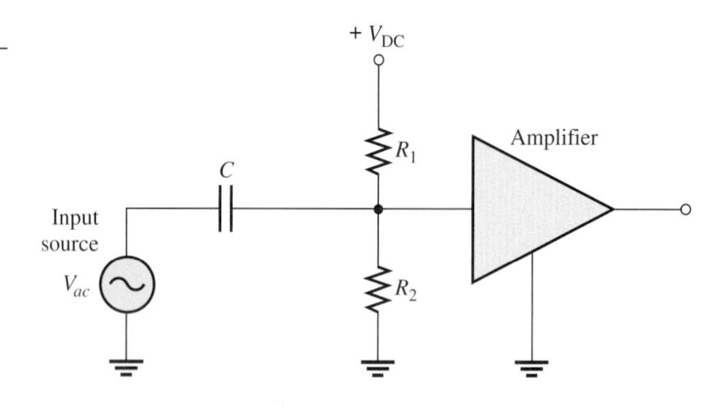

Figure 10–56 illustrates the application of the superposition principle to the circuit in Figure 10–55. In part (a), the ac source has been effectively removed from the circuit and replaced with a short to represent its ideal internal resistance. Since C is open to dc, the voltage at point A is determined by the voltage-divider action of R_1 and R_2 and the dc voltage source.

In Figure 10–56(b), the dc source has been effectively removed from the circuit and replaced with a short to represent its ideal internal resistance. Since C appears as a short at the frequency of the ac, the signal voltage is coupled directly to point A and appears across the parallel combination of R_1 and R_2. Figure 10–56(c) illustrates that the combined effect of the superposition of the dc and the ac voltages results in the signal voltage "riding" on the dc level.

FIGURE 10–56

The superposition of dc and ac voltages in an RC bias and coupling circuit.

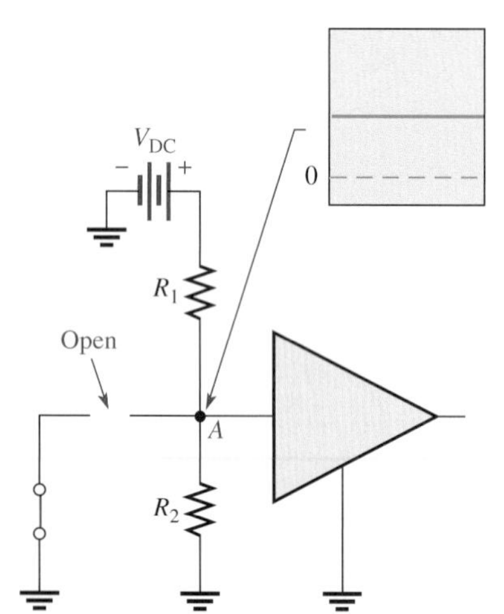

(a) dc equivalent: ac source replaced by short. C is open to dc. R_1 and R_2 act as dc voltage divider.

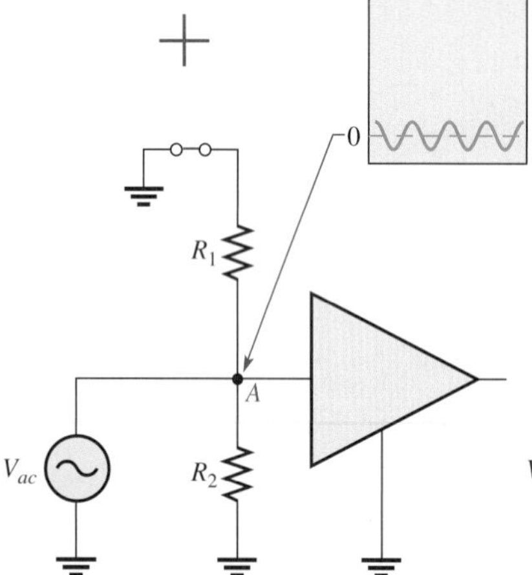

(b) ac equivalent: dc source is replaced by short. C is short to ac. All of V_{ac} is coupled to point A.

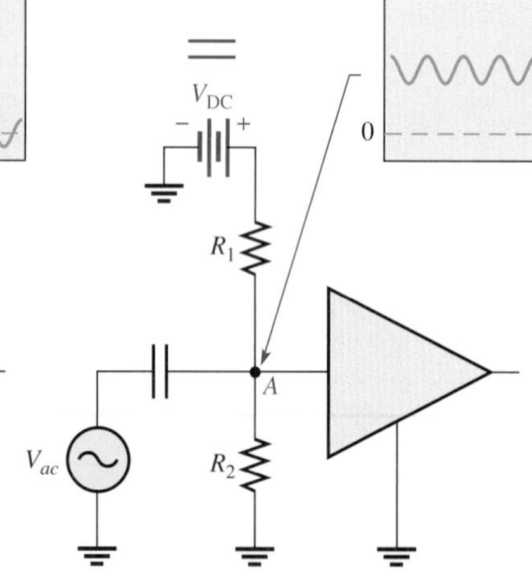

(c) dc + ac: Voltages are superimposed at point A.

SECTION 10–8 REVIEW

1. A certain RC lag network consists of a 4.7 kΩ resistor and a 0.022 μF capacitor. Determine the phase lag between the input and output voltages at a frequency of 3 kHz.
2. An RC lead network has the same component values as the lag network in Question 1. What is the magnitude of the output voltage at 3 kHz when the input is 10 V rms?
3. When an RC circuit is used as a low-pass filter, across which component is the output taken?

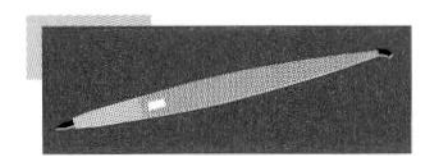

10–9 TROUBLESHOOTING

In this section, the effects that typical component failures or degradation have on the response of basic *RC* circuits are considered. An example of troubleshooting using the APM (analysis, planning, and measurement) method is also presented.

After completing this section, you should be able to

- **Troubleshoot *RC* circuits**
- Find an open resistor
- Find an open capacitor
- Find a shorted capacitor
- Find a leaky capacitor

HANDS ON TIP

Some multimeters have a relatively low frequency response of 1 kHz or less, while others can measure voltages or currents with frequencies up to about 2 MHz. Always check to make sure that your meter is capable of making accurate measurements at the particular frequency you are working with.

Effect of an Open Resistor It is easy to see how an open resistor affects the operation of a basic series *RC* circuit, as shown in Figure 10–57. Obviously, there is no path for current, so the capacitor voltage remains at zero and the total voltage, V_s, appears across the open resistor.

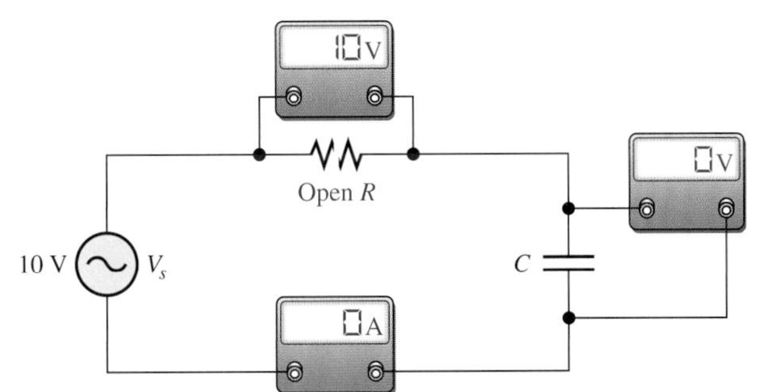

FIGURE 10–57

Effect of an open resistor.

Effect of an Open Capacitor When the capacitor is open, there is no current; thus, the resistor voltage drop is zero. The total source voltage appears across the open capacitor, as shown in Figure 10–58.

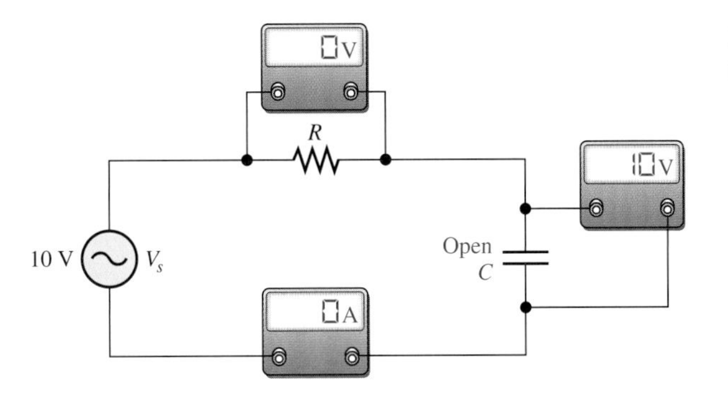

FIGURE 10–58

Effect of an open capacitor.

Effect of a Shorted Capacitor When a capacitor shorts out, the voltage across it is zero, the current equals V_s/R, and the total voltage appears across the resistor, as shown in Figure 10–59.

▶ **FIGURE 10–59**

Effect of a shorted capacitor.

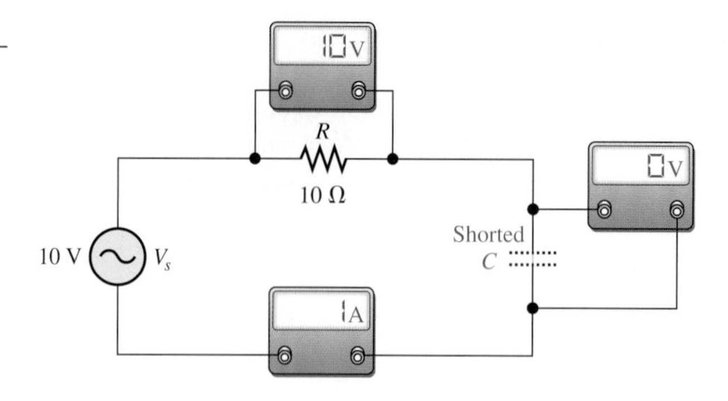

Effect of a Leaky Capacitor When a capacitor exhibits a high leakage current, the leakage resistance effectively appears in parallel with the capacitor, as shown in Figure 10–60(a). When the leakage resistance is comparable in value to the circuit resistance, R, the circuit response is drastically affected. The circuit, looking from the capacitor toward the source, can be thevenized, as shown in Figure 10–60(b). The Thevenin equivalent resistance is R in parallel with R_{leak} (the source appears as a short), and the Thevenin equivalent voltage is determined by the voltage-divider action of R and R_{leak}.

$$R_{th} = R \parallel R_{leak} = \frac{RR_{leak}}{R + R_{leak}}$$

$$V_{th} = \left(\frac{R_{leak}}{R + R_{leak}}\right)V_{in}$$

As you can see, the voltage to which the capacitor will charge is reduced since $V_{th} < V_{in}$. Also, the current is increased. The Thevenin equivalent circuit is shown in Figure 10–60(c).

▶ **FIGURE 10–60**

Effect of a leaky capacitor.

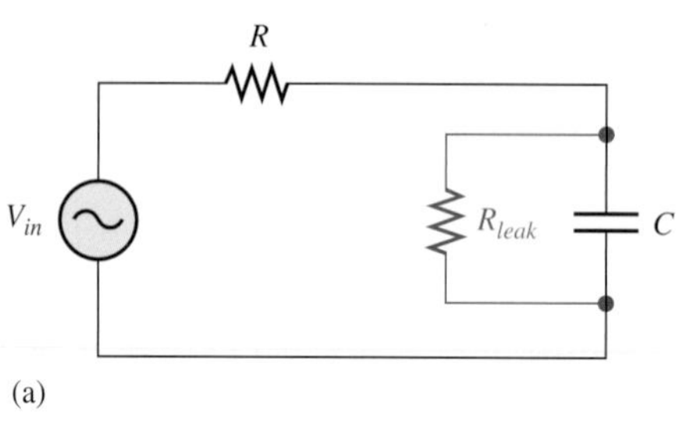

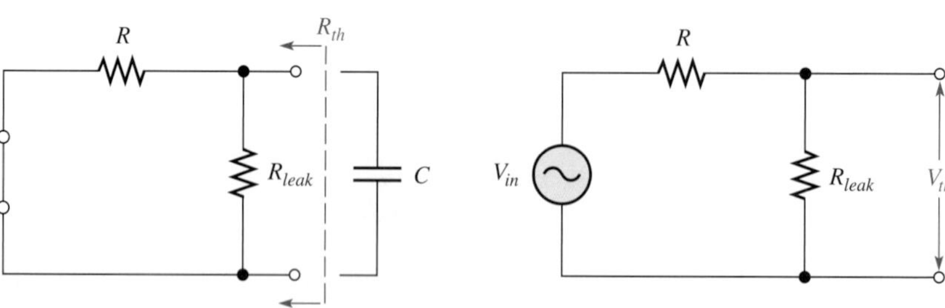

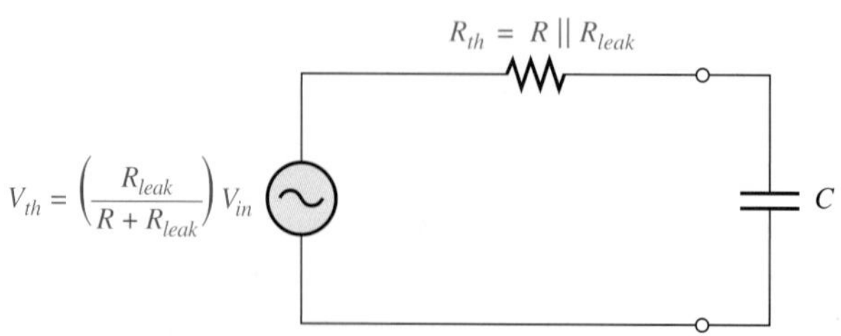

EXAMPLE 10–17

Assume that the capacitor in Figure 10–61 is degraded to a point where its leakage resistance is 10 kΩ. Determine the phase shift from input to output and the output voltage under the degraded condition.

FIGURE 10–61

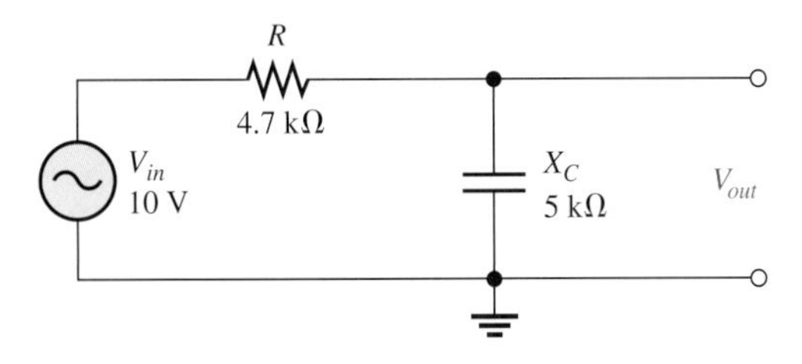

Solution The effective circuit resistance is

$$R_{th} = \frac{RR_{leak}}{R + R_{leak}} = \frac{(4.7\text{ k}\Omega)(10\text{ k}\Omega)}{14.7\text{ k}\Omega} = 3.2\text{ k}\Omega$$

The phase lag is

$$\phi = 90° - \tan^{-1}\left(\frac{X_C}{R_{th}}\right) = 90° - \tan^{-1}\left(\frac{5\text{ k}\Omega}{3.2\text{ k}\Omega}\right) = \mathbf{32.6°}$$

To determine the output voltage, first calculate the Thevenin equivalent voltage.

$$V_{th} = \left(\frac{R_{leak}}{R + R_{leak}}\right)V_{in} = \left(\frac{10\text{ k}\Omega}{14.7\text{ k}\Omega}\right)10\text{ V} = 6.80\text{ V}$$

$$V_{out} = \left(\frac{X_C}{\sqrt{R_{th}^2 + X_C^2}}\right)V_{th} = \left(\frac{5\text{ k}\Omega}{\sqrt{(3.2\text{ k}\Omega)^2 + (5\text{ k}\Omega)^2}}\right)6.8\text{ V} = \mathbf{5.73\text{ V}}$$

Related Problem What would the output voltage be if the capacitor were not leaky?

Other Troubleshooting Considerations

So far, you have learned about specific component failures and the associated voltage measurements. Many times, however, the failure of a circuit to work properly is not the result of a faulty component. A loose wire, a bad contact, or a poor solder joint can cause an open circuit. A short can be caused by a wire clipping or solder splash. Things as simple as not plugging in a power supply or a function generator happen more often than you might think. Wrong values in a circuit, such as an incorrect resistor value, the function generator set at the wrong frequency, or the wrong output connected to the circuit, can cause improper operation.

When you have problems with a circuit, always check to make sure that the instruments are properly connected to the circuits and to a power outlet. Also, look for obvious things such as a broken or loose contact, a connector that is not completely plugged in, or a piece of wire or a solder bridge that could be shorting something out.

The point is that you should consider all possibilities, not just faulty components, when a circuit is not working properly. The following example illustrates this approach with a simple circuit using the APM (analysis, planning, and measurement) method.

HANDS ON TIP

When connecting protoboard circuits, always consistently use standard colors for universal connections such as signals, power supply voltage, and ground. For example, you can use green wires for signals, red wires for power supply voltage, and black wires for ground connections. This helps you identify wiring during connecting and troubleshooting.

EXAMPLE 10–18

The circuit represented by the schematic in Figure 10–62 has no output voltage, which is the voltage across the capacitor. You expect to see about 7.4 V at the output. The circuit is physically constructed on a protoboard. Use your troubleshooting skills to find the problem.

▶ FIGURE 10–62

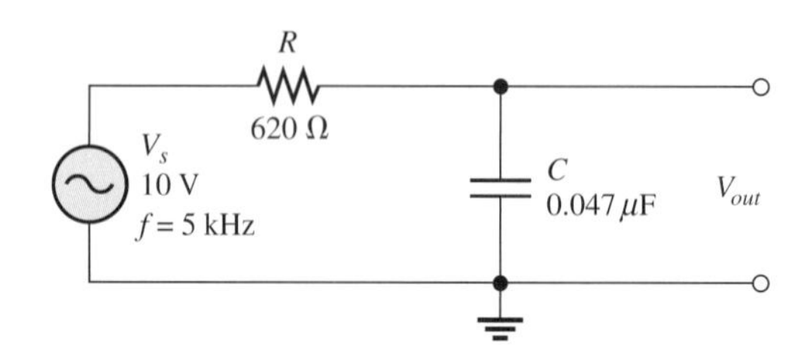

Solution Apply the APM method to this troubleshooting problem.

Analysis: First think of the possible causes for the circuit to have no output voltage.

1. There is no source voltage or the frequency is so high that the capacitor appears to be a short because its reactance is almost zero.
2. There is a short between the output terminals. Either the capacitor could be shorted, or there could be some physical short.
3. There is an open between the source and the output. This would prevent current and thus cause the output voltage to be zero. The resistor could be open, or the conductive path could be open due to a broken or loose connecting wire or a bad protoboard contact.
4. There is an incorrect component value. The resistor could be so large that the current and, therefore, the output voltage are negligible. The capacitor could be so large that its reactance at the input frequency is near zero.

Planning: You decide to make some visual checks for problems such as the function generator power cord not plugged in or the frequency set at an incorrect value. Also, broken leads, shorted leads, as well as an incorrect resistor color code or capacitor label often can be found visually. If nothing is discovered after a visual check, then you will make voltage measurements to track down the cause of the problem. You decide to use a digital oscilloscope and a multimeter to make the measurements.

Measurement: Assume that you find that the function generator is plugged in and the frequency setting appears to be correct. Also, you find no visible opens or shorts during your visual check, and the component values are correct.

The first step in the measurement process is to check the voltage from the source with the scope. Assume a 10 V rms sine wave with a frequency of 5 kHz is observed at the circuit input as shown in Figure 10–63(a). The correct voltage is present, so *the first possible cause has been eliminated.*

Next, check for a shorted capacitor by disconnecting the source and placing a multimeter (set on the ohmmeter function) across the capacitor. If the capacitor is good, an open will be indicated by an OL (overload) in the meter display after a short charging time. Assume the capacitor checks okay, as shown in Figure 10–63(b). *The second possible cause has been eliminated.*

Since the voltage has been "lost" somewhere between the input and the output, you must now look for the voltage. You reconnect the source and measure the voltage across the resistor with the multimeter (set on the voltmeter function) from one resistor lead to the other. The voltage across the resistor is zero. This means there is no current, which indicates an open somewhere in the circuit.

FIGURE 10–63

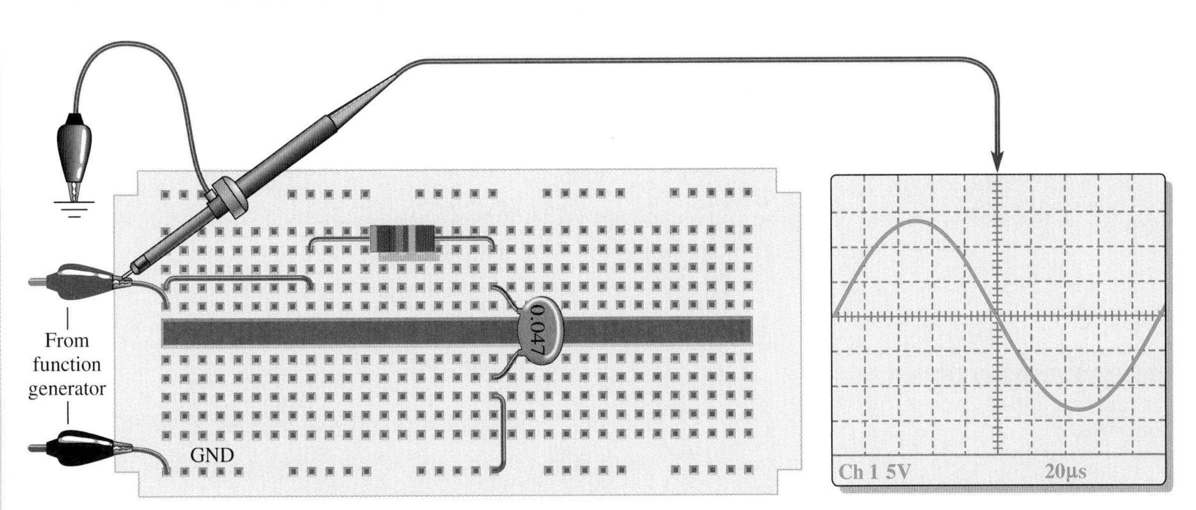

(a) Scope shows the correct voltage at the input.

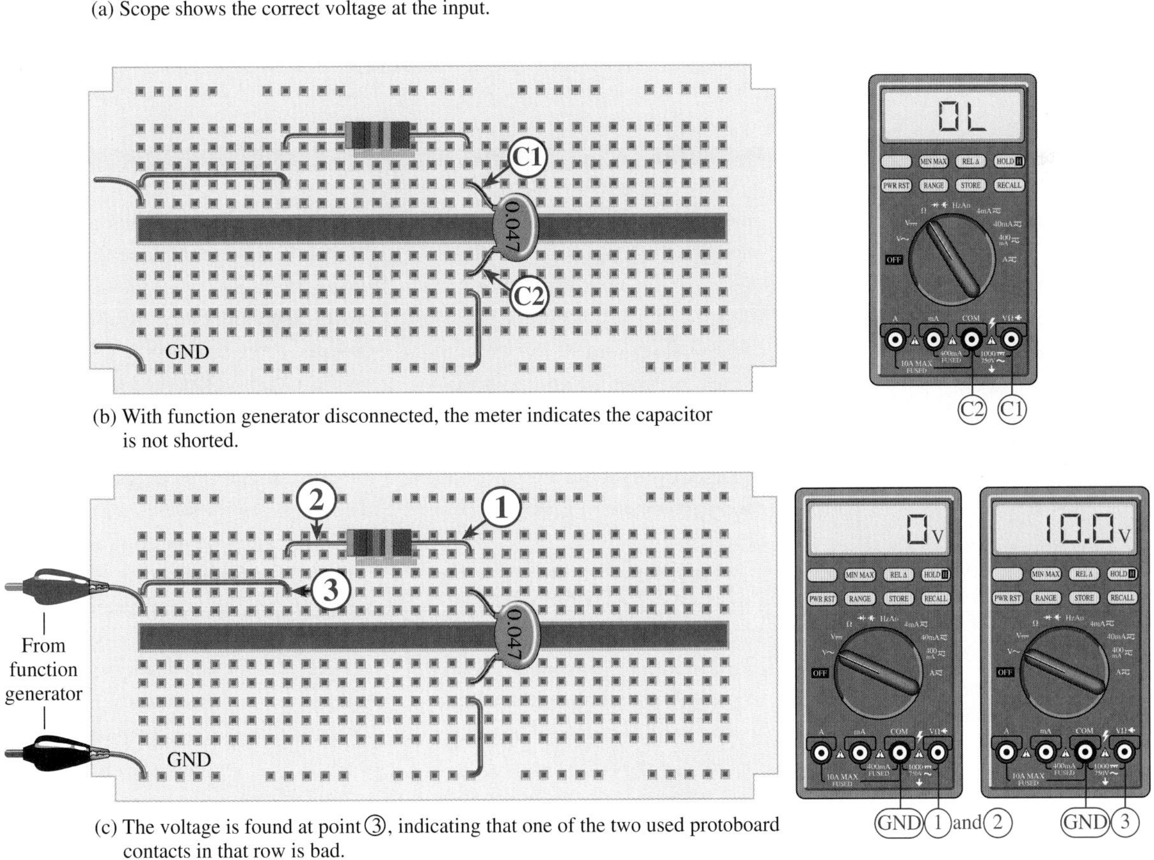

(b) With function generator disconnected, the meter indicates the capacitor is not shorted.

(c) The voltage is found at point ③, indicating that one of the two used protoboard contacts in that row is bad.

Now, you begin tracing the circuit back toward the source looking for the voltage (you could also start from the source and work forward). You can use either the scope or the multimeter, but decide to use the multimeter with one lead connected to ground and the other used to probe the circuit. As shown in Figure 10–63(c), the voltage on the right lead of the resistor, point ①, reads zero. Since you already have measured zero voltage across the resistor, the voltage on the left resistor lead at point ② must be zero as the meter indicates. Next, moving the meter probe to point ③, you read 10 V. You have found the voltage! Since there is zero volts on the left resistor lead, and there is 10 V at point ③, one of the two contacts in the protoboard hole into which the wire leads are inserted is bad. It could be that the small contacts were pushed in too far and were bent or broken so that the circuit lead does not make contact.

Move either or both the resistor lead and the wire to another hole in the same row. Assume that when the resistor lead is moved to the hole just above, you have voltage at the output of the circuit (across the capacitor).

Related Problem Suppose you had measured 10 V across the resistor before the capacitor was checked. What would this have indicated?

SECTION 10–9 REVIEW

1. Describe the effect of a leaky capacitor on the dc response of an *RC* circuit.
2. In a series *RC* circuit, if all the applied voltage appears across the capacitor, what is the problem?
3. What faults can cause 0 V across the capacitor in a series *RC* circuit?

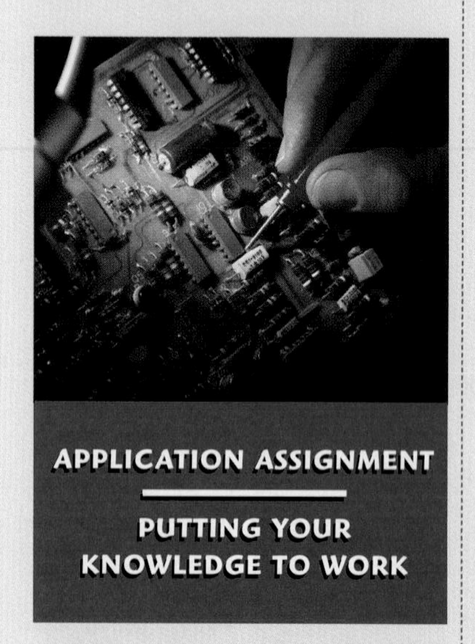

APPLICATION ASSIGNMENT

PUTTING YOUR KNOWLEDGE TO WORK

In Chapter 9, you worked with the capacitively coupled input to an amplifier with voltage-divider bias. In this assignment, you will check the voltage of a similar amplifier's input circuit to determine how it changes with the frequency of the input signal. If too much voltage is dropped across the capacitor, the overall performance of the amplifier is adversely affected. To do this assignment, you do not have to be familiar with the details of the amplifier circuit but you should review the application assignment in Chapter 9 before proceeding.

As you learned in Chapter 9, the coupling capacitor (C_1) in Figure 10–64 passes the input signal voltage to the input of the amplifier (point *A* to point *B*) without affecting the dc level at point *B* produced by the resistive voltage divider (R_1 and R_2). If the input frequency is high enough so that the reactance of the coupling capacitor is negligibly small, essentially no ac signal voltage is dropped across the capacitor. As the signal frequency is reduced, the capacitive reactance increases and more of the signal voltage is dropped across the capacitor. This lowers the output voltage of the amplifier.

The amount of signal voltage that is coupled from the input source (point *A*) to the amplifier input (point *B*) is determined by the values of the capacitor and the dc bias resistors (assuming the amplifier has no loading effect) in Figure 10–64. These components actually form a high-pass *RC* filter, as shown in Figure 10–65. The voltage-divider bias resistors are effectively in parallel with each other as far as the ac source is concerned. The lower end of R_2 goes to ground and the upper end of R_1 goes to the dc supply voltage as shown in Figure 10–65(a). Since there is no ac

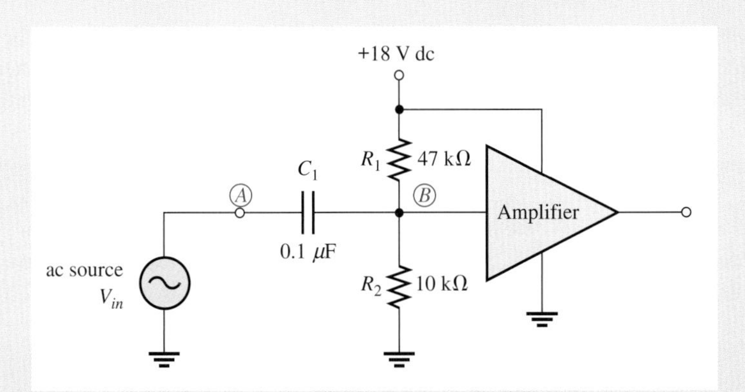

FIGURE 10–64

A capacitively coupled amplifier.

voltage at the +18 V dc terminal, the upper end of R_1 is at 0 V ac which is referred to as *ac ground.* The development of the circuit into an effective high-pass *RC* filter is shown in parts (b) and (c).

Step 1: Evaluate the Amplifier Input Circuit

Determine the value of the equivalent resistance of the input circuit. Assume the amplifier (shown inside the white dashed lines in Figure 10–66) has no loading effect on the input circuit.

Step 2: Measure the Response at Frequency f_1

Refer to Figure 10–66. The input signal voltage is applied to the amplifier circuit board and displayed on channel 1 of the oscilloscope, and channel 2 is connected to a point on the circuit

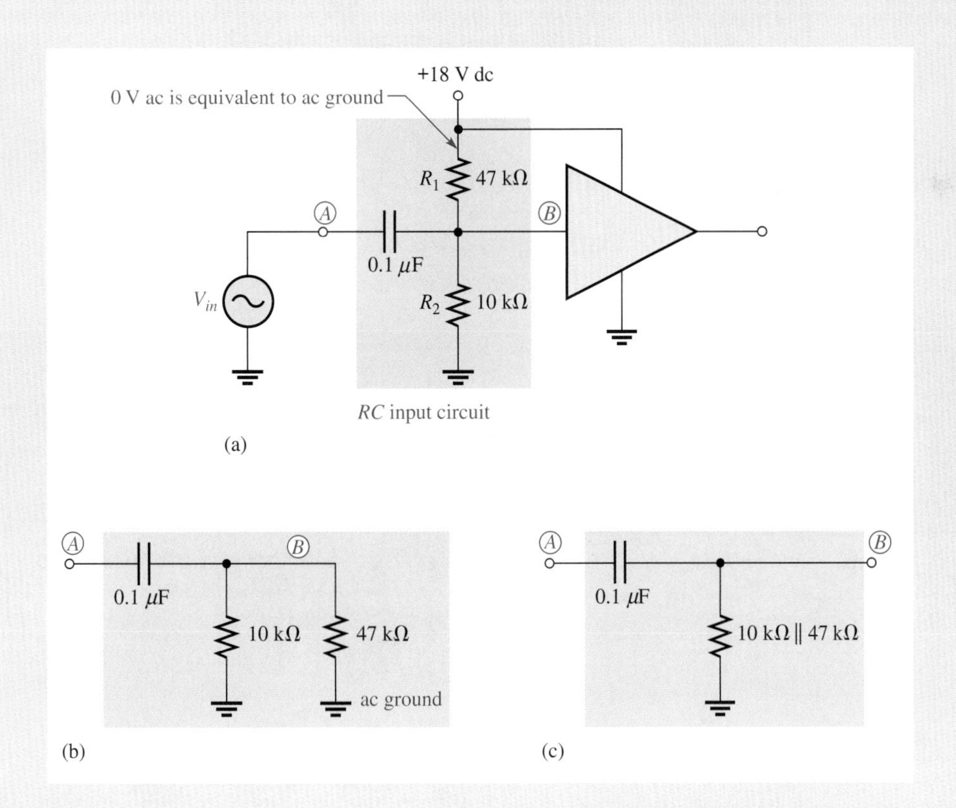

FIGURE 10–65

The *RC* input circuit acts effectively like a high-pass *RC* filter.

board. Determine to what point on the circuit the channel 2 probe is connected, the frequency, and the voltage that should be displayed.

Step 3: Measure the Response at Frequency f_2

Refer to Figure 10–67 and the circuit board in Figure 10–66. The input signal voltage displayed on channel 1 of the oscilloscope is applied to the amplifier circuit board. Determine the frequency and the voltage that should be displayed on channel 2.

State the difference between the channel 2 waveforms determined in Step 2 and Step 3. Explain the reason for the difference.

Step 4: Measure the Response at Frequency f_3

Refer to Figure 10–68 and the circuit board in Figure 10–66. The input signal voltage displayed on channel 1 of the oscilloscope is applied to the amplifier circuit board. Determine the frequency and the voltage that should be displayed on channel 2.

State the difference between the channel 2 waveforms determined in Step 2 and Step 4. Explain the reason for the difference.

Step 5: Plot a Response Curve for the Amplifier Input Circuit

Determine the frequency at which the signal voltage at point B in Figure

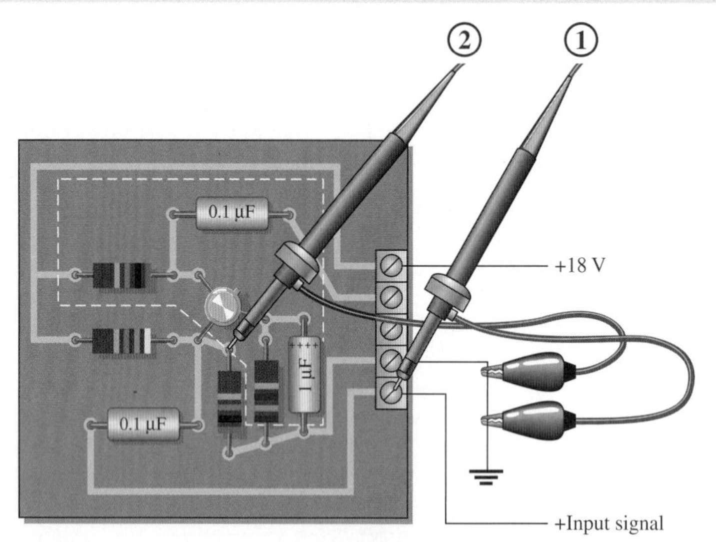

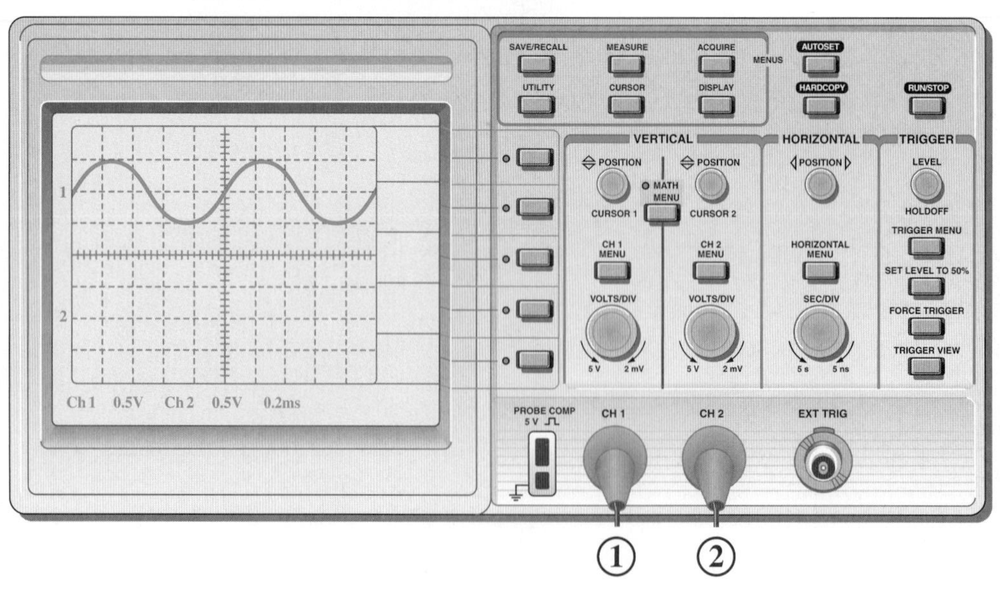

FIGURE 10–66

Measuring the input circuit response at frequency f_1. Circled numbers relate scope inputs to the probes. The channel 1 waveform is shown.

10–64 is 70.7% of its maximum value. Plot the response curve using this voltage value and the values at frequencies f_1, f_2, and f_3. How does this curve show that the input circuit acts as a high-pass filter? What can you do to the circuit to lower the frequency at which the voltage is 70.7% of maximum without affecting the dc bias voltage?

FIGURE 10–67

Measuring the input circuit response at frequency f_2. The channel 1 waveform is shown.

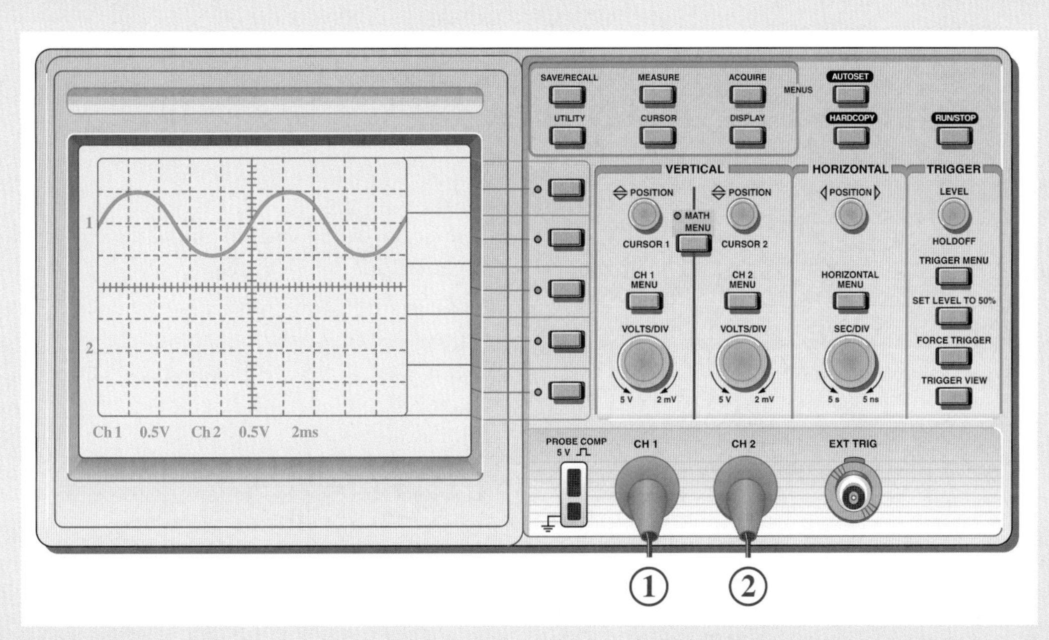

FIGURE 10–68

Measuring the input circuit response at frequency f_3. The channel 1 waveform is shown.

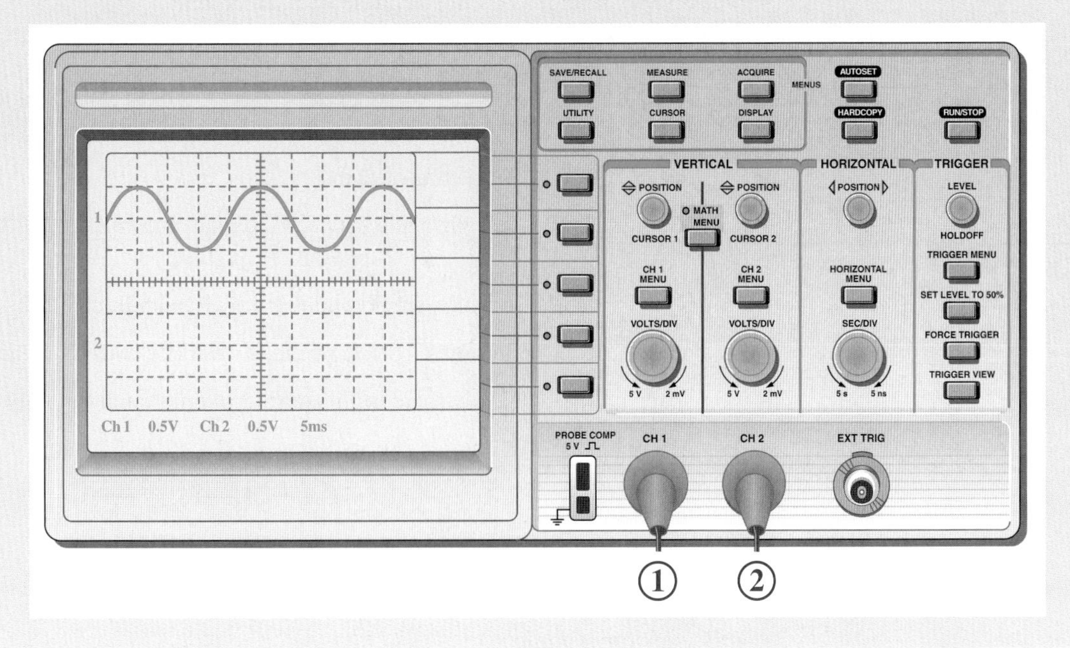

Electronics Workbench/CircuitMaker Analysis

Using Electronics Workbench or CircuitMaker, connect the equivalent circuit in Figure 10–65(b).

1. Apply an input signal voltage at the same frequency and amplitude as shown in Figure 10–66. Measure the voltage at point B with the oscilloscope and compare the result from Step 2.
2. Apply an input signal voltage at the same frequency and amplitude as shown in Figure 10–67. Measure the voltage at point B with the oscilloscope and compare the result from Step 3.
3. Apply an input signal voltage at the same frequency and amplitude as shown in Figure 10–68. Measure the voltage at point B with the oscilloscope and compare the result from Step 4.

APPLICATION ASSIGNMENT REVIEW

1. Explain the effect on the response of the amplifier input circuit of reducing the value of the coupling capacitor.
2. What is the voltage at point *B* in Figure 10–64 if the coupling capacitor opens when the ac input signal is 10 mV rms?
3. What is the voltage at point *B* in Figure 10–64 if resistor R_1 is open when the ac input signal is 10 mV rms?

SUMMARY

- When a sinusoidal voltage is applied to an *RC* circuit, the current and all the voltage drops are also sine waves.
- Total current in a series or parallel *RC* circuit always leads the source voltage.
- The resistor voltage is always in phase with the current.
- The capacitor voltage always lags the current by 90°.
- In an *RC* circuit, the impedance is determined by both the resistance and the capacitive reactance combined.
- Impedance is expressed in units of ohms.
- The circuit phase angle is the angle between the total current and the source voltage.
- The impedance of a series *RC* circuit varies inversely with frequency.
- The phase angle (θ) of a series *RC* circuit varies inversely with frequency.
- For each parallel *RC* circuit, there is an equivalent series circuit for any given frequency.
- The impedance of a circuit can be determined by measuring the source voltage and the total current and then applying Ohm's law.
- In an *RC* circuit, part of the power is resistive and part reactive.
- The phasor combination of resistive power (true power) and reactive power is called *apparent power.*

- Apparent power is expressed in volt-amperes (VA).
- The power factor (*PF*) indicates how much of the apparent power is true power.
- A power factor of 1 indicates a purely resistive circuit, and a power factor of 0 indicates a purely reactive circuit.
- In an *RC* lag network, the output voltage lags the input voltage in phase.
- In an *RC* lead network, the output voltage leads the input voltage.
- A filter passes certain frequencies and rejects others.

EQUATIONS

10–1	$Z = \sqrt{R^2 + X_C^2}$	Series *RC* impedance
10–2	$\theta = \tan^{-1}\left(\frac{X_C}{R}\right)$	Series *RC* phase angle
10–3	$V = IZ$	Ohm's law
10–4	$I = \frac{V}{Z}$	Ohm's law
10–5	$Z = \frac{V}{I}$	Ohm's law
10–6	$V_s = \sqrt{V_R^2 + V_C^2}$	Total voltage in series *RC* circuits
10–7	$\theta = \tan^{-1}\left(\frac{V_C}{V_R}\right)$	Series *RC* phase angle
10–8	$Z = \frac{RX_C}{\sqrt{R^2 + X_C^2}}$	Parallel *RC* impedance
10–9	$\theta = \tan^{-1}\left(\frac{R}{X_C}\right)$	Parallel *RC* phase angle
10–10	$G = \frac{1}{R}$	Conductance
10–11	$B_C = \frac{1}{X_C}$	Capacitive susceptance
10–12	$Y = \frac{1}{Z}$	Admittance
10–13	$Y_{tot} = \sqrt{G^2 + B_C^2}$	Total admittance
10–14	$V = \frac{I}{Y}$	Ohm's law
10–15	$I = VY$	Ohm's law
10–16	$Y = \frac{I}{V}$	Ohm's law
10–17	$I_{tot} = \sqrt{I_R^2 + I_C^2}$	Total current in parallel *RC* circuits
10–18	$\theta = \tan^{-1}\left(\frac{I_C}{I_R}\right)$	Parallel *RC* phase angle
10–19	$R_{eq} = Z \cos \theta$	Equivalent series resistance
10–20	$X_{C(eq)} = Z \sin \theta$	Equivalent series reactance

10–21	$\theta = \left(\frac{\Delta t}{T}\right)360°$	Phase angle using time measurements
10–22	$P_{true} = I_{tot}^2 R$	True power (W)
10–23	$P_r = I_{tot}^2 X_C$	Reactive power (VAR)
10–24	$P_a = I_{tot}^2 Z$	Apparent power (VA)
10–25	$P_{true} = V_s I_{tot} \cos\theta$	True power
10–26	$PF = \cos\theta$	Power factor
10–27	$\phi = 90° - \tan^{-1}\left(\frac{X_C}{R}\right)$	Phase angle of lag network
10–28	$V_{out} = \left(\frac{X_C}{\sqrt{R^2 + X_C^2}}\right)V_{in}$	Output voltage of lag network
10–29	$\phi = \tan^{-1}\left(\frac{X_C}{R}\right)$	Phase angle of lead network
10–30	$V_{out} = \left(\frac{R}{\sqrt{R^2 + X_C^2}}\right)V_{in}$	Output voltage of lead network
10–31	$f_c = \frac{1}{2\pi RC}$	Cutoff frequency of an RC filter

SELF-TEST

Answers are at the end of the chapter.

1. In a series RC circuit, the voltage across the resistance is
(a) in phase with the source voltage
(b) lagging the source voltage by 90°
(c) in phase with the current
(d) lagging the current by 90°

2. In a series RC circuit, the voltage across the capacitor is
(a) in phase with the source voltage
(b) lagging the resistor voltage by 90°
(c) in phase with the current
(d) lagging the source voltage by 90°

3. When the frequency of the voltage applied to a series RC circuit is increased, the impedance
(a) increases **(b)** decreases **(c)** remains the same **(d)** doubles

4. When the frequency of the voltage applied to a series RC circuit is decreased, the phase angle
(a) increases **(b)** decreases
(c) remains the same **(d)** becomes erratic

5. In a series RC circuit when the frequency and the resistance are doubled, the impedance
(a) doubles **(b)** is halved
(c) is quadrupled **(d)** cannot be determined without values

6. In a series RC circuit, 10 V rms is measured across the resistor and 10 V rms is also measured across the capacitor. The rms source voltage is
(a) 20 V **(b)** 14.14 V **(c)** 28.28 V **(d)** 10 V

7. The voltages in Problem 6 are measured at a certain frequency. To make the resistor voltage greater than the capacitor voltage, the frequency
   (a) must be increased **(b)** must be decreased
   (c) is held constant **(d)** has no effect
8. When $R = X_C$, the phase angle is
   (a) 0° **(b)** +90° **(c)** −90° **(d)** 45°
9. To decrease the phase angle below 45°, the following condition must exist:
   (a) $R = X_C$ **(b)** $R < X_C$ **(c)** $R > X_C$ **(d)** $R = 10X_C$
10. When the frequency of the source voltage is increased, the impedance of a parallel *RC* circuit
    (a) increases **(b)** decreases **(c)** does not change
11. In a parallel *RC* circuit, there is 1 A rms through the resistive branch and 1 A rms through the capacitive branch. The total rms current is
    (a) 1 A **(b)** 2 A **(c)** 2.28 A **(d)** 1.414 A
12. A power factor of 1 indicates that the circuit phase angle is
    (a) 90° **(b)** 45° **(c)** 180° **(d)** 0°
13. For a certain load, the true power is 100 W and the reactive power is 100 VAR. The apparent power is
    (a) 200 VA **(b)** 100 VA **(c)** 141.4 VA **(d)** 141.4 W
14. Energy sources are normally rated in
    (a) watts **(b)** volt-amperes
    (c) volt-amperes reactive **(d)** none of these
15. If the bandwidth of a certain low-pass filter is 1 kHz, the cutoff frequency is
    (a) 0 Hz **(b)** 500 Hz **(c)** 2 kHz **(d)** 1000 Hz

PROBLEMS

Answers to odd-numbered problems are at the end of the book.

BASIC PROBLEMS

SECTION 10–1 Sinusoidal Response of *RC* Circuits

1. An 8 kHz sinusoidal voltage is applied to a series *RC* circuit. What is the frequency of the voltage across the resistor? Across the capacitor?
2. What is the waveshape of the current in the circuit of Problem 1?

SECTION 10–2 Impedance and Phase Angle of Series *RC* Circuits

3. Find the impedance of each circuit in Figure 10–69.

FIGURE 10–69

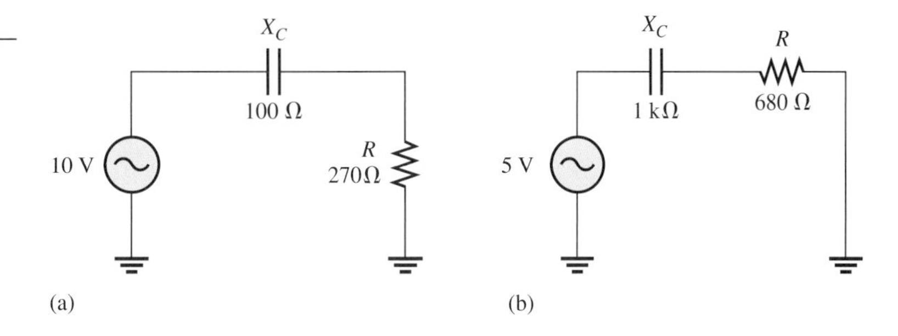

4. Determine the impedance and the phase angle in each circuit in Figure 10–70.

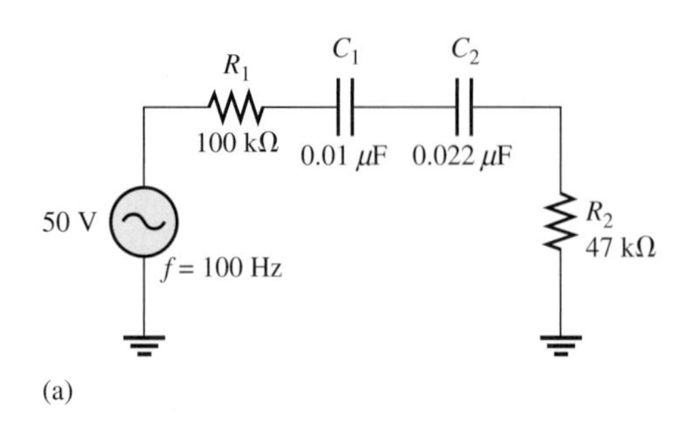

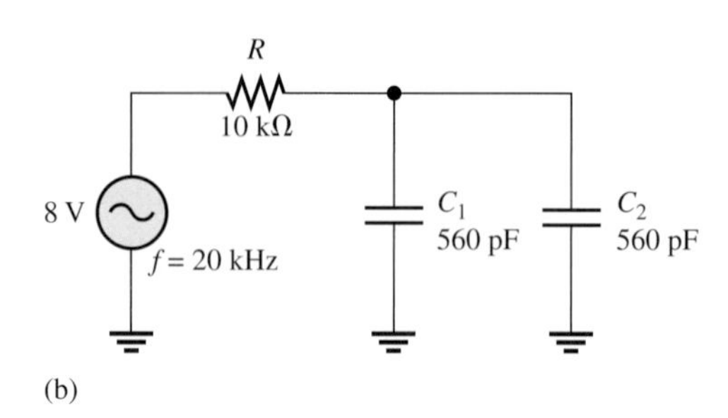

▲ FIGURE 10–70

5. For the circuit of Figure 10–71, determine the impedance for each of the following frequencies:

(a) 100 Hz **(b)** 500 Hz **(c)** 1 kHz **(d)** 2.5 kHz

▶ FIGURE 10–71

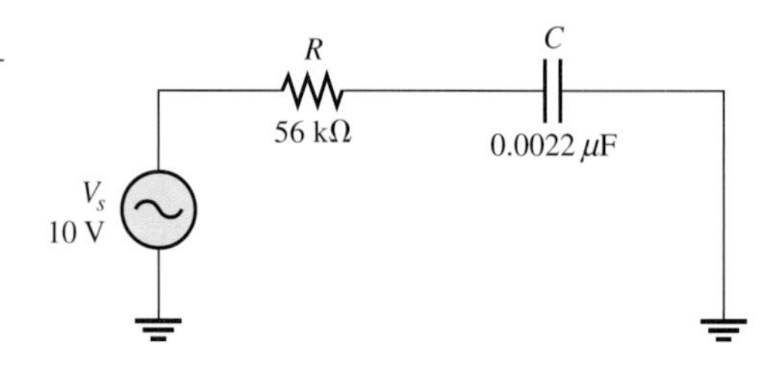

6. Repeat Problem 5 for $C = 0.0047\ \mu F$.

SECTION 10–3 **Analysis of Series *RC* Circuits**

7. Calculate the total current in each circuit of Figure 10–69.

8. Repeat Problem 7 for the circuits in Figure 10–70.

9. For the circuit in Figure 10–72, draw the phasor diagram showing all voltages and the total current. Indicate the phase angles.

10. For the circuit in Figure 10–73, determine the following:

(a) Z **(b)** I **(c)** V_R **(d)** V_C

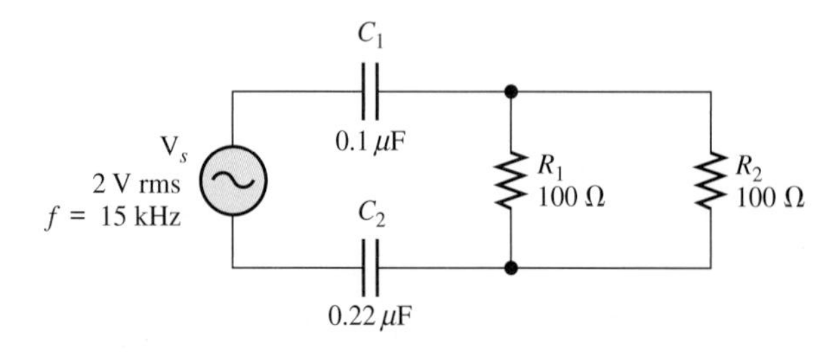

▲ FIGURE 10–72

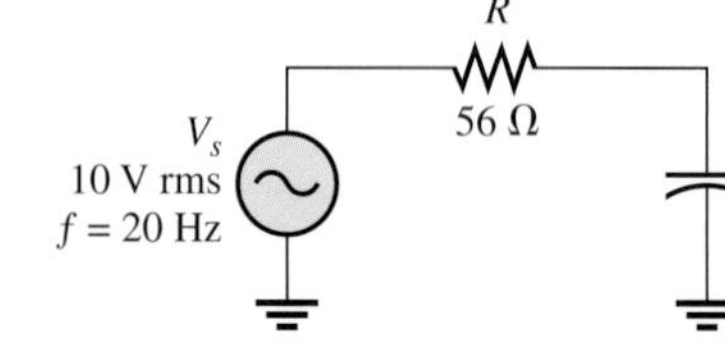

▲ FIGURE 10–73

11. To what value must the rheostat be set in Figure 10–74 to make the total current 10 mA? What is the resulting phase angle?

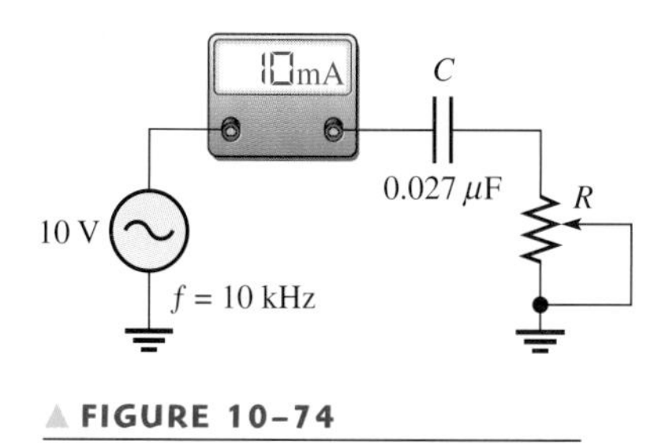

FIGURE 10–74

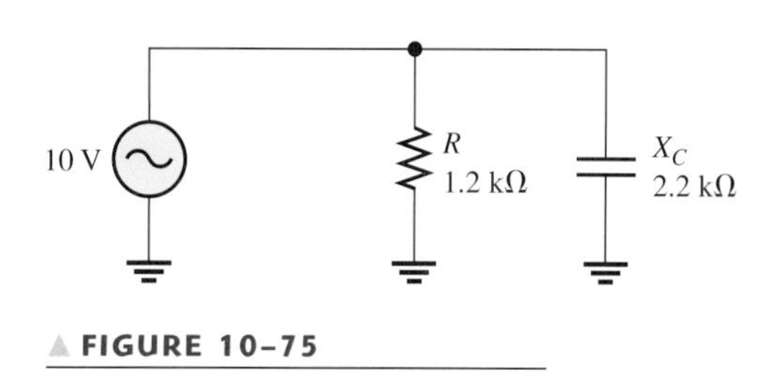

FIGURE 10–75

SECTION 10–4

Impedance and Phase Angle of Parallel *RC* Circuits

12. Determine the impedance for the circuit in Figure 10–75.

13. Determine the impedance and the phase angle in Figure 10–76.

14. Repeat Problem 13 for the following frequencies:

(a) 1.5 kHz **(b)** 3 kHz **(c)** 5 kHz **(d)** 10 kHz

15. Determine the impedance and phase angle in Figure 10–77.

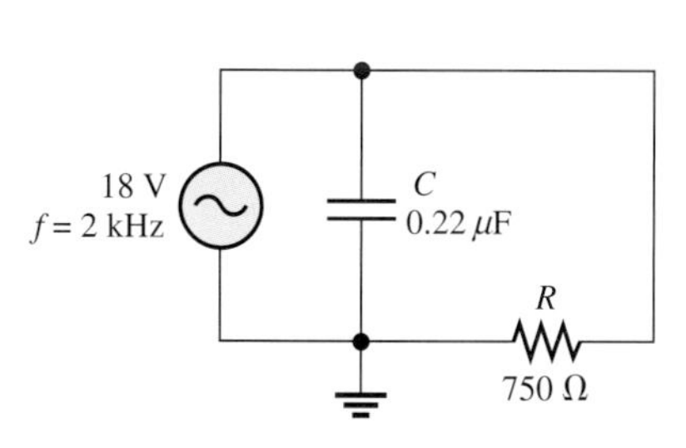

FIGURE 10–76

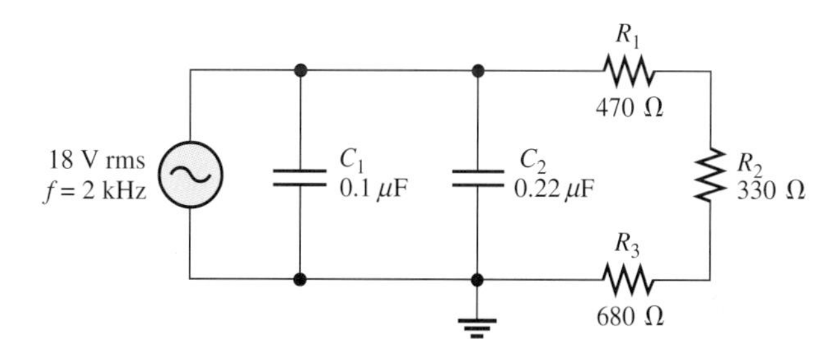

FIGURE 10–77

SECTION 10–5

Analysis of Parallel *RC* Circuits

16. For the circuit in Figure 10–78, find all the currents and voltages.

17. For the parallel circuit in Figure 10–79, find each branch current and the total current. What is the phase angle between the source voltage and the total current?

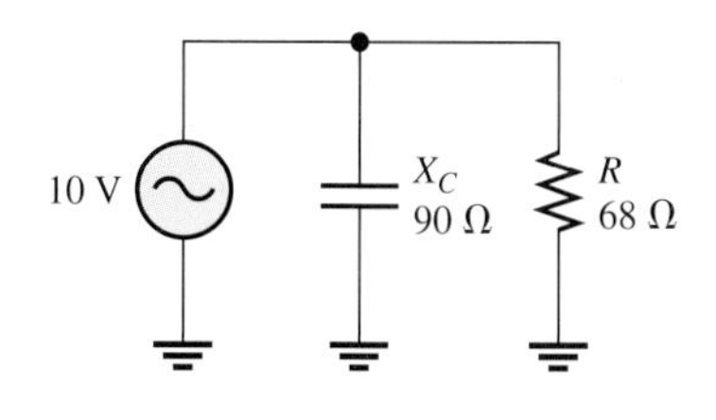

FIGURE 10–78

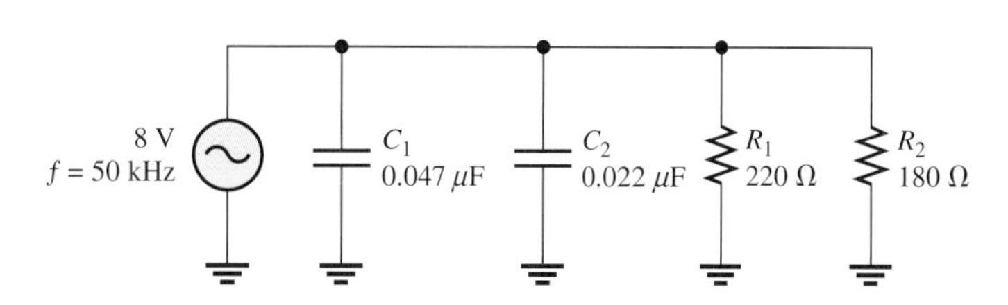

FIGURE 10–79

18. For the circuit in Figure 10–80, determine the following:
    (a) Z **(b)** I_R **(c)** I_C **(d)** I_{tot} **(e)** θ
19. Repeat Problem 18 for $R = 5\ \text{k}\Omega$, $C = 0.047\ \mu F$, and $f = 500$ Hz.
20. Convert the circuit in Figure 10–81 to an equivalent series form.

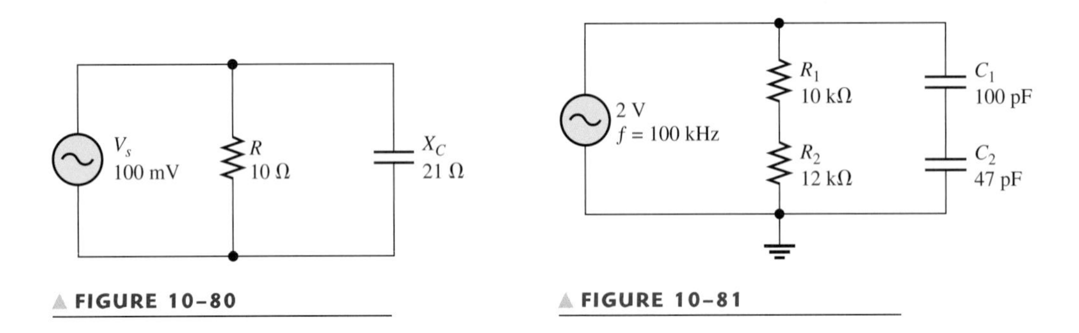

▲ FIGURE 10–80

▲ FIGURE 10–81

SECTION 10–6 **Series-Parallel Analysis**

21. Determine the voltages across each element in Figure 10–82. Draw the voltage phasor diagram.
22. Is the circuit in Figure 10–82 predominantly resistive or predominantly capacitive?
23. Find the current through each branch and the total current in Figure 10–82. Draw the current phasor diagram.
24. For the circuit in Figure 10–83, determine the following:
    (a) I_{tot} **(b)** θ **(c)** V_{R1} **(d)** V_{R2} **(e)** V_{R3} **(f)** V_C

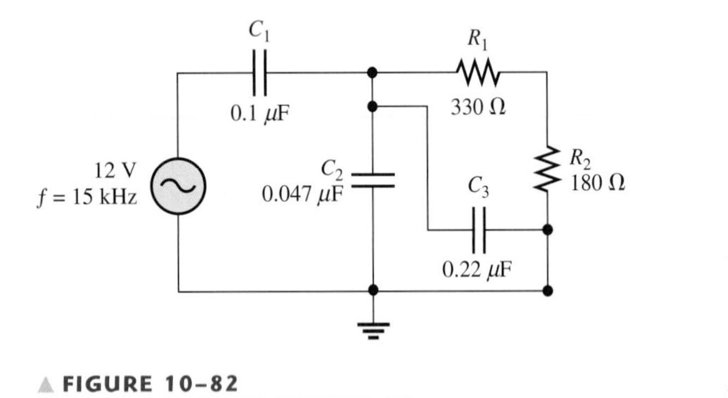

▲ FIGURE 10–82

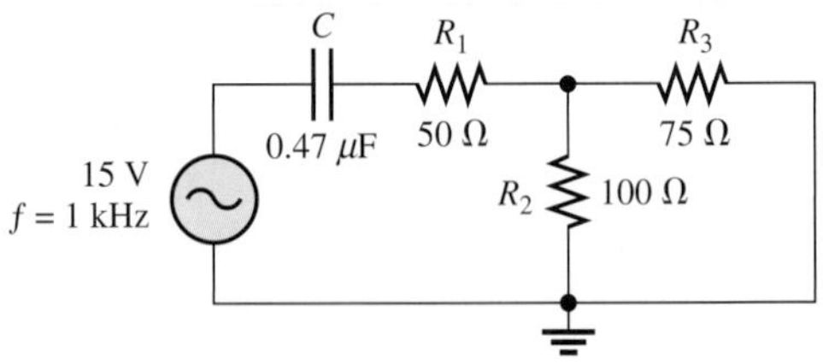

▲ FIGURE 10–83

SECTION 10–7 **Power in *RC* Circuits**

25. In a certain series *RC* circuit, the true power is 2 W, and the reactive power is 3.5 VAR. Determine the apparent power.
26. In Figure 10–73, what is the true power and the reactive power?
27. What is the power factor for the circuit of Figure 10–81?
28. Determine P_{true}, P_r, P_a, and PF for the circuit in Figure 10–83. Draw the power triangle.

SECTION 10–8 **Basic Applications**

29. For the lag network in Figure 10–84, determine the phase lag between the input voltage and the output voltage for each of the following frequencies:

(a) 1 Hz **(b)** 100 Hz **(c)** 1 kHz **(d)** 10 kHz

30. The lag network in Figure 10–84 also acts as a low-pass filter. Draw a response curve for this circuit by plotting the output voltage versus frequency for 0 Hz to 10 kHz in 1 kHz increments.

31. Repeat Problem 29 for the lead network in Figure 10–85.

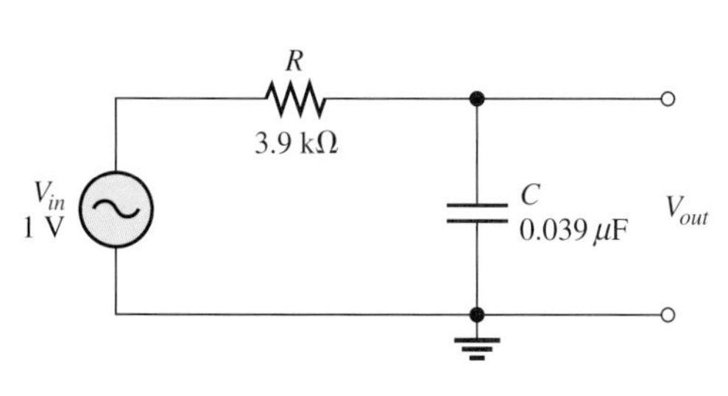

▲ FIGURE 10–84

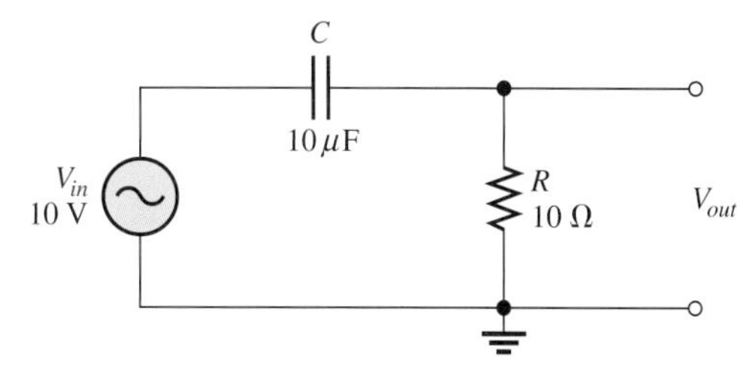

▲ FIGURE 10–85

32. Plot the frequency response curve for the lead network in Figure 10–85 for a frequency range of 0 Hz to 10 kHz in 1 kHz increments.

33. Draw the voltage phasor diagram for each circuit in Figures 10–84 and 10–85 for a frequency of 5 kHz with $V_{in} = 1$ V rms.

34. The rms value of the signal voltage out of amplifier A in Figure 10–86 is 50 mV. If the input resistance to amplifier B is 10 kΩ, how much of the signal is lost due to the coupling capacitor (C_c) when the frequency is 3 kHz?

35. Determine the cutoff frequency for each circuit in Figures 10–84 and 10–85.

36. Determine the bandwidth of the circuit in Figure 10–84.

SECTION 10–9 **Troubleshooting**

37. Assume that the capacitor in Figure 10–87 is excessively leaky. Show how this degradation affects the output voltage and phase angle, assuming that the leakage resistance is 5 kΩ and the frequency is 10 Hz.

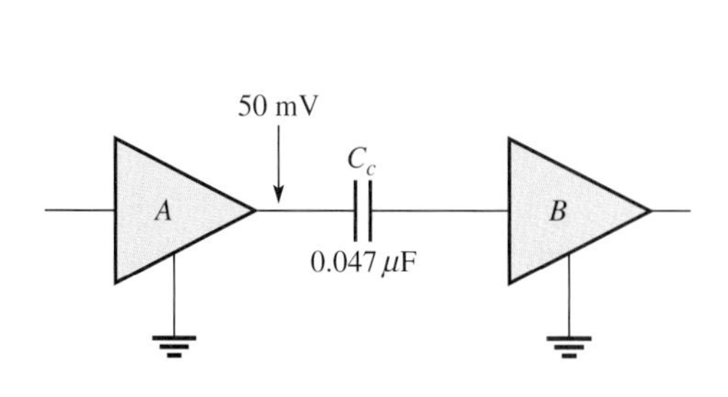

▲ FIGURE 10–86

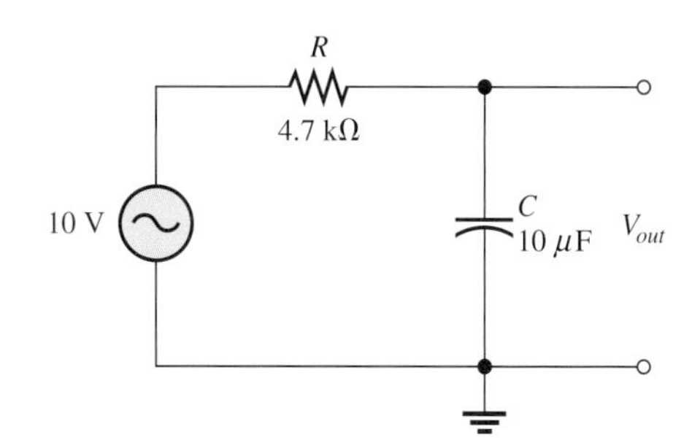

▲ FIGURE 10–87

38. Each of the capacitors in Figure 10–88 has developed a leakage resistance of 2 kΩ. Determine the output voltages under this condition for each circuit.

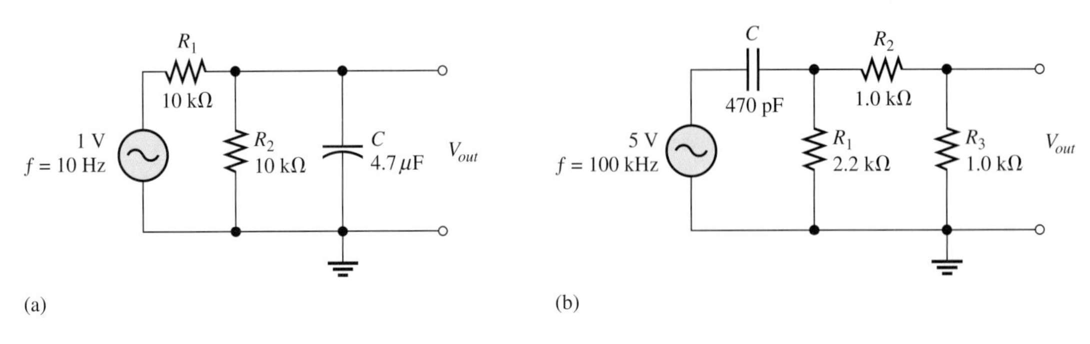

▲ FIGURE 10–88

39. Determine the output voltage for the circuit in Figure 10–88(a) for each of the following failure modes, and compare it to the correct output:

(a) R_1 open **(b)** R_2 open **(c)** C open **(d)** C shorted

40. Determine the output voltage for the circuit in Figure 10–88(b) for each of the following failure modes, and compare it to the correct output:

(a) C open **(b)** C shorted **(c)** R_1 open
(d) R_2 open **(e)** R_3 open

ADVANCED PROBLEMS

41. A single 220 V, 60 Hz source drives two loads. Load *A* has an impedance of 50 Ω and a power factor of 0.85. Load *B* has an impedance of 72 Ω and a power factor of 0.95.

(a) How much current does each load draw?

(b) What is the reactive power in each load?

(c) What is the true power in each load?

(d) What is the apparent power in each load?

(e) Which load has more voltage drop along the lines connecting it to the source?

42. What value of coupling capacitor is required in Figure 10–89 so that the signal voltage at the input of amplifier 2 is at least 70.7% of the signal voltage at the output of amplifier 1 when the frequency is 20 Hz? Disregard the input resistance of the amplifier.

▶ FIGURE 10–89

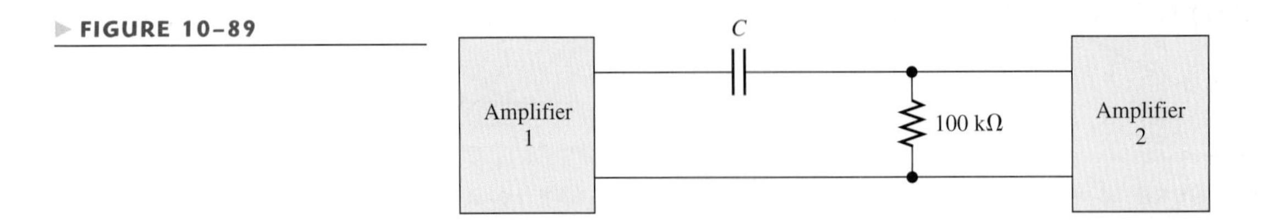

43. Determine the value of R_1 required to get a phase angle of 30° between the source voltage and the total current in Figure 10–90.

44. Draw the voltage and current phasor diagram for Figure 10–91.

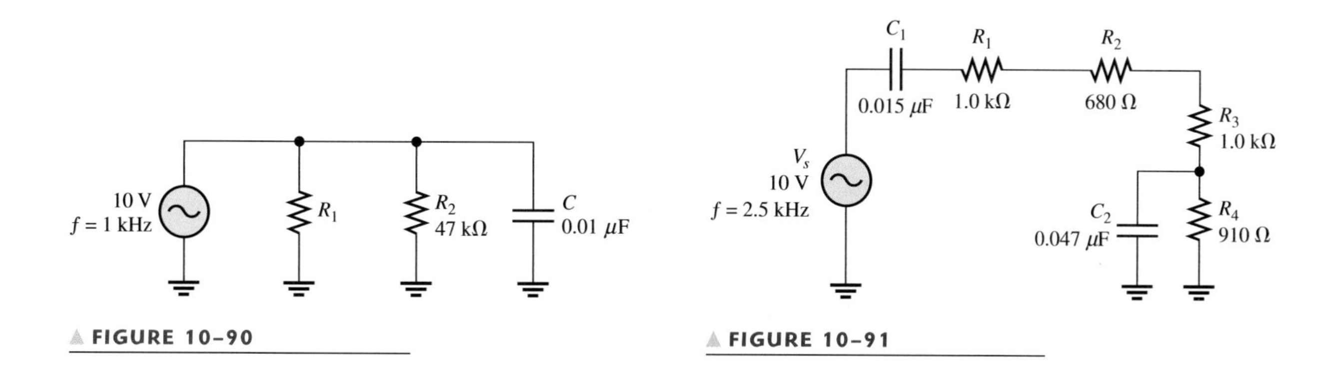

FIGURE 10–90

FIGURE 10–91

45. A certain load dissipates 1.5 kW of power with an impedance of 12 Ω and a power factor of 0.75. What is its reactive power? What is its apparent power?

46. Determine the series element or elements that are in the block of Figure 10–92 to meet the following overall circuit requirements:

(a) $P_{true} = 400$ W **(b)** leading power factor (I_{tot} leads V_s)

47. Determine the value of C_2 in Figure 10–93 when $V_A = V_B$.

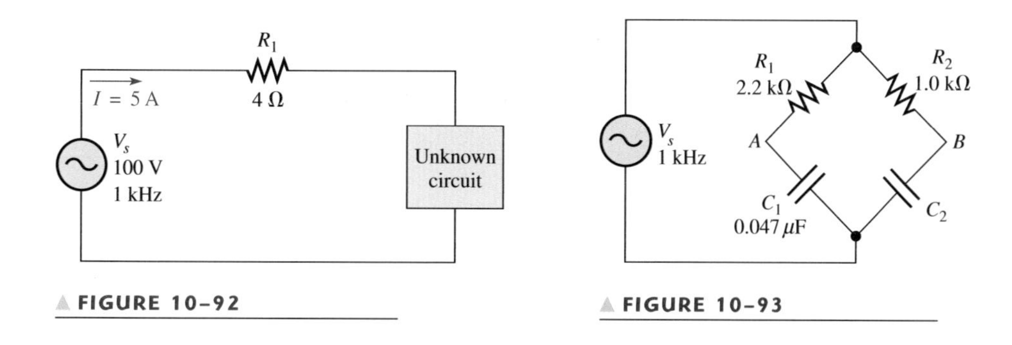

FIGURE 10–92

FIGURE 10–93

48. Draw the schematic for the circuit in Figure 10–94 and determine if the waveform on the scope is correct. If there is a fault in the circuit, identify it. After working this problem, open file P10-48 on your CD-ROM and run the setup to observe the correct operation.

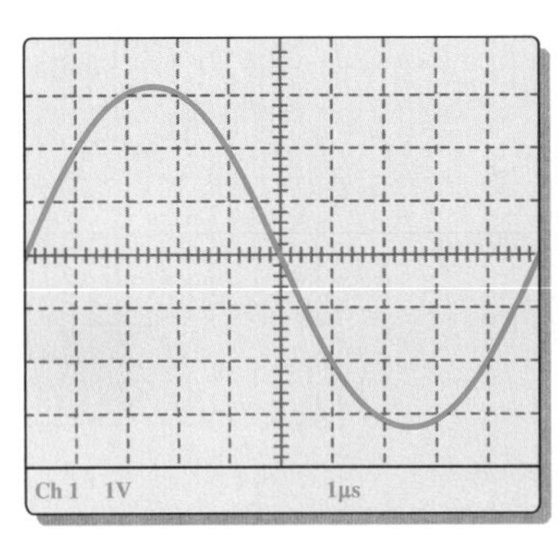

(a) Oscilloscope display

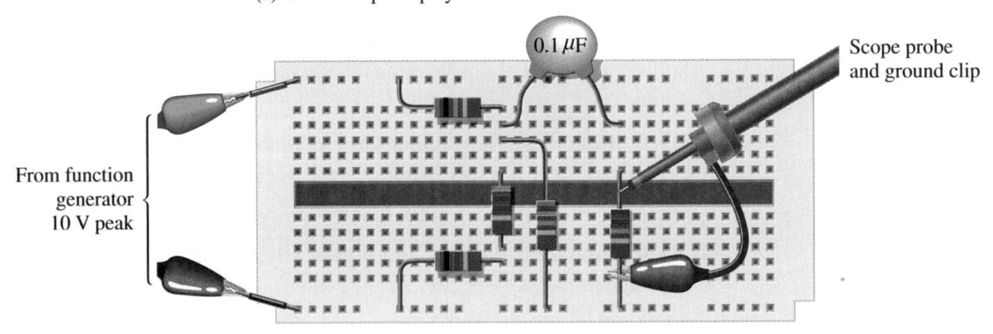

(b) Circuit with leads connected

FIGURE 10–94

ELECTRONICS WORKBENCH/CIRCUITMAKER TROUBLESHOOTING PROBLEMS

CD-ROM file circuits are shown in Figure 10–95.

49. Open file P10-49. Determine if there is a fault and, if so, identify it.

50. Open file P10-50. Determine if there is a fault and, if so, identify it.

51. Open file P10-51. Determine if there is a fault and, if so, identify it.

52. Open file P10-52. Determine if there is a fault and, if so, identify it.

53. Open file P10-53. Determine if there is a fault and, if so, identify it.

54. Open file P10-54. Determine if there is a fault and, if so, identify it.

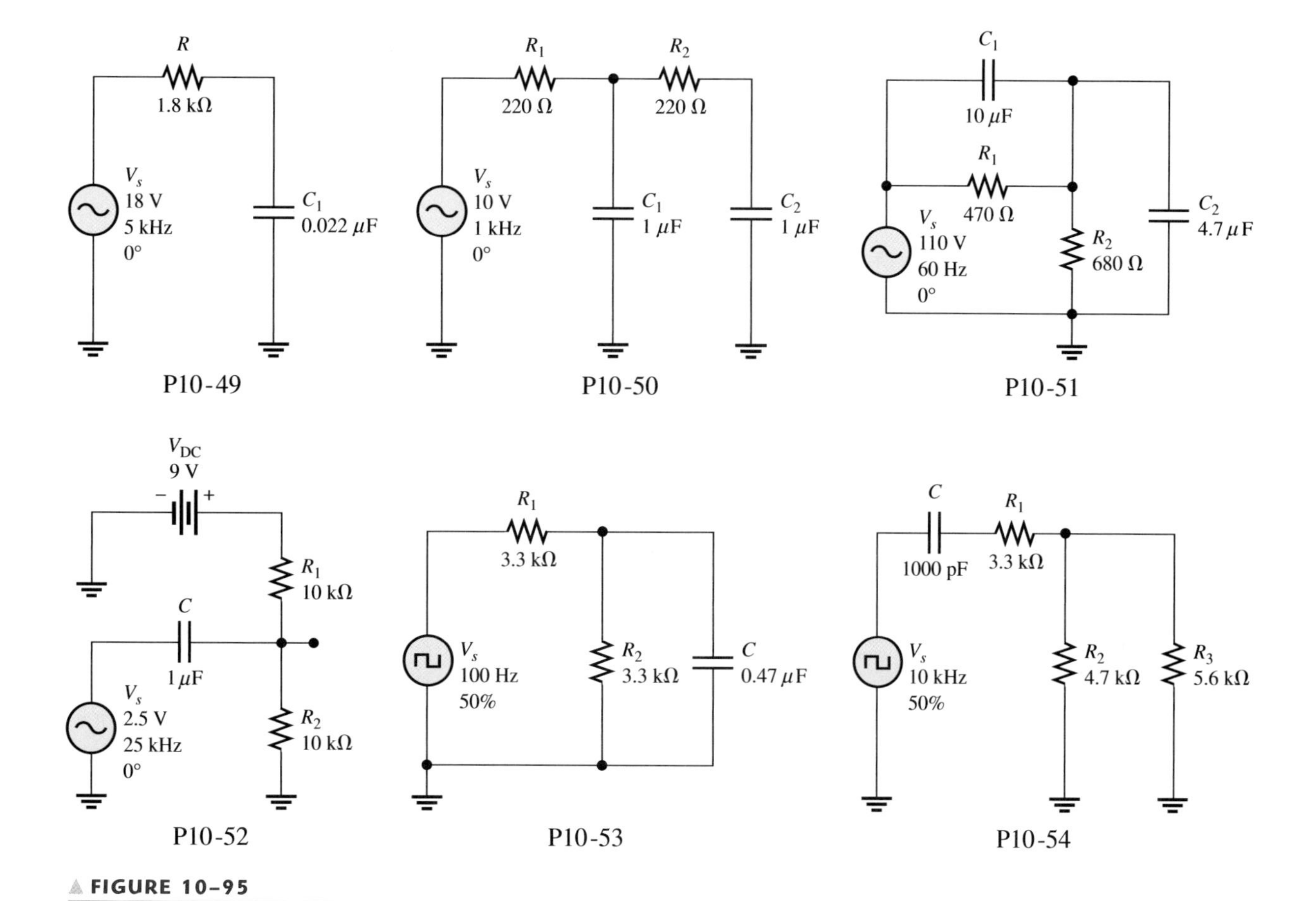

▲ FIGURE 10–95

ANSWERS

SECTION REVIEWS

SECTION 10–1 **Sinusoidal Response of *RC* Circuits**

1. V_C frequency is 60 Hz, I frequency is 60 Hz.
2. Capacitive reactance
3. θ is closer to 0° when $R > X_C$.

SECTION 10–2 **Impedance and Phase Angle of Series *RC* Circuits**

1. Impedance is the opposition to sinusoidal current.
2. V_s lags I.
3. The capacitive reactance produces the phase angle.
4. $Z = \sqrt{R^2 + X_C^2} = 59.9\ \text{k}\Omega$; $\theta = \tan^{-1}(X_C/R) = 56.6°$

SECTION 10–3 **Analysis of Series *RC* Circuits**

1. $V_s = \sqrt{V_R^2 + V_C^2} = 7.2\ \text{V}$
2. $\theta = \tan^{-1}(V_C/V_R) = 56.3°$
3. $\theta = 90°$
4. **(a)** X_C decreases when f increases. **(b)** Z decreases when f increases. **(c)** θ decreases when f increases.

SECTION 10–4 **Impedance and Phase Angle of Parallel *RC* Circuits**

1. $Z = RX_C/\sqrt{R^2 + X_C^2} = 545\ \Omega$
2. *Conductance* is the reciprocal of resistance; *capacitive susceptance* is the reciprocal of capacitive reactance; and *admittance* is the reciprocal of impedance.
3. $Y = 1/Z = 10$ mS
4. $Y = \sqrt{G^2 + B_C^2} = 24$ mS

SECTION 10–5 **Analysis of Parallel *RC* Circuits**

1. $I_{tot} = V_sY = 21$ mA
2. $\theta = \tan^{-1}(I_C/I_R) = 56.3°$; $I_{tot} = \sqrt{I_R^2 + I_C^2} = 18$ mA
3. $\theta = 90°$

SECTION 10–6 **Series-Parallel Analysis**

1. See Figure 10–96.
2. $V_1 = I_{tot}R_1 = 6.71$ V

FIGURE 10–96

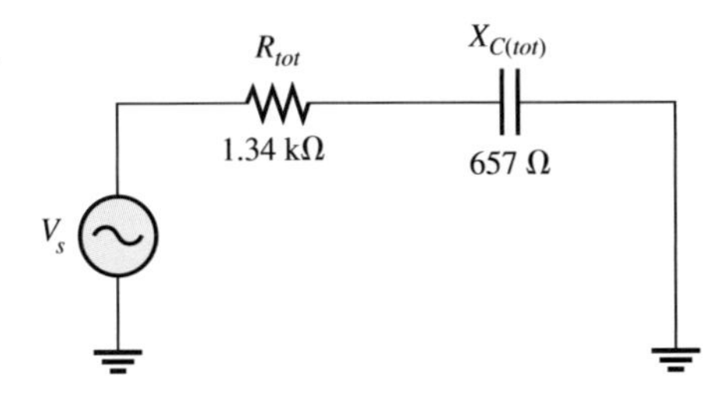

SECTION 10–7 **Power in *RC* Circuits**

1. Power dissipation is due to the resistance.
2. $PF = \cos 45° = 0.707$
3. $P_{true} = I_{tot}^2R = 1.32$ kW; $P_r = I_{tot}^2X_C = 1.84$ kVAR; $P_a = I_{tot}^2Z = 2.26$ kVA

SECTION 10–8 **Basic Applications**

1. $\phi = 90° - \tan^{-1}(X_C/R) = 62.8°$
2. $V_{out} = (R/\sqrt{R^2 + X_C^2})V_{in} = 8.9$ V rms
3. Output is across the capacitor.

SECTION 10–9 **Troubleshooting**

1. The leakage resistance acts in parallel with *C*, which alters the circuit time constant.
2. The capacitor is open.
3. A shorted capacitor, open resistor, no source voltage, or open contact can cause 0 V across the capacitor.

Application Assignment

1. A lower value coupling capacitor will increase the frequency at which a significant drop in voltage occurs.
2. $V_B = 3.16$ V dc
3. $V_B = 10$ mV rms

RELATED PROBLEMS FOR EXAMPLES

10–1 2.42 kΩ; 65.6°

10–2 2.56 V

10–3 65.5°

10–4 7.14 V

10–5 15.9 kΩ; 86.4°

10–6 24.3 Ω

10–7 4.60 mS

10–8 6.16 mA

10–9 117 mA; 31°

10–10 $R_{eq} = 8.99$ kΩ; $X_{C(eq)} = 4.38$ kΩ

10–11 $V_1 = 7.04$ V; $V_2 = 3.21$ V

10–12 0.146

10–13 990 μW

10–14 Phase lag ϕ increases.

10–15 Output voltage decreases.

10–16 Phase lead ϕ decreases; output voltage increases.

10–17 7.29 V

10–18 Resistor open

SELF-TEST

1. (c) **2.** (b) **3.** (b) **4.** (a) **5.** (d) **6.** (b) **7.** (a) **8.** (d) **9.** (c) **10.** (b) **11.** (d) **12.** (d) **13.** (c) **14.** (b) **15.** (d)

11

INDUCTORS

INTRODUCTION

Inductance is the property of a coil of wire that opposes a change in current. The basis for inductance is the electromagnetic field that surrounds any conductor when there is current through it. The electrical component designed to have the property of inductance is called an *inductor* or *coil*. Both of these terms refer to the same type of device.

In this chapter, you will study the basic inductor and its characteristics. Various types of inductors are covered in terms of their physical construction and their electrical properties. The basic behavior of inductors in both dc and ac circuits is discussed, and series and parallel combinations are analyzed. A method of testing inductors is also discussed.

CHAPTER OUTLINE

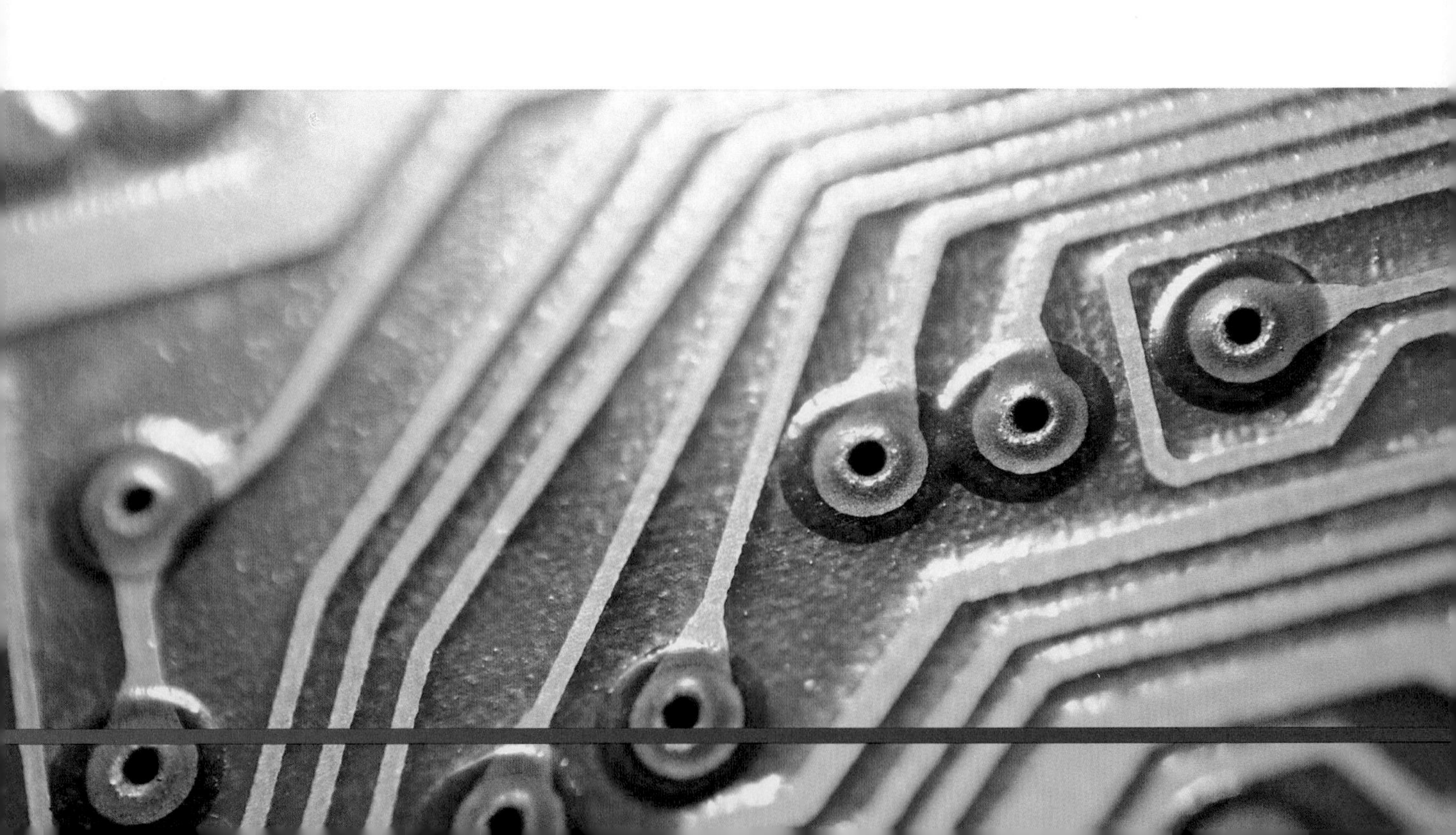

CHAPTER OBJECTIVES

- Describe the basic structure and characteristics of an inductor
- Discuss various types of inductors
- Analyze series inductors
- Analyze parallel inductors
- Analyze inductive dc switching circuits
- Analyze inductive ac circuits
- Discuss some inductor applications
- Test an inductor

KEY TERMS

- Inductor
- Winding
- Coil
- Induced voltage
- Inductance
- Winding resistance
- Faraday's law
- Lenz's law
- *RL* time constant
- Exponential
- Inductive reactance
- Quality factor

APPLICATION ASSIGNMENT PREVIEW

Assume that you are working on a piece of defective communications equipment and your supervisor asks you to check out the unmarked coils that have been removed from the system and determine their approximate inductance value by measuring the time constant. After studying this chapter, you should be able to complete the application assignment.

WWW. VISIT THE COMPANION WEBSITE

Circuit Simulation Tutorials and Other Chapter Study Tools Are Available at

http://www.prenhall.com/floyd

11–1 THE BASIC INDUCTOR

In this section, the basic construction and characteristics of inductors are examined.

After completing this section, you should be able to

- **Describe the basic structure and characteristics of an inductor**
- Explain how an inductor stores energy
- Define *inductance* and state its unit
- Discuss induced voltage
- Specify how the physical characteristics affect inductance
- Discuss winding resistance and winding capacitance
- State Faraday's law
- State Lenz's law

When a length of wire is formed into a coil, as shown in Figure 11–1, it becomes a basic inductor. Current through the coil produces an electromagnetic field. The magnetic lines of force around each loop (turn) in the winding of the coil effectively add to the lines of force around the adjoining loops, forming a strong magnetic field within and around the coil, as shown. The net direction of the total magnetic field creates a north and a south pole.

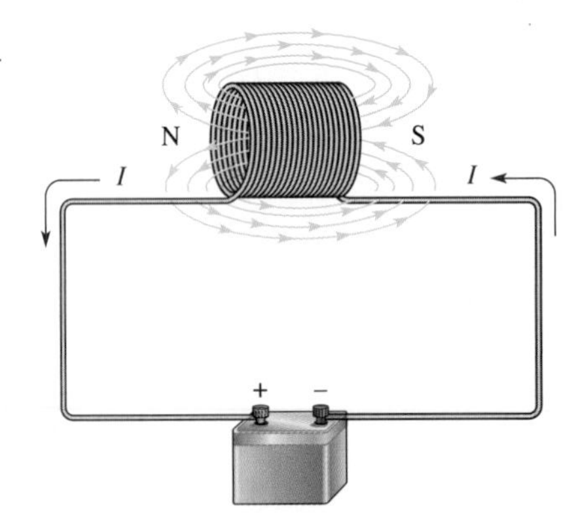

▶ FIGURE 11–1

A coil of wire forms an inductor. When there is current through it, a three-dimensional electromagnetic field is created, surrounding the coil in all directions.

Self-Inductance

When there is current through an inductor, an electromagnetic field is established. When the current changes, the electromagnetic field also changes. An increase in current expands the field, and a decrease in current reduces it. Therefore, a changing current produces a changing electromagnetic field around the inductor (also known as coil and in some applications, **choke**). In turn, the changing electromagnetic field causes an induced voltage across the coil in a direction to oppose the change in current. This property is called *self-inductance* but is usually referred to as simply inductance, symbolized by L.

Inductance is a measure of a coil's ability to establish an induced voltage as a result of a change in its current, and that induced voltage is in a direction to oppose that change in current.

The Unit of Inductance The **henry,** symbolized by H, is the basic unit of inductance. By definition, the inductance is one henry when current through the coil, *changing* at the rate of one ampere per second, induces one volt across the coil. In many practical applications, millihenries (mH) and microhenries (μH) are the more common units. A schematic symbol for the inductor is shown in Figure 11–2.

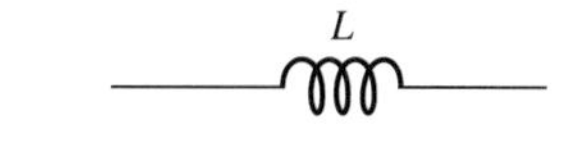

▲ **FIGURE 11–2**

Symbol for the inductor.

Energy Storage

An inductor stores energy in the magnetic field created by the current. The energy stored is expressed as

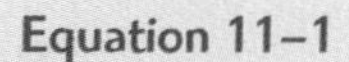

$$W = \frac{1}{2}LI^2$$

As you can see, the energy stored is proportional to the inductance and the square of the current. When current (I) is in amperes and inductance (L) is in henries, the energy (W) is in joules.

Joseph Henry 1797–1878 Henry began his career as a professor at a small school in Albany, NY, and later became the first director of the Smithsonian Institution. He was the first American since Franklin to undertake original scientific experiments. He was the first to superimpose coils of wire wrapped on an iron core and first observed the effects of electromagnetic induction in 1830, a year before Faraday, but he did not publish his findings. Henry did obtain credit for the discovery of self-induction, however. The unit of inductance is named in his honor. (Photo credit: Courtesy of the Smithsonian Institution. Photo number 52,054.)

Physical Characteristics of Inductors

The following characteristics are important in establishing the inductance of a coil: the permeability of the core material, the number of turns of wire, the core length, and the cross-sectional area of the core.

Core Material As discussed earlier, an inductor is basically a coil of wire. The material around which the coil is formed is called the **core.** Coils are wound on either nonmagnetic or magnetic materials. Examples of nonmagnetic materials are air, wood, copper, plastic, and glass. The permeabilities of these materials are the same as for a vacuum. Examples of magnetic materials are iron, nickel, steel, cobalt, or alloys. These materials have permeabilities that are hundreds or thousands of times greater than that of a vacuum and are classified as *ferromagnetic.* A ferromagnetic core provides a lower reluctance path for the magnetic lines of force and thus permits a stronger magnetic field.

As you learned in Chapter 7, the permeability (μ) of the core material determines how easily a magnetic field can be established. *The inductance is directly proportional to the permeability of the core material.*

Physical Parameters As indicated in Figure 11–3, the number of turns of wire, the length, and the cross-sectional area of the core are factors in setting the value of inductance. The inductance is inversely proportional to the length of the core and directly proportional to the cross-sectional area. Also, the inductance is directly related to the number of turns squared. This relationship is as follows:

Equation 11–2

$$L = \frac{N^2\mu A}{l}$$

where L is the inductance in henries (H), N is the number of turns, μ is the permeability in henries per meter (H/m), A is the cross-sectional area in meters squared (m^2), and l is the core length in meters (m).

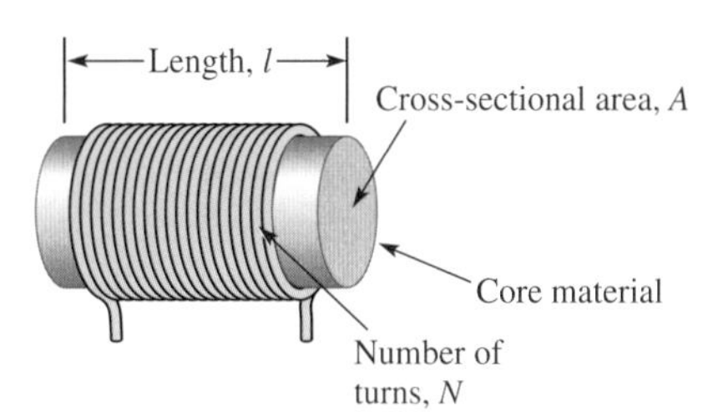

◀ **FIGURE 11–3**

Factors that determine the inductance of a coil.

EXAMPLE 11–1

Determine the inductance of the coil in Figure 11–4. The permeability of the core is 0.25×10^{-3} H/m.

▶ **FIGURE 11–4**

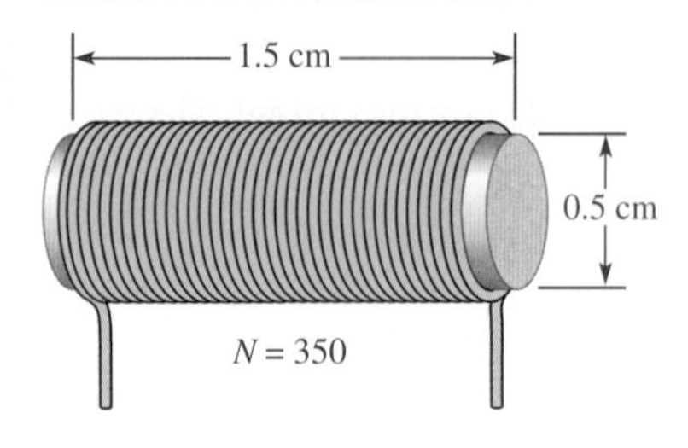

Solution 1.5 cm = 0.015 m.

$$A = \pi r^2 = \pi(0.25 \times 10^{-2}\text{ m})^2 = 1.96 \times 10^{-5}\text{ m}^2$$

$$L = \frac{N^2\mu A}{l} = \frac{(350)^2(0.25 \times 10^{-3}\text{ H/m})(1.96 \times 10^{-5}\text{ m}^2)}{0.015\text{ m}} = \mathbf{40\ mH}$$

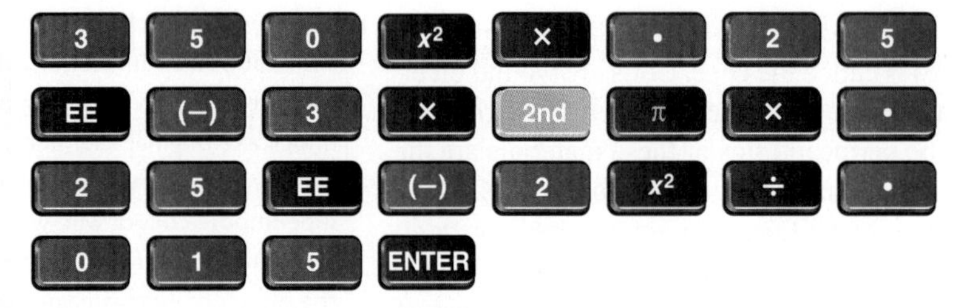

350²*.25E-3*π*.25E-2²
/.015
40.0880312567E-3

Related Problem* Determine the inductance of a coil with 400 turns on a core that is 2 cm long and has a diameter of 1 cm. The permeability is 0.25×10^{-3} H/m.

*Answers are at the end of the chapter.

Winding Resistance

When a coil is made of a certain material, for example, insulated copper wire, that wire has a certain resistance per unit of length. When many turns of wire are used to construct a coil, the total resistance may be significant. This inherent resistance is called the *dc resistance* or the winding resistance (R_W).

Although this resistance is distributed along the length of the wire, as shown in Figure 11–5(a), it is sometimes indicated in a schematic as resistance appearing in series with the inductance of the coil, as shown in Figure 11–5(b). In many applications, the winding resistance can be ignored and the coil considered as an ideal inductor. In other cases, the resistance must be considered.

▶ **FIGURE 11–5**

Winding resistance of a coil.

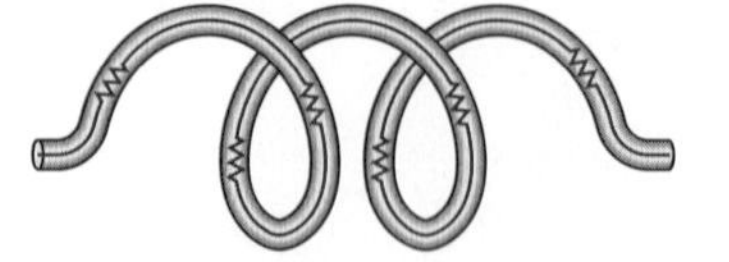

(a) The wire has resistance distributed along its length.

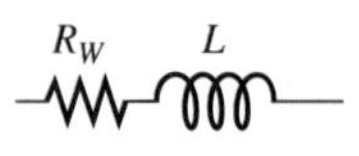

(b) Equivalent circuit

Winding Capacitance

When two conductors are placed side by side, there is always some capacitance between them. Thus, when many turns of wire are placed close together in a coil, a certain amount of stray capacitance, called *winding capacitance* (C_W), is a natural side effect. In many applications, this winding capacitance is very small and has no significant effect. In other cases, particularly at high frequencies, it may become quite important.

The equivalent circuit for an inductor with both its winding resistance (R_W) and its winding capacitance (C_W) is shown in Figure 11–6. The capacitance effectively acts in parallel. The total of the stray capacitances between each loop of the winding is indicated in a schematic as a capacitance appearing in parallel with the coil and its winding resistance, as shown in Figure 11–6(b).

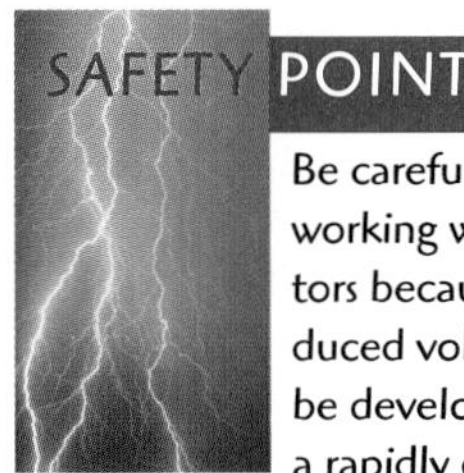

Be careful when working with inductors because high induced voltages can be developed due to a rapidly changing magnetic field. This occurs when the current is interrupted or its value abruptly changed.

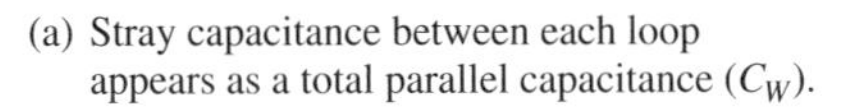

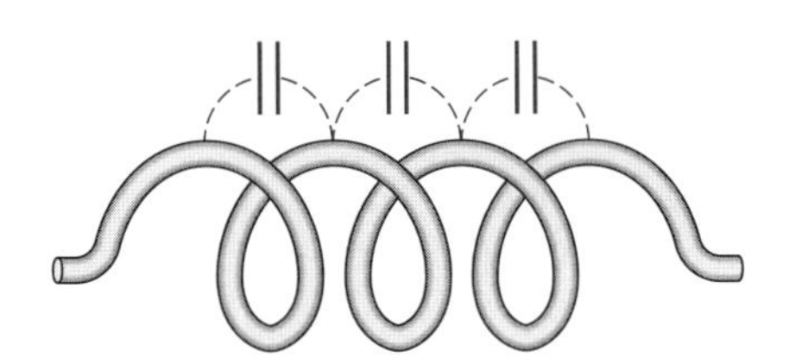

(a) Stray capacitance between each loop appears as a total parallel capacitance (C_W).

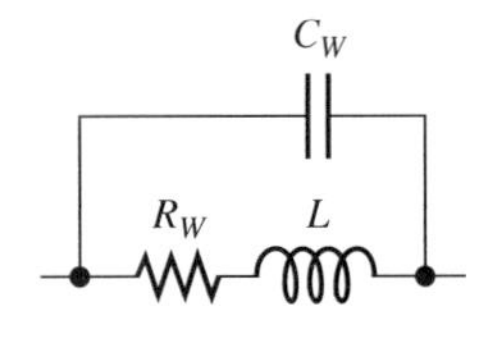

(b) Equivalent circuit

FIGURE 11–6

Winding capacitance of a coil.

Review of Faraday's Law

Faraday's law was introduced in Chapter 7 and is reviewed here because of its importance in the study of inductors. Faraday found that by moving a magnet through a coil of wire, a voltage was induced across the coil, and that when a complete path was provided, the induced voltage caused an induced current. He observed that

> **The amount of induced voltage in a coil is directly proportional to the rate of change of the magnetic field with respect to the coil.**

This principle is illustrated in Figure 11–7, where a bar magnet is moved through a coil of wire. An induced voltage is indicated by the voltmeter connected across the coil. The faster the magnet is moved, the greater the induced voltage.

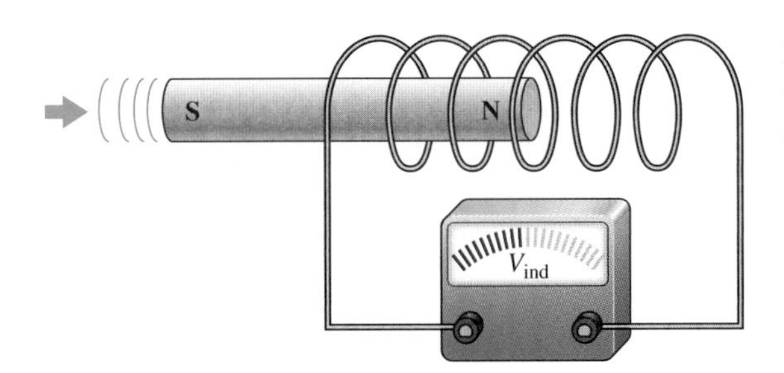

FIGURE 11–7

Induced voltage is created by a changing magnetic field.

When a wire is formed into a certain number of loops or turns and is exposed to a changing magnetic field, a voltage is induced across the coil. The induced voltage is proportional to the number of turns, *N*, of the wire in the coil and to the rate at which the magnetic field changes.

Lenz's Law

Lenz's law, also introduced in Chapter 7, adds to Faraday's law by defining the direction of induced voltage.

When the current through a coil changes and an induced voltage is created as a result of the changing magnetic field, the direction of the induced voltage is such that it always opposes the change in current.

Figure 11–8 illustrates Lenz's law. In part (a), the current is constant and is limited by R_1. There is no induced voltage because the magnetic field is unchanging. In part (b), the switch suddenly is closed, placing R_2 in parallel with R_1 and thus reducing the resistance. Naturally, the current tries to increase and the mag-

▼ FIGURE 11–8

Demonstration of Lenz's law in an inductive circuit: When the current tries to change suddenly, the electromagnetic field changes and induces a voltage in a direction that opposes that change in current.

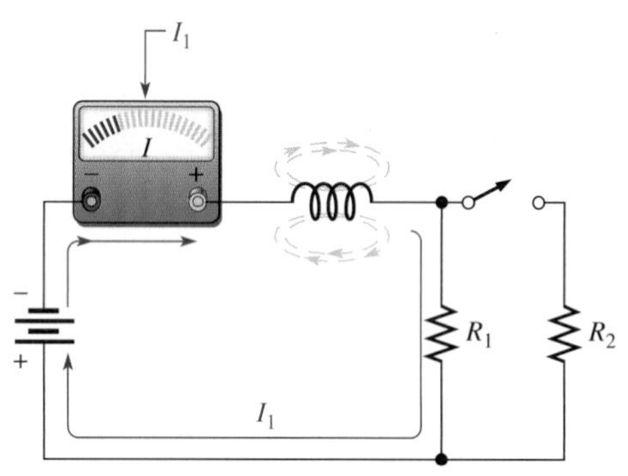

(a) Switch open: Constant current and constant magnetic field; no induced voltage.

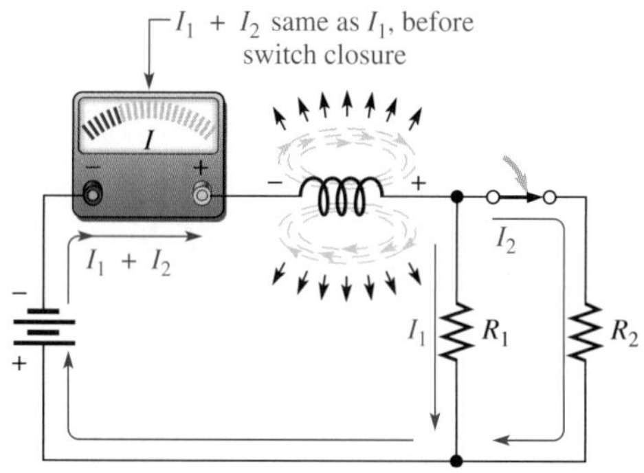

(b) At instant of switch closure: Expanding magnetic field induces voltage, which opposes increase in total current. The total current remains the same at this instant.

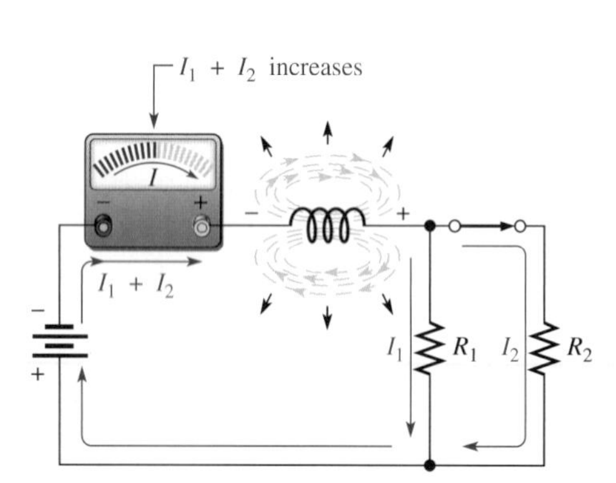

(c) Right after switch closure: The rate of expansion of the magnetic field decreases, allowing the current to increase exponentially as induced voltage decreases.

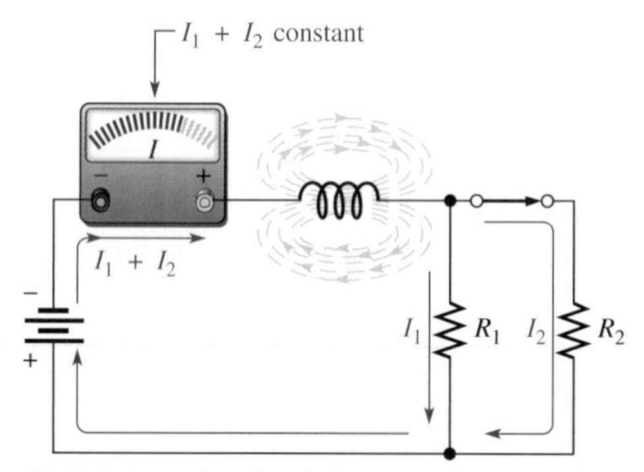

(d) Switch remains closed: Current and magnetic field reach constant value.

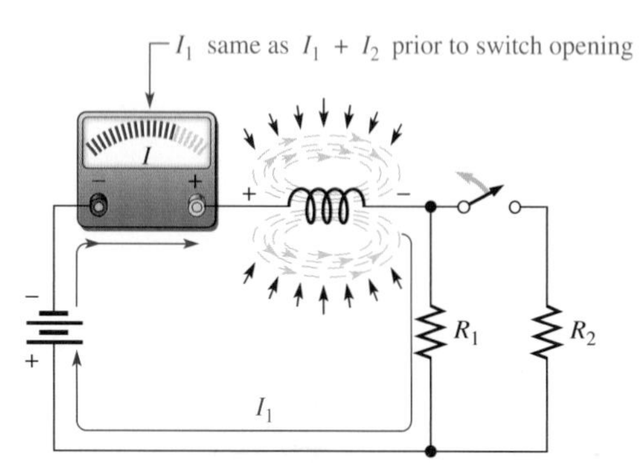

(e) At instant of switch opening: Magnetic field begins to collapse, creating an induced voltage, which opposes decrease in current.

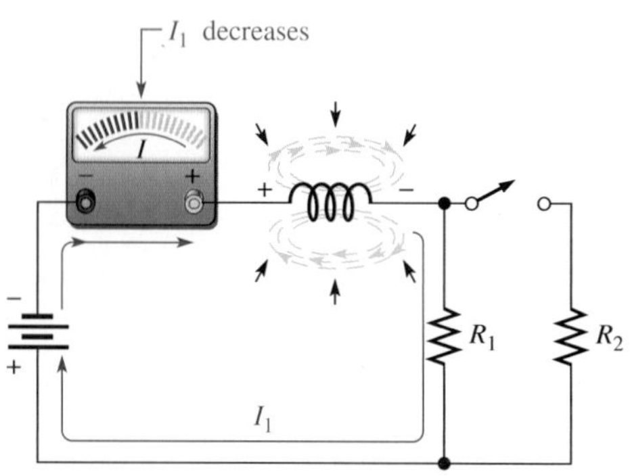

(f) After switch opening: Rate of collapse of magnetic field decreases, allowing current to decrease exponentially back to original value.

netic field begins to expand, but the induced voltage opposes this attempted increase in current for an instant.

In Figure 11–8(c), the induced voltage gradually decreases, allowing the current to increase. In part (d), the current has reached a constant value as determined by the parallel resistors, and the induced voltage is zero. In part (e), the switch has been suddenly opened, and, for an instant, the induced voltage prevents any decrease in current. In part (f), the induced voltage gradually decreases, allowing the current to decrease back to a value determined by R_1. Notice that the induced voltage has a polarity that opposes any current change. The polarity of the induced voltage is opposite that of the battery voltage for an increase in current and aids the battery voltage for a decrease in current.

SECTION 11–1 REVIEW

Answers are at the end of the chapter.

1. List the parameters that contribute to the inductance of a coil.
2. Describe what happens to L when
   (a) N is increased
   (b) the core length is increased
   (c) the cross-sectional area of the core is decreased
   (d) a ferromagnetic core is replaced by an air core
3. Explain why inductors have some winding resistance.
4. Explain why inductors have some winding capacitance.

11–2 TYPES OF INDUCTORS

Inductors normally are classified according to the type of core material. In this section, the basic types of inductors are examined.

After completing this section, you should be able to

- **Discuss various types of inductors**
- Describe the basic types of fixed inductors
- Distinguish between fixed and variable inductors

HANDS ON TIP

When breadboarding circuits which include small inductors, it is best to use encapsulated inductors for structural strength. Inductors generally have extremely fine coil wire that is connected to a much larger size of lead wires.

In unencapsulated inductors, these contact points are very vulnerable to breaking if the inductor is frequently inserted and removed from a protoboard.

Inductors are made in a variety of shapes and sizes. Basically, they fall into two general categories: fixed and variable. The standard schematic symbols are shown in Figure 11–9.

Both fixed and variable inductors can be classified according to the type of core material. Three common types are the air core, the iron core, and the ferrite core. Each has a unique symbol, as shown in Figure 11–10.

Adjustable (variable) inductors usually have a screw-type adjustment that moves a sliding core in and out, thus changing the inductance. A wide variety of inductors exists, and some are shown in Figure 11–11. Small fixed inductors are usually encapsulated in an insulating material that protects the fine wire in the coil. Encapsulated inductors have an appearance similar to a small resistor.

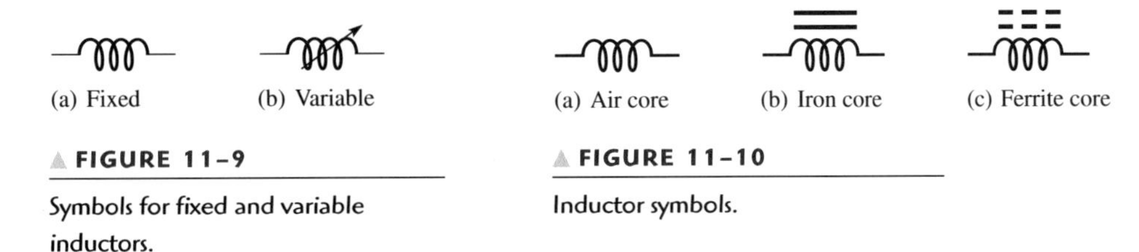

▲ FIGURE 11–9

Symbols for fixed and variable inductors.

▲ FIGURE 11–10

Inductor symbols.

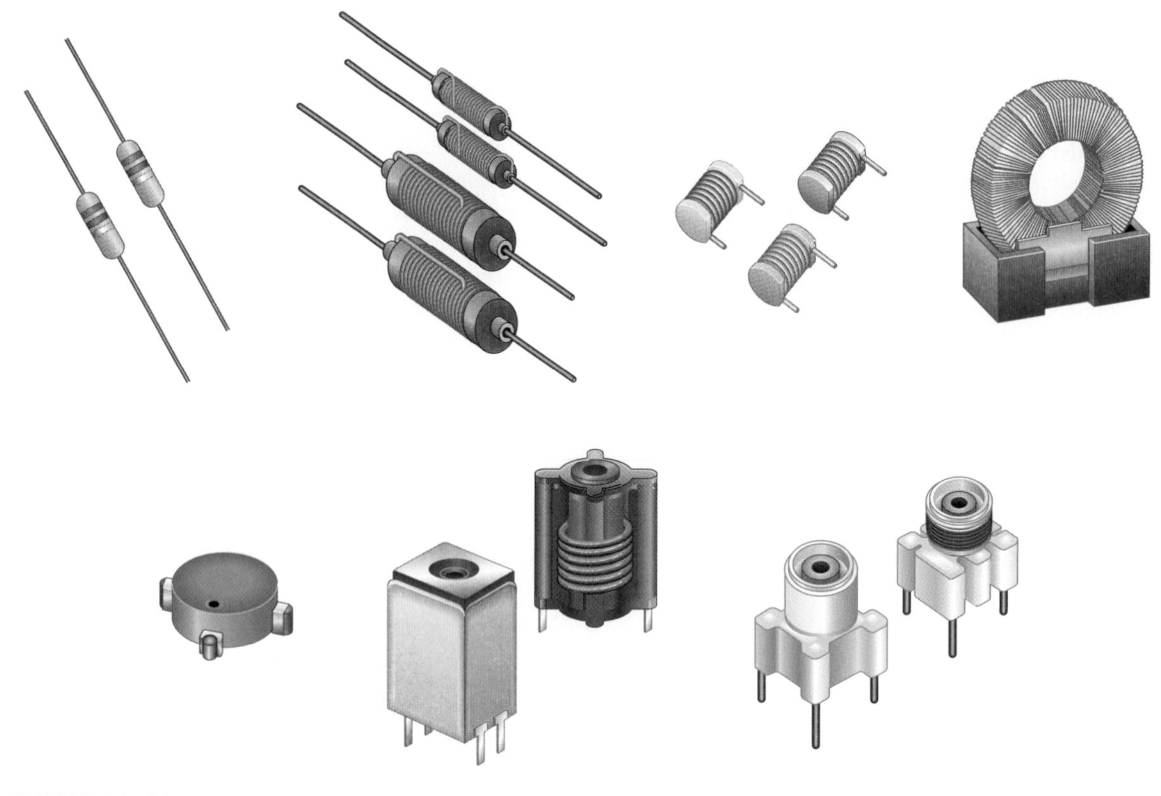

FIGURE 11–11
Typical inductors.

SECTION 11–2 REVIEW

1. Name two general categories of inductors.
2. Identify the inductor symbols in Figure 11–12.

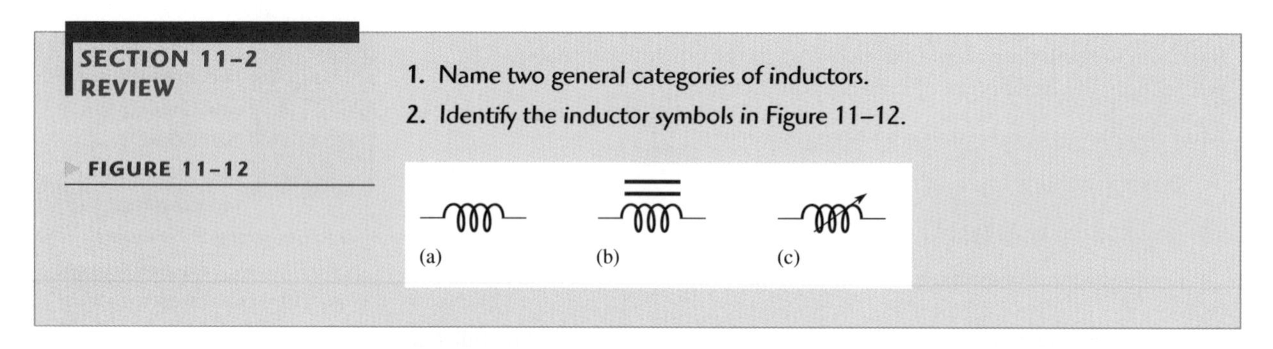

FIGURE 11–12

11–3 SERIES INDUCTORS

In this section, you will learn that when inductors are connected in series, the total inductance increases.

After completing this section, you should be able to

- **Analyze series inductors**
- Determine total inductance

When inductors are connected in series, as in Figure 11–13, the total inductance, L_T, is the sum of the individual inductances. The formula for L_T is expressed in the following equation for the general case of n inductors in series:

Equation 11–3

$$L_T = L_1 + L_2 + L_3 + \cdots + L_n$$

Notice that the formula for total inductance in series is similar to the formulas for total resistance in series (Equation 4–1) and total capacitance in parallel (Equation 9–14).

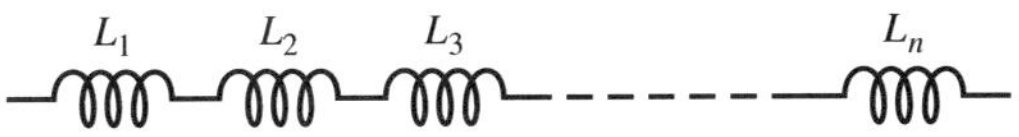

FIGURE 11–13

Inductors in series.

EXAMPLE 11–2

Determine the total inductance for each of the series connections in Figure 11–14.

FIGURE 11–14

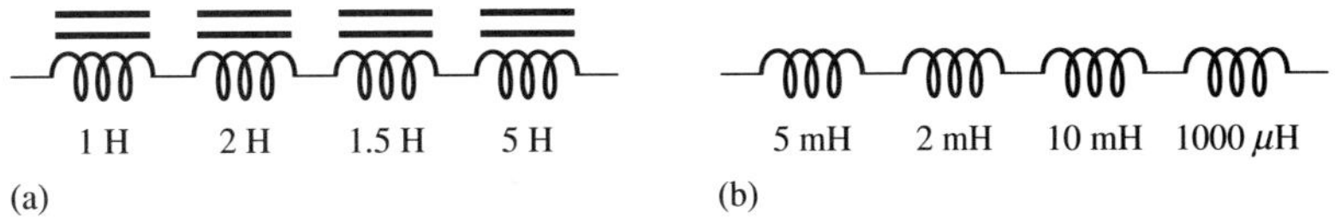

(a) 1 H, 2 H, 1.5 H, 5 H

(b) 5 mH, 2 mH, 10 mH, 1000 μH

Solution In Figure 11–14(a),

$$L_T = 1\text{ H} + 2\text{ H} + 1.5\text{ H} + 5\text{ H} = \mathbf{9.5\text{ H}}$$

In Figure 11–14(b),

$$L_T = 5\text{ mH} + 2\text{ mH} + 10\text{ mH} + 1\text{ mH} = \mathbf{18\text{ mH}}$$

Note: 1000 μH = 1 mH

Related Problem What is the total inductance of ten 50 μH inductors in series?

SECTION 11–3 REVIEW

1. State the rule for combining inductors in series.
2. What is L_T for a series connection of 100 μH, 500 μH, and 2 mH?
3. Five 100 mH coils are connected in series. What is the total inductance?

11–4 PARALLEL INDUCTORS

In this section, you will learn that when inductors are connected in parallel, total inductance is reduced.

After completing this section, you should be able to

- **Analyze parallel inductors**
- Determine total inductance

When inductors are connected in parallel, as in Figure 11–15, the total inductance is less than the smallest inductance. The general formula states that the reciprocal of the total inductance is equal to the sum of the reciprocals of the individual inductances.

$$\frac{1}{L_T} = \frac{1}{L_1} + \frac{1}{L_2} + \frac{1}{L_3} + \cdots + \frac{1}{L_n}$$

Equation 11–4

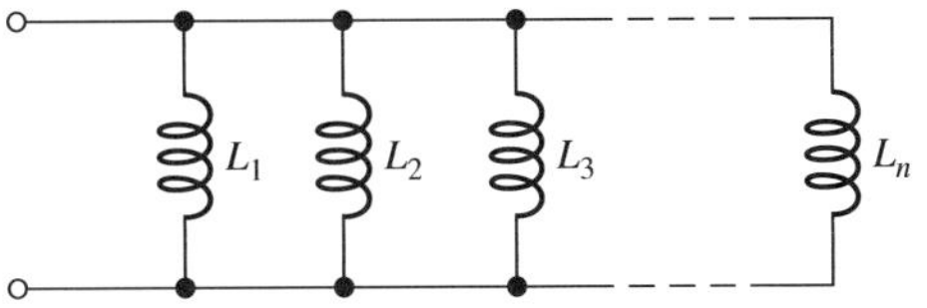

FIGURE 11–15

Inductors in parallel.

Total inductance, L_T, can be found by taking the reciprocals of both sides of Equation 11–4.

Equation 11–5

$$L_T = \frac{1}{\dfrac{1}{L_1} + \dfrac{1}{L_2} + \dfrac{1}{L_3} + \cdots + \dfrac{1}{L_n}}$$

Notice that this formula for total inductance in parallel is similar to the formulas for total parallel resistance (Equation 5–4) and total series capacitance (Equation 9–11).

EXAMPLE 11–3

Determine I_T in Figure 11–16.

FIGURE 11–16

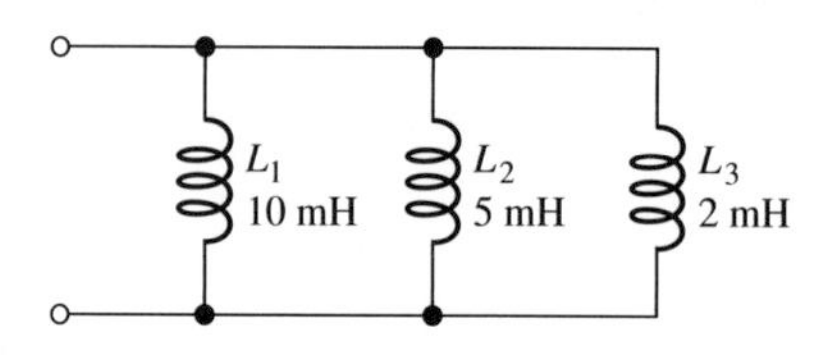

Solution

$$L_T = \frac{1}{\dfrac{1}{L_1} + \dfrac{1}{L_2} + \dfrac{1}{L_3}} = \frac{1}{\dfrac{1}{10\text{ mH}} + \dfrac{1}{5\text{ mH}} + \dfrac{1}{2\text{ mH}}} = \frac{1}{0.8}\text{ mH} = \mathbf{1.25\text{ mH}}$$

10E⁻3⁻¹+5E⁻3⁻¹+2E⁻3⁻¹
800E0
Ans⁻¹
1.25E⁻3

Related Problem Determine L_T for the following inductors in parallel: 50 μH, 80 μH, 100 μH, and 150 μH.

SECTION 11–4 REVIEW

1. Compare the total inductance in parallel with the smallest-value individual inductor.
2. The calculation of total parallel inductance is similar to that for parallel resistance. (True or False)
3. Determine L_T for each parallel combination:
   (a) 100 mH, 50 mH, and 10 mH
   (b) 40 μH and 60 μH
   (c) Ten 1 H coils

11–5 INDUCTORS IN DC CIRCUITS

An inductor will energize when it is connected to a dc voltage source. The buildup of current through the inductor occurs in a predictable manner, which is dependent on both the inductance and the resistance in a circuit. In this section, multiple switches are opened and closed simultaneously for illustration. Although this isn't practical, it serves to illustrate fundamental ideas in transient inductive circuits.

After completing this section, you should be able to

- **Analyze inductive dc switching circuits**
- Describe the energizing and deenergizing of an inductor
- Define *RL time constant*
- Relate the time constant to the energizing and deenergizing of an inductor
- Describe induced voltage
- Write the exponential equations for current in an inductor

When there is constant direct current in an inductor, there is no induced voltage. There is, however, a voltage drop due to the winding resistance of the coil. The inductance itself appears as a short to dc. Energy is stored in the magnetic field according to the formula $W = \frac{1}{2}LI^2$. The only energy conversion to heat occurs in the winding resistance ($P = I^2R_W$). This condition is illustrated in Figure 11–17.

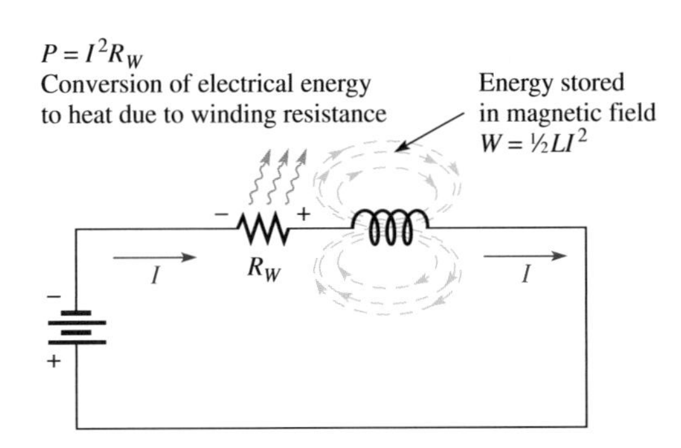

FIGURE 11–17

Energy storage and conversion to heat in an inductor.

The *RL* Time Constant

Because the inductor's basic action is to develop a voltage that opposes a change in its current, it follows that current cannot change instantaneously in an inductor. A certain time is required for the current to make a change from one value to another. The rate at which the current changes is determined by the **RL time constant**. The time constant for a series *RL* circuit is

$$\tau = \frac{L}{R}$$

Equation 11–6

where τ is in seconds when inductance (L) is in henries and resistance (R) is in ohms.

EXAMPLE 11–4

A series *RL* circuit has a resistance of 1.0 kΩ and an inductance of 1 mH. What is the time constant?

Solution

$$\tau = \frac{L}{R} = \frac{1 \text{ mH}}{1.0 \text{ k}\Omega} = \frac{1 \times 10^{-3} \text{ H}}{1 \times 10^{3} \text{ }\Omega} = 1 \times 10^{-6} \text{ s} = \mathbf{1 \text{ } \mu s}$$

Related Problem Find the time constant for $R = 2.2 \text{ k}\Omega$ and $L = 500 \text{ }\mu\text{H}$.

Energizing Current in an Inductor

In a series *RL* circuit, the current will increase to approximately 63% of its full value in one time-constant interval after the switch is closed. This buildup of current (analogous to the buildup of capacitor voltage during the charging in an *RC* circuit) follows an **exponential** curve and reaches the approximate percentage of final current as indicated in Table 11–1 and as illustrated in Figure 11–18.

TABLE 11–1

Percentage of final current after each time-constant interval during current buildup.

NUMBER OF TIME CONSTANTS	APPROXIMATE % OF FINAL CURRENT
1	63
2	86
3	95
4	98
5	99 (considered 100%)

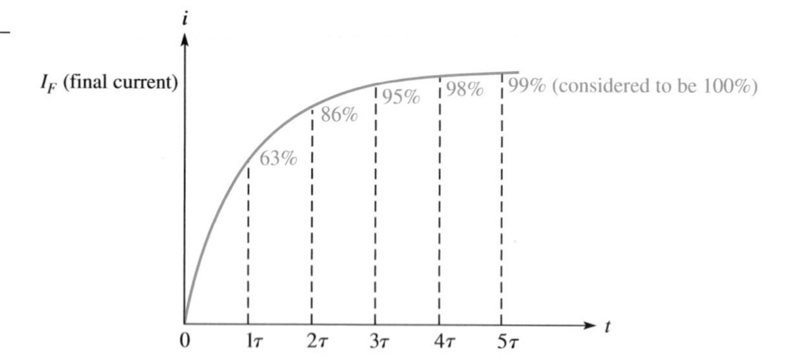

FIGURE 11–18

Energizing current in an inductor.

The change in current over five time-constant intervals is illustrated in Figure 11–19. When the current reaches its final value at approximately 5τ, it ceases to change. At this time, the inductor acts as a short (except for winding resistance) to the constant current. The final value of the current is

$$I_F = \frac{V_S}{R} = \frac{10 \text{ V}}{1.0 \text{ k}\Omega} = 10 \text{ mA}$$

(a) Initially ($i = 0$)

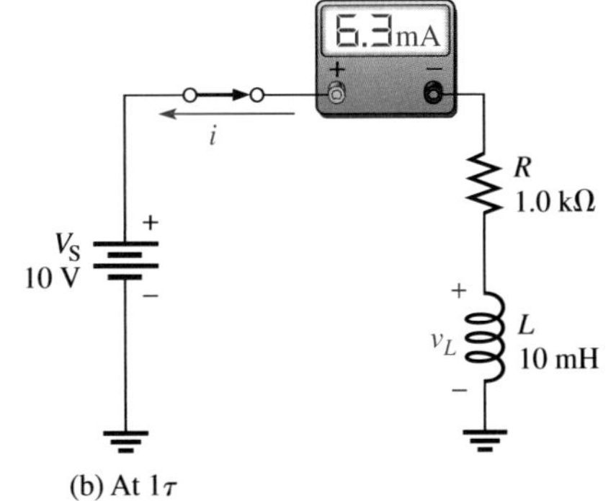

(b) At 1τ

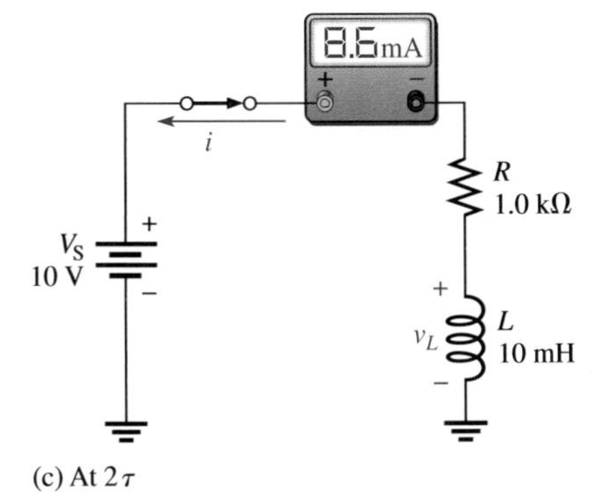

(c) At 2τ

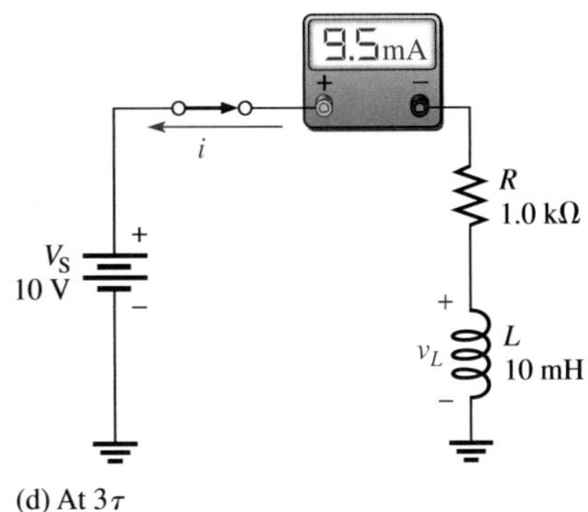

(d) At 3τ

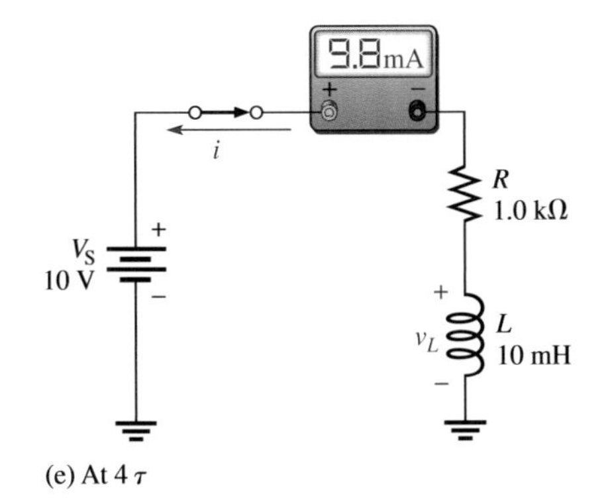

(e) At 4τ

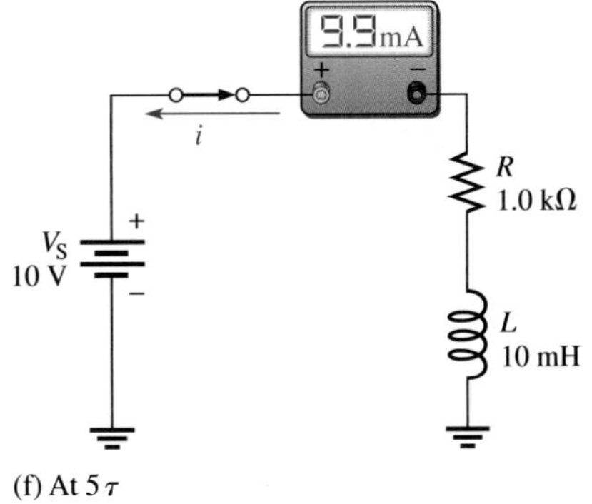

(f) At 5τ

FIGURE 11–19

Illustration of the exponential buildup of current in an inductor. The current increases approximately 63% during each time-constant interval. A voltage (v_L) is induced in the coil that tends to oppose the increase in current.

EXAMPLE 11–5

Calculate the time constant for Figure 11–20. Then determine the current and the time at each time-constant interval, measured from the instant the switch is closed.

FIGURE 11–20

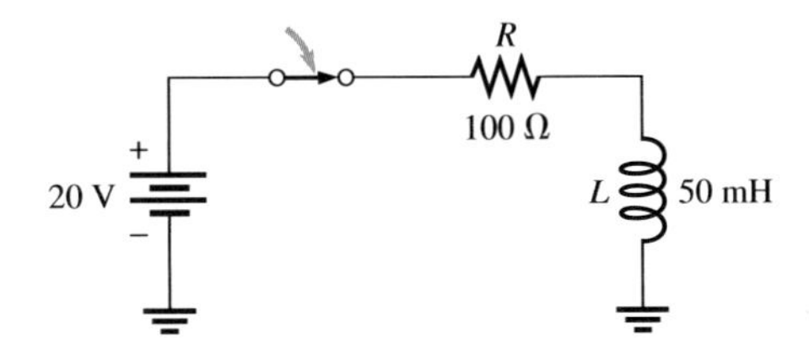

Solution The time constant is

$$\tau = \frac{L}{R} = \frac{50\text{ mH}}{100\ \Omega} = \mathbf{500\ \mu s}$$

The final current is

$$I_F = \frac{V_S}{R} = \frac{20\text{ V}}{100\ \Omega} = 200\text{ mA}$$

Use the time-constant percentage values from Table 11–1.

$$\begin{aligned}
\text{At } 1\tau = 500\ \mu\text{s:}\quad & i = 0.63(200\text{ mA}) = \mathbf{126\ mA}\\
\text{At } 2\tau = 1\text{ ms:}\quad & i = 0.86(200\text{ mA}) = \mathbf{172\ mA}\\
\text{At } 3\tau = 1.5\text{ ms:}\quad & i = 0.95(200\text{ mA}) = \mathbf{190\ mA}\\
\text{At } 4\tau = 2\text{ ms:}\quad & i = 0.98(200\text{ mA}) = \mathbf{196\ mA}\\
\text{At } 5\tau = 2.5\text{ ms:}\quad & i = 0.99(200\text{ mA}) = 198\text{ mA} \cong \mathbf{200\ mA}
\end{aligned}$$

Related Problem Repeat the calculations if $R = 680\ \Omega$ and $L = 100\ \mu$H.

Deenergizing Current in an Inductor

Current in an inductor decreases exponentially according to the approximate percentage values shown in Table 11–2 and in Figure 11–21.

TABLE 11–2

Percentage of initial current after each time-constant interval while current is decreasing.

NUMBER OF TIME CONSTANTS	APPROXIMATE % OF INITIAL CURRENT
1	37
2	14
3	5
4	2
5	1 (considered 0)

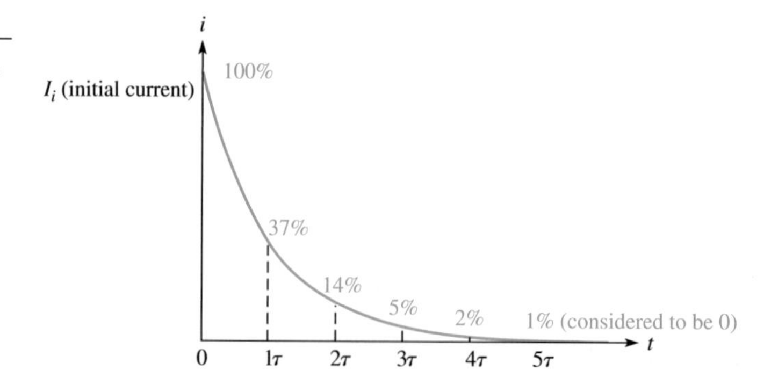

FIGURE 11–21

Deenergizing current in an inductor.

Figure 11–22(a) shows a constant current of 1 A (1000 mA) through an inductor. Then switch 1 (SW1) is opened and switch 2 (SW2) is closed simultaneously, and for an instant the induced voltage keeps the 1 A current through the inductor. During the first time-constant interval, the current decreases by 63% down to

FIGURE 11–22

Illustration of the exponential decrease of current in an inductor. The current decreases approximately 63% during each time-constant interval.

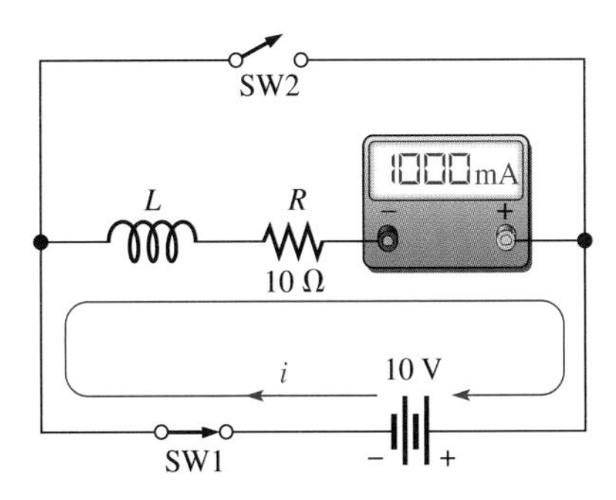

(a) Initially, there is 1 A of constant current. Then SW2 is closed and SW1 opened simultaneously ($t = 0$).

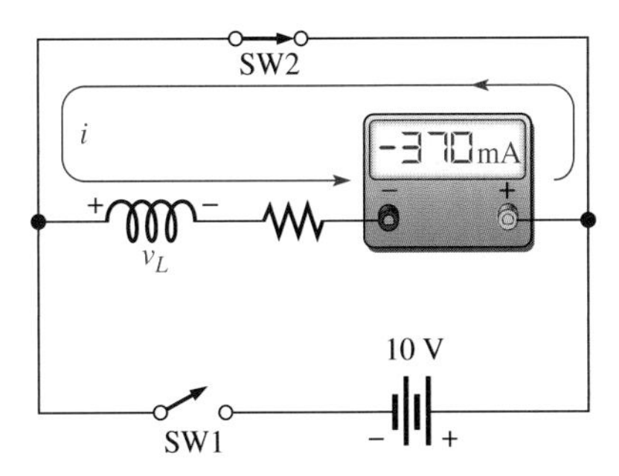

(b) At the end of one time constant ($t = 1\tau$)

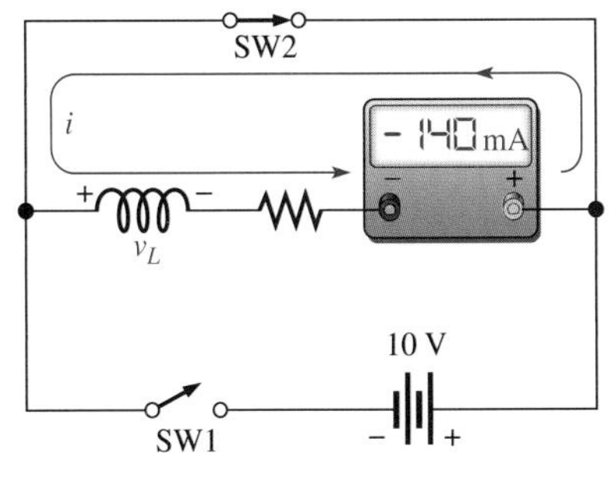

(c) At the end of two time constants ($t = 2\tau$)

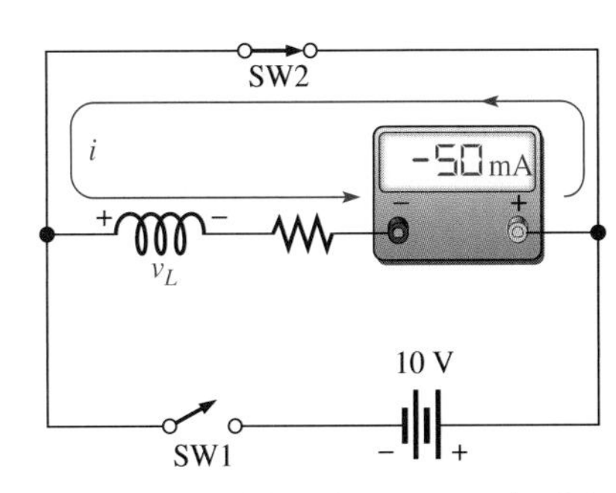

(d) At the end of three time constants ($t = 3\tau$)

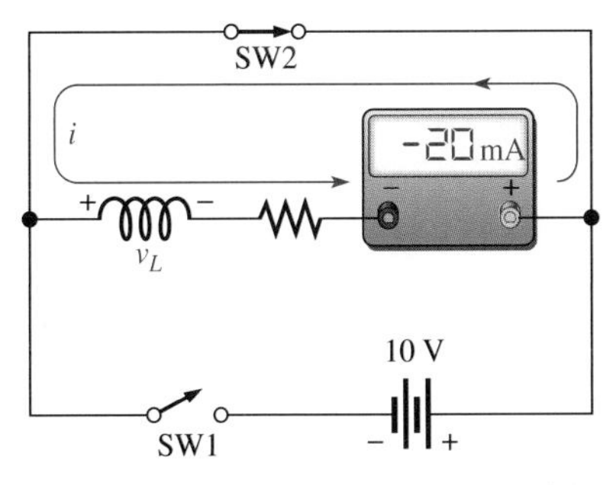

(e) At the end of four time constants ($t = 4\tau$)

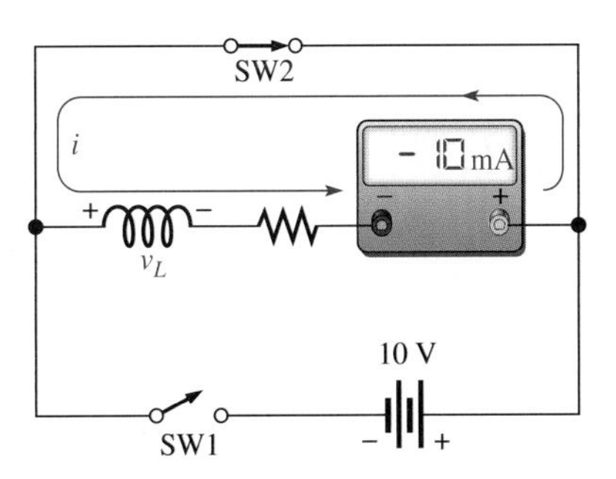

(f) At the end of five time constants ($t = 5\tau$). Since only 1% of the current is left, this value is taken as the final zero value.

370 mA (37% of its initial value), as indicated in Figure 11–22(b). During the second time-constant interval, the current decreases by another 63% to 140 mA (14% of its initial value), as shown in Figure 11–22(c). The continued decrease in the current is illustrated by the remaining parts of Figure 11–22. Part (f) shows that only 1% of the initial current is left at the end of five time constants. Traditionally, this value is accepted as the final value and is approximated as zero current. Notice that until after the five time constants have elapsed, there is an induced voltage across the coil which is trying to maintain the current. This voltage follows a decreasing exponential curve, as will be discussed later.

EXAMPLE 11–6

In Figure 11–23, switch 1 (SW1) is opened at the instant that switch 2 (SW2) is closed.

(a) What is the time constant?

(b) What is the initial coil current at the instant of switching?

(c) What is the coil current at 1τ?

Assume steady-state current through the coil prior to switch change.

▶ **FIGURE 11–23**

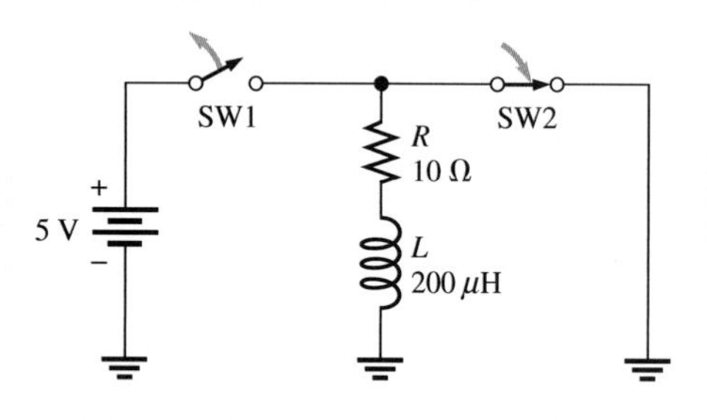

Solution **(a)** $\tau = \frac{L}{R} = \frac{200\ \mu\text{H}}{10\ \Omega} = \mathbf{20\ \mu s}$

(b) Current cannot change instantaneously in an inductor. Therefore, the current at the instant of the switch change is the same as the steady-state current.

$$I = \frac{5\text{ V}}{10\ \Omega} = \mathbf{500\ mA}$$

(c) At 1τ, the current has decreased to approximately 37% of its initial value.

$$i = 0.37(500\text{ mA}) = \mathbf{185\ mA}$$

Related Problem Change R to 47 Ω and L to 1 mH in Figure 11–23 and repeat each calculation.

Induced Voltage in a Series *RL* Circuit

As you know, when current changes in an inductor, a voltage is induced. Let's examine what happens to the voltages across the resistor and the coil in a series circuit when a change in current occurs.

Look at the circuit in Figure 11–24(a). When the switch is open, there is no current, and the resistor voltage and the coil voltage are both zero. At the instant the switch is closed, as indicated in part (b), V_R is zero and V_L is 10 V. The reason for this change is that the induced voltage across the coil is equal and opposite to the source voltage, preventing the current from changing instantaneously. Therefore, *at the instant of switch closure, the inductor effectively acts as an open with all the source voltage across it.*

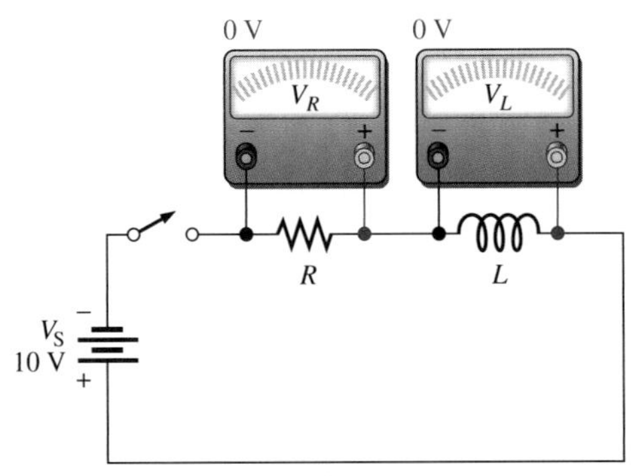

(a) Before switched is closed

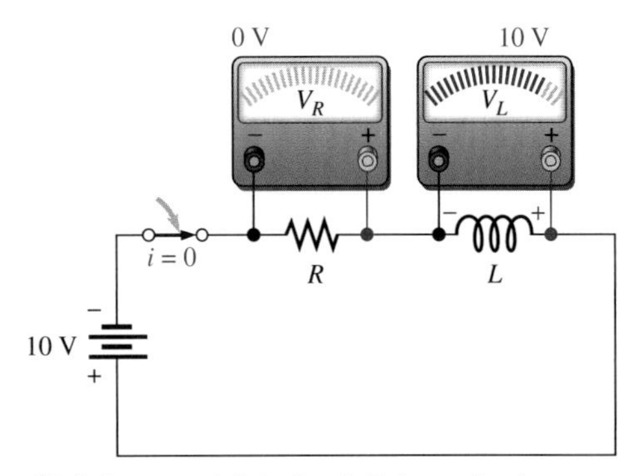

(b) At instant switch is closed, V_L is equal and opposite to V_S.

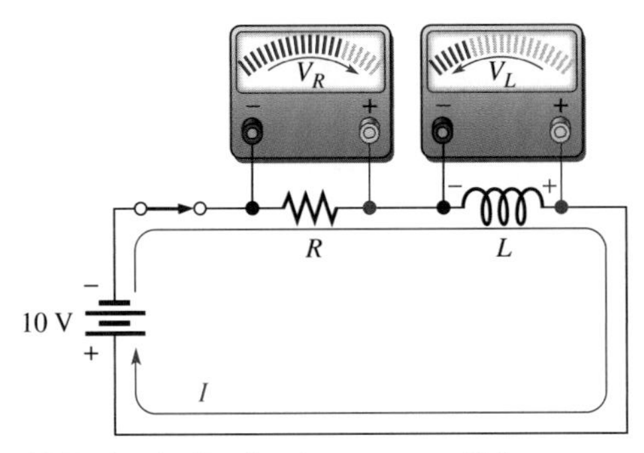

(c) During the first five time constants, V_R increases exponentially with current, and V_L decreases exponentially.

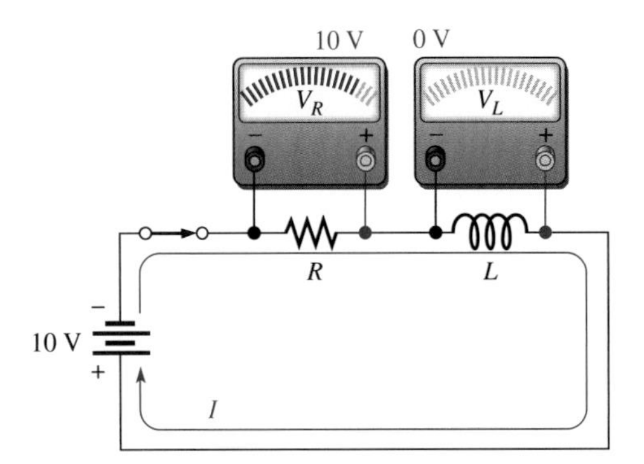

(d) After the first five time constants, $V_R = 10$ V and $V_L \cong 0$ V. The current is at a constant maximum value.

FIGURE 11–24

Illustration of how the coil voltage and the resistor voltage change in response to an increase in current. The winding resistance is neglected.

During the first five time constants, the current is building up exponentially, and the induced coil voltage is decreasing. The resistor voltage increases with the current, as Figure 11–24(c) illustrates. After five time constants have elapsed, the current has reached its final value, V_S/R. At this time, all of the source voltage is dropped across the resistor and none across the coil. Thus, the inductor effectively acts as a short to nonchanging current, as Figure 11–24(d) illustrates. Keep in mind that the inductor always reacts to a change in current by creating an induced voltage in order to counteract that change in current.

Now let's examine the case illustrated in Figure 11–25, where the current is switched out, and the inductor discharges through another path. Part (a) shows the steady-state condition, and part (b) illustrates the instant at which the source is removed by opening SW1 and the discharge path is connected by the closure of SW2. There was 1 A through the inductor prior to this. Notice that 10 V are induced in the inductor in the direction to aid the 1 A in an effort to keep it from changing. Then, as shown in Figure 11–25(c), the current decays exponentially, and so do V_R and V_L. After 5τ, as shown in Figure 11–25(d), all the energy stored in the magnetic field of the inductor is dissipated, and all values are zero.

▶ **FIGURE 11–25**

Illustration of how the coil voltage and the resistor voltage change in response to a decrease in current. The winding resistance is neglected.

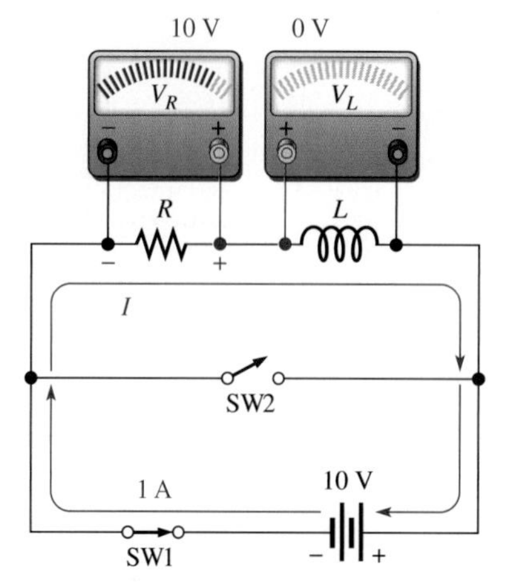

(a) Initially, there is a constant 1 A and the voltage is dropped across R.

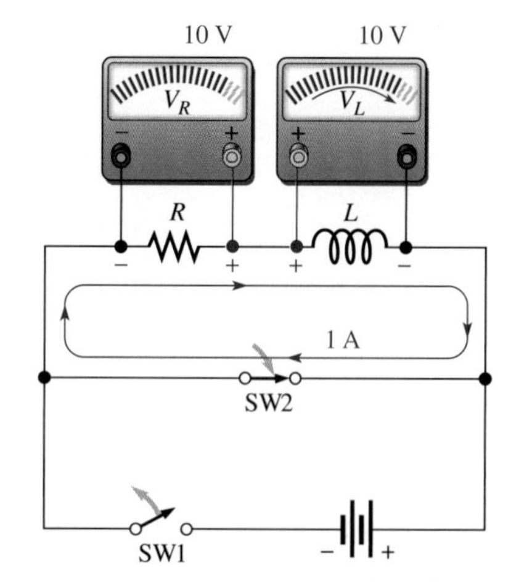

(b) At the instant that SW1 is opened and SW2 closed, 10 V is induced across L.

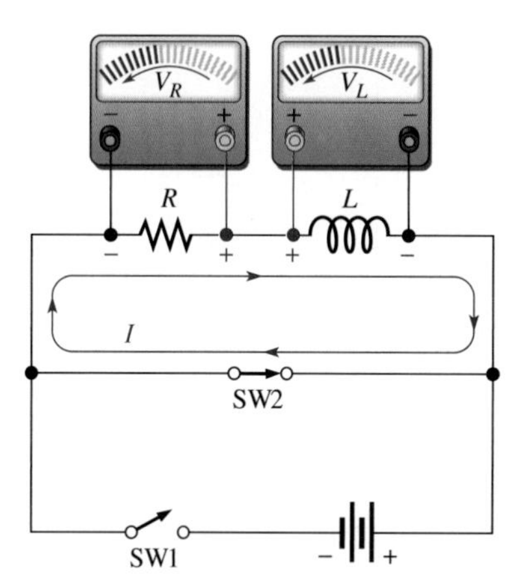

(c) During the five-time-constant interval, V_R and V_L decrease exponentially with the current.

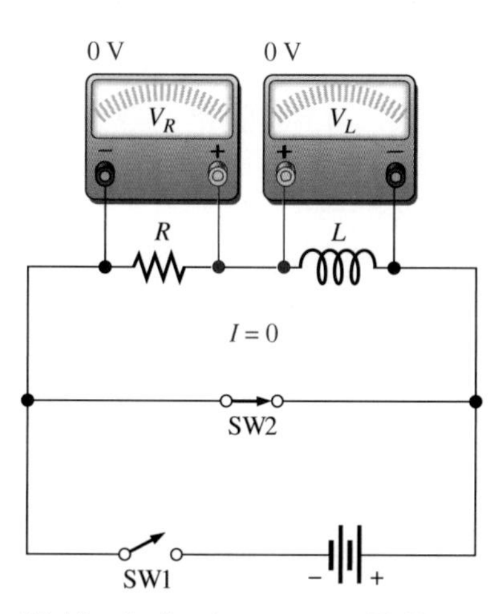

(d) After the five time constants, V_R, V_L, and I are all zero.

EXAMPLE 11–7

(a) In Figure 11–26(a), what is v_L at the instant the switch (SW) is closed? What is v_L after 5τ?

(b) In Figure 11–26(b), what is v_L at the instant SW1 opens and SW2 closes? What is v_L after 5τ?

▶ **FIGURE 11–26**

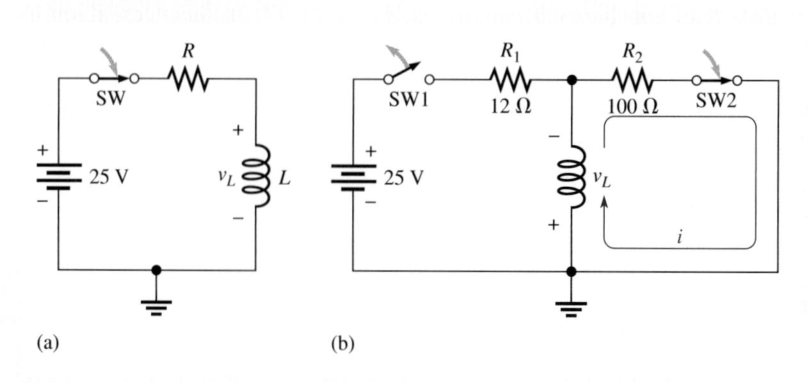

Solution **(a)** At the instant the switch is closed, all the source voltage is across the inductor. Thus, $v_L = \mathbf{25\ V}$, with the polarity as shown. After 5τ, the inductor acts as a short, so $v_L = \mathbf{0\ V}$.

(b) With SW1 closed and SW2 open, the steady-state current is

$$I = \frac{25\text{ V}}{12\ \Omega} = \mathbf{2.08\ A}$$

When the switches are thrown, an induced voltage is created across the inductor sufficient to keep this 2.08 A current for an instant. In this case, it takes an induced voltage of

$$v_L = IR_2 = (2.08\text{ A})(100\ \Omega) = \mathbf{208\ V}$$

After 5τ, the inductor voltage is **zero.**

Related Problem Repeat part (a) if the voltage source is 9 V. Repeat part (b) if R_1 is 27 Ω with the source voltage at 25 V.

The Exponential Formulas

The formulas for the exponential current and voltage in an *RL* circuit are similar to those used in Chapter 9 for the *RC* circuit, and the universal exponential curves in Figure 9–35 apply to inductors as well as capacitors. The general formulas for *RL* circuits are stated as follows:

$$v = V_F + (V_i - V_F)e^{-Rt/L}$$ **Equation 11–7**

$$i = I_F + (I_i - I_F)e^{-Rt/L}$$ **Equation 11–8**

where V_F and I_F are the final values, V_i and I_i are the initial values, and v and i are the instantaneous values of the inductor voltage or current at time t.

Increasing Current The formula for the special case in which an increasing exponential current curve begins at zero ($I_i = 0$) is

$$i = I_F(1 - e^{-Rt/L})$$ **Equation 11–9**

Using Equation 11–9, you can calculate the value of the increasing inductor current at any instant of time. The same is true for voltage by substituting V for I in Equation 11–9.

EXAMPLE 11–8

In Figure 11–27, determine the inductor current 30 μs after the switch is closed.

FIGURE 11–27

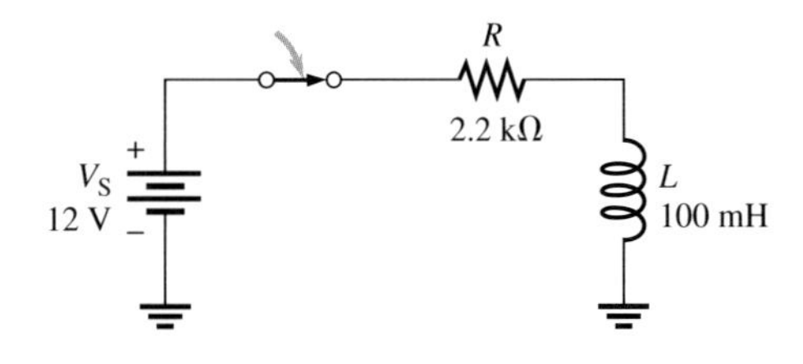

Solution The time constant is

$$\tau = \frac{L}{R} = \frac{100\text{ mH}}{2.2\text{ k}\Omega} = 45.5\ \mu\text{s}$$

The final current is

$$I_F = \frac{V_S}{R} = \frac{12\text{ V}}{2.2\text{ k}\Omega} = 5.46\text{ mA}$$

The initial current is zero. Notice that 30 μs is less than one time constant, so the current will reach less than 63% of its final value in that time.

$$i_L = I_F(1 - e^{-Rt/L}) = 5.46\text{ mA}(1 - e^{-0.66}) = \mathbf{2.64\ mA}$$

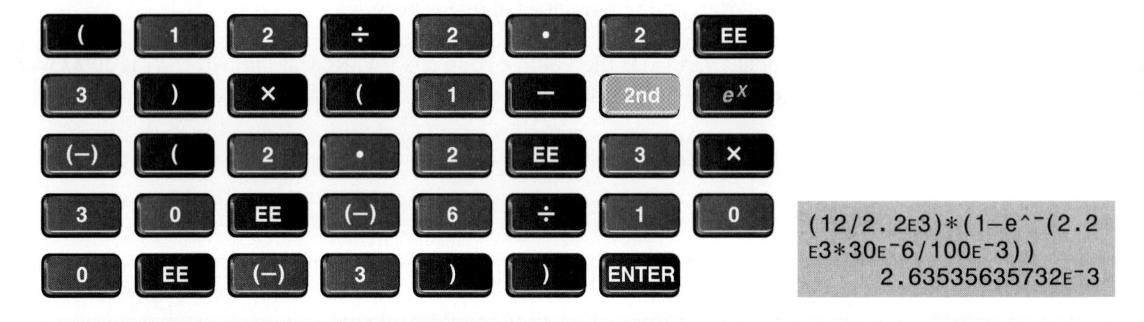

Related Problem In Figure 11–27, determine the inductor current 55 μs after the switch is closed.

Decreasing Current The formula for the special case in which a decreasing exponential current has a final value of zero is

Equation 11–10

$$i = I_i e^{-Rt/L}$$

This formula can be used to calculate the deenergizing current at any instant, as the next example shows.

EXAMPLE 11–9

Determine the inductor current in Figure 11–28 at a point in time 2 ms after the switches are thrown (SW1 opened and SW2 closed).

▶ FIGURE 11–28

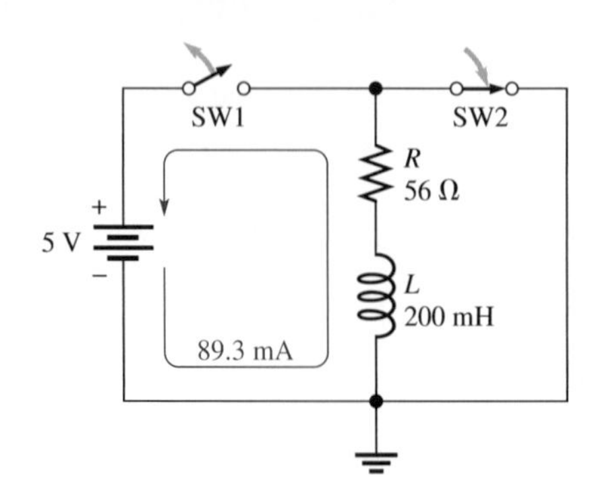

Solution The deenergizing time constant is

$$\tau = \frac{L}{R} = \frac{200\ \text{mH}}{56\ \Omega} = 3.57\ \text{ms}$$

The initial current in the inductor is 89.3 mA. Notice that 2 ms is less than one time constant, so the current will show a decrease less than 63%. Therefore, the current will be greater than 37% of its initial value at 2 ms after the switches are thrown.

$$i_L = I_i e^{-Rt/L} = (89.3\ \text{mA})e^{-0.56} = \mathbf{51.0\ mA}$$

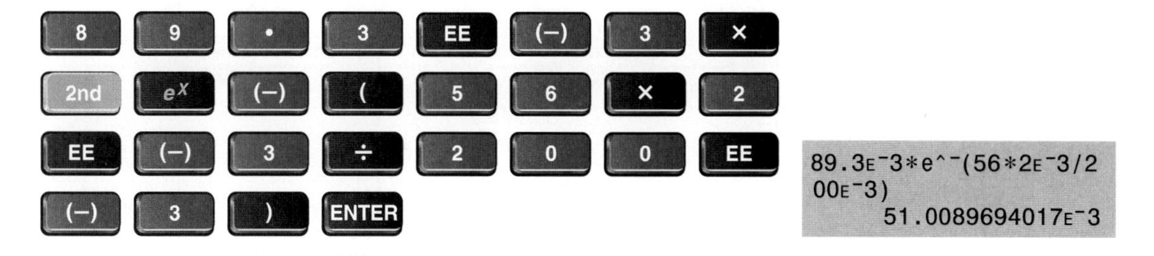

Related Problem Determine the inductor current in Figure 11–28 at a point in time 6 ms after SW1 is opened and SW2 is closed if the source voltage is changed to 10 V.

SECTION 11–5 REVIEW

1. A 15 mH inductor with a winding resistance of 10 Ω has a constant direct current of 10 mA through it. What is the voltage drop across the inductor?
2. A 20 V dc source is connected to a series *RL* circuit with a switch. At the instant of switch closure, what are the values of v_R and v_L?
3. In the same circuit as in Question 2, after a time interval equal to 5τ from switch closure, what are v_R and v_L?
4. In a series *RL* circuit where $R = 1.0\ \text{k}\Omega$ and $L = 500\ \mu\text{H}$, what is the time constant? Determine the current 0.25 μs after a switch connects 10 V across the circuit.

11–6 INDUCTORS IN AC CIRCUITS

You will learn in this section that an inductor passes ac but with an amount of opposition that depends on the frequency of the ac.

After completing this section, you should be able to

- **Analyze inductive ac circuits**
- Define *inductive reactance*
- Determine the value of inductive reactance in a given circuit
- Discuss instantaneous, true, and reactive power in an inductor

Inductive Reactance, X_L

In Figure 11–29, an inductor is shown connected to a sinusoidal voltage source. When the source voltage is held at a constant amplitude value and its frequency is increased, the amplitude of the current decreases. Also, when the frequency of the source is decreased, the current amplitude increases.

FIGURE 11–29

The current in an inductive circuit varies inversely with the frequency of the source voltage.

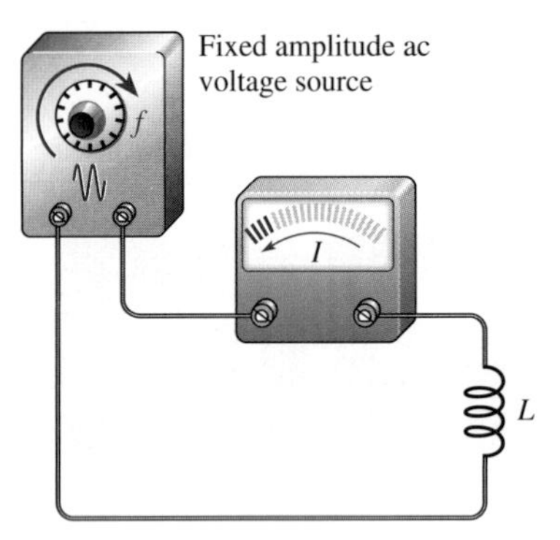

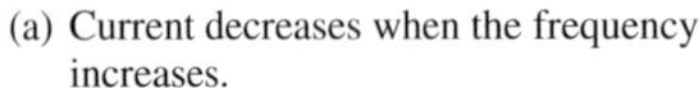

(a) Current decreases when the frequency increases.

(b) Current increases when the frequency decreases.

When the frequency of the source voltage increases, its rate of change also increases, as you already know. Now, if the frequency of the source voltage is increased, the frequency of the current also increases. According to Faraday's and Lenz's laws, this increase in frequency induces more voltage across the inductor in a direction to oppose the current and causes it to decrease in amplitude. Similarly, a decrease in frequency will cause an increase in current.

A decrease in the amount of current with an increase in frequency for a fixed amount of voltage indicates that opposition to the current has increased. Thus, the inductor offers opposition to current, and that opposition varies *directly* with frequency.

The opposition to sinusoidal current in an inductor is called inductive reactance.

The symbol for inductive reactance is X_L, and its unit is the ohm (Ω).

You have just seen how frequency affects the opposition to current (inductive reactance) in an inductor. Now let's see how the inductance, L, affects the reactance. Figure 11–30(a) shows that when a sinusoidal voltage with a fixed amplitude and fixed frequency is applied to a 1 mH inductor, there is a certain amount of alternating current. When the inductance value is increased to 2 mH, the current decreases, as shown in part (b). Thus, when the inductance is increased, the

FIGURE 11–30

For a fixed voltage and fixed frequency, the current varies inversely with the inductance value.

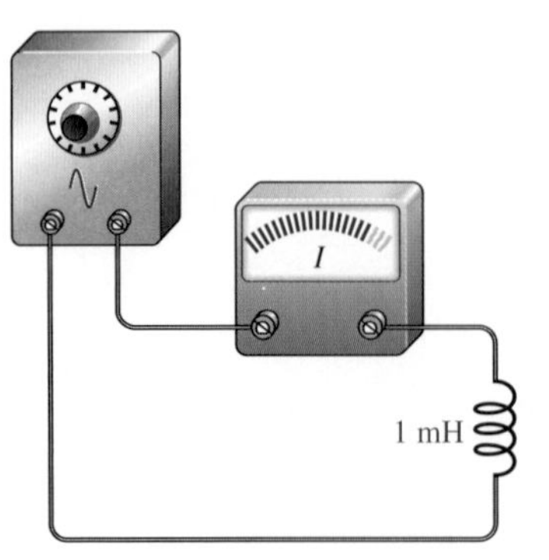

(a) Less inductance, more current.

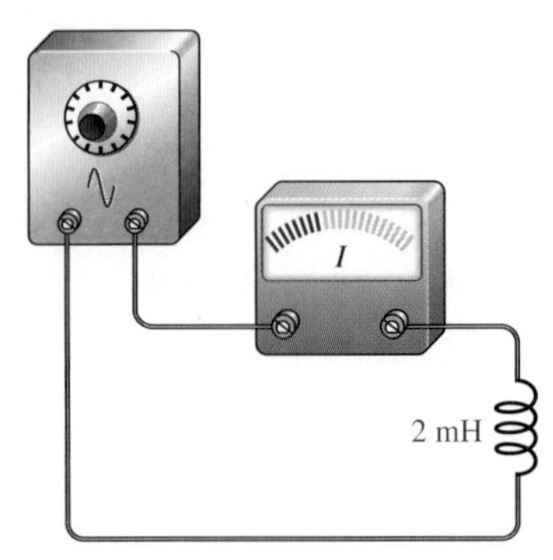

(b) More inductance, less current.

opposition to current (inductive reactance) increases. So not only is the inductive reactance *directly* proportional to frequency, but it is also *directly* proportional to inductance. This relationship can be stated as follows:

X_L is proportional to fL.

It can be proven that the constant of proportionality is 2π, so the formula for inductive reactance (X_L) is

$$X_L = 2\pi fL$$

Equation 11–11

As with capacitive reactance, the 2π term comes from the relationship of the sine wave to rotational motion. X_L is in ohms when f is in hertz and L is in henries.

EXAMPLE 11–10

A sinusoidal voltage is applied to the circuit in Figure 11–31. The frequency is 1 kHz. Determine the inductive reactance.

FIGURE 11–31

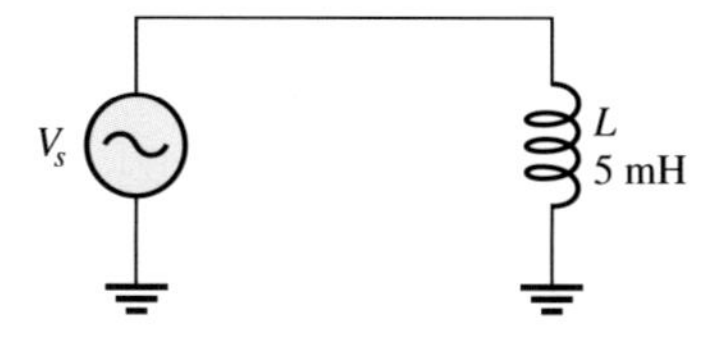

Solution Convert 1 kHz to 1×10^3 Hz and 5 mH to 5×10^{-3} H. Then, the inductive reactance is

$$X_L = 2\pi fL = 2\pi(1 \times 10^3\ \text{Hz})(5 \times 10^{-3}\ \text{H}) = \mathbf{31.4\ \Omega}$$

Related Problem What is X_L in Figure 11–31 if the frequency is increased to 3.5 kHz?

Ohm's Law in Inductive AC Circuits

The reactance of an inductor is analogous to the resistance of a resistor. In fact, X_L, just like X_C and R, is expressed in ohms. Since inductive reactance is a form of opposition to current, Ohm's law applies to inductive circuits as well as to resistive circuits and capacitive circuits; and it is stated as follows:

$$V = IX_L$$

Equation 11–12

When applying Ohm's law in ac circuits, you must express both the current and the voltage in the same way, that is, both in rms, both in peak, and so on.

EXAMPLE 11–11

Determine the rms current in Figure 11–32.

FIGURE 11–32

Solution Convert 10 kHz to 10×10^3 Hz and 100 mH to 100×10^{-3} H. Then calculate X_L.

$$X_L = 2\pi fL = 2\pi(10 \times 10^3 \text{ Hz})(100 \times 10^{-3} \text{ H}) = 6283 \ \Omega$$

Apply Ohm's law.

$$I_{rms} = \frac{V_{rms}}{X_L} = \frac{5 \text{ V}}{6283 \ \Omega} = \mathbf{796 \ \mu A}$$

Related Problem Determine the rms current in Figure 11–32 for the following values: $V_{rms} =$ 12 V, $f = 4.9$ kHz, and $L = 680 \ \mu$H.

Open file E11-11 on your EWB/CircuitMaker CD-ROM. Measure the rms current and compare to the calculated value. Change the circuit values to those given in the related problem and measure the rms current.

Current Lags Inductor Voltage by 90°

As you know, a sinusoidal voltage has a maximum rate of change at its zero crossings and a zero rate of change at the peaks. From Faraday's law (Chapter 7) you know that the amount of voltage induced across a coil is directly proportional to the rate at which the current is changing. Therefore, the coil voltage is maximum at the zero crossings of the current where the rate of change of the current is the greatest. Also, the amount of voltage is zero at the peaks of the current where the rate of change is zero. This phase relationship is illustrated in Figure 11–33. As you can see, the current peaks occur a quarter cycle after the voltage peaks. Thus, the current *lags* the voltage by 90°. Recall that in a capacitor, the current *leads* the voltage by 90°.

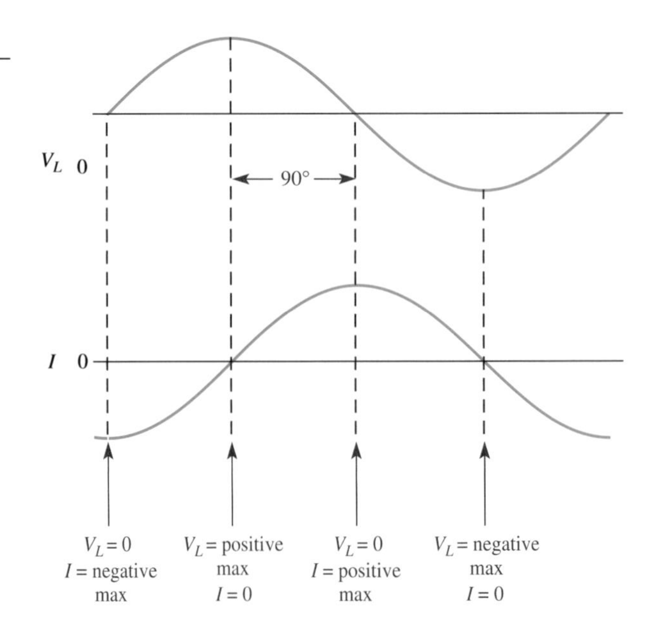

FIGURE 11–33

Current is always lagging the inductor voltage by 90°.

Power in an Inductor

As discussed earlier, an inductor stores energy in its magnetic field when there is current through it. An ideal inductor (assuming no winding resistance) does not dissipate energy; it only stores it. When an ac voltage is applied to an inductor, energy is stored by the inductor during a portion of the cycle; then the stored energy is returned to the source during another portion of the cycle. No net energy is lost in an ideal inductor due to conversion to heat. Figure 11–34 shows the power curve that results from one cycle of inductor current and voltage.

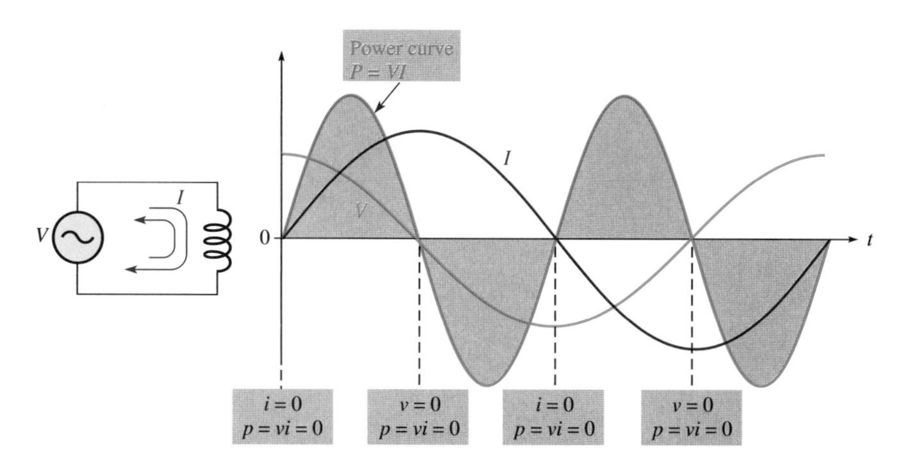

FIGURE 11–34

Power curve for an inductor.

Instantaneous Power (p) The product of instantaneous voltage, *v*, and instantaneous current, *i*, gives instantaneous power, *p*. At points where *v* or *i* is zero, *p* is also zero. When both *v* and *i* are positive, *p* is also positive. When either *v* or *i* is positive and the other negative, *p* is negative. When both *v* and *i* are negative, *p* is positive. As you can see in Figure 11–34, the power follows a sinusoidal-shaped curve. Positive values of power indicate that energy is stored by the inductor. Negative values of power indicate that energy is returned from the inductor to the source. Note that the power fluctuates at a frequency twice that of the voltage or current, as energy is alternately stored and returned to the source.

True Power (P_{true}) Ideally, all of the energy stored by an inductor during the positive portion of the power cycle is returned to the source during the negative portion. No net energy is lost due to conversion of heat in the inductance, so the power is zero. Actually, because of winding resistance in a practical inductor, some power is always dissipated, and there is a very small amount of true power, which can normally be neglected.

$$P_{true} = I^2_{rms}R_W$$

Equation 11–13

Reactive Power (P_r) The rate at which an inductor stores or returns energy is called its **reactive power,** P_r, with a unit of VAR (volt-ampere reactive). The reactive power is a nonzero quantity because at any instant in time the inductor is actually taking energy from the source or returning energy to it. Reactive power does not represent an energy loss due to conversion of heat. The following formulas apply:

$$P_r = V_{rms}I_{rms}$$

Equation 11–14

$$P_r = \frac{V^2_{rms}}{X_L}$$

Equation 11–15

$$P_r = I^2_{rms}X_L$$

Equation 11–16

EXAMPLE 11–12

A 10 V rms signal with a frequency of 1 kHz is applied to a 10 mH coil with a winding resistance of 5 Ω. Determine the reactive power (P_r) and the true power (P_{true}).

Solution First, calculate the inductive reactance and current values.

$$X_L = 2\pi fL = 2\pi(1\text{ kHz})(10\text{ mH}) = 62.8\ \Omega$$

$$I = \frac{V_S}{X_L} = \frac{10\text{ V}}{62.8\ \Omega} = 159\text{ mA}$$

Then using Equation 11–16,

$$P_r = I^2X_L = (159\text{ mA})^2(62.8\ \Omega) = \mathbf{1.59\ VAR}$$

The true power is

$$P_{true} = I^2R_W = (159\text{ mA})^2(5\ \Omega) = \mathbf{126\ mW}$$

Related Problem What happens to the reactive power if the frequency increases?

The Quality Factor (Q) of a Coil

The quality factor (Q) is the ratio of the reactive power in the inductor to the true power in the winding resistance of the coil or the resistance in series with the coil. It is a ratio of the power in L to the power in R_W. The quality factor is very important in resonant circuits, which are studied in Chapter 13. A formula for Q is developed as follows:

$$Q = \frac{\text{reactive power}}{\text{true power}} = \frac{P_r}{P_{true}} = \frac{I^2X_L}{I^2R_W}$$

In a series circuit, I is the same in L and R; thus, the I^2 terms cancel, leaving

Equation 11–17

$$Q = \frac{X_L}{R_W}$$

When the resistance is just the winding resistance of the coil, the circuit Q and the coil Q are the same. Note that Q is a ratio of like units and, therefore, has no unit itself.

SECTION 11–6 REVIEW

1. State the phase relationship between current and voltage in an inductor.
2. Calculate X_L for $f = 5$ kHz and $L = 100$ mH.
3. At what frequency is the reactance of a 50 μH inductor equal to 800 Ω?
4. Calculate the rms current in Figure 11–35.
5. An ideal 50 mH inductor is connected to a 12 V rms source. What is the true power? What is the reactive power at a frequency of 1 kHz?

FIGURE 11–35

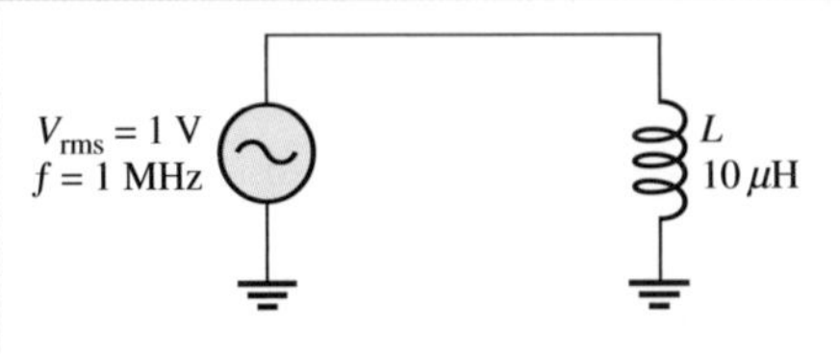

11–7 INDUCTOR APPLICATIONS

Inductors are not as versatile as capacitors and tend to be more limited in their applications due, in part, to size and cost factors. However, there are many practical uses for inductors (coils) such as those discussed in Chapter 7. Recall that relay and solenoid coils, read/write heads, speakers, and sensing elements were introduced as electromagnetic applications of coils. In this section, some additional uses of inductors are presented.

After completing this section, you should be able to

- **Discuss some inductor applications**
- Describe a power supply filter
- Explain the purpose of an RF choke
- Discuss the basics of tuned circuits

Power Supply Filter

In Chapter 9, you learned that a capacitor is used to filter the pulsating dc on the output of the rectifier in a power supply. The final output voltage was a dc voltage with a small amount of ripple. An inductor can be used in the filter, as shown in Figure 11–36(a), to smooth out the ripple voltage. The inductor, placed in series with the load as shown, tends to oppose the current fluctuations caused by the ripple voltage, and thus the voltage developed across the load is more constant, as shown in Figure 11–36(b).

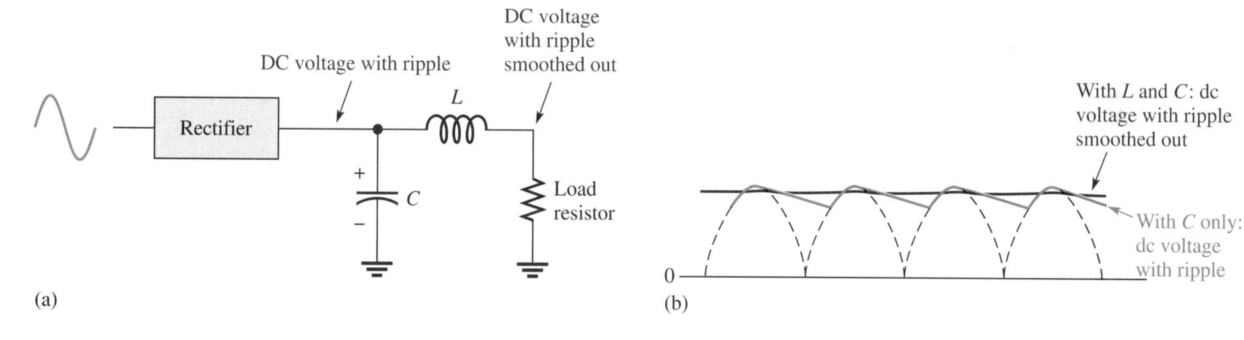

FIGURE 11–36

Basic capacitor power supply filter with a series inductor.

RF Choke

Certain types of inductors called *chokes* are used in applications where radio frequencies (RF) must be prevented from getting into parts of a system, such as the power supply or the audio section of a receiver. In these situations, an inductor is used as a series filter and "chokes" off any unwanted RF signals that may be picked up on a line. This filtering action is based on the fact that the reactance of a coil increases with frequency. When the frequency of the current is sufficiently high, the reactance of the coil becomes extremely large and essentially blocks the current. The basic illustration of an inductor used as an RF choke is shown in Figure 11–37.

Tuned Circuits

As mentioned in Chapter 9, inductors are used in conjunction with capacitors to provide frequency selection in communications systems. These tuned circuits allow a narrow band of frequencies to be selected while all other frequencies are

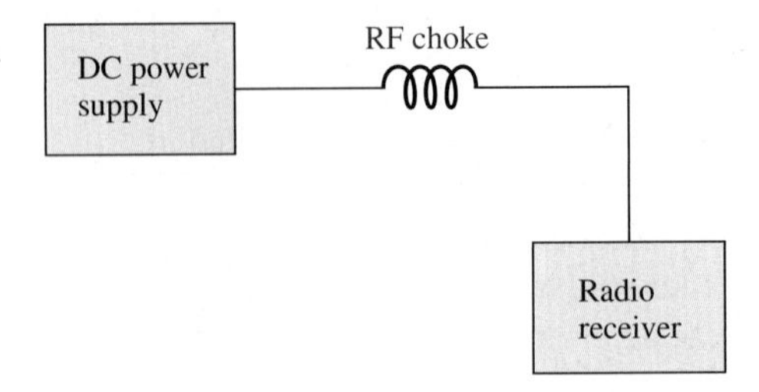

FIGURE 11–37

An inductor used as an RF choke to minimize interfering signals on the power supply line.

rejected. The tuners in your TV and radio receivers are based on this principle and permit you to select one channel or station out of the many that are available.

Frequency selectivity is based on the fact that the reactances of both capacitors and inductors depends on the frequency and on the interaction of these two components when connected in series or parallel. Since the capacitor and the inductor produce opposite phase shifts, their combined opposition to current can be used to obtain a desired response at a selected frequency. Tuned (resonant) *RLC* circuits are covered in detail in Chapter 13.

SECTION 11–7 REVIEW

1. Explain how the ripple voltage from a power supply filter can be reduced through use of an inductor.
2. How does an inductor connected in series act as an RF choke?

11–8 TESTING INDUCTORS

In this section, two basic types of failures in inductors are discussed and methods for testing inductors are introduced.

After completing this section, you should be able to

- **Test an inductor**
- Perform an ohmmeter check for an open winding
- Perform an ohmmeter check for shorted windings

The most common failure in an inductor is an open coil. To check for an open, remove the coil from the circuit. If there is an open, an ohmmeter check will indicate infinite resistance, as shown in Figure 11–38(a). If the coil is good, the ohmmeter will show the winding resistance. The value of the winding resistance depends on the wire size and length of the coil. It can be anywhere from one ohm to several hundred ohms. Figure 11–38(b) shows a good reading.

Occasionally, when an inductor is overheated with excessive current, the wire insulation will melt, and some coils will short together. This must be tested on an *LC* meter because, with two shorted turns (or even several), an ohmmeter check may show the coil to be perfectly good from a resistance standpoint. Two shorted turns occur more frequently because the turns are adjacent and can easily short across from poor insulation, voltage breakdown, or simple wear if something is rubbing on them.

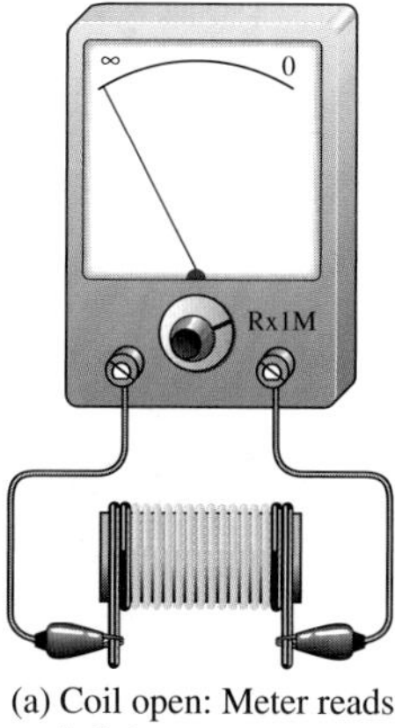

(a) Coil open: Meter reads infinity

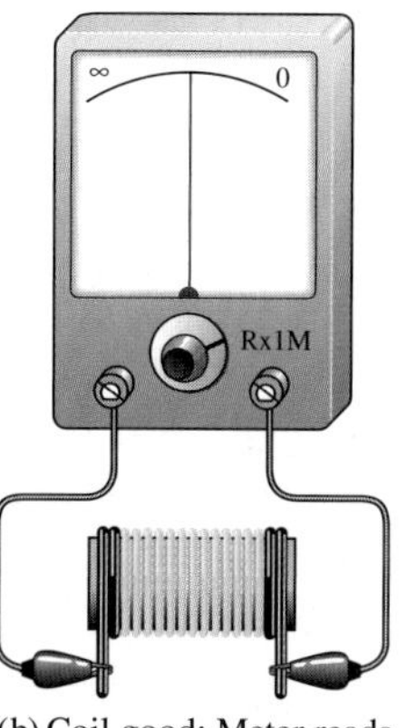

(b) Coil good: Meter reads winding resistance

FIGURE 11–38

Checking a coil by measuring the resistance.

SECTION 11–8 REVIEW

1. When a coil is checked, a reading of infinity on the ohmmeter indicates a partial short. (True or False)
2. An ohmmeter check of a good coil will indicate the value of the inductance. (True or False)

APPLICATION ASSIGNMENT

PUTTING YOUR KNOWLEDGE TO WORK

You are given two unmarked coils and asked to find their inductance values. You cannot find an inductance bridge, which is an instrument for measuring inductance directly. After scratching your head for awhile, you decide to use the time-constant characteristic of inductive circuits to determine the unknown inductances. A test setup consisting of a square wave generator and an oscilloscope is used to make the measurements.

The method is to place the coil in series with a resistor of known value and measure the time constant by applying a square wave to the circuit and observing the resulting voltage across the resistor with an oscilloscope. Knowing the time constant and the resistance value, you can calculate the inductance L.

Each time the square wave input voltage goes high, the inductor is energized; and each time the square wave goes back to zero, the inductor is deenergized. The time it takes for the exponential resistor voltage to increase to approximately its final value equals five time constants. This operation is illustrated in Figure 11–39. To make sure that the winding resistance of the coil can be neglected, it must be measured; and the value of the resistor used in the circuit must be selected to be considerably larger than the winding resistance.

FIGURE 11–39

Circuit for time-constant measurement.

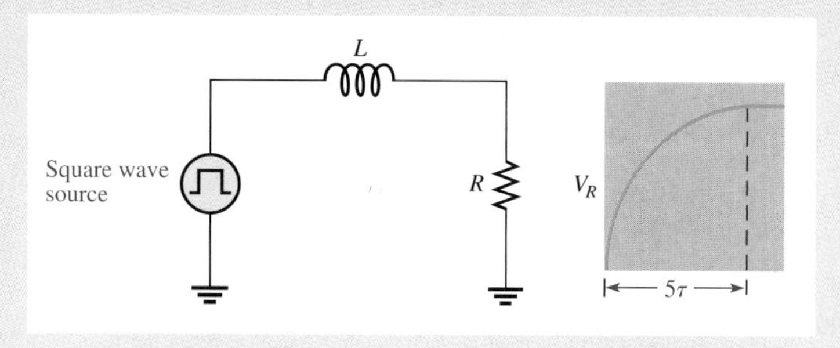

Step 1: Measure the Coil Resistance and Select a Series Resistor

Assume that the winding resistance has been measured with an ohmmeter and found to be 85 Ω. To make the winding resistance negligible, a 10 kΩ series resistor is used in the circuit.

Step 2: Determine the Inductance of Coil 1

Refer to Figure 11–40. To determine the inductance, a 10 V square wave is applied to the breadboarded circuit. The frequency of the square wave is adjusted so that the inductor has time to fully energize during each square wave pulse; the scope is set to view a complete energizing curve as shown. Determine the approximate circuit time constant from the scope display and calculate the inductance of coil 1.

Step 3: Determine the Inductance of Coil 2

Refer to Figure 11–41. To determine the inductance, a 10 V square wave is applied to the breadboarded circuit. The frequency of the square wave is adjusted so that the inductor has time to fully energize during each square wave pulse; the scope is set to view a complete energizing curve as shown. Determine the approximate circuit time constant from the scope display and calculate the inductance of coil 2. Discuss any difficulty you find with this method.

Step 4: Determine Another Way to Find Unknown Inductance

Determination of the time constant is not the only way that you can find an unknown inductance. Specify a method using a sinusoidal input voltage instead of the square wave.

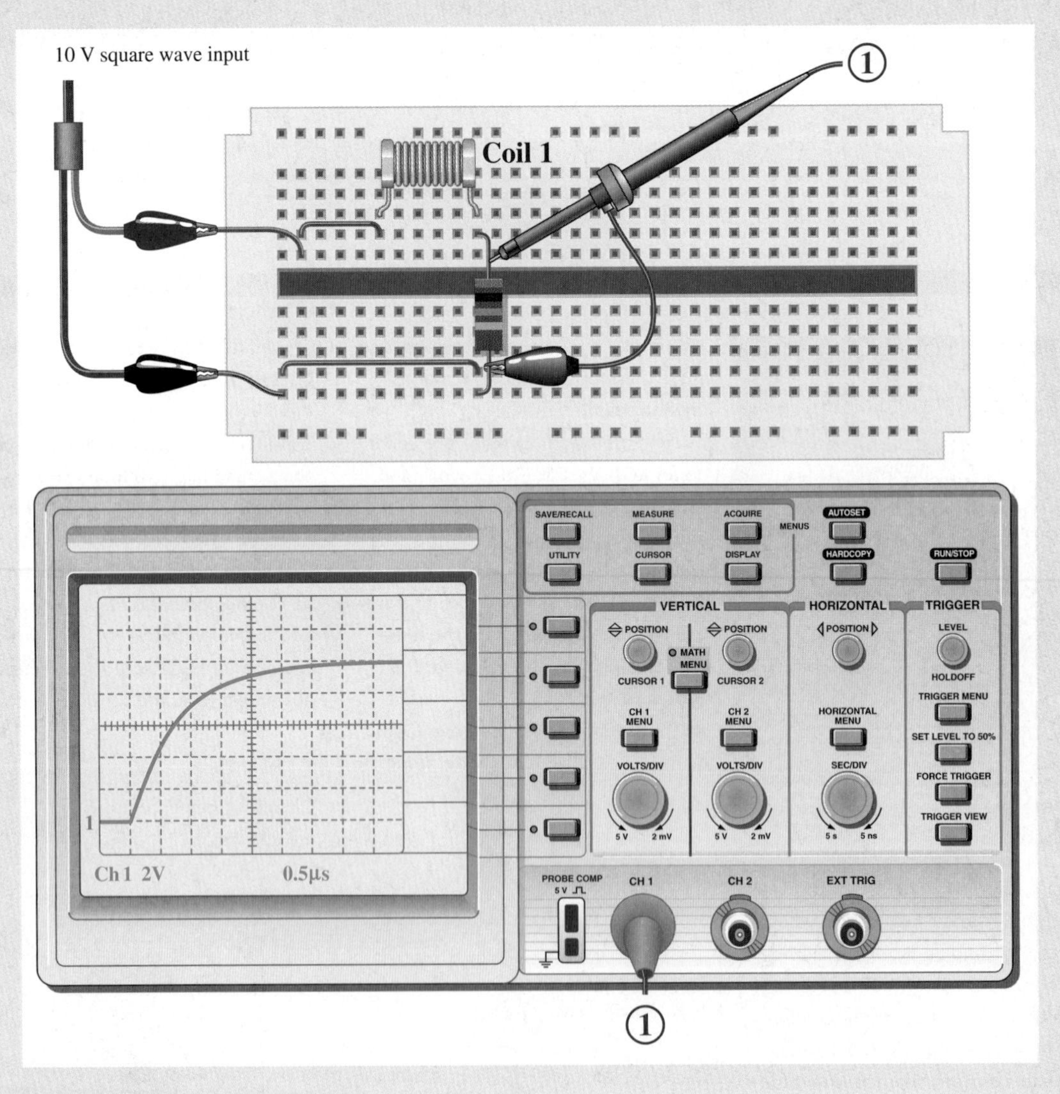

FIGURE 11–40

Testing coil 1.

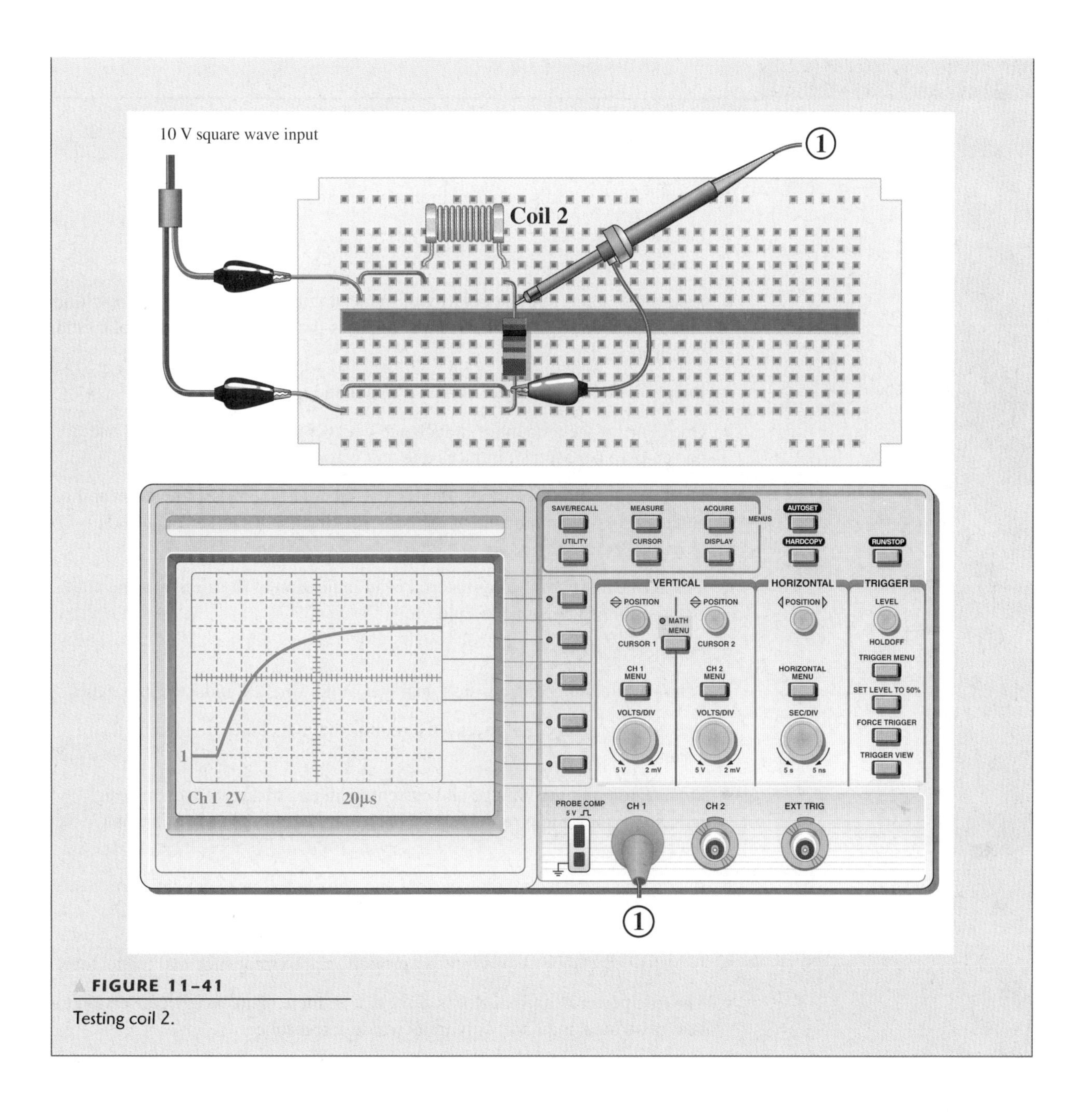

FIGURE 11–41

Testing coil 2.

APPLICATION ASSIGNMENT REVIEW

1. What is the maximum square wave frequency that can be used in Figure 11–40?
2. What is the maximum square wave frequency that can be used in Figure 11–41?
3. What happens if the frequency exceeds the maximum you determined in Questions 1 and 2? Explain how your measurements would be affected.

SUMMARY

- Self-inductance is a measure of a coil's ability to establish an induced voltage as a result of a change in its current.
- An inductor opposes a change in its own current.
- Faraday's law states that relative motion between a magnetic field and a coil induces a voltage across the coil.
- Lenz's law states that the polarity of induced voltage is such that the resulting induced current is in a direction that opposes the change in the magnetic field that produced it.
- Energy is stored by an inductor in its magnetic field.
- One henry is the amount of inductance when current, changing at the rate of one ampere per second, induces one volt across the inductor.
- Inductance is directly proportional to the square of the turns, the permeability, and the cross-sectional area of the core. It is inversely proportional to the length of the core.
- The permeability of a core material is an indication of the ability of the material to establish a magnetic field.
- Inductors add in series.
- Total parallel inductance is less than that of the smallest inductor in parallel.
- The time constant for a series *RL* circuit is the inductance divided by the resistance.
- In an *RL* circuit, the voltage and current in an energizing or deenergizing inductor make an approximately 63% change during each time-constant interval.
- Energizing and deenergizing current and voltage follow exponential curves.
- Voltage leads current by 90° in an inductor.
- Inductive reactance (X_L) is directly proportional to frequency and inductance.
- The true power in an inductor is zero; that is, there is no energy conversion to heat in an ideal inductor, only in its winding resistance.

EQUATIONS

11–1	$W = \frac{1}{2}LI^2$	Energy stored by an inductor
11–2	$L = \frac{N^2\mu A}{l}$	Inductance in terms of physical parameters
11–3	$L_T = L_1 + L_2 + L_3 + \cdots + L_n$	Series inductance
11–4	$\frac{1}{L_T} = \frac{1}{L_1} + \frac{1}{L_2} + \frac{1}{L_3} + \cdots + \frac{1}{L_n}$	Reciprocal of total parallel inductance
11–5	$L_T = \frac{1}{\frac{1}{L_1} + \frac{1}{L_2} + \frac{1}{L_3} + \cdots + \frac{1}{L_n}}$	Total parallel inductance

11–6	$\tau = \frac{L}{R}$	*RL* time constant
11–7	$v = V_F + (V_i - V_F)e^{-Rt/L}$	Exponential voltage (general)
11–8	$i = I_F + (I_i - I_F)e^{-Rt/L}$	Exponential current (general)
11–9	$i = I_F(1 - e^{-Rt/L})$	Increasing exponential current beginning at zero
11–10	$i = I_i e^{-Rt/L}$	Decreasing exponential current ending at zero
11–11	$X_L = 2\pi fL$	Inductive reactance
11–12	$V = IX_L$	Ohm's law
11–13	$P_{true} = I^2_{rms}R_W$	True power
11–14	$P_r = V_{rms}I_{rms}$	Reactive power
11–15	$P_r = \frac{V^2_{rms}}{X_L}$	Reactive power
11–16	$P_r = I^2_{rms}X_L$	Reactive power
11–17	$Q = \frac{X_L}{R_W}$	Quality factor

SELF-TEST

Answers are at the end of the chapter.

1. An inductance of 0.05 μH is larger than
   (a) 0.0000005 H **(b)** 0.000005 H
   (c) 0.000000008 H **(d)** 0.00005 mH
2. An inductance of 0.33 mH is smaller than
   (a) 33 μH **(b)** 330 μH **(c)** 0.05 mH **(d)** 0.0005 H
3. When the current through an inductor increases, the amount of energy stored in the electromagnetic field
   (a) decreases **(b)** remains constant
   (c) increases **(d)** doubles
4. When the current through an inductor doubles, the stored energy
   (a) doubles **(b)** quadruples
   (c) is halved **(d)** does not change
5. The winding resistance of a coil can be decreased by
   (a) reducing the number of turns **(b)** using a larger wire
   (c) changing the core material **(d)** either answer (a) or (b)
6. The inductance of an iron-core coil increases if
   (a) the number of turns is increased **(b)** the iron core is removed
   (c) the length of the core is increased **(d)** larger wire is used
7. Four 10 mH inductors are in series. The total inductance is
   (a) 40 mH **(b)** 2.5 mH **(c)** 40,000 μH **(d)** answers (a) and (c)
8. A 1 mH, a 3.3 mH, and a 0.1 mH inductor are connected in parallel. The total inductance is
   (a) 4.4 mH **(b)** greater than 3.3 mH
   (c) less than 0.1 mH **(d)** answers (a) and (b)

9. An inductor, a resistor, and a switch are connected in series to a 12 V battery. At the instant the switch is closed, the inductor voltage is

   (a) 0 V **(b)** 12 V **(c)** 6 V **(d)** 4 V

10. A sinusoidal voltage is applied across an inductor. When the frequency of the voltage is increased, the current

    (a) decreases **(b)** increases

    (c) does not change **(d)** momentarily goes to zero

11. An inductor and a resistor are in series with a sinusoidal voltage source. The frequency is set so that the inductive reactance is equal to the resistance. If the frequency is increased, then

    (a) $V_R > V_L$ **(b)** $V_L < V_R$ **(c)** $V_L = V_R$ **(d)** $V_L > V_R$

12. An ohmmeter is connected across an inductor and the pointer indicates an infinite value. The inductor is

    (a) good **(b)** open **(c)** shorted **(d)** resistive

PROBLEMS

Answers to odd-numbered problems are at the end of the book.

BASIC PROBLEMS

SECTION 11–1 **The Basic Inductor**

1. Convert the following to millihenries:

   (a) 1 H **(b)** 250 μH **(c)** 10 μH **(d)** 0.0005 H

2. Convert the following to microhenries:

   (a) 300 mH **(b)** 0.08 H **(c)** 5 mH **(d)** 0.00045 mH

3. How many turns are required to produce 30 mH with a coil wound on a cylindrical core having a cross-sectional area of 10×10^{-5} m^2 and a length of 0.05 m? The core has a permeability of 1.26×10^{-6}.

4. A 12 V battery is connected across a coil with a winding resistance of 12 Ω. How much current is there in the coil?

5. How much energy is stored by a 100 mH inductor with a current of 1 A?

6. The current through a 100 mH coil is changing at a rate of 200 mA/s. How much voltage is induced across the coil?

SECTION 11–3 **Series Inductors**

7. Five inductors are connected in series. The lowest value is 5 μH. If the value of each inductor is twice that of the preceding one, and if the inductors are connected in order of ascending values, what is the total inductance?

8. Suppose that you require a total inductance of 50 mH. You have available a 10 mH coil and a 22 mH coil. How much additional inductance do you need?

SECTION 11–4 **Parallel Inductors**

9. Determine the total parallel inductance for the following coils in parallel: 75 μH, 50 μH, 25 μH, and 15 μH.

10. You have a 12 mH inductor, and it is your smallest value. You need an inductance of 8 mH. What value can you use in parallel with the 12 mH to obtain 8 mH?

FIGURE 11–42

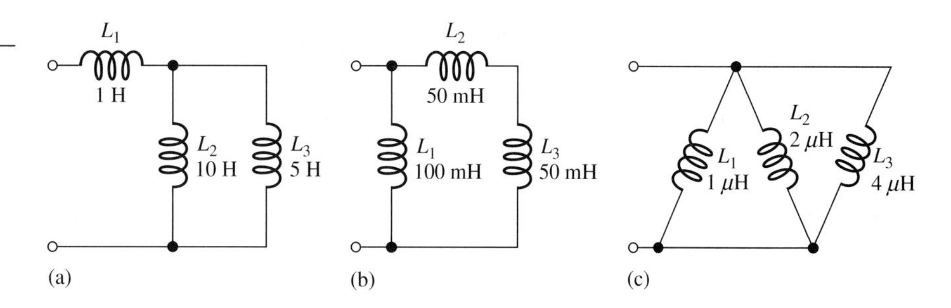

11. Determine the total inductance of each circuit in Figure 11–42.
12. Determine the total inductance of each circuit in Figure 11–43.

FIGURE 11–43

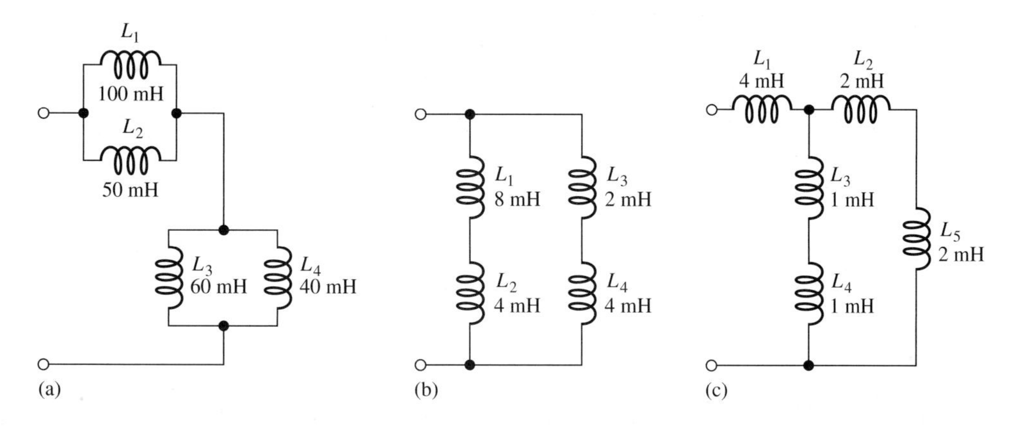

SECTION 11–5

Inductors in DC Circuits

13. Determine the time constant for each of the following series *RL* combinations:
    (a) $R = 100\ \Omega$, $L = 100\ \mu\text{H}$ **(b)** $R = 4.7\ \text{k}\Omega$, $L = 10\ \text{mH}$
    (c) $R = 1.5\ \text{M}\Omega$, $L = 3\ \text{H}$
14. In a series *RL* circuit, determine how long it takes the current to build up to its full value for each of the following:
    (a) $R = 56\ \Omega$, $L = 50\ \mu\text{H}$ **(b)** $R = 3300\ \Omega$, $L = 15\ \text{mH}$
    (c) $R = 22\ \text{k}\Omega$, $L = 100\ \text{mH}$
15. In the circuit of Figure 11–44, there is initially no current. Determine the inductor voltage at the following times after the switch is closed:
    (a) 10 μs **(b)** 20 μs **(c)** 30 μs **(d)** 40 μs **(e)** 50 μs
16. In Figure 11–45, there are 114 mA of current through the coil. When SW1 is opened and SW2 simultaneously closed, find the inductor voltage at the following times:
    (a) initially **(b)** 1.5 ms **(c)** 4.5 ms **(d)** 6 ms

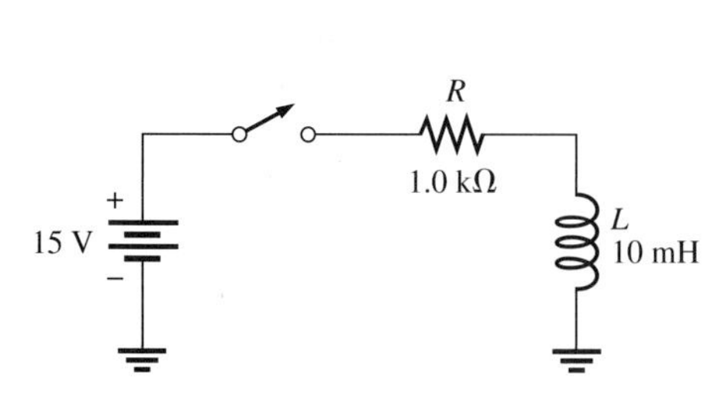

FIGURE 11–44

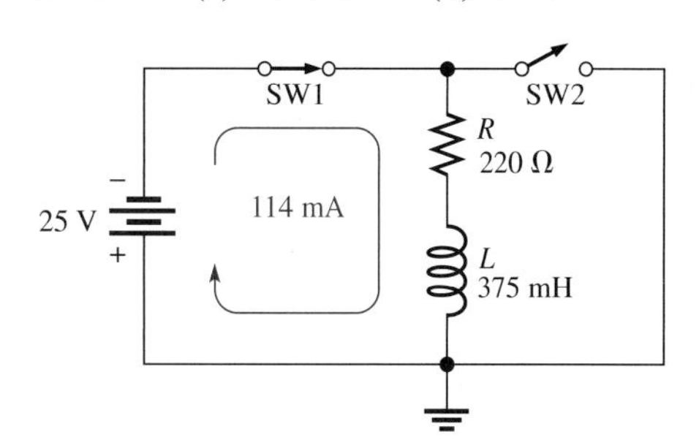

FIGURE 11–45

SECTION 11–6

Inductors in AC Circuits

17. Find the total reactance for each circuit in Figure 11–42 when a voltage with a frequency of 5 kHz is applied across the terminals.
18. Find the total reactance for each circuit in Figure 11–43 when a 400 Hz voltage is applied.
19. Determine the total rms current in Figure 11–46. What are the currents through L_2 and L_3?

FIGURE 11–46

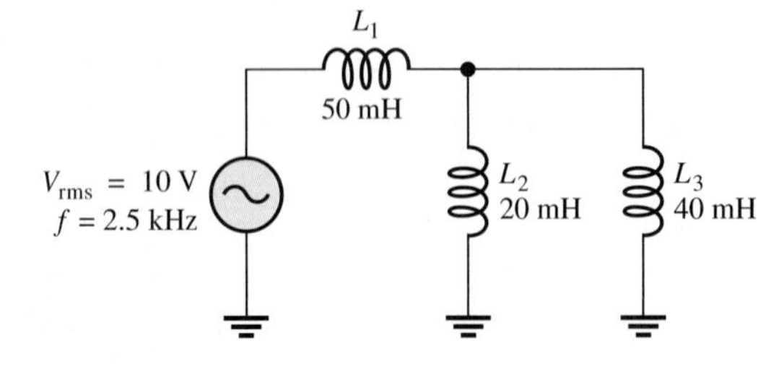

20. What frequency will produce a total rms current of 500 mA in each circuit of Figure 11–43 with an rms input voltage of 10 V?
21. Determine the reactive power in Figure 11–46 neglecting the winding resistance.

SECTION 11–8

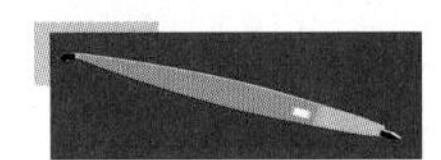

Testing Inductors

22. A certain coil that is supposed to have a 5 Ω winding resistance is measured with an ohmmeter. The meter indicates 2.8 Ω. What is the problem with the coil?
23. Determine the meter indication corresponding to each of the following failures in a coil:

    (a) open **(b)** shorted **(c)** partially shorted

ADVANCED PROBLEMS

24. Determine the time constant for the circuit in Figure 11–47.
25. Find the inductor current 10 μs after the switch is thrown from position 1 to position 2 in Figure 11–48. For simplicity, assume that the switch makes contact with position 2 at the same instant it breaks contact with position 1.

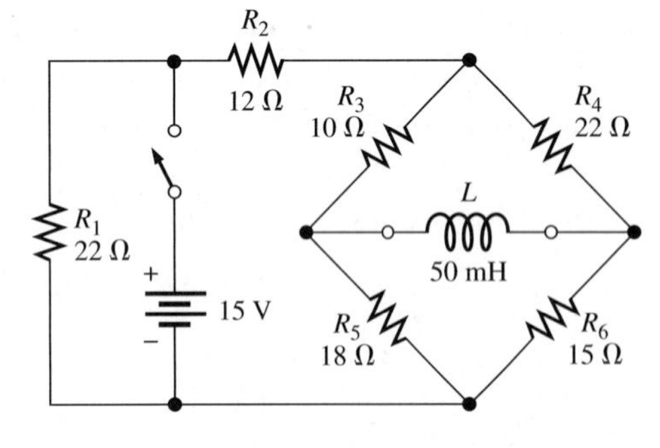

FIGURE 11–47

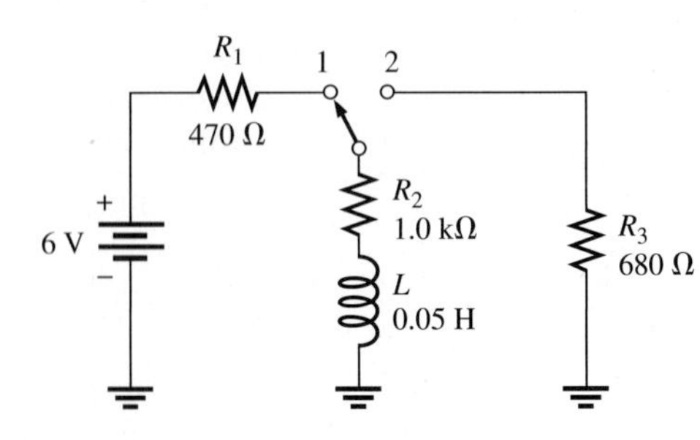

FIGURE 11–48

26. In Figure 11–49, SW1 is opened and SW2 is closed at the same instant (t_0). What is the instantaneous voltage across R_2 at t_0?
27. Determine the value of I_{L2} in Figure 11–50.
28. What is the total inductance between points A and B for each switch position in Figure 11–51?

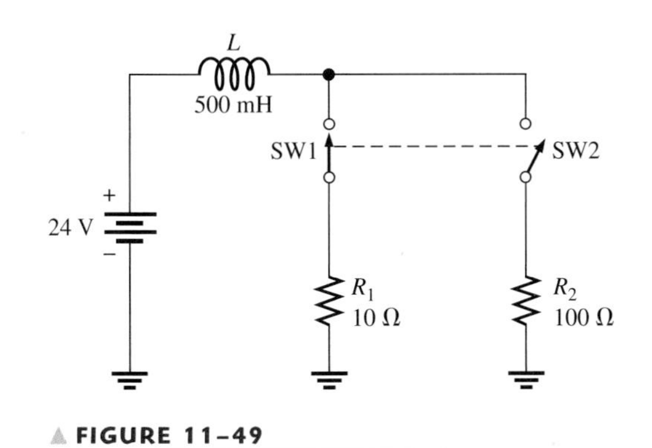

▲ FIGURE 11–49

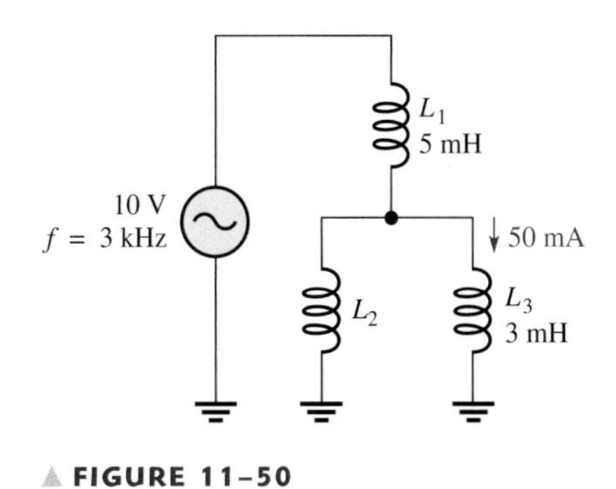

▲ FIGURE 11–50

▶ FIGURE 11–51

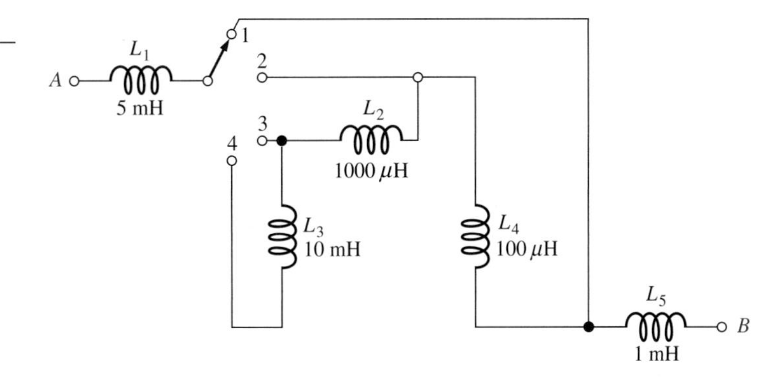

ELECTRONICS WORKBENCH/CIRCUITMAKER TROUBLESHOOTING PROBLEMS

CD-ROM file circuits are shown in Figure 11–52 on the next page.

29. Open file P11-29 and test the circuit. If there is a fault, identify it.

30. Open file P11-30 and test the circuit. If there is a fault, identify it.

31. Determine if there is a fault in the circuit in file P11-31. If so, identify it.

32. Find and specify any faulty component in the circuit in file P11-32.

33. Is there a short or an open in the circuit in file P11-33? If so, identify the component that is faulty.

ANSWERS

SECTION REVIEWS

SECTION 11–1 The Basic Inductor

1. Parameters that determine inductance are turns of wire, permeability, cross-sectional area, and core length.

2. **(a)** When N increases, L increases.

(b) When core length increases, L decreases.

(c) When area decreases, L decreases.

(d) For an air core, L decreases.

3. All wire has some resistance, and because inductors are made from turns of wire, there is always resistance in the winding.

4. Adjacent turns of wire in a coil act as plates of a capacitor and produce a small capacitance.

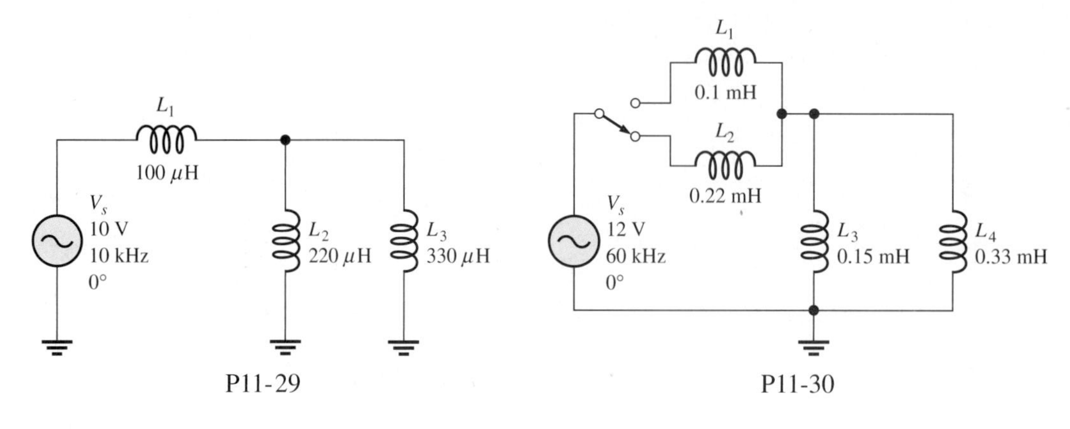

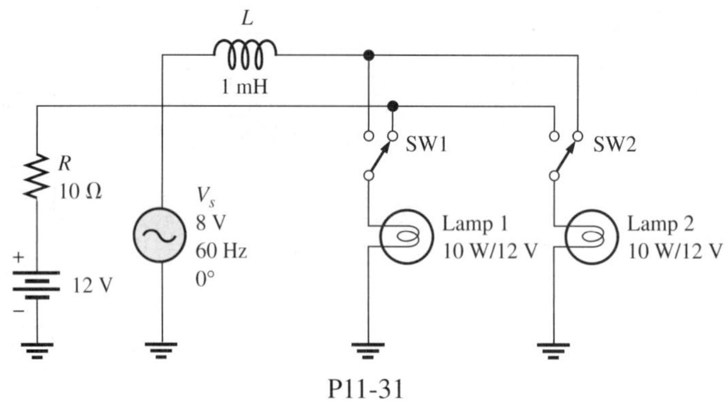

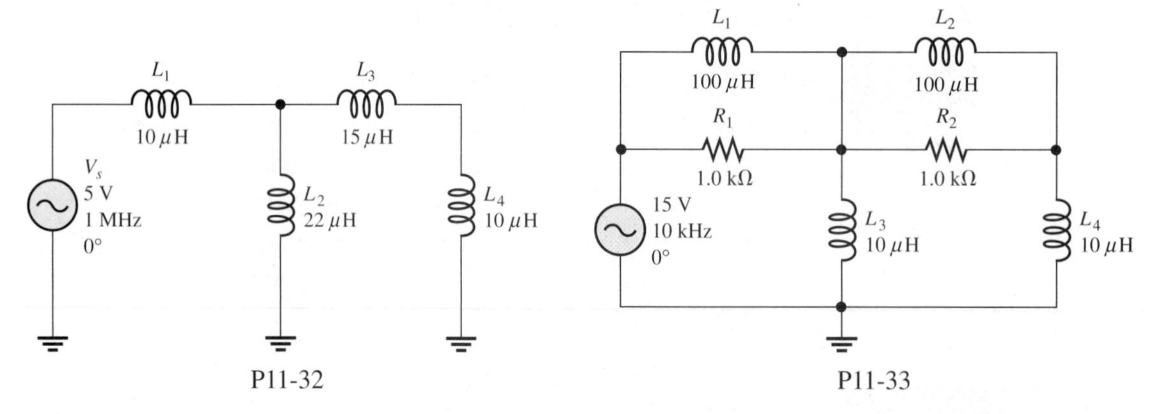

▲ **FIGURE 11–52**

SECTION 11–2 **Types of Inductors**

1. Two categories of inductors are fixed and variable.
2. **(a)** air core **(b)** iron core **(c)** variable

SECTION 11–3 **Series Inductors**

1. Inductances are added in series.
2. $L_T = 2600\ \mu H$
3. $L_T = 5 \times 100\ mH = 500\ mH$

SECTION 11–4 **Parallel Inductors**

1. The total parallel inductance is smaller than that of the smallest individual inductor in parallel.
2. True
3. **(a)** $L_T = 7.69\ mH$ **(b)** $L_T = 24\ \mu H$ **(c)** $L_T = 100\ mH$

SECTION 11–5 **Inductors in DC Circuits**

1. $V_L = (10\text{ mA})(10\ \Omega) = 100\text{ mV}$
2. Initially, $v_R = 0\text{ V}$, $v_L = 20\text{ V}$
3. After 5τ, $v_R = 20\text{ V}$, $v_L = 0\text{ V}$
4. $\tau = 500\ \mu\text{H}/1.0\text{ k}\Omega = 500\text{ ns}$, $i_L = 3.93\text{ mA}$

SECTION 11–6 **Inductors in AC Circuits**

1. Voltage leads current by 90° in an inductor.
2. $X_L = 2\pi fL = 3.14\text{ k}\Omega$
3. $f = X_L/2\pi L = 2.55\text{ MHz}$
4. $I_{rms} = 15.9\text{ mA}$
5. $P_{true} = 0\text{ W}$, $P_r = 458\text{ mVAR}$

SECTION 11–7 **Inductor Applications**

1. The inductor tends to level out the ripple because of its opposition to changes in current.
2. The inductive reactance is extremely high at radio frequencies and blocks these frequencies.

SECTION 11–8 **Testing Inductors**

1. False, a reading of infinity indicates an open.
2. False, it indicates the winding resistance.

■ Application Assignment

1. $f_{max} = 143\text{ kHz}$ ($5\tau = 3.5\ \mu\text{s}$)
2. $f_{max} = 3.57\text{ kHz}$ ($5\tau = 140\ \mu\text{s}$)
3. If $f > f_{max}$, the inductor would not fully energize because $T/2 < 5\tau$.

RELATED PROBLEMS FOR EXAMPLES

11–1 157 mH **11–2** 500 μH

11–3 20.3 μH **11–4** 227 ns

11–5 $I_F = 29.4\text{ mA}$, $\tau = 147\text{ ns}$
At 1τ, $i = 18.5\text{ mA}$
At 2τ, $i = 25.3\text{ mA}$
At 3τ, $i = 27.9\text{ mA}$
At 4τ, $i = 28.8\text{ mA}$
At 5τ, $i = 29.1\text{ mA}$

11–6 **(a)** 21.3 μs **(b)** 106.4 mA **(c)** 39.4 mA

11–7 **(a)** 9 V; 0 V **(b)** 92.6 V; 0 V

11–8 3.83 mA **11–9** 33.3 mA

11–10 110 Ω **11–11** 573 mA

11–12 P_r decreases.

SELF-TEST

1. (c) **2.** (d) **3.** (c) **4.** (b) **5.** (d) **6.** (a) **7.** (d)
8. (c) **9.** (b) **10.** (a) **11.** (d) **12.** (b)

12

RL CIRCUITS

INTRODUCTION

An *RL* circuit contains both resistance and inductance. In this chapter, basic series and parallel *RL* circuits and their responses to sinusoidal voltages are covered. In addition, series-parallel combinations are examined. Power considerations in *RL* circuits are introduced, and practical aspects of the power factor are discussed. A method of improving the power factor is presented, and two basic *RL* circuit applications are covered. Troubleshooting common faults in *RL* circuits is also covered.

The methods for analyzing reactive circuits are similar to those you studied in dc circuits. Reactive circuit problems can be solved at only one frequency at a time, and phasor math must be used.

As you study this chapter, note both the differences and the similarities in the response of *RL* circuits compared to *RC* circuits.

CHAPTER OUTLINE

CHAPTER OBJECTIVES

- Describe the relationship between current and voltage in an *RL* circuit
- Determine impedance and phase angle in a series *RL* circuit
- Analyze a series *RL* circuit
- Determine impedance and phase angle in a parallel *RL* circuit
- Analyze a parallel *RL* circuit
- Analyze series-parallel *RL* circuits
- Determine power in *RL* circuits
- Discuss some basic *RL* applications
- Troubleshoot *RL* circuits

KEY TERMS

- Impedance
- Phase angle
- Inductive susceptance
- Admittance
- Reactive power
- Apparent power
- Power factor
- *RL* lag network
- *RL* lead network
- Filter
- Frequency response
- Cutoff frequency
- Bandwidth

APPLICATION ASSIGNMENT PREVIEW

Your assignment is to identify the types of *RL* circuits contained in two sealed modules that have been removed from a communications system. You will use your knowledge of *RL* circuits and basic measurements to determine the circuit arrangement and component values. After studying this chapter, you should be able to complete the application assignment.

WWW. VISIT THE COMPANION WEBSITE

Circuit Simulation Tutorials and Other Chapter Study Tools Are Available at

http://www.prenhall.com/floyd

12–1 SINUSOIDAL RESPONSE OF *RL* CIRCUITS

As with the *RC* circuit, all currents and voltages in any type of *RL* circuit are sinusoidal when the input voltage is sinusoidal. The inductance causes a phase shift between the voltage and the current that depends on the relative values of the resistance and the inductive reactance. Because of its winding resistance, the inductor is generally not as "ideal" as a resistor or capacitor. However, it will usually be treated as ideal for purposes of illustration.

After completing this section, you should be able to

- **Describe the relationship between current and voltage in an *RL* circuit**
- Discuss voltage and current waveforms
- Discuss phase shift

In an *RL* circuit, the resistor voltage and the current lag the source voltage. The inductor voltage leads the source voltage. Ideally, the phase angle between the current and the inductor voltage is always 90°. These generalized phase relationships are indicated in Figure 12–1. Notice that they are opposite from those of the *RC* circuit that was discussed in Chapter 10.

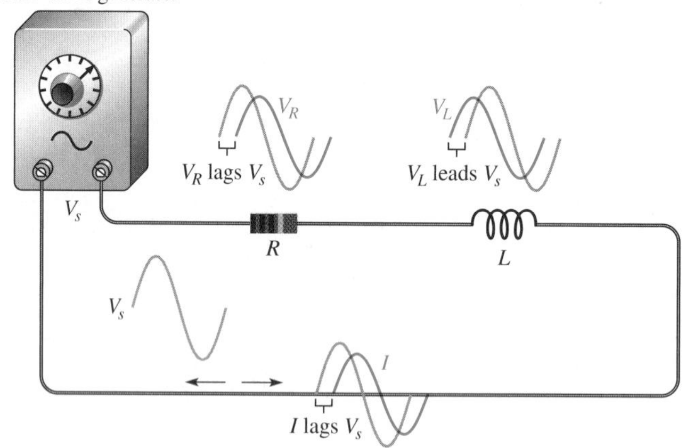

FIGURE 12–1

Illustration of sinusoidal response with general phase relationships of V_R, V_L, and I relative to the source voltage. V_R and I are in phase; V_R lags V_s; and V_L leads V_s. V_R and V_L are 90° out of phase with each other.

The amplitudes and the phase relationships of the voltages and current depend on the values of the resistance and the **inductive reactance.** When a circuit is purely inductive, the phase angle between the source voltage and the total current is 90°, with the current lagging the voltage. When there is a combination of both resistance and inductive reactance in a circuit, the phase angle is somewhere between zero and 90°, depending on the relative values of the resistance and the inductive reactance. Because all inductors have winding resistance, ideal conditions may be approached but never reached in practice.

SECTION 12–1 REVIEW

Answers are at the end of the chapter.

1. A 1 kHz sinusoidal voltage is applied to an *RL* circuit. What is the frequency of the resulting current?
2. When the resistance in an *RL* circuit is greater than the inductive reactance, is the phase angle between the source voltage and the total current closer to 0° or to 90°?

12–2 IMPEDANCE AND PHASE ANGLE OF SERIES *RL* CIRCUITS

The **impedance** of an *RL* circuit is the total opposition to sinusoidal current and its unit is the ohm. The **phase angle** is the phase difference between the total current and the source voltage.

After completing this section, you should be able to

- **Determine impedance and phase angle in a series *RL* circuit**
- Draw the impedance triangle
- Calculate impedance magnitude
- Calculate the phase angle

The impedance of a series *RL* circuit is determined by the resistance (R) and the inductive reactance (X_L), as indicated in Figure 12–2.

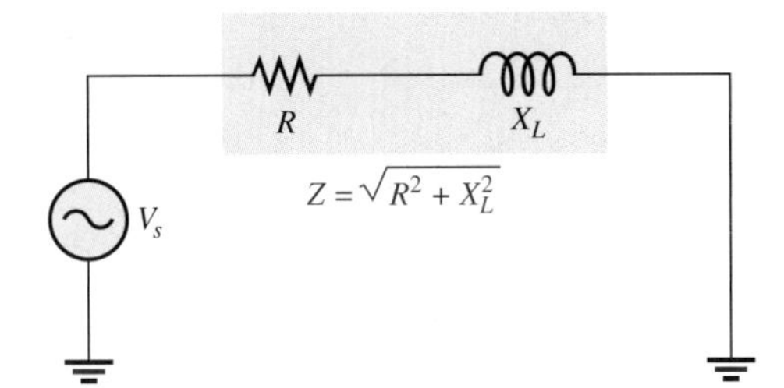

FIGURE 12–2
Impedance of a series *RL* circuit.

The Impedance Triangle

In ac analysis, both R and X_L are treated as phasor quantities, as shown in the phasor diagram of Figure 12–3(a), with X_L appearing at a $+90°$ angle with respect to R. This relationship comes from the fact that the inductor voltage leads the current, and thus the resistor voltage, by 90°. Since Z is the phasor sum of R and X_L, its phasor representation is as shown in Figure 12–3(b). A repositioning of the phasors, as shown in part (c), forms a right triangle. This formation, as you have learned, is called the *impedance triangle.* The length of each phasor represents the magnitude of the quantity, and θ is the phase angle between the source voltage and the current in the *RL* circuit.

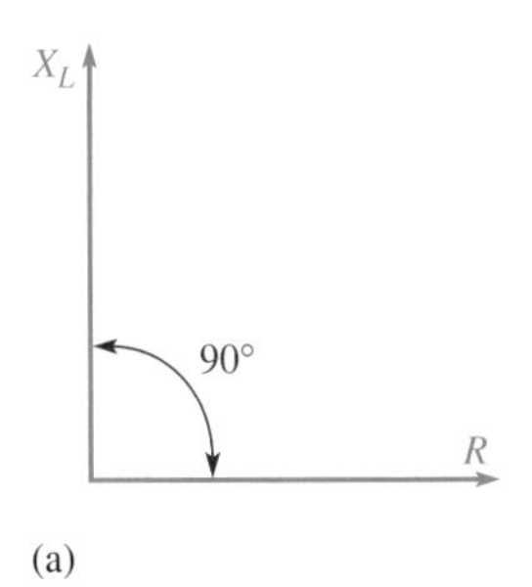

(a)

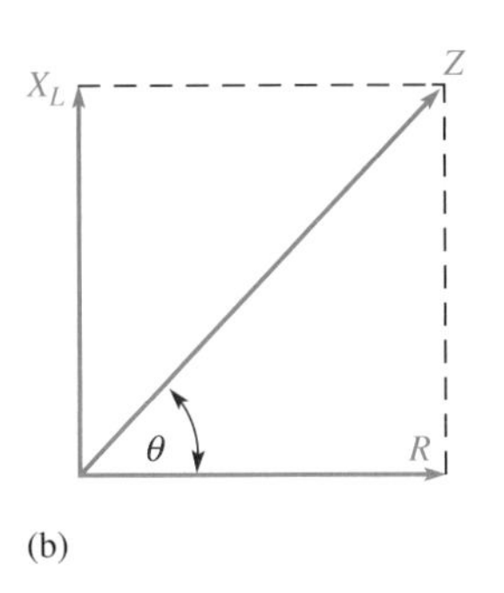

(b)

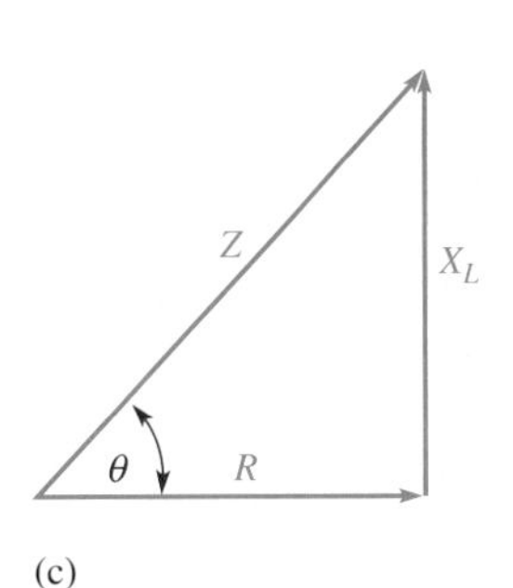

(c)

FIGURE 12–3
Development of the impedance triangle for a series *RL* circuit.

The magnitude of the impedance, Z, of the series *RL* circuit can be expressed in terms of the resistance and reactance as

$$Z = \sqrt{R^2 + X_L^2}$$

Equation 12–1

where Z is expressed in ohms.

The phase angle, θ, is expressed as

Equation 12–2

$$\theta = \tan^{-1}\left(\frac{X_L}{R}\right)$$

EXAMPLE 12–1

Determine the impedance and phase angle of the circuit in Figure 12–4. Draw the impedance triangle.

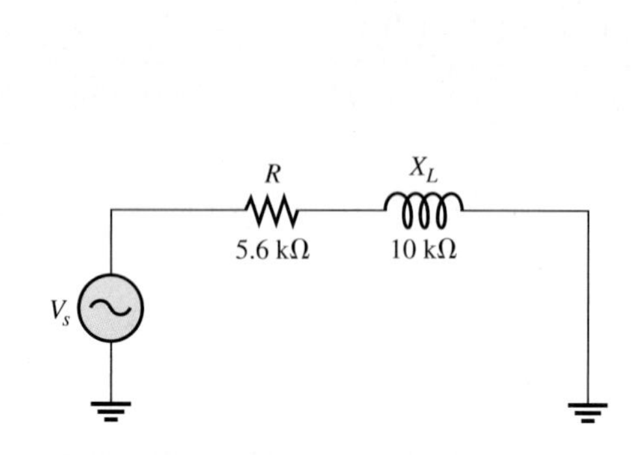

▲ **FIGURE 12–4**

$Z = 11.5\ \text{k}\Omega$

$X_L = 10\ \text{k}\Omega$

$\theta = 60.8°$

$R = 5.6\ \text{k}\Omega$

▲ **FIGURE 12–5**

Solution The impedance is

$$Z = \sqrt{R^2 + X_L^2} = \sqrt{(5.6\ \text{k}\Omega)^2 + (10\ \text{k}\Omega)^2} = \mathbf{11.5\ k\Omega}$$

The phase angle is

$$\theta = \tan^{-1}\left(\frac{X_L}{R}\right) = \tan^{-1}\left(\frac{10\ \text{k}\Omega}{5.6\ \text{k}\Omega}\right) = \mathbf{60.8°}$$

The source voltage leads the current by 60.8°. The impedance triangle is shown in Figure 12–5.

The calculator solutions for Z and θ are

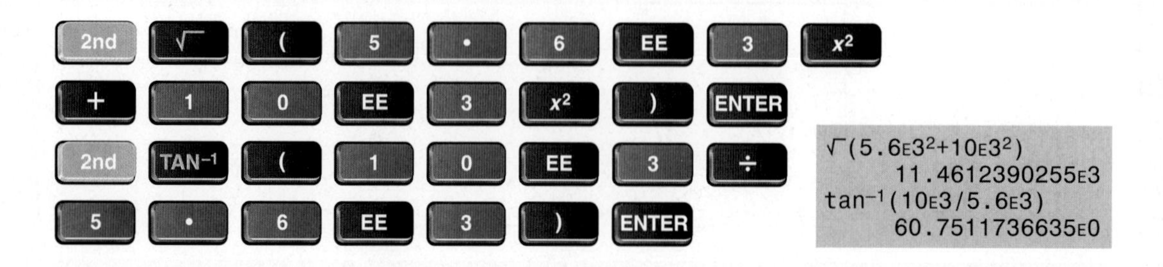

Related Problem* In a series *RL* circuit, $R = 1.8\ \text{k}\Omega$ and $X_L = 950\ \Omega$. Determine the impedance and phase angle.

*Answers are at the end of the chapter.

SECTION 12–2 REVIEW

1. Does the source voltage lead or lag the current in a series *RL* circuit?
2. What causes the phase angle in the *RL* circuit?
3. How does the response of an *RL* circuit differ from that of an *RC* circuit?
4. A series *RL* circuit has a resistance of 33 kΩ and an inductive reactance of 50 kΩ. Determine Z and θ.

12–3 ANALYSIS OF SERIES *RL* CIRCUITS

In Section 12–2, you learned how to express the impedance of a series *RL* circuit. Now, Ohm's law and Kirchhoff's voltage law are used in the analysis of *RL* circuits.

After completing this section, you should be able to

- **Analyze a series *RL* circuit**
- Apply Ohm's law and Kirchhoff's voltage law to series *RL* circuits
- Determine the phase relationships of the voltages and current
- Show how impedance and phase angle vary with frequency

Ohm's Law

The application of Ohm's law to series *RL* circuits involves the use of the quantities of Z, V, and I. The three equivalent forms of Ohm's law were stated in Chapter 10 for *RC* circuits. They apply also to *RL* circuits and are restated here:

$$V = IZ \qquad I = \frac{V}{Z} \qquad Z = \frac{V}{I}$$

The following example illustrates the use of Ohm's law.

EXAMPLE 12–2

The current in Figure 12–6 is 200 μA. Determine the source voltage.

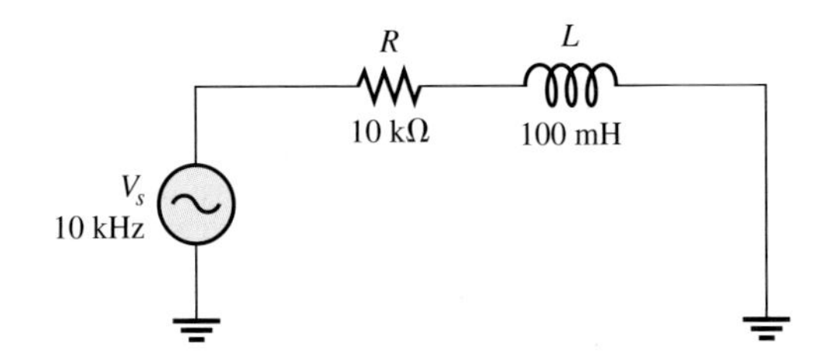

▶ FIGURE 12–6

Solution From Equation 11–11, the inductive reactance is

$$X_L = 2\pi fL = 2\pi(10\text{ kHz})(100\text{ mH}) = 6.28\text{ k}\Omega$$

The impedance is

$$Z = \sqrt{R^2 + X_L^2} = \sqrt{(10\text{ k}\Omega)^2 + (6.28\text{ k}\Omega)^2} = 11.8\text{ k}\Omega$$

Applying Ohm's law yields

$$V_s = IZ = (200\ \mu\text{A})(11.8\ \text{k}\Omega) = \mathbf{2.36\ V}$$

Related Problem If the source voltage in Figure 12–6 is 5 V, what would be the current?

Open file E12-02 on your EWB/CircuitMaker CD-ROM. Measure the current at 10 kHz, 5 kHz, and 20 kHz. Explain the results of your measurement.

Phase Relationships of the Current and Voltages

In a series *RL* circuit, the current is the same through both the resistor and the inductor. Thus, the resistor voltage is in phase with the current, and the inductor voltage leads the current by 90°. Therefore, there is a phase difference of 90° between the resistor voltage, V_R, and the inductor voltage, V_L, as shown in the waveform diagram of Figure 12–7.

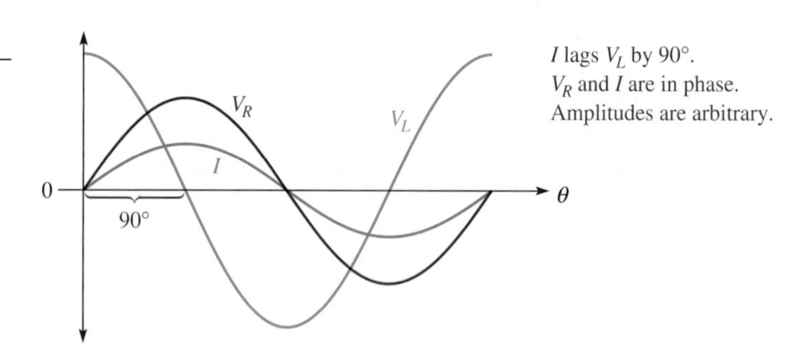

FIGURE 12–7

Phase relation of current and voltages in a series *RL* circuit.

From Kirchhoff's voltage law, the sum of the voltage drops must equal the source voltage. However, since V_R and V_L are not in phase with each other, they must be added as phasor quantities with V_L leading V_R by 90°, as shown in Figure 12–8(a). As shown in part (b), V_s is the phasor sum of V_R and V_L. This equation can be expressed as

Equation 12–3

$$V_s = \sqrt{V_R^2 + V_L^2}$$

The phase angle between the resistor voltage and the source voltage can be expressed as

Equation 12–4

$$\theta = \tan^{-1}\left(\frac{V_L}{V_R}\right)$$

Since the resistor voltage and the current are in phase, θ in Equation 12–4 also represents the phase angle between the source voltage and the current and is equivalent to $\tan^{-1}(X_L/R)$.

Figure 12–9 shows a voltage and current phasor diagram that represents the waveform diagram of Figure 12–7.

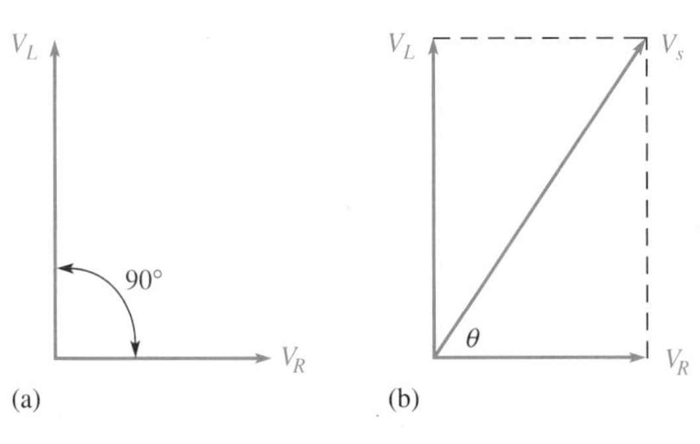

▲ **FIGURE 12–8**

Voltage phasor diagram for the waveforms in Figure 12–7.

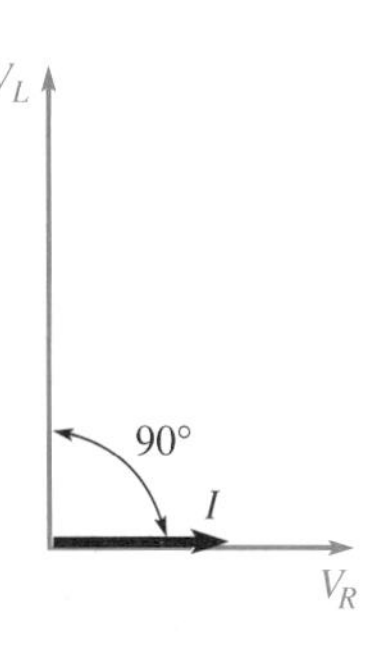

▲ **FIGURE 12–9**

Voltage and current phasor diagram for the waveforms in Figure 12–7.

EXAMPLE 12–3

Determine the source voltage and the phase angle in Figure 12–10. Draw the voltage phasor diagram.

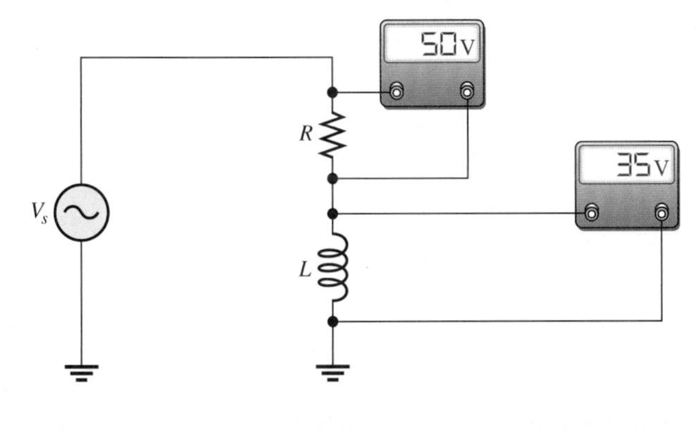

▲ **FIGURE 12–10**

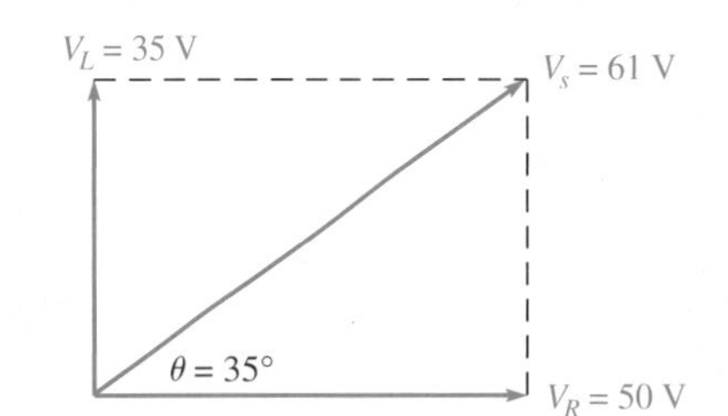

▲ **FIGURE 12–11**

Solution Since V_R and V_L are 90° out of phase, they cannot be added directly and must be added as phasor quantities. The source voltage is

$$V_s = \sqrt{V_R^2 + V_L^2} = \sqrt{(50\text{ V})^2 + (35\text{ V})^2} = \mathbf{61\text{ V}}$$

The phase angle between the current and the source voltage is

$$\theta = \tan^{-1}\left(\frac{V_L}{V_R}\right) = \tan^{-1}\left(\frac{35\text{ V}}{50\text{ V}}\right) = \mathbf{35°}$$

The voltage phasor diagram is shown in Figure 12–11.

Related Problem With the information given, can you determine the current in Figure 12–10?

Variation of Impedance with Frequency

As you know, inductive reactance varies directly with frequency. When X_L increases, the total impedance also increases; and when X_L decreases, the total impedance decreases. Thus, *in an RL circuit, Z is directly dependent on frequency.*

Figure 12–12 illustrates how the voltages and current in a series *RL* circuit vary as the frequency increases or decreases, with the source voltage held at a constant value. Part (a) shows that as the frequency is increased, X_L increases; so more of the total voltage is dropped across the inductor. Also, Z increases as X_L increases, causing the current to decrease. A decrease in current causes less voltage across R.

FIGURE 12–12

An illustration of how the variation of impedance affects the voltages and current as the source frequency is varied. The source voltage is held at a constant amplitude.

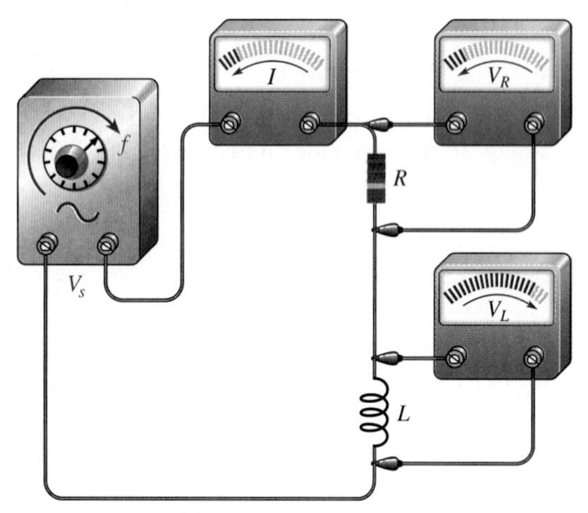

(a) As frequency is increased, I and V_R decrease and V_L increases.

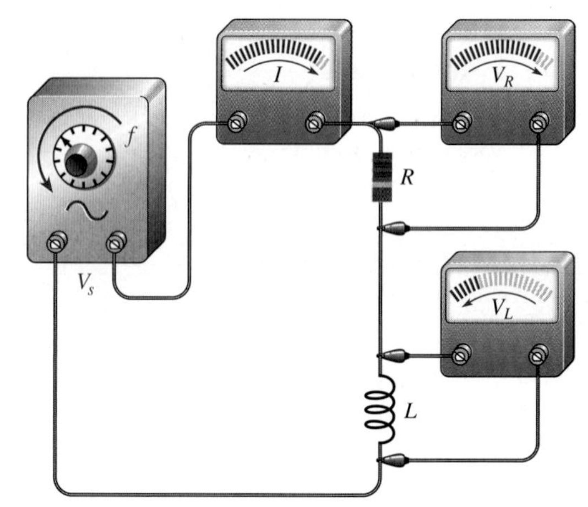

(b) As frequency is decreased, I and V_R increase and V_L decreases.

Figure 12–12(b) shows that as the frequency is decreased, X_L decreases; so less voltage is dropped across the inductor. Also, Z decreases as X_L decreases, causing the current to increase. An increase in current causes more voltage across R.

Changes in Z and X_L can be observed as shown in Figure 12–13. As the frequency increases, the voltage across Z remains constant because V_s is constant ($V_s = V_Z$), but the voltage across L increases. The decreasing current indicates that Z is increasing. It does so because of the inverse relationship stated in Ohm's law ($Z = V_Z/I$). The decreasing current also indicates that X_L is increasing. The increase in V_L corresponds to the increase in X_L.

FIGURE 12–13

Observing changes in Z and X_L with frequency by watching the meters and recalling Ohm's law.

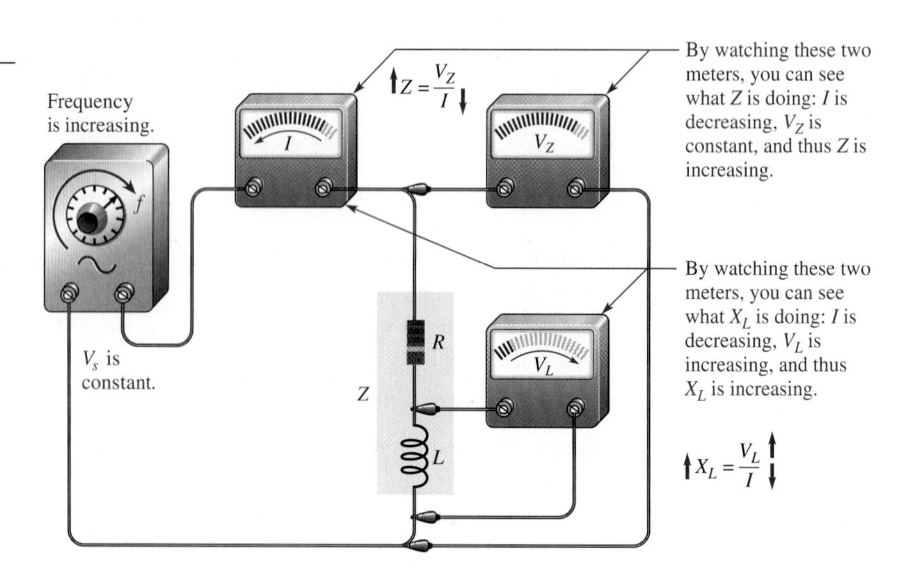

Variation of the Phase Angle with Frequency

Since X_L is the factor that introduces the phase angle in a series *RL* circuit, a change in X_L produces a change in the phase angle. As the frequency is increased, X_L becomes greater, and thus the phase angle increases. As the frequency is decreased, X_L becomes smaller, and thus the phase angle decreases. The angle between V_s and V_R is the phase angle of the circuit because I is in phase with V_R.

The variations of phase angle with frequency are illustrated with the impedance triangle as shown in Figure 12–14.

HANDS ON TIP

As you know, some multimeters have a relatively low frequency response. There are other things of which you should be aware. One is that most ac meters are accurate only if the waveform being measured is sinusoidal. Another is that accuracy when measuring small ac voltage is usually less than for dc measurements. Finally, loading can affect the accuracy of meter readings.

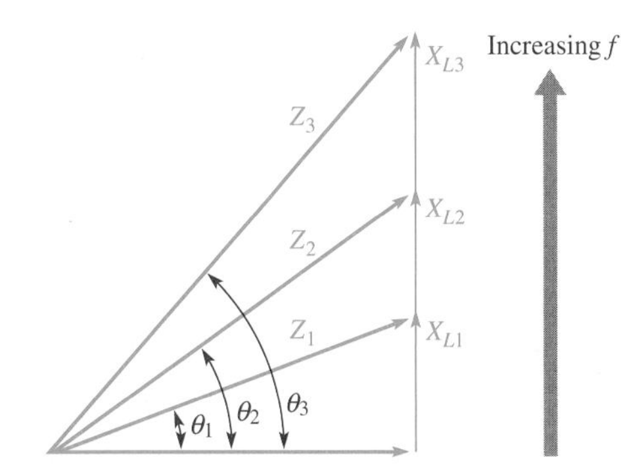

FIGURE 12–14

As the frequency increases, the phase angle θ increases.

EXAMPLE 12–4

For the series *RL* circuit in Figure 12–15, determine the impedance and the phase angle for each of the following frequencies:

(a) 10 kHz **(b)** 20 kHz **(c)** 30 kHz

FIGURE 12–15

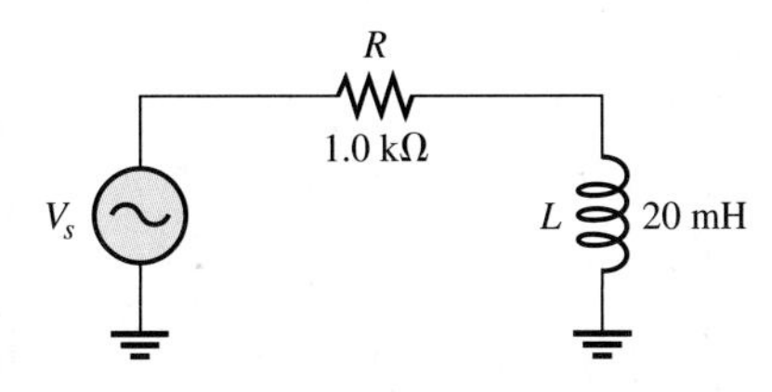

Solution **(a)** For $f = 10$ kHz, calculate the impedance as follows:

$$X_L = 2\pi fL = 2\pi(10\text{ kHz})(20\text{ mH}) = 1.26\text{ k}\Omega$$

$$Z = \sqrt{R^2 + X_L^2} = \sqrt{(1.0\text{ k}\Omega)^2 + (1.26\text{ k}\Omega)^2} = \mathbf{1.61\text{ k}\Omega}$$

The phase angle is

$$\theta = \tan^{-1}\left(\frac{X_L}{R}\right) = \tan^{-1}\left(\frac{1.26\text{ k}\Omega}{1.0\text{ k}\Omega}\right) = \mathbf{51.6^\circ}$$

(b) For $f = 20$ kHz,

$$X_L = 2\pi(20\text{ kHz})(20\text{ mH}) = 2.51\text{ k}\Omega$$

$$Z = \sqrt{(1.0\text{ k}\Omega)^2 + (2.51\text{ k}\Omega)^2} = \mathbf{2.70\text{ k}\Omega}$$

$$\theta = \tan^{-1}\left(\frac{2.51\text{ k}\Omega}{1.0\text{ k}\Omega}\right) = \mathbf{68.3^\circ}$$

(c) For $f = 30$ kHz,

$$X_L = 2\pi(30\text{ kHz})(20\text{ mH}) = 3.77\text{ k}\Omega$$

$$Z = \sqrt{(1.0\text{ k}\Omega)^2 + (3.77\text{ k}\Omega)^2} = \mathbf{3.90\text{ k}\Omega}$$

$$\theta = \tan^{-1}\left(\frac{3.77\text{ k}\Omega}{1.0\text{ k}\Omega}\right) = \mathbf{75.1^\circ}$$

Notice that as the frequency increases, X_L, Z, and θ also increase.

Related Problem Determine Z and θ in Figure 12–15 if $f = 100$ kHz.

SECTION 12–3 REVIEW

1. In a certain series *RL* circuit, $V_R = 2$ V and $V_L = 3$ V. What is the magnitude of the total voltage?
2. In Question 1, what is the phase angle?
3. When the frequency of the source voltage in a series *RL* circuit is increased, what happens to the inductive reactance? To the impedance? To the phase angle?

12–4 IMPEDANCE AND PHASE ANGLE OF PARALLEL *RL* CIRCUITS

In this section, you will learn how to determine the impedance and phase angle of a parallel *RL* circuit. Also, inductive susceptance and admittance of a parallel *RL* circuit are introduced.

After completing this section, you should be able to

- **Determine impedance and phase angle in a parallel *RL* circuit**
- Express total impedance in a product-over-sum form
- Express the phase angle in terms of R and X_L
- Determine inductive susceptance and admittance
- Convert admittance to impedance

A basic parallel *RL* circuit is shown in Figure 12–16. The expression for the impedance, using the product-over-sum rule, is

$$Z = \frac{RX_L}{\sqrt{R^2 + X_L^2}}$$

Equation 12–5

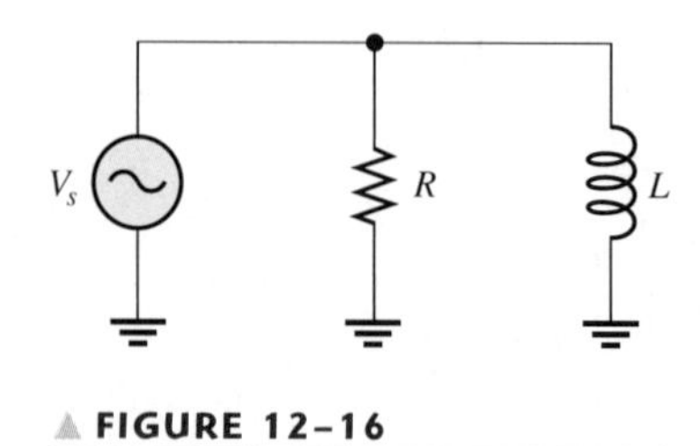

▲ **FIGURE 12–16**

Parallel *RL* circuit.

The phase angle between the source voltage and the total current can be expressed in terms of R and X_L as

$$\theta = \tan^{-1}\left(\frac{R}{X_L}\right)$$

Equation 12–6

EXAMPLE 12–5

For each circuit in Figure 12–17, determine the impedance and the phase angle.

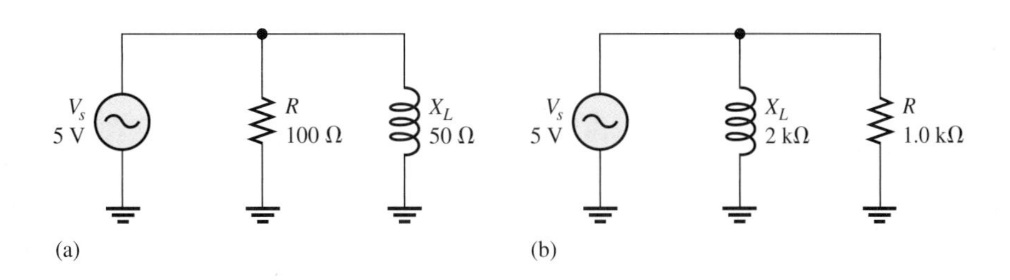

▲ **FIGURE 12–17**

Solution For the circuit in Figure 12–17(a), the impedance and phase angle are

$$Z = \frac{RX_L}{\sqrt{R^2 + X_L^2}} = \frac{(100\ \Omega)(50\ \Omega)}{\sqrt{(100\ \Omega)^2 + (50\ \Omega)^2}} = \mathbf{44.7\ \Omega}$$

$$\theta = \tan^{-1}\left(\frac{R}{X_L}\right) = \tan^{-1}\left(\frac{100\ \Omega}{50\ \Omega}\right) = \mathbf{63.4^\circ}$$

For the circuit in Figure 12–17(b),

$$Z = \frac{(1.0\ \text{k}\Omega)(2\ \text{k}\Omega)}{\sqrt{(1.0\ \text{k}\Omega)^2 + (2\ \text{k}\Omega)^2}} = \mathbf{894\ \Omega}$$

$$\theta = \tan^{-1}\left(\frac{1.0\ \text{k}\Omega}{2\ \text{k}\Omega}\right) = \mathbf{26.6^\circ}$$

The voltage leads the current, as opposed to the parallel *RC* case where the voltage lags the current.

The calculator solutions for Z and θ in Figure 12–17(a) are

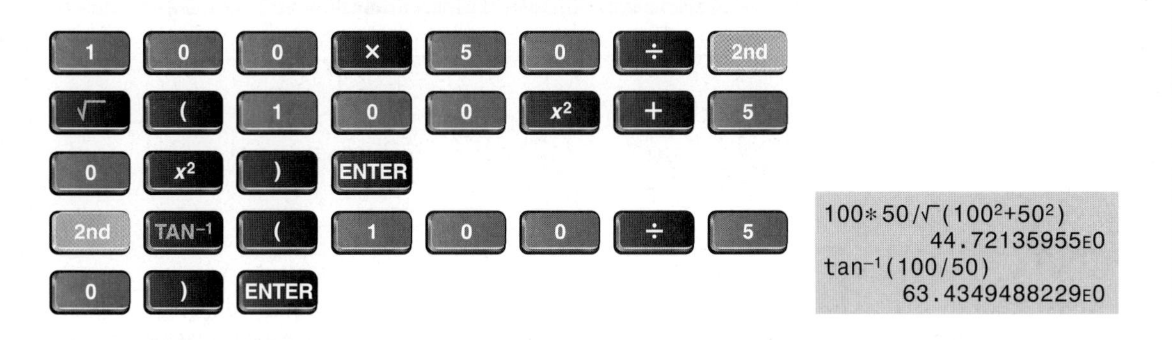

Related Problem In a parallel *RL* circuit $R = 10\ \text{k}\Omega$ and $X_L = 14\ \text{k}\Omega$. Find Z and θ.

Conductance, Susceptance, and Admittance

As you know from Section 10–4, conductance (G) is the reciprocal of resistance, susceptance (B) is the reciprocal of reactance, and admittance (Y) is the reciprocal of impedance.

For parallel RL circuits, **conductance** (G) is expressed as

Equation 12–7

$$G = \frac{1}{R}$$

Inductive susceptance (B_L) is expressed as

Equation 12–8

$$B_L = \frac{1}{X_L}$$

Admittance (Y) is expressed as

Equation 12–9

$$Y = \frac{1}{Z}$$

As with the RC circuit, the unit for G, B_L, and Y is the siemens (S).

In the basic parallel RL circuit shown in Figure 12–18, the total admittance is the phasor sum of the conductance and the inductive susceptance.

Equation 12–10

$$Y_{tot} = \sqrt{G^2 + B_L^2}$$

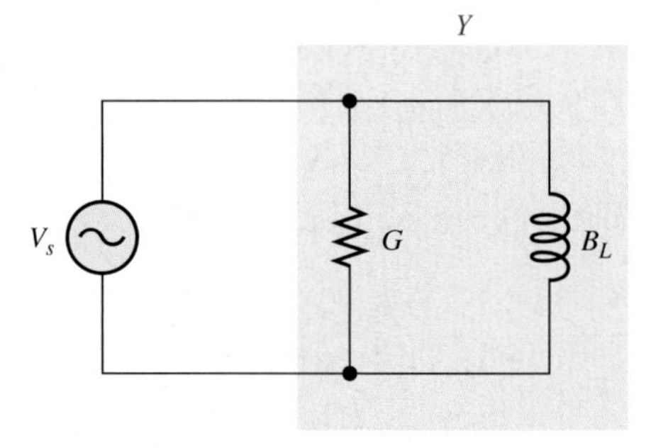

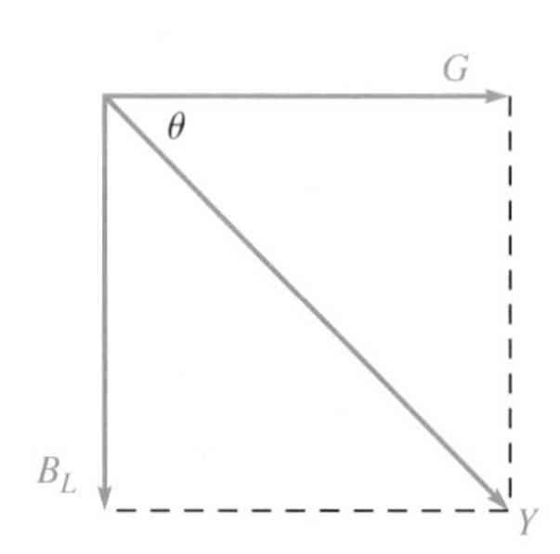

▶ **FIGURE 12–18**

Admittance in a parallel RL circuit.

EXAMPLE 12–6

Determine the total admittance in Figure 12–19; then convert it to impedance.

▶ **FIGURE 12–19**

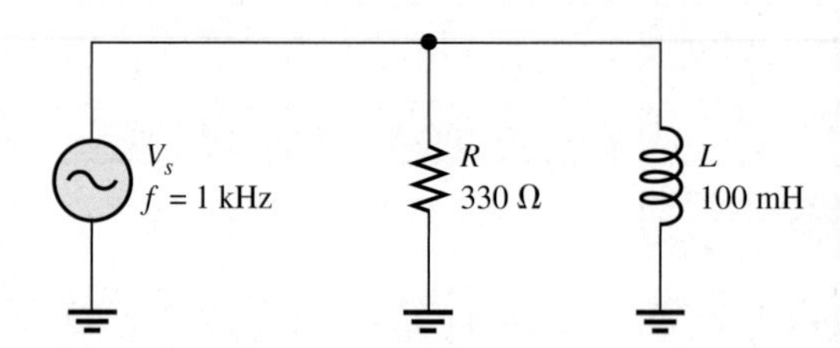

Solution To determine Y, first calculate the values for G and B_L. Since $R = 330\ \Omega$,

$$G = \frac{1}{R} = \frac{1}{330\ \Omega} = 3.03\text{ mS}$$

The inductive reactance is

$$X_L = 2\pi fL = 2\pi(1000\text{ Hz})(100\text{ mH}) = 628\ \Omega$$

The inductive susceptance is

$$B_L = \frac{1}{X_L} = \frac{1}{628\ \Omega} = 1.59\text{ mS}$$

Therefore, the total admittance is

$$Y_{tot} = \sqrt{G^2 + B_L^2} = \sqrt{(3.03\text{ mS})^2 + (1.59\text{ mS})^2} = \mathbf{3.42\text{ mS}}$$

Convert to impedance as follows:

$$Z = \frac{1}{Y_{tot}} = \frac{1}{3.42\text{ mS}} = \mathbf{292\ \Omega}$$

Related Problem What is the total admittance of the circuit in Figure 12–19 if f is increased to 2 kHz?

SECTION 12–4 REVIEW

1. If $Y = 50$ mS, what is the value of Z?
2. In a certain parallel *RL* circuit, $R = 47\ \Omega$ and $X_L = 75\ \Omega$. Determine the admittance.
3. In the circuit of Question 2, does the total current lead or lag the source voltage and by what phase angle?

12–5 ANALYSIS OF PARALLEL *RL* CIRCUITS

In the previous section, you learned how to express the impedance of a parallel *RL* circuit. Now, Ohm's law and Kirchhoff's current law are used in the analysis of *RL* circuits. Current and voltage relationships in a parallel *RL* circuit are examined.

After completing this section, you should be able to

- **Analyze a parallel *RL* circuit**
- Apply Ohm's law and Kirchhoff's current law to parallel *RL* circuits
- Determine total current and phase angle

The following example applies Ohm's law to the analysis of a parallel *RL* circuit.

EXAMPLE 12–7

Determine the total current and the phase angle in the circuit of Figure 12–20.

FIGURE 12–20

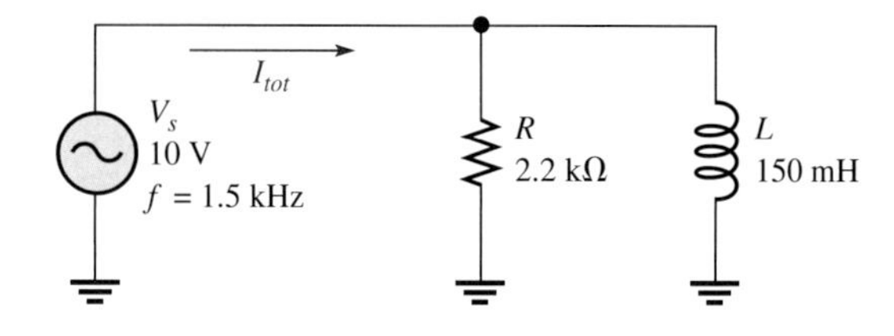

Solution First, determine the total admittance. The inductive reactance is

$$X_L = 2\pi fL = 2\pi(1.5\text{ kHz})(150\text{ mH}) = 1.41\text{ k}\Omega$$

The conductance is

$$G = \frac{1}{R} = \frac{1}{2.2\text{ k}\Omega} = 455\ \mu\text{S}$$

The inductive susceptance is

$$B_L = \frac{1}{X_L} = \frac{1}{1.41\text{ k}\Omega} = 709\ \mu\text{S}$$

Therefore, the total admittance is

$$Y_{tot} = \sqrt{G^2 + B_L^2} = \sqrt{(455\ \mu\text{S})^2 + (709\ \mu\text{S})^2} = 842\ \mu\text{S}$$

Next, use Ohm's law to get the total current.

$$I_{tot} = VY_{tot} = (10\text{ V})(842\ \mu\text{S}) = \mathbf{8.42\text{ mA}}$$

The phase angle is

$$\theta = \tan^{-1}\left(\frac{R}{X_L}\right) = \tan^{-1}\left(\frac{2.2\text{ k}\Omega}{1.41\text{ k}\Omega}\right) = \mathbf{57.3°}$$

The total current is 8.42 mA, and it lags the source voltage by 57.3°.

Related Problem Determine the total current and the phase angle if the frequency is reduced to 800 Hz in Figure 12–20.

Open file E12-07 on your EWB/CircuitMaker CD-ROM. Measure the total current and each branch current. Change the frequency to 800 Hz and measure I_{tot}.

Phase Relationships of the Currents and Voltages

Figure 12–21(a) shows all the currents and voltages in a basic parallel *RL* circuit. As you can see, the source voltage, V_s, appears across both the resistive and the inductive branches, so V_s, V_R, and V_L are all in phase and of the same magnitude. The total current, I_{tot}, divides at the junction into the two branch currents, I_R and I_L. The current and voltage phasor diagram is shown in Figure 12–21(b).

The current through the resistor is in phase with the voltage. The current through the inductor lags the voltage and the resistor current by 90°. By Kirch-

FIGURE 12–21

Currents and voltages in a parallel *RL* circuit. The current directions shown in part (a) are instantaneous and, of course, reverse when the source voltage reverses during each cycle.

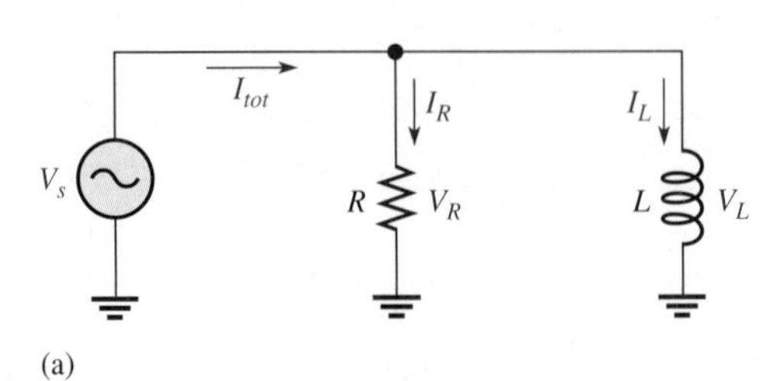

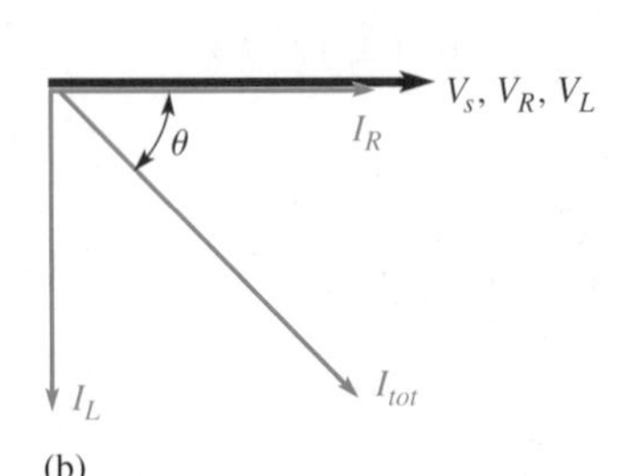

hoff's current law, the total current is the phasor sum of the two branch currents. The total current is expressed as

$$I_{tot} = \sqrt{I_R^2 + I_L^2}$$ **Equation 12–11**

The phase angle between the resistor current and the total current is

$$\theta = \tan^{-1}\left(\frac{I_L}{I_R}\right)$$ **Equation 12–12**

EXAMPLE 12–8

Determine the value of each current in Figure 12–22, and describe the phase relationship of each with the source voltage. Draw the current phasor diagram.

▶ **FIGURE 12–22**

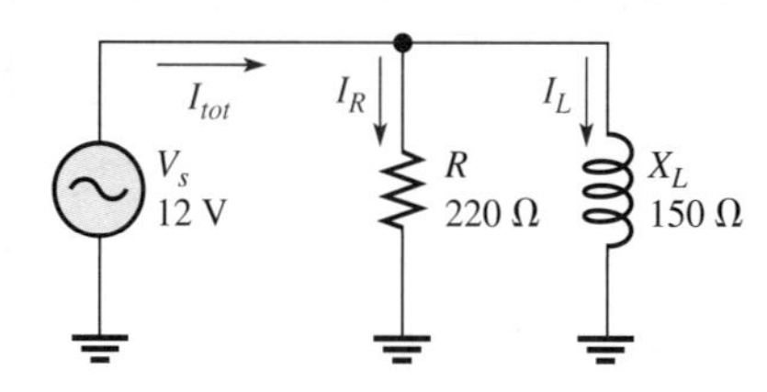

Solution Calculate the resistor current, the inductor current, and the total current.

$$I_R = \frac{V_s}{R} = \frac{12\text{ V}}{220\ \Omega} = \mathbf{54.5\ mA}$$

$$I_L = \frac{V_s}{X_L} = \frac{12\text{ V}}{150\ \Omega} = \mathbf{80\ mA}$$

$$I_{tot} = \sqrt{I_R^2 + I_L^2} = \sqrt{(54.5\text{ mA})^2 + (80\text{ mA})^2} = \mathbf{96.8\ mA}$$

The phase angle is

$$\theta = \tan^{-1}\left(\frac{R}{X_L}\right) = \tan^{-1}\left(\frac{220\ \Omega}{150\ \Omega}\right) = \mathbf{55.7°}$$

The resistor current is 54.5 mA and is in phase with the source voltage. The inductor current is 80 mA and lags the source voltage by 90°. The total current is 96.8 mA and lags the source voltage by 55.7°. The current phasor diagram is shown in Figure 12–23.

▶ **FIGURE 12–23**

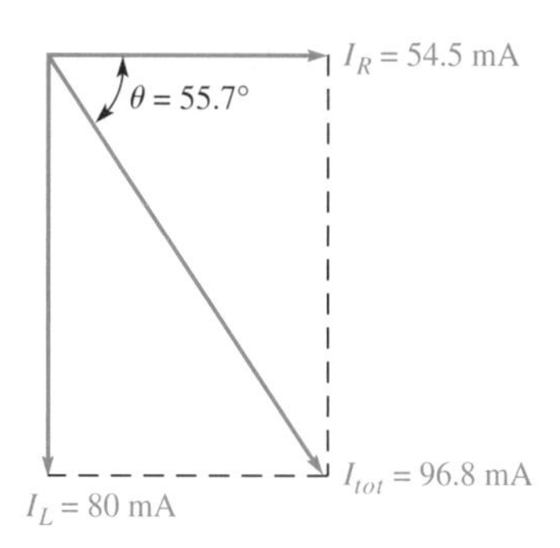

Related Problem Find I_{tot} and θ if $X_L = 300\ \Omega$ in Figure 12–22.

SECTION 12–5 REVIEW

1. The admittance of a parallel *RL* circuit is 4 mS, and the source voltage is 8 V. What is the total current?
2. In a certain parallel *RL* circuit, the resistor current is 12 mA, and the inductor current is 20 mA. Determine the phase angle and the total current.
3. What is the phase angle between the inductor current and the source voltage in a parallel *RL* circuit?

12–6 SERIES-PARALLEL ANALYSIS

In this section, the concepts studied in the previous sections are used to analyze circuits with combinations of both series and parallel *R* and *L* elements.

After completing this section, you should be able to

- **Analyze series-parallel *RL* circuits**
- Determine total impedance and phase angle
- Calculate currents and voltages

The following two examples illustrate the procedures used to analyze a series-parallel reactive circuit.

EXAMPLE 12–9

In the circuit of Figure 12–24, determine the values of the following:

(a) Z_{tot} **(b)** I_{tot} **(c)** θ

▶ FIGURE 12–24

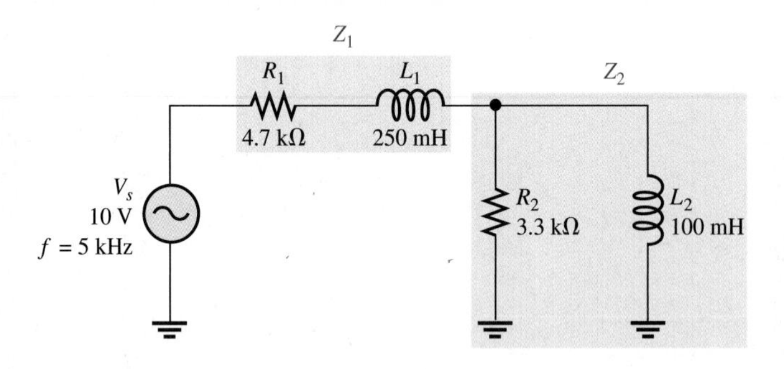

Solution **(a)** First, calculate the magnitudes of the inductive reactances.

$$X_{L1} = 2\pi f L_1 = 2\pi(5\text{ kHz})(250\text{ mH}) = 7.85\text{ k}\Omega$$

$$X_{L2} = 2\pi f L_2 = 2\pi(5\text{ kHz})(100\text{ mH}) = 3.14\text{ k}\Omega$$

One approach is to find the series equivalent resistance and inductive reactance for the parallel portion of the circuit; then add the resistances $(R_1 + R_{eq})$ to get the total resistance and add the reactances $(X_{L1} + X_{L(eq)})$ to get the total reactance. From these totals, you can determine the total impedance.

Determine the impedance of the parallel portion (Z_2) as follows:

$$G_2 = \frac{1}{R_2} = \frac{1}{3.3\ \text{k}\Omega} = 303\ \mu\text{S}$$

$$B_{L2} = \frac{1}{X_{L2}} = \frac{1}{3.14\ \text{k}\Omega} = 318\ \mu\text{S}$$

$$Y_2 = \sqrt{G_2^2 + B_L^2} = \sqrt{(303\ \mu\text{S})^2 + (318\ \mu\text{S})^2} = 439\ \mu\text{S}$$

Then

$$Z_2 = \frac{1}{Y_2} = \frac{1}{439\ \mu\text{S}} = 2.28\ \text{k}\Omega$$

The phase angle associated with the parallel portion of the circuit is

$$\theta_p = \tan^{-1}\left(\frac{R_2}{X_{L2}}\right) = \tan^{-1}\left(\frac{3.3\ \text{k}\Omega}{3.14\ \text{k}\Omega}\right) = 46.4^\circ$$

The series equivalent values for the parallel portion are found using Equations (10–19) and (10–20), adapted for parallel RL circuits as follows:

$$R_{\text{eq}} = Z_2\cos\theta_p = (2.28\ \text{k}\Omega)\cos(46.4^\circ) = 1.57\ \text{k}\Omega$$

$$X_{L(\text{eq})} = Z_2\sin\theta_p = (2.28\ \text{k}\Omega)\sin(46.4^\circ) = 1.65\ \text{k}\Omega$$

The total circuit resistance is

$$R_{tot} = R_1 + R_{\text{eq}} = 4.7\ \text{k}\Omega + 1.57\ \text{k}\Omega = 6.27\ \text{k}\Omega$$

The total circuit reactance is

$$X_{L(tot)} = X_{L1} + X_{L(\text{eq})} = 7.85\ \text{k}\Omega + 1.65\ \text{k}\Omega = 9.50\ \text{k}\Omega$$

The total circuit impedance is

$$Z_{tot} = \sqrt{R_{tot}^2 + X_{L(tot)}^2} = \sqrt{(6.27\ \text{k}\Omega)^2 + (9.50\ \text{k}\Omega)^2} = \mathbf{11.4\ k\Omega}$$

(b) Use Ohm's law to find the total current.

$$I_{tot} = \frac{V_s}{Z_{tot}} = \frac{10\ \text{V}}{11.4\ \text{k}\Omega} = \mathbf{877\ \mu A}$$

(c) To find the phase angle, view the circuit as a series combination of R_{tot} and $X_{L(tot)}$. The phase angle by which I_{tot} lags V_s is

$$\theta = \tan^{-1}\left(\frac{X_{L(tot)}}{R_{tot}}\right) = \tan^{-1}\left(\frac{9.50\ \text{k}\Omega}{6.27\ \text{k}\Omega}\right) = \mathbf{56.6^\circ}$$

Related Problem **(a)** Determine the voltage across the series part of the circuit in Figure 12–24. **(b)** Determine the voltage across the parallel part of the circuit.

Open file E12-09 on your EWB/CircuitMaker CD-ROM. Measure the current through each component. Measure the voltage across Z_1 and Z_2.

EXAMPLE 12–10

Determine the voltage across each component in Figure 12–25. Draw a voltage phasor diagram.

▶ FIGURE 12–25

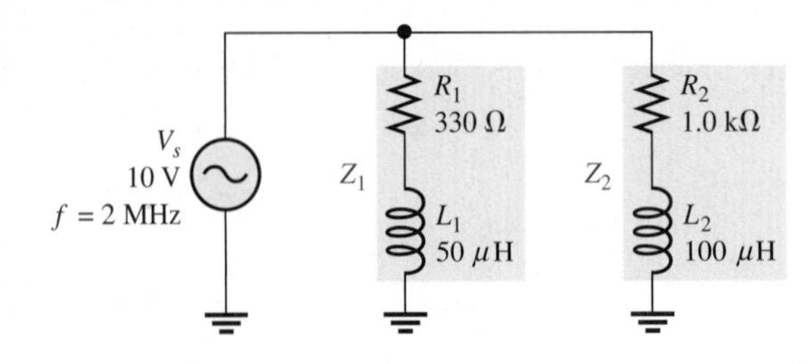

Solution First, calculate X_{L1} and X_{L2}.

$$X_{L1} = 2\pi fL_1 = 2\pi(2\text{ MHz})(50\ \mu\text{H}) = 628\ \Omega$$

$$X_{L2} = 2\pi fL_2 = 2\pi(2\text{ MHz})(100\ \mu\text{H}) = 1.26\text{ k}\Omega$$

Now, determine the impedance of each branch.

$$Z_1 = \sqrt{R_1^2 + X_{L1}^2} = \sqrt{(330\ \Omega)^2 + (628\ \Omega)^2} = 709\ \Omega$$

$$Z_2 = \sqrt{R_2^2 + X_{L2}^2} = \sqrt{(1.0\text{ k}\Omega)^2 + (1.26\text{ k}\Omega)^2} = 1.61\text{ k}\Omega$$

Calculate each branch current.

$$I_1 = \frac{V_s}{Z_1} = \frac{10\text{ V}}{709\ \Omega} = 14.1\text{ mA}$$

$$I_2 = \frac{V_s}{Z_2} = \frac{10\text{ V}}{1.61\text{ k}\Omega} = 6.21\text{ mA}$$

Now, use Ohm's law to find the voltage across each element.

$$V_{R1} = I_1R_1 = (14.1\text{ mA})(330\ \Omega) = \mathbf{4.65\ V}$$

$$V_{L1} = I_1X_{L1} = (14.1\text{ mA})(628\ \Omega) = \mathbf{8.85\ V}$$

$$V_{R2} = I_2R_2 = (6.21\text{ mA})(1.0\text{ k}\Omega) = \mathbf{6.21\ V}$$

$$V_{L2} = I_2X_{L2} = (6.21\text{ mA})(1.26\text{ k}\Omega) = \mathbf{7.82\ V}$$

Now determine the angles associated with each branch current.

$$\theta_1 = \tan^{-1}\left(\frac{X_{L1}}{R_1}\right) = \tan^{-1}\left(\frac{628\ \Omega}{330\ \Omega}\right) = 62.3°$$

$$\theta_2 = \tan^{-1}\left(\frac{X_{L2}}{R_2}\right) = \tan^{-1}\left(\frac{1.26\text{ k}\Omega}{1.0\text{ k}\Omega}\right) = 51.6°$$

Thus, I_1 lags V_s by 62.3°, and I_2 lags V_s by 51.6°, as indicated in Figure 12–26(a), where the negative signs indicate lagging angles.

The phase relationships of the voltages are determined as follows:

- V_{R1} is in phase with I_1 and therefore lags V_s by 62.3°.
- V_{L1} leads I_1 by 90°, so its angle is 90° − 62.3° = 27.7°.
- V_{R2} is in phase with I_2 and therefore lags V_s by 51.6°.
- V_{L2} leads I_2 by 90°, so its angle is 90° − 51.6° = 38.4°.

These phase relationships are shown in Figure 12–26(b).

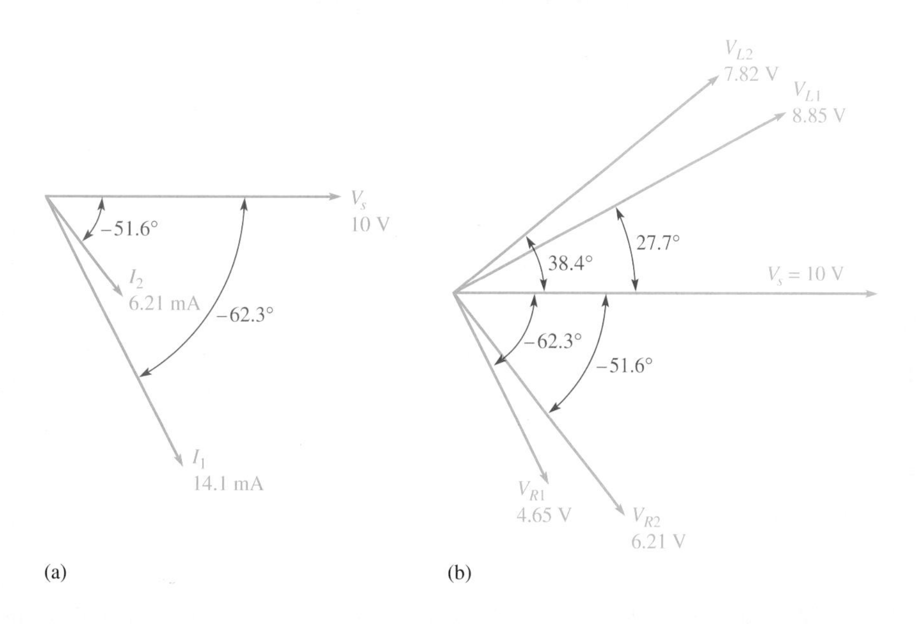

▲ **FIGURE 12–26**

Related Problem What effect does an increase in frequency have on the total current in Figure 12–25?

Open file E12-10 on your EWB/CircuitMaker CD-ROM. Measure the voltage across each component and compare to the calculated values.

SECTION 12–6 REVIEW

1. Determine the total current for the circuit in Figure 12–25. *Hint:* Find the sum of the horizontal components of I_1 and I_2 and the sum of the vertical components of I_1 and I_2. Then apply the Pythagorean theorem to get I_{tot}.
2. What is the total impedance of the circuit in Figure 12–25?

12–7 POWER IN *RL* CIRCUITS

In a purely resistive ac circuit, all of the energy delivered by the source is dissipated in the form of heat by the resistance. In a purely inductive ac circuit, all of the energy delivered by the source is stored by the inductor in its magnetic field during a portion of the voltage cycle and then returned to the source during another portion of the cycle so that there is no net energy conversion to heat. When there is both resistance and inductance, some of the energy is alternately stored and returned by the inductance and some is dissipated by the resistance. The amount of energy converted to heat is determined by the relative values of the resistance and the inductive reactance.

After completing this section, you should be able to

- **Determine power in *RL* circuits**
- Explain true and reactive power
- Draw the power triangle
- Define *power factor*
- Explain power factor correction

When the resistance is greater than the inductive reactance, more of the total energy delivered by the source is dissipated by the resistance than is stored and returned by the inductor. When the reactance is greater than the resistance, more of the total energy is stored and returned than is converted to heat.

As you know, the power dissipated in a resistor is called the true power. The power in an inductor is reactive power and is expressed as

Equation 12–13

$$P_r = I^2 X_L$$

The Power Triangle for *RL* Circuits

The generalized power triangle for the *RL* circuit is shown in Figure 12–27. The apparent power, P_a, is the resultant of the true power, P_{true}, and the reactive power, P_r.

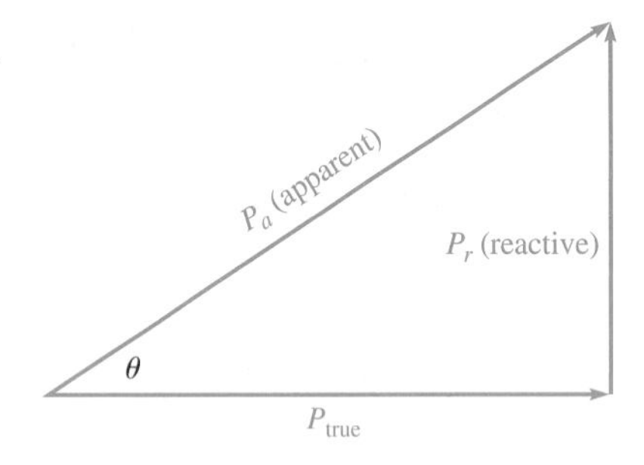

▶ **FIGURE 12–27**

Generalized power triangle for an *RL* circuit.

Recall that the power factor equals the cosine of θ ($PF = \cos\theta$). As the phase angle between the source voltage and the total current increases, the power factor decreases, indicating an increasingly reactive circuit. The smaller the power factor, the smaller the true power is compared to the reactive power. The power factor of inductive loads is called a *lagging power factor* because the current lags the source voltage.

EXAMPLE 12–11

Determine the power factor, the true power, the reactive power, and the apparent power in Figure 12–28.

▶ **FIGURE 12–28**

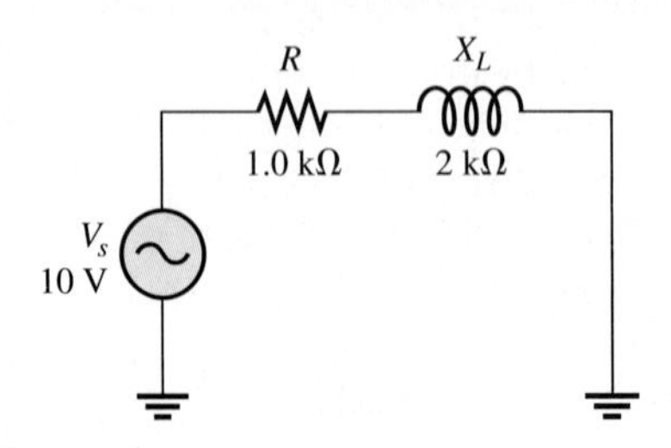

Solution The impedance of the circuit is

$$Z = \sqrt{R^2 + X_L^2} = \sqrt{(1.0\text{ k}\Omega)^2 + (2\text{ k}\Omega)^2} = 2.24\text{ k}\Omega$$

The current is

$$I = \frac{V_s}{Z} = \frac{10\text{ V}}{2.24\text{ k}\Omega} = 4.46\text{ mA}$$

The phase angle is

$$\theta = \tan^{-1}\left(\frac{X_L}{R}\right) = \tan^{-1}\left(\frac{2\text{ k}\Omega}{1.0\text{ k}\Omega}\right) = 63.4^\circ$$

Therefore, the power factor is

$$PF = \cos\theta = \cos(63.4^\circ) = \mathbf{0.448}$$

The true power is

$$P_{\text{true}} = V_sI\cos\theta = (10\text{ V})(4.46\text{ mA})(0.448) = \mathbf{20\ mW}$$

The reactive power is

$$P_r = I^2X_L = (4.46\text{ mA})^2(2\text{ k}\Omega) = \mathbf{39.8\ mVAR}$$

The apparent power is

$$P_a = I^2Z = (4.46\text{ mA})^2(2.24\text{ k}\Omega) = \mathbf{44.6\ mVA}$$

Related Problem If the frequency in Figure 12–28 is increased, what happens to P_{true}, P_r, and P_a?

Significance of the Power Factor

As you learned in Chapter 10, the power factor (*PF*) is important in determining how much useful power (true power) is transferred to a load. The highest power factor is 1, which indicates that all of the current to a load is in phase with the voltage (resistive). When the power factor is 0, all of the current to a load is 90° out of phase with the voltage (reactive).

Generally, a power factor as close to 1 as possible is desirable because then most of the power transferred from the source to the load is useful or true power. True power goes only one way—from source to load—and performs work on the load in terms of energy dissipation. Reactive power simply goes back and forth between the source and the load with no net work being done. Energy must be used in order for work to be done.

Many practical loads have inductance as a result of their particular function, and it is essential for their proper operation. Examples are transformers, electric motors, and speakers, to name a few. Therefore, inductive (and capacitive) loads are important considerations.

To see the effect of the power factor on system requirements, refer to Figure 12–29. This figure shows a representation of a typical inductive load consisting effectively of inductance and resistance in parallel. Part (a) shows a load with a relatively low power factor (0.75), and part (b) shows a load with a relatively high power factor (0.95). Both loads dissipate equal amounts of power as indicated by the wattmeters. Thus, an equal amount of work is done on both loads.

Although both loads are equivalent in terms of the amount of work done (true power), the low-power factor load in Figure 12–29(a) draws more current from the source than does the high-power factor load in Figure 12–29(b), as indicated

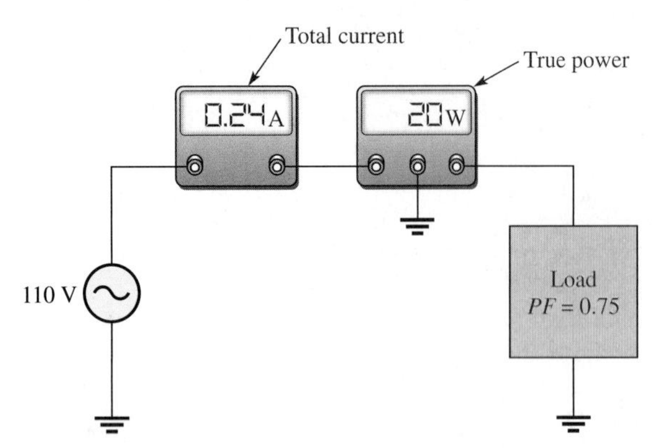

(a) A lower power factor means more total current for a given power dissipation (watts). A larger source is required to deliver the true power (watts).

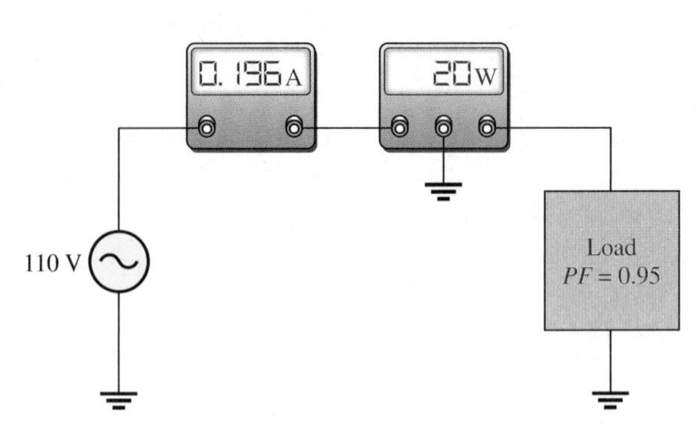

(b) A higher power factor means less total current for a given power dissipation. A smaller source can deliver the same true power (watts).

▲ **FIGURE 12–29**

Illustration of the effect of the power factor on system requirements such as source rating (VA) and conductor size.

by the ammeters. Therefore, the source in part (a) must have a higher VA rating than the one in part (b). Also, the lines connecting the source to the load must be a larger wire gauge than those in part (b), a condition that becomes significant when very long transmission lines are required, such as in power distribution.

Figure 12–29 has demonstrated that a higher power factor is an advantage in delivering power more efficiently to a load. Also, because the electric company charges for apparent power, it is less expensive to have a high power factor.

Power Factor Correction

The power factor of an inductive load can be increased by the addition of a capacitor in parallel, as shown in Figure 12–30. The capacitor compensates for the phase lag of the total current by creating a capacitive component of current that is 180° out of phase with the inductive component. This has a canceling effect and reduces the phase angle, thus increasing the power factor as well as the total current, as illustrated in the figure.

► **FIGURE 12–30**

Example of how the power factor can be increased by the addition of a compensating capacitor (C_c). As θ decreases, *PF* increases.

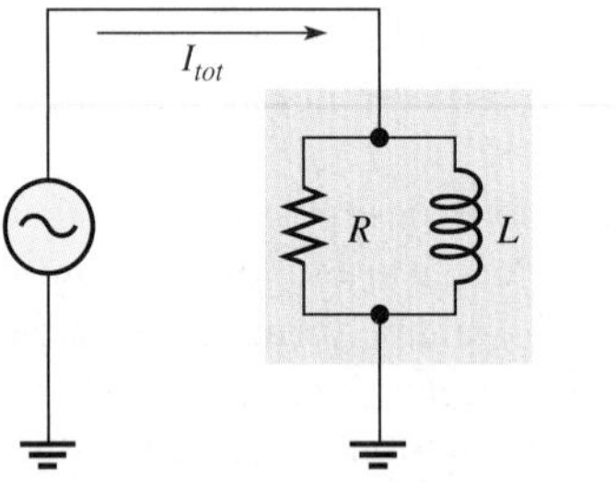

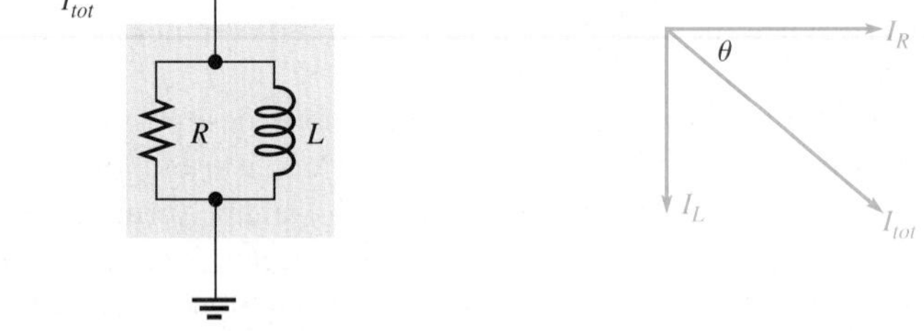

(a) Total current is the resultant of I_R and I_L.

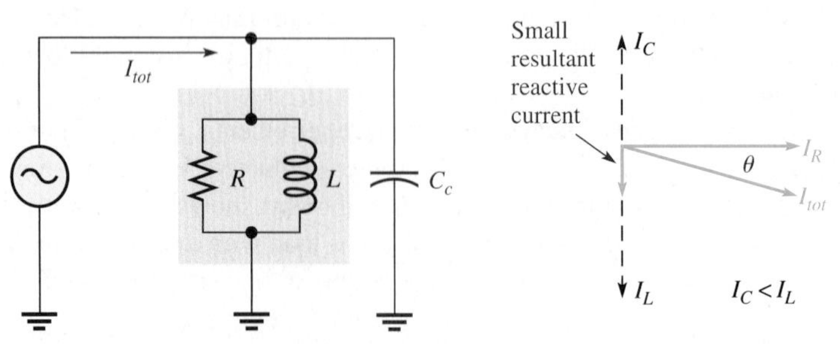

(b) I_C subtracts from I_L, leaving only a small reactive current, thus decreasing I_{tot} and the phase angle.

SECTION 12–7 REVIEW

1. To which component in an *RL* circuit is the power dissipation due?
2. Calculate the power factor when $\theta = 50°$.
3. A certain series *RL* circuit consists of a 470 Ω resistor and an inductive reactance of 620 Ω at the operating frequency. Determine P_{true}, P_r, and P_a when $I = 100$ mA.

12–8 BASIC APPLICATIONS

Two basic applications, phase shift circuits and frequency-selective circuits (filters), are covered in this section.

After completing this section, you should be able to

- **Discuss some basic *RL* applications**
- Discuss and analyze the *RL* lag network
- Discuss and analyze the *RL* lead network
- Describe how the *RL* circuit operates as a filter

RL Lag Network

The *RL* lag network is a phase shift circuit in which the output voltage lags the input voltage by a specified angle, ϕ. A basic *RL* lag network is shown in Figure 12–31(a). Notice that the output voltage is taken across the resistor and the input voltage is the total voltage applied across the circuit. The relationship of the voltages is shown in the phasor diagram in Figure 12–31(b), and a waveform diagram is shown in Figure 12–31(c). Notice that the output voltage, V_{out}, lags V_{in} by an angle, designated ϕ, that is the same as the circuit phase angle. The angles are equal, of course, because V_R and I are in phase with each other.

The phase lag, ϕ, can be expressed as

$$\phi = \tan^{-1}\left(\frac{X_L}{R}\right)$$

Equation 12–14

FIGURE 12–31

The *RL* lag network ($V_{out} = V_R$).

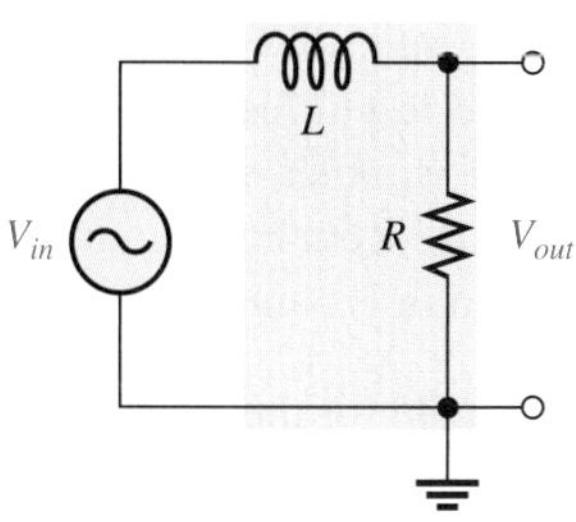

(a) A basic *RL* lag network

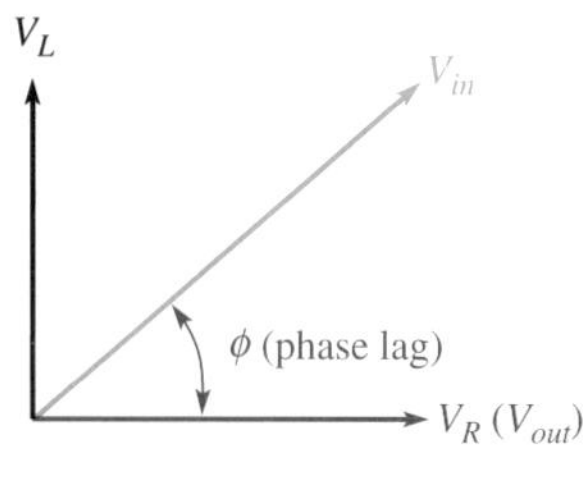

(b) Phasor voltage diagram showing phase lag between V_{in} and V_{out}

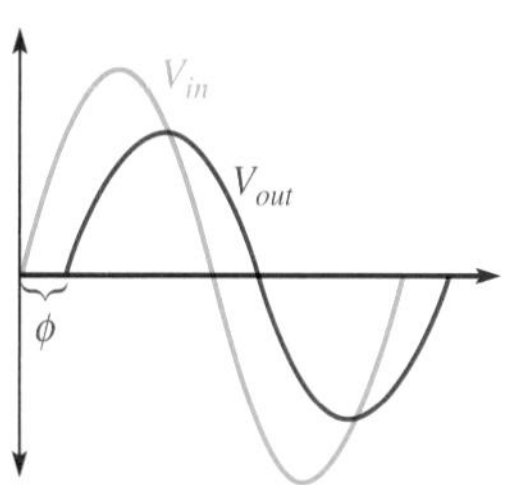

(c) Input and output waveforms

EXAMPLE 12–12

Calculate the phase lag for each circuit in Figure 12–32.

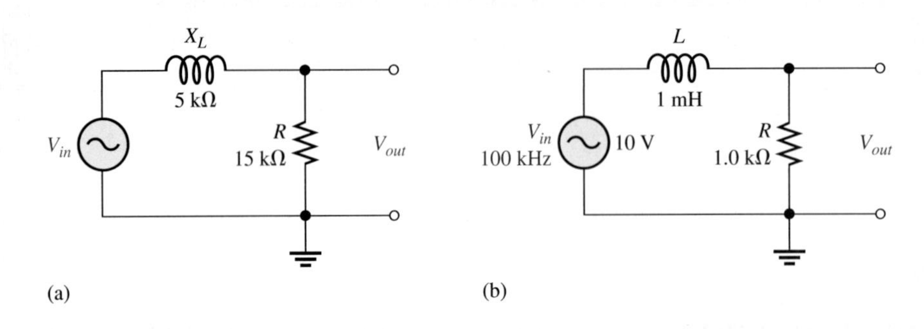

▲ FIGURE 12–32

Solution For the circuit in Figure 12–32(a),

$$\phi = \tan^{-1}\left(\frac{X_L}{R}\right) = \tan^{-1}\left(\frac{5\text{ k}\Omega}{15\text{ k}\Omega}\right) = \mathbf{18.4^\circ}$$

The output lags the input by 18.4°.

For the circuit in Figure 12–32(b), first determine the inductive reactance.

$$X_L = 2\pi fL = 2\pi(100\text{ kHz})(1\text{ mH}) = 628\ \Omega$$

The phase lag is

$$\phi = \tan^{-1}\left(\frac{X_L}{R}\right) = \tan^{-1}\left(\frac{628\ \Omega}{1.0\text{ k}\Omega}\right) = \mathbf{32.1^\circ}$$

The output lags the input by 32.1°.

Related Problem In a certain lag network, $R = 5.6\text{ k}\Omega$ and $X_L = 3.5\text{ k}\Omega$. Determine the phase lag between input and output.

The phase-lag network can be considered as a voltage divider with a portion of the input voltage dropped across L and a portion across R. The output voltage can be determined with the following formula:

Equation 12–15

$$V_{out} = \left(\frac{R}{\sqrt{R^2 + X_L^2}}\right)V_{in}$$

EXAMPLE 12–13

The input voltage in Figure 12–32(b) of Example 12–12 has an rms value of 10 V. Determine the output voltage for the lag network shown in Figure 12–32(b). Draw the waveform relationships for the input and output voltages. The phase lag (32.1°) and X_L (628 Ω) were found in Example 12–12.

Solution Use Equation 12–15 to determine the output voltage for the lag network in Figure 12–32(b).

$$V_{out} = \left(\frac{R}{\sqrt{R^2 + X_L^2}}\right)V_{in} = \left(\frac{1.0\text{ k}\Omega}{1.18\text{ k}\Omega}\right)10\text{ V} = \mathbf{8.47\ V\ rms}$$

The waveforms are shown in Figure 12–33.

▶ FIGURE 12–33

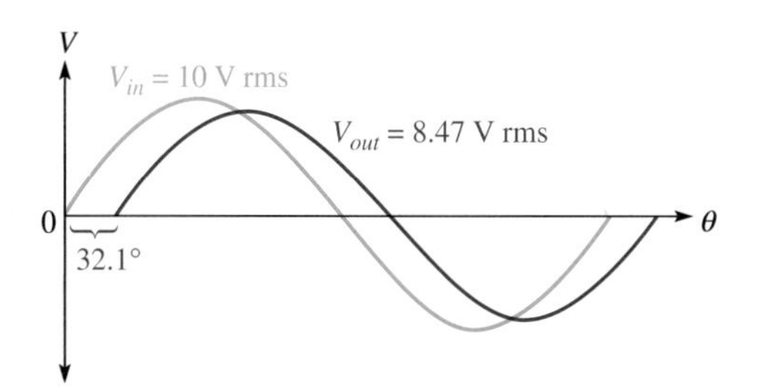

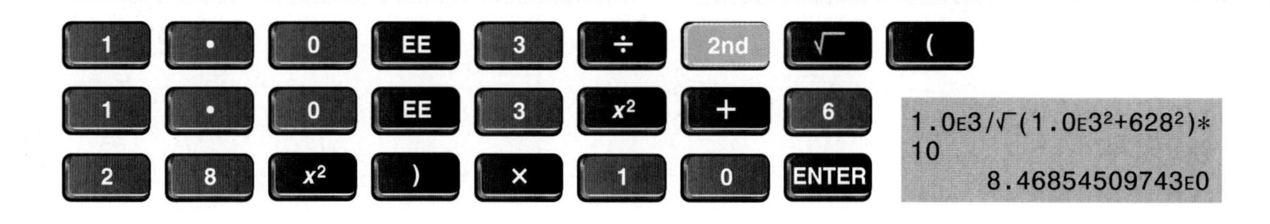

Related Problem In a lag network, $R = 4.7\text{ k}\Omega$ and $X_L = 6\text{ k}\Omega$. If the rms input voltage is 20 V, what is the output voltage?

Open file E12-13 on your EWB/CircuitMaker CD-ROM. Measure the output voltage and compare to the calculated value.

Effects of Frequency on the Lag Network Since the circuit phase angle and the phase lag are the same, an increase in frequency causes an increase in phase lag. Also, an increase in frequency causes a decrease in the magnitude of the output voltage because X_L becomes greater and more of the total voltage is dropped across the inductor and less across the resistor. Figure 12–34 illustrates the changes in phase lag and output voltage magnitude with frequency.

▼ FIGURE 12–34

Illustration of how the frequency affects the phase lag and the output voltage in an *RL* lag network with the amplitude of V_{in} held constant.

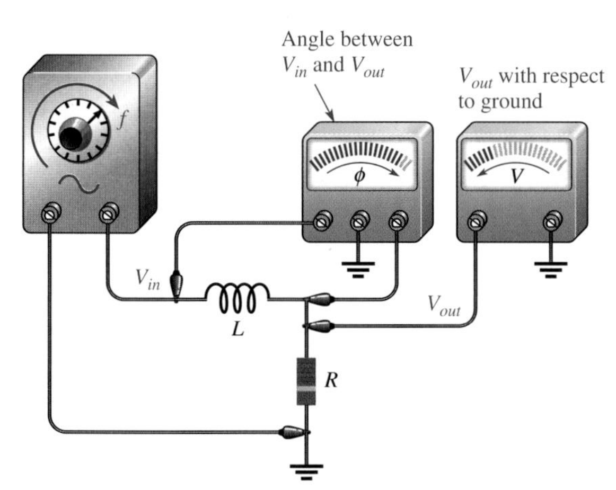

(a) When f increases, the phase lag ϕ increases and V_{out} decreases.

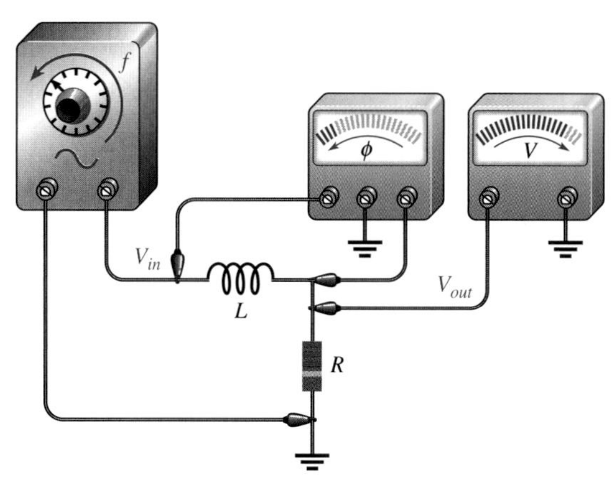

(b) When f decreases, ϕ decreases and V_{out} increases.

RL Lead Network

The **RL lead network** is a phase shift circuit in which the output voltage leads the input voltage by a specified angle, ϕ. A basic *RL* lead network is shown in Figure 12–35(a). Notice how this network differs from the lag network. Here the output voltage is taken across the inductor rather than across the resistor. The relationship of the voltages is shown in the phasor diagram of Figure 12–35(b) and in the waveform plot of Figure 12–35(c). Notice that the output voltage, V_{out}, leads V_{in} by an angle (phase lead) that is the difference between 90° and the circuit phase angle θ.

FIGURE 12–35

The *RL* lead network ($V_{out} = V_L$).

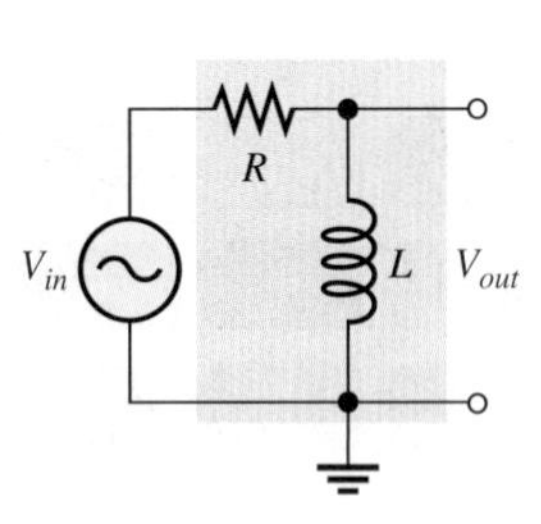

(a) A basic *RL* lead network

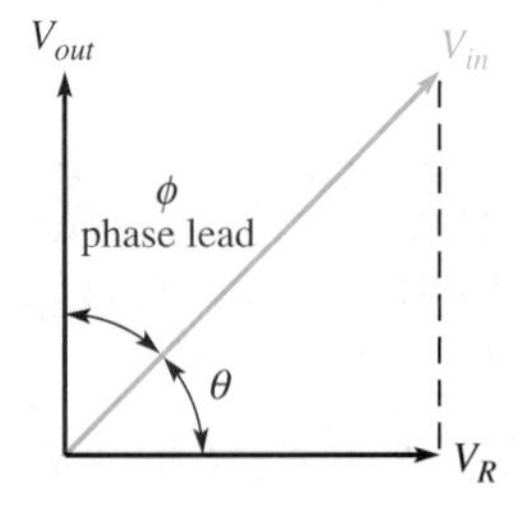

(b) Phasor voltage diagram showing V_{out} leading V_{in}

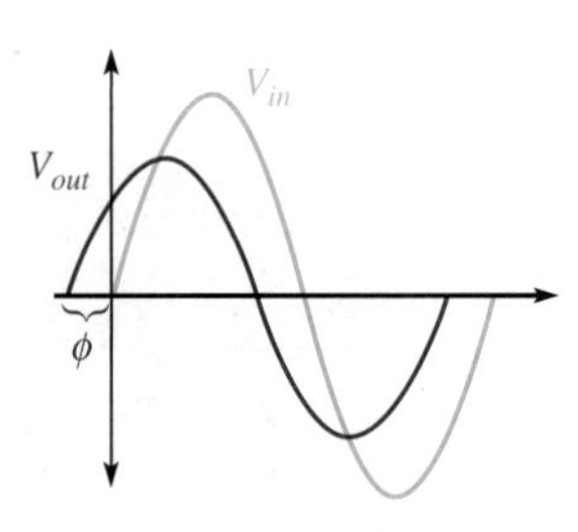

(c) Input and output waveforms

Since $\theta = \tan^{-1}(X_L/R)$, the phase lead, ϕ, can be expressed as

Equation 12–16

$$\phi = 90° - \tan^{-1}\left(\frac{X_L}{R}\right)$$

Again, the phase-lead network can be considered as a voltage divider with the voltage across *L* being the output. The expression for the output voltage is

Equation 12–17

$$V_{out} = \left(\frac{X_L}{\sqrt{R^2 + X_L^2}}\right)V_{in}$$

EXAMPLE 12–14

Determine the amount of phase lead and output voltage in the lead network in Figure 12–36.

FIGURE 12–36

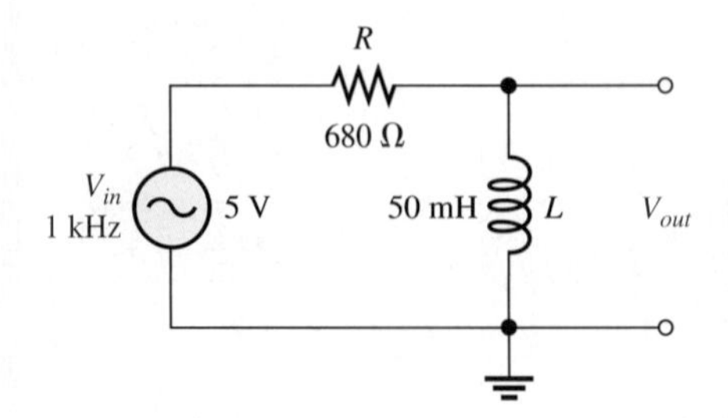

Solution First, determine the inductive reactance.

$$X_L = 2\pi fL = 2\pi(1\text{ kHz})(50\text{ mH}) = 314\ \Omega$$

The phase lead is

$$\phi = 90° - \tan^{-1}\left(\frac{X_L}{R}\right) = 90° - \tan^{-1}\left(\frac{314\ \Omega}{680\ \Omega}\right) = \mathbf{65.2°}$$

The output leads the input by 65.2°.

The output voltage is

$$V_{out} = \left(\frac{X_L}{\sqrt{R^2 + X_L^2}}\right)V_{in} = \left(\frac{314\ \Omega}{\sqrt{(680\ \Omega)^2 + (314\ \Omega)^2}}\right)5\text{ V} = \mathbf{2.1\ V}$$

Related Problem In a certain lead network, $R = 2.2\text{ k}\Omega$ and $X_L = 1\text{ k}\Omega$. What is the phase lead?

Open file E12-14 on your EWB/CircuitMaker CD-ROM. Measure the output voltage and compare to the calculated value.

Effects of Frequency on the Lead Network Since the circuit phase angle, θ, increases as frequency increases, the phase lead between the input and the output voltages decreases. You can see this relationship by examining Equation 12–16. Also, the amplitude of the output voltage increases as the frequency increases because X_L becomes greater and more of the total input voltage is dropped across the inductor. Figure 12–37 illustrates the changes in the phase lead and output voltage with frequency.

FIGURE 12–37

Illustration of how the frequency affects the phase lead and the output voltage in an *RL* lead network with the amplitude of V_{in} held constant.

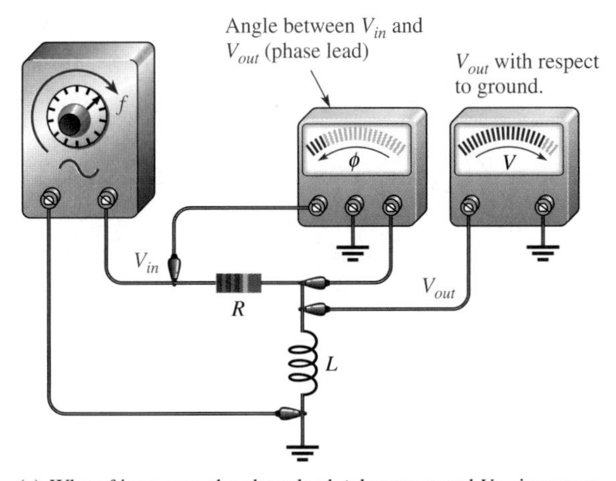

(a) When f increases, the phase lead ϕ decreases and V_{out} increases.

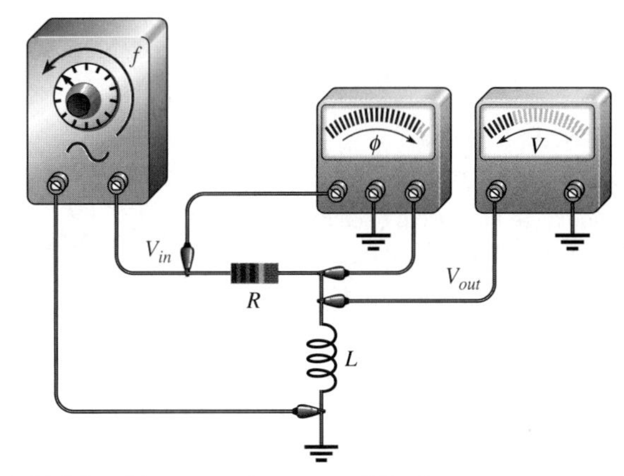

(b) When f decreases, ϕ increases and V_{out} decreases.

The *RL* Circuit as a Filter

Like *RC* circuits, series *RL* circuits also exhibit a frequency-selective characteristic and therefore act as basic filters.

Low-Pass Filter You have seen what happens to the phase angle and the output voltage in the lag network. In terms of its filtering action, the variation in the magnitude of the output voltage as a function of frequency is important.

Figure 12–38 shows the filtering action of a series *RL* circuit using a specific series of measurements in which the frequency starts at 100 Hz and is increased in increments up to 20 kHz. At each value of frequency, the output voltage is measured. As you can see, the inductive reactance increases as frequency increases, thus causing less voltage to be dropped across the resistor while the input voltage is held at a constant 10 V throughout each step. The frequency response curve for these particular values would appear similar to the response curve in Figure 10–51 for the *RC* low-pass filter.

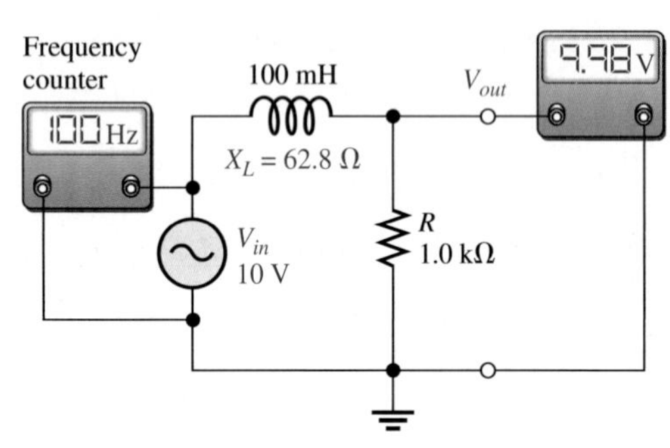

(a) $f = 100$ Hz, $X_L = 62.8\ \Omega$, $V_{out} = 9.98$ V

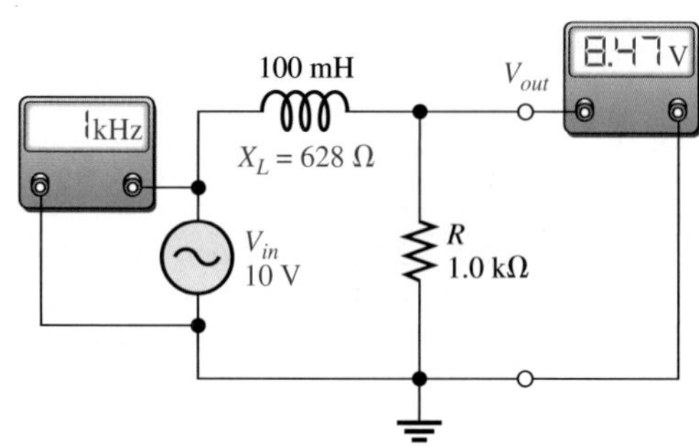

(b) $f = 1$ kHz, $X_L = 628\ \Omega$, $V_{out} = 8.47$ V

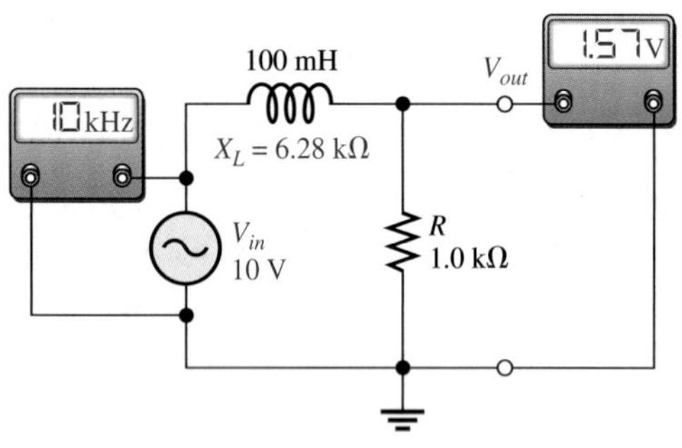

(c) $f = 10$ kHz, $X_L = 6.28\ k\Omega$, $V_{out} = 1.57$ V

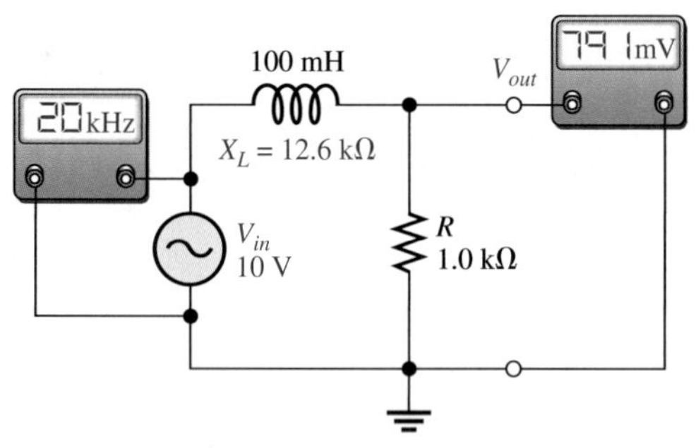

(d) $f = 20$ kHz, $X_L = 12.6\ k\Omega$, $V_{out} = 791$ mV

▲ **FIGURE 12–38**

Example of low-pass filtering action. Winding resistance has been neglected. As the input frequency increases, the output voltage decreases.

High-Pass Filter To illustrate *RL* high-pass filtering action, Figure 12–39 shows a series of specific measurements. The frequency starts at 10 Hz and is increased in increments up to 10 kHz. As you can see, the inductive reactance increases as the frequency increases, thus causing more voltage to be dropped across the inductor. Again, when the values are plotted, the response curve is similar to the one for the *RC* high-pass filter that was shown in Figure 10–53.

The Cutoff Frequency of an RL Filter The frequency at which the inductive reactance equals the resistance in a low-pass or high-pass *RL* filter is called the cutoff frequency and is designated f_c. This condition is expressed as $2\pi f_c L = R$. Solving for f_c results in the following formula:

Equation 12–18

$$f_c = \frac{R}{2\pi L}$$

As with the *RC* filter, the output voltage is 70.7% of its maximum value at f_c. In a high-pass filter, all frequencies above f_c are considered to be passed by the filter, and all those below f_c are considered to be rejected. The reverse, of course, is true for a low-pass filter. Filter bandwidth, defined in Chapter 10, applies to both *RC* and *RL* filter circuits.

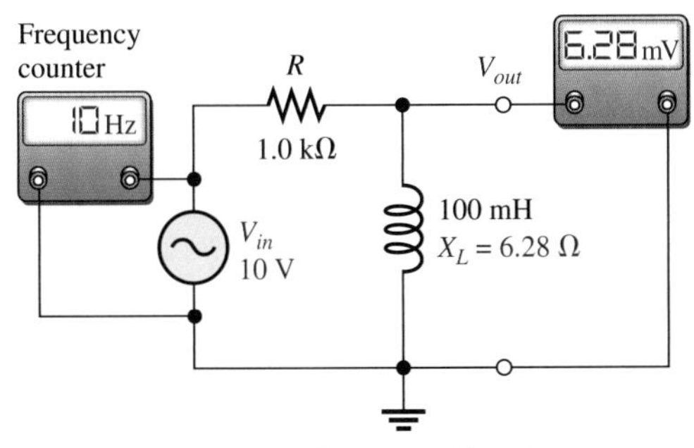

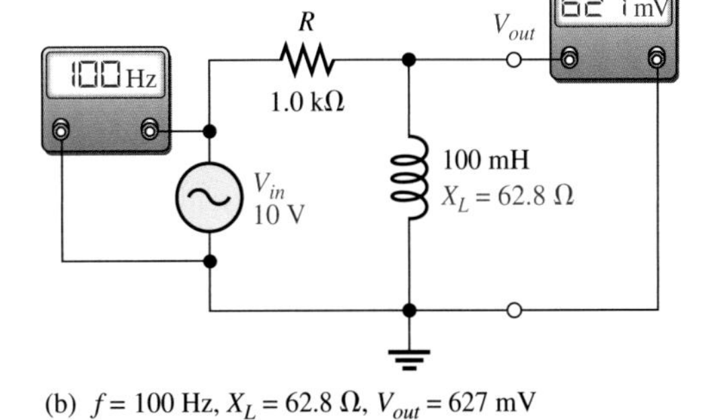

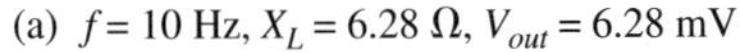

(a) $f = 10$ Hz, $X_L = 6.28\ \Omega$, $V_{out} = 6.28$ mV

(b) $f = 100$ Hz, $X_L = 62.8\ \Omega$, $V_{out} = 627$ mV

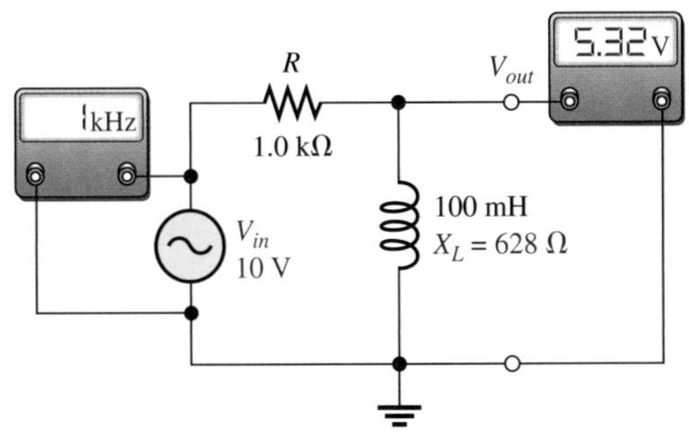

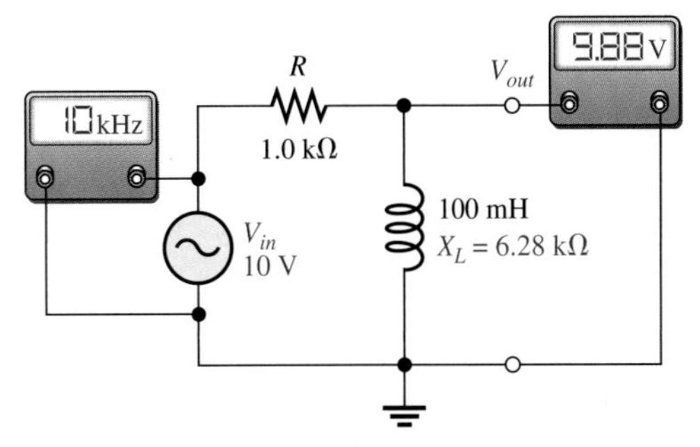

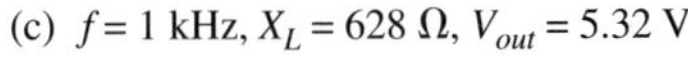

(c) $f = 1$ kHz, $X_L = 628\ \Omega$, $V_{out} = 5.32$ V

(d) $f = 10$ kHz, $X_L = 6.28$ kΩ, $V_{out} = 9.88$ V

FIGURE 12–39

Example of high-pass filtering action. Winding resistance has been neglected. As the input frequency increases, the output voltage increases.

SECTION 12–8 REVIEW

1. A certain *RL* lead network consists of a 3.3 kΩ resistor and a 15 mH inductor. Determine the phase lead between input and output at a frequency of 5 kHz.
2. An *RL* lag network has the same component values as the lead network in Question 1. What is the magnitude of the output voltage at 5 kHz when the input is 10 V rms?
3. When an *RL* circuit is used as a low-pass filter, across which component is the output taken?

12–9 TROUBLESHOOTING

In this section, the effects that typical component failures have on the response of basic *RL* circuits are considered. An example of troubleshooting using the APM (analysis, planning, and measurement) method is also presented.

After completing this section, you should be able to

- **Troubleshoot *RL* circuits**
- Find an open inductor
- Find an open resistor
- Find an open in a parallel circuit
- Find an inductor with shorted windings

Effect of an Open Inductor The most common failure mode for inductors occurs when the winding opens as a result of excessive current or a mechanical contact failure. It is easy to see how an open coil affects the operation of a basic series *RL* circuit, as shown in Figure 12–40. Obviously, there is no path for current, so the resistor voltage is zero and the total source voltage appears across the open inductor.

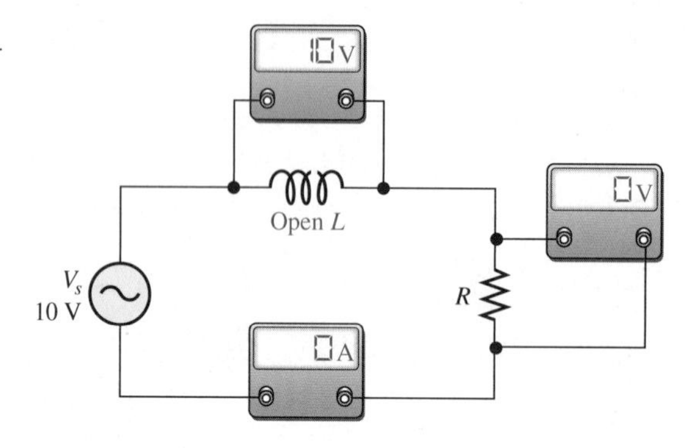

▶ **FIGURE 12–40**

Effect of an open coil.

Effect of an Open Resistor When the resistor is open, there is no current, and the inductor voltage is zero. The total input voltage appears across the open resistor, as shown in Figure 12–41.

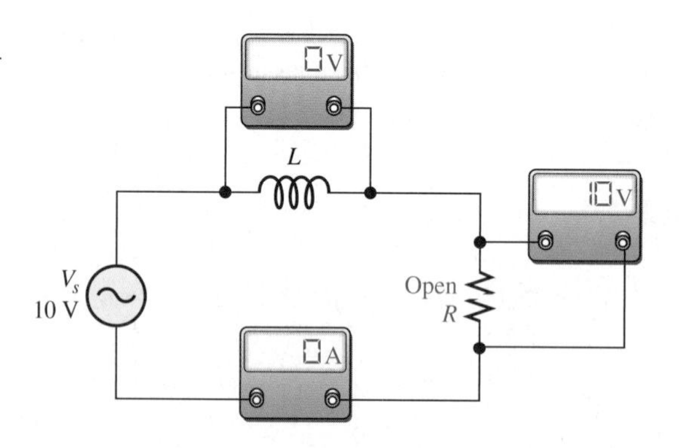

▶ **FIGURE 12–41**

Effect of an open resistor.

Open Components in Parallel Circuits In a parallel *RL* circuit, an open resistor or inductor will cause the total current to decrease because the total impedance will increase. Obviously, the branch with the open component will have zero current. Figure 12–42 illustrates these conditions.

▼ **FIGURE 12–42**

Effect of an open component in a parallel circuit with V_s constant.

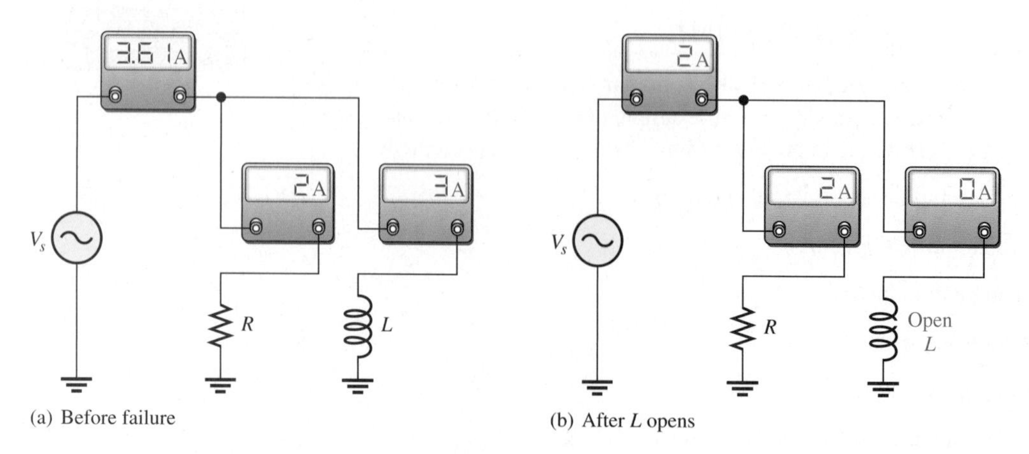

Effect of an Inductor with Shorted Windings Although a rare occurrence, it is possible for some of the windings of coils to short together as a result of damaged insulation. This failure mode is much less likely than the open coil. Shorted windings result in a reduction in inductance because the inductance of a coil is proportional to the square of the number of turns.

Other Troubleshooting Considerations

As you learned in Chapter 10, the failure of a circuit to work properly is not always the result of a faulty component. A loose wire, a bad contact, or a poor solder joint can cause an open circuit. A short can be caused by a wire clipping or solder splash. Things as simple as not plugging in a power supply or a function generator happen more often than you might think. Wrong values in a circuit, such as an incorrect resistor value, the function generator set at the wrong frequency, or the wrong output connected to the circuit, can cause improper operation.

Always check to make sure that the instruments are properly connected to the circuits and to a power outlet. Also, look for obvious things such as a broken or loose contact, a connector that is not completely plugged in, or a piece of wire or a solder bridge that could be shorting something out.

The following example illustrates a troubleshooting approach to a circuit containing inductors and resistors using the APM (analysis, planning, and measurement) method and half-splitting.

EXAMPLE 12–15

The circuit represented by the schematic in Figure 12–43 has no output voltage, which is across R_4. The circuit is physically constructed on a protoboard. Use your troubleshooting skills to find the problem.

▶ FIGURE 12–43

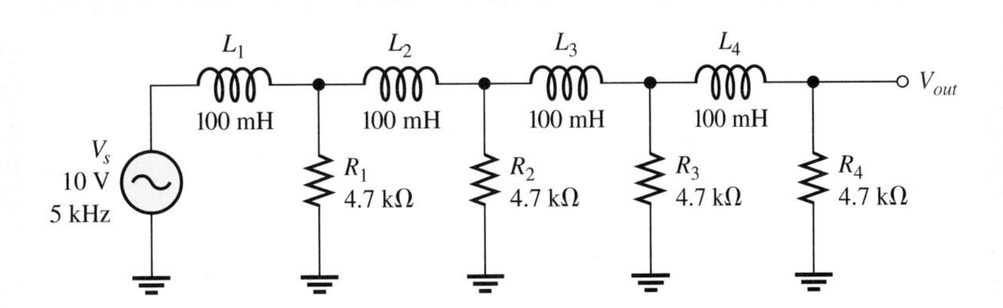

Solution Apply the APM method to this troubleshooting problem.

Analysis: First think of the possible causes for the circuit to have no output voltage.

1. There is no source voltage or the frequency is so high that the inductors appear to be open because their reactances are extremely high compared to the resistance values.
2. There is a short between one of the resistors and ground. Either a resistor could be shorted, or there could be some physical short. A shorted resistor is not a very common fault.

3. There is an open between the source and the output. This would prevent current and thus cause the output voltage to be zero. An inductor could be open, or the conductive path could be open due to a broken or loose connecting wire or a bad protoboard contact.

4. There is an incorrect component value. A resistor could be so small that the voltage across it is negligible. An inductor could be so large that its reactance at the input frequency is extremely high.

Planning: You decide to make some visual checks for problems such as the function generator power cord not plugged in or the frequency set at an incorrect value. Also, broken leads, shorted leads, as well as an incorrect resistor color code or inductor value often can be found visually. If nothing is discovered after a visual check, then you will make voltage measurements to track down the cause of the problem. You decide to use a digital oscilloscope and a multimeter to make the measurements using the half-splitting technique to more quickly isolate the fault.

Measurement: Assume that you find that the function generator is plugged in and the frequency setting appears to be correct. Also, you find no visible opens or shorts during your visual check, and the component values are correct.

The first step in the measurement process is to check the voltage from the source with the scope. Assume a 10 V rms sine wave with a frequency of 5 kHz is observed at the circuit input as shown in Figure 12–44(a). The correct ac voltage is present, so *the first possible cause has been eliminated.*

Next, check for a short by disconnecting the source and placing the multimeter (set on the ohmmeter function) across each resistor. If any resistor is shorted (unlikely), the meter will read zero or a very small resistance. Assuming the meter readings are okay, *the second possible cause has been eliminated.*

Since the voltage has been "lost" somewhere between the input and the output, you must now look for the voltage. You reconnect the source and, using the half-splitting approach, measure the voltage at point ③ (the middle of the circuit) with respect to ground. The multimeter test lead is placed on the right lead of inductor L_2, as indicated in Figure 12–44(b). Assume the voltage at this point is zero. This tells you that the part of the circuit to the right of point ③ is probably okay and the fault is in the circuit between point ③ and the source.

Now, you begin tracing the circuit back toward the source looking for the voltage (you could also start from the source and work forward). Placing the meter test lead on point ②, at the left lead of inductor L_2, results in a reading of 10 V as shown in Figure 12–44(b). This, of course, indicates that L_2 is open. Fortunately, in this case, a component, and not a contact on the board, is faulty. It is usually easier to replace a component than to repair a bad contact.

Related Problem Suppose you had measured 0 V at the left lead of L_2 and 10 V at the right lead of L_1. What would this have indicated?

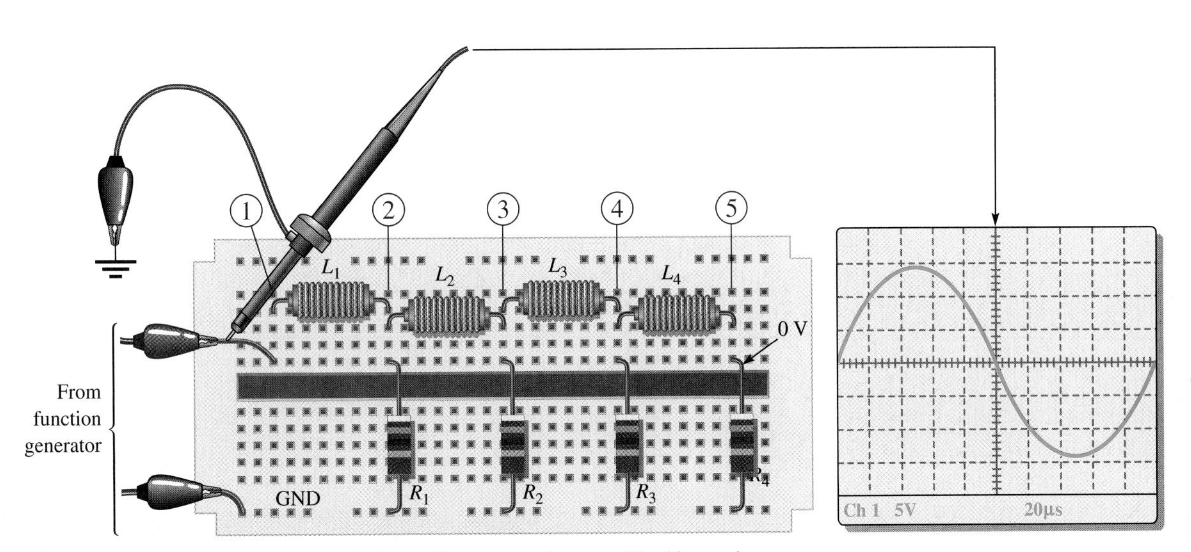

(a) Scope shows the correct voltage at the input. The scope probe ground lead is not shown.

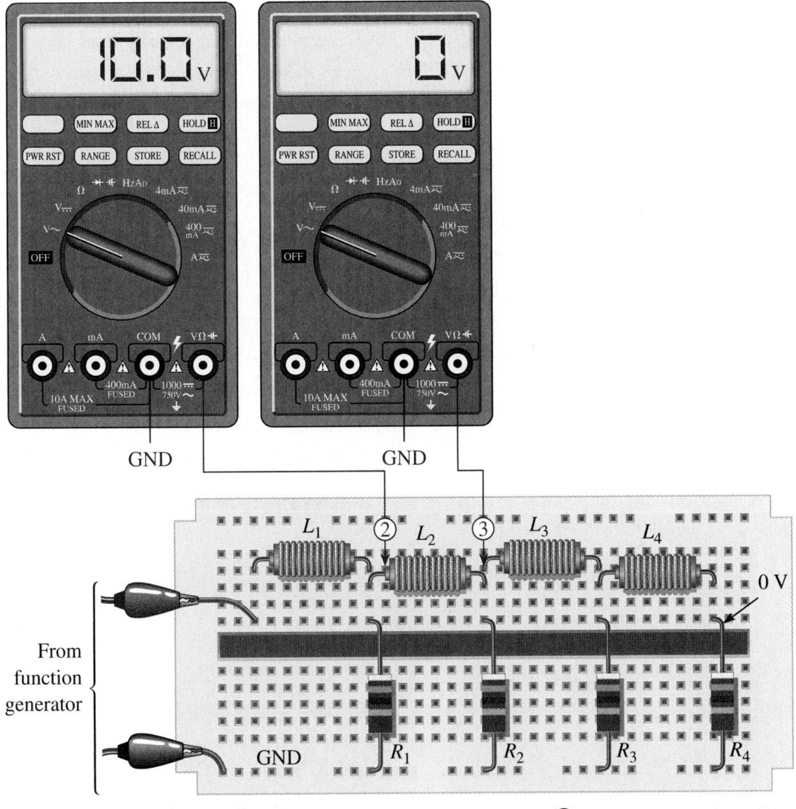

(b) A zero voltage at point ③ indicates the fault is between point ③ and the source. A reading of 10 V at point ② shows that L_2 is open.

▲ **FIGURE 12–44**

SECTION 12–9 REVIEW

1. Describe the effect of an inductor with shorted windings on the response of a series *RL* circuit.
2. In the circuit of Figure 12–45, indicate whether I_{tot}, V_{R1}, and V_{R2} increase or decrease as a result of *L* opening.

FIGURE 12–45

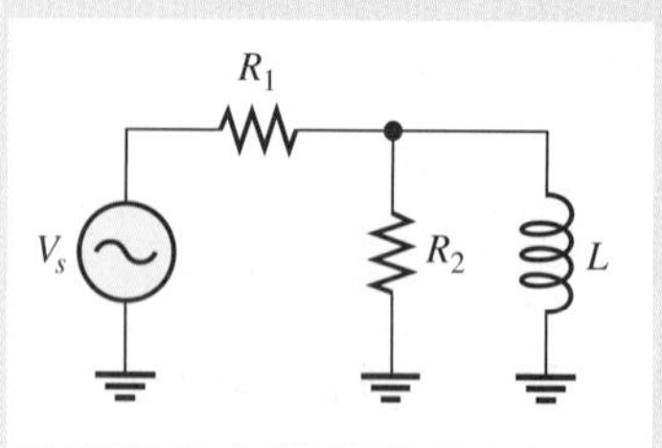

APPLICATION ASSIGNMENT

PUTTING YOUR KNOWLEDGE TO WORK

You are given two sealed modules that have been removed from a communications system that is being modified. Each module has three terminals and is labeled as an *RL* filter, but no specifications are given. Your assignment is to test the modules to determine the type of filters and the component values.

The sealed modules have three terminals labeled IN, GND, and OUT as shown in Figure 12–46. You will apply your knowledge of series *RL* circuits and some basic measurements to determine the internal circuit configuration and the component values.

Step 1: Resistance Measurement of Module 1

Determine the arrangement of two components and the values of the resistor and winding resistance for module 1 indicated by the meter readings in Figure 12–46.

Step 2: AC Measurement of Module 1

Determine the inductance value for module 1 indicated by the test setup in Figure 12–47.

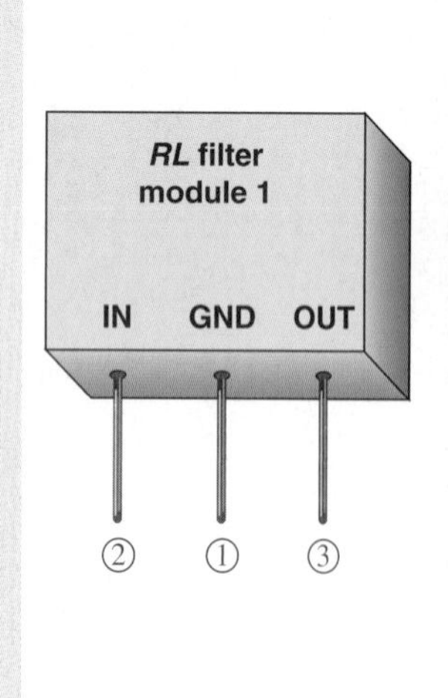

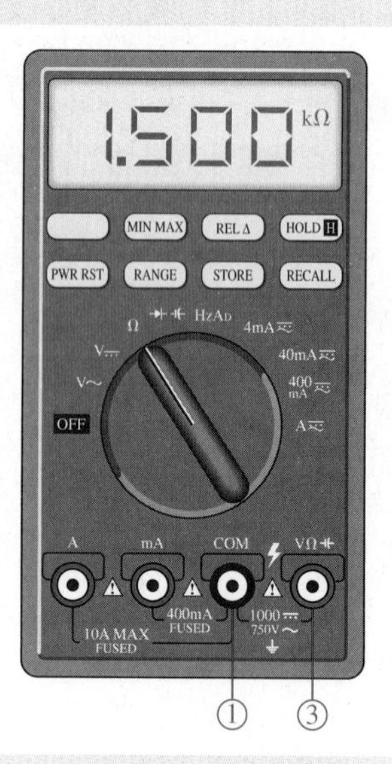

FIGURE 12–46

Resistance measurements of module 1.

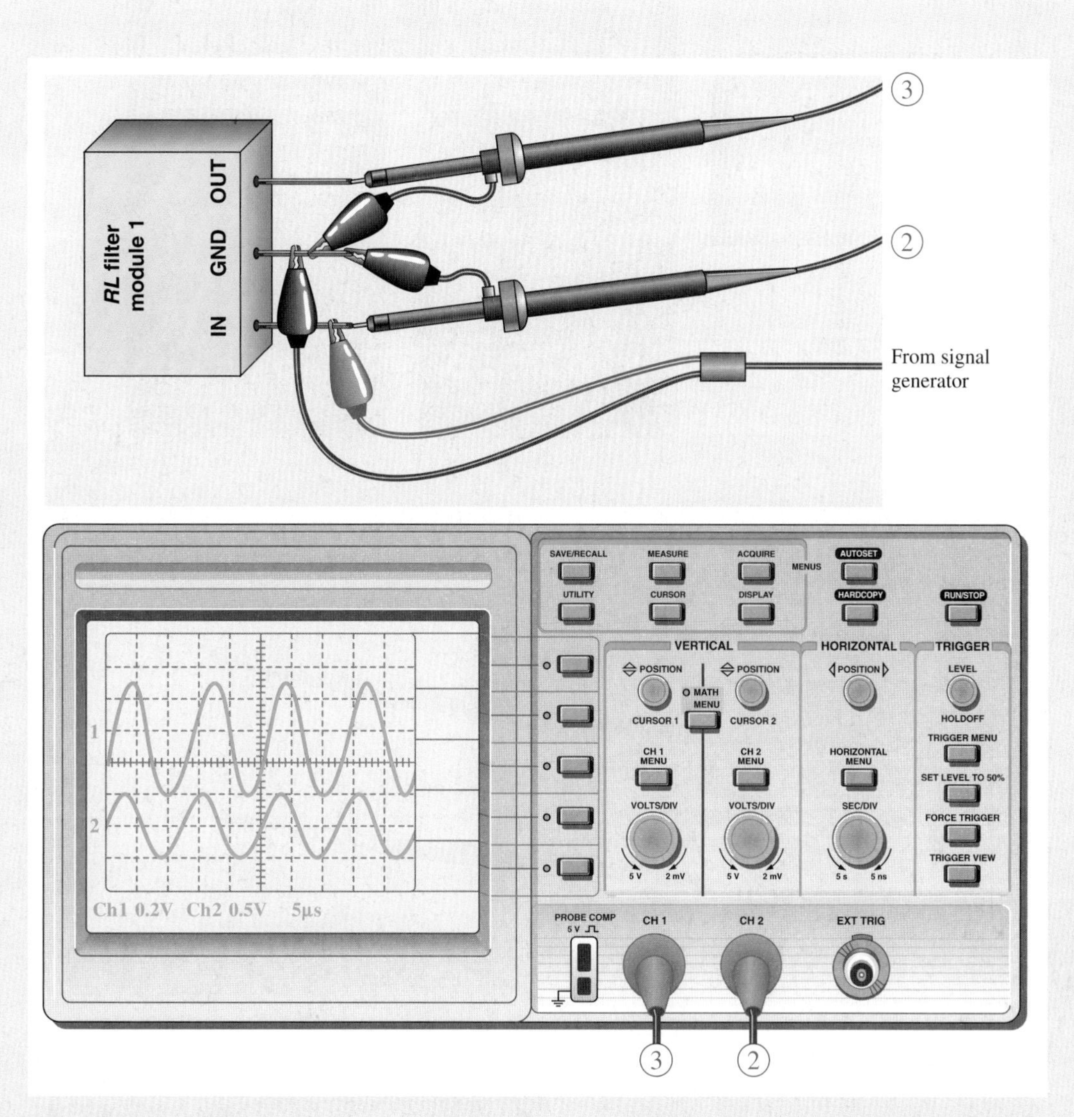

▲ **FIGURE 12–47**

AC measurements for module 1.

Step 3: Resistance Measurement of Module 2

Determine the arrangement of the two components and the values of the resistor and the winding resistance for module 2 indicated by the meter readings in Figure 12–48.

Step 4: AC Measurement of Module 2

Determine the inductance value for module 2 indicated by the test setup in Figure 12–49.

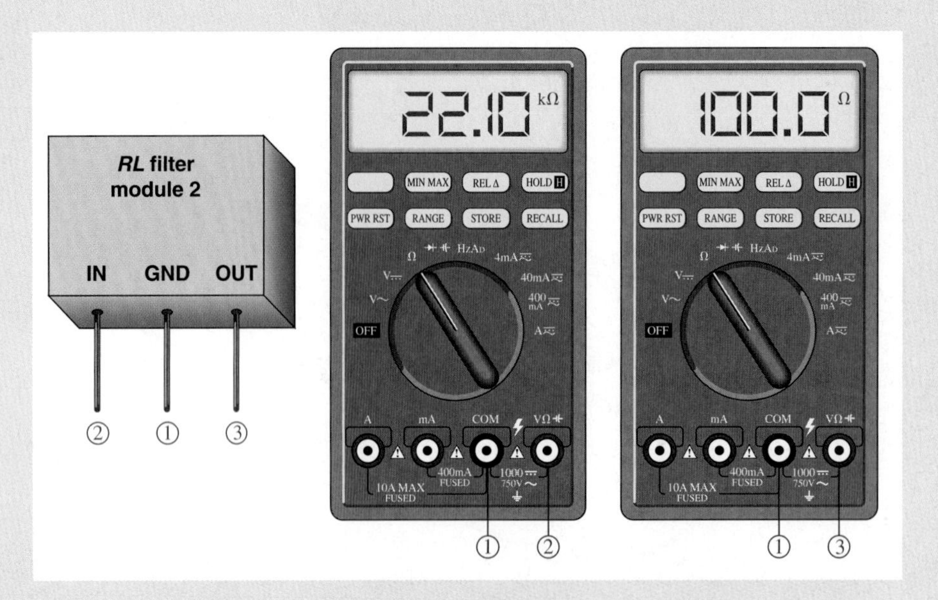

▲ **FIGURE 12–48**

Resistance measurements of module 2.

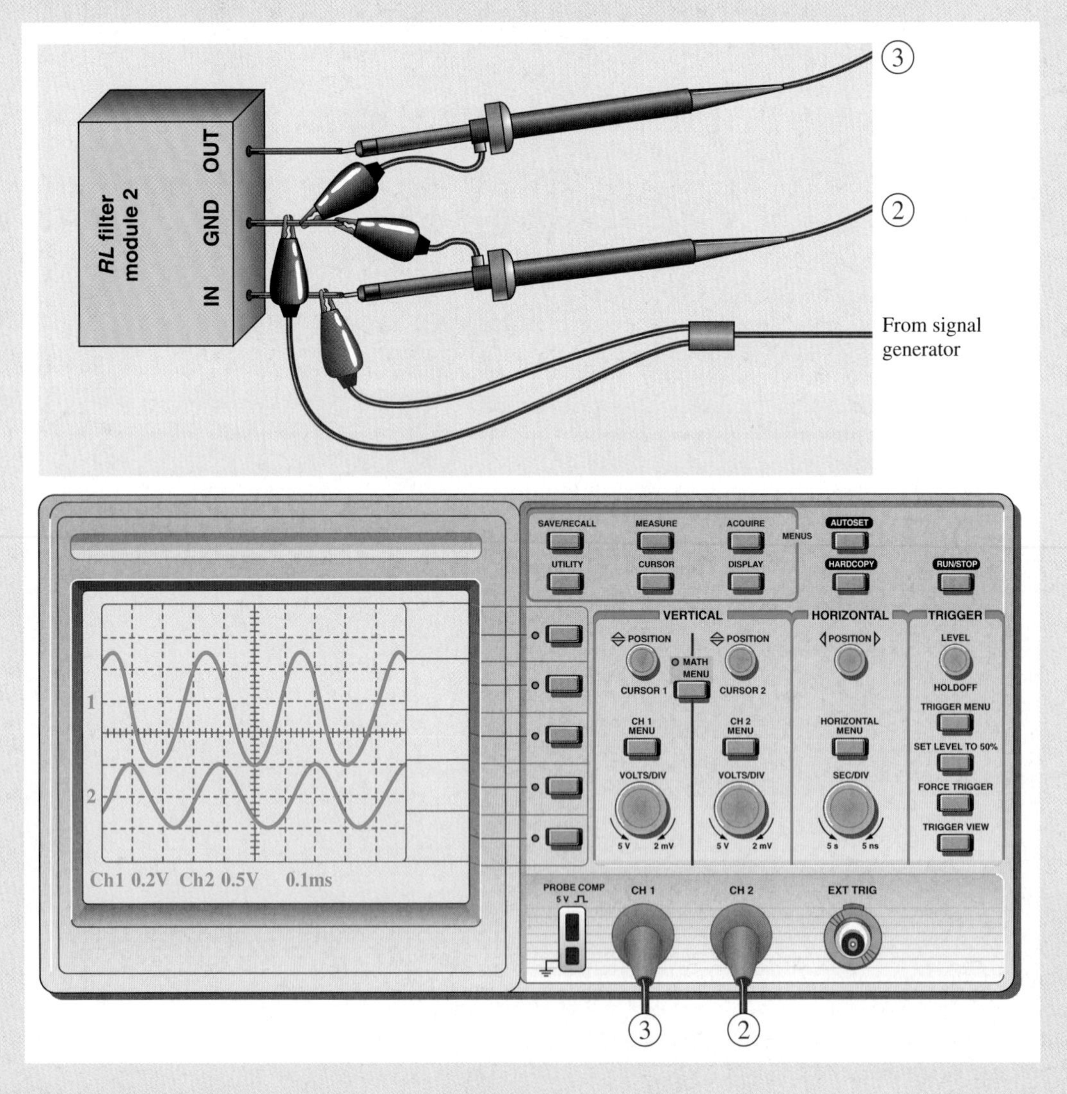

▲ **FIGURE 12–49**

AC measurements for module 2.

APPLICATION ASSIGNMENT REVIEW

1. If the inductor in module 1 were open, what would you measure on the output with the test setup of Figure 12–47?
2. If the inductor in module 2 were open, what would you measure on the output with the test setup of Figure 12–49?

SUMMARY

- When a sinusoidal voltage is applied to an *RL* circuit, the current and all the voltage drops are also sine waves.
- Total current in an *RL* circuit always lags the source voltage.
- The resistor voltage is always in phase with the current.
- In an ideal inductor, the voltage always leads the current by 90°.
- In an *RL* circuit, the impedance is determined by both the resistance and the inductive reactance combined.
- Impedance is expressed in units of ohms.
- The impedance of an *RL* circuit varies directly with frequency.
- The phase angle (θ) of a series *RL* circuit varies directly with frequency.
- You can determine the impedance of a circuit by measuring the source voltage and the total current and then applying Ohm's law.
- In an *RL* circuit, part of the power is resistive and part reactive.
- The phasor combination of resistive power (true power) and reactive power is called *apparent power.*
- The power factor indicates how much of the apparent power is true power.
- A power factor of 1 indicates a purely resistive circuit, and a power factor of 0 indicates a purely reactive circuit.
- In an *RL* lag network, the output voltage lags the input voltage in phase.
- In an *RL* lead network, the output voltage leads the input voltage in phase.
- A filter passes certain frequencies and rejects others.

EQUATIONS

12–1	$Z = \sqrt{R^2 + X_L^2}$	Series *RL* impedance
12–2	$\theta = \tan^{-1}\left(\frac{X_L}{R}\right)$	Series *RL* phase angle
12–3	$V_s = \sqrt{V_R^2 + V_L^2}$	Total voltage in a series *RL* circuit
12–4	$\theta = \tan^{-1}\left(\frac{V_L}{V_R}\right)$	Series *RL* phase angle
12–5	$Z = \frac{RX_L}{\sqrt{R^2 + X_L^2}}$	Parallel *RL* impedance
12–6	$\theta = \tan^{-1}\left(\frac{R}{X_L}\right)$	Parallel *RL* phase angle

12–7	$G = \frac{1}{R}$	Conductance
12–8	$B_L = \frac{1}{X_L}$	Inductive susceptance
12–9	$Y = \frac{1}{Z}$	Admittance
12–10	$Y_{tot} = \sqrt{G^2 + B_L^2}$	Total admittance
12–11	$I_{tot} = \sqrt{I_R^2 + I_L^2}$	Total current in parallel *RL* circuit
12–12	$\theta = \tan^{-1}\left(\frac{I_L}{I_R}\right)$	Parallel *RL* phase angle
12–13	$P_r = I^2X_L$	Reactive power
12–14	$\theta = \tan^{-1}\left(\frac{X_L}{R}\right)$	Phase angle of lag network
12–15	$V_{out} = \left(\frac{R}{\sqrt{R^2 + X_L^2}}\right)V_{in}$	Output voltage of lag network
12–16	$\phi = 90° - \tan^{-1}\left(\frac{X_L}{R}\right)$	Phase angle of lead network
12–17	$V_{out} = \left(\frac{X_L}{\sqrt{R^2 + X_L^2}}\right)V_{in}$	Output voltage of lead network
12–18	$f_c = \frac{R}{2\pi L}$	Cutoff frequency of *RL* filter

SELF-TEST

Answers are at the end of the chapter.

1. In a series *RL* circuit, the resistor voltage
   (a) leads the source voltage (b) lags the source voltage
   (c) is in phase with the source voltage (d) is in phase with the current
   (e) answers (a) and (d) (f) answers (b) and (d)
2. When the frequency of the voltage applied to a series *RL* circuit is increased, the impedance
   (a) decreases (b) increases (c) does not change
3. When the frequency of the voltage applied to a series *RL* circuit is decreased, the phase angle
   (a) decreases (b) increases (c) does not change
4. If the frequency is doubled and the resistance is halved, the impedance of a series *RL* circuit
   (a) doubles (b) halves
   (c) remains constant (d) cannot be determined without values
5. To reduce the current in a series *RL* circuit, the frequency should be
   (a) increased (b) decreased (c) constant
6. In a series *RL* circuit, 10 V rms is measured across the resistor, and 10 V rms is measured across the inductor. The peak value of the source voltage is
   (a) 14.14 V (b) 28.28 V (c) 10 V (d) 20 V
7. The voltages in Question 6 are measured at a certain frequency. To make the resistor voltage greater than the inductor voltage, the frequency is
   (a) increased (b) decreased (c) doubled (d) not a factor

8. When the resistor voltage in a series *RL* circuit becomes greater than the inductor voltage, the phase angle

   (a) increases (b) decreases (c) is not affected

9. When the frequency is increased, the impedance of a parallel *RL* circuit

   (a) increases (b) decreases (c) remains constant

10. In a parallel *RL* circuit, there are 2 A rms in the resistive branch and 2 A rms in the inductive branch. The total rms current is

    (a) 4 A (b) 5.656 A (c) 2 A (d) 2.828 A

11. You are observing two voltage waveforms on an oscilloscope. The time base (sec/div) of the scope is set to 10 μs. One half-cycle of the waveforms covers the ten horizontal divisions. The positive-going zero crossing of one waveform is at the leftmost division, and the positive-going zero crossing of the other is three divisions to the right. The phase angle between these two waveforms is

    (a) 18° (b) 36° (c) 54° (d) 180°

12. Which of the following power factors results in less energy conversion to heat in an *RL* circuit?

    (a) 1 (b) 0.9 (c) 0.5 (d) 0.1

13. If a load is purely inductive and the reactive power is 10 VAR, the apparent power is

    (a) 0 VA (b) 10 VA (c) 14.14 VA (d) 3.16 VA

14. For a certain load, the true power is 10 W and the reactive power is 10 VAR. The apparent power is

    (a) 5 VA (b) 20 VA (c) 14.14 VA (d) 100 VA

15. The cutoff frequency of a certain low-pass *RL* filter is 20 kHz. The filter's bandwidth is

    (a) 20 kHz (b) 40 kHz (c) 0 kHz (d) unknown

PROBLEMS

Answers to odd-numbered problems are at the end of the book.

BASIC PROBLEMS

SECTION 12–1

Sinusoidal Response of *RL* Circuits

1. A 15 kHz sinusoidal voltage is applied to a series *RL* circuit. Determine the frequency of *I*, V_R, and V_L.
2. What are the waveshapes of *I*, V_R, and V_L in Problem 1?

SECTION 12–2

Impedance and Phase Angle of Series *RL* Circuits

3. Find the impedance of each circuit in Figure 12–50.

FIGURE 12–50

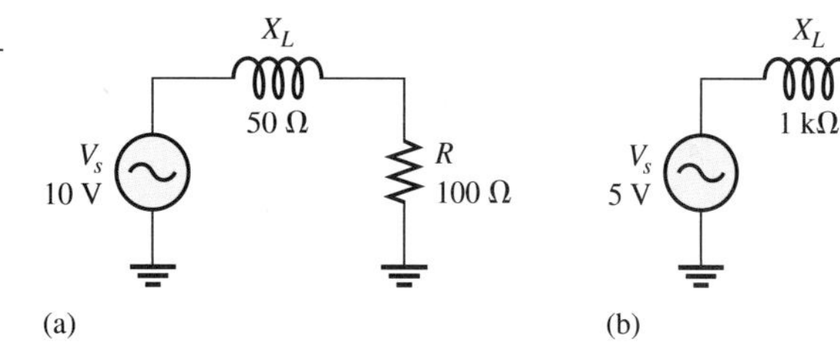

4. Determine the impedance and phase angle in each circuit in Figure 12–51.

FIGURE 12–51

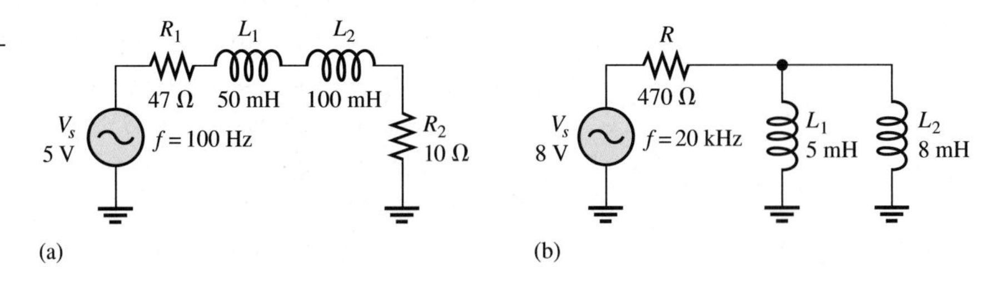

5. In Figure 12–52, determine the impedance at each of the following frequencies:

(a) 100 Hz **(b)** 500 Hz **(c)** 1 kHz **(d)** 2 kHz

FIGURE 12–52

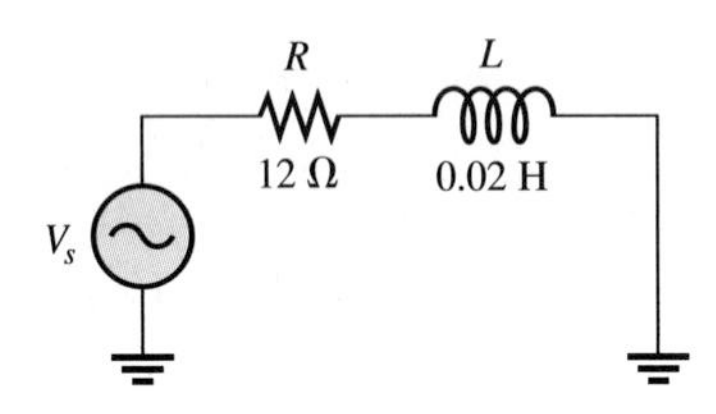

6. Determine the values of R and X_L in a series RL circuit for the following values of impedance and phase angle:

(a) $Z = 20\ \Omega, \theta = 45°$ **(b)** $Z = 500\ \Omega, \theta = 35°$

(c) $Z = 2.5\ \text{k}\Omega, \theta = 72.5°$ **(d)** $Z = 998\ \Omega, \theta = 45°$

SECTION 12–3 **Analysis of Series *RL* Circuits**

7. Find the current for each circuit of Figure 12–50.
8. Calculate the total current in each circuit of Figure 12–51.
9. Determine θ for the circuit in Figure 12–53.
10. If the inductance in Figure 12–53 is doubled, does θ increase or decrease, and by how many degrees?

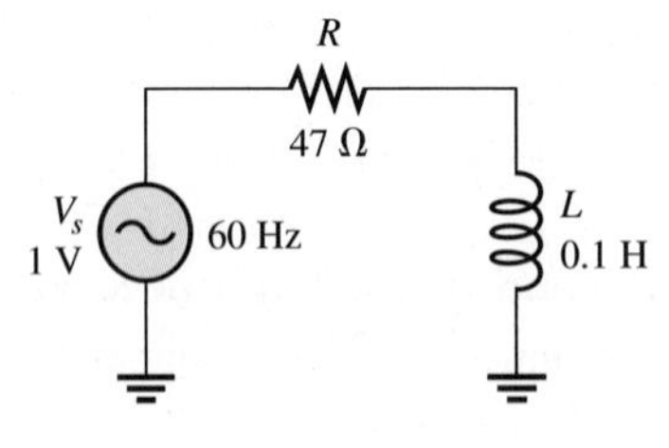

FIGURE 12–53

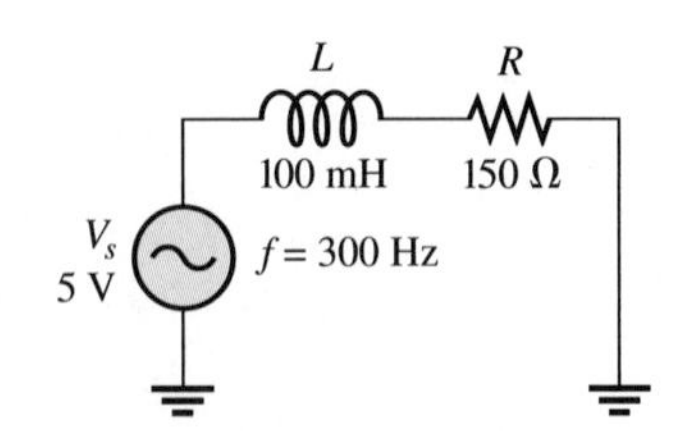

FIGURE 12–54

11. Draw the waveforms for V_s, V_R, and V_L in Figure 12–53. Show the proper phase relationships.
12. For the circuit in Figure 12–54, find V_R and V_L for each of the following frequencies:

(a) 60 Hz **(b)** 200 Hz **(c)** 500 Hz **(d)** 1 kHz

SECTION 12–4 **Impedance and Phase Angle of Parallel *RL* Circuits**

13. What is the impedance for the circuit in Figure 12–55?
14. Repeat Problem 13 for the following frequencies:
    (a) 1.5 kHz **(b)** 3 kHz **(c)** 5 kHz **(d)** 10 kHz
15. At what frequency does X_L equal R in Figure 12–55?

SECTION 12–5 **Analysis of Parallel *RL* Circuits**

16. Find the total current and each branch current in Figure 12–56.

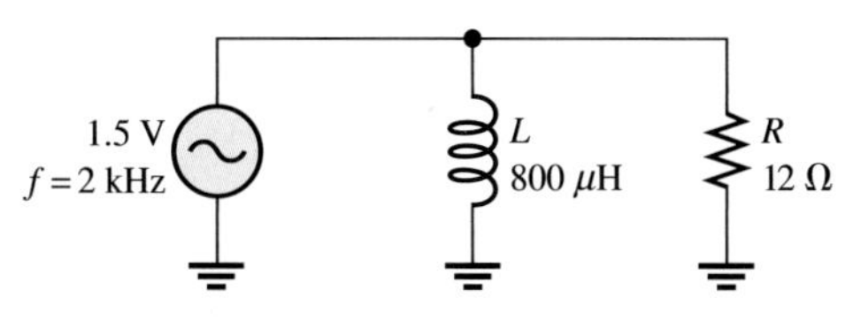

▲ FIGURE 12–55

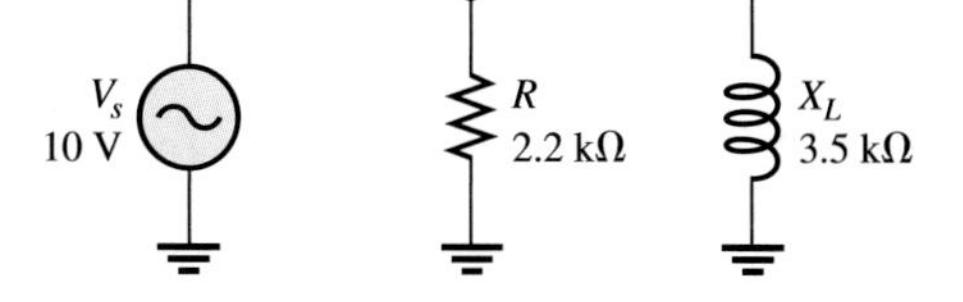

▲ FIGURE 12–56

17. Determine the following quantities in Figure 12–57:
    (a) Z **(b)** I_R **(c)** I_L **(d)** I_{tot} **(e)** θ
18. Convert the circuit in Figure 12–58 to an equivalent series form.

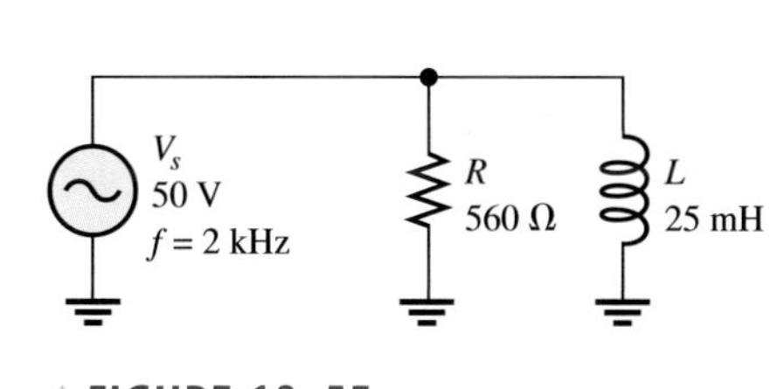

▲ FIGURE 12–57

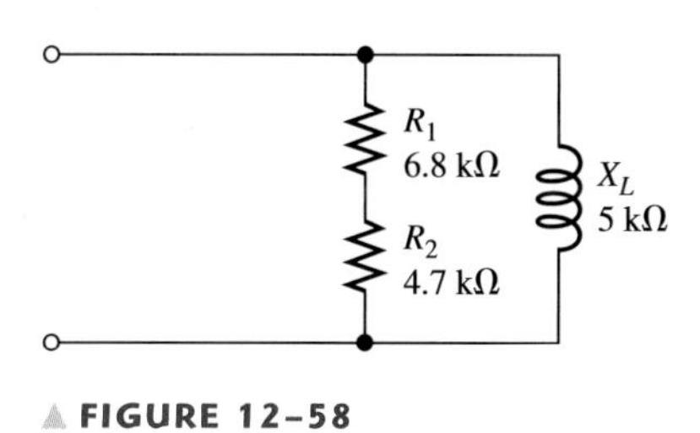

▲ FIGURE 12–58

SECTION 12–6 **Series-Parallel Analysis**

19. Determine the voltage across each element in Figure 12–59.
20. Is the circuit in Figure 12–59 predominantly resistive or predominantly inductive?
21. Find the current in each branch and the total current in Figure 12–59.

▶ FIGURE 12–59

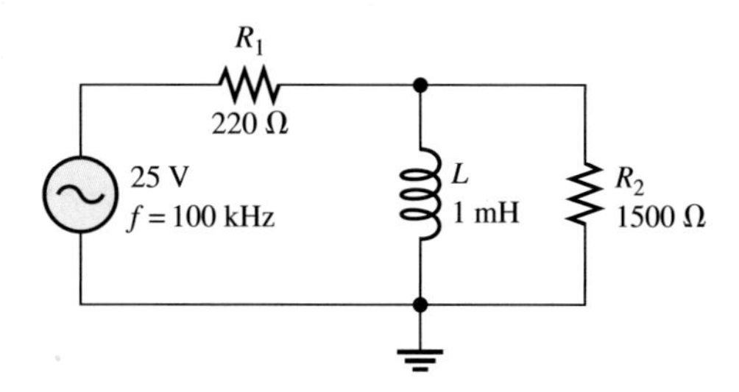

SECTION 12–7 **Power in *RL* Circuits**

22. In a certain *RL* circuit, the true power is 100 mW, and the reactive power is 340 mVAR. What is the apparent power?
23. Determine the true power and the reactive power in Figure 12–53.
24. What is the power factor in Figure 12–56?
25. Determine P_{true}, P_r, P_a, and PF for the circuit in Figure 12–59. Sketch the power triangle.

SECTION 12–8

Basic Applications

26. For the lag network in Figure 12–60, determine the phase lag of the output voltage with respect to the input for the following frequencies:

(a) 1 Hz **(b)** 100 Hz **(c)** 1 kHz **(d)** 10 kHz

27. Plot the response curve for the circuit in Figure 12–60. Show the output voltage versus frequency in 1 kHz increments from 0 Hz to 5 kHz.

28. Repeat Problem 26 for the lead network in Figure 12–61.

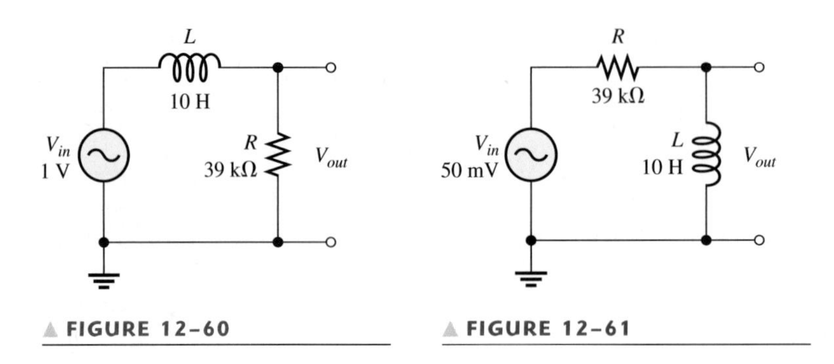

▲ FIGURE 12–60

▲ FIGURE 12–61

29. Using the same procedure as in Problem 27, plot the response curve for Figure 12–61.

30. Draw the voltage phasor diagram for each circuit in Figures 12–60 and 12–61 for a frequency of 8 kHz.

SECTION 12–9

Troubleshooting

31. Determine the voltage across each element in Figure 12–62 if L_1 were open.

▶ FIGURE 12–62

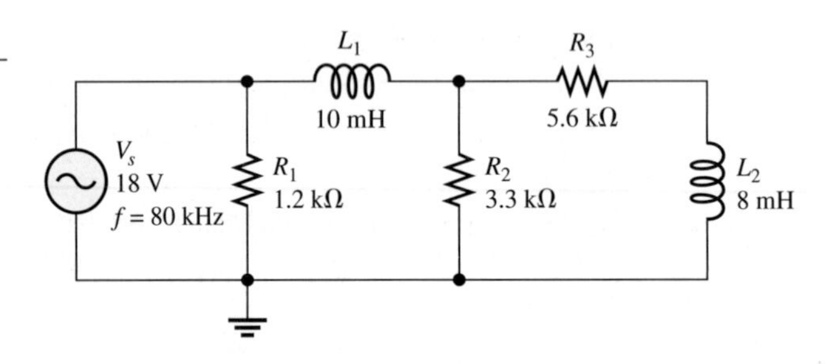

32. Determine the output voltage in Figure 12–63 for each of the following failure modes:

(a) L_1 open **(b)** L_2 open **(c)** R_1 open **(d)** a short across R_2

▶ FIGURE 12–63

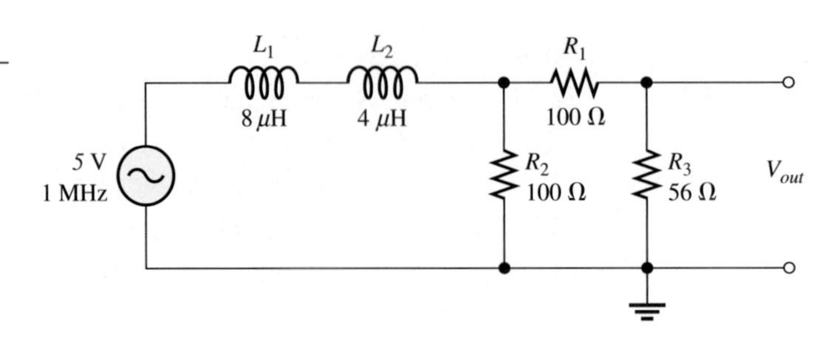

ADVANCED PROBLEMS

33. Determine the voltage across the inductors in Figure 12–64.

34. Is the circuit in Figure 12–64 predominantly resistive or predominantly inductive?

35. Find the total current in Figure 12–64.

FIGURE 12–64

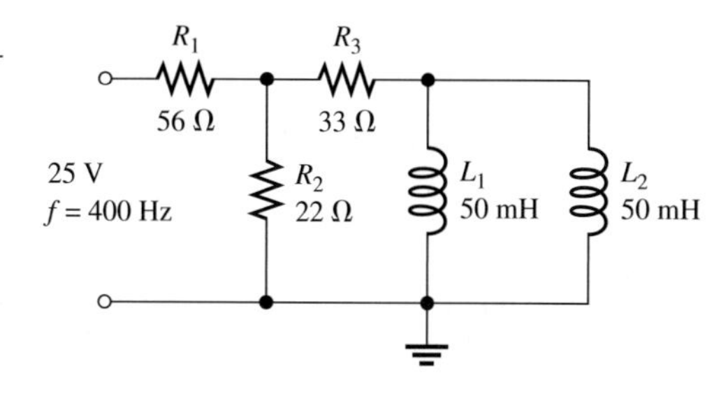

36. For the circuit in Figure 12–65, determine the following:

(a) Z_{tot} **(b)** I_{tot} **(c)** θ **(d)** V_L **(e)** V_{R3}

37. For the circuit in Figure 12–66, determine the following:

(a) I_{R1} **(b)** I_{L1} **(c)** I_{L2} **(d)** I_{R2}

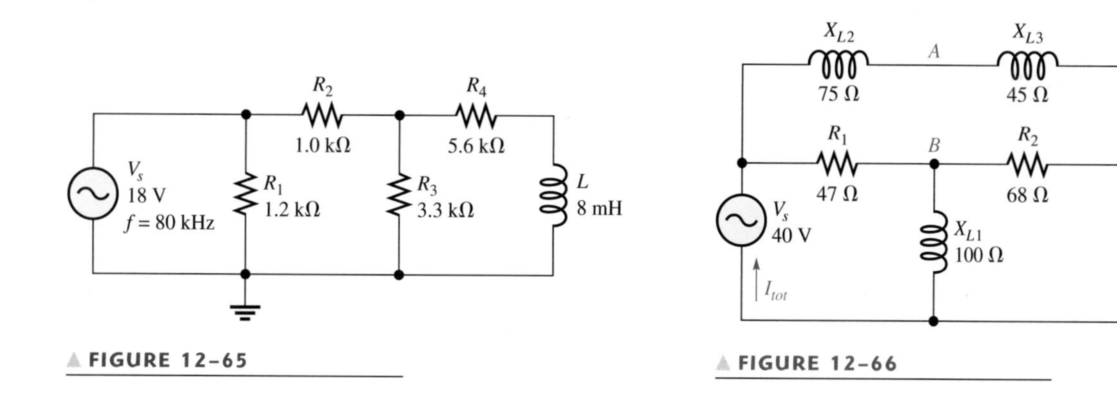

FIGURE 12–65

FIGURE 12–66

38. Determine the phase shift and attenuation (ratio of V_{out} to V_{in}) from the input to the output for the circuit in Figure 12–67.

FIGURE 12–67

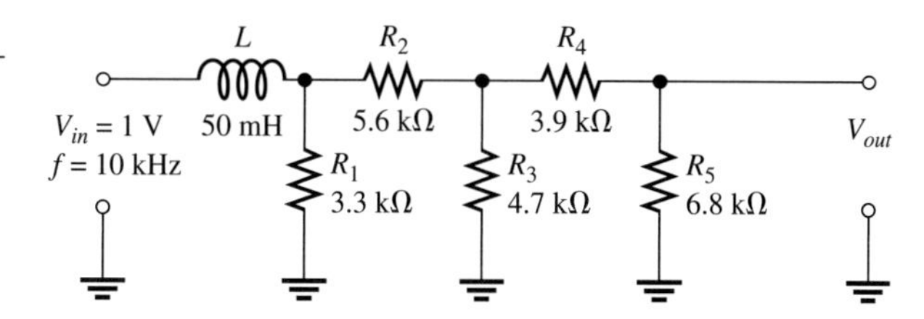

39. Determine the attenuation from the input to the output for the circuit in Figure 12–68.

FIGURE 12–68

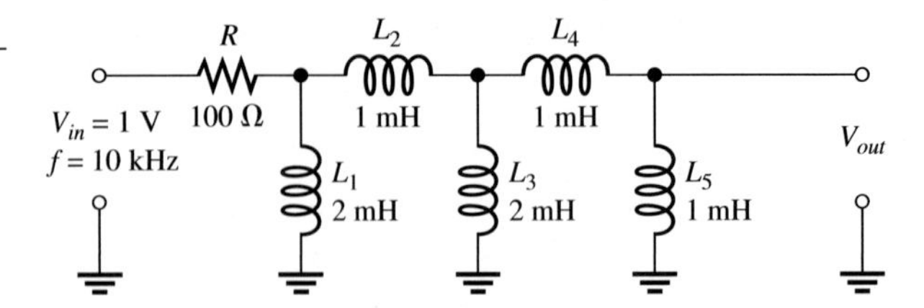

40. Design an ideal inductive switching circuit that will provide a momentary voltage of 2.5 kV from a 12 V dc source when a switch is thrown instantaneously from one position to another. The drain on the source must not exceed 1 A.

41. Draw the schematic for the circuit in Figure 12–69 and determine if the waveforms on the scope are correct. If there is a fault in the circuit, identify it. After working this problem, open file P12-41 on your CD-ROM and run the setup to observe the correct operation.

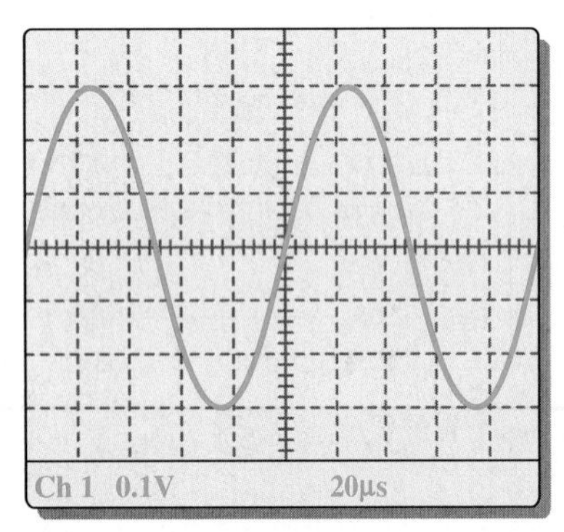

(a) Oscilloscope display

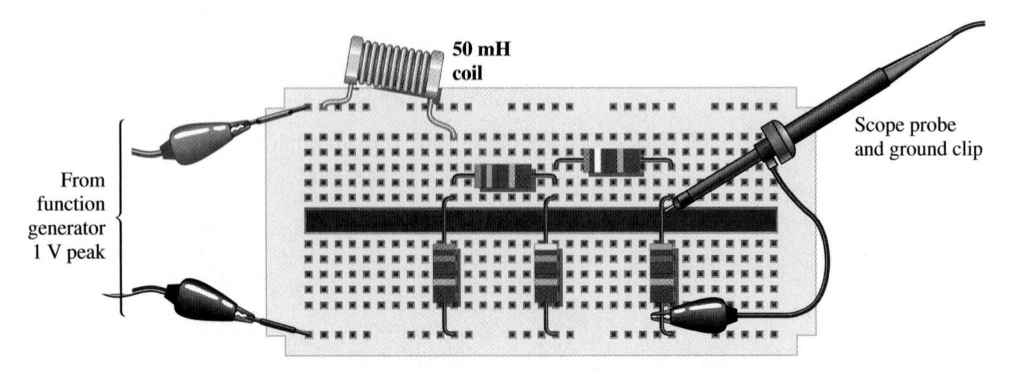

(b) Circuit with leads connected

FIGURE 12–69

ELECTRONICS WORKBENCH/CIRCUITMAKER TROUBLESHOOTING PROBLEMS

CD-ROM file circuits are shown in Figure 12–70.

42. Open file P12-42. Determine if there is a fault and, if so, identify it.

43. Open file P12-43. Determine if there is a fault and, if so, identify it.

44. Open file P12-44. Determine if there is a fault and, if so, identify it.

45. Open file P12-45. Determine if there is a fault and, if so, identify it.

46. Open file P12-46. Determine if there is a fault and, if so, identify it.

47. Open file P12-47. Determine if there is a fault and, if so, identify it.

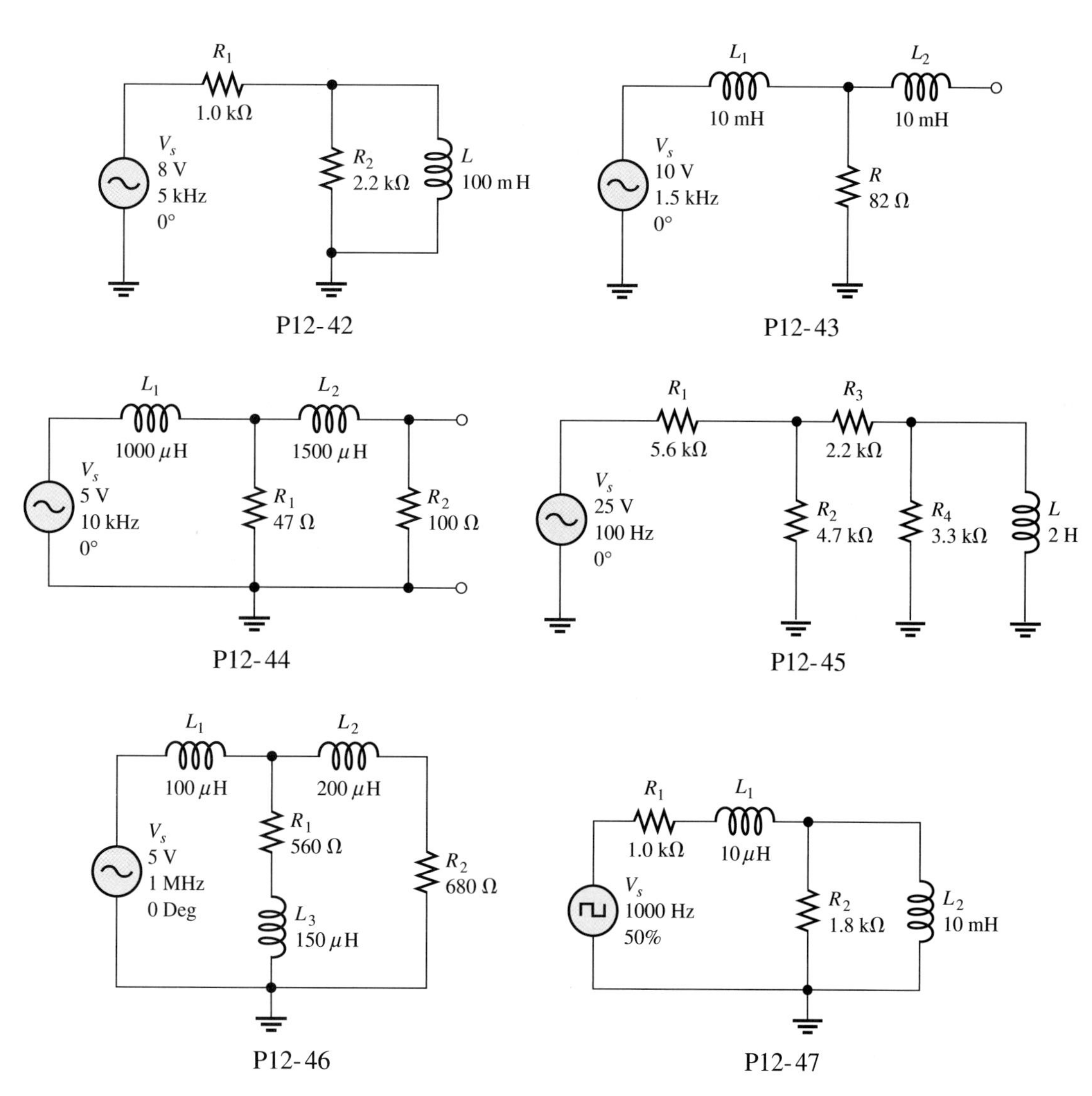

▲ **FIGURE 12–70**

ANSWERS

SECTION REVIEWS

SECTION 12-1

Sinusoidal Response of *RL* Circuits

1. The current frequency is at 1 kHz.
2. θ is closer to 0° when $R > X_L$.

SECTION 12-2

Impedance and Phase Angle of Series *RL* Circuits

1. V_s leads I.
2. The inductive reactance produces the phase angle.
3. *RL* has an opposite phase angle compared to *RC*.
4. $Z = \sqrt{R^2 + X_L^2} = 59.9\ \text{k}\Omega$; $\theta = \tan^{-1}(X_L/R) = 56.6°$

SECTION 12-3

Analysis of Series *RL* Circuits

1. $V_s = \sqrt{V_R^2 + V_L^2} = 3.61\ \text{V}$
2. $\theta = \tan^{-1}(X_L/V_R) = 56.3°$
3. When f increases, X_L increases, Z increases, and θ increases.

SECTION 12-4

Impedance and Phase Angle of Parallel *RL* Circuits

1. $Z = 1/Y = 20\ \Omega$
2. $Y = \sqrt{G^2 + B_L^2} = 25\ \text{mS}$
3. I_{tot} lags V_s by 32.1°.

SECTION 12-5

Analysis of Parallel *RL* Circuits

1. $I_{tot} = V_sY = 32\ \text{mA}$
2. $\theta = \tan^{-1}(I_L/I_R) = 59.0°$; $I_{tot} = \sqrt{I_R^2 + I_L^2} = 23.3\ \text{mA}$
3. $\theta = 90°$

SECTION 12-6

Series-Parallel Analysis

1. $I_{tot} = \sqrt{(I_1\cos\theta_1 + I_2\cos\theta_2)^2 + (I_1\sin\theta_1 + I_2\sin\theta_2)^2} = 20.2\ \text{mA}$
2. $Z = V_s/I_{tot} = 494\ \Omega$

SECTION 12-7

Power in *RL* Circuits

1. Power dissipation is in the resistor.
2. $PF = \cos 50° = 0.643$
3. $P_{\text{true}} = I^2R = 4.7\ \text{W}$; $P_r = I^2X_L = 6.2\ \text{VAR}$; $P_a = \sqrt{P_{\text{true}}^2 + P_r^2} = 7.78\ \text{VA}$

SECTION 12-8

Basic Applications

1. $\phi = 90° - \tan^{-1}(X_L/R) = 81.9°$
2. $V_{out} = \left(\dfrac{R}{\sqrt{R^2 + X_L^2}}\right)V_{in} = 9.90\ \text{V}$
3. Output is across the resistor.

SECTION 12-9

Troubleshooting

1. Shorted windings reduce L and thereby reduce X_L at any given frequency.
2. I_{tot} decreases, V_{R1} decreases, and V_{R2} increases when L opens.

Application Assignment

1. 0 V on output
2. Output same as input

RELATED PROBLEMS FOR EXAMPLES

12–1 2.04 kΩ; 27.8°

12–2 423 μA

12–3 No

12–4 12.6 kΩ; 85.5°

12–5 8.14 kΩ; 35.5°

12–6 3.13 mS

12–7 14.0 mA; 71.1°

12–8 67.6 mA; 36.3°

12–9 **(a)** 8.04 V **(b)** 2.00 V

12–10 Current decreases.

12–11 P_{true}, P_r, and P_a decrease.

12–12 32°

12–13 12.3 V rms

12–14 65.6°

12–15 Open connection between L_1 and L_2

SELF-TEST

1. (f) **2.** (b) **3.** (a) **4.** (d) **5.** (a) **6.** (d) **7.** (b) **8.** (b) **9.** (a) **10.** (d) **11.** (c) **12.** (d) **13.** (b) **14.** (c) **15.** (a)

13

RLC CIRCUITS AND RESONANCE

INTRODUCTION

In this chapter, you will learn about frequency response of *RLC* circuits—that is, circuits with combinations of resistance, capacitance, and inductance. Both series and parallel *RLC* circuits, including the concepts of series and parallel resonance will be discussed.

Because it is the basis for frequency selectivity, resonance in electrical circuits is very important to the operation of many types of electronic systems, particularly in the area of communications. For example, the ability of a radio or television receiver to select a certain frequency transmitted by a certain station and to eliminate frequencies from other stations is based on the principle of resonance.

The operation of both band-pass and band-stop filters is based on resonance of circuits containing inductance and capacitance, and these filters are discussed in this chapter. Also, certain system applications are presented.

CHAPTER OUTLINE

CHAPTER OBJECTIVES

- Determine the impedance and phase angle of a series *RLC* circuit
- Analyze series *RLC* circuits
- Analyze a circuit for series resonance
- Analyze series resonant filters
- Analyze parallel *RLC* circuits
- Analyze a circuit for parallel resonance
- Analyze the operation of parallel resonant filters
- Discuss some system applications of resonant circuits

KEY TERMS

- Series resonance
- Resonant frequency
- Band-pass filter
- Bandwidth
- Cutoff frequency
- Half-power frequency
- Decibel
- Selectivity
- Quality factor
- Band-stop filter
- Parallel resonance

APPLICATION ASSIGNMENT PREVIEW

Your application assignment in this chapter requires that you plot a frequency response curve for a resonant filter with unknown characteristics. From the frequency response measurements, you will identify the type of filter and determine the resonant frequency and the bandwidth. After studying this chapter, you should be able to complete the application assignment.

WWW. VISIT THE COMPANION WEBSITE

Circuit Simulation Tutorials and Other Chapter Study Tools Are Available at

http://www.prenhall.com/floyd

13–1 IMPEDANCE AND PHASE ANGLE OF SERIES *RLC* CIRCUITS

A series *RLC* circuit contains resistance, inductance, and capacitance. Since inductive reactance and capacitive reactance have opposite effects on the circuit phase angle, the total reactance is less than either individual reactance.

After completing this section, you should be able to

- **Determine the impedance and phase angle of a series *RLC* circuit**
- Calculate total reactance
- Determine whether a circuit is predominately inductive or capacitive

A series *RLC* circuit is shown in Figure 13–1. It contains resistance, inductance, and capacitance.

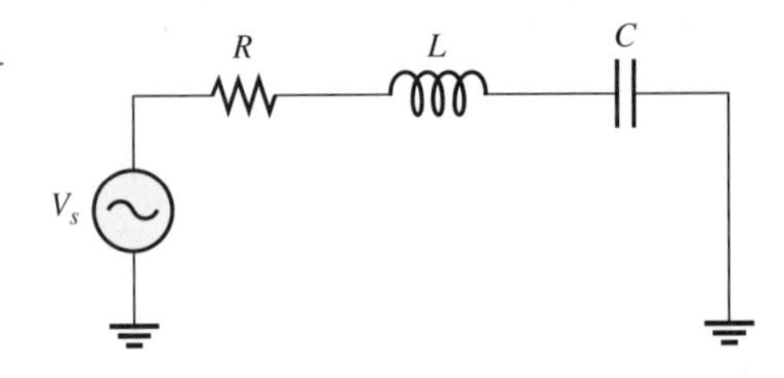

FIGURE 13–1

Series *RLC* circuit.

As you know, inductive reactance (X_L) causes the total current to lag the source voltage. Capacitive reactance (X_C) has the opposite effect: It causes the current to lead the voltage. Thus, X_L and X_C tend to offset each other. When they are equal, they cancel, and the total reactance is zero. In any case, the total reactance in a series circuit is

Equation 13–1

$$X_{tot} = |X_L - X_C|$$

The term $|X_L - X_C|$ means the absolute value of the difference of the two reactances. That is, the sign of the result is considered positive no matter which reactance is greater. For example, $3 - 7 = -4$, but the absolute value is

$$|3 - 7| = 4$$

When $X_L > X_C$, the circuit is predominantly inductive; and when $X_C > X_L$, the circuit is predominantly capacitive.

The total impedance for a series *RLC* circuit is given by

Equation 13–2

$$Z_{tot} = \sqrt{R^2 + X_{tot}^2}$$

and the phase angle between V_s and I is

Equation 13–3

$$\theta = \tan^{-1}\left(\frac{X_{tot}}{R}\right)$$

EXAMPLE 13–1

Determine the total impedance and the phase angle in Figure 13–2.

Solution First, find X_C and X_L.

$$X_C = \frac{1}{2\pi fC} = \frac{1}{2\pi(1\text{ kHz})(0.56\ \mu\text{F})} = 284\ \Omega$$

$$X_L = 2\pi fL = 2\pi(1\text{ kHz})(100\text{ mH}) = 628\ \Omega$$

FIGURE 13–2

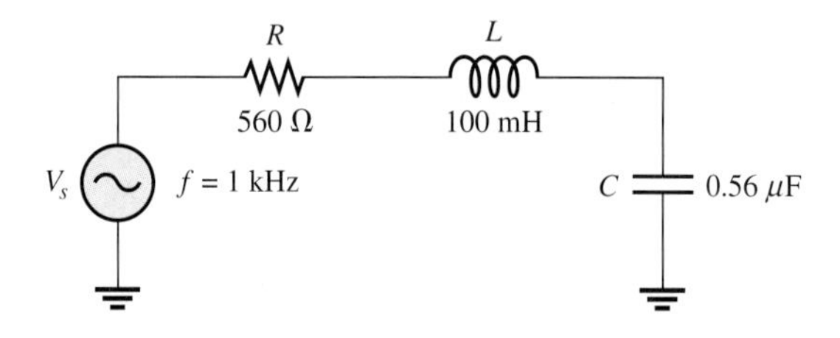

In this case, X_L is greater than X_C, and thus the circuit is more inductive than capacitive. The magnitude of the total reactance is

$$X_{tot} = |X_L - X_C| = |628\ \Omega - 284\ \Omega| = 344\ \Omega \qquad \text{(inductive)}$$

The total circuit impedance is

$$Z_{tot} = \sqrt{R^2 + X_{tot}^2} = \sqrt{(560\ \Omega)^2 + (344\ \Omega)^2} = \mathbf{657\ \Omega}$$

The phase angle (between I and V_s) is

$$\theta = \tan^{-1}\left(\frac{X_{tot}}{R}\right) = \tan^{-1}\left(\frac{344\ \Omega}{560\ \Omega}\right) = \mathbf{31.6°} \qquad \text{(current lagging } V_s\text{)}$$

Related Problem* Increase the frequency in Figure 13–2 to 2000 Hz and determine Z and θ.

*Answers are at the end of the chapter.

As you have seen, when the inductive reactance is greater than the capacitive reactance, the circuit appears to be inductive, and the current lags the source voltage. When the capacitive reactance is greater, the circuit appears to be capacitive, and the current leads the source voltage.

SECTION 13–1 REVIEW

Answers are at the end of the chapter.

1. Explain how you can determine if a series *RLC* circuit is inductive or capacitive.
2. In a given series *RLC* circuit, X_C is 150 Ω and X_L is 80 Ω. What is the total reactance in ohms? Is the circuit inductive or capacitive?
3. Determine the impedance for the circuit in Question 2 when $R = 45\ \Omega$. What is the phase angle? Is the current leading or lagging the source voltage?

13–2 ANALYSIS OF SERIES *RLC* CIRCUITS

Recall that capacitive reactance varies inversely with frequency and that inductive reactance varies directly with frequency. In this section, the combined effects of the reactances as a function of frequency are examined.

After completing this section, you should be able to

- **Analyze series *RLC* circuits**
- Determine current in a series *RLC* circuit
- Determine the voltages in a series *RLC* circuit
- Determine the phase angle

Figure 13–3 shows that for a typical series *RLC* circuit the total reactance behaves as follows: Starting at a very low frequency, X_C is high, X_L is low, and the circuit is predominantly capacitive. As the frequency is increased, X_C decreases and X_L increases until a value is reached where $X_C = X_L$ and the two reactances cancel, making the circuit purely resistive. This condition is **series resonance** and will be studied in Section 13–3. As the frequency is increased further, X_L becomes greater than X_C, and the circuit is predominantly inductive. Example 13–2 illustrates how the impedance and phase angle change as the source frequency is varied.

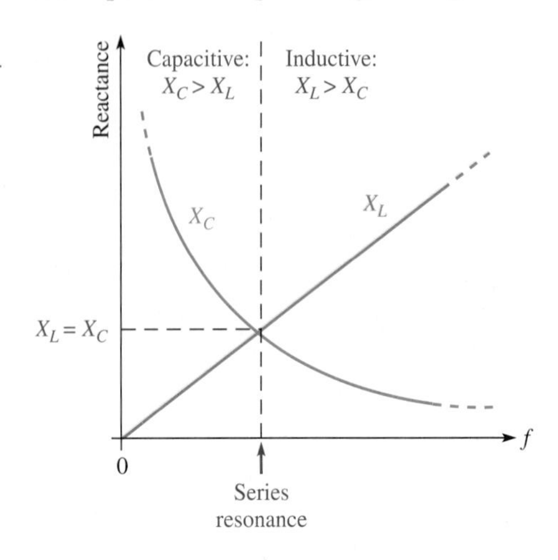

FIGURE 13–3

How X_C and X_L vary with frequency.

The graph of X_L is a straight line, and the graph of X_C is curved, as shown in Figure 13–3. The general equation for a straight line is $y = mx + b$, where m is the slope of the line and b is the y-axis intercept point. The formula $X_L = 2\pi fL$ fits this general straight-line formula, where $y = X_L$ (a variable), $m = 2\pi L$ (a constant), $x = f$ (a variable), and $b = 0$ as follows: $X_L = 2\pi Lf + 0$.

The X_C curve is called a *hyperbola,* and the general equation of a hyperbola is $xy = k$. The equation for capacitive reactance, $X_C = 1/2\pi fC$, can be rearranged as $X_C f = 1/2\pi C$ where $x = X_C$ (a variable), $y = f$ (a variable), and $k = 1/2\pi C$ (a constant).

EXAMPLE 13–2

For each of the following frequencies of the source voltage, find the impedance and the phase angle for the circuit in Figure 13–4. Note the change in the impedance and the phase angle with frequency.

(a) $f = 1$ kHz **(b)** $f = 3.5$ kHz **(c)** $f = 5$ kHz

FIGURE 13–4

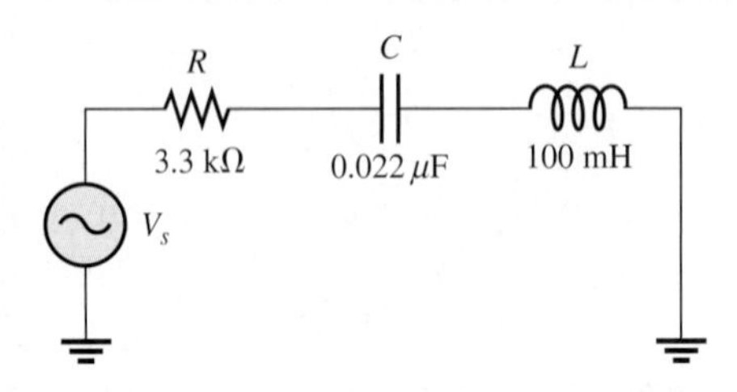

Solution **(a)** At $f = 1$ kHz,

$$X_C = \frac{1}{2\pi fC} = \frac{1}{2\pi(1\text{ kHz})(0.022\ \mu\text{F})} = 7.23\text{ k}\Omega$$

$$X_L = 2\pi fL = 2\pi(1\text{ kHz})(100\text{ mH}) = 628\ \Omega$$

The circuit is highly capacitive because X_C is much larger than X_L. The magnitude of the total reactance is

$$X_{tot} = |X_L - X_C| = |628\ \Omega - 7.23\ \text{k}\Omega| = 6.6\ \text{k}\Omega$$

The impedance is

$$Z = \sqrt{R^2 + X_{tot}^2} = \sqrt{(3.3\ \text{k}\Omega)^2 + (6.6\ \text{k}\Omega)^2} = \mathbf{7.38\ k\Omega}$$

The phase angle is

$$\theta = \tan^{-1}\left(\frac{X_{tot}}{R}\right) = \tan^{-1}\left(\frac{6.60\ \text{k}\Omega}{3.3\ \text{k}\Omega}\right) = \mathbf{63.4^\circ}$$

I leads V_s by 63.4°.

(b) At $f = 3.5$ kHz,

$$X_C = \frac{1}{2\pi(3.5\ \text{kHz})(0.022\ \mu\text{F})} = 2.07\ \text{k}\Omega$$

$$X_L = 2\pi(3.5\ \text{kHz})(100\ \text{mH}) = 2.20\ \text{k}\Omega$$

The circuit is very close to being purely resistive but is slightly inductive because X_L is slightly larger than X_C. The total reactance, impedance, and phase angle are

$$X_{tot} = |2.20\ \text{k}\Omega - 2.07\ \text{k}\Omega| = 130\ \Omega$$

$$Z = \sqrt{(3.3\ \text{k}\Omega)^2 + (130\ \Omega)^2} = \mathbf{3.30\ k\Omega}$$

$$\theta = \tan^{-1}\left(\frac{130\ \Omega}{3.3\ \text{k}\Omega}\right) = \mathbf{2.26^\circ}$$

I lags V_s by 2.26°.

(c) At $f = 5$ kHz,

$$X_C = \frac{1}{2\pi(5\ \text{kHz})(0.022\ \mu\text{F})} = 1.45\ \text{k}\Omega$$

$$X_L = 2\pi(5\ \text{kHz})(100\ \text{mH}) = 3.14\ \text{k}\Omega$$

The circuit is now predominantly inductive because $X_L > X_C$. The total reactance, impedance, and phase angle are

$$X_{tot} = |3.14\ \text{k}\Omega - 1.45\ \text{k}\Omega| = 1.69\ \text{k}\Omega$$

$$Z = \sqrt{(3.3\ \text{k}\Omega)^2 + (1.69\ \text{k}\Omega)^2} = \mathbf{3.71\ k\Omega}$$

$$\theta = \tan^{-1}\left(\frac{1.69\ \text{k}\Omega}{3.3\ \text{k}\Omega}\right) = \mathbf{27.1^\circ}$$

I lags V_s by 27.1°.

Notice how the circuit changed from capacitive to inductive as the frequency increased. The phase condition changed from the current leading to the current lagging. It is important to note that both the impedance and the phase angle decreased to a minimum and then began increasing again as the frequency went up.

Related Problem Determine Z for $f = 7$ kHz in Figure 13–4.

In a series *RLC* circuit, the capacitor voltage and the inductor voltage are always 180° out of phase with each other. For this reason, V_C and V_L subtract from each other, and thus the voltage across L and C combined is always less than the larger individual voltage across either element, as illustrated in Figure 13–5 and in the waveform diagram of Figure 13–6.

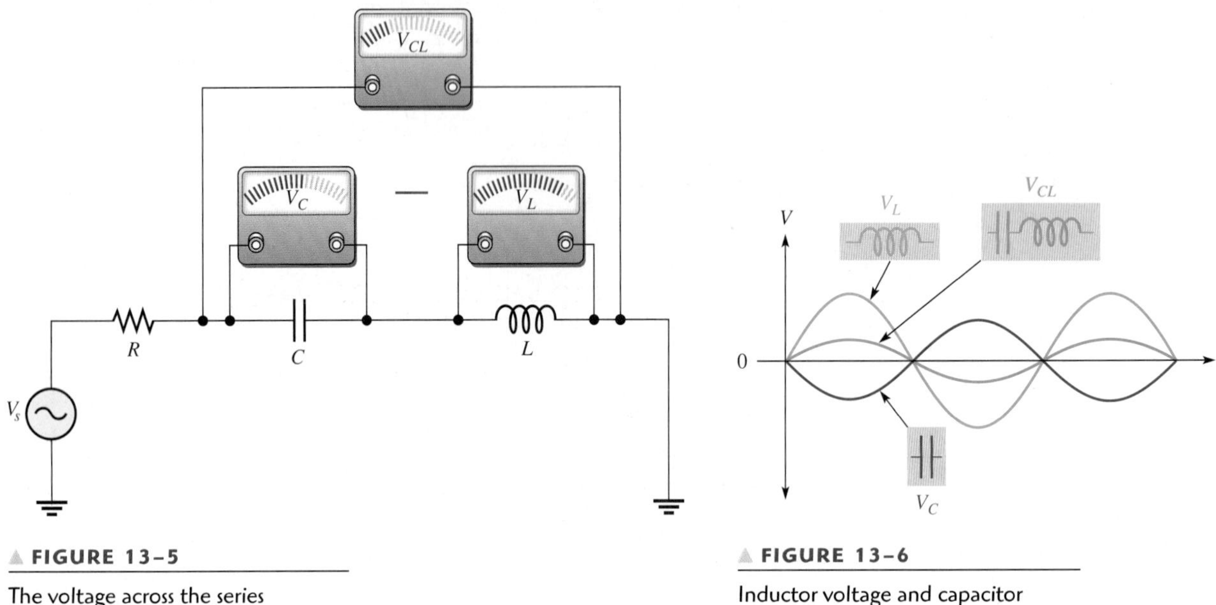

▲ **FIGURE 13–5**

The voltage across the series combination of C and L is always less than the larger individual voltage across either C or L.

▲ **FIGURE 13–6**

Inductor voltage and capacitor voltage effectively subtract because they are out of phase.

In the next example, Ohm's law is used to find the current and voltages in a series *RLC* circuit.

EXAMPLE 13–3

Find the voltage across each element in Figure 13–7, and draw a complete voltage phasor diagram. Also find the voltage across L and C combined.

▶ **FIGURE 13–7**

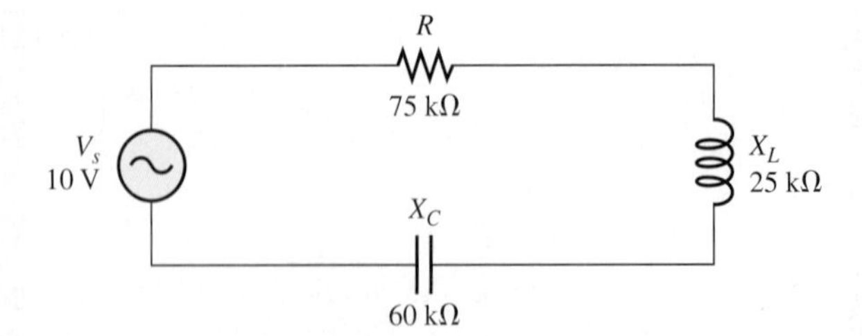

Solution First, find the total reactance.

$$X_{tot} = |X_L - X_C| = |25\text{ k}\Omega - 60\text{ k}\Omega| = 35\text{ k}\Omega$$

The total impedance is

$$Z_{tot} = \sqrt{R^2 + X_{tot}^2} = \sqrt{(75\text{ k}\Omega)^2 + (35\text{ k}\Omega)^2} = 82.8\text{ k}\Omega$$

Apply Ohm's law to find the current.

$$I = \frac{V_s}{Z_{tot}} = \frac{10\text{ V}}{82.8\text{ k}\Omega} = 121\ \mu\text{A}$$

Now, apply Ohm's law to find the voltages across *R*, *L*, and *C*.

$$V_R = IR = (121\ \mu\text{A})(75\text{ k}\Omega) = \mathbf{9.08\text{ V}}$$

$$V_L = IX_L = (121\ \mu\text{A})(25\text{ k}\Omega) = \mathbf{3.03\text{ V}}$$

$$V_C = IX_C = (121\ \mu\text{A})(60\text{ k}\Omega) = \mathbf{7.26\text{ V}}$$

The voltage across *L* and *C* combined is

$$V_{CL} = V_C - V_L = 7.26\text{ V} - 3.03\text{ V} = \mathbf{4.23\text{ V}}$$

The circuit phase angle is

$$\theta = \tan^{-1}\left(\frac{X_{tot}}{R}\right) = \tan^{-1}\left(\frac{35\text{ k}\Omega}{75\text{ k}\Omega}\right) = 25°$$

Since the circuit is capacitive ($X_C > X_L$), the current leads the source voltage by 25°.

The phasor diagram is shown in Figure 13–8. Notice that V_L is leading V_R by 90°, and V_C is lagging V_R by 90°. Also, there is a 180° phase difference between V_L and V_C. If the current phasor were shown, it would be at the same angle as V_R.

FIGURE 13–8

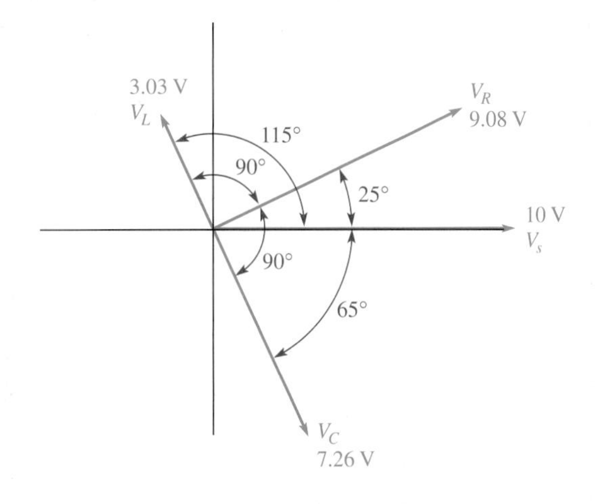

Related Problem What will happen to the current as the frequency of the source voltage in Figure 13–7 is increased?

SECTION 13–2 REVIEW

1. The following voltages occur in a certain series *RLC* circuit: $V_R = 24$ V, $V_L = 15$ V, $V_C = 45$ V. Determine the source voltage.
2. When $R = 10$ kΩ, $X_C = 18$ kΩ, and $X_L = 12$ kΩ, does the current lead or lag the source voltage? Why?
3. Determine the total reactance in Question 2.

13–3 SERIES RESONANCE

In a series *RLC* circuit, series resonance occurs when $X_L = X_C$. The frequency at which resonance occurs is called the **resonant frequency** and is designated f_r.

After completing this section, you should be able to

- **Analyze a circuit for series resonance**
- Define *series resonance*
- Determine the impedance at resonance
- Explain why the reactances cancel at resonance
- Determine the series resonant frequency
- Calculate the current, voltages, and phase angle at resonance

Figure 13–9 illustrates the series resonant condition. Since $X_L = X_C$, the reactances effectively cancel, and the impedance is purely resistive. These resonant conditions are stated in the following equations:

Equation 13–4
$$X_L = X_C$$

Equation 13–5
$$Z_r = R$$

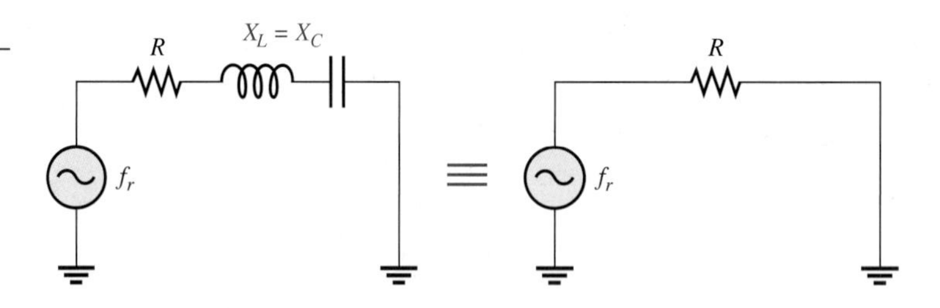

▶ **FIGURE 13–9**

At the resonant frequency (f_r), the reactances are equal in magnitude and effectively cancel, leaving $Z_r = R$.

EXAMPLE 13–4

For the series *RLC* circuit in Figure 13–10, determine X_C and Z at resonance.

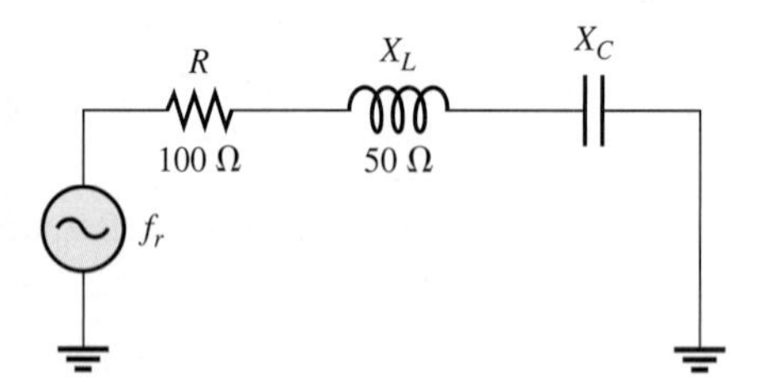

▶ **FIGURE 13–10**

Solution X_L equals X_C at resonance; therefore,

$$X_C = X_L = \mathbf{50\ \Omega}$$

Since the reactances cancel each other, the impedance equals the resistance.

$$Z_r = R = \mathbf{100\ \Omega}$$

Related Problem Just below the resonant frequency, is the circuit more inductive or more capacitive?

X_L and X_C Cancel at Resonance

At the series resonant frequency, the voltages across C and L are equal in magnitude because the reactances are equal and because the same current is through both since they are in series ($IX_C = IX_L$). Also, V_L and V_C are always 180° out of phase with each other.

During any given cycle, the polarities of the voltages across C and L are opposite, as shown in Figures 13–11(a) and 13–11(b). The equal and opposite voltages across C and L cancel, leaving zero volts from point A to point B as shown in the figure. Since there is no voltage drop from A to B but there is still current, the total reactance must be zero, as indicated in Figure 13–11(c). Also, the voltage phasor diagram in part (d) shows that V_C and V_L are equal in magnitude and 180° out of phase with each other.

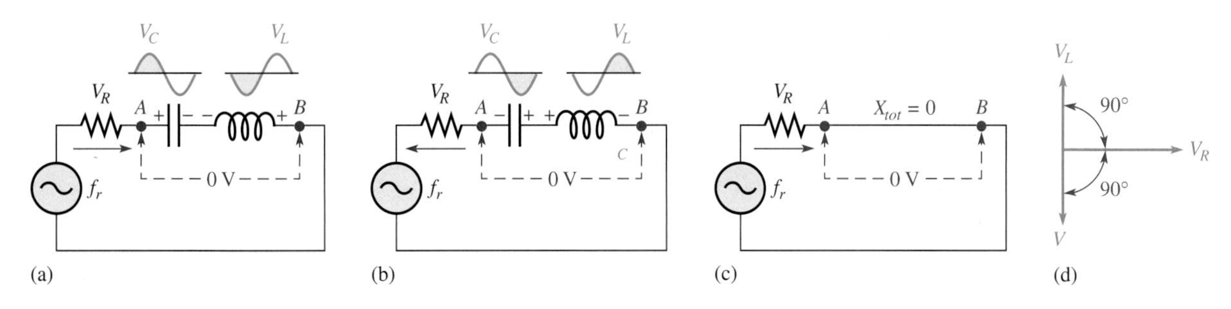

FIGURE 13–11

At the resonant frequency, f_r, the voltages across C and L are equal in magnitude. Since they are 180° out of phase with each other, they cancel, leaving 0 V across the LC combination (point A to point B). The section of the circuit from A to B effectively looks like a short at resonance (neglecting winding resistance).

The Series Resonant Frequency

For a given series RLC circuit, resonance occurs at only one specific frequency. A formula for this resonant frequency is developed as follows:

$$X_L = X_C$$

Substitute the reactance formulas, and solve for the resonant frequency (f_r).

$$2\pi f_r L = \frac{1}{2\pi f_r C}$$

$$(2\pi f_r L)(2\pi f_r C) = 4\pi^2 f_r^2 LC = 1$$

$$f_r^2 = \frac{1}{4\pi^2 LC}$$

Take the square root of both sides. The formula for resonant frequency is

$$f_r = \frac{1}{2\pi\sqrt{LC}}$$

Equation 13–6

EXAMPLE 13–5

Find the series resonant frequency for the circuit in Figure 13–12.

FIGURE 13–12

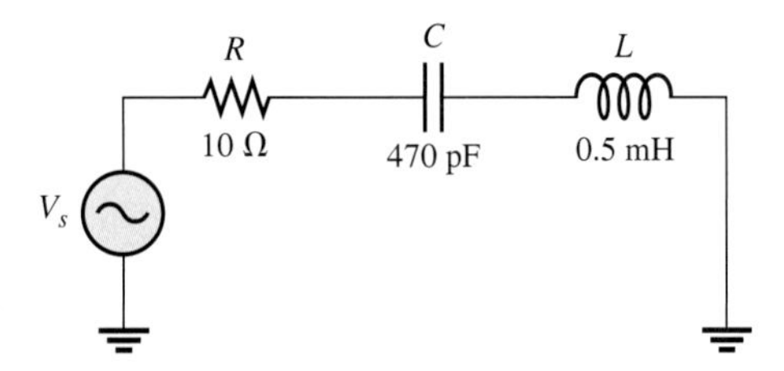

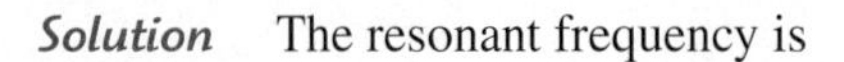

Solution The resonant frequency is

$$f_r = \frac{1}{2\pi\sqrt{LC}} = \frac{1}{2\pi\sqrt{(0.5\text{ mH})(470\text{ pF})}} = \mathbf{328\text{ kHz}}$$

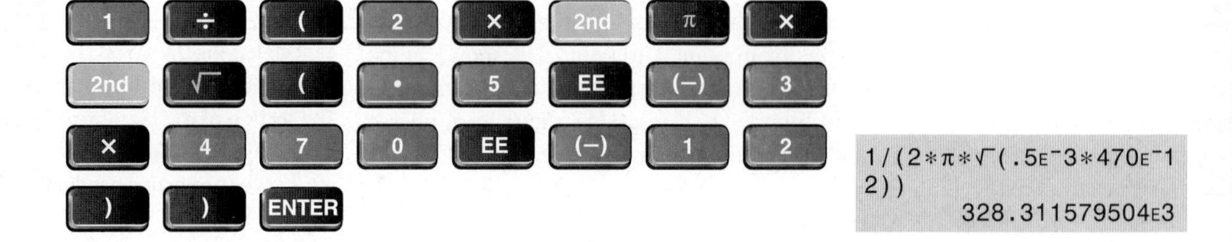

Related Problem If $C = 0.01\ \mu$F in Figure 13–12, what is the resonant frequency?

Open file E13-05 on your EWB/CircuitMaker CD-ROM. Determine the series resonant frequency by measurement.

Voltages and Current in a Series *RLC* Circuit

It is interesting to see how the current and the voltages in a series *RLC* circuit vary as the frequency is increased from below the resonant frequency, through resonance, and then above resonance. Figure 13–13 illustrates the general response of a circuit in terms of changes in the current and in the voltage drops. The Q (quality factor) of the circuit is assumed to be sufficiently high so it has no effect on the response. The Q is the ratio of reactive power to true power and is discussed later in this chapter.

Below the Resonant Frequency Figure 13–13(a) indicates the response as the source frequency is increased from zero toward f_r. At $f = 0$ Hz (dc), the capacitor appears open and blocks the current. Thus, there is no voltage across R or L, and the entire source voltage appears across C. The impedance of the circuit is infinitely large at 0 Hz because X_C is infinite (C is open). As the frequency begins to increase, X_C decreases and X_L increases, causing the total reactance, $X_C - X_L$, to decrease. As a result, the impedance decreases and the current increases. As the current increases, V_R also increases, and both V_C and V_L increase. The voltage across C and L combined decreases from its maximum value of V_s because as V_C and V_L approach the same value, their difference becomes less.

At the Resonant Frequency When the frequency reaches its resonant value, f_r, as shown in Figure 13–13(b), V_C and V_L can each be much larger than the source voltage. V_C and V_L cancel leaving 0 V across the combination of C and L because the voltages are equal in magnitude but opposite in phase. At this point the total impedance is equal to R and is at its minimum value because the total reactance is zero. Thus, the current is at its maximum value, V_s/R, and V_R is at its maximum value which is equal to the source voltage.

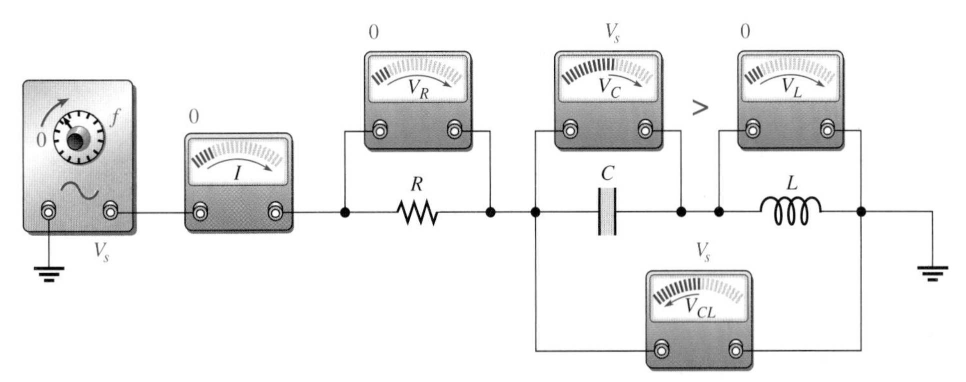

(a) As frequency is increased below resonance from 0: $X_C > X_L$, I increases from 0, V_R increases from 0, V_C increases from V_s, V_L increases from 0, and V_{CL} decreases from V_s.

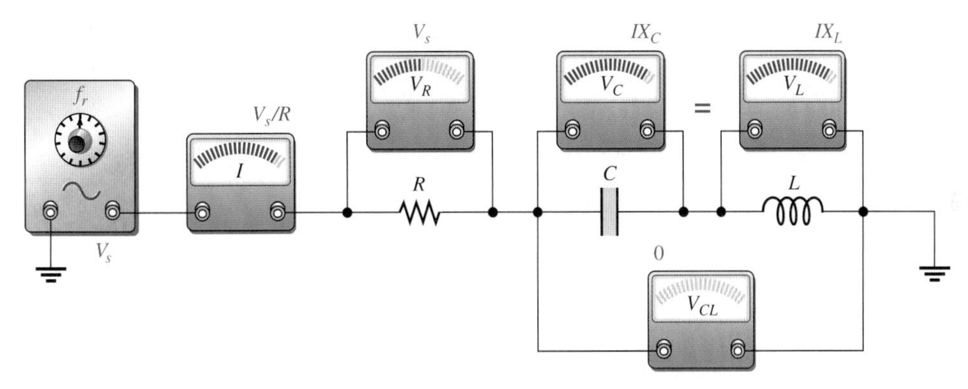

(b) At the resonant frequency, $X_C = X_L$.

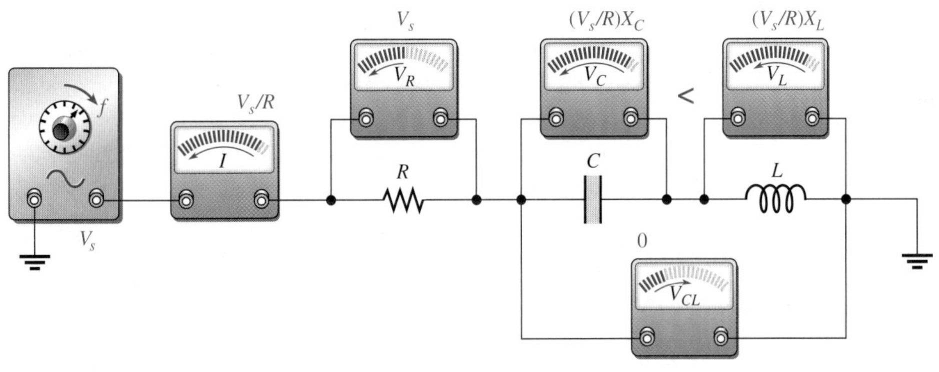

(c) As frequency is increased above resonance: $X_C < X_L$, I decreases from V_s/R, V_R decreases from V_s, V_C decreases from $(V_s/R)\,X_C$, V_L decreases from $(V_s/R)\,X_L$, and V_{CL} increases from 0.

▲ **FIGURE 13–13**

An illustration of how the voltages and the current respond in a series *RLC* circuit as the frequency is increased from below to above its resonant value. The source voltage is held at a constant amplitude.

Above the Resonant Frequency As the frequency is increased above resonance, as indicated in Figure 13–13(c), X_L continues to increase and X_C continues to decrease, causing the total reactance, $X_L - X_C$, to increase. As a result, there is an increase in impedance and a decrease in current. As the current decreases, V_R also decreases and V_C and V_L decrease. As V_C and V_L decrease, their difference becomes greater, so V_{CL} increases. As the frequency becomes very high, the current approaches zero, both V_R and V_C approach zero, and V_L approaches V_s.

Figure 13–14(a) and (b) illustrate the responses of current and voltage to increasing frequency, respectively. As frequency is increased, the current increases below resonance, reaches a peak at the resonant frequency, and then decreases above resonance. The resistor voltage responds in the same way as the current.

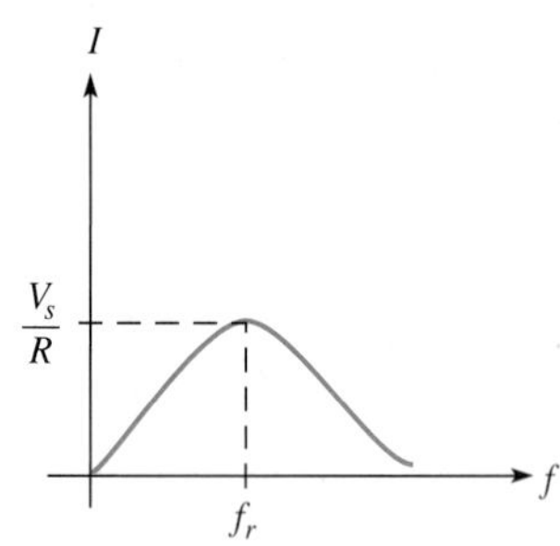

(a) Current

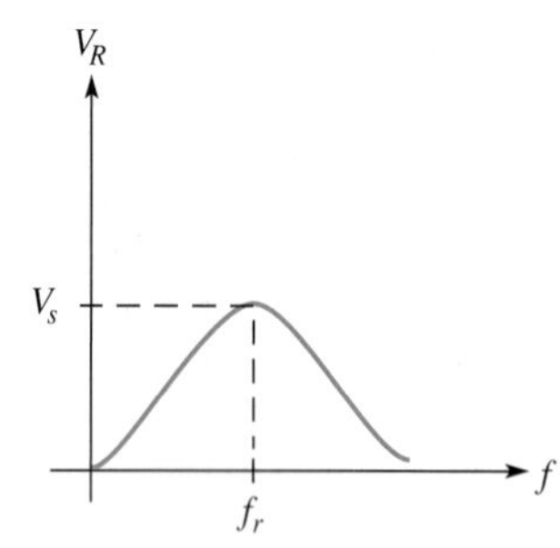

(b) Resistor voltage

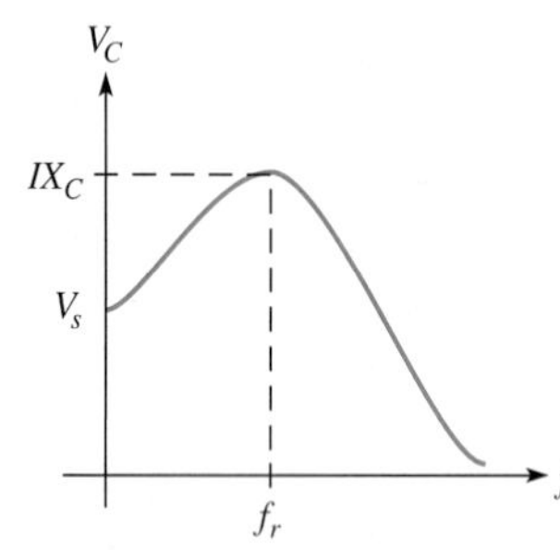

(c) Capacitor voltage

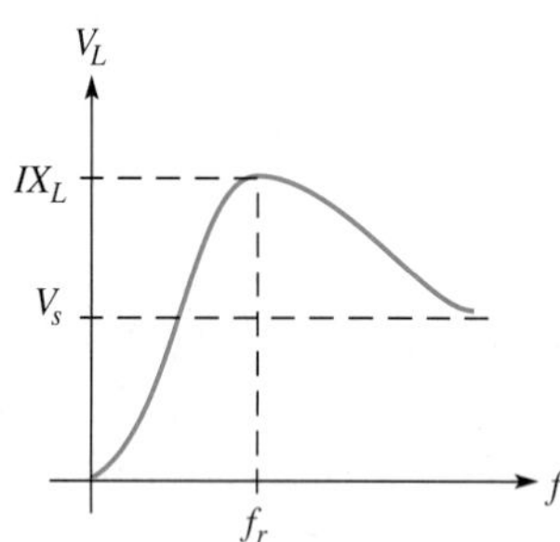

(d) Inductor voltage

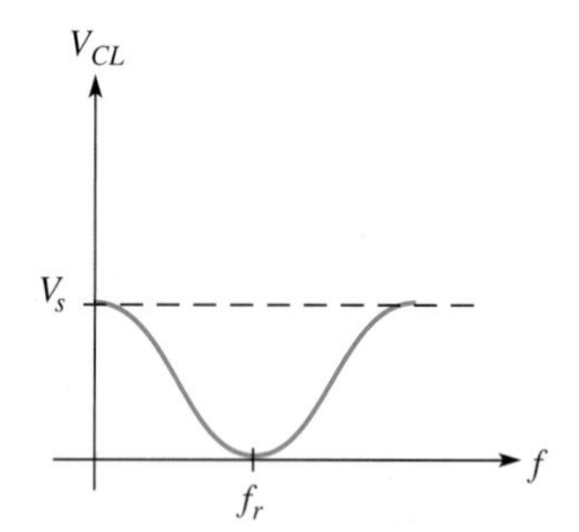

(e) Voltage across C and L combined

▲ **FIGURE 13–14**

Generalized current and voltage magnitudes as a function of frequency in a series *RLC* circuit. V_C and V_L can be much larger than the source voltage. The shapes of the graphs depend on particular circuit values.

The general slopes of the V_C and V_L curves are indicated in Figure 13–14(c) and (d). The voltages are maximum at resonance but drop off above and below f_r. The voltages across L and C at resonance are exactly equal in magnitude but 180° out of phase, so they cancel. Thus, the total voltage across both L and C is zero, and $V_R = V_s$ at resonance. Individually, V_L and V_C can be much greater than the source voltage. Keep in mind that V_L and V_C are always opposite in polarity regardless of the frequency, but only at resonance are their magnitudes equal. The voltage across the C and L combination decreases as the frequency increases below resonance, reaching a minimum of zero at the resonant frequency; then it increases above resonance, as shown in Figure 13–14(e).

EXAMPLE 13–6

Find I, V_R, V_L, and V_C at resonance in Figure 13–15.

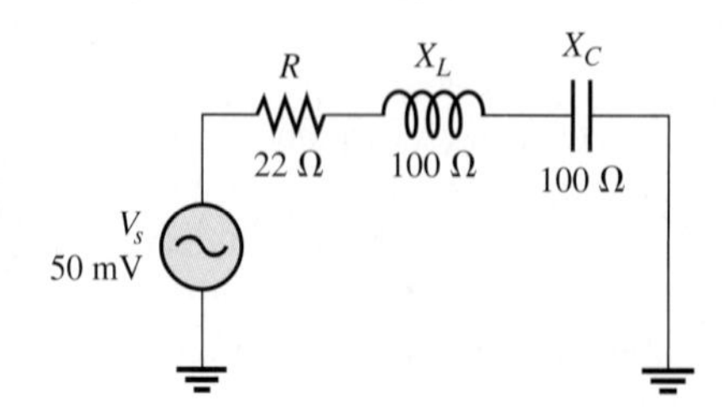

▶ **FIGURE 13–15**

Solution At resonance, I is maximum and equal to V_s/R.

$$I = \frac{V_s}{R} = \frac{50 \text{ mV}}{22\ \Omega} = \mathbf{2.27\ mA}$$

Apply Ohm's law to obtain the voltages.

$$V_R = IR = (2.27\text{ mA})(22\ \Omega) = \mathbf{50\ mV}$$

$$V_L = IX_L = (2.27\text{ mA})(100\ \Omega) = \mathbf{227\ mV}$$

$$V_C = IX_C = (2.27\text{ mA})(100\ \Omega) = \mathbf{227\ mV}$$

Notice that all of the source voltage is dropped across the resistor. Also, of course, V_L and V_C are equal in magnitude but opposite in phase. This causes the voltages to cancel, making the total reactive voltage zero.

Related Problem What is the current at resonance in Figure 13–15 if $X_L = X_C = 1\text{ k}\Omega$?

The Impedance of a Series *RLC* Circuit

Figure 13–16 shows a general graph of impedance versus frequency superimposed on the curves for X_C and X_L. At zero frequency, both X_C and Z are infinitely large and X_L is zero because the capacitor looks like an open at 0 Hz and the inductor looks like a short. As the frequency increases, X_C decreases and X_L increases. Since X_C is larger than X_L at frequencies below f_r, Z decreases along with X_C. At f_r, $X_C = X_L$ and $Z = R$. At frequencies above f_r, X_L becomes increasingly larger than X_C, causing Z to increase.

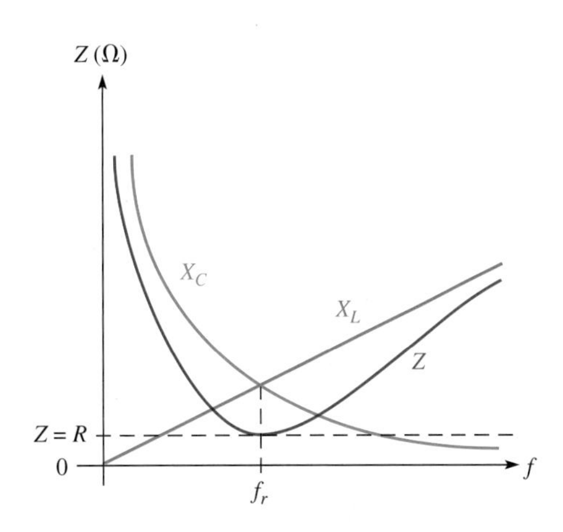

FIGURE 13–16

Series *RLC* impedance as a function of frequency.

EXAMPLE 13–7

For the circuit in Figure 13–17, determine the impedance at the following frequencies:

(a) f_r **(b)** 1 kHz below f_r **(c)** 1 kHz above f_r

FIGURE 13–17

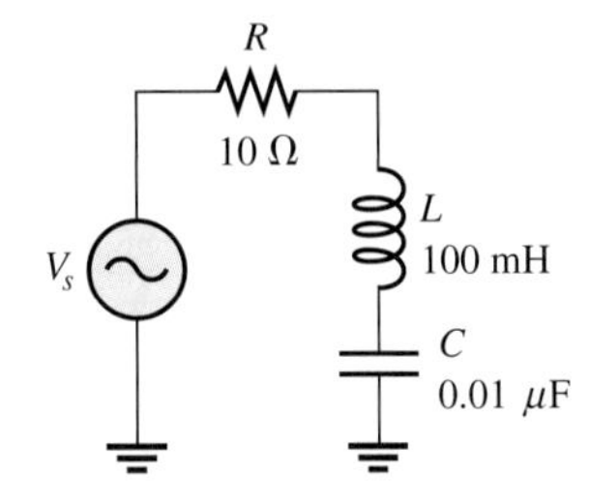

Solution **(a)** At f_r, the impedance is equal to R.

$$Z = R = \mathbf{10\ \Omega}$$

To determine the impedance above and below f_r, first calculate the resonant frequency.

$$f_r = \frac{1}{2\pi\sqrt{LC}} = \frac{1}{2\pi\sqrt{(100\text{ mH})(0.01\ \mu\text{F})}} = 5.03\text{ kHz}$$

(b) At 1 kHz below f_r, the frequency and reactances are as follows:

$$f = f_r - 1\text{ kHz} = 5.03\text{ kHz} - 1\text{ kHz} = 4.03\text{ kHz}$$

$$X_C = \frac{1}{2\pi fC} = \frac{1}{2\pi(4.03\text{ kHz})(0.01\ \mu\text{F})} = 3.95\text{ k}\Omega$$

$$X_L = 2\pi fL = 2\pi(4.03\text{ kHz})(100\text{ mH}) = 2.53\text{ k}\Omega$$

Therefore, the total reactance and the impedance at $f_r - 1$ kHz are

$$X_{tot} = |X_L - X_C| = |2.53\text{ k}\Omega - 3.95\text{ k}\Omega| = 1.42\text{ k}\Omega$$

$$Z = \sqrt{R^2 + X_{tot}^2} = \sqrt{(10\ \Omega)^2 + (1.42\text{ k}\Omega)^2} = \mathbf{1.42\text{ k}\Omega}$$

(c) At 1 kHz above f_r,

$$f = 5.03\text{ kHz} + 1\text{ kHz} = 6.03\text{ kHz}$$

$$X_C = \frac{1}{2\pi(6.03\text{ kHz})(0.01\ \mu\text{F})} = 2.64\text{ k}\Omega$$

$$X_L = 2\pi(6.03\text{ kHz})(100\text{ mH}) = 3.79\text{ k}\Omega$$

Therefore, the total reactance and the impedance at $f_r + 1$ kHz are

$$X_{tot} = |3.79\text{ k}\Omega - 2.64\text{ k}\Omega| = 1.15\text{ k}\Omega$$

$$Z = \sqrt{(10\ \Omega)^2 + (1.15\text{ k}\Omega)^2} = \mathbf{1.15\text{ k}\Omega}$$

In part (b), Z is capacitive; in part (c), Z is inductive.

Related Problem What happens to the impedance if f is decreased below 4.03 kHz? Above 6.03 kHz?

Open file E13-07 on your EWB/CircuitMaker CD-ROM. Measure the current and the voltage across each element at the resonant frequency, 1 kHz below resonance, and 1 kHz above resonance.

The Phase Angle of a Series *RLC* Circuit

At frequencies below resonance, $X_C > X_L$, and the current leads the source voltage, as indicated in Figure 13–18(a). The phase angle decreases as the frequency approaches the resonant value and is 0° at resonance, as indicated in part (b). At frequencies above resonance, $X_L > X_C$, and the current lags the source voltage, as indicated in part (c). As the frequency goes higher, the phase angle approaches 90°. A plot of phase angle versus frequency is shown in Figure 13–18(d).

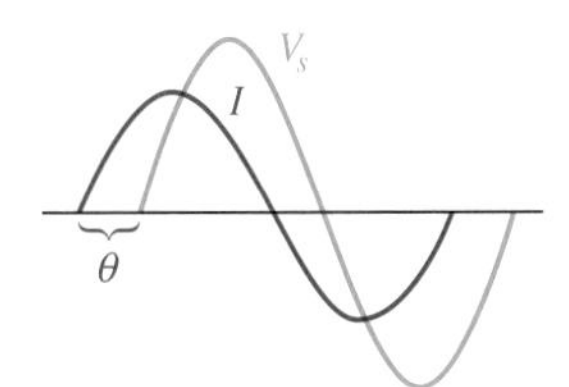

(a) Below f_r, I leads V_s.

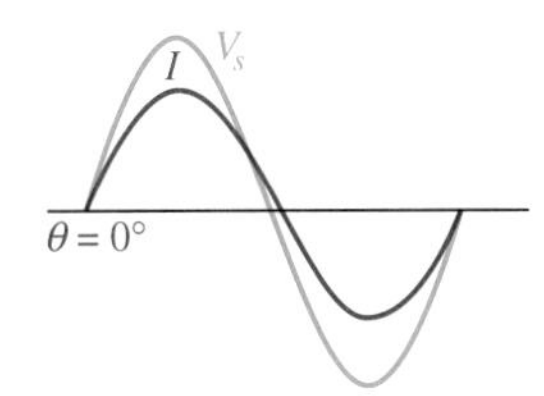

(b) At f_r, I is in phase with V_s.

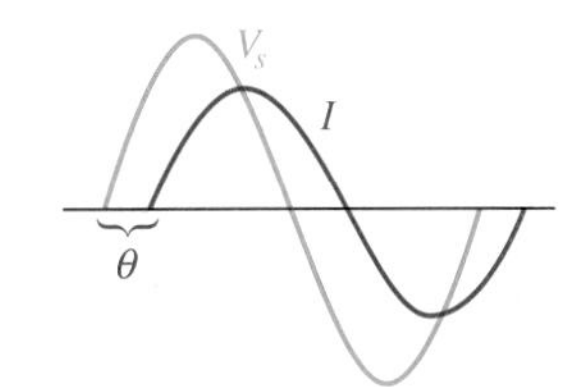

(c) Above f_r, I lags V_s.

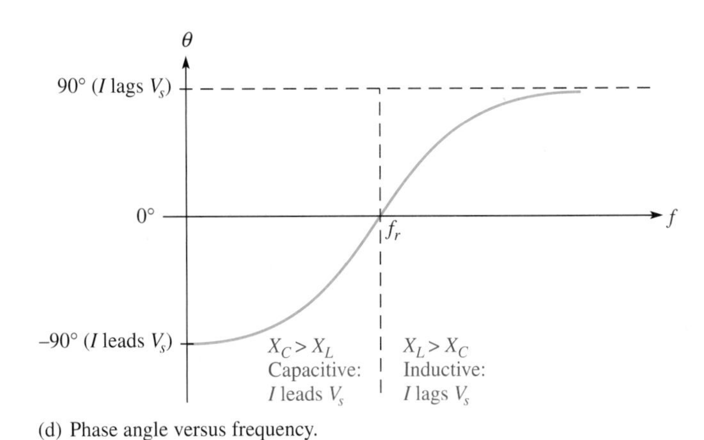

(d) Phase angle versus frequency.

FIGURE 13–18

The phase angle as a function of frequency in a series *RLC* circuit.

SECTION 13–3 REVIEW

1. What is the condition for series resonance?
2. Why is the current maximum at the resonant frequency?
3. Calculate the resonant frequency for $C = 1000$ pF and $L = 1000\ \mu$H.
4. In Question 3, is the circuit inductive, capacitive, or resistive at 50 kHz?

13–4 SERIES RESONANT FILTERS

A common use of series *RLC* circuits is in filter applications. In this section, you will learn the basic configurations for band-pass and band-stop filters and several important filter characteristics.

After completing this section, you should be able to

- **Analyze series resonant filters**
- Identify a basic series resonant band-pass filter
- Define and determine *bandwidth*
- Define the *half-power frequency*
- Discuss dB measurement
- Define *selectivity*
- Discuss filter quality factor (Q)
- Identify a series resonant band-stop filter

The Band-Pass Filter

A basic series resonant band-pass filter is shown in Figure 13–19. Notice that the series *LC* portion is placed between the input and the output and that the output is taken across the resistor.

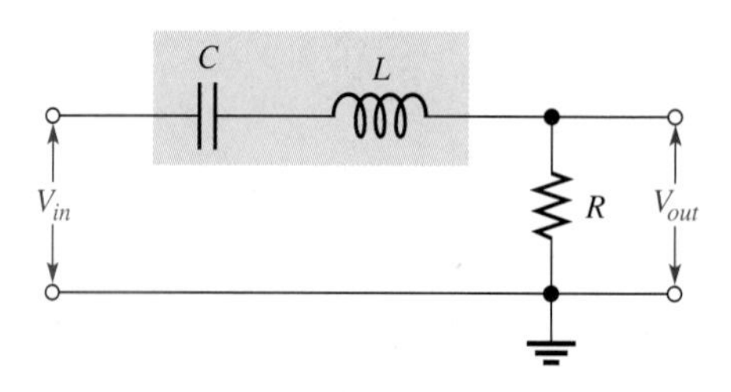

▶ **FIGURE 13–19**

A basic series resonant band-pass filter.

A **band-pass filter** allows signals at the resonant frequency and at frequencies within a certain band (or range) extending below and above the resonant value to pass from input to output without a significant reduction in amplitude. Signals at frequencies lying outside this specified band (called the **passband**) are reduced in amplitude to below a certain level and are considered to be rejected by the filter.

The filtering action is the result of the impedance characteristic of the filter. As you learned in Section 13–3, the impedance is minimum at resonance and has increasingly higher values below and above the resonant frequency. At very low frequencies, the impedance is very high and tends to block the current. As the frequency increases, the impedance drops, allowing more current and thus more voltage across the output resistor. At the resonant frequency, the impedance is very low and equal to the winding resistance of the coil. At this point there is maximum current and the resulting output voltage is maximum. As the frequency goes above resonance, the impedance again increases, causing the current and the resulting output voltage to drop. Figure 13–20 illustrates the general frequency response of a series resonant band-pass filter.

▼ **FIGURE 13–20**

Example of the frequency response of a series resonant band-pass filter with the input voltage at a constant 10 V rms. The winding resistance of the coil is neglected.

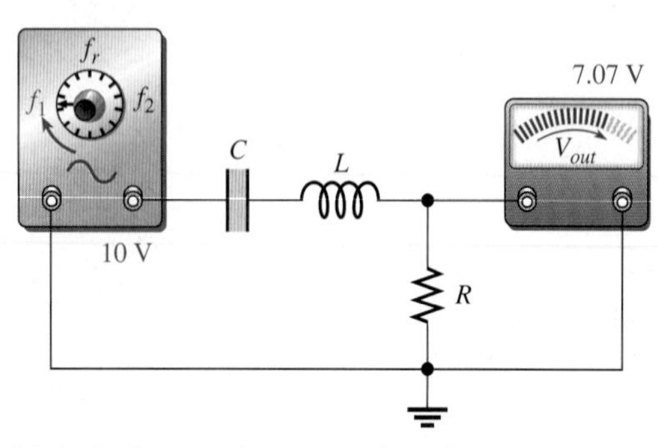

(a) As the frequency increases to f_1, V_{out} increases to 7.07 V.

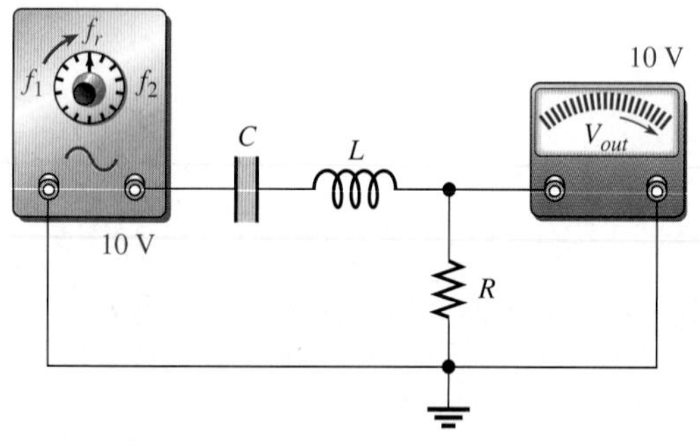

(b) As the frequency increases from f_1 to f_r, V_{out} increases from 7.07 V to 10 V.

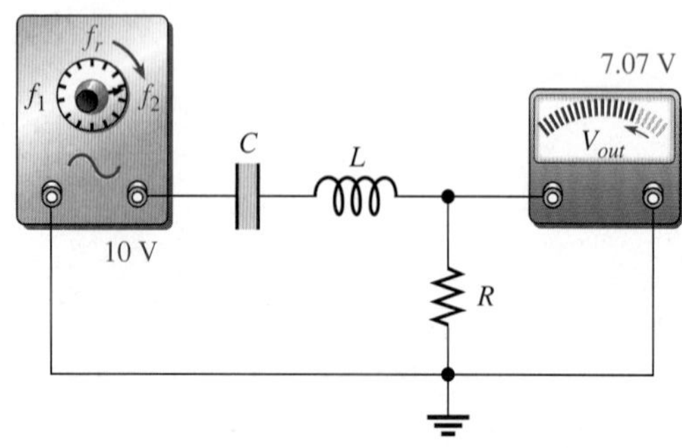

(c) As the frequency increases from f_r to f_2, V_{out} decreases from 10 V to 7.07 V.

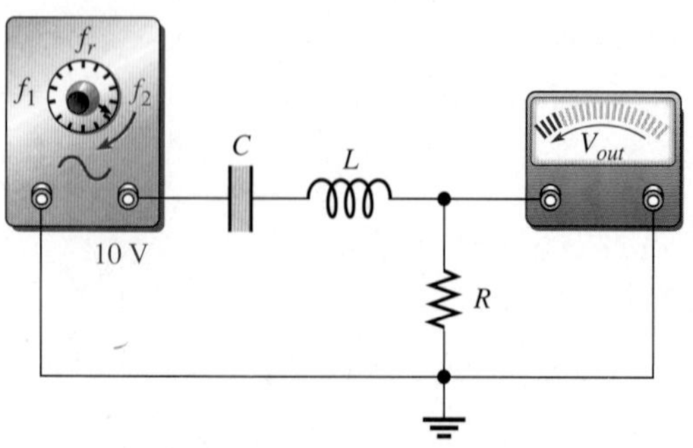

(d) As the frequency increases above f_2, V_{out} decreases below 7.07 V.

Bandwidth of the Passband

The **bandwidth** (BW) of a band-pass filter is the range of frequencies for which the current (or output voltage) is equal to or greater than 70.7% of its value at the resonant frequency. Figure 13–21 shows the bandwidth on the response curve for a band-pass filter.

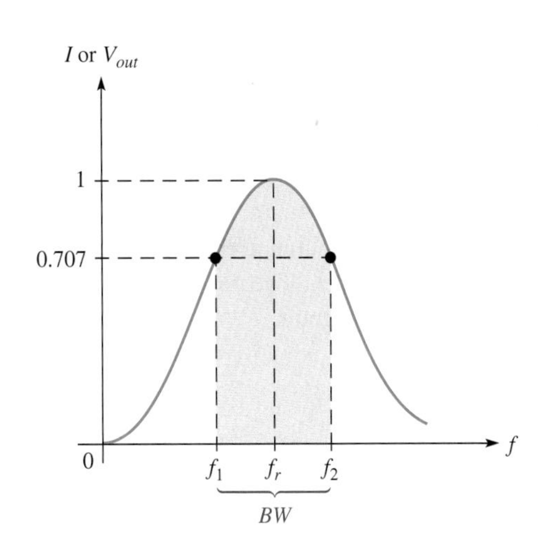

FIGURE 13–21

Generalized response curve of a series resonant band-pass filter.

The frequencies at which the output of a filter is 70.7% of its maximum are the **cutoff frequencies**. Notice in Figure 13–21 that frequency f_1, which is below f_r, is the frequency at which I (or V_{out}) is 70.7% of the resonant value (I_{max}); f_1 is commonly called the *lower cutoff frequency*. At frequency f_2, above f_r, the current (or V_{out}) is again 70.7% of its maximum; f_2 is called the *upper cutoff frequency*. Other names for f_1 and f_2 are *−3 dB frequencies, critical frequencies, band frequencies,* and *half-power frequencies.* (The term *decibel,* abbreviated dB, is defined later in this section.)

The formula for calculating the bandwidth is

$$BW = f_2 - f_1$$

Equation 13–7

The unit of bandwidth is the hertz (Hz), the same as for frequency.

EXAMPLE 13–8

A certain series resonant band-pass filter has a maximum current of 100 mA at the resonant frequency. What is the value of the current at the cutoff frequencies?

Solution Current at the cutoff frequencies is 70.7% of maximum.

$$I_{f1} = I_{f2} = 0.707I_{max} = 0.707(100\text{ mA}) = \mathbf{70.7\text{ mA}}$$

Related Problem Will a change in the cutoff frequencies affect the current at the new cutoff frequencies if the maximum current remains 100 mA?

EXAMPLE 13–9

A resonant circuit has a lower cutoff frequency of 8 kHz and an upper cutoff frequency of 12 kHz. Determine the bandwidth.

Solution

$$BW = f_2 - f_1 = 12\text{ kHz} - 8\text{ kHz} = \mathbf{4\text{ kHz}}$$

Related Problem What is the bandwidth when $f_1 = 1$ MHz and $f_2 = 1.2$ MHz?

Half-Power Points of the Filter Response

As previously mentioned, the upper and lower cutoff frequencies are sometimes called the **half-power frequencies**. This term is derived from the fact that the true power from the source at these frequencies is one-half the power delivered at the resonant frequency. The following steps show that this relationship is true for a series resonant circuit.

At resonance,

$$P_{max} = I_{max}^2 R$$

The power at f_1 (or f_2) is

$$P_{f1} = I_{f1}^2 R = (0.707 I_{max})^2 R = (0.707)^2 I_{max}^2 R = 0.5 I_{max}^2 R = 0.5 P_{max}$$

Decibel (dB) Measurement

As previously mentioned, another common term for the upper and lower cutoff frequencies is -3 *dB frequencies.* The **decibel** (dB) is a logarithmic measurement of the ratio of one voltage to another or one power to another, which can be used to express the input-to-output relationship of a filter. The following equation expresses a voltage ratio in decibels:

Equation 13–8

$$\text{dB} = 20\log\left(\frac{V_{out}}{V_{in}}\right)$$

The decibel formula for a power ratio is

Equation 13–9

$$\text{dB} = 10\log\left(\frac{P_{out}}{P_{in}}\right)$$

EXAMPLE 13–10

At a certain frequency, the output voltage of a filter is 5 V and the input is 10 V. Express the voltage ratio in decibels.

Solution

$$20\log\left(\frac{V_{out}}{V_{in}}\right) = 20\log\left(\frac{5\text{ V}}{10\text{ V}}\right) = 20\log(0.5) = \mathbf{-6.02\text{ dB}}$$

```
20*log (5/10)
      -6.02059991328E0
```

Related Problem Express the ratio $V_{out}/V_{in} = 0.2$ in decibels.

The −3 dB Frequencies The output of a filter is said to be down 3 dB at the cutoff frequencies. As you know, this frequency is the point at which the output voltage is 70.7% of the maximum voltage at resonance. We can show that the 70.7% point is the same as 3 dB below maximum (or −3 dB) as follows. The maximum voltage is the 0 dB reference.

$$20 \log\left(\frac{0.707V_{max}}{V_{max}}\right) = 20 \log(0.707) = -3 \text{ dB}$$

Selectivity of a Band-Pass Filter

The response curve in Figure 13–21 is also called a *selectivity curve.* Selectivity defines how well a resonant circuit responds to a certain frequency and discriminates against all others. *The narrower the bandwidth, the greater the selectivity.*

We ideally assume that a resonant circuit accepts frequencies within its bandwidth and completely eliminates frequencies outside the bandwidth. Such is not actually the case, however, because signals with frequencies outside the bandwidth are not completely eliminated. Their magnitudes, however, are greatly reduced. The further the frequencies are from the cutoff frequencies, the greater the reduction, as illustrated in Figure 13–22(a). An ideal selectivity curve is shown in Figure 13–22(b).

FIGURE 13–22

Generalized selectivity curve of a band-pass filter.

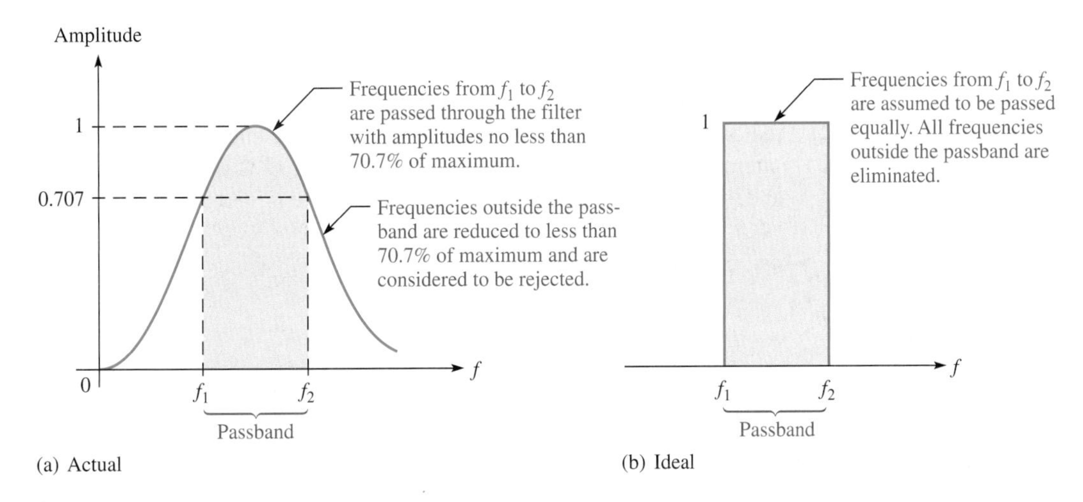

As you can see in Figure 13–23, another factor that influences selectivity is the steepness of the slopes of the curve. The faster the curve drops off at the cutoff frequencies, the more selective the circuit is because the frequencies outside the passband are more quickly reduced (attenuated).

The Quality Factor (Q) of a Resonant Circuit and Its Effect on Bandwidth

The quality factor (Q) is the ratio of the reactive power in the inductor to the true power in the winding resistance of the coil and any other resistance in series with the coil. It is a ratio of the power in L to the power in R. The quality factor is very important in resonant circuits. A formula for Q is developed as follows:

$$Q = \frac{\text{energy stored}}{\text{energy dissipated}} = \frac{\text{reactive power}}{\text{true power}} = \frac{I^2X_L}{I^2R}$$

In a series circuit, I is the same in L and R; thus, the I^2 terms cancel, leaving

$$Q = \frac{X_L}{R}$$

Equation 13–10

FIGURE 13–23

Comparative selectivity curves.

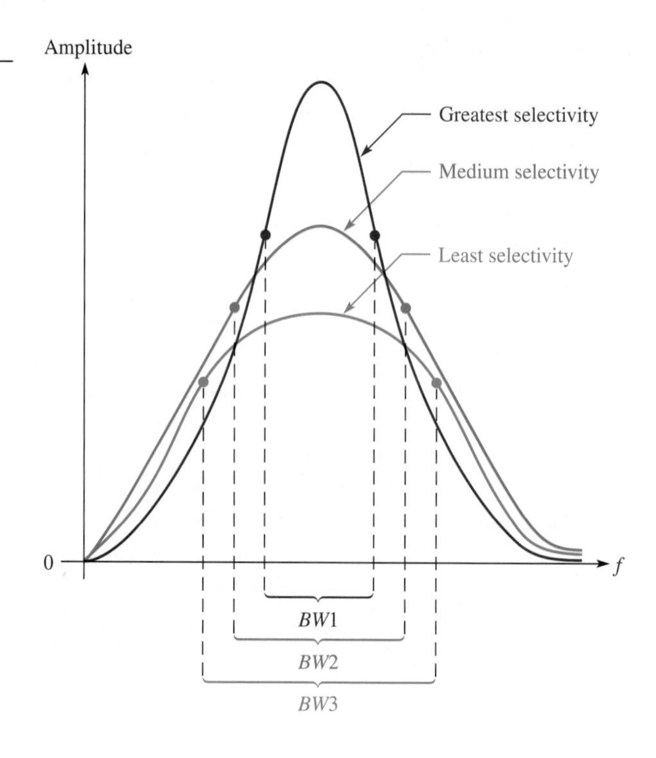

When the resistance is just the winding resistance of the coil, the circuit Q and the coil Q are the same. Since Q varies with frequency because X_L varies, we are interested mainly in Q at resonance. Note that Q is a ratio of like units (ohms) and, therefore, has no unit itself.

EXAMPLE 13–11

Determine Q at resonance for the circuit in Figure 13–24.

FIGURE 13–24

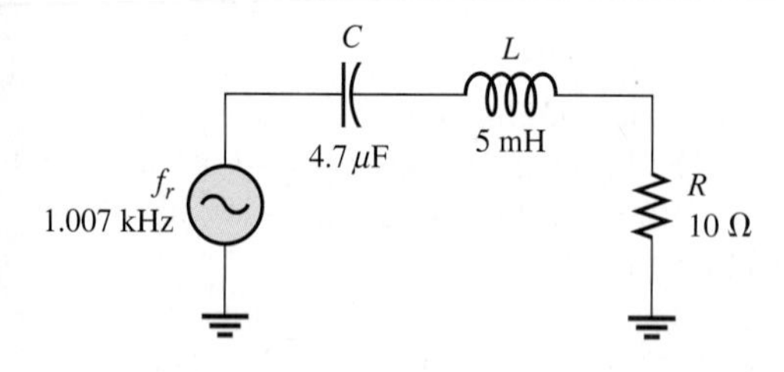

Solution Determine the inductive reactance.

$$X_L = 2\pi f_r L = 2\pi(1.007\text{ kHz})(5\text{ mH}) = 31.6\ \Omega$$

The quality factor is

$$Q = \frac{X_L}{R} = \frac{31.6\ \Omega}{10\ \Omega} = \mathbf{3.16}$$

Related Problem Calculate Q at resonance if C in Figure 13–24 is halved. The resonant frequency will increase.

How Q Affects Bandwidth A higher value of circuit Q results in a smaller bandwidth. A lower value of Q causes a larger bandwidth. A formula for the bandwidth of a resonant circuit in terms of Q is

$$BW = \frac{f_r}{Q}$$

Equation 13–11

EXAMPLE 13–12

What is the bandwidth of the filter in Figure 13–25?

FIGURE 13–25

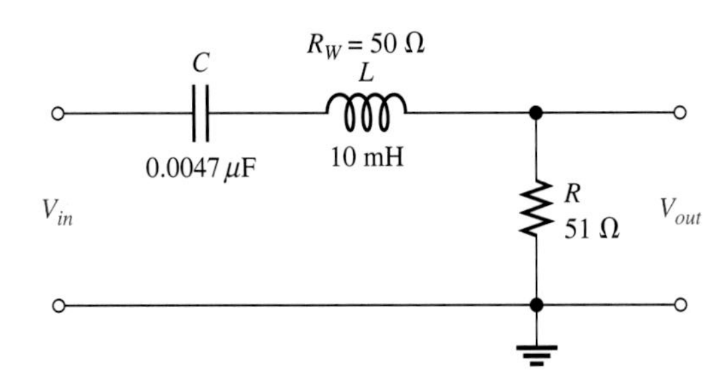

Solution The total resistance is

$$R_{tot} = R + R_W = 51\ \Omega + 50\ \Omega = 101\ \Omega$$

Determine the bandwidth as follows:

$$f_r = \frac{1}{2\pi\sqrt{LC}} = \frac{1}{2\pi\sqrt{(10\text{ mH})(0.0047\ \mu\text{F})}} = 23.2\text{ kHz}$$

$$X_L = 2\pi f_r L = 2\pi(23.2\text{ kHz})(10\text{ mH}) = 1.46\text{ k}\Omega$$

$$Q = \frac{X_L}{R_{tot}} = \frac{1.46\text{ k}\Omega}{101\ \Omega} = 14.5$$

$$BW = \frac{f_r}{Q} = \frac{23.2\text{ kHz}}{14.5} = \mathbf{1.60\text{ kHz}}$$

Related Problem Determine the bandwidth if L is changed to 50 mH in Figure 13–25.

Open file E13-12 on your EWB/CircuitMaker CD-ROM. Determine the bandwidth by measurement. How closely does this result compare to the calculated value?

The Band-Stop Filter

The basic series resonant band-stop filter is shown in Figure 13–26. Notice that the output voltage is taken across the LC portion of the circuit. This filter is still a series RLC circuit, just as the band-pass filter is. The difference is that in this case, the output voltage is taken across the combination of L and C rather than across R.

The band-stop filter rejects signals with frequencies between the lower and upper cutoff frequencies and passes those signals with frequencies below and above the cutoff values, as shown in the response curve of Figure 13–27. The

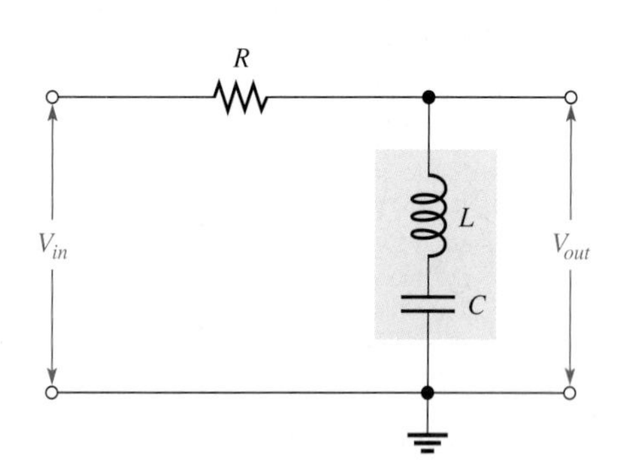

FIGURE 13–26

A basic series resonant band-stop filter.

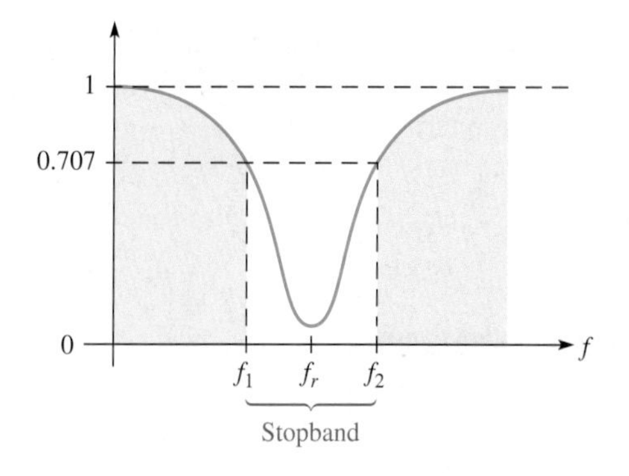

FIGURE 13–27

Generalized response curve for a band-stop filter.

range of frequencies between the lower and upper cutoff points is called the **stopband.** This type of filter is also referred to as a *band-elimination filter, band-reject filter,* or a *notch filter.*

At very low frequencies, the *LC* combination appears as a near open due to the high X_C, thus allowing most of the input voltage to pass through to the output. As the frequency increases, the impedance of the *LC* combination decreases until, at resonance, it is zero (ideally). Thus, the input signal is shorted to ground, and there is very little output voltage. As the frequency goes above its resonant value, the *LC* impedance increases, allowing an increasing amount of voltage to be dropped across it. The general frequency response of a series resonant band-stop filter is illustrated in Figure 13–28.

FIGURE 13–28

Example of the frequency response of a series resonant band-stop filter with V_{in} at a constant 10 V rms. The winding resistance is neglected.

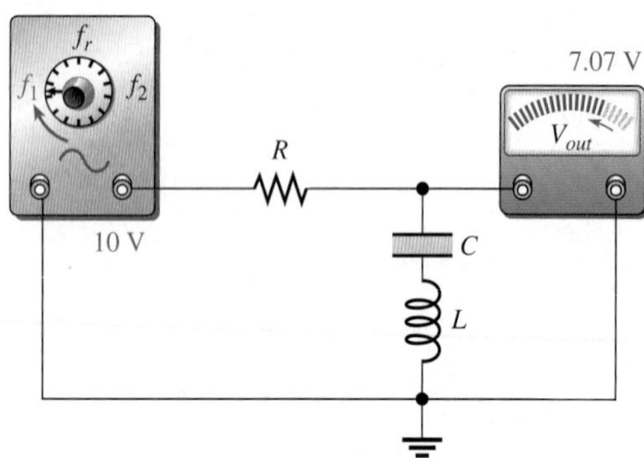

(a) As frequency increases to f_1, V_{out} decreases from 10 V to 7.07 V.

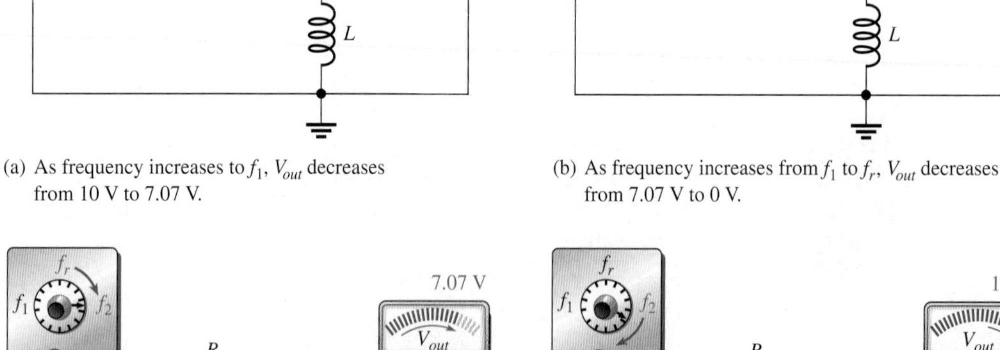

(b) As frequency increases from f_1 to f_r, V_{out} decreases from 7.07 V to 0 V.

(c) As frequency increases from f_r to f_2, V_{out} increases from 0 V to 7.07 V.

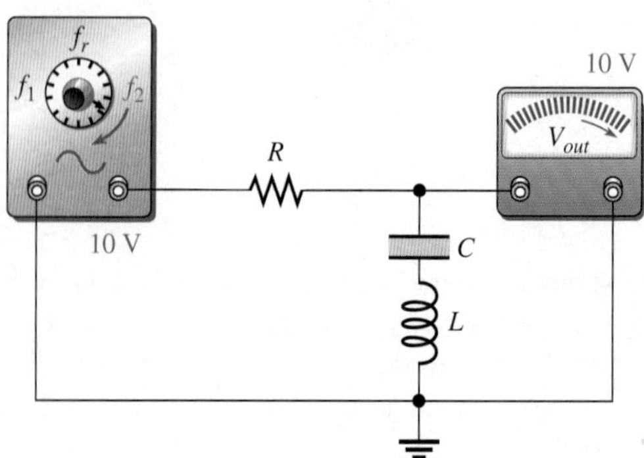

(d) As frequency increases above f_2, V_{out} increases toward 10 V.

Characteristics of the Band-Stop Filter

All the characteristics that have been discussed in relation to the band-pass filter (current response, impedance characteristic, bandwidth, selectivity, and Q) apply equally to the band-stop filter, with the exception that the response curve of the output voltage is opposite. For the band-pass filter, V_{out} is maximum at resonance. For the band-stop filter, V_{out} is minimum at resonance.

EXAMPLE 13–13

Find the output voltage at f_r and the bandwidth in Figure 13–29.

FIGURE 13–29

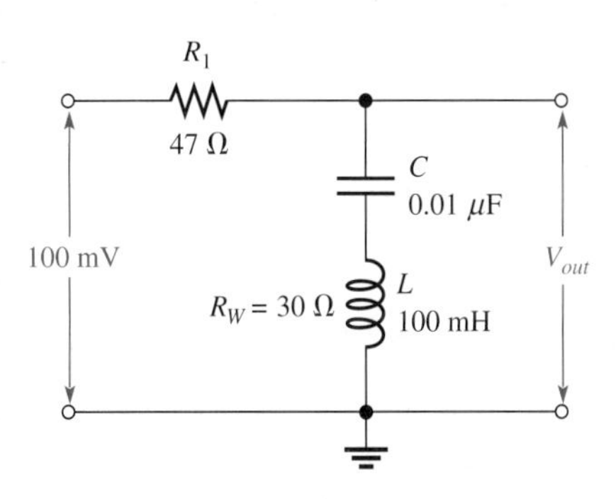

Solution Since $X_L = X_C$ at resonance, then

$$V_{out} = \left(\frac{R_W}{R_1 + R_W}\right)V_{in} = \left(\frac{30\ \Omega}{77\ \Omega}\right)100\text{ mV} = \mathbf{39.0\ mV}$$

Determine the bandwidth as follows:

$$f_r = \frac{1}{2\pi\sqrt{LC}} = \frac{1}{2\pi\sqrt{(100\text{ mH})(0.01\ \mu\text{F})}} = 5.03\text{ kHz}$$

$$X_L = 2\pi f_r L = 2\pi(5.03\text{ kHz})(100\text{ mH}) = 3160\ \Omega$$

$$Q = \frac{X_L}{R} = \frac{X_L}{R_1 + R_W} = \frac{3160\ \Omega}{77\ \Omega} = 41$$

$$BW = \frac{f_r}{Q} = \frac{5.03\text{ kHz}}{41} = \mathbf{123\ Hz}$$

Related Problem What happens to V_{out} if the frequency is increased above resonance? Below resonance?

Open file E13-13 on your EWB/CircuitMaker CD-ROM. Verify the calculated value of the resonant frequency and then measure the output voltage at resonance. Determine the bandwidth by measurement.

SECTION 13–4 REVIEW

1. The output voltage of a certain band-pass filter is 15 V at the resonant frequency. What is its value at the cutoff frequencies?
2. For a certain band-pass filter, f_r = 120 kHz and Q = 12. What is the bandwidth of the filter?
3. In a band-stop filter, is the current minimum or maximum at resonance? Is the output voltage minimum or maximum at resonance?

13–5 PARALLEL *RLC* CIRCUITS

In this section, you will learn how to determine the impedance and phase angle of a parallel *RLC* circuit. The current relationships are also covered.

After completing this section, you should be able to

- **Analyze parallel *RLC* circuits**
- Calculate the impedance
- Calculate the phase angle
- Determine all currents
- Convert from a series-parallel *RLC* circuit to an equivalent parallel form

Impedance and Phase Angle

The circuit in Figure 13–30 consists of the parallel combination of *R*, *L*, and *C*. To find the admittance, add the conductance (G) and the total susceptance (B_{tot}) as phasor quantities. B_{tot} is the difference of the inductive susceptance and the capacitive susceptance. Thus,

Equation 13–12

$$B_{tot} = |B_L - B_C|$$

$$Y = \sqrt{G^2 + B_{tot}^2}$$

The total impedance is the reciprocal of the admittance.

$$Z_{tot} = \frac{1}{Y}$$

The phase angle of the circuit is given by the following formula:

Equation 13–13

$$\theta = \tan^{-1}\left(\frac{B_{tot}}{G}\right)$$

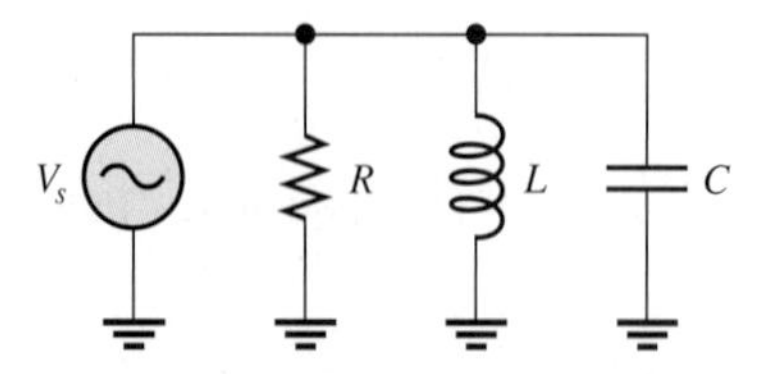

FIGURE 13–30

Parallel *RLC* circuit.

When the frequency is above its resonant value ($X_C < X_L$), the impedance of the circuit in Figure 13–30 is predominantly capacitive because the capacitive current is greater, and the total current leads the source voltage. When the frequency is below its resonant value ($X_L < X_C$), the impedance of the circuit is predominantly inductive, and the total current lags the source voltage.

EXAMPLE 13–14

Determine the total impedance and the phase angle in Figure 13–31.

FIGURE 13–31

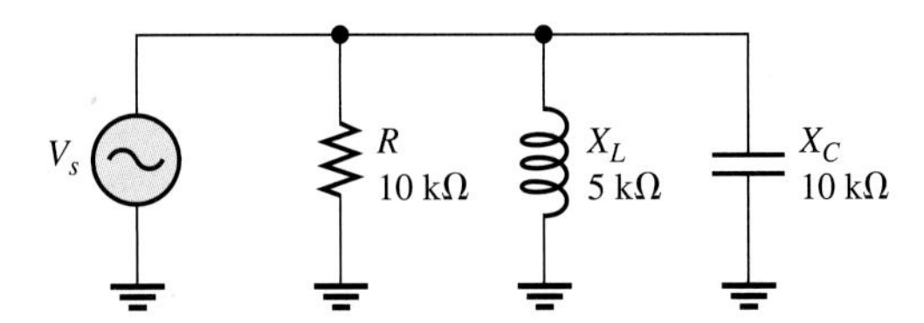

Solution First, determine the admittance as follows:

$$G = \frac{1}{R} = \frac{1}{10\text{ k}\Omega} = 100\ \mu\text{S}$$

$$B_C = \frac{1}{X_C} = \frac{1}{10\text{ k}\Omega} = 100\ \mu\text{S}$$

$$B_L = \frac{1}{X_L} = \frac{1}{5\text{ k}\Omega} = 200\ \mu\text{S}$$

$$B_{tot} = |B_L - B_C| = 100\ \mu\text{S}$$

$$Y = \sqrt{G^2 + B_{tot}^2} = \sqrt{(100\ \mu\text{S})^2 + (100\ \mu\text{S})^2} = 141.4\ \mu\text{S}$$

From Y, you can get Z_{tot}.

$$Z_{tot} = \frac{1}{Y} = \frac{1}{141.4\ \mu\text{S}} = \mathbf{7.07\ k\Omega}$$

The phase angle is

$$\theta = \tan^{-1}\left(\frac{B_{tot}}{G}\right) = \tan^{-1}\left(\frac{100\ \mu\text{S}}{100\ \mu\text{S}}\right) = \mathbf{45^\circ}$$

Related Problem Is the impedance of the circuit in Figure 13–31 predominantly inductive or predominantly capacitive?

Current Relationships

In a parallel *RLC* circuit, the current in the capacitive branch and the current in the inductive branch are always 180° out of phase with each other (neglecting any coil resistance). For this reason, I_C and I_L subtract from each other, and thus the total current into the parallel branches of L and C is always less than the largest individual branch current, as illustrated in Figure 13–32 and in the waveform diagram of Figure 13–33. Of course, the current in the resistive branch is always 90° out of phase with both reactive currents, as shown in the current phasor diagram of Figure 13–34. The total current can be expressed as

$$I_{tot} = \sqrt{I_R^2 + I_{CL}^2}$$

Equation 13–14

where I_{CL} is the absolute value of the difference of the two currents, $|I_C - I_L|$, and is the total current into the L and C branches.

The phase angle can also be expressed in terms of the branch currents as

$$\theta = \tan^{-1}\left(\frac{I_{CL}}{I_R}\right)$$

Equation 13–15

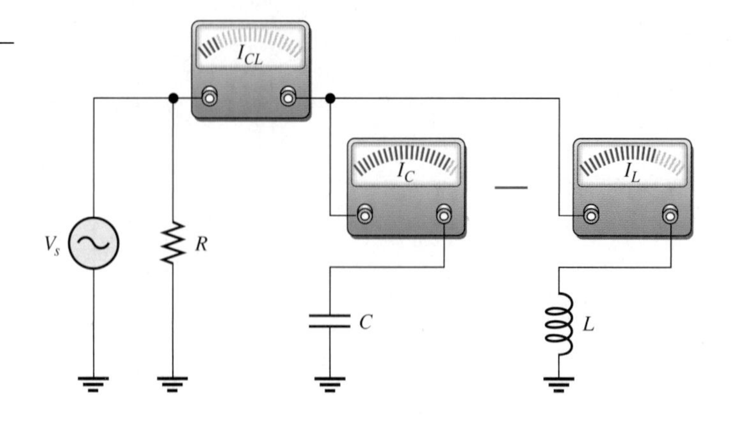

▶ FIGURE 13–32

The total current into the parallel combination of C and L is the difference of the two branch currents ($I_{CL} = |I_C - I_L|$).

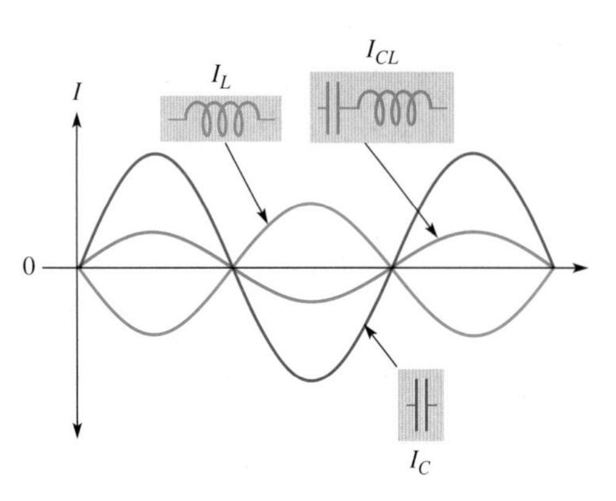

▲ FIGURE 13–33

I_C and I_L effectively subtract.

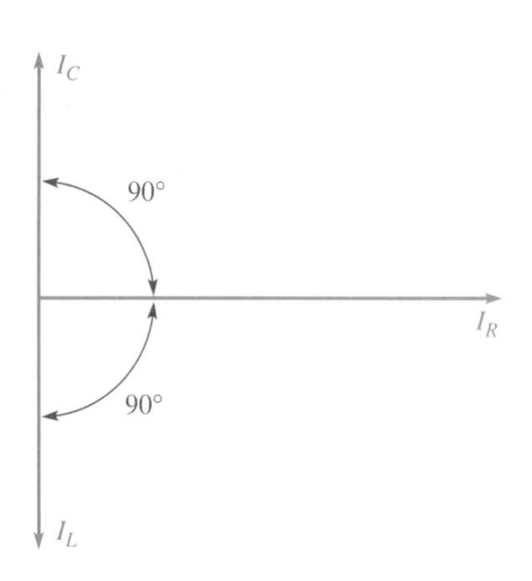

▲ FIGURE 13–34

Current phasor diagram for a parallel *RLC* circuit.

EXAMPLE 13–15

Find each branch current and the total current in Figure 13–35. Draw a diagram of their relationship.

▶ FIGURE 13–35

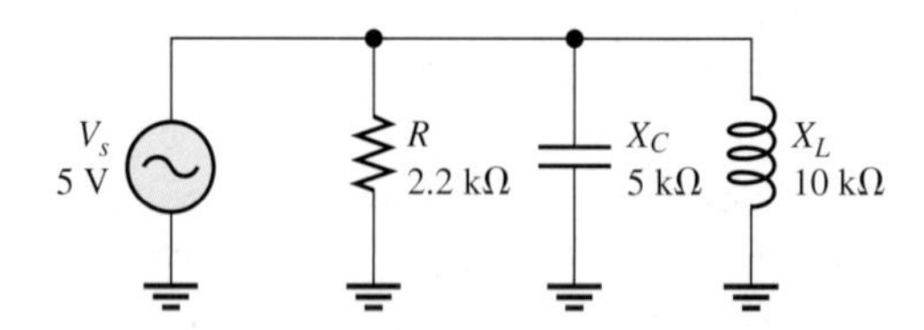

Solution Apply Ohm's law to find each branch current.

$$I_R = \frac{V_s}{R} = \frac{5\text{ V}}{2.2\text{ k}\Omega} = \mathbf{2.27\text{ mA}}$$

$$I_C = \frac{V_s}{X_C} = \frac{5\text{ V}}{5\text{ k}\Omega} = \mathbf{1\text{ mA}}$$

$$I_L = \frac{V_s}{X_L} = \frac{5\text{ V}}{10\text{ k}\Omega} = \mathbf{0.5\text{ mA}}$$

The total current is the phasor sum of the branch currents.

$$I_{CL} = |I_C - I_L| = 0.5\text{ A}$$

$$I_{tot} = \sqrt{I_R^2 + I_{CL}^2} = \sqrt{(2.27\text{ mA})^2 + (0.5\text{ mA})^2} = \mathbf{2.32\text{ mA}}$$

The phase angle is

$$\theta = \tan^{-1}\left(\frac{I_{CL}}{I_R}\right) = \tan^{-1}\left(\frac{0.5\text{ mA}}{2.27\text{ mA}}\right) = 12.4°$$

The total current is 2.32 mA, leading V_s by 12.4°. Figure 13–36 is the current phasor diagram for the circuit.

FIGURE 13–36

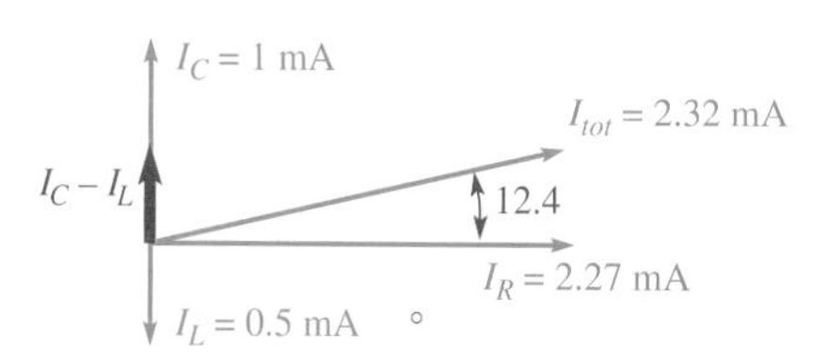

Related Problem Will the total current increase or decrease if the frequency in Figure 13–35 is increased? Why?

Conversion of Series-Parallel to Parallel

The particular series-parallel configuration shown in Figure 13–37 is important because it represents a circuit having parallel L and C branches, with the winding resistance of the coil taken into account as a series resistance in the L branch.

It is helpful to view the series-parallel circuit in Figure 13–37 in an equivalent purely parallel form, as indicated in Figure 13–38. This equivalent form will simplify the analysis of parallel resonant characteristics that will be discussed in Section 13–6.

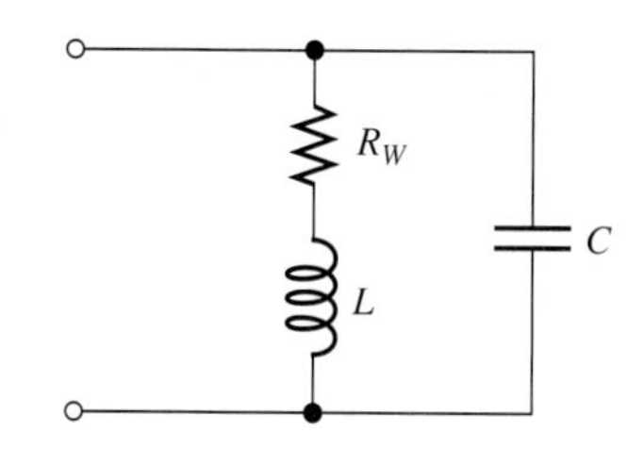

FIGURE 13–37

A series-parallel *RLC* circuit ($Q = X_L/R_W$).

The equivalent inductance, L_{eq}, and the equivalent parallel resistance, $R_{p(eq)}$, are given by the following formulas:

$$L_{eq} = L\left(\frac{Q^2 + 1}{Q^2}\right)$$

Equation 13–16

$$R_{p(eq)} = R_W(Q^2 + 1)$$

Equation 13–17

where Q is the quality factor of the coil, X_L/R_W. Derivations of these formulas are quite involved and thus are not given here. Notice in the equations that for a $Q \geq 10$, the value of L_{eq} is approximately the same as the original value of L. For example, if $L = 10$ mH and $Q = 10$, then

$$L_{eq} = 10\text{ mH}\left(\frac{10^2 + 1}{10^2}\right) = 10\text{ mH}(1.01) = 10.1\text{ mH}$$

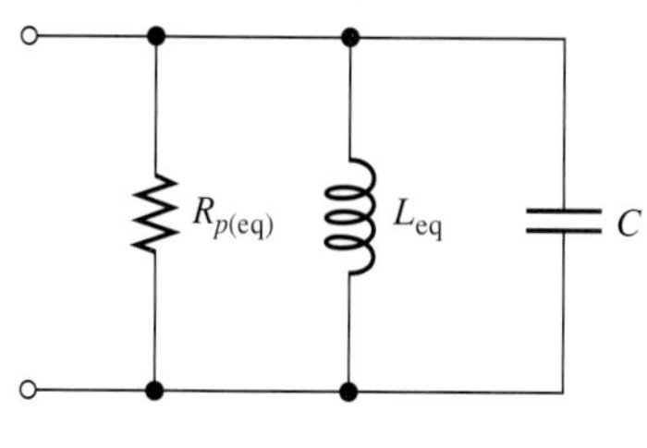

FIGURE 13–38

Parallel equivalent form of the circuit in Figure 13–37.

The equivalency of the two circuits means that at a given frequency, when the same value of voltage is applied to both circuits, the same total current is in both circuits and the phase angles are the same. Basically, an equivalent circuit simply makes circuit analysis more convenient.

EXAMPLE 13–16

Convert the series-parallel circuit in Figure 13–39 to an equivalent parallel form at the given frequency.

▶ FIGURE 13–39

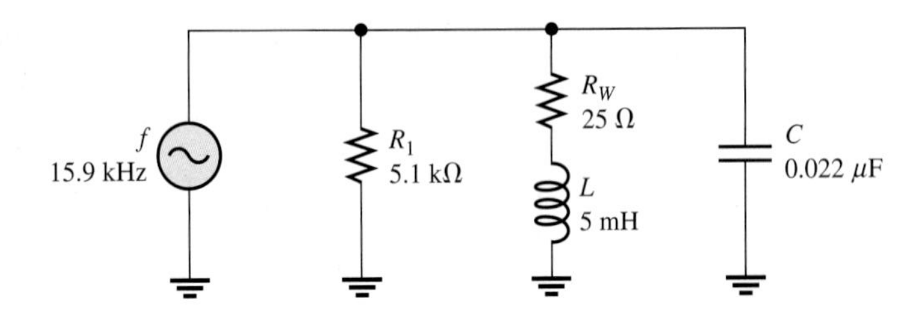

Solution Determine the inductive reactance.

$$X_L = 2\pi fL = 2\pi(15.9 \text{ kHz})(5 \text{ mH}) = 500\ \Omega$$

The Q of the coil is

$$Q = \frac{X_L}{R_W} = \frac{500\ \Omega}{25\ \Omega} = 20$$

Since $Q > 10$, then $L_{eq} \cong L = 5$ mH.

The equivalent parallel resistance is

$$R_{p(eq)} = R_W(Q^2 + 1) = (25\ \Omega)(20^2 + 1) = 10.0 \text{ k}\Omega$$

This equivalent resistance ($R_{p(eq)}$) appears in parallel with R_1 as shown in Figure 13–40(a). When combined, they give a total parallel resistance ($R_{p(tot)}$) of 3.38 kΩ, as indicated in Figure 13–40(b).

▼ FIGURE 13–40

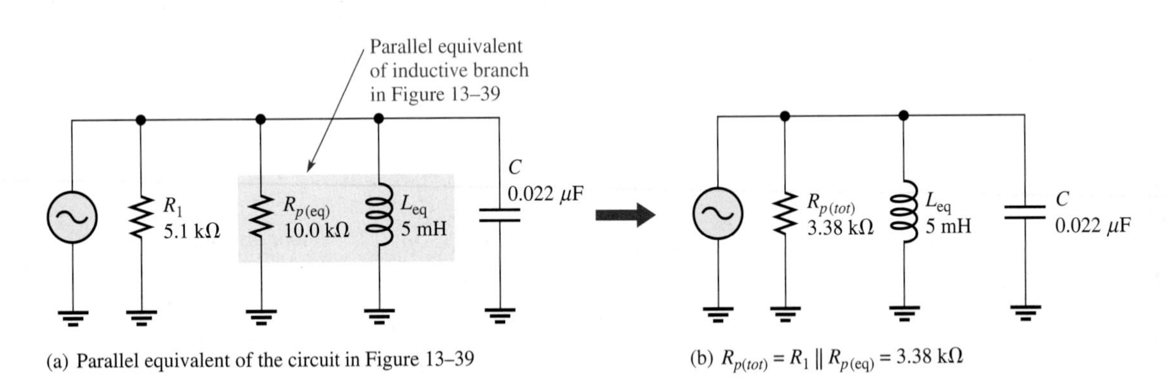

(a) Parallel equivalent of the circuit in Figure 13–39

(b) $R_{p(tot)} = R_1 \parallel R_{p(eq)} = 3.38$ kΩ

Related Problem Find the equivalent parallel circuit if $R_W = 10\ \Omega$ in Figure 13–39.

SECTION 13–5 REVIEW

1. In a three-branch parallel circuit, $R = 150\ \Omega$, $X_C = 100\ \Omega$, and $X_L = 50\ \Omega$ at a certain frequency. Determine the current in each branch when $V_s = 12$ V.
2. Is the circuit in Question 1 capacitive or inductive? Why?
3. Find the equivalent parallel inductance and resistance for a 20 mH coil with a winding resistance of 10 Ω at a frequency of 1 kHz.

13–6 PARALLEL RESONANCE

In this section, we will first look at the resonant condition in an ideal parallel *LC* circuit. Then, we will examine the more realistic case where the resistance of the coil is taken into account.

After completing this section, you should be able to

- **Analyze a circuit for parallel resonance**
- Describe parallel resonance in an ideal circuit
- Describe parallel resonance in a nonideal circuit
- Explain how impedance varies with frequency
- Determine current and phase angle at resonance
- Determine parallel resonant frequency
- Discuss the effects of loading a parallel resonant circuit

Condition for Ideal Parallel Resonance

Ideally, **parallel resonance** occurs when $X_L = X_C$. The frequency at which resonance occurs is called the *resonant frequency,* just as in the series resonant circuit. When $X_L = X_C$, the two branch currents, I_C and I_L, are equal in magnitude, and, of course, they are always 180° out of phase with each other. Thus, the two currents cancel and the total current is zero, as presented in Figure 13–41. In this ideal case, the winding resistance of the coil is assumed to be zero.

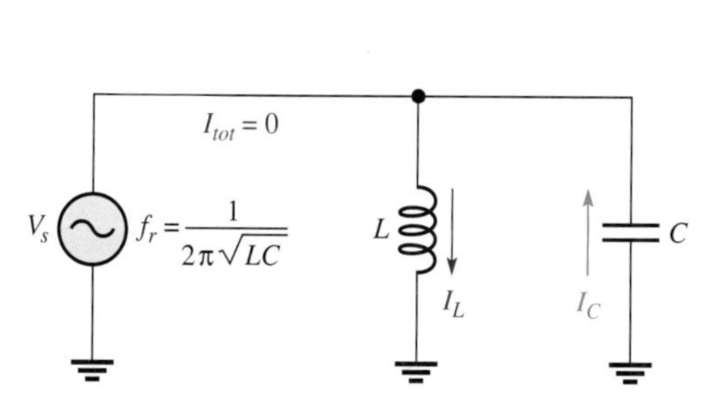

(a) Parallel circuit at resonance ($X_C = X_L$, $Z = \infty$)

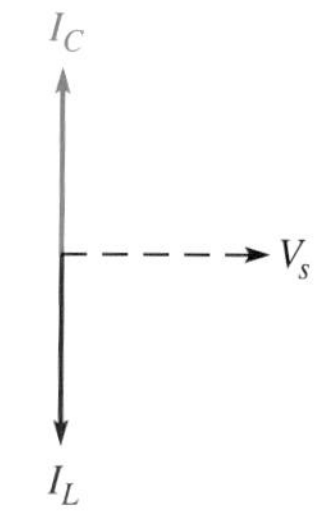

(b) Current phasors

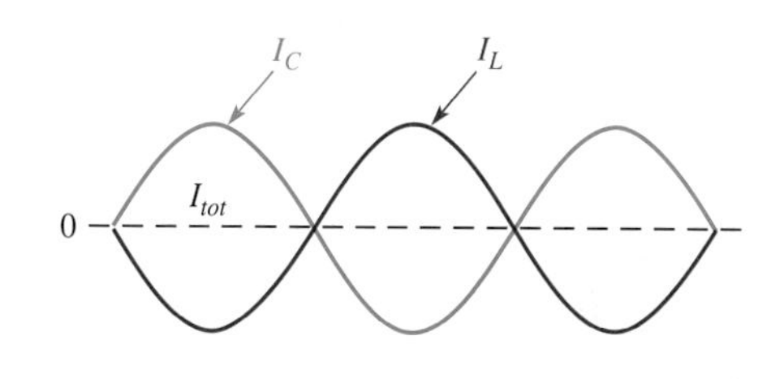

(c) Current waveforms

▼ FIGURE 13–41

An ideal parallel *LC* circuit at resonance.

Since the total current is zero, the impedance of the parallel *LC* circuit is infinitely large (∞). These ideal resonant conditions are stated as follows:

$$X_L = X_C$$

$$Z_r = \infty$$

The Parallel Resonant Frequency

For an ideal parallel resonant circuit, the frequency at which resonance occurs is determined by the same formula as in series resonant circuits.

$$f_r = \frac{1}{2\pi\sqrt{LC}}$$

Equation 13–18

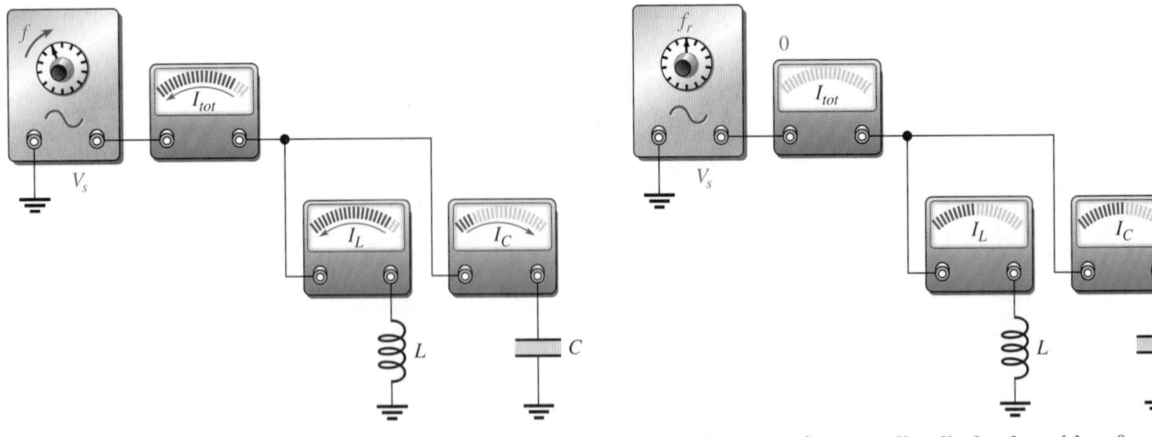

(a) As frequency is increased below resonance: $X_C > X_L$, $I_L > I_C$, I_L decreases, I_C increases, and I_{tot} decreases

(b) At the resonant frequency: $X_C = X_L$, $I_L = I_C$, and $I_{tot} = 0$

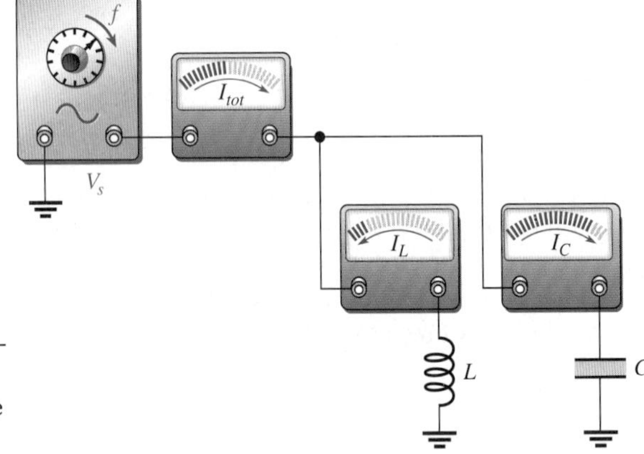

▲ **FIGURE 13–42**

An illustration of how the currents respond in a parallel *LC* circuit as the frequency is varied from below resonance to above resonance. The source voltage amplitude is constant.

(c) As frequency is increased above resonance: $X_L > X_C$, $I_L < I_C$, I_L decreases, I_C increases, and I_{tot} increases

HANDS ON TIP

When you check the resonant frequency of an *LC* circuit using a function generator, the Thevenin resistance of the generator increases the *Q* of the circuit. A generator with an internal resistance of 600 Ω can significantly broaden the bandwidth of the resonant circuit. To reduce the *Q* and obtain a narrower bandwidth, a small parallel resistor can be placed across the generator output. The drawback of this method is that the signal amplitude from the generator is reduced.

Currents in a Parallel Resonant Circuit

It is important to see how the currents in a parallel *LC* circuit vary as the frequency is increased from below the resonant value, through resonance, and then above the resonant value. Figure 13–42 illustrates the general response of an ideal circuit in terms of changes in the currents.

Below the Resonant Frequency Figure 13–42(a) indicates the response as the source frequency is increased from zero toward f_r. At very low frequencies, X_C is very high and X_L is very low, so most of the current is through *L*. As the frequency increases, the current through *L* decreases and the current through *C* increases, causing the total current to decrease. At all times, I_L and I_C are 180° out of phase with each other, and thus the total current is the difference of the two branch currents. During this time, the impedance is increasing, as indicated by the decrease in total current.

At the Resonant Frequency When the frequency reaches its resonant value, f_r, as shown in Figure 13–42(b), X_C and X_L are equal, so I_C and I_L cancel because they are equal in magnitude but opposite in phase. At this point the total current is at its minimum value of zero. Since I_{tot} is zero, *Z* is infinite. Thus, the ideal parallel *LC* circuit appears as an open at the resonant frequency.

Above the Resonant Frequency As the frequency is increased above resonance, as indicated in Figure 13–42(c), X_C continues to decrease and X_L continues to increase, causing the branch currents again to be unequal, with I_C being larger. As a result, total current increases and impedance decreases. As the frequency becomes very high, the impedance becomes very small due to the dominance of a vary small X_C in parallel with a very large X_L.

In summary, the current dips to a minimum as the impedance peaks to a maximum at parallel resonance. The expression for the total current into the L and C branches is

$$I_{tot} = |I_L - I_C|$$

Equation 13–19

EXAMPLE 13–17

Find the resonant frequency and the branch currents in the ideal (winding resistance neglected) parallel LC circuit of Figure 13–43.

▶ **FIGURE 13–43**

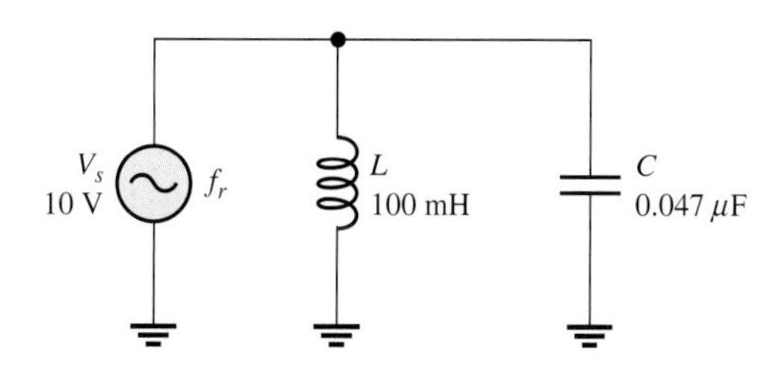

Solution The resonant frequency is

$$f_r = \frac{1}{2\pi\sqrt{LC}} = \frac{1}{2\pi\sqrt{(100\text{ mH})(0.047\ \mu\text{F})}} = \mathbf{2.32\ kHz}$$

Determine the branch currents as follows:

$$X_L = 2\pi f_r L = 2\pi(2.32\text{ kHz})(100\text{ mH}) = 1.46\text{ k}\Omega$$

$$X_C = X_L = 1.46\text{ k}\Omega$$

$$I_L = \frac{V_s}{X_L} = \frac{10\text{ V}}{1.46\text{ k}\Omega} = \mathbf{6.85\ mA}$$

$$I_C = I_L = \mathbf{6.85\ mA}$$

The total current into the parallel LC circuit is

$$I_{tot} = |I_L \quad I_C| = 0\text{ A}$$

Related Problem Find f_r and all the currents at resonance in Figure 13–43 if C is changed to 0.022 μF.

Open file E13-17 on your EWB/CircuitMaker CD-ROM. Determine the resonant frequency by measurement. Determine the currents through L and C at the resonant frequency.

Tank Circuit

The parallel resonant *LC* circuit is often called a **tank circuit.** The term *tank circuit* refers to the fact that the parallel resonant circuit stores energy in the magnetic field of the coil and in the electric field of the capacitor. The stored energy is transferred back and forth between the capacitor and the coil on alternate half-cycles as the current goes first one way and then the other when the inductor deenergizes and the capacitor charges, and vice versa. This concept is illustrated in Figure 13–44.

FIGURE 13–44

Energy storage in an ideal parallel resonant tank circuit.

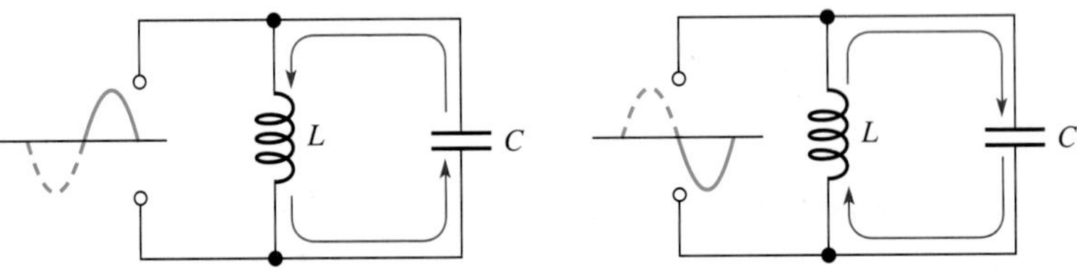

(a) The coil deenergizes as the capacitor charges.

(b) The capacitor discharges as the coil energizes.

Parallel Resonant Conditions in a Nonideal Circuit

So far, the resonance of an ideal parallel *LC* circuit has been examined. Now let's consider resonance in a tank circuit with the resistance of the coil taken into account. Figure 13–45 shows a nonideal tank circuit and its parallel *RLC* equivalent.

FIGURE 13–45

A practical treatment of parallel resonant circuits includes the coil resistance.

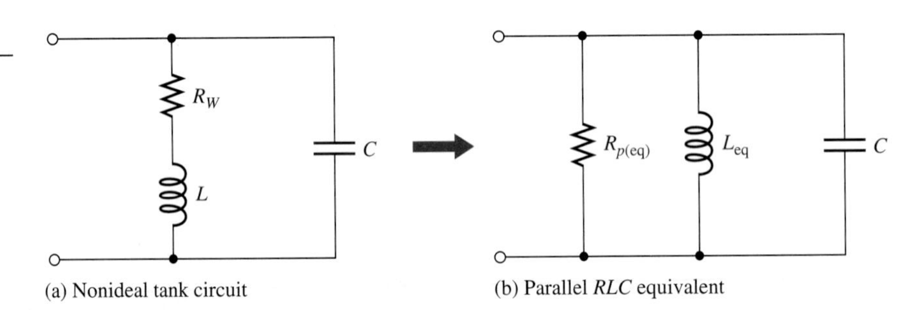

(a) Nonideal tank circuit

(b) Parallel *RLC* equivalent

If the coil's resistance is the only resistance in the circuit, the quality factor, Q, of the circuit at resonance is simply the Q of the coil.

$$Q = \frac{X_L}{R_W}$$

In terms of circuit component values, the Q can also be expressed as

$$Q = \frac{1}{R}\sqrt{\frac{L}{C}}$$

The expressions for the equivalent inductance and the equivalent parallel resistance were given in Equations 13–16 and 13–17 as

$$L_{eq} = L\left(\frac{Q^2 + 1}{Q^2}\right)$$

$$R_{p(eq)} = R_W(Q^2 + 1)$$

Recall that for $Q \geq 10$, $L_{eq} \cong L$.

At parallel resonance,

$$X_{L(eq)} = X_C$$

In the parallel equivalent circuit, $R_{p(eq)}$ is in parallel with an ideal coil and a capacitor, so the L and C branches act as an ideal tank circuit which has an infinite impedance at resonance, as shown in Figure 13–46. Therefore, the total impedance of the nonideal tank circuit at resonance can be expressed as simply the equivalent parallel resistance.

$$Z_r = R_W(Q^2 + 1)$$

Equation 13–20

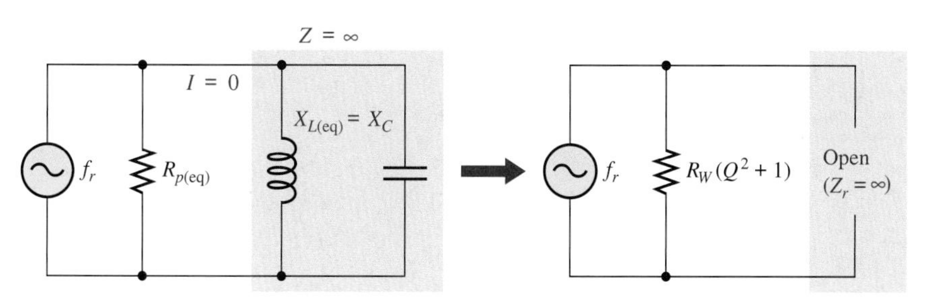

FIGURE 13–46

At resonance, the parallel LC portion appears open and the source sees only $R_{p(eq)}$, which equals $R_W(Q^2 + 1)$.

EXAMPLE 13–18

Determine the impedance of the circuit in Figure 13–47 at the resonant frequency ($f_r \cong 17{,}794$ Hz).

FIGURE 13–47

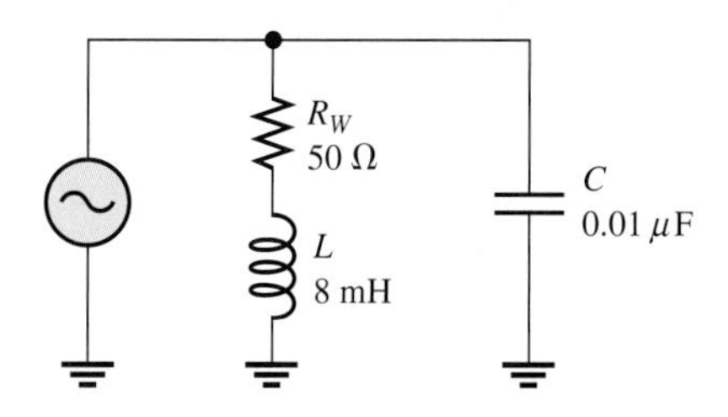

Solution Before you can calculate the impedance using Equation 13–20, you must find the quality factor. To get Q, first find the inductive reactance.

$$X_L = 2\pi f_r L = 2\pi(17{,}794 \text{ Hz})(8 \text{ mH}) = 894 \ \Omega$$

$$Q = \frac{X_L}{R_W} = \frac{894 \ \Omega}{50 \ \Omega} = 17.9$$

$$Z_r = R_W(Q^2 + 1) = 50 \ \Omega(17.9^2 + 1) = \mathbf{16.0 \ k\Omega}$$

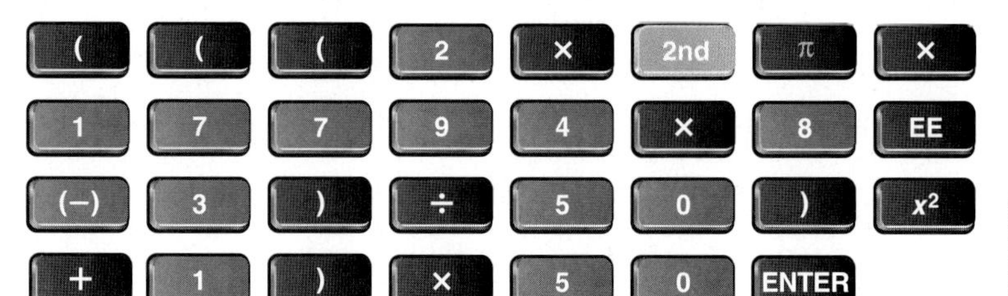

(((2∗π∗17794∗8E⁻3)/50)²+1)∗50
16.0498856512E3

Related Problem Determine Z_r in Figure 13–47 if the winding resistance is 10 Ω.

Variation of the Impedance with Frequency

The impedance of a parallel resonant circuit is maximum at the resonant frequency and decreases at lower and higher frequencies, as indicated by the curve in Figure 13–48.

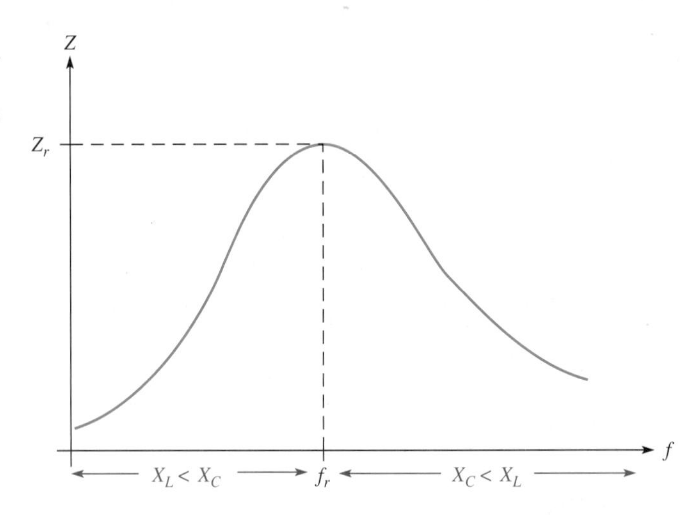

FIGURE 13–48

Generalized impedance curve for a parallel resonant circuit. The circuit is inductive below f_r, resistive at f_r, and capacitive above f_r.

At very low frequencies, X_L is very small and X_C is very high, so the total impedance is essentially equal to that of the inductive branch. As the frequency goes up, the impedance also increases, and the inductive reactance dominates (because it is less than X_C) until the resonant frequency is reached. At this point, of course, $X_L \cong X_C$ (for $Q > 10$) and the impedance is at its maximum. As the frequency goes above resonance, the capacitive reactance dominates (because it is less than X_L) and the impedance decreases.

Current and Phase Angle at Resonance

In the ideal tank circuit, the total current from the source at resonance is zero because the impedance is infinite. In the nonideal case, there is some total current at the resonant frequency, and it is determined by the impedance at resonance.

Equation 13–21

$$I_{tot} = \frac{V_s}{Z_r}$$

The phase angle of the parallel resonant circuit is 0° because the impedance is purely resistive at the resonant frequency.

Parallel Resonant Frequency in a Nonideal Circuit

As you know, when the coil resistance is considered, the resonant condition is

$$X_{L(\text{eq})} = X_C$$

which can be expressed as

$$2\pi f_r L\left(\frac{Q^2 + 1}{Q^2}\right) = \frac{1}{2\pi f_r C}$$

Solving for f_r in terms of Q yields

$$f_r = \frac{1}{2\pi\sqrt{LC}}\sqrt{\frac{Q^2}{Q^2 + 1}}$$

When $Q \geq 10$, the term with the Q factors is approximately 1.

$$\sqrt{\frac{Q^2}{Q^2 + 1}} = \sqrt{\frac{100}{101}} = 0.995 \cong 1$$

Therefore, the parallel resonant frequency is approximately the same as the series resonant frequency as long as Q is equal to or greater than 10.

$$f_r \cong \frac{1}{2\pi\sqrt{LC}} \qquad \text{for } Q \geq 10$$

For the case where the R_W of the coil is the only resistance in the circuit, a precise expression for f_r in terms of the circuit component values is

$$f_r = \frac{\sqrt{1 - (R_W^2 C/L)}}{2\pi\sqrt{LC}}$$

Equation 13–22

This precise formula is seldom necessary and for most practical situations the simpler equation $f_r = 1/(2\pi\sqrt{LC})$ is sufficient. However, the next example illustrates the use of Equation 13–22.

EXAMPLE 13–19

Find the frequency, impedance, and total current at resonance for the circuit in Figure 13–49 using Equation 13–22.

FIGURE 13–49

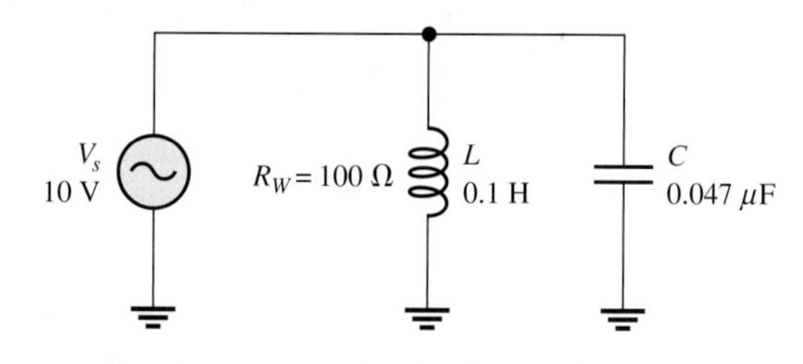

Solution The precise resonant frequency is

$$f_r = \frac{\sqrt{1 - (R_W^2 C/L)}}{2\pi\sqrt{LC}} = \frac{\sqrt{1 - [(100\ \Omega)^2(0.047\ \mu\text{F})/0.1\ \text{H}]}}{2\pi\sqrt{(0.047\ \mu\text{F})(0.1\ \text{H})}} = \mathbf{2.32\ kHz}$$

Calculate impedance as follows:

$$X_L = 2\pi f_r L = 2\pi(2.32\ \text{kHz})(0.1\ \text{H}) = 1.46\ \text{k}\Omega$$

$$Q = \frac{X_L}{R_W} = \frac{1.46\ \text{k}\Omega}{100\ \Omega} = 14.6$$

$$Z_r = R_W(Q^2 + 1) = 100\ \Omega(14.6^2 + 1) = \mathbf{21.4\ k\Omega}$$

The total current is

$$I_{tot} = \frac{V_s}{Z_r} = \frac{10\ \text{V}}{21.4\ \text{k}\Omega} = \mathbf{467\ \mu A}$$

The calculator solution for f_r is

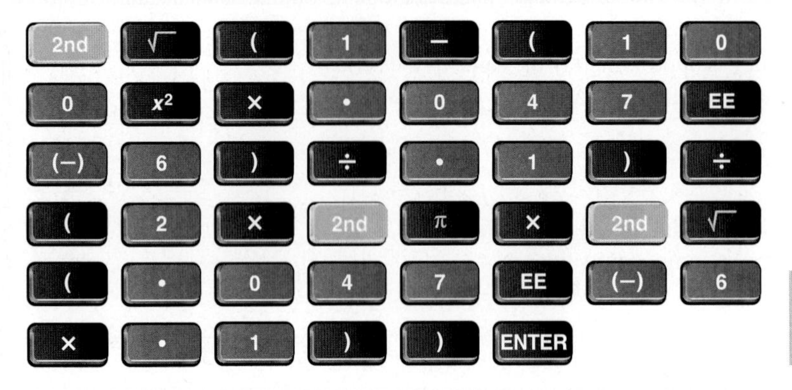

Related Problem Repeat this example using the formula $f_r = 1/(2\pi\sqrt{LC})$ and compare the results.

Open file E13-19 on your EWB/CircuitMaker CD-ROM. Determine the resonant frequency by measurement. Determine the total current and the currents through L and C at the resonant frequency. How do these results compare with the calculated values?

An External Load Resistance Affects a Tank Circuit

In most practical situations in which an external load resistance appears in parallel with a nonideal tank circuit, as shown in Figure 13–50(a). Obviously, the external resistor (R_L) will dissipate more of the energy delivered by the source and thus will lower the overall Q of the circuit. The external resistor effectively appears in parallel with the equivalent parallel resistance of the coil, $R_{p(eq)}$, and both are combined to determine a total parallel resistance, $R_{p(tot)}$, as indicated in Figure 13–50(b).

$$R_{p(tot)} = R_L \parallel R_{p(eq)}$$

FIGURE 13–50

Tank circuit with a load resistor and its equivalent circuit.

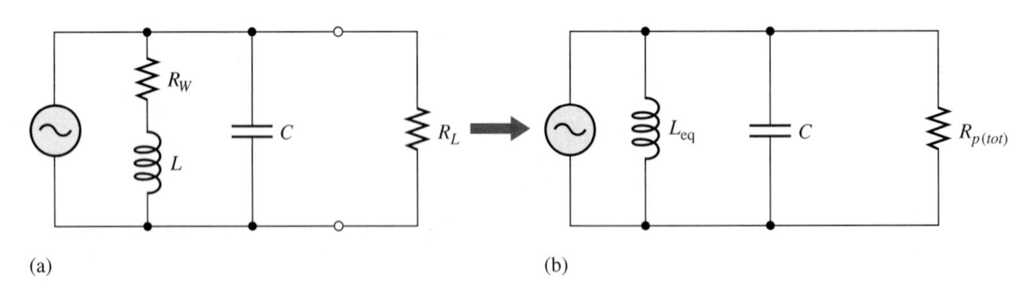

The overall Q, designated Q_O, for a parallel RLC circuit is expressed differently from the Q of a series circuit.

Equation 13–23

$$Q_O = \frac{R_{p(tot)}}{X_{L(eq)}}$$

As you can see, the effect of loading the tank circuit is to reduce its overall Q (which is equal to the coil Q when unloaded).

SECTION 13–6 REVIEW

1. Is the impedance minimum or maximum at parallel resonance?
2. Is the current minimum or maximum at parallel resonance?
3. At ideal parallel resonance, $X_L = 1.5\ \text{k}\Omega$. What is X_C?
4. A tank circuit has the following values: $R_W = 4.0\ \Omega$, $L = 50$ mH, and $C = 10$ pF. Calculate f_r and Z_r.
5. If $Q = 25$, $L = 50$ mH, and $C = 1000$ pF, what is f_r?
6. In Question 5, if $Q = 2.5$, what is f_r?
7. In a certain tank circuit, the coil resistance is $20\ \Omega$. What is the total impedance at resonance if $Q = 20$?

13–7 PARALLEL RESONANT FILTERS

Parallel resonant circuits are commonly applied to band-pass and band-stop filters. In this section, we examine these applications.

After completing this section, you should be able to

- **Analyze the operation of parallel resonant filters**
- Show how a band-pass filter is implemented
- Define *bandwidth*
- Explain how loading affects selectivity
- Show how a band-stop filter is implemented
- Determine the resonant frequency, bandwidth, and output voltage of band-pass and band-stop parallel resonant filters

The Band-Pass Filter

A basic parallel resonant band-pass filter is shown in Figure 13–51. Notice that the output is taken across the tank circuit in this application.

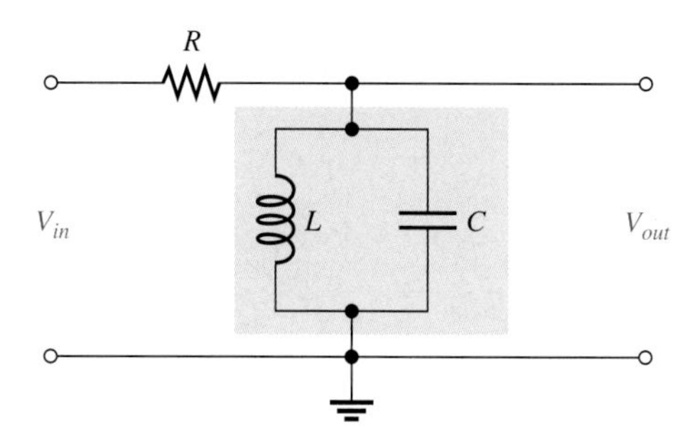

FIGURE 13–51

A basic parallel resonant band-pass filter.

The bandwidth and cutoff frequencies for a parallel resonant band-pass filter are defined in the same way as for the series resonant circuit, and the formulas given in Section 13–4 still apply. General band-pass frequency response curves showing both V_{out} and I_{tot} versus frequency are given in Figures 13–52(a) and 13–52(b), respectively.

The filtering action is as follows: At very low frequencies, the impedance of the tank circuit is very low, and therefore only a small amount of voltage is dropped across it. As the frequency increases, the impedance of the tank circuit

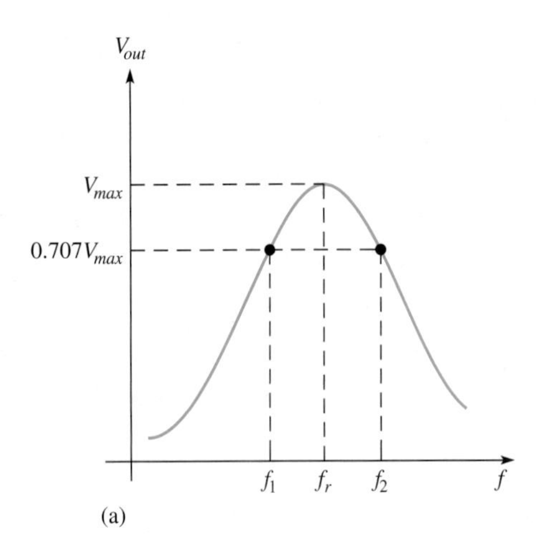

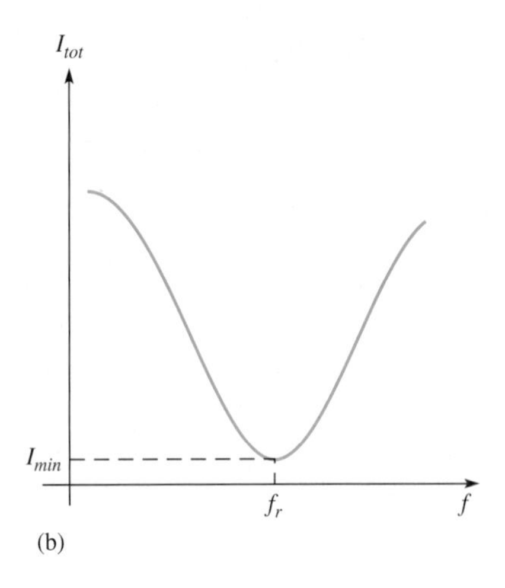

▲ **FIGURE 13–52**

Generalized frequency response curves for a parallel resonant band-pass filter.

increases, and, as a result, the output voltage increases. When the frequency reaches its resonant value, the impedance is at its maximum and so is the output voltage. As the frequency goes above resonance, the impedance begins to decrease, causing the output voltage to decrease. The general response of a parallel resonant band-pass filter is illustrated in Figure 13–53. This illustration pictorially shows how the current and the output voltage change with frequency.

▼ **FIGURE 13–53**

Example of the response of a parallel resonant band-pass filter with the input voltage at a constant 10 V rms.

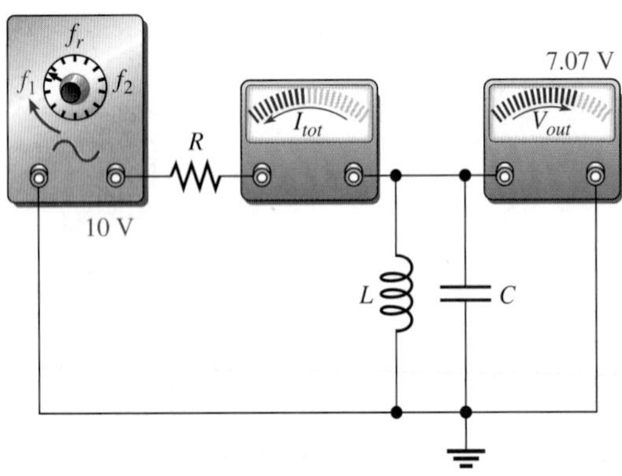

(a) As the frequency increases to f_1, V_{out} increases to 7.07 V, and I_{tot} decreases.

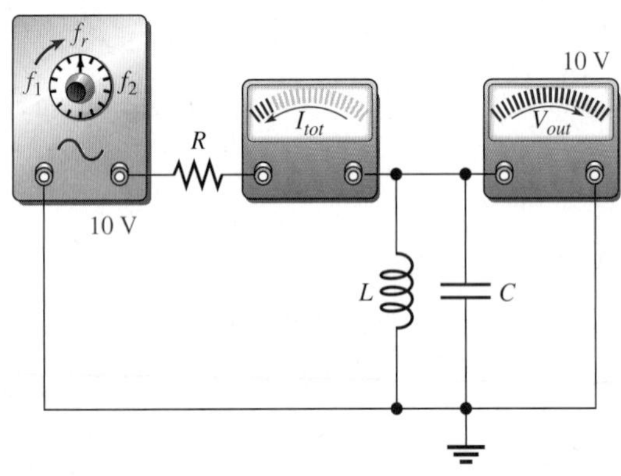

(b) As the frequency increases from f_1 to f_r, V_{out} increases from 7.07 V to 10 V, and I_{tot} decreases to its minimum value.

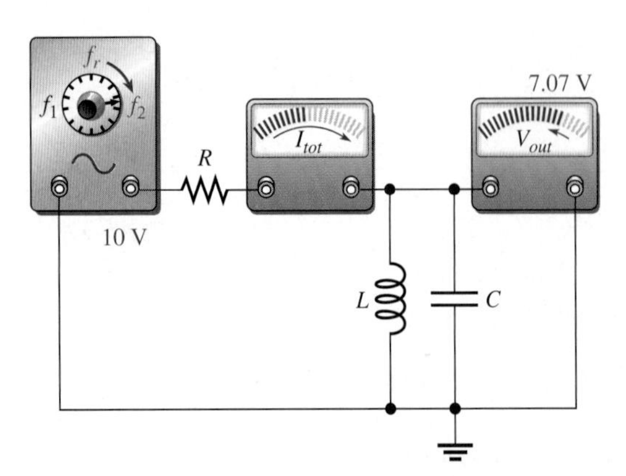

(c) As the frequency increases from f_r to f_2, V_{out} decreases from 10 V to 7.07 V, and I_{tot} increases from its minimum.

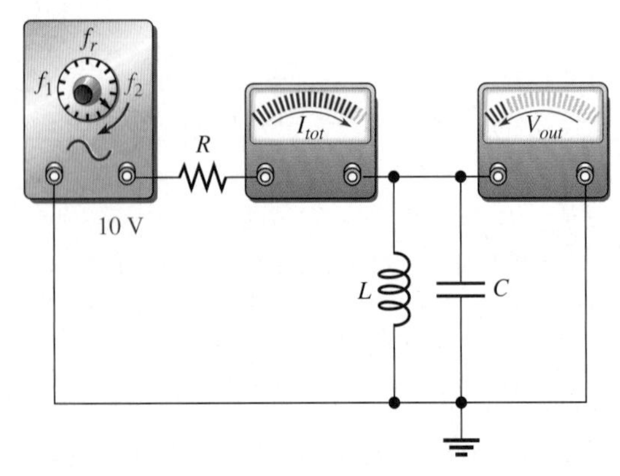

(d) As the frequency increases above f_2, V_{out} decreases below 7.07 V, and I_{tot} continues to increase.

EXAMPLE 13–20

A certain parallel resonant band-pass filter has a maximum output voltage of 4 V at f_r. What is the value of V_{out} at the cutoff frequencies?

Solution V_{out} at the cutoff frequencies is 70.7% of maximum.

$$V_{out(1)} = V_{out(2)} = 0.707V_{out(max)} = 0.707(4\text{ V}) = \mathbf{2.828\text{ V}}$$

Related Problem What is V_{out} at the cutoff frequencies if V_{out} at the resonant frequency is 10 V?

EXAMPLE 13–21

A parallel resonant circuit has a lower cutoff frequency of 3.5 kHz and an upper cutoff frequency of 6 kHz. What is the bandwidth?

Solution

$$BW = f_2 - f_1 = 6\text{ kHz} - 3.5\text{ kHz} = \mathbf{2.5\text{ kHz}}$$

Related Problem A filter has a lower cutoff frequency of 520 kHz and a bandwidth of 10 kHz. What is the upper cutoff frequency?

EXAMPLE 13–22

A certain parallel resonant band-pass filter has a resonant frequency of 12 kHz and a Q of 10. What is its bandwidth?

Solution

$$BW = \frac{f_r}{Q} = \frac{12\text{ kHz}}{10} = \mathbf{1.2\text{ kHz}}$$

Related Problem A certain parallel resonant band-pass filter has a resonant frequency of 100 MHz and a bandwidth of 4 MHz. What is its Q?

Loading Affects the Selectivity of a Parallel Resonant Band-Pass Filter

When a resistive load is connected across the output of a filter as shown in Figure 13–54(a), the Q of the filter is reduced. Since $BW = f_r/Q$, the bandwidth is increased, thus reducing the selectivity. Also, the impedance of the filter at resonance is decreased because R_L effectively appears in parallel with $R_{p(eq)}$. Thus, the maximum output voltage is reduced by the voltage-divider effect of $R_{p(tot)}$ and the internal source resistance R_s, as illustrated in Figure 13–54(b). Figure 13–54(c) shows the general effect of a load on the filter response curve.

FIGURE 13–54

Effects of loading on a parallel resonant band-pass filter.

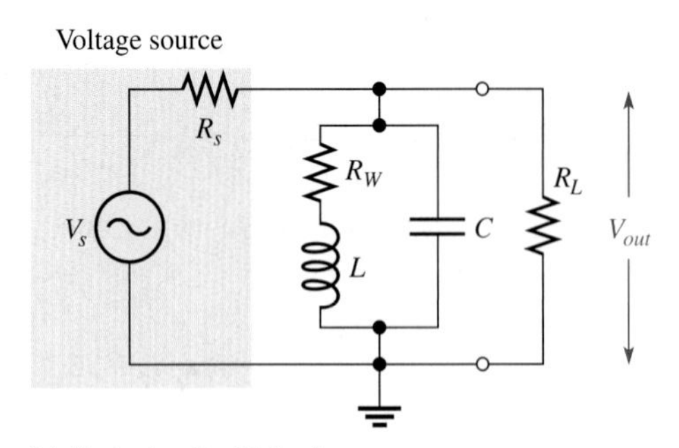

(a) Tank circuit with load

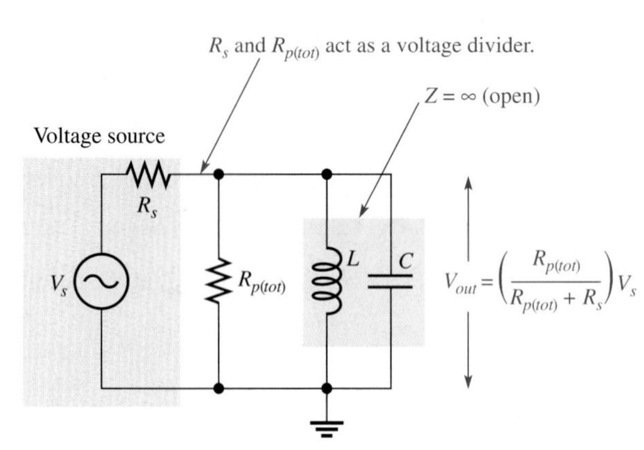

(b) Equivalent circuit

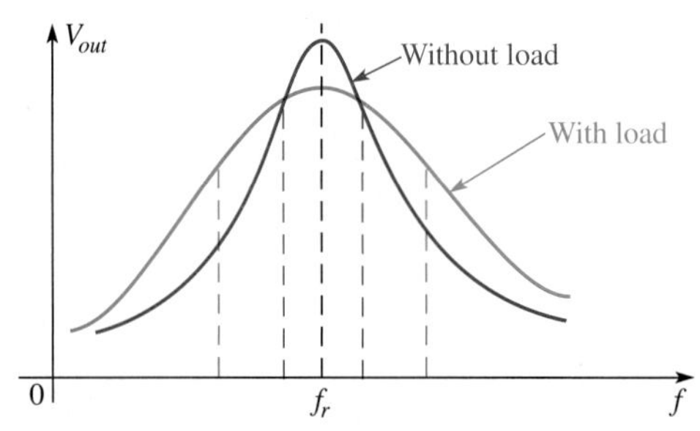

(c) Loading widens the bandwidth and reduces the output.

EXAMPLE 13–23

(a) Determine f_r, BW, and V_{out} at resonance for the unloaded filter in Figure 13–55. The source resistance is 600 Ω.

(b) Repeat part (a) when the filter is loaded with a 50 kΩ resistance, and compare the results.

FIGURE 13–55

Solution **(a)** $f_r \cong \dfrac{1}{2\pi\sqrt{LC}} = \dfrac{1}{2\pi\sqrt{(100\text{ mH})(0.1\ \mu\text{F})}} = \mathbf{1.59\ kHz}$

Calculate the bandwidth at resonance as follows:

$$X_L = 2\pi f_r L = 2\pi(1.59\text{ kHz})(100\text{ mH}) = 999\ \Omega$$

$$Q = \frac{X_L}{R_W} = \frac{999\ \Omega}{50\ \Omega} = 20$$

$$BW = \frac{f_r}{Q} = \frac{1.59\text{ kHz}}{20} = \mathbf{79.5\ Hz}$$

Calculate the output voltage.

$$R_{p(eq)} = R_W(Q^2 + 1) = 50\ \Omega(20^2 + 1) = 20.1\ \text{k}\Omega$$

$$V_{out} = \left(\frac{R_{p(eq)}}{R_{p(eq)} + R_s}\right)V_s = \left(\frac{20.1\ \text{k}\Omega}{20.7\ \text{k}\Omega}\right)5\ \text{V} = \mathbf{4.86\ V}$$

(b) When a 50 kΩ load resistance is connected, you obtain the following values. The resonant frequency is not affected.

$$R_{p(tot)} = R_{p(eq)} \parallel R_L = 20.1\ \text{k}\Omega \parallel 50\ \text{k}\Omega = 14.3\ \text{k}\Omega$$

$$Q_O = \frac{R_{p(tot)}}{X_L} = \frac{14.3\ \text{k}\Omega}{999\ \Omega} = 14.3$$

$$BW = \frac{f_r}{Q_O} = \frac{1.59\ \text{kHz}}{14.3} = \mathbf{111\ Hz}$$

$$V_{out} = \left(\frac{R_{p(tot)}}{R_{p(tot)} + R_s}\right)V_s = \left(\frac{14.3\ \text{k}\Omega}{14.9\ \text{k}\Omega}\right)5\ \text{V} = \mathbf{4.80\ V}$$

The result of adding the load resistance is an increased bandwidth and a decreased output voltage at resonance.

Related Problem How does load resistance affect the Q?

The Band-Stop Filter

A basic parallel resonant band-stop filter is shown in Figure 13–56. The output is taken across a load resistor in series with the tank circuit.

The variation of the tank circuit impedance with frequency produces the familiar current response that has been previously discussed; that is, the current is minimum at resonance and increases on both sides of resonance. Since the output voltage is across the series load resistor, the output voltage follows the current, thus creating the band-stop response characteristic, as indicated in Figure 13–57.

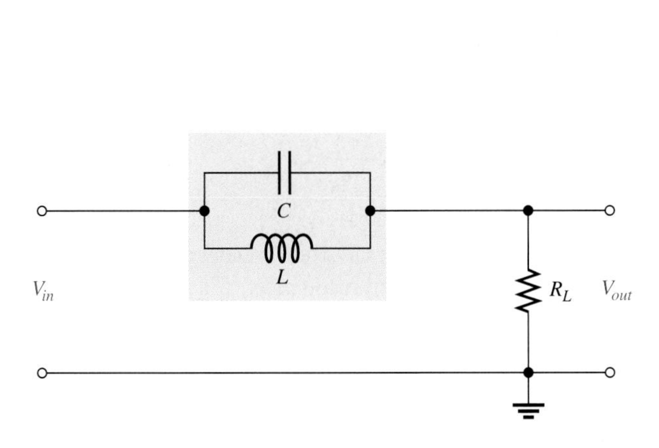

▲ FIGURE 13–56

A basic parallel resonant band-stop filter.

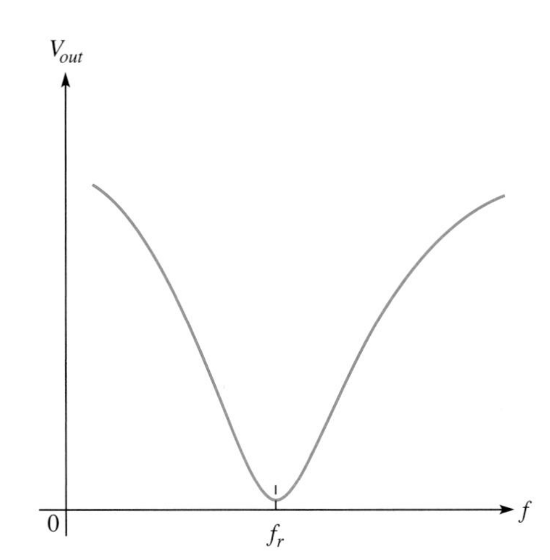

▲ FIGURE 13–57

Band-stop filter response.

Actually, the band-stop filter in Figure 13–56 can be viewed as a voltage divider created by Z_r of the tank and the load resistance. Thus, the output voltage at f_r is

$$V_{out} = \left(\frac{R_L}{R_L + Z_r}\right)V_{in}$$

EXAMPLE 13–24

Find f_r and the output voltage across R_L at resonance for the band-stop filter in Figure 13–58.

▶ FIGURE 13–58

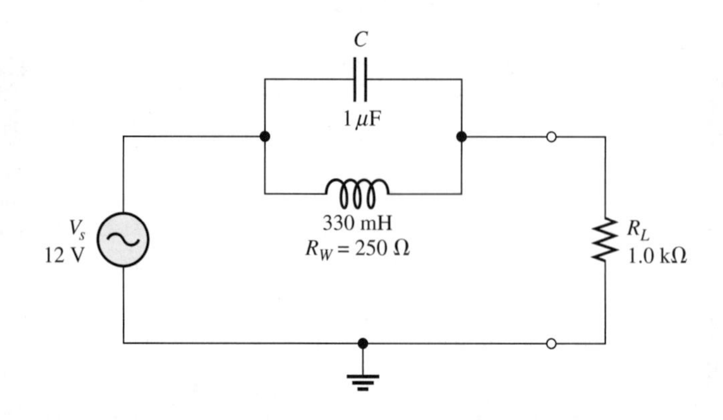

Solution The resonant frequency and impedance are calculated as follows:

$$f_r = \frac{\sqrt{1 - (R_W^2C/L)}}{2\pi\sqrt{LC}} = \mathbf{249\ Hz}$$

$$X_L = 2\pi f_r L = 516\ \Omega$$

$$Q = \frac{X_L}{R_W} = \frac{516\ \Omega}{250\ \Omega} = 2.06$$

$$Z_r = R_W(Q^2 + 1) = 1.31\ \text{k}\Omega$$

At resonance,

$$V_{out} = \left(\frac{R_L}{R_L + Z_r}\right)V_s = \left(\frac{1.0\ \text{k}\Omega}{2.31\ \text{k}\Omega}\right)12\ \text{V} = \mathbf{5.19\ V}$$

Related Problem Why wouldn't the formula for ideal resonant frequency $[f_r = 1/(2\pi\sqrt{LC})]$ work in this particular case?

Open file E13-24 on your EWB/CircuitMaker CD-ROM. Determine the resonant frequency and the voltage across R_L at resonance.

Basic filter configurations have been introduced in this section and in Section 13–4. These basic configurations are sometimes used in combination to increase filter selectivity. For example, a series resonant and a parallel resonant band-pass filter can be used together, with the series circuit connected between the input and output and the parallel circuit connected across the output.

Variable capacitors or inductors are sometimes used in certain filter applications so that the resonant circuits can be tuned over a range of resonant frequencies.

SECTION 13–7 REVIEW

1. How can the bandwidth of a parallel resonant filter be increased?
2. The resonant frequency of a certain high-Q ($Q > 10$) filter is 5 kHz. If the Q is lowered to 2, does f_r change and, if so, to what value?
3. What is the impedance of a tank circuit at its resonant frequency if $R_W = 75\ \Omega$ and $Q = 25$?

13–8 APPLICATIONS

Resonant circuits are used in a wide variety of applications, particularly in communication systems. In this section, we will look briefly at a few common communication systems applications. The purpose in this section is not to explain how the systems work, but to illustrate the importance of resonant circuits in electronic communication.

After completing this section, you should be able to

- **Discuss some applications of resonant circuits**
- Describe a tuned amplifier application
- Describe antenna coupling
- Describe tuned amplifiers
- Describe signal separation in a receiver
- Describe a radio receiver

Tuned Amplifiers

A *tuned amplifier* is a circuit that amplifies signals within a specified band. Typically, a parallel resonant circuit is used in conjunction with an amplifier to achieve the selectivity. In terms of the general operation, input signals with frequencies that range over a wide band are accepted on the amplifier's input and are amplified. The resonant circuit allows only a relatively narrow band of those frequencies to be passed on. The variable capacitor allows tuning over the range of input frequencies so that a desired frequency can be selected, as indicated in Figure 13–59.

FIGURE 13–59

A basic tuned band-pass amplifier.

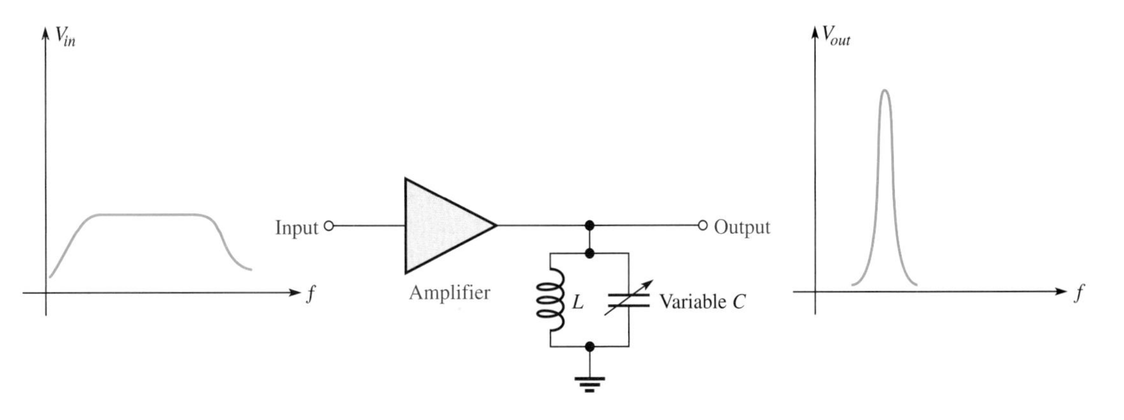

Antenna Input to a Receiver

Radio signals are sent out from a transmitter via electromagnetic waves that propagate through the atmosphere. When the electromagnetic waves cut across the receiving antenna, small voltages are induced. Out of all the wide range of electromagnetic frequencies, only one frequency or a limited band of frequencies must be extracted. Figure 13–60 shows a typical arrangement of an antenna coupled to the receiver input by a transformer. A variable capacitor is connected across the transformer secondary to form a parallel resonant circuit.

FIGURE 13–60

Resonant coupling from an antenna.

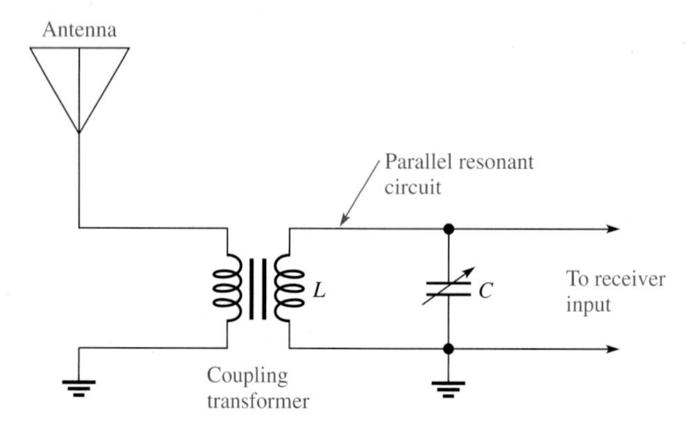

Double-Tuned Transformer Coupling in a Receiver

In some types of communication receivers, tuned amplifiers are transformer-coupled together to increase the amplification. Capacitors can be placed in parallel with the primary and secondary windings of the transformer, effectively creating two parallel resonant band-pass filters that are coupled together. This technique, illustrated in Figure 13–61, can result in a wider bandwidth and steeper slopes on the response curve, thus increasing the selectivity for a desired band of frequencies.

FIGURE 13–61

Double-tuned amplifiers.

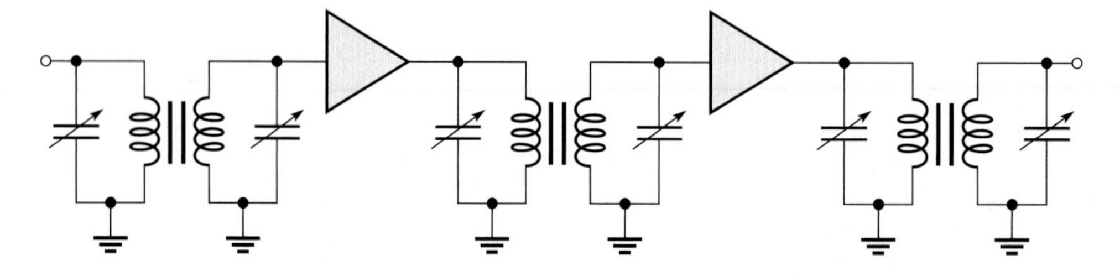

Signal Reception and Separation in a TV Receiver

A television receiver must handle both video (picture) signals and audio (sound) signals. Each TV transmitting station is allotted a 6 MHz bandwidth. Channel 2 is allotted a band from 54 MHz through 59 MHz, channel 3 is allotted a band from 60 MHz through 65 MHz, on up to channel 13, which has a band from 210 MHz through 215 MHz. You can tune the front end of the TV receiver to select any one of these channels by using tuned amplifiers. The signal output of the front end of the receiver has a bandwidth from 41 MHz through 46 MHz, regardless of the channel that is tuned in. This band, called the *intermediate frequency* (IF) band, contains both video and audio. Amplifiers tuned to the IF band boost the signal and feed it to the video amplifier.

Before the output of the video amplifier is applied to the cathode-ray tube, the audio signal is removed by a 4.5 MHz band-stop filter (called a *wave trap*), as shown in Figure 13–62. This trap keeps the sound signal from interfering with the picture. The video amplifier output is also applied to band-pass circuits that are tuned to the sound carrier frequency of 4.5 MHz. The sound signal is then processed and applied to the speaker as indicated in Figure 13–62.

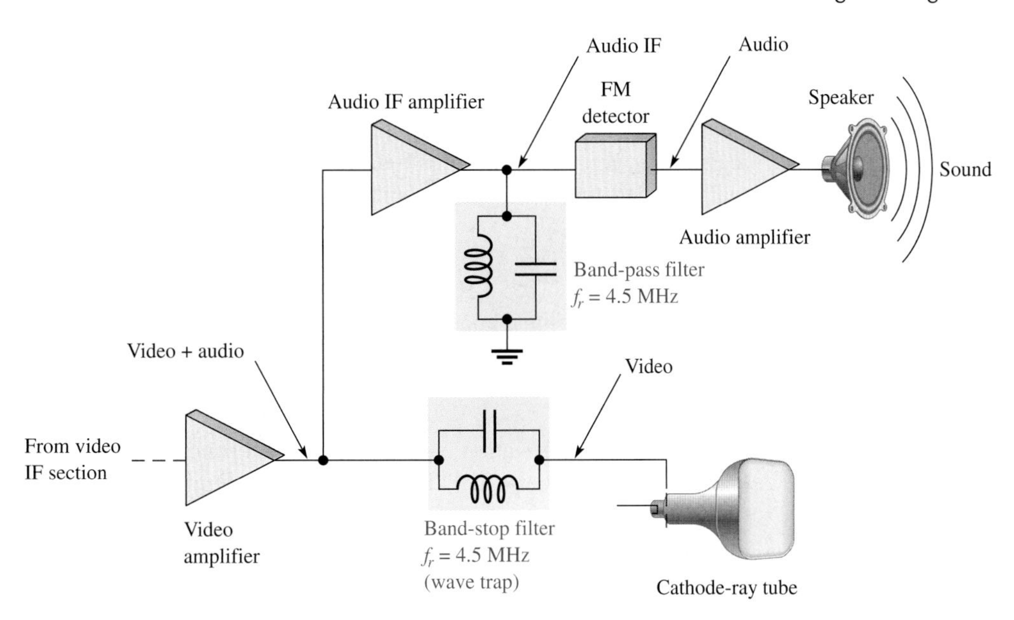

▼ FIGURE 13–62

A simplified portion of a TV receiver showing filter usage.

Superheterodyne Receiver

Another good example of filter applications is in the common AM (amplitude modulation) receiver. The AM broadcast band ranges from 535 kHz to 1605 kHz. Each AM station is assigned a certain narrow bandwidth within that range. A simplified block diagram of a superheterodyne AM receiver is shown in Figure 13–63.

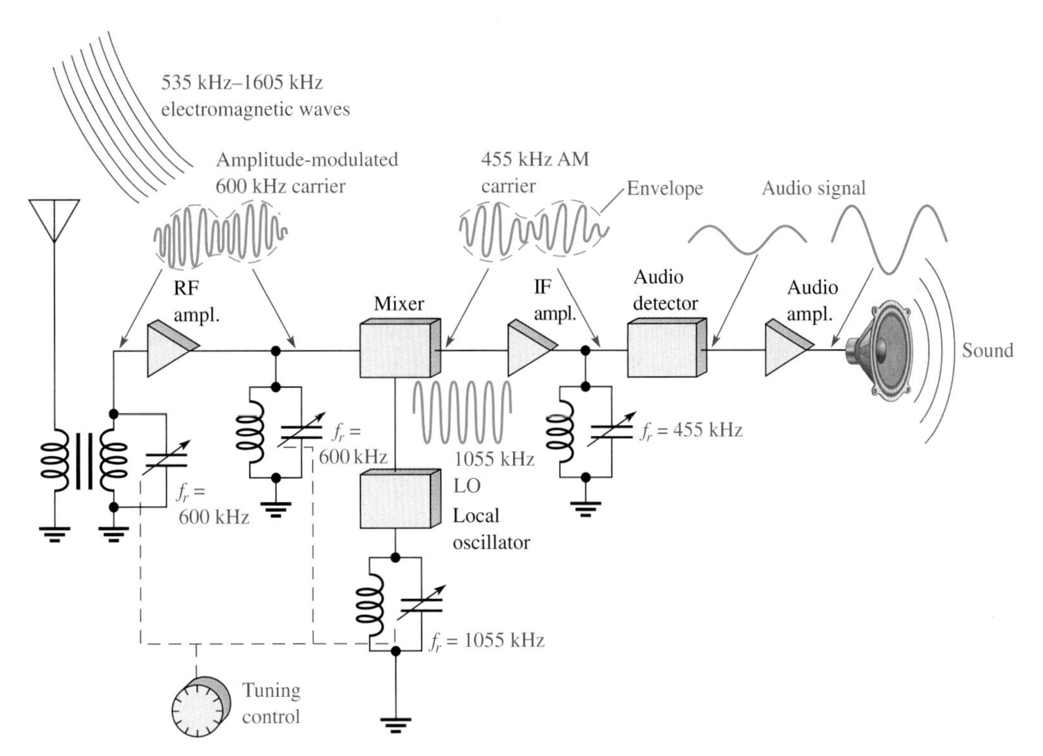

▼ FIGURE 13–63

A simplified diagram of a superheterodyne AM radio broadcast receiver showing the application of tuned resonant circuits.

There are basically three parallel resonant band-pass filters in the front end of the receiver. Each of these filters is gang-tuned by capacitors; that is, the capacitors are mechanically or electronically linked together so that they change together as a station is selected. The front end is tuned to receive a desired station, for example, one that transmits at 600 kHz. The input filter from the antenna and the RF (radio frequency) amplifier filter select only a frequency of 600 kHz out of all the frequencies crossing the antenna.

The actual audio (sound) signal is carried by the 600 kHz carrier frequency by modulating the amplitude of the carrier so that it follows the audio signal as indicated. The variation in the amplitude of the carrier corresponding to the audio signal is called the *envelope.* The 600 kHz is then applied to a circuit called the *mixer.*

The local oscillator (LO) is tuned to a frequency that is 455 kHz above the selected frequency (1055 kHz, in this case). By a process called *heterodyning* or *beating,* the AM signal and the local oscillator signal are mixed together, and the 600 kHz AM signal is converted to a 455 kHz AM signal (1055 kHz − 600 kHz = 455 kHz). The 455 kHz is the intermediate frequency (IF) for standard AM receivers.

No matter which station within the broadcast band is selected, its frequency is always converted to the 455 kHz IF. The amplitude-modulated IF is applied to an audio detector which removes the IF, leaving only the envelope or audio signal. The audio signal is then amplified and applied to the speaker.

SECTION 13–8 REVIEW

1. Generally, why is a tuned filter necessary when a signal is coupled from an antenna to the input of a receiver?
2. What is a wave trap?
3. What is meant by *ganged tuning?*

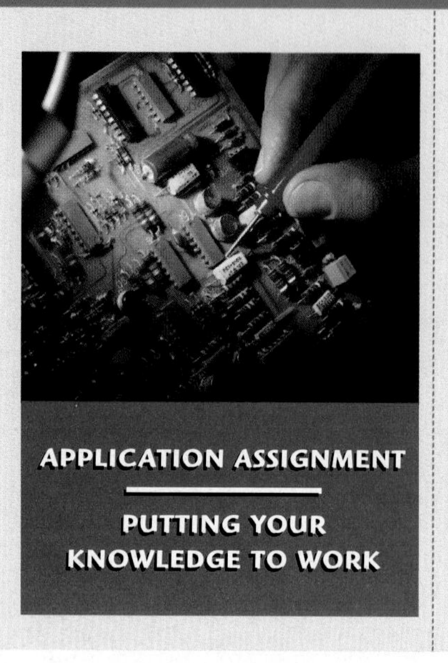

APPLICATION ASSIGNMENT

PUTTING YOUR KNOWLEDGE TO WORK

In the application assignment from the last chapter, you identified two types of filters in sealed housings. In this application assignment, you have another filter to evaluate, which is some type of resonant filter with no specifications available. Your supervisor has requested that you come up with a frequency response curve and determine the resonant frequency and the bandwidth.

In this assignment you will use oscilloscope measurements to determine the frequency response characteristics of the filter. The internal circuitry and component values are of no concern in this case.

Step 1: Measure the Frequency Response

Based on the series of five oscilloscope measurements in Figure 13–64, create a frequency response curve (plot of output voltage versus frequency) for the filter.

Step 2: Analyze the Response Curve

Specify the type of filter and determine the resonant frequency and the bandwidth.

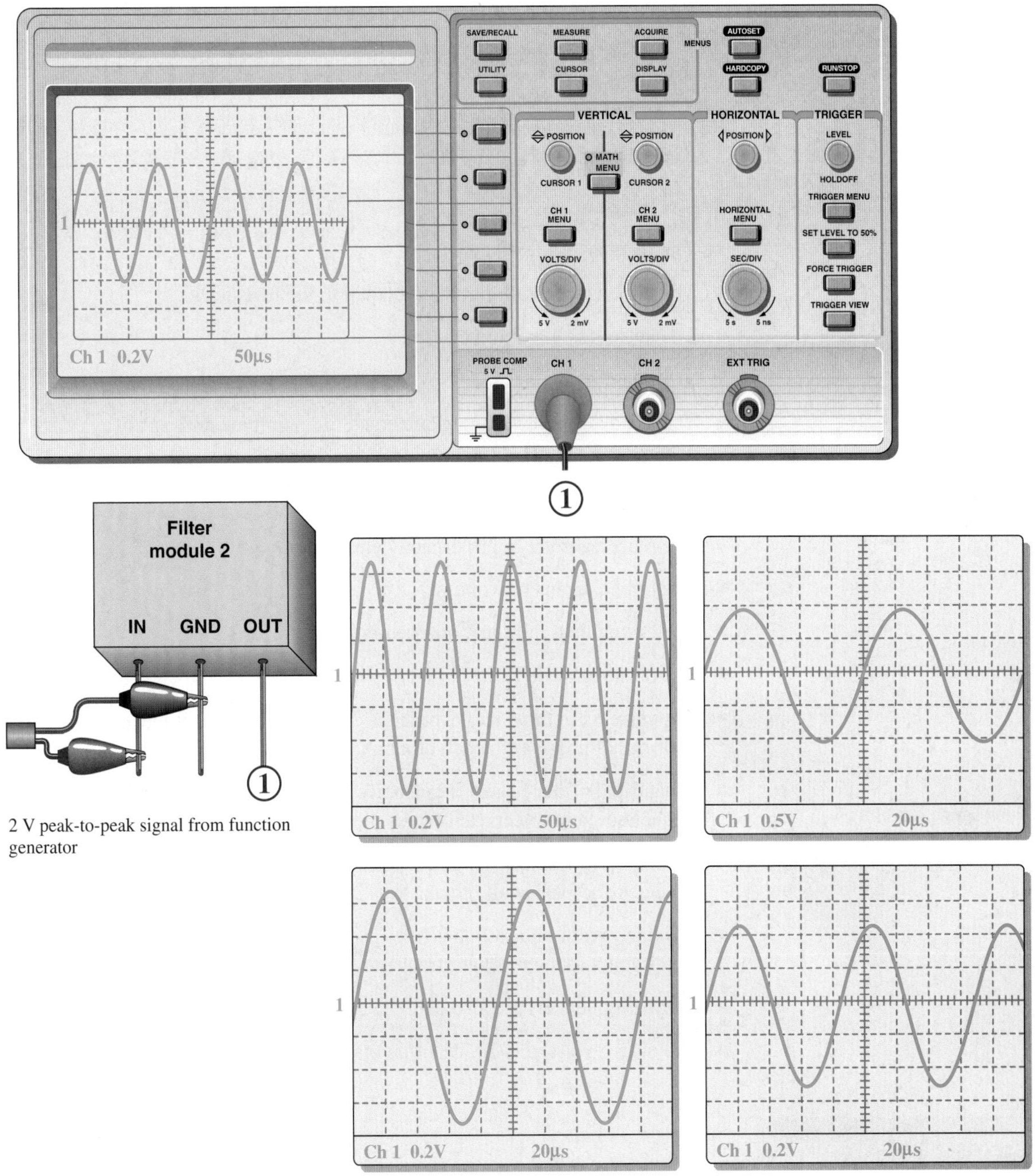

Filter module 2

IN GND OUT

2 V peak-to-peak signal from function generator

Ch 1 0.2V 50μs

Ch 1 0.2V 50μs

Ch 1 0.5V 20μs

Ch 1 0.2V 20μs

Ch 1 0.2V 20μs

FIGURE 13–64

Frequency response measurements.

APPLICATION ASSIGNMENT REVIEW

1. What are the half-power frequencies for the filter in this assignment?
2. From the measurements in Figure 13–64, can you determine the circuit arrangement or component values?

SUMMARY

- X_L and X_C have opposing effects in an *RLC* circuit.
- In a series *RLC* circuit, the larger reactance determines the net reactance of the circuit.
- In a parallel *RLC* circuit, the smaller reactance determines the net reactance of the circuit.

Series Resonance

- The reactances are equal.
- The impedance is minimum and equal to the resistance.
- The current is maximum.
- The phase angle is zero.
- The voltages across L and C are equal in magnitude and, as always, 180° out of phase with each other and thus they cancel.

Parallel Resonance

- The reactances are approximately equal for $Q \geq 10$.
- The impedance is maximum.
- The current is minimum and, ideally, equal to zero.
- The phase angle is zero.
- The currents in the L and C branches are equal in magnitude and, as always, 180° out of phase with each other and thus they cancel.
- A band-pass filter passes frequencies between the lower and upper critical frequencies and rejects all others.
- A band-stop filter rejects frequencies between its lower and upper critical frequencies and passes all others.
- The bandwidth of a resonant filter is determined by the quality factor (Q) of the circuit and the resonant frequency.
- Cutoff frequencies are also called −3 dB frequencies or critical frequencies.
- The output voltage is 70.7% of its maximum at the cutoff frequencies.

EQUATIONS

13–1	$X_{tot} = \lvert X_L - X_C \rvert$	Total series reactance (absolute value)
13–2	$Z_{tot} = \sqrt{R^2 + X_{tot}^2}$	Total series *RLC* impedance
13–3	$\theta = \tan^{-1}\left(\frac{X_{tot}}{R}\right)$	Series *RLC* phase angle
13–4	$X_L = X_C$	Condition for series resonance
13–5	$Z_r = R$	Series resonant impedance
13–6	$f_r = \frac{1}{2\pi\sqrt{LC}}$	Series resonant frequency
13–7	$BW = f_2 - f_1$	Bandwidth

13–8	$\text{dB} = 20 \log\left(\frac{V_{out}}{V_{in}}\right)$	Decibel formula for voltage ratio
13–9	$\text{dB} = 10 \log\left(\frac{P_{out}}{P_{in}}\right)$	Decibel formula for power ratio
13–10	$Q = \frac{X_L}{R}$	Series resonant quality factor
13–11	$BW = \frac{f_r}{Q}$	Bandwidth
13–12	$Y = \sqrt{G^2 + B_{tot}^2}$	Parallel *RLC* admittance
13–13	$\theta = \tan^{-1}\left(\frac{B_{tot}}{G}\right)$	Parallel *RLC* phase angle
13–14	$I_{tot} = \sqrt{I_R^2 + I_{CL}^2}$	Total parallel *RLC* current
13–15	$\theta = \tan^{-1}\left(\frac{I_{CL}}{I_R}\right)$	Parallel *RLC* phase angle
13–16	$L_{eq} = L\left(\frac{Q^2 + 1}{Q^2}\right)$	Equivalent parallel inductance
13–17	$R_{p(eq)} = R_W(Q^2 + 1)$	Equivalent parallel resistance
13–18	$f_r = \frac{1}{2\pi\sqrt{LC}}$	Ideal parallel resonant frequency
13–19	$I_{tot} = \lvert I_L - I_C \rvert$	Total parallel *LC* current (absolute value)
13–20	$Z_r = R_W(Q^2 + 1)$	Impedance at parallel resonance
13–21	$I_{tot} = \frac{V_s}{Z_r}$	Total current at parallel resonance
13–22	$f_r = \frac{\sqrt{1 - (R_W^2 C/L)}}{2\pi\sqrt{LC}}$	Parallel resonant frequency (exact)
13–23	$Q_O = \frac{R_{p(tot)}}{X_{L(eq)}}$	Parallel *RLC Q*

SELF-TEST

Answers are at the end of the chapter.

1. The total reactance of a series *RLC* circuit at resonance is
   (a) zero **(b)** equal to the resistance **(c)** infinity **(d)** capacitive
2. The phase angle of a series *RLC* circuit at resonance is
   (a) $-90°$ **(b)** $+90°$ **(c)** $0°$ **(d)** dependent on the reactance
3. The impedance at the resonant frequency of a series *RLC* circuit with $L = 15$ mH, $C = 0.015\ \mu$F, and $R_W = 80\ \Omega$ is
   (a) 15 kΩ **(b)** 80 Ω **(c)** 30 Ω **(d)** 0 Ω
4. In a series *RLC* circuit that is operating below the resonant frequency, the current
   (a) is in phase with the source voltage **(b)** lags the source voltage
   (c) leads the source voltage
5. If the value of *C* in a series *RLC* circuit is increased, the resonant frequency
   (a) is not affected **(b)** increases
   (c) remains the same **(d)** decreases

6. In a certain series resonant circuit, $V_C = 150$ V, $V_L = 150$ V, and $V_R = 50$ V. The value of the source voltage is

   (a) 150 V (b) 300 V (c) 50 V (d) 350 V

7. A certain series resonant band-pass filter has a bandwidth of 1 kHz. If the existing coil is replaced with one having a lower value of Q, the bandwidth will

   (a) increase (b) decrease
   (c) remain the same (d) be more selective

8. At frequencies below resonance in a parallel *RLC* circuit, the current

   (a) leads the source voltage (b) lags the source voltage
   (c) is in phase with the source voltage

9. The total current into the L and C branches of a parallel circuit at resonance is ideally

   (a) maximum (b) low (c) high (d) zero

10. To tune a parallel resonant circuit to a lower frequency, the capacitance should be

    (a) increased (b) decreased
    (c) left alone (d) replaced with inductance

11. The resonant frequency of a parallel circuit is the same as a series circuit using the same components when

    (a) the Q is very low (b) the Q is very high
    (c) there is no resistance

12. If the resistance in parallel with a parallel resonant filter is reduced, the bandwidth

    (a) disappears (b) decreases
    (c) becomes sharper (d) increases

PROBLEMS

Answers to odd-numbered problems are at the end of the book.

BASIC PROBLEMS

SECTION 13–1 Impedance and Phase Angle of Series *RLC* Circuits

1. A certain series *RLC* circuit operates at a frequency of 5 kHz and has the following values: $R = 10\ \Omega$, $C = 0.05\ \mu$F, and $L = 5$ mH. Determine the impedance and phase angle. What is the total reactance?
2. Find the impedance in Figure 13–65.
3. If the frequency of the source voltage in Figure 13–65 is doubled from the value that produces the indicated reactances, how does the impedance change?

▶ FIGURE 13–65

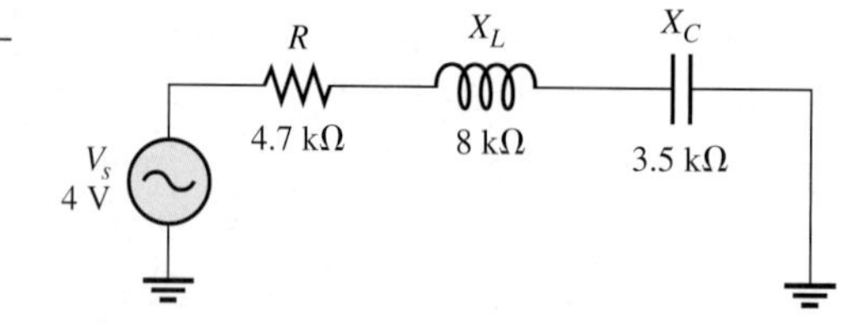

SECTION 13–2 **Analysis of Series *RLC* Circuits**

4. For the circuit in Figure 13–65, find I_{tot}, V_R, V_L, and V_C.
5. Draw the voltage phasor diagram for the circuit in Figure 13–65.
6. Analyze the circuit in Figure 13–66 for the following (f = 25 kHz):

   (a) I_{tot} **(b)** P_{true} **(c)** P_r **(d)** P_a

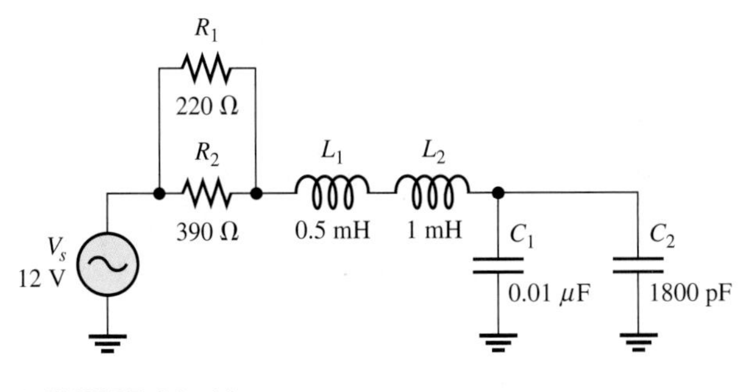

▲ FIGURE 13–66

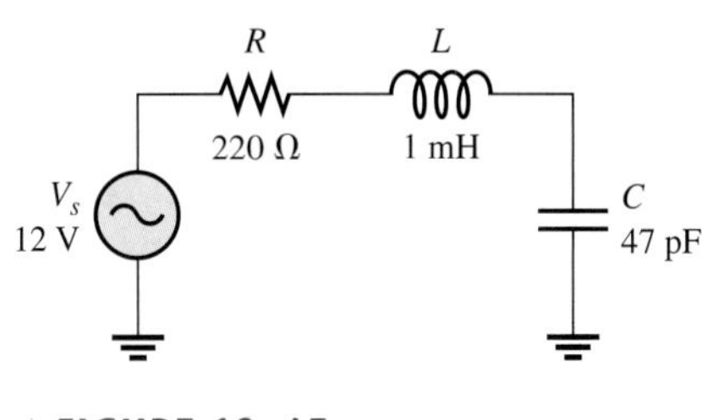

▲ FIGURE 13–67

SECTION 13–3 **Series Resonance**

7. Find X_L, X_C, Z, and I at the resonant frequency in Figure 13–67.
8. A certain series resonant circuit has a maximum current of 50 mA and a V_L of 100 V. The source voltage is 10 V. What is Z? What are X_L and X_C?
9. For the *RLC* circuit in Figure 13–68, determine the resonant frequency and the cutoff frequencies.
10. What is the value of the current at the half-power points in Figure 13–68?

SECTION 13–4 **Series Resonant Filters**

11. Determine the resonant frequency for each filter in Figure 13–69. Are these filters band-pass or band-stop types?
12. Assuming that the coils in Figure 13–69 have a winding resistance of 10 Ω, find the bandwidth for each filter.

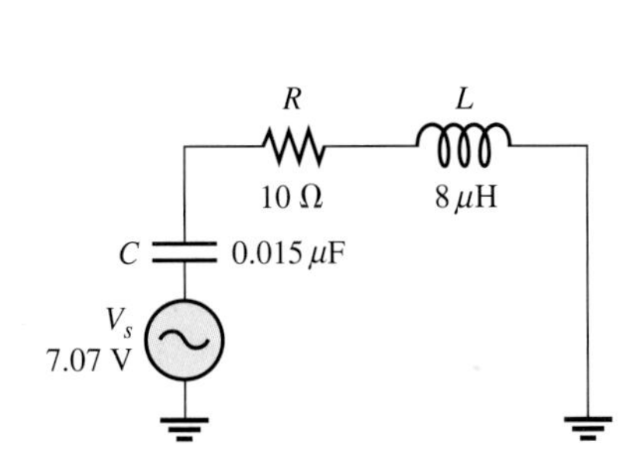

▲ FIGURE 13–68

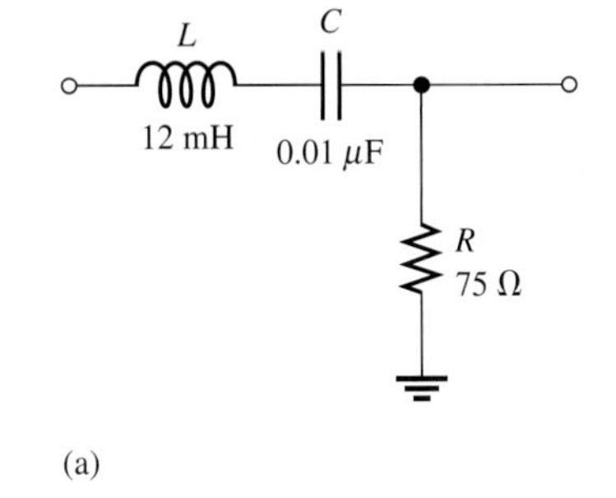

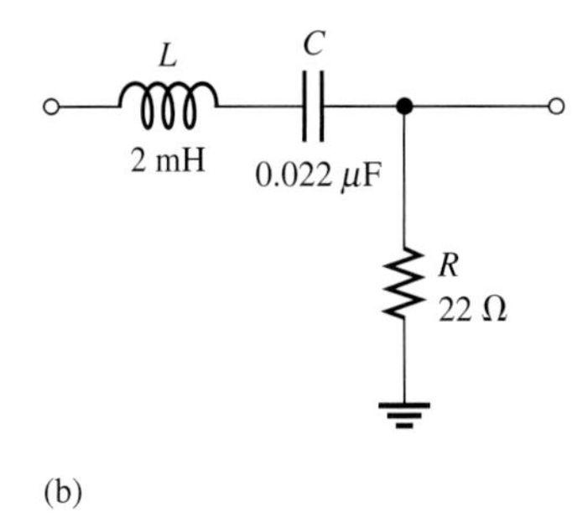

▲ FIGURE 13–69

FIGURE 13–70

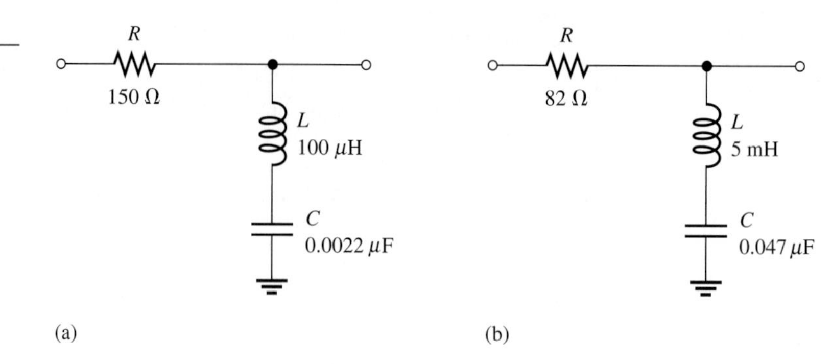

13. Determine f_r and BW for each filter in Figure 13–70.

SECTION 13–5

Parallel *RLC* Circuits

14. Find the total impedance of the circuit in Figure 13–71.
15. Is the circuit in Figure 13–71 capacitive or inductive? Explain.
16. For the circuit in Figure 13–71, find all the currents and voltages.
17. Find the total impedance for the circuit in Figure 13–72.

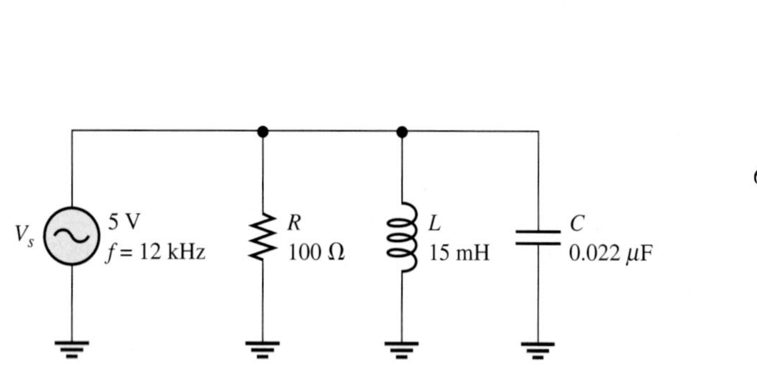

FIGURE 13–71

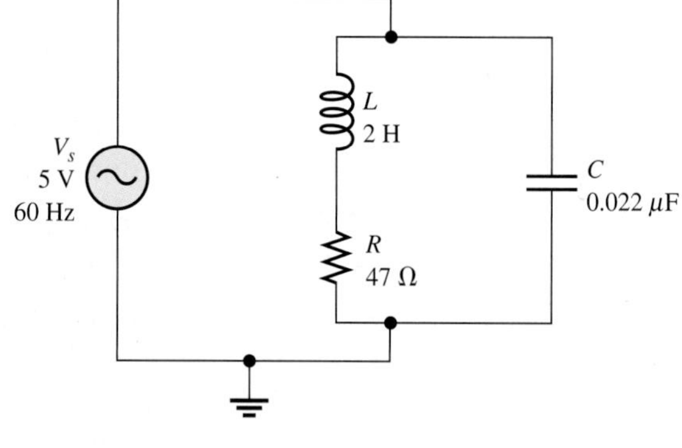

FIGURE 13–72

SECTION 13–6

Parallel Resonance

18. What is the impedance of an ideal parallel resonant circuit (no resistance in either branch)?
19. Find Z at resonance and f_r for the tank circuit in Figure 13–73.
20. How much current is drawn from the source in Figure 13–73 at resonance? What are the inductive current and the capacitive current at the resonant frequency?

FIGURE 13–73

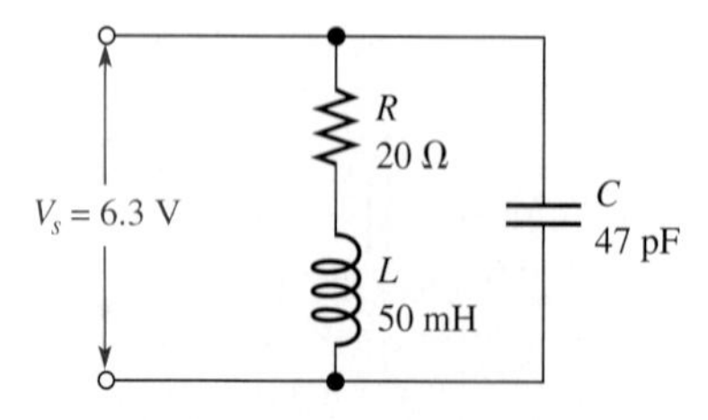

SECTION 13–7

Parallel Resonant Filters

21. At resonance, $X_L = 2\text{ k}\Omega$ and $R_W = 25\ \Omega$ in a parallel resonant band-pass filter. The resonant frequency is 5 kHz. Determine the bandwidth.
22. If the lower cutoff frequency is 2400 Hz and the upper cutoff frequency is 2800 Hz, what is the bandwidth?

23. In a certain resonant circuit, the power at resonance is 2.75 W. What is the power at the lower and upper cutoff frequencies?

24. What values of L and C should be used in a tank circuit to obtain a resonant frequency of 8 kHz? The bandwidth must be 800 Hz. The winding resistance of the coil is 10 Ω.

25. A parallel resonant circuit has a Q of 50 and a BW of 400 Hz. If Q is doubled, what is the bandwidth for the same f_r?

ADVANCED PROBLEMS

26. For each following case, express the voltage ratio in decibels:

(a) $V_{in} = 1$ V, $V_{out} = 1$ V **(b)** $V_{in} = 5$ V, $V_{out} = 3$ V

(c) $V_{in} = 10$ V, $V_{out} = 7.07$ V **(d)** $V_{in} = 25$ V, $V_{out} = 5$ V

27. Find the current through each component in Figure 13–74. Find the voltage across each component.

28. Determine whether there is a value of C that will make $V_{ab} = 0$ V in Figure 13–75. If not, explain.

29. If the value of C is 0.22 μF, how much current is through each branch in Figure 13–75? What is the total current?

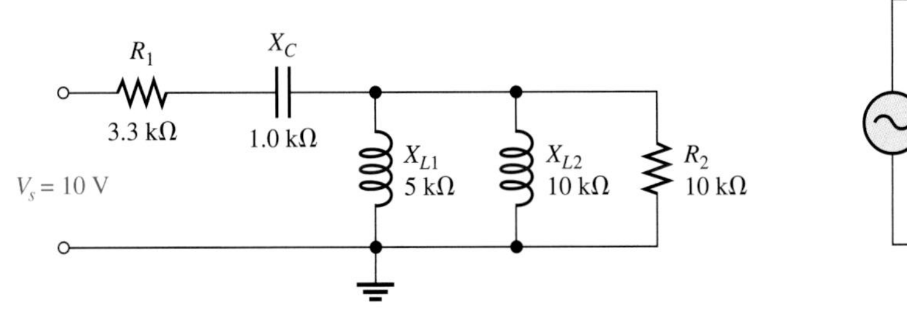

▲ FIGURE 13–74

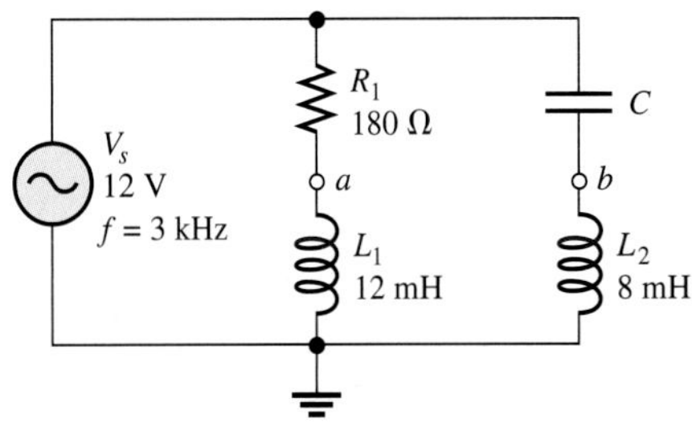

▲ FIGURE 13–75

30. Determine the resonant frequencies in Figure 13–76 and find V_{out} at each resonant frequency.

▶ FIGURE 13–76

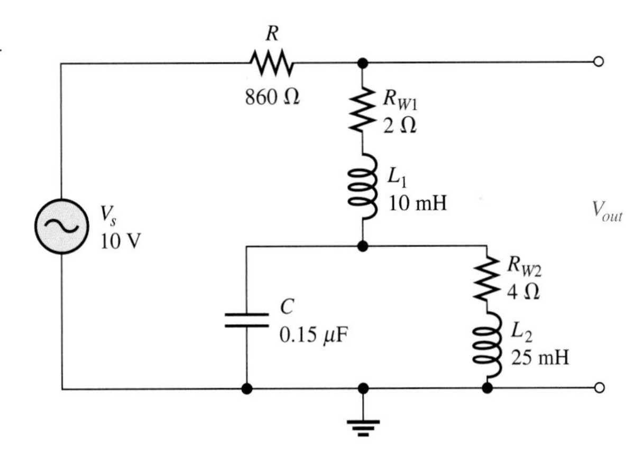

31. Design a band-pass filter using a parallel resonant circuit to meet the following specifications: $BW = 500$ Hz, $Q = 40$, $I_{C(max)} = 20$ mA, $V_{C(max)} = 2.5$ V.

32. Design a circuit in which the following series resonant frequencies are switch-selectable: 500 kHz, 1000 kHz, 1500 kHz, 2000 kHz.

33. Design a parallel-resonant network using a single coil and switch-selectable capacitors to produce the following resonant frequencies: 8 MHz, 9 MHz, 10 MHz, and 11 MHz. Assume a 10 μH coil with a winding resistance of 5 Ω.

ELECTRONICS WORKBENCH/CIRCUITMAKER TROUBLESHOOTING PROBLEMS

CD-ROM file circuits are shown in Figure 13–77.

34. Open file P13-34. Determine if there is a fault and, if so, identify it.

35. Open file P13-35. Determine if there is a fault and, if so, identify it.

36. Open file P13-36. Determine if there is a fault and, if so, identify it.

37. Open file P13-37. Determine if there is a fault and, if so, identify it.

38. Open file P13-38. Determine if there is a fault and, if so, identify it.

39. Open file P13-39. Determine if there is a fault and, if so, identify it.

▼ FIGURE 13–77

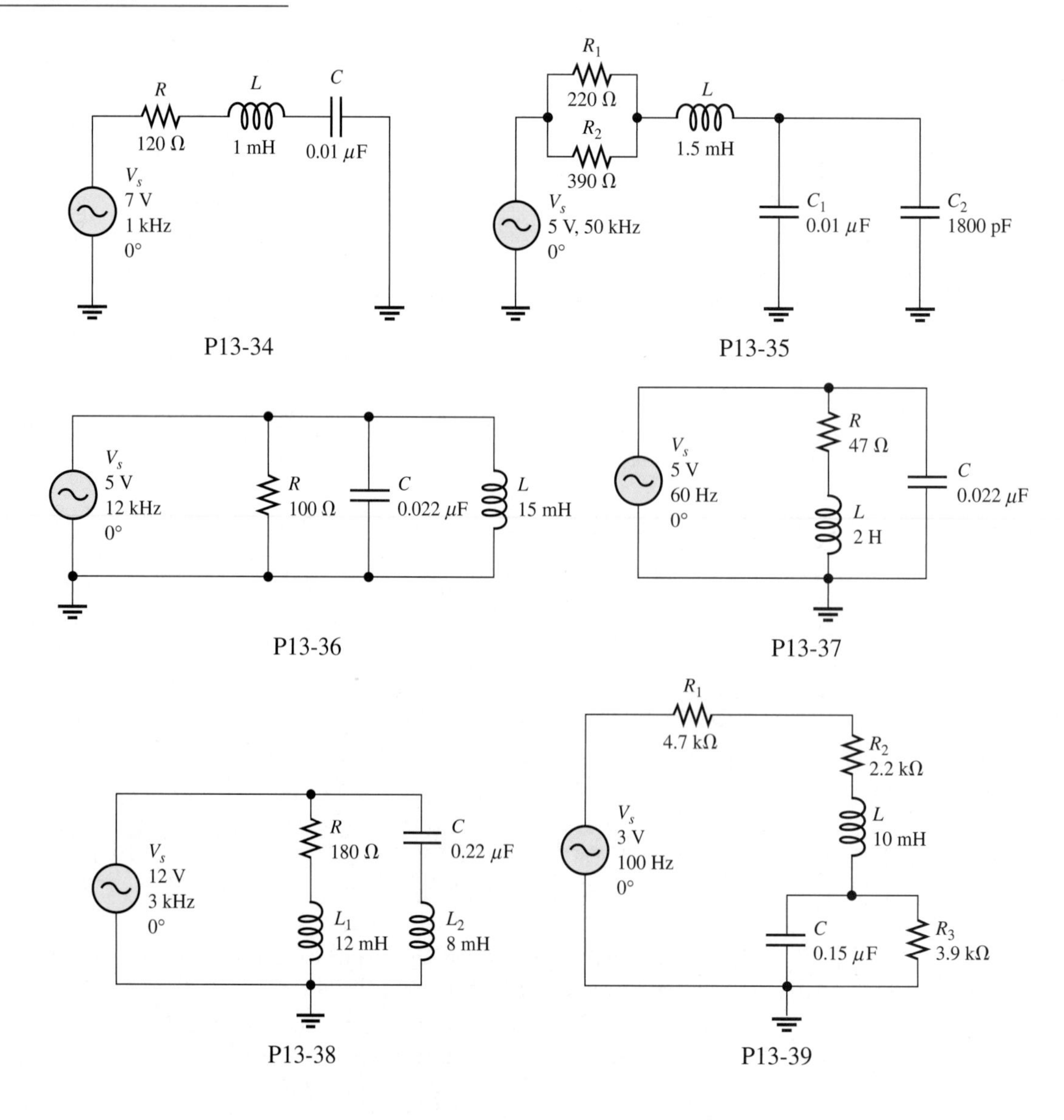

ANSWERS

SECTION REVIEWS

SECTION 13–1 **Impedance and Phase Angle of Series *RLC* Circuits**

1. The circuit is inductive if $X_L > X_C$ and capacitive if $X_C > X_L$.
2. $X_{tot} = |X_L - X_C| = 70\ \Omega$; The circuit is capacitive.
3. $Z = \sqrt{R^2 + X_{tot}^2} = 83.2\ \Omega$; $\theta = \tan^{-1}(X_{tot}/R) = 57.3°$; Current is leading V_s.

SECTION 13–2 **Analysis of Series *RLC* Circuits**

1. $V_s = \sqrt{V_R^2 + (V_C - V_L)^2} = 38.4\ \text{V}$
2. Current leads V_s because the circuit is capacitive.
3. $X_{tot} = |X_L - X_C| = 6\ \Omega$

SECTION 13–3 **Series Resonance**

1. Series resonance occurs when $X_L = X_C$.
2. The current is maximum because the impedance is minimum.
3. $f_r = 1/(2\pi\sqrt{LC}) = 159\ \text{kHz}$
4. It is capacitive because $X_C > X_L$.

SECTION 13–4 **Series Resonant Filters**

1. $V_{out} = 0.707(15\ \text{V}) = 10.6\ \text{V}$
2. $BW = f_r/Q = 10\ \text{kHz}$
3. Current is maximum; output voltage is minimum.

SECTION 13–5 **Parallel *RLC* Filters**

1. $I_R = V_s/R = 80\ \text{mA}$; $I_C = V_s/X_C = 120\ \text{mA}$; $I_L = V_s/X_L = 240\ \text{mA}$
2. The circuit is inductive ($X_L < X_C$).
3. $L_{eq} = L[(Q^2 + 1)/Q^2] = 20.1\ \text{mH}$; $R_{p(eq)} = R_W(Q^2 + 1) = 1589\ \Omega$

SECTION 13–6 **Parallel Resonance**

1. The impedance is maximum.
2. The current is minimum.
3. At ideal parallel resonance, $X_C = X_L = 1.5\ \text{k}\Omega$.
4. $f_r = \sqrt{1 - (R_W^2C/L)}/2\pi\sqrt{LC} = 225\ \text{kHz}$; $Z_r = R_W(Q^2 + 1) = 125\ \text{M}\Omega$
5. $f_r = 1/(2\pi\sqrt{LC}) = 22.5\ \text{kHz}$
6. $f_r = \sqrt{Q^2/(Q^2 + 1)}/2\pi\sqrt{LC} = 20.9\ \text{kHz}$
7. $Z_r = R_W(Q^2 + 1) = 8.02\ \text{k}\Omega$

SECTION 13–7 **Parallel Resonant Filters**

1. The bandwidth can be increased by reducing the parallel resistance.
2. f_r changes to 4.47 kHz.
3. $Z_r = R_W(Q^2 + 1) = 47.0\ \text{k}\Omega$

SECTION 13–8 **Applications**

1. A tuned filter is used to select a narrow band of frequencies.
2. A wave trap is a band-stop filter.
3. Several variable capacitors (or inductors) whose values can be varied simultaneously with a common control is an example of ganged tuning.

■ Application Assignment

1. $f_{c(1)} = 8$ kHz; $f_{c(2)} = 11.6$ kHz
2. The circuit or component values cannot be determined from the data given.

RELATED PROBLEMS FOR EXAMPLES

13–1 1.25 kΩ; 66°

13–2 4.71 kΩ

13–3 Current increases, reaches a maximum at resonance, and then decreases.

13–4 More capacitive

13–5 71.2 kHz

13–6 2.27 mA

13–7 Z increases; Z increases

13–8 No

13–9 200 kHz

13–10 −14 dB

13–11 4.61

13–12 322 Hz

13–13 V_{out} increases; V_{out} increases

13–14 Inductive

13–15 Increase. X_C will decrease and approach 0.

13–16 $R_{p(\text{eq})} = 25$ kΩ; $L_{\text{eq}} = 5$ mH; $C = 0.022$ μF, $R_{p(tot)} = 4.24$ kΩ

13–17 $f_r = 3.39$ kHz; $I_L = 4.69$ mA; $I_C = 4.69$ mA; $I_{tot} = 0$ A

13–18 80.0 kΩ

13–19 The differences are negligible.

13–20 7.07 V

13–21 530 kHz

13–22 25

13–23 Q decreases with less load resistance.

13–24 The Q is less than 10.

SELF-TEST

1. (a) **2.** (c) **3.** (b) **4.** (c) **5.** (d) **6.** (c) **7.** (a)
8. (b) **9.** (d) **10.** (a) **11.** (b) **12.** (d)

14

TRANSFORMERS

INTRODUCTION

In Chapter 11, you learned about self-inductance. In this chapter, you will study mutual inductance, which is the basis for the operation of transformers. Transformers are used in all types of applications such as power supplies, electrical power distribution, and signal coupling in communications systems.

The operation of the transformer is based on the principle of mutual inductance, which occurs when two or more coils are in close proximity. A simple transformer is actually two coils that are electromagnetically coupled by their mutual inductance. Because there is no electrical contact between two magnetically coupled coils, the transfer of energy from one coil to the other can be achieved in a situation of complete electrical isolation. This has many advantages, as you will learn in this chapter.

CHAPTER OUTLINE

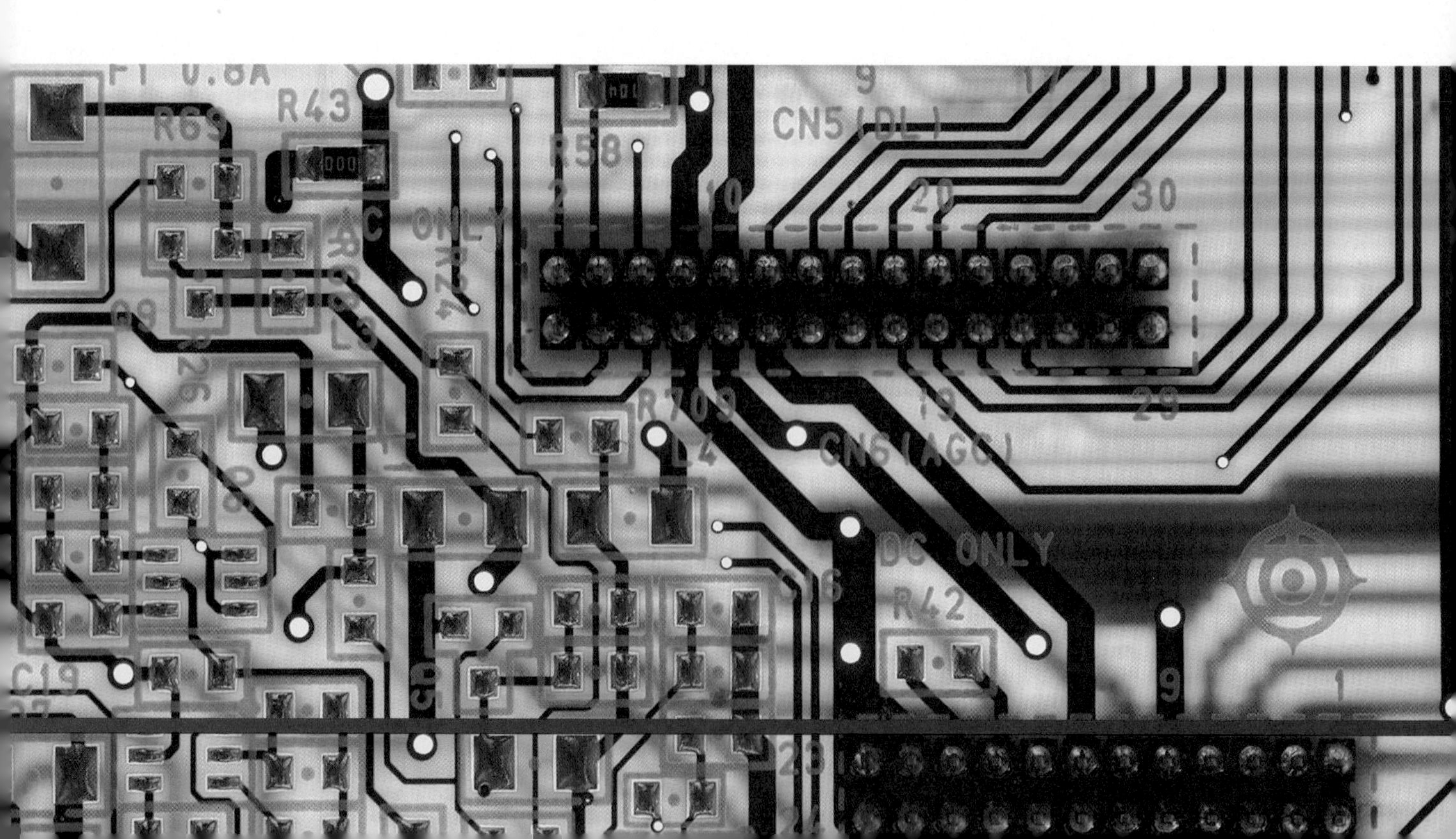

CHAPTER OBJECTIVES

- Explain mutual inductance
- Describe how a transformer is constructed and how it works
- Explain how a step-up transformer works
- Explain how a step-down transformer works
- Discuss the effect of a resistive load across the secondary winding
- Discuss the concept of a reflected load in a transformer
- Discuss impedance matching with transformers
- Explain how the transformer acts as an isolation device
- Describe a practical transformer
- Describe several types of transformers
- Troubleshoot transformers

KEY TERMS

- Mutual inductance
- Transformer
- Primary winding
- Secondary winding
- Magnetic coupling
- Turns ratio
- Reflected load
- Impedance matching
- Maximum power transfer
- Electrical isolation
- Apparent power rating
- Center tap

APPLICATION ASSIGNMENT PREVIEW

Your assignment is to troubleshoot a type of dc power supply that uses a transformer to couple the ac voltage from a standard electrical outlet. By making voltage measurements at various points, you will determine if there is a fault and be able to specify the part of the power supply that is faulty. After studying this chapter, you should be able to complete the application assignment.

WWW. VISIT THE COMPANION WEBSITE

Circuit Simulation Tutorials and Other Chapter Study Tools Are Available at

http://www.prenhall.com/floyd

14–1 MUTUAL INDUCTANCE

When two coils are placed close to each other, a changing electromagnetic field produced by the current in one coil will cause an induced voltage in the second coil because of the mutual inductance.

After completing this section, you should be able to

- **Explain mutual inductance**
- Discuss magnetic coupling
- Define *electrical isolation*
- Define *coefficient of coupling*
- Identify the factors that affect mutual inductance and state the formula

Recall that the electromagnetic field surrounding a coil expands, collapses, and reverses as the current increases, decreases, and reverses.

When a second coil is placed very close to the first coil so that the changing magnetic lines of force cut through the second coil, the coils are magnetically coupled and a voltage is induced, as indicated in Figure 14–1. When two coils are magnetically coupled, they provide electrical isolation because there is no electrical connection between them, only a magnetic link. If the current in the first coil is a sine wave, the voltage induced in the second coil is also a sine wave. The amount of voltage induced in the second coil as a result of the current in the first coil is dependent on the **mutual inductance**, L_M. The mutual inductance is established by the inductance of each coil (L_1 and L_2) and by the amount of coupling (k) between the two coils. To maximize coupling, the two coils are wound on a common core.

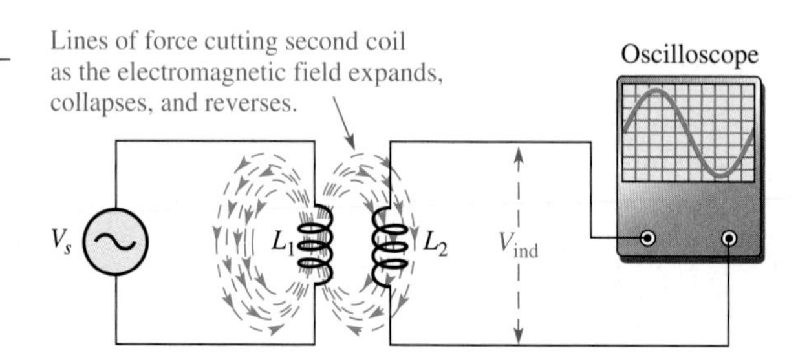

▶ **FIGURE 14–1**

A voltage is induced in the second coil as a result of the changing current in the first coil, producing a changing magnetic field that links the second coil.

Coefficient of Coupling

The **coefficient of coupling, *k*,** between two coils is the ratio of the lines of force (flux) produced by coil 1 that link coil 2 ($\phi_{1\text{-}2}$) to the total flux produced by coil 1 (ϕ_1).

Equation 14–1

$$k = \frac{\phi_{1\text{-}2}}{\phi_1}$$

For example, if half of the total flux produced by coil 1 links coil 2, then $k = 0.5$. A greater value of k means that more voltage is induced in coil 2 for a certain rate of change of current in coil 1. Note that k has no units. Recall that the unit of magnetic lines of force (flux) is the weber, abbreviated Wb.

The coefficient k depends on the physical closeness of the coils and the type of core material on which they are wound. Also, the construction and shape of the cores are factors.

Formula for Mutual Inductance

The three factors that influence mutual inductance (k, L_1, and L_2) are shown in Figure 14–2. The formula for mutual inductance is

$$L_M = k\sqrt{L_1 L_2}$$

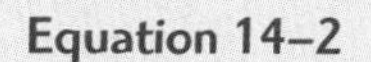

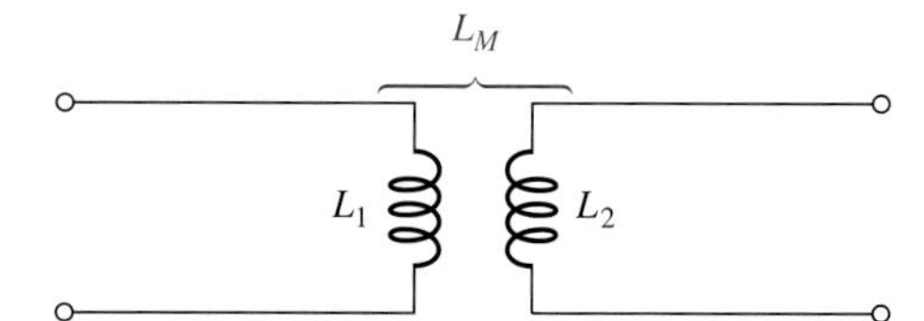

FIGURE 14–2

The mutual inductance of two coils.

EXAMPLE 14–1

One coil produces a total magnetic flux of 50 μWb, and 20 μWb link coil 2. What is k?

Solution

$$k = \frac{\phi_{1\text{-}2}}{\phi_1} = \frac{20\ \mu\text{Wb}}{50\ \mu\text{Wb}} = \mathbf{0.4}$$

*Related Problem** Determine k when $\phi_1 = 500\ \mu$Wb and $\phi_{1\text{-}2} = 375\ \mu$Wb.

*Answers are at the end of the chapter.

EXAMPLE 14–2

Two coils are wound on a single core, and the coefficient of coupling is 0.3. The inductance of coil 1 is 10 μH, and the inductance of coil 2 is 15 μH. What is L_M?

Solution

$$L_M = k\sqrt{L_1 L_2} = 0.3\sqrt{(10\ \mu\text{H})(15\ \mu\text{H})} = \mathbf{3.67\ \mu H}$$

Related Problem Determine the mutual inductance when $k = 0.5$, $L_1 = 1$ mH, and $L_2 =$ 600 μH.

SECTION 14–1 REVIEW

Answers are at the end of the chapter.

1. Define *mutual inductance*.
2. Two 50 mH coils have $k = 0.9$. What is L_M?
3. If k is increased, what happens to the voltage induced in one coil as a result of a current change in the other coil?

14–2 THE BASIC TRANSFORMER

A basic **transformer** is an electrical device constructed of two coils placed in close proximity to each other so that there is a mutual inductance.

After completing this section, you should be able to

- **Describe how a transformer is constructed and how it works**
- Identify the parts of a basic transformer

- Discuss the importance of the core material
- Define *primary winding* and *secondary winding*
- Define *turns ratio*
- Discuss how the direction of windings affects voltage polarities

A schematic of a transformer is shown in Figure 14–3(a). One coil is called the **primary winding** and the other coil is called the **secondary winding** as indicated. For standard operation, the source voltage is applied to the primary winding, and a load is connected to the secondary winding, as shown in Figure 14–3(b). The primary winding is the input winding, and the secondary winding is the output winding. It is common to refer to the side of the circuit that has the source voltage as the *primary,* and the side that has the induced voltage as the *secondary.*

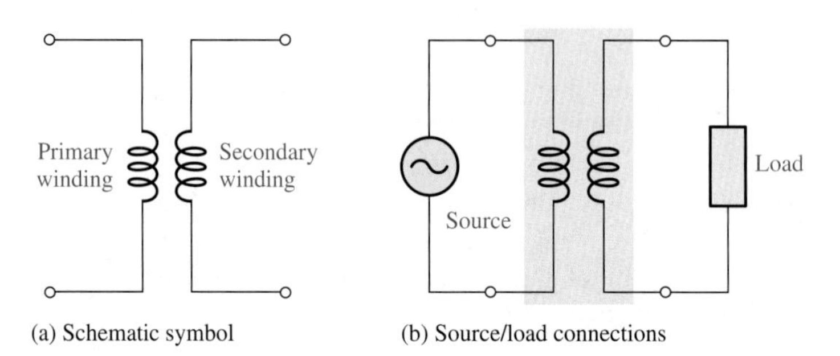

FIGURE 14–3

The basic transformer.

The windings of a transformer are formed around the core. The core provides both a physical structure for placement of the windings and a magnetic path so that the magnetic flux is concentrated close to the coils. Three general categories of core material are air, ferrite, and iron. The schematic symbol for each type is shown in Figure 14–4.

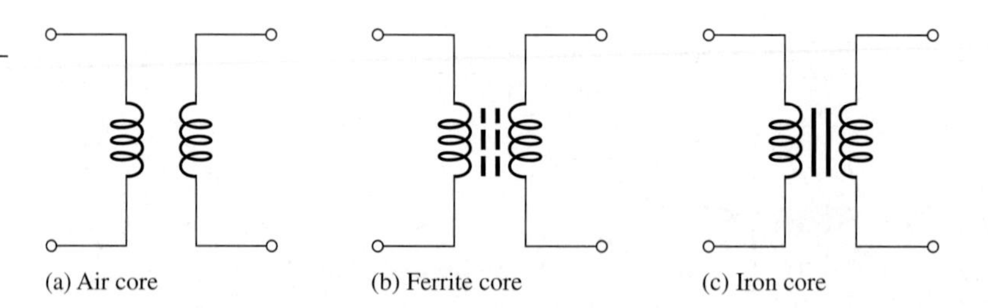

FIGURE 14–4

Schematic symbols specify the type of core.

Air-core and ferrite-core transformers generally are used for high-frequency applications and consist of windings on an insulating shell that is hollow (air) or constructed of ferrite, such as depicted in Figure 14–5. The wire is typically covered by a varnish-type coating to prevent the windings from shorting together. The amount of **magnetic coupling** between the primary winding and the secondary winding is set by the type of core material and by the relative positions of the windings. In Figure 14–5(a), the windings are loosely coupled because they are separated, and in part (b) they are tightly coupled because they are overlapping. The tighter the coupling, the greater the induced voltage in the secondary for a given current in the primary.

Iron-core transformers generally are used for audio frequency (AF) and power applications. These transformers consist of windings on a core constructed from

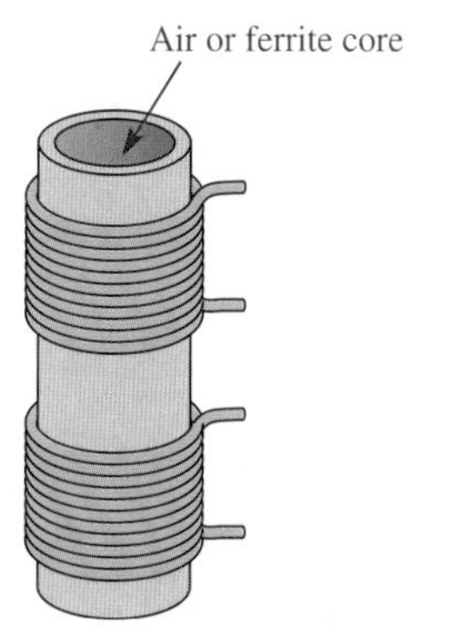

(a) Loosely coupled windings

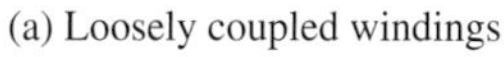

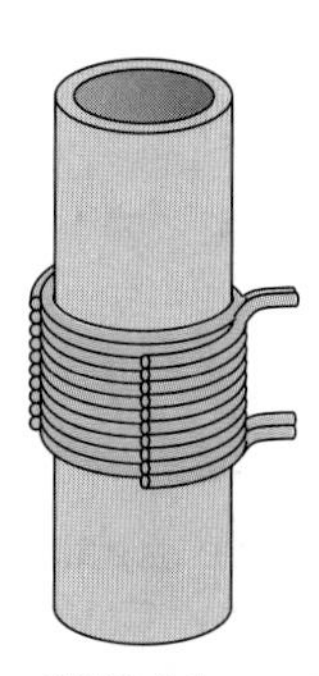

(b) Tightly coupled windings. Cutaway view shows both windings.

FIGURE 14–5

Transformers with cylindrical-shaped cores.

laminated sheets of ferromagnetic material insulated from each other, as shown in Figure 14–6. This construction provides an easy path for the magnetic flux and increases the amount of coupling between the windings. The figure also shows the basic construction of two major configurations of iron-core transformers. In the core-type construction, shown in Figure 14–6(a), the windings are on separate legs of the laminated core. In the shell-type construction, shown in part (b), both windings are on the same leg. Each type has certain advantages. In general, the core type has more room for insulation and can handle higher voltages. The shell type can produce higher core flux, so fewer turns are required.

FIGURE 14–6

Iron-core transformer construction with multilayer windings.

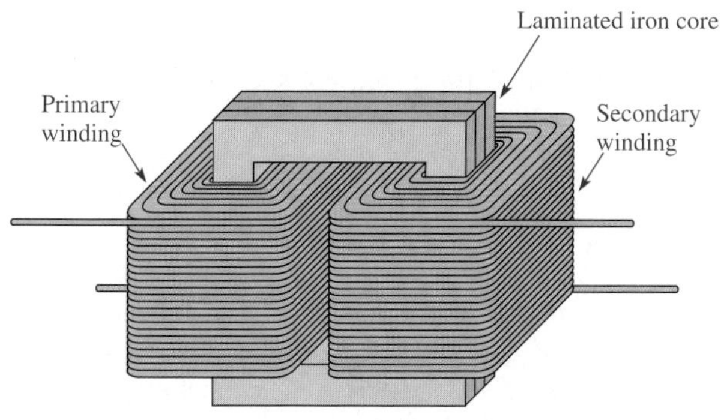

(a) Core type has each winding on a separate leg.

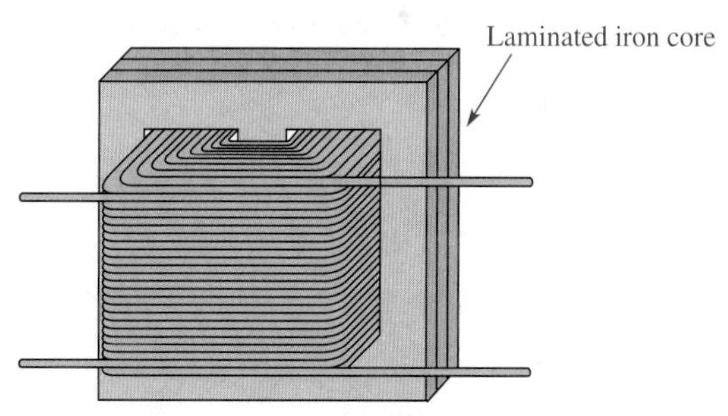

(b) Shell type has both windings on the same leg.

A variety of transformers is shown in Figure 14–7.

FIGURE 14–7

Some common types of transformers.

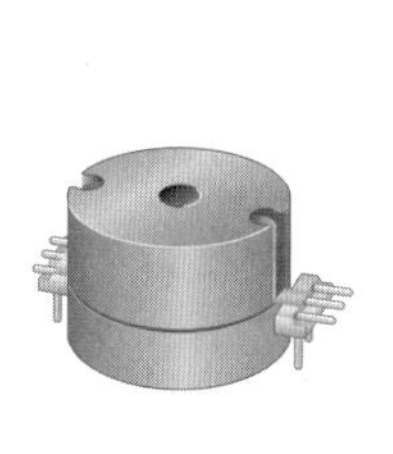

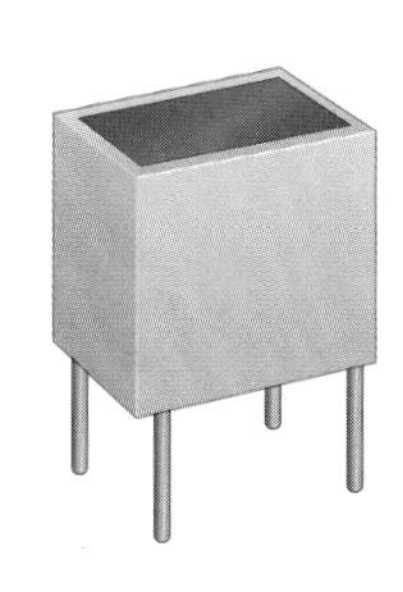

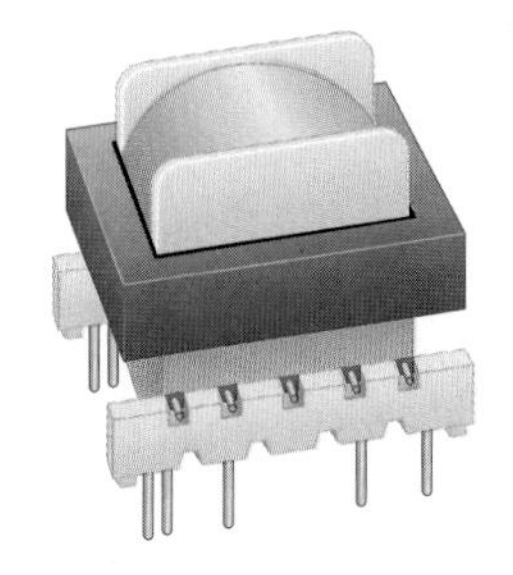

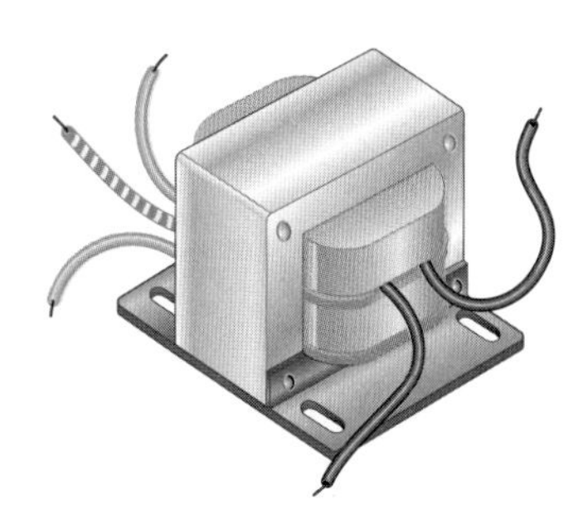

Turns Ratio

A transformer parameter that is very useful in understanding how a transformer operates is the turns ratio. In this text, the **turns ratio**, (n) is defined as the ratio of the number of turns in the secondary winding (N_{sec}) to the number of turns in the primary winding (N_{pri}).

Equation 14–3

$$n = \frac{N_{sec}}{N_{pri}}$$

Although the definition of turns ratio stated by Equation 14–3 is used in this text, there seems, unfortunately, to be no universal agreement on how the turns ratio is defined. Many sources state that the turns ratio is N_{sec}/N_{pri}, while others use N_{pri}/N_{sec}. Either definition is correct as long as it is clearly stated and used consistently. The turns ratio of a transformer is rarely if ever given as a transformer specification. Generally, the input and output voltages and the power rating are the key specifications. However, the turns ratio is useful in studying the operating principle of a transformer.

EXAMPLE 14–3

A certain transformer used in a radar system has a primary winding with 100 turns and a secondary winding with 400 turns. What is the turns ratio?

Solution $N_{sec} = 400$ and $N_{pri} = 100$; therefore, the turns ratio is

$$n = \frac{N_{sec}}{N_{pri}} = \frac{400}{100} = \mathbf{4}$$

A turns ratio of 4 can be expressed as 1:4 on a schematic.

Related Problem A certain transformer has a turns ratio of 10. If $N_{pri} = 500$, what is N_{sec}?

Direction of Windings

Another important transformer parameter is the direction in which the windings are placed around the core. As illustrated in Figure 14–8, the direction of the windings determines the polarity of the voltage across the secondary winding (secondary voltage) with respect to the voltage across the primary winding (primary voltage). Phase dots can be used to indicate polarities, as shown in Figure 14–9.

FIGURE 14–8

The direction of the windings determines the relative polarities of the voltages.

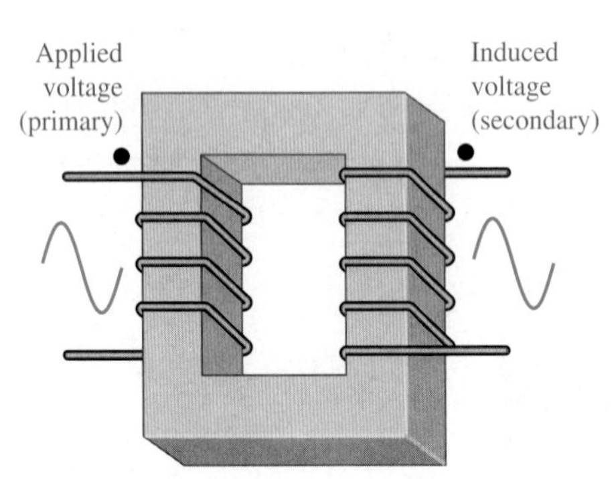

(a) The primary and secondary voltages are in phase when the windings are in the same effective direction around the magnetic path.

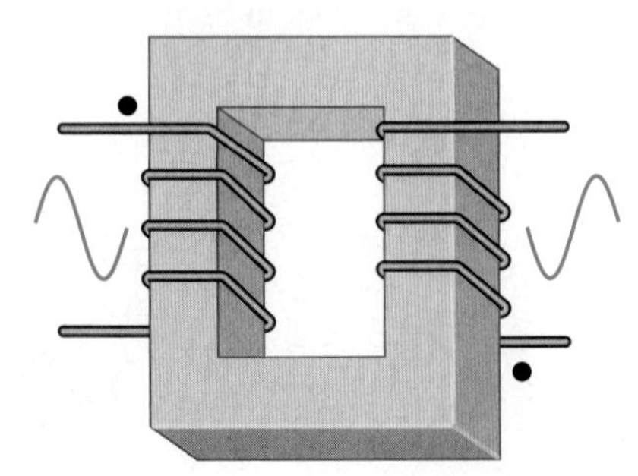

(b) The primary and secondary voltages are 180° out of phase when the windings are in the opposite direction.

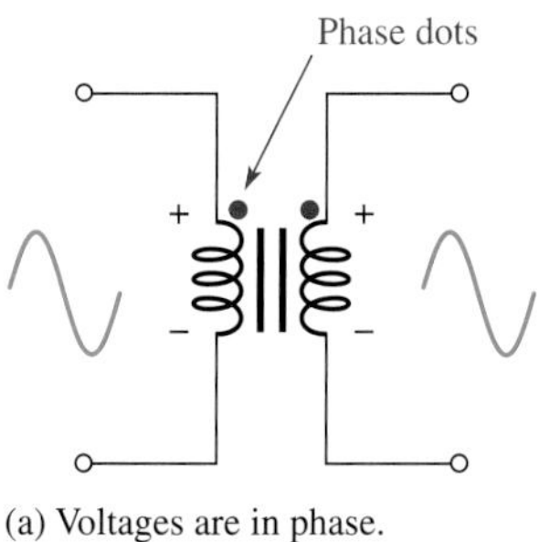

(a) Voltages are in phase.

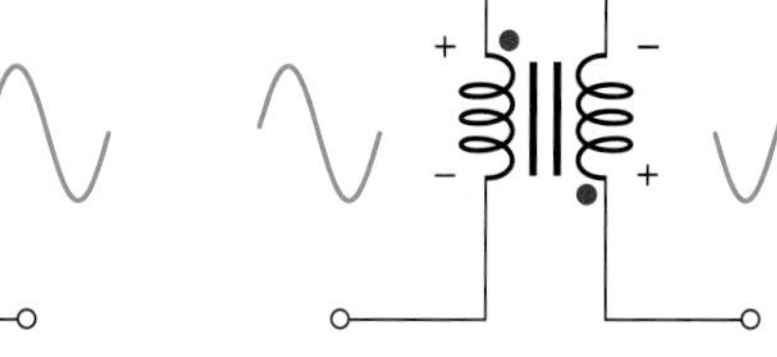

(b) Voltages are out of phase.

FIGURE 14–9

Phase dots indicate corresponding polarities of primary and secondary voltages.

SECTION 14–2 REVIEW

1. Upon what principle is the operation of a transformer based?
2. Define *turns ratio.*
3. Why are the directions of the windings of a transformer important?
4. A certain transformer has a primary winding with 500 turns and a secondary winding with 250 turns. What is the turns ratio?

14–3 STEP-UP TRANSFORMERS

A step-up transformer has more turns in its secondary winding than in its primary winding and is used to increase ac voltage.

After completing this section, you should be able to

- **Explain how a step-up transformer works**
- State the relationship between primary and secondary voltages and the turns ratio
- Identify a step-up transformer by its turns ratio

A transformer in which the secondary voltage is greater than the primary voltage is called a **step-up transformer.** The amount that the voltage is stepped up depends on the turns ratio. For any transformer,

The ratio of secondary voltage (V_{sec}) to primary voltage (V_{pri}) is equal to the ratio of the number of turns in the secondary winding (N_{sec}) to the number of turns in the primary winding (N_{pri}).

$$\frac{V_{sec}}{V_{pri}} = \frac{N_{sec}}{N_{pri}} \qquad \text{Equation 14–4}$$

Recall that N_{sec}/N_{pri} defines the turns ratio, n. Therefore, from this relationship, in Equation 14–4, V_{sec} can be expressed as

$$V_{sec} = nV_{pri} \qquad \text{Equation 14–5}$$

Equation 14–5 shows that the secondary voltage is equal to the turns ratio times the primary voltage. This condition assumes that the coefficient of coupling is 1, and a good iron-core transformer approaches this value.

The turns ratio for a step-up transformer is always greater than 1 because the number of turns in the secondary winding (N_{sec}) is always greater than the number of turns in the primary winding (N_{pri}).

EXAMPLE 14–4

The transformer in Figure 14–10 has a turns ratio of 3. What is the voltage across the secondary?

▶ FIGURE 14–10

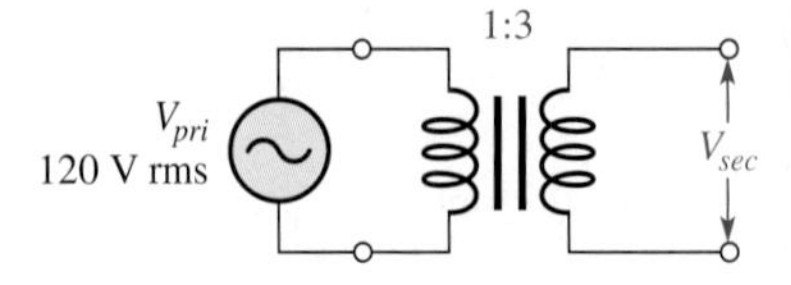

Solution The secondary voltage is

$$V_{sec} = nV_{pri} = 3(120 \text{ V}) = \mathbf{360\ V}$$

Note that the turns ratio of 3 is indicated on the schematic as 1:3, meaning that there are three secondary turns for each primary turn.

Related Problem The transformer in Figure 14–10 is changed to one with a turns ratio of 4. Determine V_{sec}.

Open file E14-04 on your EWB/CircuitMaker CD-ROM and measure the secondary voltage.

SECTION 14–3 REVIEW

1. What does a step-up transformer do?
2. If the turns ratio is 5, how much greater is the secondary voltage than the primary voltage?
3. When 240 V ac are applied to a transformer with a turns ratio of 10, what is the secondary voltage?

14–4 STEP-DOWN TRANSFORMERS

A step-down transformer has more turns in its primary winding than in its secondary winding and is used to decrease ac voltage.

After completing this section, you should be able to

- **Explain how a step-down transformer works**
- Identify a step-down transformer by its turns ratio

A transformer in which the secondary voltage is less than the primary voltage is called a **step-down transformer.** The amount by which the voltage is stepped down depends on the turns ratio. Equation 14–5 also applies to a step-down transformer.

The turns ratio of a step-down transformer is always less than 1 because the number of turns in the secondary winding (N_{sec}) is always less than the number of turns in the primary winding (N_{pri}).

EXAMPLE 14–5

The transformer in Figure 14–11 is part of a laboratory power supply and has a turns ratio of 0.2. What is the secondary voltage?

FIGURE 14–11

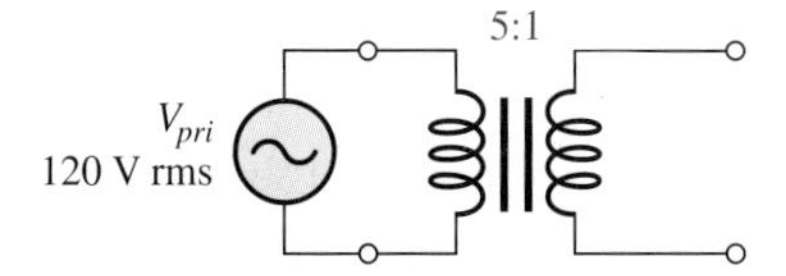

Solution The secondary voltage is

$$V_{sec} = nV_{pri} = 0.2(120\text{ V}) = \mathbf{24\text{ V}}$$

Related Problem The transformer in Figure 14–11 is changed to one with a turns ratio of 0.48. Determine the secondary voltage.

Open file E14-05 on your EWB/CircuitMaker CD-ROM and measure the secondary voltage.

SECTION 14–4 REVIEW

1. What does a step-down transformer do?
2. A voltage of 120 V ac is applied to the primary winding of a transformer with a turns ratio of 0.5. What is the secondary voltage?
3. A primary voltage of 120 V ac is reduced to 12 V ac. What is the turns ratio?

14–5 LOADING THE SECONDARY

When a resistive load is connected to the secondary winding of a transformer, the relationship of the load (secondary) current and the current in the primary circuit is determined by the turns ratio.

After completing this section, you should be able to

- **Discuss the effect of a resistive load across the secondary winding**
- Discuss power in a transformer
- Determine the current delivered by the secondary when a step-up transformer is loaded
- Determine the current delivered by the secondary when a step-down transformer is loaded

When a load is connected to the secondary winding of a transformer, the power delivered to the load can never be greater than the power delivered by the primary winding. For an ideal transformer, the power delivered by the secondary winding (P_{sec}) equals the power delivered by the primary winding (P_{pri}). When losses are considered, the power delivered by the secondary is always less.

Power is dependent on voltage and current, and there can be no increase in power in a transformer. Therefore, if the voltage is stepped up, the current is stepped down and vice versa. In an ideal transformer, the power delivered by the secondary to the load is the same as the power delivered by the primary regardless of the turns ratio.

The power delivered by the primary is

$$P_{pri} = V_{pri}I_{pri}$$

The power delivered by the secondary is

$$P_{sec} = V_{sec}I_{sec}$$

Ideally, $P_{pri} = P_{sec}$; therefore,

$$V_{pri}I_{pri} = V_{sec}I_{sec}$$

Transposing terms,

$$\frac{I_{pri}}{I_{sec}} = \frac{V_{sec}}{V_{pri}}$$

From Equation 14–4,

$$\frac{V_{sec}}{V_{pri}} = \frac{N_{sec}}{N_{pri}}$$

Therefore, since N_{sec}/N_{pri} equals the turns ratio, n, the relationship of primary current to secondary current in a transformer is

$$\frac{I_{pri}}{I_{sec}} = n$$

Equation 14–6

▼ **FIGURE 14–12**

Illustration of voltages and currents in a transformer with a loaded secondary winding.

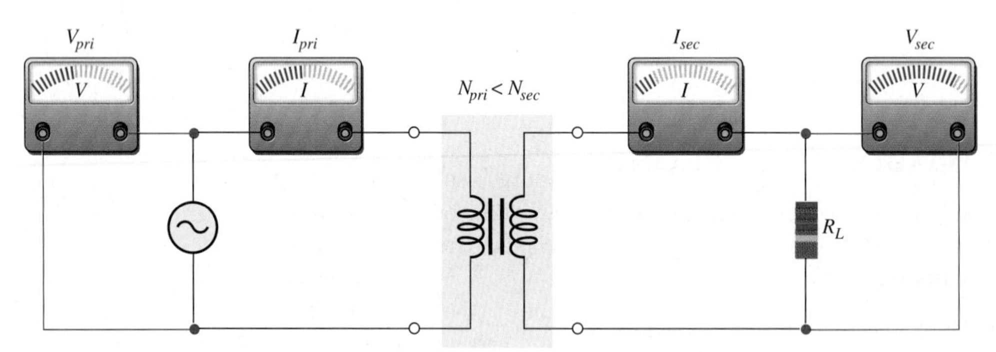

(a) Step-up transformer: $V_{sec} > V_{pri}$ and $I_{sec} < I_{pri}$

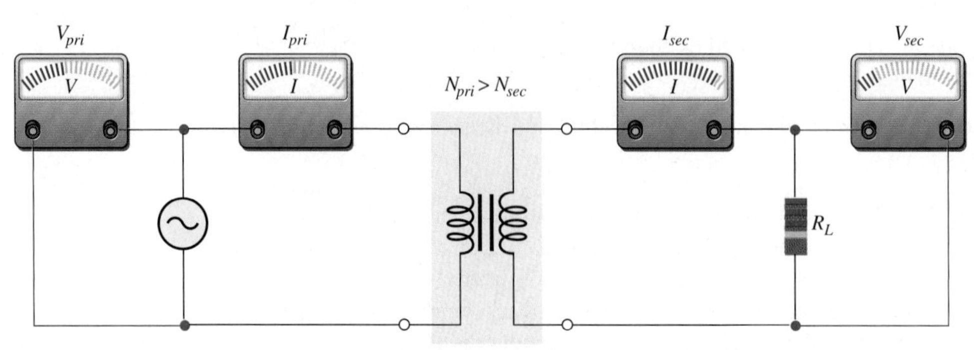

(b) Step-down transformer: $V_{sec} < V_{pri}$ and $I_{sec} > I_{pri}$

Inverting both sides of Equation 14–6 and solving for I_{sec},

$$I_{sec} = \left(\frac{1}{n}\right)I_{pri}$$

Equation 14–7

Figure 14–12 illustrates the effects of voltages and currents in a transformer. In part (a), for a step-up transformer, in which n is greater than 1, the secondary current is less than the primary current because $1/n$ is less than 1. For a step-down transformer, shown in Figure 14–12(b), n is less than 1, and I_{sec} is greater than I_{pri} because $1/n$ is greater than 1.

EXAMPLE 14–6

The two transformers shown in Figure 14–13 have loaded secondaries. If, as a result of the loaded secondary, the primary current is 100 mA in each case, what is the current through the load?

FIGURE 14–13

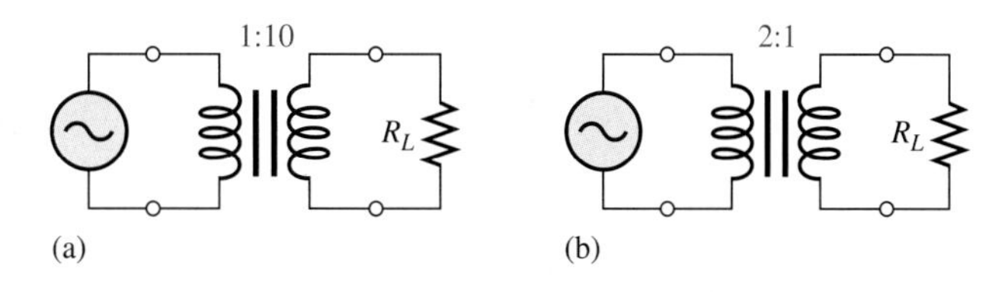

Solution In part (a), the turns ratio is 10. Therefore, the secondary load current is

$$I_L = I_{sec} = \left(\frac{1}{n}\right)I_{pri} = \left(\frac{1}{10}\right)I_{pri} = 0.1(100\text{ mA}) = \mathbf{10\text{ mA}}$$

In part (b), the turns ratio is 0.5. Therefore, the secondary load current is

$$I_L = I_{sec} = \left(\frac{1}{n}\right)I_{pri} = \left(\frac{1}{0.5}\right)I_{pri} = 2(100\text{ mA}) = \mathbf{200\text{ mA}}$$

Related Problem What is the secondary current in Figure 14–13(a) if the turns ratio is doubled? What is the secondary current in Figure 14–13(b) if the turns ratio is halved? Assume the load resistances are changed so that I_{pri} remains at 100 mA in both cases.

SECTION 14–5 REVIEW

1. If the turns ratio of a transformer is 2, is the secondary current greater than or less than the primary current? By how much?
2. A transformer has 100 primary turns and 25 secondary turns, and I_{pri} is 0.5 A. What is the turns ratio? What is the value of I_{sec}?
3. In Question 2, what is the primary current when there is a secondary load current of 10 A?

14–6 REFLECTED LOAD

From the viewpoint of the primary, a load connected across the secondary winding of a transformer appears to have a resistance that is not necessarily equal to the actual resistance of the load. The actual load is essentially "reflected" into the primary determined by the turns ratio. This reflected load is what the source effectively sees, and it determines the amount of primary current.

After completing this section, you should be able to

- **Discuss the concept of a reflected load in a transformer**
- Define *reflected resistance*
- Explain how the turns ratio affects the reflected resistance
- Calculate reflected resistance

The concept of the reflected load is illustrated in Figure 14–14. The load (R_L) in the secondary of a transformer is reflected into the primary by transformer action. The load appears to the source in the primary to be a resistance (R_{pri}) with a value determined by the turns ratio and the actual value of the load resistance. The resistance R_{pri} is called the **reflected resistance.**

FIGURE 14–14

Reflected load in a transformer circuit.

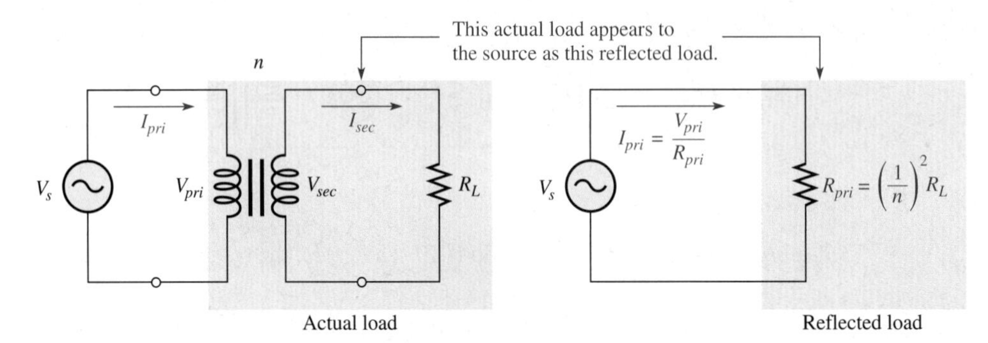

The resistance in the primary of Figure 14–14 is $R_{pri} = V_{pri}/I_{pri}$. The resistance in the secondary is $R_L = V_{sec}/I_{sec}$. From Equations 14–4 and 14–6, you know that $V_{sec}/V_{pri} = n$ and $I_{pri}/I_{sec} = n$. Using these relationships, a formula for R_{pri} in terms of R_L is determined as follows:

$$\frac{R_{pri}}{R_L} = \frac{V_{pri}/I_{pri}}{V_{sec}/I_{sec}} = \left(\frac{V_{pri}}{V_{sec}}\right)\left(\frac{I_{sec}}{I_{pri}}\right) = \left(\frac{1}{n}\right)\left(\frac{1}{n}\right) = \left(\frac{1}{n}\right)^2$$

Solving for R_{pri} yields

Equation 14–8

$$R_{pri} = \left(\frac{1}{n}\right)^2 R_L$$

Equation 14–8 shows that the resistance reflected into the primary is the square of the reciprocal of the turns ratio times the load resistance.

EXAMPLE 14–7

Figure 14–15 shows a source that is transformer-coupled to a load resistor of 100 Ω. The transformer has a turns ratio of 4. What is the reflected resistance seen by the source?

FIGURE 14–15

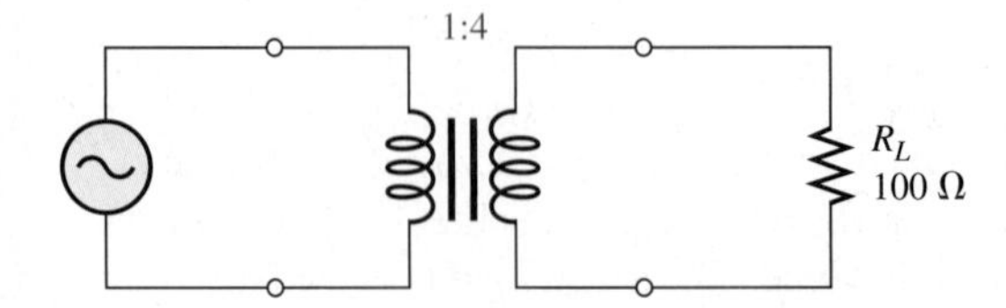

Solution The reflected resistance is

$$R_{pri} = \left(\frac{1}{n}\right)^2 R_L = \left(\frac{1}{4}\right)^2 100\ \Omega = \left(\frac{1}{16}\right) 100\ \Omega = \mathbf{6.25\ \Omega}$$

The source sees a resistance of 6.25 Ω just as if it were connected directly, as shown in the equivalent circuit of Figure 14–16.

FIGURE 14–16

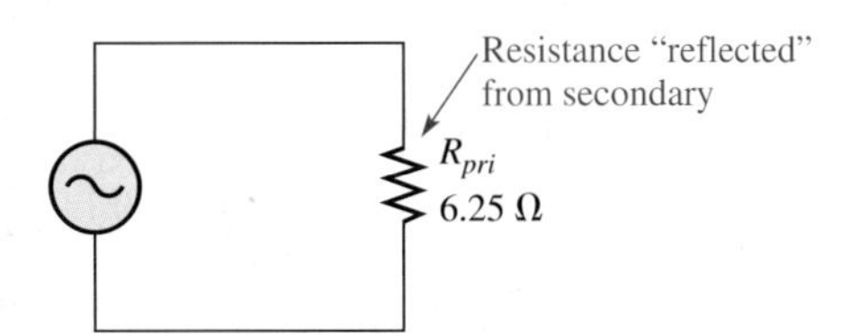

Related Problem If the turns ratio in Figure 14–15 is 10 and R_L is 600 Ω, what is the reflected resistance?

EXAMPLE 14–8

In Figure 14–15, if a transformer is used having a turns ratio of 0.25, what is the reflected resistance?

Solution The reflected resistance is

$$R_{pri} = \left(\frac{1}{n}\right)^2 R_L = \left(\frac{1}{0.25}\right)^2 100\ \Omega = (4)^2 100\ \Omega = \mathbf{1600\ \Omega}$$

Related Problem To achieve a reflected resistance of 800 Ω, what turns ratio is required in Figure 14–15?

Example 14–7 illustrated that in a step-up transformer ($n > 1$), the reflected resistance is less than the actual load resistance; Example 14–8 illustrated that in a step-down transformer ($n < 1$), the reflected resistance is greater than the load resistance.

SECTION 14–6 REVIEW

1. Define *reflected resistance.*
2. What transformer characteristic determines the reflected resistance?
3. A given transformer has a turns ratio of 10, and the load is 50 Ω. How much resistance is reflected into the primary?
4. What is the turns ratio required to reflect a 4 Ω load resistance into the primary as 400 Ω?

14–7 MATCHING THE LOAD AND SOURCE RESISTANCES

One application of transformers is in the matching of a load resistance to a source resistance in order to achieve maximum transfer of power or other results. This technique is called **impedance matching**. In audio systems, special wide-band transformers are often used to get the maximum amount of power from the amplifier to the speaker by proper selection of the turns ratio. Transformers designed specifically for impedance matching usually show the input and output impedance they are designed to match.

After completing this section, you should be able to

- **Discuss impedance matching with transformers**
- Give a general definition of impedance
- Discuss the maximum power transfer theorem
- Define *impedance matching*
- Explain the purpose of impedance matching
- Describe a practical application

As you have learned, impedance is the opposition to current, including the effects of both resistance and reactance combined. We will confine our usage in this chapter to resistance only.

The concept of **maximum power transfer** is illustrated in the basic circuit of Figure 14–17. Part (a) shows an ac voltage source with a series resistance representing its internal resistance. Some fixed internal resistance is inherent in all sources due to their internal circuitry. When the source is connected directly to a load, as shown in part (b), generally the objective is to transfer as much of the power produced by the source to the load as possible. However, a certain amount of the power produced by the source is dissipated in its internal resistance, and the remaining power goes to the load.

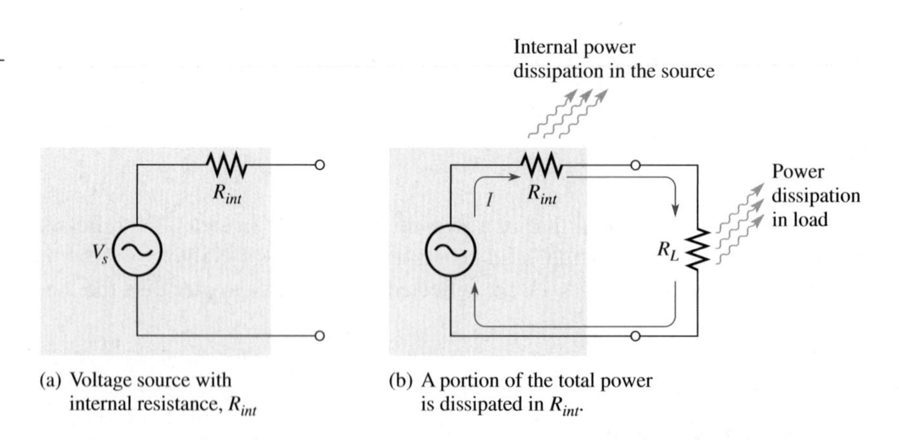

(a) Voltage source with internal resistance, R_{int}

(b) A portion of the total power is dissipated in R_{int}.

FIGURE 14–17

Power transfer from a nonideal voltage source to a load.

Maximum Power Transfer Theorem

The maximum power transfer theorem is important when you need to know the value of the load at which the most power is delivered from the source. The theorem was introduced in Chapter 6 and is restated as follows:

When a source is connected to a load, maximum power is delivered to the load when the load resistance is equal to the fixed internal source resistance.

Impedance Matching

As mentioned before, in most practical situations the internal source resistance of various types of sources is fixed and cannot be changed. Also, in many cases, the resistance of a device that acts as a load is fixed and cannot be altered. If you need to connect a given source to a given load, remember that only by chance will their resistances match. In this situation, a special type of wide-band transformer comes in handy. You can use the reflected-resistance characteristic provided by a transformer to make the load resistance appear to have the same value as the source resistance, thereby "fooling" the source into "thinking" that there is a match. This technique is called *impedance matching,* and the transformer is called an impedance-matching transformer.

Let's take a practical, everyday situation to illustrate the concept of impedance matching. The typical input resistance of a TV receiver is 300 Ω. An antenna must be connected to this input by a lead-in cable in order to receive TV signals. In this situation, the antenna and the lead-in act as the source, and the input resistance of the TV receiver is the load, as illustrated in Figure 14–18.

FIGURE 14–18

An antenna directly coupled to a TV receiver.

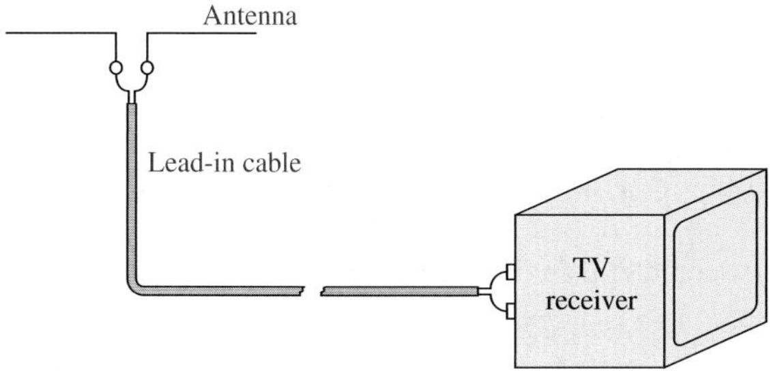

(a) The antenna/lead-in is the source; the TV input is the load.

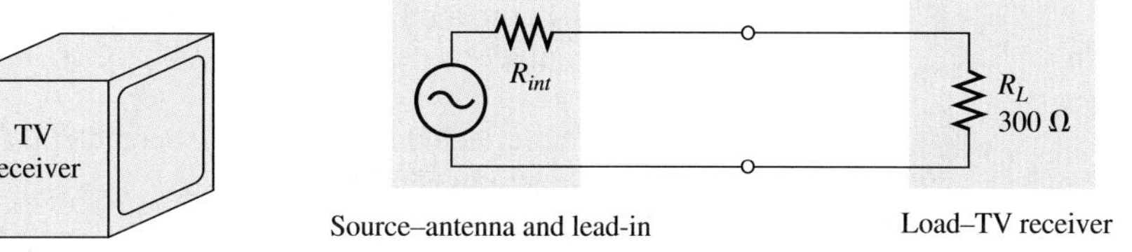

(b) Circuit equivalent of antenna and TV receiver system

It is common for an antenna system to have a characteristic resistance of 75 Ω. Thus, if the 75 Ω source (antenna and lead-in) is connected directly to the 300 Ω TV input, maximum power will not be delivered to the input of the TV, and you will have poor signal reception. The solution is to use a matching transformer, connected as indicated in Figure 14–19, in order to match the 300 Ω load resistance to the 75 Ω source resistance.

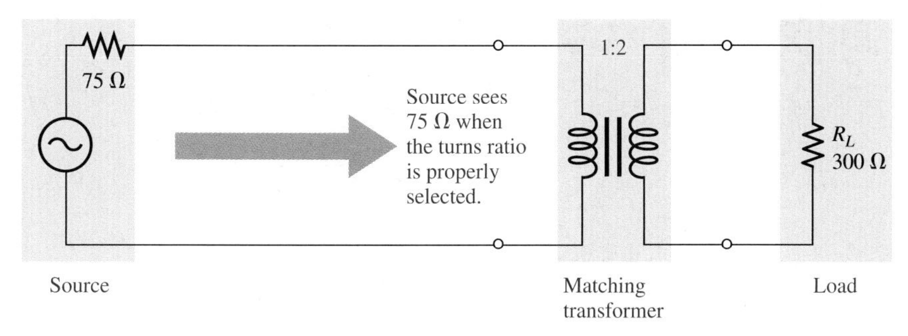

FIGURE 14–19

Example of a load matched to a source by transformer coupling for maximum power transfer.

To match the resistances, that is, to reflect the load resistance (R_L) into the primary so that it appears to have a value equal to the internal source resistance (R_{int}), you must select a proper value of turns ratio. You want the 300 Ω load resistance to look like 75 Ω to the source. Use Equation 14–8 to obtain a formula to determine the turns ratio, *n,* when you know the values for R_L and R_{pri}.

$$R_{pri} = \left(\frac{1}{n}\right)^2 R_L$$

Transpose terms and divide both sides by R_L.

$$\left(\frac{1}{n}\right)^2 = \frac{R_{pri}}{R_L}$$

Then take the square root of both sides.

$$\frac{1}{n} = \sqrt{\frac{R_{pri}}{R_L}}$$

Invert both sides to get the following formula for the turns ratio.

Equation 14–9

$$n = \sqrt{\frac{R_L}{R_{pri}}}$$

Finally, solve for the particular turns ratio to match a 300 Ω load to a 75 Ω source.

$$n = \sqrt{\frac{300\ \Omega}{75\ \Omega}} = \sqrt{4} = 2$$

Therefore, a matching transformer with a turns ratio of 2 must be used in this application.

EXAMPLE 14–9

An amplifier has an 800 Ω internal resistance. In order to provide maximum power to an 8 Ω speaker, what turns ratio must be used in the coupling transformer?

Solution The reflected resistance must equal 800 Ω. Thus, from Equation 14–9, the turns ratio can be determined.

$$n = \sqrt{\frac{R_L}{R_{pri}}} = \sqrt{\frac{8\ \Omega}{800\ \Omega}} = \sqrt{0.01} = \mathbf{0.1}$$

The diagram and its equivalent reflected circuit are shown in Figure 14–20.

▼ FIGURE 14–20

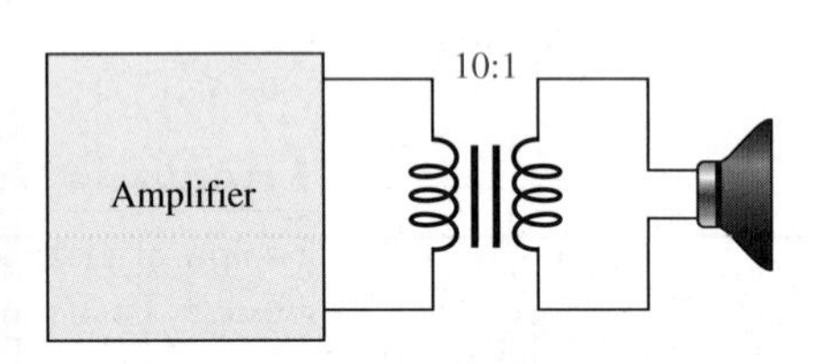

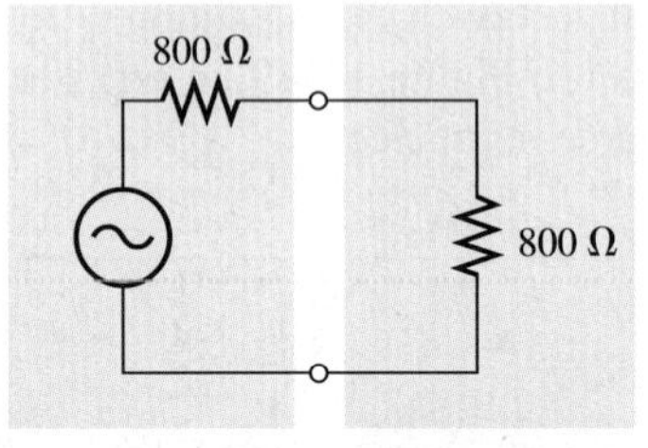

Related Problem What must be the turns ratio in Figure 14–20 to provide maximum power to two 8 Ω speakers in parallel?

SECTION 14–7 REVIEW

1. What does impedance matching mean?
2. What is the advantage of matching the load resistance to the resistance of a source?
3. A transformer has a turns ratio of 0.5. What is the reflected resistance with 100 Ω across the secondary?

14–8 THE TRANSFORMER AS AN ISOLATION DEVICE

Transformers are useful in providing electrical isolation between the primary circuit and the secondary circuit because there is no electrical connection between the two windings. In a transformer, energy is transferred entirely by magnetic coupling.

After completing this section, you should be able to

- **Explain how the transformer acts as an isolation device**
- Discuss dc isolation
- Discuss power line isolation

DC Isolation

If there is a nonchanging direct current through the primary circuit of a transformer, nothing happens in the secondary circuit, as indicated in Figure 14–21(a). A changing current in the primary winding is necessary in order to create a changing magnetic field. This will cause voltage to be induced in the secondary circuit, as indicated in Figure 14–21(b). Therefore, the transformer isolates the secondary circuit from any dc voltage in the primary circuit.

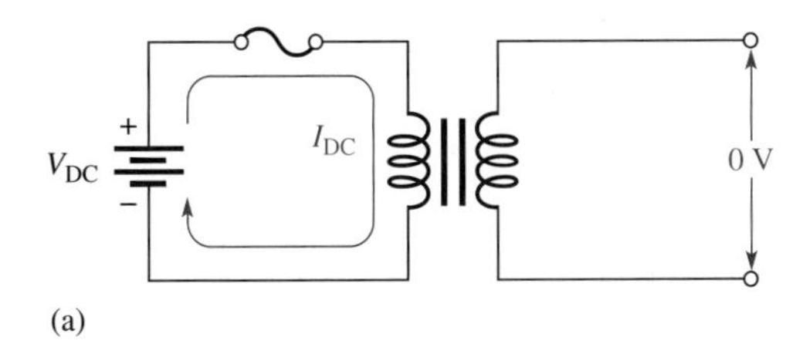

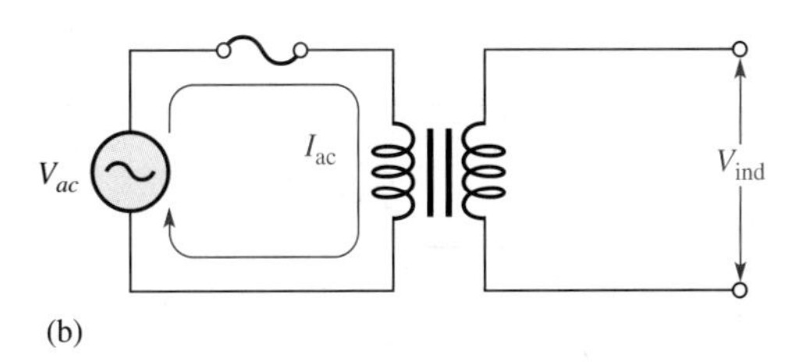

FIGURE 14–21
DC isolation and ac coupling.

In a typical application, a small transformer can be used to keep the dc voltage on the output of an amplifier stage from affecting the dc bias of the next amplifier. Only the ac signal is coupled through the transformer from one stage to the next, as Figure 14–22 illustrates.

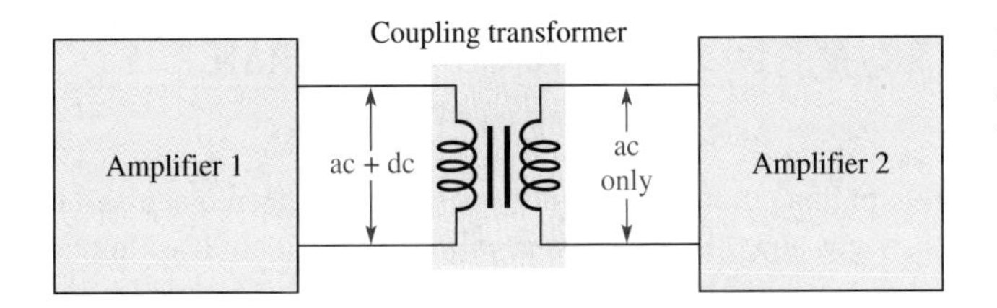

FIGURE 14–22
Audio amplifier stages with transformer coupling for dc isolation.

Power Line Isolation

Transformers are often used to electrically isolate electronic equipment from the 60 Hz, 110 V ac power line. Using an isolation transformer to couple the 60 Hz ac to an instrument prevents a possible shock hazard if the 110 V line is connected to the metal chassis of the equipment. This condition is possible if the line cord plug can be inserted into an outlet either way. Incidentally, to prevent this situation, most plugs have keyed prongs so that they can be plugged in only one way.

Figure 14–23 illustrates how a transformer can prevent the metal chassis from being connected to the 110 V line rather than to neutral (ground), no matter how the cord is plugged into the outlet. When an isolation transformer is used, the secondary circuit is said to be "floating" because it is not referenced to the power line ground. Should a person come in contact with the secondary voltage, there is no complete current path back to ground, and therefore there is a reduced shock hazard. As you know, there must be current through your body in order for you to receive an electrical shock.

▼ FIGURE 14–23

The use of an isolation transformer to reduce shock hazard.

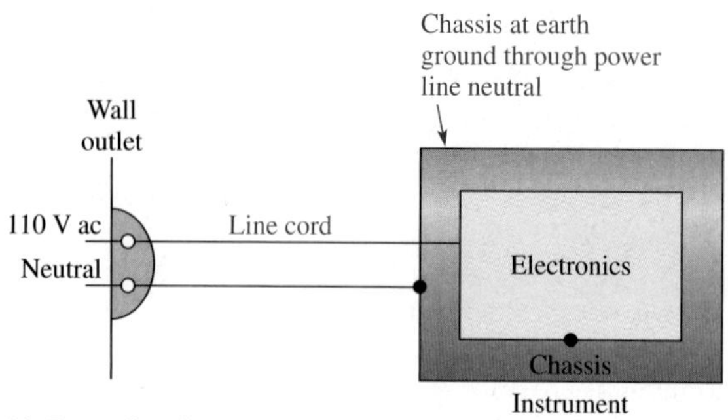

(a) Power line direct to instrument

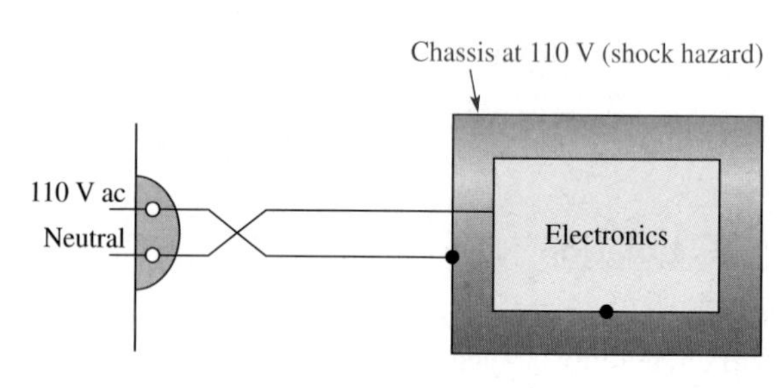

(b) Power line reversed

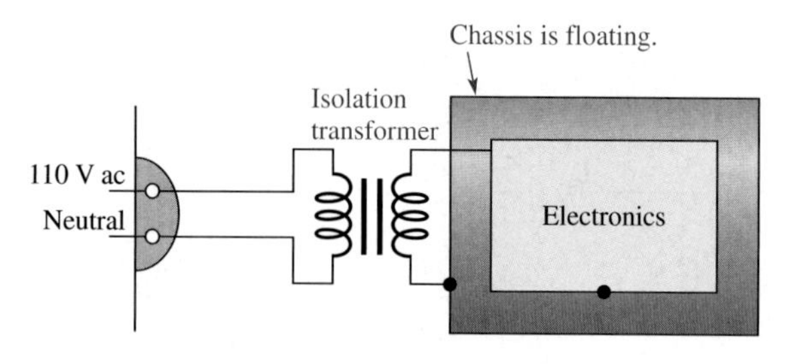

(c) Power line transformer-coupled to instrument

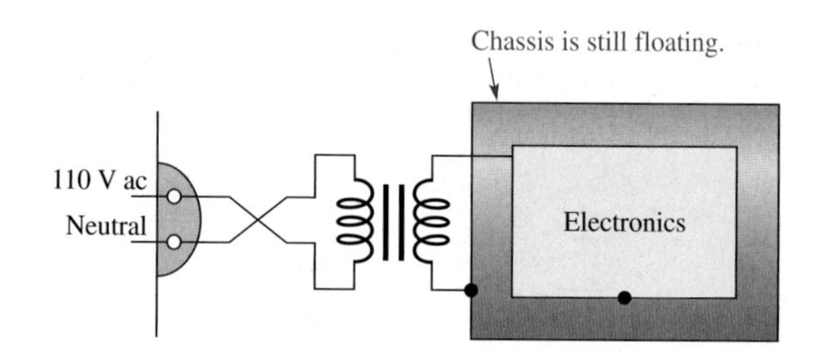

(d) Power line reversed

SECTION 14–8 REVIEW

1. What does the term *electrical isolation* mean?
2. Can a dc voltage be coupled by a transformer?

14–9 NONIDEAL TRANSFORMER CHARACTERISTICS

Up to this point, transformer operation has been discussed from an ideal point of view, and this approach is valid when you are learning new concepts. However, you should be aware of the nonideal characteristics of practical transformers and how they affect performance.

After completing this section, you should be able to

- **Describe a practical transformer**
- List and describe the nonideal characteristics
- Explain power rating of a transformer
- Define *efficiency* of a transformer

So far, the transformer has been considered as an ideal device. That is, the winding resistance, the winding capacitance, and nonideal core characteristics were all neglected and the transformer was treated as if it had an efficiency of 100%. For studying the basic concepts and in many applications, the ideal model is valid. However, the practical transformer has several nonideal characteristics of which you should be aware.

Winding Resistance

Both the primary and the secondary windings of a practical transformer have winding resistance. (You learned about the winding resistance of inductors in Chapter 11.) The winding resistances of a practical transformer are represented as resistors in series with the windings as shown in Figure 14–24.

Winding resistance in a practical transformer results in less voltage across a secondary load. Voltage drops due to the winding resistance effectively subtract from the primary and secondary voltages and result in a load voltage that is less than that predicted by the relationship $V_{sec} = nV_{pri}$. In most cases the effect is relatively small and can be neglected.

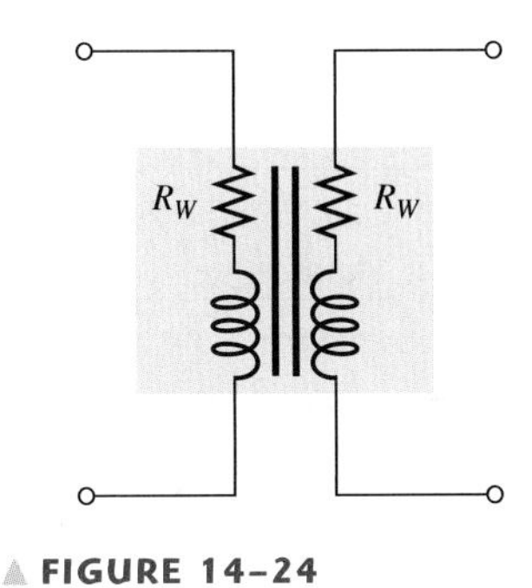

▲ FIGURE 14–24

Winding resistance in a practical transformer.

Losses in the Core

There is always some energy conversion in the core material of a practical transformer. This conversion is seen as a heating of ferrite and iron cores, but it does not occur in air cores. Part of this energy conversion is because of the continuous reversal of the magnetic field due to the changing direction of the primary current; this component of the energy conversion is called *hysteresis loss.* The rest of the energy conversion to heat is caused by eddy currents produced when voltage is induced in the core material by the changing magnetic flux, according to Faraday's law. The eddy currents occur in circular patterns in the core resistance, thus producing heat. This conversion to heat is greatly reduced by the use of laminated construction of iron cores. The thin layers of ferromagnetic material are insulated from each other to minimize the buildup of eddy currents by confining them to a small area and to keep core losses to a minimum.

Magnetic Flux Leakage

In an ideal transformer, all the magnetic flux produced by the primary current is assumed to pass through the core to the secondary winding and vice versa. In a practical transformer, some of the magnetic flux lines break out of the core and pass through the surrounding air back to the other end of the winding, as illustrated in Figure 14–25 for the magnetic field produced by the primary current. Magnetic flux leakage results in a reduced secondary voltage.

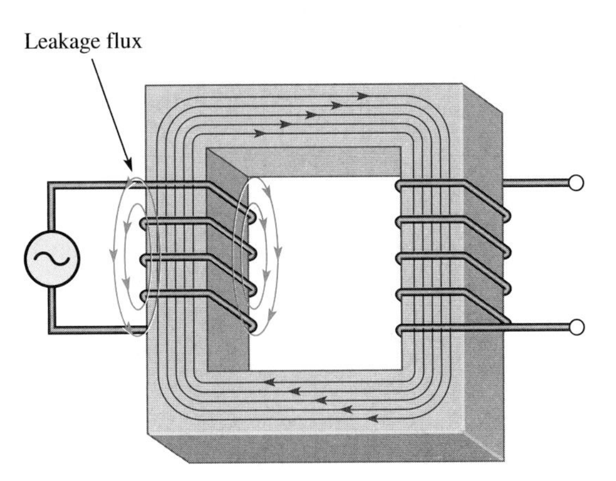

◀ FIGURE 14–25

Flux leakage in a practical transformer.

The percentage of magnetic flux that actually reaches the secondary winding determines the coefficient of coupling of the transformer. For example, if nine out of ten flux lines remain inside the core, the coefficient of coupling is 0.90 or 90%. Most iron-core transformers have very high coefficients of coupling (greater than 0.99), whereas ferrite-core and air-core devices have lower values.

Winding Capacitance

As you learned in Chapter 11, there is always some stray capacitance between adjacent turns of a winding. These stray capacitances result in an effective capacitance in parallel with each winding of a transformer, as indicated in Figure 14–26.

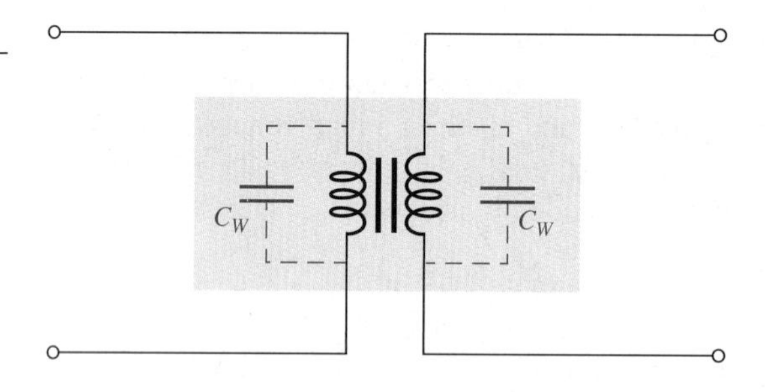

FIGURE 14–26

Winding capacitance in a practical transformer.

HANDS ON TIP

If you have an unmarked or unknown small transformer, you can use a signal generator with a relatively low output voltage to check the voltage ratio and thus the turns ratio between the input (primary) and the output (secondary). This is safer than applying 110 V ac and holding your breath. Typically, the primary wires are black, a low-voltage secondary is green, and a high-voltage secondary is red. A striped wire usually indicates a center tap. Unfortunately, all transformers do not have colored wires or the wires are not always standard colors.

These stray capacitances have very little effect on the transformer's operation at low frequencies because the reactances (X_C) are very high. However, at higher frequencies, the reactances decrease and begin to produce a bypassing effect across the primary winding and across the secondary load. As a result, less of the total primary current is through the primary winding, and less of the total secondary current is through the load. This effect reduces the load voltage as the frequency goes up.

Transformer Power Rating

A power transformer is typically rated in volt-amperes (VA), primary/secondary voltage, and operating frequency. For example, a given transformer rating may be specified as 2 kVA, 500/50, 60 Hz. The 2 kVA value is the apparent power rating. The 500 and the 50 can be either secondary or primary voltages. The 60 Hz is the operating frequency.

The transformer rating can be helpful in selecting the proper transformer for a given application. Let's assume, for example, that 50 V is the secondary voltage. In this case the load current is

$$I_L = \frac{P_{sec}}{V_{sec}} = \frac{2\text{ kVA}}{50\text{ V}} = 40\text{ A}$$

On the other hand, if 500 V is the secondary voltage, then

$$I_L = \frac{P_{sec}}{V_{sec}} = \frac{2\text{ kVA}}{500\text{ V}} = 4\text{ A}$$

These are the maximum currents that the secondary can handle in either case.

The reason that the **apparent power rating** is in volt-amperes (VA) rather than in watts (true power) is as follows: If the transformer load is purely capacitive or purely inductive, the true power (watts) delivered to the load is ideally zero. However, the current for $V_{sec} = 500$ V and $X_C = 100\ \Omega$ at 60 Hz, for example, is 5 A. This current exceeds the maximum of 4 A that the 2 kVA secondary can handle, and even though the true power is zero, the transformer may be damaged. So it is meaningless to specify power in watts for transformers.

Transformer Efficiency

Recall that the secondary power is equal to the primary power in an ideal transformer. Because the nonideal characteristics just discussed result in a power loss in the transformer, the secondary (output) power is always less than the primary (input) power. The **efficiency,** symbolized by the Greek letter eta (η), of a transformer is a measure of the percentage of the input power that is delivered to the output.

$$\eta = \left(\frac{P_{out}}{P_{in}}\right)100\%$$

Equation 14–10

Most power transformers have efficiencies in excess of 95%.

EXAMPLE 14–10

A certain type of transformer has a primary current of 5 A and a primary voltage of 4800 V. The secondary current is 90 A and the secondary voltage is 240 V. Determine the efficiency of this transformer.

Solution The input power is

$$P_{in} = V_{pri}I_{pri} = (4800\text{ V})(5\text{ A}) = 24\text{ kVA}$$

The output power is

$$P_{out} = V_{sec}I_{sec} = (240\text{ V})(90\text{ A}) = 21.6\text{ kVA}$$

The efficiency is

$$\eta = \left(\frac{P_{out}}{P_{in}}\right)100\% = \left(\frac{21.6\text{ kVA}}{24\text{ kVA}}\right)100\% = \mathbf{90\%}$$

Related Problem A transformer has a primary current of 8 A with a primary voltage of 440 V. The secondary current is 30 A and the secondary voltage is 100 V. What is the efficiency?

SECTION 14–9 REVIEW

1. Explain how a practical transformer differs from the ideal model.
2. The coefficient of coupling of a certain transformer is 0.85. What does this mean?
3. A certain transformer has a rating of 10 kVA. If the secondary voltage is 250 V, how much load current can the transformer handle?

14–10 OTHER TYPES OF TRANSFORMERS

The basic transformer has several important variations. They include tapped transformers, multiple-winding transformers, and autotransformers.

After completing this section, you should be able to

- **Describe several types of transformers**
- Describe center-tapped transformers
- Describe multiple-winding transformers
- Describe autotransformers

Tapped Transformers

A schematic of a transformer with a center-tapped secondary winding is shown in Figure 14–27(a). The **center tap** (CT) is equivalent to two secondary windings with half the total voltage across each.

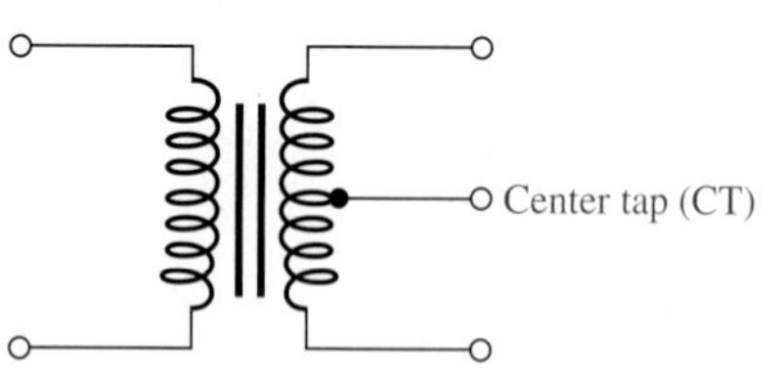

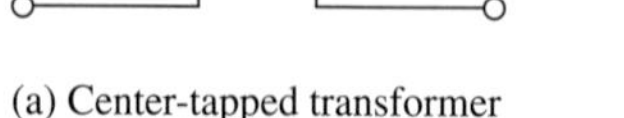

(a) Center-tapped transformer

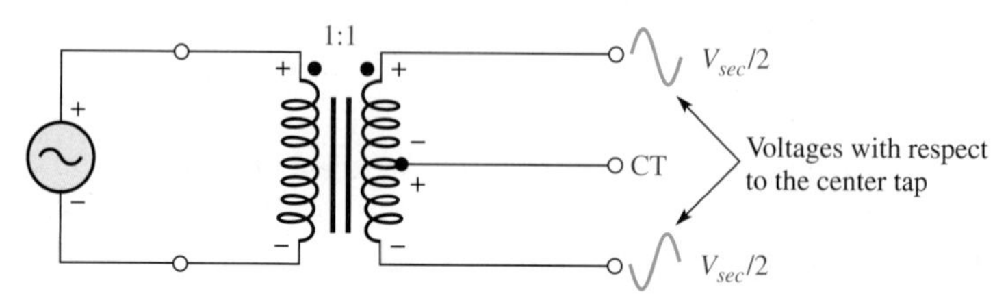

(b) Output voltages with respect to the center tap are 180° out of phase with each other and are one-half the magnitude of the secondary voltage.

▲ **FIGURE 14–27**

Operation of a center-tapped transformer.

The voltages between either end of the secondary winding and the center tap are, at any instant, equal in magnitude but opposite in polarity, as illustrated in Figure 14–27(b). Here, for example, at some instant on the sinusoidal voltage, the polarity across the entire secondary winding is as shown (top end +, bottom −). At the center tap, the voltage is less positive than the top end but more positive than the bottom end of the secondary. Therefore, measured with respect to the center tap, the top end of the secondary is positive, and the bottom end is negative. This center-tapped feature is used in some power supply rectifiers in which the ac voltage is converted to dc, as illustrated in Figure 14–28 and also in impedance-matching transformers.

▶ **FIGURE 14–28**

Application of a center-tapped transformer in ac-to-dc conversion.

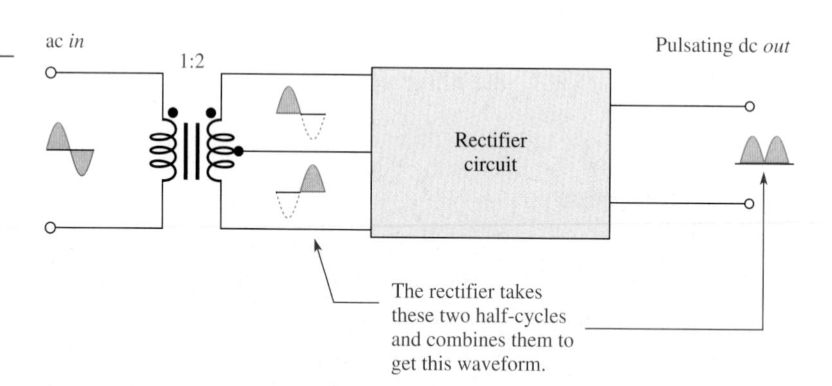

Some tapped transformers have taps on the secondary winding at points other than the electrical center. Also, multiple primary and secondary taps are sometimes used in certain applications. Examples of these types of transformers are shown in Figure 14–29.

▶ **FIGURE 14–29**

Tapped transformers.

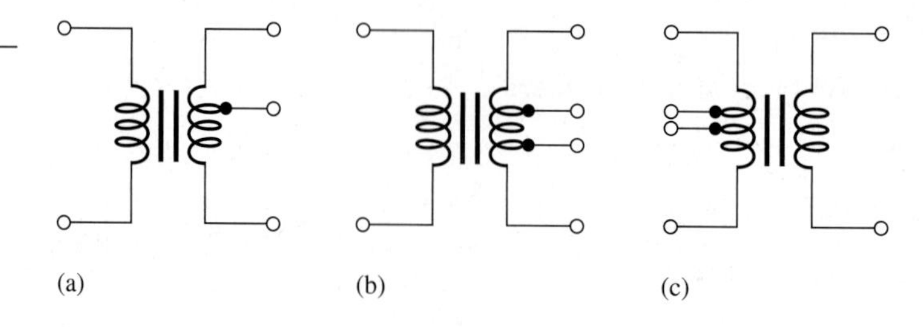

One example of a transformer with a multiple-tap primary winding and a center-tapped secondary winding is the utility-pole transformer used by power companies to step down the high voltage from the power line to 110 V/220 V service for residential and commercial customers, as shown in Figure 14–30. The multiple taps on the primary winding are used for minor adjustments in the turns ratio in order to overcome line voltages that are slightly too high or too low.

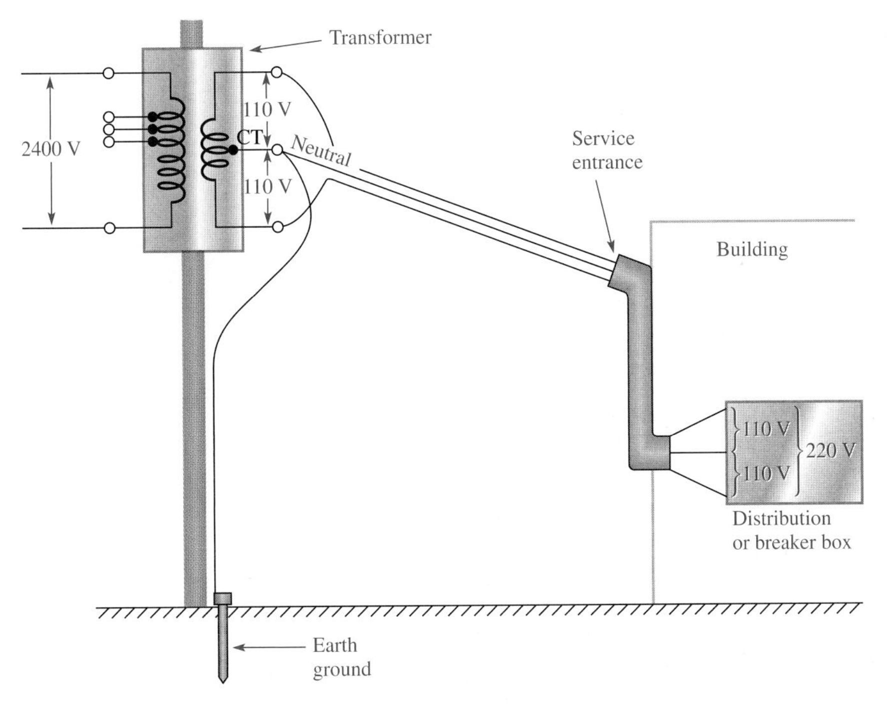

FIGURE 14–30

Utility-pole transformer in a typical power distribution system.

Multiple-Winding Transformers

Some transformers are designed to operate from either 110 V ac or 220 V ac lines. These transformers usually have two primary windings, each of which is designed for 110 V ac. When the two are connected in series, the transformer can be used for 220 V ac operation, as illustrated in Figure 14–31.

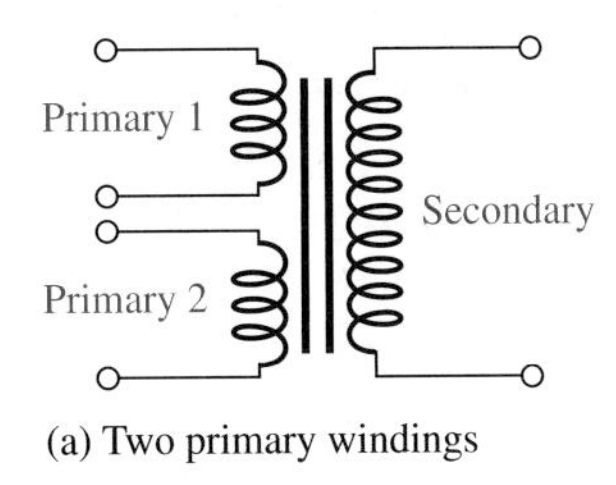

(a) Two primary windings

(b) Primary windings in parallel for 110 V ac operation

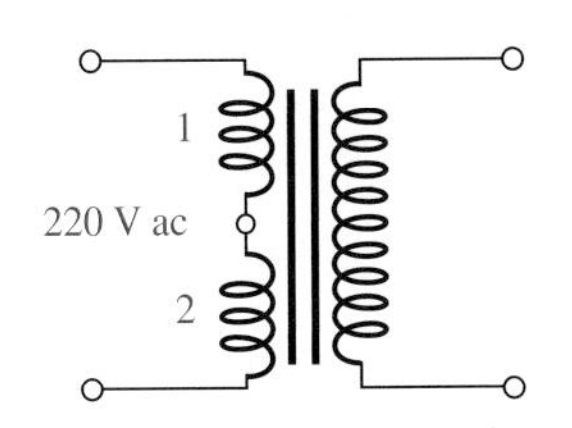

(c) Primary windings in series for 220 V ac operation

FIGURE 14–31

Multiple-primary transformer.

More than one secondary winding can be wound on a common core. Transformers with several secondary windings are often used to achieve several voltages by either stepping up or stepping down the primary voltage. These types are commonly used in power supply applications in which several voltage levels are required for the operation of an electronic instrument.

A typical schematic of a transformer with multiple secondary windings is shown in Figure 14–32; this transformer has three secondary windings. Sometimes you will find combinations of multiple primary windings, multiple secondary windings, and tapped transformers all in one unit.

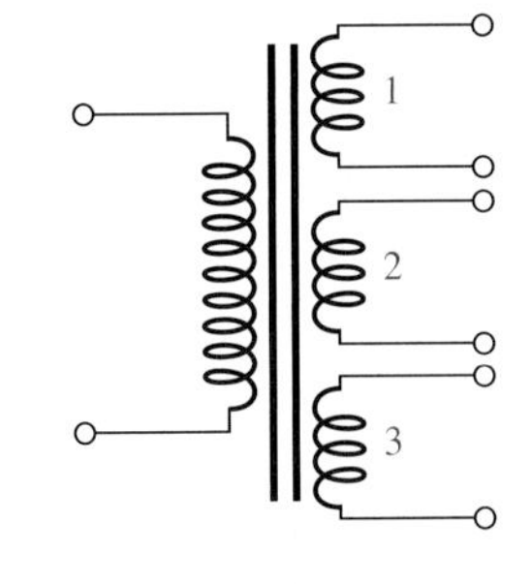

FIGURE 14–32

Multiple-secondary transformer.

EXAMPLE 14–11

The transformer shown in Figure 14–33 has the turns ratios for each secondary relative to the primary as indicated. One of the secondary windings is also center tapped. If 110 V ac are connected to the primary winding, determine each secondary voltage and the voltages with respect to CT on the middle secondary winding.

FIGURE 14–33

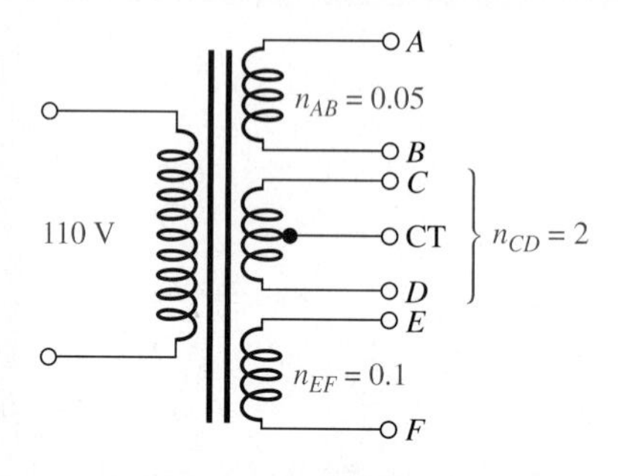

Solution

$$V_{AB} = n_{AB}V_{pri} = (0.05)110\text{ V} = \mathbf{5.5\text{ V}}$$

$$V_{CD} = n_{CD}V_{pri} = (2)110\text{ V} = \mathbf{220\text{ V}}$$

$$V_{(CT)C} = V_{(CT)D} = \frac{220\text{ V}}{2} = \mathbf{110\text{ V}}$$

$$V_{EF} = n_{EF}V_{pri}(0.1)110\text{ V} = \mathbf{11\text{ V}}$$

Related Problem Repeat the calculations if the primary winding is halved.

Autotransformers

In an **autotransformer,** one winding serves as both the primary and the secondary windings. The winding is tapped at the proper points to achieve the desired turns ratio for stepping up or stepping down the voltage.

Autotransformers differ from conventional transformers in that there is no electrical isolation between the primary and the secondary circuits because both are on one winding. Autotransformers normally are smaller and lighter than equivalent conventional transformers because they require a much lower kVA rating for a given load. Many autotransformers provide an adjustable tap using a sliding contact mechanism so that the output voltage can be varied (these are often called *variacs.*) Figure 14–34 shows schematic symbols for various types of autotransformers.

FIGURE 14–34

Variable autotransformers.

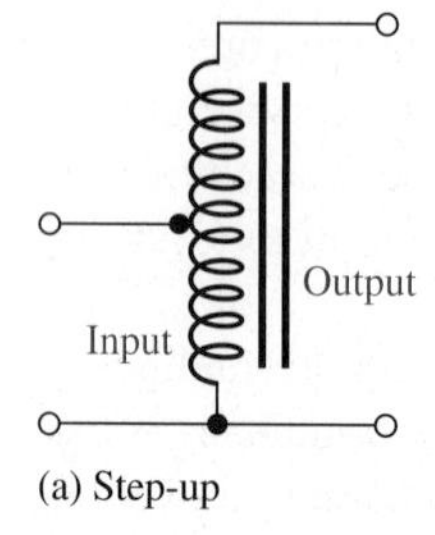

(a) Step-up

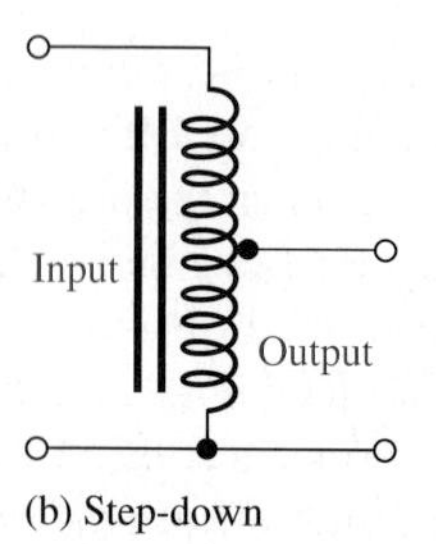

(b) Step-down

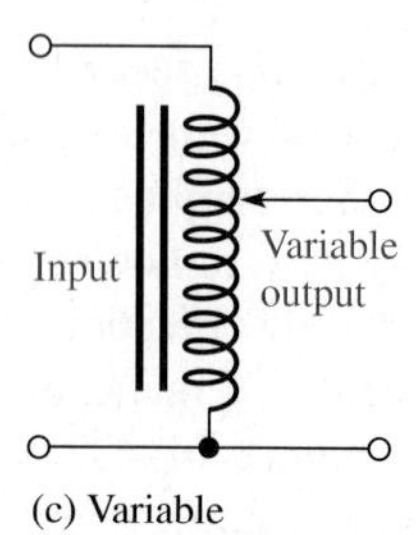

(c) Variable

SECTION 14-10
REVIEW

1. A certain transformer has two secondary windings. The turns ratio from the primary winding to the first secondary winding is 10. The turns ratio from the primary winding to the other secondary winding is 0.2. If 220 V ac are applied to the primary winding, what are the secondary voltages?
2. Name one advantage and one disadvantage of an autotransformer over a conventional transformer.

14-11 TROUBLESHOOTING

Transformers are very simple and reliable devices when operated within their specified range. The common failure in a transformer is an open in either the primary or the secondary winding. One cause of an open is the operation of the device under conditions that exceed its ratings. Shorted or partially shorted windings are possible but very rare. Transformer failure and associated symptoms are covered in this section.

After completing this section, you should be able to

- **Troubleshoot transformers**
- Find an open primary or secondary winding
- Find a shorted or partially shorted primary or secondary winding

Open Primary Winding

When there is an open primary winding, there is no primary current and, therefore, no induced voltage or current in the secondary. This condition is illustrated in Figure 14–35(a), and the method of checking with an ohmmeter is shown in part (b).

FIGURE 14-35

Open primary winding.

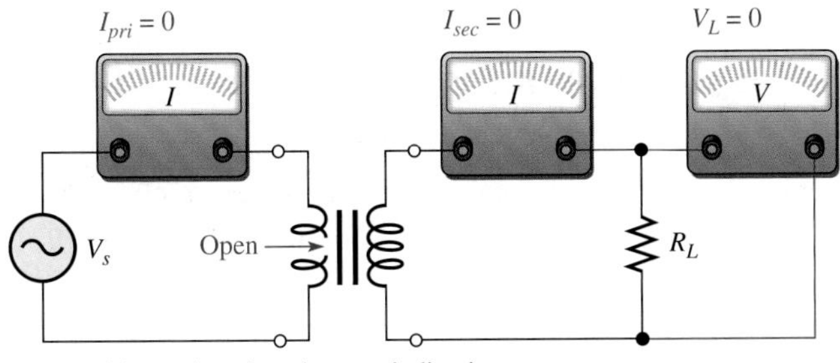

(a) Conditions when the primary winding is open

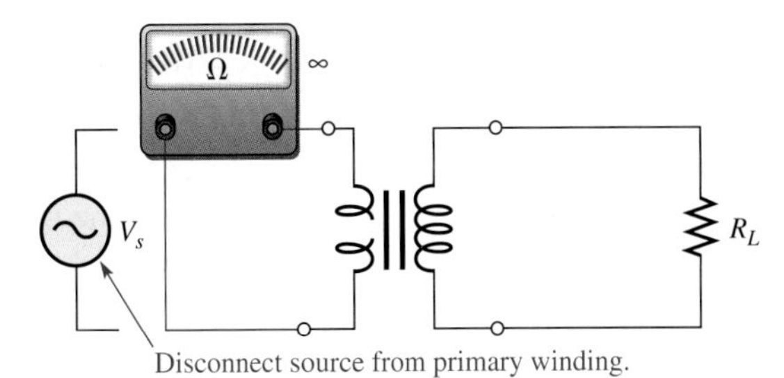

(b) Checking the primary winding with an ohmmeter

Open Secondary Winding

When there is an open secondary winding, there is no current in the secondary circuit and, as a result, no voltage across the load. Also, an open secondary winding causes the primary current to be very small (there is only a small magnetizing

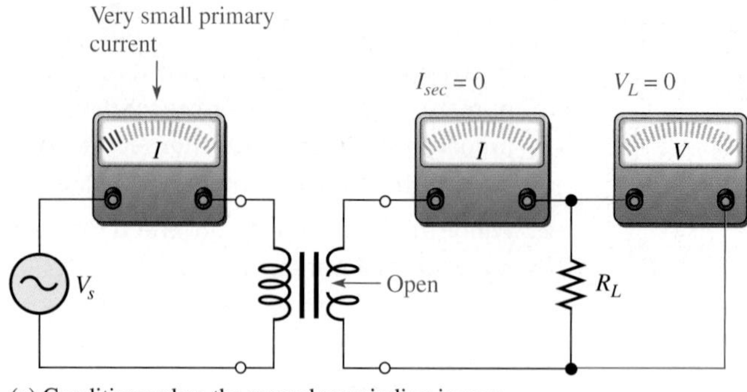

(a) Conditions when the secondary winding is open

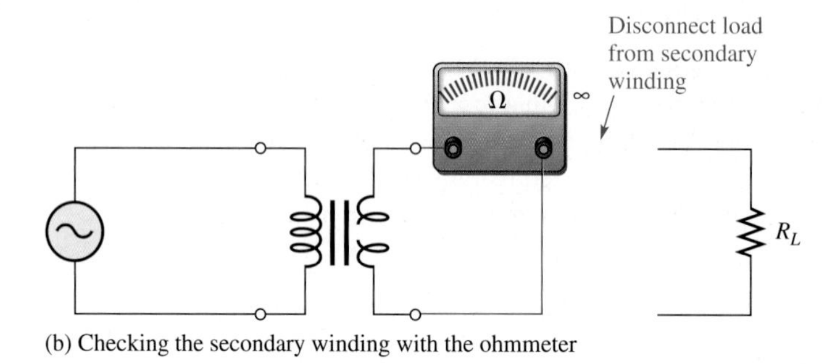

(b) Checking the secondary winding with the ohmmeter

FIGURE 14–36

Open secondary winding.

current). In fact, the primary current may be practically zero. This condition is illustrated in Figure 14–36(a), and the ohmmeter check is shown in part (b).

Shorted or Partially Shorted Primary Winding

A completely shorted primary winding will draw excessive current from the source and, unless there is a breaker or a fuse in the circuit, either the source or the transformer or both will burn out. A partial short in the primary winding can cause higher than normal or even excessive primary current. Any type of shorted winding is very uncommon.

Shorted or Partially Shorted Secondary Winding

In this case, there is an excessive primary current because of the low reflected resistance due to the short. This excessive current will cause a fuse to blow or could burn out the primary winding and result in an open. The short-circuit current in the secondary winding causes the load current to be zero (full short) or smaller than normal (partial short), as demonstrated in Figure 14–37(a) and 14–37(b). The ohmmeter check for this condition is shown in part (c).

Normally, when a transformer fails, it is very difficult to repair, and therefore the simplest procedure is to replace it.

Excessively high primary current

$I_{sec} = 0$

$V_L = 0$

Complete short

V_s

R_L

(a) Secondary winding completely shorted

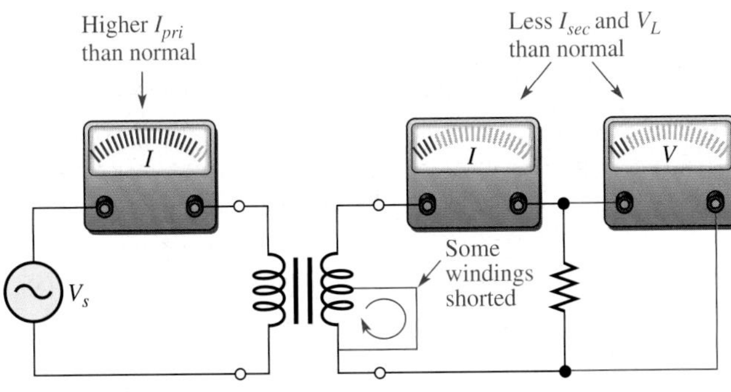

(b) Secondary winding partially shorted

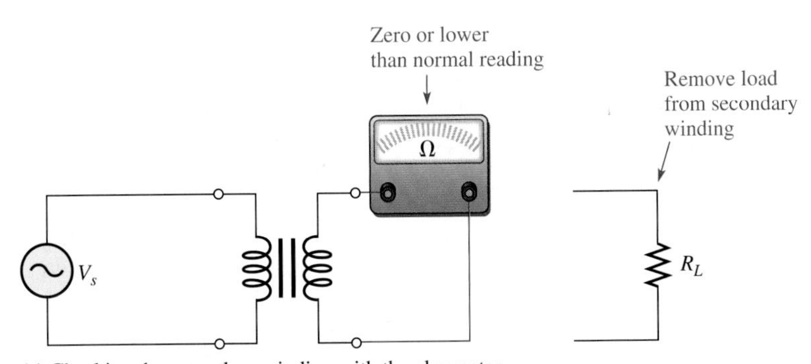

(c) Checking the secondary winding with the ohmmeter

FIGURE 14–37

Shorted secondary winding.

SECTION 14–11 REVIEW

1. What is the most probable failure in a transformer?
2. What is often the cause of transformer failure?

APPLICATION ASSIGNMENT

PUTTING YOUR KNOWLEDGE TO WORK

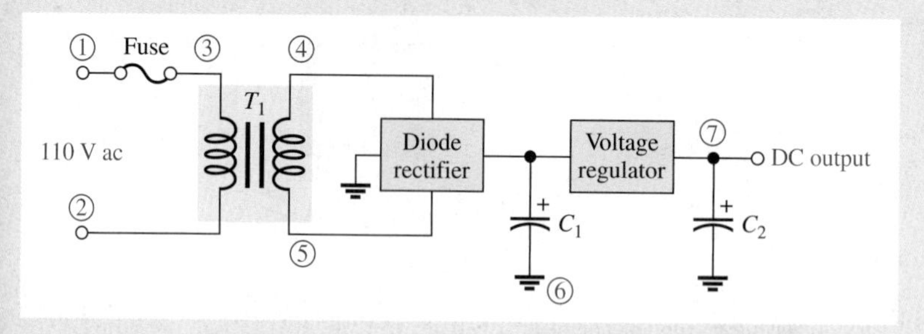

FIGURE 14–38

Basic transformer-coupled dc power supply.

A common application of the transformer is in dc power supplies. The transformer is used to couple the ac line voltage into the power supply circuitry where it is converted to a dc voltage. Your assignment is to troubleshoot four identical transformer-coupled dc power supplies and, based on a series of measurements, determine the fault, if any, in each.

The transformer (T_1) in the power supply schematic of Figure 14–38 steps the 110 V rms at the ac outlet down to 10 V rms which is converted by the diode bridge rectifier, filtered, and regulated to obtain a 6 V dc output. The diode rectifier changes the ac to a pulsating full-wave dc voltage that is smoothed by the capacitor filter C_1. The voltage regulator is an integrated circuit that takes the filtered voltage and provides a constant 6 V dc over a range of load values and line voltage variations. Additional filtering is provided by capacitor C_2. You will learn about these circuits in a later course. The circled numbers in Figure 14–38 correspond to measurement points on the power supply unit.

Step 1: Familiarization with the Power Supply

You have four identical power supply units to troubleshoot like the one shown in Figure 14–39. The power line to the primary winding of the transformer (T_1) is protected by a fuse. The secondary winding is connected to the circuit containing the rectifier, filter, and regulator. Measurement points are indicated by the circled numbers.

Step 2: Measuring Voltages on Power Supply Unit 1

After plugging the power supply into a standard wall outlet, an autoranging portable multimeter is used to measure the voltages. In an autoranging meter, the appropriate measurement range is automatically selected instead of being manually selected as in a standard multimeter.

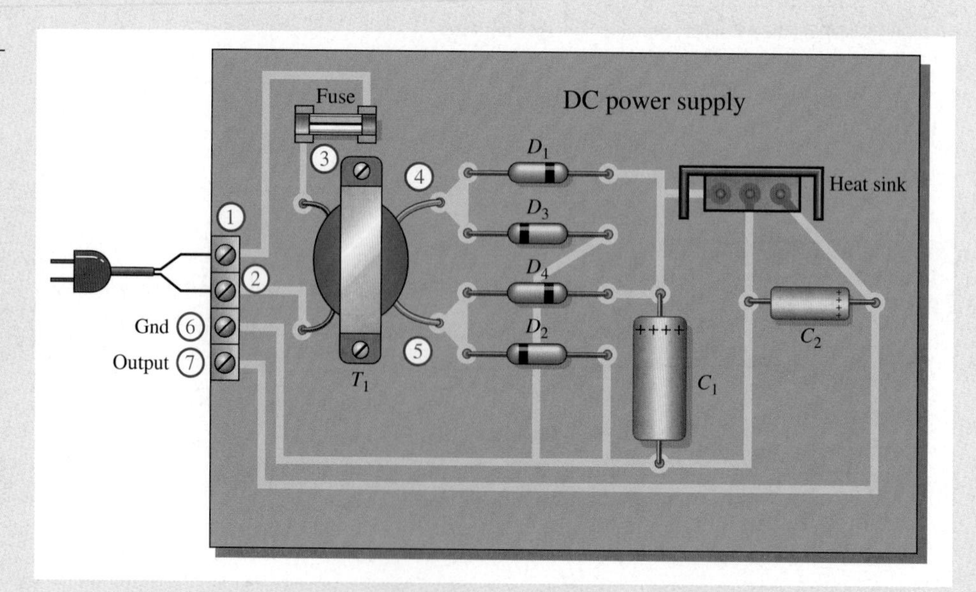

FIGURE 14–39

Power supply unit (top view).

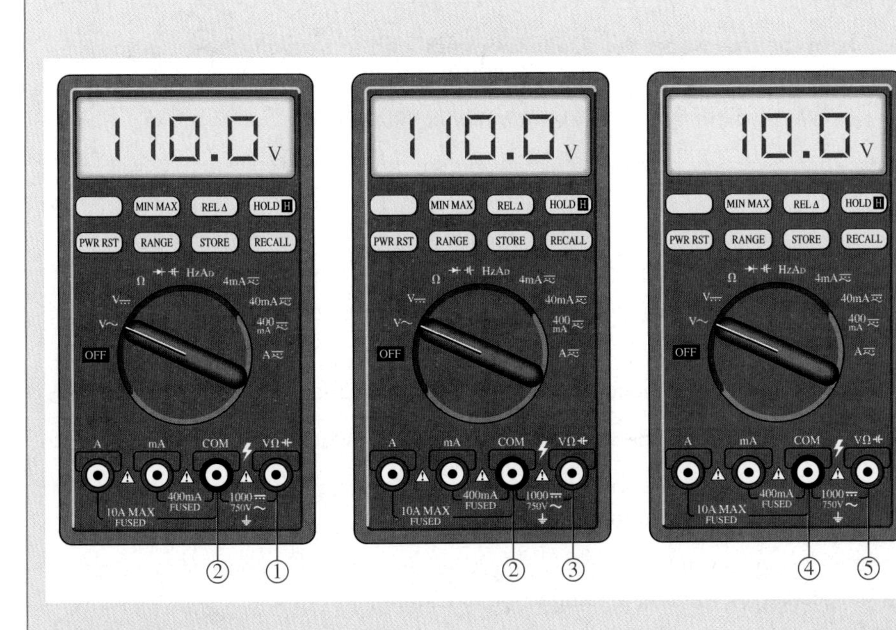

FIGURE 14–40

Voltage measurements on power supply unit 1.

Determine from the meter readings in Figure 14–40 whether or not the power supply is operating properly. If it is not, isolate the problem to one of the following: the circuit containing the rectifier, filter, and regulator; the transformer; the fuse; or the power source. The circled numbers on the meter inputs correspond to the numbered points on the power supply in Figure 14–39.

Step 3: Measuring Voltages on Power Supply Units 2, 3, and 4

Determine from the meter readings for units 2, 3, and 4 in Figure 14–41 whether or not each power supply is operating properly. If it is not, isolate the problem to one of the following: the circuit containing the rectifier, filter, and regulator; the transformer; the fuse; or the power source. Only the meter displays and corresponding measurement points are shown.

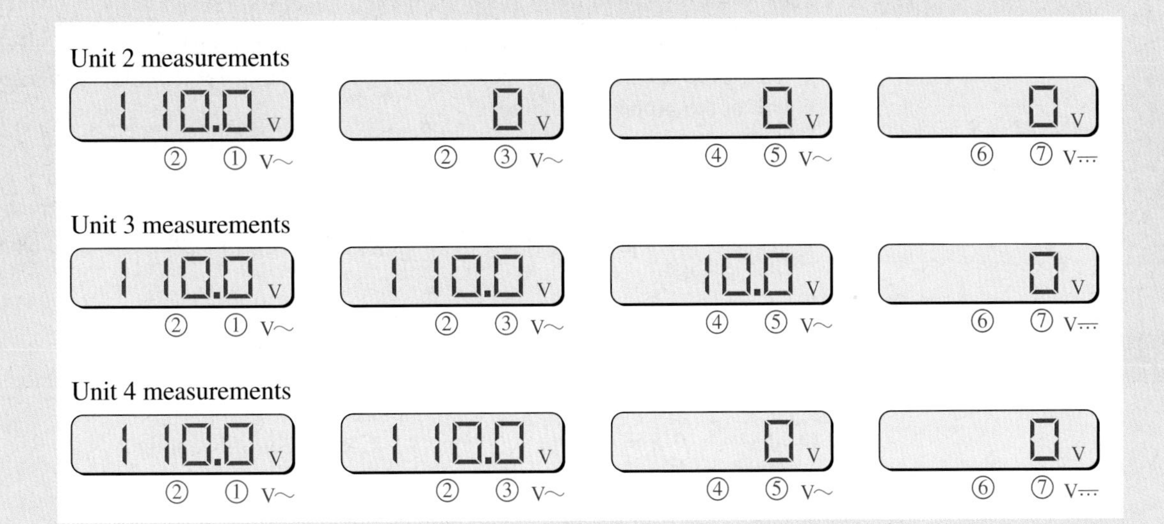

FIGURE 14–41

Measurements for power supply units 2, 3, and 4.

APPLICATION ASSIGNMENT REVIEW

1. In the case where the transformer was found to be faulty, how can you determine the specific fault (open windings or shorted windings)?
2. What type of fault in the transformer could cause the fuse to blow?

SUMMARY

- A transformer generally consists of two or more coils that are magnetically coupled on a common core.
- There is mutual inductance between two magnetically coupled coils.
- When current in one coil changes, voltage is induced in the other coil.
- The primary is the winding connected to the source, and the secondary is the winding connected to the load.
- The number of turns in the primary winding and the number of turns in the secondary winding determine the turns ratio.
- The relative polarities of the primary and secondary voltages are determined by the direction of the windings around the core.
- A step-up transformer has a turns ratio greater than 1.
- A step-down transformer has a turns ratio less than 1.
- A transformer cannot increase power.
- In an ideal transformer, the power from the source (input power) is equal to the power delivered to the load (output power).
- If the voltage is stepped up, the current is stepped down, and vice versa.
- A load connected across the secondary winding of a transformer appears to the source as a reflected load having a value dependent on the reciprocal of the turns ratio squared.
- An impedance-matching transformer can match a load resistance to an internal source resistance to achieve maximum power transfer to the load by selection of the proper turns ratio.
- A transformer does not respond to constant dc.
- Conversion of electrical energy to heat in an actual transformer results from winding resistances, hysteresis loss in the core, eddy currents in the core, and flux leakage.

EQUATIONS

14–1	$k = \dfrac{\phi_{1\text{-}2}}{\phi_1}$	Coefficient of coupling
14–2	$L_M = k\sqrt{L_1 L_2}$	Mutual inductance
14–3	$n = \dfrac{N_{sec}}{N_{pri}}$	Turns ratio
14–4	$\dfrac{V_{sec}}{V_{pri}} = \dfrac{N_{sec}}{N_{pri}}$	Voltage ratio

14–5	$V_{sec} = nV_{pri}$	Secondary voltage
14–6	$\frac{I_{pri}}{I_{sec}} = n$	Current ratio
14–7	$I_{sec} = \left(\frac{1}{n}\right)I_{pri}$	Secondary current
14–8	$R_{pri} = \left(\frac{1}{n}\right)^2 R_L$	Reflected resistance
14–9	$n = \sqrt{\frac{R_L}{R_{pri}}}$	Turns ratio for impedance matching
14–10	$\eta = \left(\frac{P_{out}}{P_{in}}\right)100\%$	Transformer efficiency

SELF-TEST

Answers are at the end of the chapter.

1. A transformer is used for
   (a) dc voltages (b) ac voltages (c) both dc and ac
2. Which one of the following is affected by the turns ratio of a transformer?
   (a) primary voltage (b) dc voltage
   (c) secondary voltage (d) none of these
3. If the windings of a certain transformer with a turns ratio of 1 are in opposite directions around the core, the secondary voltage is
   (a) in phase with the primary voltage
   (b) less than the primary voltage
   (c) greater than the primary voltage
   (d) out of phase with the primary voltage
4. When the turns ratio of a transformer is 10 and the primary ac voltage is 6 V, the secondary voltage is
   (a) 60 V (b) 0.6 V (c) 6 V (d) 36 V
5. When the turns ratio of a transformer is 0.5 and the primary ac voltage is 100 V, the secondary voltage is
   (a) 200 V (b) 50 V (c) 10 V (d) 100 V
6. A certain transformer has 500 turns in the primary winding and 2500 turns in the secondary winding. The turns ratio is
   (a) 0.2 (b) 2.5 (c) 5 (d) 0.5
7. If 10 W of power are applied to the primary winding of an ideal transformer with a turns ratio of 5, the power delivered to the secondary load is
   (a) 50 W (b) 0.5 W (c) 0 W (d) 10 W
8. In a certain loaded transformer, the secondary voltage is one-third the primary voltage. Ideally, the secondary current is
   (a) one-third the primary current (b) three times the primary current
   (c) equal to the primary current (d) less than the primary current
9. When a 1.0 kΩ load resistor is connected across the secondary winding of a transformer with a turns ratio of 2, the source "sees" a reflected load of
   (a) 250 Ω (b) 2 kΩ (c) 4 kΩ (d) 1.0 kΩ
10. In Question 9, if the turns ratio is 0.5, the source "sees" a reflected load of
    (a) 1.0 kΩ (b) 2 kΩ (c) 4 kΩ (d) 500 Ω

11. The turns ratio required to match a 50 Ω source to a 200 Ω load is
   (a) 0.25 **(b)** 0.5 **(c)** 4 **(d)** 2
12. Maximum power is transferred from a source to a load when
   (a) $R_L > R_{int}$ **(b)** $R_L < R_{int}$ **(c)** $R_L = R_{int}$ **(d)** $R_L = nR_{int}$
13. When a 12 V battery is connected across the primary winding of a transformer with a turns ratio of 4, the secondary voltage is
   (a) 0 V **(b)** 12 V **(c)** 48 V **(d)** 3 V
14. A certain transformer has a turns ratio of 1 and a 0.95 coefficient of coupling. When 1 V ac is applied to the primary, the secondary voltage is
   (a) 1 V **(b)** 1.95 V **(c)** 0.95 V

PROBLEMS

Answers to odd-numbered problems are at the end of the book.

BASIC PROBLEMS

SECTION 14–1 **Mutual Inductance**

1. What is the mutual inductance when $k = 0.75$, $L_1 = 1\ \mu H$, and $L_2 = 4\ \mu H$?
2. Determine the coefficient of coupling when $L_M = 1\ \mu H$, $L_1 = 8\ \mu H$, and $L_2 = 2\ \mu H$.

SECTION 14–2 **The Basic Transformer**

3. What is the turns ratio of a transformer having 120 turns in its primary winding and 360 turns in its secondary winding?
4. **(a)** What is the turns ratio of a transformer having 250 turns in its primary winding and 1000 turns in its secondary winding?
   (b) What is the turns ratio when the primary winding has 400 turns and the secondary winding has 100 turns?
5. Determine the phase of the secondary voltage with respect to the primary voltage for each transformer in Figure 14–42.

▼ FIGURE 14–42

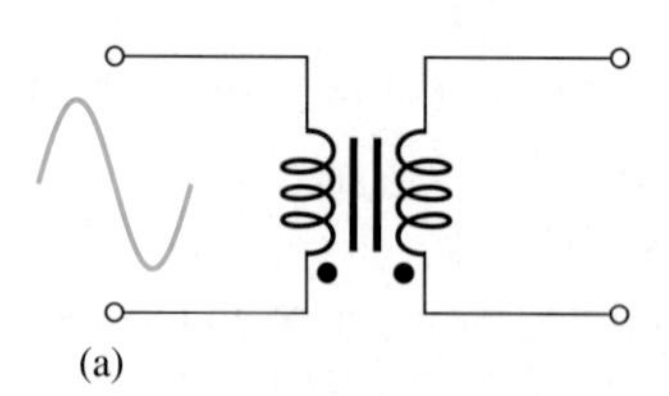

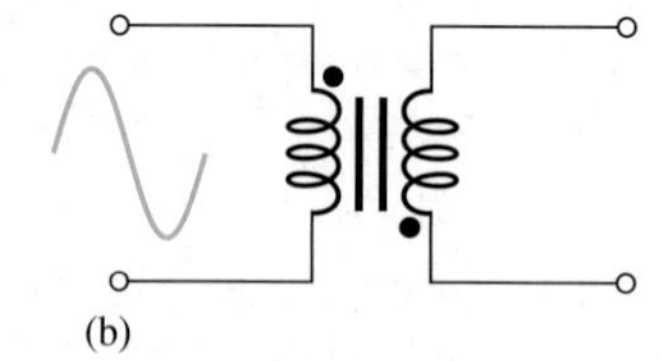

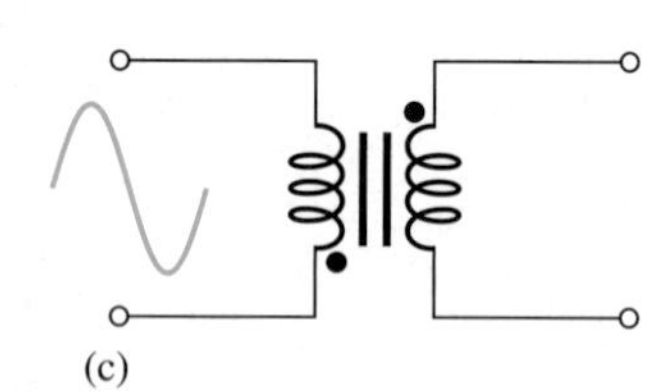

SECTION 14–3 **Step-Up Transformers**

6. If 120 V ac are connected to the primary of a transformer with a turns ratio of 1.5, what is the secondary voltage?
7. A certain transformer has 250 turns in its primary winding. In order to double the secondary voltage, how many turns must be in the secondary winding?
8. How many primary volts must be applied to a transformer with a turns ratio of 10 to obtain a secondary voltage of 60 V ac?

9. For each transformer in Figure 14–43, draw the secondary voltage showing its relationship to the primary voltage. Also indicate the amplitude.

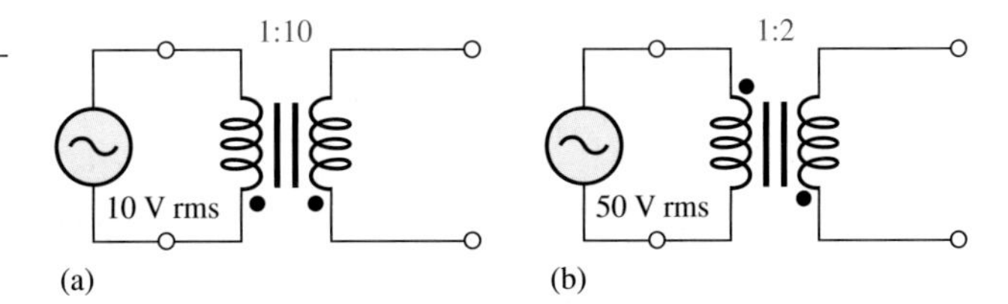

FIGURE 14–43

SECTION 14–4

Step-Down Transformers

10. To step 120 V down to 30 V, what must be the turns ratio?

11. The primary winding of a transformer has 1200 V across it. What is the secondary voltage if the turns ratio is 0.2?

12. How many primary volts must be applied to a transformer with a turns ratio of 0.1 to obtain a secondary voltage of 6 V ac?

SECTION 14–5

Loading the Secondary

13. Determine I_s in Figure 14–44.

14. Determine the following quantities in Figure 14–45:

    (a) secondary voltage **(b)** secondary current

    (c) primary current **(d)** power in the load

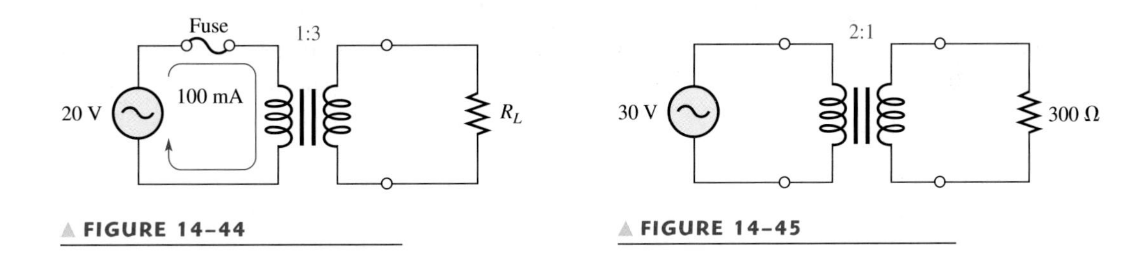

FIGURE 14–44

FIGURE 14–45

SECTION 14–6

Reflected Load

15. What is the load resistance as seen by the source in Figure 14–46?

16. What is the resistance reflected into the primary circuit in Figure 14–47?

17. What is the primary current (rms) in Figure 14–47 if the rms source voltage is 115 V?

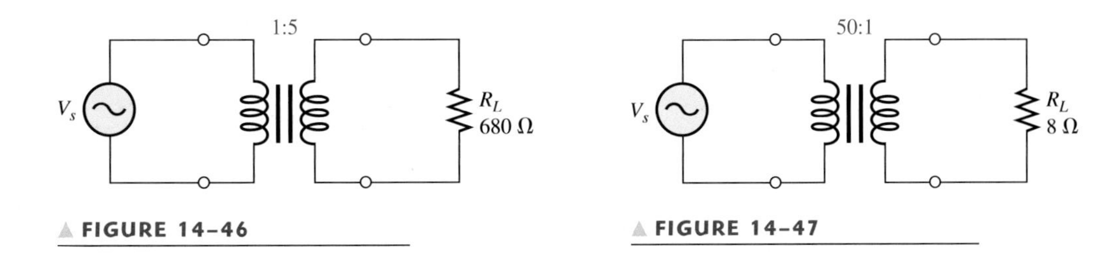

FIGURE 14–46

FIGURE 14–47

18. What must be the turns ratio in Figure 14–48 in order to reflect 300 Ω into the primary circuit?

SECTION 14–7 Matching the Load and Source Resistances

19. For the circuit in Figure 14–49, find the turns ratio required to deliver maximum power to the 4 Ω speaker.

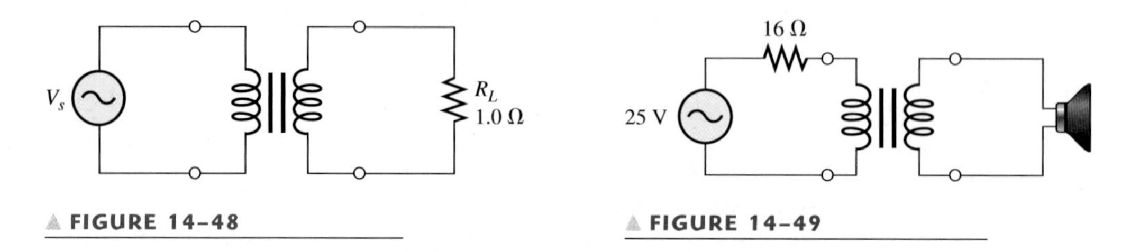

▲ FIGURE 14–48

▲ FIGURE 14–49

20. In Figure 14–49, what is the maximum power in watts delivered to the speaker?

21. Determine the value to which R_L must be adjusted in Figure 14–50 for maximum power transfer. The internal source resistance is 50 Ω.

► FIGURE 14–50

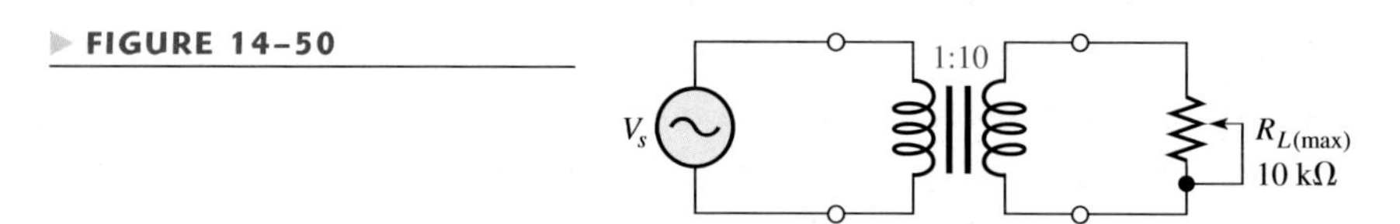

SECTION 14–8 The Transformer as an Isolation Device

22. What is the voltage across the load in each circuit in Figure 14–51?

23. If the bottom of each secondary winding in Figure 14–51 were grounded, would the values of the load voltages be changed?

▼ FIGURE 14–51

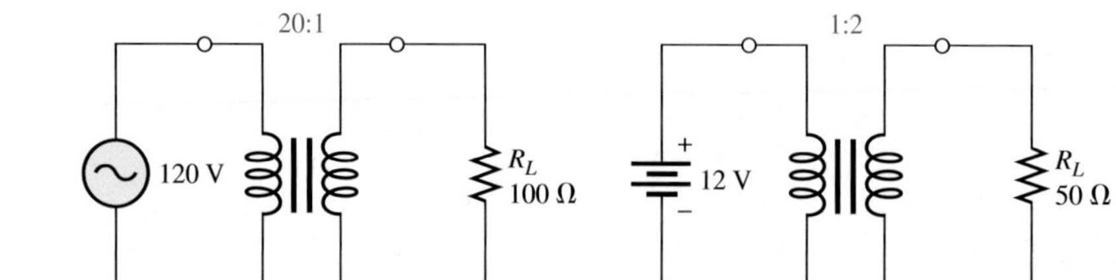

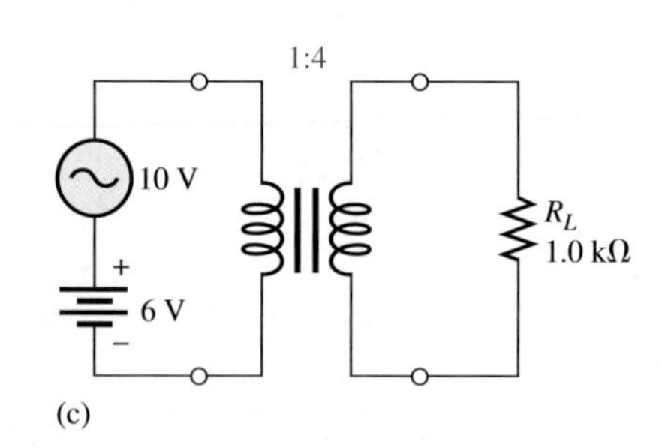

24. Determine the unspecified meter readings in Figure 14–52.

▼ FIGURE 14–52

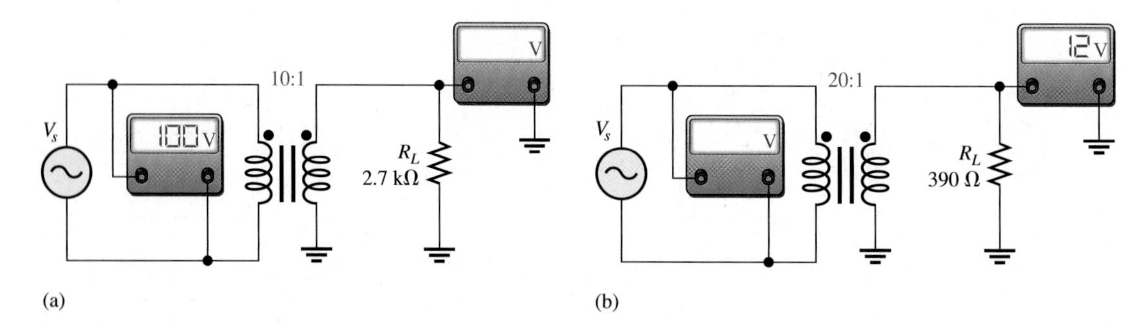

SECTION 14–9

Nonideal Transformer Characteristics

25. In a certain transformer, the input power to the primary is 100 W. If 5.5 W are dissipated in the winding resistances, what is the output power to the load, neglecting any other losses?
26. What is the efficiency of the transformer in Problem 25?
27. Determine the coefficient of coupling for a transformer in which 2% of the total flux generated in the primary does not pass through the secondary.
28. A certain transformer is rated at 1 kVA. It operates on 60 Hz, 120 V ac. The secondary voltage is 600 V.
    (a) What is the maximum load current?
    (b) What is the smallest R_L that you can drive?
    (c) What is the largest capacitor that can be connected as a load?
29. What kVA rating is required for a transformer that must handle a maximum load current of 10 A with a secondary voltage of 2.5 kV?

SECTION 14–10

Other Types of Transformers

30. Determine each unknown voltage indicated in Figure 14–53.
31. Using the indicated secondary voltages in Figure 14–54, determine the turns ratio of the primary winding to each tapped section.

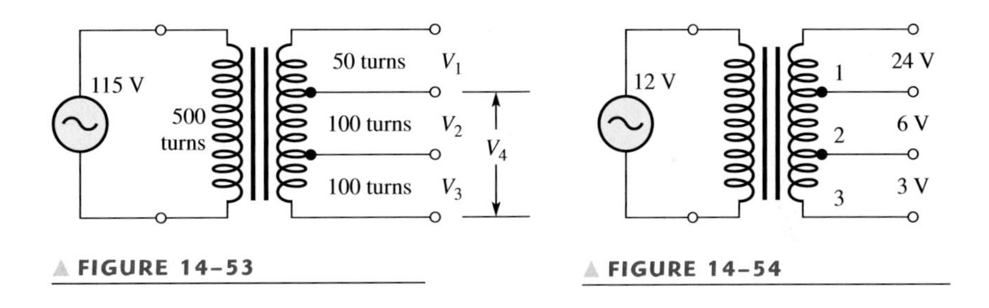

▲ FIGURE 14–53

▲ FIGURE 14–54

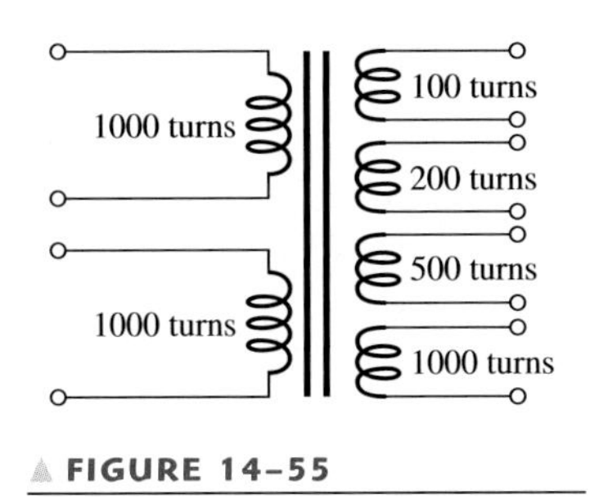

▲ FIGURE 14–55

32. In Figure 14–55, each primary winding can accommodate 120 V ac. Show how the primaries should be connected for 240 V ac operation. Determine each secondary voltage.
33. Determine the turns ratios from each primary to each secondary in Figure 14–55.

SECTION 14–11

Troubleshooting

34. When you apply 120 V ac across the primary winding of a transformer and check the voltage across the secondary winding, you get 0 V. Further investigation shows no primary or secondary current. List the possible faults. What is your next step in isolating the problem?
35. What is likely to happen if the primary winding of a transformer shorts?
36. In checking out a transformer circuit you find the secondary voltage is less than it should be, although it is not zero. What is the most likely cause?

ADVANCED PROBLEMS

37. For the loaded, tapped-secondary transformer in Figure 14–56, determine the following:

(a) all load voltages and currents

(b) the resistance looking into the primary

▶ FIGURE 14–56

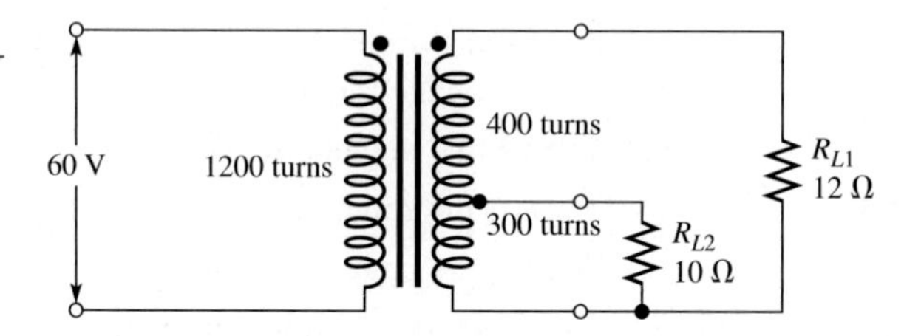

38. A certain transformer is rated at 5 kVA, 2400/120 V, at 60 Hz.

(a) What is the turns ratio if the 120 V is the secondary voltage?

(b) What is the current rating of the secondary if 2400 V is the primary voltage?

(c) What is the current rating of the primary if 2400 V is the primary voltage?

39. Determine the voltage measured by each voltmeter in Figure 14–57. The bench-type meters have one terminal that connects to ground as indicated.

▼ FIGURE 14–57

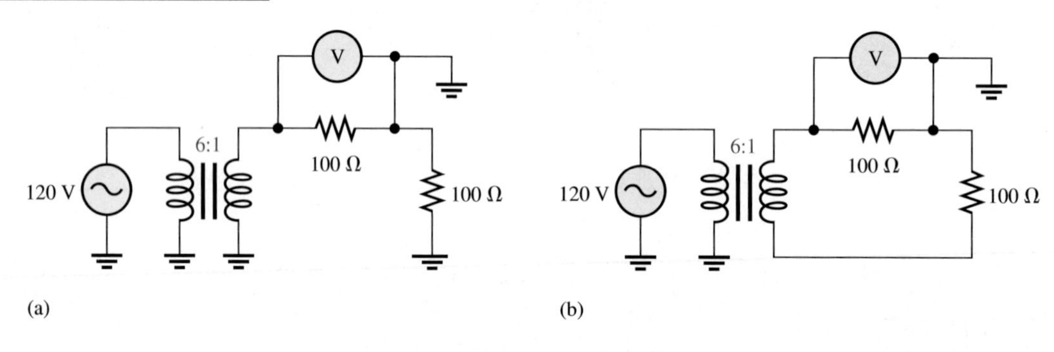

40. Find the appropriate turns ratio for each switch position in Figure 14–58 in order to transfer the maximum power to each load when the internal source resistance is 10 Ω. Specify the number of turns for the secondary winding if the primary winding has 100 turns.

▶ FIGURE 14–58

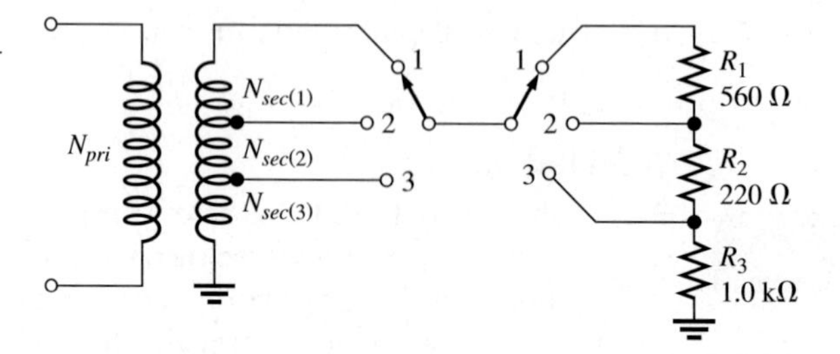

41. What must the turns ratio be in Figure 14–47 to limit the primary current to 3 mA with a source voltage of 115 V? Assume an ideal transformer and source.

42. Assume 110 V is applied to the primary of a transformer with a VA rating of 10 VA. The output voltage is 12.6 V. What is the smallest resistor that can be connected across the secondary?

43. A step-down transformer uses 110 V on the primary and 10 V on the secondary. If the secondary is rated for a maximum of 1 A, what fuse rating should be selected on the primary side?

ELECTRONICS WORKBENCH/CIRCUITMAKER TROUBLESHOOTING PROBLEMS

CD-ROM file circuits are shown in Figure 14–59.

44. Open file P14-44 and test the circuit. If there is a fault, identify it.

45. Open file P14-45 and test the circuit. If there is a fault, identify it.

46. Determine if there is a fault in the circuit in file P14-46. If so, identify it.

47. Find and specify any faulty component in the circuit in file P14-47.

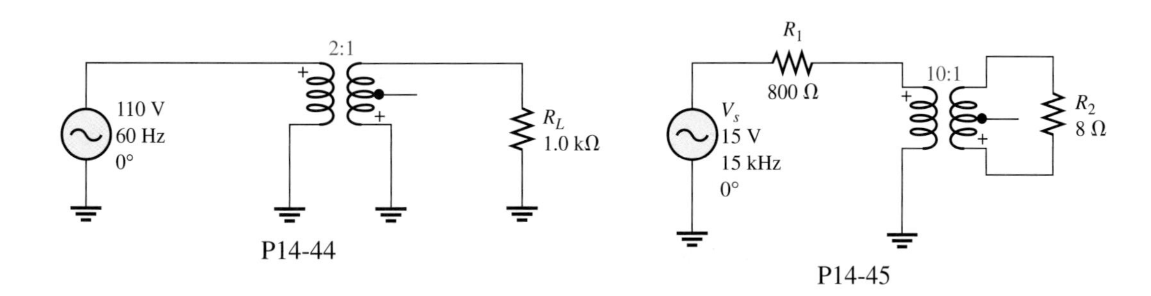

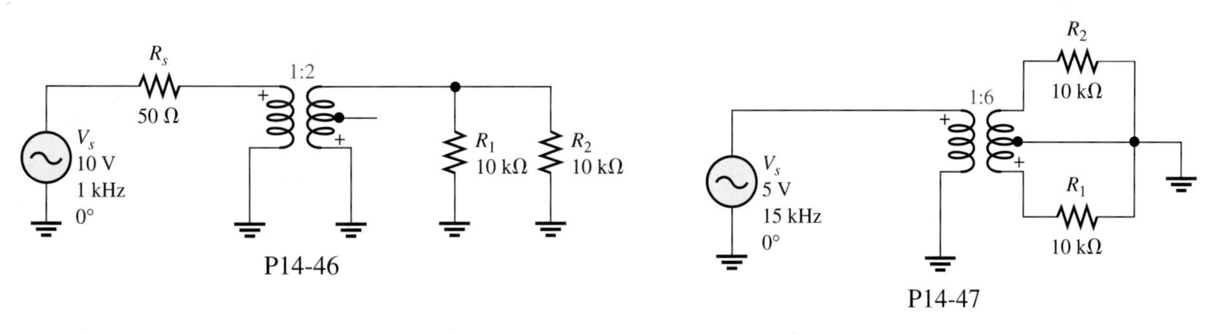

▲ FIGURE 14–59

ANSWERS

SECTION REVIEWS

SECTION 14–1 **Mutual Inductance**

1. Mutual inductance is the inductance between two coils, established by the amount of coupling between the coils.

2. $L_M = k\sqrt{L_1L_2} = 45$ mH

3. Induced voltage increases when k is increased.

SECTION 14–2 **The Basic Transformer**

1. Transformer operation is based on the principle of mutual inductance.
2. Turns ratio is the ratio of turns in the secondary to turns in the primary.
3. Directions of the windings determine the relative voltage polarities.
4. $n = N_{sec}/N_{pri} = 0.5$

SECTION 14–3 **Step-Up Transformers**

1. A step-up transformer increases voltage.
2. The secondary voltage is five times greater.
3. $V_{sec} = nV_{pri} = 2400$ V

SECTION 14–4 **Step-Down Transformers**

1. A step-down transformer decreases voltage.
2. $V_{sec} = nV_{pri} = 60$ V
3. $n = 12$ V/120 V $= 0.1$

SECTION 14–5 **Loading the Secondary**

1. The secondary current is half the primary current.
2. $n = 0.25$; $I_{sec} = (1/n)I_{pri} = 2$ A
3. $I_{pri} = nI_{sec} = 2.5$ A

SECTION 14–6 **Reflected Load**

1. Reflected resistance is the resistance in the secondary circuit as seen from the primary circuit as a function of the turns ratio.
2. The reciprocal of the turns ratio determines reflected resistance.
3. $R_{pri} = (1/n)^2R_L = 0.5\ \Omega$
4. $n = \sqrt{R_L/R_{pri}} = 0.1$

SECTION 14–7 **Matching the Load and Source Resistances**

1. Impedance matching is making the load resistance equal the source resistance.
2. Maximum power is delivered to the load when $R_L = R_{int}$.
3. $R_{pri} = (1/n)^2R_L = 400\ \Omega$

SECTION 14–8 **The Transformer as an Isolation Device**

1. Electrical isolation means there is no electrical connection between the primary and secondary circuits.
2. No, transformers do not couple dc.

SECTION 14–9 **Nonideal Transformer Characteristics**

1. In a practical transformer, conversion of electrical energy to heat reduces the efficiency. An ideal transformer has an efficiency of 100%.
2. When the coefficient of coupling is 0.85, 85% of the magnetic flux generated in the primary winding passes through the secondary winding.
3. $I_L = 10$ kVA/250 V $= 40$ A

SECTION 14–10 **Other Types of Transformers**

1. $V_{sec} = 10(220\text{ V}) = 2200\text{ V}$; $V_{sec} = 0.2(220\text{ V}) = 44\text{ V}$
2. Autotransformers are smaller and lighter for the same rating. Autotransformers provide no electrical isolation.

SECTION 14–11 **Troubleshooting**

1. The most probable failure is an open winding.
2. Operating above rated values causes transformer failure.

Application Assignment

1. Use an ohmmeter to check for open windings. Shorted windings are indicated by an incorrect secondary voltage.
2. A short will cause the fuse to blow.

RELATED PROBLEMS FOR EXAMPLES

14–1 0.75
14–2 387 μH
14–3 5000 turns
14–4 480 V
14–5 57.6 V
14–6 5 mA; 400 mA
14–7 6 Ω
14–8 0.354
14–9 0.0707 or 14.14:1
14–10 85.2%
14–11 $V_{AB} = 11\text{ V}$; $V_{CD} = 440\text{ V}$; $V_{(\text{CT})C} = V_{(\text{CT})D} = 220\text{ V}$; $V_{EF} = 22\text{ V}$

SELF-TEST

1. (b) **2.** (c) **3.** (d) **4.** (a) **5.** (b) **6.** (c) **7.** (d)
8. (b) **9.** (a) **10.** (c) **11.** (d) **12.** (c) **13.** (a) **14.** (c)

15

PULSE RESPONSE OF REACTIVE CIRCUITS

INTRODUCTION

In Chapters 10 and 12, the sinusoidal frequency response of *RC* and *RL* circuits was covered. In this chapter, the response of *RC* and *RL* circuits to pulse inputs is examined.

Before starting this chapter, you should review the material in Sections 9–5 and 11–5. Understanding exponential changes in voltages and currents in capacitors and inductors is crucial to the study of pulse response. Exponential formulas that were given in Chapter 9 are used throughout this chapter.

With pulse inputs, the time response of the circuits is of primary importance. In the areas of pulse and digital circuits, technicians are often concerned with how a circuit responds over an interval of time to rapid changes in voltage or current. The relationship of the circuit time constant to the input pulse characteristics, such as pulse width and period, determines the wave shapes of the voltages in the circuit.

Integrator and *differentiator,* terms used throughout this coverage, refer to mathematical functions that are approximated by these circuits under certain conditions. Mathematical integration is an averaging process, and mathematical differentiation is a process for establishing an instantaneous rate of change of a quantity. We will not deal with the purely mathematical aspects at this level of coverage.

CHAPTER OUTLINE

CHAPTER OBJECTIVES

- Explain the operation of an *RC* integrator
- Analyze an *RC* integrator with a single input pulse
- Analyze an *RC* integrator with repetitive input pulses
- Analyze an *RC* differentiator with a single input pulse
- Analyze an *RC* differentiator with repetitive input pulses
- Analyze the operation of an *RL* integrator
- Analyze the operation of an *RL* differentiator
- Discuss some basic applications of integrators and differentiators
- Troubleshoot *RC* integrators and *RC* differentiators

KEY TERMS

- Integrator
- Time constant
- Pulse response
- Transient time
- Steady state
- Differentiator

APPLICATION ASSIGNMENT PREVIEW

For the application assignment in this chapter, you will "breadboard" a time-delay circuit and determine the component values required to meet certain specifications. You will also determine how to test the circuit for proper operation using a pulse generator and an oscilloscope. After studying this chapter, you should be able to complete the application assignment.

WWW. VISIT THE COMPANION WEBSITE

Circuit Simulation Tutorials and Other Chapter Study Tools Are Available at

http://www.prenhall.com/floyd

15–1 THE *RC* INTEGRATOR

In terms of pulse response, a series *RC* circuit in which the output voltage is taken across the capacitor is known as an **integrator**. Recall that in terms of frequency response, this particular series *RC* circuit is a low-pass filter. The term, *integrator*, is derived from a mathematical function which this type of circuit approximates under certain conditions.

After completing this section, you should be able to

- **Explain the operation of an *RC* integrator**
- Describe how the capacitor charges and discharges
- Explain how a capacitor reacts to an instantaneous change in voltage or current
- Describe the basic output voltage waveform

Charging and Discharging of a Capacitor

When a pulse generator is connected to the input of an *RC* integrator, as shown in Figure 15–1, the capacitor will charge and discharge in response to the pulses. When the input goes from its low level to its high level, the capacitor charges toward the high level of the pulse through the resistor. This charging action is analogous to connecting a battery through a switch to the *RC* circuit, as illustrated in Figure 15–2(a). When the pulse goes from its high level back to its low level, the capacitor discharges back through the source. The resistance of the source is assumed to be negligible compared to *R*. This discharging action is analogous to replacing the source with a closed switch, as illustrated in Figure 15–2(b).

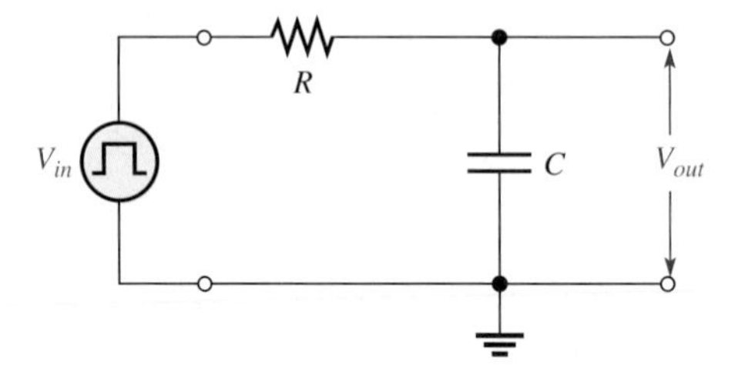

FIGURE 15–1

An *RC* integrator with a pulse generator connected.

FIGURE 15–2

The equivalent action when a pulse source charges and discharges the capacitor.

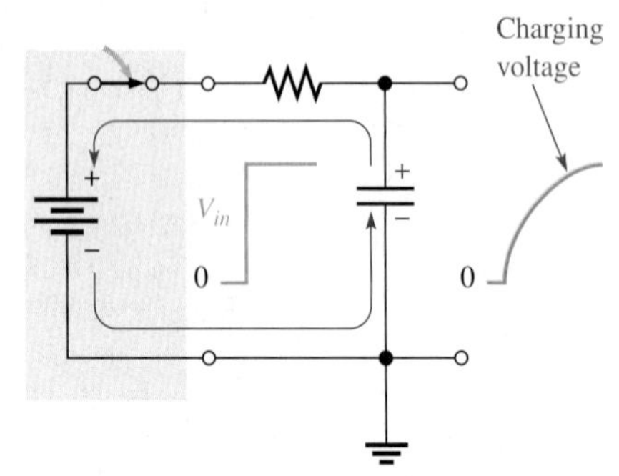

(a) When the input pulse goes HIGH, the source effectively acts as a battery in series with a closed switch, thereby charging the capacitor.

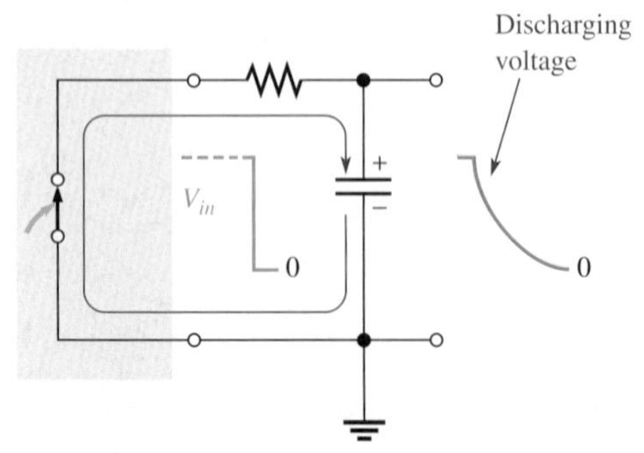

(b) When the input pulse goes back LOW, the source effectively acts as a closed switch, providing a discharge path for the capacitor.

A capacitor will charge and discharge following an exponential curve. Its rate of charging and discharging depends on the *RC* time constant ($\tau = RC$).

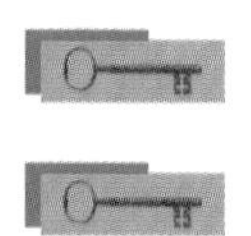

For an ideal pulse, both edges are considered to be instantaneous. Two basic rules of capacitor behavior help in understanding the pulse response of *RC* circuits.

1. The capacitor appears as a short to an instantaneous change in current and as an open to dc.
2. The voltage across the capacitor cannot change instantaneously—it can change only exponentially.

The Capacitor Voltage

In an *RC* integrator, the output is the capacitor voltage. The capacitor charges during the time that the input pulse is high. If the pulse is at its high level long enough, the capacitor will fully charge to the voltage amplitude of the pulse, as illustrated in Figure 15–3. The capacitor discharges during the time that the pulse is low. If the low time between pulses is long enough, the capacitor will fully discharge to zero, as shown in the figure. Then when the next pulse occurs, it will charge again.

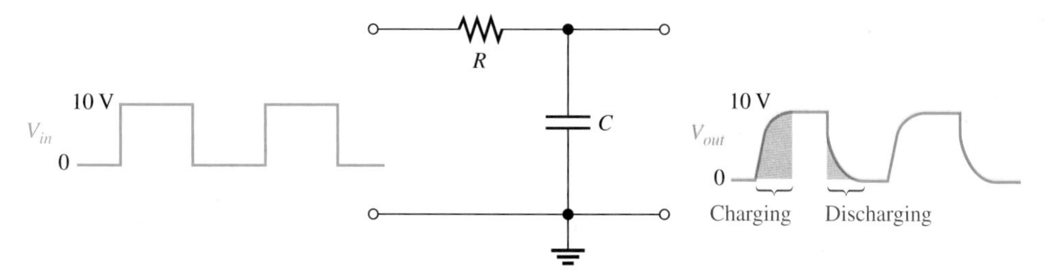

FIGURE 15–3

Illustration of a capacitor fully charging and discharging in response to a pulse input. A pulse generator is connected to the input, but the symbol is not shown, only the waveform.

SECTION 15–1 REVIEW

Answers are at the end of the chapter.

1. Define the term *integrator* in relation to an *RC* circuit.
2. What causes a capacitor in an *RC* circuit to charge and discharge?

15–2 SINGLE-PULSE RESPONSE OF *RC* INTEGRATORS

From the previous section, you have a general idea of how an *RC* integrator responds to a pulse input. In this section, the response to a single pulse is examined in detail.

After completing this section, you should be able to

- **Analyze an *RC* integrator with a single input pulse**
- Discuss the importance of the circuit time constant
- Define *transient time*
- Determine the response when the pulse width is equal to or greater than five time constants
- Determine the response when the pulse width is less than five time constants

Two conditions of pulse response must be considered:

1. When the input pulse width (t_W) is equal to or greater than five time constants ($t_W \geq 5\tau$)
2. When the input pulse width is less than five time constants ($t_W < 5\tau$)

Recall that five time constants is accepted as the time a capacitor needs to fully charge or fully discharge; this time is often called the **transient time**.

When the Pulse Width Is Equal to or Greater Than Five Time Constants

A capacitor will fully charge if the pulse width is equal to or greater than five time constants (5τ). This condition is expressed as $t_W \geq 5\tau$. At the end of the pulse, the capacitor fully discharges back through the source.

Figure 15–4 illustrates the output waveforms for various *RC* time constants and a fixed input pulse width. Notice that the shape of the output pulse approaches that of the input as the transient time is made small compared to the pulse width. In each case, the output reaches the full amplitude of the input.

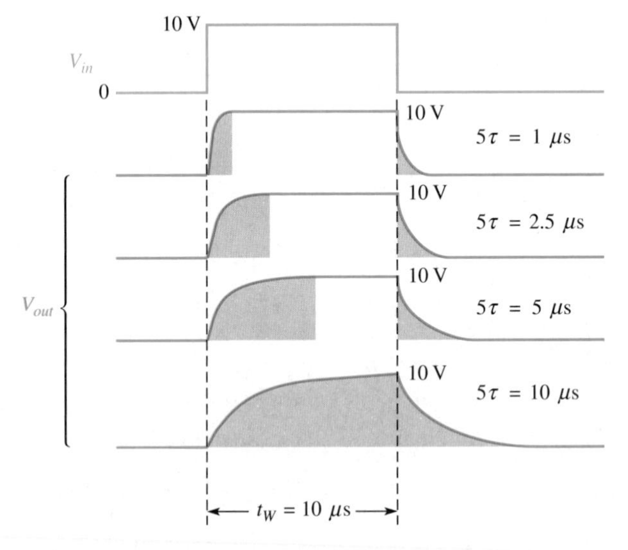

FIGURE 15–4

Variation of an integrator's output pulse shape with time constant. The shaded areas indicate when the capacitor is charging and discharging.

FIGURE 15–5

Variation of an integrator's output pulse shape with input pulse width (the time constant is fixed). Blue is input and green is output.

Figure 15–5 shows how a fixed time constant and a variable input pulse width affect the integrator output. Notice that as the pulse width is increased, the shape of the output pulse approaches that of the input because the transient time is short compared to the pulse width. The rise and fall times of the output remain constant.

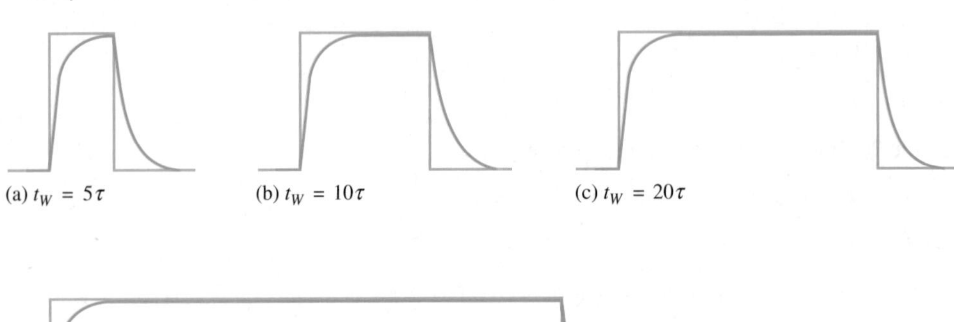

When the Pulse Width Is Less Than Five Time Constants

Now let's examine the case in which the width of the input pulse is less than five time constants of the integrator. This condition is expressed as $t_W < 5\tau$.

As you know, a capacitor charges for the duration of the pulse and the pulse width is the time it has for charging. However, because the pulse width is less than the time the capacitor needs to fully charge (5τ), the output voltage will *not* reach the full input voltage before the end of the pulse. The capacitor only partially charges, as illustrated in Figure 15–6 for several values of *RC* time constants. Notice that for longer time constants, the output reaches a lower voltage because the capacitor cannot charge as much during the pulse width. Of course, in the examples with a single-pulse input, the capacitor fully discharges after the pulse ends.

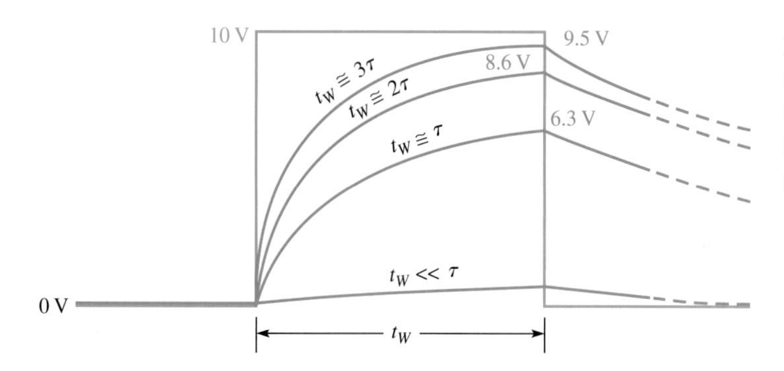

FIGURE 15–6

Capacitor voltage for various time constants that are longer than the input pulse width. Blue is input and green is output.

When the time constant is much greater than the input pulse width, the capacitor charges very little, and, as a result, the output voltage becomes a very small almost constant value, as indicated in Figure 15–6.

Figure 15–7 illustrates the effect of reducing the input pulse width for a fixed time constant value. As the width is reduced, the output voltage becomes smaller because the capacitor has less time to charge. However, it takes the capacitor the same length of time (5τ) to discharge back to zero for each condition after the pulse is removed.

FIGURE 15–7

The capacitor charges less and less as the input pulse width is reduced. The time constant is fixed.

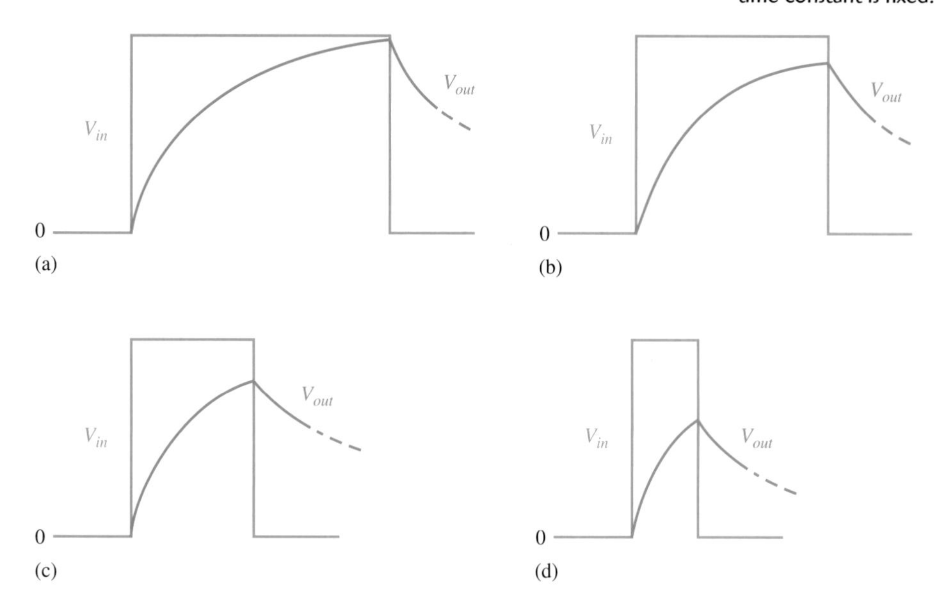

EXAMPLE 15–1

A single 10 V pulse with a width of 100 μs is applied to the integrator in Figure 15–8. The source resistance is assumed to be zero.

(a) To what voltage will the capacitor charge?

(b) How long will it take the capacitor to discharge?

(c) Show the output voltage waveform.

FIGURE 15–8

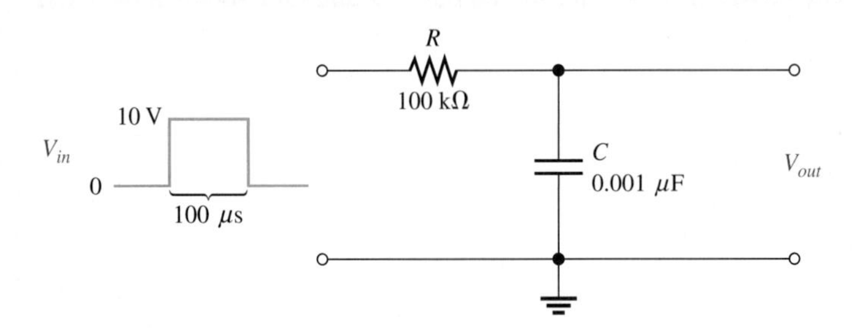

Solution **(a)** The circuit time constant is

$$\tau = RC = (100\ \text{k}\Omega)(0.001\ \mu\text{F}) = 100\ \mu\text{s}$$

Notice that the pulse width is exactly equal to one time constant. Thus, the capacitor will charge approximately 63% of the full input amplitude in one time constant, so the output will reach a maximum voltage of

$$V_{out} = (0.63)10\ \text{V} = \mathbf{6.3\ V}$$

(b) The capacitor discharges back through the source when the pulse ends. The total discharge time is

$$5\tau = 5(100\ \mu\text{s}) = \mathbf{500\ \mu s}$$

(c) The output charging and discharging curve is shown in Figure 15–9.

FIGURE 15–9

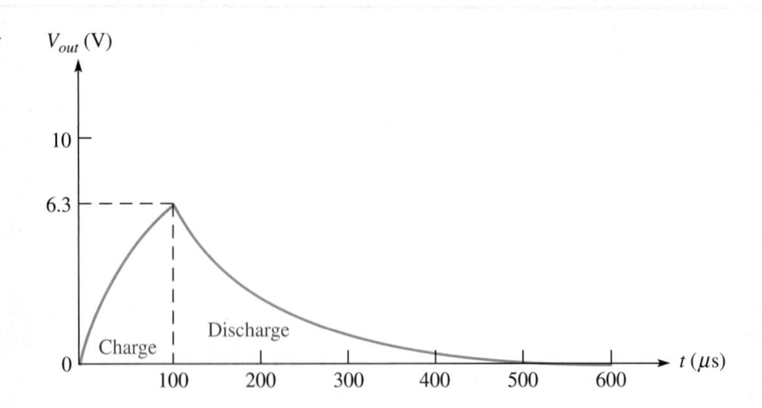

*Related Problem** If the input pulse width in Figure 15–8 is increased to 200 μs, to what voltage will the capacitor charge?

*Answers are at the end of the chapter.

EXAMPLE 15–2

Determine to what voltage the capacitor in Figure 15–10 will charge when the single pulse is applied to the input. Assume C is initially uncharged and the source resistance is zero.

FIGURE 15–10

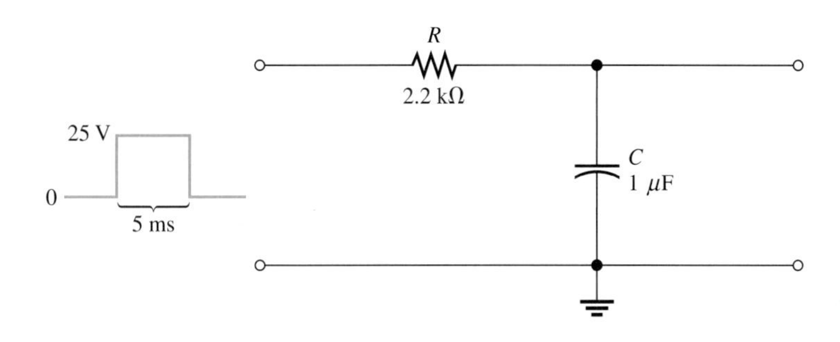

Solution Calculate the time constant.

$$\tau = RC = (2.2\ \text{k}\Omega)(1\ \mu\text{F}) = 2.2\ \text{ms}$$

Because the pulse width is 5 ms, the capacitor charges for approximately 2.27 time constants (5 ms/2.2 ms = 2.27). Use the exponential formula, Equation (9–18), to find the voltage to which the capacitor will charge. With $V_F = 25$ V and $t = 5$ ms, the calculation is done as follows:

$$v = V_F(1 - e^{t/RC}) = (25\ \text{V})(1 - e^{-5\text{ms}/2.2\text{ms}}) = \mathbf{22.4\ V}$$

These calculations show that the capacitor charges to 22.4 V during the 5 ms duration of the input pulse. It will discharge back to zero when the pulse goes back to zero.

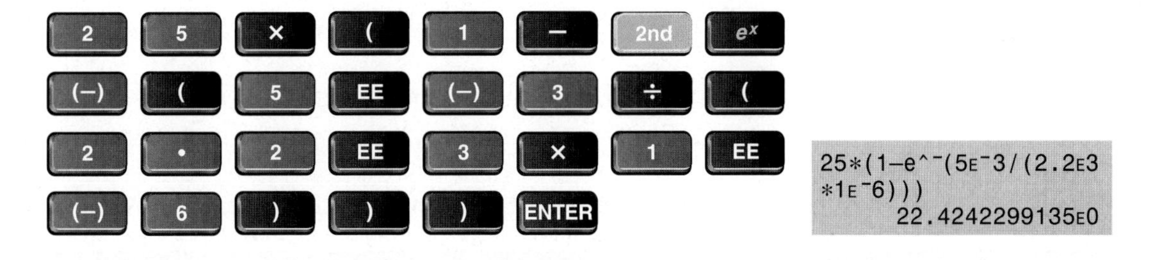

Related Problem Determine how much C will charge if the pulse width is increased to 10 ms.

SECTION 15–2 REVIEW

1. When an input pulse is applied to an *RC* integrator, what condition must exist in order for the output voltage to reach the full amplitude of the input?
2. For the circuit in Figure 15–11, which has a single input pulse, find the maximum output voltage and determine how long the capacitor will discharge.
3. For Figure 15–11, sketch the approximate shape of the output voltage with respect to the input pulse.
4. If the integrator time constant equals the input pulse width, will the capacitor fully charge?

FIGURE 15–11

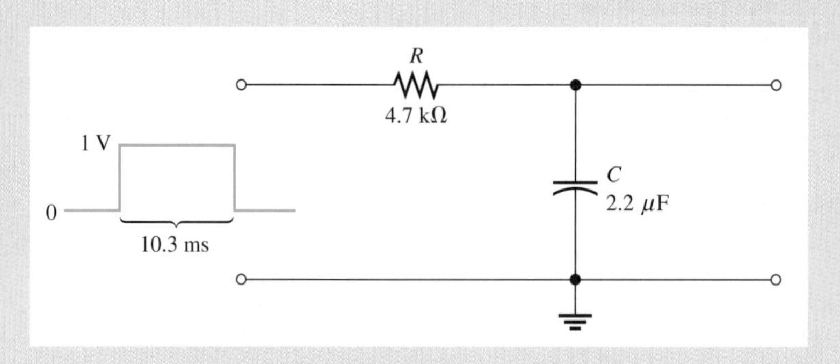

5. Describe the condition in an integrator under which the output voltage has the approximate shape of a rectangular input pulse.

15–3 REPETITIVE-PULSE RESPONSE OF *RC* INTEGRATORS

In the last section, you learned how an *RC* integrator responds to a single-pulse input. These basic ideas are extended in this section to include the integrator response to repetitive pulses. In electronic systems, you will encounter repetitive-pulse waveforms much more often than single pulses. However, an understanding of the integrator's response to single pulses is necessary in order to understand how these circuits respond to repeated pulses.

After completing this section, you should be able to

- **Analyze an *RC* integrator with repetitive input pulses**
- Determine the response when the capacitor does not fully charge or discharge
- Define *steady state*
- Describe the effect of a change in time constant on circuit response

If a periodic pulse waveform is applied to an *RC* integrator, as shown in Figure 15–12, *the output waveshape depends on the relationship of the circuit time constant and the frequency of the input pulses.* The capacitor, of course, charges and discharges in response to a pulse input. The amount of charge and discharge of the capacitor depends both on the circuit time constant and on the input frequency, as mentioned.

If the pulse width and the time between pulses are each equal to or greater than five time constants, the capacitor will fully charge and fully discharge during each period of the input waveform. This case is shown in Figure 15–12.

FIGURE 15–12

RC integrator with a repetitive-pulse waveform input.

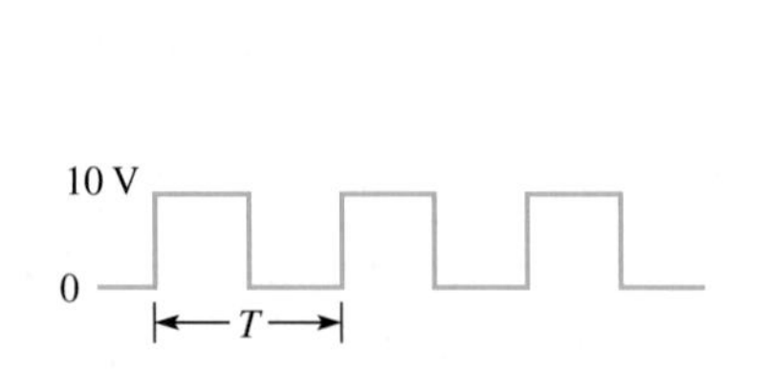

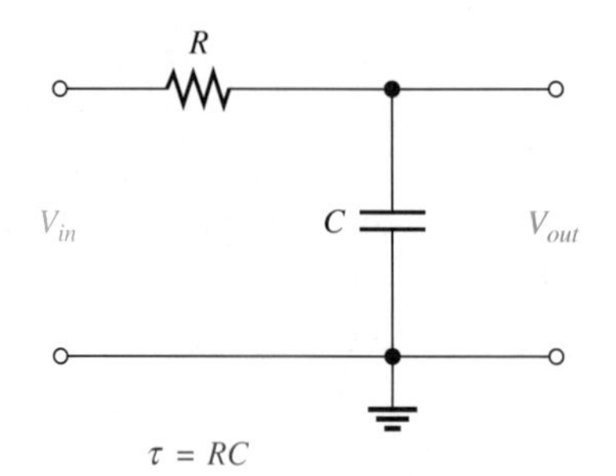

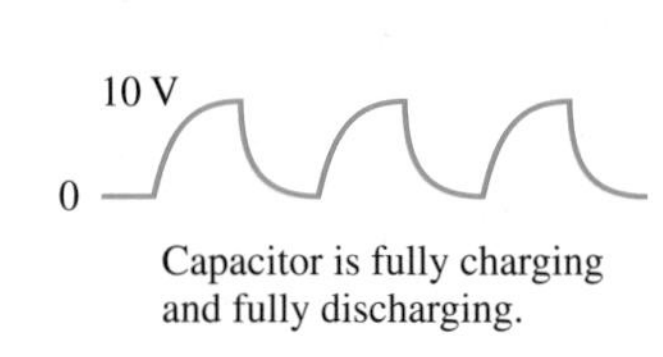

When a Capacitor Does Not Fully Charge and Discharge

When the pulse width and the time between pulses are shorter than 5 time constants, as illustrated in Figure 15–13 for a square wave, the capacitor will *not* completely charge or discharge. We will now examine the effects of this situation on the output voltage of the *RC* integrator.

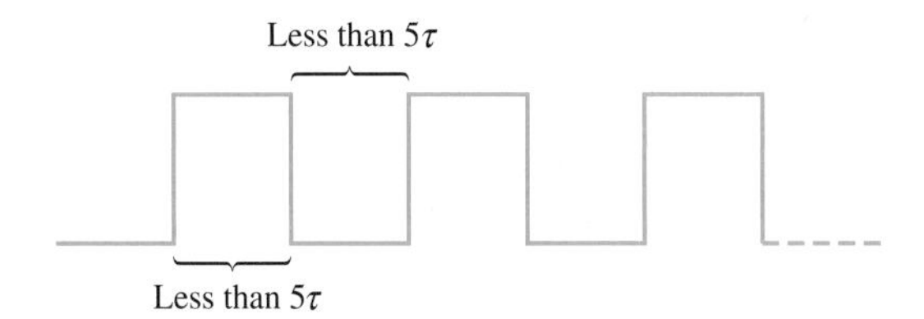

FIGURE 15–13

Input waveform that does not allow full charge or discharge of the integrator capacitor.

For illustration, let's use an *RC* integrator with a charging and discharging time constant equal to the pulse width of a 10 V square wave input, as in Figure 15–14. This choice will simplify the analysis and will demonstrate the basic action of the integrator under these conditions. At this point, you really do not care what the exact time constant value is because an *RC* circuit charges 62.3% during one time constant interval.

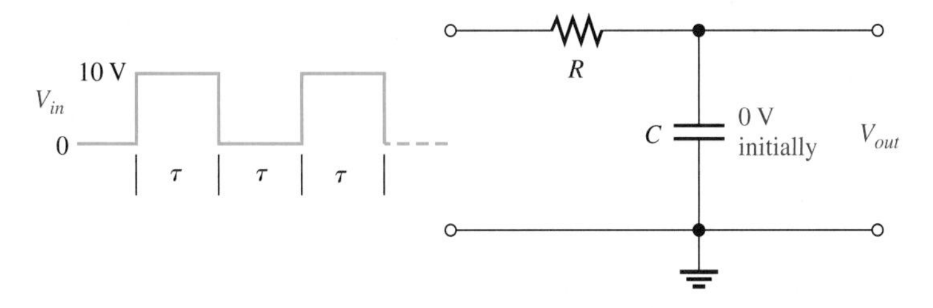

FIGURE 15–14

Integrator with a square wave input having a period equal to two time constants ($T = 2\tau$).

Let's assume that the capacitor in Figure 15–14 begins initially uncharged and examine the output voltage on a pulse-by-pulse basis. The results of this analysis are shown in Figure 15–15.

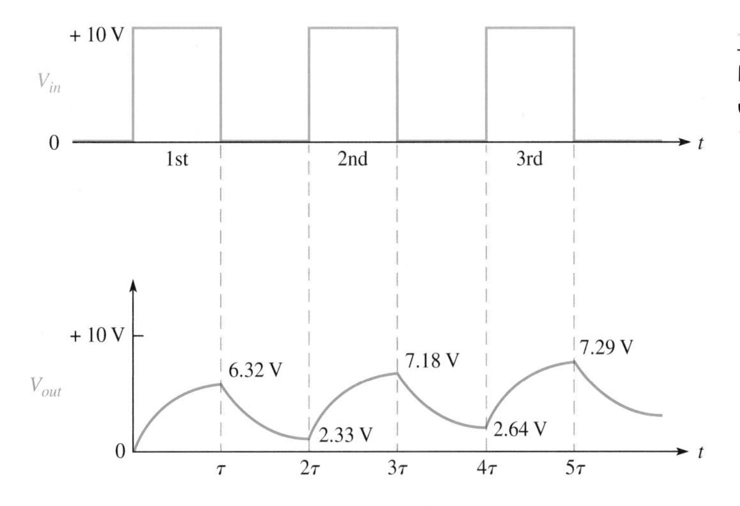

FIGURE 15–15

Input and output for the initially uncharged integrator in Figure 15–14.

First pulse During the first pulse, the capacitor charges. The output voltage reaches 6.32 V, which is 63.2% of 10 V, as shown in Figure 15–15.

Between first and second pulses The capacitor discharges, and the voltage decreases to 36.8% of the voltage at the beginning of this interval: 0.368(6.32 V) = 2.33 V.

HANDS ON TIP

When you view pulse waveforms on an oscilloscope, set the scope to be DC coupled, not AC coupled. The reason is that AC coupling places a capacitor in series and can cause distortion in low-frequency pulse signals. Also, DC coupling allows you to observe the dc level of the pulse.

Second pulse The capacitor voltage begins at 2.33 V and increases 63.2% of the way to 10 V. The total charging range is 10 V − 2.33 V = 7.67 V. The capacitor voltage will increase an additional 63.2% of 7.67 V, which is 4.85 V. Thus, at the end of the second pulse, the output voltage is 2.33 V + 4.85 V = 7.18 V, as indicated in Figure 15–15. Notice that the average is building up.

Between second and third pulses The capacitor discharges during this time, and therefore the voltage decreases to 36.8% of the initial voltage by the end of the second pulse: 0.368(7.18 V) = 2.64 V.

Third pulse At the start of the third pulse, the capacitor voltage is 2.64 V. The capacitor charges 63.2% of the way from 2.64 V to 10 V: 0.632(10 V − 2.64 V) = 4.65 V. Therefore, the voltage at the end of the third pulse is 2.64 V + 4.65 V = 7.29 V.

Steady-State Response

In the preceding discussion, the output voltage gradually built up and then began leveling off. It takes approximately 5τ for the output voltage to build up to a constant average value, regardless of the number of pulses that may occur during that time interval. This interval is the transient time of the circuit. Once the output voltage reaches the average value of the input voltage, a steady-state condition is reached, which continues as long as the periodic input continues. This condition is illustrated in Figure 15–16 based on the values obtained after five pulses.

The transient time for our example circuit is the time from the beginning of the first pulse to the end of the third pulse. The capacitor voltage at the end of the third pulse is 7.29 V.

FIGURE 15–16

Output reaches steady state after 5τ.

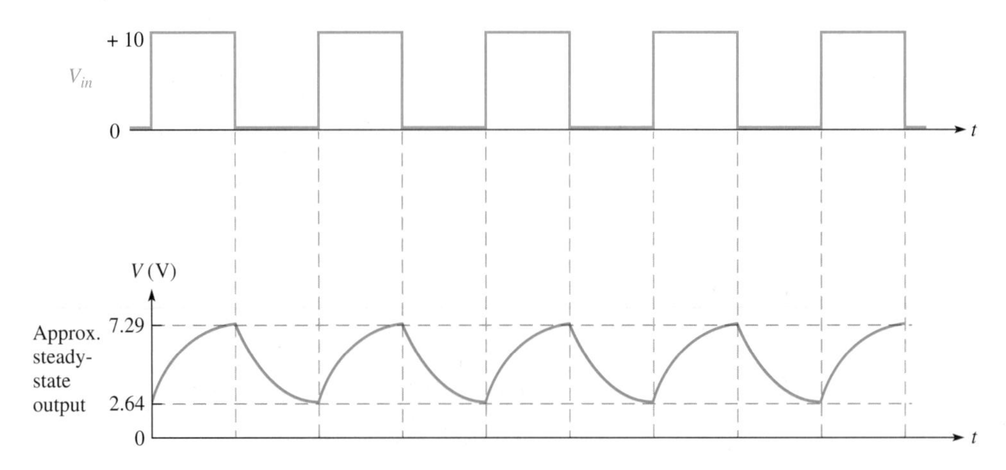

The Effect of an Increase in Time Constant

What happens to the output voltage if the RC time constant of the integrator is increased with a variable resistor, as indicated in Figure 15–17? As the time constant is increased, the capacitor charges less during a pulse and discharges less be-

FIGURE 15–17

Integrator with a variable time constant.

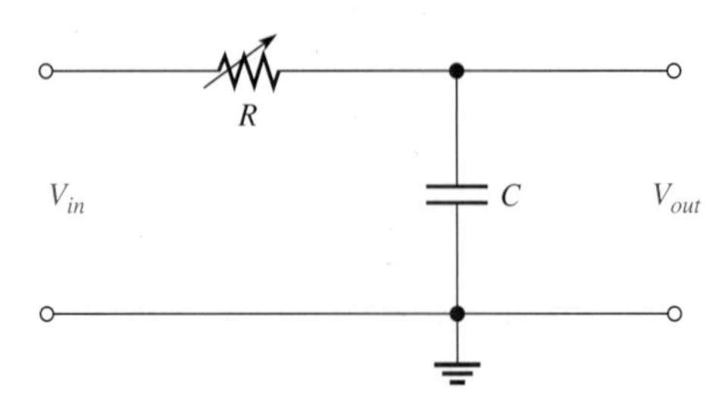

tween pulses. The result is a smaller fluctuation in the output voltage for increasing values of time constant, as shown in Figure 15–18.

As the time constant becomes extremely long compared to the pulse width, the output voltage approaches a constant dc voltage, as shown in Figure 15–18(c). This value is the average value of the input. For a square wave, it is one-half the amplitude.

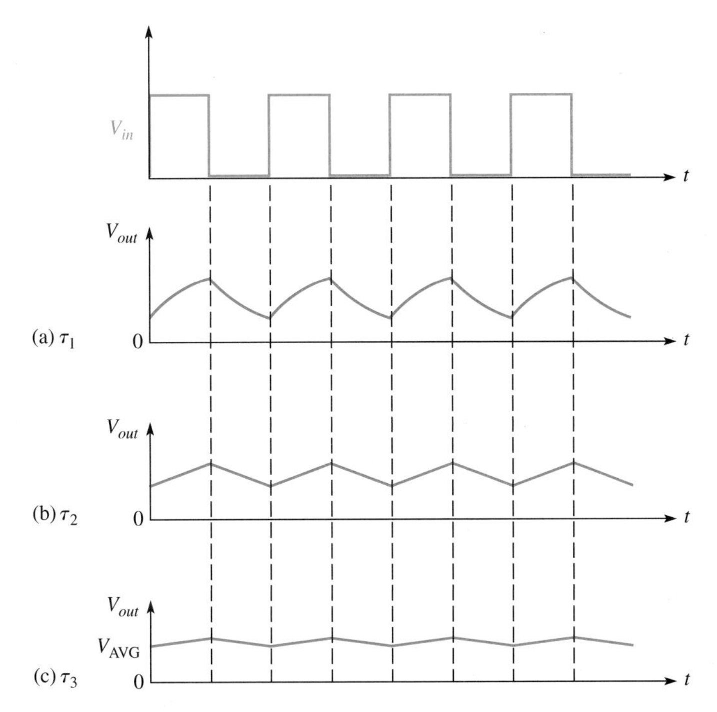

FIGURE 15–18

Effect of longer time constants on the output of an integrator ($\tau_3 > \tau_2 > \tau_1$).

EXAMPLE 15–3

Determine the output voltage waveform for the first two pulses applied to the integrator circuit in Figure 15–19. Assume that the capacitor is initially uncharged.

FIGURE 15–19

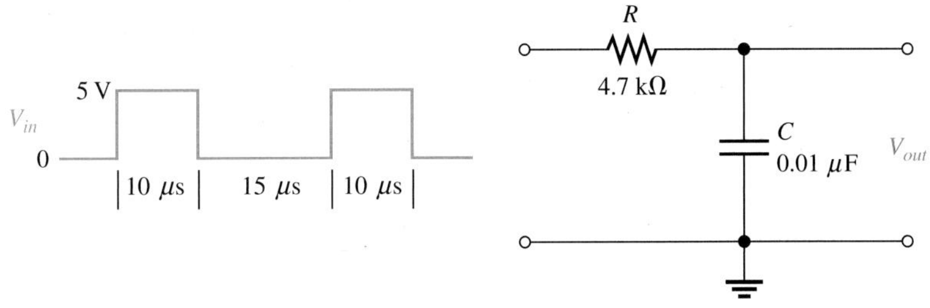

Solution First, calculate the circuit time constant.

$$\tau = RC = (4.7\ \text{k}\Omega)(0.01\ \mu\text{F}) = 47\ \mu\text{s}$$

Obviously, the time constant is much longer than the input pulse width or the interval between pulses (notice that the input is not a square wave). Thus, in this case, the exponential formulas must be applied, and the analysis is relatively difficult. Follow the solution carefully.

1. *Calculation for first pulse:* Use the equation for an increasing exponential (Eq. 9–18) because C is charging. Note that V_F is 5 V, and t equals the pulse width of 10 μs. Therefore,

$$v_C = V_F(1 - e^{-t/RC}) = (5\text{ V})(1 - e^{-10\mu s/47\mu s}) = 958\text{ mV}$$

This result is plotted in Figure 15–20(a).

FIGURE 15–20

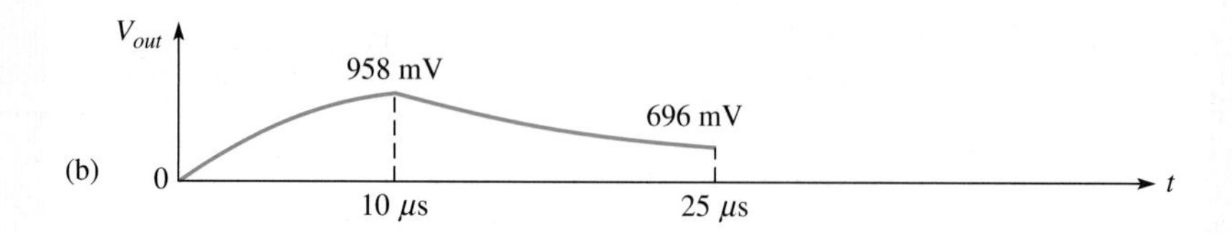

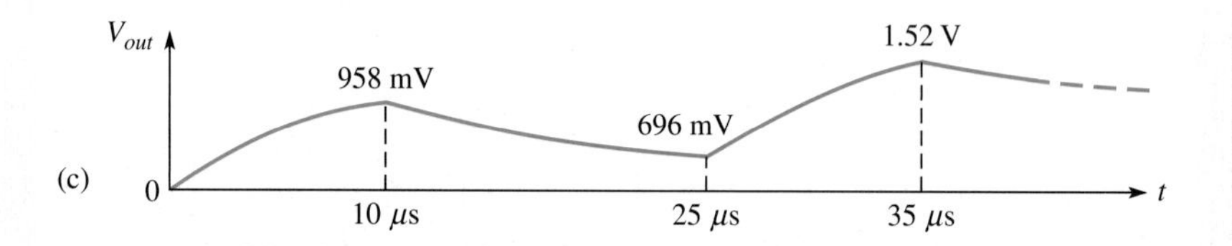

2. *Calculation for interval between first and second pulse:* Use the equation for a decreasing exponential (Eq. 9–19) because C is discharging. Note that V_i is 958 mV because C begins to discharge from this value at the end of the first pulse. The discharge time is 15 μs. Therefore,

$$v_C = V_i e^{-t/RC} = (958\text{ mV})e^{-15\mu s/47\mu s} = 696\text{ mV}$$

This result is shown in Figure 15–20(b).

3. *Calculation for second pulse:* At the beginning of the second pulse, the output voltage is 696 mV. During the second pulse, the capacitor will again charge. In this case it does not begin at zero volts. It already has 696 mV from the previous charge and discharge. To handle this situation, you must use the general exponential formula (Eq. 9–16).

$$v = V_F + (V_i - V_F)e^{-t/\tau}$$

Using this equation, you can calculate the voltage across the capacitor at the end of the second pulse.

$$v_C = V_F + (V_i - V_F)e^{-t/RC} = 5\text{ V} + (696\text{ mV} - 5\text{ V})e^{-10\mu s/47\mu s} = 1.52\text{ V}$$

This result is shown in Figure 15–20(c).

Notice that the output waveform builds up on successive input pulses. After approximately 5τ, it will reach its steady state and will fluctuate between a constant maximum and a constant minimum, with an average equal to the average value of the input. You can demonstrate this pattern by carrying the analysis in this example further.

Related Problem Determine V_{out} at the beginning of the third pulse.

Open file E15-03 on your EWB/CircuitMaker CD-ROM. Measure the steady-state output waveform in terms of its minimum, maximum, and average values.

SECTION 15–3 REVIEW

1. What conditions allow an *RC* integrator capacitor to fully charge and discharge when a periodic pulse waveform is applied to the input?
2. What will the output waveform look like if the circuit time constant is extremely small compared to the pulse width of a square wave input?
3. When 5τ is greater than the pulse width of an input square wave, the time required for the output voltage to build up to a constant average value is called ________ .
4. Define *steady-state response.*
5. What does the average value of the output voltage of an integrator equal during steady state?

15–4 SINGLE-PULSE RESPONSE OF *RC* DIFFERENTIATORS

In terms of pulse response, a series *RC* circuit in which the output voltage is taken across the resistor is known as a **differentiator**. Recall that in terms of frequency response, this particular series *RC* circuit is a high-pass filter. The term, *differentiator,* is derived from a mathematical function which this type of circuit approximates under certain conditions.

After completing this section, you should be able to

- **Analyze an *RC* differentiator with a single input pulse**
- Describe the response at the rising edge of the input pulse
- Determine the response during and at the end of a pulse for various pulse width–time constant relationships

Figure 15–21 shows an *RC* differentiator with a pulse input. The same action occurs in the differentiator as in the integrator, except the output voltage is taken across the resistor rather than the capacitor. The capacitor charges exponentially at a rate depending on the *RC* time constant. The shape of the differentiator's

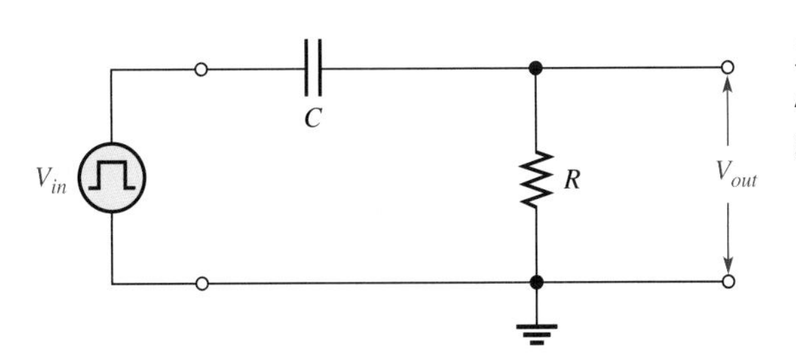

FIGURE 15–21 An *RC* differentiator with a pulse generator connected.

resistor voltage is determined by the charging and discharging action of the capacitor.

Pulse Response

To understand how the output voltage is shaped by a differentiator, you must consider the following:

1. The response to the rising pulse edge
2. The response between the rising and falling edges
3. The response to the falling pulse edge

Response to the Rising Edge of the Input Pulse Let's assume that the capacitor is initially uncharged prior to the rising pulse edge. Prior to the pulse, the input is zero volts. Thus, there are zero volts across the capacitor and also zero volts across the resistor, as indicated in Figure 15–22(a).

▼ **FIGURE 15–22**

Examples of the response of a differentiator to a single input pulse under two conditions: $t_W \geq 5\tau$ and $t_W < 5\tau$. A pulse generator is connected to the input, but the symbol is not shown, only the pulses.

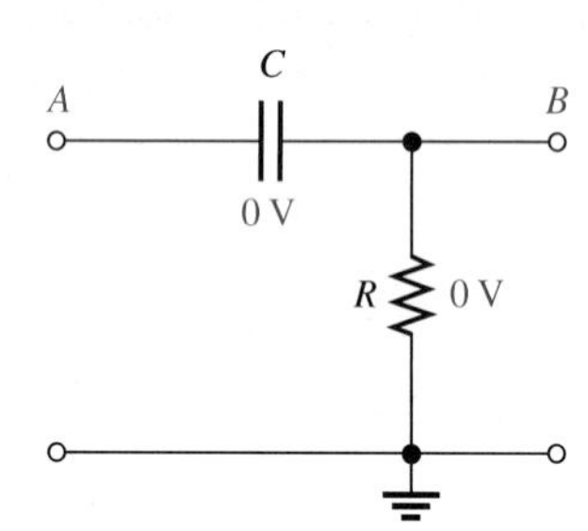

(a) Before pulse is applied

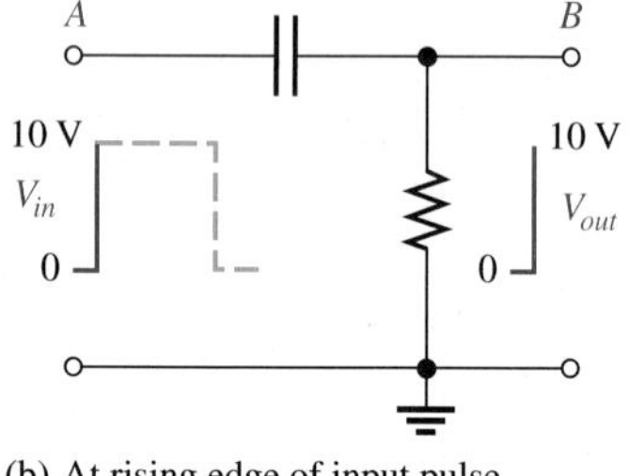

(b) At rising edge of input pulse

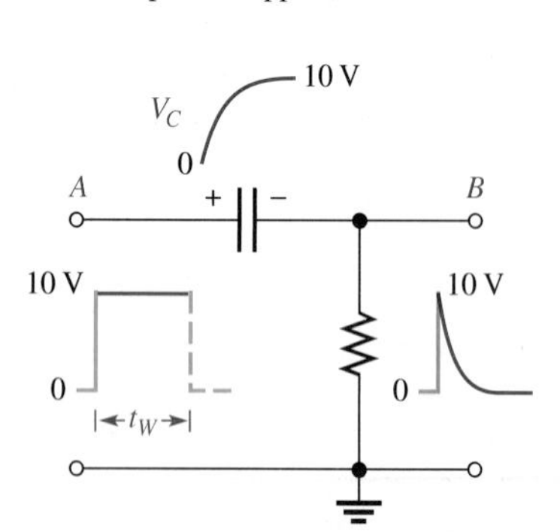

(c) During level part of pulse when $t_W \geq 5\tau$

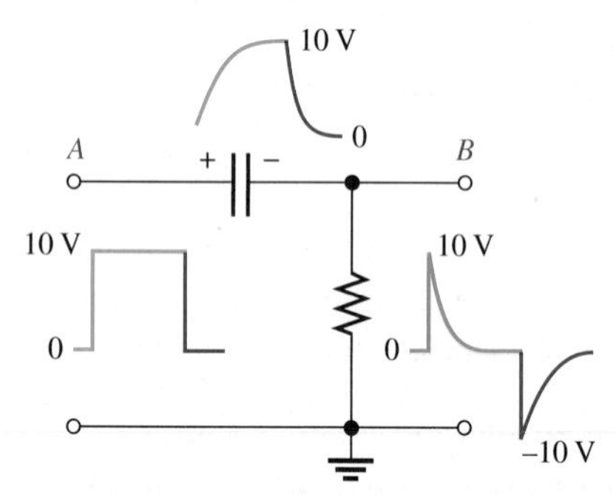

(d) At falling edge of pulse when $t_W \geq 5\tau$

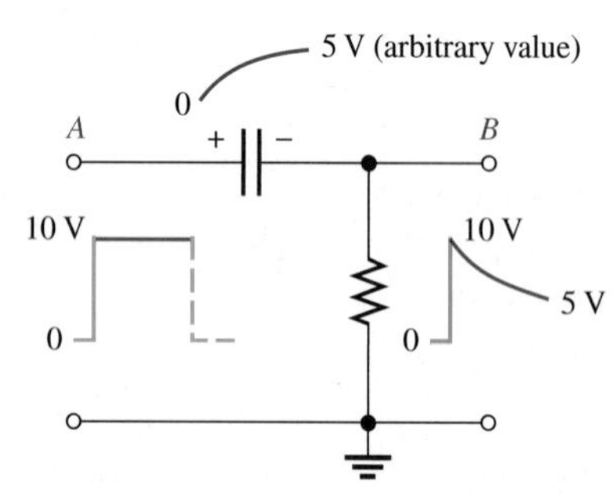

(e) During level part of pulse when $t_W < 5\tau$

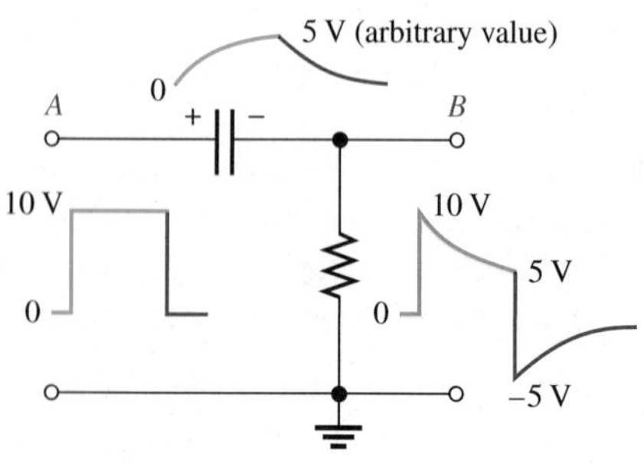

(f) At falling edge of pulse when $t_W < 5\tau$

Let's also assume that a 10 V pulse is applied to the input. When the rising edge occurs, point *A* goes to +10 V. Recall that the voltage across a capacitor cannot change instantaneously, and thus the capacitor appears instantaneously as a short. Therefore, if point *A* goes instantly to +10 V, then point *B must* also go instantly to +10 V, keeping the capacitor voltage zero for the instant of the rising edge. The capacitor voltage is the voltage from point *A* to point *B*.

The voltage at point *B* with respect to ground is the voltage across the resistor (and the output voltage). Thus, the output voltage suddenly goes to +10 V in response to the rising pulse edge, as indicated in Figure 15–22(b).

Response During Pulse When $t_W \geq 5\tau$ While the pulse is at its high level between the rising edge and the falling edge, the capacitor is charging. When the pulse width is equal to or greater than five time constants ($t_W \geq 5\tau$), the capacitor has time to fully charge.

As the voltage across the capacitor builds up exponentially, the voltage across the resistor decreases exponentially until it reaches zero volts at the time the capacitor reaches full charge (+10 V in this case). This decrease in the resistor voltage occurs because the sum of the capacitor voltage and the resistor voltage at any instant must be equal to the source voltage, in compliance with Kirchhoff's voltage law ($v_C + v_R = v_{in}$). This part of the response is illustrated in Figure 15–22(c).

Response to Falling Edge When $t_W \geq 5\tau$ Let's examine the case in which the capacitor is fully charged at the end of the pulse ($t_W \geq 5\tau$). Refer to Figure 15–22(d). On the falling edge, the input pulse suddenly goes from +10 V back to zero. An instant before the falling edge, the capacitor is charged to 10 V, so point *A* is +10 V and point *B* is 0 V. The voltage across a capacitor cannot change instantaneously, so when point *A* makes a transition from +10 V to zero on the falling edge, point *B* *must* also make a 10 V transition from zero to −10 V. This keeps the voltage across the capacitor at 10 V for the instant of the falling edge.

The capacitor now begins to discharge exponentially. As a result, the resistor voltage goes from −10 V to zero in an exponential curve, as indicated in Figure 15–22(d).

Response During Pulse When $t_W < 5\tau$ When the pulse width is less than five time constants ($t_W < 5\tau$), the capacitor does not have time to fully charge. Its partial charge depends on the relation of the time constant and the pulse width.

Because the capacitor does not reach the full +10 V, the resistor voltage will not reach zero volts by the end of the pulse. For example, if the capacitor charges to +5 V during the pulse interval, the resistor voltage will decrease to +5 V, as illustrated in Figure 15–22(e).

Response to Falling Edge When $t_W < 5\tau$ Now, let's examine the case in which the capacitor is only partially charged at the end of the pulse ($t_W < 5\tau$). For example, if the capacitor charges to +5 V, the resistor voltage at the instant before the falling edge is also +5 V because the capacitor voltage plus the resistor voltage must add up to +10 V, as illustrated in Figure 15–22(e).

When the falling edge occurs, point *A* goes from +10 V to zero. As a result, point *B* goes from +5 V to −5 V, as illustrated in Figure 15–22(f). This decrease occurs, of course, because the capacitor voltage cannot change at the instant of the falling edge. Immediately after the falling edge, the capacitor begins to discharge to zero. As a result, the resistor voltage goes from −5 V to zero, as shown.

Summary of Differentiator Response to a Single Pulse

Perhaps a good way to summarize this section is to look at the general output waveforms of a differentiator as the time constant is varied from one extreme,

▶ FIGURE 15–23

Effects of a change in time constant on the shape of the output voltage of a differentiator.

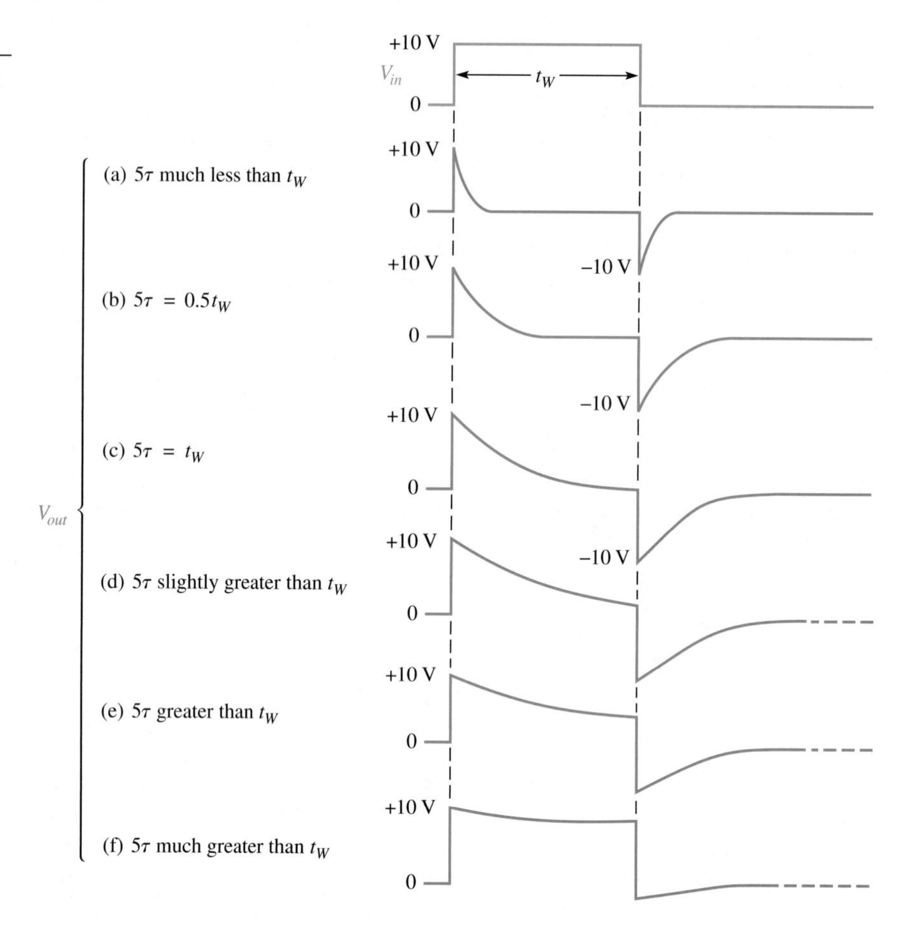

when 5τ is much less than the pulse width, to the other extreme, when 5τ is much greater than the pulse width. These situations are illustrated in Figure 15–23. In part (a), the output consists of narrow positive and negative "spikes." In part (f), the output approaches the shape of the input. Various conditions between these extremes are illustrated in parts (b) through (e).

EXAMPLE 15–4

Show the output voltage for the circuit in Figure 15–24.

▶ FIGURE 15–24

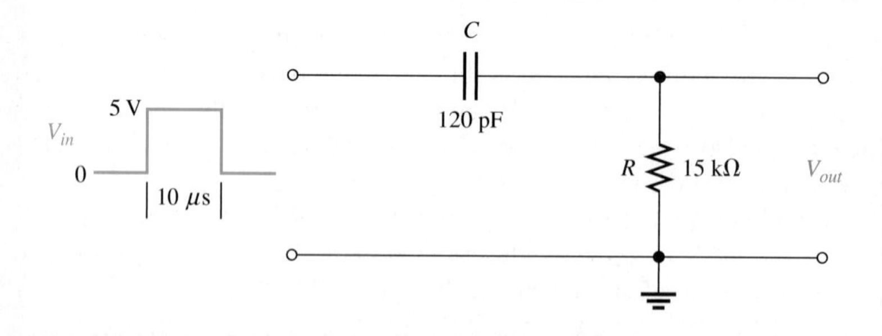

Solution First, calculate the time constant.

$$\tau = RC = (15\text{ k}\Omega)(120\text{ pF}) = 1.8\ \mu\text{s}$$

In this case, $t_W > 5\tau$, so the capacitor reaches full charge in 9 μs (before the end of the pulse).

On the rising edge, the resistor voltage jumps to +5 V and then decreases exponentially to zero before the end of the pulse. On the falling edge, the resistor voltage jumps to −5 V and then goes back to zero exponentially. The resistor voltage is, of course, the output, and its shape is shown in Figure 15–25.

▶ **FIGURE 15–25**

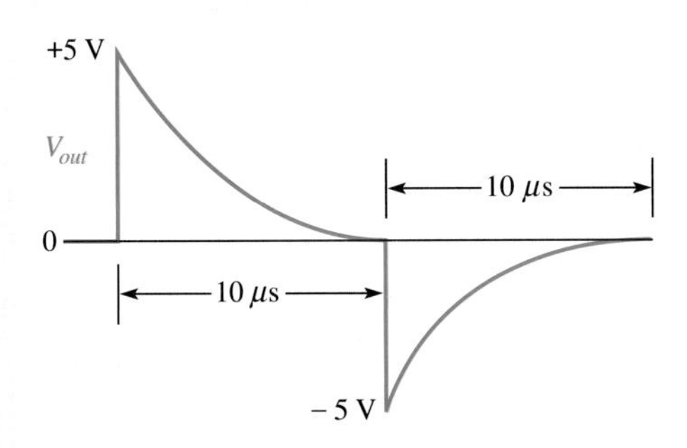

Related Problem Show the output voltage if R = 18 kΩ and C = 47 pF in Figure 15–24.

EXAMPLE 15–5

Determine the output voltage waveform for the differentiator in Figure 15–26.

▶ **FIGURE 15–26**

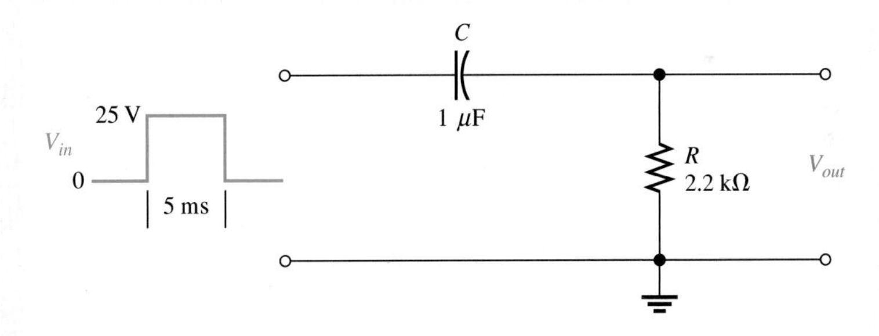

Solution First, calculate the time constant.

$$\tau = (2.2\ \text{k}\Omega)(1\ \mu\text{F}) = 2.2\ \text{ms}$$

On the rising edge, the resistor voltage immediately jumps to +25 V. Because the pulse width is 5 ms, the capacitor charges for approximately 2.27 time constants during the pulse, and therefore does not reach full charge. Thus, you must use the formula for a decreasing exponential (Eq. 9–19) in order to calculate to what voltage the output decreases by the end of the pulse.

$$v_{out} = V_i e^{-t/RC} = (25\ \text{V})e^{-5\text{ms}/2.2\text{ms}} = 2.58\ \text{V}$$

where V_i = 25 V and t = 5 ms. This calculation gives the resistor voltage at the end of the 5 ms pulse width interval.

On the falling edge, the resistor voltage immediately jumps from +2.58 V down to −22.4 V (a 25 V transition). The resulting waveform of the output voltage is shown in Figure 15–27.

FIGURE 15–27

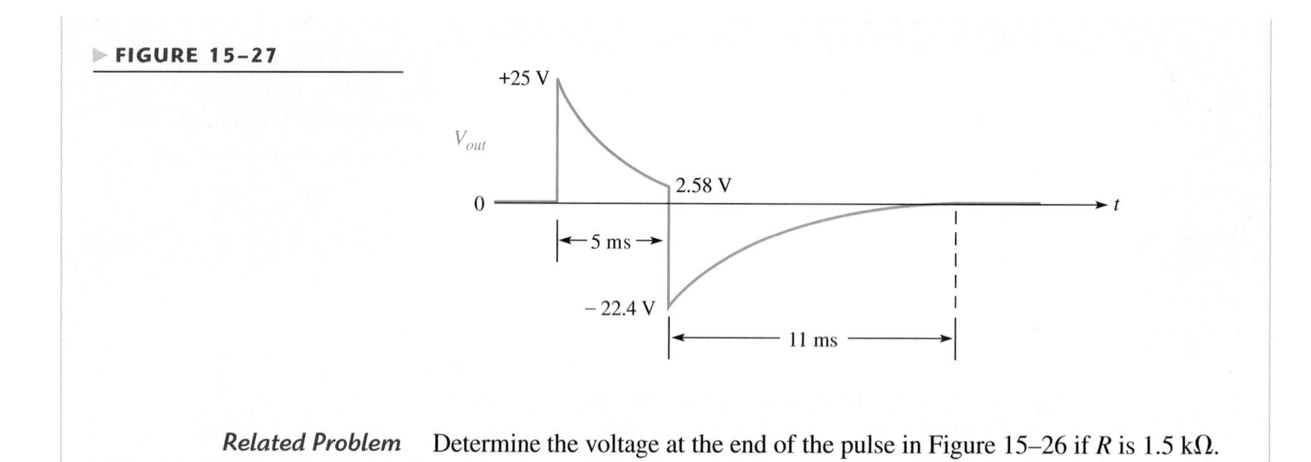

Related Problem Determine the voltage at the end of the pulse in Figure 15–26 if R is 1.5 kΩ.

SECTION 15–4 REVIEW

1. Show the output of a differentiator for a 10 V input pulse when $5\tau = 0.5t_W$.
2. Under what condition does the output pulse shape most closely resemble the input pulse for a differentiator?
3. What does the differentiator output look like when 5τ is much less than the pulse width of the input?
4. If the resistor voltage in a differentiating circuit is down to −5 V at the end of a 15 V input pulse, to what negative value will the resistor voltage go in response to the falling edge of the input?

15–5 REPETITIVE-PULSE RESPONSE OF *RC* DIFFERENTIATORS

The *RC* differentiator response to a single pulse, covered in the last section, is extended in this section to repetitive pulses.

After completing this section, you should be able to

- **Analyze an *RC* differentiator with repetitive input pulses**
- Determine the response when the pulse width is less than five time constants

If a periodic pulse waveform is applied to an *RC* differentiating circuit, two conditions again are possible: $t_W \geq 5\tau$ or $t_W < 5\tau$. Figure 15–28 shows the output when $t_W = 5\tau$. As the time constant is reduced, both the positive and the negative portions of the output become narrower. Notice that the average value of the output is zero; the waveform has equal positive and negative portions.

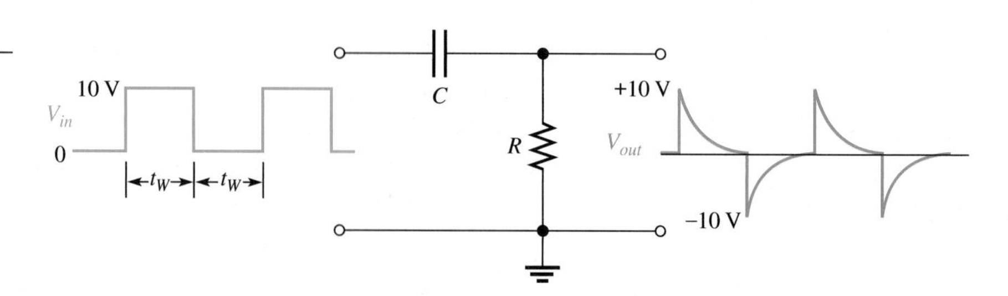

FIGURE 15–28

Example of differentiator response when $t_W = 5\tau$.

Figure 15–29 shows the steady-state output when $t_W < 5\tau$. As the time constant is increased, the positively and negatively sloping portions become flatter. For a very long time constant, the output approaches the shape of the input, but with an average value of zero. The average value of a waveform is its **dc component.** Because a capacitor blocks dc, the dc component of the input is prevented from passing through to the output.

Like the integrator, the differentiator output takes time (5τ) to reach steady state. To illustrate the response, let's take an example in which the time constant equals the input pulse width.

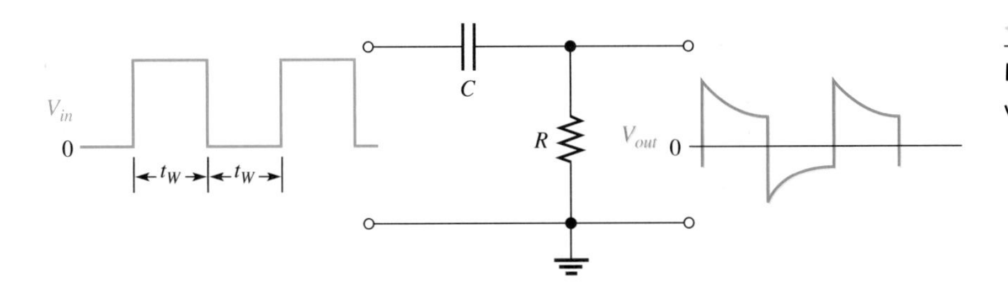

FIGURE 15–29

Example of differentiator response when $t_W < 5\tau$.

Analysis of a Repetitive Waveform

At this point, we do not care what the circuit time constant is because we know that the resistor voltage will decrease to 36.8% of its maximum value during one pulse (1τ). Let's assume that the capacitor in Figure 15–30 begins initially uncharged and examine the output voltage on a pulse-by-pulse basis. The results of this analysis are shown in Figure 15–31.

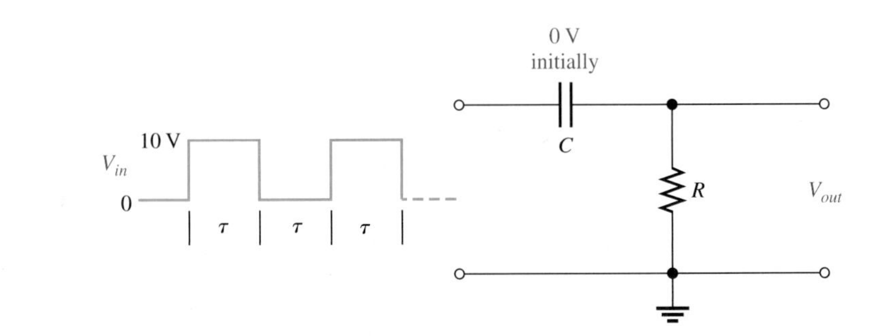

FIGURE 15–30

RC differentiator with $T = 2\tau$.

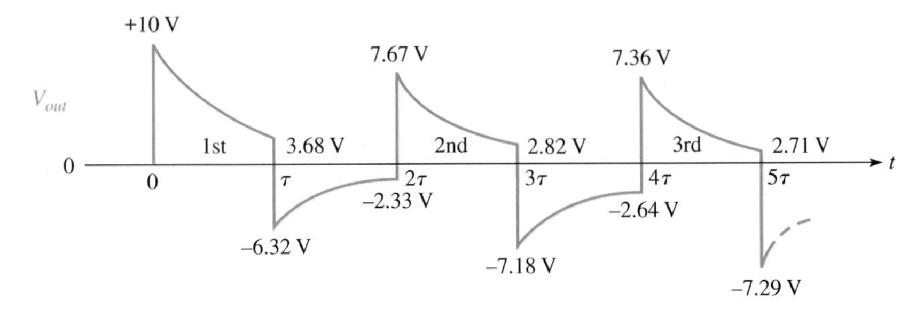

FIGURE 15–31

Differentiator output waveform during transient time for the circuit in Figure 15–30.

First pulse On the rising edge, the output instantaneously jumps to +10 V. Then the capacitor partially charges to 63.2% of 10 V, which is 6.32 V. Thus, the output voltage must decrease to 3.68 V, as shown in Figure 15–31.

On the falling edge, the output instantaneously makes a negative-going 10 V transition to −6.32 V because 3.68 V − 10 V = −6.32 V.

Between first and second pulses The capacitor discharges to 36.8% of 6.32 V, which is 2.33 V. Thus, the resistor voltage, which starts at −6.32 V, must increase to −2.33 V. Why? Because at the instant prior to the next pulse, the input voltage is zero. Therefore, the sum of v_C and v_R must be zero (+2.33 V − 2.33 V = 0). Remember that $v_C + v_R = v_{in}$ at all times, in accordance with Kirchhoff's voltage law.

Second pulse On the rising edge, the output makes an instantaneous, positive-going, 10 V transition from −2.33 V to 7.67 V. Then by the end of the pulse the capacitor charges 0.632(10 V − 2.33 V) = 4.85 V. Thus, the capacitor voltage increases from 2.33 V to 2.33 V + 4.85 V = 7.18 V. The output voltage drops to 0.368(7.67 V) = 2.82 V.

On the falling edge, the output instantaneously makes a negative-going transition from 2.82 V to −7.18 V, as shown in Figure 15–31.

Between second and third pulses The capacitor discharges to 36.8% of 7.18 V, which is 2.64 V. Thus, the output voltage starts at −7.18 V and increases to −2.64 V because the capacitor voltage and the resistor voltage must add up to zero at the instant prior to the third pulse (the input is zero).

Third pulse On the rising edge, the output makes an instantaneous 10 V transition from −2.64 V to +7.36 V. Then the capacitor charges 0.632(10 V − 2.64 V) = 4.65 V to 2.64 V + 4.65 V = +7.29 V. As a result, the output voltage drops to 0.368(7.36 V) = 2.71 V. On the falling edge, the output instantly goes from +2.71 V down to −7.29 V.

After the third pulse, five time constants have elapsed, and the output voltage is close to its steady state. Thus, the waveform in Figure 5–31 will continue to vary from a positive maximum of about +7.3 V to a negative maximum of about −7.3 V, with an average value of zero.

SECTION 15–5 REVIEW

1. What conditions allow an *RC* differentiator to fully charge and discharge when a periodic pulse waveform is applied to the input?
2. What will the output waveform look like if the circuit time constant is extremely small compared to the pulse width of a square wave input?
3. What does the average value of the differentiator output voltage equal during steady state?

15–6 PULSE RESPONSE OF *RL* INTEGRATORS

A series *RL* circuit in which the output voltage is taken across the resistor is known as an integrator in terms of pulse response.

After completing this section, you should be able to

- **Analyze the operation of an *RL* integrator**
- Determine the response to a single input pulse

Figure 15–32 shows an *RL* integrator. The output waveform is taken across the resistor and, under equivalent conditions, is the same shape as that for the *RC* integrator. Recall that in the *RC* case, the output was across the capacitor.

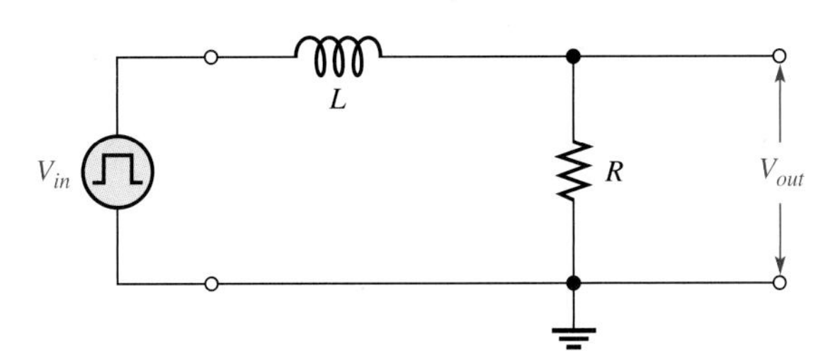

FIGURE 15–32

An *RL* integrator with a pulse generator connected.

As you know, each edge of an ideal pulse is considered to occur instantaneously. Two basic rules for inductor behavior will aid in analyzing *RL* circuit responses to pulse inputs:

1. The inductor appears as an open to an instantaneous change in current and as a short (ideally) to dc.
2. The current in an inductor cannot change instantaneously—it can change only exponentially.

Response of the Integrator to a Single Pulse

When a pulse generator is connected to the input of the integrator and the voltage pulse goes from its low level to its high level, the inductor prevents a sudden change in current. As a result, the inductor acts as an open, and all the input voltage is across it at the instant of the rising pulse edge. This situation is indicated in Figure 15–33(a).

After the rising edge, the current builds up, and the output voltage follows the current as it increases exponentially, as shown in Figure 15–33(b). The current can reach a maximum of V_p/R if the transient time is shorter than the pulse width ($V_p = 10$ V in this example where V_p is pulse amplitude).

FIGURE 15–33

Illustration of the pulse response of an *RL* integrator ($t_W > 5\tau$). A pulse generator is connected to the input but its symbol is not shown, only the pulses.

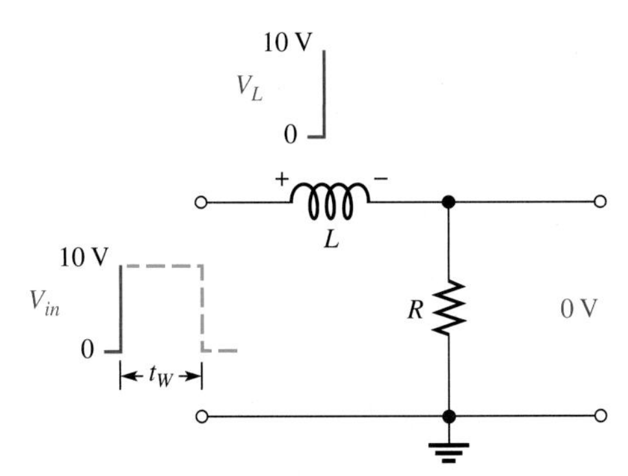

(a) At rising edge of pulse ($i = 0$)

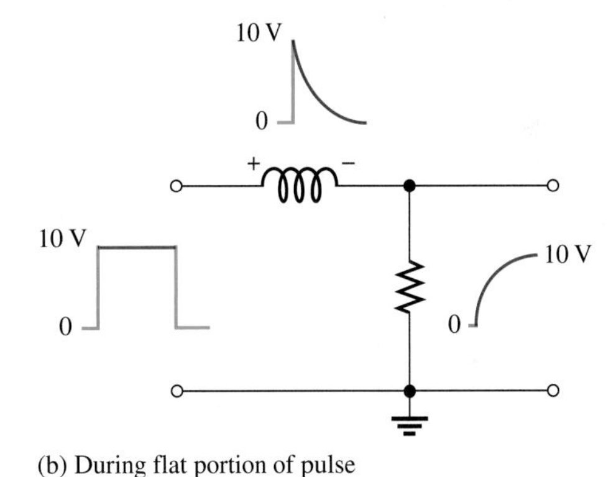

(b) During flat portion of pulse

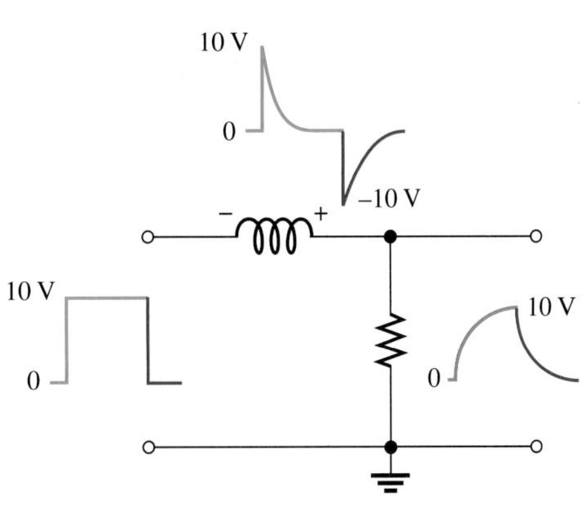

(c) At falling edge of pulse and after

When the pulse goes from its high level to its low level, an induced voltage with reversed polarity is created across the coil in an effort to keep the current equal to V_p/R. The output voltage begins to decrease exponentially, as shown in Figure 15–33(c).

The exact shape of the output depends on the L/R time constant as summarized in Figure 15–34 for various relationships between the time constant and the pulse width. You should note that the response of this RL circuit in terms of the shape of the output is identical to that of the RC integrator. The relationship of the L/R time constant to the input pulse width has the same effect as the RC time constant that we discussed earlier in this chapter. For example, when $t_W < 5\tau$, the output voltage will not reach its maximum possible value.

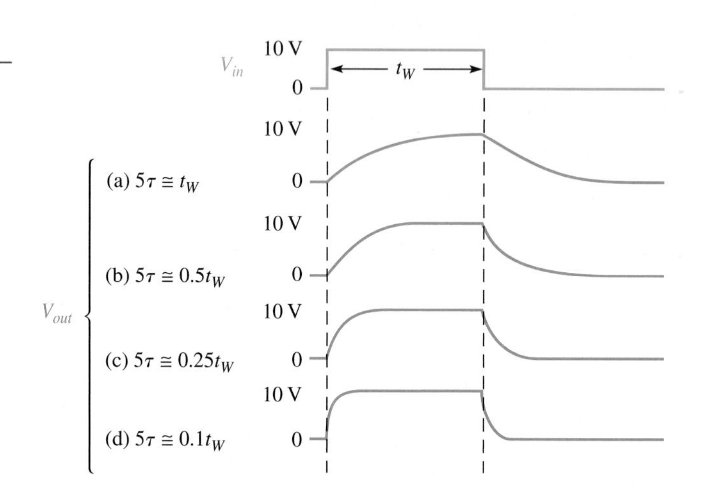

FIGURE 15–34

Illustration of the variation in integrator output pulse shape with time constant.

EXAMPLE 15–6

Determine the maximum output voltage for the integrator in Figure 15–35 when a single pulse is applied as shown.

FIGURE 15–35

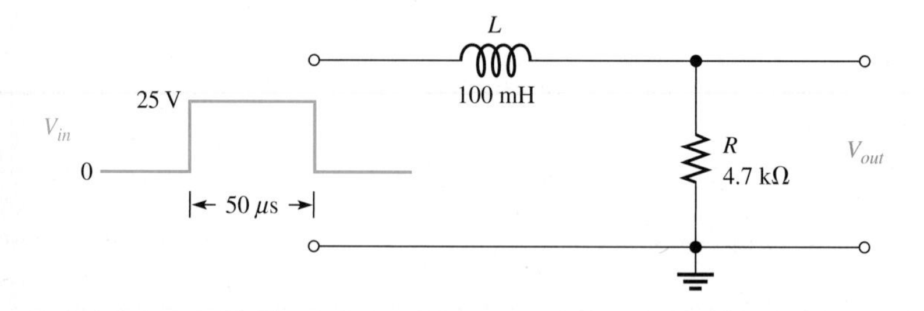

Solution Calculate the time constant.

$$\tau = \frac{L}{R} = \frac{100\text{ mH}}{4.7\text{ k}\Omega} = 21.3\ \mu\text{s}$$

Because the pulse width is 50 μs, the inductor charges for approximately 2.35τ; (50 μs/21.3 μs = 2.35). Use the exponential formula to calculate the voltage.

$$v_{out(max)} = V_F(1 - e^{-t/\tau}) = (25\text{ V})(1 - e^{-50\mu\text{s}/21.3\mu\text{s}}) = \mathbf{22.6\text{ V}}$$

Related Problem What value must R be for the output voltage to reach 25 V by the end of the input pulse in Figure 15–35?

EXAMPLE 15–7

A pulse is applied to the *RL* integrator circuit in Figure 15–36. Determine the waveforms and the values for I, V_R, and V_L.

FIGURE 15–36

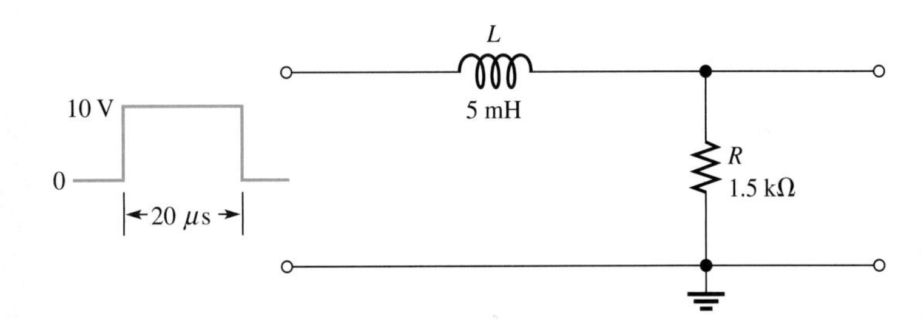

Solution The circuit time constant is

$$\tau = \frac{L}{R} = \frac{5\text{ mH}}{1.5\text{ k}\Omega} = 3.33\ \mu\text{s}$$

Since $5\tau = 16.7\ \mu$s is less than t_W, the current will reach its maximum value and remain there until the end of the pulse.

At the rising edge of the pulse,

$$i = 0\text{ A}$$

$$v_R = 0\text{ V}$$

$$v_L = 10\text{ V}$$

The inductor appears as an open, so all of the input voltage appears across L.

During the pulse,

$$i \text{ increases exponentially to } \frac{V_p}{R} = \frac{10\text{ V}}{1.5\text{ k}\Omega} = 6.67\text{ mA in } 16.7\ \mu\text{s}$$

$$v_R \text{ increases exponentially to 10 V in } 16.7\ \mu\text{s}$$

$$v_L \text{ decreases exponentially to zero in } 16.7\ \mu\text{s}$$

At the falling edge of the pulse,

$$i = 6.67\text{ mA}$$

$$v_R = 10\text{ V}$$

$$v_L = -10\text{ V}$$

After the pulse,

$$i \text{ decreases exponentially to zero in } 16.7\ \mu\text{s}$$

$$v_R \text{ decreases exponentially to zero in } 16.7\ \mu\text{s}$$

$$v_L \text{ decreases exponentially to zero in } 16.7\ \mu\text{s}$$

The waveforms are shown in Figure 15–37.

FIGURE 15–37

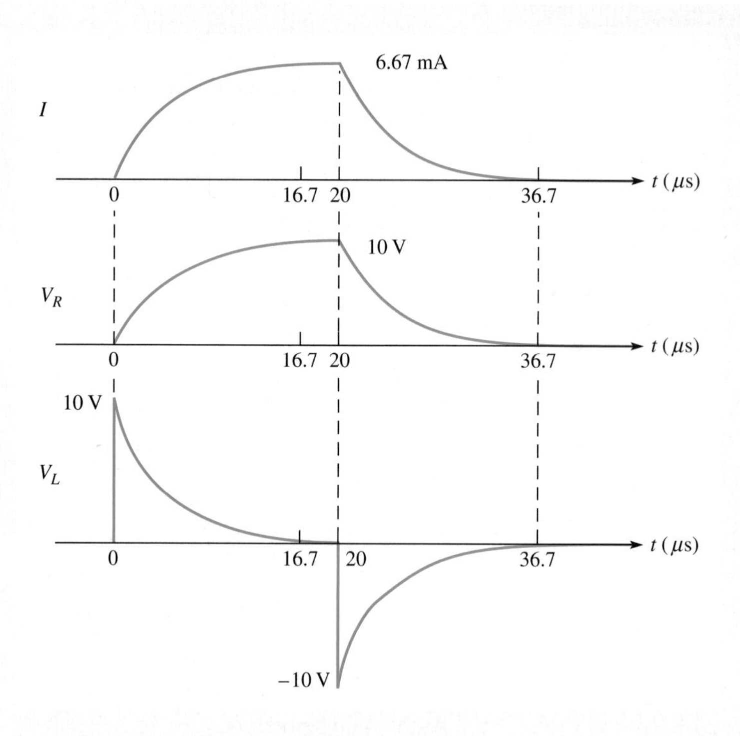

Related Problem What will be the maximum output voltage if the amplitude of the input pulse is increased to 20 V in Figure 15–36?

EXAMPLE 15–8

A 10 V pulse with a width of 1 ms is applied to the integrator in Figure 15–38. Determine the voltage level that the output will reach during the pulse. If the source has an internal resistance of 300 Ω, how long will it take the output to decay to zero? Show the output voltage waveform.

FIGURE 15–38

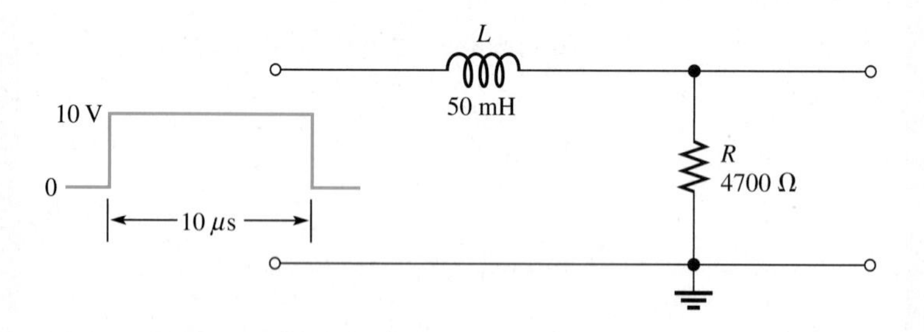

Solution The coil charges through the 300 Ω source resistance plus the 4700 Ω external resistor. The time constant is

$$\tau = \frac{L}{R_{tot}} = \frac{50\text{ mH}}{4700\ \Omega + 300\ \Omega} = \frac{50\text{ mH}}{5\text{ k}\Omega} = 10\ \mu\text{s}$$

Notice that in this case the pulse width is exactly equal to τ. Thus, the output V_R will reach approximately 63% of the full input amplitude in 1τ. Therefore, the output voltage gets to **6.3 V** at the end of the pulse.

After the pulse is gone, the inductor discharges back through the 300 Ω source and the 4700 Ω resistor. The source takes 5τ to completely discharge.

$$5\tau = 5(10\ \mu s) = \mathbf{50\ \mu s}$$

The output voltage is shown in Figure 15–39.

FIGURE 15–39

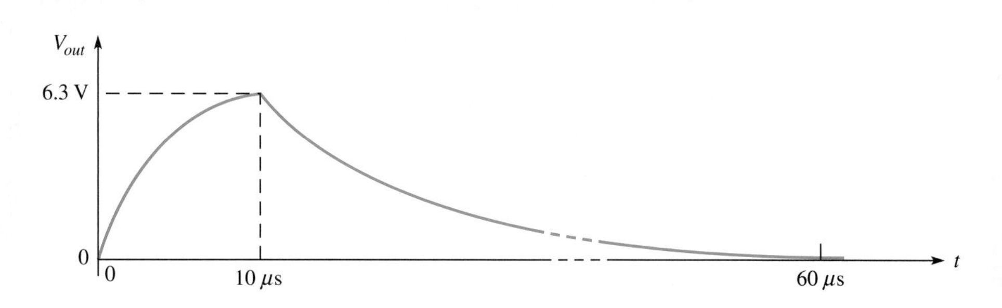

Related Problem To what approximate maximum value must *R* be changed in Figure 15–38 to allow the output voltage to reach the input level during the pulse?

SECTION 15–6 REVIEW

1. In an *RL* integrator, across which component is the output voltage taken?
2. When a pulse is applied to an *RL* integrator, what condition must exist in order for the output voltage to reach the amplitude of the input?
3. Under what condition will the output voltage have the approximate shape of the input pulse?

15–7 PULSE RESPONSE OF *RL* DIFFERENTIATORS

A series *RL* circuit in which the output voltage is taken across the inductor is known as a differentiator.

After completing this section, you should be able to

- **Analyze the operation of an *RL* differentiator**
- Determine the response to a single input pulse

Response of the Differentiator to a Single Pulse

Figure 15–40 shows an *RL* differentiator with a pulse generator connected to the input. Initially, before the pulse, there is no current in the circuit. When the input

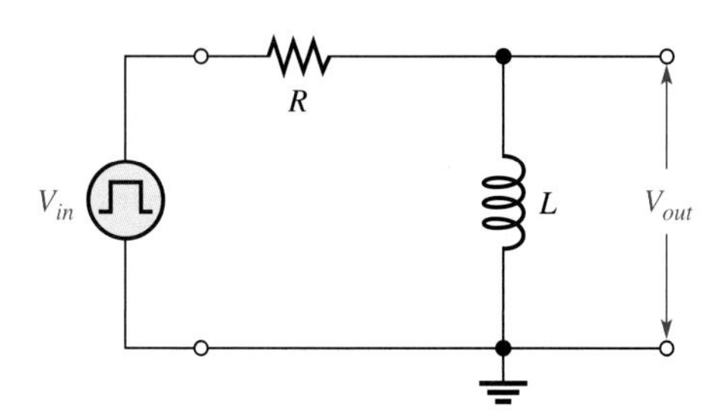

FIGURE 15–40

An *RL* differentiator with a pulse generator connected.

pulse goes from its low level to its high level, the inductor prevents a sudden change in current. It does this, as you know, with an induced voltage equal and opposite to the input. As a result, L looks like an open, and all of the input voltage appears across it at the instant of the rising edge, as shown in Figure 15–41(a) with a 10 V pulse.

During the pulse, the current exponentially builds up. As a result, the inductor voltage decreases, as shown in Figure 15–41(b). The rate of decrease, as you know, depends on the L/R time constant. When the falling edge of the input appears, the inductor reacts to keep the current as is, by creating an induced voltage in a direction as indicated in Figure 15–41(c). This reaction is seen as a sudden negative-going transition of the inductor voltage, as indicated in parts (c) and (d).

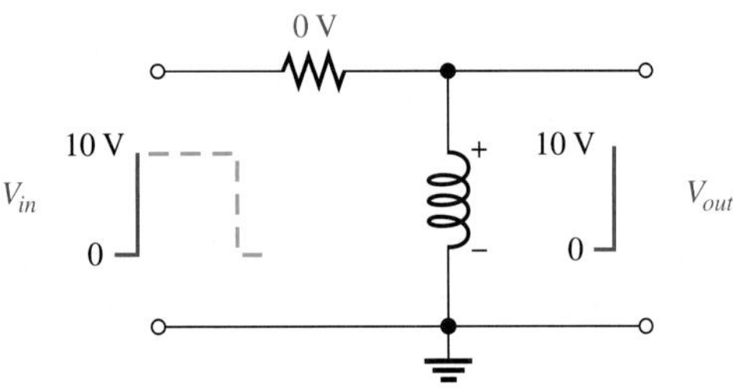

(a) At rising edge of pulse

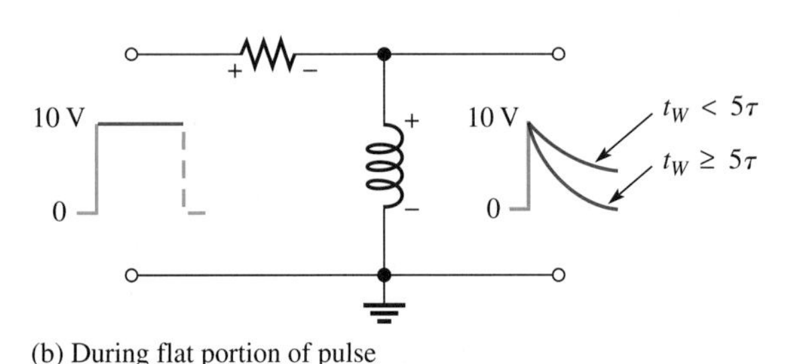

(b) During flat portion of pulse

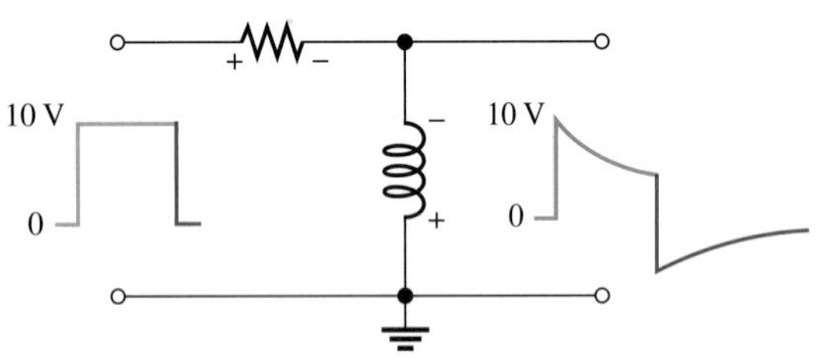

(c) At falling edge when $t_W < 5\tau$

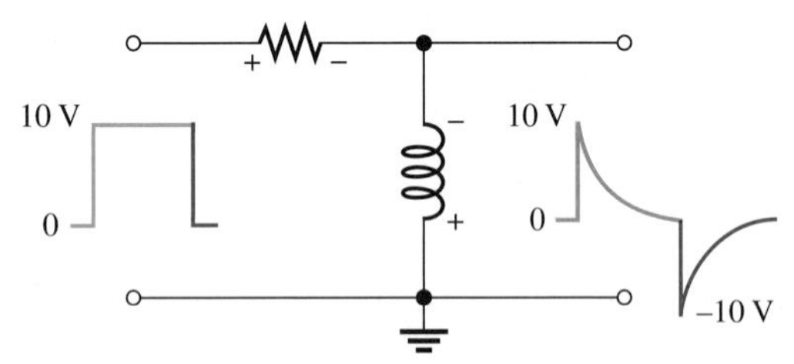

(d) At falling edge when $t_W \geq 5\tau$

FIGURE 15–41

Illustration of the response of an *RL* differentiator for both time constant conditions. The pulse generator is connected to the input, but its symbol is not shown, only the pulses.

Two conditions are possible, as indicated in Figure 15–41(c) and (d). In part (c), 5τ is greater than the input pulse width, and the output voltage does not have time to decay to zero. In part (d), 5τ is less than or equal to the pulse width, and so the output decays to zero before the end of the pulse. In this case a full -10 V transition occurs at the falling edge.

Keep in mind that as far as the input and output waveforms are concerned, the *RL* integrator and differentiator perform the same as their *RC* counterparts.

A summary of the *RL* differentiator response for relationships of various time constants and pulse widths is shown in Figure 15–42. Compare this figure to Figure 15–23 for an *RC* differentiator and you will see that they are identical.

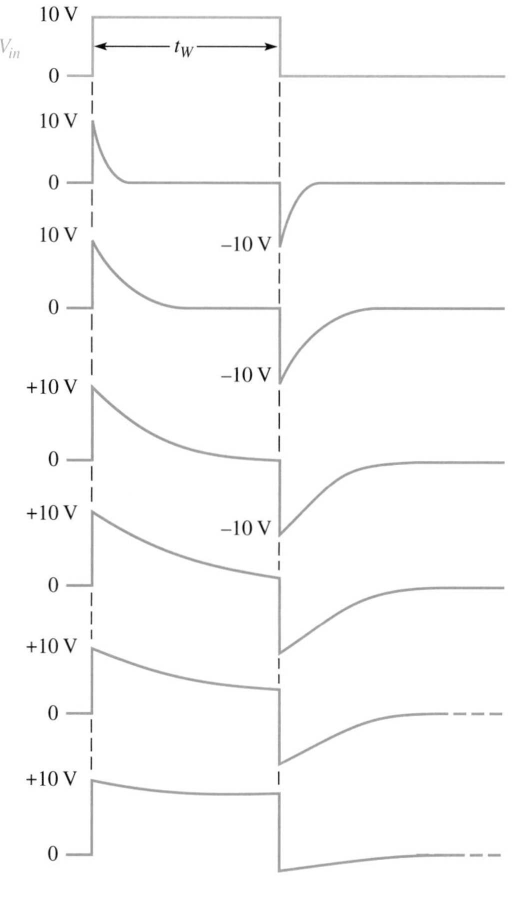

FIGURE 15–42

Illustration of the variation in output pulse shape with the time constant.

EXAMPLE 15–9

Show the output voltage waveform for the circuit in Figure 15–43.

FIGURE 15–43

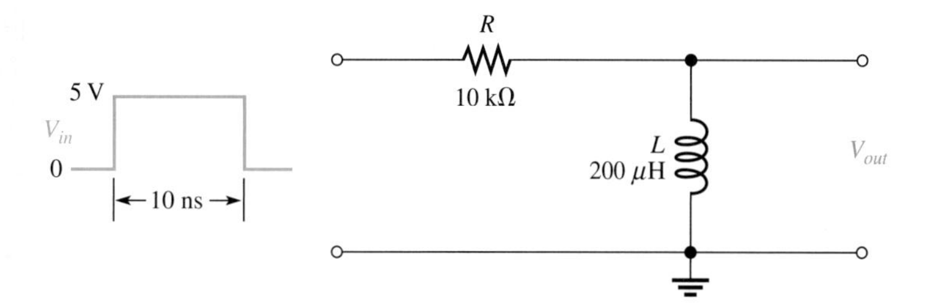

Solution First, calculate the time constant.

$$\tau = \frac{L}{R} = \frac{200\ \mu\text{H}}{10\ \text{k}\Omega} = 2\ \text{ns}$$

In this case, $t_W = 5\tau$, so the output will decay to zero at the end of the pulse.

On the rising edge, the inductor voltage jumps to +5 V and then decays exponentially to zero. It reaches approximately zero at the instant of the falling edge. On the falling edge of the input, the inductor voltage jumps to −5 V and then goes back to zero. The output waveform is shown in Figure 15–44.

▶ FIGURE 15–44

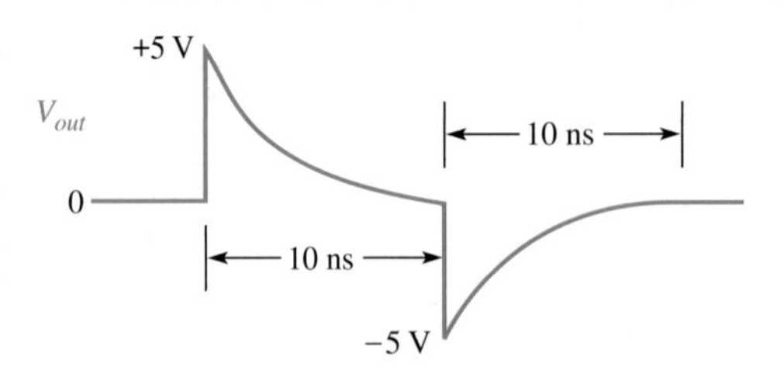

Related Problem Show the output voltage if the input pulse width is reduced to 5 ns in Figure 15–43.

EXAMPLE 15–10

Determine the output voltage waveform for the differentiator in Figure 15–45.

▶ FIGURE 15–45

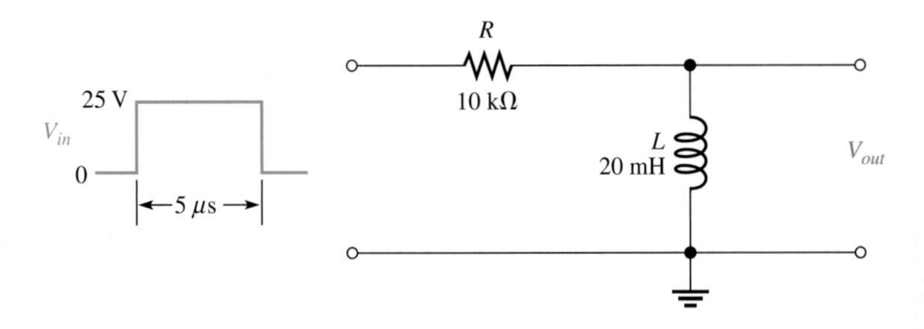

Solution First, calculate the time constant.

$$\tau = \frac{L}{R} = \frac{20 \text{ mH}}{10 \text{ k}\Omega} = 2 \ \mu\text{s}$$

On the rising edge, the inductor voltage immediately jumps to +25 V. Because the pulse width is 5 μs, the inductor charges for only two and a half time constants (2.5τ), so you must use the formula for a decreasing exponential.

$$v_L = V_i e^{-t/\tau} = (25 \text{ V})e^{-5\mu\text{s}/2\mu\text{s}} = (25 \text{ V})e^{-2.5} = 2.05 \text{ V}$$

This result is the inductor voltage at the end of the 5 μs input pulse.

On the falling edge, the output immediately drops from +2.05 V to −22.95 V (a 25 V negative-going transition). The complete output waveform is shown in Figure 15–46.

▶ FIGURE 15–46

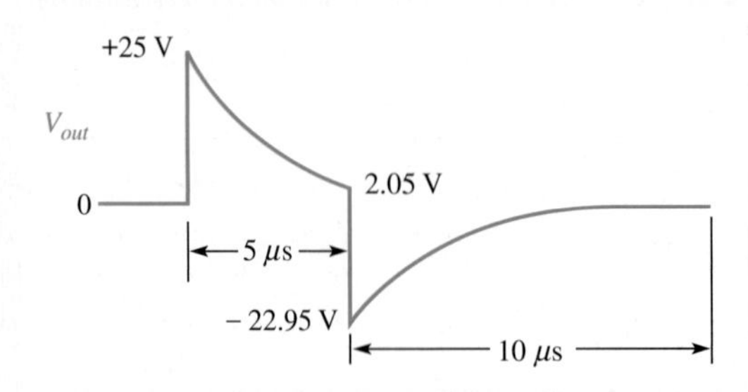

Related Problem What must be the value of R for V_{out} to reach zero by the end of the input pulse in Figure 15–45?

SECTION 15–7 REVIEW

1. In an *RL* differentiator, across which component is the output taken?
2. Under what condition does the output pulse shape most closely resemble the input pulse?
3. If the inductor voltage in an *RL* differentiator is down to +2 V at the end of a +10 V input pulse, to what negative voltage will the output voltage go in response to the falling edge of the input?

15–8 APPLICATIONS

Although integrators and differentiators are found in many different kinds of applications, only three basic ones are discussed in this section to give you a general idea of how these circuits can be used.

After completing this section, you should be able to

- **Discuss some basic applications of integrators and differentiators**
- Describe how *RC* integrators are used in timing circuits
- Describe how an integrator can be used for converting pulse waveforms to dc
- Describe how a differentiator can be used as a trigger pulse generator

Timing Circuits

RC integrators can be used in timing circuits to set a specified time interval for various purposes. By changing the time constant, the interval can be adjusted. For example, the *RC* integrator circuit in Figure 15–47(a) is used to provide a delay between the time a power-on switch is closed and the time a certain event is activated by a threshold circuit. The threshold circuit is designed to respond to the input voltage only when the input reaches a certain specified level. Don't worry

FIGURE 15–47

A basic time delay application of an *RC* integrator.

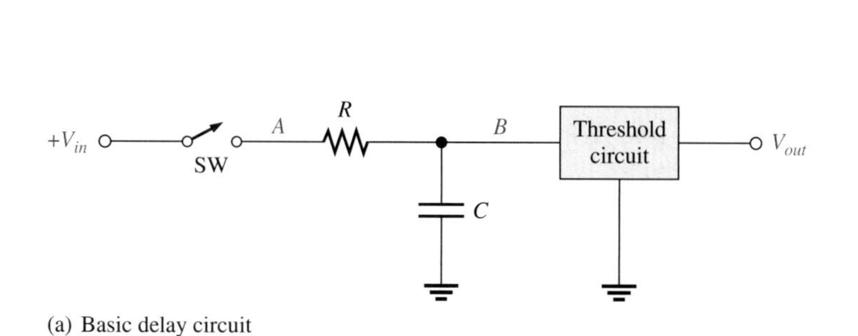

(a) Basic delay circuit

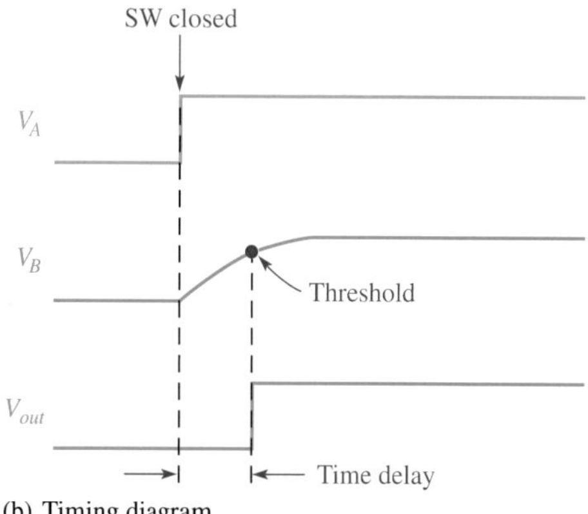

(b) Timing diagram

how the threshold circuit works at this point; you will learn about this type of circuit in a later course.

When the switch is thrown, the capacitor begins to charge at a rate set by the *RC* time constant. After a certain time, the capacitor voltage reaches the threshold value and triggers (turns on) the circuit to activate a device such as a motor, relay, or lamp, depending on the particular application. This action is illustrated by the waveforms in Figure 15–47(b).

EXAMPLE 15–11

What is the time delay in the circuit of Figure 15–47(a) with $V_{in} = 9$ V, $R = 10$ MΩ, and $C = 0.47$ μF? Assume the threshold voltage is 5 V.

Solution Use the exponential formula, Equation 9–18, and solve for time as follows:

$$v = V_F(1 - e^{-t/RC}) = V_F - V_F e^{-t/RC}$$

$$V_F - v = V_F e^{-t/RC}$$

$$e^{-t/RC} = \frac{V_F - v}{V_F}$$

Taking the natural log (ln) of both sides of the equation yields

$$-\frac{t}{RC} = \ln\left(\frac{V_F - v}{V_F}\right)$$

$$t = -RC\ln\left(\frac{V_F - v}{V_F}\right)$$

V_F is the final voltage to which C will charge and equals V_{in}. Substituting values and solving for t yields

$$t = -(10\ \text{M}\Omega)(0.47\ \mu\text{F})\ln\left(\frac{9\ \text{V} - 5\ \text{V}}{9\ \text{V}}\right) = -(10\ \text{M}\Omega)(0.47\ \mu\text{F})\ln\left(\frac{4\ \text{V}}{9\ \text{V}}\right) = \mathbf{3.8\ s}$$

Related Problem Find the time delay for the circuit in Figure 15–47 for $V_{in} = 24$ V, $R = 10$ kΩ, and $C = 1500$ pF.

A Pulse Waveform-to-DC Converter

An integrating circuit can be used to convert a pulse waveform to a constant dc value equal to the average of the waveform. This is accomplished by using a time constant that is extremely large compared to the period of the pulse waveform, as illustrated in Figure 15–48(a). Actually, there will be a slight ripple on the output as the capacitor charges and discharges a little. This ripple is less the larger the time constant and can approach a constant level, as shown in Figure 15–48(b). In terms of sinusoidal response, this circuit is a low-pass filter.

A Positive and Negative Trigger Pulse Generator

A differentiator circuit can be used to produce very short duration pulses (spikes) with both positive and negative polarities, as illustrated in Figure 15–49(a). When the time constant is small compared to the pulse width of the input, the differentiator produces a positive spike on each positive-going edge of the input and a negative spike on each negative-going edge.

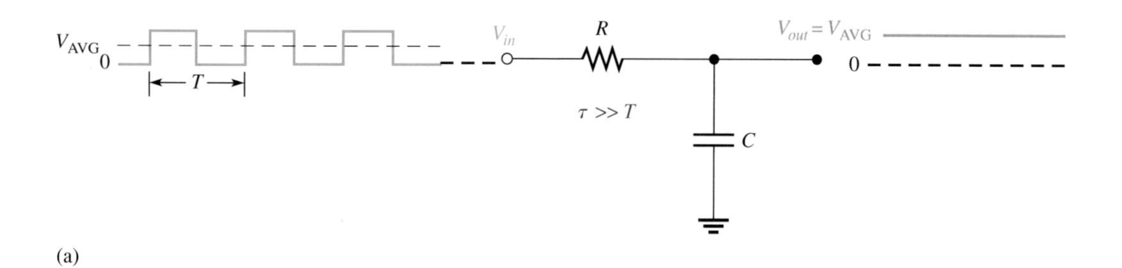

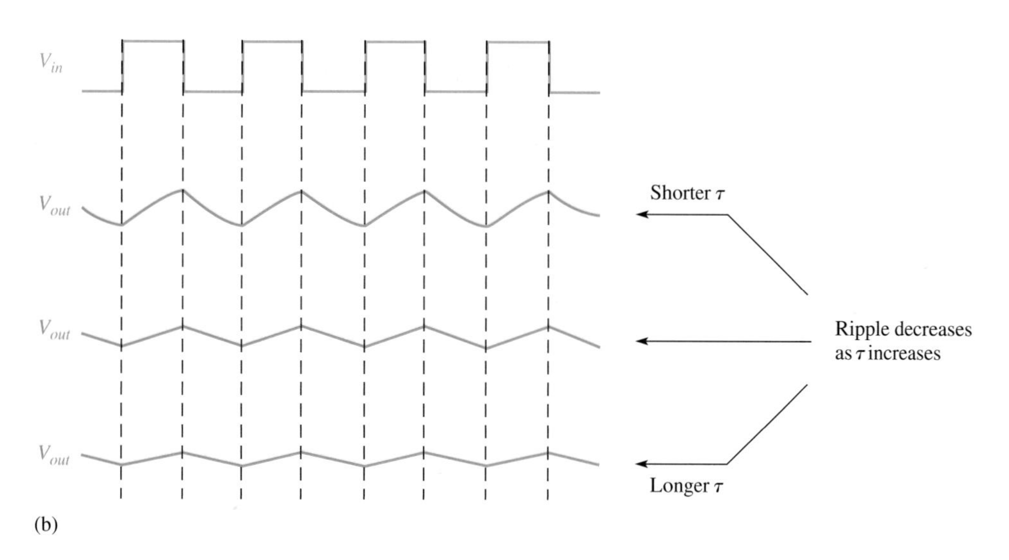

▲ **FIGURE 15–48**

A long time constant integrator acts as a pulse-to-dc converter.

▼ **FIGURE 15–49**

A very short time constant differentiator generates positive and negative spikes.

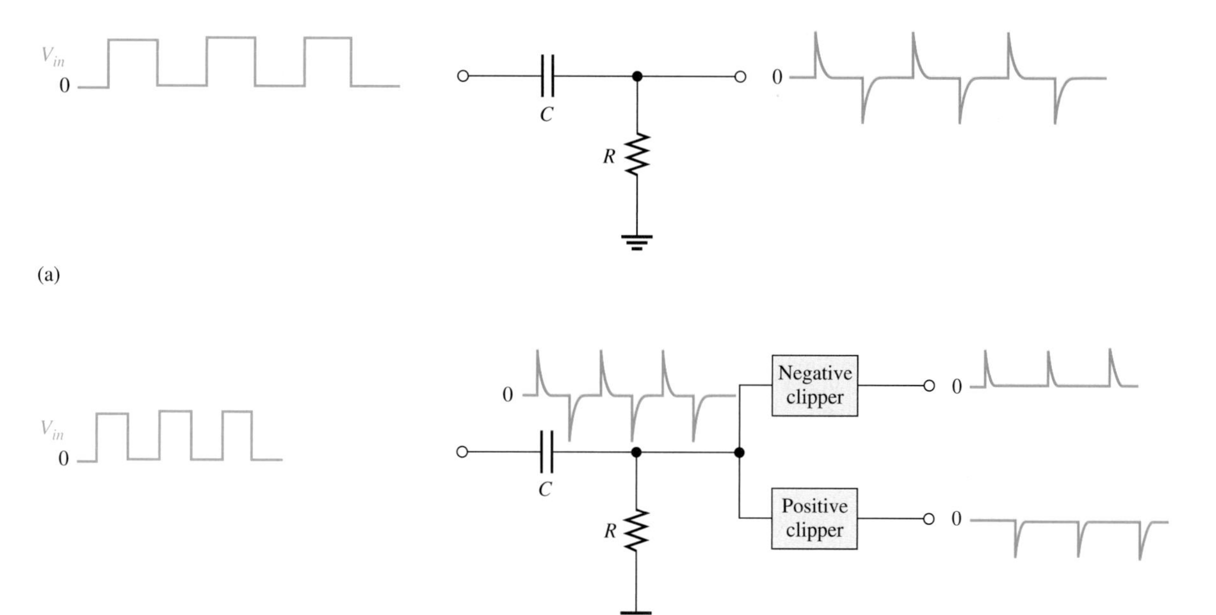

To separate the positive and negative spikes, two clipping circuits are used, as shown in Figure 15–49(b). Each clipping circuit consists of a diode, which is a semiconductor device. The negative clipper removes the negative spike and passes the positive spike. The positive clipper does just the opposite.

SECTION 15–8 REVIEW

1. How can the delay time in the circuit of Figure 15–47 be increased?
2. Reducing the value of R in Figure 15–48 will (increase, decrease) the amount of ripple on the output.
3. What characteristic must the differentiator in Figure 15–49 have in order to produce very short duration spikes on its output?

15–9 TROUBLESHOOTING

In this section, *RC* circuits with pulse inputs are used to demonstrate the effects of common component failures in selected cases. The concepts can then be easily related to *RL* circuits.

After completing this section, you should be able to

- **Troubleshoot *RC* integrators and *RC* differentiators**
- Recognize the effect of an open capacitor
- Recognize the effect of a shorted capacitor
- Recognize the effect of an open resistor

Open Capacitor

If the capacitor in an *RC* integrator opens, the output has the same waveshape as the input, as shown in Figure 15–50(a). If the capacitor in an *RC* differentiator opens, the output is zero because it is held at ground through the resistor, as illustrated in part (b).

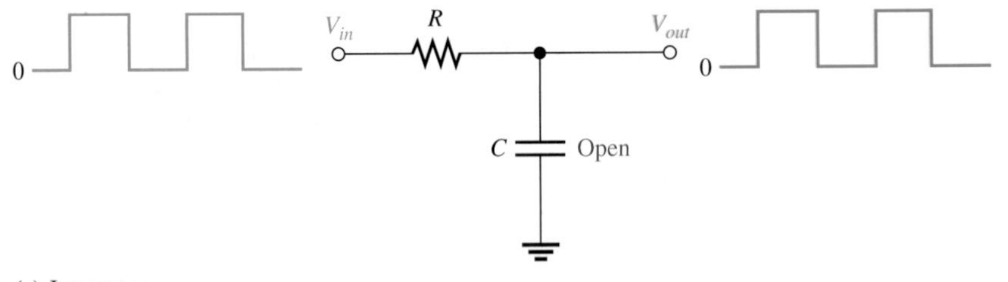

(a) Integrator

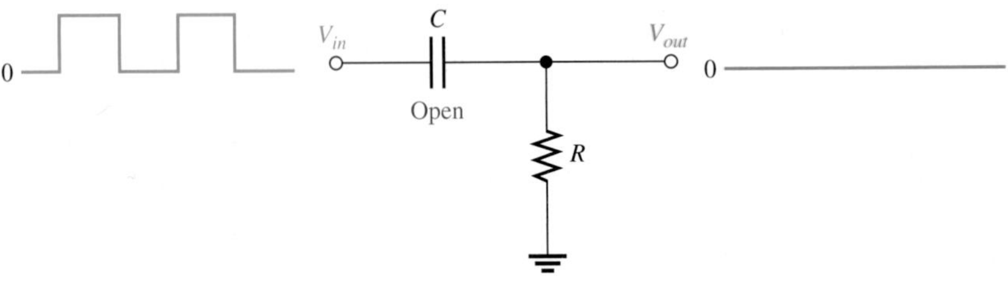

(b) Differentiator

FIGURE 15–50

Examples of the effect of an open capacitor.

Shorted Capacitor

If the capacitor in an *RC* integrator shorts, the output is at ground as shown in Figure 15–51(a). If the capacitor in an *RC* differentiator shorts, the output is the same as the input, as shown in part (b).

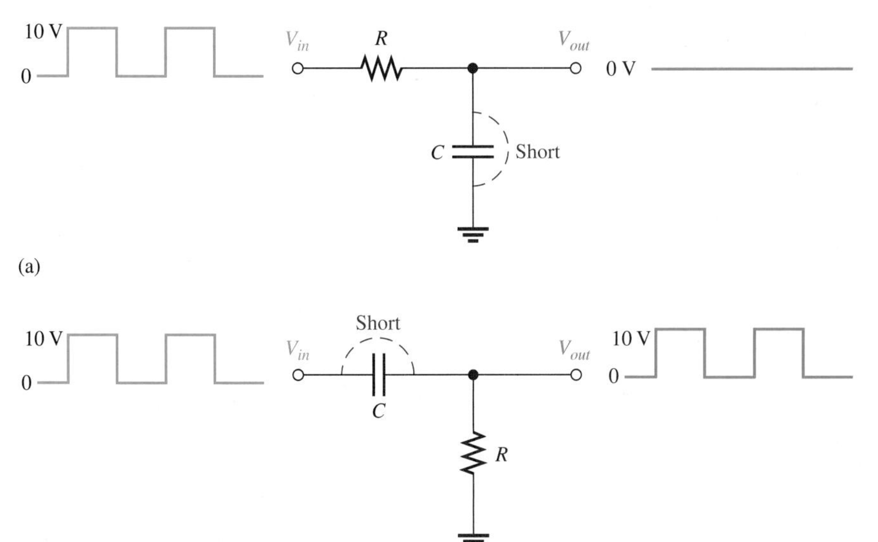

FIGURE 15–51

Examples of the effect of a shorted capacitor.

Open Resistor

If the resistor in an *RC* integrator opens, the capacitor has no discharge path, and, ideally, it will hold any charge it may have. In an actual situation, any charge will gradually leak off or the capacitor will discharge slowly through a measuring instrument connected to the output. This is illustrated in Figure 15–52(a).

If the resistor in an *RC* differentiator opens, the output looks like the input except for the dc level because the capacitor now must charge and discharge through the extremely high resistance of the oscilloscope, as shown in Figure 15–52(b).

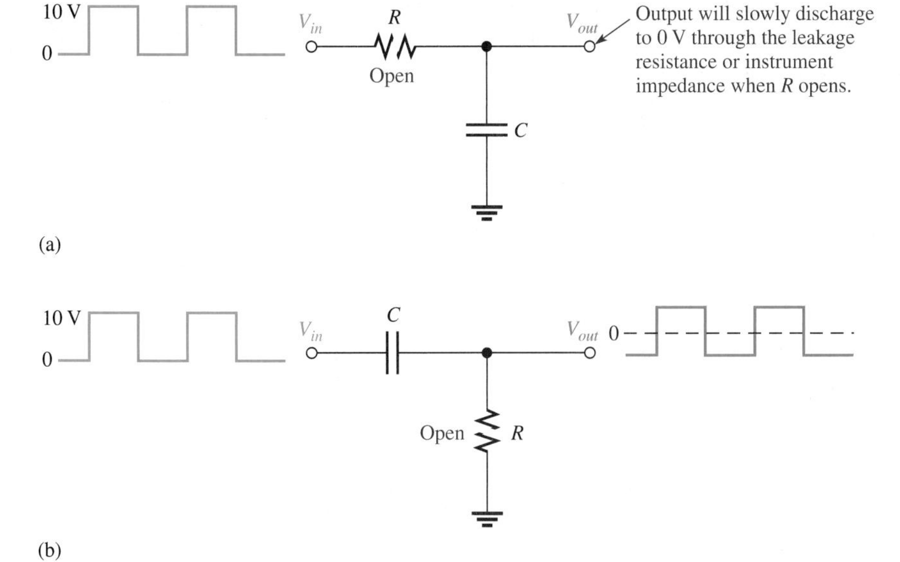

FIGURE 15–52

Examples of the effect of an open resistor.

SECTION 15-9 REVIEW

1. An *RC* integrator has a zero output with a square wave input. What are the possible causes of this problem?
2. If the integrator capacitor is open, what is the output for a square wave input?
3. If the capacitor in a differentiator is shorted, what is the output for a square wave input?

APPLICATION ASSIGNMENT

PUTTING YOUR KNOWLEDGE TO WORK

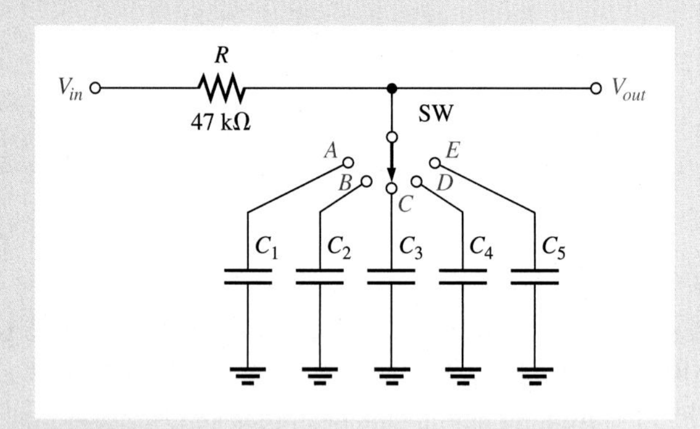

▲ **FIGURE 15-53**

Integrator delay circuit.

Suppose your supervisor has given you the assignment of breadboarding and testing a time-delay circuit that will provide five switch-selectable delay times. An *RC* integrator is selected for this application. The input is a 5 V pulse of long duration which begins the time delay. The exponential output voltage goes to a threshold trigger circuit that is used to turn the power on to a portion of a system after any one of the five selectable time delays.

A schematic of the selectable time-delay integrator is shown in Figure 15-53. The *RC* integrator is driven by a positive pulse input, and the output is the exponential voltage across the selected capacitor. The output voltage triggers a threshold circuit at the 3.5 V level, which then turns on the power to a portion of a system. The basic concept of operation is shown in Figure 15-54. In this application, the delay time of the integrator is specified as the time from the rising edge of the input pulse to the point where the output voltage reaches 3.5 V. The specified time delays are listed in Table 15-1.

Step 1: Capacitor Values

Determine a value for each of the five capacitors in the time-delay circuit that will provide a delay time within 10% of the specified delay time. Select from the following list of standard values (all are in μF): 0.1, 0.12, 0.15, 0.18, 0.22, 0.27, 0.33, 0.39, 0.47, 0.56, 0.68, 0.82, 1.0, 1.2, 1.5, 1.8, 2.2, 2.7, 3.3, 3.9, 4.7, 5.6, 6.8, and 8.2.

Step 2: Circuit Connections

Refer to Figure 15-55. The components for the time-delay circuit are assembled, but not connected, on the breadboard. Using the circled numbers, develop a point-to-point wiring list to show how the circuit and measurement instruments should be connected.

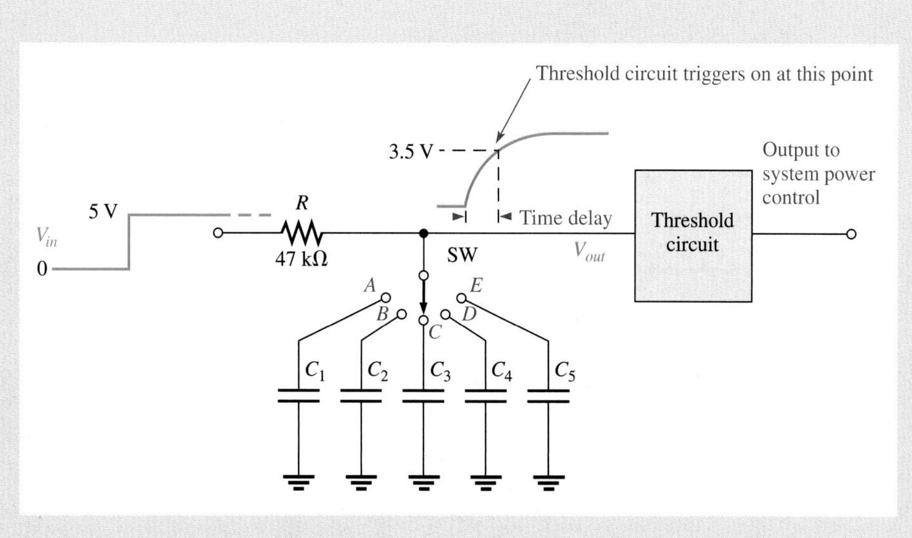

FIGURE 15–54

Illustration of the time-delay operation.

TABLE 15–1

SWITCH POSITION	DELAY TIME
A	10 ms
B	25 ms
C	40 ms
D	65 ms
E	85 ms

Step 3: Test Procedure and Instrument Settings

Develop a procedure for fully testing the time-delay circuit. Specify the amplitude, frequency, and duty cycle settings for the function generator in order to test each delay time. Specify the oscilloscope sec/div settings for measuring each of the five specified delay times.

Step 4: Measurement

Explain how you will verify that each switch setting produces the proper output delay time.

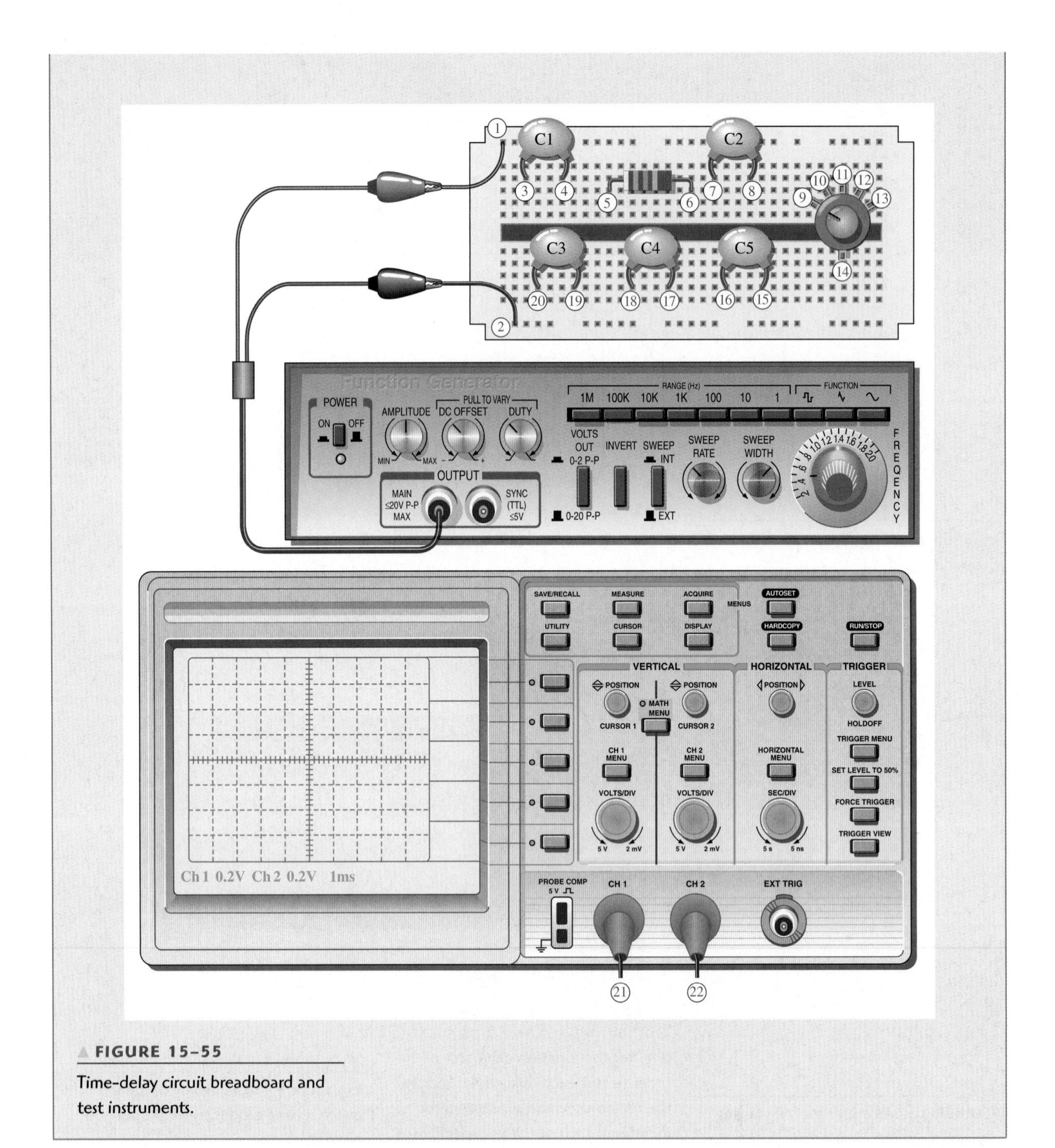

▲ **FIGURE 15–55**

Time-delay circuit breadboard and test instruments.

APPLICATION ASSIGNMENT REVIEW

1. Explain why the input pulse to the time-delay circuit must be of sufficiently long duration.
2. Indicate how you would modify the time-delay circuit so that each selected time delay had some adjustment range.

SUMMARY

- In an *RC* integrator, the output voltage is taken across the capacitor.
- In an *RC* differentiator, the output voltage is taken across the resistor.
- In an *RL* integrator, the output voltage is taken across the resistor.
- In an *RL* differentiator, the output voltage is taken across the inductor.
- In an integrator, when the pulse width (t_W) of the input is much less than the transient time, the output voltage approaches a constant level equal to the average value of the input.
- In an integrator, when the pulse width of the input is much greater than the transient time, the output voltage approaches the shape of the input.
- In a differentiator, when the pulse width of the input is much less than the transient time, the output voltage approaches the shape of the input but with an average value of zero.
- In a differentiator, when the pulse width of the input is much greater than the transient time, the output voltage consists of narrow, positive-going and negative-going spikes occurring on the leading and trailing edges of the input pulses.

SELF-TEST

Answers are at the end of the chapter.

1. The output of an *RC* integrator is taken across the
   (a) resistor **(b)** capacitor **(c)** source **(d)** coil
2. When a 10 V input pulse with a width equal to one time constant is applied to an *RC* integrator, the capacitor charges to
   (a) 10 V **(b)** 5 V **(c)** 6.3 V **(d)** 3.7 V
3. When a 10 V pulse with a width equal to one time constant is applied to an *RC* differentiator, the capacitor charges to
   (a) 6.3 V **(b)** 10 V **(c)** 5 V **(d)** 3.7 V
4. In an *RC* integrator, the output pulse closely resembles the input pulse when
   (a) τ is much larger than the pulse width
   (b) τ is equal to the pulse width
   (c) τ is less than the pulse width
   (d) τ is much less than the pulse width
5. In an *RC* differentiator, the output pulse closely resembles the input when
   (a) τ is much larger than the pulse width
   (b) τ is equal to the pulse width
   (c) τ is less than the pulse width
   (d) τ is much less than the pulse width
6. The positive and negative portions of a differentiator's output voltage are equal when
   (a) $5\tau < t_W$ **(b)** $5\tau > t_W$ **(c)** $5\tau = t_W$ **(d)** $5\tau > 0$
7. The output of an *RL* integrator is taken across the
   (a) resistor **(b)** coil **(c)** source **(d)** capacitor

8. The maximum current in an RL differentiator is

   (a) $I = \frac{V_p}{X_L}$ (b) $I = \frac{V_p}{Z}$ (c) $I = \frac{V_p}{R}$

9. The current in an RL differentiator reaches its maximum possible value when

   (a) $5\tau = t_W$ (b) $5\tau < t_W$ (c) $5\tau > t_W$ (d) $\tau = 0.5t_W$

10. If you have an RC and an RL differentiator with equal time constants sitting side-by-side and you apply the same input pulse to both,

    (a) the RC has the widest output pulse

    (b) the RL has the most narrow spikes on the output

    (c) the output of one is an increasing exponential and the output of the other is a decreasing exponential

    (d) you can't tell the difference by observing the output waveforms

PROBLEMS

Answers to odd-numbered problems are at the end of the book.

BASIC PROBLEMS

SECTION 15–1 The *RC* Integrator

1. An integrator has $R = 2.2\ \text{k}\Omega$ in series with $C = 0.047\ \mu\text{F}$. What is the time constant?

2. Determine how long it takes the capacitor in an integrating circuit to reach full charge for each of the following series RC combinations:

   (a) $R = 47\ \Omega$, $C = 47\ \mu\text{F}$ (b) $R = 3300\ \Omega$, $C = 0.015\ \mu\text{F}$

   (c) $R = 22\ \text{k}\Omega$, $C = 100\ \text{pF}$ (d) $R = 4.7\ \text{M}\Omega$, $C = 10\ \text{pF}$

SECTION 15–2 Single-Pulse Response of *RC* Integrators

3. A 20 V pulse is applied to an RC integrator. The pulse width equals one time constant. To what voltage does the capacitor charge during the pulse? Assume that it is initially uncharged.

4. Repeat Problem 3 for the following values of t_W:

   (a) 2τ (b) 3τ (c) 4τ (d) 5τ

5. Show the approximate shape of an integrator output voltage where 5τ is much less than the pulse width of a 10 V square wave input. Repeat for the case in which 5τ is much larger than the pulse width.

6. Determine the output voltage for an integrator with a single input pulse, as shown in Figure 15–56. For repetitive pulses, how long will it take this circuit to reach steady state?

SECTION 15–3 Repetitive-Pulse Response of *RC* Integrators

7. Draw the integrator output in Figure 15–57 showing maximum voltages.

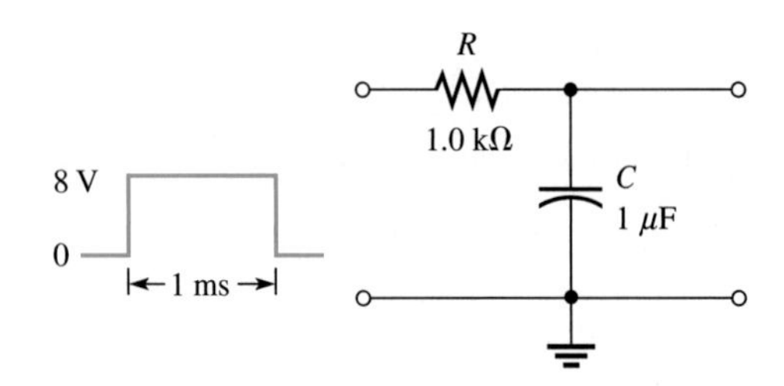

▲ FIGURE 15–56

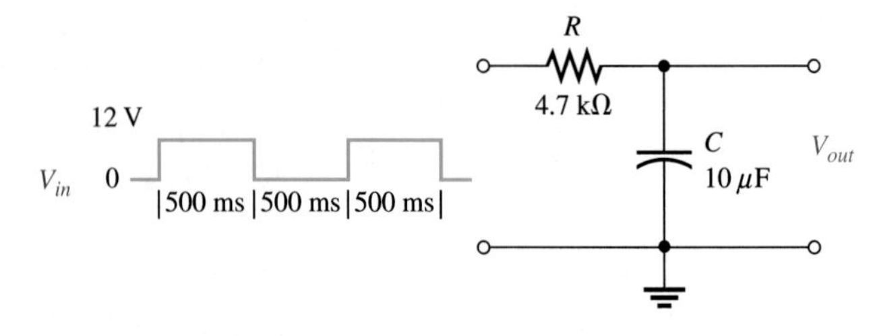

▲ FIGURE 15–57

8. A 1 V, 10 kHz pulse waveform with a duty cycle of 25% is applied to an integrator with $\tau = 25\ \mu s$. Graph the output voltage for three initial pulses. C is initially uncharged.

9. What is the steady-state output voltage of the RC integrator with a square wave input shown in Figure 15–58?

FIGURE 15–58

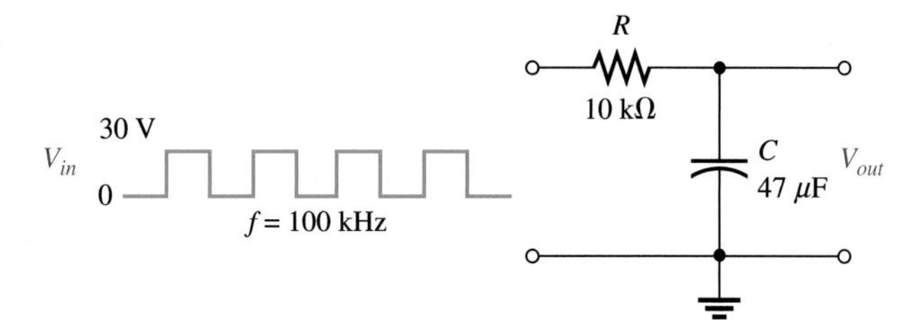

SECTION 15–4 **Single-Pulse Response of *RC* Differentiators**

10. Repeat Problem 5 for an RC differentiator.

11. Redraw the circuit in Figure 15–56 to make it a differentiator, and repeat Problem 6.

SECTION 15–5 **Repetitive-Pulse Response of *RC* Differentiators**

12. Draw the differentiator output in Figure 15–59, showing maximum voltages.

FIGURE 15–59

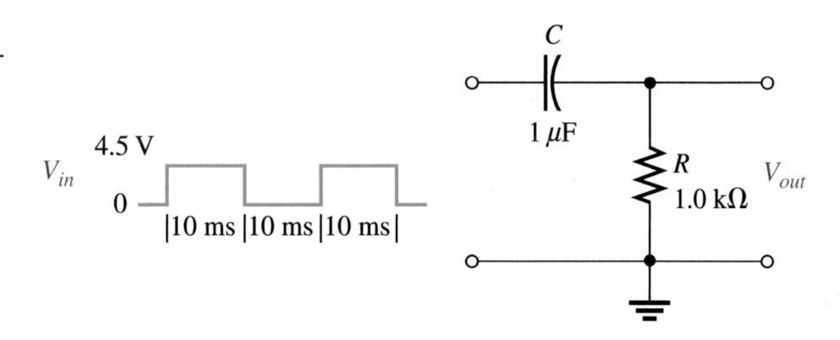

13. What is the steady-state output voltage of the differentiator with the square wave input shown in Figure 15–60?

FIGURE 15–60

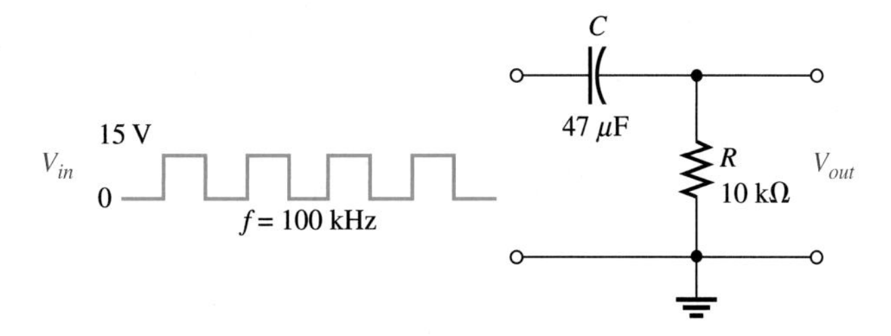

SECTION 15–6 **Pulse Response of *RL* Integrators**

14. Determine the output voltage for the circuit in Figure 15–61. A single input pulse is applied as shown.

15. Draw the integrator output in Figure 15–62, showing maximum voltages.

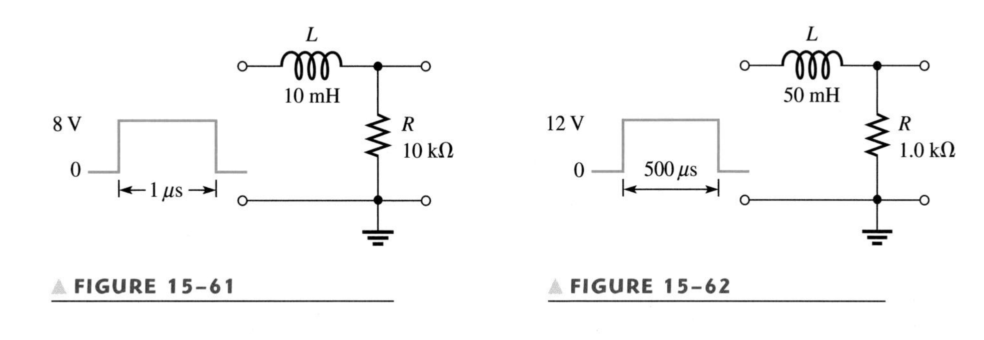

FIGURE 15–61 FIGURE 15–62

SECTION 15–7

Pulse Response of *RL* Differentiators

16. **(a)** What is τ in Figure 15–63?

(b) Draw the output voltage.

17. Draw the output waveform if a periodic pulse waveform with $t_W = 25$ ns and $T = 60$ ns is applied to the circuit in Figure 15–63.

FIGURE 15–63

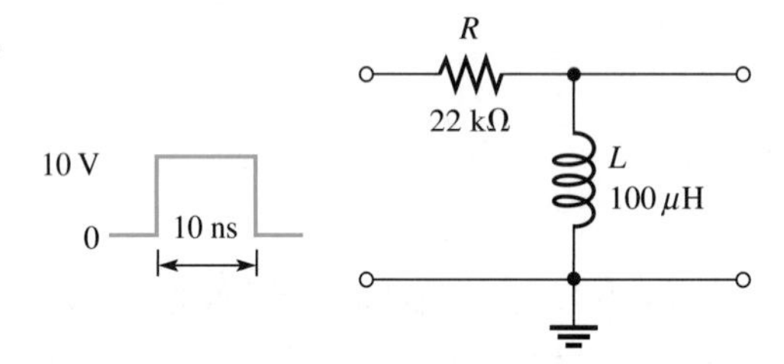

SECTION 15–8

Applications

18. What is the instantaneous voltage at point *B* in Figure 15–47 440 μs after the switch is closed for $R = 22$ kΩ and $C = 0.001$ μF? $V_{in} = 10$ V.

19. Ideally, what is the output of an *RC* integrator when the input is a square wave with an amplitude of 12 V? Assume that the time constant is much greater than the period of the input signal?

SECTION 15–9

Troubleshooting

20. Determine the most likely fault(s), in the circuit of Figure 15–64(a) for each set of waveforms in parts (b) through (d). V_{in} is a square wave with a period of 8 ms.

FIGURE 15–64

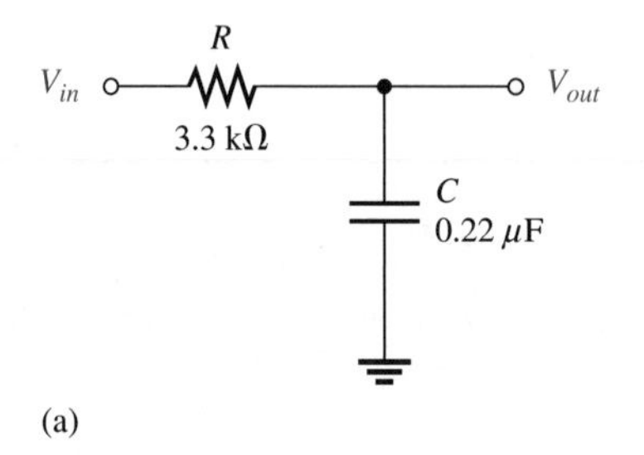

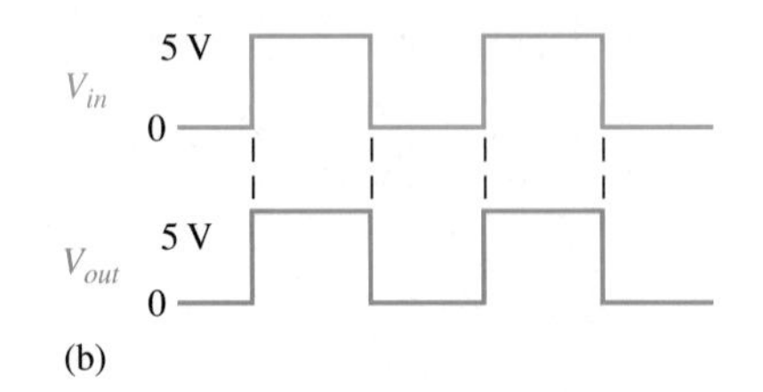

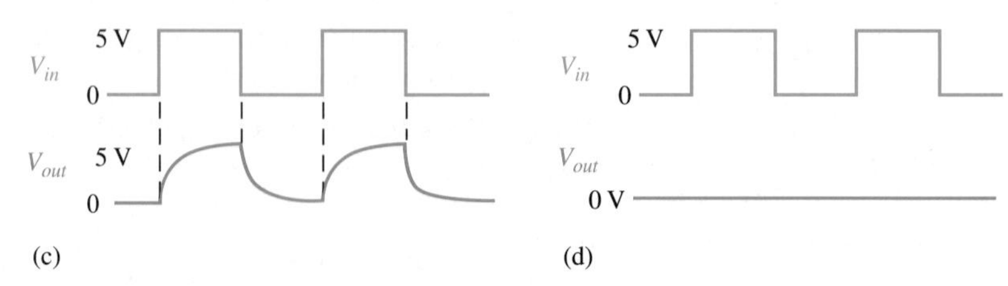

21. Determine the most likely fault(s), if any, in the circuit of Figure 15–65(a) for each set of waveforms in parts (b) through (d), V_{in} is a square wave with a period of 8 ms.

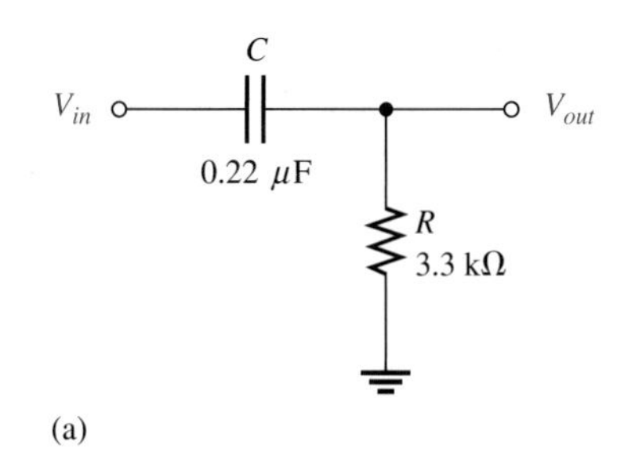

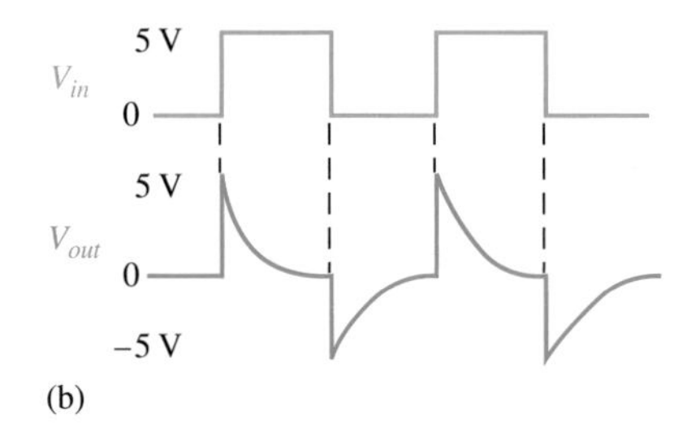

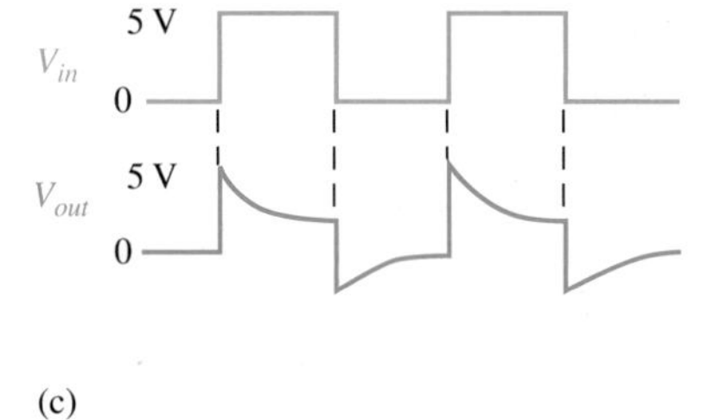

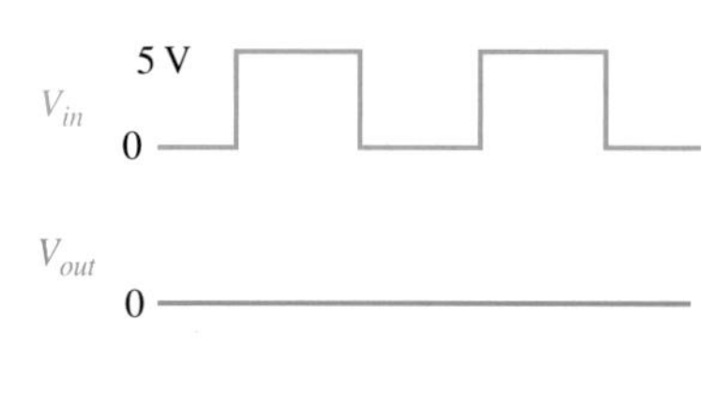

▲ FIGURE 15–65

ADVANCED PROBLEMS

22. **(a)** What is τ in Figure 15–66?

(b) Draw the output voltage.

► FIGURE 15–66

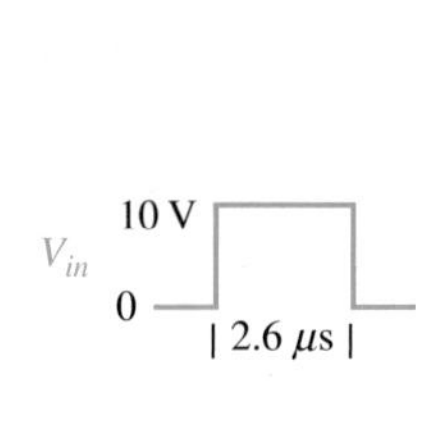

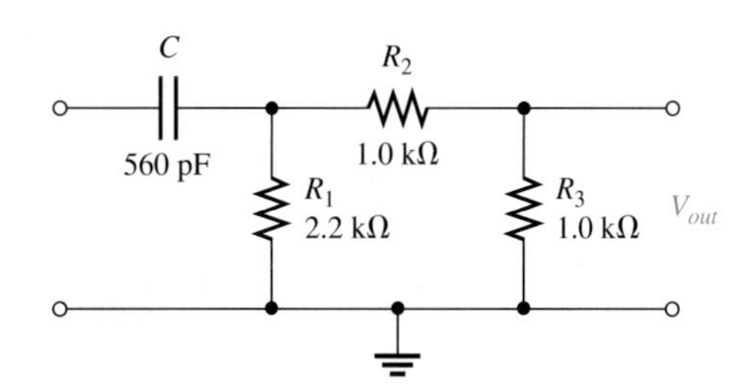

23. **(a)** What is τ in Figure 15–67?

(b) Draw the output voltage.

► FIGURE 15–67

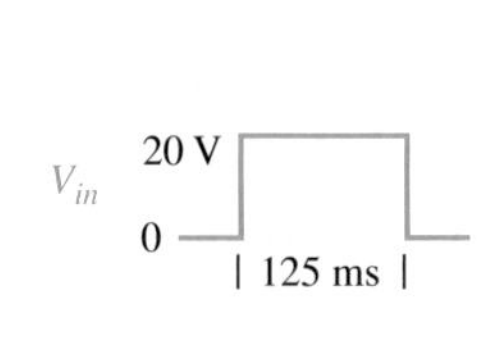

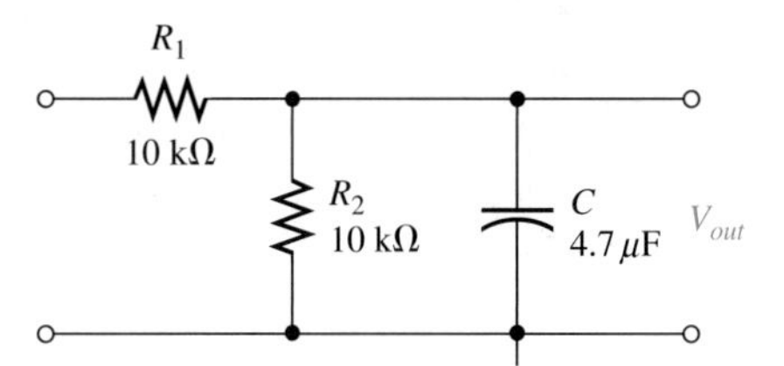

24. Determine the time constant in Figure 15–68. Is this circuit an integrator or a differentiator?

► FIGURE 15–68

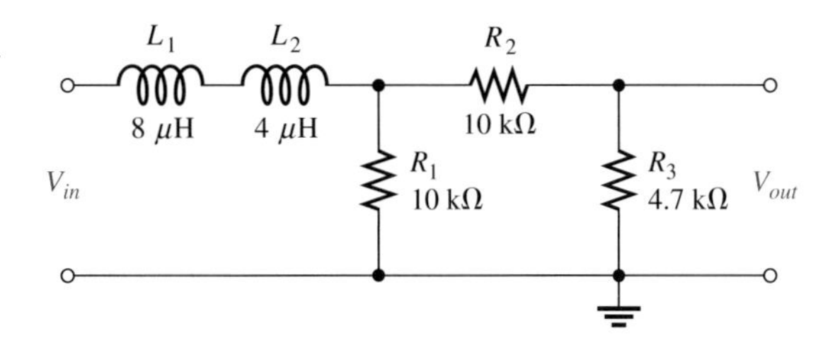

25. In a time-delay circuit like that of Figure 15–47, what time constant will produce a 1 s delay if the threshold of the circuit is 2.5 V and the amplitude of the input is 5 V?

26. Draw the schematic for the circuit in Figure 15–69 and determine if the oscilloscope presentation is correct.

FIGURE 15–69

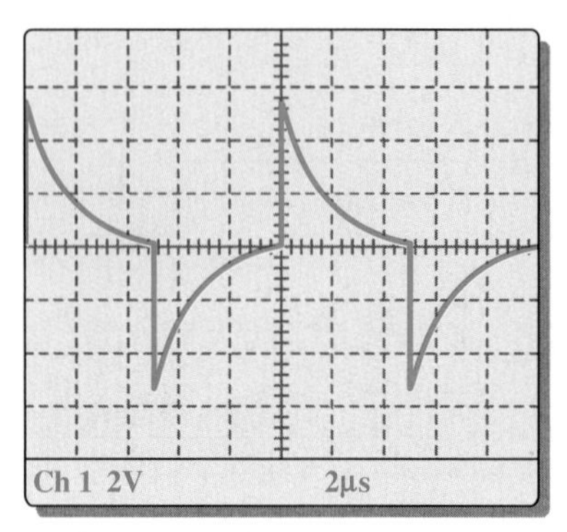

(a) Oscilloscope display

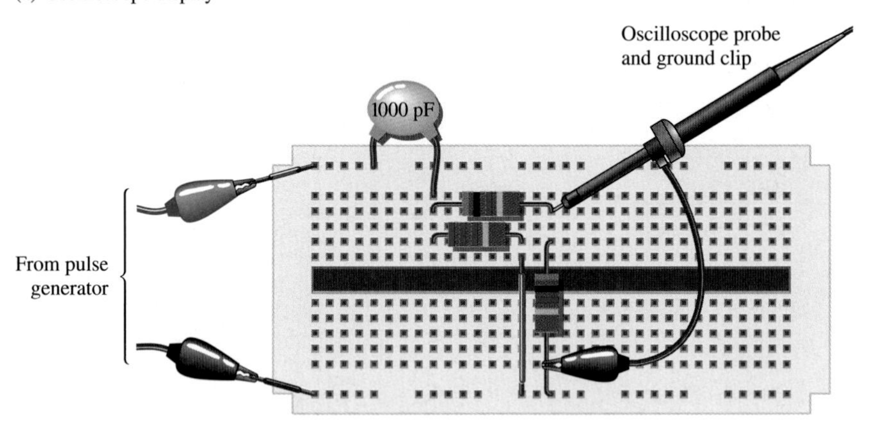

(b) Circuit board with instrument leads connected

ELECTRONICS WORKBENCH/CIRCUITMAKER TROUBLESHOOTING PROBLEMS

CD-ROM file circuits are shown in Figure 15–70.

27. Open file P15-27 and test the circuit. If there is a fault, identify it.

28. Open file P15-28 and test the circuit. If there is a fault, identify it.

29. Determine if there is a fault in the circuit in file P15-29. If so, identify it.

30. Identify any faulty component in the circuit in file P15-30.

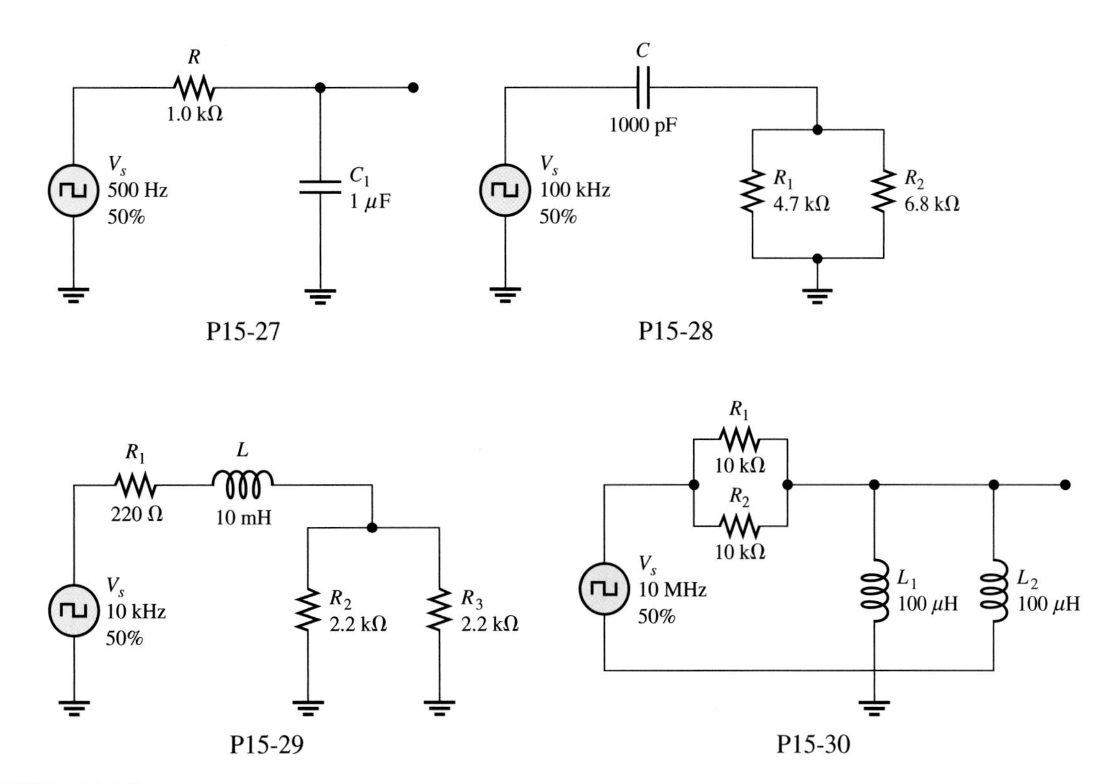

▲ FIGURE 15–70

ANSWERS

SECTION REVIEWS

SECTION 15–1 The *RC* Integrator

1. An integrator is a series *RC* circuit with a pulse input in which the output is across the capacitor.
2. A voltage applied to the input causes the capacitor to charge. Zero volts across the input causes the capacitor to discharge.

SECTION 15–2 Single-Pulse Response of *RC* Integrators

1. The output of an integrator reaches full amplitude when $5\tau \leq t_W$.
2. $V_{out} = (0.632)1\text{ V} = 0.632\text{ V}$; $t_{disch} = 5\tau = 51.7\text{ ms}$.
3. See Figure 15–71.
4. No, *C* will not fully charge.
5. The output is approximately the shape of the input when $5\tau << t_W$ (5τ much less than t_W).

▶ FIGURE 15–71

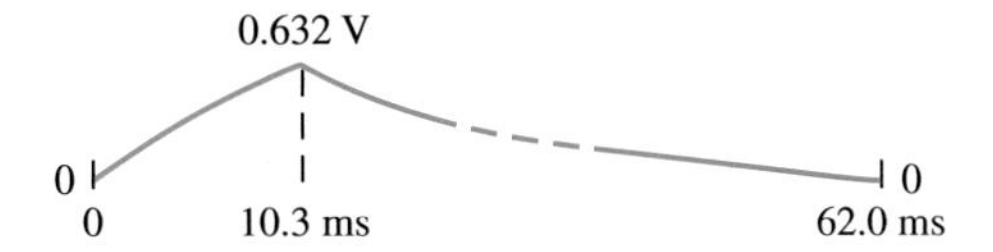

SECTION 15–3 **Repetitive-Pulse Response of *RC* Integrators**

1. An integrator capacitor will fully charge and discharge when $5\tau \leq t_W$ and $5\tau \leq$ time between pulses.
2. The output looks like the input.
3. transient time
4. Steady state is the response after the transient time has passed.
5. The average value of the output voltage equals the average value of the input voltage.

SECTION 15–4 **Single-Pulse Response of *RC* Differentiators**

1. See Figure 15–72.

▶ FIGURE 15–72

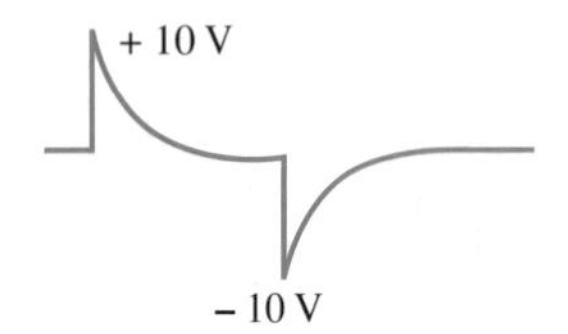

2. The output approximates the input when $5\tau >> t_W$.
3. The output consists of a positive and a negative spike.
4. $V_R = +5\text{ V} - 15\text{ V} = -10\text{ V}$

SECTION 15–5 **Repetitive-Pulse Response of *RC* Differentiators**

1. C fully charges and discharges when $5\tau \leq t_W$ and $5\tau \leq$ time between pulses.
2. The output consists of positive and negative spikes.
3. $V_{out} = 0\text{ V}$

SECTION 15–6 **Pulse Response of *RL* Integrators**

1. The output is across the resistor.
2. The output reaches the input amplitude when $5\tau \leq t_W$.
3. The output approximates the input when $5\tau << t_W$.

SECTION 15–7 **Pulse Response of *RL* Differentiators**

1. The output is across the inductor.
2. The output approximates the input when $5\tau >> t_W$.
3. $V_{out} = 2\text{ V} - 10\text{ V} = -8\text{ V}$

SECTION 15–8 **Applications**

1. Increase the time constant to increase delay time.
2. Reducing R will increase ripple.
3. A very short time constant produces very short duration spikes.

SECTION 15–9 **Troubleshooting**

1. Zero output of *RC* integrator indicates shorted capacitor, open resistor, no source voltage, or open contact.
2. An open integrator capacitor results in an output exactly like the input.
3. A shorted differentiator capacitor results in an output exactly like the input.

Application Assignment

1. The input pulse must be sufficiently long to allow the output to reach 3.5 V for the longest time-constant setting.
2. Add a series potentiometer to adjust the total resistance.

RELATED PROBLEMS FOR EXAMPLES

15–1 8.65 V
15–2 24.7 V
15–3 1.10 V
15–4 See Figure 15–73.
15–5 892 mV
15–6 10 kΩ
15–7 20 V
15–8 24.7 kΩ (assuming $R_s = 300\ \Omega$)
15–9 See Figure 15–74.
15–10 20 kΩ
15–11 3.5 μs

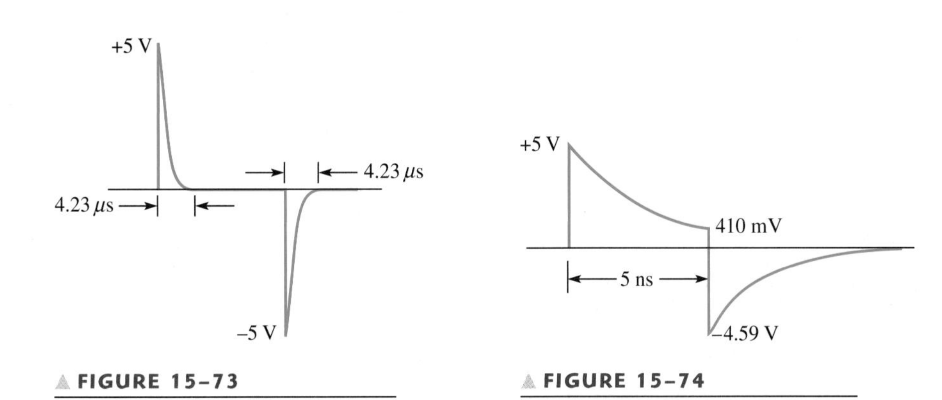

▲ FIGURE 15–73

▲ FIGURE 15–74

SELF-TEST

1. (b) **2.** (c) **3.** (a) **4.** (d) **5.** (a) **6.** (a) **7.** (a)
8. (c) **9.** (b) **10.** (d)

PART THREE

DEVICES

16

INTRODUCTION TO SEMICONDUCTORS

INTRODUCTION

To understand electronic devices you need a basic knowledge of the structure of atoms, so this chapter begins with a review from Chapter 2. The semiconductive materials used in manufacturing both **discrete devices** (such as diodes and transistors) and integrated circuits are discussed. An important concept introduced in this chapter is that of the *pn* junction that is formed when two different types of semiconductive material are joined. The *pn* junction is fundamental to the operation of devices such as the diode and certain types of transistors. The function of the *pn* junction is an essential factor in making electronic circuits operate properly. Also, diode characteristics are introduced, and you will learn how to properly use a diode in a circuit.

CHAPTER OUTLINE

CHAPTER OBJECTIVES

- Discuss the basic structure of semiconductors
- Discuss covalent bonding in silicon
- Explain how current occurs in a semiconductor
- Describe the properties of *n*-type and *p*-type semiconductors
- Describe the characteristics of a *pn* junction
- Explain how to bias a *pn* junction
- Describe the basic diode characteristics

KEY TERMS

- Valence
- Free electron
- Ion
- Silicon
- Hole
- Intrinsic semiconductor
- Doping
- Majority carriers
- Minority carriers
- *PN* Junction
- Barrier potential
- Forward bias
- Reverse bias
- Reverse breakdown
- Diode

APPLICATION ASSIGNMENT PREVIEW

Diode semiconductors are widely used in many types of electronic applications. They are of great importance in power supplies, where they are used as rectifiers to convert ac to dc. Suppose your supervisor has given you four power supply units that have failed the final assembly test and has asked you to check the units with an ohmmeter to identify the faults. After studying this chapter, you should be able to complete the application assignment.

WWW. VISIT THE COMPANION WEBSITE

Circuit Simulation Tutorials and Other Chapter Study Tools Are Available at

http://www.prenhall.com/floyd

16–1 ATOMIC STRUCTURE AND SEMICONDUCTORS

The basic structure of an atom was studied in Chapter 2, a portion of which is reviewed here. In this chapter, atomic theory is extended to semiconductors that are used in electronic devices such as diodes and transistors. This coverage lays the foundation for a good understanding of how semiconductive devices function.

After completing this section, you should be able to

- **Discuss the basic structure of semiconductors**
- Review the importance of electron shells and orbits in an atom
- Describe a valence electron and explain ionization
- Describe the atomic structure of silicon and germanium

Structure of Atoms Review

Electrons orbit the nucleus of an atom at certain distances from the nucleus. Electrons near the nucleus have less energy than those in more distant orbits. Electrons orbit only at discrete distances from the nucleus. Each discrete distance (orbit) from the nucleus corresponds to a certain energy level. In an atom, orbits are grouped into energy bands known as *shells.* A given atom has a fixed number of shells. Each shell has a fixed maximum number of electrons at permissible energy levels (orbits). Electrons in orbits farther from the nucleus have higher energy and are less tightly bound to the atom than those closer to the nucleus. The force of attraction between the positively charged nucleus and the negatively charged electron decreases with increasing distance from the nucleus.

Electrons with the highest energy levels exist in the outermost shell of an atom and are relatively loosely bound to the atom. This outermost shell is known as the

valence shell and electrons in this shell are called *valence electrons.* These valence electrons contribute to chemical reactions and bonding within the structure of a material and determine its electrical properties.

When an atom absorbs energy from a heat source or from light, for example, the energy levels of the electrons are raised. When an electron gains a certain amount of energy, it moves to an orbit farther from the nucleus. Since the valence electrons possess more energy and are more loosely bound to the atom than inner electrons, they can jump to higher orbits more easily when external energy is absorbed.

If a valence electron acquires a sufficient amount of energy, it can actually escape from the outer shell and the atom's influence. The departure of a valence electron leaves a previously neutral atom with an excess of positive charge (more protons than electrons). The process of losing a valence electron is known as *ionization.*

The escaped valence electron is called a free electron. When a free electron falls into the outer shell of a neutral atom, the atom becomes negatively charged (more electrons than protons). An ion is an atom that has gained or lost a valence electron, resulting in a net positive charge (positive ion) or a net negative charge (negative ion).

Silicon and Germanium Atoms

Two types of semiconductive materials are silicon and germanium. Both the silicon and the germanium atoms have four valence electrons. These atoms differ in

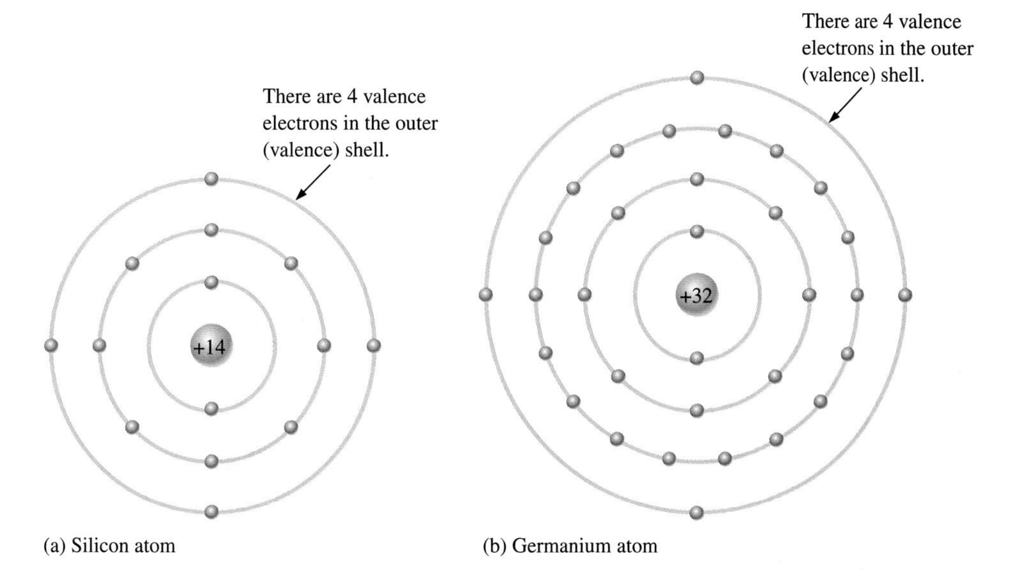

▲ **FIGURE 16–1**

Diagrams of the silicon and germanium atoms.

that silicon has 14 protons in its nucleus and germanium has 32. Figure 16–1 shows a representation of the atomic structure for both materials.

The valence electrons in germanium are in the fourth shell while the ones in silicon are in the third shell, closer to the nucleus. This means that the germanium valence electrons are at higher energy levels than those in silicon and, therefore, require a smaller amount of additional energy to escape from the atom. This property makes germanium more unstable than silicon at high temperatures, which is a main reason silicon is the most widely used semiconductive material.

SECTION 16–1 REVIEW

Answers are at the end of the chapter.

1. Define *orbit.*
2. Explain what a shell in an atom is.
3. What is a valence electron?
4. How is a positive ion created?

16–2 ATOMIC BONDING

When certain atoms combine into molecules to form a solid material, they arrange themselves in a fixed pattern called a **crystal.** The atoms within the crystal structure are held together by **covalent** bonds, which are created by the interaction of the valence electrons of each atom. A solid chunk of silicon is a crystalline material.

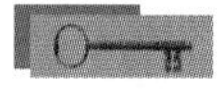

After completing this section, you should be able to

- **Discuss covalent bonding in silicon**
- Explain how covalent bonds produce a crystal

Figure 16–2 shows how each silicon atom positions itself with four adjacent atoms to form a silicon crystal. A silicon atom with its four valence electrons shares an electron with each of its four neighbors. This effectively creates eight valence electrons for each atom and produces a state of chemical stability. Also,

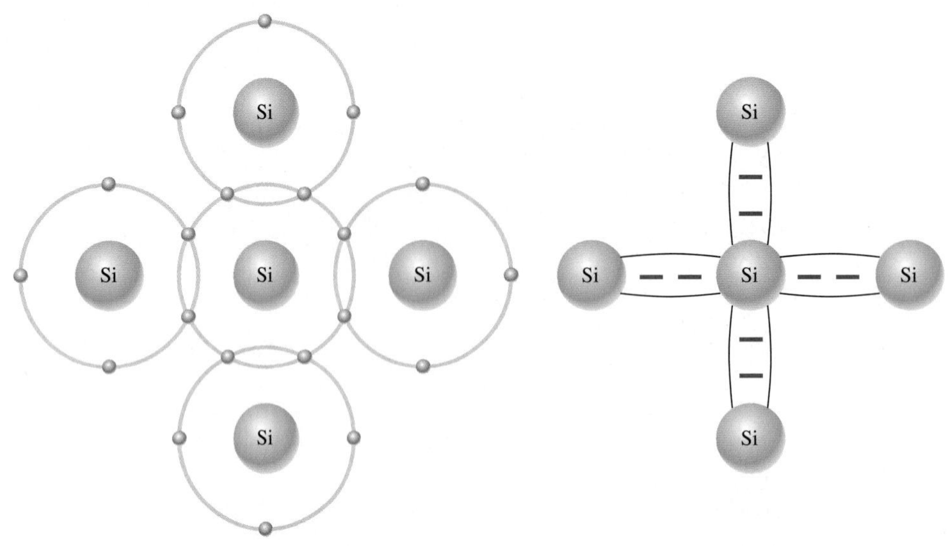

(a) The center atom shares an electron with each of the four surrounding atoms creating a covalent bond with each. The surrounding atoms are in turn bonded to other atoms, and so on.

(b) Bonding diagram. The red negative signs represent the shared valence electrons.

▲ **FIGURE 16–2**

Covalent bonds in a three-dimensional silicon crystal.

this sharing of valence electrons produces the covalent bonds that hold the atoms together; each shared electron is attracted equally by two adjacent atoms which share it. Covalent bonding in an intrinsic silicon crystal is shown in Figure 16–3. An **intrinsic** crystal is one that has no impurities. Covalent bonding for germanium is similar because it also has four valence electrons.

▶ **FIGURE 16–3**

Covalent bonds form a crystal structure.

SECTION 16–2 REVIEW

1. How are covalent bonds formed?
2. What is meant by the term *intrinsic?*
3. What is a crystal?
4. Effectively, how many valence electrons are there in each atom within a silicon crystal?

16–3 CONDUCTION IN SEMICONDUCTORS

As you have seen, the electrons of an atom can exist only within prescribed energy bands. Each shell around the nucleus corresponds to a certain energy band and is separated from adjacent shells by energy gaps, in which no electrons can exist.

After completing this section, you should be able to

- **Explain how current occurs in a semiconductor**
- Define *hole*
- Compare conduction properties of germanium and silicon
- Explain how electrons produce current
- Explain how holes are related to current
- Compare conductors, semiconductors, and insulators

An energy band diagram is shown in Figure 16–4 for a silicon crystal with only unexcited atoms (no external energy such as heat). This condition occurs *only* at a temperature of absolute 0 K.

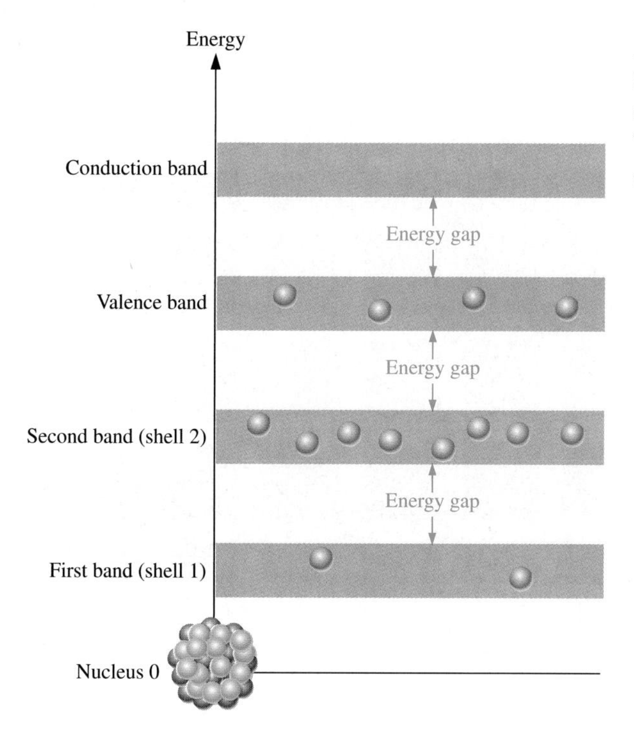

FIGURE 16–4

Energy band diagram for a pure silicon crystal with unexcited atoms. There are no electrons in the conduction band.

Conduction Electrons and Holes

An intrinsic (pure) silicon crystal at room temperature has sufficient heat (thermal) energy for some valence electrons to jump the gap from the valence band into the conduction band, becoming free electrons. This situation is illustrated in the energy diagram of Figure 16–5(a) and in the bonding diagram of Figure 16–5(b).

When an electron jumps to the conduction band, a vacancy is left in the valence band within the crystal. This vacancy is called a **hole**. For every electron raised to the conduction band by external energy, there is one hole left in the va-

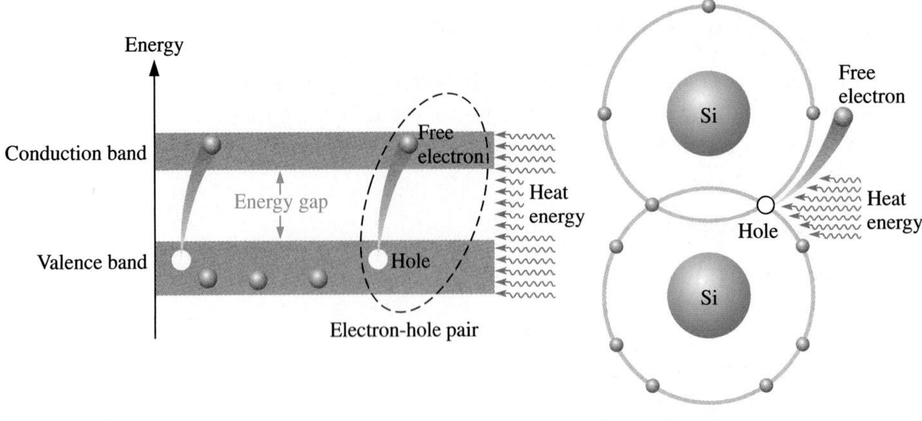

(a) Energy diagram

(b) Bonding diagram

▲ FIGURE 16–5

Creation of electron-hole pairs in a silicon crystal. An electron in the conduction band is a free electron.

lence band, creating what is called an *electron-hole pair.* **Recombination** occurs when a conduction-band electron loses energy and falls back into a hole in the valence band.

To summarize, a piece of intrinsic silicon at room temperature has, at any instant, a number of conduction-band (free) electrons that are unattached to any atom and are essentially drifting randomly throughout the material. Also, an equal number of holes are created in the valence band when these electrons jump into the conduction band. This is illustrated in Figure 16–6.

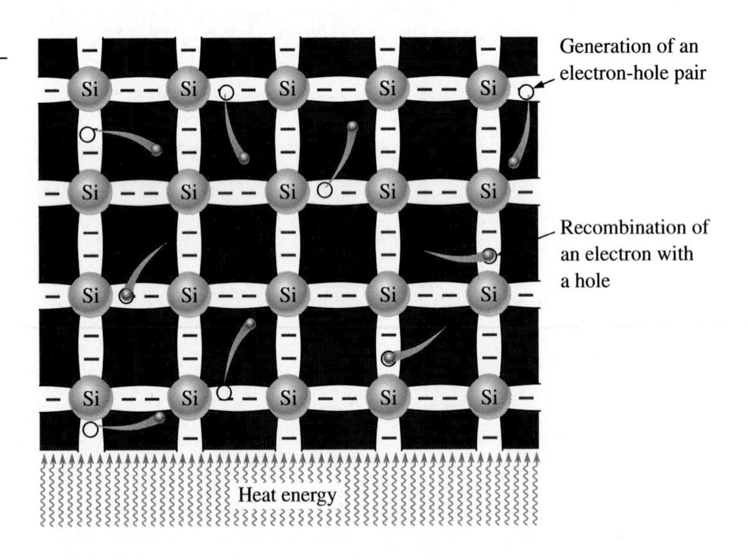

▶ FIGURE 16–6

Electron-hole pairs in a silicon crystal. Free electrons are being generated continuously while some recombine with holes.

Electron and Hole Current

When a voltage is applied across a piece of silicon, as shown in Figure 16–7, the thermally generated free electrons in the conduction band, which are free to move randomly in the crystal structure, are now easily attracted toward the positive end. This movement of free electrons is one type of current in a semiconductive material and is called *electron current.*

Another type of current occurs at the valence level, where the holes created by the free electrons exist. Electrons remaining in the valence band are still attached to their atoms and are not free to move randomly in the crystal structure. How-

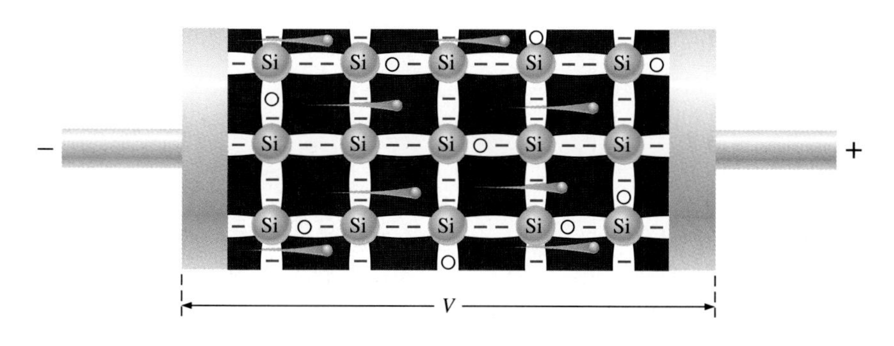

FIGURE 16–7

Free electron current in intrinsic silicon is produced by the movement of thermally generated free electrons.

ever, a valence electron can move into a nearby hole, with little change in its energy level, thus leaving another hole where it came from. The hole has effectively, although not physically, moved from one place to another in the crystal structure, as illustrated in Figure 16–8. This current is called *hole current.*

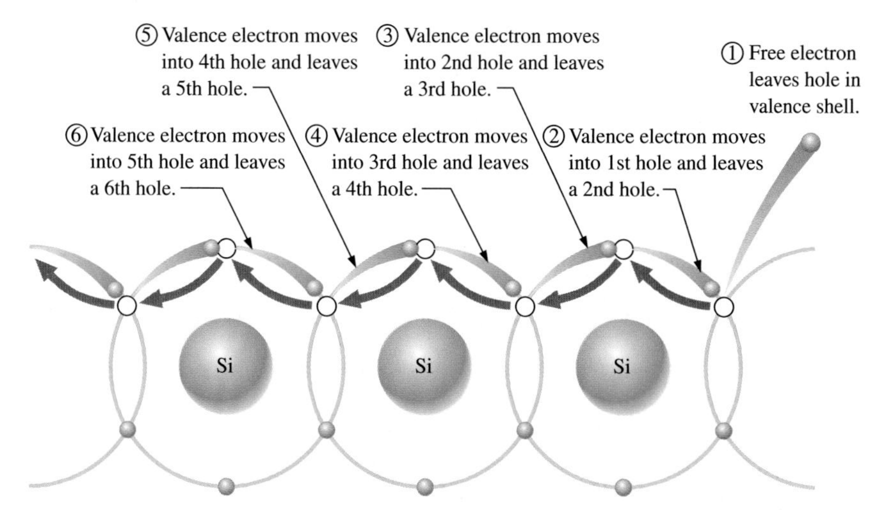

When a valence electron moves left to right to fill a hole while leaving another hole behind, a hole has effectively moved from right to left. Gray arrows indicate effective movement of a hole.

FIGURE 16–8

Hole current in intrinsic silicon.

Comparison of Semiconductors to Conductors and Insulators

In an intrinsic semiconductor, there are relatively few free electrons, so neither silicon nor germanium is very useful in its intrinsic state. Pure semiconductive materials are neither insulators nor good conductors because current in a material depends directly on the number of free electrons.

A comparison of the energy bands in Figure 16–9 for insulators, semiconductors and conductors shows the essential differences among them regarding conduction. The energy gap for an insulator is so wide that hardly any electrons acquire enough energy to jump into the conduction band. The valence band and the conduction band in a conductor (such as copper) overlap so that there are always many conduction electrons, even without the application of external energy. A semiconductor, as Figure 16–9(b) shows, has an energy gap that is much narrower than that in an insulator.

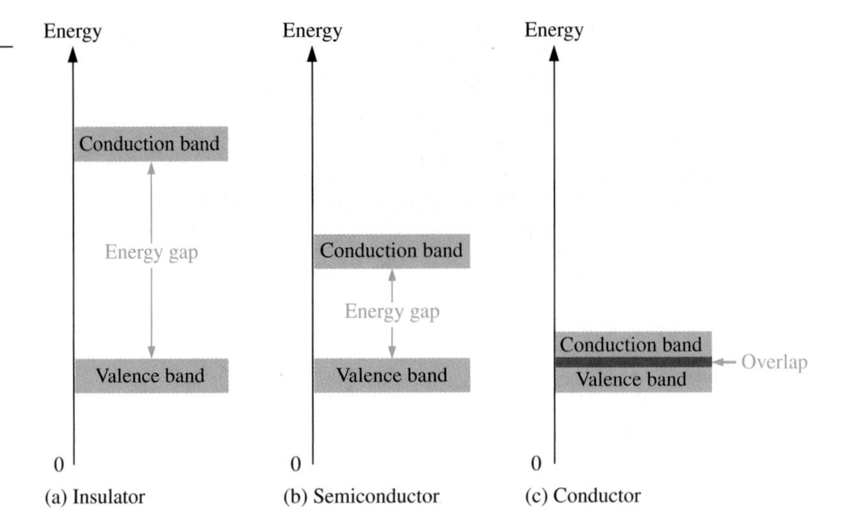

FIGURE 16–9

Energy diagrams for the three types of solids.

SECTION 16–3 REVIEW

1. In the atomic structure of a semiconductor, within which energy band do free electrons exist? Within which energy band do valence electrons exist?
2. How are holes created in an intrinsic semiconductor?
3. Why is current established more easily in a semiconductor than in an insulator?

16–4 *N*-TYPE AND *P*-TYPE SEMICONDUCTORS

Semiconductive materials do not conduct current well and are of little value in their intrinsic state. This is because of the limited number of free electrons in the conduction band and holes in the valence band. Intrinsic silicon (or germanium) must be modified by increasing the free electrons and holes to increase its conductivity and make it useful in electronic devices. This is done by adding impurities to the intrinsic material as you will learn in this section. Two types of extrinsic (impure) semiconductive materials, *n*-type and *p*-type, are the key building blocks for all types of electronic devices.

After completing this section, you should be able to

- **Describe the properties of *n*-type and *p*-type semiconductors**
- Define *doping*
- Explain how *n*-type semiconductors are formed
- Explain how *p*-type semiconductors are formed
- Define *majority carrier* and *minority carrier*

Doping

The conductivities of silicon and germanium can be drastically increased and controlled by the addition of impurities to the intrinsic (pure) semiconductive material. This process, called doping, increases the number of current carriers (electrons or holes). The two categories of impurities are *n-type* and *p-type.*

N-Type Semiconductor

To increase the number of conduction-band electrons in intrinsic silicon, **pentavalent** impurity atoms are added. These are atoms with five valence electrons, such as arsenic (As), phosphorus (P), and antimony (Sb) and are known as donor atoms because they provide an extra electron to the semiconductor's crystal structure.

As illustrated in Figure 16–10, each pentavalent atom (antimony, in this case) forms covalent bonds with four adjacent silicon atoms. Four of the antimony atom's valence electrons are used to form the covalent bonds with silicon atoms, leaving one extra electron. This extra electron becomes a conduction electron because it is not attached to any atom. The number of conduction electrons can be controlled by the number of impurity atoms added to the silicon.

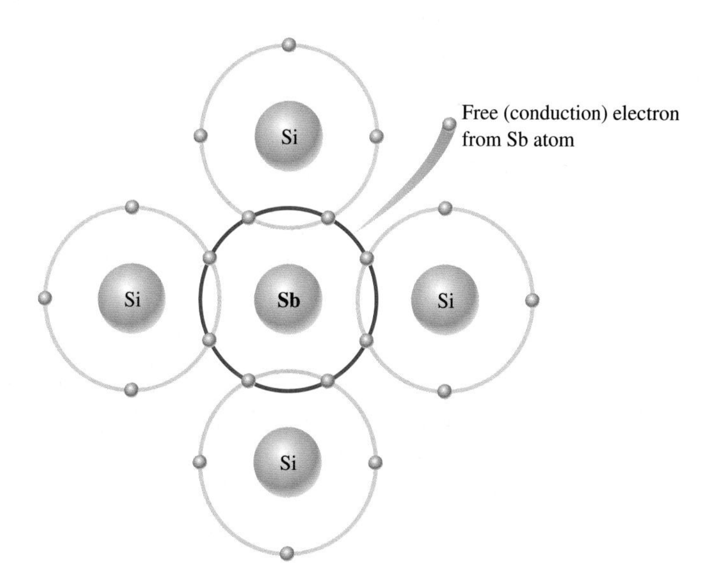

FIGURE 16–10

Pentavalent impurity atom in a silicon crystal. An antimony (Sb) impurity atom is shown in the center. The extra electron from the Sb atom becomes a free electron.

Majority and Minority Carriers Since most of the current carriers are electrons, silicon (or germanium) doped in this way is an *n*-type semiconductor (the *n* stands for the negative charge on an electron). The electrons are called the majority carriers in *n*-type material. Although the majority of current carriers in *n*-type material are electrons, there are some holes. These holes are *not* produced by the addition of the pentavalent impurity atoms. Holes in an *n*-type material are called minority carriers.

P-Type Semiconductor

To increase the number of holes in intrinsic silicon, trivalent impurity atoms are added. These are atoms with three valence electrons, such as aluminum (Al), boron (B), and gallium (Ga) and are known as acceptor atoms because they leave a hole in the semiconductor's crystal structure.

As illustrated in Figure 16–11, each trivalent atom (boron, in this case) forms covalent bonds with four adjacent silicon atoms. All three of the boron atom's valence electrons are used in the covalent bonds; and, since four electrons are required, a hole is formed with each trivalent atom. The number of holes can be controlled by the amount of trivalent impurity added to the silicon.

Majority and Minority Carriers Since most of the current carriers are holes, silicon (or germanium) doped with trivalent atoms is a *p*-type semiconductor. Holes can be thought of as positive charges. The holes are the majority carriers in *p*-type

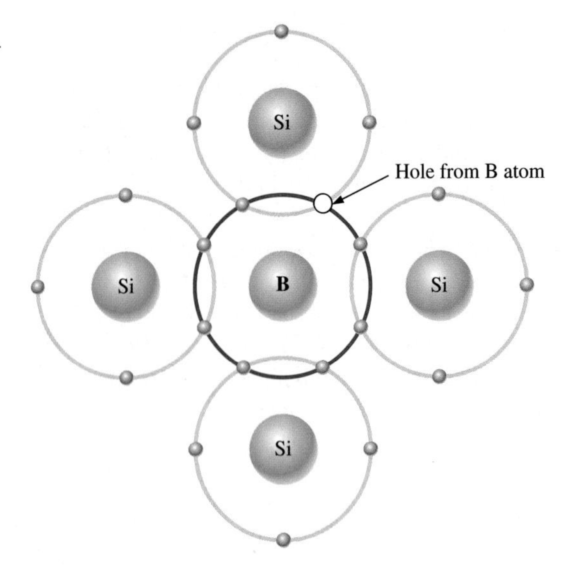

FIGURE 16–11

Trivalent impurity atom in a silicon crystal. A boron (B) impurity atom is shown in the center.

material. Although the majority of current carriers in *p*-type material are holes, there are also a few free electrons that are created when electron-hole pairs are thermally generated. These few free electrons are *not* produced by the addition of trivalent impurity atoms. Electrons in *p*-type material are the minority carriers.

SECTION 16–4 REVIEW

1. How is an *n*-type semiconductor formed?
2. How is a *p*-type semiconductor formed?
3. What is a majority carrier?

16–5 THE *PN* JUNCTION

If you take a block of silicon and dope half of it with a trivalent impurity and the other half with a pentavalent impurity, a boundary called the ***pn* junction** is formed between the resulting *p*-type and *n*-type portions. The *pn* junction is the feature that allows diodes, transistors, and other devices to work. This section and the next one will provide a basis for the discussion of the diode in Section 16–7.

After completing this section, you should be able to

- **Describe the characteristics of a *pn* junction**
- Define *pn* junction
- Discuss the depletion region in a *pn* junction
- Define *barrier potential*
- Explain the energy diagram of a *pn* junction

Formation of the Depletion Region

A *pn* junction is illustrated in Figure 16–12. The *n* region has many conduction electrons, and the *p* region has many holes. With no external voltage, the conduction electrons in the *n* region are randomly drifting in all directions. At the instant

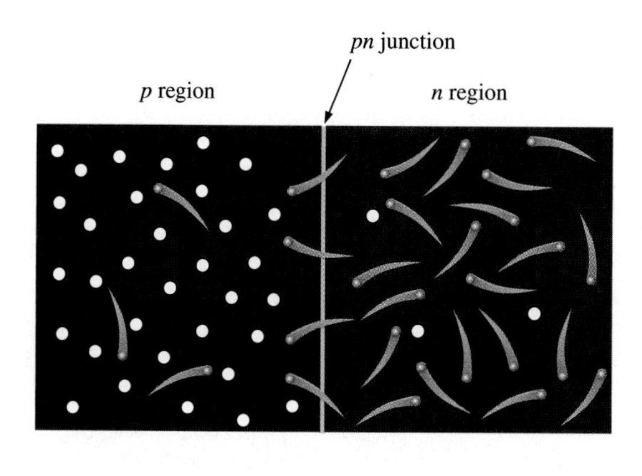

(a) At the instant of junction formation, free electrons in the *n* region near the *pn* junction begin to diffuse across the junction and fall into holes near the junction in the *p* region.

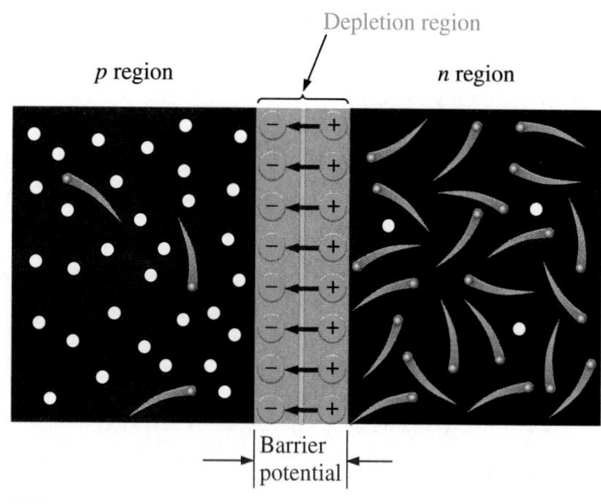

(b) For every electron that diffuses across the junction and combines with a hole, a positive charge is left in the *n* region and a negative charge is created in the *p* region, forming a barrier potential. This action continues until the voltage of the barrier repels further diffusion.

FIGURE 16–12

Formation of the depletion region in a *pn* junction.

of junction formation, some of the electrons near the junction drift across into the *p* region and recombine with holes near the junction as shown in part (a).

For each electron that crosses the junction and recombines with a hole, a pentavalent atom is left with a net positive charge in the *n* region near the junction, making it a positive ion. Also, when the electron recombines with a hole in the *p* region, a trivalent atom acquires net negative charge, making it a negative ion.

As a result of this recombination process, a large number of positive and negative ions builds up near the *pn* junction. As this buildup occurs, the electrons in the *n* region must overcome both the attraction of the positive ions and the repulsion of the negative ions in order to migrate into the *p* region. Thus, as the ion layers build up, the area on both sides of the junction becomes essentially depleted of any conduction electrons or holes and is known as the *depletion region.* This condition is illustrated in Figure 16–12(b). When an equilibrium condition is reached, the depletion region has widened to a point where no more electrons can cross the *pn* junction.

The existence of the positive and negative ions on opposite sides of the junction creates a barrier potential across the depletion region, as indicated in Figure 16–12(b). The **barrier potential**, V_B, is the amount of voltage required to move electrons through the electric field. At 25°C, it is approximately 0.7 V for silicon and 0.3 V for germanium. As the junction temperature increases, the barrier potential decreases, and vice versa.

Energy Diagram of the *PN* Junction

Now, let's look at the operation of the *pn* junction in terms of its energy level. First consider the *pn* junction at the instant of its formation. The energy bands of the trivalent impurity atoms in the *p*-type material are at a slightly higher level than those of the pentavalent impurity atom in the *n*-type material, as shown in the graph of Figure 16–13(a). They are higher because the core attraction for the valence electrons (+3) in the trivalent atom is less than the core attraction for the valence electrons (+5) in the pentavalent atom. Thus, the trivalent valence electrons are in a slightly higher orbit and, thus, at a higher energy level.

Notice in Figure 16–13(a) that there is some overlap of the conduction bands in the *p* and *n* regions and also some overlap of the valence bands in the *p* and *n* regions. This overlap permits the electrons of higher energy near the top of the

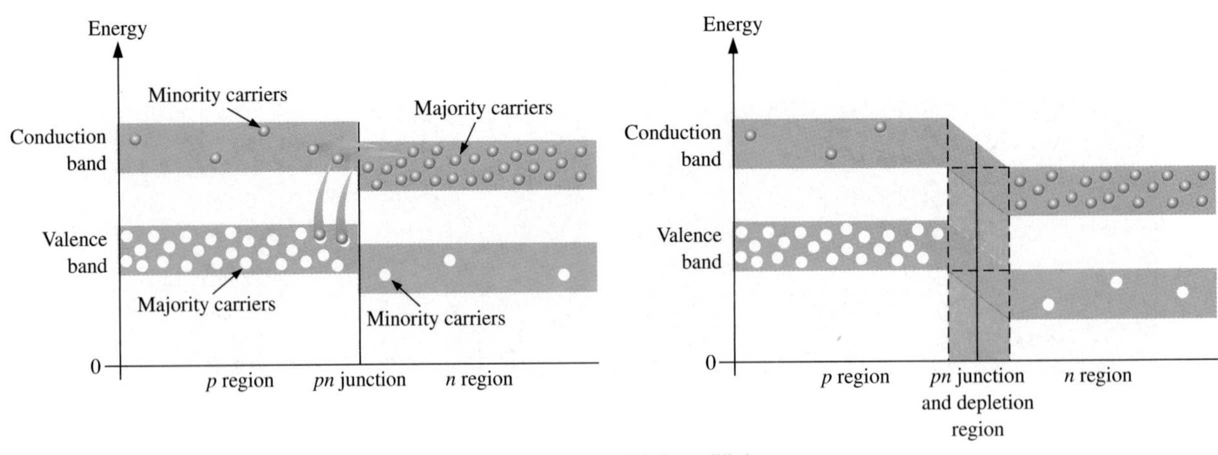

(a) At the instant of junction formation

(b) At equilibrium

▲ **FIGURE 16–13**

Energy diagrams illustrating the formation of the *pn* junction and depletion region.

n-region conduction band to begin diffusing across the junction into the lower part of the *p*-region conduction band. As soon as an electron diffuses across the junction, it recombines with a hole in the valence band. As diffusion continues, the depletion region begins to form. Also, the energy bands in the *n* region "shift" down as the electrons of higher energy are lost to diffusion. When the top of the *n*-region conduction band reaches the same level as the bottom of the *p*-region conduction band, diffusion ceases and the equilibrium condition is reached. This condition is shown in terms of energy levels in Figure 16–13(b). There is an energy gradient across the depletion region rather than an abrupt change in energy level.

SECTION 16–5 REVIEW

1. What is a *pn* junction?
2. When *p* and *n* regions are joined, a depletion region forms. Describe the characteristics of the depletion region.
3. The barrier potential for silicon is greater than for germanium. (True or False)
4. What is the barrier potential for silicon at 25°C?

16–6 BIASING THE *PN* JUNCTION

There is no movement of electrons (current) through a *pn* junction at equilibrium. The primary usefulness of the *pn* junction is its ability to allow current in only one direction and to prevent current in the other direction as determined by the bias. In electronics, **bias** refers to the use of a dc voltage to establish certain operating conditions for an electronic device. There are two practical bias conditions for a *pn* junction: forward and reverse. Either of these conditions is created by application of a sufficient external voltage of the proper polarity across the *pn* junction.

After completing this section, you should be able to

- **Explain how to bias a *pn* junction**
- Discuss forward bias
- Discuss reverse bias
- Define *reverse breakdown*

Forward Bias

Forward bias is the condition that permits current through a *pn* junction. Figure 16–14 shows a dc voltage connected in a direction to forward-bias the junction. Notice that the negative **terminal** of the V_{BIAS} source is connected to the *n* region, and the positive terminal is connected to the *p* region.

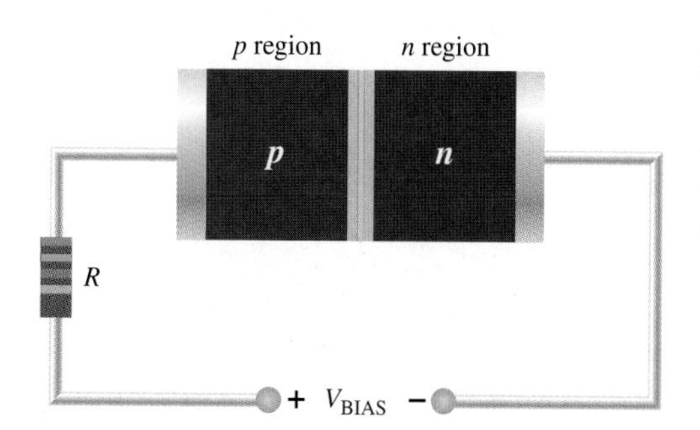

FIGURE 16–14

Forward-bias connection. The resistor limits the forward current in order to prevent damage to the semiconductor.

A discussion of the basic operation of forward bias follows: The negative terminal of the bias-voltage source pushes the conduction-band electrons in the *n* region toward the junction, while the positive terminal pushes the holes in the *p* region also toward the junction. Recall from Chapter 2 that like charges repel each other.

When it overcomes the barrier potential (V_B), the external voltage source provides the *n*-region electrons with enough energy to penetrate the depletion region and move through the junction, where they combine with the *p*-region holes. As electrons leave the *n*-region, more flow in from the negative terminal of the bias-voltage source. Thus, current through the *n* region is formed by the movement of conduction electrons (majority carriers) toward the junction.

Once the conduction electrons enter the *p* region and combine with holes, they become valence electrons. Then they move as valence electrons from hole to hole toward the positive connection of the bias-voltage source. The movement of these valence electrons is the same as the movement of holes in the opposite direction. Thus, current in the *p* region is formed by the movement of holes (majority carriers) toward the junction. Figure 16–15 illustrates current in a forward-biased *pn* junction.

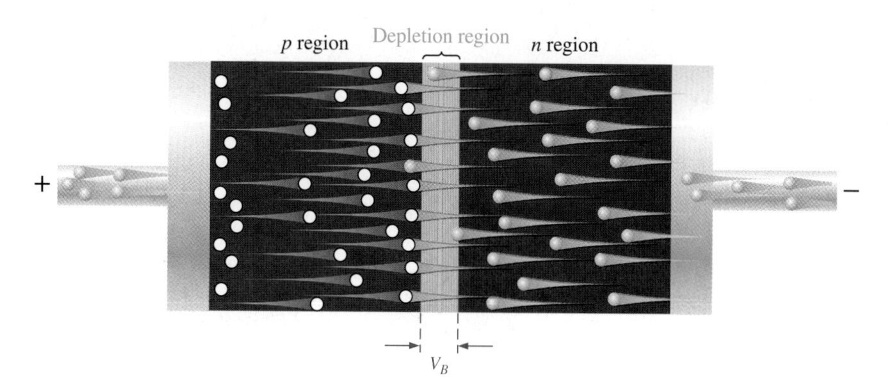

FIGURE 16–15

Current in a forward-biased *pn* junction.

The Effect of the Barrier Potential on Forward Bias The effect of the barrier potential in the depletion region is to oppose forward bias. This is because the negative ions near the junction in the *p* region tend to prevent electrons from moving through the junction into the *p* region. You can think of the barrier potential effect as simulating a small battery connected in a direction to oppose the forward-bias voltage, as shown in Figure 16–16. The resistances R_p and R_n represent the dynamic resistances of the *p* and *n* materials.

FIGURE 16–16

Barrier potential and dynamic resistance equivalent for a *pn* junction.

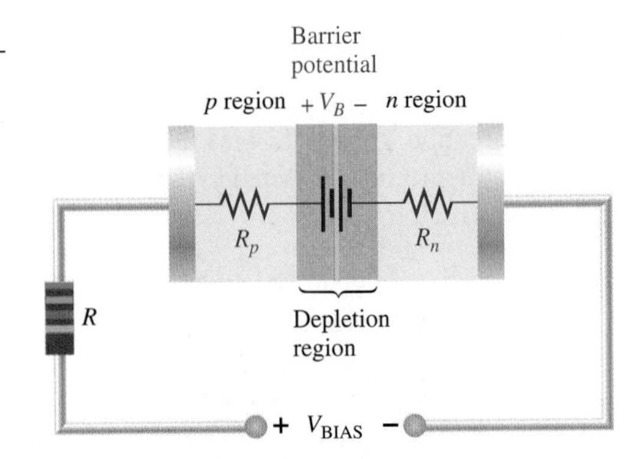

The external bias voltage must overcome the effect of the barrier potential before the *pn* junction conducts, as illustrated in Figure 16–17. Conduction occurs at approximately 0.7 V for silicon and 0.3 V for germanium. Once the *pn* junction is conducting in the forward direction, the voltage drop across it remains at approximately the barrier potential and changes very little with changes in forward current (I_F), as illustrated in Figure 16–17.

FIGURE 16–17

Illustration of *pn* junction operation under forward-bias conditions.

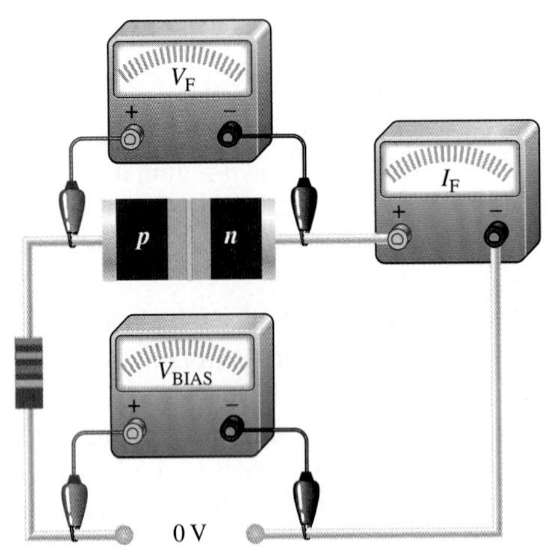

(a) No bias voltage. *PN* junction is at equilibrium.

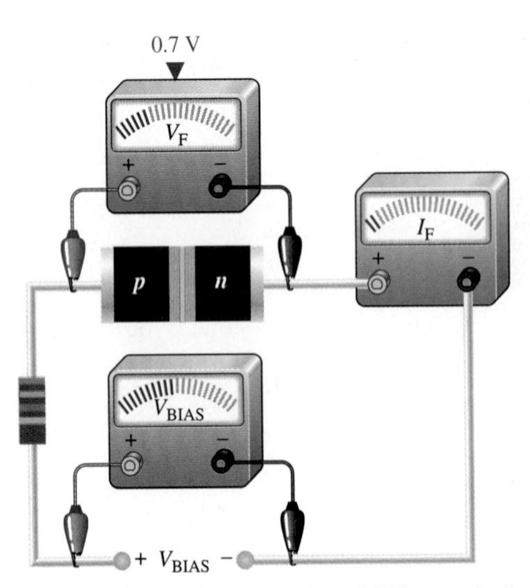

(b) Small forward-bias voltage ($V_F < 0.7$ V), very small forward current.

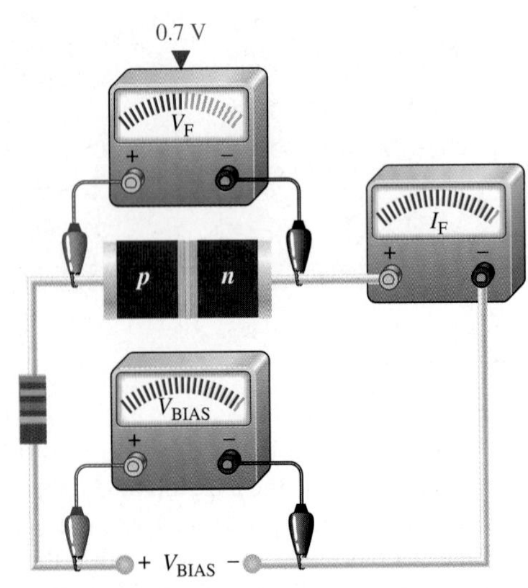

(c) Forward voltage reaches and remains at approximately 0.7 V. Forward current continues to increase as the bias voltage is increased.

Energy Diagram for Forward Bias When a *pn* junction is forward-biased, the *n*-region conduction band is raised to a higher energy level that overlaps with the *p*-region conduction band. Then large numbers of free electrons have enough energy to climb the "energy hill" and enter the *p* region where they combine with holes in the valence band. Forward bias is illustrated by the energy diagram in Figure 16–18.

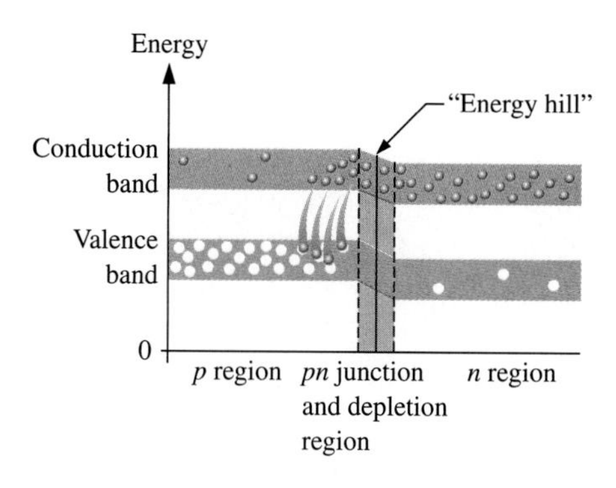

FIGURE 16–18

Energy diagram for forward bias, showing recombination in the *p* region as conduction electrons move across the junction.

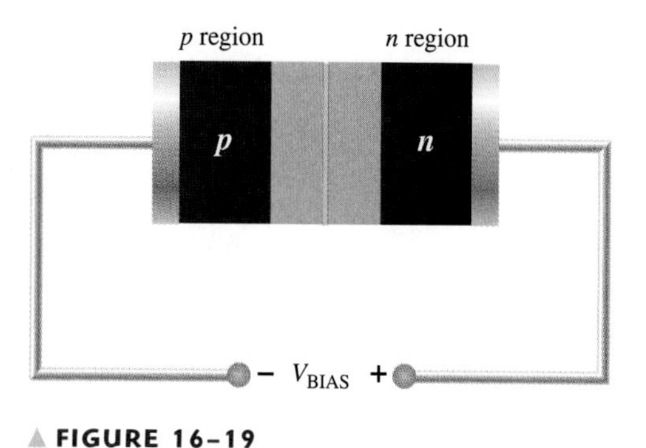

FIGURE 16–19

Reverse-bias connection.

Reverse Bias

Reverse bias is the condition that prevents current through the *pn* junction. Figure 16–19 shows a dc voltage source connected to reverse-bias the diode. Notice that the negative terminal of the V_{BIAS} source is connected to the *p* region, and the positive terminal is connected to the *n* region.

A discussion of the basic operation for reverse bias follows: The negative terminal of the bias-voltage source attracts holes in the *p* region away from the *pn* junction, while the positive terminal also attracts electrons away from the junction. As electrons and holes move away from the junction, the depletion region widens; more positive ions are created in the *n* region, and more negative ions are created in the *p* region, as shown in Figure 16–20(a). The initial flow of majority carriers away from the junction is called *transient current* and lasts only for a very short time upon application of reverse bias.

The depletion region widens until the potential difference across it equals the external bias voltage. At this point, the holes and electrons stop moving away from the junction, and majority current ceases, as indicated in Figure 16–20(b).

When the diode is reverse-biased, the depletion region effectively acts as an insulator between the layers of oppositely charged ions, forming an effective capacitance. Since the depletion region widens with increased reverse-biased voltage, the capacitance decreases, and vice versa. This internal capacitance is called the *depletion-region capacitance.*

A special type of diode that makes use of the junction capacitance is the *varactor.* It is used as a variable capacitor in which the capacitance is controlled by the reverse-bias voltage. By increasing or decreasing the reverse voltage, you can increase or decrease the capacitance.

Reverse Current As you have learned, majority current very quickly becomes zero when reverse bias is applied. There is, however, a very small current produced by minority carriers during reverse bias. Germanium, as a rule, has a greater reverse current than silicon. This current is typically in the μA or nA range. A relatively small number of thermally produced electron-hole pairs exist

FIGURE 16–20

Illustration of reverse bias.

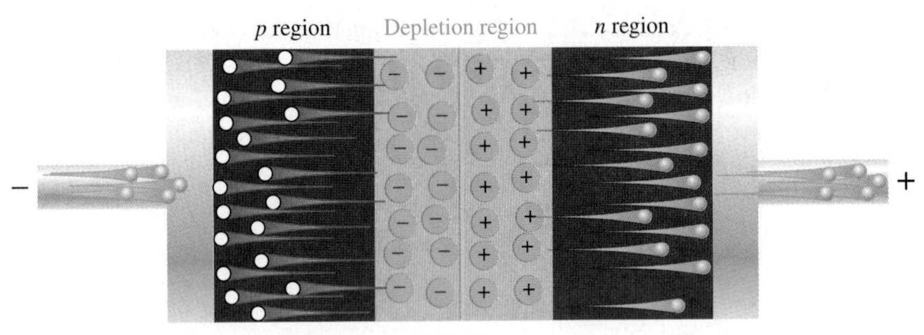

(a) There is transient current as depletion region widens.

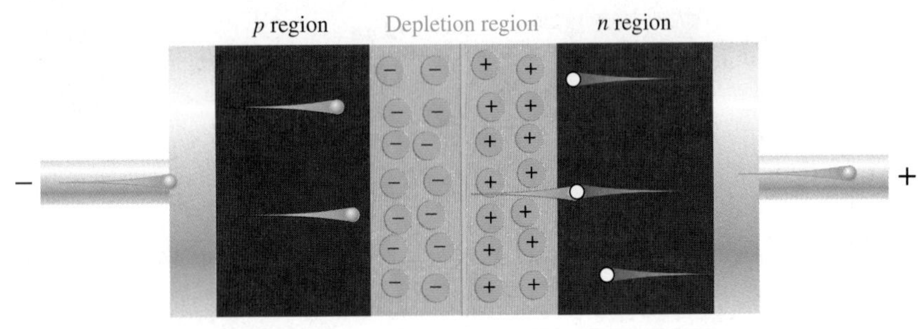

(b) Majority current ceases when barrier potential equals bias voltage. There is an extremely small reverse current due to minority carriers.

in the depletion region. Under the influence of the external voltage, some electrons manage to diffuse across the *pn* junction before recombination. This process establishes a small minority carrier current throughout the material.

The reverse current is dependent primarily on the junction temperature and not on the amount of reverse-biased voltage. A temperature increase causes an increase in reverse current.

Energy Diagram for Reverse Bias When a *pn* junction is reverse-biased, the *n*-region conduction band remains at an energy level that prevents the free electrons from crossing into the *p* region. There are a few free minority electrons in the *p*-region conduction band that can easily flow down the "energy hill" into the *n* region where they combine with minority holes in the valence band. This reverse current is extremely small compared to the current in forward bias and can normally be considered negligible. Figure 16–21 illustrates reverse bias.

FIGURE 16–21

Energy diagram for reverse bias.

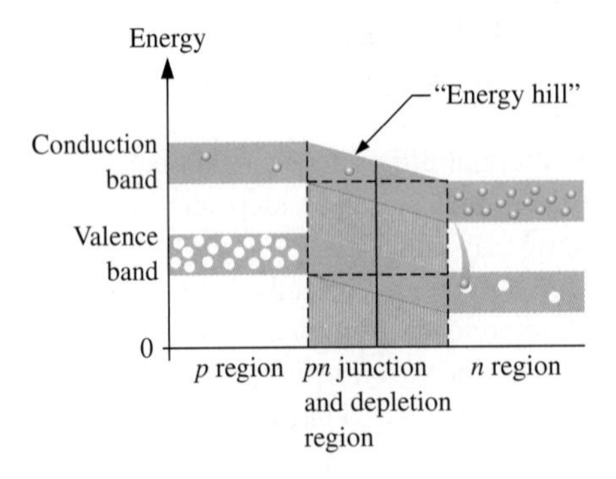

Reverse Breakdown

If the external reverse-bias voltage is increased to a large enough value, **reverse breakdown** occurs. The following describes what happens: Assume that one minority conduction-band electron acquires enough energy from the external source to accelerate it toward the positive end of the *pn* junction. During its travel, it collides with an atom and imparts enough energy to knock a valence electron into the conduction band. There are now two conduction-band electrons. Each will collide with an atom, knocking two more valence electrons into the conduction band. There are now four conduction-band electrons which, in turn, knock four more into the conduction band. This rapid multiplication of conduction-band electrons, known as an *avalanche effect,* results in a rapid buildup of reverse current.

A single *pn* junction device is called a **diode**. Most diodes normally are not operated in reverse breakdown and can be damaged if they are. However, a particular type of diode known as a zener diode (discussed in Chapter 17) is specially designed for reverse-breakdown operation.

SECTION 16–6 REVIEW

1. Name the two bias conditions.
2. Which bias condition produces majority carrier current?
3. Which bias condition produces a widening of the depletion region?
4. Minority carriers produce the current during reverse breakdown. (True or False)

16–7 DIODE CHARACTERISTICS

As you learned in the last section, a diode is a semiconductive device made with a single *pn* junction. A diode conducts current when it is forward-biased when the bias voltage exceeds the barrier potential. A diode prevents current when it is reverse-biased at less than the breakdown voltage.

After completing this section, you should be able to

- **Describe the basic diode characteristics**
- Explain the diode characteristic curve
- Recognize the standard diode symbol
- Test a diode with an ohmmeter
- Describe three diode approximations

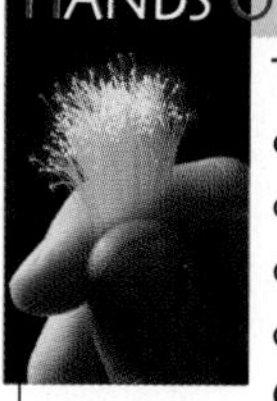

HANDS ON TIP

The diode forward characteristic curve can be plotted on the oscilloscope using the circuit shown below. Channel 1 senses the voltage across the diode and channel 2 senses a signal that is proportional to the current. The scope must be in the *X-Y* mode. The signal generator provides a 5 Vpp sawtooth or triangular waveform at 50 Hz, and its ground must not be the same as the scope ground. Channel 2 must be inverted for the displayed curve to be properly oriented.

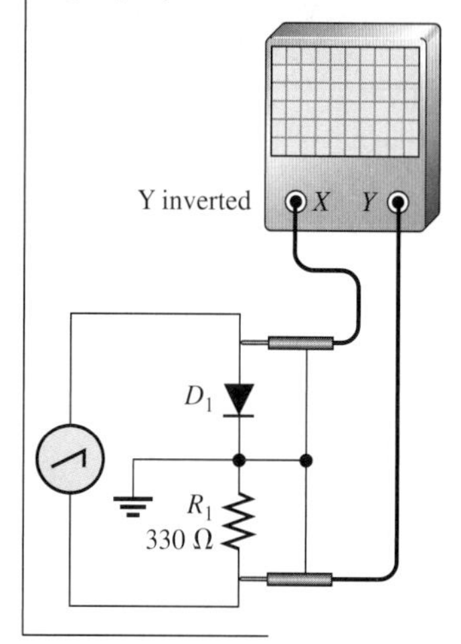

Diode Characteristic Curve

Figure 16–22 is a graph of diode current versus voltage. The upper right quadrant of the graph represents the forward-biased condition. As you can see, there is very little forward current (I_F) for forward voltages (V_F) below the barrier potential. As the forward voltage approaches the value of the barrier potential (0.7 V for silicon and 0.3 V for germanium), the current begins to increase. Once the forward voltage reaches the barrier potential, the current increases drastically and must be limited by a series resistor. *The voltage across the forward-biased diode remains approximately equal to the barrier potential.*

The lower left quadrant of the graph represents the reverse-biased condition. As the reverse voltage (V_R) increases to the left, the current remains near zero until the breakdown voltage (V_{BR}) is reached. When breakdown occurs, there is a large reverse current which, if not limited, can destroy the diode. Typically, the breakdown voltage is greater than 50 V for most rectifier diodes. Remember that most diodes should not be operated in reverse breakdown.

▶ FIGURE 16–22

Diode characteristic curve.

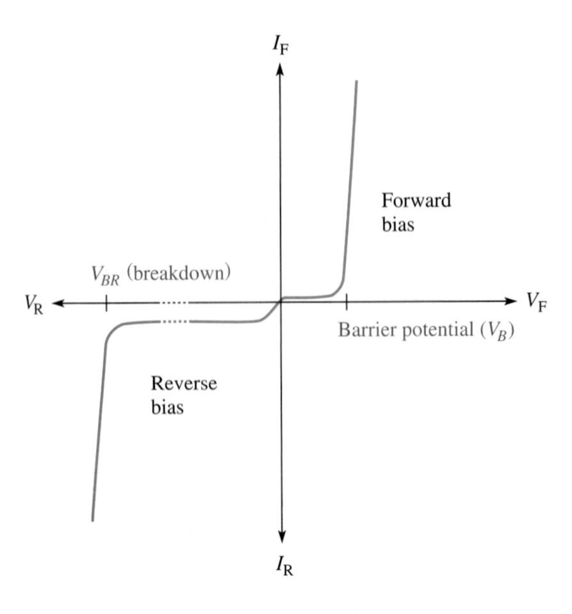

Diode Symbol

Figure 16–23(a) shows the basic diode structure and the standard schematic symbol for a general-purpose diode. The "arrowhead" in the symbol points in the direction opposite the electron flow. The two terminals of the diode are the anode (A) and cathode (K). The **anode** is the *p* region and the **cathode** is the *n* region.

When the anode is positive with respect to the cathode, the diode is forward-biased and current (I_F) is from cathode to anode, as shown in Figure 16–23(b). Remember that when the diode is forward-biased, the barrier potential, V_B, always appears between the anode and cathode, as indicated in the figure. When the anode is negative with respect to the cathode, the diode is reverse-biased, as shown in Figure 16–23(c) and there is no current. A resistor is not necessary in reverse bias but it is shown for consistency.

▼ FIGURE 16–23

Diode structure, schematic symbol, and bias circuits. V_{BIAS} is the bias voltage, and V_B is the barrier potential.

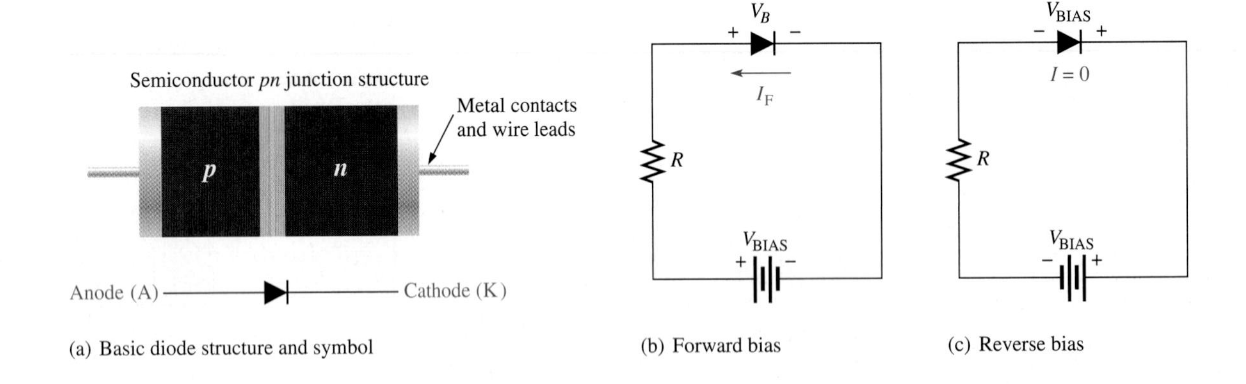

(a) Basic diode structure and symbol (b) Forward bias (c) Reverse bias

Some typical diodes and their terminal identifications are shown in Figure 16–24 to illustrate the variety of physical structures.

Diode Approximations

The Ideal Diode Model The simplest way to visualize diode operation is to think of it as a switch. When forward-biased, the diode ideally acts as a closed (on) switch, and when reverse-biased, it acts as an open (off) switch, as shown in

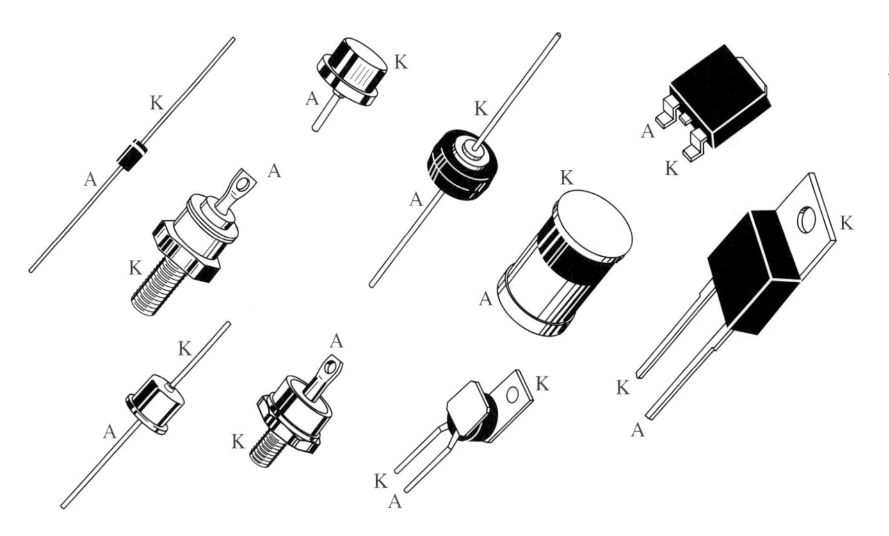

FIGURE 16–24

Typical diode packages and terminal identification. A is anode and K is cathode.

Figure 16–25. The characteristic curve for this model is also shown in part (c). Note that the forward voltage (V_F) and the reverse current (I_R) are always zero in the ideal case.

FIGURE 16–25

Ideal model of the diode as a switch.

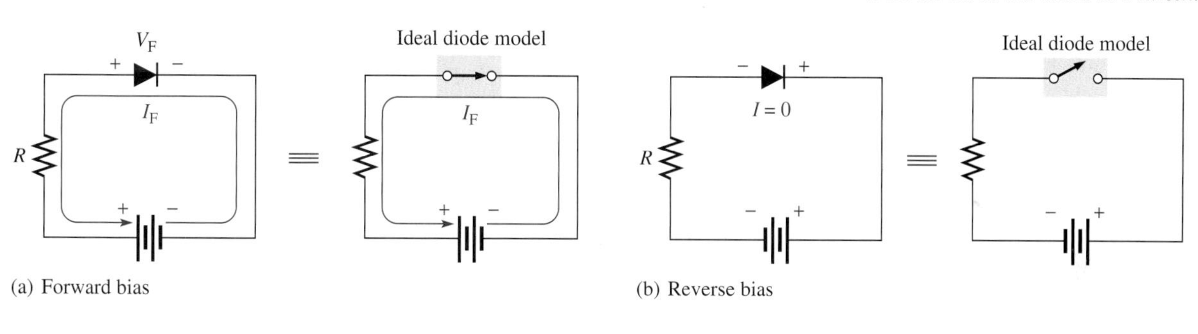

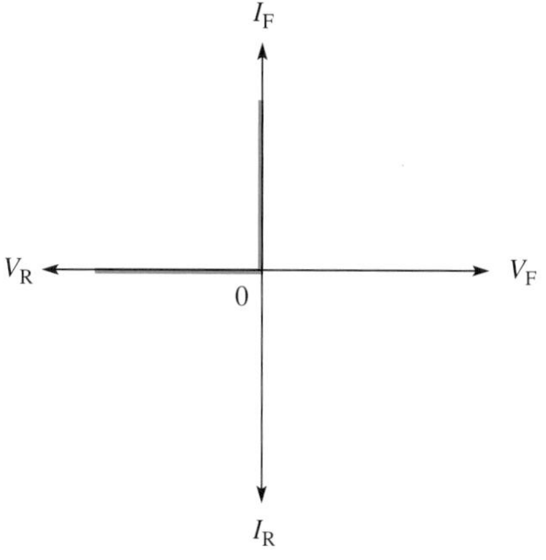

(c) Ideal characteristic curve (blue)

This ideal model, of course, neglects the effect of the barrier potential, the internal resistances, and other parameters. You may want to use the ideal model when you are troubleshooting or trying to figure out the operation of a circuit and are not concerned with more exact values of voltage or current.

The Practical Diode Model The next higher level of accuracy includes the barrier potential in the diode model. In this approximation, the forward-biased diode is represented as a closed switch in series with a small "battery" equal to the barrier potential V_B (0.7 V for Si and 0.3 V for Ge), as shown in Figure 16–26(a). The positive end of the battery is toward the anode. *Keep in mind that the barrier potential is not a voltage source and cannot be measured with a voltmeter; rather it only has the effect of a battery when forward bias is applied because the forward bias voltage, V_{BIAS}, must overcome the barrier potential before the diode begins to conduct current.* The reverse-biased diode is represented by an open switch, as in the ideal case, because the barrier potential does not affect reverse bias, as shown in Figure 16–26(b). The characteristic curve for this model is shown in Figure 16–26(c).

FIGURE 16–26

The practical model of a diode.

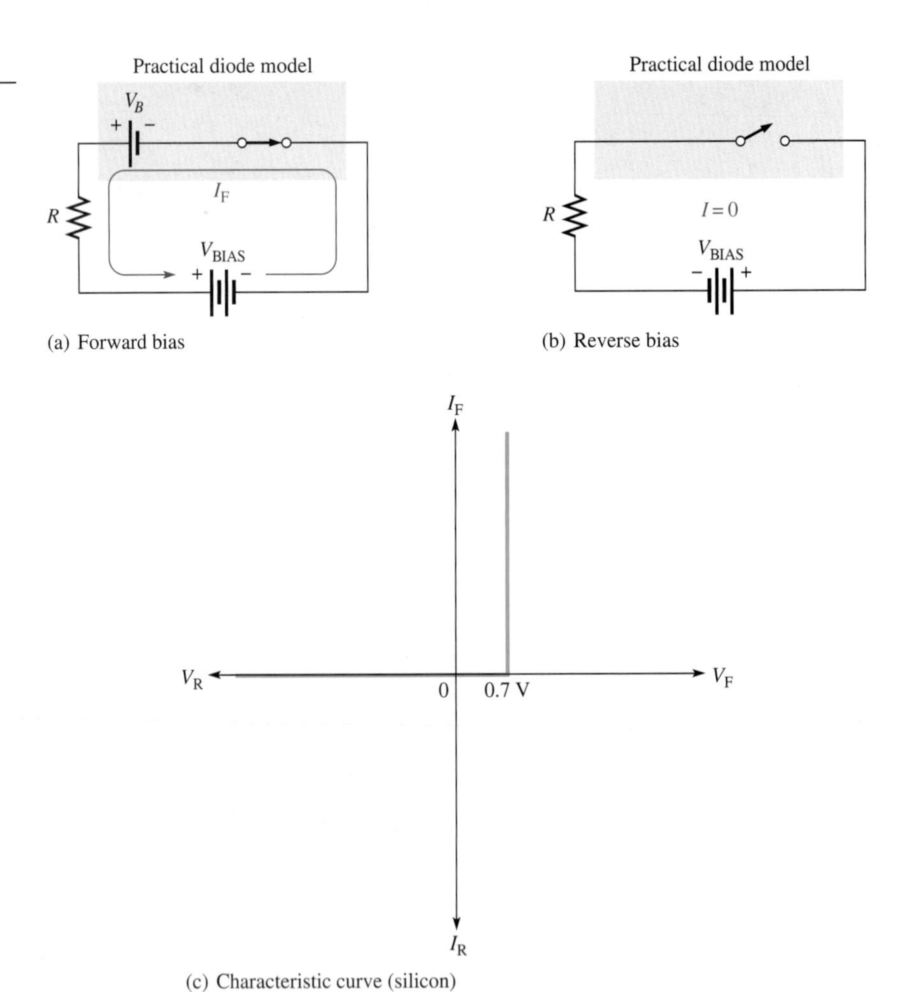

The Complex Diode Model One more level of accuracy will be considered at this point. Figure 16–27(a) shows the forward-biased diode model with both the barrier potential and the low forward dynamic resistance. Figure 16–27(b) shows how the high internal reverse resistance affects the reverse-biased model. The characteristic curve is shown in Figure 16–27(c).

Other parameters such as junction capacitance and breakdown voltage become important only under certain operating conditions and will be considered only where appropriate.

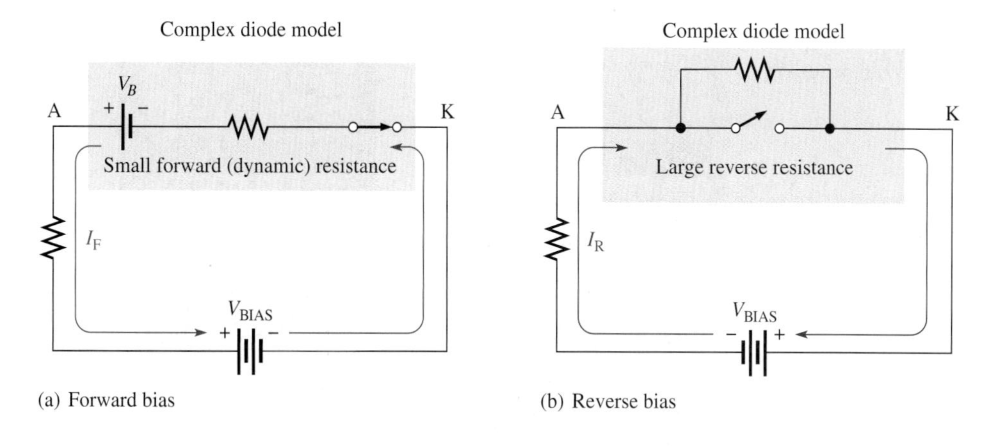

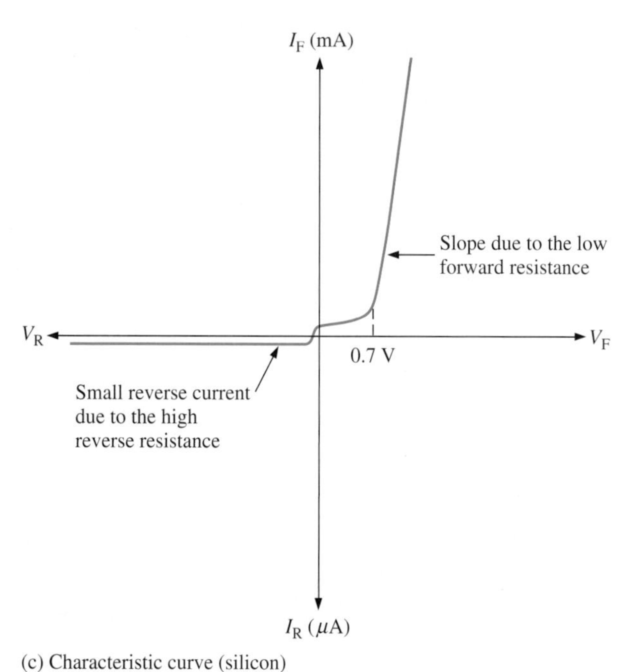

▲ **FIGURE 16–27**

The complex diode model includes barrier potential, forward resistance, and reverse resistance.

Testing a Diode

A multimeter can be used as a fast and simple way to check a diode. As you know, a good diode will show an extremely high resistance (or open) with reverse bias and a very low resistance with forward bias. A defective open diode will show an extremely high resistance (or open) for both forward and reverse bias. A defective shorted or resistive diode will show zero or a low resistance for both forward and reverse bias. An open diode is the most common type of failure.

The DMM Diode Test Position Many digital multimeters (DMMs) have a diode test position that provides a convenient way to test a diode. A typical DMM, as shown in Figure 16–28, has a small diode symbol to mark the position of the function switch. When set to *diode test,* the meter provides an internal voltage sufficient to forward bias and reverse bias a diode. This internal voltage may vary among different makes of DMM, but 2.5 V to 3.5 V is a typical range of values. The meter provides a voltage reading or other indication to show the condition of the diode under test.

FIGURE 16–28

Diode test on a properly functioning silicon diode.

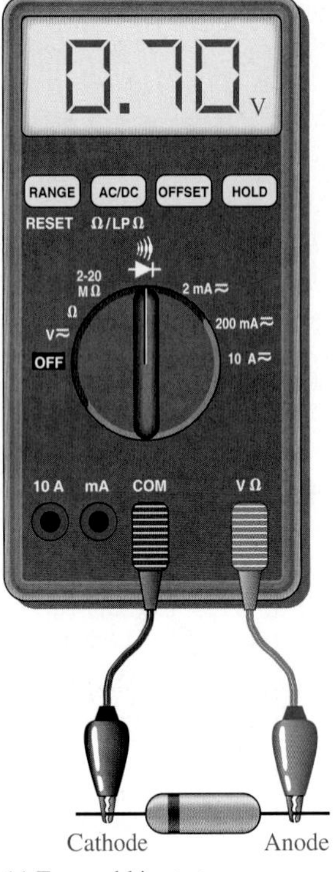

(a) Forward-bias test

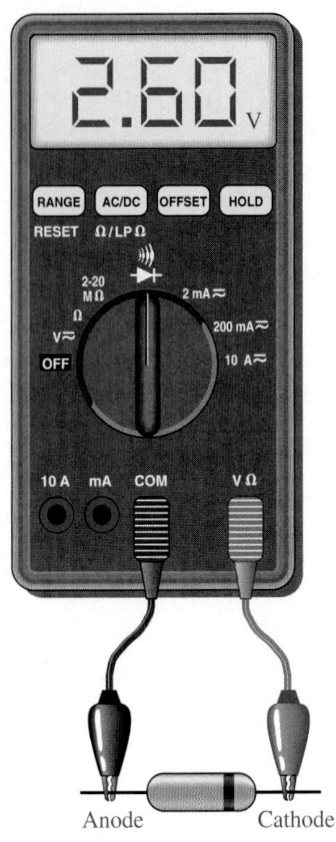

(b) Reverse-bias test

FIGURE 16–29

Testing a defective diode.

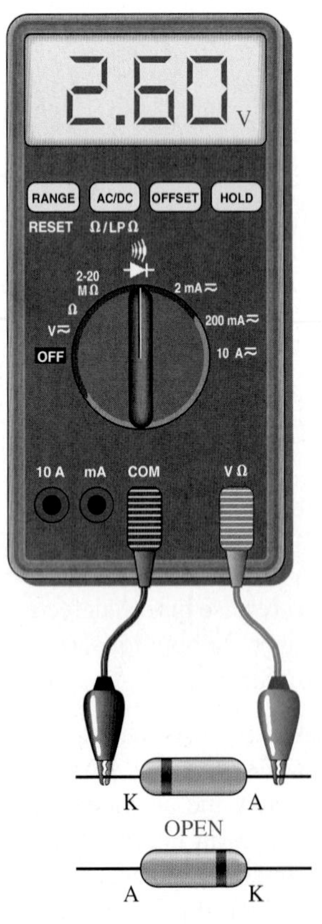

(a) Forward- and reverse-bias tests for an open diode give same indication. Some meters will display "OL."

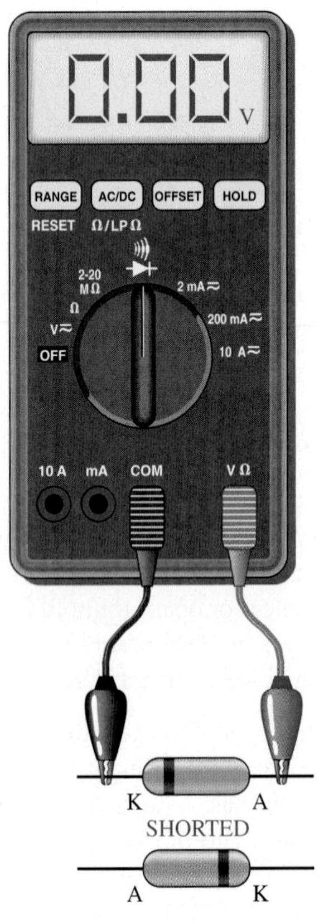

(b) Forward- and reverse-bias tests for a shorted diode give same 0 V reading. If the diode is resistive, the reading is less than 2.6 V.

When the Diode Is Working In Figure 16–28(a), the red (positive) lead of the meter is connected to the anode and the black (negative) lead is connected to the cathode to forward bias the diode. If the silicon diode is good, you will get a reading of between 0.5 V and 0.9 V, with 0.7 V being typical for forward bias.

In Figure 16–28(b), the diode is turned around to reverse bias the diode as shown. If the diode is working properly, you will get a voltage reading based on the meter's internal voltage source. The 2.6 V shown in the figure represents a typical value and indicates that the diode has an extremely high reverse resistance with essentially all of the internal voltage appearing across it. Some meters will indicate "OL" instead of a voltage.

When the Diode Is Defective When a diode has failed open, you get an open circuit voltage reading (2.6 V is typical) or "OL" indication for both the forward-bias and the reverse-bias condition, as illustrated in Figure 16–29(a). If a diode is shorted, the meter reads 0 V in both forward-bias and reverse-bias tests, as indicated in part (b). Sometimes, a failed diode may exhibit a small resistance for both bias conditions rather than a pure short. In this case, the meter will show a small voltage much less than the correct open voltage. For example, a resistive diode may result in a reading of 1.1 V in both directions rather than the correct readings of 0.7 V forward and 2.6 V reverse.

SECTION 16–7 REVIEW

1. Draw a rectifier diode symbol and label the terminals.
2. For a normal diode, the forward resistance is quite low and the reverse resistance is very high. (True or False)
3. An open switch ideally represents a ________-biased diode. A closed switch ideally represents a ________-biased diode.

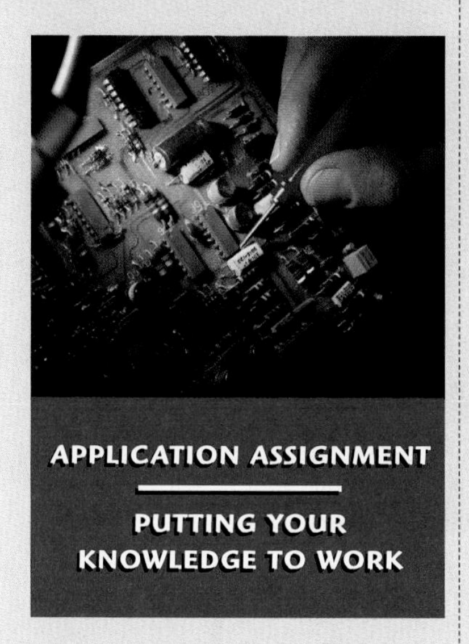

APPLICATION ASSIGNMENT

PUTTING YOUR KNOWLEDGE TO WORK

You have been assigned to check four production power supply units that have failed the final assembly test. Although you basically understand how transformers and individual diodes work, you know nothing about the overall operation of the power supply circuit or how diodes are applied. "No problem," says your supervisor. "Use a 3½-digit DMM set on the diode test function to check for open diodes or transformer windings. You don't need to understand the power supply operation for this assignment. You will learn that later."

The dc power supply and its schematic are shown in Figure 16–30. The circuit consists of a power transformer, rectifier diodes, filter capacitors, and an integrated circuit voltage regulator. Isometric views of the capacitors and the voltage regulator attached to a finned heat sink are shown.

Step 1: Identification of Components

Locate the fuse, the transformer, and the diodes in the power supply in Figure 16–30. Also identify the anode and the cathode of each diode.

Step 2: Basic Information

Before making DMM measurements, make sure the circuit is not connected to a power source (it must be unplugged). Assume that you have obtained the following specific information:

- The DMM will indicate a resistance of roughly 500 Ω when the diode is forward-biased. This reading will vary from one diode to the next.
- The DMM will indicate the very high reverse resistance of a diode by a se-

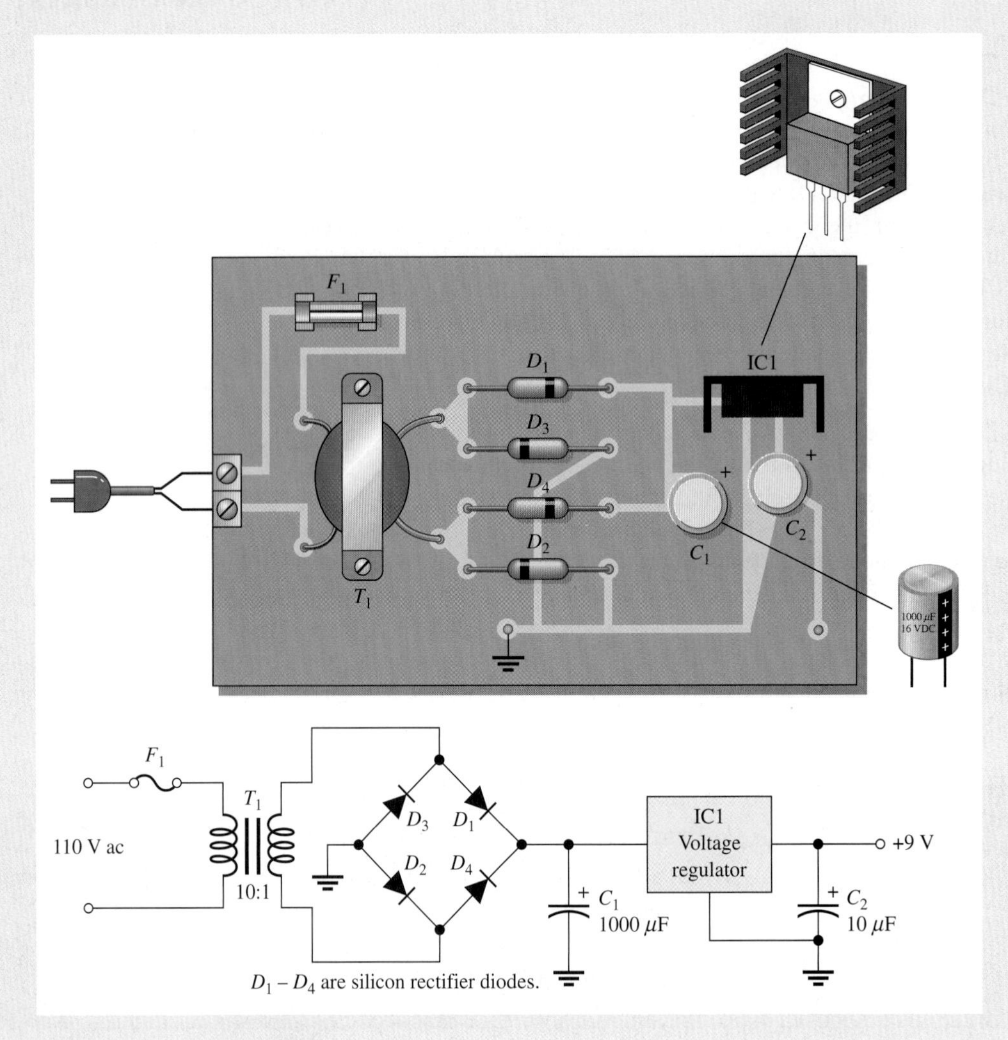

▲ **FIGURE 16–30**

DC power supply and schematic.

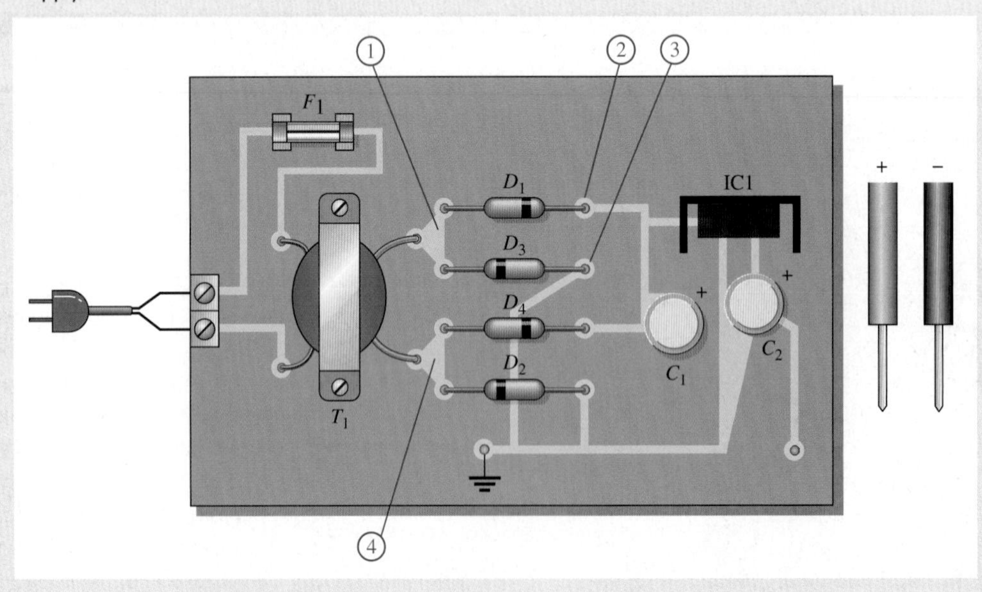

▲ **FIGURE 16–31**

Power supply showing the numbered test points to which the DMM leads are to be connected.

ries of dashes meaning that the measured value is out of the meter's range.

- The secondary winding resistance of the transformer is 10 Ω.

Step 3: Troubleshooting

Test points to which the leads are to be connected are numerically labeled in Figure 16–31. Determine the fault in each of the four power supply units based on the indicated DMM readings at the various combinations of test points shown in Figure 16–32. Each of the four diodes provides a slightly different forward reading.

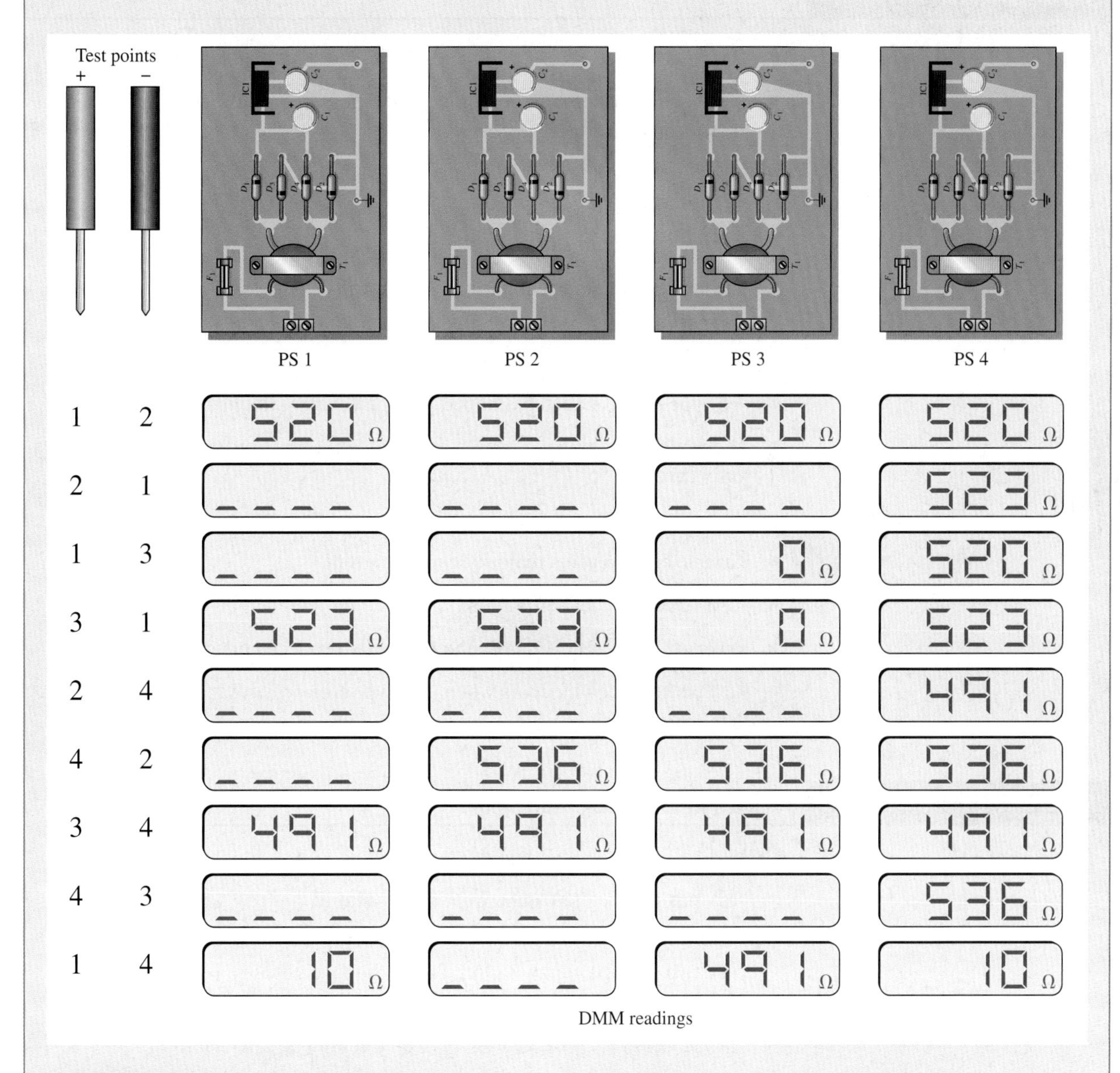

FIGURE 16–32

Results of the DMM check of each of the four power supply units.

APPLICATION ASSIGNMENT REVIEW

1. What faults could occur in the primary circuit of the power supply transformer and how would these faults be indicated by a DMM?
2. When measuring across a diode with a DMM, why is the polarity of the leads important?

SUMMARY

- The outermost shell of an atom containing electrons is the valence shell.
- Silicon and germanium are semiconductors. Silicon is the more predominant in electronics.
- Atoms within a semiconductor crystal structure are held together with covalent bonds.
- Electron-hole pairs are thermally produced.
- The process of adding impurities to an intrinsic (pure) semiconductor to increase and control conductivity is called *doping*.
- A *p*-type semiconductor is doped with trivalent impurity atoms.
- A *n*-type semiconductor is doped with pentavalent impurity atoms.
- The depletion region is a region adjacent to the *pn* junction containing no majority carriers.
- Forward bias permits majority carrier current through the *pn* junction.
- Reverse bias prevents majority carrier current.
- A *pn* structure is called a *diode*.
- Reverse leakage current is due to thermally produced electron-hole pairs.
- Reverse breakdown occurs when the reverse-biased voltage exceeds a specified value.

SELF-TEST

Answers are at the end of the chapter.

1. The charge on electrons is
   (a) positive (b) negative (c) neutral (d) variable
2. The number of valence electrons in both silicon and germanium is
   (a) two (b) eight (c) four (d) eighteen
3. When an atom loses or gains a valence electron, the atom becomes
   (a) covalent (b) neutral (c) a crystal (d) an ion
4. Atoms within a semiconductor crystal are held together by
   (a) atomic glue (b) subatomic particles
   (c) covalent bonds (d) the valence band

5. Free electrons exist in the

(a) valence band **(b)** conduction band
(c) lowest band **(d)** recombination band

6. A hole is

(a) a vacancy in the valence band left by an electron
(b) a vacancy in the conduction band
(c) a positive electron
(d) a conduction band electron

7. The widest energy gap between the valence band and the conduction band occurs in

(a) semiconductors **(b)** insulators
(c) conductors **(d)** a vacuum

8. The process of adding impurity atoms to a pure semiconductor is called

(a) recombination **(b)** crystalization
(c) bonding **(d)** doping

9. The two types of current in a semiconductor are

(a) positive current and negative current
(b) electron current and conventional current
(c) electron current and hole current
(d) forward current and reverse current

10. The majority carriers in an *n*-type semiconductor are

(a) electrons **(b)** holes **(c)** positive ions **(d)** negative ions

11. The *pn* junction is found in

(a) diodes **(b)** transistors
(c) all semiconductive material **(d)** answers (a) and (b)

12. In a diode, the region near the *pn* junction consisting of positive and negative ions is called the

(a) neutral zone **(b)** recombination region
(c) depletion region **(d)** diffusion area

13. A fixed dc voltage that sets the operating condition of a semiconductive device is called the

(a) bias **(b)** depletion voltage **(c)** battery **(d)** barrier potential

14. In a diode, the two bias conditions are

(a) positive and negative **(b)** blocking and nonblocking
(c) open and closed **(d)** forward and reverse

15. When a diode is forward-biased, it is

(a) blocking current **(b)** conducting current
(c) similar to an open switch **(d)** similar to a closed switch
(e) answers (a) and (c) **(f)** answers (b) and (d)

16. The voltage across a forward-biased silicon diode is approximately

(a) 0.7 V **(b)** 0.3 V **(c)** 0 V **(d)** dependent on the bias voltage

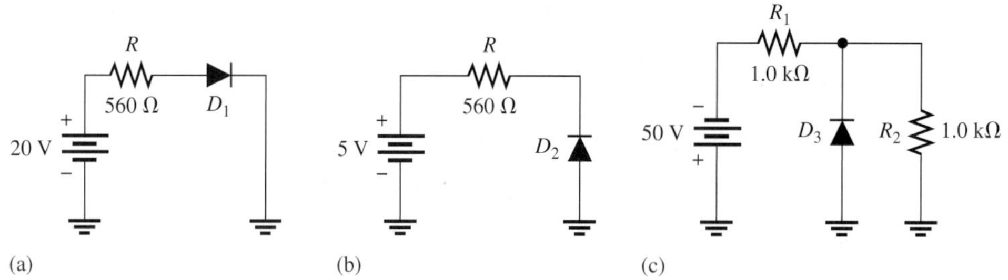

▲ FIGURE 16–33

17. In Figure 16–33, identify the forward-biased diode(s).
    (a) D_1 **(b)** D_2 **(c)** D_3 **(d)** D_1 and D_3 **(e)** D_2 and D_3

18. When the positive lead of an ohmmeter is connected to the cathode of a diode and the negative lead is connected to the anode, the meter reads
    (a) a very low resistance
    (b) an extremely high resistance
    (c) a high resistance initially, decreasing to about 100 Ω
    (d) a gradually increasing resistance

PROBLEMS

Answers to odd-numbered problems are at the end of the book.

BASIC PROBLEMS

SECTION 16–1 **Atomic Structure and Semiconductors**

1. List two semiconductive materials.
2. How many valence electrons do semiconductors have?

SECTION 16–2 **Atomic Bonding**

3. In a silicon crystal, how many covalent bonds does a single atom form?

SECTION 16–3 **Conduction in Semiconductors**

4. What happens when heat is added to silicon?
5. Name the two energy levels at which current is produced in silicon.
6. For each of the energy diagrams in Figure 16–34, determine the class of material based on relative comparisons.

► FIGURE 16–34

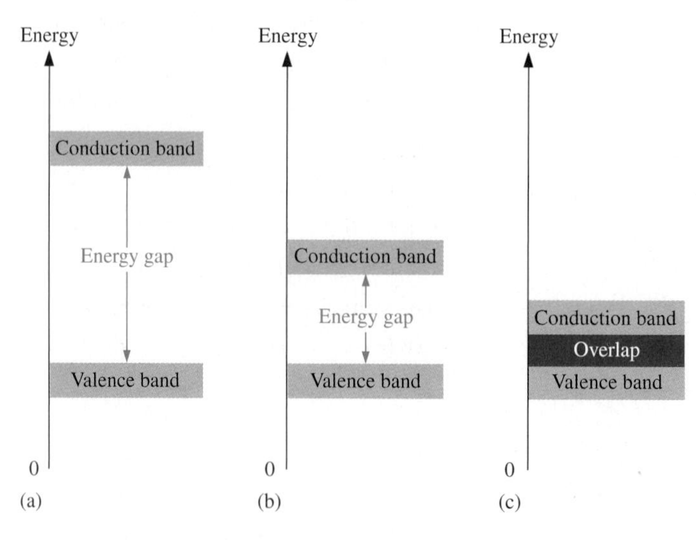

SECTION 16–4 **N-Type and P-Type Semiconductors**

7. Describe the process of doping and explain how it alters the atomic structure of silicon.
8. What type of impurity is antimony? What type of impurity is boron?

SECTION 16–5 **The PN Junction**

9. How is the electric field across the *pn* junction created?
10. Because of its barrier potential, can a *pn* junction device be used as a voltage source? Explain.

SECTION 16–6 **Biasing the PN Junction**

11. To forward-bias a *pn* junction, to which region must the positive terminal of a voltage source be connected?
12. Explain why a series resistor is necessary when a *pn* junction is forward-biased.

SECTION 16–7 **Diode Characteristics**

13. Explain how to generate the forward-bias portion of the characteristic curve.
14. What would cause the barrier potential to decrease from 0.7 V to 0.6 V?
15. Determine whether each diode in Figure 16–35 is forward-biased or reverse-biased.
16. Determine the voltage across each diode in Figure 16–35.

▶ FIGURE 16–35

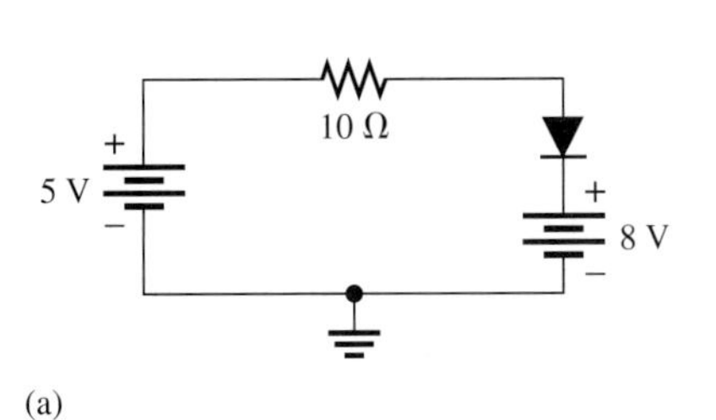

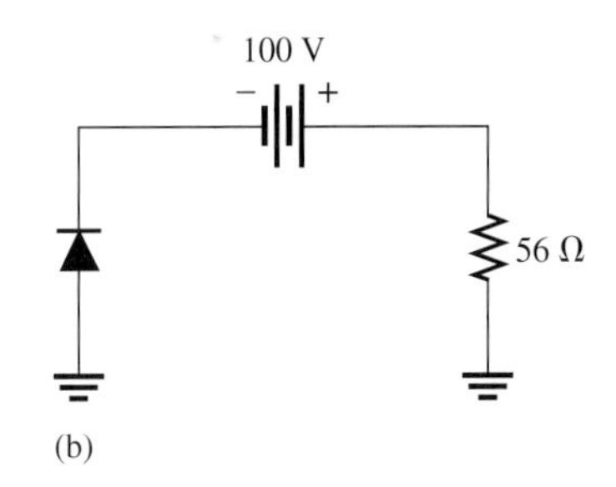

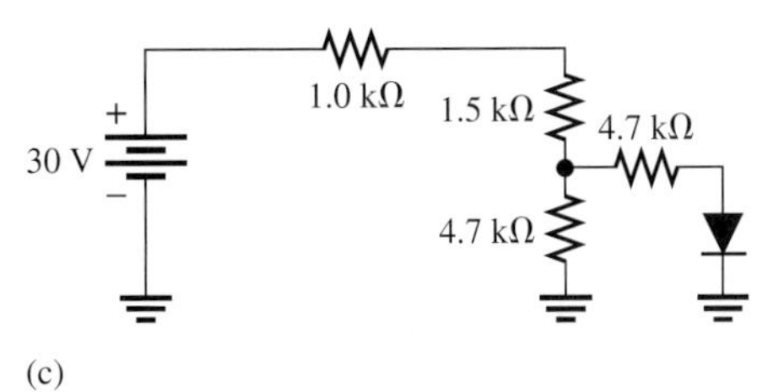

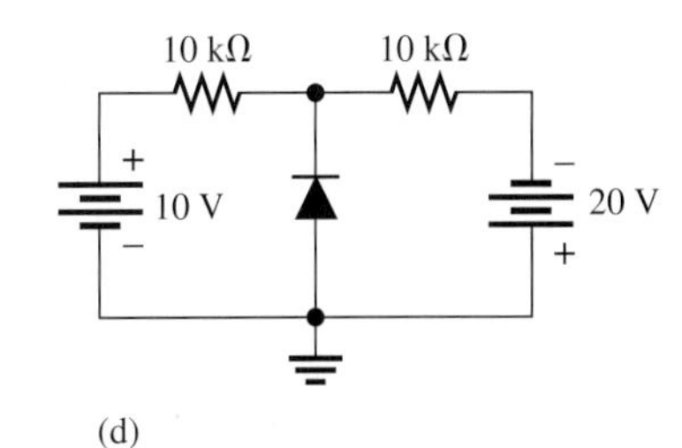

FIGURE 16–36

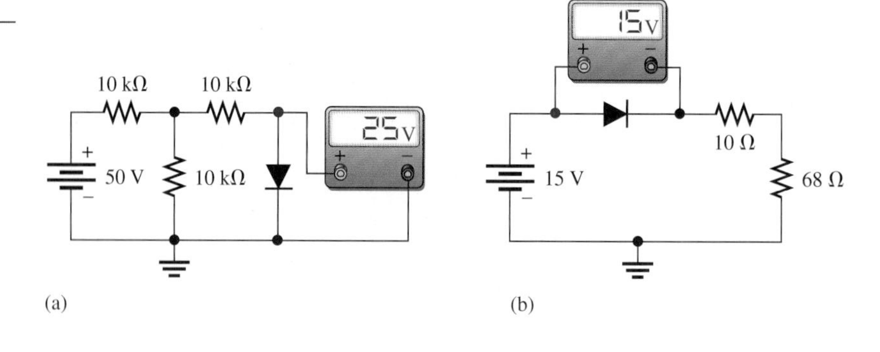

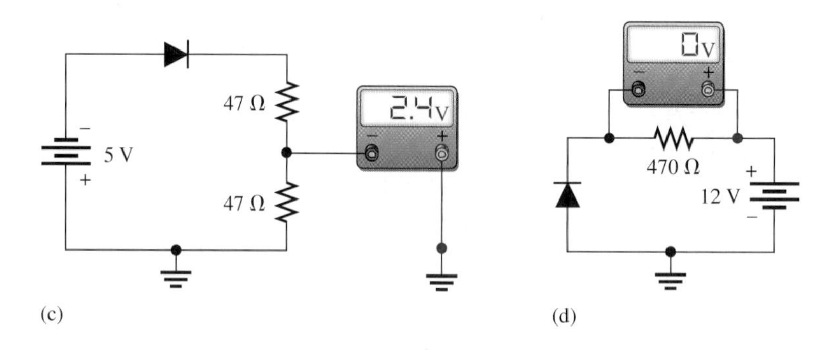

17. Examine the meter indications in each circuit of Figure 16–36, and determine whether the diode is functioning properly, or whether it is open or shorted.

18. Determine the voltage with respect to ground at each point in Figure 16–37. (The diodes are silicon.)

FIGURE 16–37

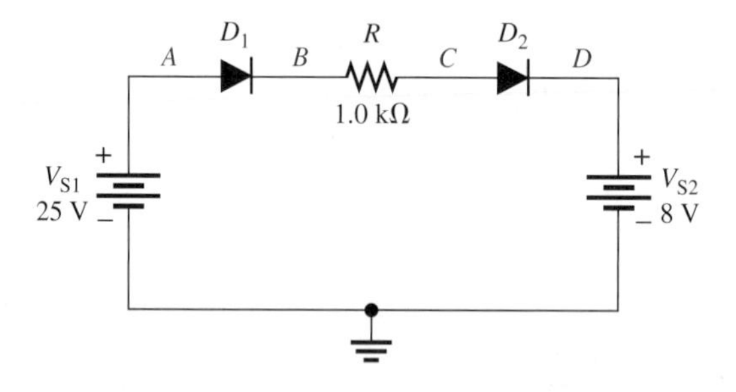

ELECTRONICS WORKBENCH/CIRCUITMAKER TROUBLESHOOTING PROBLEMS

CD-ROM file circuits are shown in Figure 16–38.

19. Open file P16-19 and determine if there is a fault. If so, identify it.

20. Open file P16-20 and determine if there is a fault. If so, identify it.

21. Open file P16-21 and determine if there is a fault. If so, identify it.

22. Open file P16-22 and determine if there is a fault. If so, identify it.

23. Open file P16-23 and determine if there is a fault. If so, identify it.

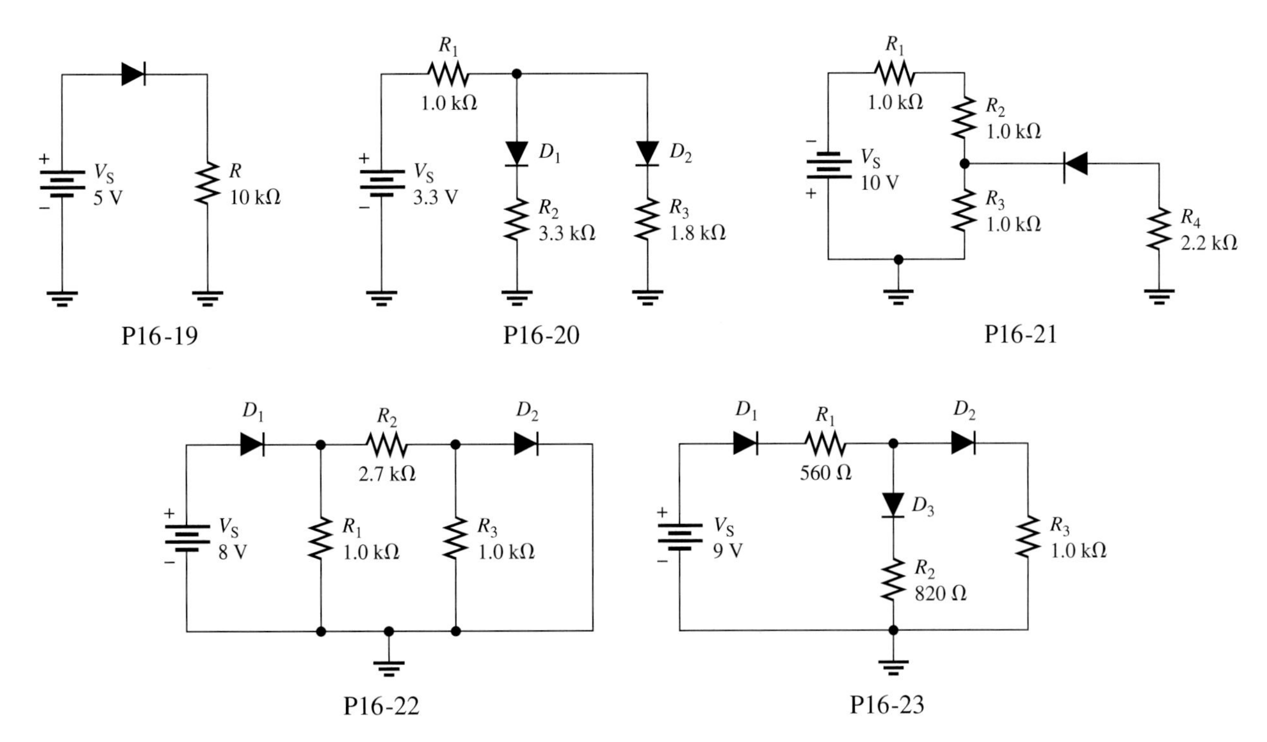

▲ FIGURE 16–38

ANSWERS

SECTION REVIEWS

SECTION 16–1 Atomic Structure and Semiconductors

1. An orbit is the path of an electron around the nucleus of an atom.
2. A shell is an energy band into which orbits are grouped.
3. A valance electron is an electron in the outermost shell (valence band).
4. A positive ion is created when an atom loses a valence electron.

SECTION 16–2 Atomic Bonding

1. Covalent bonds are formed by the sharing of valence electrons with neighboring atoms.
2. An intrinsic material is one that is in a pure state.
3. A crystal is a solid material formed by atoms bonding together in a fixed pattern.
4. There are eight shared valence electrons in each atom of a silicon crystal.

SECTION 16–3 Conduction in Semiconductors

1. Free electrons exist in the conduction band; valence electrons exist in the valence band.
2. A hole is created when an electron is thermally raised to the conduction band, leaving a hole in the valence band.
3. The energy gap between the valence band and the conduction band is narrower for a semiconductor than for an insulator.

SECTION 16–4 **_N_-Type and _P_-Type Semiconductors**

1. An n-type semiconductor is formed by the addition of pentavalent atoms to the intrinsic semiconductor.
2. A p-type semiconductor is formed by the addition of trivalent atoms to the intrinsic semiconductor.
3. A majority carrier is the particle in greatest abundance: electrons in n-type material and holes in p-type material.

SECTION 16–5 **The _PN_ Junction**

1. A pn junction is the boundary between n-type and p-type materials.
2. The depletion region is devoid of majority carriers and contains only positive and negative ions.
3. True
4. The barrier potential for silicon is 0.7 V.

SECTION 16–6 **Biasing the _PN_ Junction**

1. Two bias conditions are forward and reverse.
2. Forward bias produces majority carrier current.
3. Reverse bias causes the depletion area to widen.
4. True

SECTION 16–7 **Diode Characteristics**

Anode Cathode

▲ FIGURE 16–39

1. See Figure 16–39.
2. True
3. reverse; forward

■ Application Assignment

1. Open primary winding or open fuse; a typical DMM set on the ohms function will indicate an overload.
2. The polarity is important because the internal ohmmeter battery provides the bias voltage.

SELF-TEST

1. (b) **2.** (c) **3.** (d) **4.** (c) **5.** (b) **6.** (a) **7.** (b)
8. (d) **9.** (c) **10.** (a) **11.** (d) **12.** (c) **13.** (a) **14.** (d)
15. (f) **16.** (a) **17.** (d) **18.** (b)

17

DIODES AND APPLICATIONS

INTRODUCTION

In this chapter, the applications of diodes in converting ac to dc by the process known as *rectification* are discussed. Half-wave and full-wave rectification are introduced, and you will study the basic circuits. The limitations of diodes used in rectifier applications are examined, and you will learn about diode limiting circuits and dc restoring (clamping) circuits.

In addition to rectifier diodes, zener diodes and their applications in voltage regulation are introduced. Varactor diodes, light-emitting diodes, and photodiodes and their applications also are discussed.

CHAPTER OUTLINE

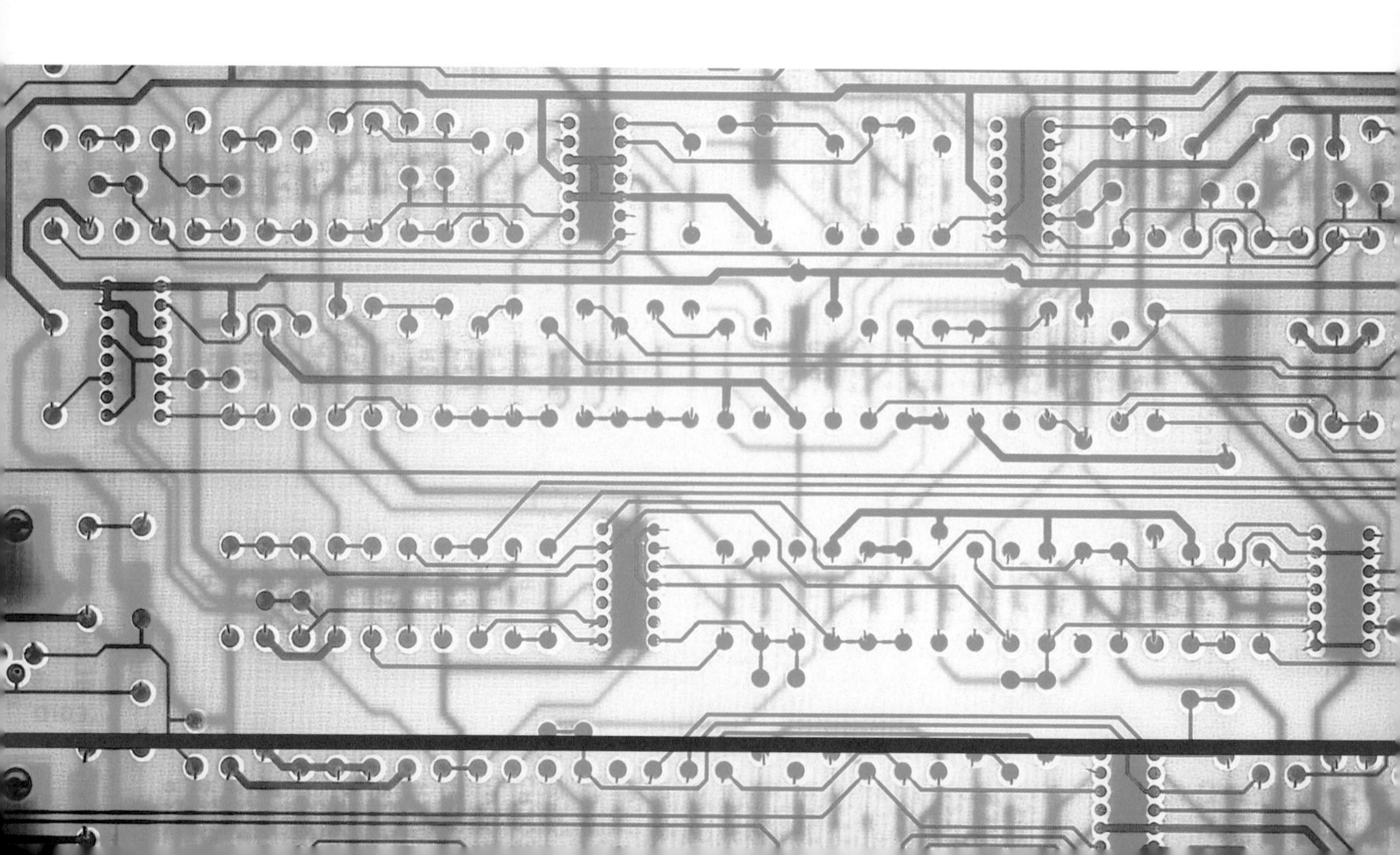

CHAPTER OBJECTIVES

- Analyze the operation of a half-wave rectifier
- Analyze the operation of a full-wave rectifier
- Describe the operation of power supply filters
- Analyze the operation of diode limiters and clampers
- Explain the characteristics and applications of a zener diode
- Describe the basic operation of a varactor diode
- Discuss the operation and application of LEDs and photodiodes
- Interpret a typical data sheet
- Troubleshoot power supplies and diode circuits, using the APM approach

KEY TERMS

- DC power supply
- Regulator
- Half-wave rectifier
- PIV
- Full-wave rectifier
- Bridge rectifier
- Capacitor-input filter
- Integrated circuit
- Line regulation
- Load regulation
- Limiter
- Clamper
- Zener diode
- Varactor
- LED
- Photodiode

APPLICATION ASSIGNMENT PREVIEW

Picture yourself as a technician in an industrial manufacturing facility responsible for maintaining and troubleshooting all of the automated production equipment. One particular system is used to count objects on a conveyor for control and inventory purposes. In order to check out and troubleshoot this system you must have a knowledge of power supply rectifiers, zener diodes, light-emitting diodes (LEDs), and photodiodes. After studying this chapter, you should be able to complete the application assignment.

WWW. VISIT THE COMPANION WEBSITE

Circuit Simulation Tutorials and Other Chapter Study Tools Are Available at

http://www.prenhall.com/floyd

17–1 HALF-WAVE RECTIFIERS

Because of their ability to conduct current in one direction and block current in the other direction, diodes are used in circuits called rectifiers that convert ac voltage into dc voltage. Rectifiers are found in all dc power supplies that operate from an ac voltage source. Power supplies are an essential part of all electronic systems from the simplest to the most complex.

After completing this section, you should be able to

- **Analyze the operation of a half-wave rectifier**
- Describe a basic dc power supply
- Explain the process of half-wave rectification
- Describe how the diode functions in a half-wave rectifier
- Determine the average value of a half-wave voltage
- Define *peak inverse voltage (PIV)*

The Basic DC Power Supply

The dc power supply converts the standard 110 V, 60 Hz ac available at wall outlets into a constant dc voltage. It is one of the most common electronic circuits that you will find. The dc voltage produced by a power supply is used to power all types of electronic circuits, such as television receivers, stereo systems, VCRs, CD players, and laboratory equipment.

A **rectifier** can be either a half-wave rectifier or a full-wave rectifier (covered in Section 17–2). The rectifier converts the ac input voltage to a pulsating dc voltage, which is half-wave rectified, as shown in the block diagram in Figure 17–1(a). A block diagram for a power supply is shown in part (b). When connected to the rectifier, the capacitor **filter** eliminates the fluctuations in the rectified voltage and produces a relatively smooth dc voltage. Power supply filters are

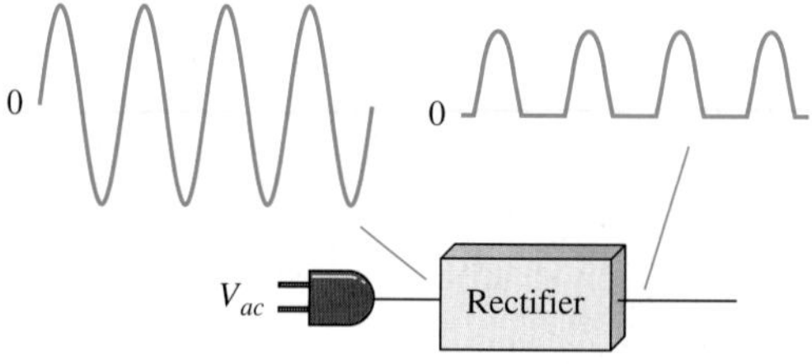

(a)

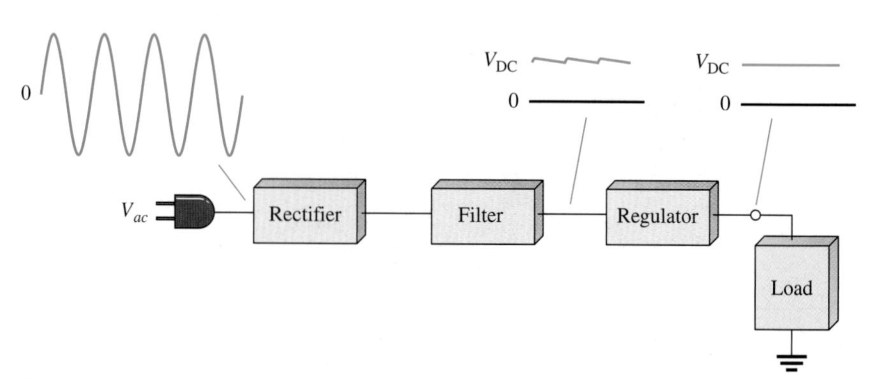

(b)

FIGURE 17–1

Block diagrams showing basic operation of a rectifier and of a regulated dc power supply.

covered in Section 17–3. The regulator is a circuit that maintains a constant dc voltage for variations in the input line voltage or in the load. Regulators vary from a single device to more complex circuits. The load block is usually a circuit for which the power supply is producing the dc voltage and load current.

The Half-Wave Rectifier

Figure 17–2 illustrates the process called *half-wave rectification.* In part (a), a diode is connected to an ac source that provides the input voltage, V_{in}, and to a load resistor, R_L, forming a half-wave rectifier. Keep in mind that all ground symbols represent the same point electrically. Let's examine what happens during one cycle of the input voltage using the ideal model for the diode. When the sinusoidal input voltage goes positive, the diode is forward-biased and conducts current through the load resistor, as shown in part (b). The current produces an output voltage across the load, which has the same shape as the positive half-cycle of the input voltage.

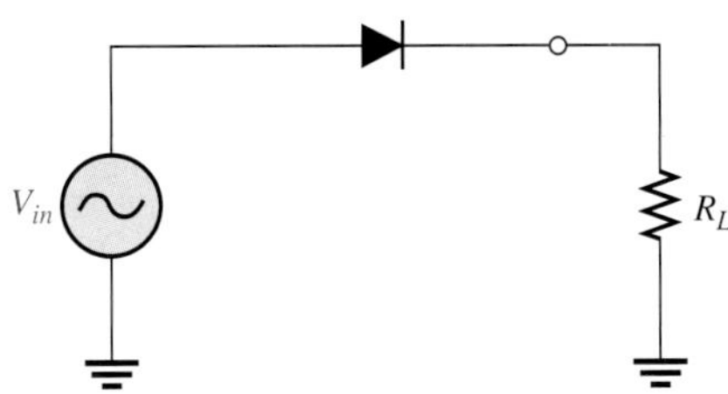

(a) Half-wave rectifier circuit

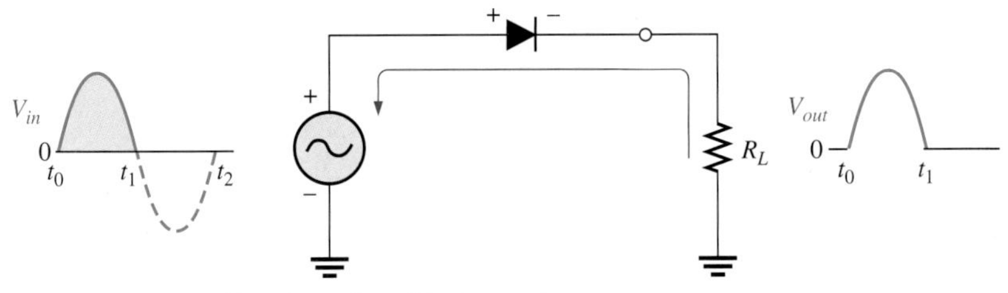

(b) Operation during positive alternation of the input voltage

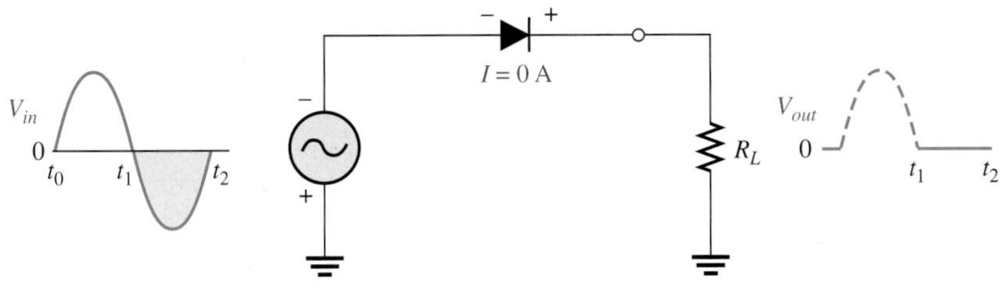

(c) Operation during negative alternation of the input voltage

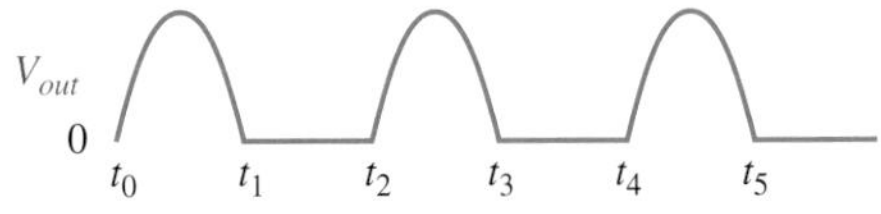

(d) Half-wave output voltage for three input cycles

FIGURE 17–2

Operation of a half-wave rectifier.

When the input voltage goes negative during the second half of its cycle, the diode is reverse-biased. There is no current, so the voltage across the load resistor is zero, as shown in Figure 17–2(c). The net result is that only the positive half-cycles of the ac input voltage appear across the load. Since the output does not change polarity, it is a pulsating dc voltage, as shown in part (d).

Average Value of the Half-Wave Output Voltage The average value of a half-wave output voltage is the value you would measure on a dc voltmeter. It can be calculated with the following equation where $V_{p(out)}$ is the peak value of the half-wave output voltage:

Equation 17–1

$$V_{AVG} = \frac{V_{p(out)}}{\pi}$$

Figure 17–3 shows the half-wave voltage with its average value indicated by the red dashed line.

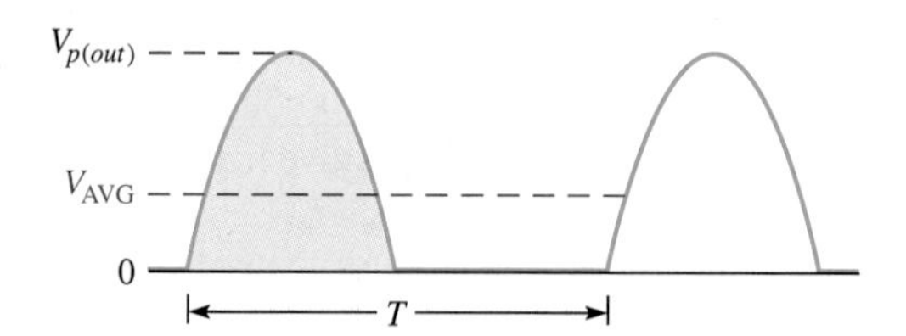

FIGURE 17–3

Average value of the half-wave rectified signal.

EXAMPLE 17–1

What is the average (dc) value of the half-wave rectified output voltage waveform in Figure 17–4?

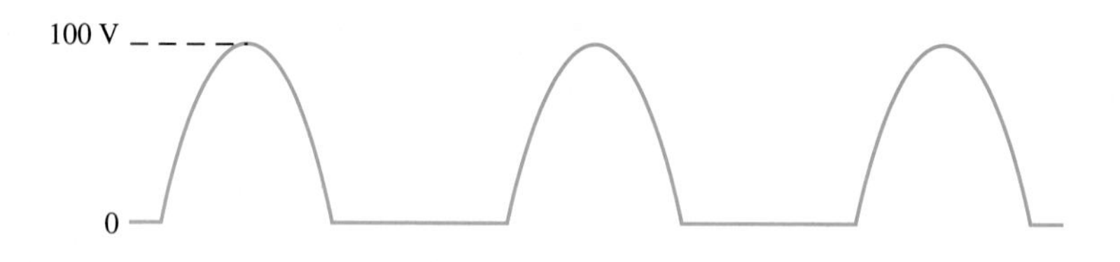

FIGURE 17–4

Solution

$$V_{AVG} = \frac{V_{p(out)}}{\pi} = \frac{100\text{ V}}{\pi} = \mathbf{31.8\text{ V}}$$

*Related Problem** Determine the average value of the half-wave output voltage if its peak amplitude is 12 V.

*Answers are at the end of the chapter.

Effect of Diode Barrier Potential on Half-Wave Rectifier Output Voltage

In the previous discussion, the diode was considered ideal. When the diode barrier potential is taken into account as in the practical model discussed in Chapter 16, here is what happens: During the positive half-cycle, the input voltage must overcome the barrier potential before the diode becomes forward-biased. This results in a half-wave output voltage with a peak value that is 0.7 V less than the peak value of the input voltage (0.3 V less for a germanium diode), as shown in Figure 17–5. The expression for peak output voltage is

Equation 17–2

$$V_{p(out)} = V_{p(in)} - 0.7\text{ V}$$

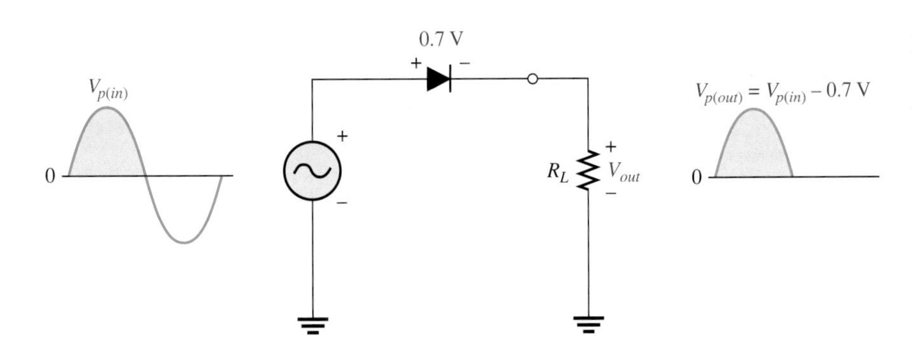

FIGURE 17–5

Effect of barrier potential on half-wave rectified output voltage (silicon diode shown).

When you work with diode circuits, it is often practical to neglect the effect of barrier potential when the peak value of the applied voltage is much greater (at least ten times) than the barrier potential.

EXAMPLE 17–2

Determine the peak output voltage and the average value of the output voltage of the rectifier in Figure 17–6 for the indicated input voltage.

FIGURE 17–6

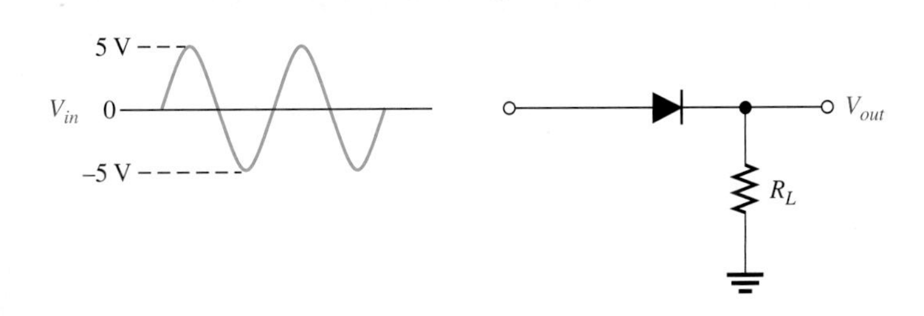

Solution

$$V_{p(out)} = V_{p(in)} - 0.7\text{ V} = 5\text{ V} - 0.7\text{ V} = \mathbf{4.3\text{ V}}$$

$$V_{AVG} = \frac{V_{p(out)}}{\pi} = \frac{4.3\text{ V}}{\pi} = \mathbf{1.37\text{ V}}$$

Related Problem Determine the peak output voltage for the rectifier in Figure 17–6 if the peak input is 3 V.

Peak Inverse Voltage (PIV)

The maximum value of reverse voltage, sometimes designated as PIV (peak inverse voltage), occurs at the peak of each negative alternation of the input cycle when the diode is reverse-biased. This condition is illustrated in Figure 17–7. The PIV equals the peak value of the input voltage, and the diode must be capable of withstanding this amount of repetitive reverse voltage.

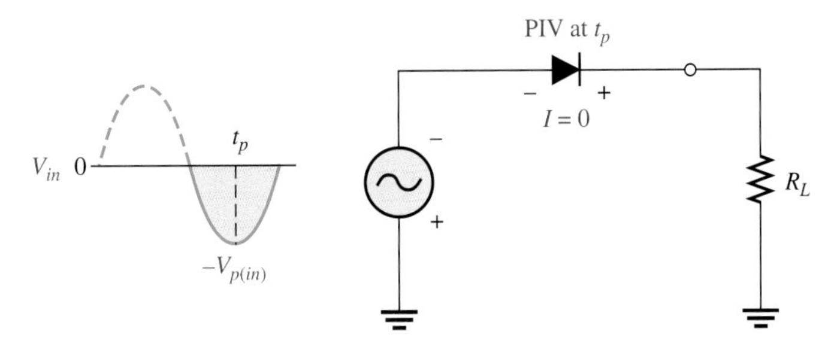

FIGURE 17–7

The PIV occurs at the peak of each half-cycle of the input voltage when the diode is reverse-biased. In this circuit, the PIV occurs at the time (t_p) of the peak of each negative half-cycle.

SECTION 17–1 REVIEW

Answers are at the end of the chapter.

1. At what point on the input cycle of a half-wave rectifier does the PIV occur?
2. For a half-wave rectifier, there is current through the load for approximately what percentage of the input cycle?
3. What is the average value of the output voltage shown in Figure 17–8?

FIGURE 17–8

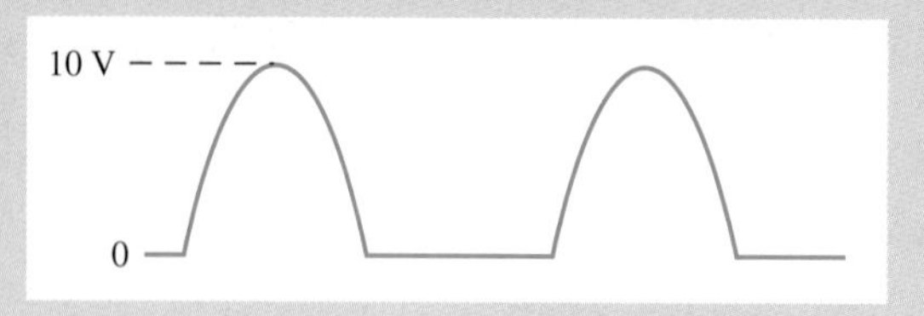

17–2 FULL-WAVE RECTIFIERS

Although half-wave rectifiers have some applications, the full-wave rectifier is the most commonly used type in dc power supplies. In this section, you will use what you learned about half-wave rectification and expand it to full-wave rectifiers. You will learn about two types of full-wave rectifiers: center-tapped and bridge.

After completing this section, you should be able to

- **Analyze the operation of a full-wave rectifier**
- Explain the process of full-wave rectification
- Analyze the center-tapped full-wave rectifier
- Analyze the full-wave bridge rectifier
- Determine the PIV for each type of rectifier

The difference between full-wave and half-wave rectification is that a **full-wave rectifier** allows unidirectional current to the load during the entire input cycle and the half-wave rectifier allows this only during one-half of the cycle. The result of full-wave rectification is a dc output voltage that pulsates every half-cycle of the input, as shown in Figure 17–9.

FIGURE 17–9

Full-wave rectification.

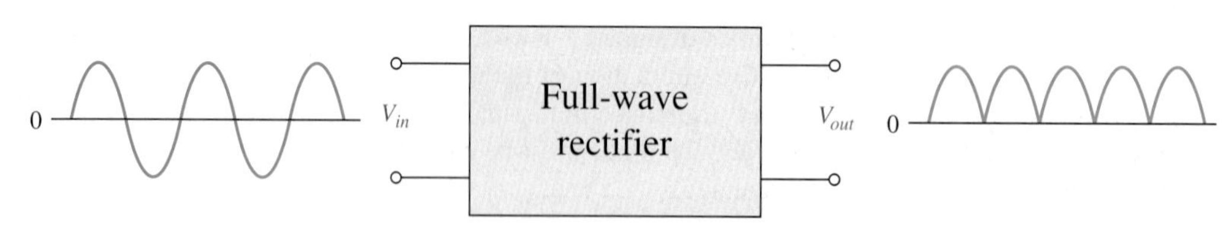

The average value for a full-wave rectified output voltage is twice that of the half-wave, expressed as follows:

Equation 17–3

$$V_{\text{AVG}} = \frac{2V_{p(out)}}{\pi}$$

Since $2/\pi = 0.637$, you can calculate V_{AVG} as $0.637V_{p(out)}$.

EXAMPLE 17–3

Find the average value of the full-wave rectified output voltage in Figure 17–10.

▶ FIGURE 17–10

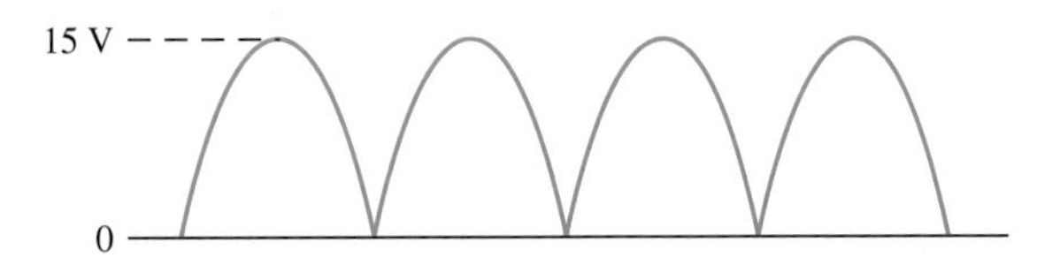

Solution

$$V_{AVG} = \frac{V_{p(out)}}{\pi} = \frac{2(15\text{ V})}{\pi} = \mathbf{9.55\ V}$$

Related Problem Find V_{AVG} for the full-wave rectified voltage if its peak is 155 V.

Center-Tapped Full-Wave Rectifier

The center-tapped (CT) full-wave rectifier uses two diodes connected to the secondary of a center-tapped transformer, as shown in Figure 17–11. The input signal is coupled through the transformer to the secondary. Half of the secondary voltage appears between the center tap and each end of the secondary winding as shown.

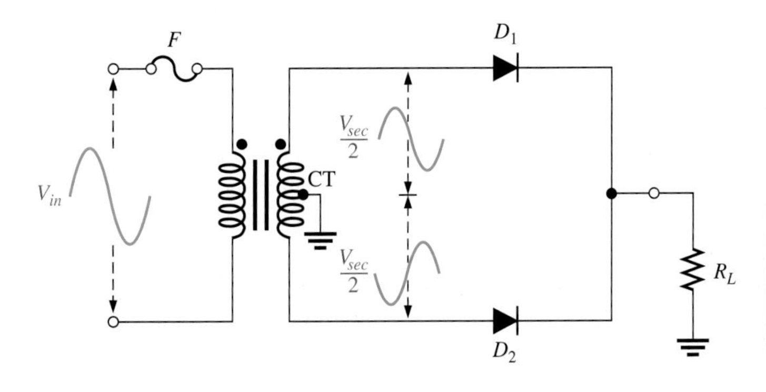

◀ FIGURE 17–11

A center-tapped full-wave rectifier.

A diode is sometimes used to isolate two power sources such as in the case of a battery backup for a computer system. The power from the power supply is connected through a series diode to the power input point on the system. A battery backup system is connected to the same point through its own series diode. The battery backup is set for a slightly lower voltage so that it does not discharge the battery during normal operation. When a power failure occurs, the battery backup system takes over and the diodes prevent the two power sources from affecting each other.

For a positive half-cycle of the input voltage, the polarities of the secondary voltages are as shown in Figure 17–12(a). This condition forward-biases the upper diode D_1 and reverse-biases the lower diode D_2. The current path is through D_1 and the load resistor R_L, as indicated. For a negative half-cycle of the input voltage, the voltage polarities on the secondary are as shown in Figure 17–12(b). This condition reverse-biases D_1 and forward-biases D_2. The current path is through D_2 and R_L, as indicated. Because the current during both the positive and the negative portions of the input cycle is in the same direction through the load, the output voltage developed across the load is a full-wave rectified dc voltage.

Effect of the Turns Ratio on Full-Wave Output Voltage If the turns ratio of the transformer is 1, the peak value of the rectified output voltage equals half the peak value of the primary input voltage. This value occurs because half of the input voltage appears across each half of the secondary winding.

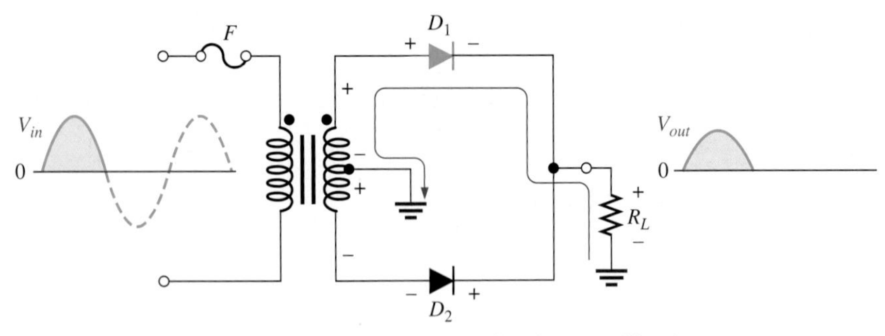

(a) During positive half-cycles, D_1 is forward-biased and D_2 is reverse-biased.

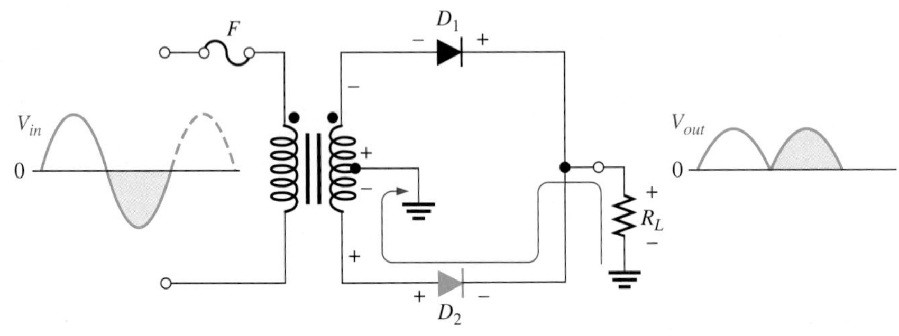

(b) During negative half-cycles, D_2 is forward-biased and D_1 is reverse-biased.

FIGURE 17–12

Basic operation of a center-tapped full-wave rectifier. Note that the current through the load resistor is in the same direction during the entire input cycle.

In order to obtain an output voltage approximately equal to the input, a step-up transformer with a turns ratio of 1 to 2 (1:2) must be used. In this case, the total secondary voltage is twice the primary voltage, so the voltage across each half of the secondary is equal to the input.

Peak Inverse Voltage (PIV) For simplicity, ideal diodes are used to illustrate PIV in a full-wave rectifier. Each diode in the full-wave rectifier is alternately forward-biased and then reverse-biased. The maximum reverse voltage that each diode must withstand is the peak value of the total secondary voltage ($V_{p(sec)}$), as illustrated in Figure 17–13. When the total secondary voltage $V_{p(sec)}$ has the polarity shown, the anode of D_1 is $+V_{p(sec)}/2$ and the anode of D_2 is $-V_{p(sec)}/2$. Since D_1 is forward-biased, its cathode is the same as its anode ($V_{p(sec)}/2$); this is also the voltage on the cathode of D_2. The total reverse voltage across D_2 is

$$V_{D2} = \frac{V_{p(sec)}}{2} - \frac{-V_{p(sec)}}{2} = V_{p(sec)}$$

FIGURE 17–13

Diode D_1 is shown forward-biased and D_2 is reverse-biased with PIV across it. The PIV across either diode is twice the peak value of the output voltage.

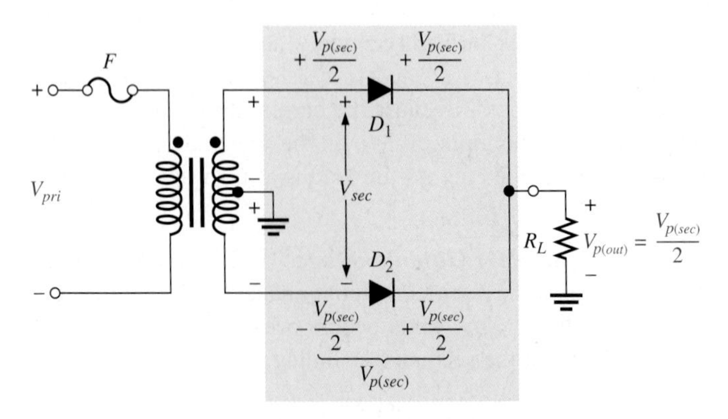

Since $V_{p(out)} = V_{p(sec)}/2$,

$$V_{p(sec)} = 2V_{p(out)}$$

For either diode, the PIV in terms of the peak secondary voltage is

$$\text{PIV} = V_{p(sec)}$$

Combining the two preceding equations, the peak inverse voltage across either diode in the center-tapped full-wave rectifier in terms of the output voltage is

$$\text{PIV} = 2V_{p(out)}$$

Equation 17–4

EXAMPLE 17–4

(a) For ideal diodes, show the voltage waveforms across the secondary winding and across R_L when a 25 V peak sine wave is applied to the primary winding in Figure 17–14.

(b) What minimum PIV rating must the diodes have?

FIGURE 17–14

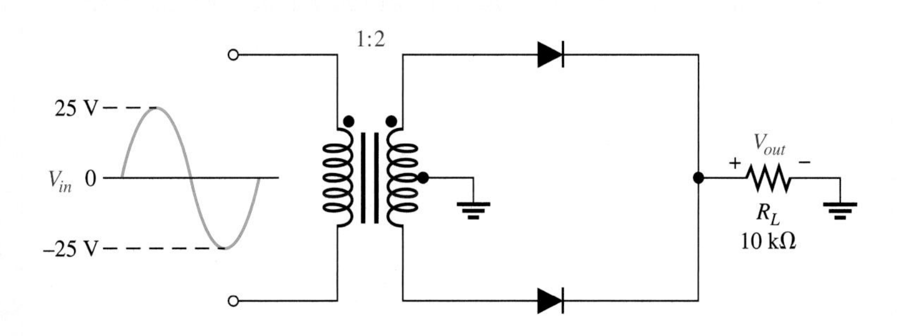

Solution **(a)** The waveforms are shown in Figure 17–15.

FIGURE 17–15

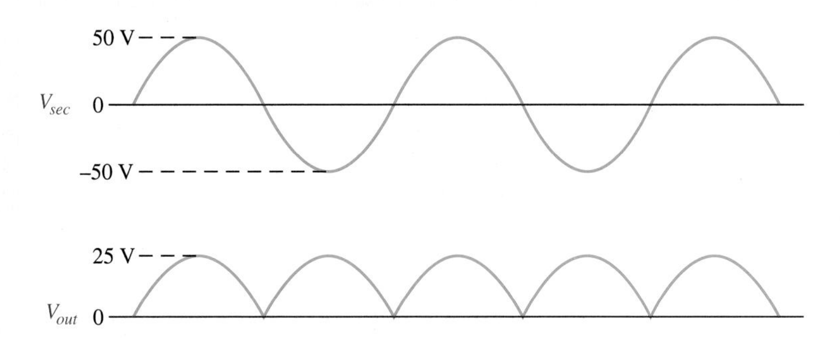

(b) The minimum PIV rating for each diode is calculated as follows. The turns ratio $n = 2$.

$$\text{PIV} = V_{p(sec)} = nV_{p(in)} = (2)25\text{ V} = \mathbf{50\text{ V}}$$

Related Problem What diode PIV rating is required to handle a peak input of 160 V in Figure 17–14?

Open file E17-04 on your EWB/CircuitMaker CD-ROM. Use the oscilloscope to display the output voltage waveform and compare to the output waveform in Figure 17–15.

Full-Wave Bridge Rectifier

The full-wave **bridge rectifier** uses four diodes, as shown in Figure 17–16. When the input cycle is positive as in part (a), diodes D_1 and D_2 are forward-biased and conduct current in the direction shown. A voltage is developed across R_L which looks like the positive half of the input cycle. During this time, diodes D_3 and D_4 are reverse-biased.

FIGURE 17–16

Operation of full-wave bridge rectifier.

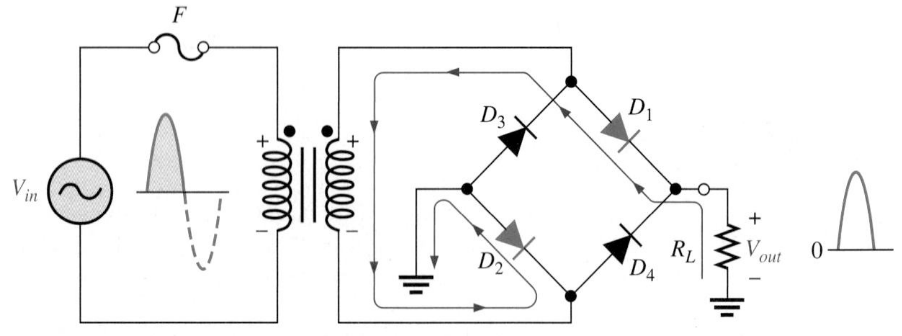

(a) During positive half-cycle of the input, D_1 and D_2 are forward-biased and conduct current. D_3 and D_4 are reverse-biased.

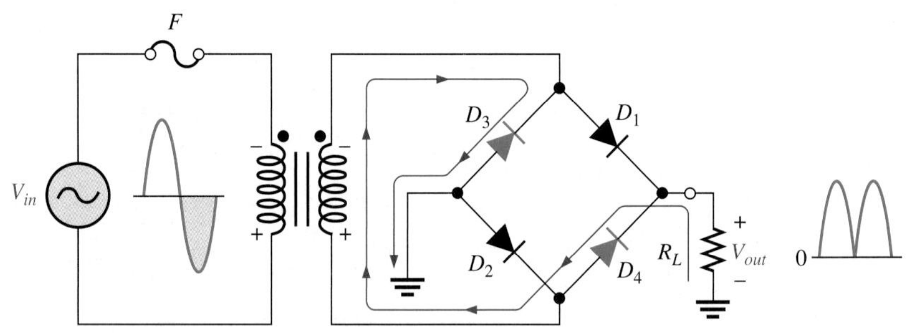

(b) During negative half-cycle of the input, D_3 and D_4 are forward-biased and conduct current. D_1 and D_2 are reverse-biased.

When the input cycle is negative, as in Figure 17–16(b), diodes D_3 and D_4 are forward-biased and conduct current in the same direction through R_L as during the positive half-cycle. During the negative half-cycle, D_1 and D_2 are reverse-biased. A full-wave rectified output voltage appears across R_L as a result of this action.

Bridge Output Voltage As you can see in Figure 17–16, two diodes are always in series with the load resistor during both the positive and the negative half-cycles. Neglecting the barrier potentials of the two diodes, the output voltage is a full-wave rectified voltage with a peak value equal to the peak secondary voltage.

Equation 17–5

$$V_{p(out)} = V_{p(sec)}$$

Peak Inverse Voltage (PIV) Let's assume that the input is in its positive half-cycle when D_1 and D_2 are forward-biased and examine the reverse voltage across D_3 and D_4. In Figure 17–17, you can see that D_3 and D_4 have a peak inverse voltage equal to the peak secondary voltage, $V_{p(sec)}$. Since the peak secondary voltage is equal to the peak output voltage, the peak inverse voltage is

Equation 17–6

$$\text{PIV} = V_{p(out)}$$

The PIV rating of the bridge diodes is half that required for the center-tapped configuration for the same output voltage.

FIGURE 17–17

PIV across diodes D_3 and D_4 in a bridge rectifier during the positive half-cycle of the input voltage.

EXAMPLE 17–5

(a) Determine the peak output voltage for the bridge rectifier in Figure 17–18.

(b) What minimum PIV rating is required for the diodes?

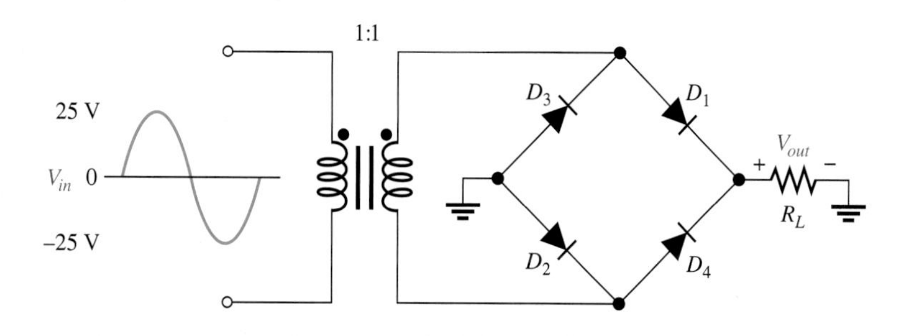

FIGURE 17–18

Solution **(a)** The peak output voltage is

$$V_{p(out)} = V_{p(sec)} = nV_{p(in)} = (1)25\text{ V} = \mathbf{25\text{ V}}$$

(b) The PIV for each diode is

$$\text{PIV} = V_{p(out)} = \mathbf{25\text{ V}}$$

Related Problem Determine the peak output voltage for the bridge rectifier in Figure 17–18 if the peak primary voltage is 160 V. What is the PIV rating for the diodes?

Open file E17-05 on your EWB/CircuitMaker CD-ROM. Measure the peak output voltage using the oscilloscope and compare to the calculated value.

SECTION 17–2 REVIEW

1. What is the average value of a full-wave rectified voltage with a peak value of 60 V?
2. Which type of full-wave rectifier has the greater output voltage for the same input voltage and transformer turns ratio?
3. For a given output voltage, the PIV for bridge rectifier diodes is less than for center-tapped rectifier diodes. (True or False)

17–3 POWER SUPPLY FILTERS AND REGULATORS

A power supply filter greatly reduces the fluctuations in the output voltage of a half-wave or full-wave rectifier and produces a nearly constant-level dc voltage. Filtering is necessary because electronic circuits require a constant source of dc voltage and current to provide power and biasing for proper operation. Filtering is accomplished using capacitors, as you will see in this section. **Voltage regulation** is usually accomplished with integrated circuit voltage regulators. A voltage regulator prevents changes in the filtered dc voltage due to variations in line voltage or load.

After completing this section, you should be able to

- **Describe the operation of power supply filters**
- Explain the operation of a capacitor-input filter
- Determine the PIV for the diode in a filtered rectifier
- Define *ripple voltage* and discuss its cause
- Calculate the ripple factor
- Discuss surge current in a capacitor-input filter
- Discuss voltage regulation
- Describe a typical IC regulator

In most power supply applications, the standard 60 Hz ac power line voltage must be converted to a nearly constant dc voltage. The 60 Hz pulsating dc output of a half-wave rectifier or the 120 Hz pulsating output of a full-wave rectifier must be filtered to reduce the large voltage variations. Figure 17–19 illustrates the filtering concept showing a relatively smooth dc output voltage. The full-wave rectifier voltage is applied to the input of the filter, and, ideally, a constant dc level appears on the output. In practice, there is usually a small voltage variation called the ripple.

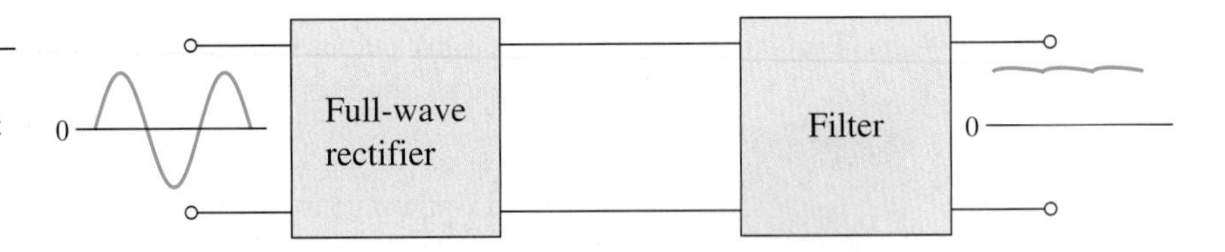

FIGURE 17–19

Basic block diagram of a power supply with a rectifier and filter that converts 60 Hz ac to dc.

Capacitor-Input Filter

A half-wave rectifier with a capacitor-input filter is shown in Figure 17–20. The half-wave rectifier is used to illustrate the filtering principle; then the concept is expanded to the full-wave rectifier.

During the positive first quarter-cycle of the input, the diode is forward-biased, allowing the capacitor to charge to approximately the diode drop of the input peak, as illustrated in Figure 17–20(a). When the input begins to decrease below its peak, as shown in part (b), the capacitor retains its charge and the diode becomes reverse-biased. During the remaining part of the cycle, the capacitor can discharge only through the load resistance at a rate determined by the R_LC time constant. The larger the time constant, the less the capacitor will discharge.

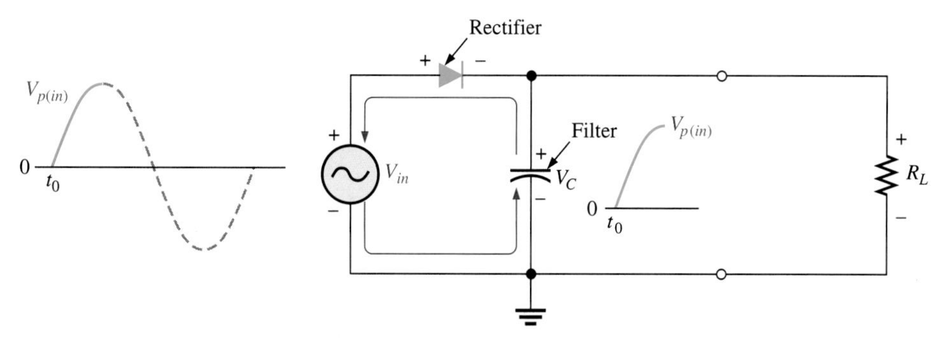

(a) Initial charging of capacitor (diode is forward-biased) happens only once when power is turned on.

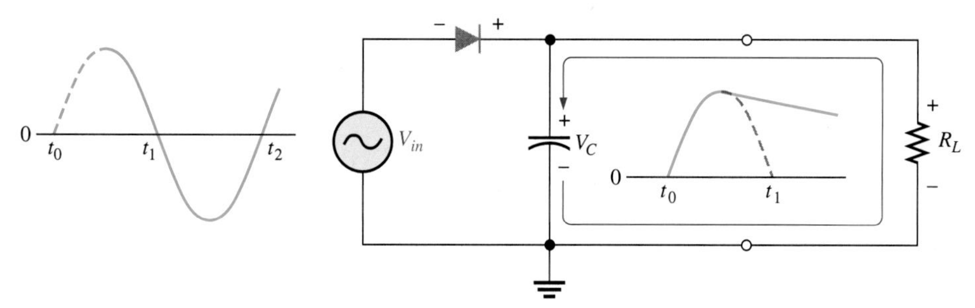

(b) The capacitor discharges through R_L after peak of positive alternation when the diode is reverse-biased. This discharging occurs during the portion of the input voltage indicated by the solid blue curve.

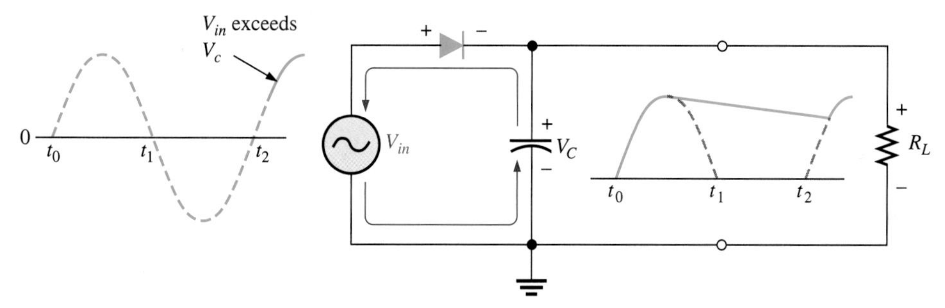

(c) The capacitor charges back to peak of input when the diode becomes forward-biased. This charging occurs during the portion of the input voltage indicated by the solid blue curve.

▲ **FIGURE 17–20**

Operation of a half-wave rectifier with a capacitor-input filter.

Because the capacitor charges to a peak value of $V_{p(in)}$, the peak inverse voltage of the diode in this application is

$$PIV = 2V_{p(in)}$$

Equation 17–7

During the first quarter of the next cycle, as illustrated in Figure 17–20(c), the diode again will become forward-biased when the input voltage exceeds the capacitor voltage.

Ripple Voltage As you have seen, the capacitor quickly charges at the beginning of a cycle and slowly discharges after the positive peak (when the diode is reverse-biased). The variation in the output voltage due to the charging and discharging is called the **ripple voltage.** The smaller the ripple, the better the filtering action, as illustrated in Figure 17–21.

For a given input frequency, the output frequency of a full-wave rectifier is twice that of a half-wave rectifier. As a result, a full-wave rectifier is easier to fil-

(a) Greater ripple means less effective filtering.

(b) Smaller ripple means more effective filtering.

▲ **FIGURE 17–21**

Half-wave ripple voltage (blue line).

ter. When filtered, the full-wave rectified voltage has a smaller ripple than does a half-wave signal for the same load resistance and capacitor values. A smaller ripple occurs because the capacitor discharges less during the shorter interval between full-wave pulses, as shown in Figure 17–22. A good rule of thumb for effective filtering is to make $R_LC \geq 10T$, where T is the period of the rectified voltage.

▶ **FIGURE 17–22**

Comparison of ripple voltages for half-wave and full-wave signals with same filter and same input frequency.

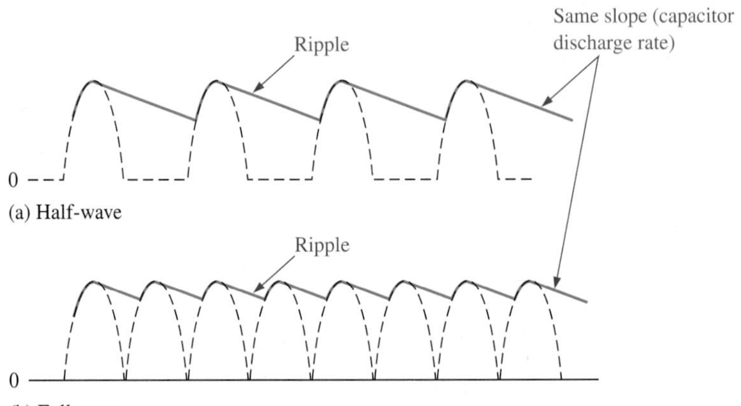

The ripple factor (r) is an indication of the effectiveness of the filter and is defined as the ratio of the ripple voltage (V_r) to the dc (average) value of the filter output voltage (V_{DC}). These parameters are illustrated in Figure 17–23.

Equation 17–8

$$r = \left(\frac{V_r}{V_{DC}}\right)100\%$$

The lower the ripple factor, the better the filter. The ripple factor can be decreased by increasing the value of the filter capacitor.

▶ **FIGURE 17–23**

V_r and V_{DC} determine the ripple factor.

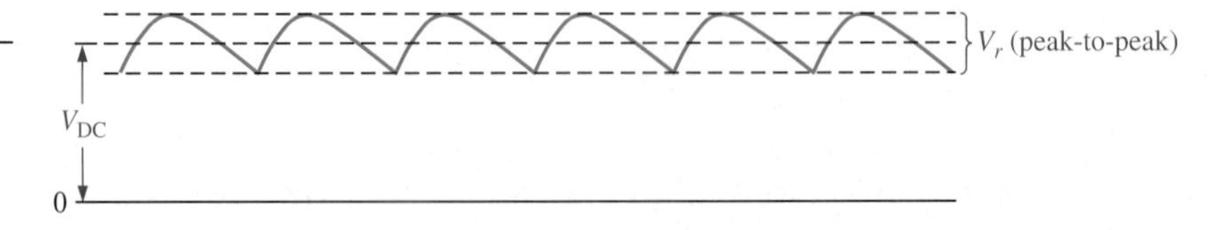

Surge Current in the Capacitor-Input Filter Before the switch in Figure 17–24(a) is closed, the filter capacitor is uncharged. At the instant the switch is closed, voltage is connected to the bridge and the capacitor appears as a short, as shown. An initial surge of current is produced through the two forward-biased diodes. The worst-case situation occurs when the switch is closed at a peak of the input voltage and a maximum surge current, $I_{surge(max)}$, is produced, as illustrated in the figure.

It is possible that the surge current could destroy the diodes, and for this reason a surge-limiting resistor R_{surge} is sometimes connected, as shown in Figure 17–24(b). The value of this resistor must be small compared to R_L. Also, the diodes must have a forward current rating such that they can withstand the momentary surge of current.

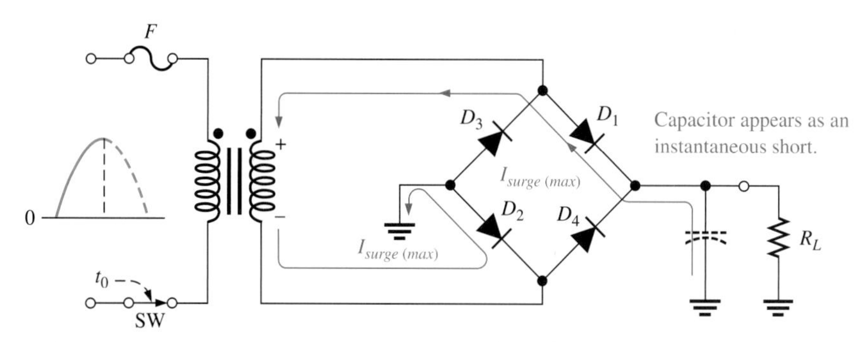

(a) Maximum surge current occurs when switch is closed at peak of input cycle.

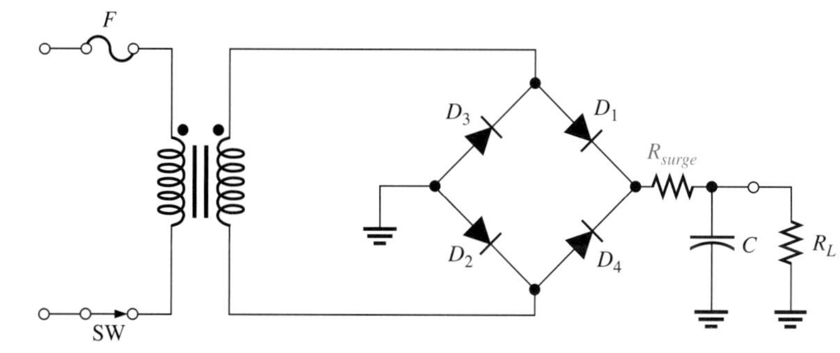

(b) A series resistor (R_{surge}) limits the surge current.

FIGURE 17–24

Surge current in a capacitor-input filter.

IC Regulators

While filters can reduce the ripple from power supplies to a low value, the most effective filter is a combination of a capacitor-input filter used with an integrated circuit (IC) voltage regulator. An integrated circuit regulator is a device that is connected to the output of a filtered rectifier and maintains a constant output voltage despite changes in the input, the load current, or the temperature. The capacitor-input filter reduces the input ripple to the regulator to an acceptable level. The combination of a large capacitor and an IC regulator is inexpensive and helps produce an excellent small power supply.

The most popular IC regulators have three terminals—an input terminal, an output terminal, and a reference (or adjust) terminal. The input to the regulator is first filtered with a capacitor to reduce the ripple to less than 10%. The regulator further reduces the ripple to a negligible amount. In addition, most regulators have an internal voltage reference, short-circuit protection, and thermal shutdown circuitry. They are available in a variety of voltages, including positive and negative outputs, and can be designed for variable outputs with a minimum of external components. Typically, IC regulators can furnish a constant output of one or more amps of current with high ripple rejection. IC regulators are available that can supply load currents of over 5 A.

Three-terminal regulators designed for a fixed output voltage require only external capacitors to complete the regulation portion of the power supply, as shown in Figure 17–25. Filtering is accomplished by a large-value capacitor between the input voltage and ground. Sometimes a second smaller-value input capacitor is

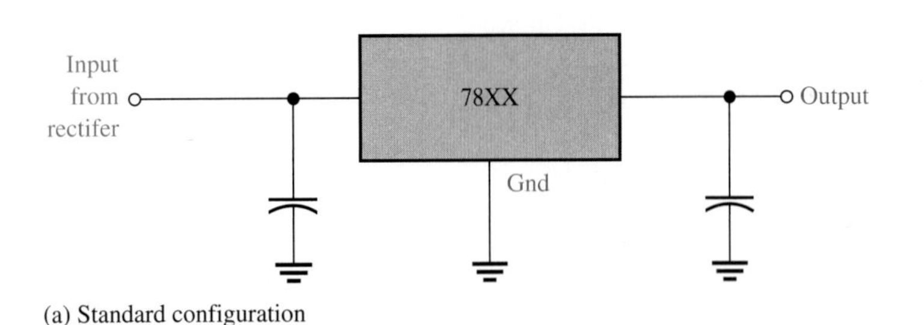

(a) Standard configuration

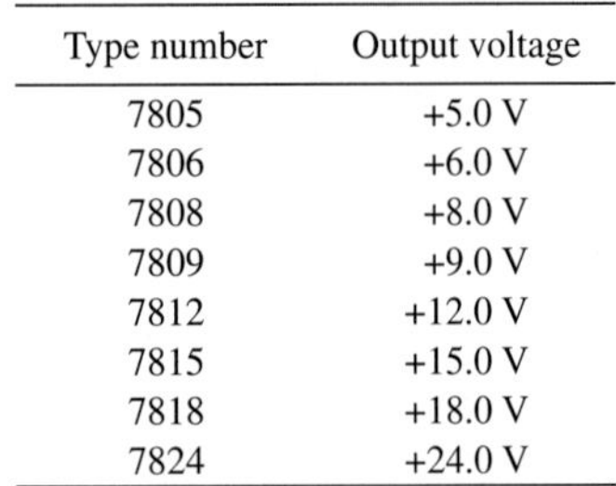

Type number	Output voltage
7805	+5.0 V
7806	+6.0 V
7808	+8.0 V
7809	+9.0 V
7812	+12.0 V
7815	+15.0 V
7818	+18.0 V
7824	+24.0 V

(b) The 7800 series

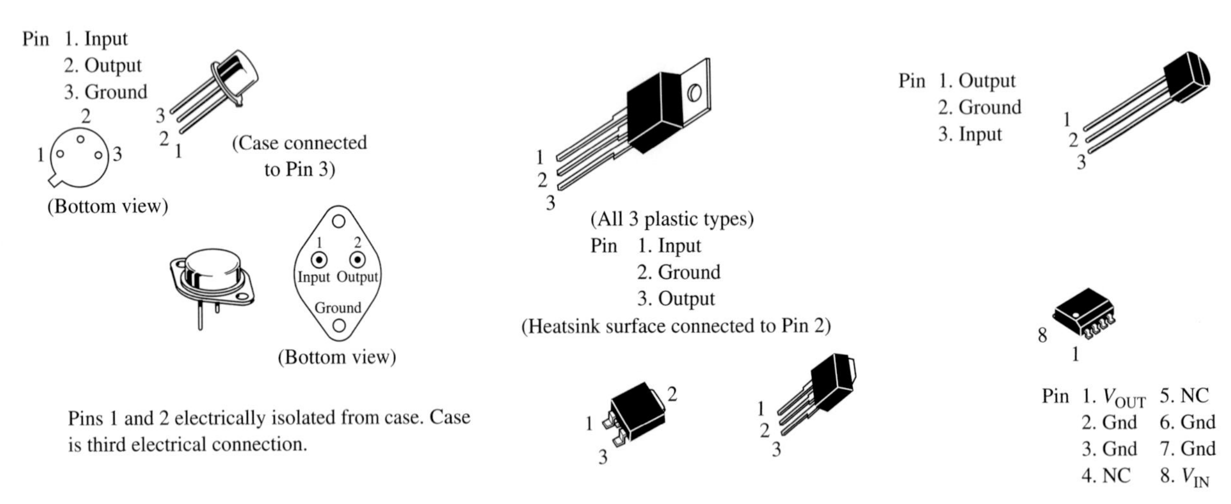

(c) Typical metal and plastic packages

▲ **FIGURE 17–25**

The 7800 series three-terminal fixed positive voltage regulators.

connected in parallel, especially if the filter capacitor is not close to the IC, to prevent oscillation. This capacitor needs to be located close to the IC. Finally, an output capacitor (typically 0.1 μF to 1.0 μF) is placed in parallel with the output to improve the transient response.

Examples of fixed three-terminal regulators are the 78XX series of regulators that are available with various output voltages and can supply up to 1 A of load current (with adequate heat sinking). The last two digits of the number stand for the output voltage; thus, the 7812 has a +12 V output. The negative output versions of the same regulator are numbered as the 79XX series; the 7912 has a −12 V output. The output voltage from these regulators is within 3% of the nominal value but will hold a nearly constant output despite changes in the input voltage or output load. A basic fixed +5 V regulated power supply with a 7805 regulator is shown in Figure 17–26.

▼ **FIGURE 17–26**

A basic +5.0 V regulated power supply.

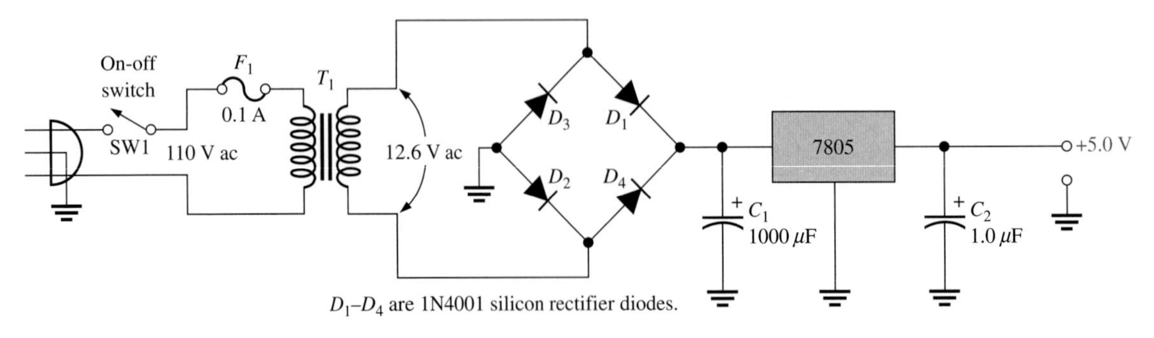

Another type of three-terminal regulator is an adjustable regulator. Figure 17–27 shows a power supply circuit with an adjustable output, controlled by the variable resistor, R_2. Note that R_2 is adjustable from zero to 1.0 kΩ. The LM317 regulator keeps a constant 1.25 V between the output and adjust terminals. This produces a constant current in R_1 of 1.25 V/240 Ω = 52 mA. Neglecting the very small current through the adjust terminal, the current in R_2 is the same as the current in R_1. The output is taken across both R_1 and R_2 and is found from the equation,

$$V_{out} = 1.25\text{ V}\left(\frac{R_1 + R_2}{R_1}\right)$$

Notice that the output voltage from the power supply is the regulator's 1.25 V multiplied by a ratio of resistances. For the case shown in Figure 17–27, when R_2 is set to the minimum (zero) resistance, the output is 1.25 V. When R_2 is set to the maximum, the output is nearly 6.5 V.

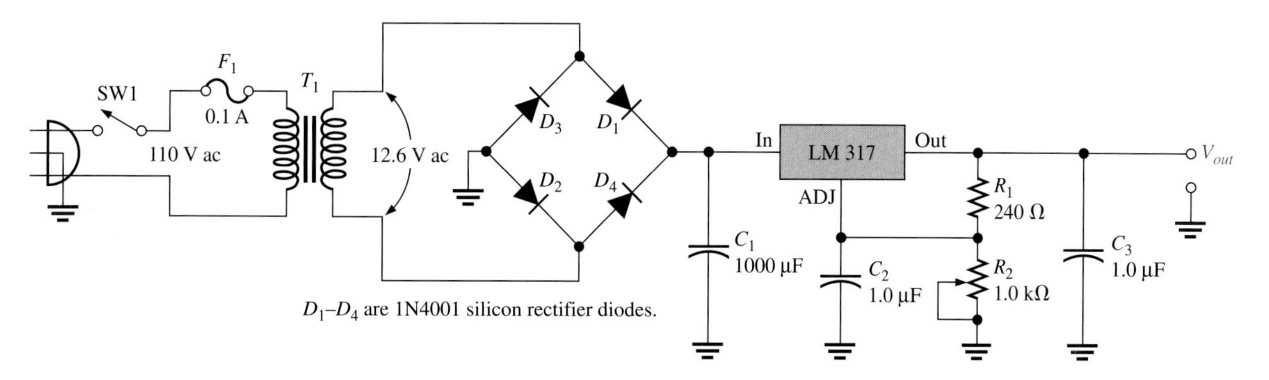

FIGURE 17–27

A basic power supply with a variable output voltage (from 1.25 V to 6.5 V).

Percent Regulation

The regulation expressed as a percentage is a figure of merit used to specify the performance of a voltage regulator. It can be in terms of input (line) regulation or load regulation. **Line regulation** specifies how much change occurs in the output voltage for a given change in the input voltage. It is typically defined as a ratio of a change in output voltage for a corresponding change in the input voltage expressed as a percentage.

$$\text{Line regulation} = \left(\frac{\Delta V_{\text{OUT}}}{\Delta V_{\text{IN}}}\right)100\%$$

Equation 17–9

Load regulation specifies how much change occurs in the output voltage over a certain range of load current values, usually from minimum current (no load, NL) to maximum current (full load, FL). It is normally expressed as a percentage and can be calculated with the following formula:

$$\text{Load regulation} = \left(\frac{V_{\text{NL}} - V_{\text{FL}}}{V_{\text{FL}}}\right)100\%$$

Equation 17–10

where V_{NL} is the output voltage with no load and V_{FL} is the output voltage with full (maximum) load.

EXAMPLE 17–6

Assume a certain 7805 regulator has a measured no-load output voltage of 5.185 V and a full-load output of 5.152 V. What is the load regulation expressed as a percentage?

Solution

$$\text{Load regulation} = \left(\frac{V_{NL} - V_{FL}}{V_{FL}}\right)100\% = \left(\frac{5.185\text{ V} - 5.152\text{ V}}{5.152\text{ V}}\right)100\% = \mathbf{0.64\%}$$

Related Problem If the no-load output voltage of a regulator is 24.8 V and the full-load output is 23.9 V, what is the load regulation expressed as a percentage?

The preceding discussion concentrated on the popular three-terminal regulators. Three-terminal regulators can be adapted to a number of specialized applications or requirements such as current sources or automatic shutdown, current limiting, and the like. For certain other applications (high current, high efficiency, high voltage), more complicated regulators are constructed from integrated circuits and discrete transistors. Chapter 21 discusses some of these applications in more detail.

SECTION 17–3 REVIEW

1. What causes the ripple voltage on the output of a capacitor-input filter?
2. The load resistance of a capacitor-filtered full-wave rectifier is reduced. What effect does this reduction have on the ripple voltage?
3. What advantages are offered by a three-terminal regulator?
4. What is the difference between input (line) regulation and load regulation?

17–4 DIODE LIMITING AND CLAMPING CIRCUITS

Diode circuits, called limiters or clippers, are sometimes used to clip off portions of signal voltages above or below certain levels. Another type of diode circuit, called a clamper, is used to restore a dc level to an electrical signal. Both of these diode circuits will be examined in this section.

After completing this section, you should be able to

- **Analyze the operation of diode limiters and clampers**
- Describe how both unbiased and biased limiters operate
- Discuss a common clamper application

Diode Limiters

Diode limiters **(clippers)** cut off voltage above or below specified levels. Figure 17–28(a) shows a diode circuit that limits or clips off the positive part of the input signal. As the input signal goes positive, the diode becomes forward-biased. Since the cathode is at ground potential (0 V), the anode cannot exceed 0.7 V. This analysis uses the diode barrier potential. Thus, the output voltage across the diode is clipped at +0.7 V when the input exceeds this value.

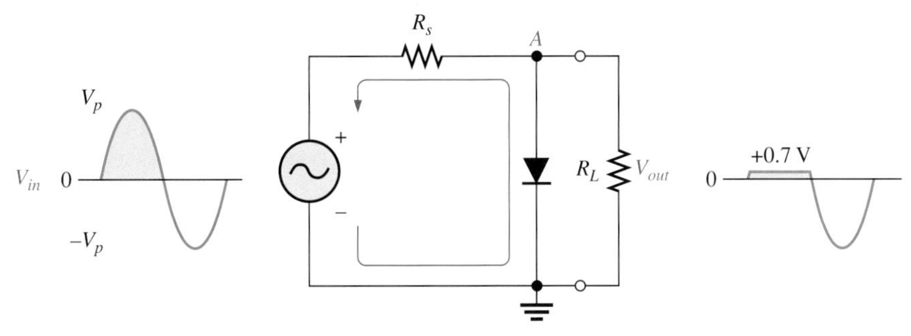

(a) Limiting of the positive alternation

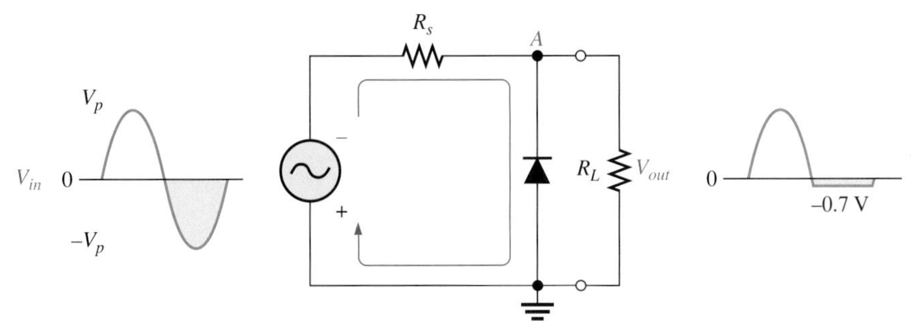

(b) Limiting of the negative alternation

FIGURE 17–28

Diode limiting (clipping) circuits.

When the input goes back below 0.7 V, the diode reverse-biases and appears as an open. The output voltage looks like the negative part of the input, but with a magnitude determined by the R_s and R_L voltage divider as follows:

$$V_{out} = \left(\frac{R_L}{R_s + R_L}\right)V_{in}$$

If R_s is small compared to R_L, then $V_{out} \cong V_{in}$. If the limiter circuit is unloaded, $V_{out} = V_{in}$.

Turn the diode around, as in Figure 17–28(b), and the negative part of the input is clipped off. When the diode is forward-biased during the negative part of the input, the output voltage across the diode is held at -0.7 V by the diode drop. When the input goes above -0.7 V, the diode is no longer forward-biased and a voltage appears across R_L proportional to the input.

EXAMPLE 17–7

What would you expect to see displayed on an oscilloscope connected as shown in Figure 17–29? Assume the time base on the scope is set to show one and one-half cycles.

FIGURE 17–29

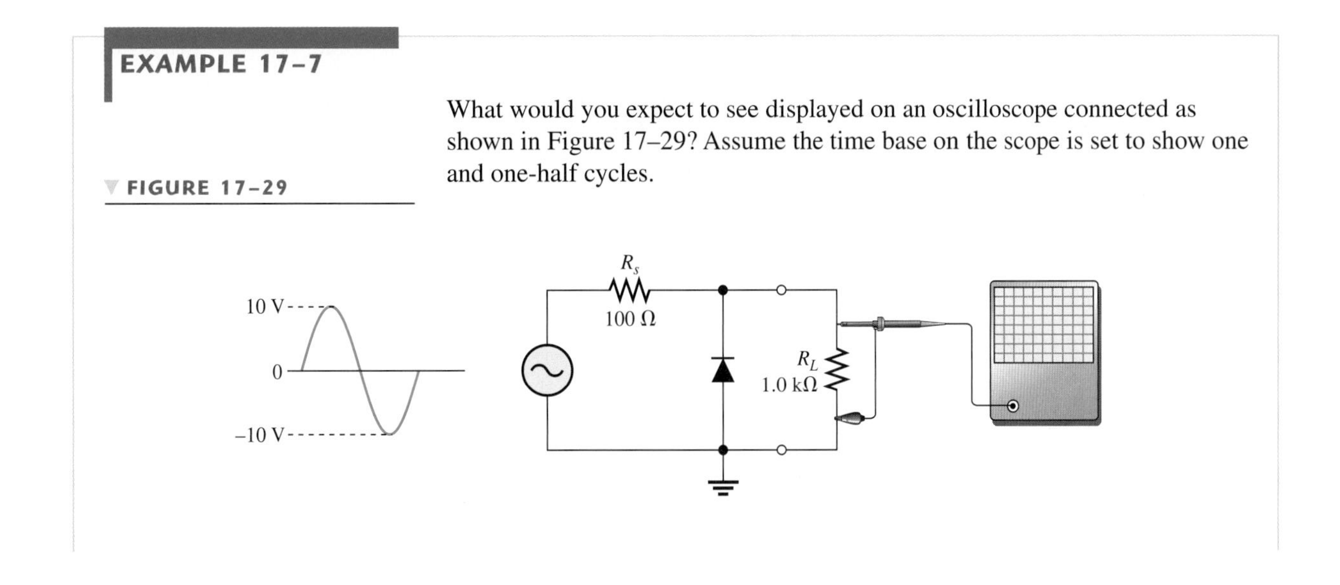

Solution The diode conducts when the input voltage goes below −0.7 V. Thus, you have a negative limiter with a peak output voltage calculated as

$$V_{p(out)} = \left(\frac{R_L}{R_s + R_L}\right)V_{p(in)} = \left(\frac{1.0\ \text{k}\Omega}{1.1\ \text{k}\Omega}\right)10\ \text{V} = 9.09\ \text{V}$$

The scope will display an output waveform as shown in Figure 17–30.

FIGURE 17–30

Waveforms for Figure 17–29.

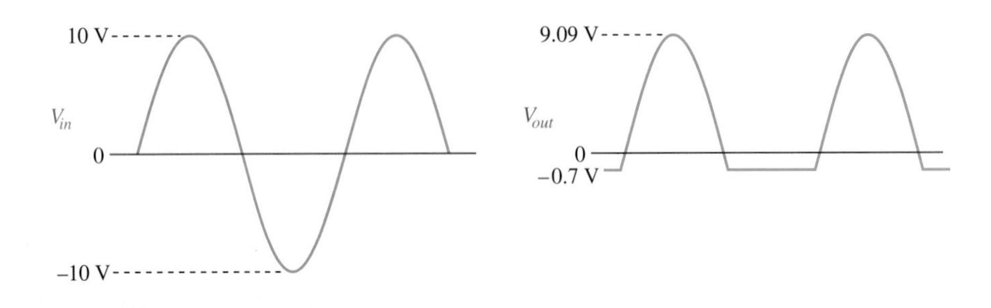

Related Problem Describe the output waveform for Figure 17–29 if the diode is germanium and R_L is changed to 680 Ω.

Open file E17-07 on your EWB/CircuitMaker CD-ROM. Display the voltage across R_L and compare to the waveform in Figure 17–30.

Adjustment of the Limiting Level To adjust the level at which a signal voltage is limited, add a bias voltage in series with the diode, as shown in Figure 17–31. The voltage at point *A* must equal $V_{\text{BIAS}} + 0.7$ V before the diode will conduct. Once the diode begins to conduct, the voltage at point *A* with respect to ground is limited to $V_{\text{BIAS}} + 0.7$ V so that the voltage above this level is clipped off, as shown in the figure.

FIGURE 17–31

Positively biased diode limiter.

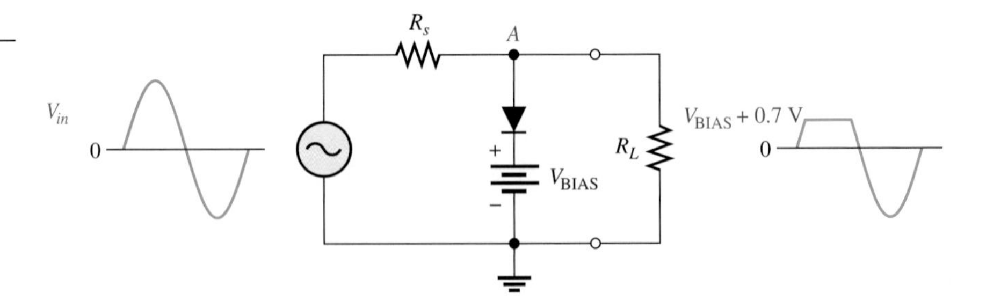

If the bias voltage is varied up or down, the limiting level changes correspondingly, as shown in Figure 17–32.

If it is necessary to limit voltage below a specified negative level, then the diode and bias battery must be connected as in Figure 17–33. In this case, the voltage at point *A* with respect to ground must go more negative than $-V_{\text{BIAS}} - 0.7$ V to forward-bias the diode and initiate limiting action, as shown.

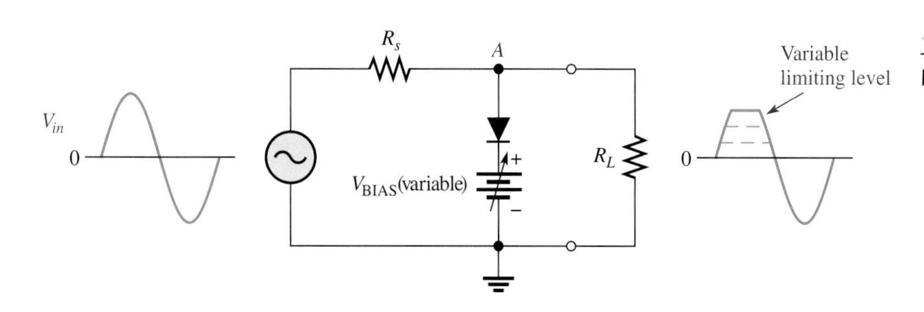

FIGURE 17–32

Positive limiter with variable bias.

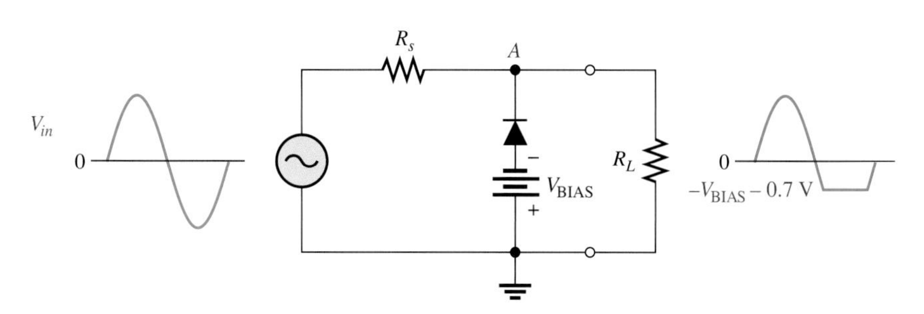

FIGURE 17–33

Negatively biased diode limiter.

EXAMPLE 17–8

Figure 17–34 shows an unloaded circuit combining a positive-biased limiter with a negative-biased limiter. Determine the output waveform.

FIGURE 17–34

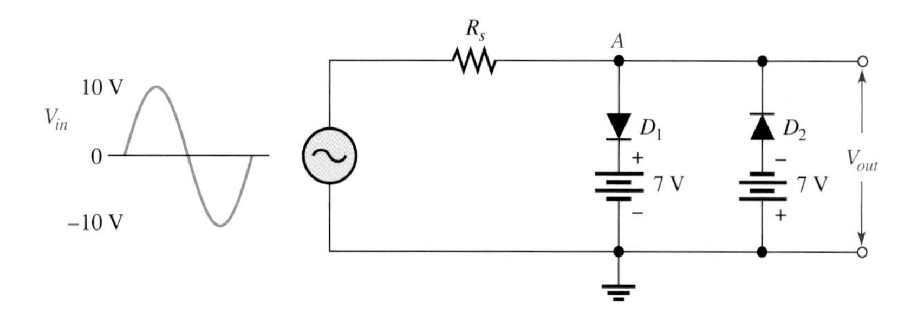

Solution When the voltage at point A with respect to ground reaches +7.7 V, diode D_1 conducts and limits the waveform at +7.7 V. Diode D_2 does not conduct until the voltage reaches −7.7 V. Therefore, positive voltages above +7.7 V and negative voltages below −7.7 V are clipped off. The resulting output waveform is shown in Figure 17–35.

FIGURE 17–35

Output waveform for Figure 17–34.

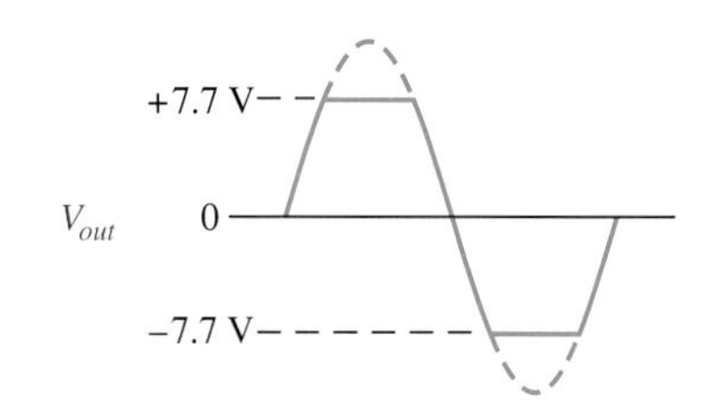

Related Problem Determine the output voltage in Figure 17–34 if both dc bias sources are 10 V and the input has a peak value of 20 V.

Open file E17-08 on your EWB/CircuitMaker CD-ROM. Display the output waveform and compare to the waveform in Figure 17–35.

Diode Clampers

A diode **clamper**, sometimes known as a *dc restorer,* adds a dc level to an ac signal. Figure 17–36 shows a diode clamper that inserts a positive dc level. To understand the operation of this circuit, start with the first negative half-cycle of the input voltage. When the input initially goes negative, the diode is forward-biased, allowing the capacitor to charge to near the peak of the input ($V_{p(in)} - 0.7$ V), as shown in Figure 17–36(a). Just past the negative peak, the diode becomes reverse-biased because the cathode is held near $V_{p(in)} - 0.7$ V by the charge on the capacitor.

FIGURE 17–36

Positive clamping operation.

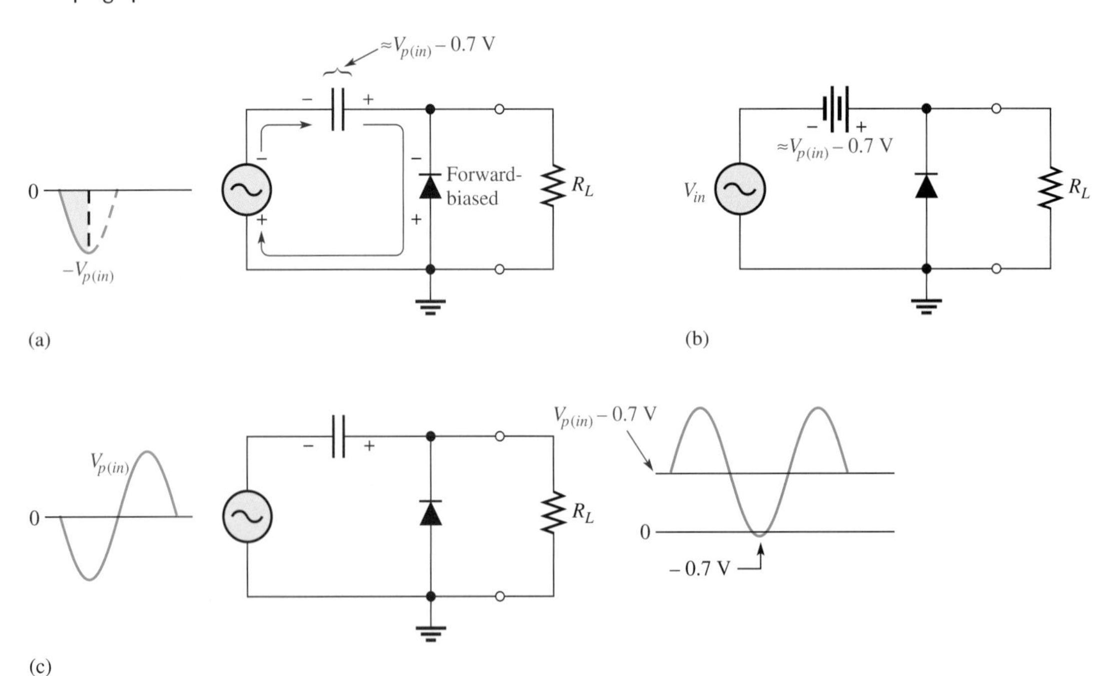

The capacitor can discharge only through the high resistance of R_L. Thus, from the peak of one negative half-cycle to the next, the capacitor discharges very little. The amount that is discharged, of course, depends on the value of R_L and the period of the input signal. For good clamping action, the *RC* time constant should be at least ten times the period of the input frequency.

The net effect of the clamping action is that the capacitor retains a charge approximately equal to the peak value of the input less the diode drop. The capacitor voltage acts essentially as a battery in series with the input signal, as shown in Figure 17–36(b). The dc voltage of the capacitor adds to the input voltage by superposition, as shown in Figure 17–36(c).

If the diode is turned around, a negative dc voltage is added to the input signal, as shown in Figure 17–37. If necessary, the diode can be biased to adjust the clamping level.

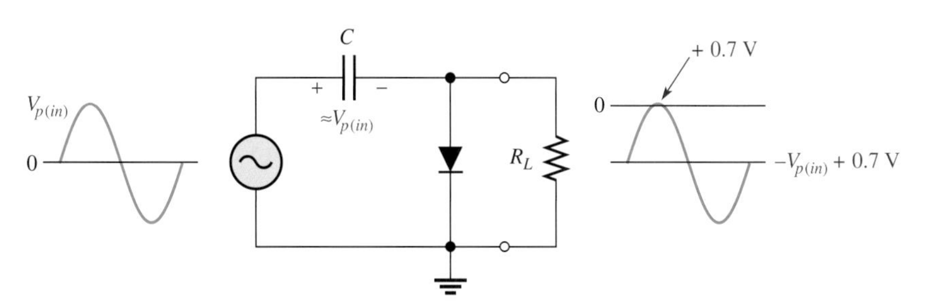

FIGURE 17–37

Negative clamping.

A Clamper Application A clamping circuit is often used in television receivers as a dc restorer. The incoming composite video signal is normally processed through capacitively coupled amplifiers that eliminate the dc component, thus losing the black and white reference levels and the blanking level. Before being applied to the picture tube, these reference levels must be restored. Figure 17–38 illustrates this process in a general way.

FIGURE 17–38

Clamping circuit (dc restorer) in a TV receiver.

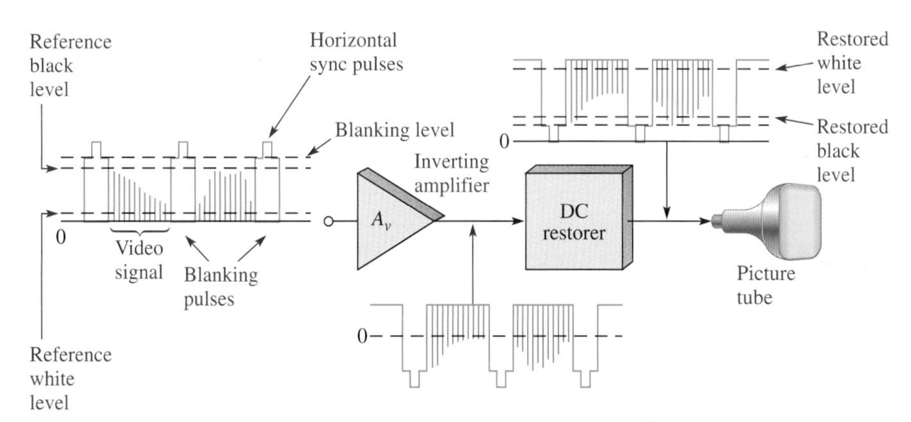

EXAMPLE 17–9

What is the output voltage that you would expect to observe across R_L in the clamping circuit of Figure 17–39? Assume that RC is large enough to prevent significant capacitor discharge.

FIGURE 17–39

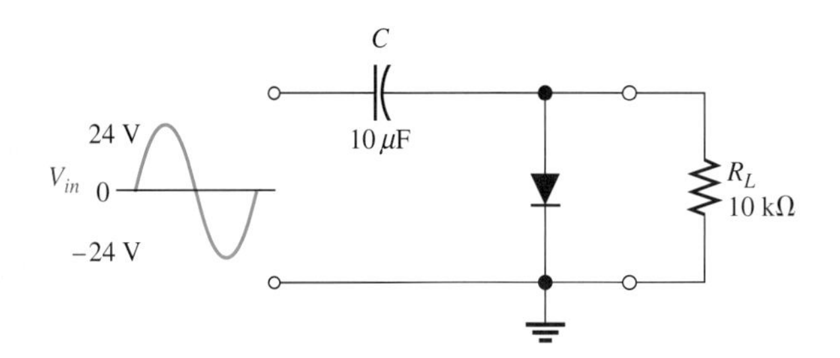

Solution Ideally, a negative dc value equal to the input peak less the diode drop is inserted by the clamping circuit.

$$V_{DC} = -(V_{p(in)} - 0.7\text{ V}) = -(24\text{ V} - 0.7\text{ V}) = \mathbf{-23.3\text{ V}}$$

Actually, the capacitor will discharge slightly between peaks, and, as a result, the output voltage will have an average value of slightly less negative than that calculated above.

The output waveform goes to +0.7 V, as shown in Figure 17–40.

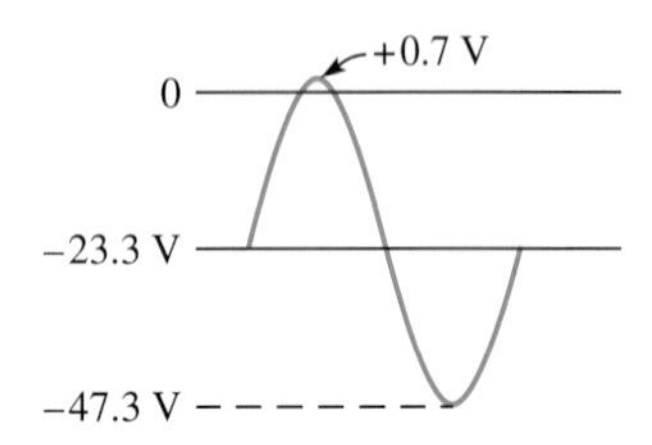

FIGURE 17–40

Output waveform across R_L for Figure 17–39.

Related Problem What is the output voltage that you would observe across R_L in Figure 17–39 if $C = 22\ \mu$F and $R_L = 18$ kΩ?

Open file E17-09 on your EWB/CircuitMaker CD-ROM. Measure the output voltage waveform across R_L and compare with the waveform in Figure 17–40.

SECTION 17–4 REVIEW

1. Discuss how diode limiters and diode clampers differ in terms of their function.
2. What is the difference between a positive limiter and a negative limiter?
3. What is the maximum voltage across an unbiased positive diode limiter during the positive alternation of the input voltage?
4. To limit the output of a positive limiter to 5 V when a 10 V peak input is applied, what value must the bias voltage be?
5. What component in a clamper circuit effectively acts as a battery?

17–5 ZENER DIODES

A major application for the zener diode is to provide an output reference voltage that is stable despite changes in input voltage. Reference voltages are used in power supplies, voltmeters, and other instruments. In this section, you will see how the zener diode maintains a nearly constant dc voltage under the proper operating conditions. You will learn the conditions and limitations for properly using the zener diode and the factors that affect its performance.

After completing this section, you should be able to

- **Explain the characteristics and applications of a zener diode**

- Discuss zener breakdown
- Describe a zener equivalent circuit
- Discuss zener voltage regulation

Figure 17–41 shows the schematic symbol for a zener diode. The zener diode is a silicon *pn* junction device that differs from the rectifier diode in that it is designed for operation in the reverse breakdown region. The breakdown voltage of a zener diode is set by carefully controlling the doping level during the manufacturing process. From the discussion of the diode characteristic curve in the last chapter, recall that when a diode reaches reverse breakdown, its voltage remains almost constant even though the current may change drastically. This volt-ampere (*V-I*) characteristic is shown again in Figure 17–42 with normal operating regions for rectifier diodes and for zener diodes shown as shaded areas. If a zener diode is forward-biased, it operates the same as a rectifier diode.

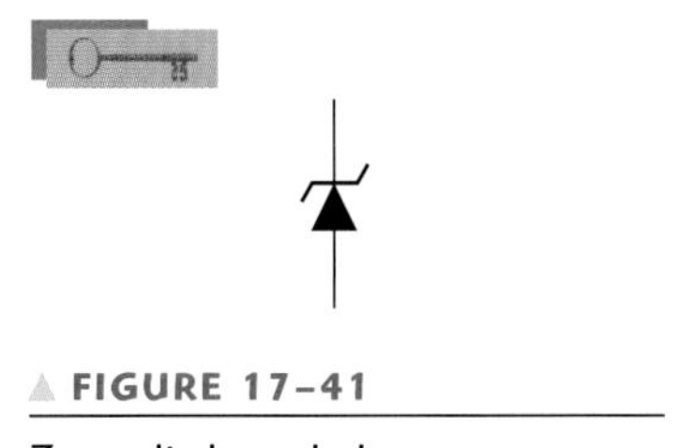

▲ **FIGURE 17–41**

Zener diode symbol.

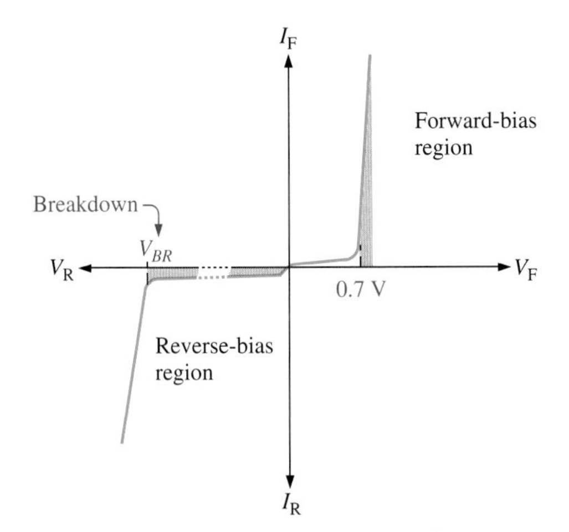

(a) The normal operating regions for a rectifier diode are shown as shaded areas.

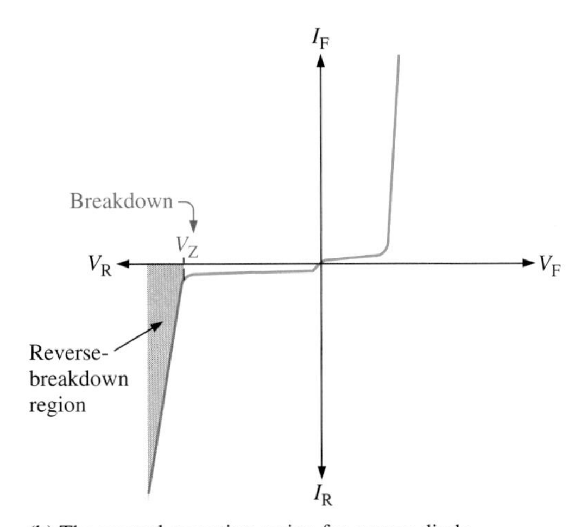

(b) The normal operating region for a zener diode is shaded.

▲ **FIGURE 17–42**

General diode *V-I* characteristic.

Zener Breakdown

Two types of reverse breakdown in a zener diode are *avalanche* and *zener.* The avalanche breakdown also occurs in rectifier diodes at a sufficiently high reverse voltage. **Zener breakdown** occurs in a zener diode at low reverse voltages. A zener diode is heavily doped to reduce the breakdown voltage, causing a vary narrow depletion region. As a result, an intense electric field exists within the depletion region. Near the breakdown voltage (V_Z), the field is intense enough to pull electrons from their valence bands and create current.

Zener diodes with breakdown voltages of less than approximately 5 V operate predominantly in zener breakdown. Those with breakdown voltages greater than approximately 5 V operate predominantly in avalanche breakdown. Both types, however, are called *zener diodes.* Zeners with breakdown voltages of 1.8 V to 200 V are commercially available.

Breakdown Characteristics Figure 17–43 shows the reverse portion of the characteristic curve of a zener diode. Notice that as the reverse voltage (V_R) is increased, the reverse current (I_R) remains extremely small up to the "knee" of the curve. At this point, the breakdown effect begins; the zener impedance (Z_Z) begins to decrease as the current (I_Z) increases rapidly. From the bottom of the knee,

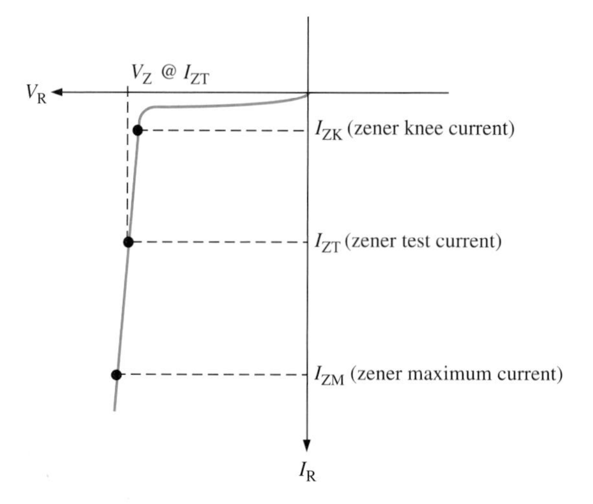

FIGURE 17–43

Reverse characteristic of a zener diode. V_Z is usually specified at the zener test current, I_{ZT}, and is designated V_{ZT}.

the breakdown voltage (V_Z) remains essentially constant. This regulating ability is the key feature of the zener diode: *A zener diode maintains an essentially constant voltage across its terminals over a specified range of reverse current values.*

A minimum value of reverse current, I_{ZK}, must be maintained in order to keep the diode in regulation. You can see on the curve that when the reverse current is reduced below the knee of the curve, the voltage changes drastically and regulation is lost. Also, there is a maximum current, I_{ZM}, above which the diode may be damaged.

Thus, basically, the zener diode maintains a nearly constant voltage across its terminals for values of reverse current ranging from I_{ZK} to I_{ZM}. A nominal zener test voltage, V_{ZT}, is usually specified on a data sheet at a value of reverse current called the *zener test current,* I_{ZT}.

Zener Equivalent Circuit

Figure 17–44(a) shows the ideal model of a zener diode in reverse breakdown. It acts simply as a battery having a value equal to the zener voltage. Figure 17–44(b) represents the practical equivalent of a zener, where the zener impedance (Z_Z) is included. The zener impedance is actually an ac resistance because it is dependent

FIGURE 17–44

Zener equivalent circuits.

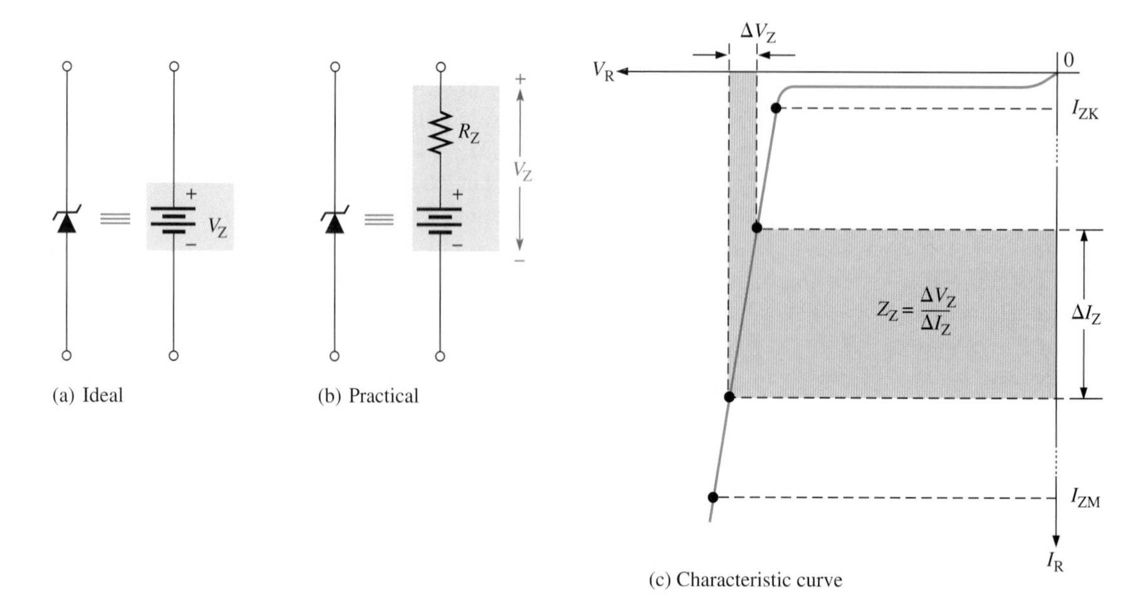

on the ratio of a change in voltage to a change in current and can be different for different portions of the characteristic curve. Since the voltage curve is not ideally vertical, a change in zener current (ΔI_Z) produces a small change in zener voltage (ΔV_Z), as illustrated in Figure 17–44(c).

The ratio of ΔV_Z to ΔI_Z is the zener impedance, expressed as follows:

$$Z_Z = \frac{\Delta V_Z}{\Delta I_Z}$$

Equation 17–11

Normally, Z_Z is specified at I_{ZT}, the zener test current. In most cases, this value of Z_Z is approximately constant over the full range of reverse-current values.

EXAMPLE 17–10

A certain zener diode exhibits a 50 mV change in V_Z for a 2 mA change in I_Z. What is the zener impedance?

Solution

$$Z_Z = \frac{\Delta V_Z}{\Delta I_Z} = \frac{50 \text{ mV}}{2 \text{ mA}} = \mathbf{25\ \Omega}$$

Related Problem Calculate the zener impedance if the zener voltage changes 100 mV for a 20 mA change in zener current.

EXAMPLE 17–11

A certain zener diode has an impedance of 5 Ω. The data sheet gives V_{ZT} = 6.8 V at I_{ZT} = 20 mA, I_{ZK} = 1 mA, and I_{ZM} = 50 mA. What is the voltage across its terminals when the current is 30 mA? What is the voltage when I = 10 mA?

Solution Figure 17–45 represents the diode.

FIGURE 17–45

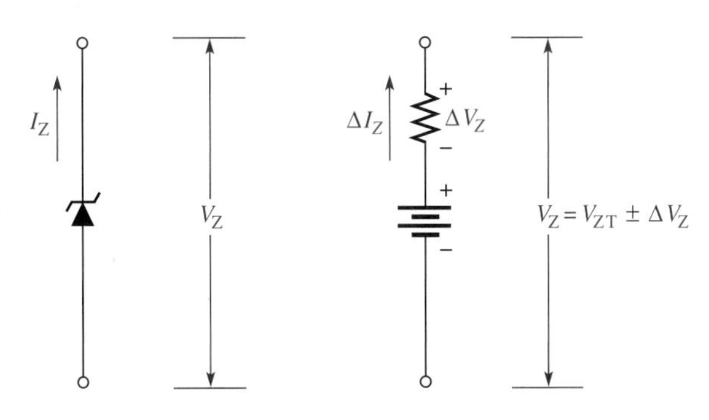

The 30 mA current is a 10 mA increase (ΔI_Z) above I_{ZT} = 20 mA.

$$\Delta I_Z = +10 \text{ mA}$$

$$\Delta V_Z = \Delta I_Z Z_Z = (10 \text{ mA})(5\ \Omega) = +50 \text{ mV}$$

The change in voltage due to the increase in current above the I_{ZT} value causes the zener terminal voltage to increase. The zener voltage for I_Z = 30 mA is

$$V_Z = 6.8 \text{ V} + \Delta V_Z = 6.8 \text{ V} + 50 \text{ mV} = \mathbf{6.85\ V}$$

The 10 mA current is a 10 mA decrease below $I_{ZT} = 20$ mA.

$$\Delta I_Z = -10 \text{ mA}$$

$$\Delta V_Z = \Delta I_Z Z_Z = (-10 \text{ mA})(5\ \Omega) = -50 \text{ mV}$$

The change in voltage due to the decrease in current below I_{ZT} causes the zener terminal voltage to decrease. The zener voltage for $I_Z = 10$ mA is

$$V_Z = 6.8 \text{ V} - \Delta V_Z = 6.8 \text{ V} - 50 \text{ mV} = \mathbf{6.75\ V}$$

Related Problem Repeat the analysis for a current of 20 mA and for a current of 80 mA using a diode with $V_{ZT} = 12$ V at $I_{ZT} = 50$ mA, $I_{ZK} = 0.5$ mA, $I_{ZM} = 100$ mA, and $Z_Z = 20\ \Omega$.

Zener Voltage Regulation

Zener diodes can be used for voltage regulation in noncritical low-current applications and for providing a known reference voltage. Figure 17–46 illustrates how a zener diode can be used to regulate a varying dc voltage to keep it at a relatively constant level. As you learned earlier, this process is called *input* or *line regulation* (see Section 17–3).

FIGURE 17–46

Zener regulation of a varying input voltage.

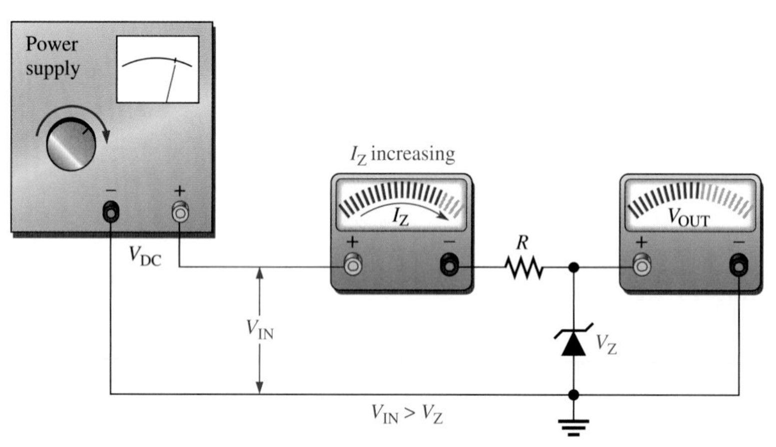

(a) As the input voltage increases, the output voltage remains constant ($I_{ZK} < I_Z < I_{ZM}$).

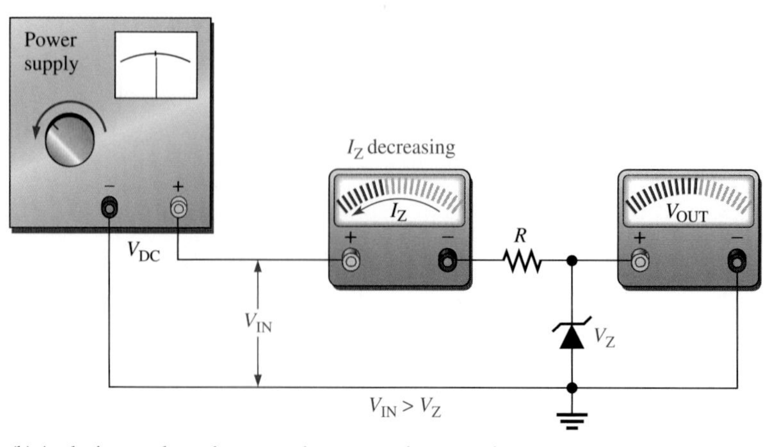

(b) As the input voltage decreases, the output voltage remains constant ($I_{ZK} < I_Z < I_{ZM}$).

As the input voltage varies (within limits), the zener diode maintains an essentially constant voltage across the output terminals. However, as V_{IN} changes, I_Z will change proportionally, and therefore the limitations on the input variation are set by the minimum and maximum current values within which the zener can operate and on the condition that $V_{IN} > V_Z$. Resistor R is the series current-limiting resistor.

For example, suppose that the zener diode in Figure 17–47 can maintain regulation over a range of current values from 4 mA to 40 mA. For the minimum current, the voltage across the 1.0 kΩ resistor is

$$V_R = (4 \text{ mA})(1.0 \text{ k}\Omega) = 4 \text{ V}$$

Since

$$V_R = V_{IN} - V_Z$$

then

$$V_{IN} = V_R + V_Z = 4 \text{ V} + 10 \text{ V} = 14 \text{ V}$$

For the maximum current, the voltage across the 1.0 kΩ resistor is

$$V_R = (40 \text{ mA})(1.0 \text{ k}\Omega) = 40 \text{ V}$$

Therefore,

$$V_{IN} = 40 \text{ V} + 10 \text{ V} = 50 \text{ V}$$

As you can see, this zener diode can regulate an input voltage from 14 V to 50 V and maintain approximately a 10 V output. The output will vary slightly because of the zener impedance.

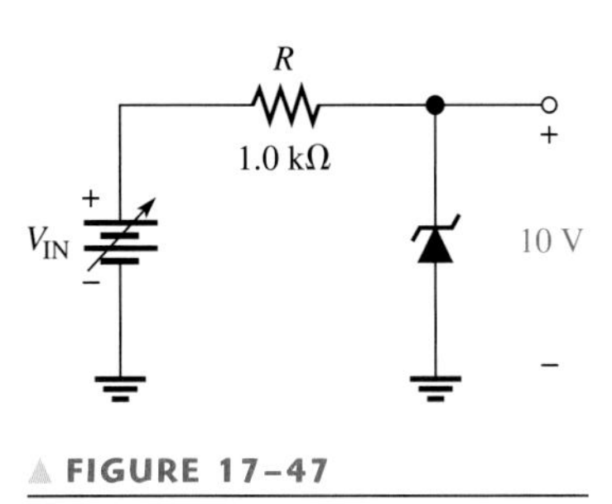

FIGURE 17–47

EXAMPLE 17–12

Determine the minimum and the maximum input voltages that can be regulated by the zener diode in Figure 17–48. Assume that $I_{ZK} = 1$ mA, $I_{ZM} = 15$ mA, $V_{ZT} = 5.1$ V at $I_{ZT} = 7$ mA, and $Z_Z = 10\ \Omega$.

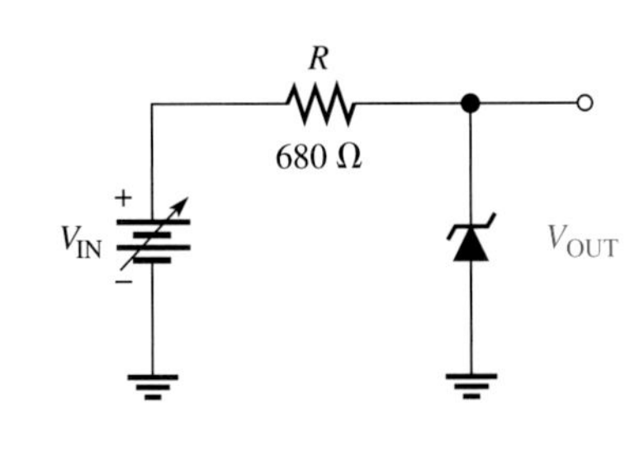

FIGURE 17–48

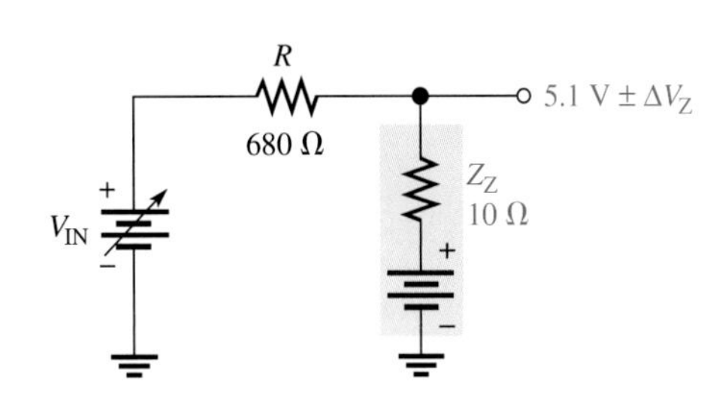

FIGURE 17–49

Equivalent of the circuit in Figure 17–48.

Solution The equivalent circuit is shown in Figure 17–49.

At $I_{ZK} = 1$ mA, the output voltage is

$$V_{OUT} = V_{ZT} - \Delta V_Z = 5.1 \text{ V} - \Delta I_Z Z_Z = 5.1 \text{ V} - (I_{ZT} - I_{ZK})Z_Z$$
$$= 5.1 \text{ V} - (7 \text{ mA})(10\ \Omega) = 5.1 \text{ V} - 0.07 \text{ V} = 5.03 \text{ V}$$

Therefore,

$$V_{IN(min)} = I_{ZK}R + V_{OUT} = (1\text{ mA})(680\ \Omega) + 5.03\text{ V} = \mathbf{5.71\text{ V}}$$

At $I_{ZM} = 15$ mA, the output voltage is

$$V_{OUT} = 5.1\text{ V} + \Delta V_Z = 5.1\text{ V} + (7\text{ mA})(10\ \Omega) = 5.1\text{ V} + 0.07\text{ V} = 5.17\text{ V}$$

Therefore,

$$V_{IN(max)} = I_{ZM}R + V_{OUT} = (15\text{ mA})(680\ \Omega) + 5.17\text{ V} = \mathbf{15.4\text{ V}}$$

Related Problem Determine the minimum and maximum input voltages that can be regulated by the zener diode in Figure 17–48. $I_{ZK} = 0.8$ mA, $I_{ZM} = 60$ mA, $V_{ZT} = 6.8$ V at $I_{ZT} = 30$ mA, and $Z_Z = 18\ \Omega$.

Zener Regulation with a Varying Load Figure 17–50 shows a zener regulator with a variable load resistor across the terminals. The zener diode maintains a constant voltage across R_L as long as the zener current is greater than I_{ZK} and less than I_{ZM}. This process is called *load regulation* and was defined in Section 17–3.

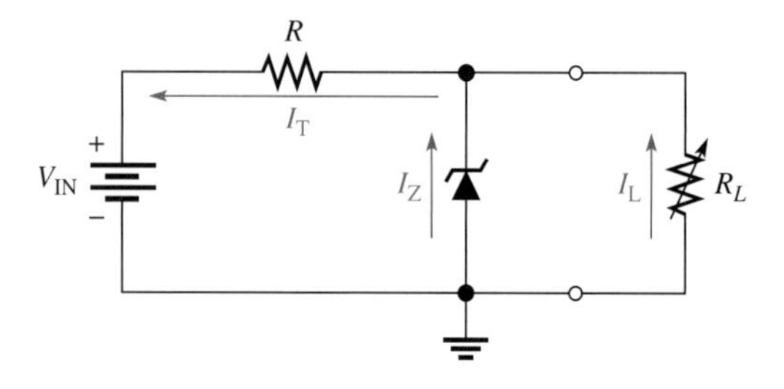

FIGURE 17–50

Zener regulation with a variable load.

From No Load to Full Load When the output terminals are open ($R_L = \infty$), the load current is zero and all of the current is through the zener. When a load resistor is connected, part of the total current is through the zener and part through R_L.

As R_L is decreased, I_L goes up and I_Z goes down. The zener diode continues to regulate until I_Z reaches its minimum value, I_{ZK}. At this point the load current is maximum. The following example illustrates.

EXAMPLE 17–13

Determine the minimum and the maximum load currents for which the zener diode in Figure 17–51 will maintain regulation. What is the minimum R_L that can be used? $V_Z = 12$ V, $I_{ZK} = 3$ mA, and $I_{ZM} = 90$ mA. Assume that $Z_Z = 0\ \Omega$ and V_Z remains a constant 12 V over the range of current values for simplicity.

FIGURE 17–51

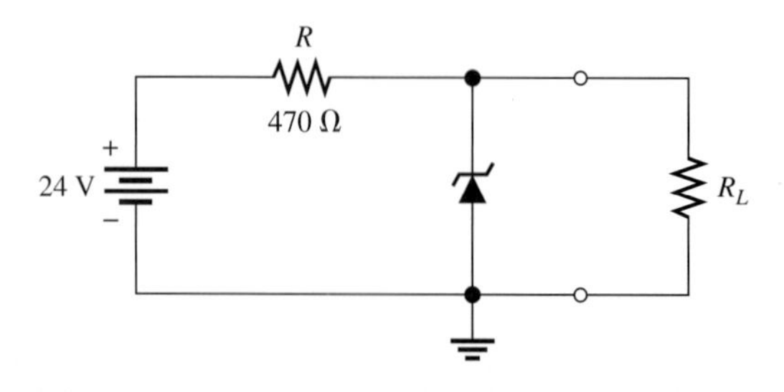

Solution When $I_L = 0$ A, I_Z is maximum and equal to the total circuit current I_T.

$$I_Z = \frac{V_{IN} - V_Z}{R} = \frac{24\text{ V} - 12\text{ V}}{470\ \Omega} = 25.5\text{ mA}$$

Since this is much less than I_{ZM}, 0 A is an acceptable minimum for I_L. That is,

$$I_{L(min)} = \mathbf{0\ A}$$

The maximum value of I_L occurs when I_Z is minimum (equal to I_{ZK}), so solve for $I_{L(max)}$ as follows:

$$I_{L(max)} = I_T - I_{Z(min)} = 25.5\text{ mA} - 3\text{ mA} = \mathbf{22.5\ mA}$$

The minimum value of R_L is

$$R_{L(min)} = \frac{V_Z}{I_{L(max)}} = \frac{12\text{ V}}{22.5\text{ mA}} = \mathbf{533\ \Omega}$$

Related Problem Find the minimum and maximum load currents for which the circuit in Figure 17–51 will maintain regulation. Determine the minimum R_L that can be used. $V_Z = 3.3$ V (constant), $I_{ZK} = 2$ mA, $I_{ZM} = 75$ mA, and $Z_Z = 0\ \Omega$.

Percent Regulation Recall that the voltage regulation expressed as a percentage is a figure of merit used to specify the performance of a voltage regulator. It can be in terms of input (line) regulation or load regulation. The *line regulation* specifies how much change occurs in the output voltage for a given change in input voltage. The *load regulation* specifies how much change occurs in the output voltage over a certain range of load current values, usually from minimum (zero) current (no load, NL) to maximum current (full load, FL).

EXAMPLE 17–14

A certain regulator has a no-load output voltage of 6 V and a full-load voltage of 5.82 V. Use Equation 17–10 to determine the load regulation expressed as a percentage.

Solution

$$\text{Load regulation} = \left(\frac{V_{NL} - V_{FL}}{V_{FL}}\right)100\% = \left(\frac{6\text{ V} - 5.82\text{ V}}{5.82\text{ V}}\right)100\% = \mathbf{3.09\%}$$

Related Problem

(a) If the no-load output voltage of a regulator is 24.8 V and the full-load output is 23.9 V, what is the load regulation expressed as a percentage?

(b) If the output voltage changes 0.035 V for a 2 V change in the input voltage, what is the line regulation expressed as a percentage? (Refer to Equation 17–9.)

SECTION 17–5 REVIEW

1. Zener diodes are normally operated in the breakdown region. (True or False)
2. A certain 10 V zener diode has a resistance of 8 Ω at 30 mA. What is the terminal voltage?

3. In a zener diode regulator, for what value of load resistance is the zener current a maximum?
4. A zener regulator has an output voltage of 12 V with no load and 11.9 V with full load. What is the load regulation expressed as a percentage?

17–6 VARACTOR DIODES

Varactor diodes are also known as variable-capacitance diodes because the junction capacitance varies with the amount of reverse-bias voltage. Varactors are specifically designed to take advantage of this variable-capacitance characteristic. The capacitance can be changed by changing the reverse voltage. These devices are commonly used in electronic tuning circuits used in communications systems.

After completing this section, you should be able to

- **Describe the basic operation of a varactor diode**
- Explain how a varactor produces a variable capacitance characteristic
- Discuss some varactor applications

A **varactor** is basically a reverse-biased *pn* junction that utilizes the inherent capacitance of the depletion region. The depletion region, created by the reverse bias, acts as a capacitor dielectric because of its nonconductive characteristics. The *p* and *n* regions are conductive and act as the capacitor plates, as illustrated in Figure 17–52.

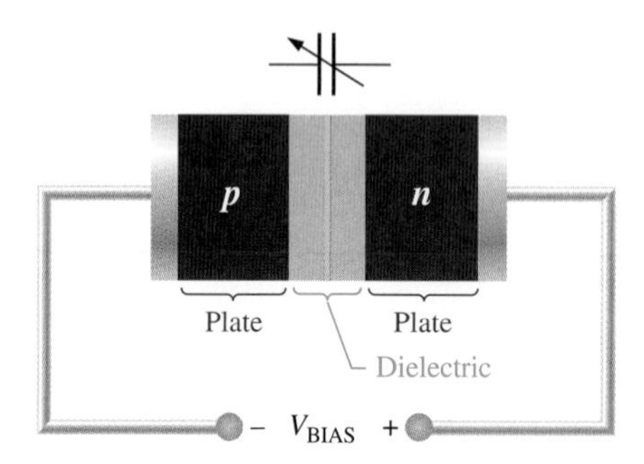

FIGURE 17–52

The reverse-biased varactor diode acts as a variable capacitor.

When the reverse-bias voltage increases, the depletion region widens, effectively increasing the dielectric thickness (d) and thus decreasing the capacitance. When the reverse-bias voltage decreases, the depletion region narrows, thus increasing the capacitance. This action is shown in Figure 17–53(a) and 17–53(b). A general curve of capacitance versus voltage is shown in Figure 17–53(c).

Recall that capacitance is determined by the plate area (A), dielectric constant (ϵ), and dielectric thickness (d), as expressed in the following formula:

Equation 17–12

$$C = \frac{A\epsilon}{d}$$

In a varactor diode, the capacitance parameters are controlled by the method of doping in the depletion region and the size and geometry of the diode's construc-

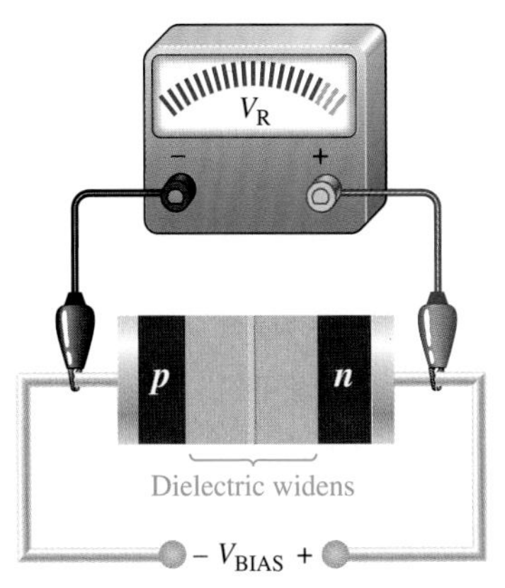

(a) Greater reverse bias, less capacitance

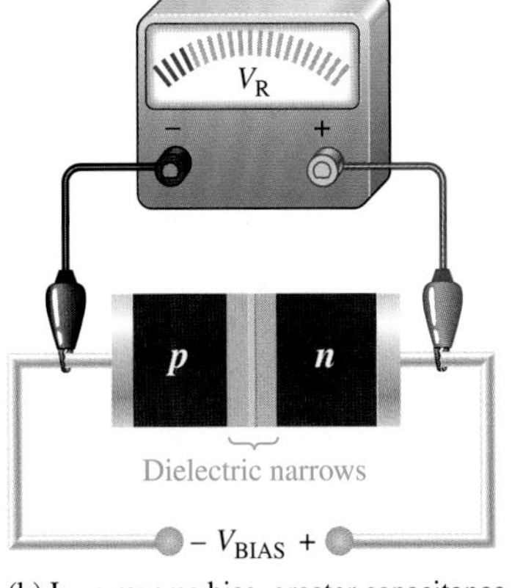

(b) Less reverse bias, greater capacitance

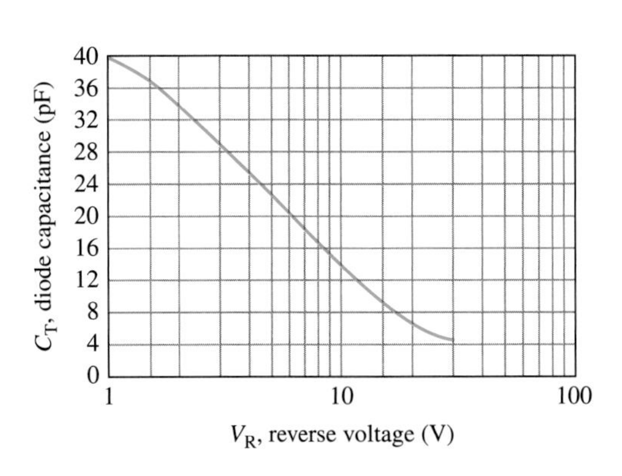

(c) Graph of diode capacitance versus reverse voltage

▲ **FIGURE 17–53**

Varactor diode capacitance varies with reverse voltage.

tion. Varactor capacitances typically range from a few picofarads to a few hundred picofarads.

Figure 17–54(a) shows a common symbol for a varactor, and part (b) shows a simplified equivalent circuit. R_S is the reverse series resistance, and C_V is the variable capacitance.

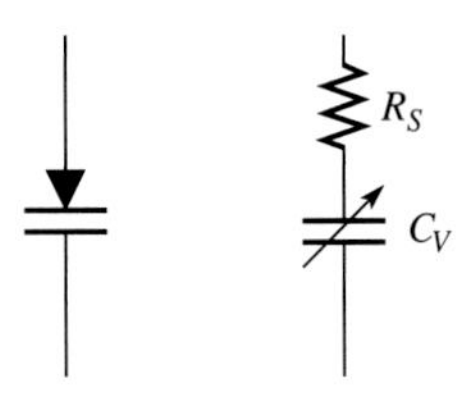

(a) Symbol (b) Equivalent circuit

▲ **FIGURE 17–54**

Varactor diode.

Applications

A major application of varactors is in tuning circuits. For example, electronic tuners in TV and other commercial receivers utilize varactors as one of their elements.

When used in a resonant circuit, the varactor acts as a variable capacitor, thus allowing the resonant frequency to be adjusted by a variable voltage level, as illustrated in Figure 17–55 where two varactor diodes provide the total variable capacitance in a parallel resonant (tank) circuit. A variable dc voltage (V_C) controls the reverse bias and therefore the capacitance of the diodes.

Recall that the resonant frequency of a tank circuit is

$$f_r = \frac{1}{2\pi\sqrt{LC}}$$

This approximation is valid for $Q > 10$.

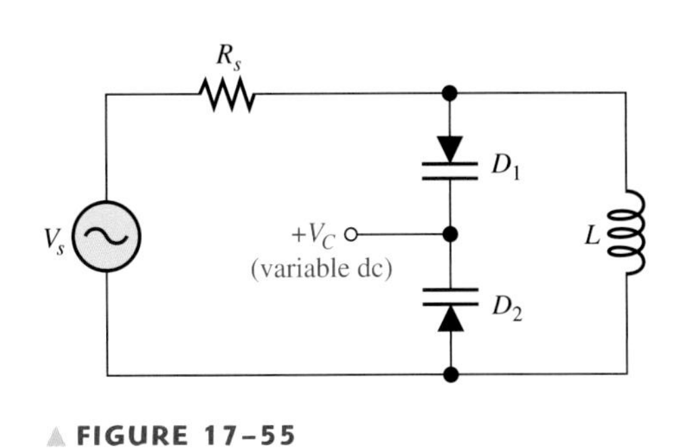

▲ **FIGURE 17–55**

Varactors in a resonant circuit.

EXAMPLE 17–15

The capacitance of a certain varactor can be varied from 5 pF to 50 pF. The diode is used in a tuned circuit similar to that shown in Figure 17–55. Determine the tuning range ($f_{r(min)}$ and $f_{r(max)}$) for the circuit if L = 10 mH.

Solution The circuit is shown in Figure 17–56. Notice that the varactor capacitances are in series.

▶ **FIGURE 17–56**

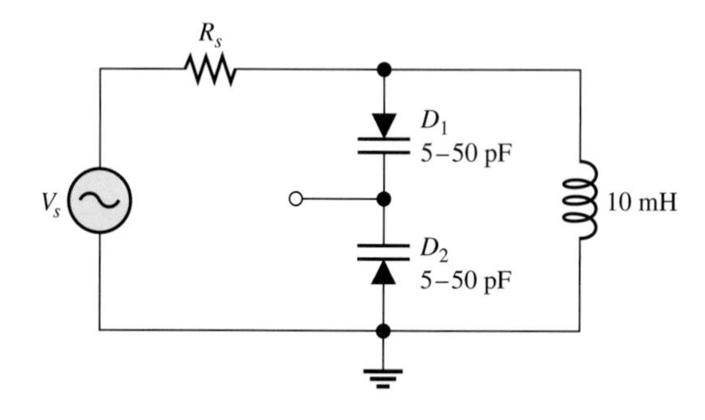

The maximum total capacitance is

$$C_{tot(max)} = \frac{C_{D1(max)}C_{D2(max)}}{C_{D1(max)} + C_{D2(max)}} = \frac{(50\text{ pF})(50\text{ pF})}{100\text{ pF}} = 25\text{ pF}$$

The minimum resonant frequency, therefore, is

$$f_{r(min)} = \frac{1}{2\pi\sqrt{LC_{tot(max)}}} = \frac{1}{2\pi\sqrt{(10\text{ mH})(25\text{ pF})}} = \mathbf{318\text{ kHz}}$$

The minimum total capacitance is

$$C_{tot(min)} = \frac{C_{D1(min)}C_{D2(min)}}{C_{D2(min)} + C_{D2(min)}} = \frac{(5\text{ pF})(5\text{ pF})}{10\text{ pF}} = 2.5\text{ pF}$$

The maximum resonant frequency, therefore, is

$$f_{r(max)} = \frac{1}{2\pi\sqrt{LC_{tot(min)}}} = \frac{1}{2\pi\sqrt{(10\text{ mH})(2.5\text{ pF})}} = \mathbf{1\text{ MHz}}$$

Related Problem Determine the tuning range for Figure 17–58 if L = 100 mH.

SECTION 17–6 REVIEW

1. What is the purpose of a varactor diode?
2. Based on the general curve in Figure 17–53(c), what happens to the diode capacitance when the reverse voltage is increased?

17–7 LEDs AND PHOTODIODES

In this section, two types of optoelectronic devices—the light-emitting diode (LED) and the photodiode—are introduced. As the name implies, the LED is a light emitter. The photodiode, on the other hand, is a light detector. We will examine the characteristics of both devices, and you will see an example of their use in the application assignment.

After completing this section, you should be able to

- **Discuss the operation and application of LEDs and photodiodes**
- Describe electroluminescence
- List the semiconductive materials used in LEDs
- Explain how the photodiode can be used as a variable resistance device

The Light-Emitting Diode (LED)

The basic operation of a LED (light-emitting diode) is as follows: When the device is forward-biased, electrons cross the *pn* junction from the *n*-type material and recombine with holes in the *p*-type material. Recall that these free electrons are in the conduction band and at a higher energy level than the holes in the valence band. When recombination takes place, the recombining electrons release energy in the form of heat and light. A large exposed surface area on one layer of the semiconductor permits the photons to be emitted as visible light. Figure 17–57 illustrates this process, called *electroluminescence.*

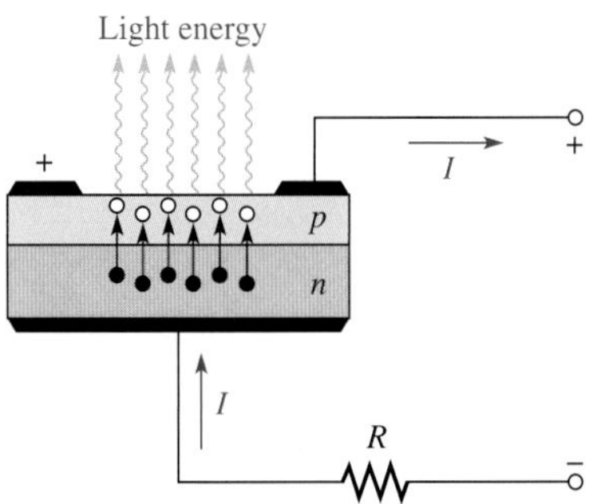

FIGURE 17–57

Electroluminescence in an LED.

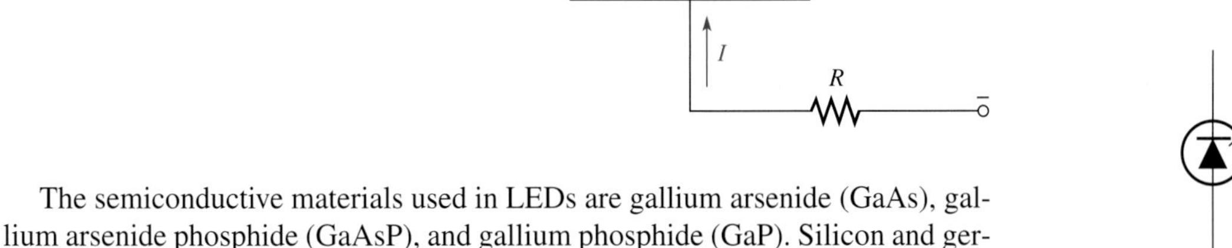

FIGURE 17–58

Symbol for an LED.

The semiconductive materials used in LEDs are gallium arsenide (GaAs), gallium arsenide phosphide (GaAsP), and gallium phosphide (GaP). Silicon and germanium are not used because they are essentially heat-producing materials and are very poor at producing light. GaAs LEDs emit infrared (IR) radiation, GaAsP produces either red or yellow visible light, and GaP emits red or green visible light. The symbol for an LED is shown in Figure 17–58.

The LED emits light in response to a sufficient forward current, as shown in Figure 17–59(a). The amount of light output is directly proportional to the forward current, as indicated in Figure 17–59(b). Typical LEDs are shown in part (c).

FIGURE 17–59

Light-emitting diodes (LEDs).

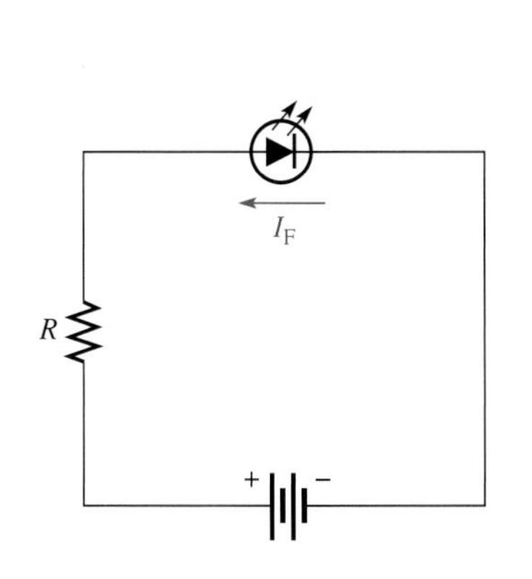

(a) Forward-biased operation

Radiated light (mW)

0

I_F (mA)

(b) Typical light output versus forward current

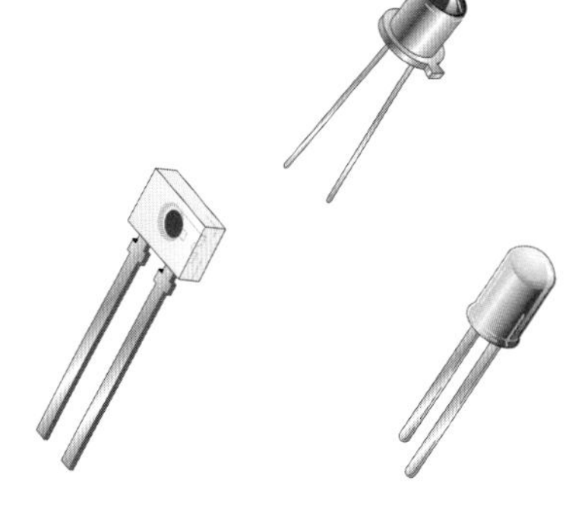

(c) Typical LEDs

Applications LEDs are commonly used for indicator lamps and readout displays on a wide variety of instruments, ranging from consumer appliances to scientific apparatus. A common type of display device using LEDs is a 7-segment display. Combinations of the segments form the ten decimal digits, as illustrated in Figure 17–60. Each segment in the display is an LED. By forward-biasing selected combinations of segments, any decimal digit and a decimal point can be formed. Two types of LED circuit arrangements are the common anode and the common cathode, as shown. Also, IR-emitting diodes are used in optical coupling applications, often in conjunction with fiber optics, in TV remote control, and other applications.

FIGURE 17–60

The 7-segment LED display.

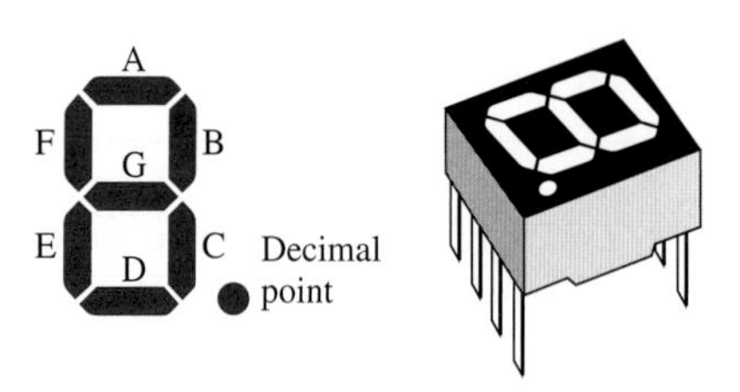

(a) LED segment arrangement and typical device

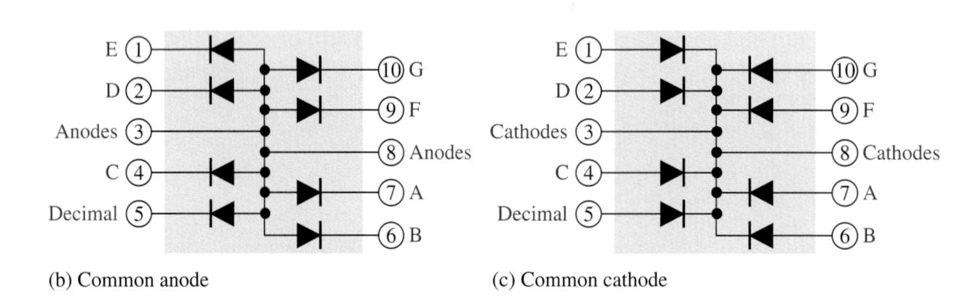

(b) Common anode (c) Common cathode

The Photodiode

The **photodiode** is a *pn* junction device that operates in reverse bias, as shown in Figure 17–61(a). Note the schematic symbol for the photodiode. The photodiode has a small transparent window that allows light to strike the *pn* junction. Typical photodiodes are shown in Figure 17–61(b), and an alternate symbol is shown in part (c).

FIGURE 17–61

Photodiode.

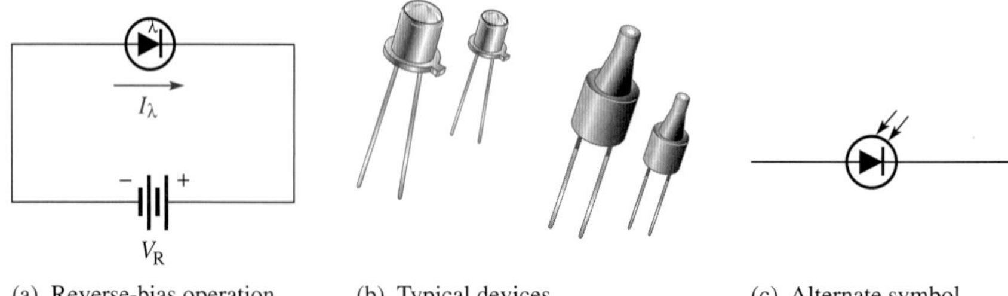

(a) Reverse-bias operation (b) Typical devices (c) Alternate symbol

Recall that when reverse-biased, a rectifier diode has a very small reverse leakage current. The same is true for the photodiode. The reverse-biased current is produced by thermally generated electron hole pairs in the depletion region, which are swept across the junction by the electric field created by the reverse voltage. In a rectifier diode, the reverse current increases with temperature due to an increase in the number of electron hole pairs.

In a photodiode, the reverse current increases with the light intensity at the exposed *pn* junction. When there is no incident light, the reverse current (I_λ) is almost negligible and is called the *dark current*. An increase in the amount of light intensity, expressed as irradiance (mW/cm^2), produces an increase in the reverse current, as shown by the graph in Figure 17–62(a).

FIGURE 17–62

Typical photodiode characteristics.

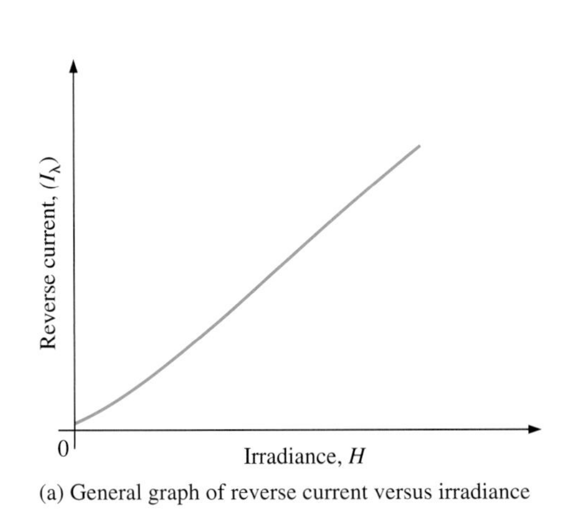

(a) General graph of reverse current versus irradiance

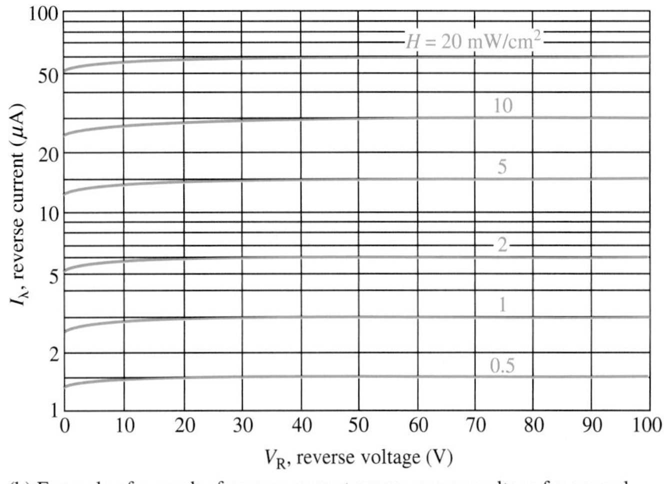

(b) Example of a graph of reverse current versus reverse voltage for several values of irradiance, H

From the graph in Figure 17–62(b), the reverse current for this particular device is approximately 1.4 μA at a reverse-bias voltage of 10 V. Therefore, the reverse resistance of the device with an irradiance of 0.5 mW/cm^2 is

$$R_R = \frac{R_R}{I_\lambda} = \frac{10\text{ V}}{1.4\ \mu\text{A}} = 7.14\text{ M}\Omega$$

At 20 mW/cm^2, the reverse current is approximately 55 μA at $V_R = 10$ V. The reverse resistance under this condition is

$$R_R = \frac{V_R}{I_\lambda} = \frac{10\text{ V}}{55\ \mu\text{A}} = 182\text{ k}\Omega$$

These calculations show that the photodiode can be used as a variable-resistance device controlled by light intensity.

Figure 17–63 illustrates that the photodiode allows essentially no reverse current (except for a very small dark current) when there is no incident light. When a

FIGURE 17–63

Operation of a photodiode.

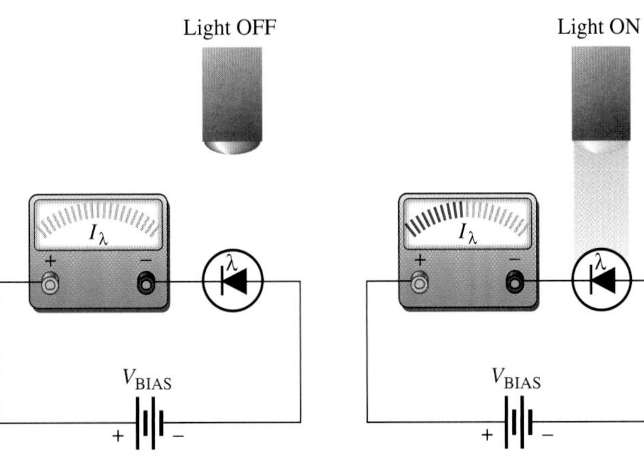

(a) No light, no current except negligible dark current

(b) Where there is incident light, resistance decreases and there is reverse current.

light beam strikes the photodiode, it conducts an amount of reverse current that is proportional to the light intensity.

An Application A simple photodiode application is depicted in Figure 17–64. Here a beam of light continuously passes across a conveyor belt and into a transparent window behind which is a photodiode circuit. When the light beam is interrupted by an object passing by on the conveyor belt, the sudden reduction in diode current activates a control circuit that advances a counter by one. The total count of objects that have passed that point is displayed by the counter. This basic concept can be extended and used for production control, shipping, and monitoring of activity on production lines.

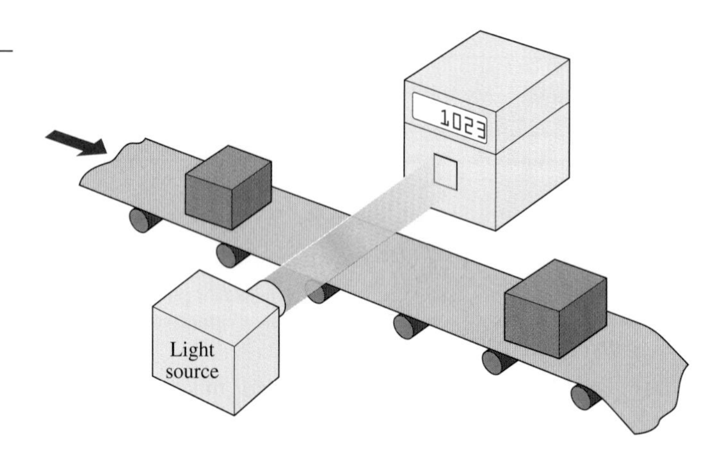

FIGURE 17–64

A photodiode circuit used in a system that counts objects as they pass on a conveyor belt.

SECTION 17–7 REVIEW

1. How does an LED differ from a photodiode?
2. List the semiconductive materials used in LEDs.
3. There is a very small reverse current in a photodiode under no-light conditions. What is this current called?

17–8 THE DIODE DATA SHEET

A manufacturer's data sheet gives detailed information on a device so that it can be used properly in a given application. A typical data sheet provides maximum ratings, electrical characteristics, mechanical data, and graphs of various parameters. A specific example is used to illustrate a typical data sheet.

After completing this section, you should be able to

- **Interpret a typical data sheet**
- Determine maximum ratings
- Determine electrical characteristics

Table 17–1 shows the maximum ratings for a certain series of rectifier diodes (1N4001 through 1N4007). These are the absolute maximum values under which the diode can be operated without damage to the device. For greatest reliability

RATING	SYMBOL	1N4001	1N4002	1N4003	1N4004	1N4005	1N4006	1N4007	UNIT
Peak repetitive reverse voltage Working peak reverse voltage DC blocking voltage	V_{RRM} V_{RWM} V_R	50	100	200	400	600	800	1000	V
Nonrepetitive peak reverse voltage	V_{RSM}	60	120	240	480	720	1000	1200	V
rms reverse voltage	$V_{R(rms)}$	35	70	140	280	420	560	700	V
Average rectified forward current (single-phase, resistive load, 60 Hz, $T_A = 75°C$)	I_O	1.0							A
Nonrepetitive peak surge current (surge applied at rated load conditions)	I_{FSM}	30 (for 1 cycle)							A
Operating and storage junction temperature range	T_J, T_{stg}	−65 to +175							°C

▲ **TABLE 17–1**

Maximum ratings.

and longer life, the diode should always be operated well under these maximum ratings. Generally, the maximum ratings are specified at 25°C and must be adjusted downward for higher temperatures.

An explanation of some of the parameters from Table 17–1 follows.

V_{RRM} The maximum reverse peak voltage that can be applied repetitively across the diode. Notice that in this case, it is 50 V for the 1N4001 and 1 kV for the 1N4007.

V_R The maximum reverse dc voltage that can be applied across the diode.

V_{RSM} The maximum reverse peak value of nonrepetitive (one cycle) voltage that can be applied across the diode.

I_O The maximum average value of a 60 Hz full-wave rectified forward current.

I_{FSM} The maximum peak value of nonrepetitive (one cycle) forward current. The graph in Figure 17–65 expands on this parameter to show values for more than one cycle at temperatures of 25°C and 175°C. The dashed lines represent values where typical failures occur.

T_A Ambient temperature (temperature of surrounding air).

T_J The operating junction temperature.

T_{stg} The storage junction temperature.

HANDS ON TIP

It is a mistake to parallel two diodes with the idea of doubling the forward current rating. Theoretically, each diode should handle half the current, but it does not work that way. One of the parallel diodes will generally be slightly warmer than the other. The warmer diode has more conductivity than the cooler one and will conduct more current. This causes the diode to become even warmer and conduct even more because of thermal runaway (also known as *current hogging*). Once thermal runaway begins, the process continues until the diode self-destructs; then the other diode will have all of the current and overheat also. So, never parallel two diodes to achieve a higher forward current rating. Always choose a diode that has a sufficient rating for the current it has to conduct.

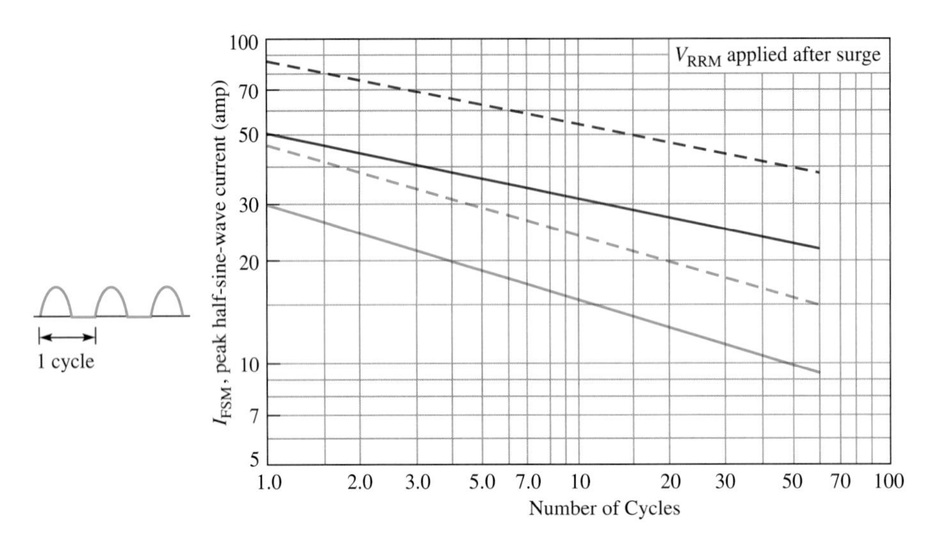

- - - Typical failures when surge applied at no-load conditions $T_J = 25°C$

—— Design limits when surge applied at no-load conditions $T_J = 25°C$

- - - Typical failures when surge applied at rated-load conditions $T_J = 175°C$

—— Design limits when surge applied at rated-load conditions $T_J = 175°C$

▲ **FIGURE 17–65**

Nonrepetitive forward surge current capability.

CHARACTERISTICS AND CONDITIONS	SYMBOL	TYPICAL	MAXIMUM	UNIT
Maximum instantaneous forward voltage drop ($I_F = 1$ A, $T_J = 25°C$)	v_F	0.93	1.1	V
Maximum full-cycle average forward voltage drop ($I_O = 1$ A, $T_L = 75°C$)	$V_{F(avg)}$	—	0.8	V
Maximum reverse current (rated dc voltage)	I_R			μA
$T_J = 25°C$		0.05	10.0	
$T_J = 100°C$		1.0	50.0	
Maximum full-cycle average reverse current ($I_O = 1$ A, $T_L = 75°C$)	$I_{R(avg)}$	—	30.0	μA

▲ **TABLE 17–2**

Electrical characteristics.

Table 17–2 lists typical and maximum values of certain electrical characteristics. These items differ from the maximum ratings in that they are not selected by design but are the result of operating the diode under specified conditions. A brief explanation of these parameters follows.

v_F The instantaneous voltage across the forward-biased diode when the forward current is 1 A at 25°C. Figure 17–66 shows how the forward voltages vary with forward current.

$V_{F(avg)}$ The maximum forward voltage drop averaged over a full cycle.

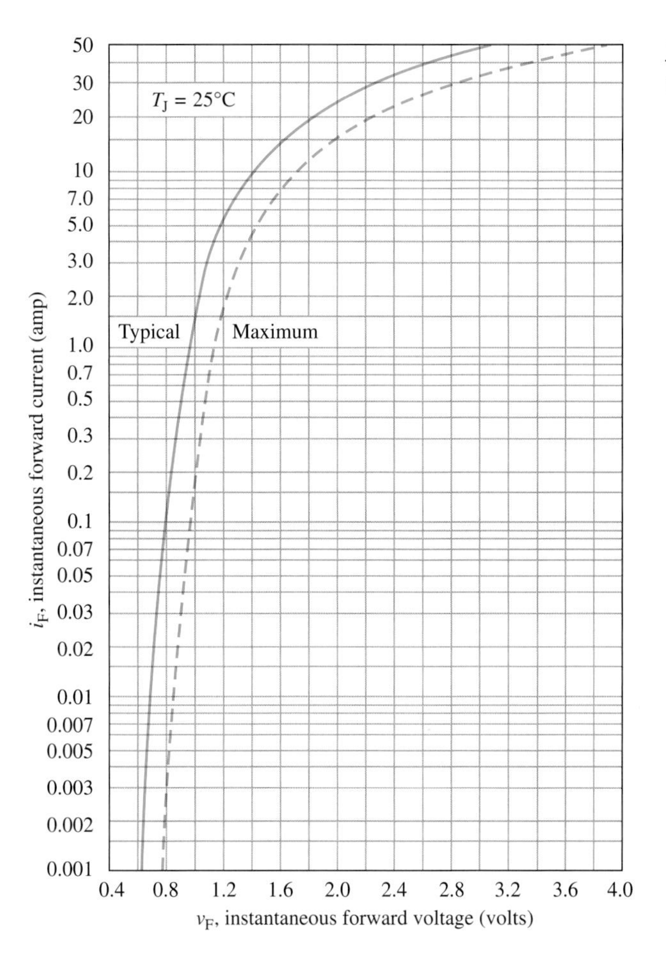

FIGURE 17–66

Forward voltage.

I_R The maximum current when the diode is reverse-biased with a dc voltage.

$I_{R(avg)}$ The maximum reverse current averaged over one cycle (when reverse-biased with an ac voltage).

T_L The lead temperature.

Figure 17–67 shows a selection of rectifier diodes arranged in order of increasing I_O, I_{FSM}, and V_{RRM} ratings.

V_{RRM} (Volts)	I_O, Average Rectified Forward Current (Amperes)					
	1.0	1.5	3.0			6.0
	59-03 (DO-41) Plastic 	59-04 Plastic 	60-01 Metal 	267-03 Plastic 	267-02 Plastic 	194-04 Plastic 
50	1N4001	1N5391	1N4719	MR500	1N5400	MR750
100	1N4002	1N5392	1N4720	MR501	1N5401	MR751
200	1N4003	1N5393 MR5059	1N4721	MR502	1N5402	MR752
400	1N4004	1N5395 MR5060	1N4722	MR504	1N5404	MR754
600	1N4005	1N5397 MR5061	1N4723	MR506	1N5406	MR756
800	1N4006	1N5398	1N4724	MR508		MR758
1000	1N4007	1N5399	1N4725	MR510		MR760
I_{FSM} (Amps)	30	50	300	100	200	400
T_A @ Rated I_O (°C)	75	$T_L = 70$	75	95	$T_L = 105$	60
T_C @ Rated I_O (°C)						
T_J (Max) (°C)	175	175	175	175	175	175

V_{RRM} (Volts)	I_O, Average Rectified Forward Current (Amperes)										
	12	20	24	25	30		40	50	25	35	40
	245A-02 (DO-203AA) Metal 		339-02 Plastic 	193-04 Plastic 	43-02 (DO-21) Metal 		42A-01 (DO-203AB) Metal 	43-04 Metal 	309A-03 	309A-02 	
50	MR1120 1N1199,A,B	MR2000	MR2400	MR2500	1N3491	1N3659	1N1183A	MR5005	MDA2500	MDA3500	
100	MR1121 1N1200,A,B	MR2001	MR2401	MR2501	1N3492	1N3660	1N1184A	MR5010	MDA2501	MDA3501	
200	MR1122 1N1202,A,B	MR2002	MR2402	MR2502	1N3493	1N3661	1N1186A	MR5020	MDA2502	MDA3502	MDA4002
400	MR1124 1N1204,A,B	MR2004	MR2404	MR2504	1N3495	1N3663	1N1188A	MR5040	MDA2504	MDA3504	MDA4004
600	MR1126 1N1206,A,B	MR2006	MR2406	MR2506			1N1190A		MDA2506	MDA3506	MDA4006
800	MR1128	MR2008		MR2508					MDA2508	MDA3508	MDA4008
1000	MR1130	MR2010		MR2510					MDA2510	MDA3510	
I_{FSM} (Amps)	300	400	400	400	300	400	800	600	400	400	800
T_A @ Rated I_O (°C)											
T_C @ Rated I_O (°C)	150	150	125	150	130	100	150	150	55	55	35
T_J (Max) (°C)	190	175	175	175	175	175	190	195	175	175	175

▲ **FIGURE 17–67**

A selection of rectifier diodes based on maximum ratings of I_O, I_{FSM}, and V_{RRM}.

SECTION 17–8 REVIEW

1. List the three rating categories typically given on all diode data sheets.
2. Define each of the following parameters: V_F, I_R, I_O.
3. Define I_{FSM}, V_{RRM}, and V_{RSM}.
4. From Figure 17–67, select a diode to meet the following specifications: $I_O = 3$ A, $I_{FSM} = 300$ A, and $V_{RRM} = 100$ V.

17–9 TROUBLESHOOTING

This section provides a general review and application of the APM (analysis, planning, and measurement) approach to troubleshooting. Specific troubleshooting examples of the power supply and diode circuits are covered.

After completing this section, you should be able to

- **Troubleshoot power supplies and diode circuits, using the APM approach**
- Use analysis to evaluate the problem based on symptoms
- Eliminate basic problems that can be detected by observation
- Plan an approach to determining what the fault is in a circuit or system
- Make appropriate measurements to isolate a fault
- Recognize symptoms caused by certain types of component failures

As you know, you can approach the troubleshooting of a defective circuit or system using the APM method. A defective circuit or system is one with a known good input but with no output or an incorrect output. For example, in Figure 17–68(a), a properly functioning dc power supply is represented by a single block with a known input voltage and a correct output voltage. A defective dc power supply is represented in part (b) as a block with an input voltage but no output voltage.

FIGURE 17–68

Block representations of functioning and nonfunctioning power supplies.

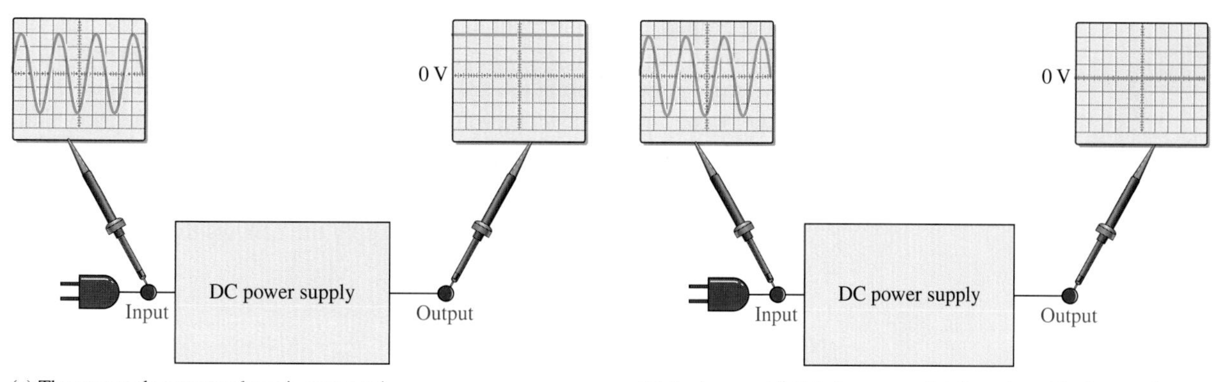

(a) The correct dc output voltage is measured.

(b) An incorrect 0 V dc is measured at the output. The input voltage is correct.

Analysis

The first step in troubleshooting a defective circuit or system is to analyze the problem, which includes identifying the symptom and eliminating as many causes as possible. In the case of the power supply example illustrated in Figure

17–68(b), the symptom is that there is no output voltage. The absence of an output voltage from the power supply does not tell you much about what the specific cause may be. In other situations, however, a particular symptom may point to a given area where a fault is most likely.

The first thing you should do in analyzing the problem is to try to eliminate any obvious causes. Start by making sure the power cord is plugged into an active outlet and that the fuse is not open. In the case of a battery-powered system, make sure the battery is good. Something as simple as this is sometimes the cause of the problem.

Beyond the power check, use your senses to detect obvious defects, such as a burned resistor, broken wire, loose connection, or an open fuse. Since some failures are temperature dependent, you can sometimes find an overheated component by touch. However, be very cautious in a live circuit to avoid possible burn or shock. For intermittent failures, the circuit may work properly for awhile and then fail due to heat buildup. As a rule, you should always do a sensory check as part of the analysis phase before proceeding.

Planning

In this phase, you must consider how you will attack the problem. There are three possible approaches to troubleshooting most circuits or systems.

1. Start at the input where there is a known input voltage and work toward the output until you get an incorrect measurement. When you find no voltage or an incorrect voltage, you have narrowed the problem to the part of the circuit between the last test point where the voltage was good and the present test point. In all troubleshooting approaches, you must know what the voltage is supposed to be at each point in order to recognize an incorrect measurement when you see it.

2. Start at the output of a circuit and work toward the input. Check for voltage at each test point until you get a correct measurement. At this point, you have isolated the problem to the part of the circuit between the last test point and the current test point where the voltage is correct.

3. Use the half-splitting method and start in the middle of the circuit. If this measurement shows a correct voltage, you know that the circuit is working properly from the input to that test point. This means that the fault is between the current test point and the output point, so begin tracing the voltage from that point toward the output. If the measurement in the middle of the circuit shows no voltage or an incorrect voltage, you know that the fault is between the input and that test point. Therefore, begin tracing the voltage from the test point toward the input.

For illustration, let's say that you decide to apply the half-splitting method using an oscilloscope.

Measurement

The half-splitting method is illustrated in Figure 17–69. Let's assume that your measurements are as indicated in the figure. At test point 3 (TP3) you observe a correct voltage which indicates that the transformer and rectifier are working properly. At test point 4 (TP4) there is no voltage, which indicates that the filter is faulty or the input to the regulator is shorted.

In this particular case, the half-splitting method took two measurements to isolate the fault to the filter circuit or regulator input. If you had started from the power supply input, it would have taken four measurements; and if you had

▲ **FIGURE 17–69**

Example of the half-splitting approach.

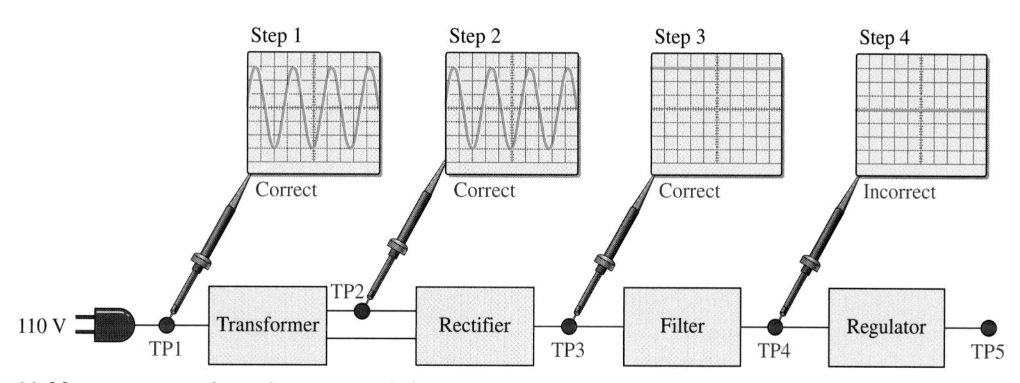

(a) Measurements starting at the power supply input.

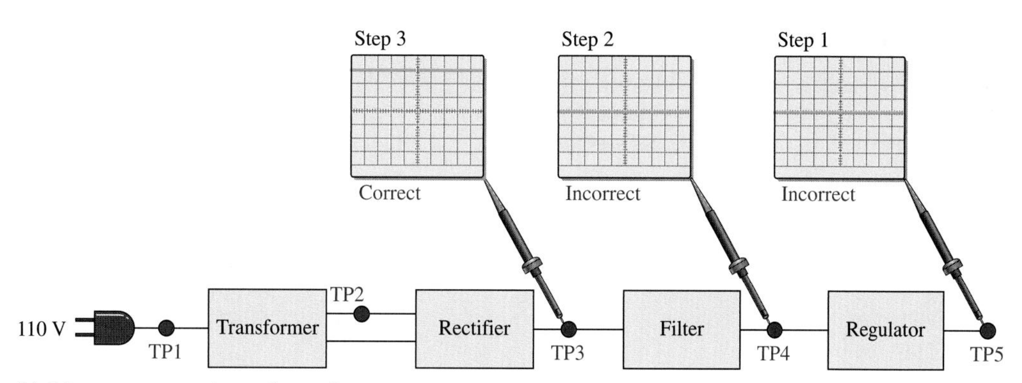

(b) Measurements starting at the regulator output.

▲ **FIGURE 17–70**

In this particular case, the two other approaches require more measurements than the half-splitting approach in Figure 17–69.

started at the output, it would have taken three measurements, as illustrated in Figure 17–70.

Next, you must determine if one of the two components in the filter is faulty or if the regulator input is shorted. Since the output voltage is zero, either the surge resistor could be open or the filter capacitor shorted. Open resistors are more common than shorted capacitors, so it is more likely the resistor that is bad. Measure the resistor with a DMM to see if it is open. If the resistor is good, then either the capacitor or the regulator input is shorted.

Fault Analysis

In some cases, after isolating a fault to a particular circuit, it may be necessary to isolate the problem to a single component in the circuit. In this event, you have to apply your knowledge of the symptoms caused by certain component failures. Some typical component failures and the symptoms they produce are now discussed.

Effect of an Open Diode in a Half-Wave Rectifier A half-wave filtered rectifier with an open diode is shown in Figure 17–71. The resulting symptom is zero output voltage as indicated. This is obvious because the open diode breaks the current path from the transformer secondary winding to the filter and load resistor and there is no load current.

Other faults that will cause the same symptom in this circuit are an open transformer winding, an open fuse, or no input voltage.

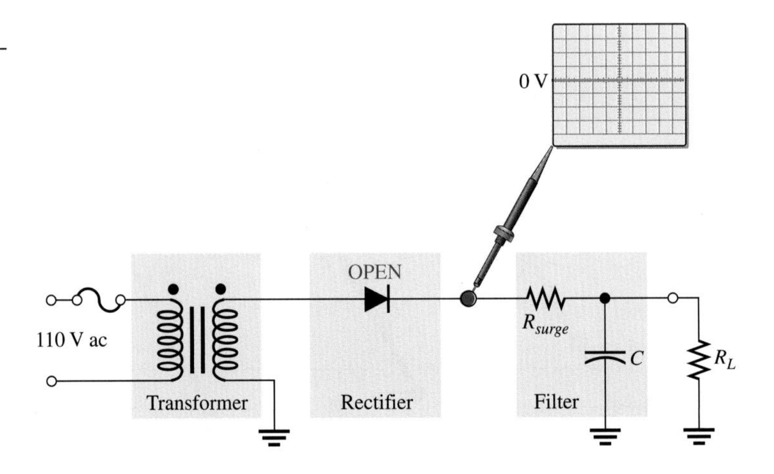

FIGURE 17–71

The effect of an open diode in a half-wave rectifier is an output of 0 V.

Effect of an Open Diode in a Full-Wave Rectifier A full-wave center-tapped filtered rectifier is shown in Figure 17–72. If either of the two diodes is open, the output voltage will have a larger than normal ripple voltage at 60 Hz rather than at 120 Hz, as indicated.

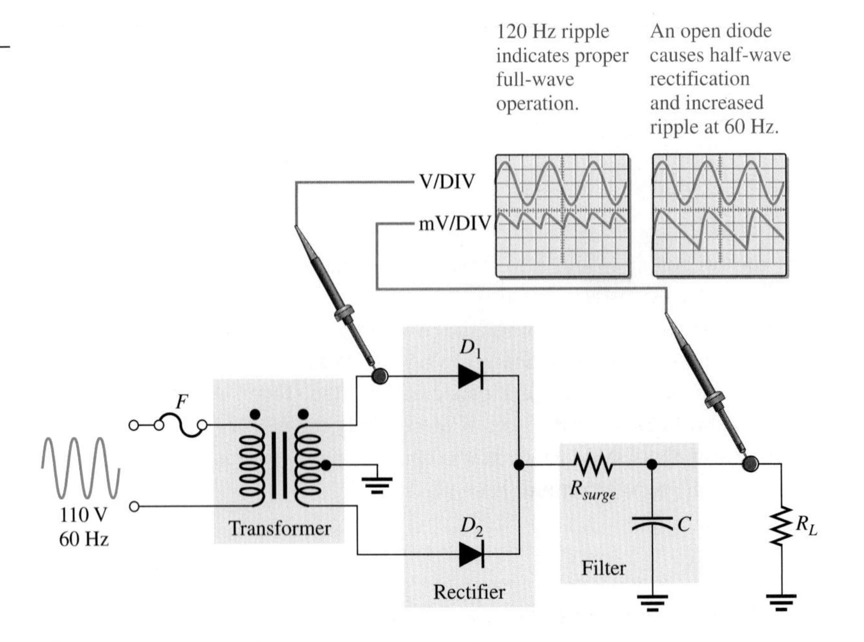

FIGURE 17–72

The effect of an open diode in a center-tapped rectifier is half-wave rectification and a larger ripple voltage at 60 Hz.

Another fault that will cause the same symptom is an open in one of the halves of the transformer secondary winding.

The reason for the increased ripple at 60 Hz rather than at 120 Hz is as follows. If one of the diodes in Figure 17–72 is open, there is current through R_L only during one half-cycle of the input voltage. During the other half-cycle of the input, the open path caused by the open diode prevents current through R_L. The result is half-wave rectification, as shown in Figure 17–72, which produces the larger ripple voltage with a frequency of 60 Hz.

An open diode in a full-wave bridge rectifier will produce the same symptom as in the center-tapped circuit, as shown in Figure 17–73. The open diode prevents current through R_L during half of the input voltage cycle. The result is half-wave rectification, which produces an increase in ripple voltage at 60 Hz.

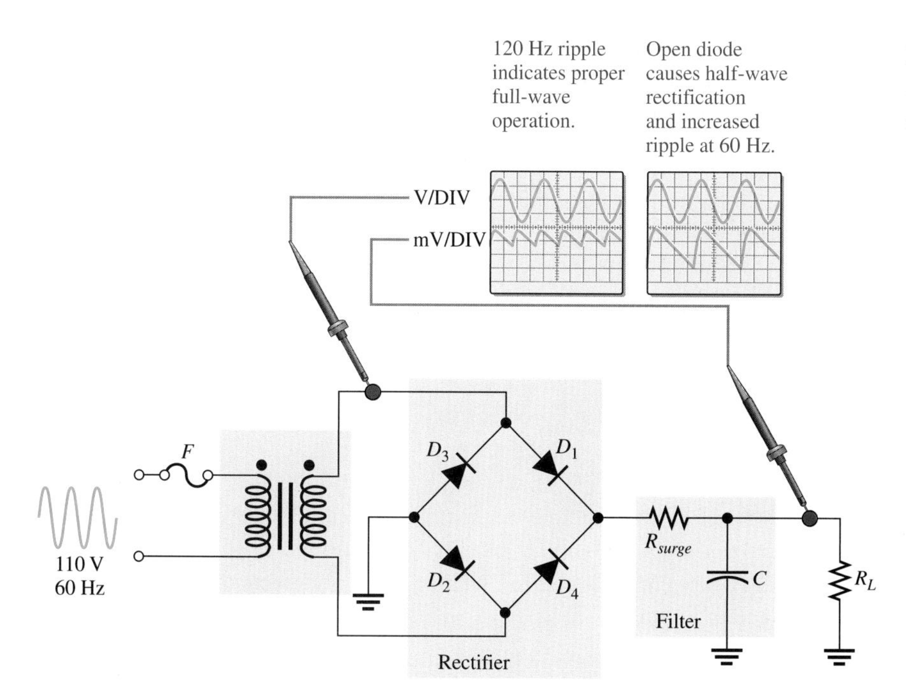

FIGURE 17–73

Effect of an open diode in a bridge rectifier.

Effect of a Shorted Diode in a Full-Wave Rectifier A shorted diode is one that has failed such that it has a very low resistance in both directions. If a diode suddenly shorts in a bridge rectifier, there will be an excessively high current during one-half of the input voltage cycle, as illustrated in Figure 17–74 with diode D_1 shorted. This condition will most likely burn open one of the diodes if the circuit is not properly fused. During the positive half-cycle of the input voltage as shown in Figure 17–74(a), the path of the load current is through the shorted diode D_1 just the same as if it were forward-biased. During the negative half-cycle of the input, the current is shorted through D_1 and D_4 as shown in part (b). This is a very low resistance path because there is a forward-biased diode and a shorted diode in series. It is likely that the excessive current will burn either or both of the diodes open if the circuit is not properly fused.

If only one of the diodes, D_1 or D_4, should burn open, the circuit will operate as a half-wave rectifier. If both diodes should burn open or if the excessive current causes the fuse to blow (which it should), the output voltage is zero.

FIGURE 17–74

Effect of a shorted diode in a bridge rectifier.

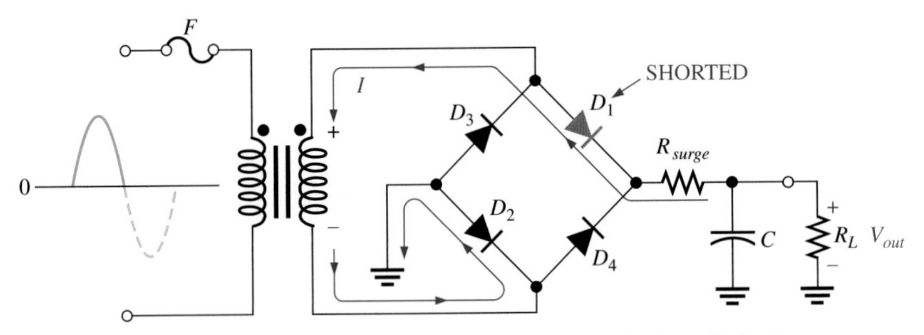

(a) Positive half-cycle: The shorted diode acts as a forward-biased diode, so the load current is normal.

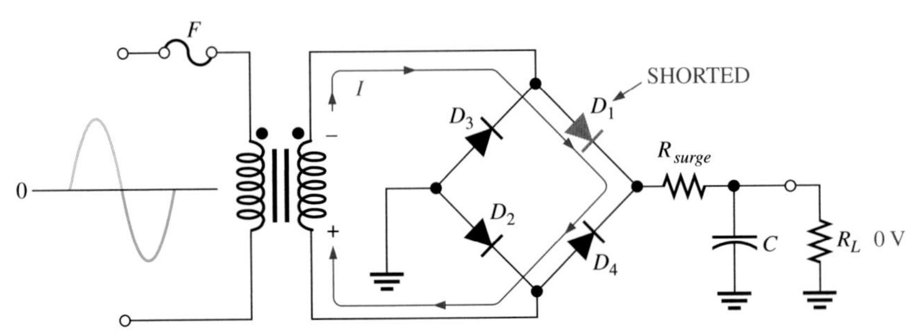

(b) Negative half-cycle: The shorted diode produces a short circuit across the source. As a result the fuse should blow. If not properly fused, D_1, D_4, or the transformer secondary will probably burn open.

Effects of a Faulty Filter Capacitor Three types of defects of a filter capacitor are illustrated in Figure 17–75.

- *Open* If the filter capacitor for a full-wave rectifier opens, the output is a full-wave rectified voltage.
- *Shorted* If the filter capacitor shorts, the output is 0 V. A shorted capacitor may cause some or all of the diodes in the rectifier to burn open due to excessive current if there is no surge resistor and if the rectifier is not properly fused. If the rectifier is properly fused, the fuse should open and prevent damage to the circuit. In any event, the output is 0 V.
- *Leaky* A leaky filter capacitor is equivalent to a capacitor with a parallel leakage resistance. The effect of the leakage resistance is to reduce the time constant and allow the capacitor to discharge more rapidly than normal. This results in an increase in the ripple voltage on the output. This fault is rare.

Effects of a Faulty Transformer An open primary or secondary winding of a power supply transformer results in an output of 0 V, as mentioned before.

A partially shorted primary winding (which is much less likely than an open) results in an increased rectifier output voltage because the turns ratio of the transformer is effectively increased. A partially shorted secondary winding results in a decreased rectifier output voltage because the turns ratio is effectively decreased.

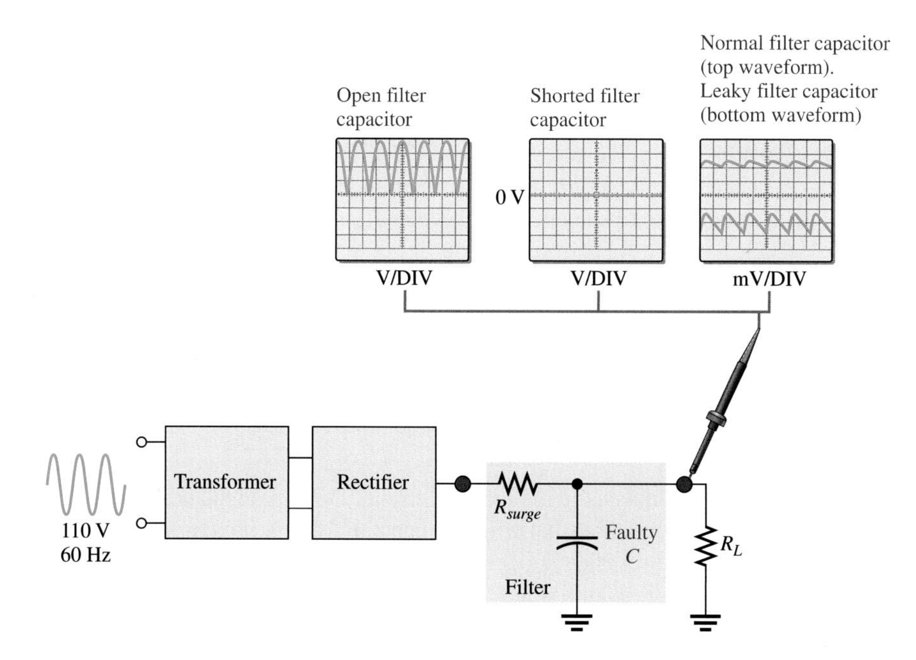

FIGURE 17–75

Effects of a faulty filter capacitor.

EXAMPLE 17–16

You are troubleshooting the power supply shown in the block diagram of Figure 17–76. You have found in the analysis phase that there is no output voltage from the regulator, as indicated. Also, you have found that the unit is plugged into the outlet and have verified the input to the transformer, as indicated. You decide to use the half-splitting method using the scope. What is the problem?

FIGURE 17–76

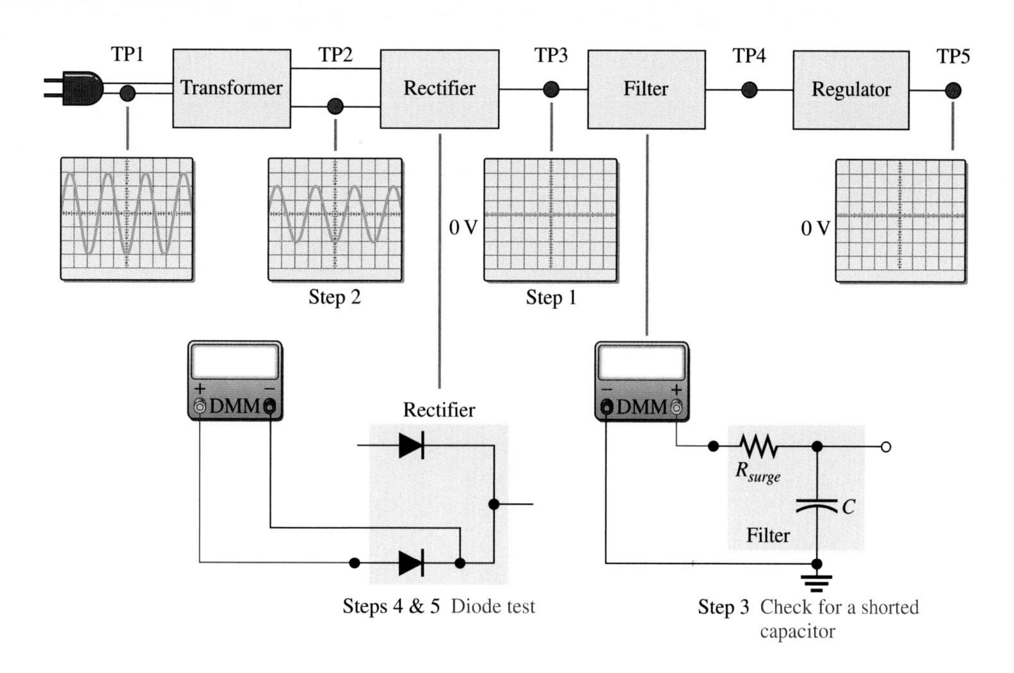

Solution The step-by-step measurement procedure is illustrated in the figure and described as follows.

Step 1: There is no voltage at test point 3 (TP3). This indicates that the fault is between the input to the transformer and the output of the rectifier. Most likely, the problem is in the transformer or in the rectifier, but there may be a short from the filter input to ground.

Step 2: The voltage at test point 2 (TP2) is correct, indicating that the transformer is working. So, the problem must be in the rectifier or a shorted filter input.

Step 3: With the power turned off, use a DMM to check for a short from the filter input to ground. Assume that the DMM indicates no short. The fault is now isolated to the rectifier.

Step 4: Apply fault analysis to the rectifier circuit. Determine the component failure in the rectifier that will produce a 0 V input. If only one of the diodes in the rectifier is open, there should be a half-wave rectified output voltage, so this is not the problem. In order to have a 0 V output, both of the diodes must be open.

Step 5: With the power off, use the DMM in the diode test mode to check each diode. Replace the defective diodes, turn the power on, and check for proper operation. Assume this corrects the problem.

Related Problem Suppose you had found a short in Step 3, what would have been the logical next step?

SECTION 17–9 REVIEW

1. What effect does an open diode have on the output voltage of a half-wave rectifier?
2. What effect does an open diode have on the output voltage of a full-wave rectifier?
3. If one of the diodes in a bridge rectifier shorts, what are some possible consequences?
4. What happens to the output voltage of a rectifier if the filter capacitor becomes very leaky?
5. The primary winding of the transformer in a power supply opens. What will you observe on the rectifier output?
6. The dc output voltage of a filtered rectifier is less than it should be. What may be the problem?

The optical counting system in this assignment incorporates the zener diode, the LED, and the photodiode, as well as other types of devices. Of course, at this point, you are interested in only how the zener and the optical diodes are used. The system power supply is similar to the one you worked with in the last chapter except that there is a zener regulator in this one rather than the integrated circuit regulator.

Description of the System

The system is used to count objects as they pass on a moving conveyor and can be used in production line monitoring, loading, shipping, inventory control, and the like. The system consists of a power supply, an infrared (IR) emitter circuit, an infrared detector, a threshold circuit, and counter/display circuits, as shown in Figure 17–77.

The power supply and infrared emitter are housed in a single unit and positioned on one side of the conveyor. The IR detector/counter unit is positioned directly opposite on the other side. The dc power supply provides power to both units. The emitter produces a continuous pattern of infrared light directed toward the detector. As long as the detector is receiving light, nothing happens. When an object passes between the two units and blocks the infrared light, the detector reacts. This reaction is sensed by the

FIGURE 17–77

The optical counting system showing the system concept, block diagram, and circuit boards.

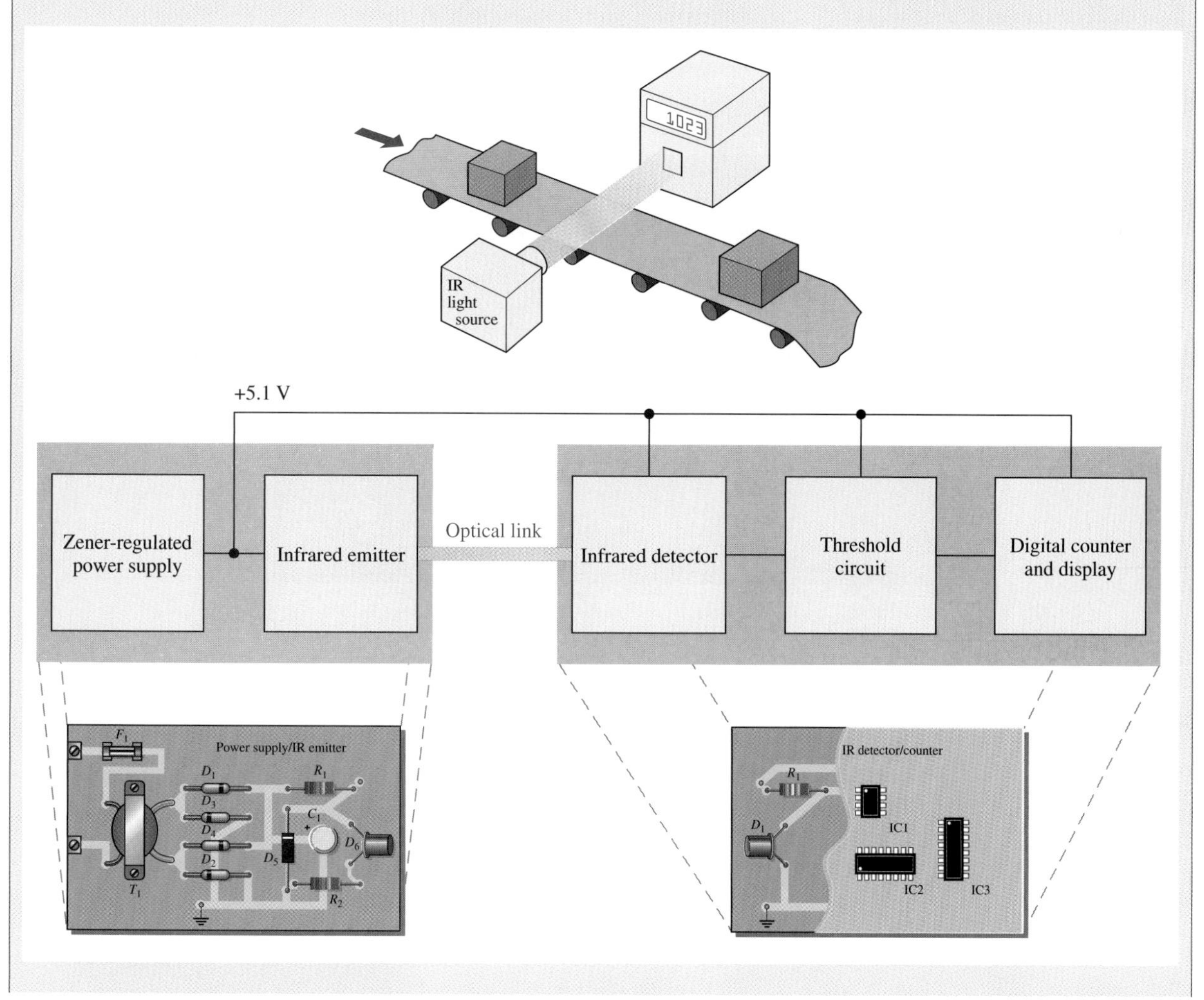

threshold circuit, which sends a signal to a digital counter, advancing its count by one. The total count can be displayed and/or sent to a central computer.

In this section, you are not concerned with the details of the threshold circuit or the digital counter and display circuit. In fact, they are beyond the scope of this book and perhaps beyond your background at this time. They are included only to make a complete system and to show that the circuits you are concerned with must **interface** with other types of circuits in order to perform the required system function. At this point, you are strictly interested in how some of the devices covered in this chapter function in the system.

Physically, the power supply/IR emitter unit consists of one printed circuit board that contains the regulated dc power supply and the infrared emitter circuit. The IR detector/counter unit consists of one printed circuit board that contains the infrared detector circuit, a threshold circuit, and the digital counter and display circuits. All of the circuits on this board except the infrared detector are covered in another course, and you do not need to know the details of how they work for the purposes of this assignment.

Step 1: Identify the Components

As you can see in Figure 17–77, the power supply/IR emitter circuit board is very similar to the power supply board that you worked with in the last chapter except for the regulator and the infrared emitter circuit. The dc voltage regulator is made up of a resistor R_1 and zener diode D_5. The infrared emitter circuit consists of R_2 and D_6, which is an infrared LED. The infrared detector circuit on the IR detector/counter board is formed by R_1 and infrared photodiode D_1. As you can see, the LED and the photodiode are identical in appearance and can only be distinguished by a part number stamped on the case or on the package from which they are taken. The digital portion of this board is shown in the light green area without detail. The digital integrated circuits, IC1, IC2, and IC3, are in SOIC (Small Outline Integrated Circuit) packages. SOIC is a surface-mount technology packaging configuration. Although digital circuits are studied in other courses, you see here an example of a common situation in many systems where linear (analog) and digital circuits work together.

FIGURE 17–78

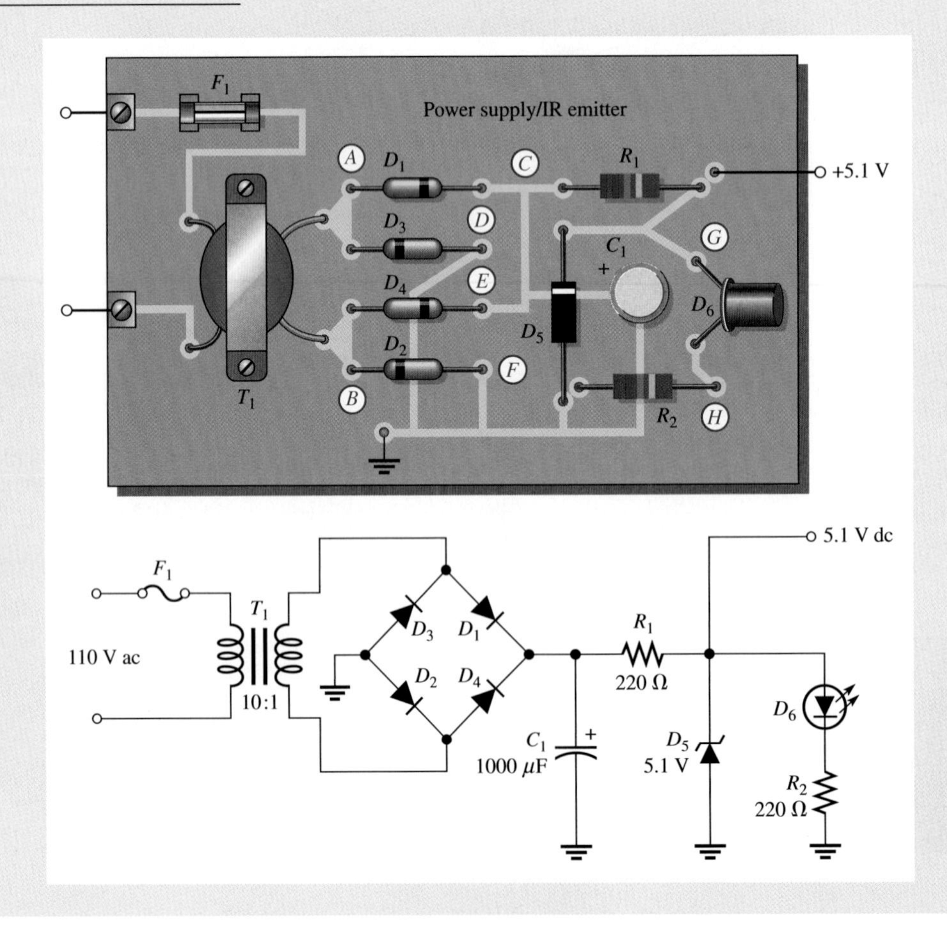

12. A diode-limiting circuit
    (a) removes part of a waveform
    (b) inserts a dc level
    (c) produces an output equal to the average value of the input
    (d) increases the peak value of the input
13. A clamping circuit is also known as a(an)
    (a) averaging circuit (b) inverter
    (c) dc restorer (d) ac restorer
14. The zener diode is designed to operate in
    (a) zener breakdown (b) forward bias
    (c) saturation (d) cutoff
15. Zener diodes are widely used as
    (a) current limiters (b) power distributors
    (c) voltage references (d) variable resistors
16. Varactor diodes are used as variable
    (a) resistors (b) current sources (c) inductors (d) capacitors
17. LEDs are based on the principle of
    (a) forward bias (b) electroluminescence
    (c) photon sensitivity (d) electron-hole recombination
18. In a photodiode, light produces
    (a) reverse current (b) forward current
    (c) electroluminescence (d) dark current

PROBLEMS

Answers to odd-numbered problems are at the end of the book.

BASIC PROBLEMS

SECTION 17–1 **Half-Wave Rectifiers**

1. Calculate the average value of a half-wave rectified voltage with a peak value of 200 V.
2. Draw the waveforms for the load current and voltage for Figure 17–82. Show the peak values.
3. Can a diode with a PIV rating of 50 V be used in the circuit of Figure 17–82?
4. Determine the peak power delivered to R_L in Figure 17–83.

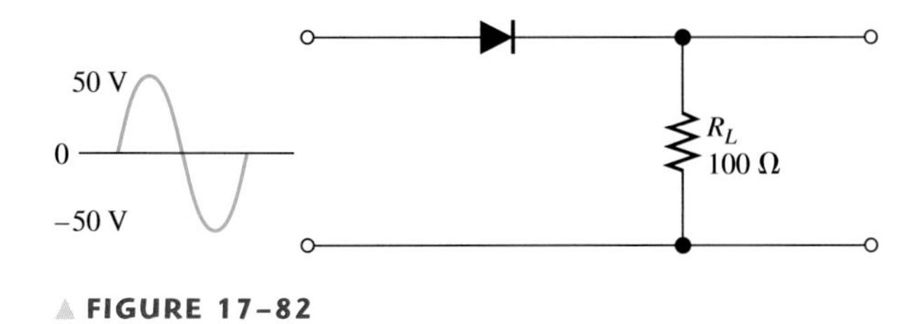

FIGURE 17–82

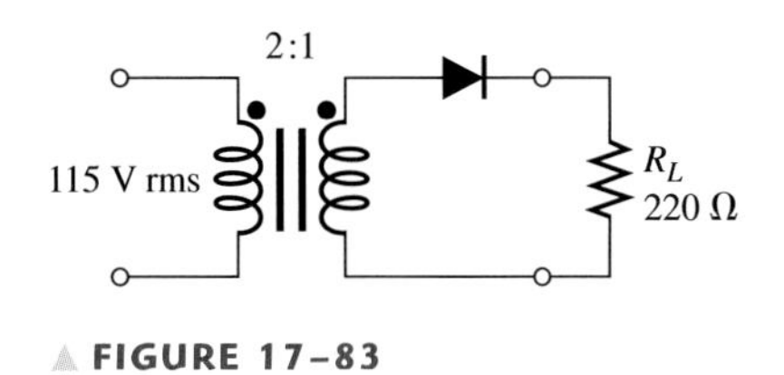

FIGURE 17–83

SECTION 17–2 **Full-Wave Rectifiers**

5. Calculate the average value of a full-wave rectified voltage with a peak value of 75 V.

6. Consider the circuit in Figure 17–84.

(a) What type of circuit is this?

(b) What is the total peak secondary voltage?

(c) Find the peak voltage across each half of the secondary.

(d) Sketch the voltage waveform across R_L.

(e) What is the peak current through each diode?

(f) What is the PIV for each diode?

▶ FIGURE 17–84

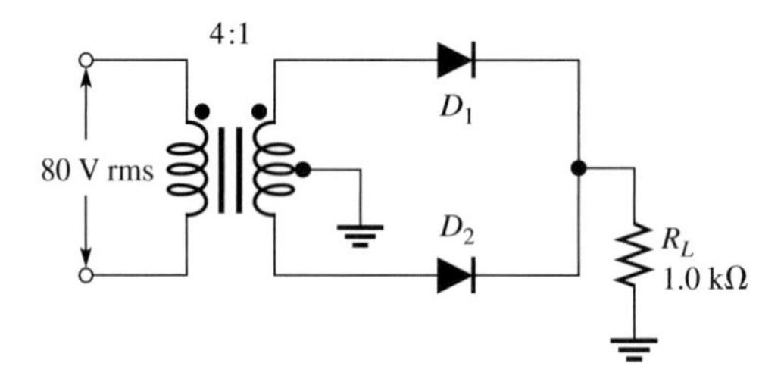

7. Calculate the peak voltage rating of each half of a center-tapped transformer used in a full-wave rectifier that has an average output voltage of 110 V.

8. Show how to connect the diodes in a center-tapped rectifier in order to produce a negative-going full-wave voltage across the load resistor.

9. What PIV rating is required for the diodes in a bridge rectifier that produces an average output voltage of 50 V?

SECTION 17–3 **Power Supply Filters and Regulators**

10. The ideal dc output voltage of a capacitor-input filter is the (peak, average) value of the rectified input.

11. Refer to Figure 17–85 and sketch the following voltage waveforms in relationship to the input waveform: V_{AB}, V_{AC}, and V_{BC}.

▶ FIGURE 17–85

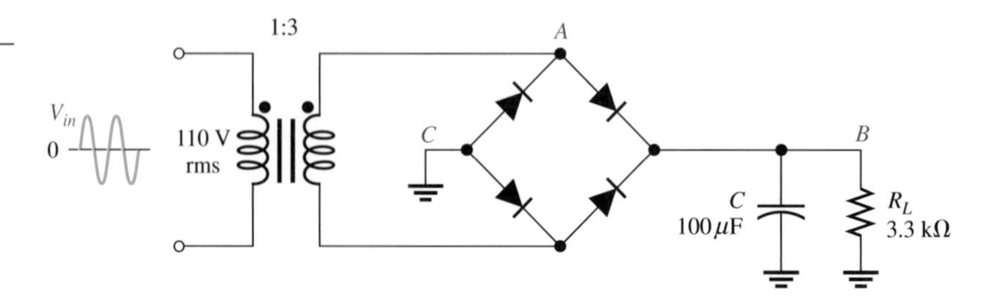

SECTION 17–4 **Diode Limiting and Clamping Circuits**

12. Draw the output waveforms for each circuit in Figure 17–86.

FIGURE 17–86

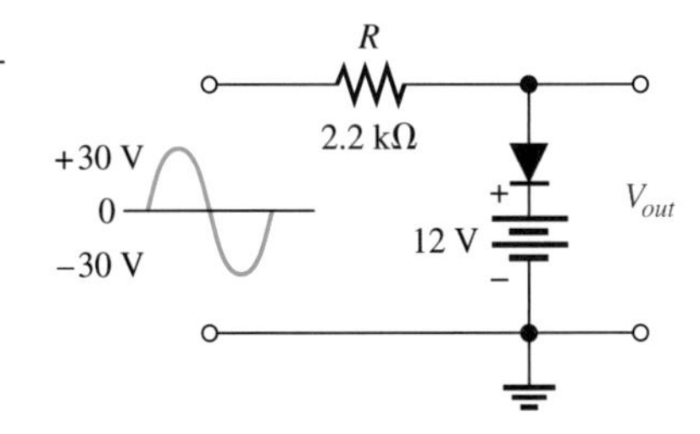

(a)

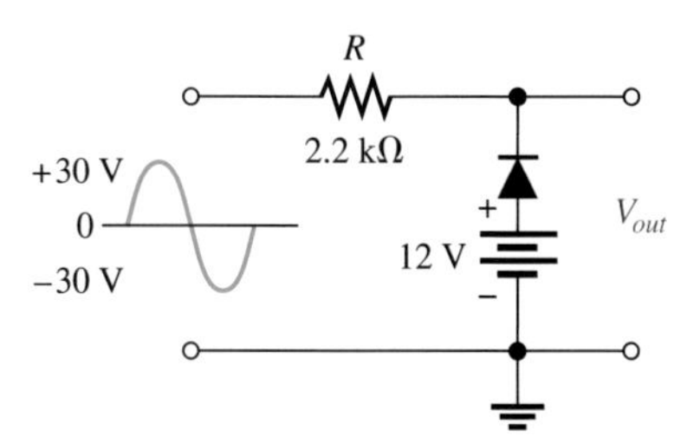

(b)

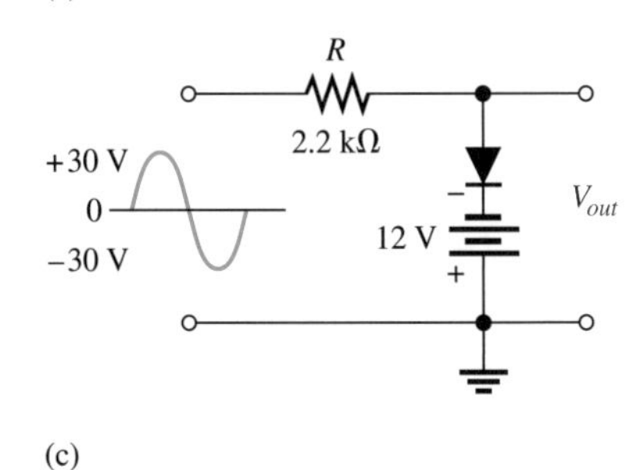

(c)

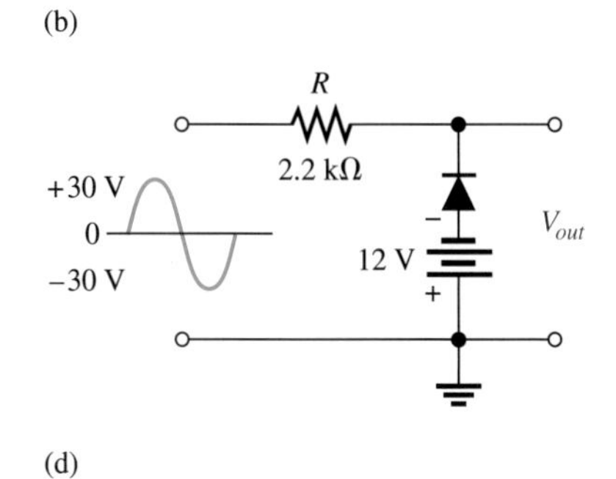

(d)

13. Describe the output waveform of each circuit in Figure 17–87. Assume that the R_LC time constant is much greater than the period of the input.

FIGURE 17–87

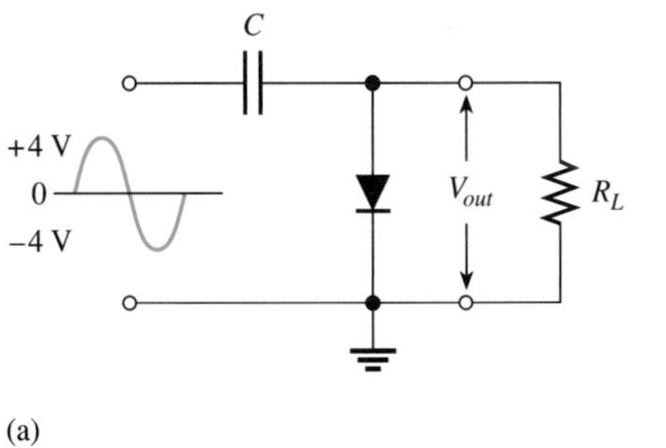

(a)

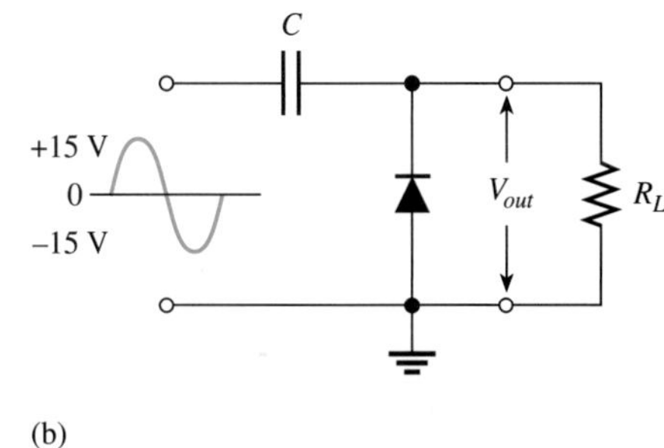

(b)

SECTION 17–5

Zener Diodes

14. A certain zener diode has a $V_Z = 7.5$ V and a $Z_Z = 5\ \Omega$. Draw the equivalent circuit.

15. Determine the minimum input voltage required for regulation to be established in Figure 17–88. Assume an ideal zener diode with $I_{ZK} = 1.5$ mA and $V_Z = 14$ V.

16. To what value must the total series resistance be adjusted in Figure 17–89 to make $I_Z = 40$ mA? Assume that $V_Z = 12$ V at 30 mA and $Z_Z = 30\ \Omega$.

17. A loaded zener regulator is shown in Figure 17–90. For $V_Z = 5.1$ V at 35 mA, $I_{ZK} = 1$ mA, $I_{ZM} = 70$ mA, and $Z_Z = 12\ \Omega$, determine the minimum and maximum permissible load currents.

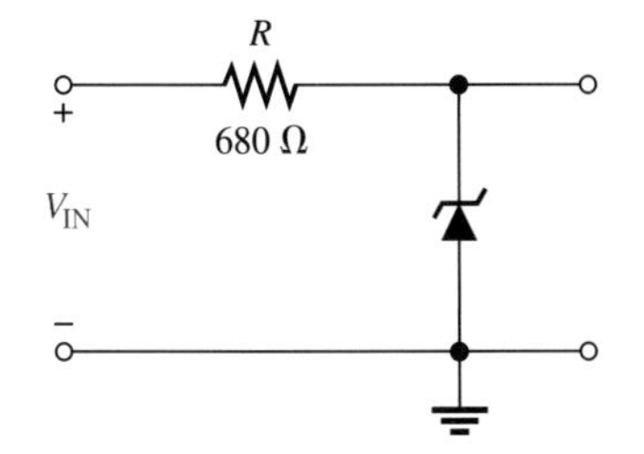

FIGURE 17–88

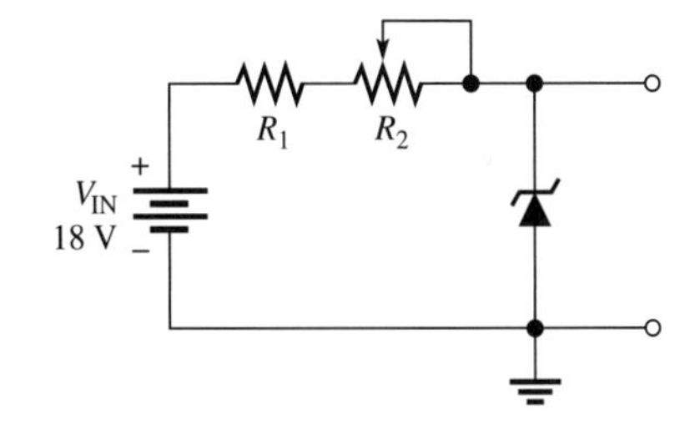

FIGURE 17–89

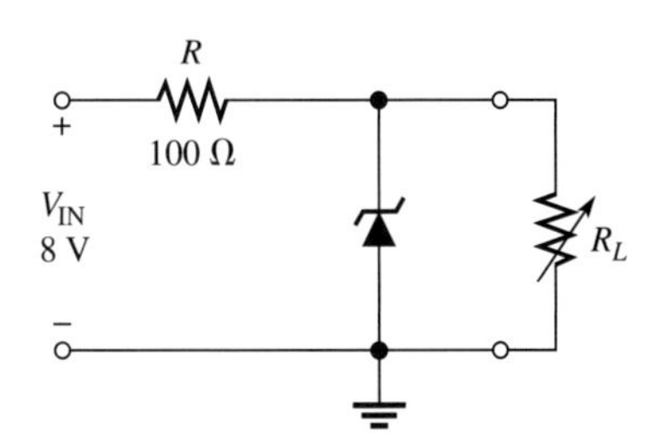

FIGURE 17–90

18. Find the percent load regulation in Problem 17.
19. For the circuit of Problem 17, assume that the input voltage is varied from 6 V to 12 V. Determine the percent line regulation with no load and with maximum load current.
20. The no-load output voltage of a certain zener regulator is 8.23 V, and the full-load output is 7.98 V. Calculate the percent load regulation.

SECTION 17–6 **Varactor Diodes**

21. Figure 17–91 is a curve of reverse voltage versus capacitance for a certain varactor. Determine the change in capacitance if V_R varies from 5 V to 20 V.
22. Refer to Figure 17–91 and determine the value of V_R that produces 25 pF.
23. What capacitance value is required for each of the varactors in Figure 17–92 to produce a resonant frequency of 1 MHz?

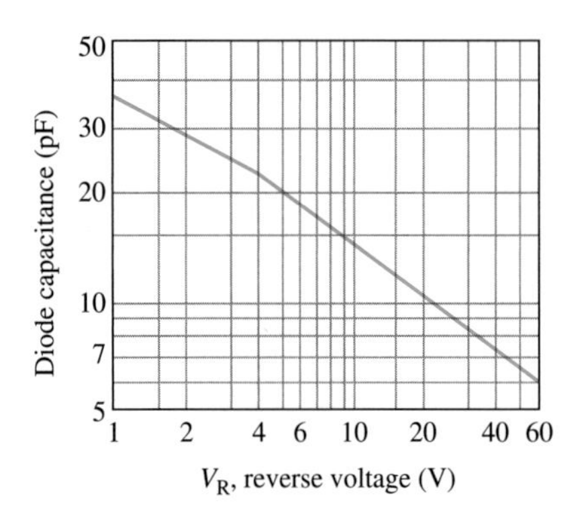

▲ FIGURE 17–91

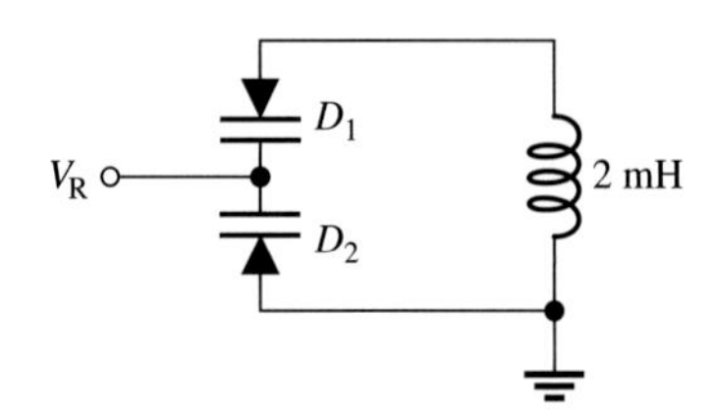

▲ FIGURE 17–92

24. At what value must the control voltage be set in Problem 23 if the varactors have the characteristic curve in Figure 17–91?

SECTION 17–7 **LEDs and Photodiodes**

25. When the switch in Figure 17–93 is closed, will the microammeter reading increase or decrease? Assume that D_1 and D_2 are optically coupled.

▶ FIGURE 17–93

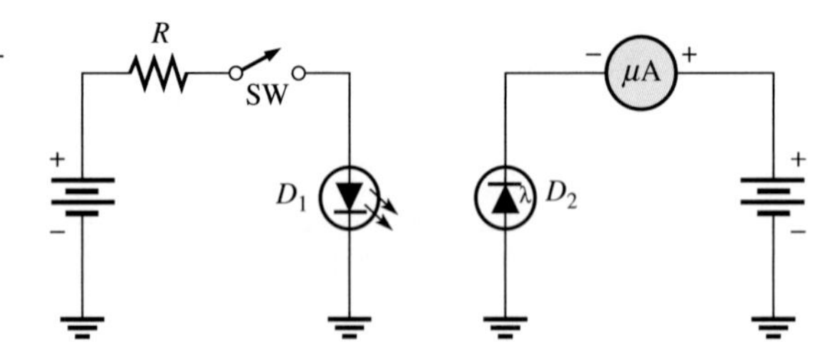

26. With no incident light, there is a certain amount of reverse current in a photodiode. What is this current called?

SECTION 17–8 **The Diode Data Sheet**

27. From the data sheet in Appendix E, determine how much peak inverse voltage that a 1N4001 diode can withstand.
28. Repeat Problem 27 for a 1N4004.

29. If the peak output voltage of a full-wave bridge rectifier is 50 V, determine the minimum value of the surge limiting resistor required when 1N4001 diodes are used.

SECTION 17–9

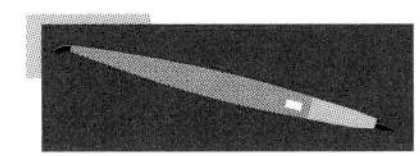

Troubleshooting

30. From the meter readings in Figure 17–94, determine if the rectifier is functioning properly. If it is not, determine the most likely failure(s). DMM1 is an ac voltmeter; DMM2 and DMM3 are dc voltmeters.

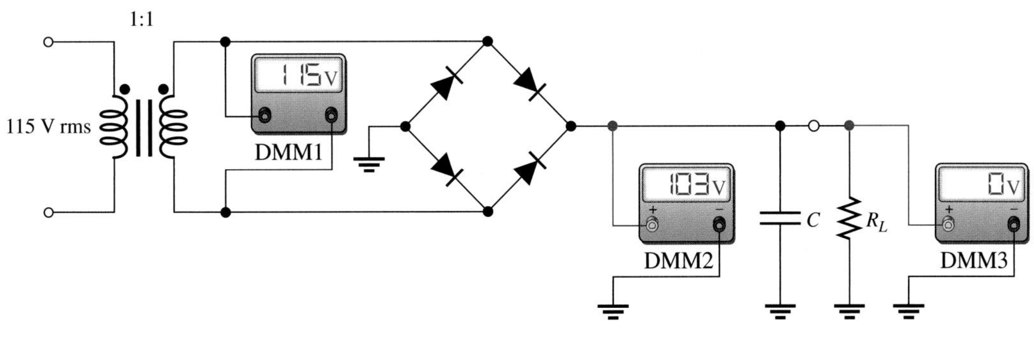

FIGURE 17–94

31. Each part of Figure 17–95 shows oscilloscope displays of rectifier output voltages. In each case, determine whether or not the rectifier is functioning properly and, if it is not, identify the most likely failure(s).

FIGURE 17–95

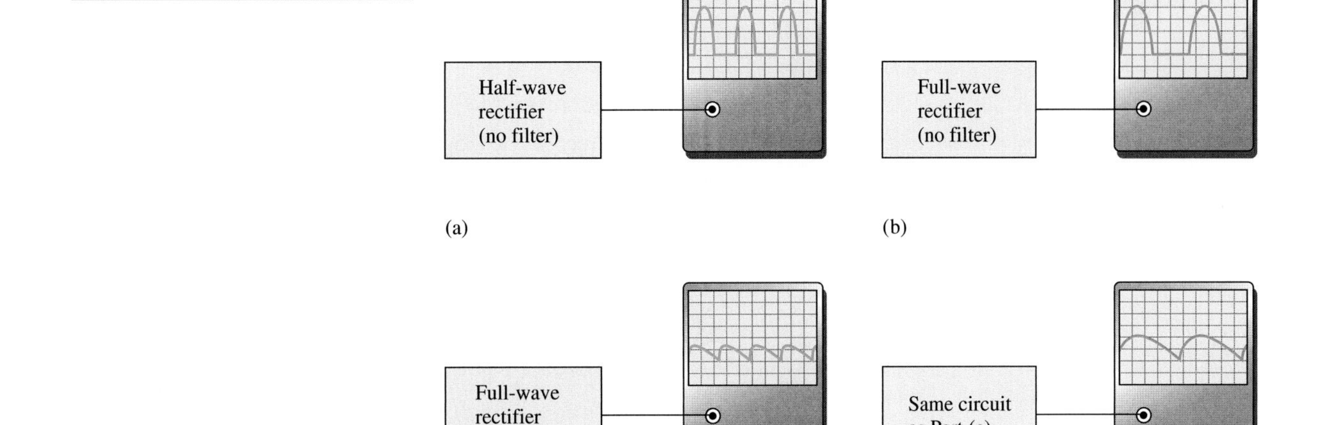

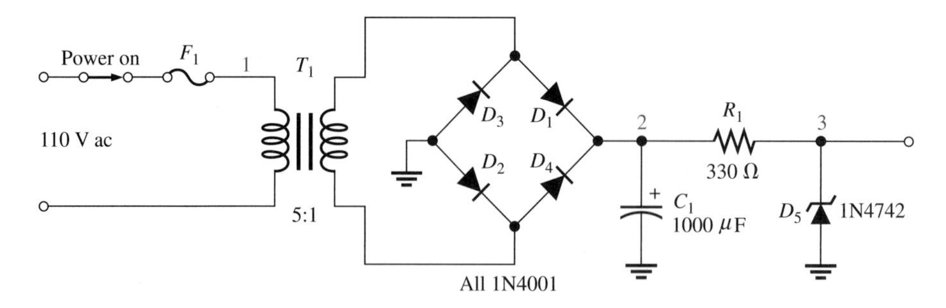

▲ FIGURE 17–96

32. For each set of measured voltages at the points (1, 2, and 3) indicated in Figure 17–96, determine if they are correct, and if not, identify the most likely fault(s). State what you would do to correct the problem once it is isolated.

(a) $V_1 = 110$ V rms, $V_2 \cong 30$ V dc, $V_3 \cong 12$ V dc

(b) $V_1 = 110$ V rms, $V_2 \cong 30$ V dc, $V_3 \cong 30$ V dc

(c) $V_1 = 0$ V, $V_2 = 0$ V, $V_3 = 0$ V

(d) $V_1 = 110$ V rms, $V_2 \cong 30$ V peak full-wave 120 Hz voltage, $V_3 \cong 12$ V 120 Hz pulsating voltage

(e) $V_1 = 100$ V rms, $V_2 = 0$ V, $V_3 = 0$ V

33. Determine the most likely failure in the circuit board of Figure 17–97 for each of the following symptoms. State the corrective action you would take in each case. The transformer has a turns ratio of 1.

(a) No voltage across the primary

(b) No voltage at point 2 with respect to ground; 110 V rms across the primary

(c) No voltage at point 3 with respect to ground; 110 V rms across the primary

(d) 150 V rms at point 2 with respect to ground; input is correct at 110 V rms

(e) 68 V rms at point 3 with respect to ground; input is correct at 110 V rms

(f) A pulsating full-wave rectified voltage with a peak of 155.5 V at point 4 with respect to ground

▶ FIGURE 17–97

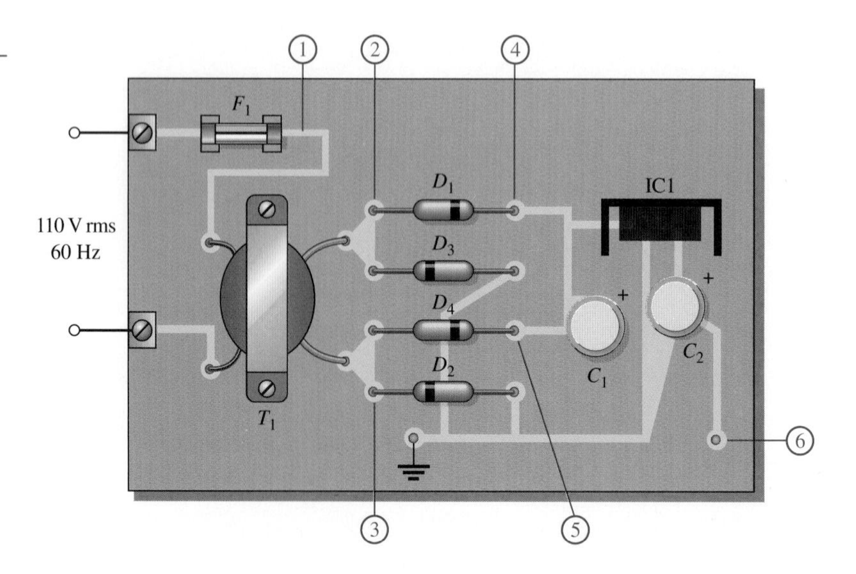

(g) Excessive 120 Hz ripple voltage at point 5 with respect to ground

(h) Ripple voltage has a frequency of 60 Hz at point 4 with respect to ground

(i) No voltage at point 6 with respect to ground

ELECTRONICS WORKBENCH/CIRCUITMAKER TROUBLESHOOTING PROBLEMS

CD-ROM file circuits are shown in Figure 17–98.

34. Open file P17-34 and determine if there is a fault. If so, identify it.

35. Open file P17-35 and determine if there is a fault. If so, identify it.

36. Open file P17-36 and determine if there is a fault. If so, identify it.

37. Open file P17-37 and determine if there is a fault. If so, identify it.

38. Open file P17-38 and determine if there is a fault. If so, identify it.

FIGURE 17–98

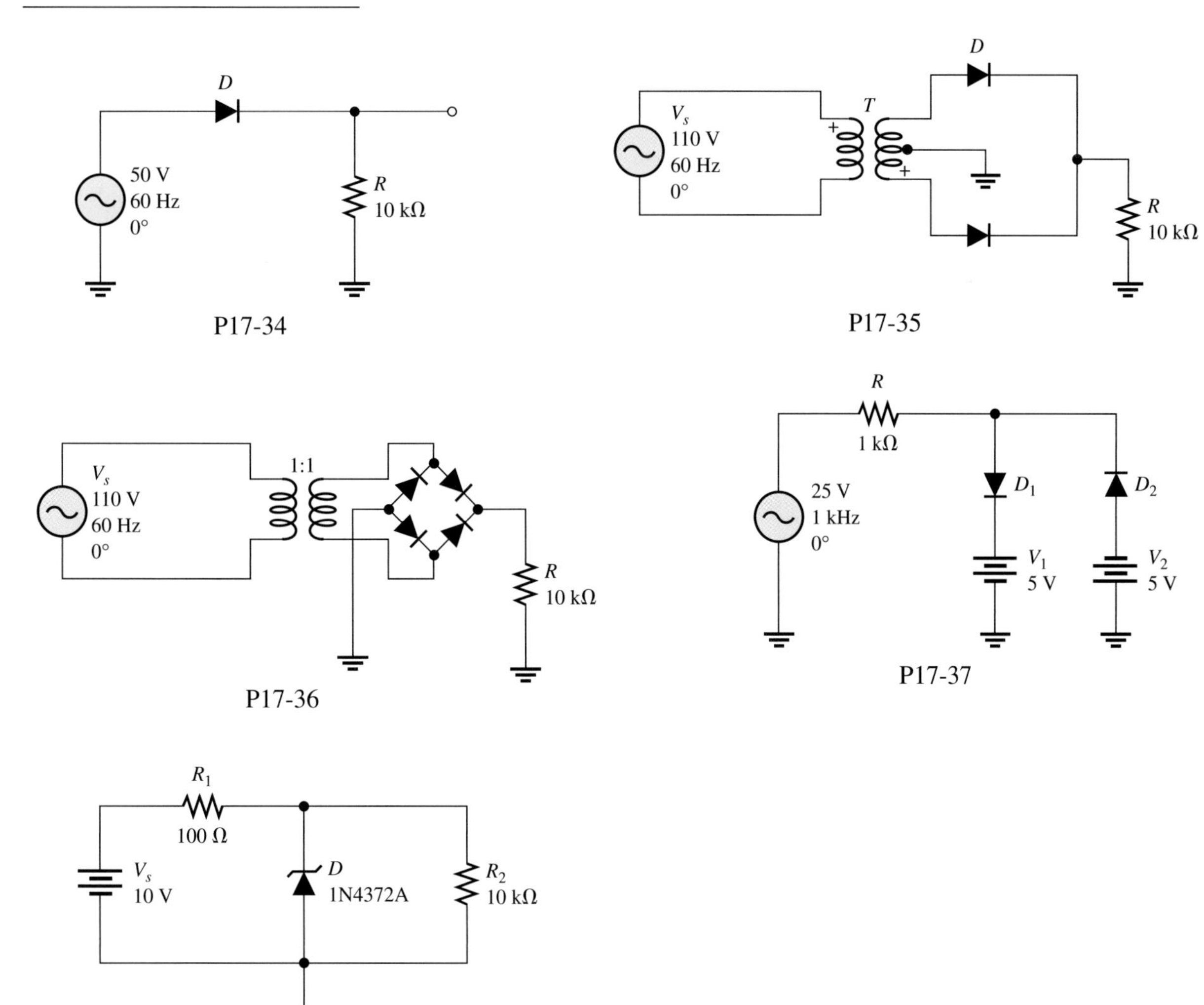

ANSWERS

SECTION REVIEWS

SECTION 17–1

Half-Wave Rectifiers

1. The PIV occurs at 270°, the peak of the negative alternation of the input cycle.
2. There is load current for 50% of the input cycle.
3. $V_{AVG} = V_{p(out)}/\pi = 3.18$ V

SECTION 17–2

Full-Wave Rectifiers

1. $V_{AVG} = 2V_{p(out)}/\pi = 38.2$ V
2. A bridge rectifier has the greater V_{out}.
3. True

SECTION 17–3

Power Supply Filters and Regulators

1. Ripple voltage is caused by the capacitor charging and discharging slightly.
2. Reducing load resistance increases ripple voltage.
3. Better ripple rejection, line and load regulation, thermal protection.
4. *Line regulation:* Constant output voltage with varying input voltage.
   Load regulation: Constant output voltage with varying load current.

SECTION 17–4

Diode Limiting and Clamping Circuits

1. Limiters clip off or remove portions of a waveform. Clampers insert a dc level.
2. A positive limiter clips off positive voltages. A negative limiter clips off negative voltages.
3. 0.7 V appears across the diode in the limiter during the positive alternation.
4. The bias voltage must be 5 V − 0.7 V = 4.3 V.
5. The capacitor acts as a battery in a clamper.

SECTION 17–5

Zener Diodes

1. True
2. $V_Z = 10\text{ V} + (30\text{ mA})(8\ \Omega) = 10.2\text{ V}$
3. I_Z is maximum for maximum R_L.
4. % load reg. = $[(V_{NL} - V_{FL})/V_{FL}] \times 100\% = 0.833\%$

SECTION 17–6

Varactor Diodes

1. A varactor acts as a variable capacitor.
2. Diode capacitance decreases when reverse voltage is increased.

SECTION 17–7

LEDs and Photodiodes

1. LEDs give off light when forward-biased; photodiodes respond to light when reverse-biased.
2. Gallium arsenide, gallium arsenide phosphide, and gallium phosphide are LED materials.
3. Dark current is the small photodiode reverse current with no light.

SECTION 17–8

The Diode Data Sheet

1. The three rating categories on a diode data sheet are maximum ratings, electrical characteristics, and mechanical data.

2. V_F is forward voltage, I_R is reverse current, and I_O is peak average forward current.
3. I_{FSM} is maximum forward surge current, V_{RRM} is maximum reverse peak repetitive voltage, and V_{RSM} is maximum reverse peak nonrepetitive voltage.
4. The 1N4720 has an $I_O = 3.0$ A, $I_{FSM} = 300$ A, and $V_{RRM} = 100$ V.

SECTION 17–9 **Troubleshooting**

1. An open diode results in no output voltage.
2. An open diode produces a half-wave output voltage.
3. The shorted diode may burn open. Transformer will be damaged. Fuse will blow.
4. The amplitude of the ripple voltage increases with a leaky filter capacitor.
5. There will be no output voltage when the primary opens.
6. The problem may be a partially shorted secondary winding.

■ Application Assignment

1. Rectifier, zener, light-emitting, photo
2. R_2 limits the LED current.
3. Use an 8.2 V zener diode.
4. 1N4738
5. The LED emits a continuous IR beam.
6. The photodiode senses the absence of the IR beam.

RELATED PROBLEMS FOR EXAMPLES

17–1 3.82 V

17–2 2.3 V

17–3 98.7 V

17–4 320 V

17–5 $V_{p(out)} = 160$ V; PIV $= 160$ V

17–6 3.7%

17–7 A positive peak of 8.72 V and clipped at -0.3 V

17–8 A sine wave clipped at $+10.7$ V and -10.7 V

17–9 Same as Figure 17–40

17–10 5 Ω

17–11 $V_Z = 11.4$ V at 20 mA; $V_Z = 12.6$ V at 80 mA

17–12 $V_{IN(min)} = 6.81$ V; $V_{IN(max)} = 48.1$ V

17–13 $I_{L(min)} = 0$ A; $I_{L(max)} = 42$ mA; $R_{L(min)} = 79$ Ω

17–14 **(a)** 3.77% **(b)** 1.75%

17–15 101 kHz to 318 kHz

17–16 Verify C is shorted and replace it.

SELF-TEST

1. (c)	**2.** (b)	**3.** (a)	**4.** (a)	**5.** (c)	**6.** (d)	**7.** (b)
8. (d)	**9.** (a)	**10.** (d)	**11.** (b)	**12.** (a)	**13.** (c)	**14.** (a)
15. (c)	**16.** (d)	**17.** (b)	**18.** (a)			

18

TRANSISTORS AND THYRISTORS

INTRODUCTION

Two basic types of transistors will be discussed in this chapter: the bipolar junction transistor (BJT) and the field-effect transistor (FET). The two major application areas of amplification and switching are introduced. Also, the unijunction transistor (UJT) will be introduced and common types of thyristors presented.

CHAPTER OUTLINE

CHAPTER OUTLINE

CHAPTER OBJECTIVES

- Describe the basic structure and operation of bipolar junction transistors
- Analyze a transistor circuit with voltage-divider bias
- Explain the operation of a bipolar junction transistor as an amplifier
- Analyze a transistor switching circuit
- Discuss important transistor parameters and maximum ratings
- Describe the basic structure and operation of a JFET
- Explain the operation of a JFET in terms of its characteristics
- Explain the basic structure and operation of MOSFETs
- Analyze the common biasing arrangements for JFETs and MOSFETs
- Discuss the operation and application of the UJT
- Explain the basic structure and operation of the SCR, the triac, and the diac
- Identify various types of transistor package configurations
- Troubleshoot transistors

KEY TERMS

- Bipolar junction transistor (BJT)
- Emitter
- Base
- Collector
- Bipolar
- Saturation
- Cutoff
- Q-point
- Amplification
- Junction field-effect transistor (JFET)
- Drain
- Source
- Gate
- Metal-oxide semiconductor field-effect transistor (MOSFET)
- Depletion mode
- Enhancement mode
- Unijunction transistor (UJT)
- Silicon-controlled rectifier (SCR)
- Triac
- Diac

APPLICATION ASSIGNMENT PREVIEW

You have been assigned to work on a proportional heat-control system that maintains a liquid in a tank at a precise temperature for a certain industrial production process. You are asked to analyze and troubleshoot a particular transistor circuit in the system that produces a dc output voltage proportional to the temperature of the substance in the tank. This circuit uses the variable resistance of a temperature sensor as part of its bias circuit. The output of this temperature detector circuit is converted to digital form for precise control of the burner that heats the tank. If the temperature of the substance tends to increase above a certain value, the circuitry causes less fuel to be fed to the burner and thus reduces the temperature. If the temperature tends to decrease, the circuitry causes more fuel to be fed to the burner and thus raises the temperature. This type of system maintains the substance at a nearly constant temperature and is typical of many industrial process-control systems. After studying this chapter, you should be able to complete the application assignment.

WWW. VISIT THE COMPANION WEBSITE

Circuit Simulation Tutorials and Other Chapter Study Tools Are Available at

http://www.prenhall.com/floyd

18–1 BIPOLAR JUNCTION TRANSISTORS (BJTs)

The basic structure of the bipolar junction transistor, BJT, determines its operating characteristics. In this section, you will see how semiconductive materials are joined to form a transistor, and you will learn the standard transistor symbols. Also, you will see how important dc bias is to the operation of transistors in terms of setting up proper currents and voltages in a transistor circuit. Finally, two important parameters, α_{DC} and β_{DC}, are introduced.

After completing this section, you should be able to

- **Describe the basic structure and operation of bipolar junction transistors**
- Explain the difference between *npn* and *pnp* transistors
- Explain transistor biasing
- Discuss transistor currents and their relationships
- Define β_{DC}
- Define α_{DC}
- Discuss the transistor voltages
- Calculate dc bias currents and voltages in a basic transistor circuit

The **bipolar junction transistor (BJT)** is constructed with three doped semiconductor regions separated by two *pn* junctions, as shown in the epitaxial planar structure in Figure 18–1(a). The three regions are called **emitter**, **base**, and **collector**. Representations of the two types of bipolar transistors are shown in Figure 18–1(b) and (c). One type consists of two *n* regions separated by a *p* region (*npn*), and the other consists of two *p* regions separated by an *n* region (*pnp*).

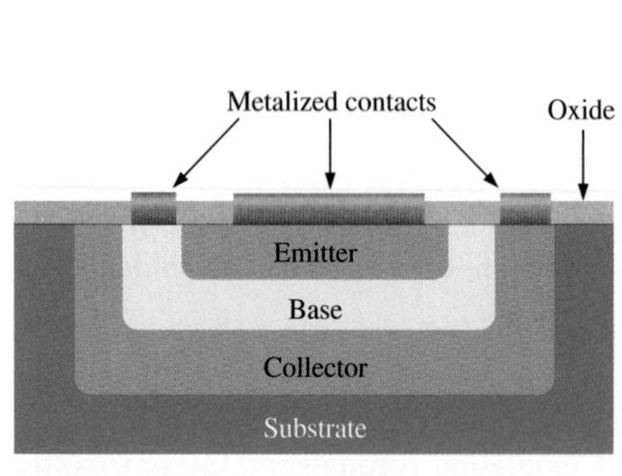

(a) Basic epitaxial planar structure

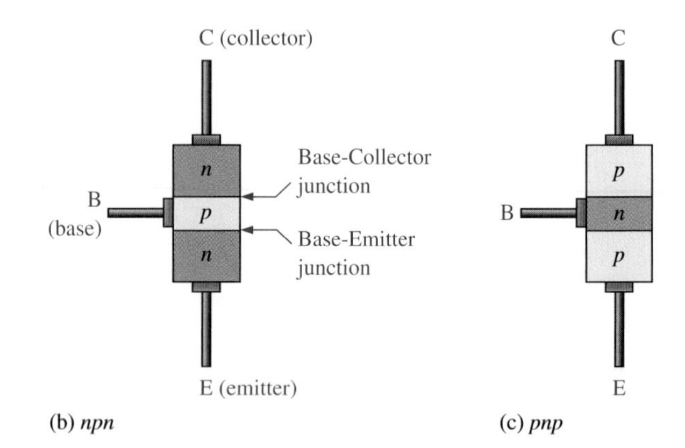

(b) *npn* (c) *pnp*

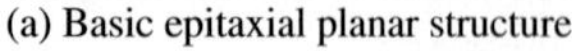

FIGURE 18–1

Basic construction of bipolar junction transistors.

The *pn* junction joining the base region and the emitter region is called the *base-emitter junction.* The junction joining the base region and the collector region is called the *base-collector* junction, as indicated in Figure 18–1(b). A wire lead connects to each of the three regions, as shown. These leads are labeled E, B, and C for emitter, base, and collector, respectively. The base material is lightly doped and very narrow compared to the heavily doped emitter and collector materials. The reason for this is discussed in the next section.

Figure 18–2 shows the schematic symbols for the *npn* and *pnp* bipolar transistors. Notice that the emitter terminal has an arrow. The term **bipolar** refers to the use of both holes and electrons as charge carriers in the transistor structure.

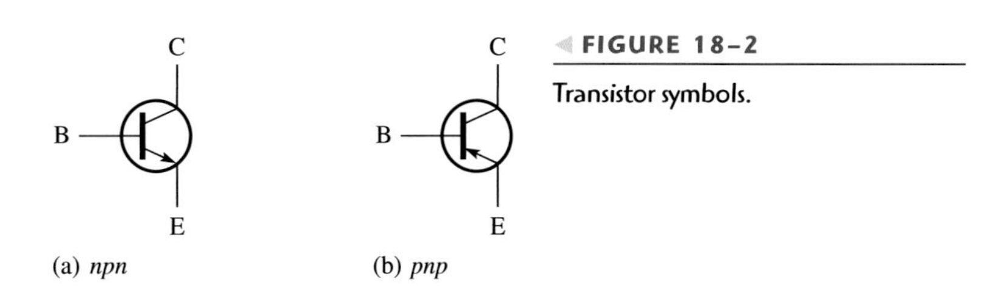

FIGURE 18–2

Transistor symbols.

Transistor Biasing

In order for the **transistor** to operate properly as an amplifier, the two *pn* junctions must be correctly biased with external dc voltages. We will use the *npn* transistor to illustrate transistor biasing. The operation of the *pnp* is the same as for the *npn* except that the roles of the electrons and holes, the bias voltage polarities, and the current directions are all reversed. Figure 18–3 shows the proper bias arrangement for both *npn* and *pnp* transistors. Notice that in both cases the base-emitter (BE) junction is forward-biased and the base-collector (BC) junction is reverse-biased. This is called *forward-reverse bias.*

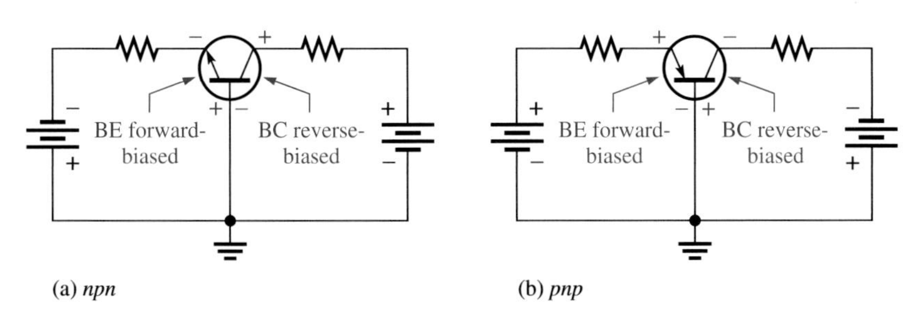

FIGURE 18–3

Forward-reverse bias of a bipolar junction transistor.

To illustrate transistor action, let's examine what happens inside the *npn* transistor when the junctions are forward-reverse biased. The forward bias from base to emitter narrows the BE depletion region, and the reverse bias from base to collector widens the BC depletion region, as depicted in Figure 18–4. The heavily doped *n*-type emitter region is teeming with conduction-band (free) electrons that easily diffuse through the forward-biased BE junction into the *p*-type base region where they become minority carriers, just as in a forward-biased diode. The base region is lightly doped and very thin so that it has a very limited number of holes. Thus, only a small percentage of all the electrons flowing through the BE junction can combine with the available holes in the base. These relatively few recombined electrons flow out of the base lead as valence electrons, forming the small base electron current, as shown in Figure 18–4.

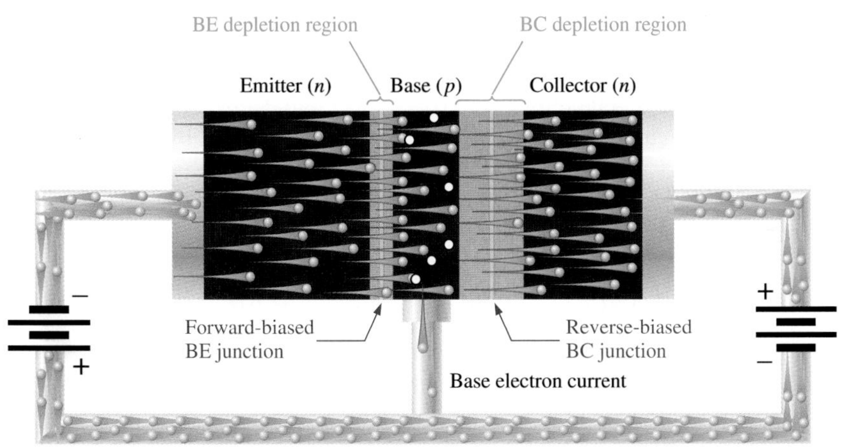

FIGURE 18–4

Pictorial representation of transistor operation.

Most of the electrons flowing from the emitter into the thin, lightly doped base region do not recombine but diffuse into the BC depletion region. Once in this region they are pulled through the reverse-biased BC junction by the electric field set up by the force of attraction between the positive and negative ions. Actually, you can think of the electrons as being pulled across the reverse-biased BC junction by the attraction of the collector supply voltage. The electrons now move through the collector region, out through the collector lead, and into the positive terminal of the collector voltage source. This forms the collector electron current, as shown in Figure 18–4.

Transistor Currents

The directions of current in an *npn* and a *pnp* transistor are as shown in Figure 18–5(a) and 18–5(b), respectively. An examination of these diagrams shows that the emitter current is the sum of the collector and base currents, expressed as follows:

Equation 18–1

$$I_E = I_C + I_B$$

As mentioned before, I_B is very small compared to I_E or I_C. The capital-letter subscripts indicate dc values.

FIGURE 18–5

Transistor currents.

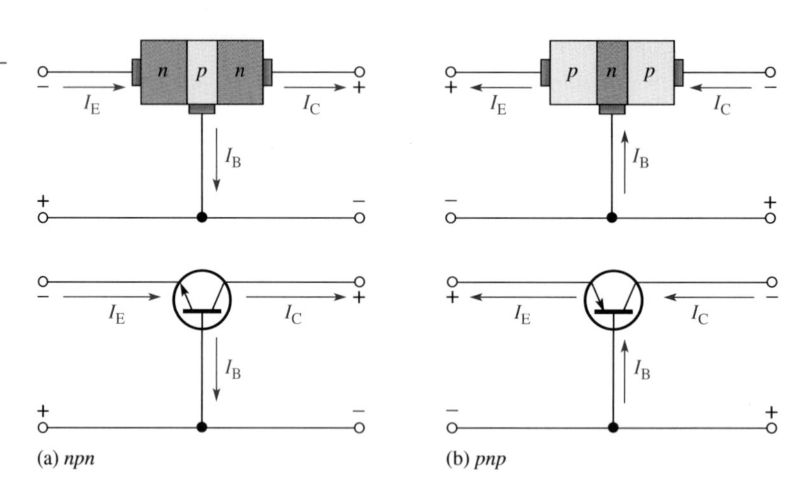

These direct currents (emitter, base, and collector) are also related by two parameters: the dc **alpha** (α_{DC}), which is the ratio I_C/I_E and the dc **beta** (β_{DC}), which is the ratio I_C/I_B. β_{DC} is the direct current gain and is usually designated as h_{FE} on the transistor data sheets.

The collector current is equal to α_{DC} times the emitter current.

Equation 18–2

$$I_C = \alpha_{DC} I_E$$

where α_{DC} typically has a value between 0.950 and 0.995. Generally α_{DC} can be considered to be 1 and therefore $I_C = I_E$.

The collector current is equal to the base current multiplied by β_{DC}.

Equation 18–3

$$I_C = \beta_{DC} I_B$$

where β_{DC} typically has a value between 20 and 200.

Transistor Voltages

The three dc voltages for the biased transistor in Figure 18–6 are the emitter voltage (V_E), the collector voltage (V_C), and the base voltage (V_B). These voltages are with respect to ground. The collector voltage is equal to the dc supply voltage, V_{CC}, less the drop across R_C.

Equation 18–4

$$V_C = V_{CC} - I_C R_C$$

The base voltage is equal to the emitter voltage plus the base-emitter junction barrier potential (V_{BE}), which is about 0.7 V for a silicon transistor.

$$V_B = V_E + V_{BE}$$ Equation 18–5

In the configuration of Figure 18–6, the emitter is the common terminal, so $V_E =$ 0 V and $V_B = 0.7$ V.

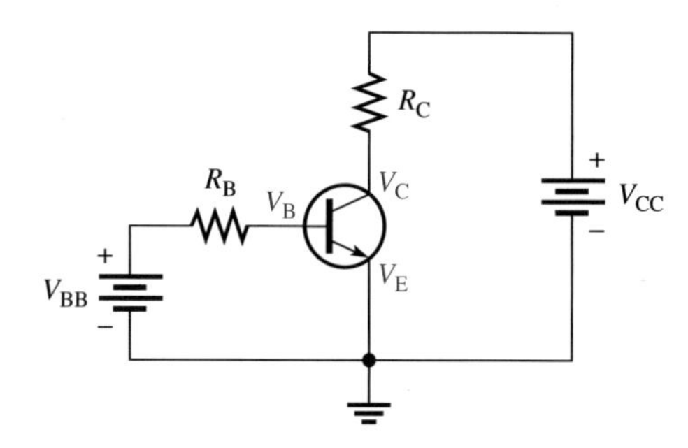

FIGURE 18–6

Bias voltages.

EXAMPLE 18–1

Find currents I_B, I_C, and I_E and voltages with respect to ground V_B and V_C in Figure 18–7. β_{DC} is 50.

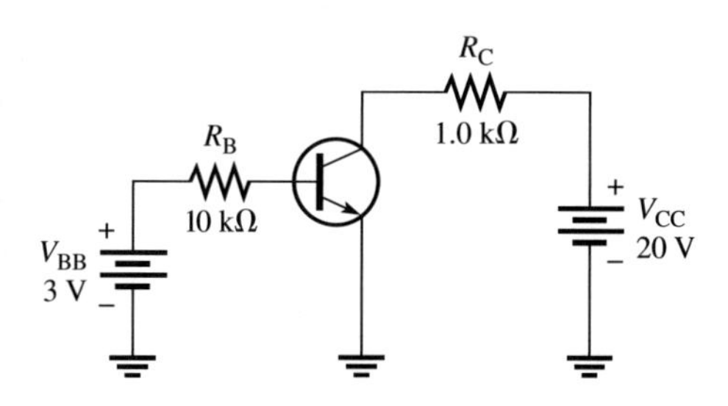

FIGURE 18–7

Solution Since the emitter is at ground, $V_B =$ **0.7 V**. The drop across R_B is $V_{BB} - V_B$, so I_B is calculated as follows:

$$I_B = \frac{V_{BB} - V_B}{R_B} = \frac{3\text{ V} - 0.7\text{ V}}{10\text{ k}\Omega} = \mathbf{230\ \mu A}$$

Now you can find I_C, I_E, and V_C.

$$I_C = \beta_{DC} I_B = 50(230\ \mu\text{A}) = \mathbf{11.5\ mA}$$

$$I_E = I_C + I_B = 11.5\text{ mA} + 230\ \mu\text{A} = \mathbf{11.7\ mA}$$

$$V_C = V_{CC} - I_C R_C = 20\text{ V} - (11.5\text{ mA})(1.0\text{ k}\Omega) = \mathbf{8.5\ V}$$

*Related Problem** Determine I_B, I_C, I_E, V_B, and V_C in Figure 18–7 for the following values: $R_B = 22$ kΩ, $R_C = 220$ Ω, $V_{BB} = 6$ V, $V_{CC} = 9$ V, and $\beta_{DC} = 90$.

Open file E18-01 on your EWB/CircuitMaker CD-ROM. Measure all of the transistor currents and terminal voltages. Change all values as specified in the related problem and repeat the measurements.

*Answers are at the end of the chapter.

SECTION 18–1 REVIEW

Answers are at the end of the chapter.

1. Name the three transistor terminals for a bipolar junction transistor.
2. Define *forward-reverse bias.*
3. What is β_{DC}?
4. If I_B is 10 μA and β_{DC} is 100, what is the collector current?

18–2 VOLTAGE-DIVIDER BIAS

In this section, you will study a method of biasing a transistor for linear operation using a resistive voltage divider. Although there are other bias methods, this one is the most widely used.

After completing this section, you should be able to

- **Analyze a transistor circuit with voltage-divider bias**
- Determine base input resistance
- Determine base, emitter, and collector voltages
- Determine base, emitter, and collector currents

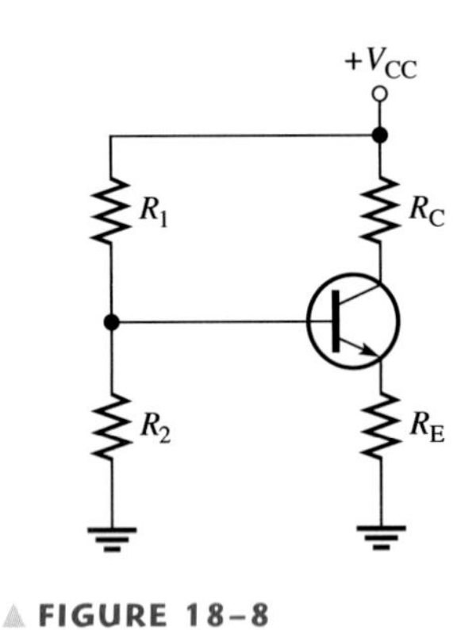

FIGURE 18–8

Voltage-divider bias.

The voltage-divider bias configuration uses only a single dc source to provide forward-reverse bias to the transistor, as shown in Figure 18–8. Resistors R_1 and R_2 form a voltage divider that provides the base bias voltage. Resistor R_E allows the emitter to rise above ground potential.

The voltage divider is loaded by the resistance as viewed from the base of the transistor. In some cases, this loading effect is significant in determining the base bias voltage. Let's examine this arrangement in more detail.

Input Resistance at the Base

The approximate input resistance of the transistor, viewed from the base of the transistor in Figure 18–9, is derived as follows:

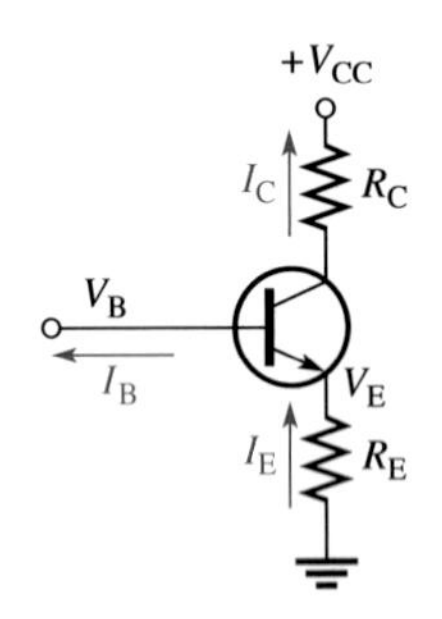

FIGURE 18–9

Circuit for deriving input resistance.

$$R_{IN} = \frac{V_B}{I_B}$$

Neglecting the V_{BE} of 0.7 V,

$$V_B \cong V_E = I_E R_E$$

Since $I_E \cong I_C$, then

$$I_E \cong \beta_{DC} I_B$$

Substituting yields

$$V_B \cong \beta_{DC} I_B R_E$$

$$R_{IN} \cong \frac{\beta_{DC} I_B R_E}{I_B}$$

The I_B terms cancel, leaving

Equation 18–6

$$R_{IN} \cong \beta_{DC} R_E$$

Base Voltage

Now, using the voltage-divider formula, you get the following equation for the approximate base voltage for the circuit in Figure 18–8:

$$V_B = \left(\frac{R_2 \parallel R_{IN}}{R_1 + R_2 \parallel R_{IN}}\right)V_{CC}$$

Equation 18–7

If R_{IN} is at least ten times greater than R_2, Equation 18–7 can be simplified.

$$V_B \cong \left(\frac{R_2}{R_1 + R_2}\right)V_{CC}$$

Equation 18–8

Once you have determined the base voltage, you can determine the emitter voltage V_E (for an *npn* transistor).

$$V_E = V_B - 0.7\text{ V}$$

Equation 18–9

EXAMPLE 18–2

Determine V_B, V_E, V_C, V_{CE}, I_B, I_E, and I_C in Figure 18–10.

▶ FIGURE 18–10

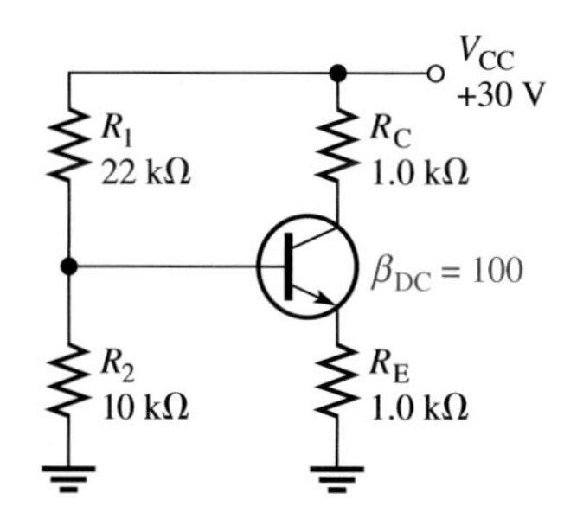

Solution The input resistance at the base is

$$R_{IN} \cong \beta_{DC}R_E = 100(1.0\text{ k}\Omega) = 100\text{ k}\Omega$$

Since R_{IN} is ten times greater than R_2, you can neglect it and the base voltage is approximately

$$V_B \cong \left(\frac{R_2}{R_1 + R_2}\right)V_{CC} = \left(\frac{10\text{ k}\Omega}{32\text{ k}\Omega}\right)30\text{ V} = \mathbf{9.38\text{ V}}$$

Therefore, $V_E = V_B - 0.7\text{ V} = \mathbf{8.68\text{ V}}$.

Now that you know V_E, you can find I_E by Ohm's law.

$$I_E = \frac{V_E}{R_E} = \frac{8.68\text{ V}}{1.0\text{ k}\Omega} = \mathbf{8.68\text{ mA}}$$

Since α_{DC} is so close to 1 for most transistors, it is a good approximation to assume that $I_C \cong I_E$. Thus,

$$I_C \cong \mathbf{8.68\text{ mA}}$$

Use $I_C = \beta_{DC}I_B$ and solve for I_B.

$$I_B = \frac{I_C}{\beta_{DC}} = \frac{8.68\text{ mA}}{100} = \mathbf{86.8\ \mu A}$$

Since you know I_C, you can find V_C.

$$V_C = V_{CC} - I_C R_C = 30\text{ V} - (8.68\text{ mA})(1.0\text{ k}\Omega) = 30\text{ V} - 8.68\text{ V} = \mathbf{21.3\text{ V}}$$

Since V_{CE} is the collector-to-emitter voltage, it is the difference of V_C and V_E.

$$V_{CE} = V_C - V_E = 21.3\text{ V} - 8.68\text{ V} = \mathbf{12.6\text{ V}}$$

Related Problem Determine V_B, V_E, V_C, V_{CE}, I_B, I_E, and I_C taking into account R_{IN} at the base in Figure 18–10.

Open file E18-02 on your EWB/CircuitMaker CD-ROM. Measure all of the transistor currents and the terminal voltages.

SECTION 18–2 REVIEW

1. How many dc voltage sources are required for voltage-divider bias?
2. In Figure 18–10, if R_2 is 4.7 kΩ, what is the value of V_B?
3. If V_{CC} is reduced to 20 V in Figure 18–10 with no other changes, what is I_C?

18–3 THE BIPOLAR JUNCTION TRANSISTOR AS AN AMPLIFIER

DC bias allows a transistor to operate as an amplifier. Thus, a transistor can be used to produce a larger signal using a smaller signal as a "pattern." In this section, we will discuss how a transistor acts as an amplifier. The subject of amplifiers is covered in more detail in the next chapter.

After completing this section, you should be able to

- **Explain the operation of a bipolar junction transistor as an amplifier**
- Discuss the collector characteristic curves
- Define *cutoff* and *saturation*
- Analyze the dc load line operation of an amplifier
- Discuss the significance of the Q-point
- Explain the signal operation of an amplifier
- Calculate the voltage gain of an amplifier

Let's examine the parameters and dc operating conditions important in the operation of a transistor **amplifier.**

Collector Characteristic Curves

Using a circuit like that shown in Figure 18–11(a), you can generate a set of *collector characteristic curves* that show how the collector current, I_C, varies with the collector-to-emitter voltage, V_{CE}, for specified values of base current, I_B. Notice in the circuit diagram that both V_{BB} and V_{CC} are variable sources of voltage.

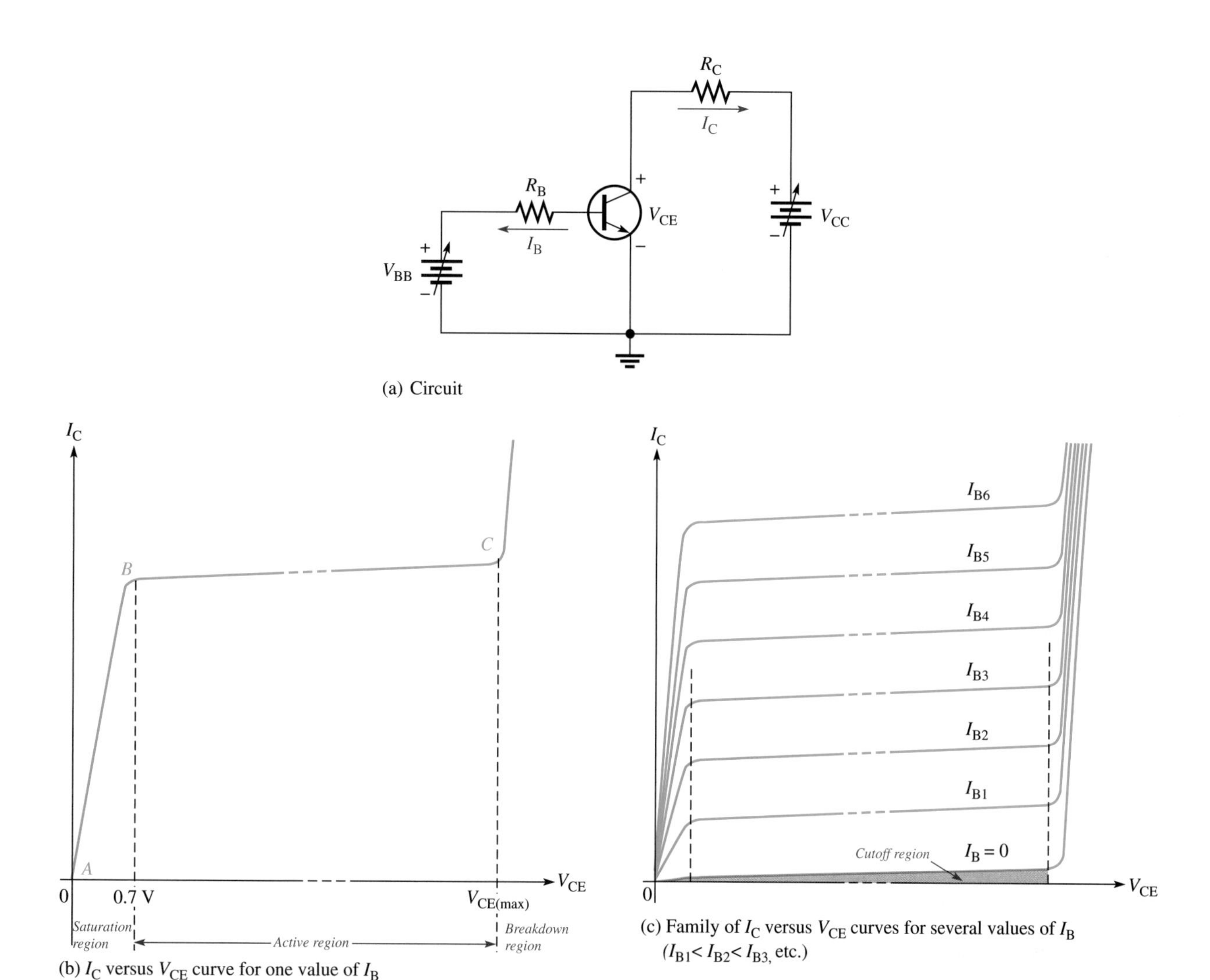

▲ **FIGURE 18–11**

Collector characteristic curves.

Assume that V_{BB} is set to produce a certain value of I_B and V_{CC} is zero. For this condition, both the base-emitter junction and the base-collector junction are forward-biased because the base is at approximately 0.7 V while the emitter and the collector are at 0 V. The base current is through the base-emitter junction because of the low impedance path to ground and, therefore, I_C is zero. When both junctions are forward-biased, the transistor is in the saturation region of its operation.

As V_{CC} is increased, V_{CE} increases gradually as the collector current increases. This is indicated by the portion of the characteristic curve between points *A* and *B* in Figure 18–11(b). I_C increases as V_{CC} is increased because V_{CE} remains less than 0.7 V due to the forward-biased base-collector junction.

Ideally, when V_{CE} exceeds 0.7 V, the base-collector junction becomes reverse-biased and the transistor goes into the *active* or **linear** region of its operation. Once the base-collector junction is reverse-biased, I_C levels off and remains essentially constant for a given value of I_B as V_{CE} continues to increase. Actually, I_C increases very slightly as V_{CE} increases due to widening of the base-collector depletion region. This results in fewer holes for recombination in the base region which effectively causes a slight increase in β_{DC}. This is shown by the portion of the characteristic curve between points *B* and *C* in Figure 18–11(b). For this portion of the characteristic curve, the value of I_C is determined only by the relationship expressed as $I_C = \beta_{DC} I_B$.

When V_{CE} reaches a sufficiently high voltage, the reverse-biased base-collector junction goes into breakdown; and the collector current increases rapidly as indicated by the part of the curve to the right of point C in Figure 18–11(b). A transistor should never be operated in this breakdown region.

A family of collector characteristic curves is produced when I_C versus V_{CE} is plotted for several values of I_B, as illustrated in Figure 18–11(c). When $I_B = 0$, the transistor is in the cutoff region although there is a very small collector leakage current as indicated. The amount of collector leakage current for $I_B = 0$ is exaggerated on the graph for purposes of illustration.

EXAMPLE 18–3

Draw the family of collector characteristic curves for the circuit in Figure 18–12 for $I_B = 5\ \mu A$ to $25\ \mu A$ in $5\ \mu A$ increments. Assume that $\beta_{DC} = 100$.

FIGURE 18–12

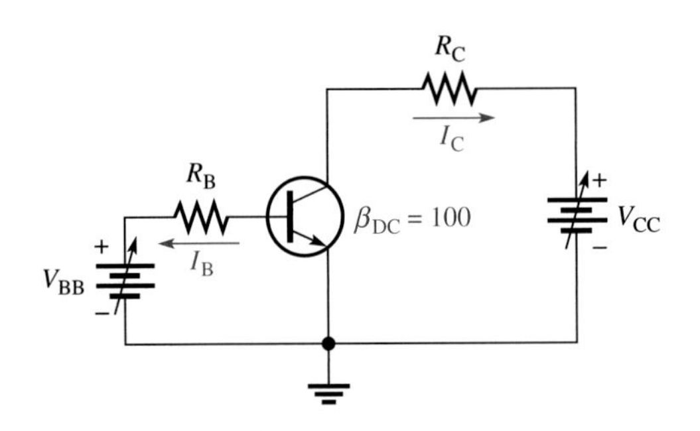

Solution Using the relationship $I_C = \beta_{DC} I_B$, you can calculate the ideal values of I_C and tabulate as in Table 18–1.

The resulting curves are plotted in Figure 18–13. Generally, characteristic curves have a slight upward slope as shown. This indicates that, for a given value of I_B, I_C increases slightly as V_{CE} increases.

TABLE 18–1

I_B	I_C
5 μA	0.5 mA
10 μA	1.0 mA
15 μA	1.5 mA
20 μA	2.0 mA
25 μA	2.5 mA

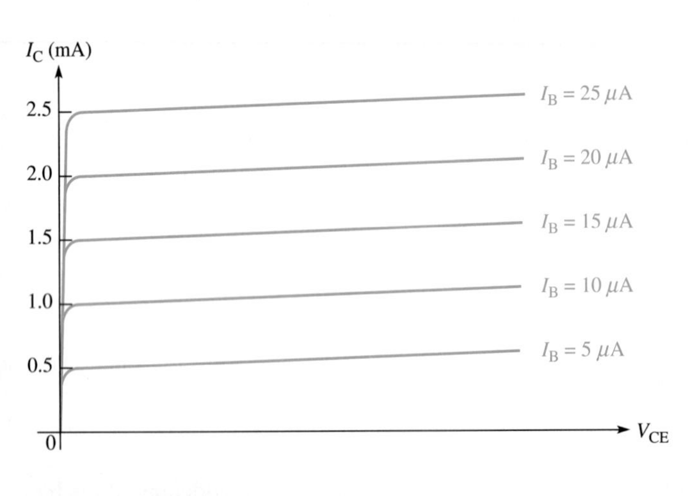

FIGURE 18–13

Related Problem Where would the curve for $I_B = 0$ appear on the graph in Figure 18–13?

Cutoff and Saturation

As previously mentioned, when $I_B = 0$, the transistor is in cutoff. Under this condition, there is a very small amount of collector leakage current, I_{CEO}, due mainly to thermally produced carriers. In cutoff, both the base-emitter and the base-collector junctions are reverse-biased.

Now let's consider the condition known as saturation. When the base current is increased, the collector current also increases and V_{CE} decreases as a result of more drop across R_C. When V_{CE} reaches a value called $V_{CE(sat)}$, the base-collector junction becomes forward-biased and I_C can increase no further even with a continued increase in I_B. At the point of saturation, the relation $I_C = \beta_{DC} I_B$ is no longer valid.

For a transistor, $V_{CE(sat)}$ occurs somewhere below the knee of the collector curves and is usually only a few tenths of a volt for silicon transistors and is often assumed to be zero for analysis purposes.

Load Line Operation

A straight line drawn on the collector curves between the cutoff and saturation points of the transistor is called the *load line.* Once set up, the transistor always operates along this line. Thus, any value of I_C and the corresponding V_{CE} will fall on this line. Notice that the load line is determined by the collector circuit resistance and V_{CC} and not by the transistor itself.

Now let's set up a dc load line for the circuit in Figure 18–14. First determine the cutoff point on the load line. When the transistor is in cutoff, there is essentially no collector current. Thus, the collector-emitter voltage, V_{CE}, is equal to V_{CC}. In this case, $V_{CE} = 30$ V.

Next, determine the saturation point on the load line. When the transistor is saturated, V_{CE} is approximately zero. (Actually, it is usually a few tenths of a volt, but zero is a good approximation.) Therefore, all the V_{CC} voltage is dropped across $R_C + R_E$. From this you can determine the saturation value of collector current, $I_{C(sat)}$. This value is the maximum value for I_C. You cannot possibly increase it further without changing V_{CC}, R_C, or R_E. In Figure 18–14, the value of $I_{C(sat)}$ is $V_{CC}/(R_C + R_E)$, which is 93.8 mA.

Finally, the cutoff and saturation points are plotted on the assumed curves in Figure 18–15, and a straight line, which is the dc load line, is drawn between them.

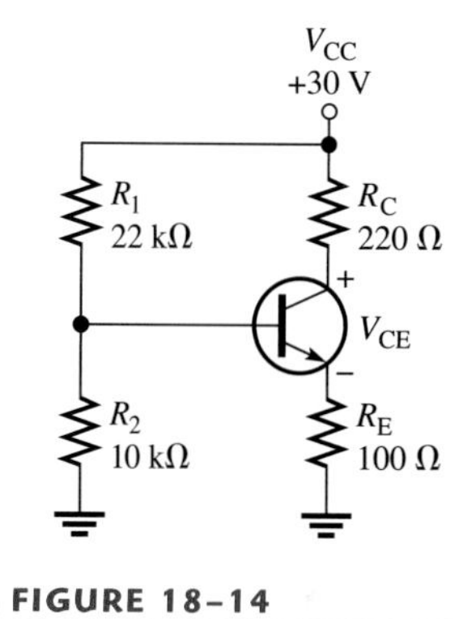

▲ FIGURE 18–14

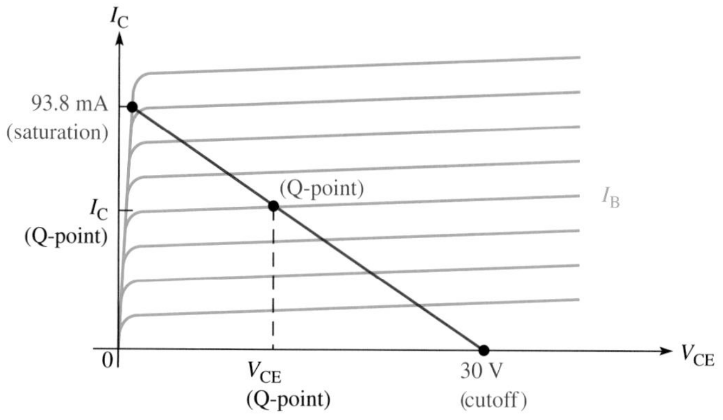

◀ FIGURE 18–15

DC load line (red) for the circuit in Figure 18–14.

Q-Point

The base current, I_B, is established by the base bias. The point at which the base current curve intersects the dc load line is the *quiescent* or Q-point for the circuit. The coordinates of the Q-point are the values for I_C and V_{CE} at that point, as illustrated in Figure 18–15.

Now we have completely described the dc operating conditions for the circuit. In the following paragraphs we discuss ac signal conditions.

Signal (ac) Operation of an Amplifier

The circuit in Figure 18–16 produces an output signal with the same waveform as the input signal but with a greater amplitude. This increase is called amplification. The figure shows an input signal, V_{in}, capacitively coupled to the base. The collector voltage is the output signal, as indicated. The input signal voltage causes the base current to vary at the same frequency above and below its dc value. This variation in base current produces a corresponding variation in collector current. However, the variation in collector current is much larger than the variation in base current because of the current **gain** through the transistor. The ratio of the ac collector current (I_c) to the ac base current (I_b) is designated β_{ac} (the ac beta) or h_{fe}.

Equation 18–10

$$\beta_{ac} = \frac{I_c}{I_b}$$

The value of β_{ac} normally differs slightly from that of β_{DC} for a given transistor. Remember that lowercase subscripts distinguish ac currents and voltages from dc currents and voltages.

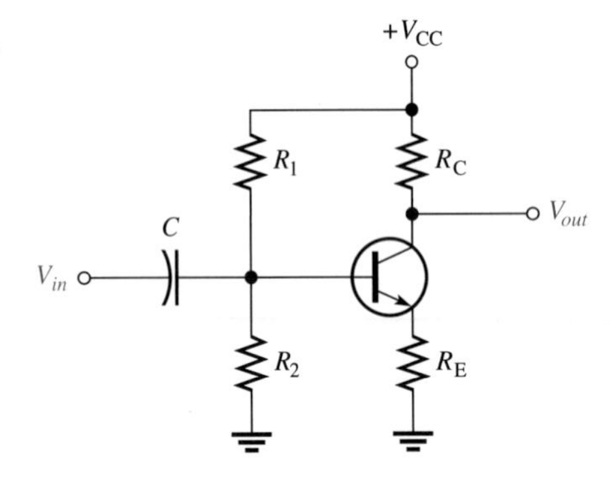

FIGURE 18–16

An amplifier with voltage-divider bias with capacitively coupled input signal. V_{in} and V_{out} are with respect to ground.

Signal Voltage Gain of an Amplifier

Now let's take the amplifier in Figure 18–16 and examine its voltage gain with a signal input. The output voltage is the collector voltage. The variation in collector current produces a variation in the voltage across R_C and a resulting variation in the collector voltage, as shown in Figure 18–17.

As the collector current increases, the I_cR_C drop increases. The increase produces a decrease in collector voltage because $V_c = V_{CC} - I_cR_C$. Likewise, as the collector current decreases, the I_cR_C drop decreases and produces an increase in collector voltage. Therefore, there is a 180° phase difference between the collector current and the collector voltage. The base voltage and collector voltage are also 180° out of phase, as indicated in Figure 18–17. This 180° phase difference between input and output is called an *inversion.*

The voltage gain A_v of the amplifier is V_{out}/V_{in}, where V_{out} is the signal voltage at the collector and V_{in} is the signal voltage at the base. Because the base-emitter

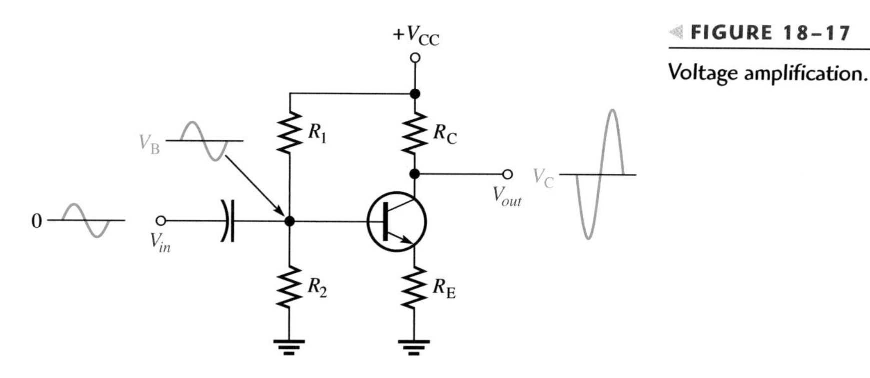

FIGURE 18–17

Voltage amplification.

junction is forward-biased, the signal voltage at the emitter is approximately equal to the signal voltage at the base. Thus, since $V_b \cong V_e$, the gain is approximately $V_c/V_e = I_c R_C / I_e R_E$. Because the α_{ac} is close to 1, I_c and I_e are very close to the same value. Therefore, they cancel, giving the voltage gain formula for the amplifier in Figure 18–17:

$$A_v \cong \frac{R_C}{R_E}$$

Equation 18–11

A negative sign on A_v is often used to indicate the inversion between input and output.

EXAMPLE 18–4

In Figure 18–18, a signal voltage of 50 mV rms is applied to the base.

(a) Determine the output signal voltage for the amplifier.

(b) Find the dc collector voltage on which the output signal voltage is riding.

(c) Draw the output waveform.

FIGURE 18–18

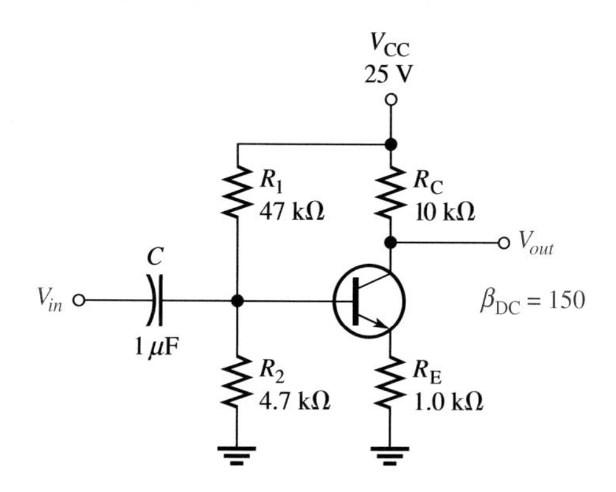

Solution **(a)** The signal voltage gain is

$$A_v \cong \frac{R_C}{R_E} = \frac{10\ \text{k}\Omega}{1.0\ \text{k}\Omega} = 10$$

The output signal voltage is the input signal voltage times the voltage gain.

$$V_{out} = A_v V_{in} = (10)(50\ \text{mV}) = \mathbf{500\ mV\ rms}$$

(b) Next find the dc collector voltage.

$$R_{IN} \cong \beta_{DC}R_E = (150)(1.0\text{ k}\Omega) = 150\text{ k}\Omega$$

R_{IN} can be neglected since it is more than ten times R_2. Thus,

$$V_B \cong \left(\frac{R_2}{R_1 + R_2}\right)V_{CC} = \left(\frac{4.7\text{ k}\Omega}{51.7\text{ k}\Omega}\right)25\text{ V} = 2.27\text{ V}$$

$$I_C \cong I_E = \frac{V_E}{R_E} = \frac{V_B - 0.7\text{ V}}{1.0\text{ k}\Omega} = 1.57\text{ mA}$$

$$V_C = V_{CC} - I_CR_C = 25\text{ V} - (1.57\text{ mA})(10\text{ k}\Omega) = \mathbf{9.3\text{ V}}$$

This value is the dc level of the output. The peak value of the output signal is

$$V_p = 1.414(500\text{ mV}) = 707\text{ mV}$$

(c) Figure 18–19 shows the output waveform with the signal voltage riding on the 9.3 V dc level. Remember, the output waveform is inverted compared to the input.

FIGURE 18–19

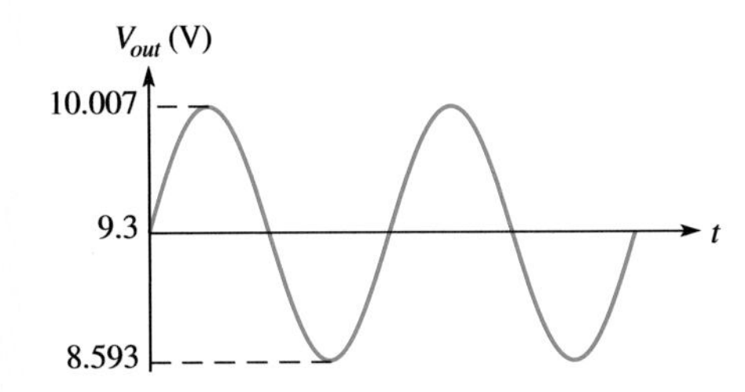

Related Problem If R_C is changed to 12 kΩ in Figure 18–18, what is the output waveform?

Open file E18-04 on your EWB/CircuitMaker CD-ROM. Measure the output signal voltage and the dc level with the oscilloscope. Compare with the waveform in Figure 18–19.

Signal Operation on the Load Line

You can obtain a graphical picture of an amplifier's operation by showing an example of signal variations on a set of collector curves with a load line, as shown in Figure 18–20. Let's assume that the dc Q-point values are as follows: I_B = 40 μA, I_C = 4 mA, and V_{CE} = 8 V. The input signal varies the base current from a maximum of 50 μA to a minimum of 30 μA. The resulting variations in collector current and V_{CE} are shown on the graph. The operation is linear as long as the variations do not reach cutoff at the lower end of the load line or saturation at the upper end.

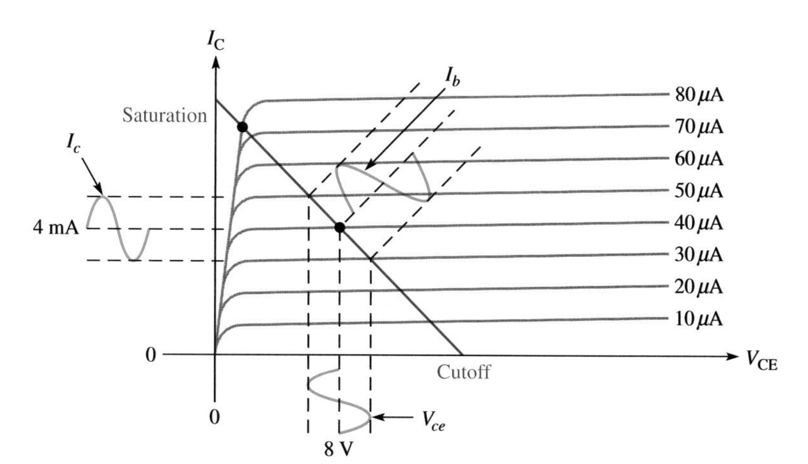

FIGURE 18–20

Signal operation on the load line.

SECTION 18–3 REVIEW

1. What does the term *amplification* mean?
2. What is the Q-point?
3. A certain transistor circuit has $R_C = 47\text{ k}\Omega$ and $R_E = 2.2\text{ k}\Omega$. What is the approximate voltage gain?

18–4 THE BJT AS A SWITCH

In the previous section, we discussed the transistor as a linear amplifier. The second major application area is switching applications. When used as an electronic switch, a transistor normally is operated alternately in cutoff and saturation.

After completing this section, you should be able to

- **Analyze a transistor switching circuit**
- Describe the conditions in cutoff
- Describe the conditions in saturation
- Calculate the collector-emitter cutoff voltage
- Calculate the collector saturation current
- Calculate the minimum base current to produce saturation

Figure 18–21 illustrates the basic operation of a transistor as a switching device. In part (a), the transistor is in the cutoff region because the base-emitter junction is not forward-biased. In this condition, there is, ideally, an open between collector and emitter, as indicated by the switch equivalent. In part (b), the transistor is in the saturation region because the base-emitter junction and the base-collector junction are forward-biased, and the base current is made large enough to cause the collector current to reach its saturated value. In this condition there is, ideally, a short between collector and emitter, as indicated by the switch equivalent. Actually, a voltage drop of up to a few tenths of a volt normally occurs, which is the saturation voltage, $V_{CE(sat)}$.

FIGURE 18–21

Ideal switching action of a transistor.

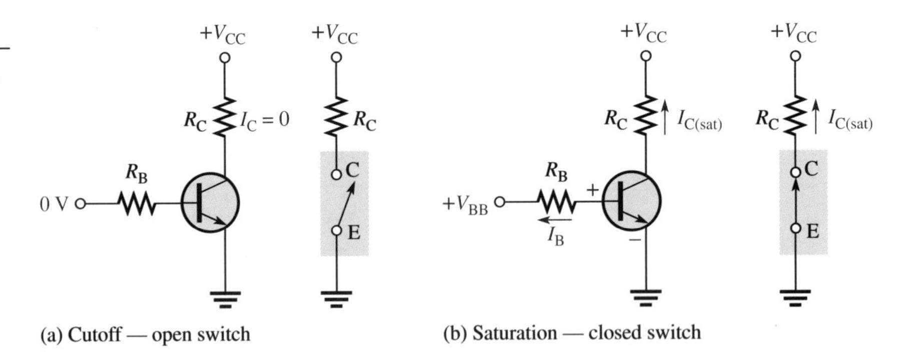

Conditions in Cutoff

As mentioned before, a transistor is in cutoff when the base-emitter junction is *not* forward-biased. Neglecting leakage current, all of the currents are approximately zero, and V_{CE} is approximately equal to V_{CC}.

Equation 18–12

$$V_{CE(cutoff)} \cong V_{CC}$$

Conditions in Saturation

When the base-emitter junction is forward-biased and there is enough base current to produce a maximum collector current, the transistor is saturated. Since $V_{CE(sat)}$ is very small compared to V_{CC}, it can usually be neglected, so the collector current is

Equation 18–13

$$I_{C(sat)} \cong \frac{V_{CC}}{R_C}$$

The minimum value of base current needed to produce saturation is

Equation 18–14

$$I_{B(min)} = \frac{I_{C(sat)}}{\beta_{DC}}$$

EXAMPLE 18–5

(a) For the transistor switching circuit in Figure 18–22, what is V_{CE} when $V_{IN} = 0$ V?

(b) What minimum value of I_B will saturate this transistor if the β_{DC} is 200?

(c) Calculate the maximum value of R_B when $V_{IN} = 5$ V.

FIGURE 18–22

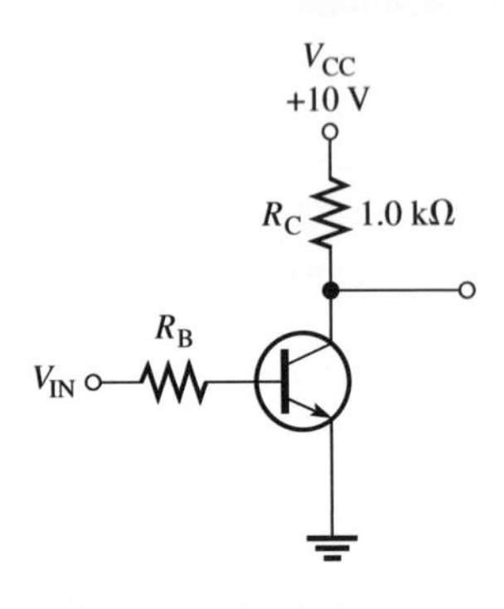

Solution **(a)** When $V_{IN} = 0$ V, the transistor is in cutoff and $V_{CE} = V_{CC} =$ **10 V.**

(b) When the transistor is saturated, $V_{CE} \cong 0$ V, so

$$I_{C(sat)} \cong \frac{V_{CC}}{R_C} = \frac{10\text{ V}}{1.0\text{ k}\Omega} = 10\text{ mA}$$

$$I_{B(min)} = \frac{I_{C(sat)}}{\beta_{DC}} = \frac{10\text{ mA}}{200} = \mathbf{50\ \mu A}$$

This is the value of I_B necessary to drive the transistor to the point of saturation. Any further increase in I_B will drive the transistor deeper into saturation but will not increase I_C.

(c) When the transistor is saturated, $V_{BE} = 0.7$ V. The voltage across R_B is

$$V_{R_B} = V_{IN} - 0.7\text{ V} = 4.3\text{ V}$$

The maximum value of R_B that will allow a minimum I_B of 50 μA is calculated by Ohm's law. The actual value used should be much less.

$$R_{B(max)} = \frac{V_{R_B}}{I_B} = \frac{4.3\text{ V}}{50\ \mu\text{A}} = \mathbf{86\ k\Omega}$$

Related Problem Determine the minimum value of I_B that will saturate the transistor in Figure 18–22 if β_{DC} is 125 and $V_{CE(sat)}$ is 0.2 V.

SECTION 18–4 REVIEW

1. When a transistor is used as a switching device, in what two states is it operated?
2. When does the collector current reach its maximum value?
3. When is the collector current approximately zero?
4. Name the two conditions that produce saturation.
5. When is V_{CE} equal to V_{CC}?

18–5 BJT PARAMETERS AND RATINGS

Two parameters, β_{DC} and α_{DC}, were introduced in Section 18–1. Although these parameters are related, β_{DC} is the more useful, and we will discuss it further in this section. Maximum transistor ratings are also discussed because they are important in establishing the limits within which a transistor must operate.

After completing this section, you should be able to

- **Discuss important transistor parameters and maximum ratings**
- Explain how β_{DC} varies with I_C and temperature
- Calculate I_C and V_{CE} based on maximum power dissipation

More about β_{DC}

The β_{DC} is an important bipolar transistor parameter that we need to examine further. β_{DC} varies with both collector current and temperature. Keeping the junction temperature constant and increasing I_C causes β_{DC} to increase to a maximum. A further increase in I_C beyond this maximum point causes β_{DC} to decrease.

If I_C is held constant and the temperature is varied, β_{DC} changes directly with the temperature. If the temperature goes up, β_{DC} goes up, and vice versa. Figure 18–23 shows an example of the variation of β_{DC} with I_C and junction temperature (T_J) for a typical transistor.

FIGURE 18–23

Variation of β_{DC} with I_C for several temperatures.

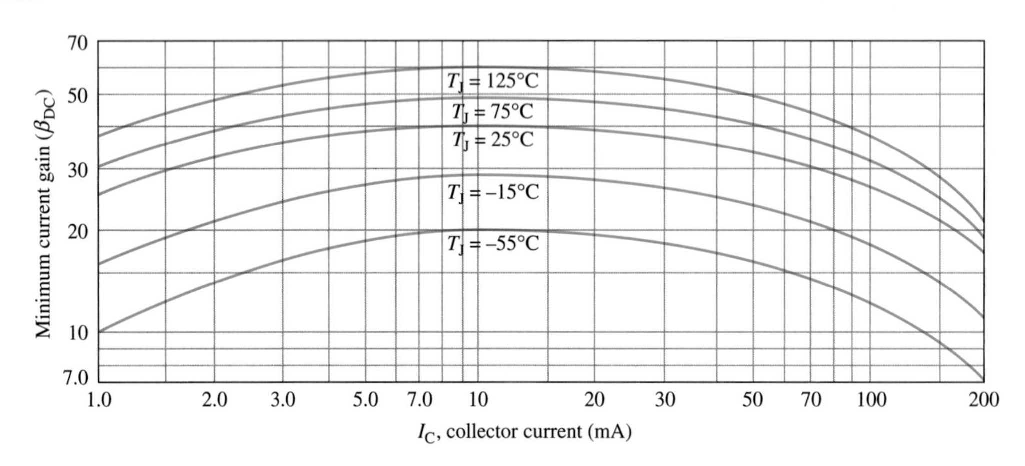

A transistor data sheet usually specifies β_{DC} (h_{FE}) at specific I_C values. Even at fixed values of I_C and temperature, β_{DC} varies from device to device for a given transistor due to inconsistencies in the manufacturing process that are unavoidable. The β_{DC} specified at a certain value of I_C is usually the minimum value, $\beta_{DC(min)}$, although the maximum and typical values are also sometimes specified. A typical transistor data sheet is shown in Appendix E.

Maximum Transistor Ratings

Like any other electronic device, the transistor has limitations on its operation. These limitations are stated in the form of maximum ratings and are normally specified on the manufacturer's data sheet. Typically, maximum ratings are given for collector-to-base voltage (V_{CB}), collector-to-emitter voltage (V_{CE}), emitter-to-base voltage (V_{EB}), collector current (I_C), and power dissipation (P_D).

The product of V_{CE} and I_C must not exceed the maximum power dissipation. Both V_{CE} and I_C cannot be maximum at the same time. If V_{CE} is maximum, I_C can be calculated as

Equation 18–15

$$I_C = \frac{P_{D(max)}}{V_{CE}}$$

If I_C is maximum, V_{CE} can be calculated by rearranging the terms in Equation 18–15.

Equation 18–16

$$V_{CE} = \frac{P_{D(max)}}{I_C}$$

For any given transistor, a maximum power dissipation curve can be plotted on the collector curves, as shown in Figure 18–24(a). These values are tabulated in part (b). For this transistor, $P_{D(max)}$ is 500 mW, $V_{CE(max)}$ is 20 V, and $I_{C(max)}$ is 50 mA. The curve shows that this particular transistor cannot be operated in the

FIGURE 18–24

Maximum power dissipation curve.

(a)

$P_{D(max)}$	V_{CE}	I_C
500 mW	5 V	100 mA
500 mW	10 V	50 mA
500 mW	15 V	33 mA
500 mW	20 V	25 mA

(b)

shaded portion of the graph. $I_{C(max)}$ is the limiting rating between points *A* and *B*, $P_{D(max)}$ is the limiting rating between points *B* and *C*, and $V_{CE(max)}$ is the limiting rating between points *C* and *D*.

EXAMPLE 18–6

The transistor in Figure 18–25 has the following maximum ratings: $P_{D(max)} =$ 800 mW, $V_{CE(max)} = 15$ V, $I_{C(max)} = 100$ mA, $V_{CB(max)} = 20$ V, and $V_{EB(max)} =$ 10 V. Determine the maximum value to which V_{CC} can be adjusted without exceeding a rating. Which rating would be exceeded first?

FIGURE 18–25

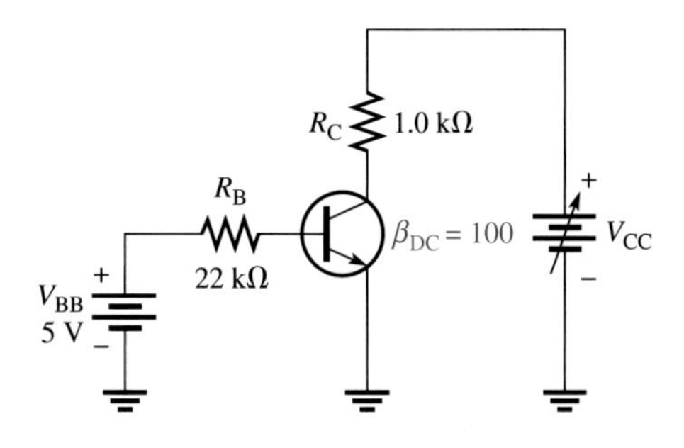

Solution First, find I_B so that you can determine I_C.

$$I_B = \frac{V_{BB} - V_{BE}}{R_B} = \frac{5\text{ V} - 0.7\text{ V}}{22\text{ k}\Omega} = 195\ \mu\text{A}$$

$$I_C = \beta_{DC} I_B = (100)(195\ \mu\text{A}) = 19.5\text{ mA}$$

I_C is much less than $I_{C(max)}$ and will not change with V_{CC}. It is determined only by I_B and β_{DC}.

The voltage drop across R_C is

$$V_{R_C} = I_C R_C = (19.5\text{ mA})(1.0\text{ k}\Omega) = 19.5\text{ V}$$

Now you can determine the maximum value of V_{CC} when $V_{CE} = V_{CE(max)} =$ 15 V.

$$V_{R_C} = V_{CC} - V_{CE}$$

Thus,

$$V_{CC(max)} = V_{CE(max)} + V_{R_C} = 15\text{ V} + 19.5\text{ V} = \mathbf{34.5\ V}$$

V_{CC} can be increased to 34.5 V, under the existing conditions, before $V_{CE(max)}$ is exceeded. However, at this point it is not known if $P_{D(max)}$ has been exceeded. The power dissipation is

$$P_D = V_{CE(max)}I_C = (15\text{ V})(19.5\text{ mA}) = 293\text{ mW}$$

Since $P_{D(max)}$ is 800 mW, it is *not* exceeded when $V_{CE} = 34.5$ V. Thus, $V_{CE(max)}$ is the limiting rating in this case.

If the base current is removed, causing the transistor to turn off, $V_{CE(max)}$ will be exceeded because the entire supply voltage, V_{CC}, will be dropped across the transistor.

Related Problem The transistor in Figure 18–25 has the following maximum ratings: $P_{D(max)} =$ 500 mW, $V_{CE(max)} = 25$ V, $I_{C(max)} = 200$ mA, $V_{CE(max)} = 30$ V, $V_{EB(max)} =$ 15 V. Determine the maximum value to which V_{CC} can be adjusted without exceeding a rating. Which rating would be exceeded first?

SECTION 18–5 REVIEW

1. Does the β_{DC} of a transistor increase or decrease with temperature?
2. Generally, what effect does an increase in I_C have on the β_{DC}?
3. What is the allowable collector current in a transistor with $P_{D(max)} = 32$ mW when $V_{CE} = 8$ V?

18–6 THE JUNCTION FIELD-EFFECT TRANSISTOR (JFET)

Recall that the bipolar junction transistor (BJT) is a current-controlled device; that is, the base current controls the amount of collector current. The **field-effect transistor (FET)** is different; it is a voltage-controlled device in which the voltage at the gate terminal controls the amount of current through the device. Also, compared to the BJT, the FET has a very high input resistance, which makes it superior in certain applications.

After completing this section, you should be able to

- **Describe the basic structure and operation of a JFET**
- Identify the standard JFET symbols
- Explain the difference between *n*-channel and *p*-channel JFETs
- Label the terminals of a JFET

The **junction field-effect transistor (JFET)** is a type of FET that operates with a reverse-biased junction to control current in the channel. Depending on their structure, JFETs fall into either of two categories: *n* channel or *p* channel. Figure 18–26(a) shows the basic structure of an *n* channel JFET. Wire leads are connected to each end of the *n* channel; the **drain** is at the upper end and the **source** is at the lower end. Two *p*-type regions are diffused in the *n*-type material to form a channel, and both *p*-type regions are connected to the **gate** lead. For simplicity, the gate lead is shown connected to only one of the *p* regions. A *p*-channel JFET is shown in Figure 18–26(b).

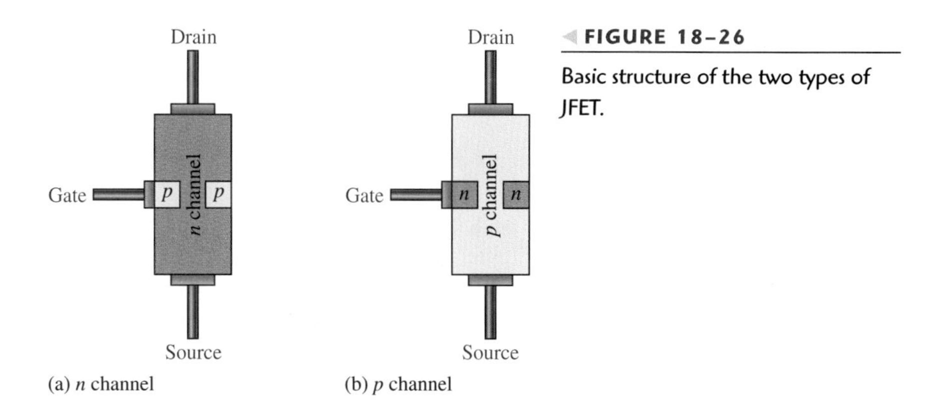

(a) n channel (b) p channel

FIGURE 18–26

Basic structure of the two types of JFET.

Basic Operation

To illustrate the operation of a JFET, Figure 18–27(a) shows bias voltages applied to an n-channel device. V_{DD} provides a drain-to-source voltage and supplies current from drain to source. V_{GG} sets the reverse-biased voltage between the gate and the source. The white areas surrounding the p material of the gate represent the depletion region created by the reverse bias. This depletion region is wider toward the drain end of the channel because the reverse-biased voltage between the gate and the drain is greater than that between the gate and the source.

The JFET is always operated with the gate-to-source pn junction reverse-biased. Reverse-biasing of the gate-source junction with a negative gate voltage produces a depletion region in the n channel and thus increases its resistance. The

FIGURE 18–27

Effects of V_{GG} on channel width and drain current ($V_{GG} = V_{GS}$).

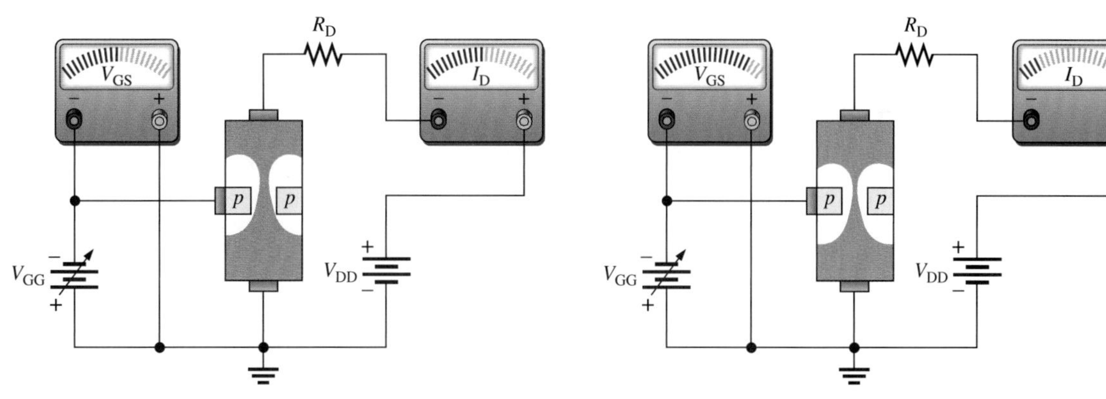

(a) JFET biased for conduction

(b) Greater V_{GG} narrows the channel (between white areas) which increases the resistance of the channel and decreases I_D.

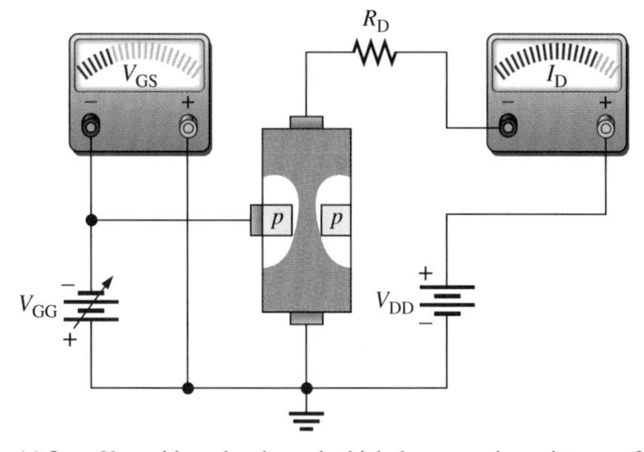

(c) Less V_{GG} widens the channel which decreases the resistance of the channel and increases I_D.

channel width can be controlled by varying the gate voltage, and thereby the amount of drain current, I_D, can also be controlled. This concept is illustrated in Figure 18–27(b) and (c).

JFET Symbols

The schematic symbols for both *n*-channel and *p*-channel JFETs are shown in Figure 18–28. Notice that the arrow on the gate points "in" for *n* channel and "out" for *p* channel.

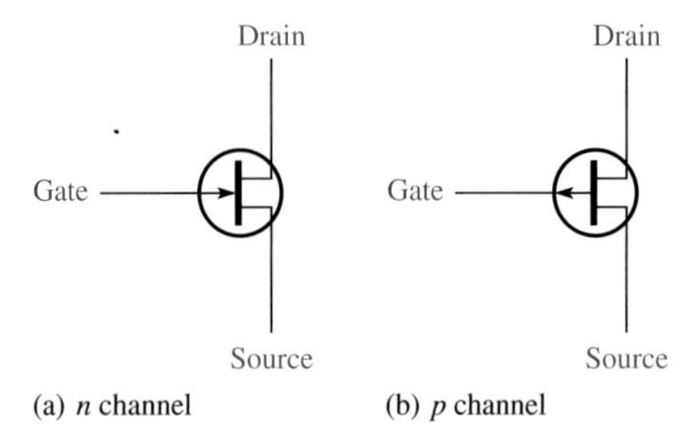

FIGURE 18–28

JFET schematic symbols.

SECTION 18–6 REVIEW

1. Name the three terminals of a JFET.
2. An *n*-channel JFET requires a (positive, negative, or 0) V_{GS}.
3. How is the drain current controlled in a JFET?

18–7 JFET CHARACTERISTICS

In this section, you will see how the JFET operates as a voltage-controlled, constant-current device. You will also learn about cutoff and pinch-off as well as JFET input resistance and capacitance.

After completing this section, you should be able to

- **Explain the operation of a JFET in terms of its characteristics**
- Define *pinch-off voltage*
- Describe how V_{GS} controls I_D
- Discuss cutoff and how it differs from pinch-off

Let's consider the case where the gate-to-source voltage is 0 ($V_{GS} = 0$ V). This voltage is produced by shorting the gate to the source, as in Figure 18–29(a) where both are grounded. As V_{DD} (and thus V_{DS}) is increased from 0 V, I_D will increase proportionally, as shown in the graph of Figure 18–29(b) between points *A* and *B*. In this region, the channel resistance is essentially constant because the depletion region is not large enough to have significant effect. This region is called the *ohmic region* because V_{DS} and I_D are related by Ohm's law.

At point *B* in Figure 18–29(b), the curve levels off and I_D becomes essentially constant. As V_{DS} increases from point *B* to point *C*, the reverse-bias voltage from gate to drain (V_{GD}) produces a depletion region large enough to offset the increase in V_{DS}, thus keeping I_D relatively constant.

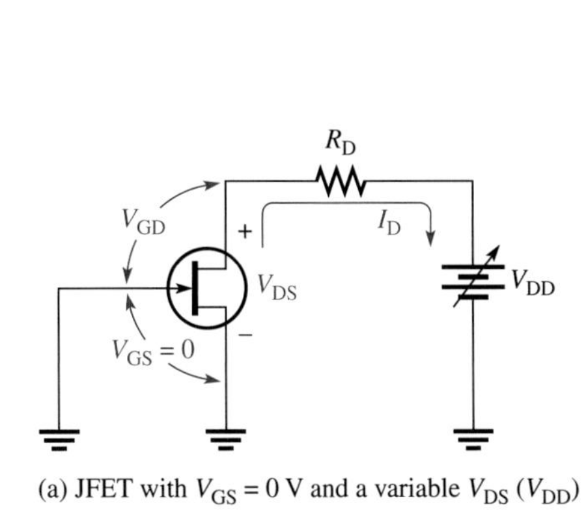

(a) JFET with $V_{GS} = 0$ V and a variable V_{DS} (V_{DD})

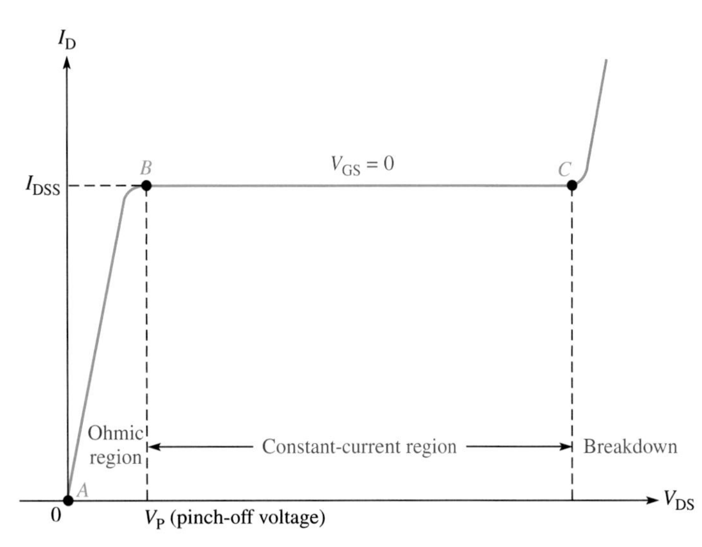

(b) Drain characteristic

▲ **FIGURE 18–29**

The drain characteristic curve of a JFET for $V_{GS} = 0$ V, showing pinch-off.

Pinch-Off Voltage

For $V_{GS} = 0$ V, the value of V_{DS} at which I_D becomes essentially constant (point B on the curve in Figure 18–29(b)) is the **pinch-off voltage,** V_P. For a given JFET, V_P has a fixed value. As you can see, a continued increase in V_{DS} above the pinch-off voltage produces an almost constant drain current. This value of drain current is I_{DSS} (*D*rain to *S*ource current with gate *S*horted) and is always specified on JFET data sheets. I_{DSS} is the *maximum* drain current that a specific JFET can produce regardless of the external circuit, and it is always specified for the condition, $V_{GS} = 0$ V.

As shown in the graph in Figure 18–29(b), breakdown occurs at point C when I_D begins to increase very rapidly with any further increase in V_{DS}. Breakdown can result in irreversible damage to the device, so JFETs are always operated below breakdown and within the constant-current region (between points B and C on the graph).

V_{GS} Controls I_D

Let's connect a bias voltage, V_{GG}, from gate to source as shown in Figure 18–30(a). As V_{GS} is set to increasingly more negative values by adjusting V_{GG}, a family of drain characteristic curves is produced as shown in Figure 18–30(b). Notice that I_D decreases as the magnitude of V_{GS} is increased to larger negative values. Also notice that, for each increase in V_{GS}, the JFET reaches pinch-off (where constant current begins) at values of V_{DS} less than V_P. So, the amount of drain current is controlled by V_{GS}.

Cutoff Voltage

The value of V_{GS} that makes I_D approximately zero is the cutoff voltage, $V_{GS(off)}$. The JFET must be operated between $V_{GS} = 0$ V and $V_{GS(off)}$. For this range of gate-to-source voltages, I_D will vary from a maximum of I_{DSS} to a minimum of almost zero.

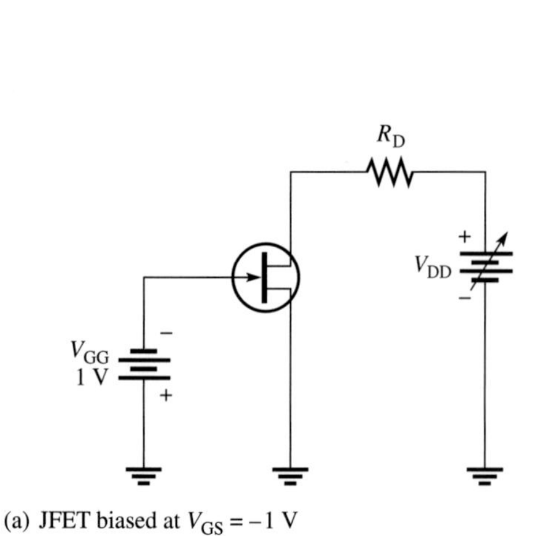

(a) JFET biased at $V_{GS} = -1$ V

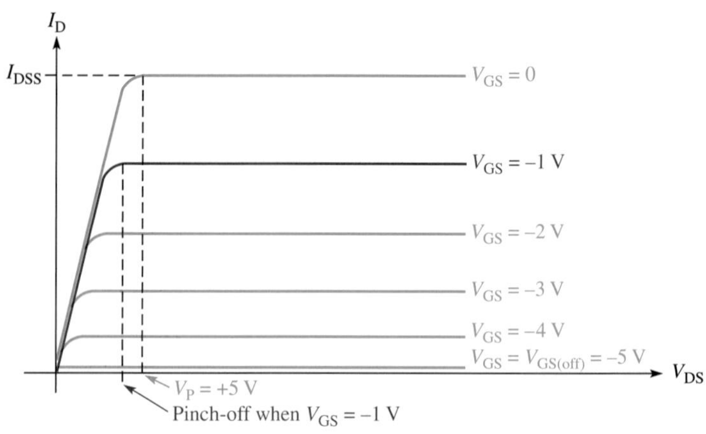

(b) Family of drain characteristic curves

▲ **FIGURE 18–30**

Pinch-off occurs at a lower V_{DS} as V_{GS} is increased to more negative values for a specific JFET.

As you have seen, for an *n*-channel JFET, the more negative V_{GS} is, the smaller I_D becomes in the constant-current region. When V_{GS} has a sufficiently large negative value, I_D is reduced to zero. This cutoff effect is caused by the widening of the depletion region to a point where it completely closes the channel as shown in Figure 18–31.

▶ **FIGURE 18–31**

JFET at cutoff.

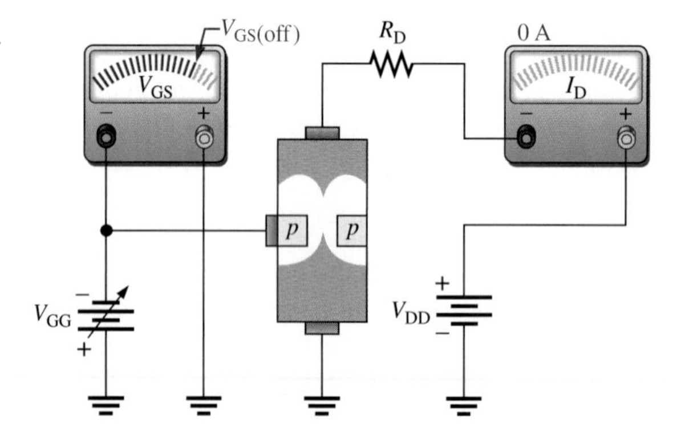

The basic operation of a *p*-channel JFET is the same as for an *n*-channel device except that it requires a negative V_{DD} and a positive V_{GS}.

Comparison of Pinch-Off and Cutoff

As you have seen, there is definitely a difference between pinch-off and cutoff. There is also a connection. V_P is the value of V_{DS} at which the drain current becomes constant and is always measured at $V_{GS} = 0$ V. However, pinch-off occurs for V_{DS} values less than V_P when V_{GS} is nonzero. So, although V_P is a constant, the minimum value of V_{DS} at which I_D becomes constant varies with V_{GS}.

$V_{GS(off)}$ and V_P are always equal in magnitude but opposite in sign. A data sheet usually will give either $V_{GS(off)}$ or V_P, but not both. However, when you know one, you have the other. For example, if $V_{GS(off)} = -5$ V, then $V_P = +5$ V as shown in Figure 18–30(b).

EXAMPLE 18–7

For the JFET in Figure 18–32, $V_{GS(off)} = -4$ V and $I_{DSS} = 12$ mA. Determine the minimum value of V_{DD} required to put the device in the constant-current region of operation.

▶ FIGURE 18–32

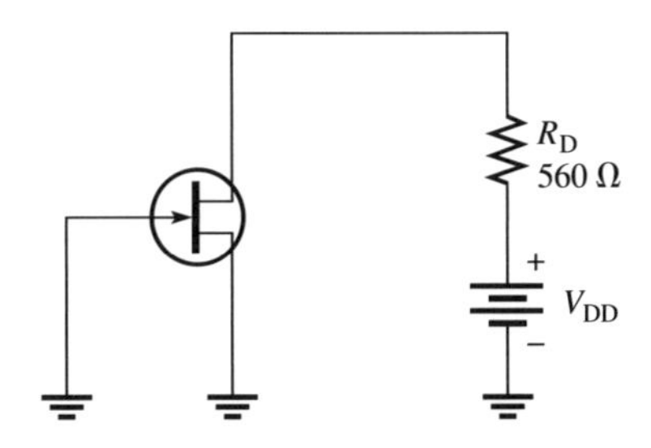

Solution Since $V_{GS(off)} = -4$ V, $V_P = 4$ V. The minimum value of V_{DS} for the JFET to be in its constant-current region is

$$V_{DS} = V_P = 4\text{ V}$$

In the constant-current region with $V_{GS} = 0$ V,

$$I_D = I_{DSS} = 12\text{ mA}$$

The drop across the drain resistor is

$$V_{R_D} = (12\text{ mA})(560\ \Omega) = 6.72\text{ V}$$

Apply Kirchhoff's law around the drain circuit.

$$V_{DD} = V_{DS} + V_{R_D} = 4\text{ V} + 6.72\text{ V} = \mathbf{10.7\text{ V}}$$

This is the value of V_{DD} to make $V_{DS} = V_P$ and put the device in the constant-current region.

Related Problem If V_{DD} is increased to 15 V in Figure 18–32, what is the drain current?

JFET Input Resistance and Capacitance

A JFET operates with its gate-source junction reverse-biased. Therefore, the input resistance at the gate is very high. This high input resistance is one advantage of the JFET over the bipolar transistor. (Recall that a bipolar transistor operates with a forward-biased base-emitter junction.)

JFET data sheets often specify the input resistance by giving a value for the gate reverse current, I_{GSS}, at a certain gate-to-source voltage. The input resistance can then be determined using the following equation. The vertical lines indicate an absolute value (no sign).

$$R_{IN} = \left|\frac{V_{GS}}{I_{GSS}}\right|$$

Equation 18–17

For example, the 2N3970 data sheet lists a maximum I_{GSS} of 250 pA for $V_{GS} = -20$ V at 25°C. I_{GSS} increases with temperature, so the input resistance decreases.

The input capacitance C_{iss} of a JFET is considerably greater than that of a bipolar transistor because the JFET operates with a reverse-biased *pn* junction.

Recall that a reverse-biased *pn* junction acts as a capacitor whose capacitance depends on the amount of reverse voltage. For example, the 2N3970 has a maximum C_{iss} of 25 pF for $V_{GS} = 0$ V.

EXAMPLE 18–8

A certain JFET has an I_{GSS} of 1 nA for $V_{GS} = -20$ V. Determine the input resistance.

Solution

$$R_{IN} = \left|\frac{V_{GS}}{I_{GSS}}\right| = \frac{20\text{ V}}{1\text{ nA}} = \mathbf{20{,}000\ M\Omega}$$

Usually, a resistance this large cannot be measured.

Related Problem Determine the minimum input resistance for the 2N3970 JFET.

SECTION 18–7 REVIEW

1. The drain-to-source voltage at the pinch-off point of a particular JFET is 7 V. If the gate-to-source voltage is 0 V, what is V_P?
2. The V_{GS} of a certain *n*-channel JFET is increased negatively. Does the drain current increase or decrease?
3. What value must V_{GS} have to produce cutoff in a *p*-channel JFET with a $V_P = -3$ V?

18–8 THE METAL-OXIDE SEMICONDUCTOR FET (MOSFET)

The metal-oxide semiconductor field-effect transistor (MOSFET) is the second category of field-effect transistor. The MOSFET differs from the JFET in that it has no *pn* junction structure; instead, the gate of the MOSFET is insulated from the channel by a silicon dioxide (SiO_2) layer. The two basic types of MOSFETs are depletion (D) and enhancement (E).

After completing this section, you should be able to

- **Explain the basic structure and operation of MOSFETs**
- Discuss the D-MOSFET
- Discuss the E-MOSFET
- Recognize the symbols for both types of MOSFETs
- Describe the proper handling of MOSFETs

Depletion MOSFET (D-MOSFET)

One type of MOSFET is the depletion MOSFET (D-MOSFET), and Figure 18–33 illustrates its basic structure. The drain and source are diffused into the substrate material and then connected by a narrow channel adjacent to the insulated gate. Both *n*-channel and *p*-channel devices are shown in the figure. We will

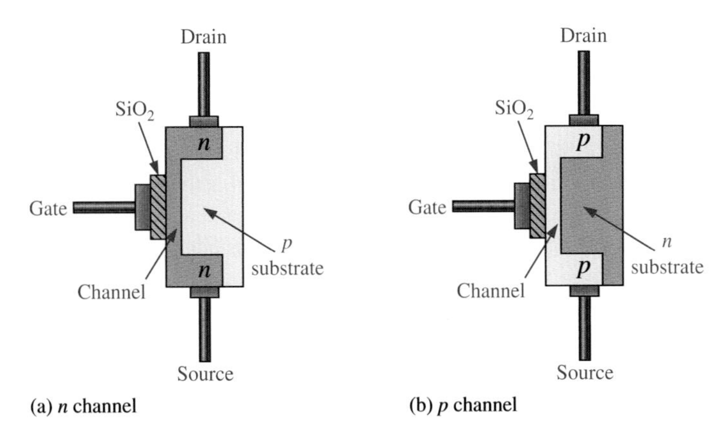

(a) *n* channel (b) *p* channel

FIGURE 18–33

Representation of the basic structure of D-MOSFETs.

use the *n*-channel device to describe the basic operation. The *p*-channel operation is the same, except the voltage polarities are opposite those of the *n*-channel device.

The D-MOSFET can be operated in either of two modes—the depletion mode or the enhancement mode—and is sometimes called a *depletion/enhancement MOSFET.* Since the gate is insulated from the channel, either a positive or a negative gate voltage can be applied. The *n*-channel MOSFET operates in the depletion mode when a negative gate-to-source voltage is applied and in the enhancement mode when a positive gate-to-source voltage is applied.

Depletion Mode Visualize the gate as one plate of a parallel plate capacitor and the channel as the other plate. The silicon dioxide insulating layer is the dielectric. With a negative gate voltage, the negative charges on the gate repel conduction electrons from the channel, leaving positive ions in their place. The *n*-channel is depleted of some of its electrons, so the channel conductivity is decreased. The greater the negative voltage on the gate, the greater the depletion of *n*-channel electrons. At a sufficiently negative gate-to-source voltage, $V_{GS(off)}$, the channel is totally depleted and the drain current is zero. This depletion mode is illustrated in Figure 18–34(a).

Like the *n*-channel JFET, the *n*-channel D-MOSFET conducts drain current for gate-to-source voltages between $V_{GS(off)}$ and 0 V. In addition, the D-MOSFET conducts for values of V_{GS} above 0 V.

FIGURE 18–34

Operation of *n*-channel D-MOSFET.

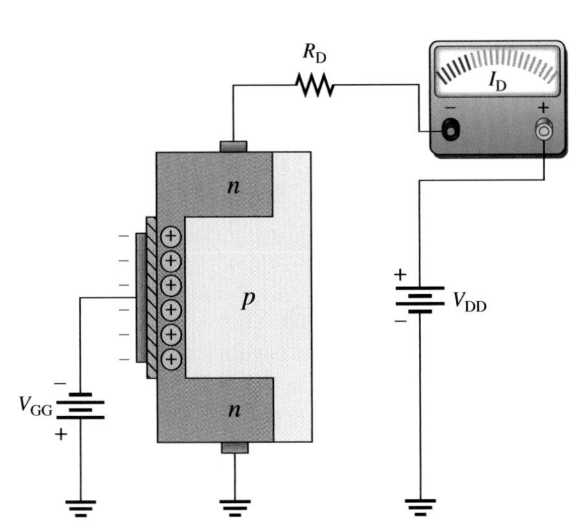

(a) Depletion mode: V_{GS} negative and less than $V_{GS(off)}$

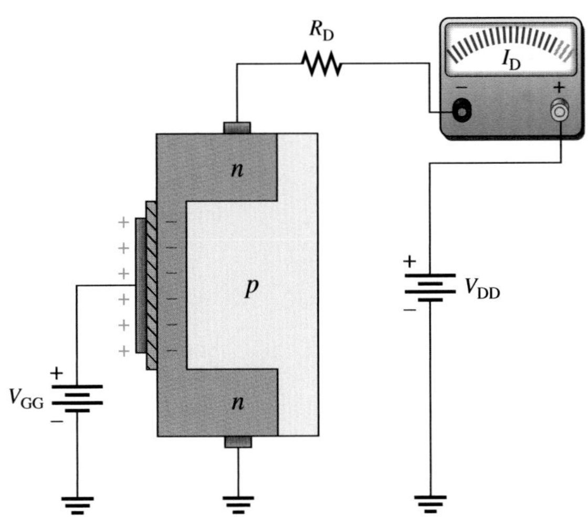

(b) Enhancement mode: V_{GS} positive

Enhancement Mode With a positive gate voltage, more conduction electrons are attracted into the channel, thus increasing (enhancing) the channel conductivity, as illustrated in Figure 18–34(b).

D-MOSFET Symbols The schematic symbols for both the *n*-channel and the *p*-channel depletion/enhancement MOSFETs are shown in Figure 18–35. The substrate, indicated by the arrow, is normally (but not always) connected internally to the source. An inward substrate arrow is for *n*-channel, and an outward arrow is for *p*-channel.

FIGURE 18–35

D-MOSFET schematic symbols.

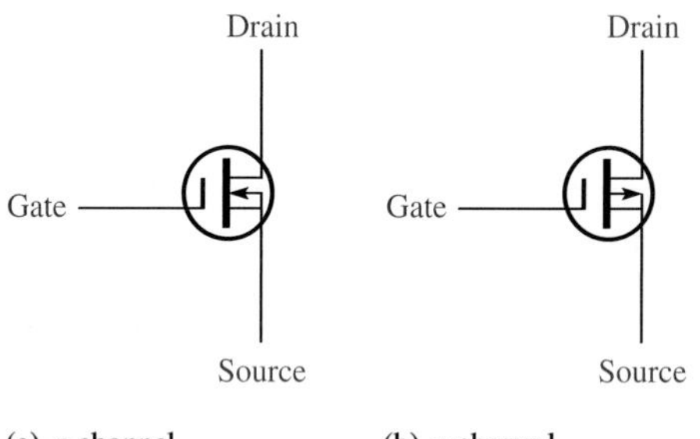

Enhancement MOSFET (E-MOSFET)

This type of MOSFET operates *only* in the enhancement mode and has no depletion mode. It differs in construction from the D-MOSFET in that it has no structural channel. Notice in Figure 18–36(a) that the substrate extends completely to the SiO_2 layer.

FIGURE 18–36

E-MOSFET construction and operation.

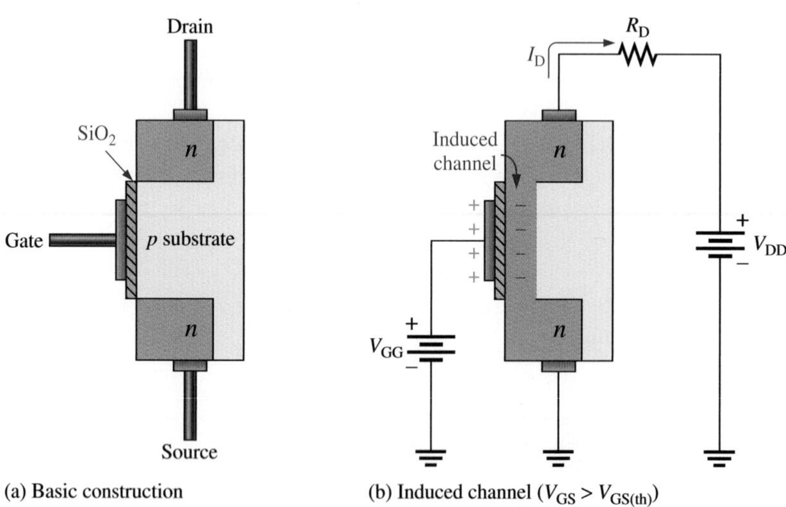

For an *n*-channel device, a positive gate voltage above a threshold value, $V_{GS(th)}$, *induces* a channel by creating a thin layer of negative charges in the substrate region adjacent to the SiO_2 layer, as shown in Figure 18–36(b). The conductivity of the channel is enhanced by increasing the gate-to-source voltage, thus pulling more electrons into the channel. For any gate voltage below the threshold value, there is no channel.

The schematic symbols for the *n*-channel and *p*-channel E-MOSFETs are shown in Figure 18–37. The broken lines symbolize the absence of a structural channel.

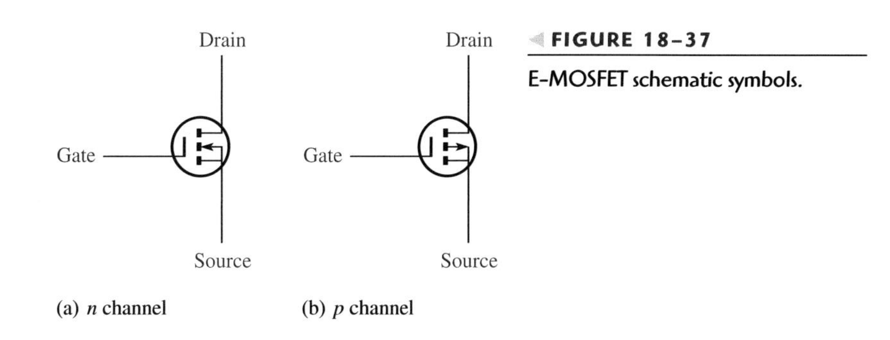

FIGURE 18–37
E-MOSFET schematic symbols.

Handling Precautions

Because the gate of a MOSFET is insulated from the channel, the input resistance is extremely high (ideally infinite). The gate leakage current, I_{GSS}, for a typical MOSFET is in the pA range, whereas the gate reverse current for a typical JFET is in the nA range.

The input capacitance, of course, results from the insulated gate structure. Excess static charge can accumulate because the input capacitance combines with the very high input resistance and can result in damage to the device as a result of electrostatic discharge (ESD). To avoid ESD and possible damage, the following precautions should be taken:

1. MOS devices should be shipped and stored in conductive foam.
2. All instruments and metal benches used in assembly or test should be connected to earth ground (round prong of wall outlets).
3. The assembler's or handler's wrist should be connected to earth ground with a length of wire and a high-value series resistor.
4. Never remove a MOS device (or any other device, for that matter) from the circuit while the power is on.
5. Do not apply signals to a MOS device while the dc power supply is off.

SECTION 18–8 REVIEW

1. Name two types of MOSFETs, and describe the major difference in construction.
2. If the gate-to-source voltage in a depletion/enhancement MOSFET is 0 V, is there current from drain to source?
3. If the gate-to-source voltage in an E-MOSFET is 0 V, is there current from drain to source?

18–9 FET BIASING

Using some of the FET parameters discussed in the previous sections, we will now describe how to dc bias FETs. Biasing is used to select a proper dc gate-to-source voltage to establish a desired value of drain current.

After completing this section, you should be able to

- **Analyze the common biasing arrangements for JFETs and MOSFETs**
- Analyze a self-biased JFET circuit
- Analyze a zero-biased D-MOSFET circuit
- Analyze drain-feedback and voltage-divider biased E-MOSFET circuits

Self-Biasing a JFET

Recall that a JFET must be operated such that the gate-source junction is always reverse-biased. This condition requires a negative V_{GS} for an *n*-channel JFET and a positive V_{GS} for a *p*-channel JFET. This can be achieved using the self-bias arrangements shown in Figure 18–38.

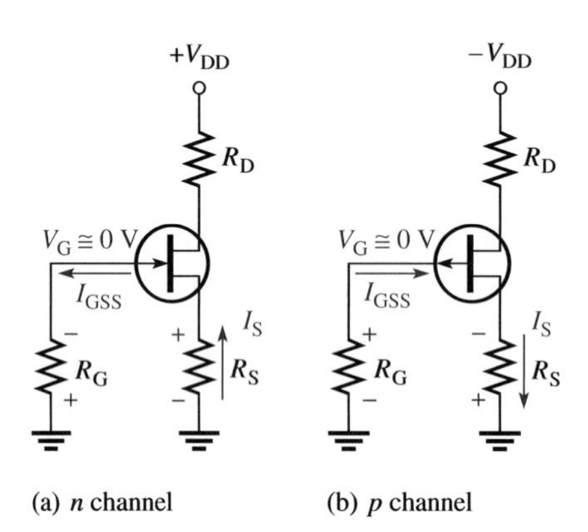

FIGURE 18–38

Self-biased JFETs ($I_S = I_D$ in all FETs).

Notice that the gate is biased at approximately 0 V by resistor R_G connected to ground. The reverse leakage current, I_{GSS}, does produce a very small voltage across R_G, but this can be neglected in most cases. We will assume that R_G has no voltage drop across it.

For the *n*-channel JFET in Figure 18–38(a), I_S produces a voltage drop across R_S and makes the source positive with respect to ground. Since $I_S = I_D$ and $V_G = 0$, then $V_S = I_D R_S$. The gate-to-source voltage is

$$V_{GS} = V_G - V_S = 0\text{ V} - I_D R_S$$

Thus,

Equation 18–18

$$V_{GS} = -I_D R_S$$

For the *p*-channel JFET shown in Figure 18–38(b), the current through R_S produces a negative voltage at the source, making the gate positive with respect to the source. Therefore, since $I_S = I_D$,

Equation 18–19

$$V_{GS} = +I_D R_S$$

In the following analysis, the *n*-channel JFET is used for illustration. Keep in mind that analysis of the *p*-channel JFET is the same except for opposite polarity voltages.

The drain voltage with respect to ground is determined as follows:

Equation 18–20

$$V_D = V_{DD} - I_D R_D$$

Since $V_S = I_D R_S$, the drain-to-source voltage is

$$V_{DS} = V_D - V_S$$

Equation 18–21

$$V_{DS} = V_{DD} - I_D(R_D + R_S)$$

EXAMPLE 18–9

Find V_{DS} and V_{GS} in Figure 18–39, given that $I_D \cong 5$ mA.

FIGURE 18–39

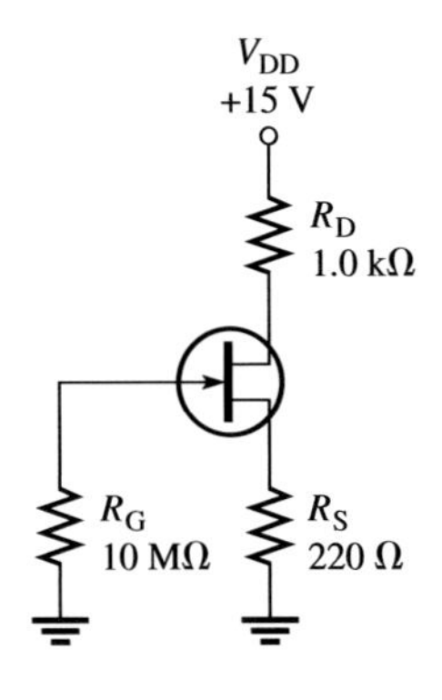

Solution

$$V_S = I_D R_S = (5 \text{ mA})(220\ \Omega) = 1.1 \text{ V}$$

$$V_D = V_{DD} - I_D R_D = 15 \text{ V} - (5 \text{ mA})(1.0 \text{ k}\Omega) = 15 \text{ V} - 5 \text{ V} = 10 \text{ V}$$

Therefore,

$$V_{DS} = V_D - V_S = 10 \text{ V} - 1.1 \text{ V} = \mathbf{8.9\ V}$$

Since $V_G = 0$ V,

$$V_{GS} = V_G - V_S = 0 \text{ V} - 1.1 \text{ V} = \mathbf{-1.1\ V}$$

Related Problem Determine V_{DS} and V_{GS} in Figure 18–39 when $I_D = 8$ mA. Assume that $R_D = 860\ \Omega$, $R_S = 390\ \Omega$, and $V_{DD} = 12$ V.

Open file E18-09 on your EWB/CircuitMaker CD-ROM. Measure V_{DS} and V_{GS}.

D-MOSFET Bias

Recall that depletion/enhancement MOSFETs can be operated with either positive or negative values of V_{GS}. A simple bias method is to set $V_{GS} = 0$ V so that an ac signal at the gate varies the gate-to-source voltage above and below this bias point. A MOSFET with zero bias is shown in Figure 18–40. Since $V_{GS} = 0$ V, $I_D = I_{DSS}$ as indicated. The drain-to-source voltage is expressed as

$$V_{DS} = V_{DD} - I_{DSS} R_D$$

Equation 18–22

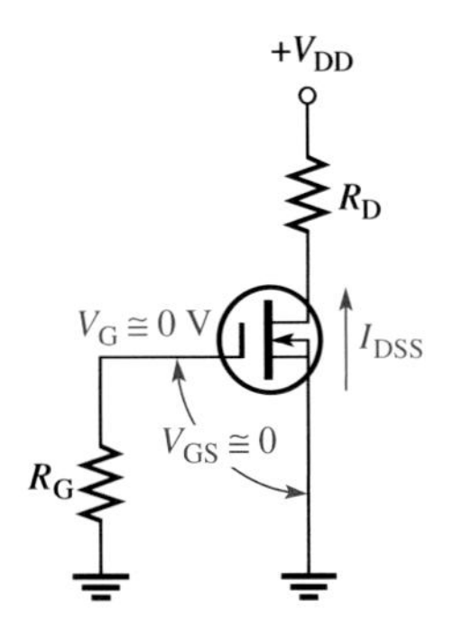

FIGURE 18–40

A zero-biased D-MOSFET.

EXAMPLE 18–10

Determine the drain-to-source voltage in the circuit of Figure 18–41. The MOSFET data sheet gives $V_{GS(off)} = -8$ V and $I_{DSS} = 12$ mA.

▶ FIGURE 18–41

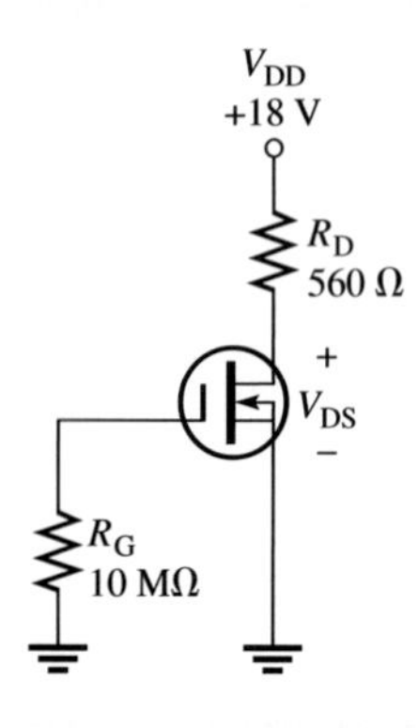

Solution Since $I_D = I_{DSS} = 12$ mA, the drain-to-source voltage is

$$V_{DS} = V_{DD} - I_{DSS}R_D = 18\text{ V} - (12\text{ mA})(560\ \Omega) = \mathbf{11.3\ V}$$

Related Problem Find V_{DS} in Figure 18–41 when $V_{GS(off)} = -10$ V and $I_{DSS} = 20$ mA.

E-MOSFET Bias

Recall that enhancement-only MOSFETs must have a V_{GS} greater than the threshold value, $V_{GS(th)}$. Figure 18–42 shows two ways to bias an E-MOSFET where an *n*-channel device is used for illustration. In either bias arrangement, the purpose is to make the gate voltage more positive than the source by an amount exceeding $V_{GS(th)}$.

▶ FIGURE 18–42

E-MOSFET biasing arrangements.

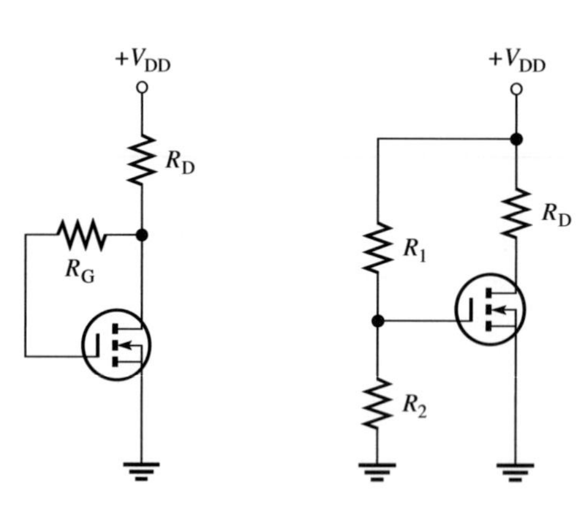

In the drain-feedback bias circuit in Figure 18–42(a), there is negligible gate current and, therefore, no voltage drop across R_G. As a result, $V_{GS} = V_{DS}$.

Equations for the voltage-divider bias in Figure 18–42(b) are as follows:

Equation 18–23

$$V_{GS} = \left(\frac{R_2}{R_1 + R_2}\right)V_{DD}$$

Equation 18–24

$$V_{DS} = V_{DD} - I_DR_D$$

EXAMPLE 18–11

Determine the amount of drain current in Figure 18–43. The MOSFET has a $V_{GS(th)}$ of 3 V.

FIGURE 18–43

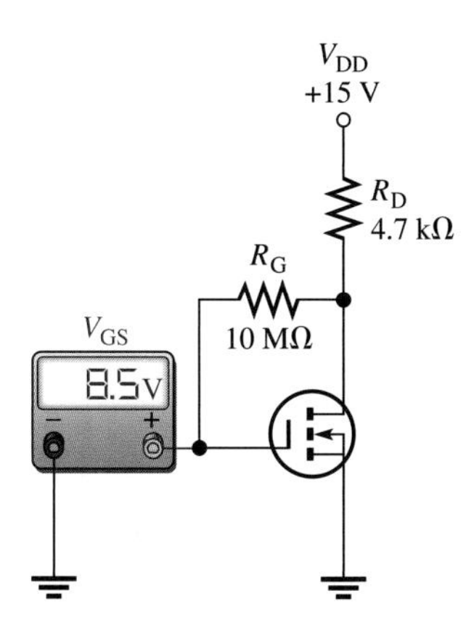

Solution The meter indicates that $V_{GS} = 8.5$ V. Since this is a drain feedback configuration, $V_{DS} = V_{GS} = 8.5$ V. Use Ohm's law to find the current.

$$I_D = \frac{V_{DD} - V_{DS}}{R_D} = \frac{15\text{ V} - 8.5\text{ V}}{4.7\text{ k}\Omega} = \mathbf{1.38\text{ mA}}$$

Related Problem Determine I_D if the meter in Figure 18–43 reads 5 V.

SECTION 18–9 REVIEW

1. A *p*-channel JFET must have a (positive, negative) V_{GS}.
2. In a certain self-biased *n*-channel JFET circuit, $I_D = 8$ mA and $R_S = 1.0$ kΩ. Determine V_{GS}.
3. For a D-MOSFET biased at $V_{GS} = 0$ V, is the drain current equal to 0 V, I_{GSS}, or I_{DSS}?
4. For an *n*-channel E-MOSFET with $V_{GS(th)} = 2$ V, V_{GS} must be in excess of what value in order to conduct?

18–10 UNIJUNCTION TRANSISTORS (UJTs)

The unijunction transistor (UJT) consists of an emitter and two bases. It has no collector. The UJT is used mainly in switching and timing applications.

After completing this section, you should be able to

- **Discuss the operation and application of the UJT**
- Recognize the UJT symbol
- Draw an equivalent circuit
- Describe standoff ratio
- Explain how a UJT oscillator works

Figure 18–44(a) shows the construction of a **unijunction transistor (UJT)**. The base contacts are made to the n-type bar. The emitter lead is connected to the p region. The UJT schematic symbol is shown in Figure 18–44(b). Do not confuse this symbol with that of a JFET; *the difference is that the arrow is at an angle for the UJT.*

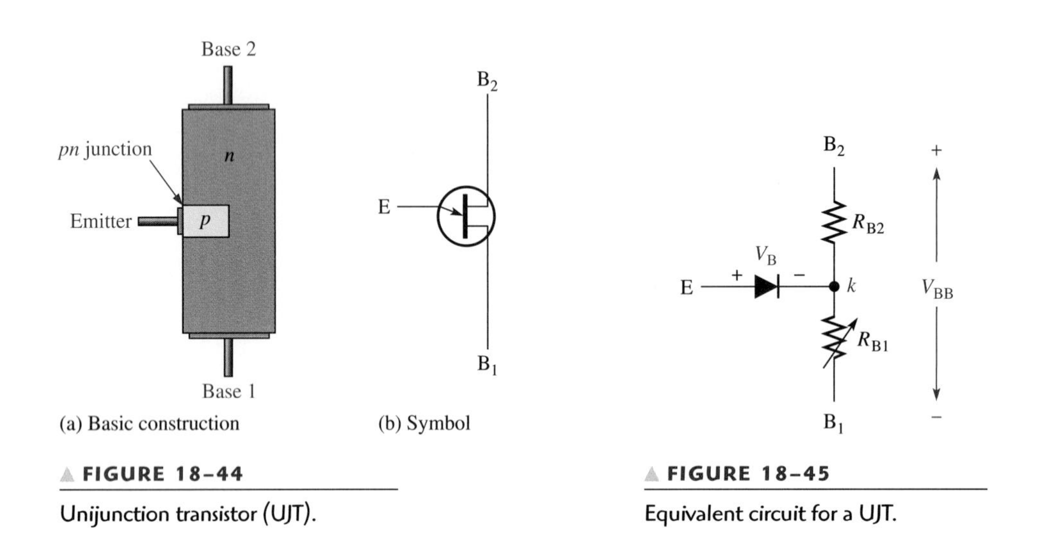

FIGURE 18–44
Unijunction transistor (UJT).

FIGURE 18–45
Equivalent circuit for a UJT.

UJT Operation

In normal UJT operation, base 2 (B_2) and the emitter are biased positive with respect to base 1 (B_1). Figure 18–45 will illustrate the operation. This equivalent circuit represents the internal UJT characteristics. The total resistance between the two bases, R_{BB}, is the resistance of the n-type material. R_{B1} is the resistance between point k and base 1. R_{B2} is the resistance between point k and base 2. The sum of these two resistances makes up the total resistance, R_{BB}. The diode represents the pn junction between the emitter (p-type material) and the n-type material.

The ratio R_{B1}/R_{BB} is designated η (the Greek letter eta) and is defined as the *intrinsic standoff ratio.* It takes an emitter voltage of $V_B + \eta V_{BB}$ to turn the UJT on. This voltage is called the *peak voltage.* Once the device is on, resistance R_{B1} drops in value. Thus, as emitter current increases, emitter voltage decreases because of the decrease in R_{B1}. This characteristic is the negative resistance characteristic of the UJT.

As the emitter voltage decreases, it reaches a value called the *valley voltage.* At this point, the pn junction is no longer forward-biased, and the UJT turns off.

An Application

UJTs are commonly used in oscillator circuits. Figure 18–46(a) shows a typical circuit. Its operation is as follows: Initially the capacitor is uncharged and the UJT is off. When power is applied, the capacitor charges up exponentially. When it reaches the peak voltage, the UJT turns on and the capacitor begins to discharge, as indicated. When the emitter reaches the valley voltage, the UJT turns off and the capacitor begins to charge again. The cycle repeats. Waveforms of the capacitor voltage and the R_2 voltage are shown in Figure 18–46(b).

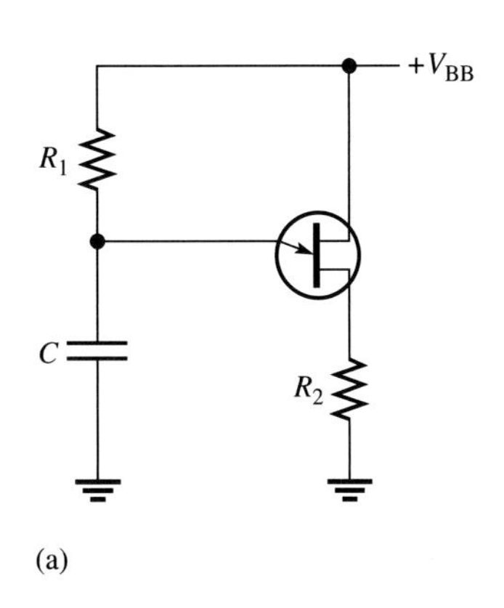

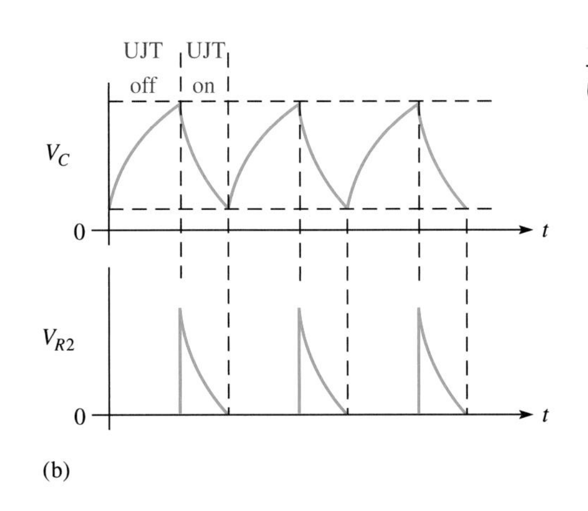

FIGURE 18–46

UJT oscillator circuit.

SECTION 18–10 REVIEW

1. What does "UJT" stand for?
2. What are the terminals of a UJT?
3. What is the intrinsic standoff ratio?

18–11 THYRISTORS

Thyristors are devices constructed of four layers of semiconductive material. The types of thyristors covered in this section are the silicon-controlled rectifier (SCR), the diac, and the triac. These thyristor devices share certain characteristics. They act as open circuits capable of withstanding a certain rated voltage until triggered. When turned on (triggered), they become low-resistance current paths and remain so, even after the trigger is removed, until the current is reduced to a certain level or until they are turned off, depending on the type of device. Thyristors can be used to control the amount of ac power to a load and are used in lamp-dimming circuits, motor speed control, ignition systems, and charging circuits, to name a few. UJTs are often used as trigger devices for thyristors.

After completing this section, you should be able to

- **Explain the basic structure and operation of the SCR, the triac, and the diac**
- Recognize the SCR, triac, and diac symbols
- Use a transistor analogy to describe SCR operation
- Describe triac operation
- Describe diac operation
- Discuss typical thyristor applications

A (anode)

G (gate)

K (cathode)

FIGURE 18–47

SCR symbol.

Silicon-Controlled Rectifiers (SCRs)

The silicon-controlled rectifier (SCR) has three terminals, as shown in the symbol in Figure 18–47. Like a rectifier diode, it is a unidirectional device, but the conduction of current is controlled by the gate, G. Current is from the cathode K to the anode A when the SCR is on. A positive voltage on the gate will turn the

SCR on. The SCR will remain on as long as the current from the cathode to the anode is equal to or greater than a specified value called the *holding current*. When the current drops below the holding value, the SCR will turn off. It can be turned on again by a positive voltage on the gate.

SCR Construction The SCR is a four-layer device. It has two *p* regions and two *n* regions, as shown in Figure 18–48. This construction can be thought of as an *npn* and a *pnp* transistor interconnected. The upper *pnp* layers act as a transistor, Q_1, and the lower *npn* layers act as a transistor, Q_2. Notice that the two middle layers are "shared."

FIGURE 18–48

Construction of SCR and its transistor equivalent.

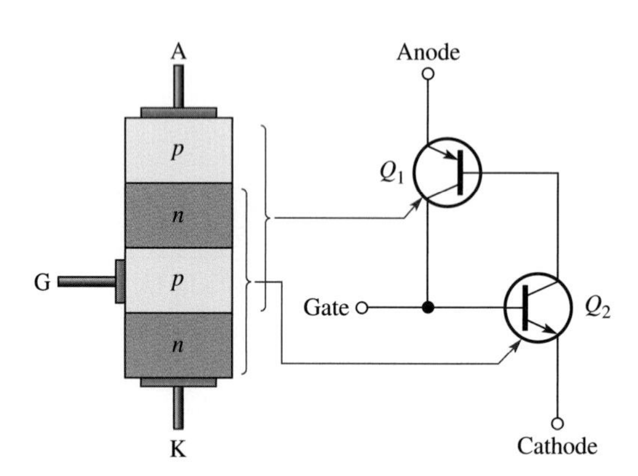

An Application Figure 18–49 shows an SCR used to rectify and control the average power delivered to a load. The SCR rectifies the 60 Hz ac just as a conventional rectifier diode. It can conduct only during the positive half-cycle. The phase controller produces a trigger pulse at the SCR gate in order to turn it on during any portion of the positive half-cycle of the input. If the SCR is turned on earlier in the half-cycle, more average power is delivered to the load. If the SCR is fired later in the half-cycle, less average power is delivered.

FIGURE 18–49

SCR controls average power to load.

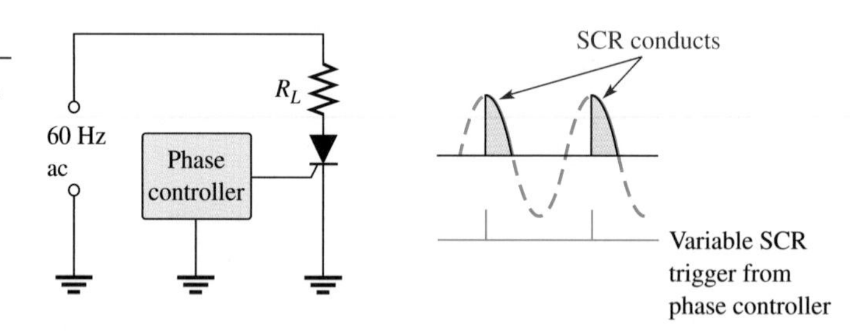

Triacs

The **triac** is also known as a *bidirectional triode thyristor.* It is equivalent to two SCRs connected to allow current in either direction. The SCRs share a common gate. The symbol for a triac is shown in Figure 18–50.

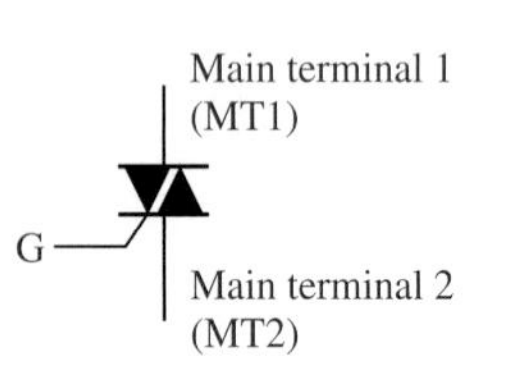

FIGURE 18–50

Triac symbol.

Applications Like the SCR, triacs are also used to control average power to a load by the method of phase control. The triac can be triggered such that the ac power is supplied to the load for a controlled portion of each half-cycle. During each positive half-cycle of the ac, the triac is off for a certain interval, called the *delay angle* (measured in degrees). Then it is triggered on and conducts current through the load for the remaining portion of the positive half-cycle, called the *conduction angle.* Similar action occurs on the negative half-cycle except that, of

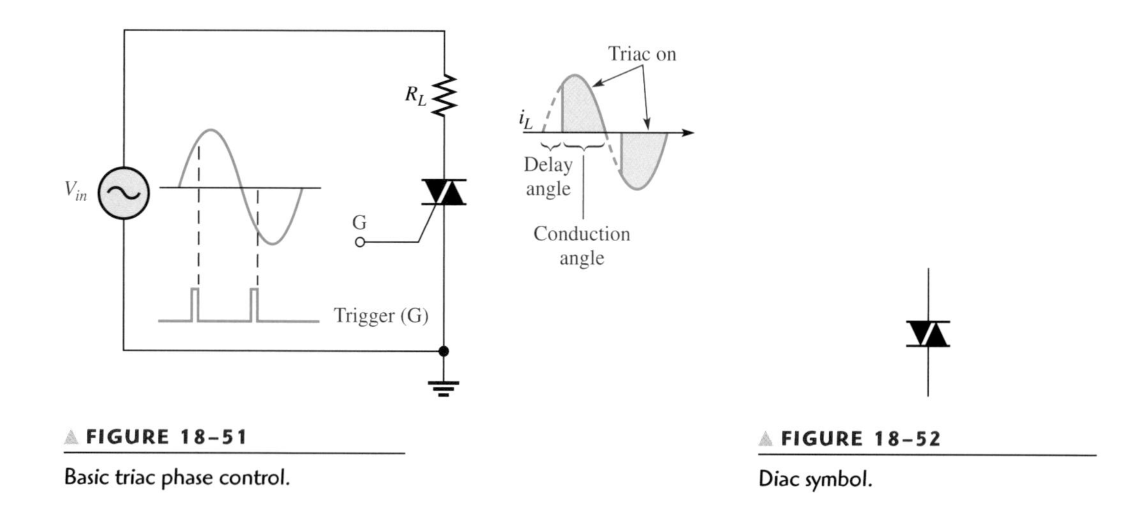

FIGURE 18–51

Basic triac phase control.

FIGURE 18–52

Diac symbol.

course, current is conducted in the opposite direction through the load. Figure 18–51 illustrates this action.

Diacs

The diac is a bidirectional device that does not have a gate. It conducts current in either direction when a sufficient voltage, called the *breakover potential,* is reached across the two terminals. The symbol for a diac is shown in Figure 18–52.

SECTION 18–11 REVIEW

1. What does "SCR" stand for?
2. Explain how an SCR basically works.
3. Name a typical application of a triac.
4. A diac is gate-controlled. (True or False)

18–12 TRANSISTOR PACKAGES AND TERMINAL IDENTIFICATION

Transistors are available in a wide range of package types for various applications. Those with mounting studs or heat sinks are usually power transistors. Low-power and medium-power transistors are usually found in smaller metal or plastic cases. Still another package classification is for high-frequency devices. You should be familiar with common transistor packages and be able to identify the emitter, base, and collector terminals.

After completing this section, you should be able to

- **Identify various types of transistor package configurations**
- List three broad categories of transistors
- Recognize various types of cases and identify the pin configurations

Transistor Categories

Manufacturers generally classify transistors into three broad categories: general-purpose/small-signal devices, power devices, and RF (radio frequency/microwave) devices. Although each of these categories, to a large degree, has its own unique package types, you will find certain types of packages used in more than one device category. While keeping in mind there is some overlap, we will look at transistor packages for each of the three categories, so that you will be able to recognize a transistor when you see one on a circuit board and have a good idea of what general category it is in.

General-Purpose/Small-Signal Transistors General-purpose/small-signal transistors are generally used for low- or medium-power amplifiers (less than 1 W) or switching circuits. The packages are either plastic or metal cases. Certain types of packages contain multiple transistors. Figure 18–53 illustrates common plastic cases, Figure 18–54 shows packages called *metal cans,* and Figure 18–55 shows multiple-transistor packages. Some of the multiple-transistor packages such as the dual-in-line (DIP) and the small-online (SO) are the same as those used for many integrated circuits. Typical pin connections are shown so you can identify the emitter, base, and collector.

Power Transistors Power transistors are used to handle large currents (typically more than 1 A) and/or large voltages. For example, the final audio stage in a

FIGURE 18–53

Plastic cases for general-purpose/small-signal transistors. Both old and new JEDEC TO numbers are given. Pin configurations may vary. Always check the data sheet.

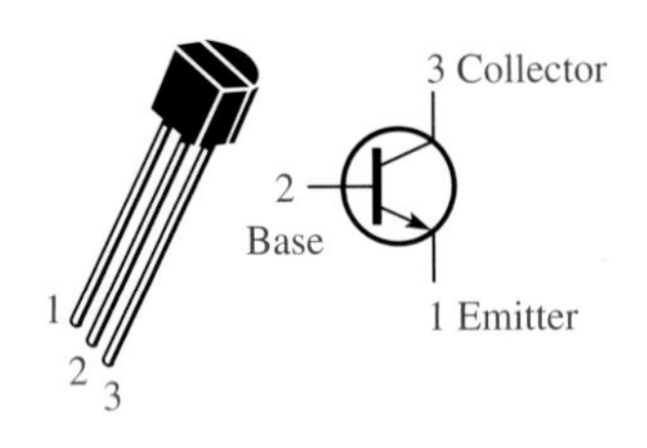

(a) TO-92 or TO-226AA

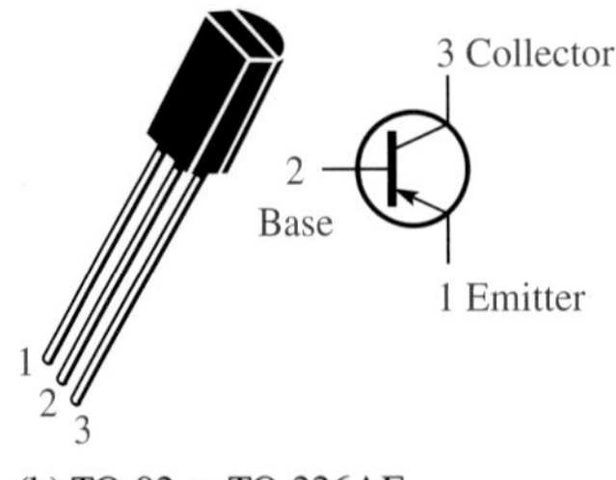

(b) TO-92 or TO-226AE

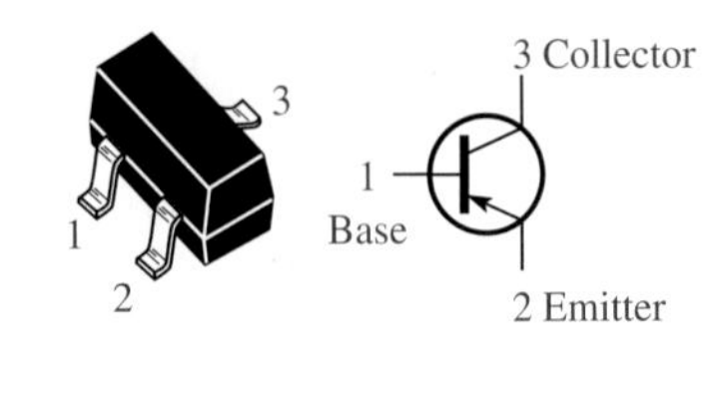

(c) SOT-23 or TO-236AB

FIGURE 18–54

Metal cases for general-purpose/small-signal transistors.

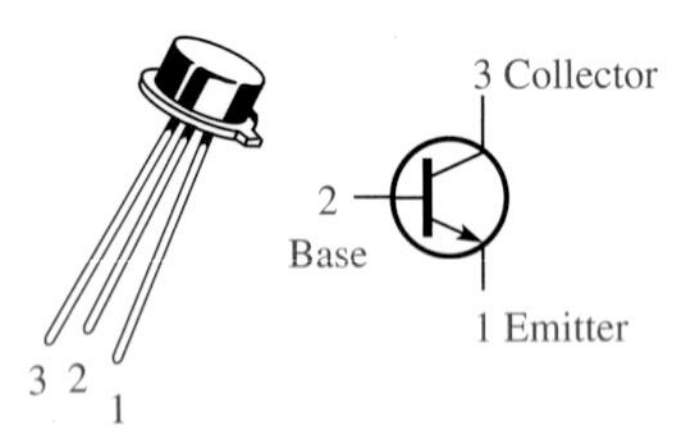

(a) TO-18 or TO-206AA

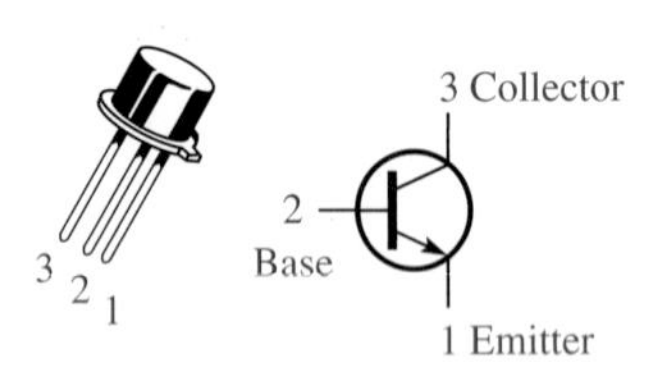

(b) TO-39 or TO-205AD

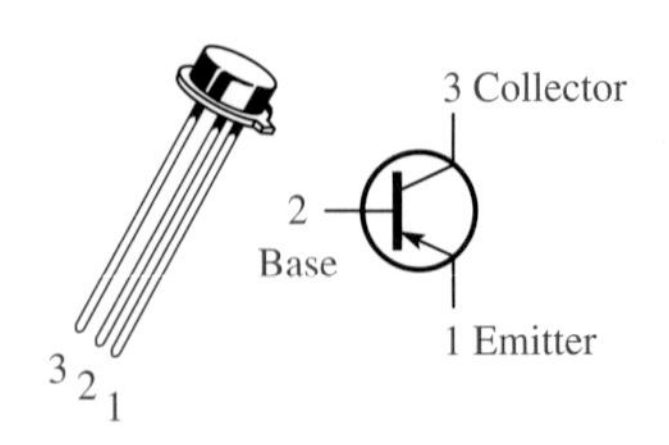

(c) TO-46 or TO-206AB

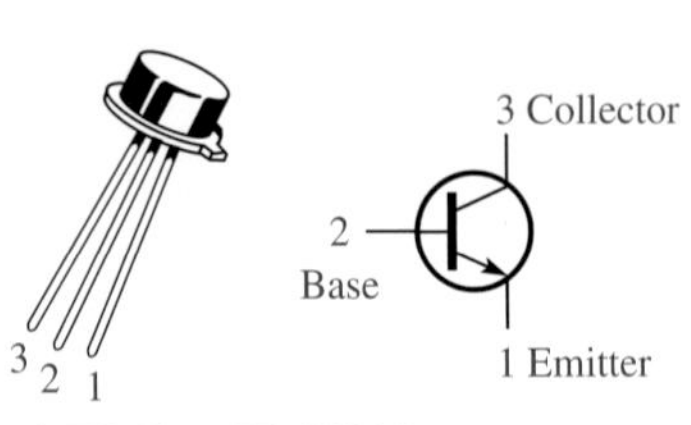

(d) TO-52 or TO-206AC

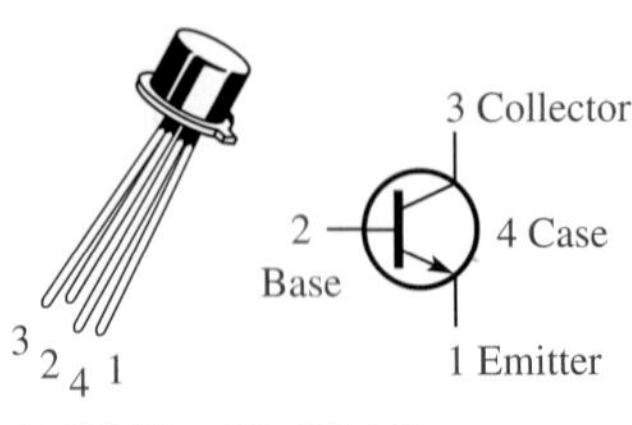

(e) TO-72 or TO-206AF

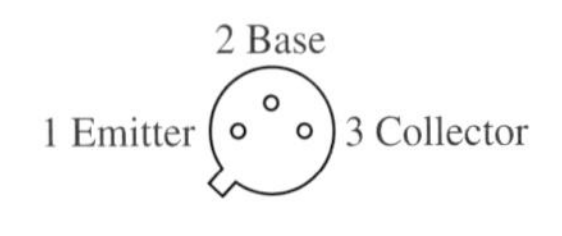

(f) Pin configuration (bottom view). Emitter is closest to tab.

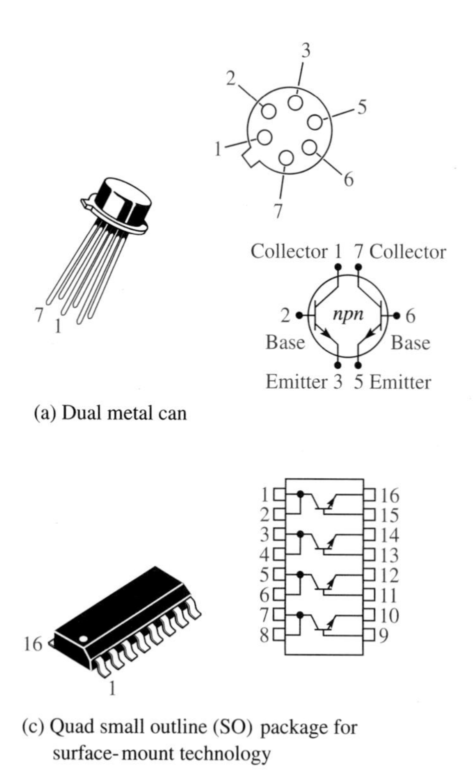

(a) Dual metal can

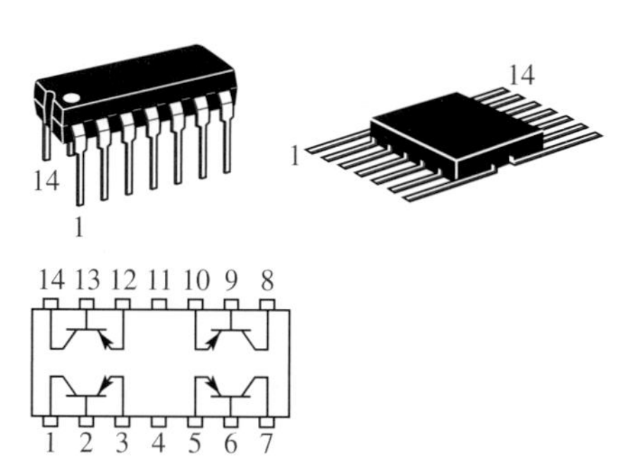

(b) Quad dual in-line (DIP) and quad flat-pack. Dot indicates pin 1.

(c) Quad small outline (SO) package for surface-mount technology

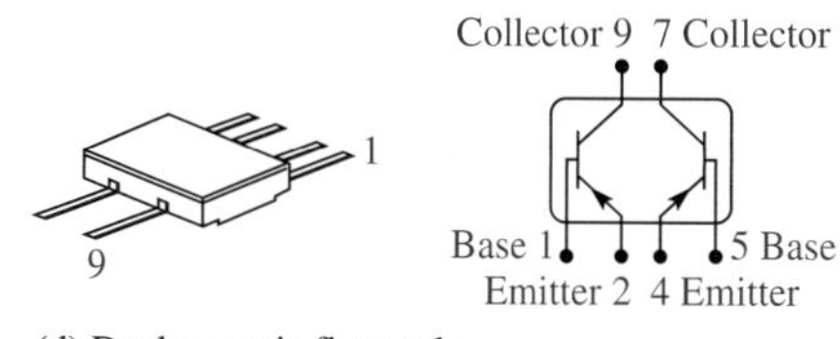

(d) Dual ceramic flat-pack

▲ **FIGURE 18–55**

Typical multiple-transistor packages.

stereo system uses a power transistor amplifier to drive the speakers. Figure 18–56 shows some common package configurations. In most applications, the metal tab or the metal case is common to the controller and is thermally connected to a heat sink for heat dissipation. Notice in part (g) how the small transistor chip is mounted inside the much larger package.

▼ **FIGURE 18–56**

Typical power transistors.

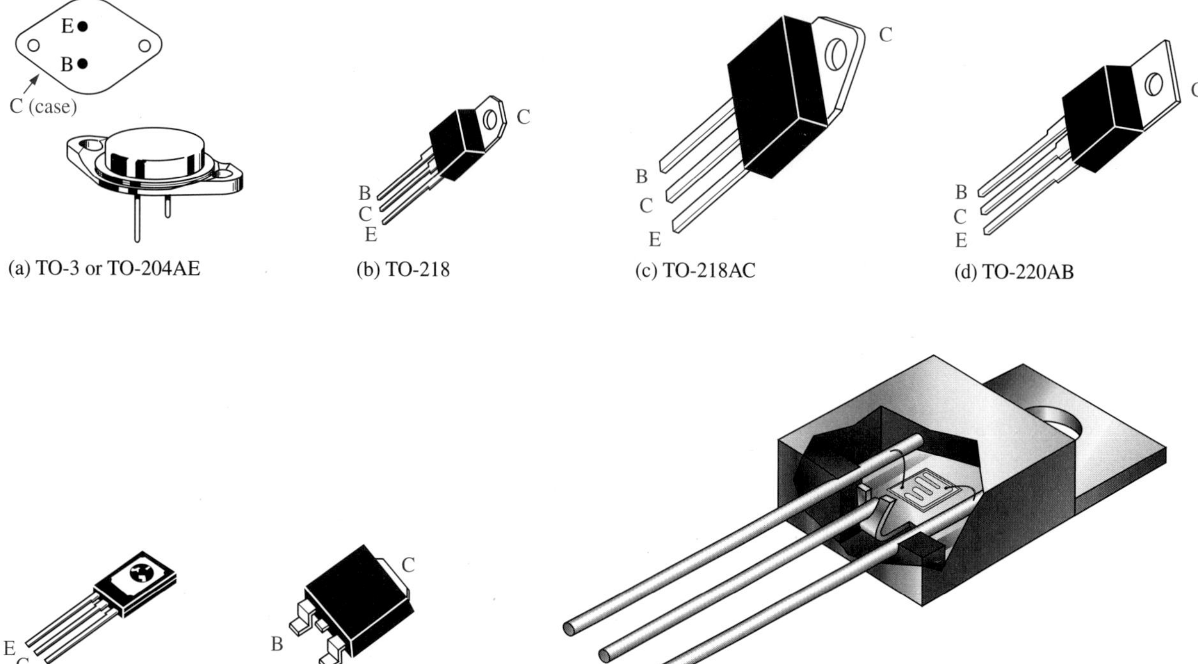

(a) TO-3 or TO-204AE

(b) TO-218

(c) TO-218AC

(d) TO-220AB

(e) TO-225AA

(f) Surface-mount technology

(g) Cutaway view of tiny transistor chip mounted in the encapsulated package

RF Transistors RF transistors are designed to operate at extremely high frequencies and are commonly used for various purposes in communications systems and other high-frequency applications. Their unusual shapes and lead configurations are designed to optimize certain high-frequency parameters. Figure 18–57 shows some examples.

FIGURE 18–57

Examples of RF transistors.

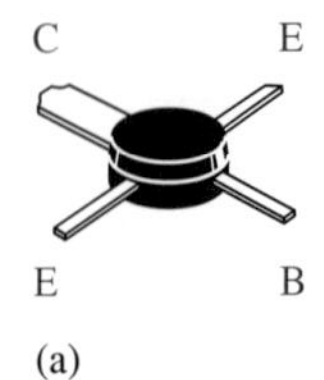

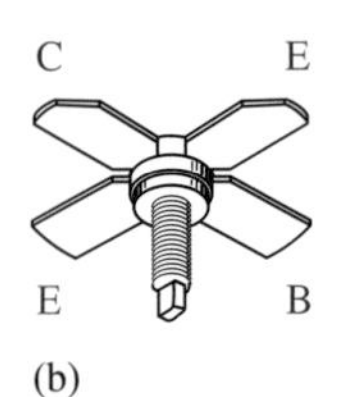

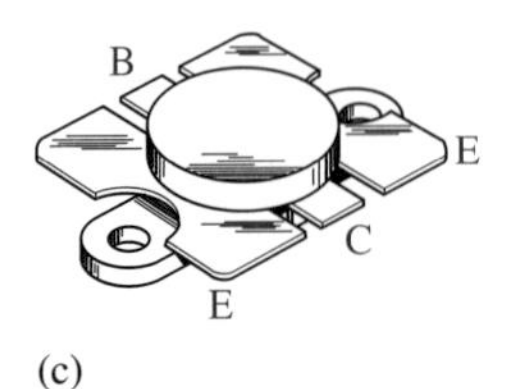

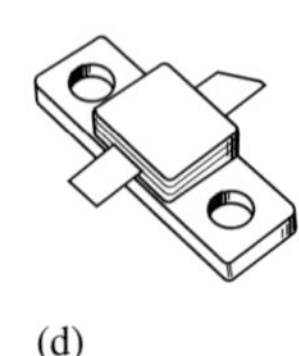

SECTION 18–12 REVIEW

1. List the three broad categories of transistors.
2. In a single-transistor metal case, how do you identify the leads?
3. In power transistors, the metal mounting tab or case is connected to which transistor region?

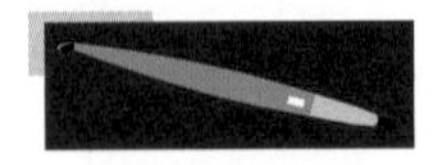

18–13 TROUBLESHOOTING

As you already know, a critical skill in electronics is the ability to identify a circuit malfunction and to isolate the failure to a single component if possible. In this section, the basics of troubleshooting transistor bias circuits and testing individual transistors are covered.

After completing this section, you should be able to

- **Troubleshoot transistors**
- Troubleshoot a transistor bias circuit
- Test a transistor in-circuit and out-of-circuit
- Discuss gain measurement

Troubleshooting a Transistor Bias Circuit

Several faults can occur in a transistor bias circuit. An initial check of the collector-emitter voltage will show if the transistor is saturated or cut off. The most common faults are open base or collector resistor, loss of base or collector bias voltage, open transistor junctions, and open contacts or interconnections on the printed circuit board. Figure 18–58 illustrates a basic four-point check that will help isolate one of these faults in the circuit. The term *floating* refers to a point that is effectively disconnected from a "solid" voltage or ground. A very small, erratically fluctuating voltage is sometimes observed.

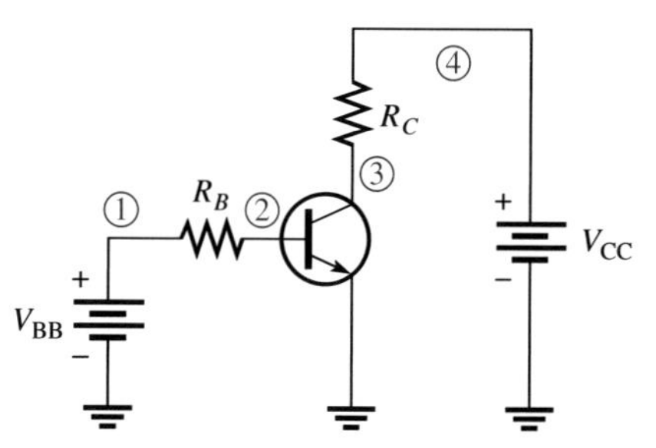

Symptom	Test point voltages 1	2	3	4
A. No V_{BB}				
Source shorted	0 V	0 V	V_{CC}	V_{CC}
Source open	Floating	Floating	V_{CC}	V_{CC}
B. No V_{CC}				
Source shorted	V_{BB}	0.7 V	0 V	0 V
Source open	V_{BB}	0.7 V	Floating	Floating
C. R_B open	V_{BB}	Floating	V_{CC}	V_{CC}
D. R_C open	V_{BB}	0.7 V	Floating	V_{CC}
E. Base-to-emitter junction open	V_{BB}	V_{BB}	V_{CC}	V_{CC}
F. Base-to-collector junction open	V_{BB}	0.7 V	V_{CC}	V_{CC}
G. Low β	V_{BB}	0.7 V	Higher than normal	V_{CC}

FIGURE 18–58

Test point troubleshooting guide for a basic transistor bias circuit.

EXAMPLE 18–12

From the voltage measurements in Figure 18–59, what is the most likely problem?

FIGURE 18–59

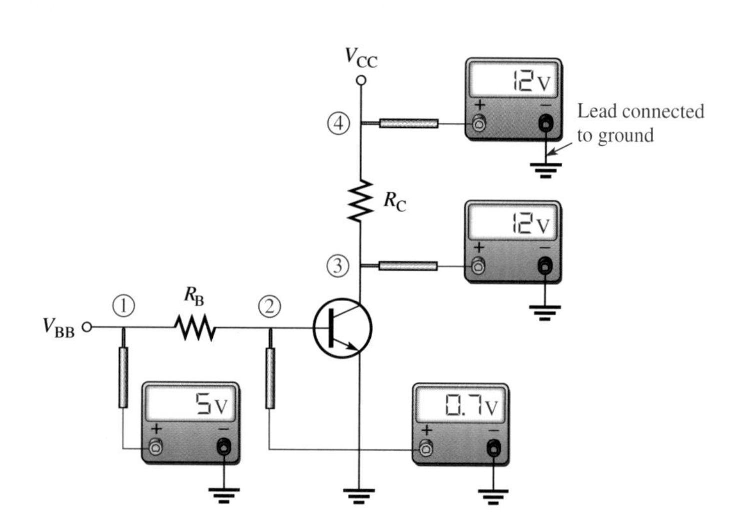

Solution Since $V_{CE} = 12$ V, the transistor is obviously in cutoff. The V_{BB} measurement appears OK and the voltage at the base is correct. Therefore, the collector-base junction of the transistor must be open. Replace the transistor and check the voltages again.

Related Problem What is the voltage at point 2 if the BE junction opens in Figure 18–59?

Open file E18-12 on your EWB/CircuitMaker CD-ROM and determine the fault, if any.

Transistor Testers

▲ **FIGURE 18–60**

Transistor tester. (Photography courtesy of B&K Precision Corp.)

An individual transistor can be tested either in-circuit or out-of-circuit. For example, let's say that an amplifier on a particular printed circuit (PC) board has malfunctioned. You perform a check and find that the symptom is the same as Item F in the table of Figure 18–58, indicating an open from base to collector. This can mean either a bad transistor or an open in an external circuit connection. Your next step is to find out which one.

Good troubleshooting practice requires that you do not remove a component from a circuit board unless you are reasonably sure that it is bad or you simply cannot isolate the problem down to a single component. When components are removed, there is a risk of damage to the PC board contacts and traces.

The next step is to do an in-circuit check of the transistor using a transistor tester similar to the one shown in Figure 18–60. The three clip-leads are connected to the transistor terminals and the tester gives a GOOD/BAD indication.

Case 1 If the transistor tests BAD, it should be carefully removed and replaced by a known good one. An out-of-circuit check of the replacement device is usually a good idea, just to make sure it is OK. The transistor is plugged into the socket on the transistor tester for out-of-circuit tests.

Case 2 If the transistor tests GOOD in-circuit, examine the circuit board for a poor connection at the collector pad or for a break in the connecting trace. A poor solder joint often results in an open or a highly resistive contact. The physical point at which you actually measure the voltage is very important in this case. If you measure on the collector side of an open that is external to the transistor, you will get the indication listed in Item D of the table in Figure 18–58. If you measure on the side of the external open closest to R_C, you will get the indication in Item F of the table. This situation is illustrated in Figure 18–61.

Importance of Point-of-Measurement in Troubleshooting

In Figure 18–61, if you had taken the initial measurement on the transistor lead itself and the open were internal to the transistor, you would have measured V_{CC}, as illustrated in Figure 18–62. This would have indicated a bad transistor even before the tester was used. This simple concept emphasizes the importance of point-of-measurement in certain troubleshooting situations. This is also valid for opens at the base and emitter of a transistor.

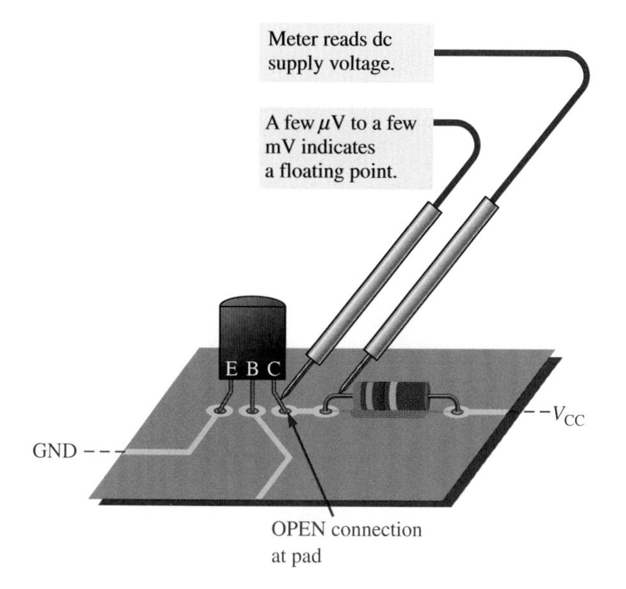

▲ **FIGURE 18–61**

The indication of an open, when it is in the external circuit, depends on where you measure. The other meter probe is connected to ground.

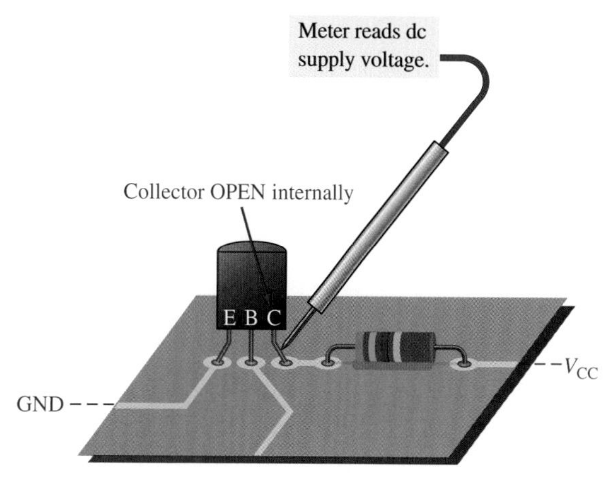

▲ **FIGURE 18–62**

Illustration of an internal transistor open. Compare with Figure 18–61. The other meter probe is connected to ground.

EXAMPLE 18–13

What fault do the measurements in Figure 18–63 indicate? The other meter probe is touching ground.

▼ **FIGURE 18–63**

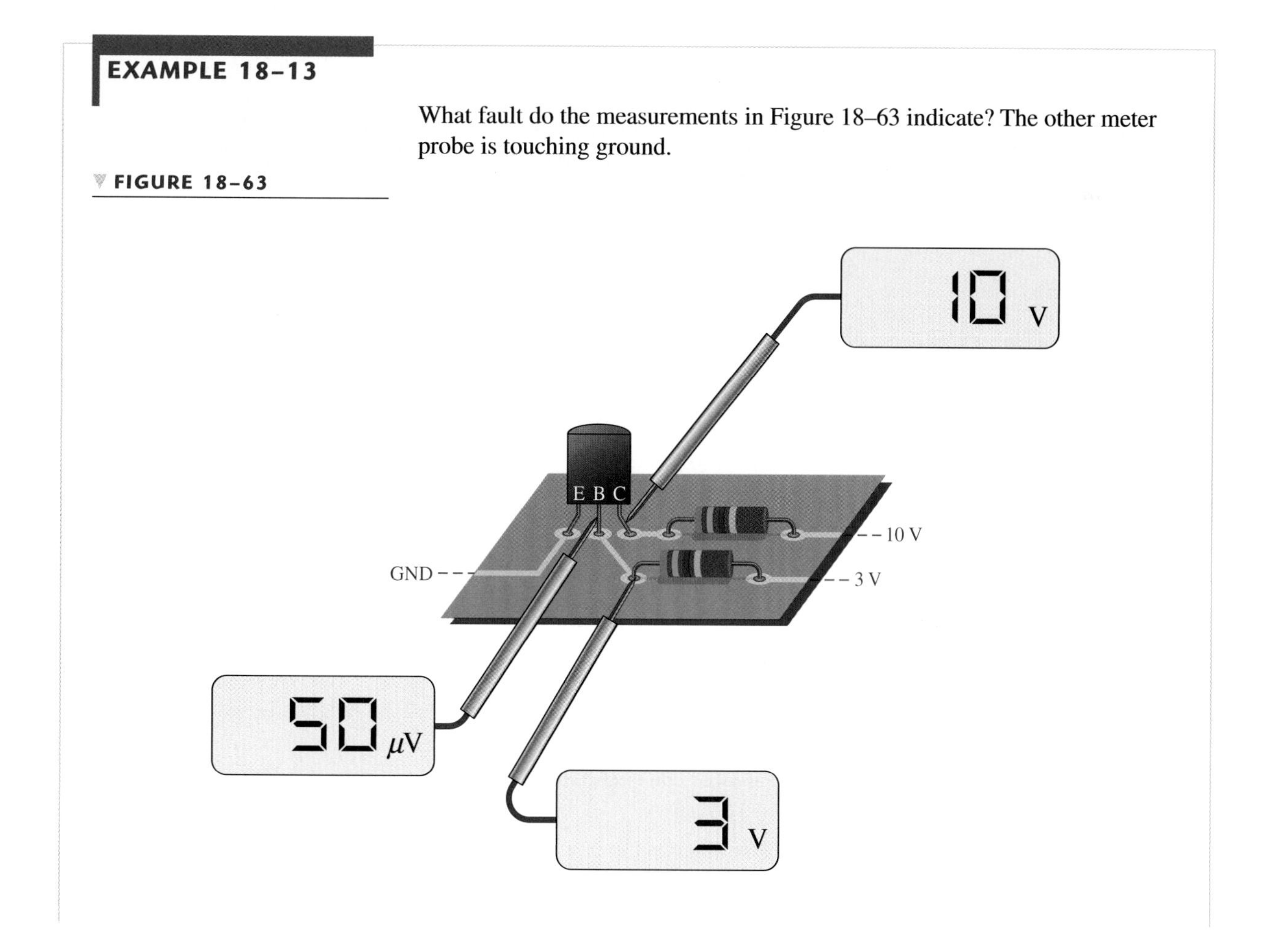

Solution The transistor is in cutoff, as indicated by the 10 V measurement on the collector lead. The base bias voltage of 3 V appears on the PC board contacts but not on the transistor lead. This indicates that there is an open external to the transistor between the two measured points. Check the solder joint at the base contact on the PC board. If the open were internal, there would be 3 V on the base lead.

Related Problem If the meter in Figure 18–63 that now reads 3 V indicates a floating point, what is the most likely fault?

Out-of-Circuit Multimeter Test

If a transistor tester is not available, an analog multimeter can be used to perform a basic diode check on the transistor junctions for opens or shorts. For this test, a transistor can be viewed as two back-to-back diodes as shown in Figure 18–64. The BC junction is one diode and the BE junction is the other. Also many DMMs have a transistor test feature.

FIGURE 18–64

The transistor is viewed as two diodes for a junction check with a multimeter.

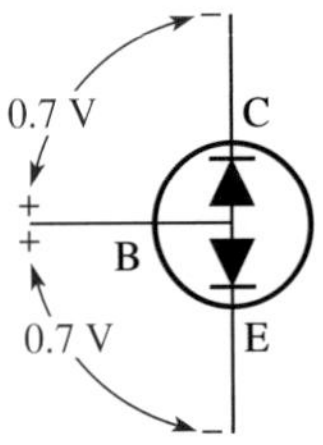

(a) Both junctions should read approximately 0.7 V when forward-biased

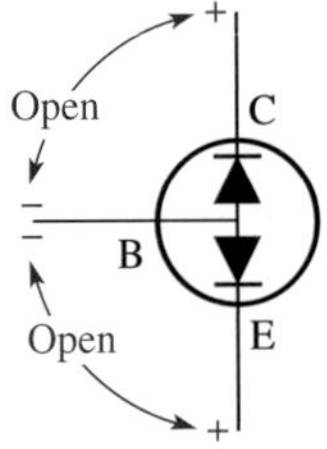

(b) Both junctions should read open (ideally) when reverse-biased

Gain Measurement

Some transistor testers check the β_{DC}. A known value of I_B is applied and the resulting I_C is measured. The reading will indicate the value of the I_C/I_B ratio, although in some units only a relative indication is given. Most testers provide for an in-circuit β_{DC} check, so that a suspected device does not have to be removed from the circuit for testing.

Curve Tracers

The curve tracer is an oscilloscope type of instrument that can display transistor characteristics such as a family of collector curves. In addition to the measurement and display of various transistor characteristics, diode curves can also be displayed.

SECTION 18–13 REVIEW

1. If a transistor on a circuit board is suspected of being faulty, what should you do first?
2. In a transistor bias circuit, such as the one in Figure 18–58, what happens if R_B opens?
3. In a circuit such as the one in Figure 18–58, what are the base and collector voltages if there is an open between the emitter and ground?

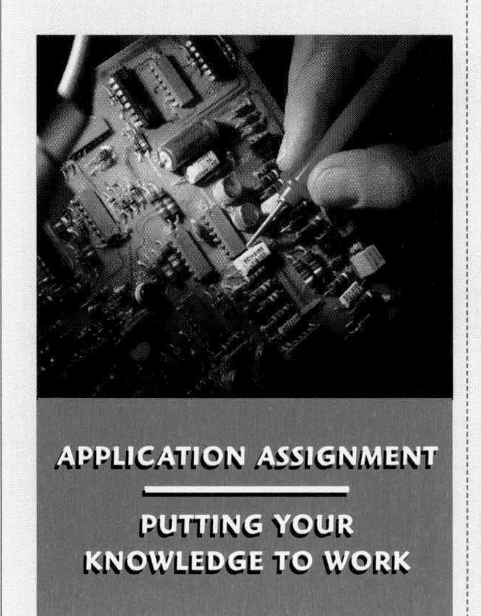

APPLICATION ASSIGNMENT

PUTTING YOUR KNOWLEDGE TO WORK

As stated in the chapter opener, you will work on an industrial temperature-control system. The system has several parts, but in this assignment you are concerned with the transistor circuit that monitors the temperature changes in the fluid in a tank and provides a proportional output that is used to precisely control the temperature. In this application, the transistor is used as a dc amplifier.

Description of the System

The system in Figure 18–65 is an example of a process control system that uses the principle of closed-loop feedback to maintain a substance at a constant temperature of 50°C ± 1°C. The temperature sensor in the tank is a **thermistor,** which is a temperature-sensitive resistor whose resistance has a negative temperature coefficient. The thermistor forms part of the bias circuit for the transistor detector.

A small change in temperature causes the resistance of the thermistor to change, which results in a proportional change in the output voltage from the transistor detector. Since thermistors are typically nonlinear devices, the purpose of the digital portions of the system is to precisely compensate for the nonlinear characteristic of the temperature detector in order to provide a continuous linear adjustment of the fuel flow to the burner to offset any small shift in temperature away from the desired value.

The temperature detector board (highlighted in color) that you are concerned with in this application assignment is obviously only a small part of the total system. It is interesting, however, to see how the variations in a transistor's bias point can be used in a situation such as this. Although the digital portions of the system are not covered, they show that typical systems have various types of elements.

The temperature detector circuit board and the thermistor probe that mounts into the side of the tank are shown in Figure 18–66. The thermistor symbol and the case configuration of the *npn* transistor are included.

Step 1: Relate the PC Board to the Schematic

Develop a schematic of the circuit board in Figure 18–66 including the thermistor. Arrange the schematic into a form that is familiar to you and identify the type of biasing used.

Step 2: Analyze the Circuit

With a dc supply of +15 V, analyze the temperature detector circuit to determine the output voltage at 50°C and at 1°C above and below 50°C. Use 0.7 V for the base-emitter voltage. Assume $\beta_{DC} = 100$ for the transistor. Since accuracy on the graph is difficult to achieve for this small range, assume that the thermistor resistances for the controlled temperature range are as follows:

At $T = 50°C$, $R = 2.75\ k\Omega$

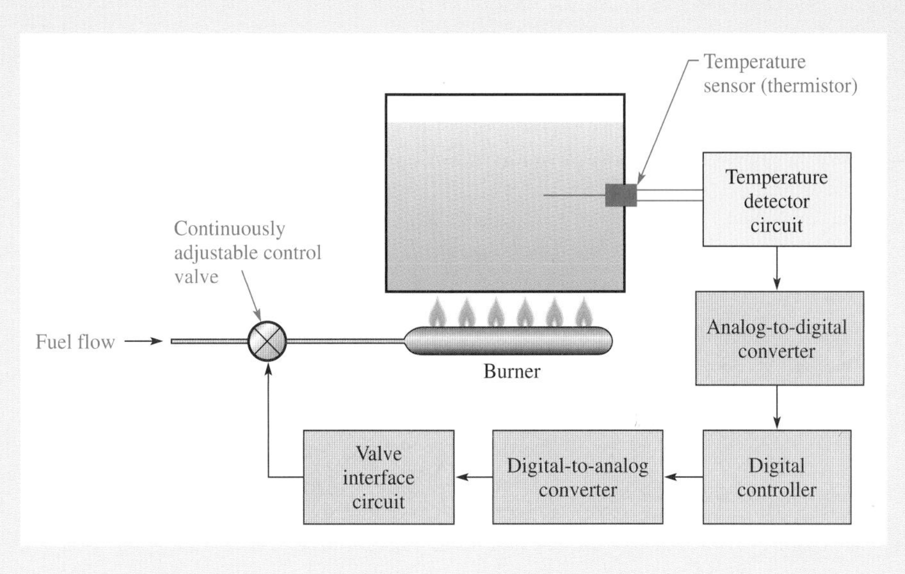

FIGURE 18–65

Block diagram of the industrial temperature-control system.

FIGURE 18–66

The temperature detector circuit.

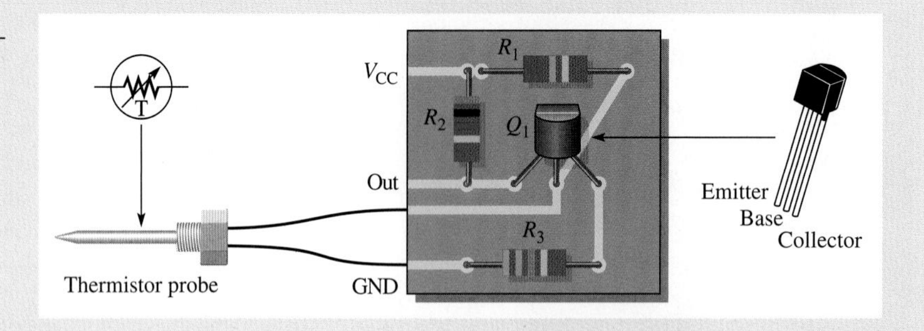

At $T = 49°C$, $R = 3.1\ k\Omega$

At $T = 51°C$, $R = 2.5\ k\Omega$

Will your values change significantly if you take into account the 1.0 MΩ input resistance of the analog-to-digital converter?

Step 3: Check Output over Temperature Range

With the circuit in a controlled-temperature environment, determine the output voltage over a range from 30°C to 110°C at 20° intervals. Use the graph in Figure 18–67.

Step 4: Troubleshoot the Circuit Board

For each of the following problems, state the probable cause or causes in each case (Refer to Figure 18–66):

1. V_{CE} is approximately 0.1 V and V_C is 3.8 V.
2. Collector of Q_1 remains at 15 V.
3. For each of the above conditions, if there is more than one possible fault, describe how you will go about isolating the problem.

FIGURE 18–67

Graph of thermistor resistance versus temperature.

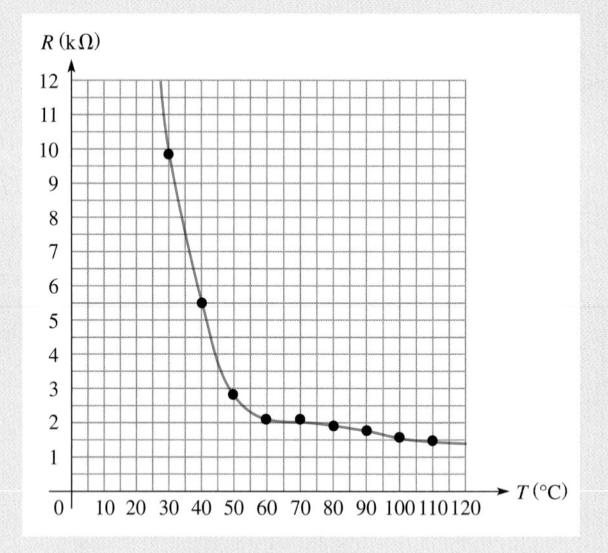

APPLICATION ASSIGNMENT REVIEW

1. What is the basic purpose of the detector circuit board in the process control system?
2. In Step 2, does the transistor go into saturation and/or cutoff?
3. Calculate the output voltage at 40°C.
4. Calculate the output voltage at 60°C.

SUMMARY

- A bipolar junction transistor (BJT) consists of three regions: emitter, base, and collector.
- The three regions of a BJT are separated by two *pn* junctions.
- The two types of bipolar transistor are the *npn* and the *pnp*.
- The term *bipolar* refers to two types of current: electron current and hole current.
- A field-effect transistor (FET) has three terminals: source, drain, and gate.
- A junction field-effect transistor (JFET) operates with a reverse-biased gate-to-source *pn* junction.
- JFETs have very high input resistance due to the reverse-biased gate-source junction.
- JFET current between the drain and the source is through a channel whose width is controlled by the amount of reverse bias on the gate-source junction.
- The two types of JFETs are *n*-channel and *p*-channel.
- Metal-oxide semiconductor field-effect transistors (MOSFETs) differ from JFETs in that the gate of a MOSFET is insulated from the channel.
- A depletion/enhancement MOSFET (D-MOSFET) can operate with a positive, negative, or zero gate-to-source voltage.
- The D-MOSFET has a physical channel between the drain and the source.
- An enhancement-only MOSFET (E-MOSFET) can operate only when the gate-to-source voltage exceeds a threshold value.
- The E-MOSFET has no physical channel.
- Transistors are used as either amplifying devices or switching devices.
- Transistor and thyristor symbols are shown in Figure 18–68.

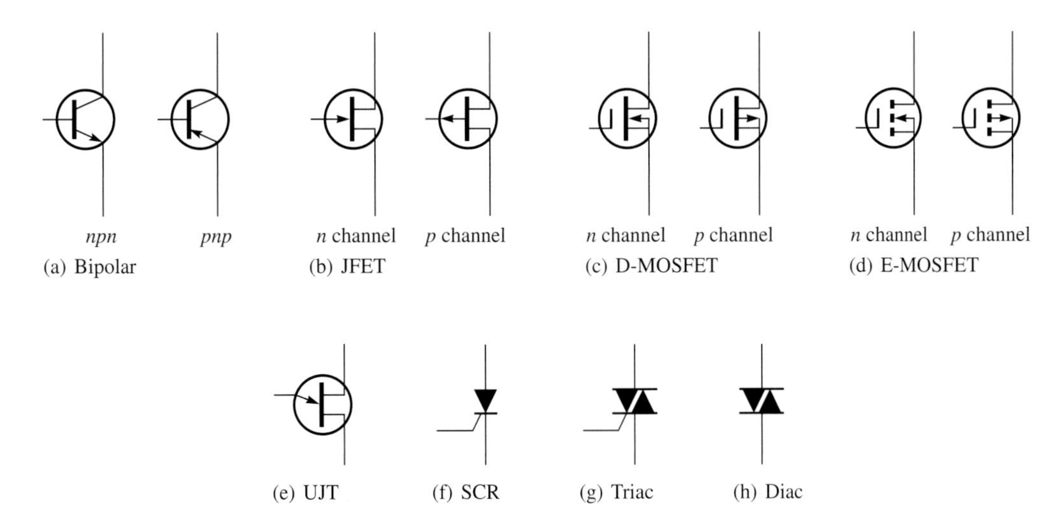

FIGURE 18–68

Transistor and thyristor symbols.

EQUATIONS

18–1	$I_E = I_C + I_B$	Bipolar transistor currents
18–2	$I_C = \alpha_{DC} I_E$	Relationship of collector and emitter currents
18–3	$I_C = \beta_{DC} I_B$	Relationship of collector and base currents
18–4	$V_C = V_{CC} - I_C R_C$	Collector voltage
18–5	$V_B = V_E + V_{BE}$	Base voltage
18–6	$R_{IN} \cong \beta_{DC} R_E$	Input resistance at base
18–7	$V_B = \left(\frac{R_2 \parallel R_{IN}}{R_1 + R_2 \parallel R_{IN}}\right) V_{CC}$	Base voltage with voltage-divider bias
18–8	$V_B \cong \left(\frac{R_2}{R_1 + R_2}\right) V_{CC}$	Approximate base voltage ($R_{IN} >> R_2$)
18–9	$V_E = V_B - 0.7\text{ V}$	Emitter voltage
18–10	$\beta_{ac} = \frac{I_c}{I_b}$	ac beta
18–11	$A_v \cong \frac{R_C}{R_E}$	Voltage gain (Amplifier in Figure 18–17)
18–12	$V_{CE(cutoff)} \cong V_{CC}$	V_{CE} at cutoff
18–13	$I_{C(sat)} \cong \frac{V_{CC}}{R_C}$	Collector saturation current
18–14	$I_{B(min)} = \frac{I_{C(sat)}}{\beta_{DC}}$	Minimum base current for saturation
18–15	$I_C = \frac{P_{D(max)}}{V_{CE}}$	I_C for maximum V_{CE}
18–16	$V_{CE} = \frac{P_{D(max)}}{I_C}$	V_{CE} for maximum I_C
18–17	$R_{IN} = \left\lvert \frac{V_{GS}}{I_{GSS}} \right\rvert$	JFET input resistance
18–18	$V_{GS} = -I_D R_S$	Self-bias voltage for *n*-channel JFET
18–19	$V_{GS} = +I_D R_S$	Self-bias voltage for *p*-channel JFET
18–20	$V_D = V_{DD} - I_D R_D$	Drain voltage
18–21	$V_{DS} = V_{DD} - I_D(R_D + R_S)$	Drain-to-source voltage
18–22	$V_{DS} = V_{DD} - I_{DSS} R_D$	Drain-to-source voltage D-MOSFET
18–23	$V_{GS} = \left(\frac{R_2}{R_1 + R_2}\right) V_{DD}$	Gate-to-source voltage E-MOSFET
18–24	$V_{DS} = V_{DD} - I_D R_D$	Drain-to-source voltage E-MOSFET

SELF-TEST

Answers are at the end of the chapter.

1. The *n*-type regions in an *npn* bipolar junction transistor are
   (a) collector and base **(b)** collector and emitter
   (c) base and emitter **(d)** collector, base, and emitter
2. The *n*-region in a *pnp* transistor is the
   (a) base **(b)** collector **(c)** emitter **(d)** case
3. For normal operation of an *npn* transistor, the base must be
   (a) disconnected
   (b) negative with respect to the emitter
   (c) positive with respect to the emitter
   (d) positive with respect to the collector
4. The three currents in a BJT are
   (a) forward, reverse, and neutral **(b)** drain, source, and gate
   (c) alpha, beta, and sigma **(d)** base, emitter, and collector
5. Beta (β) is the ratio of
   (a) collector current to emitter current
   (b) collector current to base current
   (c) emitter current to base current
   (d) output voltage to input voltage
6. Alpha (α) is the ratio of
   (a) collector current to emitter current
   (b) collector current to base current
   (c) emitter current to base current
   (d) output voltage to input voltage
7. If the beta of a certain transistor operating in the linear region is 30 and the base current is 1 mA, the collector current is
   (a) 0.33 mA **(b)** 1 mA **(c)** 30 mA **(d)** unknown
8. If the base current is increased,
   (a) the collector current increases and the emitter current decreases
   (b) the collector current decreases and the emitter current decreases
   (c) the collector current increases and the emitter current does not change
   (d) the collector current increases and the emitter current increases
9. When an *n*-channel JFET is biased for conduction, the gate is
   (a) positive with respect to the source
   (b) negative with respect to the source
   (c) positive with respect to the drain
   (d) at the same voltage as the drain
10. When the gate-to-source voltage of an *n*-channel JFET is increased, the drain current
    (a) decreases **(b)** increases
    (c) stays constant **(d)** becomes zero

11. When a negative gate-to-source voltage is applied to an n-channel MOSFET, it operates in the
    (a) cutoff state **(b)** saturated state
    (c) enhancement mode **(d)** depletion mode
12. A UJT consists of
    (a) two emitters **(b)** a collector and two emitters
    (c) an emitter and two bases **(d)** an emitter, base, and collector
13. An SCR is like a rectifier diode except that
    (a) it can be triggered into conduction by a gate signal
    (b) it can conduct current in both directions
    (c) it can handle more power
    (d) it has four terminals
14. A triac is a device that
    (a) can be triggered into conduction
    (b) can conduct current in both directions
    (c) answers (a) and (b)
    (d) none of these answers

PROBLEMS

Answers to odd-numbered problems are at the end of the book.

BASIC PROBLEMS

SECTION 18–1

Bipolar Junction Transistors (BJTs)

1. What is the exact value of I_C for $I_E = 5.34$ mA and $I_B = 475\ \mu$A?
2. What is the α_{DC} when $I_C = 8.23$ mA and $I_E = 8.69$ mA?
3. A certain transistor has an $I_C = 25$ mA and an $I_B = 200\ \mu$A. Determine the β_{DC}.
4. In a certain transistor circuit, the base current is 2% of the 30 mA emitter current. Determine the collector current.
5. Find I_B, I_E, and I_C in Figure 18–69 given that $\beta_{DC} = 49$.

FIGURE 18–69

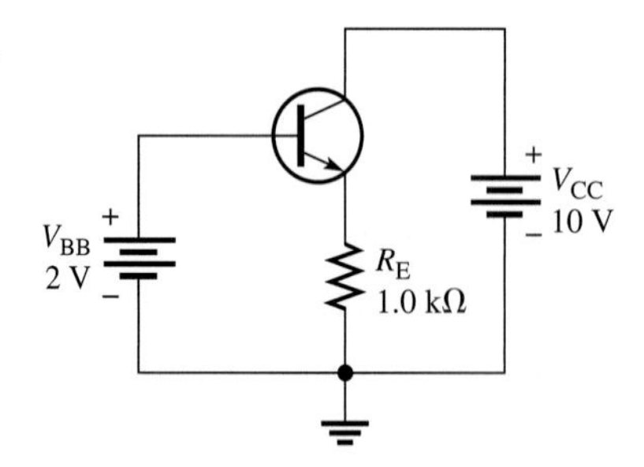

6. Determine the terminal voltages of each transistor with respect to ground for each circuit in Figure 18–70. Also determine V_{CE}, V_{BE}, and V_{BC}. $\beta_{DC} = 50$.

FIGURE 18–70

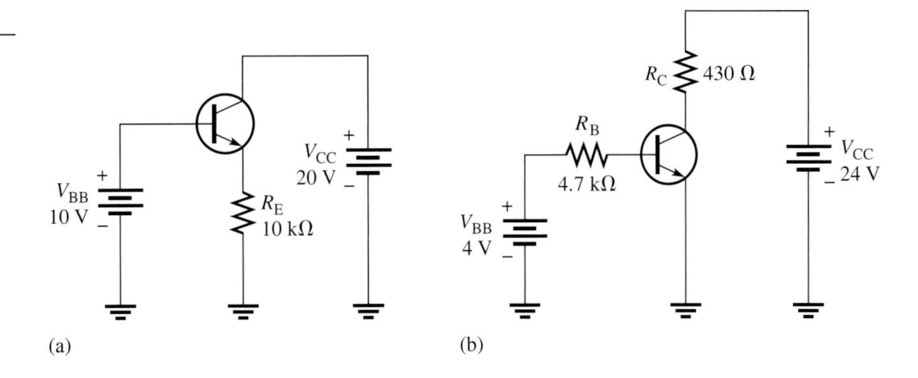

SECTION 18–2 **Voltage-Divider Bias**

7. Determine I_B, I_C, and V_C in Figure 18–71.

FIGURE 18–71

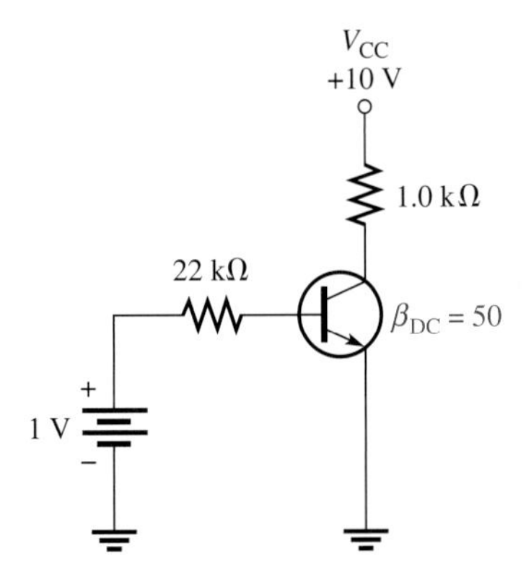

8. For the circuit in Figure 18–72, find V_B, V_E, I_E, I_C, and V_C.

9. In Figure 18–72, what is V_{CE}? What are the Q-point coordinates?

FIGURE 18–72

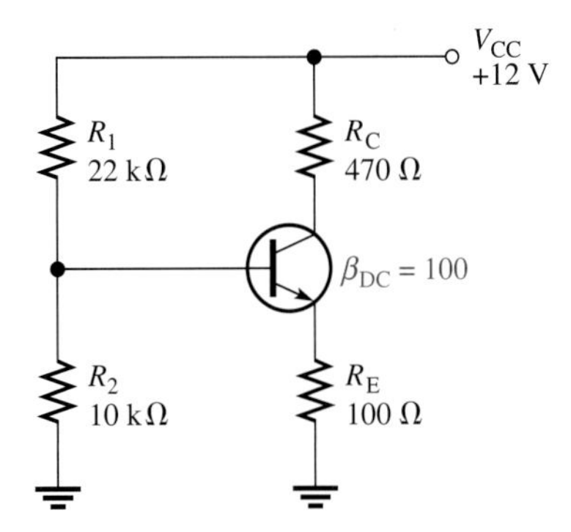

SECTION 18–3 **The Bipolar Junction Transistor as an Amplifier**

10. A transistor amplifier has a voltage gain of 50. What is the output voltage when the input voltage is 100 mV?

11. To achieve an output of 10 V with an input of 300 mV, what voltage gain is required?

12. A 50 mV signal is applied to the base of a properly biased transistor with $R_E = 100\ \Omega$ and $R_C = 500\ \Omega$. Determine the signal voltage at the collector.

SECTION 18–4 **The BJT as a Switch**

13. Determine $I_{C(sat)}$ for the transistor in Figure 18–73. What is the value of I_B necessary to produce saturation? What minimum value of V_{IN} is necessary for saturation?

14. The transistor in Figure 18–74 has a β_{DC} of 150. Determine the value of R_B required to ensure saturation when V_{IN} is 5 V. What must V_{IN} be to cut off the transistor?

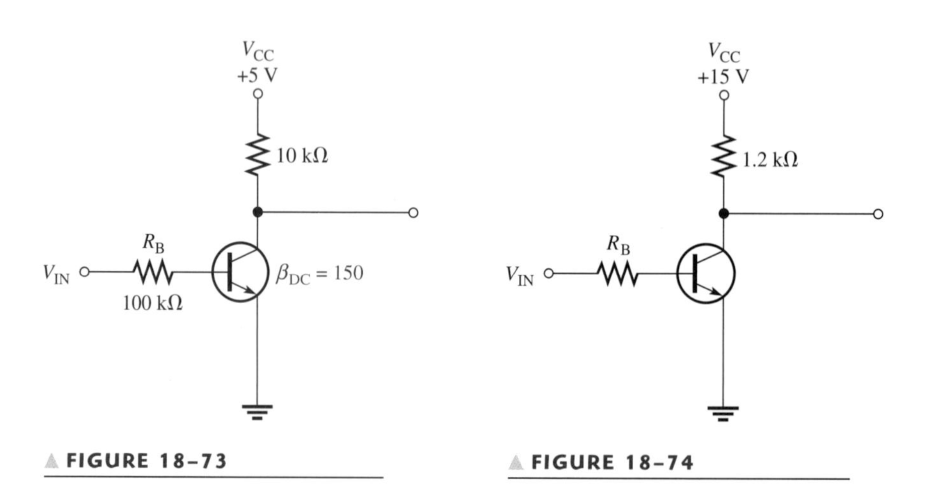

▲ FIGURE 18–73

▲ FIGURE 18–74

SECTION 18–5 **BJT Parameters and Ratings**

15. If the β_{DC} in Figure 18–70(a) changes from 100 to 150 due to a temperature increase, what is the change in collector current?

16. A certain transistor is to be operated at a collector current of 50 mA. How high can V_{CE} go without exceeding a $P_{D(max)}$ of 1.2 W?

SECTION 18–6 **The Junction Field-Effect Transistor (JFET)**

17. The V_{GS} of a *p*-channel JFET is increased from 1 V to 3 V.

(a) Does the depletion region narrow or widen?

(b) Does the resistance of the channel increase or decrease?

18. Why must the gate-to-source voltage of an *n*-channel JFET always be either 0 V or negative?

SECTION 18–7 **JFET Characteristics**

19. A JFET has a specified pinch-off voltage of 5 V. When $V_G = 0$, what is V_{DS} at the point where I_D becomes constant?

20. A certain JFET is biased such that $V_{GS} = -2$ V. If V_{GS} is changed to -3 V, how do V_{DS} at pinch-off and I_D change?

21. For a certain JFET, $I_{DSS} = 10$ mA and $V_{GS(off)} = -8$ V. If V_{GS} is varied from 0 V to -8 V, what will be the maximum and minimum values of I_D?

22. A certain *p*-channel JFET has a $V_{GS(off)} = 6$ V. What is I_D when $V_{GS} = 8$ V?

23. The JFET in Figure 18–75 has a $V_{GS(off)} = -4$ V. Assume that you increase the supply voltage, V_{DD}, beginning at 0 until the ammeter reaches a steady value. What does the voltmeter read at this point?

FIGURE 18–75

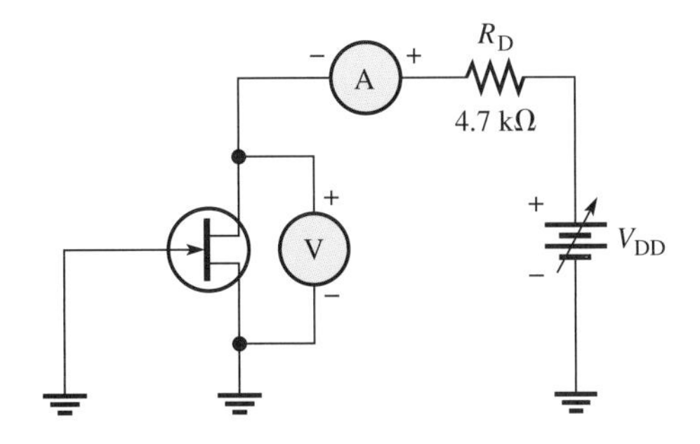

SECTION 18–8 **The Metal-Oxide Semiconductor FET (MOSFET)**

24. Draw the schematic symbols for n-channel and p-channel D-MOSFETs and E-MOSFETs. Label the terminals.
25. Explain why both types of MOSFETs have an extremely high input resistance at the gate.
26. In what mode is an n-channel D-MOSFET operating with a positive V_{GS}?
27. A certain E-MOSFET has a $V_{GS(th)} = 3$ V. What is the minimum V_{GS} for the device to turn on?

SECTION 18–9 **FET Biasing**

28. For each circuit in Figure 18–76, determine V_{DS} and V_{GS}.

FIGURE 18–76

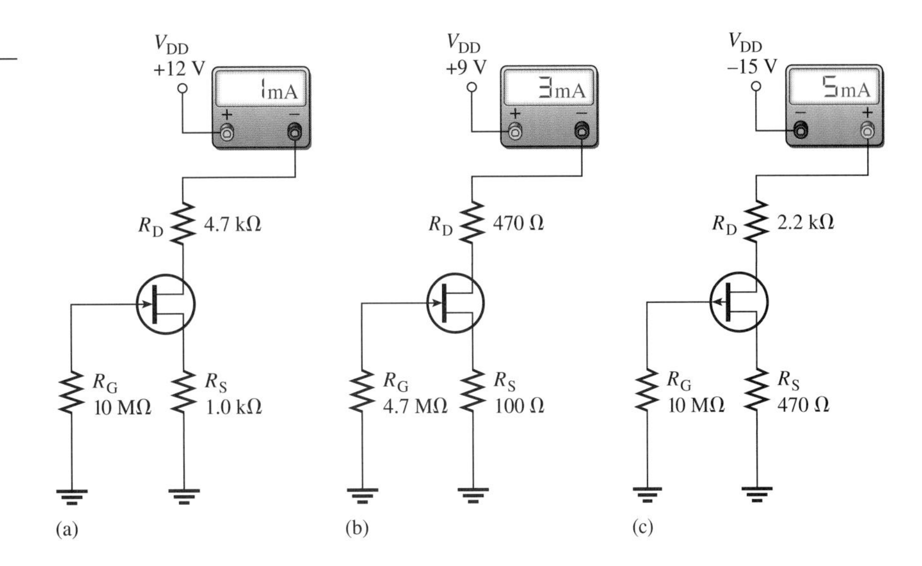

29. Determine in which mode (depletion or enhancement) each D-MOSFET in Figure 18–77 is biased.

FIGURE 18–77

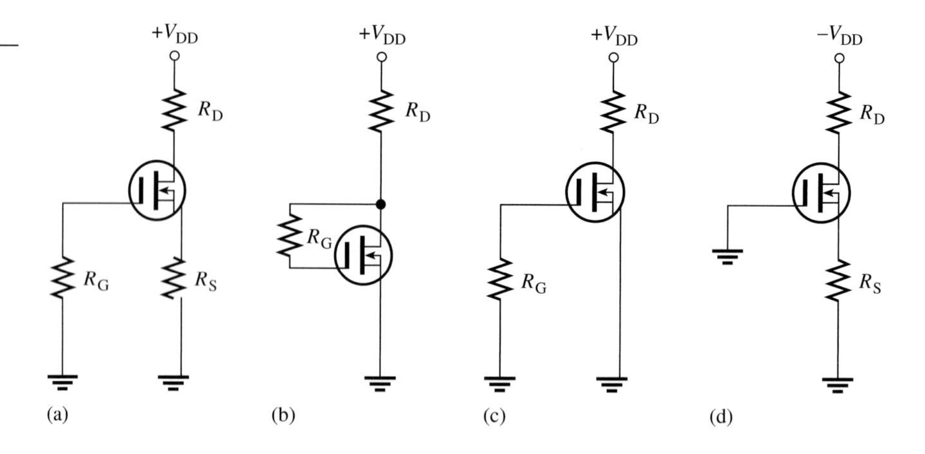

30. Each E-MOSFET in Figure 18–78 has a $V_{GS(th)}$ of +5 V or −5 V, depending on whether it is an *n*-channel or a *p*-channel device. Determine whether each MOSFET is *on* or *off*.

FIGURE 18–78

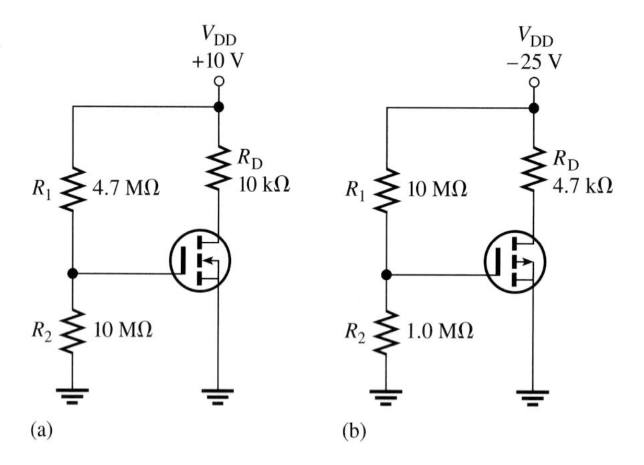

SECTION 18–10

Unijunction Transistors (UJTs)

31. In a UJT, if R_{BB} is 5 kΩ and R_{B1} is 3 kΩ, what is η?

32. What is the charging time constant in the oscillator circuit of Figure 18–79?

33. How long will the capacitor in Figure 18–79 initially charge if the peak voltage of the UJT is 6.3 V?

34. If the valley voltage of the UJT in Figure 18–79 is 2.23 V, what is the peak-to-peak value of the voltage across the capacitor, if the peak voltage is the same as in Problem 33? Draw its general shape.

FIGURE 18–79

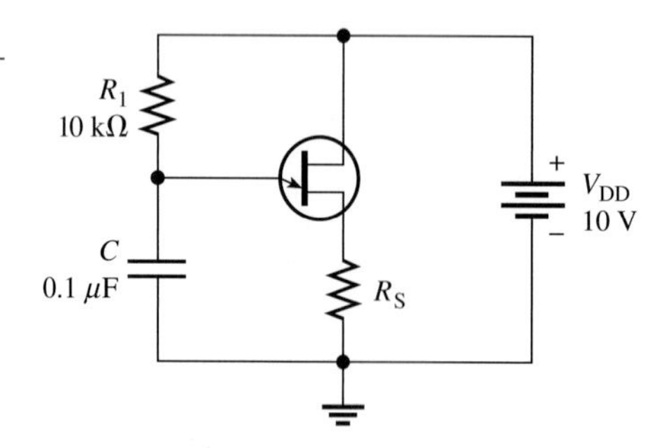

SECTION 18–11

Thyristors

35. Assume that the holding current for a particular SCR is 10 mA. What is the maximum value of R in Figure 18–80 necessary to keep the SCR in conduction once it is turned on? Neglect the drop across the SCR.

36. Repeat Problem 35 for a holding current of 500 μA.

FIGURE 18–80

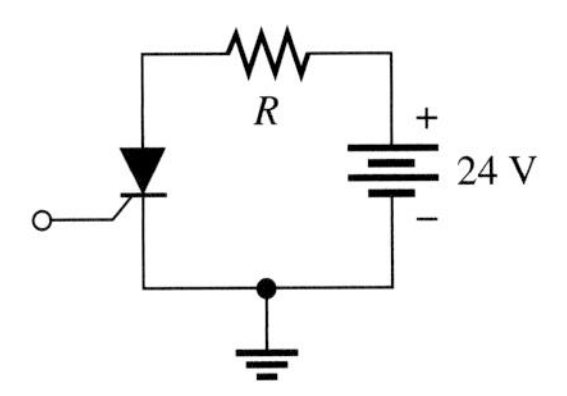

SECTION 18–13

Troubleshooting

37. In an out-of-circuit test of a good *npn* transistor, what should an analog ohmmeter indicate when its positive probe is touching the emitter and the negative probe is touching the base? When its positive probe is touching the base and the negative probe is touching the emitter?

38. What is the most likely problem, if any, in each circuit of Figure 18–81? Assume β_{DC} of 75.

FIGURE 18–81

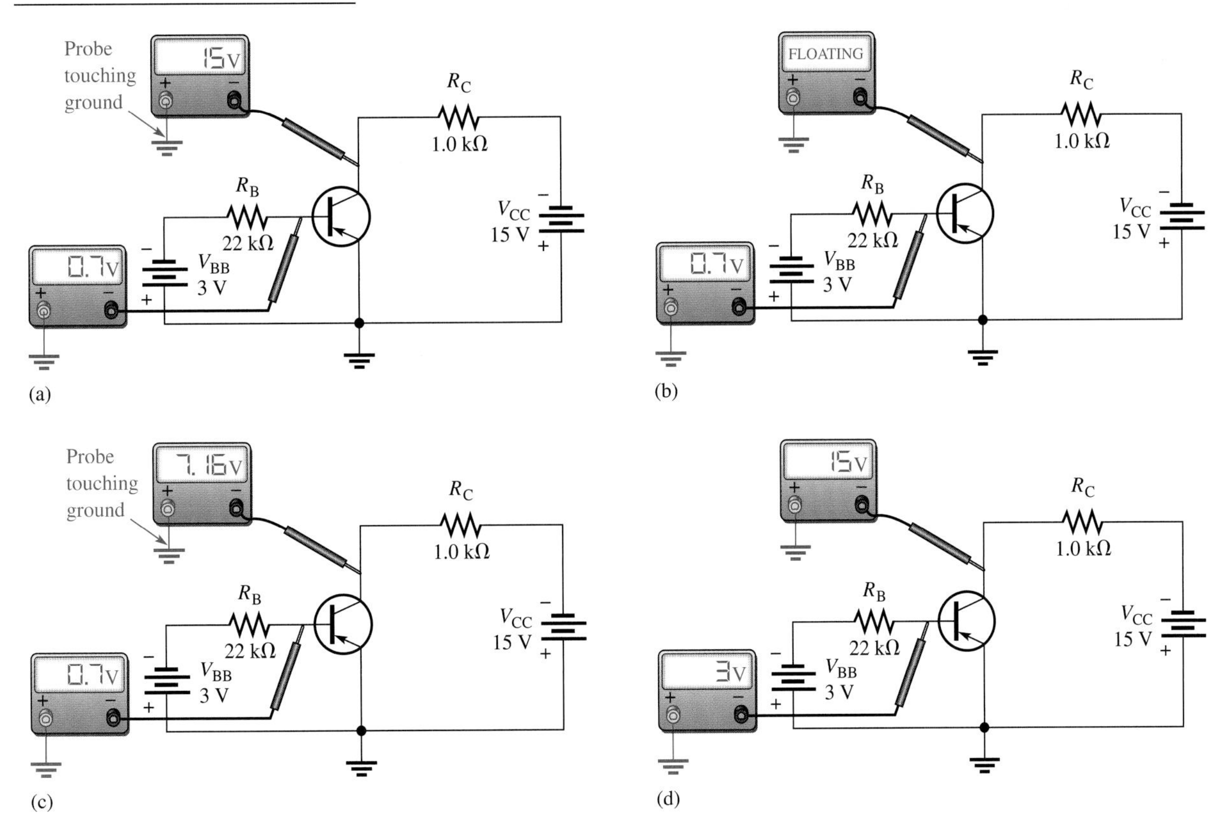

39. What is the value of the dc beta of each transistor in Figure 18–82?

FIGURE 18–82

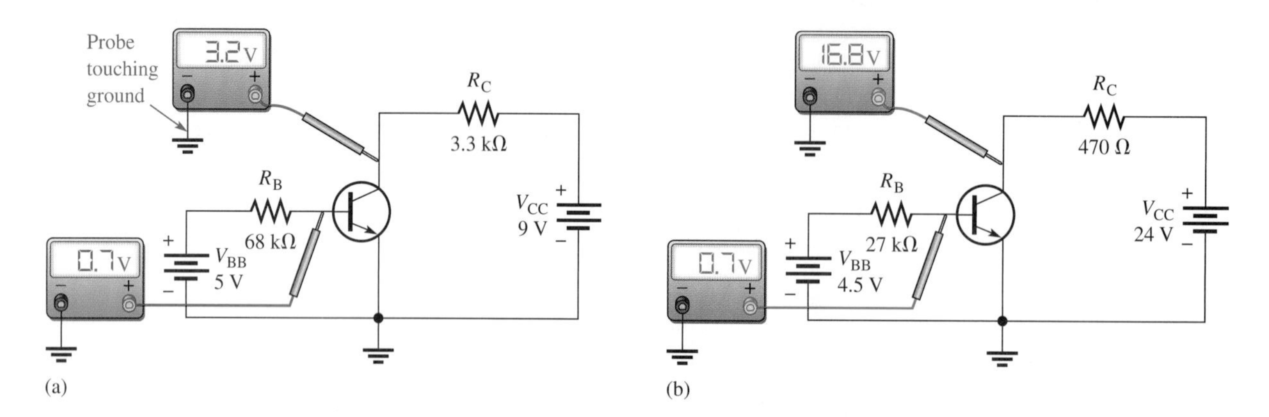

ELECTRONICS WORKBENCH/CIRCUITMAKER TROUBLESHOOTING PROBLEMS

CD-ROM file circuits are shown in Figure 18–83.

40. Open file P18-40 and test the circuit. If there is a fault, identify it.
41. Open file P18-41 and test the circuit. If there is a fault, identify it.
42. Determine if there is a fault in the circuit in file P18-42. If so, identify it.
43. Identify any faulty component in the circuit in file P18-43.

FIGURE 18–83

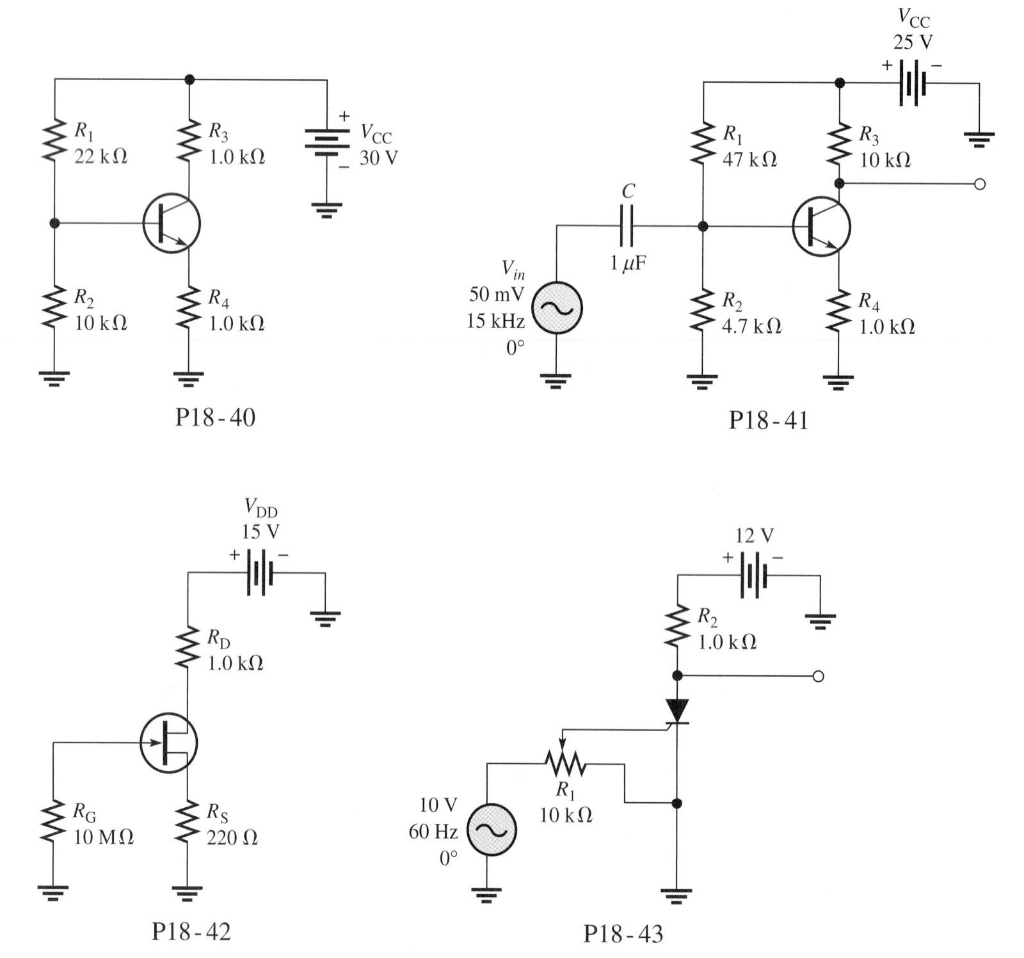

ANSWERS

SECTION REVIEWS

SECTION 18–1

Bipolar Junction Transistors (BJTs)

1. The three transistor terminals for a BJT are emitter, base, and collector.
2. Forward-reverse bias means that the base-emitter junction is forward-biased and the base-collector junction is reverse-biased.
3. β_{DC} is direct current gain.
4. $I_C = \beta_{DC} I_B = 1$ mA

SECTION 18–2

Voltage-Divider Bias

1. One source is required for voltage-divider bias.
2. $V_B = (4.7\ \text{k}\Omega/26.7\ \text{k}\Omega)(30\ \text{V}) = 5.28\ \text{V}$
3. $I_C \cong I_E = V_E/R_E = 5.55\ \text{V}/1.0\ \text{k}\Omega = 5.55\ \text{mA}$

SECTION 18–3

The Bipolar Junction Transistor as an Amplifier

1. Amplification is the process of increasing the amplitude of an electrical signal.
2. The Q-point is the dc operating point.
3. $A_v = R_C/R_E = 21.4$

SECTION 18–4

The BJT as a Switch

1. A switching transistor operates in saturation and cutoff.
2. I_C is maximum at saturation.
3. $I_C \cong 0$ A at cutoff.
4. Saturation occurs when the base-emitter junction and base-collector junction are forward-biased, and there is sufficient base current.
5. $V_{CE} = V_{CC}$ at cutoff.

SECTION 18–5

BJT Parameters and Ratings

1. β_{DC} increases with temperature.
2. β_{DC} increases with I_C to a certain value, and then decreases.
3. $I_C = P_{D(max)}/V_{CE} = 40$ mA

SECTION 18–6

The Junction Field-Effect Transistor (JFET)

1. The three JFET terminals are drain, source, and gate.
2. An n-channel JFET requires a negative V_{GS}.
3. I_D is controlled by V_{GS}.

SECTION 18–7

JFET Characteristics

1. $V_P = 7$ V
2. I_D decreases when V_{GS} is increased negatively.
3. $V_{GS} = +3$ V

SECTION 18–8 **The Metal-Oxide Semiconductor FET (MOSFET)**

1. Depletion/enhancement MOSFET; enhancement only MOSFET. The D-MOSFET has a structured channel; the E-MOSFET does not.
2. Yes, there is I_{DS} when $V_{GS} = 0$ V.
3. No, there is no I_{DS} when $V_{GS} = 0$ V.

SECTION 18–9 **FET Biasing**

1. A *p*-channel JFET must have a positive V_{GS}.
2. $V_{GS} = -I_D R_S = -8$ V
3. For $V_{GS} = 0$ V, $I_D = I_{DSS}$.
4. $V_{GS} > 2$ V

SECTION 18–10 **Unijunction Transistors (UJTs)**

1. UJT stands for unijunction transistor.
2. The terminals of a UJT are emitter, base 1, and base 2.
3. The intrinsic standoff ratio is R_{B1}/R_{BB}.

SECTION 18–11 **Thyristors**

1. SCR stands for silicon-controlled rectifier.
2. A positive voltage on the gate turns the SCR on, and the SCR conducts current from cathode to anode. If the current drops below the holding value, the SCR turns off.
3. A triac can be used in a lamp dimmer.
4. False

SECTION 8–12 **Transistor Packages and Terminal Identification**

1. Three categories of transistor are small signal/general purpose, power, and RF.
2. Clockwise from tab: emitter, base, collector (bottom view)
3. The metal mounting tab or case in power transistors is the collector.

SECTION 18–13 **Troubleshooting**

1. First test the transistor in-circuit.
2. If R_B opens, the transistor is in cutoff.
3. The base voltage is V_{BB} and the collector voltage is V_{CC}.

■ Application Assignment

1. The detector circuit senses a change in temperature and produces a proportional change in output voltage.
2. No.
3. 9.09 V
4. 13.5 V

RELATED PROBLEMS FOR EXAMPLES

18–1 $I_B = 241\ \mu A$; $I_C = 21.7$ mA; $I_E = 21.9$ mA; $V_B = 0.7$ V; $V_C = 4.23$ V

18–2 $V_B = 8.77$ V; $V_E = 8.07$ V; $V_C = 21.9$ V; $V_{CE} = 13.9$ V; $I_B = 80.7\ \mu A$; $I_E = 8.07$ mA; $I_C = 8.07$ mA

18–3 Along the horizontal axis

18–4 An 850 mV peak sine wave riding on a 6.16 V dc level

18–5 78.4 μA

18–6 $V_{CC(max)} = 44.6$ V; $V_{CE(max)}$ is exceeded first.

18–7 $I_D = 12$ mA

18–8 80,000 MΩ

18–9 $V_{DS} = 2$ V; $V_{GS} = -3.12$ V

18–10 6.8 V

18–11 2.13 mA

18–12 5 V

18–13 R_B open

SELF-TEST

1. (b)	**2.** (a)	**3.** (c)	**4.** (d)	**5.** (b)	**6.** (a)	**7.** (c)
8. (d)	**9.** (b)	**10.** (a)	**11.** (d)	**12.** (c)	**13.** (a)	**14.** (c)

19

AMPLIFIERS AND OSCILLATORS

INTRODUCTION

As you learned in the previous chapter, the biasing of a transistor is purely a dc operation. The purpose of biasing, however, is to establish an operating point (Q-point) about which variations in current and voltage can occur in response to an ac input signal.

When very small signal voltages must be amplified, such as from an antenna in a receiver, variations about the Q-point of an amplifier are relatively small.

Amplifiers designed to handle these small ac signals are called *small-signal amplifiers.* When large swings or variations in voltage and current about the Q-point are required for power amplification, *large-signal amplifiers* are used. An example is the power amplifier that drives the speakers in a stereo system.

Regardless of whether an amplifier is in the small-signal or large-signal category, it will be in one of three configurations of BJT amplifier circuits: common-emitter, common-collector, and common-base. For FETs the configurations are common-source, common-drain, and common-gate. Also, for any of the configurations, there are three basic modes of operation possible—class A, class B, and class C—which are covered in this chapter.

Oscillators are also introduced in this chapter. An oscillator is a circuit that produces a sustained sinusoidal output without an input signal. Oscillators operate on the principle of positive feedback.

CHAPTER OUTLINE

CHAPTER OBJECTIVES

- Analyze a common-emitter amplifier
- Analyze a common-collector amplifier
- Analyze a common-base amplifier
- Analyze three types of FET amplifiers
- Analyze a multistage amplifier
- Explain class A amplifier operation
- Analyze class B amplifiers
- Explain the basic operation of a class C amplifier
- Discuss the theory and analyze the operation of several types of oscillators
- Troubleshoot amplifier circuits

KEY TERMS

- Common-emitter (CE)
- Amplifier
- Voltage gain
- Current gain
- Power gain
- Common-collector (CC)
- Common-base (CB)
- Transconductance
- Common-source (CS)
- Common-drain (CD)
- Decibel
- Class A
- Class B
- Class C
- Oscillator
- Feedback

APPLICATION ASSIGNMENT PREVIEW

You have been assigned to check out and troubleshoot an audio preamplifier board that is thought to be malfunctioning. This circuit board consists of transistor amplifiers used in a superheterodyne AM receiver being manufactured by your company. The circuit amplifies the audio signal before it goes to the power amplifier that drives the speaker. After studying this chapter, you should be able to complete the application assignment.

WWW. VISIT THE COMPANION WEBSITE

Circuit Simulation Tutorials and Other Chapter Study Tools Are Available at

http://www.prenhall.com/floyd

19–1 COMMON-EMITTER AMPLIFIERS

The common-emitter (CE) is a type of bipolar junction transistor amplifier configuration in which the emitter is at ac ground. The other two types of amplifier configuration, the common-collector and common-base, are covered in the following sections.

After completing this section, you should be able to

- **Analyze a common-emitter amplifier**
- Identify a common-emitter configuration
- Explain how a bypass capacitor increases voltage gain
- Calculate voltage gain
- Discuss phase inversion
- Calculate input resistance
- Calculate current gain
- Calculate power gain

Figure 19–1 shows a typical common-emitter (CE) amplifier. The one shown has voltage-divider bias, although other types of bias methods are possible. C_1 and C_2 are coupling capacitors used to pass the signal into and out of the amplifier such that the source or load will not affect the dc bias voltages. C_3 is a bypass capacitor that shorts the emitter signal voltage (ac) to ground without disturbing the dc emitter voltage. Because of the bypass capacitor, the emitter is at signal ground (but not dc ground), thus making the circuit a common-emitter amplifier. The bypass capacitor increases the signal voltage gain. (The reason why this increase occurs is discussed next.) Notice that the input signal is applied to the base, and the output signal is taken from the collector. All capacitors are assumed to have a reactance of approximately zero at the signal frequency.

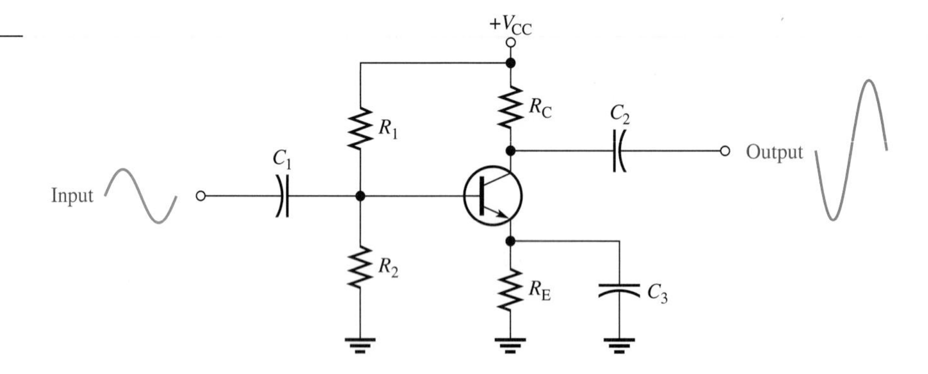

FIGURE 19–1

Typical common-emitter (CE) amplifier.

A Bypass Capacitor Increases Voltage Gain

The bypass capacitor shorts the signal around the emitter resistor, R_E, in order to increase the voltage gain. To understand why, let's consider the amplifier without the bypass capacitor and see what the voltage is. The CE amplifier with the bypass capacitor removed is shown in Figure 19–2.

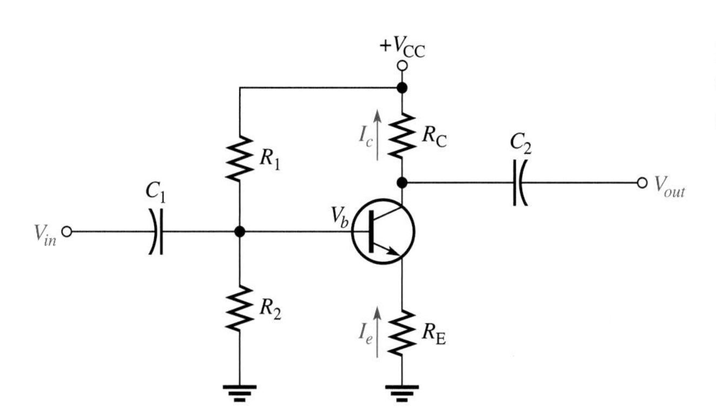

FIGURE 19–2
CE amplifier with bypass capacitor removed.

As before, lowercase italic subscripts indicate signal (ac) voltages and signal (alternating) currents. The voltage gain of the amplifier is V_{out}/V_{in}. The output signal voltage is

$$V_{out} = I_c R_C$$

The signal voltage at the base is approximately equal to

$$V_b \cong V_{in} \cong I_e(r_e + R_E)$$

where r_e is the internal emitter resistance of the transistor (not shown in the schematic). The voltage gain, A_v, can now be expressed as

$$A_v = \frac{V_{out}}{V_{in}} = \frac{I_c R_C}{I_e(r_e + R_E)}$$

Since $I_c \cong I_e$, the currents cancel and the gain is the ratio of the resistances.

$$A_v = \frac{R_C}{r_e + R_E}$$

Equation 19–1

Keep in mind that this formula is for the CE configuration without the bypass capacitor. If R_E is much greater than r_e, then $A_v \cong R_C/R_E$ as given earlier in Equation 18–11.

If the bypass capacitor is connected across R_E, it effectively shorts the signal to ground, leaving only r_e in the emitter. Thus, the voltage gain of the CE amplifier with the bypass capacitor shorting R_E is

$$A_v = \frac{R_C}{r_e}$$

Equation 19–2

The transistor parameter r_e is important because it determines the voltage gain of a CE amplifier in conjunction with R_C. A formula for estimating r_e is given without derivation in the following equation:

$$r_e \cong \frac{25\text{ mV}}{I_E}$$

Equation 19–3

EXAMPLE 19–1

Determine the voltage gain of the amplifier in Figure 19–3 both with and without a bypass capacitor in normal and in decibel form.

FIGURE 19–3

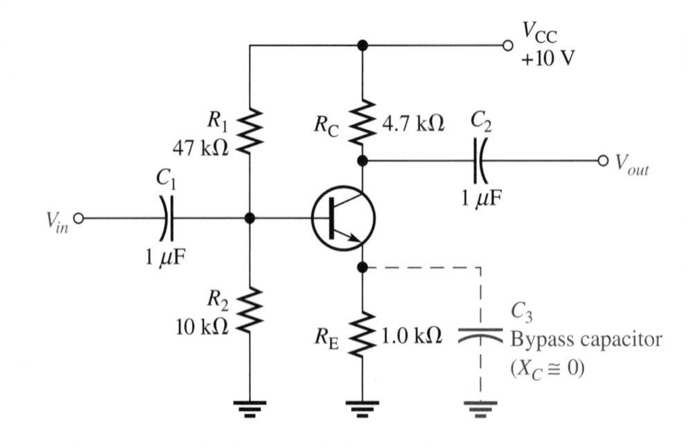

Solution First, determine r_e. To do so, you need to find I_E. Thus,

$$V_B \cong \left(\frac{R_2}{R_1 + R_2}\right)V_{CC} = \left(\frac{10\ \text{k}\Omega}{47\ \text{k}\Omega + 10\ \text{k}\Omega}\right)10\ \text{V} = 1.75\ \text{V}$$

$$V_E = V_B - 0.7\ \text{V} = 1.05\ \text{V}$$

$$I_E = \frac{V_E}{R_E} = \frac{1.05\ \text{V}}{1.0\ \text{k}\Omega} = 1.05\ \text{mA}$$

$$r_e \cong \frac{25\ \text{mV}}{I_E} = \frac{25\ \text{mV}}{1.05\ \text{mA}} = 23.8\ \Omega$$

The voltage gain without a bypass capacitor is

$$A_v = \frac{R_C}{r_e + R_E} = \frac{4.7\ \text{k}\Omega}{1023.8\ \Omega} = \mathbf{4.59}$$

In decibel form,

$$A_v = 20\log(4.59) = \mathbf{13.24\ dB}$$

The voltage gain with the bypass capacitor installed is

$$A_v = \frac{R_C}{r_e} = \frac{4.7\ \text{k}\Omega}{23.8\ \Omega} = \mathbf{197}$$

As you can see, the voltage gain is greatly increased by the addition of the bypass capacitor. In terms of decibels (dB), the voltage gain is

$$A_v = 20\log(197) = \mathbf{45.9\ dB}$$

*Related Problem** What is the voltage gain with the bypass capacitor if $R_C = 5.6\ \text{k}\Omega$?

Open file E19-01 on your EWB/CircuitMaker CD-ROM. Measure the voltage gain of the amplifier without a bypass capacitor. Connect a 10 μF bypass capacitor and measure the gain. Compare both measurements to the calculated values.

*Answers are at the end of the chapter.

Phase Inversion

As we discussed in Chapter 18, the output voltage at the collector is 180° out of phase with the input voltage at the base. Therefore, the CE amplifier is characterized by a phase inversion between the input and the output. As mentioned before, this inversion is sometimes indicated by a negative voltage gain.

AC Input Resistance

The dc input resistance (R_{IN}), viewed from the base of the transistor, was developed in Section 18–2. The input resistance "seen" by the signal at the base is derived in a similar manner when the emitter resistor is bypassed to ground.

$$R_{in} = \frac{V_b}{I_b}$$

$$V_b = I_e r_e$$

$$I_e \cong \beta_{ac} I_b$$

$$R_{in} \cong \frac{\beta_{ac} I_b r_e}{I_b}$$

The I_b terms cancel, leaving

$$R_{in} \cong \beta_{ac} r_e$$

Equation 19–4

Total Input Resistance of a CE Amplifier

Viewed from the base, R_{in} is the ac resistance. The actual resistance seen by the source includes that of bias resistors. We will now develop an expression for the total input resistance. The concept of ac ground was mentioned earlier. At this point it needs some additional explanation because it is important in the development of the formula for total input resistance, $R_{in(tot)}$.

You have already seen that the bypass capacitor effectively makes the emitter appear as ground to the ac signal because the X_C of the capacitor is nearly zero at the signal frequency. Of course, to a dc signal the capacitor looks like an open and thus does not affect the dc emitter voltage.

In addition to seeing ground through the bypass capacitor, the signal also sees ground through the dc supply voltage source, V_{CC}. It does so because there is zero signal voltage at the V_{CC} terminal. Thus, the $+V_{CC}$ terminal effectively acts as ac ground. As a result, the two bias resistors, R_1 and R_2, appear in parallel to the ac input because one end of R_2 goes to actual ground and one end of R_1 goes to ac ground (V_{CC} terminal). Also, R_{in} at the base appears in parallel with $R_1 \parallel R_2$. This situation is illustrated in Figure 19–4.

▼ FIGURE 19–4

Total input resistance.

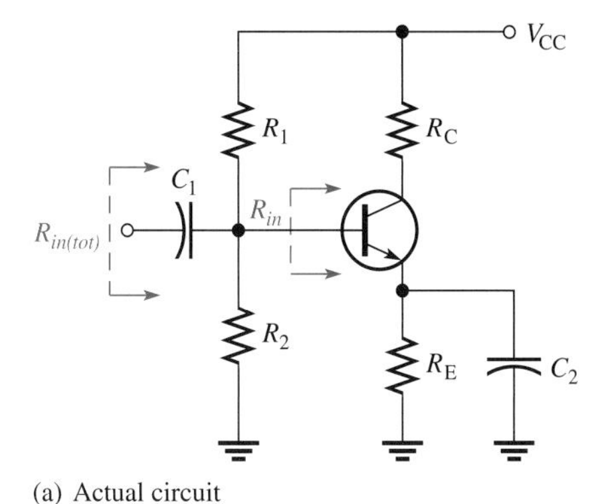

(a) Actual circuit

(b) ac equivalent circuit seen by ac source (V_{in})

The expression for the total input resistance to the CE amplifier as seen by the ac source is

Equation 19–5

$$R_{in(tot)} = R_1 \parallel R_2 \parallel R_{in}$$

R_C has no effect because of the reverse-biased, base-collector junction.

EXAMPLE 19–2

Determine the total input resistance seen by the signal source in the CE amplifier in Figure 19–5. $\beta_{ac} = 150$.

FIGURE 19–5

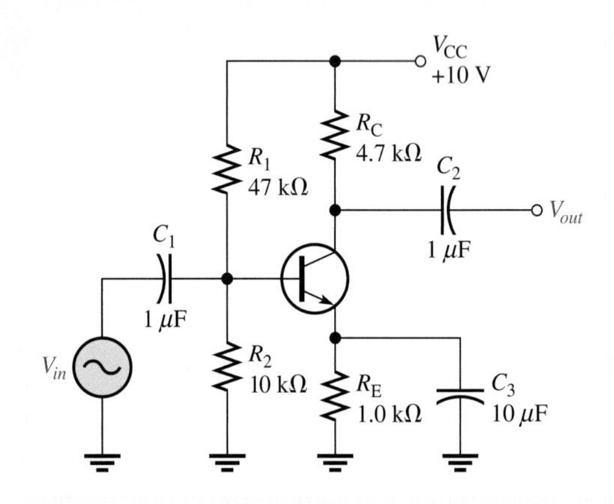

Solution In Example 19–1, you found r_e for the same circuit. Thus,

$$R_{in} = \beta_{ac} r_e = 150(23.8\ \Omega) = 3.57\ \text{k}\Omega$$

$$R_{in(tot)} = R_1 \parallel R_2 \parallel R_{in} = 47\ \text{k}\Omega \parallel 10\ \text{k}\Omega \parallel 3.57\ \text{k}\Omega = \mathbf{2.49\ k\Omega}$$

Related Problem What is the total input resistance see by the signal source in Figure 19–5 if C_3 is removed?

Current Gain

The signal current gain of a CE amplifier is

Equation 19–6

$$A_i = \frac{I_c}{I_s}$$

where I_s is the source current and is calculated by $V_{in}/R_{in(tot)}$.

Power Gain

The power gain of a CE amplifier is the product of the voltage gain and the current gain.

Equation 19–7

$$A_p = A_v A_i$$

EXAMPLE 19–3

Determine the voltage gain, current gain, and power gain for the CE amplifier in Figure 19–6. $\beta_{DC} = \beta_{ac} = 100$. Also, express the voltage and power gains in decibels.

FIGURE 19–6

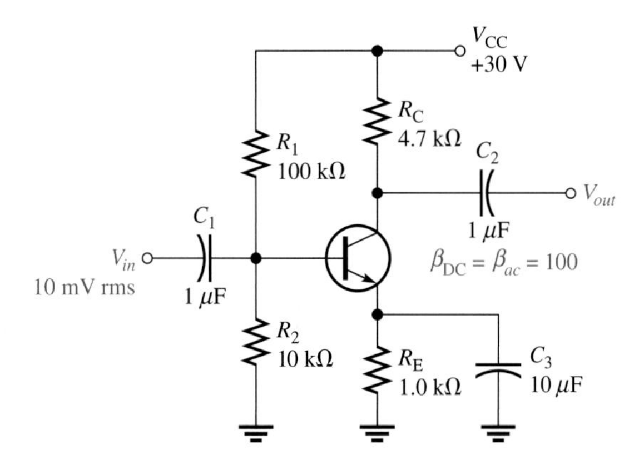

Solution First find r_e. To do so, you must find I_E. Begin by calculating V_B. Since $R_{IN} = \beta_{DC}R_E = 100\ \text{k}\Omega$ is ten times greater than R_2, it can be neglected. Thus,

$$V_B \cong \left(\frac{R_2}{R_1 + R_2}\right)V_{CC} = \left(\frac{10\ \text{k}\Omega}{110\ \text{k}\Omega}\right)30\ \text{V} = 2.73\ \text{V}$$

$$I_E = \frac{V_E}{R_E} = \frac{V_B - 0.7\ \text{V}}{R_E} = \frac{2.03\ \text{V}}{1.0\ \text{k}\Omega} = 2.03\ \text{mA}$$

$$r_e \cong \frac{25\ \text{mV}}{I_E} = \frac{25\ \text{mV}}{2.03\ \text{mA}} = 12.3\ \Omega$$

The ac voltage gain is

$$A_v = \frac{R_C}{r_e} = \frac{4.7\ \text{k}\Omega}{12.3\ \Omega} = \mathbf{382}$$

Determine the signal current gain by first finding $R_{in(tot)}$ to get I_s.

$$R_{in(tot)} = R_1 \parallel R_2 \parallel \beta_{ac}r_e = 100\ \text{k}\Omega \parallel 10\ \text{k}\Omega \parallel 1.23\ \text{k}\Omega = 1.08\ \text{k}\Omega$$

$$I_s = \frac{V_{in}}{R_{in(tot)}} = \frac{10\ \text{mV}}{1.08\ \text{k}\Omega} = 9.26\ \mu\text{A}$$

Next determine I_c.

$$I_c = \frac{V_{out}}{R_C} = \frac{A_vV_{in}}{R_C} = \frac{(382)(10\ \text{mV})}{4.7\ \text{k}\Omega} = 813\ \mu\text{A}$$

$$A_i = \frac{I_c}{I_s} = \frac{813\ \mu\text{A}}{9.26\ \mu\text{A}} = \mathbf{87.8}$$

The power gain is

$$A_p = A_vA_i = (382)(87.8) = \mathbf{33{,}540}$$

The voltage gain and the power gain in decibels are as follows:

$$A_v = 20\log(382) = \mathbf{51.6\ dB}$$

$$A_p = 10\log(33{,}540) = \mathbf{45.3\ dB}$$

Related Problem Determine A_v, A_i, and A_p in this example when R_{IN} is taken into account.

Open file E19-03 on your EWB/CircuitMaker CD-ROM. Measure the voltage gain and compare with the calculated value.

SECTION 19–1 REVIEW

Answers are at the end of the chapter.

1. What is the purpose of the bypass capacitor in a CE amplifier?
2. How is the voltage gain of a CE amplifier determined?
3. If A_v is 50 and A_i is 200, what is the power gain of a CE amplifier?

19–2 COMMON-COLLECTOR AMPLIFIERS

The common-collector (CC) amplifier, commonly referred to as an emitter-follower, is the second of the three basic BJT amplifier configurations. The input is applied to the base and the output is at the emitter. There is no collector resistor. The voltage gain of a CC amplifier is approximately 1.

After completing this section, you should be able to

- **Analyze a common-collector amplifier**
- Calculate voltage gain
- Calculate input resistance
- Calculate current gain
- Calculate power gain
- Describe the darlington pair configuration

Figure 19–7 shows a common-collector (CC) circuit with a voltage-divider bias. Notice that the input is applied to the base and the output is taken from the emitter.

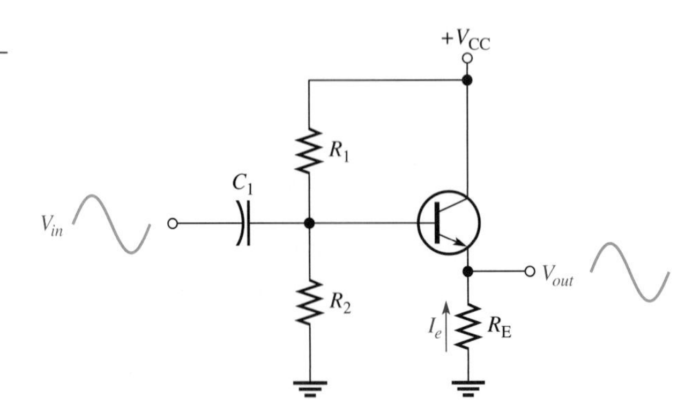

FIGURE 19–7

Typical emitter-follower (common-collector, CC) amplifier.

Voltage Gain

As in all amplifiers, the voltage gain in a CC amplifier is $A_v = V_{out}/V_{in}$. For the emitter-follower, V_{out} is I_eR_E, and V_{in} is $I_e(r_e + R_E)$. Therefore, the gain is $I_eR_E/I_e(r_e + R_E)$. The currents cancel, and the gain expression simplifies to

$$A_v = \frac{R_E}{r_e + R_E}$$

Equation 19–8

It is important to notice here that *the gain is always less than 1.* Because r_e is normally much less than R_E, then a good approximation is $A_v = 1$.

Since the output voltage is the emitter voltage, it is in phase with the base or the input voltage. As a result, and because the voltage gain is close to 1, the output voltage follows the input voltage—thus the common-collector amplifier is also known as the **emitter-follower.**

Input Resistance

The emitter-follower is characterized by a high input resistance, which makes it a very useful circuit. Because of the high input resistance, the emitter-follower can be used as a buffer to minimize loading effects when one circuit is driving another.

The derivation of the input resistance viewed from the base is similar to that for the CE amplifier. In this case, however, the emitter resistor is not bypassed.

$$R_{in} = \frac{V_b}{I_b} = \frac{I_e(r_e + R_E)}{I_b} \cong \frac{\beta_{ac}I_b(r_e + R_E)}{I_b} = \beta_{ac}(r_e + R_E)$$

If R_E is at least ten times larger than r_e, then the input resistance at the base is

$$R_{in} \cong \beta_{ac}R_E$$

Equation 19–9

In Figure 19–7, the bias resistors appear to the input signal to be in parallel with R_{in}, just as in the voltage-divider biased CE amplifier. The total ac input resistance is

$$R_{in(tot)} = R_1 \parallel R_2 \parallel R_{in}$$

Equation 19–10

Since R_{in} can be made large with the proper selection of R_E, a much higher input resistance results for this configuration than for the CE circuit.

Current Gain

The signal current gain for the emitter-follower is I_e/I_s where I_s is the signal current and can be calculated as $V_s/R_{in(tot)}$. If the bias resistors are large enough to be neglected so that $I_s = I_b$, then the current gain of the amplifier is equal to the current gain of the transistor, β_{ac}. Of course, the same was also true for the CE amplifier. β_{ac} is the maximum achievable current gain in both types of amplifiers.

$$A_i = \frac{I_e}{I_s}$$

Equation 19–11

Since $I_e = V_{out}/R_E$ and $I_s = V_{in}/R_{in(tot)}$, A_i can also be expressed as $R_{in(tot)}/R_E$, as will be shown in Example 19–4.

Power Gain

The power gain is the product of the voltage gain and the current gain. For the emitter-follower, the power gain is approximately equal to the current gain because the voltage gain is approximately 1.

$$A_p \cong A_i$$

Equation 19–12

EXAMPLE 19–4

Determine the input resistance of the emitter-follower in Figure 19–8. Also find the voltage gain, current gain, and power gain.

FIGURE 19–8

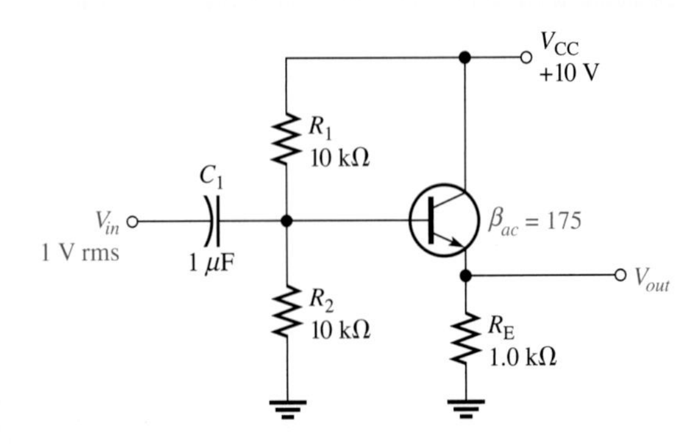

Solution The approximate input resistance viewed from the base is

$$R_{in} \cong \beta_{ac}R_E = (175)(1.0\text{ k}\Omega) = 175\text{ k}\Omega$$

The total input resistance is

$$R_{in(tot)} = R_1 \parallel R_2 \parallel R_{in} = 10\text{ k}\Omega \parallel 10\text{ k}\Omega \parallel 175\text{ k}\Omega = \mathbf{4.86\text{ k}\Omega}$$

The voltage gain is, neglecting r_e,

$$A_v \cong \mathbf{1}$$

The current gain is

$$A_i = \frac{I_e}{I_s}$$

$$I_e = \frac{V_{out}}{R_E}$$

$$I_s = \frac{V_{in}}{R_{in(tot)}}$$

$$A_i = \frac{V_{out}/R_E}{V_{in}/R_{in(tot)}}$$

Since $V_{in} \cong V_{out}$,

$$A_i = \frac{R_{in(tot)}}{R_E} = \frac{4.86\text{ k}\Omega}{1.0\text{ k}\Omega} = \mathbf{4.86}$$

The power gain is

$$A_p \cong A_i = \mathbf{4.86}$$

Related Problem If R_E in Figure 19–8 is decreased to 820 Ω, what is the power gain?

Open file E19-04 on your EWB/CircuitMaker CD-ROM. Verify that the voltage gain is close to 1.

The Darlington Pair

As you have seen, β is a major factor in determining the input resistance. The β of the transistor limits the maximum achievable input resistance you can get from a given emitter-follower circuit.

One way to boost input resistance is to use a **darlington pair,** as shown in Figure 19–9. The collectors of two transistors are connected together, and the emitter of the first drives the base of the second. This configuration achieves β multiplication as shown in the following steps. The emitter current of the first transistor is

$$I_{e1} \cong \beta_1 I_{b1}$$

This emitter current becomes the base current for the second transistor, producing a second emitter current of

$$I_{e2} \cong \beta_2 I_{e1} = \beta_1 \beta_2 I_{b1}$$

Therefore, the effective current gain of the darlington pair is

Equation 19–13

$$\beta = \beta_1 \beta_2$$

The input resistance is

Equation 19–14

$$R_{in} = \beta_1 \beta_2 R_E$$

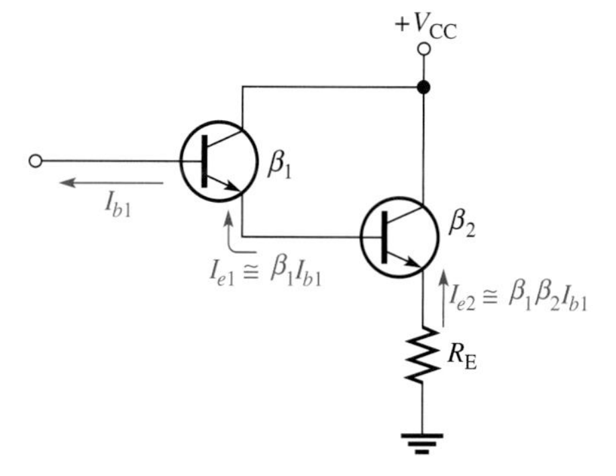

FIGURE 19–9

Darlington pair.

SECTION 19–2 REVIEW

1. What is a common-collector amplifier called?
2. What is the ideal maximum voltage gain of a CC amplifier?
3. Can the current gain be higher than the voltage gain in the CC amplifier?

19–3 COMMON-BASE AMPLIFIERS

The third basic BJT amplifier configuration is the common-base (CB). It provides high voltage gain with unity current gain. Since it has a low input resistance, the CB amplifier is the most appropriate type for certain high-frequency applications where sources tend to have very low output resistances.

After completing this section, you should be able to

- **Analyze a common-base amplifier**
- Calculate voltage gain
- Calculate input resistance
- Calculate current gain

- Calculate power gain
- Compare the three basic amplifier configurations: CE, CC, and CB

A typical common-base (CB) circuit is pictured in Figure 19–10. The base is at signal (ac) ground, and the input is applied to the emitter. The output is taken from the collector and is in phase with the input.

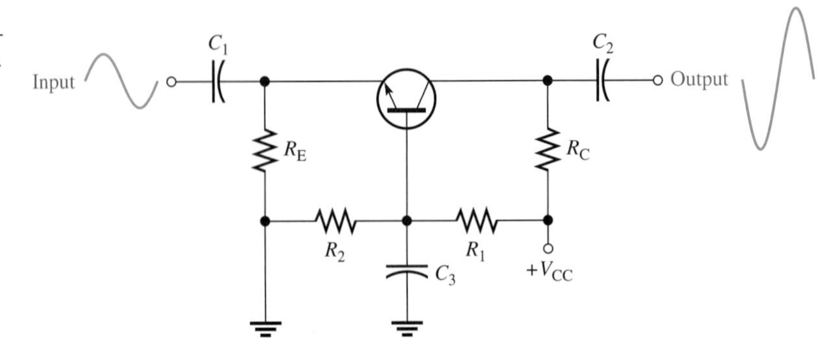

FIGURE 19–10

Typical common-base (CB) amplifier.

Voltage Gain

The input voltage is the emitter voltage V_e. The output voltage is the collector voltage V_c. With this in mind, we develop the voltage gain formula as follows:

$$A_v = \frac{V_c}{V_e} = \frac{I_c R_C}{I_e r_e} \cong \frac{I_e R_C}{I_e r_e}$$

Equation 19–15

$$A_v = \frac{R_C}{r_e}$$

Notice that the gain expression is the same as that for the CE amplifier (when R_E is bypassed).

Input Resistance

The resistance viewed from the emitter appears to the input signal as follows:

$$R_{in} = \frac{V_{in}}{I_{in}} = \frac{V_e}{I_e} = \frac{I_e r_e}{I_e}$$

Equation 19–16

$$R_{in} = r_e$$

Viewed from the source, R_E appears in parallel with R_{in}. However, r_e is normally so small compared to R_E that Equation 19–16 is also valid for the total input resistance, $R_{in(tot)}$.

Current Gain

The current gain is the output current I_c divided by the input current I_e. Since $I_c \cong I_e$, the signal current gain is approximately 1.

Equation 19–17

$$A_i \cong 1$$

Power Gain

Since the current gain is approximately 1 for the CB amplifier, the power gain is approximately equal to the voltage gain.

Equation 19–18

$$A_p \cong A_v$$

EXAMPLE 19–5

Find the input resistance, voltage gain, current gain, and power gain for the CB amplifier in Figure 19–11.

FIGURE 19–11

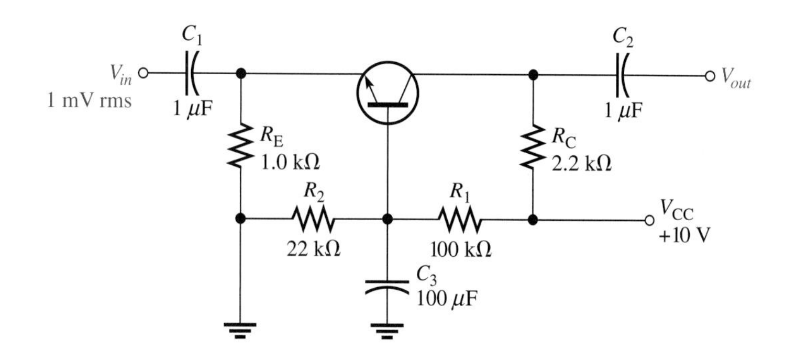

Solution First, find I_E so that you can determine r_e. Then $R_{in} = r_e$.

$$V_B \cong \left(\frac{R_2}{R_1 + R_2}\right)V_{CC} = \left(\frac{22\text{ k}\Omega}{122\text{ k}\Omega}\right)10\text{ V} = 1.80\text{ V}$$

$$V_E = V_B - 0.7\text{ V} = 1.80\text{ V} - 0.7\text{ V} = 1.10\text{ V}$$

$$I_E = \frac{V_E}{R_E} = \frac{1.10\text{ V}}{1.0\text{ k}\Omega} = 1.10\text{ mA}$$

$$R_{in} = r_e \cong \frac{25\text{ mV}}{1.10\text{ mA}} = \mathbf{22.7\ \Omega}$$

The signal voltage gain is

$$A_v = \frac{R_C}{r_e} = \frac{2.2\text{ k}\Omega}{22.7\ \Omega} = \mathbf{96.9}$$

Thus,

$$A_i \cong \mathbf{1}$$

$$A_p \cong \mathbf{96.9}$$

Related Problem If R_E is increased to 1.5 kΩ in Figure 19–11, what is the voltage gain?

Open file E19-05 on your EWB/CircuitMaker CD-ROM. Measure the voltage gain and compare to the calculated value.

Summary

Table 19–1 summarizes the important characteristics of each of the three amplifier configurations. Also, relative values are indicated for general comparison of the amplifiers.

TABLE 19–1

Comparison of BJT amplifier configurations. The current gains and the input resistance are the maximum achievable values, with the bias resistors neglected.

	CE	CC	CB
Voltage gain, A_v	High R_C/r_e	Low $\cong 1$	High R_C/r_e
Current gain, $A_{i(max)}$	High β_{ac}	High β_{ac}	Low $\cong 1$
Power gain, A_p	Very high A_iA_v	High $\cong A_i$	High $\cong A_v$
Input resistance, $R_{in(max)}$	Low $\beta_{ac}r_e$	High $\beta_{ac}R_E$	Very low r_e
Phase of output relative to input	180° (inversion)	0° (no inversion)	0° (no inversion)

SECTION 19–3 REVIEW

1. Can the same voltage gain be achieved with a CB as with a CE amplifier?
2. Is the input resistance of a CB amplifier very low or very high?

19–4 FET AMPLIFIERS

Field-effect transistors, both JFETs and MOSFETs, can be used as amplifiers in any of three circuit configurations similar to those for the bipolar junction transistor. The FET configurations are common-source, common-drain, and common-gate. These are similar to the BJT configurations of common-emitter, common-collector, and common-base, respectively.

After completing this section, you should be able to

- **Analyze three types of FET amplifiers**
- Calculate the transconductance of a FET
- Analyze a common-source amplifier
- Analyze a common-drain amplifier
- Analyze a common-gate amplifier
- Compare the three FET amplifier configurations

Transconductance of a FET

Recall that in a bipolar junction transistor, the base current controls the collector current, and the relationship between these two currents is expressed as $I_c = \beta_{ac}I_b$. In a FET, the gate voltage controls the drain current. An important FET parameter is the **transconductance**, g_m, which is defined as

Equation 19–19

$$g_m = \frac{I_d}{V_{gs}}$$

The transconductance is one factor that determines the voltage gain of a FET amplifier. On data sheets, the transconductance is sometimes called the *forward transadmittance* and is designated y_{fs} with units of Siemens (S). You will still find some data sheets using the older unit mho for y_{fs}.

Common-Source (CS) Amplifiers

A self-biased n-channel common source (CS) JFET amplifier with an ac source capacitively coupled to the gate is shown in Figure 19–12. The resistor R_G serves two purposes: (1) It keeps the gate at approximately 0 V dc (because I_{GSS} is extremely small), and (2) its large value (usually several megohms) prevents loading of the ac signal source. The bias voltage is created by the drop across R_S. The bypass capacitor, C_3, keeps the source of the FET effectively at ac ground.

The signal voltage causes the gate-to-source voltage to swing above and below its Q-point value, causing a swing in drain current. As the drain current increases, the voltage drop across R_D also increases, causing the drain voltage (with respect to ground) to decrease.

The drain current swings above and below its Q-point value in phase with the gate-to-source voltage. The drain-to-source voltage swings above and below its Q-point value 180° out of phase with the gate-to-source voltage, as illustrated in Figure 19–12.

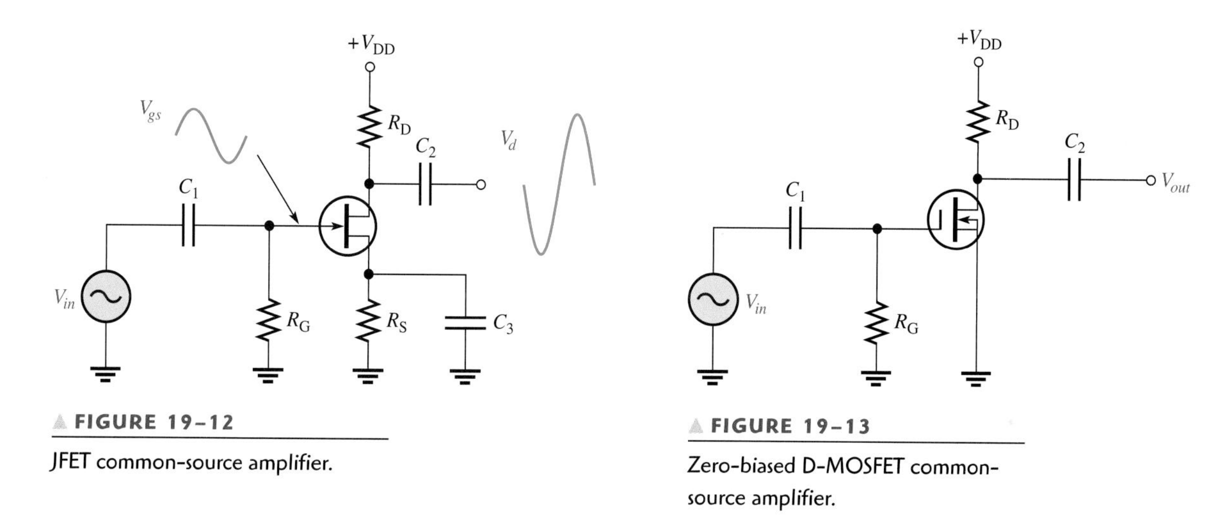

FIGURE 19–12

JFET common-source amplifier.

FIGURE 19–13

Zero-biased D-MOSFET common-source amplifier.

D-MOSFET A zero-biased n-channel D-MOSFET with an ac source capacitively coupled to the gate is shown in Figure 19–13. The gate is at approximately 0 V dc and the source terminal is at ground, thus making $V_{GS} = 0$ V.

The signal voltage causes V_{gs} to swing above and below its 0 value, producing a swing in I_d. The negative swing in V_{gs} produces the depletion mode, and I_d decreases. The positive swing in V_{gs} produces the enhancement mode, and I_d increases.

E-MOSFET Figure 19–14 shows a voltage-divider-biased, n-channel E-MOSFET with an ac signal source capacitively coupled to the gate. The gate is biased with a positive voltage such that $V_{GS} > V_{GS(th)}$, where $V_{GS(th)}$ is the threshold value.

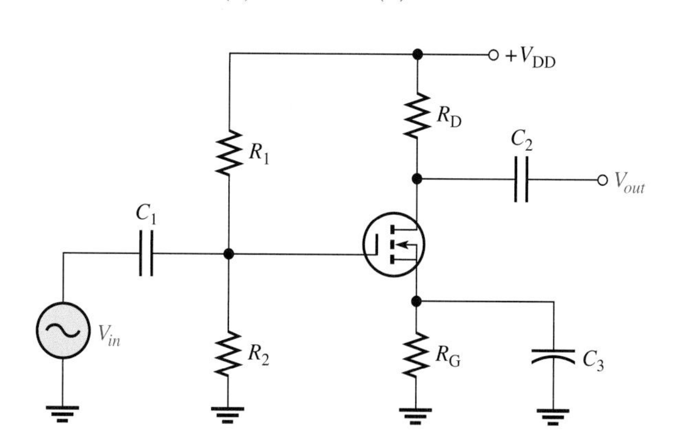

FIGURE 19–14

Common-source E-MOSFET amplifier with voltage-divider bias.

As with the JFET and D-MOSFET, the signal voltage produces a swing in V_{gs} above and below its Q-point value. This swing, in turn, causes a swing in I_d. Operation is entirely in the enhancement mode.

Voltage Gain Voltage gain, A_v, of an amplifier always equals V_{out}/V_{in}. In the case of the CS amplifier, V_{in} is equal to V_{gs}, and V_{out} is equal to the signal voltage developed across R_D, which is I_dR_D. Thus,

$$A_v = \frac{I_d R_D}{V_{gs}}$$

Since $g_m = I_d/V_{gs}$, the common-source voltage gain is

Equation 19–20

$$A_v = g_m R_D$$

Input Resistance Because the input to a CS amplifier is at the gate, the input resistance is extremely high. Ideally, it approaches infinity and can be neglected. As you know, the high input resistance is produced by the reverse-biased *pn* junction in a JFET and by the insulated gate structure in a MOSFET.

The actual input resistance seen by the signal source is the gate-to-ground resistor R_G in parallel with the FET's input resistance, V_{GS}/I_{GSS}. The reverse leakage current I_{GSS} is typically given on the data sheet for a specific value of V_{GS} so that the input resistance of the device can be calculated.

EXAMPLE 19–6

(a) What is the total output voltage (dc + ac) of the amplifier in Figure 19–15? The g_m is 1800 μS, I_D is 2 mA, $V_{GS(off)}$ is −3.5 V, and I_{GSS} is 15 nA.

(b) What is the input resistance seen by the signal source?

FIGURE 19–15

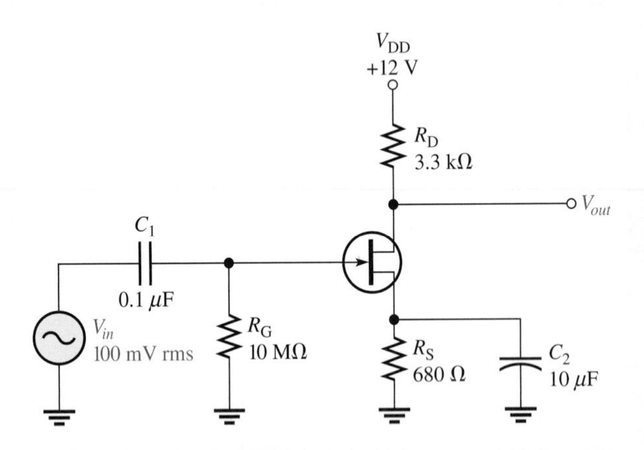

Solution **(a)** First, find the dc output voltage.

$$V_D = V_{DD} - I_D R_D = 12\text{ V} - (2\text{ mA})(3.3\text{ k}\Omega) = 5.4\text{ V}$$

Next, find the ac output voltage by using the gain formula.

$$A_v = \frac{V_{out}}{V_{in}} = g_m R_D$$

$$V_{out} = g_m R_D V_{in} = (1800\ \mu\text{S})(3.3\text{ k}\Omega)(100\text{ mV}) = 594\text{ mV rms}$$

The total output voltage is an ac signal with a peak-to-peak value of 594 mV × 2.828 = **1.67 V, riding on a dc level of 5.4 V.**

(b) The input resistance is determined as follows (since $V_G = 0$ V):

$$V_{GS} = I_D R_S = (2 \text{ mA})(680 \ \Omega) = 1.36 \text{ V}$$

The input resistance at the gate of the JFET is

$$R_{IN(gate)} = \frac{V_{GS}}{I_{GSS}} = \frac{1.36 \text{ V}}{15 \text{ nA}} = 91 \text{ M}\Omega$$

The input resistance seen by the signal source is

$$R_{in} = R_G \parallel R_{IN(gate)} = 10 \text{ M}\Omega \parallel 91 \text{ M}\Omega = \mathbf{9.0 \ M\Omega}$$

Related Problem What is the total output voltage in the amplifier of Figure 19–15 if V_{DD} is changed to 15 V? Assume the other parameters are the same.

Common-Drain (CD) Amplifier

A common-drain (CD) JFET amplifier is shown in Figure 19–16 with voltages indicated. Self-biasing is used in this circuit. The input signal is applied to the gate through a coupling capacitor, and the output is at the source terminal. There is no drain resistor. This circuit, of course, is analogous to the bipolar emitter-follower and is sometimes called a *source-follower.*

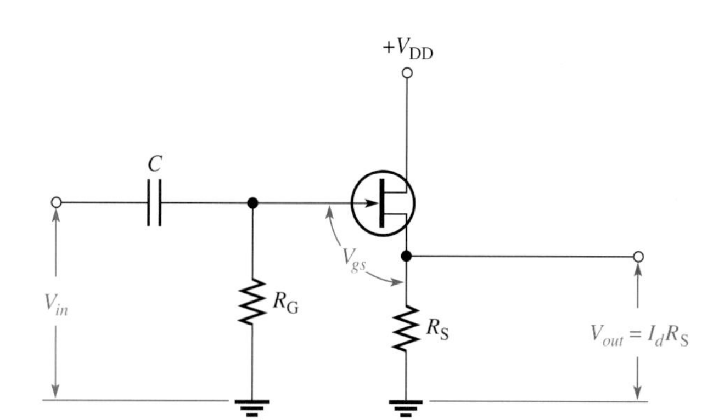

FIGURE 19–16

JFET common-drain amplifier (source-follower).

Voltage Gain As in all amplifiers, the voltage gain is $A_v = V_{out}/V_{in}$. For the source-follower, V_{out} is $I_d R_S$ and V_{in} is $V_{gs} + I_d R_S$, as shown in Figure 19–16. Therefore, the gate-to-source voltage gain is $I_d R_S/(V_{gs} + I_d R_S)$. Substituting $I_d = g_m V_{gs}$ into the expression gives the following result:

$$A_v = \frac{g_m V_{gs} R_S}{V_{gs} + g_m V_{gs} R_S}$$

Canceling V_{gs} yields

$$A_v = \frac{g_m R_S}{1 + g_m R_S}$$

Equation 19–21

Notice here that *the gain is always slightly less than 1.* If $g_m R_S >> 1$, then a good approximation is $A_v \cong 1$. Since the output voltage is at the source, it is in phase with the gate (input) voltage.

Input Resistance Because the input signal is applied to the gate, the input resistance seen by the input signal source is extremely high, just as in the CS amplifier configuration. The gate resistor R_G, in parallel with the input resistance looking in at the gate, is the total input resistance.

EXAMPLE 19–7

(a) Determine the voltage gain of the amplifier in Figure 19–17(a) using the data sheet information in Figure 19–17(b).

(b) Also determine the input resistance. Assume minimum data sheet values where available.

▼ **FIGURE 19–17**

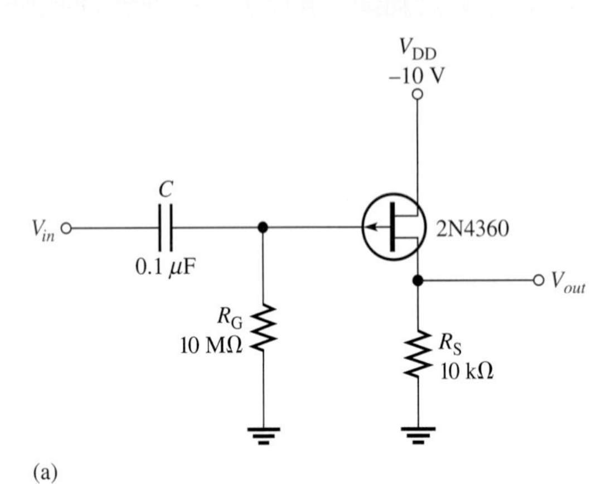

(a)

*ELECTRICAL CHARACTERISTICS (T_A = 25°C unless otherwise noted)

Characteristic	Symbol	Min	Max	Unit
OFF CHARACTERISTICS				
Gate-Source Breakdown Voltage (I_G = 10 μAdc, V_{DS} = 0)	$V_{(BR)GSS}$	20	–	Vdc
Gate-Source Cutoff Voltage (V_{DS} = −10 Vdc, I_D = 1.0 μAdc)	$V_{GS(off)}$	0.7	10	Vdc
Gate Reverse Current (V_{GS} = 15 Vdc, V_{DS} = 0) (V_{GS} = 15 Vdc, V_{DS} = 0, T_A = 65°C)	I_{GSS}	– –	10 0.5	nAdc μAdc
ON CHARACTERISTICS				
Zero-Gate Voltage Drain Current (Note 1) (V_{DS} = −10 Vdc, V_{GS} = 0)	I_{DSS}	3.0	30	mAdc
Gate-Source Breakdown Voltage (V_{DS} = −10 Vdc, I_D = 0.3 mAdc)	V_{GS}	0.4	9.0	Vdc
SMALL SIGNAL CHARACTERISTICS				
Drain-Source "ON" Resistance (V_{GS} = 0, I_D = 0, f = 1.0 kHz)	$r_{ds(on)}$	–	700	Ohms
Forward Transadmittance (Note 1) (V_{DS} = −10 Vdc, V_{GS} = 0, f = 1.0 kHz)	$\vert y_{fs}\vert$	2000	8000	μmhos
Forward Transconductance (V_{DS} = −10 Vdc, V_{GS} = 0, f = 1.0 MHz)	$Re(y_{fs})$	1500	–	μmhos
Output Admittance (V_{DS} = −10 Vdc, V_{GS} = 0, f = 1.0 kHz)	$\vert$	–	100	μmhos
Input Capacitance (V_{DS} = −10 Vdc, V_{GS} = 0, f = 1.0 MHz)	C_{iss}	–	20	pF
Reverse Transfer Capacitance (V_{DS} = −10 Vdc, V_{GS} = 0, f = 1.0 MHz)	C_{rss}	–	5.0	pF
Common-Source Noise Figure (V_{DS} = −10 Vdc, I_D = 1.0 mAdc, R_G = 1.0 Megohm, f = 100 Hz)	NF	–	5.0	dB
Equivalent Short-Circuit Input Noise Voltage (V_{DS} = −10 Vdc, I_D = 1.0 mAdc, f = 100 Hz, BW = 15 Hz)	E_n	–	0.19	$\mu V/\sqrt{Hz}$

*Indicates JEDEC Registered Data.

Note 1: Pulse Test: Pulse Width ≤ 630 ms, Duty Cycle ≤ 10%.

(b)

Solution **(a)** From the data sheet, $g_m = y_{fs} = 2000\ \mu S$ minimum. The gain is

$$A_v \cong \frac{g_m R_S}{1 + g_m R_S} = \frac{(2000\ \mu S)(10\ k\Omega)}{1 + (2000\ \mu S)(10\ k\Omega)} = \mathbf{0.952}$$

(b) From the data sheet, $I_{GSS} = 10$ nA at $V_{GS} = 15$ V. Therefore,

$$R_{IN(gate)} = \frac{15\text{ V}}{10\text{ nA}} = 1500\text{ M}\Omega$$

$$R_{IN} = R_G \parallel R_{IN(gate)} = 10\text{ M}\Omega \parallel 1500\text{ M}\Omega = \mathbf{9.93\text{ M}\Omega}$$

Related Problem If the g_m of the JFET in the source-follower of Figure 19–17 is doubled, what is the voltage gain?

Common-Gate (CG) Amplifier

A typical common-gate amplifier is shown in Figure 19–18. The gate is effectively at ac ground because of the bypass capacitor C_3. The input signal is applied at the source terminal through C_1. The output is coupled through C_2 from the drain terminal. This configuration is seldom used mainly because it gives up the major advantage of JFETs, which is high input resistance.

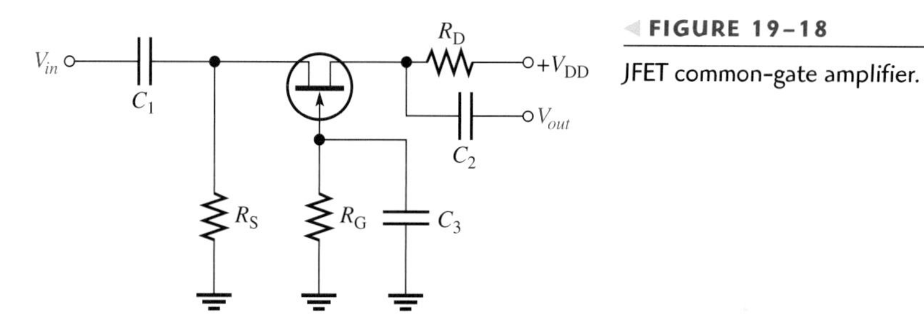

FIGURE 19–18
JFET common-gate amplifier.

Voltage Gain The voltage gain from source to drain is developed as follows:

$$A_v = \frac{V_{out}}{V_{in}} = \frac{V_d}{V_{gs}} = \frac{I_d R_D}{V_{gs}} = \frac{g_m V_{gs} R_D}{V_{gs}}$$

$$A_v = g_m R_D$$

Equation 19–22

Notice that the gain expression is the same as for the CS JFET amplifier.

Input Resistance As you have seen, both the CS and the CD configurations have extremely high input resistances because the gate is the input terminal. In contrast, the common-gate configuration has a low input resistance, as shown in the following steps.

First, the input current (source current) is equal to the drain current.

$$I_{in} = I_d = g_m V_{gs}$$

The input voltage equals V_{gs}.

$$V_{in} = V_{gs}$$

The input resistance at the source terminal is, therefore,

$$R_{in(source)} = \frac{V_{in}}{I_{in}} = \frac{V_{gs}}{g_m V_{gs}}$$

$$R_{in(source)} = \frac{1}{g_m}$$

Equation 19–23

Summary

Table 19–2 summarizes the gain, input resistance, and phase characteristics for the three FET amplifier configurations.

TABLE 19–2

Comparison of FET amplifier configurations.

	CS	CD	CG
Voltage gain, A_v	$g_m R_D$	$\dfrac{g_m R_S}{1 + g_m R_S}$	$g_m R_D$
Input resistance, R_{in}	$\left(\dfrac{V_{GS}}{I_{GSS}}\right) \parallel R_G$	$\left(\dfrac{V_{GS}}{I_{GSS}}\right) \parallel R_G$	$\left(\dfrac{1}{g_m}\right) \parallel R_G$
Phase of output relative to input	180° (inversion)	0° (no inversion)	0° (no inversion)

SECTION 19–4 REVIEW

1. What factors determine the voltage gain of a CS FET amplifier?
2. A certain CS amplifier has an $R_D = 1.0\ k\Omega$. When a load resistance of $1.0\ k\Omega$ is capacitively coupled to the drain, how much does the gain change?
3. What is a major difference between a CG amplifier and the other two configurations?

19–5 MULTISTAGE AMPLIFIERS

Several amplifiers can be connected in a cascaded arrangement with the output of one amplifier driving the input of the next. Each amplifier in the cascaded arrangement is known as a stage. The purpose of a multistage arrangement is to increase the overall gain.

After completing this section, you should be able to

- **Analyze a multistage amplifier**
- Determine the multistage voltage gain
- Convert the voltage gain to decibels
- Determine the effects of loading on the gain of each stage and on the overall gain

Multistage Gain

The overall gain, $A_{v(tot)}$, of cascaded amplifiers as in Figure 19–19 is the product of the individual loaded voltage gains.

Equation 19–24

$$A_{v(tot)} = A_{v1}A_{v2}A_{v3} \cdots A_{vn}$$

where n is the number of stages and A_{v1}, A_{v2}, . . . are the gains taking into account the loading effects of the following stages.

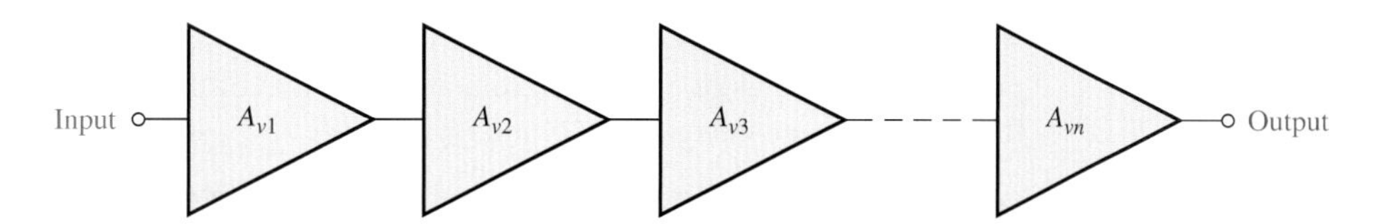

▲ **FIGURE 19–19**

Cascaded amplifiers.

Decibel Voltage Gain

Amplifier voltage gain is often expressed in decibels (dB) as follows:

$$A_v\ (\text{dB}) = 20 \log A_v$$ Equation 19–25

This formula is particularly useful in multistage systems because the overall dB voltage gain is the *sum* of the individual dB gains.

$$A_{v(tot)}\ (\text{dB}) = A_{v1}\ (\text{dB}) + A_{v2}\ (\text{dB}) + \cdots + A_{vn}\ (\text{dB})$$ Equation 19–26

EXAMPLE 19–8

A given cascaded amplifier arrangement has the following loaded voltage gains: $A_{v1} = 10$, $A_{v2} = 15$, and $A_{v3} = 20$. What is the overall gain? Also express each gain in decibels and determine the total decibel voltage gain.

Solution

$$A_{v(tot)} = A_{v1}A_{v2}A_{v3} = (10)(15)(20) = \mathbf{3000}$$

$$A_{v1}\ (\text{dB}) = 20 \log 10 = \mathbf{20.0\ dB}$$

$$A_{v2}\ (\text{dB}) = 20 \log 15 = \mathbf{23.5\ dB}$$

$$A_{v3}\ (\text{dB}) = 20 \log 20 = \mathbf{26.0\ dB}$$

$$A_{v(tot)}\ (\text{dB}) = 20\ \text{dB} + 23.5\ \text{dB} + 26.0\ \text{dB} = \mathbf{69.5\ dB}$$

Related Problem In a certain multistage amplifier, the individual stages have the following voltage gains: $A_{v1} = 25$, $A_{v2} = 5$, and $A_{v3} = 12$. What is the overall gain? Express each stage gain in dB and determine the total dB gain.

Multistage Analysis

We will use the two-stage amplifier in Figure 19–20 to illustrate multistage analysis. Notice that both stages are identical CE amplifiers with the output of the first

▼ **FIGURE 19–20**

Two-stage common-emitter amplifier.

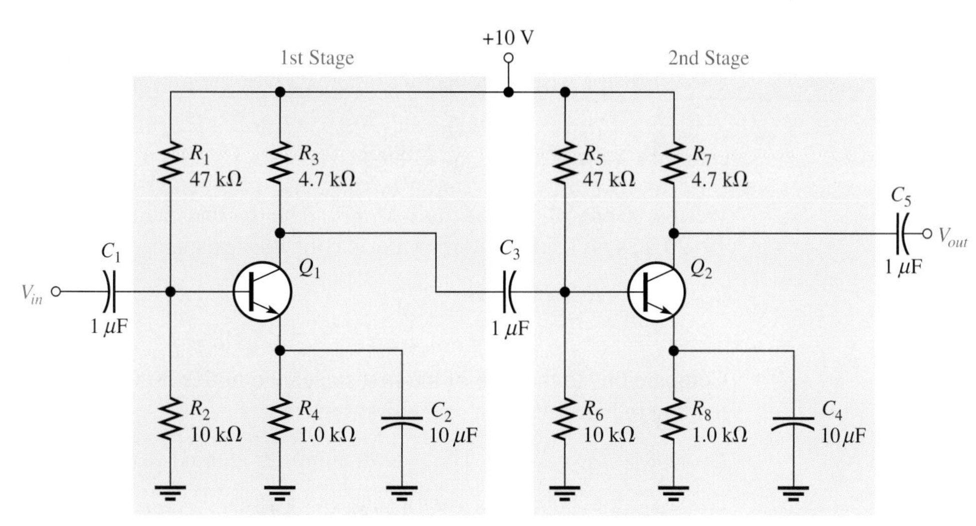

stage capacitively coupled to the input of the second stage. Capacitive coupling prevents the dc bias of one stage from affecting that of the other. Also notice that the transistors are designated Q_1 and Q_2.

Loading Effects In determining the gain of the first stage, you must consider the loading effect of the second stage. Because the coupling capacitor C_3 appears as a short to the signal frequency, the total input resistance of the second stage presents an ac load to the first stage.

Looking from the collector of Q_1, the two biasing resistors, R_5 and R_6, appear in parallel with the input resistance at the base of Q_2. In other words, the signal at the collector of Q_1 "sees" the collector resistor R_3 and R_5, R_6, and $R_{in(base2)}$ of the second stage all in parallel to ac ground. Thus, the effective ac collector resistance of Q_1 is the total of all these in parallel, as Figure 19–21 illustrates.

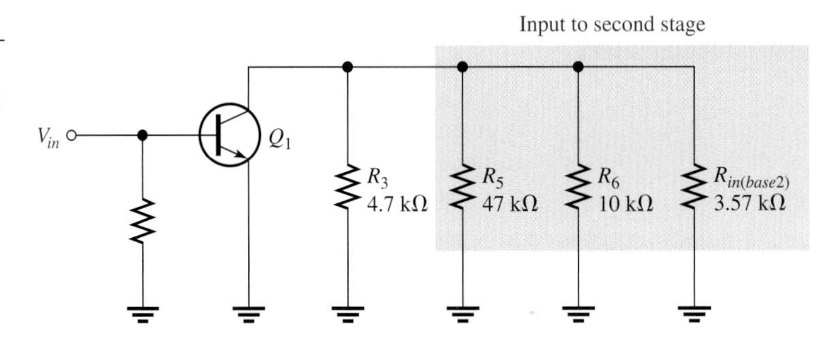

FIGURE 19–21

AC equivalent of first stage in Figure 19–20, showing loading from second stage.

The voltage gain of the first stage is reduced by the loading of the second stage because the effective ac collector resistance of the first stage is less than the actual value of its collector resistor, R_3. Remember that $A_v = R_C/r_e$ for an unloaded amplifier.

Voltage Gain of the First Stage The ac collector resistance of the first stage, as shown in Figure 19–20, is

$$R_{c1} = R_3 \parallel R_5 \parallel R_6 \parallel R_{in(base2)}$$

Keep in mind that lowercase italic subscripts denote ac quantities such as for R_c.

You can verify that $I_E = 1.05$ mA, $r_e = 23.8\ \Omega$, and $R_{in(base2)} = 3.57\ \text{k}\Omega$. The effective ac collector resistance of the first stage is

$$R_{c1} = 4.7\ \text{k}\Omega \parallel 47\ \text{k}\Omega \parallel 10\ \text{k}\Omega \parallel 3.57\ \text{k}\Omega = 1.63\ \text{k}\Omega$$

Therefore, the base-to-collector voltage gain of the first stage is

$$A_v = \frac{R_{c1}}{r_e} = \frac{1.63\ \text{k}\Omega}{23.8\ \Omega} = 68.5$$

Voltage Gain of the Second Stage The second stage, as shown in Figure 19–20, has no load resistor, so the ac collector resistance is R_7, and the gain is

$$A_v = \frac{R_7}{r_e} = \frac{4.7\ \text{k}\Omega}{23.8\ \Omega} = 198$$

Compare this to the gain of the first stage, and notice how much the second-stage loading reduced the gain of the first stage.

Overall Voltage Gain The overall amplifier gain is

$$A_{v(tot)} = A_{v1}A_{v2} = (68.5)(198) = 13{,}563$$

If an input signal of, say, 100 μV, is applied to the first stage and if the **attenuation** or reduction in gain of the input base circuit is neglected, an output from the second stage of (100 μV)(13,563) $\cong$ 1.36 V will result. The overall gain can be expressed in decibels as follows:

$$A_{v(tot)}\ (\text{dB}) = 20 \log(13{,}563) = 82.6\ \text{dB}$$

SECTION 19–5 REVIEW

1. What does the term *stage* mean?
2. How is the overall gain of a multistage amplifier determined?
3. Express a voltage gain of 500 in decibels.

19–6 CLASS A OPERATION

When an amplifier, whether it is one of the three configurations of BJT or a FET type, is biased such that it always operates in the linear region where the shape of the output signal is an amplified replica of the input signal, it is a class A amplifier. The discussion and formulas in the previous sections apply to class A operation.

After completing this section, you should be able to

- **Explain class A amplifier operation**
- Define Q-point
- Use a load line to analyze class A operation
- Calculate power gain of a common-emitter class A amplifier
- Determine dc quiescent power
- Determine output power
- Determine maximum efficiency of class A amplifiers

When the output signal takes up only a small percentage of the total load line excursion, the amplifier is a **small-signal amplifier.** When the output signal is larger and approaches the limits of the load line, the amplifier is a **large-signal amplifier.** Amplifiers are typically operated as large-signal devices when power amplification is the major objective. Figure 19–22 illustrates class A operation of an inverting amplifier.

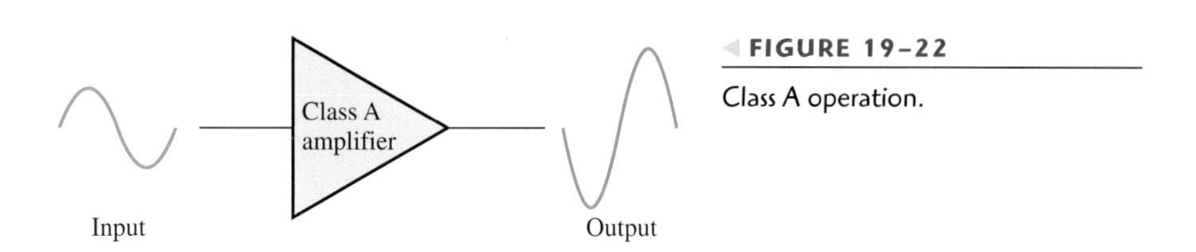

FIGURE 19–22 Class A operation.

The Q-Point Is Centered for Maximum Output Signal

When the dc operating point (Q-point) is at the center of the load line, a maximum class A signal can be obtained. You can see this concept by examining the graph of the load line for a given amplifier in Figure 19–23(a). This graph shows

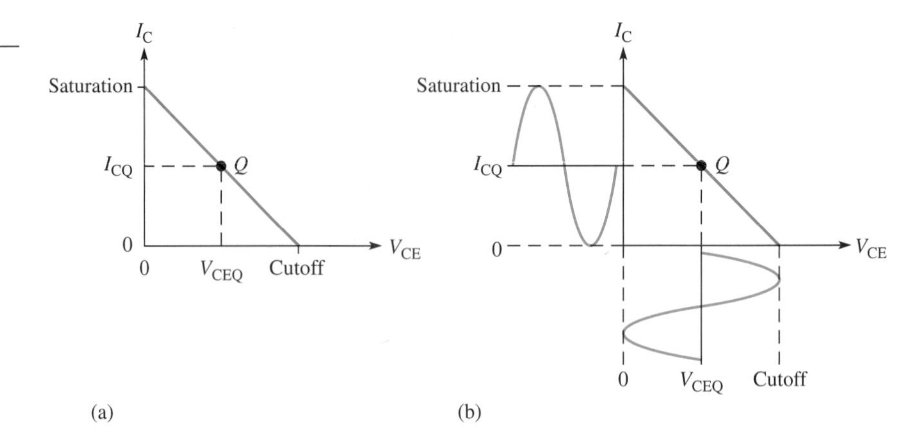

FIGURE 19–23

Maximum class A output occurs when the Q-point is centered on the load line.

the load line with the Q-point at its center. The collector current can vary from its Q-point value, I_{CQ}, up to its saturation value and down to its cutoff value of zero. Likewise, the collector-to-emitter voltage can swing from its Q-point value, V_{CEQ}, up to its cutoff value and down to its saturation value of near zero. This operation is indicated in Figure 19–23(b).

The peak value of the collector current is I_{CQ}, and the peak value of the collector-to-emitter voltage is V_{CEQ} in this case. This signal is the maximum that can be obtained from the class A amplifier. Actually, saturation or cutoff cannot be reached, so the practical maximum is slightly less.

A Noncentered Q-Point Limits Output Swing

If the Q-point is not centered, the output signal is limited. Figure 19–24 shows a load line with the Q-point moved away from the center toward cutoff. The output variation is limited by cutoff in this case. The collector current can only swing down to near zero and an equal amount above I_{CQ}. The collector-to-emitter voltage can only swing up to its cutoff value and an equal amount below V_{CEQ}. This situation is illustrated in Figure 19–24(a). If the amplifier is driven any further than this, it will "clip" at cutoff, as shown in Figure 19–24(b).

FIGURE 19–24

Q-point closer to cutoff.

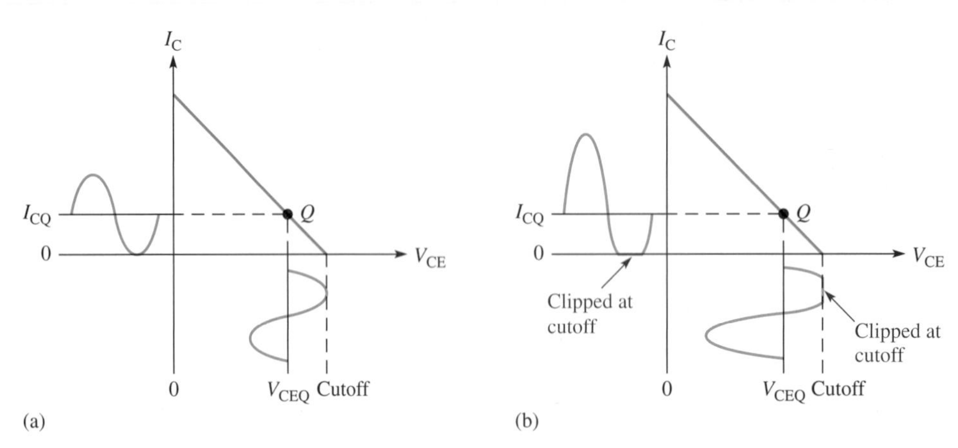

Figure 19–25 shows a load line with the Q-point moved away from center toward saturation. In this case, the output variation is limited by saturation. The collector current can only swing up to near saturation and an equal amount below I_{CQ}. The collector-to-emitter voltage can only swing down to its saturation value

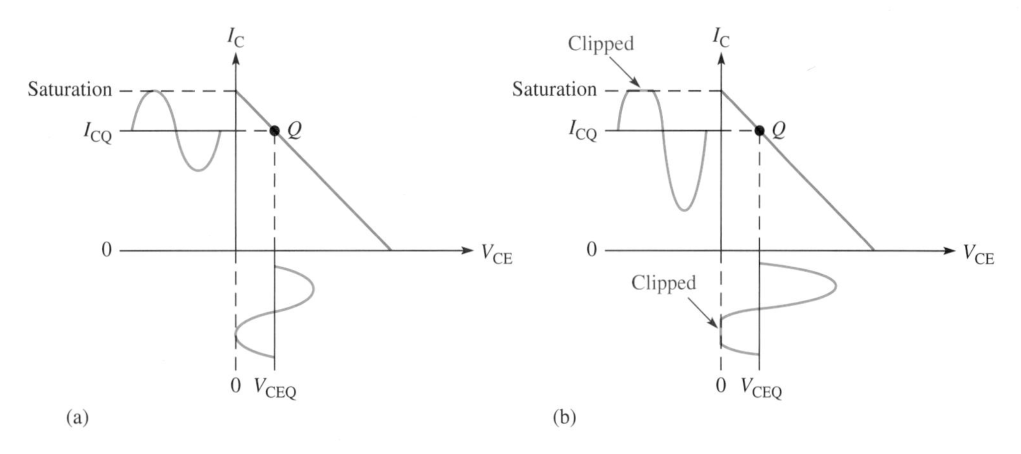

▲ **FIGURE 19–25**

Q-point closer to saturation.

and an equal amount above V_{CEQ}. This situation is illustrated in Figure 19–25(a). If the amplifier is driven any further, it will "clip" at saturation, as shown in Figure 19–25(b).

Power Gain

The main purpose of a large-signal amplifier is to achieve power gain. If we assume that the large-signal current gain A_i is approximately equal to β_{DC}, then the power gain for a common-emitter amplifier is

$$A_p \cong \beta_{DC} A_v$$

Equation 19–27

DC Quiescent Power

The power dissipation of a transistor with no signal input is the product of its Q-point current and voltage.

$$P_{DQ} = I_{CQ} V_{CEQ}$$

Equation 19–28

The quiescent power is the maximum power that the class A transistor must handle; therefore, its power rating should exceed this value.

Output Power

In general, for any Q-point location, the output power of a CE amplifier is the product of the rms collector-to-emitter voltage and the rms collector current.

$$P_{out} = V_{ce} I_c$$

Equation 19–29

Q-Point Centered When the Q-point is centered, the maximum collector current swing is I_{CQ}, and the maximum collector-to-emitter voltage swing is V_{CEQ}, as was shown in Figure 19–23(b). The output power therefore is

$$P_{out} = (0.707 V_{CEQ})(0.707 I_{CQ})$$

$$P_{out(max)} = 0.5 V_{CEQ} I_{CQ}$$

Equation 19–30

This is the maximum ac output power from a class A amplifier under signal conditions. Notice that it is one-half the quiescent power dissipation.

Efficiency

Efficiency of an amplifier is the ratio of ac output power to dc input power. The dc input power is the dc supply voltage times the current drawn from the supply.

$$P_{DC} = V_{CC} I_{CC}$$

The average supply current I_{CC} equals I_{CQ}, and the supply voltage V_{CC} is twice V_{CEQ} when the Q-point is centered. Therefore, ideally the maximum efficiency is

$$\text{eff}_{max} = \frac{P_{out}}{P_{DC}} = \frac{0.5V_{CEQ}I_{CQ}}{V_{CC}I_{CC}} = \frac{0.5V_{CEQ}I_{CQ}}{2V_{CEQ}I_{Q}} = \frac{0.5}{2}$$

Equation 19–31

$$\text{eff}_{max} = 0.25$$

Thus 25% is the highest possible efficiency available from a class A amplifier and is approached only when the Q-point is at the center of the load line (power in the bias resistors is ignored).

EXAMPLE 19–9

Determine the following values for a class A amplifier operated with a centered Q-point with $I_{CQ} = 50$ mA and $V_{CEQ} = 7.5$ V:

(a) minimum transistor power rating **(b)** ac output power **(c)** efficiency

Solution **(a)** The maximum power that the transistor must be able to handle is the minimum rating that you would use. Thus,

$$P_{DQ} = I_{CQ}V_{CEQ} = (50 \text{ mA})(7.5 \text{ V}) = \mathbf{375\ mW}$$

(b) $P_{out} = 0.5V_{CEQ}I_{CQ} = 0.5(7.5 \text{ V})(50 \text{ mA}) = \mathbf{188\ mW}$

(c) Since the Q-point is centered, the efficiency is at its maximum possible value of **25%.**

Related Problem Explain what happens to the output voltage of the class A amplifier in this example if the Q-point is shifted to $V_{CE} = 3$ V.

SECTION 19–6 REVIEW

1. What is the optimum Q-point location for class A amplifiers?
2. What is the maximum efficiency of a class A amplifier?
3. A certain amplifier has a centered Q-point of $I_{CQ} = 10$ mA and $V_{CEQ} = 7$ V. What is the maximum ac output power?

19–7 CLASS B PUSH-PULL AMPLIFIER OPERATION

When an amplifier is biased such that it operates in the linear region for 180° of the input cycle and is in cutoff for 180°, it is a class B amplifier. The primary advantage of a class B amplifier over a class A is that the class B is more efficient; you can get more output power for a given amount of input power. A disadvantage of class B is that it is more difficult to implement the circuit in order to get a linear reproduction of the input waveform. As you will see in this section, the term **push-pull** refers to a common type of class B amplifier circuit in which the input wave shape is approximately reproduced at the output.

After completing this section, you should be able to

- **Analyze class B amplifiers**
- Explain class B operation

- Discuss the meaning of push-pull amplifier operation
- Define crossover distortion
- Describe how a class B push-pull amplifier is biased
- Determine the maximum output power
- Determine the efficiency of a class B amplifier

Class B amplifier operation is illustrated in Figure 19–26, where the output waveform is shown relative to the input.

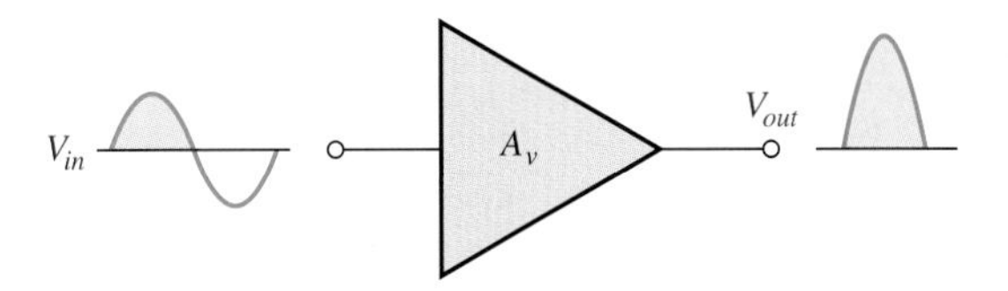

FIGURE 19–26

Class B amplifier (noninverting).

The Q-Point Is at Cutoff

The class B amplifier is biased at cutoff so that $I_{CQ} = 0$ and $V_{CEQ} = V_{CE(cutoff)}$. It is brought out of cutoff and operates in its linear region when the input signal drives it into conduction. This is illustrated in Figure 19–27 with an emitter-follower circuit.

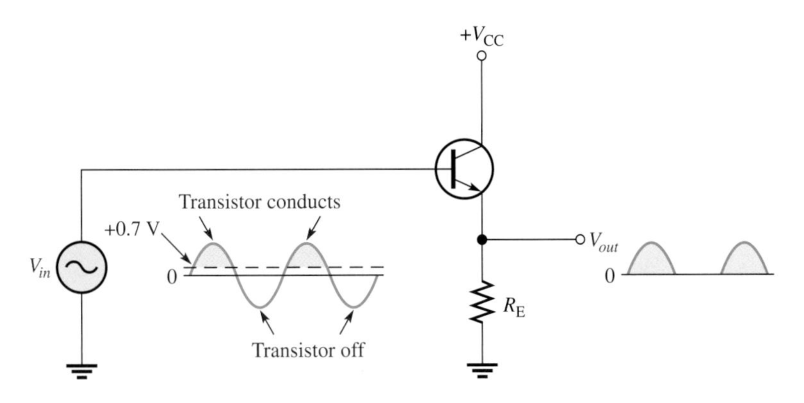

FIGURE 19–27

Common-collector class B amplifier.

Push-Pull Operation

Figure 19–28 shows one type of push-pull class B amplifier using two emitter-followers. This is a complementary amplifier because one emitter-follower uses an *npn* transistor and the other a *pnp,* which conducts on opposite alternations of the input cycle. Notice that there is no dc base bias voltage ($V_B = 0$). Thus, only the signal voltage drives the transistors into conduction. Q_1 conducts during the positive half of the input cycle, and Q_2 conducts during the negative half.

Crossover Distortion

When the dc base voltage is zero, the input signal voltage must exceed V_{BE} before a transistor conducts. As a result, there is a time interval between the positive and negative alternations of the input when neither transistor is conducting, as shown in Figure 19–29. The resulting distortion in the output waveform is quite common and is called *crossover distortion.*

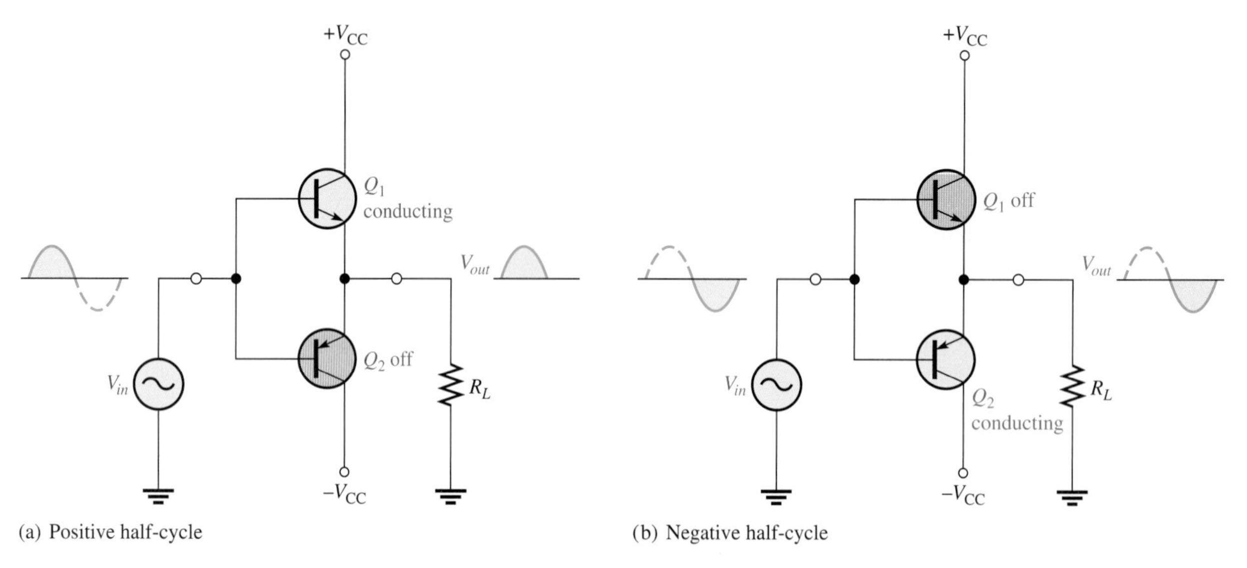

▲ **FIGURE 19–28**

Class B push-pull operation.

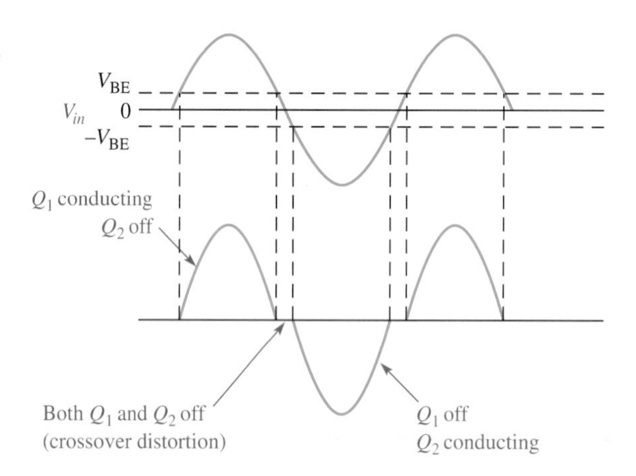

▶ **FIGURE 19–29**

Illustration of crossover distortion in a class B push-pull amplifier.

Biasing the Push-Pull Amplifier

To eliminate crossover distortion, both transistors in the push-pull arrangement must be biased slightly above cutoff when there is no signal. This can be done with a voltage divider and diode arrangement, as shown in Figure 19–30. When the diode characteristics of D_1 and D_2 are closely matched to the characteristics of the transistor base-emitter junctions, a stable bias is maintained.

Since R_1 and R_2 are of equal value, the voltage with respect to ground at point A between the two diodes is $V_{CC}/2$. Assuming that both diodes and both transistors are identical, the drop across D_1 equals the V_{BE} of Q_1, and the drop across D_2 equals the V_{BE} of Q_2. As a result, the voltage at the emitters is also $V_{CC}/2$, and therefore, $V_{CEQ1} = V_{CEQ2} = V_{CC}/2$, as indicated. Because both transistors are biased near cutoff, $I_{CQ} \cong 0$.

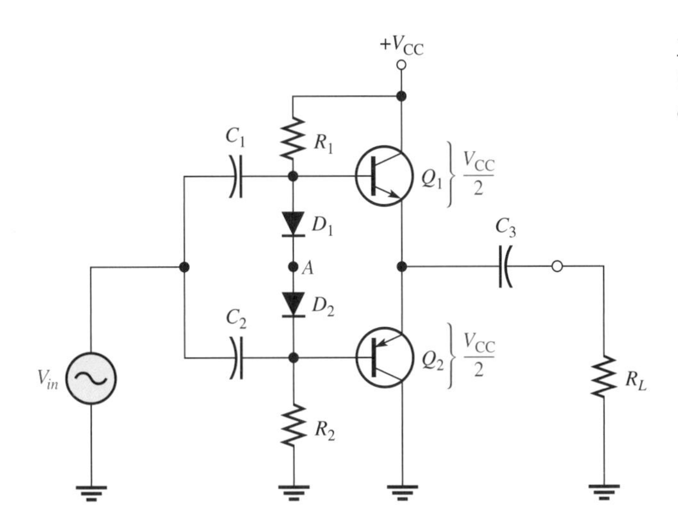

FIGURE 19–30

Biasing the push-pull amplifier to eliminate crossover distortion.

AC Operation

Under maximum conditions, transistors Q_1 and Q_2 are alternately driven from near cutoff to near saturation. During the positive alternation of the input signal, the Q_1 emitter is driven from its Q-point value of $V_{CC}/2$ to near V_{CC}, producing a positive peak voltage approximately equal to V_{CEQ}. At the same time, the Q_1 current swings from its Q-point value near zero to near-saturation value, as shown in Figure 19–31(a).

FIGURE 19–31

AC push-pull operation. Capacitors are assumed to be shorts at the signal frequency, and the dc source is at ac ground.

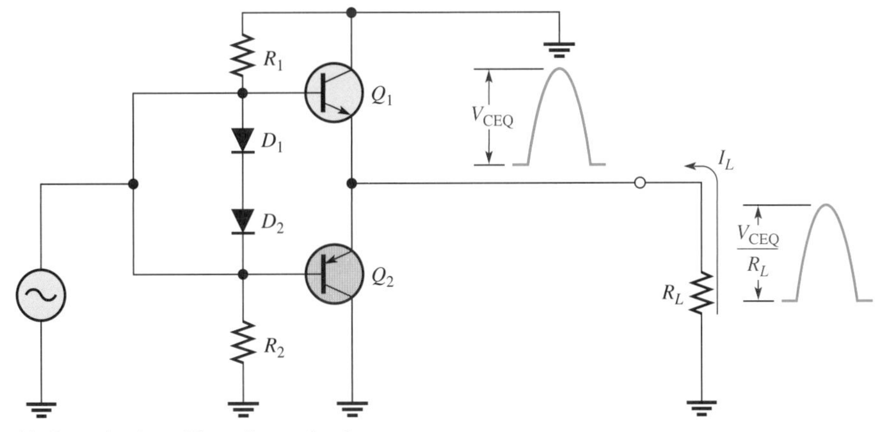

(a) Q_1 conducting with maximum signal output

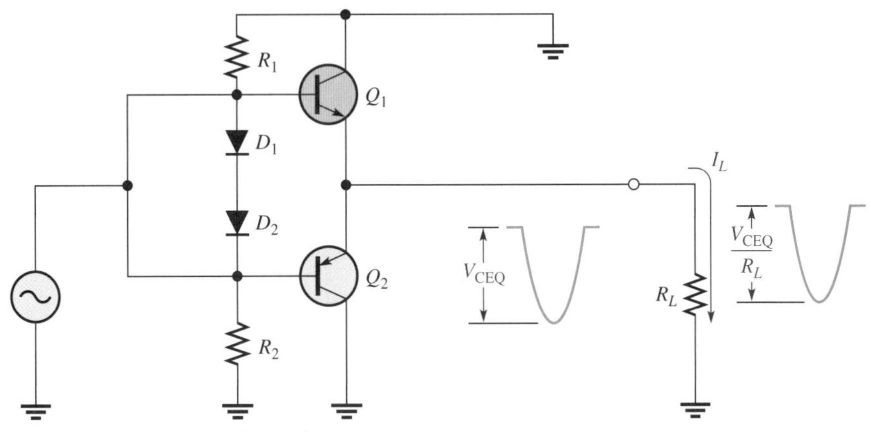

(b) Q_2 conducting with maximum signal output

During the negative alternation of the input signal, the Q_2 emitter is driven from its Q-point value of $V_{CC}/2$ to near zero, producing a negative peak voltage approximately equal to V_{CEQ}. Also, the Q_2 current swings from near zero to near-saturation value, as shown in Figure 19–31(b).

Because the peak voltage across each transistor is V_{CEQ}, the ac saturation current is

Equation 19–32

$$I_{c(sat)} = \frac{V_{CEQ}}{R_L}$$

Since $I_e \cong I_c$ and the output current is the emitter current, the peak output current is also V_{CEQ}/R_L.

EXAMPLE 19–10

Determine the maximum peak values for the output voltage and current in Figure 19–32.

FIGURE 19–32

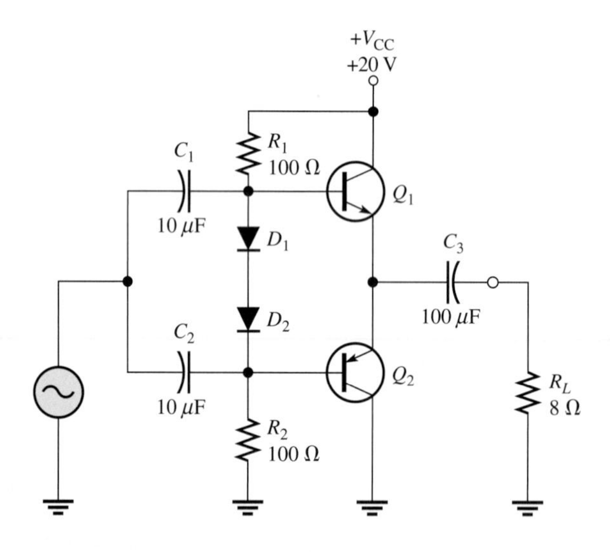

Solution The maximum peak output voltage is

$$V_{p(out)} \cong V_{CEQ} = \frac{V_{CC}}{2} = \frac{20\text{ V}}{2} = \mathbf{10\text{ V}}$$

The maximum peak output current is

$$I_{p(out)} \cong I_{c(sat)} = \frac{V_{CEQ}}{R_L} = \frac{10\text{ V}}{8\ \Omega} = \mathbf{1.25\text{ A}}$$

Related Problem Find the maximum peak values for the output voltage and current in Figure 19–32 if V_{CC} is lowered to 15 V and the load resistor is changed to 16 Ω.

Maximum Output Power

It has been shown that the maximum peak output current is approximately $I_{c(sat)}$, and the maximum peak output voltage is approximately V_{CEQ}. The maximum average output power therefore is

$$P_{out} = V_{rms(out)} I_{rms(out)}$$

Since

$$V_{rms(out)} = 0.707 V_{p(out)} = 0.707 V_{CEQ}$$

and

$$I_{rms(out)} = 0.707 I_{p(out)} = 0.707 I_{c(sat)}$$

then

$$P_{out} = 0.5 V_{CEQ} I_{c(sat)}$$

Substituting $V_{CC}/2$ for V_{CEQ},

$$P_{out(max)} = 0.25 V_{CC} I_{c(sat)}$$

Equation 19–33

Input Power

The input power comes from the V_{CC} supply and is

$$P_{DC} = V_{CC} I_{CC}$$

Since each transistor draws current for a half-cycle, the current is a half-wave signal with an average value of

$$I_{CC} = \frac{I_{c(sat)}}{\pi}$$

Thus,

$$P_{DC} = \frac{V_{CC} I_{c(sat)}}{\pi}$$

Equation 19–34

Efficiency

The main advantage of push-pull class B amplifiers over class A is that a class B amplifier has a much higher efficiency. This advantage usually overrides the difficulty of biasing the class B push-pull amplifier to eliminate crossover distortion.

The efficiency is again defined as the ratio of ac output power to dc input power.

$$\text{eff} = \frac{P_{out}}{P_{DC}}$$

The maximum efficiency for a class B amplifier is designated eff_{max} and is developed as follows, using Equations 19–33 and 19–34:

$$P_{out(max)} = 0.25 V_{CC} I_{c(sat)}$$

$$\text{eff}_{max} = \frac{P_{out(max)}}{P_{DC}} = \frac{0.25 V_{CC} I_{c(sat)}}{V_{CC} I_{c(sat)}/\pi} = 0.25\pi$$

$$\text{eff}_{max} = 0.785$$

Equation 19–35

Therefore, ideally the maximum efficiency is 78.5%. Recall that the maximum ideal efficiency for class A is 0.25 (25%). Actual maximum efficiencies are always lower.

EXAMPLE 19–11

Find the maximum ac output power and the dc input power of the amplifier in Figure 19–32 of Example 19–10.

Solution In Example 19–10, $I_{c(sat)}$ was found to be 1.25 A. Thus,

$$P_{out(max)} = 0.25V_{CC}I_{c(sat)} = 0.25(20\text{ V})(1.25\text{ A}) = \mathbf{6.25\ W}$$

$$P_{DC} = \frac{V_{CC}I_{c(sat)}}{\pi} = \frac{(20\text{ V})(1.25\text{ A})}{\pi} = \mathbf{7.96\ W}$$

Related Problem What is the maximum ac output power in Figure 19–32 if $R_L = 4\ \Omega$?

SECTION 19–7 REVIEW

1. Where is the Q-point for a class B amplifier?
2. What causes crossover distortion?
3. What is the maximum ideal efficiency of a push-pull class B amplifier?
4. Explain the purpose of the push-pull configuration for class B.

19–8 CLASS C OPERATION

Class C amplifiers are biased so that conduction occurs for much less than 180°. Class C amplifiers are more efficient than either class A or push-pull class B. Thus, more output power can be obtained from class C operation. Because the output waveform is severely distorted, class C amplifiers are normally limited to applications as tuned amplifiers at radio frequencies (RF).

After completing this section, you should be able to

- **Explain the basic operation of a class C amplifier**
- Discuss power in a class C amplifier
- Discuss a tuned amplifier
- Determine maximum output power
- Determine class C efficiency

Basic Operation

Class C amplifier operation is illustrated in Figure 19–33. A basic common-emitter class C amplifier with a resistive load is shown in Figure 19–34(a). It is biased below cutoff with the $-V_{BB}$ supply. The ac source voltage has a peak value that is slightly greater than $V_{BB} + V_{BE}$ so that the base voltage exceeds the barrier potential of the base-emitter junction for a short time near the positive peak of each cycle, as illustrated in Figure 19–34(b). During this short interval, the transistor is turned on. When the entire load line is used, the maximum collector current is approximately $I_{C(sat)}$, and the minimum collector voltage is approximately $V_{CE(sat)}$.

FIGURE 19–33

Class C amplifier (inverting).

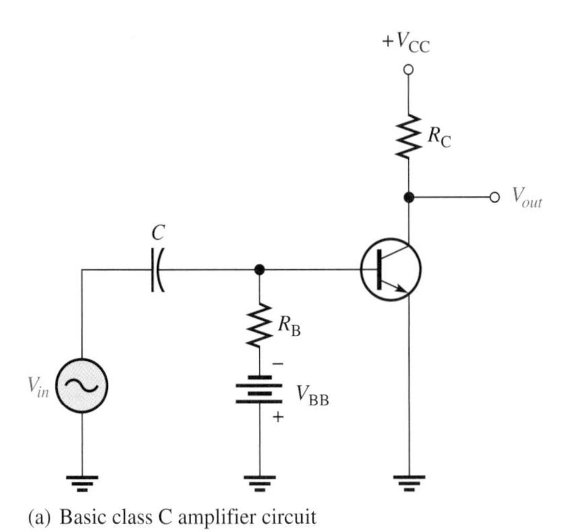

(a) Basic class C amplifier circuit

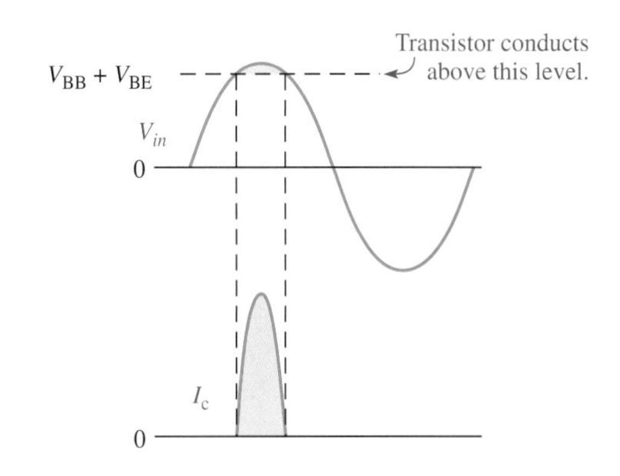

(b) Input voltage and output current waveforms

FIGURE 19–34

Class C operation.

Power Dissipation

The power dissipation of the transistor in a class C amplifier is low because it is on for only a small percentage of the input cycle. Figure 19–35 shows the collector current pulses. The time between pulses is the period (T) of the ac input voltage.

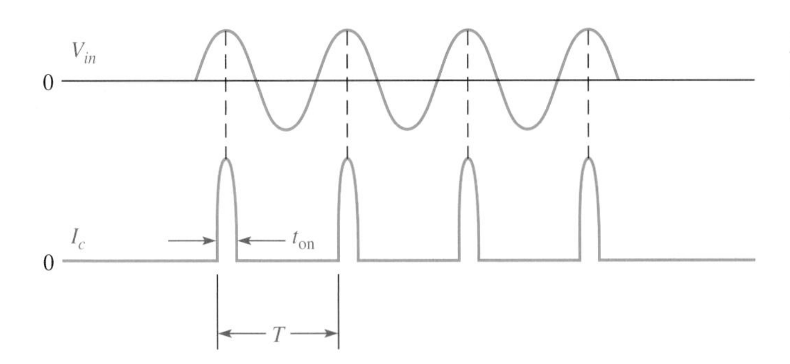

FIGURE 19–35

Collector current pulses in a class C amplifier.

The transistor is on for a short time, t_{on}, and off for the rest of the input cycle. Since the power dissipation averaged over the entire cycle depends on both the ratio of t_{on} to T and the power dissipation during t_{on}, it is typically very low.

Tuned Operation

Because the collector voltage (output) is not a replica of the input, the resistively loaded class C amplifier is of no value in linear applications. Therefore, it is necessary to use a class C amplifier with a parallel resonant circuit (tank), as shown in Figure 19–36(a). The resonant frequency of the tank circuit is determined by the formula $f_r = 1/(2\pi\sqrt{LC})$.

The short pulse of collector current on each cycle of the input initiates and sustains the oscillation of the tank circuit so that an output sinusoidal voltage is produced, as illustrated in Figure 19–36(b).

FIGURE 19–36

Tuned class C amplifier.

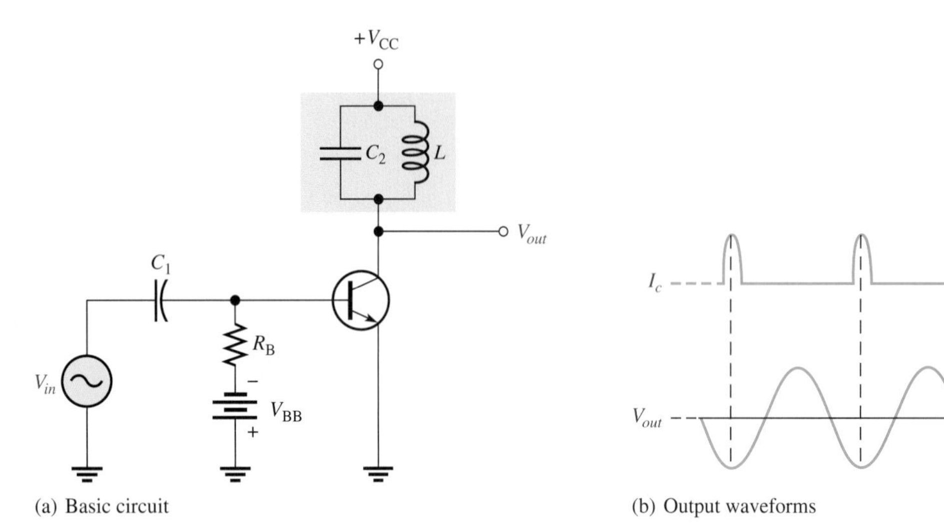

(a) Basic circuit

(b) Output waveforms

The amplitude of each successive cycle of the oscillation would be less than that of the previous cycle because of energy conversion in the resistance of the tank circuit, as shown in Figure 19–37(a), and the oscillation would eventually die out. However, the regular recurrences of the collector current pulse re-energizes the resonant circuit and sustains the oscillations at a constant amplitude. When the tank circuit is tuned to the frequency of the input signal, re-energizing occurs on each cycle of the tank voltage, as shown in Figure 19–37(b).

FIGURE 19–37

Tank circuit oscillations. V_r is the voltage across the tank circuit.

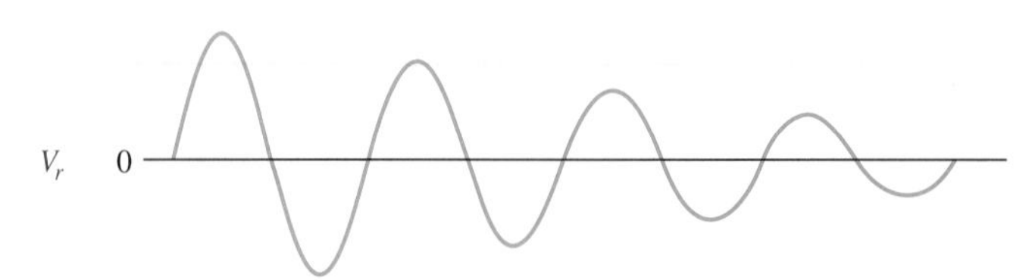

(a) Oscillation gradually dies out due to energy conversion (loss) in the tank circuit.

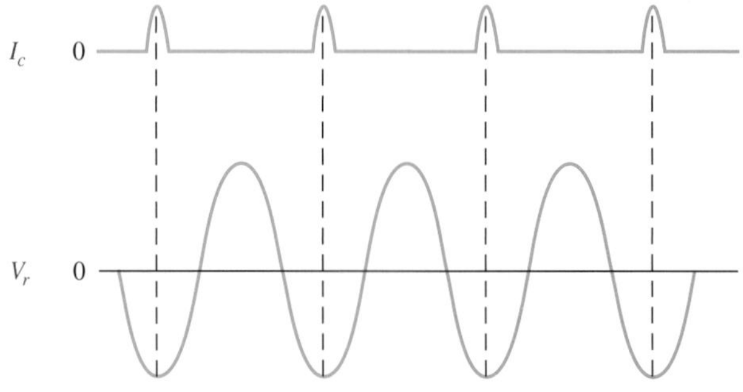

(b) Oscillation can be sustained by short pulses of collector current.

Maximum Output Power

Since the voltage developed across the tank circuit has a peak-to-peak value of approximately $2V_{CC}$, the maximum output power can be expressed as

$$P_{out} = \frac{V_{rms}^2}{R_c} = \frac{(0.707V_{CC})^2}{R_c}$$

$$P_{out(max)} = \frac{0.5V_{CC}^2}{R_c}$$

Equation 19–36

where R_c is the equivalent parallel resistance of the collector tank circuit and represents the parallel combination of the coil resistance and the load resistance. It usually has a low value.

The total power that must be supplied to the amplifier is

$$P_T = P_{out} + P_{D(avg)}$$

where $P_{D(avg)}$ is the average power dissipation. Therefore, the efficiency is

$$\text{eff} = \frac{P_{out}}{P_{out} + P_{D(avg)}}$$

Equation 19–37

When $P_{out} >> P_{D(avg)}$, the class C efficiency closely approaches 100%.

EXAMPLE 19–12

A certain class C amplifier has a $P_{D(avg)}$ of 100 mW, a V_{CC} equal to 24 V, and an R_c of 100 Ω. Determine the efficiency.

Solution

$$P_{out(max)} = \frac{0.5V_{CC}^2}{R_c} = \frac{0.5(24\text{ V})^2}{100\ \Omega} = 2.88\text{ W}$$

Therefore,

$$\text{eff} = \frac{P_{out}}{P_{out} + P_{D(avg)}} = \frac{2.88\text{ W}}{2.88\text{ W} + 100\text{ mW}} = \mathbf{0.966}$$

or

$$\%\text{eff} = 96.6\%$$

Related Problem What happens to the efficiency of the amplifier if R_c is increased?

SECTION 19–8 REVIEW

1. How is a class C amplifier normally biased?
2. What is the purpose of the tuned circuit in a class C amplifier?
3. A certain class C amplifier has an average power dissipation of 100 mW and an output power of 1 W. What is its efficiency?

19–9 OSCILLATORS

An oscillator is a circuit that produces a repetitive waveform on its output with only the dc supply voltage as an input. A repetitive input signal is not required. The output voltage can be either sinusoidal or nonsinusoidal, depending on the type of oscillator. Generally, oscillator operation is based on the principle of positive feedback. In this section, we will examine this concept, look at the general conditions required for oscillation to occur, and introduce several basic oscillator circuits. Additional coverage of oscillators is provided in Chapter 21.

After completing this section, you should be able to

- **Discuss the theory and analyze the operation of several types of oscillators**
- Explain what an oscillator is
- Discuss positive feedback
- Describe the conditions for oscillation
- Identify an *RC* oscillator and discuss its basic operation
- Identify a Colpitts oscillator and discuss its basic operation
- Identify a Hartley oscillator and discuss its basic operation
- Identify a Clapp oscillator and discuss its basic operation
- Identify a crystal oscillator and discuss its basic operation
- Describe a basic quartz crystal

Oscillator Principles

The basic concept of an oscillator is illustrated in Figure 19–38. Essentially, an oscillator converts electrical energy in the form of dc to electrical energy in the form of ac. A basic oscillator consists of a transistor amplifier for gain (op-amp oscillators are covered in Chapter 21) and a positive feedback circuit that produces phase shift and provides attenuation, as shown in Figure 19–39.

FIGURE 19–38

The basic oscillator concept showing three possible types of output waveforms.

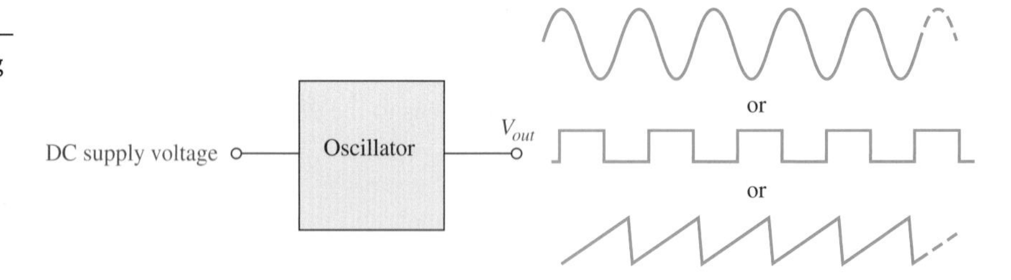

Positive Feedback **Positive feedback** is characterized by the condition wherein a portion of the output voltage of an amplifier is fed back to the input with no net phase shift, resulting in a reinforcement of the output signal. The basic idea is illustrated in Figure 19–40. As you can see, the in-phase feedback voltage is amplified to produce the output voltage, which in turn produces the feedback voltage. That is, a loop is created in which the signal sustains itself with no input signal and a continuous sinusoidal output is produced. This phenomenon is called *oscillation.*

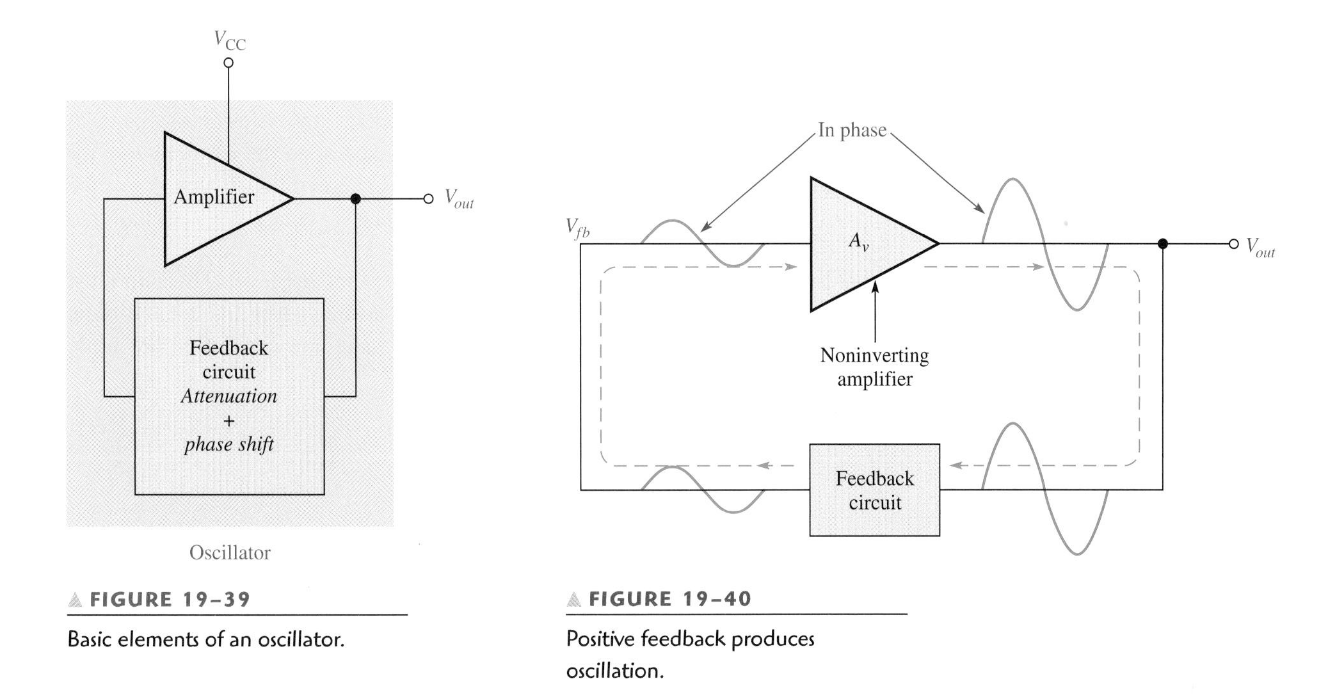

▲ **FIGURE 19–39**

Basic elements of an oscillator.

▲ **FIGURE 19–40**

Positive feedback produces oscillation.

Conditions for Oscillation Two conditions, illustrated in Figure 19–41, are required for a sustained state of oscillation:

1. The phase shift around the feedback loop must be 0°.
2. The loop gain (voltage gain, A_{cl}, around the closed feedback loop) must be at least 1 **(unity gain).**

▼ **FIGURE 19–41**

Conditions for oscillation.

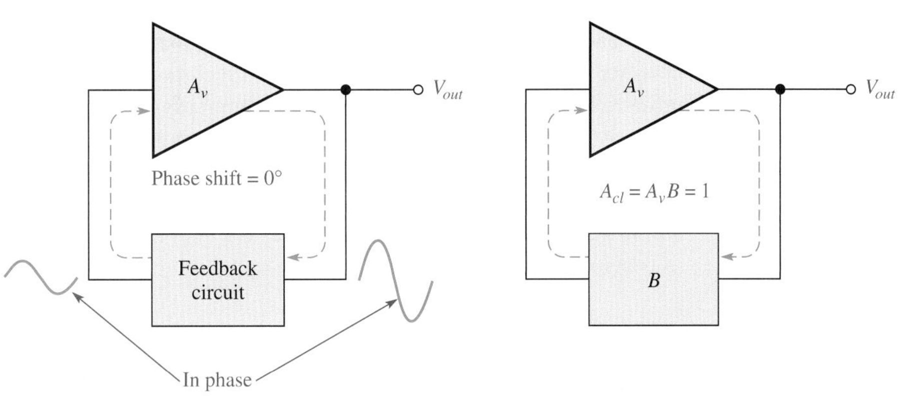

(a) The phase shift around the loop is 0°.

(b) The closed loop gain is 1.

The voltage gain around the closed feedback loop (A_{cl}) is the product of the amplifier gain (A_v) and the attenuation (B) of the feedback circuit.

$$A_{cl} = A_v B$$

Equation 19–38

If a sinusoidal wave is the desired output, a loop gain greater than 1 will rapidly cause the output to saturate at both peaks of the waveform, producing unacceptable distortion. To avoid this, some form of gain control must be used to keep the loop gain at exactly 1, once oscillations have started. For example, if the attenuation of the feedback network is 0.01, the amplifier must have a gain of exactly 100 to overcome this attenuation and not create unacceptable distortion

(0.01 × 100 = 1.0). An amplifier gain of greater than 100 will cause the oscillator to limit both peaks of the waveform.

Start-Up Conditions So far, you have learned what it takes for an oscillator to produce a continuous sinusoidal output. Now let's examine the requirements for the oscillation to start when the dc supply voltage is turned on. As you know, the unity-gain condition must be met for oscillation to be sustained. For oscillation to begin, the voltage gain around the positive feedback loop must be greater than 1 so that the amplitude of the output can build up to a desired level. The gain must then decrease to 1 so that the output stays at the desired level and oscillation is sustained. The conditions for both starting and sustaining oscillation are illustrated in Figure 19–42.

FIGURE 19–42

When oscillation starts at t_0, the condition $A_{cl} > 1$ causes the sinusoidal output voltage amplitude to build up to a desired level, where A_{cl} decreases to 1 and maintains the desired amplitude.

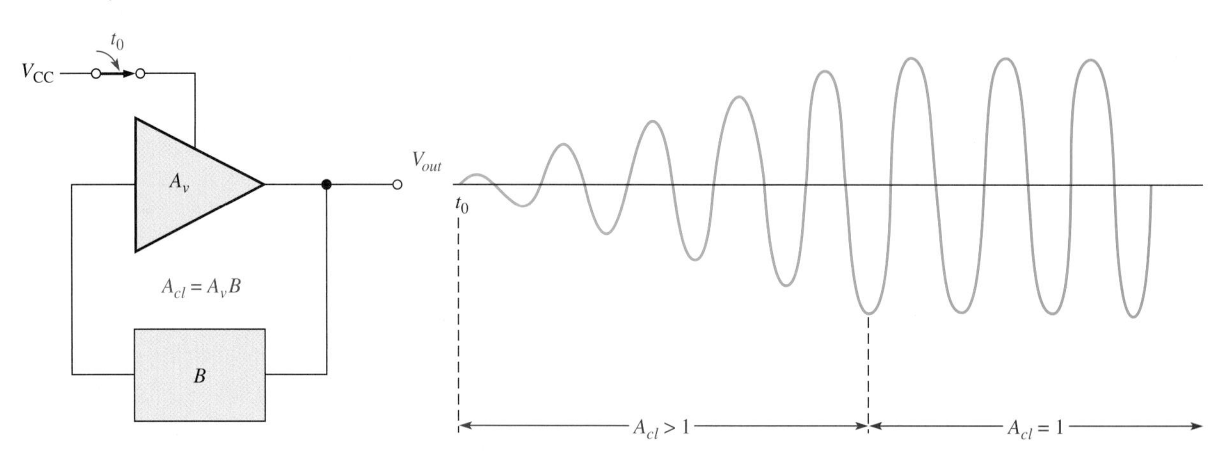

A question that normally arises is this: If the oscillator is off (no dc voltage) and there is no output voltage, how does a feedback signal originate to start the positive feedback build-up process? Initially, when power is turned on, a small positive feedback voltage develops from thermally produced broad-band noise in the resistors or other components or from turn-on transients. The feedback circuit permits only a voltage with a frequency equal to the selected oscillation frequency to appear in phase on the amplifier's input. This initial feedback voltage is amplified and continually reinforced, resulting in a buildup of the output voltage as previously discussed.

The *RC* Oscillator

The basic *RC* oscillator shown in Figure 19–43 uses an *RC* circuit as its feedback circuit. In this case, three *RC* lag networks have a total phase shift of 180°. The common-emitter transistor contributes a 180° phase shift. The total phase shift through the amplifier and feedback circuit therefore is 360°, which is effectively 0° (no phase shift). The attenuation of the *RC* circuit and the gain of the amplifier must be such that the overall gain around the feedback loop is equal to 1 at the frequency of oscillation. This circuit will produce a continuous sinusoidal output.

The Colpitts Oscillator

One basic type of tuned oscillator is the Colpitts, named after its inventor. As shown in Figure 19–44, this type of oscillator uses an *LC* circuit in the feedback loop to provide the necessary phase shift and to act as a filter that passes only the specified frequency of oscillation. The approximate frequency of oscillation is es-

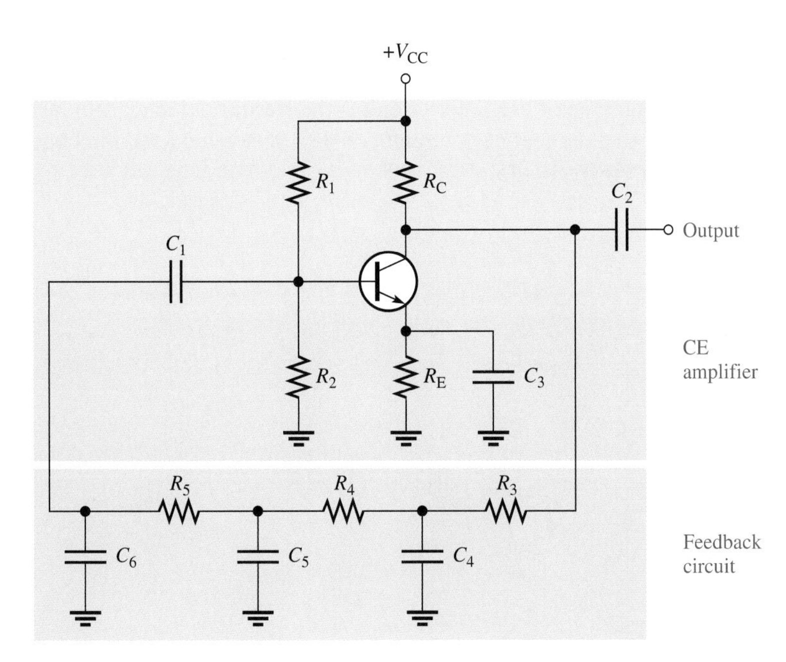

FIGURE 19–43

A basic *RC* oscillator.

tablished by the values of C_1, C_2, and L according to the following familiar formula:

$$f_r \cong \frac{1}{2\pi\sqrt{LC_T}}$$

Equation 19–39

Because the capacitors effectively appear in series around the tank circuit, the total capacitance is

$$C_T = \frac{C_1C_2}{C_1 + C_2}$$

Equation 19–40

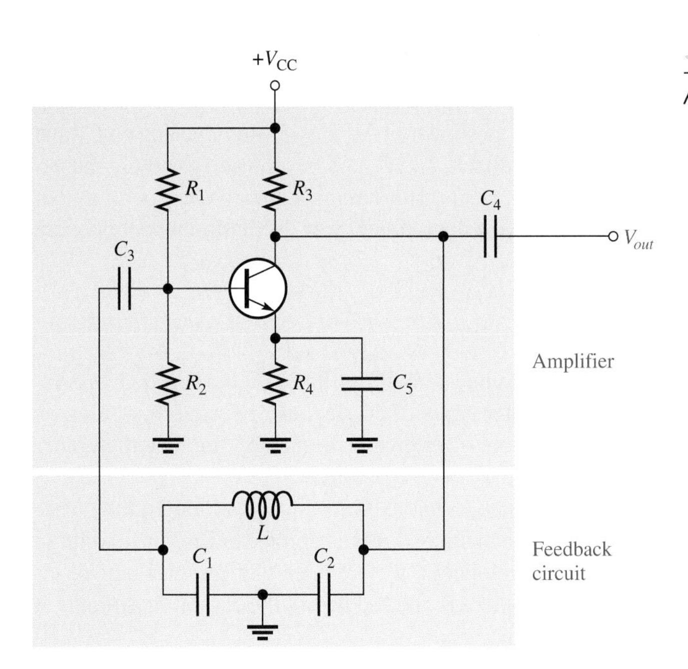

FIGURE 19–44

A basic Colpitts oscillator.

The Hartley Oscillator

Another basic type of oscillator circuit is the Hartley, which is similar to the Colpitts except that the feedback circuit consists of two inductors and one capacitor, as shown in Figure 19–45.

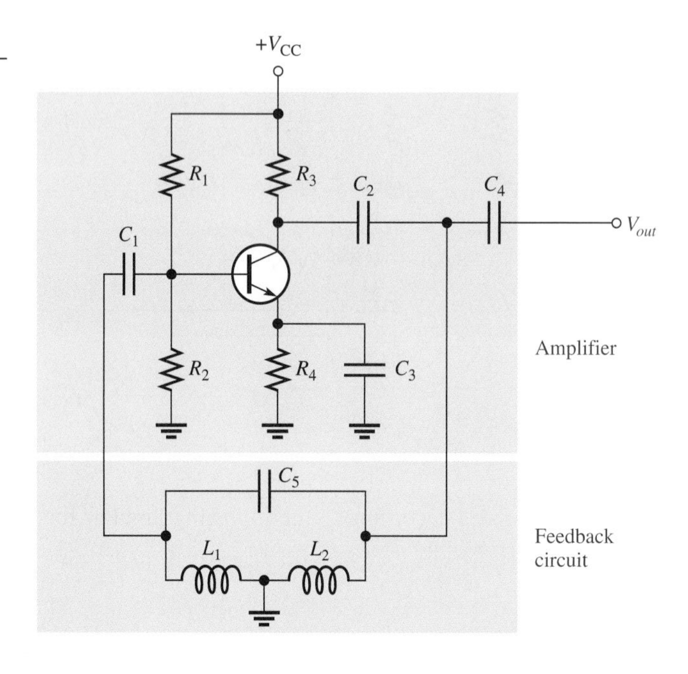

FIGURE 19–45

A basic Hartley oscillator.

The frequency of oscillation for the Hartley oscillator is

Equation 19–41

$$f_r \cong \frac{1}{2\pi\sqrt{L_T C_5}}$$

The total induction is the series combination of L_1 and L_2.

The Clapp Oscillator

The Clapp oscillator is similar to the Colpitts except that there is an additional capacitor in series with the inductor, as shown in Figure 19–46. Capacitors C_1 and C_2 can be selected for optimum feedback, and C_3 can be adjusted to obtain the desired frequency of oscillation. Also, a capacitor having a negative temperature coefficient can be used for C_3 to stabilize the frequency of oscillation when there are temperature changes.

The Crystal Oscillator

A crystal oscillator is essentially a tuned-circuit oscillator that uses a quartz crystal as the resonant tank circuit. Other types of crystals can be used, but quartz is the most prevalent. Crystal oscillators offer greater frequency stability than other types.

Quartz is a substance found in nature that exhibits a property called the *piezoelectric effect.* When a changing mechanical stress is applied across the crystal to cause it to vibrate, a voltage is developed at the frequency of the mechanical vibration. Conversely, when an ac voltage is applied across the crystal, it vibrates at the frequency of the applied voltage.

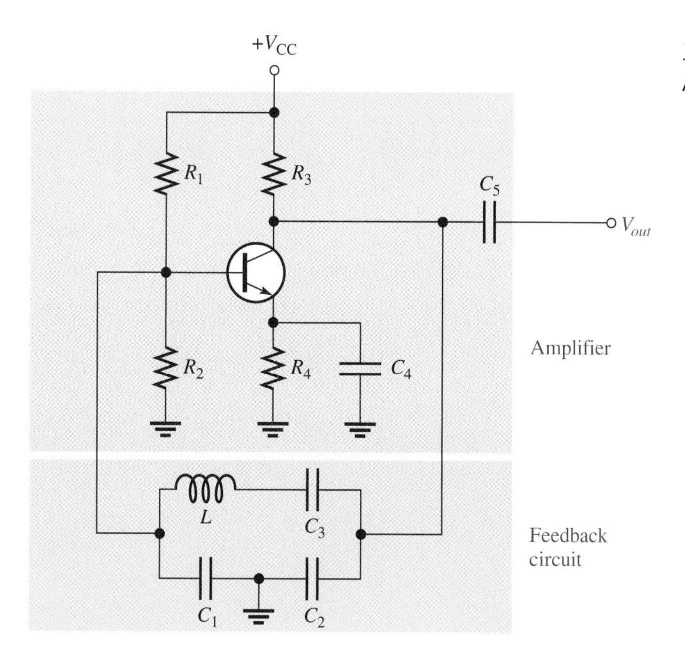

FIGURE 19–46

A basic Clapp oscillator.

The symbol for a crystal is shown in Figure 19–47(a), the electrical equivalent is shown in part (b), and a typical mounted crystal is shown in part (c). In construction, a slab of quartz is mounted as shown in Figure 19–47(d).

FIGURE 19–47

Quartz crystal.

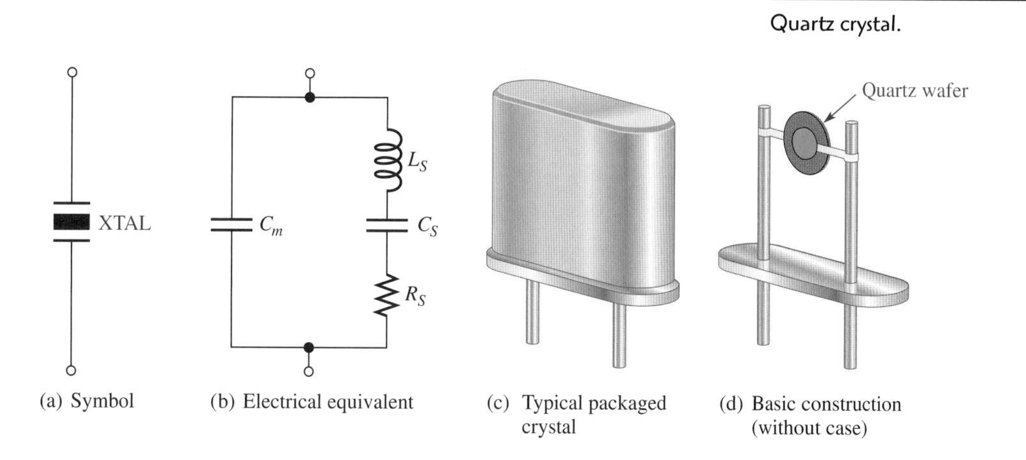

Series resonance occurs in the crystal when the reactances in the series branch are equal. Parallel resonance occurs, at a higher frequency, when the inductive reactance of L_S equals the reactance of the parallel capacitor C_m.

A crystal oscillator using the crystal as a series resonant tank circuit is shown in Figure 19–48(a). The impedance of the crystal is *minimum* at the series resonance, thus providing maximum feedback. The crystal tuning capacitor, C_C, is used to "fine-tune" the oscillator frequency by "pulling" the resonant frequency of the crystal slightly up or down.

A modified Colpitts configuration, shown in Figure 19–48(b), uses the crystal in its parallel resonant mode. The impedance of the crystal is maximum at parallel resonance, thus developing the maximum voltage across both C_1 and C_2. The voltage across C_2 is fed back to the input.

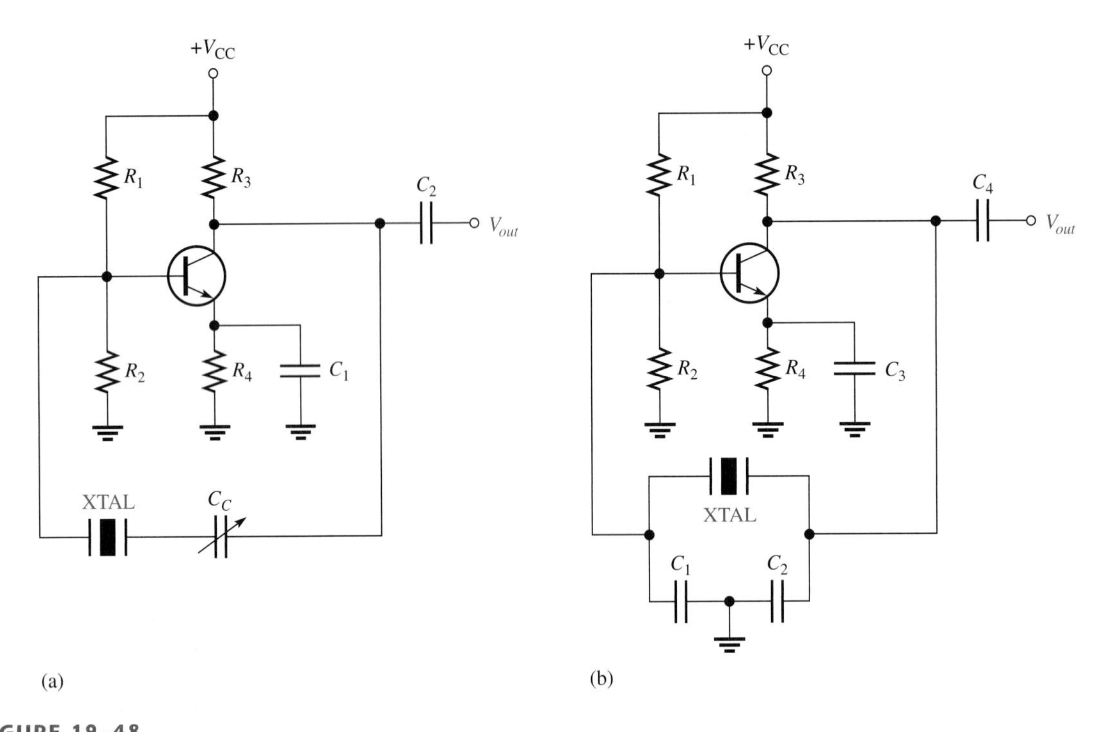

FIGURE 19–48

Basic crystal oscillators.

SECTION 19–9 REVIEW

1. What is an oscillator?
2. What type of feedback does an oscillator use and what is the purpose of the feedback circuit?
3. What are the conditions required for a circuit to oscillate?
4. What are the start-up conditions for an oscillator?
5. Name four types of oscillators.
6. Describe the basic difference between the Colpitts and Hartley oscillators.
7. What is the main advantage of a crystal oscillator?

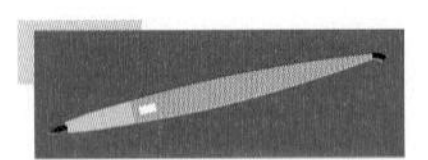

19–10 TROUBLESHOOTING

In working with any circuit, you must first know how it is supposed to work before you can troubleshoot it for a failure. The two-stage capacitively coupled amplifier discussed in Section 19–5 is used to illustrate a typical troubleshooting procedure.

After completing this section, you should be able to

- **Troubleshoot amplifier circuits**
- Describe the process of signal tracing

The correct signal levels and dc voltage levels (approximate) for the capacitively coupled two-stage amplifier are shown in Figure 19–49.

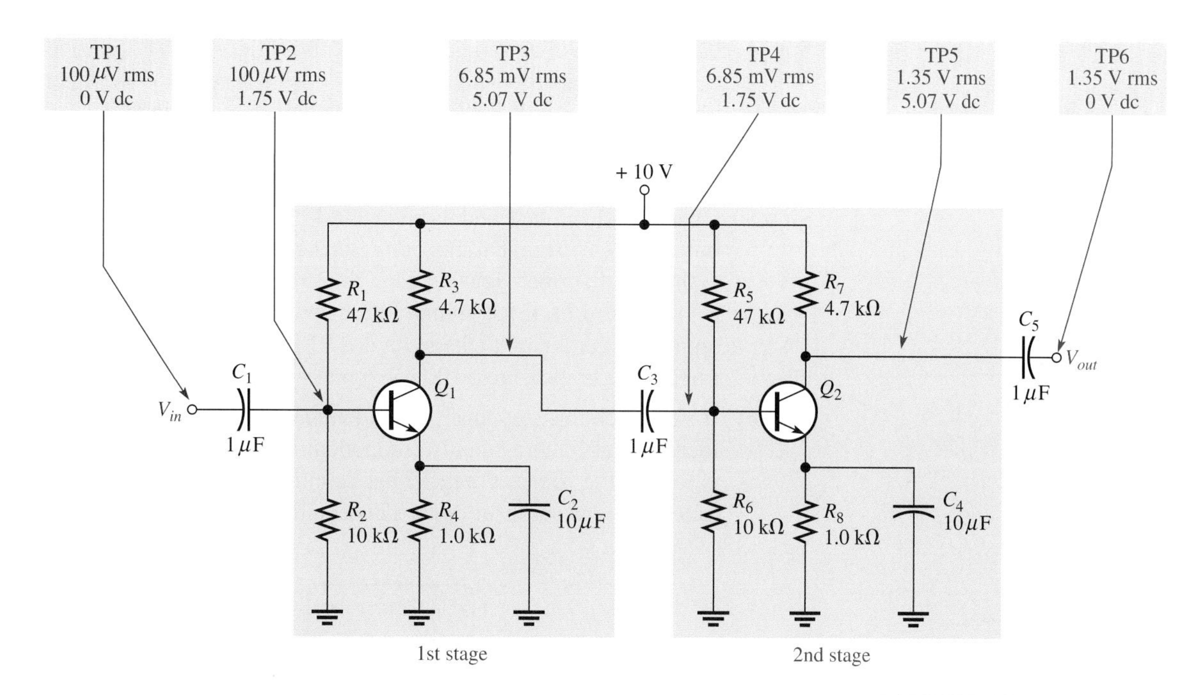

▲ **FIGURE 19–49**

Two-stage amplifier with correct ac and dc voltage levels indicated.

Troubleshooting Procedure

The APM (analysis, planning, and measurement) approach to troubleshooting will be used.

Analysis It has been found that there is no output voltage, V_{out}. You have also determined that the circuit did work properly and then failed. A visual check of the circuit board or assembly for obvious problems such as broken or poor connections, solder splashes, wire clippings, or burned components turns up nothing. You conclude that the problem is most likely a faulty component in the amplifier circuit or an open connection. Also, the dc supply voltage may not be correct or may be missing.

Planning You decide to use an oscilloscope to check the dc levels and the ac signals (some may prefer to use a DMM to measure the dc voltages). Also, you decide to apply the half-splitting method to trace the voltages in the circuit and use an in-circuit transistor tester if a transistor is suspected of being faulty.

Measurement The following steps indicate the measurements and the reasoning process in isolating the fault.

Step 1: Check the dc supply voltage. If it is 10 V, the problem is in the amplifier circuit. If the dc voltage is absent or has an incorrect value, check the dc source and related connections for the problem.

Step 2: Check the dc voltage and the signal at the input to the second stage (TP4). If the correct dc voltage and signal are present at this point, the fault is in the second stage or the coupling capacitor C_5 is open. Go to Step 3.

If the dc voltage is incorrect or if there is no signal at TP4, the fault is in the first stage, C_1, or C_3, so go to Step 4.

Step 3: Check the dc voltage and the signal at the collector of Q_2 (TP5). If the dc voltage is correct and the signal is present at this point, the coupling capacitor C_5 is open. If the dc level is incorrect or there is no signal at TP5, the second stage is defective. Either the transistor Q_2, one of the resistors, or the bypass capacitor C_4 is faulty. Go to Step 6.

Step 4: Check the dc voltage and the signal at the collector of Q_1 (TP3). If the dc voltage is correct and the signal is present at this point, the coupling capacitor C_3 is faulty. If the dc voltage is incorrect or there is no signal at TP3, the first stage is defective or coupling capacitor C_1 is faulty. Go to Step 5.

Step 5: Check the dc voltage and the signal at TP2. If the dc voltage is correct but there is no signal at this point, the coupling capacitor C_1 is open. If there is a correct signal at TP2, the first stage is faulty. Go to Step 6. *Note:* To check the signal TP2, it will probably be necessary to apply a larger signal to the input at TP1 because it is very difficult or impossible to measure a 100 μV signal with typical test equipment.

Step 6: If you reach this step, one of the identical amplifier stages have been identified as having a faulty component or connection. The next step is to find the specific fault. If the dc collector voltage is at V_{CC}, the transistor is in cutoff or open. This could be caused by a bad transis-

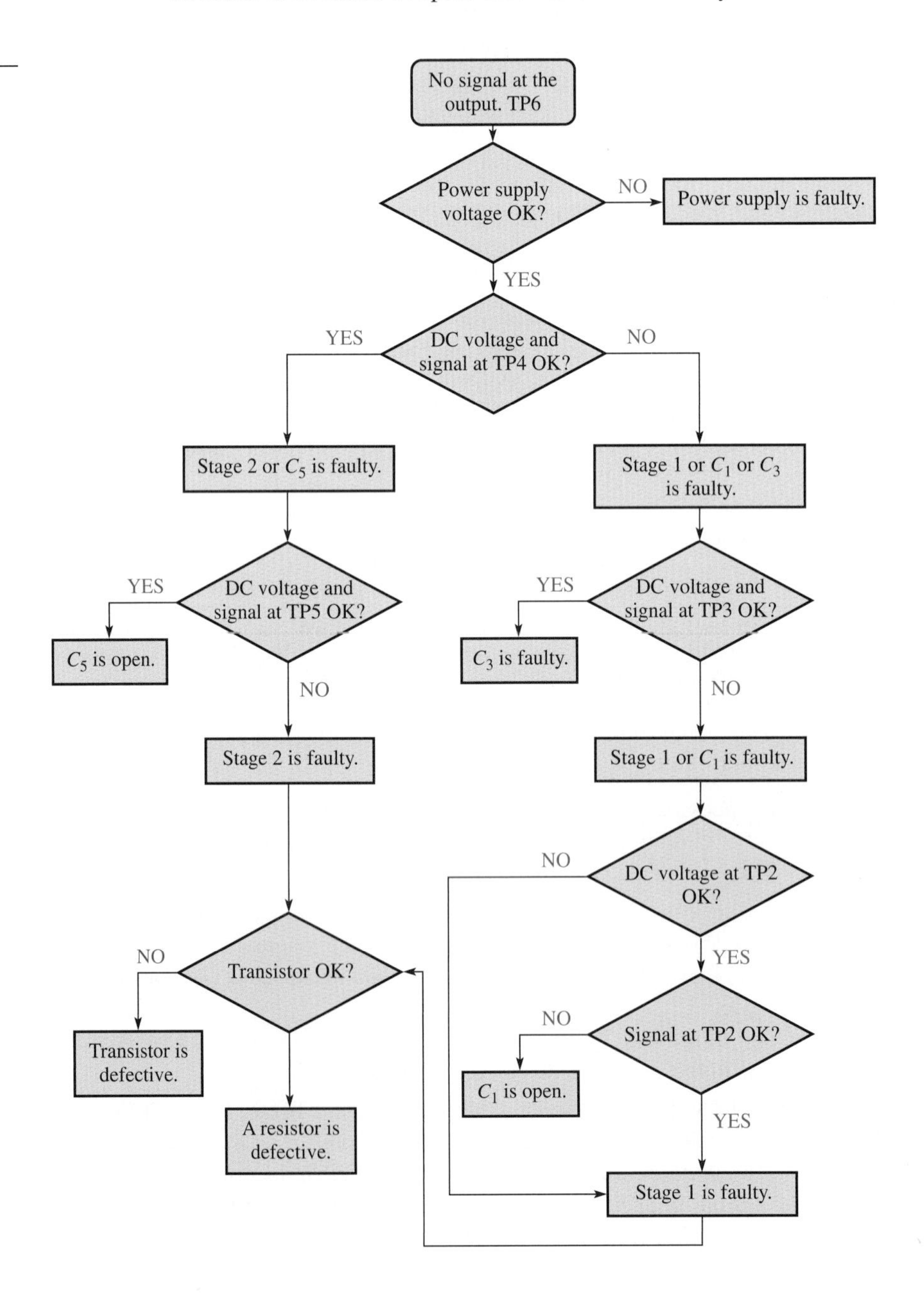

FIGURE 19–50

Troubleshooting flowchart.

tor, an open bias resistor circuit (either R_1 or R_5, depending on the stage), an open emitter resistor, or open connection in the collector. If the dc collector voltage is approximately equal to the dc emitter voltage, the transistor is in saturation or shorted. This could be caused by a bad transistor, an open bias resistor (either R_2 or R_6, depending on the stage), an open collector resistor (either R_3 or R_7, depending on the stage), or a shorted contact. An in-circuit check of the transistor should be done at this point. If the transistor checks okay, begin checking for open resistors, or bad contacts.

A troubleshooting flowchart illustrates this process in Figure 19–50. The chart is based on the assumption that there is a correct input signal.

Fault Analysis

As an additional example of isolating a component failure in a circuit, let's use a class A amplifier with the output monitored by an oscilloscope, as shown in Figure 19–51. As shown, the amplifier has a normal sinusoidal output when a sinusoidal input signal is applied.

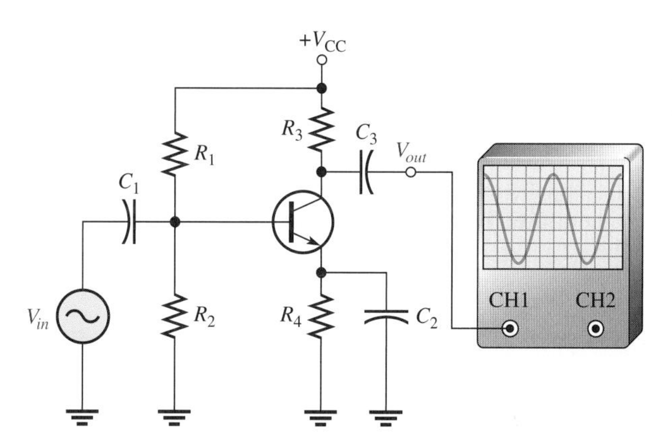

FIGURE 19–51

Class A amplifier with correct output display.

Now, several incorrect output waveforms will be considered and the most likely causes discussed. In Figure 19–52(a), the scope displays a dc level equal to the dc supply voltage, indicating that the transistor is in cutoff. The two possible causes of this condition are (1) the transistor is open from collector to emitter, or (2) R_4 is open, preventing collector and emitter current.

FIGURE 19–52

Oscilloscope displays of output voltage for the amplifier in Figure 19–51, illustrating several types of failures.

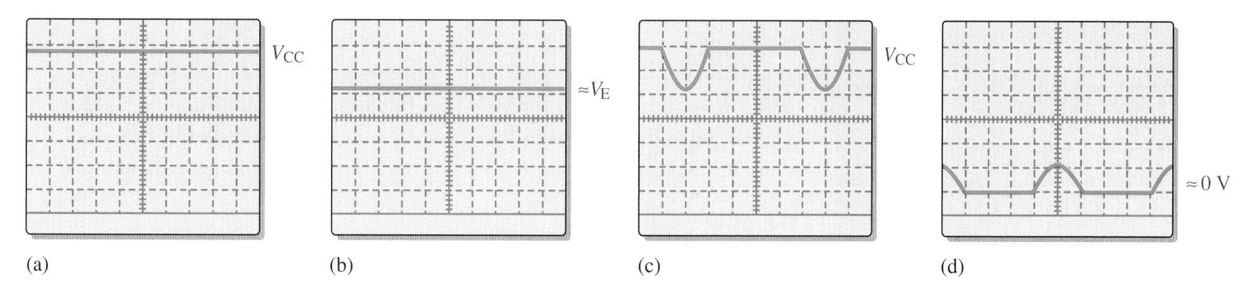

In Figure 19–52(b), the scope displays a dc level at the collector approximately equal to the emitter voltage. The two possible causes of this indication are (1) the transistor is shorted from collector to emitter, or (2) R_2 is open, causing the transistor to be biased in saturation. In the second case, a sufficiently large input signal can bring the transistor out of saturation on its negative peaks, resulting in short pulses on the output.

In Figure 19–52(c), the scope displays an output waveform that is clipped at cutoff. Possible causes of this indication are (1) the Q-point has shifted down due to a wrong resistor value, or (2) R_1 is open, biasing the transistor in cutoff. In the second case, the input signal is sufficient to bring it out of cutoff for a small portion of the cycle.

In Figure 19–52(d), the scope displays an output waveform that is clipped at saturation. Again, it is possible that a wrong resistance value has caused a drastic shift in the Q-point up toward saturation, or R_2 is open, causing the transistor to be biased in saturation, and the input signal is bringing it out of saturation for a small portion of the cycle.

SECTION 19–10 REVIEW

1. If C_4 in Figure 19–49 were open, how would the output signal be affected? How would the dc level at the collector of Q_2 be affected?
2. If R_5 in Figure 19–49 were open, how would the output signal be affected?
3. If the coupling capacitor C_3 in Figure 19–49 shorted out, would any of the dc voltages in the amplifier be changed? If so, which ones?
4. Assume that the base-emitter junction of Q_2 in Figure 19–49 shorts.
   (a) Will the ac signal at the base of Q_2 change? In what way?
   (b) Will the dc level at the base of Q_2 change? In what way?
5. What would you check for if you noticed clipping at both peaks of the output waveform in Figure 19–49?
6. A significant loss of gain in the amplifier of Figure 19–51 would most likely be caused by what type of failure?

APPLICATION ASSIGNMENT

PUTTING YOUR KNOWLEDGE TO WORK

The audio preamplifier in the receiver system that you have been assigned to work on accepts a very small audio signal out of the detector circuit and amplifies it to a level for input to the power amplifier that provides audio power to the speaker. In this application, the transistors are used as common-emitter small-signal linear amplifiers.

Description of the System

The block diagram for the AM radio receiver is shown in Figure 19–53. A very basic system description follows. The antenna picks up all radiated signals that pass by and feeds them into the RF (radio frequency) amplifier. The voltages induced in the antenna by the electromagnetic radiation are extremely small.

The RF amplifier is tuned to select and amplify a desired frequency within the AM broadcast band (535 kHz to 1605 kHz). Since it is a frequency-selective circuit, the RF amplifier eliminates essentially all but the selected frequency band. The output of the RF amplifier goes to the mixer where it is combined with the output of the local oscillator which is 455 kHz above the selected RF frequency.

In the mixer, a nonlinear process called *heterodyning* takes place and produces one frequency that is the sum of the RF and local oscillator frequencies and another frequency that is the difference (always 455 kHz). The sum frequency is filtered out and only the 455 kHz difference frequency is used. This frequency is amplitude modulated just like the much higher RF frequency and, therefore, contains the audio signal.

The IF (intermediate frequency) amplifier is tuned to 455 kHz and amplifies the mixer output. The detector takes the amplified 455 kHz AM signal and recovers the audio from it while eliminating the intermediate frequency. The output of the detector is a small audio signal that goes to the

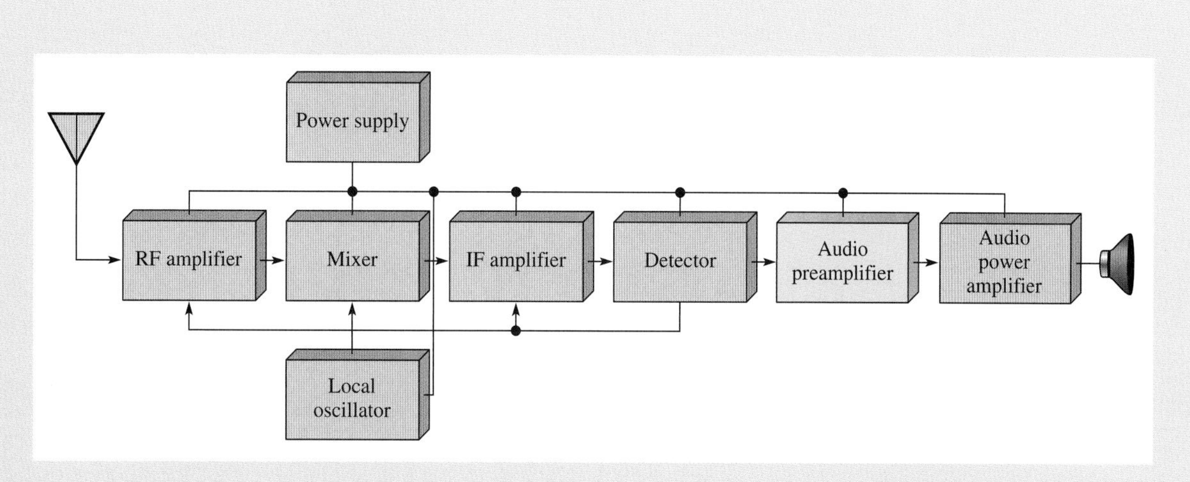

FIGURE 19–53

Block diagram of AM radio receiver.

audio preamplifier and then to the power amplifier, which drives the speaker to convert the electrical audio signal into sound.

Step 1: Relate the PC Board to the Schematic

Notice that there are no component labels on the PC board shown in Figure 19–54. Label the components using the schematic in Figure 19–55 as a guide. That is, find and label R_1, R_2, and so forth on the PC board and make the board and schematic agree. Figure 19–54 may be copied for this purpose.

Step 2: Analyze the Amplifier

Assume $\beta_{ac(min)} = 175$.

1. Calculate all of the dc voltages in the audio preamp.
2. Determine the voltage gain of each stage and the overall voltage gain of the two-stage amplifier. Don't forget to take into account the loading effect of the input resistance of the second stage. Assume that the second stage is unloaded and all capacitive reactances are zero.
3. Specify what you would expect to happen as the frequency is reduced to a very low value. Why?

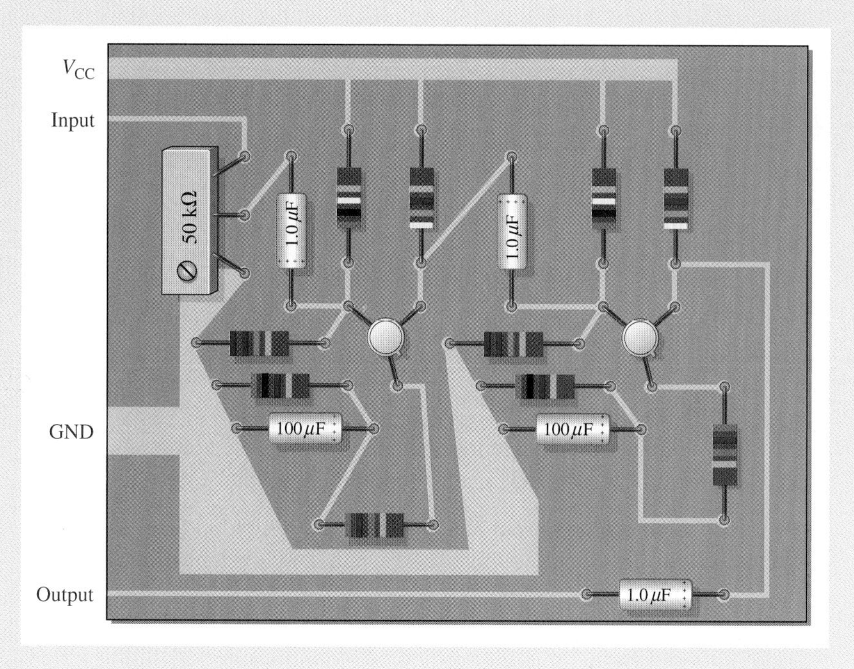

FIGURE 19–54

Audio preamplifier circuit board.

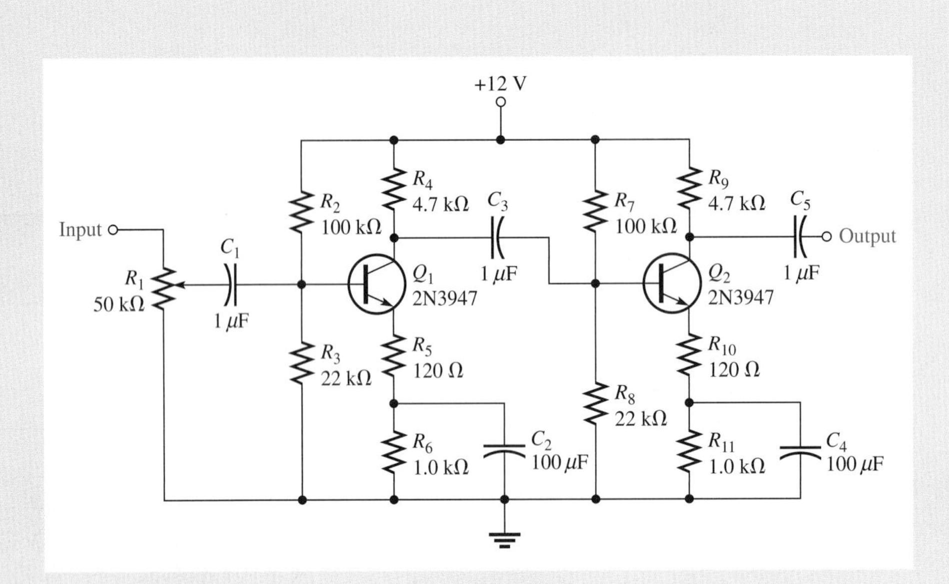

FIGURE 19–55

Audio preamplifier schematic.

4. Compute the voltage gain for the case of a 15 kΩ load resistor connected to the output of the second stage.

Step 3: Troubleshoot the Circuit

For each of the following problems state the probable cause or causes in each case.

1. Proper ac and dc voltages at base of Q_2, but signal voltage at collector is less than it should be.
2. Proper signal at collector of Q_1, but no signal at base of Q_2.
3. Amplitude of output signal much less than it should be.
4. The collector of Q_2 at +12 V dc and no signal voltage at the collector of Q_2 with a verified signal at the base.

APPLICATION ASSIGNMENT REVIEW

1. Explain why R_1 is in the circuit of Figure 19–55.
2. What happens if C_2 opens?
3. What happens if C_3 opens?
4. How can you reduce the gain of each stage without changing the dc voltages?

SUMMARY

- The three bipolar transistor amplifier configurations are common-emitter (CE), common-collector (CC), and common-base (CB).
- The general characteristics of a common-emitter amplifier are high voltage gain, high current gain, very high power gain, and low input resistance.
- The general characteristics of a common-collector amplifier are a voltage gain of approximately 1, high current gain, power gain, and high input resistance.

- The general characteristics of a common-base amplifier are high voltage gain, a current gain of approximately 1, high power gain, and very low input resistance.
- The three FET amplifier configurations are common-source (CS), common-drain (CD), and common-gate (CG).
- The overall voltage gain of a multistage amplifier is the product of the loaded gains of all the individual stages.
- Any of the bipolar or FET configurations can be operated as class A, class B, or class C amplifiers.
- The class A amplifier conducts for the entire 360° of the input cycle.
- The class B amplifier conducts for 180° of the input cycle.
- The class C amplifier conducts for a small portion of the input cycle.
- Sinusoidal oscillators operate with positive feedback.
- The two conditions for positive feedback are the phase shift around the feedback loop must be 0° and the voltage gain around the feedback loop must be at least 1.
- For initial start-up, the voltage gain around the feedback loop must be greater than 1.
- The feedback signal in a Colpitts oscillator is derived from a capacitive voltage divider in the *LC* circuit.
- The Clapp oscillator is a variation of the Colpitts with a capacitor added in series with the inductor.
- The feedback signal in a Hartley oscillator is derived from an inductive voltage divider in the *LC* circuit.
- Crystal oscillators are the most stable type.

EQUATIONS

19–1	$A_v = \dfrac{R_C}{r_e + R_E}$	CE voltage gain (unbypassed)
19–2	$A_v = \dfrac{R_C}{r_e}$	CE voltage gain (bypassed)
19–3	$r_e \cong \dfrac{25\text{ mV}}{I_E}$	Internal emitter resistance
19–4	$R_{in} \cong \beta_{ac} r_e$	CE input resistance
19–5	$R_{in(tot)} = R_1 \parallel R_2 \parallel R_{in}$	CE total input resistance
19–6	$A_i = \dfrac{I_c}{I_s}$	CE current gain
19–7	$A_p = A_v A_i$	CE power gain
19–8	$A_v = \dfrac{R_E}{r + R_E}$	CC voltage gain
19–9	$R_{in} \cong \beta_{ac} R_E$	CC input resistance
19–10	$R_{in(tot)} = R_1 \parallel R_2 \parallel R_{in}$	CC total input resistance

19–11	$A_i = \dfrac{I_e}{I_s}$	CC current gain
19–12	$A_p \cong A_i$	CC power gain
19–13	$\beta = \beta_1\beta_2$	Beta for a darlington pair
19–14	$R_{in} = \beta_1\beta_2 R_E$	Darlington input resistance
19–15	$A_v = \dfrac{R_C}{r_e}$	CB voltage gain
19–16	$R_{in} = r_e$	CB input resistance
19–17	$A_i \cong 1$	CB current gain
19–18	$A_p \cong A_v$	CB power gain
19–19	$g_m = \dfrac{I_d}{V_{gs}}$	FET transconductance
19–20	$A_v = g_m R_D$	CS voltage gain
19–21	$A_v = \dfrac{g_m R_S}{1 + g_m R_S}$	CD voltage gain
19–22	$A_v = g_m R_D$	CG voltage gain
19–23	$R_{in(source)} = \dfrac{1}{g_m}$	CG input resistance
19–24	$A_{v(tot)} = A_{v1}A_{v2}A_{v3} \cdots A_{vn}$	Multistage gain
19–25	$A_v\,(dB) = 20 \log A_v$	Voltage gain in dB
19–26	$A_{v(tot)}\,(dB) = A_{v1}\,(dB) + A_{v2}\,(dB) + \cdots + A_{vn}\,(dB)$	Multistage dB gain
19–27	$A_p \cong \beta_{DC} A_v$	CE large-signal power gain
19–28	$P_{DQ} = I_{CQ} V_{CEQ}$	Transistor power dissipation
19–29	$P_{out} = V_{ce} I_c$	CE output power
19–30	$P_{out\,(max)} = 0.5 V_{CEQ} I_{CQ}$	CE output power with centered Q-point (maximum)
19–31	$\text{eff}_{max} = 0.25$	Class A ideal maximum efficiency
19–32	$I_{c(sat)} = \dfrac{V_{CEQ}}{R_L}$	Class B ac saturation current
19–33	$P_{out(max)} = 0.25 V_{CC} I_{c(sat)}$	Class B output power (maximum)
19–34	$P_{DC} = \dfrac{V_{CC} I_{c(sat)}}{\pi}$	Class B input power
19–35	$\text{eff}_{max} = 0.785$	Class B ideal maximum efficiency
19–36	$P_{out\,(max)} = \dfrac{0.5 V_{CC}^2}{R_c}$	Class C output power (maximum)
19–37	$\text{eff} = \dfrac{P_{out}}{P_{out} + P_{D(avg)}}$	Class C efficiency
19–38	$A_{cl} = A_v B$	Closed loop gain
19–39	$f_r \cong \dfrac{1}{2\pi\sqrt{LC_T}}$	Colpitts frequency of oscillation
19–40	$C_T = \dfrac{C_1 C_2}{C_1 + C_2}$	Colpitts total feedback capacitance
19–41	$f_r \cong \dfrac{1}{2\pi\sqrt{L_T C_5}}$	Hartley frequency of oscillation

SELF-TEST

Answers are at the end of the chapter.

1. In a common-emitter (CE) amplifier, the capacitor from emitter to ground is called the
   (a) coupling capacitor (b) decoupling capacitor
   (c) bypass capacitor (d) tuning capacitor
2. If the capacitor from emitter to ground in a CE amplifier is removed, the voltage gain
   (a) increases (b) decreases
   (c) is not affected (d) becomes erratic
3. When the collector resistor in a CE amplifier is increased in value, the voltage gain
   (a) increases (b) decreases
   (c) is not affected (d) becomes erratic
4. The input resistance of a CE amplifier is affected by
   (a) α and r_e (b) β and r_e (c) R_c and r_e (d) R_e, r_e, and β
5. The output signal of a CE amplifier is always
   (a) in phase with the input signal
   (b) out of phase with the input signal
   (c) larger than the input signal
   (d) equal to the input signal
6. The output signal of a common-collector amplifier is always
   (a) in phase with the input signal (b) out of phase with the input signal
   (c) larger than the input signal (d) exactly equal to the input signal
7. The largest *theoretical* voltage gain obtainable with a CC amplifier is
   (a) 100 (b) 10 (c) 1 (d) dependent of β
8. Using a darlington pair has the advantage of
   (a) increasing the overall voltage gain (b) less cost
   (c) decreasing the input resistance (d) increasing the overall β
9. Compared to CE and CC amplifiers, the common-base (CB) amplifier has
   (a) a lower input resistance (b) a much larger voltage gain
   (c) a larger current gain (d) a higher input resistance
10. In terms of higher power gain, the amplifier of choice is
    (a) CB (b) CC (c) CE (d) any of these
11. The most efficient amplifier configuration is
    (a) class A (b) class B (c) class C
12. When a FET with a lower transconductance is submitted into a FET amplifier circuit,
    (a) the voltage gain increases (b) the voltage gain decreases
    (c) the input resistance decreases (d) nothing changes
13. When amplifiers are cascaded,
    (a) the gain of each amplifier is increased
    (b) each amplifier has to work less
    (c) a lower supply voltage is required
    (d) the overall gain is increased

14. If three amplifiers, each with a voltage gain of 30, are connected in a multistage arrangement, the overall voltage gain is
    (a) 27,000 (b) 90 (c) 10 (d) 30
15. In a class A amplifier, the output signal is
    (a) distorted (b) clipped
    (c) the same shape as the input (d) smaller in amplitude than the input
16. Oscillators operate on the principle of
    (a) signal feedthrough (b) positive feedback
    (c) negative feedback (d) attenuation

PROBLEMS

Answers to odd-numbered problems are at the end of the book.

BASIC PROBLEMS

SECTION 19–1 Common-Emitter Amplifiers

1. Determine the voltage gain for Figure 19–56.
2. Determine each of the dc voltages, V_B, V_C, and V_E, with respect to ground in Figure 19–56.
3. Determine the following dc values for the amplifier in Figure 19–57:
   (a) V_B (b) V_E (c) I_E (d) I_C (e) V_C (f) V_{CE}

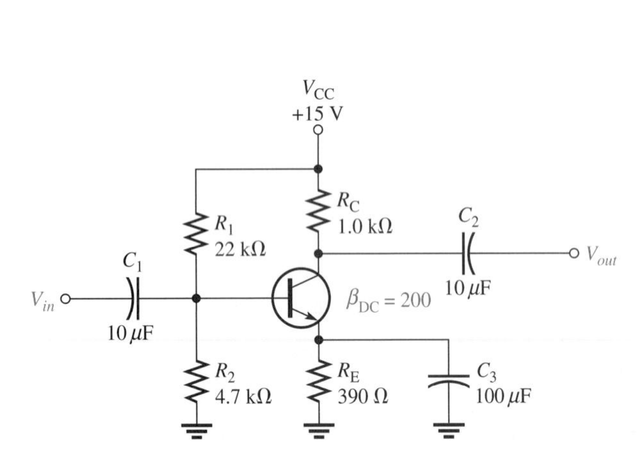

▲ FIGURE 19–56

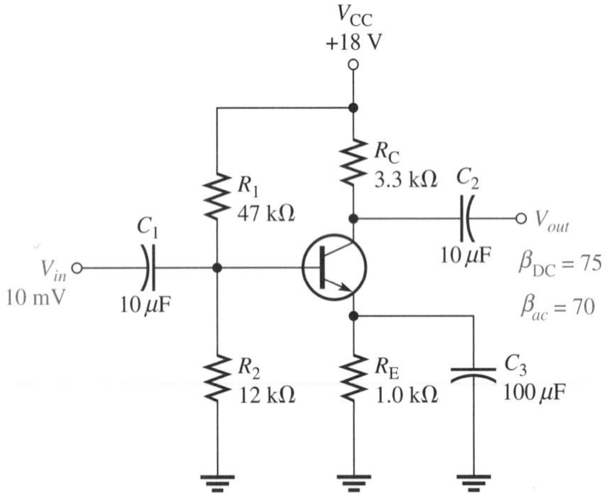

▲ FIGURE 19–57

4. Determine the following ac values for the amplifier in Figure 19–57:
   (a) R_{in} (b) $R_{in(tot)}$ (c) A_v (d) A_i (e) A_p
5. The amplifier in Figure 19–58 has a variable gain control, using a 100 Ω potentiometer for R_E with the wiper ac grounded. As the potentiometer is adjusted, more or less of R_E is bypassed to ground, thus varying the gain. The total R_E remains constant to dc, keeping the bias fixed. Determine the maximum and minimum gains for this amplifier.
6. If a load resistance of 600 Ω is placed on the output of the amplifier in Figure 19–58, what is the maximum gain?

SECTION 19–2 Common-Collector Amplifiers

7. Determine the *exact* voltage gain for the emitter-follower in Figure 19–59.
8. What is the total input resistance in Figure 19–59? What is the dc output voltage when $V_{in} = 10$ mV?

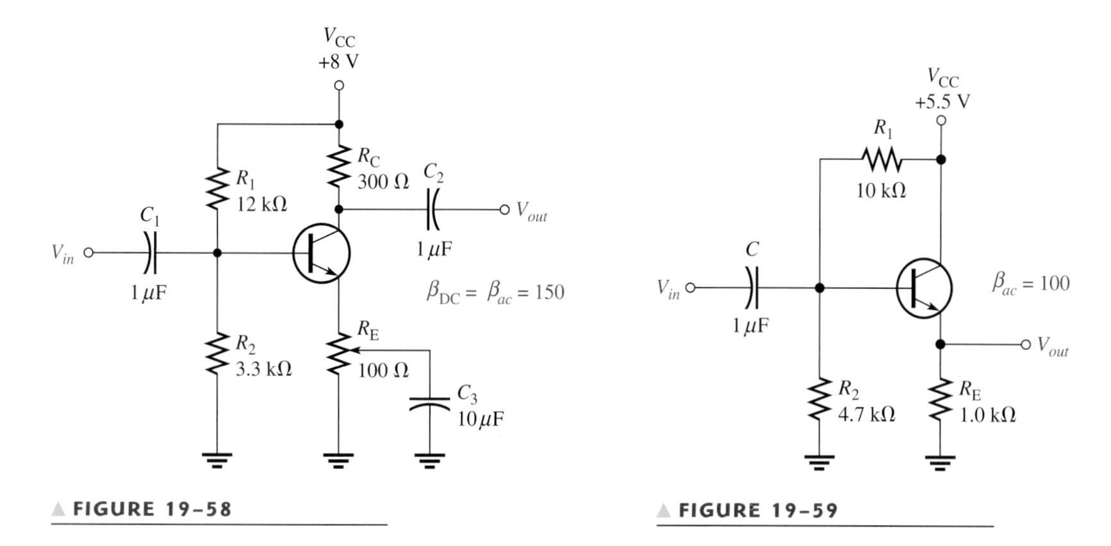

▲ **FIGURE 19–58**

▲ **FIGURE 19–59**

9. A load resistance is capacitively coupled to the emitter in Figure 19–59. In terms of signal operation, the load appears in parallel with R_E and reduces the effective emitter resistance. How does this affect the voltage gain?

SECTION 19–3 Common-Base Amplifiers

10. What is the main disadvantage of the CB amplifier compared to the CE and the emitter-follower?

11. Find R_{in}, A_v, A_i, and A_p for the amplifier in Figure 19–60.

▶ **FIGURE 19–60**

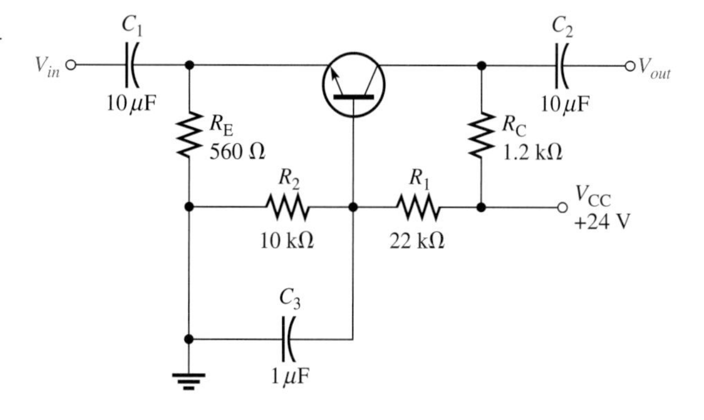

SECTION 19–4 FET Amplifiers

12. Determine the voltage gain of each CS amplifier in Figure 19–61.

▼ **FIGURE 19–61**

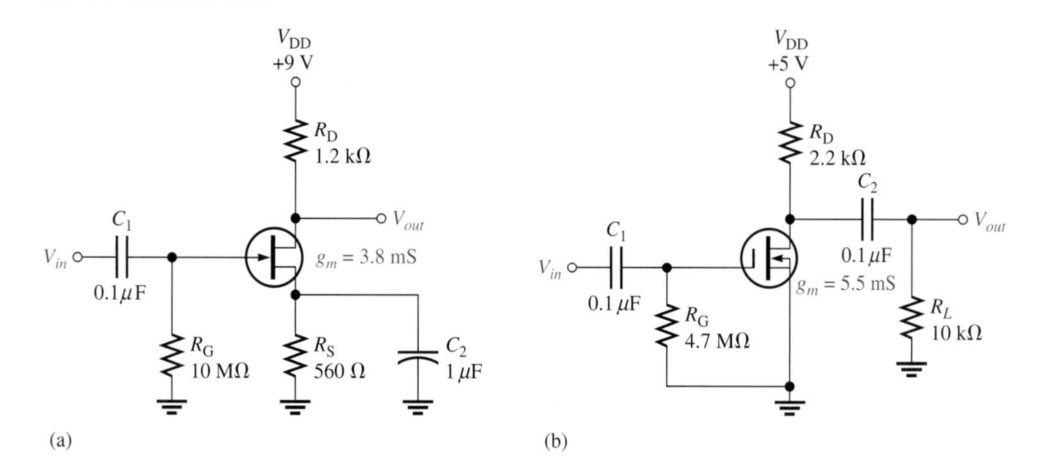

13. Find the gain of each amplifier in Figure 19–62.

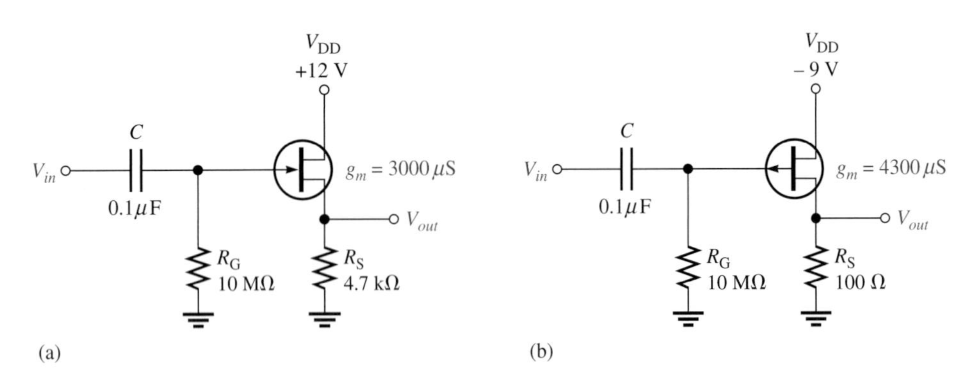

FIGURE 19–62

14. Determine the gain of each amplifier in Figure 19–62 when a 10 kΩ load is capacitively coupled from source to ground.

SECTION 19–5 **Multistage Amplifiers**

15. Each of three cascaded amplifier stages has a decibel voltage gain of 10. What is the overall decibel voltage gain? What is the actual overall voltage gain?

16. For the two-stage, capacitively coupled amplifier in Figure 19–63, find the following values:

(a) Voltage gain of each stage

(b) Overall voltage gain

(c) Express the gains found above in decibels

FIGURE 19–63

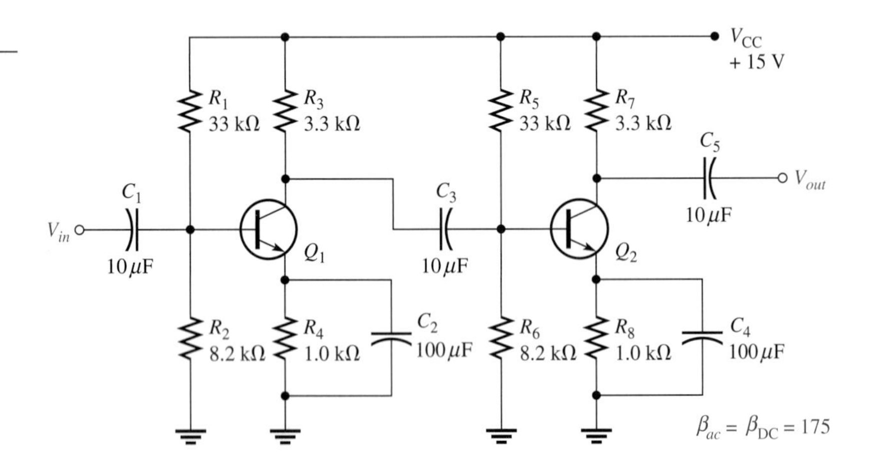

SECTION 19–6 **Class A Operation**

17. Determine the minimum power rating for each of the transistors in Figure 19–64.

SECTION 19–7 **Class B Push-Pull Operation**

18. Determine the dc voltages at the bases and emitters of Q_1 and Q_2 in Figure 19–65. Also determine V_{CEQ} for each transistor.

FIGURE 19–64

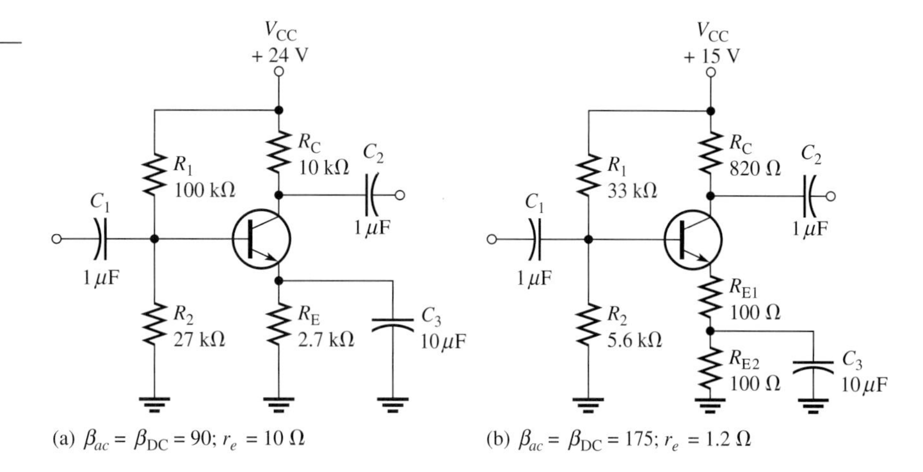

FIGURE 19–65

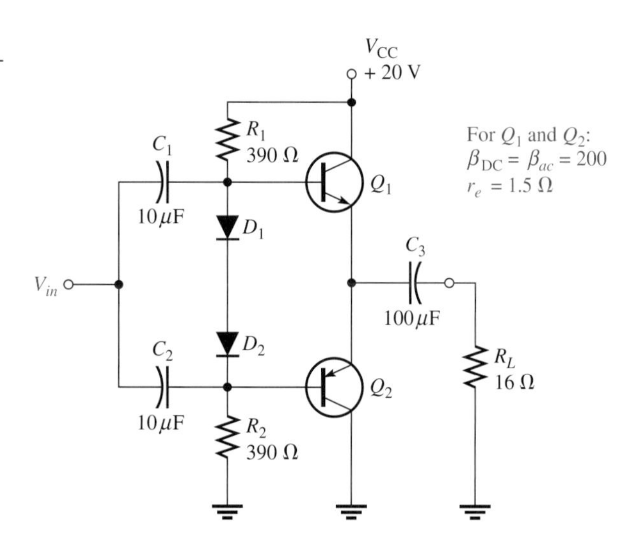

19. Determine the maximum peak output voltage and peak load current for the circuit in Figure 19–65.

20. The efficiency of a certain class B push-pull amplifier is 0.71, and the dc input power is 16.3 W. What is the ac output power?

SECTION 19–8

Class C Operation

21. What is the resonant frequency of the tank circuit in a class C amplifier with $L = 10$ mH and $C = 0.001\ \mu$F?

22. Determine the efficiency of the class C amplifier when $P_{D(avg)} = 10$ mW, $V_{CC} = 15$ V, and the equivalent parallel resistance in the collector tank circuit is 50 Ω.

SECTION 19–9

Oscillators

23. If the voltage gain of the amplifier portion of an oscillator is 75, what must be the attenuation of the feedback circuit to sustain the oscillation?

24. Generally describe the change required to the oscillator of Problem 23 in order for oscillation to begin when the power is initially turned on.

25. Calculate the frequency of oscillation for each circuit in Figure 19–66, and identify each type of oscillator.

FIGURE 19–66

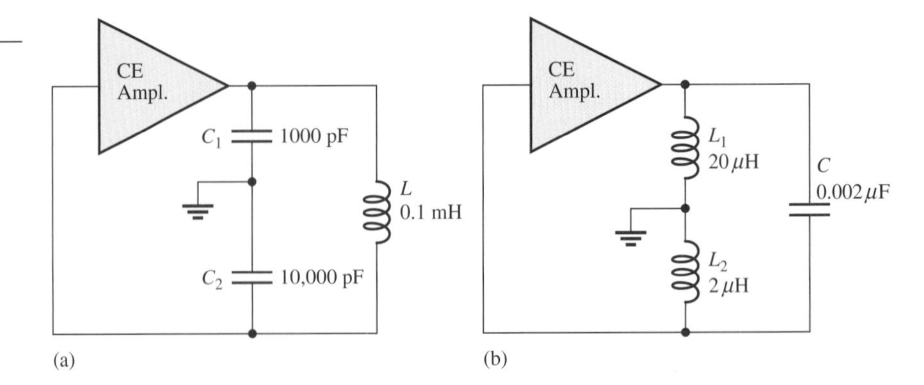

SECTION 19–10

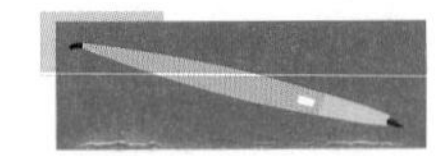

Troubleshooting

26. Assume that the coupling capacitor C_3 is shorted in Figure 19–20. What dc voltage will appear at the collector of Q_1? $\beta_{DC} = 125$.
27. Assume that R_5 opens in Figure 19–20. Will Q_2 be in cutoff or in conduction? What dc voltage will you observe at the Q_2 collector?
28. Refer to Figure 19–63 and determine the general effect of each of the following failures:
    (a) C_2 opens **(b)** C_3 opens **(c)** C_4 opens **(d)** C_2 shorts
    (e) Base-collector junction of Q_1 opens **(f)** Base-emitter of Q_2 opens
29. Assume that you must troubleshoot the amplifier in Figure 19–63. Set up a table of test point values, input, output, and all transistor terminals that include both dc and rms values that you expect to observe when a 300 Ω test signal source with a 25 μV rms output is used.
30. What symptom(s) would indicate each of the following failures under signal conditions in Figure 19–67?
    (a) Q_1 open from drain to source **(b)** R_3 open **(c)** C_2 shorted
    (d) C_3 shorted **(e)** Q_2 open from drain to source

FIGURE 19–67

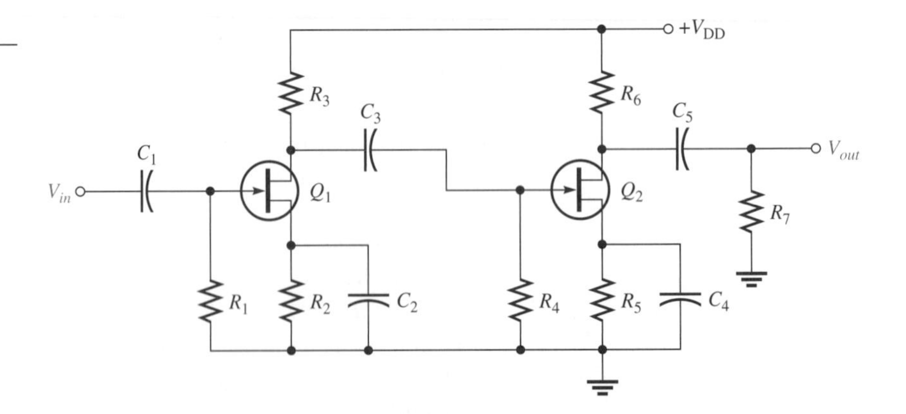

ELECTRONICS WORKBENCH/CIRCUITMAKER TROUBLESHOOTING PROBLEMS

CD-ROM file circuits are shown in Figure 19–68.

31. Open file P19-31. Determine whether or not the circuit is operating properly. If not, identify the fault.

32. Open file P19-32. Determine whether or not the circuit is operating properly. If not, identify the fault.
33. Open file P19-33. Determine whether or not the circuit is operating properly. If not, identify the fault.
34. Open file P19-34. Determine whether or not the circuit is operating properly. If not, identify the fault.
35. Open file P19-35. Determine whether or not the circuit is operating properly. If not, identify the fault.

▼ FIGURE 19–68

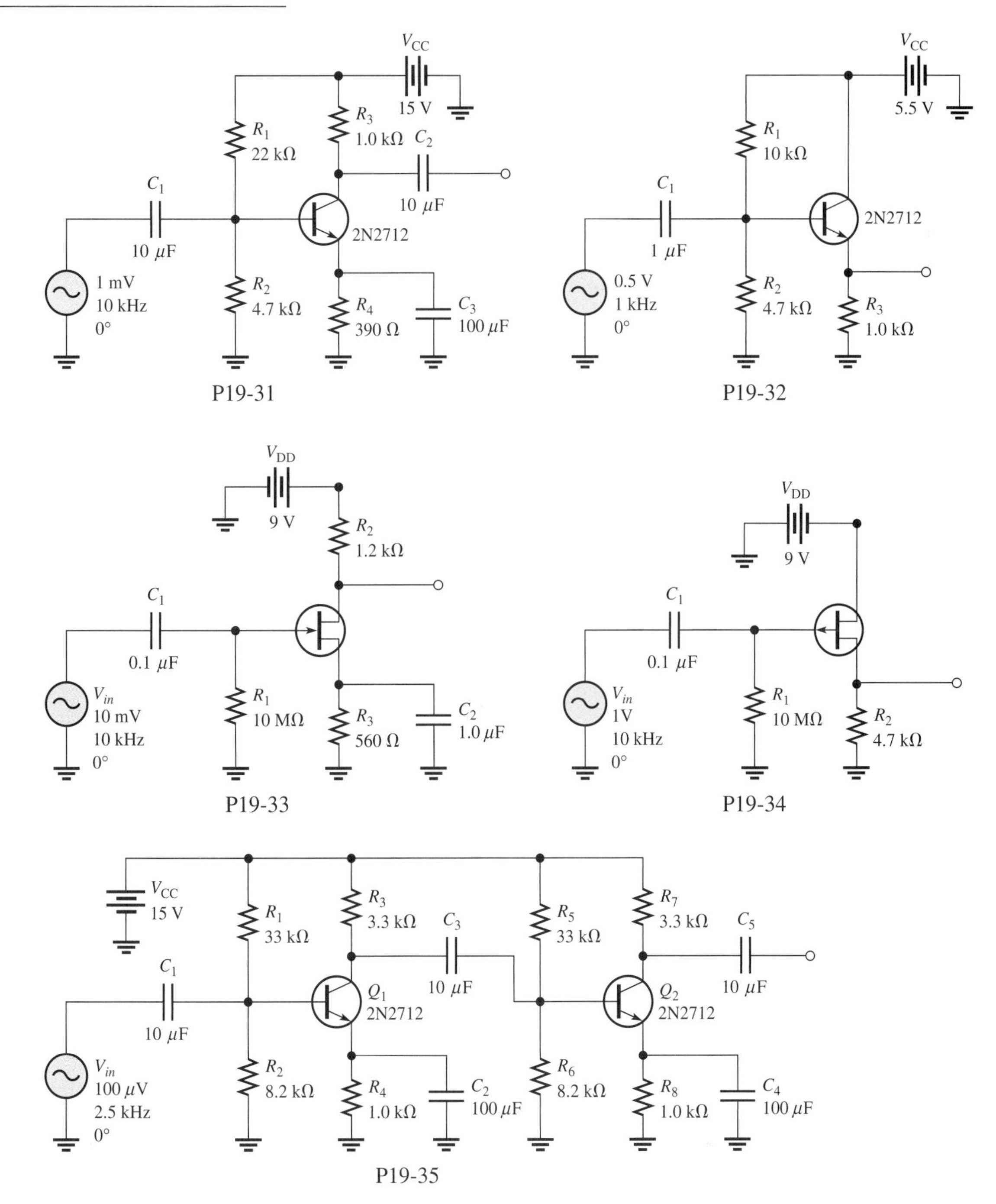

ANSWERS

SECTION REVIEWS

SECTION 19–1

Common-Emitter Amplifiers

1. The bypass capacitor increases voltage gain.
2. The voltage gain of a CE amplifier is the ratio of collector resistance to total emitter resistance.
3. $A_p = A_v A_i = 10{,}000$

SECTION 19–2

Common-Collector Amplifiers

1. A common-collector amplifier is an emitter-follower.
2. $A_{v(max)} = 1$
3. Yes

SECTION 19–3

Common-Base Amplifiers

1. Yes
2. The input resistance is very low.

SECTION 19–4

FET Amplifiers

1. The A_v of a CS FET amplifier is determined by g_m and R_D.
2. A_v is halved because R_D is halved.
3. Common-gate has low input resistance; CS and CD do not.

SECTION 19–5

Multistage Amplifiers

1. A stage is one amplifier in a cascaded arrangement.
2. The overall gain is the product of individual loaded voltage gains.
3. $20 \log(500) = 54$ dB

SECTION 19–6

Class A Operation

1. The Q-point is centered on the load line.
2. $\text{Eff}_{(max)} = 25\%$
3. $P_{out(max)} = 0.5V_{CEQ}I_{CEQ} = 35$ mW

SECTION 19–7

Class B Push-Pull Operation

1. The Q-point is at cutoff.
2. The barrier potential of the base-emitter junction causes crossover distortion.
3. $\text{Eff}_{max} = 78.5\%$
4. The push-pull configuration reproduces both positive and negative alternations of the input signal with greater efficiency.

SECTION 19–8

Class C Operation

1. A class C amplifier is biased in cutoff.
2. The tuned circuit produces a sinusoidal output.
3. $P_{out}/(P_{out} + P_{D(avg)}) = 90.9\%$

SECTION 19–9

Oscillators

1. An oscillator is a circuit that produces a repetitive output waveform with no input signal.

2. Positive feedback; it provides attenuation and phase shift.
3. Conditions for oscillation are zero phase shift and unity voltage gain around the closed loop.
4. Start-up conditions are loop gain at least 1 and zero phase shift.
5. Four types of oscillators are *RC* phase-shift, Colpitts, Hartley, and crystal.
6. Colpitts uses two capacitors, center-tapped to ground in parallel with an inductor. Hartley uses two coils, center-tapped to ground in parallel with a capacitor.
7. A crystal oscillator has a greater frequency stability.

SECTION 19–10 **Troubleshooting**

1. If C_4 opens, the gain drops. The dc level would not be affected.
2. Q_2 would be biased in cutoff.
3. The collector voltage of Q_1 and the base voltage of Q_2 would change. A change in V_{B2} will also cause V_{E2}, I_{E2}, and V_{C2} to change.
4. **(a)** The ac signal will disappear because it is shorted to ground through the base-emitter junction and C_4.
   (b) Yes, the dc level will decrease.
5. Check for excess input signal voltage.
6. Open bypass capacitor C_2 will cause a loss of gain.

Application Assignment

1. R_1 adjusts the volume by increasing or decreasing the input voltage.
2. The gain of the first stage decreases if C_2 opens.
3. There will be no output if C_3 opens because the signal is not coupled through to the second stage.
4. The gain can be reduced by partially bypassing the emitter resistors.

RELATED PROBLEMS FOR EXAMPLES

19–1 235 **19–2** 7.83 kΩ

19–3 $A_v = 338$; $A_i = 77.6$; $A_p = 26{,}228$

19–4 5.89 **19–5** 64.5

19–6 1.67 V peak-to-peak signal riding on a dc level of 8.4 V

19–7 0.976

19–8 $A_{v(tot)} = 1500$, $A_{v1}\ (\text{dB}) = 28\ \text{dB}$, $A_{v2}\ (\text{dB}) = 14\ \text{dB}$, $A_{v3}\ (\text{dB}) = 21.6\ \text{dB}$, $A_{v(tot)}\ (\text{dB}) = 63.6\ \text{dB}$

19–9 The peak unclipped output voltage decreases to 3 V.

19–10 7.5 V; 469 mA **19–11** 12.5 W

19–12 Efficiency decreases.

SELF-TEST

1. (c) **2.** (b) **3.** (a) **4.** (d) **5.** (b) **6.** (a) **7.** (c)
8. (d) **9.** (a) **10.** (c) **11.** (c) **12.** (b) **13.** (d) **14.** (a)
15. (c) **16.** (b)

20

OPERATIONAL AMPLIFIERS (OP-AMPS)

INTRODUCTION

In the previous chapters of this book, several electronic devices were introduced. These devices, such as the diode and the transistor, are individually packaged and are connected in a circuit with other individual devices to form a functional unit. Individually packaged devices are referred to as *discrete components.*

We will now introduce the topic of linear integrated circuits (ICs), in which many transistors, diodes, resistors, and capacitors are fabricated on a single silicon chip and packaged in a single case to form an operational amplifier. The manufacturing process for ICs is complex and beyond the scope of this coverage.

In our study of ICs, we will treat the entire circuit as a single device. That is, we will be more concerned with what the circuit does from an external point of view and will be less concerned about the internal, component-level operation.

CHAPTER OUTLINE

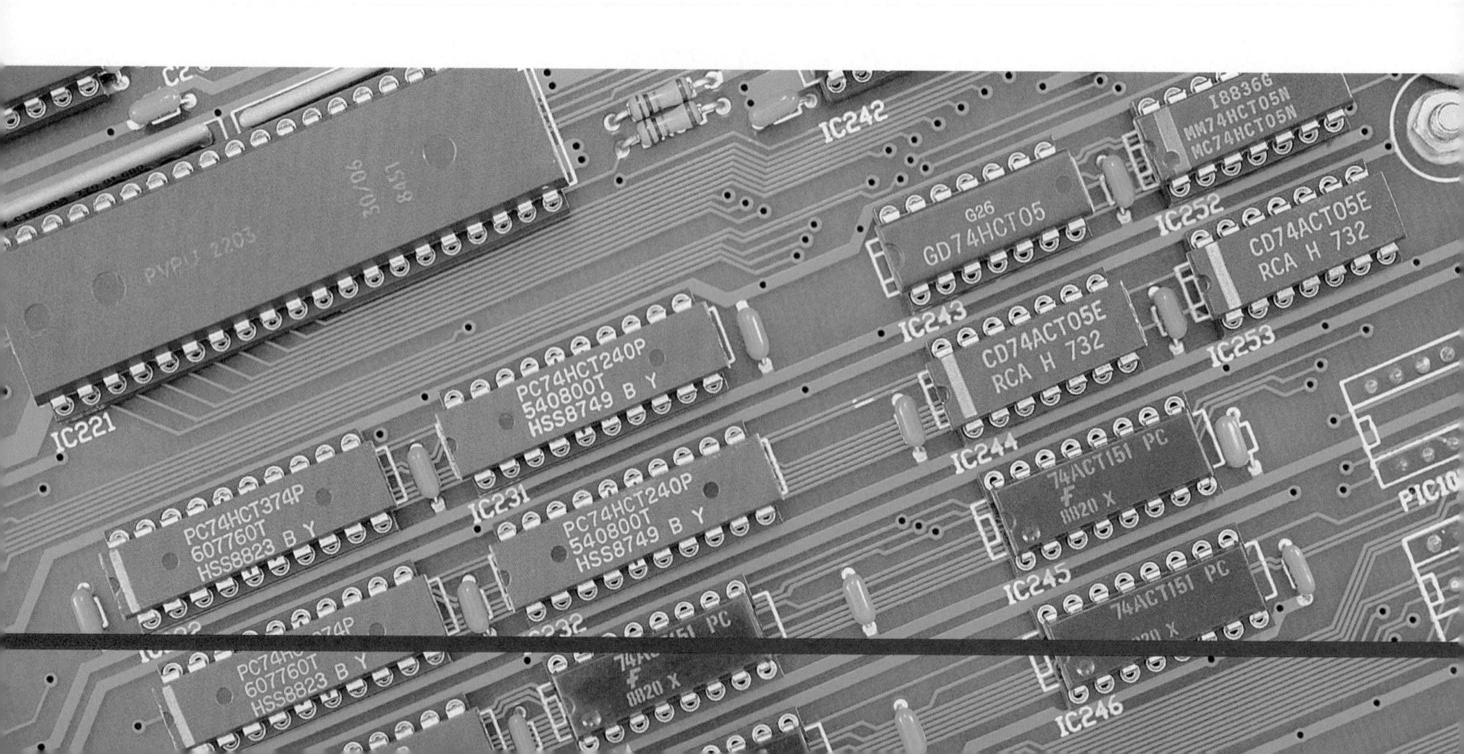

CHAPTER OBJECTIVES

- Discuss the basic op-amp
- Explain the basic operation of a differential amplifier
- Discuss several op-amp parameters
- Explain negative feedback in op-amp circuits
- Analyze three op-amp configurations
- Describe the effects of negative feedback on the three basic op-amp configurations
- Troubleshoot op-amp circuits

KEY TERMS

- Operational amplifier
- Differential amplifier
- Common-mode rejection ratio
- Open-loop voltage gain
- Negative feedback
- Closed-loop voltage gain
- Noninverting amplifier
- Voltage-follower
- Inverting amplifier

APPLICATION ASSIGNMENT PREVIEW

As an electronics technician in a chemical laboratory, you are assigned responsibility for the maintenance and troubleshooting of a spectrophotometer system that is used to analyze chemical solutions to determine their contents. The system combines electronic, mechanical, and optical technology to accomplish its task. The electronics portion uses a photocell and an op-amp to detect the wavelength of light passing through the solution and to amplify the resulting signal prior to sending it to a processor for analysis and display. After studying this chapter, you should be able to complete the application assignment.

WWW. VISIT THE COMPANION WEBSITE

Circuit Simulation Tutorials and Other Chapter Study Tools Are Available at

http://www.prenhall.com/floyd

20–1 INTRODUCTION TO OPERATIONAL AMPLIFIERS

Early operational amplifiers (op-amps) were used primarily to perform mathematical operations such as addition, subtraction, integration, and differentiation—hence the term *operational.* These early devices were constructed with vacuum tubes and worked with high voltages. Today's op-amps are linear integrated circuits (ICs) that use relatively low dc supply voltages and are reliable and inexpensive.

After completing this section, you should be able to

- **Discuss the basic op-amp**
- Recognize an op-amp symbol
- Identify terminals on op-amp packages
- Describe an ideal op-amp
- Describe a practical op-amp

Symbol and Terminals

The standard operational amplifier symbol is shown in Figure 20–1(a). It has two input terminals, the inverting input (−) and the noninverting input (+), and one output terminal. The typical op-amp requires two dc supply voltages, one positive and the other negative, as shown in Figure 20–1(b). Usually these dc voltage terminals are left off the schematic symbol for simplicity, but they are always understood to be there. Typical IC packages are shown in Figure 20–1(c).

FIGURE 20–1

Op-amp symbols and packages.

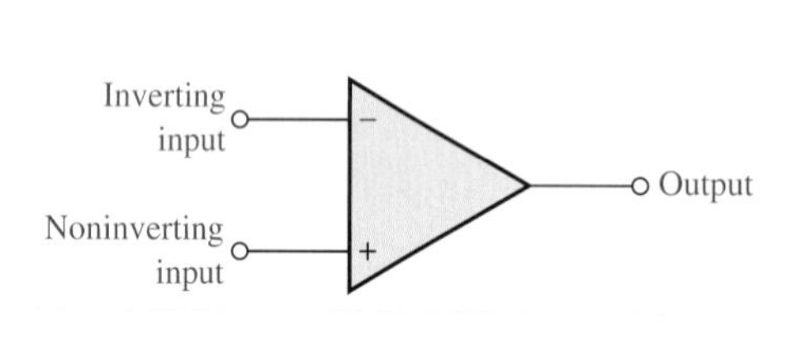

(a) Symbol

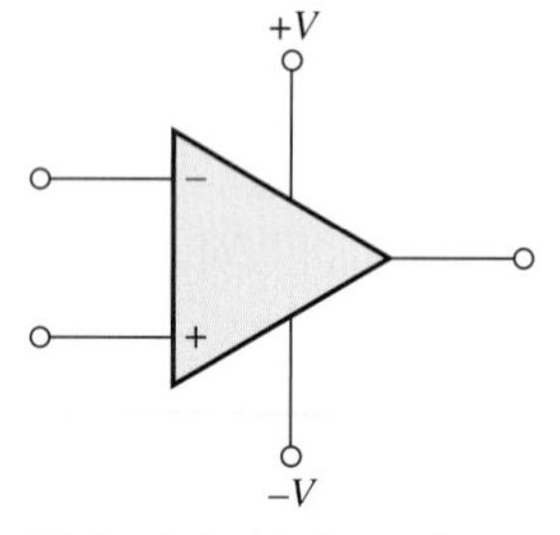

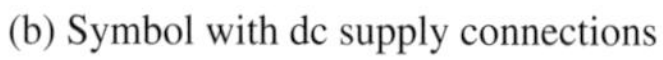

(b) Symbol with dc supply connections

DIP

DIP

SOIC

PLCC

(c) Typical packages. Looking from the top, pin 1 always is to the left of the notch or dot on the DIP and SOIC packages. The dot indicates pin 1 on the plastic-leaded chip carrier (PLCC) package.

The Ideal Op-Amp

To illustrate what an op-amp is, let's consider its *ideal* characteristics. A practical op-amp, of course, falls short of these ideal standards, but it is much easier to understand and analyze the device from an ideal point of view.

The ideal op-amp has *infinite voltage gain, infinite bandwidth,* and an *infinite input impedance* (open), so that it does not load from the driving source. Also, it has a *zero output impedance.* These characteristics are illustrated in Figure 20–2. The input voltage, V_{in}, appears between the two input terminals, and the output voltage is A_vV_{in}, as indicated by the symbol for internal voltage source. The concept of infinite input impedance is a particularly valuable analysis tool for the various op-amp configurations, which will be discussed in Section 20–5.

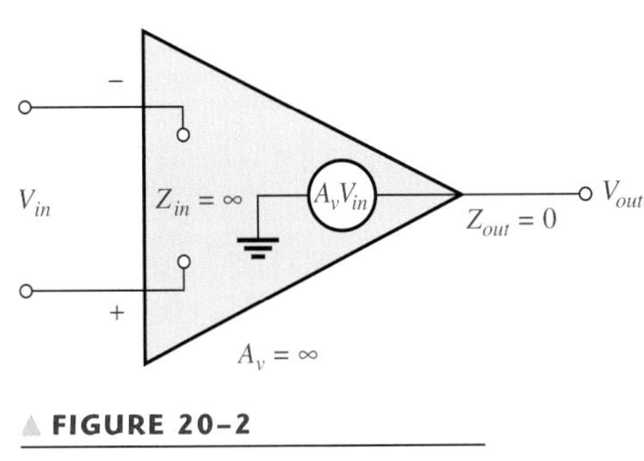

FIGURE 20–2

Ideal op-amp representation.

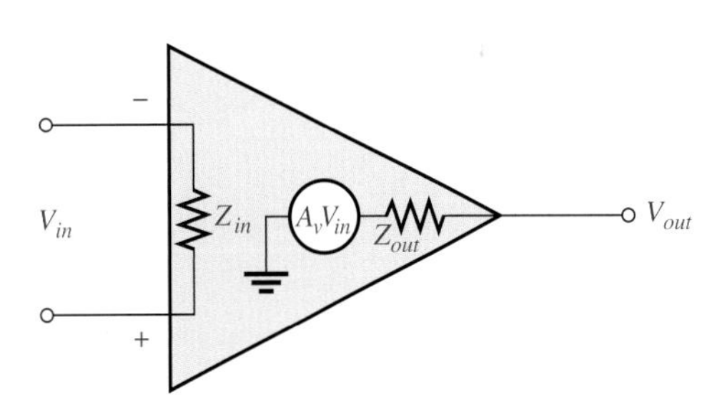

FIGURE 20–3

Practical op-amp representation.

The Practical Op-Amp

Although modern IC op-amps approach parameter values that can be treated as ideal in many cases, the ideal device can never be made. Any device has limitations, and the IC op-amp is no exception. Op-amps have both voltage and current limitations. Peak-to-peak output voltage, for example, is usually limited to slightly less than the two supply voltages. Output current is also limited by internal restrictions such as power dissipation and component ratings.

Characteristics of a practical op-amp are *high voltage gain, high input impedance, low output impedance,* and *wide bandwidth,* as illustrated in Figure 20–3.

SECTION 20–1 REVIEW

Answers are at the end of the chapter.

1. Sketch the symbol for an op-amp.
2. Describe the ideal op-amp.
3. Describe some of the characteristics of a practical op-amp.

20–2 THE DIFFERENTIAL AMPLIFIER

The op-amp, in its basic form, typically consists of two or more differential amplifier stages. Because the differential amplifier (diff-amp) is fundamental in the op-amp's internal operation, it is useful to have a basic understanding of this type of circuit.

After completing this section, you should be able to

- **Explain the basic operation of a differential amplifier**
- Describe single-ended input operation
- Describe differential input operation

- Describe common-mode input operation
- Define *common-mode rejection ratio* (CMRR)
- Describe how diff-amps are used to implement an op-amp

A basic **differential amplifier** circuit and its symbol are shown in Figure 20–4. The diff-amp stages that make up part of the op-amp provide high voltage gain and common-mode rejection (defined later in this section). Notice that the differential amplifier has two outputs where the op-amp has only one output.

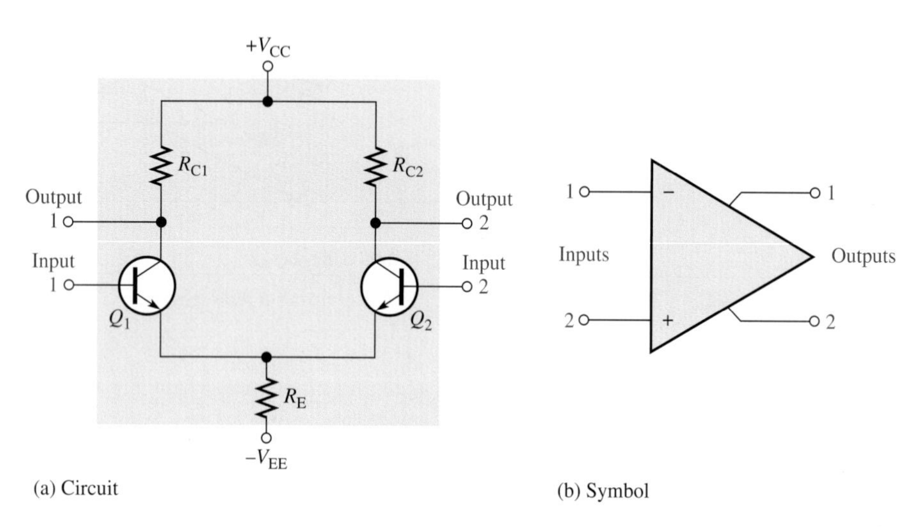

▲ **FIGURE 20–4**

Basic differential amplifier.

Basic Operation

Although an op-amp typically has more than one diff-amp stage, we will use a single diff-amp to illustrate the basic operation. The following discussion is in relation to Figure 20–5 and consists of a basic dc analysis of the diff-amp's operation.

First, when both inputs are grounded (0 V), the emitters are at −0.7 V, as indicated in Figure 20–5(a). It is assumed that the transistors are identically matched by careful process control during manufacturing so that their dc emitter currents are the same when there is no input signal. Thus,

$$I_{E1} = I_{E2}$$

Since both emitter currents combine through R_E,

$$I_{E1} = I_{E2} = \frac{I_{R_E}}{2}$$

where

$$I_{R_E} = \frac{V_E - V_{EE}}{R_E}$$

Based on the approximation that $I_C \cong I_E$, it can be stated that

$$I_{C1} = I_{C2} \cong \frac{I_{R_E}}{2}$$

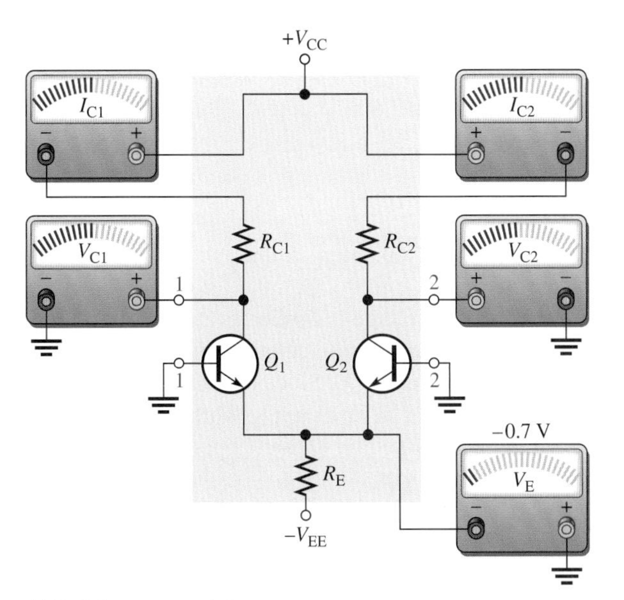

(a) Both inputs grounded.

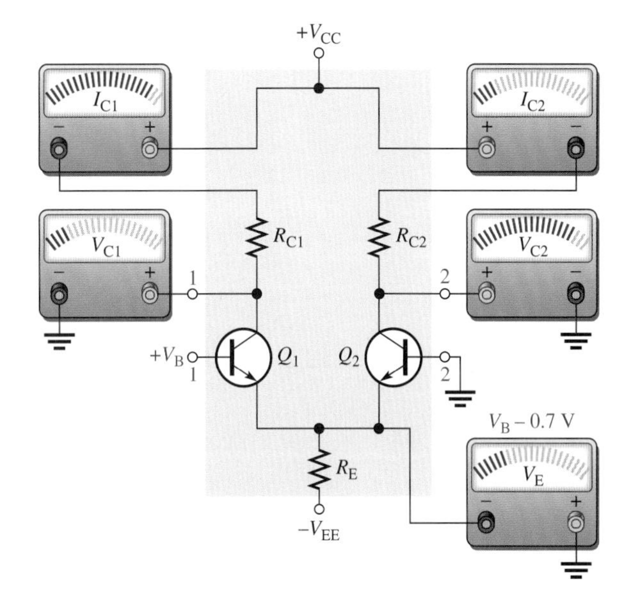

(b) Bias voltage on input 1 with input 2 grounded.

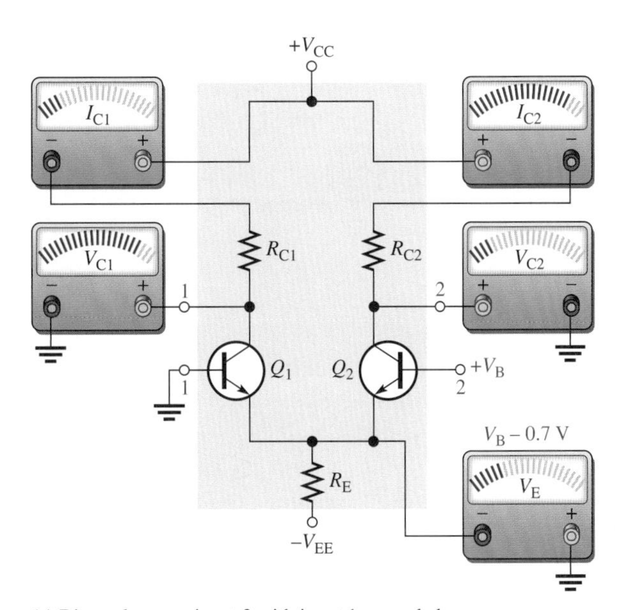

(c) Bias voltage on input 2 with input 1 grounded.

FIGURE 20–5

Basic operation of a differential amplifier (ground is 0 V) showing relative changes in currents and voltages.

Since both collector currents and both collector resistors are equal (when the input voltage is zero),

$$V_{C1} = V_{C2} = V_{CC} - I_{C1}R_{C1}$$

This condition is illustrated in Figure 20–5(a).

Next, input 2 remains grounded, and a positive bias voltage is applied to input 1, as shown in Figure 20–5(b). The positive voltage on the base of Q_1 increases I_{C1} and raises the emitter voltage. This action reduces the forward bias (V_{BE}) of Q_2 because its base is held at 0 V (ground), thus causing I_{C2} to decrease, as indicated in Figure 20–5(b). The net result is that the increase in I_{C1} causes a decrease in V_{C1}, and the decrease in I_{C2} causes an increase in V_{C2}, as shown.

Finally, input 1 is grounded and a positive bias voltage is applied to input 2, as shown in Figure 20–5(c). The positive bias voltage causes Q_2 to conduct more, thus increasing I_{C2}. Also, the emitter voltage is raised. This reduces the forward bias of Q_1 because its base is held at ground and causes I_{C1} to decrease. The result is that the increase in I_{C2} produces a decrease in V_{C2}, and the decrease in I_{C1} causes V_{C1} to increase as shown.

Modes of Signal Operation

Single-Ended Input When a diff-amp is operated in this mode, one input is grounded and the signal voltage is applied only to the other input, as shown in Figure 20–6. In the case where the signal voltage is applied to input 1, as in part (a), an inverted, amplified signal voltage appears at output 1 as shown. Also, a signal voltage appears in phase at the emitter of Q_1. Since the emitters of Q_1 and Q_2 are common, this emitter signal becomes an input to Q_2, which functions as a common-base amplifier. The signal is amplified by Q_2 and appears, noninverted, at output 2. This action is illustrated in part (a).

FIGURE 20–6

Single-ended input operation of a differential amplifier.

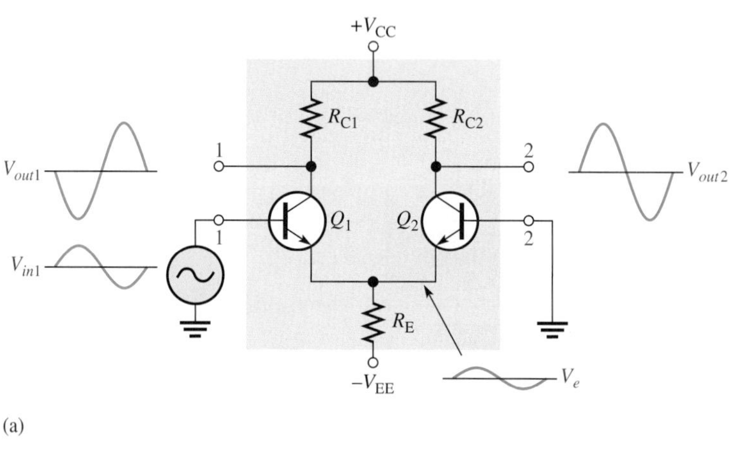

(a)

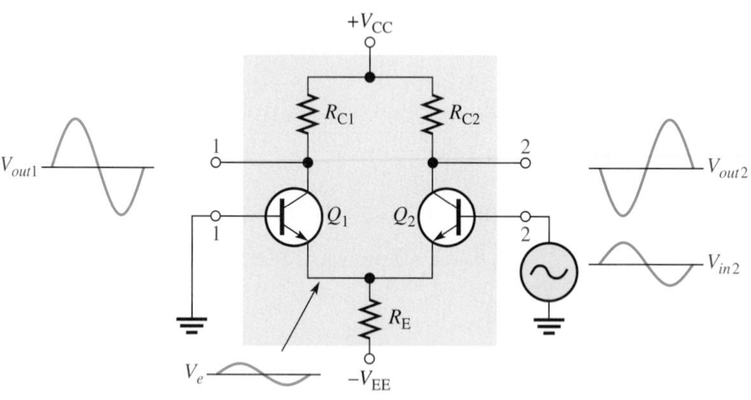

(b)

In the case where the signal is applied to input 2 with input 1 grounded, as in Figure 20–6(b), an inverted, amplified signal voltage appears at output 2. In this situation, Q_1 acts as a common-base amplifier, and a noninverted, amplified signal appears at output 1.

Differential Input In this mode, two signals of opposite polarity (out of phase) are applied to the inputs, as shown in Figure 20–7(a). This type of operation is also referred to as *double-ended.* Each input affects the outputs, as you will see in the following discussion.

Figure 20–7(b) shows the output signals due to the signal on input 1 acting alone as a single-ended input. Figure 20–7(c) shows the output signals due to the signal on input 2 acting alone as a single-ended input. In parts (b) and (c), notice that the signals on output 1 are of the same polarity. The same is also true for output 2. By superimposing both output 1 signals and both output 2 signals, we get the total differential operation, as pictured in Figure 20–7(d).

FIGURE 20–7

Differential operation of a differential amplifier.

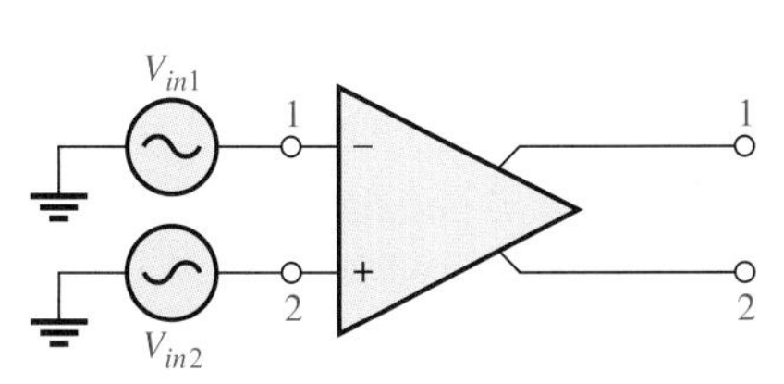

(a) Differential inputs

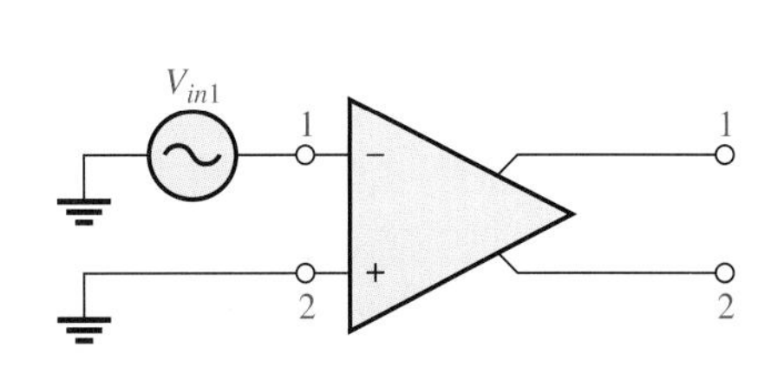

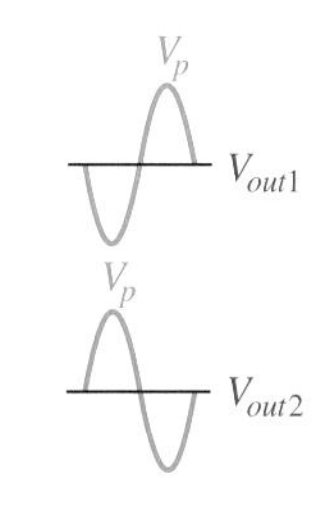

(b) Outputs due to V_{in1}

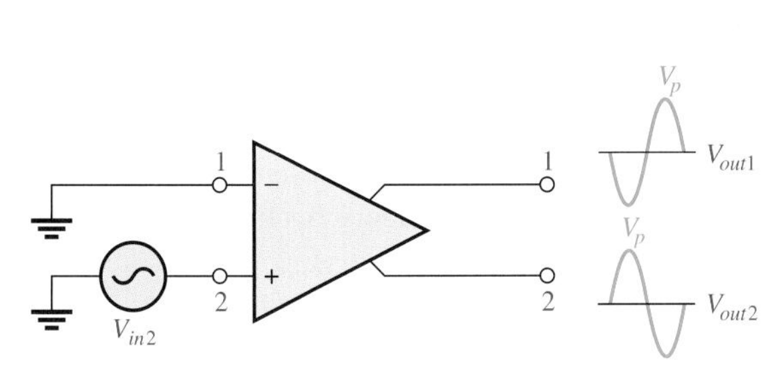

(c) Outputs due to V_{in2}

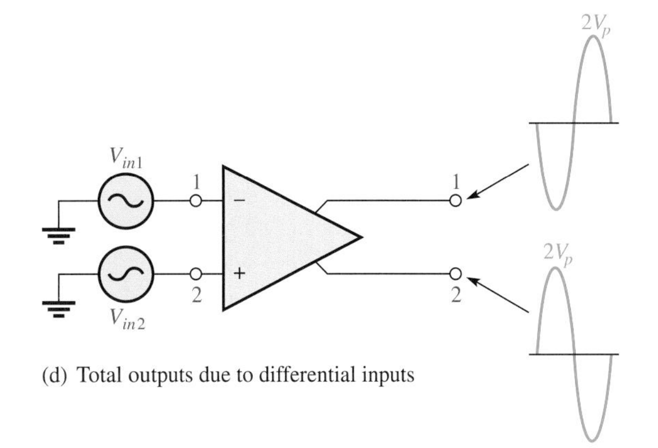

(d) Total outputs due to differential inputs

Common-Mode Input One of the most important aspects of the operation of a diff-amp can be seen by considering the common-mode condition where two signal voltages of the same phase, frequency, and amplitude are applied to the two inputs, as shown in Figure 20–8(a). Again, by considering each input signal as acting alone, you can understand the basic operation.

Figure 20–8(b) shows the output signals due to the signal only on input 1, and part (c) shows the output signals due to the signal on only input 2. Notice in parts (b) and (c) that the signals on output 1 are of the opposite polarity, and so are those on output 2. When the input signals are applied to both inputs, the outputs are superimposed and they cancel, resulting in a near zero output voltage, as shown in Figure 20–8(d).

FIGURE 20–8

Common-mode operation of a differential amplifier.

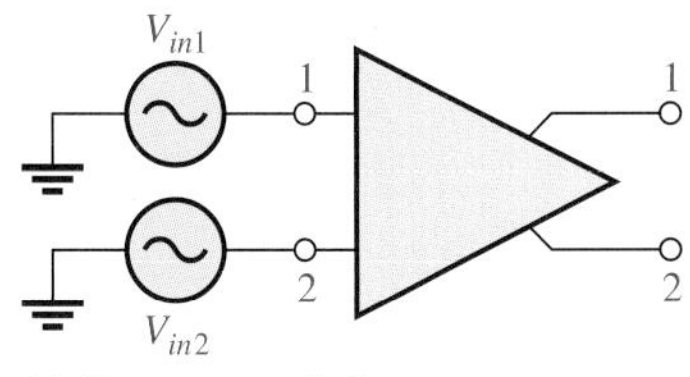

(a) Common-mode inputs

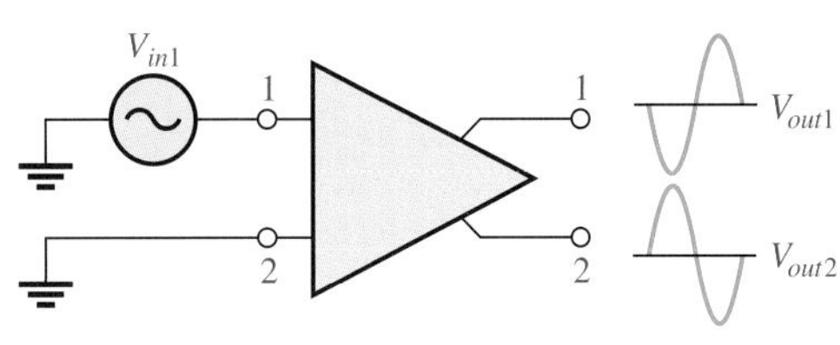

(b) Outputs due to V_{in1}

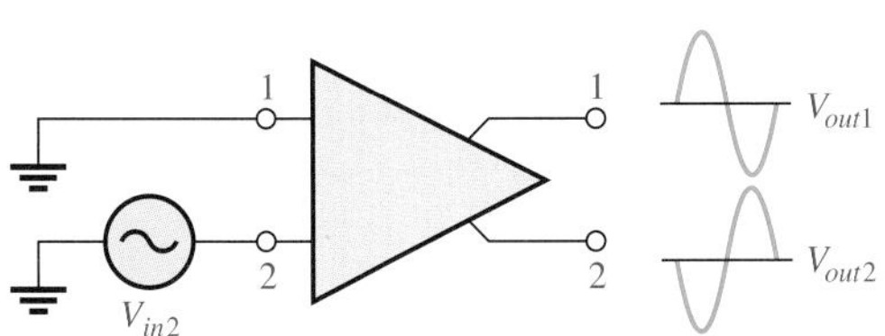

(c) Outputs are due to V_{in2}

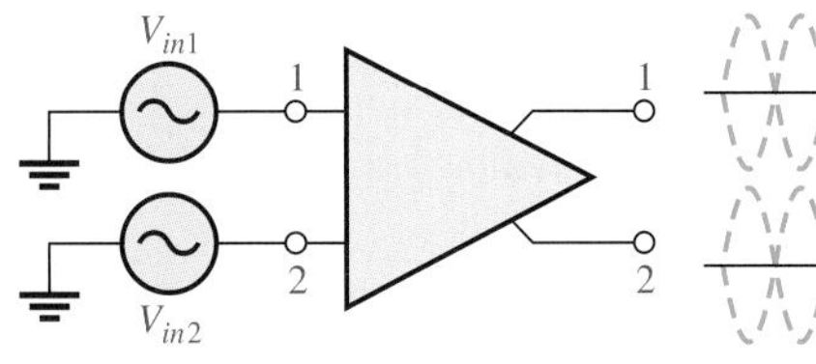

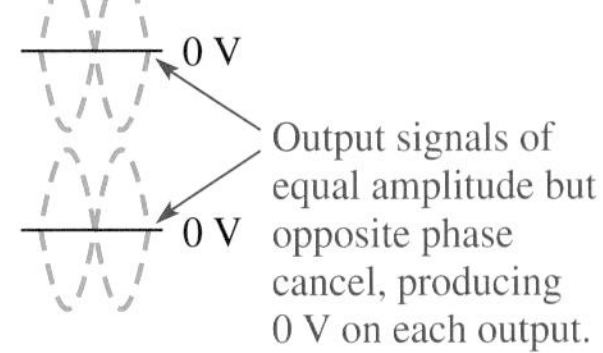

(d) Ouputs cancel when common-mode signals are applied

This action is called *common-mode rejection.* Its importance can be seen in the situation where an unwanted signal appears commonly on both diff-amp inputs. Common-mode rejection means that this unwanted signal will not appear on the outputs to distort the desired signal. **Common-mode signals** (noise) generally are the result of the pick-up of radiated energy on the input lines, from adjacent lines, or the 60 Hz power line, or other sources.

Common-Mode Rejection Ratio

Desired signals appear on only one input or with opposite polarities on both input lines. These desired signals are amplified and appear on the outputs as previously discussed. Unwanted signals (noise) appearing with the same polarity on both input lines are essentially canceled by the diff-amp and do not appear on the outputs. The measure of an amplifier's ability to reject common-mode signals is a parameter called the **common-mode rejection ratio (CMRR)**.

Ideally, a diff-amp provides a very high gain for desired signals (single-ended or differential), and zero gain for common-mode signals. Practical diff-amps, however, do exhibit a very small common-mode gain (usually much less than 1), while providing a high differential voltage gain (usually several thousand). The higher the differential gain with respect to the common-mode gain, the better the performance of the diff-amp in terms of rejection of common-mode signals. This suggests that a good measure of the diff-amp's performance in rejecting unwanted common-mode signals is the ratio of the differential gain, $A_{v(d)}$, to the common-mode gain, A_{cm}. This ratio is the common-mode rejection ratio, CMRR.

Equation 20–1

$$\text{CMRR} = \frac{A_{v(d)}}{A_{cm}}$$

The higher the CMRR, the better. A very high value of CMRR means that the differential gain $A_{v(d)}$ is high and the common-mode gain A_{cm} is low.

The CMRR is often expressed in decibels (dB) as

Equation 20–2

$$\text{CMRR} = 20\log\left(\frac{A_{v(d)}}{A_{cm}}\right)$$

EXAMPLE 20–1

A certain diff-amp has a differential voltage gain of 2000 and a common-mode gain of 0.2. Determine the CMRR and express it in decibels.

Solution $A_{v(d)} = 2000$ and $A_{cm} = 0.2$. Therefore,

$$\text{CMRR} = \frac{A_{v(d)}}{A_{cm}} = \frac{2000}{0.2} = \mathbf{10{,}000}$$

Expressed in decibels,

$$\text{CMRR} = 20\log(10{,}000) = \mathbf{80\ dB}$$

*Related Problem** Determine the CMRR and express it in dB for an amplifier with a differential voltage gain of 8500 and a common-mode gain of 0.25.

*Answers are at the end of the chapter.

A CMRR of 10,000, for example, means that the desired input signal (differential) is amplified 10,000 times more than the unwanted noise (common-mode). So, as an example, if the amplitudes of the differential input signal and the common-mode noise are equal, the desired signal will appear on the output 10,000 times greater in amplitude than the noise. Thus, the noise or interference has been essentially eliminated.

Example 20–2 illustrates further the idea of common-mode rejection and the general signal operation of the diff-amp.

EXAMPLE 20–2

The diff-amp shown in Figure 20–9 has a differential voltage gain of 2500 and a CMRR of 30,000. In part (a), a single-ended input signal of 500 μV rms is applied. At the same time a 1 V, 60 Hz common-mode interference signal appears on both inputs as a result of radiated pick-up from the ac power system. In part (b), differential input signals of 500 μV rms each are applied to the inputs. The common-mode interference is the same as in part (a).

(a) Determine the common-mode gain.

(b) Express the CMRR in dB.

(c) Determine the rms output signal for parts (a) and (b).

(d) Determine the rms interference voltage on the output.

▼ **FIGURE 20–9**

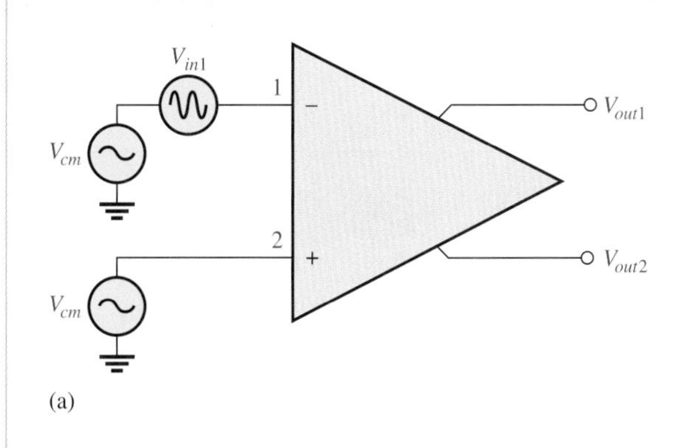

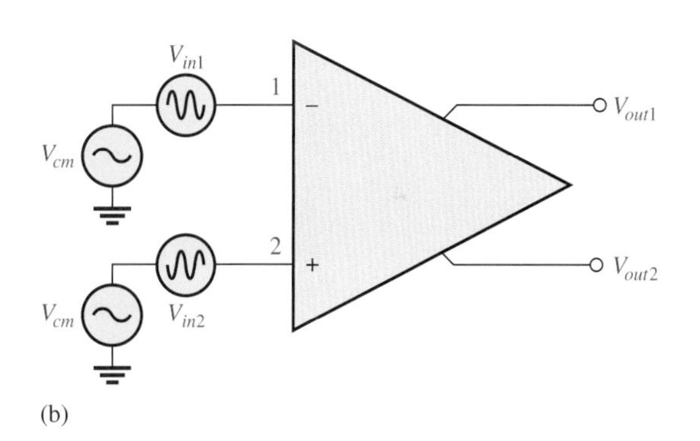

Solution **(a)** CMRR $= A_{v(d)}/A_{cm}$. Therefore,

$$A_{cm} = \frac{A_{v(d)}}{\text{CMRR}} = \frac{2500}{30{,}000} = \mathbf{0.083}$$

(b) CMRR $= 20\log(30{,}000) = \mathbf{89.5\ dB}$

(c) In Figure 20–9(a), the differential input voltage, $V_{in(d)}$, is the difference between the voltage on input 1 and that on input 2. Since input 2 is grounded, the voltage is zero. Therefore,

$$V_{in(d)} = V_{in1} - V_{in2} = 500\ \mu\text{V} - 0\ \text{V} = 500\ \mu\text{V}$$

The output signal voltage in this case is taken at output 1.

$$V_{out1} = A_{v(d)}V_{in(d)} = (2500)(500\ \mu\text{V}) = \mathbf{1.25\ V\ rms}$$

In Figure 20–9(b), the differential input voltage is the difference between the two opposite-polarity, 500 μV signals.

$$V_{in(d)} = V_{in1} - V_{in2} = 500\ \mu\text{V} - (-500\ \mu\text{V}) = 1000\ \mu\text{V} = 1\ \text{mV}$$

The output signal voltage is

$$V_{out1} = A_{v(d)}V_{in(d)} = (2500)(1\text{ mV}) = \mathbf{2.5\ V\ rms}$$

This shows that a differential input (two opposite-polarity signals) results in a gain that is double that for a single-ended input.

(d) The common-mode input is 1 V rms. The common-mode gain A_{cm} is 0.083. The interference voltage on the output, therefore, is

$$A_{cm} = \frac{V_{out(cm)}}{V_{in(cm)}}$$

$$V_{out(cm)} = A_{cm}V_{in(cm)} = (0.083)(1\text{ V}) = \mathbf{83\ mV}$$

Related Problem The amplifier in Figure 20–9 has a differential voltage gain of 4200 and a CMRR of 25,000. For the same single-ended and differential input signals as described in the example: **(a)** Find A_{cm}. **(b)** Express the CMRR in dB. **(c)** Determine the rms output signal for parts (a) and (b) of the figure. **(d)** Determine the rms interference (common-mode) voltage appearing on the output.

A Simple Op-Amp Arrangement

Figure 20–10 shows two diff-amp stages and an emitter-follower connected to form a simple op-amp. The first stage can be used with a single-ended or a differential input. The differential outputs of the first stage feed into the differential in-

FIGURE 20–10

Simplified internal circuitry of a basic op-amp.

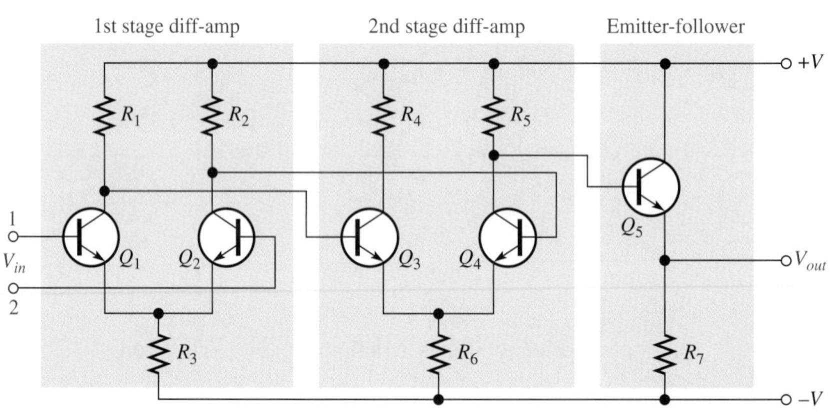

(a) Circuit

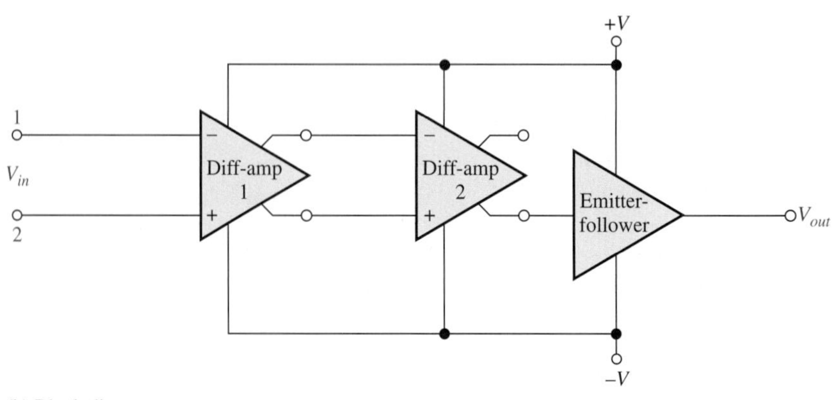

(b) Block diagram

puts of the second stage. The output of the second stage is single-ended to drive an emitter-follower to achieve a low output impedance. Both differential stages together provide a high voltage gain and a high CMRR.

SECTION 20–2 REVIEW

1. Distinguish between differential and single-ended inputs.
2. Define *common-mode rejection.*
3. For a given value of differential gain, does a higher CMRR result in a higher or lower common-mode gain?

20–3 OP-AMP PARAMETERS

In this section, several important op-amp parameters are defined. (These are listed in the objectives that follow.) Also, several popular IC op-amps are compared in terms of these parameters.

After completing this section, you should be able to

- **Discuss several op-amp parameters**
- Define *input offset voltage*
- Discuss input offset voltage drift with temperature
- Define *input bias current*
- Define *input impedance*
- Define *input offset current*
- Define *output impedance*
- Discuss common-mode input voltage range
- Discuss open-loop voltage gain
- Define *common-mode rejection ratio*
- Define *slew rate*
- Discuss frequency response
- Compare the parameters of several types of IC op-amps

Input Offset Voltage

The ideal op-amp produces zero volts out for zero volts in. In a practical op-amp, however, a small dc voltage, $V_{OUT(error)}$, appears at the output when no differential input voltage is applied. Its primary cause is a slight mismatch of the base-emitter voltages of the differential input stage, as illustrated in Figure 20–11(a).

The output voltage of the differential input stage is expressed as

$$V_{OUT} = I_{C2}R_C - I_{C1}R_C$$ **Equation 20–3**

A small difference in the base-emitter voltages of Q_1 and Q_2 causes a small difference in the collector currents. This results in a nonzero value of V_{OUT}. (The collector resistors are equal.)

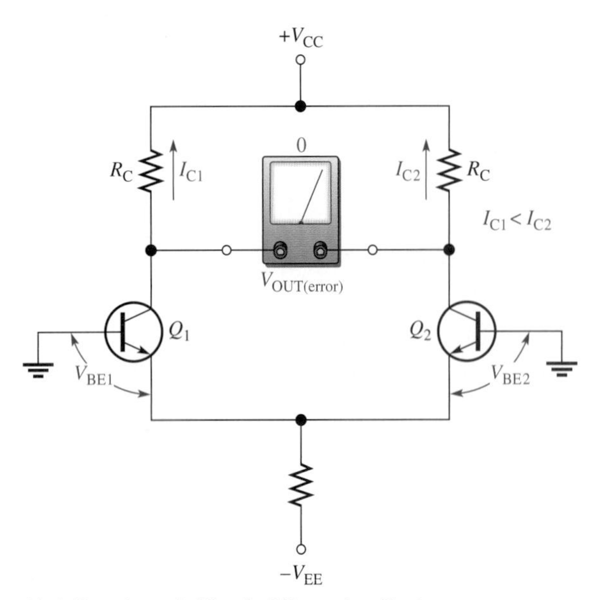

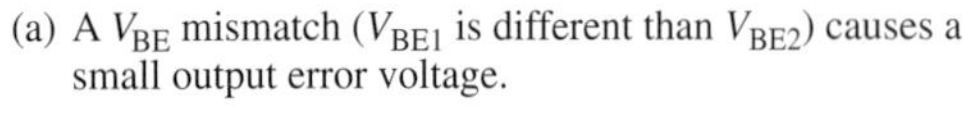

(a) A V_{BE} mismatch (V_{BE1} is different than V_{BE2}) causes a small output error voltage.

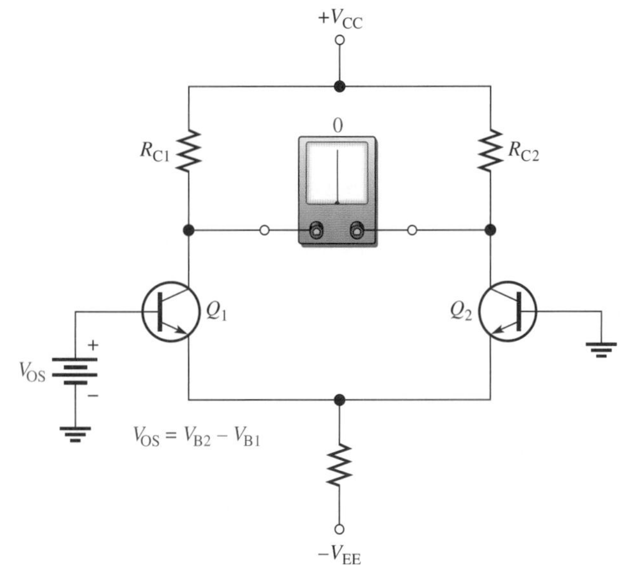

(b) The input offset voltage is the difference in the voltage between the inputs that is necessary to eliminate the output error voltage (makes $V_{OUT} = 0$)

▲ **FIGURE 20–11**

Input offset voltage, V_{OS}.

As specified on an op-amp data sheet, the *input offset voltage,* V_{OS}, is the differential dc voltage required between the inputs to force the differential output to zero volts, as demonstrated in Figure 20–11(b). Typical values of input offset voltage are in the range of 2 mV or less. In the ideal case, it is 0 V.

Input Offset Voltage Drift with Temperature

The *input offset voltage drift* is a parameter related to V_{OS} that specifies how much change occurs in the input offset voltage for each degree change in temperature. Typical values range anywhere from about 5 μV per degree Celsius to about 50 μV per degree Celsius. Usually, as op-amp with a higher nominal value of input offset voltage exhibits a higher drift.

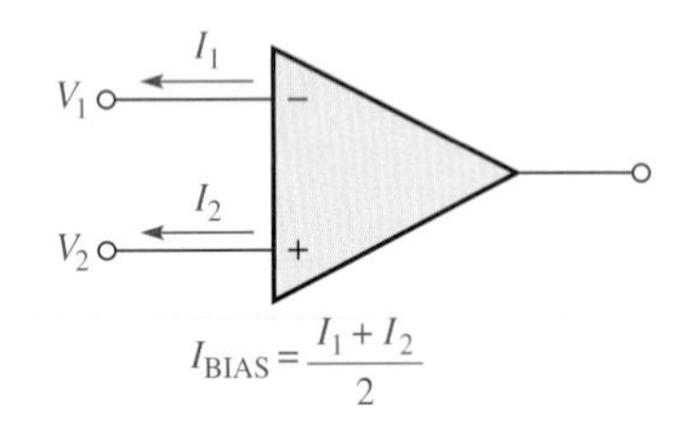

▲ **FIGURE 20–12**

Input bias current is the average of the two op-amp input currents.

Input Bias Current

You have seen that the input terminals of a diff-amp are the transistor bases and, therefore, the input currents are the base currents.

The *input bias current* is the direct current required by the inputs of the amplifier to properly operate the first stage. By definition, the input bias current is the average of both input currents and is calculated as follows:

Equation 20–4

$$I_{BIAS} = \frac{I_1 + I_2}{2}$$

The concept of input bias current is illustrated in Figure 20–12.

Input Impedance

Two basic ways of specifying the input impedance of an op-amp are the differential and the common mode. The *differential input impedance* is the total resistance between the inverting and the noninverting inputs, as illustrated in Figure 20–13(a). Differential impedance is measured by determining the change in bias current for a given change in differential input voltage. The *common-mode input impedance* is the resistance between each input and ground and is measured by determining the change in bias current for a given change in common-mode input voltage. It is depicted in Figure 20–13(b).

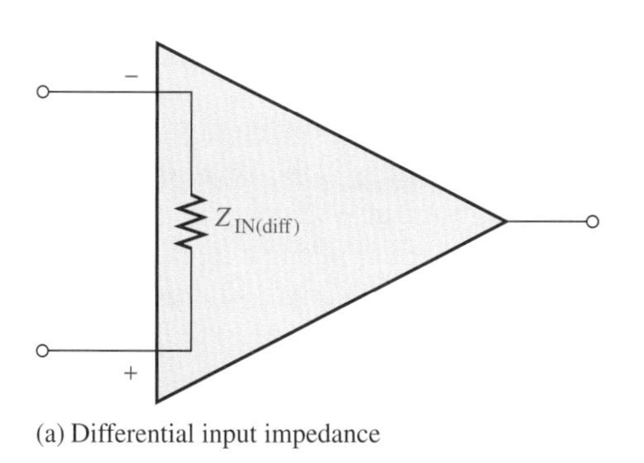

(a) Differential input impedance

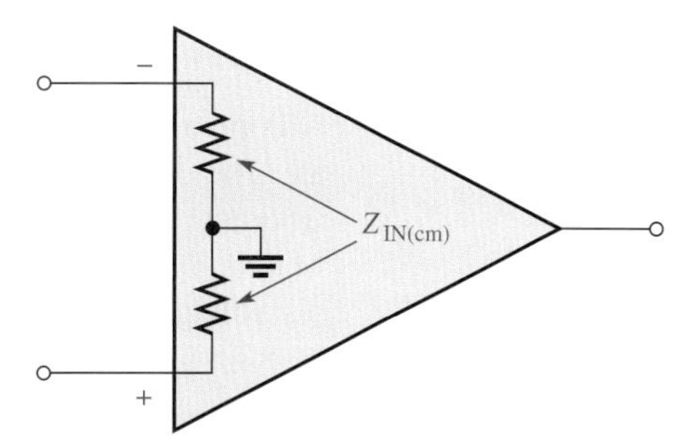

(b) Common-mode input impedance

▲ **FIGURE 20–13**

Op-amp input impedance.

Input Offset Current

Ideally, the two input bias currents are equal, and thus their difference is zero. In a practical op-amp, however, the bias currents are not exactly equal.

The *input offset current* is the difference of the input bias currents, expressed as an absolute value.

$$I_{OS} = |I_1 - I_2| \qquad \text{Equation 20–5}$$

Actual magnitudes of offset current are usually at least an order of magnitude (ten times) less than the bias current. In many applications, the offset current can be neglected. However, high-gain, high-input impedance amplifiers should have as little I_{OS} as possible because the difference in currents through large input resistances develops a substantial offset voltage, as shown in Figure 20–14.

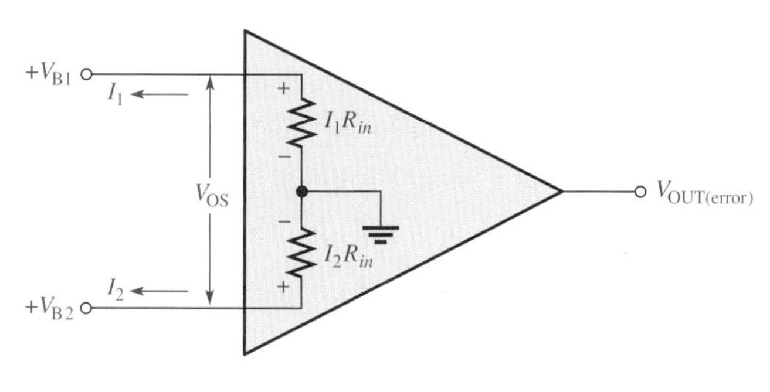

▲ **FIGURE 20–14**

Effect of input offset current.

▲ **FIGURE 20–15**

Op-amp output impedance.

The offset voltage developed by the input offset current is

$$V_{OS} = I_1R_{in} - I_2R_{in} = (I_1 - I_2)R_{in}$$

$$V_{OS} = I_{OS}R_{in} \qquad \text{Equation 20–6}$$

The error created by I_{OS} is amplified by the gain A_v of the op-amp and appears in the output as

$$V_{OUT(error)} = A_vI_{OS}R_{in} \qquad \text{Equation 20–7}$$

A change in offset current with temperature affects the error voltage. Values of temperature coefficient for the offset current in the range of 0.5 nA/°C are common.

Output Impedance

The *output impedance* is the resistance viewed from the output terminal of the op-amp, as indicated in Figure 20–15.

Common-Mode Input Voltage Range

All op-amps have limitations on the range of voltages over which they will operate. The *common-mode input voltage range* is the range of input voltages which, when applied to both inputs, will not cause clipping or other output distortion. Many op-amps have common-mode input voltage ranges of ± 10 V with dc supply voltages of ± 15 V.

Open-Loop Voltage Gain, A_{ol}

The open-loop voltage gain of the op-amp is the internal voltage gain of the device and represents the ratio of output voltage to input voltage when there are no external components. The open-loop voltage gain is set entirely by the internal design. Open-loop voltage gain can range up to 200,000 and is *not a well-controlled parameter.* Data sheets often refer to the open-loop voltage gain as the *large-signal voltage gain.*

Common-Mode Rejection Ratio

The *common-mode rejection ratio* (CMRR), as discussed in conjunction with the diff-amp, is a measure of an op-amp's ability to reject common-mode signals. An infinite value of CMRR means that the output is zero when the same signal is applied to both inputs (common-mode).

An infinite CMRR is never achieved in practice, but a good op-amp does have a very high value of CMRR. As previously discussed, common-mode signals are undesired interference voltages such as 60 Hz power-supply ripple and noise voltages due to pickup of radiated energy. A high CMRR enables the op-amp to virtually eliminate these interference signals from the output.

The accepted definition of CMRR for an op-amp is the open-loop voltage gain (A_{ol}) divided by the common-mode gain.

Equation 20–8

$$\text{CMRR} = \frac{A_{ol}}{A_{cm}}$$

It is commonly expressed in decibels as follows:

Equation 20–9

$$\text{CMRR} = 20\log\left(\frac{A_{ol}}{A_{cm}}\right)$$

EXAMPLE 20–3

A certain op-amp has an open-loop voltage gain of 100,000 and a common-mode gain of 0.25. Determine the CMRR and express it in decibels.

Solution

$$\text{CMRR} = \frac{A_{ol}}{A_{cm}} = \frac{100{,}000}{0.25} = \mathbf{400{,}000}$$

$$\text{CMRR} = 20\log(400{,}000) = \mathbf{112\ dB}$$

Related Problem If a particular op-amp has a CMRR of 90 dB and a common-mode gain of 0.4, what is the open-loop voltage gain?

Slew Rate

The maximum rate of change of the output voltage in response to a step input voltage is the *slew rate* of an op-amp. The slew rate is dependent upon the frequency response of the amplifier stages within the op-amp.

Slew rate is measured with an op-amp connected as shown in Figure 20–16(a). This particular op-amp connection is a unity-gain, noninverting configuration that will be discussed later. It gives a worst-case (slowest) slew rate. As shown in part (b), a pulse is applied to the input, and the ideal output voltage is measured. The width of the input pulse must be sufficient to allow the output to "slew" from its lower limit to it upper limit, as shown. As you can see, a certain time interval, Δt, is required for the output voltage to go from its lower limit $-V_{max}$ to its upper limit $+V_{max}$, once the input step is applied. The slew rate is expressed as

Equation 20–10

$$\text{Slew rate} = \frac{\Delta V_{out}}{\Delta t}$$

where $\Delta V_{out} = +V_{max} - (-V_{max})$. The unit of slew rate is volts per microsecond (V/μs).

FIGURE 20–16

Measurement of slew rate.

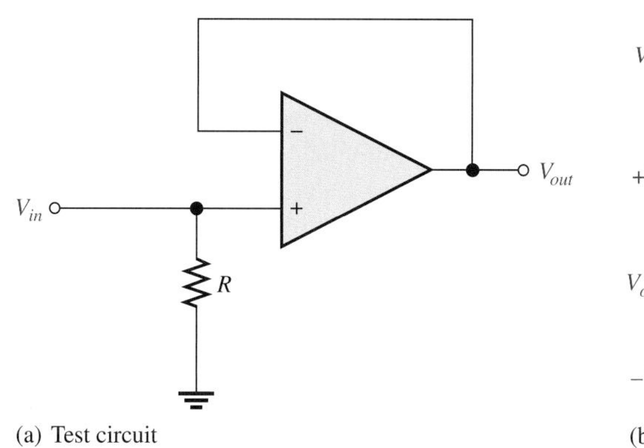

(a) Test circuit

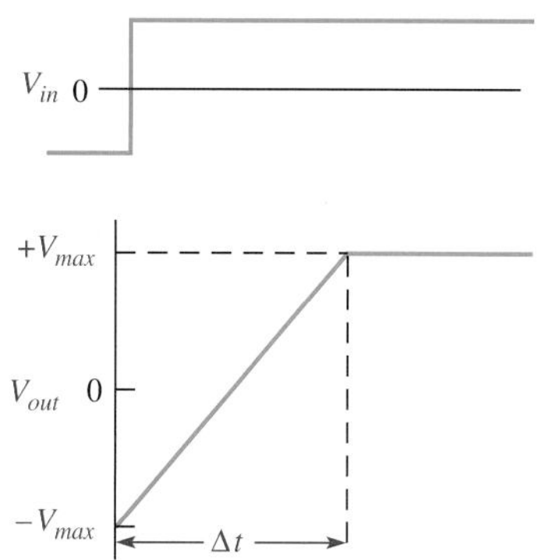

(b) Step input voltage and the resulting output voltage

EXAMPLE 20–4

The output voltage of a certain op-amp appears as shown in Figure 20–17 in response to a step input. Determine the slew rate.

FIGURE 20–17

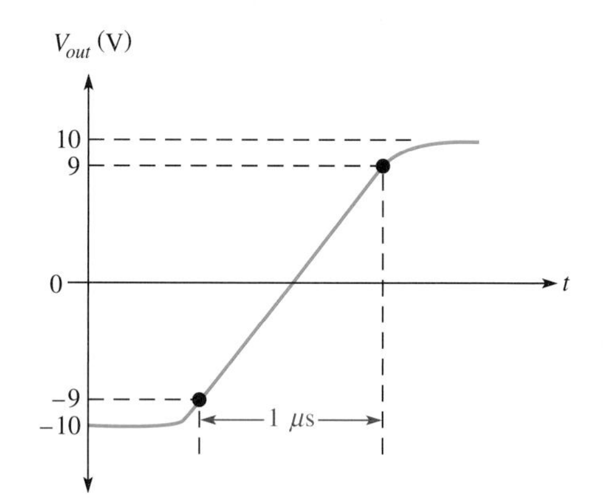

Solution The output goes from the lower to the upper limit in 1 μs. Since this response is not ideal, the limits are taken at the 90% points, as indicated. Thus, the upper limit is $+9$ V and the lower limit is -9 V. The slew rate is

$$\text{Slew rate} = \frac{\Delta V_{out}}{\Delta t} = \frac{+9\text{ V} - (-9\text{ V})}{1\ \mu\text{s}} = \mathbf{18\ V/\mu s}$$

Related Problem When a pulse is applied to an op-amp, the output voltage goes from -8 V to $+7$ V in 0.75 μs. What is the slew rate?

Frequency Response

The internal amplifier stages that make up an op-amp have voltage gains limited by junction capacitances. Although the diff-amps used in op-amps are somewhat different from the basic amplifiers discussed earlier, the same principles apply. An op-amp has no internal coupling capacitors, however; therefore, the low frequency response extends down to dc (0 Hz).

Comparison of Op-Amp Parameters

Table 20–1 provides a comparison of values of some of the parameters just described for several popular IC op-amps. Any values not listed were not given on the manufacturer's data sheet.

TABLE 20–1

OP-AMP	INPUT OFFSET VOLTAGE (mV)(MAX)	INPUT BIAS CURRENT (nA)(MAX)	INPUT IMPEDANCE (MΩ)(MIN)	OPEN-LOOP GAIN (TYP)	SLEW RATE (V/μs) (TYP)	CMRR (dB) (MIN)	COMMENT
LM741C	6	500	0.3	200,000	0.5	70	Industry standard
LM101A	7.5	250	1.5	160,000	—	80	General-purpose
OP113E	0.075	600	—	2,400,000	1.2	100	Low noise, low drift
OP177A	0.01	1.5	26	12,000,000	0.3	130	Ultra precision
OP184E	0.065	350	—	240,000	2.4	60	Precision, rail-to-rail*
AD8009AR	5	150	—	—	5500	50	*BW* = 700 MHz, ultra fast, low distortion, current feed-back
AD8041A	7	2000	0.16	56,000	160	74	*BW* = 160 MHz, rail-to-rail
AD8055A	5	1200	10	3500	1400	82	Very fast voltage feedback

*Rail-to-rail means that the output voltage can go as high as the supply voltages.

Other Features

Most available op-amps have three important features: short-circuit protection, no latch-up, and input offset nulling. Short-circuit protection keeps the circuit from being damaged if the output becomes shorted. The no-latch-up feature prevents the op-amp from hanging up in one output state (high or low voltage level) under certain input conditions. Input offset nulling is achieved by an external potentiometer that sets the output voltage at precisely zero with zero input.

SECTION 20–3 REVIEW

1. List ten or more op-amp parameters.
2. List two parameters, not including frequency response, that are frequency dependent.

20–4 NEGATIVE FEEDBACK

Negative feedback is one of the most useful concepts in electronics, particularly in op-amp applications. Negative feedback is the process whereby a portion of the output voltage of an amplifier is returned to the input with a phase angle that opposes (or subtracts from) the input signal.

After completing this section, you should be able to

- **Explain negative feedback in op-amp circuits**
- Describe the effects of negative feedback
- Discuss why negative feedback is used

Negative feedback is illustrated in Figure 20–18. The inverting input (−) effectively makes the feedback signal 180° out of phase with the input signal. The op-amp has extremely high gain and amplifies the *difference* in the signals applied to the inverting and noninverting inputs. A very tiny difference in these two signals is all the op-amp needs to produce the required output. *When negative feedback is present, the noninverting and inverting inputs are nearly identical.* This concept can help you figure out what signal to expect in many op-amp circuits.

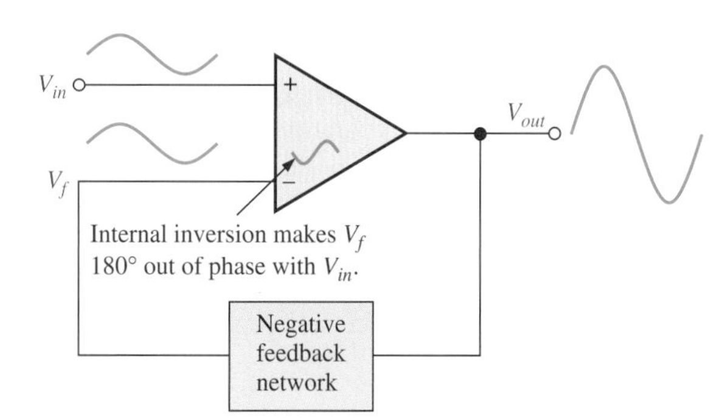

FIGURE 20–18

Illustration of negative feedback.

Now let's review how negative feedback works and why the signals at the inverting and noninverting terminals are nearly identical when negative feedback is used. Assume a 1.0 V input signal is applied to the noninverting terminal and the open-loop gain of the op-amp is 100,000. The amplifier responds to the voltage at its noninverting input terminal and moves the output toward saturation. Immediately, a fraction of this output is returned to the inverting terminal through the feedback path. But if the feedback signal ever reaches 1.0 V, there is nothing left for the op-amp to amplify! Thus, the feedback signal tries (but never quite succeeds) in matching the input signal. The gain is controlled by the amount of feedback used. When you are troubleshooting an op-amp circuit with negative feedback present, remember that the two inputs will look identical on a scope but in fact are very slightly different.

Now suppose something happens that reduces the internal gain of the op-amp. This causes the output signal to drop a small amount, returning a smaller signal to the inverting input via the feedback path. This means the difference between the signals is larger than it was. The output increases, compensating for the original drop in gain. The net change in the output is so small, it can hardly be measured. The main point is that any variation in the amplifier is immediately compensated for by the negative feedback, resulting in a very stable, predictable output.

Why Use Negative Feedback?

As you have seen, the inherent open-loop gain of a typical op-amp is very high (usually greater than 100,000). Therefore, an extremely small difference in the two input voltages drives the op-amp into its saturated output states. In fact, even the input offset voltage of the op-amp can drive it into saturation. For example, assume $V_{in} = 1$ mV and $A_{ol} = 100{,}000$. Then,

$$V_{in}A_{ol} = (1 \text{ mV})(100{,}000) = 100 \text{ V}$$

Since the output level of an op-amp can never reach 100 V, it is driven into saturation and the output is limited to its maximum output levels, as illustrated in Figure 20–19 for both a positive and a negative input voltage of 1 mV.

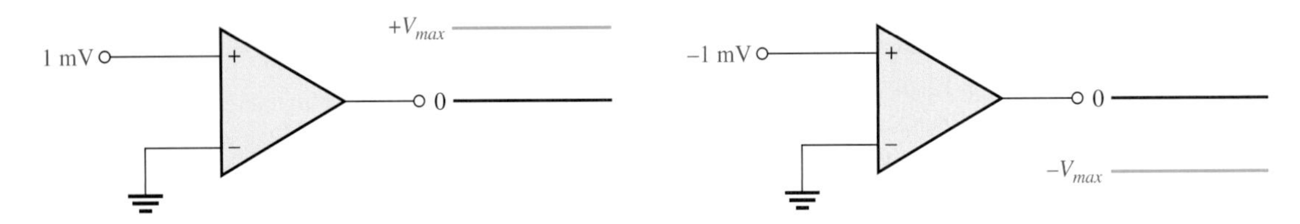

FIGURE 20–19

Without negative feedback, an extremely small difference in the two input voltages drives the op-amp to its output limits and it becomes nonlinear.

The usefulness of an op-amp operated in this manner is severely restricted and is generally limited to comparator applications (to be studied in Chapter 21). With negative feedback, the overall closed-loop voltage gain (A_{cl}) can be reduced and controlled so that the op-amp can function as a linear amplifier. In addition to providing a controlled, stable voltage gain, negative feedback also provides for control of the input and output impedances and amplifier bandwidth. Table 20–2 summarizes the general effects of negative feedback on op-amp performance.

TABLE 20–2

	VOLTAGE GAIN	INPUT Z	OUTPUT Z	BANDWIDTH
Without negative feedback	A_{ol} is too high for linear amplifier applications	Relatively high (see Table 20–1)	Relatively low	Relatively narrow (because the gain is so high)
With negative feedback	A_{cl} is set to desired value by the feedback network	Can be increased or reduced to a desired value depending on type of circuit	Can be reduced to a desired value	Significantly wider

SECTION 20–4 REVIEW

1. What are the benefits of negative feedback in an op-amp circuit?
2. Why is it necessary to reduce the gain of an op-amp from its open-loop value?
3. When troubleshooting an op-amp circuit in which negative feedback is present, what do you expect to observe on the input terminals?

20–5 OP-AMP CONFIGURATIONS WITH NEGATIVE FEEDBACK

In this section, we will discuss three basic ways in which an op-amp can be connected using negative feedback to stabilize the gain and increase frequency response. As mentioned, the extremely high open-loop gain of an op-amp creates an unstable situation because a small noise voltage on the input can be amplified to a point where the amplifier is driven out of its linear region. Also, unwanted oscillators can occur. In addition, the open-loop gain parameter of an op-amp can vary greatly from one device to the next. Negative feedback takes a portion of the output and applies it back out of phase with the input, creating an effective reduction in gain. This closed-loop gain is usually much less than the open-loop gain and independent of it.

After completing this section, you should be able to

- **Analyze three op-amp configurations**
- Identify the noninverting amplifier configuration
- Determine the voltage gain of a noninverting amplifier
- Identify the voltage-follower configuration
- Identify the inverting amplifier configuration
- Determine the voltage gain of an inverting amplifier

Closed-Loop Voltage Gain, A_{cl}

The closed-loop voltage gain is the voltage gain of an op-amp with negative feedback. The amplifier configuration consists of the op-amp and an external feedback network that connects the output to the inverting input. The closed-loop voltage gain is then determined by the component values in the feedback network and can be precisely controlled by them.

Noninverting Amplifier

An op-amp connected in a closed-loop configuration as a noninverting amplifier is shown in Figure 20–20. The input signal is applied to the noninverting input (+). A portion of the output is applied back to the inverting input (−) through the feedback network. This constitutes negative feedback. The feedback fraction, B, is determined by R_f and R_i, which form a voltage-divider network. The attenuation of the feedback network is the portion of the output returned to the inverting

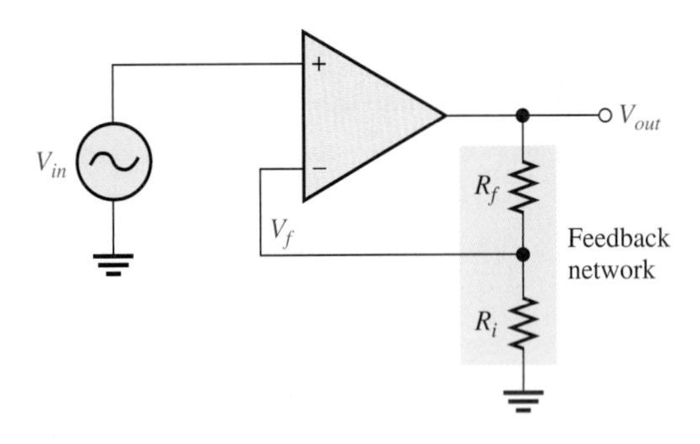

FIGURE 20–20

Noninverting amplifier.

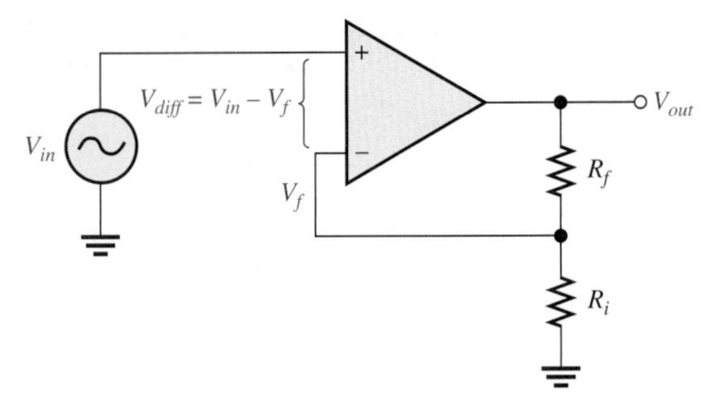

FIGURE 20–21

Differential input, $V_{in} - V_f$.

input and determines the gain of the amplifier as you will see. This smaller feedback voltage, V_f, can be written

$$V_f = \left(\frac{R_i}{R_i + R_f}\right)V_{out} = BV_{out}$$

The differential voltage, V_{diff}, between the op-amp's input terminals is illustrated in Figure 20–21 and can be expressed as

$$V_{diff} = V_{in} - V_f$$

This input differential voltage is forced to be very small as a result of the negative feedback and the high open-loop gain, A_{ol}. Therefore, a close approximation is

$$V_{in} \cong V_f$$

By substitution,

$$V_{in} \cong BV_{out}$$

Rearranging,

$$\frac{V_{out}}{V_{in}} \cong \frac{1}{B}$$

The ratio of the output voltage to the input voltage is the closed-loop gain. This result shows that the closed-loop gain for the noninverting amplifier, $A_{cl(\text{NI})}$, is approximately

$$A_{cl(\text{NI})} = \frac{V_{out}}{V_{in}} \cong \frac{1}{B}$$

The fraction of the output voltage, V_{out}, that is returned to the inverting input is found by applying the voltage-divider rule to the feedback network.

$$V_{in} \cong BV_{out} \cong \left(\frac{R_i}{R_f + R_i}\right)V_{out}$$

Rearranging,

$$\frac{V_{out}}{V_{in}} = \left(\frac{R_f + R_i}{R_i}\right)$$

which can be expressed as follows:

Equation 20–11

$$A_{cl(\text{NI})} = \frac{R_f}{R_i} + 1$$

Equation 20–11 shows that the closed-loop voltage gain, $A_{cl(NI)}$, of the noninverting (NI) amplifier is not dependent on the op-amp's open-loop gain but can be set by selecting values of R_i and R_f. This equation is based on the assumption that the open-loop gain is very high compared to the ratio of the feedback resistors, causing the input differential voltage, V_{diff}, to be very small. In nearly all practical circuits, this is an excellent assumption.

For those rare cases where a more exact equation is necessary, the output voltage can be expressed as

$$V_{out} = V_{in}\left(\frac{A_{ol}}{1 + A_{ol}B}\right)$$

The following formula gives the exact solution of the closed-loop gain:

$$A_{cl(NI)} = \frac{V_{out}}{V_{in}} = \left(\frac{A_{ol}}{1 + A_{ol}B}\right)$$

EXAMPLE 20–5

Determine the closed-loop voltage gain of the amplifier in Figure 20–22. The open-loop voltage gain of the op-amp is 100,000.

▶ **FIGURE 20–22**

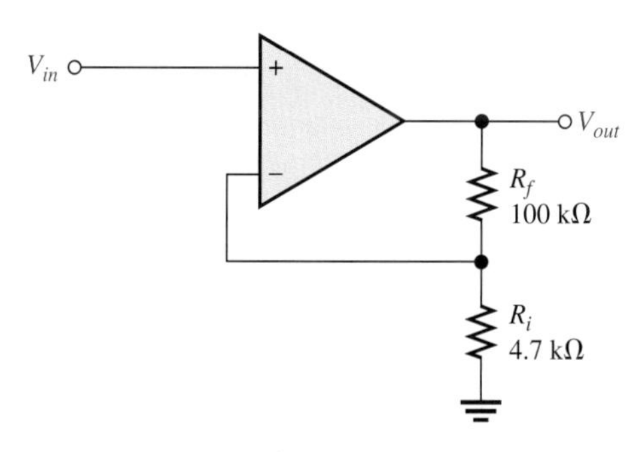

Solution This is a noninverting op-amp configuration. Therefore, the closed-loop voltage gain is

$$A_{cl(NI)} = \frac{R_f}{R_i} + 1 = \frac{100\ \text{k}\Omega}{4.7\ \text{k}\Omega} + 1 = \mathbf{22.3}$$

Related Problem If R_f in Figure 20–22 is increased to 150 kΩ, determine the closed-loop gain.

Open file E20-05 on your EWB/CircuitMaker CD-ROM. Measure the closed-loop gain of the amplifier. Change R_f to 150 kΩ and measure the gain.

Voltage-Follower The voltage follower configuration is a special case of the noninverting amplifier where all of the output voltage is fed back to the inverting input (−) by a straight connection, as shown in Figure 20–23. As you can see, the straight feedback connection has a voltage gain of approximately 1. The closed-

loop voltage gain of a noninverting amplifier is $1/B$ as previously derived. Since $B = 1$, the closed-loop gain of the voltage-follower is

Equation 20–12

$$A_{cl(\mathrm{VF})} = 1$$

The most important features of the voltage-follower configuration are its very high input impedance and its very low output impedance. These features make it a nearly ideal buffer amplifier for interfacing high-impedance sources and low-impedance loads. This is discussed further in Section 20–6.

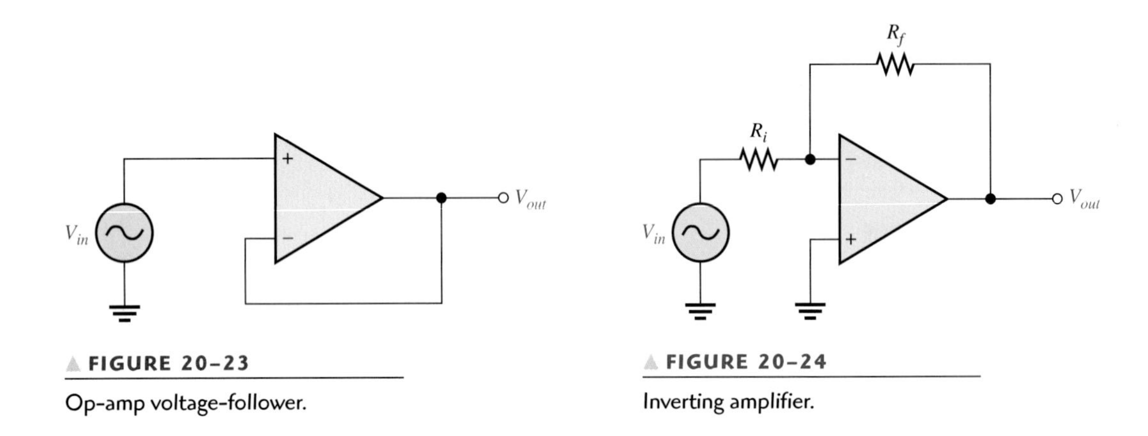

▲ **FIGURE 20–23**

Op-amp voltage-follower.

▲ **FIGURE 20–24**

Inverting amplifier.

Inverting Amplifier

An op-amp connected as an **inverting amplifier** with a controlled amount of voltage gain is shown in Figure 20–24. The input signal is applied through a series input resistor (R_i) to the inverting input (−). Also, the output is fed back through R_f to the inverting input. The noninverting input (+) is grounded.

At this point, the ideal op-amp parameters mentioned earlier are useful in simplifying the analysis of this circuit. In particular, the concept of infinite input impedance is of great value. An infinite input impedance implies that there is *no* current out of the inverting input. If there is no current through the input impedance, then there must be *no* voltage drop between the inverting and noninverting inputs. This means that the voltage at the inverting input (−) is zero because the noninverting input (+) is grounded. This zero voltage at the inverting input terminal is referred to as *virtual ground.* This condition is illustrated in Figure 20–25(a).

▼ **FIGURE 20–25**

Virtual ground concept and closed-loop voltage gain development for the inverting amplifier.

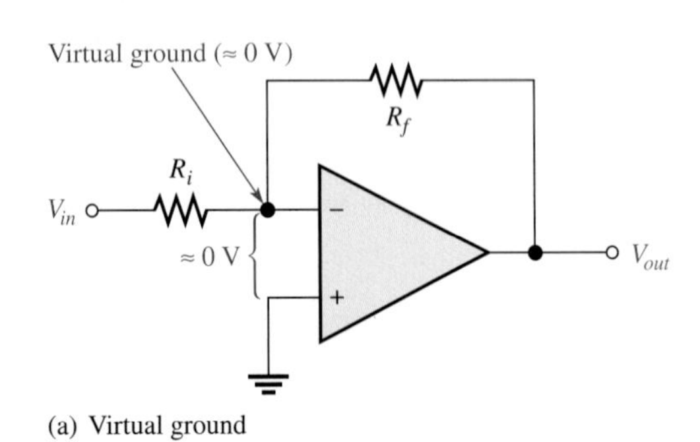

(a) Virtual ground

I_f
+ R_f −
I_{in}
0 V
V_{in}
+ R_i −
$I_i = 0$
V_{out}

(b) $I_{in} = I_f$ and current at the inverting input, $I_i = 0$

Since there is no current at the inverting input, the current through R_i and the current through R_f are equal, as shown in Figure 20–25(b).

$$I_{in} = I_f$$

The voltage across R_i equals V_{in} because of virtual ground on the other side of the resistor. Therefore,

$$I_{in} = \frac{V_{in}}{R_i}$$

Also, the voltage across R_f equals $-V_{out}$ because of virtual ground, and therefore,

$$I_f = \frac{-V_{out}}{R_f}$$

Since $I_f = I_{in}$,

$$\frac{-V_{out}}{R_f} = \frac{V_{in}}{R_i}$$

Rearranging the terms,

$$\frac{V_{out}}{V_{in}} = -\frac{R_f}{R_i}$$

Of course, V_{out}/V_{in} is the overall gain of the inverting amplifier.

$$A_{cl(\text{I})} = -\frac{R_f}{R_i}$$

Equation 20–13

Equation 20–13 shows that the closed-loop voltage gain $A_{cl(\text{I})}$ of the inverting amplifier is the ratio of the feedback resistance R_f to the input resistance R_i. *The closed-loop gain is independent of the op-amp's internal open-loop gain.* Thus, the negative feedback stabilizes the voltage gain. The negative sign indicates inversion.

EXAMPLE 20–6

Given the op-amp configuration in Figure 20–26, determine the value of R_f required to produce a closed-loop voltage gain of -100.

▶ **FIGURE 20–26**

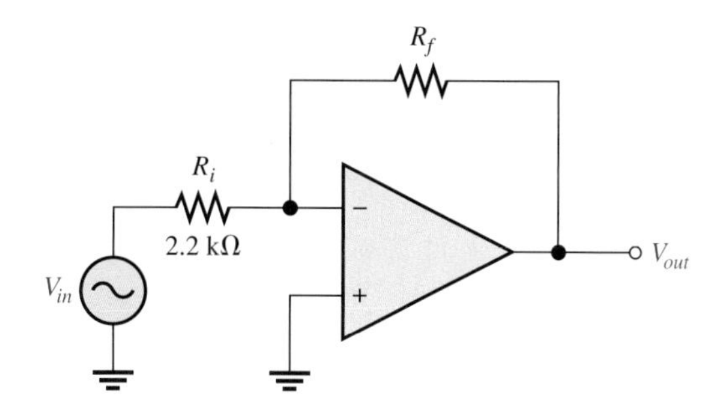

Solution Knowing that $R_i = 2.2\ \text{k}\Omega$ and $A_{cl(\text{I})} = -100$, calculate R_f as follows:

$$A_{cl(\text{I})} = -\frac{R_f}{R_i}$$

$$R_f = -A_{cl(\text{I})}R_i = -(-100)(2.2\ \text{k}\Omega) = \mathbf{220\ k\Omega}$$

Related Problem **(a)** If R_i is changed to 2.7 kΩ in Figure 20–26, what value of R_f is required to produce a closed-loop gain of −25?

(b) If R_f failed open, what would you expect to see at the output?

Open file E20-06 on your EWB/CircuitMaker CD-ROM. Set the value of R_f to the value calculated in the example and measure the closed-loop gain.

SECTION 20–5 REVIEW

1. What is the main purpose of negative feedback?
2. The closed-loop voltage gain of each of the op-amp configurations discussed is dependent on the internal open-loop voltage gain of the op-amp. (True or False)
3. The attenuation (B) of the negative feedback circuit of a noninverting op-amp configuration is 0.02. What is the closed-loop gain of the amplifier?

20–6 OP-AMP IMPEDANCES

In this section, you will see how a negative feedback connection affects the input and output impedances of an op-amp. The effects on both inverting and noninverting amplifiers are examined.

After completing this section, you should be able to

- **Describe effects of negative feedback on the three basic op-amp configurations**
- Determine the input and output impedances for the noninverting amplifier
- Determine the input and output impedances for the voltage-follower
- Determine the input and output impedances for the inverting amplifier

Input and Output Impedances of a Noninverting Amplifier

The input impedance of a noninverting amplifier configuration, $Z_{in(NI)}$, shown in Figure 20–27, is greater than the internal input impedance of the op-amp itself (without feedback) by a factor of $1 + A_{ol}B$, as expressed by the following equation:

Equation 20–14

$$Z_{in(IN)} = (1 + A_{ol}B)Z_{in}$$

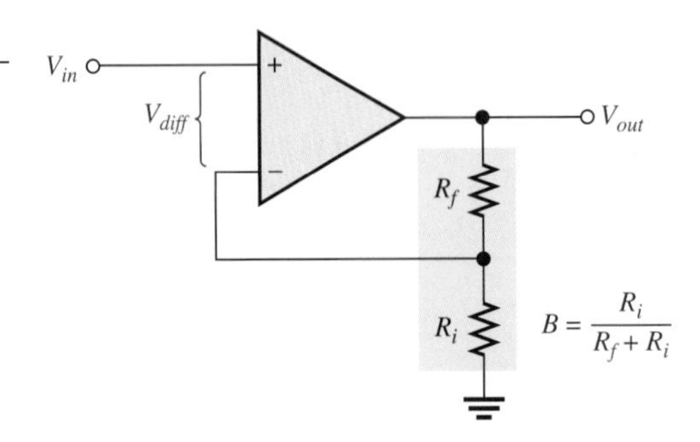

▶ **FIGURE 20–27**

Noninverting amplifier.

The output impedance with the negative feedback, $Z_{out(NI)}$, is less than the op-amp output impedance by a factor of $1/(1 + A_{ol}B)$, expressed as follows:

$$Z_{out(NI)} = \frac{Z_{out}}{1 + A_{ol}B}$$

Equation 20–15

In summary, the negative feedback in a noninverting configuration increases the input impedance and decreases the output impedance.

EXAMPLE 20–7

(a) Determine the input and output impedances of the amplifier in Figure 20–28. The op-amp data sheet gives $Z_{in} = 2\ \text{M}\Omega$, $Z_{out} = 75\ \Omega$, and $A_{ol} = 200{,}000$.

(b) Find the closed-loop voltage gain.

FIGURE 20–28

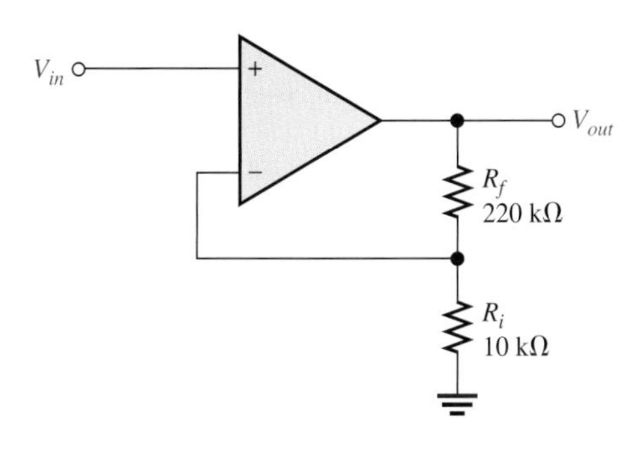

Solution **(a)** The attenuation, B, of the feedback circuit is

$$B = \frac{R_i}{R_i + R_f} = \frac{10\ \text{k}\Omega}{230\ \text{k}\Omega} = 0.0435$$

$$Z_{in(NI)} = (1 + A_{ol}B)Z_{in} = [1 + (200{,}000)(0.0435)](2\ \text{M}\Omega) = \mathbf{17.4\ G\Omega}$$

$$Z_{out(NI)} = \frac{Z_{out}}{1 + A_{ol}B} = \frac{75\ \Omega}{1 + 8700} = \mathbf{8.6\ m\Omega}$$

(b) $$A_{cl(NI)} = \frac{1}{B} = \frac{1}{0.0435} = \mathbf{23.0}$$

Related Problem **(a)** Determine the input and output impedances in Figure 20–28 for op-amp data sheet values of $Z_{in} = 3.5\ \text{M}\Omega$, $Z_{out} = 82\ \Omega$, and $A_{ol} = 135{,}000$.

(b) Find $A_{cl(NI)}$.

Open file E20-07 on your EWB/CircuitMaker CD-ROM. Measure the closed-loop gain.

Voltage-Follower Impedances

Since the voltage-follower is a special case of the noninverting configuration, the same impedance formulas are used, but with $B = 1$.

$$Z_{in(VF)} = (1 + A_{ol})Z_{in}$$

Equation 20–16

$$Z_{out(VF)} = \frac{Z_{out}}{1 + A_{ol}}$$

Equation 20–17

As you can see, the voltage-follower input impedance is greater for a given A_{ol} and Z_{in} than for the noninverting configuration with the voltage-divider feedback circuit. Also, its output impedance is much smaller because B is normally much less than 1 for a noninverting configuration.

EXAMPLE 20–8

The same op-amp in Example 20–7 is used in a voltage-follower configuration. Determine the input and output impedances.

Solution Since $B = 1$,

$$Z_{in(\text{VF})} = (1 + A_{ol})Z_{in} = (1 + 200{,}000)2\text{ M}\Omega = \mathbf{400\text{ G}\Omega}$$

$$Z_{out(\text{VF})} = \frac{Z_{out}}{1 + A_{ol}} = \frac{75\ \Omega}{1 + 200{,}000} = \mathbf{375\ \mu\Omega}$$

Notice that $Z_{in(\text{VF})}$ is much greater than $Z_{in(\text{NI})}$, and $Z_{out(\text{VF})}$ is much less than $Z_{out(\text{NI})}$ from Example 20–7.

Related Problem If the op-amp in this example is replaced with one having a higher open-loop gain, how are the input and output impedances affected?

Input and Output Impedances of an Inverting Amplifier

For the inverting amplifier configuration shown in Figure 20–29, the output impedance, $Z_{in(\text{I})}$, approximately equals the external input resistance, R_i.

Equation 20–18

$$Z_{in(\text{I})} \cong R_i$$

The output impedance, $Z_{out(\text{I})}$, approximately equals the internal output impedance of the op-amp, Z_{out}.

Equation 20–19

$$Z_{out(\text{I})} \cong Z_{out}$$

FIGURE 20–29

Inverting amplifier.

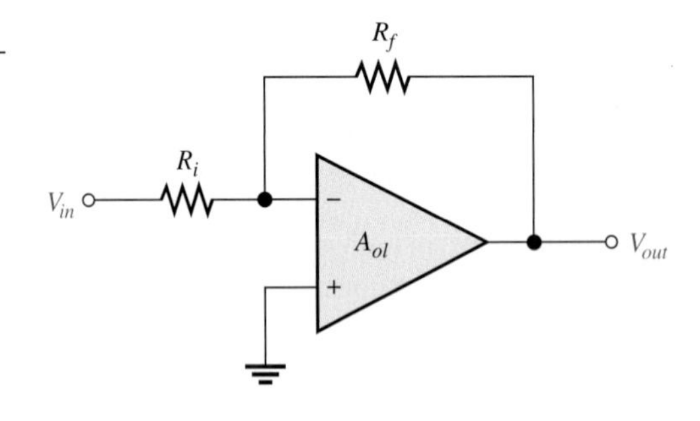

EXAMPLE 20–9

The op-amp in Figure 20–30 has the following parameters: $A_{ol} = 50{,}000$, $Z_{in} = 4\text{ M}\Omega$, and $Z_{out} = 50\ \Omega$.

(a) Find the values of the input and output impedances.

(b) Determine the closed-loop voltage gain.

FIGURE 20–30

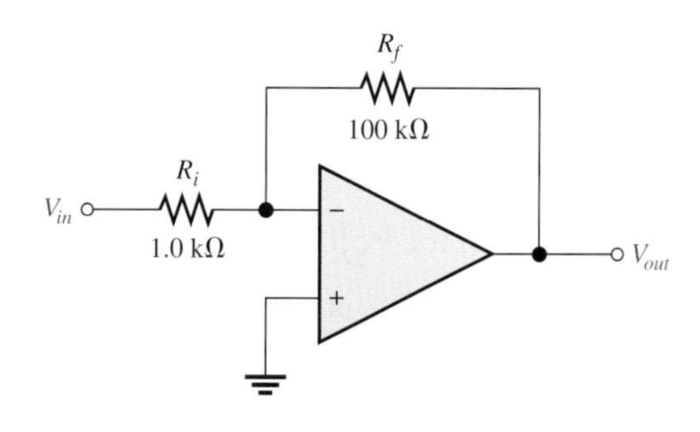

Solution (a) $Z_{in(I)} \cong R_i = \mathbf{1.0\ k\Omega}$

$Z_{out(I)} \cong Z_{out} = \mathbf{50\ \Omega}$

(b) $A_{cl(I)} = -\dfrac{R_f}{R_i} = -\dfrac{100\ \text{k}\Omega}{1.0\ \text{k}\Omega} = \mathbf{-100}$

Related Problem Determine $Z_{in(I)}$, $Z_{out(I)}$, and A_{cl} in Figure 20–30 for the following values: $A_{ol} = 100{,}000$; $Z_{in} = 5\ \text{M}\Omega$; $Z_{out} = 75\ \Omega$; $R_i = 560\ \Omega$, $R_f = 82\ \text{k}\Omega$.

Open file E20-09 on your EWB/CircuitMaker CD-ROM. Measure the closed-loop gain.

SECTION 20–6 REVIEW

1. How does the input impedance of a noninverting amplifier configuration compare to the input impedance of the op-amp itself?
2. Connecting an op-amp in a voltage-follower configuration (increases, decreases) the input impedance.
3. Given that $R_f = 100\ \text{k}\Omega$, $R_i = 2\ \text{k}\Omega$, $A_{ol} = 120{,}000$, $Z_{in} = 2\ \text{M}\Omega$, and $Z_{out} = 60\ \Omega$, determine $Z_{in(I)}$ and $Z_{out(I)}$ for an inverting amplifier configuration.

20–7 TROUBLESHOOTING

As a technician, you will encounter situations in which an op-amp or its associated circuitry has malfunctioned. The op-amp is a complex integrated circuit with many types of internal failures possible. However, since you cannot troubleshoot the op-amp internally, you treat it as a single device with only a few connections to it. If it fails, you replace it just as you would a resistor, capacitor, or transistor.

After completing this section, you should be able to

- **Troubleshoot op-amp circuits**
- Compensate for input offset voltage
- Troubleshoot for faulty op-amps, open resistors, and incorrect offset compensation

In the basic op-amp configurations, there are only a few external components that can fail. These are the feedback resistor, the input resistor, and the potentiometer used for offset voltage compensation. Also, of course, the op-amp itself can fail or there can be faulty contacts in the circuit. Let's examine the three basic configurations for possible faults and the associated symptoms.

Input Offset Voltage Compensation

Most integrated circuit op-amps provide a means of compensating for offset voltage. This is usually done by connecting an external potentiometer to designated pins on the IC package, as illustrated in Figure 20–31(a) and (b) for a 741 op-amp. The two terminals are labeled *offset null*. With no input, the potentiometer is simply adjusted until the output voltage reads 0, as shown in Figure 20–31(c).

FIGURE 20–31

Input offset voltage compensation for a 741 op-amp.

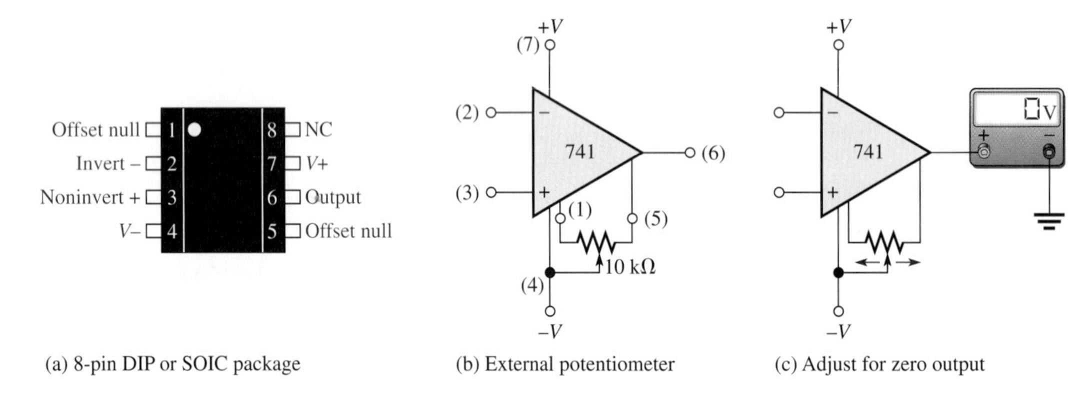

(a) 8-pin DIP or SOIC package (b) External potentiometer (c) Adjust for zero output

Faults in a Noninverting Amplifier

The first thing to do when you suspect a faulty circuit is to check for the proper supply voltage and ground. After you have done that, troubleshoot for other possible faults.

Open Feedback Resistor If the feedback resistor, R_f, in Figure 20–32 opens, the op-amp is operating with its very high open-loop gain, which causes the input signal to drive the device into nonlinear operation and results in a severely clipped output signal as shown in part (a).

Open Input Resistor In this case, you still have a closed-loop configuration. But, since R_i is open and effectively equal to infinity (∞), the closed-loop gain from Equation 20–11 is

$$A_{cl(NI)} = \frac{R_f}{R_i} + 1 = \frac{R_f}{\infty} + 1 = 0 + 1 = 1$$

This shows that the amplifier acts like a voltage-follower. You would observe an output signal that is the same as the input, as indicated in Figure 20–32(b).

Open or Incorrectly Adjusted Offset Null Potentiometer In this situation, the output offset voltage will cause the output signal to begin clipping on only one peak as the input signal is increased to a sufficient amplitude. This is indicated in Figure 20–32(c).

Faulty Op-Amp As mentioned, many things can happen to an op-amp. In general, an internal failure will result in a loss or distortion of the output signal. The best approach is to first make sure that there are no external failures or faulty conditions. If everything else is good, then the op-amp must be bad.

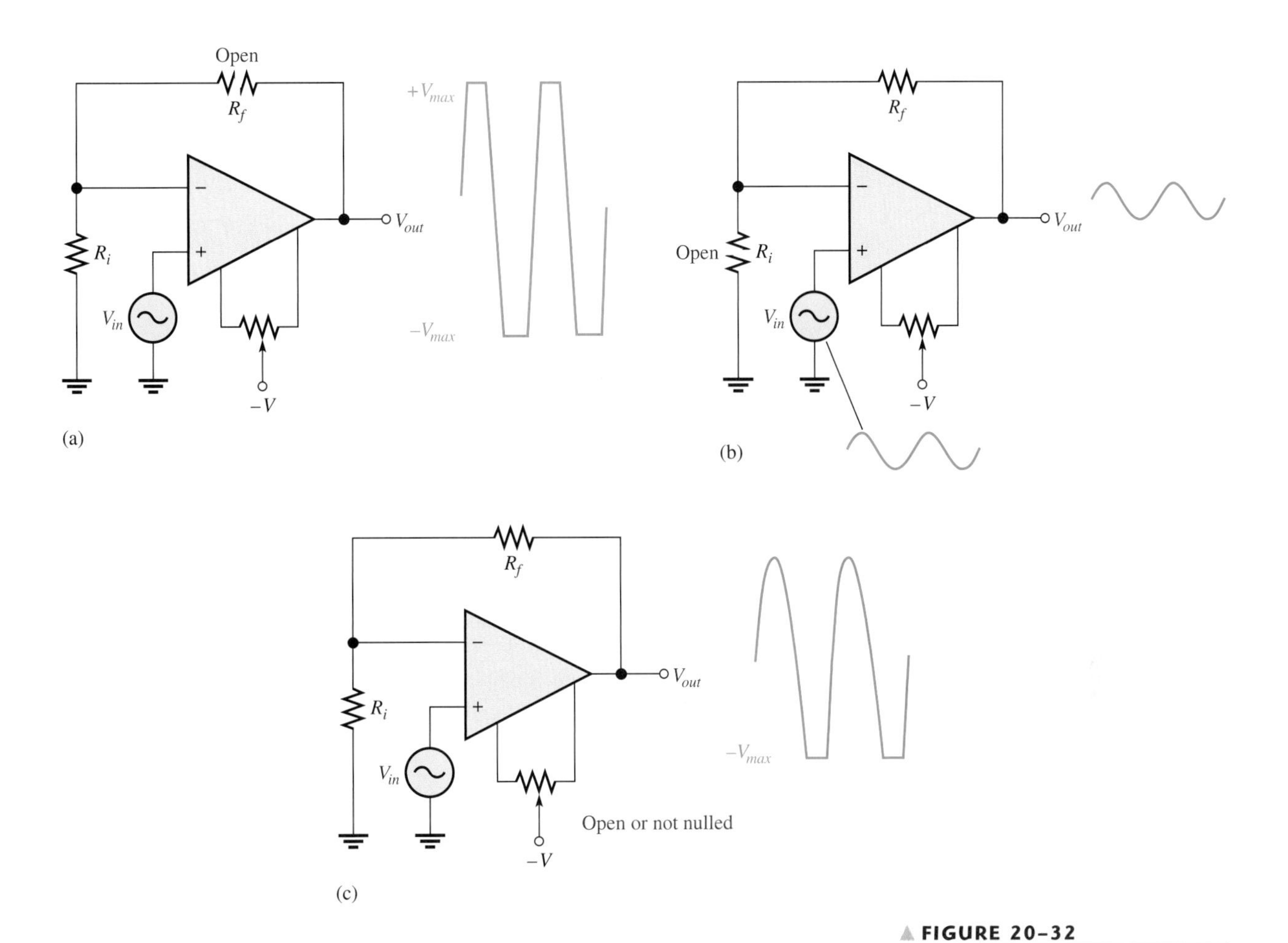

▲ **FIGURE 20–32**

Faults in a noninverting amplifier.

Faults in a Voltage-Follower

The voltage-follower is a special case of a noninverting amplifier. Except for a bad op-amp, a bad external connection, or a problem with the offset null potentiometer, about the only thing that can happen in a voltage-follower circuit is an open feedback loop. This would have the same effect as an open feedback resistor as previously discussed.

Faults in an Inverting Amplifier

Open Feedback Resistor If R_f opens as indicated in Figure 20–33(a), the input signal still feeds through the input resistor and is amplified by the high open-loop gain of the op-amp. This forces the device to be driven into nonlinear operation, and you will see an output something like that shown. This is the same result as in the noninverting configuration.

Open Input Resistor This prevents the input signal from getting to the op-amp input, so there will be no output signal, as indicated in Figure 20–33(b).

Failures in the op-amp itself or the offset null potentiometer have the same effects as previously discussed in the noninverting amplifier.

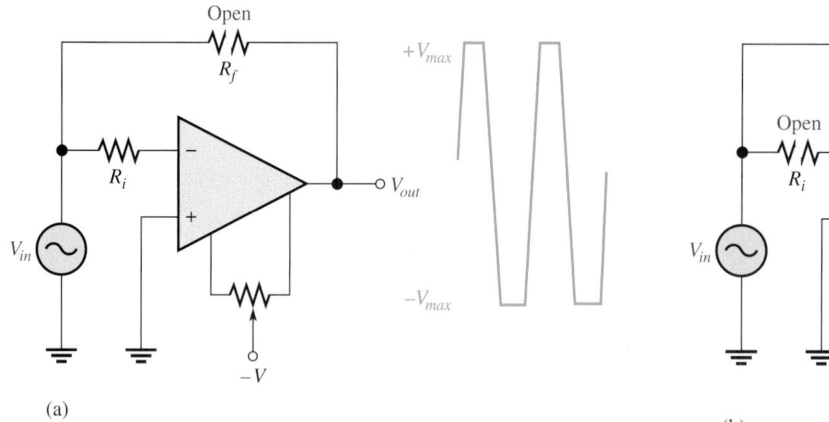

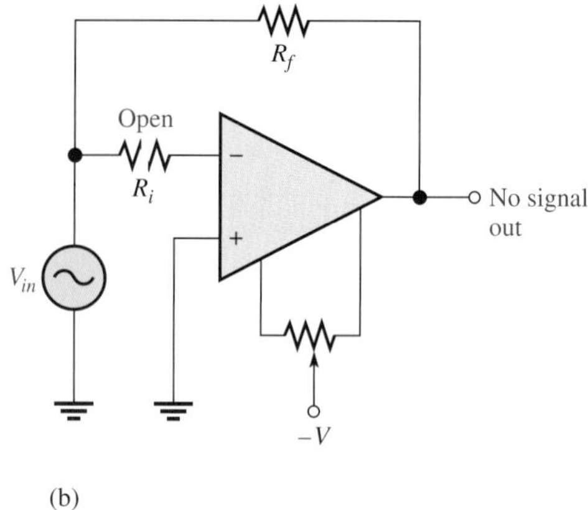

FIGURE 20–33

Faults in an inverting amplifier.

SECTION 20–7 REVIEW

1. If you notice that the op-amp output signal is beginning to clip on one peak as you increase the input signal, what should you check first?
2. If there is no op-amp output signal when there is a verified input signal, what would you suspect as being faulty?

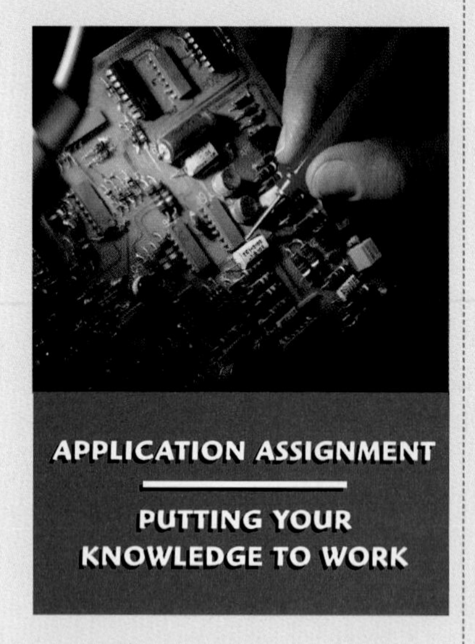

APPLICATION ASSIGNMENT

PUTTING YOUR KNOWLEDGE TO WORK

As an electronics technician in a chemical lab, you have been assigned to work on a spectrophotometer system which combines light optics with electronics to analyze the chemical makeup of various solutions. This type of system is common in chemical and medical laboratories as well as many other areas. It is another example of mixed systems in which electronic circuits interface with other types of systems, such as mechanical and optical, to accomplish a specific function.

Description of the System

The light source in Figure 20–34 produces a beam of visible light containing a wide spectrum of wavelengths. Each component wavelength in the beam of light is refracted at a different angle by the prism, as indicated. Depending on the angle of the platform as set by the pivot angle controller, a certain wavelength passes through the narrow slit and is transmitted through the solution under analysis. By precisely pivoting the light source and prism, a selected wavelength can be transmitted. Every chemical and compound absorbs different wavelengths of light in different ways, so the resulting light coming through the solution has a unique "signature" that can be used to define the chemicals in the solution.

The photocell on the circuit board produces a voltage that is proportional to the amount of light and wavelength. The op-amp circuit amplifies the photovoltaic cell output and sends the resulting signal to the processing and display unit where the type of chemical(s) in the solution is identified. This is usually a microprocessor-based digital system. Although these other system blocks are interesting, our focus in this section is the photocell and op-amp circuit board.

Step 1: Relate the PC Board to a Schematic

Develop a complete schematic by carefully following the conductive traces on the PC board shown in Figure 20–34 to see how the components are interconnected. The middle pin on the potentiometer is the variable. Some of the interconnecting traces are on the reverse side of the board, but if you are familiar with basic op-amp configurations, you should have no trouble figuring out the connections. Refer to Figure 20–31 for the pin layout of the 741 op-amp. This op-amp is housed in a surface-mount SOIC-8 package.

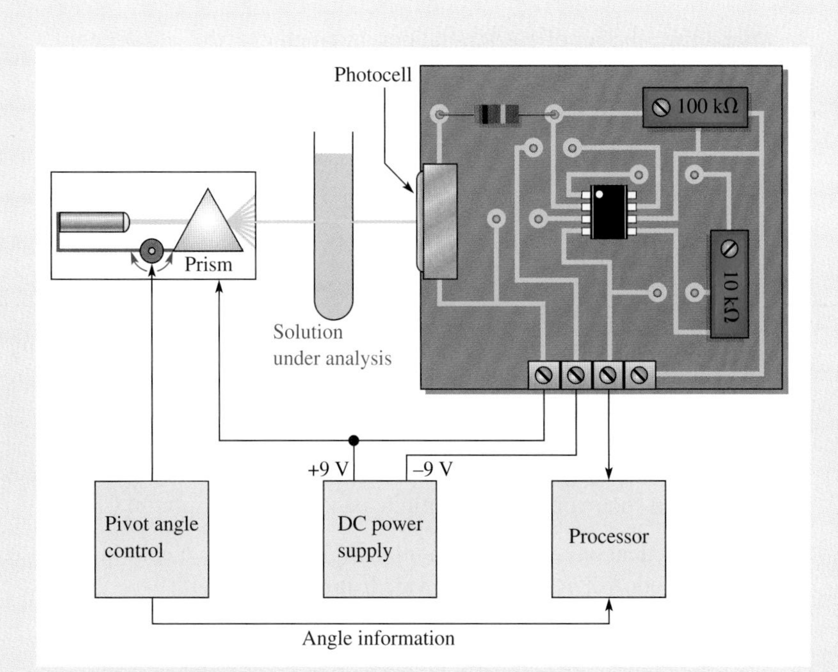

FIGURE 20–34

Spectrophotometer system.

Step 2: Analyze the Circuit

1. Determine the resistance value to which the feedback rheostat must be adjusted for a voltage gain of 10.
2. Assume the maximum linear output of the op-amp is 1 V less than the supply voltage. Determine the voltage gain required and the value to which the feedback resistance must be set to achieve the maximum linear output. The system light source produces constant light output from 400 nm to 700 nm, which is approximately the full range of visible light from violet to red. The maximum voltage from the photocell is 0.5 V at 800 nm.
3. Using the gain previously found, determine the op-amp output voltage over the range of wavelengths from 400 nm to 700 nm in 50 nm intervals and plot a graph of the results. Refer to the photocell response curve in Figure 20–35.

Step 3: Troubleshoot the Circuit

For each of the following problems, state the probable cause or causes:

1. Zero voltage at the op-amp output. List three possible causes.
2. Output of op-amp stays at approximately −8 V.
3. A small dc voltage on the op-amp output under no-light conditions.

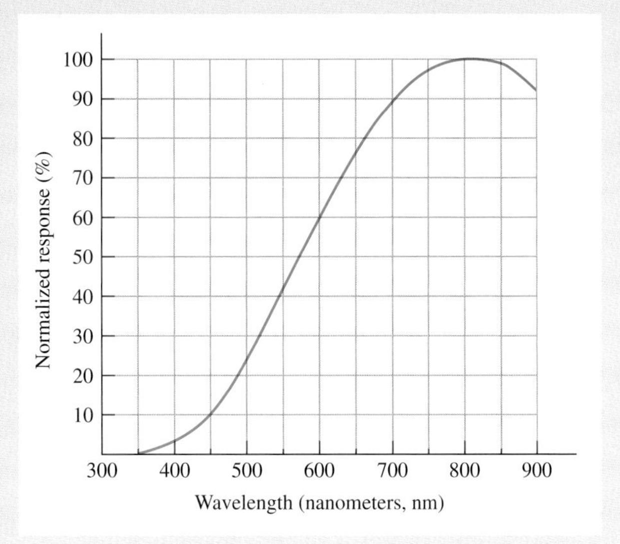

FIGURE 20–35

Photocell response curve.

APPLICATION ASSIGNMENT REVIEW

1. What is the purpose of the 100 kΩ potentiometer on the circuit board?
2. What is the purpose of the 10 kΩ potentiometer?
3. Explain why the light source and prism must be pivoted.

SUMMARY

- The op-amp has three terminals not including power and ground: inverting input (−), noninverting input (+), and output.
- Most op-amps require both a positive and a negative dc supply voltage.
- The ideal (perfect) op-amp has infinite impedance, zero output impedance, infinite open-loop voltage gain, infinite bandwidth, and infinite CMRR.
- A good practical op-amp has high input impedance, low output impedance, high open-loop voltage gain, and wide bandwidth.
- A diff-amp is normally used for the input stage of an op-amp.
- A differential input voltage appears between the inverting and noninverting inputs of a diff-amp.
- A single-ended input voltage appears between one input and ground (with the other inputs grounded).
- A differential output voltage appears between two output terminals of a diff-amp.
- A single-ended output voltage appears between the output and ground of a diff-amp.
- Common mode occurs when equal, in-phase voltages are applied to both input terminals.
- Input offset voltage produces an output error voltage (with no input voltage).
- Input bias current also produces an output error voltage (with no input voltage).
- Input offset current is the difference between the two bias currents.
- Open-loop voltage gain is the gain of the op-amp with no external feedback connections.
- Slew rate is the rate (in volts per microsecond) that the output voltage of an op-amp can change in response to a step input.
- Negative feedback occurs when a portion of the output voltage is connected back to the inverting input such that it subtracts from the input voltage, thus reducing the voltage gain but increasing the stability and bandwidth.
- There are three basic op-amp configurations: inverting, noninverting, and voltage-follower.
- All op-amp configurations (except comparators, covered in Chapter 21) employ negative feedback.
- A noninverting amplifier configuration has a higher impedance and a lower output impedance than the op-amp itself (without feedback).

- An inverting amplifier configuration has an input impedance approximately equal to the input resistor R_i and an output impedance approximately equal to the internal output impedance of the op-amp itself.
- The voltage-follower has the highest input impedance and the lowest output impedance of the three configurations.

EQUATIONS

20–1	$\mathbf{CMRR} = \dfrac{A_{v(d)}}{A_{cm}}$	Common-mode rejection ratio (diff-amp)
20–2	$\mathbf{CMRR} = 20 \log\left(\dfrac{A_{v(d)}}{A_{cm}}\right)$	Common-mode rejection ratio (dB)
20–3	$V_{\mathrm{OUT}} = I_{\mathrm{C2}}R_{\mathrm{C}} - I_{\mathrm{C1}}R_{\mathrm{C}}$	Differential output
20–4	$I_{\mathrm{BIAS}} = \dfrac{I_1 + I_2}{2}$	Input bias current
20–5	$I_{\mathrm{OS}} = \lvert I_1 - I_2 \rvert$	Input offset current
20–6	$V_{\mathrm{OS}} = I_{\mathrm{OS}}R_{\mathrm{in}}$	Offset voltage
20–7	$V_{\mathrm{OUT(error)}} = A_v I_{\mathrm{OS}}R_{\mathrm{in}}$	Output error voltage
20–8	$\mathbf{CMRR} = \dfrac{A_{ol}}{A_{cm}}$	Common-mode rejection ratio (op-amp)
20–9	$\mathbf{CMRR} = 20 \log\left(\dfrac{A_{ol}}{A_{cm}}\right)$	Common-mode rejection ratio (dB)
20–10	$\textbf{Slew rate} = \dfrac{\Delta V_{out}}{\Delta t}$	Slew rate
20–11	$A_{cl(\mathrm{NI})} = \dfrac{R_f}{R_i} + 1$	Voltage gain (noninverting)
20–12	$A_{cl\,(\mathrm{VF})} = 1$	Voltage gain (voltage-follower)
20–13	$A_{cl\,(\mathrm{I})} = -\dfrac{R_f}{R_i}$	Voltage gain (inverting)
20–14	$Z_{in(\mathrm{NI})} = (1 + A_{ol}B)Z_{in}$	Input impedance (noninverting)
20–15	$Z_{out(\mathrm{NI})} = \dfrac{Z_{out}}{1 + A_{ol}B}$	Output impedance (noninverting)
20–16	$Z_{in(\mathrm{VF})} = (1 + A_{ol})Z_{in}$	Input impedance (voltage-follower)
20–17	$Z_{out(\mathrm{VF})} = \dfrac{Z_{out}}{1 + A_{ol}}$	Output impedance (voltage-follower)
20–18	$Z_{in(\mathrm{I})} \cong R_i$	Input impedance (inverting)
20–19	$Z_{out\,(\mathrm{I})} \cong Z_{out}$	Output impedance (inverting)

SELF-TEST

Answers are at the end of the chapter.

1. Which characteristic does not necessarily apply to an op-amp:

(a) high gain **(b)** low power

(c) high input impedance **(d)** low output impedance

2. In selecting an op-amp suppose you have several choices. Of the CMRR values listed, the most desirable is

   (a) 10 dB (b) 20 dB (c) 50 dB (d) 100 dB

3. The output voltage of a particular op-amp increases 8 V in 12 μs in response to a step voltage on the input. The slew rate is

   (a) 0.667 V/μs (b) 1.5 V/μs (c) 96 V/μs (d) 0.75 V/μs

4. A noninverting op-amp configuration has an R_i of 1.0 kΩ and an R_f of 100 kΩ. If V_{out} is 5 V, the value of V_f is

   (a) 50 mV (b) 49.5 mV (c) 495 mV (d) 500 mV

5. In the amplifier described in Question 4, the value of B is

   (a) 0.01 (b) 0.1 (c) 0.0099 (d) 101

6. The closed-loop gain of the amplifier in Question 4 is

   (a) 0.0099 (b) 1 (c) 99 (d) 101

7. One characteristic of a voltage-follower is

   (a) $A_{cl} > 1$ (b) inversion (c) high Z_{out} (d) noninversion

8. An inverting amplifier has the following circuit values: $R_f = 220$ kΩ, $R_i = 2.2$ kΩ, and $A_{ol} = 25{,}000$. The closed-loop gain is

   (a) −100 (b) −0.01 (c) 100 (d) −250

9. If you know an op-amp's open-loop gain and nothing else, you can determine the closed-loop gain of

   (a) an inverting amplifier

   (b) a noninverting amplifier

   (c) a voltage-follower

   (d) none of the amplifier configurations without additional information

10. The feedback attenuation of a voltage-follower is

    (a) unity (b) less than unity

    (c) greater than unity (d) variable

11. The value of B in a certain noninverting amplifier is 0.025. The closed-loop gain is

    (a) unity (b) 40 (c) 0.025 (d) undeterminable

12. The highest possible input impedance is achieved with the

    (a) inverting amplifier (b) noninverting amplifier

    (c) differential amplifier (d) voltage-follower

PROBLEMS

Answers to odd-numbered problems are at the end of the book.

BASIC PROBLEMS

SECTION 20–1 **Introduction to Operational Amplifiers**

1. Compare a practical op-amp to an ideal op-amp.
2. Two IC op-amps are available to you. Their characteristics are as follows:

   Op-amp 1: $Z_{in} = 5$ MΩ, $Z_{out} = 100$ Ω, $A_{ol} = 50{,}000$

   Op-amp 2: $Z_{in} = 10$ MΩ, $Z_{out} = 75$ Ω, $A_{ol} = 150{,}000$

   Choose the one you think is generally more desirable.

SECTION 20–2 **The Differential Amplifier**

3. Identify the type of input and output configuration for each diff-amp in Figure 20–36.

▼ FIGURE 20–36

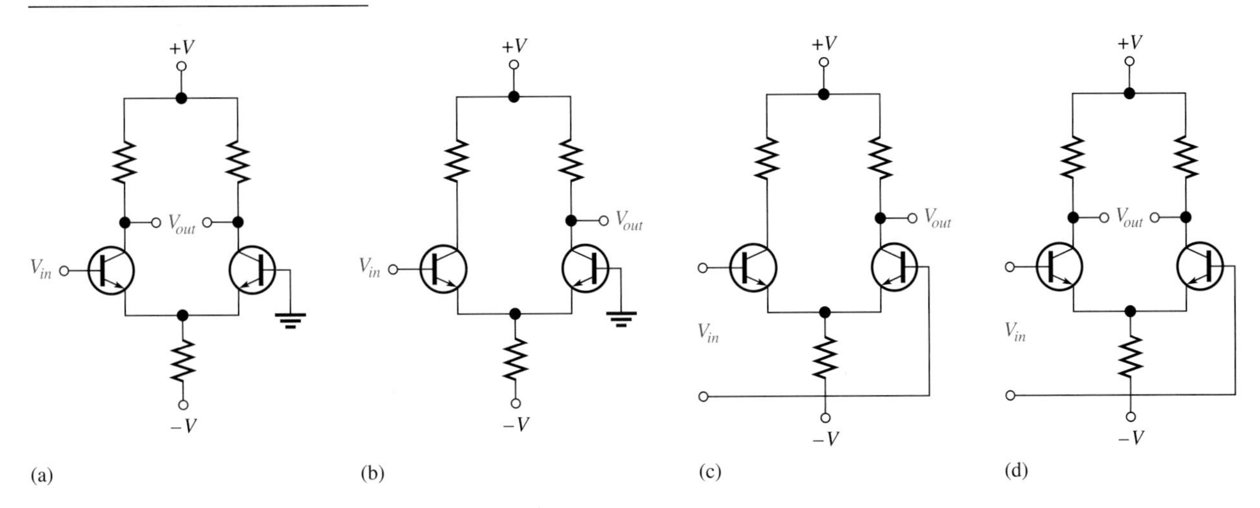

4. The dc base voltages in Figure 20–37 are zero. Using your knowledge of transistor analysis, determine the dc differential output voltage. Assume that Q_1 has an $\alpha = 0.98$ and Q_2 has an $\alpha = 0.975$.

▶ FIGURE 20–37

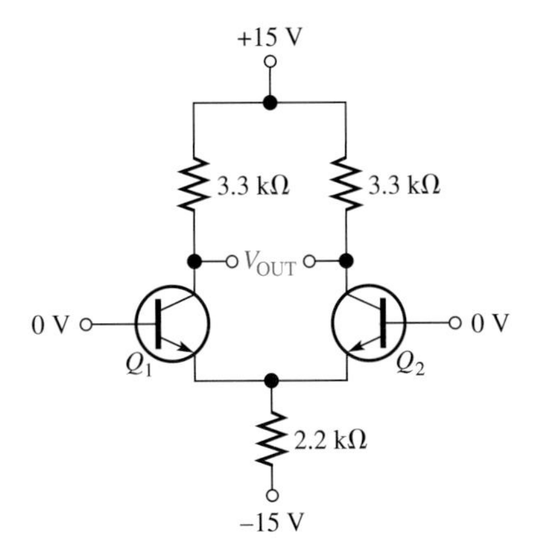

SECTION 20–3 **Op-Amp Parameters**

5. Determine the bias current, I_{BIAS}, given that the input currents to an op-amp are 8.3 μA and 7.9 μA.
6. Distinguish between *input bias current* and *input offset current,* and then calculate the input offset current in Problem 5.
7. A certain op-amp has a CMRR of 250,000. Convert this to decibels.
8. The open-loop gain of a certain op-amp is 175,000. Its common-mode gain is 0.18. Determine the CMRR in decibels.
9. The op-amp data sheet specifies a CMRR of 300,000 and an A_{ol} of 90,000. What is the common-mode gain?

10. Figure 20–38 shows the output voltage of an op-amp in response to a step input. What is the slew rate?

FIGURE 20–38

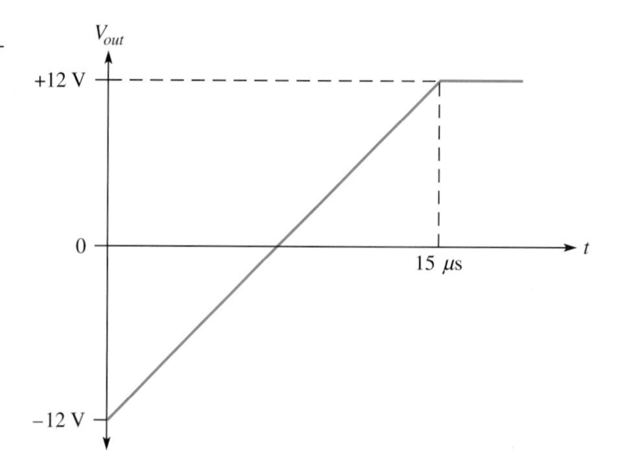

11. How long does it take the output voltage of an op-amp to go from −10 V to +10 V if the slew rate is 0.5 V/μs?

SECTION 20–5

Op-Amp Configurations with Negative Feedback

12. Identify each of the op-amp configurations in Figure 20–39.

FIGURE 20–39

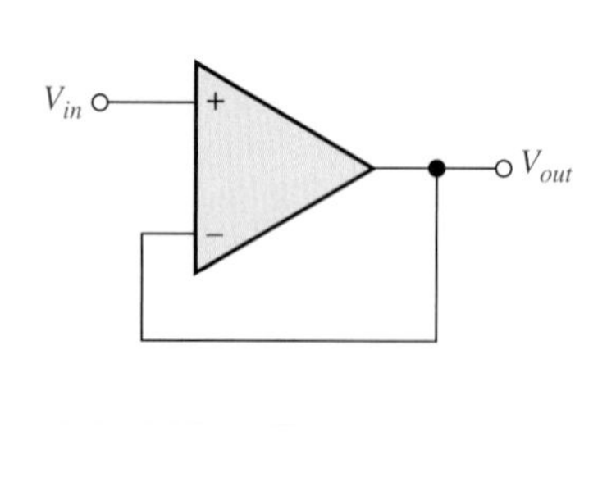

(a)

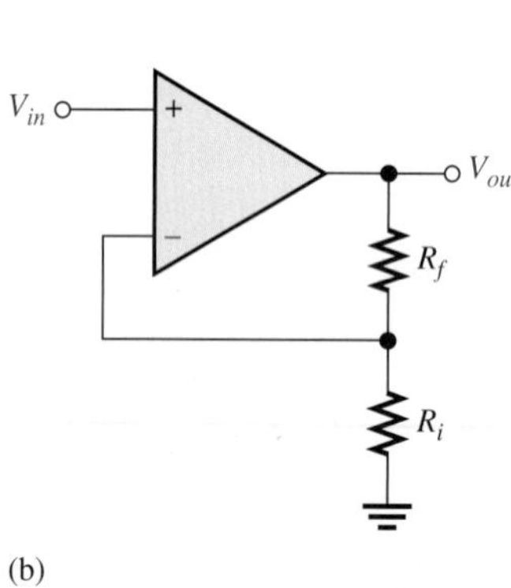

(b)

(c)

13. For the amplifier in Figure 20–40, determine the following:

(a) $A_{cl(NI)}$ **(b)** V_{out} **(c)** V_f

FIGURE 20–40

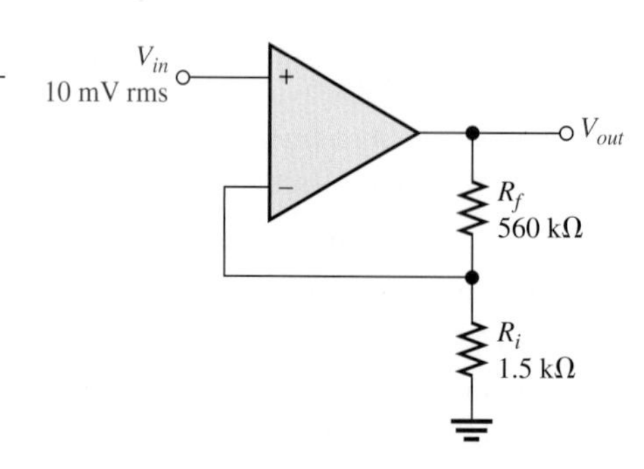

14. Determine the closed-loop gain of each amplifier in Figure 20–41.

▼ FIGURE 20–41

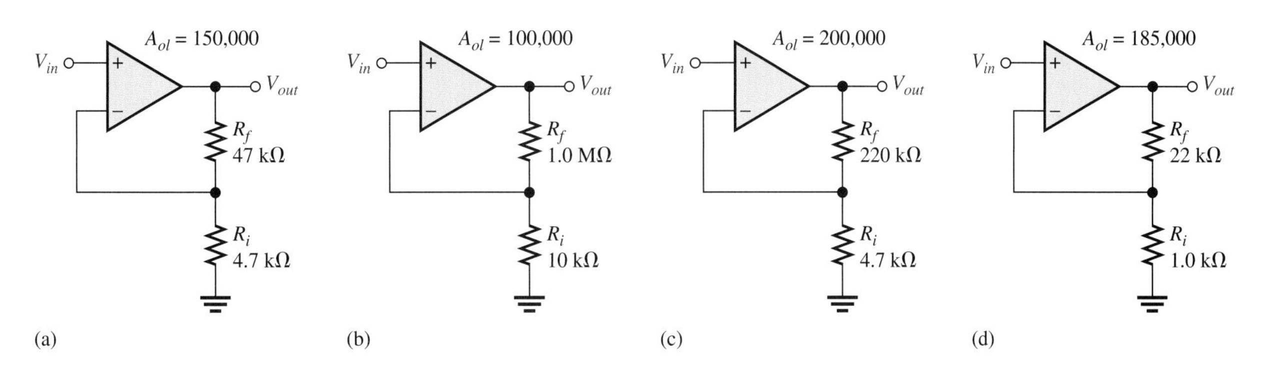

15. Find the value of R_f that will produce the indicated closed-loop gain in each amplifier in Figure 20–42.

▼ FIGURE 20–42

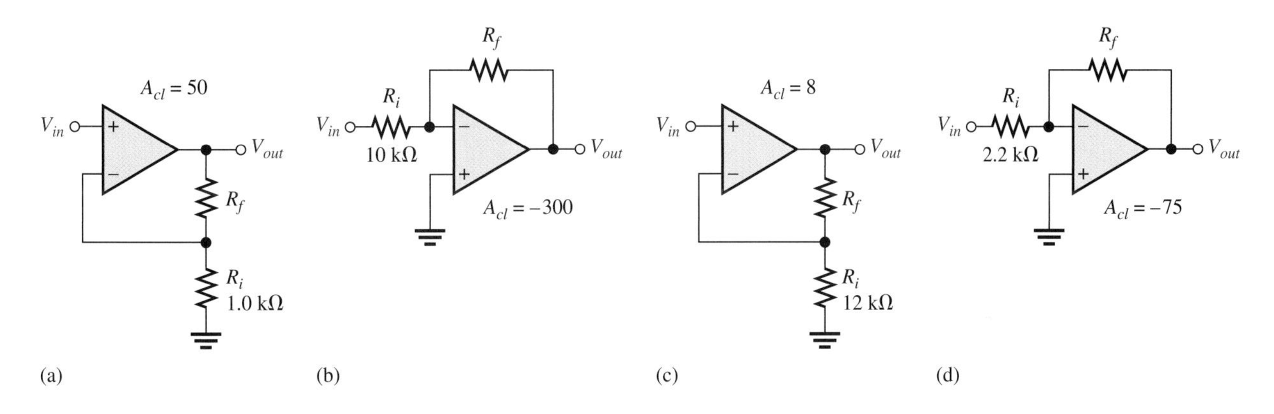

16. Find the gain of each amplifier in Figure 20–43.

▼ FIGURE 20–43

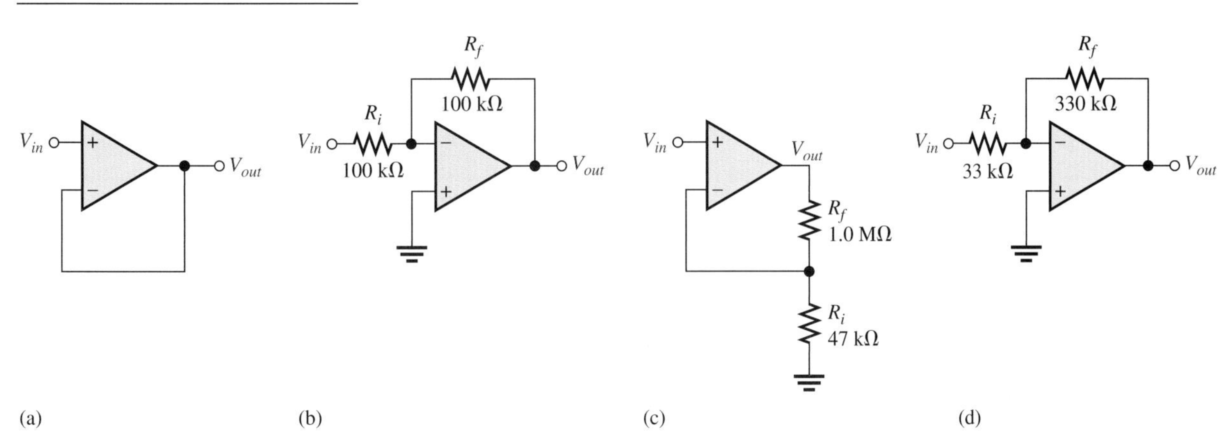

17. If a signal voltage of 10 mV rms is applied to each amplifier in Figure 20–43, what are the output voltages and what is their phase relationship with inputs?

18. Determine the approximate values for each of the following quantities in Figure 20–44:

(a) I_{in} **(b)** I_f **(c)** V_{out} **(d)** closed-loop gain

FIGURE 20–44

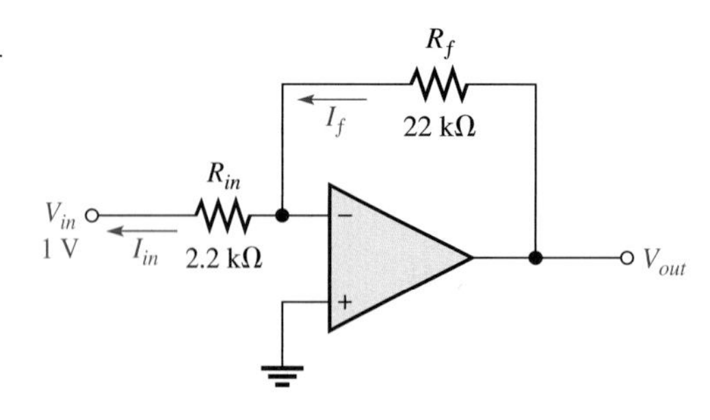

SECTION 20–6

Op-Amp Impedances

19. Determine the input and output impedances for each amplifier configuration in Figure 20–45.

FIGURE 20–45

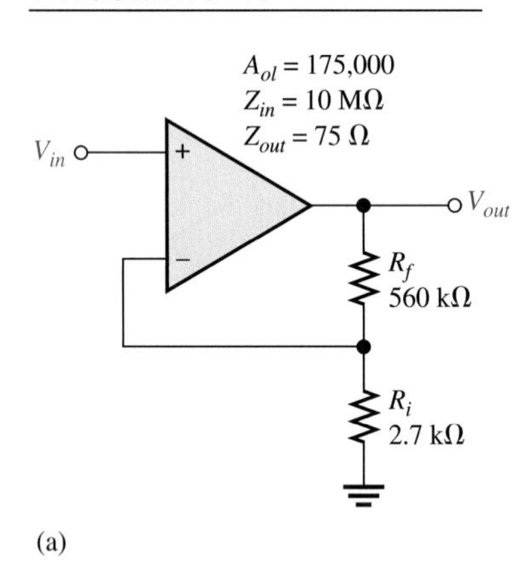

(a)

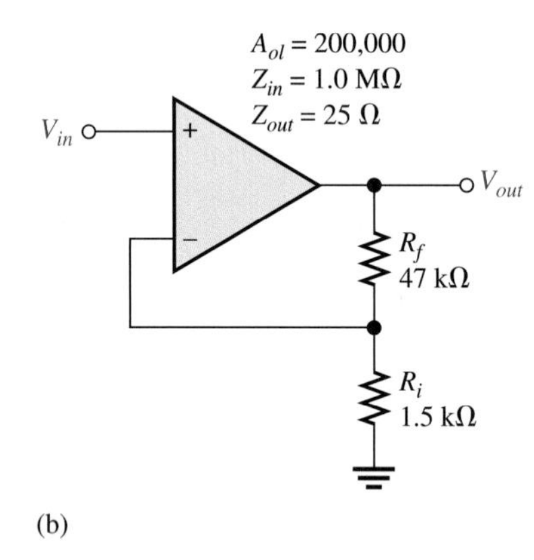

(b)

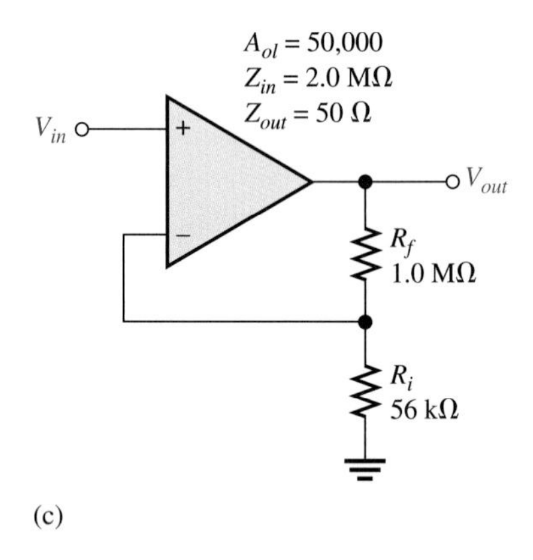

(c)

20. Repeat Problem 19 for each circuit in Figure 20–46.

FIGURE 20–46

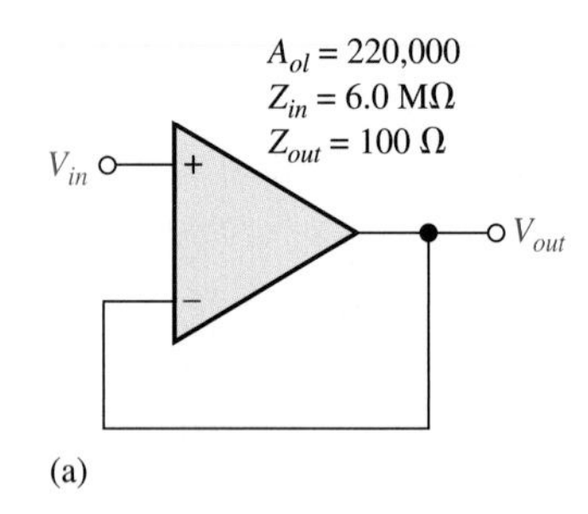

(a)

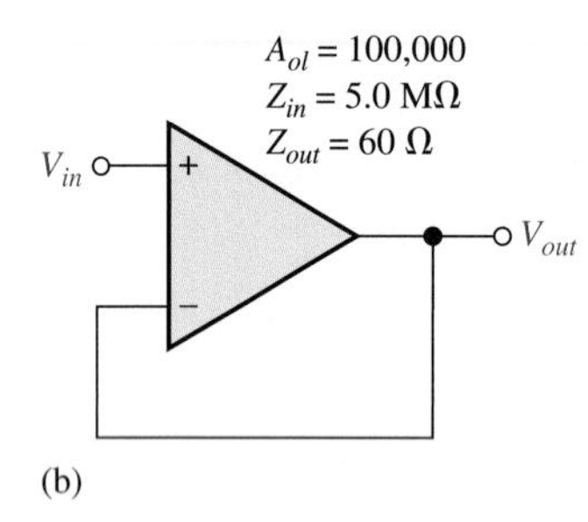

(b)

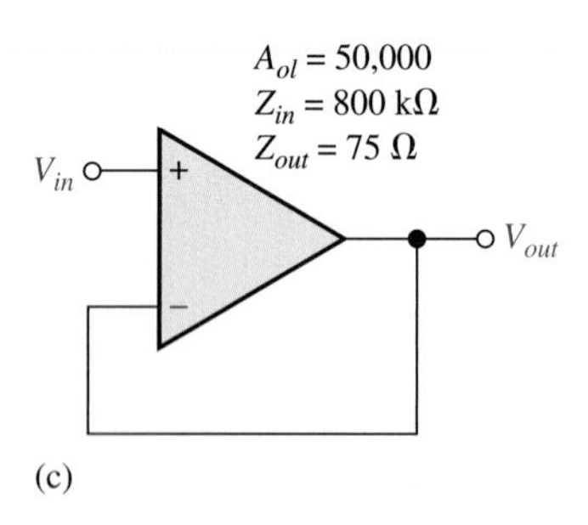

(c)

21. Repeat Problem 19 for each circuit in Figure 20–47.

SECTION 20–7

Troubleshooting

22. Determine the most likely fault(s) for each of the following symptoms in Figure 20–48 with a 100 mV signal applied.

(a) No output signal

(b) Output severely clipped on both positive and negative swings

(c) Clipping on only positive peaks when input signal is increased to a certain point

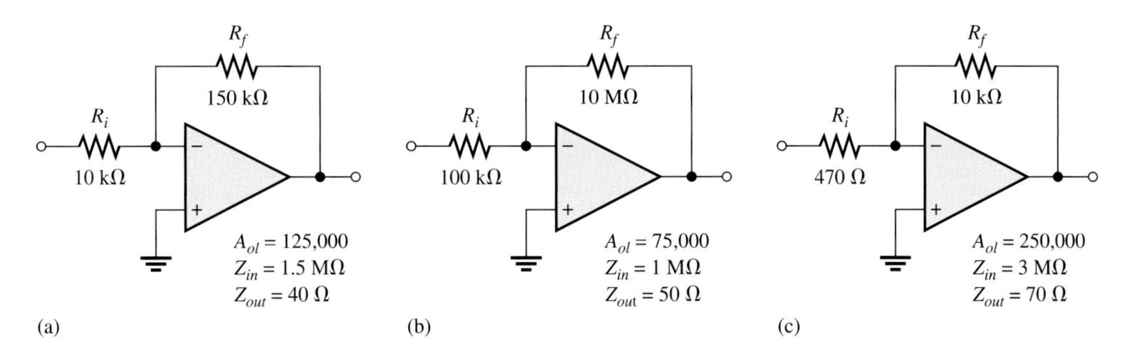

▲ FIGURE 20–47

▶ FIGURE 20–48

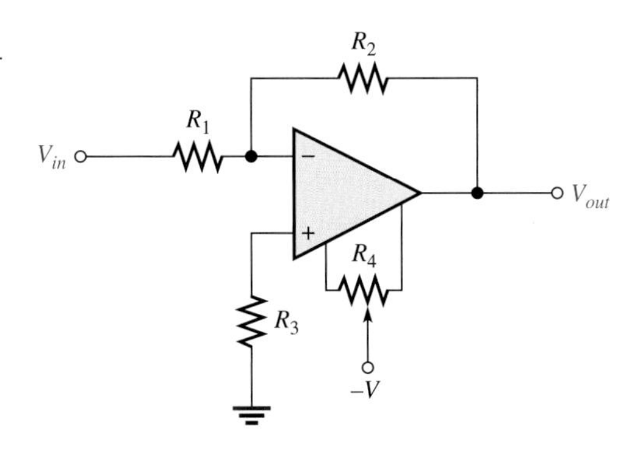

23. On the circuit board in Figure 20–34, what happens if the middle lead (wiper) of the 100 kΩ potentiometer is broken (open)?

ELECTRONICS WORKBENCH/CIRCUITMAKER TROUBLESHOOTING PROBLEMS

CD-ROM file circuits are shown in Figure 20–49 on the next page.

24. Open file P20-24 and determine if there is a fault. If so, identify it.

25. Open file P20-25 and determine if there is a fault. If so, identify it.

26. Open file P20-26 and determine if there is a fault. If so, identify it.

27. Open file P20-27 and determine if there is a fault. If so, identify it.

28. Open file P20-28 and determine if there is a fault. If so, identify it.

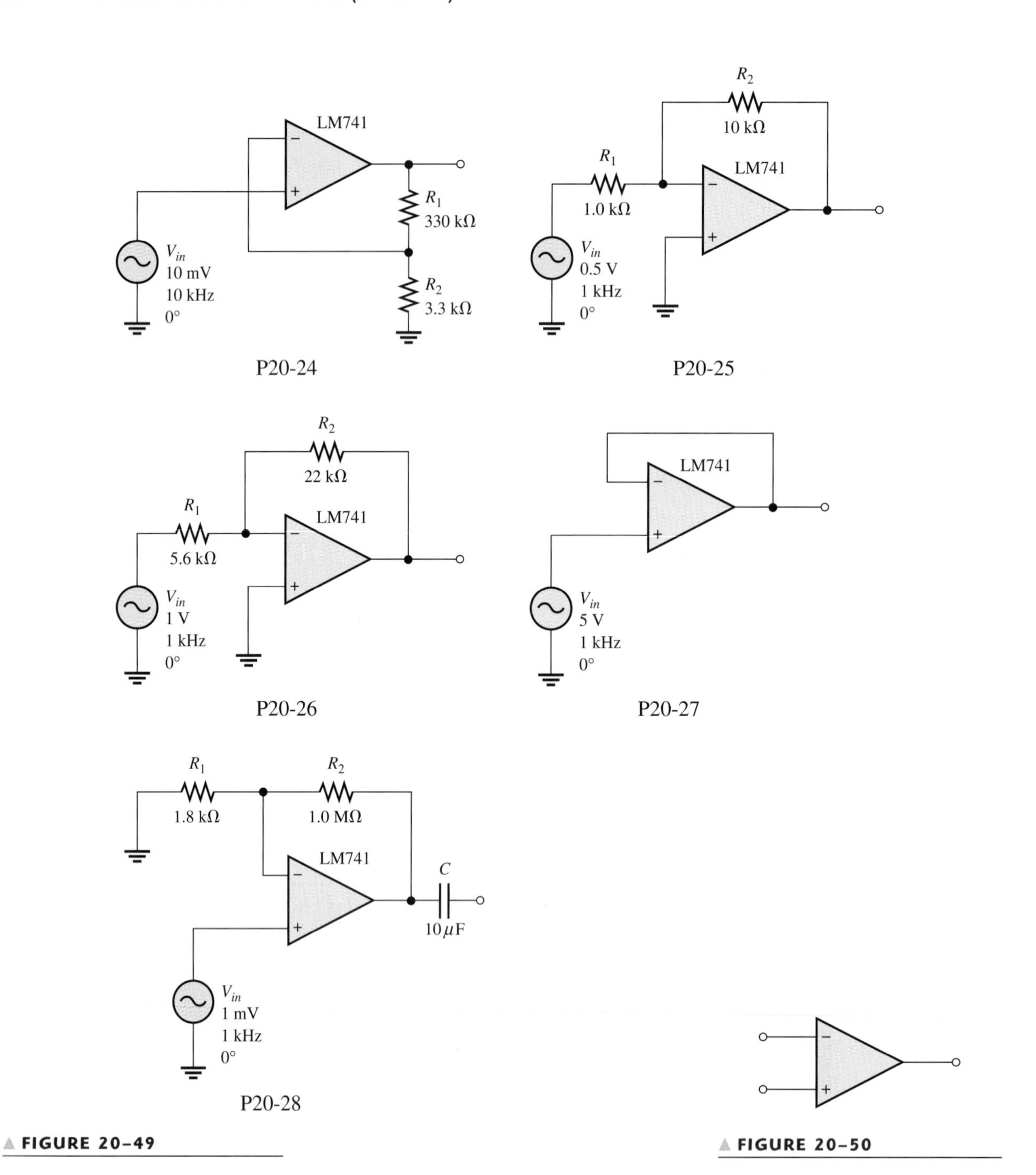

▲ FIGURE 20–49

▲ FIGURE 20–50

ANSWERS

SECTION REVIEWS

SECTION 20–1 Introduction to Operational Amplifiers

1. See Figure 20–50.
2. Infinite Z_{in}, zero Z_{out}, infinite voltage gain, infinite bandwidth
3. High Z_{in}, low Z_{out}, high voltage gain, wide bandwidth.

SECTION 20–2 **The Differential Amplifier**

1. Differential input is between two input terminals; single-ended input is from one input terminal to ground (with the other inputs grounded).
2. Common-mode rejection is the ability of an op-amp to reject common-mode signals.
3. A higher CMRR results in a lower A_{cm}.

SECTION 20–3 **Op-Amp Parameters**

1. Op-amp parameters are input bias current, input offset voltage, drift, input offset current, input impedance, output impedance, common-mode input voltage range, CMRR, open-loop voltage gain, slew rate, and frequency response.
2. Slew rate and voltage gain are frequency-dependent parameters.

SECTION 20–4 **Negative Feedback**

1. Negative feedback provides a stable controlled voltage gain, control of input and output impedances, and wider bandwidth.
2. The open-loop gain is so high that a very small signal on the input will drive the op-amp into saturation.
3. Both inputs will be the same.

SECTION 20–5 **Op-Amp Configurations with Negative Feedback**

1. Negative feedback stabilizes gain. **2.** False **3.** $A_{cl} = 1/B = 50$

SECTION 20–6 **Op-Amp Impedances**

1. Z_{in} of the noninverting amplifier is higher than that of the op-amp itself.
2. The voltage-follower configuration increases Z_{in}.
3. $Z_{in(I)} = 2\ \text{k}\Omega$; $Z_{out(I)} = 60\ \Omega$

SECTION 20–7 **Troubleshooting**

1. Check the output null adjustment. **2.** The op-amp.

■ Application Assignment

1. The 100 kΩ potentiometer is the feedback resistor.
2. the 10 kΩ potentiometer is for nulling the output.
3. The light source and the prism must be pivoted to allow different wavelengths of light to pass through the slit.

RELATED PROBLEMS FOR EXAMPLES

20–1 34,000; 90.6 dB

20–2 **(a)** 0.168 **(b)** 88 dB **(c)** 2.1 V rms, 4.2 V rms **(d)** 0.168 V

20–3 12,649 **20–4** 20 V/μs **20–5** 32.9

20–6 **(a)** 67.5 kΩ **(b)** The amplifier would have an open-loop gain, producing a square wave output.

20–7 **(a)** 20.6 GΩ, 14 mΩ **(b)** 23

20–8 Z_{in} increases; Z_{out} decreases.

20–9 $Z_{in(I)} = 560\ \Omega$; $Z_{out(I)} = 75\ \Omega$; $A_{cl} = -146$

SELF-TEST

1. (b) **2.** (d) **3.** (a) **4.** (b) **5.** (c) **6.** (d) **7.** (d)
8. (a) **9.** (c) **10.** (a) **11.** (b) **12.** (d)

21

BASIC APPLICATIONS OF OP-AMPS

INTRODUCTION

Op-amps are used in such a wide variety of applications that it is impossible to cover all of them in one chapter or even in one book. Therefore, in this chapter we examine some of the more fundamental applications to illustrate the versatility of the op-amp.

CHAPTER OUTLINE

CHAPTER OBJECTIVES

- Explain the basic operation of a comparator circuit
- Analyze summing amplifiers, averaging amplifiers, and scaling amplifiers
- Explain the operation of op-amp integrators and differentiators
- Discuss the operation of several types of op-amp oscillators
- Recognize and evaluate basic op-amp filters
- Describe the operation of basic series and shunt voltage regulators

KEY TERMS

- Comparator
- Summing amplifier
- Averaging amplifier
- Scaling adder
- Integrator
- Differentiator
- Wien-bridge oscillator
- Triangular-wave oscillator
- Voltage-controlled oscillator (VCO)
- Relaxation oscillator
- Active filter
- Series regulator
- Shunt regulator

APPLICATION ASSIGNMENT PREVIEW

You are the owner of an electronics repair service that handles all kinds of consumer electronic equipment. In this particular case, a dead FM stereo receiver system has come into your shop. After a preliminary check, you decide that the dual-polarity power supply that provides ± 12 V to all the op-amps in both channels of the receiver is faulty. The power supply uses positive and negative integrated circuit voltage regulators. After studying this chapter, you should be able to complete the application assignment.

WWW. VISIT THE COMPANION WEBSITE

Circuit Simulation Tutorials and Other Chapter Study Tools Are Available at

http://www.prenhall.com/floyd

21–1 COMPARATORS

Operational amplifiers are often used to compare the amplitude of one voltage with another. In this application, the op-amp is used in the open-loop configuration, with the input voltage on one input and a reference voltage on the other.

After completing this section, you should be able to

- **Explain the basic operation of a comparator circuit**
- Discuss zero-level detection
- Discuss nonzero-level detection

Zero-Level Detection

One application of the op-amp used as a **comparator** is to determine when an input voltage exceeds a certain level. Figure 21–1(a) shows a zero-level detector. Notice that the inverting input (−) is grounded and the input signal voltage is applied to the noninverting input (+). Because of the high open-loop voltage gain, a very small difference voltage between the two inputs drives the amplifier into saturation, causing the output voltage to go to its limit. For example, consider an op-amp having $A_{ol} = 100{,}000$. A voltage difference of only 0.25 mV between the inputs could produce an output voltage of (0.25 mV)(100,000) = 25 V if the op-amp were capable. However, since most op-amps have output voltage limitations of less than ±15 V, the device would be driven into saturation. For many comparison applications, special op-amp comparators are selected.

FIGURE 21–1

The op-amp as a zero-level detector.

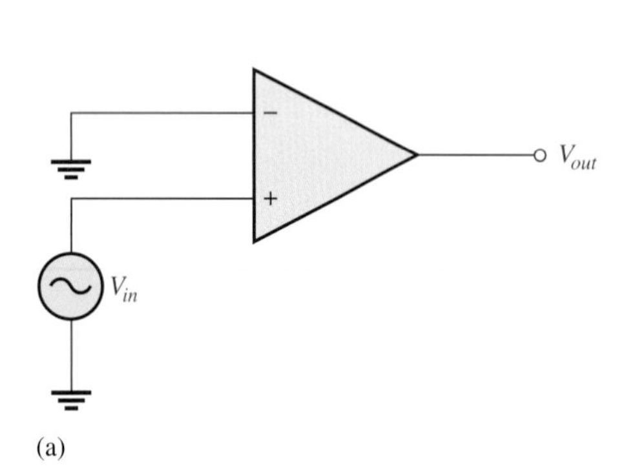

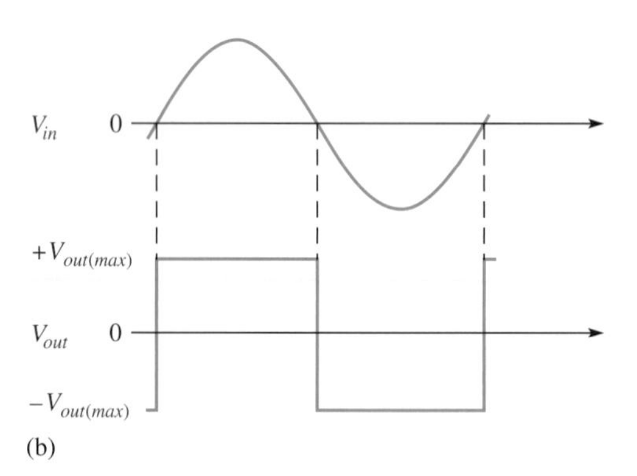

Figure 21–1(b) shows the result of a sinusoidal input voltage applied to the noninverting input of the zero-level detector. When the sine wave is negative, the output is at its maximum negative level. When the sine wave crosses 0, the amplifier is driven to its opposite state and the output goes to its maximum positive level, as shown.

As you can see, the zero-level detector can be used as a squaring circuit to produce a square wave from a sine wave.

Nonzero-Level Detection

The zero-level detector in Figure 21–1 can be modified to detect voltages other than zero by connecting a fixed reference voltage to the inverting input (−), as shown in Figure 21–2(a). A more practical arrangement is shown in part (b) using a voltage divider to set the reference voltage as follows:

$$V_{\text{REF}} = \frac{R_2}{R_1 + R_2}(+V)$$

Equation 21–1

where $+V$ is the positive op-amp supply voltage. As long as the input voltage (V_{in}) is less than V_{REF}, the output remains at the maximum negative level. When the input voltage exceeds the reference voltage, the output goes to its maximum positive state, as shown in Figure 21–2(c) with a sinusoidal input voltage.

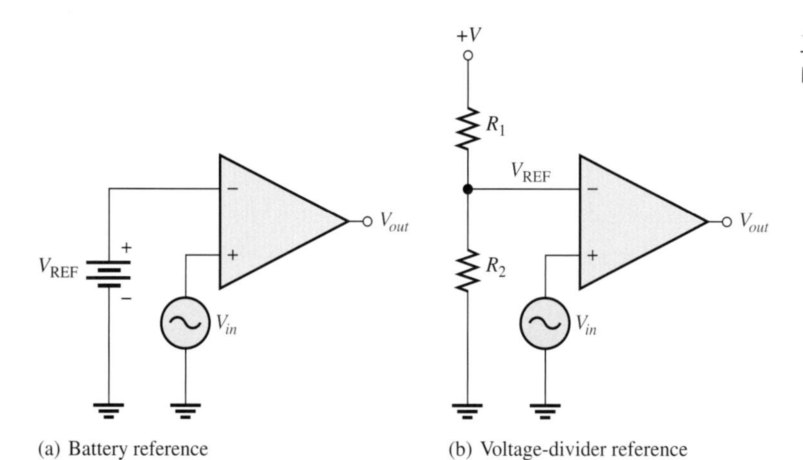

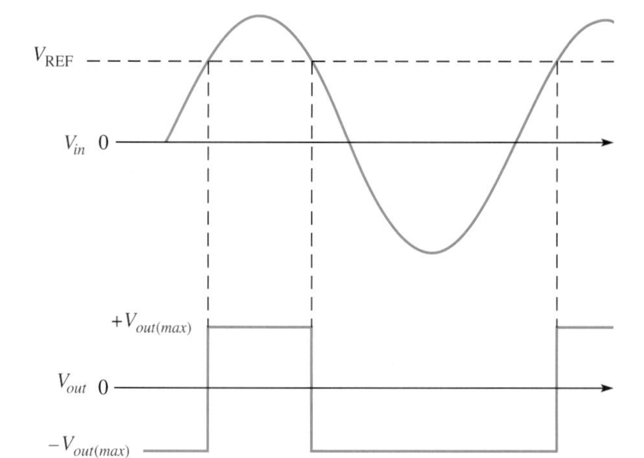

FIGURE 21–2

Nonzero-level detectors.

EXAMPLE 21–1

The input signal in Figure 21–3(a) is applied to the comparator circuit in part (b). Draw the output showing its proper relationship to the input signal. Assume that the maximum output levels of the op-amp are ±12 V.

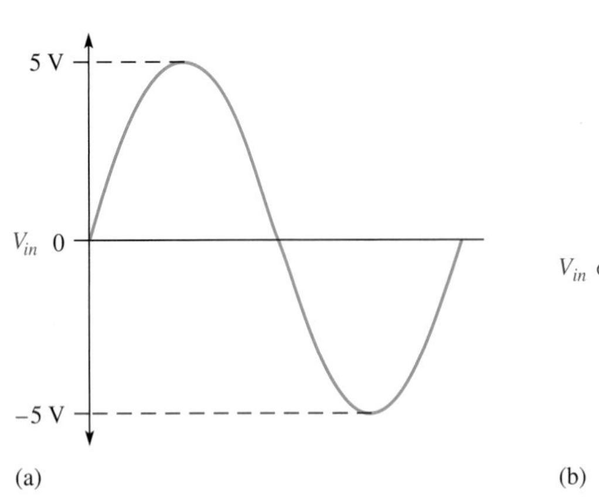

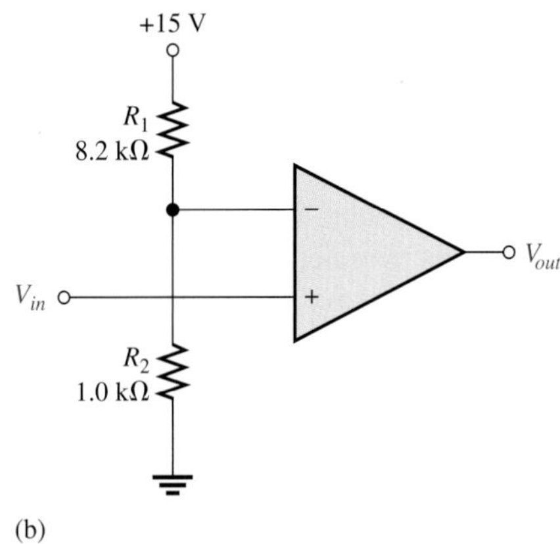

▲ **FIGURE 21–3**

Solution The reference voltage is set by R_1 and R_2.

$$V_{\text{REF}} = \frac{R_2}{R_1 + R_2}(+V) = \frac{1.0\text{ k}\Omega}{8.2\text{ k}\Omega + 1.0\text{ k}\Omega}(+15\text{ V}) = 1.63\text{ V}$$

As shown in Figure 21–4, each time the input exceeds +1.63 V, the output voltage switches to its +12 V level, and each time the input goes below +1.63 V, the output switches back to its −12 V level.

▶ **FIGURE 21–4**

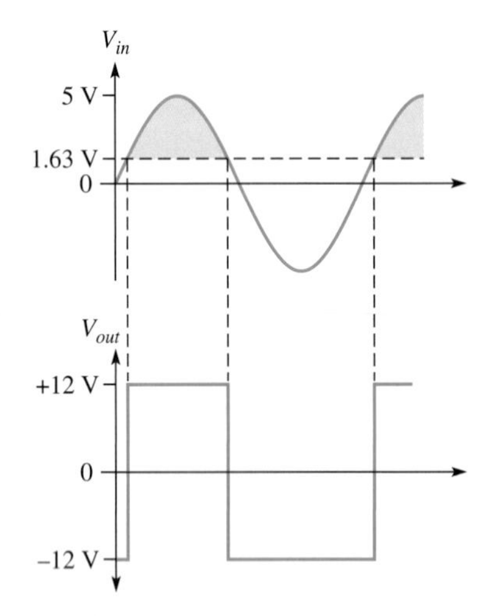

Related Problem* Determine the reference voltage in Figure 21–3 if R_1 = 22 kΩ and R_2 = 3.3 kΩ.

Open file E21-01 on your EWB/CircuitMaker CD-ROM. Measure the output voltage waveform and determine if it matches the waveform in Figure 21–4.

*Answers are at the end of the chapter.

SECTION 21–1 REVIEW

Answers are at the end of the chapter.

1. What is the reference voltage for the comparator in Figure 21–5?
2. Draw the output waveform for Figure 21–5 when a sine wave with a 5 V peak is applied to the input.

FIGURE 21–5

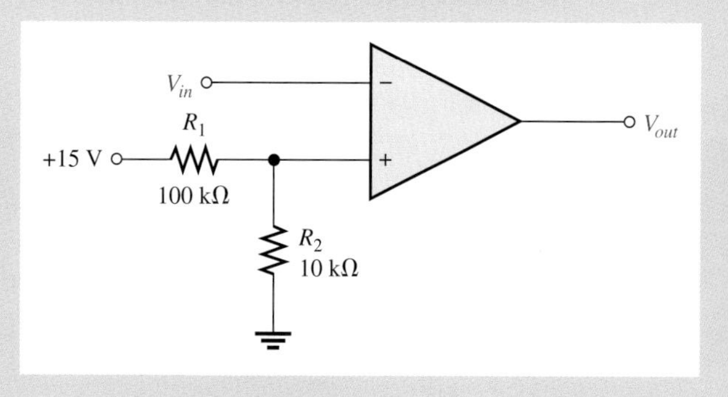

21–2 SUMMING AMPLIFIERS

The summing amplifier is a variation of the inverting op-amp configuration covered in Chapter 20. The **summing amplifier** has two or more inputs, and its output voltage is proportional to the negative of the algebraic sum of its input voltages. In this section, you will see how a summing amplifier works, and you will learn about the averaging amplifier and the scaling amplifier, which are variations of the basic summing amplifier.

After completing this section, you should be able to

- **Analyze summing amplifiers, averaging amplifiers, and scaling amplifiers**
- Calculate summing amplifier output voltage for given inputs for both unity gain and nonunity gain conditions
- Calculate the output voltage for an averaging amplifier
- Calculate the output voltage for a scaling adder
- Explain how a scaling adder can be used in a digital-to-analog converter

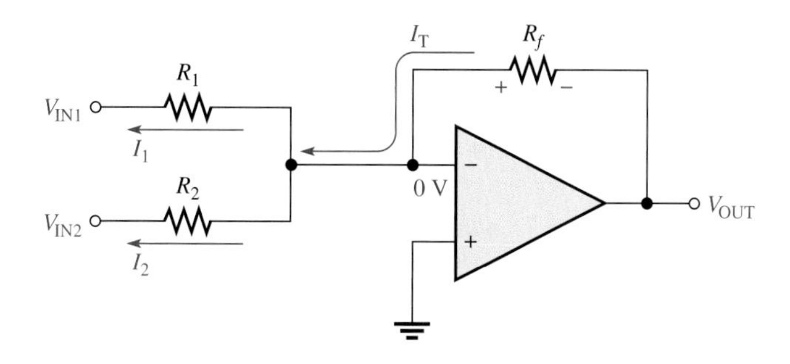

FIGURE 21–6

Two-input, inverting summing amplifier.

Figure 21–6 shows a two-input summing amplifier. Two voltages, V_{IN1} and V_{IN2}, are applied to the inputs and produce currents I_1 and I_2, as shown. Using the concepts of infinite input impedance and virtual ground, you can see that the inverting input of the op-amp is approximately 0 V, and there is no current from the input. Therefore, the total current, which is the sum of I_1 and I_2, is through R_f.

$$I_T = I_1 + I_2$$

Since $V_{OUT} = -I_T R_f$,

$$V_{OUT} = -(I_1 + I_2)R_f = -\left(\frac{V_{IN1}}{R_1} + \frac{V_{IN2}}{R_2}\right)R_f$$

If all three of the resistors are equal ($R_1 = R_2 = R_f = R$), then

$$V_{OUT} = -\left(\frac{V_{IN1}}{R} + \frac{V_{IN2}}{R}\right)R$$

Equation 21–2

$$V_{OUT} = -(V_{IN1} + V_{IN2})$$

Equation 21–2 shows that the output voltage is the sum of the two input voltages. A general expression is given in Equation 21–3 for a summing amplifier with n inputs, as shown in Figure 21–7, where all the resistors are equal in value.

Equation 21–3

$$V_{OUT} = -(V_{IN1} + V_{IN2} + V_{IN3} + \cdots + V_{INn})$$

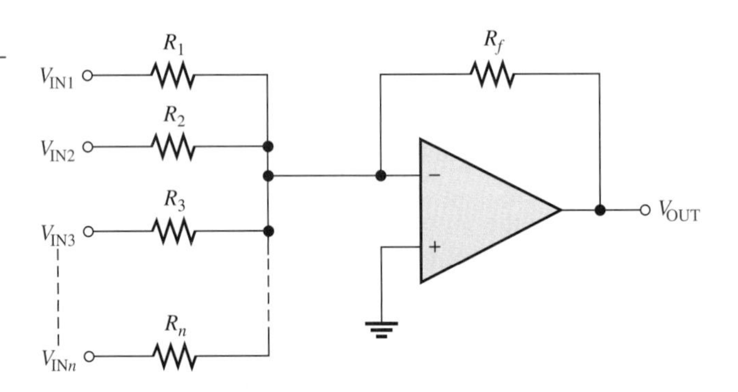

FIGURE 21–7

Summing amplifier with n inputs.

EXAMPLE 21–2

Determine the output voltage in Figure 21–8.

FIGURE 21–8

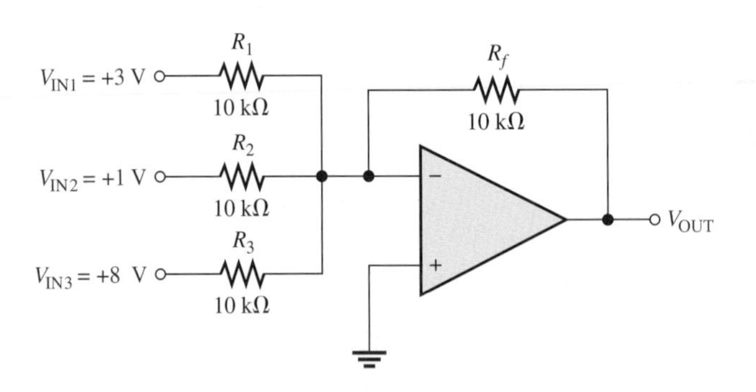

Solution

$$V_{OUT} = -(V_{IN1} + V_{IN2} + V_{IN3}) = -(3\text{ V} + 1\text{ V} + 8\text{ V}) = \mathbf{-12\text{ V}}$$

Related Problem If a fourth input of +0.5 V is added to Figure 21–8 with a 10 kΩ resistor, what is the output voltage?

Open file E21-02 on your EWB/CircuitMaker CD-ROM. Measure the output voltage and verify that it is the sum of the input voltages.

Summing Amplifier with Gain Greater Than Unity

When R_f is larger than the input resistors, the amplifier has a gain of R_f/R, where R is the value of each input resistor. The general expression for the output is

$$V_{OUT} = -\frac{R_f}{R}(V_{IN1} + V_{IN2} + \cdots + V_{INn})$$

Equation 21–4

As you can see, the output is the sum of all the input voltages multiplied by a constant determined by the ratio R_f/R.

EXAMPLE 21–3

Determine the output voltage for the summing amplifier in Figure 21–9.

▶ **FIGURE 21–9**

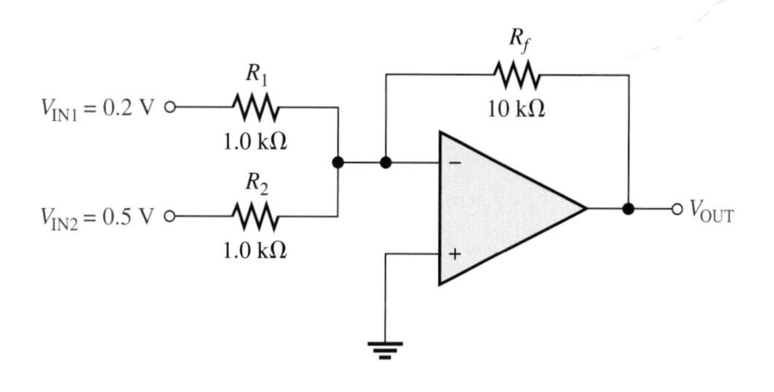

Solution $R_f = 10\text{ k}\Omega$ and $R = R_1 = R_2 = 1.0\text{ k}\Omega$. Therefore,

$$V_{OUT} = -\frac{R_f}{R}(V_{IN1} + V_{IN2}) = -\frac{10\text{ k}\Omega}{1.0\text{ k}\Omega}(0.2\text{ V} + 0.5\text{ V}) = -10(0.7\text{ V}) = \mathbf{-7\text{ V}}$$

Related Problem Determine the output voltage in Figure 21–9 if the two input resistors are 2.2 kΩ and the feedback resistor is 18 kΩ.

Open file E21-03 on your EWB/CircuitMaker CD-ROM. Measure the output voltage and verify that it matches the calculated value.

Averaging Amplifier

An averaging amplifier, which is a variation of a summing amplifier, can produce the mathematical average of the input voltages. This is done by setting the ratio R_f/R equal to the reciprocal of the number of inputs. You know that the average of several numbers is obtained by first adding the numbers and then dividing by the quantity of numbers you have. Examination of Equation 21–4 and a little thought will convince you that a summing amplifier will do the same. Example 21–4 will illustrate.

EXAMPLE 21–4

Show that the amplifier in Figure 21–10 produces an output whose magnitude is the average of the input voltages.

▶ **FIGURE 21–10**

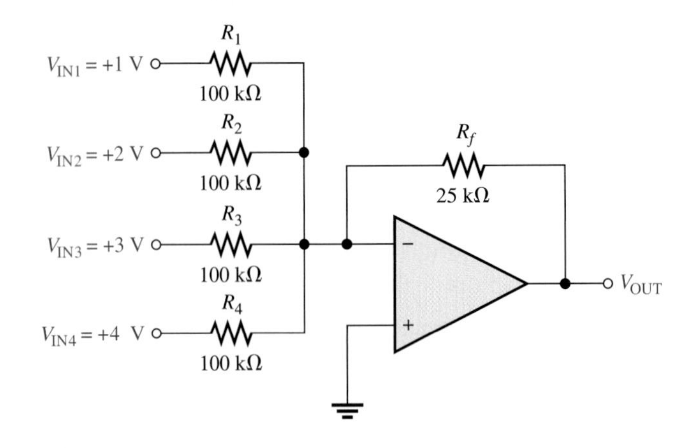

Solution Since the input resistors are equal, $R = 100\text{ k}\Omega$. The output voltage is

$$V_{OUT} = -\frac{R_f}{R}(V_{IN1} + V_{IN2} + V_{IN3} + V_{IN4})$$

$$= -\frac{25\text{ k}\Omega}{100\text{ k}\Omega}(1\text{ V} + 2\text{ V} + 3\text{ V} + 4\text{ V}) = -\frac{1}{4}(10\text{ V}) = \mathbf{-2.5\text{ V}}$$

A simple calculation shows that the average of the four input values is the same as V_{OUT} but of opposite sign.

$$V_{IN(avg)} = \frac{1\text{ V} + 2\text{ V} + 3\text{ V} + 4\text{ V}}{4} = \frac{10\text{ V}}{4} = 2.5\text{ V}$$

Related Problem Specify the changes required in the averaging amplifier in Figure 21–10 in order to handle five inputs.

Open file E21-04 on your EWB/CircuitMaker CD-ROM. Measure the output voltage and verify that it is the average of the input voltages.

Scaling Adder

A different weight can be assigned to each input of a summing amplifier, forming a scaling adder by simply adjusting the values of the input resistors. As you have seen, the output voltage can be expressed as

Equation 21–5

$$V_{OUT} = -\left(\frac{R_f}{R_1}V_{IN1} + \frac{R_f}{R_2}V_{IN2} + \cdots + \frac{R_f}{R_n}V_{INn}\right)$$

The weight of a particular input is set by the ratio of R_f to the resistance for that input. For example, if an input voltage is to have a weight of 1, then $R = R_f$. Or, if a weight of 0.5 is required, $R = 2R_f$. The smaller the value of the input resistance R, the greater the weight, and vice versa.

EXAMPLE 21–5

For the scaling adder in Figure 21–11, determine the weight of each input voltage and find the output voltage.

FIGURE 21–11

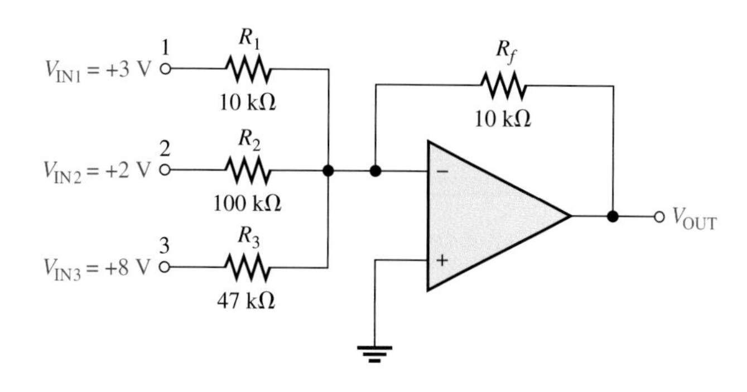

Solution Weight of input 1: $\dfrac{R_f}{R_1} = \dfrac{10\text{ k}\Omega}{10\text{ k}\Omega} = \mathbf{1}$

Weight of input 2: $\dfrac{R_f}{R_2} = \dfrac{10\text{ k}\Omega}{100\text{ k}\Omega} = \mathbf{0.1}$

Weight of input 3: $\dfrac{R_f}{R_3} = \dfrac{10\text{ k}\Omega}{47\text{ k}\Omega} = \mathbf{0.213}$

The output voltage is

$$V_{OUT} = -\left(\frac{R_f}{R_1}V_{IN1} + \frac{R_f}{R_2}V_{IN2} + \frac{R_f}{R_3}V_{IN3}\right) = -[1(3\text{ V}) + 0.1(2\text{ V}) + 0.213(8\text{ V})]$$

$$= -(3\text{ V} + 0.2\text{ V} + 1.7\text{ V}) = \mathbf{4.9\text{ V}}$$

Related Problem Determine the weight of each input voltage in Figure 21–11 if $R_1 = 22\text{ k}\Omega$, $R_2 = 82\text{ k}\Omega$, $R_3 = 56\text{ k}\Omega$, and $R_f = 10\text{ k}\Omega$. Also find V_{OUT}.

Open file E21-05 on your EWB/CircuitMaker CD-ROM. Measure the output voltage and compare to the calculated value.

SECTION 21–2 REVIEW

1. Define *summing point.*
2. What is the value of R_f/R for a five-input averaging amplifier?
3. A certain scaling adder has two inputs, one having twice the weight of the other. If the resistor value for the lower weighted input is 10 kΩ, what is the value of the other input resistor?

21–3 INTEGRATORS AND DIFFERENTIATORS

An op-amp integrator simulates mathematical integration, which is basically a summing process that determines the total area under the curve of a function. An op-amp differentiator simulates mathematical differentiation, which is a process of determining the instantaneous rate of change of a function. The integrators and differentiators shown in this section are idealized to show basic principles. Practical integrators often have an additional resistor or other circuitry in parallel with the feedback capacitor to prevent saturation. Practical differentiators may include a series resistor to reduce high frequency noise.

After completing this section, you should be able to

- **Explain the operation of op-amp integrators and differentiators**
- Recognize an integrator
- Determine the rate of change of the integrator output voltage
- Recognize a differentiator
- Determine the differentiator output voltage

The Op-Amp Integrator

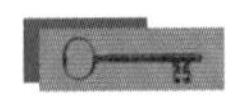

An ideal integrator is shown in Figure 21–12. Notice that the feedback element is a capacitor that forms an *RC* circuit with the input resistor. Although a large-value resistor is normally used in parallel with the capacitor to limit the gain, it does not affect the basic operation and is not shown for purposes of this analysis.

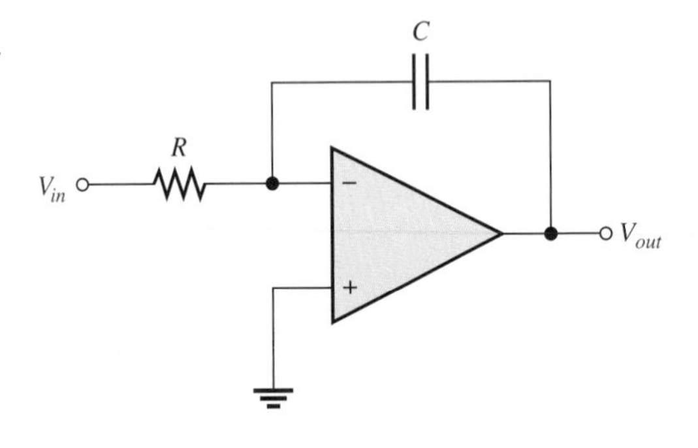

FIGURE 21–12

An ideal op-amp integrator.

How a Capacitor Charges To understand how the integrator works, it is important to review how a capacitor charges. Recall that the charge Q on a capacitor is proportional to the charging current and the time.

$$Q = I_C t$$

Also, in terms of the voltage, the charge on a capacitor is

$$Q = CV_C$$

From these two relationships, the capacitor voltage can be expressed as

$$V_C = \left(\frac{I_C}{C}\right)t$$

This expression has the form of an equation for a straight line beginning at zero with a constant slope of I_C/C. Remember from algebra that the general formula for a straight line is $y = mx + b$. In this case, $y = V_C$, $m = I_C/C$, $x = t$, and $b = 0$.

Recall that the capacitor voltage in a simple *RC* circuit is not linear but is exponential. This is because the charging current continuously decreases as the capacitor charges and causes the rate of change of the voltage to continuously decrease. The key thing about using an op-amp with an *RC* circuit to form an integrator is that the capacitor's charging current is made constant, thus producing a straight-line (linear) voltage rather than an exponential voltage. Now let's see why this is true.

In Figure 21–13, the inverting input of the op-amp is at virtual ground (0 V), so the voltage across R_i equals V_{in}. Therefore, the input current is

$$I_{in} = \frac{V_{in}}{R_i}$$

If V_{in} is a constant voltage, then I_{in} is also a constant because the inverting input always remains at 0 V, keeping a constant voltage across R_i. Because of the very high input impedance of the op-amp, there is negligible current from the inverting input. All of the input current is through the capacitor, as indicated in Figure 21–13, so

$$I_C = I_{in}$$

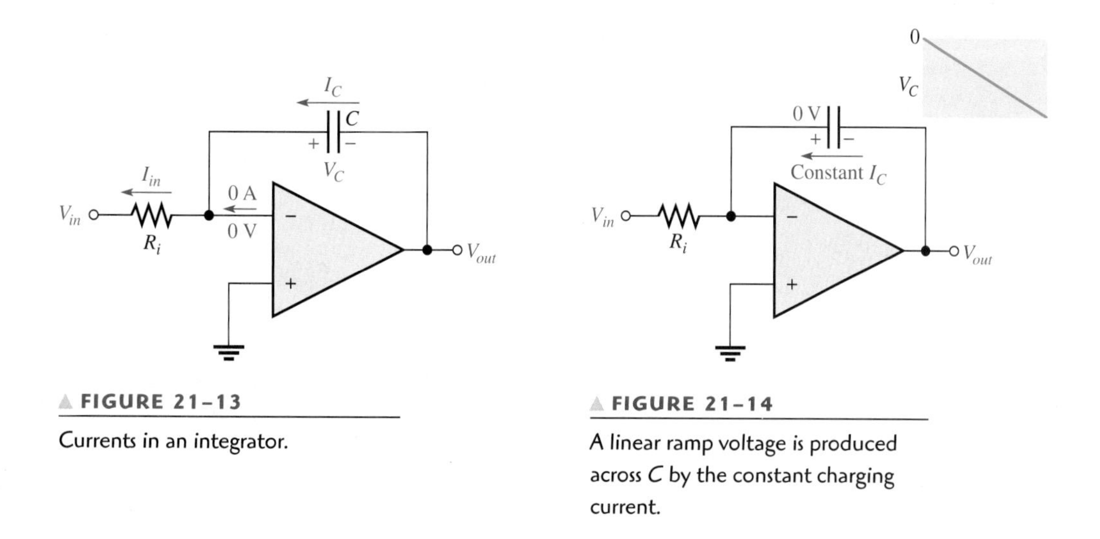

FIGURE 21–13

Currents in an integrator.

FIGURE 21–14

A linear ramp voltage is produced across *C* by the constant charging current.

The Capacitor Voltage Since I_{in} is constant, so is I_C. The constant I_C charges the capacitor linearly and produces a linear voltage across *C*. The positive side of the capacitor is held at 0 V by the virtual ground of the op-amp. The voltage on the negative side of the capacitor decreases linearly from zero as the capacitor charges, as shown in Figure 21–14. This voltage is called a *negative ramp*.

The Output Voltage V_{out} is the same as the voltage on the negative side of the capacitor. When a constant input voltage in the form of a step or pulse (a pulse has a constant amplitude when high) is applied, the output ramp decreases negatively until the op-amp saturates at its maximum negative level. This is indicated in Figure 21–15.

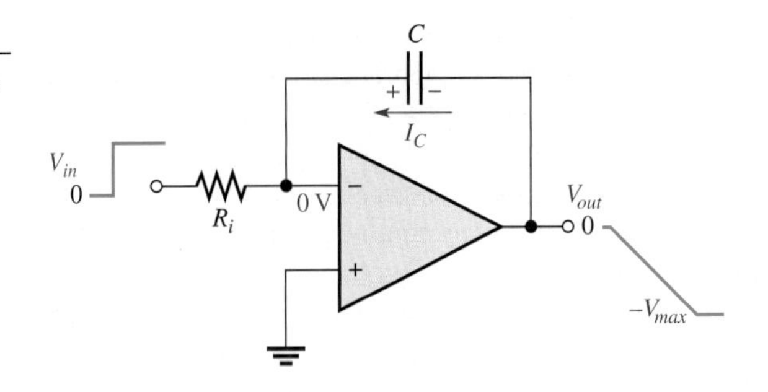

FIGURE 21–15

A constant input voltage produces a ramp on the output.

Rate of Change of the Output The rate at which the capacitor charges, and therefore the slope of the output ramp, is set by the ratio I_C/C, as you have seen. Since $I_C = V_{in}/R_i$, the rate of change or slope of the integrator's output voltage is $\Delta V_{out}/\Delta t$.

Equation 21–6

$$\frac{\Delta V_{out}}{\Delta t} = -\frac{V_{in}}{R_i C}$$

EXAMPLE 21–6

(a) Determine the rate of change of the output voltage in response to the first input pulse in a pulse waveform, as shown for the integrator in Figure 21–16(a). The output voltage is initially zero.

(b) Describe the output after the first pulse. Draw the output waveform.

FIGURE 21–16

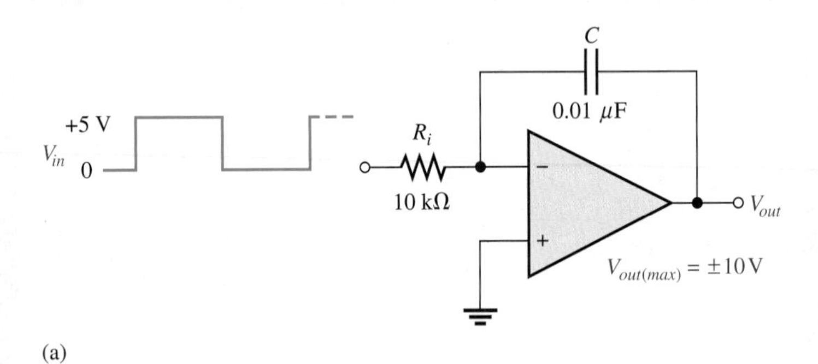

(a)

+5 V
V_{in} 0
100 μs
100 μs
100 μs
0
V_{out}
−5 V
−10 V

(b)

Solution

(a) The rate of change of the output voltage during the time that the input pulse is HIGH is

$$\frac{\Delta V_{out}}{\Delta t} = -\frac{V_{in}}{R_iC} = -\frac{5\text{ V}}{(10\text{ k}\Omega)(0.01\ \mu\text{F})} = -50\text{ kV/s} = \mathbf{-50\ mV/\mu s}$$

(b) The rate of change was found to be -50 mV/μs in part (a). When the input is at $+5$ V, the output is a negative-going ramp. When the input is at 0 V, the output is a constant level. In 100 μs, the voltage decreases.

$$\Delta V_{out} = (-50\text{ mV/}\mu\text{s})(100\ \mu\text{s}) = -5\text{ V}$$

Therefore, the negative-going ramp reaches -5 V at the end of the pulse. The output voltage then remains constant at -5 V for the time that the input is zero. When the next pulse occurs, the output again becomes a negative-going ramp which reaches -10 V at the end of the second pulse. The output will remain at -10 V thereafter because it has reached its maximum negative limit. The waveforms are shown in Figure 21–16(b).

Related Problem

Modify the integrator in Figure 21–16 to make the output change from 0 to -5 V in 50 μs with the same input.

Open file E21-06 on your EWB/CircuitMaker CD-ROM. Observe the output voltage waveform and compare to the waveform in Figure 21–16(b).

The Op-Amp Differentiator

An ideal **differentiator** is shown in Figure 21–17. Notice how the placement of the capacitor and resistor differs from their placement in the integrator. The capacitor is now the input element. A differentiator produces an output that is proportional to the rate of change of the input voltage. Although a small-value resistor is normally used in series with the capacitor to limit the gain, it does not affect the basic operation and is not shown for purposes of this analysis.

To see how the differentiator works, let's apply a positive-going ramp voltage to the input as indicated in Figure 21–18. In this case, $I_C = I_{in}$ and the voltage across the capacitor is equal to V_{in} at all times ($V_C = V_{in}$) because of virtual ground on the inverting input.

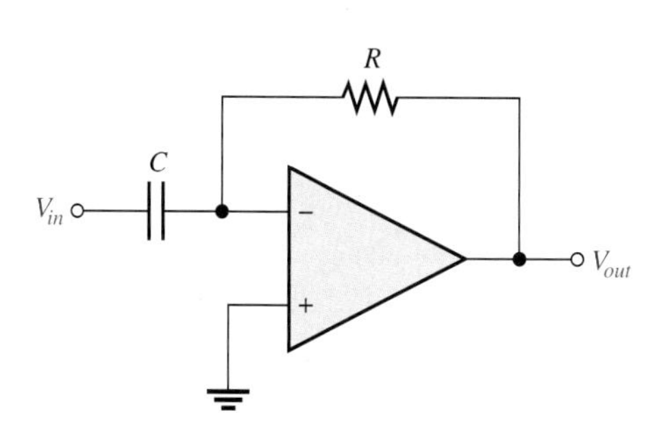

FIGURE 21–17

An ideal op-amp differentiator.

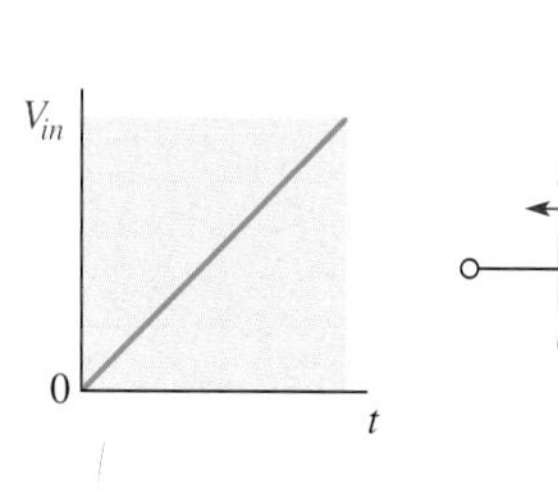

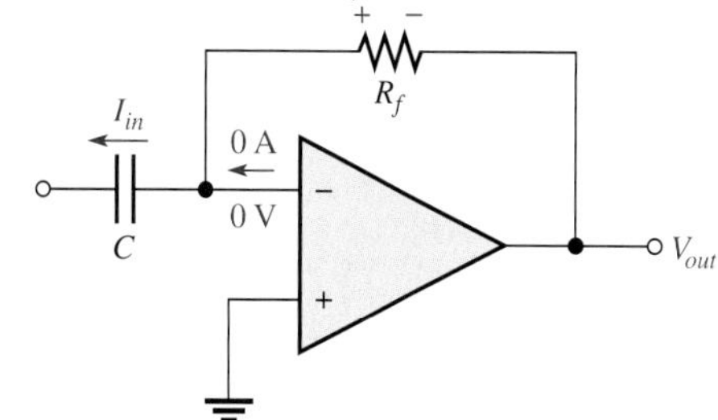

FIGURE 21–18

A differentiator with a ramp input.

From the basic formula, $V_C = (I_C/C)t$, the capacitor current is

$$I_C = \left(\frac{V_C}{t}\right)C$$

Since the current at the inverting input is negligible, $I_R = I_C$. Both currents are constant because the slope of the capacitor voltage (V_C/t) is constant. The output voltage is also constant and equal to the voltage across R_f because one side of the feedback resistor is always 0 V (virtual ground).

$$V_{out} = I_R R_f = I_C R_f$$

Equation 21–7

$$V_{out} = -\left(\frac{V_C}{t}\right)R_f C$$

The output is negative when the input is a positive-going ramp and positive when the input is a negative-going ramp, as illustrated in Figure 21–19. During this positive slope of the input, the capacitor is charging from the input source and the constant current through the feedback resistor is in the direction shown. During the negative slope of the input, the current is in the opposite direction because the capacitor is discharging.

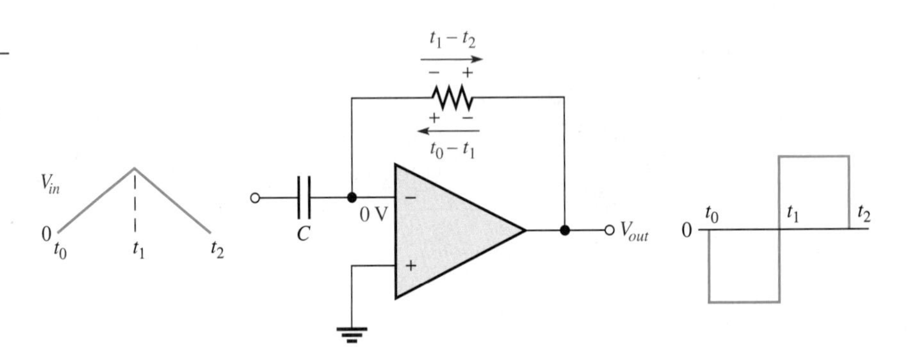

FIGURE 21–19

Output of a differentiator with a series of positive and negative ramps (triangle wave) on the input.

Notice in Equation 21–7 that the term V_C/t is the slope of the input. If the slope increases, V_{out} increases. If the slope decreases, V_{out} decreases. So, the output voltage is proportional to the slope (rate of change) of the input. The constant of proportionality is the time constant, $R_f C$.

EXAMPLE 21–7

Determine the output voltage of the op-amp differentiator in Figure 21–20 for the triangular-wave input shown.

FIGURE 21–20

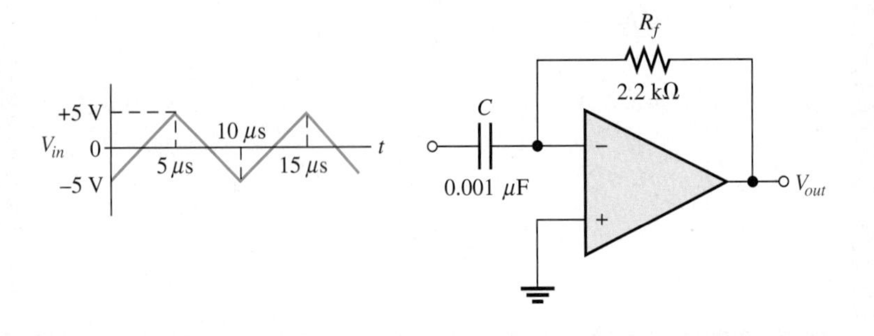

Solution Starting at $t = 0$, the input voltage is a positive-going ramp ranging from -5 V to $+5$ V (a $+10$ V change) in 5 μs. Then it changes to a negative-going ramp ranging from $+5$ V to -5 V (a -10 V change) in 5 μs.

The time constant is

$$R_fC = (2.2\ \text{k}\Omega)(0.001\ \mu\text{F}) = 2.2\ \mu\text{s}$$

Determine the slope or rate of change (V_C/t) of the positive-going ramp and calculate the output voltage as follows:

$$\frac{V_C}{t} = \frac{10\ \text{V}}{5\ \mu\text{s}} = 2\ \text{V}/\mu\text{s}$$

$$V_{out} = -\left(\frac{V_C}{t}\right)R_fC = -(2\ \text{V}/\mu\text{s})2.2\ \mu\text{s} = \mathbf{-4.4\ V}$$

Likewise, the slope of the negative-going ramp is -2 V/μs. Calculate the output voltage.

$$V_{out} = -\left(\frac{V_C}{t}\right)R_fC = -(-2\ \text{V}/\mu\text{s})2.2\ \mu\text{s} = \mathbf{4.4\ V}$$

Finally, Figure 21–21 shows the output voltage waveform relative to the input.

FIGURE 21–21

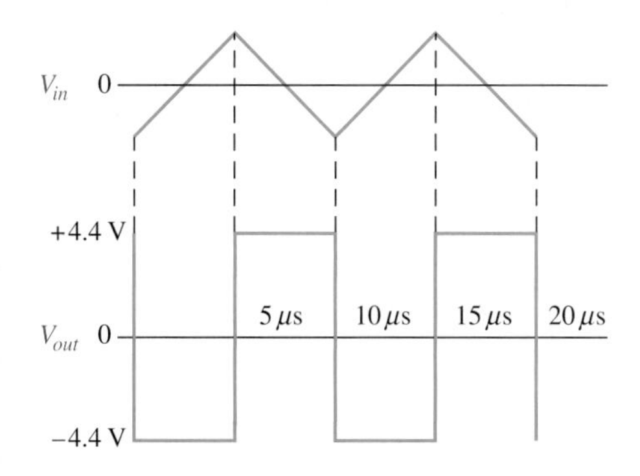

Related Problem What would the output voltage be if the feedback resistor in Figure 21–20 is changed to 3.3 kΩ?

SECTION 21–3 REVIEW

1. What is the feedback element in an op-amp integrator?
2. For a constant input voltage to an integrator, why is the voltage across the capacitor linear?
3. What is the feedback element in an op-amp differentiator?
4. How is the output of a differentiator related to the input?

21–4 OSCILLATORS

Oscillators were introduced in Chapter 19 and the principles of operation were discussed. Also, several types of oscillator circuits that use discrete transistors were covered. In this section, several types of oscillators implemented with op-amps are introduced. The Hartley, Colpitts, and Clapp oscillators discussed in Chapter 19 can also be implemented with op-amps.

After completing this section, you should be able to

- **Discuss the operation of several types of op-amp oscillators**
- Identify a Wien-bridge oscillator and analyze its operation
- Identify a triangular-wave oscillator and analyze its operation
- Identify a voltage-controlled oscillator (VCO) and analyze its operation
- Identify a relaxation oscillator and analyze its operation

The Wien-Bridge Oscillator

One type of sinusoidal oscillator is the **Wien-bridge oscillator**. A fundamental part of the Wien-bridge oscillator is a lead-lag circuit like that shown in Figure 21–22(a). R_1 and C_1 together form the lag portion of the circuit; R_2 and C_2 form the lead portion. The operation of this circuit is as follows. At lower frequencies, the lead circuit dominates due to the high reactance of C_2. As the frequency increases, X_{C2} decreases, thus allowing the output voltage to increase. At some specified frequency, the response of the lag circuit takes over, and the decreasing value of X_{C1} causes the output voltage to decrease.

FIGURE 21–22

A lead-lag circuit and its response curve.

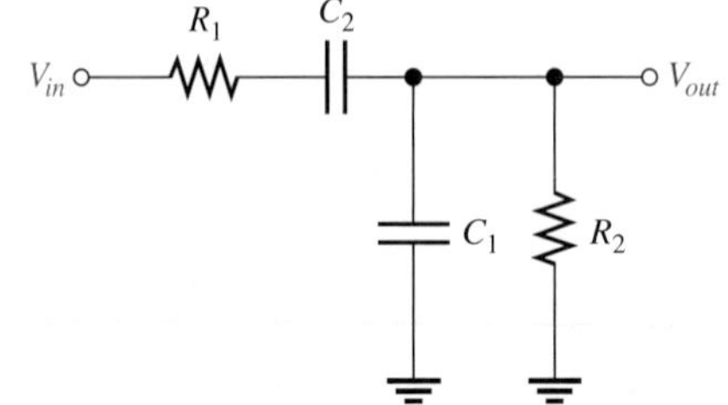

(a) Circuit

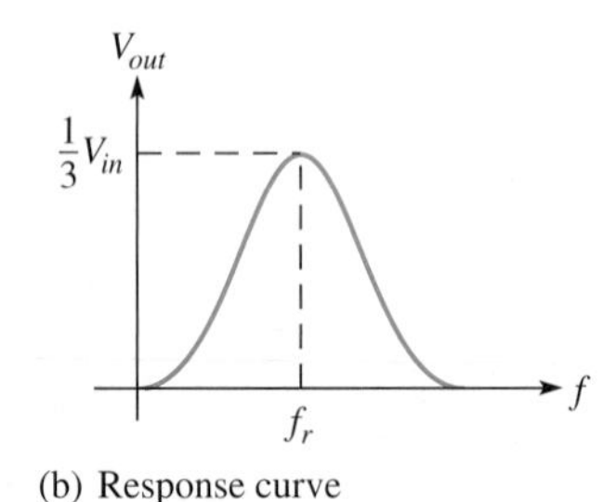

(b) Response curve

The response curve for the lead-lag circuit shown in Figure 21–22(b) indicates that the output voltage peaks at a frequency f_r. At this point, the attenuation (V_{out}/V_{in}) of the circuit is ⅓ if $R_1 = R_2$ and $X_{C1} = X_{C2}$ as stated by the following equation:

Equation 21–8

$$\frac{V_{out}}{V_{in}} = \frac{1}{3}$$

The formula for the resonant frequency is

Equation 21–9

$$f_r = \frac{1}{2\pi RC}$$

To summarize, the lead-lag circuit has a resonant frequency, f_r, at which the phase shift through the circuit is 0° and the attentuation is ⅓. Below f_r, the lead circuit dominates and the output leads the input. Above f_r, the lag circuit dominates and the output lags the input.

The Basic Circuit The lead-lag circuit is used in the positive feedback loop of an op-amp, as shown in Figure 21–23(a). A voltage divider is used in the negative feedback loop. The Wien-bridge oscillator circuit can be viewed as a noninverting amplifier configuration with the input signal fed back from the output through the lead-lag circuit. Recall that the closed-loop gain of the amplifier is determined by the voltage divider.

$$A_{cl} = \frac{1}{B} = \frac{1}{R_2/(R_1 + R_2)} = \frac{R_1 + R_2}{R_2}$$

The circuit is redrawn in Figure 21–23(b) to show that the op-amp is connected across the bridge circuit. One leg of the bridge is the lead-lag circuit, and the other is the voltage divider.

FIGURE 21–23

The Wien-bridge oscillator schematic shown in two equivalent forms.

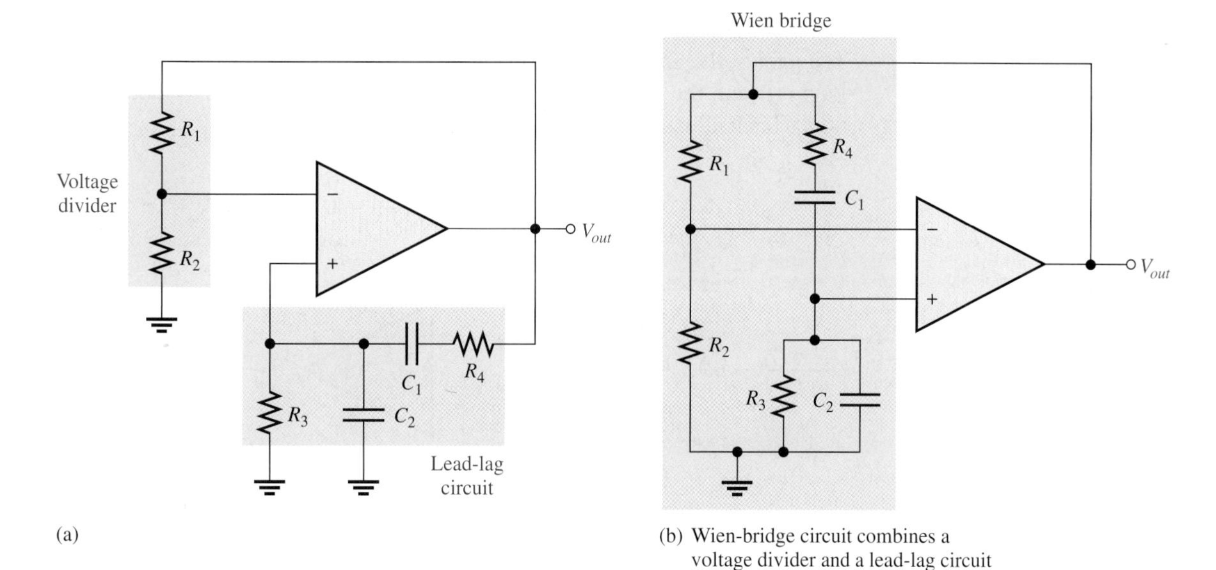

(b) Wien-bridge circuit combines a voltage divider and a lead-lag circuit

Positive Feedback Conditions for Oscillation As you know, for the circuit to produce a sustained sinusoidal output (oscillate), the phase shift around the positive feedback loop must be 0° and the gain around the loop must be unity (1). The 0° phase-shift condition is met when the frequency is f_r because the phase shift through the lead-lag circuit is 0° and there is no inversion from the noninverting input (+) of the op-amp to the output. This is shown in Figure 21–24(a).

The unity-gain condition in the feedback loop is met when

$$A_{cl} = 3$$

This offsets the ⅓ attenuation of the lead-lag circuit, thus making the total gain around the positive feedback loop equal to 1, as depicted in Figure 21–24(b). To achieve a closed-loop gain of 3,

$$R_1 = 2R_2$$

Then

$$A_{cl} = \frac{R_1 + R_2}{R_2} = \frac{2R_2 + R_2}{R_2} = \frac{3R_2}{R_2} = 3$$

FIGURE 21–24

Conditions for oscillation in the Wien-bridge circuit.

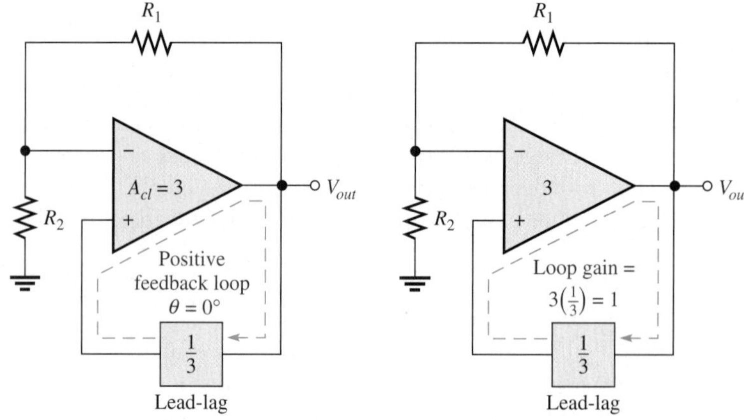

(a) The phase shift around the loop is 0°. (b) The voltage gain around the loop is 1.

Start-Up Conditions Initially, the closed-loop gain of the amplifier itself must be more than three ($A_{cl} > 3$) until the output signal builds up to a desired level. The gain of the amplifier must then decrease to 3 so that the total gain around the loop is 1 and the output signal stays at the desired level, thus sustaining oscillation. This is illustrated in Figure 21–25.

FIGURE 21–25

Oscillator start-up conditions.

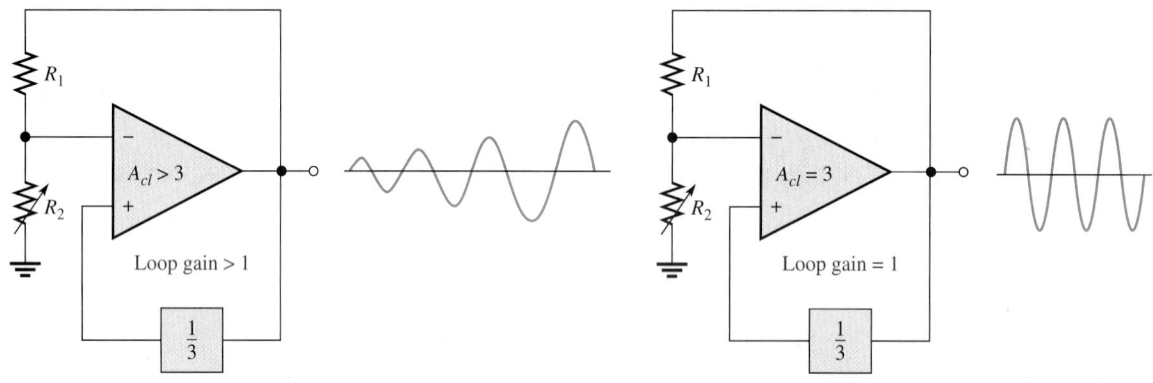

(a) Initially, loop gain greater than 1 causes output to build up. (b) Loop gain of 1 causes a sustained constant output.

The circuit in Figure 21–26 illustrates a basic method for achieving the condition described above. Notice that the voltage-divider circuit has been modified to include an additional resistor, R_3, in parallel with a back-to-back zener diode arrangement. When dc power is first applied, both zener diodes appear as opens. This places R_3 in series with R_1, thus increasing the closed-loop gain of the amplifier ($R_1 = 2R_2$).

$$A_{cl} = \frac{R_1 + R_2 + R_3}{R_2} = \frac{3R_2 + R_3}{R_2} = 3 + \frac{R_3}{R_2}$$

Initially when the power is turned on, a small positive feedback signal develops from noise or turn-on **transients.** The lead-lag circuit permits only a signal with a frequency equal to f_r to appear in phase on the noninverting input. This feedback signal is amplified and continually reinforced, resulting in a buildup of the output voltage. When the output signal reaches the zener breakdown voltage, the zeners conduct and effectively short out R_3. This lowers the amplifier's closed-loop gain to 3. At this point the total loop gain is 1 and the output signal levels off, sustaining the oscillation. All practical methods to achieve stability for

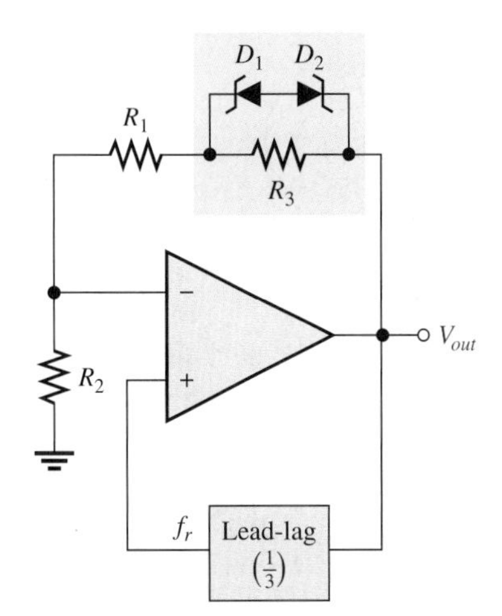

◀ **FIGURE 21–26**

Self-starting Wien-bridge oscillator using back-to-back zener diodes.

feedback oscillators require the gain to be self-adjusting. This requirement is a form of automatic gain control (AGC). The zener diodes in this example limit the gain at the onset of a nonlinearity, in this case, zener conduction.

Another method to control the gain uses a JFET as a voltage-controlled resistor in a negative feedback path. A JFET operating with a small or zero V_{DS} is operating in the ohmic region. As the gate voltage increases, the drain-source resistance increases. If the JFET is placed in the negative feedback path, automatic gain control can be achieved because of this voltage-controlled resistance.

A JFET stabilized Wien-bridge oscillator is shown in Figure 21–27. The gain of the op-amp is controlled by the components shown in the blue box, which include the JFET. The JFET's drain-source resistance depends on the gate voltage. With no output signal, the gate is at zero volts, causing the drain-source resistance to be at the minimum. With this condition, the loop gain is greater than 1. Oscillations begin and rapidly build to a large output signal. Negative excursions of the output signal forward-bias D_1, causing capacitor C_3 to charge to a negative voltage. This voltage increases the drain-source resistance of the JFET and reduces the gain (and hence the output). This is classic negative feedback at work. With the proper selection of components, the gain can be stabilized at the required level.

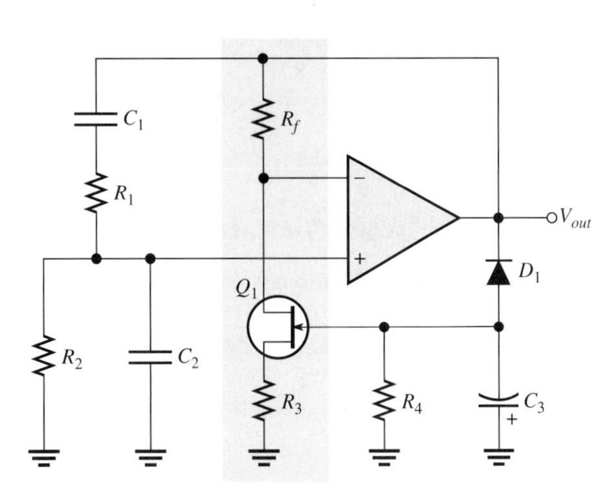

◀ **FIGURE 21–27**

Self-starting Wien-bridge oscillator using a JFET in the negative feedback loop.

EXAMPLE 21–8

Determine the frequency of oscillation for the Wien-bridge oscillator in Figure 21–28. Also, verify that oscillations will start and then continue when the output signal reaches 5.4 V.

▶ **FIGURE 21–28**

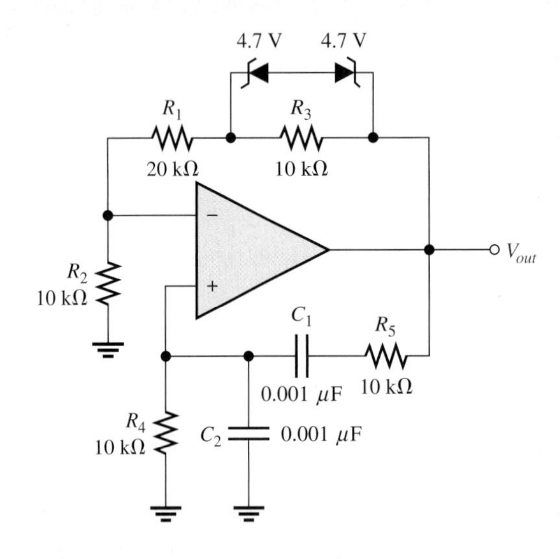

Solution For the lead-lag circuit, $R_4 = R_5 = R = 10\text{ k}\Omega$ and $C_1 = C_2 = C = 0.001\ \mu\text{F}$. The resonant frequency is

$$f_r = \frac{1}{2\pi RC} = \frac{1}{2\pi(10\text{ k}\Omega)(0.001\ \mu\text{F})} = \mathbf{15.9\text{ kHz}}$$

Initially, the closed-loop gain is

$$A_{cl} = \frac{R_1 + R_2 + R_3}{R_2} = \frac{40\text{ k}\Omega}{10\text{ k}\Omega} = 4$$

Since $A_{cl} > 3$, the start-up condition is met.

When the output reaches 4.7 V + 0.7 V = 5.4 V, the zeners conduct (their forward resistance is assumed small, compared to 10 kΩ), and the unity closed-loop gain is reached. Thus, oscillation is sustained.

$$A_{cl} = \frac{R_1 + R_2}{R_2} = \frac{30\text{ k}\Omega}{10\text{ k}\Omega} = 3$$

Related Problem What change is required in the oscillator in Figure 21–28 to produce an output with an amplitude of 6.8 V?

A Triangular-Wave Oscillator

One practical implementation of a **triangular-wave oscillator** utilizes an op-amp comparator as shown in Figure 21–29. The operation is as follows. To begin, assume that the output voltage of the comparator is at its maximum negative level. This output is connected to the inverting input of the integrator through R_1, producing a positive-going ramp on the output of the integrator. When the ramp voltage reaches the upper trigger point (UTP), the comparator switches to its

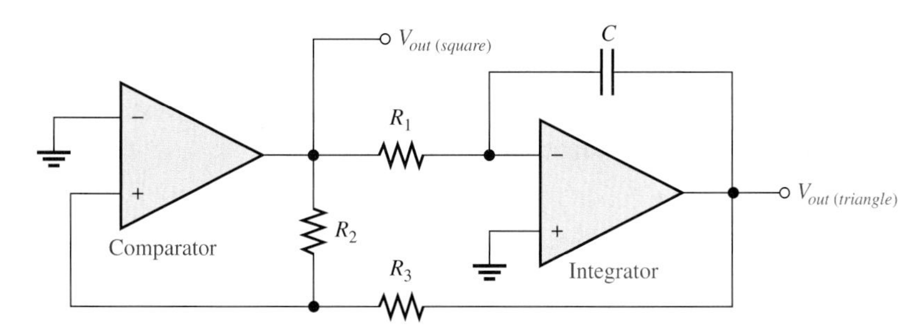

FIGURE 21–29

A triangular-wave oscillator using two op-amps.

maximum positive level. This positive level causes the integrator ramp to change to a negative-going direction. The ramp continues in this direction until the lower trigger point (LTP) of the comparator is reached. At this point, the comparator output switches back to the maximum negative level and the cycle repeats. This action is illustrated in Figure 21–30.

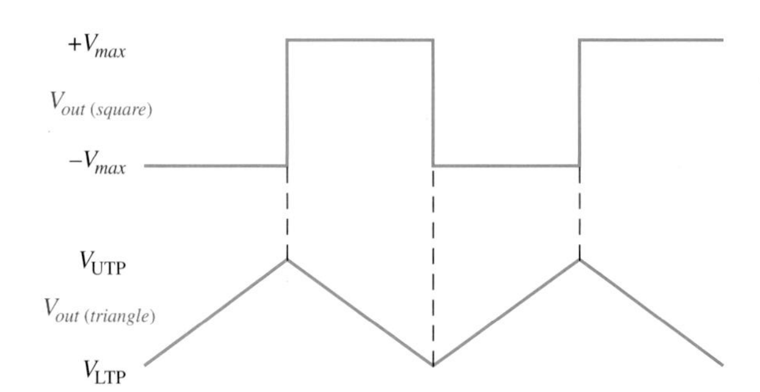

FIGURE 21–30

Waveforms for the circuit in Figure 21–29.

Since the comparator produces a square-wave output, the circuit in Figure 21–29 can be used as both a triangular-wave oscillator and a square-wave oscillator. Devices of this type are commonly known as *function generators* because they produce more than one output function. The output amplitude of the square wave is set by the output swing of the comparator, and the resistors R_2 and R_3 set the amplitude of the triangular output by establishing the UTP and LTP voltages according to the following formulas:

$$V_{\mathrm{UTP}} = +V_{max}\left(\frac{R_3}{R_2}\right)$$

Equation 21–10

$$V_{\mathrm{LTP}} = -V_{max}\left(\frac{R_3}{R_2}\right)$$

Equation 21–11

where the comparator output levels, $+V_{max}$ and $-V_{max}$, are equal. The frequency of both waveforms depends on the R_1C time constant as well as the amplitude-setting resistors, R_2 and R_3. By varying R_1, the frequency of oscillation can be adjusted without changing the output amplitude.

$$f = \frac{1}{4R_1C}\left(\frac{R_2}{R_3}\right)$$

Equation 21–12

EXAMPLE 21–9

Determine the frequency of the circuit in Figure 21–31. To what value must R_1 be changed to make the frequency 20 kHz?

▶ **FIGURE 21–31**

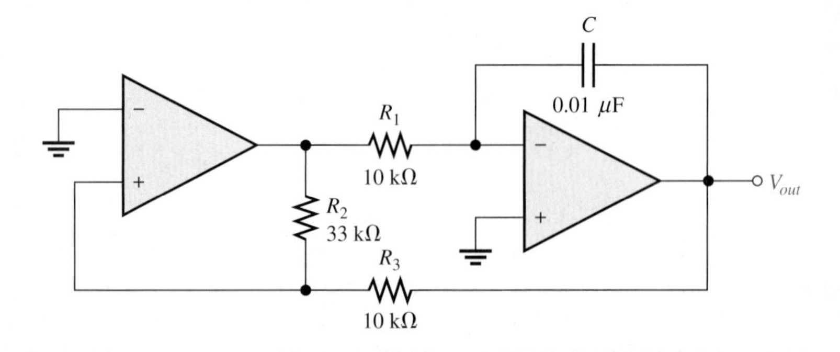

Solution

$$f = \frac{1}{4R_1C}\left(\frac{R_2}{R_3}\right) = \left(\frac{1}{4(10\text{ k}\Omega)(0.01\ \mu F)}\right)\left(\frac{33\text{ k}\Omega}{10\text{ k}\Omega}\right) = \mathbf{8.25\ kHz}$$

To make $f = 20$ kHz,

$$R_1 = \frac{1}{4fC}\left(\frac{R_2}{R_3}\right) = \left(\frac{1}{4(20\text{ kHz})(0.01\ \mu\text{F})}\right)\left(\frac{33\text{ k}\Omega}{10\text{ k}\Omega}\right) = \mathbf{4.13\ k\Omega}$$

Related Problem What is the amplitude of the triangular wave in Figure 21–31 if the comparator output is ±10 V?

A Voltage-Controlled Sawtooth Oscillator (VCO)

The voltage-controlled oscillator (VCO) is an oscillator whose frequency can be changed by a variable dc control voltage. VCOs can be either sinusoidal or nonsinusoidal. One way to build a voltage-controlled sawtooth oscillator is with an op-amp integrator that uses a switching device (PUT) in parallel with the feedback capacitor to terminate each ramp at a prescribed level and effectively "reset" the circuit. Figure 21–32(a) shows the implementation.

The PUT is a programmable unijunction transistor with an anode, a cathode, and a gate terminal. The gate is always biased positively with respect to the cathode. When the anode voltage exceeds the gate voltage by approximately 0.7 V, the PUT turns on and acts as a forward-biased diode. When the anode voltage falls below this level, the PUT turns off. Also, the current must be above the holding value to maintain conduction.

The operation of the sawtooth oscillator begins when the negative dc input voltage, $-V_{IN}$, produces a positive-going ramp on the output. During the time that the ramp is increasing, the circuit acts as a regular integrator. The PUT triggers on when the output ramp (at the anode) exceeds the gate voltage by 0.7 V. The gate is set to the approximate desired sawtooth peak voltage. When the PUT turns on, the capacitor rapidly discharges, as shown in Figure 21–32(b). The capacitor does not discharge completely to zero because of the PUT's forward voltage, V_F. Discharge continues until the PUT current falls below the holding value. At this point, the PUT turns off and the capacitor begins to charge again, thus generating a new output ramp. The cycle continually repeats, and the resulting output is a repetitive sawtooth waveform, as shown. The sawtooth amplitude and period can be adjusted by varying the PUT gate voltage.

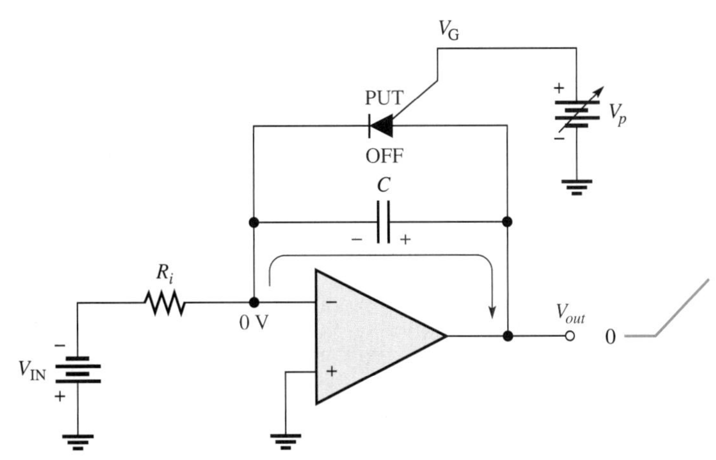

(a) Initially, the capacitor charges, the output ramp begins, and the PUT is off.

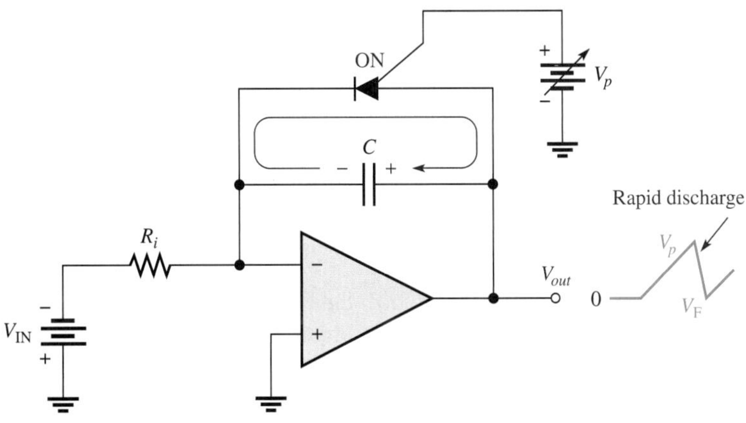

(b) The capacitor rapidly discharges when the PUT momentarily turns on.

FIGURE 21–32

Voltage-controlled sawtooth oscillator operation.

The frequency of oscillation is determined by the R_iC time constant of the integrator and the peak voltage set by the PUT. Recall that the charging rate of the capacitor is V_{IN}/R_iC. The time it takes the capacitor to charge from V_F to V_p is the period, T, of the sawtooth (neglecting the rapid discharge time).

$$T = \frac{V_p - V_F}{|V_{IN}|/R_iC}$$

Equation 21–13

From $f = 1/T$,

$$f = \frac{|V_{IN}|}{R_iC}\left(\frac{1}{V_p - V_F}\right)$$

Equation 21–14

EXAMPLE 21–10

(a) Find the peak-to-peak amplitude and frequency of the sawtooth output in Figure 21–33. Assume that the forward PUT voltage, V_F, is approximately 1 V.

(b) Draw the output waveform.

FIGURE 21–33

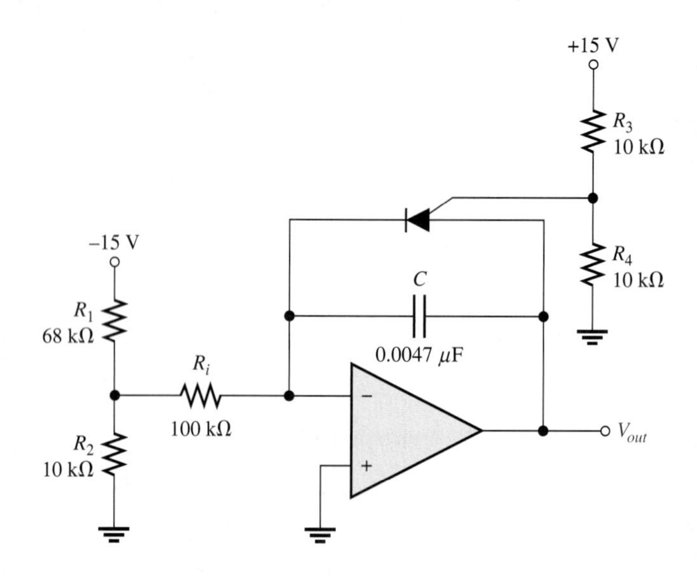

Solution **(a)** First, find the gate voltage in order to establish the approximate voltage at which the PUT turns on.

$$V_G = \frac{R_4}{R_3 + R_4}(+V) = \frac{10\text{ k}\Omega}{20\text{ k}\Omega}(15\text{ V}) = 7.5\text{ V}$$

This voltage plus 0.7 V by which the PUT anode must exceed the gate voltage sets the approximate maximum peak value of the sawtooth output.

$$V_p = 7.5\text{ V} + 0.7\text{ V} = 8.2\text{ V}$$

The minimum peak value (low point) is

$$V_F \cong 1\text{ V}$$

So the peak-to-peak amplitude is

$$V_{pp} = V_p - V_F = 8.2\text{ V} - 1.0\text{ V} = \mathbf{7.2\text{ V}}$$

The frequency is determined as follows:

$$V_{IN} = \frac{R_2}{R_1 + R_2}(-V) = \frac{10\text{ k}\Omega}{78\text{ k}\Omega}(-15\text{ V}) = -1.92\text{ V}$$

$$f = \frac{|V_{IN}|}{R_iC}\left(\frac{1}{V_p - V_F}\right) = \left(\frac{1.92\text{ V}}{(100\text{ k}\Omega)(0.0047\ \mu\text{F})}\right)\left(\frac{1}{8.2\text{ V} - 1\text{ V}}\right) = \mathbf{567\text{ Hz}}$$

(b) The output waveform is shown in Figure 21–34, where the period is determined as follows:

$$T = \frac{1}{f} = \frac{1}{567\text{ Hz}} = 1.76\text{ ms}$$

FIGURE 21–34

Output of the circuit in Figure 21–33.

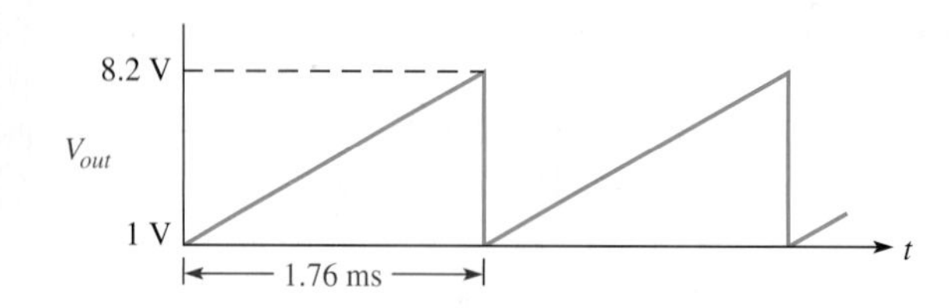

Related Problem If R_i is changed to 56 kΩ in Figure 21–33, what is the frequency?

A Square-Wave Relaxation Oscillator

The basic square-wave oscillator shown in Figure 21–35 is a type of relaxation oscillator because its operation is based on the charging and discharging of a capacitor. Notice that the op-amp's inverting input is the capacitor voltage and the noninverting input is a portion of the output fed back through resistors R_2 and R_3. When the circuit is first turned on, the capacitor is uncharged, and thus the inverting input is at 0 V. This makes the output a positive maximum, and the capacitor begins to charge toward V_{out} through R_1. When the capacitor voltage reaches a value equal to the feedback voltage on the noninverting input, the op-amp switches to the maximum negative state. At this point, the capacitor begins to discharge from $+V_f$ toward $-V_f$. When the capacitor voltage reaches $-V_f$, the op-amp switches back to the maximum positive state. This action continues to repeat, as shown in Figure 21–36, and a square-wave output voltage is obtained.

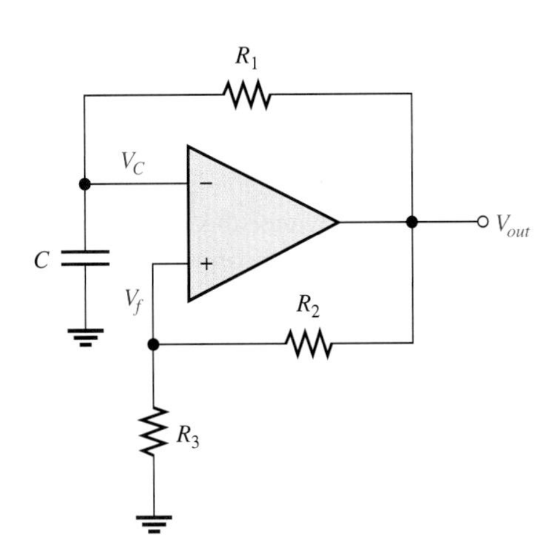

FIGURE 21–35

A square-wave relaxation oscillator.

FIGURE 21–36

Waveforms for the square-wave relaxation oscillator.

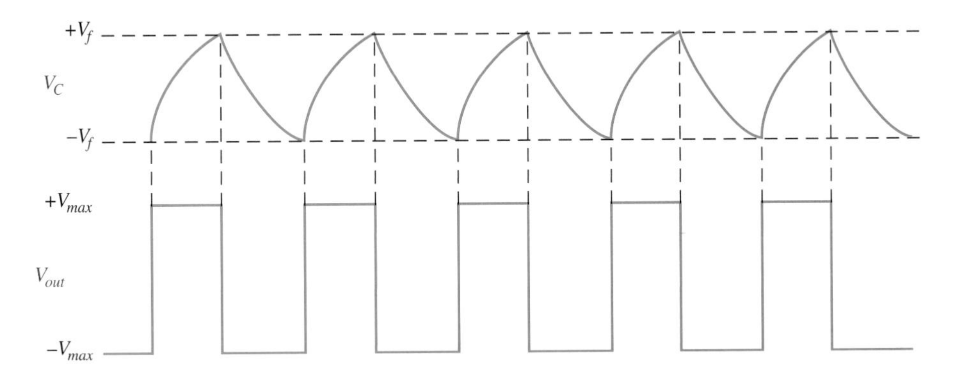

SECTION 21–4 REVIEW

1. There are two feedback loops in a Wien-bridge oscillator. What is the purpose of each?
2. What is a VCO, and basically what does it do?
3. Upon what principle does a relaxation oscillator operate?

21–5 ACTIVE FILTERS

Filters are usually categorized by the manner in which the output voltage varies with the frequency of the input voltage. The categories of active filters that we will examine in this section are low-pass, high-pass, and band-pass.

After completing this section, you should be able to

- **Recognize and evaluate basic op-amp filters**
- Evaluate single-pole and two-pole low-pass filters
- Evaluate single-pole and two-pole high-pass filters
- Determine the resonant frequency of a certain type of band-pass filter

Low-Pass Active Filters

Figure 21–37 shows a basic **active filter** and its response curve. Notice that the input circuit is a single low-pass *RC* circuit, and unity gain is provided by the op-amp with a negative feedback loop. Simply stated, this is a voltage-follower with an *RC* filter between the input signal and the noninverting input.

FIGURE 21–37

Single-pole, active low-pass filter and response curve.

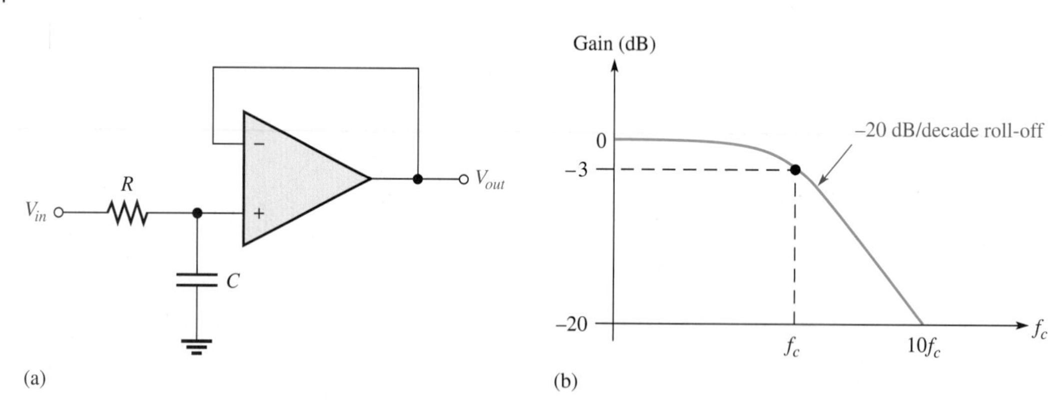

The voltage at the noninverting input, $V+$, is as follows:

$$V+ = \left(\frac{X_C}{\sqrt{R^2 + X_C^2}}\right)V_{in}$$

Since the gain of the op-amp is 1, the output voltage is equal to $V+$.

Equation 21–15

$$V_{out} = \left(\frac{X_C}{\sqrt{R^2 + X_C^2}}\right)V_{in}$$

A filter with one *RC* circuit that produces a −20 dB/decade **roll-off** beginning at f_c is said to be a *single-pole* or *first-order filter.* The term "−20 dB/decade" means that the voltage gain decreases by ten times (−20 dB) when the frequency increases by ten times **(decade).**

Low-Pass Two-Pole Filters There are several types of active filters and they can have varying numbers of **poles,** but we will use a two-pole filter to illustrate. Figure 21–38(a) shows a two-pole (second-order) low-pass filter. Since each *RC* circuit in a filter is considered to be one-pole, the two-pole filter uses two *RC* circuits to produce a roll-off rate of −40 dB/decade, as indicated in Figure 21–38(b). The active filter in Figure 21–38 has unity gain below f_c because the op-amp is connected as a voltage-follower.

FIGURE 21–38

Two-pole, active low-pass filter and its ideal response curve.

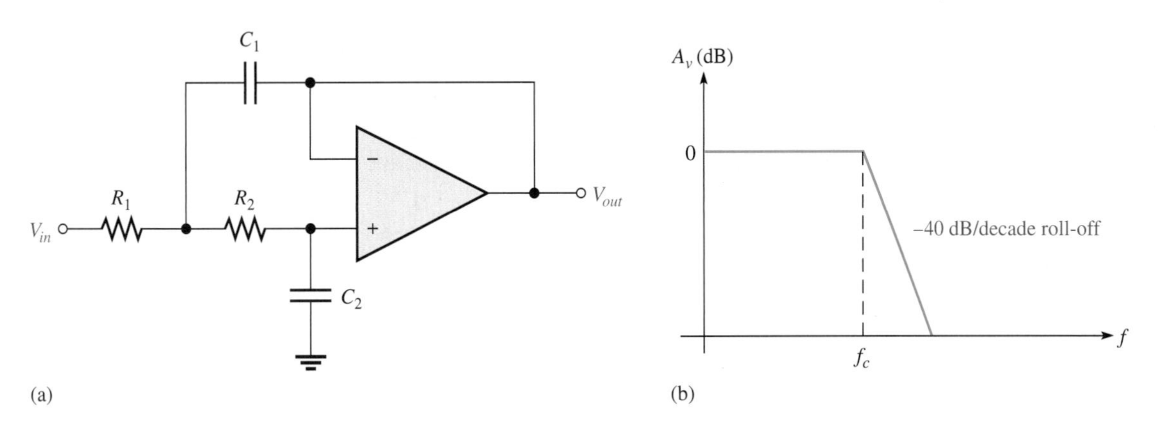

One of the *RC* circuits in Figure 21–38(a) is formed by R_1 and C_1, and the other by R_2 and C_2. The critical frequency of this filter can be calculated using the following formula:

$$f_c = \frac{1}{2\pi\sqrt{R_1R_2C_1C_2}}$$

Equation 21–16

Figure 21–39(a) shows an example of a two-pole low-pass filter with values chosen to produce a response with a critical frequency of 1 kHz. Note that $C_1 = 2C_2$ and $R_1 = R_2$ because these relationships result in a gain of 0.707 (−3 dB) at f_c. For critical frequencies other than 1 kHz, the capacitance values can be scaled inversely with the frequency. For example, as shown in Figure 21–39(b) and (c), to get a 2 kHz filter, halve the values of C_1 and C_2; for a 500 Hz filter, double the values.

FIGURE 21–39

Examples of low-pass filters (two-pole).

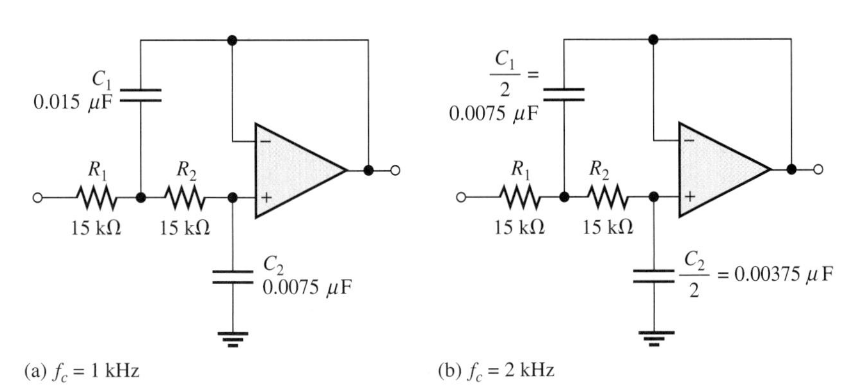

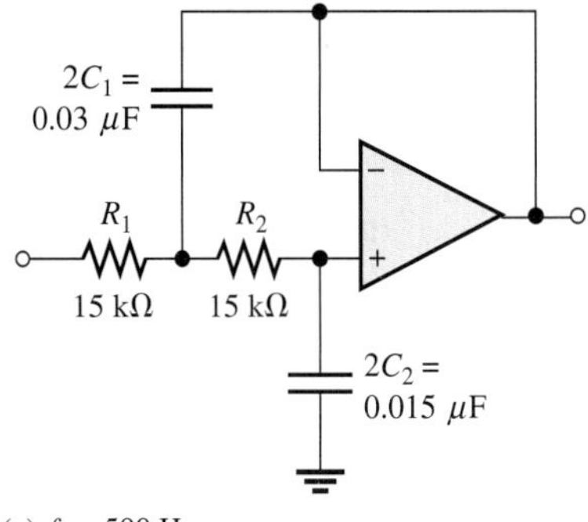

EXAMPLE 21–11

Calculate the capacitance values required to produce a 3 kHz critical frequency in the low-pass filter of Figure 21–40.

FIGURE 21–40

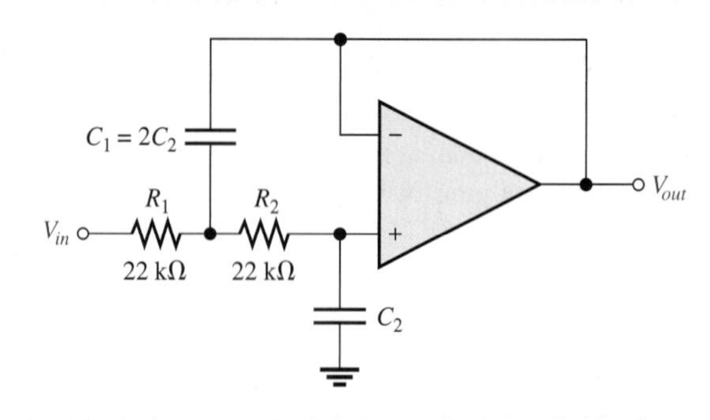

Solution The resistor values have already been set at 22 kΩ each. Since these differ from the 1 kHz reference filter, you cannot use the scaling method to get the capacitance values. Use Equation 21–16.

$$f_c = \frac{1}{2\pi\sqrt{R_1R_2C_1C_2}}$$

Then square both sides.

$$f_c^2 = \frac{1}{4\pi^2R_1R_2C_1C_2}$$

Since $C_1 = 2C_2$ and $R_1 = R_2 = R$,

$$f_c^2 = \frac{1}{4\pi^2R^2(2C_2^2)}$$

Solve for C_2, and then determine C_1.

$$C_2^2 = \frac{1}{8\pi^2R^2f_c^2}$$

$$C_2 = \frac{1}{\sqrt{2}\,2\pi Rf_c} = \frac{0.707}{2\pi Rf_c} = \frac{0.707}{2\pi(22\text{ k}\Omega)(3\text{ kHz})} = \mathbf{0.0017\ \mu F}$$

$$C_1 = 2C_2 = 2(0.0017\ \mu\text{F}) = \mathbf{0.0034\ \mu F}$$

Related Problem Determine f_c in Figure 21–40 for $R_1 = R_2 = 27$ kΩ, $C_1 = 0.001$ μF, and $C_2 = 500$ pF.

Open file E21-11 on your EWB/CircuitMaker CD-ROM. Verify that the critical frequency is 3 kHz.

High-Pass Active Filters

In Figure 21–41(a), a high-pass active filter with a −20 dB/decade roll-off is shown. Notice that the input circuit is a single high-pass *RC* circuit and that unity

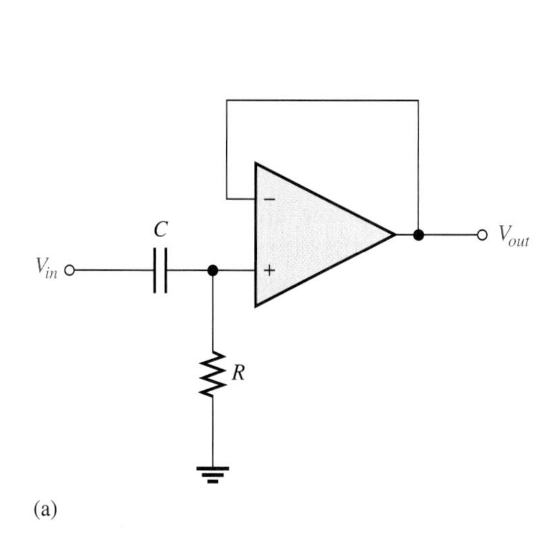

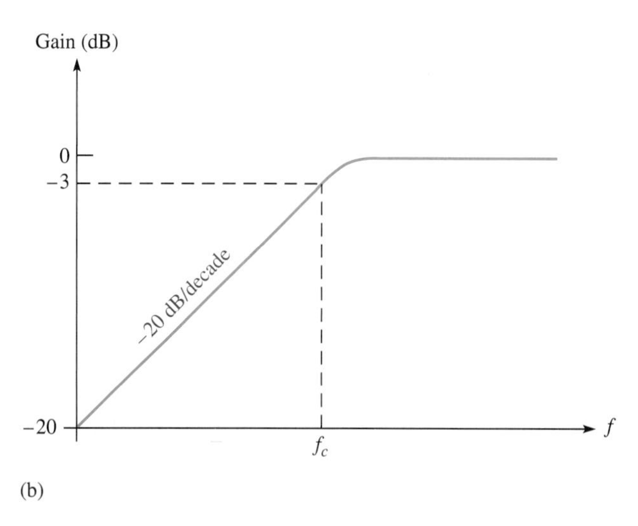

▲ **FIGURE 21–41**

Single-pole, active high-pass filter and response curve.

gain is provided by the op-amp with negative feedback. The response curve is shown in Figure 21–41(b).

Ideally, a high-pass filter passes all frequencies above f_c without limit, as indicated in Figure 21–42(a). In practice, of course, such is not the case. All op-amps inherently have internal *RC* circuits that limit the amplifier's response at high frequencies. Such is the case with the active high-pass filter. There is an upper frequency limit to its response, which, in effect, makes this type of filter a band-pass filter with a very wide bandwidth rather than a true high-pass filter, as indicated in Figure 21–42(b). In many applications, the internal high-frequency cutoff is so much greater than the filter's critical frequency that the internal high-frequency cutoff can be neglected.

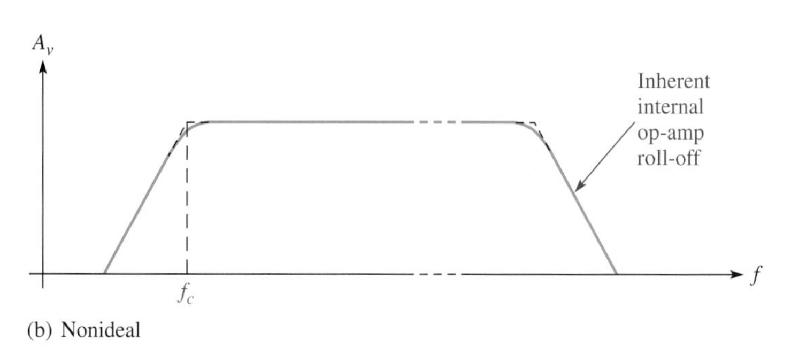

◀ **FIGURE 21–42**

High-pass filter response.

The voltage at the noninverting input is as follows:

$$V+ = \left(\frac{R}{\sqrt{R^2 + X_C^2}}\right)V_{in}$$

Since the op-amp is connected as a voltage-follower with unity gain, the output voltage is the same as $V+$.

Equation 21–17

$$V_{out} = \left(\frac{R}{\sqrt{R^2 + X_C^2}}\right)V_{in}$$

If the internal critical frequencies of the op-amp are assumed to be much greater than the desired f_c of the filter, the gain will roll off at -20 dB/decade as shown in Figure 21–42(b). This is a single-pole filter because it has one *RC* circuit.

High-Pass Two-Pole Filters Figure 21–43 shows a two-pole active high-pass filter. Notice that it is identical to the corresponding low-pass type, except for the positions of the resistors and capacitors. This filter has a roll-off rate of -40 dB/decade below f_c, and the critical frequency is the same as for the low-pass filter given in Equation 21–16.

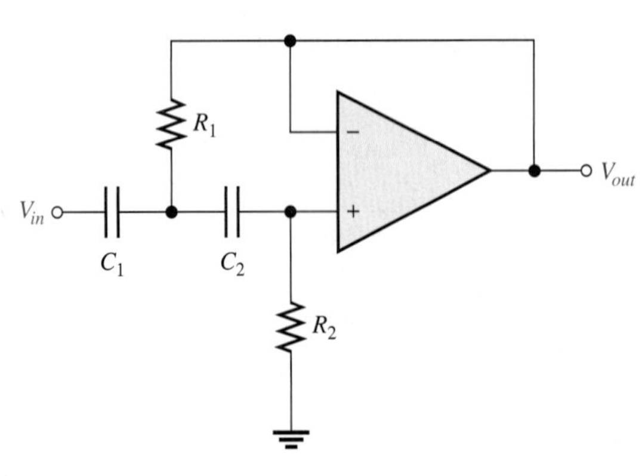

FIGURE 21–43

Two-pole, active high-pass filter.

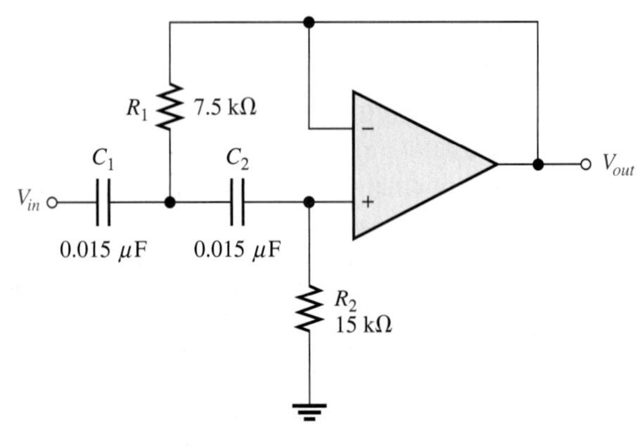

FIGURE 21–44

Two-pole, high-pass filter (f_c = 1 kHz).

Figure 21–44 shows a two-pole high-pass filter with values chosen to produce a response with a critical frequency of 1 kHz. Note that $R_2 = 2R_1$ and $C_1 = C_2$ because these relationships result in a gain of 0.707 (-3 dB) at f_c. For frequencies other than 1 kHz, the resistance values can be scaled inversely, as was done with the capacitors in the low-pass case.

EXAMPLE 21–12

For the filter of Figure 21–45, calculate the resistance values required to produce a critical frequency of 5.5 kHz.

Solution The capacitor values have been preselected to be 0.0022 μF each. Since these differ from the 1 kHz reference filter, you cannot use the scaling method to get the resistor values. Use Equation 21–16 and square both sides.

$$f_c = \frac{1}{2\pi\sqrt{R_1R_2C_1C_2}}$$

$$f_c^2 = \frac{1}{4\pi^2R_1R_2C_1C_2}$$

FIGURE 21–45

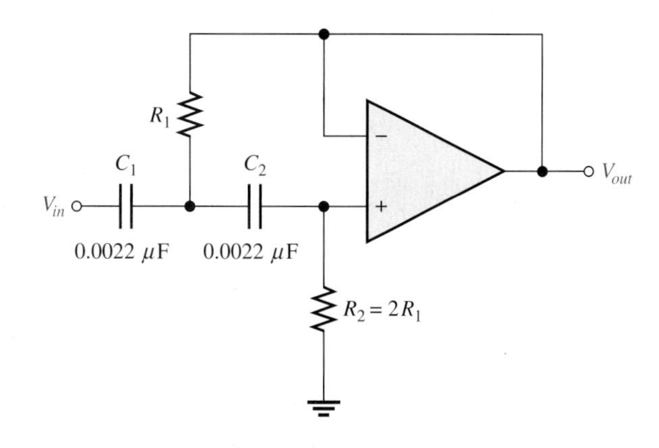

Since $R_2 = 2R_1$ and $C_1 = C_2 = C$,

$$f_c^2 = \frac{1}{4\pi^2(2R_1^2)C^2}$$

Solve for R_1, and then determine R_2.

$$R_1^2 = \frac{1}{8\pi^2 C^2 f_c^2}$$

$$R_1 = \frac{1}{\sqrt{2}\,2\pi C f_c} = \frac{0.707}{2\pi C f_c} = \frac{0.707}{2\pi(0.0022\ \mu\text{F})(5.5\ \text{kHz})} = \mathbf{9.3\ k\Omega}$$

$$R_2 = 2R_1 = 2(9.3\ \text{k}\Omega) = \mathbf{18.6\ k\Omega}$$

Related Problem Determine f_c in Figure 21–45 for $R_1 = 9.3$ kΩ and $R_2 = 18.6$ kΩ if C_1 and C_2 are changed to 4700 pF.

Open file E21-12 on your EWB/CircuitMaker CD-ROM. Verify that the critical frequency is 5.5 kHz.

Band-Pass Filter Using a High-Pass/Low-Pass Combination

One way to implement a band-pass filter is to use a cascaded arrangement of a high-pass filter followed by a low-pass filter, as shown in Figure 21–46(a). Each of the filters shown is a two-pole configuration so that the roll-off rates of the response curve are −40 dB/decade, as indicated in the composite response curve of part (b). The critical frequency of each filter is chosen so that the response curves overlap, as indicated. The critical frequency of the high-pass filter is lower than that of the low-pass filter.

The lower frequency, f_{c1}, of the passband is set by the critical frequency of the high-pass filter. The upper frequency, f_{c2}, of the passband is the critical frequency of the low-pass filter. Ideally, the center frequency, f_r, of the passband is the

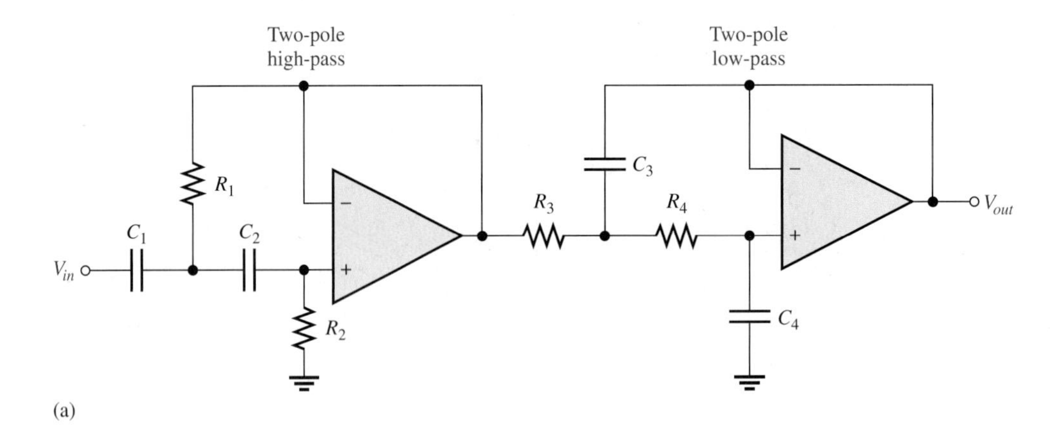

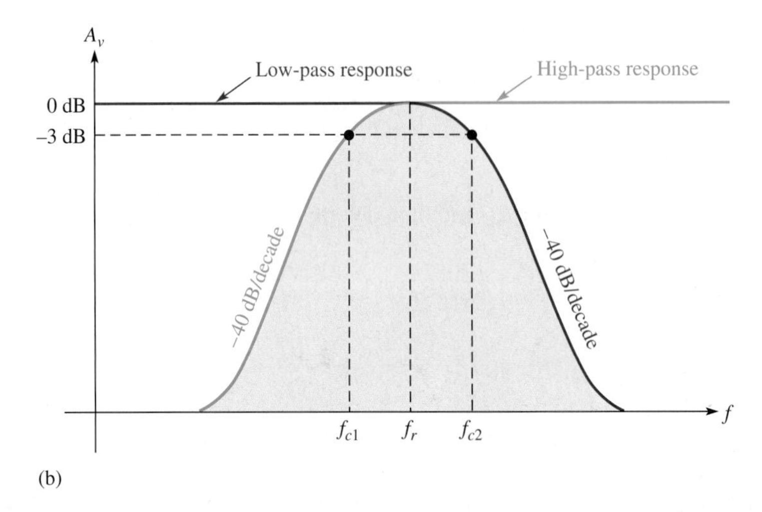

FIGURE 21–46

Band-pass filter formed by combining two-pole, high-pass filter with two-pole, low-pass filter. (It does not matter in which order the filters are cascaded.)

geometric average of f_{c1} and f_{c2}. The following formulas express the three frequencies of the band-pass filter in Figure 21–46.

Equation 21–18

$$f_{c1} = \frac{1}{2\pi\sqrt{R_1R_2C_1C_2}}$$

Equation 21–19

$$f_{c2} = \frac{1}{2\pi\sqrt{R_3R_4C_3C_4}}$$

Equation 21–20

$$f_r = \sqrt{f_{c1}f_{c2}}$$

EXAMPLE 21–13

(a) Determine the bandwidth and center frequency for the filter in Figure 21–47.

(b) Draw the response curve.

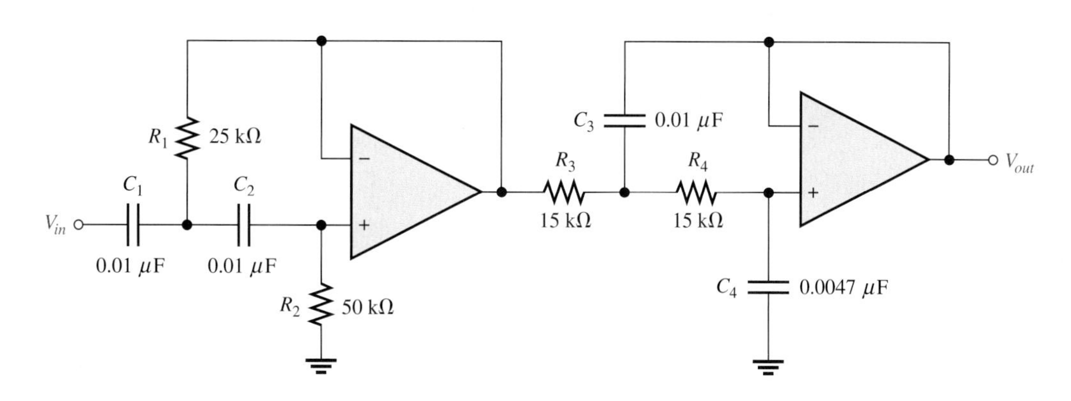

▲ FIGURE 21–47

Solution **(a)** The critical frequency of the high-pass filter is

$$f_{c1} = \frac{1}{2\pi\sqrt{R_1R_2C_1C_2}} = \frac{1}{2\pi\sqrt{(25\text{ k}\Omega)(50\text{ k}\Omega)(0.01\ \mu\text{F})(0.01\ \mu\text{F})}} = 450\text{ Hz}$$

The critical frequency of the low-pass filter is

$$f_{c2} = \frac{1}{2\pi\sqrt{R_3R_4C_3C_4}} = \frac{1}{2\pi\sqrt{(15\text{ k}\Omega)(15\text{ k}\Omega)(0.01\ \mu\text{F})(0.0047\ \mu\text{F})}} = 1.55\text{ kHz}$$

$$BW = f_{c2} - f_{c1} = 1.55\text{ kHz} - 450\text{ Hz} = \mathbf{1.1\ kHz}$$

$$f_r = \sqrt{f_{c1}f_{c2}} = \sqrt{(450\text{ Hz})(1.55\text{ kHz})} = \mathbf{822\ Hz}$$

(b) The response curve is shown in Figure 21–48.

▶ FIGURE 21–48

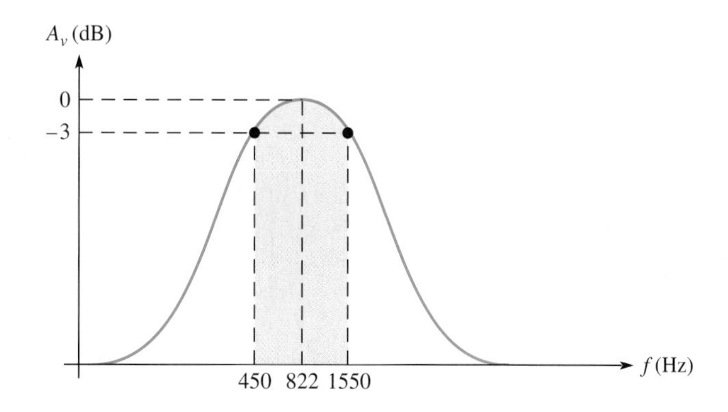

Related Problem Describe how you can increase the bandwidth of the filter in Figure 21–47 without changing f_{c1}.

SECTION 21–5 REVIEW

1. In terms of circuit components, what does the term *pole* refer to?
2. What type of response characterizes the single-pole low-pass filter?

3. How does a high-pass filter differ in its implementation from a corresponding low-pass version?
4. If the resistance values of a high-pass filter are doubled, what happens to the critical frequency?

21–6 VOLTAGE REGULATOR FUNDAMENTALS

Two fundamental types of linear voltage regulators are introduced in this section. One is the series regulator and the other is the shunt regulator.

After completing this section, you should be able to

- **Describe the operation of basic series and shunt voltage regulators**
- Explain how a basic op-amp series regulator works
- Discuss short circuit and overload protection
- Explain how a basic op-amp shunt regulator works

Basic Series Regulator

A simple representation of a linear **series regulator** is shown in Figure 21–49(a), and the basic components are shown in the block diagram in part (b). Notice that the control element is in series with the load between input and output. The output sample circuit senses a change in the output voltage. The error detector compares the sample voltage with a reference voltage and causes the control element to compensate in order to maintain a constant output voltage.

FIGURE 21–49

Block diagrams of a three-terminal, series voltage regulator.

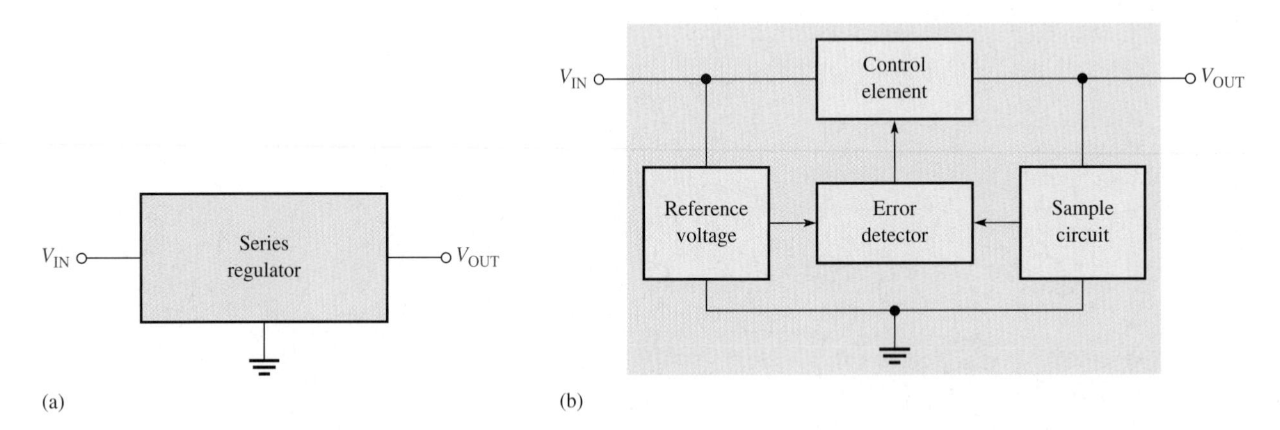

Regulating Action A basic op-amp series regulator circuit is shown in Figure 21–50. Its operation is illustrated in Figure 21–51 and is as follows. The resistive voltage divider formed by R_2 and R_3 senses any change in the output voltage. When the output tries to decrease, as shown in part (a), because of a decrease in V_{IN} or because of an increase in I_L (decrease in R_L), a proportional voltage decrease is applied to the op-amp's inverting input by the voltage divider. Since the zener diode, D_1, holds the other op-amp input at a nearly constant reference voltage (V_{REF}), a small difference voltage (error voltage) is developed across the op-amp's inputs. This difference voltage is amplified, and the op-amp's output

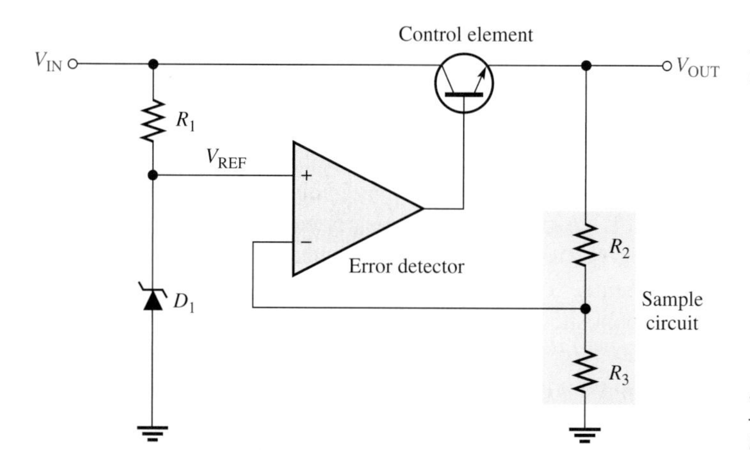

FIGURE 21–50

Basic op-amp series regulator.

FIGURE 21–51

Illustration of series regulator action that keeps V_{OUT} constant when V_{IN} or R_L changes.

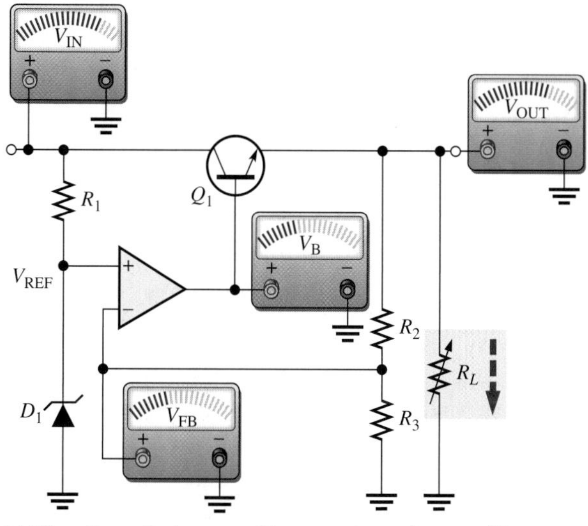

(a) When V_{IN} or R_L decreases, V_{OUT} attempts to decrease. The feedback voltage, V_{FB}, also attempts to decrease, and as a result, the op-amp's output voltage V_B attempts to increase, thus compensating for the attempted decrease in V_{OUT} by increasing the Q_1 emitter voltage. Changes in V_{OUT} are exaggerated for illustration.

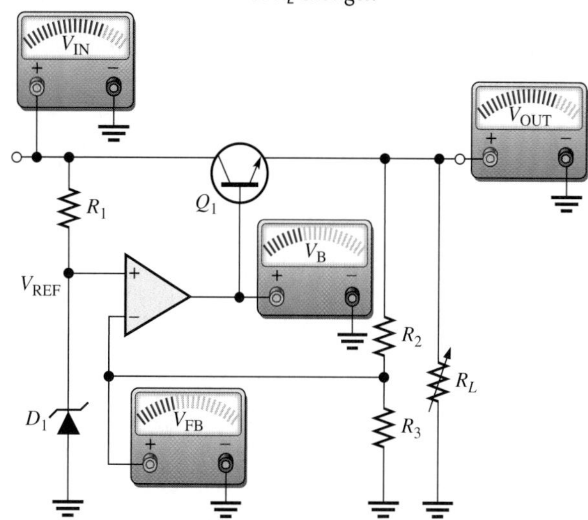

(b) When V_{IN} (or R_L) stabilizes at its new lower value, the voltages return to their original values, thus keeping V_{OUT} constant as a result of the negative feedback.

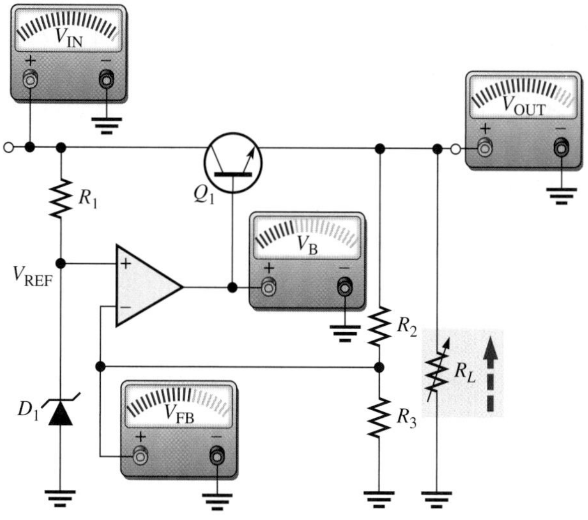

(c) When V_{IN} or R_L increases, V_{OUT} attempts to increase. The feedback voltage, V_{FB}, also attempts to increase, and as a result, V_B, applied to the base of the control transistor, attempts to decrease, thus compensating for the attempted increase in V_{OUT} by decreasing the Q_1 emitter voltage.

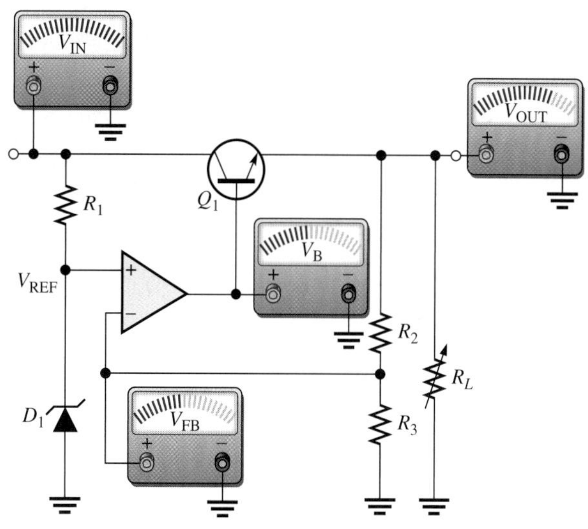

(d) When V_{IN} (or R_L) stabilizes at its new higher value, the voltages return to their original values, thus keeping V_{OUT} constant as a result of the negative feedback.

voltage (V_B) increases. This increase is applied to the base of Q_1, causing the emitter voltage, V_{OUT}, to increase until the voltage to the inverting input again equals the reference (zener) voltage. This action offsets the attempted decrease in output voltage, thus keeping it nearly constant, as shown in part (b). The power transistor, Q_1, is usually used with a heat sink because it must handle all of the load current.

The opposite action occurs when the output tries to increase, as illustrated in Figure 21–51(c) and (d). Percent regulation was discussed in Chapter 17.

The op-amp in Figure 21–50 is actually connected as a noninverting amplifier in which the reference voltage, V_{REF}, is the input at the noninverting terminal, and the R_2/R_3 voltage divider forms the negative feedback circuit. The closed-loop voltage gain is

Equation 21–21

$$A_{cl} = \frac{R_2}{R_3} + 1$$

Therefore, the regulated output voltage is

Equation 21–22

$$V_{OUT} = \left(\frac{R_2}{R_3} + 1\right)V_{REF}$$

From this analysis you can see that the output voltage is determined by the zener voltage and the resistors R_2 and R_3. It is relatively independent of the input voltage, and therefore, regulation is achieved (as long as the input voltage and load current are within specified limits).

EXAMPLE 21–14

Determine the output voltage for the regulator in Figure 21–52.

▶ FIGURE 21–52

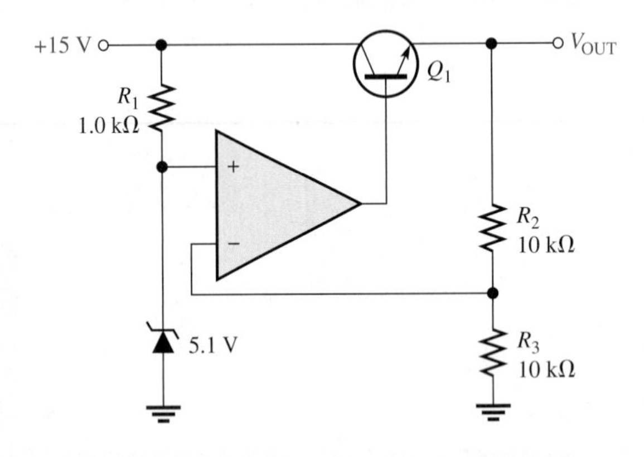

Solution $V_{REF} = 5.1$ V, the zener voltage. Therefore,

$$V_{OUT} = \left(\frac{R_2}{R_3} + 1\right)V_{REF} = \left(\frac{10\text{ k}\Omega}{10\text{ k}\Omega} + 1\right)5.1\text{ V} = (2)5.1\text{ V} = \mathbf{10.2\text{ V}}$$

Related Problem The following changes are made in the circuit of Figure 21–52: A 3.3 V zener replaces the 5.1 V zener, $R_1 = 1.8$ kΩ, $R_2 = 22$ kΩ, and $R_3 = 18$ kΩ. What is the output voltage?

Short-Circuit or Overload Protection If an excessive amount of load current is drawn, the series-pass transistor can be quickly damaged or destroyed. Most regulators use some type of protection from excess current in the form of a current-limiting mechanism.

Figure 21–53 shows one method of current limiting to prevent overloads called *constant-current limiting.* The current-limiting circuit consists of transistor Q_2 and resistance R_4.

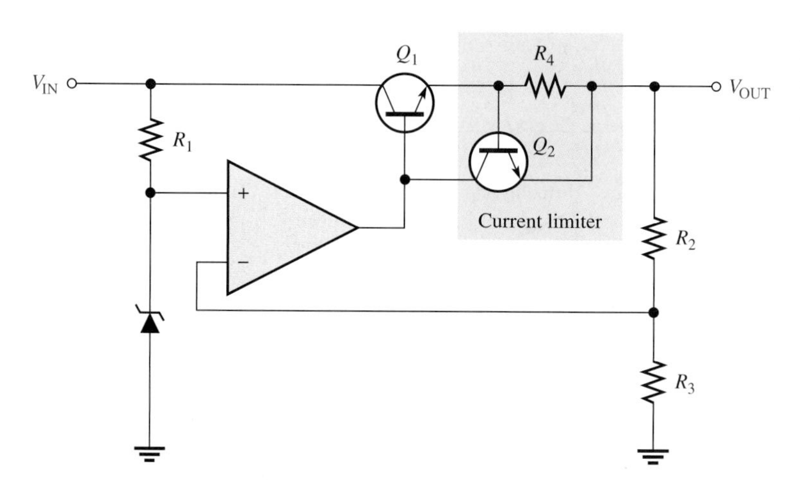

FIGURE 21–53

Series regulator with constant-current limiting.

The load current through R_4 creates a voltage from base to emitter of Q_2. When I_L reaches a predetermined maximum value, the voltage drop across R_4 is sufficient to forward-bias the base-emitter junction of Q_2, thus causing it to conduct. Enough Q_1 base current is diverted into the collector of Q_2 so that I_L is limited to its maximum value $I_{L(max)}$. Since the base-to-emitter voltage of Q_2 cannot exceed about 0.7 V for a silicon transistor, the voltage across R_4 is held to this value, and the load current is limited to

$$I_{L(max)} = \frac{0.7\text{ V}}{R_4}$$

Equation 21–23

Basic Shunt Regulator

As you have seen, the control element in the series regulator is the series-pass transistor. A simple representation of a shunt type of linear regulator is shown in Figure 21–54(a), and the basic components are shown in the block diagram in part (b).

FIGURE 21–54

Block diagrams of a three-terminal shunt regulator.

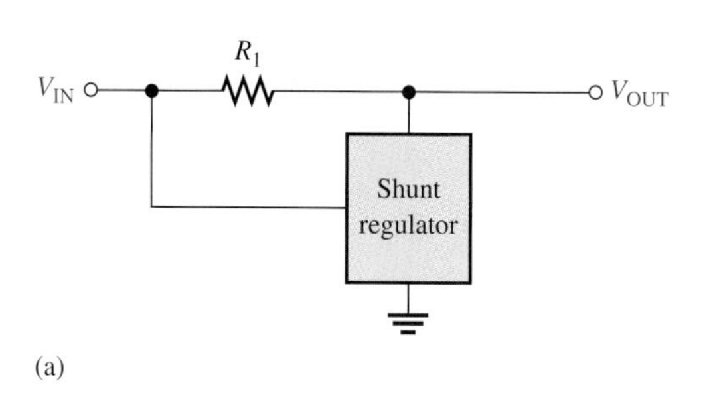

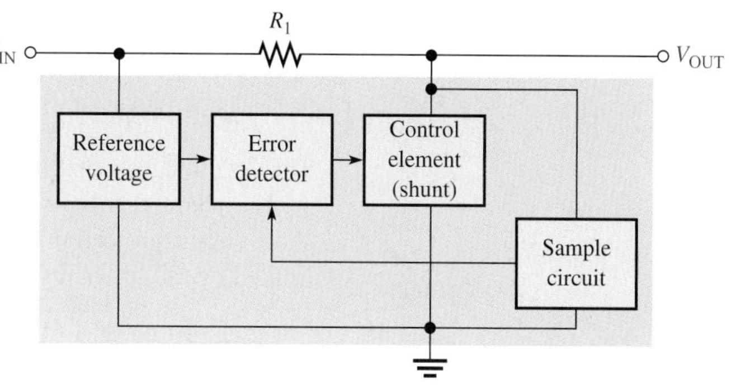

In the basic **shunt regulator**, the control element is a transistor (Q_1) in parallel with the load, as shown in Figure 21–55. A series resistor (R_1) is in series with the load. The operation of the circuit is similar to that of the series regulator, except that regulation is achieved by controlling the current through the parallel transistor Q_1.

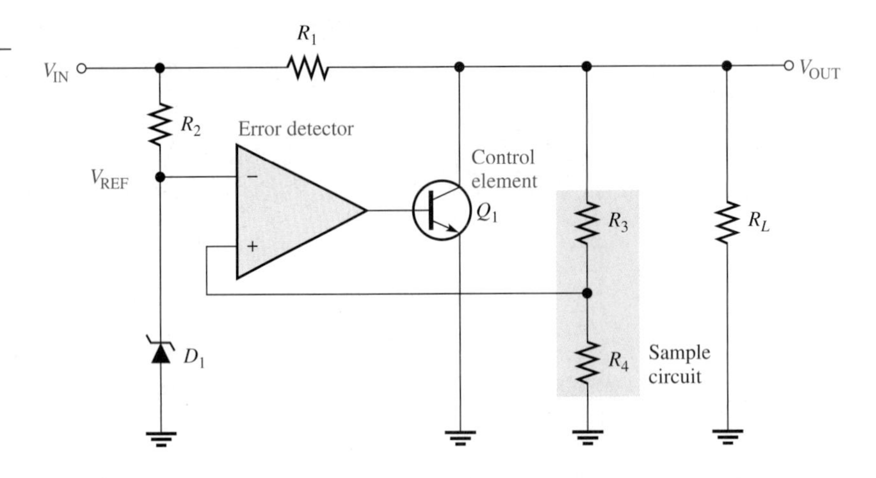

▶ **FIGURE 21–55**

Basic op-amp shunt regulator.

When the output voltage tries to decrease due to a change in input voltage or load current, as shown in Figure 21–56(a), the attempted decrease is sensed by R_3 and R_4 and applied to the op-amp's noninverting input. The resulting difference in voltage reduces the op-amp's output (V_B), driving Q_1 less, thus reducing its collector current (shunt current) and increasing its internal collector-to-emitter resistance, r_{ce}. Since r_{ce} acts as a voltage divider with R_1, this action offsets the attempted decrease in V_{OUT} and maintains it at an almost constant level.

The opposite action occurs when the output tries to increase, as shown in Figure 21–56(b). With I_L and V_{OUT} constant, a change in the input voltage produces a change in shunt current (I_S) as follows:

$$\Delta I_S = \frac{\Delta V_{IN}}{R_1}$$

With a constant V_{IN} and V_{OUT}, a change in load current causes an opposite change in shunt current.

$$\Delta I_S = -\Delta I_L$$

This formula says that if I_L increases, I_S decreases, and vice versa.

The shunt regulator is less efficient than the series type but offers inherent short-circuit protection. If the output is shorted ($V_{OUT} = 0$), the load current is limited by the series resistor, R_1, to a maximum value ($I_S = 0$).

$$I_{L(max)} = \frac{V_{IN}}{R_1}$$

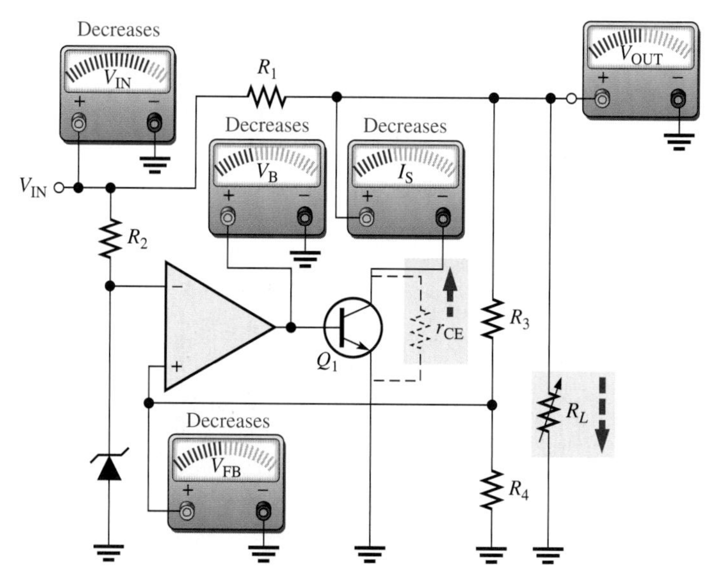

(a) Response to a decrease in V_{IN} or R_L

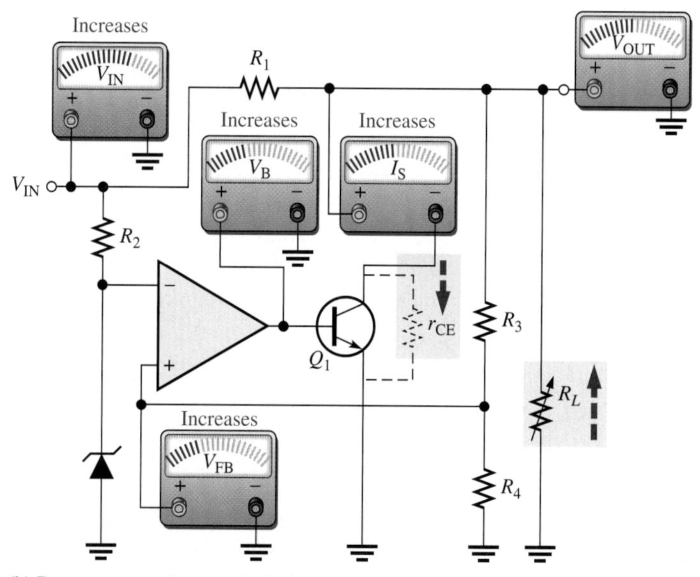

(b) Response to an increase in V_{IN} or R_L

FIGURE 21–56

Sequence of responses when V_{OUT} tries to decrease as a result of a decrease in V_{IN} or R_L.

EXAMPLE 21–15

In Figure 21–57, what power rating must R_1 have if the maximum input voltage is 12.5 V?

Solution The worst-case power dissipation in R_1 occurs when the output is short-circuited and $V_{OUT} = 0$. When $V_{IN} = 12.5$ V, the voltage dropped across R_1 is

$$V_{IN} - V_{OUT} = 12.5\text{ V}$$

FIGURE 21–57

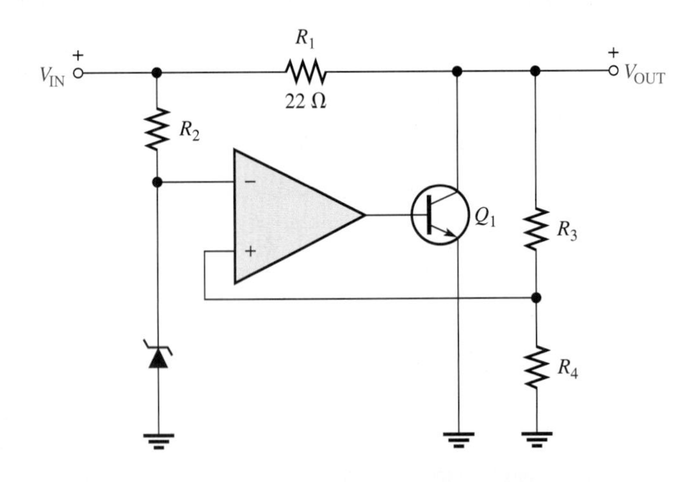

The power dissipation in R_1 is

$$P_{R1} = \frac{V_{R1}^2}{R_1} = \frac{(12.5\text{ V})^2}{22\ \Omega} = \mathbf{7.1\ W}$$

Therefore, a resistor of at least 10 W should be used.

Related Problem In Figure 21–57, R_1 is changed to 33 Ω. What must be the power rating of R_1 if the maximum input voltage is 24 V?

SECTION 21–6 REVIEW

1. How does the control element in a shunt regulator differ from that in a series regulator?
2. Name one advantage and one disadvantage of a shunt regulator over a series type.

APPLICATION ASSIGNMENT

PUTTING YOUR KNOWLEDGE TO WORK

In this application assignment, you will focus on the regulated power supply which provides the FM stereo receiver with dual polarity dc voltages. The op-amps in the channel separation circuits and the audio amplifiers operate from ±12 V. Both positive and negative voltage regulators are used to regulate the rectified and filtered voltages from a bridge rectifier. You should review section 17–3.

This power supply utilizes a full-wave bridge **rectifier** with both the positive and negative rectified voltages taken off the bridge at the appropriate points and filtered by electrolytic capacitors. Integrated circuit voltage regulators (7812 and 7912) provide regulation for the positive and negative voltages. A block diagram is shown in Figure 21–58.

Step 1: Relate the PC Board to a Schematic

Develop a schematic for the power supply in Figure 21–59. Add any missing labels and include the IC pin numbers. The rectifier diodes are 1N4001s, the filter capacitors C_1 and C_2 are 100 μF, and the transformer has a turns ratio of 5 : 1. Determine the backside PC board connections as you develop the schematic.

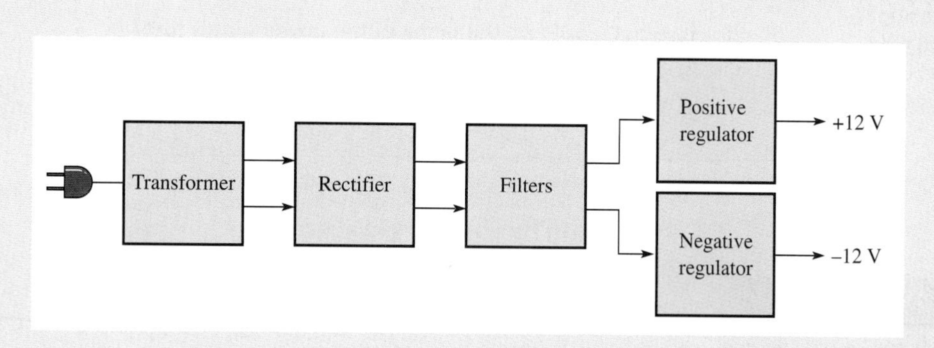

FIGURE 21–58

Block diagram of the dual-polarity power supply.

Step 2: Analyze the Power Supply Circuits

1. Determine the approximate voltage with respect to ground at each of the four "corners" of the bridge. State whether each voltage is ac or dc.
2. Calculate the peak inverse voltage of the rectifier diodes.
3. Show the voltage waveform across D_1 for a full cycle of the ac input.

Step 3: Troubleshoot the Power Supply

State the probable cause or causes for the following:

1. Both positive and negative output voltages are zero.
2. Positive output voltage is zero, and the negative output voltage is −12 V.
3. Negative output voltage is zero, and the positive output voltage is +12 V.
4. There are radical voltage fluctuations on the output of the positive regulator.

Indicate the voltages you should measure at the four corners of the diode bridge for the following faults:

1. Diode D_1 open
2. Capacitor C_2 open

FIGURE 21–59

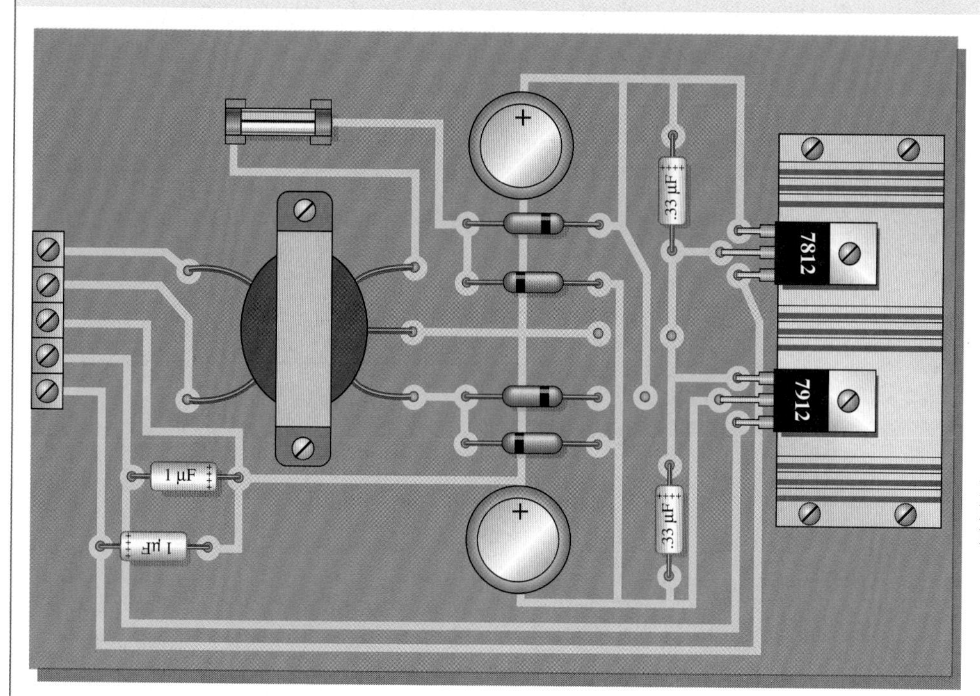

APPLICATION ASSIGNMENT REVIEW

1. What should be the rating of the power supply fuse?
2. What purpose do the 0.33 μF capacitors serve?
3. Which regulator provides the negative voltage?

SUMMARY

- In an op-amp comparator, when the input voltage exceeds a specified reference voltage, the output changes state.
- The output voltage of a summing amplifier is proportional to the sum of the input voltages.
- An averaging amplifier is a summing amplifier with a closed-loop gain equal to the reciprocal of the number of inputs.
- In a scaling adder, a different weight can be assigned to each input, thus making the input contribute more or contribute less to the output.
- The integral of a step is a ramp.
- The derivative of a ramp is a step.
- In a Wien-bridge oscillator, the closed-loop gain must be equal to 3 in order to have unity gain around the positive feedback loop.
- In filter terminology, a single *RC* circuit is called a *pole.*
- Each pole in a filter causes the output to roll off (decrease) at a rate of −20 dB/decade.
- Two-pole filters roll off at a maximum rate of −40 dB/decade.
- In a series voltage regulator, the control element is a transistor in series with the load.
- In a shunt voltage regulator, the control element is a transistor (or zener diode) in parallel with a load.
- The terminals on a three-terminal voltage regulator are input voltage, output voltage, and ground.

EQUATIONS

21–1 $$V_{\text{REF}} = \frac{R_2}{R_1 + R_2}(+V)$$ Comparator reference

21–2 $$V_{\text{OUT}} = -(V_{\text{IN1}} + V_{\text{IN2}})$$ Two-input adder

21–3 $$V_{\text{OUT}} = -(V_{\text{IN1}} + V_{\text{IN2}} + V_{\text{IN3}} + \cdots + V_{\text{IN}n})$$ *n*-input adder

21–4 $$V_{\text{OUT}} = -\frac{R_f}{R}(V_{\text{IN1}} + V_{\text{IN2}} + \cdots + V_{\text{IN}n})$$ Adder with gain

21–5 $$V_{\text{OUT}} = -\left(\frac{R_f}{R_1}V_{\text{IN1}} + \frac{R_f}{R_2}V_{\text{IN2}} + \cdots + \frac{R_f}{R_n}V_{\text{IN}n}\right)$$ Adder with gain

21–6 $$\frac{\Delta V_{out}}{\Delta t} = -\frac{V_{in}}{R_iC}$$ Rate of change in integrator

21–7	$V_{out} = -\left(\frac{V_C}{t}\right)R_f C$	Differentiator output with ramp input
21–8	$\frac{V_{out}}{V_{in}} = \frac{1}{3}$	Lead-lag attenuation at f_r
21–9	$f_r = \frac{1}{2\pi RC}$	Lead-lag resonant frequency
21–10	$V_{\text{UTP}} = +V_{max}\left(\frac{R_3}{R_2}\right)$	Upper trigger point, triangular-wave oscillator
21–11	$V_{\text{LTP}} = -V_{max}\left(\frac{R_3}{R_2}\right)$	Lower trigger point, triangular-wave oscillator
21–12	$f = \frac{1}{4R_1C}\left(\frac{R_2}{R_3}\right)$	Frequency of oscillation, triangular-wave oscillator
21–13	$T = \frac{V_p - V_{\text{F}}}{\lvert V_{\text{IN}}\rvert /RC}$	Sawtooth period, VCO
21–14	$f = \frac{\lvert V_{\text{IN}}\rvert}{RC}\left(\frac{1}{V_p - V_{\text{F}}}\right)$	Sawtooth frequency, VCO
21–15	$V_{out} = \left(\frac{X_C}{\sqrt{R^2 - X_C^2}}\right)V_{in}$	Output of one-pole filter (low-pass only)
21–16	$f_c = \frac{1}{2\pi\sqrt{R_1R_2C_1C_2}}$	Critical frequency of two-pole filter (low-pass)
21–17	$V_{out} = \left(\frac{R}{\sqrt{R^2 + X_C^2}}\right)V_{in}$	Output of one-pole filter (high-pass only)
21–18	$f_{c1} = \frac{1}{2\pi\sqrt{R_1R_2C_1C_2}}$	Lower critical frequency of a band-pass filter
21–19	$f_{c2} = \frac{1}{2\pi\sqrt{R_3R_4C_3C_4}}$	Upper critical frequency of a band-pass filter
21–20	$f_r = \sqrt{f_{c1}f_{c2}}$	Center frequency of a band-pass filter
21–21	$A_{cl} = \frac{R_2}{R_3} + 1$	Closed-loop voltage gain, voltage regulator
21–22	$V_{\text{OUT}} = \left(\frac{R_2}{R_3} + 1\right)V_{\text{REF}}$	Series regulator output voltage
21–23	$I_{\text{L(max)}} = \frac{0.7\text{ V}}{R_4}$	For constant-current limiting

SELF-TEST

Answers are at the end of the chapter.

1. The purpose of a comparator is to

(a) amplify an input voltage

(b) detect the occurrence of a changing input voltage

(c) produce a change in output when an input voltage equals a reference voltage

(d) maintain a constant output when the dc input voltage changes

2. To use a comparator for zero-level detection, the inverting input is connected to
   (a) ground **(b)** the dc supply voltage
   (c) a positive reference voltage **(d)** a negative reference voltage
3. In a 5 V level detector circuit,
   (a) the noninverting input is connected to +5 V
   (b) the inverting input is connected to +5 V
   (c) the input signal is limited to a 5 V peak value
   (d) the input signal must be riding on a +5 dc level
4. In a certain four-input summing amplifier, all the input resistors are 2.2 kΩ and the feedback resistor in 2.2 kΩ. If all the input voltages are 2 V, the output voltage is
   (a) 2 V **(b)** 10 V **(c)** 2.2 V **(d)** 8 V
5. The gain of the amplifier in Question 4 is
   (a) 1 **(b)** 2.2 **(c)** 4 **(d)** unknown
6. To convert a summing amplifier to an averaging amplifier,
   (a) all input resistors must be a different value
   (b) the ratio R_f/R must equal to the reciprocal of the number of inputs
   (c) the ratio R_f/R must equal to the number of inputs
   (d) answers (a) and (b)
7. In a scaling adder,
   (a) the input resistors are all the same value
   (b) the input resistors are all different values
   (c) the input resistors have values that depend on the assigned weight of each input
   (d) the ratio R_f/R must be the same for each input
8. The feedback path in an op-amp integrator consists of
   (a) a resistor
   (b) a capacitor
   (c) a resistor and a capacitor in series
   (d) a resistor and a capacitor in parallel
9. The feedback path in an op-amp differentiator consists of
   (a) a resistor
   (b) a capacitor
   (c) a resistor and a capacitor in series
   (d) a resistor and a capacitor in parallel
10. The op-amp comparator circuit uses
    (a) positive feedback **(b)** negative feedback
    (c) regenerative feedback **(d)** no feedback
11. Unity gain and zero phase shift around the feedback loop are conditions that describe
    (a) an active filter **(b)** a comparator
    (c) an oscillator **(d)** an integrator or differentiator
12. The input frequency of a single-pole, low-pass active filter increases from 1.5 kHz to 150 kHz. If the critical frequency is 1.5 kHz, the gain decreases by
    (a) 3 dB **(b)** 20 dB **(c)** 40 dB **(d)** 60 dB

PROBLEMS

Answers to odd-numbered problems are at the end of the book.

BASIC PROBLEMS

SECTION 21–1

Comparators

1. Determine the output level (maximum positive or maximum negative) for each comparator in Figure 21–60.

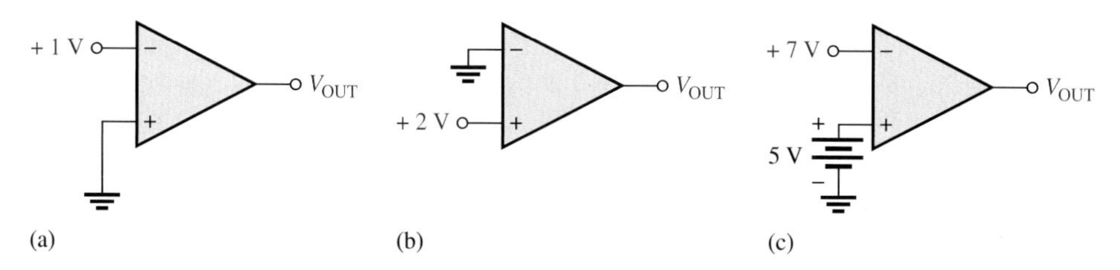

▲ FIGURE 21–60

2. A certain op-amp has an open-loop gain of 80,000. The maximum saturated output levels of this particular device are ±12 V when the dc supply voltages are ±15 V. If a differential voltage of 0.15 mV rms is applied between the inputs, what is the peak-to-peak value of the output?
3. Draw the output voltage waveform for each circuit in Figure 21–61 with respect to the input. Show voltage levels.

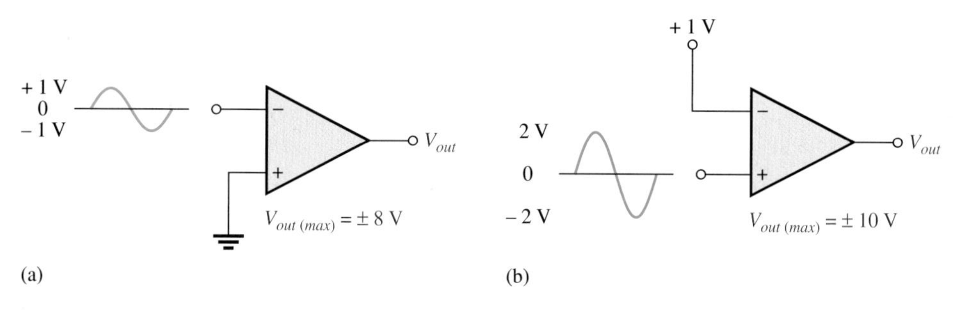

▲ FIGURE 21–61

SECTION 21–2

Summing Amplifiers

4. Determine the output voltage for each circuit in Figure 21–62.

▼ FIGURE 21–62

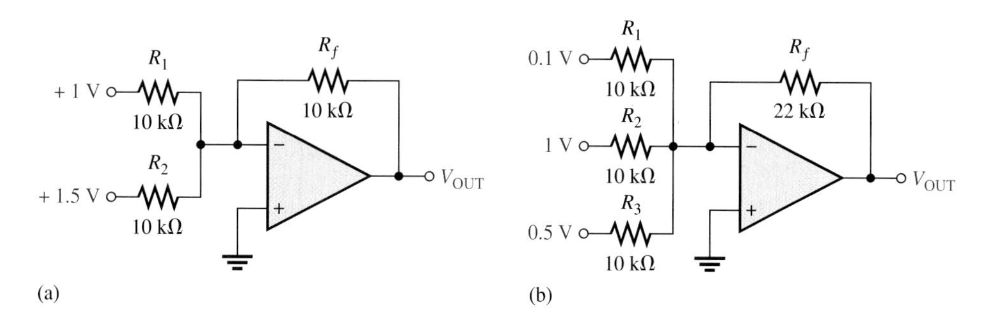

5. Determine the following in Figure 21–63:

   (a) V_{R1} and V_{R2} (b) current through R_f (c) V_{OUT}

6. Find the value of R_f necessary to produce an output that is 5 times the sum of the inputs in Figure 21–63.

▶ FIGURE 21–63

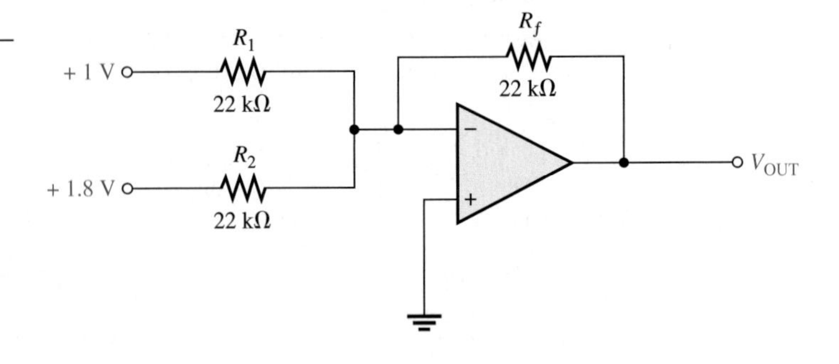

7. Find the output voltage when the input voltages shown in Figure 21–64 are applied to the scaling adder. What is the current through R_f?

▶ FIGURE 21–64

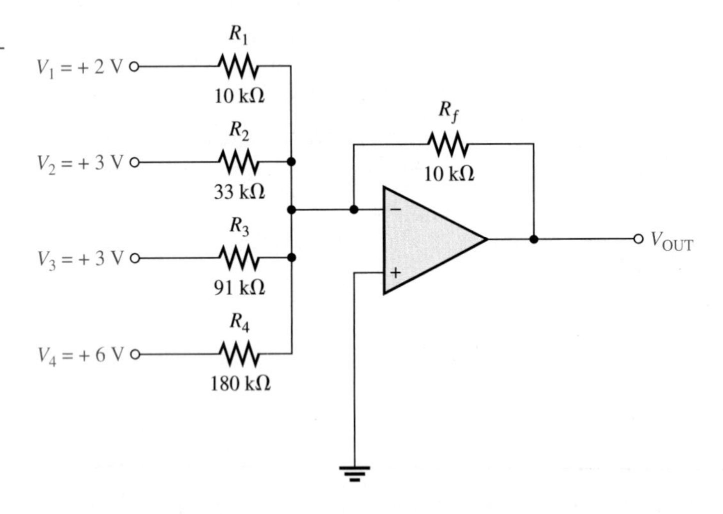

8. Determine the values of the input resistors required in a six-point scaling adder so that the lowest weighted input is 1 and each successive input has a weight *twice* the previous one. Use $R_f = 100\text{ k}\Omega$.

SECTION 21–3 **Integrators and Differentiators**

9. Determine the rate of change of the output voltage in response to the step input to the ideal integrator in Figure 21–65.

▶ FIGURE 21–65

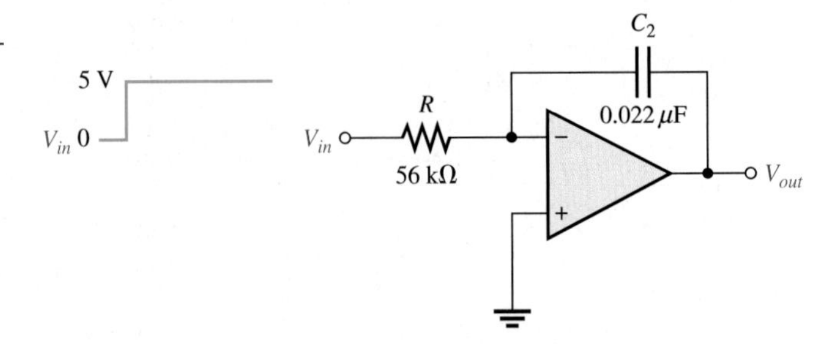

10. A triangular waveform is applied to the input of the ideal differentiator in Figure 21–66 as shown. Determine what the output should be, and draw its waveform in relation to the input.

FIGURE 21–66

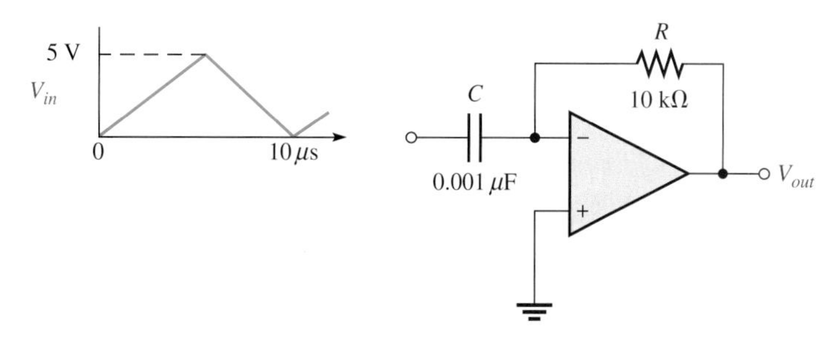

SECTION 21–4

Oscillators

11. A certain lead-lag circuit has a resonant frequency of 3.5 kHz. What is the rms output voltage if an input signal with a frequency equal to f_r and with an rms value of 2.2 V is applied to the input?

12. Calculate the resonant frequency of a lead-lag circuit with the following values: $R_1 = R_2 = 6.2\text{ k}\Omega$, and $C_1 = C_2 = 0.02\ \mu\text{F}$

13. Determine the necessary value of R_2 in Figure 21–67 so that the circuit will oscillate. Neglect the forward resistance of the zener diodes.

FIGURE 21–67

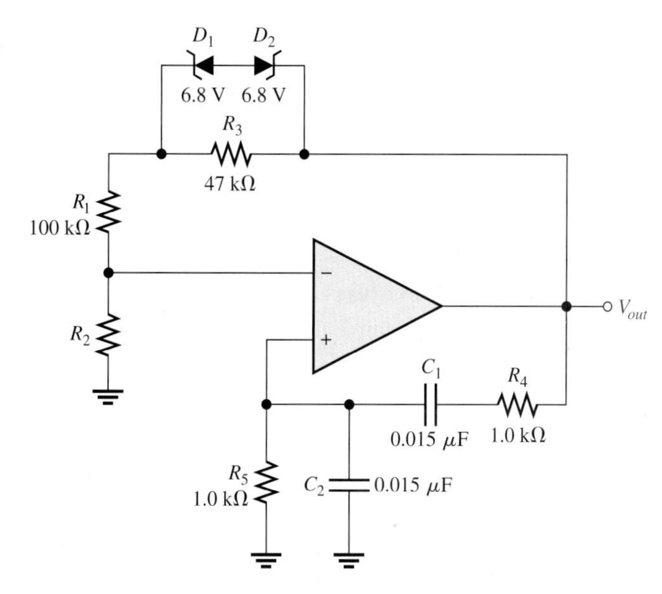

14. Explain the purpose of R_3 in Figure 21–67.

15. Find the frequency of oscillation for the Wien-bridge oscillator in Figure 21–67.

16. What type of signal does the circuit in Figure 21–68 produce? Determine the frequency of the output.

FIGURE 21–68

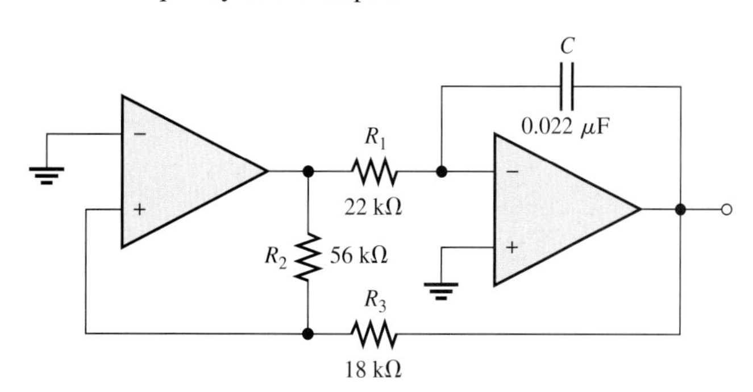

17. Show how to change the frequency of oscillation in Figure 21–68 to 10 kHz.
18. Determine the amplitude and frequency of the output voltage in Figure 21–69. Use 1 V as the forward PUT voltage.

▶ FIGURE 21–69

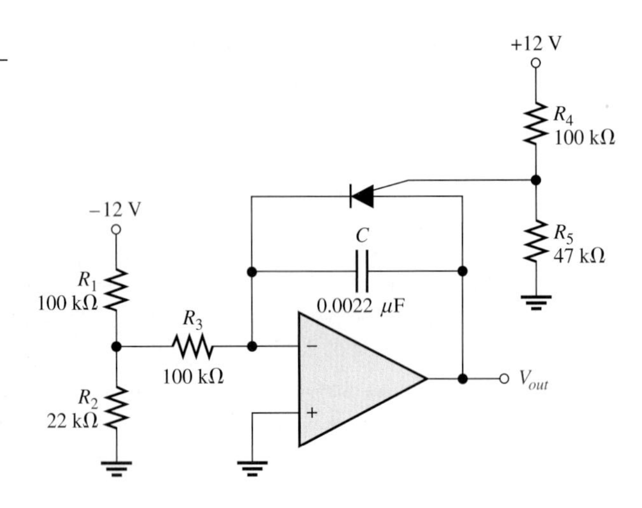

19. Modify the sawtooth oscillator in Figure 21–69 so that its peak-to-peak output is 4 V.
20. A certain sawtooth oscillator has the following parameter values: $V_{IN} = 3$ V, $R = 4.7$ kΩ, $C = 0.001$ μF, and V_F for the PUT is 1.2 V. Determine its peak-to-peak output voltage if the period is 10 μs.

SECTION 21–5 **Active Filters**

21. Determine the number of poles in each active filter in Figure 12–70, and identify its type.

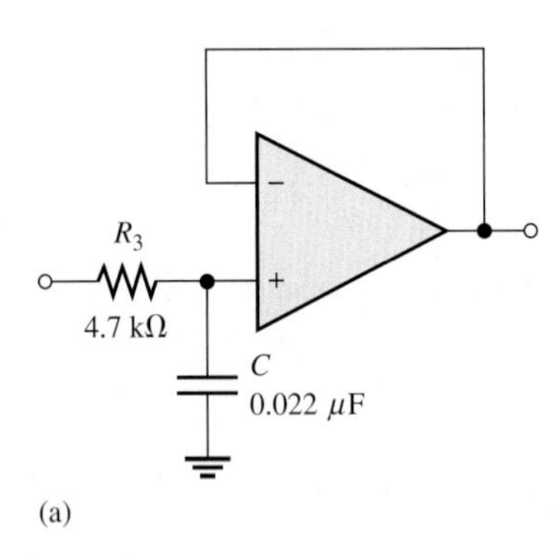

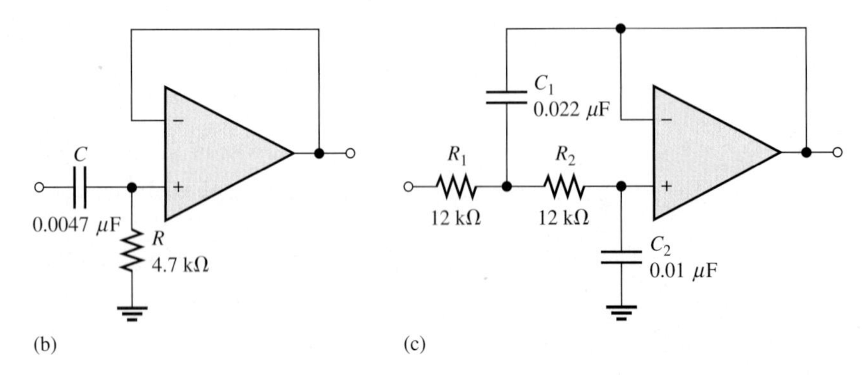

▲ FIGURE 21–70

22. Calculate the critical frequencies for the filters in Figure 21–70.
23. Determine the bandwidth and center frequency of each filter in Figure 21–71.

FIGURE 21–71

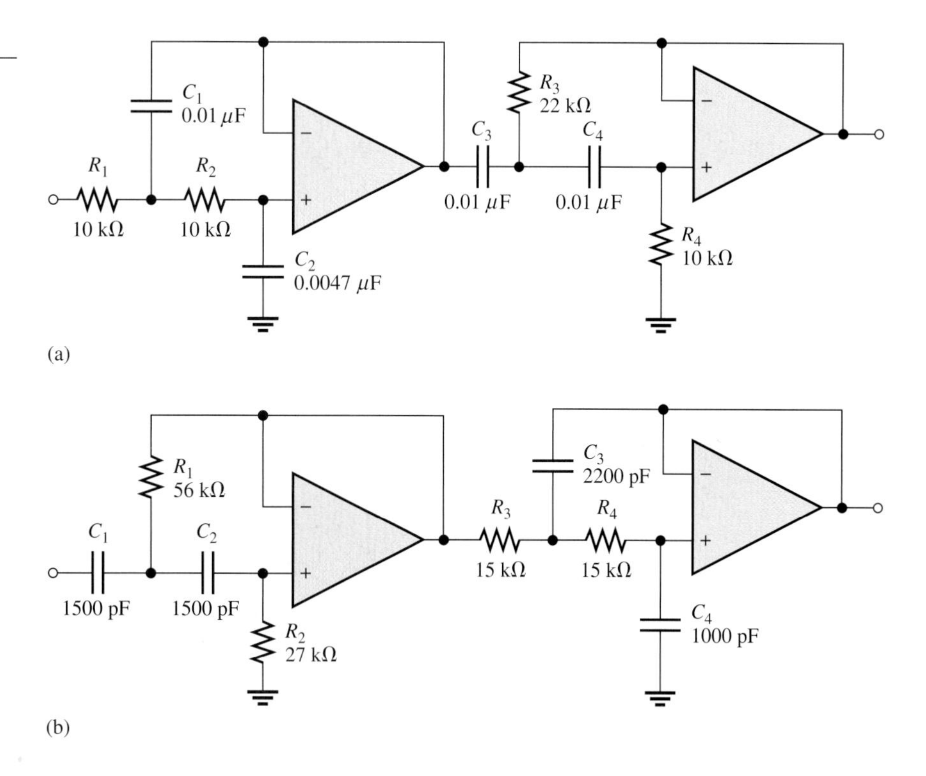

SECTION 21–6

Voltage Regulator Fundamentals

24. Determine the output voltage for the series regulator in Figure 21–72.

25. If R_3 in Figure 21–72 is doubled, what happens to the output voltage?

26. If the zener voltage is 2.7 V instead of 2 V in Figure 21–72, what is the output voltage?

27. A series voltage regulator with constant current limiting is shown in Figure 21–73. Determine the value of R_4 if the load current is to be limited to a maximum value of 250 mA. What power rating must R_4 have?

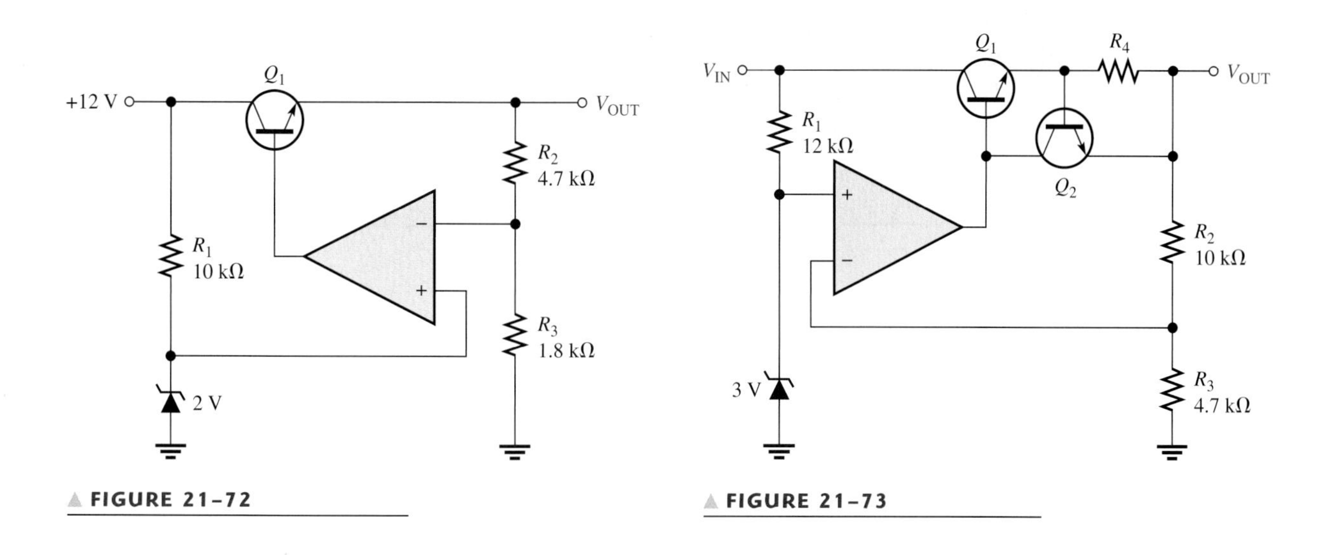

FIGURE 21–72

FIGURE 21–73

28. If R_4 (determined in Problem 27) is halved, what is the maximum load current?

29. In the shunt regulator of Figure 21–74, when the load current increases, does Q_1 conduct more or less? Why?

30. Assume that I_L remains constant and V_{IN} increases by 1 V in Figure 21–74. What is the change in the collector current of Q_1?

▶ FIGURE 21–74

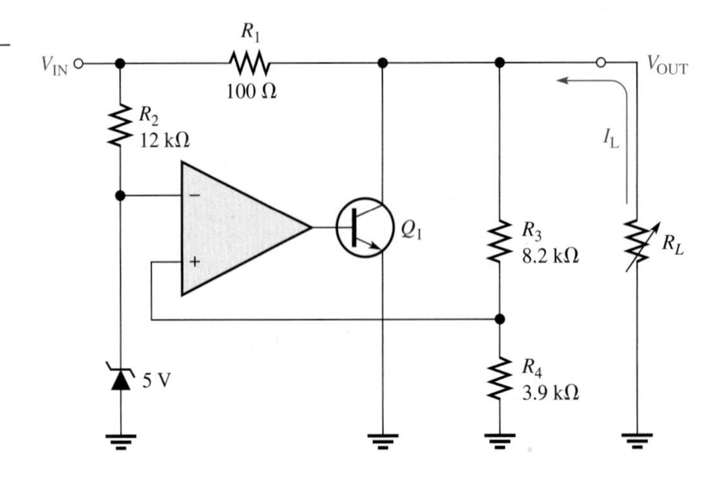

31. With a constant input voltage of 18 V, the load resistance in Figure 21–74 is varied from 1.0 kΩ to 1.2 kΩ. Neglecting any change in output voltage, how much does the shunt current through Q_1 change?

ELECTRONICS WORKBENCH/CIRCUITMAKER TROUBLESHOOTING PROBLEMS

CD-ROM file circuits are shown in Figure 21–75.

32. Open file P21-32 and determine if there is a fault. If so, identify it.

33. Open file P21-33 and determine if there is a fault. If so, identify it.

34. Open file P21-34 and determine if there is a fault. If so, identify it.

35. Open file P21-35 and determine if there is a fault. If so, identify it.

36. Open file P21-36 and determine if there is a fault. If so, identify it.

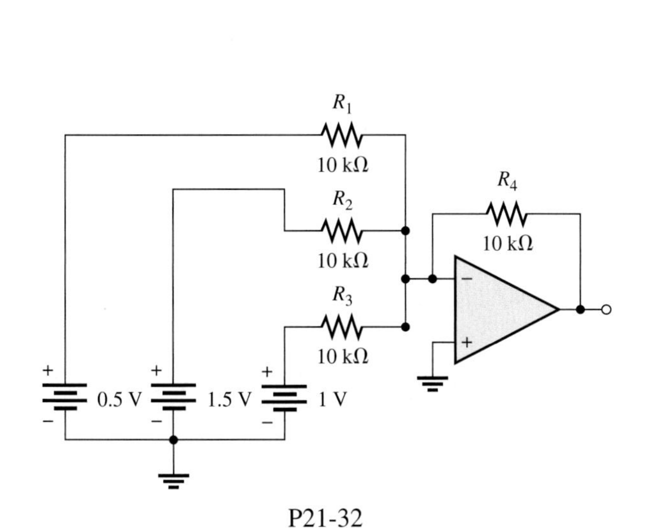

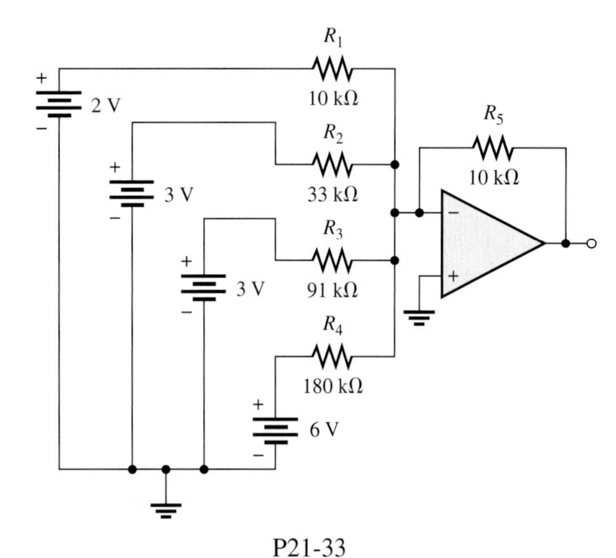

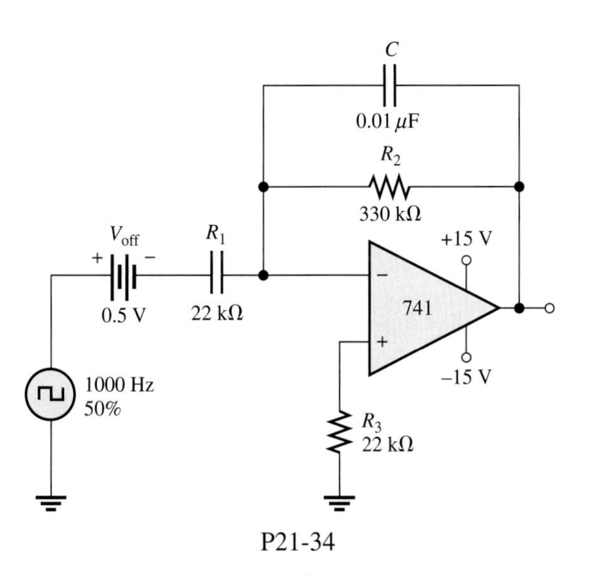

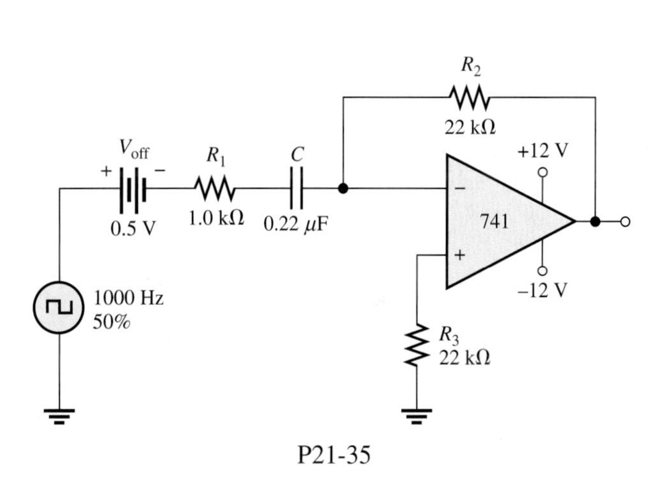

▲ **FIGURE 21–75**

ANSWERS

SECTION REVIEWS

SECTION 21–1 **Comparators**

1. $(10\ k\Omega/110\ k\Omega)15\ V = 1.36\ V$
2. See Figure 21–76.

► **FIGURE 21–76**

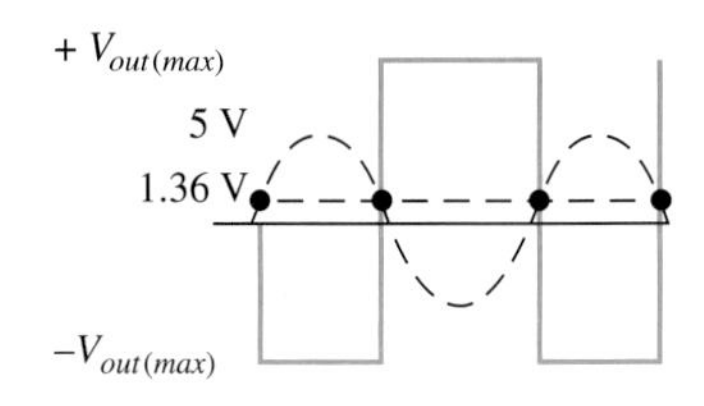

SECTION 21–2 **Summing Amplifiers**

1. The summary point is the terminal of the op-amp where the input resistors are commonly connected.
2. $R_f/R = 1/5 = 0.2$
3. 5 kΩ

SECTION 21–3 **Integrators and Differentiators**

1. The feedback element in an integrator is a capacitor.
2. The capacitor voltage is linear because the capacitor current is constant.
3. The feedback element in a differentiator is a resistor.
4. The output of a differentiator is proportional to the rate of change of the input.

SECTION 21–4 **Oscillators**

1. The negative feedback loop sets the closed-loop gain; the positive feedback loop sets the frequency.
2. A VCO exhibits a frequency that can be varied with a dc control voltage.
3. The basis of a relaxation oscillator is the charging and discharging of a capacitor.

SECTION 21–5 **Active Filters**

1. A pole is a single *RC* circuit.
2. A single-pole low-pass frequency response is flat from dc to the critical frequency.
3. The *R* and *C* positions are interchanged.
4. The critical frequency is halved.

SECTION 21–6 **Voltage Regulator Fundamentals**

1. In a shunt regulator, the control element is in parallel with the load rather than in series with the load.
2. A shunt regulator has inherent current limiting, but it is less efficient than a series regulator because part of the load current must be bypassed through the control element.

■ Application Assignment

1. The fuse rating should be 1 A.
2. These are optional capacitors that prevent oscillations.
3. The 7912 is a negative-voltage regulator.

RELATED PROBLEMS FOR EXAMPLES

21–1 1.96 V

21–2 −12.5 V

21–3 −5.73 V

21–4 Add a 100 kΩ input resistor and change R_f to 20 kΩ.

21–5 0.45; 0.12; 0.18; $V_{OUT} = -3.03$ V

21–6 Change C to 5000 pF.

21–7 Same waveform but with amplitude of 6.6 V

21–8 Use 6.1 V zeners.

21–9 6.06 V peak-to-peak

21–10 952 Hz

21–11 8.34 kHz

21–12 2.57 kHz

21–13 Increase f_{c2} by reducing the resistor and/or capacitor values.

21–14 7.33 V

21–15 17.5 W

SELF-TEST

1. (c) **2.** (a) **3.** (b) **4.** (d) **5.** (a) **6.** (b) **7.** (c)
8. (b) **9.** (a) **10.** (d) **11.** (c) **12.** (c)

A: Table of Standard Resistor Values

Resistance Tolerance (±%)

0.1% 0.25% 0.5%	1%	2% 5%	10%	0.1% 0.25% 0.5%	1%	2% 5%	10%	0.1% 0.25% 0.5%	1%	2% 5%	10%	0.1% 0.25% 0.5%	1%	2% 5%	10%	0.1% 0.25% 0.5%	1%	2% 5%	10%	0.1% 0.25% 0.5%	1%	2% 5%	10%
10.0	10.0	10	10	14.7	14.7	—	—	21.5	21.5	—	—	31.6	31.6	—	—	46.4	46.4	—	—	68.1	68.1	68	68
10.1	—	—	—	14.9	—	—	—	21.8	—	—	—	32.0	—	—	—	47.0	—	47	47	69.0	—	—	—
10.2	10.2	—	—	15.0	15.0	15	15	22.1	22.1	22	22	32.4	32.4	—	—	47.5	47.5	—	—	69.8	69.8	—	—
10.4	—	—	—	15.2	—	—	—	22.3	—	—	—	32.8	—	—	—	48.1	—	—	—	70.6	—	—	
10.5	10.5	—	—	15.4	15.4	—	—	22.6	22.6	—	—	33.2	33.2	33	33	48.7	48.7	—	—	71.5	71.5	—	—
10.6	—	—	—	15.6	—	—	—	22.9	—	—	—	33.6	—	—	—	49.3	—	—	—	72.3	—	—	—
10.7	10.7	—	—	15.8	15.8	—	—	23.2	23.2	—	—	34.0	34.0	—	—	49.9	49.9	—	—	73.2	73.2	—	—
10.9	—	—	—	16.0	—	16	—	23.4	—	—	—	34.4	—	—	—	50.5	—	—	—	74.1	—	—	—
11.0	11.0	11	—	16.2	16.2	—	—	23.7	23.7	—	—	34.8	34.8	—	—	51.1	51.1	51	—	75.0	75.0	75	—
11.1	—	—	—	16.4	—	—	—	24.0	—	24	—	35.2	—	—	—	51.7	—	—	—	75.9	—	—	—
11.3	11.3	—	—	16.5	16.5	—	—	24.3	24.3	—	—	35.7	35.7	—	—	52.3	52.3	—	—	76.8	76.8	—	—
11.4	—	—	—	16.7	—	—	—	24.6	—	—	—	36.1	—	36	—	53.0	—	—	—	77.7	—	—	—
11.5	11.5	—	—	16.9	16.9	—	—	24.9	24.9	—	—	36.5	36.5	—	—	53.6	53.6	—	—	78.7	78.7	—	—
11.7	—	—	—	17.2	—	—	—	25.2	—	—	—	37.0	—	—	—	54.2	—	—	—	79.6	—	—	—
11.8	11.8	—	—	17.4	17.4	—	—	25.5	25.5	—	—	37.4	37.4	—	—	54.9	54.9	—	—	80.6	80.6	—	—
12.0	—	12	12	17.6	—	—	—	25.8	—	—	—	37.9	—	—	—	56.2	—	—	—	81.6	—	—	—
12.1	12.1	—	—	17.8	17.8	—	—	26.1	26.1	—	—	38.3	38.3	—	—	56.6	56.6	56	56	82.5	82.5	82	82
12.3	—	—	—	18.0	—	18	18	26.4	—	—	—	38.8	—	—	—	56.9	—	—	—	83.5	—	—	—
12.4	12.4	—	—	18.2	18.2	—	—	26.7	26.7	—	—	39.2	39.2	39	39	57.6	57.6	—	—	84.5	84.5	—	—
12.6	—	—	—	18.4	—	—	—	27.1	—	27	27	39.7	—	—	—	58.3	—	—	—	85.6	—	—	—
12.7	12.7	—	—	18.7	18.7	—	—	27.4	27.4	—	—	40.2	40.2	—	—	59.0	59.0	—	—	86.6	86.6	—	—
12.9	—	—	—	18.9	—	—	—	27.7	—	—	—	40.7	—	—	—	59.7	—	—	—	87.6	—	—	—
13.0	13.0	13	—	19.1	19.1	—	—	28.0	28.0	—	—	41.2	41.2	—	—	60.4	60.4	—	—	88.7	88.7	—	—
13.2	—	—	—	19.3	—	—	—	28.4	—	—	—	41.7	—	—	—	61.2	—	—	—	89.8	—	—	—
13.3	13.3	—	—	19.6	19.6	—	—	28.7	28.7	—	—	42.2	42.2	—	—	61.9	61.9	62	—	90.9	90.9	91	—
13.5	—	—	—	19.8	—	—	—	29.1	—	—	—	42.7	—	—	—	62.6	—	—	—	92.0	—	—	—
13.7	13.7	—	—	20.0	20.0	20	—	29.4	29.4	—	—	43.2	43.2	43	—	63.4	63.4	—	—	93.1	93.1	—	—
13.8	—	—	—	20.3	—	—	—	29.8	—	—	—	43.7	—	—	—	64.2	—	—	—	94.2	—	—	—
14.0	14.0	—	—	20.5	20.5	—	—	30.1	30.1	30	—	44.2	44.2	—	—	64.9	64.9	—	—	95.3	95.3	—	—
14.2	—	—	—	20.8	—	—	—	30.5	—	—	—	44.8	—	—	—	65.7	—	—	—	96.5	—	—	—
14.3	14.3	—	—	21.0	21.0	—	—	30.9	30.9	—	—	45.3	45.3	—	—	66.5	66.5	—	—	97.6	97.6	—	—
14.5	—	—	—	21.3	—	—	—	31.2	—	—	—	45.9	—	—	—	67.3	—	—	—	98.8	—	—	—

NOTE: These values are generally available in multiples of 0.1, 1, 10, 100, 1 k, and 1 M.

B: Batteries

Batteries are an important source of dc voltage. They are available in two basic categories: the wet cell and the dry cell. A battery generally is made up of several individual cells.

A cell consists basically of two electrodes immersed in an electrolyte. A voltage is developed between the electrodes as a result of the chemical action between the electrodes and the electrolyte. The electrodes typically are two dissimilar metals, and the electrolyte is a chemical solution.

Simple Wet Cell

Figure B–1 shows a simple copper-zinc (Cu-Zn) chemical cell. One electrode is made of copper, the other of zinc. These electrodes are immersed in a solution of water and hydrochloric acid (HCl), which is the electrolyte.

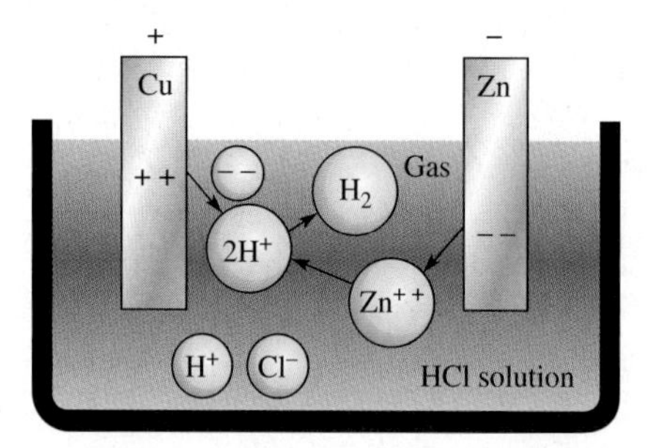

FIGURE B–1

Simple chemical cell.

Positive hydrogen ions (H^+) and negative chlorine ions (Cl^-) are formed when the HCl ionizes in the water. Since zinc is more active than hydrogen, zinc atoms leave the zinc electrode and form zinc ions (Zn^{++}) in the solution. When a zinc ion is formed, two excess electrons are left on the zinc electrode, and two hydrogen ions are displaced from the solution. These two hydrogen ions will migrate to the copper electrode, take two electrons from the copper, and form a molecule of hydrogen gas (H_2). As a result of this reaction, a negative charge develops on the zinc electrode, and a positive charge develops on the copper electrode, creating a potential difference or voltage between the two electrodes.

In this copper-zinc cell, the hydrogen gas given off at the copper electrode tends to form a layer of bubbles around the electrodes, insulating the copper from the electrolyte. This effect, called *polarization,* results in a reduction in the voltage produced by the cell. Polarization can be remedied by the addition of an agent to the electrolyte to remove hydrogen gas or by the use of an electrolyte that does not form hydrogen gas.

Lead-Acid Cell The positive electrode of a lead-acid cell is lead peroxide (PbO_2), and the negative electrode is spongy lead (Pb), as indicated in Figure B–2. The electrolyte is sulfuric acid (H_2SO_4) in water. Thus, the lead-acid cell is classified as a wet cell.

Two positive hydrogen ions ($2H^+$) and one negative sulfate ion (SO_4^{--}) are formed when the sulfuric acid ionizes in the water. Lead ions (Pb^{++}) from both electrodes displace the hydrogen ions in the electrolyte solution. When the lead ion from the spongy lead electrode enters the solution, it combines with a sulfate ion (SO_4^{--}) to form lead sulfate ($PbSO_4$), and it leaves two excess electrons on the electrode.

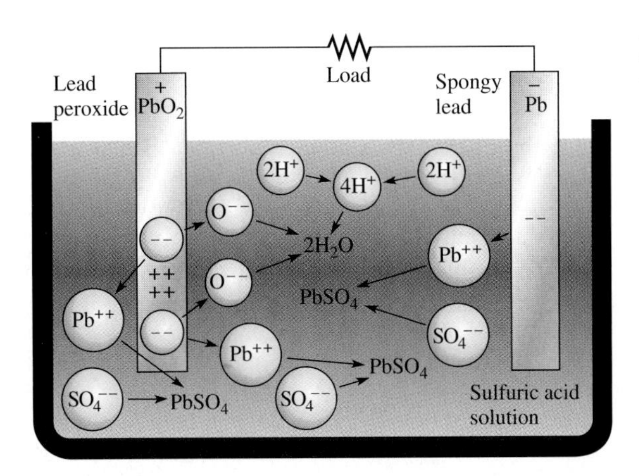

FIGURE B–2

Chemical reaction in a discharging lead-acid cell.

When a lead ion from the lead peroxide electrode enters the solution, it also leaves two excess electrons on the electrode and forms lead sulfate in the solution. However, because this electrode is lead peroxide, two free oxygen atoms are created when a lead atom leaves and enters the solution as a lead ion. These two oxygen atoms take four electrons from the lead peroxide electrode and become oxygen ions (O^{--}). This process creates a deficiency of two electrons on this electrode (there were initially two excess electrons).

The two oxygen ions ($2O^{--}$) combine in the solution with four hydrogen ions ($4H^{+}$) to produce two molecules of water ($2H_2O$). This process dilutes the electrolyte over a period of time. Also, there is a buildup of lead sulfate on the electrodes. These two factors result in a reduction in the voltage produced by the cell and necessitate periodic recharging.

As you have seen, for each departing lead ion, there is an excess of two electrons on the spongy lead electrode, and there is a deficiency of two electrons on the lead peroxide electrode. Therefore, the lead peroxide electrode is positive, and the spongy lead electrode is negative. This chemical reaction is pictured in Figure B–2.

As mentioned, the dilution of the electrolyte by the formation of water and lead sulfate requires that the lead-acid cell be recharged to reverse the chemical process. A chemical cell that can be recharged is called a *secondary cell.* One that cannot be recharged is called a *primary cell.*

The cell is recharged by the connection of an external voltage source to the electrodes, as shown in Figure B–3. The formula for the chemical reaction in a lead-acid cell is

$$Pb + PbO_2 + 2H_2SO_4 \rightarrow 2PbSO_4 + 2H_2O$$

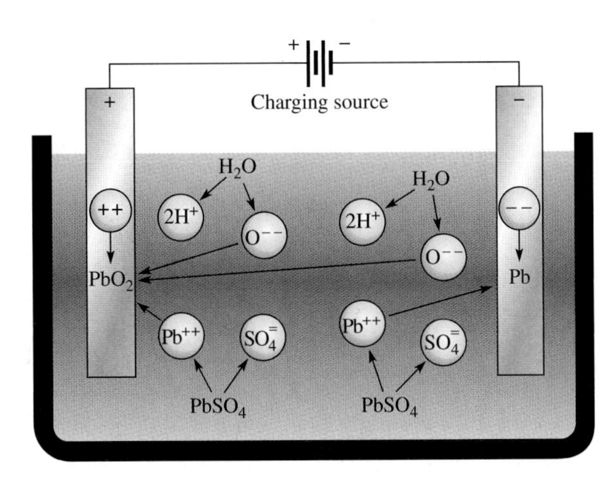

FIGURE B–3

Recharging a lead-acid cell.

Dry Cell

In a dry cell, some of the disadvantages of a liquid electrolyte are overcome. Actually, the electrolyte in a typical dry cell is not dry but rather is in the form of a moist paste. This electrolyte is a combination of granulated carbon, powdered manganese dioxide, and ammonium chloride solution.

A typical carbon-zinc dry cell is illustrated in Figure B–4. The zinc container or can is dissolved by the electrolyte. As a result of this reaction, an excess of electrons accumulates on the container, making it the negative electrode.

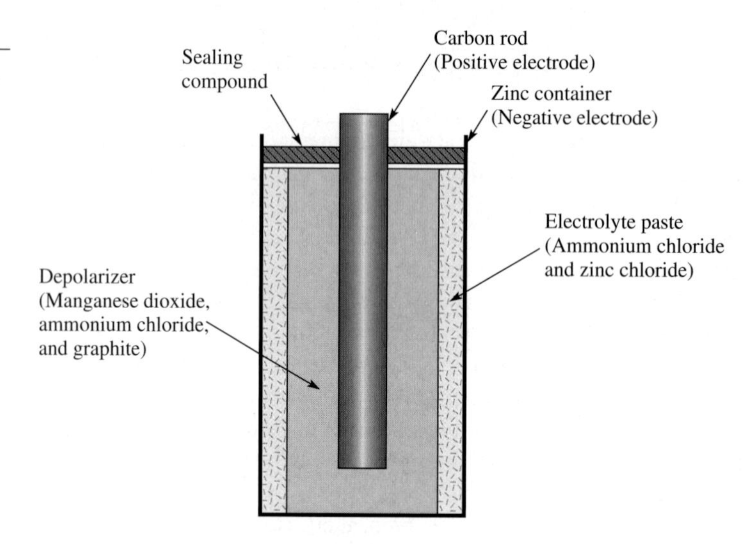

FIGURE B–4

Simplified construction of a dry cell.

The hydrogen ions in the electrolyte take electrons from the carbon rod, making it the positive electrode. Hydrogen gas is formed near the carbon electrode, but this gas is eliminated by reaction with manganese dioxide (called a *depolarizing agent*). This depolarization prevents bursting of the container due to gas formation. Because the chemical reaction is not reversible, the carbon-zinc cell is a primary cell.

Types of Chemical Cells

Although only two common types of battery cells have been discussed, there are several types, listed in Table B–1.

TABLE B–1

Types of battery cells.

TYPE	+ ELECTRODE	– ELECTRODE	ELECTROLYTE	VOLTS	COMMENTS
Carbon-zinc	Carbon	Zinc	Ammonium and zinc chloride	1.5	Dry, primary
Lead-acid	Lead peroxide	Spongy lead	Sulfuric acid	2.0	Wet, secondary
Manganese-alkaline	Manganese dioxide	Zinc	Potassium hydroxide	1.5	Dry, primary or secondary
Mercury	Zinc	Mercuric oxide	Potassium hydroxide	1.3	Dry, primary
Nickel-cadmium	Nickel	Cadmium hydroxide	Potassium hydroxide	1.25	Dry, secondary
Nickel-iron (Edison cell)	Nickel oxide	Iron	Potassium hydroxide	1.36	Wet, secondary

C: Capacitor Color Coding and Marking

Capacitor Color

Some capacitors have color-coded designations. The color code used for capacitors is basically the same as that used for resistors. Some variations occur in tolerance designation. The basic color codes are shown in Table C–1, and some typical color-coded capacitors are illustrated in Figure C–1.

TABLE C–1

Typical composite color codes for capacitors (picofarads).

COLOR	DIGIT	MULTIPLIER	TOLERANCE
Black	0	1	20%
Brown	1	10	1%
Red	2	100	2%
Orange	3	1000	3%
Yellow	4	10000	
Green	5	100000	5% (EIA)
Blue	6	1000000	
Violet	7		
Gray	8		
White	9		
Gold		0.1	5% (JAN)
Silver		0.01	10%
No color			20%

NOTE: EIA stands for Electronic Industries Association, and JAN stands for Joint Army-Navy, a military standard.

FIGURE C–1

Typical color-coded capacitors.

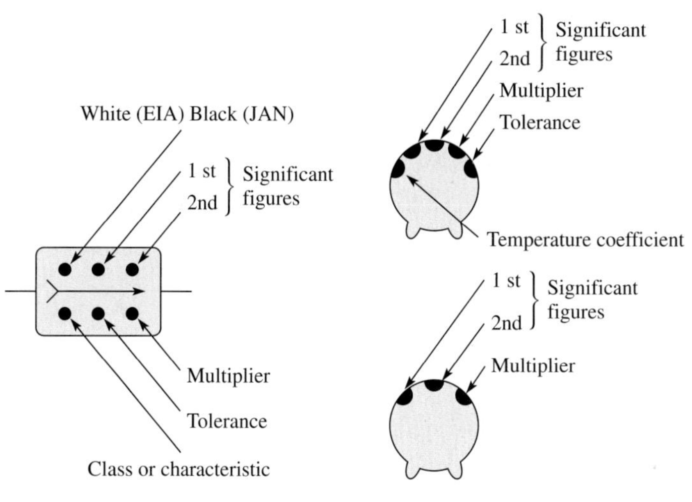

(a) Molded mica

(b) Disc ceramic

1 st
2nd
Significant figures
Multiplier
Tolerance

Indicates outer foil. May be on either end. May also be indicated by other methods such as typographical marking or black stripe.

1 st
2nd
Significant voltage figures

Add two zeros to significant voltage figures. One band indicates voltage ratings under 1000 volts.

(c) Molded tubular

Marking Systems

A capacitor, as shown in Figure C–2, has certain identifying features.

- Body of one solid color (off-white, beige, gray, tan or brown).
- End electrodes completely enclose ends of part.
- Many different sizes:
  1. Type 1206: 0.125 inch long by 0.063 inch wide (3.2 mm × 1.6 mm) with variable thickness and color.
  2. Type 0805: 0.080 inch long by 0.050 inch wide (2.0 mm × 1.25 mm) with variable thickness and color.
  3. Variably sized with a single color (usually translucent tan or brown). Sizes range from 0.059 inch (1.5 mm) to 0.220 inch (5.6 mm) in length and in width from 0.032 inch (0.8 mm) to 0.197 inch (5.0 mm).
- Three different marking systems:
  1. Two place (letter and number only).
  2. Two place (letter and number or two numbers).
  3. One place (letter of varying color).

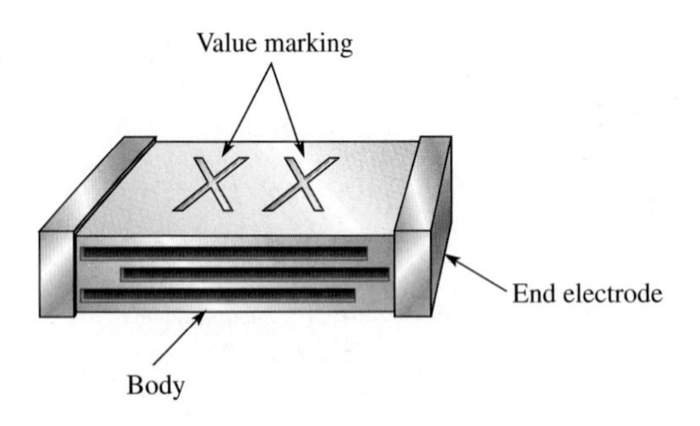

FIGURE C–2

Capacitor marking.

Standard Two-Place Code

Refer to Table C–2.

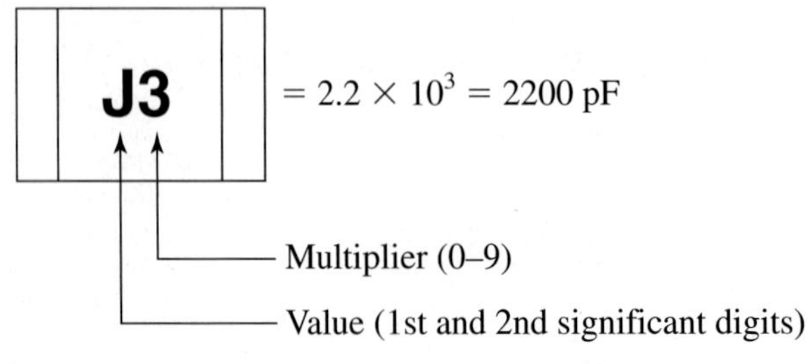

Examples: S2 = 4.7 × 100 = 470 pF
b0 = 3.5 × 1.0 = 3.5 pF

TABLE C–2

VALUE*						MULTIPLIER
A	1.0	L	2.7	T	5.1	0 = ×1.0
B	1.1	M	3.0	U	5.6	1 = ×10
C	1.2	N	3.3	m	6.0	2 = ×100
D	1.3	b	3.5	V	6.2	3 = ×1000
E	1.5	P	3.6	W	6.8	4 = ×10000
F	1.6	Q	3.9	n	7.0	5 = ×100000
G	1.8	d	4.0	X	7.5	etc.
H	2.0	R	4.3	t	8.0	
J	2.2	e	4.5	Y	8.2	
K	2.4	S	4.7	y	9.0	
a	2.5	f	5.0	Z	9.1	

*Note uppercase and lowercase letters.

Alternate Two-Place Code

Refer to Table C–3.

- Values below 100 pF—Value read directly

05 = 5 pF   82 = 82 pF

- Values 100 pF and above—Letter/Number code

A1 = 10 × 10 = 100 pF

N3 = 33 × 1000
= 33000 pF = .033 μF

Multiplier (1–9)

Value (1st and 2nd significant digits)

TABLE C–3

VALUE*						MULTIPLIER
A	10	J	22	S	47	1 = ×10
B	11	K	24	T	51	2 = ×100
C	12	L	27	U	56	3 = ×1000
D	13	M	30	V	62	4 = ×10000
E	15	N	33	W	68	5 = ×100000
F	16	P	36	X	75	
G	18	Q	39	Y	82	etc.
H	20	R	43	Z	91	

*Note uppercase letters only.

Standard Single-Place Code

Refer to Table C–4.

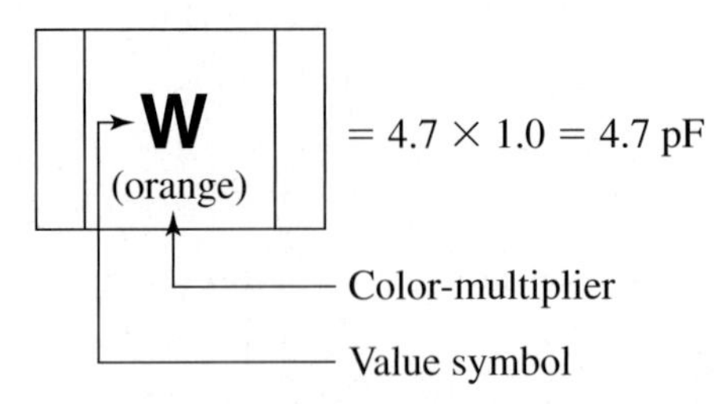

Examples: R (Green) = 3.3 × 100 = 330 pF
7 (Blue) = 8.2 × 1000 = 8200 pF

▶ TABLE C–4

VALUE						MULTIPLIER (COLOR)
A	1.0	K	2.2	W	4.7	Orange = ×1.0
B	1.1	L	2.4	X	5.1	Black = ×10
C	1.2	N	2.7	Y	5.6	Green = ×100
D	1.3	O	3.0	Z	6.2	Blue = ×1000
E	1.5	R	3.3	3	6.8	Violet = ×10000
H	1.6	S	3.6	4	7.5	Red = ×100000
I	1.8	T	3.9	7	8.2	
J	2.0	V	4.3	9	9.1	

D: The Current Source, Norton's Theorem, and Millman's Theorem

THE CURRENT SOURCE

The current source is another type of energy source that ideally provides a constant current to a load even when the resistance of the load varies. The concept of the current source is important in certain types of transistor circuits.

Figure D–1(a) shows a symbol for the ideal current source. The arrow indicates the direction of current, and I_S is the value of the source current. An ideal current source produces a constant value of current through a load, regardless of the value of the load. This concept is illustrated in Figure D–1(b), where a load resistor is connected to the current source between terminals *A* and *B*. The ideal current source has an infinitely large internal parallel resistance.

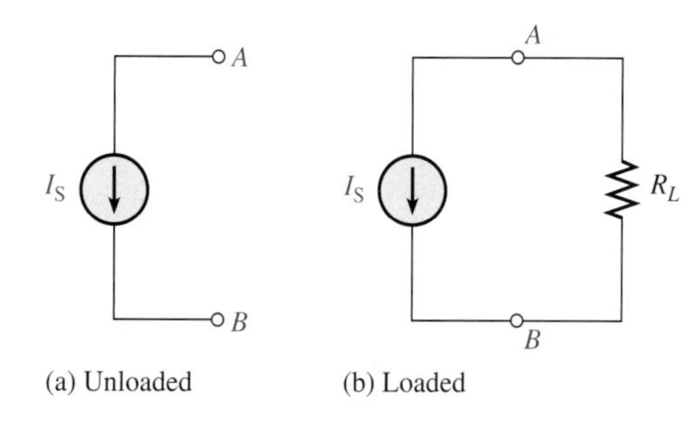

FIGURE D–1

Ideal current source.

Transistors act basically as current sources, and for this reason, knowledge of the current source concept is important. You will find that the equivalent model of a transistor does contain a current source.

Although the ideal current source can be used in most analysis work, no actual device is ideal. A practical current source representation is shown in Figure D–2. Here the internal resistance appears in parallel with the ideal current source.

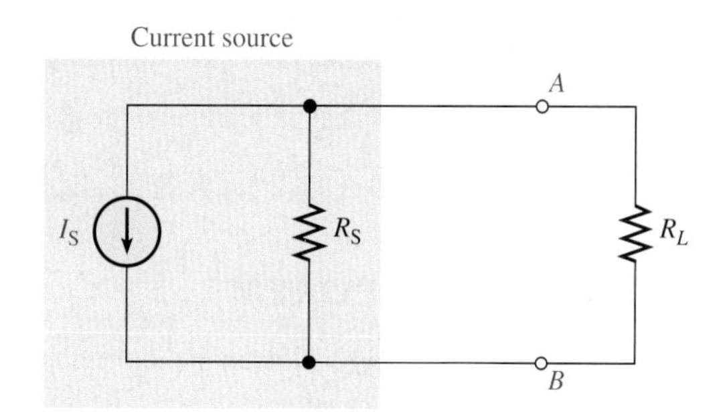

FIGURE D–2

Practical current source with load.

If the internal source resistance, R_S, is much larger than a load resistor, the practical source approaches ideal. The reason is illustrated in the practical current source shown in Figure D–2. Part of the current, I_S, is through R_S, and part is through R_L. Resistors R_S and R_L act as a current divider. If R_S is much larger than R_L, most of the current is through R_L and very little through R_S. As long as R_L remains much smaller than R_S, the current through it will stay almost constant, no matter how much R_L changes.

If there is a constant-current source, you can normally assume that R_S is so much larger than the load that R_S can be neglected. This simplifies the source to ideal, making the analysis easier.

Example D–1 illustrates the effect of changes in R_L on the load current when R_L is much smaller than R_S. Generally, R_L should be at least ten times smaller than R_S ($10R_L \leq R_S$).

EXAMPLE D–1

Calculate the load current (I_L) in Figure D–3 for the following values of R_L: 100 Ω, 560 Ω, and 1.0 kΩ.

FIGURE D–3

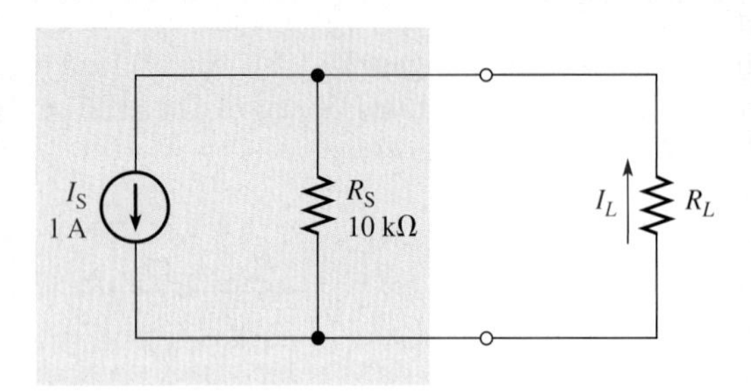

Solution For $R_L = 100\ \Omega$, the load current is

$$I_L = \left(\frac{R_S}{R_S + R_L}\right)I_S = \left(\frac{10\text{ k}\Omega}{10.1\text{ k}\Omega}\right)1\text{ A} = \mathbf{990\text{ mA}}$$

For $R_L = 560\ \Omega$,

$$I_L = \left(\frac{10\text{ k}\Omega}{10.56\text{ k}\Omega}\right)1\text{ A} = \mathbf{947\text{ mA}}$$

For $R_L = 1.0\text{ k}\Omega$,

$$I_L = \left(\frac{10\text{ k}\Omega}{11\text{ k}\Omega}\right)1\text{ A} = \mathbf{909\text{ mA}}$$

Notice that the load current, I_L, is within 10% of the source current for each value of R_L because R_L is at least ten times smaller than R_S.

NORTON'S THEOREM

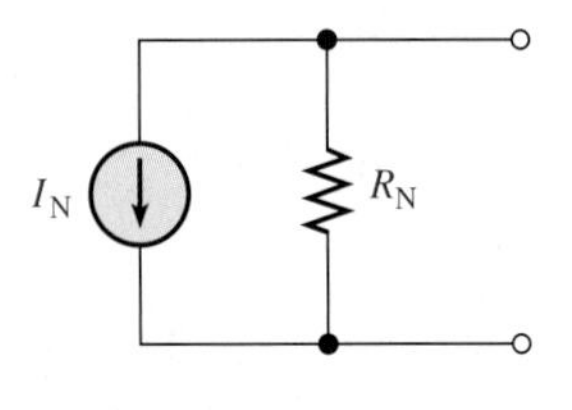

FIGURE D–4

The form of Norton's equivalent circuit.

Like Thevenin's theorem, Norton's theorem provides a method of reducing a more complex circuit to a simpler form. The basic difference is that Norton's theorem gives an equivalent current source in parallel with an equivalent resistance. The form of Norton's equivalent circuit is shown in Figure D–4. Regardless of how complex the original circuit is, it can always be reduced to this equivalent form. The equivalent current source is designated I_N, and the equivalent resistance, R_N.

To apply Norton's theorem, you must know how to find the two quantities I_N and R_N. Once you know them for a given circuit, simply connect them in parallel to get the complete Norton circuit.

Norton's Equivalent Current (I_N)

As stated, I_N is one part of the complete Norton equivalent circuit; R_N is the other part. I_N is defined to be the short circuit current between two points in a circuit. Any component connected between these two points effectively "sees" a current source of value I_N in parallel with R_N.

To illustrate, suppose that a resistive circuit of some kind has a resistor connected between two points in the circuit, as shown in Figure D–5(a). We wish to find the Norton circuit that is equivalent to the one shown as "seen" by R_L. To find I_N, calculate the current between points A and B with these two points shorted, as shown in Figure D–5(b). Example D–2 demonstrates how to find I_N.

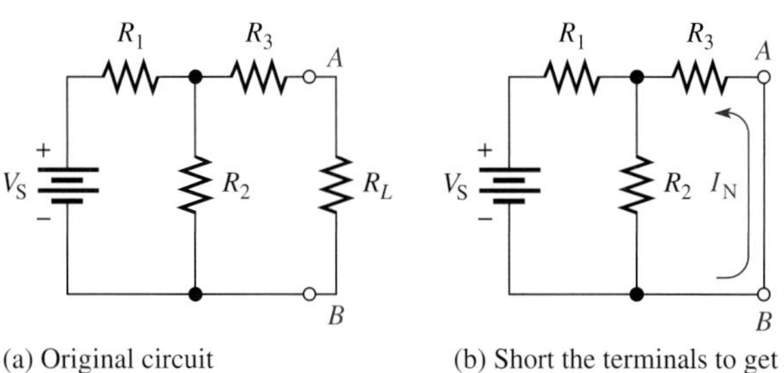

FIGURE D–5

Determining the Norton equivalent current, I_N.

EXAMPLE D–2

Determine I_N for the circuit within the shaded area in Figure D–6(a).

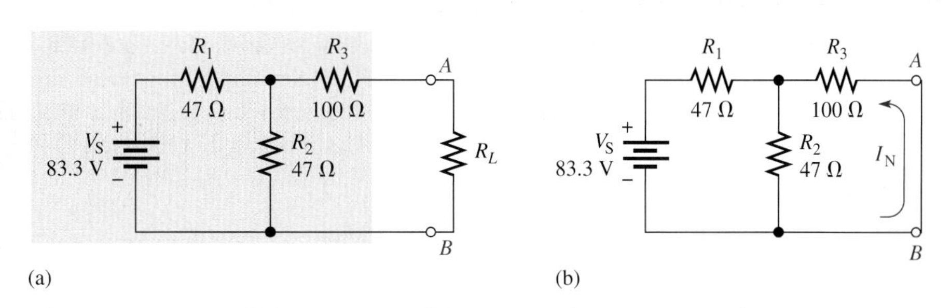

FIGURE D–6

Solution Short terminals A and B as shown in Figure D–6(b). I_N is the current through the short and is calculated as follows: First, the total resistance seen by the voltage source is

$$R_T = R_1 + \frac{R_2 R_3}{R_2 + R_3} = 47\ \Omega + \frac{(47\ \Omega)(100\ \Omega)}{147\ \Omega} = 79\ \Omega$$

The total current from the source is

$$I_T = \frac{V_S}{R_T} = \frac{83.3\ \text{V}}{79\ \Omega} = 1.05\ \text{A}$$

Now apply the current-divider formula to find I_N (the current through the short):

$$I_N = \left(\frac{R_2}{R_2 + R_3}\right) I_T = \left(\frac{47\ \Omega}{147\ \Omega}\right) 1.05\ \text{A} = \mathbf{336\ mA}$$

This is the value for the equivalent Norton current source.

Norton's Equivalent Resistance (R_N)

We define R_N in the same way as R_{TH}: it is the total resistance appearing between two terminals in a given circuit with all sources replaced by their internal resistances. Example D–3 demonstrates how to find R_N.

EXAMPLE D–3

Find R_N for the circuit within the shaded area of Figure D–6 (see Example D–2).

Solution First reduce V_S to zero by shorting it, as shown in Figure D–7.

Looking in at terminals *A* and *B*, you can see that the parallel combination of R_1 and R_2 is in series with R_3. Thus,

$$R_N = R_3 + \frac{R_1}{2} = 100\ \Omega + \frac{47\ \Omega}{2} = \mathbf{124\ \Omega}$$

FIGURE D–7

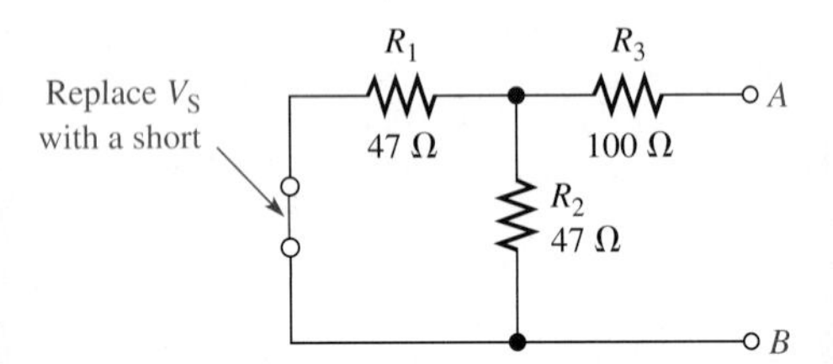

The two examples have shown how to find the two equivalent components of a Norton equivalent circuit, I_N and R_N. Keep in mind that these values can be found for any linear circuit. Once these are known, they must be connected in parallel to form the Norton equivalent circuit, as illustrated in Example D–4.

EXAMPLE D–4

Draw the complete Norton circuit for the original circuit in Figure D–6 (Example D–2).

Solution In Examples D–2 and D–3 you found that $I_N = 336$ mA and $R_N = 124\ \Omega$. The Norton equivalent circuit is shown in Figure D–8.

FIGURE D–8

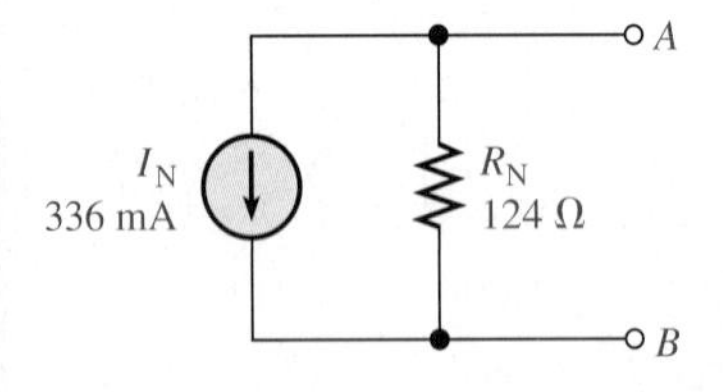

Summary of Norton's Theorem

Any load resistor connected between the terminals of a Norton equivalent circuit will have the same current through it and the same voltage across it as if it were connected to the terminals of the original circuit. A summary of steps for theoretically applying Norton's theorem is as follows:

1. Short the two terminals between which you want to find the Norton equivalent circuit.

2. Determine the current (I_N) through the shorted terminals.
3. Determine the resistance (R_N) between the two terminals (opened) with all voltage sources shorted and all current sources opened ($R_N = R_{TH}$).
4. Connect I_N and R_N in parallel to produce the complete Norton equivalent for the original circuit.

Norton's equivalent circuit can also be derived from Thevenin's equivalent circuit by use of the source conversion method.

MILLMAN'S THEOREM

Millman's theorem permits any number of parallel voltage sources to be reduced to a single equivalent voltage source. It simplifies finding the voltage across or current through a load. Millman's theorem gives the same results as Thevenin's theorem for the special case of parallel voltage sources. A conversion by Millman's theorem is illustrated in Figure D–9.

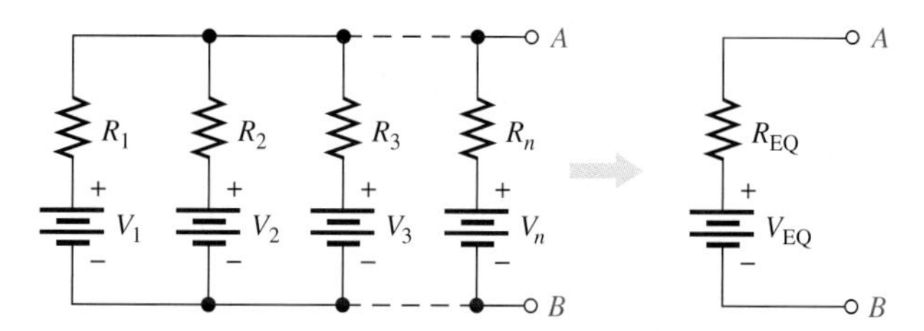

FIGURE D–9

Reduction of parallel voltage sources to a single equivalent voltage source.

Millman's Equivalent Voltage (V_{EQ}) and Equivalent Resistance (R_{EQ})

Millman's theorem gives us a formula for calculating the equivalent voltage, V_{EQ}. To find V_{EQ}, convert each of the parallel voltage sources into the current sources, as shown in Figure D–10.

FIGURE D–10

Parallel voltage sources converted to current sources.

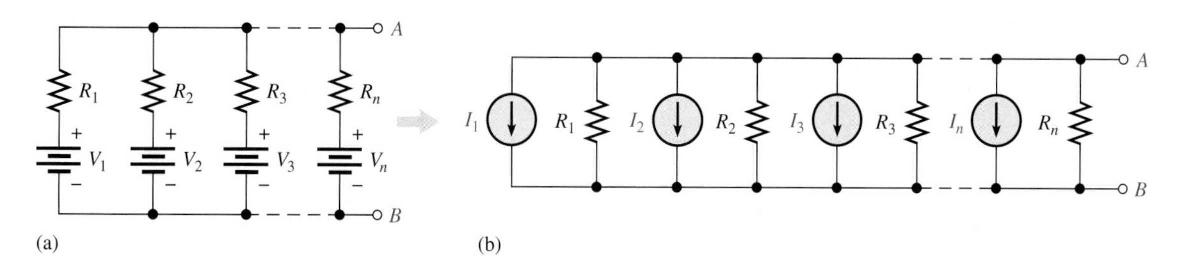

In Figure D–10(b), the total current from the parallel current sources is

$$I_T = I_1 + I_2 + I_3 + \cdots + I_n$$

The total conductance between terminals A and B is

$$G_T = G_1 + G_2 + G_3 + \cdots + G_n$$

where $G_T = 1/R_T$, $G_1 = 1/R_1$, and so on. Remember, the current sources are effectively open. Therefore, by Millman's theorem, the equivalent resistance, R_{EQ}, is the total resistance R_T.

$$R_{EQ} = \frac{1}{G_T} = \frac{1}{(1/R_1) + (1/R_2) + (1/R_3) + \cdots + (1/R_n)}$$

Equation D–1

By Millman's theorem, the equivalent voltage is $I_T R_{EQ}$, where I_T is expressed as follows:

$$I_T = I_1 + I_2 + I_3 + \cdots + I_n = \frac{V_1}{R_1} + \frac{V_2}{R_2} + \frac{V_3}{R_3} + \cdots + \frac{V_n}{R_n}$$

The following is the formula for the equivalent voltage:

Equation D–2

$$V_{EQ} = \frac{(V_1/R_1) + (V_2/R_2) + (V_3/R_3) + \cdots + (V_n/R_n)}{(1/R_1) + (1/R_2) + (1/R_3) + \cdots + (1/R_n)}$$

Equations D–1 and D–2 are the two Millman formulas. The equivalent voltage source has a polarity such that the total current through a load will be in the same direction as in the original circuit.

EXAMPLE D–5

Use Millman's theorem to find the voltage across R_L and the current through R_L in Figure D–11.

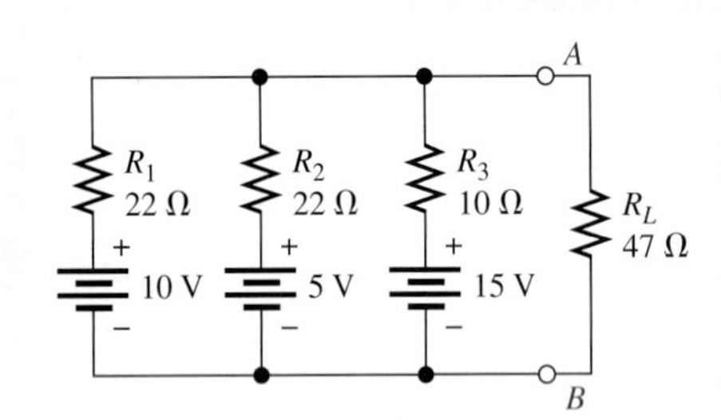

FIGURE D–11

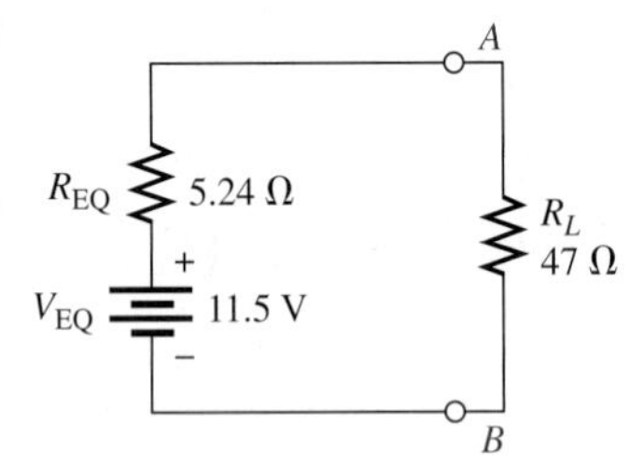

FIGURE D–12

Solution Apply Millman's theorem as follows:

$$R_{EQ} = \frac{1}{(1/R_1) + (1/R_2) + (1/R_3)}$$

$$= \frac{1}{(1/22\ \Omega) + (1/22\ \Omega) + (1/10\ \Omega)} = \frac{1}{0.19} = 5.24\ \Omega$$

$$V_{EQ} = \frac{(V_1/R_2) + (V_2/R_2) + (V_3/R_3)}{(1/R_1) + (1/R_2) + (1/R_3)}$$

$$= \frac{(10\text{ V}/22\ \Omega) + (5\text{ V}/22\ \Omega) + (15\text{ V}/10\ \Omega)}{(1/22\ \Omega) + (1/22\ \Omega) + (1/10\ \Omega)} = \frac{2.18\text{ A}}{0.19\text{ S}} = 11.5\text{ V}$$

The single equivalent voltage source is shown in Figure D–12.

Now calculate I_L and V_L for the load resistor:

$$I_L = \frac{V_{EQ}}{R_{EQ} + R_L} = \frac{11.5\text{ V}}{52.2\ \Omega} = \mathbf{220\ mA}$$

$$V_L = I_L R_L = (220\text{ mA})(47\ \Omega) = \mathbf{10.3\ V}$$

E: Devices Data Sheets

1N4001 thru 1N4007

Designers▲Data Sheet

"SURMETIC"▲ RECTIFIERS

. . . subminiature size, axial lead mounted rectifiers for general-purpose low-power applications.

Designers Data for "Worst Case" Conditions

The Designers▲ Data Sheets permit the design of most circuits entirely from the information presented. Limit curves — representing boundaries on device characteristics — are given to facilitate "worst case" design.

LEAD MOUNTED
SILICON RECTIFIERS

50-1000 VOLTS
DIFFUSED JUNCTION

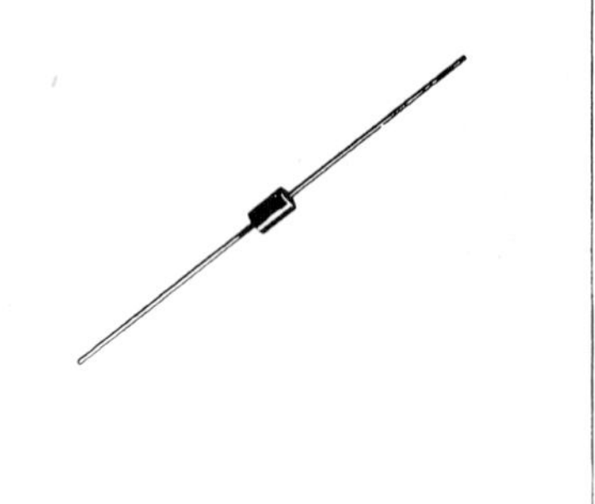

*MAXIMUM RATINGS

Rating	Symbol	1N4001	1N4002	1N4003	1N4004	1N4005	1N4006	1N4007	Unit
Peak Repetitive Reverse Voltage Working Peak Reverse Voltage DC Blocking Voltage	V_{RRM} V_{RWM} V_R	50	100	200	400	600	800	1000	Volts
Non-Repetitive Peak Reverse Voltage (halfwave, single phase, 60 Hz)	V_{RSM}	60	120	240	480	720	1000	1200	Volts
RMS Reverse Voltage	$V_{R(RMS)}$	35	70	140	280	420	560	700	Volts
Average Rectified Forward Current (single phase, resistive load, 60 Hz, see Figure 8, $T_A = 75^oC$)	I_O	1.0							Amp
Non-Repetitive Peak Surge Current (surge applied at rated load conditions, see Figure 2)	I_{FSM}	30 (for 1 cycle)							Amp
Operating and Storage Junction Temperature Range	T_J, T_{stg}	-65 to +175							^oC

*ELECTRICAL CHARACTERISTICS

Characteristic and Conditions	Symbol	Typ	Max	Unit
Maximum Instantaneous Forward Voltage Drop ($i_F = 1.0$ Amp, $T_J = 25^oC$) Figure 1	v_F	0.93	1.1	Volts
Maximum Full-Cycle Average Forward Voltage Drop ($I_O = 1.0$ Amp, $T_L = 75^oC$, 1 inch leads)	$V_{F(AV)}$		0.8	Volts
Maximum Reverse Current (rated dc voltage) $T_J = 25^oC$	I_R	0.05	10	µA
$T_J = 100^oC$		1.0	50	
Maximum Full-Cycle Average Reverse Current ($I_O = 1.0$ Amp, $T_L = 75^oC$, 1 inch leads	$I_{R(AV)}$	—	30	µA

*Indicates JEDEC Registered Data

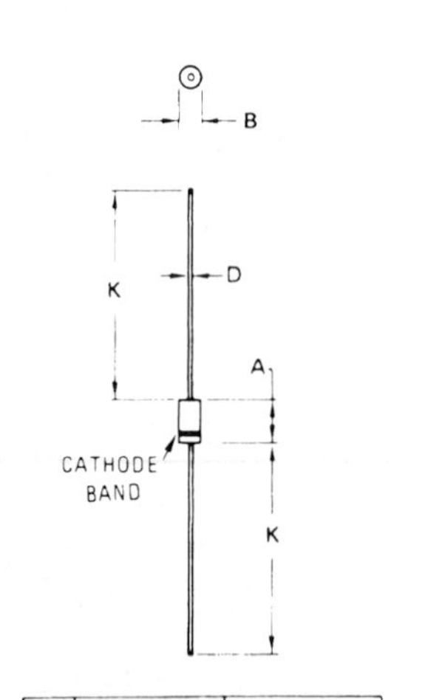

DIM	MILLIMETERS MIN	MILLIMETERS MAX	INCHES MIN	INCHES MAX
A	5.97	6.60	0.235	0.260
B	2.79	3.05	0.110	0.120
D	0.76	0.86	0.030	0.034
K	27.94	–	1.100	–

CASE 59-04
Does Not Conform to DO-41 Outline.

MECHANICAL CHARACTERISTICS

CASE: Transfer Molded Plastic
MAXIMUM LEAD TEMPERATURE FOR SOLDERING PURPOSES: 350°C, 3/8" from case for 10 seconds at 5 lbs. tension
FINISH: All external surfaces are corrosion-resistant, leads are readily solderable
POLARITY: Cathode indicated by color band
WEIGHT: 0.40 Grams (approximately)

▲Trademark of Motorola Inc.

DS 6015 R3

1N4728, A thru 1N4764, A

Designers▲ Data Sheet

1.0 WATT

ZENER REGULATOR DIODES

3.3–100 VOLTS

ONE WATT HERMETICALLY SEALED GLASS SILICON ZENER DIODES

- Complete Voltage Range – 2.4 to 100 Volts
- DO-41 Package – Smaller than Conventional DO-7 Package
- Double Slug Type Construction
- Metallurgically Bonded Construction
- Nitride Passivated Die

Designer's Data for "Worst Case" Conditions

The Designers▲ Data sheets permit the design of most circuits entirely from the information presented. Limit curves – representing boundaries on device characteristics – are given to facilitate "worst case" design.

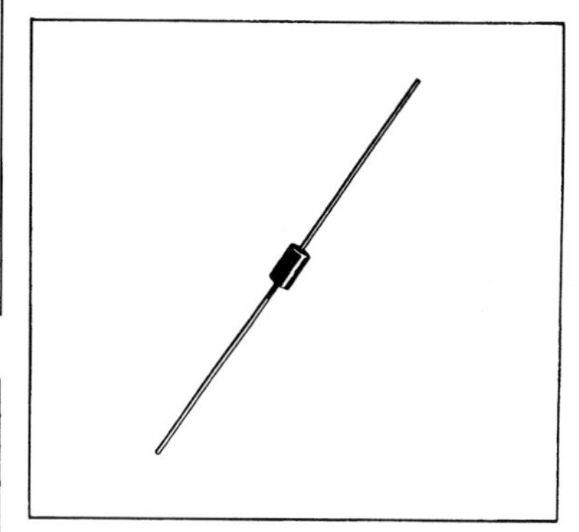

*MAXIMUM RATINGS

Rating	Symbol	Value	Unit
DC Power Dissipation @ $T_A = 50^oC$ Derate above 50^oC	P_D	1.0 6.67	Watt $mW/^oC$
Operating and Storage Junction Temperature Range	T_J, T_{stg}	-65 to +200	^oC

MECHANICAL CHARACTERISTICS

CASE: Double slug type, hermetically sealed glass

MAXIMUM LEAD TEMPERATURE FOR SOLDERING PURPOSES: 230°C, 1/16" from case for 10 seconds

FINISH: All external surfaces are corrosion resistant with readily solderable leads.

POLARITY: Cathode indicated by color band. When operated in zener mode, cathode will be positive with respect to anode.

MOUNTING POSITION: Any

FIGURE 1 – POWER TEMPERATURE DERATING CURVE

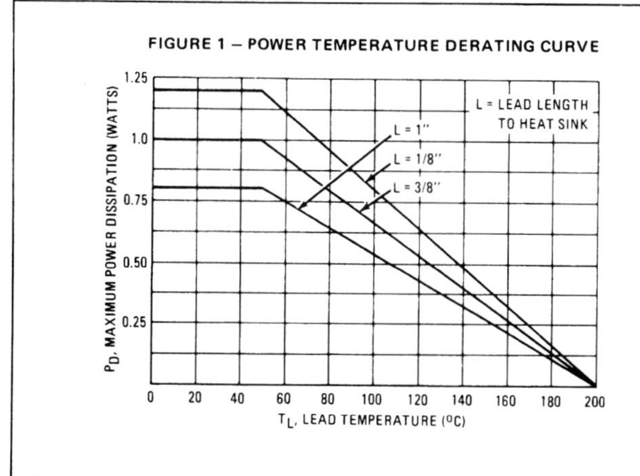

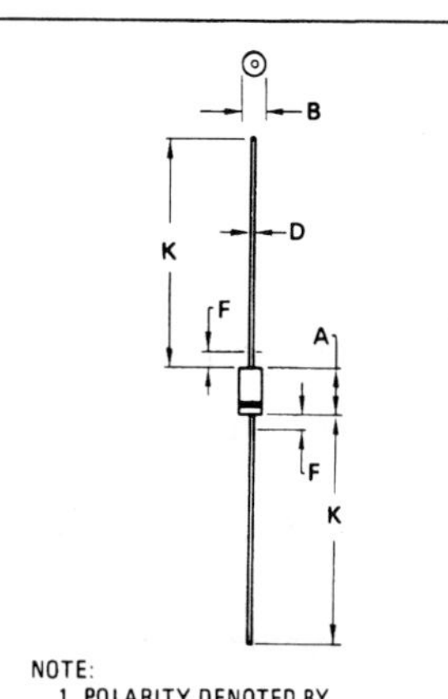

NOTE:
1. POLARITY DENOTED BY CATHODE BAND
2. LEAD DIAMETER NOT CONTROLLED WITHIN "F" DIMENSION.

DIM	MILLIMETERS		INCHES	
	MIN	MAX	MIN	MAX
A	4.07	5.20	0.160	0.205
B	2.04	2.71	0.080	0.107
D	0.71	0.86	0.028	0.034
F	–	1.27	–	0.050
K	27.94	–	1.100	–

All JEDEC dimensions and notes apply.

CASE 59-03 (DO-41)

*Indicates JEDEC Registered Data
▲Trademark of Motorola Inc.

DS 7039 R1

***ELECTRICAL CHARACTERISTICS** (T_A = 25°C unless otherwise noted) V_F = 1.2 V max, I_F = 200 mA for all types.

JEDEC Type No. (Note 1)	Nominal Zener Voltage V_Z @ I_{ZT} Volts (Notes 2 and 3)	Test Current I_{ZT} mA	Maximum Zener Impedance (Note 4)			Leakage Current		Surge Current @ T_A = 25°C i_r – mA (Note 5)
			Z_{ZT} @ I_{ZT} Ohms	Z_{ZK} @ I_{ZK} Ohms	I_{ZK} mA	I_R µA Max	V_R Volts	
1N4728	3.3	76	10	400	1.0	100	1.0	1380
1N4729	3.6	69	10	400	1.0	100	1.0	1260
1N4730	3.9	64	9.0	400	1.0	50	1.0	1190
1N4731	4.3	58	9.0	400	1.0	10	1.0	1070
1N4732	4.7	53	8.0	500	1.0	10	1.0	970
1N4733	5.1	49	7.0	550	1.0	10	1.0	890
1N4734	5.6	45	5.0	600	1.0	10	2.0	810
1N4735	6.2	41	2.0	700	1.0	10	3.0	730
1N4736	6.8	37	3.5	700	1.0	10	4.0	660
1N4737	7.5	34	4.0	700	0.5	10	5.0	605
1N4738	8.2	31	4.5	700	0.5	10	6.0	550
1N4739	9.1	28	5.0	700	0.5	10	7.0	500
1N4740	10	25	7.0	700	0.25	10	7.6	454
1N4741	11	23	8.0	700	0.25	5.0	8.4	414
1N4742	12	21	9.0	700	0.25	5.0	9.1	380
1N4743	13	19	10	700	0.25	5.0	9.9	344
1N4744	15	17	14	700	0.25	5.0	11.4	304
1N4745	16	15.5	16	700	0.25	5.0	12.2	285
1N4746	18	14	20	750	0.25	5.0	13.7	250
1N4747	20	12.5	22	750	0.25	5.0	15.2	225
1N4748	22	11.5	23	750	0.25	5.0	16.7	205
1N4749	24	10.5	25	750	0.25	5.0	18.2	190
1N4750	27	9.5	35	750	0.25	5.0	20.6	170
1N4751	30	8.5	40	1000	0.25	5.0	22.8	150
1N4752	33	7.5	45	1000	0.25	5.0	25.1	135
1N4753	36	7.0	50	1000	0.25	5.0	27.4	125
1N4754	39	6.5	60	1000	0.25	5.0	29.7	115
1N4755	43	6.0	70	1500	0.25	5.0	32.7	110
1N4756	47	5.5	80	1500	0.25	5.0	35.8	95
1N4757	51	5.0	95	1500	0.25	5.0	38.8	90
1N4758	56	4.5	110	2000	0.25	5.0	42.6	80
1N4759	62	4.0	125	2000	0.25	5.0	47.1	70
1N4760	68	3.7	150	2000	0.25	5.0	51.7	65
1N4761	75	3.3	175	2000	0.25	5.0	56.0	60
1N4762	82	3.0	200	3000	0.25	5.0	62.2	55
1N4763	91	2.8	250	3000	0.25	5.0	69.2	50
1N4764	100	2.5	350	3000	0.25	5.0	76.0	45

*Indicates JEDEC Registered Data.

NOTE 1 — Tolerance and Type Number Designation. The JEDEC type numbers listed have a standard tolerance on the nominal zener voltage of ±10%. A standard tolerance of ±5% on individual units is also available and is indicated by suffixing "A" to the standard type number.

NOTE 2 — Specials Available Include:

A. Nominal zener voltages between the voltages shown and tighter voltage tolerances,

B. Matched sets.

For detailed information on price, availability, and delivery, contact your nearest Motorola representative.

NOTE 3 — Zener Voltage (V_Z) Measurement. Motorola guarantees the zener voltage when measured at 90 seconds while maintaining the lead temperature (T_L) at 30°C ± 1°C, 3/8″ from the diode body.

NOTE 4 — Zener Impedance (Z_Z) Derivation. The zener impedance is derived from the 60 cycle ac voltage, which results when an ac current having an rms value equal to 10% of the dc zener current (I_{ZT} or I_{ZK}) is superimposed on I_{ZT} or I_{ZK}.

NOTE 5 — Surge Current (i_r) Non-Repetitive. The rating listed in the electrical characteristics table is maximum peak, non-repetitive, reverse surge current of 1/2 square wave or equivalent sine wave pulse of 1/120 second duration superimposed on the test current, I_{ZT}, per JEDEC registration; however, actual device capability is as described in Figures 4 and 5.

APPLICATION NOTE

Since the actual voltage available from a given zener diode is temperature dependent, it is necessary to determine junction temperature under any set of operating conditions in order to calculate its value. The following procedure is recommended:

Lead Temperature, T_L, should be determined from

$$T_L = \theta_{LA} P_D + T_A$$

θ_{LA} is the lead-to-ambient thermal resistance (°C/W) and P_D is the power dissipation. The value for θ_{LA} will vary and depends on the device mounting method. θ_{LA} is generally 30 to 40°C/W for the various clips and tie points in common use and for printed circuit board wiring.

The temperature of the lead can also be measured using a thermocouple placed on the lead as close as possible to the tie point. The thermal mass connected to the tie point is normally large enough so that it will not significantly respond to heat surges generated in the diode as a result of pulsed operation once steady-state conditions are achieved. Using the measured value of T_L, the junction temperature may be determined by:

$$T_J = T_L + \Delta T_{JL}.$$

ΔT_{JL} is the increase in junction temperature above the lead temperature and may be found as follows:

$$\Delta T_{JL} = \theta_{JL} P_D$$

θ_{JL} may be determined from Figure 3 for dc power conditions. For worst-case design, using expected limits of I_Z, limits of P_D and the extremes of $T_J(\Delta T_J)$ may be estimated. Changes in voltage, V_Z, can then be found from:

$$\Delta V = \theta_{VZ} \Delta T_J$$

θ_{VZ}, the zener voltage temperature coefficient, is found from Figure 2.

Under high power-pulse operation, the zener voltage will vary with time and may also be affected significantly by the zener resistance. For best regulation, keep current excursions as low as possible.

Surge limitations are given in Figure 5. They are lower than would be expected by considering only junction temperature, as current crowding effects cause temperatures to be extremely high in small spots resulting in device degradation should the limits of Figure 5 be exceeded.

2N4877

MEDIUM-POWER NPN SILICON TRANSISTOR

. . . designed for switching and wide band amplifier applications.

- Low Collector-Emitter Saturation Voltage – $V_{CE(sat)}$ = 1.0 Vdc (Max) @ I_C = 4.0 Amp
- DC Current Gain Specified to 4 Amperes
- Excellent Safe Operating Area
- Packaged in the Compact TO-39 Case for Critical Space-Limited Applications.

4 AMPERE
POWER TRANSISTOR

NPN SILICON
60 VOLTS
10 WATTS

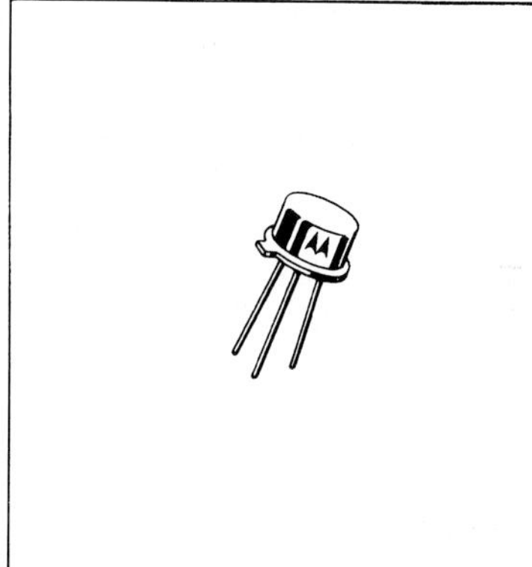

* MAXIMUM RATINGS

Rating	Symbol	Value	Unit
Collector-Emitter Voltage	V_{CEO}	60	Vdc
Collector-Base Voltage	V_{CB}	70	Vdc
Emitter-Base Voltage	V_{EB}	5.0	Vdc
Collector Current – Continuous	I_C	4.0	Adc
Base Current	I_B	1.0	Adc
Total Device Dissipation @ T_C = 25°C Derate above 25°C	P_D	10 57.2	Watts mW/°C
Operating and Storage Junction Temperature Range	T_J, T_{stg}	-65 to +200	°C

*Indicates JEDEC Registered Data

THERMAL CHARACTERISTICS

Characteristic	Symbol	Max	Unit
Thermal Resistance, Junction to Case	θ_{JC}	17.5	°C/W

FIGURE 1 – POWER-TEMPERATURE DERATING CURVE

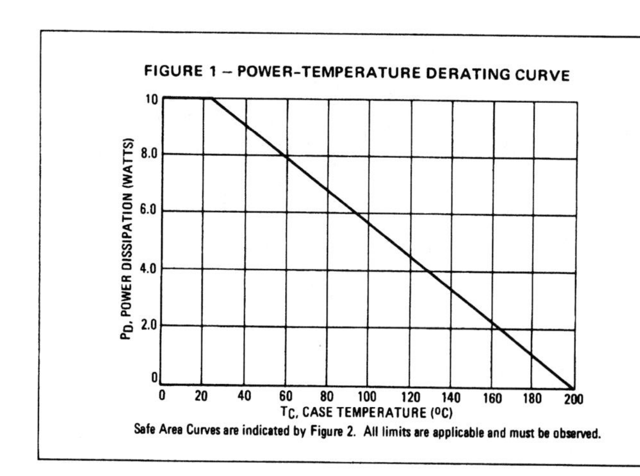

Safe Area Curves are indicated by Figure 2. All limits are applicable and must be observed.

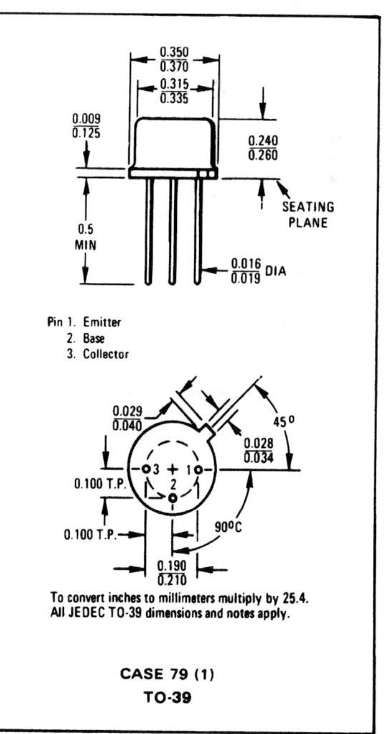

CASE 79 (1)
TO-39

*ELECTRICAL CHARACTERISTICS (T_C = 25°C unless otherwise noted)

Characteristic	Symbol	Min	Max	Unit
OFF CHARACTERISTICS				
Collector-Emitter Sustaining Voltage (1) (I_C = 200 mAdc, I_B = 0)	$V_{CEO(sus)}$	60	–	Vdc
Collector Cutoff Current (V_{CE} = 70 Vdc, $V_{EB(off)}$ = 1.5 Vdc) (V_{CE} = 70 Vdc, $V_{EB(off)}$ = 1.5 Vdc, T_C = 100°C)	I_{CEX}	– –	100 1.0	μAdc mAdc
Collector Cutoff Current (V_{CB} = 70 Vdc, I_E = 0)	I_{CBO}	–	100	μAdc
Emitter Cutoff Current (V_{BE} = 5.0 Vdc, I_C = 0)	I_{EBO}	–	100	μAdc
ON CHARACTERISTICS(1)				
DC Current Gain (I_C = 1.0 Adc, V_{CE} = 2.0 Vdc) (I_C = 4.0 Adc, V_{CE} = 2.0 Vdc)	h_{FE}	30 20	– 100	–
Collector-Emitter Saturation Voltage (I_C = 4.0 Adc, I_B = 0.4 Adc)	$V_{CE(sat)}$	–	1.0	Vdc
Base-Emitter Saturation Voltage (I_C = 4.0 Adc, I_B = 0.4 Adc)	$V_{BE(sat)}$	–	1.8	Vdc
DYNAMIC CHARACTERISTICS				
Current-Gain-Bandwidth Product (I_C = 0.25 Adc, V_{CE} = 10 Vdc, f = 1.0 MHz) (I_C = 0.25 Adc, V_{CE} = 10 Vdc, f = 10 MHz)**	f_T	4.0 30	– –	MHz
SWITCHING CHARACTERISTICS				
Rise Time (V_{CC} = 25 Vdc, I_C = 4.0 Adc, I_{B1} = 0.4 Adc)	t_r	–	100	ns
Storage Time (V_{CC} = 25 Vdc, I_C = 4.0 Adc, I_{B1} = I_{B2} = 0.4 Adc)	t_s	–	1.5	μs
Fall Time (V_{CC} = 25 Vdc, I_C = 4.0 Adc, I_{B1} = I_{B2} = 0.4 Adc)	t_f	–	500	ns

*Indicates JEDEC Registered Data.
**Motorola guarantees this value in addition to JEDEC Registered Data.
Note 1: Pulse Test: Pulse Width ≤ 300 μs, Duty Cycle ≤ 2.0%.

FIGURE 2 – ACTIVE-REGION SAFE OPERATING AREA

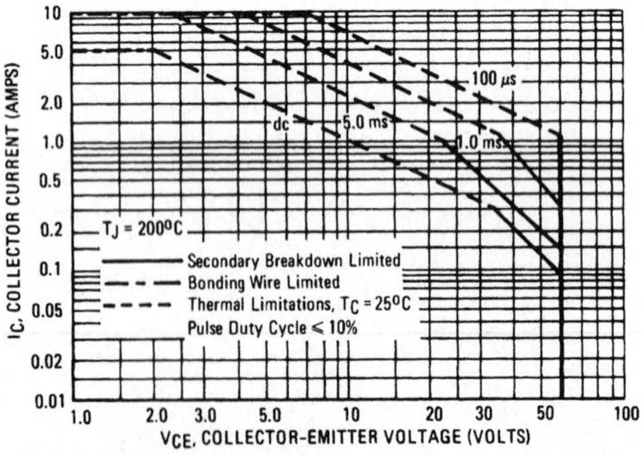

100 μs
dc
5.0 ms
1.0 ms
T_J = 200°C
Secondary Breakdown Limited
Bonding Wire Limited
Thermal Limitations, T_C = 25°C
Pulse Duty Cycle ≤ 10%
I_C, COLLECTOR CURRENT (AMPS)
V_{CE}, COLLECTOR-EMITTER VOLTAGE (VOLTS)

FIGURE 3 – SWITCHING TIME TEST CIRCUIT

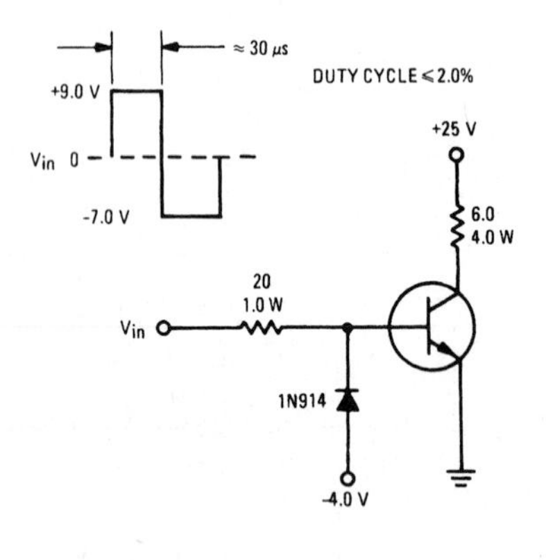

There are two limitations on the power handling ability of a transistor: average junction temperature and second breakdown. Safe operating area curves indicate I_C–V_{CE} limits of the transistor that must be observed for reliable operation; i.e., the transistor must not be subjected to greater dissipation than the curves indicate.

The data of Figure 2 is based on $T_{J(pk)}$ = 200°C; T_C is variable depending on conditions. Second breakdown pulse limits are valid for duty cycles to 10% provided $T_{J(pk)} \leq$ 200°C. At high case temperatures, thermal limitations will reduce the power that can be handled to values less than the limitations imposed by second breakdown. (See AN-415)

MOTOROLA Semiconductor Products Inc.

BOX 20912 • PHOENIX, ARIZONA 85036 • A SUBSIDIARY OF MOTOROLA INC.

6025-2 PRINTED IN USA 3-71 IMPERIAL LITHO 821668 10M DS 3189

2N1595 thru 2N1599

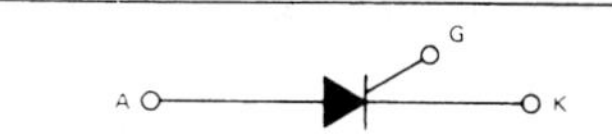

SILICON CONTROLLED RECTIFIERS

1.6 AMPERE RMS
50 thru 400 VOLTS

REVERSE BLOCKING TRIODE THYRISTORS

These devices are glassivated planar construction designed for gating operation in mA/μA signal or detection circuits.

- Low-Level Gate Characteristics –
  I_{GT} = 10 mA (Max) @ 25°C
- Low Holding Current
  I_H = 5.0 mA (Typ) @ 25°C
- Glass-to-Metal Bond for Maximum Hermetic Seal

***MAXIMUM RATINGS** (T_J = 125°C unless otherwise noted)

Rating	Symbol	Value	Unit
Repetitive Peak Reverse Blocking Voltage	V_{RRM}		Volts
2N1595		50	
2N1596		100	
2N1597		200	
2N1598		300	
2N1599		400	
Repetitive Peak Forward Blocking Voltage	V_{DRM}		Volts
2N1595		50	
2N1596		100	
2N1597		200	
2N1598		300	
2N1599		400	
RMS On-State Current (All Conduction Angles)	$I_{T(RMS)}$	1.6	Amps
Peak Non-Repetitive Surge Current (One Cycle, 60 Hz, T_J = -65 to +125°C)	I_{TSM}	15	Amps
Peak Gate Power	P_{GM}	0.1	Watt
Average Gate Power	$P_{G(AV)}$	0.01	Watt
Peak Gate Current	I_{GM}	0.1	Amp
Peak Gate Voltage Forward	V_{GFM}	10	Volts
Reverse	V_{GRM}	10	
Operating Junction Temperature Range	T_J	-65 to +125	°C
Storage Temperature Range	T_{stg}	-65 to +150	°C

*Indicates JEDEC Registered Data

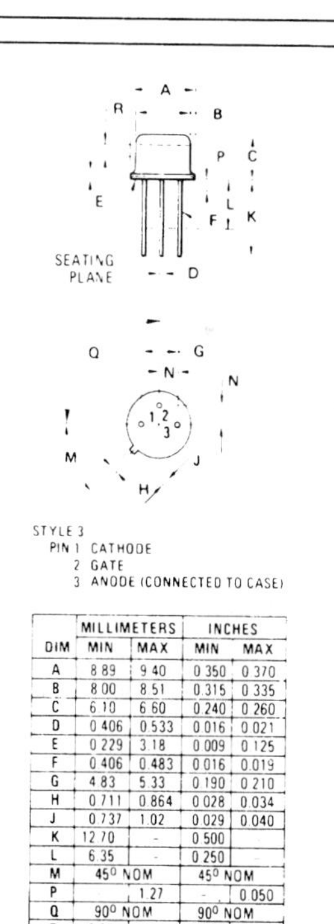

DIM	MILLIMETERS MIN	MILLIMETERS MAX	INCHES MIN	INCHES MAX
A	8.89	9.40	0.350	0.370
B	8.00	8.51	0.315	0.335
C	6.10	6.60	0.240	0.260
D	0.406	0.533	0.016	0.021
E	0.229	3.18	0.009	0.125
F	0.406	0.483	0.016	0.019
G	4.83	5.33	0.190	0.210
H	0.711	0.864	0.028	0.034
J	0.737	1.02	0.029	0.040
K	12.70	-	0.500	
L	6.35	-	0.250	-
M	45° NOM		45° NOM	
P		1.27	-	0.050
Q	90° NOM		90° NOM	
R	2.54	-	0.100	-

All JEDEC dimensions and notes apply.

CASE 79-02
TO-39

2N1595 thru 2N1599

ELECTRICAL CHARACTERISTICS (T_C = 25°C unless otherwise noted).

Characteristic	Symbol	Min	Typ	Max	Unit
*Peak Reverse Blocking Current (Rated V_{RRM}, T_J = 125°C)	I_{RRM}	–	–	1000	μA
*Peak Forward Blocking Current (Rated V_{DRM}, T_J = 125°C)	I_{DRM}	–	–	1000	μA
*Peak On-State Voltage (I_F = 1.0 Adc, Pulsed, 1.0 ms (Max), Duty Cycle ≤ 1%)	V_{TM}		1.1	2.0	Volts
*Gate Trigger Current (V_{AK} = 6.0 V, R_L = 12 Ohms)	I_{GT}	–	2.0	10	mA
*Gate Trigger Voltage	V_{GT}				Volts
(V_{AK} = 6.0 V, R_L = 12 Ohms)		–	0.7	3.0	
(V_{AK} = 6.0 V, R_L = 12 Ohms, T_J = 125°C)		0.2	–	–	
Reverse Gate Current (V_{GK} = 10 V)	I_{GR}	–	17	–	mA
Holding Current (V_{AK} = 12 V)	I_H	–	5.0	–	mA
Turn-On Time	t_{gt}				μs
(I_{GT} = 10 mA, I_F = 1.0 A)		–	0.8	–	
(I_{GT} = 20 mA, I_F = 1.0 A)		–	0.6	–	
Turn-Off Time (I_F = 1.0 A, I_R = 1.0 A, dv/dt = 20 V/μs, T_J = 125°C)	t_q		10	–	μs

*Indicates JEDEC Registered Data.

CURRENT DERATING

FIGURE 1 – CASE TEMPERATURE REFERENCE

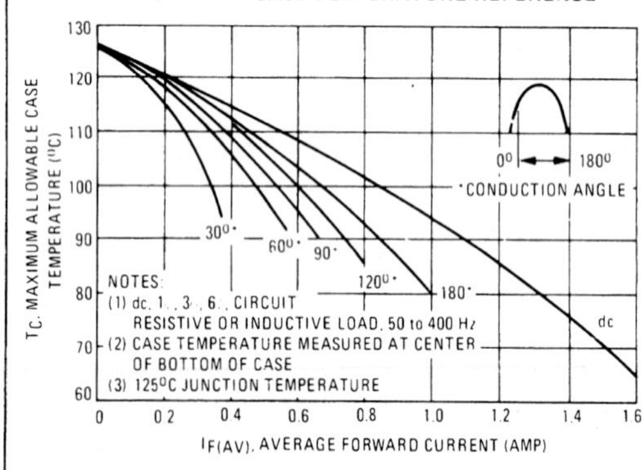

FIGURE 2 – AMBIENT TEMPERATURE REFERENCE

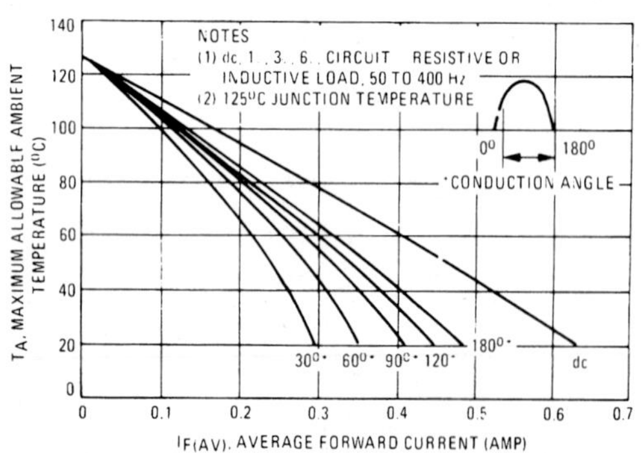

MOTOROLA Semiconductor Products Inc.
BOX 20912 • PHOENIX ARIZONA 85036 • A SUBSIDIARY OF MOTOROLA INC
10137 PRINTED IN USA (8 77) MPS 10M

MC1741, MC1741C
MC1741N, MC1741NC

OPERATIONAL AMPLIFIER
SILICON MONOLITHIC
INTEGRATED CIRCUIT

INTERNALLY COMPENSATED, HIGH PERFORMANCE OPERATIONAL AMPLIFIERS

. . . designed for use as a summing amplifier, integrator, or amplifier with operating characteristics as a function of the external feedback components.

- No Frequency Compensation Required
- Short-Circuit Protection
- Offset Voltage Null Capability
- Wide Common-Mode and Differential Voltage Ranges
- Low-Power Consumption
- No Latch Up
- Low Noise Selections Offered – N Suffix

MAXIMUM RATINGS (T_A = +25°C unless otherwise noted)

Rating	Symbol	MC1741C	MC1741	Unit
Power Supply Voltage	V_{CC} V_{EE}	+18 -18	+22 -22	Vdc Vdc
Input Differential Voltage	V_{ID}	±30		Volts
Input Common Mode Voltage (Note 1)	V_{ICM}	±15		Volts
Output Short Circuit Duration (Note 2)	t_S	Continuous		
Operating Ambient Temperature Range	T_A	0 to +70	-55 to +125	°C
Storage Temperature Range Metal, Flat and Ceramic Packages Plastic Packages	T_{stg}	 -65 to +150 -55 to +125		°C

Note 1. For supply voltages less than ±15 V, the absolute maximum input voltage is equal to the supply voltage.

Note 2. Supply voltage equal to or less than 15 V.

EQUIVALENT CIRCUIT SCHEMATIC

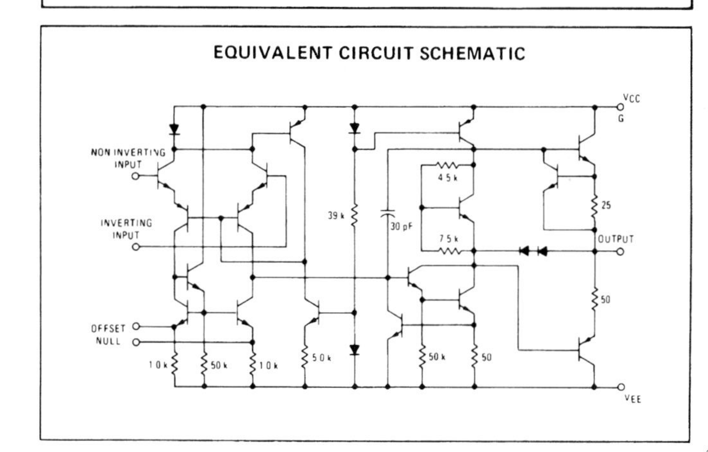

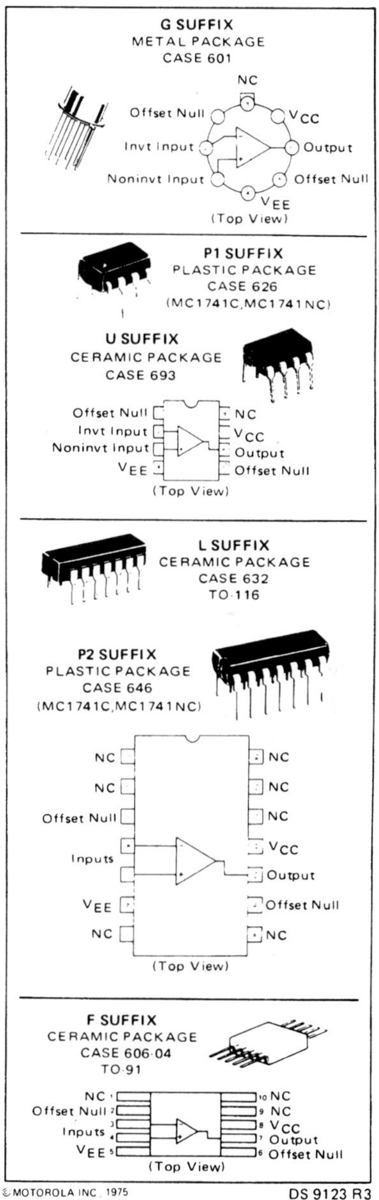

MC1741 • MC1741C • MC1741N • MC1741NC

ELECTRICAL CHARACTERISTICS (V_{CC} = 15 V, V_{EE} = 15 V, T_A = 25°C unless otherwise noted).

Characteristic	Symbol	MC1741			MC1741C			Unit
		Min	Typ	Max	Min	Typ	Max	
Input Offset Voltage ($R_S \leq$ 10 k)	V_{IO}	–	1.0	5.0	–	2.0	6.0	mV
Input Offset Current	I_{IO}	–	20	200	–	20	200	nA
Input Bias Current	I_{IB}	–	80	500	–	80	500	nA
Input Resistance	r_i	0.3	2.0	–	0.3	2.0	–	MΩ
Input Capacitance	C_i	–	1.4	–	–	1.4	–	pF
Offset Voltage Adjustment Range	V_{IOR}	–	±15	–	–	±15	–	mV
Common Mode Input Voltage Range	V_{ICR}	±12	±13	–	±12	±13	–	V
Large Signal Voltage Gain (V_O = ±10 V, $R_L \geq$ 2.0 k)	A_v	50	200	–	20	200	–	V/mV
Output Resistance	r_o	–	75	–	–	75	–	Ω
Common Mode Rejection Ratio ($R_S \leq$ 10 k)	CMRR	70	90	–	70	90	–	dB
Supply Voltage Rejection Ratio ($R_S \leq$ 10 k)	PSRR	–	30	150	–	30	150	μV/V
Output Voltage Swing	V_O							V
($R_L \geq$ 10 k)		±12	±14	–	±12	±14	–	
($R_L \geq$ 2 k)		±10	±13	–	±10	±13	–	
Output Short-Circuit Current	I_{os}	–	20	–	–	20	–	mA
Supply Current	I_D	–	1.7	2.8	–	1.7	2.8	mA
Power Consumption	P_C	–	50	85	–	50	85	mW
Transient Response (Unity Gain – Non-Inverting)								
(V_I = 20 mV, $R_L \geq$ 2 k, $C_L \leq$ 100 pF) Rise Time	t_{TLH}	–	0.3	–	–	0.3	–	μs
(V_I = 20 mV, $R_L \geq$ 2 k, $C_L \leq$ 100 pF) Overshoot	os	–	15	–	–	15	–	%
(V_I = 10 V, $R_L \geq$ 2 k, $C_L \leq$ 100 pF) Slew Rate	SR	–	0.5	–	–	0.5	–	V/μs

ELECTRICAL CHARACTERISTICS (V_{CC} = 15 V, V_{EE} = 15 V, T_A = *T_{high} to T_{low} unless otherwise noted.)

Characteristic	Symbol	MC1741			MC1741C			Unit
		Min	Typ	Max	Min	Typ	Max	
Input Offset Voltage ($R_S \leq$ 10 kΩ)	V_{IO}	–	1.0	6.0	–	–	7.5	mV
Input Offset Current	I_{IO}							nA
(T_A = 125°C)		–	7.0	200	–	–	–	
(T_A = -55°C)		–	85	500	–	–	–	
(T_A = 0°C to +70°C)		–	–	–	–	–	300	
Input Bias Current	I_{IB}							nA
(T_A = 125°C)		–	30	500	–	–	–	
(T_A = -55°C)		–	300	1500	–	–	–	
(T_A = 0°C to +70°C)		–	–	–	–	–	800	
Common Mode Input Voltage Range	V_{ICR}	±12	±13	–	–	–	–	V
Common Mode Rejection Ratio ($R_S \leq$ 10 k)	CMRR	70	90	–	–	–	–	dB
Supply Voltage Rejection Ratio ($R_S \leq$ 10 k)	PSRR	–	30	150	–	–	–	μV/V
Output Voltage Swing	V_O							V
($R_L \geq$ 10 k)		±12	±14	–	–	–	–	
($R_L \geq$ 2 k)		±10	±13	–	±10	±13	–	
Large Signal Voltage Gain ($R_L \geq$ 2 k, V_{out} = ±10 V)	A_v	25	–	–	15	–	–	V/mV
Supply Currents	I_D							mA
(T_A = 125°C)		–	1.5	2.5	–	–	–	
(T_A = -55°C)		–	2.0	3.3	–	–	–	
Power Consumption (T_A = +125°C)	P_C	–	45	75	–	–	–	mW
(T_A = -55°C)		–	60	100	–	–	–	

*T_{high} = 125°C for MC1741 and 70°C for MC1741C
T_{low} = -55°C for MC1741 and 0°C for MC1741C

Answers to Odd-Numbered Problems

Chapter 1

1. **(a)** 3×10^3 **(b)** 7.5×10^4 **(c)** 2×10^6

3. **(a)** 8.4×10^3 **(b)** 9.9×10^4 **(c)** 2×10^5

5. **(a)** 0.0000025 **(b)** 500 **(c)** 0.39

7. **(a)** 4.32×10^7 **(b)** 5.00085×10^3
(c) 6.06×10^{-8}

9. **(a)** 2.0×10^9 **(b)** 3.6×10^{14}
(c) 1.54×10^{-14}

11. **(a)** 89×10^3 **(b)** 450×10^3
(c) 12.04×10^{12}

13. **(a)** 345×10^{-6} **(b)** 25×10^{-3}
(c) 1.29×10^{-9}

15. **(a)** 7.1×10^{-3} **(b)** 101×10^6
(c) 1.50×10^6

17. **(a)** 22.7×10^{-3} **(b)** 200×10^6
(c) 848×10^{-3}

19. **(a)** 345 μA **(b)** 25 mA **(c)** 1.29 nA

21. **(a)** 3 μF **(b)** 3.3 MΩ **(c)** 350 nA

23. **(a)** 5000 μA **(b)** 3.2 mW
(c) 5 MV **(d)** 10,000 kW

25. **(a)** 50.68 mA **(b)** 2.32 MΩ
(c) 0.0233 μF

Chapter 2

1. 80×10^{12} C

3. **(a)** 10 V **(b)** 2.5 V **(c)** 4 V

5. 20 V

7. **(a)** 75 A **(b)** 20 A **(c)** 2.5 A

9. 2 s

11. **(a)** 6800 Ω ± 10% **(b)** 33 Ω ± 10%
(c) 47,000 Ω ± 5%

13. **(a)** Red, violet, brown
(b) (b) 330 Ω, (d) 2.2 kΩ, (a) 39 kΩ, (e) 56 kΩ, (f) 100 kΩ

15. **(a)** 22 Ω **(b)** 4.7 kΩ **(c)** 82 kΩ
(d) 3.3 kΩ **(e)** 56 Ω **(f)** 10 MΩ

17. There is current through lamp 2.

19. Ammeter in series with resistors, with its negative terminal to the negative terminal of source and its positive terminal to one side of R_1. Voltmeter placed across (in parallel with) the source (negative to negative, positive to positive).

21. Position 1: V1 = 0, V2 = V_S
Position 2: V1 = V_S, V2 = 0

23. **(a)** 0.25 V **(b)** 250 V

25. **(a)** 200 Ω **(b)** 150 MΩ **(c)** 4500 Ω

27. 33.3 V

29. AWG #27

31. Circuit (b)

33. One ammeter in series with battery. One ammeter in series with each resistor (seven total).

35. See Figure P–1.

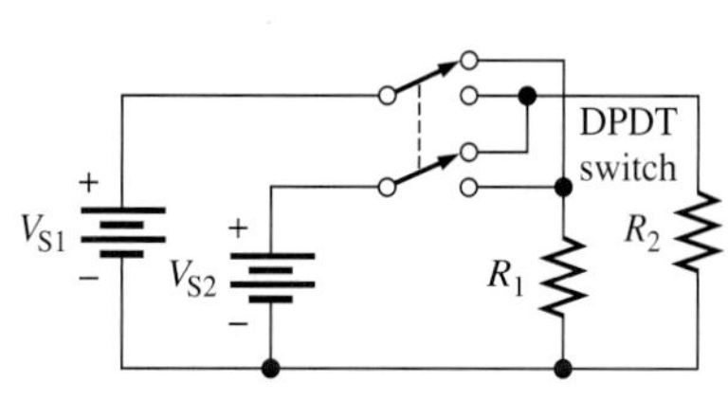

FIGURE P–1

Chapter 3

1. **(a)** 3 A **(b)** 0.2 A **(c)** 1.5 A

3. 15 mA

5. **(a)** 3.33 mA **(b)** 550 μA (c) 588 μA
(d) 500 mA **(e)** 6.60 mA

7. **(a)** 2.50 mA **(b)** 2.27 μA **(c)** 8.33 mA

9. **(a)** 10 mV **(b)** 1.65 V **(c)** 14.1 kV
(d) 3.52 V **(e)** 250 mV **(f)** 750 kV
(g) 8.5 kV **(h)** 3.53 mV

11. **(a)** 81 V **(b)** 500 V **(c)** 125 V

13. **(a)** 2 kΩ **(b)** 3.5 kΩ **(c)** 2 kΩ
(d) 100 kΩ **(e)** 1 MΩ

15. **(a)** 4 Ω **(b)** 3 kΩ **(c)** 200 kΩ

17. 417 mW

19. **(a)** 1 MW **(b)** 3 MW
(c) 150 MW **(d)** 8.7 MW

21. **(a)** 2,000,000 μW **(b)** 500 μW
(c) 250 μW **(d)** 6.67 μW

23. 16.5 mW

25. 1.18 kW

27. 5.81 W

29. 25 Ω

31. 0.00186 kWh

33. 1 W

35. **(a)** positive at top **(b)** positive at bottom **(c)** positive at right

37. 36 Ah

39. 13.5 mA

41. 4.25 W

43. Five

45. 150 Ω

47. $V = 0$ V, $I = 0$ A; $V = 10$ V, $I = 100$ mA; $V = 20$ V, $I = 200$ mA; $V = 30$ V, $I = 300$ mA; $V = 40$ V, $I = 400$ mA; $V = 50$ V, $I = 500$ mA; $V = 60$ V, $I = 600$ mA; $V = 70$ V, $I = 700$ mA; $V = 80$ V, $I = 800$ mA; $V = 90$ V, $I = 900$ mA; $V = 100$ V, $I = 1$ A

49. $R_1 = 0.5\ \Omega$; $R_2 = 1\ \Omega$; $R_3 = 2\ \Omega$

51. 10 V; 30 V

53. $I_{MAX} = 3.83$ mA; $I_{MIN} = 3.46$ mA

55. 216 kWh

57. 12 W

59. 2.5 A

61. No fault

63. Lamp 4 is shorted.

Chapter 4

1. See Figure P–2.

3. 0.1 A

5. 138 Ω

7. **(a)** 7.9 kΩ **(b)** 33 Ω **(c)** 13.24 MΩ

FIGURE P–2

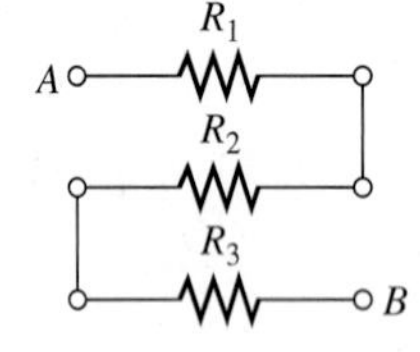

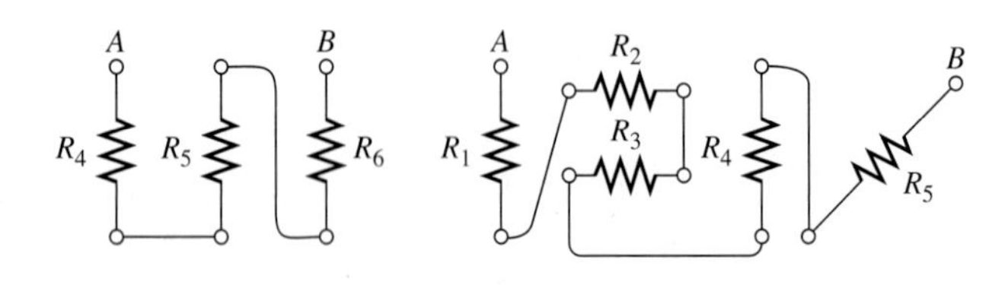

The series circuit is disconnected from the source, and the ohmmeter is connected across the circuit terminals.

9. 1126 Ω

11. **(a)** 170 kΩ **(b)** 50 Ω **(c)** 12.4 kΩ **(d)** 1.97 kΩ

13. **(a)** 625 μA **(b)** 4.26 μA. The ammeter is connected in series.

15. 355 mA

17. 14 V

19. 26 V

21. **(a)** $V_2 = 6.8$ V **(b)** $V_R = 8$ V, $V_{2R} = 16$ V, $V_{3R} = 24$ V, $V_{4R} = 32$ V

The voltmeter is connected across (in parallel with) each resistor for which the voltage is unknown.

23. **(a)** 3.84 V **(b)** 6.77 V

25. $V_{5.6k\Omega} = 10$ V; $V_{1k\Omega} = 1.79$ V; $V_{560\Omega} = 1$ V; $V_{10k\Omega} = 17.9$ V

27. 55 mW

29. Measure V_A and V_B individually with respect to ground; then $V_{R2} = V_A - V_B$.

31. **(a)** R_4 open **(b)** R_4 and R_5 shorted

33. 780 Ω

35. $V_A = 10$ V; $V_B = 7.72$ V; $V_C = 6.68$ V; $V_D = 1.81$ V; $V_E = 0.57$ V; $V_F = 0$ V

37. 500 Ω

39. **(a)** 19.1 mA **(b)** 45.8 V **(c)** R(⅛ W) = 343 Ω, R(¼ W) = 686 Ω, R(½ W) = 1371 Ω

41. See Figure P–3.

43. $R_1 + R_7 + R_8 + R_{10} = 4.23$ kΩ; $R_2 + R_4 + R_6 + R_{11} = 23.6$ kΩ; $R_3 + R_5 + R_9 + R_{12} = 19.9$ kΩ

45. *A:* 5.45 mA; *B:* 6.06 mA; *C:* 7.95 mA; *D:* 12 mA

47. *A:* $V_1 = 6.03$ V, $V_2 = 3.35$ V, $V_3 = 2.75$ V, $V_4 = 1.88$ V, $V_5 = 4.0$ V;

B: $V_1 = 6.71$ V, $V_2 = 3.73$ V, $V_3 = 3.06$ V, $V_5 = 4.5$ V;

C: $V_1 = 8.1$ V, $V_2 = 4.5$ V, $V_5 = 5.4$ V;

D: $V_1 = 10.8$ V, $V_5 = 7.2$ V

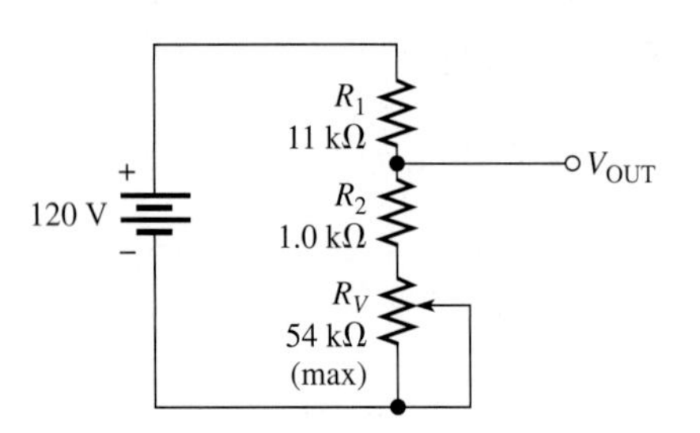

FIGURE P–3

49. Yes, R_3 and R_5 are shorted.

51. **(a)** R_{11} has burned open due to excessive power.
(b) Replace R_{11} (10 kΩ). **(c)** 338 V

53. R_6 is shorted.

55. Lamp 4 is open.

57. The 82 Ω resistor is shorted.

Chapter 5

1. See Figure P–4.

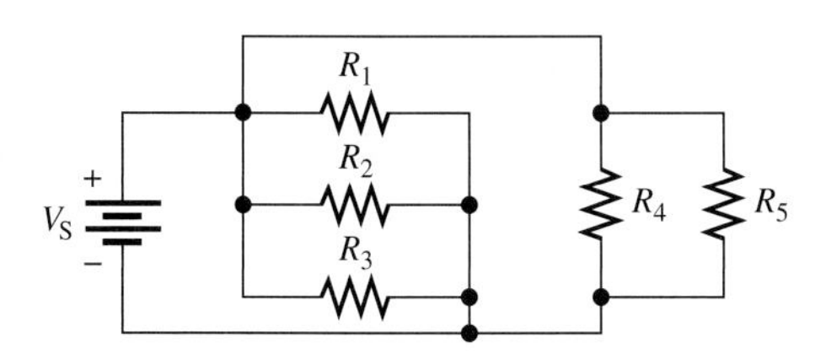

FIGURE P–4

3. 12 V; 5 mA

5. 1350 mA

7. $I_2 = I_3 = 7.5$ mA. An ammeter in series with each resistor in each branch.

9. **(a)** 25.6 Ω **(b)** 359 Ω **(c)** 819 Ω
(d) 996 Ω

11. 2 kΩ

13. **(a)** 909 mA **(b)** 76 mA

15. Circuit (a)

17. $I_1 = 2.19$ A; $I_2 = 811$ mA

19. 200 mW

21. 682 mA; 4.09 A

23. The 1.0 kΩ resistor is open.

25. R_2 is open.

27. $R_2 = 25\ \Omega$; $R_3 = 100\ \Omega$; $R_4 = 12.5\ \Omega$

29. $I_R = 4.8$ A; $I_{2R} = 2.4$ A; $I_{3R} = 1.6$ A; $I_{4R} = 1.2$ A

31. **(a)** $R_1 = 100\ \Omega$, $R_2 = 200\ \Omega$, $I_2 = 50$ mA
(b) $I_1 = 125$ mA, $I_2 = 74.9$ mA, $R_1 = 80\ \Omega$, $R_2 = 134\ \Omega$, $V_S = 10$ V
(c) $I_1 = 253$ mA, $I_2 = 147$ mA, $I_3 = 100$ mA, $R_1 = 395\ \Omega$

33. 53.7 Ω

35. 3.92 A; 1.08 A

37. $R_1 \parallel R_2 \parallel R_5 \parallel R_9 \parallel R_{10} \parallel R_{12} = 100\text{ k}\Omega \parallel 220\text{ k}\Omega \parallel 560\text{ k}\Omega \parallel 390\text{ k}\Omega \parallel 1.2\text{ M}\Omega \parallel 100\text{ k}\Omega = 33.6\text{ k}\Omega$

$R_4 \parallel R_6 \parallel R_7 \parallel R_8 = 270\text{ k}\Omega \parallel 1.0\text{ M}\Omega \parallel 820\text{ k}\Omega \parallel 680\text{ k}\Omega = 135.2\text{ k}\Omega$

$R_3 \parallel R_{11} = 330\text{ k}\Omega \parallel 1.8\text{ M}\Omega = 278.9\text{ k}\Omega$

39. $R_2 = 750\ \Omega$; $R_4 = 423\ \Omega$

41. The 4.7 kΩ resistor is open.

43. **(a)** One of the resistors has opened due to excessive power dissipation. **(b)** 30 V
(c) Replace the 1.8 kΩ resistor.

45. **(a)** 940 Ω **(b)** 518 Ω **(c)** 518 Ω
(d) 422 Ω

47. R_3 is open.

49. **(a)** R from pin 1 to pin 4 agrees with calculated value.
(b) R from pin 2 to pin 3 agrees with calculated value.

Chapter 6

1. R_2, R_3, and R_4 are in parallel, and this parallel combination is in series with both R_1 and R_5.

3. See Figure P–5.

5. 2003 Ω

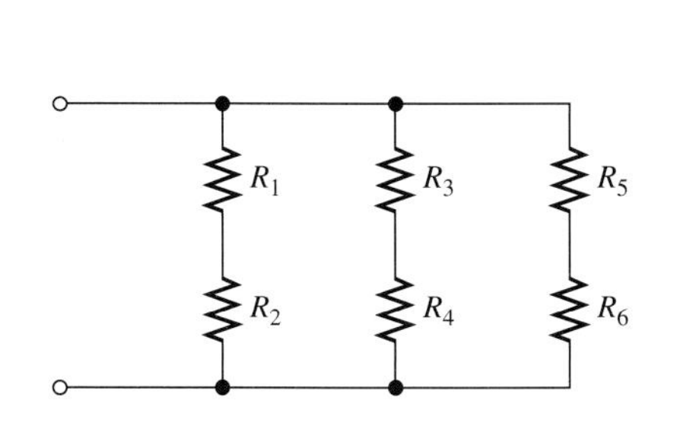

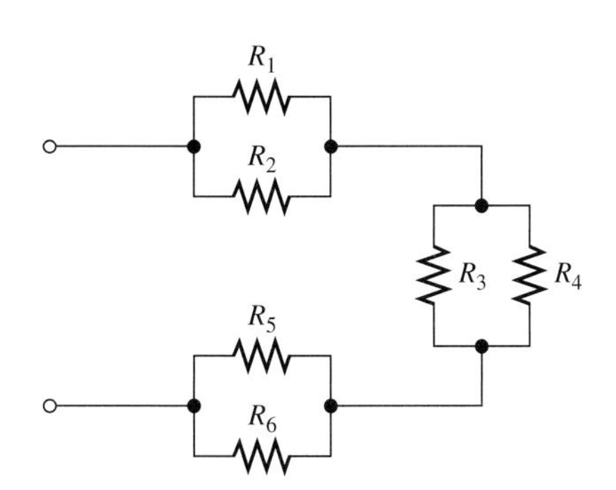

FIGURE P–5

7. **(a)** 128 Ω **(b)** 791 Ω

9. **(a)** $I_1 = I_4 = 11.7$ mA, $I_2 = I_3 = 5.85$ mA; $V_1 =$ 655 mV, $V_2 = V_3 = 585$ mV, $V_4 = 257$ mV

(b) $I_1 = 3.8$ mA, $I_2 = 618\ \mu$A, $I_3 = 1.27$ mA, $I_4 = 1.91$ mA; $V_1 = 2.58$ V, $V_2 = V_3 =$ $V_4 = 420$ mV

11. 7.5 V unloaded; 7.29 V loaded

13. 56 kΩ load

15. 22 kΩ

17. 2 V

19. 360 Ω

21. 7.33 kΩ

23. $R_{TH} = 18$ kΩ; $V_{TH} = 2.7$ V

25. 1.06 V; 226 μA

27. 21 mA

29. No, the meter should read 4.39 V. The 680 Ω is open.

31. The 7.62 V and 5.24 V readings are incorrect, indicating that the 3.3 kΩ resistor is open.

33. **(a)** $V_1 = -10$ V, all others 0 V

(b) $V_1 = -2.33$ V, $V_4 = -7.67$ V, $V_2 = -7.67$ V, $V_3 = 0$ V

(c) $V_1 = -2.33$ V, $V_4 = -7.67$ V, $V_2 = 0$ V, $V_3 = -7.67$ V

(d) $V_1 = -10$ V, all others 0 V

35. See Figure P–6.

37. $R_T = 5.76$ kΩ; $V_A = 3.3$ V; $V_B = 1.7$ V; $V_C = 850$ mV

39. $V_1 = 1.61$ V; $V_2 = 6.77$ V; $V_3 = 1.72$ V; $V_4 =$ 3.33 V; $V_5 = 378$ mV; $V_6 = 2.57$ V; $V_7 = 378$ mV; $V_8 = 1.72$ V; $V_9 = 1.61$ V

41. 110 Ω

43. $R_1 = 180$ Ω; $R_2 = 60$ Ω. Output across R_2.

45. 845 μA

47. 11.7 V

49. See Figure P–7.

51. Pos 1: $V_1 = 88.0$ V, $V_2 = 58.7$ V, $V_3 = 29.3$ V

Pos 2: $V_1 = 89.1$ V, $V_2 = 58.2$ V, $V_3 = 29.1$ V

Pos 3: $V_1 = 89.8$ V, $V_2 = 59.6$ V, $V_3 = 29.3$ V

53. One of the 12 kΩ resistors is open.

55. The 2.2 kΩ resistor is open.

57. $V_A = 0$ V; $V_B = 11.1$ V

59. R_2 is shorted.

61. No fault

63. R_4 is shorted.

65. R_5 is shorted.

Chapter 7

1. Decrease

3. 37.5 μWb

FIGURE P–6

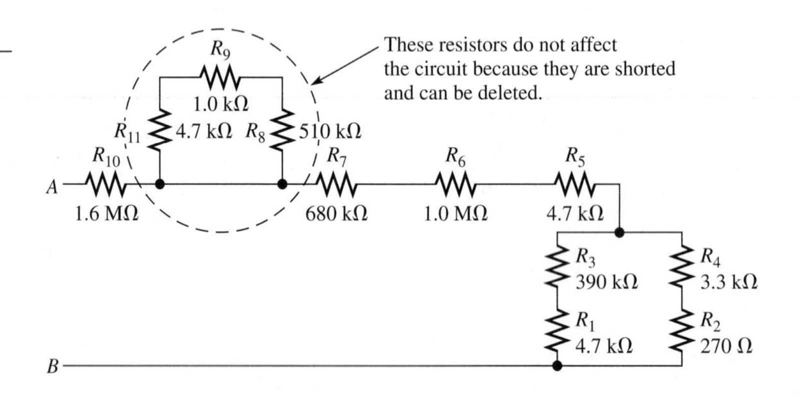

FIGURE P–7

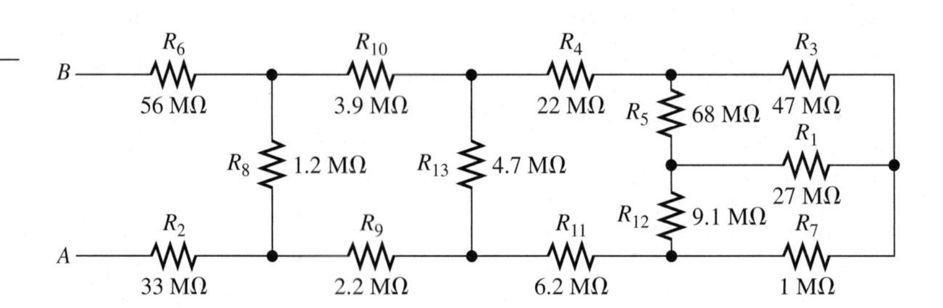

5. 597
7. 1500 At
9. **(a)** Electromagnetic force **(b)** Spring force
11. Electromagnetic force
13. Change the current.
15. Material A
17. 1 mA
19. To electrically connect the loop to the external circuit
21. The output voltage appears as the solid blue curve in Figure 7–39 with a frequency of 120 Hz and a peak of 10 V.
23. No fault

Chapter 8

1. **(a)** 1 Hz **(b)** 5 Hz **(c)** 20 Hz **(d)** 1 kHz **(e)** 2 kHz **(f)** 100 kHz
3. 2 μs
5. 250 Hz
7. 200 rps
9. **(a)** 7.07 mA **(b)** 4.5 mA **(c)** 14.14 mA
11. 15°, *A* leads.
13. See Figure P–8.

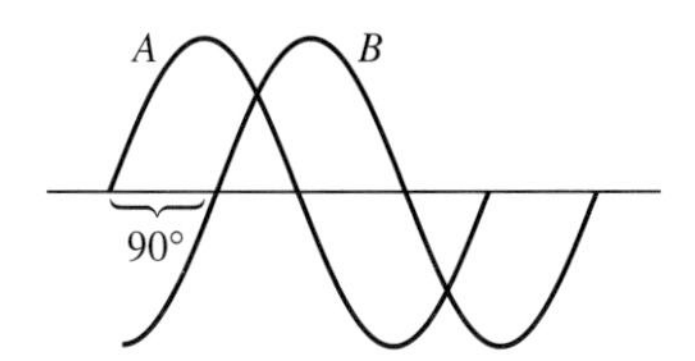

FIGURE P–8

15. **(a)** 22.5° **(b)** 60° **(c)** 90° **(d)** 108° **(e)** 216° **(f)** 324°
17. **(a)** 57.4 mA **(b)** 99.6 mA **(c)** −17.4 mA **(d)** −57.4 mA **(e)** −99.6 mA **(f)** 0 mA
19. 30°: 13.0 V; 45°: 14.5 V; 90°: 13.0 V; 180°: −7.5 V; 200°: −11.5 V; 300°: −7.5 V
21. **(a)** 7.07 mA **(b)** 0 A **(c)** 10 mA **(d)** 20 mA **(e)** 10 mA
23. 7.38 V
25. 4.24 V
27. $t_r \cong 3.5$ ms; $t_f \cong 3.5$ ms; $t_W \cong 12.5$ ms; $V = 5$ V
29. **(a)** −0.375 V **(b)** 3.0 V
31. **(a)** 50 kHz **(b)** 10 Hz
33. 25 kHz
35. 0.424 V; 2 Hz
37. 1.4 V; 120 ms; 30%
39. $I_{max} = 2.38$ A; $V_{avg} = 136$ V; See Figure P–9.

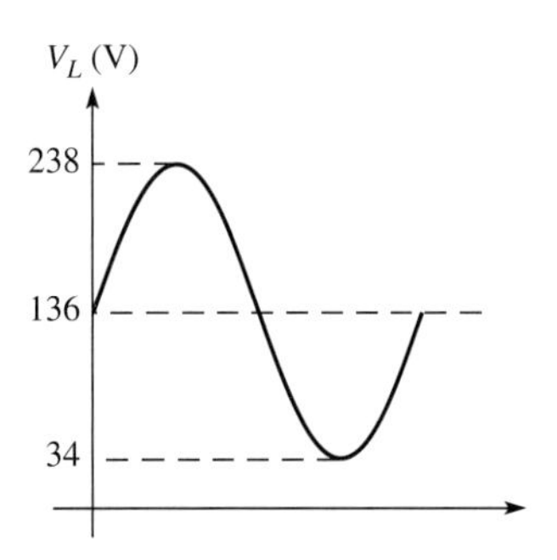

FIGURE P–9

41. **(a)** 2.5 **(b)** 3.96 V **(c)** 12.5 kHz
43. See Figure P–10.

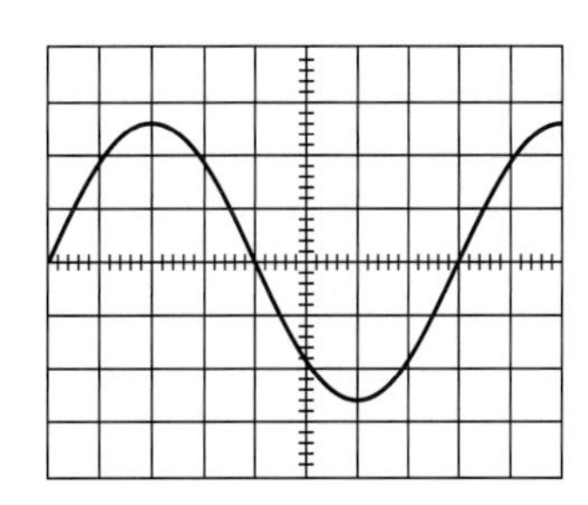

FIGURE P–10

45. $V_{p(in)} = 4.44$ V; $f_{in} = 2$ Hz
47. R_3 is open.
49. 5 V; 1 ms

Chapter 9

1. **(a)** 5 μF **(b)** 1 μC **(c)** 10 V
3. **(a)** 0.001 μF **(b)** 0.0035 μF **(c)** 0.00025 μF
5. 2 μF
7. 88.5 pF
9. A 12.5 pF increase
11. Ceramic
13. **(a)** 0.02 μF **(b)** 0.047 μF **(c)** 0.001 μF **(d)** 22 pF
15. **(a)** Encapsulation **(b)** Dielectric (ceramic disk) **(c)** Plate (metal disk) **(d)** Leads

17. **(a)** 0.69 μF **(b)** 69.7 pF **(c)** 2.6 μF

19. $V_1 = 2.13$ V; $V_2 = 10$ V; $V_3 = 4.55$ V; $V_4 = 1$ V

21. **(a)** 2.62 μF **(b)** 689 pF **(c)** 1.6 μF

23. **(a)** 100 μs **(b)** 560 μs **(c)** 22.1 μs **(d)** 15 ms

25. **(a)** 9.48 V **(b)** 13.0 V **(c)** 14.3 V **(d)** 14.7 V **(e)** 14.9 V

27. **(a)** 2.72 V **(b)** 5.90 V **(c)** 11.7 V

29. **(a)** 339 kΩ **(b)** 13.5 kΩ **(c)** 677 Ω **(d)** 33.9 Ω

31. **(a)** 30.4 Ω **(b)** 115 kΩ **(c)** 49.7 Ω

33. 200 Ω

35. $P_{true} = 0$ W; $P_r = 3.39$ mVAR

37. 0 Ω

39. Shorted capacitor

41. 3.18 ms

43. 3.24 μs

45. **(a)** Charges to 3.32 V in 10 ms, then discharges to 0 V in 215 ms.
(b) Charges to 3.32 V in 10 ms, then discharges to 2.96 V in 5 ms, then charges toward 20 V.

47. 0.0056 μF

49. $V_1 = 7.25$ V; $V_2 = 2.76$ V; $V_3 = 0.79$ V; $V_5 = 1.19$ V; $V_6 = 0.79$ V; $V_4 = 1.98$ V

51. C_2 is open.

53. No fault

Chapter 10

1. 8 kHz; 8 kHz

3. **(a)** 288 Ω **(b)** 1209 Ω

5. **(a)** 726 kΩ **(b)** 155 kΩ **(c)** 91.5 kΩ **(d)** 63.0 kΩ

7. **(a)** 34.7 mA **(b)** 4.14 mA

9. $I_{tot} = 12.3$ mA; $V_{C1} = 1.31$ V; $V_{C2} = 0.595$ V; $V_R = 0.616$ V; $\theta = 72.0°$ (V_s lagging I_{tot})

11. 808 Ω; $-36.1°$

13. 326 Ω; 64.3°

15. 245 Ω; 80.5°

17. $I_{C1} = 118$ mA; $I_{C2} = 55.3$ mA; $I_{R1} = 36.4$ mA; $I_{R2} = 44.4$ mA; $I_{tot} = 191$ mA; $\theta = 65.0°$ (V_s lagging I_{tot})

19. **(a)** 4.02 kΩ **(b)** 20 μA **(c)** 14.8 μA **(d)** 24.9 μA **(e)** 36.4° (V_s lagging I_{tot})

21. $V_{C1} = 8.74$ V; $V_{C2} = 3.26$ V; $V_{C3} = 3.26$ V; $V_{R1} = 2.11$ V; $V_{R2} = 1.15$ V

23. $I_{tot} = 82.4$ mA; $I_{C2} = 14.4$ mA; $I_{C3} = 67.6$ mA; $I_{R1} = I_{R2} = 6.39$ mA

25. 4.03 VA

27. 0.915

29. **(a)** 0.0548° **(b)** 5.46° **(c)** 43.7° **(d)** 84.0°

31. **(a)** 90° **(b)** 86.4° **(c)** 57.8° **(d)** 9.04°

33. See Figure P–11.

35. Figure 10–84: 1.05 kHz; Figure 10–85: 1.59 kHz

37. No leakage: $V_{out} = 3.21$ V; $\theta = 18.7°$;
Leakage: $V_{out} = 2.83$ V, $\theta = 33.3°$

39. **(a)** 0 V **(b)** 0.321 V **(c)** 0.5 V **(d)** 0 V

41. **(a)** $I_{L(A)} = 4.4$ A; $I_{L(B)} = 3.06$ A
(b) $P_{r(A)} = 509$ VAR; $P_{r(B)} = 211$ VAR
(c) $P_{true(A)} = 823$ W; $P_{true(B)} = 641$ W
(d) $P_{a(A)} = 968$ VA; $P_{a(B)} = 675$ VA
(e) Load A

43. 11.4 kΩ

45. $P_r = 1.32$ kVAR; $P_a = 2$ kVA

47. 0.103 μF

49. C is leaky.

51. No fault

53. R_2 is open.

FIGURE P–11

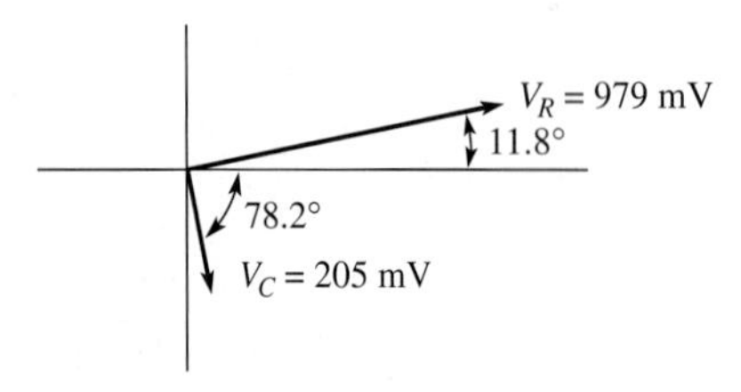

(a) For Figure 10–84

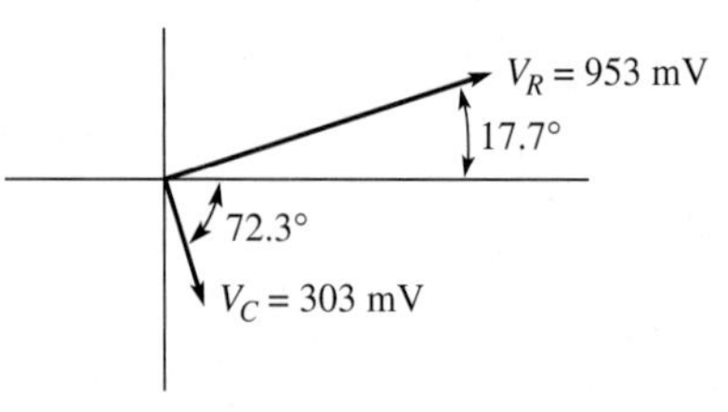

(b) For Figure 10–85

Chapter 11

1. **(a)** 1000 mH **(b)** 0.25 mH **(c)** 0.01 mH **(d)** 0.5 mH

3. 3450 turns

5. 50 mJ

7. 155 μH

9. 7.14 μH

11. **(a)** 4.33 H **(b)** 50 mH **(c)** 0.57 μH

13. **(a)** 1 μs **(b)** 2.13 μs **(c)** 2 μs

15. **(a)** 5.52 V **(b)** 2.03 V **(c)** 0.747 V **(d)** 0.275 V **(e)** 0.101 V

17. **(a)** 136 kΩ **(b)** 1.57 kΩ **(c)** 17.9 mΩ

19. $I_{tot} = 10.1$ mA; $I_{L2} = 6.7$ mA; $I_{L3} = 3.35$ mA

21. 101 mVAR

23. **(a)** Infinite resistance **(b)** Zero resistance **(c)** Lower R_W

25. 2.92 mA

27. 26.1 mA

29. L_3 is open.

31. No fault

33. L_3 is shorted.

Chapter 12

1. 15 kHz

3. **(a)** 112 Ω **(b)** 1.8 kΩ

5. **(a)** 17.4 Ω **(b)** 64 Ω **(c)** 127 Ω **(d)** 251 Ω

7. **(a)** 89.3 mA **(b)** 2.78 mA

9. 38.7°

11. See Figure P–12.

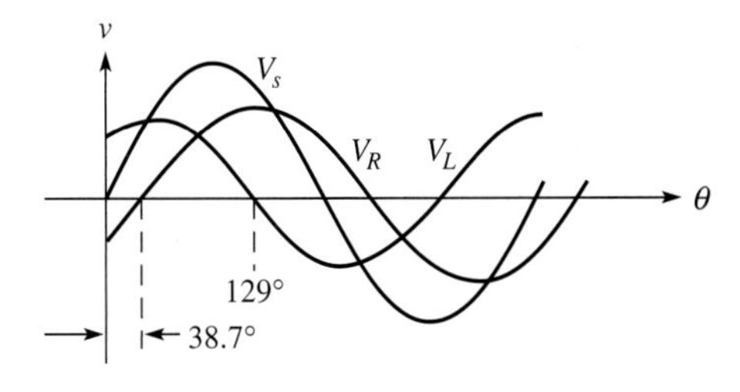

FIGURE P–12

13. 7.69 Ω

15. 2.39 kHz

17. **(a)** 274 Ω **(b)** 89.3 mA **(c)** 159 mA **(d)** 183 mA **(e)** 60.7° (I_{tot} lagging V_s)

19. $V_{R1} = 7.92$ V; $V_{R2} = V_L = 20.8$ V

21. $I_{tot} = 36$ mA; $I_L = 33.2$ mA; $I_{R2} = 13.9$ mA

23. 13.0 mW; 10.4 mVAR

25. $PF = 0.386$; $P_{true} = 347$ mW; $P_r = 692$ mVAR; $P_a = 900$ mVA

27. See Figure P–13.

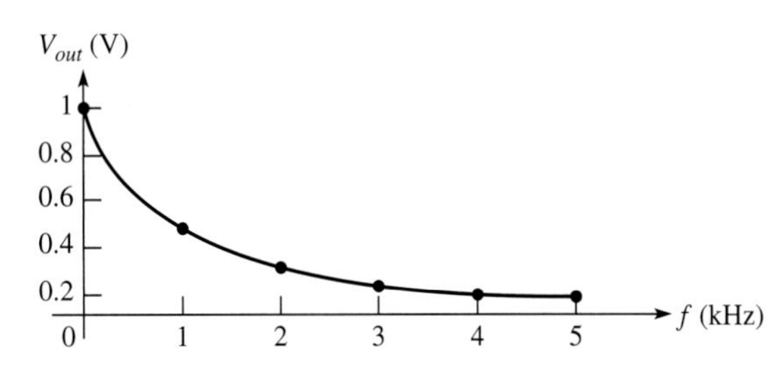

FIGURE P–13

29. See Figure P–14.

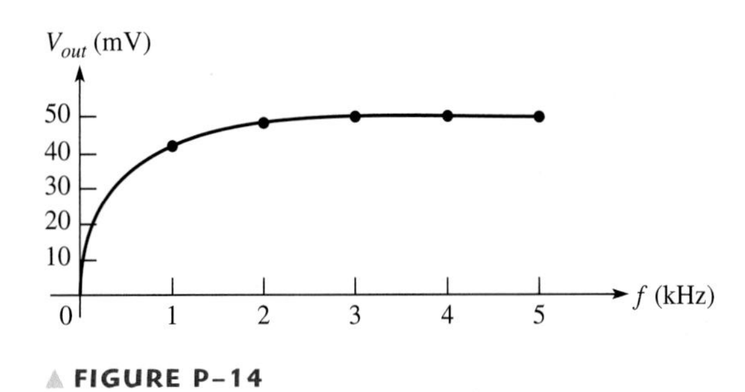

FIGURE P–14

31. $V_{R1} = V_{L1} = 18$ V; $V_{R2} = V_{R3} = V_{L2} = 0$ V

33. 5.57 V

35. 88.7 mA

37. **(a)** 405 mA **(b)** 228 mA **(c)** 333 mA **(d)** 335 mA

39. 0.133

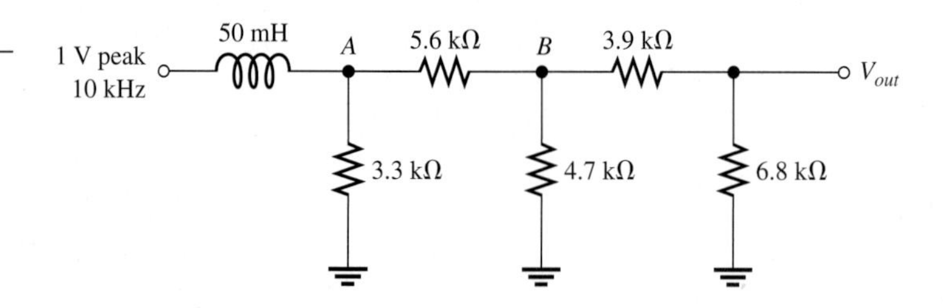

FIGURE P–15

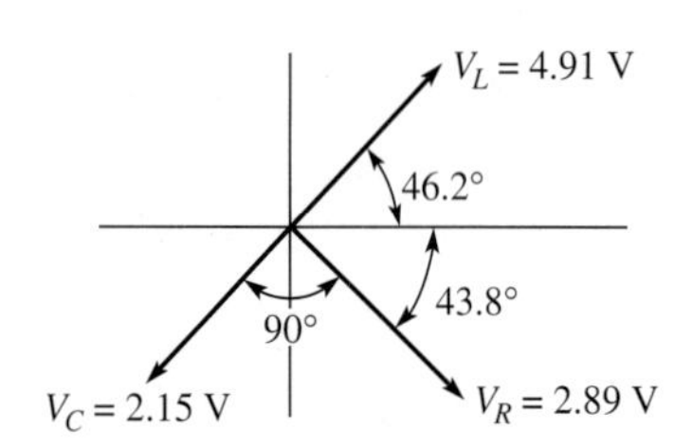

FIGURE P–16

41. See Figure P–15.
43. L_2 is open.
45. R_2 is open.
47. L_1 is shorted.

Chapter 13

1. 480 Ω, 88.8° (V_s lagging I); 480 Ω capacitive
3. Impedance increases.
5. See Figure P–16.
7. $X_L = 4.61$ kΩ; $X_C = 4.61$ kΩ; $Z = 220$ Ω; $I = 54.5$ mA
9. $f_r = 459$ kHz; $f_1 = 360$ kHz; $f_2 = 559$ kHz
11. **(a)** 14.5 kHz; band-pass
    (b) 24.0 kHz; band-pass
13. **(a)** $f_r = 339$ kHz, $BW = 239$ kHz
    (b) $f_r = 10.4$ kHz, $BW = 2.61$ kHz
15. Capacitive, $X_C < X_L$.
17. 758 Ω
19. 53.1 MΩ; 104 kHz
21. 62.5 Hz
23. 1.38 W
25. 200 Hz
27. $I_{R1} = I_C = 2.11$ mA; $I_{L1} = 1.33$ mA; $I_{L2} = 667$ μA; $I_{R2} = 667$ μA; $V_{R1} = 6.96$ V; $V_C = 2.11$ V; $V_{L1} = V_{L2} = V_{R2} = 6.67$ V
29. $I_{R1} = I_{L1} = 41.5$ mA; $I_C = I_{L2} = 133$ mA; $I_{tot} = 104$ mA
31. $L = 989$ μH; $C = 0.064$ μF
33. 8 MHz; $C = 40$ pF; 9 MHz: $C = 31$ pF; 10 MHz: $C = 25$ pF; 11 MHz: $C = 21$ pF
35. No fault
37. L is shorted.
39. L is shorted.

Chapter 14

1. 1.5 μH
3. 3
5. **(a)** Positive at top **(b)** Positive at bottom
   (c) Positive at bottom
7. 500 turns
9. **(a)** Same polarity, 100 V rms
   (b) Opposite polarity, 100 V rms
11. 240 V
13. 33.3 mA
15. 27.2 Ω
17. 5.75 mA
19. 0.5
21. 5 kΩ
23. No
25. 94.5 W
27. 0.98
29. 25 kVA
31. Secondary 1: 2; Secondary 2: 0.5; Secondary 3: 0.25
33. Top secondary: $n = 100/1000 = 0.1$
    Next secondary: $n = 200/1000 = 0.2$
    Next secondary: $n = 500/1000 = 0.5$
    Bottom secondary: $n = 1000/1000 = 1$
35. Excessive primary current is drawn, potentially burning out the source and/or the transformer, unless the primary is protected by a fuse.
37. **(a)** $V_{L1} = 35$ V, $I_{L1} = 2.92$ A, $V_{L2} = 15$ V, $I_{L2} = 1.5$ A
    (b) 28.9 Ω
39. **(a)** 20 V **(b)** 10 V
41. 0.0145 (69.2:1)
43. 90 mA or less
45. Primary is shorted.
47. Primary is open.

Chapter 15

1. 103 μs

3. 12.6 V

5. See Figure P–17.

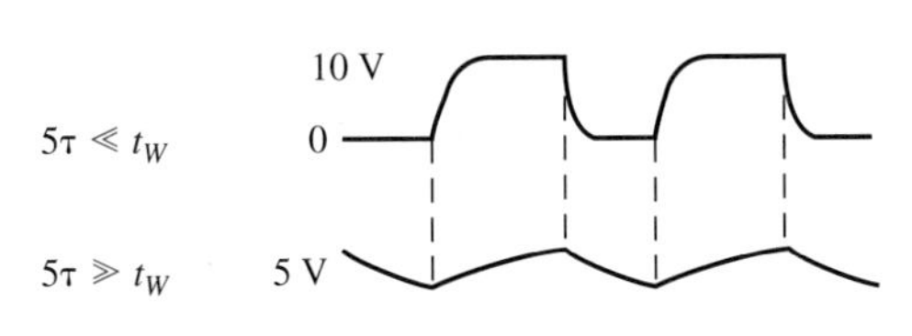

FIGURE P–17

7. See Figure P–18.

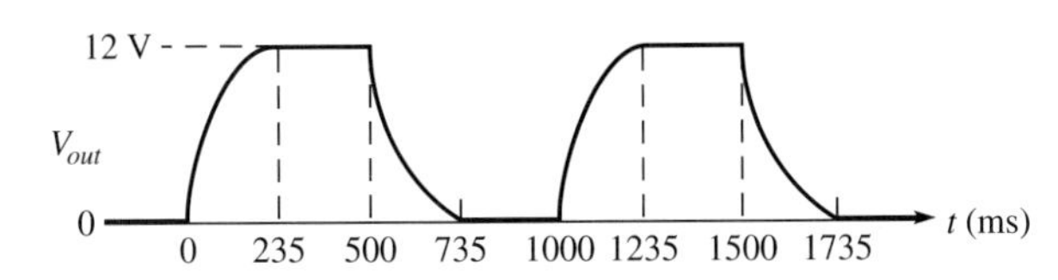

FIGURE P–18

9. 15 V dc level with a very small charge/discharge fluctuation

11. Exchange positions of R and C. See Figure P–19 for the output voltage. $5\tau = 5$ ms (repeating Problem 6)

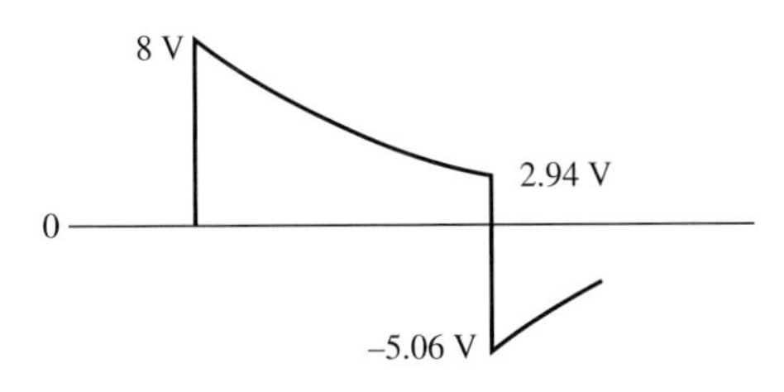

FIGURE P–19

13. Approximately the same shape as input, but with an average value of 0 V

15. See Figure P–20.

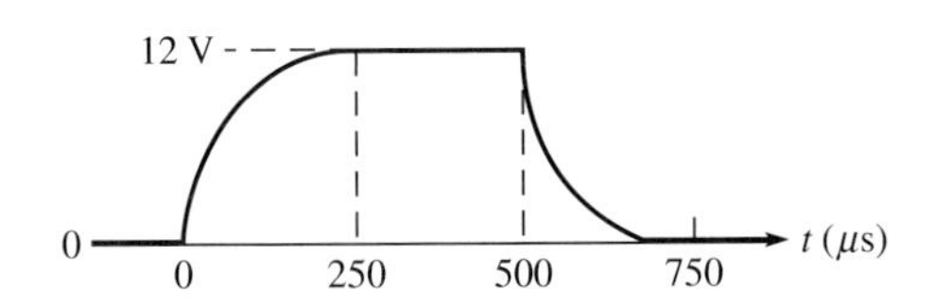

FIGURE P–20

17. See Figure P–21.

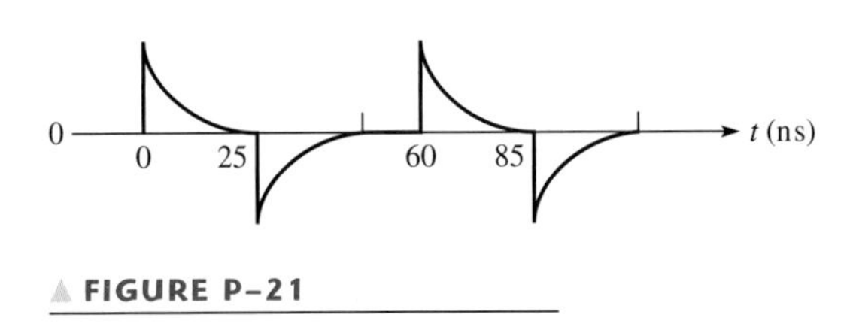

FIGURE P–21

19. A 6 V dc level

21. **(b)** No fault **(c)** Leaky capacitor
(d) Open C or shorted R

23. **(a)** 23.5 ms **(b)** See Figure P–22.

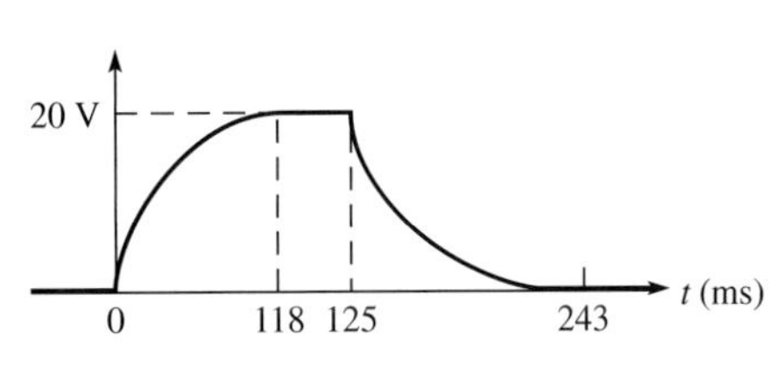

FIGURE P–22

25. 1.44 s

27. Capacitor is leaky.

29. No fault

Chapter 16

1. Silicon, germanium

3. Four

5. Conduction band, valence band

7. Doping is the process of adding trivalent or pentavalent impurity atoms into an intrinsic semiconductor to increase the effective number of free electrons.

9. The electric field across a pn junction is created by the diffusion of free electrons from the n-material across the barrier and their recombination with holes in the p-material. This results in a net negative charge on the p-side of the junction and net positive charge on the n-side of the junction, forming an electric field between the opposite charges.

11. The positive terminal must be connected to the p region.

FIGURE P-23

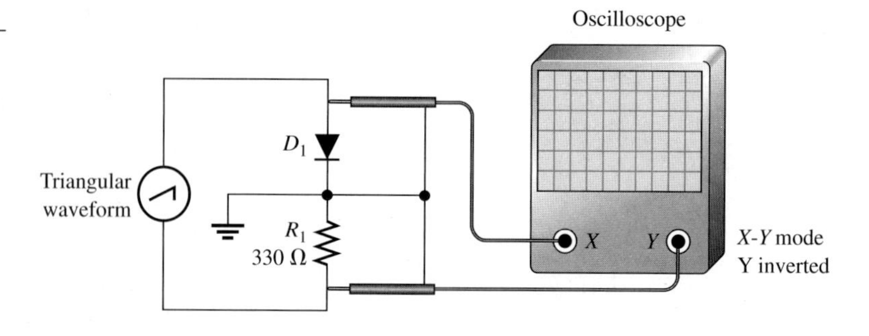

FIGURE P-24

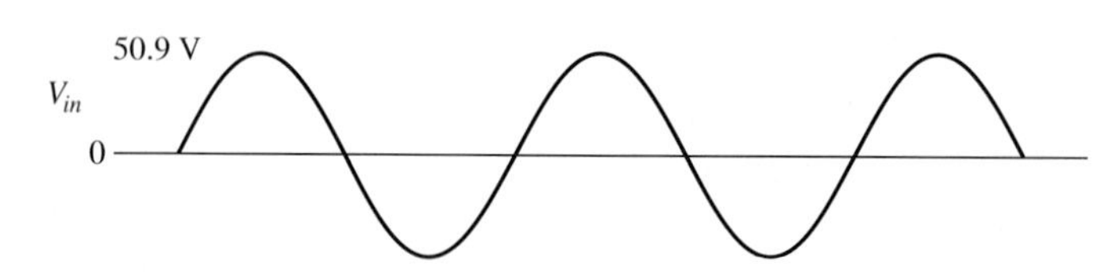

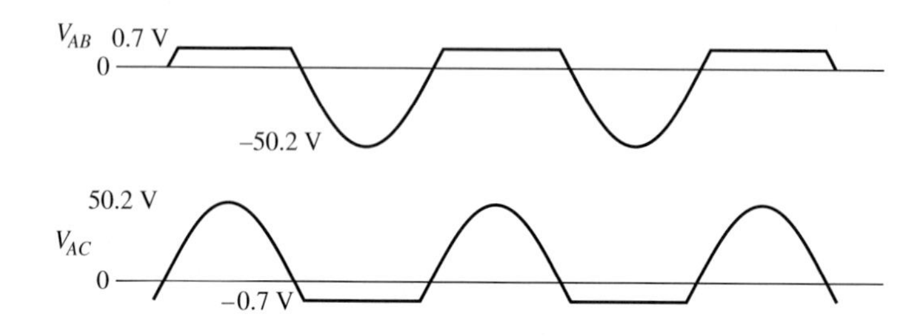

50.2 V

V_{BC}

0

13. See Figure P–23.
15. (a) reverse (b) forward
    (c) forward (d) forward
17. (a) open (b) open (c) shorted
    (d) open or okay
19. open diode
21. No fault
23. D_1 is leaky.

Chapter 17

1. 63.7 V
3. Yes
5. 47.7 V
7. 173 V
9. 78.5 V
11. See Figure P–24.
13. (a) A sine wave with a positive peak at +0.7 V, a negative peak at −7.3 V, and a dc value of −3.3 V
    (b) A sine wave with a positive peak at +29.3 V, a negative peak at −0.7 V, and a dc value of +14.3 V
15. 15.0 V
17. $I_{L(min)} = 0$ A; $I_{L(max)} = 28$ mA
19. 12%; 8%
21. A decrease of 9 pF
23. 25.4 pF each
25. Increase
27. 50 V
29. 1.67 Ω
31. (a) Correct (b) Incorrect, open diode
    (c) Correct (d) Incorrect, open diode
33. (a) Blown fuse, replace
    (b) Open transformer winding or connection
    (c) Open transformer winding or connection
    (d) Some primary windings shorted

(e) Some secondary windings shorted

(f) C_1 open (g) C_1 leaky

(h) A diode open (i) C_2 shorted

35. Lower diode (D_2) is open.

37. No fault

Chapter 18

1. 4.87 mA

3. 125

5. $I_B = 25.9\ \mu A$; $I_E = 1.3$ mA; $I_C = 1.27$ mA

7. $I_B = 13.6\ \mu A$; $I_C = 680\ \mu A$; $V_C = 9.32$ V

9. $V_{CE} = 3.34$ V; Q-point: $I_C = 15.2$ mA, $V_{CE} = 3.34$ V

11. 33.3

13. 0.5 mA; 3.33 μA; 1.03 V

15. 3 μA

17. (a) Narrows (b) Increases

19. 5 V

21. $I_{D(max)} = 10$ mA; $I_{D(min)} \cong 0$

23. 4 V

25. The gate is insulated from the channel by an SiO_2 layer.

27. 3 V

29. (a) Depletion (b) Enhancement (c) Zero bias (d) Enhancement

31. 0.6

33. 1 ms

35. 2.4 kΩ

37. Very high resistance; very low resistance

39. (a) 28 (d) 109

41. No fault

43. SCR gate is open.

Chapter 19

1. 199

3. (a) 3.25 V (b) 2.55 V (c) 2.55 mA (d) $\cong$ 2.55 mA (e) 9.59 V (f) 7.04 V

5. $A_{v(max)} = 92.3$; $A_{v(min)} = 2.91$

7. 0.976

9. A_v is reduced slightly.

11. $R_{in} = 2.07\ \Omega$; $A_v = 580$; $A_i \cong 1$; $A_p = 580$

13. (a) 0.934 (b) 0.301

15. 30 dB; 31.6

17. (a) 7.71 mW (b) 53.6 mW

19. 10 V; 625 mA

21. 50.3 kHz

23. 0.0133

25. (a) 528 kHz, Colpitts (b) 759 kHz, Hartley

27. Cutoff, 10 V

29. See tables below.

DC VALUES

V_{B1}	V_{B2}	V_{E1}	V_{E2}	V_{C1}	V_{C2}	V_{IN}	V_{OUT}
2.88 V	2.88 V	2.18 V	2.18 V	7.84 V	7.84 V	0 V	0 V

AC VALUES (RMS)

V_{b1}	V_{b2}	V_{e1}	V_{e2}	V_{c1}	V_{c2}	V_{in}	V_{out}
20.9 μV	1.96 mV	0 V	0 V	1.96 mV	594 mV	20.9 μV	594 mV

31. R_2 is open.

33. C_2 is open.

35. C_3 is open.

Chapter 20

1. *Practical op-amp:* High open-loop gain, high input impedance, low output impedance, large bandwidth, high CMRR

Ideal op-amp: Infinite open-loop gain, infinite input impedance, zero output impedance, infinite bandwidth, infinite CMRR

3. (a) Single-ended input, differential output

(b) Single-ended input, single-ended output

(c) Differential input, single-ended output

(d) Differential input, differential output

FIGURE P–25

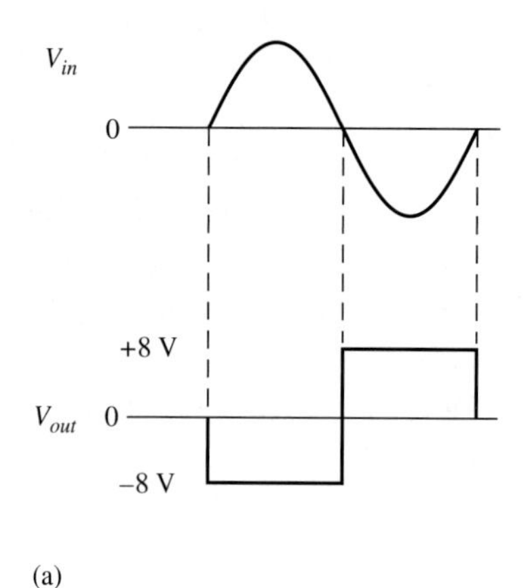

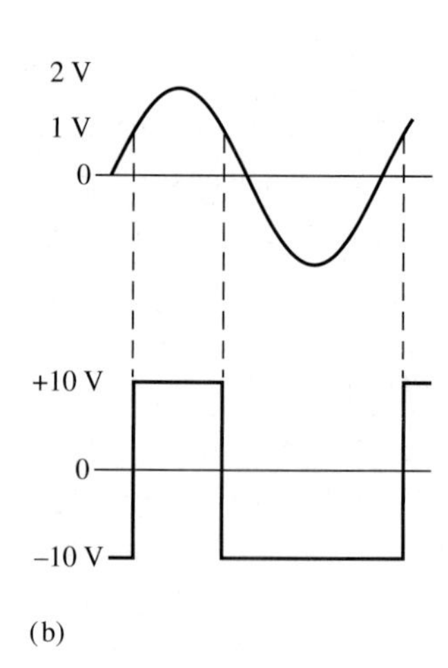

5. 8.1 μA

7. 108 dB

9. 0.3

11. 40 μs

13. **(a)** 374 **(b)** 3.74 V **(c)** 10 mV

15. **(a)** 49 kΩ **(b)** 3 MΩ **(c)** 84 kΩ
(d) 165 kΩ

17. **(a)** 10 mV, in-phase
(b) −10 mV, 180° out-of-phase
(c) 223 mV, in-phase
(d) −100 mV, 180° out-of-phase

19. **(a)** $Z_{in(NI)} = 8410$ MΩ, $Z_{out(NI)} = 89.2$ mΩ
(b) $Z_{in(NI)} = 6181$ MΩ, $Z_{out(NI)} = 4.04$ mΩ
(c) $Z_{in(NI)} = 5302$ MΩ, $Z_{out(NI)} = 18.9$ mΩ

21. **(a)** $Z_{in(I)} = 10$ kΩ, $Z_{out(I)} = 40$ Ω
(b) $Z_{in(I)} = 100$ kΩ, $Z_{out(I)} = 50$ Ω
(c) $Z_{in(I)} = 470$ Ω, $Z_{out(I)} = 70$ Ω

23. A_{cl} increases to 100.

25. R_1 is open.

27. Op-amp faulty

Chapter 21

1. **(a)** Maximum negative
(b) Maximum positive
(c) Maximum negative

3. See Figure P–25.

5. **(a)** $V_{R1} = 1$ V, $V_{R2} = 1.8$ V **(b)** 127 μA
(c) −2.8 V

7. −3.57 V; 357 μA

9. −4.06 mV/μs

11. 733 mV

13. 50 kΩ

15. 10.6 kHz

17. Change R_1 to 3.54 kΩ.

19. Change R_4 to 65.8 kΩ.

21. **(a)** 1 pole, low pass **(b)** 1 pole, high pass
(c) 2 poles, band pass

23. **(a)** $BW = 1.25$ kHz; $f_r = 1.58$ kHz
(b) $BW = 4.42$ kHz; $f_r = 4.42$ kHz

25. Output voltage decreases.

27. 2.8 Ω; > 175 mW

29. Q_1 conducts more.

31. Increases by 2.6 mA

33. R_5 is shorted.

35. C_1 is open.

Glossary

active filter A frequency-selective circuit consisting of active devices such as transistors or op-amps combined with reactive (*RC*) circuits.

admittance (Y) A measure of the ability of a reactive circuit to permit current; the reciprocal of impedance. The unit is the siemens (S).

alpha (α) The ratio of collector current to emitter current in a bipolar junction transistor.

alternating current (ac) Current that reverses direction in response to a change in source voltage polarity.

ammeter An electrical instrument used to measure current.

ampere (A or amp) The unit of electrical current.

ampere-hour rating A number given in ampere-hours determined by multiplying the current in amps (A) times the length of time in hours (h) a battery can deliver that current to a load.

ampere-turn The unit of magnetomotive force (mmf).

amplification The process of producing a larger voltage, current, or power using a smaller input signal as a pattern.

amplifier An electronic circuit having the capability of amplification and designed specifically for that purpose.

amplitude The maximum value of a voltage or current.

anode The most positive terminal of a diode or other electronic device.

apparent power (P_a) The product of the voltage times the current, expressed in volt-amperes (VA). In a purely resistive circuit it is the same as the true power.

apparent power rating The method of rating transformers in which the power capability is expressed in volt-amperes (VA).

atom The smallest particle of an element possessing the unique characteristics of that element.

atomic number The number of protons in the nucleus of an atom.

attenuation A gain less than unity.

autotransformer A transformer in which the primary and secondary windings are in a single winding.

average value The average of a sine wave over one half-cycle. It is 0.637 times the peak value.

averaging amplifier An amplifier with several inputs that produces an output voltage that is the mathematical average of the input voltages.

AWG (American wire gauge) A standardization based on wire diameter.

balanced bridge A bridge circuit that is in the balanced state as indicated by zero volts across the bridge.

band-pass filter A filter that passes a range of frequencies lying between two cutoff frequencies and rejects frequencies above and below the range.

band-stop filter A filter that rejects a range of frequencies lying between two cutoff frequencies and passes frequencies above and below the range.

bandwidth (BW) The characteristic of certain electronic circuits that specifies the usable range of frequencies for which signals pass from input to output without significant reduction in amplitude.

barrier potential The inherent voltage across the depletion region of a *pn* junction.

base One of the semiconducting regions in a bipolar junction transistor.

baseline The normal level of a pulse waveform; the voltage level in the absence of a pulse.

battery An energy source that uses a chemical reaction to convert chemical energy into electrical energy.

beta (β) The ratio of collector current to base current in a bipolar junction transistor.

bias The application of a dc voltage to a diode or other electronic device to produce a desired mode of operation.

bipolar Characterized by two *pn* junctions.

bipolar junction transistor (BJT) A transistor constructed with three doped semiconducting regions separated by two *pn* junctions.

bleeder current The current left after the total load current is subtracted from the total current into the circuit.

branch One current path in a parallel circuit.

bridge rectifier A type of full-wave rectifier consisting of diodes arranged in a four-cornered configuration.

bypassing The method of connecting a capacitor from a point to ground to remove the ac signal without affecting the dc voltage. A special case of decoupling.

capacitance The ability of a capacitor to store electrical charge.

capacitive reactance The opposition of a capacitor to sinusoidal current. The unit is the ohm (Ω).

capacitive susceptance (B_C) The ability of a capacitor to permit current; the reciprocal of capacitive reactance. The unit is the siemens (S).

capacitor An electrical device consisting of two conductive plates separated by an insulating material and possessing the property of capacitance.

capacitor-input filter A power supply filter which uses a capacitor from the rectifier output to ground to eliminate most of the variation in the rectifier output voltage.

cathode The more negative terminal of a diode or other electronic device.

center tap (CT) A connection at the midpoint of the secondary winding of a transformer.

charge An electrical property of matter that exists because of an excess or a deficiency of electrons. Charge can be either positive or negative.

charging The process in which a current removes charge from one plate of a capacitor and deposits it on the other plate, making one plate more positive than the other.

choke An inductor. The term is used more commonly in connection with inductors used to block or choke off high frequencies.

circuit An interconnection of electrical components designed to produce a desired result. A basic circuit consists of a source, a load, and an interconnecting current path.

circuit breaker A resettable protective device used for interrupting excessive current in an electric circuit.

circuit ground A method of grounding whereby the metal chassis that houses the assembly or a large conductive area on a printed circuit board is used as the common or reference point; also called *chassis ground.*

circular mil (CM) The unit of the cross-sectional area of a wire.

clamper A circuit that adds a dc level to an ac signal; a dc restorer.

class A A category of amplifier circuit that conducts for the entire input cycle and produces an output signal that is a replica of the input signal in terms of its waveshape.

class B A category of amplifier circuit that conducts for half of the input cycle.

class C A category of amplifier circuit that conducts for a very small portion of the input cycle.

closed circuit A circuit with a complete current path.

closed loop voltage gain (A_{cl}) The overall voltage gain of an op-amp with negative feedback.

coefficient of coupling (k) A constant associated with transformers that is the ratio of secondary magnetic flux to primary magnetic flux. The ideal value of 1 indicates that all the flux in the primary winding is coupled into the secondary winding.

coil A common term for an inductor.

collector One of the semiconducting regions in a BJT.

color code A system of color bands or dots that identify the value of a resistor or other component.

common base (CB) A BJT amplifier configuration in which the base is the common (grounded) terminal.

common-collector (CC) A BJT amplifier configuration in which the collector is the common (grounded) terminal.

common-drain (CD) A FET amplifier configuration in which the drain is the common terminal.

common-emitter (CE) A BJT amplifier configuration in which the emitter is the common (grounded) terminal.

common-mode rejection ratio (CMRR) A measure of an op-amp's ability to reject signals that appear the same on both inputs; the ratio of open-loop gain to common-mode gain.

common-mode signals Signals that appear the same on both inputs of an op-amp.

common-source (CS) A FET amplifier configuration in which the source is the common terminal.

comparator A circuit which compares two input voltages and produces an output in either of two states indicating the greater or less than relationship of the inputs.

conductance (G) The ability of a circuit to allow current; the reciprocal of resistance. The unit is the siemens (S).

conductor A material in which electrical current is established with relative ease. An example is copper.

core The structure around which the winding of an inductor is formed. The core material influences the electromagnetic characteristics of the inductor.

coulomb (C) The unit of electrical charge.

Coulomb's law A physical law that states a force exists between two charged bodies that is directly proportional to the product of the two charges and inversely proportional to the square of the distance between them.

coupling The method of connecting a capacitor from between two points in a circuit to allow ac to pass from one point to the other while blocking dc.

covalent Related to the bonding of two or more atoms by the interaction of their valence electrons.

crystal The pattern or arrangement of atoms forming a solid material.

current The rate of flow of charge (electrons).

current divider A parallel circuit in which the total current divides among the branches.

current gain The ratio of output current to input current.

cutoff The nonconducting state of a transistor.

cutoff frequency (f_c) The frequency at which the output voltage of a filter is 70.7% of the maximum output voltage.

cycle One repetition of a periodic waveform.

darlington pair A two-transistor arrangement that produces a multiplication of current gain.

dc component The average value of a pulse waveform.

dc power supply An electronic instrument that produces voltage, current, and power from the ac power line or batteries in a form suitable for use in powering electronic equipment.

decade A tenfold change in the value of a quantity. When a quantity becomes ten times less or ten times greater, it has changed a decade.

decibel (dB) The unit of logarithmic expression of a ratio, such as a power ratio or a voltage ratio.

decoupling The method of connecting a capacitor from one point, usually the dc power supply line, to ground to short ac to ground without affecting the dc voltage.

degree The unit of angular measure corresponding to 1/360 of a complete revolution.

depletion mode The condition in a MOSFET when the channel is depleted of conduction electrons.

diac A semiconductive device that can conduct current in either of two directions when properly activated.

dielectric The insulating material between the plates of a capacitor.

dielectric constant A measure of the ability of a dielectric material to establish an electric field.

dielectric strength A measure of the ability of a dielectric material to withstand voltage without breaking down.

differential amplifier An amplifier that produces an output proportional to the difference of two inputs.

differentiator A circuit that produces an output that approaches the mathematical derivative of the input, which is the rate of change.

digital multimeter An electronic instrument that combines meters for the measurement of voltage, current, and resistance.

diode An electronic device that permits current in only one direction.

discrete device An individual electrical or electronic component that must be used in combination with other components to form a complete functional circuit.

doping The process of imparting impurities to an intrinsic semiconductive material in order to control its conduction characteristics.

drain One of the three terminals of a field-effect transistor.

duty cycle A characteristic of a pulse waveform that indicates the percentage of time that a pulse is present during a cycle; the ratio of pulse width to period.

effective value A measure of the heating effect of a sine wave; also known as the *rms* (root mean square) *value.*

efficiency The ratio of the output power to the input power, expressed as a ratio.

electrical Related to the use of electrical voltage and current to achieve desired results.

electrical isolation The condition that exists when two coils are magnetically linked but have no electrical connection between them.

electrical shock The physical sensation resulting from electrical current through the body.

electromagnetic field A formation of a group of magnetic lines of force surrounding a conductor created by electrical current in the conductor.

electromagnetic induction The phenomenon or process by which a voltage is produced in a conductor when there is relative motion between the conductor and a magnetic or electromagnetic field.

electron The basic particle of electrical charge in matter. The electron possesses negative charge.

electronic Related to the movement and control of free electrons in semiconductors or vacuum devices.

element One of the unique substances that make up the known universe. Each element is characterized by a unique atomic structure.

emitter One of the three semiconducting regions in a BJT.

emitter-follower A popular term for a common-collector amplifier.

energy The fundamental capacity to do work. The unit is the joule (J).

engineering notation A system for representing any number as a one-, two-, or three-digit number times a power of ten with an exponent that is a multiple of 3.

enhancement mode The condition of a MOSFET when the channel has an abundancy of conduction electrons.

exponent The number to which a base number is raised.

exponential A mathematical function described by a natural logarithm (base). The charging and discharging of a capacitor are described by an exponential function.

falling edge The negative-going transition of a pulse.

fall time (t_f) The time interval required for a pulse to change from 90% to 10% of its amplitude.

farad (F) The unit of capacitance.

Faraday's law A law stating that the voltage induced across a coil of wire equals the number of turns in the coil times the rate of change of the magnetic flux.

feedback The process of returning a portion of a circuit's output signal to the input in such a way as to create certain specified operating conditions.

field-effect transistor (FET) A type of transistor that uses an induced electric field within its structure to control current.

filter A type of circuit that passes certain frequencies and rejects all others.

forward bias The condition in which a *pn* junction conducts current.

free electron A valence electron that has broken away from its parent atom and is free to move from atom to atom within the atomic structure of a material.

frequency A measure of the rate of change of a periodic function; the number of cycles completed in 1 s. The unit of frequency is the hertz.

frequency response In electrical circuits, the variation in the output voltage (or current) over a specified range of frequencies.

full-wave rectifier A circuit that converts an alternating sine wave into a pulsating dc consisting of both halves of a sine wave for each input cycle.

function generator An electronic instrument that produces electrical signals in the form of sine waves, triangular waves, and pulses.

fundamental frequency The repetition rate of a waveform.

fuse A protective device that burns open when there is excessive current in a circuit.

gain The amount by which an electrical signal is increased or decreased; the ratio of output to input; the amount of amplification.

gate One of the three terminals of a FET.

gauss A CGS unit of flux density.

generator An energy source that produces electrical signals.

germanium A semiconductive material.

ground In electronic circuits, the common or reference point.

half-power frequency The frequency at which the output of a filter is 70.7% of maximum.

half-splitting A troubleshooting procedure where one starts in the middle of a circuit or system and, depending on the first measurement, works toward the output or toward the input to find the fault.

half-wave rectifier A circuit that converts an alternating sine wave into a pulsating dc consisting of one-half of a sine wave for each input cycle.

harmonics The frequencies contained in a composite waveform, which are integer multiples of the repetition frequency (fundamental).

henry (H) The unit of inductance.

hertz (Hz) The unit of frequency. One hertz equals one cycle per second.

high-pass filter A certain type of filter whereby higher frequencies are passed and lower frequencies are rejected.

hole The absence of an electron in the valence band of an atom.

hypotenuse The longest side of a right triangle.

hysteresis A characteristic of a magnetic material whereby a change in magnetization lags the application of a magnetic force.

impedance (Z) The total opposition to sinusoidal current expressed in ohms.

impedance matching A technique used to match a load resistance to an internal source resistance in order to achieve a maximum transfer of power.

induced current A current induced in a conductor when the conductor moves through a magnetic field.

induced voltage Voltage produced as a result of a changing magnetic field.

inductance (*L*) The property of an inductor that produces an opposition to any change in current.

inductive reactance (X_L) The opposition of an inductor to sinusoidal current. The unit is the ohm.

inductive susceptance (B_L) The reciprocal of inductive reactance. The unit is the siemens.

inductor An electrical device formed by a wire wound in a coil around a core having the property of inductance and the capability to store energy in its electromagnetic field; also known as a coil or, in some applications, a choke.

input The voltage, current, or power applied to an electrical circuit to produce a desired result.

instantaneous power The value of power in a circuit at any given instant of time.

instantaneous value The voltage or current value of a waveform at a given instant in time.

insulator A material that does not allow current under normal conditions.

integrated circuit (IC) A type of circuit in which all the components are constructed on a single tiny chip of silicon.

integrator A circuit that produces an output that approaches the mathematical integral of the input.

interface To make the output of one type of circuit compatible with the input of another so that they can operate properly together.

intrinsic The pure or natural state of a material.

intrinsic semiconductor A pure semiconductive material with relatively few free electrons.

inverting amplifier An op-amp closed-loop configuration in which the input signal is applied to the inverting input.

ion An atom that has gained or lost a valence electron resulting in a net positive or negative charge.

ionization The removal or addition of an electron from or to a neutral atom so that the resulting atom (called an ion) has a net positive or negative charge.

joule (J) The unit of energy.

junction A point at which two or more components are connected.

junction field-effect transistor (JFET) A type of FET that operates with a reverse-biased junction to control current in a channel.

kilowatt-hour (kWh) A common unit of energy used mainly by utility companies.

Kirchhoff's current law A law stating that the total current into a junction equals the total current out of the junction.

Kirchhoff's voltage law A law stating that (1) the sum of the voltage drops around a closed loop (path) equals the source voltage or (2) the sum of all the voltages (drops and sources) around a closed loop is zero.

lag To be behind; describes a condition of the phase or time relationship of waveforms in which one waveform is behind the other in phase or time.

large-signal amplifier An amplifier that produces an output signal that approaches the limits of the load line.

lead To be ahead; describes a condition of the phase or time relationship of waveforms in which one waveform is ahead of the other in phase or time; also, a wire or cable connection to a device or instrument.

leading edge The first step or transition of a pulse.

LED (light-emitting diode) A type of diode that emits light when there is forward current.

Lenz's law A physical law that states when the current through a coil changes, the polarity of the induced voltage created by the changing magnetic field is such that it always opposes the change that caused it. The current cannot change instantaneously.

limiter A circuit that removes part of a waveform above or below a specified level; a clipper.

linear Characterized by a straight-line relationship.

line regulation The change in output voltage for a given change in line (input) voltage, normally expressed as a percentage.

lines of force Magnetic flux lines that exist in a magnetic field radiating from the north pole to the south pole.

load An element (resistor or other component) connected across the output terminals of a circuit that draws current from the circuit.

load current The output current of a circuit supplied to a load.

loading The effect on a circuit when an element that draws current from the circuit is connected across the output terminals.

load regulation The change in output voltage for a given change in load current, normally expressed as a percentage.

low-pass filter A certain type of filter in which lower frequencies are passed and higher frequencies are rejected.

magnetic coupling The magnetic connection between two coils as a result of the changing magnetic flux lines of one coil cutting through the second coil.

magnetic field A force field radiating from the north pole to the south pole of a magnet.

magnetic flux The lines of force between the north and south poles of a permanent magnet or an electromagnet.

magnetic flux density The number of lines of force per unit area perpendicular to a magnetic field.

magnetizing force The amount of mmf per unit length of magnetic material.

magnetomotive force The force that produces a magnetic field.

magnitude The value of a quantity, such as the number of volts of voltage or the number of amperes of current.

majority carrier The most numerous charge carrier in a doped semiconductive material (either free electrons or holes.)

maximum power transfer The condition, when the load resistance equals the source resistance, under which maximum power is transferred from source to load.

maximum power transfer theorem A theorem that states the maximum power is transferred from a source to a load when the load resistance equals the internal source resistance.

metric prefix A symbol that is used to replace the power of ten in numbers expressed in engineering notation.

minority carrier The least numerous charge carrier in a doped semiconductive material (either free electrons or holes).

MOSFET Metal-oxide semiconductor field-effect transistor.

multimeter An instrument that measures voltage, current, and resistance.

mutual inductance (L_M) The inductance between two separate coils, such as a transformer.

negative feedback The return of a portion of the output signal to the input such that it is out of phase with the input signal.

neutron An atomic particle having no electrical charge.

noninverting amplifier An op-amp closed-loop configuration in which the input signal is applied to the noninverting input.

nucleus The central part of an atom containing protons and neutrons.

ohm (Ω) The unit of resistance.

ohmmeter An instrument for measuring resistance.

Ohm's law A law stating that current is directly proportional to voltage and inversely proportional to resistance.

open A circuit condition in which the current path is interrupted.

open circuit A circuit in which there is not a complete current path.

open-loop gain (A_{ol}) The internal voltage gain of an op-amp without feedback.

operational amplifier (op-amp) A special type of amplifier exhibiting very high open-loop gain, very high input impedance, very low output impedance, and good rejection of common-mode signals.

orbit The path an electron takes as it circles around the nucleus of an atom.

oscillator An electronic circuit consisting of an amplifier and a phase-shift network connected in a positive feedback loop that produces a time-varying output signal without an external input signal using positive feedback.

oscilloscope A measurement instrument that displays signal waveforms on a screen.

parallel The relationship in electric circuits in which two or more current paths are connected between the same two points.

parallel resonance In a parallel *RLC* circuit, the condition where the impedance is maximum and the reactances are equal.

passband The range of frequencies passed by a filter.

peak-to-peak value The voltage or current value of a waveform measured from its minimum to its maximum points.

peak value The voltage or current value of a waveform at its maximum positive or negative points.

pentavalent Describes an atom with five valence electrons.

period (T) The time interval of one complete cycle of a given sine wave or any periodic waveform.

periodic Characterized by a repetition at fixed time intervals.

permeability The measure of ease with which a magnetic field can be established in a material.

phase The relative displacement of a time-varying waveform in terms of its occurrence with respect to a reference.

phase angle The angle between the source voltage and the total current in a reactive circuit.

phasor A representation of a sine wave in terms of both magnitude and phase angle.

photoconductive cell A type of variable resistor that is light-sensitive.

photodiode A diode whose reverse resistance changes with incident light.

pinch-off voltage The value of the drain-to-source voltage of a FET at which the drain current becomes constant when the gate-to-source voltage is zero.

PIV (peak inverse voltage) The maximum value of reverse voltage which occurs at the peak of the input cycle when the diode is reverse-biased.

***PN* junction** The boundary between *n*-type and *p*-type semiconductive materials.

pole In practical terms, a single *RC* circuit in a filter or amplifier that causes the response to change at a 20 dB per decade rate above or below a certain frequency.

positive feedback The return of a portion of the output signal to the input such that it is in phase with the input signal.

potentiometer A three-terminal variable resistor.

power The rate of energy usage.

power factor The relationship between volt-amperes and true power or watts. Volt-amperes multiplied by the power factor equals true power.

power gain The ratio of output power to input power; the product of voltage gain and current gain.

power of ten A numerical representation consisting of a base of 10 and an exponent; the number 10 raised to a power.

power rating The maximum amount of power that a resistor can dissipate without being damaged by excessive heat buildup.

primary winding The input winding of a transformer; also called *primary*.

probe An accessory used to connect a voltage to the input of an oscilloscope or other instrument.

proton A positively charged atomic particle.

pulse A type of waveform that consists of two equal and opposite steps in voltage or current separated by a time interval.

pulse repetition frequency The fundamental frequency of a repetitive pulse waveform; the rate at which the pulses repeat expressed in either hertz or pulses per second.

pulse response The reaction of a circuit to a given input.

pulse width (t_W) The time interval between the opposite steps of an ideal pulse. For a nonideal pulse, the time between the 50% points on the leading and trailing edges.

push-pull A type of class B amplifier in which one output transistor conducts for one half-cycle and the other conducts for the other half-cycle.

Q-point The dc operating (bias) point of an amplifier.

quality factor (Q) The ratio of reactive power to true power in a coil or a resonant circuit.

radian A unit of angular measurement. There are 2π radians in one complete revolution. One radian equals 57.3°.

ramp A type of waveform characterized by a linear increase or decrease in voltage or current.

***RC* lag network** A phase shift circuit in which the output voltage, taken across the capacitor, lags the input voltage by a specified angle.

***RC* lead network** A phase shift circuit in which the output voltage, taken across the resistor, leads the input voltage by a specified angle.

***RC* time constant** A fixed time interval set by the values of *R* and *C* that determines the time response of a series *RC* circuit. It equals the product of the resistance and the capacitance.

reactive power (P_r) The rate at which energy is stored and alternately returned to the source by a capacitor or inductor. The unit is the VAR.

recombination The process of a free electron falling into a hole in the valence band of an atom.

rectifier An electronic circuit that converts ac into pulsating dc; one part of a power supply.

reflected load The load as it appears to the source in the primary of a transformer.

reflected resistance The resistance in the secondary circuit reflected into the primary circuit.

regulator An electronic circuit that maintains an essentially constant output voltage with a changing input voltage or load.

relaxation oscillator A type of oscillator, generally nonsinusoidal, whose operation is based on the charging and discharging of a capacitor.

relay An electromagnetically controlled mechanical device in which electrical contacts are open or closed by a magnetizing current.

reluctance The opposition to the establishment of a magnetic field in a material.

resistance Opposition to current. The unit is the ohm (Ω).

resistor An electrical component designed specifically to provide resistance.

resolution The smallest increment of a quantity that a meter can measure.

resonant frequency The frequency at which a resonant condition occurs in a series or parallel *RLC* circuit.

retentivity The ability of a material, once magnetized, to maintain a magnetized state without the presence of a magnetizing force.

reverse bias The condition in which a *pn* junction prevents current.

reverse breakdown The condition of a diode in which excessive reverse-bias voltage causes a rapid buildup of reverse current.

rheostat A two-terminal variable resistor.

right triangle A triangle with a 90° angle.

ripple voltage The small variation in the dc voltage on the output of a filtered rectifier caused by the slight charging and discharging action of the filter capacitor.

rise time (t_r) The time interval required for a pulse to change from 10% to 90% of its amplitude.

rising edge The positive-going transition of a pulse.

***RL* lag network** A phase shift circuit in which the output voltage, taken across the resistor, lags the input voltage by a specified angle.

***RL* lead network** A phase shift circuit in which the output voltage, taken across the inductor, leads the input voltage by a specified angle.

***RL* time constant** A fixed time interval set by the values of *R* and *L* that determines the time response of a circuit.

rms value The value of a sine wave that indicates its heating effect, also known as the *effective value.* It is equal to 0.707 times the peak value. RMS stands for root mean square.

roll-off The decrease in the response of a filter below or above a critical frequency.

saturation The state of a BJT in which the collector current has reached a maximum and is independent of the base current.

sawtooth waveform A type of electrical waveform composed of ramps; a special case of a triangular waveform in which one ramp is much shorter than the other.

scaling adder A special type of summing amplifier with weighted inputs.

schematic A symbolized diagram of an electrical or electronic circuit.

scientific notation A system for representing any number as a number between 1 and 10 times an appropriate power of ten.

secondary winding The output winding of a transformer; also called *secondary.*

selectivity A measure of how effectively a filter passes certain frequencies and rejects others. The narrower the bandwidth, the greater the selectivity.

semiconductor A material that has a conductance value between that of a conductor and that of an insulator. Silicon and germanium are examples.

sensitivity factor The ohms-per-volt rating of a voltmeter.

series In an electrical circuit, a relationship of components in which the components are connected such that they provide a single current path between two points.

series-aiding An arrangement of two or more series voltage sources with polarities in the same direction.

series-opposing An arrangement of two series voltage sources with polarities in the opposite direction.

series regulator A type of voltage regulator with the control element in series between the input and output.

series resonance In a series *RLC* circuit, the condition where the impedance is minimum and the reactances are equal.

shell An energy band in which electrons orbit the nucleus of an atom.

short circuit A zero or abnormally low resistance between two points; usually an inadvertent condition.

shunt regulator A type of voltage regulator with the control element between the output and ground.

siemens The unit of conductance.

silicon A semiconductive material used in diodes and transistors.

silicon-controlled rectifier (SCR) A device that can be triggered on to conduct current in one direction.

sine wave A type of electrical waveform that is based on the mathematical sine function.

small-signal amplifier An amplifier that produces an output signal that takes up a small percentage of the total load line excursion.

solenoid An electromagnetically controlled device in which the mechanical movement of a shaft or plunger is activated by a magnetizing current.

source Any device that produces energy; one of the three terminals of a FET.

speaker An electromagnetic device that converts electrical signals to sound waves.

steady state The equilibrium condition of a circuit that occurs after an initial transient time.

step-down transformer A transformer in which the secondary voltage is less than the primary voltage.

step-up transformer A transformer in which the secondary voltage is greater than the primary voltage.

stopband The range of frequencies between the upper and lower cutoff points.

summing amplifier An amplifier with several inputs that produces an output voltage proportional to the algebraic sum of the input voltages.

superposition A method for analyzing circuits with two or more sources by examining the effects of each source by itself and then combining the effects.

switch An electrical or electronic device for opening and closing a current path.

tank circuit A parallel resonant circuit.

tapered Nonlinear, such as a tapered potentiometer.

temperature coefficient A constant specifying the amount of change in the value of a quantity for a given change in temperature.

terminal An external contact point on an electronic device.

terminal equivalency A condition that occurs when two circuits produce the same load voltage and load current where the same value of load resistance is connected to either circuit.

tesla The unit of flux density.

thermistor A type of variable resistor that is temperature-sensitive.

Thevenin's theorem A circuit theorem that provides for reducing any two-terminal resistive circuit to a single equivalent voltage source in series with an equivalent resistance.

thyristor A class of four-layer semiconductive devices.

time constant A fixed-time interval set by *R* and *C*, or *R* and *L* values, that determines the time response of a circuit.

tolerance The limits of variation in the value of a component.

trailing edge The second step or transition of a pulse.

transconductance (g_m) The ratio of drain current to gate-to-source voltage in a FET.

transformer A device formed by two or more windings that are magnetically coupled to each other and provide a transfer of power electromagnetically from one winding to another.

transient A temporary passing condition in a circuit; a sudden or temporary change in circuit conditions.

transient time An interval equal to approximately five time constants.

transistor A semiconductive device used for amplification and switching applications in electronic circuits.

triac A bidirectional thyristor that can be triggered into conduction.

triangular waveform A type of electrical waveform that consists of two ramps.

triangular wave oscillator A type of oscillator that produces a triangular wave output voltage.

trigger The activating mechanism of some electronic devices or instruments.

trimmer A small variable capacitor.

troubleshooting A systematic process of isolating, identifying, and correcting a fault in a circuit or system; the application of logical thinking combined with a thorough knowledge of circuit or system operation to find and correct a malfunction.

true power The power that is dissipated in a circuit, usually in the form of heat.

turns ratio The ratio of turns in the secondary winding to turns in the primary winding.

unbalanced bridge A bridge circuit that is in the unbalanced state as indicated by a voltage across the bridge proportional to the amount of deviation from the balanced state.

unijunction transistor (UJT) A type of transistor consisting of an emitter and two bases.

unity gain A gain of 1.

valence Related to the outer shell or orbit of an atom.

valence electron An electron that is present in the outermost shell of an atom.

VAR (volt-ampere reactive) The unit of reactive power.

varactor A diode that is used as a voltage-variable capacitor.

volt The unit of voltage or electromotive force.

voltage The amount of energy available to move a certain number of electrons from one point to another in an electric circuit.

voltage-controlled oscillator A type of oscillator whose frequency can be varied over a specified range with a dc control voltage.

voltage divider A circuit consisting of series resistors across which one or more output voltages are taken.

voltage drop The difference in the voltage at two points due to energy conversion.

voltage-follower A closed-loop, noninverting op-amp with a voltage gain of 1.

voltage gain The ratio of output voltage to input voltage.

voltage regulation The process of maintaining an essentially constant output voltage over variations in input voltage or load.

voltmeter An instrument used to measure voltage.

watt (W) The unit of power.

Watt's law A law that states the relationships of power to current, voltage, and resistance.

waveform The pattern of variations of a voltage or current showing how the quantity changes with time.

weber The unit of magnetic flux.

Wheatstone bridge A 4-legged type of bridge circuit with which an unknown resistance can be accurately measured using the balanced state of the bridge. Deviations in resistance can be measured using the unbalanced state.

Wien-bridge oscillator A sinusoidal oscillator that employs a lead-lag circuit in the feedback loop.

winding The loops or turns of wire in an inductor.

winding resistance The resistance of the length of wire that makes up a coil.

wiper The sliding contact in a potentiometer.

zener breakdown The lower voltage breakdown in a zener diode.

zener diode A type of diode that operates in reverse breakdown (called zener breakdown) to provide voltage regulation.

Index